# NORTH AMERICAN TUNNELING

## 2014 PROCEEDINGS

# NORTH AMERICAN TUNNELING

## 2014 PROCEEDINGS

Edited by Gregg Davidson, Alan Howard, Lonnie Jacobs, Robert Pintabona, and Brett Zernich

PUBLISHED BY THE
SOCIETY FOR MINING, METALLURGY & EXPLORATION

**Society for Mining, Metallurgy & Exploration Inc. (SME)**
12999 E. Adam Aircraft Circle
Englewood, Colorado, USA 80112
(303) 948-4200 / (800) 763-3132
www.smenet.org

SME advances the worldwide mining and minerals community through information exchange and professional development. With members in more than 70 countries, SME is the world's largest association of mining and minerals professionals.


Electronic edition published 2014.



ISBN 978-0-87335-400-4
Ebook 978-0-87335-402-8

# Contents

# Preface

The program chairs, program committee, executive committee, and SME staff extend a warm welcome to all authors, exhibitors, students, and other attendees of the 2014 North American Tunneling Conference (NAT 2014) in Los Angeles, California.

The program committee has selected a theme for this year's conference that it believes best reflects where we as an industry find ourselves and, more importantly, where we are heading: "Mission Possible." It is not surprising that each time we assemble, our commitment to the industry compels us to share new theories, novel innovations, and the latest tools that make what once may have been perceived as impossible, now possible. This year, we continue this tradition, and the papers assembled herein reflect the persistent progress our industry has made as it strives to drive value and experience new possibilities.

The success of this conference is due to the efforts of authors who have taken time from their demanding schedules to share with us the successes and failures they've experienced. Their dedication is testament to our theme, and the industry as a whole benefits; and for that, these individuals have our sincerest appreciation and gratitude. Further, the 2014 NAT Conference would not be complete without the participation of enterprise owners, engineers, contractors, suppliers, and manufacturers, and their willingness to engage. Their involvement continues to provide a foundation that allows our industry to move into the future.

We thank the session chairs, co-chairs, authors, and members of the NAT 2014 Executive Committee for their contributions and dedication to this conference. Additionally, the chairs express appreciation to the SME staff for their hard work, patience, and enthusiastic support. Lastly, we thank all the participants for joining us and making this conference a success that will drive value for the benefit of all.

Gregg Davidson
Alan Howard
Lonnie Jacobs
Robert Pintabona
Brett Zernich

## NAT 2014 Chairs

Gregg Davidson, Jacobs Associates
Alan Howard, Brierley Associates
Lonnie Jacobs, Frontier-Kemper
Robert Pintabona, MWH Global
Brett Zernich, Traylor Brothers

## NAT 2014 Session Chairs

John Bednarski
MWD

Mike Bruen
MWH Global

Peter Chou
Parsons Corporation

William Dean
Frontier-Kemper

Christopher Fleming
Michels Pipeline

Brian Harris
Drill Tech

Sean Harvey
Brierley Associates

David Haug
LACSD

Ivan Hee
Arup

Harold Leiendecker
Halcrow

Mike McKenna
LACMTA

Axel Nitschke
Gall Zeidler Consultants

Derek Penrice
Hatch Mott MacDonald

Scott Rand
King Packaging

Brett Robinson
Traylor Brothers

Mike Stokes
Skanska

Joel Theodore
Sound Transit

Michael Torsiello
Jacobs Associates

James Wonneberg
EPC Consultants

Brian Zalenko
Parsons Brinckerhoff

## UCA of SME 2014 Executive Committee

Bill Edgerton
Jacobs Associates

Lester Bradshaw
Bradshaw Construction Corporation

Judy Cochran
Jacobs Associates

Robert Goodfellow
Aldea Services

Gregory Hauser
Dragados USA-Seattle Tunnel Partners

Heather Ivory
ITunnel

Marcus Jensen
Southern Nevada Water Authority

Colin Lawrence
Hatch Mott MacDonald

Rick Lovat
L2 Advisors

Pamela Moran
Wisko America

Nasri Munfah
HNTB Corporation

Krishniah Murthy
LACMTA

Jeffrey Petersen
Kiewit Infrastructure Company

Michael Rispin
Normet Americas

Michael Roach
Traylor Brothers

David Rogstad
Frontier Kemper Constructors

Kellie Rotunno
Northeast Ohio Reg Sewer District

Arthur Silber
Hatch Mott McDonald

Leonard Worden
CSI-Tunnel Systems Division

## NAT 2014 Executive Committee

Bradford F. Townsend
Parsons Corporation

Brett Campbell
J.F. Shea Construction

Gregg Davidson
Jacobs Associates

Robert R. Freeman
Cellular Concrete

Dave Haug
Sanitation Districts of Los Angeles County

Alan Howard
Brierley Associates

Jon Hurt
Arup

Heather Ivory
ITunnel

Lonnie Jacobs
Frontier-Kemper

Marc R. Kritzer
Northeast Ohio Reg Sewer District

Colin Lawrence
Hatch Mott MacDonald

John Morgan
Citizens Energy Group

Derek Penrice
Hatch Mott MacDonald

Robert Pintabona
MWH

Jim Rush
Tunnel Business Magazine

Mike Smithson
Skanska USA

Brett Zernich
Traylor Brothers

# TRACK 1: TECHNOLOGY

## Session 1: Operation Monitoring and Control

Mike Bruen, Chair

# Global Monitoring and Data Management Application to the Alaskan Way Tunnel Project

**Boris Caro Vargas**
Soldata

**ABSTRACT:** Urban tunnel construction can impact the whole City environment as it generates settlement of the surrounding ground, the adjacent buildings, utilities and structures. The goal of all the "geotechnical players" involved in design and construction of these tunnel projects is to mitigate the impacts related to ground settlement and reduce risks associated with geotechnical hazards to an acceptable level. As all the geotechnical parameters are linked together the only way to achieve the lowest global settlement impact is to monitor in real time and integrate all the different parameters into a single database. This article will define some key concepts of automated global data management related to settlement control and then focus on how the global data management from design phase through construction has been implemented in the monitoring program of the SR-99 Tunnel Project.

## INTRODUCTION

The Alaskan Way Viaduct Replacement Program will replace the 1950s seismically vulnerable viaduct structure that blights the waterfront with a deep bored tunnel under the City of Seattle. The design and build contract, currently under construction by Seattle Tunnel Partners, STP, a joint Venture of Dragados and Tutor Perini, for the Washington State Department of Transport, WSDOT, is due for completion in 2016. Soldata as a subcontractor to the Contractor is providing third party instrumentation and monitoring services of surface infrastructure. The needs for instrumentation monitoring and Global Data Management need to be considered at an early stage in the project development and adequately provided for in design and construction. The future infrastructure instrumentation needs are dependent upon the design assumptions made and in particular to the assumptions of TBM performance. Within the construction phase accurate real time data acquisition, review and analysis plays an important part in managing the risks associated with ground movement. Without this, the owner is unable to accurately manage his risk profile and carry out required ongoing assessments throughout the project life. The recent advances in the ability to obtain large quantities of data and to analyze this in a short duration have led to as step change in the quality control process during construction. Through ongoing daily communication with key staff, increased awareness of the impact of TBM operation on ground settlement control, the risks to both the Owner and the Contractor are reduced.

## DATA MANAGEMENT REQUIREMENTS IN THE TUNNELING INDUSTRY

### Definition and Differences Between Data Collection and Data Reporting

Over the last 20 years progress in data processing and data transmission have had a tremendous impact on the way the monitoring of the instrumentation on a tunnel project is specified and performed. As a first step it is worth reminding that, for automated monitoring systems there are several time steps between the moment the data is collected from a sensor (or more generally a monitoring point) and the moment when the data is actually reported to the user interface. To simplify things we'll refer to:

- data collection time, or how fast the data can be recorded from a monitoring point
- data reporting time, or how fast information can be posted on an end user interface.

The "standard" automated instruments, such as vibrating wire sensors, 4–20mA electrical sensors or more recently Micro Electro Mechanical Systems (MEMS) have acquisition times measured in fractions of a second. The total data collection time for a set of sensors (like a MEMS In-place inclinometer chain or sets of strain gages) rarely exceed a few seconds and is governed by the way the data acquisition box is setup and by how many monitoring points are connected to the same datalogger (with or without multiplexers). The data reporting time depends on how many operations are performed on the raw set of data and on the path the data has to follow from the acquisition unit to the final location of the database. This time is usually measured in seconds or a few minutes. For remote methods for displacement monitoring (Dunnicliff, 2012) or what we can call "modern" automated instruments, the time range is broader. Data collection times can be measured in seconds (for Automatic Motorized Total Stations (AMTS) or terrestrial radar), minutes (Scanner) or even weeks (Satellite interferometry). However, there is no point in this case to compare data acquisition times per monitoring point as the quantity of monitoring points per acquisition units (except from GPS) ranges from dozens to several millions of points. Although recent progress has been made, the data processing times can still be much longer than the ones for standard sensors, and can have a limit to the use of certain types of instruments (Tamagnan and Beth, 2012), mainly due to the amount of data recorded at each acquisition. Some of these techniques heavily rely on algorithm calculation iterations to reach an acceptable accuracy (AMTS, GPS). In the case of remote deformation monitoring techniques, the data collection and reporting times are a function of the project specification and the design of the monitoring program. Many recent project specifications require *continuous* data acquisition frequency for automated instruments and do not specifically differentiate between the data collection and the data reporting times mentioned in this section, and do not take into account that these times can be a function of the monitoring program, the instrumentation layout or the accuracy. What is usually more important for risk management is to ensure that accurate and reliable data is provided at the right time to allow the project players to make informed decisions. Other recent specifications require faster data acquisition and reporting when the construction activity is closer to the instrument, as fast as 2 minutes for automated instruments and one hour for AMTS readings; however, unlike the monitoring of manual instruments, there is minimal additional cost involved in having a higher frequency of monitoring for automated instruments, so once the instrument is installed, the monitoring frequency (data collection + reporting time) should be defined by:

- How sensitive the structure is to settlement
- What is the layout of monitoring points in relation to the acquisition units
- What are the limits of the system to reach an acceptable accuracy and repeatability.

On the Alaskan Way Tunnel project, manual monitoring points are generally monitored as frequently as daily (depending on the location of the TBM excavation face) and automated points have to be monitored as frequently as hourly (tiltmeters, AMTS).

### When Should the Monitoring Program Be Started

When should we start monitoring? This question is posed on every tunnel project and more generally for every construction monitoring plan. The common answer is that if the construction has not started there is no point in monitoring! There are several reasons why this answer is wrong. First of all, there are several factors that can generate movements in the area of influence of a tunnel project before any construction activity start:

- Weather changes during the year, the season, or the day (temperatures, humidity and pressure)
- Tidal changes
- Changes in groundwater levels
- Impacts of other construction projects
- Other activities in the area, related or not, to the project (excavation or utility relocation projects, traffic)
- Local or global previous existing settlement 'trends' of buildings
- Earthquakes etc.

All these factors can have an impact greater than the total anticipated settlement due to the tunnel excavation, and therefore generate "false alarms" during construction. If this is the case, how is it possible to differentiate between the impact of the tunnel construction and external non-related factors? How to assign responsibilities and liability if damages occur, if action levels are reached? Many tunneling projects still tend to limit the duration of the baseline monitoring to the minimum (one or two weeks) while on other recent projects baseline monitoring periods of 6 months have been established (Sowers and Caro Vargas, 2013) In addition to detecting any preconstruction potential impact, the longer monitoring periods allow a better calibration of the instruments and allow the adjustment in the setup in data processing routines to get more accurate and repeatable data. It is clear that longer monitoring periods can only be practically implemented on projects where the monitoring firm is involved at early stages, such as design-build projects or PPPs. Although for conventional Design-Bid-Build Projects, monitoring programs can be implemented directly by the Owner during design, however the continuity between *pre*-construction and actual construction monitoring programs is hard to achieve. At another level, similar

**Figure 1. Overview of radar interferometry satellite coverage overlapping the ZOI**

questions have to be resolved for the post construction period.

### How Should the Extent of the Monitoring Area Be Defined

Another key question is what should be the limits of the area to be monitored. This area is usually a function of the zone (or area) of influence of the tunnel excavation defined during the initial settlement analysis. Similar questions for the timing are raised for "space," not only do we have to answer the question: When should the monitoring period start? What should the limits be for the monitoring area?

The design analysis usually defines the settlement profile due to TBM excavation. The theoretical width of the settlement profile is usually defined as a function of the depth of the tunnel, the ground parameters and the tunnel diameter (Peck, 1969). The zone of influence of the TBM excavation (ZOI) is usually defined based on this settlement profile and extends from the tunnel centerline to the points where the settlement is considered to be "negligible." All the buildings and structures located within this ZOI are usually instrumented. The degree of instrumentation of each structure depends on how sensitive it is to absolute and differential settlement. The problem of this binary approach (structure to be monitored or NOT to be monitored) is that buildings just outside the ZOI might settle. In this case, if damage occurs, how are responsibility and liability determined? On the other hand, it might be very costly to monitor a larger ZOI, when compared to the lower probability of the associated risk.

Some recent monitoring techniques allow monitoring settlement over large areas beyond standard definition of the ZOI of the tunnel impact (Sowers and Caro Vargas, 2013) without a complete modification of the monitoring plan and instrumentation layout. This technique allows to monitor thousands of points over large urban areas (10,000/km$^2$) by processing the radar images taken by the satellite going over a specific location. The frequency of monitoring depends on the return of the satellite over the same area (generally around two weeks), and so it cannot be considered as a "real-time" monitoring technique, but its coverage allows to mitigate any litigation related to "non-monitored buildings" outside the ZOI at a lower cost than traditional monitoring techniques. This technique has been implemented and used on the Alaskan Way Tunnel Project successfully and allowed to separate the potential impact of the SR-99 Tunnel Project from other construction project and already existing settlement trends in some areas of the City. An example of the Satellite interferometry output is shown on Figure 1.

## DEFINITION OF GLOBAL MONITORING

Monitoring geotechnical, structural and environmental sensors is not a goal per se. The objectives of a monitoring program are generally to understand the correlation between a construction activity and the impact on the surrounding environment (ground, structures, water, air, etc.) in order to mitigate any "excessive" impact. Since the early implementation of monitoring plans, manual survey data has been combined with standard geotechnical data retrieved from underground sensors. Movements of surface monitoring points, near surface monitoring, structure monitoring points and other excavation monitoring points are still measured with traditional survey equipment. However, the accuracy and data collection frequency (see section: "Definition and Differences between Data Collection and Data Reporting" above) often requires the surveying to be automated using AMTS in order to be compared with structural and geotechnical sensors with submillimiter accuracy. Using AMTS systems with an accuracy of 1mm at 300' distances allows the combination of traditional geotechnical sensors and AMTS into the same monitoring plan. Most tunneling projects today require the geotechnical engineer and the surveyor to work hand in hand. It is not uncommon for current projects to specify requirements for data managers or instrumentation specialists with experience in both standard geotechnical instruments (piezometers, extensometers, strain gages) and AMTS, with the expertise to collect, verify, manage, analyze and present data from both types of sensors. Vibration and noise monitoring data (although different in its dynamic nature) using geophones and microphones is also often part of the construction monitoring program, as well as utility monitoring remote techniques, such as acoustic leak detection or video inspections. Most of the monitoring firms have expertise or at least are familiar with all these different data sources and types and know how to process the raw data into the geotechnical instrumentation database and interpret the results. But when it comes to monitoring actual construction parameters, the monitoring firm does not usually have access to the source of data, simply because it has no control over the installation of sensors to monitor the construction process itself. Construction processes using automated sensors are increasing; compensation grouting (Thurlow et al., 1999) and mechanized tunneling with tunnel boring machines have set up the standard of the industry in that regard. To integrate the data (pressure and volumes are the most common) from the sensors installed to monitor these activities, both the contractor (specialty contractor or general contractor) and the monitoring firm have to collaborate closely. This collaboration is needed at early stages. The following sections will present an example of global data integration and monitoring on the SR-99 tunnel project, starting from the design phase with the definition of the settlements to the integration of the tunneling parameters of the current world's largest TBM.

### Settlement Control Process in a Design-Build Project and Definition of Key Parameters During the Design Phase

During the design phase of the project, the potential impact on surface structures and buried utilities is critical information in not only determining the impact but also accurately assessing the cost of mitigating the impact and ensuring that the risk profile is accommodated within the budget. The approach taken is iterative in nature and as the project develops and both the ground conditions and the conditions of structures becomes better known, the risk profile can be adjusted accordingly. Surface structures can be analyzed and studied to a high level of certainty to determine their response to ground movement. Buried utilities are often more difficult to assess as the age and condition is often an unknown. As data becomes available and more analysis is undertaken the design team has to revise the degree of impact and the extent of the zone of influence. Coordinating this information with the civil design team allows for a coordinated approach to be established and risks to be managed at the earliest opportunity through such factors as a change in alignment or deepening of the project alignment in order to reduce the predicted impact. The generic design process followed is set out in Figure 2.

On the Alaskan Way Project the initial assessment of buildings that might potentially be impacted yielded a number in the region of 280. This number was then reduced to 158 through ongoing assessment during the design process. The determination of the ground loss figure to be used for conceptual design has a great significance on the potential impact on 3rd parties and thus the viability of the project as a whole. A sensitivity study should be carried out to determine the potential for impact beyond that which is considered appropriate. Only in this way will the risk to the project be able to be determined and quantified, if at some time in the future the ground loss changes from what was assumed initially, either theoretically or based on actual conditions. For planning purposes a 0.5% face loss figure is generally reasonable.

Projects that envisage the use of large diameter machines requires careful assessment of the base parameters to be used in order to correctly assess the potential face loss that will occur. Utilizing a common figure of 0.5% face loss and then applying it to a 17m ID TBM will yield a trough width approximately only double that of a 5m ID TBM with similar cover in the same conditions. More significant is the magnitude of potential absolute settlement. With a 10 times greater factor in potential settlement over a 5m TBM, in order to control the settlement to that

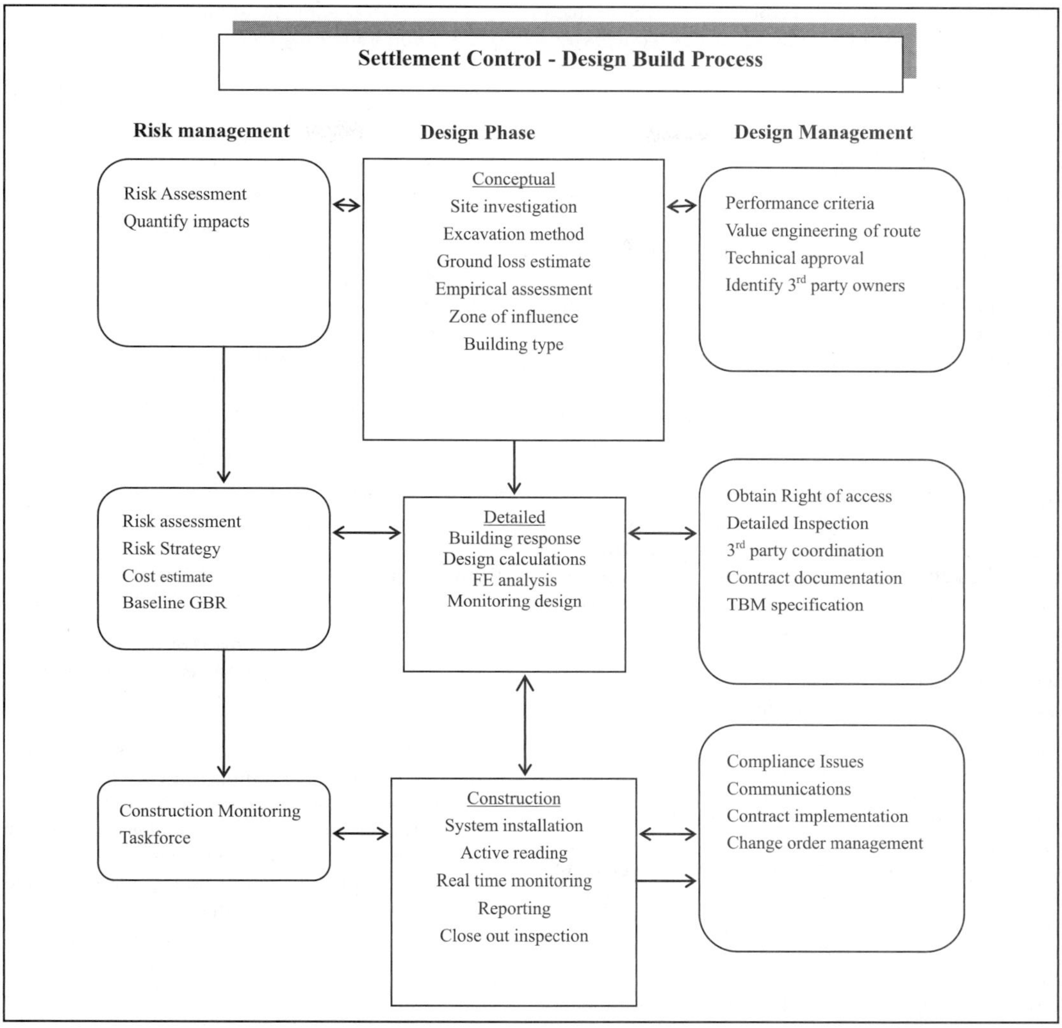


**Figure 2. Generic design process on a design-build project**

which could be considered "normal" or "acceptable" to 3rd parties, the face loss needs to be controlled by the same factor, i.e., face loss needs to be controlled to a value of 0.05%! This figure would appear to be very tight and difficult to achieve. However this is the sort of number that is theoretically required to be achieved if large diameter TBMs are to become accepted in a dense urban environment. The simple application of face loss alone as project criteria needs careful consideration, and yet placing an absolute value of movement as contract criteria may lead to an unrealistic constraint on the Contractor. For contractual purposes, the assessment and allocation of risk responsibility between the Contractor and WSDOT, 1% face loss and 1 inch criteria for absolute vertical movement of buildings were set.

Settlement on a TBM Project can materialize based on the following factors:

- Overexcavation
- Ground loss along the skin of the TBM
- Ground loss at backfill grouting location
- Machine stoppages
- Human error in operating the machine
- TBM design (TBM type, cutterwheel, etc.)
- Lack of control of face pressure.

These factors are in part able to be analyzed through 3D and 4D FE type analysis and the resultant need for enhanced pressurized face TBM systems can be specified to manage the potential impact. Mitigation measures developed on the Alaskan Way TBM include TBM cutter replacement from within the spokes of the machine, dual TBM key systems to prevent downtime, bentonite injection along the TBM skin, dual component (A+B) tail skin grouting system, comprehensive real time data collection allowing continuous monitoring and analysis by the project team. The human factor is potentially the weak link in the system, yet an over reliance on automation can have a negative effect as well. A judicious use of sensors with predetermined alarms allowing manual enquiry and override as required is the effective way to manage the operation with daily reviews through Task Force meetings involving all parties from TBM Operator to Construction Manager and with surface monitoring staff and geotechnical engineers being present.

In the design phase of the project it was identified that 158 buildings might be impacted by tunnel construction and 20 of the buildings could be impacted with slight damage arising. The RFP that was issued mandated risk mitigation measures to be applied to these buildings including foundation strengthening and/or structural strengthening and in one case demolition. In the early stages of the contract through a collaborative approach, the building that was to be demolished was able to be saved through an extensive structural remodeling; other buildings were able to have the risk of induced damage reduced by carrying out minor structural improvements. In order to incentivize the contractor's performance, an allowance has been set aside. Buildings are categorized into either Group A or Group B structures. Group A structures are required to be protected by the Contractor and costs covered within his contract price. Any damage occurring to Group B structures, the majority, will be compensated through the use of the Deformation and Mitigation repair fund which covers costs up to a value of $20M. The Contractor is responsible for costs above this value. In order to incentivize best practice, any funds that remain within the fund will be shared on a 75:25 basis in favor of the Contractor (Nielsen et al., 2011).

## Monitoring Program

The monitoring program on the Alaskan Way tunnel has already been discussed in several articles (Sowers and Caro Vargas, 2013). The figures presented below include the sensors installed in the vicinity of the portal areas (up to 200' around launch and exit shafts), as well as along the tunnel alignment within the zone of influence of tunneling as shown in Figure 3. All the instruments listed below are automatically monitored unless mentioned otherwise:

- 120 Manual Inclinometer casings
- 8 in-place inclinometers with MEMS serial chains
- 138 Vibrating wire strain gages
- 59 vibrating wire load cells
- 178 vibrating wire piezometers
- 120 extensometers with three to five vibrating wire sensors
- 143 MEMs tiltmeters
- 54 pressure liquid level cells
- 55 electrolevel beam with MEMs tilt sensors
- 10 Vibrating Wire Crack gages
- 37 Motorized Total Stations monitoring 710 3D prisms
- 200 reflectorless settlement points (Tamagnan and Beth, 2011)
- 928 Manual levelling structure monitoring points
- 150 manual near surface settlement points

Baseline monitoring for all these instruments was started six months prior to the tunneling activity at each location.

## Tunneling Parameters Integration

The increase in the availability of data from the TBM and the ability to directly correlate this data with the impact that the TBM has on the surface allows operations to be reviewed and controlled. Many of the TBM data acquisition points are related to the direct operation of the machine itself, in this category are the on/off sensors determining valve operation on TBM systems and the status of the operation of TBM systems. Data acquired from the head of the machine recording pressure, density, injection volume and muck weights give a direct relationship to ground movement and it is becoming routine to analyze this data to determine where excess face loss may have occurred. The former category provides important indirect data which can, if properly understood and analyzed by a skilled team, indicate incipient reasons for reduced ability to control face pressure which in turn will manifest itself in fluctuations in the volume, pressure and weight parameters and again in turn face loss and surface settlement. Primary and secondary data sets are shown in Table 1. Secondary data has a lower potential influence on settlement but is worthy of review in order to validate any impact.

The importance of the integration of the available systems cannot be overstated in a modern tunneling project in an urban environment. The consequences of not adequately understanding the implications arising from actions taken are significant and can be managed and mitigated against at low cost through the deployment of an adequately skilled and experienced oversight team. Since the beginning of the discussion of the monitoring program, and as specified in the initial specifications, all project players emphasized the need to incorporate the TBM parameters into the geotechnical monitoring database

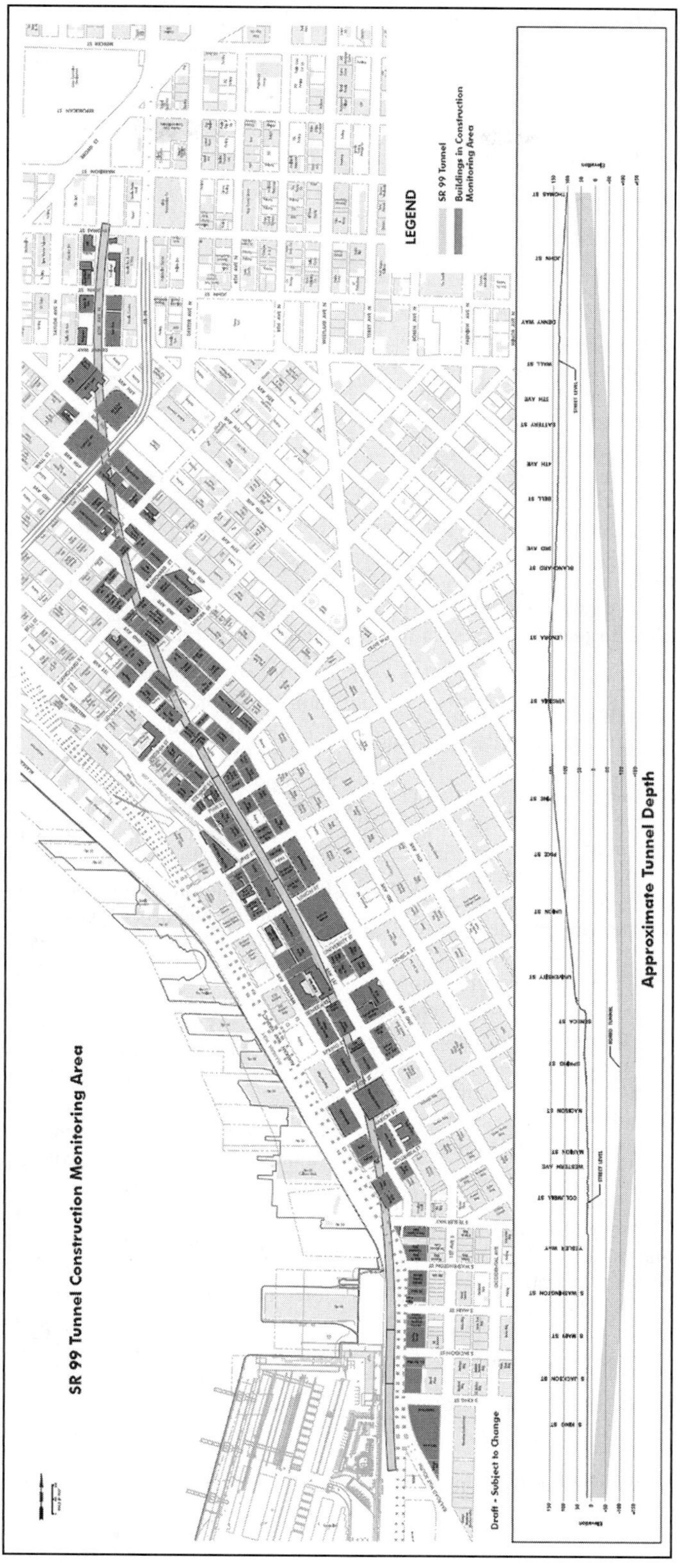


Figure 3. SR-99 tunnel construction monitoring area

**Table 1. Key TBM data influencing settlement**

| Primary TBM Data | Secondary TBM Data |
|---|---|
| Muck weight on belt | Conditioner injection to face |
| Muck volume on belt | Shove pressure |
| Face pressure | Advance rate |
| Liner grout pressure | TBM torque |
| Liner grout volume | Power usage |
| TBM skin injection pressure | |
| Screw pressure sensors | |

and allow the TBM operators to access the settlement monitoring data and the monitoring engineers to visualize on a single platform the TBM parameters and the geotechnical data. This was only achievable because of the early involvement of the monitoring firm and the TBM manufacturer in the design process. A working group composed of representatives of the TBM manufacturer (Hitachi), the monitoring firm (Soldata), the Contractor (STP JV) and the Owner's Engineer (HMMD) met regularly to define which information from the machine would need to be directly reported into the monitoring database.

The parameters that the task force decided to include were:

- Face pressure at 12 different locations (given the diameter of the machine and the potential differences along the face
- TBM ring installation progress (ring number)
- TBM Station
- Muck volume and weight
- Injected Grout volume and grout pressure
- Bentonite pressure and bentonite volume
- Shield pressure at two different locations

The second challenge the task force had to resolve was to find a way to have both data management systems "communicate" together and make sure the data from the TBM sensors would flow into the global monitoring database in a compatible way. As the monitoring firm did not have access to the machine and was not supposed to install any sensor, the work was to connect the TBM database to the Global monitoring database and send the TBM sensors' data in a compatible format. The Global monitoring software would inquiry the TBM database and retrieve the information every 5 minutes and incorporate the data into the global database.

## CONTINGENCY AND MITIGATION PLANS

Every monitoring program of a tunnel project is divided into 4 phases:

1. Definition of the Monitoring Plan during the design process
2. Installation period, when all the sensors specified in the monitoring plan are commissioned (including setup into the Global Monitoring database)
3. Monitoring period, including contingency plans
4. Retrieval, abandonment and restoration

Phases 1, 2 and 4 follow classical paths and are usually well defined in the project specifications. Phase 3, definitely the most important step for a project in providing accurate and reliable data at the right time is what makes the value and the success of a monitoring program. Several questions have to be answered during the design of the monitoring program, and a contingency plan, as well as a mitigation plan, have to be incorporated into the monitoring as essential parts. The answers to those questions are often underestimated, as they are clearly the most difficult scope activities to price, and that is where the expertise and past experience of the monitoring firms plays a big role.

### Global Monitoring and Contingency Plan

A monitoring program is not perfect. There is a common belief that an automated system where data is not acquired, processed and reported manually cannot fail. But it can. A contingency plan for a monitoring program defines the process to follow to minimize the impact of those failures. There are several types of system failures, but they can be classified in two categories, whether they involve hardware or software. On an urban tunnel project, most of the instruments and data acquisition units are located outdoor or in the ground. They can be exposed to extreme conditions, weather or construction operations related. The first questions the contingency plan has to answer is: How fast a failing or damaged sensor has to be repaired? In case of major system failure, what is the "back-up" plan? Do we have redundancy for critical sensors that are vital for the construction progress? The maximum period of time from the moment the failure is noticed until the problem is fixed or an alternate solution is implemented is generally set between 24 and 72 hours. The other kind of failure is related to software and data management. Between the instant the data is sent from the field to the moment it is reported to the end user, several steps can go wrong. IT related glitches are far more common than sensor failures and their extent usually impacts the whole monitoring program. In the case of the Alaskan Way Tunnel, the specifications only required a period of 72 hours to repair failing or damaged sensors. However, after discussions between the project players, a more detailed contingency plan was defined to solve both hardware and software problems, mainly

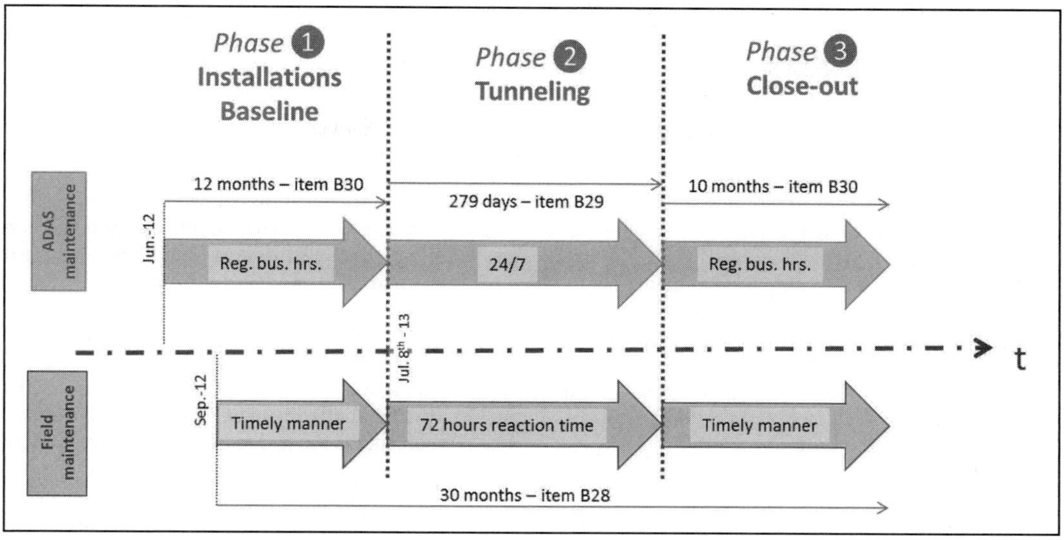


**Figure 4. Scheme of the 2 maintenance regimes on the Alaskan Way Tunnel Project**

through proactive maintenance operations. It is then composed of 2 distinct regimes:

- The field maintenance, requiring proactive or reactive work at the sensor physical location
- The database management, requiring mostly proactive work and regular checks by a data manager

The field maintenance activities have to be defined in order of priorities, usually the maintenance of sensors closer to the tunnel face are the top priority. Maintenance activities range from regular checks or calibration procedures (proactive maintenance) to full replacement of a defective installation (sensors + data acquisition unit) (reactive maintenance). One of the main constraints of the reactive measures, because they are unexpected by nature, is to perform them in a timely manner, especially when restrictive access protocols need to be followed. Everything has to be done to avoid them, starting with a well planned and executed installation. To optimize costs, as the reactive measures cannot be predicted but *WILL* happen, it is usual to define an "on-call" protocol outside of regular business hours, when field technicians or engineers might be alerted. This protocol is made possible only when the right database maintenance regime is in place. On the Alaskan Way Tunnel project, this regime is implemented 24/7 during the most sensitive period of tunneling. A data manager is constantly watching key parameters of the global monitoring system. If a major dysfunction in the hardware or software on the critical sensors is observed and cannot be resolved remotely, the data manager will inform the on-call technician, who will perform a site-inspection at the physical location of the sensor or data acquisition units.

The general maintenance regime is presented in Figure 4.

## Mitigation Plan, Construction Monitoring Task Force, and Alert Management

As seen in the previous section, a series of measures have to be operational to ensure the right data is provided at the right time. Once the data is correctly reported, it is compared with the threshold level. On the Alaskan Way Tunneling Project, each sensor type has 2 or 3 different threshold levels defined. If the data is greater than the threshold level, an automated alert is sent through email to the construction monitoring task force members. The Construction Monitoring Task Force (CMTF) is defined as follow in the contract's Technical Requirements: group of individuals responsible for the planning, implementing, and processing monitoring data; evaluating results; and making recommendations to mitigate settlement/ground deformation. With executive participation by WSDOT and Design-Builder, the Task Force has authority to direct rapid and effective changes in construction to achieve settlement/ground deformation mitigation.

The CMTF is composed of the following members:

- Construction Manager or TBM Equipment Superintendent
- Geotechnical Instrumentation Engineer/Geologist

- Monitoring Project Manager or Superintendent
- WSDOT Representative(s)
- Design Builder Engineering Design Representative.

Each of the task force members may have one or 2 backups to guarantee its functioning. Their role is to meet daily (during active tunneling) or weekly (outside the active tunneling period), review the data and evaluate the actions to implement in the next 24 hours period to mitigate any non-anticipated behavior or excessive settlement. Most of the active actions that can be implemented are related to the TBM operations. In practice, the monitoring firm presents the relevant data of the past 24 hours to the CMTF members, who will, in addition to interpreting the monitoring data, discuss:

- TBM location and how current operation is impacting structure and utility
- Current ground conditions and likely changes that may occur
- Face pressures to be used. Provide Task Force and TBM operator, a tabulation of the face support pressures by station, based upon soil, groundwater and tidal conditions that the Design-Builder has determined, will be needed for that day's Tunnel drive.
- Current deformation impacts and status of Alert and Maximum levels on affected facilities
- Any mechanical or operational issues that could impact tunnel progress
- A statement of maintenance works to be carried out
- A review of TBM data to determine trends that could indicate changes in ground conditions, TBM wear over excavation of material, and inadequate backfill grouting and gap injection
- Increase in data collection frequency
- Installation of additional instruments

## CONCLUSION

Monitoring the risks related to settlement control on a tunneling project generates a considerable amount of data. These risks are usually interrelated and in order to make informed decisions in a timely manner, the project team needs simple and easy access to the data. As the amount of data is growing, and the time allowed to make decisions is shrinking, the role of the monitoring firm is key to process and extract the relevant information from all the different sources. Preventing the project decision makers to be "swamped" by the data flow is imperative. The design process does not end with the definition of the monitoring program, and the monitoring program should not end with the project completion. In many other project areas, and especially in cost and schedule control, feedback from past projects is used to prepare better the following one. This approach based on experience reduces the risk of cost overruns and construction schedule delays. In the geotechnical field, a similar approach should be followed and ground settlement monitoring data should be used to optimize design models of future projects and mitigate the risks related to geotechnical hazards. Some project with specific construction activities are already using what is called the observational method as a standard (Caro Vargas, 2010), but this could also be applied to TBM Projects. The global data management processes described in this article should be used as a tool to optimize the design, and define more precisely key parameters, such as ground loss. Ultimately, a better connection between design and actual monitoring data could allow a reduction the global safety factors used in the geotechnical industry.

## REFERENCES


John Dunnlicliff, Remote Monitoring of deformation. Introduction, Geotechnical Instrumentation News, p. 26 March 2012.

Damien Tamagnan and Martin Beth, State of the art of recent monitoring topographic techniques, 2012.

Peck, R.B., 1969, Deep Excavations and Tunneling in Soft Ground, State of the Art Volume, 7th International Conference on Soil Mechanics and Foundation Engineering, Mexico City, pages 225 to 290.

David Sowers and Boris Caro Vargas, Instrumentation approach for the Alaskan Way Bored Tunnel, RETC 2013.

Paul Thurlow, Martin Beth and Anthony Lucas, A controlled approach to providing protection to structures located above tunnel construction works, Tunnel Construction and Piling—International Symposium, September 1999.

Damien Tamagnan and Martin Beth, Monitoring of Surface Deformation with Robotic Total Stations Using Reflectorless Measurement, Geotechnical Instrumentation News, p. 21 September 2011.

Brian Nielsen, Patrick McCormick, Eldon Abbott, Chris Bambridge, Contracting for the SR-99 Bored Tunnel, RETC 2011.

Boris Caro Vargas and Martin Beth, Observational method using real time surface settlement monitoring—The South Toulon Tunnel Project, ORVSS 42, Cincinnati OH, October 21, 2011.

# Quality and Process Control at Tunnelling Jobsites: Digital Communication Systems, Examples for Real-Time Information

**Nod Clarke-Hackston and Mathias Knoll**
VMT

**ABSTRACT:** Data is being generated in increasing quantities on most major machine bored tunnelling projects. Transmitting this data to where it can be meaningfully processed into useful information is of paramount importance to both the contractor and the owner.

Creation of data by the processes of segmentally lined, machine bored tunnel construction and the effects of that process on the ground through which it is taking place, including any structures located thereabouts, is currently the task of numerous suppliers who have monitoring capability on their individual equipment.

Being able to collect and collate these data in all of its various formats and to produce useful real-time information for all concerned is essential for a safe and successful project.

## INTRODUCTION

If information is needed then you need a method of producing the data at source, transferring this data to a storage area where it can be maintained and then processed into useful information. The key factors to consider are the information transfer infrastructure itself, the transducers that are installed in and around the project that produce the data and what that data is to be used for and how this is presented to the end user. It is necessary to migrate from the "Lots of Data; lack of information" syndrome, to a philosophy where "meaningful information" is the keyword.

## DATA COLLECTION

### Information Transfer Infrastructure

On large tunnel construction sites, a reliable and integrated communication system is not only a question of technical and economic efficiency, but also significantly helps to provide safety for all parties involved. VMT's High Adaptability Data and Emergency System (HADES) which provides an autonomous, consistent, and integrated infrastructure that is independent of, the public power supply, the domestic telephone network, or commercial mobile radio coverage but can also interface with, the mobile phone network, and telephone network and the internet is one such system. The redundant Gigabit (TCP/IP) network provides interfaces for all system components in any construction site section and ensures consistent coverage, even in difficult to access areas. The entire tunnel infrastructure can be covered using a single industrial fiber optic data line.

The base unit allows the network to be set up in a ring or tree configuration using preconfigured certified fiber-optic cable assemblies and incorporates uninterruptable power supply with surge protection where redundant connection to the network is provided by means of up to three fiber-optic ports. The key aspects of this infrastructure are sensor data flow, telecommunications, video control, access control, tracking, emergency alerts, office communication and fire-brigade response plan. The HADES software provides easy-to-handle network management using a web browser.

### Integrated Communication System

The HADES communication telephone module enables wireless communication on site with consistent radio coverage both above and below ground. It also includes a link to the mobile GSM network or to the local public telephone network using analogue, ISDN, or DSL connection which allows communication to and from external users when required. All modules and components are designed to be used in harsh environments. HADES Mobile telephone, access control, and tracking are based on proven DECT (Digital Enhanced Cordless Telecommunications) technology. Various telephone modules are available including special hard-shell mobile phones designed for use with long standby and short recharging times and may be combined with electro-larynx for use in a respirator mask. Wall-mounted IP phones have heavy-duty cord, high calling tone volume with separate calling relay contact, and visual/acoustic secondary signalling device.

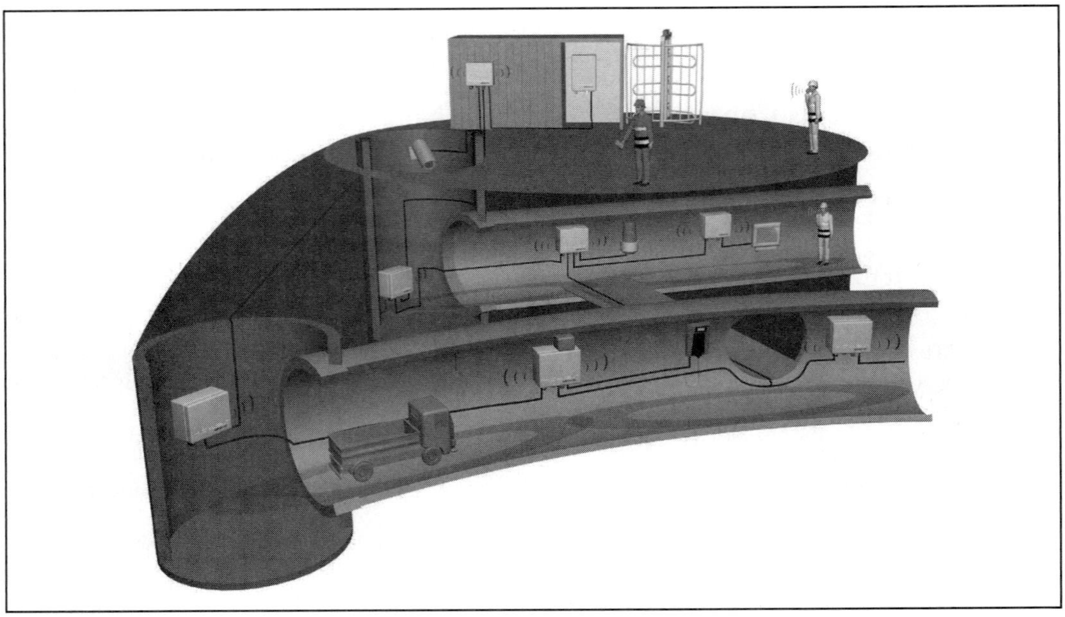

**Figure 1. Information transfer infrastructure schematic**

### Video Monitoring

Using the HADES video module, the construction site or key locations therein can be visually monitored in real-time or recorded for later review using VGA IP colour cameras or high resolution cameras for poor lighting conditions. The video software allows flexible and scalable camera management and offers database connectivity.

All services are provided using a single infrastructure giving a broadband network throughout the entire tunnel and on the TBM. PCs, cameras, and safety sensors may be positioned as desired and connected to the broadband network. At each monitored workplace within the network, camera images may be viewed and by using secure VPN access, authorized personnel have access to the construction site network from anywhere in the world.

### PLC

All data from the PLC on the machine can be transmitted to the site office, machine manufactured head office, the IRIS data base or other location for viewing, storage or control.

### Sensors

The HADES sensorics module is used to gather measured values from the machine PLC and from any individual (e.g., earth pressure) sensors. It also includes network interfaces to fire detecting equipment (and to existing 3rd party systems), such as gas detectors, and monitoring equipment.

### Access Control

Access control is achieved by either ID cards or tags that are worn on the body or by using mobile phones and HADES Messenger. Access gates, barriers, or turnstiles are of necessity incorporated into the access control system.

### Tracking

Mobile phones or HADES Messenger using proven DECT technology provide consistent tracking based on each individual cell. Each person or vehicle may be tracked in each cell and shown in the HADES software at any time. This helps to improve the safety aspect, particularly in cases of an emergency.

### Emergency

An emergency button on the HADES Messenger or mobile phone may be used to trigger an alert from any place on the construction site. Conversely, dedicated notifications may be sent from the operations centre to mobile terminals or emergency intercom stations. The UPS ensures stand-alone operation, when mains power is not available.

### Office

Office communication such as data transfer, fax, telephone switching is done using dedicated interfaces within the network.

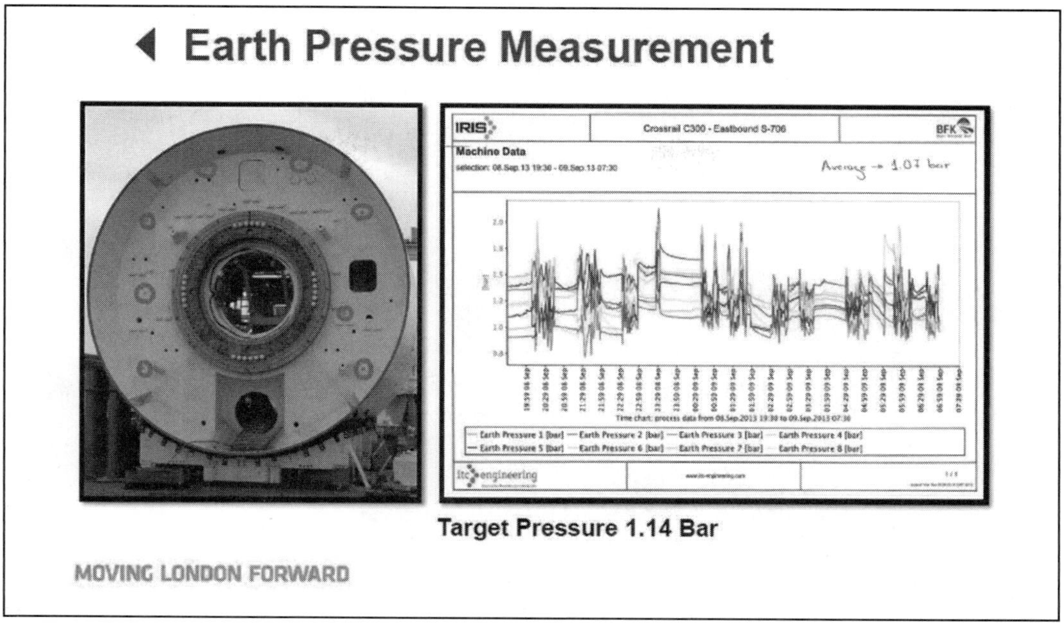


**Figure 2. PLC Data presentation ex CrossRail**

### Remote Access

Remote Desktop Virtual Private Network provides easy, secure, and flexible remote access to defined segments of the construction site network. Enabling employees and service technicians have access to the site network services and data from anywhere in the world. Remote desktop sessions via Web File Access are used to adapt configurations and provide access to defined data.

## DATA USAGE

Whilst excavation by a Tunnel Boring Machine is often considered to be a continuous process it should be more correctly described as a cyclical one where the processes of excavation, spoil removal and tunnel support should all be coordinated as they are both interrelated and interdependent. Keeping track of all these activities, how they affect both the ground and the structures located above the tunnelling activities and maintaining a traceable record are common prerequisites.

As it is unlikely that any one supplier will be responsible for all the monitoring and recording activities taking place on a given project, the method and format of the data produced will typically vary according to the specifications of the sensors and the systems of the individual manufacturers. In order to make full use of this a Data Management System for Tunnelling, IRIS (Integrated Risk and Information System) has been created. The structure of the system is a sophisticated data exchange facility that enables data types of all common formats to be integrated in to a common "web-based" database where there are manifold applications for the correlation of data from the differing data sources. Correlation of all these activities is determined by time, ring number and chainage. Clock synchronization ensures that the correct time stamp for all data is applied and navigational data and monitoring results give the relevant positional information.

There are modules available that cover all aspects of data on a tunnelling project including the machine data, navigation system, tunnel supply and equipment logistics including complete Segment Documentation System (SDS), measurement systems both inside the tunnel and on surface structures along the course of the tunnel, geological and geotechnical investigations, as well as project design.

Dynamic Information Systems such as IRIS are the most important part in judging the entire situation—where correlation between events in the past, the present and the future can be undertaken.

During TBM tunnel construction for example, the ability to make fast decisions in potentially critical situations are very important; these decisions are dependent on the amount and quality of available data and information. IRIS greatly assists project management in the provision of these. It is however of paramount importance that the system is fully established at the commencement of the project so that all parties "buy into the system from day one." The objective of the system is to improve the overall performance of the project, to be able to access and interpret the

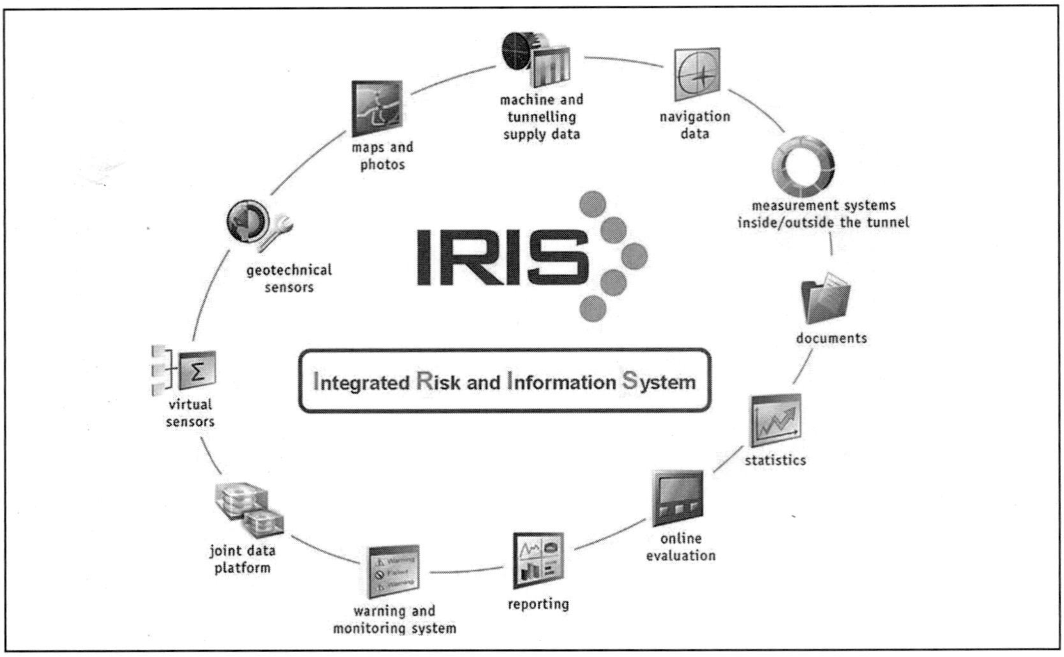


**Figure 3. Data management system for tunnelling**

data in real time, not a means for each side to score points of the other. An agreement should be made between the contractor and the owner client at the beginning of any project where it should be decided what information shall be collected and to whom this information is available—job specific access should be made for those who will benefit from access such as Mechanic, Electrician, Surveyor, TBM operator, project engineer, project manager, etc.

Data Management System for Tunnelling are able to integrate the data provided and give composite displays and reports of the information and trends to those directly responsible for controlling the tunnelling and to give suitable warnings when preset limits are approached or exceeded. These warnings can take the form of simple screen messages through to email and SMS messages sent to the relevant engineers or managers. All warning messages are recorded and acknowledgment that the warning has been received is logged, thus giving complete transparency to the tunnelling operations.

On projects where multiple sites are all working simultaneously, links to a central control room (manned 24 hours per day) would be implemented to enable the owner's management team to have an overall view of all activities throughout the project and to respond to any unforeseen situation that may arise. Furthermore, areas at risk of settlement are detected at an early stage and are displayed perspicuously, in order that corresponding measures can be initiated if necessary.

All relevant data can be inserted either automatically by an electronic interface or by manual input either during the tunnelling process on the TBM, in the tunnel, from the office or from remote locations off site. The option to have handwritten notes or photos that can be introduced into the data base is also included.

The systematic storage and automated reporting provide reliable continuous documentation of the entire tunnelling process and the ability to use this information in planning future projects will prove to be very beneficial.

## SEGMENT DOCUMENTATION

One subset of the Data Management system for tunnelling is the Segment Documentation System (SDS). This is used for the production and installation of segments and other ready-mixed concrete parts on site. All production methods, industrial goods, and production tools are identified and effectively documented using barcode labels or RFID tags. The SDS can use a production forecast to minimize the necessary storage capacities and thus enables the user to reduce storage and labour costs.

Consistent logging is carried out from all manufacturing steps, for quality assurance purposes. Thanks to automatic reporting and statistics

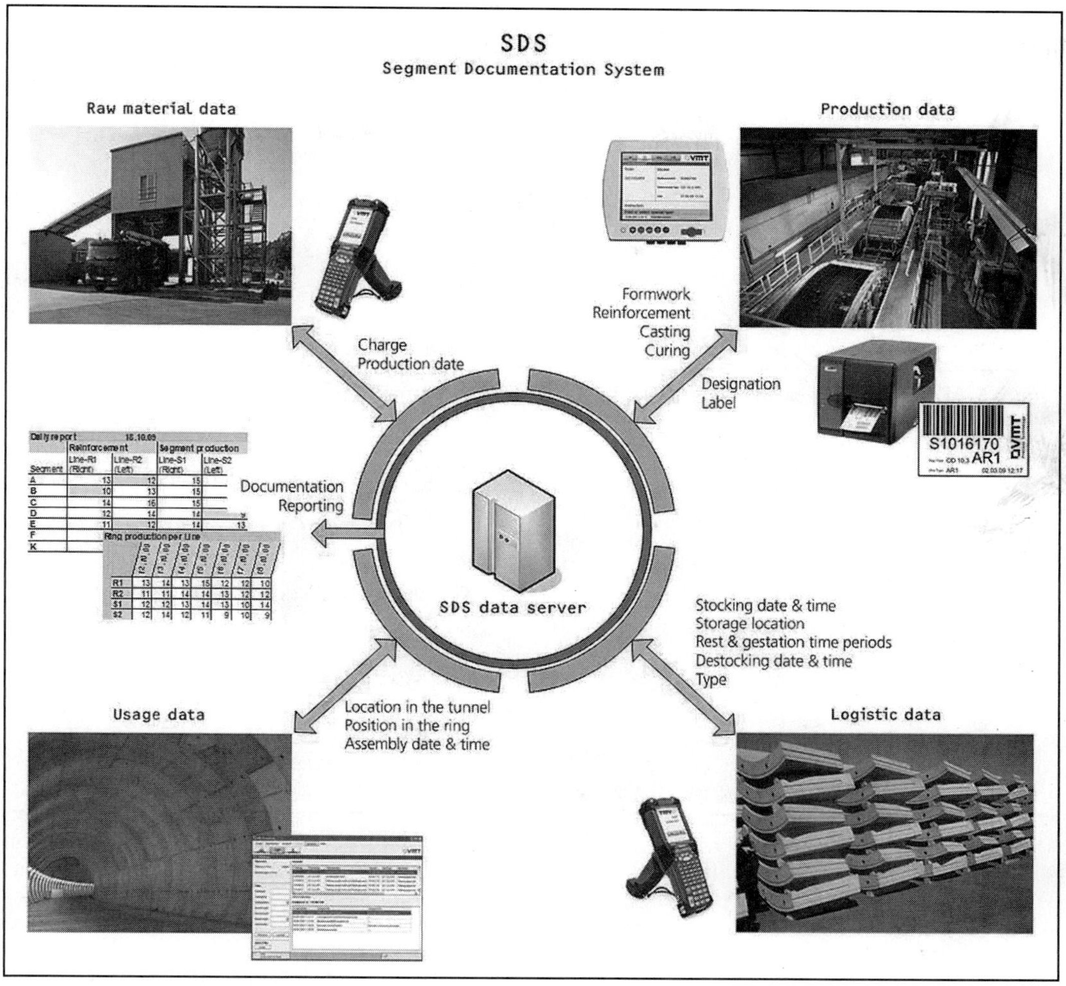


**Figure 4. Segment documentation system**

capabilities, the end user gets an overview of the entire manufacturing process and can track deadlines and technical specifications. Similarly, SDS provides long-term archiving and document management. If required, production control can be accessed at any time to make modifications.

SDS is modularly structured, has a multilingual menu, and may be individually configured. The system is able to handle both stationary as well as carousel production arrangements.

SDS supports various storage options. In the basic version, the inventory is not managed, i.e., the SDS does not know individual storage locations. In a customized version, the inventory is managed so that the SDS defines and records all inventory activities. To achieve this, the SDS knows all storage locations and rules and is able to clearly inform on storage locations, contents, and utilization. Using this configuration, the SDS may also be used to control the gantry cranes. In addition to the production module, the site module combines and enhances the previously recorded data by key figures from transporting and inventory management on site. Furthermore, the database records storage times and locations of the segments and their transportation routes up to their installation in the tunnel.

An example of a SDS that has been integrated with the Web based "Data Management System for Tunnelling" is the Koralm tunnel. The Koralm tunnel is part of a major project in Europe the "Baltic–Adriatic axis" where between 2013 and 2016 two 10m Aker Wirth Doubleshield TBM's will excavate 17.1km and 15.6km of the twin tubes within the central section of the Koralm tunnel totalling 33km in length. The lining will be precast concrete segments of a 6 + 0 type plus one additional invert

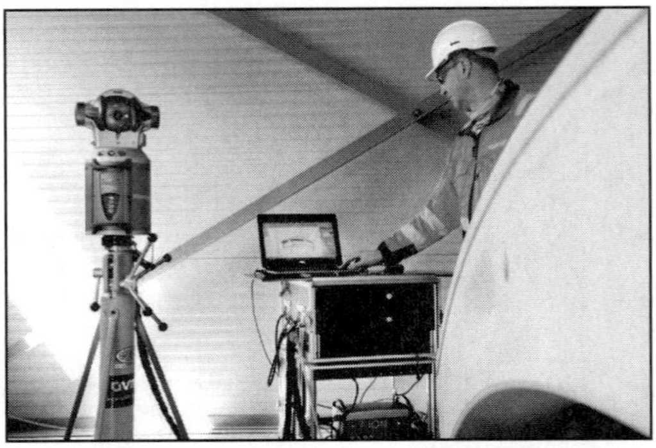

Figure 5. 3D full plane measurement using LaserTracker instrument

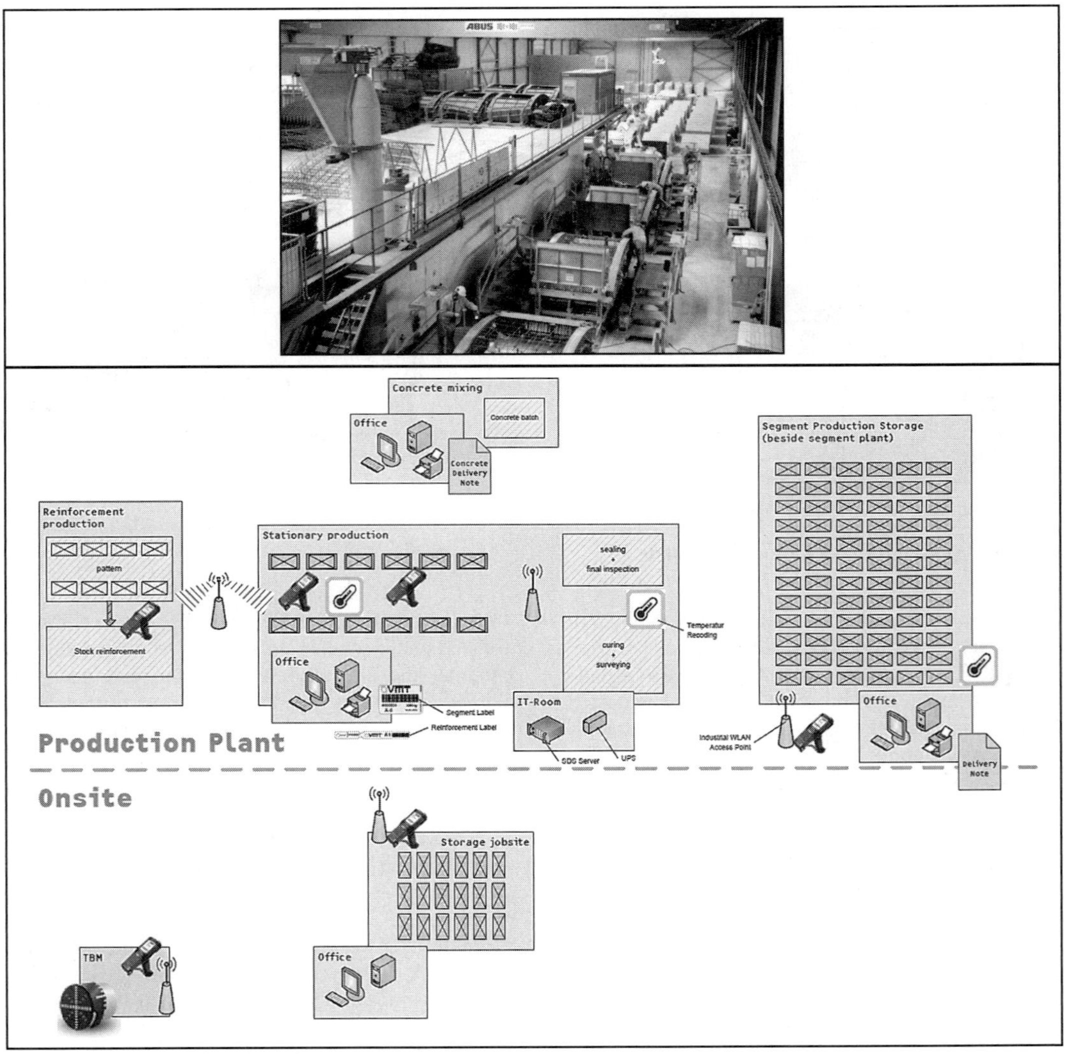


Figure 6. Full traceability of the entire production process

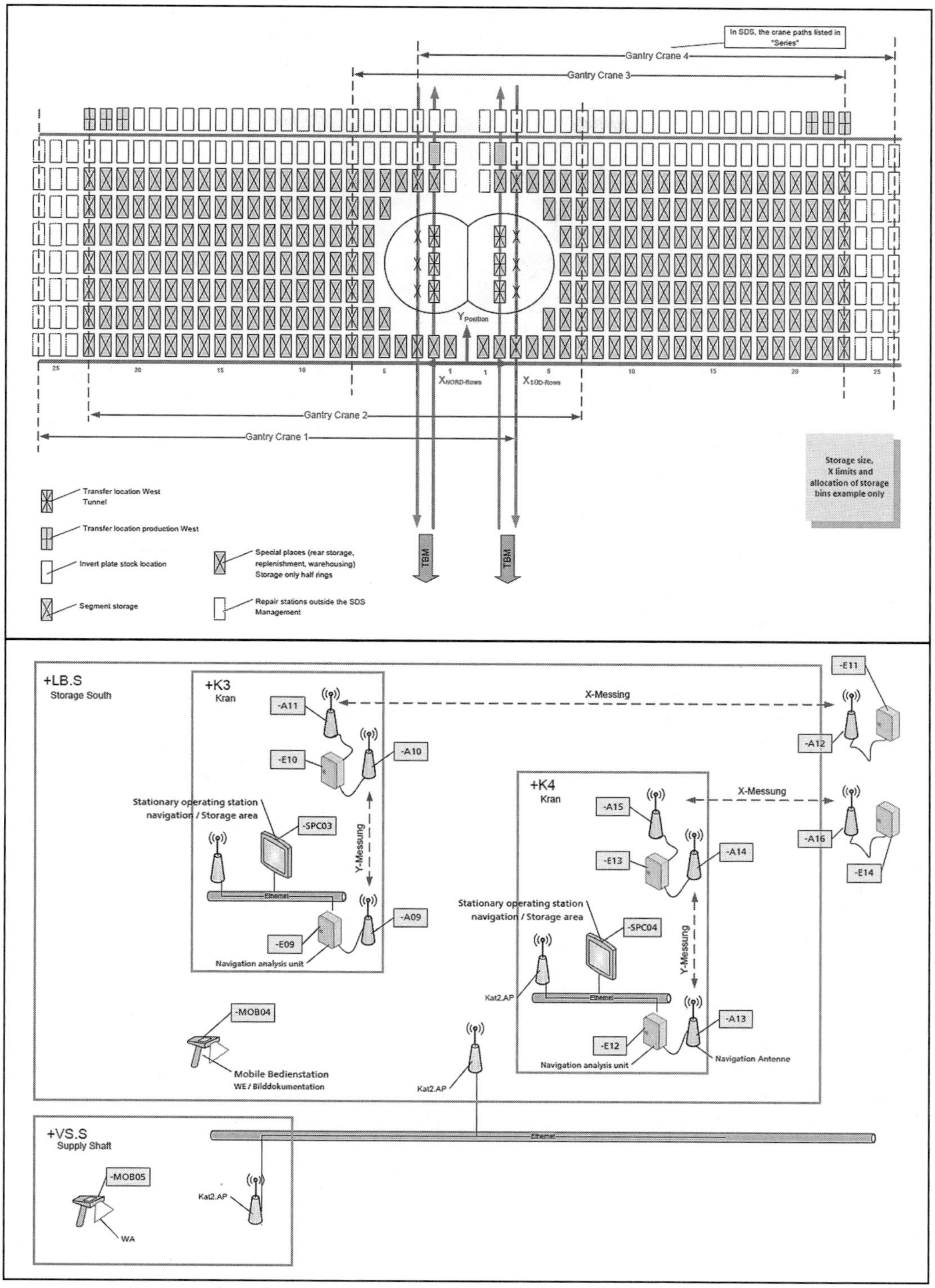


**Figure 7. Traceability within the stock yard**

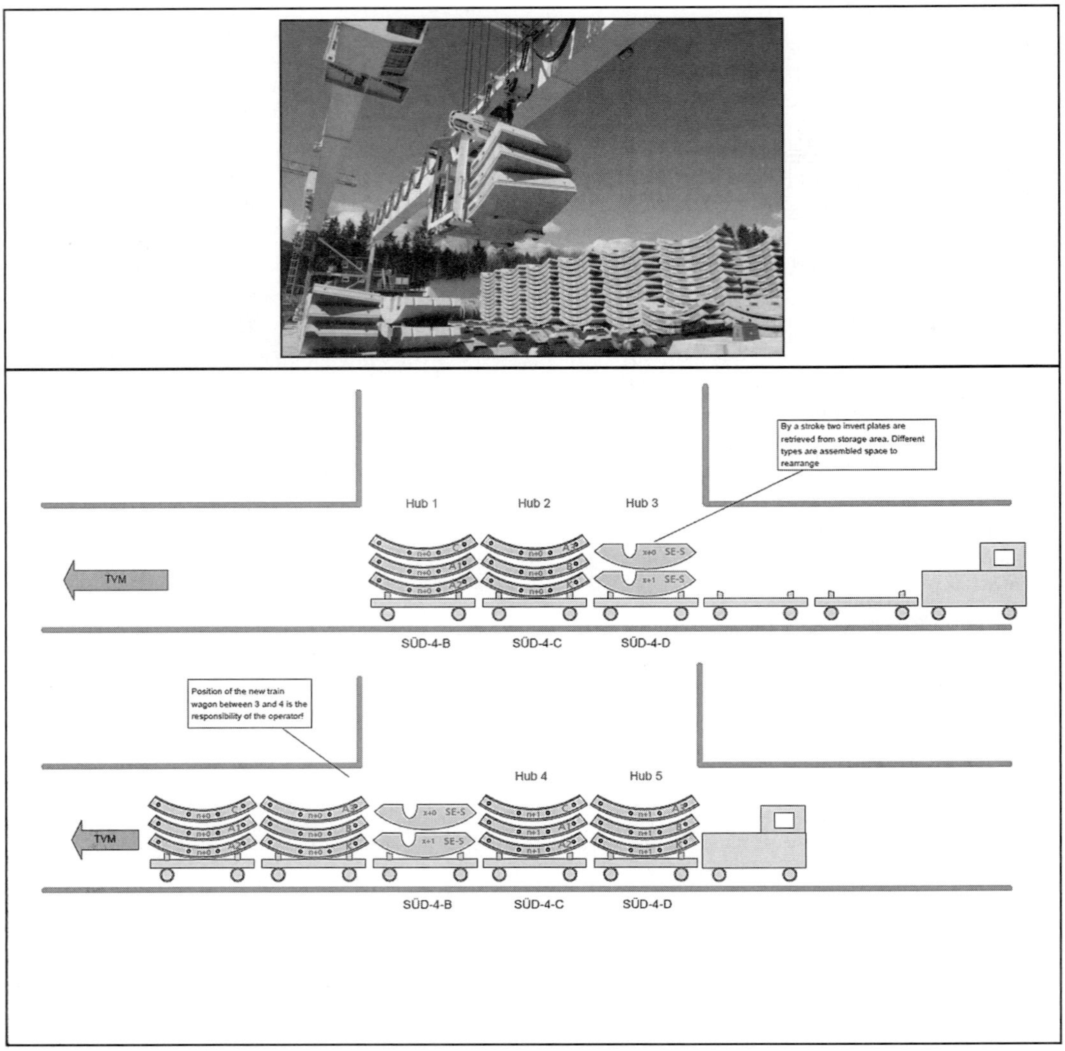


**Figure 8. Precise segment selection and loading**

plate. Interestingly due to the variable geological and hydrological conditions there are 16 differing types of reinforcement and 3 different types of concrete mixes. Production is by carousel and the reinforcement cages are manufactured on site. There are 2 carousels each with 7 sets of moulds in use and a complete set in the measurement area for exchange whenever it is needed. 3D Full plane segment dimensional checks are carried out according to the following criteria: 1st 10 segments from each mould then every 50th segment from each mould however in case of out of tolerance values are discovered the measuring frequency re commences with 1st 10 segments then every 50th after the mould has been adjusted.

The precast site storage area has a very limited footprint and access to the tunnel is via an onsite 60m deep shaft, however in order to supply the most appropriate ring type to the ring-build crew there is an involved process to go through. Firstly geophysical probing attempts to determine the geology ahead of the TBM which in turn enables the rebar and concrete type to be determined. The ring type is then chosen through the ring selection programme and this is forwarded via the IRIS Data Management System to the SDS which accesses its database and directs the gantry cranes to precise location in the stockyard. The complete ring of segments is then loaded in the correct build order and sent to the TBM in time for the Ringbuild.

## CONCLUSIONS

The combination of a comprehensive Information Transfer Infrastructure, a full tunnelling Data Management System and the specific segment documentation system enables the on time delivery of the correct ring type and the full traceability including the final installed location of each and every segment on complex projects and helps to maximise the efficiency of the tunnel construction and for the long term data storage for the design life of the tunnel.

## REFERENCES

Clarke-Hackston. N. (2001). "Geometric Assessment of Steel Moulds & Concrete segments" Presentation at International Centre for Geotechnics & Underground Construction. Sargans. December 2001.

Messing. M. (2002). "Lasertracker für die Vrmessung von Tübbingen" Interdisziplinäre Messaufgabe im Bauwesen. Beiträge zum 56. VW-Fortbildungseminar im September 2002 in der Bauhaus-Universität, Weimar.

Katja Remus; & Nod Clarke-Hackston BSc (Hons) C. Eng. MIMMM "The part of Tunnelling Information Systems in Mechanized Tunnelling. Article for Tunnels & Tunnelling 2009.

# Innovative Infrastructure Inspection Technologies

**Andre Hawks**
ILF Consultants Inc.

**Bernhard Spoerr**
DIBIT Measuring Technique USA Inc.

**ABSTRACT:** With aging infrastructure becoming a common problem throughout the United States, time efficient and innovative inspection technologies are needed in order to facilitate infrastructure inspections that can be done rapidly and in-depth with minimal impact to infrastructure operation as cost effective as possible. Innovative infrastructure inspection technologies will be discussed. Technologies examined will include the recently studied high speed nondestructive testing (NDT) and tunnel scanning methods to facilitate mapping defects on the surface, within, and behind tunnel linings.

## INTRODUCTION

Infrastructure in the United States is in desperate need of repair and upgrading. ASCE's Report Card For America's Infrastructure released in March 2013 gave an overall grade of D+ (p. 3, DiLoreto) across 16 categories in comparison to a D given in ASCE's 2009 Report Card. Although an improvement, a D+ is nothing the current richest and most powerful country in the world should brag about. And yet with the consistent low grading of America's infrastructure systems, owners are specifying old technologies in their Tunnel Inspection RFP's including recent examples in Pennsylvania and Oregon.

Like the personal computer so has infrastructure inspection technology advanced and yet these new technologies are not being specified by owners. This paper will examine innovative inspection technologies and hopes to open a dialogue with infrastructure owners and government agencies like the FHWA and it's National Tunnel Inspection Standards (NTIS) which are currently being developed and are close to becoming required. And not only should new technologies provide more in depth information, but inspection systems should also be developed that have minimal impact to the infrastructures operation. The inability to shut down infrastructure is often why infrastructure inspection is delayed due to the inability to close down a Tunnel or a Bridge for more than 5 hours in the middle of the night or the inability to shut down at all such as critical drinking water supply lines like the San Francisco Public Utilities Commission's Old Irvington Tunnel.

The key aspects new inspection technologies need to provide are as follows:

- Minimizes impacts to operation
- Provides comprehensive condition assessment
- Allows comparison to future inspections to measure the change in condition
- Allows inspections to be safely performed
- Cost effective

## NON DESTRUCTIVE TESTING & RAPID SCAN SYSTEMS

In recognition of the issues discussed above the Transportation Research Board's SHRP 2 program has funded a research program titled "R06(G) Mapping Voids, Debonding, Delaminations, Moisture, and Other Defects Behind Or Within Tunnel Linings" being headed by Texas A&M University. This paper will not cover this study in depth as this studies own reports, but will give the reader of this paper valuable information on the inspection technologies being evaluated and studied under the R06(G) research project.

One of the main goals of the SHRP 2 study is the development of both rapid screening and in-depth inspection whereby rapid screening is first completed which helps identify areas of greatest concern which are then inspected with the slower indepth methods. The performance criteria that was set by the expert panel for both the rapid screening and indepth methods established that the NDT methods should detect a defect within or immediately behind the tunnel linings that have a minimum surface area of 1 $ft^2$ and any defect needs to be located within 1 ft. of the actual location on the tunnel lining. Additionally, NDT methods should identify delaminated area voids up to 4 inches deep as measured from the lining surface with an accuracy of within 0.25 inches.

To determine which NDT methods meet this criteria the SHRP 2 research team conducted research on the Eisenhower Memorial Tunnel, Hanging Lakes Tunnel, and No Name Tunnel in Colorado, the Chesapeake Channel Tunnel in Virginia, and the Washburn Tunnel in Texas to study tunnels constructed with different construction methods. All three tunnels in Colorado were built using drill & blast methods while the Chesapeake Channel Tunnel was built via the cut and cover method and the Washburn Tunnel was constructed via the immersed tube method. Therefore their subway covered all but a TBM tunnel and therefore represents a majority of the tunneling methods typically used. All five tunnels use tile over their final lining also known as tile concrete which is typical of highway tunnels.

The following indepth and rapid screening techniques were able to meet the criteria set in the SHRP 2 study and are briefly described below:

- **Air-Coupled Ground Penetrating Radar (GPR):** Uses discrete electromagnetic pulses sent into a structure and then captures the reflections from layer interfaces in the structure. At each interface within the structure, part of the energy is reflected and part of the energy is transmitted. This difference is used to calculate the layer thickness and dielectrics. Defects can only be detected if significant moisture or air pockets are within the defects.
- **Thermography (handheld thermal camera):** Uses significantly improved thermal cameras to capture temperature differences within the lining possibly indicating defects within or behind the lining.
- **Tunnel Scanner (discussed in detail below):** Uses photogrammetry and or terrestrial laser scanning to capture surface conditions and insitu geometry.
- **Ground Coupled GPR:** Similar physical phenomena as Air Coupled GPR except needs to be in contact or very close to the lining surface when data is being collected. Difference is Ground Coupled GPR can detect more accurately and can also detect rebar.
- **Ultrasonic Tomography:** Uses an array of ultrasonic transducers to transmit and receive acoustic stress waves which are then converted into a three dimensional volume with a digitally focused algorithm. The intensities of the returned waves are color coated to show discontinuities (voids, cracks, delaminations,etc) with distinct wave speeds.
- **Ultrasonic Echo:** Uses an ultrasonic transducer to send and receive ultrasonic pulses from the same side of the test object by the same or two separate transducers which is then measured to locate discontinuities and the thickness of the object.
- **Portable Seismic Property Analyzer (PSPA) Ultrasonic Surface Waves & Impact Echo:** Ultrasonic Surface Waves uses the time difference in surface wave propagation to determine the modulus. The differences in velocity with wavelength is then used to generate a dispersion curve whose variation in velocity identifies discontinuities. Impact Echo uses the stress waves generated by impact then converted into a frequency domain by a fast Fourier transform algorithm to detect changes in amplitude and shape of the frequency curve to indicate discontinuities. Both these methods are simultaneously conducted with the PSPA using a solenoid-type impact hammer and two high-frequency accelerometers.

Some of the methods above are for use in rapid scanning and some for more indepth scanning. Therefore based upon the goal to perform a rapid initial scan then followed by an indepth scan of specific areas identified, the following procedures were suggested by the R06(G) research team (p. 5, Wimsatt et al.):

1. "Collect thermal images and air coupled GPR data on the tunnel lining. Air coupled GPR data should be collected every foot along the tunnel lining. Thermal images can be collected every foot as well; however, the equipment covered in this report can collect data at a spacing determined by the camera operator or tunnel inspector. This data should be collected ideally on the same day; however, it can be collected separately. The thermal images should be collected when the air temperature is rising or falling; areas of possible defects may show up better in the thermal images. The data from any of these devices can be obtained at a walking pace (around 1 mph or 1.61 kmh). Air coupled GPR data can be obtained at much higher speeds, but the geometry and features in tunnels may make it difficult to operate the equipment at speeds much greater than 1 mph.
2. Analyze the data from the scanning devices above. Select areas for indepth testing based on the GPR surface dielectric results, thermal images, and observed surface distresses that are of concern to tunnel inspectors.

3. Conduct indepth testing with the ground coupled GPR and either the ultrasonic tomography, ultrasonic echo, or portable seismic property analyzer device. The choice of equipment could be based on the cost and the type of defect to be detected (tile debonding, delamination, and voids) The ultrasonic tomography and ultrasonic echo devices may be more appropriate for measuring and mapping defects greater than two inches from the tunnel lining surface. The ultrasonic tomography device is more expensive than the other two devices; however, it has the capability to provide more information in the field about such defects. The portable seismic property analyzer may be more appropriate for determining the limits of shallow defects.
4. Evaluate the data collected from these devices."

Additionally, Table 1 was provided by the R06(G) research team (p. 3–4, Wimsatt et al.) for each of the

**Table 1. Summary of NDT devices**

| Device | Accuracy | Detection Depth | Deterioration Mechanisms Detected | Tunnel Lining Types | Other Information |
|---|---|---|---|---|---|
| Air-coupled ground penetrating radar (GPR) | Locates defects within 1 foot of its actual location | Does not measure depth, but indicates areas of high moisture or low density (high air voids). Such areas may represent problems within or behind the tunnel lining | Tile the bonding delaminations air-filled voids, water-filled voids, moisture intrusion | Concrete, tile-lined concrete, and shotcrete | This is scanning tool that can indicate where to conduct testing with in depth devices. |
| Thermography (handheld thermal camera) | Locates defect within 1 foot of its actual location | Does not measure depth, but can indicate tile bonding delaminations up to 1 inch, voids up to 3 inches | Tile the bonding delaminations air-filled voids, water-filled voids, moisture intrusion | Concrete, tile-lined concrete, and shotcrete | This is scanning tool that can indicate where to conduct testing with in depth devices. |
| SPACETEC scanner | Locates defect within 1 foot of its actual location | Does not measure depth but can indicate tile bonding, possibly delaminations up to 1 inch, and possibly voids up to 3 inches | Tile the bonding delaminations air-filled voids, water-filled voids, moisture intrusion | Concrete, tile-lined concrete, and shotcrete | This is scanning tool that can indicate where to conduct testing with in depth devices. Testing can only be conducted through a service contract. |
| Ground-coupled (GPR) | Can determine defect within 10% of the actual depth without reference cores, 5% of cores are available | The device can possibly detect defects at any depth within or immediately behind tunnel linings. However, specimen testing indicates it cannot locate 1 square foot voids in steel plates behind tunnel linings | Delaminations, air-filled voids, water-filled voids, moisture intrusion | Concrete, tile-lined concrete, and shotcrete | Experienced personnel are needed to interpret defect locations and depths from the GPR scans. Specimen testing indicates it cannot locate 1 square foot voids in steel plates behind tunnel linings |

(table continues)

**Table 1. Summary of NDT devices (continued)**

| Device | Accuracy | Detection Depth | Deterioration Mechanisms Detected | Tunnel Lining Types | Other Information |
|---|---|---|---|---|---|
| Ultrasonic tomography | Concrete: voids within 0.5 in. shallow delaminations within 0.75 in. Shotcrete: air-filled voids within 0.7 in. water-filled voids within 1.21 in. shallow delaminations within 1.88 in. | Can detect defects up to 8 in. deep based on specimen test. Tunnel tests indicate it can detect possible defects up to 20 in. deep | Delaminations and voids | Concrete, tile-lined concrete, and shotcrete | May not be effective for measuring the defects that are 2 in. or less from the lining surface. May not be accurate enough for measuring defect depths in shotcrete |
| Ultrasonic echo | Comparable to ultrasonic tomography system based on tunnel testing with both devices. Past experience indicates it also can measure tunnel lining thickness within 3% of the actual thickness | Comparable to ultrasonic tomography system based on tunnel testing with both devices | Delaminations and voids | Concrete, and shotcrete | May not be effective for measuring the defects that are 2 in. or less from the lining surface. May not be accurate enough for measuring defect depths in shotcrete. Tunnel tests indicate problems with using this device on tiles. |
| Portable seismic property analyzer (PSPA) ultrasonic surface waves and impact echo | Ultrasonic surface waves: about 15% of the actual depth for defects up to 6 in. deep.<br><br>Impact echo 10% for deep delaminations greater than 6 in. deep | Ultrasonic surface waves to 6 in. deep<br><br>Impact echo up to 18 in. deep | Delaminations and voids | Concrete, tile-lined concrete, and shotcrete | Maybe difficult to quantify the depth of defects that are shallow or extensive. May not get good results when testing and very rough concrete surfaces, oily surfaces, and severely curved surfaces |

Notes:

1. Detection depth is the maximum depth a discontinuity is able to be detected.
2. SPACETEC Scanner is the specified brand of tunnel scanner used by the R06(G) research team. The available manufacturers are DIBIT, SPACETEC, Amberg Technologies, and Geodata Group. The use of the tunnel scanners is mainly procured through a service contract where the equipment is rented.
3. The detection depth would not be that different in comparing cast in place concrete vs shotcrete as long as the quality of placement is similar.

methods tested outlining key information for each method.

## TUNNEL SCANNING SYSTEM

3D imaging in tunnels is becoming a valuable NDT tool for gaining accurate and comprehensive information of the condition of the tunnel lining surface. Although this technology does not allow seeing what is behind or within the tunnel lining, it does provide high resolution 3D images of common deficiencies such as cracks, spalls, or leakage on the surface, in relation to the tunnel alignment and stations. By performing a tunnel scan as a first inspection measure, the overall condition of the tunnel surface can be documented and the data used to locate visible deficiencies on the surface. These detected deficiencies and their locations are stored in a deficiency database and can be effectively used as a baseline for further NDT. The advantage of performing a tunnel scan over a manual visual inspection lies in the quality of information gained from the scan data as well as overall time and personnel savings. Both of these advantages relate to aging infrastructure because often high quality information on the aging infrastructure is lacking or not available and shutdown time and inspection costs are always needed to be kept to a minimum.

Terrestrial laser scanning is on its way to become one of the standard technologies for object acquisition in surveying engineering. The possibility to obtain a dense three-dimensional point cloud of the surface of the object under investigation immediately excels other traditional surveying techniques. Combining the high spatial resolution of photogrammetric imaging with the excellent capability of measuring 3D space by laser scanning bear great potential for both data acquisition and compilation. With the help of such hybrid scanning system which combines lidar and photogrammetric technology, it is possible to obtain an accurate point cloud and high resolution digital images simultaneously resulting in a true color rendered 3D scan model. This technology can be very well utilized for tunnel inspections, as it delivers a comprehensive and dense illustration of the tunnel surface. State-of-Art software allows performing a virtual inspection from the desktop generating tunnel maps and inspection reports in such detailed way then never before.

Two types of tunnel scanning systems that utilize hybrid sensor technology have proven to be very effective for road and rail tunnel inspections. The more traditional way of scanning is the stop & go or more commonly known as static scanning method. The heart of this system consists of a terrestrial laser scanner utilizing the time-of-flight measurement principles. The scanner emits an infrared laser beam which is moved in a plane by a rotating mirror resulting in a dense cloud of points. The success rate of laser scanning depends on the reflectivity of the surface. Common materials used for tunnel lining (i.e., concrete, shotcrete, tiles) or just rock tunnels have proven to be well suited for laser sensors. The instrument as shown in Figure 1 as example is the Riegl LMS Series which is tilted 90 degrees to enable more efficient data collecting in tunnels. By aligning the scanner axis with the tunnel axis, data can be collected 360 degrees in form of individual rings or sections rather than spheres. The static scanning system is positioned close to tunnel center line and moved along it, scanning sections of the tunnel with each set up. At the end, multiple scans from each set up are merged together to form the 3D point cloud. Scan data registration and geo-referencing occurs through reflective targets that are positioned on the instrument and in the view of the sensor. This method is called indirect geo-referencing. Reflective targets are detected on the scan and assigned with their coordinates in a local scanner based coordinate system. A total station is then used to measure every target of each set up and obtain their spatial coordinates in the external coordinate system. Through a procedure called geo-referencing, the registered scans are transformed from the scanner based local coordinate system to an external (geodetic) coordinate system. This allows the scan data to be integrated into other geospatial data.

The stop & go system has been used very successfully in road tunnels but can also be adapted to rail tunnels with a platform that can be pushed along rail tracks. The data acquisition rate depends on how much tunnel can be scanned with one set up and how fast the targets can be measured for each scan. Due to a fully automated target measurement process and a robotic total station which remotely obtains the coordinates of each of the scanner targets, the time for one scan can be as fast as 3 minutes. A point accuracy of 5 mm can be achieved using this method of measured points on the tunnel surface in relation to tunnel control points.

The second hybrid scanning system used for tunnel inspection is a more recent development with a different and more advanced data collection approach then the stop & go scanning method. This system is based on the kinematic laser scanning method as opposed to static terrestrial laser scanning. During kinematic laser scanning, the system changes its position during the data acquisition and therefore scans the tunnel surface from a moving position.

The system is installed on a rail vehicle, hand pushed by the operator. A dense array of data is

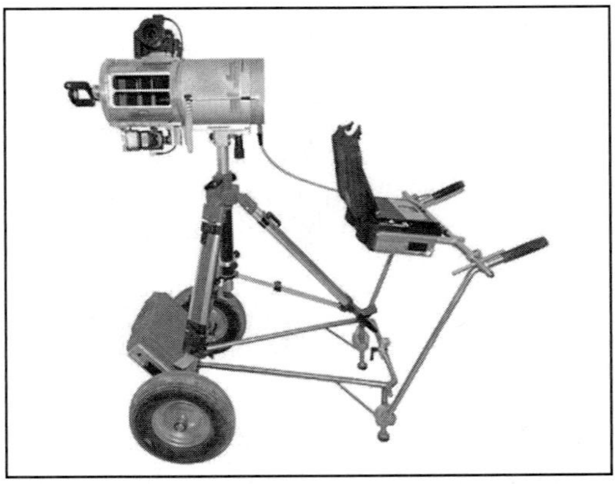

**Figure 1. Example of stop & go (static) scanning system**

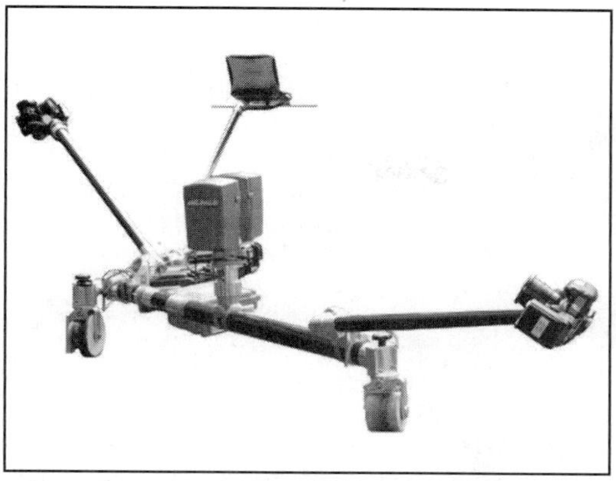

**Figure 2. Example of kinematic scanning system**

acquired as the scanner is walked along the tracks recording detailed information of visible defects on tunnel linings. The heart of the kinematic system is a phase shift laser scanner with a 360 degrees sensor view. The instrument as shown in Figure 2 as example is the FARO scanner. The scanner emits an infrared laser beam which is moved in a plane by a rotating mirror. The results are transversal profiles. From the forward motion of the rail vehicle, a 3D point cloud forms in the appearance of a line rather than a sphere, hence the term line-scanner. This method of scanning differs in a way that there are no single scans that have to be registered in the end to complete the scan model. The line-scanner collects one continuous data stream during a steady movement of the platform. The number of data points is dependent on the acquisition rate of the laser scanner, the rotation rate of the deflection system and the velocity of the platform. Additionally, a signal intensity value provides information about the reflectivity of the surface at each target point. The kinematic system also includes odometer to measure the trajectory of the moving platform and a tilt sensor for the determination of the systems 3D state. The data of these sensors are stored in a data logger which is used for scan orientation. The scans of a kinematic system are oriented relative to the existing tracks and can be accurately referenced to tunnel stations. Therefore no control points and reference targets need to be installed, which eliminates additional time spent for target measurement by a total station. Instead, the rail track centerline serves as a baseline and for scan registration. The absolute achievable accuracy after adjustments is 10 mm.

The obvious advantage of the kinematic system is its high scanning speed. This comes into play in tunnels with high frequency traffic such as subway tunnels in which complete shutdown times have to be held at a minimum. Past projects have proven that

**Figure 3. Scan data processing stages, form point mesh to laser intensity image to true color rendered scan model**

a distance of 1 mile can be scanned in just 3 hours which is significantly faster than using a static scanning system.

In addition to the laser scanner, digital cameras are installed on the scanner or in the case of the kinematic system on strategic locations on the platform. This enhances the quality of the final scan model in terms of rendering the point cloud using high resolution imagery obtained during scanning. The digital

**Figure 4. True color rendered scan model of a road tunnel**

SLR cameras are configured to capture overlapping pictures automatically during scanning.

### Data Processing

The type of data acquisition method of hybrid scanning systems provides dense three-dimensional point clouds and high resolution images. The first step to recreate the tunnel digitally is to build a reference model based on design drawings including alignment, profile and the typical cross section of the tunnel. The 3D registered point cloud and surface reflectivity values from the laser scanner are projected onto the reference model. For the image registration, stereo matching occurs. The registered images are then aligned with the laser scanner data and refined through data point matching between the laser reflectivity texture and the digital images. The result of this process is a complete 3D model of the tunnel surface, rendered high resolution images. See Figure 3.

### Data Analysis and Mapping

Increasingly faster computer technology enables software to handle high density data and load the rendered scan model in real-time. This allows the engineer to virtually walk through the tunnel and observe, investigate and analyze the state of the tunnel lining surface in a 3D environment. This means that a tunnel inspection can be done form the desktop and pre-inspection overview can be made. The user can virtually walk along the tunnel alignment or in a free 3D orbit mode always in reference to the actual tunnel stations. It is also possible to compare initial and follow up scans and determine deficiency changes over time (i.e.: crack growth) or overlay them and check for deformations. The software and scan model is also a great visualization tool to present the state of the infrastructure to their owners. See Figure 4.

Data mapping takes place in interactive digital mapping software. The software offers the ability to display the full colored digital tunnel model with real time cross section display and dimensions at various locations including statistical values such as volume. The final generated digital tunnel model is imported into the software for data manipulation and analysis in a 2D mode. The 2D viewer mode unfolds the tunnel image and displays it on a plane with an orthogonal viewing point. This allows for easier data analysis. Data reduction can be performed to filter or mask distorted and excessive data to ensure a quality digital tunnel model for further inspection and mapping.

The goal of a tunnel scanning survey is to be able to observe the tunnel surface, detect visible defects and record it using digital mapping tools. TIS (Tunnel Information System) is a tool that allows mapping of deficiencies such as cracks, spalls or moisture. TIS provides a flexible object database that enables a 3D localization of objects within the tunnel. All mapped objects can be assigned with the respective inspection terminology key. Each defect was stored into the database including an identification number, dimension, description and location. See Figure 5.

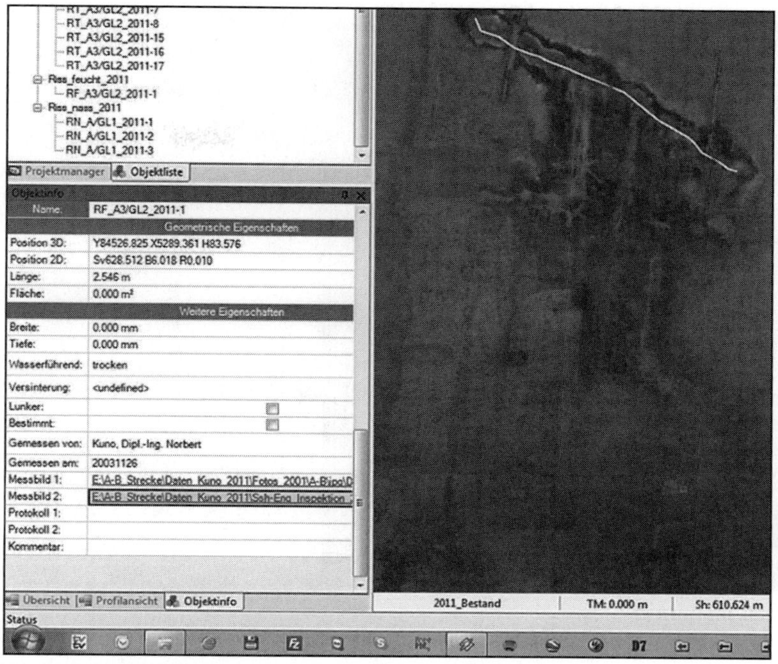

**Figure 5. Categorizing of deficiencies**

For crack detection, the TIS module has a semi-automated crack detection function for precise localization and dimensions. Initiated by user-given estimations of crack start and end point, the software initiates an automated crack tracing on the basis of local line fitting and mathematical observations in both directions of the crack. Several restrictions, rules and optimization criteria to find the correct crack trajectory are taken into account. The crack tracing process creates polygons of the extracted cracks and feeds them into TIS. This method is applicable to various types of surface texture. See Figure 6.

### Tunnel Maps

Tunnel maps are common types of reports to portray the existing condition the tunnel lining. Tunnel maps can be generated from the scan model illustrating the rendered point cloud with the tunnel invert, side-walls and crown "folded out" to enable representation of the tunnel perimeter in a flat map, on which all documented features (i.e., deficiencies, joints) are recorded with respect to both their longitudinal and radial position. See Figure 7.

### Future Tunnel Scans

The result from a tunnel scan is a comprehensive digital representation of the tunnel in its existing condition. The data obtained through the scanning system reflects the state of the tunnel in a modern and very detailed way. The true color rendered scan model can be visualized on current computers with very powerful software and a variety of results can be derived from it. What makes this hybrid scanning technology stand out to other terrestrial laser scanner is the integration of high resolution imagery, which enhances the quality of the scan model to a point, which makes it efficient for investigating tunnel lining surfaces. It is also a great tool for monitoring deficiencies over time. As soon as a baseline scan is done, follow up scan data can be compared against each other to determine timely changes. This technology has been used in conjunction with tunnel condition assessments in Europe for the past 10 years and its use continues to grow there on rail, subway, road and water tunnels. Examples of recent tunnel scanning for US projects includes the Boston, San Francisco and St Louis subway tunnels.

All technology presented in this paper are products of Dibit Measuring Technique USA, (DMT) Inc. DMT is a worldwide operating tunnel surveying and scanning firm specializing in 3D imaging of tunnel surfaces.

Other companies which provide similar products and solutions are: SPACETEC Datengewinnung GmbH, Amberg Technologies and Geodata Group.

Figuure 6. Crack mapping

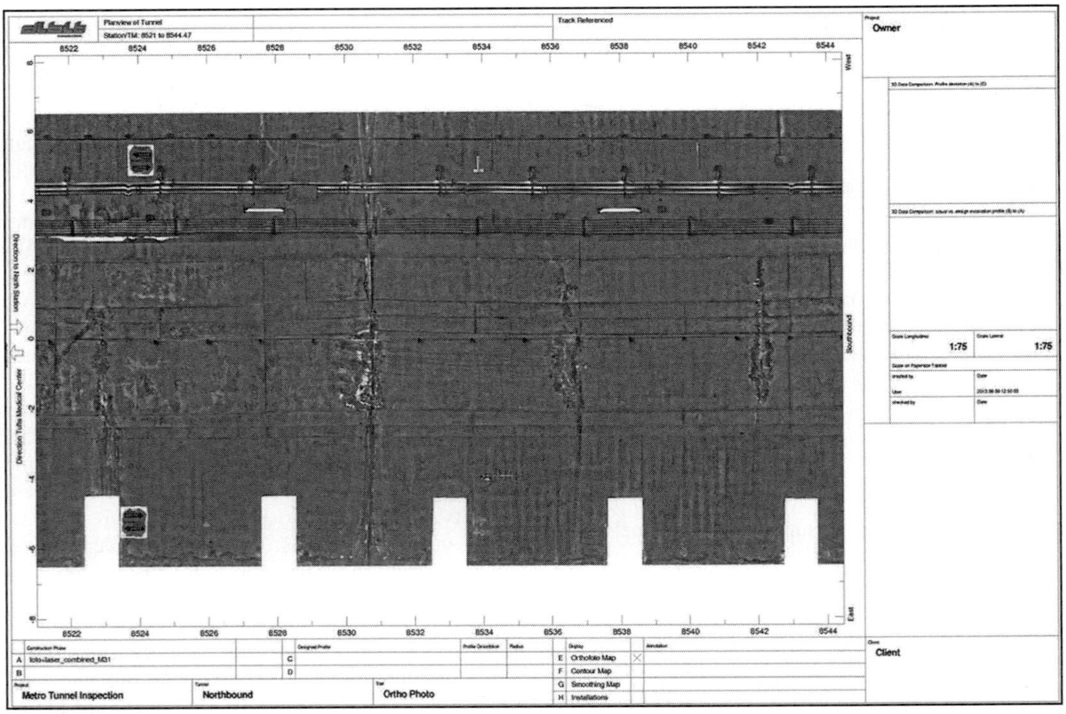


Figure 7. Tunnel map

## CONCLUSIONS

As Engineers, we should always strive to use the latest and most efficient technologies for our clients who are tasking us to rehabilitate or inspect a piece of infrastructure that is likely severely overdue as demonstrated by ASCE's overall infrastructure rating of a D+. Every category that states the needed expenditures but one in ASCE's report outlines expenditures in the billions with Drinking Water over a Trillion. These numbers are overwhelming to say the least, but if we do nothing our nation will begin to digress in quality of life and economic impacts will be felt. In fact ASCE states that if nothing is done disposable income will be impacted \$3,100 a year for each American family.

Each of the inspection technologies discussed meet all of the key aspects new inspection technologies need to provide in order to gain the interest of infrastructure owners and the engineers implementing the inspections so badly needed. The current standard of sending in large teams of inspectors that rely on the accuracy of the human eye often in miserable conditions is slow, costly, and does not provide all of the digitized and quantitative data provided by the methods discussed. The drawback of using these technologies is that their use in tunnel inspections are in their infancy and therefore are foreign to most owners and tunnel professionals, but with exposure to the benefits of these technologies it is hoped that the methods discussed above become the norm.

With the technologies discussed efficient and indepth infrastructure inspections can be performed efficiently and cost effective to begin to thoroughly analyze and document the state of critical pieces of infrastructure so that repair and upgrade schemes can be developed and implemented. Our current approach of "fix it later" is not working and is putting our society on the edge of economic disaster that once felt, is likely insurmountable to overcome.

## REFERENCES

DiLoreto, G., *2013 Report Card For America's Infrastructure, American Society of Civil Engineers*, March 2013.

Wimsatt et al., *Mapping Voids, Debonding, Delaminations, Moisture, and Other Defects Behind Or Within Tunnel Linings*, The Strategic Highway Research Program 2 Transportation Research Board of the National Academies, December 2012.

# Influence of Geological Conditions on Measured TBM Vibration Frequency

**Michael Mooney, Bryan Walter, John Steele, and Daniel Cano**
Colorado School of Mines

**ABSTRACT:** This paper examines TBM vibration as a source of information about geological conditions. An EPB TBM was outfitted with accelerometers to monitor vibration during excavation of the University Link light rail tunnel project (U230) in Seattle Washington. Impact-response testing of the TBM indicated that significant signal over a wide range of frequencies transfers from the cutterhead where vibration due to ground interactions emanate, to the bulkhead where sensors can be installed. Analysis of the vibration data collected during excavation indicates that both amplitude and frequency content appear to be influenced by TBM operating parameters and by geological conditions.

## INTRODUCTION

The vibration characteristics of a system when subjected to external stimulus have long been used as a way to monitor the system itself. For example, the field of vibration based condition or health monitoring uses the measured vibration response and its changes to identify wear and damage of rotating shafts, wind turbines, hydroelectric turbines, bridges and buildings. The vibration characteristics of earth construction equipment have been used in the same manner and also to interrogate the condition of the ground with which the equipment is interacting. Examples include intelligent vibratory soil compactors and smart drilling (Mooney & Rinehart 2009, Richard et al. 2002). In the former case, the elastic stiffness and compacted state of the soil is estimated based on measured drum vibration. In the latter case, the rock hardness and strength is estimated based on measured drill bit vibration. The goal of the study described in this paper is to develop a similar approach where TBM vibration can be used to assess geological conditions.

The underlying premise is that the measurable vibration characteristics of equipment interacting with the ground will be influenced by the ground properties under certain operating conditions. The ground conditions, therefore, can be estimated by back-analysis that uses either statistical or physical models of the ground/machine interaction. For example, a vibratory drum operating on soil can be physically modeled with lumped masses, springs and dashpots (see Figure 1). The model predicts the contact force vs. deflection response that is a function of the roller parameters, operating frequency and amplitude, as well as the assumed ground stiffness and damping. Through a process of matching experimentally measured vibration response to modeled response, the ground stiffness and damping are estimated. This estimation of soil stiffness and compacted state is provided continuously and in real time, and is now routinely performed in earthwork construction practice.

If a physical model is difficult to develop, statistical approaches can also be used. For example, in smart drilling, empirical relationships between measured vibration amplitudes and frequencies with rock types and stress conditions have been developed through statistical analysis. The rock type can then be estimated based on these empirical models.

With this previous work in mind, there is considerable potential benefit to using TBM vibration as a continuous means of characterizing both the ground conditions and the TBM condition. One significant limitation to TBM tunneling, particularly in pressurized face conditions, is the inability to log or otherwise document the geology through which the TBM is excavating. There is currently no way to catalog the as-built geological conditions that could be very beneficial for lifecycle engineering of the tunnel. Further, the lack of documentation of excavated geology makes it impossible to assess the accuracy of the geotechnical baseline report. This has significant implications on differing site conditions, contractor claims, resolution of disputes, etc. Finally, real time characterization of geological conditions could serve to optimize TBM performance and avoid damage, e.g., through identification of boulder impacts, changing ground conditions that require different ground conditioning, face support, etc.

This paper examines the potential for TBM vibration based assessment of geology and ground conditions in general. An experimental program was

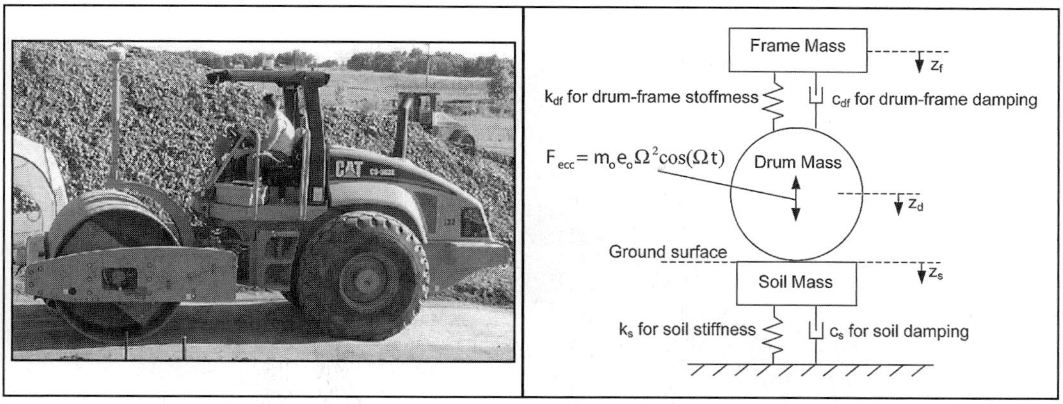


**Figure 1. Dynamic model of vibratory drum roller interacting with the soil**

conducted wherein a 6.44 m diameter Hitachi Zosen earth pressure balance (EPB) TBM was outfitted with accelerometers. Vibration data was collected during TBM excavation of the University Link Light Rail Tunnel project (U230) in Seattle Washington. In addition, the vibratory response of the TBM was explored prior to tunneling through impact—response testing to explore the transfer of vibration from the cutterhead through to the bulkhead.

## PROJECT BACKGROUND

Vibration was monitored during EPB TBM operations on the University Link Light Rail Tunnel project (U230) in Seattle, Washington. Monitoring was carried out during southbound tunneling over approximately 1 km from the Capitol Hill Station to Pine Street Stub Tunnel (right to left on Figure 2). The complex soft ground geology (see Figure 1) is divided into fluvial deposits, glacial deposits, lacustrine and glaciolacustrine deposits, all of which have been glacially over-ridden and are therefore very overconsolidated (Irish 2009). The overburden varied from a minimum of 4.2 m under Interstate 5 (Station 1046) to over 40 m (Station 1060). The entire tunnel alignment has a steep, curved downhill grade (4.8%).

The 6.44 m diameter Hitachi EPB TBM was instrumented with four triaxial accelerometers mounted on the bulkhead of the TBM near the main bearings (Figure 3a). Reliable instrumentation and high bandwidth data acquisition are not currently feasible at the ideal location of the cutterhead. The bulkhead was selected with the assumption that vibration occurring at the cutterhead would travel through to the bulkhead. The orientation of the triaxial axes and their relationship to the TBM are shown in Figure 3b. The accelerometers have a bandwidth of 0–600 Hz and were sampled at 2 kHz. TBM operating parameter (OP) data was accessed from the Hitachi PLC every 10 seconds. The OPs used in the study include cutterhead torque (T), axial thrust (F), cutterhead rotational speed (N), average face pressure ($\sigma$), and advance rate (AR). Further details about the instrumentation and data acquisition system deployed can be found in Walter (2013).

## TBM IMPACT-RESPONSE BEHAVIOR

Impact-response testing was performed on the TBM prior to excavation to characterize vibration signatures and their transmission from the cutterhead to the bulkhead inside the TBM. Testing was performed by striking the cutterhead at locations 0–7 shown in Figure 4 with a hammer. A roving accelerometer (denoted $a_1$) was magnetically mounted next to the impact location (shown in Figure 4) to capture the input signal to the cutterhead. Vibration was recorded at fixed cutterhead locations $a_2$ and $a_3$ as well as fixed bulkhead locations $a_1$–$a_4$ (Figure 3). The aim behind this impact-response testing was to quasi simulate boulder interactions at the cutterhead (cutting tools impacting boulders during advance and rotation), and then assess how vibration carries through the main bearing to the bulkhead where sensors can practically be mounted.

One advantage of hammer impact testing is that the stiff hammer contacting a steel cutterhead introduces broadband vibration frequency content. The analysis of response observed provides an indication of which frequencies pass through and which are mechanically filtered due to the makeup of the TBM frame. An example set of bulkhead response time histories from five sequential hammer impacts on the cutterhead (same position) is shown in Figure 5. The signals are clear, repeatable and convey vibration response that extends for 200 ms.

All time domain signals were analyzed via discrete Fourier transform to explore the frequency content of the cutterhead vibration and bulkhead

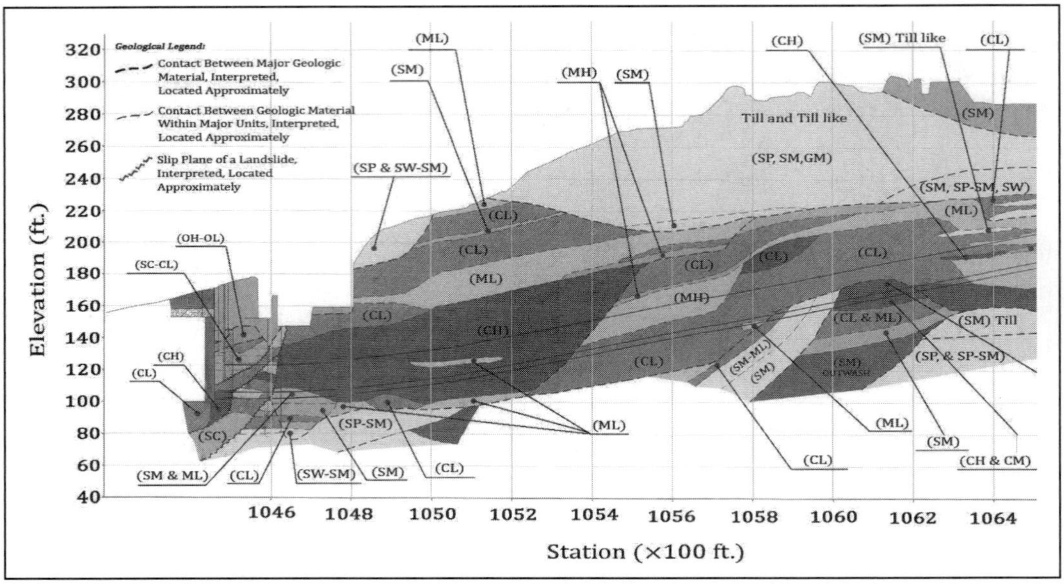


**Figure 2. Estimated geological conditions along U230 alignment (based on geotechnical data report by Irish 2009)**

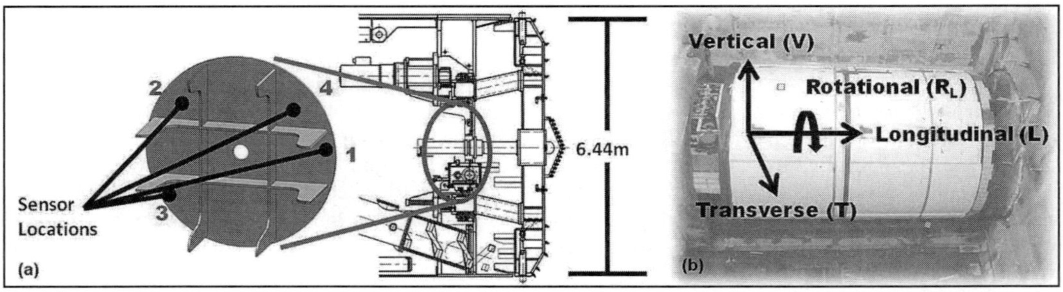


**Figure 3. Schematic of accelerometer locations: (a) side view of the TBM, highlighting the bulkhead of the machine where triaxial accelerometers 1–4 were mounted; (b) accelerometer coordinate system used in the study**

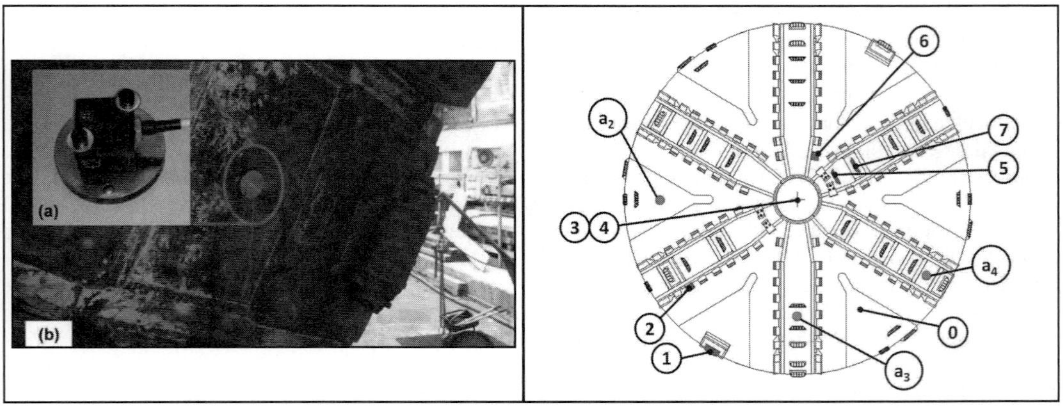


**Figure 4. Schematic of Hitachi 6.44 m diameter cutterhead and locations of impacts (0–7) as well as locations of accelerometers ($a_2$–$a_4$)**

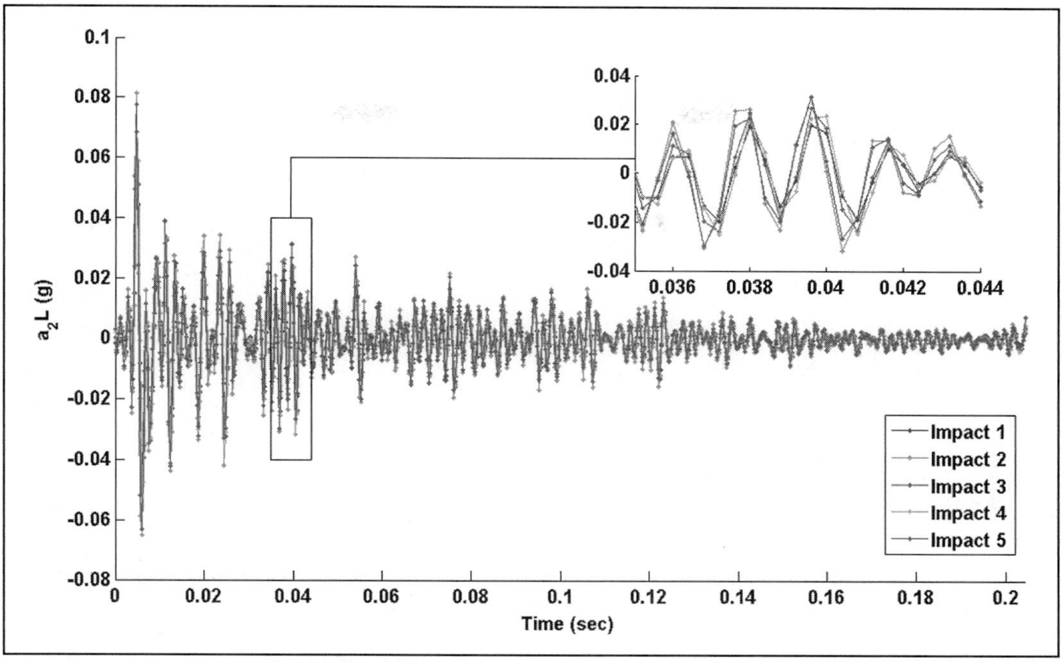


**Figure 5. Measured longitudinal bulkhead vibration ($a_2L$) resulting from five impacts at position 7 of cutterhead**

vibration. As one example, Figure 6a shows the amplitude portion of the frequency response spectrum of longitudinal cutterhead vibration due to an impact at position 4 while Figure 6b shows the amplitude spectrum of longitudinal vibration at the bulkhead. The noise floor amplitude of the accelerometers was found to be 0.1 mg across the frequency spectrum. Therefore, the majority of the signals shown in Figure 6 are well above the noise floor. Figure 6a illustrates some observed resonant modes of the cutterhead. The largest magnitudes occur at 260 and 380 Hz, while smaller magnitudes occur at 180, 340, 400, 460 Hz and 580 Hz. These are considered natural or resonant modes of the cutterhead wherein any excitation signal amplitude (from the impact) would be amplified. Figure 6b illustrates a number of frequencies where significant vibration amplitude was measured at the bulkhead, e.g., at 260, 340, 380, 400, 460, 530 and 580 Hz. Some of these bulkhead frequencies are similar to the cutterhead peak frequencies and some are not. A comparison of cutterhead and bulkhead vibration magnitudes at similar frequencies reveals the amount of signal attenuation or amplification.

An informative way to assess what frequencies pass from the bulkhead through the TBM main bearing to the bulkhead is through calculation of a transfer function, the ratio of bulkhead (output) FFT to cutterhead (input) FFT. The transfer function amplitudes for the transverse, vertical and longitudinal bulkhead vibration due to impact at cutterhead position 4 are shown in Figure 7. The amplitude at each frequency in Figure 7 is reported in dB where the reference signal is the cutterhead vibration. Therefore the amplitude of the transfer function quantifies how the input signal has changed from cutterhead to bulkhead. For interpretation, –10 dB and –20 dB imply that the vibration amplitude has decreased by a factor of 3 and 10, respectively from cutterhead to bulkhead. Conversely, +10 dB and +20 dB imply that the vibration amplitude has been amplified by a factor of 3 and 10, respectively, from cutterhead to bulkhead. An amplitude of 0 dB implies that the input and output amplitudes are the same. The noise floor for the transfer function is –110 dB; therefore, all amplitudes measured are well above the noise.

Figure 7 shows a variety of responses with vibration reduction and amplification depending on direction (T, V or L) and frequency. Transverse vibrations are either amplified or remain unchanged for many frequency bands up until 500 Hz. A number of frequency bands in the vertical and longitudinal directions also show negligible reduction or significant amplification. While the transfer functions are complex, the main takeaway is that considerable and measurable vibration signal generated by cutterhead

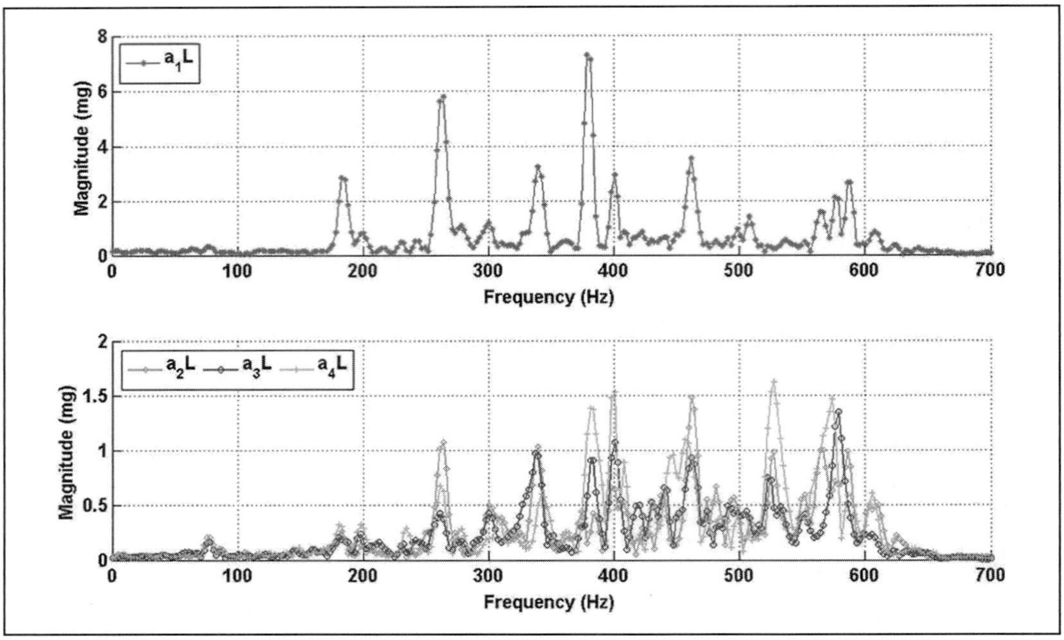


**Figure 6. (a) Amplitude response spectrum of cutterhead position 4 longitudinal vibration (impact at 4); (b) the resulting amplitude response spectra at the bulkhead in longitudinal direction**

interactions with the ground can reach the bulkhead where accelerometers can be easily mounted.

## VIBRATION RESPONSE DURING U230 EXCAVATION

TBM vibration was continuously recorded at bulkhead sensors $a_1$–$a_4$ throughout the southbound drive. To first paint the big picture, the measured response over the 1 km long drive is shown in Figure 8. Here, the root mean square (rms) amplitude of time domain bulkhead vibration recorded by sensor $a_1$ is reported in the L, T and V directions. To provide some perspective regarding tunneling conditions and performance, the key TBM operating parameters (OP) measured are also shown, namely the advance rate (AR), cutterhead torque (T), cutterhead rotation rate (N), average face pressure ($\sigma$), and axial/thrust force (F). The geology and overburden along the alignment are also shown (see Figure 2 for a geology legend). Figure 8 illustrates that the measured bulkhead vibration amplitude clearly varies along the alignment as do the measured OPs. In fact, a visual assessment of Figure 8 shows that vibration amplitudes often change when OPs change. The broader question is whether vibration characteristics change with ground conditions.

To more closely examine the frequency content of TBM vibration measured, data from individual rings were investigated. Figure 9a illustrates the recorded OP data and Figure 9b the $a_1$ vibration data recorded during excavation of ring 502 (Station 1057+13). According to the geological profile (Figure 2), the TBM was mining through full face low plasticity clay during ring 502. Both the raw vibration records and rms amplitude are shown in Figure 9b. OP parameters indicate excavation with a constant advance rate = 100 mm/min through homogeneous ground as evidenced by the fairly constant torque, thrust and face pressure. The vibration response echoes this homogeneous behavior and constant operation over the 21 minutes of ring excavation. Figure 9c presents the results of a joint time-frequency of $a_1$V vibration. Here, FFT analysis was performed on 10 second segments and stitched together throughout the 21 minutes of excavation. The resulting spectrogram presents both the frequency content and amplitudes at each frequency as a function of time. Figure 9c reveals distinct frequencies in the bulkhead vibration response, with the largest amplitudes present at 460–470, 390 and 290 Hz. Consistent with the OP data and time history vibration data for ring 502, the frequencies and amplitudes remain constant throughout excavation. These frequencies are in the range of those observed during impact-response testing presented earlier but do not match them exactly. This is understandable given that impact-response testing was performed on the TBM without soil pressing on the face, an empty chamber, and with no pressure on the shield or cutterhead.

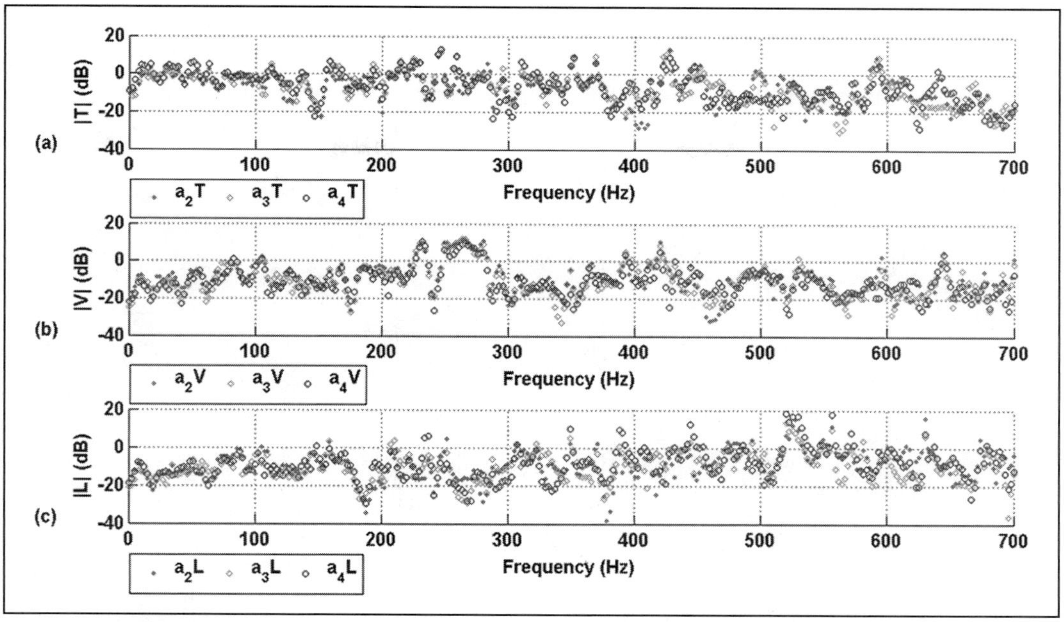


**Figure 7. Transfer function amplitudes vs. frequency for determined from cutterhead impact/vibration at position 4 (input) and bulkhead vibration (output). Plots (a), (b) and (c) show the transfer function amplitudes in the transverse, vertical and longitudinal directions, respectively.**

Figure 10 explores the amplitude and frequency content more closely by zooming in on the TBM start-up phase at the beginning of ring 502 excavation. The distinct frequencies of 470, 390 and 300 Hz are achieved after 12 sec when the cutterhead rotation speed has reached its constant 2.2 rpm. Prior to this, it is clear that the dominant frequencies increase with cutterhead rotation speed. A similar behavior is evident during the cutterhead rotation ramp down of ring 507 excavation (Station 1056+88). Though not shown here, frequency content and TBM OP values were similar from ring 502 excavation through ring 507 excavation.

Bulkhead vibration frequency domain analysis was carried out for data along the southbound alignment. Ten rings were selected in different geological conditions for analysis (see Figure 8 for locations). Ramp up and ramp down operations were not included so that the analysis could focus on steady state data. The dominant frequencies and their amplitudes of $a_1V$ vibration observed during excavation of these ten rings are presented in Figure 11. Vibration frequency is normalized by cutterhead rotation speed to enable direct comparison of observed frequencies. Figure 11 shows at a broad scale that the dominant frequencies are consistent throughout data during each ring excavation. These frequencies are evident in primarily three clusters with central frequencies of approximately 280, 180 and 130 Hz/rpm. Individual dominant frequencies and their amplitudes within these clusters were found to be variable across the rings and appear to be sensitive to soil type. For example, TBM vibration while excavating in CH material (rings 665, 586, 425 partial) did not exhibit dominant frequencies within the three cluster areas while TBM vibration in other soil types exhibited clear dominant frequencies. The nature of these frequencies and their relationship with soil type requires further study and verification. Further, the influence of TBM OP values on these frequencies and their amplitude also require more complete characterization.

## DISCUSSION

While there is significant measurable TBM vibration during excavation, the nature of the vibration, particularly the frequency content, is quite complicated. The observation of dominant frequencies in the 300–500 Hz range, the range that is consistent with harmonic modes of the cutterhead and bulkhead based on impulse-response testing and modeling efforts (see Walter 2013 for finite element analysis results), suggests that the TBM vibration is influenced by free vibration response during excavation. The strong influence of cutterhead rotation speed on the dominant frequencies, however, suggests that the TBM is exhibiting forced vibration response, where the cutterhead rotation speed serves as the root forcing frequency. That the observed frequencies are five

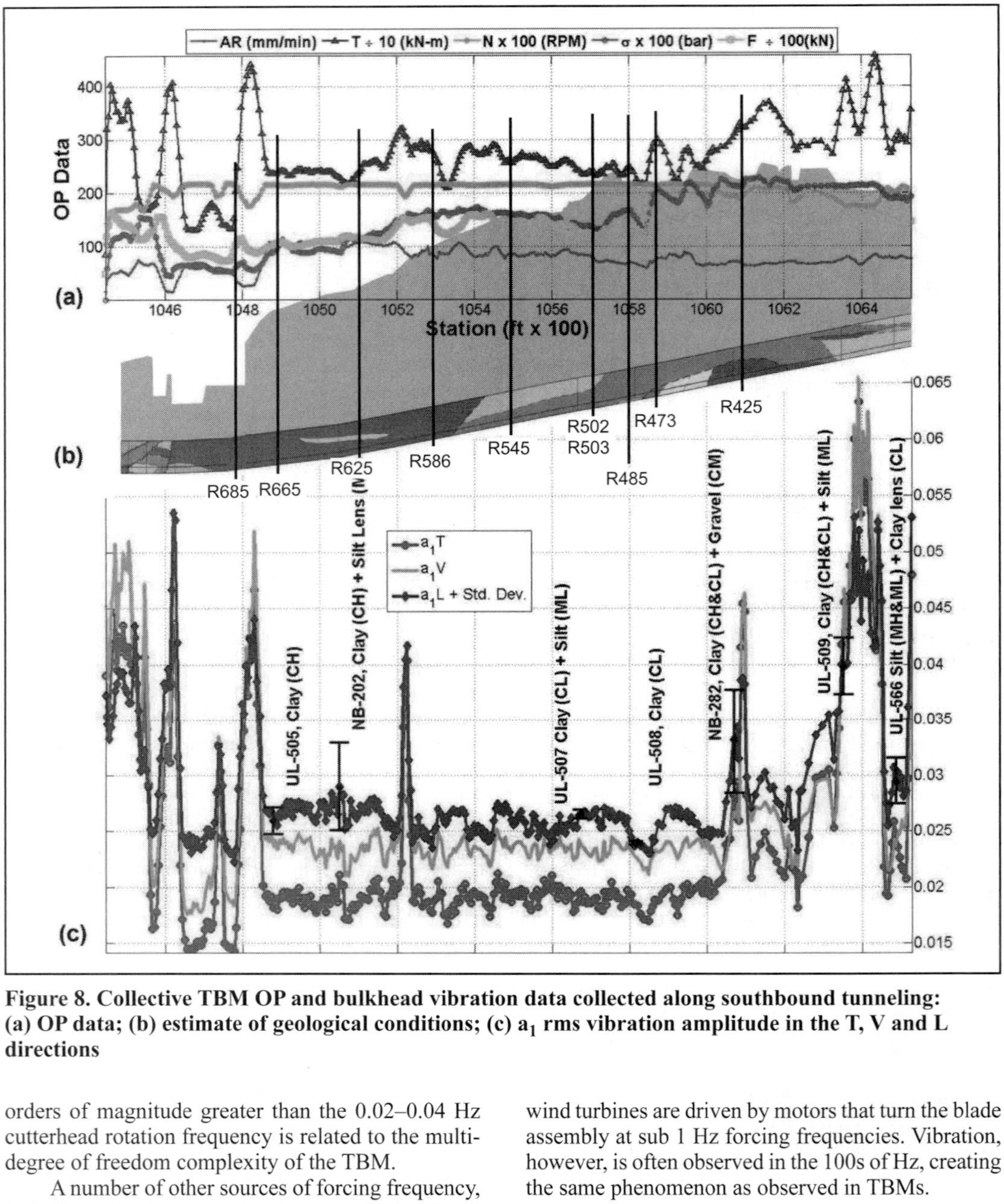


**Figure 8. Collective TBM OP and bulkhead vibration data collected along southbound tunneling: (a) OP data; (b) estimate of geological conditions; (c) $a_1$ rms vibration amplitude in the T, V and L directions**

orders of magnitude greater than the 0.02–0.04 Hz cutterhead rotation frequency is related to the multi-degree of freedom complexity of the TBM.

A number of other sources of forcing frequency, including electric motors turning the cutterhead and the main thrust bearing, may also influence TBM vibration. A detailed assessment of these sources, however, revealed that these sub-100 Hz frequencies are evident in the TBM bulkhead response but at amplitudes much less than those observed between 300–500 Hz (Walter 2013). The complex characteristics of TBM vibration and the desire to use TBM vibration as a continuous monitoring approach are likened to wind turbine vibration monitoring. Modern wind turbines are driven by motors that turn the blade assembly at sub 1 Hz forcing frequencies. Vibration, however, is often observed in the 100s of Hz, creating the same phenomenon as observed in TBMs.

## CONCLUSIONS

An experimental program was conducted to relate measurable TBM bulkhead vibration to geological conditions. A used during the Seattle University Link light rail project was outfitted with triaxial accelerometers. Vibration data was collected from a 6.44 m diameter Hitachi-Zosen EPB TBM during the 1 km long southbound excavation. Impact-response

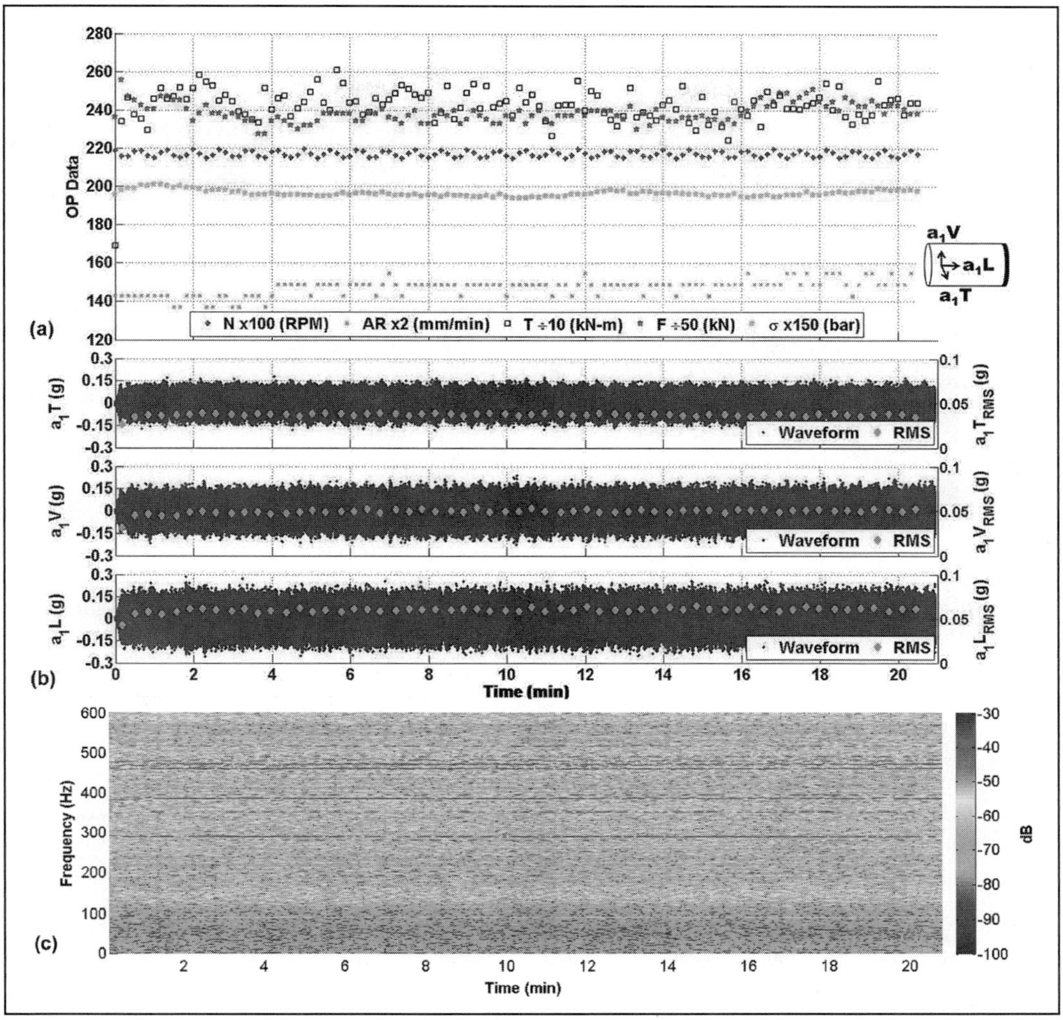


**Figure 9. (a) OP data, (b) bulkhead vibration time history data, and (c) joint time-frequency response of $a_1V$ during excavation of ring 502 (Station 1057+13)**

testing revealed that measurable vibration signature up to 500 Hz propagated from the cutterhead where geology-related vibrations would originate to the bulkhead where accelerometers can be easily mounted. For a number of frequency bands, cutterhead vibration amplitudes were either amplified at the bulkhead or attenuated only slightly. This is important because to date accelerometers cannot reliably be placed at the cutterhead. The results show that the bulkhead is a suitable surrogate location for ground-cutterhead vibration measurements. The analysis of bulkhead vibration data collected revealed clearly measurable vibration signals with amplitude and frequency content changing along the alignment. Vibration response was found to be influenced by TBM OPs, namely cutterhead rotation speed, that varied considerably throughout the alignment. When normalized by cutterhead rotation speed, changes in the resulting frequency content can be related to geological conditions. Vibration amplitude was also found to vary with geological conditions.

## ACKNOWLEDGMENTS

This research was partially funded by an IGERT grant from the National Science Foundation (DGE-0801692). The authors would like to acknowledge Jay Dee Contractors, Inc. for their support in this research.

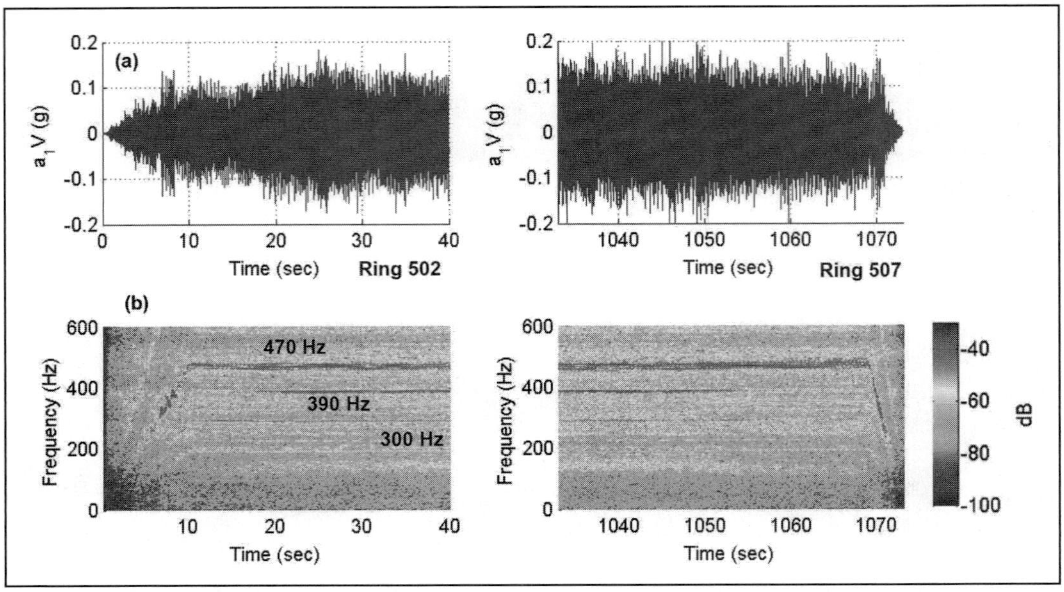


**Figure 10. (a) Bulkhead vibration time history and (b) joint time-frequency response at the beginning of ring 502 excavation (Station 1057+13) through the end of ring 507 (Station 1056+88) excavation**

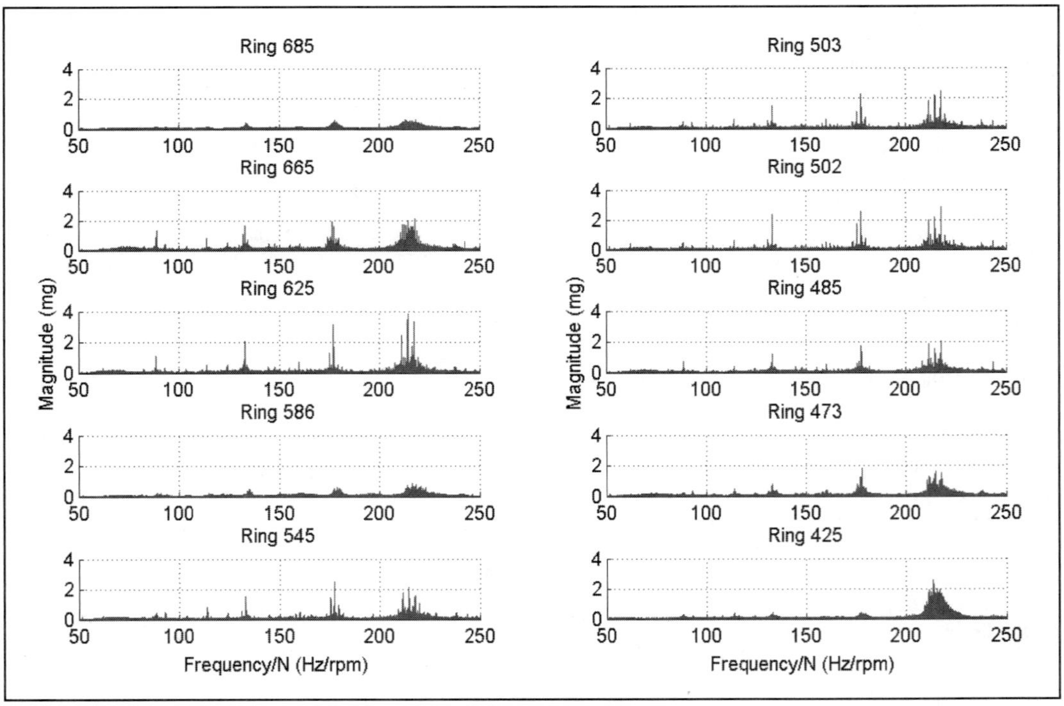


**Figure 11. FFT amplitude spectra of vertical acceleration ($a_1V$) from ten different rings. Note that the x-axis is normalized by the rms cutterhead rotation (N) of each ring.**

## REFERENCES

Carden, E.P. (2004) "Vibration based condition monitoring: A review," Structural Health Monitoring, 3(4), 355–377.

Cooper, G. Lesage, M. Sheppard, M. and Wand, P. (1987) "The interpretation of tricone drill bit vibrations for bit wear and rock type," Proc. RETC, June, 14–18.

Farrar, C.R and Worden, K. (2007) "An introduction to structural health monitoring," Phil. Trans. R. Soc. A, 365(1851), 303–315.

Hameed, Z., Hong, Y.S., Cho, Y.M., Ahn, S.H. and Song, C.K. (2009) "Condition monitoring and fault detection of wind turbines and related algorithms: A review," Renewable and Sustainable Review, 13(1), 1–39.

Irish, R.J. (2009) "Pre-Bid Engineering Geologic Evaluation of Subsurface Conditions for the University Link Light Rail Tunnels, Capitol Hill Station to Pine Street Stub Tunnel, and Capitol Hill Station Excavation and Supprt, Contract U230 Seattle, Washington."

Ledgerwood, L.W., Hoffmann, O.J., Jain, J.R., Herbig, C., and Spencer, R.W. (2010) "Downhole Vibration Measurement, Monitoring and Modeling Reveal Stick-Slip as a Primary Cause of PDC Bit Damage in Today's Applications," SPE Annual Technical Conference, 10.

Mooney, M.A. and Rinehart, R.V. (2009) "In Situ Soil Response to Vibratory Loading and Its Relationship to Roller-Measured Soil Stiffness," J. Geotechnical & Geoenvironmental Engineering, ASCE, 1,022–1,031.

Walter, B. (2013) Detecting changing geologic conditions with tunnel boring machines by using passive vibration measurements, PhD Dissertation, Colorado School of Mines, 163 pp.

# Urban Tunnel Monitoring: What's Next?

**Boris Caro Vargas**
Soldata

**ABSTRACT:** In the past 15 years the tools and techniques used to monitor the performance of tunnel construction in relation to settlement have improved significantly. From manual intrusive instruments the trend is now to use more automatic, remote sensing, real time and reflectorless techniques. Optical and radar signals are now commonly used for a wide range of high accuracy movement detection and measurement. The improvement of data processing methods and capabilities help to manage the risk. This paper illustrates this evolution with worldwide tunnel case studies and references. Mitigating the risks related to settlement has always been one of the top priorities of the geotechnical players of the tunneling industry. While other sectors have adopted smart technology long ago into their decision making processes to manage the risks, the civil engineering industry is now understanding the benefits of instrumentation and monitoring applied to all phases of a construction project.

## RECENT DEVELOPMENTS AND BIG INNOVATION STEPS IN THE CONSTRUCTION MONITORING

### Data Management Evolution

Before the 1990s all the monitoring of geotechnical parameters and related impacts of a tunneling project were recorded through manually read instruments. The only applications of automated or semi-automated geotechnical sensors were found in dam monitoring. Most of the manual instrument types and sensor technology are still widely used today, from observation wells to manual inclinometers. Only a few new sensor technologies have emerged, like Micro Electro Mechanical Systems (MEMS) or fiber optics, and most of the evolution is due to improvements in data collection, processing and management. As in many other industries, the geotechnical monitoring has been carried into the PC revolution and has widely benefited from the exponential increase in data collection and processing capabilities of computers and the birth of internet. Managing databases of several gigabytes is now common but was impossible in the 1990s. At the time, where underground construction challenges were continuously pushed back thanks to an improvement in the drilling, excavating and grouting equipment, the use of computers to control some of the most sensitive construction processes (such as compensation grouting or mechanized tunneling), gave birth to a new era of risk management, with "real time" data management. The Jubilee Line Project in London between 1994 and 1997, was constructed under the scrutiny of hundreds of electrolevel beams connected to Automatic Data Acquisition Systems, and of various other instruments (Thurlow et al. 1999). The first monitoring databases were also developed and put in place at that time, running in DOS environment and soon in Windows 3.1 (Beth et al., 2011). The computerized grout control process, combined with the automatic data acquisition and processing of geotechnical instrumentation, was the first implementation of real time risk management for an urban tunnel project.

Similar projects, in other parts of the World, especially in Hong Kong and Singapore, under the influence of the British "pioneers," emerged with similar applications. The first large scale monitoring project in the US, the Boston Big Dig, was mainly performed using "armies" of field engineers to manually collect data from thousands of instruments in the vicinity of the excavation.

The first significant use of automated geotechnical instruments on a tunnel project was performed on the East Side access in NYC from 1999, as well as in Puerto Rico in 1998 on the Rio Piedras Station (Beth et al., 1999). The real time monitoring in the US and the application of latest data collection / processing techniques was lagging behind compared to Southeast Asia and Europe. Database sizes of several gigabytes (up to 120Gbs in Barcelona L5) have not been seen until very recently in the US. However, due to recent global economic changes and the complexity of some recent urban tunnel projects, the US has closed their "technological" gap and has seen similar data collection applications (Subway Line 7 Extension in NYC—Seattle SR-99 Tunnel Project). Today many different software programs

**Figure 1. Layout of AMTS and prisms on the south end of the alignment for the SR-99 Tunnel Project**

are available to collect, store, manage and report geotechnical instrumentation data and their use is common on most large-scale urban tunnel project.

### The Emergence of Remote Sensing with Automatic Motorized Total Stations

Apart from the use of data management capabilities that followed the PC hardware and software evolution, a new approach to collect "geotechnical" parameters or "structural" deformation data has emerged. In the construction monitoring field, the "geotechnical engineer" approach has followed a parallel path to the "Professional Land Surveyor" approach. But are these different? How do these professionals help to manage the risks on a tunnel project? The answer is clearly by predicting, evaluating, accepting, transferring or avoiding the impact of construction related settlement. This impact can be monitored by installing sensors in the ground that *directly* measure the change of a parameter and translates it into a signal at a specific location underground (or above ground), or by measuring the displacements of specific points using *standard* survey techniques. These two approaches share the same goal, although they use different expertise. The technological computer revolution that transformed the tools used by surveyors has had a profound impact on both the hardware and the software (whereas the change for geotechnical instruments mostly affected the "software" or the way to collect and process data). These changes are best illustrated in the use of specific equipment: Automatic Motorized Total Stations (AMTS) AKA Robotic Total Stations (RTS). This equipment has the ability to monitor absolute 3D movements 24/7 automatically. The equipment hardware has improved significantly over the last 15 years and today is more precise, robust, reliable, quiet and can be totally controlled remotely. It is now possible to automatically monitor fifty 3D precise targets at a 300' distance with an accuracy better than 1mm in each of the three dimensions in less than one hour. In terms of software, the improvement in computer processing speed has reduced the calculation time to obtain absolute monitoring 3D displacements in "near" real time and has allowed more complex calculations using networks of AMTS instead of single AMTS. Figure 1 shows the layout of the AMTS on the SR-99 Tunnel Project in Seattle. However, the AMTS and all other recent "remote sensing" techniques cannot totally replace the "underground" sensors as they cannot monitor below the surface. Pure geotechnical instruments still provide us with a wealth of critical information that cannot be found by any other method.

## EVOLUTION IS NOTHING WITHOUT CONTROL: TODAY'S CHALLENGES FOR LARGE-SCALE MONITORING PROGRAMS

### Data Is Different from Information

During the 2000s the enthusiasm for the improvements in data acquisition and processing described in the first section, combined with the obligation to provide data in real time to all the different project stakeholders, the lower cost of the sensors and the difficulties of tunneling in always more congested urban areas resulted in an exponential growth of geotechnical data collection. On each urban tunnel project where poor soils conditions were one of the main risk factors, the common idea was to install as many geotechnical sensors as possible. As the acquisition and processing capabilities seem limitless, the design engineers thought they should collect as much data as possible. However, no data management system is self-sufficient. There is always a need to convert the *data* into *information* and this process cannot be 100% automated.

In large tunnel construction fields, where very different soils and groundwater conditions are often encountered along the tunnel drive, the interpretation of the data retrieved from sensors requires a high level of local geotechnical expertise. Also, the correlation between the construction sequence and its impact on existing surrounding structures require deep knowledge of pre-design and post-construction data analysis. The geotechnical engineer cannot be a slave to the data, the data and the related instrumentation program must be a tool at the service of the Engineer. Since the middle of the 2000s, and after a slight decrease in tunnel collapses from the previous decade, there was an illusion that real time data management could prevent any major collapse from happening and that the more sensors and data that could be collected, the safer the project would be.

This proved to be wrong: In Cologne in 2009, during the construction of one of the crossover passages of a new light rail underground project, an extensive monitoring program had been implemented. However, 2 buildings collapsed in the vicinity of the construction of a shaft on March 3rd 2009, killing 2 people. In Amsterdam, during the construction of the North/South subway line and despite the largest monitoring program implemented up to that date (recently outnumber by the CrossRail Project), several buildings sank several centimeters in September 2008 during the construction of the Vijzelgracht Station. In both cases water seemed to have played a major role and both buildings, ground, and groundwater were heavily scrutinized. One of the main lessons learned from those events was that data has to be processed into information and fed into early warning systems in order to be an efficient tool for managing the risk of settlement control. After two decades of the limitless increase in the amount of data collected (quantity), one of the major challenges is to extract the relevant information (quality) sifting through and interpreting the data.

## Selection of the Monitoring Company

Since the early publications about geotechnical instrumentation and monitoring (Dunnicliff, 1993, 2011), the benefits and drawbacks of "who should be responsible for monitoring and instrumentation during construction" are clear. The conclusion is: "*The people who have the greatest interest in the monitoring and instrumentation data should be given direct responsibility for obtaining the data.*" This is in theory. In practice it is almost never the case, especially in the US on Design-Build–Bid Projects. In most cases the General Contractor is the one responsible for the monitoring, but he is clearly the one with the least interest of obtaining the data. In most contract specifications if the data reaches certain thresholds the construction has to be done differently, sometimes even stop, whereas the interest of the Contractor is obviously to never stop the construction, as the consequences of a delay are often their major concern and risk for extra costs. The Owner is logically the one with an interest in building a risk-free project with the least impact to the City stakeholders, but in reality the Contractor has to assume all the liability related to construction, with little incentive to provide a high quality monitoring program.

With increasing competition in the monitoring industry, and a selection process based exclusively on the "lowest bid," it is difficult to expect an improvement in the quality of the instrumentation data and the monitoring plan. The cost of the monitoring plan after the installation phase is mostly based on the evaluation of contingency and proactive measures to be implemented to ensure the right information is delivered in due time during all construction steps to all interested parties and stakeholders (Contractor, Subcontractors, Owner, public, 3rd party utilities, building owners, engineers).

It is sometimes difficult to understand the *value* of a monitoring program: on a project where the construction has no impact and no major deformation is observed, the cost of the monitoring *could have been* reduced to the minimum! It can be hard to realize that the value of the programs resides in the *awareness* that nothing was happening *during* the construction. A monitoring program also indirectly benefits the overall Project, by improving safety and helping to meet the schedule.

Competition has increased in the monitoring industry because the selection criteria was never based on qualifications, basically almost anybody with some geotechnical expertise could name himself a "monitoring specialist." However, there has been a healthy evolution in the recent years. With always bigger tunnel projects being built and the spread of "design-build" projects (as opposed to design-bid-build projects), a pre-qualification process based on *qualitative* criteria is now a common rule when a General Contractor has to make a decision for each of the "key" subcontractors. As monitoring is now one of the visible parts of the project and with the possibility for the contractor to optimize the monitoring program for design-build projects (as he is responsible for the final design), there is a clear benefit for the Contractor to choose from experienced monitoring firms capable of both handling the challenges of mega monitoring programs associated with their mega projects and share their experience and expertise to provide the latest innovative techniques. On design-bid-build projects the specifications are often hard to meet, and to avoid the "low-cost low-quality monitoring" Engineers had a tendency to make the specifications always stricter; however monitoring specifications are sometimes hard to enforce.

On design-build projects the design is finalized in a collaborative approach between the Owner, the General Contractor and the monitoring firm (Sowers and Caro Vargas, 2013). In this case, if the prequalification and filtering of the capable monitoring firms were handled properly by the General Contractor in due time, the risk of having a low cost (high risk) monitoring program is minimized. Some of the criteria for prequalification of the monitoring firm are:

- Experience in similar (environment, size, geology, Risk) recent large scale monitoring programs
- In-house data management capabilities (including software, programming, IT security)
- In-house data acquisition, processing and transmission expertise
- Experience in implementing Contingency Plans to ensure the continuity of the data reporting
- Metrological interpretation expertise (to ensure the reliability of the data)

## WHAT SHOULD BE A MONITORING SYSTEM

Tunneling in urban areas faces a myriad of challenges. Downtown areas are more and more crowded, with layers of tunnels overlapping each other, (sometimes within a few feet from each other, like in the recent construction of the Central Subway line in San Francisco), existing utilities and buildings' basements (the Grand Wilshire Tower in Los Angeles are built within a few feet of the existing Red Line). In addition to the technical challenges associated with such presence of nearby underground (and above ground) structures, monitoring the impact of a tunnel excavation involves the installation of sensors on these structures or their vicinity. Apart from the drilling itself, the additional constraints to install an underground sensor (right of way, access, traffic control, utility obstructions, relocations or temporary shutdown, specific equipment that complies with City regulations for dust, noise and clearance, protection of the instruments, safety of the instrumentation technicians, trenching, restoration) increases the cost of the installation significantly, when the installation is even possible. These costs are extremely difficult to evaluate at bid stage, and the design is usually done without taking these constraints into account. Even when the work can be performed, the efforts and time necessary to overcome these challenges often result in delays and design changes. The costs have pushed the monitoring firms to look for less intrusive solutions, either by adapting existing techniques or inventing new tools.

### Wireless Communication and Power Consumption

Data communication is always specific to an instrument location and every new installation brings a specific challenge. With the high cost of cabling, trenching, and restoration the radio communication was the preferred solution for data transmission between a sensor location and the base station. However, the lack of reliability of such signals and the limitation in distance, especially in heavily urbanized areas, limited the use of radios. On large projects, specific Wi-Fi networks have been implemented (Amsterdam, Toulon), but this solution is costly for medium size tunnel projects. The decrease in costs in 3G communications and the increase of coverage in most of the urban areas make it the most reliable solution for data transmission. For large data transfers, 4G is now available and being used successfully. Remote communication is now also installed for underground instrumentation without the need of trenching with autonomous installation for each instrumented borehole.

The second area of improvement not related to the sensor technology itself is related to power consumption. Even though it is relatively easy to get access to permanent power in an urban environment, it might be a challenge in specific project areas, especially in ever changing construction sites or for the underground instrumentation installed below streets. The loss of power is a risk that can cause loss of data, something unacceptable for any monitoring firm. That is why there have been significant improvements in reducing the power consumptions of most of the control units (dataloggers, industrial PCs etc.) so that autonomous power sources could be used as a primary source of power or as a back-up. Additional components are used to switch between active and passive modes, depending on the data collection frequency. The use of solar panels is common in most of the areas, even with low sunlight. Future applications and optimization will reduce even more the need of permanent power and smaller size more efficient batteries and charging systems will reduce the presence of cables and improve the autonomy and reliability of the monitoring systems.

**Non-Intrusive Monitoring Techniques**

Besides data transmission, the trend is to reduce the intrusiveness and to use techniques to monitor structures without the need to access them at any time. The use of AMTS was the first remote sensing tool where the data acquisition was totally separated from the monitoring point. In this case, at the monitoring location, no specific sensor is installed, only a prism that reflects the laser signal sent by the AMTS. Networks of AMTS are today the preferred monitoring tool on many European urban tunneling projects, where most of the monitoring points are 3D prisms, placing the traditional underground sensors (inclinometers, extensometers and piezometers) installed in boreholes or structural instruments (tiltmeters, strain gages, liquid level systems, crack gages…) on a secondary level. This trend is accelerating since the AMTS can now be used to monitor surface settlement without the need to install any prism, by simply using the reflective properties of the surface This use of the reflectorless properties of the instrument now allows a full coverage of the surface and structures along a tunnel alignment (Tamagnan and Beth, 2011) with an accuracy similar to more traditional AMTS use with prisms (±1mm), even though the range of such application is limited and depends on the reflective properties of the road surface. The combined use of AMTS capabilities allows a full coverage of both the buildings and the road surface, as shown in Figure 2.

Another "survey" tool, laser scanners (or LiDAR) are being used to monitor deformation. Mostly used in the mining industry, their limitations in terms of accuracy (1cm @ 100m) and data processing times, as well as data output size make it a non-suitable tool *yet* for the real time and accuracy needs of the urban tunneling industry. Another technology, using radar interferometry, is a tool that has been widely used for slope stability. However, the scale each image is converging and the accuracy (3 to 5mm) make it a suitable complement to monitor beyond the traditional zone of influence of a tunneling project in time and space. The capability to process images from the past and the size each image can cover (5 km × 5 km with a density of 10,000 points/km$^2$ or more) make it a very usable tool to complete existing monitoring programs at a lower cost than traditional techniques (Sowers and Caro Vargas, 2013).

These 3 techniques and their variations are the present and future of non-intrusive monitoring tools. The last technique worth mentioning is the use of GPS sensors, however their principle is closer to standard sensors where each monitoring *point* is a sensor. Although there has been a clear improvement in accuracy and data processing times, their cost and intrusiveness is comparable to former structural sensors. All these tools will not replace standard instrumentation, but, as data processing times decrease and accuracy increases, they will reach the breaking points to be used in the automatic monitoring programs for tunneling projects. Additional technological improvements worth mentioning are the MEMs sensors and the fiber optics, but this is not the aim of

**Figure 2. Combined use of AMTS for prism and reflectorless monitoring applications**

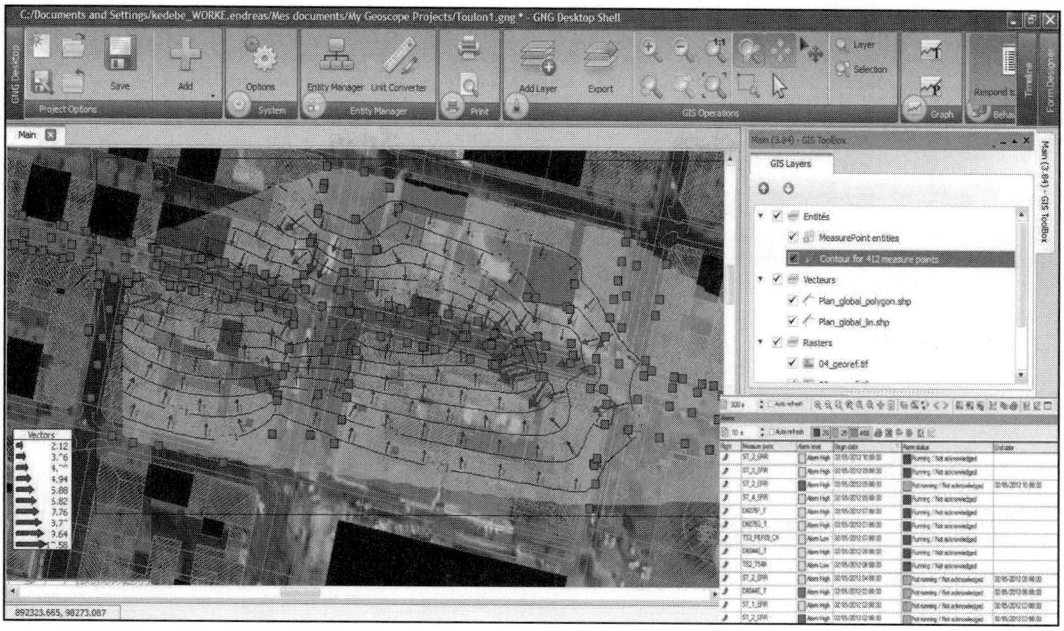


**Figure 3. Global data integration and alert management**

this article, as they relate to more traditional sensors improvements.

### Global Data Integration and Decision Making Tool

The software revolution has impacted all of the construction process. Most of the heavy construction equipment, and especially the TBMs, now include a personal computer or black box that records all the important parameters necessary to understand the mechanical behavior and the performance of the machine. In addition to these internal dataloggers and the standard geotechnical instrumentation data recording programs already discussed in this article, the whole impact of the construction has to be monitored, recorded, and compared to acceptable standards. Apart from vibration limits, each City has now noise and dust limits that construction operations cannot exceed. Monitoring should be a requirement of all the urban tunneling projects, including:

- Real time geotechnical instrumentation sensors
- Equipment performance parameters
- Environmental data

All three systems can live independently of each other; however they all have a clear interaction and understanding how one affects the other and is key to mitigate the impacts of all three. This global integration has started with the first real time monitoring project where compensation grouting data (grout volume and pressure at each port location) had to be combined and compared with real time settlement or heave at each specific monitoring point location, but is now widely used on tunneling projects where TBM parameters, such as face pressure and muck volumes, are combined to settlement data, to allow for a better adjustment of construction parameters in real time (Figure 3).

The next step of this global integration is to finally be able to compare the anticipated ground behavior from design analysis with the construction monitoring data as easily and as fast as possible. For some very specific sensitive techniques, such as NATM (Caro Vargas and Beth, 2011) there is an obligation to use the actual monitoring data to adjust the design during construction. But this obligation could be turned into real time experience more widely on most of the tunneling projects to compare the face loss calculation, draw a “live” zone of influence and more generally feed the construction models with actual parameters. Until there is no real incentive for the Contractors or the Engineers to do so and talk together, the Design and the Monitoring Industry will not share what they are both looking for: reliable and accurate geotechnical information.

## CONCLUSION

It is now clear that the monitoring industry for urban tunneling projects is a “work in progress.” Mistakes have been made along the way in making sure the data is automated, reliable and accurate, and accessible in real time. Priority should be emphasized on the quality of the information, not the largest amount of data. After a period of exponential growth in data collection, the focus is now on making sure that the right information is communicated to the right person at the right time. After being focused in integrating large amounts of data, the industry needs to focus on making sure the message goes through, and that early warning systems filter the non-critical parameters from the "absolutely vital" information. The work of the Engineer is still to interpret and correlate the data, and to identify construction impacts at the earliest possible time, as he has no time to decipher and filter between signals. The upfront work between the Engineer, the contractor, the monitoring firm and the Owner is a mandatory step so that all the players understand their interest in obtaining the best monitoring and instrumentation data, as this is the best way to limit the impact of ground settlement on the construction and the public.

## REFERENCES


Paul Thurlow, Martin Beth and Anthony Lucas, A controlled approach to providing protection to structures located above tunnel construction works, Tunnel Construction and Piling—International Symposium, September 1999.

Martin Beth, Giorgio Pezzetti and Clemens Schmuck, Urban Excavation Instrumentation: Analysis and Reflexions, FMGM May 2011, Berlin.

Martin Beth, Stephane Carayol and Laurent Lavene, Ground Treatment and ground and structures movement monitoring in Puerto Rico, International Congress on Urban Heritage and Building Maintenance, October 25th 1999, Delft (NL).

Federico Valdemarin, Jean-Christophe Monneret and Henning Schwarz, Una red de estaciones totales robotizadas para auscultar la construcción de la Línea 9 del Metro de Barcelona: el caso del doble túnel de la Sagrera, Jornada Tecnica: Instrumentaction de tuneles y Excavaciones, Barcelona, October 21st 2010.

Damien Tamagnan and Martin Beth, Monitoring of Surface Deformation with Robotic Total Stations Using Reflectorless Measurement, Geotechnical Instrumentation News, p. 21 September 2011.

John Dunnicliff, Geotechnical Instrumentation for Monitoring Field Performance, 1993.

John Dunnlicliff, Remote Monitoring of deformation. Introduction, Geotechnical Instrumentation News, p. 23–24 June 2011.

David Sowers and Boris Caro Vargas, Instrumentation approach for the Alaskan Way Bored Tunnel, RETC 2013.

Boris Caro Vargas and Martin Beth, Observational method using real time surface settlement monitoring—The South Toulon Tunnel Project, ORVSS 42, Cincinnati OH, October 21, 2011.

# TRACK 1: TECHNOLOGY

## Session 2: Ground Support, Final Lining, and Design

Harold Leiendecker, Chair

# Ground Characterization While Drilling Roofbolters in Tunneling Operations

**A. Naeimipour, S. Bahrampour, and J. Rostami**
The Pennsylvania State University

**C. Dogruoz**
Dumlupinar University

**ABSTRACT:** Characterization of the ground surrounding the tunnel can improve safety and allow for optimizing the ground support design during the tunnel construction. Measurement while drilling (MWD) systems offers valuable information about the ground condition in probe drilling as well as drilling of blastholes and roof bolts. This paper will discuss the current efforts underway at Penn State Univ. in collaboration with J.H. Fletcher & Co for developing instrumented roof bolters which can detect bedding and discontinuities and provide a rough estimate on the rock strength along the borehole. Recent improvements in performance of MWD for roofbolters as an effective way to obtain geological information about the rock and 3D visualization of this data as a prelude to development of rock mass classification and hazard mapping of the roof and walls will also be presented.

## INTRODUCTION

In underground construction, maintaining a safe work environment often involves installation of suitable ground support system. It is very common to use rock bolts in tunneling projects due to the relatively simple installation, low cost, and substantial support capacity. Proper selection of rock bolts depends on correct identification of the geological conditions, which can dramatically change even within a short distance (Gu, 2003). However, there is often limited information about the surrounding rock and the exploration borehole measures are few and far in between. Preparing geological maps while tunnel excavation operation is proceeding is not always practically possible and safe. The application of MWD for roof characterizations has been recently attracted many researchers because it can potentially provide the needed geological information in a short time without the need for any operational interruptions. In this approach roof bolt drilling system collects additional information such as torque, thrust and penetration rate which can be used for estimating the rock strength and evaluating rock mass conditions. This information will be utilized for development of geological map of the ground around the tunnel. Such maps will in turn allow for development of hazard-maps and 3D data visualization of the ground. Moreover, the information can potentially be employed to evaluate the suitability of the applied ground support for a given section. A series of studies has demonstrated the potential for analyzing drilling parameters from roof bolters to estimate rock properties and to identify discontinuities (Itakura et al., 2001; Itakura et al., 2008; LaBelle, 2001). West Virginia University (WVU) in collaboration with J.H. Fletcher & Co. also worked on this subject between 1999 to 2006 and have developed the first generation of the void detection system for Fletcher roofbolters (Finfinger, 2003; Gu, 2003; Mirabile, 2003; Tang, 2006; Finfinger et al., 2000; Luo, Peng, and Wilson, 2003; Luo et al., 2002; Peng et al., 2003). Additional studies are being conducted by Penn State University and J.H. Fletcher & Co. to continue the work performed at WVU. This project is sponsored by National Institute of Occupational Safety and Health (NIOSH). This paper will discuss the testing and related analysis underway in this project to improve the void detection capabilities of the roofbolters as well as rock strength measurement and implementation of the obtained results into a 3D visualization of the ground and development of hazard maps for tunnel wall. Available and applicable borehole logging systems will be explained and their use in detection of discontinuities, Rock mass strength estimation and ground conditions will be discussed. Data visualization will also be briefly covered and its potential for use in tunneling and underground construction will be evaluated.

## BACKGROUND

Itakura et al. (1997) employed a portable pneumatic roof bolter with the ability to record torque, thrust, revolution, and stroke. Torque and thrust were monitored by using strain gauges installed on the surface of the drilling rod, while penetration and rotation rate were kept constant during the tests. They manufactured blocks included sandstone, sandy shale, and coal samples with three different discontinuity angles of 0 degrees, 30 degrees, and 60 degrees. The average value of torque and/or thrust was found to be an indicative index to allow for classification of the rock layers along the borehole. Furthermore, it has been proposed that patterns of thrust or torque along with neural network algorithms may be used to categorize the discontinuities, but the resulting error was rather large (Itakura, 1998; Itakura et al., 2001). Itakura et al. (2008) reported that the roof bolter examined in an underground coal mine in Queensland, Australia, successfully showed the distribution of discontinuities and layer boundaries using the ratio of recorded parameters of torque and thrust.

In addition, J.H. Fletcher & Co. as a pioneer in developing instrumented roof bolters, developed a system that monitors drilling operations and drilling parameters, including thrust, torque, rotation speed (rpm), and bit position (Gu, 2003). This system has been employed to detect rock discontinuities including voids, fractures, and bed separations and to estimate the relative hardness of the rock mass (Finfinger et al., 2000). Variation of thrust or feed pressure had been found to be the most suitable identifier of discontinuities (Finfinger, 2003; Peng et al., 2003). Based on Finfinger (2003) concept of thrust valley, thrust decreases rapidly after reaching a void and increases rapidly again when it goes through the discontinuity to maintain constant penetration rate. A drop of more than 50% was then considered as an index to detect discontinuity. The distance between the two sides of the valley was also used to measure the discontinuity aperture. Although the instrumented system has been improved to a great extent, there are still some inaccuracies in detecting the location and, especially, the size of discontinuities. Collins, et al. (2004) explained that some major voids could not initially be detected by the system during a series of field experiments in a limestone mine, mainly, because of the difference between the hardness of concrete used in the laboratory used for training the drill and the limestone at the roof of the mine. In this situation, the parameters of the roof mapping algorithm needed to be updated constantly. It was also found that unlike the usual pattern observed in the laboratory, in which both thrust and torque would drop simultaneously, a sudden rise in the rotation torque happened just before encountering the voids. Meanwhile, the thrust did not have a consistent reaction. New theories were developed later to describe the observed trends. Another problem was reported by Anderson and Prosser (2007), in which the hairline and vertical cracks along with layers of the rocks were not correctly identified. Moreover, Tang (2006) elaborated that the applicability of the developed system is limited to voids with size of ⅛ inch or larger.

In a more recent study, it has been shown that vibration and acoustic measurements can also be used to improve the accuracy of the void detection and rock characterization algorithms (Bahrampour et al. 2013). For this purpose, it was observed that valuable information can be extracted from the high frequency components of the vibration and acoustic signals that can be subsequently used for void detection by modeling the problem. Moreover, the author suggested that combinations of the new measurements with the drilling parameters such as torque and thrust can potentially further improve the accuracy of the rock characterization algorithms and provide robustness by adding redundancy in the measurements.

However, further developments of the measurement while drilling systems is dependent on extensive field tests and how fast and accurate the properties of rock mass inside the drilled boreholes can be identified by means of sampling or probing. This allows the borehole drilled by the instrumented roof bolter to be logged for training and verification of the results by the related algorithms and pattern recognition programs. Borehole logging or well logging is a conventional practice in oil and gas, groundwater, and mining industries, which continuously records the information related to variation in targeted physical properties of the rocks in bore hole. There are many different logging methods namely electric, radiation, sonic or acoustic, and optical probes. Each of these methods consists of varied sub-approaches that are suitable to measure a physical parameter in a particular situation such as lateral resistivity, neutron and gamma, caliper, optical or TV logs and sonic or acoustic methods. Well logging in oil industry utilizes probes with relatively large size and much heavier apparatuses, whereas, in mining and civil operations simpler and lighter devices should be used.

In this project, slimmer, lighter probes are needed so that they can be run easily in the roof bolt holes, which are typically about 1–1½ inch in diameter, and limited space in underground environment. Unfortunately, very limited attempts have been made to employ borehole logging methods to verify the validity of the results from the instrumented roof bolter in the field tests. Gu (2003) and Tang (2006) mentioned application of a simple borehole camera system to be used in addition to coring to verify the roof bolter results during the underground tests. However, many researchers have studied geophysical methods in order to be able to estimate different properties of the rock mass especially the strength (McNally 1990; Payne and Ward 2002; Zhou et al. 2005; Oyler et al. 2010). Also, borehole televiewers and cameras are developed to take continuous picture of the borehole wall so discontinuities could be detected and analyzed (Unrug 1994; Ellenberger 2009; Bae et al., 2011). Unlike these studies that most of the logging runs were performed in downhole boreholes, in this project upward boreholes which are usually dry need to be logged. This is a logistic issue that should be solved in order to pave the way for further investigation of the application of borehole logging in training the instrumented roof bolter. The only system that is readily compatible with this condition is optical televiewer or bore-cams. Obviously, this information can also be obtained from coring into the roof or walls and testing the core samples, but this could take a substantial amount of time and cause interruption in the operation, the results will be available with a time lag related to testing, and require additional equipment and setting that is not readily available at the site. Despite these issues, some cores will be retrieved from the formations where the coring operations are deemed not to have the least impact on normal operations. These cores will be used to validate and adjust the measurements made by various probing devices.

## VOID DETECTION

Void detection is a feature that could help identify the discontinuities and joints in various rock mass conditions. These features are known to dominate rock mass behavior. The term *void* typically refers to a joint with open aperture or other conditions in the rock mass that represents open space or area filled with weak deposits. However, in this study, void refer to any discontinuity that could weaken the rock mass, including but not limited to bedding, joints, cracks, fractures, fissure, etc. in any type of rock. A series of full scale experimental tests were conducted at J.H. Fletcher & Co. facilities in Huntington WV. Figure 1 shows the picture of the test setup. In these tests concrete blocks with various strength as well as rock samples from the sedimentary layers of mid-Atlantic region casted in concrete were drilled using an instrumented roof bolter. Test results are consistent with previous studies by WVU and Fletcher and it has been observed that the feed pressure, which can be translated to thrust, drops within the void. This pattern is used for void detection using a new algorithm as it will be explained. The pattern of dropping thrust force has been previously studied (Collins et al., 2004), but no adaptive algorithm yet exists to model this behavior and correctly detect the voids independent of the rock strength. In other words, feed pressure signal is tightly correlated to the strength of the rock being drilled. Consequently, the drop in the feed pressure is dependent on the strength of the rock. In this study, the void detection problem was modeled as a mean change problem and an efficient mean change detection algorithm, known as cumulative sum (CUSUM) algorithm (Basseville et al., 1993), was used to detect the voids. Preliminary results of using this algorithm for void detection has proven to be effective. The CUSUM algorithm has an adaptive threshold that does not need careful fine tuning when dealing with different rock strengths and various drilling parameters such as desired penetration rates and rpm.

**Figure 1. J.H. Fletcher & Co. instrumented roof bolter setup**

### Pattern of Feed Pressure for Void Detection

Figure 2 shows a typical sensory data collected while drilling into a stack of two concrete blocks with a void at the intersection of the two concrete blocks. The measured attributes are rotation pressure, feed pressure, rpm, position, bite rate (penetration per revolution), and vacuum pressure. It is seen here that feed pressure (thrust) has a sudden decrease when the drilling bit approaches the void while the rotation pressure (torque) is almost constant during the drilling. This pattern was observed in almost all the experiments performed. Having this in mind, void detection can be mathematically formulated as a mean change detection problem. In this formulation, it is assumed that the mean of the stochastic

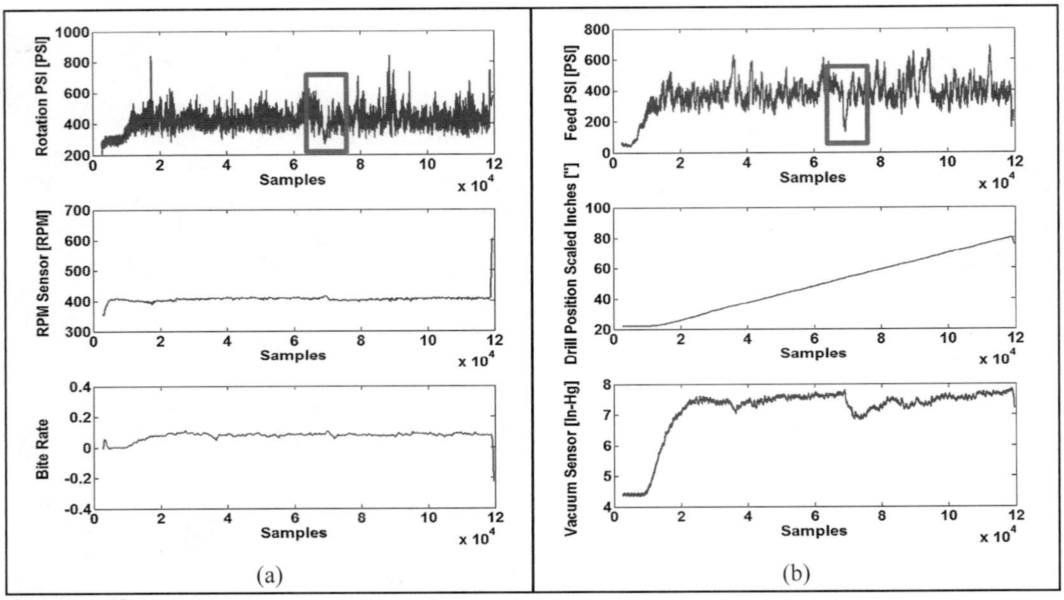


**Figure 2. Typical sensory data collected while drilling into a stack of two concrete blocks with a void at the intersection of the two concrete blocks. Samples represent the time scale, where every 10 samples is 1 sec.**

signal (i.e., feed pressure) is almost constant when there is no void and it undergoes a change (decrease here) when a void appears. The goal was to detect the change as quickly as possible with high detection rate and small false alarm rate.

For solving this problem, the CUMSUM algorithm which is a well-known change detection algorithm will be used. In this real time change detection algorithm, the initial mean of the signal is estimated using the initial samples and a cumulative sum is computed and monitored at each time step. Deviation from the initial mean is then detected by comparing the cumulative sum to an adaptive threshold. As soon as a change is detected, the CUSUM algorithm will be restarted and will initialize to detect the possible changes in following samples. The details of the CUSUM algorithm are omitted here and interested readers are referred to (Basseville et al., 1993) for detailed derivation of the algorithm. In the above formulation, it is assumed that feed pressure drops when the drill bit approaches a void. For this purpose, the void detection algorithm is turned off during the first 5 inches of the drilling and the initial time series is used to estimate mean and variance. Information about the magnitude change is usually not known a-priori. One good choice is to replace mean with minimum possible magnitude of the jump. In this paper, or change in amplitude is selected to be 40% of the mean . In other words, the algorithm would be sensitive to all change greater than 40% of the mean value of the time series. More sophisticated algorithms can be used to estimate the parameter such as the generalized likelihood ratio method but this will significantly add the computational costs.

In order to evaluate the suitability of this algorithm, data from full scale testing experiments were used to detect the interface between concrete blocks. The experiments consisted of a set of 16 concrete blocks with different strength. The blocks are approximately 0.5 × 0.5 × 0.75 m (~20 × 20 × 30 inches), and the concrete mix was designed for various strengths: low (~20 MPa or 3,000 psi), medium (50 MPa or 7,500 psi), and high (70 MPa or 10,000 psi). Different combinations of concrete blocks were used to test robustness of the algorithm to deal with different setups. For examples, a hard concrete block on top of a soft concrete block (H-S) or a hard concrete block on top of another hard concrete block (H-H). The gap between the blocks were less than a 1–2 of millimeters and was considered to represent a "void" in this study. Additional tests were performed in samples of rock from several different mining operations in PA and WV that were case in the concrete block to simulate the variation of strata in the roof and walls of an underground openings. These samples were subjected to various rock mechanics testing to measure their strength and other related properties.

Table 1 summarizes the different concrete combinations studied in this paper with the number of holes drilled in each setup along with void detection rate and false alarm rate where 'S', 'M', and 'H' letters stand for soft, medium, and hard concrete, respectively. Different values of penetration rate

**Table 1. Results of the void detection algorithm on different combinations of concrete soft (S), medium (M), and hard (H) blocks**

| Concrete combinations | S-H | H-S | M-H | H-H | H-M | M-S | S-M | M-M | S-S |
|---|---|---|---|---|---|---|---|---|---|
| Number of holes | 17 | 17 | 17 | 18 | 21 | 18 | 18 | 18 | 17 |
| Detection rate | 82.3% | 88.2% | 100% | 94.4% | 100% | 100% | 88.8% | 88.8% | 100% |
| Number of false alarms | 1 | 1 | 0 | 1 | 0 | 2 | 2 | 1 | 2 |

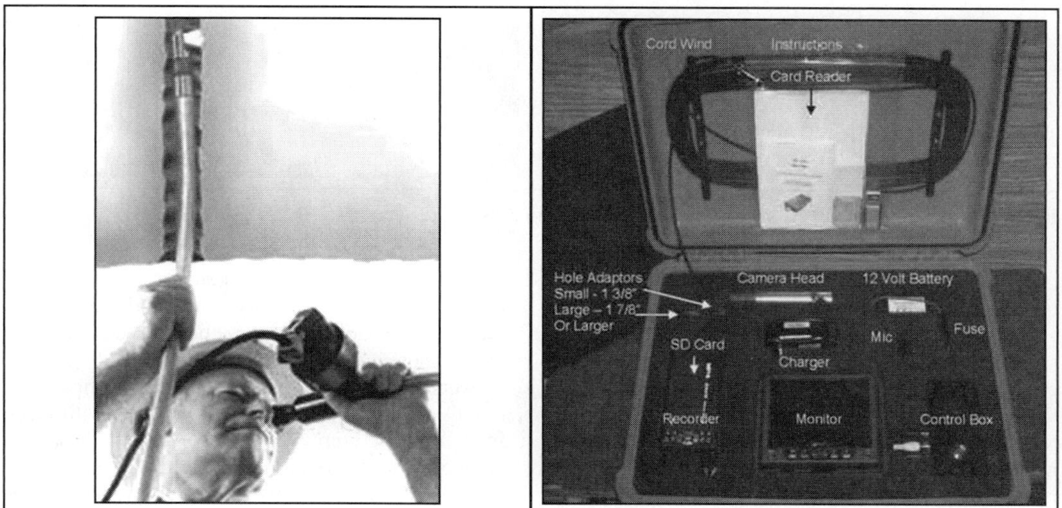


(a) (b)

**Figure 3. (a) Strata-scope and its schematic application in mining industry("Optim Stratascope," 2013); (b) typical bore-scope package and its different components ("Borehole Inspection Systems," 2013)**

and rpm are used in the tests to evaluate the performance of the algorithm with respect to various drilling parameters. As it can be seen, the voids have been detected with high detection rate and a few false alarms. Overall, the detection rate of 93.3% is obtained with false alarm rate of 6.2% in 161 holes. It should be emphasized that the above algorithm used only feed pressure for detection of the voids. Using other sensory data such as vacuum pressure and rotation pressure can further improve the results which will be discussed more as future research topic in next section.

## BOREHOLE LOGGING

As mentioned earlier, further improvement in the capability of the drilling system to characterize the ground is largely dependent on extensive field tests and how fast and accurately the rock masses surrounding the drilled boreholes can be analyzed. The conventional method of coring/testing requires considerable time and budget. Borehole logging is an effective alternative to address this issue in which variation in some physical properties of rocks in the bore hole is continuously recorded and later would be related to rock type and rock mass mechanical and physical properties. This is because usually accessing the target area is not practical due to their relative depth from the surface. The only exception is borescoping which is commonly used in mining applications where the side wall of the boreholes are inspected for joints and voids. However, the information generated from borescoping is very limited, has high chances of missing some features, is unable to verify the direction of the joints, and cannot be used in additional analysis of the spatial distribution of the joints in the ground. Uniaxial Compressive Strength of intact Rock (UCS) and condition of the discontinuities are among the most important parameters of the rock mass classification to be defined for stability evaluation. In the following, estimation and detection of these features by means of borehole probing will be elaborated in more details.

### Evaluating the Conditions of Discontinuities

Condition of discontinuities, such as frequency, orientation, roughness, filling and etc., is one of the main factors that control stability of the underground spaces. There are methods which employ borescope

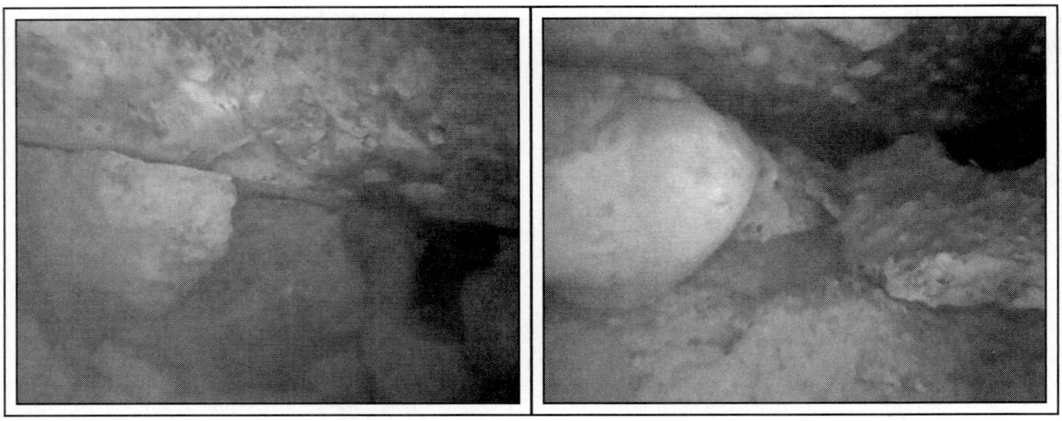

**Figure 4. Sample photos of borescoping operation in J.H. Fletcher & Co. R&D laboratory**

or endoscope as the device for evaluation of the fractures and other discontinuities (Ellenberger, 2009). Figure 3a shows a strata-scope. This device is a simple monitoring tool for checking the mine roof condition mainly to see if there is any fracture near the opening boundary or if any fracture or bed separation is initiated because of the mining operations. A more advanced tool with almost the same application is bore-scope. Different parts of a bore-scope package are shown in Figure 2b.

With bore-scope, unlike the strata-scope, the picture of the borehole can be recorded for future reference. In some cases, it is also possible to record the operator voice so the main features observed and their depths can be mentioned to make the video more informative. Moreover, the real-time picture can be seen in a LCD monitor. Both strata-scope and borescope are designed to be used in slim boreholes with the diameter of one inch or more. Also, one operator with a short-time training can conduct the inspection and carry the equipment. These tools are generally inexpensive (< ~$10k). However, since these types of instruments provide a narrow directional view and not a full 360 degree image, picture of borehole wall is of limited application in analysis of discontinuities. It should also be highlighted that even if the recorded video is available for future evaluations, it is not convenient to review the information and compare the data from multiple boreholes. Furthermore, some features, such as hair cracks, cannot be detected easily with these tools. Figure 4 shows two sample photos related to borescoping operation in R&D laboratory of J.H. Fletcher & Co. as part of instrumented roof bolter development project. The aim of this operation was to investigate the condition and location of the contact area between the concrete blocks, stacked on top of each other after drilling.

More advanced and expensive tools for borehole imaging are resistivity, sonic and optical televiewers. For each of these methods several products are manufactured. Both resistivity and acoustic viewers need a fluid-filled borehole. Optical TVs (OPTVs) can be used in borehole filled with clear, fresh fluid or empty borehole. Acoustic televiewers work based on amplitude and travel time of the reflected acoustic signals. Resistivity tools provide the image of the strata by means of the sensor pads which record the difference in resistivity between various layers on the borehole wall. Generally, the main components in the head of OPTV tools are a fish-eye mirror, LED light ring, and image sensor or camera. The obtained images are particularly suited to fracture and fault analysis and can also be used for interpretation of the near-wellbore stress field from borehole breakouts and drilling-induced fractures. Furthermore, unlike the borescopes they produce an unwrapped 360 degree picture, and not a pointed video of the borehole wall; therefore, comparison of different logs is more convenient. The depth relative to the collar of the borehole is also recorded automatically. The orientation of boreholes can be determined by the built-in 3-axis magnetometer or three accelerometers.

Optical televiewer or OPTV seems to be a better option for this study since it is dealing with empty boreholes. However, it also should be checked if the geometry and physical features of the probe are also suitable for this application. In addition, they may need to be explosion proofed in case they are to be used in coal mines or gassy tunnels. Figure 5a and (b) show the OPTV probe heads of ALT-Mt Sopris and DMT, respectively which are the two systems considered for application in our current study.

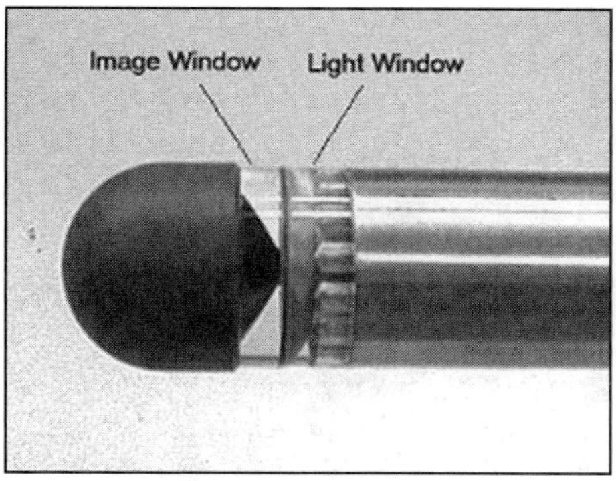


(a)

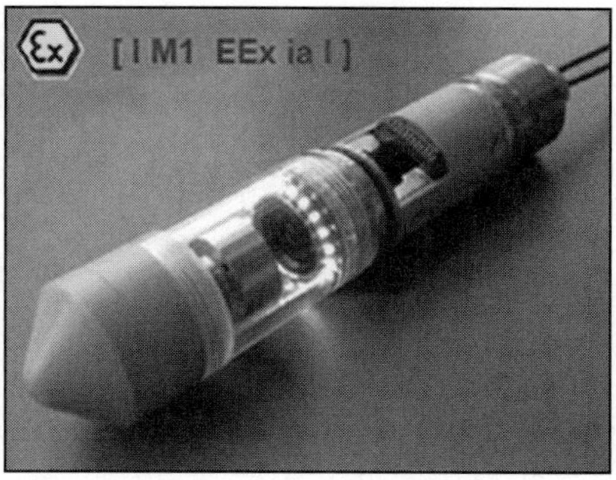


(b)

**Figure 5. OPTV probe head of (a) ALT-Mt Sopris Q40OBI-1000 ("Q40 OBI-1000." 2013); (b) DMT Slim Borehole Scanner ("DMT SlimBoreholeScanner," 2013) tools**

### Estimation of Rock Strength

For intact rock strength (UCS) estimation, sonic logging seem to be the most commonly used method (McNally, 1990; Guo & Zhou, 2011). Full waveform sonic probes and acoustic TVs can provide useful information related to the strength of the rock in addition to the conventional sonic tools (Guo & Zhou, 2011). As mentioned earlier, McNally (1990) and Oyler et al. (2010) developed equations to estimate UCS of intact rocks in the immediate roof of mines in Australia and the U.S., respectively. However, most of the studies were done on the surface and in down-hole boreholes. In these cases, filling the borehole with a fluid is not an issue; but these systems cannot be applied to this research since the boreholes are upward and most likely dry. Employing rubber packer system for filling the upward dry hole with a fluid such as water or gel to conduct sonic logging through the packer can be a simple solution. In addition, the size of the existing acoustic and sonic probes do not match short and slim borehole about 10 ft. long and 1" in diameter commonly used for roof bolting. The height restriction of the underground spaces especially in small tunnels can also be a problem for inserting the probes, which are typically around 2 m (6 ft) in the borehole. To get a more accurate data from the closely bedded layers, a high vertical resolution is needed which needs employment of more receivers in the sonic log. This will leads to a longer probe which is already unfavorable for the logging operation in an underground space. Overall, these issues make it very unlikely that the conventional sonic televiewers could be used for tunneling applications. However, employment of Acoustic TV can still be considered in conjunction with fluid-filled upward/slim/short drill-holes. The advantage of this method is that the transmitter works as the receiver, which maximizes the covered area. However, the initial processed data from this log is qualitative and just indicates the rock strength relative to the adjacent layers and feature. More studies should be done to produce localized or general quantitative information about the UCS of the intact rock.

As an alternative, mechanical probes could be used or developed to measure UCS. Stamp test and borehole penetrometer test are among the methods that could be used in the borehole to estimate the strength of the rock (Wagner and Schümann 1971; Wijk 1989; von Unrug 1999). However, these methods will be unable to provide continuous information about rock strength as the stamp penetration are performed at certain intervals. On the other hand, scratch test showed to be a relatively accurate and reliable approach (Roberto and Fabrice 2002; Schei and Detournay 2000). This method can provide continuous information about the strength of the rock along the borehole by making a scratch on its surface and measuring the forces and subsequent analysis of data to estimate rock properties. Extensive studies are needed to develop this system for boreholes. Efforts are underway to design and deploy a mechanical borehole logging unit as part of the current study which will pave the way for further investigation of the application of borehole logging in rock characterization and in training the instrumented roof bolter.

## DATA VISUALIZATION

Since this project involves dealing with a large amount of data, there is a need to be able to visualize the recorded data from various boreholes in a 3D so that it could be interpreted and presented easier and in a more useful manner for practical applications. Initial efforts in developing a 3D visualization of the borehole ground characterization data has been done by using commercial software packages, namely

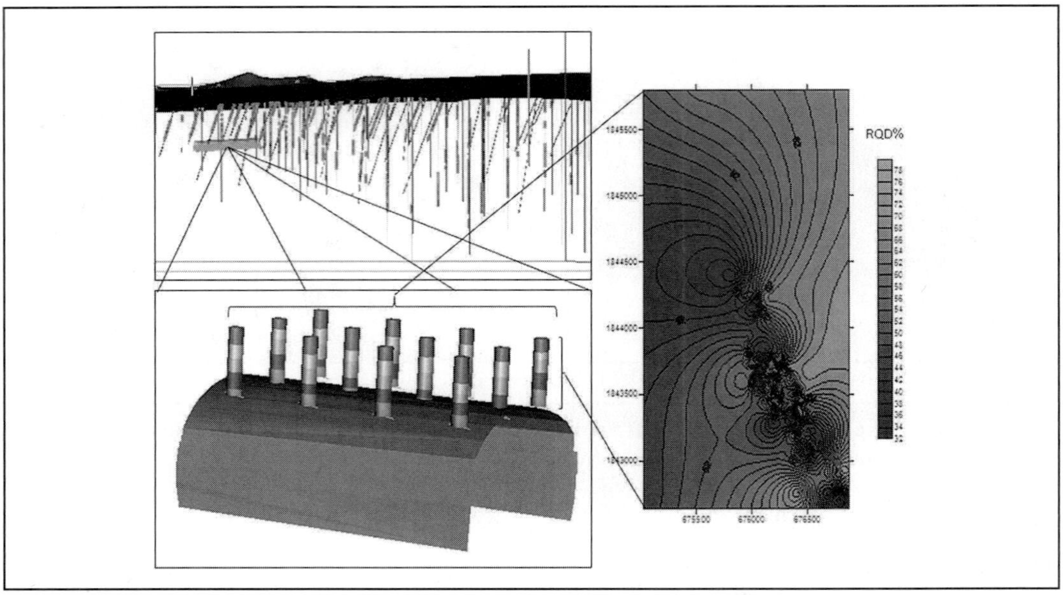


**Figure 6. 3D visualization of an unreal situation of boreholes drilled in a tunnel and the developed contour map for RQD (rock quality designation) of those boreholes, by Gems and Surfer software programs**

GEMCOM by Dassault Systemes Geovia Inc. and Surfer by Golden Software Inc. The objective of this exercise was to illustrate the encountered rocks and possibly voids and bed separation detected in a borehole around the underground opening in 3D. By detecting the position, distance, and frequency of the discontinuities in boreholes using the instrumented roof bolter, it is possible to calculate the RQD of each borehole and further plot the RQD variation in different sections. Figure 6 shows the result of preliminary trials using GEMCOM software package. Additional efforts are underway to work with other visualization software to select a platform for future developments and to enable the programs to show the stratification around the opening and to develop an algorithm for ground support evaluation and mapping of the failure risks.

## CONCLUSION

The use of data obtained from roof bolter to characterize the ground around an opening can be a substantial resource in understanding the ground conditions without interruption of the tunneling operations and related activities. This includes evaluation of the rock strength as well as detection of voids and joints using various operational parameters of the drills such as feed pressure, rotation pressure, drilling rate, and RPM. The CUMSUM algorithm was used to identify joints and bed separation from the full scale drilling test data collected from J.H. Fletcher test unit. The algorithm enjoys an adaptive threshold that does not need fine tuning and works well in different studied scenarios. Moreover, the algorithm has a recursive formulation which facilitates real-time computations. The results demonstrate the suitability of the proposed algorithm to achieve high detection rate with small false alarm rate.

For training of the drilling unit, a variety of borehole probing and logging systems will be used. This allows for quantifying rock properties and rock mass characteristics in shortest possible time so that pertinent data needed to estimate rock properties for correlation with drilling information can be generated. The rock strength and joint and discontinuity information can be used to develop a real time rock mass classification that can subsequently be used in related analysis to evaluate optimize ground support design. Ultimately, the generated information will be used to develop a 3D image of the geology of the ground and a hazard map for the roof and walls using a commercial program for data visualization.

## ACKNOWLEDGMENTS

This project has been funded by National Institute of Occupational Safety and Health (NIOSH) ground control group under the contract no. 211-2011-41138. Authors would like to express their gratitude for this funding and acknowledge the active support of NIOSH project manager, John Elenberger in this study.

## REFERENCES


Bae, D., Kim, K., Koh, Y., and Kim, J. 2011. Characterization of Joint Roughness in Granite by Applying the Scan Circle Technique to Images from a Borehole Televiewer. *Rock Mechanics and Rock Engineering* 44 (4) (February 5): 497–504.

Bahrampour, S., Rostami, J., Naeimipour, A., Collins, C., and Kahraman S., Instrumentation of a Roof Bolter Machine for Void Detection and Rock Characterization. *In: Proceedings of the 32th International Conference on Ground Control in Mining.* Morgantown, WV.

Borehole Inspection Systems. 2013. Retrieved October 21, 2013, from http://www.emsmining.com/productspage.html.

DMT SlimBoreholeScanner. 2013. Retrieved October 21, 2013, from http://www.dmt.de/fileadmin/PDF/Broschueren/Produkte/DMT_SlimBoreholeScanner_1109.pdf.

Ellenberger, John. 2009. A Roof Quality Index for Stone Mines Using Borescope Logging. *28th International Conference on Ground Control in Mining* (Class I): Morgantown, WV. 143–148.

Gu, Q., 2003. Geological Mapping of Entry Roof in Mines, *Dissertation* (PhD in Mining Engineering), West Virginia University, 190 pp.

Guo, H., & Zhou, B. 2011. Geotechnical Characterization from Geophysical Logs for Longwall Coal Mining. *73rd EAGE Conference & Exhibition*, (May 2011), 23–26.

Itakura, K., Sato, K., Ichihara, Y., Deguchi, G., Matsumoto, H., and Eguchi, H. 1997. Development of a roof-logging system by rock-bolt drilling. *Transactions of the Institution of Mining and Metallurgy, Section A: Mining Industry*. 106: A118–A123.

Itakura, K. 1998. Current drill monitoring system using mechanical data of drilling machine and estimation of roof rock structure. *In: Proceedings of the Aust.-Japan Tech. Exchange Workshop in Coal Mining 98.* Brisbane, Australia, 8–10.

Itakura, K., Sato, K., Deguchi, G., Ichihara, Y., and Matsumoto H. 2001. "Visualization of geo-structure by mechanical data logging of rock-bolt drilling and its accuracy." *In: Proceedings of the 20th International Conference on Ground Control in Mining.* Morgantown, WV: West Virginia University, 184–190.

Itakura, K., Goto, T., Yoshida, Y., Tomita, S., Iguchi, S., Ichihara, Y., and Mastalir, P. 2008. "Portable intelligent drilling machine for visualizing roof-rock geostructure." *In: Proceedings of the Rockbolting in Mining & Injection Technology and Roadway Support Systems Conference.* Aachen, Germany, 597–609.

McNally, G.H. 1990. The Prediction of Geotechnical Rock Properties from Sonic and Neutron Logs. *Exploration Geophysics* 21, 65–71.

Optim Stratascope. 2013. Retrieved October 21, 2013, from http://www.deterco.com/products/Optim/optim_stratascope.htm.

Oyler, D.C., Mark, C. and Molinda, G.M. 2010.In situ estimation of roof rock strength using sonic logging. *International Journal of Coal Geology* 83: 484–490.

Payne, D. and Ward, B. 2002. Strata Management in Weak Roof Conditions at Crinum Mine, in Aziz, N (ed), *Coal 2002: Coal Operators' Conference, University of Wollongong & the Australasian Institute of Mining and Metallurgy*, 126–134.

Q40 OBI-1000, Optical Televiewer. 2013. Retrieved October 21, 2013, from http://mountsopris.com/products/downhole-probes/item/q40obi-1000?category_id=12.

Roberto, S.R., Jørn, S.T., & Fabrice, D. 2002, October. Continuous scratch testing on core allows effective calibration of log-derived mechanical properties for use in sanding prediction evaluation. *In SPE/ISRM Rock Mechanics Conference.*

Schei, G., Fjær, E., and Detournay E. 2000. The Scratch Test: An Attractive Technique for Determining Strength and Elastic Properties of Sedimentary Rocks. *SPE Annual Technical.*

Tang, X. 2006.Development of Real Time Roof Geology Detection System Using Drilling Parameters During Roof Bolting Operation. *Dissertation* (PhD in Dept. of Mining Engineering), West Virginia University, 357 p.

Unrug, K.F. 1994. Realistic Design of Ground Control Based on Geotechnical Data Obtained During Mine Development. *Proc. 13th Int. Conf. Ground Control in Mining*, August 2–4, S. Peng and Charles T Holland (Eds), WV, pp. 227–232.

Unrug, K.F. 1999. Characterization of rock strength and jointing 'in situ'. *The 37th U.S. Symposium on Rock Mechanics (USRMS)*, June 7–9, 1999, Vail, CO.

Wagner, H, and EHR Schümann. 1971. The Stamp-load Bearing Strength of Rock an Experimental and Theoretical Investigation. *Rock Mechanics* 3: 185–207.

Wijk, G. 1989. The Stamp Test for Rock Drillability Classification. *International Journal of Rock Mechanics and Mining...* 26 (I): 0–7.

Zhou, B., Fraser, S., Borsaru, M., Aizawa, T., Sliwa, R. and Hashimoto, T. 2005. New Approaches for Rock Strength Estimation from Geophysical Logs. *Proceedings of Bowen Basin Symposium 2005–The future for coal–Fuel for thought.* Yeppoon, Queensland, pp. 151–164 (published by GSA Coal Geology Group).

# Black River Tunnel Phase 1: Innovative Quality Control Approach for Installation and Testing of Rock Bolts in Shafts

**J. Edberg and J. Brown**
NTH Consultants, Ltd.

**S. Pearson**
ARCADIS US, Inc.

## INTRODUCTION

This paper details the use of an alternative testing method to the 2004 "Post-Tensioning Institute (PTI) Recommendations for Prestressed Rock and Soil Anchors" manual's recommendations for temporary and permanent rock anchors for applications in shafts and tunnels. This method was developed and used for the Black River Tunnel (BRT) Project in Lorain, OH as a means to reduce testing time of rock bolts during shaft construction while maintaining a similar level of quality control and assurance. The following sections will detail the project background, subsurface characteristics, shaft construction, the 2004 PTI testing method and associated difficulties, the alternative testing method and conclusions reached.

## PROJECT BACKGROUND

The City of Lorain (City) is located in north central Ohio on the south shore of Lake Erie, approximately 25 miles west of Cleveland and 74 miles east of Toledo. The City chose to build a large diameter storage tunnel and pump station to meet their requirements to reduce sanitary sewer overflows to the nearby Black River. NTH Consultants, Ltd. (NTH) was retained by the City's prime consultant, Malcolm Pirnie, Inc. (now the water division of ARCADIS), to perform a geotechnical investigation, develop design and construction bid documents for the tunnel and shaft liner systems and ultimately provide oversight assistance during the construction phase of the project.

The BRT is an approximately 5,560 feet long, 23-foot diameter rock tunnel to be lined with a 19-foot inside diameter secondary concrete lining (see Figure 1). The tunnel is designed to accept flow at a new drop shaft (Shaft 3) located on the west side of the Black River, across from the City of Lorain Black River Waste Water Treatment Plant (BRWWTP). The tunnel will be dewatered at the south end by a large diameter pump station (Shaft 1)

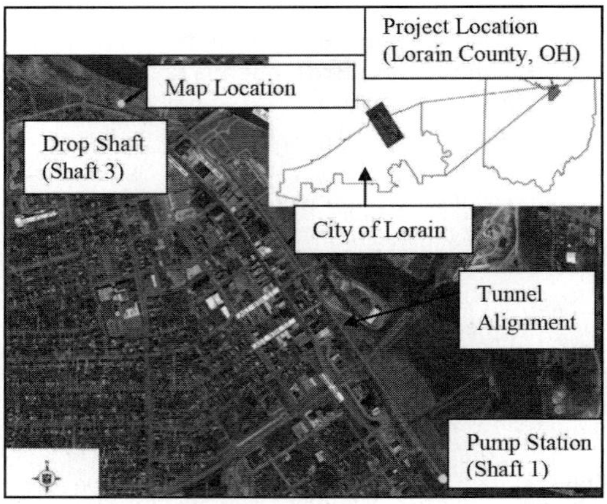


**Figure 1. Project plan view**

located in the existing Lorain Port Authority Public Boat Launch. Flow from the new pump station will be discharged into an existing shallow interceptor sewer and directed back to the BRWWTP.

## SUBSURFACE CHARACTERISTICS

The subsurface characteristics at the shaft sites generally consisted of overburden soils underlain by horizontally bedded shale bedrock which varies from highly weathered to fresh (unweathered). The shale is characterized as the upper Devonian-aged Ohio Shale Formation of Northeastern and Southeastern Ohio, according to the Ohio Department of Natural Resources Division of Geologic Survey's "Generalized Column of Bedrock Units in Ohio." (State 2004) In general, the top of weathered rock slopes downward from south to north with depths 20 and 41 feet below existing ground surface for Shaft 3 and Shaft 1 respectively. The highly weathered rock zone may vary from several inches to approximately 25 feet in thickness. Vertical joint sets are present in the rock formation. Average RQD values from top of

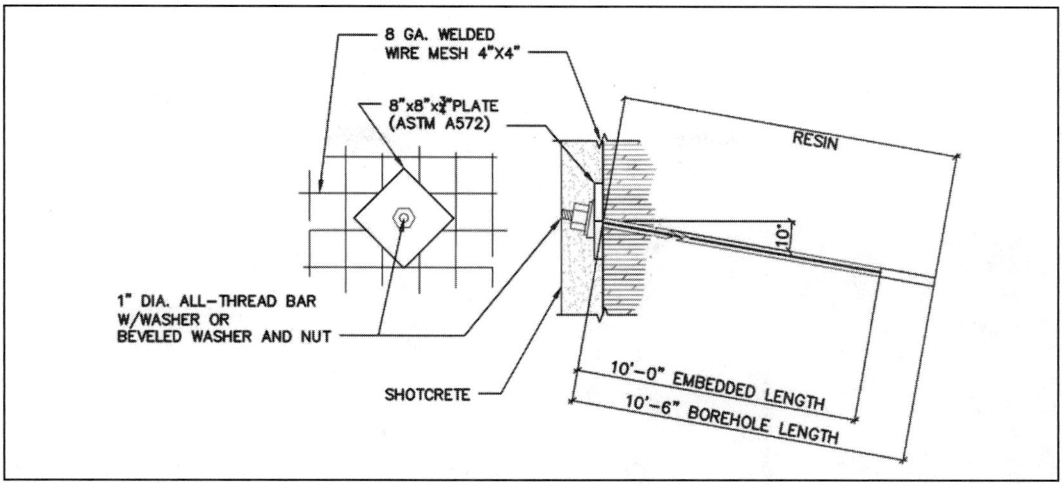


**Figure 2. Rock bolt detail**

bedrock to approximately 80 to 110 feet are 55 percent, with fractures per foot of 2.0 (ranged from 0.5 to 5.9 fractures per foot). Below this level, RQD values average 86 percent. For design of the shaft liners, the unconfined compressive strength was chosen as 1,500 psi. Other pertinent parameters for shale in the tunnel influence zone include the following: average indirect tensile test strength of 340 psi, average Cerchar abrasivity index of 0.3, and average slake durability of 76%. These numbers indicate the shale formation can be characterized as a soft rock with medium durability (Gamble 1971) that becomes less weathered and is of higher quality as depth increases. Groundwater infiltration was expected to be in the range of 13 gallons per minute (GPM) for Shaft 1 and 44 GPM for Shaft 3.

## SHAFT CONSTRUCTION

The temporary earth retention systems for the shafts were constructed as a two-tiered system consisting of a steel rib and liner plate upper portion through the soil overburden and highly weathered shale, followed by a grouted steel thread bar rock bolt with wire mesh and shotcrete temporary support system for the lower portion, through the less weathered to intact shale. Shaft 1 had a 46-foot diameter upper portion with a 42-foot diameter lower portion and an overall depth of 184 feet. Shaft 3 had a 53-foot diameter offset upper portion (to accommodate an influent chamber) and a 35-foot diameter lower portion with an overall depth of 117 feet. For both shafts, the overburden was excavated using an excavator, bucket and muck bins. The lower portion of Shaft 1 was excavated using blasting techniques, whereas the lower portion of Shaft 3 was mechanically excavated using a rock ripper and hydraulic hammer. The excavation methods at each of the shafts left the rock face uneven, particularly so with the use of blasting at Shaft 1. Additionally, both shafts had infiltration from the surrounding rock mass that was light but consistent. Both the uneven rock face and presence of water proved to be challenging to rock bolt testing as discussed later in this paper.

The rock bolts used for the lower portions of both shafts were 1-inch diameter, 10.5-foot long A615 all threaded steel bolts that were to be fully resin-grouted into the rock on a 3.14 (pi) ft × 3 to 3.5 ft grid spacing. They not only functioned as the primary liner system of the shaft, they also were designed to provide uplift resistance for the final concrete liner system.

The installation sequence involved drilling a 1⅜-inch diameter bore hole 10 feet, 6 inches deep on a 10 degree downward angle (see Figure 2), blowing out the hole with pressurized air, inserting the two-part resin cartridges into the hole, and then driving and spinning the rock bolt through the resin cartridges to thoroughly mix the resin. The resin was set within 30 minutes. The contractor used a 4-wheel hydraulic, self-propelled drilling unit to both drill the bore hole and insert the rock anchor. After installation, 4-inch × 4-inch, 8-gage wire mesh was placed over the rock face and then an 8-inch × 8-inch × ¾-inch thick steel plate and nut were installed over the mesh and locked off at a load of 25 kips according to design. The contract documents specified that each rock bolt be "proof tested" according to the "2004 PTI Recommendations for Prestressed Rock and Soil Anchors" manual to verify it was capable of holding the design load. The rock bolt was then be pre-tensioned to a required load of 25 kips. Finally, unreinforced shotcrete was applied to a thickness of

5-inches to protect the shale from water and temperature exposure.

## PTI TESTING PROCEDURE

The 2004 PTI manual proof test recommends that each rock bolt first be subjected to an incremental loading sequence, starting at a small "alignment load" typically between 5% and 10% of the lock-off load, followed by loadings of 25%, 50%, 75%, 100%, 120% and 133% of the design load (25 kips). A 10-minute hold at the 133% load is then performed at the end of the incremental loading and is referred to as a "creep test."

Figure 3. Rock bolt testing setup

The test loads were applied through the use of a calibrated center-hole jack that was set up over a cribbing system. The cribbing system allowed for a rod extension to be put on the end of rock bolt to accommodate the jack. The system also provided access to the nut in order to lock off the nut against the plate at the required design load upon completion of the test. See Figure 3 for an illustration of the test setup.

For the incremental load test, a deflection gauge is fixed to the jacking plate prior to beginning the loading sequence. The rock bolt is then subjected to each loading increment established by a jack pressure that is correlated to an axial load through a calibration procedure. Upon each loading increment, deflection readings are taken from the gauge. During the creep test, readings are taken at the 1, 2, 3, 4, 5, 6 and 10 minute time intervals with the load maintained at the 133% level.

The deflection readings taken during the incremental loading phase are then plotted against predicted theoretical deflection under the same axial load using a bar length consisting of the unbonded zone (free bar beyond the rock face) with 20% of the resin zone and a length consisting of the unbonded zone with 50% of the resin zone. An example of such a plot from the initial testing of Shaft 1 is shown in Figure 4.

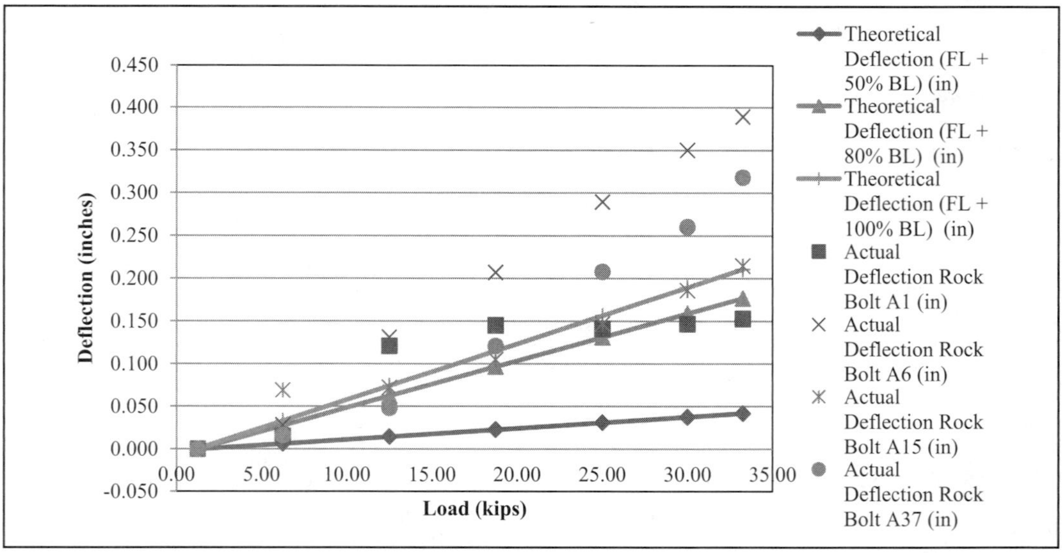


Figure 4. Deflection vs. Load for select initial shaft 1 rock bolts

In typical practice, Figure 4 provides an indication of how much resin is being mobilized to resist the applied load. For the bolt to be acceptable by PTI standards, the actual deflection should fall within the theoretical ranges in the plot. In reviewing Figure 4, it is apparent that the deflection falls outside the deflection parameters allowed by the PTI manual for most of the rock bolts, with the exception of Rock Bolt A1. Therefore only rock bolt A1 would be considered a passing bolt. The remaining rock bolts, in fact, yielded values that exceeded the theoretical deflection using the entire bolt length. This would indicate that the bolt should have pulled out of the wall. In reviewing these tests, and acquiring other results and observations from the initial testing, it became apparent that on many of the tests, the baseplate was locally crushing and ultimately embedding into the shale. This was due to the unevenness of the rock face and the softening of the rock caused by exposure to water. The movement of the baseplate ultimately resulted in a certain amount of angular distortion of the cribbing, which, in turn, would then artificially inflate the deflection readings. The angular distortion of the cribbing impacted testing in several ways:

- Proved difficult to determine how much of the resin was being mobilized to develop the load. Therefore, the criteria outlined in the PTI manual for the incremental load phase could not definitively used to pass or fail a bolt.
- Necessitated more frequent reliance of the creep test to verify that the rock bolt was satisfactorily holding the load. The PTI criterion of 0.04 inches was assigned as the threshold for a passing rock bolt. If the rock bolt deflected less than this value during the ten-minute hold, the rock bolt was deemed acceptable. For this project, since the rock bolts were used for uplift resistance in which the rock bolt would be sheared rather than pulled, the long term pullout performance of the bar was of less importance. This essentially allowed focus to be placed more so on the rock bolts ability to hold the load, rather than how it is exactly holding it.
- Resulted in increased test time as a result of constantly resetting the deflection gauge due to the cribbing movement.

The contractor attempted to mitigate the cribbing movement by using a pneumatic drill to "pre-torque" the bolts so they were seated better for testing. However a single test still could take between 20 and 30 minutes to perform. For a given row in Shaft 1 (42 bolts), this required as much as 41 man-hours using a two-man crew to complete. It became apparent that the contractor had not fully accounted for the schedule impacts the testing regime would have on the project. In order to maintain schedule, the contractor asked if the NTH/ARCADIS team could develop an alternative procedure that would save schedule while maintaining the required level of testing quality.

## ALTERNATIVE TESTING METHOD

The engineering team developed a procedure to maintain full testing of the rock bolts while significantly reducing the amount of time required for testing. The method involved transitioning from a predominately quality control approach solely through PTI testing to a more proactive quality assurance approach supplemented with quality control PTI and torque wrench testing. It should be noted that there are other rock bolt tests, such as pull-out testing (ASTM 2007) and electronic non-destructive testing (Hartman et.al. 2010); however, torque wrench testing was selected for its ease and familiarity of use, as well as its ability to test the strength of the rock bolt while simultaneously allowing the rock bolt to be locked off and used as a production rock bolt. To specify the correct torque, the manufacturer provided a correlation chart between axial load and torque, as shown in Figure 5. The prescribed torque for the 25 kip load was approximately 830 ft-lbs.

The alternative test method generally followed these steps:

1. Performed PTI testing of all rock bolts within the first two rows while maintaining full-time observation of installation by the engineer. Installation observation verified that the rock bolts were installed according to the contract drawings/specification as well as the manufacturer's recommendations. It also provided a measure of installation consistency between rock bolts. In particular, it was important to take note the following:

   a. Length of borehole
   b. Borehole cleaning
   c. Resin cartridges used
   d. Rotation and spin time of the rock bolt

The initial PTI testing allowed an understanding of the performance characteristics of the rock bolt, and verified that the contractor's rock bolt installation practices resulted in a rock bolt that produced a passing bolt. As stated previously, due to rock face conditions, emphasis was placed more on the rock bolt's ability to pass the creep test than a review of the incremental loading data. For this project,

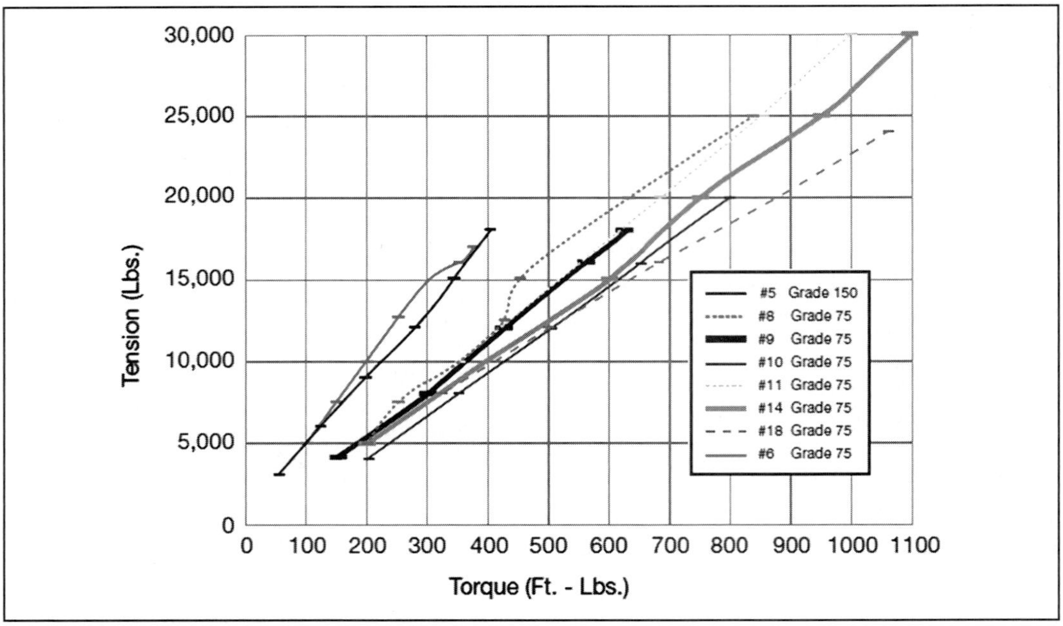


**Figure 5. Correlation chart between applied torque and axial load**

initial testing of the first row yielded a 29% failure rate (12/42). Failures ranged from immediately pulling out of the wall to failing during the 10-minute creep test at the 133% load. It is worth noting that the initial installation procedures varied between rock bolts, where some rock bolts were overspun (spinning transcended into gel time), had insufficient spinning to mix the resin, or in one particular instance, did not install the resin cartridges. Knowing that the installation practices were producing failing bolts, the contractor then established uniform, proper procedures within the second row to produce rock bolt that passed PTI testing.

2. Once acquiring an installation procedure that resulted in passing rock bolts, continued full time installation observation of subsequent rows to verify installation procedures were consistent with the initial passing rows.
3. Incrementally reduced the amount of PTI to 10% of the rock bolts. The approach on this project was to perform 10 PTI tests on the third row, with subsequent rows then reduced to 10% PTI testing. The10% PTI testing was continued in order to verify that performance and behavior of the rock bolts were consistent with previous rows and were satisfactorily passing according to PTI standards.
4. Tested all remaining rock bolts within the row with a torque wrench to ensure the rock bolt could carry the design load. The torque wrench was utilized in lieu of the PTI testing based on the following considerations:

   a. The design load of 25 kips was chosen on the basis that it was not only the required lock-off load, but during the initial testing of the rock bolts, the majority of failures occurred below this load. The rationale was that if the torque wrench successfully locked off to the design load, it was probable that the rock bolt would pass a PTI test. In future applications, it may be more prudent to lock off at the 133% level (highest PTI level), then back the load off to 100% for lock-off.
   b. Installation observation verified consistency between rock bolts within a given row that were PTI tested and rock bolts that were torque tested. Similar to the above rationale, provided the PTI tested rock bolts passed and the torque tested rock bolts were installed in the same manner, they also should pass a PTI test.

5. In the event of a failure whether it be through the PTI test or the inability to lock off the rock bolt with the torque wrench, the PTI testing is increased to restore confidence in the installation procedure and verify performance metrics that ensure the rock bolts satisfy the design and PTI criteria.

This change in testing procedure resulted in a reduction in test time per row from 41 man-hours to approximately 8 man-hours per row (using a two-man crew) while still maintaining a similar testing quality. However, there are some specific limitations that must be considered prior to implementing this procedure.

### Alternative Test Limitations

#### *Maintaining Lock-off and Creep Considerations*

The rock bolts utilized did not have an unbonded zone. Without an unbonded zone that is post-grouted after testing, the tensioning is essentially not "locked-in." This deviates somewhat from the PTI recommendations and can result in additional creep and loss of tensioning. For this project, since there was a small free length, the rock bolt would only need to mobilize (creep) 1/25th of an inch to regain the design load. This was considered negligible. For other applications in which creep could generate excessive movement (>1-inch) that may be detrimental to a wall system, a bond zone should be introduced. The PTI test or torque lock-off should be implemented and the rock bolt locked off before the unbonded zone resin gels. Additionally, it may be prudent to develop more long term time-load-creep relationships through the use of extended creep tests. Depending on the results of the testing, pre-tensioning loads may be increased to accommodate for the creep potential.

#### *Bonded Zone Penetration Considerations*

As previously stated, moving of the cribbing made it difficult to ascertain from the incremental loading phase of the PTI test how much of the bonded zone was penetrated to develop the load resistance. Again, for this case, since the rock bolts were used for uplift resistance in which the rock bolt would be sheared rather than pulled, the long term pullout performance of the bar was of less importance. In instances where long term conditions and the bond zone performance are critical, shotcrete could first be applied prior to placing the plate and nut and testing the rock bolt. This would provide a more stable surface that would allow bar deflection to be measured more accurately. If this is not possible, it may be more prudent to test the rock bolts to failure according to ASTM D4435 and determine an appropriate factor of safety.

## CONCLUSIONS

This paper presented a testing method alternative to utilizing 100% testing according to the 2004 "Post-Tensioning Institute (2004 PTI) Recommendations for Prestressed Rock and Soil Anchors" manual's recommendations for temporary and permanent rock anchors for applications in shafts and tunnels. Full testing of the rock bolts using the 2004 PTI technique proved to be difficult due to the jaggedness and softening of the rock face after excavation. This resulted in an increase in testing time that the contractor had not accounted for. The proposed alternative testing method involved a sequence of full PTI testing of the rock bolts to establish performance characteristics and verify the installation technique produced a passing bolt, followed by a transition to only 10% PTI testing. The remaining rock bolts were tested through a hand torque wrench. The testing method essentially allowed for a transition from a predominately quality control approach through full PTI testing, to a combination of heightened quality assurance through full time oversight with appropriate levels of quality control using PTI and torque wrench testing. This effectively maintained a similar quality of testing while reducing the overall test time. However, use of the technique must consider the effects of creep and the necessity for maintaining an appropriate lock-off load. The introduction of a bond zone and the development of a firm surface, such as shotcreting, prior to application of the pre-tensioning force, would allow for a more accurate understanding of creep and allow the pre-tensioning force to be more effectively maintained at the lock-off load.

## REFERENCES

ASTM Standard D4435, 2008, "Standard Test Method for Rock Anchor Pull Tests," ASTM International, West Conshohocken, PA, 2003, DOI: 10.1520/D4435-08, www.astm.org.

Gamble, J.C., 1971, Durability-Plasticity Classification of Shales and Other Argillaceous Rocks, Ph.D. Thesis, University of Illinois.

Hartman, W, Lecinq, B, Higgs, J and Tongue, D, "Non Destructive Integrity Testing of Rock Reinforcement Elements in Australian Mines," in Aziz, N (ed), 10th Underground Coal Operators' Conference, University of Wollongong & the Australian Institute of Mining and Metallurgy, 2010, pgs. 161–170.

Post-Tensioning Institute (PTI), 2004, Recommendations for Prestressed Rock and Soil Anchors.

State of Ohio Department of Natural Resources: Division of Geological Survey, "Generalized Column of Bedrock Units in Ohio," revised 2004. 15 Jan. 2015. http://www.dnr.state.oh.us/portals/10/pdf/stratcol.pdf.

# Myth and Reality: Bolts in Modern Concrete Segmental Tunnel Linings

**Anthony Harding, Malcolm Chappell, and Mike King**
CH2M HILL

**ABSTRACT:** Despite being a small component of concrete segmental linings the bolts are usually the topic of much debate. What should they be designed for? Do they serve a long-term function? Can they be removed at the back of the TBM backup or even omitted from the design altogether? This paper attempts to answer these questions by looking at the development of segmental linings and the reasons bolts were originally included, and then looking at some of the functional requirements commonly specified by designers and clients today. Myths are exposed and guidance for setting and fulfilling realistic functional requirements is provided.

## INTRODUCTION

The majority of concrete segmental linings being constructed around the world today use steel spear or curved bolts for the connections between segments. While recent years have seen steel bolts on the circumferential joints replaced with push-fit dowels in many projects, most of those projects retain steel bolts for connecting the longitudinal joints. While there are plenty of examples of linings that do not use bolts at all, they are the exception rather than the rule, and there is little guidance in the literature as to how segmental linings can be installed successfully without bolts in a conventional TBM. For this reason many clients, designers, and contractors with no experience of bolt-free segmental linings are unwilling to discard the tried and tested bolt system.

The authors of this paper have many years of experience in the design, specification, manufacture and erection of segmental tunnel linings and wish to share their views and experiences with the reader. The aim of the paper is to provide some guidance as to what functions the bolts perform in segmental linings. It first provides background to the historical reasons for the inclusion of bolts, noting the implications of the changes on concrete segments that has occurred in recent years. Then it makes reference to an industry consultation that the authors have undertaken, to understand what practitioners understand the bolts to do, and also what they cannot do. These functions are tested with reference to current design practice, providing three useful reference points for clients, designers and practitioners:

- Functions that the bolts normally perform in a typical lining
- Functions that are sometimes quoted, but that the bolts are unable to perform under normal circumstances ('mythical' functions)
- Criteria that may assist in omitting bolts from the design altogether

This paper does not comment on the use of bolts in cast iron linings, which are usually designed differently, although some of the concepts discussed may transfer to such linings. The paper also does not consider special concrete linings for which the bolting system has been designed to provide full moment transfer across the joint, as they typically have a much more structurally robust connection detail than one or two spear or curved bolts across the longitudinal joint.

## HISTORICAL BACKGROUND

The history of segmental linings in the UK, and the place of bolts in their development, is relatively easy to trace thanks to a couple of comprehensive papers on the development of precast concrete segmental linings in the UK. The first of these is TRRL Supplementary Report 335, by Craig and Muir Wood (1978). This report provides a state-of-the-art of UK tunnel lining practice in the late 1970s, with some reference to how the various technologies developed. It notes three very common types of precast concrete (PCC) segmental lining:

- Grouted smooth bore PCC segmental linings, which contained no bolts and required a former (circular steel support used for erection purposes)
- Bolted PCC linings which originally took the same profile as cast iron and were bolted with straight bolts across concrete flanges

- Expanded linings, which require no bolts

The report is quite clear that the reason for the inclusion of the bolts is to facilitate the build, which at that time was still predominantly lifted into place by hand or using roller bolts and winches. Indeed, it cites research undertaken by the Building Research Establishment that demonstrated that bolts make little or no difference to the long-term stress distributions in the ring, confirming the prevailing wisdom of the time.

The second paper was by Winterton (1994), and reports on some developments in the 15 years since from the TRRL report. These developments included:

- Mechanically jointed linings with dowels on the circumferential joints and steel connectors or hidden bolts in the longitudinal joints
- Trapezoidal smoothbore linings with curved bolts

The paper states that the bolts are required for erection and that if bolts are not provided then alternative support must be provided until the ring is fully grouted and the grout has reached adequate strength to support the ring. It also mentions that steel bolts perform no long-term structural function, and that engineers were seeking to develop a non-ferrous solution to avoid corrosion problems. However, it also noted that significant corrosion problems with this kind of bolt had not been reported. The authors think that this may be due to one-pass bolted lining systems seldom being used in corrosive environments, although the paper does not comment on this.

The paper also reports on the increase in segment sizes driven by a desire for reduced erection times. It is important to note that since the publication of the review this trend has continued. The size of segments has grown as the number of segments in the ring has reduced and ring lengths have increased in the endeavour to reduce connections and to increase productivity, with the attendant cost savings. This development has run in parallel with improved hydraulic engineering in TBM erectors.

Ring lengths in the UK at the time of publication of Winterton's paper would likely have been in the region of 1.0m long, which was a big increase from the original 2 ft (610 mm). Today segments are typically 1.2–1.5 m (2–4 ft) for smaller rings and can be in excess of 2 m (6.5 ft) for larger rings. Similarly the number of segments for a 6 m (20 ft) internal diameter ring in 1994 would normally have been in the region of 8 segments plus key or more, while today 5 segments plus key is routinely specified. Thus the weight per unit has increased significantly since 1994. Conversely, the bolts have not shown a comparable increase in size. Indeed, if anything the number and size of bolts has reduced slightly since the mid-1990s (some linings even only employ 1 bolt on the longitudinal joint), and are routinely employed in conjunction with short sockets that have less capacity than the bolts. If the bolts retained their original function, to provide structural support to the segments of a partially built ring, then the bolt sizes and numbers would have been expected to increase with the increasing segment size. The fact that this has not happened suggests that the bolts have become dissociated from this original function.

**Tunnel Failures**

One useful indication of a structural element's function is what happens when it is no longer able to fulfil its function. Therefore a short review of past tunnel failures has been undertaken, primarily based on a study by the Civil Engineering and Development Department of Hong Kong (CEDD 2008) into 42 tunnel collapses around the world, and a further 6 in Hong Kong itself. Of these collapses only 5 of these cases involved the failure of a segmental lining. In none of the projects were the bolts implicated. The bolts were only mentioned in two projects.

In the Humbercare tunnel in the UK, the fact that the bolts had been removed prior to collapse was noted in Grose et al. (1999). However, the report notes that high deformations appear to have been the cause. These deformations were reported in the press to have been as much as 0.5m (New Civil Engineer, 1999)—deformations that the bolts would have been unable to resist. The analysis showed that the differential pressures were high enough to cause shear failure of the contact pads—a force much higher than the bolt capacity. Therefore it is unlikely that the bolts would have prevented the initial water ingress that ultimately led to the collapse. It should be noted that independent review by the Closed Face Working Party of the British Tunnelling Society also did not implicate the omission of the bolts in the failure (Tunnels and Tunnelling International 2003).

In Cairo, failure of the lining occurred just behind the TBM. Reportedly (TunnelTalk, 2009), a segment at about the shoulder point in the previous ring that fell when the last segment of the subsequent ring was being inserted. The segment was in a ring just leaving the tail shield and its fall allowed water and soil to pour into the tunnel under the tailseal, filling the interior of the TBM and the tunnel and generating the ground loss that created a sinkhole crater on the surface. This lining was bolted and the bolts failed to prevent the collapse, suggesting that whatever the cause of the failure any additional structural capacity provided by the bolts was insufficient to resist the forces involved. As an aside the authors note that the size of keys has increased significantly since the early days of concrete segmental linings, as have grout pressures, and this means that the shape of the key must be carefully engineered such that the grout pressures are resisted by the lining in compression as bolts cannot perform this function.

Overall, an analysis of tunnel failures is consistent with the view that the bolts perform no structural function in the long term.

## STRUCTURAL BEHAVIOR

To assist in understanding the function bolts might serve in the long term it is important to understand the structural behavior of a tunnel lining. Segmentally lined tunnels are similar to brick and stone arch bridges, which are proven to work without bolts (some for over a thousand years even when subject to modern traffic loading today). Indeed, modern expanded concrete segmental linings work very well with no bolts, as do the brick arch linings constructed in many parts of the world. The shear forces in circular linings are typically very small, and the shear resistance provided by friction under the compression loads is much higher. Furthermore, in the unlikely event that shear displacements were to occur, they would typically arise from a load imbalance on either side of the joint. The displacement would result in the higher loaded side of the joint displacing inwards, reducing the load on that side of the joint and hence the shear force. In all but the softest of soils this would require only very small displacements to reach a stable equilibrium.

This fact is reflected in much design guidance available in different countries. Design guidance from the British Tunnelling Society (BTS 2004), French underground construction society (AFTES 2005), and German unneling committee (DAUB 2000) state that the bolts assist in assuring build accuracy. The BTS and DAUB guidance do not suggest any other functions, but the AFTES guidance states that the bolts will hold the gasket closed, and mentions some other functions that the bolts can perform in certain circumstances:

- Assist segment stability for accidental load cases during build
- Help the ring retain its shape prior to ground load coming onto the ring
- Provide resistance to internal pressures in pressure tunnels

Design guidance from the US Federal Highway Administration (FHWA 2009) only mentions that the bolts are provided hold the gasket closed, noting that bolts are often omitted from the circumferential joint as the TBM rams exert sufficient force to hold the gasket closed.

The Japan Society of Civil Engineers (JSCE 2009) guidance for segmental linings provides methods for the design of joints and bolts to ensure moment resistance across the joint (a special case that is outside the scope of this paper), and also provides methods for bolt-less systems. Similarly the International Tunnelling Association (ITA 2000) guidance states that if bolts are left in the lining then they may be considered as reinforcement, and ideally designed for the worst moment. It further states that if bolts are not included or removed then the joint must be designed to transfer an eccentric normal force at the joint. The authors note that the methods to design for moment resistance typically require more robust connections than one or two spear bolts that are usually observed in segmental tunnel linings. For this reason the joints many or most segmental linings around the world are designed as if no bolts are present irrespective of whether bolts are retained or removed, which is consistent with the guidance from BTS, DAUB, AFTES, and FWHA.

Therefore it is clear that under normal design conditions (once the ground load has come on the tunnel) a well-engineered concrete segmental lining can perform without any requirement for bolts. The remainder of this paper will discuss the requirement for bolts prior to normal ground loads being established on the tunnel, and try to clarify circumstances where bolts may assist in the permanent condition.

## INDUSTRY CONSULTATION

The authors have undertaken an industry canvassing exercise to determine what practitioners consider to be the function of bolts, and also what functions they believe bolts cannot perform. The pool of individuals canvassed covered clients, contractors and designers. Most meetings were face to face or telephone conversations, but some were by filling in a questionnaire and views were also provided through a discussion thread on the social networking site LinkedIn. The following functions were generally cited:

- Hold the rings together
- Historical—"If it has worked before why change it?"
- Keeps the ring from relaxing in the tail shield
- Holds the gasket closed

In addition, a number of other functions that the authors have encountered in the past were discussed, and generally agreed not to be credible functions:

- Pulling the joints closed
- Restricting rotation/moments in service
- Preventing displacements in service

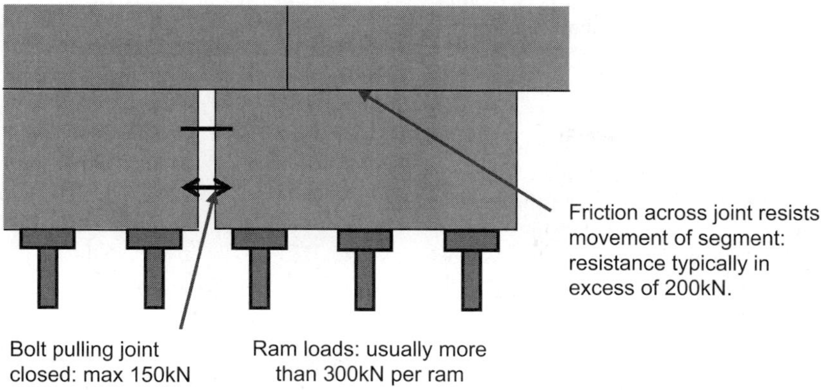


**Figure 1. Pulling joints closed**

This section discusses the functions discussed above and makes other observations pertinent to the functionality of connection systems for the longitudinal joint.

### Holding Rings Together During Ring Build

Bolts between segments (and between rings if bolts are employed rather than dowels) are typically installed immediately after each segment has been placed, although one of the authors is aware of a tunnel where the bolts were installed at the end of the build prior to advancing the TBM. The most commonly cited reason for including the bolts was to maintain the shape or build of the ring once erected. In some cases this was expressed as holding rings together, keeping the ring from relaxing in that tailshield, or simply holding the gasket closed. The important factor in all of these cases is that before the ring is grouted the ring is not firmly held in place and small forces can cause significant deformations, which can become 'locked in' to the lining once it is grouted.

Perhaps chief among these forces is the force from the gasket, which when fully compressed can exert forces of 30–50 kN per m (2.1–3.4 kips/ft). These loads are eccentric to the joint, thereby also creating the potential for rotations if movement is not adequately restrained.

It was also noted that the action of the tail shield brushes (which compress the ring) can often force the ring into a more round shape, as can correctly executed tail shield grouting. This was evidenced by the observation that some bolts around the ring are often found to have loosened during the shove and need to be subsequently re-tightened.

A second function of the bolts was to hold the segments of partially built rings together. One contractor cited an instance where the wrong ram had been removed from the ring, resulting in an unbolted segments becoming unsupported and falling onto an operative. It is recommended that rings should be designed for the removal of rams from a segment in a partially built ring (incorrect TBM operation).

It should be noted that some modern TBMs are provided with ring adjusters that hold and/or adjust the ring to a circle as it passes through the tail seals and whilst it is grouted.

### Pulling the Joints Closed

Some of those interviewed raised the point that the bolts cannot be used to pull the gasket closed, only to hold an already compressed gasket closed. One person mentioned an example when closing the joint using the bolt had been attempted, with the consequence that the bolt threads were stripped and the joints remained open. The possibility of pulling the joints closed was considered by the authors to be unlikely due to the large forces that would be required. In many interviews the restraint of the previous rings was discussed. For the joints to close the two segments either side of the joint must move together, but are restrained from doing so by the friction with the previous ring. This is illustrated in Figure 1, which uses likely minimum ram forces and frictions for a typical project. As the friction created by the ram forces is higher than the bolt capacity, the bolts are not able to close the joint and will fail.

This highlights the important of ensuring that the joints are closed with gaskets compressed upon placement of each segment, which was mentioned by many interviewees.

As a slight aside, in the authors' experience it is often specified that bolts are not used to compress the gasket, but only designed to hold it closed once compressed by the TBM erector. It is also frequently specified that the segment erector have sufficient power to compress the gaskets in all expected build configurations. The authors believe that this is a prudent thing to specify.

### Restricting Rotation/Moments in Service

None of the individuals canvassed considered that the bolts might offer much resistance to movements in service. Their ability to resist rotation is limited. In a typical 300 mm (12 in) thick lining the joint contact area will be around 170 mm (around 7 inches), so the maximum lever arm for moment development will be less than 100 mm (4 in). Given that the capacity of a spear bolt is usually around 100 kN (22 kips), this gives a moment capacity of less than 10 kNm (7 kipft)—substantially less than the moments required to restrain significant movements, or to close open joints, in most ground conditions.

This suggests that designers should be very careful not to consider the bolts as providing extra security against rotation. The bolts cannot perform this function in a normal design with spear or curved bolts, so if extra security is required for any reason then an alternative means of restraint must be provided.

### Preventing Displacements in Service

This issue primarily relates to lateral slipping of one segment relative to another of the kind that results in lipping. The prevailing view seems to be that the bolts offer little restraint in this regard once the ring is grouted. The first reason for this is that the hoop load across the longitudinal joint is typically at least 1000kN/m. Therefore the force required to overcome the friction across the joint is several times greater than the bolt strength. Furthermore, the force that would be required to cause such a displacement is much larger than what would usually be designed for. However, given that the bolt will only increase the capacity of the joint by around 10%, it offers little additional security against such an unusual force.

The second consideration with regard to this potential function is the detailing of the typical bolted connection, which is usually fluted towards the joint to facilitate demoulding and help guide the bolt into the socket. Initially the bolt is more-or-less centered in the hole, so if there are any shear movements the bolt bends along the entire length between socket and bolt head. The resulting loads in the bolt are very small due to the long length. Once the bolt touches the side of the bolt hole, typically at around 5 mm (0.4 in) the bolt acts in bending between the joint face and the socket. Over this distance the bending moments are much higher and the plastic capacity of the bolts is reached rather quickly. Analyses using typical arrangements and bolt sizes suggest that this plastic deformation provides 15–30% of

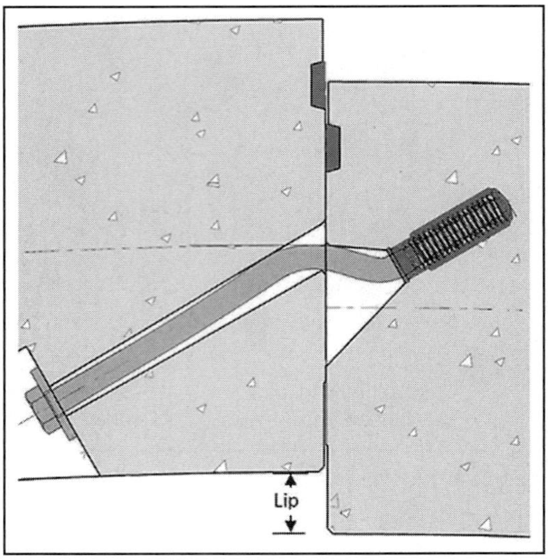


**Figure 2. Bolt deformation**

the bolt's shear capacity depending on the precise configuration. Only after significant movement has occurred does the bolt truly engage the full shear capacity of the bolt, as illustrated in Figure 2. The movements that would be required to reach this situation would be significant—potentially in excess of 50mm (2 inches).

No one interviewed could recall seeing lipping of this magnitude, so it would appear that in most normal tunnels the bolts are not significantly engaged in shear, and do not really contribute to the shear resistance of the joint. Furthermore, if the bolts were to become engaged in shear the deformations would compromise the waterproofing as the gaskets would no longer be in contact with each other—resulting in significant water ingress and the likely attendant consequences. It was also reported that after the Northridge earthquake in Los Angeles, a tunnel under construction that was affected suffered no movement of this kind despite the fact that it had no bolts. The fact that this kind of movement has not been observed demonstrates that the bolts would not have contributed to the structural resistance to shear movements even if they had been installed.

The above observations are relevant to spear bolts. However, it should be noted that curved bolts will probably engage in shear at 10–15 mm (0.4–0.6 in) displacement, providing little or no resistance prior to that.

As with the joint rotations, designers should be extremely careful before relying on the bolts to restrain shear forces in any permanent and serviceable manner, as they will be unsuitable for that function in most circumstances. As was mentioned by a number of those consulted, bolts should certainly not be relied upon to provide shear resistance between rings at cross passage openings.

### Seismic Design

While no-one interviewed stated that bolts were required for seismic design, reference was made to the fact that bolts are sometimes specified as remaining in the lining to provide additional redundancy. This is also mentioned by Dean et al. (2006), who reviewed a number of concrete segmental linings subject to significant earthquakes, citing only one instance of minor damage being sustained by a segmental lining. Hashash (2001) mentions that ensuring relative flexibility of linings is an important aspect of seismic design, which suggests that a component that has the potential to decrease flexibility (albeit only marginally) might be best removed. Furthermore, if rotations are significant then bolt failures—which could take the form of concrete cone failure—could result from the movements, thereby creating damage. However, as no instances of bolts causing damage have been noted it is likely that the argument as to whether bolts are best left in or out will remain moot for some time to come.

Nevertheless, the authors note that the shear resistance that bolts offer can equally be provided by guide rods, which have the triple benefit of engaging in shear at lower deflections, providing a more linear increase in resistance with displacement, and not being subject to the same durability concerns as bolts.

### 'Shove' Rings

When launching from a shaft tunnel boring machines typically build a number of rings that remain within the shaft. As these rings carry no load but their own self weight and do not benefit from the supporting effects of grout and ground, a support system is usually required. In such a low load situation bolts can offer benefits to the support system. Therefore while not normally quantified in the design of the ring support, the bolts are usually installed to provide a little redundancy.

### When Might Bolts Not Be Required?

Examples were cited of rings, other than expanded rings, that had been built without bolts. Such rings were fully grouted with two-component tail shield grouting such that rings were fully grouted as they exited the tail shield. Guide rods are usually used to facilitate correct location of the segments at the longitudinal connections, and push-fit dowels that provide tensile resistance were used on the circumferential joint.

Before the ring has passed through the tail shield brushes the segments are initially held in place by build bars in the invert of the tail shield, their own self weight, and friction between them and the previous ring created by the TBM rams. The previous ring is usually at least partially grouted in and therefore offers significant resistance to deformation. The friction resistance generated by the TBM rams would generally be several times higher than the forces that the bolts are capable of, but may not necessarily be uniformly distributed around the segment even with small imperfections in ring plane. Furthermore, the restraint offered is only at the trailing edge of the ring that has just been built, while the gaskets on the longitudinal joints exert forces all the way to the leading edge. For these reasons, and also simply knowing that bolting offers a degree of redundancy to the system, many felt that having the bolts present offered additional assurance of build quality.

When tail shield grouting is not employed, however, the retraction of the rams to build the next ring can leave the segments without restraint against movement. While dowels will restrain longitudinal movement, the gaskets would tend to push the segments out of position radially, particularly at the leading edge of the ring. Therefore in this circumstance bolts should certainly be provided.

Among the contractors interviewed there was no consistent view as to whether removing bolts from the design would save time or money, apart from the materials saving. It is the authors' view that the issue depends very much on how the TBM crews are organized, and how the ring build is undertaken. Furthermore, the successful building of a ring with no bolts will depend on workmanship. Therefore, while the decision to pursue a bolt-less solution should be led by the designer to ensure that there are no unintended consequences, full buy-in from the contractor is also required to overcome any issues during construction.

## OTHER ISSUES

As well as functional requirements for the bolts a number of other issues were raised and discussed as part of the canvassing exercise which are also relevant considerations for segment design.

### Durability Concerns

While both Craig and Muir Wood (1978) and Winterton (1994) mention that in many tunnels the bolts remain in place without excessive corrosion for decades, there are examples of bolts falling out due to corrosion. Furthermore, designers may be asked to demonstrate that their solution provides durability for the tunnel design life, usually 100–120 years.

Perhaps the simplest method of dealing with this issue is to specify corrosion resistant materials such as galvanized or even stainless steel bolts. However, this can add extra cost, and in the case of galvanized bolts can be subject to damage of the coating during installation. Even if the coating is maintained galvanizing may still not be sufficient in some environments.

A second option is to fill the bolt pockets with grout or concrete, thereby protecting the bolt head, which is most exposed to the environment. However, this does not prevent galvanic cells forming in the joint (if still water is present) and corroding the bolt shaft, which may be a risk. This can be overcome by filling the joint area around the bolt with grout, but this is a costly exercise and can be difficult to demonstrate full encapsulation of the bolt shaft. Shrinkage of the material in the pocket can also result in cracks sufficiently large to permit corrosion. Furthermore, the authors are aware of instances of the pocket infill itself falling out. Therefore it is recommended that if designers specify bolt pocket filling then full consideration of the above issues should be made in the design and specification of the solution. Consideration should also be given to whether the bolts are truly required in the permanent design before specifying durability measures, as they may be adding unnecessary cost to the project.

### Bolt Removal Post Erection

Bolt removal post-construction is commonly undertaken in many projects around the world. This is an obvious solution to the durability issue, but in response to this a number of issues were raised by a number of those interviewed.

Firstly, while more than 50% of the contractors interviewed thought that taking the bolts out would save cost, a significant minority believed that the cost of taking the bolts out would be more than the cost of the bolts saved. The authors believe that this probably hinges on whether operatives have 'down time' during the normal TBM cycle to remove bolts, as well as the cost of local labor in comparison to the cost of sourcing the additional bolts. If additional operatives are required to remove the bolts then removal is unlikely to be cost effective.

One pitfall raised by an individual with considerable experience of removing bolts is that it is often assumed that bolts can be re-used indefinitely. He stated that this was not the case, and that around 10 re-uses was a practical limit. Therefore if the contractor has not re-used bolts before the designer or client may consider stipulating criteria for re-use with reference to the contractor maintaining appropriate stock levels.

Another issue raised was that not all bolts come out. If bolts are to be removed purely to save cost and there are no safety or durability concerns with leaving them in then the easiest option is simply to leave them in. However, there can be safety concerns with them falling out, such as in transportation tunnels. The authors are aware of metro and rail tunnels where loose bolts have been observed and the risk of a bolt falling out on a train is a very real operational concern. Therefore it is recommended not only the bolts are removed, but that a mitigation is required for bolts that cannot be removed with an air spanner. Cutting of bolt heads was mentioned as a mitigation by some of those consulted, but there are other solutions. If portions of the bolt are left in then their durability must be addressed in the same manner as where bolts are retained.

### Bolt Design—Failure Modes

Most of the practitioners who referred to witnessing bolt failure referred to 'thread stripping'. This is not surprising as in the experience of the authors this is often the point in the system with the lowest capacity. However, this is an issue that is often missed by inexperienced segmental lining designers. The capacity of the bolt itself is not the only consideration: there are five ways that the bolting system can fail under tensile loading.

1. Failure of the bolt in tension
2. Failure of the interface between the bolt and the socket
3. Failure of the interface between the socket and the concrete
4. Concrete cone pull-out
5. Bearing failure of the seating under the bolt head

Items 1, 4 and 5 can usually be checked in accordance with relevant codes of practice, albeit that cone failure requires careful interpretation as the joint face is not perpendicular to the bolt shaft, and the intrados face of the segment can bisect the failure cone. Items 2 and 3 can only be checked with reference to test data for pull-out of the socket embedded in concrete. This information is usually provided by the manufacturers for a 'normal' concrete mix but tests with the actual segment mix can be undertaken if required.

### Providing "That Little Extra"

A couple of those consulted in the exercise mentioned that they had undertaken assessments of existing tunnels for which the actual load conditions differed from those originally designed for, including a case where there was insufficient hoop load to hold the gasket closed. In most of those instances the small (typically circa 5%) difference that the bolts made was the difference between a pass and fail. However, one of those interviewed cited another instance where bolts were removed (at great expense) to prevent damage arising from settlement induced deformations. This fits with the authors' own experience with London Underground Limited, where bolts are routinely removed from existing cast-iron linings to prevent damage from new tunneling works in the vicinity. Therefore while there may be unanticipated cases where bolts might provide a benefit, leaving them in can also risk damage or other unintended consequences.

### Designing Rings for Post Construction Cases

The fact that the bolts are unsuitable for use in restraining ring to ring movements at cross passages was raised by some of those consulted. This is because bolts provide little resistance until large displacements are observed, larger than permissible steps and lips between segments. Stiffer cross-joint solutions are therefore recommended for this purpose.

Where tunnels are required to resist small internal pressures (or small differences between internal and external pressures where the net effect is outwards), it is possible to design the bolts for tension if the durability of the bolts can be assured. However, the impact of one or more bolts failing should be assessed as an ultimate limit state or accidental limit state case, to verify that unexpected failure will not result in unacceptable effects such as collapse, excessive lipping, or excessive leakage. Particular attention should be paid to the fact that a pair of bolts will typically act across the joint, and that if one bolt fails then all the load will be transferred to the adjacent bolt, doubling the force.

## CONCLUSIONS

### Myths

The discussion in this paper aimed to uncover the function of the bolts in conventional precast concrete segmental linings. Perhaps the only definitive function that has emerged from the research is to hold the joint closed under gasket loads, which is what designers usually design for and most design guidance specifies. However, though exploring the responses to questions on the current use and function of bolts, and examining some of the issues with bolts performing other functions, it has been possible to expose some myths about the normal application of bolts in segmental linings. Specific observations are:

- Bolts rarely perform any structural function in the long term
- Bolts cannot be used to pull the joints closed
- Bolts have a limited capacity to improve lining performance by restricting rotation/ moments in service
- Bolts cannot practically prevent displacements in service

Of course, these only apply to the normal application of segmental lining designs. Nevertheless, it is clear from the research is that normal spear or curved bolt design is unlikely to be adequate to secure any of the above functions, so if any of these functions are actually required then careful design and specification of the system will be needed.

### Reality

Whilst the authors don't universally recommend the removal of bolts from all segmental linings it is recognized that their removal will improve the carbon footprint of the lining. With the global proliferation of major tunneling projects this is an important issue for all parties involved in the project to consider carefully and profoundly.

The exercise has also raised some useful advice to designers for non-expanded linings, which is presented in Table 1.

Table 1 is not intended to be exhaustive, but to highlight some of the key issues that should be considered when designing with or without bolts under different construction or operational cases. It has been noted in a number of places in the paper that robust engineering is required.

This suggests that, as with all design advice, the advice is useful only if paired with the practical application of sound engineering principles and a clear understanding of the specific problem in hand.

## ACKNOWLEDGMENTS

The authors would like to acknowledge that much information shared in this paper has been provided by the significant number of individuals in the industry who took the time to share their understanding of the issues with us. Without their input this paper would not have been possible.

Particular thanks is due to Weidong Zeng of CH2M HILL for his assistance in sourcing and translating the design guidance from JSCE.

**Table 1. Design advice**

| Issue | Design Advice |
|---|---|
| Bolt design | • Design for retaining gasket compression forces.<br>• Design for partially built case unless alternative safeguards are available.<br>• Ensure that all modes of bolt failure are assessed: failure of bolt/socket interface or concrete cone-pull-out are often more critical than failure of the bolt itself. |
| Designing without bolts | • Two component continuous tail shield grouting should be used. Grout must have sufficient strength to restrain the ring before the rams are withdrawn to build the segments of the next ring.<br>• The decision to pursue bolt-less solution should be led by the designer to ensure that there are no unintended consequences, but with full buy-in from the contractor.<br>• Dowels should be provided on the circumferential joints.<br>• Guide rods should be considered to assist in the build.<br>• The restraint provided by the tailshield brushes, while real, should usually be ignored as the brushes can become worn or damaged.<br>• Good control of ring plane is required.<br>• Ensure that mitigation measures are provided for the possibility of a segment of a partially built ring falling in the event that support from the TBM rams is accidentally withdrawn. |
| Bolt removal and re-use | • Bolt removal and re-use is acceptable in most normal design situations, but both designer and contractor should be mindful that bolts cannot be re-used indefinitely, and that a mitigation is required for bolts that cannot be removed. |
| Reliance on bolts post-construction | • The bolts on the longitudinal joint should not routinely be relied on to provide any restraint against slipping or rotation under normal post-erection load cases. Any attempt to rely on this resistance must take into account the displacement required to mobilize the resistance of the bolts, and the forces required to generate such movements (which are likely to be much larger than the bolt capacity).<br>• Conventional spear or curved bolts should not be used to provided shear restraint between rings at temporary cross passage openings.<br>• Bolts may be designed for tension, but with care.<br>• Bolts may improve lining performance in a seismic event, but this is hard to prove. Designers should consider carefully whether shear resistance is required, and if so consider the use of guide rods as they may perform this function better.<br>• Any reliance on bolts in the long term should be subject to a thorough review of bolt durability, and the implications of single bolt failing must be considered. |
| Bolt pocket filling | • If durability is to be provided by filling the bolt pocket then the design of the filling should ensure that cracks from shrinkage of the material in the pocket cannot occur, and the fill material is securely fixed to the pocket or bolt. Consideration should also be given to the potential for corrosion of the shaft. |

## REFERENCES


Association Française des Tunnels et de l'Espace Souterrain (AFTES), The design, sizing and construction of precast concrete segments installed at the rear of a tunnel boring machine (TBM), GT18R1A1, AFTES, France, 2005.

British Tunnelling Society (BTS) and Institution of Civil Engineers, Tunnel lining design guide, London, 2004.

Craig, R N; Miur-Wood, A M; A Review of Tunnel Lining Practice in the United Kingdom; TRRL Supplementary Report 335.

Dean, A, Young, D.J., Kramer, G.J.E., The use and performance of precast concrete tunnel linings in seismic Areas. IAEG2006, Nottingham, United Kingdom 6–10 September 2006.

Deutscher Ausschuss für unterirdisches Bauen (DAUB), Concrete Linings for Tunnel built by underground Construction, Germany, 2001.

Grose, B; Benton, L; Hull wastewater flow transfer tunnel: tunnel collapse and causation investigation; Proceedings of the Institution of Civil Engineers, Geotechnical Engineering 158, October 2005 Issue GE4, Pages 179–185.

Hashash, Y.M.A., Hook, J.J., Schmidt, B., Yao, J.I.-C., 2001. Seismic design and analysis of underground structure. T Tunnelling and Underground Space Technology 16, 247–293, 2001.

Japan society of Civil Engineers (JSCE), Design of Segment for Shield Tunnel [in Japanese], Tunnel Library 23, Japan, 2009.

Mainland East Division, Geotechnical Engineering Office; Catalogue of Notable Tunnel Failure Case Histories (up to December 2008); Civil Engineering and Development Department, Hong Kong.

Winterton, T R; Developments in precast concrete tunnel linings in the United Kingdom, London 1994, Tunnelling '94, pp. 601–633.

Magazine article: Workers fought to save collapsing tunnel; New Civil Engineer; 25 November 1999.

Magazine article: UK EPBM collapse comparisons; Tunnels and Tunnelling international, 1 March 2003.

TunnelTalk, Cairo Metro tunnel collapse; Internet article, http://tunneltalk.com/Cairo-Metro-Sep09-tunnel-collapse.php, 2009.

US Department of Transport Federal Highway Administration (FHWA), Technical Manual for Design and Construction of Road Tunnels, FWHA-NHI-10-034, USA, 2009.

Working Group No. 2, International Tunnelling Association (ITA), Guidelines for the Design of Shield Tunnel Lining, Tunnelling and Underground Space Technology, Vol. 15, No. 3, pp. 303–331, 2000.

# Development of Sprayed Concrete Linings for London Underground Bond Street Station Upgrade Project, London, UK

**B. Lyons and S. Kumpfmueller**
Dr. Sauer & Partners Ltd.

**M. Sleath**
Atkins Ltd.

**D. Terry**
Costain Laing O'Rourke Joint Venture

**ABSTRACT:** Bond Street station is undergoing a major upgrade to increase capacity, improve accessibility and create interchange with the new Crossrail station. It is one of the most complex sprayed concrete lining (SCL) tunnel design projects undertaken by London Underground. The design has had to address challenging connections to and interfaces with the existing infrastructure requiring the use of binocular sections, an underpass tunnel and a large cruciform junction. This has required significant design effort, maximizing innovation and utilisation of the latest technology including the use of fibre reinforced sprayed concrete for both the primary and secondary linings.

## OVERVIEW

London Underground's Bond Street Station is an important destination in the Tube network, serving tourists, residents and businesses in the heart of the West End. It is also a key interchange between the Central and Jubilee lines. The station suffers severe congestion, and passenger numbers, currently at 155,000 per day, are predicted to rise to 225,000 once Crossrail services arrive in 2018. The £300 million Bond Street Station Upgrade scheme will provide essential congestion relief by means of a new station entrance and satellite ticket hall, and additional escalators, lifts and passageways. Apart from congestion relief, the project will also allow step-free access throughout the station, improved fire safety, and provision for interchange with the new Crossrail station directly adjacent.

To enable these works, London Underground (LU) in 2007 commissioned Atkins, supported by Dr. Sauer and Partners, to carry out Outline design of the station upgrade. Following this, LU in 2010 let a Design & Build contract to a joint venture of Costain Laing O'Rourke (CoLOR). CoLOR in turn let the multidisciplinary detailed design to a joint venture of Atkins and Halcrow (HAT) with Dr. Sauer and Partners as specialist subconsultant responsible for all design of sprayed concrete lined (SCL) tunnelling works.

## HISTORICAL BACKGROUND

Bond Street underground station was first opened in 1900 as part of the Central London Railway. Access to the platforms was via a lift shaft under the Ticket Hall at the junction of Oxford Street to Davies Street. The running tunnels and platform tunnels oriented east-west with bolted segmental linings were constructed in London Clay. The first major upgrade came in the 1920s, with the installation of twin escalators under Oxford Street and decommissioning of the lifts. This was soon followed by enlargement of the Central line running tunnels and extension of the platforms.

A further major upgrade was carried out in the 1970s with the construction of the Fleet Line (now Jubilee line) running north-south, 13m deeper than the Central line. This required substantial expansion of the station with enlargement of the Ticket Hall, now subsurface, installation of two banks each of three escalators, and construction of associated passages and staircases. Again, all tunnelling works were in bolted cast iron, up to 10m in diameter. Compressed air was used in the construction of the shallower tunnels in an attempt to minimise surface settlement.

The proposed Crossrail scheme, now under construction, includes a Bond Street station immediately to the south of the LU station. The two stations will be integrated and will be linked by a paid-side passenger tunnel. Modelling indicates that interchange between Crossrail and the LU lines will greatly increase passenger numbers within Bond Street Station. The Crossrail works were given the necessary legal powers by the Crossrail Act 2008, which also provide LU the necessary powers to construct the Bond Street Station Upgrade.

## GEOLOGY

The geology at the Bond Street site is typical of central London, with superficial deposits (Made Ground, River Terrace Deposits) overlying London Clay (principally units A2 and A3). Under the London Clay is the Upper Mottled Clay of the Lambeth Group strata, with frequent sand channels, potentially extensive and water-bearing. The majority of the tunnelling works required for the Bond Street Station Upgrade lie within the London Clay, with the deepest tunnels at Jubilee line level reaching into the Lambeth Group strata.

Ground level is at approximately 122.5m LUD (London Underground Datum, being 100m below Ordnance Survey Datum) with a gentle fall from north to south. The London Clay starts at a depth of approximately 7m, again with a gentle north-south fall. A shallow aquifer within the superficial deposits sits at 117m LUD whilst the deep aquifer under the clay aquitard has been historically drawn down by groundwater abstraction and is now maintained at a level of approximately 60m LUD by the GARDIT (General Aquifer, Research, Development and Investigation team) programme. The piezometric profile within the London Clay is therefore close to hydrostatic in the upper part of the stratum, but shows variation in pressure at different locations at the base of the clay. There is also evidence of pore water pressure drop adjacent to cast iron tunnel linings.

The lowest unit of the London Clay, A2, is around 10m thick and consists of alternating sandy clays and silty clays, with a high percentage of silt, occasional wood fragments and pyrite nodules. This layer does not generally contain claystones. Unit A3 is about 12m thick with a base layer of homogenous silty clay, followed by silty clay with three or four layers of claystones. Silt and sand partings become more common towards the top of the unit. Both units A2 and A3 are described as stiff to very stiff fissured clays. Insitu "greenfield" Ko was measured in the range 1.0–1.3, but it is noted that the Station Upgrade works are largely within ground that has been disturbed by previous tunnelling works.

Apart from the normal hazards associated with tunnelling in London Clay and the Lambeth Group materials, the Outline design identified a risk of unexploded ordnance in the superficial deposits, especially towards the north end of the site. In addition, historical borehole locations were logged, however there remains a risk of uncharted boreholes and wells intercepting the tunnel alignment.

## THE TUNNELLING WORKS

All tunnelling works will be carried out from a very small site on the footprint of Nos 354–358 Oxford Street. The tunnelling works can be divided into three geographical areas (refer to Figure 1):

1. Northern Tunnels—comprising the chambers for new escalators 9&10, associated passages, overbridge and stairs, and a binocular cross passage connecting to the north end of the Jubilee line platforms. Construction access is via Shaft 3, which will house plant and provide "back of house" access to the new escalator machine chamber in the permanent condition.
2. Southern Tunnels—comprising lift shafts, stairs, overbridge and associated passages connecting the new satellite Ticket Hall to the Central line and the south end of the Jubilee line. Construction access is via Shaft 1, which will subsequently contain a lift and emergency access/egress stairs.
3. Crossrail link passage—comprising the tunnelled connection from the end of the Crossrail works to the LU station passages. The works are shallower than the above tunnel sections and require an overbridge over the Central line as well as Lift 3 down to Jubilee line level.

The design of the tunnelling works faced several severe challenges, which demanded a high level of expertise in developing the solution:

- Severe spatial constraints, due to proximity of existing LU tunnels and the Post Office mail rail tunnel, the Limits of Deviation defined in the Crossrail Act, and also the requirement to provide public passages fully compliant with LU standards. This has resulted in some very complex tunnel geometry
- High ground movements risks, with tunnelling taking place close under sensitive infrastructure such as LU escalators and track, Grade-listed buildings, cast iron water mains and strategic sewers
- Presence of asbestos in the caulking of existing tunnel linings
- Requirement to maintain the LU station operational throughout the works.

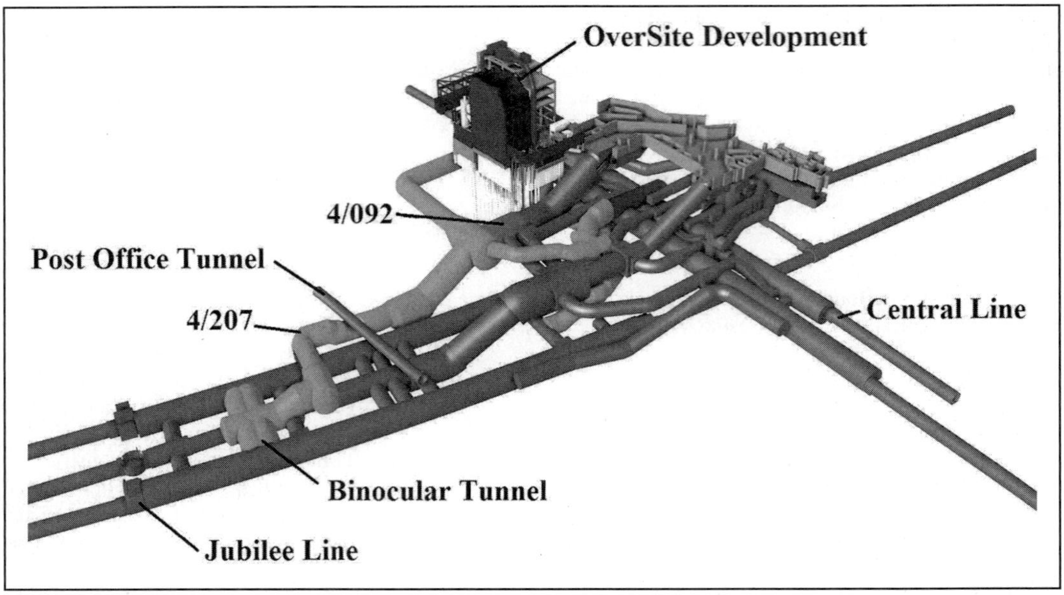


**Figure 1. Overall view of Bond Street Station Upgrade**

At Outline design stage the decision was taken to maximise the use of SCL, rather than rely on traditional segmental linings. This allowed the works to be mechanised as far as possible, maximised flexibility in geometry, and increased confidence in the control of ground movements. Where available space does not permit the use of SCL, hand-mined "squareworks" with insitu concrete and steel sections is employed.

## SCL DESIGN CONCEPT

Dr. Sauer and Partners, HAT and CoLOR undertook the detailed design in a collaborative manner exploiting the benefits of the Contractor's input during detailed design. The Dr. Sauer and Partners design team split the detailed design into the following stages:

- Identification of value engineering opportunities, safety improvements and critical design issues requiring early confirmation
- A conceptual phase including Initial sequencing/face division assumptions, lining thickness calculation through 2D analyses and confirmation of internal spaceproofing through coordination with the wider Halcrow-Atkins design team
- Detailed design progressed based on the "frozen" concept including 3D FE and CAD modelling plus staged constructability/plant co-ordination with the Contractor, and coordination with the compensation grouting designer.

During the concept phase a number of value engineering initiatives were introduced including:

- Revised escalator chamber sizing to suit "HD Metro" type escalator hardware currently being introduced across LU's network
- A blind lift shaft which was to be constructed partially upwards using hand mining methods was repositioned to sit below a SCL chamber allowing the lift shaft to be sunk safely from the chamber using SCL techniques
- A tunnel which directly underpasses the existing Post Office tunnel was changed from traditional hand mining methods to SCL providing programme and safety benefits.

During this period the sequencing for each tunnel was developed with the Contractor to provide early ring closure and divide the face to minimise open face area but also allow plant access. In some areas the face division was influenced by the presence of existing tunnels which must be dismantled as part of the tunnel excavation and in others by the need for compensation grouting during the tunnelling process.

Also the waterproofing and lining design concept was confirmed during the concept phase. The typical tunnel support in this project consists of primary and final lining formed of steel-fibre-reinforced sprayed concrete with a sprayed waterproofing membrane between the linings.

The primary lining is designed to carry short-term loads and the final lining to sustain the long term loads assuming that the primary lining deteriorates in time. Since the sprayed water proofing system lies between the primary and final lining, the final lining is considered to be fully tanked and designed to carry the full water pressure and ground loads. In addition to that, a uniform 75 kPa surcharge load on the ground surface has been taken into account as part of the long term loads for the final lining. For the purpose of the design a suite of comprehensive 3D analyses was developed which is detailed in the next section.

The residual flexural strength offered by a fibre reinforced lining has been exploited to minimise the use of bar reinforcement at areas where tensile stresses may be induced e.g., junctions to other tunnels. The Designer proposed this approach in order to provide cost/programme savings and avoid issues with spraying around large diameter bars. The Project includes a number of complex junction geometries including cruciform junctions, junctions with thin pillars between adjacent openings and junctions to squareworks tunnels. At each location the Designer has succeeded in reducing the bar reinforcement quantities and diameters to a practical amount. The Specification has been written to define tight criteria for the testing of the concrete including the flexural strength parameters.

The waterproofing design has been progressed on a risk-based philosophy targeting those areas judged to be particularly susceptible to differential movement (e.g., shaft to tunnel junctions,) connections to existing tunnels and those tunnels in the potentially water-bearing strata of the Lambeth Group.

## FINITE ELEMENT ANALYSIS (FEA)

Sprayed concrete has been modelled by the "concrete damage plasticity" model which takes into account the tensile and compressive strength limits of the sprayed concrete in order to perform nonlinear analysis of the tunnel linings. However, in the analysis only the tensile strength of the lining material is limited to its residual flexural resistance and the compression part remains elastic to avoid curvature limit check required by EC2. Knowing that steel-fibre-reinforced shotcrete grade C30/37 is used for the construction of primary lining, a design residual flexural strength of 0.4MPa has been considered in the analysis.

All excavation stages have been modelled and a check on equilibrium and stability performed at each stage. Stress concentrations at the breakouts can be readily identified in the 3D model. Considering the tensile and compressive capacity of sprayed concrete, the lining has been thickened or reinforced around the large openings. Based on the Austrian Guideline (2008) for steel fibre concrete, the allowable stresses and the strains in the steel fibre sprayed concrete have been checked throughout the model in order to demonstrate the support adequacy in the design. The maximum allowable tensile and compressive strains in the SCL tunnels were adopted as –0.01 and +0.002 respectively.

Due to the complexity of the existing Tube system and the proposed tunnelling scheme for Bond Street Station Upgrade, conducting a comprehensive 3D analysis was more efficient than 2D modelling. This type of analysis was performed at the final design stage (RIBA F) to achieve the following:

- Obtaining a good estimate of ground and lining movement during the tunnel construction allowing the setting of trigger values for SCL convergence monitoring
- Dimensioning the new SCL tunnels and providing a basis for calculation of reinforcement
- Face stability during excavation
- Stress concentrations at junctions, curves, underpass, etc.

In total three separate large 3D models were developed to cover all the tunnels in the station. The yellow, blue and green transparent blocks depicted in Figure 2 indicate the boundary of the separate FE models. ABAQUS which is a general purpose finite element software was employed to perform the 3D FE analyses.

As an example FE mesh of one of the models that comprises a large connection chamber, passenger lift shaft and connection to future Crossrail station at Bond Street station is shown in Figure 3. The model contains approximately 450,000 solid tetrahedral linear elements and 17,000 triangular shell elements to model soil layers and sprayed concrete lining respectively. It is noted that the models were considered successful but depended on the work carried out at the concept stage to freeze the geometry/sequence allowing the 3D models to be constructed during the detailed design period.

All excavation sequences have been modelled with some simplifications. Excavation is simulated mostly full face with 1-metre round length, and in parts where a temporary invert was required, the excavation is divided into Top-Heading and Bench-Invert sequences.

The Mohr-Coulomb failure criterion was utilised for the elements forming the ground. The numerical analyses have been undertaken on the basis of a total stress analysis using undrained soil parameters and no water pressure is produced during the analysis. It should be noted that both un-drained shear strength (Cu) and Young modulus (Eu) of London Clay stratums have been linearly increased with depth. As an example a contour plot of the

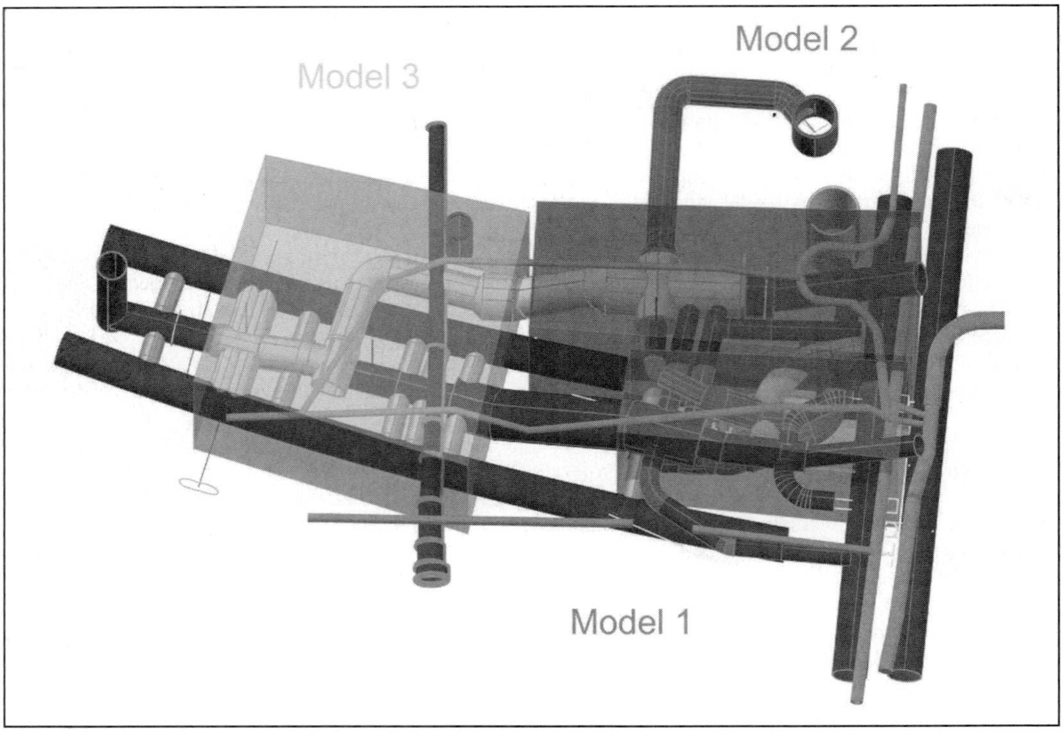


**Figure 2. Project was covered by three separate 3D FE models**

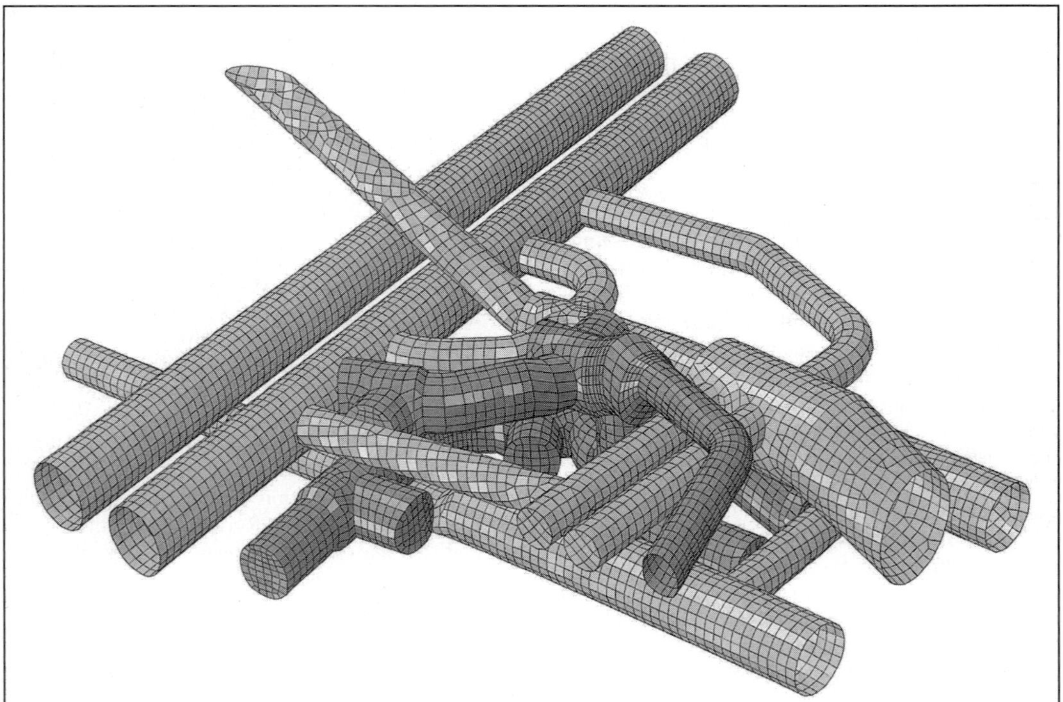

**Figure 3. Illustration of the 3D FE mesh of a connection chamber at Bond Street Station**

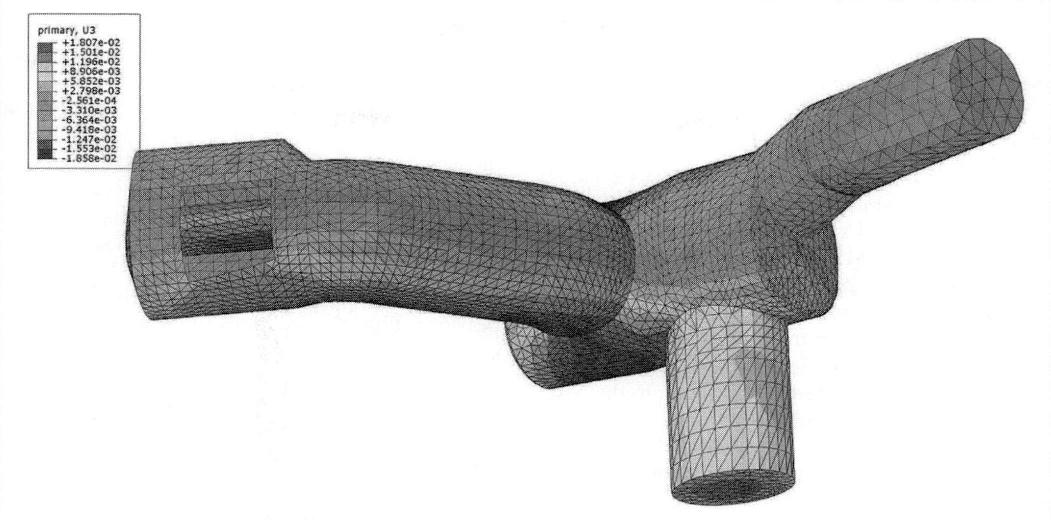


**Figure 4. Contour plot of vertical displacement at the connection chamber and the lift shaft**

vertical displacement at one of the critical tunnel structures in the project (connection chamber 2) has been presented in Figure 4.

## INTERFACE BETWEEN SCL AND COMPENSATION GROUTING WORKS

As previously noted, the tunnelling works pose significant ground movement risks, including risks to Eighteenth Century properties in Stratford Place of high Heritage value. Halcrow-Atkins carried out potential damage assessments in accordance with the requirements of the Crossrail Act 2008 and determined damage risks accordingly. In order to mitigate settlement risks to properties the primary measure is best tunnelling techniques to minimise ground movements at source. The secondary measure selected for the Bond Street Station Upgrade is use of compensation grouting. However this poses a number of issues for the design of the new tunnelled works:

- Risks of grouting above or close to the advancing tunnel face, causing instability of the face and damage to immature shotcrete
- Risk of raised ground pressures over completed tunnels during grouting operations.

These issues required pro-active management in the design of the SCL tunnels, as tunnelling design was carried out in advance of the detailed design of the compensation grouting scheme. There is limited information available as to the effect of compensation grouting on tunnels in London Clay, however reference was made to contemporaneous design work ongoing for the Crossrail project, and to experience from the 1990s Jubilee Line Extension. Good information is available regarding typical pump pressures for combinations of equipment and ground conditions similar to those expected at Bond Street. It should be noted that the very high pressure required to initiate hydrofracture of the London Clay was disregarded in the design assessment, as it applies only to a small area of the ground and thus is not a significant load at a distance of some metres away. However the lower pressure required to propagate the fracture to full extent is typically in the order of 8 bar at the pump, and if replicated in the ground this would present a highly onerous load on a nearby tunnel. A number of measures were therefore specified in order to manage risks:

- Specification of a moving grouting exclusion zone around and above the advancing tunnelling face and immature shotcrete;
- Development of a design grouting load case defined by an additional vertical pressure of 140kPa applied over an area equal to the width of the affected tunnel, at a distance no less than 3m above the tunnel;
- Requirement for active verification of actual grouting pressures;
- Specification of drilling and grouting exclusion zones for tubes-a-manchettes in close proximity to new and existing tunnels.

The result of this was to constrain the compensation grouting design, and in one area the SCL design had to be amended to allow an onerous grouting impact, lest protection of the building above be severely compromised. The design amendment consisted of provision of mesh reinforcement and specification of an enhanced monitoring regime in recognition of lower design partial safety factors applied in the temporary condition.

## CONSTRUCTABILITY

A key aspect in the development of the design was the input of buildability. In line with the agreed assurance regime, formal staged submissions were used, at nominally 50%, 80% and 100% complete, for formal review and comment by both CoLOR JV and the client, London Underground. In addition there was a continuous ongoing informal review and input process throughout the design period between designer and contractor. To enable this to happen to the best level resourcing of the tunnel team was important in that it enabled the correct level of review in advance of the works. Detailed construction sequences were included in the design submission as part of this review and to provide assurance to the client that the scheme was buildable as designed.

## KEY CHALLENGES DURING CONSTRUCTION

The site of the works on the UK's busiest shopping street presents a challenge in that the only available site area is within the footprint of the building, providing a very constrained site (refer to Figure 5). The overlying building is designed, in the temporary condition, to accommodate welfare, M & E workshops and cranage facilities. The SCL equipment is set up within the basement and sub-basement areas within the new building. This requires extensive temporary works to make the installation fit, especially given the required redundancy in the SCL supply which means an installation with multiple silos and multiple shotcrete pumps.

Below Oxford Street is a dense network of utilities consisting of sewers, gas and water mains, electricity and telecommunication cables. Many of these date from the early or mid-19th century and are of a cast iron construction.

All the utilities have been assessed as being extremely sensitive to ground movements and have therefore been subject to a range of damage mitigation measures. These measures include lining of the larger diameter pipes using modern materials and methods. An ongoing programme of remediation has been agreed with the asset owners and undertaken prior to commencement of the tunnelling works.

The majority of the tunnels are constructed in, around and adjacent to the station which must remain operational during the works. To fit the works into

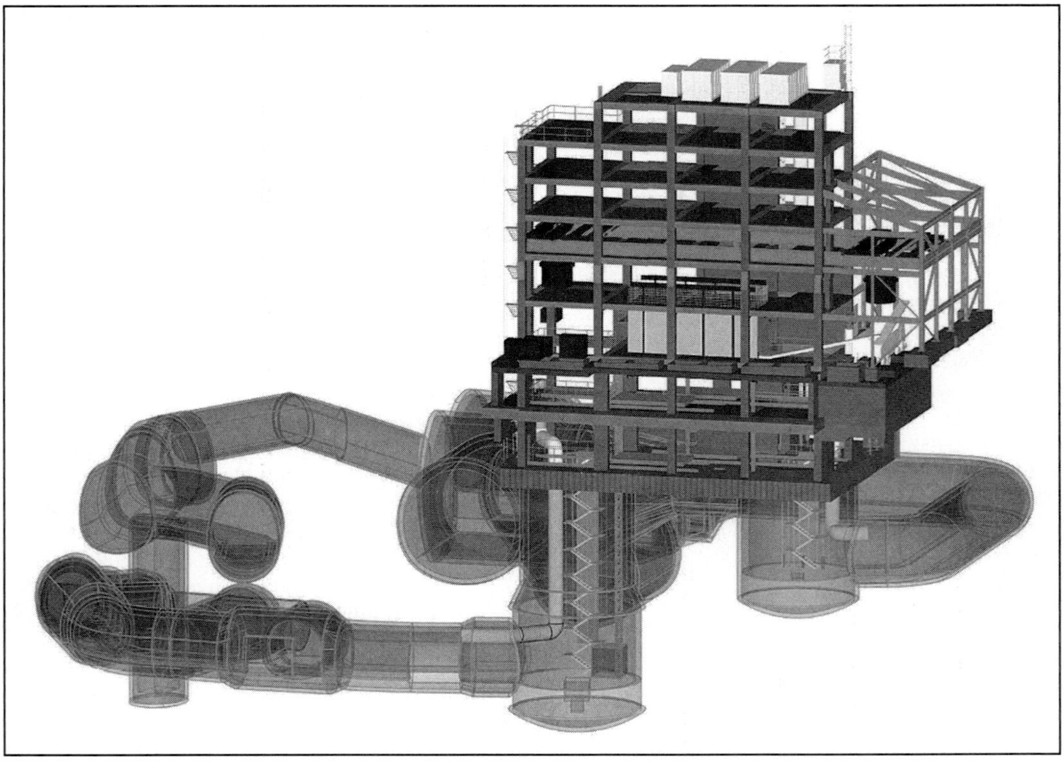

**Figure 5. Challenging construction site setup**

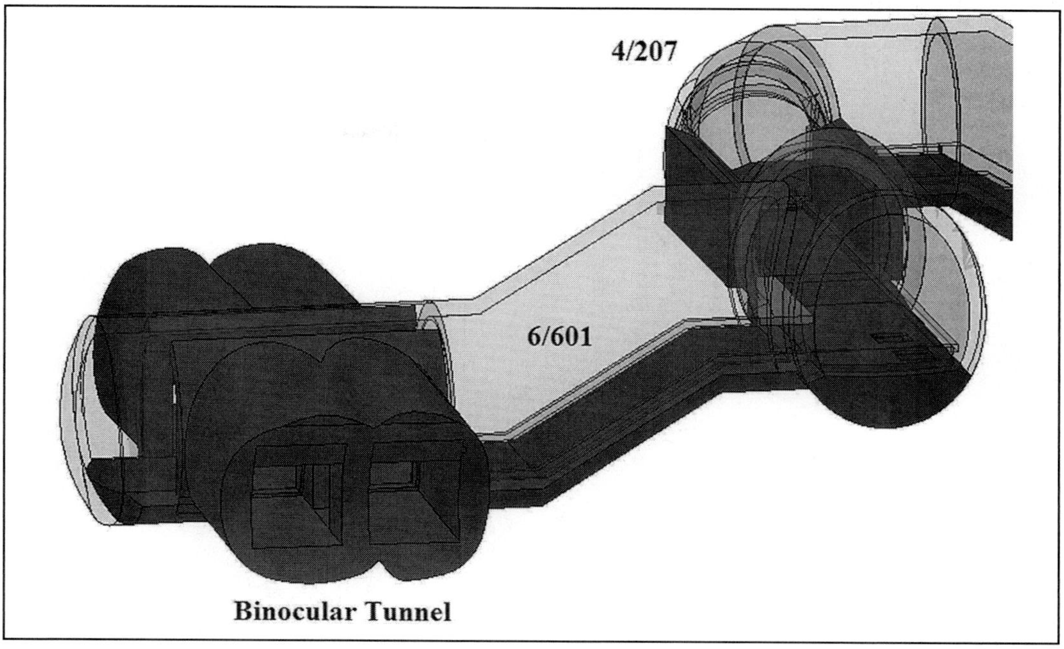


**Figure 6. 3D CAD model of the binocular tunnel**

the available underground space, numerous tunnel cross-sections and geometries have been designed. Of particular note in terms of key construction challenges are the following, discussed below:

- Binocular tunnels 6/209
- Concourse 4/092
- Post Office Tunnel Underpass
- Tight corners—notably passage 4/207

At the proposed connections to the Jubilee line the passenger flow analysis requires two passages. However, due to the restrictions created by the existing infrastructure there is insufficient space for two separate passageways. A binocular structure was designed to connect to the existing platform tunnel. This binocular tunnel (refer to Figure 6), 9m width × 6.5m height, incorporates two intersecting SCL sections. After construction of the primary lining of the first tunnel, a cast in-situ secondary lining formed. The second tunnel is subsequently excavated with the primary lining bearing onto the completed first tunnel. The principal design issues with this are the waterproofing connection between the two tunnels and the structural connection between the two primary linings.

The solution devised requires a niche to be formed in the secondary lining of the first tunnel which facilitates a section of the sprayed waterproof membrane to be exposed for an overlap onto the membrane of the second tunnel. During the excavation of the second tunnel, the redundant section of the first tunnel is broken out to permit plant access. Once primary lining is completed, a wire cutting operation provides a profile which permits the junction of the waterproof membranes and allows sufficient space for the final cast insitu lining to be formed.

One of the principal benefits of the BSSU project is the provision of two new escalators down to Jubilee line level. To connect the new escalators to the existing station, an extension to the lower concourse of existing escalators 6,7 & 8 was designed. This tunnel denoted 4/092, 10m width × 8m height, is driven from a connection chamber (Conc1), 10m width × 10.5m height (refer to Figure 7). The construction of 4/092 is further complicated as two existing curved passageways constructed from segmental cast iron linings, internal diameter 3.85m, will need to be demolished within the footprint of the new tunnel and subsequently reconnected to the new lining of the proposed 4/092 tunnel. The construction of 4/092 must be completed in a planned 6 month closure of the Jubilee line.

The chosen solution includes preliminary works to replace the segmental lining of the two passageways, with sprayed concrete. This work is preceded by the remediation of any asbestos caulking within the old tunnels. The subsequent construction of 4/092 is progressed using a sidewall drift and enlargement with provision for the connections into the part demolished cast iron passageways. The structural

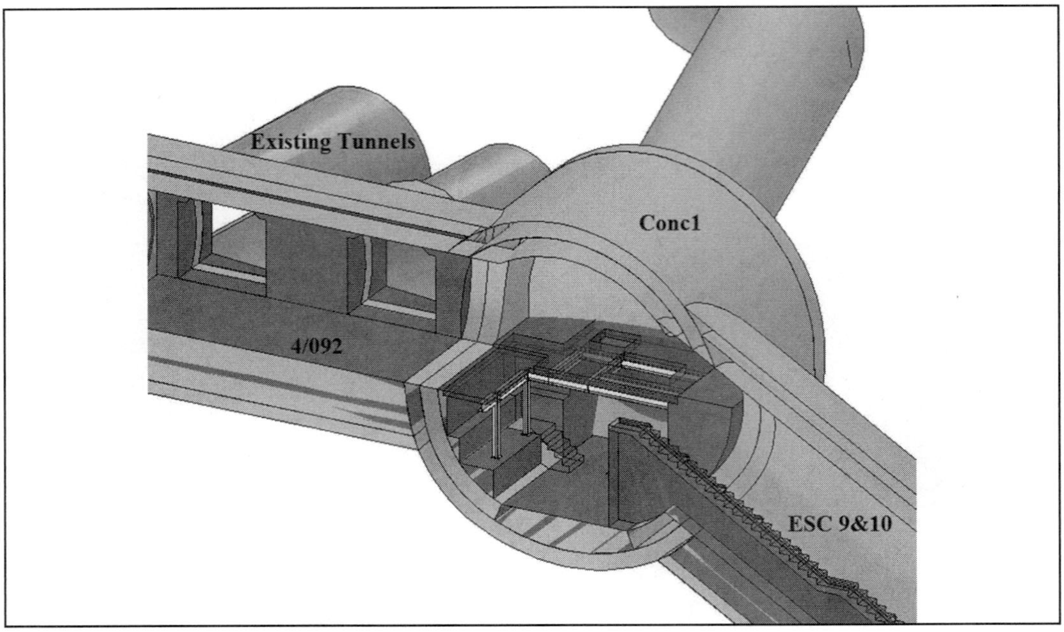


**Figure 7. Longitudinal section through 4/092 and escalator 9&10**

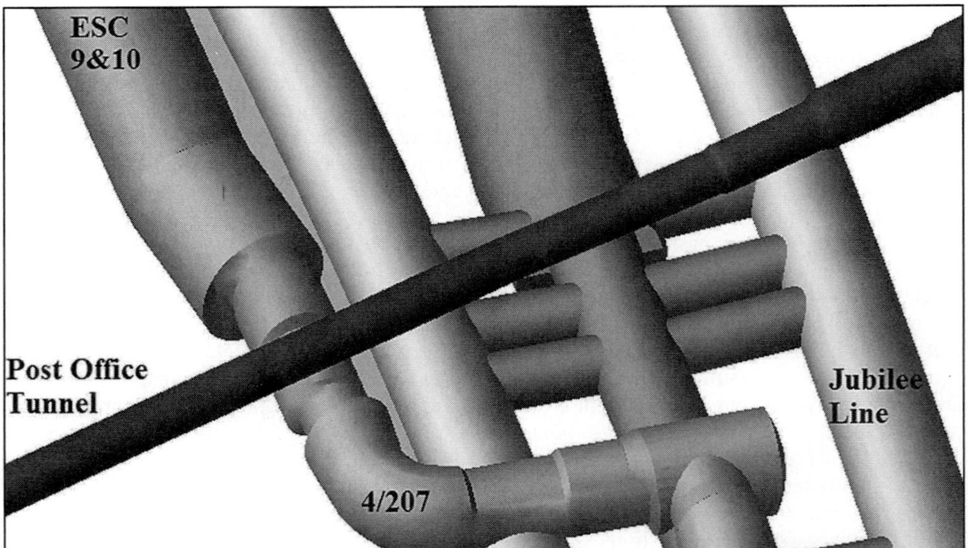


**Figure 8. The Post Office tunnel running diagonally from bottom left to top right**

connections and waterproofing details are unique for each connection to the existing tunnels. Based on the successful design by Dr. Sauer and installation of a similar profile at another recent project, the Tottenham Court Station Upgrade, no lattice girders will be used for the sidewall / enlargement process.

The Post Office Tunnel (refer to Figure 8) runs directly above the access tunnel to the Jubilee line works and impinges onto the works such that the excavation for the SCL tunnel exposes the lowermost 3rd of the existing tunnel. To maintain the requisite passenger envelope and to ensure that the

Post Office Tunnel remains operational for maintenance purposes a "squashed" SCL profile has been designed to allow the new tunnel to pass beneath the Post Office Tunnel.

To enable the under-crossing to be undertaken safely, a true real-time monitoring regime is being installed, which entails a data link to be established between the existing Post Office Tunnels and LU tunnels.

To comply with the allowable limits of deviation, several sharp corners with tight radii have to be negotiated. Of special note is the corner in passage 4/207 where a 'lobster back' profile has been adopted to balance the advance length in the crown against a dimension on the inside of the bend which permits efficient sprayed concrete application. In this instance, due to the sharpness of the corner and the resultant dimension on the outside of the bend, a sequential advance sequence has been adopted to mitigate a large open excavation.

## CONCLUSION

Bond Street Station Upgrade is one of the most complex SCL design projects undertaken by London Underground which has required significant design effort including complex modelling. Close integration between designers, contractor and client have been key in minimising construction risks and optimizing the design. In combination with the adjacent Crossrail station the upgrade of the LU Bond Street Station will create a state-of-the-art transport hub with minimal impact on the surrounding infrastructure.

## ACKNOWLEDGMENTS

The authors would like to thank London Underground for granting permission for preparation of this paper. We also wish to thank Costain Laing O'Rourke JV, Halcrow-Atkins JV and Dr. Sauer & Partners for their assistance in the production of this paper.

# Corrosion—The Scourge of Steel Rock Reinforcement

**Sean Tokarz and Greg Raines**
MWH

**Jason Moral**
Corrpro

## INTRODUCTION

Steel rock reinforcements such as friction type anchors, mechanical anchor bolts and grouted dowels or bolts are common to the support of underground structures such as roadways, conveyance and pumping stations not to mention the widespread use in underground mining. Some of these reinforcements are placed as "temporary" support, meant to provide for safety against rockfall during construction. Many of these support types are specified for use as "permanent" support and often these application are intended to maintain an opening for the structures useful life, which could be 100 years or more. Yet there is little understanding of the true service life of these reinforcements and the existing case studies do not go back nearly long enough. One of the most important considerations in the longevity of a steel reinforcement is the corrosivity of the environment in which the bolt or dowel is being installed (Figure 1).

The parameters involved with steel and grout degradation over time are often not well understood and as a consequence the true corrosive nature of the environment is often overlooked prior to support installation. Even when testing of the environment is performed the currently available design guides provide only an indication of whether it is corrosive or non-corrosive and the level at which it may be an issue is largely left to the designer's discretion.

## IMPORTANCE & DESIGN STANDARDS

In the mining industry the use of steel rock reinforcements is widespread. Often these applications are for temporary support while ore body is being extracted from a particular adit. Even for longer term applications the miners and engineers maintain access to the tunnel or cavern and additional reinforcements can be added as necessary. In the civil underground industry steel rock supports are often specified as either temporary rock reinforcement during construction, for permanent rock support or both. In the case of most civil underground works the life of the permanent rock reinforcement is intended to match or exceed the life of the roadway, conveyance, pump station or other civil structure being housed within the opening. It is typical in a pressure tunnel to install the permanent reinforcement, never to be inspected again unless there is an issue.

The design standard in the USA is the Post Tensioning Institute's Recommendations for Prestressed Rock and Soil Anchors (2004). It should be noted that there is an updated document planned to be issued in 2014. The document is geared to high capacity, pretensioned anchors and does not specifically address passive anchorages commonly used for support. The standard specifies two classes of corrosion protection consisting of two barriers of corrosion protection (Class I) or one barrier over the bond length (Class II) including the drill hole grout. The document mentions epoxy coating as providing additional corrosion protection but does not count it as an additional barrier. Class II protection is specified for temporary supports in "aggressive" ground or permanent supports in "non-aggressive" environments where the consequence of failure is low. All other permanent applications require a Class I level of protection. The document provides a definition of aggressive ground in terms of several quantitative measurements including pH, resistivity, and presence of sulfides or stray currents.

The international standard commonly referenced includes British Standard 8081:1989 "British Standard Code of Practice for Ground Anchorages" which has been partially superseded by BS EN 1537 "Execution of Special Geotechnical Work, Ground Anchors." The original 1989 code is meant as a comprehensive guide to the design, installation and testing of ground anchorages including corrosion protection design. The EN 1537 report was published in 2000 and updated in 2013 and is meant as a revised guide to the installation of ground anchorages with sections devoted to the topic of corrosion protection. The standard calls for a double corrosion

**Figure 1. Installing "permanent" anchor bolts next to "temporary" friction anchors and mine straps already in advanced stages of corrosion**

protection systems independent of the drillhole grout for permanent installations, which is similar to the US standard. The standard also stipulates a single measure of corrosion protection is acceptable in some applications provided electrical isolation of the anchor from the ground can be determined.

## CASE STUDIES

Although steel rock reinforcements, grouted and ungrouted friction types, have been in use for underground applications for more than 50 years, little is known about their true service life. Littlejohn (1987) did a study of over 30 corrosion failures reported in the literature and found that most of the failures could be divided equally into issues with the free stressing length and issues at the bolt head. Baxter (1996) provided a review of rock reinforcements used primarily in hydro applications from around the world. His conclusions were that the industry typically relies on grout encapsulation and passivation for corrosion protection, but issues arising from poor installation procedures can severely limit reinforcement service life. He was particularly critical of the use of resin cartridges as a means of providing encapsulation for corrosion protection.

In the studies available there is little information on the aggressiveness of the ground conditions and the effect on reinforcement service life. The Army Core of Engineers (1980) mentions the encapsulation of bolts with grout as a means of preventing corrosion. Charette et al. (2004) investigated the long term effects of corrosion in North American mines on hydraulically expanded, friction type bolts using pull out tests. The environments in which the bolts were installed were grouped into low to moderate and moderate to high corrosive level environments. The observed rates of corrosion for the highly corrosive environments were found to be two to ten times that of the low. It was also observed that the onset of corrosion after installation was approximately 18 months and 12 months for low and high corrosive environments respectively.

## CORROSIVE ENVIRONMENTS

Corrosion of steel in conductive electrolytes, such as soil or water, occurs as the loss of metal ions due to the electrochemical interaction between anodic and cathodic areas at the metal surface. Corrosion can be in the form of ferrous-oxides, ferric salts or dissolved metal ions. In various soil and water applications, most commonly observed corrosion mechanics are galvanic corrosion, uniform corrosion and stray current corrosion. In uniform corrosion, the metal loss is relatively uniform throughout the surface. In galvanic corrosion, the corrosion occurs due to interaction between dissimilar metals or the interaction of anodic and cathodic areas on the same metal surface. Stray current corrosion takes place when the external AC or DC currents discharge at an unintended structure.

Corrosion can take place in variety of environments, however, a High Risk of Corrosion environment has been defined per PTI (1996) as an environment where one or more of the following factors are present:

- Electrolyte Resistivity under 3000 ohm-cm
- pH less than 5.0

**Table 1. Corrosivity ratings chart from resistivity**

| Resistivity (ohm-cm) | Corrosivity Rating |
|---|---|
| <1,000 | Extremely corrosive |
| 1,000 to 3,000 | Highly corrosive |
| 3,000 to 5,000 | Corrosive |
| 5,000 to 10,000 | Moderately corrosive |
| 10,000 to 20,000 | Mildly corrosive |
| >20,000 | Progressively less corrosive |

- Presence of Moisture and Chlorides
- Presence of Sulfides
- Presence of Stray Currents

Resistivity is the inverse of conductivity and is measured in the units of ohm-centimeters. The resistivity of a soil or rock mass is a function of moisture and the concentrations of ionic soluble salts and is considered to be the most comprehensive indicator of corrosivity. Variations in resistivity indicate variations in composition which are conducive to galvanic corrosion. Since ionic current flow is associated with steel-rock corrosion reactions, high resistivity will slow down corrosion reactions. It is important to note that there is no universally accepted criterion for resistivity and corrosivity ratings, since the occurrence of corrosion is the result of multiple factors. Table 1 represents an approach used by Corrpro.

pH is the negative logarithm of the hydrogen ion concentration. For ferrous materials used in construction, pH in the range of 6 to 10 has little effect on the rate of corrosion under oxidizing conditions at ambient temperatures. At pH values above 10, the steel readily polarizes which tends to passivate the steel greatly reducing the potential for formation of corrosion cells. On the other end of the pH scale, acidic environments are commonly associated with heavy corrosion to iron alloys. In acidic conditions, the hydrogen ions present act as cathode depolarizers, hence increasing the corrosion reaction rate. At pH values below 4, the rate of corrosion accelerates rapidly. It is estimated that for each pH level change, the corrosion rate increases tenfold.

Chloride ions are depolarizing agents and cause pitting of many common materials of construction. They also break the crystalline passivizing ferrous-oxide layer on metal surfaces and expose the surface to other types of corrosion. As the chloride concentration increases, the rate of corrosion increases progressively. Concentrations over 50 ppm are significant and may cause depolarization and corrosion fatigue problems on steel. Corrosion fatigue is a phenomenon where the existing fatigue cracks which occur due to operating stresses and material defects propagate exponentially due to corrosion.

Sulfide ions present in a material, if any, are indicative of anaerobic conditions. Under these conditions, sulfate-reducing bacteria can greatly accelerate the rate of corrosion of ferrous materials. The bacteria reduce sulfates to sulfides and in the process oxidize iron, hence causing heavy corrosion on metal surfaces. Detectable concentrations of sulfide ions can be indicative of anaerobic conditions, however, several other factors are also required to provide a suitable environment for the bacteria to exist, such as temperature, pH and moisture.

Contrary to the common conception, availability of sulfates do not possess an immediate threat to metal. Depending on the concentration of sulfates, heavy corrosion can take place on mortar, grout and other cementitious bodies. Sulfates are naturally present in many soils and natural waters. They may be developed by bacterial action or introduced by industrial pollution, as well as being present in coal or related natural compounds. Sulfates are the ubiquitous corrosive species for concrete and they affect cementitious bodies such as grout, backfill and concrete-based coatings. In some areas, gypsum-bearing ground waters exist, which are extremely corrosive to concrete. Specific cements should be specified based on known levels of sulfate contamination.

Stray DC currents through the earth can emanate from the operation of DC transit systems, foreign cathodic protection rectifiers, welding and DC motors. When discharged from the surface of ferrous piping, these currents may cause significant corrosion on steel structures. It is important to note that stray current effects are nearly impossible to predict prior to construction due to their very complicated nature.

As previously mentioned, most metals corrode in low pH environments by dissolving in the electrolyte. This principle is at the heart of cathodic protection, the pH level at the metal and electrolyte interface is increased to a level where the metal is passive, by increasing the electrical polarization. However, zinc and aluminum are amphoteric metals, which can be described as metals that are soluble at high pH environments. Particularly zinc, depending on the application, may become highly active at pH levels around 12 and more, as shown in Figure 2 (Xiaoge Gregory Zhang).

When using galvanized rock reinforcements, it is imperative that the long-term pH level is considered especially if the bolt will be placed inside grout, where the pH levels may become critical.

Cathodic protection is a proven method of providing corrosion protection. At conditions where it may be possible to electrically bond a number of rock bolts, cathodic protection can be economical and practical. An impressed current system placed in the vicinity of the application can provide corrosion protection against galvanic, uniform, stray current and microbiologically induced corrosion. Also, at some applications, cathodic protection may provide protection

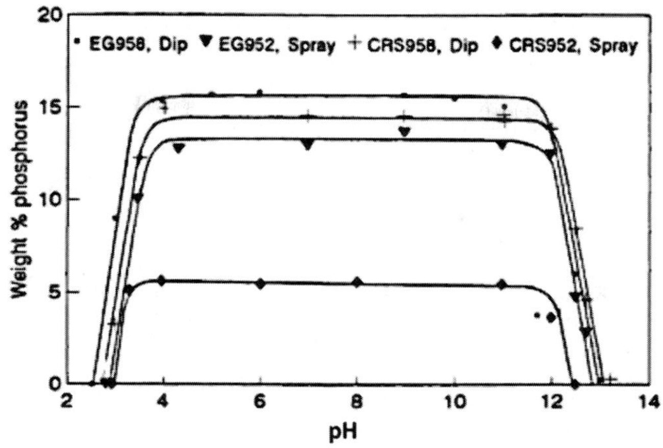


**Figure 2. Dissolution of Phosphate coatings obtained by different methods after immersion is stirred HCI or NaOH solutions of different pH values for 30 min at 25°C (Source: Xiaoge Gregory Zhang, Corrosion and Electrochemistry of Zinc, page 173)**

against corrosion fatigue. Adequate cathodic protection yields to 1 mpy or less corrosion rates.

## ROCK REINFORCEMENT TYPES

The purpose of solid steel rock reinforcements is not only to support the rockmass but to aid the rockmass in supporting itself. The benefits of using solid steel over other support systems include; speed of application, versatility, ease of installation and relatively low cost. Generally, all steel rock reinforcing systems will corrode. The important questions to be answered are the design service life of the system, followed by the corrosivity of the environment. Although the service life of a steel reinforcement system cannot be separated from the types and concentrations of corrosive elements in general terms we can rank steel reinforcement systems by their level of protection as shown in Table 2.

The primary corrosion protection role of grout backfills is to prevent the movement of ions particularly within migrating groundwater. The benefits of cement grouting over resin is the alkaline environment which the cement creates in the area in contact with and surrounding the steel anchorage. This high pH environment tends to passivate the steel slowing or stopping the process of corrosion. Resin on the other hand is inert. The key to protection however is assurance of complete encapsulation and limiting the size and frequency of cracks after curing.

In addition to the measures shown in the table most manufacturers offer options for polymer or epoxy coatings to extend the service lives of the steel reinforcements. The issue with coatings is the opportunity for scratches to occur during the installation process. Scratches in the coating would not only create an opening for degrading reactions to occur but could actually speed up the corrosion process by concentrating the reaction on a single location in the steel. Another measure commonly employed is the addition of a galvanizing layer by a hot dipped process. The issue with galvanization is that it is sacrificial in nature and corrosion tends to concentrate in discrete locations. Additionally as previously mentioned, galvanizing layers are susceptible to corrosion at high ph, alkaline environments and therefore should not be coupled with cement-grout backfill.

As shown by Littlejohn (1987) nearly half of all incidences of corrosion induced bolt failure were due to issues at the bolt head. Similar to protection of the steel support shaft, protection of the bolt head and face plate is typically achieved by encapsulation. For most installations this can be easily achieved by installing a PVC "Trumpet" in the rock nearest the surface exposure and covering the bolt head with concrete or shotcrete. In instances where the bolt head is not covered by additional concrete reinforcement a grease filled protective cap can be installed. One benefit of using the cap is the ability to come back at a later date, remove the cap and pull test the bolt.

## CONSIDERATIONS AFFECTING SERVICE LIFE

Galvanization may be better than other coatings in instances where scratches during installation are likely because it is sacrificial in nature and has self-healing properties. Simply oversizing the reinforcing

**Table 2. Steel rock reinforcements and corrosion susceptibility**

| Most Susceptible | Defining Characteristics | Benefits | Corrosion Concerns |
|---|---|---|---|
| Steel Friction Type Anchorages | Either steel C-shaped slotted forced into undersized hole or hydraulically expanded in oversized hole. Steel grips rock over entire length | Inexpensive, Fast installation, immediate support, bolt provides some support after failure (slip) | Steel in contact with rock over entire length |
| Steel Mechanical Anchors | Expansion shell anchor at bolt head grips rock and creates tension between bolt head and face plate | Relatively inexpensive, fast installation, immediate support, can be designed for high loads | Anchor in direct contact with rock. Post installation grouting extends service life. |
| Resin Grouted Steel Bolts | Polyester resin and catalyst cartridges placed in open drill hole and mixed by bolt spinning, grout transfers load carrying capacity to steel | Fast installation, capable of high load carrying within several minutes, can be post tensioned after initial grout set | Steel encapsulation is difficult to achieve |
| Cement Grouted Bolts or Dowels | Grout filled hole either pre (better) or post bar installation, grout transfers load capacity to steel | Capable of carrying high loads, can be post tensioned after grout set | Steel encapsulation may be difficult in highly fractured rockmass or with high groundwater infiltration |
| Poly Sheathed, Cement Grouted Bolts | Similar to cement grouted rockbolts except poly (or equiv.) grease or grout filled sheath provides additional layer of protection | Capable of carrying high loads, can be post tensioned after grout set | Only in areas not protected by sheathing, same concerns as cement grouted bolts |
| **Least Susceptible to Corrosion** | | | |

steel bar would be another sacrificial means of providing additional service life for the steel. Additionally the zinc within the galvanic coating would react with the cement within any backfill and eventually corrode leaving a pathway for water to get to the steel. Galvanization or oversizing is recommended for anchors directly in contact with the formation such as friction type or (ungrouted) mechanical anchors providing additional service life for reinforcements used for short term applications.

One of the primary issues for grouted rockbolts in caused by grout flowing into the voids of a fractured rock mass, preventing complete bolt encapsulation. For this reason it should be clear which grouts are to be employed for bonding to the rockmass and which are for encapsulation and passivation. Grouts used for these two very different purposes should be kept separate for long term applications. Breather tubes typically used to backfill around a bolt can become tangled and damaged during the installation process. For grouting applications hollow core bolts are more expensive than solid core bolts or simple grouted rebar but the ability to gout the bolt from the exterior (more reliable than from center) and allow for grout return through the center of the bolt eliminates this issue.

The use of fully resign grouted rockbolts as a means of "permanent" support has been recommended by the Army Corps and others. The combination of fast set bonding cartridges in the bonding zone and slow set cartidges in the free stressing zone is used as a means of corrosion protection. Additionally the bolts can be tensioned after the fast set and prior to the slow set to lock in post tensioning. The issue with the use of cartridges is the poor level of encapsulation that has been found on several studies to be as low as 10 to 25% of the bar length (Baxter 1996). The cartridges themselves can prove to be an issue once they are broken they remain in the grout as it sets, potentially creating a void large enough for groundwater to infiltrate. Resin cartridges also have a shelf life that should be verified with the supplier prior to installation. It is also recommended to use one resin cartridge from each package as a test to ensure shelf life has not been exceeded.

Where backfill grouting is used as the primary form of corrosion protection of steel reinforcements the installation process and quality control procedures are paramount. The size of the hole should be large enough to ensure complete encapsulation and bar centralizers should be placed per manufacturer's instructions. For resin cartridges a hole that is drilled too far into the rock formation will cause issues as the resin may collect at the bottom of the hole and not around the reinforcement. If the hole is not properly cleaned prior to installation the grout may not penetrate to the base or may not provide encapsulation. Groundwater flowing from the hole is always an issue for backfill grouting, particularly when using portland cement. In this case an initial

**Table 3. Generalized service life for common steel supports**

| Corrosivity of Environment | | Mild to Moderate | Moderate to High | High to Severe |
|---|---|---|---|---|
| Example Indicators | Resistivity | 10,000–≥20,000 | 3,000–10,000 | <1,000–3,000 |
| | pH | >11 | 5–11 | <5 |
| **Recommended Service Life for Steel Supports** | | | | |
| Friction type & mechanical anchors | | Years | <1 Year | Immediate use only |
| Resin grouted | | Decades | Years | <1 Year |
| Cement grouted | | Several decades | Decades (+years) | Years–decades* |
| Poly-sheathed, Cement Grouted (Per PTI & BS 8081) | | Several decades | Several decades | Several decades* |

\+ Presence of sulfates.
* Monitoring recommended.

pressure grouting technique may be necessary to fill the discontinuities followed by the steel reinforcement installation and backfill grouting. Also an area of concern is just beneath the bolt head which is often left unbackfilled because workers didn't want to let grout flow from out of the drillhole onto the ground. Of course the most accurate procedure for achieving encapsulation is to first fill the drillhole with grout and then place the steel bar into the filled hole.

## RECOMMENDATIONS FOR SERVICE LIFE

The available design guidelines are vague in the discussions of support service life, typically referring to either "permanent" or "temporary" support types and the level of corrosiveness of an environment is typically defined as being either "corrosive" or "non-corrosive." Table 3 is an initial attempt to "fill in the gaps" left by the available guidelines. The purpose is to provide some general guidance on the use of certain support types. The basis of the table is the available case histories of support performance which is usually limited to locations where issues were encountered and recommendations for service life are likely on the conservative side. Since most of the case histories are based on installations within the previous few decades no attempt has been made to make recommendations beyond that duration. The indicators shown on the table are for general reference to corrosion rates only, for design all parameters effecting the corrosion of steel and backfill grouts, discussed previously, should be considered.

One important but commonly overlooked step in the life of a supporting system is the long term maintenance and monitoring. Various schemes for monitoring corrosion have been identified but few provide examples of case histories. Permanent reference electrodes can provide low IR-drop polarized potential readings if installed correctly. By recording and historically keeping track of potentials, corrosion may be monitored. Various types and shapes of coupons are commercially available. By utilizing a coupon, actual free corrosion rates can be estimated. Any installed coupons should be replaced at least semi-annually, and sent to a lab for further analysis without contamination. The most accurate method of estimating corrosion rates remains pull testing or overcoring of the rock bolt and visual inspection. The bolt head and face plate, when accessible, can be also visually inspected without removing the bolt.

## CONCLUSIONS

One of the most important considerations in the longevity of a steel reinforcement is the corrosivity of the environment in which the bolt or dowel is being installed. Understanding the parameters which cause the degradation of steel or backfill grouts is important in the design of underground supports. Although each case is different some general guidelines for support types and service lives have been presented for conceptual level understanding. Additional measures to extend and monitor the useful life of steel supports were also discussed.

The authors would like to thank the representatives from several underground support suppliers who provided recommendations and insight that aided in the development of this material.

## REFERENCES

Army Corps of Engineers, "Rock Reinforcement" Engineering Manual 1110-1-2907, 1980.

Baxter, D., "Do All Rock Bolts Rust? Can Q.A. help? Does it Matter?" Proceedings of the 9th Australian Tunneling Conf., IE Aust, Sydney, 1996.

Baxter, D., "Rock Solid" International Water Power and Dam Construction, 1996, V. 50, n. 11, pg. 38.

British Standards "BS8081 Code of Practice for Ground Anchorages," 1989.

British Standards "BS EN 1537 Execution of Special Geotechnical Work, Ground Anchors.," 2003.

Charette, F., Germain, P., Hadijgeoriou, J. "Corrosion Behavoir of Swellex Bolts in Underground Mining Environments" Candian Institue of Mining, Annual General Meeting, 2004, Edmonton, pg. 9.

Littlejohn, G.S. "Ground Anchorages: corrosion performance" Proc. Instn Civ. Engrs., 1987, Part 1, 82, pp. 645–662.

Post Tensioning Institute (PTI). "Recommendations for Prestressed Rock and Soil Anchors," 1996.

Zhang X.G., "Corrosion and Electrochemistry of Zinc," 1996.

# Structural Repair of Cross-Passage 5, Interstate 70 Hanging Lakes Tunnel

**Ralph Trapani and Jeremy Wehner**
Parsons Corporation

**Rex Goodrich**
Colorado Department of Transportation

**ABSTRACT:** The twin bore Hanging Lake Tunnels (HLT) in Colorado on Interstate 70 (I-70) are operated by the Colorado Department of Transportation (CDOT). The tunnels include eight (8) cross-passages that house sensitive equipment and provide emergency egress. Structural integrity inspections were conducted in 2009. Cracked shotcrete was observed in Cross-Passage 5 (CP-5).

Areas in CP-5 had substantial de-lamination of the shotcrete. A ground characterization plan was accomplished based on exploratory drilling and testing. An innovative repair was developed including Polyurethane (PUR) injection, and installation of tensioned rock bolts. The repair was successfully completed in late 2010 and monitoring of the cross-passage is ongoing.

## INTRODUCTION

The Hanging Lake Tunnels convey I-70 traffic through the southern wall of Glenwood Canyon at M.P. 125.3 to M.P. 126.0, for a length of approximately 4000 Linear Feet (LF). The original construction project occurred from 1988–1992. The innovative I-70 Glenwood Canyon Tunnels (including Hanging Lake) were the first major vehicle tunnels constructed in the USA, using rock reinforcement and steel-fiber reinforced shotcrete as a permanent excavation support solution (Trapani, et al., 2003).

In late September of 2009, CDOT issued a Notice to Proceed to Parsons Transportation Group (PTG) for a Baseline Tunnel Structural Integrity Inspection, under PTG's Statewide Tunnel and Complex Structure Non-Project Specific (NPS) contract. This inspection was reported in Trapani, et al., 2010. The inspection activities revealed cracked shotcrete in Cross-Passage 5 (CP-5). The cross-passages house sensitive electrical and communications equipment and provide an emergency pedestrian exit from one vehicular bore to the other in the event of a fire emergency.

Areas in CP-5 were found to have substantial de-lamination of the 4-inch thick shotcrete layer from the underlying rock. The distressed shotcrete in the north end of CP-5 had moved to within ½ inch of an electrical switchgear unit that takes a 4,160-volt main feed to a transformer that converts the power into usable 480-volt electricity (Figure 1). As originally constructed, the distance between the switchgear and the cross passage wall was approximately 12-inches.

The bulging mass of rock behind the cracked shotcrete, if dislodged with sufficient inertia to overturn the switchgear, could result in an instantaneous release of energy (electrical explosion) within the tunnel cross passage.

Initially, a full inspection and sounding of the east wall and ceiling were not possible due to conflicts with electrical equipment and conduits (Figure 2). To immediately resolve any safety and access issues, CDOT personnel rerouted circuits, and removed all equipment from the cross passage to a safe off-site storage location.

After this equipment and conduit was removed, complete soundings and measurements were completed on April 13, 2010. A man-lift and ladder were used to provide safe access to all walls and the cross passage ceiling. Cracked shotcrete was identified by visual inspection. Shotcrete was sounded using hammers. Shotcrete that had delaminated from the surrounding rock gave a drummy or hollow sound. Intact shotcrete gave a solid, ringing sound.

In summary, 3090.75 square feet of shotcrete was noted to be in a distressed condition. This represented about 70% of the total surface area of shotcrete (4429 square feet) in CP-5.

Based on interviews with parties involved in the original interstate tunnel construction, the cross passages were top headed just like the tunnel and were driven half way from each tunnel bore as

Figure 1.

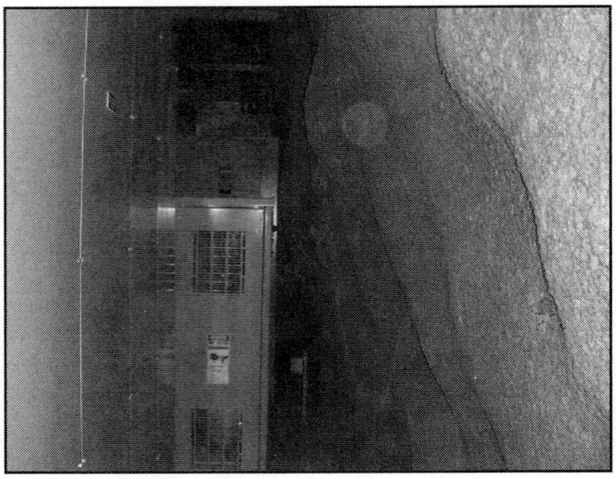

Figure 2.

construction progressed. CP-5 was the first cross passage excavated. The main westbound tunnel (North bore) had a major roof fall out at the intersection of the no. 5 cross passage. The over break extended up at least 10 feet above the crown of the tunnel. There was difficulty holding the ground during the cross passage excavation.

## SUBSURFACE EXPLORATION

As a first step, a comprehensive program of subsurface exploration and ground characterization was directed by CDOT. Parsons Transportation Group, and it's subconsultant, Agapito Associates, Inc. (AAI) were contracted to conduct studies and assist CDOT in the subsurface investigation, and repair design for the CP-5 rehabilitation. PTG/Agapito team conducted subsurface investigations and ground characterization.

The ground characterization plan was developed based on core drilling, geologic and geotechnical logging, and on-site point load testing. To characterize the rock mass surrounding CP-5, a series of 5 holes was cored at each of three sections. Holes were drilled from inside the cross-passage (Figure 3) and depths ranged for 27.7 feet to 45.0 feet in length. The core was logged onsite, and lithology, core recovery percentage, and fracture spacing were noted.

Point load tests were conducted, and continuous photo logs of the core hole walls were generated by CoLog Inc.

Rock mass quality for each length of recovered core was estimated using the Norwegian Geotechnical Institute (NGI) Q-System. The rock mass consisted of highly altered granitic rock. It was determined that the majority of the rock surrounding CP-5 falls into the Q-System classification of "very poor."

Rock quality did not appreciably improve with distance into the rock. The rock quality did appear to be highest in the crown, and lower in the sidewalls. The southern section was slightly more competent than the northern and center sections of the cross passage. Ground water inflow is present, but characterized as minor.

## REHABITATION RECOMMENDATIONS

The rehabilitation design was developed by Agapito Associates, and independently checked by Parsons Transportation Group. Since the rehabilitation work had to be done from a single closed lane with the tunnel in operation, the plan sought to minimize the amount of shotcrete and wall rock material removed.

Given the rock quality, and the desire to maintain clearance between the cross passage walls and installed equipment, the recommended rehabilitation scheme focused on the development of an internal rock reinforcement system. A combination of Polyurethane (PUR) injection (to tie the broken rock mass together to create a self-supporting shell around the opening) and tensioned rock bolts (to further reinforce the shell, limit future deformation, and to provide skin control for the damaged shotcrete) were used.

PUR has a long history of use in mining applications, as described in Molinda, 2008. In addition, CDOT (in partnership with the Federal Highway Administration) had conducted full scale demonstration projects (Arndt, et al. 2008) using PUR repairs to stabilize damaged rock structures. The benefits of choosing PUR for this repair included its ability to inject under pressure and naturally flow to open fractures and discontinuities in need of reinforcement/stabilization. PUR also has a low viscosity to penetrate into the fractures in the rock mass surrounding CP-5. The slightly expansive quality of PUR enhanced its ability to penetrate into the small fractures.

Based on the ground characterization results there were slight amounts of water noted in the jointing. The PUR was specified to be a hydrophobic

**Figure 3.**

to mildly hydrophilic material. This was intended to allow for permeation into the moist voids and fractures without requiring significant pumping pressures.

The rehabilitation scheme included tensioned rock bolts so the shell could be further reinforced. The rock bolts were incorporated to limit further convergence of the cross passage crown and walls, and constrain any shotcrete and rock that remains detached near the excavation surface following PUR injection.

This rehabilitation scheme only required that shotcrete and wall rock be removed where additional clearance was needed, for example near critical electrical switchgear. Other damaged and displaced shotcrete would be stabilized by the PUR injection and bolting, so removal and disposal was minimized.

Agapito designers determined bolt spacing and length using the Q-System. From design charts, a circumferential and lateral bolt spacing of 1 m (3.3 ft) was recommended. This was a conservative recommendation, as it used the pre-PUR injection $Q$ value of 0.5. Bolt length was determined from design equations included in Barton et al., 1974.

The equation used included an Excavation Support Ratio (ESR) parameter. ESR is an attempt to quantify the final use, and thus the required safety factor, of an opening. The lower the ESR, the higher the safety factor. It was judged that CP-5 falls into the category of ESR = 1.3, which includes storage rooms and minor road and railway tunnels. With a roof span of 15.7 ft (4.8 m) and a sidewall height of 10.9 ft (3.3 m), the recommended crown bolt length was 2.55 m (8.4 ft) for the roof and 2.38 m (7.8 ft) for the sidewalls. Using these values as guidelines, and standardizing for ease of application, an 8-ft bolt length for both the crown and sidewalls was recommended. Based on anticipated anchorage factors, a 3-ft length of fast set resin was recommended to provide adequate anchorage, leaving 5 ft to be tensioned and resin-encapsulated.

Based on the desire to apply a significant tensioning preload in the bolts, and on the guideline that bolts should be tensioned to approximately 60% of their minimum yield, a bolt with a minimum yield of 20,000 pounds force (lbf) was specified. This allowed the bolts to be tensioned to 12,000 lbf, and allowed for a capacity far in excess of that required using the conservative assumption of suspension loading. For a 3.3 ft by 3.3 ft area, 5 ft long (the non-anchor portion of the resin column), the suspended load is approximately 8,700 lbs, assuming a rock density of 160 pounds per cubic ft (pcf). The 20,000-lbf minimum yield specification could be met with #6 (¾-in) Grade 60 rebar. Recommended hole size for #6 rebar is 1-in, allowing for a ⅛-in annulus.

Standard rock bolt plates are commonly 6 inches by 6 inches. To provide extra skin control, an 8-in by 8-in plate was recommended. Plate profile and grade was to be as recommended by the bolt manufacturer.

A summary of the recommended bolting system for CP-5 is listed here:

- Bolt spacing 3.3 ft
- Bolt length 8 ft
- Bolt minimum yield 20,000 lbf
- Pre-load tension 12,000 lbf
- Bolt hole diameter 1 in
- Fast set resin 3 ft at back of hole
- Normal set resin from collar to 5 ft
- Plate size 8 in by 8 in

The installed bolt pattern was as described above, with rows were staggered so that bolts in the current row lie between bolts in the previous row. Schematically, bolting begins at the crown centerline and extends down the crown and sidewalls. Following the 3.3 ft spacing, the nearest bolt to the floor was approximately 3.7 ft from the floor line.

Although the detailed design of the PUR injection was left to the successful contractor, it was suggested that the injection effort would utilize the existing core holes (nominal 3 inches diameter) that were drilled by Agapito Associates during the exploration. It would be most cost-effective to limit the injection to a zone from the excavation surface to about 15 ft (one excavation diameter) into the rock mass. This could be accomplished by packing off deeper portions of the existing 30-ft to 45-ft core holes. The hole wall images and core photographs provided in the drill logs were used to target injection areas along the lengths of the holes. To maximize coverage, a new series of holes between the existing sections could be allowed if necessary.

After the PUR injection effort is complete, the recommended rock bolting was performed. A stiff system was desirable, one that can develop load in the rock bolts without significant additional deformation of the cross passage crown and walls. To accomplish that, a fully resin-grouted, torque-tensioned rebar system was recommended.

## CONSTRUCTION

Based on the AAI-recommended repair, Parsons was directed by CDOT to assist in the preparation of plans, specifications, and estimates (PS&E) to develop a construction project that would execute the repair. The project was bid on August 26, 2010, and Mayes Concrete Specialties (MCS) of Grand Junction, Colorado was the successful low bidder.

The contractor mobilized quickly, as the project completion was required by winter 2010. The completion date was based on CDOT's desire to open all lanes of the tunnel and replace sensitive electrical switchgear into a safe, stabile cross passage to allow full operations before the winter season.

MCS began work to stabilize the cross passage as per the PS&E in September 2010. As PUR injection proceeded, other adjacent tunnel elements, such as the mainline concrete lining and tiled areas in the roadway space, were monitored. No additional distress was noted during or after the PUR injection work. MCS injected grout during the period of early October, until early November (Figure 4). MCS injected a total of 28,000 lbs of Stratathane Lock-Roc Strata-Tech Soil Stabilizer™. Samples of the PUR were tested for compressive strength after 2 days. The compressive strength of the PUR tested was over 10,700 PSI.

The actual amount of PUR that MCS injected was about 50% less than estimates. CDOT ordered additional borehole photography, using special lighting techniques to reveal the presence of the injected PUR in the cracks and fissures. This was accomplished by COLOG Inc.. That, along with post grouting core-drilling verified the migration of grout into cracks and fissures. On that basis, CDOT accepted the PUR stabilization.

MCS's installation of rock reinforcement proceeded until the end of November. CP-5 final repair

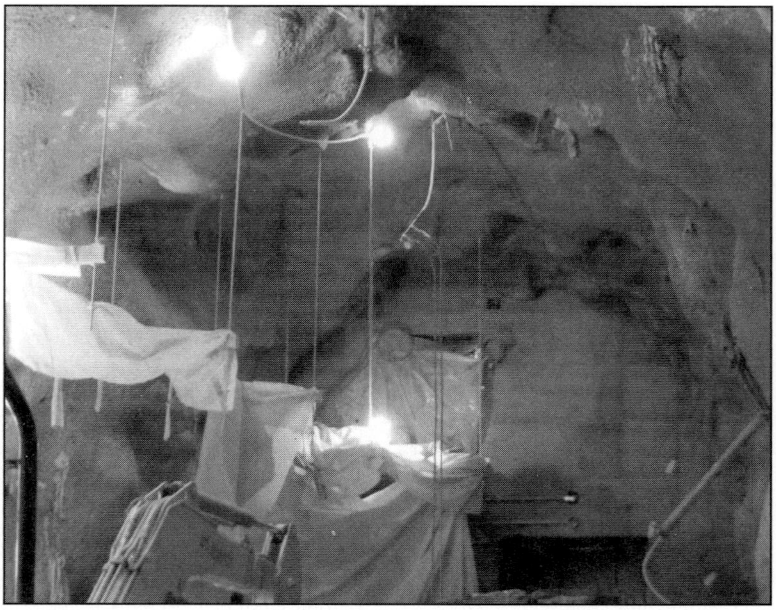

**Figure 4.**

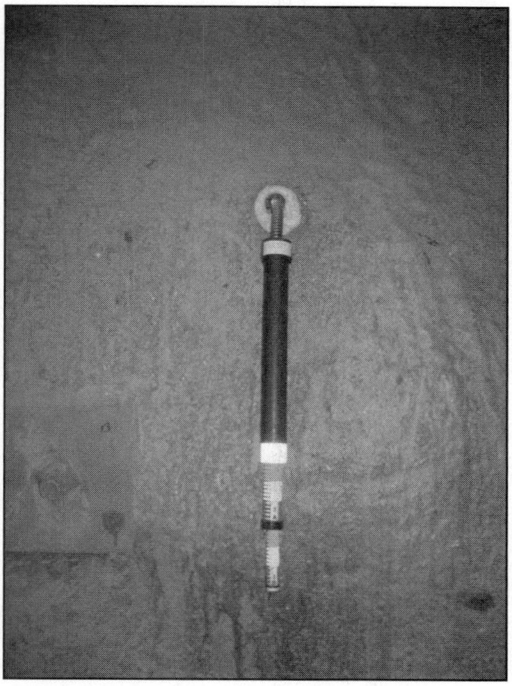

**Figure 5.**

work and cleanup continued until the first week in December.

A final walk-through occurred on December 6, 2010, and the project was accepted by CDOT. After final clean-up, the original electrical equipment was re-installed in mid-December 2010. The original electrical circuits were re-established, and the tunnel electrical operation was restored to its normal operating configuration.

To provide long-term monitoring of the rock mass around CP-5, direct reading tell-tale instruments were installed at 13 locations in the sidewalls and roof (Figure 5).

These were monitored and recorded on a monthly basis by CDOT tunnel maintenance staff and by Parsons Engineers during a summer 2013 structural integrity re-inspection. Tell tales showed movements of about ¼ inch maximum as of January 2014.

## SUMMARY

The original construction of the I-70 Glenwood Canyon tunnels brought modern, state of the art techniques to the American tunneling scene. To continue this legacy of innovation, this successful repair allowed for the continued safe operation of these tunnels while minimizing construction time and cost. At this time, the opening has stabilized, and the travelling public will enjoy a safe passage through the Colorado Rocky Mountains for many years to come.

## ACKNOWLEDGMENTS

The authors wish to thank their respective employers and the Colorado Department of Transportation for the permission to publish this paper. Important project participants from CDOT included Steve Quick-Tunnel Superintendent, Joe Elsen-Program Engineer, Roland Wagner-Resident Engineer, and Damian Leyba-Project Engineer. From the construction contractor, Mayes Concrete Specialties, the hard work of Dwain Cook, Gordon McFerron, and Kyle Vanderberg is recognized. Many thanks to the engineers and drillers at Agapito Associates including Tom Vandergrift, Tim Ross, Leo Gilbride, and Jim Barrix. Portions of this paper are derived from their design report for the project. From Parsons Transportation Group, the contributions by John Braaksma, Brad Doyle, Larry Eckert, and Jon Kaneshiro greatly contributed to the success of the project. The diligent construction inspection and documentation by Steve Foster (Yeh and Associates) and Joe Bair (PTG) ensured a quality outcome.

## REFERENCES

Arndt, Ben, Matthew DeMarco, Richard Andrew, Polyurethane Resin (PUR) Injection for Rock Mass Stabilization, *Publication no. FHWA-CFL/TD-08-004*, September 2008.

Barton, N.R., R. Lien, and J. Lunde. Engineering Classification of Rock Masses for the Design of Tunnel Support." *Rock Mechanics,* 6(4):189–239, (1974).

Molinda, G., "Reinforcing Coal Mine Roof with Polyurethane Injection:4 Case Studies," *Geotechnical and Geologic Engineering,* October 26(5):553–566,(2008).

Trapani, Ralph, R. Andrew, and R. Pihl, "Innovation by Teamwork: Engineering Geology of Interstate 70 through Glenwood Canyon.," Chapter in *Engineering Geology of Colorado*, Association of Engineering Geologists (AEG), Omnipress, (2003).

Trapani, Ralph, J. Kaneshiro, et al., "Structural Inspections of Colorado's Eisenhower Johnson Memorial Tunnel, Hanging Lake Tunnel and Reverse Curve Tunnel.," *Proceedings of the North American Tunneling Conference,* paper P122, (2010).

# TRACK 1: TECHNOLOGY

## Session 3: TBM Technology and Selection

Brett Robinson, Chair

# The Greatest Challenges in TBM Tunneling: Experiences from the Field

**Jim Clark and Steve Chorley**
The Robbins Company

**ABSTRACT:** TBM tunneling is an ever-increasing prospect for underground construction, and with each new tunnel bored there are unknown elements. When boring through the earth, even extensive Geotechnical Baseline Reports can miss fault lines, water inflows, squeezing ground, rock bursting, and other types of extreme conditions. This paper will draw on the considerable field service experience within Robbins to analyze successful methods of dealing with the most challenging conditions encountered.

## INTRODUCTION

Many tunnel projects are located in areas with relatively poor access along the tunnel alignment and bored under extremely high overburden. These two factors often result in limited geological information. It would be reasonable to state that the deeper the tunnel, the greater the level of uncertainties. When faced with these uncertainties everyone involved with project including the owner, the contractor, and the machine supplier must be prepared to tackle geological surprises. This paper describes the problematic geological conditions and associated difficulties faced on three separate projects and the measures that were taken to overcome these difficulties.

## KARGI KIZILIRMAK HYDROELECTRIC PROJECT

### Background

The Kargi Kizilirmak Hydroelectric Project is located on the Sakarya River, near the Beypazarı district of Ankara province in Turkey. The Robbins Company supplied a 9.84 meter diameter Double Shield TBM and continuous conveyor system to Gülermak for excavation of the 11.8 kilometer head race tunnel (see Figure 1). The tunnel is being driven through a mountainside with up to 600 m of overburden. The geology consists of volcanic rock and softer limestone for the first 3.0 kilometers, followed by harder rock including marble and basalt for the remainder of the tunnel alignment. Due to the variation in geology the ground support regimes range from pre-cast segmental lining for the first 3.0 kilometers transitioning into ring beams, rock bolts and shotcrete as the tunnel moves into more competent geology. Several unique features were incorporated into the TBM design to facilitate installation of the various ground support regimes.

### Issues Encountered (Trapped Cutterhead)

The machine was launched in the spring of 2012 and almost immediately encountered geology that was substantially more problematic than was described in the geological reports. The geology consisted of blocky rock, sand and clays. As a countermeasure that was immediately put into place to avoid the cutterhead becoming stuck in the blocky material, crews began boring half strokes and half resets. This ensured that there was always the option of rapidly retracting the cutterhead in the event that torque reached critical levels. After boring through 80 meters of these difficult ground conditions, the machine encountered a section of extremely loose running ground with high clay content. A collapse occurred in front of the cutterhead and the cathedral effect resulted in a cavity forming that extended more than 10 m above the crown of the tunnel. The weight of the collapsed material trapped the cutterhead. After several unsuccessful attempts to clean out and restart the cutterhead, consolidation of the ground above and in front of the machine was carried out. Injection of polyurethane resins via lances inserted through the cutter housings and muck buckets was the method utilized for consolidation operations; however, injection locations were restricted to the available openings and subsequent attempts to restart the cutterhead proved to be unsuccessful.

### Bypass Tunnel

After assessing all the available options it was decided that a bypass tunnel would be required. Robbins Field Service assisted Gülermak with bypass tunnel design and work procedures to free the cutterhead and stabilize the disturbed ground. Blasting techniques were ruled out due to concern over further collapses caused by blast induced

**Figure 1. Double shield TBM for Kargi**

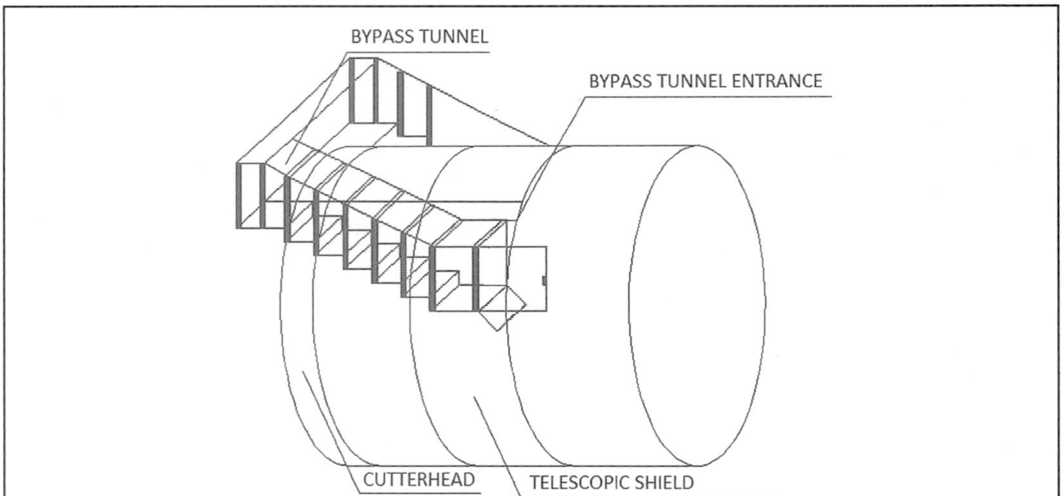


**Figure 2. Bypass tunnel**

vibration; hence, the excavation was undertaken using pneumatic hand held breakers. Details of the bypass tunnel can be seen in Figures 2 and 3.

Upon completion of the bypass tunnel, further stabilization of the collapsed material above the machine and the ground ahead of the machine was carried out. The injection process this time was far more comprehensive due to the vastly improved access provided by the bypass tunnel. The area around the cutterhead was able to be cleared of material and the cutterhead was freed, allowing boring to recommence.

At this point in time it was believed that the collapse was an isolated event and that the geology would improve as the overburden increased; however, material for a second bypass tunnel was stored at site. Unfortunately this measure proved to be prudent planning. Although the machine passed through several weak zones successfully, a further five bypass tunnels were required to free the cutterhead during the first 2 kilometers of boring. Robbins and Gülermak analyzed the bypass tunnel excavation procedures and implemented improvements that resulted in a reduction in the time taken for bypass operations from 28 days to 14 days. One of the main aspects of the improved procedures was the implementation of breaking out for the bypass tunnel through the telescopic shield area of the TBM rather than the accepted norm of breaking out from the tail shield. This modification resulted in reducing the length of each bypass tunnel by over 4 meters.

## Pipe Roof Canopy

The possibility of installing ground support such as fore-poles or a pipe roof canopy ahead of the tunnel face was investigated and after consultation with Gülermak a custom design canopy drill was installed in the forward shield for installation of a tube canopy (See Figures 4 and 5). The space in the forward shield area is limited; hence, the extension section of each tube is only 1.0 m in length. However the advantages of drilling closer to the tunnel face more than compensates for the time spent adding extensions to the tube length. The location of the canopy drill reduces the length of each canopy tube by more than 3 meters when compared to installation using the main TBM probe drills. Apart from the obvious savings in drilling time, the extra 3 meters of drilling length can result in a significant increase in hole deviation. The diameter of the canopy tubes is 90 mm, each canopy typically extends up to 10 m from the tunnel face and the drill positioner, carriage and slew ring provide 130 degrees of coverage.

**Figure 3. Bypass tunnel excavation**

## Squeezing Ground

The time dependency of ground behavior is due to the creep and consolidation processes taking place around the tunnel (Anagnostou & Kovári 2005). In many cases the convergence can be a gradual process taking place over a period of days, weeks or even months. On several stretches of the Kargi tunnel, rapid convergences occurred in the space of a few hours. The geology at the time of these rapid convergences consisted of Serpentine with high content of swelling clay. The convergence was of a radial nature, and distributed relatively evenly around the profile of the TBM.

Probe drilling ahead of the tunnel face identified the majority of the areas considered to be at risk from squeezing conditions. As it is generally accepted that there is a direct relationship between TBM advance rates and problems caused by squeezing ground it was essential that TBM downtime was minimized while boring through these stretches. On the occasions that squeezing ground had been

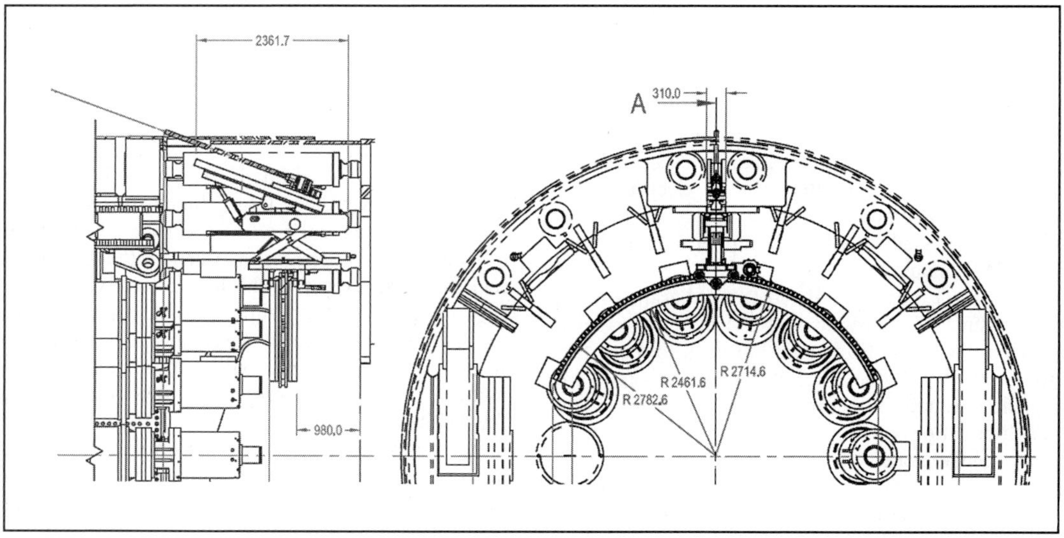


**Figure 4. Custom canopy drill**

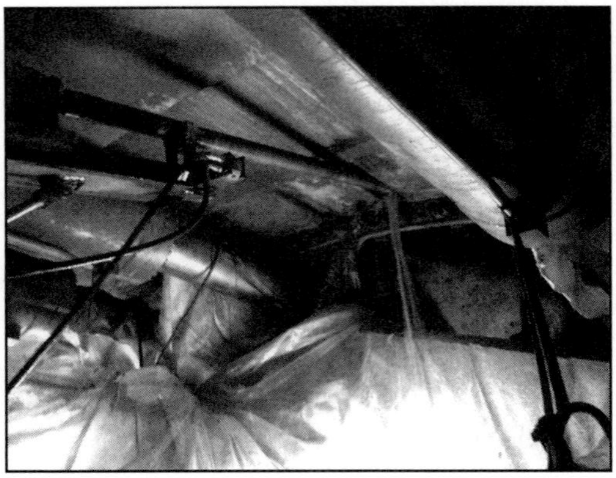

**Figure 5. Canopy tube drilling**

identified all outstanding maintenance works, repairs and replacement of worn cutters was completed before boring through the zone of concern commenced. Inevitably, even after taking these precautions there were unscheduled stoppages. On many occasions the only successful means of restarting the machine after stoppages in convergence zones was to utilize single shield mode boring. In this mode the TBM gripper shoes are retracted, the main thrust cylinders are closed up and the auxiliary thrust cylinders are utilized to propel the machine forward by thrusting off the segmental lining. The typical thrust force for standard boring operations using the main thrust cylinders on the Kargi machine is approximately 21,000 kN. On several occasions thrust force up to 136,000 kN was applied through the auxiliary thrust system before the machine could be freed from squeezing ground. Generally after boring one or two meters in single shield mode the TBM was freed and it was possible to return to double shield mode.

On several stretches of tunnel the rate of convergence coupled with the comparative softness of the ground caused the gripper shield to act as a plough and force muck into the telescopic shield area. The buildup of material became so severe that a mucking system had to be installed in the telescopic shield area. The system consisted of two electric hoists mounted on a running beam that allowed muck kibbles to be placed, lifted, and emptied directly onto the TBM conveyor.

Another measure utilized to combat the effects of the squeezing ground was the application of a polymer based biodegradable lubricant to the extrados of the TBM shields. Eight injection ports were installed around the perimeter of the forward shield and lubrication was injected when boring through convergence zones. It is difficult to quantify the advantage obtained as there was very little consistency in ground conditions and associated thrust pressures; however, it is clear that the application of lubrication reduced the frictional forces between the shields and converging ground.

### Solution (Gear Reduction)

To further mitigate the effects of squeezing ground or collapses, custom-made gear reducers were ordered and retrofitted to the cutterhead motors as a solution. They were installed between the drive motor and the primary two-stage planetary gearboxes. During standard boring operations the gear reducers operate at a ratio of 1:1, offering no additional reduction and allowing the cutterhead to reach design speeds for hard rock boring. When the machine encounters loose or squeezing ground the reducers are engaged, which results in a reduction in cutterhead speed but the available torque is increased. Figure 6 shows the torque curves for both standard and reduced gearing.

Since the installation of the canopy drill and the increase in available cutterhead torque, the TBM has traversed several sections of adverse geology including stretches of severe convergence without becoming trapped. As of November 2013 over 4,250 m of boring has been completed.

## LOS OLMOS

The Los Olmos tunnel is a 12.5 km long water transfer tunnel that was bored through the Andes Mountains in Peru. Odebrecht was the main contractor and the tunnel was driven using a 5.3 m diameter Robbins main beam TBM. It is the World's second deepest civil works tunnel after the Gotthard base tunnel with overburden of up to 2000 meters. The tunnel alignment is through complex geology consisting of quartz porphyry, andesite, and tuff with rock strengths ranging from 60 to 225 MPa. The machine crossed over 400 fault lines including two major faults of approximately 50 m wide.

The machine was launched in March 2007 and by February 2008 it had bored over four kilometers. The geology over the first 4,000 m of boring was far more challenging than was anticipated. As the height of the overburden increased, the geological conditions became gradually more severe and long stretches of extremely loose, blocky ground were encountered. The rock stresses caused by the high overburden also resulted in over 16,000 recorded rock bursting events. TBM utilization was as low as 18.7% of working time because rock support installation was requiring a very high 43.5% of the working time (Roby & Willis 2008). One of the main problems faced was ground deterioration and the resulting falls of blocky ground. The majority of these events occurred during the time taken for the newly excavated bore to pass behind the rear fingers of the roof shield, where ring beams and mesh are installed.

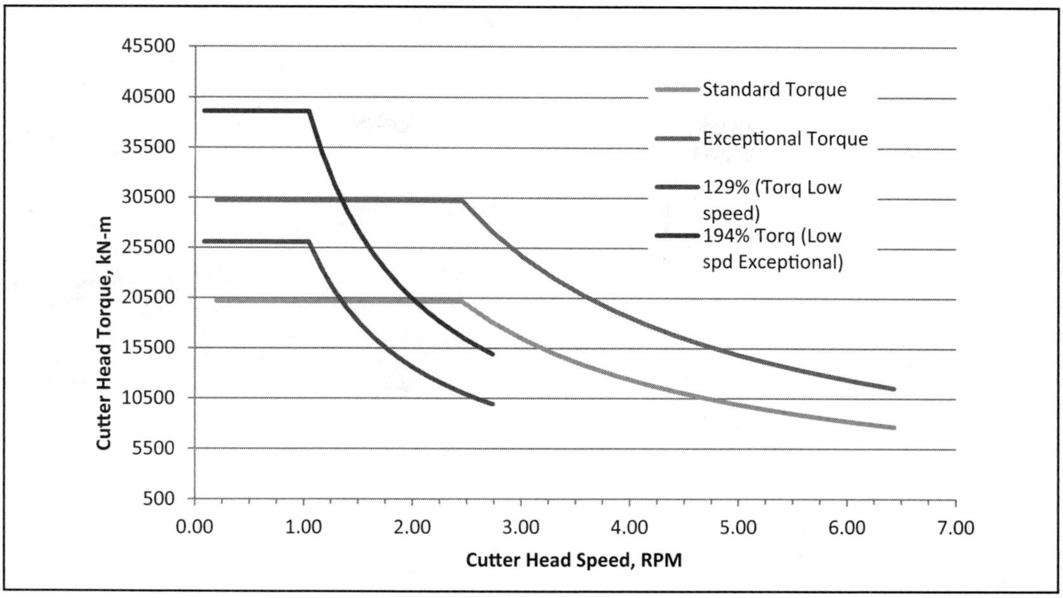


**Figure 6. Cutterhead torque curves**

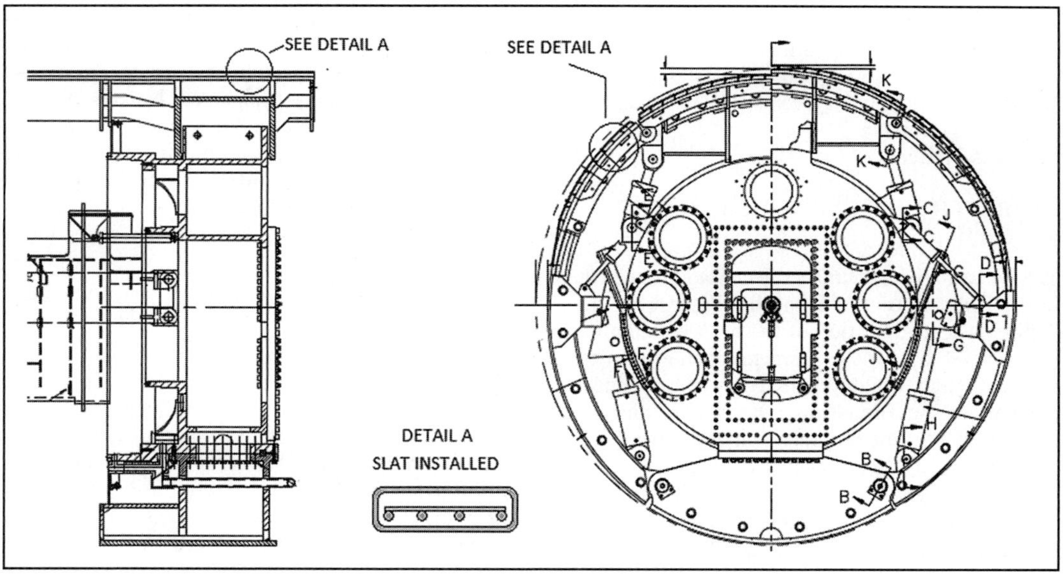


**Figure 7. McNally support system**

### McNally Roof Supports System

During consultations between Robbins and Odebrecht, a decision was taken to modify the machine to facilitate the installation of the McNally roof support system, which allows support to be installed directly behind the main roof shield. The main components of the initial modification consisted of removing the shield roof fingers and forming rectangular pockets with a length of 1.4 m. The pockets run from the rear side of the cutterhead to the trailing edge of the roof support. At a later stage when the ground conditions worsened these pockets were extended to cover the profile of the side supports. Figure 7 shows details of the modifications that were implemented to enable use of the McNally System.

The procedures for installation of the McNally Roof Support System were as follows:

1. Two slats, formed from 6 mm rebar are loaded into each of the pockets.
2. The upper slats in each pocket are drawn from the pocket and pinned to the tunnel wall by means of ring beams or rock bolts.
3. As the machine advances the slats are held in place and extruded from the pockets.
4. When the leading edge of the upper slat is completely withdrawn it is fixed to the trailing edge of the lower slat, with an overlap of 200 mm. Additional slats are then loaded.

The Main advantage of the McNally support system is that is installed closer to the face than other ground support methods used on TBMs, which reduces the required standup time of the excavation. It holds loose rock in place (see Figure 8) which in turn helps to mobilize the strength of the rock mass and maintain the inherent strength of the tunnel arch. When used correctly the system can significantly reduce the time taken to provide adequate support and can also offer reductions in the level of support required.

Incorporation of the McNally support system and various other modifications to the TBM resulted in a steady increase in production rates in spite of continuous rock bursting events. The machine broke though in December 2011 having achieved production rates in excess of 670 m a month.

**Figure 8. Loose rock held in place by McNally system**

## PARBATI HYDROELECTRIC PROJECT STAGE II

The Parbati Hydroelectric Project Stage II is located in the Kullu district of Himachal Pradesh in India. A nine kilometer section of the head race tunnel is being driven by a 6.8 m diameter open gripper type TBM, through a highly stressed mountain range at the foot of the Himalayas. Overburden along the TBM section of the headrace tunnel reaches as high as 1400 m. The geology consists of granite/gneissose, and quartzite with bands of biotite schist and talc. Rock strengths are expected to exceed 270 MPa.

The contractor, Himachal Joint Venture (HJV), purchased a refurbished Robbins-Atlas Jarva TBM from Norwegian Company NCC. The machine was launched in May 2004 and after the completion of 500 m of boring NCC handed over the machine to HJV. HJV operated the machine up to Chainage 1300 m but due to technical difficulties associated with the machine and relatively slow progress they approached Robbins for assistance. Robbins provided a field service team to supervise repairs, maintenance and operation of the TBM. Repairs were carried out, the machine restarted and despite crossing several minor fault zones operations went smoothly with productions rates of up to 526 m a month. Tunnel support ranged from spot bolting through to complete ring beams, mesh, shotcrete and rock bolts.

### Rock Bursts

By mid-October 2006 with over four kilometers of boring completed and overburden of over 1,100 m, several major rock bursting events occurred. The rock bursting was accompanied by moderate to severe loss of ground so the support regime was upgraded to include ring beams, rock bolts, lagging sheets and concrete backfilling. During the following 50 m of boring the incidences of rock bursting events increased to the point that at times they were almost continuous.

### Probe Drilling

The Parbati project is typical of many hydroelectric projects in that it is located in a mountainous area where there is limited access and high overburden above the alignment of the tunnel. These factors resulted in limited availability of detailed geological information. Bearing this in mind geological investigation ahead of the tunnel face was essential and was achieved by maintaining a strict regime of probe drilling.

A routine probe hole (P1) was drilled at chainage 4056 m at the 11 o'clock position on the face. The depth of the hole was 27 m and minor ingress of water and silt was observed from probe chainage 4066.5 m up to 4077.3 m. A decision was made to drill a second probe hole (P2) at the 1 o'clock face position in order to gain further information on the geology/hydrology ahead of the face. During the night shift of the 18th November 2006 the P2 probe drilling operations were underway when the crew heard several cracking sounds emanating from the surrounding rock mass. Shortly after these events the initial probe hole (P1) was observed to be discharging water and silt under high pressure. It took the crew almost 2½ hours to seal the 51 mm hole using a mechanical packer attached to the probe drill. During these 2½ hours approximately 180 $m^3$ of silt and 125,000 liters of water were discharged, and continuous rock bursting was occurring.

### Inundation

Due to the high pressure and high volume of the discharge it was decided that the best course of action would be to drill drainage holes to relieve the pressure ahead of the tunnel face, before a programme of consolidation grouting could be undertaken. Both drainage holes and grout holes were to be drilled via standpipes. The design of the standpipe arrangement consisted of drilling a 75 mm hole 5.0 m deep, inserting a 6.0 m long, 64 mm steel pipe with a threaded section on the trailing end, and anchoring the pipe in place by cement grouting. A ball valve and pressure gauge were attached to the threaded end of the pipe.

A third probe hole (P3) was drilled utilizing the standpipe arrangement, to a depth of 38 meters. Although the location of the P3 probe hole was adjacent to the P1 probe hole location at the 10 o'clock face position, it did not encounter silt or high pressure water. The next course of action was to attempt drilling a fourth hole that would intersect probe hole P1 to facilitate drainage operations. The hole was drilled though a standpipe which was subsequently fitted with a valve to enable regulation of flow, a pressure gauge and a length of 75 mm hose to allow drainage of material directly into the tunnel muck cars (see Figure 8).

On the 24th November probe hole P1 was successfully intersected and drainage operations were underway when several rock bursting events occurred. The pressure in probe hole P1 gradually increased until it exceeded the 25 bar capacity of the pressure gauge and minor inflows of silt and water began to flow through fissures in the rock mass close to the face. Further rock bursting fractured the rock mass surrounding the collar of probe hole P1 causing the rock to fall away and expose the hole behind resulting in an inrush of water and silt under massive pressure. The crew tried unsuccessfully for several hours to insert a packer into P1 to stem the flow of material but at 7:00 am with silt levels rising rapidly and rock bursting continually occurring, the tunnel was evacuated for safety reasons.

**Figure 8. Drilling through stand pipe at Parbati**

During the 25th November it was deemed impractical and unsafe to enter the tunnel. Water ingress was measured at the portal throughout the day and flow rates gradually increased until they exceeded 7000 liters/min. On the 26th November flow rates stabilized so a team entered the tunnel to assess the situation. They observed that the inundation had almost completely buried the TBM (see Figure 9) and that silt and water were still flowing from the probe hole. However the pressure of the discharge had reduced and a crew was mobilized and managed to seal the probe hole by inserting a mechanical packer. The total amount of silt deposited during this event was over 14,000 $m^3$ and the cleanup operation took over 2 months.

### TBM Refurbishment & Modification

Robbins was awarded a refurbishment contract for the TBM as many parts and assemblies had been damaged due to being submerged for a prolonged period of time. Once the refurbishment was complete, cement grouting with OPC was carried out to consolidate the ground in front of the TBM. The project was then held up due to contractual issues until January 2010 when Robbins was awarded a contract to modify the TBM. The main components of the modifications included installation of pockets for the McNally support system, upgrading the cutterhead support system, and an improved probe drilling system. The existing probe drilling system accommodated drilling from two fixed positions only. The modified system provides 110 degrees of coverage.

After the modifications were completed further consolidation grouting was carried out before the machine advanced. A system of boring in increments of 8.0 m advances interspersed by extensive consolidation grouting proved to be successful and the machine successfully crossed the geological feature that had caused the inundation. 50 m of boring was completed before the project was again held up due to contractual issues. The project was retendered early 2013 and works resumed in November 2013, although boring will not commence immediately as remedial works to ground support are required in several sections of the tunnel.

**Figure 9. Parbati TBM buried in silt**

## CONCLUSIONS

TBMs are often the only viable option for the excavation of long tunnels with high overburden, due to the impracticalities of opening several faces via adits to enable the application of traditional tunneling methods. As with the three case studies outlined in this paper geological surprises are frequently encountered in long and deep tunnels. Due to cost constraints contractors often decide to procure a TBM that is suited to the geological baseline reports rather than opting for additional features that insure against geological anomalies. It is more often than not possible to retrofit additional features but TBM down time for preparatory works, installation, and component lead times is usually substantial. The actual cost of the additional features applied to the machines described in this paper would have been a fraction of the costs involved had they been installed during the manufacturing process. When compared to the overall cost of a project, additional features installed during manufacturing become almost insignificant.

Technical features on the TBM are not the only insurance required when faced with geological uncertainties. The contractor should have an action plan in place to cover all eventualities. Ground treatment materials and equipment, as well as bypass tunnel materials and equipment should be available at site. Again the cost of these items is almost insignificant when compared to the cost of the project, and their availability will provide substantial reductions in project delays should they be required.

## REFERENCES

Anagnostou, G. & Kovári, K. 2005. Tunnelling through geological fault zones. Proceedings of International symposium on design, construction and operation of long tunnels (Taipei).

Roby, J. & Willis D. 2008. Coping with difficult ground and 2000m of cover in Peru. Proceeding of World Tunnel Congress, Underground Facilities for Better Environment and Safety (India).

# Tunnel Boring Machine Selection for the Baltimore Red Line Downtown Tunnel

**Harald Cordes, Richard Flanagan, and Mala Ciancia**
Parsons Brinckerhoff

**ABSTRACT:** The Baltimore Red Line Project by the Maryland Transit Administration (MTA) involves design and construction of a three mile long tunnel through downtown Baltimore, enabling light rail service between residential and business areas. This paper focuses on the considerations for the selection of an appropriate Tunnel Boring Machine methodology for the Downtown Tunnel contract procurement. Design challenges of tunneling beneath downtown Baltimore include: a tight right-of-way; close proximity to historic and modern high-rise buildings; pre-existing underground tunnels and utilities; and extremely challenging ground conditions consisting of soil and rock in both full-face and mixed-face conditions. A comprehensive assessment and comparison of state-of-the-art tunnel face support, including innovative hybrid machine techniques with earth pressure balance, slurry face control and open-mode capabilities was performed to determine the most suitable approach for the project.

## RED LINE PROJECT DESCRIPTION

The Baltimore Red Line Project is a new 14.1-mile-long east-west light rail transit (LRT) line, connecting the areas of Woodlawn, Edmondson Village, West Baltimore, downtown Baltimore, Inner Harbor East, Fells Point, Canton and the Johns Hopkins Bayview Medical Center Campus. As shown in Figure 1, the alignment includes a 3-mile-long underground section through downtown Baltimore.

The so-called Downtown Tunnel will be excavated by Tunnel Boring Machines (TBMs) between five cut-and-cover stations. It is expected to utilize two 23 ft diameter TBMs with segmental liners for the dual running tunnels. The horizontal alignment is constrained by the existing street grid with the intent to locate the tunnels within the right-of-way, with a minimum curvature limited to 650 ft. The vertical alignment is constrained by a maximum operational grade that will not exceed 5% in the mined tunnel sections, the location and depth of the cut-and-cover stations, the presence of adjacent and overlying structures and subsurface conditions.

The tunnel alignment will pass beneath and directly adjacent to multiple buildings, including both, partially-historic row houses and older and recently constructed mid- and high-rise buildings. Some of the structures are located in former offshore areas or along the shoreline. Buildings along the alignment are located on a mix of either deep foundations or spread footings. It is currently envisioned that the TBMs will mine through four out of five unexcavated stations with subsequent station excavation and demolishing of the segmental liner within the station footprint. Only the Inner Harbor Station will be fully excavated prior to TBM arrival.

The Red Line General Engineering Consultant (RL GEC) was tasked in July 2013 by Maryland Transit Administration (MTA) with Final Design services for the Baltimore Red Line. According to the schedule at the time of this writing, the Downtown Tunnel is proposed to be constructed under a Design-Bid-Build Contract with contract procurement in mid 2015. The RL GEC is working closely with the Red Line Program Management Consultant (RL PMC) and MTA to complete the design and develop contract documents. For the purpose of this paper, the RL GEC, RL PMC, and the MTA, together, are the RL Team.

## GROUND CONDITIONS

### Geologic Setting

The geologic structure, lithology, and stratigraphy of rock and soils in the Baltimore Red Line Project area are complex and reflect a comprehensive sequence of tectonic, erosional, and depositional events. The Baltimore Red Line Project is located within the Piedmont Upland Piedmont Plateau and Coastal Plain Physiographic Provinces. Two general types of rock materials underlie the Baltimore region: the crystalline rocks of the Piedmont Plateau and the wedge of unconsolidated sediments overlying the east-sloping surface of crystalline rocks in the Coastal Plain.

Overburden in the Downtown Tunnel area consists of loose, non-lithified material lying above

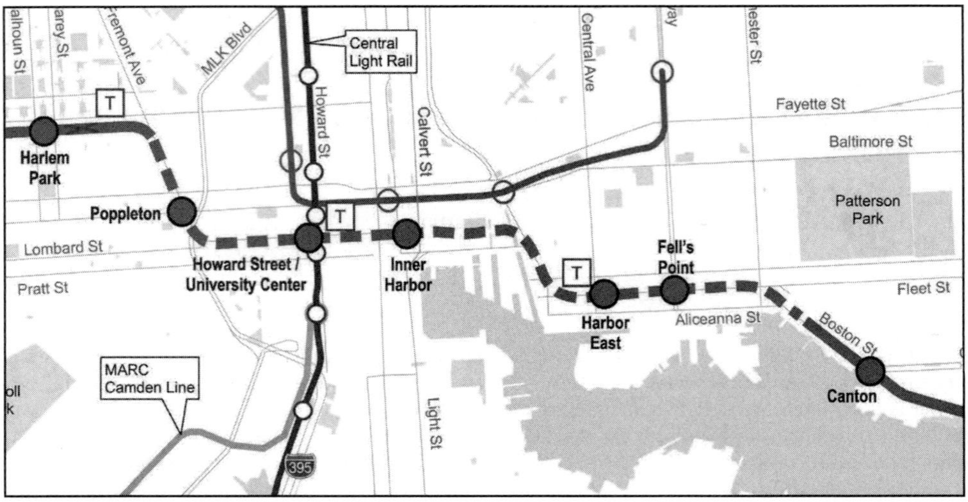


**Figure 1. Downtown tunnel alignment**

weathered rock. The overburden comprises Fill, Post-Cretaceous and Cretaceous deposits, and Residual Soil. A transition zone of highly weathered to completely weathered rock is present between rock and overburden. The transition zone generally reflects the character of the parent rock, and its thickness varies with the lithology of the underlying rock and its drainage and erosional history.

**Red Line Stratigraphy and Ground Classification System**

A project-specific ground classification system was developed for Red Line Project. This ground classification system considers the geologic setting, the nature and variability of rock and soil to be encountered, and the probable construction methods to be used. The unweathered to moderately weathered rock classes are linked to International Society for Rock Mechanics (ISRM) weathering grades, fracture spacing, strength, number of sets of slickensided fractures, number and thickness of planar weakness zones, and presence/absence of inherently weak rock types. For highly to completely weathered rock and residual soil, the classification is also linked to ISRM weathering grade criteria, including decomposition and disintegration, and behavior when agitated in water. For soils, two natural soil groups were defined above residual soil, each of which is further divided on the basis of USCS classification: the Cretaceous Group and the Post-Cretaceous Group. Ground Class Groups occur in the following general stratigraphic sequence from the ground surface down.

The **Fill** is highly heterogeneous. Much of the fill placement along the proposed alignment took place before the twentieth century and was uncontrolled, as fill was placed in former marsh and in open water areas. Fill includes buried shoreline structures, such as timber cribs containing miscellaneous uncompacted fill and debris, and remains of stone masonry walls, quays, and piers supported on timber piles.

The **Post-Cretaceous Group** consists of terrace and/or alluvial deposits of interbedded clean sands and gravels, sand and gravel with fines, soft clays and silts, and organic deposits. The organic deposits, which accumulated in marshes adjacent to estuaries, are moderately to highly compressible. Results of field testing available to date indicate estimated in-situ permeability of some Post-Cretaceous ground classes to be as high as $10^{-2}$ cm/sec.

The **Cretaceous Group** sediments to be encountered by proposed Downtown Tunnel excavations consists of intercalated sands and gravels, sands, silts, and clays of the Patuxent and Arundel Clay Formations. The Cretaceous Group is locally present in thicknesses up to about 25 feet at the western end of the alignment and increases in thickness to the east to more than 120 feet.

The coarse-grained soils of the Cretaceous Group are all very dense, with mean N-values near 75. For fine-grained soils of the Cretaceous Group, consistency based on N-values ranges from stiff to hard, but is mostly within the hard range, with mean and median N-values greater than 50. Cretaceous Group ground classes are heavily overconsolidated and generally have low compressibility and high strength. Results of field testing available to date indicate estimated in-situ permeability of $10^{-5}$ to $10^{-2}$ cm/sec for some Cretaceous Group sand and sand and gravel ground classes. One of the authors experience, during Baltimore Metro Section C construction, indicated permeability as high as 1.5 ×

$10^{-2}$ cm/sec in this type of material. This value was back-calculated based on data from long term dewatering some 1000 ft away from the Baltimore Red Line alignment. Historically, these units are known to have the producing capacity of major water-supply aquifers. Although separated by lenses of fine-grained sediments, they essentially function as a single hydrologic unit.

**Residual Soil** of Ground Class VI is often absent along the Downtown Tunnel alignment, but where present it is generally thicker at the western end of the alignment. Residual Soils are a medium to very dense mix of sand and fines derived from weathered rock. No crystalline texture or rock mass structure is visible. Residual soils in the Downtown Tunnel area generally grade downward into completely weathered rock.

The **Transition Group** consists of highly weathered to completely weathered rock (ISRM Weathering Grades IV and V) with relict rock structure. In completely weathered rock, chemical weathering has progressed to a point where there are more void spaces and fine-grained materials and fewer and less stable mineral aggregates. The completely weathered rock is typically a mixture of sand, silt, and clay. It is dense to very dense or stiff to hard and becomes more granular with increasing proximity to the top of Ground Class III (GC III) rock. In highly weathered rock, the aggregates of sound mineral grains are larger and more strongly bonded and can comprise a volume nearly equal to that of the fine-grained matrix. The highly weathered rock generally consists of less than about 50 percent rock irregularly distributed in a soil matrix. Transition Group thickness was found to be variable in the Downtown Tunnel area. Maximum thickness is in the vicinity of a former stream channel, where was found to be about 60 feet thick. Transition Group ground classes are unstable below the groundwater table and are expected to exhibit raveling to flowing behavior during excavation. Much of the groundwater flow in Transition Group ground classes occurs along fractures or fissures. Permeability is expected to be generally low to moderate but locally much higher, up to $10^{-3}$ to $10^{-2}$ cm/sec, at open fractures which could produce significant inflows.

The top of **Rock** ground classes in the Downtown Tunnel area is not a well-defined surface but a gradational zone, the thickness and nature of which depend on parent material, erosion, drainage history, and other factors. The ground classification approach for the Baltimore Red Line Project established the top of GC III or better rock as the level below which material behavior will be primarily rock-like. Rock of GC III is typically of fair to poor quality based on Rock Quality Designation (RQD). It is slightly to moderately weathered with fracture spacing less than 2 feet and has multiple planar weakness zones less than 6 inches thick or a single planar weakness zone greater than 6 inches thick. Below a vertical distance of 5 to 20 feet below the top of GC III or better rock, rock quality improves, but quality varies depending on lithology and fault proximity. Below the upper 20 feet, rock is typically slightly weathered to unweathered and of good to excellent quality based on RQD but with localized zones of fair to poor quality rock. Rock types and their properties vary both vertically and laterally along the alignment.

The Amphibolite Group is the most common Downtown Tunnel rock, constituting about one-third of the rock to be excavated, mostly at the western end of the alignment. The unconfined compressive strength of Amphibolite Group rocks is likely to be high, up to 34,800 pounds per square inch (psi).

Granite Group rocks are typically intermixed with other rock types and constitute about 15 percent of the rock in the Downtown Tunnel area. Based on laboratory test results available to date, the unconfined compressive strength of Granite Group rocks may be up to 29,600 psi. The Granite Group rocks appear to be among the most abrasive rock types along the Downtown Tunnel alignment.

Tonalite Group rocks locally constitute 30 to 70 percent of the rock mass volume at the western end of the alignment but overall constitute only 5 percent of the rock in the Downtown Tunnel area. Quartz content in the Tonalite Group is high, ranging from 20 to 55 percent. The unconfined compressive strength of Tonalite Group rocks may be up to 34,100 psi.

The Granite Gneiss Group makes up about 25 percent of the rock in the Downtown Tunnel area and generally occurs with intermixed amphibolite in the eastern half of the alignment. Granite Gneiss rocks show a wide range of unconfined compressive strengths, from about 1,700 psi to 31,000 psi. Mica Schist Group rocks appear to occur only locally, in one central portion of the alignment, and have the lowest unconfined compressive strength, less than about 10,000 psi. The Pegmatite Group includes pegmatite, vein quartz, aplite, and migmatite and constitutes about 10 percent of the rock in the Downtown Tunnel area, scattered throughout the length of the alignment. Available data are limited, but unconfined compressive strength appears moderate. Hard mineral content is high due to high quartz content, and the Cerchar Abrasivity Index of 6.1 is the highest of all rock types in the Downtown Tunnel area. A group of Marble/Calcareous Rocks, including dolomitic marble and calc-silicate schist, and a group of Cataclastic Rocks, including mylonite and cataclasite, are also present at several locations along the alignment.

**Table 1. Anticipated ground conditions in the tunnel excavation (west to east)**

| Face Condition | Length | Comment |
|---|---|---|
| Mixed-ground | 900 ft | Mixed ground consists of Cretaceous Group and Residual Soil over Transition Group |
| Mixed-face | 11,00 ft | Mixed face is Transition Group above rock (mostly Amphibolite and Tonalite) |
| Full-face rock | 900 ft | Rock, mostly Amphibolite and Granite |
| Mostly mixed-face | 3,000 ft | Mixed face is Transition Group above rock (mostly Amphibolite and Granite) |
| Mixed-ground | 1,000 ft | Mixed ground on either side of Inner Harbor Station. Mixed ground consists of Cretaceous Group above Transition Group |
| Full-face transition zone | 600 ft | Mostly Transition Group |
| Full-face rock | 2,200 ft | Rock is Granite, Amphibolite, Granite Gneiss, Mica Schist, and Pegmatite |
| Mixed-face | 1,200 ft | Mixed face is Transition Group above rock (mostly Granite Gneiss and Amphibolite, some Granite, Pegmatite, and Marble) |
| Mixed-ground | 2,000 ft | Mixed ground consists of Cretaceous Group and Residual Soils above Transition Group |
| Full-face Cretaceous | 3,000 ft | Cretaceous Group includes full-face sections with clean sands and sections with sands, silt, and clay |

### Anticipated Ground and Groundwater Conditions in the Tunnel Excavation

With the complex intercalation of overburden, Transition Group, and rock, the tunnel excavation face is likely to contain multiple ground classes at most locations. Sudden variations in ground behavior and groundwater inflow associated with variations in ground classes are likely to occur during construction. Highly variable groundwater conditions due to natural variability of subsurface materials, as well as possible localized artesian and perched water conditions create the potential for very large groundwater inflows with infinite recharge from the nearby Inner Harbor. All tunnel excavation is expected to be below the groundwater table.

The tunnel excavation will encounter sections of full face soil (soft ground), full face rock, mixed face, and mixed ground. Mixed-face conditions were defined as instances where Rock is overlain within the excavation face by Transition Group and/or locally Residual Soil or Cretaceous Group ground classes. Mixed-face conditions are anticipated over a total distance of about 5,300 feet along the alignment. A mixed-ground conditions were defined as instances where Transition Group is overlain within the excavation face by Residual Soil, Cretaceous Group, or Post-Cretaceous Group ground classes. Mixed-ground conditions are anticipated over a total distance of about 3,900 feet. Table 1 includes the currently anticipated ground conditions in the tunnel face and the approximate lengths of these conditions along the alignment from west to east.

## TUNNEL CONSTRUCTION CONSIDERATIONS

The ground conditions expected for the Downtown Tunnel are primarily mixed-face conditions (33%), followed by mixed-ground conditions (28%), full face rock (20%) and full face Cretaceous soils (19%), with high groundwater pressures throughout the alignment. In soft ground tunneling beneath the groundwater table, especially in the Cretaceous Group granular materials, and the Transition Group materials which will break down into gravel/sand/silt size particles along with some cobble size material, successful ground control is highly dependent on effective groundwater control. This control must be maintained at all times, especially in granular soils with relatively high permeability, where infiltrating water carries material with it. Loss of groundwater control under these conditions invariably leads to loss of ground control, resulting in face instabilities. Both ground control and groundwater control are achievable with current state-of-the-art TBM equipment technology, but the most appropriate technology will be a function of the anticipated ground conditions. TBM operations will need to be continually adjusted to control face stability and minimize ground loss. Similar adjustments will be necessary when the excavation face includes the extensive length of mixed-face conditions and hard rock fragments in the Transition Group. The tunneling technology for these conditions should be a pressurized-face TBM with additional open-mode and/or semi-open mode capabilities in the full face rock sections.

The mechanized tunneling technology offers basically two different types of TBMs: open-mode and closed-mode. The open-mode TBMs are typical hard rock TBMs, where the face is sufficiently stable and groundwater control is not an issue. Closed-mode TBM operation requires a pressurized face to support unstable ground and groundwater pressure. These pressurized face TBMs are usually divided into Earth Pressure Balance and Slurry Face Machines. Each of these proven methods have advantages in their special range of application. Recent technological advances have also enhanced the respective geological range of implementation over each mode and the combination of these types have been further explored:

- A closed earth pressure balance (EPB) mode would be most appropriate for fine-grained and unstable water bearing soils such as fine-grained Cretaceous Group ground classes, Residual Soil, fine-grained Transition Group, mixed-ground, and some mixed-face conditions. In the EPB mode, the excavation chamber is filled entirely with excavated ground under a pre-determined pressure. The face pressure is controlled by balancing the rate of advance of the TBM with the rate of discharge of the excavated material. A screw conveyor will provide the mechanism to adjust pressure in the chamber at the muck discharge point. Conditioning of the excavated ground will typically be provided by adding foam and or polymers into the chamber, to improve workability and reduce wear/torque of the machine.
- The closed slurry face (SF) mode would be most appropriate for coarse grained and unstable water bearing soils such as the Cretaceous C1/2 Group, coarse Transition Group and mixed-face conditions. The face pressure would be applied by means of pressurized bentonite slurry, with or without additives, depending upon the soil conditions encountered. Slurry is contained in the excavation chamber and would allow for removal of excavated material, suspended in slurry, by pumping from the chamber at approximately the same rate slurry is introduced into the chamber. At the surface, a separation plant would be set up to remove excavated material from the loaded slurry. The treated slurry would then be returned to the pressure chamber.
- The open single shield mode would be most appropriate for stable ground/rock that is not or just slightly water bearing. This would allow the excavation under atmospheric conditions to facilitate cutting tool changes and dry muck removal with a belt conveyer.

It is the unique challenge for this project that none of the above modes on their own are considered to be most suitable for the entire Downtown Tunnel alignment, for the following reasons.

First, full face, highly permeable coarse-grained granular Cretaceous Group ground classes in the east section of the alignment that are overlain by compressible Post–Cretaceous soils, in combination with a high groundwater table, would be at the limit of the typical EPB mode application range. In particular, the groundwater drawdown will need to be controlled and limited to minimize drops in preconstruction pore pressures in the overlying compressible materials, which could induce potentially damaging ground displacements.

Construction experience from EPB tunneling in clean sands at the Botlek Tunnel in the Netherlands [5] revealed process technology difficulties when applying high support pressures with related sedimentation process and an air bubble at the tunnel face. It was concluded that the limit of the EPB application range is defined by permeable non-cohesive grounds under water pressure over 2.5 bar. The earth pressure balance procedure will reach its applicable limit in these conditions when restrictive requirements are set to the admissible surface settlements [5]. It should be noted that since the Botlek Tunnel, the operation range for EBP modes has been increased. However, there still remain unresolved issues with the air bubble, as encountered most recently at an EPB tunnel excavation in Seattle, Washington [2]. The air bubble accumulation in the crown of the excavation chamber needed to be frequently removed manually through ports by bleeding out the air using, simple hand-operated valves in the crown of the bulkhead. The formation of such air bubbles can lead to a variety of problems, such as: excessive drop in EPB pressure between tunnel advances, less stable face support and the possibility of water and material flowing into the excavation chamber, over-excavation and possibility of surface settlements and potential blowouts through the screw or to the surface [2]. However, even though these adverse ground conditions may be encountered only along relatively smaller parts of the Downtown Tunnel alignment, this needs to be factored into the TBM selection.

Mixed-face conditions with rock and unstable material typically present two excavation problems. The tunnel face consists of both stable rock that takes time to excavate while the remaining upper face area is unstable that displaces quickly into any excavation cavity. The vibration associated with machine excavation also increases the mobility of the unstable material. Such conditions are, for example, anticipated in the area of the Poppleton Station on a westerly mixed-face stretch where very strong and massive Amphibolite Group and Tonalite Group rock is overlain by mostly sand and gravel. In these conditions it is expected to be difficult to mix the excavated material into a suitable pulp with "body" that is needed to consistently support the face in EPB mode. In this environment, an SF mode would have advantages despite higher cost for slurry processing. However, having more fine-grained materials and Residual Soil with a high silt and clay content overlying the rock, as on many other mixed-face stretches along the alignment, would reverse this and would make the EPB mode generally more desirable and cost effective by avoiding slurry processing costs. There is also the need for higher power demand, torque and forces on the segmental lining by EPB mode compared to SF mode. However, this appears to be manageable with the anticipated 23 ft TBM diameter under consideration that nowadays the world largest TBM will be a 57 ft diameter EPB. It is also noted that approximately 90% of pressurized face TBM applications are now EPB TBMs [7]. This is likely due to the fact that the initial capital and operating costs for EPB are lower than those for SF since the latter requires a large slurry processing plant. However, as discussed above, EPB mode may entail significantly greater construction risks for certain ground conditions that are present in the Downtown Tunnel in comparison to the SF mode.

Open-mode is only applicable for limited alignment sections. Even when fully in rock, a nominal face pressure may need to be applied due to water pressure build up through joints, faults, shear zones and fractures in rock. An open-mode or semi-open-mode excavation with an EPB would potentially expose the screw conveyor to intense wear, and would require additional protection measures. However, in comparison to an EPB, an open-mode would not be possible with a typical SF TBM.

Any cutterhead design should be, independent of the applied mode, equipped to excavate mixed-face ground conditions with a configuration of both soil rippers and rock disc cutters. Minor adjustment may be made based on the ground conditions encountered.

It should be noted that the five stations along the Downtown Tunnel alignment are not conveniently positioned to allow for a change of TBM mode to match the anticipated ground conditions described herein. Additionally, it is anticipated that four out of the five stations will be excavated after the TBMs pass through, due to right-of-way constraints. Therefore, a mode change would not be possible at these locations. Also, the approach of utilizing multiple TBMs, with different modes that would divide the alignment into multiple short sections appropriate for specific TBM modes, is considered to be uneconomical e.g., it would increase the overall schedule and costs.

Recent advances in mechanized tunneling technology allow tunnel face stability to be applied to the exposed ground in one TBM by either slurry (SF) or conditioned native soil material (EPB) with additional open-mode capabilities. These so-called Hybrid machines can generally provide a safer work environment, mitigating ground loss potential and the resulting displacements of the ground adjacent to the tunnel. Projects that were previously not technically possible are now feasible. Such TBMs can also be equipped to handle rock conditions and provide ground support with the same tunnel lining as used in soft ground. Tunneling Boring Machine selection for use on the Downtown Tunnel should account for these innovative Hybrid TBM types.

## STATE-OF-THE-ART HYBRID TBM TECHNOLOGY

Technical and commercial limits of mechanized tunneling with TBMs are often reached when variable ground conditions become too extensive. Multi-mode or Hybrid machines incorporate the possibility to operate in different modes and therefore adapt the excavation technology in the tunnel to the actual ground conditions encountered [1]. Hybrid TBMs are designed to incorporate the most favorable attributes of different types of machines in order to excavate mixed geology in a more efficient manner [3]. Hybrid TBMs include combinations of traditional muck handling systems and the capability to apply combinations of ground support systems. Design features of Hybrid TBMs often include mixed-face cutterheads with a combination of disk cutters and carbide bits [3].

These Hybrid TBMs are also called Dual-Mode, Multi-Mode, Variable Density TBMs or All-Condition Tunneler depending on their manufacturer and can have very little or even no functional compromises compared to conventional single-mode TBMs. However, this type of TBM is a substantial investment, and if not used properly in the ground type each mode was designed for, advance rates will be less than desired. The requirement for smooth and efficient mode changes is thus essential [4].

### Dual-Mode TBMs (closed/open-mode)

The combination of closed-mode EPB and open-mode has already been well established by several TBM manufacturers and project applications. A classic earth pressure balance machine with a screw conveyor at the invert can be easily operated in open-mode with only partially filled excavation chamber. However, the mode change may involve cutterhead adjustments, and using a more robust screw conveyor that can excavate rock. If a complete conversion is desired, the screw conveyor can be replaced by a conveyor belt from within the tunnel or an intermediate shaft. The more technically complex system is an integrated machine concept with a parallel installation for both mucking systems, the screw and the belt conveyor.

The combination of SF and open-mode usually requires the installation of a secondary mucking system, because the slurry mode is based on hydraulic muck removal. Even with a partially filled excavation chamber, the transport system still needs to be a hydraulic circuit with almost no benefits derived from operating the Slurry TBM in open-mode.

The conversion to an open-mode operation with desired "dry" muck removal would require either replacing the suction grid and stone crusher (if installed) by a conveyor belt or, alternatively, utilizing an integrated machine concept with parallel installed retractable screw conveyor and/or belt in the center. The cutterhead concept needs to be modified with the installation of additional muck buckets, channels or guide plates to transfer the muck to the center mounted screw/conveyor belt.

The integrated concept for either EPB and SF in combination with open-mode has the advantage of a quicker mode change with minimized downtimes and labor costs, but larger TBM investments. Other factors to consider: slurry is very expensive, and the size of the slurry plant and its requirements can be large. The advances in EPB machine technology are closing the gap between Slurry and EPB, and the hybrid design lends itself better to EPB conversion [6].

### Multi-Mode TBMs (EPB/SF/open-mode)

Muck transport and muck handling systems as well as the properties of the muck are significantly different for Slurry and EPB modes. In addition, the mechanism of pressure control is fundamentally different between these types. That makes the combination of EPB and Slurry mode in one machine relatively complex.

The solution of interchangeable systems still requires a free air chamber access, which for most projects is a very difficult and time consuming challenge [1]. An alternative solution is to install both systems parallel in the invert area. However, this requires a relatively large TBM diameter to avoid functional compromises. The stone crusher in front of the suction grid, which could be mandatory for the slurry mode, will create additional constraints.

The newest concept, the so-called Variable Density Machine, can be operated in the classic slurry-mode, incorporating an air bubble system for face pressure control, as well as in a full or open EPB mode. The transfer between the operational modes can be done gradually under permanent and full control of the face pressure and without any need for chamber interventions [1]. The cutterhead is designed to limit the particle size that can enter the excavation chamber. In both modes, the muck would be extracted with a screw conveyor from the excavation chamber and may require a second screw arrangement and subsequent hydraulic circuit with stone crusher.

This type of machine requires two parallel muck transport systems in the tunnel, a closed hydraulic circuit for the SF mode and a dry system with conveyor belt for EPB mode. Based on the project condition, these systems can be designed to perform equally or alternatively the conveyor belt can be downgraded to muck cars based on the utilization of the primary mode.

A real world example of the flexibility of this type of machine concept is the OSIS Augmentation and Relief Sewer project (OARS) in Columbus, OH [1]. Since June 2013, a 23 ft diameter Herrenknecht machine has been utilized to excavate mostly good quality dolomite and limestone rock. Due to anticipated bedding planes as well as possible faults and voids, including karstic and solution features with high water pressure of up to 5.2 bar, the machine is designed to operate in different modes from fully open to semi-closed and fully closed pressurized face mode. The excavated material will either pass through a screw conveyor extending from the invert of the excavation chamber through the pressure bulkhead onto a continuous conveyor, or extracted as a slurry and pumped with an hydraulic circuit to a surface treatment plant.

A relatively similar setting was utilized for the Port of Miami Tunnel with the so-called Water Control Process (WCP) mode. With a hydraulic mucking process, water was pumped into the excavation chamber and muck was discharged through the screw conveyor with a subsequent so-called slurrifier box with integrated rock crusher, then pumped to a separation plant located on the surface. The treated water was re-circulated to the TBM. However, the WCP mode was used only where ground was stabilized by pre-grouting. In all the other areas, the TBM operated in closed EPB mode.

A similar configuration with an additional high density slurry supply system is currently utilized in Kuala Lumpur [1]. Several 22 ft diameter variable density Herrenknecht TBMs are being utilized for the Klang Valley Mass Rapid Transit. These machines are designed to deal with the complex geology of a 9.5 km underground section of Kuala Lumpur's Mass Rapid Transit Sungai Buloh-Kajang Line when it traverses the city centre. The TBMs will break through seven stations along the alignment. The MRT Sungai Buloh-Kajang Line's underground alignment will cut through 2 different geological formations, namely the Kenny Hill Formation which consists of sedimentary rocks such as mudstone, shale, phyllite and sandstone, and the Kuala Lumpur Limestone Formation with erratic Karstic features comprising eroded limestone rock beneath a layer of top soil. A Variable Density TBM was developed which enables the density and viscosity of the slurry to be varied. This prevents the slurry from escaping into cavities or blowing out from fissures leading to the surface. With this method, the face pressure of the TBM is preserved, and the overburden will be kept stable during the excavation process.

A 20.5 ft diameter Hybrid Robbins EPB/Slurry TBM will be soon utilized for the Baku Metro in Azerbaijan. This new hybrid machine will further extend the Baku Metro excavation in 2014. The machine will excavate through 3.5 miles of mixed ground including silt, clay, sandstone, and limestone at pressures up to 5 bar. The hybrid TBM will break through three stations along the alignment. To tackle the difficult ground conditions, the machine will be built using standard EPB features including a screw conveyor, which can be switched over to slurry pipes within the machine shield for muck removal. Both systems will be installed on the machine before launch, so that a simple switchover in just a few hours can be performed as the geology evolves [6].

Even though the geologic conditions of the projects described above are different than the anticipated ground conditions for the Downtown Tunnel, such innovative hybrid technology could mitigate project risk and would likely provide benefits for the project.

## RED LINE APPROACH FOR DOWNTOWN TUNNEL TBM PROCUREMENT

The design of a hybrid machine must be customized for the particular geology in order to be the most effective. Hybrid machines have the potential to lower risk and make difficult excavations possible, as long as accurate geologic information is available [6]. Therefore a Geotechnical Baseline Report (GBR) is planned, as part of the Contract Documents, which will provide information and guidance for selecting a TBM.

The Specification Section for the Tunnel Boring Machine will be equally as important as the GBR for TBM selection. The Specification will include mandatory functional requirements for the machine concept to make it most suitable for the Downtown Tunnel. The development of such a Specification Section requires input from the ongoing site investigations, input from TBM manufacturers, and published experience from similar applications.

Based on these inputs, the RL Team will develop a detailed profile and requirements for the TBM (without limiting potential bidders) to incorporate additional innovations. Based on the current level of design and knowledge at the end the Preliminary Engineering design phase, the RL Team anticipates the following requirements for the Downtown Tunnel TBM selection:

- Provide TBM capable of active face support (closed-mode) by pressurized slurry mode and/or earth pressure balance mode at all times in areas specified in the GBR, to limit volume loss to less than 0.5%. Adapt pressurized face mode based on actual ground conditions encountered.
- Provide TBM capable of operating in both closed-mode (primary mode) and open-mode (secondary mode) without compromising either mode. The design of the machine must be customized for the geologic conditions described in the GBR. Provide TBM with integrated system installation (parallel) for dual-mode and/or multi-mode with smooth and efficient mode changes from within the tunnel without the need for interventions.
- The muck handling system should be designed for high water inflow in weak and unstable ground with groundwater pressure of up to 3 bar. It should efficiently collect the excavated material under all conditions.
- The TBM should have the thrust and torque capacity to overcome high ground loads and free the cutterhead if it gets jammed by unstable ground. Recommended cutterhead drives are electric, variable frequency drives that can operate at variable torque and optimize excavation for a wide variety of ground conditions [4].
- The TBM shield should be tapered and configured to minimize the effects of ground and material interaction on the TBM operation. The shield length should be minimized to allow 650 ft radius alignment curvature.
- The cutterhead should provide a stable face when the TBM is operated in open-mode and allow sufficient material flow through the cutterhead into the excavation chamber in closed-mode.
- A customized mixed ground cutterhead design should be provided, featuring heavy duty knife edge bits that can be interchanged with disc cutters with access from behind the cutterhead.
- Full wear protection for abrasive mixed-ground and mixed-face conditions should be provided. Wear plate should cover the entire exposed front surface of the cutterhead that is not shared with a cutting tool location or injection ports.
- Cutterhead wear protection monitoring systems shall be provided.
- If a screw conveyor is utilized, it should be furnished with a replaceable inner liner and carbide bits for abrasion protection.

It should be noted that this is not a complete list regarding the TBM selection, and additional requirements will be identified as the design advances and will be included in the finalized TBM Specification. However, the requirements stated above should give the reader a good indication of the type of TBM that will be required for the Downtown Tunnel. This level of detail will be included in the Contract Documents, because it is the RL Team's understanding that a Hybrid TBM has the potential to lower project risks and make excavations in difficult ground conditions possible, as long as determining geologic information is included in the GBR and appropriate requirements and guidance for the TBM configuration are included in the Specifications.

## CONCLUSION AND PREVIEW

Conventional pressurized face Tunnel Boring Machines reach their technical and economic limits with their specific method when they have to mine through highly variable ground conditions. Along the Baltimore Red Line Downtown Tunnel alignment, the ground changes from stable rock to soft water-bearing ground and mixed-face conditions. The alignment poses the most demanding challenges for the selection of appropriate, innovative TBMs in order to mitigate project risks and to allow the cost-effective construction of the Baltimore Red Line.

Based on the RL Team's ongoing design and planning, the utilization of Hybrid Machines is being considered during design as a viable and flexible option for tunneling in the adverse and challenging conditions along the Downtown Tunnel alignment. The use of state-of-the-art hybrid technology may be included in the preparation of the bid documents. The benefits of such techniques would be to mitigate excavation impacts on adjacent structures, reduce cutterhead interventions, reduce quantities and costs for ground treatment and avoid the utilization of several TBM types at the same time. These benefits are anticipated to outweigh the negative factor of additional costs for the hybrid technology. However, this is only the case if the specified and offered hybrid technology can be tailored to the ground conditions encountered.

An effective hybrid machine design is about minimizing complications, and the designed machine must be able to mine efficiently in the prescribed conditions while minimizing cost to everyone involved [6].

## ACKNOWLEDGMENT

The authors want to thank Werner Burger/Herrenknecht and Lok Holm/Robbins for their valuable feedback and information regarding state-of-the-art Hybrid TBM technology. We also want to thank the GEC and the PMC team for their contribution. We want to thank Brett Zernich/Traylor for review of this paper.

## REFERENCES

[1] Burger, W., Multi-Mode TBMs—State of the Art and Recent Development, RETC 2013, Washington DC, USA.

[2] Gharahbagh E.A., Diponio M.A., Raleigh P., Hagan B., Keeping the chamber full: Managing the "air bubble" in EPB Tunneling, RETC 2013, Washington DC, USA.

[3] Grothen B., Clark. J, Hybrid TBM Design Selection, World Tunnel Congress 2012, Bangkok, Thailand.

[4] Jordan D., The Next Generation of Large Diameter, Mixed Ground Tunnel Boring Machines, World Tunnel Congress 2013, Geneva, Switzerland.

[5] Maidl U. and Cordes H., Active Earth Pressure Control with Foam, World Tunneling Congress, International Tunneling Association (ITA), 2003, Amsterdam, Netherlands.

[6] Robbins Company, One Machine, Two Modes—Hybrid TBMs Gain Ground in the Industry, Tunnels & Tunnelling International, December 2013.

[7] Thewes M., TBM Tunnelling Challenges—Refining the State-of-the-Art, International Tunnel Association (ITA), 2007, Prague, Czech Republic.

# The Next Level: Why Deeper Is Better for TBMs in Mining

**Ryan Gratias, Craig Allan, and Desiree Willis**
The Robbins Company

**ABSTRACT:** Diminishing surficial mineral deposits, increasing environmental regulation and advanced geological exploration techniques are ushering in a new era of mining. Unconventional technology must be adopted to ensure that safe, efficient and responsible access to minerals is possible as prospecting continues to push the mining industry deeper. This paper discusses why competitive mining operations will become increasingly dependent on Tunnel Boring machines (TBMs) for mine development and expansion, and explores the implications of TBMs in a drill and blast dominated industry.

## INTRODUCTION

Like all industries, mining is constantly changing or evolving. The quality and types of materials being mined, the methods to extract those materials, the geographical location of the materials, the manner in which materials are accessed, and the social and political climate in which mines operate are all changing at a rapid pace. Narrowing the focus to mining minerals in ore bodies, the changes are no less significant. Current trends include a global reduction of surface deposits and continued increased awareness of environmental impact from mining.

## GEOLOGICAL EXPLORATION

### Ore Genesis

Ore genesis in its broadest sense determines how mineral deposits form within the earth's crust. Locating and classifying these ore bodies is a complex endeavor. Geologic exploration techniques range from conventional prospecting to the use of airborne and satellite imagery. Once a prospective site is identified, geophysical prospecting allows for surveying and mapping the ore deposit. Ore bodies are formed by a variety of geological processes and therefore can be found in a range of formations. Methods to survey and map the formation include remote sensing, aeromagnetic surveying, regional gravity surveying, and airborne radiometric methods.

### Ore Evaluation

After an ore body is located, it must be evaluated to determine the content and concentration of the ore mineral in order to assess the economic viability of extraction. As demand for minerals found in ore deposits continues to rise, and as environmental concerns continue to grow, mining operations are forced to extract minerals from more complex locations and with less environmental impact. As surface ore deposits are being depleted and environmental concerns over surface displacement grow, underground mining operations will continue to become more prevalent. For subsurface mines, core drilling provides mineral samples and helps narrow the specific boundaries between materials.

The time and financial investment required to locate, identify, map and evaluate the ore, all critical components to successful mining operations, also contribute to the cost of mining. Yet, when well executed, these significant investments can help reduce cost to access and harvest minerals.

## ACCESSING MINERAL DEPOSITS

### Surface Versus Subsurface Ore Deposits

Underground mines will continue to be the primary mining method in the future—this fact is evidenced by the global reduction of available surface ore beds, the cost and environmental implications of overburden removal, the environmental impact and the public relation implications of large surface mining operations, and the increased global demand for minerals and metals.

### Access Methods

Subsurface ore bodies are accessed by shafts and declines. The primary methods of accessing subsurface ore bodies are drill and blast—by far the most common—shaft boring machines and tunnel boring machines. In this paper we are discussing only decline access tunnels and are discussing TBM and drill and blast methods. As previously mentioned, ore deposits are found in a variety of formations. The depth and type of formation, as well as the type and quality of the material to be mined determine the

**Table 1. Comparison of TBM and drill and blast methods**

| Factor | Drill and Blast | Tunnel Boring Machine |
|---|---|---|
| Site prep time | Requires less start up time | Requires 3 to 12 months |
| Equipment storage | Requires explosive storage permits | Requires slightly larger foot print |
| Length of the tunnel | Slower excavation rate (typically 3 to 9 meters per day averaging 180m/month with three shifts) | Significantly faster excavation rates from 15 meters to 50 meters per day, 450+/month) |
| Shape of the tunnel | Typically horseshoe-shaped but can be other shapes | Uniformly round |
| Length and depth of required tunnel | Difficult in low overburden settings<br>Substantially slower in longer access tunnels (over 2 km) | Not comparable to drill and blast for short tunnels (less than 2 km)<br>Minimum 30 m turn radius<br>Faster for long, straight tunnels<br>Can be used in low or high overburden |
| Ore body orientation/mining method used | Can be used with any ore body orientation | Best for use with deep or long ore bodies |
| Removal, disposal or reuse of spoils | Can be reused but spoil size and consistency is highly variable. Removal due to variable size of rocks can be difficult. | Can be reused; uniformly sized muck chips. Uniform rock also makes for easier removal by continuous conveyor |
| Means for removing mined material | Continuous conveyor; muck cars | Continuous conveyor; muck cars |
| Ground vibration | High | Low |
| Existence of explosive and/or hazardous gases | Mitigation possible | Mitigation Possible |
| Populated or unpopulated area | Typically unpopulated, or in populated areas with restrictions | Populated or unpopulated |
| Access to skilled labor | Requires unique skill sets and certification | Primarily mechanics |

method in which the ore is extracted. These and other factors also play an important role in determining the most cost effective and efficient way to access the ore body itself.

### Factors Affecting Sub-Surface Access Methods

Table 1 provides a partial list of factors that affect the type of access method employed.

## COMPARISON OF MINING METHODS FOR A DEEP ORE BODY

For the purposes of this paper we will use an example that is likely to become more and more prevalent in the future—that of a deep ore body. To extend the life of a hypothetical mine, an access bore must be excavated to a depth of 750 m below the surface (see Figure 1). Assuming a 15% grade, the bore will need to be approximately 5km in length. Because this is an existing mine, there is minimal site prep, logistics and permitting and therefore excavation can begin in six months.

### Surface Mining

Surface mining for such a deep ore body, while possible, is unlikely. Removing hundreds of meters of overburden would probably not be financially viable, and would definitely have negative environmental implications. The PR implications associated with surface mining to such a depth are also likely to be negative.

### Drill and Blast

While Drill and Blast (D&B) is likely to be favorable over surface mining at such a depth, the method has advantages and disadvantages. No overburden must be removed and the method is considered much better for the environment than surface mining. The D&B method can be mobilized fairly quickly, starting immediately after the site prep is complete and can excavate short radius turns in tunnels. However, the 5-km tunnel length exposes the drill and blast method's major weakness—advance rate. The excavation rate of a drill and blast operation may average out to 6 m per day (Tarkoy & Byram, 1991). At this rate, and assuming 6 months for site prep, logistics and permitting, it will take about 2¾ years to finish the access tunnel.

### TBM Tunneling

Through decades of experience in tunnels around the world, it has been observed that in tunnels over 2 km in length, TBMs are the most effective tunneling method (see Figure 2).

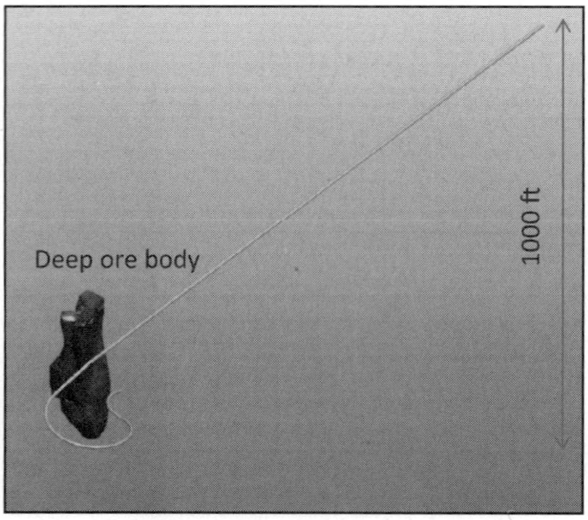


**Figure 1. Deep ore body with access tunnel (OZ Minerals, 2013)**

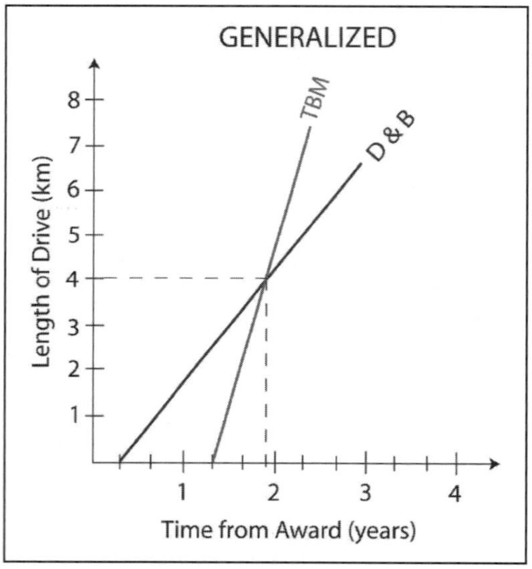


**Figure 2. Generalized graph comparing advantages and disadvantages of TBMs vs. D&B**

In comparison with D&B methods, TBMs have many advantages. TBMs are a more automated form of construction, requiring fewer workers. It has been shown that less ground support is needed in comparison with drill and blast. This can be attributed to the smooth excavation profile. The type of ground support is also more widely varying for TBMs—from wire mesh to ring beams, rock bolts, and steel slats using the McNally Support System. Installation of these types of ground support from within the machine shield, paired with the absence of explosive materials for excavation, also makes TBM tunneling safer in general than Drill and Blast.

Time is both the main advantage and disadvantage of TBMs. The advantage comes in the form of advance rate whereas the disadvantage is due to delivery/setup time. TBMs average speeds of 20 m per day which means it will take a TBM only 250 days to excavate the access tunnel, as opposed to the 830 days needed for D&B. However delivery and setup for a new, custom TBM is about 1 year. This means that the TBM will start six months after the D&B operations would. Despite the six month latency, using a TBM will still beat D&B to the finish by nearly a year. Furthermore, a TBM can be reused, so if a mining operation were to own one then the lead time for startup could be reduced from one year to a couple of months.

The addition of a continuous conveyor for muck removal can further increase TBM advance rates over long distances, with typical conveyor system availability rates of 90% or higher observed. Ventilation is also much better in TBM tunnels using conveyors, as there is a substantial reduction in exhaust from locomotives. Continuous conveyors could also be used with drill and blast operations, with the same effect of speeding up advance rates over rail car haulage.

### Chosen Method

Given the advantages offered by a TBM in a longer access tunnel scenario, paired with modern TBMs' unique abilities to excavate in conditions such as decline tunnels, make this the obvious choice. Modern TBMs can be designed with shorter main beams to bore in reduced radii curves, be outfitted with core drills and other ancillary equipment for ore body exploration, and can be specially designed for muck haulage on a decline.

## MAJOR MINING PROJECTS

### Stillwater Mine, Montana, USA

Examples of successful mining projects using TBMs are available worldwide. The Stillwater mine is perhaps the best example of TBMs being used over a significant period of time to extend the life of a mine and access a longitudinal ore body.

The Stillwater Mining Company (SMC) is the largest producer of platinum group metals (PGMs) in North America and the only producer in the United States. Its J-M Reef lies under southern Montana's Stillwater, Sweet Grass and Park Counties and is located approximately 30 miles north of Yellowstone National Park. Discovered in the early 1970s, the 28-mile-long J-M Reef is part of the Stillwater Complex, a layered succession of ultramafic to mafic rocks in the earth's crust. Its uniform layers of mineral concentrations and proximity to the earth's surface make the J-M Reef a world-class ore body for Platinum Group Metals.

SMC has selected TBMs for mine development because of the benefits they offer over conventional mining methods. The mine has found that TBMs have increased advance rates over traditional mining methods. While the capital cost for TBMs is approximately 1.5 times that of conventional mining fleets, they only have 33% of the operating costs. SMC has used four TBMs for mining in the past, with the first TBM used at the Stillwater mine in 1988. Table 2 shows a list of TBM drives completed or started at SMC since 1988.

SMC's latest TBM bore is the Blitz Tunnel, a 7.1 km (4.4 mi) mine development tunnel, which will map the location of the reef in the Eastern portion of the mine where there is limited drilling data. SMC ordered a 5.5 m (18.0 ft) Main Beam TBM manufactured by The Robbins Company for the job (see Figures 3 and 4).

In order to detect the reef in relation to the TBM, careful analysis is required during drilling. Diamond core drills on the TBM, in addition to probe drills, take samples above, ahead, and alongside the machine every 150 m (500 ft). The cores are logged and interpreted on the spot, concurrent with boring. Based on the data, the TBM is then readjusted so that it stays on the correct bore path—near but not intersecting the reef.

### Magma Copper Mine, Arizona

The San Manuel Mine is one of the largest underground mines in the world, but projections before the tunnel was built estimated its reserves would be depleted by 1998. The tunnel allowed the

**Table 2. List of TBM drives at SMC since 1988**

| Mine | Machine | Drive | Start Date | Finish Date | Length (m) |
|---|---|---|---|---|---|
| Stillwater | Robbins MB 146-193-1 | 5000 East FWL | March 1988 | July 1988 | 975 |
| Stillwater | Robbins MB 146-193-1 | 5900West FWL | May 1989 | August 1990 | 3,390 |
| Stillwater | Robbins MB 146-193-1 | 5700West FWL | October 1990 | January 1991 | 1,405 |
| Stillwater | Robbins MB 146-193-1 | 5500West FWL | February 1991 | June 1991 | 7,500 |
| East Boulder | CTS | Access #1 | July 1998 | July 2000 | 2,286 |
| East Boulder | Robbins MB 156-275 | Access #2 | March 1999 | September 2000 | 5,530 |
| East Boulder | Robbins MB 156-275 | West FWL | September 2000 | September 2008 | 2,200 |
| East Boulder | Robbins MB 156-275 | Graham Creek | January 2011 | 2012 | 2,590 |
| Stillwater | Robbins MB244-313-2 | Blitz 5000 East | May 2012* | | 6,858* |
| | | | **Total** | | **32,734** |

* Indicates project in progress as of January 2014.

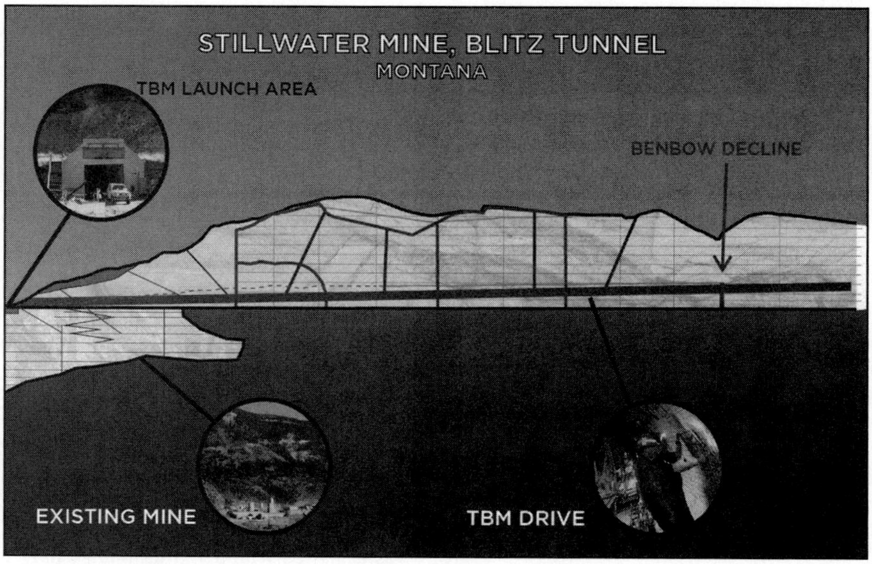


**Figure 3. Blitz tunnel diagram**

**Figure 4. Main beam TBM for Stillwater Mine**

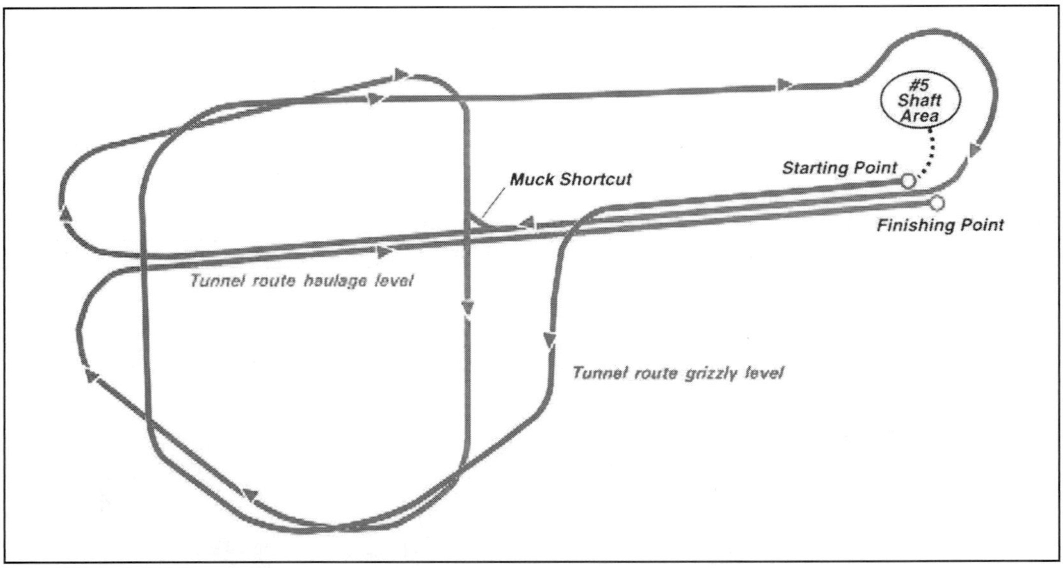


**Figure 5. Route of Magma copper tunnel**

development of the Lower Kalamazoo ore body, in the vicinity of dwindling ore bodies that had already been tapped. As a result the mine was able to stay open until 2003.

The project owner, Magma Copper Company, awarded the construction contract to a joint venture of Frontier-Kemper Constructors Inc. and Deilmann-Haniel GmbH. The joint venture chose a 4.6 m Main Beam Robbins TBM to bore the 10.5 km mining tunnel (see Figure 5).

The Lower Kalamazoo geology is quite complex, consisting of porphyry, and granodiorite. The tunnel route includes numerous faults and dikes—it passes through the San Manuel fault six times and the Virgin Fault five times. Much of the rock has been weakened by hydrothermal metamorphosis.

The cutterhead of the 4.6 m diameter machine could reverse rotational direction to prevent jamming when it encountered fractured rock. The machine was designed with a shorter main beam, allowing it to excavate reduced radii curves in the tunnel. Boring

**Figure 6. Breakthrough ceremony at Magma Copper Mine**

began on November 11, 1993 in a specially prepared concrete chamber. There were no major problems crossing the San Manuel Fault, but wet clay at the Virgin Fault slowed boring. The TBM continued to encounter soft clay and crumbling ground.

Robbins and the contractors added several features to the machine to optimize performance. They increased muck flow through the cutterhead, increased cutterhead torque, and added additional rock support to the tunnel. After the initial modifications, TBM performance greatly improved. Daily advances tripled to 22.94 m per day for the first 15 months of boring and the machine averaged more than 30 m per day for the rest of the project. The TBM stayed on schedule and holed through on December 4, 1995 (see Figure 6).

### Grosvenor Decline Tunnel, Australia

A unique tunnel has just begun excavation near Moranbah, Australia at the Anglo American Coal Mine. An access tunnel is required for deep coal drifts. Two decline tunnels, at grades of 1:6 and 1:8, will be used for the mine access to new coal seams. An 8.0 m hybrid EPB/rock machine was supplied for mixed ground conditions ranging from sand and clay to varying grades of hard rock up to 120 MPa UCS, as well as coal seams. Methane gas is expected to be present throughout the tunnel, so the machine has been designed as Explosion Proof Compliant to ERZ-1. The TBM was launched in December 2013 (see Figure 7).

Only about 300 m of ground are expected to require EPB mode, while the rest will be bored in hard rock mode. Thus, the design was optimized towards hard rock excavation. In EPB Mode, the machine utilizes a two-stage, center-mounted screw, with a replaceable inner liner and carbide bits for abrasion protection. A mixed ground cutterhead is fitted with interchangeable knife bits and Trimay abrasion resistant wear plates for abrasion protection. To keep the mixing chamber spark-safe in the presence of methane, the chamber is filled with water, foam, and other additives. To deal with the resulting watery muck, the first screw conveyor is run faster while the second screw conveyor is run slower, creating a muck plug in the beginning of screw conveyor #2, which keeps most of the water in front of the machine.

The machine essentially uses its EPB technology to deal with any methane gas safely. If any methane leakage is detected, an evacuation system called a "snuffer box" will draw methane out of the end of the screw conveyor and directly into the ventilation system.

To convert to hard rock mode, a hydraulically operated muck chute is deployed around the screw. The muck is then picked up by paddles in the muck chamber to load the screw. Interchangeable EPB knife bits must be replaced with disc cutters on the cutterhead, and the EPB scrapers on the cutterhead must be replaced with hard rock bucket lips.

A skew ring twists the thrust cylinders in order to react the torque of the machine in hard rock, allowing for more efficient single direction cutterhead excavation and muck pickup. Mini grippers on the rear shield allow the machine to bore 400 to 600 mm forward, then be retracted for cutter changes (see Figure 8).

A final unique aspect of the machine is a specially designed "Quick Removal System." As no

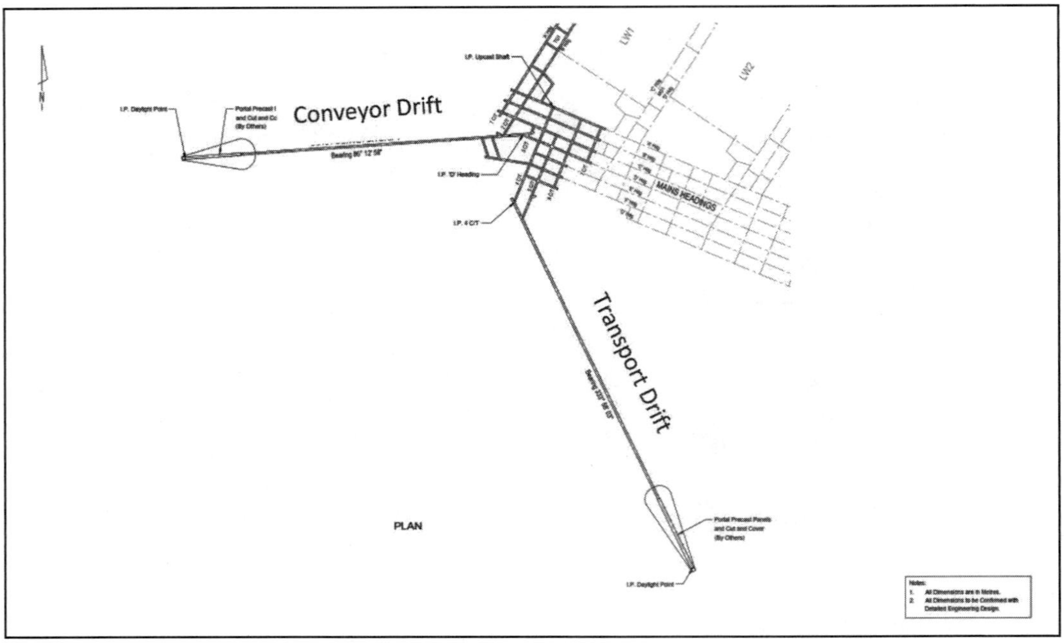


**Figure 7. Layout of tunneling at Anglo American Coal Mine**

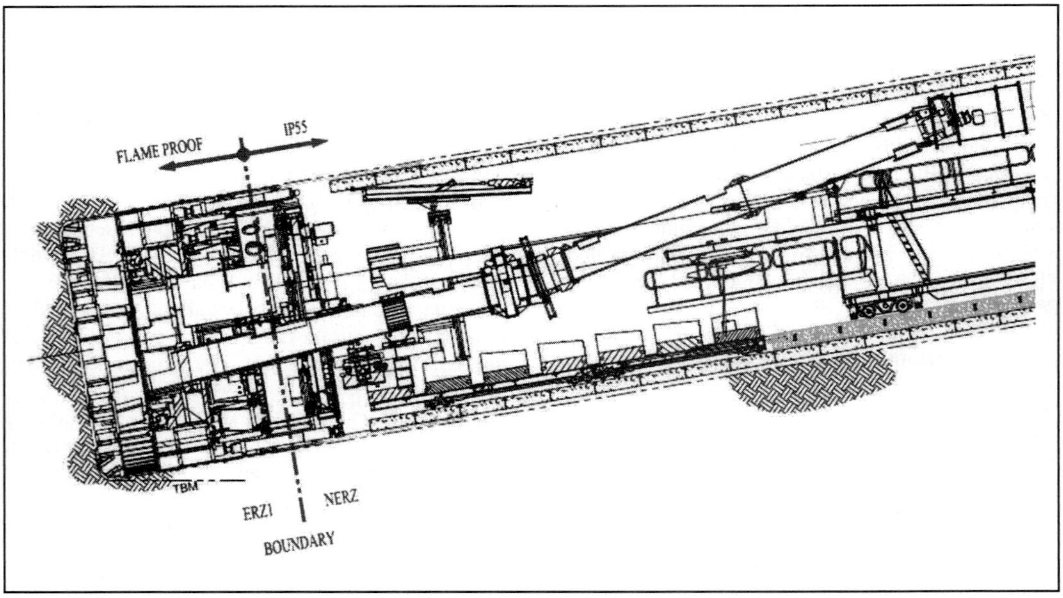


**Figure 8. Explosion-proof TBM on a decline**

ground in Australia can be left unsupported and the machine is boring a blind tunnel, it is designed to be retracted in one piece from its shield, leaving the shield in place. The core of the machine is a bolted design and separates from the shield, in a process that does not require a cutting torch. The machine will then be walked up the decline tunnel on a set of specially designed transport dollys and sent by rail to the second decline tunnel, where another shield will be waiting for machine assembly prior to launch.

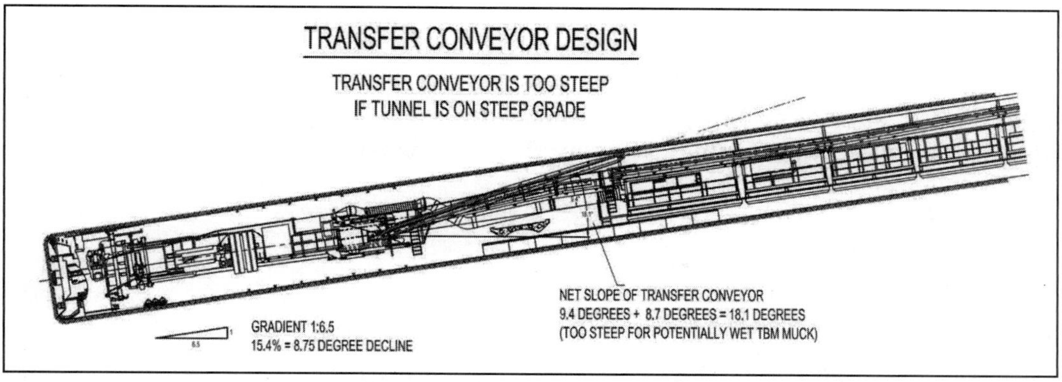


**Figure 9. Typical conveyor design does not provide correct angle for muck removal**

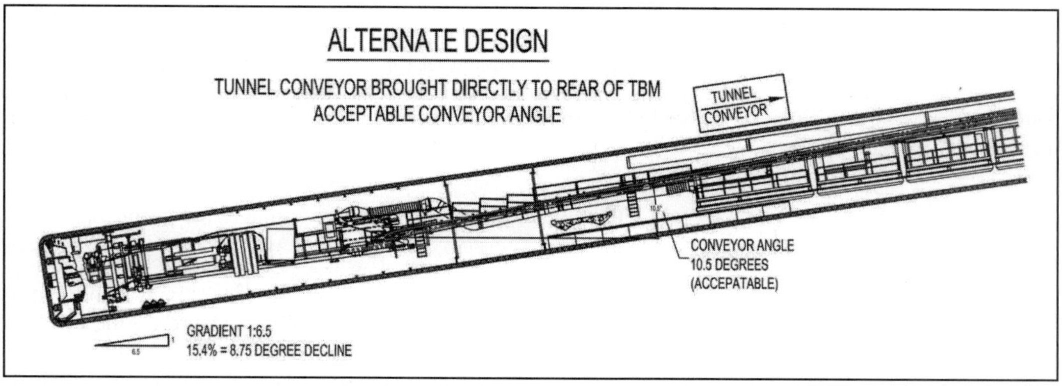


**Figure 10. Alternate conveyor design brings tunnel conveyor directly to the rear of the TBM**

### Carrapateena Decline Tunnel, Australia

Another decline tunnel, yet to begin excavation is located at the OZ Minerals copper and gold mine in southern Australia. A high grade, cylindrical ore deposit has been identified 500 to 1,500 m below the ground. To excavate the ore body, a TBM access tunnel 1,000 m deep is required. A 5.8 m diameter Main Beam TBM was procured to excavate a 7 km access tunnel at 15.4% grade. The angle of decline requires the TBM and continuous conveyor to be uniquely designed to maintain an acceptable angle for conveyor muck removal (see Figures 9 and 10).

The TBM is currently being assembled at Robbins' manufacturing facility in Shanghai, China, and will be delivered in early 2014. The project is on hold and has an unknown start date.

## CONCLUSIONS

Looking at only TBMs manufactured by Robbins, 29 have been used in mining applications over the years and mining use is accelerating. Given the various aspects that these projects have demonstrated—boring longitudinal ore bodies, curved tunnel drives, steep declines, and in gaseous conditions—modern TBMs have what it takes to make mine development rapid, efficient and economical. For deep ore bodies requiring drives over 2 km in length, TBMs should be seriously considered for their higher advance rates, improved range of ground support, and safety.

With the global demand for minerals increasing, mines can only be pushed in one direction—deeper. As the location of deposits change, the excavation must necessarily evolve with it. Those mines embracing mechanized tunneling, and more specifically TBMs, will experience a paradigm shift in their mining operations. Ore bodies which were once considered inaccessible will finally be within reach. Early adopters of the TBM method will be able to better meet the increased demand and/or extend the life of the mine—a result every miner hopes for.

## REFERENCES

OZ Minerals 2013. Carrapateena Decline Excavation by TBM. Powerpoint presentation.

Tarkoy, P.J., Byram J.E. 1991. The advantages of tunnel boring: a qualitative/quantitative comparison of D&B and TBM excavation. Hong Kong Engineer, January 1991.

# Urban EPB Tunneling in Limited Space: A Case Study of the San Francisco Central Subway Project

**Noah Johnson**
The Robbins Company

**ABSTRACT:** The San Francisco Central Subway project is a challenging modern example of urban tunneling in limited space conditions. Two 6.3 m diameter Earth Pressure Balance Machines (EPBs) are excavating parallel 2.5 km long tunnels under low cover and in mixed ground conditions. The small launch site situated between an interstate and an off-ramp, highly curved tunnel alignment, and geology are particular challenges. These elements required customized tunnel and machine design, from TBM shipment and assembly, to launch and excavation.

This paper discusses the project challenges and solutions at the Central Subway project, with a focus on TBM and continuous conveyor logistics. Requirements of the project include explosion-proof electrical components, laser-guided survey, rubber-tired supply vehicles, and machine and back-up solutions for steep inclines and tight curves.

## INTRODUCTION: CHALLENGES OF URBAN TUNNELING

Tunneling in urban areas lends itself to certain challenges to overcome during all stages of design and construction. While there are also many political, environmental, financial and logistical concerns with tunneling in urban areas, this paper will focus on the issues seen from a TBM manufacturer's standpoint. The main issue faced is that of space at the site level. More frequently, urban tunneling jobsites are relatively small. In the ideal situation there would be enough room at the jobsite for staging of all the TBM parts, services, and tunnel excavation support equipment. This is not the case in most urban environments as the ground level is usually congested with other infrastructure. Because of this, tunnel launching and reception shafts need to be squeezed into existing unused or underused areas. The TBM must be able to be broken down into smaller pieces in order to fit and navigate these small sites. The TBM must then be assembled, launched, operated, and disassembled under these same constraints. A large range of sizes of tunnels are used in urban environments, from micro utility tunnels to multiple lane highway mega-tunnels. This paper will focus on challenges specific for light-rail-sized tunnels. There are also considerations to be made in order to protect the surrounding infrastructure. Settlement monitoring and excavation volume monitoring are two such measures that can be used to ensure minimal disturbance at the ground level.

An efficient, inexpensive, and reliable transportation network is an important part of any urbanization plan as it moves the people and goods inside, and through these urban areas. Generally, as cities grow their need for these networks increases, but the space required for them is in high demand from other forms of infrastructure. In the past and ever more so in the present and future, cities are utilizing underground space to create and expand their transportation systems.

### Surrounding Infrastructure

Urban tunneling is a unique challenge for a number of reasons—one being that the surrounding infrastructure already exists. Designing and constructing a tunnel near other buildings, roads, highways, utilities and other tunnels must be done in a way to minimize risk. Usually this infrastructure needs to be kept in operation during all phases of tunneling. Caution must be taken to either design the tunnel in a way to avoid this infrastructure or to have proper mitigation plans in place in case something does go wrong. Also, due to the surface infrastructure that is in place, space is often limited for assembly, launching, operation, and recovery of the TBM and its related systems. It is of vital importance to the success of a project that specifics of the assembly, launch and reception sites are shared with the TBM manufacturer as early on in the design process as possible. The site requirements can greatly affect the design of the machine as it might require that the TBM be customized to meet special assembly/disassembly requirements.

Detailed assembly plans can also be created as the machine design progresses in order to confirm

**Figure 1. Lowering of machine shield sections, MX12 Line, Mexico City, Mexico**

the logistics of part delivery to the site, crane and lifting requirements, or special tools or procedures that need to be created (see Figure 1, a limited space assembly in Mexico City).

### Often Difficult Soft or Mixed Ground Geology

One of the biggest under-defined variables in most tunneling projects is the geology of the tunnel alignments. Many people have theorized that you can never fully determine all the parameters of a particular geology, but only estimate them by sampling discrete areas and extrapolating the results. This is where the experience of the contractor and the history of tunnels in the area are invaluable. The past performance of tunneling projects near the proposed alignment can give huge insights into the predicted geology and performance of the current project. Another must is working with the TBM manufacturer to make certain that the TBM specifications are properly set so the machine can effectively operate in all expected conditions, and beyond.

The geology of urban sites, just like every part of the world, is greatly varied, but many urban areas are located in alluvial zones. People have historically settled near water and many urban areas have grown from these settlements over the centuries. It is imperative that detailed geological and hydrological studies be done when urban tunneling. The risk and consequences of something going wrong due to encountering unexpected geological conditions is greater in urban areas due to the proximity of vulnerable infrastructure. A thorough geological study should identify the location of all the expected ground types and water table levels throughout the tunnel alignment. This information must be shared with the TBM supplier from the initiation of the design as well, to ensure that the machine is capable of operating under those conditions. Because of the alluvial geology of most urban sites, the majority of recent urban tunneling projects have been completed with Earth Pressure Balance or Slurry TBMs. The acceptance and performance of these machines has grown greatly over the past few decades. The risk for settlement is greatly reduced with the technology of these machines if they are properly operated. They have become much more industrialized over the years, increasing the speed and efficiency of the boring cycle (see Figures 2 and 3, showing the variety of cutterhead setups).

### Tunnel Alignment

The alignment of urban tunnels can also be quite curvy both horizontally and vertically. In most cases the tunnel profile will follow along with the contour of the terrain. Also, the alignment may follow the path of surface streets in order to connect between stations placed along those streets. These streets are rarely arranged in straight lines; so consequently, the overall metro route can have complex curves in the design. The TBM must be designed to be able to handle these curves plus some margin of error to allow for steering corrections.

### Low Cover

Many urban tunnels are designed with low cover along their alignment. An average of 1.5 to 3 times the diameter is commonplace. With this amount of cover, settlement due the operation of the TBM is a huge risk. Settlement monitoring and mitigation plans, like ground improvement or ground freezing, must be in place to cover this risk. Operating the TBM with the correct parameters is also a huge factor in minimizing the amount of settlement seen at the surface. If these parameters are not vigilantly monitored and maintained then surface ground disturbances due to over excavation, blowouts or incomplete tail void filling can occur. Over excavation at the beginning of the bore must also be addressed. Every machine and geology has a learning curve and this must be expected and appropriate countermeasures implemented. Ground freezing, pre-excavation grouting, jet grouting, etc. can be implemented both at the start of boring and at certain key areas along the alignment to mitigate the risk of settlement. During the

**Figure 2. Mixed ground cutterhead**

**Figure 3. Soft ground cutterhead**

bore, instruments and controls such as ground settlement monitors, belt scales, annular grout volume calculations, etc. can and should be in place to identify when unacceptable settlement is occurring and when the mitigation plans need to be put in place.

### Cost and Project Schedule

Cost and project duration are always concerns in any project, but especially so in urban environments. Since urban tunnels have a higher visibility and they affect more people and infrastructure, the timeframe for construction must be kept in mind and usually minimized as well. The decision must be made early on if it is more efficient to operate multiple machines and reduce the duration of the project or bore with fewer machines and increase the duration. In shorter parallel tunnel runs, one machine may be a better option, but as the tunnel length increases more machines becomes the efficient solution. The project owner or contractor must study the tunnel plan and determine the optimum solution for the situation. One of the biggest factors is weighing the additional cost and complexity to operate multiple machines against the increased duration of the project with fewer TBMs.

## URBAN TUNNELING CASE STUDY: SAN FRANCISCO CENTRAL SUBWAY

San Francisco's Central Subway extension project will add a vital north-south link to the city, connecting commerce, tourist and historical locations with fast and efficient public transport. The project will create a twin tube extension to the Third Street T line, extending it from the 4th street Cal Tran station to Chinatown. The planned opening date for the extension is in 2019. JV Contractor Barnard Impregilo Healy was selected to complete the major tunneling works by the project owner San Francisco Mass Transit Authority, or SFMTA. The contractor selected and purchased two 6.3 m Robbins Earth Pressure Balance machines to bore the 2.5 km long parallel tunnels. Both machines were launched sequentially from 4th and Bryant Street north towards Chinatown. They will be removed from a reception shaft situated across the street from Washington Square Park.

### Project Challenges

- Densely populated urban setting in a historic city.
- Small launch pit with low head height and mostly covered with steel plates, situated below an interstate freeway that allows assembly of only one TBM at a time.
- Steep grades—Sections of ± 7% do not allow the use of conventional locomotives.
- Curvy alignment—there is a 137 m radius curve followed by a 260 m curve
- Complex geology—753 m of the alignment will be in the "Franciscan Complex": Abrasive sandstone that is tough on the cutterhead, disc cutters, and the screw conveyor.
- Classified as "Potentially Gassy."
- Cal/OSHA and numerous regulatory bodies with complicated, contradictory and extremely strict safety and environmental codes.

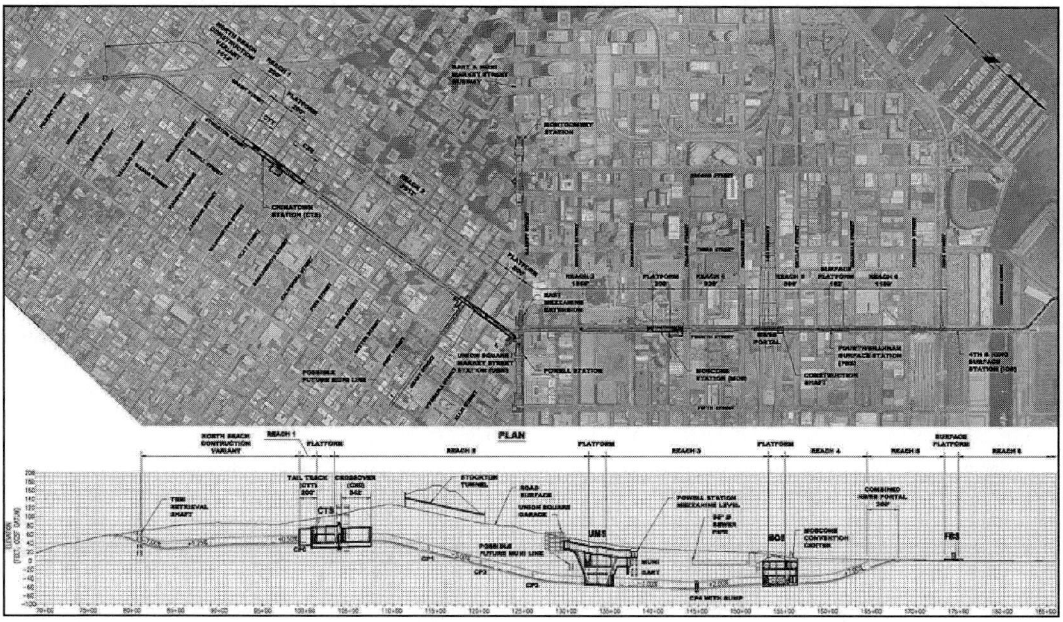


**Figure 4. Tunnel and station alignment (Image credit: SFMTA Pre-bid Presentation)**

### Alignment

Much like the city of San Francisco itself, the tunnel alignment has nearly continuous elevation and grade changes (fix this)The alignment begins at the intersection of 4th and Bryant streets. The machines will be launched heading to the northwest below 4th street, diving immediately. The tunnels will level out then take a sharp right turn (R 137 m) towards the north at Market Street, where they will pass under two other operational subway lines. In order to ensure the safety of those lines during boring operations, a liquid level system working in concert with longitudinal and transverse strain gauges and other instrumentation will be used to monitor ground disruption. The system will be used under the live tracks and determine if settlement mitigation measures must be executed.

After leaving the Union Square/Market Street Station the TBMs continue along below Stockton St. at a steep 7% climb for approximately 800 m through Franciscan bedrock. The tunnels travel under Nob Hill and the Stockton St. Tunnel, which will present more settlement monitoring activities and the additional challenge of increased overburden. After the Chinatown Station there is another long R 850 m curve just before the final, approximately 100 m, 7% climb to the retrieval shaft. The exit shaft for the two machines is located at Columbus Ave. and Union St. across from Washington Square Park (see Figure 4).

### Geologic Setting

The geology of the tunnel alignment is mixed and it ranges from mud deposits, sand and clay to sandstone, mudstone, and shale. The TBMs will encounter three disparate geological formations: the Colma, Old Bay and Franciscan. However, stating that there are only three types of ground is a vast oversimplification. There is certain to be a wide range of diverse and fluctuating ground conditions between these three generalized formations.

It is anticipated that the three formations will be distributed roughly as follows:

- Colma Formation (Surficial soils): ~ 1,750 m
- Undifferentiated Old Bay Deposits: >500 m Interspersed
- Franciscan Bedrock (sandstone): ~ 750 m
- Maximum rock strength (UCS): 27 MPa

### Machine and Back-Up Specifications

The TBMs were designed with a number of special features to efficiently manage the varied geology, navigate the steep and turning alignment, and bore in what has been rated as "Potentially Gassy with Special Conditions" by Cal/OSHA.

A mixed face cutterhead was selected and designed to excavate a wide variety of ground ranging from soft soils to hard rock, as well as diaphragm walls. The wear surfaces of the cutterhead are clad in a combination of chromium carbide plating, hard facing, and tungsten carbide bits to ensure the life of the head in the abrasive environment. The TBMs can be equipped with either a full dress of soft ground tools (picks, rippers, scrapers, etc.) or a mixed dress that incorporates 17-inch, pressure compensated disc cutters when the machine will encounter rock or concrete. Forty specially designed housings have the ability to mount either disc cutter or soft ground tools. The opening ratio is 31% to allow efficient and controlled muck flow through the head. Grizzly bars are also incorporated to prevent boulders that are too large to pass through the screw conveyor from entering the mixing chamber. The cutterhead also features 5 foam and 2 water injection ports for soil conditioning and a programmable copy cutter to create additional overcut in order to negotiate tight turns. In order to detect the need for an intervention there are also 3 sets of wear detection bits to automatically detect when the wear of the cutters gets down to unacceptable levels.

The cutterhead is driven by five, 210 kW VFD-controlled electric motors that transmit power through multi-stage planetary gear reducers and a large diameter bull gear. This setup is integral to a high capacity three-axis main bearing design modeled on hard rock TBMs. The VFD motor control allows infinitely adjustable cutterhead speed.

Like all EPBMs the Central Subway TBMs are fitted with a screw conveyor. Fortunately the relatively low hydrostatic pressures on this project only necessitate the use of a single-stage screw. However, due to the abrasive quality of the muck, the owner has specified replaceable wear protection on both the flights of the screw as well as the screw casing. To accommodate this, the shell of the screw conveyor is built of multiple replaceable sections.

Due to the complex geometry of the alignment, steering the TBMs accurately through the tight curves (min R 137 m) is one of the key challenges of the project. To accomplish this it was necessary to articulate the TBM shields. An active articulation system was integrated as it allows the thrust cylinders to remain parallel to the tail skin and react evenly with the segments. This feature mitigates the risk of segment damage, ring deformation, or settlement during boring.

Like all highly urban tunnel projects, another key challenge is ground loss or settlement, especially where the alignment crosses under live metro rail tunnels. As noted above the owner and contractor have highly instrumented, key areas where settlement could pose a serious risk to existing property or infrastructure. To prevent ground loss a precise system of control and measurement of the excavated material must also be implemented to eliminate over-excavation. The machine is fitted with two electronic scales and a laser system that constantly monitors the weight and volume of the tunnel muck travelling down the backup conveyor. These measurements are then compared to theoretical values to determine if over-excavation is occurring. The machines also have an active face support system that can detect if rapid pressure loss is taking place at the excavation face. The system will automatically inject pressurized Bentonite slurry into the mixing chamber to restore the lost pressure. The operator will then close the guillotine gates on the screw until face pressure is restored and it is safe to resume operation.

The geology of the alignment has the potential to contain ignitable concentrations of flammable gasses. For this reason, a requirement of the electrical design for both machines is to be ANSI/NFPA Class 1 Div 2 compliant. This was achieved using a combination of both inherently safe electrical designs and sealed or purged cabinets in all areas of the machines. The machines are also equipped with a gas detection system in order to identify the presence of multiple gasses in the tunnel.

Both machines and their respective backups were factory assembled and tested in Robbins' Pudong facility in Shanghai, China. The machines were then disassembled and shipped to San Francisco. Both machines were shipped in the largest sub-assemblies possible in order to reduce the assembly time at the site (see Figure 5).

### Site Setup

The launching and service portal presents a unique challenge to the project. Most of the project site is located within the busy on-ramp/off-ramp interchange of Interstate 80 and 4th street. The launching pit (137 m × 11 m) for the machine is actually directly below 4th street with a small access window

**Figure 5. First of two San Francisco Central Subway EPBs with mixed ground tool configuration**

**Figure 6. Small jobsite passes under Interstate highway 80 (Image credit: SFMTA)**

(11 m × 11 m) to the pit on the southeast side of the intersection. All the tunnel services and operations are squeezed into the available space between and below the highway and ramps. When both machines were launched and boring became fully operational the tunnel service vehicles had to navigate a tight course through the jobsite and down to the tunnel portal. This includes a sharp 90 degree turn down a ramp to the launching pit (see Figures 6 and 7).

### Continuous Conveyors

To get the muck out of the tunnels the contractor made the wise decision to source Robbins continuous conveyors designed and manufactured per the specific requirements of the site. In addition to its ability to transport muck on grades not serviceable by traditional locomotives, the system provides other benefits. These include greater system availability, less down time and more importantly, greatly simplified logistics when compared to rail-based muck cars. The material/personnel "supply trains" are free to come and go at any time.

Each TBM's conveyor discharges muck onto the trailing-gear-mounted advancing tailpiece that contains and aligns the tail pulley of the tunnel conveyor. The tailpiece has an "installation window" through which the carrying structure is assembled and troughing idlers are attached. As the TBM advances, the structure and troughing idlers emerge from the rear of the tailpiece, and the side rail supports and return rollers are installed. From there the muck is deposited onto the continuous tunnel conveyor.

Both of the extensible conveyor systems are equipped with a 500 m capacity belt storage cassette and splicing stand to allow the TBMs to bore approximately 250 m before more of the fabric-reinforced belt needs to be added. Belting is added during regularly scheduled TBM maintenance shifts. The continuous tunnel conveyors deposit onto a single stationary overland conveyor located in the aft end of the launch box that feeds a nearby radial stacker conveyor (see Figure 8).

### Supply of Materials and Personnel

Supply of tunnel lining segments, utility extension supplies, other materials and personnel is handled using specially designed rubber tired vehicles (RTVs).

The RTVs are engineered to drive on the curved tunnel invert with automated self-centering, as well as flat ground with seamless transition from one to the other without operator input. To negotiate the tight curves and fit through the trailing gear of the TBM, the RTVs are articulated. Further insurance to prevent collision is provided by an auto pilot system that is engaged when the RTV approaches the backup and safely guides it through.

The single most important feature is that the RTVs have operator's cabins on each end, allowing them to be driven in and out of the tunnel without being turned around or needing to be "backed out" (see Figure 9).

### Project Status as of Press Time—Mid-December 2013

Presently the first TBM has been fully commissioned, has built 428 rings (652 m) and is boring at a rate of 16–20 rings per day. The typical operating "day" is 20 hours of operation comprised of 2 × 10 hour long excavation shifts and 4 hours for machine maintenance. The continuous conveyor system is fully installed on this drive and muck flows from it to the overland conveyor onto the radial stacker.

The second TBM has also been assembled and launched with the full trailing gear installed. After advancing 113 m (74 rings) using piston pumps to discharge muck, the initial section was completed in mid-November 2013. Boring is presently suspended while the contractor removes the piston

**Figure 7. Lowering a section of machine shield (Image credit: SFMTA)**

**Figure 8. Horizontal conveyors going around a curve**

**Figure 9. RTV transporting segments (Image credit: sfmta.com)**

pump mucking system, thrust frame and temporary free standing rings. Concurrently they will begin the installation of main drive, belt storage cassette and remaining components of the continuous conveyor system to complete the commissioning process and resume production.

## CONCLUSIONS

Urban tunneling can be a difficult and risky endeavor, but working closely with the TBM manufacturer can greatly alleviate many of these difficulties. By having assurance that the most current technology is being used much of the risk of urban tunneling can be reduced. Also, having detailed assembly, startup, operational and settlement mitigation procedures in place at the outset of a project can greatly increase its chances of success.

## REFERENCES


centralsubwaysf.com

Mechanized Tunneling in Urban Areas, design methodology and construction control.

Tyler Sandell—Urban EPB Tunneling in Mixed Ground Conditions: A Case Study of the San Francisco Central Subway Project. Arabian Tunneling Conference, 2013.

PB Telemon JV—Central Subway Geotechnical Baseline Report—Report CS-155-1 Task DP1.04.40.

Arthor Wong, John Funghi—SFMTA Central Subway Pre-Bid Conference Presentation.

# King Road Grade Separation: Accelerated Underpass Construction by Jacking

**Jeb Pittsinger**
Hatch Mott MacDonald

**Robert Doucet**
Western Mechanical Electrical Millwright Services

**ABSTRACT:** The project team designed an innovative strategy for the King Road Grade Separation which involved constructing a reinforced concrete tunnel structure 'off-line' of a major rail corridor, and subsequently sliding the completed four-track tunnel into place using an open-cut tunnel-jacking methodology. The slide utilized a one of a kind low friction, sliding base assembly. The structure was pushed into place under a fraction of the structure's weight using a system of hydraulic jacks. This methodology allowed for a significantly abbreviated shutdown of mainline tracks over a single 72 hour period, minimizing operational impact to CN, Metrolinx and VIA trains.

## THE CHALLENGE

There are tens of thousands of road/rail grade crossings in North America. At best, these are often the cause of vehicular traffic delays and associated air pollution from vehicle idling. At worst, they present safety hazards where vehicles or pedestrians can find themselves on the rail tracks while fast moving trains are approaching.

Additionally, traditional grade separation construction typically requires a great deal of work to be done within the rail Right-of-Way during short duration work windows with significant disruption to the rail services and vehicular traffic. This type of work not only increases operational risk to the rail users but can also create safety risks for the contractor throughout the duration of the project.

The specific challenge from rail and road owners has been to create practical, constructible, cost-effective solutions that can be safely constructed while minimizing the operational risks to the rail and road users. The accelerated construction techniques developed and utilized on the King Road Grade Underpass to overcome these challenges can become a model for grade separations throughout North America.

## BACKGROUND

The project is located in the City of Burlington, Ontario within what is known as the "The Golden Horseshoe" surrounding the Greater Toronto and Hamilton area. Like many boroughs along southern Ontario, through the years, the City of Burlington (herein referred to as "the City") has grown from a cluster of villages and townships into a full-fledged City. Many arterial roads have been established along the Canadian National Railway (CN) corridor that now serve the community for much broader demands, connecting communities and freeway corridors. Approximately 100 trains per day, utilize this rail corridor servicing freight, passenger and commercial, inter-modal needs. Through its urbanization and master planning, numerous grade separations have been constructed in the City over the years as shown in Figure 1, with several more planned over the next 10 to 15 years. To this point, more traditional construction techniques had been utilized to construct the grade separations. However, both the city and railroad desired a more efficient method that caused less disruption to rail operations and motorists, as well as accommodating more challenging site constrains. Hatch Mott MacDonald (HMM) was engaged to investigate grade separation options for the King Road level crossing. At the onset HMM was given several project constraints:

- Maintain continuous freight rail service throughout the duration of construction
- Minimize construction work within the rail ROW
- Maximize safety for workers, rail users and motorists
- Contain temporary railway alignments to the existing rail ROW

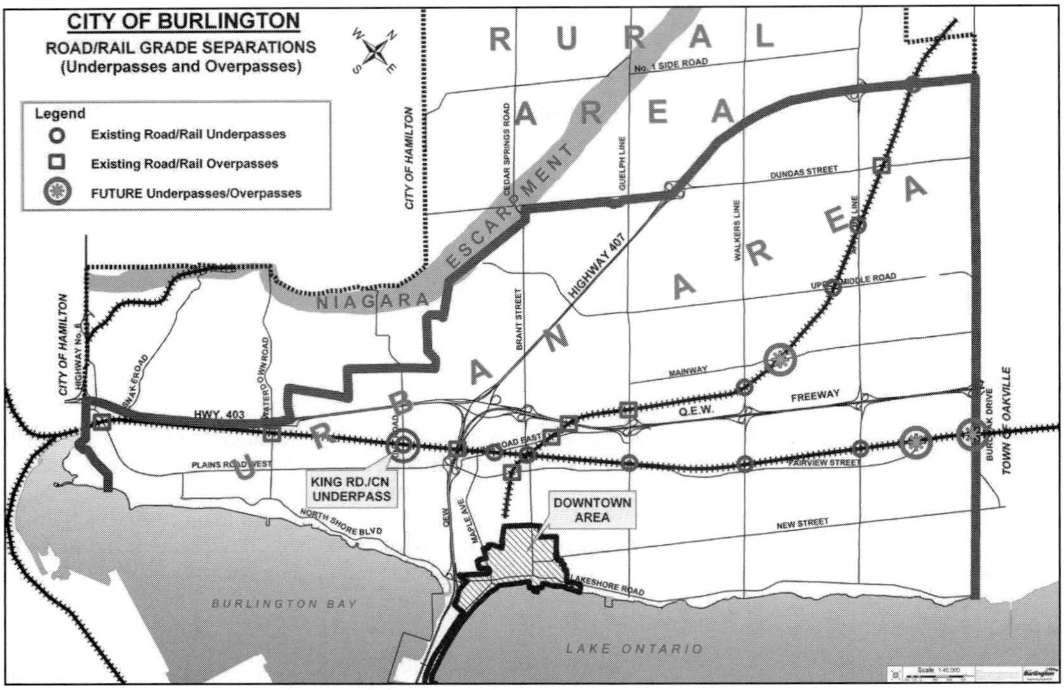


**Figure 1. Rail corridors through the City of Burlington**

- Minimize roadway closures
- Protect adjacent stream and wetlands
- Provide a low maintenance, aesthetically appealing structure

In addition to the project constrains, there were also several project stakeholders that utilized the rail ROW that had to be coordinated with including Metrolinx commuter trains, VIA and Amtrak inter-city passenger traffic as well as the transcontinental and international gateway freight traffic of Canadian Pacific Rail. By working closely with the City and CN, HMM was able to develop a plan centered around minimizing railway and roadway closures, preserving the project site, and maximizing safety while creating a durable, pleasing structure.

## THE SOLUTION

The solution developed for the King Road Grade Separation was an accelerated construction technique called Open Cut Box/Bridge Jacking. In principal, the plan was relative simple:

- Excavate a jacking pit adjacent to the rail tracks
- Construct a reinforced concrete "box structure" within the jacking pit
- Install a headwall and diversion track for freight service
- Initiate a three day rail closure and within that time frame
  - Remove the tracks and excavate the ROW
  - Push the box structure into final position
  - Backfill then reinstate the rail tracks

However, in practice the solution required a great deal of careful planning and attention to detail to ensure the structure could be successfully installed within the scheduled work window.

## SITE LAYOUT

A residential community is located on the southern side of the project site and several commercial retailers are toward the north. King Road is a north-south roadway and is intersected by four mainline rail tracks running east-west. The underpass structure was constructed south of the tracks in a pit excavated to the depth of the final vertical roadway alignment. To allow for continued use of the existing roadway during construction, by commuters and emergency vehicles, the underpass was constructed just west of roadway and a curve was introduced into the final horizontal roadway alignment. Due to the strict limitation on railroad Right-of-Way, the entire underpass could not be cast then jacked into position. Instead, the underpass was constructed to allow three rail lines to be reinstated immediately. Then it was extended to carry the forth northern most track after

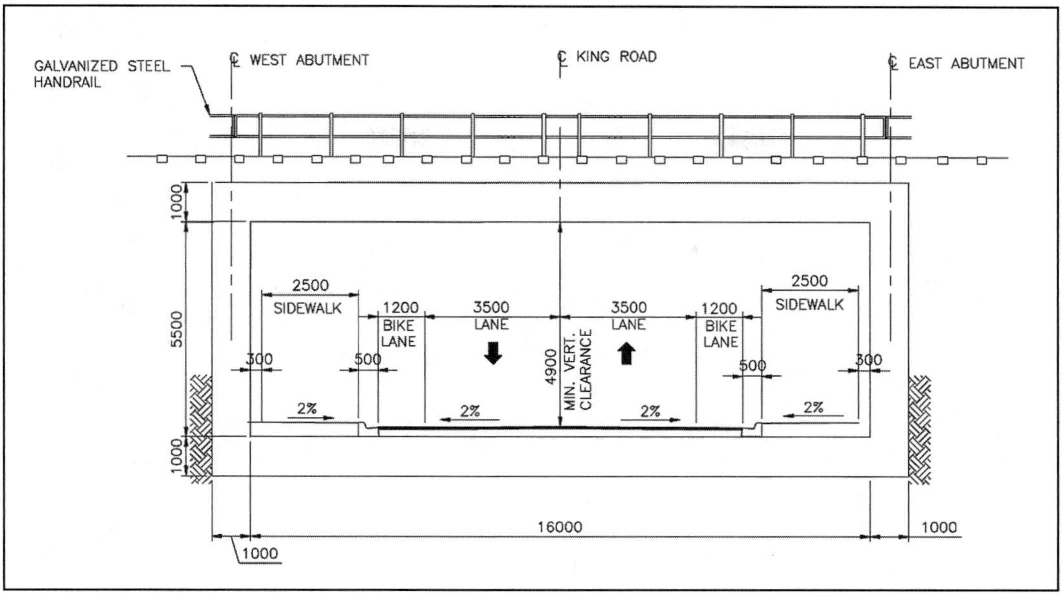


**Figure 2. Cross section of King Road underpass**

it has been jacked into the final position. To maintain continuous freight traffic during the weekend excavation and underpass installation a by-pass track was constructed on the north side of the rail ROW. To support the by-pass track and facilitate the installation of the underpass a secant pile headwall was constructed between the two northern most tracks. The box structure was jacked up to the head wall and the freight rail was transferred from the by-pass track on to the box structure. Then the secant pile head wall was demolished and the final 2 meters of the underpass was completed insitu. The portion of the underpass that was jacked into position was approximately 7.5 meters in height, 18 meters in width and 20 meters in length. The structure was installed so that the top of that rood slab was approximately 450 millimeters below the elevation of the rail ties and would be topped with ballast between parapet walls.

## GROUND CONDITIONS

The existing rail corridor is located on a fill embankment approximately 2 to 3 meters in height with the roadway ramping up to meet the grade crossing on either side. The embankment was granular fill predominantly consisting of sand and ballast. The embankment was underlain by approximately 5 meters of dense silty sands above firm to very firm clay and clayey silty till. The bedrock was shale which was encountered at approximately 12 meters below ground surface. Due to clearance requirements the underpass would be founded on the clay stratum. The properties of the clay were reviewed and the underpass was designed to be a compensated structure with a base slab that performed as a single large spread footing so no significant settlement was predicted. Groundwater was recorded at approximately 4.8 meters below ground surface so temporary dewatering and permanent gravity drains were installed to locally lower the groundwater table.

## THE BOX

The reinforced concrete underpass structure was designed as a single monolithic structure to maximize structural efficiency and minimize long term maintenance. The underpass was designed to carry two 3.5 meter wide vehicular traffic lanes, as well as two 1.2 meter wide dedicated bike lanes and two 2.5 meter wide sidewalks. The box was constructed with a 6 degree skew to meet the roadway alignment and several utility chases were constructed integrally to the slabs. A trainman's walkway was also cast integrally to the structure and also served as a parapet to retain ballast. Additionally, wing walls were added on the trailing end of the box to simplify the backfilling process. Finally, an aesthetic finish was added to the interior of the box to create a uniform appearance with the approach walls. See Figure 2.

## THE JACKING SYSTEM

The jacking system utilized a very unique low friction slide rail system that had never been utilized for an open cut box jack. The system reduced the jacking forces from almost 50% of the box weight down to

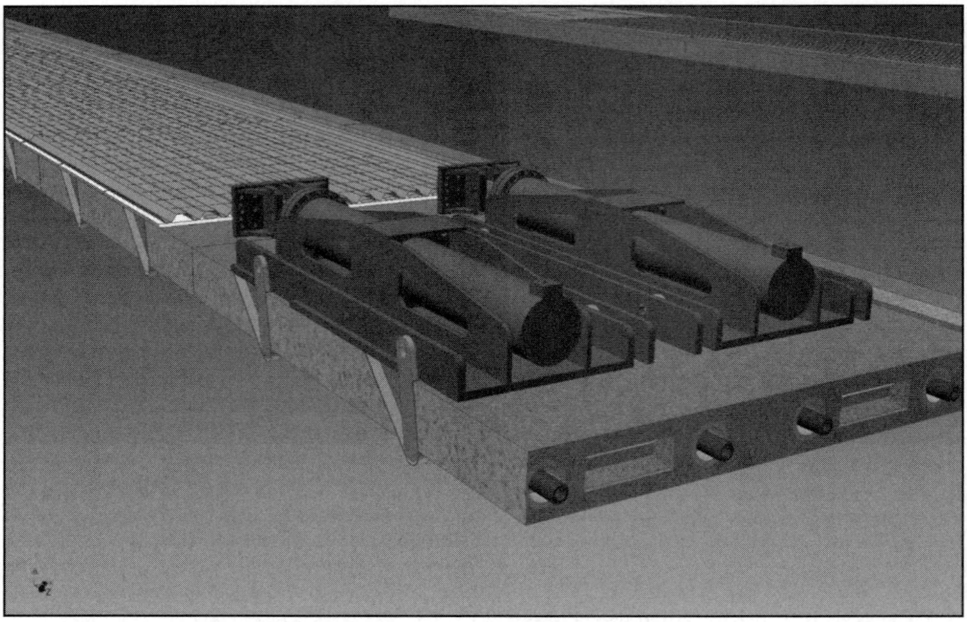

**Figure 3. Rendering of precast slide track and jacking sled**

only 12% or from 1,250 tons to 300 tons. The system was designed and constructed by Western Mechanical Contractors (WM). WM brought significant value to the project because of their previous experience of installing pedestrian tunnels along the CN corridor. The jacking system included precast concrete slide rails, low friction slide materials, a pneumatic system to lift the structure and self-advancing hydraulic jacking sleds. Each of the precast slide rails were approximately 3.6 meters by 3.6 meters and were all mechanically connected to create two continuous slide rails on each side of the excavation. A durable low friction material was placed between the concrete slide rails and the underside of the box to create the sliding surface. Pneumatics were devised to transfer the weight of the structure to the low friction material placed on the slide rails through a system of pressurized hoses. The structure was then advanced into the final position by extending the hydraulic rams mounted on the jacking sleds. Once the hydraulics were fully extended they would begin retracting pulling the sled forward until it reset cambing devises into the slide rails to begin advancing again. The approach meant that external thrust blocks or reaction pads were not needed. Instead the forces were completely internal to the underpass structure, jacking sleds and slide rails. As part of the design two test pushes were conducted on the box at different stages of construction to ensure the system performed as expected. Total thrust for each of the initial pushes was less than 6% of the box weight. See Figure 3.

## PREPARATION

The work leading up to the jacking of the box was just as critical as the work done over the long weekend. Advanced contracts were let prior to the construction of the structure to relocate utilities and install piping for future drainage. Work for the temporary headwall supporting the by-pass track consisted of 40 linear meters of secant piling with 55 interlocking caissons 1,180 millimeters in diameter. A deadman wall and a series of horizontal threaded bars were utilized to stabilize the headwall. A deadman wall was chosen over other options to minimize disruption to the railway and allow for installation with the rail ROW.

Eight hundred millimeter diameter secant piles with a single row of grouted soil anchors were utilized to support King Road while the jacking pit was excavated. Soldier piles and lagging were installed to complete the support of excavation for the jacking pit. Once the support of excavation was constructed the contract for the construction of the box structure could be executed. Reinforced concrete footings were then constructed to create a foundation for the slide rails at the base of the excavation. The jacking system was prepared and installed then covered with corrugated metal decking to separate the box structure from the slide rail. The reinforcing for the box structure was placed and concrete was poured in three phases: the base, the walls and the roof structure. While each of the phases was curing, the test pushes were conducted. Parapets, water proofing and

**Figure 4. Underpass structure completed prior to the jacking operation**

temporary handrails were all installed to prepare the structure to be pushed into place.

During construction of the box structure, the design and construction team were developing a very detailed schedule and resource list for the three day track closure and box structure installation which was to be from late Friday evening to early Tuesday morning. The schedule included every task that would have to be completed as well as manpower and equipment requirements. Due to the critical nature of the work, additional manpower and equipment were brought to the site to ensure work would be able to continue without delaying construction. Contingency scenarios were also developed to ensure the tracks would be operating by the Tuesday morning rush hour even if the jacking did not proceed as planned. Additionally, CN prepared teams for track cutting and replacement, repairing signals and communications, and providing flagging. Furthermore, a system of communication was developed to maintain contact with key stakeholders so they knew of any delays as soon as possible.

Immediately prior to the weekend of the push the construction crews began dismantling the support of excavation between the box structure and the rail corridor to limit the amount of work to be done over the weekend.

## WEEKEND CLOSURE

### Friday

The box structure was scheduled to be pushed into place over a three day weekend, beginning at 20:00 hours on Friday, October 5, 2012 with the passage of the last commuter train. From this point on all rail traffic utilized the diversion track on a slow work order. As scheduled, CN crews began removing cutting rails and removing track in sections with excavators and front end loaders pushing them down the rail corridor, clear of the worksite. See Figure 4.

### Saturday

By early Saturday morning the tracks had been completely cleared and construction crews were beginning to excavate the area where the box was to be pushed. Up to four excavators were working simultaneously to remove spoil and load it into dump trucks to stockpile it in designated locations near the work zone. The movements of the dump trucks needed to be coordinated with the train movements through the site which was up to the CN flagmen. Once excavation became too deep for the excavators to pull material from existing grade they were moved into the excavation and dump trucks were backed through the

**Figure 5. Underpass structure in its final position with the rail tracks being replaced Monday afternoon**

box structure to be loaded. By late in the evening the excavation was at the target elevation and the focus changed to preparing the base of the excavation and installing the slide track. Granular fill was placed and carefully leveled then the precast concrete slide track was placed and connected together to create a continuous surface. This procedure continued through the night.

### Sunday

The slide track was finished by late morning and the pneumatic jacking system was pressurized to lift the box. Just after noon the hydraulic system was engaged and the jacking sleds began to push the box into place. By extending the hydraulics to advance the box structure then retracting them to pull the jacking sleds forward the box was jacked into final position within five hours. Immediately after the jacking was completed the backfilling of the excavation began and continued through the night.

### Monday

By Monday morning the areas around the box structure had been backfilled and temporary retaining walls had been placed. Ballast was installed and track segments were tracked in with excavators and placed into their permanent positions. Simultaneously, track signaling was also reestablished. As the evening approached tamping trains were run over the area and all systems were tested. By midnight the three tracks had been completed and returned to service for the morning commute. See Figure 5.

## CONTINUED WORK

With the installation of the underpass structure the most critical elements of the construction had been completed with respect to track operations. In the following months many other aspects of the project were completed including the cast-in-place extension for the fourth track, installation of the approach walls, construction of an aqueduct structure to allow an adjacent stream to cross over the newly depressed roadway, utilities grading and pavement. Each of these aspects of the project was able to be completed with no effect to the rail ROW and with short closures to King Road.

## CONCLUSION

The project was completed on schedule despite only nine months between design NTP and the weekend closure. Additionally, final construction costs for the accelerated construction were shown to be equivalent to preliminary cost estimates for traditional construction. With attention to detail and careful planning, the unique accelerated construction techniques developed for the King Road Grade Separation Project can provide a model to create practical, constructible, and cost effective grade separations that minimize operational risks to rail and road users while maintaining the highest level of safety.

# Tunneling Under the Hudson and East Rivers in the Early 1900s: Risk Identification and Management Lessons That Are Still Useful Today

**Vincent Tirolo Jr.**
Skanska USA Civil Inc.

**ABSTRACT:** In New York City more than a dozen subaqueous tunnels were mined between 1900 and 1920 under both the Hudson and East Rivers. This was a spectacular achievement because it was less than 30 years since Peter Barlow and James Henry Greathead introduced the first circular tunnel shield with a cast iron segmental liner on Tower Subway project in London in 1869 and only 14 years since the first successful use of compressed air in tunneling again by Greathead on the City and South London Railway in 1886. The most spectacular of the tunnel achievements during this period were the six subaqueous tunnels built by Pennsylvania Railroad Company (PRR) under the Hudson (North) and East Rivers to expand the PRR system into New York and New England. These six tunnels were mined with 12 shields using compressed air through highly variable and difficult ground. The maximum depth to tunnel invert was 30 meters (98 ft.) below mean high water. Average compressed air pressures exceeded 2 bar (200 KPa) with a maximum compressed air of 3 bar (300 KPa). The risks involved with this work included compressed air sickness; blasting near the shield; mixed face tunneling; tunneling through man-made obstructions such as piers and bulkheads; compressed air blows; maintaining the tunnel alignment and profile; tunnel settlement and lateral movement; and cracked segments. These risks were overcome by a combination of assembling an international team of tunneling experts, early risk identification, the use of observation method during and after construction, in-situ testing, and the ongoing input from an engineering review board. These six tunnels were constructed, outfitted and brought into service within six years. In today's dollars their cost would be two and a half to three billion dollars.

## INTRODUCTION

In 1904 the Pennsylvania Railroad Company (PRR) started an ambitious program to expand into New York and New England. This program included the mining of two tunnels under the Hudson (North) River and four tunnels under the East River, land tunnels and the Pennsylvania Station in Manhattan, railroad yards in the New Jersey Meadowlands and Sunnyside, New York and finally the Hellgate Bridge connected New England with Long Island. The cost of this project in 2014 dollars would between 2.5 and 3 billion dollars. The project was entirely privately funded and completed on schedule in less than ten years.

The paper focuses on the risks of mining subaqueous tunnels under the North and East Rivers in the early 1900s and how these risks were identified, managed and overcome. It is a story of hard work, innovation, triumph and sometimes tragedy which provides lessons for recognizing and reducing risk that endure into our current era.

The engineers involved on the project were legendary figures at the time but almost are forgotten today. That is a shame because as a group they part of our engineering heritage.

A number of these legendary engineers are shown in Figure 1. The photograph was made at a "holing through" when two of the four North River shields met under the North River in 1906. Standing on the top step of the stairs on the left is Charles M. Jacobs, chief engineer of the North River Division. Immediately to Jacobs right is Alfred Noble, chief engineer of the East River Division. B.H.M. Hewett, John O'Rourke and James Forgie are standing to the right of the stairs. Figure 2 shows Alexander Cassett, President of the PRR and Samuel Rea of the Vice President of the PRR and the executive in charge of the project. Forty years earlier, Cassett and Rea began their careers at the PRR as engineers.

## CONSTRUCTION METHODOLOGY

### Tunnel Shield

The "tunnel shield" was conceived and patented by Marc Isambard Brunel in 1818. Interestingly the Brunel's patent was for a circular shield and completely different from the rectangular compartmentalize shield he used to tunnel in free air across of the Thames River in London in 1825. After many delays, that tunnel was completed in 1843. Twelve years after Brunel's shield patent, Thomas Cochrane

Figure 1. Shields meet under the North River (1906)

Figure 2. Alexander Cassett (left) Samuel Rea (right)

submitted a patent for the use of compressed air in tunnel and shaft construction. In 1864 W.P. Barlow patented a circular tunnel shield. Copperthwaite (1906) quoted from Barlow's patent:

> "...The earth is continuously removed from within this cylinder (the shield), and the cylinder is from time to time forced forward a short distance to admit a ring of iron being put together within the inner end of the cylinder, such iron rings being of a strength suitable for forming a permanent lining within the tunnel. It is desirable that the thickness of the iron of the cylinder should be as little as may be, in order that the space between the outer surfaces of the ring and the earth which surrounds them may not produce any subsidence of the land above.... The cylinder is from time to time forced forward by screws. The space, as it is left between the earth and the exterior of the tunnel, may be filled by injecting or running in fluid cement."

Copperthwaite (1906) further discuses improvements to Barlow's shield:

> "In 1866, R. Morton took out a provisional patent for a segmental cast iron liner erected with a shield and propelled forward by hydraulic jacks reacting against previously erected liners. In 1868, Barlow obtained another patent for a shield with transverse diaphragms and a partially closed tunnel face. Therefore by 1868, all the features of compressed air shield tunneling were in place: cylindrical shield; cast iron liner; hydraulic propulsion jacks; and compressed. What was needed was a masterful tunnel engineer to economically combine them all and projects to demonstrate their applicability. That engineer was James Henry Greathead. The projects were the Tower Subway in 1869 (shield, cast iron liner) in City and South London Railway in 1886 (shield, cast iron liner, hydraulic propulsion jacks and compressed air)."

In the United States, A.E. Beach designed and constructed a shield in 1869 to mine a short tunnel under Broadway in New York City. This shield was propelled with hydraulic jacks.

In 1879, DeWitt Clinton Haskins attempted to build a tunnel under the Hudson River using compressed air *but without a shield.* The tunnel liner was masonry. The work was abandoned in 1882. It was restarted in 1889 by the English Contractor S. Pearson and Son (Pearson) using compressed air, a shield, cast iron tunnel liners, and hydraulic propelling jacks. E.W. Moir, who managing the project for Pearson, greatly improved the safety of the miners (sandhogs) by developed the first medical lock in 1889. Due to lack of funds this work was also abandoned in 1891. This tunnel was eventually completed in 1905 as the H&M Tunnels (McAdoo Tunnels) by J.V. Davies and Charles Jacobs.

The same Charles Jacobs, an Englishman, in 1894 had completed Ravenswood Gas Tunnel the first subaqueous crossing in New York City. Jacobs became the PRR Chief Engineer of the North River Division. Pearson became the tunneling contractor for the PRR crossings of the East River.

## GEOLOGIC CONDITIONS

The geological conditions under the North and East Rivers are highly variable. On the New Jersey, Manhattan and Long Island City shores lines the bedrock that drops often off rapidly. The rock is the vicinity of the Weehawken, New Jersey shaft is a contact zone between the hard igneous rock of Palisades Sill diabase and the softer sedimentary sandstone and shale strata of the Newark Formation. The east and west shorelines of Manhattan primarily consist of the Manhattan Formation mica and hornblende schists. Quartz pegmatite intrusions are common. The rock at the Long Island City shaft is Ravenswood granodiorite gneiss and Fordham gneiss. The East River contains shallow submerged ridges of Inwood marble, Hellgate dolomite and Fordham gneiss.

The soft ground in the North River consists primarily of sensitive black (near shoreline) and grey inorganic silts and plastic clays. Saxena and Smirnoff (1973) indicated the terminology of "Hudson River silts" was based on the tunneling behavior of these soils which, being highly sensitive, behave like silt when remolded as tunnel muck. The soft ground in the East River consisted glacial sands and gravels, numerous boulders and occasional pockets of very fine sand and silt.

The shorelines of New Jersey, Manhattan and Long Island City contained active piers, rail yards, bulkheads and rip rap. Tunnel cover at the shorelines was often less than one tunnel diameter.

## CONTRACTUAL ARRANGEMENTS

### Board of Engineers

A "Board of Engineers" (the Board) was organized by Alexander Cassatt the President of the PRR and Samuel Rea, Vice President of the PRR. Rea was in charge of all matters related to the project. The Board reported directly to and made recommendations to Rea on all engineering and construction issues on the project. When there were disagreements, Rea made the final decision.

The Board also advised the PRR on high risk engineering and construction activities. The Board met for the first time in January 11, 1902. Meeting continued on a regular and special session basis until the Board was dissolved at the effective completion of the project in April 1909.

The composition of the board varied over its 8 year life but when first organized consisted of General Charles W. Raymond, as chair, Charles Jacob, member and Chief Engineer of the North River Division, Alfred Noble, member and Chief Engineer of the East River Division, Gustav Lindenthal, William Brown, Chief Engineer of the Meadows Division, and George Gibbs, Chief Engineer of Electric Traction and New York Station Division.

These individuals were selected by the PRR because of their experience in both mega projects and large highly complex railroad construction projects (Raymond, Noble and Brown) or because they had specialized expertise in structural engineering (Lindenthal), compressed air tunneling (Jacobs) and electric power traction (Gibbs). The Board was a mixture of individuals directly involved in executing the work and others who were not. The contractors involved in the work were not directly represented on the Board.

The Board also monitored overall design and construction consistency throughout the project. The work was divided into four Divisions: The Meadows Division; the North River Division; the New York Station; and the East River Division. Each river division produced its own contract documents and specifications under the uniform guidelines of the Board. The cast iron liners for the river crossings were designed by each division (essentially the same design). Both sets of contract documents specified that the river tunnels be constructed by the use of shields, hydraulic propelling jacks and compressed air. A shield design was provided in both sets of contract documents. Contractors could propose alternative means and methods but the final decision was always with the Board.

Many decisions of the Board were not unanimous. At times, the Board's discussions were heated. The final decision was always made by Rea. In a least one case, which will be discussed later, Rea's decided against the majority opinion of the Board.

## Contractors

Since the tunnels were being built by a private corporation and not a public agency there was greater flexibility in contract procurement. In general work was awarded on the basis of the lowest qualified bid but there were negotiations on specific work items and construction risk.

Geologic risk was greater in the East River where the shield would encounter mixed face conditions in the form of submerged rock ridges. The East River also contained strata sand and gravel strata not conducive to "holding" compressed air. There were also four East River Tunnels totaling 4755 m (15,600 ft.) of bored tunnel compared with two North River tunnels totaling 4023 m (13,200 ft.) of bored tunnel. Also, eight shields were used for the East River Tunnels compared with four for the North River Tunnels.

The soft ground conditions with the North River were more conducive to compressed air shield tunneling. However, maintaining the tunnels' alignment and profile while mining though the "Hudson River Silts" were more challenging in the North River than the East River.

### *North River*

The contracts were awarded in May 1904 for the two North River tunnels to the O'Rourke Construction Company (O'Rourke). The bid was a unit price. The O'Rourke Construction Company was an American company. Its owner John F. O'Rourke, a son of Irish immigrants, was a graduate of Cooper Union in New York City with a BS in both civil and mechanical engineering. Prior to being awarded the contract O'Rourke Construction Company experience with compressed air was limited to the construction of caisson foundations in New York City (New York Times 1904). With the exception of the shield's hydraulic jacking system which was modified by the contractor, the shield used under the North River was the shield provided in the contract documents. The shield was designed by James Forgie, the Chief Assistant Engineer of the North River Division, under the supervision of Charles Jacobs. Forgie, a native of Scotland, came to America with Jacobs.

To minimize schedule risk during the design of the tunnels, a separate contract was let in June 1903 to the United Engineering & Construction Company for construction of both the Manhattan and Weehawken shafts. These shafts were completed in September 1904. The shields were launched in May 1905 and the last ring of cast iron was erected in the tunnels, one year ahead of schedule, in November 1906.

The decision by the Board and Rea to let an early-on shaft construction contract prior to the completion of the tunnel design proved critical to maintaining the construction schedule for the North River tunnels.

### *East River*

The contract for the East River Tunnels was awarded in May 1904 to S. Pearson & Son Limited (Pearson). Pearson was an international company based in London, England with extensive tunnel experience in both London and New York City. Pearson completed the Blackwall Tunnel in London. The Blackwall Tunnel was discussed in Rea's 1892 report to the then president of the PRR G.B. Roberts (Jonnes 2007). Rea had been sent to London by Roberts to study the feasibility of using shields and compressed to mine subaqueous heavy rail tunnels. The Blackwall Tunnel was a large, 8 m (27 ft.) external diameter, pedestrian/vehicular tunnel circular with a cast iron final liner. The crown of the tunnel was within 1.5 m (5 ft.) of the Thames. The maximum air pressure used in the Blackwall tunnel was 2.6 bar (37 psi). The PRR selected Pearson as contractor for the East River Tunnels because of their experience on the Blackwall Tunnel.

The Board and Rea always considered the tunneling under the East River to be more hazardous that tunneling under the North River. Therefore it was absolutely necessary that the contractor selected for this work we experience in compressed air tunneling. Also, to attract competent contractors to this work it was important that the contractual relationship limit the contractor's risk. Therefore, there was a limit set on the monetary losses Pearson could sustain in the work. A stated by Noble (1913):

> "A fixed amount was named as contractor's profit. If the actual cost of the work when completed, including this sum named as contractor's profit, should be less than a certain estimated amount named in the contract, the contractor should have one-half of the saving. If, on the other hand, the actual cost of the completed work, including the fixed sum for contractor's profit, should exceed the estimated cost named in the contract, the contractor should pay one-half the excess and the railroad company the other half; the contractor's liability was limited, however, to the amount named for profit plus $1,000,000; or, in other words, his maximum money loss would be $1,000,000. Any further excess of cost was to be borne wholly by the railroad company. The management of the work, with some unimportant restrictions, was placed with the contractor; the relations of the engineer, as to plans, inspection, etc., were the same as in ordinary work, and the interest of the contractor to reduce cost was the same in kind as in ordinary work."

The Pearson's costs were audited by the PRR. Costs were generally divide into Unit Labor costs for work within the tunnel directly chargeable to a specific unit of work, e.g., erecting the cast iron liner, and "Top Charges" to including the contractor's staff, yardmen, plant operators, electricians, pipemen and other support staff (Brace, et al. 1910).

## RISKS AND RISK REDUCTION METHODS

### Risk Identification and Reduction

Tunneling involves a considerable amount of risk. Certain risks can be managed others while can only be recognized. The PRR subaqueous tunnels were mined through rock, soft soils, and mixed face conditions (rock and soil). Liner distortion was always a problem in rock and mixed face conditions. The tunnels encountered obstructions near the shorelines including concrete bulkheads, nests of timber piles and riprap. The tunnels were shallow often with less than one tunnel diameter below the mudline. Even with the use of temporary clay blankets in the river, blows of compressed air through the tunnel face were common. Compressed air sickness was common as was injury and death (more than 30 sandhogs died in these tunnels). Within this environment, the engineers and sandhogs focused on expeditiously erecting the liner and completing the tunnel on schedule. Part of this focus included the early identification risks. These risks included:

1. Schedule Risk
2. Mixed Face and Rock Tunneling with Shields and Liner Distortion
3. Tunnel Alignment, Settlement and/or Tunnel Buoyancy
4. Compressed Air Blows and Compressed Air Sickness

### Schedule Risk

#### *Shafts*

As previously discussed on the North River tunnels, the Weehawken and Manhattan (Eleventh Avenue) shafts were let as part of a separate contract prior to awarding the tunnel work. This had a major factor in keeping the North River tunnels on and even ahead of schedule.

The early on construction of the shafts was particularly significant in Weehawken. The shaft site was located in the contact of the igneous rock of Palisades Sill diabase and the sedimentary sandstone and shale strata of the Newark Formation. The rock in the contact zone was highly altered and fractured. This required the Weehawken Shaft sidewalls to be carried down through the rock to the invert of the North River tunnels.

Forgie (1906) described the situation as:

> "However, during excavation and notwithstanding the exploratory borings which indicated good self-supporting material, the rock was found to be rotten and disintegrated (the junction of the trap of the Palisades and the sandstone country-rock occurring at the one-third of the length of the shaft from the west end. It was decided to carry the concrete walls down the whole depth of the shaft and also to excavate a length of 40 ft. of the tunnels both east and west of it, to insure the safety of the end walls."

Rather than seat the shaft's concrete walls at the top of rock, it was decided to carry the shaft walls down to entire depth of the shaft, over 21 m (70 ft.). This delayed the completion of shaft construction until September 1904, more than six months, *but did not delay tunneling*. The Manhattan Shaft which was started concurrently with the Weehawken Shaft in June 1903 was completed in December 1903 without incident.

Because of the special contractual issues, an early on contract was not let for the First Avenue and Long Island City Shafts for the East River Tunnels. Fortunately, both shafts were constructed without incident.

### *Shields*

Another schedule risk mitigation measure was the decision that the tunnels in both the North River and East River Division be driven from both shorelines concurrently with a mid-river "Holing Through." This required the use of four shields in the Hudson River and eight shields in the East River. However, simultaneously mining with multiple shields would trigger unforeseen consequences which will be discussed later.

This decision to mine from both shorelines was based on the assumption that the rate of tunneling would be slow and using multiple shields would improve daily progress. Multiple shields also permitted greater labor flexible. Crews could be shifted from one shield to another if equipment or mining problems halted progress in one or more of the bores. However, this schedule risk mitigation measure required greater upfront (mobilization) capital costs not only for the additional shields and their trailing gear but also multiple and higher capacity compressed air plants to provide high (tools) and low (sandhog) compressed air requirements at each bore. The PRR assumed the additional capital costs.

Risk can only be addressed using the data, local experience and expertise available at the time of the risks was analyzed. In hindsight, driving multiple shields for mid-river tunnel junctions in both the North and East River tunnels were not necessary, but for different reasons in each tunnel.

According to Forgie (1907), tunnel driving in the North River through full face "Hudson River Silt" averaged more 4.4 m (14.4 ft.) in 24 hours. The slowest progress was in mixed face which averaged less than 0.6 m (2 ft.) in 24 hours. Tunneling progress was continuous from erection of the first ring within the tunnel shield on May 12, 1905 to the mid-river "holing through" occurred on September 12, 1906, one year ahead of schedule.

The experience in the East River Tunnels was entirely different. Pearson installed bulkheads and the first tunnel put under air pressure (Tunnel D) on October 5, 1905. Subsequently, Tunnels C, B, and A were put under air pressure. Work was finished four years later in May 1909 two years behind schedule. For various reasons related to either equipment problems, ground conditions and limited compressor plant capacity, work was suspended in one or more of the tunnels for months at a time. Work was suspended on Tunnel A for a year. Blows were so common in the East River, the work was suspended in one or more tunnels to allow higher volumes of compressed air to flow into the working tunnels.

## Mixed Face and Rock Tunneling with a Shield

Shields at the time (early 1900s) were not intended to bore through rock or mixed face ground. Excavating through mixed face ground conditions involved the loss face stability; compressed air blows; blasting generating flying muck and toxic gases injuring sandhogs and damaging the shield, cast iron liner distortion, damage to the shield skin plates rubbing against irregular rock surfaces, etc.

### *Full Face Rock*

To reduce the risk when the shield encountered rock or mixed face, engineers required that the shield be moved forward on rails embedded into a cast in place invert concrete cradle. Headings were maintained in front of the shield and the rock removed by bench blasting whenever possible (Figure 3). Liner distortion was common in full rock and mixed face tunnels because of the unbalanced load on the tunnel liners and distortion of the liners under its own weight. Cross-diameter temporary tie-rods keep tight with turnbuckles reduced or eliminated the risk of both liner distortion and cracked segments. Promptly filling the voids between the liner and the rock were filled 1:1 sand cement grout or large stone which

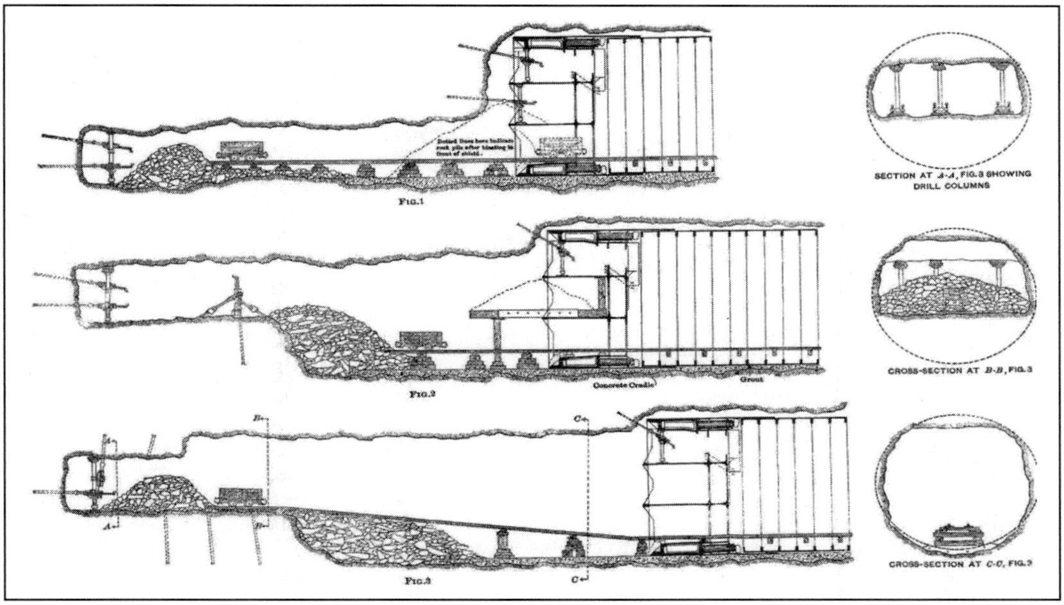


**Figure 3. Full rock face tunneling—rock removed ahead of shield (Brace, et al. 1910)**

were the grouted into place facilitated the early removal of the tie-rods.

### *Mixed-Face*

A number of different options for mining through mixed face conditions were tried during the course of the work. Poling and breasting were the primary measures used reduce the risk in the soft ground portion of tunnel face instability. Poling systems were kept as far in front of the face as possible. Another method used to reduce risk was equipping the shield with sliding forward platforms to provide additional support to breasting and breasting screw jacks.

Forgie (1907) said it was almost impossible to keep the shield from being damaged in either mixed face or full face rock because of the inaccessibility at the bottom of the shield. The concrete cradle helped in this regard to reduce risk of damaging the shield. These cradles were constructed of cast-in-place concrete installed in drifts mined ahead of the tunnel face. Figure 3 illustrates a shield mining through full face rock tunneling face. Figure 4 illustrates mixed face conditions.

The rails in the concrete cradle were set so that the top of the rails were at the elevation of the tunnel invert. Therefore as the shield passed over the cradle, the liner erected within the shield was high by the thickness of the shield skin. As the liner passed out the back of the shield, it would settle down on the cradle at the correct invert elevation. However, as the rock dropped off and the cradle eliminated there was a risk that the liner will crack unless the invert soils were stiffened by continuously grouting.

Highly variable and rapidly changing ground conditions made driving the East River tunnels very difficult. For example, the average drive eastward from Manhattan shaft to the mid-river junction was 537 m (1,760 ft.) while the average tunnel drive from the Long island City Shaft westward was 653 m (2,142 ft.). The Manhattan shields encountered, 23% full face rock, 37% mixed face and 40% soft ground. The rock and mixed face conditions were not continuous but occurred throughout the drive. Often the mixed stratum of boulders and cobbles covered the top of rock surface. The Long Island City drives were less variable with a higher percentage of soft ground.

The soft ground was typically coarse to fine grained sands, gravels and boulders. Fine grained silts and clays typical for the North River Tunnel occurred only in pockets within the East River. It was difficult to prevent the loss of compressed air through the tunnel face. The use of clay blankets was essential for the completion of the East River Tunnels. In addition to soft ground geology, ground cover was shallow and obstructions at the shorelines extensive. As discussed previously, even with the extensive use of clay blankets, air losses at time were so severe that the Long Island City compressed air plant could not provide enough air to maintain operations in four tunnels simultaneously. Even though this risk losing air during mining had been anticipated, the volume of loss air was much greater than projected.

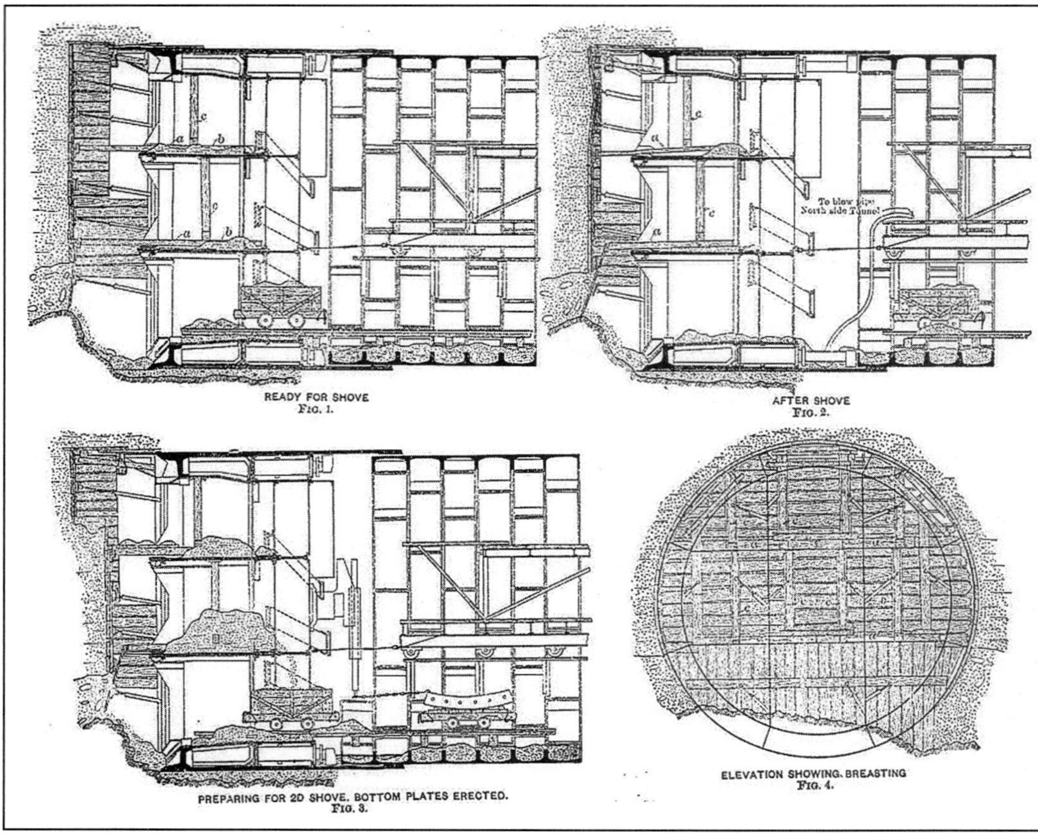


**Figure 4. Mixed face tunneling—poling and breasting aided by the shield's sliding platforms supported the soft ground in portions of tunnel face above rock (Brace, et al. 1910)**

## Alignment, Settlement, and/or Tunnel Buoyancy

### *Alignment*

A major risk throughout the mining operation was maintaining tunnel alignment. A number of tunneling operations increased the risk of the tunnels losing their alignment. Most of these were unavoidable and "field" corrections were always necessary. As discussed earlier, cast iron liners erected with a shield supported on a concrete cradle were erected higher than the design profile. Also, completed cast iron liners tend to sag when the shield passes from rock into loose sands and gravels or soft silts and clays. This sag was a result of liner self-weight and soil disturbance due to movement of the shield in these transition zones. If the sag was not corrected quickly, the brittle cast iron liners cracked. Cast steel liners were substituted for cast iron liner in the transition zones between rock and soil to minimize the possibly of cracked segments.

Leaving a hard support and moving into soft ground could also only cause the shield to move to the left or right. Forgie (1907) describes the method of "leads" that was used to correct the alignment of the shield to reduce this alignment risks.

Another unforeseen risk occurred from the decision to reduce schedule risk with multiple shields. Based on their experience on the smaller H&M Tunnels (McAdoo Tunnels), Jacobs and Forgie initially attempted to shove the North River shields through the soft Hudson River silts "blind" (all shield doors closed). Unfortunately, the larger North River shields displaced, "mud waved," a significantly larger volume of soft soils. These mud waves pushed against the cast iron liners of the adjacent previously mined adjacent bore causing the liners to both distort and move laterally. Shoving blind also made it more difficult to maintain the shield's alignment. To mitigate these risks shoving "blind" was eliminated. The maximum amount of soil displaced was limited to about 50% (less than half the shield doors remained open during a shove). This change plus learning to stage adjacent shields movements significantly reduced the of alignment deviations.

The tunnels were constantly being surveyed and the orientation of the shield modified in same

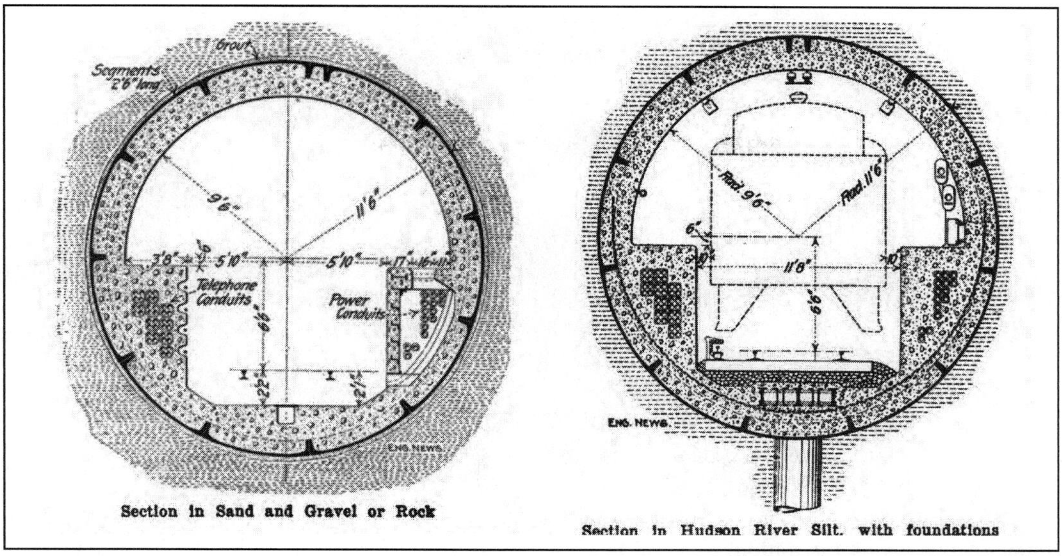


**Figure 5. Tunnel support options: sand and gravel or rock, or Hudson River silts**

increments to change the tunnels line and grade. Checks were made on alignment and profile every four cast iron ring. Since the segments had a length of 30 inches that means to tunnel alignment and profile was checked every 3 m (10 ft.). When necessary, tapered rings were used to correct the tunnel alignment.

### *Settlement and Buoyancy*

The greatest concern for the Board of Engineers and the PRR was the fear the tunnels would settle excessively and/or rise and fall with the tide. The tunnels did both.

Calculations using dead load only indicated that the tunnels were buoyant and therefore the in-situ soil was unloaded. However, if live load was included in the computations, in the form of heavy rail traffic, the soft in-situ soils were required to partially support the tunnel.

Since the PRR tunnels were much larger and would carry heavier rail traffic than the previously construction H&M Tunnels, these risks were anticipated prior to construction. Movements due to buoyancy were monitored and determined to be too small to impact rail operations. However, "By mid-April 1906 Cassatt decided to make the North River tunnels' cast iron ring linings much heavier, increasing the liners weight of each lineal foot from 4,205 kilograms (9,272 lbs) to 5,260 kilogram (11,594 lbs), hoping the additional weight would better settle the tunnel" (Jonnes, 2007) e.g., reduce buoyant movements Increasing the weight of the liner delayed to completion of the tunnel since mining had to be suspended from time to time due to unavailable of segments.

The tunnel settlement was a major concern for years after the tunnel was completed. The fear was that excessive tunnel settlement would distort the tunnel liner enough to infringe on the train's dynamic envelope and even prevent clear passage of the trains. There was also a concern settlement would open the joints in the cast iron liner and cause significant water leakage. In either case, train traffic through the tunnels would need to be suspended. To address this risk, a special tunnel segments (called "Bore Segments" Hewett and Brown 1910) were fabricated and placed in every 4.6 m (15 ft.) through the North River tunnels to accommodate a screw piles. See Figure 5 and 6. Test screw piles were installed and testing in various locations. Screw piles were to extend over 60 meters (200 ft.) to rock.

After the North River Tunnels were completed test screw piles were installed in a number of locations to test the feasibility of the methodology. Surveying readings were taken daily to monitor liner movement. These readings were taken for years after the tunnels were in full operation. The screw piles were never installed thanks to Rea. Going against the majority opinion of the Board of Engineers, Rea made the heroic decision to adopt an Observation Method when it came to the installation of piles. The tunnel segments were designed to accept a pile support if it was necessary. Piles were available. The piles could be installed at night through track outages or by closing one tunnel at a time (single track operations). Therefore, Rea decided to monitor the tunnel

**Figure 6. "Bore segment" with cover after decision was made not to install screw piles (Forgie personal papers, Smithsonian Institution)**

over time and then make a determination whether piles were necessary when more data was available.

Rea reasoned that movement of the tunnels with the tide was real and measurable while data on "possible" excessive settlements were yet to be determined. Small tidal movements of the tunnel did not damage the tunnel or impact rail operations. However, a discontinuous "rigid" pile system would limit the liners free buoyant movements and potentially damage the liner. Finally, the test pile program showed that the installation of screw piles disturbed the tunnel subgrade and increased localized settlements.

It was the right decision. Measurements taken since the tunnels were completed indicate only minor settlements and the tunnels remain dry. Rea's decision not to support the North River tunnels on piles was not only correct but considering the circumstances heroic and an excellent example of not only the Observational Method (60 years before Peck's 1969 paper) and true example risk management.

### Compressed Air Blows and Compressed Air Sickness

These subaqueous tunnels were mined through rock, soft soils, and mixed face conditions (rock and soil). The tunnels often encountered obstructions near the shorelines including concrete bulkheads, nests of timber piles and riprap. The tunnels were shallow often with less than one tunnel diameter below the mudline. To reduce the risk of compressed air blows temporary clay blankets in the river were used wherever necessary. The Corps of Engineers allowed these temporary blankets, even if they obstructed navigation, as long as the clay blankets were removed promptly when the tunnel had passed through the area.

Even with the use of clay blankets compressed air sickness was common as was injury and death were common (more than 30 sandhogs died in these tunnels). This was particularly true in the East River Tunnels. Average compressed air pressures exceeded 2 bar (200 KPa) with a maximum compressed air of 3 bar (300 KPa).

The tunnels were two years behind schedule because of difficult mining conditions and high levels of compressed air sickness. This changed when E.W. Moir was assigned to the project by Pearson. Moir was an expert of compressed air and compressed air sickness. In 1890 he invented an installed the first medical air lock during Pearson's unsuccessful attempt to complete Hudson and Manhattan Tubes in the Hudson River. The installation of medical locks and Moir's insistence on the more effective use of clay blankets were instrumental in the completion of the PRR east river Tunnels in 1908.

An article in the April 3, 1908 magazine Railway Way sums up the importance of Moir to the successful completion of the project:

> "On March 20 the workmen and staff of S. Pearson & Son, Inc. who have just completed the excavation of the East river tunnels for the Pennsylvania Railroad, presented to E.W. Moir, vice-president of the company, an interesting model of a medical air lock bearing the inscription: 'presented to Mr. E.W. Moir, the maker of the first medical lock on the Hudson river Tunnel, 1890, by grateful "Sand Hogs" on

the Pennsylvania East River Tunnels, New York 1908."

The effective use of medical locks together with a dedicated medical staff significantly reduced the risk associated with compressed air tunnel and lead an renaissance in tunneling in New York City and beyond for decades to come.

## CONCLUSIONS AND LESSON LEARNED

The Pennsylvania Railroad tunnels build under the Hudson and East Rivers from 1904 through 1908. Over sixteen miles of tunnel, including seven miles of subaqueous tunnels, four permanent ventilation shafts, a major terminal and two major rail yards all conceived constructed and placed into revenue service with 10 years at a cost of $150,000,000. The completion of these tunnels and the extension of the PRR into New York, Long Island and New England made the PRR more competitive with its rival the New York Central Railroad. But it was also works done in the public interest by the largest and most powerful railroad company in America with no government monies. We will never see the likes of this again. However, their approach to identifying and reducing risks has applicability today.

The PRR prequalify and selected contractors only with the requisite experience to perform the work. Contractors were offered agreed upon an equitable means of payment based on the risks of the project. The PRR assumed the risk of the ground in the East River Tunnels. The PRR assumed costs of capital expenditures (multiple shields and compressed air plants) when necessary to move both the North River and East River Tunnels forward. Key individuals were identified early (Charles Jacobs, James Forgie, Alfred Noble, etc.) and assigned to the project both in the field and in the office. Belated, but still ahead of their time, they focused on safety with the insistence that E.W. Moir be assigned to the East River Tunnels. Rea and the Board focused on early identification of risks. They carefully considered on options. They tried alternative methods to reduce risk and adopted the most effective measures as early in the project as possible. Rea and the Board invested in obtaining and analyzing field data. They pioneered the observational approach to help with tough decisions (Rea).

Most importantly, the PRR make decisions in real time. Rea spoke for the railroad and had final authority on all engineering and business matters.

We also can learn something about have the courage to make tough engineering decisions in real time even. Everyone involved in the tunnel work should be aware and proud of the engineers and sandhogs who involved in the project. Starting with the vision of Alexander Cassatt, who died before his dream was accomplished, to Samuel Rea would make tough decisions in the face of real uncertainty, to E.W. Moir who did everything in his power to protect the sandhogs and engineers executing the work in the field. Forgie, Jacobs, Noble, Japp, Brace, Hewett and Brown who not only worked on the project but recorded the details of the construction for future generations in journals such as the Transactions of ASCE, Engineering News and the Franklin Institute. Finally, let us acknowledge the sandhogs themselves who accomplished this monumental project under very difficult conditions.

## REFERENCES

Brace, James H., Francis Mason, and S.H. Woodard (1910), "The New York Tunnel Extension of the Pennsylvania Railroad—The East River Tunnels," *Transactions, ASCE, Vol. LXVIII*, Paper No. 1159, September.

Copperthwaite (1906), *Tunnel Shields and the Use of Compressed Air in Subaqueous Works*, D. Van Nostrand Company, New York.

Cudahy, Brian J. (2002), *Rails under the Mighty Hudson*, Fordham University Press, New York.

Forgie, James (1906), "The Construction of the Pennsylvania R.R. Tunnels Under the Hudson River at New York City"—Tunnel Design, Shields and Plant," Engineering News, Vol. 56, No. 24, December 13, 2006.

Forgie, James (1907), "The Construction of the Pennsylvania R.R. Tunnels Under the Hudson River at New York City—Driving the Tunnels," Engineering News, Vol. 57, No. 9, February 28, 1907.

Hewett, B.H.M. and Brown, W.L. (1910), The New York Extension of the Pennsylvania Railroad. The North River Tunnels," *Transactions, ASCE, Vol. LXVIII*, Paper No. 1155, Sept. 1910.

Hewett, B.H.M. and S. Johannesson (1922), *Shield and Compressed Air Tunneling*, McGraw-Hill, 1st Edition.

Jacobs, Charles M. (1910), "The New York Extension of the Pennsylvania Railroad. The North River Division," *Transactions, ASCE, Vol. LXVIII*, Paper No. 1151, Sept. 1910.

Japp, Henry (1910), "The New York Extension of the Pennsylvania Railroad. East River Tunnels—Contractors Plant for East River Tunnels," *Transactions, ASCE, Vol. LXIX*, Paper No. 1160, October 1910.

Jenny, Robert J. (1983) "Compressed Air Used in Soft Ground Tunneling," ASCE, *Journal of Construction Engineering and Management*, Vol. 109, No. 2, June, p. 206–213.

Jonnes, Jill (2007), *Conquering Gotham—A Gilded Age Epic: The Construction of Penn Station and Its Tunnels*, Viking Penguin, New York.

Middleton, William D. (1996), *Manhattan Gateway New York's Pennsylvania Station*, Kalmbach Books, Waukesha, WI.

New York Times (1904), "O'Rourke Will Build Pennsylvania Tunnel," March 12.

Noble, Alfred (1910), "The New York Tunnel Extension of the Pennsylvania Railroad—East River Division," *Transactions, ASCE*, Vol. LXVIII, Sept, Paper No. 1152.

Noble, Alfred (1913), "The Railroad Tunnels of New York City," *Journal of the Franklin Institute, Vol. CLXXV, No. 4*, April 1913.

Peck, R.B. (1969), "Advantages and Limitations of the Observational Method in Applied Soil Mechanics," *Geotechnique*, Vol. 19, No. 2, pp. 171–187.

Prelini, Charles (1912), *Tunneling—A Practical Treatise*, 6th Edition, D. Van Nostrand Company, New York.

Raymond, Charles W. (1910), "The New York Tunnel Extension of the Pennsylvania Railroad," *Transactions, ASCE, Vol. LXVIII*, Paper No. 1150.

Railway Age (1906), "Meeting of the Pennsylvania Tunnel Shields," Vol. 42, No. 12.

Railway Age (1908), "Medical Air Lock," Vol. 45, No. 14, p. 520.

Rea, Samuel (1909), "Pennsylvania Railroad; New York Tunnel Extension. Historical Outline" .Remarks of Mr. Samuel Rea at the annual dinner given by the Board of Directors of the Pennsylvania Railroad Co. December 15, 1909.

Saxena, S. and Smirnoff, T. (1973), "Geotechnical Properties of Hudson River Silts," *Journal of the Soil Mechanics and Foundations Division*, ASCE, Vol. 99, No. SM10, October, pp. 912–917.

Singstad, O. (1944), "The Queens Midtown Tunnel—Under the East River Between Mid-Manhattan and Queens, New York City, *Transactions, ASCE*, Vol. 109, pp. 679–762.

# The Multilayer Wedge Method for Estimating TBM Face Pressures in Stratified Soils

**D. Scott Kieffer**
Graz University of Technology

**Zixin Zhang**
Tongji University

**Xinyu Hu**
Shanghai Institute of Technology

**ABSTRACT:** The current tendency toward ever-larger TBMs is accompanied by a higher probability of significant mixed-face conditions. The unprecedented 17.45 m diameter earth pressure balance TBM for the Alaskan Way Viaduct Replacement Program (AWVRP) is a contemporary example. The TBM is being driven through complexly layered glacial sands and clays, wherein several alternating soil units may comprise the tunnel face. Analyzing the required TBM face pressure under such conditions is a non-trivial task and has been facilitated by the recently developed multilayer wedge method (MWM) (Hu et al., 2012). This paper introduces key aspects of the MWM and its application at the AWVRP. Results obtained with the MWM are compared to classical methods for predicting required face pressure.

## INTRODUCTION

Adequate internal face support pressure prevents a face collapse when driving a TBM shield through soft ground. The minimum required (critical) support pressure can be calculated using the classical wedge (Horn, 1961) and trapezoid wedge (Broere, 2001; Wei, 2005) limit equilibrium analogies shown in Figure 1. In the wedge analysis, the vertical force above the tunnel at the interface BCEF must be estimated. Anagnostou et al. (1994) and Jancsecz & Steiner (1994) propose using soil arching theory to predict the minimum support pressure at collapse, rather than presuming full overburden pressure. The classical wedge model is applicable only in homogeneous ground conditions. Belter et al. (1999) introduced a two-dimensional wedge model to handle stratified soil conditions within the critical failure volume above the tunnel crown, but not ahead of the TBM. Broere (2001) proposed a three-dimensional wedge model to consider the effect of layered soils in front of the TBM, and Wei (2005) idealized the failure geometry as a tapered (rather than orthogonal) wedge and prism.

Centrifuge tests and field observations of face failures confirm that failure characteristics in sand and clay are notably different (e.g., Mair & Taylor, 1997). As shown in Figure 2, face failures in sand typically have a silo-like failure geometry defined by near-vertical sidewalls of the critical failure surface (a condition well suited to the wedge stability models described above). However, face failures in clay exhibit significantly broader failure volumes and the sidewalls of the critical failure surface are inclined significantly from vertical.

In stratified soils, the critical shear surface takes on different orientations according to soil type, and the failure geometry is resultantly more complex (Figure 3).

## THE MULTILAYER WEDGE METHOD

The multilayer wedge method (MWM) for face pressure analysis (Hu et al., 2012) was developed to handle horizontally stratified soil conditions in front and above the TBM face, and considers general failure geometry characteristics depicted in Figure 3. As enumerated below, the MWM analysis includes a modified wedge stability model and modified Terzaghi soil arching model.

### Modified Wedge Stability Model

The modified wedge stability model for accommodating stratified soils is depicted in Figure 4. For the two-layer condition shown in Figure 4a, the critical failure surface geometry in the upper and lower soil layer are defined by the angles $\alpha_l$, $\theta_l$ and $\alpha_0$, $\theta_0$, respectively. More generally, the potential failure

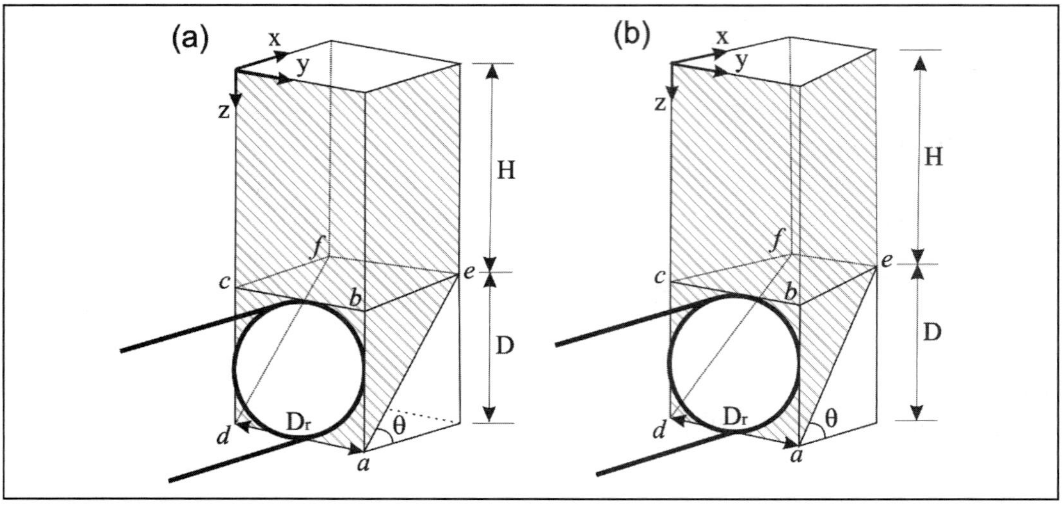


**Figure 1. Classical models for estimating face pressure: (a) wedge analogy (Horn, 1961) and (b) and trapezoid wedge analogy (modified after Broere, 2001; Wei, 2005)**

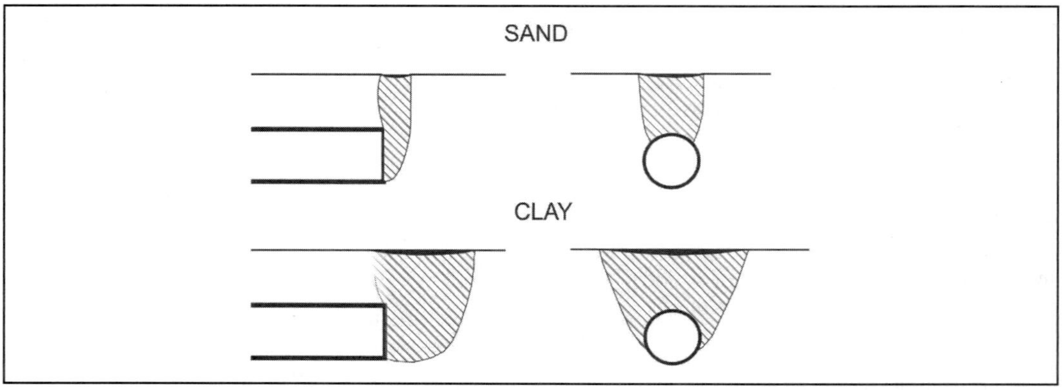


**Figure 2. Failure geometry characteristics in sand and clay (modified after Mair & Taylor, 1997)**

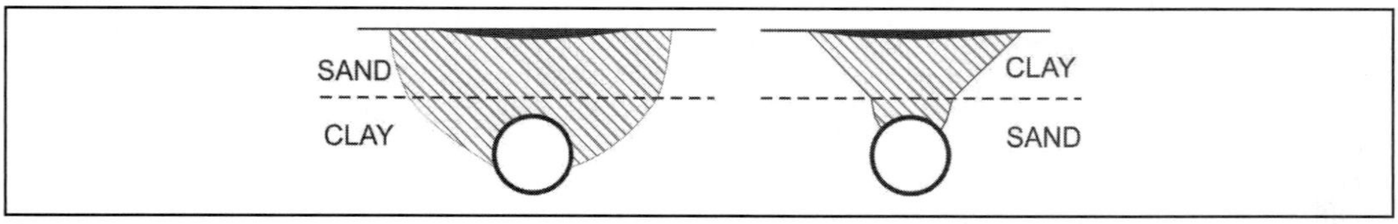


**Figure 3. Failure geometry of two-layer soils (modified after Selby, 1988)**

volume may include *n*-layers of soil above or in front of the TBM face (Figure 4b), where the failure geometry of the $i^{th}$ layer is defined by the angles $\alpha_i$ and $\theta_i$.

As shown in Figure 5, each soil layer *i* is subjected to the overburden force $G_{(i)}$ and the effective weight of the slice itself $W_{(i)}$, together with the reactive loading $R_{i-1(i)}$ and $R_{i+1(i)}$ from the overlying ($i$–1) and underlying ($i$+1) layers, respectively. Acting along the potential failure plane within the $i^{th}$ layer are the shear and normal forces $T_{(i)}$ and $N_{(i)}$, where $T_{(i)} = C_{(i)+} N_{(i)} \tan\phi_i$, and $C_{(i)}$ and $\phi_i$, represent the layer cohesion and angle of internal friction, respectively. Not depicted in the two-dimensional representation of Figure 5 are the out-of plane shear force $S_{(i)}$ and normal force $N_{(i)}$, acting on the side planes delimiting the $i^{th}$ layer.

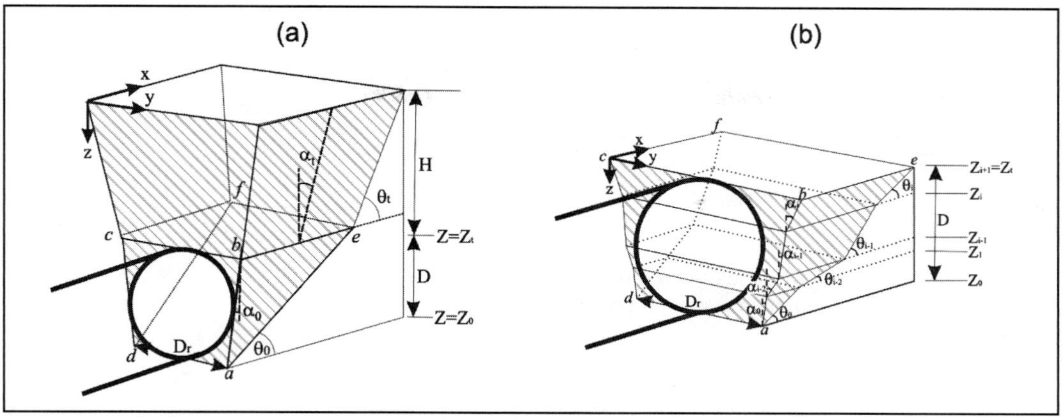


**Figure 4. Stratified soil conditions: (a) above and (b) in front of TBM face**

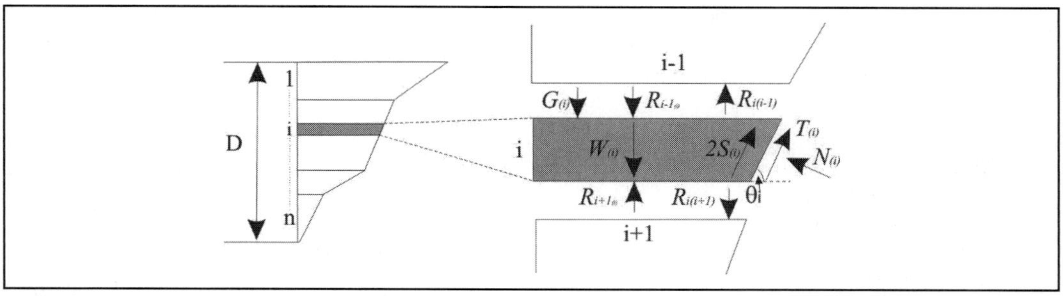


**Figure 5. Two-dimensional representation of loading applied to $i^{th}$ layer of $n$-layer system**

From horizontal and vertical force equilibrium, combined with the continuity condition $R_{i-1(i)} = R_{i(i-1)}$ and boundary condition $R_{0(1)} = 0$ and $R_{n+1(n)} = 0$, the minimum required support pressure can be calculated as:

$$2\sum_{i=1}^{N} S_{(i)}\frac{1}{\eta_{(i)}} + \sum_{i=1}^{N} P_{(i)}\frac{\eta'_{(i)}}{\eta_{(i)}} + \sum_{i=1}^{N} C_{(i)}\frac{1}{\eta_{(i)}} - \sum_{i=1}^{N} G_{(i)} - \sum_{i=1}^{N} W_{(i)} = 0$$

where

$$\begin{cases} \eta_{(i)} = \sin\theta_i - \cos\theta_i \tan\varphi_{(i)} \\ \eta'_{(i)} = \cos\theta_i + \sin\theta_i \tan\varphi_{(i)} \end{cases}$$

The forces acting on the $i^{th}$ layer are given by:

$$G = \sigma_v(z_t)\cdot 2y(z_t)\cdot x(z_t)$$

$$W_{(i)} = \int_{z_i}^{z_{i+1}} \gamma_{(i)}\cdot x(z)\cdot 2y(z)dz$$

$$C_{(i)} = \int_{z_i}^{z_{i+1}} \frac{2y(z)}{\sin\theta_i}\cdot c_{(i)}dz$$

$$S_{(i)} = \int_{z_i}^{z_{i+1}} x(z)\cdot[c_{(i)} + K_{y(i)}\cdot\sigma_z(z)\cdot\cos\alpha_i\cdot\tan\varphi_i]$$

$$\sigma_z(z) = \frac{\int_{z_t}^{z}(\sigma_v(z_t) + \gamma_{(i)}\cdot(z - z_t))\cdot x(z)\cdot 2y(z)dz}{\int_{z_t}^{z} x(z)\cdot 2y(z)\,dz}$$

Where the wedge dimensions $x(z)$ and $y(z)$ as a function of depth can be solved as:

$$\begin{aligned} x(z) &= (z_i - z)\cot\theta_i + x(z_i) \\ &= (z_i - z)\cot\theta_i + (z_0 - z_1)\cot\theta_0 \\ &\quad + (z_1 - z_2)\cot\theta_1 + \ldots + (z_{i-1} - z_i)\cot\theta_{i-1} \\ &= (z_i - z)\cot\theta_i + \sum_{k=1}^{i}(z_{k-1} - z_k)\cot\theta_{k-1} \end{aligned}$$

$$y(z) = (z_i - z)\tan\alpha_i + y(z_i)$$
$$= (z_i - z)\tan\alpha_i + D_r/2 + (z_0 - z_1)\tan\alpha_0$$
$$+(z_1 - z_2)\tan\alpha_1 + \ldots + (z_{i-1} - z_i)\tan\alpha_{i-1}$$
$$= \frac{(z_i - z)\tan\alpha_i + D_r}{2 + \sum_{k=1}^{i}(z_{k-1} - z_k)\tan\alpha_{k-1}}$$

**Two soil layers.** For the case of two soil layers (Figure 5a), the minimum required support pressure is expressed as:

$$P_{cr(1)}\frac{\eta'_{(1)}}{\eta_{(1)}} + P_{cr(2)}\frac{\eta'_{(2)}}{\eta_{(2)}} = G_{(1)} + G_{(2)} + W_{(1)} + W_{(2)}$$
$$-2S_{(1)}\frac{1}{\eta_{(1)}} - 2S_{(2)}\frac{1}{\eta_{(2)}}$$
$$-C_{(1)}\frac{1}{\eta_{(1)}} - C_{(2)}\frac{1}{\eta_{(2)}}$$

$$P_{cr} = P_{cr(1)} + P_{cr(2)}$$

where

$$G = G_{(1)} + G_{(2)}$$
$$= \begin{bmatrix}(z_0 - z_l)\cot\theta_0 \\ +(z_l - z_t)\cot\theta_t\end{bmatrix} \cdot \begin{bmatrix}D_r + 2(z_0 - z_l)\tan\alpha_0 \\ +2(z_l - z_t)\tan\alpha_t\end{bmatrix} \cdot \sigma_v(z_t)$$

$$W = W_{(1)} + W_{(2)}$$
$$= \frac{1}{2}\gamma_0 \cdot D_r(z_0 - z_l)^2\cot\theta_0 + \frac{2}{3}\gamma_0$$
$$\cdot(z_0 - z_l)^3\tan\alpha_0\cot\theta_0$$
$$+(z_l - z_t)\gamma_t\begin{bmatrix}\frac{1}{2}(z_l - z_t)(D_r \\ +2(z_0 - z_l)\tan\alpha_0)\cot\theta_t \\ +\frac{2}{3}(z_l - z_t)^2\tan\alpha_t\cot\theta_t \\ +(z_0 - z_l)(D_r \\ +2(z_0 - z_l)\tan\alpha_0)\cot\theta_0 \\ +(z_0 - z_l)(z_l - z_t)\tan\alpha_t\cot\theta_0\end{bmatrix}$$

$$C = C_{(1)} + C_{(2)}$$
$$= \frac{(D_r + (z_0 - z_l)\tan\alpha_0)(z_0 - z_l)\cdot c_0}{\sin\theta_0}$$
$$+ \frac{\begin{pmatrix}D_r + 2(z_0 - z_l)\tan\alpha_0 \\ +(z_l - z_t)\tan\alpha_t\end{pmatrix}(z_l - z_t)\cdot c_t}{\sin\theta_t}$$

$$S = S_{(1)} + S_{(2)}$$
$$= \frac{1}{2}(z_0 - z_l)^2\cot\theta_0\begin{pmatrix}c_0 + K_{y0} \\ \cdot\sigma_{z0}\cdot\cos\alpha_0\cdot\tan\varphi_0\end{pmatrix}$$
$$+\frac{1}{2}(2(z_0 - z_l)\cot\theta_0$$
$$+(z_l - z_t)\cot\theta_t)(z_l - z_t)\begin{pmatrix}c_t + K_{yt} \\ \cdot\sigma_{zt}\cdot\cos\alpha_t\cdot\tan\varphi_t\end{pmatrix}$$

$$\sigma_{zt} = \frac{\int_{z_t}^{z_l}\begin{bmatrix}(\sigma_v(z_t) + \gamma_t\cdot(z - z_t)) \\ \cdot(D_1\cot\theta_0 + (z_l - z)\cot\theta_1) \\ \cdot(D_r + 2D_1\tan\alpha_0 + 2(z_l - z)\tan\alpha_t)\end{bmatrix}dz}{\int_{z_t}^{z_l}(D_1\cot\theta_0 + (z_l - z)\cot\theta_1)\cdot(D_r + 2D_1\tan\alpha_0 + 2(z_l - z)\tan\alpha_t)dz}$$

$$= \frac{\begin{bmatrix}\begin{pmatrix}D_r \\ +2D_1\tan\alpha_0\end{pmatrix}\begin{pmatrix}\sigma_v(z_t)D_2\begin{pmatrix}D_1\cot\theta_0 \\ +\frac{1}{2}D_2\cot\theta_1\end{pmatrix} \\ +\gamma_t D_2^2\begin{pmatrix}\frac{1}{2}D_1\cot\theta_0 \\ +\frac{1}{6}D_2\cot\theta_1\end{pmatrix}\end{pmatrix} \\ +2\tan\alpha_t\begin{pmatrix}\sigma_v(z_t)D_2^2\begin{pmatrix}\frac{1}{2}D_1\cot\theta_0 \\ +\frac{1}{3}D_2\cot\theta_1\end{pmatrix} \\ +\gamma_t D_2^3\begin{pmatrix}\frac{1}{6}D_1\cot\theta_0 \\ +\frac{1}{12}D_2\cot\theta_1\end{pmatrix}\end{pmatrix}\end{bmatrix}}{D_2\begin{bmatrix}\frac{1}{2}D_2(D_r + 2D_1\tan\alpha_0)\cot\theta_t \\ +\frac{2}{3}D_2^2\tan\alpha_t\cot\theta_t \\ +D_1(D_r + 2D_1\tan\alpha_0)\cot\theta_0 \\ +D_1D_2\tan\alpha_t\cot\theta_0\end{bmatrix}}$$

$$\sigma_{z0} = \frac{\int_{z_l}^{z_0} \begin{bmatrix} \left(\sigma_{zt} + \gamma_0 \cdot (z - z_l)\right) \\ \cdot\left((z_0 - z)\cot\theta_0\right) \\ \cdot\left(D_r + 2(z_0 - z)\tan\alpha_0\right) \end{bmatrix} dz}{\int_{z_l}^{z_0} (z_0 - z)\cot\theta_0 \cdot \left(D_r + 2(z_0 - z)\tan\alpha_0\right) dz}$$

$$= \frac{\cot\theta_0 \cdot \begin{bmatrix} \frac{1}{6} D_1^2 \cdot \sigma_{zt} \cdot (3D_r + 4D_1 \tan\alpha_0) \\ + \frac{1}{6} D_1^3 \cdot \gamma_0 \cdot (D_r + D_1 \tan\alpha_0) \end{bmatrix}}{\frac{1}{2} D_r D_1^2 \cot\theta_0 + \frac{2}{3} D_1^3 \tan\alpha_0 \cot\theta_0}$$

and $\gamma_0$, $c_0$, $\varphi_0$, and $\gamma_t$, $c_t$, $\varphi_t$ are the soil parameters for the lower and upper layers, respectively.

**Single soil layer.** For the case of a single homogenous soil layer, the critical failure geometry is defined by the angles $\theta$ *and* $\alpha$. The minimum required support pressure is then expressed as:

$$P_{cr} = \frac{1}{\eta}\left[\eta(G + W) - (2S + C)\right]$$

where

$$G = (D_r + 2D\tan\alpha)D\cot\theta \cdot \sigma_v(z_t)$$

$$W = \frac{1}{2}\gamma \cdot D_r D^2 \cot\theta + \frac{2}{3}\gamma \cdot D^3 \cot\theta \tan\alpha$$

$$S = \frac{1}{2} D^2 \cot\theta (c + K_y \cdot \sigma_z \cdot \cos\alpha \cdot \tan\varphi)$$

$$C = \frac{D}{\sin\theta}(D\tan\alpha + D_r) \cdot c$$

$$\sigma_z = \frac{\int_{z_t}^{z_0} \begin{bmatrix} \left(\sigma_v(z_t) + \gamma \cdot (z - z_t)\right) \\ \cdot (z_0 - z)\cot\theta \\ \cdot\left(D_r + 2(z_0 - z)\tan\alpha\right) \end{bmatrix} dz}{\int_{z_t}^{z_0} \begin{bmatrix} (z_0 - z)\cot\theta \\ \cdot (D_r + 2(z_0 - z)\tan\alpha \end{bmatrix} dz}$$

$$= \frac{\cot\theta \cdot \begin{bmatrix} \frac{1}{6} D^2 \cdot \sigma_v(z_t) \cdot (3D_r + 4D\tan\alpha) \\ + \frac{1}{6} D^3 \cdot \gamma \cdot (D_r + D\tan\alpha) \end{bmatrix}}{\frac{1}{2} D_r D^2 \cot\theta + \frac{2}{3} D^3 \cot\theta \tan\alpha}$$

## Modified Terzaghi Soil Arching Model—2D

Soil arching is an important aspect of wedge-based models for calculating face pressure (Broere, 2001). As enumerated below and depicted in Figure 6, a modified Terzaghi soil-arching model is proposed as part of the MWM of analysis. Main assumptions of the modified arching model are:

- The arch width is not constant as assumed in Terzaghi's classical silo analogy, but gradually increases toward the ground surface according to the angle $\alpha_a$; and
- Thin rectangular strips are assumed to comprise a trapezoidal failure geometry.

A two-dimensional representation of the modified arching model is shown in Figure 6. In this simplified case (a soil layer of any unit length at depth $z$) the width of the corresponding rectangle is $2b + 2(H - z)\tan\alpha_a$. The vertical stress for modified Terzaghi arching of a homogenous soil in two-dimensions to satisfy equilibrium conditions results in:

$$\sigma_v = \left( \frac{\gamma(b + (H - z)\tan\alpha_a)}{K\cos\alpha_a \tan\varphi - \tan\alpha_a} - \frac{c}{K\cos\alpha_a \tan\varphi} \right) + \left( \begin{array}{l} \left(b + (H - z)\tan\alpha_a\right)^{\frac{K\cos\alpha_a \tan\varphi}{\tan\alpha_a}} \\ \cdot\left(b + H\tan\alpha_a\right)^{-\frac{K\cos\alpha_a \tan\varphi}{\tan\alpha_a}} \end{array} \right) \cdot \begin{bmatrix} q_0 - \frac{\gamma(b + H\tan\alpha_a)}{K\cos\alpha_a \tan\varphi - \tan\alpha_a} \\ + \frac{c}{K\cos\alpha_a \tan\varphi} \end{bmatrix}$$

where $K$ is the coefficient of lateral earth pressure.

In stratified soils (Figure 6b) a similar result is obtained for each soil layer, which is then integrated from top to bottom and using the effective stress at the bottom of the previous layer as the continuity condition for the following layer. Step by step the vertical stress at the base of the soil silo is calculated as:

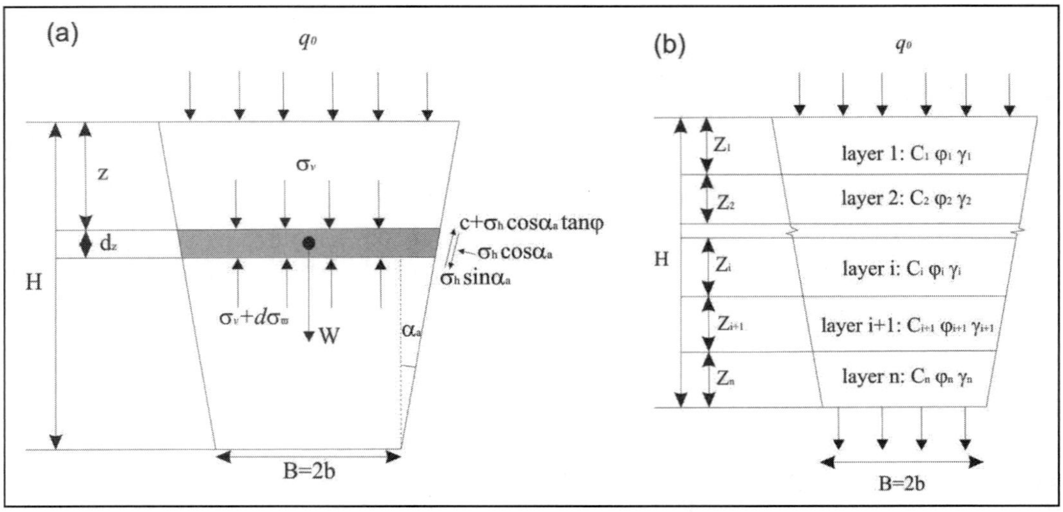


**Figure 6. Simplified two-dimensional representation of the modified Terzaghi soil arching model: (a) homogenous soil and (b) stratified soil**

$$\begin{aligned}
\sigma_v = P_n = {} & \left(\frac{\gamma_n(b+(H-z_n)\tan\alpha_{an})}{K_n\cos\alpha_{an}\tan\varphi_n-\tan\alpha_{an}}-\frac{c_n}{K_n\cos\alpha_{an}\tan\varphi_n}\right)+\left(b+(H-z_n)\tan\alpha_{an}\right)^{\frac{K_n\cos\alpha_{an}\tan\varphi_n}{\tan\alpha_{an}}} \\
& \cdot(b+H\tan\alpha_{an})^{-\frac{K_n\cos\alpha_{an}\tan\varphi_n}{\tan\alpha_{an}}}\cdot\left(\frac{c_n}{K_n\cos\alpha_{an}\tan\varphi_n}-\frac{\gamma_n(b+H\tan\alpha_{an})}{K_n\cos\alpha_{an}\tan\varphi_n-\tan\alpha_{an}}\right) \\
& +\sum_{i=n-1}^{1}\left[\begin{aligned}
& \left(\frac{\gamma_i(b+(H-z_i)\tan\alpha_{ai})}{K_i\cos\alpha_{ai}\tan\varphi_i-\tan\alpha_{ai}}-\frac{c_i}{K_i\cos\alpha_{ai}\tan\varphi_i}\right) \\
& +\left(b+(H-z_i)\tan\alpha_{ai}\right)^{\frac{K_i\cos\alpha_{ai}\tan\varphi_i}{\tan\alpha_{ai}}}\cdot\left(b+H\tan\alpha_{ai}\right)^{-\frac{K_i\cos\alpha_{ai}\tan\varphi_i}{\tan\alpha_{ai}}} \\
& \cdot\left(\frac{c_i}{K_i\cos\alpha_{ai}\tan\varphi_i}-\frac{\gamma_i(b+H\tan\alpha_{ai})}{K_i\cos\alpha_{ai}\tan\varphi_i-\tan\alpha_{ai}}\right) \\
& \cdot\prod_{j=n}^{i+1}\left(b+(H-z_j)\tan\alpha_{aj}\right)^{\frac{K_j\cos\alpha_{aj}\tan\varphi_j}{\tan\alpha_{aj}}}\cdot\left(b+H\tan\alpha_{aj}\right)^{-\frac{K_j\cos\alpha_{aj}\tan\varphi_j}{\tan\alpha_{aj}}}
\end{aligned}\right] \\
& +q_0\prod_{k=1}^{n}\left(b+(H-z_k)\tan\alpha_{ak}\right)^{\frac{K_k\cos\alpha_{ak}\tan\varphi_k}{\tan\alpha_{ak}}}\cdot\left(b+H\tan\alpha_{ak}\right)^{-\frac{K_k\cos\alpha_{ak}\tan\varphi_k}{\tan\alpha_{ak}}}
\end{aligned}$$

### Modified Terzaghi Soil Arching Model—3D

The preceding model considers an infinitely long or two dimensional wedge, wherein shear stresses act only along the two delimiting surfaces. For three-dimensional soil arching (Figure 7) the following assumptions are made:

- The arch width is not constant, but gradually increases toward the ground surface according to the angles $\alpha_a$ and $\theta_a$; and
- Thin rectangular slabs are assumed to comprise a trapezoidal-prism failure geometry.

At any depth z, the width and the length of the corresponding rectangular slab are $B + 2(H - z)\tan\alpha_a$ and $L + 2(H - z)\tan\theta_a$, respectively. By satisfying equilibrium conditions, the vertical stress for modified Terzaghi arching in three-dimensions is expressed as:

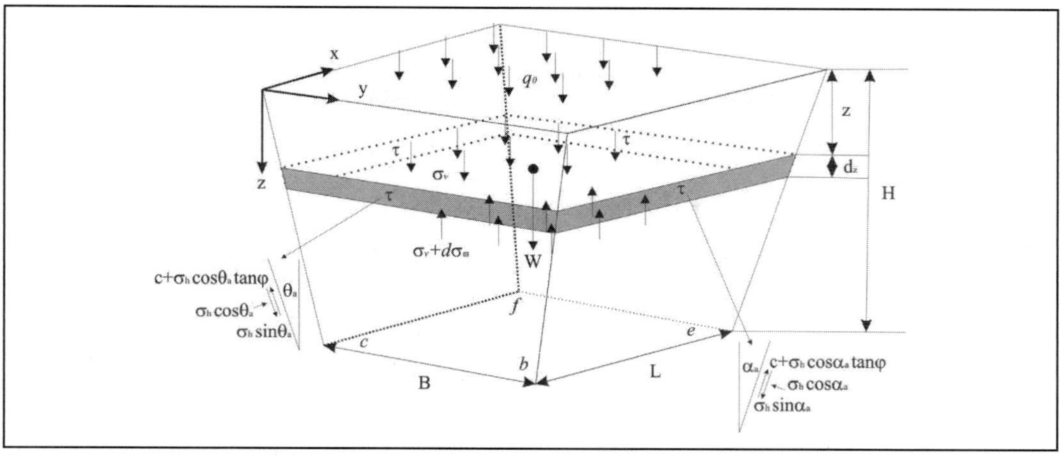


**Figure 7. Modified Terzaghi soil arching model in three dimensions**

$$\sigma_v = \left(B+2H\tan\alpha_a\right)^{n-m}\cdot\left(L+2H\tan\theta_a\right)^{-(n+m)}\cdot\left(B+2(H-z)\tan\alpha_a\right)^{m-n}\cdot\left(L+2(H-z)\tan\theta_a\right)^{m+n}$$
$$\cdot\left[\begin{array}{l}\left(B+2H\tan\alpha_a\right)^{m-n}\cdot\left(L+2H\tan\theta_a\right)^{n+m}\cdot\int_0^z\left(\gamma-2c\left(\frac{1}{(L+2(H-z)\tan\theta_a)}+\frac{1}{(B+2(H-z)\tan\alpha_a)}\right)\right.\\ \left.\cdot(B+2(H-z)\tan\alpha_a)^{n-m}\cdot(L+2(H-z)\tan\theta_a)^{-(m+n)}dz+q_0\right.\end{array}\right]$$

where

$$\begin{cases} m=\dfrac{K}{2}\left(\dfrac{\cos\theta_a\tan\varphi}{\tan\theta_a}+\dfrac{\cos\alpha_a\tan\varphi}{\tan\alpha_a}\right) \\ n=\dfrac{K}{2}\left(\dfrac{\cos\theta_a\tan\varphi}{\tan\theta_a}-\dfrac{\cos\alpha_a\tan\varphi}{\tan\alpha_a}\right)\end{cases}$$

## THE MULTILAYER WEDGE METHOD APPLIED TO THE AWVRP

With a diameter of 17.45 m, the earth pressure balance TBM commissioned for the ongoing Alaskan Way Viaduct Replacement Program (AWVRP) sets a new record in terms of size. The unprecedented tunnel diameter in combination with complex stratified glacial sands and clays create a high probability for encountering significant mixed face conditions. As an example Figure 8 shows a simplified cross section of the ground conditions at Sta 210+00, based on interpretation of information provided in the project's Geotechnical Baseline Report (GBR) (WSDOT, 2010).

Six engineering soil units comprise the cross section of Figure 8, and the maximum hydraulic head associated with Puget Sound is one meter above the existing ground surface. With an overburden-to-diameter ratio of 1.3, in a heavily urbanized area, face pressure control is one of the project's key factors.

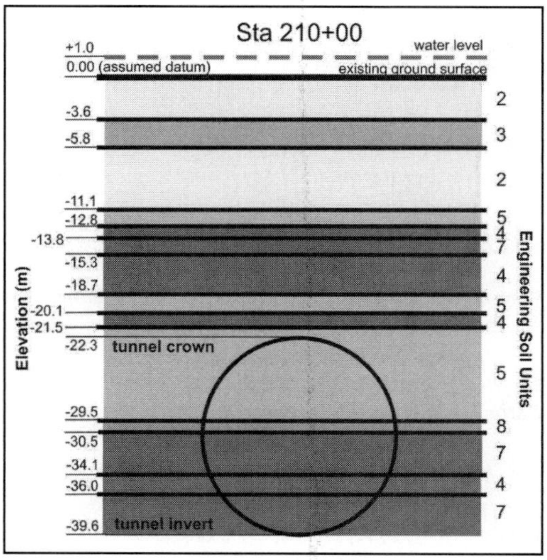


**Figure 8. Cross-sectional idealization at Sta 210+00**

The MWM analysis for stratified soils is applied to the cross section and compared to established methods for calculating face support pressure. According to baseline values and ranges given in the project's GBR, the geotechnical parameters summarized in Table 1 have been assigned to the soil layers at Sta 210+00. As engineering soil units 5 and 7 comprise

**Table 1. Geotechnical parameters assigned to soil units at Sta 210+00**

| Engineering Soil Unit* | Moist Unit wt. ($kN/m^3$) | Effective Cohesion c' (kPa) | Effective Friction Angle, $\phi$' | Undrained Shear Strength $S_u$ (kPa) | At-Rest Earth Pressure, $K_o$ | Hydraulic Conductivity, Horiz (cm/sec) | Hydraulic Conductivity, Vert (cm/sec) |
|---|---|---|---|---|---|---|---|
| 2 | 18.5 | 0 | 32 | — | 0.45 | $5 \times 10^{-3}$ | $3 \times 10^{-3}$ |
| 3 | 18.5 | 5 | 24 | 45 | 0.55 | $3 \times 10^{-4}$ | $3 \times 10^{-5}$ |
| 4 | 23.5 | 240 | 40 | 620 | 0.6 | $1 \times 10^{-5}$ | $1 \times 10^{-7}$ |
| 5 | 21.0 | 0 | 39 | — | 0.8 | $5 \times 10^{-3}$ | $1 \times 10^{-3}$ |
| 7 | 19.0 | 60 | 25 | 330 | 1.4 | $1 \times 10^{-5}$ | $1 \times 10^{-6}$ |
| 8 | 23.0 | 0 | 40 | — | 1.0 | $1 \times 10^{-4}$ | $1 \times 10^{-4}$ |

* As defined in project GBR (WSDOT, 2010).

the majority of the tunnel face, the following calculations correspond to these soil units.

Methods for estimating face support pressure for comparison to the MWM described herein include the following.

***Chinese Code***

$$p_f = p_w + k\sum \gamma h + \text{fluctuating pressure}$$

where $p_f$ is the face pressure, $p_w$ is the water pressure, k is the coefficient of lateral earth pressure (ranging from active to passive states), $\gamma$ is the unit weight of soil, and $h$ is the dimension from ground surface to depth where $p_f$ is calculated.

***Japanese Code***

$$p_f = \sum \gamma h$$

***Broms & Bennermark (1967) Empirical Method***

$$p_f = \sum \gamma h - \text{N}\, c_u$$

where N is the stability number and $c_u$ is undrained shear strength.

***Davis et al. (1980) Empirical Method for Estimating N***

$$\text{N} = 2 + 2\ln(C/R + 1)$$

where $C$ and $R$ represent the overburden depth and tunnel radius, respectively. Alternatively, simply assume that N = 6 based on case studies (Broms & Bennermark, 1967). In many cases the Broms & Bennermark and Davis approaches provide representative results in clay, but tend toward unreasonable values in soils with little or no cohesion.

**Undrained Loading Condition**

During normal excavating conditions, it is reasonable to assume that the soils behave according to undrained loading conditions, thus water pressure and earth pressure are calculated together to determine the face pressure. Results obtained from the multilayer wedge method and comparative analyses are summarized in Figure 9. The reported face pressure corresponds to the depth-averaged value obtained for engineering soil units 5 and 7, which comprise the majority of the tunnel face and failure volume.

Due to the arch effect of soil above the tunnel the multilayer wedge method (MWM) provides the minimum face pressure. Considering the relatively small overburden-to-diameter ratio of 1.3, it is justifiably conservative to assume that no soil arching occurs.

**Drained Loading Condition**

During significant TBM downtime, pore pressures will tend to dissipate during shearing. Therefore it is reasonable to assume that the soils behave like under drained loading condition, meaning that water pressure and earth pressure should be calculated separately to determine the face pressure. Results obtained from the multilayer wedge method and comparative analyses are also summarized in Figure 9. The reported face pressure corresponds to the depth-averaged value obtained for engineering soil units 5 and 7, which comprise the majority of the tunnel face and failure volume.

## CONCLUSIONS

The multilayer wedge method (Hu et al., 2012) summarized herein allows computation of the minimum face support pressure in horizontally stratified soil conditions and is consistent with general characteristics of failure geometries observed in centrifuge tests and in the field. The multilayer wedge method includes a modified wedge stability model and modified Terzaghi soil arching model. Results obtained with this method provide additional insight to the potential range of face pressures to be expected during TBM driving through significant mixed face conditions.

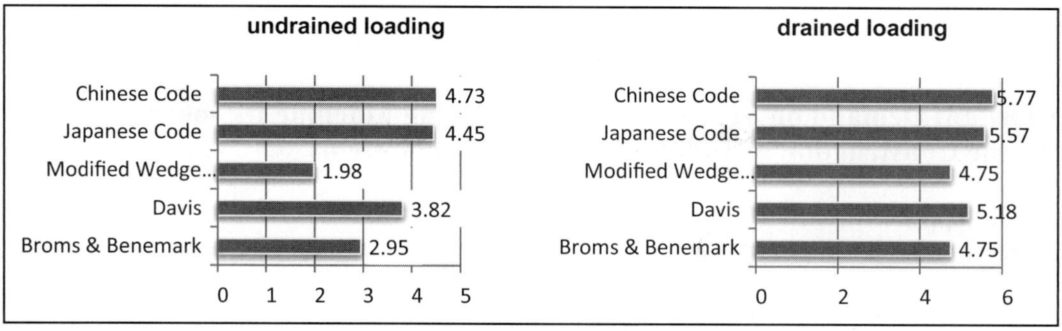


**Figure 9. Minimum face support pressures for undrained and drained loading conditions at Sta 210+00**

## REFERENCES


Belter, B., Heiermann, W., Katzenbach, R., Maidl, B., Quick, H, and Wittke, W., 1999, NBS Köln-Rhein/Main–Neue Wege bei der Umsetzung von Verkehrsprojekten. Bauingenieur, 74(1): 1–7.

Broere, W., 2001, Tunnel Face Stability & New CPT Applications: [PhD thesis], Delft University of Technology, Delft.

Broms, B.B., and Bennermark, H., 1967, Stability of Clay at Vertical Openings, ASCE Journal of Soil Mech. and Found. Div., 93.

Davis, E.H., Gunn, M.J., Mair, R.J., and Seneviratne, H.N., 1980, The stability of shallow tunnels and underground openings in cohesive material, Geotechnique, 304, 397–416.

Horn, M., 1961, Horizontal earth pressure on perpendicular tunnel face, Hungarian National Conference of the Foundation Engineer Industry, Budapest, Hungary.

Hu, X., Zhang, Z., and Kieffer, D.S., 2012, A real-life stability model for a large shield-driven tunnel in heterogeneous soft soils, Front. Struct. Civ. Eng., 6(2): 176–187.

Jancsecz, S., and Steiner, W., 1994, Face support for a large mix-shield in heterogeneous ground conditions. Tunnelling, 94: 531–550.

Mair, R.J., and Taylor, R.N., 1997, Bored tunnelling in the urban environment. Proceedings of the 14th International Conference on Soil Mechanics and Foundation Engineering, Rotterdam: 2,353–2,385.

Qin, J., 2005, Study on face deformation and collapse of earth pressure shield tunnel: [PhD thesis], HoHai University, Nanjing, China.

Selby, A.R., 1988, Surface movements caused by tunnelling in two-layer soil. Engineering Geology of Underground Movements Bell, F.G. et al. (eds.), Geological Society Engineering Geology Special Publication, 5: 71–77.

Washington State Department of Transportation (WSDOT), 2010, Alaskan Way Viaduct and Seawall Replacement Program, Appendix G1, Revised SR 99 Bored Tunnel Alternative Design-Build Project, Geotechnical Baseline Report, June, 2010.

Wei, G., 2005, Theoretical study on properties of soil and structure during pipe jacking construction: [PhD thesis], Zhejiang University, Huangzhou, China (in Chinese).

# Constant Demand on Very Large Tunnel Boring Machine Diameters for the Construction of Today's Infrastructure Systems

**Karin Bäppler**
Herrenknecht AG

**ABSTRACT:** The project demands of today's infrastructure systems are focused on a high level of technology and realization where mechanized tunnelling technology displays one of the solutions becoming more and more predominant today by creating conditions for environmentally friendly mobility of people and goods, by creating sustainable management of precious resources as well as the targeted development of raw materials and energy sources. This article is focused on the mechanized tunnelling of very large diameter tunnel boring machines (TBMs) for underground infrastructures. TBMs with diameters exceeding 15 meters are able to construct tunnels safe, reliable, within the given time frame and fulfilling also the demand on high quality standards. The dimensions the projects take is shown with the successful completion of projects such as the M30 highway tunnel in Madrid (EPB Shield Ø15.2m) Shanghai Changijiang Under River Tunnel Project in China (Mixshield Ø15.43m), and the Galleria Sparvo Tunnel in Italy (EPB Shield Ø15.6m). These are only a few mentioned very large diameter tunnel projects of today. The article addresses on TBMs with very large diameter which are currently in operation and design and have to deal with the complexity of geology, side constraints and tight time schedules. The complexity to fulfill highest technology standards with TBMs will be addressed which starts already in the design phase of a machine and is followed by the demand, input and involvement of all parties in the project.

## INTRODUCTION

The successful completion of large diameter tunnelling projects in the past recent years supports the ongoing trend of further design of tunnels with very large profiles and exceeding also the already secured experiences in the diameter ranges of 15 to 16 meters. In May and September 2008 the two largest Mixshields with diameters of 15.43 meters finished their excavation of 7.2km long drives crossing beneath the Yangtze and this 10 and 12 months earlier than planned. In July 2013 one of the largest EPB Shields with a diameter of 15.6 meters finished the excavation of the twin-bore road Tunnel Sparvo in Italy in about two years matching with the program of the project. For both projects the, at that time, largest machines ever built had been put underground to excavate the very large tunnel profiles used for road traffic. The machine that excavated the Tunnel Sparvo faced challenging conditions such as instable rock conditions with presence of methane gas; the machines that crossed beneath the Yangtze faced long tunnel drives and high hydrostatic pressures. Apart from the large tunnelling diameter, the projects are to be excavated in more challenging subsurface conditions and environment. Geology can bear issues and uncertainties which all parties involved in the project have to consider and which should be addressed early in the TBM design to have backup solutions and to be able to react with a certain kind of flexibility in case of occurrence.

In 2013 a 13.6-m-diameter Mixshield was manufactured in Germany with destination Istanbul. The machine will excavate and line within the next years a large diameter road tunnel crossing beneath the strait of Istanbul connecting Asia with Europe. The tunnel construction is one of the most challenging tunnelling projects. The alignment of the Istanbul Strait Road Tube Crossing Project is designed with clearances of 100 meters below sea level at its deepest point. 3.34 kilometers of the tunnel with a total length of 5.4 kilometers is going to be built by means of mechanized tunnelling technology starting on the Asian side. Another project that was contracted in autumn 2013 is focused on a short tunnel drive with a machine diameter of 17.6 meters in diameter being part of a road tunnel project in Asia. The paper focuses on the TBM design issues for the 13.6m- and 17.6m-diameter Mixshields.

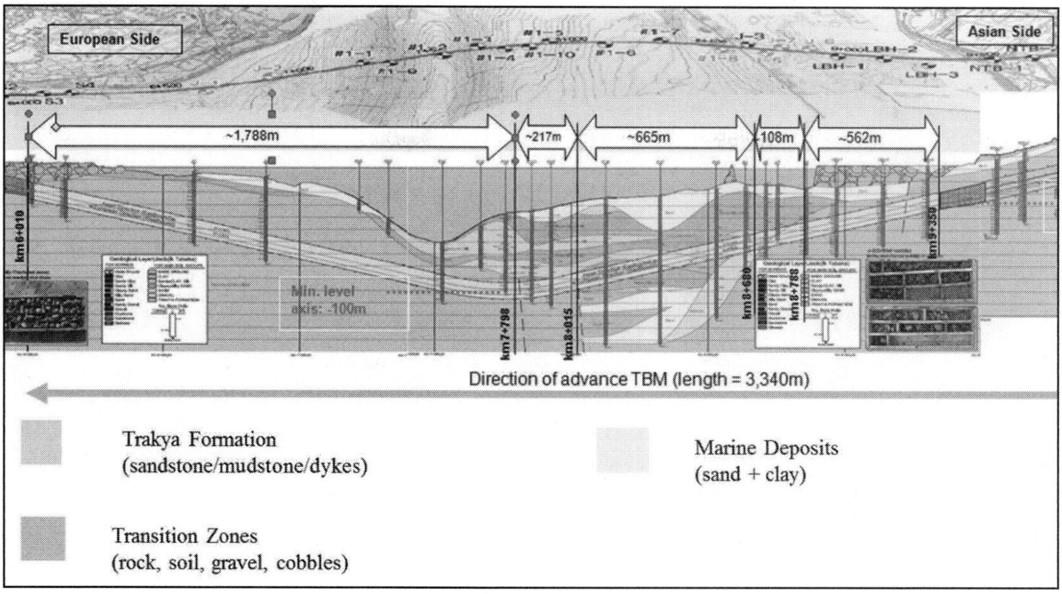


**Figure 1. Overview about TBM tunnelling section with predicted geological conditions**

## TBM DESIGN AND CHALLENGES FOR THE EXCAVATION OF THE FIRST ROAD TUNNEL CROSSING BENEATH THE BOSPORUS IN ISTANBUL

The main part of the Istanbul Strait Road Tunnel Crossing Project will be a 3.34 kilometer long tunnel having an internal diameter of 12.0 meters. The tunnel profile will take two road levels, each with two traffic lanes and a shoulder. The Strait Road Crossing tunnel will be excavated from the Asian side of Istanbul towards the European side by means of mechanized tunnelling technology and this in a complex geological and hydrogeological environment. It will be the first road tunnel with a total length of 5.4km beneath the Bosporus connecting both sides of Istanbul. See Figure 1.

The TBM design is mainly focused on varying subsurface conditions of hard and soft rock, mixed face conditions and high hydrostatic pressures. Apart from these conditions the alignment will likely impact existing utilities and structures of the harbor and dock.

The predicted subsurface conditions to be faced with TBM excavation will comprise about 70% hard rock of the Trakya formation (mudstone, sandstone and magmatic rock dykes) and about 20% of marine deposits (clay, silt, sand, gravel and cobbles). The dykes are out of diabase, andesite or diorite rocks of high strength (UCS of up to 250MPa) and high abrasiveness. Also the granular material, sands and cobbles are supposed to be very abrasive. The alignment of the tunnel with gradients of ±5% leads to a maximum depth of the tunnel to tunnel invert of 105 meters below sea level. High hydrostatic pressures of about 10bars at tunnel axis are expected. These issues from geology had been considered in the design of the applied technology. The 120 meter long TBM that will excavate beneath the Bosporus is a Mixshield with a diameter of 13.66m. It is designed for 12bars maximum face pressure on tunnel axis. One of the main tasks is the ability to change the excavation tools quickly and safely even with high pressures. Therefore a cutting wheel concept was developed which was first implemented in the design of the 4th tube Elbe tunnel project in a 14.2m-diameter Mixshield in 1997. The cutting wheel is accessible from the rear of the machine in free air with the possibility to change safely all disc cutters and a large number of the assembled cutting knives. With this design time and cost consuming accesses for tool change work under pressurized, hyperbaric conditions or even saturation can be avoided taking the health and safety of the personnel into account. In addition, the Mixshield is equipped with the possibility of executing saturation diving comparable to the system applied for the construction of the Westerschelde road tunnel project in the Netherlands.

Access to the pressurized chamber is possible through two separate man locks. Consumables and tools required in the working chamber for maintenance or tool change can be supplied through three material locks.

The cutting wheel is designed with 6 main arms which are accessible in free air and 6 auxiliary spokes. Due to the prevailing conditions of soft ground, hard

**Figure 2. Accessible cutting wheel in free air for tool change**

rock and mixed face conditions it is equipped with 35 cutters and 192 cutting knives and 12 buckets. Stones, rocks and boulders can be crushed by means of hydraulic jaw crusher handling grain sizes of 1,200mm. The cutters are 19" 2-ring monoblock disc cutters with cutter rings and hubs consisting of a single steel body. This design is advantageous in mixed face conditions and blocky tunnel faces and in case bigger stones get stacked between the cutter rings no damage will appear due to pushed apart rings. Additionally the 19" cutter bearings are prepared to resist high impact loads reducing the amount of blocked disc cutters considerably.

The full set of disc cutters (number of 35) and scrapers (number of 48) can be changed in free air; the remaining 144 scrapers and 12 buckets are changeable according to the prevailing pressures under hyperbaric conditions or under saturation. For these interventions (access into the excavation chamber for maintenance and tool change during a working pressure over approximately 4.5bars) the involvement of divers is required with the need of a shuttle transport from surface to the machine. Interventions under saturation with the involvement of shuttles and habitats on surface were the first time used for the Westerschelde project in the Netherlands in 1999 where two Mixshields with diameters of 11.34m had been applied with prevailing support pressures of 7.5bars. See Figure 2.

Based on the information that the basic Trakya formation is composed of mixed abrasive siltstone and mudstone with Cerchar Abrasivity Index (CAI) values in the range of 1 to 3 and sandstone with CAI values of 2 to 3.5, thus classified as abrasive to very abrasive, special attention was taken on the detection of possible wear on the cutting tools and cutting wheel structure. Wear detection pins are integrated in each disc cutter to detect possible wear at the hub and to be informed about the conditions of the tools and to be ready for necessary maintenance accesses in a targeted manner. Installed face and rear face wear detectors as well as rim wear detectors will give early information about possible wear at the steel construction of the cutting wheel. Each disc cutter is equipped with a Disc Cutter Rotation Monitoring (DCRM) system developed by Herrenknecht with signal transfer into the control cabin. The DCRM system provides data about the rotational movement and the temperature of the disc cutters in real time. For wear detection of the cutting wheel structure eight radial hydraulically lines are integrated in the structure of which six radial sensors are installed in the front plate and two radial sensors at the backside. To protect the cutting wheel structure from possible wear the cutting wheel is designed with two rows of grillbars, hardox plates and in the gauge area with composite wear plates.

The excavation of the tunnel by means of a Mixshield that is designed for maximum face pressures of 12bars still bears uncertainties of what will be faced during tunnelling in complex geological and hydrogeological environment of the Strait Road Crossing. Therefore the TBM is equipped with probe hole drilling and injection systems to be able to execute forward ground investigations or ground treatments from within the machine. This foresees 17 inclined injection lines for crown injections with angles of 11° and nominal diameters of 100mm and 3 horizontal drill pipes in the face area with nominal diameters of 100mm. The installation of drilling equipment for rotary percussive rock drills can be installed on the erector (1 temporary drill rig) and on

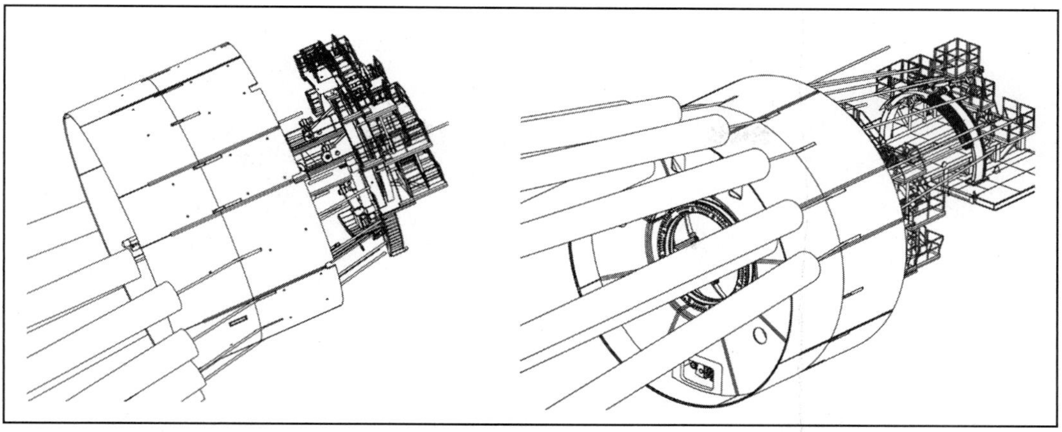

**Figure 3. Permanent drill rig on the back up bridge and temporary drill rig on the erector**

the back up bridge (1 permanent drill rig). Systematic probe drilling is planned in the mixed face conditions and when approaching dike intercepts. See Figure 3.

The Mixshield has an excavation diameter of 13.71m and an installed cutting wheel power of 4,900kilowatts. It is equipped with a 6m-diameter electric drive and an installed nominal torque of 23,290kNm. The cutting wheel is able to rotate with maximum 3.2 revolutions per minute.

In total 51 thrust cylinders arranged in 17 triple units allow a maximum thrust force of 247,300kN. They thrust forward on the installed segmental lining ring. The tunnel lining consists of 2m-long reinforced precast concrete segments comprising six standard segments, two counter-key segments and one key-segment. Each segment is fitted with a double compressible gasket. The annular gap will be backfilled using a two component grout that is injected through 8 grout lines arranged at the shield periphery. Three rows of brushes and one inflatable emergency sealing plus 1 row of tailskin sealing spring plates seal the shield from possible earth and water inflows.

The Mixshield was manufactured in Germany and was named Yildrim Bayezid. This is the name of a sultan who drove the expansion of the Ottoman Empire successfully forward at the end of the 14th Century.

The machine was transported on site. The assembly of the TBM on site was focused on limited availability of surface space. The launch box was built on the Asian side and was designed with nominal dimensions of 15–40m in width, 168m in length and 40m in depths. TBM assembly took place in the trapezoidal launching box. Start of tunnelling work in Istanbul was scheduled for spring 2014.

After completion of tunnelling works and following opening of the tunnel, the new link between Europe and Asia will initially be operated for 26 years by the Joint Venture "Avrasya Tüneli İşletme İnşaat ve Yatırım A.Ş. (ATAŞ)" and will be subsequently handed over to the government of Istanbul.

## OVERVIEW OF PROJECT AND TBM DESIGN FOR THE LARGEST TBM WITH A DIAMETER OF 17.6 METERS

In Hong Kong the largest TBM with a diameter of 17.6m will be used for the excavation of an about 480m long section. The tunnelling section is part of a 5km-long subsea tunnel link that will be excavated by means of mechanized tunnelling technology. Tunnelling works will be executed by Dragages-Bouygues Joint Venture.

The project in focus in this paper is the subsea tunnel of the Tuen Mun-Chek Lap Kok Link (TM-CLKL). It comprises the excavation of two parallel subsea tunnels. The tunnel profile of the subsea tunnels will take two traffic lanes for each TBM tunnel tube. The design carriageway width in each tube is planned to be 7.3m wide for 2 lanes with 0.5m wide marginal strip at both sides of the carriageway.

The sub-sea tunnel is located north of Lantau Island across the Urmston Road, a busy navigation channel. This north-south underground link between Pillar Point in Tuen Mun (north) and Hong Kong—Zhuhai—Macao Bridge Hong Kong Boundary Crossing Facilities near the Hong Kong International Airport at Chek Lap Kok (south) will be excavated using apart from two 13.6m-diameter Mixshields also a 17.6m-diameter TBM. See Figure 4.

The sub-sea tunnel can be generally divided into three major sections:

a. Northern Landfall where reclamation of approximately 16.5 hectares and associated seawall need to be built with the need of the construction of TBM launching shafts, cut &

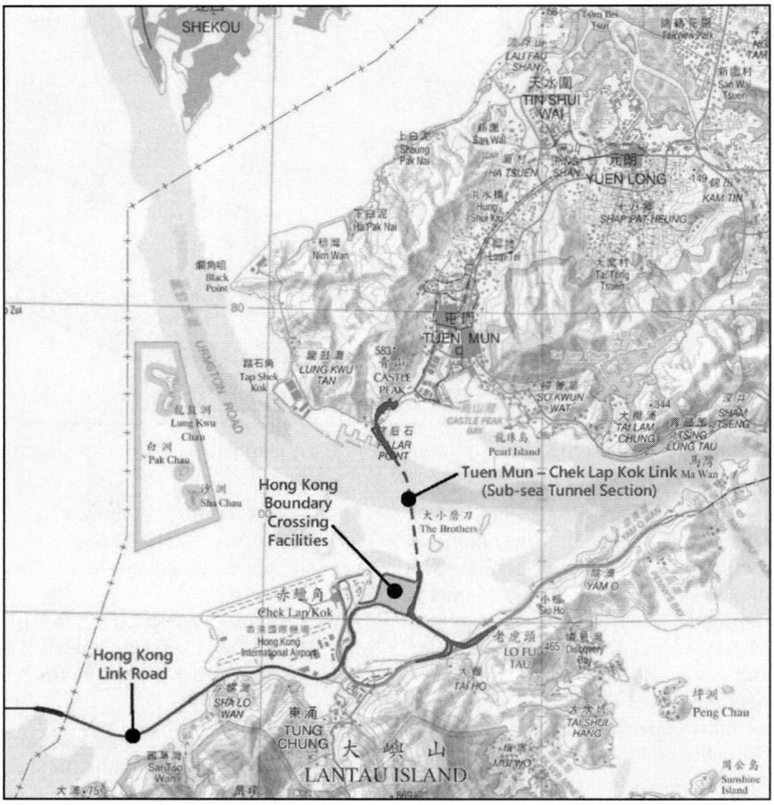


**Figure 4. Overview of the Tuen Mun-Chek Lap Kok Link (TM-CLKL) sub-sea tunnels**

cover boxes and U-shape ramps connecting the tunnel to the adjoining road network

b. TBM Bored Tunnels comprising two parallel sections of about 4.2km with shield diameters of 17.6m and 13.6m. The tunnels will be connected by cross passages at every 100m interval
c. Southern Landfall comprising TBM arrival shafts, cut & cover boxes and U-shape ramps connecting the tunnel to the adjoining road network

The project demands for the twin bored subsea tunnels of the TM-CLKL are tunnelling drives of 4.2km, high hydrostatic pressures exceeding 5 bars associated with highly unstable ground conditions. The subsurface conditions are mainly characterized by alluvium comprising mainly sand with alterations of clay and silt, completely to highly decomposed granite, slightly decomposed to fresh granite and marine deposits comprising sand and clay. Mixed face conditions will be encountered on about 50% of the tunnelling drives and about 50% full face in alluvium. The rock strength of the granite was estimated to be in the range of 70 to 170MPa. Further demands on the TBM drives are sections with an overburden of only 1D and apart from high support pressures also expected high wear in the granite, clogging potential in the clayey soils as well as pockets of biogenic gas within the marine and/or alluvial deposits.

For the bored tunnelling section three TBMs will be used. Based on the geological and hydrogeological conditions Mixshields were specified. One Mixshield is designed with a diameter of 17.6m and two Mixshields with diameters of 13.6m. Due to the fact that the machines are currently still in the design stage and the final design of the Mixshields is not finished yet, this article will focus only on the demand on the TBM designs for the TM-CLKL project.

Figure 5 gives an overview on the planned application of the three large diameter TBMs.

The 17.6m-diameter Mixshield will start excavation for the Northbound ramp tunnel from within the launching shaft for an about 480m long section towards the next shaft.

Due to the demands from geology and specific project conditions, the TBMs are specified and will be designed to minimize routine maintenance providing maximum reliability and good access to all components. In respect of cutterhead intervention for

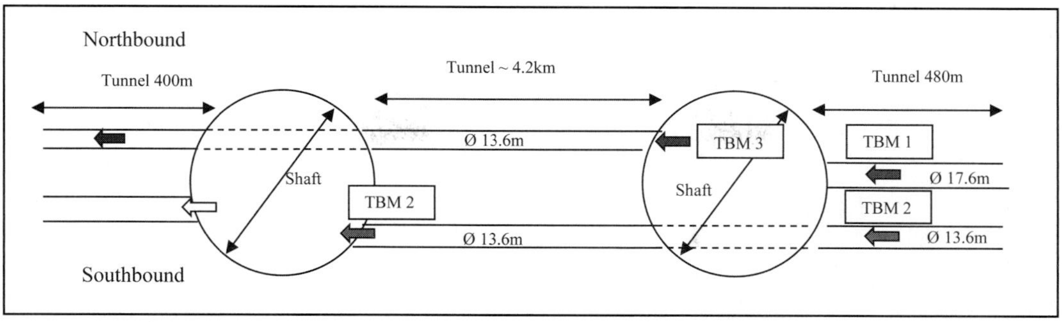


**Figure 5. Planned application of tunnelling sections for large diameter TBMs for TM-CLKL**

regular inspection and maintenance of cutting tools and TBM cutterhead redundant systems for inspection and replacement are required especially when focusing on hyperbaric intervention. In order to be prepared to inspect the cutterhead in the worst case conditions with high water pressure and unstable face, the machine is equipped with all necessary basic installation for chamber access in saturation mode. Besides piping and connections required for saturation access a permanent pre-chamber will be installed in the shield to which a transport shuttle can be connected. This transport shuttle is available on site. All means of transport and passage of the shuttle through the backup to the pre-chamber are foreseen in the design.

## CONCLUSION

Numerous very large diameter tunnelling projects exceeding TBM diameters of 14 meters were completed successfully in the past and generally with acceptable and also outstanding progress in challenging conditions such as for example the completion of the Galleria Sparvo road tunnel project in Italy with a 15.6m-diameter EPB Shield in gaseous conditions. The very large tunnel bores are accepted and trusted by the public which is shown with the increasing number of designed large to very large diameter infrastructure tunnels.

Mechanized tunnelling technology is well advanced initiated by requirements of numerous large scale projects worldwide. Experiences from past tunnelling projects are permanently implemented in the TBM design of upcoming projects with features such as e.g., advanced wear detection systems for cutting tools and structures, accessibility of cutting wheels in free air, advanced ground improvement scenarios for pressurized tunnelling drives from within the machine and with focus on river crossing projects such as stated with the two examples highlighted in this article of the crossing of the Bosporus Strait or the TM-CLKL subsea tunnels, also the ability to handle high water pressures. These are just a few technical features in TBM design which show the state of the art TBM technology and which open the mind for future large to very large diameter tunnel design.

# TRACK 1: TECHNOLOGY

## Session 4: Innovative "Toolbox"

Christopher Fleming, Chair

# Impact of the 2010–2011 La Niña Weather Phenomenon on Terrain Stability and Utilization of Long Tunnels for Mitigation Along the Ruta del Sol Project Alignment in Colombia, South America

**Vojtech Gall, Axel G. Nitschke, Andres Giraldo, and Richard Blessing IV**
Gall Zeidler Consultants

**ABSTRACT:** The Ruta del Sol Road Project is located in the seismically active Eastern Cordillera of Colombia, South America and includes approximately 3.2 kilometers of tunnels between the towns of Villeta and Guaduas through an area extremely prone to landslides. Between 2010 and 2011, extended periods of rainfall associated with the La Niña weather phenomenon affected the region and extensive slope failures occurred along the Project alignment. Gall Zeidler Consultants analyzed the correlation between La Niña on the geotechnical conditions and the possible need for a redesign of the alignment, for which additional long tunnels may present the most preferable alternative.

## INTRODUCTION

The Ruta del Sol Road Project is a massive infrastructure project that, once completed, will offer a safe, reliable transportation route between Colombia's capital city of Bogotá and the Caribbean Coast via a 1,000-kilometer long, double-lane highway. This project also includes the construction, rehabilitation, and expansion of several bridges, viaducts, tunnels, high cuts and embankments. The Project is split into three Sectors, which are each further divided into smaller subsections, called Tramos. Sector 1 is 78 kilometers long and, although it is the shortest of the three Sectors, it runs through some of the most technically challenging terrain of the entire Project. Sector 1 mostly constitutes the complete construction of an entirely new road, while Sectors 2 and 3 focus on the rehabilitation and expansion of existing roads. Sector 1 is further subdivided into three Tramos; Tramo 1 starts at K0+000 and ends at K21+600, crossing through the Eastern Cordillera mountain range and connecting the municipalities of Villeta and Guaduas in Cundinamarca Department. An area 500 meters wide on either side of the alignment (1 kilometer total) was set aside as an alignment corridor in which the actual alignment may move on an as-needed basis. Tramos 2 and 3, which total approximately 56 kilometers long, are currently under construction, while construction on Tramo 1 has yet to begin. The contractual partners of the Ruta del Sol Road Project are the owner, Agencia Nacional de Infraestructura (ANI), previously known as Instituto Nacional de Concesiones (INCO), and the contractor, Consorcio Vial Helios (HELIOS). ANI awarded Sector 1 of Ruta del Sol in December 2009.

### Scope

During the last six months of 2010 and the first four months of 2011, the country of Colombia was impacted by the seasonal weather anomaly known as La Niña, which refers to the periodic cooling of ocean surfaces temperatures in the central and east-central equatorial Pacific Basin; on the Multivariate El Niño Southern Oscillation (ENSO) Index, or MEI, La Niña events correspond to a negative value and are generally characterized by wetter-than-average conditions in Colombia. According to the MEI, approximately nine La Niña events have occurred since 1950, which corresponds to approximately one La Niña event occurring every six to seven years.

The 2010–2011 La Niña event ranks as one of the most pronounced since 1950. Landslides, mass movements and flooding were reported throughout Colombia, including along the Project alignment, to such a degree that the government of Colombia declared a State of National Catastrophe. In the Project area, the historic San Francisco Landslide was reactivated, while smaller landslides and slope failures were also reported throughout Cundinamarca Department. Flooding in the many tributary streams of the Río San Francisco led to the destruction of at least one bridge (which was later rebuilt), as well as slope failures and excessive erosion along streambeds. These events resulted in a technical argument between ANI and HELIOS over the influence of La Niña events on the design, construction, operational reliability, sustainability, and safety of Sector 1, Tramo 1 of the Ruta del Sol Project.

Gall Zeidler Consultants (GZ) was retained by ANI and HELIOS as an independent, objective arbitrator to review the current design of the alignment of Tramo 1 in relation to the geomorphological conditions before and after the 2010–2011 La Niña event in order to determine the consequences, if any, of this event on the long-term performance of Tramo 1. ANI and HELIOS jointly provided GZ with six specific questions that were to be answered in order to settle the dispute between the two entities. Specifically, ANI and HELIOS wanted to know whether the 2010–2011 La Niña event affected the geomorphological characteristics throughout the Project area to a degree that warranted the Design Sector 2 alignment to be changed either within or, if deemed necessary, outside of the designate 1-kilometer wide alignment corridor. Additionally, the two parties were interested in knowing if the vulnerability and risks of the Project had increased, if the stability and resistance of planned slopes higher than 70 meters along the Project alignment had been affected, if the constructability of the alignment near the San Francisco Landslide was feasible and sustainable over the design life of the Project (20 years), and if a specific date that these geomorphological affectations could be determined with scientific accuracy. ANI and HELIOS asked GZ to provide technical reasons for any and all conclusions made during the arbitration process.

Based on years of global experience on projects with similar geomorphological and geotechnical characteristics, GZ provided expertise as an independent engineering consultant in order to successfully settle the dispute between ANI and HELIOS. GZ assigned a team of highly qualified experts in the fields of geology, meteorology, hydrology, and transportation, as well as civil, structural and geotechnical engineering, to review the Design Sector 2 design of Sector 1, Tramo 1, including all available geological, geomorphological and geotechnical information provided by ANI and HELIOS, in order to successfully answer the proposed six questions in an unambiguous manner in order to avoid possible misinterpretations.

GZ determined that a thorough understanding of the Project's geotechnical baseline, which was used as a basis for the design of the structural elements of the Project, was essential in determining the effects, if any, that the 2010–2011 La Niña event may have had on the geotechnical and geomorphological parameters of the Project area. This investigation would be carried out using two different perspectives: retrospectively, by forensically evaluating the impact on the Project area and thus establishing a "Past Baseline," and prospectively, by providing recommendations on how to address future weather extremes for the engineered lifetime of the Ruta del Sol Road Project and thus establishing a "Present Baseline." Using this approach, GZ would be able to accurately assign a risk-based characterization of individual structural elements along the alignment of Tramo 1 in regards to both the current conditions of the Project area as well as any future conditions that might be brought about by future La Niña events as they pertained to the constructability and sustainability of the Project (Gall Zeidler Consultants, 2013).

**July 2013 Site Visit**

In order to gain a first-hand understanding of the physical properties of the Project area and to establish the Present Baseline, several members of the GZ Project Team, along with representatives from both ANI and HELIOS, visited the project site in early July 2013. For two days, the group visited multiple locations along the alignment of Sector 1, Tramo 1 between K0+000 and K21+600, ending the visit near the municipality of Guaduas. The trip concluded with a helicopter flyover of the entire alignment of Sector 1, with special emphasis on Tramo 1, before heading back to Bogotá.

Some key observations that were made during the Site Visit included the common occurrence of several geomorphological features throughout the Project area, including triangular facets and fault scarps, which indicate geologic creep and the active erosion of the mountainsides. Upon failure, these features can create landslides, slope failures and other mass movements; these mass movement events have been affecting National Road 50, currently the only transportation artery connecting Villeta and Guaduas, for years. The underlying geology of the Project area is dominated by marine sedimentary rocks, including shales, mudstones and siltstones, with interlayered sandstones and conglomerates of terrestrial origin. In particular, dark grey to black marine sediments located throughout the Project area were especially fissile, and were easily broken into several pieces in hand sample. A combination of tectonic activity, hydrology, climate (during both La Niña and non-La Niña years) and land use are also contributing to the weathering, erosion, and eventual slope failure of the mountainsides throughout the Project area. Finally, large historic landslides, including El Cune Landslide in Villeta (K0+000, circa 2004) and the San Francisco Landslide (K17+00 to K18+000) that was reactivated after the 2010–2011 La Niña event, indicate that geomorphological processes throughout the Project area are constantly evolving and changing. Thus, it can be expected that landslides and slope failures will continue to dominate the landscape throughout the area of Sector 1, Tramo 1 during both La Niña and non-La Niña years.

Following the site visit the team interviewed members of various governmental agencies in order to gather as much information on the Project area as possible. After the 2010–2011 La Niña event, several government agencies began to dedicate more resources to studying and investigating the impacts that this event had throughout Colombia, including the Project area, and thus the amount and accuracy of available information has increased. This information became critical in order to establish both the Past and Present baselines, as well as how the two differ from one another as a result of the 2010–2011 La Niña event. The government agencies that were interviewed during the July 2013 Site Visit include: Unidad Nacional de Gestión del Riesgo y Desastres (UNGRD), Colombia Humanitaria, Instituto de Hidrología, Meteorología y Estudios Ambientales (IDEAM), Instituto Geográfico Agustín Codazzi (IGAC), Sociedad Colombiana de Geología y Mineria (SCG), Agencia Nacional de Licencias Ambientales (ANLA), the Ministry of Transportation, and Instituto Nacional de Vías (INVIAS) (Gall Zeidler Consultants, 2013).

## PROJECT AREA

### Regional Tectonic Setting

The alignment of the Ruta del Sol Road Project lies within the Eastern Cordillera, or Cordillera Oriental, which is an intracontinental orogenic belt in the northern Andes Mountains of central Colombia. Stretching for 750 kilometers from Ecuador to Venezuela, the Eastern Cordillera is the longest and widest of the three branches of the Andes (Hudson 2010) and is composed primarily of deformed Mesozoic and Cenozoic sedimentary rocks overlying a polymetamorphic basement (Taboada et al. 2000).

The geodynamics of Colombia are highly complex and dominated by the convergence of Nazca, Caribbean and South American tectonic plates, as well as the Chocó block (also referred to as the Baudó-Panama Arc or the Panama-Chocó block) (Pulido 2003). Off the Pacific coast of Colombia, the Nazca plate is subducting beneath the Paleo-Caribbean plate (the remnants of the collision between the Chocó block and the South American plate) and, subsequently, the South American plate at a rate of approximately 60 millimeters per year. This subduction is responsible for the deep seismic activity along the Colombian Trench, as well as the active volcanism seen today in the Western and Central cordilleras. Conversely, the shallow, non-magmatic southeastward subduction of the Caribbean plate beneath the South American plate at a rate of approximately 20 millimeters per year is related to the shallow but moderate to high levels of seismicity seen in the Eastern Cordillera. The subduction of the Caribbean plate, combined with the southwestward movement of the South American plate against the North Andean block (NAB)—bounded by the Santa Marta-Bucaramanga Fault (SMB) in the north and the Ibagué Fault (IF) in the south—results in the thrusting of the Eastern Cordillera over the more stable Precambrian South American craton, also known as the Llanos shield (Paris et al. 2000). This dynamic, multi-plate interaction is illustrated below in Figure 1.

This complex multi-plate interaction scheme has produced extensive folding and faulting throughout the Eastern Cordillera, and, thus, the Project area. Major fault systems generally trend north-south throughout the region, such as the Salinas Fault System (SFS) and the Eastern Cordillera Frontal Fault System (FFS) that bound the western and eastern margins of the cordillera, respectively. In the Project area, the Sistema de Fallas de Bituima, Falla de Bituima, and Falla de Alto del Trigo are all a part of the larger SFS. Additionally, compressive forces related to the movement of the South American plate in relation to the NAB have created oblique and strike-slip components to the regional dynamics, resulting in a secondary set of east-west to northwest-southeast trending faults throughout the Project area (Pulido 2003). Several smaller east-west and northwest-southeast striking faults and suspected faults, or lineaments, affect the majority of the project alignment by running parallel or obliquely to it, or traversing it. Named faults and lineaments include: Falla la Masata, Falla Honda, Falla San Isidro, Falla Tibayes, Falla Don Joaco, and Falla Columpio. The two tunnels located along the alignment of Sector 1, Tramo 1, El Trigo Tunnel (2,248 meters) and La Cumbre Tunnel (978 meters) run nearly parallel to Falla Tibayes for their entire length. Additionally, several smaller unnamed faults, suspected faults, and lineaments come into either direct or indirect contact with the alignment multiple times along its entire length. Table 1 summarizes major tectonic structures, including major faults and fault systems, throughout the Project area of Sector 1, Tramo 1. The structures listed are in order from east (K0+000) to west (K21+600).

Seismicity in the mountainous areas of the Project area is typically shallow, where intracontinental deformation tends to occur along the reactivated north-northeast trending faults. Some of these faults have yet to be identified or fully studied. The Bucaramanga Seismic Nest is an area of concentrated seismic activity approximately 275 kilometers northeast of the Project area, where the activity is clustered in a small volume of approximately 2,800 cubic kilometers at a depth of 160 kilometers (Pulido 2003). The potential of earthquakes is a major design consideration for Sector 1, Tramo 1 of the Ruta del Sol Project (Gall Zeidler Consultants, 2013).

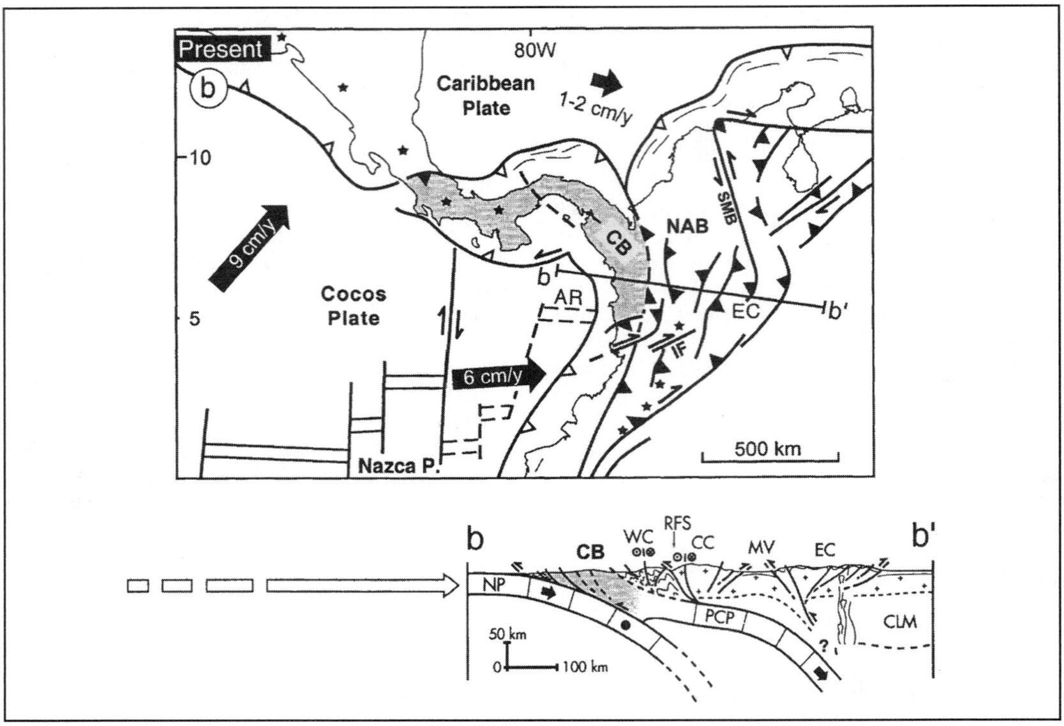


**Figure 1. The Geodynamics of Northwestern South America depicting the Project Area of the Eastern Cordillera, EC (Pulido, 2003)**

### Geology

The dominant materials along the Project alignment are sedimentary rocks with bedding planes generally striking north-south or north-northeast and dipping to the west; the only exception is the sedimentary rocks of the Bituima Syncline (Guaguaquí and Olini groups), which dip to the east. The rock units east of the Falla de Bituima are believed to be primarily marine in origin, while the units to the west of this fault are thought be terrestrial. The sequence of lithologies, beginning with the oldest, is detailed in Table 2. Figure 2 depicts the alignment of Sector 1, Tramo 1 in relation to the geologic units of the Project area (Gall Zeidler Consultants, 2013).

### Geotechnics

Slope failure along Sector 1, Tramo 1 is of great concern to the safety, constructability and reliability of the alignment. The majority of geologic units found within the Project area contain large amounts of fine-grained sedimentary rocks, which are generally extremely susceptible to weathering and erosion in their undisturbed state. However most, if not all, of the lithologic units in the Project area have been deformed by active faulting throughout the region, further increasing their permeability and thus, their susceptibility to weathering and erosion. Additionally, though generally strong in their intact, undisturbed state, the coarser-grained terrestrial conglomerates and sandstones in the western area of the alignment have also undergone extensive deformation throughout geologic time; permeability has been increased in these units as well. The dip angle and orientation of bedding planes will also impact the risk of slope failure along the alignment, particularly where the dip angle of the geologic units is downhill and in areas through which high cuts (up to 80 meters) are to be excavated. In areas that are overlain by several to tens of meters of alluvial and colluvial materials (refer to Table 2), including the San Francisco Landslide (K17+500 to K18+500), slope failure and mass movements are again of great concern as these materials become water saturated, enabling them to slide easily over the beds of more competent materials (sandstones and conglomerates) that they overlie (Gall Zeidler Consultants, 2013).

### Hydrology

As the alignment of Sector 1, Tramo 1 winds through the steep, mountainous terrain, a total of 81 small catchments will be crossed. These crossings are designed for 1-in-100 year storms with durations

**Table 1. Major structural and geomorphological features along the project alignment of Sector 1, Tramo 1**

| Name | Structure | Station | Trend | Dip | Geomorphological Features | Geologic Units |
|---|---|---|---|---|---|---|
| El Cune Landslide | Landslide | K0+000 | North-south | — | Scarps, transverse ridges and cracks | Trincheras and Útica formations overlain by several tens of meters of colluvium |
| Sistema de Fallas de Quebrada Negra | Thrust fault system | K0+000 to K5+500 | North-south | — | — | Útica Formation |
| Villeta Anticlinorium | Anticlinorium | K0+000 to K11+700 | North-south | — | — | Útica, Trincheras, Socotá, Capotes, Hiló formations |
| Sistema de Fallas de Bituima | Thrust fault system; part of the Salinas Fault System | K10+500 to K11+700 | North-south | — | — | Socotá, Capotes, Hiló formations |
| Falla de Bituima | Thrust fault; part of the Salinas Fault System | K10+700 to K11+700 | North-south | East; steep | Ridges, fault-controlled streams; movement of less than 0.2 millimeters per year, recurrence interval between 3,000 and 30,000 years | Bounded by the Capotes and Hiló formations to the east and the Guaguaquí Formation to the west |
| Bituima Syncline | Syncline | K11+700 to K15+000 | North-south | — | — | Guaguaquí, Olini groups |
| Falla de Alto del Trigo | Thrust fault; part of the Salinas Fault System | K14+700 to K15+000 | North-south | East; moderate to steep | Well-defined fault line with old scarps, saddles, triangular facets, aligned drainages, ponded (confined) alluvial deposits; movement of less than 0.2 millimeters per year, recurrence between 10,000 and 400,000 years | Bounded by the Guaguaquí Group to the east and the Seca Formation to the west |
| Falla la Magdalena | Strike-slip fault | Approximately K18+200 | East-west | — | — | Cuts perpendicularly through all formations |
| San Francisco Landslide | Landslide | K17+500 to K18+500 | East-west | — | Scarps, transverse ridges and cracks | Seca, Hoyón, San Juan de Río Seco formations overlain by several tens of meters of colluvium |
| Guaduas Syncline | Syncline | K15+900 to K21+600 | North-south | — | — | Seca, Hoyón, San Juan de Río Seco formations |

**Table 2. The geologic units of Sector 1, Tramo 1 of Ruta del Sol**

| Geologic Unit | Symbol | Description | Station |
|---|---|---|---|
| Útica Formation | Kiuh | Arkosic sandstones, fine-grained to conglomeritic sublitharenites, interbedded mudstones and siltstones; black biomicrites | K5+500 to K7+600 |
| Trincheras Formation | Kitr | Black siliceous and calcareous mudstones with sporadic intercalations of limestones | K0+000 to K5+500; K7+000 to K9+700 |
| Socotá Formation | Kis | Black siliceous and calcareous mudstones with sporadic intercalations of limestones; interbedded graded layers of fine-grained to conglomeritic quartz sandstones with calcareous cement | K9+700 to K10+400 |
| Capotes Formation | Kic | Black siltstone and mudstones with concretions | K10+400 to K10+700; K11+500 to K11+700 |
| Hiló Formation | Kih | Calcareous and siliceous siltstones and mudstones with layers of sandy limestones; limestones and fine-grained calcareous sandstones with mudstones | K10+700 to K11+500 |
| Guaguaquí Group | Kgg | Black, calcareous mudstones with calcareous concretions and intercalations of siliceous siltstone, chert and limestone; occasional coal | K11+700 to K12+400; K14+500 to K14+900 |
| Olini Group (Kso) | Ksli | Thin, undulating layers of chert and siliceous silt | K12+400 to K14+500 |
| | Ksl | Siliceous siltstones, interbedded fine-grained quartz arenites | |
| | Ksls | Fine- to medium-grained sandstones with interbeds of siliceous siltstone | |
| Seca Formation | KPgs | Mudstones and claystones with caliche and interbedded fine- to medium-grained quartz arenite | K14+900 to K15+500 |
| Hoyón Formation (Pgh) | Pgh-1 | Thick to very thick layers of polymictic conglomerates and quartz sandstones | K15+500 to K18+300 |
| | Pgh-2 | Claystones with sporadic layers of polymictic pebble conglomerates | — |
| | Pgh-3 | Polymictic conglomerates with interbedded quartz sandstones and mudstones | — |
| San Juan de Río Seco Formation (Pgs) | Pgs-1 | Conglomeritic sandstones with agglomerated nodules | K18+300 to K19+200 |
| | Pgs-2 | Alternating layers of claystones and mudstones with varying thicknesses, medium to thick interbeds of fine- to coarse-grained sublitharenites | K19+200 to K21+600 |
| | Pgs-3 | Thin layers of medium- to coarse-grained quartz sandstones and sublitharenites | — |
| | Pgs-4 | Claystones with thin, interbedded layers of medium- to coarse-grained quartz arenites and sublitharenites | — |
| Colluvium | Qc | Weathered sediments of bedrock units, outlining tributary stream and river beds; colluvial materials attributed to weaknesses in fine-grained sedimentary rocks of the Útica, Trincheras, Socotá, Capotes, and Hiló formations | K2+200 to K10+500; K13+000 to K14+800 |
| Alluvium | Qal | | K2+200 to K10+500 |

of 10 minutes to match the catchment sizes. For the ease of maintenance and modular construction, standard box culverts with standardized cross sections have been proposed. The alignment also crosses a few larger catchments, which include the Quebrada El Cune (K2+054, K9+121), the Quebrada Honda (K11+800), the Río San Francisco (K14+900, K18+308) and the Río Honda (K21+161), which will all be crossed with viaducts so that the flood capacity of the rivers are largely irrelevant. However, bridge foundations and piers will need to take flood levels into account. River flow records dating back 30 years come from Río Villeta at Tobia and using the data collected at this location, it was observed that the longer-duration (weeks to months) flood events between 2010 and 2011 associated with La Niña appear to be more significant than shorter events, due to multiple days of intense rainfall.

The permeability (saturated hydraulic conductivity) of the geologic units in the Project are span a wide range of values, but indicate permeability in soils and sandy rocks between $5\times10^{-5}$ and $1\times10^{-3}$ meters per second, and very high values for the alluvial and colluvial materials ranging between $1\times10^{-3}$

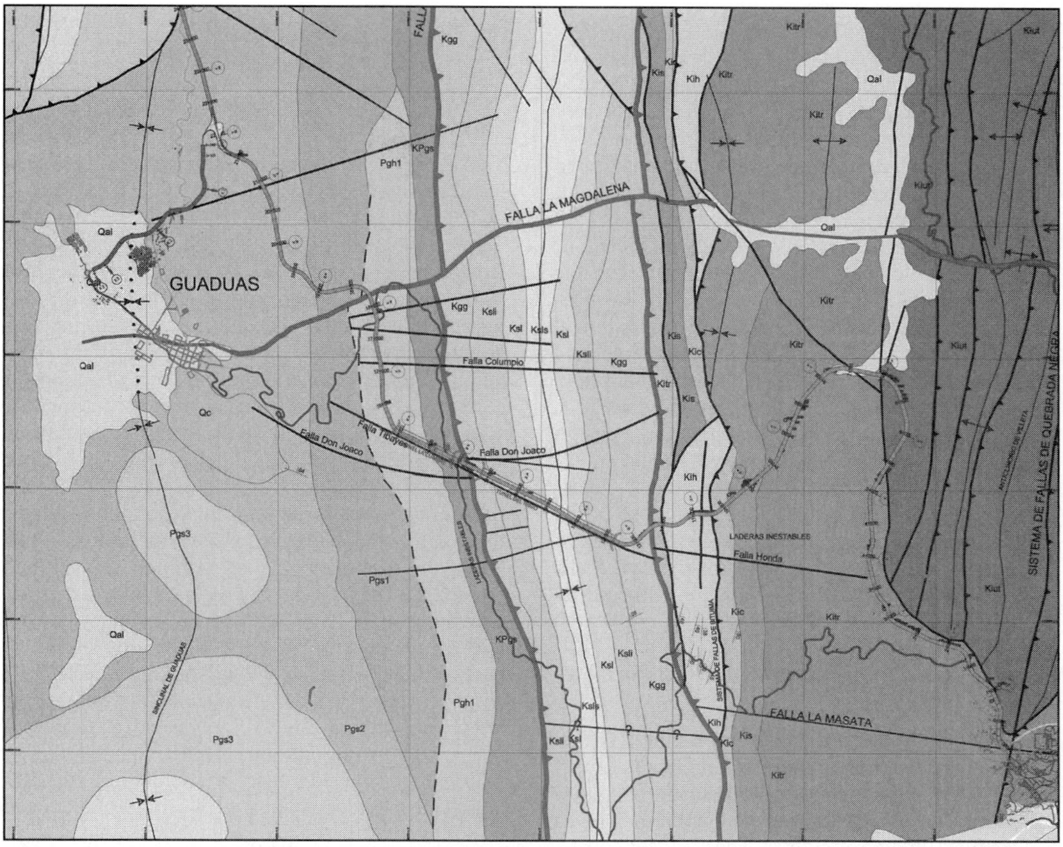


**Figure 2. Geologic units of the Project Area and the alignment of Sector 1, Tramo 1 in green (Gall Zeidler Consultants, 2013)**

and $3 \times 10^{-2}$ meters per second. No pumping tests have been carried out. Groundwater measurements taken during borehole drilling indicate perched water conditions throughout the Project area, with water within a few meters of the ground surface. There is no other monitoring of groundwater levels (Gall Zeidler Consultants, 2013).

## Land Use

Land use in the Project area is primarily agricultural, including 940 hectares of pastures and 865 hectares of crops. 520 hectares of the Project area include forested land. The watershed of the Río San Francisco, as well as the ecosystem of the northern Andes, is protected in the 2,872-hectare San Francisco National Natural Forest Reservation, or NNFR (K13+600 to K19+500). Although not an official reservation as designated by the Colombian government, the area surrounding K12+600 to K13+100—known as the La Esmeralda Forest Reservation—is considered to be another environmentally important area by the local people, who work to conserve La Esmeralda's natural resources. All activities related to the construction of Tramo 1 should be heavily monitored so as not to affect these environmentally important areas (Gall Zeidler Consultants, 2013).

## Climate

The climate of the Project area is classified as mostly "temperate semi-humid" by IDEAM, with a mean temperature between 18°C and 25°C. Average precipitation in the area has a mean annual value of 1,000 and 2,000 millimeters. Two pluviometric stations near the Project site, one at Útica and one at El Tuscolo, have precipitation records dating from 1981 to 2012. These two datasets indicate that precipitation in the project area has a bimodal distribution, with one peak in precipitation levels during the months of April and May, and the second peak occurring around October and November (Gall Zeidler Consultants, 2013).

## STRUCTURAL OVERVIEW OF THE PROJECT

The structural elements of Sector 1, Tramo 1 include bridges, viaducts, and tunnels, as well as earthwork (embankments, cuts and fills) and roadway construction elements. The alignment is highly affected by the mountainous terrain and has many tight, horizontal curves and maximum grades to keep the alignment mainly on the surface by means of the structural elements listed above. Critical to the investigation into the sustainability and constructability of Sector 1, Tramo 1 is the understanding and analysis of the interaction of the proposed structures along the alignment with both current and future geotechnical and geomorphological conditions. The current Design Sector 2 alignment includes a design speed of 80 kilometers per hour, a maximum allowable gradient of 7%, and a minimum radius of horizontal curves of 235 meters. For the purpose of the arbitration investigation and review, the alignment of Sector 1, Tramo 1 was subdivided into three sections. An overview of the structural elements of the alignment in each of these sections is summarized below.

### Part 1—Ascent from Villeta to the East Portal of El Trigo Tunnel

This section of the alignment begins at K0+000 and ends at the East Portal of El Trigo Tunnel (K12+382), and includes sharp horizontal curvature with deep cuts and fills, as well as numerous bridges and viaducts on both straight and curved alignments. Six of these structures are post-tensioned I-girder type bridges, six are cast-in-place, post-tensioned box beam viaducts, and one is a steel arch structure. All abutments are supported grade beams on caissons with spill-through slopes, and the piers and columns are integral with the foundational elements. Foundations are to be drilled caissons embedded into shale, sandstone, or claystone. These materials are very fissile and break easily along bedding planes, particularly when they encounter excessive water flow. Additionally, this section of the alignment lies in extremely close proximity to several faults and fault systems (refer to Table 1). There is also a high risk of landslides and mass movements in this area, both of which have previously occurred on high, steep hills where deep cuts and fills are planned along the Design Sector 2 alignment; these risks are highest in areas when the bedding of the underlying rock units dips in excess of 45 degrees towards these planned cuts and/or excavations. Anticipated cut slopes in this section of the alignment have a maximum height between 13 and 62 meters with slope gradients ranging from 0.25:1 to 2.3:1 (vertical height to horizontal width, or V:H). At least 37 landslides or mudflows have been reported in this area of the Project (Gall Zeidler Consultants, 2013).

### Part 2—El Trigo and La Cumbre Tunnels

This section of the alignment includes the area between the East Portal of El Trigo Tunnel (K12+382) and the West Portal of La Cumbre Tunnel (K16+118); the Río San Francisco Alto Bridge, which spans between the two tunnels, is also included in this section and is a single 140-meter long steel arch span with a separate 28-meter high arch for a twin superstructure. Currently, the Design Sector 2 design has one bore in each direction for each tunnel, with each bore capable of carrying two 4.15-meter wide lanes as well as a curb and sidewalk section on either side (1.25 meters and 1.15 meters, respectively). The finished tunnels will be 12 meters wide and 9 meters high, with a horseshoe shape. Steel piles and concrete walls with two or three levels of long tiebacks and reinforced slopes will protect the tunnels' portals where appropriate. The twin bores of El Trigo Tunnel will be connected with five 2.3-meter wide cross passages, while the twin bores of La Cumbre Tunnel will be connected by two cross passages of the same width. The tunnels are to be excavated in four subsequent drifts: top heading, left and right side drift, and bench/invert, and will feature sprayed concrete (shotcrete) initial linings, lattice girders spaced between 0.5 meters and 2 meters apart based on ground conditions, 6-centimeter diameter drilled and grouted arch pipes/spiling spaced at 0.5 meters where selected (total 40 pipes), and reinforced concrete final linings. These structures run through the La Esperanza Natural Reservation between K12+600 and K13+100 and also lie entirely within the San Francisco NNFR. The Falla Tibayes runs parallel to this section for its entire length, and the alignment crosses the Falla de Alto del Trigo at approximately K14+900 (refer to Table 1). This section of the alignment is bounded on both sides by the suspected fault, Falla Don Joaco. Additionally, the proximity of the structures of this part of the alignment to these various geologic faults indicates that the surrounding ground which is to support the structures may be highly disturbed and fractured, increasing their susceptibility to water infiltration and subsequent erosion in the form of landslides and mudslides. With the proximity to confirmed and suspected faults aside, this section of the alignment, and more specifically, the two tunnels, is the least susceptible to the geotechnical risks of landslides and mass movements than the other more vulnerable structures of Sector 1, Tramo 1 (Gall Zeidler Consultants, 2013).

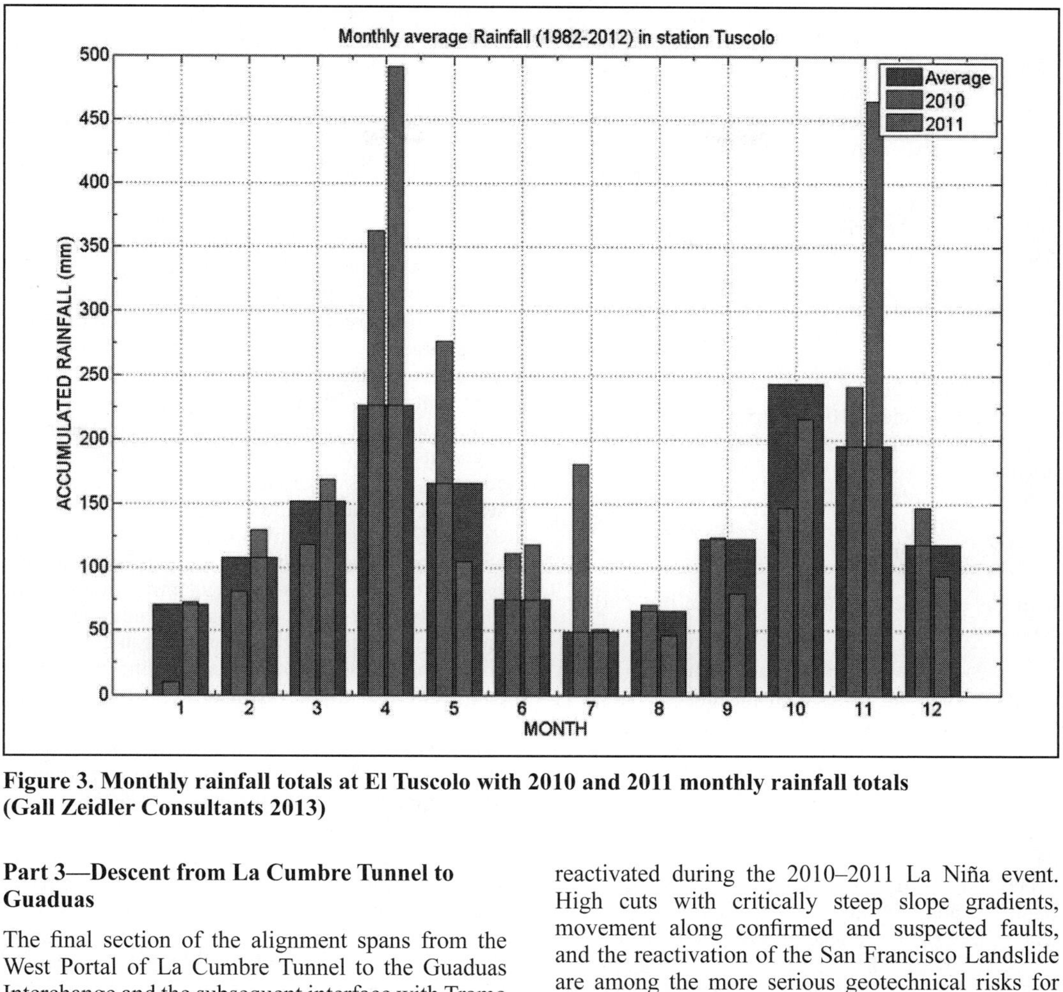


**Figure 3. Monthly rainfall totals at El Tuscolo with 2010 and 2011 monthly rainfall totals (Gall Zeidler Consultants 2013)**

### Part 3—Descent from La Cumbre Tunnel to Guaduas

The final section of the alignment spans from the West Portal of La Cumbre Tunnel to the Guaduas Interchange and the subsequent interface with Tramo 2 (K21+600). This section includes the construction of four bridges and viaducts, and has several high cuts and fills on both sides of the alignment. One of the highest cut planned along the alignment of Sector 1, Tramo 1, which is 78 meters high, is located in this section, as well as several other high cuts and fills. Some of these high cuts and fills, with critical slope gradients ranging from 2.2:1 to 2.4:1 (V:H), are located either directly in or adjacent to the San Francisco Landslide. As with the rest of the alignment, high cuts in areas where the bedding planes of the underlying geological formations are dipping out of the proposed cuts are of particular concern in this section. Additionally, this section includes the perpendicular crossing of four faults, including the Falla Columpio (K17+150) and three unnamed, suspected faults at K17+650, K17+950, and K20+950. Further, the alignment crosses the large strike-slip Falla La Magdalena at K18+200, as well as the massive San Francisco Landslide (K19+100), which was reactivated during the 2010–2011 La Niña event. High cuts with critically steep slope gradients, movement along confirmed and suspected faults, and the reactivation of the San Francisco Landslide are among the more serious geotechnical risks for this part of the alignment. At least 12 landslides or mudflows have been reported in this section of the Project area (Gall Zeidler Consultants, 2013).

## 2010–2011 LA NIÑA EVENT

According to records from the Útica and El Tuscolo pluviometric stations, periods of heavy rain began in April 2010—when the ENSO MEI was still positive—with a total of 362.2 millimeters of precipitation. This is the second largest value for precipitation levels in April since 1982; the positive ENSO MEI does not qualify as "La Niña" conditions. In May 2010, the ENSO MEI values became negative—indicating La Niña conditions—and another peak of rainfall was recorded with the largest value for May (276 millimeters). The largest value for a 24-hour period was recorded in June 2010 (51.1 millimeters). July 2010 saw the largest number of days with rain (13), as well as the highest value of recorded rain for the month of July since 1982 (180.7 millimeters).

This is significant, as July typically has the lowest precipitation values of the year in the Project area, averaging around 50 millimeters between 1981 and 2012. Precipitation values for the rest of 2010 and into 2011 were close to the historical average, though November and December were above average. Precipitation values for February, March and April 2011 were all above average, with April seeing both the highest recorded value of monthly rainfall (421 millimeters) and the most days with rain (24). November 2011 also saw the largest values of recorded precipitation for that month (464.3 millimeters) and the most days with rain (24). This data is illustrated in Figure 3. According to the ENSO MEI, the 2010–2011 La Niña event extended from May through June 2010 and from February through March 2011, and is ranked as one of the most pronounced events since 1950. In the 31 years of precipitation records at El Tuscolo, 2011 was the wettest year and 2010 was the fifth wettest year on record (Gall Zeidler Consultants, 2013).

## Impacts of the 2010–2011 La Niña in the Project Area

Due to local differences in climate and topography, different areas of the alignment, and Colombia in general, experienced effects from the 2010–2011 La Niña at different times. However, these effects were widespread. Landslides, mass movements and flooding were the main outcomes experienced throughout the country, with over 313 fatalities and over 345,000 homes destroyed or damaged; over 807,000 hectares flooded throughout the country, and 751 roadways and 66 bridges were also affected in some way (CEPAL, 2012). According to IDEAM (2011), 23 of the 39 landslides recorded in Colombia in 2011 were in Cundinamarca Department, where the Project alignment lies. The Colombian government created the agency Colombia Humanitaria in order to raise and manage funds in order to remedy the various natural disasters associated with the 2010–2011 La Niña. Colombia Humanitaria has invested about COP $5.38 billion (approximately US $2.83 billion) to deal with the effects of this historic catastrophe. Further, the government also defined clear requirements for the development of future infrastructure projects in Colombia, which the GZ Project Team carefully analyzed throughout the arbitration process while evaluating the Project's constructability and sustainability. During the July 2013 Site Visit to the Project site, the team interviewed a local family whose home sat within the San Francisco Landslide, which became reactivated after months of heavy rain in April 2011, the wettest month ever recorded in the Project area. According to these locals, their house moved a distance of approximately 250 to 300 meters over an estimated time period of between 12 and 24 hours, suggesting an average velocity on the order of 10 to 20 meters per hour (Gall Zeidler Consultants, 2013).

## Vulnerability and Risk of Sector 1, Tramo 1

Vulnerability to extreme weather events is typically split into three components: exposure, the degree to which the system is exposed to significant climatic variation; sensitivity, the degree to which the system is positively or negatively affected by climate-related stimuli, and adaptive capacity, or the ability of the system to adjust to climate change (including climate variability and extremes) to moderate potential damages, to take advantage of opportunities, or to cope with the consequences. The adaptive capacity of infrastructure projects during the planning phase is given; however, once a project has been built, the adaptive capacity becomes minimal. Similarly, the sensitivity of large infrastructure projects such as Ruta del Sol can be investigated and more robust structures can be designed to mitigate any potential risks. However, similar to the adaptive capacity, the sensitivity can only be manipulated minimally after the Project is built. For linear projects in general, the alignment selection is of crucial significance for the entire life of the infrastructure project, whereas operability has the highest priority. It is therefore of utmost importance to examine the vulnerability of the planned route, identify its sensitivity, and adapt the structures with a robust design as much as possible during the planning phase. However, it should be noted that even with the greatest care, it is not always possible to build a roadway that is sustainable enough to cope with extreme weather events, as well as landslides, mass movements, and other changes to the geomorphological and geotechnical properties of the project area.

Risk is a situation involving exposure to threats, and within the scope of this Project, risk along the alignment of Sector 1, Tramo 1 arises from natural causes, triggered by extreme rainfall events. GZ was tasked with evaluating the vulnerability of the alignment in great detail in order to assess the risk of threats, including mass movements, landslides, and flooding, from extreme weather events. In order to accomplish this task, the Project Team utilized a comprehensive Risk Register as a central repository of all risks to the Project alignment identified in order to provide information regarding the probability, impact, counter measures and risk ownership. Specifically, the probability and impact of risks triggered by extreme rainfall events were the main focus throughout the arbitration process. The Risk Register provided a qualitative assessment for risk probability and impact for the identified threats. This information was used to develop a risk profile along the length of the Project alignment. Criteria were developed to

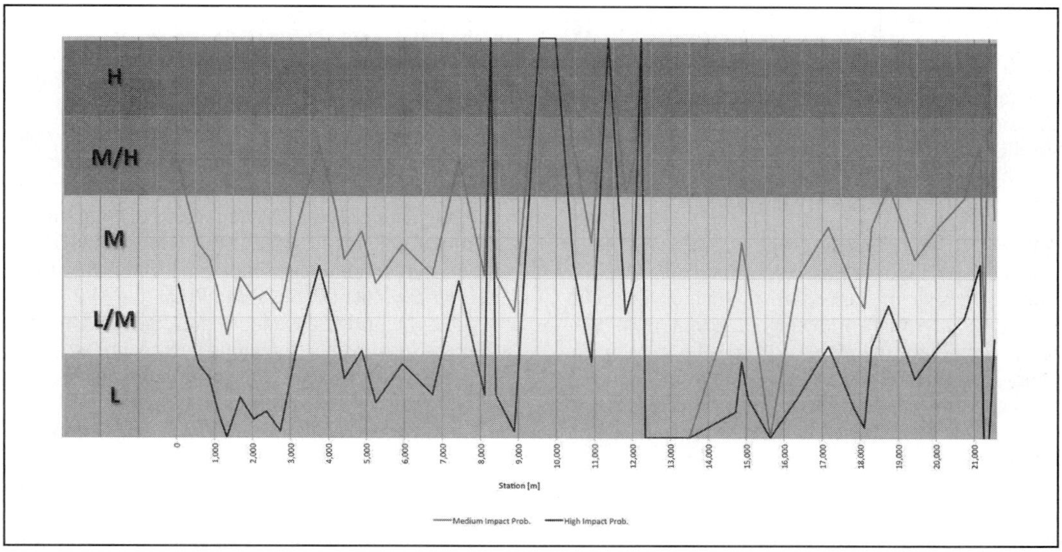


**Figure 4. Risk probability profile for the threat of mass movement along the alignment of Sector 1, Tramo 1 of the Ruta del Sol Road Project (Gall Zeidler Consultants 2013)**

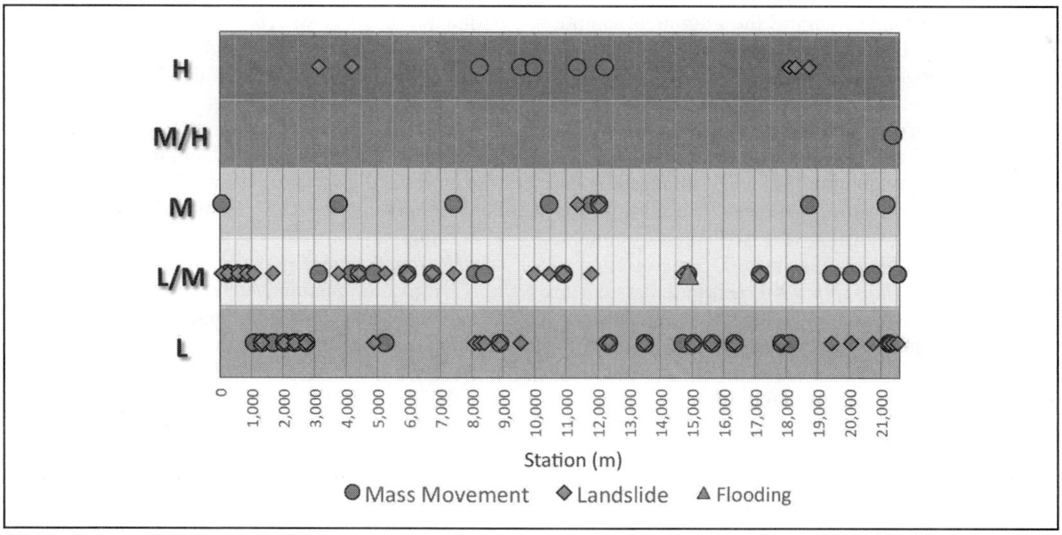


**Figure 5. Risk profile of Sector 1, Tramo 1 of the Ruta del Sol Road Project for extreme weather events (Gall Zeidler Consultants)**

identify the vulnerability of sections and structures, which was used to quantitatively assess the probability of threats materializing into events. The criteria for the vulnerability were divided into two groups: natural features of the topography and terrain along specific sections of the alignment of the Project, and design features of the structural elements of Sector 1, Tramo 1. These criteria were evaluated separately, and then merged into a joint criterion to both qualitatively assess the probability of each threat for each particular alignment section or structure. Structural elements that were deemed to possess moderate to high vulnerability were the main focus of the arbitration process, and were the baseline for the development of the risk profile along the entire alignment length. Using this approach, the Project Team developed a risk probability profile for the threat of mass movements, which can be seen in Figure 4, while the risk profile for all of Sector 1, Tramo 1 can be seen in Figure 5. Using this risk-based analytical approach, the Project Team was able to succinctly answer the questions proposed by ANI and HELIOS in order to successfully settle the dispute between the two contractual partners by providing reasons for their decisions with a strong technical background (Gall Zeidler Consultants, 2013).

### Answering the Proposed Questions

Many geomorphological changes, such as weathering, are typically slow processes that take place over longer periods of time. Other changes, including landslides and mass movements, result from long-term weathering processes but may be triggered by short-term changes in external conditions. The GZ Project Team's investigation into the impact of the 2010–2011 La Niña event on the current Design Sector 2 alignment illustrated that the excessive rains associated with this event were most certainly a trigger event for several slope failures throughout the Project area. The cyclical nature of La Niña events and the continuously-occurring geomorphological process that shape the terrain in the Project area interact in such a way that an exact date on which the alignment experienced sufficient geomorphological changes to warrant changing the alignment cannot be determined. However, the month of April 2011—the wettest month on record at the Project site—saw multiple days of excessive rainfall and several landslides were documented throughout the area during this time, offering a strong indication for an accumulation of extreme geomorphological changes in the Project area during the 2010–2011 La Niña event. As La Niña is a cyclical atmospheric phenomenon, it can be assumed that La Niña events, possibly comparable to that of 2010–2011, will continue to occur in the future. It was clear to the team that certain areas along the alignment, including those where critically high cuts in excess of 70 meters are planned, areas where the bedding planes of the underlying geologic units are be dipping towards proposed cuts, or where large, historically active landslides are present, such as the San Francisco Landslide are extremely vulnerable to future La Niña events. Additionally, the underlying geology of the project area and the alignment's proximity to several major faults and suspected faults led the team to conclude that the proposed 1-kilometer alignment corridor did not provide enough flexibility for movement in order to avoid these obstacles. Considering these factors, and taking into account guidelines defined by the Colombian government after the 2010–2011 La Niña regarding the sustainability of infrastructure projects in the country, GZ believes that the current Design Sector 2 alignment of Sector 1, Tramo 1 is not capable of providing sustainable and reliable infrastructure during future extreme weather events (Gall Zeidler Consultants, 2013).

## ADDITIONAL LONG TUNNELS AS POSSIBLE MITIGATION MEASURES

GZ's scope of work throughout this Project did not include suggesting alternative alignments for Tramo 1, Sector 1, and thus no such suggestions were made to ANI or HELIOS either throughout or after the arbitration process. However, for similar geological and geotechnical conditions, the use of long tunnels to mitigate constructability and vulnerability issues often offers attractive solutions, which is equally the case for the Ruta del Sol Road Project.

Unlike bridges, viaducts, cuts, and fills, tunnels offer the most protection against exposure along the Project alignment, as well as offer a drastic decrease in the sensitivity of specific structures to landslides and mass movements. This is seen in Section 2 of Tramo 1, where the El Trigo and La Cumbre tunnels are located. The construction of tunnels is often met with resistance, as they are generally initially more expensive than most other structures. Regarding sustainability of Sector 1, Tramo 1 of the Ruta del Sol Road Project, however, tunnels are superior to other structures under the current and future conditions along the Project alignment, as their long-term maintenance costs will be much lower than other structures; using tunnels, high maintenance costs relating to road closures, landslide removal, and landslide prevention measures can be decreased or avoided throughout the entire operational life of the Project. The decrease in partial and full road closures will increase the general availability of the roadway. For a concessionaire, this means less risk to the flow of revenue, and for the users, this means decreased average travel times.

During the planning phase of the Project, tunnels would provide a very high flexibility for the alignment selection. For example, as is with the alignment of Sector 1, Tramo 1, there is typically the need to gain elevation when crossing mountainous regions. Long sections with close-to-maximum slopes are common in such crossings and generally increase travel times, especially for the truck traffic that is the main source of revenue. These sections typically follow the natural topography of the terrain in order to gain the needed elevation, which greatly limits the flexibility for the alignment. Such alignments require the need for very high cuts and bridges, which is also seen with Sector 1, Tramo1. The use of long tunnels will greatly decrease the required elevation needed to cross the mountains, shortening the alignment and allowing for the avoidance of steep grades. In turn, these factors increase the average travel speed along the alignment, reduce travel time, and increase transportation revenue, as more commuters are able to travel the alignment. Due to these factors that the authors believe the use of tunnels is the most attractive option for Sector 1, Tramo 1 of Ruta del Sol (Gall Zeidler Consultants, 2013).

## CONCLUSION

The 2010–2011 La Niña had a profound effect on the geomorphology of the project area, as well as a change on the geotechnical properties of the materials underlying the mountains and valleys between Villeta and Guaduas. These changes have increased the vulnerability and risk of landslides and mass movements in an area already susceptible to such events, especially in areas where water-saturated fine-grained marine sediments and disturbed, highly permeable coarse-grained sandstones and conglomerates are more susceptible to erosion and weathering than under normal conditions, areas where critically high cuts are planned along the alignment, particularly in areas where the bedding planes of the geologic units are dipping towards such proposed high cuts, and where historically active landslides are located. Increased precipitation over extended periods of time, as seen with the 2010–2011 La Niña event, dramatically increased the rate at which these materials weather and erode, increasing the vulnerability to the proposed structures of the alignment of Sector 1, Tramo 1, as well as increasing the risk of mass movement events. Future La Niña events will most certainly continue to occur and affect the geomorphology along the Project area, although the timing and intensity of such events can only be speculated. The new alignment of Sector 1, Tramo 1 of Ruta del Sol should be designed accordingly, and should be able to mitigate these risks (Gall Zeidler Consultants, 2013). For example, increasing the amount of tunneling along the alignment, though initially more expensive than other structures, will provide protection against future landslide and mass movements events, dramatically decrease the time required for road closures and repairs, decrease the alignment length and thus, overall travel times, and offer long term sustainability and cost savings throughout the life of the Project alignment.

## ACKNOWLEDGMENTS

The authors of this paper would like to thank ANI and HELIOS for their ongoing work on Sector 1, Tramo 1 of the Ruta del Sol Road Project, as well as for their contribution to the tunneling and underground construction industry. The authors would also like to thank John Rudolf and Nikos Syrtariotis at GZ for their assistance in the fields of structural and geotechnical engineering during the completion of this Project, and also for their contributions to the writing of the Final Report. The authors would also like to thank Fernando Oropeza (Fugro México), and Ernesto Vicente and Kathy Reyes (Fugro Consultants, Inc.), who lent their expertise in the fields of meteorology, climatology and civil, structural, and geotechnical engineering, as well as David Hartwell, who offered his expertise to the team in the field of hydrology as it pertained to the geometrical design of the alignment. Finally, the authors would like to thank Mark Urban and Bill Wiedelman (HNTB) for their insight regarding the alignment and geometrical design of the Project.

## REFERENCES

Banco Interamericano de Desarrollo (BID), Comisión Económica Para América Latina y el Caribe (CEPAL), 2012. Ola Invernal en Colombia 2010–2011. 240 pages.

Gall Zeidler Consultants, 2013. Impact on the Geomorphological Conditions Caused by the 2010–2011 La Nina Phenomenon in the Project Area of Ruta del Sol, Sector 1, Tramo 1. Final Report. *Unpublished document.* 214 pages, 4 appendices.

Hudson, R.A., 2010. *Colombia: A Country Study.* Pages 70–76.

IDEAM, 2011. S ervicio de Seguimiento y Pronóstico de le Amenaza por Deslizamientos, *Boletín número 104.*

Paris, G., Machette, M.N., Dart, R.L., and Haller, K.M., May 2000. Map and Database of Quaternary Faults and Folds in Colombia and its Offshore Regions. *U.S. Geological Survey Open-File Report 00-0284.*

Pulido, N., 2003. Seismotectonics of the Northern Andes (Colombia) and the Development of Seismic Networks. *Bulletin of the International Institute of Seismology and Earthquake Engineering.* Special Edition. Pages 69–76.

Taboada, A., Rivera, L.A., Fuenzalida, A., Citernas, A., Philip, H., Bijwaard, H., Olaya, J., and Rivera, C., 2000. Geodynamics of the northern Andes: Subductions and intracontinental deformation (Colombia). *Tectonics, Vol. 19, No. 5.* Pages 787–813.

# Achieving Fast EPB Advance in Mixed Ground: A Study of Contributing Factors

**Joe Roby and Desiree Willis**
The Robbins Company

**ABSTRACT:** Earth Pressure Balance (EPB) tunneling in mixed ground conditions is a challenging prospect, as it often includes excavation in boulder fields, sections of rock, and/or sticky clay, under high water pressure or changing water pressure. Maintaining a rapid advance rate in such conditions is a function of many factors—from adequate cutting tools to cutterhead design, pre-planning and execution of an appropriate ground conditioning regime as well as proper maintenance and operation of the TBM. This paper will analyze recent record-breaking and high-performing projects seeking to identify factors that contribute to fast machine advance. These factors will then be discussed and an effort made to form simple, high level guidelines for optimal TBM excavation in mixed ground conditions.

## INTRODUCTION

Labor costs for tunnels excavated with hard rock TBM and soft ground EPB machines (EPBMs) typically are 30 to 50 plus percent of total project cost. Reduction of the time for tunnel construction without increasing staffing results in a substantial savings in total project cost. Finding methods by which we can safely reduce total tunnel construction time has a clear cost benefit to project owners and generally to the tax paying public. It also has the benefit of bringing needed infrastructure online sooner, which never meets with public disapproval.

In this paper the authors attempt to find commonalities among EPBMs operating in mixed ground conditions that achieved higher than average advance rates within a given sample of projects. By mixed ground we mean that the tunnel alignment contains some fairly easy to excavate material for an EPBM, which typically implies soils, sands, gravel & clays in some combination, as well as material that is not easily excavated by an EPB machine, which typically implies:

- Coarse sands and gravels, below the water table, with insufficient fines to form a plug in the screw conveyor
- Large boulders requiring disc cutters to break
- Competent rock
  - Above the water table
  - Impermeable rock below the water table
  - Permeable rock below the water table

Each of these geological types imposes somewhat unique challenges when excavated with an EPBM.

## MIXED GROUND CHALLENGES

Following is a brief discussion of some of the challenges each of the above mentioned geological types presents when excavation is attempted with an EPB machine.

### Coarse Sand and Gravels

When EPBs are below the water table and contain insufficient fines to form a plug in the screw, it is necessary to add foams, polymers or fine material to form the plug.

In addition, sands and gravels can be extremely abrasive and it is usually prudent to add friction-reducing foams and polymers. This addition reduces the rate of wear on the cutterhead, screw conveyor and other components. Reducing wear is essential to high performance because it reduces the number of interventions likely to be required for maintenance of cutters, cutterhead and other wearing components forward of the pressure bulkhead. In all of the mixed ground conditions we are discussing, the importance of reducing wear is paramount.

### Large Boulders

When large boulders are expected the cutterhead is typically fitted with disc cutters. However, when the tunnel also passes through more traditional EPB materials, it is important to maintain the cutterhead face opening ratio. Disc cutters take up a lot of precious cutterhead space compared to EPB picks and bits. The design of the cutters and cutterhead take on great importance for mixed ground tunnels with a probability of large boulders, as the appropriate EPB cutterhead opening ratio for excavating traditional EPB materials must still be maintained. Restricting the size of rock pieces that may pass through the cutterhead is important to reducing the risk of blockage of the screw conveyor. Also, in such situations, minimizing wear is imperative.

## Competent Rock

### *Competent Rock Above the Water Table*

Generally, in this condition, we have the same concern as mentioned above for large boulders. In addition, we have a muck flow issue and a potentially extreme EPB wear issue. EPBs depend upon a combination of face pressure and the always full mixing chamber to charge the screw conveyor with muck. When cutting solid rock above the water table, there is no face pressure and so the mixing chamber will not naturally fill, meaning the screw conveyor will also not fill naturally. In practice, if no extraordinary measures are taken, the flow of material through the screw happens cyclically:

- Machine bores rock until a sufficient amount of material is in the mixing chamber
- The rock in the mixing chamber, finally at sufficient height, under its own weight, will flow into the screw conveyor
- The screw conveyor will then discharge the muck, and the cycle repeats.

Of course, with practice, it is sometimes possible to balance the screw conveyor drive speed to the EPBM advance speed to maintain a charged mixing chamber to maintain flow to the screw. However, this requires the rock to break in consistent ways to provide a smooth, almost fluid flow of the excavated material, which rarely happens.

In reality, machine operators generally must resort to injecting material into the chamber to mix with the cut rock in order to create a mix of materials that will flow in a more fluid-like manner. Generally, the material injected into the mixing chamber includes a volume of water along with foams, polymers or other materials. Often, the mixing chamber may have to be artificially pressurized with compressed air in order to help the material flow into the screw conveyor.

Depending on the abrasivity of the rock being excavated, anti-wear, torque-reducing foams and polymers will likely be required.

### *Competent, Impermeable Rock Below the Water Table*

The challenges of this condition are essentially the same as described above, for solid rock above the water table.

### *Competent, Permeable Rock Below the Water Table*

This situation is essentially the same as that described for the previously mentioned two solid rock sections, except that the rate of water injection into the mixing chamber to achieve a properly flowing material will be affected by the natural flow rate of water into the cutting chamber. It remains highly likely that it will require the injection of foams, polymers or other fines in order to form a plug in the screw conveyor.

Again, depending on on the abrasivity of the rock being excavated, anti-wear, torque-reducing foams and polymers will likely be required. In addition abrasive wear on the cutters due to water injection and the presence of rock is a challenge.

## THE PROJECT DATABASE

For this paper the authors reviewed 25 projects in 10 different countries which employed 40 different EPBMs on projects for which we deemed the geology to be "mixed." Obviously, the geology of some of these projects was decidedly more challenging than others but all contained at least some sections of geology that included coarse sands and gravels that wouldn't form a plug, or they contained large boulders or hard rock. Many of the tunnels contained some combination of these "difficult to excavate with an EPB" geologies.

We were looking for machines that had achieved high advance rates relative to the other machines in our sample. But, it would not be sufficient to have merely had a world record "best day" or "best month." We were looking for projects on which the EPBM performance over the entire tunnel excavation was significantly better than others operating in similarly difficult geology. For this purpose, we elected to use "average weekly meterage" as our measure of total productivity. One caveat to the reader: contractors and consultants are loathe to share complete information on their projects because it is hard-won intellectual property that enables them to more accurately tender future work. In some cases, we were not given accurate data regarding total working hours per week, holidays and other information which would have allowed us to normalize the data completely (i.e., providing an average advance per working hour). We were forced to look at the total length of the tunnel versus the weeks required for excavation and assume that a similar number of hours were worked each week on average. Of course, in industry publications and on the internet we also sought and found additional data regarding each project (e.g., confirmed dates, additional geological data, additional EPBM specifications, etc.) These data helped to ensure a more complete and objective data set.

The basic data set for each project / EPBM included:

- Project name
- Country
- Length of tunnel
- Average weekly advance in meters
- Geological description
- Water / face pressure
- Diameter of machine
- Cutterhead drive type (e.g., hydraulic, VFD electric)
- Cutterhead power
- Cutting tools fitted to cutterhead and quantity
- Muck removal system (e.g., muck cars / rail, conveyor)
- Ground conditioning (e.g., existence of pre-project ground conditioning planning and coordination with machine manufacturer and chemical supplier and/or near continuous use of ground conditioning agents, and a list of chemicals employed)

## THE PROJECTS, THE EPBMs, AND THEIR PERFORMANCE

For the 40 EPBMs reviewed the diameter ranged from 5.9 to 10.2 meters, though the vast majority were in the 6 to 6.5 m range. Thirty-one of the machines were employed on metro projects, eight on sewerage projects and one on a train tunnel. They were supplied by three different manufacturers. The face pressure under which they worked ranged from 0 to 13.5 bar with an average of 3.6 bar, with seven projects not reporting the ground pressure. Forty-seven percent of the machines were fitted with variable frequency electric cutterhead drives and the balance were driven hydraulically. The geology on which the machines operated varied widely from sedimentary rock and weathered rock through glacial till, gravel, sands, soils and clays, however all had encountered mixed conditions.

Fifty-four percent of the projects gave information regarding ground conditioning employed. Several projects gave detailed information regarding ground conditioning, or that information is publicly available in articles published in industry periodicals and conference papers. Unfortunately, no ground conditioning information was forthcoming or could be found in searches for nearly 40% of the projects. Given the apparent importance of this subject, and the currently fast growing knowledge on the subject of ground conditioning and its importance, it would be beneficial to have more details in this area for better statistical analysis of performance between machines employing state of the art ground conditioning and those that do not.

Thirty percent (12 machines) had average weekly advance rates exceeding 100m/week. Forty-five percent or 18 projects had average weekly advance rates exceeding the average of 85m/week (see Table 1, a summary of EPB data set).

**Table 1. EPB data set summary**

| | |
|---|---|
| Number of EPBMs | 40 |
| Diameter range | 5.9 to 10.2m |
| Face pressure range | 0 to 13.5 bar,<br>3.6 bar average |
| Average weekly advance rate | 85.4m/week |
| Maximum advance rate | 178.5m/week |
| Minimum advance rate | 32.6m/week |
| Standard deviation | 36.0m/week |

## WHAT DID THE HIGH-PERFORMING EPBMs HAVE IN COMMON?

We sorted the data several ways looking for data which had a close correlation with high average weekly advance. Against the following data we found only weak correlation:

- Machine diameter
- Cutter configuration
- Cutterhead drive type (electric and hydraulic)
- Face pressure
- Mucking system
- Tunnel length
- Country of project, and developed / developing nations

For example, Canada had two of the top 10 performers, but it also had 2 of the bottom 10 performers. The top 10 performers were about equally divided between developed and developing countries with the top performer being on the Moscow Metro Line 3 project.

There was no correlation between performance and face pressure and, in fact, four machines with very high average weekly advances of 120 to 179m/week were working at 6 to 8 bar on the Abu Dhabi STEP project.

Perhaps not surprisingly, contractor experience does have some correlation with machine performance. All of the contractors operating machines that had average weekly advance rates in excess of 100m/week had previously excavated at least three prior EPB tunnels with some of them having excavated many. With one exception, the bottom 40% of performers was operated by contractors very new to EPB operations.

Conveyor mucking systems were used on seven of the projects, but there was no correlation with performance with conveyors being used on top, mid and bottom performers. Obviously perhaps, conveyors can help set the stage for high performance but are not alone sufficient to guarantee high performance. Neither did tunnel length strongly correlate though longer tunnels trended toward higher average weekly advance rates, as one would expect.

High performance appears to be at least lightly linked to a mixed ground EPBM being fitted with a cutterhead designed and fitted for mixed ground (i.e., fitted with disc cutters as well as soft ground tools). Perhaps more to the point, machines that started and had to be stopped one or more times in the tunnel to have the cutting head redressed, from soft ground tools to full disc cutters, under pressure often lost so much time for the retrofit as to make it impossible to achieve a rapid tunnel excavation. Clearly, accurate geological mapping must be made available in the tendering stage if the contractor and machine manufacturer are to agree to the correct design and cutting tool selection prior to the start of excavation.

The single factor that had the strongest correlation to machine performance appears to be ground conditioning. The best performers nearly all had soils tested in a laboratory in advance of the start of boring and had established an initial ground conditioning regime in coordination with the contractor, the machine manufacturer and the chemical supplier. Even those projects that merely brought in the chemical supplier at the start of boring had more success than those who did not employ chemicals or did so only late in the project. There seems to be sufficient evidence to support the avocation for laboratory testing and coordination between contractor, machine manufacturer and chemical supplier in order to insure the best machine design for chemical injections and provide the best basis for early high performance of the EPBM.

## THE IMPORTANCE OF GROUND CONDITIONING

While perhaps such a strong correlation between EPBM performance and a quality ground conditioning regime may not have been anticipated by all, those who have been heavily involved in the EPBM excavation of difficult geological conditions may not be surprised in the least. Most of those who have been involved in the use of ground conditioning for EPBMs operating in coarse gravel have known for years about the efficacy of using foams to form a plug in the screw. This method allows EPBMs to excavate material previously considered the sole domain of the slurry TBM.

A good ground conditioning regime can be equally as important as the machine design and logistical aspects on any EPB project. Additives are used to consolidate ground and maintain a smooth flow of muck through the cutterhead, thereby maintaining consistent earth pressure.

The use of ground conditioning at the cutterhead has further been shown to reduce wear and increase advance rates. The type of additive used, and indeed whether or not additive is needed at all, is determined by soil permeability, ground water pressure, and the risk of clogging/adhesion (Langmaack, 2006).

Japan, the country that truly created the modern EPBM, has been well aware of the importance of ground conditioning additives for many years and is a leader in the development of foam additives. Table 2 is a 1996 recommendation on the use of additives for EPBMs from the Japanese Society of Civil Engineers. According to the Shield Tunneling Association of Japan (established in 1985), the first EPB with a foam GC system was delivered in 1980 and a total of 431 EPBs fitted with foam GC systems have been delivered in Japan through 2007.

Over the decades we have seen the use and function of ground conditioning additives broaden substantially. From providing a method to form a plug in the screw conveyor in coarse materials, ground conditioning additives now provide a method by which to increase the cohesiveness of material, reduce the adhesiveness of material, reduce the friction of material (i.e., reduce the torque on cutterheads and screw conveyors) and more.

Soil consistence is described in 4 states: solid, semi-solid, plastic and liquid. To this standard description of "soil," on a mixed ground project we add the possibility of boulders, hard rock above and below the water table, etc. EPBMs are not capable of safely, efficiently and economically excavating materials at the extremes of these states, especially so when under the water table. However, when we change the characteristics of these materials through the use of ground conditioning agents, and when the EPBM design has been done with full knowledge of the ground conditions, we substantially broaden the range of materials that can successfully excavated by EPBMs.

## ESTABLISHING A GROUND CONDITIONING REGIME

A good place to start an understanding of the basics of ground conditioning is the *Specifications and Guidelines for the Use of Specialist Products for Mechanised Tunnelling* published in 2001 by EFNARC, the European federation focused on specialist construction chemicals and concrete systems. In 2005 the document was updated to include hard rock TBMs as well. EFNARC engages with the European Commission and the CEN technical committees as well as other groups participating in the European Harmonization of Specifications and Standards. We recommend the EFNARC document to our readers for its considerable valuable information (see Figure 1).

Geotechnical Baseline Reports (GBRs) for most projects will define the geological and hydrological conditions anticipated along the tunnel

**Table 2. Table from Japanese Society of Civil Engineers (1996) with recommendations regarding use of additives for EPB applications**

| Shield Type | | | EPBM | | |
|---|---|---|---|---|---|
| | Soil Type | SPT N | Without Additives | With Additives | Slurry |
| Alluvial cohesive soil | Silk and clay | 0–2 | Y | Y | Y |
| | Sandy silt, sandy clay | 0–5 | Y | Y | Y |
| | Sandy silt, sandy clay | 5–10 | Y | Y | Y |
| Pleistocene cohesive soil | Loam and clay | 10–20 | N | Y | Y |
| | Sandy loam, sandy clay | 15–25 | N | Y | Y |
| | Sandy loam, sandy clay | over 25 | N | Y | Y |
| Sandy soil | Sandy with silty clay | 10–15 | Y | Y | Y |
| | Loose sandy soil | 10–30 | N | Y | Y |
| | Consolidated sand | over 30 | N | Y | Y |
| Gravel with boulders | Loose gravel | 10–40 | N | Y | Y |
| | Consolidated gravel | over 40 | N | Y | Y |
| | Gravel with boulders | — | N | Y | N |
| | Boulder gravel, boulders | — | N | N | N |

alignment including photographs, in situ test results and laboratory test results including particle size distributions, presence of boulders, rock types and strengths, ground water information, permeability, moisture content of clays, etc. With the GBR information and the EFNARC recommendations one can form a very rough idea of the ground conditioning that might be appropriate. Further consultation with the ground conditioning chemical supplier will result in a more well-defined initial ground conditioning plan. Further coordination with the EPBM supplier will insure that the EPBM is delivered with foam, polymer and other systems designed for the best implementation of the ground conditioning regime immediately upon launch of the EPBM. It is, however, recommended to take the ground conditioning planning a step further, to the laboratory.

## SPECIAL LABORATORY TESTING FOR EPB SOIL CONDITIONING SPECIFICATION

Today there are a growing number of laboratories, in private companies and at universities, which can perform a number of tests aimed specifically at defining a ground conditioning regime for an EPB project. Typically, these laboratories mix actual soil samples from the job site, at their in situ moisture content, with various foams and polymers and then test the treated samples (see Figure 2). One such simple test is a slump test, such as is typically performed on wet concrete to determine its workability. (This test can also be done on the job site, if the correct equipment is made available at the site). As written in the paper *Characterization of Soil Conditioning for Mechanized Tunneling:* "…the carried out tests show that the slump test is a good indicator to define the global behavior of a conditioned soil and due to its simplicity, can be used in the preliminary design stage but in particular on the job site to keep the conditioning development under control during excavation" (Borio 2007).

Other tests include permeability testing of the sample to determine the probability of the material forming a plug in the screw conveyor. Other lab testing done today includes wear testing and even scale model screw conveyance of the material under pressure.

Professionally performed specialist laboratory testing can give us a much better recommendation for an initial soil conditioning regime to be employed at EPBM launch, including recommended foam and polymer types along with specifying the important parameters for use, including:

- **Cf**—the concentration of foam product in water. Generally this will be in the 0.1 to 4.0% range, though it is dependent upon the ground condition and the specific foam product selected.
- **FER**—the Foam Expansion Ratio. Values are typically ×5 to ×30, being expressed as the ratio of air to foam, where ×18 will be 17 parts air and 1 part foam/water solution. The larger the expansion ratio the dryer the foam. Generally, the wetter the soil is, the dryer the foam should be.
- **FIR**—the Foam Injection Ratio. This is the ratio of foam injected into the cutting head and the in situ volume of soil being excavated. This is typically in the range of 30 to 60% per EFNARC guidelines, but in the Japanese standard goes beyond 100% up to 130% foam/insitu soil volume. (The reader should bear in mind that the actual ratio of

foam to soil in the chamber will be dependent upon the pressure in the chamber, as the air in the foam compresses under pressure, hence the ability to go above 100% and still excavate material.)

- **Cp**—the concentration of Polymer product in water, typically in the 0.1 to 2.0% range, but can go to 5% according to EFNARC.

Many foam products are provided with polymers so that only the foam guidelines need be followed.

If wear tests are provided they can aid the contractor in making a better estimate of wear of the EPBM and cutting elements thereby assisting with both the cost estimate and estimation of down time for interventions for repairs. While the wear tests won't provide definitive numbers, if the wear tests

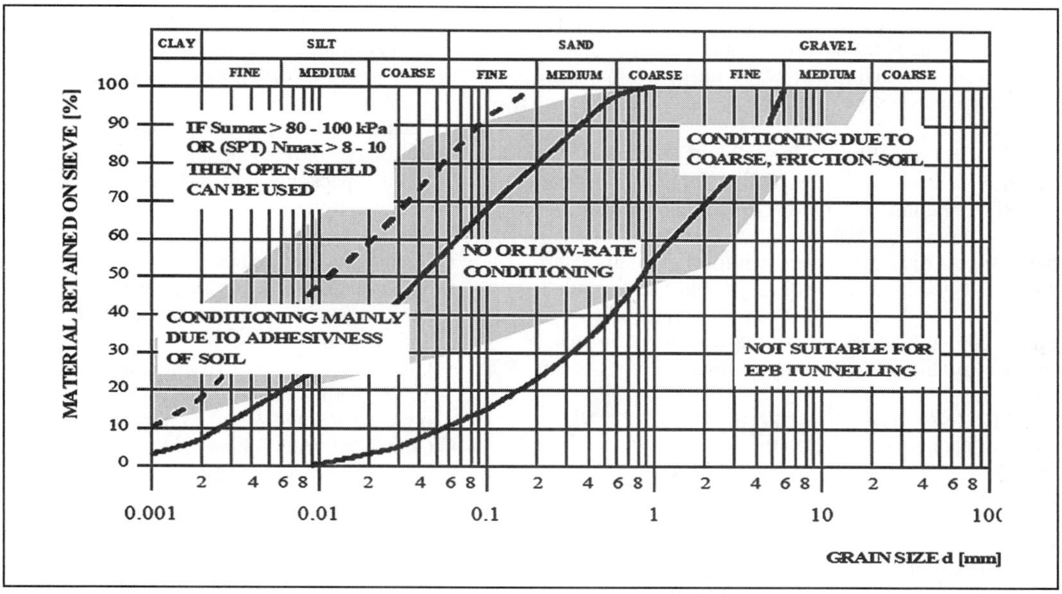


**Figure 1. EFNARC guideline for particle size distribution in which EPBs can be employed, as well as soil conditioning needs in different ground types (boundaries are only indicative)**

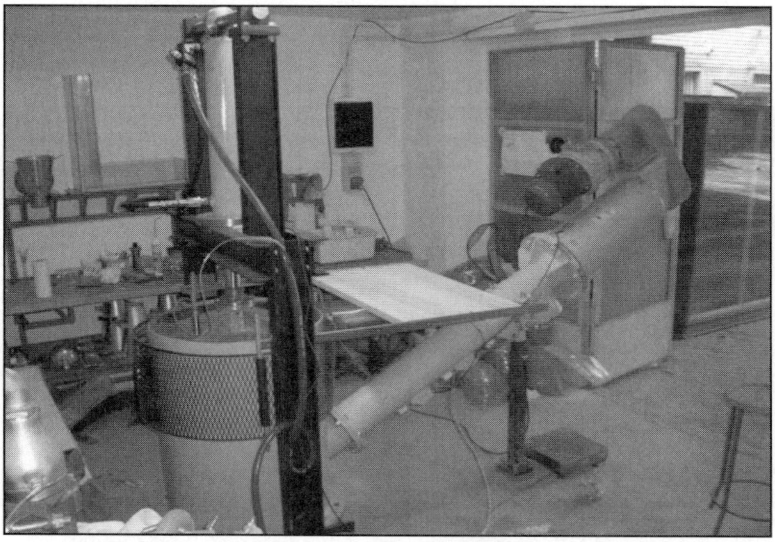

**Figure 2. Testing fixture. Treated sample is placed in barrel on left and subjected to pressure and extracted from barrel through screw conveyor on right (Photo courtesy of Mapei-UTT).**

show a reduction in wear of 25% with the use of additives, it provides some indication of the savings one might reasonably expect to see in the field. Given the danger, downtime and cost of hyperbaric interventions, reduction in wear may well prove to be one of the higher motivations for the use of ground conditioning / wear reduction agents.

Specialist laboratory testing has proven its worth. Speaking of one of the higher performing projects in our data base, it was stated, "The average ground conditioning parameters used at the job site are comparable with the values found after the laboratory tests.... This confirms the utility of making laboratory tests before the TBM launch" (Dal Negro et al. 2013).

In 2011 the Shield Tunneling Association of Japan issued a technical guideline for use of foam in EPB tunneling. The guideline includes a formula for calculating the FIR based on the results of the particle size distribution curve information and can provide a good starting point, thought the formula does not consider ground pressure, permeability or pore volume. Unfortunately, the document is currently available officially only in Japanese.

## DESIGNING THE EPBM FOR THE GROUND CONDITIONING REGIME

It is imperative that the EPBM manufacturer is aware of the GC regime plan and that appropriate foam generators, polymer plant, air compressors and bentonite systems are included, as well as proper distribution and injection points on the cutterhead, into the cutting chamber and into the screw conveyor. Results from the 40 EPBMs reviewed and anecdotal evidence points to this being an area of coordination which is often overlooked or under emphasized and where a little effort early in the EPBM design can result in vastly improved performance on the project.

A properly designed EPBM GC system requires input from the contractor and the GC additives supplier (see Figure 3).

The team must agree to the GC plan and ensure that the EPBM design and GC equipment supply will fully support the GC plan. Some things that must be considered:

- Probable quantities of foam agent, polymers and bentonite (or other fine material) to be consumed, consumption rates and estimated TBM production rates
- Package sizes to be used for each GC agent
- Logistics; movement and handling of GC agents / packages into and out of the tunnel
- Specification of the dosing units
- Specification of the foam generator
- Specification of dedicated air compressor
- Specification of bentonite plants
- Locations of the above systems on the TBM and back-up
- Quantity and location of injection nozzles for all GC additives and water (cutterhead, mixing chamber and screw conveyor)
- Control systems for manual, semi-automatic and fully automatic control
- Location of system adjustment controls and ability to "lockout" to prevent unauthorized adjustments
- Quantity and placement of additional water lines into mixing chamber

Regarding this last point, yes, it is important to have the capability to inject water into the chamber in addition to GC agents. When the ground is too dry, it

**Figure 3. Silty clay prior to and following GC treatment (Photo courtesy of Condat)**

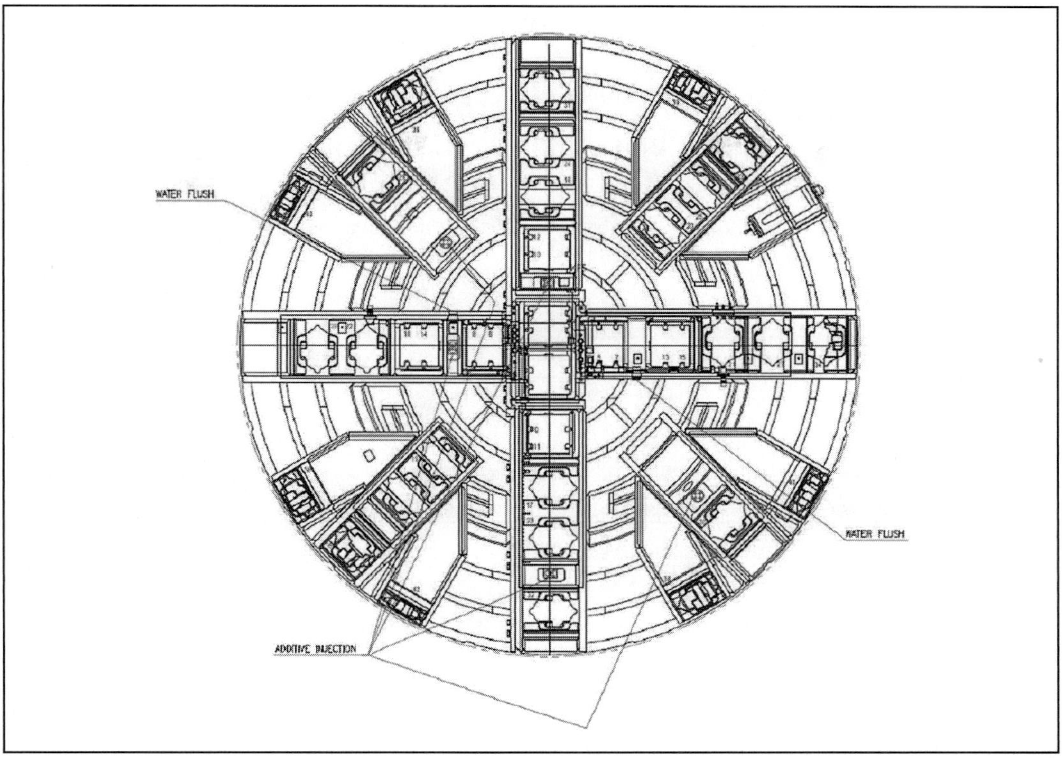


**Figure 4. Ø6.6 meter EPB cutterhead with five additive injection ports and two water injection ports to prevent clogging**

is most effective to use water to wet the soil and GC agents to condition the soil.

In general, it is best to inject all GC agents from the cutterhead because this provides the best possibility for GC agents to flow with and become thoroughly mixed with the excavated material. However, there are times when it might be advantageous to inject GC agents into the mixing chamber. For example it is prudent to inject bentonite during a machine stoppage because foam will collapse, eventually leaving an air bubble in the top of the chamber and water in the bottom. Under certain conditions it might be necessary to inject directly into the screw conveyor to form a plug, or to reduce friction and torque at the screw conveyor. When designing the EPBM for GC use, it is important that the systems be designed for flexibility and with redundancy. A properly designed EPBM will offer the user opportunities to employ all of the GC agents (water, foam, polymers and bentonite) in any combination and at an array of injection points on the cutterhead, into the mixing chamber and into the screw conveyor. In addition, because of the danger and difficulty associated with effecting repairs beyond the pressure bulkhead, distribution line redundancy is advisable.

### Cutterhead Foam Injection Ports

EPB cutterheads should be designed with certain port sizes and locations and minimum quantities. Figure 4 shows an example of additive injection port locations on a Ø6.6m EPB cutterhead. These injection ports should be capable of injecting foam, polymer, bentonite, or any mix of these and should be located with the first port as close to the center of the cutterhead as possible. Remaining ports should be located with decreased radial spacing as they near the outer periphery of the cutterhead. It is not necessary for the ports to reach the outermost radius of the cutterhead, this being the area of fastest motion and therefore best mixing. For "metro sized" cutterheads 6 to 7m in diameter, a minimum of five injection ports is standard, with all piping having an internal diameter of about 1.5 inches (38mm). For each injection port on EPB cutterheads, protection bits with tungsten carbide inserts and hard facing should be placed on both sides of the port for protection in both directions of cutterhead rotation.

As EPB cutterheads get larger, more ports are of course needed. For example, in the Ø9m and Ø10m range EPB cutterheads, seven additive injection

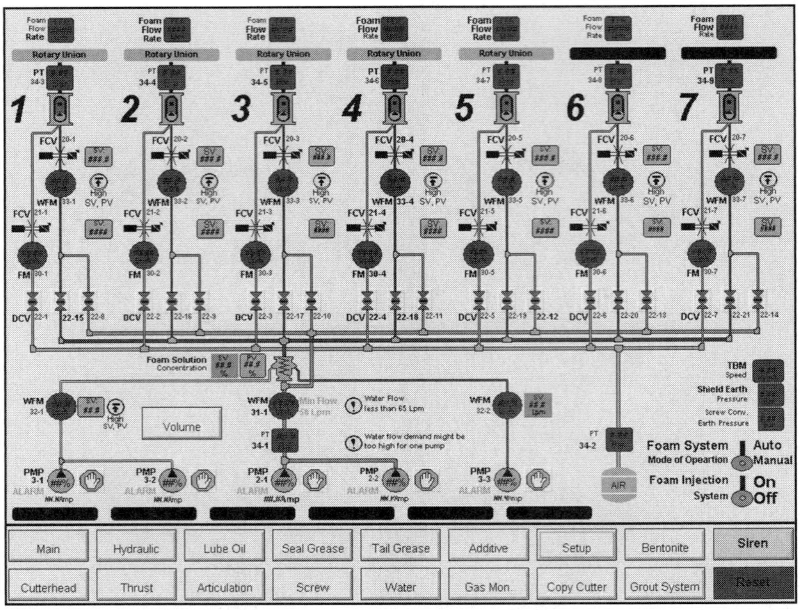


**Figure 5. Foam and polymer system setup screen on EPBM operator's Human-Machine Interface**

ports are used, with piping having an internal diameter of about 2 inches (50mm).

It is advisable to fit the screw conveyor with a minimum of three 50 to 100mm diameter injection ports with one located as near the pressure bulkhead as possible and the others located along the conveyor. The pressure bulkhead should have a minimum of ten 50mm diameter injection ports with at least one located immediately each side of the screw conveyor intake and the remaining distributed roughly evenly around the bulkhead.

It should be noted that GC systems (foam generators, polymer pumps, bentonite pumps and water lines) will not be connected to all of the ports fitted to the EPBM. There will be a substantial surplus of ports when the quantity is compared to the quantity of GC injection lines. What is important is, again, flexibility and redundancy so the contractor can make adjustments to the ground treatment as needed to achieve success based on actual results.

### Operator Station and Software

The operator's station for the EPBM, with the usual Human Machine Interface (HMI) touch screens, typically has several screens dedicated to GC systems. The foam system will generally have one screen for setup (to set Cf, FIR and FER) and one screen for operation where the operator can monitor status in automatic mode, or control the system in manual mode.

FIR, again, is the ratio of foam injected as a percent of the in situ volume of soil being excavated. Since the rate of volume of soil being excavated is dependent upon the EPBM's advance rate, the rate at which the foam is injected must vary with the EPB advance rate in order to maintain a constant FIR, that is, the same proportion of foam to soil at all times. This being the case, it is advantageous to operate in automatic mode in order to maintain a consistent state of soil conditioning.

Of course, there are similar options on the operator's control screens for setting the parameters for polymer. The HMI may have an additional screen which shows the total volumes of air, water, foam and polymers that have been injected over some period of time which can, of course, be reset (see Figure 5).

The geology anticipated on a project affects the final design of a number of components of an EPBM: cutterhead, cutting tools, screw conveyor(s), ground conditioning systems, grout systems, etc. However, it is worth noting that if the contractor, the GC chemical supplier and TBM designers work together, the design of cutterheads and conveyors can be positively impacted for improved TBM performance and reduced component wear (see Figure 6).

## OTHER CONTRIBUTING FACTORS

Other factors contributing to high advance rate in mixed ground are many, yet one of the most compelling is proper cutterhead and screw conveyor design. In mixed ground conditions, EPB cutterheads must balance an optimal cutterhead opening ratio for smooth muck flow with a robust cutterhead structure and the adequate number of disc cutters and cutting

**Figure 6. Well-conditioned clay leaving the screw conveyor onto the belt conveyor (photo courtesy of Mapei UTT)**

tools. Screw conveyors must be designed with the knowledge of the maximum face pressure to be encountered, the probable presence of boulders and the maximum boulder size which will be allowed to pass through the cutterhead.

### Cutting Tools

The optimal primary protection for EPB cutterheads is the replaceable knife bit. These come in standard duty and heavy duty, but standard duty is only recommended for geology that is very easy to excavate. In a mixed face application, these bits are interchangeable with disc cutters. Cutterhead spokes are designed to alternate between primary and secondary cutting tools. It has been found that a radial spacing of these primary cutting tools at about 3.5 in (89mm) apart is efficient in the breaking up of soft ground. When hard rock or boulders are encountered and these tools are replaced by disc cutters, this same spacing allows the discs to break up the rock and allows the cracked rock in-between cutters to fall out.

### Abrasion-Resistant Wear Plate

The optimal design for EPB cutterheads includes full protection with an outer cladding of abrasion resistant wear plate. There are greatly varying grades of abrasion resistant wear plate available, and the selection of this plate is usually project specific, based on balancing cost with sufficient hardness and wear resistance. There is wear plate available that can resist the wear of nearly all types of ground conditions, including very abrasive rock and long tunnels, but the cost and workability varies quite considerably.

Wear plate should cover the entire exposed front surface of the cutterhead that is not shared with a cutting tool location or a chemical injection port. Figure 7 gives an example of the type of coverage that should be given by cutterhead wear plates.

### Screw Conveyors

Screw conveyors can be designed with replaceable bolt on sections and hard facing on each turn of the screw to withstand abrasive ground. The screw conveyor casings can be lined with abrasion resistant plate as well. Again, the actual abrasion resistant material selected can have a dramatic impact on cost.

Screws may have a shaft or no shaft (a "ribbon" conveyor). Shafted screws have a greater pressure drop across each flight and therefore can be made shorter than a ribbon screw to achieve the same total pressure drop across the conveyor. However, ribbon screws can pass a larger boulder within the same casing diameter compared to a shafted screw. Often times two screw conveyors are used in series to achieve the required pressure drop and these are often a combination of a ribbon screw for the first conveyor and a shafted screw for the second conveyor.

Screw conveyors can also be designed to be disassembled within the tunnel, even with the face under pressure, to make it possible to more safely

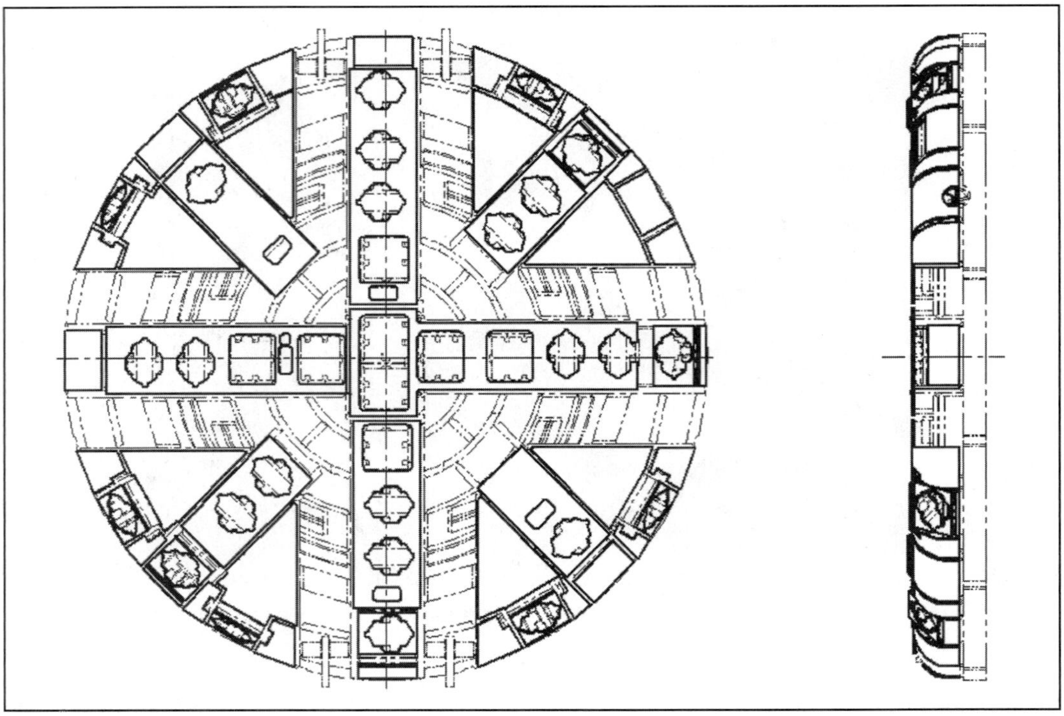

**Figure 7. Drawing showing coverage of wear plate material, which is not always obvious on the EPB as wear plate and structure are often the same color**

and rapidly repair and maintain worn screw flights and casings. However, this necessarily requires dividing the casing and screws into smaller pieces with bolted joints, etc., all of which increases the manufacturing complexity and cost but saves time in the tunnel.

With all of the variables available in selecting a properly designed screw conveyor, or conveyors, for the EPBM it is again imperative to have good information on the full range of geology, hydrology and pressures to be encountered in the tunnel.

As important as a well planned and executed ground condition conditioning regime is, in most cases the best GC plan cannot overcome a poorly designed EPBM.

## CONCLUSIONS

It was our intention at the outset to attempt to derive some simple, high-level guidelines that if followed would provide the highest probability of an EPBM reaching the best possible performance in a mixed ground tunnel. Following are those guidelines, some of which are simply common sense, known already by experienced EPBM users and some of which have been suggested by several other recent authors on the subject of ground conditioning:

1. **Geological samples:** Prior to tendering, the project owner should engage an experienced geological / hydrological testing firm to perform as many hydrological tests and obtain test samples from as many points as reasonably possible along the tunnel alignment, and if possible from the actually tunnel depth. Sufficient sample quantities should be obtained to provide the tendering contractors to perform laboratory testing on the samples prior to bid. If that is not possible, then the owner or their consultants should have such laboratory testing performed, which can establish a base-line initial ground conditioning recommendation by one or several chemical suppliers. This will allow the tendering contractors to make adjustments in their commercial budgets and schedules for the improvement in performance they may reasonably expect to see on the project with the proper use of ground conditioning.
2. **Laboratory testing for ground conditioning specification:** Should the owner not provide the contractors with laboratory test results of the geological sample testing, then the contractor would be well advised to have such tests carried out at their own expense in

order to obtain a recommended ground conditioning regime from an experienced EPB chemicals supplier. The results of such tests will go far toward providing the best possibility of high performance on the project, as well as giving the tendering contractor much information regarding probable costs for ground conditioning agents.

3. **EPBM design:** Though ground conditioning is extremely important, it is equally important on mixed ground projects that the contractor and machine manufacturer review the probable geology, hydrology and face pressures of the project in detail and discuss the impact on the EPBM design, which might include:
    - Dress of cutterhead: disc cutters, scrapers, picks, bits, etc.
    - Opening ratio of cutterhead
    - Type of screw conveyors: ribbon or shafted
    - Quantity and length of screw conveyors
    - Abrasion-resistant cladding requirements: cutterhead, mixing chamber, mixing bars, screw conveyor flights and casing, etc.
    - Face pressure related design: pressure bulkhead, thrust ram sizing, articulation ram sizing, tail shield seals, main bearing seals, man-lock and tool-lock, breathable air design, air compressors, etc.
    - Ground conditioning foam, polymer and bentonite systems, air compressors, etc.
4. **Coordination and equipment specification for ground conditioning:** Early in the EPBM procurement / design phase, the contractor, chemical supplier and EPBM supplier should meet and discuss the results of the ground conditioning laboratory results. There should be agreement regarding the systems required on the EPBM to properly inject the agreed upon chemicals into the proper locations on the EPBM (e.g., cutterhead, pressure bulkhead / mixing chamber, screw conveyor points, etc.). There should be agreement on foam generation plant specifications, probable ranges for Cf, Cp, FER, FIR, and it should be ensured that those calculations for the sizing of plants (e.g., air compressors) consider the likely face pressures under which the EPBM will be working.
5. **On-site ground conditioning testing:** The job site should have the ability to do on-site testing of ground conditioning agents in order to make adjustments throughout the tunnel drive without undue downtime for the machine. At minimum this should include:
    - A laboratory scale foam generator
    - A 5 liter heavy duty mixer with 3 speeds and standard paddles
    - DIN flow table (30cm table) with standard mortar cone (slump test)
    - A graduated container of 1 or 2 liters capacity (plastic or non-breaking)
    - Weighing balance accurate to 0.1 gram
    - Stop watch
    - Calibrated glass or clear plastic cylinder, with perforated base, 1 liter capacity
    - Various calibrated plastic containers up to 2 liters
    - A 50ml graduated cylinder
    - A filter–funnel of 1 liter capacity with non-absorbent filter
6. **EPBM launch, ground conditioning adjustment and site lab setup:** At the start of boring, on the job site, there should be representatives from the chemical supplier and the EPBM supplier to work with the contractor to make any adjustments to the ground conditioning regime to obtain optimal EPBM performance. In addition, this time can be used to ensure that the ground conditioning testing that is done on site is done properly, including the training of personnel as may be required.

Ground conditioning, as the main factor explored here affecting advance rate, is the first line of influence for the contractor/additive supplier/equipment supplier to influence how material is excavated. The GC plan, implemented in front of the cutterhead, impacts the entire operation as the material must flow through the machine, out the heading, over the surface and off the site. It affects every part of the job from the number of tool changes required to the amount of cleanup in the heading and on the surface due to spillage. When this global impact of ground conditioning is taken into account, it makes good sense that advance rates are closely correlated. The authors believe that it is this overarching influence that makes a good GC plan, in combination with an EPBM properly designed for executing the plan, one of the most powerful tools available in achieving good project success.

## REFERENCES

L. Borio, D. Peila, C. Oggeri, S. Pelizza, 2007, *Characterization of Soil Conditioning for Mechanized Tunneling,* Mediterranean NO-DIG XXV International Conference and Exhibition, Roma, Italy. www.iattmed.com/pdf/s6_02_Borio.pdf.

EFNARC, April 2005, Specifications and Guidelines for the use of specialist products for Mechanized Tunnelling (TBM) in Soft Ground and Hard Rock. www.enfarc.org.

Shield Tunnelling Association of Japan, 2011, Technical Guideline for Foam Shield Tunnelling, 7th edition.

E. Dal Negro, A. Boscaro, D. Michelis, C. Campinoti, D. Nebbia, 2013, *Ground conditioning: STEP Abu Dhabi sewer Project*, International Tunnelling Association, Switzerland, proceedings of the World Tunnel Congress 2013 Geneva, Switzerland. www.tunnelbuilder.it/uploads/CMS/Documents/CH248.pdf.

# Automatic Soil Conditioning Through Clay

**Richard McLane**
Traylor Bros., Inc.

**ABSTRACT:** Unique soil conditioning techniques were developed during the Sound Transit University Link Light Rail U220 tunnel project in Seattle, WA. The project consists of twin bore, 21.5 foot diameter EPB TBM tunnels extending 11,400 feet under urban environments and through varying soil formations including gravels, sands, silts and clays. Earth pressures ranged from 2.2 bar to 6.1 bar. This paper outlines the development of the automatic soil conditioning system used on the TBMs through the clay formations.

## INTRODUCTION

The University Link project (Figure 1) extends Seattle's light rail from the downtown Pine Street Stub Tunnel (PSST) to University of Washington Station (UWS). The 5 km (3.15 mi) of twin-bore, 6.56 m (21.5 feet) excavated diameter tunnels were constructed under two separate contracts, U220 and U230. The tunnels from UWS to the Capitol Hill Station (CHS) are part of the U220 Contract, and utilized the UWS site for tunneling operations.

The U220 project was awarded to a joint venture of Traylor Bros., Inc. and Frontier-Kemper Constructors (TFK JV) in March of 2009. Notice to proceed was given January 4, 2010. The tunnels were mined with two Herrenknecht EPB TBMs. The first drive commenced May 31, 2011, with the second drive starting about five weeks later. The first machine holed through into the Capitol Hill station on March 21, 2012, and the second TBM holed through on April 2, 2012.

## LOCAL GEOLOGY

The U220 tunnels were constructed in an area where the present day land surface reflects glacial and non-glacial sediments. The Quaternary geologic history of the Puget Sound region is dominated by a succession of at least six continental glaciations. Because of the significant ice thickness, the soils in the Puget Sound region are generally overconsolidated.

Pre-construction exploration for the project encountered deposits from at least three glacial cycles and three non-glacial cycles. The deposits consist of clays, silts, sands, and gravels in various combinations, relative densities, and consistencies. In many areas, the borings showed soils comprised of glacial tills overlying a glacial lacustrine clay. The hard glacial lacustrine clay deposits are overconsolidated as a result of the glaciations.

The Geotechnical Baseline report (GBR) divided the tunnel into five soil types, four of which were encountered on the project:

- Blue SG represents overconsolidated fine-grained, plastic soils
- Turquoise SG represents overconsolidated fine-grained, non-plastic soils
- Yellow SG represents overconsolidated fine to coarse sand, with varying amounts of gravel, silt, and clay
- Red SG represents overconsolidated coarse sand and gravel, with varying amounts of fine to coarse sand, silt, and clay
- Purple SG represents normally consolidated fine to coarse sand, with varying amounts of gravel, silt, and clay.

Nearly 75% of the drive consisted of the Blue Soil Group.

## SOIL TESTING

Prior to the TBMs being launched, TFK carried out testing of foam conditioning agents for use on the project. It was clear from the GBR, the stickiness of the Blue SG could potentially be a problem, especially for the long screw conveyors on the TBMs, and could lead to clogging of the cutterhead and/or screw conveyors.

TFK tested three different soil conditioners in a small scale laboratory. The test apparatus could be used to test bubble size and life at various pressures, but because the testing regime focused on stickiness, the tests were done at atmospheric pressure.

The three conditioning agents tested were:

- Boraid–Soilax S (surfactant)
- Condat–CL B4/AC (surfactant + anti-clay)
- BASF–Rheosoil 127 (surfactant + anti-clay)

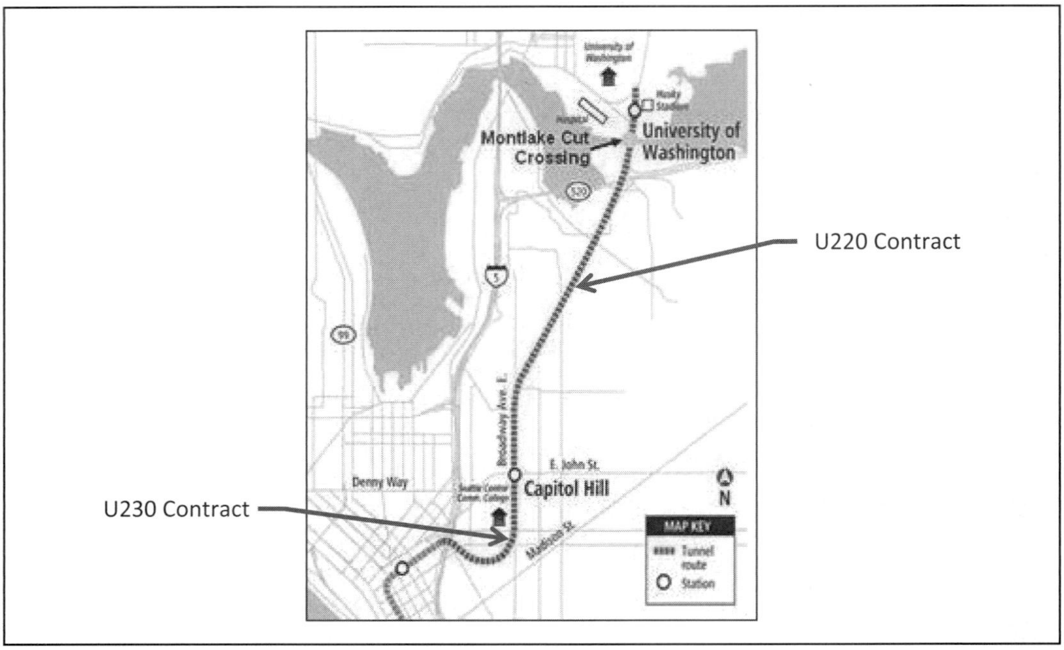


**Figure 1. University link alignment**

Test results concluded that the Condat and BASF products, both of which contain surfactant and anti-clay agents, greatly reduced the stickiness of the muck. Boraid's Soilax-S had no effect on the stickiness.

During the testing, TFK used traditional settings for clay and set the Foam injection Ratio (FIR) at 30%, and the Foam Expansion Ratio was set at 8:1. Concentrations of the conditioning agents were set at 2% by volume.

Test results indicated the stickiness could be managed and its effects minimized through the use of the anti-clay soil conditioners.

## DEVELOPMENT

Because of the semi-short startup configuration at the UWS launch box, muck was removed initially from the TBMs using several 18 cyd muck boxes. Once the TBMs had tunneled far enough (900 feet), the muck removal system was switched over to a 26 inch wide continuous conveyor belt due to steep grades (up to 4.5%) along the alignment. The use of a conveyor belt was mandated by the Contract Specifications.

Beginning at the headwall, TFK encountered the Blue SG overly consolidated lacustrine clay. It was clear that the material was difficult to handle and, as feared, it easily plugged the cutterhead and the screws. The early attempts to condition the material with anti-clay foaming agents with traditional FIR and FER settings proved difficult because the Blue SG material which had such low permeability and high unit weights, that it would collapse the foam bubbles. The collapse of the foam bubbles at 2.3 bar (Earth pressure at launch), meant that difficulties certainly would occur at the highest anticipated EPB pressure of 6.1 bar. Furthermore, the collapsed foam bubbles would collect at the top of the excavation chamber, and initially this air had to be "burped" by opening a two inch ball valve in the crown of the bulkhead. The "burping" of the air at the top of the chamber, although effective, would drastically increase the likelihood of a collapsed face, and thus cause settlement. TFK quickly moved to reduce FER as low as possible with the intent of keeping the chamber full at all times and not relying on human involvement to relieve the air bubble. The reduction of FER to 0.5 was extremely effective in reducing the air bubble at the top of the chamber.

Because the initial portion of the drive utilized muck boxes, controlling the behavior or consistency of the material, although difficult, was not hindering TBM advance. TBM operators made the material as wet as possible to facilitate it moving through the cutterhead, the three screw conveyors and into the muck boxes. In basic terms, they took the material past the liquid limit on the water content continuum and turned it into a slurry. Because the material flowed as a liquid, boxes were easily filled, but because the operators were using so much surfactant to get the material flowing, the muck was too wet

**Table 1. U220 Blue soil group properties**

| Soil Group | Wet Unit Weight (pcf) | Dry Unit Weight (pcf) | Water Content (%) | Liquid Limit (%) | Plasticity Index (%) |
|---|---|---|---|---|---|
| Blue | 128 ±4 | 100 ±6 | 28 ± 8 | 48 ±14 | 24 ±5 |

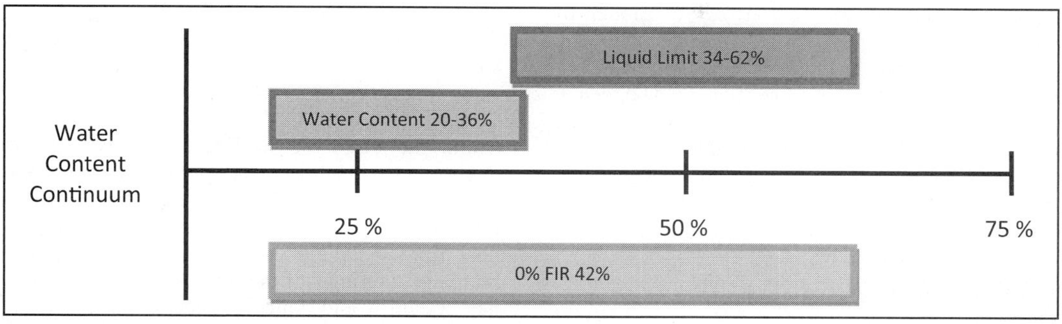


**Figure 2. Graphical representation of Blue SG properties**

and consumption of the concentrate was nearly 20 times normal use. As a result, TFK decided to cut off the use of soap concentrate and allowed only water for the TBMs until a better solution could be found.

As mentioned above, once the advance had reached about 900 feet, the mucking was switched over to the continuous conveyor belt. At first, the TBM operators were given full reign over the soil conditioning. Their operational thought process was the same as with the boxes: get the material wet enough to pass through the cutterhead and screws. Because the clay was being taken past the liquid limit, the material would not stay on the belt to be transferred out of the tunnel, up the 15 degree incline at the shaft and to the muck bin. Machine advance rates went from 20 min/push to 2–3 hours/push. Additionally, the material consistency proved challenging when using the conveyor weigh scales on the TBM, as outlined in the 2013 RETC paper titled, "TBM Conveyor Belt Scales: The University Link Project Experience" (Banerjee et. al).

It was clear to TFK management that something had to change. The final solution was developed through several iterations. Clearly, to keep material on the belt it had to be dry, yet plastic enough to move through the cutterhead and screw conveyors. The material had to be taken as close to the liquid limit without going over and turning to slurry.

## WATER CONTENT, LIQUID LIMIT & FIR

Using the data from the GBR, TFK was able to estimate the required FIR over the rest of the drive. Table 1 outlines soil properties for the blue soil group for the U220 project.

From the graphic in Figure 2, it can be seen that an estimate can be made of the FIR needed to reach the liquid limit of the material. Because the lower limit of the in situ water content could range between 20–36%, and the material Liquid Limit could be 34–62%, a wide band of FIR exists to transform the material into a flowable plastic. For U220, TFK experienced FIRs anywhere from 4%–30%, and a general average was between 17–25%.

Since the in situ water content and the material's liquid limit properties could change over a short distance, it was important that the ground conditioning system could react quickly enough to maintain a uniform consistency.

TFK tunnel engineers limited the amount of conditioners, both liquid and air, that operators could inject into the excavation chamber by setting limits with the TBM's Programmable Logic Controller (PLC) soil conditioner module. This narrowed the range of operational parameters the TBM operators could adjust during an advance helping to refine the correct setpoints. Additionally, the FIR was tied directly to TBM advance rate such that the injection rate matched the excavation rate as is done on nearly every EPB TBM and is the typical "Automatic" foam mode found on most TBMs.

## ANTI-CLAY ADDITIVE

TFK discovered that there existed a very narrow operational FIR band within which the material would turn from a pliable plastic to a liquid when using only water as a conditioning agent. Additionally, re-amalgamation of the clay would happen once the material exited the screw conveyor. The addition of a gridded bar, a "grizzly bar," helped minimize the re-amalgamation, but the introduction of a small amount of the anti-clay additive greatly increased the of the operational bandwidth. The use of the agent was not intended to form foam bubbles in the traditional foam sense, but rather to act as a clay solvent. By using a small fraction of the concentrate, 0.2%–0.4% by volume, the range of plasticity would widen and prevent the re-amalgamation of the clay as it exited the screw. The percentage was found, like most foam parameters, through trial and error. Not enough of the anti-clay agent and the material would make long logs, labeled "tuna" by the miners. With too much of the additive, a soap coating, or sheen could be seen in the material. When dosed properly, the material has a dull complexion, while the material coming out of the screw had a crumbly, "gorgonzola cheese" like consistency.

## SYSTEM FEEDBACK

The final step in conquering the Blue SG was to fully control the FIR by the TBM PLC with minimal operator input. In order to determine the plasticity of the clay, feedback information was needed. Through direct observation, TFK discovered that the hydraulic drive pressures of the TBM cutterhead and number one screw provided the direct feedback needed to indicate the material's plasticity. The hydraulic drive pressures directly equate to torque of the cutterhead and screw conveyor, therefore, material consistency could be controlled via a feedback loop analyzing the cutterhead and screw torques and drive pressures within the TBM PLC.

*System parameters within the PLC:*

- Cutterhead Drive Pressure above 190 bar = Material too dry, at the Plastic Limit
- Cutterhead Drive Pressure below 125 bar = Material too wet, at the Liquid Limit
- Screw Drive Pressure above 125 bar = Material too dry, at the Plastic Limit
- Cutterhead System Reaction Time = 45 seconds
- Screw System Reaction Time = 90 Seconds
- Anti-Clay concentration = 0.2–0.4%
- PLC Controlled Foam Injection Ratio (FIR) = 0–30% (variable)
- Foam Expansion Ratio (FER) = 0.5 (fixed)
- TBM advance rate = 0–100 mm/min
- Cutterhead speed = 2 RPM (stage 1)

*System logic within the TBM PLC:*

- Every 45 seconds, system snapshot, evaluate
  - If Cutterhead Drive Pressure above 190 bar, then Increase FIR by 0.1%
  - If Cutterhead Drive Pressure below 125 bar, then Decrease FIR by 0.1%
- Every 90 seconds, system snapshot, evaluate
  - If Screw Drive Pressure above 125 bar, then Increase FIR by 0.25%
- Adjust foam pump flow based on TBM advance rate to maintain FIR
- FER is fixed at 0.5, but the volume of air introduced is calculated based on EPB pressures.

Within the PLC, the feedback loop executed at 45 seconds and 90 seconds for the cutterhead and number one screw conveyor, respectively. These intervals relate to the reaction time of an adjustment from the previous executed adjustment and the realized change. If the time was too short, then they system would typically inject either too much or not enough of the soil conditioner and adjust FIR too quickly. These time intervals would be different based on various TBM configurations and ground conditions.

## TYPE C FOAM

In the EPB Tunneling industry, there are two generally accepted types of foam, 'Type A' and 'Type B.' These two types of foam were developed for various grain size distributions and are discussed in depth in the 1999 RETC paper titled, "Soil Conditioning for EPB Shield Tunneling of the South Bay Ocean Outfall" (Williamson et al. 1999). During the project's development of the automatic soil conditioning through the clay, TFK developed a new type of foam for used in ultra-fine grain soils.

Figure 3 shows the traditional Grain Size Distribution graph with four zones; each of the zones show when to use 'Type A' or 'Type B' foam, depending on the bore log grain size distribution.

As mentioned previously, TFK found that traditional foam did not work in the Blue SG. Using typical FIR and FER settings, as well as surfactant concentrations, did not transform the material in to a flowable material, and negatively impacted the face stability by creating the air bubble at the top of the excavation chamber. As a reference, typical FIR and FER settings would be 20–30% and 8:1 respectively, and surfactant dosage is around 2–4%.

Through the development of the automatic soil conditioning and experimenting with all of the foam settings to transform the in situ material to an earth paste that can support the face, pass through the cutterhead and screws, and stay on the conveyor belt, TFK considers the foam used as a new type foam: 'Type C' foam. Figure 4 shows a typical grain size distribution for the Blue SG. Also shown is a new 'Zone V' where the majority of ultra-fine grained material passes a #200 sieve.

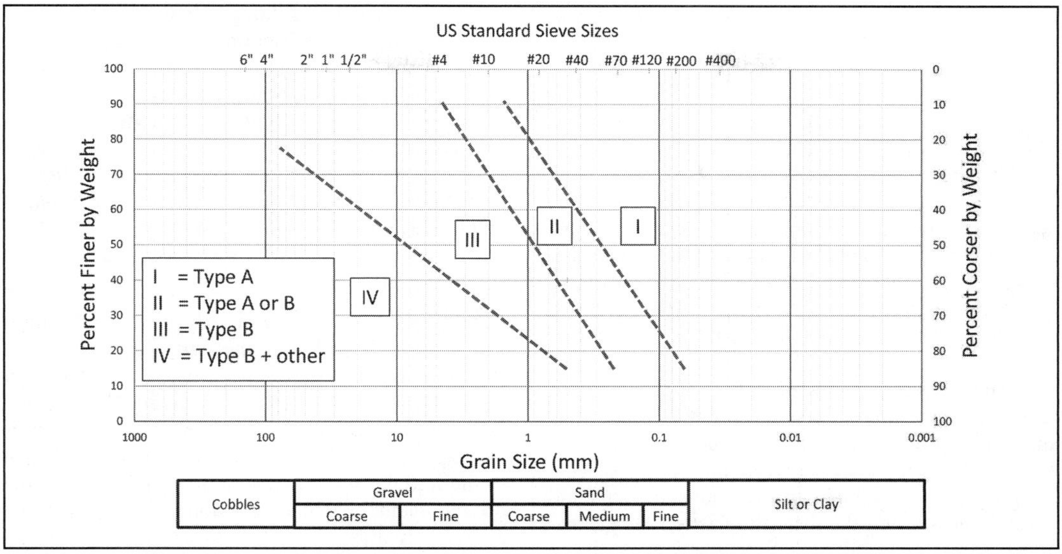


**Figure 3. Traditional foam type vs. grain size**

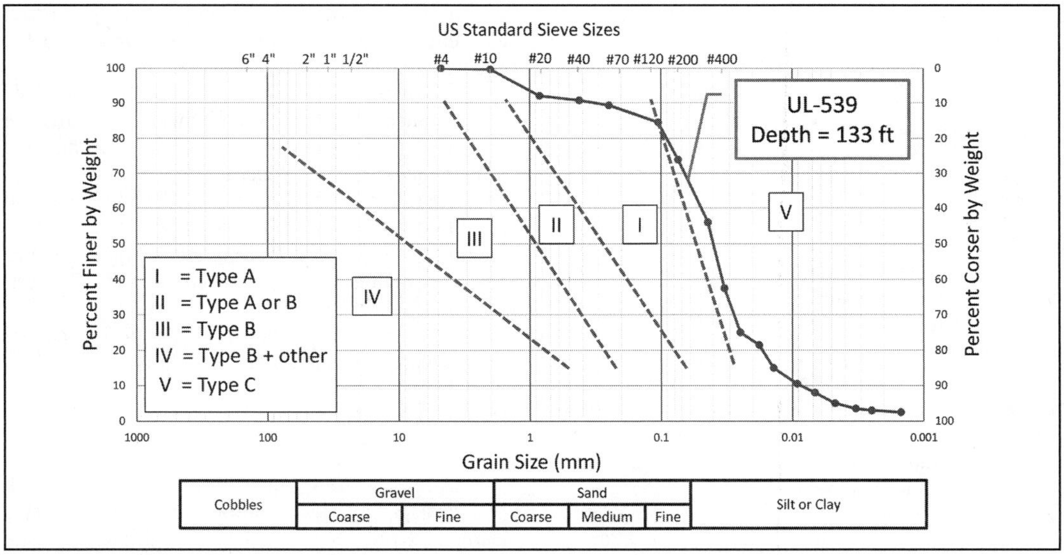


**Figure 4. Foam types vs. grain size**

As mentioned previously, the FER and FIR, as well as surfactant concentrations were drastically different than typical foam usage. TFK used FIRs between 4–30%, FER at 0.5:1, and surfactant with anti-clay agents at 0.2–0.4%. Note that the 'Type C' foam doesn't actually even look like traditional foam; visually it looks like soda water with large amounts of liquid and very small amounts of air.

## SYSTEM MODELING

As stated previously, TFK used the drive pressures between 125 and 190 bar, and the cutterhead running in Stage 1. It should also be noted that the feed pressure is nominally 25 bar. Subtracting feed pressure, and converting, cutterhead drive pressures are converted to cutterhead torque:

- 125 bar – 25 bar = 100 bar = 1,945 kNm Lower Limit
- 190 bar – 25 bar = 165 bar = 3,210 kNm Upper Limit

These torque values directly correlated with material consistency.

In order to use these values for various cutterhead sizes, modeling of the interaction of the cutterhead and the ground, as well as the earth paste between the cutterhead and the bulkhead in the excavation chamber can be done. The equation that most closely matches the interaction is that of a rotating multiple disk friction clutch.

The torque capacity of a friction clutch is a function of the normal pressure applied, the inner and outer radii of the disk, the number of surfaces, and the friction factor between the surfaces. The equation below outlines the relationship between Torque, pressure on the disk, the friction factor, the outer radius of the disk, and the number of disks; note that inner radius has been removed from the final equation:

$$T = N \cdot \int_{r_i}^{r_o} 2 \cdot \pi \cdot r \cdot p \cdot f \cdot r \cdot dr$$

$$T = \frac{2}{3} \pi p f r_O^{\,3} N$$

where

$T$ = torque, Nm
$r_O$ = cutterhead outer radius, m
$p$ = earth pressure, kPa
$f$ = friction factor
$N$ = number of disks

To use the equation, solve for the friction factor at both the lower limit and the upper limit using the calculated torque from above, average EPB pressure, Cutterhead radius, and setting N=2.

$$1{,}945{,}000 \text{ Nm} = \frac{2}{3} \cdot \pi \cdot 3.75 \text{ bar} \cdot \frac{100 \text{ kPa}}{1 \text{ bar}} \cdot f_{lower} \cdot \left(\frac{6.59}{2}\right)^3 \cdot 2$$

$$f_{lower} = 35$$

$$3{,}210{,}000 \text{ Nm} = \frac{2}{3} \cdot \pi \cdot 3.75 \text{ bar} \cdot \frac{100 \text{ kPa}}{1 \text{ bar}} \cdot f_{upper} \cdot \left(\frac{6.59}{2}\right)^3 \cdot 2$$

$$f_{upper} = 57$$

Coincidentally, although seemingly unrelated, the upper and lower friction factors are very close to the in situ liquid limits for the material shown in Table 1. This could be explained by the fact that as liquid is being injected into the cutterhead, the friction against the cutterhead is reduced and thus a reduction in torque is seen. Therefore, it is possible to determine initial operational torque limits based on GBR established liquid limits.

## CONCLUSIONS

Conditioning the clay with only water and dumping it into a muck box at the start of tunneling was straightforward. Challenges arose when the clay needed to be placed it on a belt moving 600 feet/min for disposal outside of the tunnel. The use of anti-clay agents helpful, but not in the traditional form as the initial threat of stickiness did not become the driving factor in the excavation, whereas material consistency did.

The conditioning of the clay to a consistency that was workable with the conveyor belt mucking system and compatible with the TBM Belt scales lead to new developments in both foam type, usage, as well as how the foam is proportioned into the excavation chamber during excavation.

Production rates: The first TBM finished three months ahead of schedule with an average advance rate of 62 feet/day and the second machine finished four months ahead of schedule with an average advance rate of 64 feet/day. Once the automatic soil condition system was fully operational, average production rates were approximately 75 feet/day. The best production from one machine was 175 feet/ 24 hours.

Further reading about the U220 project can be found in previous RETC and NAT proceedings.

## REFERENCES

Banerjee, McLane, Shorey, Gasson, "TBM Conveyor Belt Scales: The University Link Project Experience," Proceedings, RETC 2013.

Burdick, Krulc, McLane, Brandt, "University Link Light Rail TBM Tunnel UWS to CHS Contract U220: Case History," Proceedings, RETC 2013.

Menke, McLane, Salvati, Sleavin, "Design and Construction of Light Rail Tunnels Under Montlake Cut: Sound Transit University Link," Proceedings, NAT 2012.

Williamson, Traylor, Higuchi, "Soil Conditioning for EPB Shield Tunneling on the South Bay Ocean Outfall," Proceedings, RETC 1999.

# Relationship Between Cutterhead Torque and Tool Wear: A Laboratory-Scale Study

**Ehsan Alavi Gharahbagh**
Jay Dee Contractors, Inc.

**Jamal Rostami and Kaveh Talebi**
The Pennsylvania State University

**ABSTRACT:** During the last 50 years, soft ground mechanized tunneling has witnessed significant improvements in machinery and methods. Despite advancements in the use of soft ground Tunnel Boring Machines (TBM) such as slurry and Earth Pressure Balance (EPB), prediction and quantifying the primary wear of cutting tool and secondary wear on other components of the machines is an issue that has remained challenging. This is due to the complex tribological system at the face and along the path of the muck which include two and three body wear, presence of water, various hardness of the tool and other machine parts, mineralogy of the soil, grain size/shape, deposition and consolidation of the soil, the use of soil conditioning and anti-abrasion agents, face pressure, and finally machine operating parameters. Limited amount of research has been performed to characterize tool wear in soft ground tunneling environment. This paper offers a brief review of the ongoing research on this topic around the world and introduces the Penn State Soil Abrasion Index (PSAI) to provide a basis for soil abrasion measurement for geotechnical investigations. The results of testing in various soil types and the impact of different parameters on the measured wear and torque in the testing device will be discussed. In addition a parametric study of the effect of parameters such as grain size distribution, water content, and soil conditioning on torque and wear will be presented. The paper will offer some recommendations relative to the practical implications of the results of the testing program and related analysis.

## INTRODUCTION

The demand for underground structures and tunnels in urban areas has been on the rise around the world. These structures are often built in soft ground (i.e., soil) and are mainly bored using Tunnel Boring Machines (TBMs). For TBMs, the issue of primary wear on cutting tools and secondary wear on other components that come to contact with soil are crucial since in many cases tool inspection, maintenance, and replacement are done under extremely difficult conditions.

Many factors influence soil abrasivity. These factors include in-situ shear strength, heterogeneity, unit weight, porosity, mineral composition, grain size distribution, sphericity and roundness, cementation, and moisture content. The Soil Abrasion Test (SAT) developed by the Norwegian University of Science and Technology (NTNU) and the LCPC Test developed by the Laboratoire Central des Ponts et Chaussées, France (Thuro and Plinninger 2007) have been developed during the last decade to address the soil abrasivity. These tests have been examined in several publications by the authors (e.g., Alavi Gharahbagh et al., 2010 & 2011) and inherently have some short comings that will limit their use for application in soft ground tunneling. Some of these short-comings are as follows:

- The tests are performed in dry conditions, whereas the field conditions are in variable moisture conditions, most often saturated soils.
- In-situ soil conditions, including grain size distribution, are altered during the sample preparation.
- Grain shapes are disturbed during the sample preparation for testing.
- The tool and the soil pressure/contact stress are not comparable between the field conditions and the test conditions.
- Other parameters such as the presence of high ambient water pressure (which could reach up to 17 bars) are not accounted for.
- Both tests are limited by the size of the largest grain that the testing equipment can handle (4 m and 6.3 mm), but in the field, it is possible to have gravel and cobbles (<300 mm or 12 in).
- Tunnel practitioners often utilize various types of soil conditioners to improve cutting

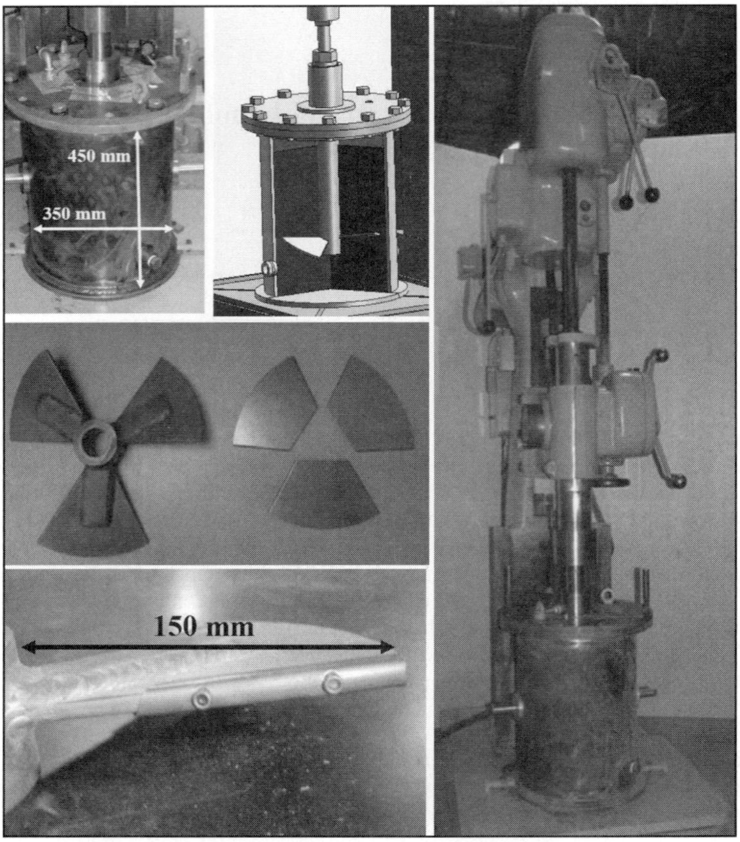


**Figure 1. Penn State soil abrasion testing device: cylindrical chamber, device overview; propeller and metal covers**

efficiency. The existing tests cannot measure such working conditions and their impact on soil abrasion.

Study of soil abrasion is started in 2009 at Penn State University which will be discussed in the following section. The testing included a close simulation of working conditions of soft ground tunneling machines in a small chamber. The preliminary results were very promising and the concept has been considered for use by other research groups. There are some simultaneous research activities by various researchers including the work underway at Torino Italy (Peila 2012), at SINTEF (Jakobsen 2010), and at Switzerland (Barzegari et al., 2012) that concur with the Penn State soil abrasion testing research.

## PENN STATE SOIL ABRASION TESTING SYSTEM

The testing system consists of a cylindrical chamber, where the soil can be loaded without any changes to its nature and grain size distribution. The diameter and length of the chamber are 350 mm and 450 mm (14×18 in), respectively. The unit is mounted on a 5-hp drill press which delivers the rotational torque at a rotational speed of 60–1100 rpm to the drive unit which in turn transfers the motion to the propeller through a solid shaft. The chamber is partially filled with soil samples to the height of 300 mm (12 in) and a propeller with three blades is lowered into the chamber. The final position of the propeller during the test is approximately in the middle of the soil column with 150 mm (6 in) soil above and below the propeller blades. The wear specimen is a cover that is mounted on each blade. For each test, the covers are weighed before and after the test; the total weight loss of the covers is the tool wear and can be considered as a measure of soil abrasion. The device and its various components are shown in Figure 1.

The testing system can accommodate various moisture conditions including dry, wet, and saturated (i.e., submerged) and under up to 10 bars of ambient pressure. Tests were performed on various soil samples with known properties including grain

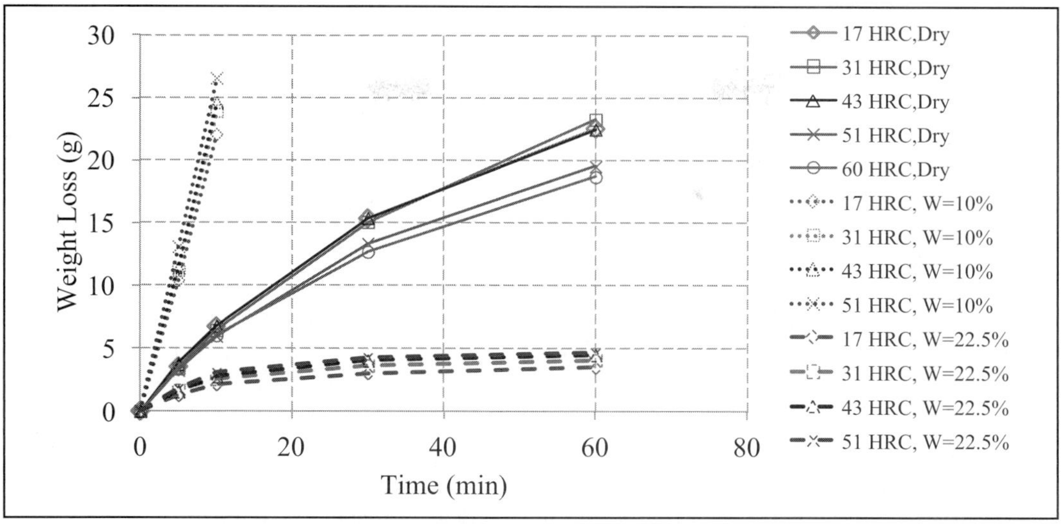


**Figure 2. Test results of Silica sand for dry, moist, and saturated conditions using various covers**

size distribution, mineral content, and grain sphericity and roundness. The samples included Silica sand with high quartz content, Limestone sand, ASTM standard sands, and Silty sand.

In addition, several tests were performed on a series of soil samples from several ongoing tunneling projects around the US (e.g., WSSC tunnel in Washington DC area, University Link and Brightwater tunnel projects in Seattle, WA). The test matrix included testing these soils under dry, wet and saturated conditions and also with various ambient pressures. The rotational speed of the propellers is also a variable that can be changed if needed. While majority of tests were performed at 60 rpm, some higher speed tests up to 180 rpm were also performed. The test duration was a function of observed wear and varied from 5–60 minutes. The propeller can be arranged with various pitch angles (10, 20, 30 degrees). The hardness of the metal covers was also varied by using different heat-treated high-grade steel covers made of AISI 4130 steel alloy with Rockwell hardness (HRC) of 17, 31, 43, 51, and 60. In addition, the issue of tool wear under high ambient pressure is being studied and the results indicated that by increasing the ambient pressure, the amount of weight loss on the covers increases (Rostami et al., 2012). This confirms the actual phenomena that happens in the field and has been observed by various machine manufacturers and operators. Over 200 sets of tests were performed at various settings. Each set included several stops to measure progression of wear on the covers at various time intervals. The figures are presented to demonstrate typical test results. Figure 2 shows typical soil abrasion test results for a Silica sand in dry, 10% water content, and saturated conditions using various hardness. This figure demonstrates the effect of water content and metal cover hardness on soil abrasion and tool wear. To address the effect of the relative hardness of metal cover and soil on tool wear, a series of tests were performed where mixtures of Silica sand (i.e., higher hardness) and Limestone sand (i.e., lower hardness) at controlled proportions were used to create variable ratios of tool/mineral hardness (Mosleh et al. 2012). Figure 3 shows the weight loss on the covers verses the hardness ratio (tool/mineral). Rostami et al. 2012, Mosleh et al., 2012, and Alavi Gharahbagh et al., 2012 discussed the testing parameters in more details. In the following section, the effect of torque on several parameters of interest in the experimental setting is presented.

## EFFECT OF TORQUE ON SOIL ABRASION

Torque is one of the main operational parameters in soft ground mechanized tunneling. In theory, the force that is required to turn the cutterhead is directly related to the anticipated wear on the tools. Therefore, if there is better understanding of what kind and amount of wear takes place over given operational time frame, then a reliable estimate of the time that tools need to be changed can be calculated rather than simply guessing.

In order to accommodate torque in this experimental study, a direct torque-measuring system was designed and installed on the testing device. This system contains a 4500 N (~0.5 ton) capacity round turn table with a diameter of 310 mm (~12 in) that is bolted to the base of the drill press (Figure 4a). The chamber is centered and secured on top of the turn table (Figure 4b). The turn table allows for the

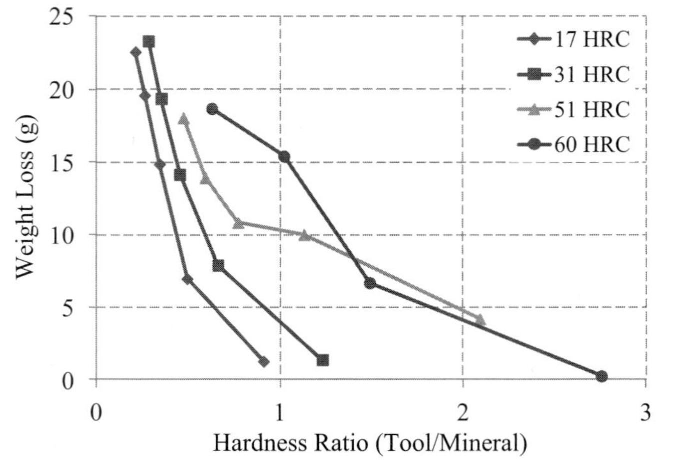


**Figure 3. Weight loss of covers versus hardness ratio (tool/mineral) in dry sand mixtures**

rotation of the chamber during the test. Two arms instrumented by using individual S-shape, 100-kg (200 lbs) capacity load cells are attached to the chamber in order to measure the applied force for rotation of the chamber and also to secure the chamber from rotation (Figure 4c). By considering the clockwise rotation of the propeller during the test, the arm on the left side of the chamber registers compression forces while the arm on the right side of the chamber registers tension forces. The data from two S-shape force meters are monitored by using the computer-based data acquisition system. By considering the distance between the center of propeller inside the chamber and the location of the attached arm to the chamber (22 cm or ~9 in), torque can be calculated.

Several key tests were performed in order to investigate the relationship between torque and different parameters of interest such as rotational speed of the propeller, water content, pitch angle, and weight loss of the covers using the direct torque measurement system. Figure 5 displays the measured torque on Silica sand samples in dry condition and by using different rotational speeds during the first 5 minutes of testing. As it can be seen in Figure 5, by increasing the rotational speed, starting value of torque increases as it was expected. In addition, the applied torque to the sample by using 60 and 105 rpm is more stable in compare to the applied torque by using 180 rpm set-up.

Figure 6 shows the measured torque for individual tests by using 60 rpm, 105 rpm, and 180 rpm in 5 and 15 minute time steps in addition to the maximum measured torque with respect to rotational speed and weight loss after 15 minutes of testing. As it can be seen from parts (a) to (c) of Figure 6, in contrast to 105 rpm and 180 rpm set-ups in which, as the test progressed, the amount of applied torque decreases, in 60 rpm set-up the amount of applied torque increases as the test progressed. One of the explanations can be that, at higher rotational speeds, while the overall applied torque is higher, the amount of dynamic compaction of the sample is less and thus torque gradually decreases as the test proceeds. Meanwhile part (d) of Figure 6, confirms a linear relationship between rotational speed and mean value of torque as well as rotational speed and weight loss in the Silica sand samples.

Figure 7 summarizes the results of testing on dry clay samples (sample is obtained from U230 project tunnel in Seattle, WA) at 60 and 180 rpm. Figure 7a displays the comparison of the testing results in the first 5 minutes of testing by using 60 and 180 rpm. As it can be seen, during the first 5 minutes of the test at 60 rpm, the amount of torque continuously increases as the test progressed until it reaches a stable compaction level. After this stage, the amount of applied torque remains constant during the next 10 minutes of testing (Figure 7b). In contrast, at 180 rpm, the amount of applied torque continuously decreases as the test progressed. The amount of weight loss decreases by increasing the rpm (Figure 7d).This result is consistent with the results of testing in Silty sand samples as discussed by Rostami et al., 2012.

In addition, the effect of pitch angle on torque was studied by performing a set of tests using 10, 20, and 30 degrees pitch angle on Silica sand samples. Figure 8 shows that despite the increase in torque at higher pitch angles, the weight loss of the covers decreases in both 60 and 180 rpm tests.

**Figure 4. Direct torque measurement system set-up: (a) installed turn table on the base of the testing device, (b) secured chamber by using four rollers, (c) S-shape force-meters installed on both sides of the chamber**

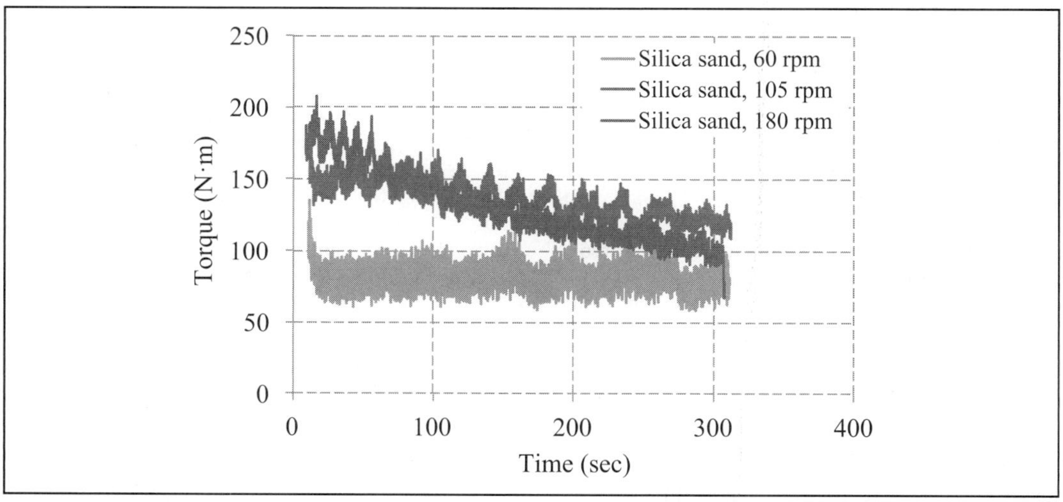


**Figure 5. Measured torque on dry Silica sand by using 10° pitch angle propeller with different rotational speeds**

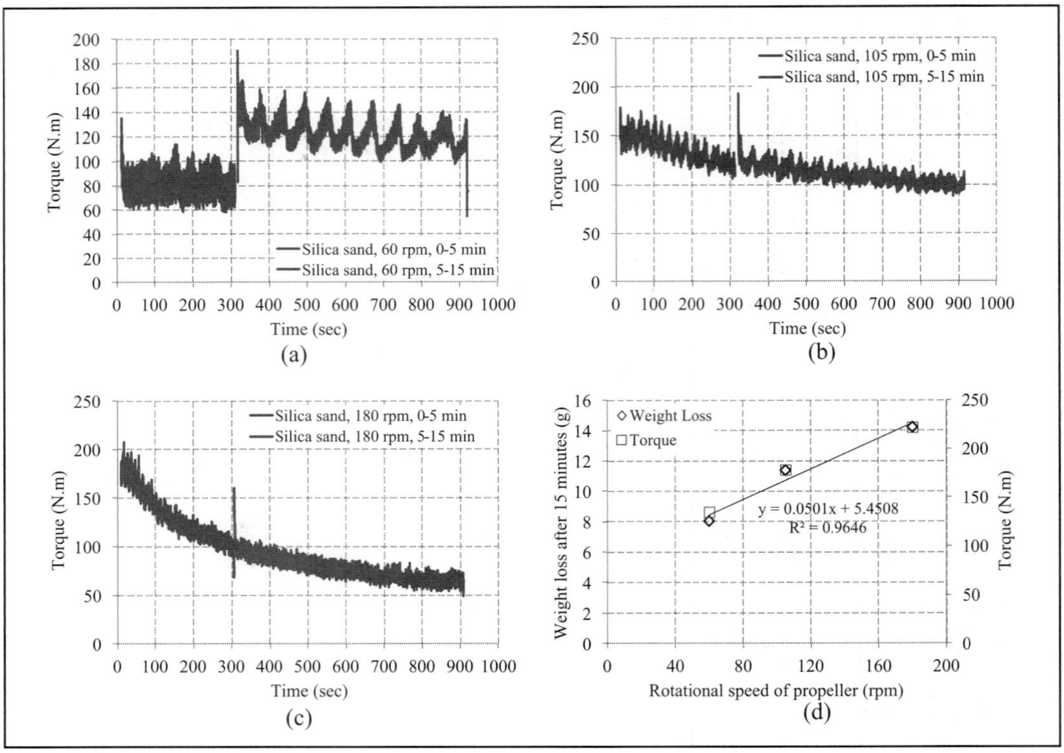


**Figure 6. Measured torque on dry Silica sand by using 10° pitch angle propeller with (a) 60 rpm, (b) 105 rpm, (c) 180 rpm, and (d) rotational speed of propeller versus torque and weight loss after 15 minutes of testing**

Furthermore, the effect of water content on torque is studied by performing a set of tests at 12.5% and 22.5% water content Silica sand and Limestone sand by using 10° pitch angle propellers and 60 rpm rotational speed. The result of this study is summarized in Figures 9 and 10.

Water content plays a significant role on abrasion and wear as well as torque requirement of the machine. As one can see in Figure 9 the water contents close to dry of optimum in compaction test (W = 12.5%), cause much higher torque compared to dry and saturated (W = 22.5%) conditions. It should be noted that due to a very high amount of torque applied to the Limestone sand sample, when testing at 12.5% water content, the safety shear pin was cut off and the test stopped. It is interesting to consider that the amount of applied torque at 12.5% water content test was much higher in Limestone sand as compared to Silica sand, but the amount of weight loss of covers or wear was much higher in Silica sand samples. This shows that high torque values individually cannot be used as a measure of abrasion and a combination of different variables such as mineral content, sphericity and roundness, grain size distribution, etc. should be considered in the wear analysis.

The results of torque studies in addition to the performed parametric study were used to study the impact of different variables on wear and hence examine the sensitivity of the results to each testing parameter. The analysis of available data has allowed the research team to select a special test setting as the basis for standard soil abrasion index test. This means that these operational settings will be kept constant, so that the wear can be measured for various soil samples and thus the results be used as a quantitative measure of abrasion properties for the given soil sample.

## PENN STATE SOIL ABRASION INDEX

The standard setting for the testing device is obtained based on the parametric study on different parameters. A detailed study about the parameters of interest can be found in Rostami et al., 2012, and Alavi Gharahbagh et al., 2012. This setting includes the rotational speed of 60 rpm, propeller pitch angle of 10°, and cover hardness of 17 HRC. The test will be performed at various moisture contents including a dry sample, soil with water content dry of optimum compaction, and saturated soil. The Penn State soil

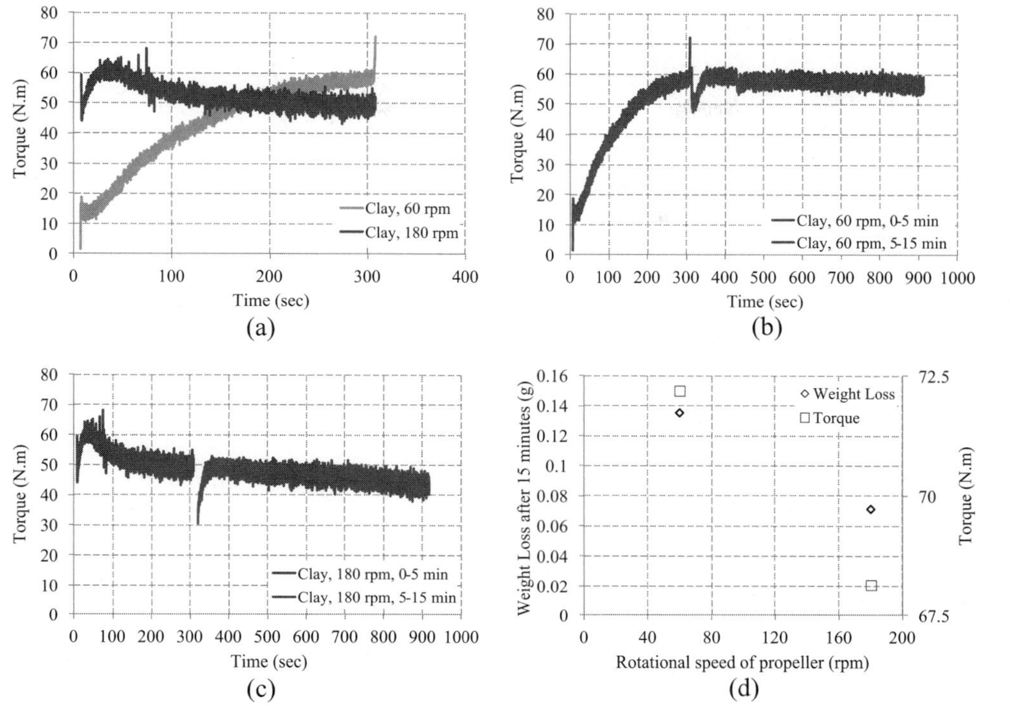


**Figure 7. Measured torque on dry clay by using 10° pitch angle propeller with (a) 60 and 180 rpm in 5 minutes, (b) 60 rpm in 5 and 15 minutes time steps, (c) 180 rpm in 5 and 15 minutes time steps, and (d) rotational speed of propeller versus torque and weight loss after 15 minutes of testing**

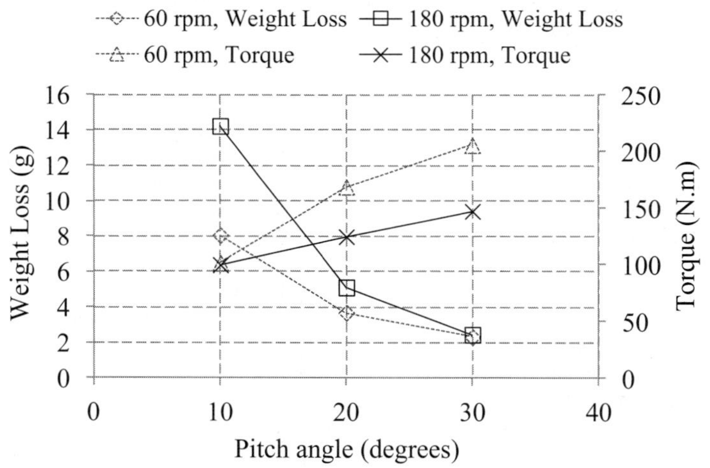


**Figure 8. Effect of pitch angle on torque and weight loss of covers in 60 and 180 rpm set-ups**

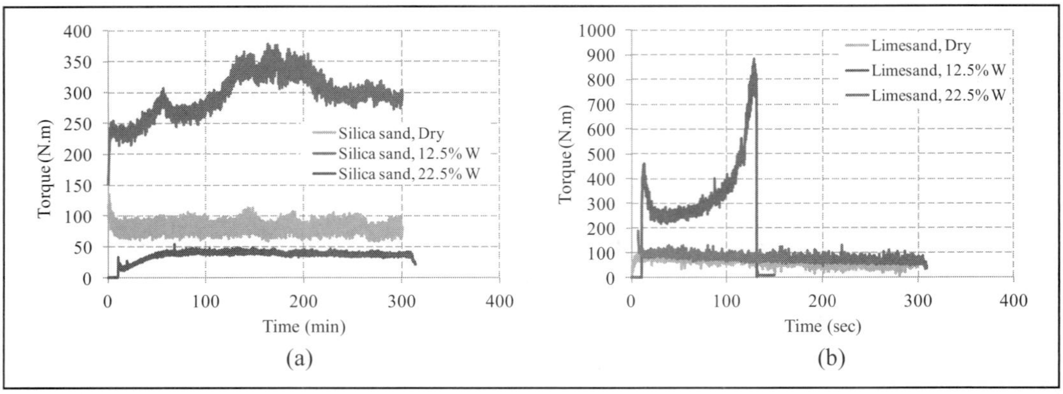


**Figure 9. (a) Comparison of torque with different water contents in Silica sand and (b) comparison of torque with different water contents in Limestone sand**

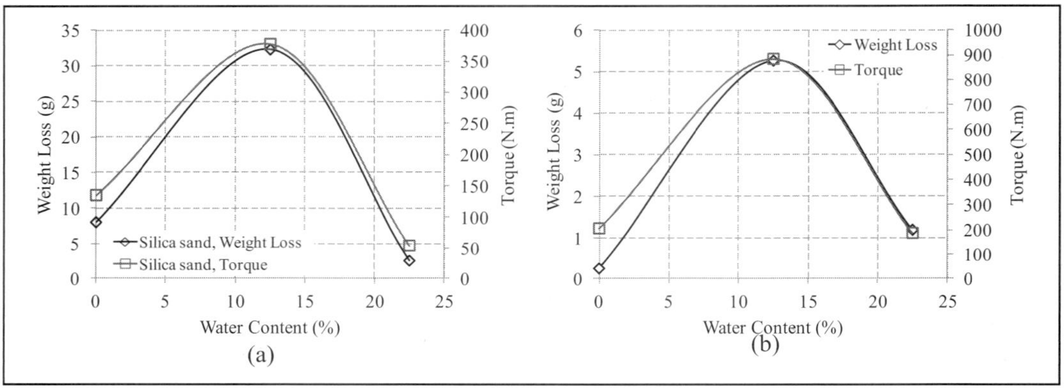


**Figure 10. (a) Effect of water content on torque and weight loss in Silica sand and (b) effect of water content on torque and weight loss in Limestone sand (the reported weight loss is for 15 minutes of testing)**

abrasion index (PSAI) which is introduced in this section is the result of wear measurements using the defined standard test setting on the developed soil abrasion testing device. The observations during testing at various conditions show that at wear rates of above 20 gram, the blade covers do experience some losses in the surface area towards the lower side of the propeller around the outer edge (Figure 11). This refers to actual removal of the corners and sometimes in small pieces.

This means that the propeller blades will get exposed to highly abrasive soils and will wear and incur permanent damage if the tests continue beyond this point. In other words, there are cases that the testing had to be stopped within a few minutes to avoid permanent damage to the propeller or the equipment. In addition, the accuracy of the testing results could be compromised since instead of weight loss on the covers; it would be registered on the propeller blades which are not measured. This is while in some other cases, the covers hardly show any sign of wear even after 60 minutes of testing. Therefore, there was a need to develop a procedure to allow for comparing the weight loss of covers in different soil types based on the same testing time under the same testing conditions to define a standard abrasion index.

To achieve this goal, the test results for various test durations could be combined in a graph and a best-fit curve will be developed to estimate the characteristics of the soil. Given the shape of the wear curves when plotted against time (refer to Figure 2, performed tests in dry and saturated conditions after 60 minutes of testing), a power function seems to offer the best fit. Thus the variation of the wear on the cover as a function of time can be expressed as follows:

$$W = AT^b \quad (1)$$

where W is wear (gram), T is time (minutes), and "A" and "b" are constants defining the shape of the curve. With the use of "A" and "b" in this study, one

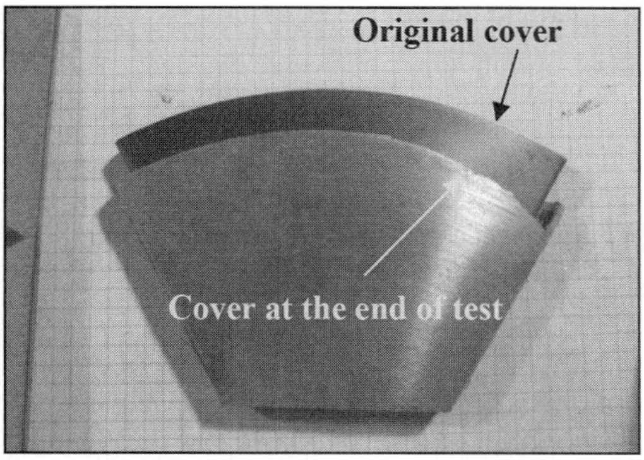


**Figure 11. Significant damage to the covers at the end of the test**

**Table 1. Criteria for the PSAI soil abrasion index**

| Classification | Weight Loss (g) After 60 Minutes Based on $W=AT^b$ |
|---|---|
| Non to very low abrasivity | <2 |
| Low abrasivity | 2–5 |
| Medium abrasivity | 5–10 |
| High abrasivity | 10–15 |
| Extremely high abrasivity | >15 |

can extrapolate the measured wear on the covers for shorter tests. This facilitates estimation of the anticipated wear for a given time frame. For example, with only two measurement of wear within the first 5–10 minutes and using the origin (0,0) as a starting point, a best-fit curve can be developed for the available data to estimate the anticipated wear at 30 and 60 minutes. The measured or calculated wear at 60 minutes (from the formula) is defined as the Soil Abrasion Index or "PSAI." Moisture content can be noted with the subscript to indicate the testing conditions. As such $PSAI_D$, $PSAI_S$, $PSAI_{10\%}$ would represent the abrasion index in Dry, Saturated, and at 10% water contents, respectively. This approach allows for expansion of the application of the soil abrasion index. While "A" signifies the magnitude and intensity of wear, the power "b" shows the long-term effect of continued testing. "b" values of over one shows the increasingly aggressive abrasion while values below one show somewhat slower wear as the testing continues. A more detailed analysis of the PSAI is discussed in Alavi Gharahbagh et al., 2013.

## SOIL CLASSIFICATION BASED ON PSAI

Given the ability to measure the soil abrasion index using the new testing device at Penn State University, a preliminary classification has been developed to offer a qualitative description of abrasivity of different soil types. Table 1 shows the suggested soil abrasion classification based on the PSAI soil abrasion index. This is an introductory classification and will be subject to future modification based on comparison of the measured index and the wear on various machine components observed in the field.

This preliminary classification for soil abrasion based on the PASI can be used for classification of different soil types for selection of cutting tools and hard facing as well as quantification of wear on various machine components. This requires comparison of field data with measured PSAI values and regression analysis of the results to obtain a correlation between these parameters. One should also note that wear on the cutterhead of tunneling machines is an operational parameter and cannot be simply estimated by using this or any other index. This is to say that while PSAI can offer a measure of soil abrasion in different conditions, in the real-life application, the soil is modified by using various soil conditioners to reduce its abrasivity and required cutterhead torque. Therefore, the actual primary and secondary wear of the cutterhead and machine components are in fact a function of these operational parameters that are different from site to site, and at various times in the same tunneling operation. This includes using the same machine in the same soil, but changing soil conditioners or various foam injection or expansion ratios (FIRs, FERs), face pressure, or advance rates.

The results of testing and study of wear and torque of the soil abrasion device in various soil types in the laboratory scale is that the behavior of soil at various point along the tunnel could be different depending on the soil types present at cross sections of the tunnel. Practical implications of this study is that while the amount of wear in the same soil type is directly related to the amount of torque, in low abrasion soils, machine can experience extremely high torques at low moisture contents. In general, application of water will reduce the wear and torque in a given soil if no conditioning is used, and if the foam is expected to be dry or unstable, it is preferred to increase the water content to lower the torque and wear. Lab testing has shown that the application of soil conditioning can reduce the wear by a factor of over 200 times as compared to soils with moisture content close to their optimum proctor's test water content. While the use of harder steel can improve the life of various machine components that are in contact with the soil, the improved wear properties of the hardened steel is not surely proportional to the nominal hardness values in Rockwell or Brinel scale. Additional testing and analysis are underway to quantify the impact of various soil conditioning measures on wear and torque and results will be published in upcoming publications.

## CONCLUSION

The Penn State Soil Abrasion Testing device as well as the developed Soil Abrasion Index can be used as a standard for measuring soil abrasivity in the design and construction phases of the soft ground tunnels. The test is designed in such a way that simulates the actual working conditions of the soft ground mechanized TBMs as much as possible and fulfill several shortcomings of the other developed tests. Furthermore in case of tribology, the tool wear mechanism developed in this test, follows the actual mechanism that causes tool wear in the field. To date over 250 sets of tests have been performed. The testing system can discriminate between various working conditions, including the grain mineralogy, shape, size distribution, and water content. The test results could be included in geotechnical site investigations and for prediction of wear on various soft ground tunneling machines in the future. The main findings of this paper are summarized as follows:

- Moist samples (W = 12.5%) caused higher torque and wear compared to dry and saturated samples (W = 22.5%) in Silica sand and in general in the same soil type. The amount of torque in Limestone sand was much higher than the Silica sand sample in moist conditions (W = 12.5%). Despite the higher amount of torque in Limestone sand, the weight loss was lower than Silica sand.
- Increasing the rotational speed resulted in higher torque and higher weight loss of covers in sand-size samples but had the opposite effects in the fine grain soils.
- A formula was introduced to estimate the anticipated wear at various testing times for given soil under various testing conditions. This is to allow for shorter tests in moist/abrasive samples to be extrapolated to desired time scale and therefore the impact of time on tool wear can be evaluated.

A preliminary classification for soil abrasion is offered based on PASI index in this paper that can be used for classification of different soil types to provide a qualitative description of soil for selection of cutting tools and hard facing as well as quantification of wear on various machine components.

## REFERENCES

Alavi Gharahbagh, E., Rostami, J., Gilbert, M. 2010. Tool Wear Issue in Soft Ground Tunneling, Developing a Reliable Soil Abrasivity Index. Presented at North American Tunneling conference, 19–23 June, Portland, OR (SME publication available in www. Onemine.org).

Alavi Gharahbagh, E., Rostami, J., Palomino, A.M., 2011. New soil abrasion testing method for soft ground tunneling applications. Tunneling and Underground Space Technology Journal, 26 (5), 604–613.

Alavi Gharahbagh, E., Qiu, T., Rostami, J. 2012. Evaluation of Granular Soil Abrasivity for Wear on Cutting Tools in Excavation and Tunneling Equipment, Journal of Geotechnical and Geoenvironmental Engineering. http://ascelibrary.org/doi/abs/10.1061/%28ASCE%29GT.1943-5606.0000897.

Alavi Gharahbagh, E., Rostami, J., Talebi, K., 2013. Introducing the PennState soil abrasion index (PSAI) for application in soft ground mechanized tunneling, World Tunneling Congress, Geneva, Switzerland.

Barzegari, G., Uromeihy, A., Zhao, J., 2012. A newly developed soil abrasion testing method for tunneling using shield machines, Quarterly Journal of Engineering Geology and Hydrogeology, 46, 63–74.

Jakobsen, P.D., Dahl, F., 2010. Soil Abrasion in TBM Tunneling, KTA International Symposium on Mechanized Tunneling Technology/November 5/Seoul/Korea.

Mosleh, M., Alavi Gharahbagh, E., Rostami, J., 2012. Effects of relative hardness and moisture on tool wear in soft ground tunneling, Journal of Wear. http://dx.doi.org/10.1016/j.wear.2012.11.041.

Peila, D., Picchio, A., Chieregato, A., Barbero, M., Dal Negro, E., Boscaro, A., 2012, Test procedure for assessing the influence of soil conditioning for EPB tunneling on the tool wear, WTC, Bangkok, Thailand.

Rostami, J., Alavi Gharahbagh, E., Palomino, A.M., Mosleh, M., 2012. Development of Soil Abrasivity Testing for Soft Ground Tunneling Using Shield Machines, Tunneling and Underground Space Technology Journal, 28, 245–256.

Thuro, K. Plinninger, R. 2007. Determining abrasivity with the LCPC test. Proceedings of the 1st Canada—U.S. Rock Mechanics Symposium, p. 8.

# Improvement of Testing Accuracy by a New Generation of Cerchar Abrasivity Testing Device

**Farzad Sotoudeh and Hossein Memarian**
School of Mining Engineering, University of Tehran

**Mohammad Taghi Hamzaban**
School of Mining Engineering, Sahand University of Technology

**Jamal Rostami**
The Pennsylvania State University

**ABSTRACT:** Rock abrasivity is one of the most important factors in estimating the wear of rock cutting tools. Various methods and indices have been used to evaluate rock abrasivity in mechanized excavation applications. Cerchar abrasivity index (CAI) is one of the simplest and most commonly used methods to assess the wear of cutting tools of various mechanized excavation machines. In this paper, a new generation of the Cerchar abrasivity testing device is introduced and the advantages of the proposed configuration and testing system are discussed. The device is capable for automatic control of the length and speed of scratch while measuring the applied forces for pin movement and amount of tip wear. Studies showed that the precise control of testing length has considerable effect on the quality of the results especially in abrasive samples. The initial results indicate that while CAI values are not very sensitive to the length of the scratch in soft to medium rocks, they increase with the length of the scratch in the hard to very hard and abrasive rocks. The measurement of pin forces also allows for determination of the amount of pin penetration into the rock surface that can be used to distinguish rock strength better than standard CAI testing.

## INTRODUCTION

Mechanized excavators used in underground construction and mining projects are generally very expensive and their productivity is sensitive to site conditions. In addition, high operating costs of these machines increase the importance of machine utilization rate and reducing the downtime, including the cutter change time which is a function of rock abrasivity (Valantin, 1973). Rock abrasivity can also be used in selection of appropriate cutting tools and preliminary assessment of drilling costs. One of the most common methods for evaluating of rock abrasivity, in association with mechanized drilling machines is the Cerchar abrasivity test. This test is a simple, fast and effective way to evaluate and compare the various rock types and their strength and abrasivity. The test was proposed for the first time by the CERCHAR* in France (Cerchar, 1973).

The test involves the use of a steel stylus having a 90° conical tip, which is placed perpendicular to the surface of the specimen under a constant load of 70N. The stylus is scratched in a direction parallel to the rock surface over a distance of 10mm. Figure 1 shows the details of the Cerchar testing apparatus commonly used in rock mechanics laboratories. After the test, the width of the wear flat on the tip of the cone is measured in units of 0.1mm as the Cerchar Abrasion Index. The Cerchar test is repeated on 3 to 6 pins and the average of the measured diameter of wear flat in two perpendicular direction is reported as the Cerchar Abrasivity Index (CAI) (ASTM, 2010). See Table 1.

Based on numerous experiments conducted by various authors, about 70% of the stylus wear occur in the first millimeter of the scratch (Al-Ameen& Waller, 1994 and Plinninger et al., 2003). In the Cerchar test, some parameters have very significant influence on the results. These parameters can be expressed as follows:

1. Surface condition of the specimen which can be rough (natural surface) or smooth (sawn)
2. Sliding distance
3. Pin hardness
4. Test speed while scratching the rock

* Laboratoire du Centre Et Recherches des Charbonnages (CERCHAR) de France.

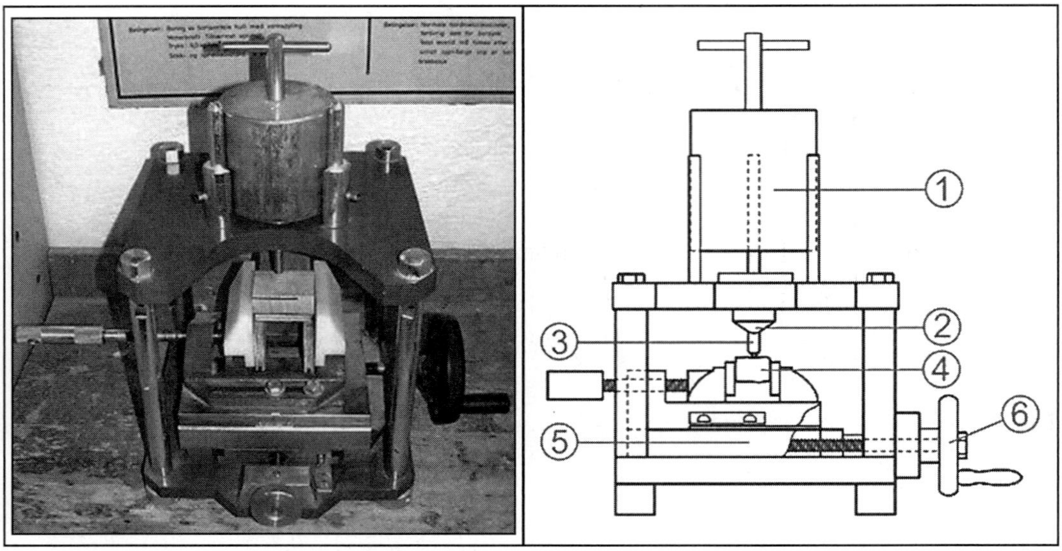


**Figure 1. The Cerchar testing machine: 1. the constant weight (7kg), 2. pin guide, 3. testing pin, 4. rock sample, 5. vice, 6. hand crank (West 1989)**

**Table 1. Rock abrasivity classification based on the CAI values (Cerchar 1973)**

| Abrasivity Classification | Cerchar Abrasivity Index (CAI) |
|---|---|
| Non-abrasive | 0.3–0.5 |
| Slightly abrasive | 0.5–1.0 |
| Medium abrasivity | 1.0–2.0 |
| Abrasive | 2.0–4.0 |
| Highly abrasive | 4.0–6.0 |
| Extremely abrasive | 6.0–7.0 |

5. The method used to measure the wear flat of the pin tip (Ghasemi, 2010)

In commonly used devices, the 2nd and 4th parameters, are somewhat uncontrollable because of the manual mechanism of moving the pin and operator errors. Rostami (2005) conducted the Cerchar tests on some rock samples at various laboratories and the results of these tests on rough and sawn surfaces of rock samples can be seen in Figure 2 and show high dispersion. Thus, in spite of widespread use of the Cerchar test, there are still some major issues effecting the reliability and repeatability of the test results. This becomes more important when the results are used in design and selection of TBMs or other excavation machines with millions of dollars investment and high operating costs hanging in the balance with the cutting tools and daily production rate of the machine.

To improve the accuracy and repeatability of Cerchar test and eliminate some of the sources of error discussed earlier, a new electromechanical apparatus was designed and built that provides automatic control of testing parameters. Also, a new method for measuring wear flat of the pin tip has been introduced that greatly increase the accuracy of wear flatness measurement. Tests performed on different rock samples showed that the results of cerchar abrasivity index measurement on new apparatus are more reliable and can offer additional information that can be used for distinguishing rock strength and abrasivity.

## NEW CERCHAR TESTING APPARATUS

The new device was designed for two main purposes (Hamzaban et al., 2013):

1. To reduce the effect of operator errors on the test results
2. To generate more information about the interaction between the rock and pin and wear phenomenon that occur during the test

To achieve these objectives, the new apparatus was designed with the capability to control pin movement on rock surface at a steady predetermined speed and more accurate control of the sliding distance.

To achieve these objectives, the following components were implemented in the design of the new device (Figure 3):

1. An electric motor with variable speed, to move the pin with a steady and desired speed;
2. A horizontal displacement sensor for measuring the sliding length of the pin on the sample;

3. A data acquisition system to collect and store data from various sensors;
4. A computer program to monitor data from various sensors and to send feedback commands to the device;
5. Mechanical structure to enable performing tests by electric motor and with various sensors.

Figure 3 shows the overview of the new Cerchar testing apparatus.

## MEASUREMENT OF WEAR FLAT

Use of a microscope with minimum magnification factor of 30× to measure wear flatness of pin tips is recommended (ASTM, 2010). Measuring can be done in two ways: from top view and from the side view. During the test, usually a small bulge in the downstream of pin tip is formed. The effect of this bulge should be removed from test results (West, 1989). In measuring from top view, detection and removal of the bulge is quite difficult and subjective

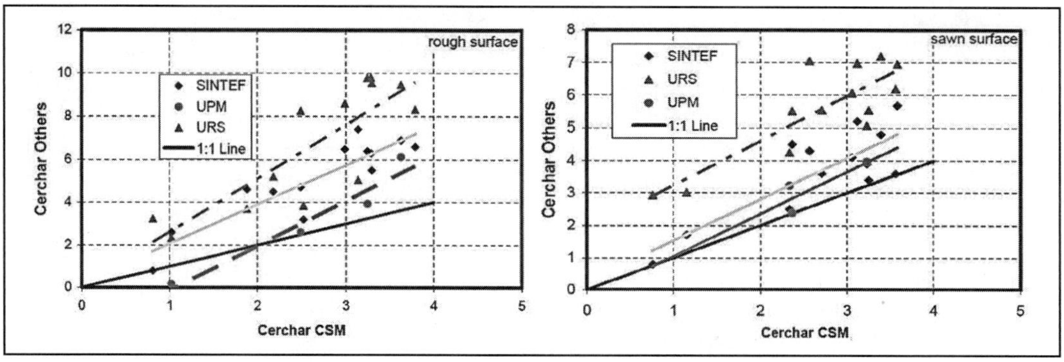


**Figure 2. Results of CAI tests in different laboratories (Rostami et al., 2005)**

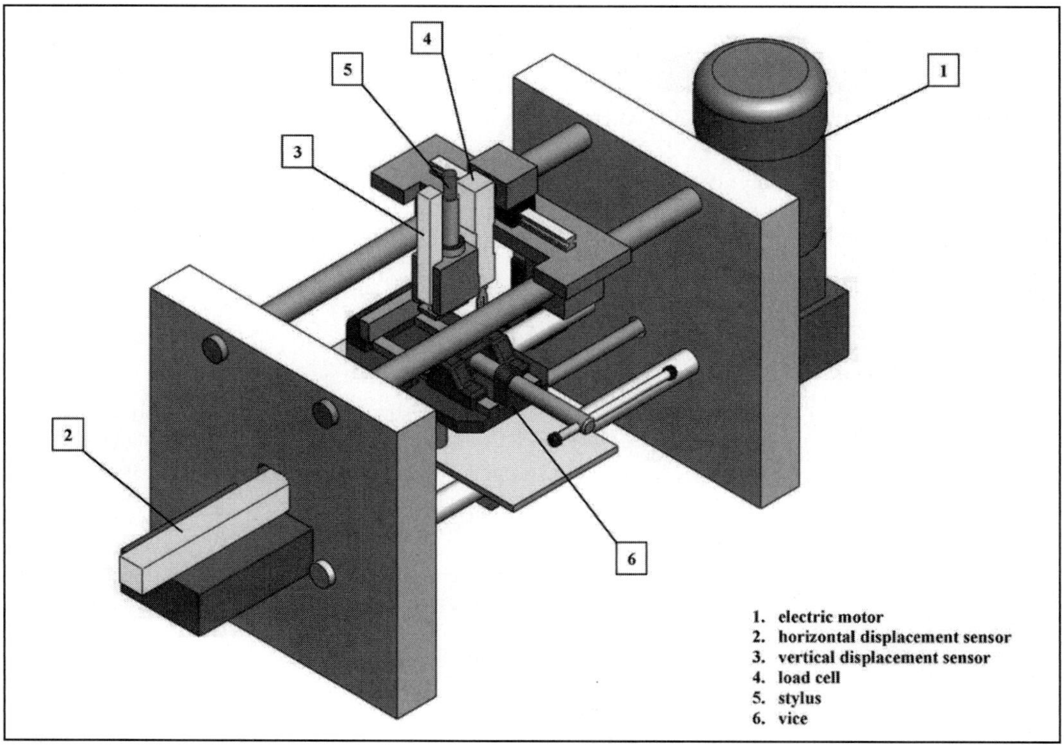


**Figure 3. The new Cerchar testing apparatus (Hamzaban et al., 2013)**

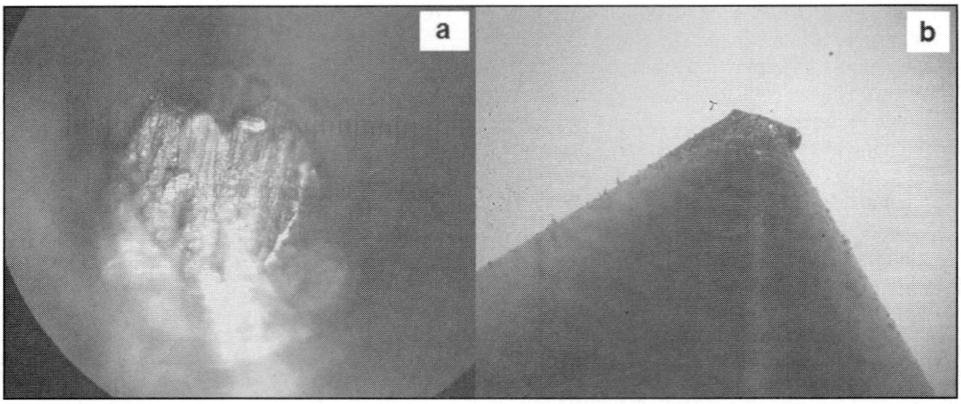


**Figure 4. (a) Measurement of pin tip wear from top view and (b) bulge created at the pin tip during sliding**

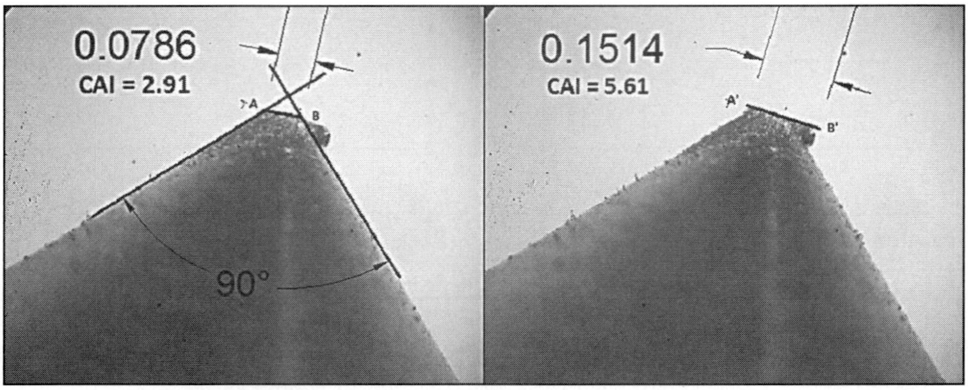


**Figure 5. Comparison of the CAI results when reading from the side view and from the top view**

(Figure 4), this issue can be easily addressed and rectified while measuring from a side view. Therefore the side view measurement is more accurate and the dispersion of the results is much lower, meaning less operator sensitivity/dependency.

In this study, a microscope with a magnification factor of 32× was used to measure the diameter of wear flat from side view.

To determine Cerchar Index, measurements should be done in two perpendicular directions and their average should be reported. This is done by taking two pictures of each pin after 90 degrees pin rotation relative to the first image. Difference of results when measuring from top view and the side view is shown in Figure 5. In this case, CAI is 5.61 when measured from top view, while its correct value is 2.91 in side view.

## TESTING PROGRAM

A total of 24 various rock samples from the Aras hydropower plant project (IWP, 2011) were tested to compare CAI results obtained by the new device and standard testing device. All tests were carried out according to ASTM and CERCHAR standards on sawn surfaces of rock samples. Wear flat on pin tips was measured by measuring from the side view according to the procedure described in previous section.

## DISCUSSION

Figure 6, shows the values of CAI obtained by the new device versus to the values from standard device (Cerchar testing machine used by West (1989)). Also, 1:1 line of $CAI_{New}$ and standard Cerchar or $CAI_{West}$ is drawn in this graph. There is a strong linear relationship between the results of two devices by a correlation coefficient of 0.9363. However, by comparing the best fit trend line on data point with 1:1 line, it can be concluded that the new device gives smaller values than the standard device.

A total of five pins were used on every sample and ten readings were made for each sample. Figure 7 shows the values of standard deviation of ten measurements and the average values which was

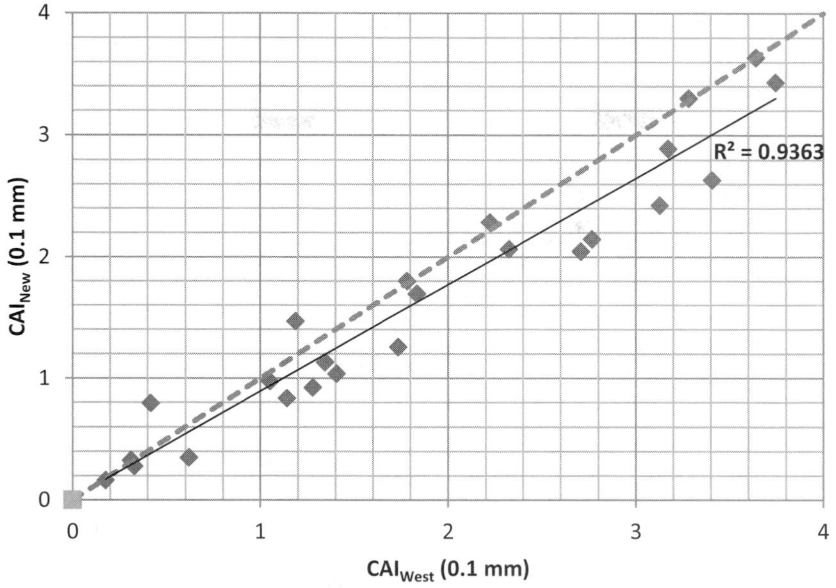


**Figure 6. The CAI values from the New device ($CAI_{New}$) and the West device ($CAI_{West}$)**

considered as CAI for these samples. Dispersion of data obtained by testing the standard device increases with increase in rock abrasivity. There is a weak correlation between the standard deviation and abrasivity of samples. However, the increase in standard deviation with abrasivity is smaller in the data obtained by using the new device. The fitted trend lines in Figure 7 can be compared to the lines in Figure 8.

The reduced dispersion of the results in abrasive samples can be attributed to accurate control of testing length and speed by the new testing device. The difference of the length of scratch can be considered to be the main reason for the dispersion of the data in abrasive samples. This factor has a minor effect on the less abrasive samples. This is contrary to the common conclusion by previous researchers when it was commonly believed that major part of the wear on pin tip is occurs in the first 1–2 millimeters of scratch (Al-Ameen & Waller, 1994 and Plinninger et al., 2003). It was often believed that wear flat on pin tip does not change in the remaining part of sliding path. So, variations of ±0.5 mm in the length of the sliding path was deemed acceptable without any noticeable effect on the CAI results. However, according to Figure 8 this is true only in non to low abrasive samples and in more abrasive samples the size of the wear flat changes with the length of the scratch, meaning more dispersion in the CAI data due to variations in sliding lengths. Accurate measurement of the length of scratch on tested samples showed that often the length of scratches created by the standard apparatus is greater than 1cm.

## CONCLUSION

In this study, a new version of the Cerchar abrasivity testing device was introduced. More accurate control of the testing parameters can be achieved with the new device. This allows for reducing the impact of operator's skill and accuracy on test results. Comparison of the CAI test results obtained by using new device and those of standard Cerchar testing device showed that more accurate control of test parameters, especially the sliding length, has considerable effect on the quality and accuracy of the results. The importance of the testing length is increased in more abrasive samples, meaning that inaccurate control of length of scratch in abrasive samples can lead to a high dispersion in test results. Therefore, it is recommended that length of the sample should be closely controlled in abrasive rock samples and the wear flat should preferably be measured by looking at the pin under the microscope in side view to minimize the operator sensitivity and improve the accuracy of the test results.

## ACKNOWLEDGMENTS

The authors would like to express their gratitude to Mr. Reza Sanajian and Mrs. Mahdieh Nasiri in Folowrd Industrial Plans Co. for their unsparing technical supports in designing and manufacturing of

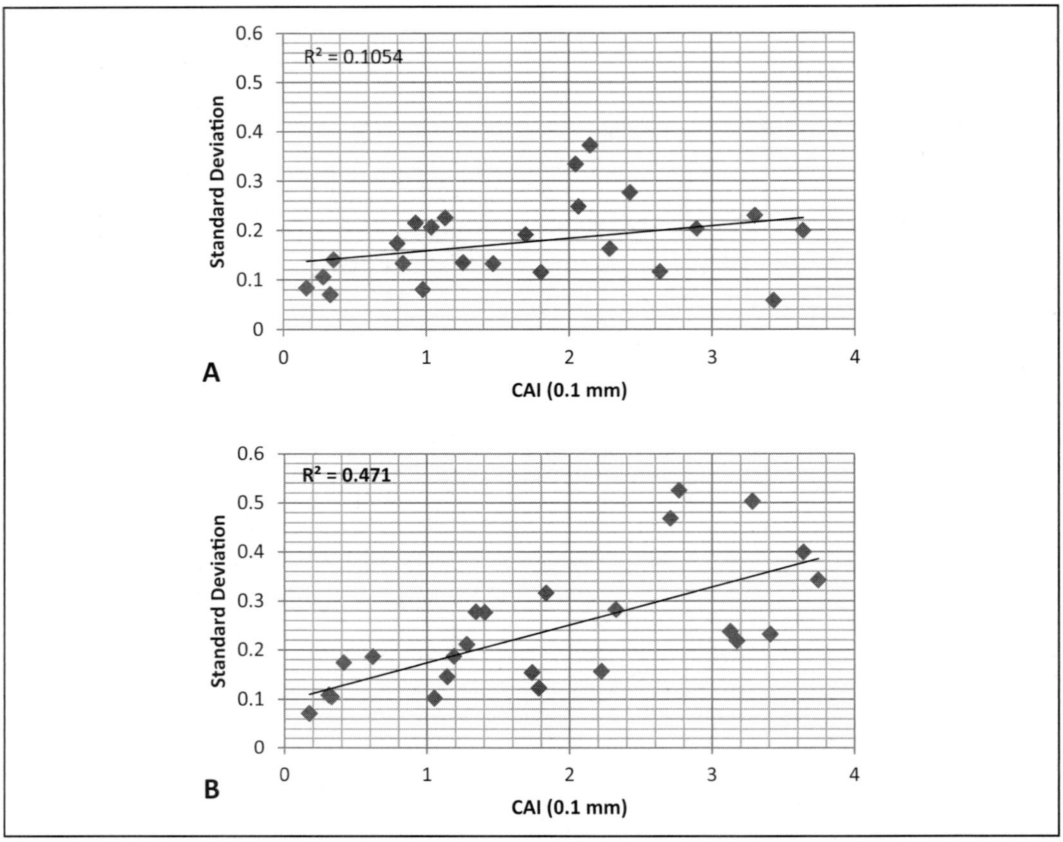


**Figure 7. The standard deviations of readings versus to CAI values obtained by: (a) the West device and (b) the new device**

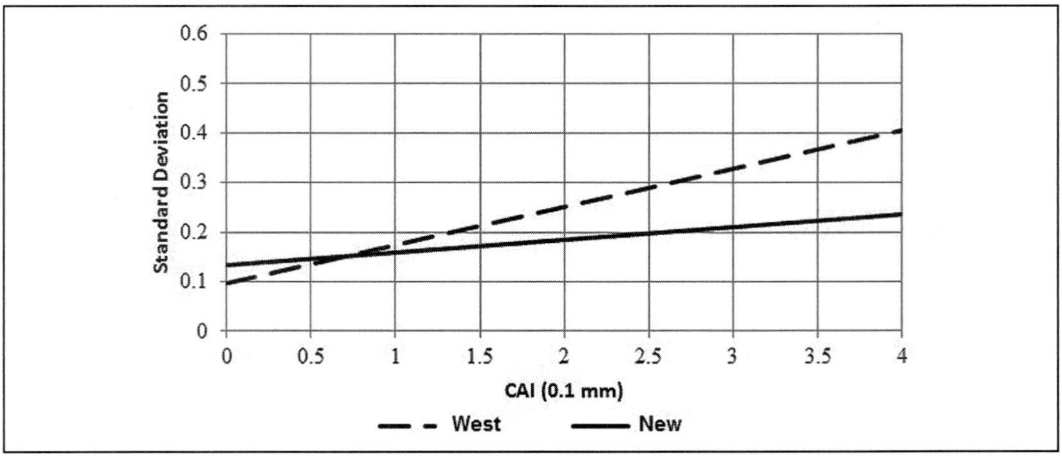


**Figure 8. Comparison of standard deviation trends in the West device and the new one**

the testing device. Also, the authors wish to express their appreciation to Iran Water & Power Resources Development Co., Mr. Banihashemi of MahabGhods Consultant Engineers Co. and Mr. Mahinraad of Moshanir Consultant Engineers Co. for providing the samples used in testing and additional information necessary to conduct experiments and subsequent analysis.

## REFERENCES

Al-Ameen, S.I., and Waller, M.D. 1994. The influence of rock strength and abrasive mineral content on the Cerchar abrasive index. Eng Geol 36:293–301.

ASTM D7625-10 ASTM. 2010. Standard test method for laboratory determination of abrasiveness of rock using the CERCHAR method. ASTM International.

Cerchar. 1973. Cerchar tests for the measurement of hardness and abrasivity of rocks. Centre Study Research, French Coal Industry, Document 73-59:1–10.

Ghasemi, A. 2010. Study of CERCHAR abrasivity index and potential modifications for more consistent measurement of rock abrasion. Dissertation, Pennsylvania State University.

Hamzaban, M.T., and Memarian, H. 2011. Methods of evaluating rock abresivity, Iranian Mining council Journal. No. 12, pp. 14–22.

Hamzaban, M.T., Memarian, H., and Rostami, J. 2013. Continuous Monitoring of Pin Tip Wear and penetration into Rock Surface Using a New Cerchar Abrasivity Testing Device, Rock Meck Rock Eng, DOI 10.1007.

Iran Water & Power Resources Development Co (IWPRD). 2011. Aras hydroelectric power plan. Report of phase 2 site investigation Studies.

Käsling, H., and Thuro, K. Determining rock abrasivity in the labrotary. Engineering Geology, Technische Universität München, Germany.

Michalakopoulos, T.N., Anagnostou, V.G., Bassanou, M.E., and Panagiotou, G.N. 2006. The influence of styli hardness on the Cerchar abrasiveness index value. Int J of Rock Mech & Min Sci 43:321–327.

Plinninger, R.J., Kasling, H., Thuro, K., and Spaun, G. 2003. Testing conditions and geomechanical properties influencing the cerchar abrasivity index (CAI) value. Int J of Rock Mech & Min Sci 40:259–263.

Plinninger, R.J., Kasling, H., and Thuro, K. 2004. Wear prediction in hard rock excavation using the Cerchar abrasiveness index (CAI). In: Proceedings of the ISRM regional Symposium EUROCK 2004 & 53rd Geomechanics Colloquy, Salzburg, pp. 599–604.

Rostami, J. 2005. CAI testing and its implications. Tunn &Tunn Int 37:43–45.

Rostami, J. 2012. Study of pressure distribution within the crushed zone in the contact area between rock and disc cutters, Int J Rock Mech Min Sci 9.

Rostami, J., Ozdemir, L., Bruland, A., and Dahl, F. 2005. Review of issues related to Cerchar Abrasivity testing and their implications on geotechnical investigations and cutter cost estimates. In: Proceedings of rapid excavation and tunneling conference (RETC). Seattle, pp. 738–751.

Stanford, J., and Hagan, P. The impact of stylus metallurgy on the Cerchar Abrasivity Index value. The University of New South Wales (UNSW), Sydney.

West, G. 1989. Technical Note—rock abrasiveness testing for tunneling. Int J of Rock Mech & Min Sci Geomech 26:151–160.

# Keeping Your Cool—Backfill Grout Placement Around Carrier Pipes

**Rory P.A. Ball and Craig Camp**
Hatch Mott MacDonald

**ABSTRACT:** The placement of backfill grout between carrier pipes and tunnel linings involves several major risks. For a successful project, the designer must understand the scope of these risks and consider the possible mitigation measures to be addressed in the design. These risks include: unwanted pipe flotation, voids from improper grout injection, detrimental long term movement, and pipe damage from grout pressure or excessive heat during grout cure. This paper evaluates these risks and offers mitigation measures for backfill grouting in tunnels and shafts. It focuses primarily on tunnels over two meters in diameter. Relevant project experience is offered in terms of design, risk management, and construction considerations.

## BACKFILL PURPOSE

Backfill grout fills the annular space located between the carrier pipe(s) and the casing pipe or the initial ground support system; referred to herein as the initial support system. The initial support system is installed by trenching, trenchless methods or tunneling, and the carrier pipe(s) is subsequently installed. The purpose of placing backfill around carrier pipes is to provide long-term stability and to lock it in place instead of it resting permanently on saddles. Seismic induced damage, flotation deflection, damage from river scour, or other sources of instability can be prevented when a properly designed grout mix and installation system protects the carrier pipe(s). Backfill grout can also protect overlying utilities and surface structures if the initial support system were to fail. The entire backfill grouting system may also be designed to prevent migration of groundwater and/or contaminants or to allow thermal conductance away from the carrier pipe in the case of electric transmission lines. Backfill grout can provide corrosion protection due to the cement content of the mix and the insulation from initial support system.

## METHODS AND EQUIPMENT

Initial support systems include casings, steel liner plates, concrete segments, and spray-on liners. In some cases the existing support may be the old pipeline being slip-lined with a new carrier pipe. Steel casings may be required by some right-of-way (ROW) owners in case of carrier pipe failure so that fluids are contained within the casing and drain away from the ROW owner's facilities or to provide separation to prevent cross contamination as in potable water lines. The casing may be required to extend beyond the sensitive zones (longitudinally) in order to protect the ROW owner's property which typically includes railroads, state department of transportation, and other utility owners. Casings might also be required so the carrier pipe can be reset to line and grade tolerances which are tighter than the construction methods capability in the anticipated ground conditions. This is typical in ground containing cobbles and boulders or a mixed face. In these cases, the tunnel needs to be surveyed and an acceptable design submitted to re-establish line and grade before the carrier pipe installation commences.

After the initial support system is installed, the carrier pipe(s) can be placed. The entire carrier pipe installation and backfilling process must be clearly discussed in the contract documents and detailed in contractor submittals. Some of the necessary topics include: the materials used and methods employed; the backfill volume estimate and measurement; the required number of lifts; a description of how loadings will be balanced; void monitoring and remediation; installation tolerances; and the process for keeping both the grout and pipe cool.

The carrier pipe can be installed on systems such as: a cast invert, rails, skids, casing centralizers, casing spacers, and blocking. Each option has its advantage(s) and disadvantage(s):

- A cast invert is where the contractor pours an invert and screeds the surface to the desired elevation. The carrier pipe is then set on the invert and secured to prevent unwanted pipe movement. In some cases the carrier pipe is placed on engineered saddles to hole the pipe off the invert before grouting. Depending on the project specific requirements, the carrier pipe is then grouted in-place. This tends to be a low cost option with the invert being a concrete pour and the backfill grout being customized based upon the project requirements.

- Rails, similar to railroad tracks, are securely placed to line and grade. The carrier pipe is then installed and secured as previously described. This tends to be a low cost option with the invert and the backfill grout being customized based upon the project requirements.
- Blocking requires man entry as each pipe is carried into the casing and the pipe string is assembled within the tunnel. The carrier pipe(s) can be brought into the tunnel using a small pipe carrier, depending on the internal diameter. Blocking design is very similar to the other designs except for man entry and assembly within the tunnel. This allows each pipe to be individually set to line and grade. Depending on the space between the casing and carrier pipe(s), this can lead to a serious safety issue if the ease of egress is impacted.
- Skids are a low cost option consisting of dimensional lumber strapped to the carrier pipe. They allow for the carrier pipe to be slid into the casing (without manned entry) and, depending on the backfill material, prevent the carrier pipe from floating and coming in direct contact with the casing pipe. Careful attention must be paid to be sure they stay in place while installing the pipe. Often skids tend to slide or skew which may lead to carrier pipe damage. Runners can be attached to the bottom of the casing to help center the carrier pipe(s). The ends of the skids should be dog-eared and not square cut, allowing the skids to ride up and over any irregularity within the casing. Skids tend to be a low cost option.
- Casing centralizers are designed to place the carrier pipe in the center of the casing, both horizontally and vertically, and may provide dielectric isolation. The design and installation process are very similar to skidding. Centralizers force the carrier pipe to match the orientation of the casing pipe which may be an issue if reverse grades occur on a gravity line. Casing centralizers need to be designed by a professional engineer for the dynamic installation loads and detailed on sealed drawings. Cost is slightly greater than casing spacers. Centralizers and spacers should have an even number of legs of symmetrical design about the vertical centerline.
- Casing spacers are similar to casing centralizers except that the carrier pipe sits lower in the casing which reduces the dynamic installation load on the spacers. Some spacers are adjustable and can help ease grade changes associated with the casing pipe orientation. Spacers typically carry a higher cost than skids.

Once the carrier pipe is installed, interior bulkheads are constructed as needed to separate lifts and provide neat joints and appropriate piping installed for the grouting process. Interior bulkheads are commonly constructed of wood while the end bulkheads are normally constructed using bricks and mortar along with a waterproofing component to prevent the flow of fluids. The ends of the tunnel reach are sealed using bulkheads or end seals. The material selection can be specified based upon the ROW or property owner's requirements or other design criteria. It is critical for these bulkheads/end seals to be engineered to withstand the loading imposed on them during the backfill grout placement.

## MIXES

Grout mixes vary from low to high strength cement mixes with variable unit weights. Because the grout completely surrounds the carrier pipe, the whole system is permanent and must be designed for both short term and long term loading conditions.

Backfill grout options range from simple mixes of cement and water to proprietary formulations such as self-consolidating concrete mixes. Some of the mixes options are the following:

- Neat Grout—A mix of cement and water. Additives or pozzolans may also be added. Neat grout has great flowability characteristics, self-levels, and is capable of filling voids other mixes may miss. The downside of neat grout can be high heat of hydration temperatures, shrinkage, cracking, bleed, easy dilution when contacting standing water, and a limited range of final unit weights.
- Flowable Fill Grout—A mix of cement, water, and sand. Additives or pozzolans may also be added. Similar characteristics to neat grout, but with the added benefit of more options for adjusting the unit weight of the final grout due to the granular content. Shorter pumping distances may be required to prevent segregation and bleed. The mix will have less flowability (depending on viscosity) and may not fill all the voids.
- Low Density Cellular Concrete (LDCC) Grout—A mix of cement, water, and preformed foam. Additives or pozzolans, sometimes aggregate, may also be added. LDCC grout is a highly flowable, self-levels, and provides the designer with the option for a

variety of final unit weights (generally from 0.32 to 1.92 g/cm$^3$). Having air bubbles in the grout mix also provides greater resistance to the damaging effects in a freeze/thaw environment. Typically, strength rises as the unit weight rises. Careful consideration must be paid to the integrity of the bubbles as they travel from the surface through curing in their final position.
- Plain Concrete—A mix of cement, water, and aggregates. Additives or pozzolans may also be added. Advantages include adjustable strengths and unit weights, readily available mix designs and tests, and readily available local suppliers. Plain concrete does not have the long distance pumpability necessary for many backfilling operations and typically requires higher grouting pressures to sustain flow. Segregation of the granular content can lead to clogging. Additionally, it is not self-leveling and may not fill all the voids in the annular space.
- Self-Consolidating Concrete (SCC)—A specially designed mix of cement, water, aggregates, pozzolans, and additives. SCC is a self-leveling fluid mixture capable of being pumped longer distances (if properly designed) while preventing segregation. It is more capable to flow into tight spaces compared to plain concrete. SCC has a lower heat of hydration and shrinkage than neat grout. However, the availability of SCC testing may be limited.
- Additives and pozzolans—There are numerous ingredient options available to mix designers for a range of desired effects. For example:
  - Fly ash—reduces heat of hydration, delays strength rise, improves pumpability
  - Slag—reduces heat of hydration, delays strength rise
  - Bentonite—reduces bleed, increase pumpability
  - Retarders—increases set time
  - Fluidizer—increases flowability, self-leveling, pumpability

## TYPICAL RISKS

The installation and backfill grouting of carrier pipes unfortunately involves numerous risks. The severity of the consequences associated with these risks range from perhaps tolerable (e.g., minor ponding in the carrier pipe) to catastrophic (e.g., carrier pipe collapse due to grout over-pressurization). Regardless, once a carrier pipe has been grouted, there is little that can be done to remediate any of these consequences. Chipping out a failed carrier pipe in grout is cost prohibitive. For this reason, the installation and backfill grouting of the carrier pipe must be done right the first time. This section addresses some of the risks and possible mitigations for successful installations. Topics include the following:

- Carrier pipe transportation
- Grout mix issues
- Grout system issues
- Flotation risks and unbalanced loading risks
- Pressure concerns
- Grout cure concerns
- Volume issues

### Carrier Pipe Transportation

The road from the factory to the carrier pipe's final position within the tunnel can be a lengthy, bendy, and treacherous journey depending on a project's unique characteristics. Performance requirements are critical on the delivery, storage, handling, joining, placement, and testing of the pipe. Contractors must submit a clear plan for review that unquestionably conforms to the contract and the approved manufacturer's requirements. Common construction related issues include:

- Length and weight of each segment of pipe
- Alignment and placement of the pipe within the tunnel
- Welding or joining of pipe segments
- Stulling, interior pipe support, and removal
- Repair or placement of linings and coatings
- Excessive exposure to weather
- Gouges during movement of the pipe
- Excessive bending of the pipe during transport from the surface to the tunnel

In addition to performance requirements on transporting the pipe, designers should include qualification requirements on the contractor and their personnel. These requirements should include an installation certification from the manufacturer qualifying the contractor to be able to properly join, lay and handle the pipe. Designers must review construction submittals in detail alongside contract and manufacturer requirements. Construction managers must provide training to their inspectors to watch for critical areas of possible damage.

The carrier pipe needs to be installed to acceptable line and grade. In order to ensure proper alignment, the tunnel should be surveyed as required and the alignment plotted to a scale that demonstrates the constructed alignment will accommodate the design requirements. If the as-built tunnel does not meet the requirement, then an acceptable alternative approach can be developed before commencing with carrier pipe installation. This may require the use of adjustable spacers instead of centralizers.

Pipe deflection and the maximum allowable joint deflection should always be less than the maximum allowed deflection by the design. 50% of the maximum allowed deflection value is reasonable. If more deflection is required, the contractor should consider using shorter pipe joint segments.

### Grout Mix Issues

Five critical properties of backfill grout are the following: flowability, survivability, density, strength, and heat of hydration. Details on these properties are as follows:

- Flowability is the viscosity of the grout. The grout must be able to travel from the mixing plant to the final position. It must be able to flow through small openings and fill all voids.
- Survivability is how well the grout survives the installation process without detrimental effects. If the grout collapses (e.g., bubble collapse in low density cellular grout), then the backfill operation will require more grout than anticipated. Segregation, bleed, dilution, and other unwanted effects must be minimized.
- Density is the design weight as mixed in the field. The grout density should be measured before pumping, at the injection point, incrementally, and upon completion as the grout overflows. When the grout overflow comes through the drain lines, it should be approximately the same density as the grout injected.
- Strength refers to the unconfined compressive strength (UCS). The designed strength may be verified by testing test samples taken during injection at the point of injection. Depending on the mixing method and batching size, several samples need to be collected during the installation process as well as the overflow. The strength helps ensure the cement content as installed is acceptable to the design.

The contract should clearly state the requirements of the grout mix in terms of the following: unit weight, strength, shrinkage, allowable ingredients, flowability, survivability, maximum mix curing temperature, any testing obligations, and cement content for corrosion resistance. It is critical for the designer to understand tradeoffs for specifying each of the criteria. For example, specifying a minimum strength may be necessary to support the carrier pipe and surrounding ground, but too high of a strength may cause problems with pumpability. As another example, neat grout has the maximum flowability of any grout. However, the heat of hydration and shrinkage associated with neat grout may not be acceptable to some installations. Neat grout may also dilute more than other mixes if there is excessive groundwater in the tunnel.

If a unique grout mix is required, it is common for contractors to request early approval of the grout mix to allow early testing to begin well ahead of construction. This was the case for the long distance pumping required for the Seymour-Capilano Project (Yanagisawa et al. 2013) where many rounds of mix and equipment testing were implemented.

Mix testing is important to monitor many factors. For example, if a large amount of fly ash or other retarders are used, strength testing may need to be performed on 56 day samples instead of 28 day samples. For testing the heat of hydration, a test pit may be required for the specific grout mix proposed for the project. It is recommended this pit be at least 25% larger in all directions than diameter of the initial support system to be grouted. The top of the pit can be covered in plastic and burlap sacks to help contain some of the heat to prevent excessive atmospheric losses. Temperature probes should then be installed within the mass and electronically recorded during the cure. The results of this test can be used to design a mix meeting the allowable temperature gain requirements. In some cases, an acceptable cooling system can be designed to dissipate the heat and protect the carrier pipe.

Long pumping distances can lead to undesirable consequences to backfill grout mixes. The specification must require the grout be able to stand the maximum anticipated pumping distance and pressures as determined by the contractor's setup. If large volumes are necessary to pump, the grout must be able to sit at a standstill for a period of time and be able to restart pumping. Minimal segregation and bleeding must also be required in this case. Long distance pumping must not adversely affect the distribution of aggregate in mixes containing sand, pea gravel, or other granular ingredients. Pumping lines must be rust free and debris free to maximize the potential for smooth flow. Pre-cleaning of all feed lines may be required even if they are new. If cellular grout is used, the integrity of the air bubbles must be maintained from batching through cure of the mix. High pressures and excessive transport distances may adversely affect the integrity of the air bubbles. At a certain height (depending on the mix), the weight of the cellular grout reaches a point where the grout above collapses the air bubbles below. The maximum allowable lift height should never be exceeded. Surface pumpability tests may be necessary to determine the robustness of the grout mix for the specific project scenario.

The engineer must specify the type of permissible joint between pours and especially for the crown pour. Multiple pours can create vertical joints which require special care. Bird beaks may form where the crown-pour pinches out due to the slope of the initial support system; thus creating a pour that resembles a bird's beak. The crown pour may need to be done in a single pour to completely fill the top space.

### Grout System Issues

The backfill procedure must be well thought-out and planned. Plans should include materials, manpower, equipment, and durations. The plan should include calculations and estimates where possible for each of the following items.

Proper planning should include contingency plans to account for problems anywhere within the grout system. Grouting equipment may need to have backup components at the critical stages such as final crown pours. Contractors should have on-hand methods for patching any bulkhead leaks. Volume and pressure gauges sometimes fail and therefore need readily available backups. If delays in the pour occur, how can grout lines be cleaned out to prevent unwanted blockages?

The entire rout mixing system must be design so as to mix and deliver the proper mix at the correct pressure and volume to the correct location and for the mix to perform as intended. The entire system must be robust so as to deliver the product. The system must also be designed to dispose of the grout material if and when needed. Some of the items to consider include the following:

- Proposed mix(es)
- Mix water source, capacity and chemistry
- Grout batching capacity
- Grout source from on-site batch plant or transit mixed
- Time in process before wasting materials
- Disposal options for the waste

The grout placement system must be capable of the following:

- Circulation and re-circulation of grout
- Monitoring injection pressure
- Monitoring volumetric measurements
- Disposal of grout from within the lines
- Cleaning of grout lines
- The capacity to vent air and drain nuisance water

Standing or flowing water within the tunnel are serious concerns for the backfill grouting process. These conditions can wash away or dilute grout leading to undesirable consequences such as the following: ungrouted zones, lower unit weights/strengths, higher flotation forces, groundwater contamination and issues with water handing systems. Active groundwater infiltration should be remediated if possible. Foam or chemical grout injection may stop water flows. Alternatively, a water handling system consisting of panning, drains and sumps is an option for some grouting scenarios.

Grout can clog water handling systems because grout will mix with any nuisance water in the tunnel. The normal process is to collect water into a sump, pump it to the surface, and treat as required. Cement mixing into water will change the water pH which may negatively impact the purification process. In some cases, the inflow of grout may overwhelm the process or bring it to a halt. The sump system design must be robust to work as intended and prevent damage such as shaft and tunnel flooding.

### Flotation and Unbalanced Loading Risks

Assuming the initial support system is restrained from floating or sinking before carrier pipe installation, the grouting plan must consider how to prevent movement of the carrier pipe(s). First, the carrier pipe(s) must be secured in all directions and properly supported to prevent sagging if cooling water is added within the pipe(s). Afterwards, depending on the driving force from the unit weight of the grout mix and the resisting force from the tunnel weight, the carrier pipe will tend to float or sink within the grout matrix. This unit weight influences the long-term floatation or sinking which must be evaluated for the completed initial support system/grout/carrier pipe configuration. Any strapping used must be designed to distribute the loading across the pipe surface and prevent damage. Sand bags are sometimes used to weight down the liner, but can also lead to problems if not properly designed. The range of allowable grout unit weights must be monitored to prevent excessive forces if the unit weight becomes too high. Any heat softening of the material must be taken into account. Preventing movement of the pipe may be critical to minimize ponding, reverse grades, sedimentation, and the loss of design clearances to the initial support system. To help the contractor estimate the spacing between saddles/straps, specifying the allowable movement can be helpful. Also, high definition video of the completed carrier pipe helps to evaluate the end product.

Whether the grout is injected through feed lines in the crown or through ports in the tunnel, buoyancy at every stage must be evaluated. The worst case (maximum) value of unit weight of the grout must be estimated to create a buoyancy mitigation plan. An undesired value of the grout's unit weight may occur when the grout is improperly batched, excessive bleed occurs, or entrained bubbles collapse. This is one reason why the mixture must be designed and tested for durability specific to the aspects of the planned grouting scenario. A licensed professional engineer must take the geometry, material types, and grout lift heights to plan the blocking needed to support the carrier pipe. The blocking must safely distribute the forces on the carrier pipe without excessive point loading. Multiple grout lifts help to minimize the stress on linings if each lift is allowed to cure before subsequent lifts. The design must include a way to ensure damage will not occur if subsequent grout lifts penetrate previous lifts at the carrier pipe interface. A sample scenario of lift heights may be the following:

- Lift #1—from the invert up high enough to place the carrier pipe in compression against the blocking on the crown
- Lift #2 &3—increases the fill height while minimizing excessive flotation forces
- Lift #4—brings grout above the springline and places a downward force the carrier pipe; fills the remaining void in the crown

Unbalanced loads may occur if grout is filled unevenly from either side of the annular space outside the carrier pipe. These loads can deflect pipe segments or deform thin pipe sections out of round. Whether the grout is injected by crown feed lines or ports, the injection points must be staggered on either side. Volumes injected on each side must be monitored and balanced. Timber stulls may need to be added within the tunnel if an impact to roundness is anticipated.

Monitoring grout injection levels and pressures can be done at multiple ports via standpipes. This requires more ports, staggered at different elevations and stations from the springline to the crown. These ports are preferably the more expensive 75-mm (3-in.) diameter threaded design. They can help monitor for pressure, grout communication, and release air. They can also be used for injection additional grout into remaining voids after the final lift is completed. Voids can be identified by simply tapping a hammer on the lining and listening for the change in sound. This may be necessary for a steel carrier pipe if it contracts from cooling once the heat of hydration dissipates. If this void were to be left behind, the pipe may be subject to uneven external loading in the future.

Any ports penetrating the carrier pipe must be repaired and sealed before the line is commissioned.

### Pressure Concerns

Fluid pressure develops when injecting the grout into the enclosed annular space. The allowable pressure to prevent damage to the carrier pipe(s) is critical to understand and monitor. In some scenarios, the pressure can be directly measured at several points along the carrier pipe(s) at the grout placement locations when drop holes are used. However, injection systems from bulkheads only allow the operators to measure pressure within the feed-lines before they disappear into the bulkhead. This "bulkhead" pressure may be higher than the placement location when friction causes line losses. Any change in elevation between measurement and placement must also be considered.

Grout pressure may lead to damage of the carrier pipe if injection is not properly managed. Supply pumps must provide constant pressure and flow rates to help prevent overpressurization. The pressure must be monitored directly at the point of injection and evaluated against the allowable pressure. As allowable pressures are reached, the valve operator must be able to immediately stop or divert the flow of grout. In some scenarios, grout communication between ports can be monitored. The grouting crew must be mindful of elevation differences longitudinally along the tunnel if grouting proceeds uphill from port to port. In this scenario, the uncured grout may be at an acceptable pressure at the port of injection, but may be overpressurized at recent downhill ports zones.

The following are some suggestions when it comes to preventing damage to carrier pipes from grout pressure:

- Evaluate the allowable pressure before buckling of the pipe occurs
- Use a grout pump that supplies a constant pressure
- Maintain direct communication between the pump operator and pressure monitor/valve operator
- Use multiple lifts if possible that cure and protect the carrier pipe before a final lift is placed in the crown
- Prevent blockages in vent pipes which may lead to excessive pressures
- If multiple feed lines are used at different length intervals, monitor exhaust air in each feed line to determine when grout passes each location (as determined by air no longer flowing through the pipe)
- Install a manifold system that automatically bypasses the feed line if pressures go above limits
- Monitor drain and vent lines for any slowdown in water or grout flow
- Carefully monitor grout volumes

The fluid pressures from backfill grout can place undesirable strain on the carrier pipe if the restraint and injection system are not properly engineered. The cause for the strain can be several factors; these include: buoyancy forces, point loading, unbalanced loading, and grout overpressurization.

### Grout Cure Concerns

The heat of hydration is the heat caused by the cement hydrating. This complex process includes a number of exothermic chemical reactions which can be influenced by several factors. These factors include: water/cement ratio, air entrainment, chemical admixtures, cement type, additional cementitious materials, cement fineness, sulfate content, initial grout/aggregate temperatures, ambient temperatures, and geometric or environmental factors. Therefore, each heat of hydration scenario for a specific grout mix is distinctive under various mix and environmental scenarios. Consequently, the properties of the mix can be affected. These properties include the rate of strength gain, workability, pore development, and curing behavior. Most importantly, excessive heat may cause damage to the carrier pipes and cause unwanted sagging in plastic grout feed lines.

If curing temperatures reach over 71°C (160°F), unstable hydration products develop in some concretes due to a change in the cement hydration reactions (Gajda 2006). This is referred to as Delayed Ettringite Formation (DEF) and may happen many months or years later. DEF may lead to cracking and expansion of the grout which may adversely affect the carrier pipe(s). If the contractor does not believe temperatures can be maintained below 71°C (160°F), testing of the mix should be performed ahead of time to evaluate the potential for DEF. High fly ash mixtures can help to prevent DEF.

Mass concrete is defined by the American Concrete Institute (ACI) as "any volume of concrete with dimensions large enough to require that measures be taken to cope with generation of heat from hydration from cement and attendant volume change to minimize cracking." Because of the inhomogeneous hydration within the grout arrangement and the inhomogeneous loss of heat to the surrounding environment, temperature differences will occur throughout the grouted zone. These temperature differences can induce thermal strains and stresses that could potentially initiate cracking if they exceed the early age tensile strength of the grout. Although cracking is not ideal, it may not much of a concern depending on the project specific design issues.

Neat grouts and some cellular grout tend to have fast and hot hydrations while plain concrete may have a slow and low hydration. Heat can cause steel pipe to undergo elastic deformation and can cause undesired deformations in other pipe types. Contractors must submit a plan to address the heat of hydration and provide a monitoring method. These plans commonly require upfront mix testing, field trials, and sometimes advanced thermodynamic analyses of the proposed tunnel configuration. Thermodynamic analyses can also be an aid in determining what cooling may be necessary. Active cooling options include circulating cold water (not air) through the carrier pipe(s) and possibly inserting extra embedded cooling pipes in the grout mix. These cooling systems must be able to remove heat from within their own system and have the ability to adjust the flow depending on the temperature rise.

The following is a list of some, but not all, of the further options for reducing the heat of hydration:

- Require grout transportation and placement when the ambient temperature is the most favorable (e.g., nighttime or early morning)
- Use low-heat cement such as Type II (not Type I/II)
- Minimize the total content of cement and/or use larger and better graded aggregates
- Include Class F fly ash or slag cement
- Minimize the water-to-cement (W/C) ratio to as low as possible
- Add shaved or chipped ice to replace some of the mix water
- Add dry ice or liquid nitrogen to precool the grout mix (note: this is extremely expensive)
- Precool the mix water and aggregate
- Place multiple lifts of grout at separate intervals to allow more of the heat to escape
- Utilize different mix designs for lower lifts (such as plain concrete) before injecting grout with better flowability characteristics to fill voids (such as neat grout)
- Precool the cooling water flowing through the carrier pipe(s) or embedded pipes

For projects where heat of hydration is a concern, the authors suggest the industry consider using fiber optic lines to monitor temperatures nearly continuously along the entire length of the carrier pipe(s). There currently exist fiber optic products for monitoring cast in place pipe (CIPP) curing temperatures along one to two foot longitudinal zones of the host pipes. Products like this could be fastened to the exterior of the carrier pipe(s) before backfill grouting and help monitor the interface heat until the heat of hydration subsides.

### Volume Issues

The backfill grout plan should allow for complete filling of the voids around the carrier pipe(s) and within the initial support system. Monitoring the volume of grout is also important to help the contractor reduce the potential for over-pressurizing the grout and warn of a possible void. The plan and monitoring can benefit from the following:

- Calculate the estimated grout volumes ahead of construction and subsequently adjust it based on actual field conditions. (Use this information to plan batch sizes and mix availability.)
- Include contingency plans if supplies run low during backfill grouting.
- Consider possible trucking issues—city noise/traffic concerns; plant hours; restrictions on truck movements during rush hour.
- Install digital volume meters at the point of injection (typically the bulkhead).
- Include a way to check that air and water are purged. (The consistency and unit weight of the mix must be carefully considered if there is water that needs to be offset in the tunnel.)

## QUALITY CONTROL AND RECORDS

Quality control (QC) during execution of the pipe installation and backfill grouting work is crucial to supporting a successful operation. Construction QC includes material review, process review, field control tests, laboratory testing, detailed records, volume estimates, carrier pipe surveying and post installation testing.

Grout testing typically includes unit weight and strength measurements. The samples used for these tests should be pulled from the closest possible point relative to the location the grout is injected. As mentioned previously, injection samples should be compared to overflow samples from drains.

Inspectors must carefully review approved submittals and be familiar with contract requirements. The elevations of the carrier pipe must be verified before bulkheads are constructed. Any deviations to the planned means and methods must be brought to the attention of the contractor's superintendent, construction manager and design engineer for review and comment. Modifying the old carpenter's mantra slightly provides us with good advice—check twice and backfill grout once.

The quality control and quality assurance inspectors must monitor and record a variety of data. This data includes surveying information, batch volumes, injection volume, grout temperatures, cooling water flow, cooling water temperatures, grout pressures, unit weight testing, and sample identifications (for lab testing). If possible, this data should be recorded both manually and electronically. Temperature rises in the cooling water or in the grout will require higher cooling water flow rates to prevent detrimental effects. For this reason, it is important that the operators do not leave the system unattended during the critical cure times.

## RECENT PROJECTS

### Project #1—Santa Ana River Interceptor (SARI) Relocation

The SARI Relocation project in Yorba Linda, California, included over a mile of backfill grouting in carrier pipes placed within casing installed by trenching or trenchless methods. The backfill grout scenarios include:

- Low density cellular concrete (LDCC)
  - 328 m (1,078 ft) of 1,372-mm (54-in.) ID Hobas within a 2,134-mm (84-in.) ID corrugated metal pipe trenched into place
  - 856 m (2,810 ft) of 1,372-mm (54-in.) Hobas within a 2,134-mm (84-in.) ID reinforced concrete microtunnel casing
- Neat cement grout above plain concrete (inverted siphon crossings)
  - 333 m (1,092 ft) of three 864-mm (34-in.) HDPE pipe within 2,527-mm (99.5-in.) ID Permalok steel microtunnel casing
  - 241 m (794 ft) of two 305-mm (12-in.) and one 406-mm (16-in.) HDPE pipes within a 1,918-mm (75.5-in.) ID welded steel microtunnel casing

For the LDCC backfill grouting operations, the Hobas pipes need blocks at the top to resist flotation and saddles at the bottom to provide clearance for grout to the casing invert. The saddles were designed for holding a completely water-filled pipe without damage from point loading. The water was required by the contract to be continuously flowed through the pipe from before grout placement until the heat of hydration subsided.

To prevent uneven loading on the Hobas pipe, grout feed pipes had injection points in the crown that were staggered left to right. This setup was needed to balance the loading from the grout fluid pressure to minimize strain on the pipe joints. The PVC feed lines were also secured on tight intervals to prevent sagging of the lines due to the heat of grout curing.

### Project #2—(Anonymous) 1,524-mm Steel Carrier Pipe in 2,438-mm Steel Casing

The project installed one reach of welded steel pipe in a 2438-mm (96-in.) steel casing installed using microtunneling. The single reach was approximately 305 m (1,000 ft) long. The contractor installed the carrier pipe to the design grade and meeting the design requirements. The contractor self-performed the cellular backfill grouting. The project owner raised several questions regarding the contractor's plan during the submittal process. Most of the questions were seeking clarification of the work plan. The contractor's response to most questions was that it was their problem. The contractor made one set-up at the downstream end to fill from the low end to the high end. The high end vent assembly was mounted on the crown and extended approximately 0.3 m (1 ft) above the ground surface. There was no valve in the vent assembly. Injection pressure never exceeded the maximum allowable pressure. The cellular backfill grout was at 0.48–0.56 g/cm$^3$ (30–35 pcf) with a minimum unconfined compressive strength of 21 kg/cm$^2$ (300 lb/in$^2$). Samples were taken for testing at the injection point and the overflow. All pre-injection samples exceeded the minimum density and the break strength met the design requirements. The overflow purged the annular space of water and the overflow grout density was substantially greater than the injected density. The actual volume of cellular grout injected exceeded the estimated volume which required additional truck loads ordered at the end of the day.

## CONCLUSIONS

Owners desire quality finished products that meet or exceed the intended design life. No one gains when a backfill operation goes wrong and quality falls short of expectations. To have the most successful backfill grouting operations possible, team members must work collaboratively and thoughtfully to address every foreseeable risk. Once the grout injection is underway, time becomes the enemy as the grout curing process begins and access to the annulus outside the carrier pipe may be completely off-limits. This paper highlights some of the major risks involved with the installation and backfill of carrier pipes. Designers should evaluate these risks based on their project specific conditions and contemplate additional risk factors.

## REFERENCES

Gajda, J. and Alsamsam, E. (2006)—Engineering Mass Concrete Structures, Portland Cement Association, USA.

Yanagisawa, S., Lowson, A., and Banek, K. (2013)—Massive Annular Grout and Long-Distance Pumping at the Seymour-Capilano Project, Rapid Excavation and Tunneling Conference (RETC), SME, USA.

# TRACK 1: TECHNOLOGY

## Session 5: Fresh Approach on Performance

James Wonneberg, Chair

# New Advancements in Dry Mix Shotcrete Using Rapid Set Cement in Lieu of Accelerator Admixtures in Tunnels, Shafts, and Pipe Liners

**Simon Reny**
King Packaged Materials Company

**Mike Ballou**
CTS Cement Manufacturing Corporation

**ABSTRACT:** In the mining and tunneling industries time is critical, and as the mining/tunneling cycle becomes shorter production increases. Shotcrete is often used for ground support when using the drill and blast method or other tunneling methods. But before re-opening access for the next phase of the underground heading, the applied shotcrete is required to reach a minimum compressive strength in order to ensure the safety of the workers going into the heading. This article demonstrates the possibility of speeding up significantly mining and tunnelling process using calcium sulfo-aluminate cement in dry-mix process shotcrete mixes.

## INTRODUCTION

In the mining and tunneling industries, time is critical, and as the mining/tunneling cycle becomes shorter production increases. Shotcrete is often used for ground support when using the drill and blast method or other tunneling methods. But before re-opening access for the next phase of the underground heading, the applied shotcrete is required to reach a minimum compressive strength in order to ensure the safety of the workers going into the heading. In order to speed up the mining and tunneling process, King Packaged Materials Company, Burlington, ON, Canada, (King), a specialized company in pre-packaged, pre-blended dry cementitious material, has been working together with mines in Eastern Canada to develop a shotcrete mix design which meets the minimum required compressive strength as fast as possible. Using high early strength cement (Type III or Type HE) and a high accelerator dosage, it is possible to provide a shotcrete mix design capable of reaching early age compressive strengths of up to 7 MPa (1000 psi) at 4 hours. But to go over this previous limit, the cement technology needed to be reviewed. Working with calcium sulfo-aluminate (CSA) cement, the Rapid Set® Cement technology from CTS Cement, King has developed a research program to bring early age compressive strength gain to another level.

CSA cement sets very quickly and generates high early age compressive strengths. The main challenge when working with CSA cement is to mix and place the shotcrete or concrete mix before it sets. When water is mixed with any shotcrete or concrete mix made with CSA cement, the cement hydration process starts abruptly wherein the workability of the mixture decreases quickly but strength gain begins immediately after final set. By using a chemical retarder, it is possible to increase open time and workability but it also increases set time and delays strength gain. From the shotcrete technology perspective, the dry-mix process is the ideal choice. With technology used in the dry-mix shotcrete process, the shotcrete mixture is only mixed with water at the end of the transportation hose, at the nozzle, just prior to being shot onto the receiving surface. The mixture sets almost instantly in place and begins to gain strength. This makes the combination of the CSA cement technology with the dry-mix shotcrete process an ideal solution for reducing the mining and tunneling cycle.

The testing program included a first phase where the cement paste was optimized with the use of different pozzolans. Following this initial testing the target final set time was established to be 10 minutes after shooting. The rapid strength gain dry-mix shotcrete went through several levels of testing prior to being available for commercial use. Initially the rapid strength gain dry-mix shotcrete was tested internally by King in both winter and summer conditions on surface. Following that, the rapid strength gain dry-mix shotcrete was tested in both a mine training facility (to observe the effect of underground conditions) and at Laval University (Quebec City, QC, Canada) where all parameters of the shotcrete application could be controlled. The third portion of the testing protocol involved testing the rapid strength gain dry-mix shotcrete underground at a

**Figure 1. Aliva 252 dry-mix shotcrete machine being loaded with a dry, pre-mixed, pre-packaged shotcrete mix in a bulk bag**

mining facility in Northern Ontario, Canada under a cemented sand-fill section. The latest segment of the testing program was to test the effect on energy absorption of this particular cement matrix when used in conjunction with steel fibers. The steel fiber dosage used was the same dosage used in dry-mix shotcrete for ground support in mining applications in Northern Ontario, Canada.

## TEST METHODS

Shooting operations were conducted using both the Aliva 246 and Aliva 252 dry-mix shotcrete machines (Figure 1).

Regular shooting procedures were followed as described in the ACI 506 Guide to Shotcrete.[1] During the first two phases of testing a standard mining shotcrete (produced by King) was used as a control mix to make sure all of the different parameters were typical to normal shotcrete operations. The control mix results met the usual standard, and therefore these results are not presented in the article as they are not relevant to the topic. Set time was determined using a hand-held penetrometer in accordance with ASTM C1117, "Standard Test Method for Time of Setting of Shotcrete Mixtures by Penetration Resistance (Withdrawn 2003)." Early age compressive strength was determined using the end-beam test method, adapted from ASTM C116 "Test Method for Compressive Strength of Concrete Using Portions of Beams Broken in Flexure (Withdrawn 1999)"[2], (Figure 2). This method requires shooting directly in a mold to produce 75×75×100 mm (3×3×14 in.) beams specimens. Each end of each beam is then tested at a specific age using the end beam testing apparatus (Figure 2). This method is the most reliable and practical method to evaluate shotcrete early age compressive strength between 0.1 to 10 MPa (15 to 1,450 psi)[3]. Coring shotcrete when the compressive strength is low produces unreliable results or is logistically challenging.

Later age compressive strength was determined in accordance with ASTM C 1604 "Standard Test Method for Obtaining and Testing Drilled Cores of Shotcrete." The boiled absorption and volume of permeable voids were determined in accordance with ASTM C642 "Standard Test Method for Density, Absorption and Voids in Hardened Concrete." Also the shotcrete nozzleman was asked to provide comments regarding the evaluation of the material (including rebound) based on his experience. The rapid strength gain dry-mix shotcrete was tested for all of the properties listed above at different material and ambient temperatures in order to observe the effects of temperature on the shotcrete mix. The toughness of the steel fiber-enhanced rapid strength gain dry-mix shotcrete was measured in accordance with ASTM C1550 "Standard Test Method for Flexural Toughness of Fiber Reinforced Concrete (Using Centrally Loaded Round Panel)."

**Table 1. Set time results**

| Date | Jan-12 | Jun-12 | Dec-12 | Feb-13 |
|---|---|---|---|---|
| Final set time | 4 minutes | 4 minutes | 6 minutes | n/a |

## RESULTS

The results of the different tests are presented in this section. Set time results are provided in Table 1.

Early age compressive strength and later age compressive strength development curves with

**Figure 2. End beam testing apparatus**

relation to material temperature are shown respectively in Figure 3 and Figure 4. The target compressive strengths in both graphs are typical for a mining shotcrete specification.

The flexural strength results are presented in Table 2 and the volume of permeable voids and boiled absorption results are given in Table 3. Only one series of flexural strength tested was conducted to evaluate if the results were to differ from regular mining shotcrete.

Different nozzleman, who shot the material in the lab or on surface, commented that the rebound levels were as low as or even lower than typical silica fume enhanced dry-mix shotcrete. They also added that the water was easier to adjust for the proper consistency and the rapid strength gain dry-mix shotcrete seemed to be less sensitive to water fluctuation than the control mix.

The testing underground in situ brings minor daily challenges that a reliable product should be able to overcome. In this case when testing was conducted in March 2013, the equipment was in poor condition and required maintenance prior to shooting. Even after emergency maintenance was performed on the equipment, the material feeding rate was not as consistent as usually expected. On the second in situ testing conducted in May 2013, typical underground conditions such has minimal lighting and poor equipment maintenance were observed. Results from the in situ underground testing program with respect to set time, early age compressive strength, later age compressive strength, ambient temperature, volume of permeable voids and boiled absorption are shown in Table 4.

Different nozzleman who shot the material underground provided the same comments as the nozzleman who shot the material in lab/surface conditions, which were that the rebound levels were as low as or even lower than typical silica fume enhanced dry-mix shotcrete.

The toughness of the steel fiber-enhanced rapid strength requires shooting round determined panels (RDP). The RDP were shot on surface to simplify the logistic of testing, (Figure 5 and Figure 6). The results are presented in Table 5.

## DISCUSSION

The early age compressive strength curves presented in Figure 3 indicate that the material temperature had the largest effect on the time taken to reach compressive strengths in excess of 20 MPa (2,900 psi). It should be noted that even with an initial material temperature of 5°C (41°F) it was possible to reach compressive strengths in excess of 20 MPa (2,900 psi) within 4 hours after shooting. The later age compressive strength curves presented in Figure 4 indicate that the material temperature did not have a major impact on later age compressive strengths, and all of the samples tested were shown to exceed the target compressive strengths of 10 MPa (1,450 psi) at 24 hours, 20 MPa (2,900 psi) at 3 days, 30 MPa (4,350 psi) at 7 days and 40 MPa (5,800 psi) at 28 days. It should be noted that the lower compressive strength results for the "Indoor Lab Test" (tested Feb. 2013) in Figure 4, can be attributed to the fact that the material was shot at the wettest possible consistency without sloughing. The flexural

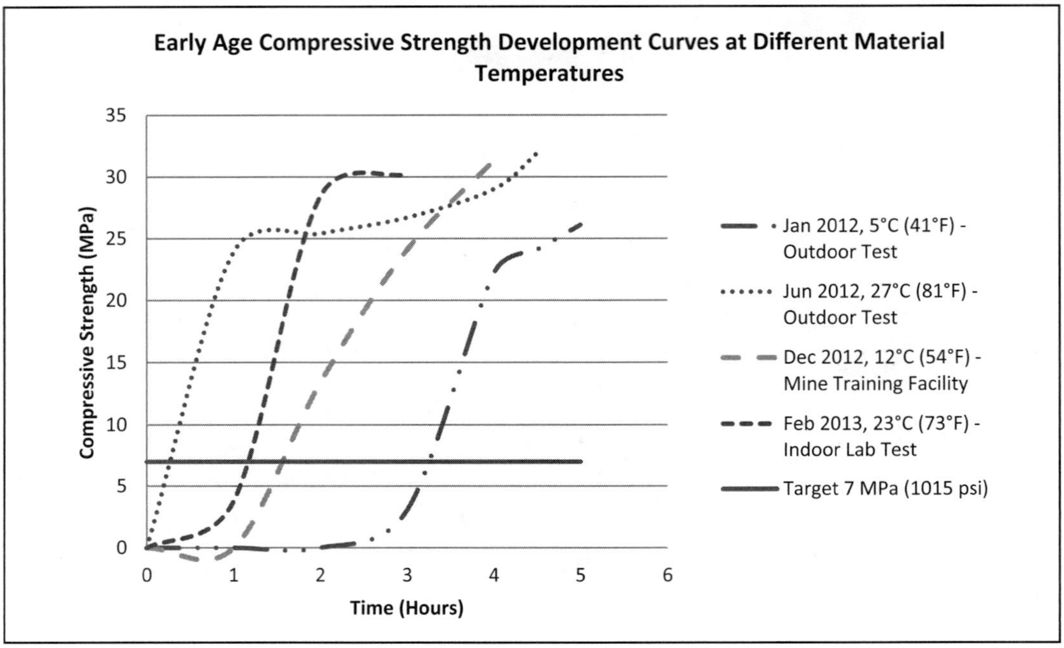


**Figure 3. Early age compressive strength of end beam specimens**

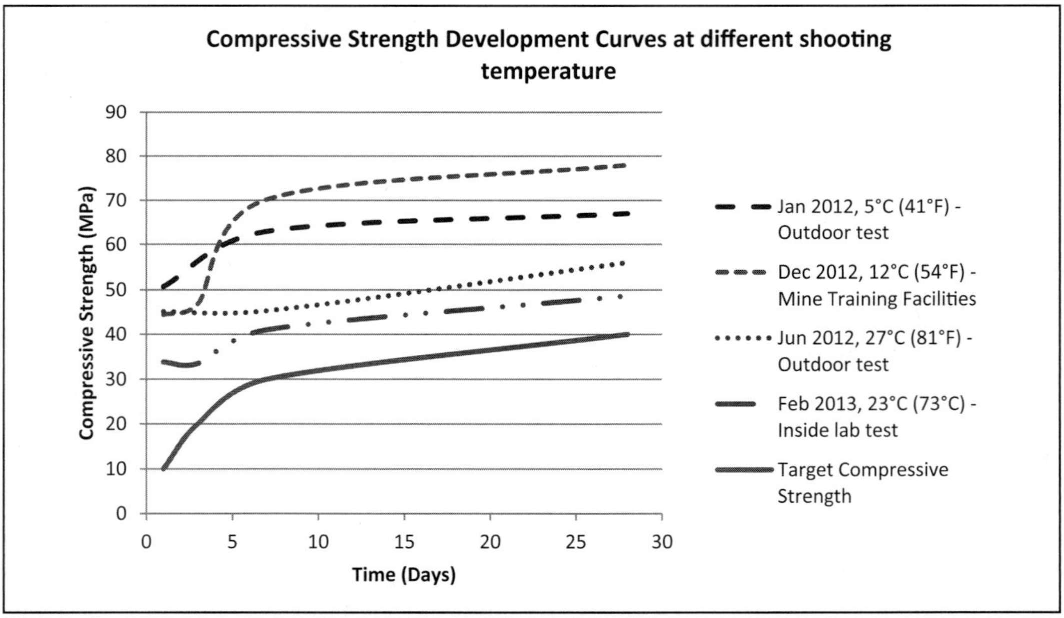


**Figure 4. Later age compressive strength of shotcrete cores**

strength results presented in Table 2 are very similar to results that would be expected from a normal Portland cement-based, silica fume enhanced dry-mix shotcrete.

When comparing the early age compressive strength results between values obtained in lab/surface conditions to underground conditions, it is apparent that the same level of strength development has not been shown to be present in underground conditions. It is possible that this could have been caused by a higher water/cement ratio being used underground as it can be more difficult to visually attain the proper consistency in underground conditions. It is also possible that sand lenses or entrapped rebound could have been present in end beam samples due to the poor condition of the shotcrete equipment used underground. Future testing will help provide values that can be expected for early age compressive strength in underground conditions.

Temperature has a big impact on early age compressive strength (same as with Portland cement), but since the goal of the testing program was to show it is possible to re-open the heading when compressive strengths reach at least 7 MPa or 1,000 psi (around 1 or 2 hours) the slight reduction in early age compressive strength in underground conditions was not considered an issue. Therefore, ambient temperature as well as temperatures of the dry material and water must be controlled and monitored to ensure safety. The set time results were found to meet the requirements of the testing program and therefore satisfactory.

The absorption values are higher than usual but the commonly used guidelines that are proposed in literature, and generally accepted in the industry[4], were all obtained using Portland cement based shotcrete, so the values available in this test program must be taken as data to be collected for further development. These higher values could possibly be related to shooting with too high of a water/cement ratio or poor compaction/consolidation which could also explain the lower compressive strength results.

The acceptable or lower rebound level can be explained by the combination of the different pozzolans and the fineness of the CSA cement. CSA

**Table 2. Flexural strength results**

| Time | Jan-12 |
|---|---|
| Initial material temperature | 5°C (41°F) |
| **Day** | **MPa** |
| 7 | 5.6 |
| 28 | 6 |

**Table 3. Volume of permeable voids and boiled absorption results**

| Time | Jan-12 | Jun-12 | Feb-13 |
|---|---|---|---|
| Initial mix temperature | 5°C (41°F) | 27°C (81°F) | 23°C (73°F) |
| Volume of permeable voids (%) | 15.8 | 15.0 | 15.9 |
| Boiled absorption (%) | 7.1 | 7.0 | 7.1 |

**Table 4. In situ test results from underground testing**

| Properties | Mar-13 | May-13 |
|---|---|---|
| Ambient temperature | 20°C (68°F) | 23°C (73°F) |
| Set time | 10 minutes | N/A |
| Compressive strength at 1 hour | 0.7 MPa (101 psi) | 12.4 MPa (1798 psi) |
| Compressive strength at 1.5 Hours | 9.6 MPa (1392 psi) | N/A |
| Compressive strength at 2 hours | 15.6 MPa (2263 psi) | 14 MPa (2031 psi) |
| Compressive strength at 3 days | 30.1 MPa (4366 psi) | 41.3 MPa (5990 psi) |
| Compressive strength at 7 days | 45.4 MPa (6584 psi) | 42.4 MPa (6150 psi) |
| Compressive strength at 28 days | 47.5 MPa (6889 psi) | 51.2 MPa (7426 psi) |
| Volume of permeable voids | 19.0% | 19.3% |
| Boiled absorption | 8.7% | 9.0% |

**Table 5. Round determined panel results**

| | | | Average Toughness as a Function of Deflection | | | | |
|---|---|---|---|---|---|---|---|
| | Average Peak Applied Load | | 5 mm | 10 mm | 20 mm | 30 mm | 40 mm |
| **Sample Age** | **(kN)** | **(lb·f)** | **(J)** | **(J)** | **(J)** | **(J)** | **(J)** |
| 24 hours | 26.3 | 5918.1 | 103.8 | 188.3 | 301.1 | 370.8 | 418.6 |
| 28 days | 36.0 | 8081.9 | 124.8 | 219.9 | 345.7 | 422.0 | 464.5 |

**Figure 5. Nozzleman shooting a round determinate panel**

cements are usually finer than normal Portland cement, causing increased adhesion and compaction. All nozzlemen pointed out the fact that the water was easier to adjust and seemed to fluctuate less than conventional Portland cement based dry-mix shotcrete.

The toughness results obtained at 24 hours are slightly higher than results normally obtained at 28 days with a Portland cement based steel fiber reinforced dry-mix shotcrete used for ground support in mining applications. Typically specifications call for 400 Joules at 28 days for a 40 mm deflection. Moreover, there is a noticeable increase of the average peak load applied and toughness between 24 hours and 28 days results. Figure 7 shows the fiber content in a section of a broken RDP sample after being tested in accordance with ASTM C1550.

## CONCLUSIONS

1. It is possible to obtain 20 MPa or even higher compressive strengths at 2 hours in the right conditions using dry-mix shotcrete with CSA cement for mining applications.
2. The rapid strength gain dry-mix shotcrete should be considered a very robust product that is suitable for regular mining and tunneling operations.
3. Early age strength development seems to be sensitive to ambient and material temperatures.
4. Absorption results are higher than usual when compared to Portland cement based, silica fume enhanced shotcrete mixes.
5. In situ testing provided sufficient confidence to the mine in order to include this new product in the mining cycle. Since being introduced into the mining process, the mine requested that a pigment be added for safety reasons to differentiate where this mix is used instead of their regular dry-mix shotcrete.
6. It is possible to combine the CSA cement and steel fiber technologies into a dry-mix shotcrete material to provide similar toughness earlier than normally obtained with a steel fiber reinforced Portland cement based shotcrete used for ground support in underground applications.

Looking forward, the next step is to improve the formulation for higher strengths at earlier ages, which there is sufficient information on CSA cement technology to believe it is possible. The boiled absorption level must be monitored and investigated, in order to evaluate if it is the nature of the new technology or results from poor compaction. The early age toughness will have to be evaluated, as early as 3 hours after application. This technology is very promising and brings new possibilities for mining and tunneling methods in the future.

**Figure 6. Surface finish of a round determinate panel**

**Figure 7. Steel fiber content of a broken RDP after being tested**

## REFERENCES


[1] ACI Committee 506, "Guide to Shotcrete (ACI 506R-05)," American Concrete Institute, Farmington Hills, Mich., 2005, 40 pp.

[2] Heere, R., Morgan, D.R. 2002, "Determination of Early-Age Compressive Strength of Shotcrete," Shotcrete Magazine, Vol. 4, No. 2, pp. 28–31.

[3] Bernard, S., Geltinger, C., 2007, "Determination of Early-Age Compressive Strength for FRS," Shotcrete Magazine, Vol. 9, No. 4, pp. 22–27.

[4] Austin, S.A., Robins, P.J., 1995, "Sprayed Concrete: Properties, Design and Application," Department of Civil & Building Engineering, Loughborough University of Technology, Loughborough, UK, 382 pages.

# TBM Performance and Tool Wear Prediction Along Two Lots of Dyaaba Headrace Tunnel (Uma-Oya Project, Sri Lanka)

**J. Hassanpour**
University of Tehran

**S. Tarigh Azali**
Ferdowsi University

**J. Rostami**
The Pennsylvania State University

**ABSTRACT:** Predicting TBM performance and cutter consumption in tunneling projects plays an important role on project schedule and cost. To predict these parameters, it is necessary to have reliable input parameters and suitable predictive models. In this paper, a new procedure used to anticipate TBMs performance parameters and cutter consumption along two lots of Dyaaba headrace tunnel (Uma-Oya project at Sri Lanka) have been explained. Applied empirical prediction models are simple equations and charts that use usual and easily-available rock mass parameters as input data and can be applied simply. They are based on experiences gained from actual TBM performance and disc cutters life measurements along similar tunneling projects.

## INTRODUCTION

Predicting TBM performance and cutter consumption in tunneling projects plays an important role on project schedule and cost. To predict these parameters, it is necessary to have reliable input parameters and suitable predictive models. Since there is no suitable prediction method or due to lack of required input data for existing models, their application in special condition of different projects is very difficult.

In this study, TBMs performance parameters and cutter consumption along two lots of Dyaaba headrace tunnel have been anticipated according to experiences gained from actual TBM performance and disc cutters life measurements along similar tunneling projects.

Applied empirical prediction models which recently have been developed and reported in literature are simple equations and charts that use usual and available rock mass parameters as input data and can be applied easily.

## PROJECT DESCRIPTION

Uma-Oya multipurpose project in Sri Lanka consists of four major parts including Power Plant, two Dams, Water Conveying tunnels with total length of about 23km and a 700m long vertical shaft which will generate hydro power and irrigate farm lands in the Moneragala, Badulla and Ampara districts in the south east of Sri Lanka. The most important part of this project is a long tunnel will be excavated by mechanized and drill & blast methods. As illustrated in Table 1, this tunnel has been divided into three main sections according to excavation method selected for boring each part. Two similar double shield machines manufactured by Herrenknecht (Germany) have been selected for excavating the main section of tunnel namely Dyraaba headrace tunnel with total length of about 15km. The first machine will excavate 6200m total length of lot A and the second lot (Lot B) with total length of 9700m will be excavated by the second machine.

## ENGINEERING GEOLOGICAL PROPERTIES OF ROCK MASSES

Based on petrographic analyses done on the samples taken from boreholes and also field observations, it is possible to divide tunnel route into four different petrographic units. These main petrographic units identified along two lots of tunnel, include (1) quartz-feldspathic gneisses partly with frequent garnet and biotite minerals (Pmgga and Pmgqf); (2) undifferentiated charnockitic (greenish, medium to coarse grained) biotite gneisses (Pmgk); (3) pure coarse grained quartzites (Pmq) and (4) calc-gneisses and marbles (Pmc). Table 2 lists these units and their distribution along the two lots of tunnel.

**Table 1. Three main sections of water conveying tunnel**

| Structure | Description | Excavation Method |
|---|---|---|
| Puhulpola Reservoirs link tunnel | 3,750m long free flow tunnel to Dyraaba Dam | Drill & Blast |
| Dyraaba headrace tunnel | 15,150m long headrace tunnel to the Vertical Pressure shaft | Mechanized method |
| Tailrace tunnel | 3,600m long tailrace tunnel | Drill & Blast |

**Table 2. Engineering geological properties of identified stratigraphic units along two sections of the tunnel**

| No. | | Chainage (m) Start | Finish | Section Length (m) | Symbol | UCS (Mpa) | RQD (%) | RMR |
|---|---|---|---|---|---|---|---|---|
| 1 | Lot A | 0 | 171 | 171 | Pmgk | 120 | 60 | 58 |
| 2 | | 171 | 293 | 122 | Pmc | 75 | 90 | 55 |
| 3 | | 293 | 447 | 154 | Pmq | 90 | 60 | 55 |
| 4 | | 447 | 1,514 | 1,067 | Pmgk | 120 | 75 | 60 |
| 5 | | 1,514 | 2,558 | 1044 | Pmgqf | 90 | 75 | 58 |
| 6 | | 2,558 | 3,536 | 978 | Pmgga | 120 | 80 | 60 |
| 7 | | 3,536 | 4,147 | 611 | Pmgqf | 90 | 100 | 64 |
| 8 | | 4,147 | 4,754 | 607 | Pmgk | 120 | 85 | 64 |
| 9 | | 4,754 | 5,072 | 318 | Pmgqf | 90 | 100 | 64 |
| 10 | | 5,072 | 5,298 | 226 | Pmq | 90 | 90 | 64 |
| 11 | | 5,298 | 5,383 | 85 | Pmgqf | 50 | 25 | 30 |
| 12 | | 5,383 | 5,617 | 234 | Pmgqf | 90 | 50 | 50 |
| 13 | | 5,617 | 6,200 | 583 | Pmgk | 120 | 90 | 65 |
| 1 | Lot B | 6,200 | 7,738 | 1,538 | Pmgk | 120 | 90 | 65 |
| 2 | | 7,738 | 7,952 | 214 | Pmc | 75 | 90 | 65 |
| 3 | | 7,952 | 8,640 | 688 | Pmgqf | 90 | 100 | 65 |
| 4 | | 8,640 | 8,899 | 259 | Pmq | 90 | 25 | 40 |
| 5 | | 8,899 | 10,502 | 1,603 | Pmgqf | 90 | 50 | 45 |
| 6 | | 10,502 | 10,650 | 148 | Pmq | 90 | 85 | 65 |
| 7 | | 10,650 | 11,165 | 515 | Pmgqf | 90 | 100 | 65 |
| 8 | | 11,165 | 11,767 | 602 | Pmgk | 120 | 80 | 60 |
| 9 | | 11,767 | 12,112 | 345 | Pmc | 40 | 25 | 30 |
| 10 | | 12,112 | 13,068 | 956 | | 80 | 50 | 55 |
| 11 | | 13,068 | 13,289 | 221 | Pmc | 40 | 25 | 40 |
| 12 | | 13,289 | 13,650 | 361 | | 40 | 25 | 30 |
| 13 | | 13,650 | 14,189 | 539 | Pmgqf | 90 | 50 | 55 |
| 14 | | 14,189 | 14,806 | 617 | Pmgga | 120 | 25 | 30 |
| 15 | | 14,806 | 15,260 | 454 | | 90 | 80 | 55 |

There are several inferred regional tectonic structures crossing the tunnel, such as faults or thrust planes whose appearance on the terrain is as lineaments.

The main discontinuities systems along the Dyraaba headrace tunnel are foliation and joints. The trends of foliations in the metamorphic rocks along the tunnel route are variable and their trends are slightly oblique to the tunnel axis with medium dip angles.

According to exploratory boreholes (at the section of tunnel and near it) the rock mass quality is good to excellent (R.Q.D=84–100%) along the major part of tunnel and poor to medium (R.Q.D=40–65%) due to highly to completely weathered rock or fractured zones along minor parts of tunnel. Variations of rock quality designation along Dyraaba headrace tunnel are shown in Table 2.

According to field estimations and regarding the laboratory test results, the tunnel will be excavated in strong to very strong rocks with a usual range of UCS between 75 and 120MPa (Table 2).

The weathering of the natural bedrock surface varies much, from slightly weathered rocks to reddish brown residual soil. The fresh bedrock has usually a high content of feldspar, which alters to the reddish clay minerals in weathered units. The weathering degree in depth reduced considerably and it is

**Table 3. Summary results of petrographic analyses and abrasiveness estimation using different methods**

| | Mineral Content (%) | | | | | | Ave. | Ave. UCS | | | |
|---|---|---|---|---|---|---|---|---|---|---|---|
| Petrog. Unit | C | Q | F | G | B | O | VHNR | (MPa) | ABI | CAI | ABR |
| Garnet Gneiss | | 40–50 | 20–50 | 10–40 | <10 | | 915 | 90 | 824 | | |
| Biotite Gneiss | | 15 | 15 | < 20 | 20–30 | 20–30 | 635 | 120 | 762 | | |
| Quartzite | | 70–95 | 5–30 | | | | 1,030 | 90 | 924 | | |
| Marble | 50–60 | 20–30 | | 10 | | | 500 | 75 | 374 | | |

C: Calcite (Dolomite), Q: Quartz, G: Garnet, B: Biotite, F: Feldspare, O: Orthopyroxene

**Table 4. Classification of rock abrasiveness based on new geotechnical index of ABI (Hassanpour and Rostami, 2010)**

| No. | Description | ABI | Examples |
|---|---|---|---|
| 1 | Not abrasive | <75 | Very weak sedimentary rocks like Mudstone, Marlstone |
| 2 | Slightly abrasive | 75–200 | Limestone, Marble |
| 3 | Abrasive | 200–400 | Sandstone, Tuff |
| 4 | Very abrasive | 400–900 | Strong igneous and metamorphic rocks like Granite and Schist |
| 5 | Extremely abrasive | >900 | Very strong metamorphic rocks like Gneiss and Quartzite |

expected that the rock at tunnel level will be fresh or slightly weathered except for fault zones which due to existence of water the rock can be potentially weathered.

In this study a number of samples were taken from boreholes and sent to Switzerland (EPFL rock mechanics laboratory) for estimation of abrasiveness using two methods of Cerchar and LCPC. Table 3 summarizes the obtained results. As shown, the values of CAI parameter (Cerchar Abrasivity Index) for four main types of lithologies including (1) Granite, (2) Biotite gneiss, (3) Garnet gneiss and (4) Quartzite are different. Results of LCPC tests more or less accord the results obtained from Cerchar tests.

In addition to laboratory test procedures some geotechnical indices are used as a measure for evaluating and classifying rock abrasiveness. One of the most important indices usually used in practice is Vickers Hardness Number of the Rock (VHNR). This index can be simply measured based on data obtained from microscopic analyses of thin sections (Table 3). Also, ABI or Abrasiveness Index which introduced by Hassanpour and Rostami (2010) can be considered as a new index for assessing rock abrasiveness. This index is derived by combination of two important parameters of VHNR and UCS as follow:

$$ABI = VHNR \cdot \left( \frac{UCS}{100} \right) \qquad (1)$$

Based on experience gained from different projects a classification of rock abrasiveness using new geotechnical index of ABI is proposed and introduced in Table 4.

Table 3 lists summary results of petrographic analyses and abrasiveness evaluation of the main rock types identified along two lots of Dyaraab tunnel.

**Table 5. Main specifications of selected machines for Uma-Oya project**

| | Value in Two Selected Machines | |
|---|---|---|
| Parameter | Lot A | Lot B |
| Machine type | D.S. TBM | D.S. TBM |
| Machine diameter | 4.3m | 4.3m |
| Cutters diameter | 432mm (17 inch) | 432mm (17 inch) |
| Number of disc cutters | 27 | 27 |
| Max. operating cutterhead thrust | 21,287KN | 21,287KN |
| Cutterhead power | 1,250KW | 1250KW |
| Cutterhead speed | 0 to 11 rpm | 0 to 11 rpm |
| Cutterhead torque (nominal) | 1,725kN·m (11rpm) | 1,725kN·m (11rpm) |
| Thrust cylinder stroke | 1,300mm | 1,300mm |
| TBM weight | 247 ton | 247 ton |

## MACHINES SPECIFICATIONS

Two similar double shield hard rock TBMs used in this project are manufactured by Herrenknecht. The cutterhead is laced with 27, each 17 inch or 432mm diameter cutters with a load capacity of 250KN. Other main technical specifications of the machines are summarized in Table 5.

## PERFORMANCE PREDICTION OF SELECTED MACHINES

TBM performance prediction models should only be used with a reasonable understanding of their basis and considering of their limitations. An improper use of a model to a ground condition that is different than the original set of data used for development of that model will leads to inaccurate and even wrong results in predicting TBM performance and preparing construction schedule for a tunnel project. Meanwhile, application of experiences gained from similar projects is a practical way to predict TBM performance in projects with similar ground conditions.

Hassanpour et al. (2011a) have compiled geological and machine performance data from Manapouri tailrace tunnel project at New Zealand and performed an extensive statistical analysis on the available information from this project. Based on the statistical analyses, they have proposed following equation to predict TBM performance:

$$FPI = \exp(0.005UCS + 0.020RQD + 1.644) \quad (2)$$

where

$$FPI = \frac{Th}{ROP} \quad (3)$$

and

$$PR = \frac{ROP \cdot RPM \cdot 60}{1000} \quad (4)$$

In the above equations: FPI = Field penetration index (KN/cutter/mm/rev); ROP = Rate of penetration (mm/rev), PR = Penetration rate (m/h), Th = cutter thrust (kN/cutter), RPM = cutterhead revolution speed (rev/min), UCS = Uniaxial compressive strength (MPa); and RQD = Rock Quality Designation (%).

Equation 2 is proposed for jointed, blocky and massive igneous and metamorphic rocks (with a UCS range of 50 to 250MPa) which are very similar to rocks exposed in Uma-Oya project. The advantage of this equation is that it is based on FPI, allowing for the calculation of penetration per revolution for given level of cutter load. Therefore, given the compressive strength of the rock specimen and the estimated RQD, it allows for estimation of FPI and with a given set of machine specifications, it allows for estimation of ROP.

Hassanpour et al. (2011b) has proposed some other predictive equations which are obtained based on data from different tunneling projects excavated in different geological conditions. They also combined data from different projects and developed a general FPI prediction chart (Figure 1) a boreability classification system (Table 6) for estimating TBM performance in different geological conditions.

In this study, FPI and ROP and PR values are estimated using Equations 2, 3 and 4 and listed in Table 7. As shown, variation range of FPI is between 10 (in fault zones) and 62kN/cutter/mm/rev. (in massive and strong rocks). It means that identified rock units along two lots of tunnel can be categorized in four boreability classes according to Table 6 and Figure 1.

It must be emphasized that for estimating ROP in different geological units it is necessary to have an estimate of operating machine thrust and RPM. These parameters were assumed based on experiences gained from previous similar projects and in general from machine specifications.

To prepare construction schedule, it is necessary to know TBM advance rate in each unit. Advance rate can be calculated using following equation:

$$AR = \frac{PR \cdot U \cdot h_s \cdot n_s}{100} \quad (5)$$

where U = Utilization factor (%), $n_s$ = number of shifts per day and $h_s$ = number of hours per shift. In this project the values of $n_s$ and $h_s$ were assumed 2 shifts per day and 10 hours per shift, respectively.

Number of days that machine will be excavating through each geological unit ($N_{day}$) can be estimated using Equation 6:

$$N_{day} = \frac{L_{sec}}{AR} \quad (6)$$

where $L_{sec}$ is length of given geological unit along the tunnel (m) and AR is advance rate of machine (m/day) at that geological unit. $N_{day}$ is the most important parameter for preparing tunnel construction schedule. Knowing this parameter, it is possible to calculate TBM advance during construction period. Table 8 shows results of calculation of this parameter based on geo-mechanical characteristics, unit length and assumed utilization factor for tunnel construction in each section. Figure 2 shows predicted histograms of monthly TBM advance and accumulated advance of machines for Lots A and B respectively.

As shown in these graphs, it is predicted that total required times for completing excavation and installation of lining in lots A and B are about 15 and 21 months, respectively. The estimates indicate that maximum and minimum monthly advance in Lot A of tunnel will be 649 and 150m respectively. In lot B, maximum and minimum monthly advance are higher and exceed 1,100 and 200m, respectively.

Low anticipated advance in 12th and 13th months of operation in lot A and first 4 months in lot B is due to adverse geological condition predicted in identified fault zones. In these sections due to occurrence of probable instabilities in tunnel, utilization factor is assumed to be 5% or less.

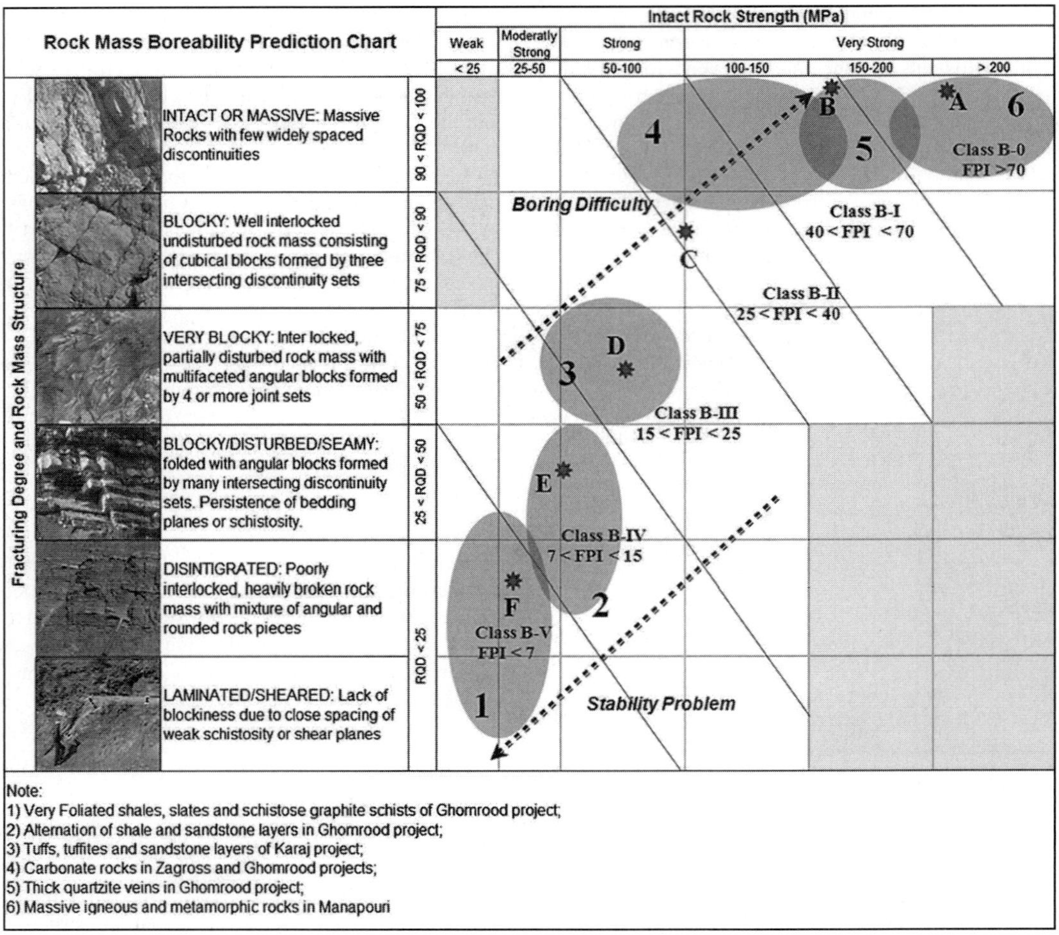


**Figure 1. General rock mass boreability prediction chart (Hassanpour et al., 2011b)**

**Table 6. Summary of ground conditions for various boreability classes (Hassanpour et al., 2011b)**

| Boreability Class | FPI Range (kN/mm/rev) | Rock Mass Boreability | Stability Condition | TBM Excavatability | Example |
|---|---|---|---|---|---|
| B-0 | >70 | Tough | Completely stable | Tough | Very strong and massive intrusive and metamorphic rocks |
| B-I | 40–70 | Fair-tough | Stable | Fair | Massive igneous and metamorphic rocks |
| B-II | 25–40 | Good-fair | Minor instabilities | Good | Blocky and jointed Tuff, limestone |
| B-III | 15–25 | good | Only local structural instabilities | Very good | Alternations of sandstone, limestone and shale |
| B-IV | 7–15 | Very good | Some major instabilities | Good | Alternations of thin bedded Shale and Sandstone layers |
| B-V | <7 | Excellent | Collapse, gripper problems, squeeze, etc. | May be problematic | Highly foliated and schistose metamorphic rocks (Slate, Phyllite, Graphite schist), Shale, Marlstone, thick fault zones |

**Table 7. Input parameters (geological units and their geo-mechanical properties) and results of ROP estimation at each rock unit**

| Zone | | Unit Name | Chainage (m) Start | Chainage (m) Finish | Section Length (m) | UCS (Mpa) | RQD (%) | FPI (kN/c./mm/rev) | ROP (mm/rev) | PR (m/h) |
|---|---|---|---|---|---|---|---|---|---|---|
| 1 | Lot A | Pmgk | 0 | 171 | 171 | 120 | 60 | 31.1 | 7.7 | 3.25 |
| 2 | | Pmc | 171 | 293 | 122 | 75 | 90 | 46.7 | 5.2 | 2.17 |
| 3 | | Pmq | 293 | 447 | 154 | 90 | 60 | 26.8 | 9.0 | 3.77 |
| 4 | | Pmgk | 447 | 1,514 | 1,067 | 120 | 75 | 42.6 | 5.6 | 2.37 |
| 5 | | Pmgqf | 1,514 | 2,558 | 1,044 | 90 | 75 | 36.7 | 6.6 | 2.75 |
| 6 | | Pmgga | 2,558 | 3,536 | 978 | 120 | 80 | 47.4 | 5.1 | 2.13 |
| 7 | | Pmgqf | 3,536 | 4,147 | 611 | 90 | 100 | 62.1 | 3.9 | 1.63 |
| 8 | | Pmgk | 4,147 | 4,754 | 607 | 120 | 85 | 52.6 | 4.6 | 1.92 |
| 9 | | Pmgqf | 4,754 | 5,072 | 318 | 90 | 100 | 62.1 | 3.9 | 1.63 |
| 10 | | Pmq | 5,072 | 5,298 | 226 | 90 | 90 | 50.3 | 4.8 | 2.01 |
| 11 | | Pmgqf | 5,298 | 5,383 | 85 | 50 | 25 | 10.5 | 22.9 | 9.61 |
| 12 | | Pmgqf | 5,383 | 5,617 | 234 | 90 | 50 | 21.7 | 11.1 | 4.66 |
| 13 | | Pmgk | 5,617 | 6,200 | 583 | 120 | 90 | 58.4 | 4.1 | 1.73 |
| 1 | Lot B | Alt. | 14,806 | 15,260 | 454 | 90 | 80 | 40.8 | 5.9 | 2.48 |
| 2 | | Pmgga | 14,189 | 14,806 | 617 | 120 | 25 | 14.9 | 16.1 | 6.77 |
| 3 | | Pmgqf | 13,650 | 14,189 | 539 | 90 | 50 | 21.7 | 11.1 | 4.66 |
| 4 | | Alt. | 13,289 | 13,650 | 361 | 40 | 25 | 10.0 | 24.1 | 10.11 |
| 5 | | Pmc | 13,068 | 13,289 | 221 | 40 | 25 | 10.0 | 24.1 | 10.11 |
| 6 | | Alt. | 12,112 | 13,068 | 956 | 80 | 50 | 20.7 | 11.7 | 4.90 |
| 7 | | Pmc | 11,767 | 12,112 | 345 | 40 | 25 | 10.0 | 24.1 | 10.11 |
| 8 | | Pmgk | 11,165 | 11,767 | 602 | 120 | 80 | 47.4 | 5.1 | 2.13 |
| 9 | | Pmgqf | 10,650 | 11,165 | 515 | 90 | 100 | 62.1 | 3.9 | 1.63 |
| 10 | | Pmq | 10,502 | 10,650 | 148 | 90 | 85 | 45.3 | 5.3 | 2.23 |
| 11 | | Pmgqf | 8,899 | 10,502 | 1,603 | 90 | 50 | 21.7 | 11.1 | 4.66 |
| 12 | | Pmq | 8,640 | 8,899 | 259 | 90 | 25 | 12.8 | 18.7 | 7.87 |
| 13 | | Pmgqf | 7,952 | 8,640 | 688 | 90 | 100 | 62.1 | 3.9 | 1.63 |
| 14 | | Pmc | 7,738 | 7,952 | 214 | 75 | 90 | 46.7 | 5.2 | 2.17 |
| 15 | | Pmgk | 6,200 | 7,738 | 1,538 | 120 | 90 | 58.4 | 4.1 | 1.73 |

## PREDICTING CUTTER CONSUMPTION

One of the main cost items in mechanized tunneling projects in rock and soil is the cost of changing of damaged or worn cutting tools. In addition, disc cutter change is a time consuming operation which can have a negative influence on TBM performance. So, to estimate exact cost of project and TBM performance it is required that disc cutter wear or life as well as TBM penetration rate evaluated using appropriate methods.

At present, there are a few common prediction models for the calculation of the cutter life. Among them two models developed in NTNU (Bruland, 1998) and CSM (Colorado School of Mines, Rostami, 1997) are more important ones. Although, these methods compute the total cost of replacing the cutters, the delays due to replacements and the number of cutters broken and replaced, but have some shortcomings when applied to real TBM projects. The main limitation is that they are based on some parameters like CLI (cutter life index) and CAI (Cerchar abrasivity index) that usually are not available especially in preliminary phases of studies.

In recent years Hassanpour and Rostami (2010) based on data collected from different mechanized tunneling projects in Iran and abroad have developed some empirical equations for predicting cutter wear. They have used ABI index as the main input geological parameter in their prediction models. One of the tunneling cases they used for developing their equations is Manapouri tailrace tunnel in New Zealand constructed in similar geological condition. As mentioned before, this tunnel has been excavated in strong to very strong igneous and metamorphic rocks which lithologically consist of biotite gneiss, hornblende gneiss, calc-silicate gneiss, pegmatite and gabbro. Based on experiences gained from this particular project they have proposed following equation for estimating cutter life and cutter consumption in similar geological conditions:

**Table 8. Calculation results of construction schedule of the tunnel (Lot A)**

| Zone | | Unit Name | Chainage (m) Start | Finish | Section Length (m) | PR (m/h) | Predicted Utilization Factor (%) | Advance Rate (m/day) | Required Days for Tunnel Excavation | Accumu-lated No. of Days | Accumu-lated No. of Months |
|---|---|---|---|---|---|---|---|---|---|---|---|
| 1 | Lot A | Pmgk | 0 | 171 | 171 | 3.25 | 10 | 6.5 | 26 | 26 | 0.9 |
| 2 | | Pmc | 171 | 293 | 122 | 2.17 | 40 | 17.3 | 7 | 33 | 1.1 |
| 3 | | Pmq | 293 | 447 | 154 | 3.77 | 40 | 30.2 | 5 | 38 | 1.3 |
| 4 | | Pmgk | 447 | 1,514 | 1,067 | 2.37 | 40 | 19.0 | 56 | 95 | 3.2 |
| 5 | | Pmgqf | 1,514 | 2,558 | 1,044 | 2.75 | 40 | 22.0 | 47 | 142 | 4.7 |
| 6 | | Pmgga | 2,558 | 3,536 | 978 | 2.13 | 40 | 17.1 | 57 | 199 | 6.6 |
| 7 | | Pmgqf | 3,536 | 4,147 | 611 | 1.63 | 40 | 13.0 | 47 | 246 | 8.2 |
| 8 | | Pmgk | 4,147 | 4,754 | 607 | 1.92 | 40 | 15.4 | 39 | 286 | 9.5 |
| 9 | | Pmgqf | 4,754 | 5,072 | 318 | 1.63 | 40 | 13.0 | 24 | 310 | 10.3 |
| 10 | | Pmq | 5,072 | 5,298 | 226 | 2.01 | 40 | 16.1 | 14 | 324 | 10.8 |
| 11 | | Pmgqf | 5,298 | 5,383 | 85 | 9.61 | 5 | 9.6 | 9 | 333 | 11.1 |
| 12 | | Pmgqf | 5,383 | 5,617 | 234 | 4.66 | 5 | 4.7 | 50 | 383 | 12.8 |
| 13 | | Pmgk | 5,617 | 6,200 | 583 | 1.73 | 40 | 13.8 | 42 | 425 | 14.2 |
| 1 | Lot B | Alt. | 14,806 | 15,260 | 454 | 2.48 | 20 | 9.9 | 46 | 46 | 1.5 |
| 2 | | Pmgga | 14,189 | 14,806 | 617 | 6.77 | 5 | 6.8 | 91 | 137 | 4.6 |
| 3 | | Pmgqf | 13,650 | 14,189 | 539 | 4.66 | 40 | 37.3 | 14 | 151 | 5.0 |
| 4 | | Alt. | 13,289 | 13,650 | 361 | 10.11 | 5 | 10.1 | 36 | 187 | 6.2 |
| 5 | | Pmc | 13,068 | 13,289 | 221 | 10.11 | 5 | 10.1 | 22 | 209 | 7.0 |
| 6 | | Alt. | 12,112 | 13,068 | 956 | 4.90 | 40 | 39.2 | 24 | 233 | 7.8 |
| 7 | | Pmc | 11,767 | 12,112 | 345 | 10.11 | 5 | 10.1 | 34 | 267 | 8.9 |
| 8 | | Pmgk | 11,165 | 11,767 | 602 | 2.13 | 40 | 17.1 | 35 | 303 | 10.1 |
| 9 | | Pmgqf | 10,650 | 11,165 | 515 | 1.63 | 40 | 13.0 | 40 | 342 | 11.4 |
| 10 | | Pmq | 10,502 | 10,650 | 148 | 2.23 | 40 | 17.9 | 8 | 350 | 11.7 |
| 11 | | Pmgqf | 8,899 | 10,502 | 1603 | 4.66 | 40 | 37.3 | 43 | 394 | 13.1 |
| 12 | | Pmq | 8,640 | 8,899 | 259 | 7.87 | 5 | 7.9 | 33 | 426 | 14.2 |
| 13 | | Pmgqf | 7,952 | 8,640 | 688 | 1.63 | 40 | 13.0 | 53 | 479 | 16.0 |
| 14 | | Pmc | 7,738 | 7,952 | 214 | 2.17 | 40 | 17.3 | 12 | 492 | 16.4 |
| 15 | | Pmgk | 6,200 | 7,738 | 1,538 | 1.73 | 40 | 13.8 | 111 | 603 | 20.1 |

$$H_f(\mathrm{m}^3/\mathrm{cutter}) = -173.67\,\mathrm{Ln(ABI)} + 1379.19 \tag{7}$$

and

$$W_f(\mathrm{cutter/m}) = \frac{\pi D^2}{4H_f} \tag{8}$$

where $H_f$ is instantaneous cutter life in $\mathrm{m}^3$/cutter, $W_f$ is instantaneous cutter consumption in cutter/m (Bruland, 1998), $D$ is tunnel diameter and ABI is abrasivity index calculated using Equation 1.

In this study, due to geological similarity of two projects, the Equations 7 and 8 were applied to predict cutter consumption and cutter life parameters along the tunnel. Table 9 summarizes the results of computation of two parameters of $H_f$ and $W_f$ in different geological units identified along the Uma-Oya tunnel.

By multiplying $W_f$ and $L_{sec}$ (tunnel section length), total required number of disc cutters in each section can be obtained (last column of Table 9). So, it is anticipated that a total number of about 500 and 655 disc cutters will be required for completing lots A and B of Uma-Oya project, respectively.

## CONCLUSION

In this paper new empirical models for predicting TBM performance and cutter consumption were introduced. To explain the application procedure of the model, TBM performance and cutter consumption along an under-construction tunnel in Sri Lanka were calculated.

As illustrated in this paper, applied empirical prediction models are simple equations and charts that use usual and easily-available rock mass parameters as input data and can be applied very easily for predicting time schedule and cutter consumption. They are based on experiences gained from actual

**Table 9. Results of computation of cutter life, cutter consumption, and total number of disc cutters required for completing each section of tunnel**

| Zone | | Chainage (m) Start | Finish | Section Length (km) | Unit | VHNR | UCS (MPa) | ABI | $H_f$ ($m^3/c$) | $W_f$ (c/m) | Required No of Disc Cutters |
|---|---|---|---|---|---|---|---|---|---|---|---|
| Lot A | 1 | 0 | 171 | 171 | Pmgk | 800 | 120 | 960 | 186.6 | 0.085 | 15 |
| | 2 | 171 | 293 | 122 | Pmc | 500 | 75 | 375 | 349.9 | 0.045 | 6 |
| | 3 | 293 | 447 | 154 | Pmq | 1,030 | 90 | 927 | 192.7 | 0.083 | 13 |
| | 4 | 447 | 1,514 | 1,067 | Pmgk | 800 | 120 | 960 | 186.6 | 0.085 | 91 |
| | 5 | 1,514 | 2,558 | 1,044 | Pmgqf | 850 | 90 | 765 | 226.0 | 0.070 | 73 |
| | 6 | 2,558 | 3,536 | 978 | Pmgga | 915 | 120 | 1098 | 163.3 | 0.097 | 95 |
| | 7 | 3,536 | 4,147 | 611 | Pmgqf | 850 | 90 | 765 | 226.0 | 0.070 | 43 |
| | 8 | 4,147 | 4,754 | 607 | Pmgk | 800 | 120 | 960 | 186.6 | 0.085 | 52 |
| | 9 | 4,754 | 5,072 | 318 | Pmgqf | 850 | 90 | 765 | 226.0 | 0.070 | 22 |
| | 10 | 5,072 | 5,298 | 226 | Pmq | 1,030 | 90 | 927 | 192.7 | 0.083 | 19 |
| | 11 | 5,298 | 5,383 | 85 | Pmgqf | 850 | 50 | 425 | 328.1 | 0.048 | 4 |
| | 12 | 5,383 | 5,617 | 234 | Pmgqf | 850 | 90 | 765 | 226.0 | 0.070 | 16 |
| | 13 | 5,617 | 6,200 | 583 | Pmgk | 800 | 120 | 960 | 186.6 | 0.085 | 50 |
| | **Sum** | | | | | | | | | | **499** |
| Lot B | 1 | 15,260 | 14,806 | 454 | Alt. | 900 | 90 | 810 | 216.1 | 0.074 | 33 |
| | 2 | 14,806 | 14,189 | 617 | Pmgga | 915 | 120 | 1098 | 163.3 | 0.097 | 60 |
| | 3 | 14,189 | 13,650 | 539 | Pmgqf | 850 | 90 | 765 | 226.0 | 0.070 | 38 |
| | 4 | 13,650 | 13,289 | 361 | Alt. | 900 | 40 | 360 | 357.0 | 0.045 | 16 |
| | 5 | 13,289 | 13,068 | 221 | Pmc | 500 | 40 | 200 | 459.0 | 0.035 | 8 |
| | 6 | 13,068 | 12,112 | 956 | Alt. | 900 | 80 | 720 | 236.6 | 0.067 | 64 |
| | 7 | 12,112 | 11,767 | 345 | Pmc | 500 | 40 | 200 | 459.0 | 0.035 | 12 |
| | 8 | 11,767 | 11,165 | 602 | Pmgk | 800 | 120 | 960 | 186.6 | 0.085 | 51 |
| | 9 | 11,165 | 10,650 | 515 | Pmgqf | 850 | 90 | 765 | 226.0 | 0.070 | 36 |
| | 10 | 10,650 | 10,502 | 148 | Pmq | 1030 | 90 | 927 | 192.7 | 0.083 | 12 |
| | 11 | 10,502 | 8,899 | 1,603 | Pmgqf | 850 | 90 | 765 | 226.0 | 0.070 | 113 |
| | 12 | 8,899 | 8,640 | 259 | Pmq | 1,030 | 90 | 927 | 192.7 | 0.083 | 21 |
| | 13 | 8,640 | 7,952 | 688 | Pmgqf | 850 | 90 | 765 | 226.0 | 0.070 | 48 |
| | 14 | 7,952 | 7,738 | 214 | Pmc | 500 | 75 | 375 | 349.9 | 0.045 | 10 |
| | 15 | 7,738 | 6,200 | 1,538 | Pmgk | 800 | 120 | 960 | 186.6 | 0.085 | 131 |
| | **Sum** | | | | | | | | | | **653** |

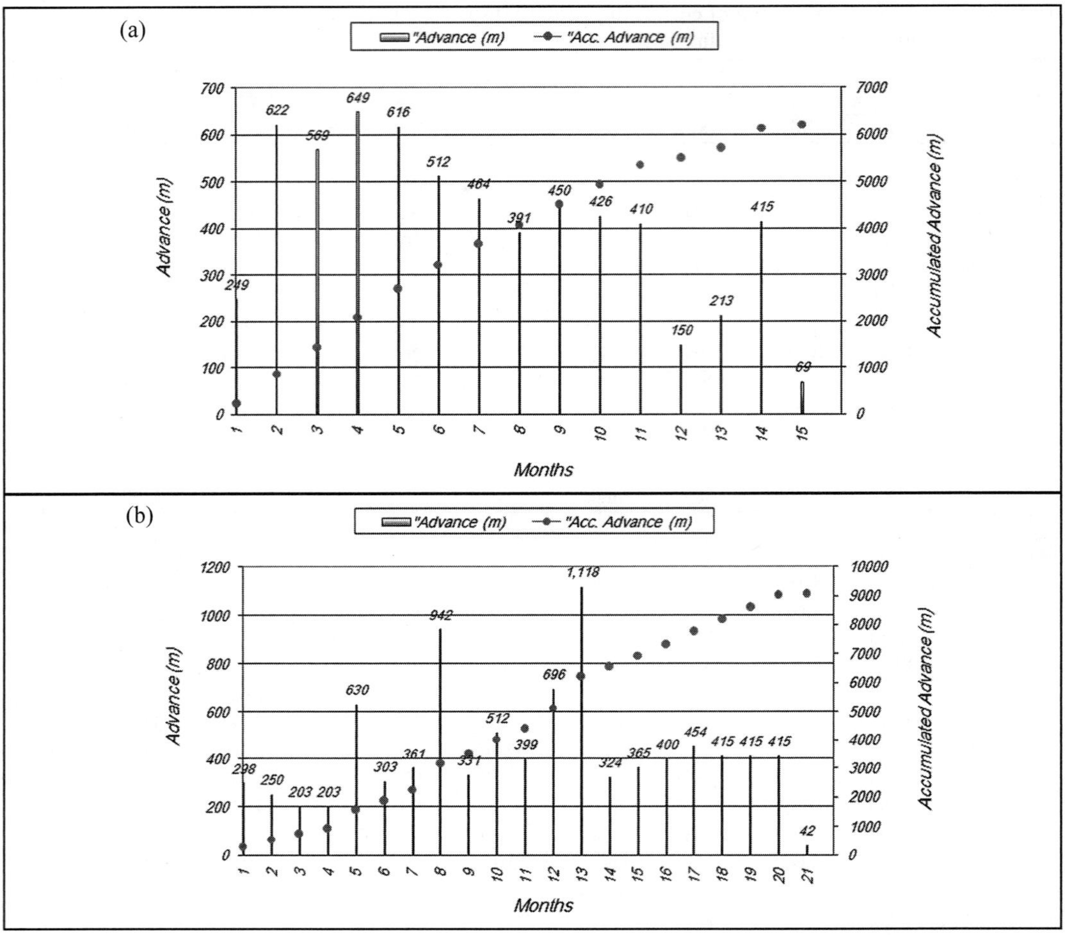


**Figure 2. Summary of predicted tunnel construction schedule, including monthly and accumulative advance rates in two lots of tunnel: (a) Lot A and (b) Lot B**

TBM performance and disc cutters life measurements along similar tunneling projects.

Results of analyses using newly developed model show that total required time for completing excavation in two lots of Uma-Oya tunnel are about 15 and 21 months, respectively. It is also predicted that a total number of about 500 and 655 disc cutters will be required for constructing these two lots of tunnel.

## REFERENCES

Bruland, A. 1998. Hard rock tunnel boring, NTNU, Trondheim, Norway.

Hassanpour, J. and Rostami, J. 2010. TBM performance in hard rock. Tehran: Nashr Fan Aria, p. 558.

Hassanpour, J., Rostami, J. and Tarigh Azali, S. 2011a. Predicting TBM penetration rate in igneous and high grade metamorphic rocks: a case history of Manapouri 2nd tailrace tunnel, proc. of 4th Iranian rock mechanics symposium, (article in Persian).

Hassanpour, J., Rostami, J. and Zhao, J. 2011b. A new hard rock TBM performance prediction model for project planning. Tunneling and Underground Space Technology, Vol. 26: 595–603.

Rostami, J. 1997. Development of a force estimation model for rock fragmentation with disc cutters through theoretical modeling and physical measurement of crushed zone pressure. Ph.D. thesis, Colorado School of Mines, Golden, Colorado, USA, p. 249.

# Shrinkage Compensating Concrete for Use in Underground Concrete Structures

**Matt Jabbari and Ken Vallens**
CTS Cement Manufacturing Corp.

**ABSTRACT:** Tunnels and underground structures are regularly specified with 100-year design lifetime concrete. In order to achieve this we must eliminate cracking in the concrete. Cracks are pathways for the migration of water, chemicals and associated ions that can corrode structural steel, and eventually affect the integrity of the structure. The most common type of cracking is drying shrinkage cracking.

The use of Shrinkage-Compensating Concrete is an effective way to minimize the cracking caused by drying shrinkage. By designing & producing controlled compressive stresses in the concrete, Shrinkage Compensating Concrete reduces shrinkage and the associated detrimental tensile forces which lead to shrinkage cracking. Shrinkage Compensating Concrete is popular in wastewater and water treatment infrastructure design, where liquid or chemical penetration/escape is strictly prohibited. Shrinkage Compensating Concrete is also widely used in industrial floors where the reduction of joints and elimination of cracking is highly desirable.

Shrinkage Compensating Concrete has lower permeability, higher durability & abrasion resistance, higher freeze-thaw resistance, and higher resistance to sulfate attack than ordinary portland cement concrete.

## INTRODUCTION

While Shrinkage Compensating Concrete has been used in various structures since the 1960s, this type of concrete is generally not in the curriculum of the colleges and universities that teach engineering. Normally, an engineering graduate is not familiar with this type of concrete. However, the specification and use of Shrinkage Compensating Concrete is growing rapidly, and more and more infrastructures are being built with this type of concrete.

Shrinkage Compensating Concrete has been used in numerous underground applications going back to the 1960s. 500 First Street NW in Washington D.C. is an eight story building with two basement levels. Similar to many sites in the Washington D.C. area, there were significant groundwater issues to contend with in constructing the foundation. In 1967, a post-tensioned mat concrete foundation was constructed using Shrinkage Compensating Concrete. Forty-five years later this mat foundation is still solid and watertight (Thornton, Chusid, Miller et al. 2009).

Shrinkage Compensating Concrete was used for the waffle ceilings in many of Washington D.C.'s yellow and green metro line subterranean stations (Figure 1; Sullivan, Horwitz-Bennett et al. 2013).

The State of Nevada used Shrinkage Compensating Concrete in the construction of the River Mountain Tunnel #2 in Henderson, NV.

Shrinkage Compensating Concrete can be used in tunneling and underground construction, including but not limited to cut and cover tunnels, tunnel inverts, pre-cast concrete segmental liners, and cast-in place tunnel lining.

## CRACK AND SHRINKAGE OVERVIEW

### Why Does Concrete Crack?

Concrete cracks can occur because of shrinkage, external effects, and detrimental internal expansion. Shrinkage cracking can be either a result of plastic or drying shrinkage. The most common type of cracks are drying shrinkage cracks.

External effects causing cracking can be in the form of thermal stresses, differential settlement, differential movement, or damage due to freezing and thawing. Internal expansion can result from corrosion of reinforcement or chemical reaction between the components of the concrete, such as Alkali-Silica Reaction (ASR) or Delayed Ettringite Formation (DEF). On top of all of these factors, errors in design and detailing, poor construction practices (including construction overloads, excessive water addition during mixing or finishing, and inadequate curing), or overloading during use can also cause cracking in concrete. Thus, the first challenge to anyone trying to sort out the cause(s) of concrete cracking is to attempt to determine the source of cracks (Coleman et al. 2013).

The end use of the concrete application will determine the extent to which cracking is acceptable or unacceptable. For instance, cracks are not

**Figure 1.**

acceptable in tunneling construction, where they might be acceptable in a slab on grade (SOG) application.

The measures used to control cracking depend, to a large extent, on the economics of the situation and the seriousness of cracking if not controlled. Cracks are objectionable where their size and spacing compromise the strength, stability, serviceability, function, or appearance of the structure (Coleman et al. 2013).

A concrete tunnel liner placed below the water table may be subject to attack from sulfate bearing groundwater. In conventional portland cements, hydrated calcium aluminate (C3A) will react with sulfate ions to form detrimental expansive compounds. The consequence of this reaction is that the newly formed substance takes up a larger volume than the reactants causing expansion and cracking. Long-term exposure causes continual expansion leading to extensive deterioration.

One of the contributing factors to cracking is high water content in the concrete's cement paste. Portland cement needs a W/C ratio of 0.25 to hydrate (25 pounds of water is needed to hydrate 100 pounds of portland cement). At this W/C ratio, the concrete is very stiff and not workable (you cannot even get this concrete out of a ready mix truck). Higher W/C ratios are used to make the concrete workable. The extra water added is "water of convenience," which will end up on the concrete surface as bleed water. Excessive water is one of the culprits in shrinkage cracking. The bleed water escapes to the surface through small capillaries. These capillaries reduce the durability of the concrete. Also, bleed water on the surface of concrete causes the W/C ratio to change. The W/C ratio on the surface (present as bleed water), is higher than the W/C ratio in the bulk of the concrete, contributing to lower durability, lower abrasion resistance and higher shrinkage.

Engineers constantly struggle to reduce the drying shrinkage, and cracking associated with drying shrinkage in portland cement concrete. In this effort, they try to lower the W/C ratio, use gap-graded aggregates, and even reduce the amount of cement in the mix (lowering the paste content). There are drawbacks associated with any of these measures, such as: lack of workability, material availability, and strength loss, to name a few.

However, by replacing a small portion of portland cement with a mineral expansive additive (Komponent), one can convert a high shrinkage mix to a Shrinkage Compensating mix (Type-K) conforming to ASTM C 845; and consequently reduce shrinkage cracking. All of this can be done while using local portland cement and aggregates without major changes the mix design. The added cost is very minimal when you look at the overall benefits associated with the project.

## SHRINKAGE COMPENSATING CONCRETE

Shrinkage Compensating concrete made with Type-K cement, is an effective way to minimize the cracking caused by drying shrinkage. Shrinkage Compensating Concrete expands during the first part of curing process (7 days of wet curing). Expansion will induce tension in the reinforcement and compression in the concrete. Shrinkage cracks are eliminated if the Shrinkage Compensating Concrete's

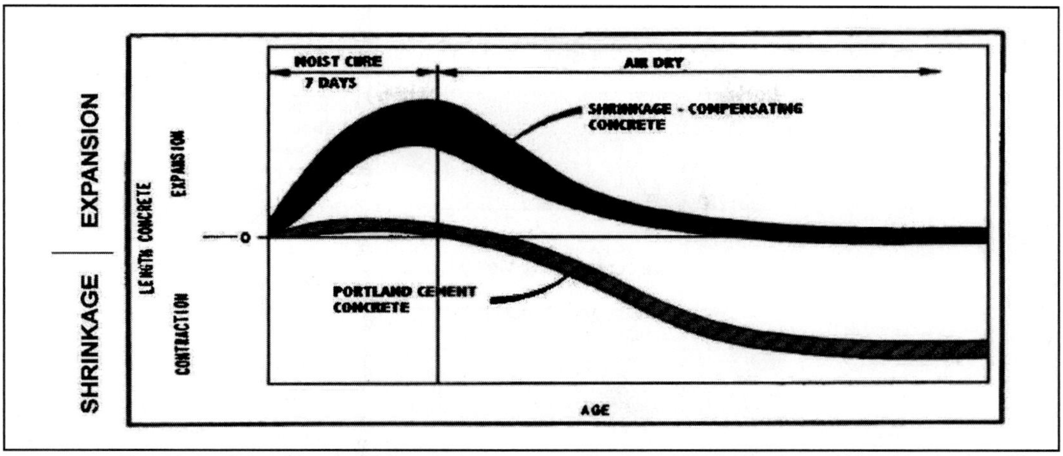


**Figure 2. Typical length change characteristics of shrinkage-compensating concrete and portland-cement concrete (ACI 223-98)**

expansion is greater than its anticipated shrinkage (Figure 2).

By designing and producing controlled compressive stresses in the concrete, Shrinkage Compensating Concrete reduces the detrimental tensile forces which lead to shrinkage cracking. The concrete mass will remain in compression as long as the compressive stresses are more than the tensile stresses.

As known in the industry, concrete is about 10 times stronger in compression than it is in tension. As long as the concrete mass is in compression, it won't crack!

There are no added compressive stresses in structural members, and the designer need not make any design adjustments. All other design parameters are unchanged and should be in accordance with good engineering practices, standards, and code requirements.

There are three characteristics of Shrinkage Compensating Concrete that make it the product of choice for tunneling & underground construction: ability to design and construct large monolithic placements, absence of shrinkage cracks, and greatly reduced permeability (Valentine et al. 2000).

A tunneling design engineer looks for durable, water tight, low permeability, 100-year life concrete for the underground structures he/she designs. Shrinkage Compensating Concrete has lower permeability, higher durability & abrasion resistance, higher freeze-thaw resistance, and higher resistance to sulfates than ordinary portland cement concrete.

## COMPARING SHRINKAGE COMPENSATING CONCRETE TO ORDINARY PORTLAND CEMENT CONCRETE

Shrinkage Compensating Cement consumes more water to hydrate, therefore less water "bleeding" occurs with this type of concrete (Figure 3).

Figure 3 shows a comparison of hydrated portland cement concrete to hydrated shrinkage-compensating concrete by mass and volume.

## ANTI-ASR AND ACR

Calcium Sulfoaluminate cement, which one of the main ingredients in Type-K cement, is relatively inactive to Alkali-Silica Reaction (ASR) and Alkali-Carbonate Reaction (ACR) in comparison with portland cement. This is due likely to:

1. The ettringite ($3CaO \cdot Al_2O_3 \cdot 3CaSO_4 \cdot 32H_2O$) hydration product of Calcium Sulfoaluminate cement, with 32 crystalline water molecules, decreases the porosity of hardened Calcium Sulfoaluminate cement dramatically, and
2. Lower PH-values in the liquid phase of hydration products of Calcium Sulfoaluminate cement in comparison with portland cement (Valentine et al. 1994, Yanjun, Yongmo, Chunlei, et al. 2012).

## CASE STUDIES

Type K cement has been available since the 1960s and has exhibited an excellent track record.

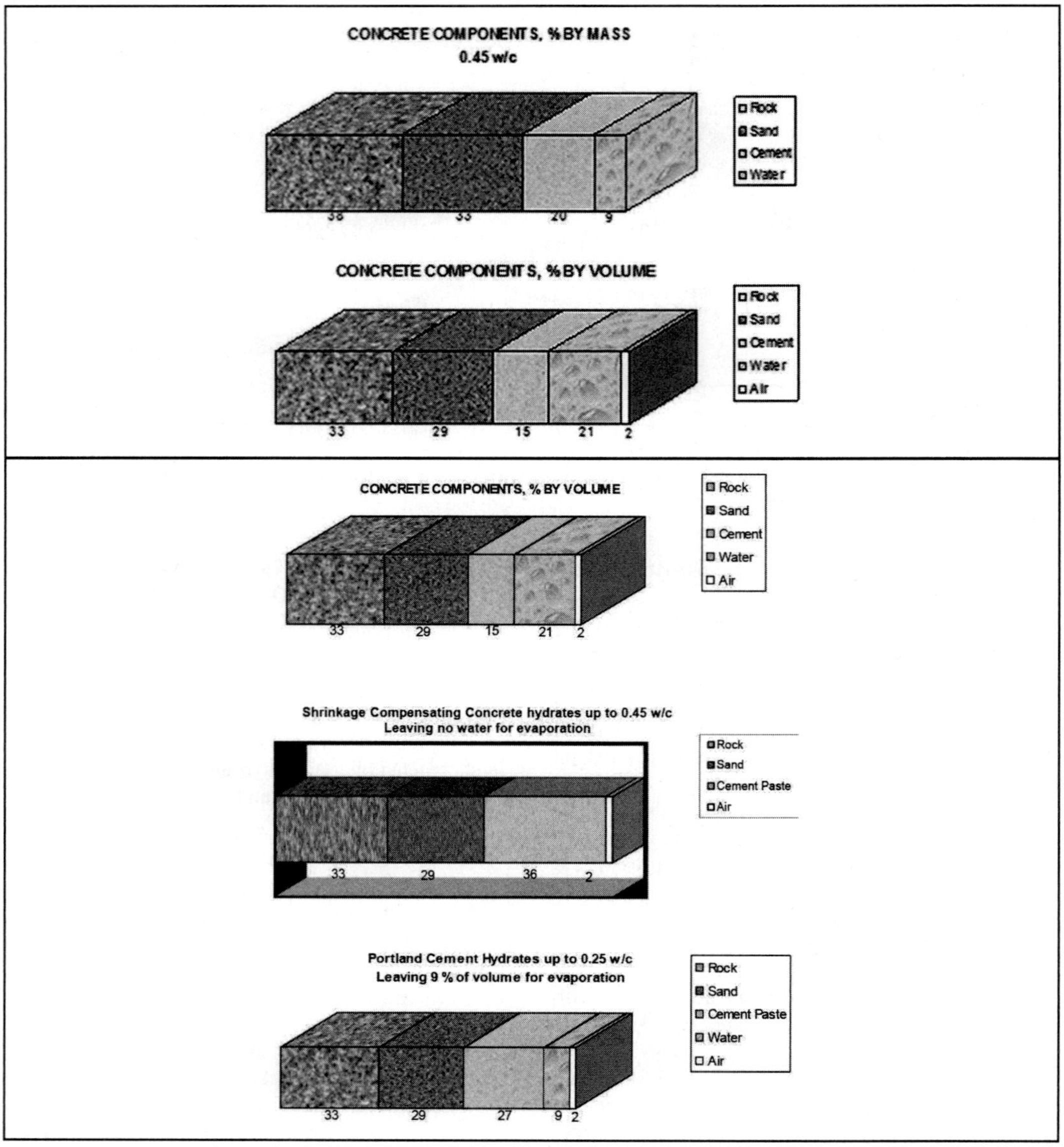


**Figure 3.**

In the early 1990s at Orange County California's John Wayne airport, portland cement concrete was used on level one of the parking structure resulting in approximately one mile of cracks. The engineer switched to Shrinkage Compensating Concrete for the placement of level 2 of the same parking structure which resulted in no cracks. As a matter of fact, after 14 years of heavy traffic, the finish of the Shrinkage Compensating Concrete is still crisp and looking new, due to the materials superior abrasion resistance (Chusid et al. 2007). See Figure 4.

Other construction industries have used Shrinkage Compensating Concrete successfully.

Shrinkage Compensating Concrete is popular in wastewater and water treatment infrastructure design, where liquid or chemical penetration/escape is strictly prohibited.

Shrinkage Compensating Concrete is also popular in rock anchoring, soil nailing, and roof bolting operations, as well as grouting of post-tensioned structures. It is used due to its expansive characteristics and its ability to compensate for shrinkage.

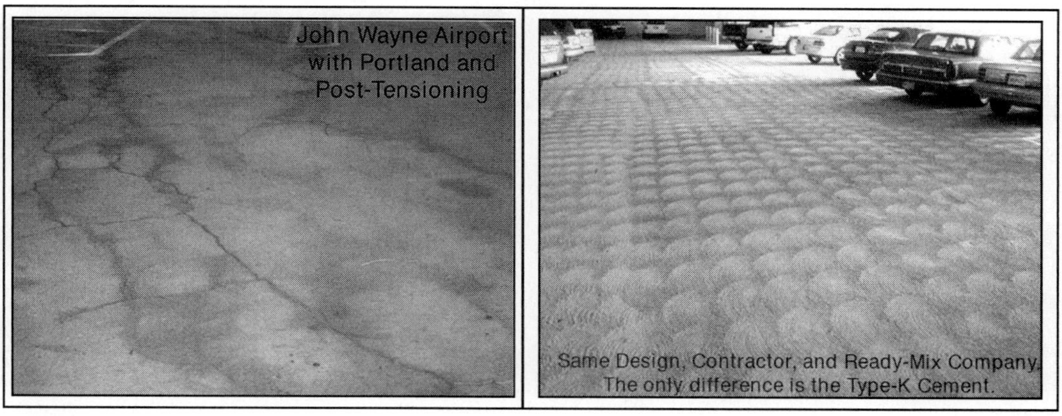


**Figure 4.**

**Figure 5.**

In warehouses and distribution centers, Shrinkage Compensating Concrete is used to reduce the number of joints, typically placing floors in excess of 50,000 square feet with no joints.

The Turnpike Authorities of Michigan, Ohio, New Jersey, and Pennsylvania use Shrinkage Compensating Concrete toppings in their bridge rehabilitation operations to eliminate decking cracks, therefore preventing the bridge's steel reinforcement from exposure to chloride ions.

Recently a monolithic roof (64 feet × 40 feet) of a custom residence in the Eastern Mountains of San Diego (Julian, CA), using Shrinkage Compensating Concrete was placed with no roofing membrane. This residence is located 4000 feet above sea level. The area is known to have a quite high accumulated snow fall in the winter. No leakage was reported by the owner. As a matter of fact Shrinkage Compensating Concrete was used to construct the entire house, including the subterranean garage and the tunnel connecting the garage to the observation tower. Please see Figures 5, 6, and 7, which show Type K concrete monolithic roof placements.

Edward K Rice; PE, F.ASCE, F.ACI; one of the original developers of Shrinkage Compensating Concrete built his house in 1963 using Type K concrete. To this day, he lives in the same house and its exposed concrete roof has not leaked (Chusid et al. 2006).

## CONCLUSION

Shrinkage Compensating Concrete technology is ideal for, and should be adopted extensively in tunnels and underground structures.

**Figure 6.**

**Figure 7.**

While the cost of Shrinkage Compensating Concrete can be slightly more than conventional portland cement concrete, the cost reduction associated with lesser joints and larger pours makes using Shrinkage Compensating Concrete a cost benefit. Also, savings resulting from extended life, reduced shrinkage cracking, reduced leakage, and reduced need for repairs make the overall life cycle cost significantly lower than conventional portland cement concrete construction. The true benefit is derived from having a leak-proof, structurally sound and environmentally safe underground structure.

## REFERENCES

ACI Committee. 1998. *Standard Practice for the Use of Shrinkage-Compensating Concrete* (ACI 223-98). Chapter 1.4. Farmington Hills, MI: American Concrete Institute. December.

Bondy, Kenneth. President of the Post-Tensioned Institute and was inducted into the PTI Hall of Fame in 2005.

Chusid, M. 2006. All-Concrete House Turns 40 Years Old. *Concrete International* March.

Chusid, M. 2007. A Perfect Match: Post Tensioning and Shrinkage Compensating Concrete Form a Durable Union at John Wayne Airport. *PTI Journal* July.

Chusid, M. Principle and founder of Chusid Associates, a consultant to building product manufacturer, based in Tarzana, CA.

Coleman, J. 2013. Cracking...Defect or Normal? When is concrete cracking a construction defect? *Concrete International* September.

Sullivan, C.C., and Horwitz-Bennett, B. 2013. Building with Concrete-Design and Construction Techniques. *Concrete Construction AIA Journal.*

Thornton, K., and Chusid, M. 2009. Waterproofing Without Membrane. *PTI Journal* August.

Valentine, L. 1994. Environmental Containment Structures. *Concrete International* July.

Valentine, L. 2000. Containment Structures in the Chemical Industry. *Concrete International* January.

Yanjun, L., Yongmo, X., and Chunlei, G. 2012. Sulfoaluminate Cement: An Alternative to Portland Cement. *Advanced Materials Research* Vols. 368–373, pp. 478–484. Trans Tech Publications, Switzerland.

# Geoelectrics-While-Tunneling: Methodology and Performance Aspects

**K. Schaeffer**
Department of Civil & Environmental Engineering, Colorado School of Mines

**ABSTRACT:** This paper presents a detailed investigation of electric geophysical techniques that 'look ahead' of the tunnel face during TBM operations for both hard rock and soft ground conditions, called geoelectrics-while-tunneling. These methods utilize integrated components of the TBM as electrodes and use current injection to 'illuminate' the ground ahead of the TBM, identifying changing ground conditions, faults, groundwater inflows, boulders, cavities, utilities etc. Classical earth based methodology, called geoelectrics, is well established and is briefly presented, where many of its aspects can be translated to geoelectrics-while-tunneling. An overview of geoelectrics-while-tunneling is presented pertaining to its geometry, influences on performance, and its allowable electrical current injection limit, so as to better establish these methods for use in industry.

## INTRODUCTION

In order for tunnel boring machines (TBMs) to maintain precision,* especially for complex and large diameter excavations, methods that can monitor ground conditions are valuable in identifying geologic changes or manmade structures.† If a TBM encounters an unanticipated change in ground conditions, precision may decrease as the original TBM excavation settings (cutterhead torque, advance rate, cutterhead RPM etc.) could be used. As an additional consequence of improper TBM settings, unnecessary wear and costs in cutting tools (rippers, disc cutters, and scrappers) may be incurred. In order to increase precision and reduce unnecessary costs, research has given attention to the adaptation of established geophysical techniques which non-destructively evaluate subsurface properties (material type, porosity, water content, salinity, temperature etc.), termed geoelectrics. This paper discusses geoelectrics application in a TBM tunneling environment, termed geoelectrics-while-tunneling.

To date, two commercially available systems have attempted geoelectrics-while-tunneling; the BEAM (Kaus & Boening 2008) and the BEAM4 (Kopp 2012). While these systems have been implemented on numerous projects (e.g., CCS Project, Ecuador; YinTao T7 & T9, China; JinPing II Diversion Tunnel, China), there are limited published results of their successes, failures, challenges, etc.. Many applications of geoelectrics are well established and have been successfully implemented for over a century (e.g., surface and borehole based methods including: oil/gas, mineral exploration, vadoze zone detection, hydrology) (Reynolds 2011).

This paper first summarizes (a) surface and (b) borehole based geoelectrics, and emphasizes the salient features that enable their success. Building on the successful approaches of geoelectrics, the paper then examines how certain aspects (i.e., borehole effects) of borehole-based geoelectrics can be translated into geoelectrics-while-tunneling. Next, this paper discusses how the performance of geoelectrics-while-tunneling is partially dependent upon the TBM type‡ and the electrode configuration. Finally, this paper details how to maximize the level of injected current while maintaining the safety of staff onboard the TBM.

## GEOELECTRICS

Geoelectrics involve the injection and retrieval of either direct current (DC) or alternating current (AC) in the earth, and the simultaneous measurement of the induced electrical field at discrete locations. The objective of geoelectrics is to evaluate the electrical conductivity field, σ (S/m), and then to predict ground conditions that exhibit similar characteristics as the evaluated σ. The magnitude of current injection/retrieval (ranges from 1 mA to 1 A) is a function

*Precision of excavation may include the precision of thrust, precision of steering, or the minimization of surface settlement (especially in urban environments).

†Structures may include foundation elements, tie-backs, or abandoned wells.

‡TBM types include open face, and closed face (EPB or slurry).

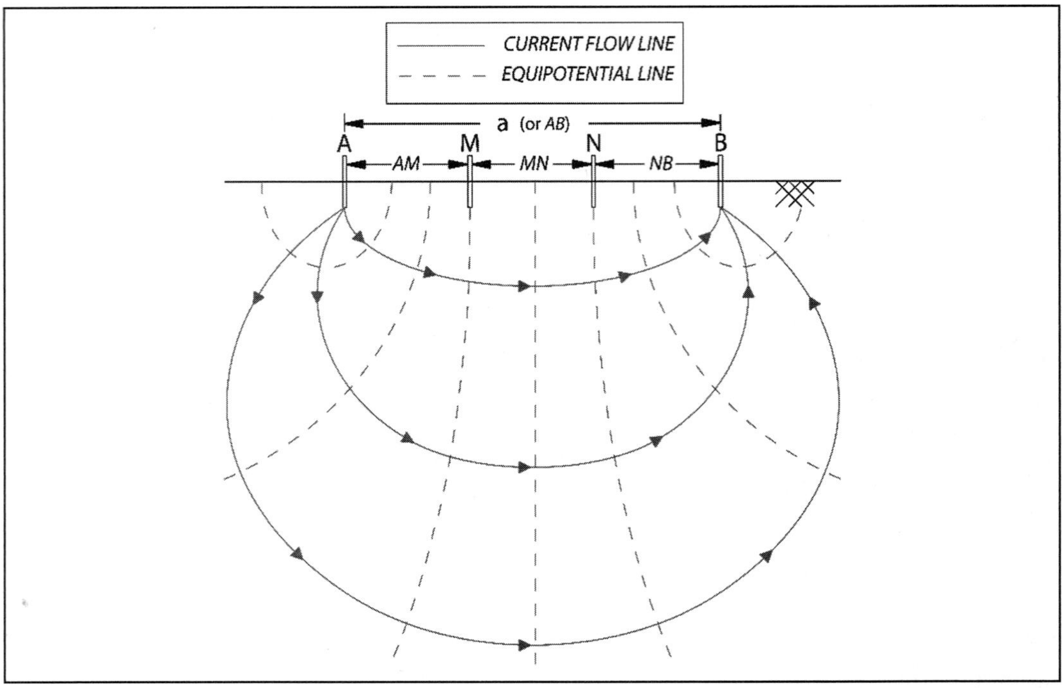


**Figure 1. Surface-based geoelectrics set-up with current injection/retrieval electrodes A/B, potential measuring electrodes M and N, and current flow lines and induced electrical field**

of the application's physical size, electrical current type (AC or DC), extent of observation, and ambient electrical noise. The flow of current and the induced electrical field (e.g., see Figure 1) are dependent upon the electrode configuration and the electrical resistivity/conductivity* field of the earth.

The electrical conductivity of the earth at any point in space is primarily influenced by the characteristics of the pore space. Pore space characteristics include, but are not limited to, salinity, temperature, porosity, clay content and rock fracture density. The soil and rock grains act as insulating material. Samouëlian et al. (2005) summarizes contemporary theories for the influence of pore space characteristics on electrical conductivity. Since σ is more dependent upon the pore space electrical conductivity, geoelectrics is routinely used in petroleum, mining, and hydrological studies including oil/gas location, subsurface cavity location, mineral exploration, down-hole logging, vadoze zone hydrology (Reynolds 1997). However, with prior knowledge of the site and with all else being equal, different types of soil (clay, sand, silt etc.) or rock (sandstone, limestone, claystone etc.) can present site specific ranges in electrical conductivity, and therefore, may be distinguished and identified using geoelectrics. Many studies have attempted to provide ranges of electrical conductivity for various earth materials, where most earth materials fall within a range of 0.0001 to 5 S/m (Samouëlian et al. 2005).

Electrical conductivity is a complex variable and denoted as $\sigma^*$ (S/m) in Equation 1.

$$\sigma^* = \sigma' + i\sigma'' \tag{1}$$

The real part of complex conductivity, σ', is directly related to the electrical conductivity of the pore space fluid as well as the presence/concentration of electrically conductive minerals, e.g., metallic ore, sulfides (Kemna 2000, Revil 2013). As the salinity, temperature, degree of saturation, or porosity of the pore space is increased, σ' is also increased. For example, in highly fractured rock, the porosity is increased and generally results in an increased σ'.

The imaginary part of the complex conductivity, σ'', reflects the ability of the medium to store an electrical charge. Much like a capacitor, as current flows through the material in one direction, charge accumulates to some maximum and eventually current can no longer flow. After current flow is ceased,

*Resistivity and conductivity are used interchangeably in literature and are simply the reciprocal of one another. This paper will use conductivity to describe how well electrical current is able to flow through a material. Materials with a higher conductivity will pass current more easily than materials with a lower conductivity.

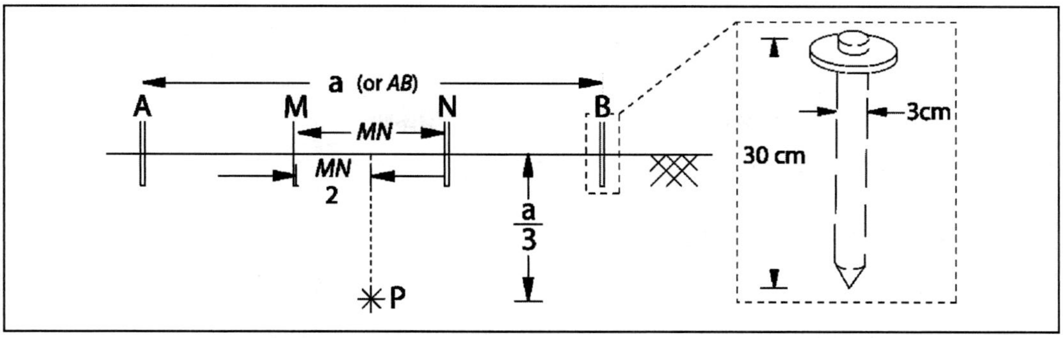


**Figure 2. Electrical conductivity point estimation and typical dimensions of a steel spike electrode**

the medium can then release this charge over time even though the current is shut off. This is called the induced polarization effect (IP effect). The IP effect has been attributed to the presence of the electrical double layer most significantly found in clays. Therefore, $\sigma''$ is directly proportional to the clay content (Sumner 1984).

The following two sections briefly comment on surface based and borehole based geoelectrics. These comments in no way attempt to summarize the field, and only aim to outline features implementation aspects that assist in later geoelectrics-while-tunneling discussion.

## Surface-Based Applications

Surface-based geoelectrics employ an array of four metallic electrodes for a single measurement of $\sigma^*$ at a point P in the subsurface. Figure 2 shows a rule-of-thumb estimation for the location of P. The electrodes are inserted into the earth in a configuration defined by their relative spacing (Figure 1). Figure 1 (see also Figure 2) shows a general electrode configuration, however, many unique electrode configurations exist and are well defined by the separation distances between the current injection/retrieval electrodes (A and B) and the potential measuring electrodes (M and N). Some well-known configurations include the Wenner, Dipole-Dipole, and Schlumberger (Reynolds 2011, Samouëlian et al. 2005). A typical electrode separation may be on the order of 1 to 100 meters. The depth of current penetration is directly proportional to the spacing between electrodes A and B. the spatial resolution of electrical conductivity measurements is inversely proportional to the relative spacing between electrodes A and B, as well as M and N. In other words, the availability of electrical conductivity information decreases with depth. At best, current surface based geoelectrics can achieve a spatial resolution on the centimeter scale (Samouëlian et al. 2005).

In theory, the four electrode array is idealized by four points with no physical properties or influence on the electrical physics of the surrounding environment. In reality, the four electrodes do have material properties and do have influence on the electrical physics of the surrounding environment.

When the electrodes are made of iron alloy, such as steel, the electrodes themselves can become polarized and demonstrate an IP Effect during the measurement process. As the steel electrode is polarized its contribution to the IP Effect is far greater than that of the surrounding earth. Therefore, when measuring the IP effect of the surrounding earth, the measurements will be inaccurate as the polarization of both the electrode and the earth are captured. A non-polarizing metal, such as silver/silver-chloride is used to eliminate the contribution from the electrodes on the measured IP Effect. However, this type of electrode can be expensive to implement. Instead, steel spikes (Figure 2) are used with the injection of DC current only, where the IP effect is neglected and not captured for either the electrodes or the earth. Understandably, this approach is not ideal, as it cannot fully characterize $\sigma^*$.

## Borehole Techniques

The need for high spatial resolution (cm) data at greater depths (>50 meters) motivated the oil and gas industry in developing borehole based techniques in the early 1900s using the Schlumberger methods for well logging (Schlumberger 1933, Spies 1996, Slater 2000). In contrast to surface based methods, borehole based methods can achieve continued high spatial resolutions at greater depths because of the ability to position sensors at any interval along the depth of the borehole. Some of the earliest borehole based methods utilized a four electrode array similar to surface based methods. Figure 3a shows a traditional (called unfocused methods; Spies 1996) A, B, M and N configuration applied to a vertical borehole, where three electrodes (A, M and N) are confined to a housing unit and are restricted in movement relative to one another. The fourth electrode (B) is located at a fixed point on the earth surface. The electrode housing is

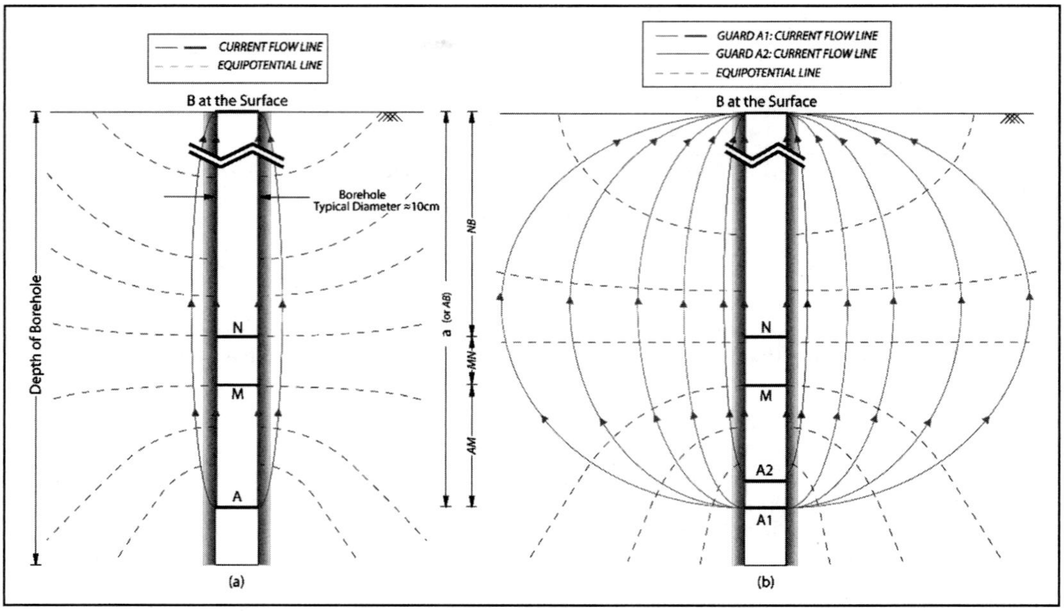


**Figure 3. Borehole electric geophysical methods: (a) unfocused, four electrode array, (b) focused array with two injection guard electrodes**

lowered into a previously drilled borehole, where it is successively moved to image different horizons.

Using so-called 'current focusing', the four-electrode implementations were soon advanced to increase the spatial resolution of measurements as well as the current penetration distance (Figure 3b). Current focusing uses guard injection electrodes (Figure 3b: A1 and A2) instead of a single injection electrode (Figure 3a: A). Each guard electrode injects a separate stream of current. The additional current from the upper guard injection electrode (Figure 3b: A2), forces the stream of current from the lower guard injection electrode (Figure 3b: A1) to propagate further into the medium before it can return to the retrieval electrode (Figure 3b: B).

Current focusing is primarily used to mitigate adverse borehole affects. The most significant borehole effects include the undesired absorption of electrical current from the addition of (a) clay to increase the contact between the earth and the electrodes, and from (b) metallic borehole casings used to increase borehole structural stability. These two effects will be discussed in the following paragraphs.

In a case where a dry borehole exists (no borehole fluid, e.g., bentonite slurry), the electrical contact between an electrode and the earth (borehole sidewall in this case) can be poor, and as a consequence current injection into the earth is substantially reduced. To improve current injection, clay-mud is injected into the borehole to help improve the electrode contact with the earth. In unfocused methods (Figure 3a), this solution can lead to additional problems, especially when mud with higher salinity (and therefore higher conductivity) is used (Anderson 2001). In these cases, current may actually remain confined within the more conductive mud and travel up the borehole sidewall (Figure 3a). When current is injected in equal magnitudes from two closely spaced injection electrodes (guard electrodes; Figure 3b), the potential between both electrodes is reduced, consequently reducing the flow of current between the electrodes, and forcing the current from the lower guard injection electrode into the earth.

Traditionally, borehole based geoelectrics are used in uncased boreholes, where direct electrical contact between the earth and the electrodes is possible. In many applications over recent years, boreholes are cased to prevent collapse of the borehole. Due to the large difference in conductivity between steel casing and the earth (on the order of $10^6$ S/m), it is very difficult to propagate current into the earth without it flowing directly through the casing and back to the retrieval electrode (Figure 3a). Early research has explored the idea of measuring very small electrical potential changes due to a very small amount of current propagation into the earth, whenever a metallic casing is present (Alpin 1939). Early applications were not yet feasible until recent advances in modeling and measurement sensitivity. Numerical and experimental studies have shown success in geoelectrics using cased boreholes (Kaufman 1990, Kaufman and Wightman 1993, Klein and Martin 1993, Schnenkel and Morrison 1994).

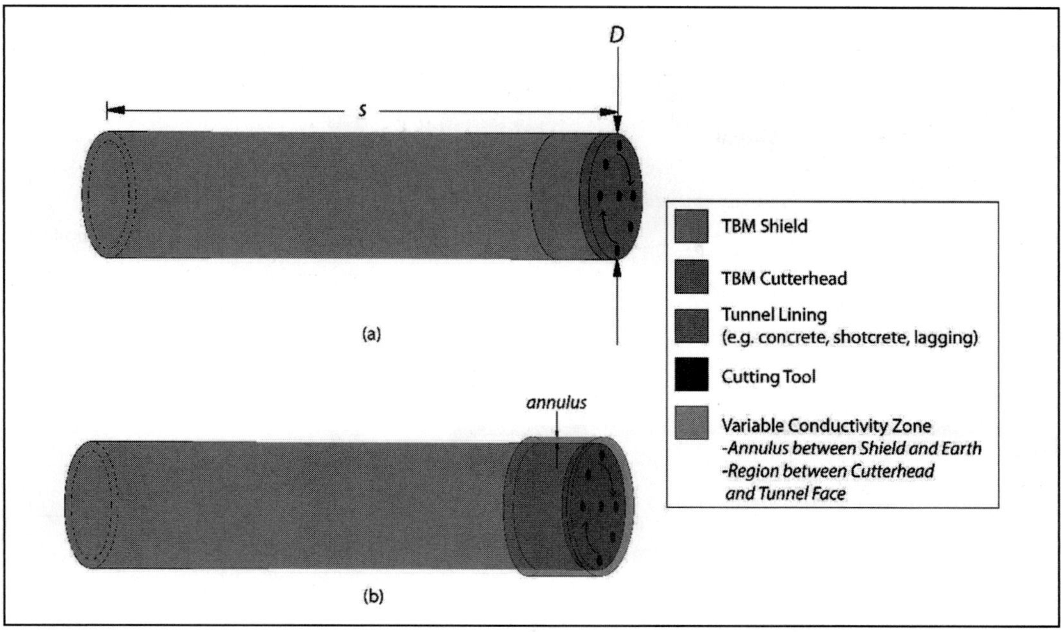


**Figure 4. TBM tunneling geometry: (a) dimensions of TBM and tunnel cylinder, (b) region of variable contact at TBM cutterhead and shield**

## GEOELECTRICS-WHILE-TUNNELING

The following sections discuss, (a) the TBM tunneling environment, (b) a TBM-mounted geoelectrics-while-tunneling electrode array, (c) how the performance of geoelectrics-while-tunneling is defined, (d) the effect of TBM type and mode of operation on the performance of geoelectrics-while-tunneling, and (e) how to maximize the level of current injection while maintaining the safety of onboard crew members.

### Tunneling Environment and Geoelectrics-While-Tunneling

The environment (material types and geometry) for geoelectrics-while-tunneling is shown in Figure 4. A TBM of diameter D (2–17 m) excavates through the ground creating an air-filled cylinder (length *s* in Figure 4a). Mounted at multiple locations on the cutterhead, cutting tools (e.g., rippers, scrapers, disk cutter) are used to fracture or dig into the tunnel face (Figure 4a). Cutting tools generally have good contact with the soil and/or rock. The TBM itself is made up largely of a metallic structure with high electrical conductivity.* The concrete lining and the air-filled cylinder have relatively low electrical conductivity* compared to the TBM and the surrounding earth material. The electrical conductivity of the earth is variable, and its value likely lies between the lining and the TBM. A region of variable electrical conductivity surrounds the TBM (Figure 4b). This region includes the space a) between the cutterhead and the tunnel face, and (b) between the shield and the tunnel wall where the TBM over-excavates an annulus. These regions vary in size and electrical conductivity depending upon the TBM type and mode of operation (discussed in more detail later).

Geoelectrics-while-tunneling consists of a current injection electrode (Figure 5b: A; ideally an isolated cutterhead or cutting tool), a current retrieval electrode (Figure 5b: B; fixed anchor outside of tunnel lining), electrical potential measuring electrodes (Figure 5b: M and N), and a computer unit (not shown) that records and processes incoming electrical measurements. Figure 5b shows a potential measurement electrode, N, at a location similar to the retrieval electrode, B. This location is optimal as it gives a reference voltage measurement that is relatively constant as the TBM moves through differing ground conditions. Therefore, the potential between M and N is directly related to the changing ground conditions ahead of the TBM, measured by electrode M. Lines of equipotential are not shown in Figure 5, but exist in reality.

Under ideal conditions, current is injected into the earth via an injection electrode, where current will flow through the earth and around the TBM.

*High conductivity (steel): $10^6$ S/m, Low conductivity (concrete): $10^{-4}$ S/m

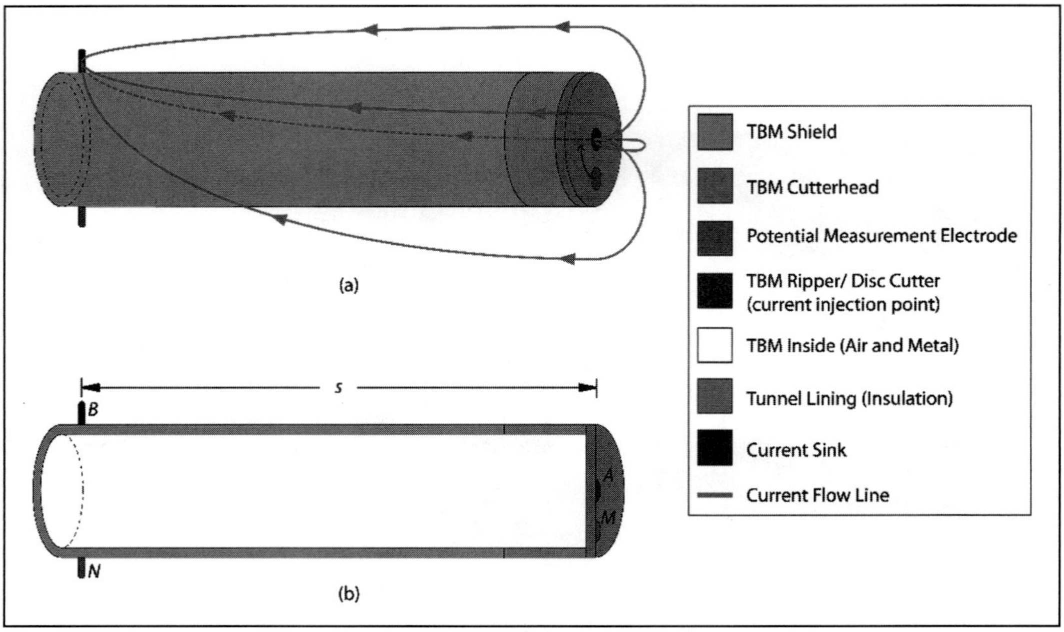


**Figure 5. Idealized current flow for TBM mounted electrode array: (a) current flow around TBM to sink, (b) cross section of the TBM/Tunnel, and locations of A-B-M-N electrodes**

Figure 5a shows current injected from an electrically insulated cutting tool mounted on the front of the cutterhead. However, other options of current injection may exist including the cutterhead, the shield, and/or a recessed dummy cutting tool*. The retrieval electrode is located in the ground, outside of the tunnel lining, at some distance, *s*, behind the front of the cutterhead (Figure 5). Note that *s* increases as the TBM excavates the tunnel because electrode B is a fixed point. Similar to borehole-based approaches, geoelectrics-while-tunneling utilizes an array of electrodes that move relative to one another. This is advantageous as the electrode array will be automatically moved as the TBM advances. Therefore, electrical conductivity information can be continuously acquired throughout the length of the tunnel excavation.

If the injection electrode is not electrically isolated from the rest of the TBM, current will likely pass from the injection electrode back into the TBM without much propagation into the surrounding earth. This result is analogous to a situation where a metal casing is used in a vertical borehole, as previously discussed. For the remainder of discussion it is assumed that all electrodes are electrically isolated from the rest of the TBM. Still, there may be some implementation complications in the electrical isolation of TBM components and should be investigated.

## Look-Ahead Distance and Performance Optimization

The performance of geoelectrics-while-tunneling is based upon its ability to detect changing ground conditions ahead of the TBM. The location of the changing ground conditions will be defined here by the horizontal distance, *x*, in front of the TBM cutterhead (Figure 6). Since, no actual observation can be made of the ground change, detection relies on geoelectrics-while-tunneling to alert the TBM operational staff through continuous monitoring of electrical potential. Electrode M and N perform this monitoring by reporting changes in electrical potential at defined time intervals. As the measurement of electrical potential is already a difference in voltage between electrodes M and N, it should be noted that the term 'change in electrical potential' refers to the measurement of potential at two different times. At one point in time, $t_1$, the TBM can be considered in a completely homogeneous medium with no geologic change anywhere in its vicinity. The first electrical potential measurement is taken at $t_1$, called the far-field measurement, and is held constant for all changes in electrical potential. At a defined time interval, $n \cdot \Delta t$, potential measurements are taken continuously (n = 1, 2, 3…) after $t_1$ and are referenced to

*A dummy tool would be mounted onto the cutterhead much like a regular cutting tool, however, would be recessed slightly to avoid use and wear in the actual excavation process.

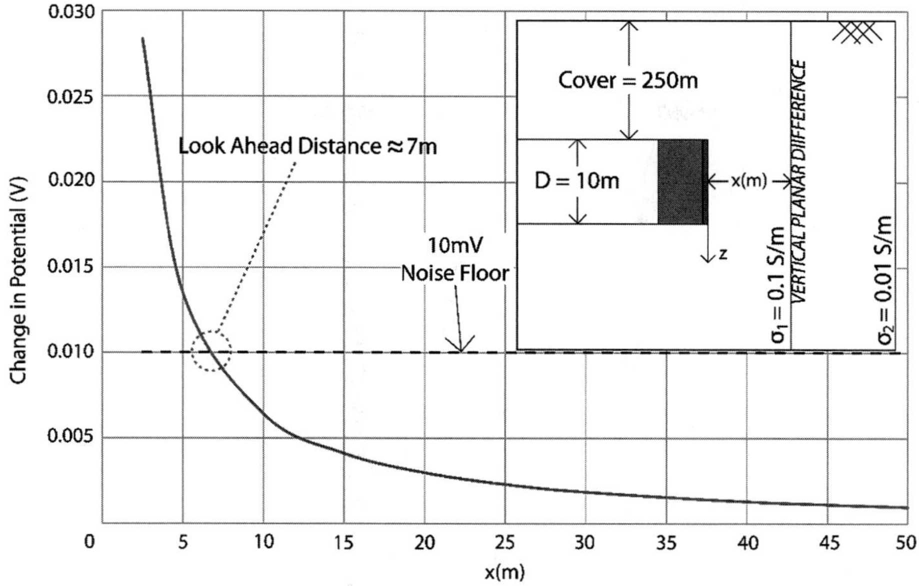


**Figure 6. Example look ahead distance, *x*, for a vertical planar difference in geology**

the far-field electrical potential. The look-ahead distance for geoelectrics-while-tunneling is dependent upon measured changes in electrical potential and will be discussed next.

Figure 6 shows a case (from Schaeffer and Mooney 2014b) where the TBM excavates through a homogeneous medium of some electrical conductivity and advances toward a homogenous medium of a different electrical conductivity (moving from $\sigma_1$ to $\sigma_2$). The different conductivities in Figure 6 may represent the case where the TBM moves from clay to sand. Figure 6 shows a plot of the measured change in electrical potential reported by electrodes M and N as the TBM moves closer (*x* decreasing) to the geologic change. According to this plot, the change in electrical potential clearly increases from as far as 50 meters away. In reality, environmental electrical noise is present and may result from the TBM or ambient electric signals in the earth. Therefore, in this paper, the look-ahead distance is defined as the distance, x, between the front of the TBM and an incoming geologic difference, where the measured change in electrical potential exceeds a signal to noise ratio (SNR) greater than unity. The value of electrical noise is not easily approximated, and is likely site specific. Current modeling approaches use a change in electrical potential noise floor of 10mV, which is believed to be fairly conservative (Schaeffer et al. 2014a, 2014b).

The performance of geoelectrics-while-tunneling is optimized when the look-ahead distance is maximized. The earlier the geologic change is detected, the more time is given to adjust TBM operations or mitigate poor ground conditions ahead of the TBM. Two major factors, (a) TBM type/mode, and (b) electrical current magnitude influence the look-ahead distance, and will discussed in the following two sections.

### Influence of TBM Type and Mode of Operation

In hard rock conditions, open face TBMs are often used for excavation. The hard rock at the tunnel face is crushed into smaller fragments by cutting tools. The region of variable conductivity as shown in Figure 4b will made up of pieces of crushed rock and air, and will have a lower electrically conductivity than the virgin rock at the tunnel face. Even though the cutting tools may make good contact with the rock, it can be reasonably assumed that poor contact exists between the cutterhead and the tunnel face. Therefore, in order to inject current into the earth in front of the TBM, an injection electrode would need to be a at least one cutting tool (Figure 7a). Dummy mounted tools would not work in this case as they would be recessed slightly and their contact with the earth may be too variable.

In soft ground conditions requiring pressurized face tunneling, earth pressure balance (EPB) and slurry shield TBMs are employed. Both modes provide very different electrical conductivity environments. In EPB mode, water and/or soil conditioning agents, most commonly in foam form, are mixed with

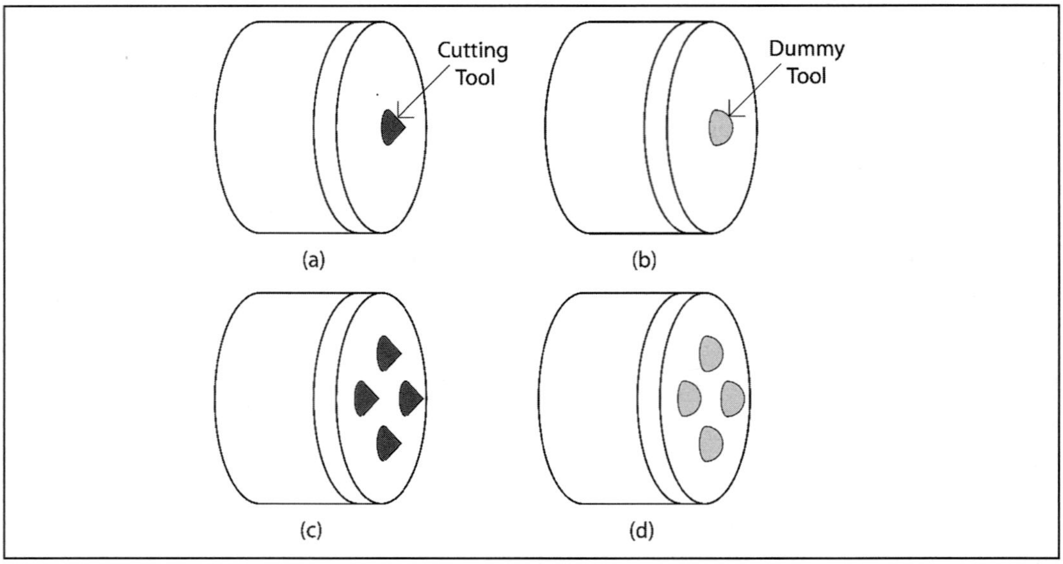


**Figure 7. Configuration of current injection electrodes for each TBM type: (a) hard rock open face, (b) soft ground EPB, (c) soft ground EPB, (d) soft ground slurry**

the soil at the cutterhead. The conditioned soil within the region from the tips of the rippers and scrapers to the cutterhead face (10–30 cm thickness) has a significantly reduced electrical conductivity than the virgin soil. To inject current into the soil ahead of the TBM, injection electrodes need to be embedded within active ripper/s or scraper/s (Figure 7a) or in dummy tools protected by rippers or scrapers (Figure 7b). The conditioned soil between the cutting tool tips and the cutterhead forms an insulating layer that minimizes current flow into the TBM body. And though difficult to verify, it is commonly assumed that the annulus around the forward shield is filled with conditioned soil. The annulus is expected to insulate the virgin soil from the metallic shield. The path of least resistance for current flow is, therefore, ahead of the TBM face and into the earth. To this end, soil conditioning produces a side effect that improves the performance of geoelectrics-while-tunneling.

In slurry shield TBM tunneling, a bentonite clay mixed with water is injected in the same manner as soil conditioner. In contrast, slurry and water are highly conductive and likely more conductive than the virgin soil. Electrical current injected through a cutting tool is likely to flow directly back to the TBM through the slurry. This leaves a reduced level of current to propagate ahead of the TBM face into the virgin earth. This deleterious side effect is analogous to borehole geoelectrics where mud is used to increase contact between the electrodes and borehole wall. This can be addressed by current focusing used in borehole geoelectrics. If current is injected in equal magnitudes through closely spaced cutting tools simultaneously (Figure 7c) or dummy tools (Figure 7d), then little to no current is allowed to travel between them, which steers current into the earth. Field testing is needed to confirm this adaptation on current focusing, as it may have some complications in complex TBM environments.

**Electrical Current Limit**

The look-ahead distance is directly proportional to the depth of current penetration, where the depth of current penetration is directly proportional to the electrode spacing, *s*, and also the magnitude of injected current, I (Amps). Since, *s* is controlled, the magnitude of electrical current must, therefore, be maximized. However, for safety considerations, it is essential to limit the magnitude of electrical current such that no individual is harmed on board the TBM via current leaked to the TBM in the case of an electrical shortage. Based upon safety limits of current flow through a human being, an analysis on the total limit of available current is performed.

According to OSHA Title 29 Code of Federal Regulations (CFR) Part 1910.302 through 1910.308—Design Safety Standards for Electrical Systems, the absolute limit of DC current which can pass through a human without physical damage is 3mA. This limit varies substantially when AC current is introduced in IP methods at different frequencies, but is only increased in these cases. When soaking wet (worst case scenario), OSHA estimates the average human electrical resistance to be on the order of 1,000 Ohms (1K Ohm).

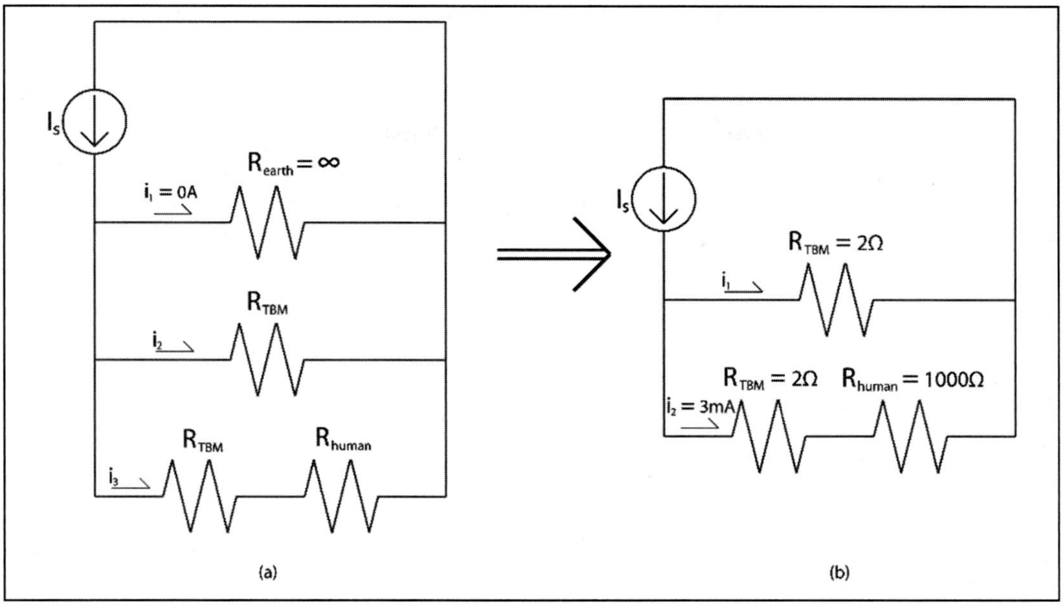


**Figure 8. Simple circuit analysis of TBM-mounted electrical system**

A theoretical analysis of a TBM, earth, and human simple circuit, is performed. Figure 8a shows the circuit diagram for three parallel legs of resistance. Current injected from a cutterhead electrode has three simple, parallel paths that it can take:

1. Through the earth
2. Through the TBM
3. Through an equivalent series resistor equal to the summed resistance of the TBM and a human

The resistance of the TBM, $R_{TBM}$ ($\Omega$), can be reasonably calculated as $2\Omega$. However, the resistance of the earth proves to be a difficult task as the earth is highly variable and anisotropic. Therefore, there is no single, unique resistance that can be reasonably used for the earth. The worst case scenario is where the earth has an infinite resistance, and effectively all of the current short circuits directly back into the TBM. The simple circuit is now reduced to Figure 8b.

It is now possible to calculate a value for the magnitude of total allowable current injected into the circuit, $I_s$ (Amps), using Equation 2:

$$I_s = i_2 \frac{R_{TBM} + R_{human}}{R_{TBM}} = (0.003A)\frac{(2\Omega + 1000\Omega)}{2\Omega} \approx 15A \quad (2)$$

This provides an allowable limit for $I_s$ of 15A. To comply with appropriate voltage limits (<55V), this solution also solves for the associated voltage seen by the equivalent resistance in our circuit (~$2\Omega$) using Ohm's Law (Equation 3):

$$V = IR = 15A * 2\Omega = 30V \quad (3)$$

It is shown that the voltage observed across this theoretical circuit remains at 30V, well below the allowable 55V limit.

The only time during which a problem may occur is if someone were to make direct contact with a current injection/retrieval electrode. In this situation, all of the current would short circuit through that person. This event would only occur during a time where cutting tools are being replaced or repaired, and will be avoided by shutting off power to the geoelectrics-while-tunneling system. It is certain that OSHA 'Lock-Out-Tag-Out' regulations would avoid this occurrence.

## CONCLUSIONS

Classic geoelectrics are well established in methodology. Many theories are presented in geoelectrics literature which attempt to estimate the material type (e.g., soil: clay, sand, silt; rock: sandstone, limestone, claystone) and pore space characteristics (e.g., porosity, permeability, water content, salinity, temperature) from a measured electrical conductivity field, $\sigma$. Geoelectrics present practical implemenetation obstacles, the solutions for which can be translated to geoelectrics-while-tunneling. The need for geoelectrics-while-tunneling is becoming more evident as tunnels are increasingly excavated by TBMs in complex ground conditions (e.g., mixed face with soil and rock) and in urban environments where remnants of underground entities may be present. Research is underway, which investigates the sensitivity of geoelectrics-while-tunneling to its various geometrical aspects and various geologic changes.

## REFERENCES


Alpin, L.M., 1939, The method of the electric logging in the borehole with casing: U.S. patent N56026.

Anderson, Barbara Ina. "Modeling and inversion methods for the interpretation of resistivity logging tool response." (2001).

Binley, Andrew, and Andreas Kemna. "DC resistivity and induced polarization methods." Hydrogeophysics. Springer Netherlands, 2005. 129–156.

Davies, D.H., et al. "Azimuthal resistivity imaging: A new generation laterolog."SPE Formation Evaluation 9.3 (1994): 165–174.

Hoopes, J. Inversion Cross Section of Stromboli Volcano: Data Provided by Revil Et Al. (2011). Digital image, 2013.

Kaufman, Alexander A. "The electrical field in a borehole with a casing."Geophysics 55.1 (1990): 29–38.

Kaufman, Alexander A., and W. Edward Wightman. "A transmission-line model for electrical logging through casing." Geophysics 58.12 (1993): 1,739–1,747.

Kaus, A., Boening, W. 2008. BEAM–Geoelectrical Ahead Monitoring for TBM-Drives, Geomechanics and Tunneling, 1(5), 442–450.

Kemna, Andreas. "Tomographic inversion of complex resistivity." Ruhr-Universität Bochum (2000).

Klein, J.D., P.R. Martin, and A.E. Miller. "Cement resistivity and implications for measurement of formation resistivity through casing." SPE Annual Technical Conference and Exhibition. 1993.

Kopp, T. 2012. Real-Time Monitoring of Geological Conditions During Mechanized Tunneling by Means of Beam4 Method, Proc. 1st Eastern European Tunneling Conference, Budapest, Hungary, Aug. 18–21, 2012.

Kovári, Kalman, and Maorco Ramoni. "Urban tunnelling in soft ground using TBM's." Proceedings of the International Congress on Mechanized Tunneling: Challenging Case Histories, Politecnico Ditorino, Italy. 2004.

Kulenkampff, J., 1994, Die komplexe elektrische Leitfahigkeit poroser Gesteine im Frequenzbereich von 10 Hz bis 1 Mhz: Einflusse von Porenstruktur and Porenfullung: Ph.D. dissertation, Tech. Univ. Clausthal.

Linde, Niklas, et al. "Improved hydrogeophysical characterization using joint inversion of cross-hole electrical resistance and ground-penetrating radar traveltime data." Water Resources Research 42.12 (2006): W12404.

Milligan, George. "Lubrication and soil conditioning in tunnelling, pipe jacking and microtunnelling: A state-of-the-art review." Geotechnical Consulting Group, London, UK (2000).

Mooney, M.A., M. Karaoulis, and A. Revil. "Investigation of Geoelectric-while-tunneling Methods through Numerical Modeling." WTC 2013 (2013).

Revil, André, et al. "Hydrogeology of Stromboli volcano, Aeolian Islands (Italy) from the interpretation of resistivity tomograms, self-potential, soil temperature and soil $CO_2$ concentration measurements." Geophysical Journal International186.3 (2011): 1,078–1,094.

Revil, A. "Spectral induced polarization of shaly sands: Influence of the electrical double layer." *Water Resources Research* 48.2 (2012).

Revil, A., et al. "Low-frequency complex conductivity of sandy and clayey materials." Journal of colloid and interface science (2013).

Reynolds, John M. "Seven: Electrical Resistivity Methods." An Introduction to Applied and Environmental Geophysics. Chichester: John Wiley, 1997. 418–19.

Samouëlian, Anatja, et al. "Electrical resistivity survey in soil science: a review." Soil and Tillage Research 83.2 (2005): 173–193.

Schaeffer, K.P., M.A. Mooney, and M. Karaoulis. "3D Computational Evaluation of Electrical Imaging during Tunnel Construction, A Geometrical and Implementation Review." Tunnelling and Underground Space Technology (2014a). *Note: we expect the paper to be accepted by the time we submit the final paper. Title subject to change.*

Schaeffer, K.P., M.A. Mooney, M. Karaoulis. "3D Computational Geoelectric Tunnel Look Ahead Using Resistivity Sensitivity Analysis. Tunnelling and Underground Space Technology (2014b). *Note: we expect the paper to be accepted by the time we submit the final paper. Title subject to change.*

Schenkel, Clifford J., and H. Frank Morrison. "Electrical resistivity measurement through metal casing." Geophysics 59.7 (1994): 1,072–1,082.

Schlumberger, Conrad. "Conrad Schlumberger." U.S. Patent No. 1,913,293. 6 Jun. 1933.

Siegel, H.O., 1959, Mathematical formulation and type curves for in-duced polarization: Geophysics, 24, 547–565.

Spies, B.R. "Electrical and electromagnetic borehole measurements: A review." Surveys in Geophysics 17.4 (1996): 517–556.

Slater, Lee, et al. "Cross-hole electrical imaging of a controlled saline tracer injection." Journal of Applied Geophysics 44.2 (2000): 85–102.

Snyder, Larry L. "Tunnel boring machine." U.S. Patent No. 4,494,799. 22 Jan. 1985.

Sumner, John S. Principles of induced polarization for geophysical exploration. Elsevier Science, 1984.

Ward, Stanley H. "The resistivity and induced polarization methods." 1st EEGS Symposium on the Application of Geophysics to Engineering and Environmental Problems. 1988.

Wenner, F., 1912, The four-terminal conductor and the Thomson bridge: U.S. Bur. Standards Bull. p. 559–610. Wenner, F., 1915, A method of measuring earth resistivity US Bureau of Standards Bulletin 12 469–78.

# TRACK 2: DESIGN

## Session 1: Design of Underground Spaces

Peter Chou, Chair

# 3D Analysis of Precast Segmental Liner and Induced Settlement for EPB Excavation

**Pooyan Asadollahi and Jon Y. Kaneshiro**
Parsons Corporation

**ABSTRACT:** The Anacostia River Tunnel with a length of 12,300 feet (2.33 miles) and internal diameter of 23 feet will be excavated using an Earth Pressure Balance Tunnel Boring Machine (EPBM) in the hard/stiff clays and silts/sands of the Potomac Formation. This paper presents the design of the one-pass precast steel fiber reinforced segmental liner and the analysis conducted to predict surface settlement due to the tunnel excavation. The liner was designed using closed-form solutions, as well as 2D and 3D finite element analyses. The tunnel was analyzed using Plaxis 3D to simulate the excavation, the segmental liner, the annular grout pressure, EPBM earth pressure, and the thrust of EPBM jacks.

## INTRODUCTION

The Anacostia River Tunnel (ART) Project is part of the DC Clean Rivers Project (DCCR) through which the District of Columbia Water and Sewer Authority (DC Water) is implementing a Long Term Control Plan (LTCP) for Combined Sewer Overflow (CSO) discharges to the Potomac and Anacostia rivers and Rock Creek within the District of Columbia. In June 2013, the ART Design-Build contract was awarded to Impregilo-Healy-Parsons (IHP) Joint Venture with Parsons Corporation as the lead designer.

The ART, also known as Division H of DCCR, includes an approximately 12,300-foot-long tunnel with a 23-foot inside diameter. The tunnel extends from the CSO 019 North Drop Shaft, located near the southern end of the RFK Stadium parking lot, at the upstream end to the Poplar Point Junction Shaft, which is designed and constructed under a separate contract (Division A), located in the median of South Capitol Street, at the downstream end. Six drop shafts and four adits will be constructed along the ART alignment.

The tunnel will be excavated using an Earth Pressure Balance Tunnel Boring Machine (EPBM). This paper presents the analyses preformed to design the steel fiber reinforced (SFR) precast segmental liner and evaluates anticipated tunneling-induced surface settlements. The analyses were conducted using empirical, analytical, and finite element 2D/3D numerical methods. In addition, the following studies and analyses, not presented in this paper, were performed:

- Durability analysis considering 100-year design life and severe exposure environment;
- Seismic analysis for the ordinary and maximum design earthquake events; and
- Design of segments accessories and joints.

## GEOLOGY

From the surface downward or from youngest to oldest, the encountered geologic units in the project area include: recent Fill; Quaternary-age Alluvium; Cretaceous-age Patapsco/Arundel Formation (undivided) of the Potomac Group (P/A); and Cretaceous-age Patuxent Formation of the Potomac Group (PTX).

The main ART tunnel extends from CSO 19 North Shaft to the Poplar Point Shaft and is expected to be excavated through the P/A and PTX Formations. Figure 1 illustrates the subsurface ground layers along the tunnel axis. Table 1 summarizes the characteristics of the P/A and PTX Formations.

## SUBSURFACE GROUND CONDITIONS AT CRITICAL STATIONS

The ART passes through two formations. For analysis and design of the bored tunnel, the deepest and shallowest stations at each formation were selected as the critical stations, which are as follows:

- STA 132+50, which is the deepest cross section and passes through the P/A Formation;
- STA 92+00, where the tunnel will be excavated through mixed face of the P/A and PTX Formations;

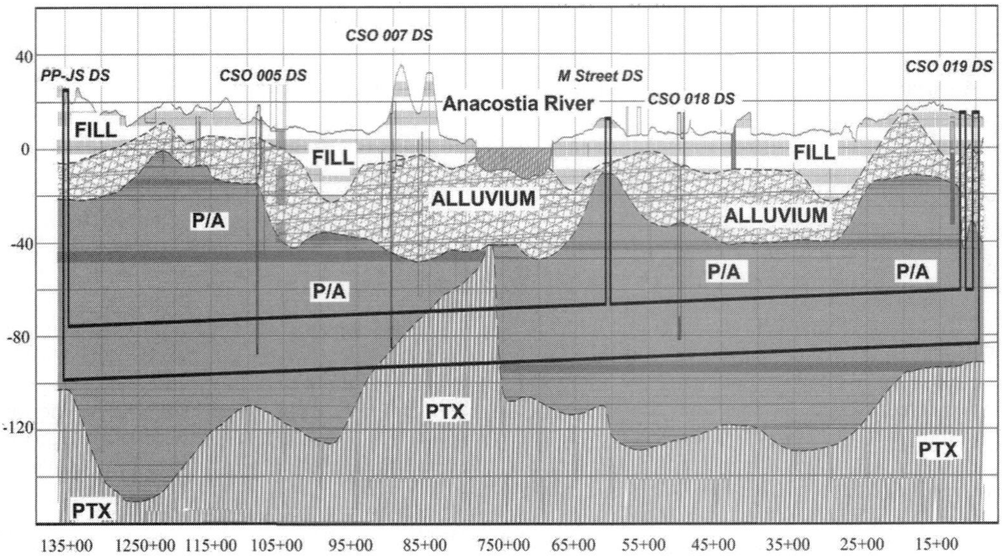


**Figure 1. Geological profile along the Anacostia River Tunnel (ART)**

**Table 1. Characteristics of the P/A and PTX Formations**

| Formation | Description | Permeability (cm/sec) | Shear Strength | $K_0$ |
|---|---|---|---|---|
| P/A | • Hard/stiff clay or silt<br>• Plastic Index between 18 and 70 Extremely low to low abrasivity<br>• High potential for stickiness | $10^{-7}$~$10^{-4}$ | $S_u = 750+0.24\sigma'_v$ psf | 1.1~1.3 |
| PTX | • Silty/clayey sand<br>• Medium to extremely high abrasivity<br>• High artesian groundwater pressure | $10^{-4}$~$10^{-1}$ | $\Phi' = 36°$ | 0.9~1.1 |

- STA 77+00, which is the shallowest section and passes through the PTX Formation underneath the Anacostia River; and
- STA 71+00, where the tunnel is to be excavated through the P/A Formation underneath the Anacostia River.

Table 2 summarizes the ground properties, geometry, and ground layer thicknesses at the critical stations.

## EPBM CHARACTERISTICS

The tunnel will be excavated using an EPBM with a front shield diameter of 25 feet, 11.42 inches. The TBM has a maximum thrust force of 12,990 kips, which will be applied through 38 cylinders, two at each position/ram. The TBM jacking shoes have been designed so that their load is applied on the centerline of 12-inch segments and does not cause any eccentric loading. However, since this may not be reached during construction an eccentricity of $1/12$ of the thickness was assumed to account for jacking shoe misalignment.

## PRECAST SEGEMENTAL LINER LAYOUT AND PROPERTIES

The precast segmental liner that was designed for the ART is a universal ring configuration consisting of six segments and a key. The general segment layout is depicted in Figure 2. The segmental liner will be 12 inches thick and will be reinforced by steel fibers. The minimum compressive strength of concrete will be 6,000 psi. However, typical cylinder breaks are greater than 7,000 psi.

## INDUCED SETTLEMENT: EMPIRICAL METHOD

The empirical method presented in the literature (Mair et al. 1993; Schmidt 1974; Schmidt 1979) was used to estimate the induced surface settlement due to tunneling. This approach requires an assumption for volume loss, which typically comes from previous experience in similar projects.

For "good practice in firm ground; tight control face pressure with closed face machine in slowly raveling and squeezing ground," the FHWA (2009) suggests 0.5% volume loss. In addition, Wang et al. (2000) gave the same range of volume loss for hard/stiff clay.

Based on Impregilo's experience on similar projects—such as Thessaloniki Metro; Strategic Tunnel Enhancement Program (STEP) T-02, Abu Dhabi; Naples Metro—the following volume losses were assumed for estimating the settlement:

- 0.4 percent for the ART in clay of the P/A Formation; and
- 0.5 percent for the ART in sand of the PTX Formation.

Using above-mentioned empirical method and volume losses, the following vertical maximum surface settlements were predicted:

- 0.22 inch, at STA 132+50;
- 0.29 inch, at STA 92+00;
- 0.42 inch, at STA 77+00; and
- 0.35 inch, at STA 71+00.

**Table 2. Subsurface ground properties at the critical stations**

| Stations | Tunnel Invert Elevation [ft] | Soil Layers | Elevation of Top of the Layer [ft] | Undrained Shear Strength $S_u$ [psf] | Drained Shear Strength $\Phi$ [°]; C = 0 | Young's Modulus E [ksf] | Lateral Pressure Coefficient $K_0$ |
|---|---|---|---|---|---|---|---|
| STA 132+50 | –98.3 | Fill | +21 | N/A | 30 | 300 | 0.5 |
| | | Alluvium | –3 | 730 | 24 | 300 | 0.59 |
| | | P/A | –21 | 2025 | 22 | 750 | 1.3 |
| | | PTX | –121 | N/A | 36 | 1,000 | 0.9 |
| STA 92+00 | –93.4 | Fill | +10 | N/A | 30 | 300 | 0.5 |
| | | Alluvium | –5 | 885 | 24 | 300 | 0.59 |
| | | P/A | –45 | 1625 | 22 | 650 | 1.3 |
| | | PTX | –85 | N/A | 36 | 1,000 | 0.9 |
| STA 77+00 | –91.6 | Alluvium | –10 | 150 | 24 | 100 | 0.47 |
| | | PTX | –40 | N/A | 36 | 1,000 | 0.9 |
| STA 71+00 | –90.8 | Alluvium | –12 | 150 | 24 | 100 | 0.47 |
| | | P/A | –44 | 1530 | 22 | 610 | 1.3 |
| | | PTX | –105 | N/A | 36 | 1000 | 0.9 |

- Unit weight of soil layers: $\gamma_{Fill}$ = 115 pcf; $\gamma_{Alluvium}$ = 100 pcf; $\gamma_{P/A}$ = 130 pcf; $\gamma_{PTX}$ = 130 pcf.
- Groundwater level is at an elevation of 0 feet or is conservatively assumed to be so.

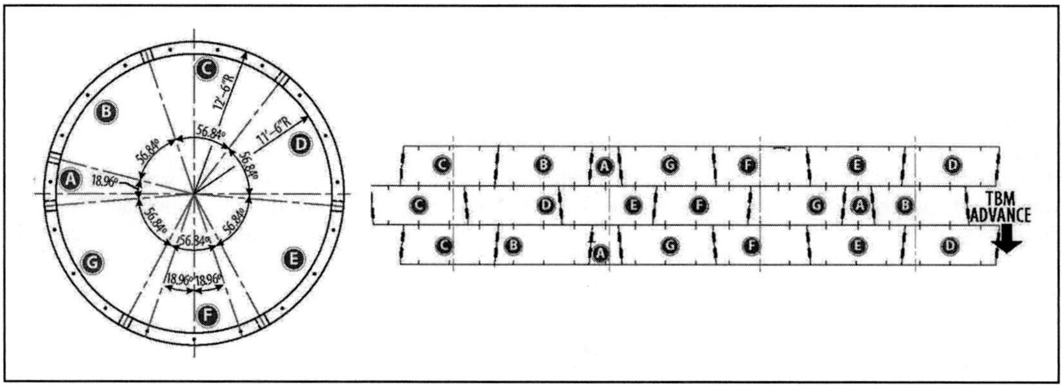


**Figure 2. General layout of precast segmental liner**

**Table 3. Maximum thrust and bending moment from Ranken et al. (1978) method**

| | Full Slippage | | No Slippage | |
|---|---|---|---|---|
| | Thrust (kip/ft) | Moment (kip-ft/ft) | Thrust (kip/f) | Moment (kip-ft/ft) |
| STA 132+50 | 234.8 | 37.02 | 221.42 | 32.27 |
| STA 92+00 | 195.27 | 30.46 | 184.32 | 26.58 |
| STA 77+00 | 125.87 | 10.14 | 128.13 | 9.71 |
| STA 71+00 | 130.06 | 17.93 | 124.01 | 15.78 |

## RANKEN ET AL. CLOSED FORM SOLUTION

Ranken et al. (1978) introduced an analytical closed form solution for ground-support interaction for a tunnel in soil based on two dimensional, plane strain, linear elasticity assumptions, where the tunnel is assumed to be deep and in contact with the ground.

The Ranken et al. method was employed to determine the maximum thrust force and bending moment developed in segmental liners at each critical station. Table 3 summarizes the results of the Ranken et al. analyses considering load factors of 1.2 and 1.6, for groundwater and ground loads, respectively. These forces have been compared to the segment strength in Section "M-N Interaction Chart."

## JACKING AND HANDLING LOADS

To ensure that the ART segments are not damaged because of jacking loads, the segments are assumed to be subjected to the full thrust of the jacks (i.e., 12,990 kips), which is applied to the segments through 19 rams (i.e., 684 kips per ram). Even though the jacking shoe and shield articulation are designed for near zero eccentricity, an eccentricity of 1 inch for the jack thrust was assumed, which causes a bending moment of 57 kip-ft. Each ram (jacking shoe) has an area of more than 2.17 ft$^2$. Thus, TBM jacks, for the worst case scenario, will apply axial force of 316 kips/ft and a bending moment of 26.4 kip-ft/ft. This combination is checked versus the segment capacity in Section "M-N Interaction Chart."

Handling and stacking can cause an ultimate bending moment of 3.7 kip-ft/ft. This is a relatively small bending moment; however, no axial force is applied. In addition, based on codes alone, the design of plain concrete is not to take any tensile stress or bending moment. In theory, however, fiber reinforcement is essential to ensure safe handling/stacking. The handling bending moment is compared with the capacity of the steel fiber reinforced segments in Section "M-N Interaction Chart."

## FINITE ELEMENT 2D ANALYSIS

All four critical cross sections were simulated using Plaxis 2D version 2011.02. Considering the length of the tunnel, plane strain concepts properly simulate the tunnel behavior and its surrounding ground. The ground layers were modeled with the given properties, and the analyses were performed both for undrained and drained conditions. The segmental liner was simulated using structural monolithic elastic plate element. The effects of joints in the precast segmental liner were taken into account using Muir Wood's (1975) equation, which reduces the effective moment of inertia of the liner by a factor of $(4/n)^2$, where $n$ is the number of joints, which is 7 for the ART.

In order to make sure that Muir Wood's equation can properly determine the effect of joints on segmental liner stiffness, Plaxis 2D and Phase2 (Version 8) analyses were performed on the undrained model at STA 132+50 and the results were compared. The ring moment of inertia was not reduced in Phase2 analysis, but hinges were simulated at joints assuming ideally no moment transfer at the joints. All input parameters and construction sequences were the same in Phase2 and Plaxis 2D models.

Numerical analyses were performed in two stages: (1) initial stage, which simulates the ground condition before construction; and (2) tunnel excavation and liner installation. In order to account for all existing and future surcharge loads, a uniformly distributed load of 1 ksf was applied to the ground surface. For the EPBM excavation, ground relaxation before installation of tunnel support is assumed to be zero. This assumption is conservative in structural analysis of the liner but may underestimate the ground deformation. To evaluate settlement more appropriately, it is prudent to perform analysis by allowing for ground relaxation before installation of the support. However, the percentage of ground relaxation is unknown for EPBM excavation in soft ground (as compared to conventional excavation, e.g., Vlachopoulos and Diederichs, 2009). Any empirical approach to ground relaxation in two dimensions does not provide a better estimation of the maximum surface settlement compared to settlement calculated using the volume loss approach described in this paper. Consequently, 2D Finite Element (FE) settlement analyses were not performed.

The axial and shear forces, as well as the bending moment were determined from undrained and drained FE analyses performed at all critical stations. After applying appropriate load factors, all axial forces and bending moments were plotted versus the capacity of segmental liner.

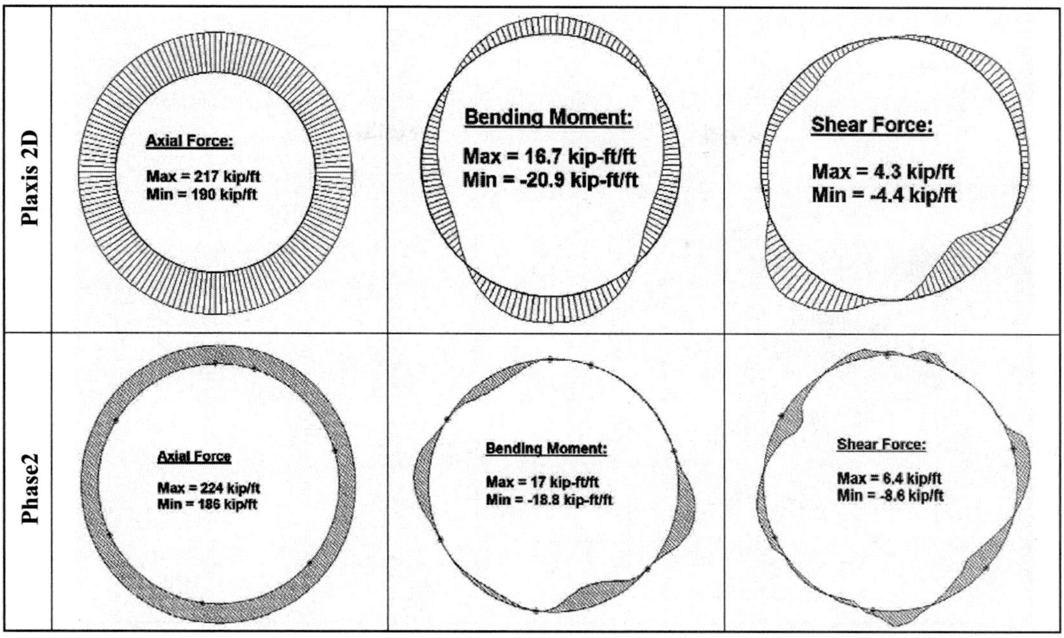


**Figure 3. Axial force, bending moment, and shear diagram; STA 132+50, undrained**

Figure 3 presents of the forces and bending moment diagrams obtained using FE Analyses for the undrained model at STA 132+50, which was found to be the most critical station. It can be seen that the results of axial forces and bending moments calculated using Plaxis 2D and Phase2 analyses are very close to each other. This also shows that the Muir Wood's equation appropriately determines the stiffness of segmental liner by taking into account joints' effect. However, the maximum shear force at joints has been underestimated by a factor of two in the Plaxis 2D analysis, where Muir Wood's equation was used.

## DOSAGE OF STEEL FIBER REINFORCEMENT

Handling and stacking causes an ultimate bending moment of 3.7 kip-ft/ft which is equivalent to a tensile stress of 154 psi in 12-inch segments. Steel fiber was designed to ensure a minimum tensile strength of 154 psi. Moreover, steel fiber or other kinds of reinforcement is essential to ensure durability of the liner for 100-year life time of the ART. Studies by Singh and Singhal (2010) have shown that with increasing fiber content, there is significant increase in compressive strength and hence increase in durability.

The same as rebar, fibers will become effective after development of some [micro-] cracks. After cracking, the tensile strength of SFR concrete would be a fraction of the modulus of rupture of plain concrete. The ratio can be obtained using the following equation (DBV, 1992):

$$\boldsymbol{Re} = \frac{180 \times \text{Dosage} \times \dfrac{\text{Length}}{\text{Diameter}} \times \sqrt[3]{\text{Diameter}}}{180 \times 20 + \text{Dosage} \times \dfrac{\text{Length}}{\text{Diameter}} \times \sqrt[3]{\text{Diameter}}}$$

During stripping of the molds, the compressive strength can be as low as 2,000 psi, which has modulus of rupture of 335 psi. For 60 pounds per cubic yard (lb/cy) of steel fiber with a length of 1.97 inches and a diameter of 0.03 inch, ***Re*** would be equal to 0.67. This means that tensile strength of the SFR segment after applying a strength reduction factor of 0.9 is 202 psi.

## M-N INTERACTION CHART

Figure 4 presents an Bending Moment-Axial Force (M-N) interaction chart of plain and SFR concrete versus the combinations of axial force and bending moment determined using different analytical and numerical methods as well as jacking and handling loads. It can be seen that a 12-inch segment with a compressive strength of 6,000 psi and reinforced with 60 lb/cy of steel fiber can carry all applied loads.

## FACE PRESSURE: CLOSED-FORM SOLUTION

The face pressure at tunnel axis was calculated using a closed-form solution presented in Broere and van

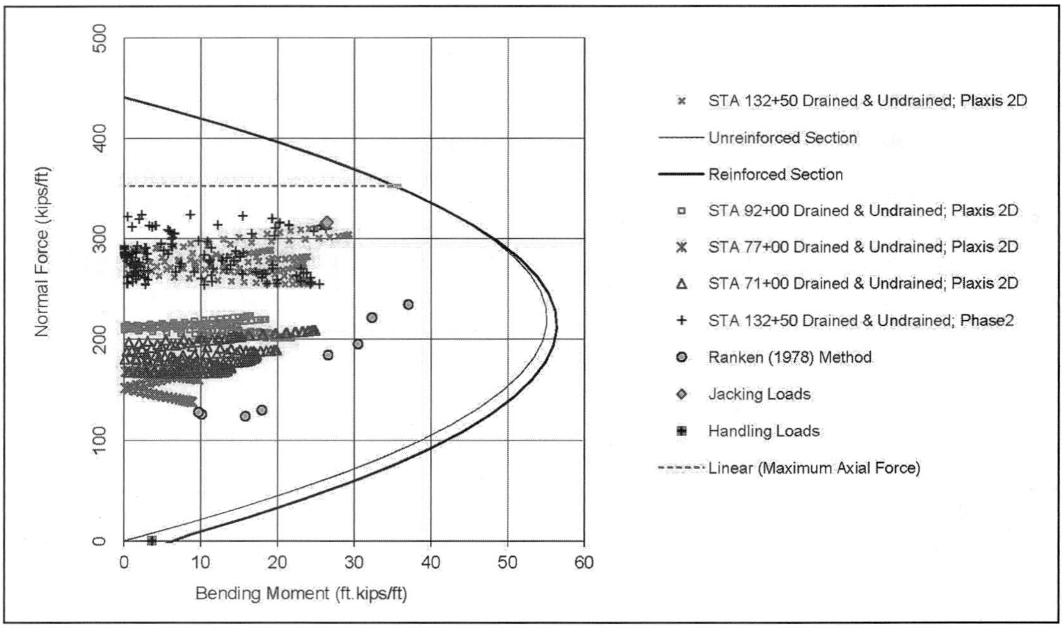


**Figure 4. Bending moment—Axial force interaction chart**

**Table 4. Face pressure at tunnel axis determined using the Broere method**

| | Face Stabilizing Pressure (bar) |
|---|---|
| STA 132+50 | 2.9 |
| STA 92+00 | 2.5 |
| STA 77+00 | 2.8 |
| STA 71+00 | 2.6 |

Tol (2001) in order to be utilized as an input of Plaxis 3D analysis. Broere and van Tol (2001) modified some important limitations of current analytical methods such as the heterogeneity of the ground at the face, and suggested an approach based on wedge and silo theory to evaluate required pressure at the tunnel face. Table 4 summarizes face pressure calculated at each station.

## FINITE ELEMENT 3D ANALYSIS

The three-dimensional finite element numerical analysis was performed using Plaxis 3D version 2012. All four critical stations were simulated. In the models, only one symmetric half was included. The models were 75 feet wide and expanded 300 feet along the tunnel axis. The ground (mesh) was carried to more than 100 feet below the invert of the tunnel, which is sufficient to capture any failure mechanism and to avoid any influence from the model boundaries.

After the initial phase, which generates in situ stresses before any construction activities, the second phase simulated where the TBM has advanced 90 feet into the ground. Then, four rounds of excavation were simulated. The ground excavation and the tunnel construction were modeled in stages. At each stage:

- The ground in front of the TBM was excavated.
- The support pressure at the tunnel axis in Table 4 was applied at the tunnel face. The vertical increment from the tunnel crown to the invert was modeled.
- The TBM shield was activated and the shape of the shield was simulated.
- At the back of the TBM, the pressure due to grouting was applied. The grout pressure at each station was determined based on the depth and the ground condition.
- The thrust forces (on the hydraulic jacks) driving the TBM against the tunnel liner were applied.
- The next segmental ring was installed.

Figure 5a and 5b presents "extrusion" (i.e., vertical deformation of the tunnel face) and vertical ground deformation at STA 132+50, respectively. The maximum surface settlement was determined to be 0.162 inch, which corresponds to a volume loss of 0.295%.

Figure 6 presents the radial and longitudinal stresses in the segmental liner. It can be seen that the

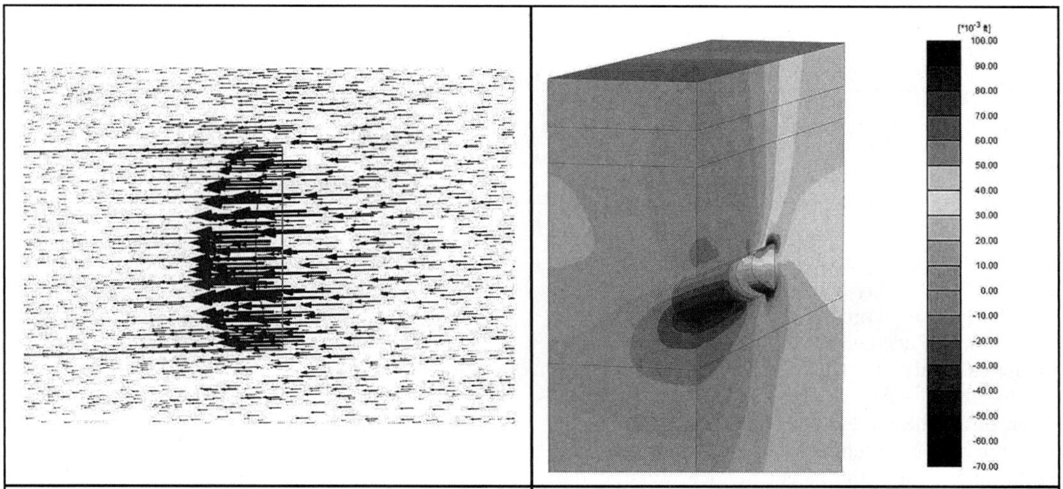


| (a) Extrusion of face<br>(Max = 0.46 inch) | (b) Vertical ground deformation<br>(Max at Surface = 0.162 inch) |
|---|---|

**Figure 5. Ground deformation; Plaxis 3D; STA 132+50**

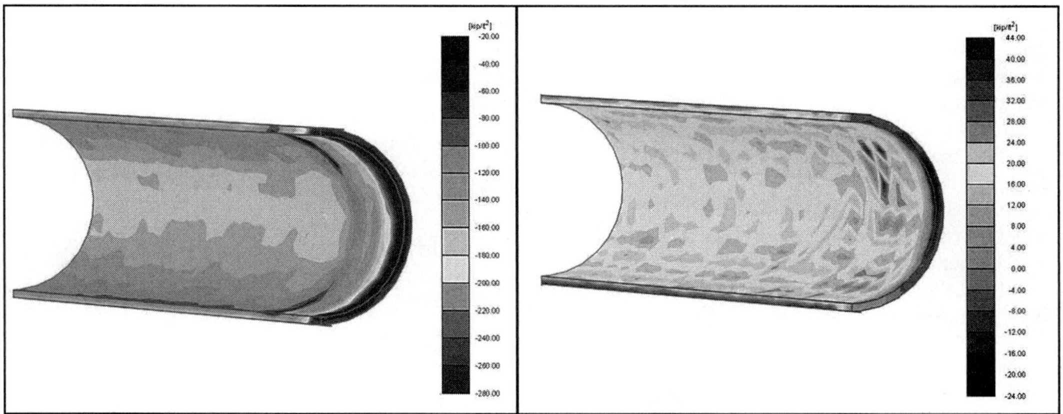


| (a) Radial stress in liner<br>(Max Compression = 237 ksf) | (b) Longitudinal stress in liner<br>(Tension = 15.57 ksf; Compression = 19.76 ksf) |
|---|---|

**Figure 6. Ground deformation; Plaxis 3D; STA 132+50**

maximum compressive stress is 237 ksf (1.646 ksi) and the maximum tensile stress is 15.57 ksf (108 psi). It should be noted that this tensile stress is due to bursting of EPMB jacks on the segment. Considering load factors of 1.2 for the groundwater loads and 1.6 for the ground loads, the factored compressive and tensile stresses would be 2.30 ksi and 0.151 ksi, respectively. Following ACI 350 requirements, the allowable compressive strength of 6 ksi concrete would be 2.65 ksi. With 60 pounds per cubic yard of steel fiber reinforcement, the tensile strength of segments would be 464 psi. Consequently, 12-inch segments can properly carry all loads with adequate factors of safety.

The same 3D analysis was performed at other critical stations. The results are summarized in Table 5.

It can be seen that the results of 3D analyses demonstrate that the designed 12-inch steel fiber reinforced segmental liner can properly support the bored tunnel. On average, the Plaxis 3D models predict the following approximate volume loss for the EPB excavation:

- 0.3 percent for the ART in clay of the P/A Formation; and
- 0.4 percent for the ART in sand of the PTX Formation.

**Table 5. Summary of the results of Plaxis 3D analysis**

| | Max Surface Settlement (inch) | Ground Loss (%) | Max Ultimate Liner Stress (ksi) | |
|---|---|---|---|---|
| | | | Compressive Stress | Tensile Stress |
| STA 132+50 | 0.162 | 0.295 | 2.3 | 0.151 |
| STA 92+00 | 0.219 | 0.34 | 1.63 | 0.197 |
| STA 77+00 | 0.33 | 0.4 | 1.24 | 0.112 |
| STA 71+00 | 0.242 | 0.277 | 1.82 | 0.143 |

At STA 92+00, the tunnel will be excavated through a mixed face of clay and sand. The Broere and van Tol (2001) method predicted that a 2.6 bar face pressure would be adequate at this station. However, the numerical analysis was stopped by soil body failure due to the lack of sufficient face pressure. The existence of a cohesionless soil with high groundwater pressure below the tunnel axis requires a higher supporting pressure at the face. The face pressure was increased to 2.9 bar to prevent failure at the face.

## SUMMARY AND CONCLUSIONS

The precast segmental liner was designed for the ART using the closed-form solution as well as 2D and 3D finite element analyses. The universal ring configuration comprises six segments and a key, 12-inch segments with a minimum compressive strength of 6,000 psi and steel fiber reinforcement of 60 pounds per cubic yard proves to be satisfactory.

The maximum surface settlement was calculated. It was found that the EPB excavation in stiff clay of the Potomac Formation causes less than 0.3% volume loss while excavation through sandy layer has a higher volume loss of 0.4%.

The following general conclusions can be made based on the analyses presented in this paper:

- Muir Wood's equation properly reduces the stiffness of segmental liners considering the effect of the joints. However, the maximum shear force in joints may be underestimated using Muir Wood's equation. This may be attributed to a small amount of moment transfer at the joints referred to as joint stickiness (see for example, Iftimie, 1994). Thus, other method should be considered when designing joints and accessories.
- A closed-form solution (e.g., presented in Boere and van Tol, 2001) properly determines face pressure at a tunnel face excavated through clayey and sandy layers; however, the approach should be used in mixed faces with caution. For mixed face excavations, it is recommended that the face pressure be calculated by conservatively assuming the weakest layer for the whole face.
- In plane strain 2D analysis of EPB excavation, the ground relaxation before installation of the liner can be disregarded. This is because the result of the 2D plane strain closely approximates the stresses in the liner and induced ground deformations compared to the 3D analyses.

## ACKNOWLEDGMENTS

The authors thank their employer, Parsons Corporation, as well as IHP JV and DC Water for permission to publish this paper. It should be acknowledged that the paper summarizes the results of Parsons' independent analyses of the design performed by HSA-Jenny Engineering, the Engineer of Record, who employed three dimensional shell elements to simulate three rings.

## REFERENCES

ACI Committee 350, 2006, Code requirements for environmental engineering concrete structures and commentary (ACI 350-06).

Broere W., and van Tol, A.F. 2001. Time-dependant infiltration and groundwater flow in a face stability analysis. In *Modern Tunneling Science and Technology*. Kyoto, Japan.

DBV. 1992. Technology of Steel Fiber Reinforced Concrete and Shotcrete, Deutscher Beton-Verein E.V. DBV Merkblatter Faserbeton, Fassung. In German.

Federal Highway Administration (FHWA). 2009. Technical Manual for Design and Construction of Road Tunnels—Civil Elements, Publication No. FHWA-NHI-10-034.

Iftimie, T., 1994, "Prefabricated lining, conceptual analysis and comparative studies for optimal solution," in Abel Salam, M.E., ed., Proceedings of the Int'l Congress on Tunnelling and Ground Conditions, Cairo. April 3–7.

Mair, R.J., Taylor, R.N., and Bracegirdle, A. 1993. Subsurface settlement profiles above tunnels in clays, *Geotechnique*, 43(2), 315–320.

Muir Wood, A. M. 1975. The circular tunnel in elastic ground. *Geotechnique*, 1, 115–127.

Ranken, R.E., Ghaboussi, J., and Hendron, A.J. 1978. Analysis of Ground-Liner Interaction for Tunnels. Report No. UMTA-IL-06-0043-78-3, US Department of Transportation, Washington, DC.

Singh, A.P. and D. Singhal, 2010, Strength and Permeability Relationship of Steel Fibrous Concrete, *35th Conf. on Our World in Concrete & Structures*, August, Singapore.

Schmidt, B. 1979. Settlements and ground movements associated with tunneling in soil. PhD thesis, University of Illinois.

Schmidt, B. 1974. Prediction of settlement due to tunneling in soil, Three case histories. In *Rapid Excavation and Tunneling Conference, San Francisco, California, June 1974.*

Vlachopoulos, N., and Diederichs, M.S. 2009. Improved longitudinal displacement profiles for convergence-confinement analysis of deep tunnels, *Rock Mechanics and Rock Engineering*, 42(2): 131–146.

Wang, Z.W, Sampaco, K.L., Fischer, G.R., Kucker, M.S., Godlewski, P.M., and Robinson, R.A. 2000. Models for predicting surface settlements due to soft ground tunneling. In *Proceedings of the North American Tunneling Conference, 2000.*

# Review of International Practice on Critical Aspects of Segmental Tunnel Lining Design

**Mehdi Bakhshi and Verya Nasri**
AECOM

**ABSTRACT:** Methods for analysis of segmental tunnel linings are presented in conformance with standards and guidelines from various countries in Europe, Asia and America. Effect of embedment loads on segments is analyzed using elastic equations, beam-spring model, FEM and DEM. Analysis of segments are studied against load case of jack thrust forces and cross section changes in joints using ACI, DAUB, Iyengar diagram, and 3D/2D FEM. Resulting forces on segments are used in order to design concrete strength and reinforcement. Best methods of practice are recommended and fiber reinforcement is presented as one of the latest developments in segmental lining design.

## INTRODUCTION

Segmental tunnel linings are designed as initial ground support and final lining in TBM-bored tunnels. Procedures to design concrete lining for embedment loads; cross section changes in joints; and checks against construction loads such as segment demolding, stacking, handling, TBM jack thrust forces, and grouting pressure have been presented elsewhere (Bakhshi and Nasri, 2013a, b, c).

Several recommendations, guidelines, and standards are available for analyses and design of precast concrete segmental linings. In this paper, special attention is given to recommendations and guidelines for analyses of segments for determining embedment loads on the one hand, and tensile stresses in joints due to jack thrust forces and cross section changes on the other. Standard design methods for precast concrete segments are presented. Best methods of practice for analysis and design of these elements are recommended. Finally, a fiber reinforcement system is presented as one of the latest developments in segmental tunnel lining systems.

## METHODS OF ANALYSIS FOR EMBEDMENT LOADS

### Elastic Equations Method

The Elastic Equations Method, recommended by JSCE (2007) and ITA (2000), is a simple method for calculating member forces of circular tunnels. The load distribution model consists of uniform vertical soil and water pressures, a triangularly distributed horizontal soil reaction between 45° and 135° from the crown on both sides in addition to a linearly varying lateral earth pressure, and dead weight of the lining. Distribution of loads used in this method is shown in Figure 1. Member forces are calculated using elastic equations available in the literature (JSCE, 2007; and ITA, 2000). In this method, a uniform bending rigidity is assumed for the lining, which cannot accurately represent the staggered geometry of segmental lining.

### Beam-Spring Method

In the Beam-Spring Method—recommended by JSCE (2007), ÖVBB (2011), and FHWA (2009)—the lining is modeled in the cross-sectional plane perpendicular to the longitudinal direction of the tunnel as a series of beam elements spanning between longitudinal joints of segments. The interaction between the ground and the lining is modeled by linear translational springs in radial, tangential, and longitudinal directions. Since the lining and ground are represented by a series of beams and springs, this method is referred to as the Beam-Spring Method.

The method of calculation of the springs' stiffness can be found elsewhere (Bakhshi and Nasri, 2013b). Various 2D approaches have been developed in order to evaluate the effect of the segment joints, including models that assume the segmental ring as a solid ring with fully bending rigidity, solid ring with reduced bending rigidity (Muir Wood, 1975), ring with multiple hinged joints, and ring with rotational springs. However, 2D models cannot represent circumferential joints and the staggered arrangement of segments in adjoining rings. As shown in Figure 2, a so-called "2½-dimensional" multiple hinged segmented double ring beam-spring model can be used to evaluate the reduction of bending rigidity and effects of staggered geometry by modeling segments as curved beams, longitudinal joints as rotational

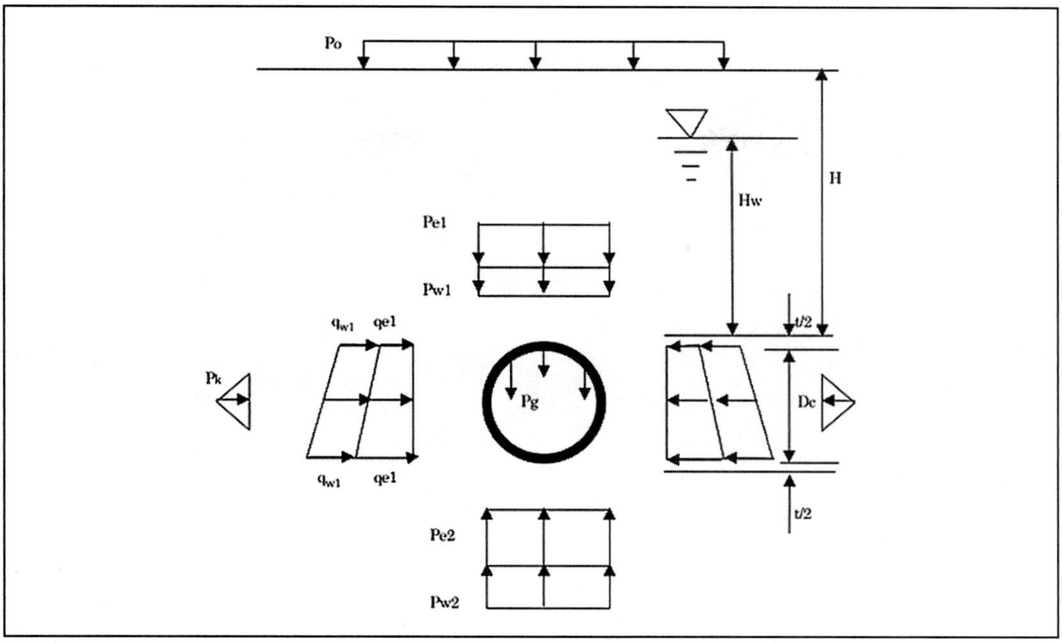


**Figure 1. Distribution of loads used in the Elastic Equations Method (ITA, 2000)**

springs (Janßen joints), and circumferential joints as shear springs. Equations and estimations regarding rotational spring stiffness and the shear spring constant of the joints have been presented in previous publications (Bakhshi and Nasri, 2013a, b, c). Two rings are used in this analysis in order to evaluate the coupling of rings; however, only half of the segment width is considered in this model to include only the influence zone of longitudinal and circumferential joints of one ring. Considering the dead weight of the lining, and applying ground and water pressures as distributed member loads projected along the beam direction, member forces are calculated using a structural analysis package.

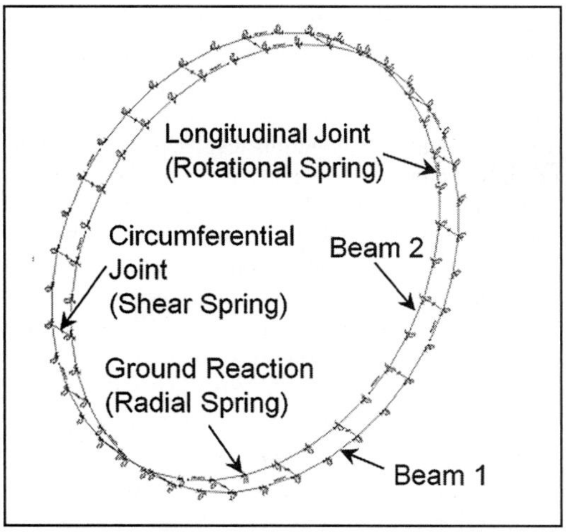


**Figure 2. Multiple-hinged, segmented, double-ring beam-spring model**

### Finite Element and Finite Difference Methods

The two-dimensional Finite Element Method (FEM) or the Finite Difference Method (FDM) is recommended by ÖVBB (2011) and AFTES (2005) for calculation of tunnel lining forces in soft ground, loose rock, and in solid rock classified as partly homogeneous. A two-dimensional approach is sufficient for a continuous linear structure without sudden changes in cross section or concentrated load intensities, while three-dimensional approaches are generally recommended for areas of intersection between crosscuts and the main tunnel (ÖVBB, 2011). In FEM, as shown in Figure 3, the surrounding ground is modeled as a continuum medium discretized into a limited number of smaller elements connected at nodal points. This method of analysis has the advantage of taking into account the deformability of the ground and in particular, its behavior after failure, the redistribution of loads resulting from lining deformation, and excavation stages (ÖVBB, 2011). This numerical method of analysis is also valid for non-uniform and anisotropic initial stresses—i.e., when a dissymmetrical feature is present in the surrounding ground because of several different formations or in

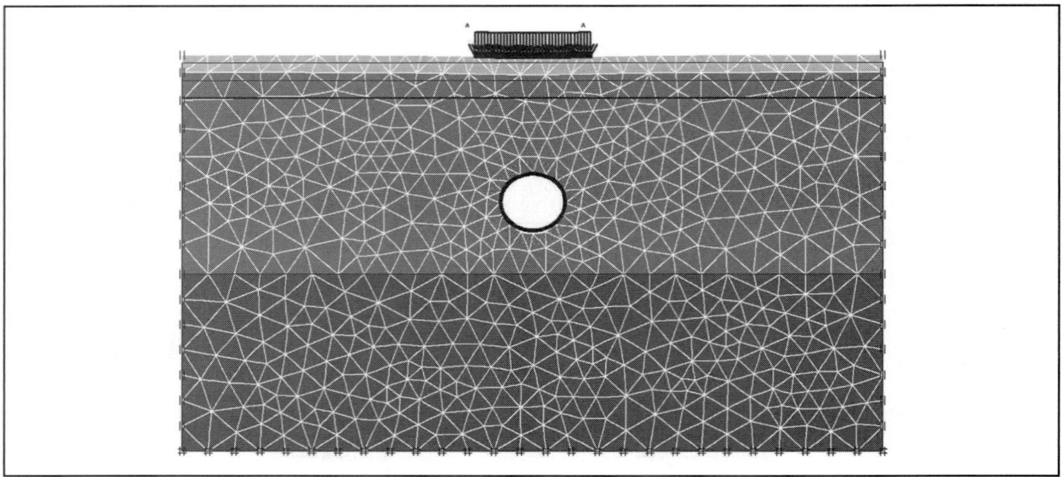

**Figure 3. FEM model for tunnel excavation in soft ground**

the external loads because of nearby existing structures (AFTES, 2005). By means of FEM, complex underground conditions and tunnel characteristics can be analyzed.

## METHODS OF ANALYSIS FOR BURSTING AND SPLITTING TENSILE FORCES

After assembly of a complete ring, the TBM moves forward by pushing its jacks on the bearing pads placed on the circumferential joints of the newest assembled ring. This action results in development of high compression stresses under the jack pads, as well as bursting tensile stresses deep in the segment and splitting tensile forces between the pads. Similar to the effect of jack thrust forces in circumferential joints, bursting tensile stresses are present at the longitudinal joints that are due to change of cross section because of the gasket and the stress relief grooves.

### ACI Simplified Equations Methods

ACI 318 section 18.13 (ACI, 2008) specifies simplified equations to determine the magnitude of the bursting force, $T_{burst}$, and its centroidal distance from the face of the segment, $d_{burst}$ as:

$$T_{burst} = 0.25P_{pu}\left(1-\frac{h_{anc}}{h}\right);\ d_{burst} = 0.5\left(h-2e_{anc}\right) \quad (1)$$

As shown in Figure 4, for the case of jack thrust forces applied on the circumferential joints, $P_{pu}$ is the maximum extraordinary jacking force applied on each jack pad, $h_{anc}$ is the length of contact area between jack pads and reduced depth of cross section on the segment face, $h$ is the depth of cross section, and $e_{anc}$ is the maximum possible eccentricity of jack pads with respect to the centroid of the cross section. For the case of cross section change at the longitudinal joints, $P_{pu}$ is the maximum normal force due to permanent embedment loads, and $e_{anc}$ is the maximum total eccentricity consists of normal force eccentricity (M/N) and eccentricity of load transfer area.

### DAUB Simplified Equations Methods

Similar to ACI, DAUB (2013) recommends simplified equations for bursting and splitting tensile stresses in the joints based on the assumption that force transfers by means of a tension block.

$$F_{sd} = 0.25 \cdot N_{Ed} \cdot \left(1 - d_1 / d_s\right) \quad (2)$$

$$F_{sd,R} = N_{Ed} \cdot \left(\frac{e}{d} - \frac{1}{6}\right);\ F_{sd,2} = 0.3F_{sd,R} \quad (3)$$

where $F_{sd}$, $F_{sd,R}$, and $F_{sd,2}$ are bursting, splitting, and secondary tensile stresses developed close to the segment face, and $N_{Ed}$ is the maximum normal force due to jack thrust force or embedment loads.

As shown in Figure 5, for the case of cross section change at the longitudinal joints, $e$ is the total eccentricity consisting of eccentricity of normal force and the hinge neck ($e = e_l + e_k = M/N + e_k$), $d_k$ is the width of the hinge neck, $d_l$ is the length of load transfer zone on the face of segment ($d_l = d_k - 2e$), $d_s$ is the distributed width of tension block inside the segment ($d_s = 2e' = d - 2e_l$), and $d$ is the total width of the segment cross section. Note that DAUB recommends splitting and secondary tensile reinforcement for only highly eccentric normal force conditions ($e > d/6$). According to DAUB (see Figure 5), bursting tensile reinforcement are placed at a distance of $0.4d_s$ from the face of segments,

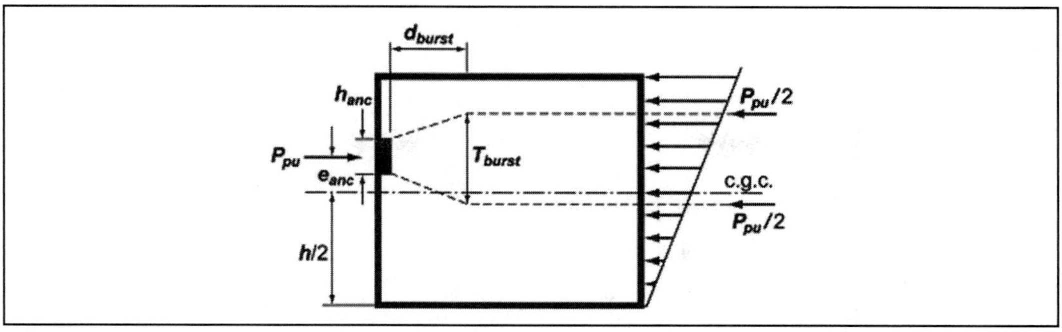


**Figure 4. Bursting tensile forces and associated parameters recommended by ACI R18.13**

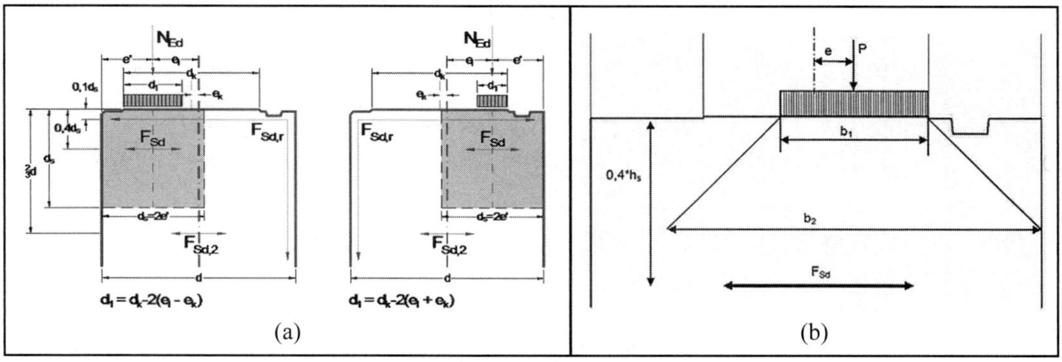


**Figure 5. Force transfer recommended by DAUB in: (a) longitudinal joints using a tension block concept, (b) circumferential joints under an eccentric jack thrust force load case (e = 50 mm)**

while splitting and secondary tensile reinforcements, if necessary, are placed at $0.1d_s$ and $⅔d$ from the face of segment, respectively.

### Iyengar Diagram Method

The analytical method of the Iyengar Diagram (Iyengar 1962) for calculation of bursting tensile stresses has been used in design of tunnels in the Netherlands (Groeneweg, 2007). Similar to previous methods, the extent of the spreading and therefore the magnitude of the tensile stresses, as shown in Figure 6, depend on the dimensions of the introduction surfaces (β) and final spreading surfaces ($a$). According to this diagram, bursting tensile stresses ($\sigma_{cx}$), which vary significantly from the face toward inside the segment, are determined as a fraction of the fully spread compressive stress ($\sigma_{cm} = F/ab$).

### Finite Element Methods

As shown in Figure 7, the effect of jack thrust force is simulated by modeling typical segments of two adjoining rings. The jack thrust forces are applied on the net contact area of the jack pads and segment face on the front circumferential joint. The recess (due to the gasket and the stress relief grooves) is modeled on the connection between two segments to simulate force transfer through a reduced cross section through the middle circumferential joint. Compressive forces of the gasket in the early hours of installation are simulated by applying maximum reaction force of the gasket. Solid elements are used for this analysis. The translational degrees of freedom are fixed in all directions at the back of behind segment, which is part of a previously installed ring. As shown in Figure 8, typical analytical results consist of transversal and radial bursting and spalling tensile stresses developed under the jack pad and in the areas between the pads.

Bursting stresses at the vicinity of the longitudinal joints are analyzed for the case of maximum normal force and gasket pressure. The two-dimensional FE model used to simulate the longitudinal joint consists of small end parts of two adjacent segments in a ring (curvature of elements are neglected) modeled with recess of the gasket and the stress relief grooves. The contact zone is modeled as a discontinuity between two adjacent segments. Nonlinear compression-only springs attach segment faces in the longitudinal joint, simulating behavior of the

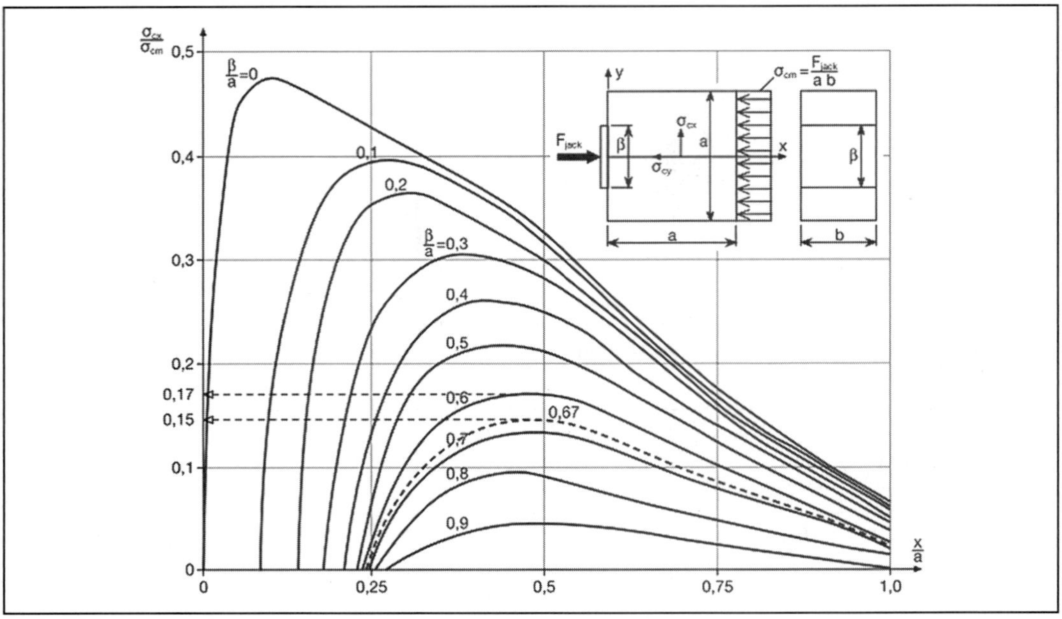


**Figure 6. Iyengar Diagram (1962) for determining bursting tensile stresses**

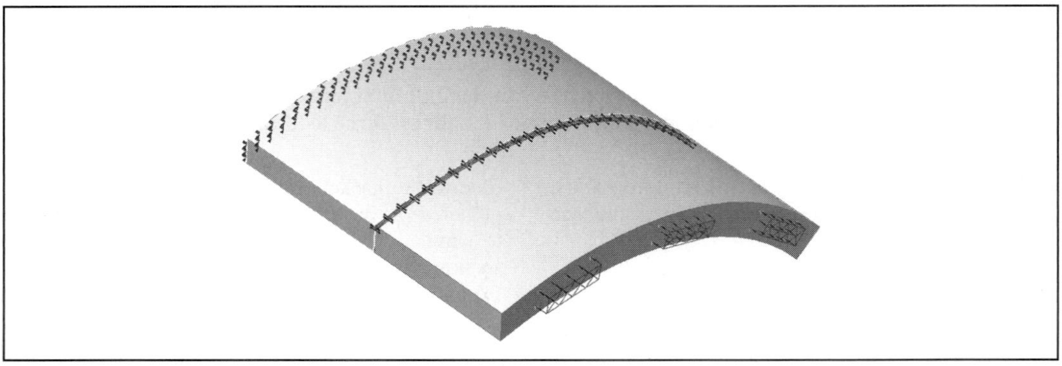

**Figure 7. 3D FEM model for case of jack thrust force**

plywood material. Translational degrees of freedom along the farthest vertical face of one of the segments are fixed in both directions, while the vertical face of the other segment is loaded with the uniformly distributed pressure of maximum normal force. Figure 9 shows typical analytical results including bursting tensile and compressive stresses in the area around longitudinal joints.

## BEST METHODS OF PRACTICE

Among different methods of analysis for determining embedment loads, the Elastic Equations Method gives the largest member forces since a uniform bending rigidity is assumed for the lining. The multiple hinged segmented double ring Beam-Spring model gives reasonable forces as a result of analysis, especially for the transferred bending moment in the longitudinal joints. Finite Element and Finite Difference Methods are superior methods when a asymmetrical feature is present in the structure, in the surrounding ground, or in the external loads.

Simplified equations methods of analysis for determining jack thrust forces and cross section changes in the joints result in a more conservative and uniformly distributed reinforcement plan. Analytical and numerical methods of analysis such as Iyengar Diagram and FEM may result in a more cost-effective and nonuniform reinforcement design.

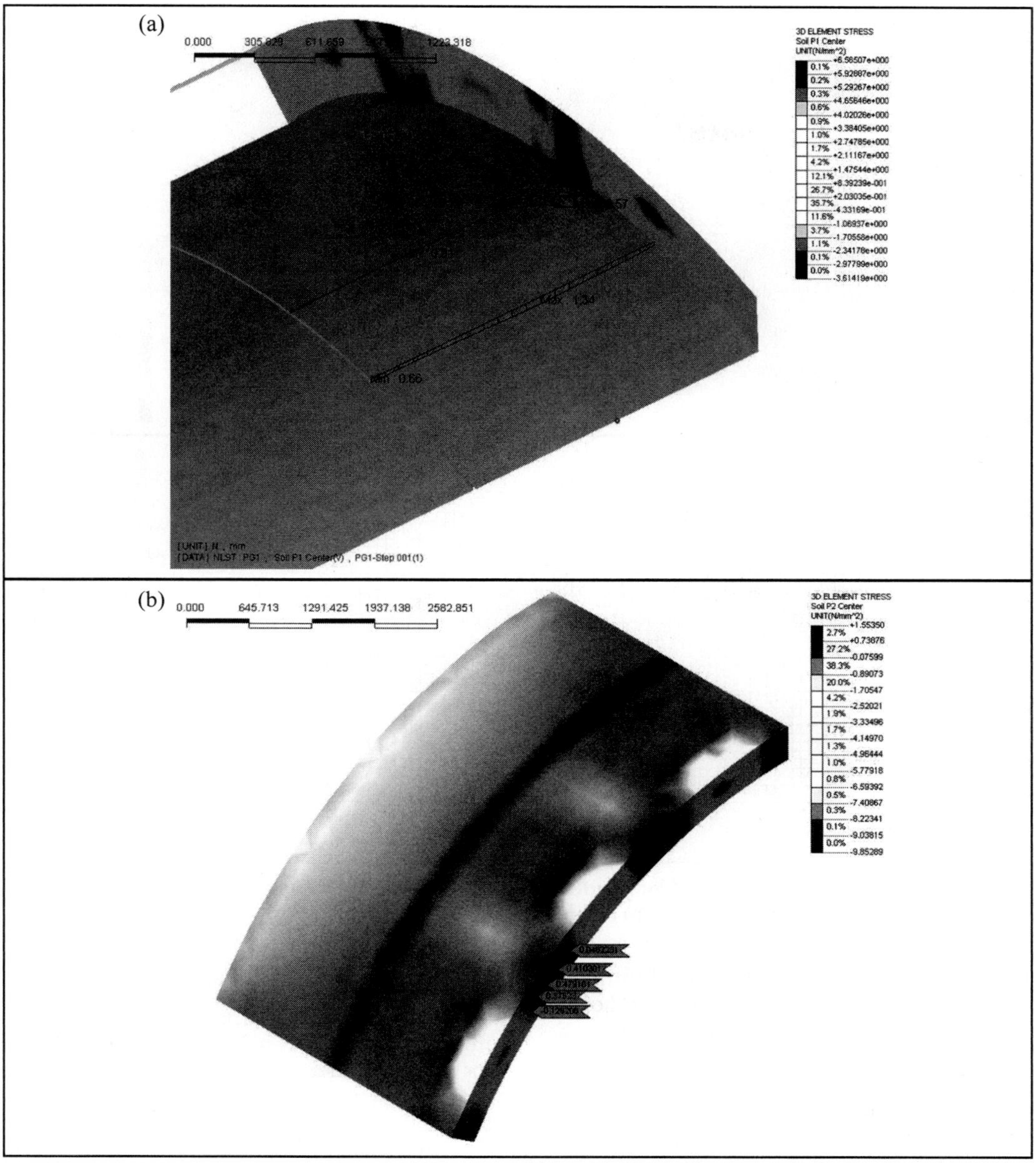


**Figure 8. Bursting and spalling tensile stresses developed in segments due to TBM jack thrust forces and gasket pressure: (a) transversal stresses, (b) radial stresses**

## LATEST TECHNOLOGIES IN SEGMENTAL TUNNEL LINING

### Fiber Reinforced Concrete Segments

Among structural applications of Fiber Reinforced Concrete (FRC), there is a growing interest in precast tunnel segments, where fibers may be substituted, partially or totally, for conventional steel bars (Plizzari and Tiberti, 2006). FRC segments have several advantages over conventional reinforced concrete segments, among which are the potential cost saving and longer service life of tunnels (Angerer and Chappell, 2008; Hilar and Beno, 2012). Fibers are uniformly dispersed through the segment, so their presence in the cover zone is very advantageous, especially with high bursting and splitting stresses developed in this zone induced by jack thrust forces.

Current codes and standards propose stress-crack opening or stress-strain constitutive laws for

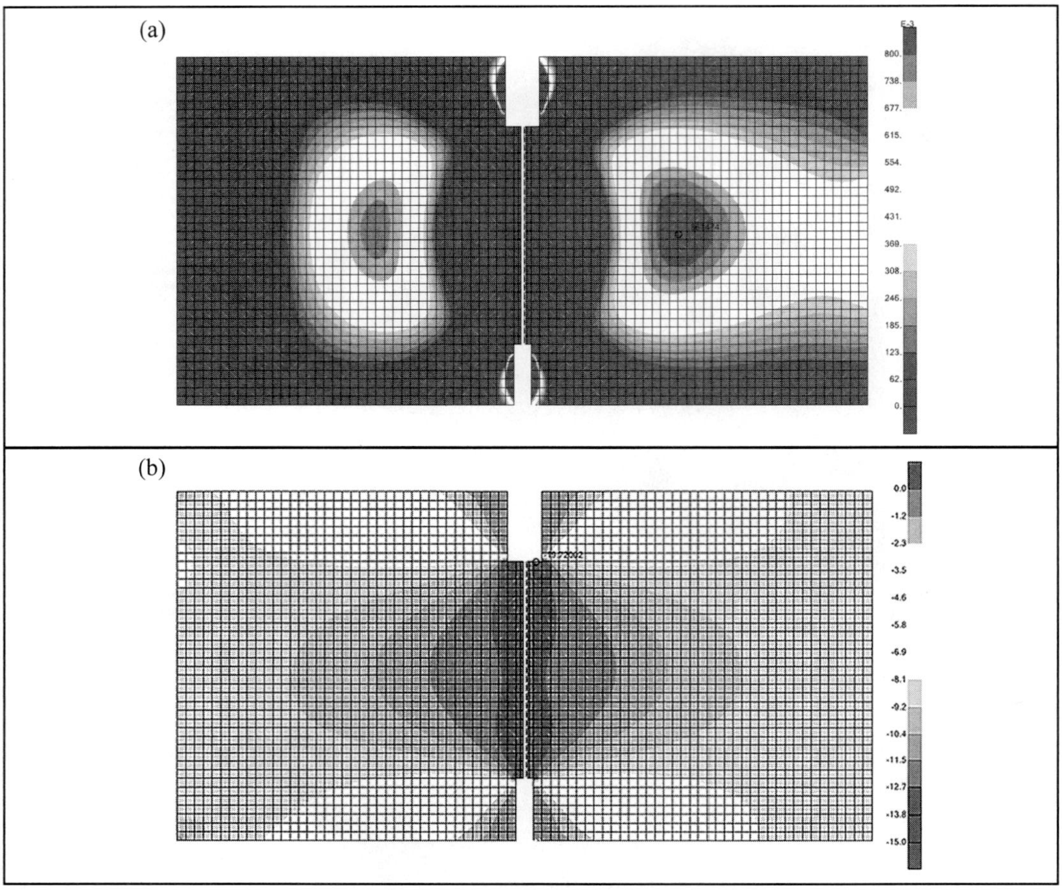


**Figure 9. Developed stresses around longitudinal joints due to maximum normal force and gasket pressure: (a) bursting tensile stresses, (b) compressive stresses**

FRC as a linear post-cracking behavior (hardening or softening) or as a plastic rigid behavior, as shown in Figures 10a and 10b. Plastic rigid constitutive law has been used successfully in calculation of resistance moment of segmental FRC tunnel linings (Angerer and Chappell, 2008; Caratelli et al. 2012). This constitutive model is based on residual tensile strength, which can be obtained from standard tests (ASTM C1399, 2010; ASTM C1609, 2010; EN14651, 2003; RILEM TC 162-TDF, 2002; JCI-SF4, 1984), and considering a correction factor of ⅓ for correlation with the design tensile strength (Bakhshi et al. 2013; fib Model Code 2010). Using such an approach, moment-axial force interaction diagrams can be obtained. These diagrams—such as the ones shown in Figure 10c for a 400 mm deep, 1,700 mm wide section reinforced with fiber dosages of 20 to 50 kg/m$^3$—are used to check against applied forces. FRC resistance diagrams are comparable with a bar reinforcement system—e.g., 28 steel bars with a cross sectional area of approximately 100 mm$^2$.

## CONCLUSION

Existing recommendations, guidelines, and standards from various countries in Europe, Asia and America for precast concrete segmental tunnel linings were evaluated. Among the different methods of analysis for determining embedment loads, the Elastic Equations Method gives the largest member forces since a uniform bending rigidity is assumed for the whole lining. The multiple hinged segmented double ring Beam-Spring model gives more reasonable internal forces as results of analysis, especially for the transferred bending moment in the longitudinal joints. Finite Element, Finite Difference and Discrete Element Methods are superior when an asymmetrical feature is present in the structure, in the surrounding ground, or in the external loads. On the other hand, simplified equations methods of analysis for determining jack thrust forces and cross section changes in the joints result in a more conservative and uniformly distributed reinforcement plan. Analytical

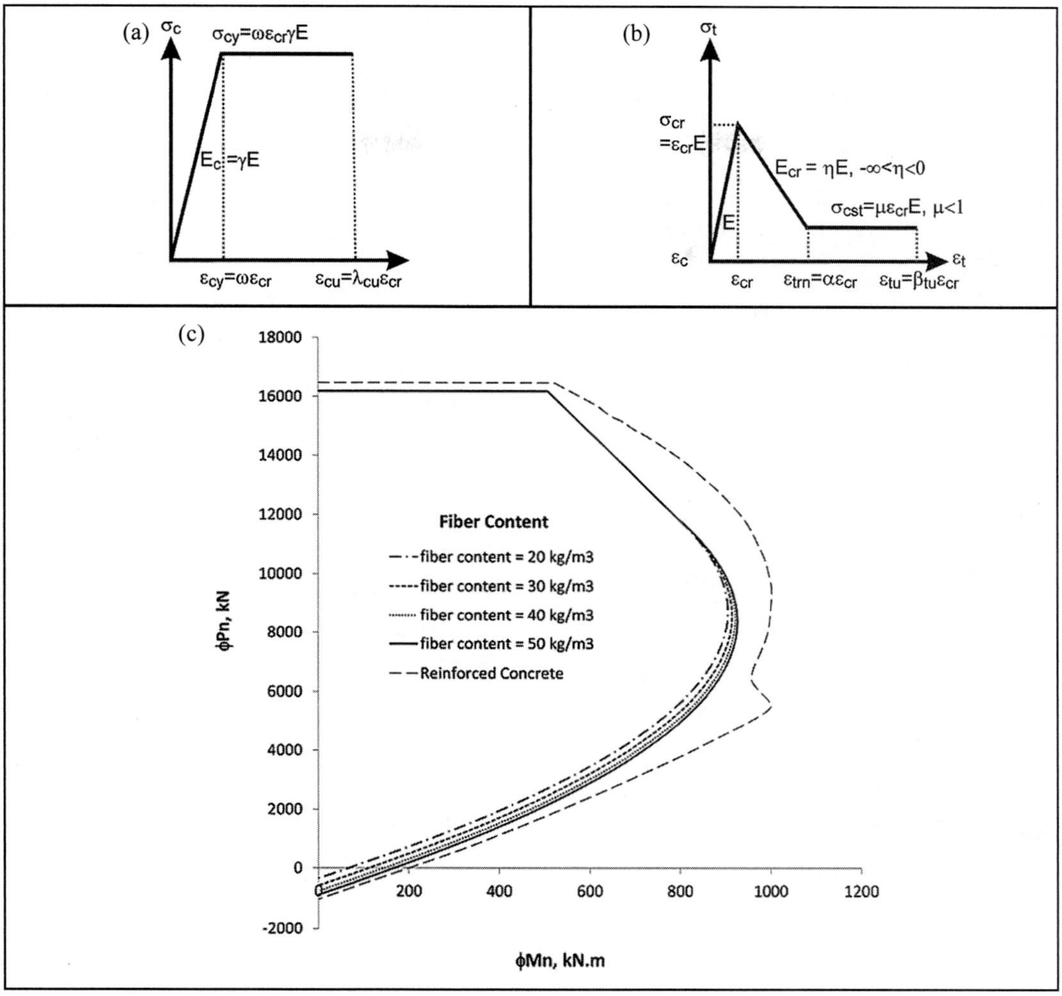


**Figure 10. Design of precast FRC segmental tunnel lining: (a) constitutive law in compressive, (b) constitutive law in tension, (c) resulting moment-axial force resistance diagrams**

and numerical methods of analysis such as Iyengar Diagram and FEM may result in a nonuniform and more cost-effective reinforcement design. FEM based approaches are recommended for the analysis of the internal forces in the proximity of the joints due to embedment loads or jack thrust forces, especially when segments are made of FRC materials.

## REFERENCES


ACI 318-08. 2008. Building Code Requirements for Structural Concrete and Commentary. American Concrete Institute Committee 318.

Angerer, W., and Chappell, M. 2008. Design of steel fibre reinforced segmental lining for the gold coast desalination tunnels. In *Proceedings of the 13th Australian Tunnelling Conference*, 4–7.

ASTM C1399-10. 2010. Standard test method for obtaining average residual-strength of fiber-reinforced concrete. American Society for Testing and Materials, West Conshohocken.

ASTM C1609-10. 2010. Standard Test Method for Flexural Performance of Fiber-Reinforced Concrete (using beam with third-point loading). American Society for Testing and Materials, West Conshohocken.

Bakhshi, M., Barsby, C., and Mobasher, B. 2013. Comparative evaluation of early age toughness parameters in fiber reinforced concrete. *Materials and Structures*, 1–20.

Bakhshi, M., and Nasri, V. 2013a. Structural design of segmental tunnel linings. In Proceedings of 3rd International Conference on Computational Methods in Tunnelling and Subsurface

Engineering: EURO: TUN 2013. Ruhr University Bochum, April 2013, 17–19.

Bakhshi, M., and Nasri, V. 2013b. Latest developments in design of segmental tunnel linings. In Proceedings of the Canadian Society for Civil Engineering General Conference (CSCE 2013), Montréal, Québec. May 29 to June 1, 2013.

Bakhshi, M. and Nasri, V. 2013c. Practical aspects of segmental tunnel lining design. Underground—The way to the future: In *Proceedings of the World Tunnel Congress 2013. Geneva. May 31–June 7, 2013*. G. Anagnostou & H. Ehrbar (eds).

Caratelli, A., Meda, A., and Rinaldi, Z. 2012. Design according to MC2010 of a fibre reinforced concrete tunnel in Monte Lirio, Panama. *Structural Concrete*. 13(3), 166–173.

DAUB (German Tunnelling Committee) Working Group "Lining Segment Design." 2013. Recommendations for the Design, Production, and Installation of Segmental Rings.

EN 14651. 2003. Precast Concrete Products—Test method for metallic fiber concrete—Measuring the flexural tensile strength. *European Standards*, Brussels, Belgium.

FHWA. 2009. Technical Manual for Design and Construction of Road Tunnels—Civil Elements. US Department of Transportation, Federal Highway Administration.

fib Model Code. 2010. Bulletin 55: Model Code 2010—First complete draft. Fédération internationale du béton/the International Federation for Structural Concrete. Lausanne, Switzerland.

Groeneweg, T. 2007. Shield Driven Tunnels in Ultra High Strength Concrete: Reduction of the Tunnel Lining Thickness. Thesis (MSc in Civil Engineering), Delft University of Technology, The Netherlands.

Hilar, M.; Beno, J. 2012. Steel fibre reinforced segmental tunnel linings. *Czech and Slovak ITA-AITES magazine Tunel*. 3, 30–36.

International Tunneling Association (ITA) Working Group No. 2. 2000. Guidelines for the design of shield tunnel lining. *Tunneling and Underground Space Technology*. 15(3), 303–331.

Iyengar, K.T. 1962. Two-dimensional theories of anchorage zone stresses in post-tensioned beams. *Journal of the American Concrete Institute (ACI)*. 59(10), 1443–1466.

JSCE. 2007. Standard Specifications for Tunneling: Shield Tunnels. Japan Society of Civil Engineers.

JCI-SF4. 1984. Methods of Tests for flexural Strength and flexural Toughness of fiber Reinforced Concrete (using beam with third-point loading). Japanese Concrete Institute, 45–56.

Muir Wood, A.M. 1975. The circular tunnel in elastic ground. *Géotechnique*, 25(1), 115–127.

ÖVBB Guideline. 2011. Guideline for Concrete Segmental Lining Systems. Austrian Society for Concrete and Construction Technology. February 2011.

Plizzari, G.A., and Tiberti, G. 2006. Steel fibers as reinforcement for precast tunnel segments. *Tunnelling and Underground Space Technology*, 21(3), 438–439.

RILEM TC 162-TDF. 2002. Test and design methods for steel fibre reinforced concrete: bending test (final recommendation). *Materials and Structures*, 35, 579–582.

# Tunneling Under the Sydney Opera House: The Vehicle Access and Pedestrian Safety Project

**Seth Pollak, Nik Sokol, and Mark Adams**
Arup

**Greg McTaggart**
Sydney Opera House

**Ben Clarke**
John Holland Group

**ABSTRACT:** The current traffic arrangements for servicing the Sydney Opera House (SOH) necessitates that delivery vehicles share the open forecourt and boardwalk areas with visitors, complicating logistics and resulting in several incidents per year. The Vehicle Access and Pedestrian Safety (VAPS) project was initiated in order separate vehicle and pedestrians by diverting freight traffic via a mined tunnel into a new sub-level loading dock. New access tunnels and lifts link the dock to the existing SOH basement and ground level facilities. The design and construction of the tunnels are discussed with particular attention paid to the risks involved in excavating under a UNESCO World Heritage Listed Site. The project was further complicated by low rock cover, high surcharge loads, a high horizontal stress regime, and a skewed, 17m spanning, flat roof intersection of the tunnel and loading dock. Observations from the robust ground and structure instrumentation program are compared against design assumptions.

## INTRODUCTION

The Sydney Opera House (SOH) is one of the world's most iconic structures. With 8.3 million visitors each year and 1,000 vehicles arriving and departing each week, the original delivery access routes shared between pedestrians and traffic can no longer cope with this combined volume. The AUD $152M Vehicle Access and Pedestrian Safety (VAPS) project is being constructed to separate pedestrians and delivery vehicles for safety and efficiency while maintaining operation of six performance spaces and seven restaurants within the SOH complex. VAPS is also the first phase of a proposed and notably more ambitious renewal plan to be completed over the next decade. Planning for the next series of projects will commence in 2014.

The VAPS project is being managed by the Sydney Opera House Building Development & Maintenance team and is the largest building project to be undertaken at Sydney Opera House since the building opened in 1973. Construction and closure of the Forecourt area commenced in February 2011. The Forecourt was partially reopened in September 2013, with rotating occupation of work areas to proceed until project completion later in 2014. Figures 1 and 2 illustrate some of the major facilities to be constructed as part of the VAPS project, including an 11m span access tunnel under the Forecourt to the SOH, a 15m deep, 45m by 45m underground loading dock and truck turning bay built by cut and cover means under a renovated vehicle concourse, an egress passage from the loading dock to the cut and cover portion of the Forecourt Tunnel, access tunnels/ corridors to both the Concert Hall and Opera Theatre, a cross passage between these two tunnels, and several new lift shafts and stairwells which interface with the existing operational areas of the SOH basement and ground level.

Tunneling and excavation work underneath the SOH has proceeded without disturbance or interruption to pedestrian access and site services or a performance schedule which averages more than 40 events per week. Since work began onsite in early 2011, underground work has proceeded almost entirely behind the scenes—all planned performances and tours have continued and for the busy shops, cafes and restaurants it has been business as usual.

### Project Participants

The VAPS project was procured as a fully designed project through a tender process which considered the construction methodology proposed, company

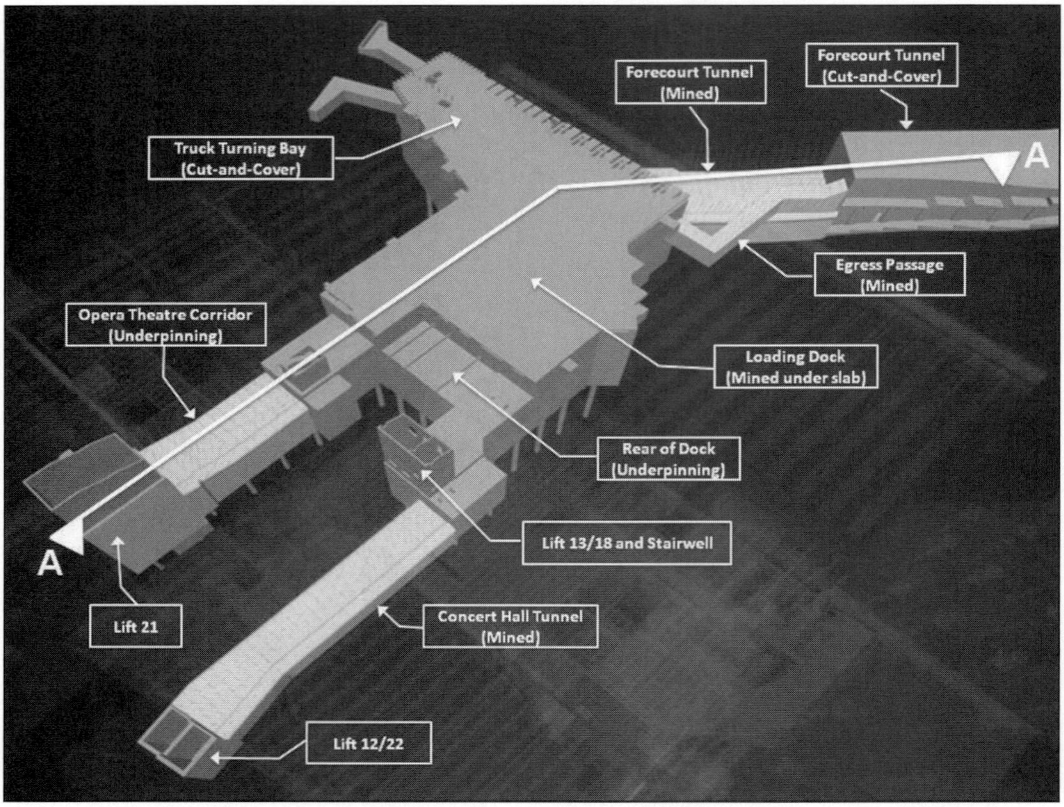


**Figure 1. Three-dimensional view looking southeast at the major features of the Sydney Opera House VAPS project**

**Figure 2. Section A-A from Figure 1, architectural rendering looking west. From left to right: cut-and-cover access ramp, mined tunnel under Forecourt, loading dock under vehicle concourse, access tunnels and lifts. Image courtesy of Scott Carver.**

**Table 1. VAPS project team members**

| Discipline | Company |
|---|---|
| Client | Sydney Opera House Trust |
| Contractor | John Holland Group |
| Tunnel, geotechnical, civil, structural, fire, theatre, acoustics, traffic, security | Arup |
| Architect | Scott Carver |
| Vertical transport | Norman Disney Young |
| Electrical, mechanical | Hyder + Steensen Varming |
| Hydraulics | Hyder + Warren Smith & Partners |
| Survey, instrumentation | Lynton Surveys |
| Resident engineers—structural, geotechnical & tunneling | Arup |

experience, expertise of the team, project program, price and quality. The VAPS project participants are presented in Table 1.

## TUNNEL DESIGN

The design of the temporary support on the VAPS project is founded on principles developed and applied in Hawkesbury Sandstone over the past 30 years, beginning with the adjacent Opera House Underground Parking Station. Making use of the Voussoir beam rock mechanics model, calculations were verified by 2D and 3D discrete element analyses. The timing of the support installation was closely tied to the excavation sequence in order to limit ground movements under structures. A rigorous instrumentation and monitoring program measured sidewall deflection and tunnel convergence, both of which are compared to the design assumptions later in the paper.

### Temporary Support Design Methodology

The temporary support design methodology adopted for VAPS mined tunnels included analytical beam-arch methods and numerical (discrete element) methods. Based on the structural uniformity and specific behavior mechanisms observed in Hawkesbury Sandstone, an empirical approach such as the Q system or RMR was not considered appropriate to arrive at a safe and reliable temporary support design. The adopted methods account for site specific joint orientations and properties as derived from both the ground investigation and adjacent excavations, ensuring against an oversimplification of the rock mass as may be true by use of empirical methods.

Failure mechanisms considered and designed against in the analyses included:

- Reinforced rock beam failure through lateral stress induced bedding shear displacement;
- Vertical slip of the reinforced rock beam under surcharge and self-weight loading along joints at the abutments;
- Compressive crushing of the rock at the top of the beam midspan under deflection;
- Build-up of localized water pressure in bedding joints due to laminate seam or aquitard;
- Beam delamination at midspan due to vertical tension from self-weight and surcharge loads;
- Delay in support installation leads to excessive beam deflection, block rotations, and ultimate instability;
- Rock reinforcement bond failure;
- Failure of the thin shotcrete membrane through debonding, and ultimately, flexural failure; or punching shear due to a detached block between bolts;
- Instability of vertical walls, both at tunnel face and temporary drift walls.

### Voussoir Beam Analysis

The Voussoir beam, or linear arch analogue has been applied to Hawkesbury Sandstone in nearly every tunnel project since the Opera House Underground Parking Station. The method as applied to the VAPS tunnel design is that proposed by Diederichs and Kaiser (1999) and modified as suggested by Asche and Lechner (2003) for simplicity. In this "hybrid" Voussoir beam analysis, the secondary loop in the Diederichs and Kaiser method is replaced by direct calculation of abutment deflection through logic proposed by Asche and Lechner. Once the shape and stress distribution of the linear arch thrust line has been established, the component of stress along and normal to the bedding plane is calculated by considering the slope of the parabolic arch as a function of distance along the beam. The shear and normal stress are then compared to the shear strength of the bedding plane, with the bolting pattern designed to carry the excess stress.

### 2D Numerical Analysis (UDEC)

Confirmation of the analytical results, as well as incorporation of site specific geology and loading conditions, was undertaken through numerical modelling using the distinct element code UDEC (v.4.01). Joint set orientations were determined through examination of borehole optical televiewer data (RAAX). In developing the model, justification needed to be provided for the adopted joint pattern. A block size (volume) analysis (after Palmström, 2005) was carried out using the dip angle and depth of the recorded fractures in the raw RAAX data for the

**Table 2. Discontinuity shear strength parameters**

| Type | φ | Dilation | Normal Stiffness (MPa/m) | Shear Stiffness (MPa/m) |
|---|---|---|---|---|
| Bedding (clean) | 34° | 3° | 4,000 | 400 |
| Bedding (1–5mm clay infill) | 29° | 3° | 1,500 | 150 |
| Erosional plane | 20° | 0° | 200 | 20 |
| Sub vertical joint (clean) | 34° | 0° | 4,000 | 400 |
| Sub vertical joint (1–5mm clay infill) | 32° | 0° | 1,500 | 150 |

**Table 3. Forecourt mined tunnel key results comparison**

| Analysis Type | Mid-Span Deflection (mm) | Horizontal Abutment Stress (kPa/m) | Minimum Thrust Arch Horizontal Stress (kPa/m) |
|---|---|---|---|
| Voussoir beam | 1 | 599 | 336 |
| UDEC | 10 | 800 | 400 |

project boreholes. The data was subsequently cross referenced with core logs and photos to exclude any cross bedding that was accidentally logged as a fracture by the televiewer. The generated block size distribution in UDEC was then plotted after a normal distribution profile, and the estimated block size from the Palmström analysis verified to fall within the standard deviation of this profile. Once satisfied, the model was considered "calibrated" to the actual block size. Table 2 defines the discontinuity shear strength parameters used in UDEC.

A significant characteristic of the Sydney geology is the known high value of horizontal stress compared to the vertical. The effect of this high stress is to cause relief during tunnel excavation, which can manifest itself in shearing along bedding planes and opening of discontinuities. Along with increased ground movements and loss of shear strength, this dilation can result in increased joint aperture leading to increased groundwater inflow. Various results have been published with a range of stress magnitudes. For conservatism, two cases of horizontal stress were examined for the VAPS study; a lower bound stress profile which was critical to tunnel stability (minimizing confinement of the Voussoir beam abutments), and an upper bound profile which maximized shearing along bedding planes (ground movements).

## Results

The results of the analytical and numerical design analyses for the typical Forecourt Mined Tunnel are presented in Table 3.

The results show that the mid span deflection of the rock beam is greater in UDEC than in the analytical model. This is likely due to the effects of the erosional plane which has been included above the crown in UDEC.

The adopted temporary support for the tunnel consisted of 5m long RB310 rock bolts spaced at 1.25m centers and pre-tensioned to 75 kN. A 100mm thick steel fiber shotcrete layer was specified to retain any raveling pieces in between the bolt pattern. The benefit of shotcrete was not included in design calculations or models as a conservative measure.

## Special Case—Monumental Stairs

A 16.25m long section of the Forecourt tunnel runs obliquely underneath the Monumental Stairs (MS) strip footing and is subject to high surcharge loads of up to 700 kN/m. Preliminary tunnel stability analyses carried out in UDEC during the design development stage indicated that the load carried straight through the rock beam, with little spreading as this is limited by the interlayer friction between bedding planes. The result is a narrow pressure distribution bulb and increased load onto the tunnel support, manifesting itself as large vertical displacements and settlement under the stairs. The problem is compounded by the skewed angle between the tunnel axis and the trend of the stair load (approximately 45°). The 2D case of subjecting the full skew span to the surcharge load applied across the top of the model is not geometrically accurate. Therefore, it was decided to model the entire Forecourt tunnel in 3D.

Use of 3DEC (Itasca Consulting Group Pty Ltd) to develop discrete element model (discontinuum) was elected for the following reasons:

- Accounts for geological structure of Hawkesbury Sandstone;
- Provides suitable comparison with 2D UDEC models also used in design;
- Prevents defining new set of geotechnical continuum parameters characterising anisotropic rock mass as isotropic;

- Provides more conservative load spread due to rock mass bedding compared to continuum model.

The design intent was to provide a stiff and robust support (and construction sequence) which minimized movements to the overlying stairs, only 6.9m above the crown (cover/span ratio of 0.58). Full face excavation with short (1–1.5m advances) was considered preferable over sequential methods involving complex reinforcement and shotcrete overlaps.

The adopted passive support consisted of a 600mm thick shotcrete lining, reinforced with lattice girders, welded wire mesh, and additional steel reinforcing bars at critical locations. Tensioned RB550 rock bolts were designed between lattice girders to promote beam action within the horizontally bedded rock, and also to prevent shear displacement of the rock during the future loading dock excavation. For each 1.25m advance (13 total stages) the maximum axial force and maximum/minimum bending moments were extracted from the lining beam elements. Therefore, for each stage of construction, every previous 1.25m wide beam element was checked again for a change in stress. Reinforcement was then designed to meet the critical demand. The following shotcrete performance requirements were specified in the design:

- $f'c_{1\ day} = 10$ MPa
- $f'c_{3\ day} = 20$ MPa
- $f'c_{7\ day} = 30$ MPa
- $f'c_{28\ day} = 40$ MPa
- Toughness (by ASTM C1550) = 40J at 4mm deflection
- Residual flexural strength, L/600 (by ASTM C1609) = 2.5 MPa at 7 days

In a unique concept, 40mm diameter reinforcing bars at 200mm spacing were hung between lattice girders after being welded together with a 20mm diameter cross tie piece. This provided positive moment reinforcement in the correct location of the lining, and allowed for complete encapsulation by the shotcrete. The final design rock bolt spacing of 1.44m × 1.25m (longitudinal/transverse) was set to fit with the required lattice girder spacing of 720mm on center. A summary of the structural build-up of the lining under the stairs is given in Table 4.

**Table 4. Summary of layer attributes for temporary lining design case under MS**

| Layer No. | Thickness (mm) | Shotcrete Type | Reinforcement |
|---|---|---|---|
| 1 | 100 | S1* | Fiber |
| 2 | 100 | S1 | Fiber+mesh+L-bar |
| 3 | 100 | S2* | Lattice girder |
| 4 | 100 | S2 | Lattice girder |
| 5 | 100 | S2 | φ40 diam. bars |
| 6 | 100 | S1 | Fibre |

* S1 = steel fiber reinforced
* S2 = plain shotcrete

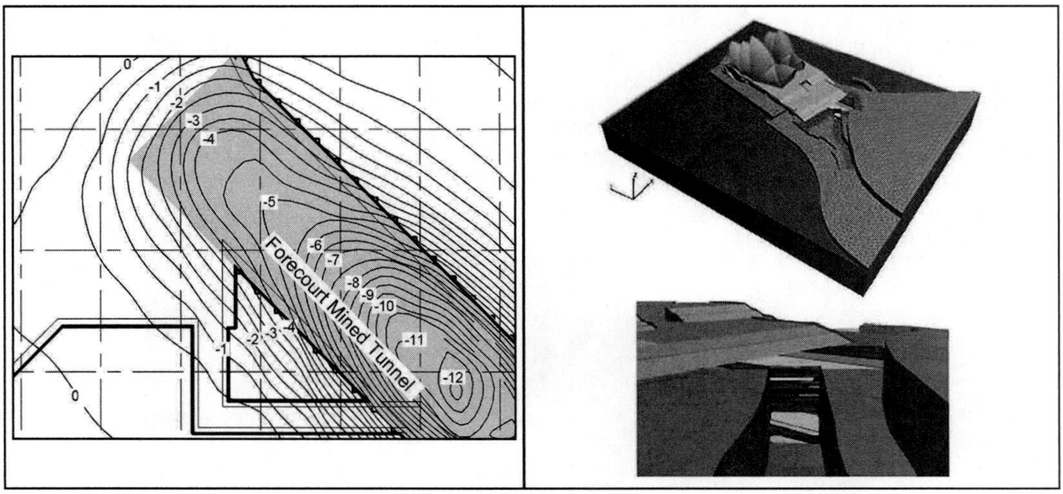


**Figure 3. Settlement contours (left) based on 2D and 3D analyses (right). Settlement of 5mm was predicted under the Monumental Stairs, compared to 12mm away from the stairs, where a less stiff support was used closer to the portal.**

### 3DEC Model Results

With the inclusion of the temporary support, the maximum top of rock displacement under the stairs was calculated as 5mm, as shown in Figure 3. In general, the 3DEC model produced similar in-tunnel convergence displacements compared to the 2D model. However, surface settlements were greater for the 3D model. This is likely a result of the major principal stress running parallel to the tunnel axis, which is largely ignored in the 2D plane strain model. In addition, the 3DEC model incorporates the Forecourt Tunnel portal, a boundary condition that introduces more relaxation than the 2D case.

Two separate in situ stress cases were evaluated in the 2D and 3D models; the first followed the recommendations of Pells (2004) and the second considered the notably high in-stress results obtained from hydrofracture tests in VAPS boreholes. Table 5 presents the sensitivity analysis results, the basis for the trigger levels used during excavation and to validate anticipated ground-structure behavior.

## CONSTRUCTION PROGRESS

### Forecourt Tunnel (Mined)

Commencing in August 2012, the ~40m section of mined tunnel was excavated using an S300 roadheader. After excavating a split heading through the first 3m of the portal area, the remainder was excavated full face with 1.5m advance lengths (Figure 4). Shotcrete was applied after mining to comply with personnel re-entry requirements (min. f'c=1 MPa) with subsequent installation of pattern rock bolts.

As discussed above, the temporary liner for the section of the Forecourt tunnel passing under MS was designed to be installed in three distinct passes, with the full 600mm thickness achieved no further than 3.75m from the face (Figure 5). Prior to commencing tunnel excavation, the Contractor requested to use a double corrosion protected DCP310 rock bolt under the MS in lieu of the RB550 in order to match the bolt with the drill rig available to the project. Reanalysis using 3DEC verified the DCP310

**Table 5. Results of sensitivity analysis of in situ stress profiles for Forecourt mined tunnel**

| Model | Stress Profile | Surface Settlement (Under stairs) | Surface Settlement (Under Forecourt) | Sv (mm) | Sh (mm) |
|---|---|---|---|---|---|
| UDEC | Pells (2004) | N/A | 5.5 | 8 | 10 |
| | VAPS hydro | | 5 | 8 | 12 |
| 3DEC | Pells (2004) | 5 | 12 | 10 | 13 |
| | VAPS hydro | 5 | 12 | 12 | 16 |

Sv = in tunnel vertical convergence
Sh = in tunnel horizontal convergence

**Figure 4. View into the Forecourt Tunnel during full-face roadheader excavation**

**Figure 5. Typical temporary support installation sequence for the section of Forecourt Tunnel under the MS**

was adequate, provided a decreased spacing of 750 × 1,000 (longitudinal/transverse) was adopted.

Under the MS through the heavily reinforced section John Holland worked with Arup to optimize the excavation and support sequence. Advance lengths were increased to 1.5m to allow a more efficient excavation cycle that reduced plant movement and interchange of activities such as shotcrete and steel installation. The 40mm diameter bars were prefabricated into panels and lifted into place. Edge boards were used to provide accurate control of the shotcrete profile and to ensure subsequent lattice girders and steel reinforcement could be installed to the design position in the concrete profile. Tunnel convergence and surface settlement over the tunnel were closely monitored and observed to be within the design predictions allowing the excavation to proceed.

### Opera Theatre Corridor

The Opera Theatre (now known as the Joan Sutherland Theatre) Corridor is an approximately 10m wide, 15m deep, 40m long excavation extending from the loading dock to Lift 21, a new lift for scenes and large goods. Excavation by hydraulic hammer began at Lift 21 from within the SOH and the top heading proceeded to the south. This required underpinning beams to be installed progressively to support the SOH ground level rooms as the excavation progressed. Flat jacks were utilized to negate movement of the existing structure. The bench was advanced from the loading dock using the S300 roadheader, which mined underneath a newly constructed stairwell and post-tensioned floor slab, back to Lift 21.

### Cross Cut and Concert Hall Tunnels

After benching the Opera Theatre Corridor, a Contractor-proposed Cross Cut access tunnel was advanced to the west with the S125 roadheader, to provide access to mine the Concert Hall Tunnel that was subsequently was mined to the south and north. These tunnels as well as the egress tunnel had a similar rectangular profile and were supported with pattern rock bolts and shotcrete.

## INSTRUMENTATION AND MONITORING

A robust instrumentation and monitoring program was specified for the VAPS project.

Instrumentation and monitoring on the project included 6 in-place (automated) inclinometers, 2 in-place borehole extensometers, 200+ surface and building monitoring points, arrays of tunnel and excavation deformation monitoring points, noise and vibration monitoring, geological mapping and crack monitoring.

An online monitoring website was maintained by John Holland's instrumentation subcontractor, Lynton Survey. Inclinometers and extensometers were automated and data was updated on the online system every 15 minutes. All other instrumentation was manually read with survey reports uploaded to the system daily. Green, amber and red trigger levels were assigned to each instrumentation point, with green representing 80%, amber 100% and red 125% of values calculated during design.

### Geological Mapping and Designation of Additional Support

Geological mapping was undertaken throughout the period and after each tunnel advance. The mapping gathered information regarding key parameters required for the validation of design or selection of rock support. Spot bolting was required on the existing rock wall adjacent to the access ramp. On rare occasion, the newly excavated vertical sidewalls of the ramp and loading dock revealed potential rock wedges/ blocks formed by an occurrence of subvertical joint intersection with bedding. Such unstable blocks were either scaled and removed, or supported

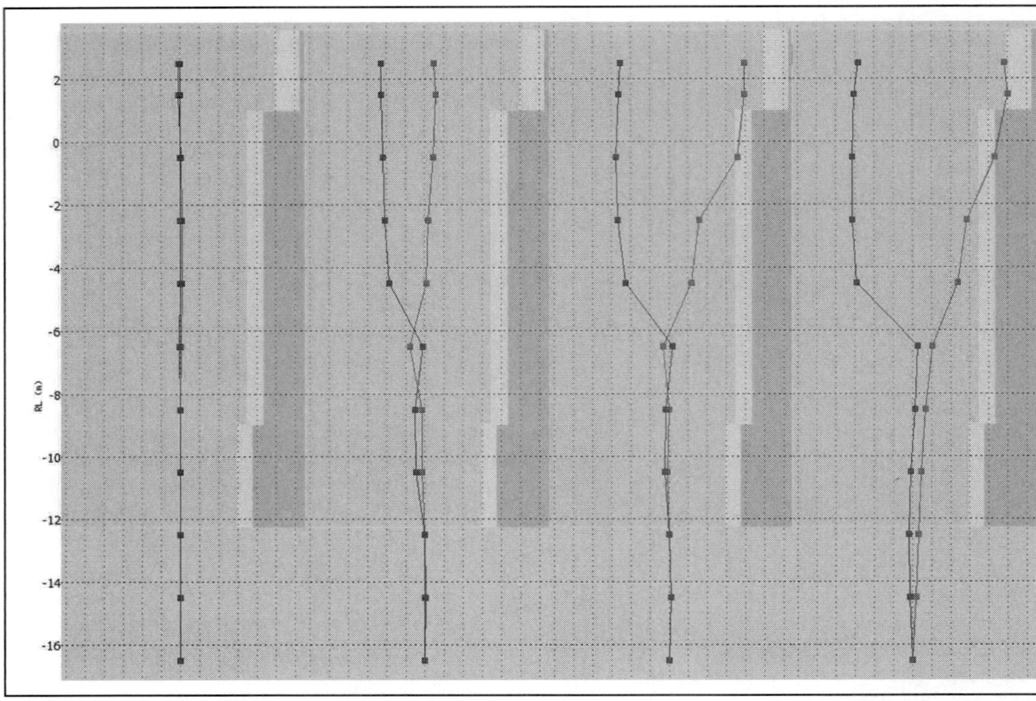


**Figure 6. Cumulative displacement profiles of Inclinometer 5 from before (July 2012) during (January and May 2013) and after (September 2013) excavation of the loading dock and TTB. Note close correlation of design (shaded area) and observed displacement. X-axis grid cell width is 2mm.**

with spot bolting. The sidewalls were in large part self-supporting as observed in similar sandstone cuts around the Sydney area.

Around the loading dock, several footings sit in close proximity to the edge of excavation. The rock subgrade was assessed and determined to meet the required bearing capacity, typically 3 or 5MPa, prior to placement of the footings. In some instances, rock bolts were installed under the footings to address the risk of parallel geological features which could potentially cause rock wedge or planar failures, or in one case to reinforce and anchor a temporary prop footing vulnerable to accidental impact.

### Management Action Team/Permit to Tunnel

A Management Action Team (MAT) with representatives from Arup, John Holland and Lynton Survey discussed the instrumentation results daily, identifying notable trends and responding to any alarm trigger breaches in accordance with plans presented in the contract documents. After a thorough review and risk assessment of instrumentation, ground behaviour, and relevant geological/structural observations, a permit to tunnel was issued. A valid permit is required for compliance with the *Workplace Health and Safety—New South Wales—Tunnelling Construction Code of Practice* and includes detailed information on the work area, construction sequence, constraints, ground support type, minimum support requirements for personnel re-entry and a summary of instrumentation observations.

### Extensometer Observations

Two borehole extensometers were installed to verify ground movements over the middle of the Forecourt tunnel crown. In both cases, ground movements observed were less than those anticipated. This result is principally attributed to two reasons: (1) as discussed above, JHG opted to install rock bolts at a spacing half that as specified by the design, stiffening the response of the rock beam; (2) tunnel face advances in the Forecourt tunnel, specifically those under the MS section, achieved higher shotcrete strengths than assumed in the design prior to subsequent excavation. Ongoing back analysis seeks to assess how actual shotcrete strengths compare to the model.

### Inclinometer Observations

Excavation of the north wall of the Truck Turning Bay (TTB) and the east wall of the loading dock created the longest sandstone wall on the VAPS project

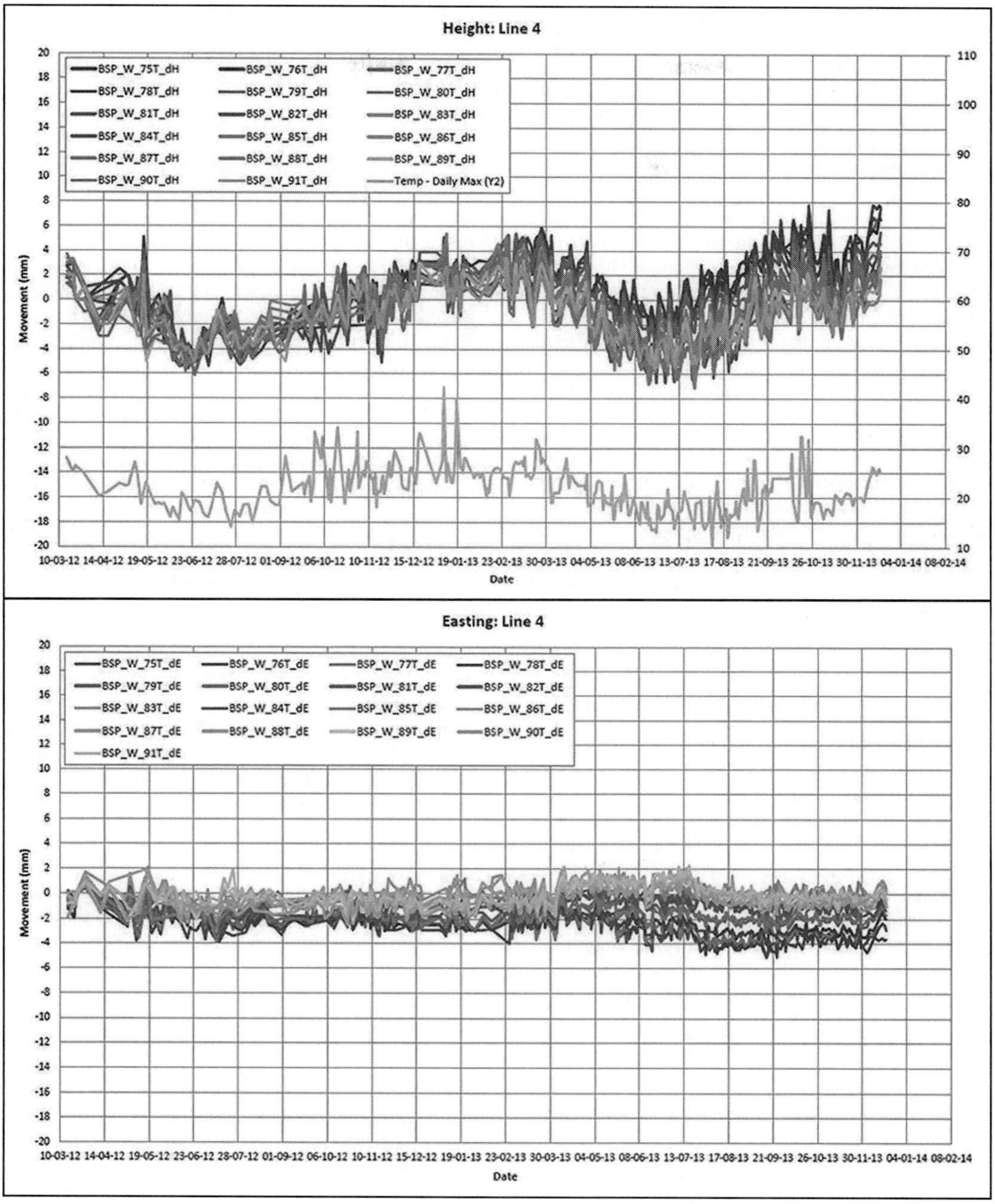


**Figure 7. Plot of change in vertical height (top, upper line), air temperature (top, lower line) and eastings (bottom) of survey targets on the underside of the Monumental Stairs from May 2012 through December 2013. A clear correlation between vertical movement and temperature is evident whereas there is none between eastings and temperature. A slight negative easting trend is observed, indicating a lateral structural response to excavation.**

and the area anticipated to experience the most lateral ground movement, as the wall runs generally perpendicular to the NE-SW oriented principal horizontal stress direction. Inclinometer IN5 was installed just beyond this wall to monitor the ground relaxation during mining. A clear response to the excavation was observed on IN5, the top sensors of which logged a deflection of 10mm towards the excavation, exactly matching the predicted value for this instrument (Figure 6). No notable movements have been recorded by IN-5 since completion of the loading dock and TTB excavations to full depth, confirming that rock excavation and immediate stress re-distribution are responsible for the observed deflection.

### Tunnel Deformation Monitoring Points

Three point arrays (sidewall, crown, sidewall) of tunnel deformation monitoring points (TDMPs) were installed progressively behind the advancing face. TDMPs did not show significant convergence of any of four tunnels monitored. The majority of sandstone relaxation and associated ground movement likely occurs soon after excavation. Consequently, despite being installed as soon as practical after advancing the tunnel face, the TDMPs likely did not capture much of the tunnel convergence that occurred.

### Excavation Deformation Monitoring Points

Arrays of EDMPs were installed progressively with the loading dock and Opera Theatre Corridor excavations. As with the TDMPs, it is believed that much of the rock relaxation had occurred by the time is was safe to install the EDMPs. However, around the loading dock excavation, the majority of EDMPs showed minor movement into the excavation. EDMPs and in-place inclinometers generally indicated similar movement directions, but varying magnitudes of deflection, within the predicted value threshold, which is solid evidence of a highly anisotropic insitu stress state.

### Structural Monitoring

Over 200 survey targets were installed on the exterior and interior portions of the SOH structure within the zone of the influence of the VAPS project. The vast majority of points on the interior of the SOH never recorded an alarm reading. For the vast majority of points that recorded a trigger alarm breach, the data were shown to be a result of survey error or negligible structural movement.

A number of target arrays were installed on the exterior of the SOH, specifically to the underside of the MS. These exterior points have almost continuously indicated movements, particularly in settlement/ heave, which breach the specified triggered levels for both positive and negative changes. From the data collected to date, clear correlation of structural movements and air temperature exists (Figure 7, top). A few structural monitoring points on the MS indicate movements not attributed to air temperature. These trends are considered to be a result of excavation activities and typically involve small lateral movements of less than 5mm (Figure 7, bottom), which is approximately 50% of those predicted. The movements are likely attributed to manifestation of stress release along low friction bedding planes. However, use of generally lower bound shear strength parameters likely played a role in overestimating movements. Further back analyses would be valuable to understand the role that joint friction and stiffness played in matching these measured movements.

## SUMMARY

The SOH is a unique environment in which to design and construct an underground excavation. The complexity and sensitivity of the existing structure, excavation shape and geotechnical conditions drove the development of robust ground support designs, innovative construction sequences and a comprehensive instrumentation program. Collaboration between the designer, constructor and client allowed optimization of the support and sequencing and successful completion of the excavation with no disruption to the Sydney Opera House performance schedule.

## REFERENCES

Asche, H.R. and Lechner, M.K. 2003. Design for the Cross City Tunnel, Sydney. *RETC.* Seattle. p. 122–139.

Diederichs, M.S. and Kaiser, P.K. 1999. Stability of large excavations in laminated hard rock masses: the voussoir analogue revisited. *Inter. Journal of Rock Mechanics and Mining Sciences.* No. 36. p. 97–117.

Palmström, A. 2005. Measurements of and correlations between block size and rock quality designation (RQD). *Tunnels and Underground Space Technology.* v. 20. p. 362–377.

Pells, P.J.N. 2004. Substance and mass properties for the design of engineering structures in Hawkesbury Sandstone. Australian Geomechanics. v. 39. No. 3. p. 1–21.

# Modeling of Precast Concrete Tunnel Lining Subjected to Fire Event

**Jimmy Susetyo, Reza Shobayry, and Tomas Gregor**
Hatch Mott MacDonald

**ABSTRACT:** Numerical simulations using FLAC3D were performed to evaluate the response of Precast Concrete Tunnel Lining (PCTL) of the Eglinton-Scarborough Crosstown project to a major fire event. The behavior of the concrete and the steel reinforcement was explicitly modeled using Eurocode 2 Part 1.2 provisions to obtain a more accurate prediction of the PCTL response. Pore-pressure build-up in the concrete was also simulated to predict the occurrence of concrete spalling.

This paper presents the methodology and assumptions used in the simulations. A verification study was performed to compare the prediction made by the model and the experimental results by Phan (2008). The model was found to produce a slightly conservative prediction of the PCTL response to the major fire event when compared to experimental results.

## INTRODUCTION

As widely known, exposure of a reinforced concrete (RC) member to an extremely high temperature will severely impact the performance of the member. The heating of the member will result in a degradation of strength and stiffness of the concrete and the reinforcing steel. Moreover, additional stresses and strains will occur in the member as a result of heating exposure, and spalling may also occur. These conditions may severely affect the structural integrity of the member.

Metrolinx (MX) is currently undertaking the construction of a new Light Rail Transit (LRT) line in Toronto, Ontario. The line, named Eglinton-Scarborough Crosstown (ESC), will be constructed using Tunnel Boring Machines (TBMs) and lined with single-pass precast concrete tunnel lining (PCTL). As part of the design of the PCTL, Hatch Mott MacDonald (HMM) performed an evaluation of the structural integrity of the PCTL in the event of a major fire in the tunnels.

The PCTL's structural performance during a major fire event was evaluated by performing numerical simulations using FLAC3D, a finite-difference program for engineering mechanics computation developed by ITASCA Consulting Group (Itasca 2009). The simulations include an explicit modeling of the concrete and the reinforcing steel. The behavior of the concrete and steel reinforcement under high temperature was simulated using the models provided in Eurocode 2 Part 1.2 (2004) in order to obtain a representative behavior of the PCTL under fire. In addition, pore-pressure build-up caused by heating of moisture in the concrete was also simulated to predict the occurrence of concrete spalling.

This paper presents a method to predict the response of PCTL during a major fire event. The methodology and assumptions are discussed, as well as the simulation results. Comparison between the model predictions and the experimental results obtained by Phan (2008) is also presented.

## SIGNIFICANCE

The method described in this paper incorporates an explicit modeling of the behavior of concrete and reinforcing steel in the PCTL when subjected to high temperature and ground load, as well as a consideration of concrete spalling due to pressure build-up in the concrete. Full ground-structure interaction was employed in order to simulate the actual stresses exerted onto the PCTL by the ground.

## EGLINTON-SCARBOROUGH CROSSTOWN (ESC)

The Eglinton-Scarborough Crosstown project is a new LRT line that will travel across Eglinton Avenue between Mount Dennis and Kennedy Station (see Figure 1). It is considered to be the largest transit expansion in Toronto, Canada, to date. The total length of the line is 19 km (11.8 miles), including a 10 km (6.2 miles) twin tunnels between Keele Street and Laird Drive.

The internal diameter of the ESC twin tunnels is 5.75 m. The PCTL rings consist of four 67.5° parallelogram segments and two 45° trapezoidal segments. The nominal thickness and width of the segments are 250 mm and 1500 mm, respectively. The PCTL is made of concrete with a compressive strength of 60 MPa. Twenty MD154 steel bars are provided in

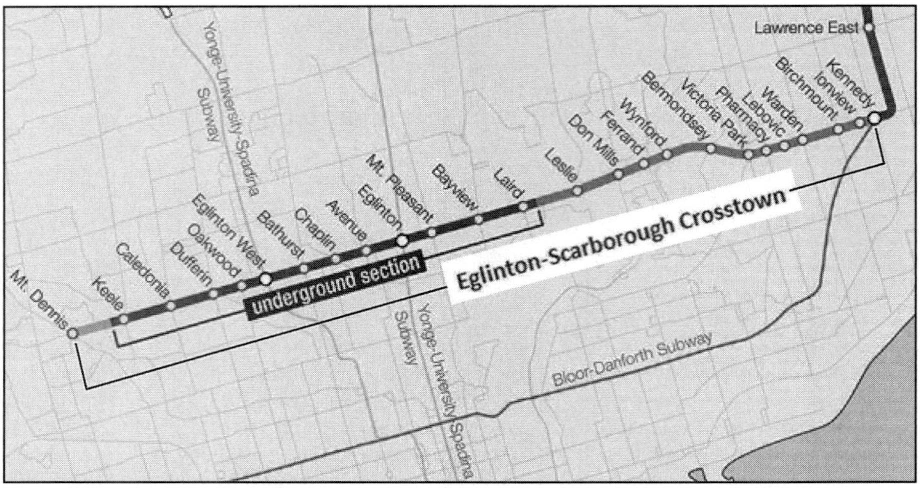


**Figure 1. Eglinton-Scarborough Crosstown (Metrolinx 2013)**

the circumferential direction and MD90.3 steel bars with an average spacing of 121 mm are provided in the longitudinal direction. In addition, two rows of MD58.1 steel ties spaced averagely at 100 mm are provided at the radial joints and the circumferential joints. The cross-sectional area of MD154, MD90.3, and MD58.1 steel bars are 154 mm$^2$, 90.3 mm$^2$, and 58.1 mm$^2$, respectively. The yield strength of all steel reinforcement is 450 MPa. In addition to the steel reinforcement, 1 kg/m$^3$ of polypropylene fibers are also added to the concrete to increase the concrete fire resistance.

## NUMERICAL SIMULATION

The FLAC3D model used in the numerical simulations is depicted in Figure 2, and represents a 6 m long portion of the tunnel. The model is mechanically restrained against translations normal to the face at its external boundaries except at the top boundary, which is unrestrained. Thermally, the model's external boundaries are set as adiabatic boundaries, while the temperature at the PCTL intrados surface is set to a pre-determined temperature. The intrados temperature will be evaluated and adjusted during the analytical run, in accordance with the temperature development over time during a major fire event.

FLAC3D solid elements were used to model the ground and the concrete, and linear-elastic cable elements were used to model the reinforcing steel. Each PCTL segment was modeled explicitly, with the PCTL rings connected one to another through sixteen dowels uniformly spaced along each circumferential joint. Compression-only frictional interface elements were provided at the extrados, the circumferential joints, and the radial joints of the PCTL to model the stress transfer between ground and the PCTL and between PCTL segments.

The tunnel springline and the groundwater table are located 33 m and 30 m below the ground surface, respectively. The stratigraphic profile of the soil is listed in Table 1, and the mechanical and thermal properties of the soil are listed in Table 2.

Coupled thermal-mechanical-fluid process was used in the simulations. Pore-pressure generation in the ground as a result of volumetric straining was considered. A linear elastic-plastic constitutive model with Mohr-Coulomb failure criterion was used to evaluate the mechanical behavior of the ground. In modeling the concrete, FLAC3D's strain-softening constitutive model was employed. The cohesion and friction angle of the concrete material were defined using a piecewise-linear hardening/softening function of the plastic shear strain, and the tensile strength was also defined using a piecewise-linear softening function of the plastic tensile strain. These strength parameters, as well as the modulus of elasticity and the thermal properties, were adjusted during the analysis runtime based on the calculated temperature, total plastic shear strain, and total plastic tensile strain, to conform to the concrete stress-strain relationships. To accurately model the behavior of the reinforcing steel, the strength parameters, the modulus of elasticity, and the thermal properties of the linear-elastic cable elements representing the steel were also adjusted during the analysis running time based on the calculated temperature. For all elements, isotropic heat conduction was used to model the heat transfer.

Loads considered in the analyses included dead loads (in-situ soil stresses and self-weight of the soil and PCTL), hydrostatic pressure, and fire loading.

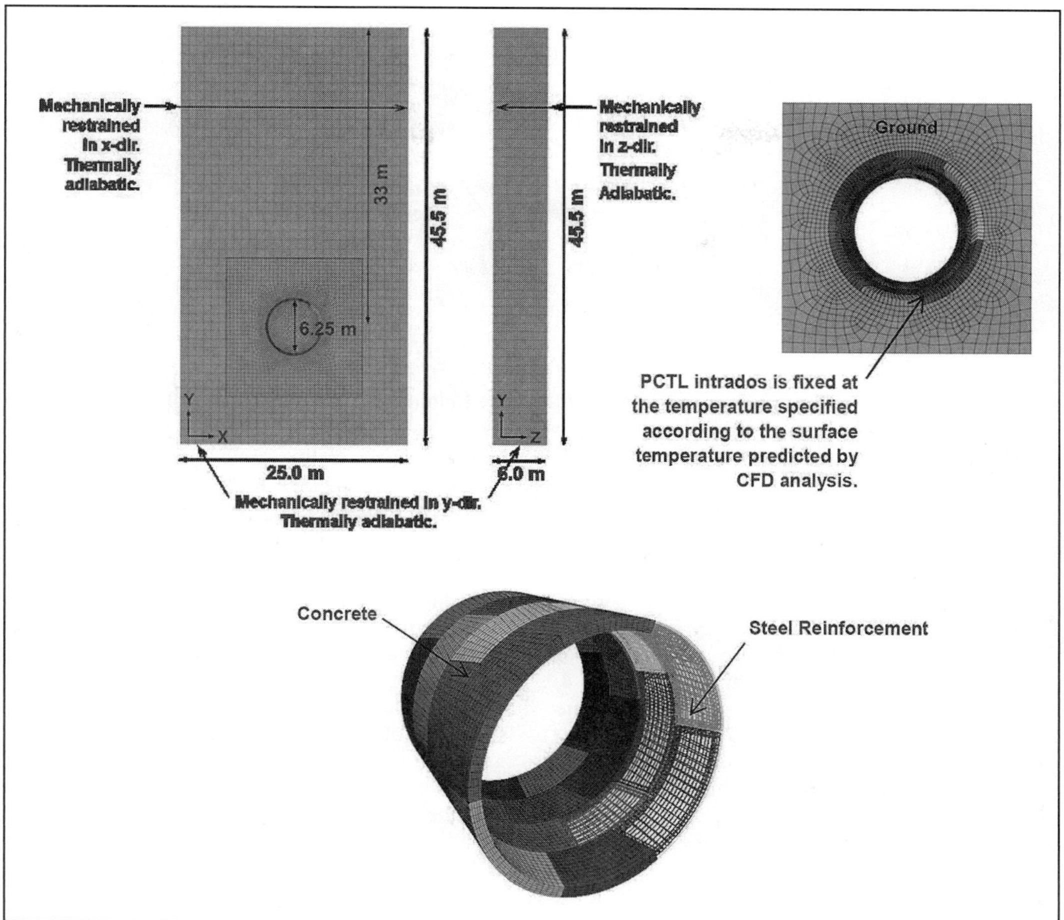


**Figure 2. FLAC3D model**

**Table 1. Soil stratigraphy**

| | Depth (m) | | | |
|---|---|---|---|---|
| **Soil Unit** | **Top** | **Bottom** | **Tunnel Springline** | **Groundwater** |
| Sand/Silt | 0.0 | –14.0 | | |
| Till | –14.0 | –29.0 | –33.0 | –30.0 |
| Sand/Silt | –29.0 | –54.0 | | |

**Table 2. Geotechnical properties**

| Properties | Till | Sand/Silt |
|---|---|---|
| Dry unit weight (kN/m$^3$) | 16.1 | 16.1 |
| Saturated unit weight (kN/m$^3$) | 21 | 20 |
| Elastic modulus (MPa) | 20 | 20 |
| Effective cohesion (kPa) | 0 | 0 |
| Effective Friction Angle (°) | 28 | 32 |
| Poisson's Ratio | 0.3 | 0.3 |
| Porosity | 0.5 | 0.4 |
| Effective horizontal/vertical stress ratio | 1.1 | 1.1 |
| Thermal conductivity (W/m.°C) | 1.3 | 1.3 |
| Specific heat (kJ/kg.°C) | 0.92 | 0.92 |
| Coefficient of thermal expansion (/°C) | $15\times10^{-6}$ | $15\times10^{-6}$ |

The fire loading was developed under a separate study using Computational Fluid Dynamic (CFD) analyses, in which two cases of with and without mechanical ventilation were considered; the case without mechanical ventilation generated the highest load on the PCTL, as expected. The fire loading was modeled as the intrados surface temperature that varied over the course of the fire event, as illustrated in Figure 3. The duration of the fire was assumed two hours.

Heating of moisture in the concrete pores may cause the moisture to undergo transformation from liquid to gaseous phase. This transformation will lead to steam pressure build-up inside the concrete. If this internal pressure build-up cannot be successfully relieved, concrete surface may spall. In order

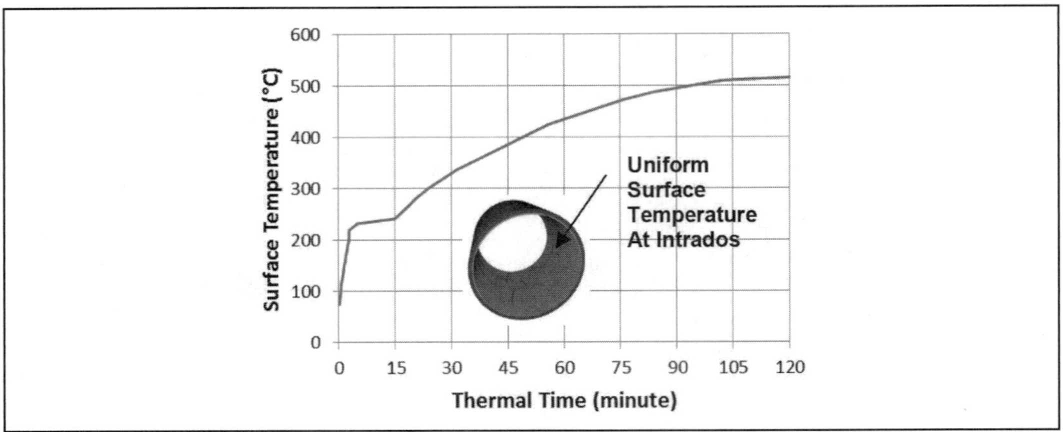


**Figure 3. Development of the PCTL Intrados temperature during a 2-hr fire**

to simulate this phenomenon, pore-pressure was applied to the concrete elements. The magnitude of the applied pore-pressure varied depending on the calculated element temperatures, and was derived from the experimental study done by Phan (2008). The variation of the pore-pressure as a function of the concrete temperature is illustrated in Figure 4. Spalling was considered to occur when the concrete element was failing in tension and the maximum shear strain in the element exceeded 0.1 (a large though rather arbitrary chosen value). The spalling element was then deleted, and the element under the spalled element would be subject to the surface temperature. Note that although Phan (2008) observed that the pore-pressure varied depending on its location with respect to the heated surface, the pore-pressure magnitude used in the model was assumed independent of its location. Note also that a concrete moisture content of 1.5% was assumed.

Provisions of Eurocode 2 Part 1.2 (2004) were used to describe the variation in the compressive and tensile stress-strain relationships and in the thermal properties of the concrete and the reinforcing steel under high temperature; the variations are illustrated in Figure 4.

## MODEL VERIFICATION

In order to evaluate the accuracy of the spalling prediction made using the methodology described in this paper, Specimen I-1.5-13-M-5 tested by Phan (2008) was modeled. Specimen I-1.5-13-M-5 was a 200×200×100 mm concrete block with a concrete compressive strength of 75.3 MPa and containing 1.5 kg/m$^3$ of 13 mm long polypropylene fibers. During the test, the specimen was insulated on all faces but the front face, to which heat was applied with a rate of 5°C/min. Details of the test setup can be found in Phan (2008).

In the study, Phan found that the specimen survived the heating exposure without spalling. The presence of the polypropylene fibers helped in providing an interconnecting network that enabled steam movement through the concrete. The pore-pressure measured during the test is illustrated in Figure 4; this pore-pressure was applied to the concrete elements during the course of the verification study. All other strength, deformation, and thermal parameters were calculated as previously described.

In contrast to what was observed during the test, the model predicted the spalling of concrete after 144 minutes of heating exposure. The variation of the concrete tensile strength, concrete principal tensile stress, and pore-pressure predicted at the concrete surface during the analysis runtime are shown in Figure 5.

As illustrated in the figure, an increase in the surface temperature resulted in an increase in the concrete principal tensile stress. As the temperature increased, the concrete tensile strength deteriorated and pore-pressure was generated in the concrete, in accordance with the prescribed behavior. At 100°C, just before the concrete tensile strength dropped to 90% of the characteristic tensile strength, the concrete tensile stress was predicted to be 2.43 MPa, 85% of the characteristics tensile strength. When the concrete tensile strength dropped, the tensile stress exceeded the strength and redistribution of stress occurred until equilibrium was attained. Redistribution of stress also occurred when the concrete became plastic (i.e., when the tensile stress exceeded the tensile strength and the tensile strength dropped to its nominal residual value). These processes occurred repeatedly throughout the duration of the heating exposure. Spalling was triggered when the sum of the pore-pressure and the tensile stress exceeded the tensile strength; equilibrium could no

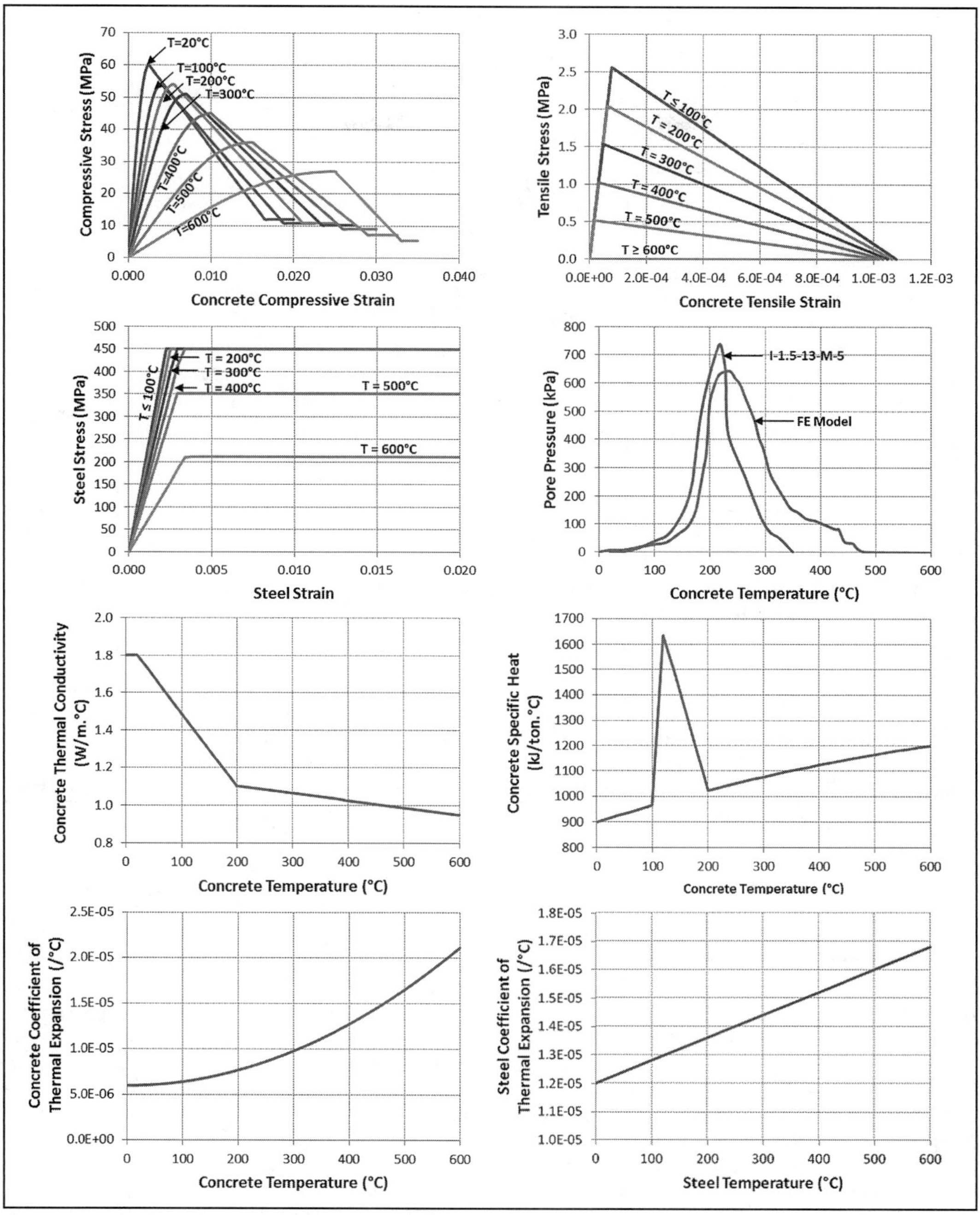


**Figure 4. Constitutive models for concrete and steel used in the analyses**

longer be maintained when this occurred. In the analysis, spalling was predicted to occur when the temperature reached 310°C.

There are three possible reasons that may explain the discrepancy between the experimental results and the analysis prediction. First, due to lack of available information, estimations were made on the tensile strength, modulus of elasticity, and thermal properties of Specimen I-1.5-13-M-5 using available empirical relationships. Second, cracks may have developed during the heating exposure. The cracks may have helped in dissipating the pressure build-up

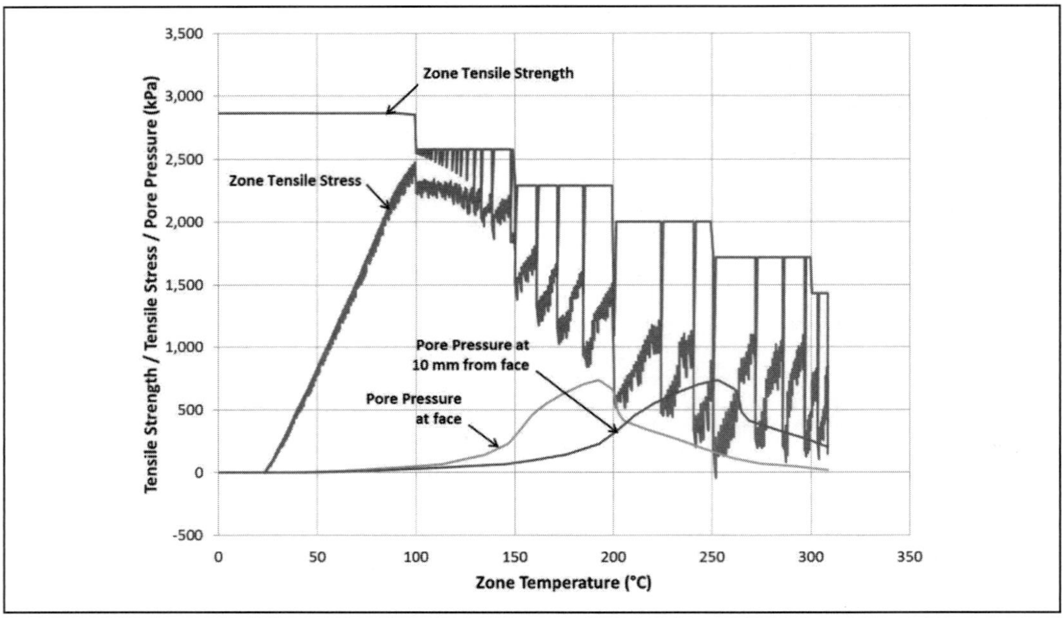


**Figure 5. Development of concrete tensile strength, principal tensile stress, and pore pressure at the heated face of specimen I-1.5-13-M-5**

**Table 3. Maximum PCTL internal forces predicted by the analysis**

| | Concrete | | | | | | | | | | | |
|---|---|---|---|---|---|---|---|---|---|---|---|---|
| Heat Time | Circ. Axial Force (kN/m) | | Circ. Bending Moment (kNm/m) | | Long. Axial Force (kN/m) | | Long. Bending Moment (kNm/m) | | Max. Shear Force | Steel Stress (MPa) | | |
| (min) | Max | Min | Max | Min | Max | Min | Max | Min | (kN/m) | Circ. | Long. | Ties |
| 0 | –1,492 | –1,963 | 65.5 | –59.1 | –190 | –313 | 14.8 | –14.7 | 105 | 68.2 | 10.1 | 4.5 |
| 30 | –1,369 | –2,049 | 125.3 | 1.4 | –624 | –1,315 | 87.7 | 39.8 | 213 | 210.4 | 33.1 | 23.3 |
| 60 | –1,065 | –2,151 | 171.9 | 51.7 | –997 | –2,684 | 172.0 | 83.6 | 425 | 331.3 | 82.2 | 64.7 |
| 90 | –1,143 | –2,618 | 176.5 | –8.3 | –991 | –2,843 | 219.9 | 20.1 | 480 | 390.1 | 188.7 | 119.1 |
| 120 | –1,029 | –2,721 | 63.0 | 53.2 | –1,108 | –3,323 | 46.0 | –42.7 | 548 | 233.2 | 237.9 | 143.2 |

faster than the rate of steam leaking through pores alone; such allowance has not been reflected in the model. Third, the pore-pressure magnitude used in the model was assumed independent of its location, in contrary to what was observed by Phan (2008). Nevertheless, it is believed that the methodology used to predict the response of concrete exposed to heat are appropriate and will yield slightly conservative results.

## DISCUSSION OF ANALYSIS RESULTS

The results of the analyses performed to evaluate the ESC PCTL structural integrity in the event of a major fire are summarized in Table 3. The PCTL forces were calculated by integrating the element stresses across the lining thickness. As indicated in the table, initially, an increase in concrete temperature resulted in an increase in the axial compression and bending moment in the PCTL. The stresses in the circumferential and longitudinal reinforcement, as well as in the ties, also increased. When the concrete has been exposed to heating for 68 minutes, the sum of the pore-pressure and the calculated concrete tensile stress in some portion of the PCTL exceeded the concrete tensile strength. This triggered concrete spalling of the portion, as indicated in Figure 6. Spalling of concrete progressed throughout the remaining duration of heating exposure. The occurrence of spalling reduced the PCTL thickness and thereby increased the flexibility of the PCTL ring. As a result, reductions in bending moment and the reinforcing steel stresses were predicted as the PCTL was further

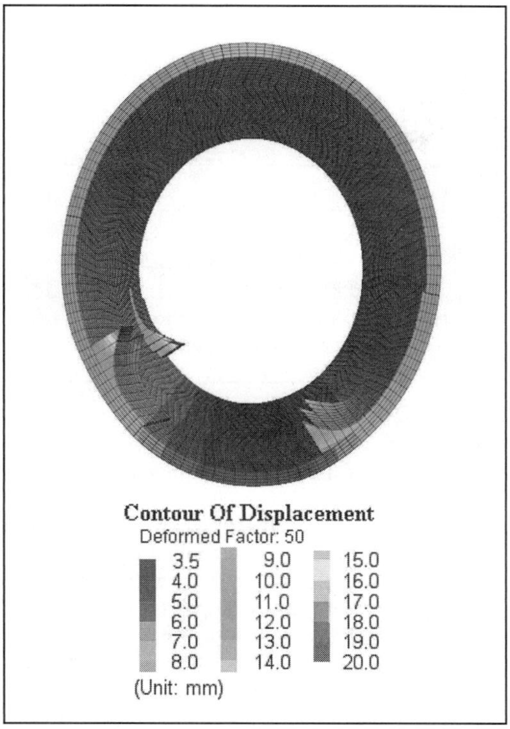


**Figure 6. Prediction of spalling at 68 minutes of heating exposure**

exposed to heating. The axial compressive force in the PCTL, however, continued to increase with further heating.

Despite the occurrence of spalling, the PCTL was predicted to be able to sustain the applied loading until the termination of heating exposure at 2 hours. Only the first intrados layer (i.e., the first 50 mm layer exposed to heating) was predicted to spall; no progressive spalling across the lining thickness was predicted. The PCTL internal forces were predicted to still lie within the concrete capacity curve, as indicated in Figure 7, and all reinforcing steel stresses were predicted to remain below the yield strength adjusted to account for the steel temperature.

The deformation of the PCTL throughout the duration of heating exposure is illustrated in Figure 8. Prior to spalling, exposure to heating resulted in expansion of the tunnels. Just before spalling occurred, the net diameter expansions (excluding deformation prior to heating exposure) across the springlines and across the crown and invert were predicted to be 1.78 mm and 2.28 mm, respectively. After the onset of spalling, an inward movement of the springline and outward movement of the crown and invert were predicted. At the end of the heating exposure, the net diameter expansion across the springline was reduced to a net diameter contraction of 0.66 mm, whereas the net diameter expansion across the crown and the invert was increased to 4.95 mm.

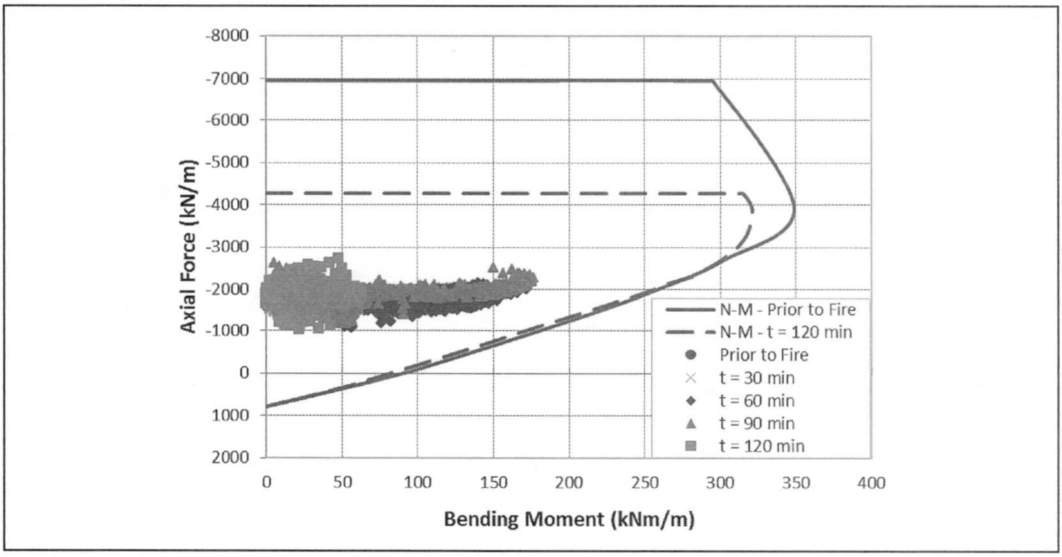


**Figure 7. PCTL predicted axial force and bending moment**

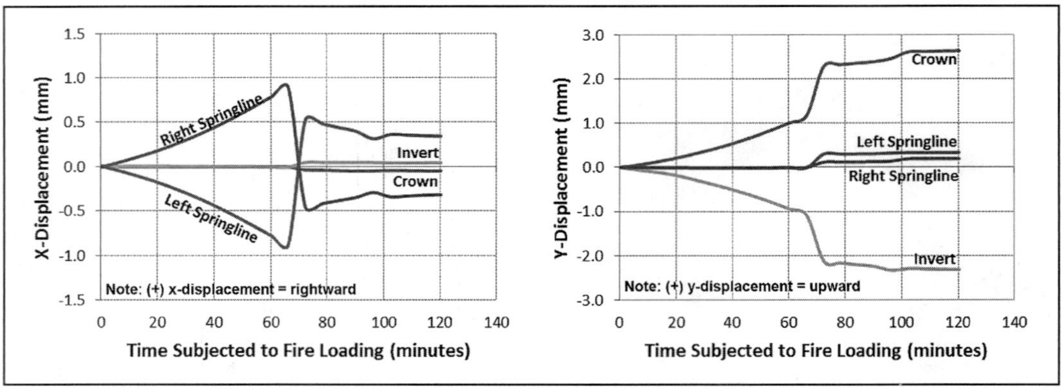


**Figure 8. Deformation of PCTL due to fire loading only**

## CONCLUSIONS AND RECOMMENDATIONS

Numerical simulations were performed to evaluate the PCTL structural capacity of the Eglinton-Scarborough Crosstown twin tunnels during a fire event. Based on the analysis results, the following conclusions can be made:

- The methodology used in the analyses is considered to yield slightly conservative results, based on the results of the verification study done on Specimen I-1.5-13-M-5 tested by Phan (2008). Whereas spalling was not observed during the experiment, the model predicted the occurrence of concrete spalling after over two hours of fire exposure.
- The discrepancy between the test result of Specimen I-1.5-13-M-5 and the model prediction may be caused by:
  - The difference between the actual concrete properties and the properties estimated using the models provided in Eurocode 2 Part 1.2 (2004).
  - The lack of consideration of crack formation during the heating exposure that may have helped in dissipating the pressure build-up faster than the rate of steam leaking through concrete pores alone.
  - The assumption that the magnitude of pressure build-up is independent of its location, in contrary to what was observed in the study performed by Phan (2008).
- In the PCTL model, spalling was predicted to occur after 68 minutes of heating exposure. However, the PCTL was predicted to be able to sustain the applied loading and to remain stable until the termination of heating exposure at 2 hours, despite the occurrence of spalling. Spalling depth was predicted to be limited to the first 50 mm layer of concrete exposed to heating.
- Although spalling was predicted to not compromise the PCTL structural integrity, the damage due to spalling would have to be repaired to ensure the long-term structural integrity and durability of the PCTL.

## REFERENCES

European Committee for Standardization, 2004. "Eurocode 2: Design of Concrete Structures, Part 1–2: Structural Fire Design," EN 1992-1-2, 97 pp.

Itasca Consulting Group, 2009. "FLAC3D—Fast Lagrangian Analysis of Continua in 3 Dimensions User's Guide."

Metrolinx, 2013. "Eglinton Crosstown," http://www.thecrosstown.ca/the-project, Last Accessed: December 2013.

Phan, L. T., 2008. "Pore Pressure and Explosive Spalling in Concrete," Materials and Structures, Vol. 41, Issue No. 10, December 2008, pp. 1623–1632.

# Second Avenue Subway Project: Design and Construction of 72nd Street Station and G3/G4 Cavern Final Linings

**Michael Voorwinde**
AECOM

**Richard Giffen**
Arup

**Robert Powers**
Schiavone Construction Company

**ABSTRACT:** This paper will focus on the structural design and construction of the final lining for a mined cavern underground subway station at 72nd Street in Phase 1 of the Second Avenue Subway Project, located in Manhattan. Major topics covered will include: the geometric design, the selection of drained versus undrained structures, consideration of rock, hydrostatic and temperature loading, modeling techniques and structural analysis, groundwater pressure relief system details, and constructability. The discussion will also explore engineering work undertaken to address changes and challenges during construction from the perspective of both the Designer and Contractor. These include reanalysis of the liner to address areas of over-break, provision of additional access, and modifications to the groundwater pressure relief drainage details to suit the Contractor's preferred construction methods.

## INTRODUCTION

The Second Avenue Subway will be the first major expansion of New York's subway system in over 50 years. To be constructed in four phases, it will run from 125th Street in Harlem to Hanover Square in the Financial District along the east side of Manhattan (see Figure 1). For most of its length, the alignment is directly under Second Avenue.

Designed by a joint venture of AECOM and Arup (AAJV) for Metropolitan Transportation Authority Capital Construction (MTACC), Phase 1 is under construction and consists of twin bored tunnels, three new stations at 96th Street, 86th Street and 72nd Street respectively, and major modifications to an existing station at 63rd Street. The new station structures are either cut-and-cover construction or mined rock caverns depending on elevation and ground conditions. There are 10 construction contracts in Phase 1, with multiple contracts for each station. This paper covers the design and construction of the final linings in mined rock cavern for the new 72nd Street Station and in the tunnels connecting this station to the existing station at 63rd Street, which are named after their rail track designations as G3/G4 (see Figure 2).

The 72nd Street Station structure, under Second Avenue between 69th and 73rd Streets, includes a main station cavern of 64'-10" internal span with a full length mezzanine housing public space and ancillary areas. At the north and south ends of the station are smaller crossover caverns. Nine smaller adits, with widths varying from 8'-0" to 32'-0", have been constructed for entrances, egress, utilities and ventilation purposes and these connect the main station cavern to adjacent cut-and-cover street entrances and ancillary structures, most off of the Second Avenue right-of-way. Tunnels on the G3 and G4 track alignments each include turnout caverns tapering in width from 45'-2½" to 32'-6" with provision for future expansion further south along Second Avenue in Phase 3 of the project.

## CONSTRUCTION OVERVIEW

Construction of the 72nd Street Station and G3/G4 Tunnels began in late 2010 with the award of Contract C-26007 to a joint venture of Schiavone, J.F. Shea and Kiewit (SSK) for $447 million. Parsons Brinkerhoff (PB) is acting as the Consultant Construction Manager. The contract schedule spanned 37 months, with 23 for excavation during which time 185,000 cubic yards of rock was excavated from the shafts, caverns and drill-and-blast tunnels, and another 14 months to complete the final lining installation in the four caverns and two running tunnels during which 12 million pounds of

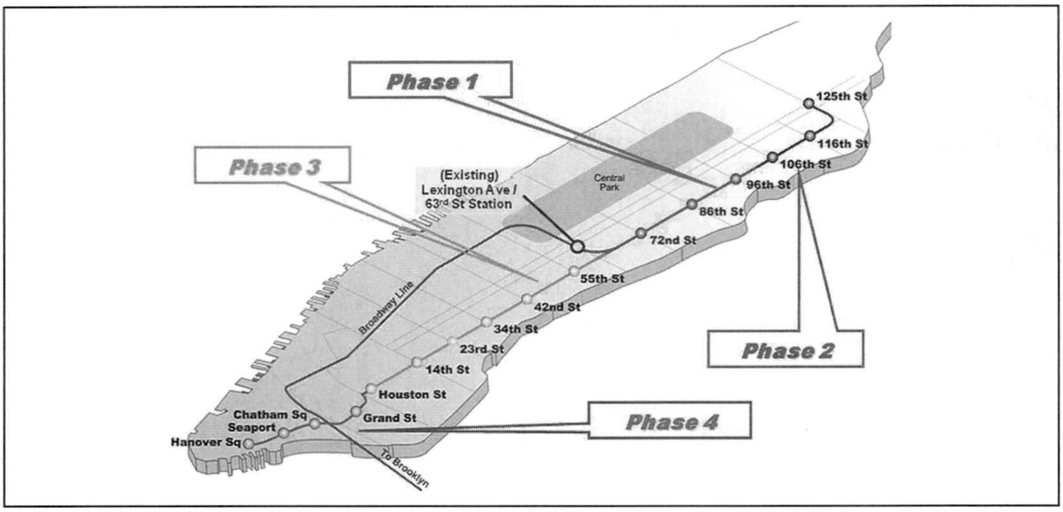


**Figure 1. Project map**

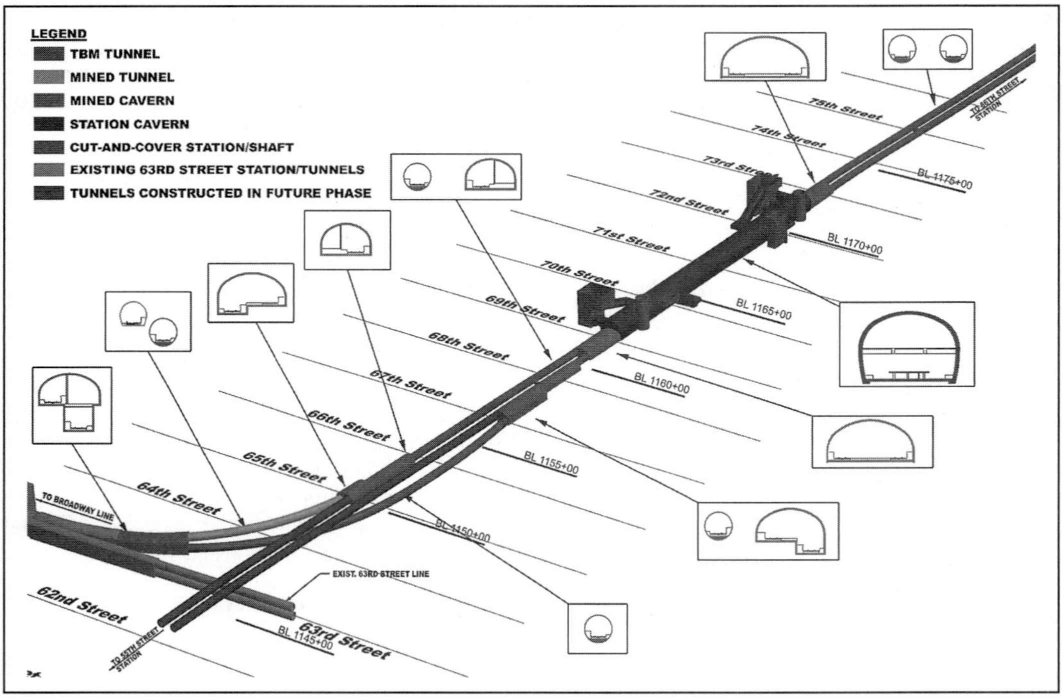


**Figure 2. Isometric of 72nd Street station and the G3/G4 tunnels**

reinforcing steel was installed with 84,000 cubic yards of concrete placed. The first Contract milestone was reached when the northern half of the station lining was completed in September 2013. The last final lining pour was in November of 2013 and substantial completion was scheduled to be achieved in mid-January 2014.

## FINAL LINING DESIGN

A common design objective for the cast-in-place reinforced concrete cavern final linings was minimizing rock excavation volume. With relatively low rock permeability, the below ground structures in rock were predominantly designed as drained, reducing excavation volumes without causing excessive

**Figure 3. Architectural rendering of station mezzanine with exposed coffers in cavern final lining**

drawdown or large water flows. Although a groundwater pressure relief system needs to be provided and maintained, this was found to be more economical than thicker concrete linings and additional rock excavation. The final lining design for the public station cavern incorporated coffers recessed into the exposed architectural concrete visible from the mezzanine below (shown in Figure 3).

### Loading

The cavern final liner arches were designed for dead load, rock loads, groundwater loads and temperature effects. Live loads, including equipment and train loading, and superimposed dead loads were also considered on the inverts and mezzanines where applicable. The geology in the proximity of the site consists of shallow fill above Manhattan Schist. Rock loading on the cavern arches was determined by extensive empirical analysis and discontinuum modeling using Universal Distinct Element Code (UDEC) and other software to mimic the rock jointing indicated in the rock core data, which also considered overburden and building surcharges. The rock load considered for each cavern section varied with rock quality, and was considered buoyant when applied with water pressure. Test wells indicated a normal high water level several feet below the existing grade, and a groundwater pressure relief system was employed to limit hydrostatic pressures. Drained liners with a drainage composite layer behind the PVC waterproofing membrane were designed for 25% of full hydrostatic groundwater pressure. Temperature gradients relative to liner thickness were also considered for both increasing and decreasing temperature load cases.

### Structural Analysis

STAAD Pro software was used for the structural analysis of the cavern final linings. Two-dimensional sections modeling a 1-foot strip were used for each cavern configuration. Compression-only springs were used for supports, with rock and the drainage composite acting in series. For sections with mezzanines, beam stiffness and loading was scaled to represent the beam spacing. These 2D models were supplemented by 3D analysis in areas with complex geometry such as adit penetrations. For the station cavern, the mezzanine will be constructed under a future contract so the lining design had to accommodate a cavern with and without the mezzanine beams in place.

### Groundwater Pressure Relief System

Groundwater pressure relief under the invert slab is provided by a layer of crushed stone with a network of perforated PVC pipes at regular intervals. Larger ductile iron pipes under each track carry the water to an ejector pit. Additionally, perforated pipes run behind the footings to collect groundwater from the drainage composite across the arch. Rather than use a floating invert slab, the cavern liner wall to invert slab connection was designed as a moment connection to gain efficiency in the wall design as well as end fixity at the base, further reducing thickness of the lining. Reinforced footings transfer liner thrust through the drainage layer, and locally thicken the invert slab to capture peak bending and shear from the moment connection at the wall.

## FINAL LINING DESIGN MODIFICATIONS DURING CONSTRUCTION

Over the course of cavern final lining construction, factors such as scheduling, access, and Contractor means and methods necessitated some changes to the final lining design.

### TBM Run Shift at Crossover Caverns

A single 22 ft diameter hard rock tunnel boring machine (TBM) was used to excavate the east and

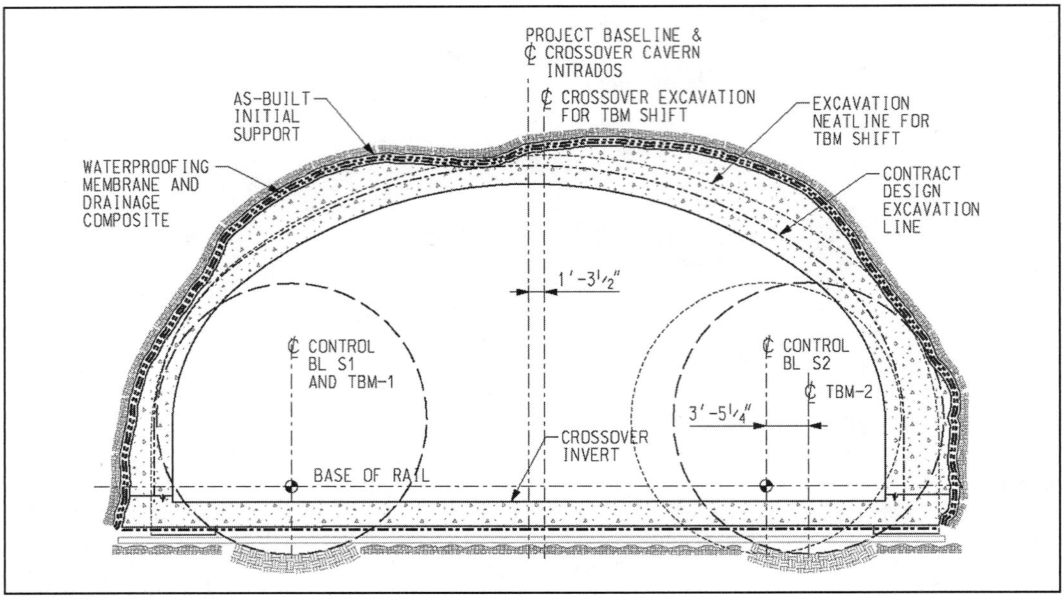


**Figure 4. Section of crossover cavern lining with shifted TBM run**

west running tunnels from a launch box at 92nd Street with the tunneling contract awarded to a joint venture of Skanska, Schiavone and J.F. Shea (S3) in early 2007. The longer east TBM tunnel drive extends to an existing 63rd Street Station bell-mouth. Initially, the east tunnel drive was to occur first, followed by the west tunnel which was only to extend to the north end of 72nd Street station, allowing all mining to proceed at 72nd Street after the first drive. Due to poorer than expected ground conditions on the east tunnel alignment near the launch box, S3 adopted ground freezing during assembly of the TBM on site. To maintain the project schedule, the west tunnel was instead bored first as ground freezing took effect over the east tunnel. Taking advantage of cheaper excavation by TBM and favorable advance rates being achieved, the west drive was extended through the 72nd Street Station to the G3/S1 turnout cavern, changing the mining sequence of the station cavern in SSK's contract. Keeping the project on schedule required station mining to begin from temporary construction shafts after the west tunnel drive, but before the east tunnel drive had passed through the station. During construction, the east TBM alignment was offset further east over the length of the station to allow bench excavation at the west side to proceed while providing additional rock pillar adjacent to the second TBM drive. Introducing S-curves in the shorter crossover caverns maximized the clearance in the larger station cavern being mined first.

TBM excavation in the shifted alignment was outside of the crossover cavern excavation profile neat line by almost 4 ft. For mined excavation and initial rock support, the width of both crossover caverns was enlarged to include the shifted east TBM tunnel excavation and maintain an optimal excavation profile (see Figure 4). The intrados of the crossover final linings was maintained per the contract geometry so as not to impact internal space requirements. Consequently, this created a final lining of tapering thickness across the arch, varying threefold in thickness from one side to the other. Three separate sections were considered for both the rock support and final lining designs as the TBM offset increased through the S-curves. The initial approach for accommodating this change in liner thickness was to introduce a third layer of reinforcing at the outer face of the thickened section, following the revised initial support profile. As discussed later in the paper, this design approach was modified to consider overbreak.

### G3 Horseshoe Tunnel

Selecting a tunneling method for connecting the new 72nd Street station and the existing 63rd Street station was largely determined by scheduling. With the TBM passing through 72nd Street station for both drives, it was hoped that the west run could continue to 63rd Street to take advantage of mechanized tunneling, but the 560 ft radius horizontal curve beyond the G3/S1 turnout cavern proved too tight for the TBM's capabilities. Drill-and-blast tunneling did not have this limitation. Another constraint was the stacked configuration of the north and southbound track alignments in the existing bell-mouth. These track alignments had to diverge sufficiently before splitting into separate running tunnels. A new stub cavern extended the existing bell-mouth for this purpose. Minimizing final liner thicknesses also reduced the length of the stub cavern as an adequate rock pillar could be developed between the adjacent tunnels more rapidly. Water pressures are high at this location, the deepest section of the Phase 1 alignment. The contract design included a reinforced concrete drained horseshoe tunnel with a 19'-5" internal diameter arch above the springline.

SSK put forward a Value Engineering (VE) proposal to change the contract horseshoe tunnel to a drill-and-blast undrained 19'-9" internal diameter circular tunnel with a Steel Fiber Reinforced Concrete (SFRC) lining, the SFRC approach being the existing project standard specified by the Designer. SSK worked with a reduced width of flat invert at tunnel invert for drill-and-blast operations (see Figure 5). This allowed a more circular lining arch to be achieved, minimizing the tunnel width where the rock pillar width was critical at the stub cavern interface, locally eliminating the need for groundwater pressure relief. By eliminating the piping and bar reinforcing, and re-using the circular lining from their TBM tunnel form system, SSK were able to share a cost saving in materials and labor for installation with the Owner. AAJV designed the modified tunnel lining as part of the VE exercise with calculations for the final liner based on as-built excavation scans.

### Addressing Overbreak in the Mined Caverns

The geometry, access and scheduling created several challenging mining advances. These included the development of the station cavern top heading from the offset circular construction shafts, the expansion

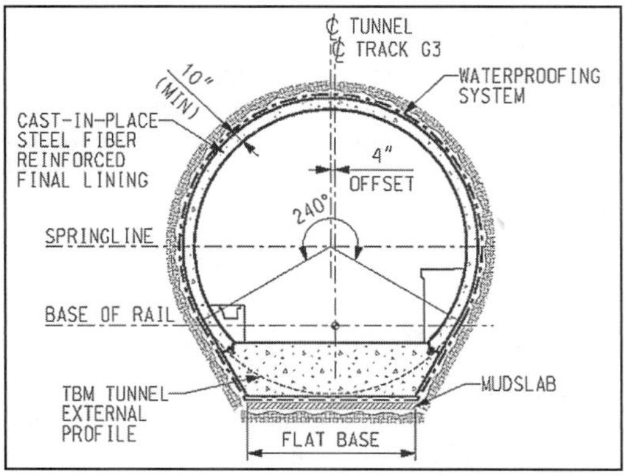


**Figure 5. Typical section—G3 horseshoe tunnel**

of the stub cavern top heading from the end of the horseshoe tunnel, and the crossover arch excavation from the west side station bench excavation. During excavation, there were some areas of considerable overbreak exceeding the limits set out in the contract specifications of 12 inches or 18 inches over 20% of the perimeter. Due to access limitations, SSK did not wish to fill the overbreak with shotcrete back to the contract limits. Instead, the Contractor proposed hanging the waterproofing membrane from the initial support shotcrete, and casting a thicker monolithic final lining. The Contractor's surveyor produced scans of the excavation profile at various stages. These made clear that further investigation was needed to determine the impact on the cavern liner design. Pushing the exterior face reinforcement outward, or introducing a third layer of reinforcement near the exterior face were not easily achievable because of the undulating initial support. In many areas, thickening of the final lining by the magnitude of overbreak surveyed was considered too large to disregard. In the crossovers, this effect magnified the asymmetry of the final lining cross section for the TBM shift as described earlier.

There were two key considerations related to the increased and varying lining thickness; the flexural behavior of the liner modified by increased and asymmetrical stiffness relative to the support, and shrinkage and temperature effects due to the additional thickness. The area and position of cavern final lining reinforcement shown in the contract drawings was no longer sufficient in all locations, and did not meet minimum reinforcement requirements.

For each zone, critical worst-case excavation profiles were selected for analysis. Dead loads and thermal gradients increased with thickness from over-excavation, while hydrostatic and rock loads were relatively unchanged. Sections were checked for combined axial compression and bending. Effects of the thickening were most pronounced in cavern sections with thinner linings and short spans with lighter contract reinforcement. Scans showed relatively constant overbreak in many sections, increasing the overall liner stiffness, attracting more load relative to the rock-springs and magnifying moments, shears and axial forces. In several other sections, thickened arch shoulders increased maximum negative moments but not effective depth, while having no overbreak at the crown to increase effective depth for maximum positive moments. When considering the overbreak, contract reinforcement was found to be insufficient in some of these areas, and the primary reinforcement was augmented to account for the magnified moments.

Based on various codes, guidelines and technical literature, the design approach provided

**Figure 6. Photograph of public station cavern final lining with exposed coffers and formwork system in view**

longitudinal temperature and shrinkage reinforcement for normally restrained section thickness up to 24 inches with half typically placed at each face. With the additional thickness from over-excavation, a minimum reinforcing ratio on the inner 12 inches was maintained, and additional bars were added to the outer face contract reinforcing such that the code requirements of ACI 318 for the full section were still met. To satisfy serviceability requirements, the SFRC mix developed for the unreinforced tunnel linings was also used in areas of larger over-break. Steel fibers increased liner robustness, improving toughness and ductility while reducing shrinkage strains and crack widths. Although free (unrestrained) shrinkage strains are only reduced slightly by the addition of steel fibers, relatively small dosages can substantially reduce shrinkage crack widths when restrained. Per the Owner's preference, steel fibers were not considered for strength design, or for replacing reinforcement required by code.

A contract requirement for sand-blasting of the exposed architectural concrete in the public cavern (shown with coffers in Figure 6) and entrance adits prevented the use of SFRC to address overbreak. Instead, longitudinal bar reinforcement was added. AAJV developed tables to indicate where additional reinforcing bars and SFRC were needed to address the predominant maximum overbreak for each pour. The detailer used these to develop shop drawings.

### Temporary Access Tunnel at Invert Level

Towards the end of mining operations in the station cavern, SSK proposed a short temporary access tunnel from the cut-and-cover excavation at Ancillary 1 station cavern invert. As this station has an island platform, the final configuration did not require any permanent adits connecting into the station invert at track level. Deepening the cut-and-cover rock excavation and mining a 14.5 feet wide by 16.5 feet high drill-and-blast tunnel was justified by the improved access to the station cavern for materials handling after both construction shafts in the right-of-way are backfilled to complete the arch of the final lining. In particular, the cavern final lining formwork systems could be removed from the station without being scrapped on-site. Although no permanent use could be found for the tunnel space, located behind the over-track exhaust plenum, this tunnel will be handed over for use by the follow-on finishes contractor. The void will be backfilled with concrete prior to completing the station cavern final lining.

AAJV analyzed the final lining with a 20 ft wide rectangular wall block out spanned by a 30 ft wide arch pour centered above the temporary tunnel (See Figure 7). The 2D STAAD model of the final lining was modified to include a pin support at the top of the block out. Strut-and-tie analysis was then used to determine the area of supplemental lintel and hanging reinforcement at the base of the liner arch to resist the resulting thrust from the arch above for this short-term condition. One of the cavern wall form systems will be handed over to the finishes contractor to complete the lining once access is no longer needed and the tunnel is backfilled.

### Elimination of Reinforced Footings

As described earlier, the contract liner and pressure relief configuration was selected to minimize rock excavation. Ultimately, the Contractor's approach was to reduce intricate labor at the expense of excavating slightly more rock to simplify the footing details. AAJV confirmed that by increasing the

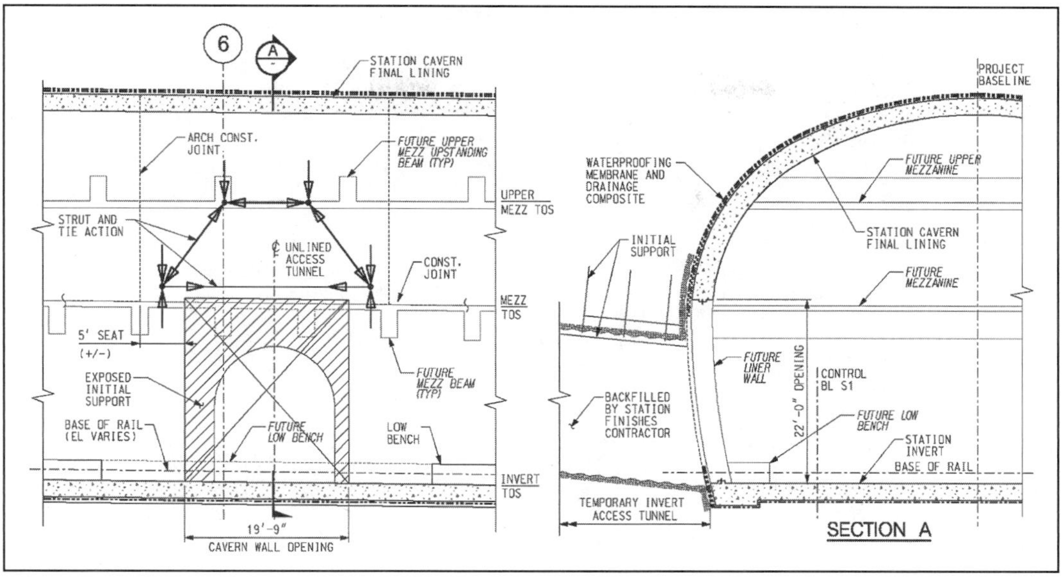


**Figure 7. Block-out in station cavern final lining for temporary access tunnel**

thickness of the station invert slab several inches to match the cavern liner wall thickness, the bending and shear capacity of the thickened slab was sufficient without additional corner bar reinforcing or shear links. An added advantage to this modification was that crews could complete all piping installation prior to waterproofing (completed by a sub-contractor), and prior to placement of reinforcing without the need for crews to revisit each area. Separating these decoupled the scheduling for each operation.

## CONSTRUCTION ENGINEERING

With restrictions on access and daily work schedule to limit the impact on the community, the Contractor developed creative engineering solutions to meet Contract milestones while preserving the design intent.

### Muck House Enclosures at Construction Access Shafts

Satisfying the needs of the community in the densely populated residential neighborhood surrounding the jobsite was a major consideration. The Contract placed strict work restrictions on activities such as blasting and surface work. For the volume of shot-rock to be removed and trucked from the site in the allotted contract time from the two planned construction access shafts, SSK needed to mitigate nuisance to the community to allow work passed curfew. The solution was to construct temporary "Muck Houses" with gantry cranes over the 69th and 72nd street construction shafts, covering two of Second Avenue's six lanes. A full proposal, including a 3D animation

**Figure 8. Screen capture from SSK's muck house animation**

of a "Muck House" in operation (see Figure 8) was presented to the Community Board who approved the scheme. J.F. Shea and Schiavone's engineering departments worked together to develop the concept and design the mucking system and steel framed enclosures. Installed on micropiles between utilities in the right-of-way, these "Muck Houses" allowed SSK to raise full muck bins from the tunnel and load trucks while limiting noise and dust. Also useful once concreting began—they allowed SSK to bring in oversized reinforcing mat deliveries and lower them into the station cavern after the surface work restriction cut-off time. These structures permitted completion of the excavation and concreting phases within the allotted time, despite the imposed work restrictions, overcoming limited access to achieve high daily production.

**Figure 9. Photograph of welded mat reinforcement installed in the G4/S2 turnout cavern**

### Welded Mat Reinforcement and Fabrications

Vitally important to SSK was maximizing productivity and efficiency of reinforcement and concreting operations to complete the work as quickly as possible. One way this was accomplished was the use of welded reinforcing mats and fabrications, improving ease of installation and quality control while reducing site installation time. This method was used for most inverts, endwalls and arches throughout the project. The maximum mat size was controlled by the shaft opening in each muck house as well as trucking restrictions for wide loads in NYC. Bars were first cut in Sayerville, NJ; bent by Local 46 reinforcing lathers in Newark, NJ; then shipped up to Fryeburg, Maine for welding into mats. Completed mats were then sent back to Sayerville for reloading onto smaller trucks, which are able to maneuver into the "Muck Houses," for ease of delivery and unloading at the job site. This journey was justified by the efficiencies gained for site work, although the lead-time for fabricating these reinforcing mats required shop drawing approval well in advance of installation for each scheduled pour.

Reinforcement detailing was mostly dictated by the installation method. In the smaller cavern arches, the reinforcing mats were hung from WA anchors in rock before the forms were assembled (See Figure 9). This allowed for fast turnarounds once concrete placements began. In the larger station and crossover caverns, a reinforcing template and gantry crane were utilized. This involved building the crown reinforcing cage of the arch on top of the template, which was then hoisted onto the arch form then raised into place with the form. This method was modified for the architectural coffers in the public portion of the cavern. Both the template and form were modified for this configuration; coffer frames were added to the template and the forms had additional hinge points installed, allowing the recessed coffers panels to be stripped first and the main arch form to be lowered vertically. Throughout the entire project, the Contractor communicated their method of installation to the Designer before submitting shop drawings for that area so the specific splice locations and other reinforcing details could be reviewed in the proper light.

### Use of 3D Computer Aided Design as a Planning Tool

There were many areas where the complexity of the design required an in-depth examination by the Contractor on how they were to be constructed. One of the more difficult details of this contract was the installation of #11 reinforcing couplers for future mezzanine beams, with hooked dowels developed into the wall and arch concrete. These couplers needed to be installed with tight quality control for use by the future contractor constructing the mezzanines. SSK utilized 3D AutoCAD modeling to develop dowel arrangements for each beam type (see example in Figure 10), varying embedment lengths and hook orientations to gain sufficient clearances. Through close collaboration with the Designer, SSK found bar configurations that met the design intent, but also allowed for efficient and accurate installation of the beam couplers. 3D CAD was also instrumental for finding other reinforcement detailing solutions at other complex portions of the job. At the complex adit miters and transitions, 3D CAD modeling allowed different sections to be cut and

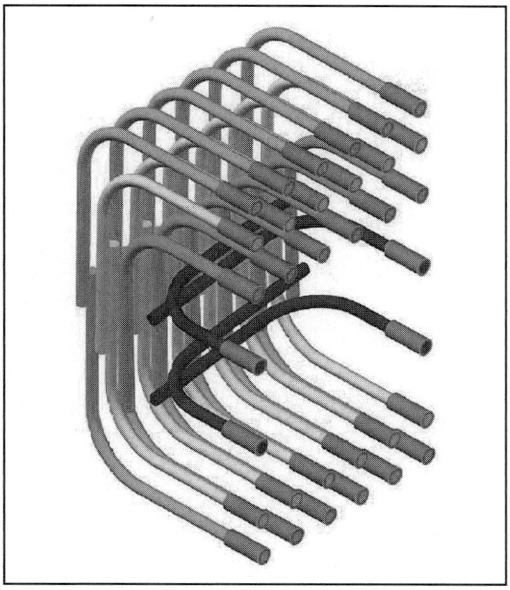

**Figure 10. Isometric of mezzanine beam dowel reinforcing cage**

the reinforcing details displayed clearly for both the Designer and Contractor to visualize.

3D CAD was useful in developing a plan for backfilling the construction shafts to top-of-rock with concrete, complicated by a large overlap between the shafts and cavern. An inclined form system was needed across each shaft opening to replicate the excavation profile in the adjacent cavern arch. After several lifts of concrete backfill at the cavern wall, steel beams were seated on the bench concrete spanning to the shotcreted rock surface above the shaft opening to support the sacrificial formwork. Using surface scans, 3D models of the shaft openings into the cavern were created to determine the length and mounting angles for each beam. Beams were set-out in the model, and fabrication drawings developed to suit the as-excavated condition.

## DESIGNER-CONTRACTOR-CONSULTANT CONSTRUCTION MANAGER COLLABORATION

Linear construction on multiple work fronts posed a challenge in scheduling, and also in preparing and approving issued-for-construction working drawings for permanent works in the caverns and tunnels to meet changing priorities, often at short notice. Regular and open communication between all parties was key. Bi-weekly technical meetings involving engineers from SSK, AAJV and PB allowed concerns, challenges and proposals to be raised as they developed. A detailed log was maintained by the contractor listing each of the topics covered and highlighting action items. Another beneficial collaboration tool was a "hot list" created by the contractor, updated and distributed weekly. It tracked the priority and status of all open submittals and technical items, allowing the designer to react to changing field conditions and refocus when needed on expediting and resolving the most critical technical issues. A pro-active, flexible and collaborative approach by all parties gave ample opportunity to resolve many critical issues before they impacted the construction schedule.

## CONCLUSION

The design of underground stations in developed urban environments requires consideration of the constructability and sequencing as part of the design process. During construction of these final linings, the Designer and Contractor worked to develop and maintain a strong working relationship and were able to quickly assess and implement changes to the permanent design to suit revised sequencing and to solve construction challenges. Overcoming obstacles together supported the completion of contract work on schedule, moved the overall project forward, and demonstrated a model of success for future projects.

# Innovative Shaft and Tunnel Linings for the Thames Water Lee Tunnel

**Peter Jewell**
CH2M HILL

**Alan Barker, Matthew Bellhouse, and Chebli Matta**
MVB Joint Venture

**Andy Sefton**
Thames Water

**ABSTRACT:** London's growth and requirements for even cleaner rivers have caused Thames Water to plan the Thames Tideway Tunnel, a 21.75mi (35km) long, 23.6 ft (7.2m) internal diameter tunnel mostly beneath the River Thames. The first section to be built is the Lee Tunnel, which runs for 4.3mi (6.9km) in East London at up to 262.47 ft (80m) depth between the Beckton Sewage Treatment Works and the Abbey Mills Pumping Station. This paper describes the project generally, and specifically the challenging geotechnical conditions and the novel approaches adopted for the design and construction of the five large diameter deep shafts and main tunnel's linings.

## INTRODUCTION

The Thames Water Lee Tunnel is the first stage of an enhancement to London's sewer system which will remove the discharge of combined storm water and sewage to the River Thames and lower River Lee. MVB, a joint venture of Morgan Sindall, Vinci Construction Grands Projets and Bachy Soletanche, commenced construction in 2010. Completion is expected at the end of 2015. In addition to construction, MVB is responsible for developing the project's Reference Design to the Detailed Design stage. Thames Water's Project Management Team, led by CH2M HILL, is working collaboratively with MVB to deliver the $1.0 billion (£ Sterling 635 million) project within budget.

The project requires the construction of five shafts and a 23.6 ft (7.2m) internal diameter tunnel running 4.3 mi (6.9km) from the Abbey Mills Pumping Station complex to the Beckton Sewage Treatment Works (STW) in East London (Figure 1). The tunnel is being constructed at depths of between 213 ft (65m) and 262 ft (80m). The scheme is required to store 12,360,100 $ft^3$ (350,000$m^3$) of combined storm water and sewage which can be pumped to the Beckton STW once rainstorms have abated.

The structural fabric of the concrete lined shafts and tunnels is critical to the integrity of the Lee Tunnel system, which has a design life of 120 years. This led to the Reference Design requiring shafts and tunnels to be provided with both primary and secondary linings, which in turn has required a number of innovative approaches. This paper describes the geotechnical and groundwater conditions and the approaches taken for the following major structural components:

- Bar reinforced concrete diaphragm walls forming the primary shaft linings;
- Steel fibre reinforced slipformed secondary shaft linings with annulus infilling; and
- Steel fibre reinforced concrete segmental primary tunnel lining.

A cast in-situ steel fibre reinforced concrete design is being developed for the secondary tunnel lining.

## GEOTECHNICAL CONDITIONS

The stratigraphy at the project site is similar to that encountered under much of east London within the London Basin. It comprises a sequence of Made Ground, Alluvium and River Terrace Gravels overlying Tertiary age clays and sands with occasional pebble beds which in turn overlie Chalk (a weak limestone) of Cretaceous age (Figure 2). There is also an industrial legacy of historic contamination.

The ground conditions at each of the five shaft locations are similar. Heterogeneous layers of Made

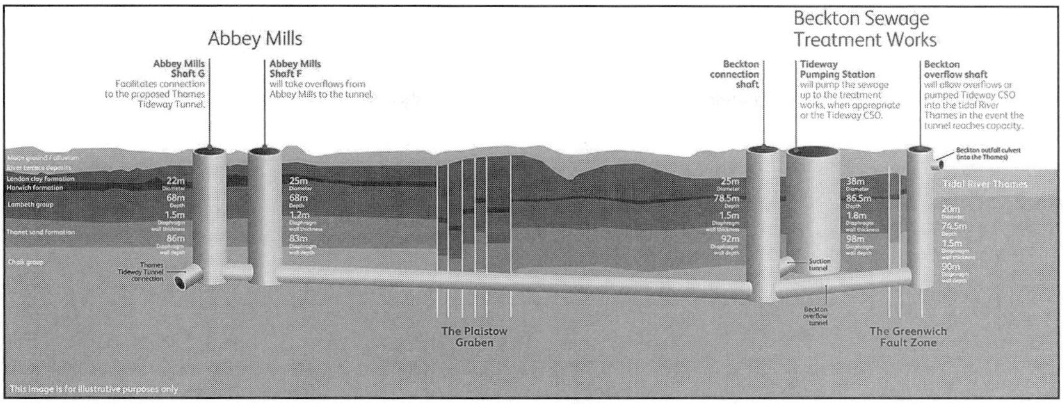


**Figure 1. The Lee Tunnel System**

Ground are typically 3.3 ft to 9.8 ft (1m to 3m) thick at the Beckton STW, but up to 26 ft (8m) thick at Abbey Mills. The Made Ground overlies very soft to soft alluvial organic clay which is up to 16 ft (5m) thick at the Beckton STW, but only around 3.3 ft (1m) thick at Abbey Mills. The alluvial clay is underlain by River Terrace Deposits comprising layers of medium dense to dense sandy gravel typically between 13 ft and 20 ft (4m and 6m) thick.

These deposits are generally underlain by London Clay, which is a stiff to very stiff fissured inorganic silty clay. The London Clay is weathered in the upper 6.6 ft (2m) leading to reductions in its strength. The thickness of London Clay at four of the shafts is between 16 ft and 26 ft (5m and 8m), but it is absent at the Beckton Overflow Shaft.

The London Clay is underlain by the Harwich Formation which is up to 9.8 ft (3m) thick, though again absent at the location of the Beckton Overflow Shaft. The Harwich formation is a mixture of sandy clay and clayey sand and contains rounded pebbles typically up to 0.4 in. (10mm) diameter.

Beneath lie the Lambeth Group and Thanet Sand, the full sequence of which was encountered at each of the shafts. The Lambeth Group comprises stiff and very stiff fissured clays interbedded with very dense sands which can also occur in channels cut into the clay layers. The Lambeth Group can also contain calcareous cemented horizons over 3.3 ft (1m) thick. The Thanet Sand generally comprises very dense silty fine sand. At the base of the Thanet Sand is a layer of re-worked flint in a clayey matrix known as the Bullhead Beds. The Thanet Sands lie unconformably upon the chalk.

Encountered towards the base of each of the shaft, the Chalk is a weak limestone with an unconfined compressive strength typically between 3 and 6MPa. The degree of fracturing of the Chalk is variable, but is typically greater nearer the interface

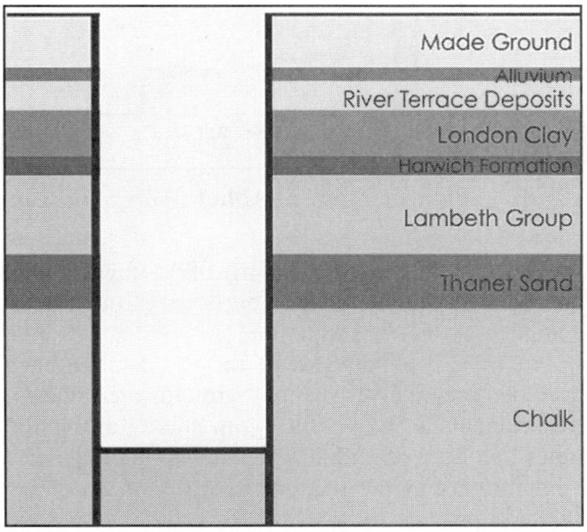


**Figure 2. Typical geological sequence**

with the Thanet Sand. The Chalk contains regular continuous and discontinuous bands and nodules of extremely strong siliceous flint.

During ground investigation, two geological structures were identified: the Greenwich Fault zone at Beckton and a newly identified 'Plaistow Graben' fault zone which is located approximately two-thirds into the tunnel drive.

## GROUNDWATER CONTROL FOR SHAFT CONSTRUCTION

At Beckton, the groundwater pressures are hydrostatic with the groundwater table a few feet below ground level.

At Abbey Mills the piezometric surface in the Thanet Sand and Chalk is around 20m (295 ft) below ground level due to historic pumping from the Chalk and there is a second observable water pressure close

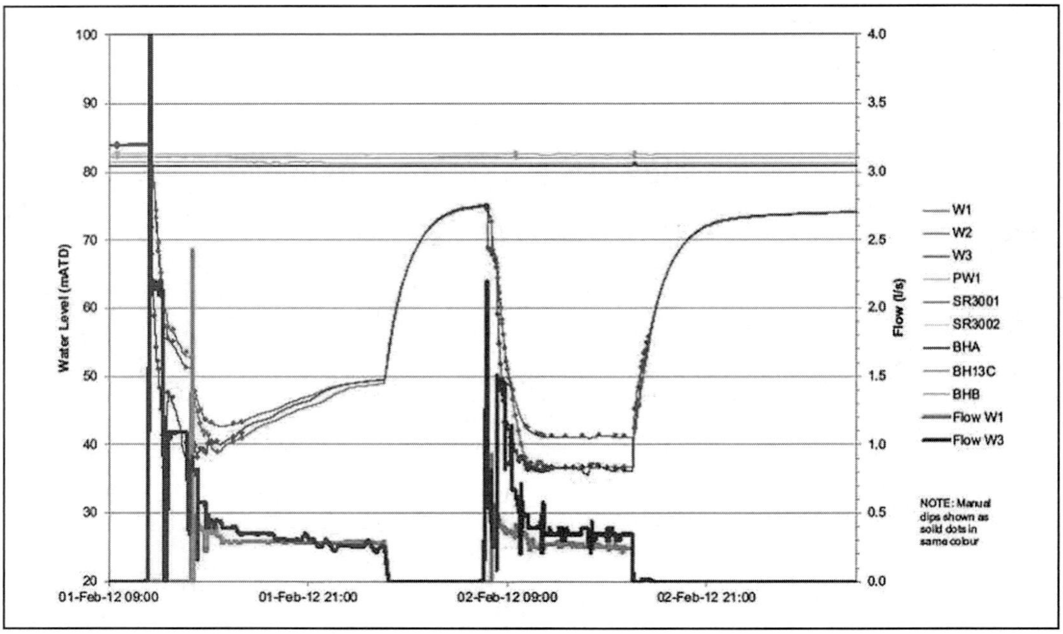


**Figure 3. Pumping tests at Abbey Mills Shaft F**

to ground level. The permeability of the unweathered chalk at depth is dominated by flows through horizontal to sub-horizontal fissures.

Borehole geophysics of the Chalk have been used to accurately ascertain groundwater control requirements at the various work sites. Reservation tubes had been specified under the contract as steel void formers in the diaphragm walls of all of the shafts to provide for possible toe grouting from the base of the diaphragm wall.

However, accurate plotting of the major water paths at the shaft sites concluded that reservation tubes could be omitted from three shafts, namely Shafts F, G and the Pumping Station Shaft. Reservation tubes were installed in the Overflow Shaft where there was concern of the impact upon permeability from faulting, and in the Connection Shaft where significant flow paths were identified close to the base level of the diaphragm wall at 295 ft (90m) below ground level.

The reservation tubes were only used at the Overflow Shaft. Here, toe grouting was used in conjunction with a pumping test, to confirm that permeabilities were sufficiently low to allow commencement of excavation under sump pumping at excavation level, utilising the pumping test wells for passive relief. At the Connection Shaft, pumping tests showed that grouting through the reservation shafts was not necessary.

Further pumping tests within the footprints of the shafts at Abbey Mills were completed to confirm design assumptions made within hydrogeological modeling, that the identified fissure had been cut-off by the diaphragm wall. By drawing the water level with shaft F to formation whilst monitoring external piezometers, it was possible to plot the zone of influence during dewatering and to demonstrate that this activity would not impact upon the known significant hydrocarbon contamination to the north.

Figure 3 demonstrates a high drawdown, of up to 164 ft (50m) with a maximum flow of 0.021 cfs (0.6L/s). The piezometric levels external to the shaft remained unaffected during the pumping trial.

## DESIGN AND CONSTRUCTION OF SHAFT LININGS

### Primary Shaft Linings

The shafts required for Lee Tunnel are some of the deepest and largest in diameter ever constructed in the UK and range in internal diameter from 66 ft (20m) to 125 ft (38m) (Stanley et al. 2012a, 2012b). The design of the walls was required to follow Eurocodes and had to consider an assessment of chalk stiffness, high hoop stresses, large multiple openings and non-axisymmetric loadings. These considerations led to the four diaphragm-wall shafts being formed of linked panels up to 5.9 ft (1.8m) wide and up to 322 ft (98m) deep, which necessitated:

- A project-specific design of concrete mix
- Demanding verticality tolerances of 1 in 300

**Figure 4. Three concrete trucks at a diaphragm wall primary lining pour**

- Design and safe and accurate placement of extremely long, spliced, reinforcement cages
- Management of bentonite support fluid
- Verification of construction tolerances through monitoring

The verticality observed following excavation of the shafts has matched the surveys undertaken during installation, and in fact has exceeded all expectations.

The concrete mixes developed for the steel bar reinforced diaphragm walls had to meet a number of requirements:

- For such large concrete pours (up to 49440 $ft^3$ (1400$m^3$)) in an individual panel) the workability had to be retained for much longer times than normal.
- However the concrete could not contain so much water that it would become susceptible to bleed.
- The design required a concrete strength of 8,700 psi (60MPa).

The diaphragm walls were constructed using concrete grade C50/60, incorporating 70% Ground Granulated Blastfurnace Slag (GGBS) cement replacement (Figure 4). The high workability requirement led to placement via tremie pipes, which ensured a good degree of self compaction, particularly at base slab level where the highest density of reinforcement was located. In order to avoid the formation of 'cold' construction joints during these very long concrete pours, the concrete mix was designed for a workability retention of 6 hours. The GGBS reduced the early temperature gain and the rate of strength gain, with the intention that the minimum compressive strength of 8700 psi (60MPa) would not be achieved until 56 days. This enabled the subsequently constructed adjacent secondary and infill panels to 'bite' into primary panels, thus ensuring an 'overlap' detail.

During the mixing trials it became apparent that the super plasticisers had to be very well mixed to coat all of the cement particles and thus gain the required workability. This prolonged the mixing process leading to concerns over the production capability of the plants to supply at the required rate of 4240 $ft^3$ (120$m^3$) per hour.

In four of the shafts, 'soft' eyes were created by replacing the steel reinforcement with glass fibre reinforcing bars, in order to facilitate easier 'break outs' and 'break ins' for the TBM during tunnelling.

The diaphragm walls were formed as a series of overlapping rectangular diaphragm wall panels forming a faceted circle. The design did not require the primary lining (the diaphragm walls) to be watertight. Primary panels were excavated at up to 23.6 ft (7.2m) wide. Secondary panels overlapped and therefore cut into the concrete of the primaries. The primary panels were constructed at a rate of one per week, with panel sequencing coordinated such that excavation could continue at a safe distance from cage placement and concreting. Excavation of the primary panels was sequenced such that concrete of similar strength existed on each side of a secondary panel.

The walls were excavated by means of conventional grabs to the top of the Lambeth Group, and "Hydrofraise" equipment (Figure 5) beneath. The Hydrofraise units (reverse circulation milling

**Figure 5. Hydrofraise unit for diaphragm wall construction**

machines) featured pairs of toothed cutter wheels to mine the ground with the excavated materials being removed through the suction lines of the bentonite circulation system. Instrumentation integral with the Hydrofraise informed the operator of the plan position, inclination and twist of the cutter body together with the rotation speeds of the cutter wheels during excavation. Hydraulic rams and pressure plates were fitted to both sides of the bodies of the Hydrofraises. The pressure plates were jacked against the side of the panels by the operator to change the inclination of the cutter wheels and therefore retain the diaphragm wall panels within the required tolerances.

Each diaphragm wall panel contained a reinforcing cage over its full depth. In the primary panels it was difficult to envisage that the cages could be shaped to suit the three separately excavated bites, and be placed up to 328 ft (100m) below ground. Accordingly it was elected to place three separate cages in the primary panels, spaced 400mm apart from each other. The maximum cage length able to be transported to site was 73.8 ft (22.5m), so a three-bite primary panel required fifteen separate pieces of reinforcing cage to be installed. Cages were connected using a combination of threaded couplers and bulldog grips, with the cage undergoing temporary works designs to check that handling, lifting, splicing and placing could all be undertaken safely.

The shape of the overlap between the panels is trapezoidal, thus the Hydrofraises had to cut an uneven profile whilst excavating the secondary panels. To even up the shape of the overcuts, polystyrene panels were connected to the cage and placed on the edge of the primary panels.

To form the Hydrofraise-excavated diaphragm walls, the bentonite acts both as a support fluid and as a medium to pump the excavated cuttings back to the surface. In common with normal practice, the bentonite was passed through a series of desanders and desilters to remove the excavated solids. Bentonite that has been through the excavation process a number of times eventually becomes too heavy to be recleaned, and is then discarded. For the Lee Tunnel project it was decided to dry out the bentonite so that it could be removed from site along with the excavated spoil, the waste water being discharged directly into the Beckton STW. Figure 6 shows the appearance of the diaphragm walls revealed during excavation.

### Secondary Shaft Linings

A radical structural rethink of the shaft lining design has eliminated more than 500 tons of reinforcing steel from each of the five shafts. Believed to be a world first, MVB chose to adopt a double-sided slipform shutter to construct what was essentially a free standing chimney within the excavated shaft. This decision, together with innovative design features for the base slab, the elimination of water paths at construction joints by the installation of re-injectable channels, and careful specification of the concrete used to fill the annulus between the diaphragm walls and the slipformed internal lining, has successfully achieved the requirements.

Not only did these design innovations yield a significant commercial saving on steel, but by largely removing the need to handle and fix heavy steelwork, they have enabled faster, safer construction and will significantly improve lifetime durability, avoiding the risks of steel corrosion and spalling.

The concrete mixes were designed to meet the particular requirements of each construction process, in addition to the normal strength and durability requirements needed to achieve the specified design life of 120 years. Laboratory and on-site testing was undertaken for each of the trial mixes to verify the mix designs, which included the construction of a full size slipform test panel at ground level prior to slipforming the inner lining wall of the first shaft to be constructed.

The slipformed inner linings (Figure 7) were constructed with a steel fibre reinforced concrete

**Figure 6. Beckton overflow shaft diaphragm walls exposed during excavation**

**Figure 7. Slipformed construction of secondary lining**

grade C50/60 incorporating 36% GGBS cement replacement, with the concrete being delivered to the slipform rig by skips supported from cranes. The mix was designed to accommodate continuous slipforming, with the ability to provide 'cold' construction joints in exceptional circumstances when halting the slipform was unavoidable. A plasticiser and retarder were therefore incorporated within the mix, with the level of retardation being monitored, and adjusted as necessary to suit the planned pour rates and formwork travel speeds achieved. The workability of the concrete was critical in allowing the formwork to travel upwards continuously, as well as enabling the surface finish to be worked upon from the trailing platform of the slipform rig. The slipform process progressed well and consistently achieved high qualities of finish.

The inclusion of steel fibres increased the ductility of the otherwise unreinforced concrete, reduced the potential for spalling damage, and assisted in controlling drying shrinkage cracking. Extensive shrinkage testing was carried out at Bekaert's laboratory in Belgium, followed up by trials on site, to determine optimum workability and doseage rates.

The mixes incorporated 1.87 lb per $ft^3$ (30kg per $m^3$) of Dramix fibers 1.4 in. (35mm) long. The fibres were introduced into the mix at the batching plant.

The design requirement for the inner lining wall to be able to move laterally under the compression pressure exerted by the annulus mass concrete was achieved by building it upon two layers of 0.08 in. (2mm) thick PVC low friction membrane with grease between. The bottom layer was fixed down to the base slab with contact adhesive, and the slip-formed inner lining wall was cast on top of the upper layer. The slip membrane methodology was verified by earlier laboratory testing.

The mass concrete used to infill the annulus between the diaphragm wall and the inner lining wall is grade C25/30 incorporating 50% GGBS cement replacement. This mix was designed as a flowable, retarded concrete to be placed in a single pour operation, in order to exert the required design pressure on the shaft inner lining. The concrete incorporated a plasticiser to ensure it had the required workability to enable it to flow evenly around the shaft perimeter from the four tremie pipes used to place the concrete, and to aid self compaction without external means. A retarder was used within the concrete to maintain its fluidity and prevent the onset of the curing process until a sufficient head of concrete had been poured to exert the required design pressure on the lining wall. The required period of retardation was thus determined by the rate at which the infill concrete could be placed. Instruments were placed in the annulus to enable concrete temperatures and pressures to be monitored in real time, in order to confirm design assumptions.

### Shaft Excavation Ground Displacement Research

The construction of Abbey Mills Shaft F provided the unique opportunity to implement a large scale monitoring scheme, which was directed by the University of Cambridge, UK. Three diaphragm wall panels were instrumented with distributed fibre optic instrumentation and inclinometers to investigate the structural performance of the shaft wall. Ground movements were measured with multi-point rod extensometers and inclinometers which were installed in two arrays radiating from the shaft.

One empirical formula (New and Bowers, 1994) is frequently used in British design practice; applying this to Abbey Mills shaft predicted large surface settlements up to 1.6 in. (40mm) next to the shaft wall. Extensometer measurements show much smaller settlements: a maximum of 0.20 in. (5mm) settlement at 8.2 ft (2.5m) distance from the shaft wall was recorded during diaphragm wall construction; shaft excavation itself caused only 0.04–0.08 in. (1–2mm) surface settlement. This research work provides valuable information for the construction of future shafts for the Thames Tunnel project.

## DESIGN AND CONSTRUCTION OF TUNNEL LININGS

Lying between 213 ft (65m) and 262 ft (80m) below ground level, the main tunnel is exceptional in being deeper than any other tunnel in London, and in being constructed very largely through a full face of chalk under full hydrostatic pressure. The ground and groundwater pressures combine to subject the tunnel linings to high external stresses at most times. However, when the tunnel becomes full of combined storm water and sewage effluent, the tunnel linings have to withstand a net internal pressure which leads to the development of significant tensile stresses.

It was considered that a single pass lining would present too high a risk to achieving these requirements and accordingly a two pass lining was decided upon. The watertightness requirements for the completed tunnel were as follows:

- There shall be no running water present on the finished surface of the lining. Occasional damp patches and staining are permitted. Total water ingress shall not exceed 0.03 gal (0.1L) per hour per linear metre of nominal bore.
- Sealing of leaks in the segmental lined tunnel shall be carried out prior to the secondary lining installation.

The use of a Slurry Tunnel Boring Machine (TBM) was a mandatory requirement from Thames Water. This methodology enabled the primary lining to be formed of pre-cast concrete segments. However, it would be impractical for the secondary lining to be segmental. Thames Water envisaged a steel bar reinforced cast in-situ concrete formed in a shutter.

For the primary lining, MVB elected to design a single 'universal' 25.6 ft (7.8m) internal diameter segmental ring, 13.8 in. (350mm) thick, of nominal width 5.6 ft (1.7m) and a 0.83 in. (21mm) taper. Each of these rings comprises seven main segments plus a key segment. The geometry was exceptionally tightly-specified.

Building on Morgan Sindall's widespread successful experiences with steel fibre reinforced concrete linings, both as sprayed concrete and as segmental concrete linings, MVB proposed an alternative design of steel fibre segmental concrete linings without any mesh or steel bar reinforcement.

The design of these segments involved an extensive testing programme, notably including included full scale compression tests undertaken at the BRE, to examine the effectiveness of the convex radial joint geometry (Figure 8). A key objective was

**Figure 8. Full-scale load testing of tunnel segmental primary linings**

to design fibre reinforced concrete segments which did not require any steel bar or mesh reinforcement. Accordingly, testing of various concrete mixes with different proportions of steel fibres was crucial to developing a concrete mix which provided the necessary strength and integrity without being difficult to place in the moulds.

The adopted concrete mix design featured 30kg/m$^3$ of Dramix RC 80/60 BN steel fibres, plus 0.0624 ft/m$^3$ (1kg/m$^3$) of Adfil monofilament polypropylene fibres to eliminate risk of spalling in the event of an accidental fire during construction. The characteristic flexural residual strength was not less than 2.2N/mm$^2$, and the characteristic tensile splitting strength was not less than 3.3N/mm$^2$.

Supply of the tunnel segments was entrusted to Ridham Precast Concrete (RPC), which is a Kent, UK subsidiary company of Morgan Sindall. RPC was already experienced in the production of precision engineered precast tunnel segments for other projects, but the demanding requirements of the Lee Tunnel specification led to RPC ordering special moulds from specialist French manufacturer CBE Group. These moulds were fabricated using laser cloud robotics technology to create the required precision.

The Reference Design specified a steel bar reinforced secondary tunnel ling, but MVB is currently developing a design based upon the use of steel fibre reinforced concrete, thus disposing of the need to fix steel in a very onerous environment.

## CONCLUSION

In view of its exceptional depth, very large shaft sizes, full hydrostatic groundwater pressures, and uncommon ground conditions, the Lee Tunnel is pushing the boundaries of tunnel and shaft design and construction in London. The five shafts have been constructed through a series of strata having varied geotechnical properties. They have terminated in fully-saturated Chalk at greater depths than ever reached previously in London. The main tunnel needed to be driven with a closed-face slurry TBM, mostly through the Chalk, but also including highly fractured zones, an unexpected drift filled hollow feature and a mixed face of Chalk with Thanet Sand. These conditions are demanding by any reasonable tunneling standard.

The geotechnical conditions and structural performance requirements led to pre-contract decisions to provide primary and secondary linings for both the shafts and the tunnels. The shafts were so deep that in practice diaphragm walls were the only available construction method. Given the practical difficulties in installing waterstops between adjacent diaphragm wall panels, it was decided not to require the diaphragm walls to be watertight. However, this decision placed an enhanced obligation on the watertightness performance of the secondary lining. The adoption of the double slipformed fibre reinforced concrete linings with the annulus concrete infilling was a novel and previously untried approach, but has produced a combination of dry shafts with high quality finishes.

The precision moulds used for the primary tunnel lining has led to the production of segments which have been installed to a very high standard of ring build. There were very few instances of lips and steps out of tolerance and only negligible water ingress noted. Without steel bar reinforcement, it was inevitable that some segments would crack during placement, but these cracks have successfully self healed and pose no concern to either the structural integrity, watertightness requirement or durability of the primary lining. The decision to adopt steel fibre reinforced concrete for the secondary lining was innovative, but is paying off in terms of reducing the costs and duration which a steel bar reinforced concrete lining would have required.

The successful use of these innovative approaches to the design and construction of the major civil components of the Lee Tunnel will help provide a robust foundation for the delivery of Thames Water's forthcoming Thames Tideway Tunnels programme. At the end of December 2013, four of the five shafts were completed, and diaphragm walls were complete for the fifth shaft, and 95% of the tunnel excavation and primary lining has been finished.

## ACKNOWLEDGMENTS

The authors acknowledge with thanks Thames Water for giving their permission to publish this paper. The assistance of colleagues on the Lee Tunnel project in the preparation of this paper is sincerely appreciated.

## REFERENCES

New, B.M. and Bowers, K. 1994. Ground Movement Model Validation at the Heathrow Express Trial Tunnel. *Tunnelling '94*. London.

Stanley, M., Gatward, J, Puller, D., Matta, C., Pound, C., Dobson, N., Jewell, P. and O'Connor, M. 2012. Design and Construction of the Thames Water Lee Tunnel Shafts, London, Part 1, Construction of the Shaft Diaphragm Walls. 3rd International Conference on the Design and Construction of Shafts, London.

Stanley, M., Gatward, J, Puller., D, Matta, C., Pound, C., Dobson., N., Jewell, P. and O'Connor, M. 2012. Design and Construction of the Thames Water Lee Tunnel Shafts, London, Part 2, Construction of the Shaft Base Plugs, Slipformed Inner Lining and Annulus Infill. 3rd International Conference on the Design and Construction of Shafts, London.

# TRACK 2: DESIGN

## Session 2: Challenging Design Issues

Joel Theodore, Chair

# Precast Segment Design for High Hydrostatic Head Conditions in Rock: State of Practice

**Paolo M. Brion**
New York City Department of Environmental Protection

**Z. Bade Sozer, Daniel Ebin, and Sam Swartz**
Jacobs Associates

**ABSTRACT:** A bolted and gasketed precast concrete segmental initial lining is being designed for the Rondout-West Branch Bypass Tunnel utilizing industry state-of-practice design procedures for extreme head loading. This paper discusses these design procedures and outlines construction recommendations for the main tunnel drive. The initial lining system is reinforced for high hydrostatic head (23 bar) and considerable TBM thrust loads in addition to other applicable loading conditions. Experience on other high head projects using precast segments in rock and lessons learned are highlighted.

## INTRODUCTION

The Rondout-West Branch Tunnel (RWBT), a segment of the Delaware Aqueduct, was built from 1937 to 1944 and provides about 50% of New York City's total water supply. The tunnel has a finished diameter of 13.5 feet and runs southeast about 45 miles from the Rondout Reservoir to the West Branch Reservoir. Monitoring and tunnel operations have shown that the RWBT is leaking up to 35 million gallons per day (MGD), mainly through two locations, Roseton and Wawarsing (as shown in Figure 1). The Wawarsing area will be repaired through an extensive grouting program. The Roseton area, on the other hand, because of the extensive nature of the leakage, has to be bypassed. To accomplish this, the Rondout-West Branch Bypass Tunnel (Bypass Tunnel) will be constructed. This paper discusses the precast concrete segment design for the Bypass Tunnel.

### Purpose of Precast Concrete Segments

One proposed method of excavation for the Bypass Tunnel is through the use of a single shield tunnel boring machine (TBM). Bolted, gasketed, precast concrete segments will form the initial lining of the proposed two-pass lining system. The segments are designed for temporary construction loads such as for stacking, lifting, transportation, and full hydrostatic loads, as well as for backfill grouting loads in the short term. Once the final lining is installed, it is anticipated that the segments will leak and transfer the hydrostatic loads to the final lining. In the long term, the segments are designed for full ground loads.

The main design challenge associated with using a precast segmental lining system for the Bypass Tunnel project is the high hydrostatic pressure, with a groundwater head ranging from 600 to 775 feet.

### Previous Projects with High Loading Conditions for Segment Design

The Lake Mead Intake No. 3 and Arrowhead Tunnels of the Inland Feeder System are two projects that employed bolted and gasketed precast concrete segmental linings with high loading conditions in rock. A comparison of key parameters of the segment linings from those two projects and the Bypass Tunnel is shown in Table 1.

The Lake Mead Intake No. 3, outside of Las Vegas, NV, taps into Lake Mead to provide an additional means of supplying water as the water level in Lake Mead lowers. The Arrowhead Tunnels Project, part of the Inland Feeder System Project, consists of two water tunnels below the San Bernardino Mountains outside of Los Angeles, CA. Part of the Arrowhead Tunnels Project included two stages of large-scale verification testing, performed for both the structural capacity of the segments and watertightness of the gaskets. Initial testing was performed during design to verify assumptions and analysis, and a second stage of testing was required during construction by the Contractor to verify the final geometry and products selected. These tests and their results were used as the basis of design for the Bypass Tunnel, as discussed below.

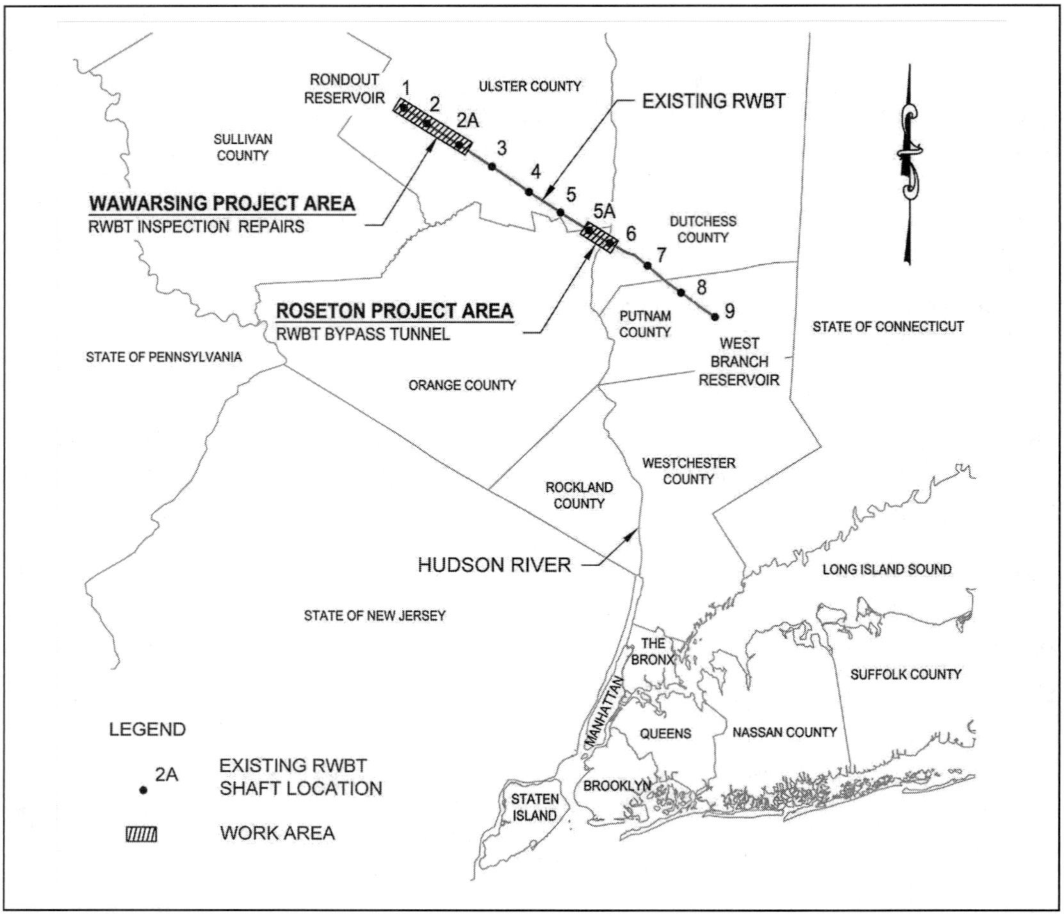


**Figure 1. Rondout-West Branch Bypass Tunnel Project**

**Table 1. Segment comparison table**

| Project | Number of Segments | Design Groundwater Head (ft) | Segment Thickness (in.) | Internal Diameter (ft) | Concrete Strength (ksi) | Reinforcement Type | Lining Type |
|---|---|---|---|---|---|---|---|
| Arrowhead | 5+1 | 900 | 13 | 16 | 8 | Rebar | Two Pass |
| Lake Meade | 5+1 | 570 | 14 | 20 | 6 | Wire bar* | One Pass |
| Bypass Tunnel | 5+1 | 775 | 13 | 18.17 | 9 | Rebar | Two Pass |

*75 ksi wire bar reinforcement used.

## LOADING CONDITIONS

The Bypass Tunnel segments are designed with five segments and a key with an internal diameter of 18.17 feet. A layout of the segments is shown in Figure 2.

As mentioned previously, the segments are designed for short- and long-term ground loads as well as the high hydrostatic loads until the final lining is installed. Ground loads were conservatively based on Terzaghi's rock loading classification table for completely crushed rock (see Table 2).

The segments are also designed for the high TBM thrust loads required to advance the TBM. Potential bending moments caused by eccentric thrust loads on the segments are also considered.

After the segmental ring is erected, the annulus between the ring and the surrounding ground must be grouted soon after the ring exits the tail shield. Two stages of grouting are anticipated. Stage I grouting ensures confinement of the segments by filling the voids between the segments and the excavated rock. Stage II grouting is used to control heading water inflows and is triggered based on the Stage I grout volume and the amount of muck excavated. Both Stage I and II grout loads were modeled for a maximum of 50 psi above the hydrostatic head applied over a 67.5 degree angle on the segment. The critical angle was determined by a sensitivity analysis.

Additionally, the segment design is checked for loading conditions during handling and transport, including segment build in the tunnel. Analysis showed that a minimum concrete strength of 2,000 psi is required before the segments can be stripped from the molds.

## SEGMENT DESIGN

The final design of the segments is left to the Contractor to fit his means and methods. However, a working design for basis-of-bid will be provided in the Contract Documents (Figure 3 is a working design of a precast segment ring). Some of these guidelines and key issues are briefly discussed in this section.

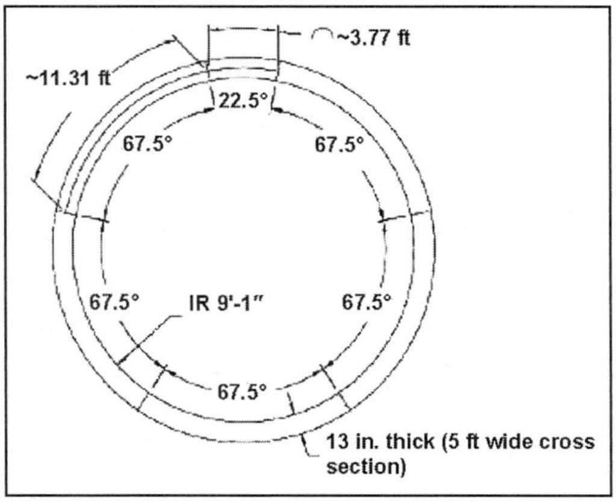


**Figure 2. Segment configuration**

### Methods of Analysis

Preliminary analyses were completed for the lining using closed form, ground-lining interaction analyses developed by Ranken et al. (1978) which were used to develop preliminary thickness, strength, and reinforcement requirements. A beam-spring model using STAAD Pro (Bentley, 2010) was then developed for final design, and is discussed herein.

### Beam-Spring Analyses

A two-dimensional, beam-spring numerical model using the structural program STAAD Pro (Bentley, 2010) was created to assess the controlling load combinations (see Figure 4). The model assumes that the surrounding ground stays in intimate contact with the lining and that any outward translation of the lining will result in a passive reaction of the ground. The passive reaction of the ground is modeled by compression-only radial springs located on each node of the lining, which release if any tension is induced in the spring.

Janssen's Method, was selected for joint behavior, as it provides an intermediate condition of the joint between a full hinge and an intact lining. The segment joints are modeled as partial hinges allowing some moment transfer through the joint where the amount of moment transfer is based on the joint's rotational stiffness. The rotational stiffness is derived by representing the joint as an equivalent concrete beam based on the contact between the adjacent segments, which is controlled by the packing material dimensions.

The program output from the STAAD model provides bending moments, axial thrusts, and shears induced in the lining. These results can be plotted

**Table 2. Rock load classifications**

| Rock Condition | Road Load $H_p$, in feet | Remarks |
|---|---|---|
| Hard and intact | Zero | Light lining, required only if spalling or popping occurs. |
| Hard stratified or schistose | 0 to 0.5 B | Light support. |
| Massive, moderately jointed | 0 to 0.25 B | Load may change erratically from point to point. |
| Moderately blocky and seamy | 0.25 B to 0.35 (B + $H_t$) | No side pressure. |
| Very block and seamy | 0.35 to 1.10 (B + $H_t$) | Little to no side pressure. |
| Completely crushed but chemically intact | 1.10 (B + $H_t$) | Considerable side pressure. Softening effect of seepage towards bottom of tunnel requires either continuous support for lower ends of ribs or circular ribs. |
| Squeezing rock, moderate depth | 1.10 to 2.10 (B + $H_t$) | Heavy side pressure, invert struts required. Circular ribs are recommended. |
| Squeezing rock, great depth | 2.10 to 4.50 (B + $H_t$) | |
| Swelling rock | Up to 250 ft irrespective of value of (B + $H_t$) | Circular ribs required. In extreme cases, use yielding support. |

From Terzaghi (1946, Table 3, figure references omitted, B = tunnel width, $H_t$ = tunnel height).

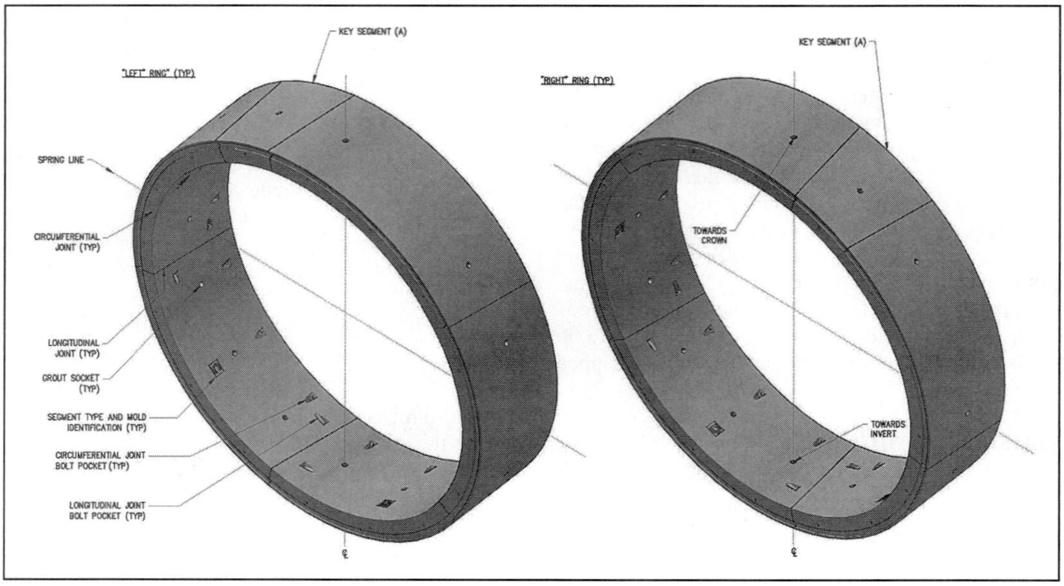


**Figure 3. Precast segment rings**

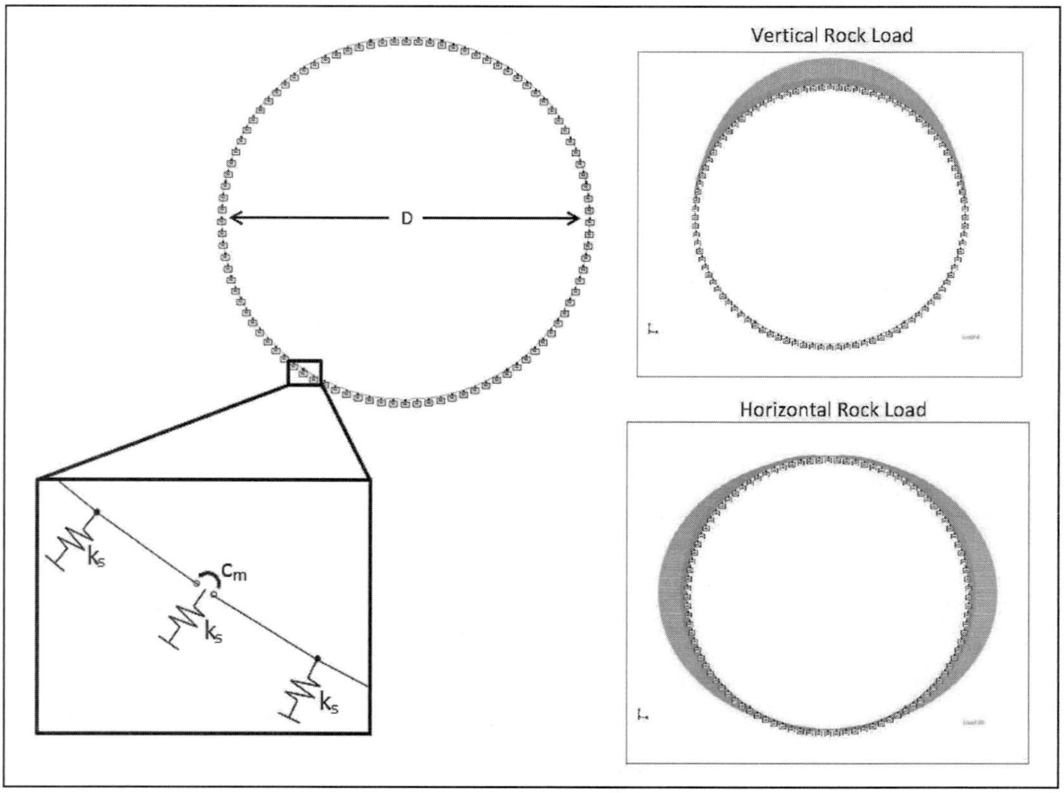


**Figure 4. STAAD Beam Spring Model (left) and example ground loading (right, top and bottom)**

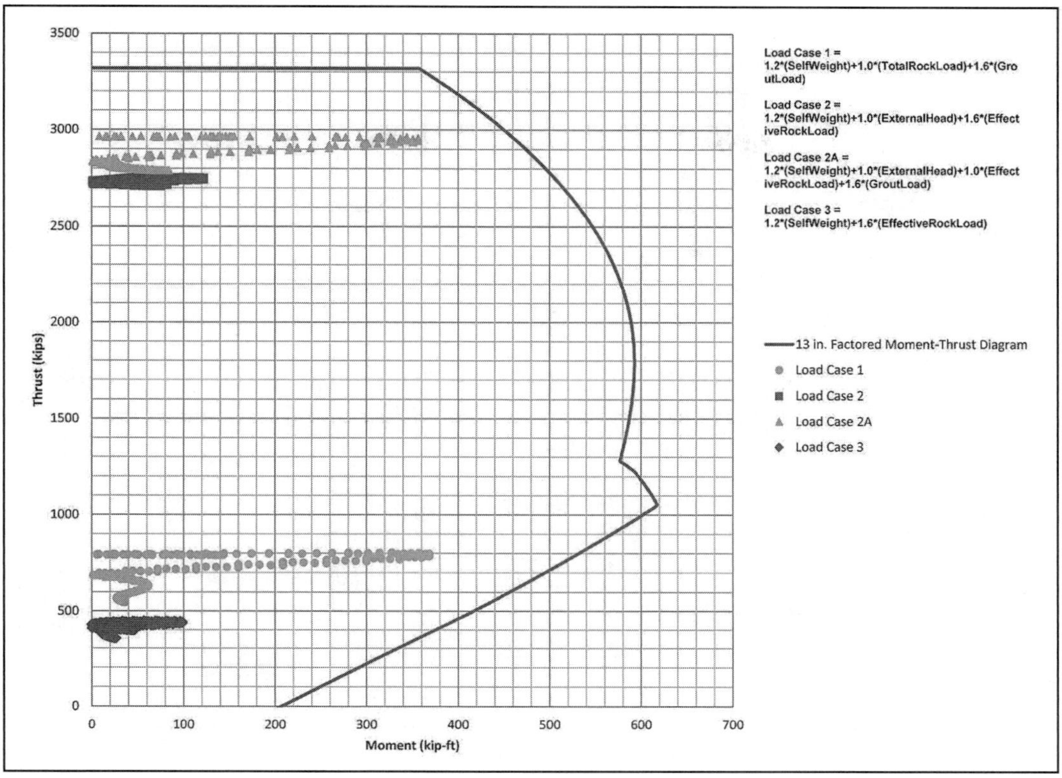


**Figure 5. Moment-thrust interaction diagram**

on moment-thrust interaction diagrams. Output from several design load cases can be superimposed on the interaction diagram and checked to see if they fall within a precast segment's capacity envelope. For short term rock and grout loading (LC1), short term rock and full hydrostatic loading (LC2), short term rock, hydrostatic and secondary grouting (LC2A), and long term rock loading (LC4), the data points are shown in Figure 5. An example of the preliminary segment reinforcing is also shown for reference in Figure 6.

## JOINT DESIGN

### Longitudinal Joint Design

In the case of high hydrostatic pressures, the segmental lining is under a uniform ring thrust with negligible or no bending moments. The primary failure mode occurs as tensile splitting at the longitudinal joints. This was observed by large-scale tests performed on prototype segment designs from the Channel Tunnel (Eves and Curtis, 1992) and the Arrowhead Tunnels Project (Swartz et al., 2002). The load transfer from segment to segment occurs over a smaller area (assumed as approximately 50% of the segment thickness because of a packer or slightly raised bearing pads). This results in tensile stresses behind the face of a joint extending to typically 80% to 85% of the segment thickness into the segment. Joint reinforcement is required to resist these tensile stresses since the concrete itself is typically assumed to have minimal capacity. Low angle shear failure across the joint, which occurs along a shear plane at approximately a 26 to 27 degree angle, is also considered. Shear failure is more critical for high eccentric loading conditions and for wider bearing surfaces.

Joint bursting capacity can be checked using two methods—as described by Eves and Curtis (1992) and Swartz et al. (2002). The first method uses empirical equations based large-scale tests and are complemented by theoretical derivations to estimate the loads that will cause splitting and shearing failures at the longitudinal segment joints. This method is valid for a reinforcement ratio between 0.375 and 1.2 (Eves and Curtis, 1992), where reinforcement ratio is the joint capacity from the steel reinforcement divided by the assumed tensile capacity of concrete. It should be noted that the amount of steel considered in their equations are for one side of the joint only for tensile splitting (as it only takes

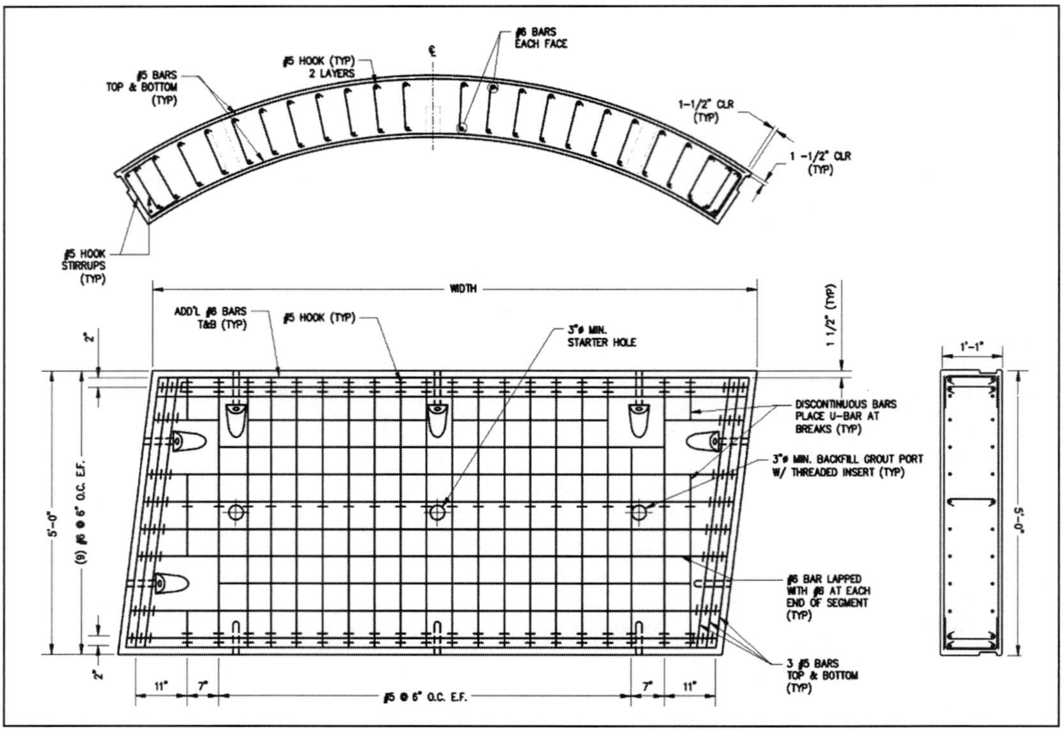


**Figure 6. Preliminary precast segment rebar layout**

one side to fail) and both sides of the joint for shearing failure.

The second method is based on Leonhardt (1964), which was originally developed for prestressed concrete. This method neglects the tensile strength of the concrete, and assumes that all load is taken by the reinforcement. This method estimates the amount of tensile force at the joints that must then be resisted by joint reinforcement. A strength reduction factor or appropriate partial factor of safety should also be applied to the resisting force from the reinforcement. A strength reduction factor of about 0.77, corresponding to a partial factor of safety of 1.3, is recommended for this method. This method gave a more conservative reinforcement value and controlled for the Bypass Tunnel segment joint calculations. See Figure 7.

### Circumferential Joint Design

The thrust cylinders of the TBM push off the segmental rings, resulting in loads at discrete points through jacking shoes on the circumferential joints. These loads can also cause shearing or splitting damage to the precast segments. Calculations and rebar placement are similar to the longitudinal joint reinforcement design, but thrust from hydrostatic pressure is replaced by TBM thrust. Consideration also has to be made for the size of the thrust ram shoes, as the thrust loads are applied over a limited portion of the circumferential joint.

For the high TBM thrust forces anticipated for this project, the circumferential joint geometry also must be designed to limit the offset between the centerline of the TBM thrust rams and the centerline of the segments. Any significant eccentricity will induce bending moments in the segments adjacent to the circumferential joints.

### Gasket Design

The segments are designed to be watertight until the final lining is installed. This is achieved by use of EPDM (ethylene propylene diene monomer) gaskets along the longitudinal and circumferential joints. The Bypass Tunnel segment gaskets are designed to withstand the maximum groundwater head conditions with a factor of safety of 1.5, resulting in a design pressure of 35 bar. The factor of safety is based on tests performed for the Arrowhead Tunnels. These tests show that the gaskets are anticipated to relax 65% for a design life of 5 years. Also considered in segment design is the maximum gap allowed to ensure that the gaskets can withstand a 35 bar hydrostatic pressure based on tightness tests. The design gap is the sum of tolerances to ensure water

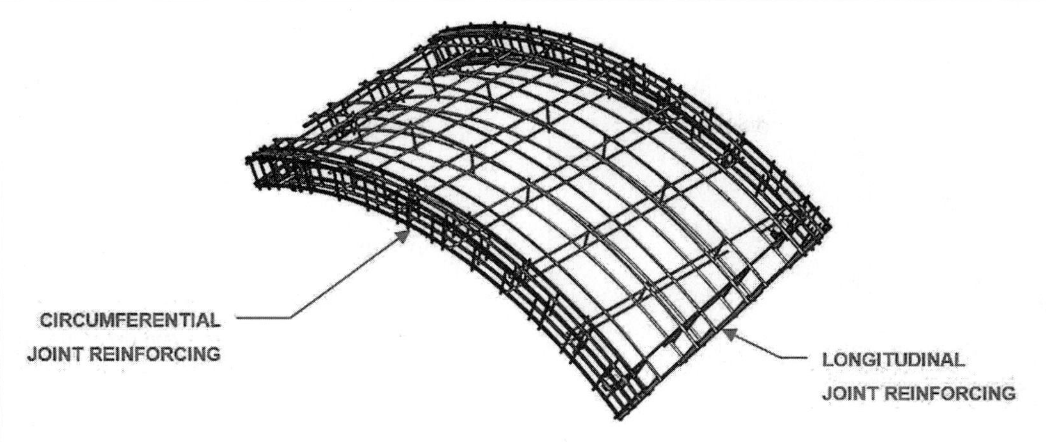


Source: Shop drawing for Brooklyn/Staten Island Siphon Replacement Tunnel (welded rebar cage)
Courtesy of CSI Tunnel Systems, NH

**Figure 7. General example of precast segment rebar**

tightness, which is verified by testing. An offset of the gaskets relative to each other is also considered, to account for lips and steps during ring build. This offset is typically set at about 10 to 15-mm. The compression of the gaskets causes line loads along the joints that are used as input for the required pull-out capacity of the connectors, which is discussed below.

Gasket compression line loads required to seal against the external hydrostatic pressures will result in an additional localized load on the segments at the extrados joint edge. This localized load will be applied as a line load along the length of the gasket. As there is little reinforcement in proximity of the gasket groove, the concrete will need to resist these loads. The magnitude of the line load will be dependent on a number factors, including the gasket material stiffness, the gasket and gasket groove geometry, and the compression force required to seal against the anticipated hydrostatic pressures.

### Packing Material Requirements

Packers are used in joints to provide more uniform distribution of high thrust loads across the segment joint surface, and to prevent concrete-to-concrete contact. Because of the ring deformation at ring build plus deformations due to external loads, longitudinal joints rotate, resulting in birdsmouthing. Birdsmouthing can cause localized high stresses and potential spalling of the concrete segments if concrete-to-concrete contact is not prevented by the use of packers.

Packers need to have certain properties to satisfy design requirements. A packer should allow initial deformation to better distribute loads. Deformations should be primarily in the direction of loading by crushing without significant deformations in other directions. In other words, packer material needs to have low Poisson's ratios. At some point the packer must be stiff enough to withstand the high stresses without excessive deformation or crushing.

A number of materials are typically used for packers. Marine grade plywood has good initial deformation properties because of the closure of voids in the plywood, the material exhibits a low Poisson's ratio, and it has strain hardening behavior to limit the amount of deformation for higher stresses. Bituminous packer boards typically have good initial deformation and crushing, but exhibit elastic behavior resulting in shear stresses in the lateral direction along the joint face. Also, the material does not typically strain-harden, so deformation continues under constant pressure, and it oozes out of the joint under high loading conditions (see Figure 8). Composite materials are also now commercially available, but experience with these materials is limited.

For the Bypass Tunnel, Marine Grade A-B plywood packers will be required in the longitudinal joints and optional in the circumferential joints. Based on the segment design and assumed joint layout, recommended dimensions for longitudinal packers are 6.5 inches wide (half segment width) and 0.25 inch thick. Alternative packer materials will be considered, but must exhibit behavior consistent with the plywood material selected.

### Taper Requirements

There are three angles considered for segment design. The first one is for the design of the ring taper required to negotiate the horizontal curves along the alignment. The required taper can be designed for a

**Figure 8. Mastic packer material test (Arrowhead project)**

single side or for both sides of the ring. The Bypass Tunnel alignment has two 1,000-foot radius curves. For assumed 5-foot-long segments, the calculated taper is approximately 0.3 degree.

The second angle is related to the segment shape. Other than the key segment and the two adjacent segments, which have to be trapezoidal, the rest of the segments can either be rectangles or parallelograms based, on this longitudinal joint angle between the segments. The Bypass Tunnel project allows for either configuration based on the Contractor's means and methods, with a minimum of 7 degrees for the parallelogram-shaped segments.

The third angle is for the key segment and is related to whether the joints are true radial or angled to allow for ease of installation. For the Bypass Tunnel design, modifying joint angles on the key segment for ease of installation is prohibited because of the extremely high loading conditions.

### Connector Requirements

Connectors between rings and segments serve two main purposes: to maintain gasket compression until grouting has locked the segments in place, and to hold the joints in alignment prior to ring grouting. The gaskets along the circumferential joints are initially compressed when the thrust of the machine pushes against the segments, and the gaskets along the longitudinal joints are initially compressed by the segment erector. Once the ring is built, connectors are required to maintain some of the compression. For the Bypass Tunnel project, bolts are required at the longitudinal joints because of the anticipated high gasket compression forces. For the circumferential joints, it is currently anticipated that performance requirements will be provided in the specifications to allow the Contractor to choose between either bolts or dowels. The circumferential joint connector should have certain deformation characteristics to ensure that gaskets are compressed enough to be watertight, and to satisfy pull-out and shear capacity based on the TBM thrust loading and installation conditions. Based upon design calculations for circumferential joints, three bolts or dowels, each with approximately 20 kips pull-out capacity, are required. This calculation was based upon the compression characteristics of the M385 73 Datwyler gasket used on the Arrowhead Tunnels Project, to seal against the design water pressure of approximately 35 bar.

## ADDITIONAL CONSTRUCTION MEASURES

For TBM excavation along the main drive, preliminary groundwater inflow analysis indicates heading inflows over 1,000 gallons per minute (gpm) for an approximately 1,500-foot fault zone if left untreated. Because of the anticipated high groundwater inflows for a portion of the alignment, two segment types are recommended: Type I segments, with grout ports that do not fully penetrate the segment; and Type II segments, with grout ports that fully penetrate the segment thickness, as well as marked locations for drain hole installation. The Contractor will be required to use Type II to deal with high groundwater inflows, which will be identified through mandatory probing and pre-excavation grouting. Type II segments will allow easier backfill grouting through the segments and lowering of the water head by the use of drain holes. It is anticipated that the Contractor will take advantage of drain hole installation to divert heading inflows into the excavation in a controlled manner, thus allowing the TBM to manage the remaining heading inflow with a maximum capacity of 800 gpm. The Bypass Tunnel segment design is also equipped with contractor-designed modified grout inserts. The Contractor has to install these modified grout inserts at locations measuring over 700 feet of groundwater head to monitor the groundwater pressure behind the segments. This is to ensure the segment design capacity is not exceeded.

## RECOMMENDATIONS AND CONCLUSIONS

Precast concrete segmental linings have been used successfully on projects with high hydrostatic loading conditions in the last 10 to 15 years, two of which have been mentioned earlier in this paper. As such, the design of the segments for the Bypass Tunnel project presented herein is based on proven and reliable segment design and construction considerations from similar past projects. The design shows that in high head conditions, thrust loads on the precast segment joints are a major concern but potential eccentricities also need to be considered for short term loading and proper reinforcement must be detailed at the joints to avoid critical splitting failures.

Additionally the proposed groundwater control plan utilized in the overall design scheme is essential for the high inflow conditions.

It is important to provide a working design on the Contract Documents for basis-of-bid. The Bypass Tunnel segment design limits certain aspects of design based on previous experiences but allows modifications as long as the alternatives do not compromise the stability of the design. At the same time, the design requirements are more performance based rather than descriptive to allow the Contractor to modify the design to fit his means and methods.

## REFERENCES

Bentley. 2010. STAAD Pro.

Blom, C.B.M. 2002. Design philosophy of concrete linings for tunnels in soft soils. PhD Thesis for Delft University of Technology.

Eves, R.C.W., and D.J. Curtis. 1992. The Channel Tunnel Part 1—Tunnels and tunnel lining design and procurement. In *Proceedings of the Institute of Civil Engineers, Special Issue,* Paper 10045, London.

Leonhardt, F. 1964. *Prestressed Concrete Design and Construction*, 2d edition. Berlin: Wilhelm Ernst and Sohn.

Muir Wood, A.M. 1975. The circular tunnel in elastic ground. *Goetechnique,* 25(1), 115–127.

Proctor R.V. and T.L. White. 1988. *Rock Tunneling with Steel Supports.* Youngstown, Ohio. Commercial Shearing and Stamping Co.

Ranken, R.E., J. Ghaboussi, and A.J. Hendron. 1978. Analysis of Ground-Liner Interaction for Tunnels. Report No. UMTA-IL-06-0043-78-3, US Department of Transportation, Washington, DC.

Swartz, S., H. Lum, M. McRae, D.J. Curtis, and J. Shamma. 2002. Structural design and testing of a bolted and gasketed pre-cast concrete segmental lining for high external hydrostatic pressure. In *Proceedings of the North American Tunneling Conference, 2002, Seattle, WA,* 151–160.

# Refurbishment of the Ross Shaft at the Sanford Underground Research Facility in Lead, South Dakota, USA

**Syd De Vries**
SURF/Lawrence Berkeley Lab

**ABSTRACT:** The Sanford Underground Research Facility (SURF) is an Underground Science Laboratory located at the old Homestake Gold Mine in Lead, South Dakota. SURF has been constructed to host sensitive particle physics experiments requiring shielding from cosmic radiation. The Ross Shaft is one of two access shafts used to access physics experiments currently located nearly 5,000 feet below the surface. Built in the early 1930s, Ross Shaft has been subject to significant deterioration and requires a complete refurbishment to support construction of future underground laboratories. This paper discusses the design and construction associated with the refurbishment of the Ross Shaft.

## ORIGINAL SHAFT

The Ross Shaft was constructed in stages with the first stage completed from the surface to 3,200 ft. depth in late 1934 and eventually completed to a depth of 5,171 ft. Levels were established at various elevations to access the ore body and were identified based on their depth in feet below shaft collar. Full hoisting and skipping operation from the 5000 Level (L) was established in 1956. The shaft consists of six compartments including Cage, Counterweight, Manway, Utilities and Skips. Figure 1 shows the layout of a typical shaft set.

## DESIGN

In 2007, a decision was made to re-open the mine to prepare it to be developed as an Underground Science Laboratory. As part of the re-opening process, Arup USA was hired to perform design engineering services. The design scope included site investigations to determine the condition of existing infrastructure to support lab development as well as facility upgrades design. Arup procured the services of SRK Consulting in Denver CO. who subcontracted to G.L. Tiley and Associates for all design activities associated with underground shafts.

Under the direction of Tiley, a series of visual inspections and NDT tests were performed on the existing steel to determine its capacity to support current operations and future Laboratory construction. The load bearing capacity of shaft steel sets needs to consider the maximum load that is applied to the structure during operation which include vertical forces applied to the Cage Guides in the event of an emergency stop or dogging event, horizontal forces applied to steel sets as conveyances ride along guides as well as the static loads of the set steel frames suspended from a bearing set as well as the connected utilities. Using the NDT data, a Finite Element Analysis (FEA) was performed on the existing steel sets to determine its capacity to support operation considering the following:

1. The total weight of a steel set is 6,100 lbs
2. Cage weight = 8,800 lbs
3. Skip weight = 16,500 lbs, Skip payload = 22,000 lbs
4. Maximum Dogging load = 2.5 G, where G = Total Weight of Conveyance
5. Horizontal Impact load = 0.5 G North/South and East/West
6. All Bearing sets are replaced with new steel and support 26 Typical sets
7. Every fourth Typical set is replaced with new Hollow Structural Steel (HSS)

The analysis revealed that the existing shaft furnishings could not support even reduced conveyance speeds and weights without significant shaft steel replacement. Table 1 shows a summary of the FEA results.

Based on the results listed in Table 1, a decision was made to perform a complete removal of old shaft furnishings and replacement with new.

## BASIS OF DESIGN

It was decided during design scoping meetings that the existing Shaft hoisting capacities for men, materials, equipment and waste rock removal were sufficient to support future Laboratory construction and operation. Therefore, the design for new shaft steel furnishings would be based on the loads generated

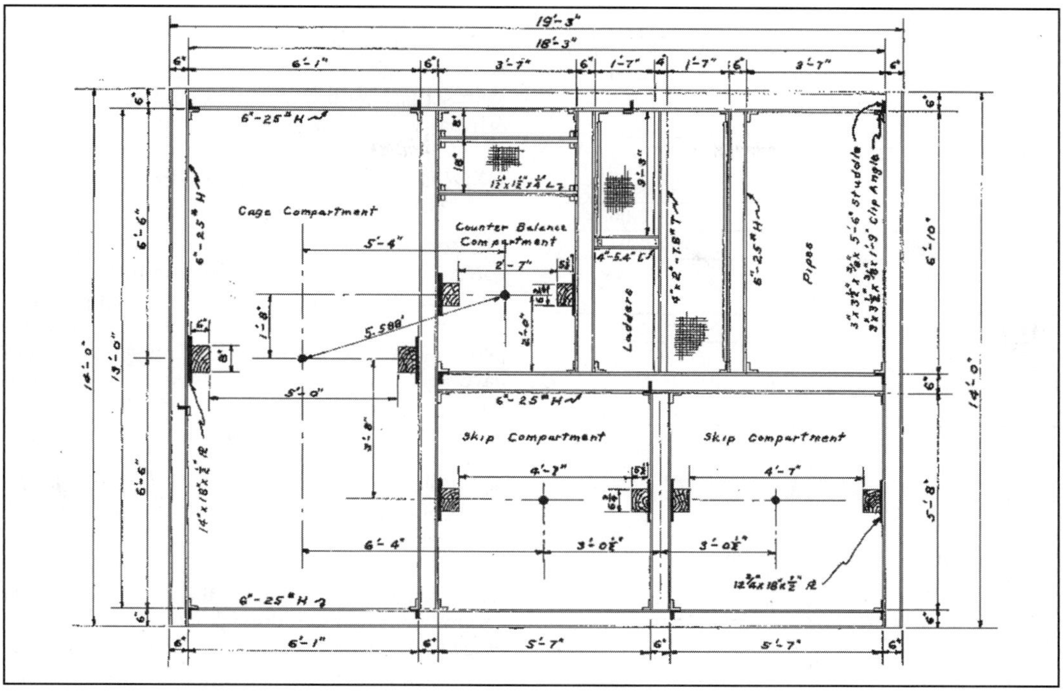


**Figure 1. Old shaft steel typical set general arrangement**

**Table 1. FEA results**

| | Cage Conveyance Payloads | | |
|---|---|---|---|
| | 8,000 lb | 6,000 lb | 4,000 lb |
| Total beams analyzed | 6,808 | 6,808 | 6,808 |
| Total beams requiring replacement | 6,037 | 5,268 | 5,136 |
| % Shaft steel replacement required | 89 | 77 | 75 |

from these hoisting capacities. Table 2 lists the shaft operating parameters used in the design which includes hoisting speeds, conveyance payloads and dimensions.

Working within the shaft operating parameters listed above, a design for shaft steel replacement was undertaken. The details of the new design are compared to the old design in Table 3.

Another design consideration was the requirement that access over the entire length of the shaft be maintained throughout the refurbishment process. Pump stations are located at the 1250, 2450, 2600, 3650, and 5000 L's and need to be accessed routinely for maintenance. A 12 in. diameter pump discharge column exists in the shaft and must continue to operate as well as electrical and communication lines that service the pump stations. In addition, the Ross Shaft is required to provide emergency egress for personnel from the underground in the event of

**Table 2. Shaft operating parameters**

| Specifications | | Ross Shaft | |
|---|---|---|---|
| | | Service Hoist | Production Hoist |
| Payload | Mass (tons) | 6 | 11 |
| | Personnel | 60 | N/A |
| Cage dimensions inside per deck | No. of decks | 2 | N/A |
| | Height | 7 ft. | N/A |
| | Width (ft) | 4 ft 8 in. | N/A |
| | Length (ft) | 12 ft 4 ½ in. | N/A |
| Hoisting speed | ft/min | 2,200 | 2,800 |
| Hoist power rated | Hp | 1,500 | 2,400 |
| Production capacity | Tons per day | N/A | 3,000 |

a failure in the Yates shaft. As a result, there could be no substantive changes to the layout of the various compartments ensuring that conveyances would be able to travel through the refurbished portion of the shaft and transition into the old shaft furnishings throughout the duration of the shaft refurbishment program. This requirement would not only impact the layout of the newly refurbished shaft but would also affect the construction schedule as periodic construction shutdowns would be required to allow for maintenance personnel to access the pump stations

**Table 3. Shaft design details**

| | Old Shaft Steel Design | New Shaft Steel Design |
|---|---|---|
| Typical set steel construction | WF6 × 25 | 5 in. × 7 in. × ½ in. HSS |
| Bearing set steel construction | 3 × W12 × 50 beams | 2 × W18 × 40 beams |
| Typical set weight | 6,100 lbs | 18,000 |
| Bearing set weight | 9,100 lbs | 19,500 |
| Utilities weight per set | Unknown | 11,500 |
| Typical set spacing | 6 ft | 18 ft |
| Bearing set spacing | 150 ft | 180 ft |
| Dogging load (2.5G) | 53,000 lbs | 53,000 lbs |
| Total bearing set load | 208,500 lbs + Utilities | 349,500 lbs |
| Cage guides | 6 in. × 8 in. Kari Wood | 5½ in. × 8 in. Fir |
| Skip/counterweight guides | 6 in. × 8 in. Kari Wood | 5 in. × 7 in. × ½ in. HSS |
| Design life | Unknown | 50 years |

for pump maintenance and shaft inspections would be required to ensure safe passage for emergency egress. As such, the general arrangement of a typical shaft set has not been altered significantly from the original as is shown in Figure 2.

## GEOLOGIC SETTING AND GROUND SUPPORT

The Ross Shaft is sited within the eroded Precambrian core of the northern Black Hills uplift of western South Dakota. Rocks exposed within the Ross Shaft include Early Proterozoic and Phanerozoic metamorphosed sediments and volcanics. This series of rocks have been regionally metamorphosed from greenschist to upper amphibolite facies and are highly deformed through multiple events which developed strong foliation and re-healed geologic discontinuities. These metamorphosed rocks were subsequently cross cut by later Tertiary-age rhyolite and phonolite intrusions.

Original ground support in the shaft consists of reinforced concrete walls for the first 300 ft and at various other locations throughout the shaft. The rest of the shaft is supported with a ½ in. layer of sprayed

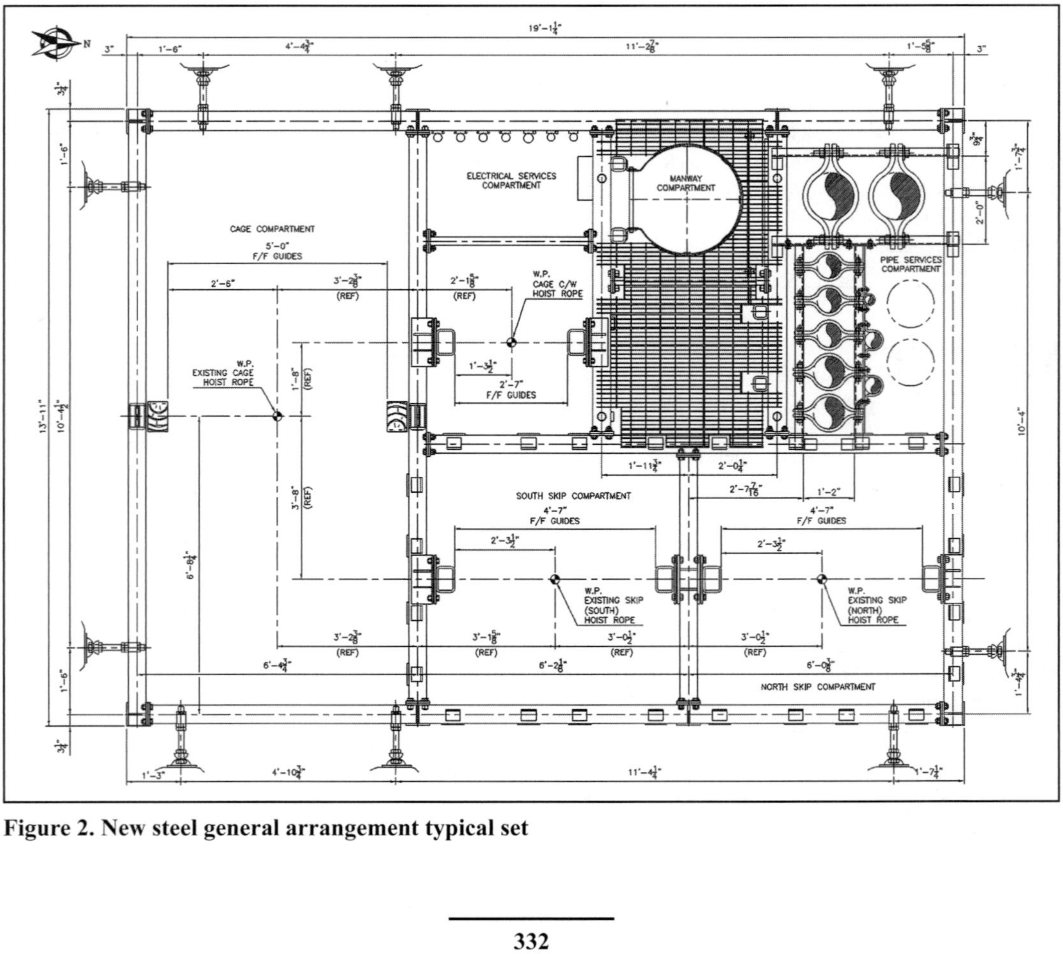


**Figure 2. New steel general arrangement typical set**

**Table 4. Types of ground support**

| Location | Type of Anchor | Dimensions | Type A | Type B | Type C |
|---|---|---|---|---|---|
| North and South Walls | Mesh Anchors | Length (ft) | 4 | 4 | 4 |
| | | Spacing (ft × ft) | 6 × 4 | 3 × 4 | 3 × 4 |
| | Primary Anchors | Length (ft) | 5 | 5 | 5 |
| | | Spacing (ft × ft) | 6 × 4 | 6 × 4 | 3 × 4 |
| East and West Walls | Mesh Anchors | Length (ft) | 4 | 4 | 4 |
| | | Spacing (ft × ft) | 6 × 4 | 3 × 4 | 3 × 4 |
| | Primary Anchors | Length (ft) | 7 | 7 | 7 |
| | | Spacing (ft × ft) | 6 × 4 | 6 × 4 | 3 × 4 |

on gunite and a combination of resin grouted rebar, split set bolts, expanded metal mesh and mats at sporadic locations. Corrugated steel lacing is provided as an additional layer of protection and was installed between sets around the perimeter of the shaft.

As part of the shaft refurbishment program, all old steel furnishings are being removed, including the existing perimeter lacing. In doing so, a form of ground support is removed. It was decided that new ground support would need to be installed to ensure that at all times, shaft refurbishment crews are working below secured ground. New ground support would consist of resin grouted ⅞ in. diameter Dywidag rebar and 3 in. × 3 in. × #6 gauge welded wire mesh. Newly installed ground support in the shaft can be seen in Figure 3.

There are essentially three Types of ground support that are used in the shaft and are listed in Table 4.

Ground support determinations are made following a Rock Mass Assessment performed by a Lab Geotechnical Engineer. The Geological Strength Index (GSI) is calculated during this assessment and typically ranges between 75 and 85. The minimum ground support required based on these values is Type A. However, recommendations for increases in ground support from Type B to Type C are based on the presence, condition and orientation of wedges, shear zones or localized discontinuities. Pull tests are performed on a frequency of 3 tests per 180 linear feet of shaft.

## CONSTRUCTION

### Work Decks

In order to execute the shaft refurbishment program, work decks are required to enable the removal of the existing steel furnishings, install new ground support, new steel furnishings and utilities. These work decks are designed to safely access both the old and the new steel.

**Figure 3. Shaft turnbuckle temporary support**

The Cage compartment is equipped with a three level work deck with an overall dimension of 41 ft high × 12 ft 6 in. long × 4 ft 11 in. wide. The upper deck is furnished with an articulated one Ton pneumatically operated jib crane for rigging old and new steel. The jib crane is able to fold into the work deck space during conveyance within the shaft but can expand to reach 11 ft in its open position.

The South Skip compartment is equipped with one of the original ore skips fitted with a work deck on top of it. This skip is located below the platform and is used as a container for waste material generated during the refurbishment process to be hoisted to the surface and dumped for final disposal.

The North Skip Compartment is equipped with a four level work deck with an overall dimension of 48 ft high × 5 ft 3 in. long × 4 ft 6 in. wide. A one-ton articulated jib crane has also been installed on this work deck and has a total reach of 6 ft.

In addition to the Cage and North Skip work decks which are operated from the Service and Production hoists, a Swing Stage is operated in the

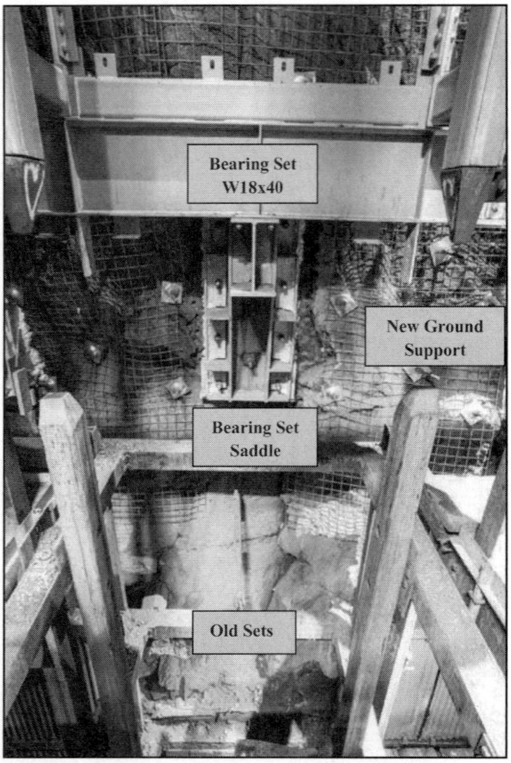


**Figure 4. New bearing set installation (photo by Steve Babbit)**

Manway compartment to provide access for steel assembly in the North West corner of the shaft. This platform is suspended from aircraft cable attached to an 8 in. × 4 in. Aluminum I-beam anchored in a completed new set above the work area and is raised and lowered with electric winches mounted on the platform. Each set up covers a vertical span of 100 ft before the unit needs to be torn down and re-set up lower in the shaft.

### Old Steel Removal

All old steel sets in the shaft are suspended from Bearing beams anchored into the shaft walls at 150 ft intervals. Since the method of steel replacement follows a top down method, temporary shoring of the old steel is required before any old Bearing set is removed. This temporary shoring consists of six × 1 in. diameter × 4 ft long turnbuckles anchored into the West and East walls spaced every eight sets or 50 ft vertically. Figure 3 shows a turnbuckle installation.

Once the old sets are secured, a temporary platform is constructed over the Counterweight, Manway and Services compartment using 3 in. × 8 in. planks. With old sets secured and platforms in place, the removal of the old steel can commence. Oxygen and Acetylene cutting torches are used by crews to cut out the old steel. Water sprays below the work area provide fire protection during the cutting process. The Jib Cranes on the Cage and North Skip platforms are used to secure the old steel during the cutting process and load the cut steel pieces onto the Cage Work deck or South skip for hoisting to the surface. Only one set is removed at a time.

## NEW STEEL INSTALLATION

### Bearing Sets

New Bearing sets are installed every 180 ft vertically in the shaft. Each Bearing set consists of two W18×40 I beams that are anchored into the east and west walls of the shaft with three support weldments or "saddles" which are bolted to the shaft walls with ten × 1 in. diameter 150 KSI threaded rods anchored a minimum of 5 ft into the rock walls. The depth of the rods and configuration of the saddles depends on the span between the rock wall and the "I" beam wall plate. Prior to saddle installation, rock walls needs to be trimmed at the saddle locations to a flat surface to minimize the gap between the saddle back plate and rock wall to no more than 3 in. Drilling and blasting is not performed as part of the saddle excavation due to the requirement to maintain access and utilities throughout the shaft during refurbishment. Consequently, a combination of drilling, cutting with rock saws and breaking with chipping hammers are used for excavation at the saddle locations.

Once the saddles are bolted to the shaft walls, the space between the back plate and the wall rock is formed and filled with a high strength, cementitious grout. The grout is allowed to set for a minimum of one day to give it a strength of 3,500 psi before the nuts on the 1 in. thread bars are torqued to 300 ft-lbs. Figure 4 shows a completed Bearing set installed as well as the transition between the new and the old steel furnishings.

### Typical Sets

Typical sets are installed every 18 ft vertically in the shaft, suspended below a Bearing set as shown in Figure 5. With Bearing sets spaced 180 ft vertically, a total of 9 typical sets are normally installed between Bearing sets. Each typical set is constructed of 5 in. × 7 in. × ½ in. Hollow Structural Steel (HSS) and is connected vertically to the next set with 16 × 3½ in. × 2 ½ in. × ½ in. angle iron hangers and eight guides. All steel pieces are bolted together with ⅞ in. diameter galvanized A325 Tension Control (TC) bolts. The specifications for steel assembly call for all set steel connection bolts to be torqued to 500 ft-lbs. TC bolts are designed with splines that break off inside a special electric wrench when the designed

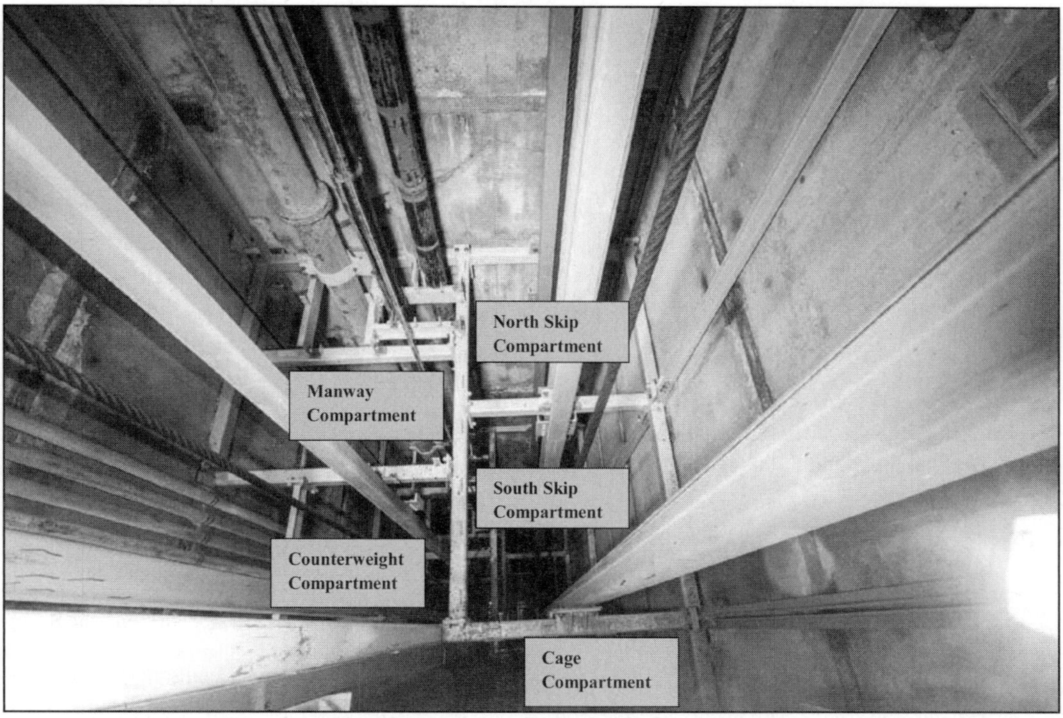


**Figure 5. Typical sets installation looking down the shaft (photo by Steve Babbit)**

torque has been reached. The use of TC bolts as compared to regular A325 Hex bolts has made steel assembly fast and efficient.

Additional components of typical sets include steel ladders and landings within the Manway compartment at specified locations as well as Expanded Metal Brattice installed around the perimeter of the Skipping compartments. These components are to be installed when the shaft refurbishment has reached predetermined locations.

### Shaft Alignment

Prior to the start of shaft refurbishment, two plumb lines were suspended the complete length of the shaft from which offsets were measured to the existing sets and wall rock locations. This information revealed that the shaft requires very little adjustment over the first 2,000 ft. Between 3,000 and 4,000 ft depth, the shaft penetrates the historic Homestake ore body. In this section of the shaft, in situ rock stresses generated from mining have resulted in as much as a 4¾ in. movement of the shaft furnishings to the south and 4¼ in. to the east. It is within this area where significant adjustments are required in shaft steel alignment.

For the shaft refurbishment, three plumb lines have been installed in the South West, South East and North East corners of the shaft. Each plumb line is constructed of 5⁄16 in. diameter aircraft cable suspended from electric winches located at the shaft collar and is long enough to be lowered the full length of the shaft. Based on the results of the initial pre-construction survey, offsets have been calculated from the plumb lines in all three corners for each set location and the set steel is adjusted to fit within these offset calculations. A shaft blocking system consisting of a total of ten 1–1½ in. diameter threaded bars are installed around the perimeter of the shaft steel set and are adjusted to obtain the required offsets. The maximum adjustment in any horizontal direction is ⅝ in. over 100 ft. Figures 6 and 7 show this blocking system as well as the shaft plumb line offset measurement.

### Key Elements of Success

Although the project is far from complete, the successful refurbishment of the first 1,250 feet of shaft can be attributed to a number of factors:

1. Early involvement of experienced shaft Construction Specialists and Infrastructure

**Figure 6. Set steel blocking installation**

**Figure 7. Shaft alignment measurements**

Techs in the design process and development of procedures,

2. Construction of a scaled model of the shaft complete with work decks and set steel pieces that were used to test the work steps and determine if there were fatal flaws in the plans,
3. Having the right equipment for the job including effective and practical work decks equipped with Cranes for rigging,
4. A simple design for steel replacement that is easily assembled in the field,
5. An engineering team that is responsive towards recommendations for improvement,
6. An experienced 3rd party review team that meets on a semiannual basis to review work progress and make recommendations for improvements,
7. A steel fabricator that has provided quality steel fabrications on schedule and been a demonstrated partner on the project, making valuable recommendations for improvement,
8. Conducting a trial fit up of two shaft sets in the fabricator's yard for inspection by Infrastructure Technicians,
9. Complete commitment to a successful project implementation by all stakeholders.

## ACKNOWLEDGMENTS

The author wishes to thank Ken Lau, Pierre Shields and their team of Engineers at G.L. Tiley and Associates for the Ross Shaft steel design. The SURF Engineering team is also thanked for their support in reviewing this paper and especially David Vardiman for his geological and geotechnical contributions. A special thanks goes out to the Construction team headed up by William McElroy and George Vandine, as well as all the shaft technicians for their support, without which the success of this project would not be possible.

## REFERENCE

Bjorge, Guy N., Ross, A.J.M., Johnson, J.D., Staple, S.J., Wiggert, J.F., 1935, Construction and Equipment of the Ross Shaft, Homestake Mining Company, A.I.M.E. New York Meeting.

# Research on the Effect of Risk Mitigation Measures Against Earthquake for Mountain Tunnel Through Static Loading Test

**Kosuke Kawata, Nobuharu Isago, Atsushi Kusaka, and Hideto Mashimo**
Road Technology Research Group, Public Works Research Institute

**ABSTRACT:** Rock tunnels are generally considered to have the required structural strength to resist earthquakes; recently, however, a number of rock tunnels in Japan have suffered severe earthquake damage to their permanent lining. Although several countermeasures are available to mitigate the risk of such damage, their mechanism and effect are not fully understood. In this study, static loading experiments assuming a mountain tunnel affected by an earthquake were carried out to clarify the mechanism of lining damage and the effect of countermeasures. The results revealed that the presence of void spaces behind the lining reduces the structural load-bearing performance, especially in tunnels constructed by conventional timber lagging method. Moreover, reinforcement such as inverted arch, rock bolts, lining rebars and carbon-fiber sheets for inner surfaces can potentially mitigate the risk of lining collapse due to earthquake.

## INTRODUCTION

A number of rock tunnels with no apparent unfavorable ground conditions suffered severe damage, including the collapse of their permanent lining, in large seismic events such as the Chuetsu Earthquake in 2004 (e.g., Mashimo 2005). Since these tunnels were considered to have the required structural strength to resist earthquakes, studies are being carried out to clarify the mechanical behavior of tunnels during earthquakes and the damage mechanism (e.g., Kusaka et al. 2009); however, the aseismic design methodology has not yet been established.

In many cases of earthquake-damaged tunnels, some defects already existed, such as improper construction and deterioration of lining. Although countermeasures are usually implemented against such problems, their earthquake-resistant effect was not fully understood.

In this study, static loading experiments were carried out assuming a mountain tunnel affected by earthquake motion to clarify the mechanism of lining damage and the effect of countermeasures.

## EXPERIMENTAL METHOD

### Experimental Apparatus and Models

Figure 1 shows both photographic and schematic views of the static loading apparatus used in the experiments. The mechanical behavior of the tunnel structure subjected to excessive force was examined. The experimental apparatus is composed of a movable wall for loading with oil pressure jacks and reaction walls. Three jacks are used, each one applying a load of up to 400kN. The soil tank is 1.3m in width, 1.3m in breadth and 0.3m in height. Displacement of the movable walls is up to 50mm. The specimen was not displaced in the longitudinal direction, because the tank was covered by a stiff plate during the experiments. The tunnel lining was made from mortar and set in the middle of the soil tank. The ground was made from a weak mortar mixture and the load was applied through the three oil pressure jacks from one side. Steel piles were set on the three planes so that the walls were perfectly fixed. Teflon sheets were used to eliminate friction between the tank and the specimen.

The tunnel was a $\frac{1}{20}$ scale model assuming a standard cross section of national road tunnel in Japan. The targeted uniaxial compressive strength of the lining was 18MPa. The ground model was prepared using a weak-mixture mortar with targeted uniaxial compressive strength of 0.5MPa. Table 1 shows the specification of the materials which were used for the tunnel lining and the ground. Figure 2 shows the load displacement curve of the weak-mixture mortar used for the ground. This curve was obtained from another test where only the weak-mixture mortar was cast in the tank and was loaded. From the test, it was found that the elastic deformation modulus was about 250MPa for the range from 0mm to 3mm displacement and 130MPa for the range from 3mm to 25mm.

The loading was stopped when the jacks reached their capacity or when the collapse of tunnel specimen occurred. The jack loads, the displacements of steel frame, the horizontal displacements of tunnel and the strains on inner and outer surfaces of the lining were recorded during experiment. Cracks

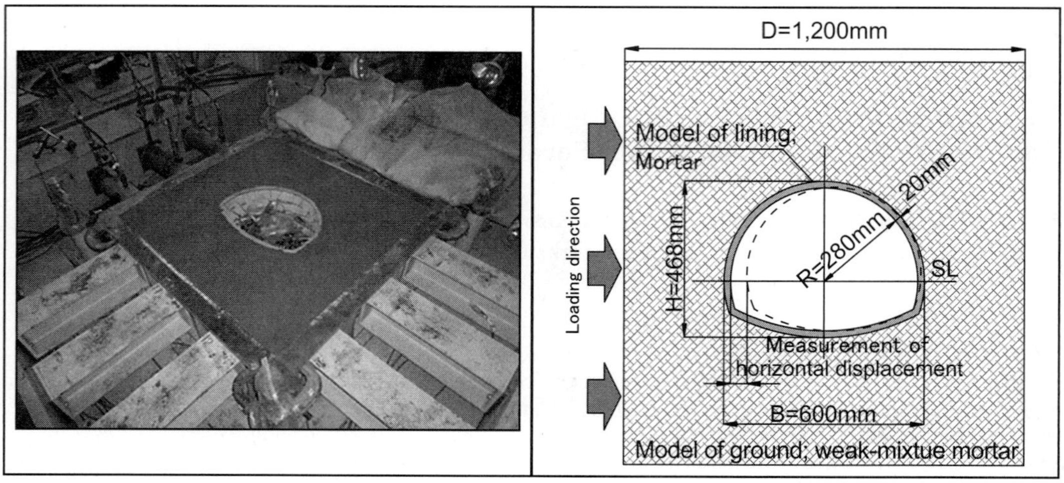


**Figure 1. Experimental apparatus**

**Table 1. The recipe of mortar and poor-mixture mortar**

| Mortar for Lining | | | | | Poor-Mixture Mortar for Ground | | | |
|---|---|---|---|---|---|---|---|---|
| | Unit Weight (kg/m$^3$) | | | | | Unit Weight (kg/m$^3$) | | |
| W/C (%) | Cement | Water | Soil | AE Water Reducing Agent | W/C (%) | Cement | Water | Soil |
| 66.8 | 352 | 235 | 1,575 | 3.52 | 191.8 | 55 | 105.5 | 1,000 |

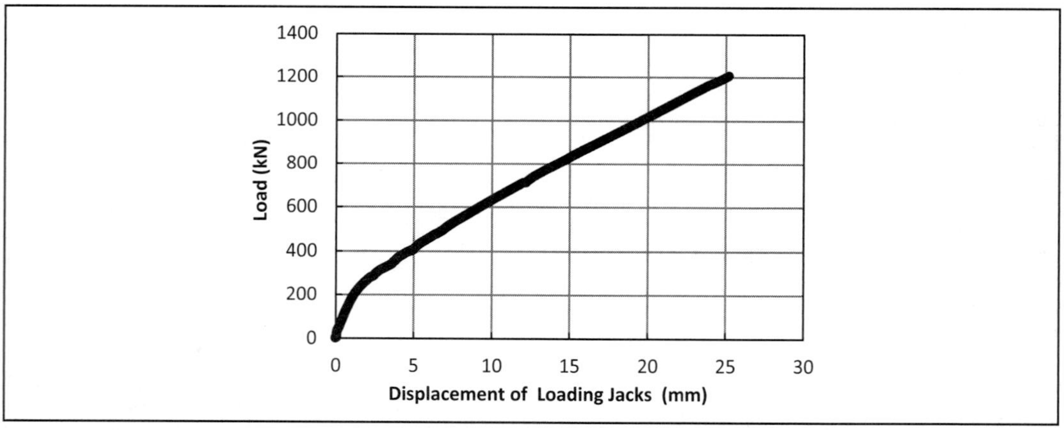


**Figure 2. Load displacement curve of the poor-mixture mortar in the tank without tunnel**

on the inner surface of the lining were also monitored through the observation window.

### Assumed Loading Mode

Based on past studies, earthquake damage to mountain tunnels is divided into three types, as shown in Figure 3. Type-I indicates flexural compression failure or flexural crack at the shoulder of the tunnel. Type-II indicates compression failure or flexural compression failure at the crown of the tunnel. Type-III indicates flexural compression failure or compression failure at the tunnel sidewalls.

These damage modes were mainly observed in tunnels constructed using conventional timber lagging method. According to Kusaka et al. (2011), the damage modes are relevant to the ground deformation mode for shearing (Type-I), horizontal compression (Type-II) and vertical compression (Type-III).

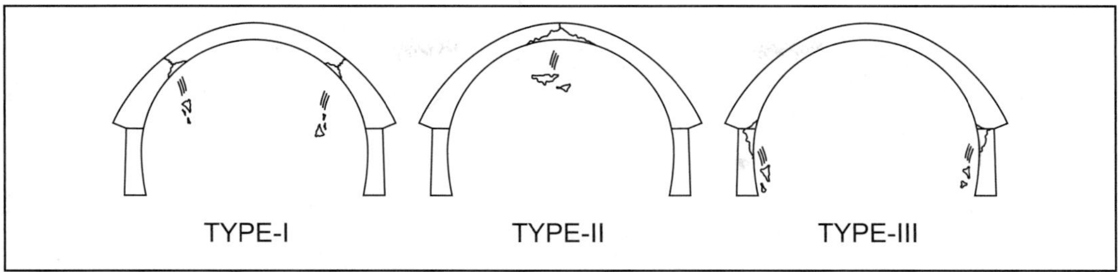


**Figure 3. Modes of seismic damage to mountain tunnels**

**Table 2. Experimental cases**

| | Risk Mitigation Measures Resisting Earthquake | | | Material Properties of Specimen | | | | |
|---|---|---|---|---|---|---|---|---|
| | | | | Ground | Mortar of Lining | | | |
| No. | Special Condition | Counter-measure | Materials and Range | UCS (MPa) | UCS (MPa) | Young's Modulus (GPa) | Poisson's Ratio | Tensile Strength (MPa) |
| 1 | Void behind lining | — | 60~120 degree | 0.434 | 26.5 | 15.7 | 0.18 | 2.53 |
| 2 | Without Inverted arch | — | — | 0.455 | 24.1 | 15.3 | 0.20 | 2.27 |
| 3 | — | — | — | 0.486 | 23.8 | 14.3 | 0.17 | 1.93 |
| 4 | — | Rebar | Wire Netting (Whole circumference) | 0.478 | 18.1 | 13.0 | 0.19 | 1.94 |
| 5 | — | Rockbolts | Brass Stick (15 degrees interval) | 0.458 | 21.8 | 14.9 | 0.19 | 1.84 |
| 6 | — | Sheet | Carbon Fiber Sheet (Inner surface of lining) | 0.478 | 24.2 | 15.8 | 0.19 | 2.12 |

Especially, Type-II potentially gives a direct injury to tunnel users.

In this study, Type-II was selected given its potential to cause direct damage in order to experimentally examine the mechanical behavior of tunnels. In other words, a ground deformation mode in which horizontal compressive strain is dominant was used in this experiment.

### Experimental Cases

Specimens simulating special conditions or risk mitigation measures are as shown in Table 2 and Figure 4. The effects of these conditions and measures were examined through comparison of Case 3, which is the basic case in this study, and other cases as described below.

Case 1 has void spaces behind the tunnel lining around the crown. These void spaces were located between 60° and 120° at a maximum depth of 5cm from the tunnel crown. For Case 2, an inverted arch was not considered. The other cases had an inverted arch. Case 3 was the general type, where the tunnel had an inverted arch and there were no void spaces behind the lining. Case 4 had rebars in the lining, which was modeled by wire mesh 1 mm in diameter at 1-cm intervals. Case 5 was reinforced by material simulating rock bolts. The bolts were made from a brass stick 3 mm in diameter and 150mm in length and placed at 15° intervals in the circumferential direction. For Case 6, carbon fiber sheets (50g/m$^2$) were attached to the inner surface of the lining.

## RESULTS AND DISCUSSION

The load-displacement relationship is shown in Figure 5, where the displacement indicates tunnel convergence in the horizontal direction and the load represents the total load of the three jacks. Failure load of the lining model is shown in Figure 6. Strain distribution of the tunnel specimens at specific steps is shown in Figure 7, where the compressive strain is indicated as a negative value. Solid symbols illustrate the inner strain of the lining and open symbols illustrate the outer strain. Recorded progress of the cracks on the inner surface of the lining is shown in Figure 8.

### Influence of Void Space Behind Lining Around Crown

Figure 5a compares the behavior of Case 1 (with void spaces) and Case 3 (without void spaces) using the

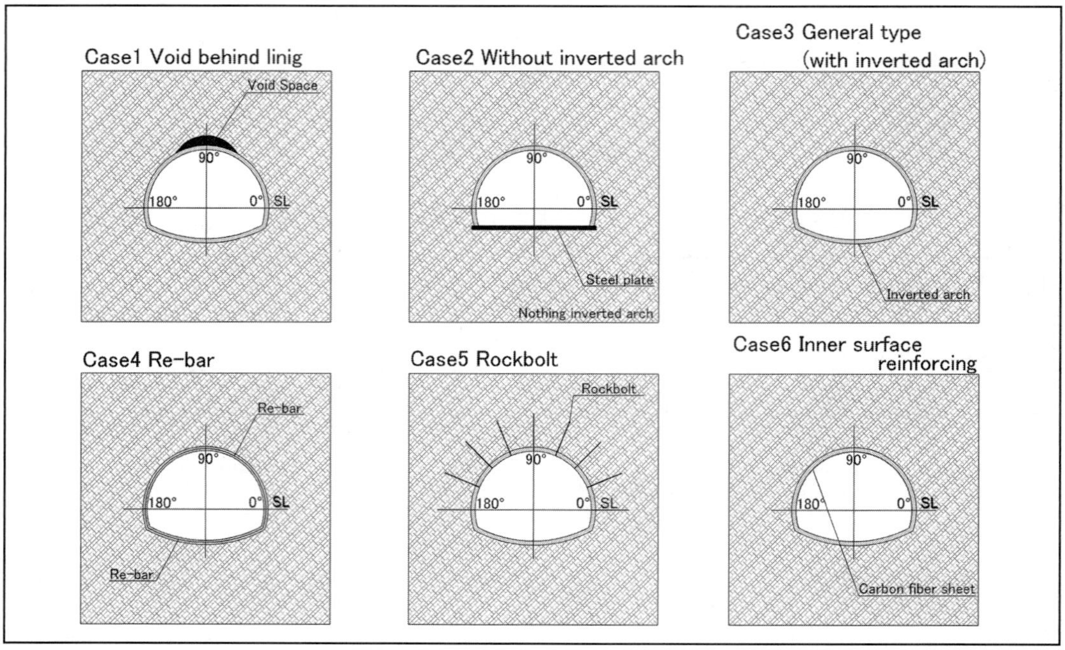


Figure 4. Overview of experimental cases

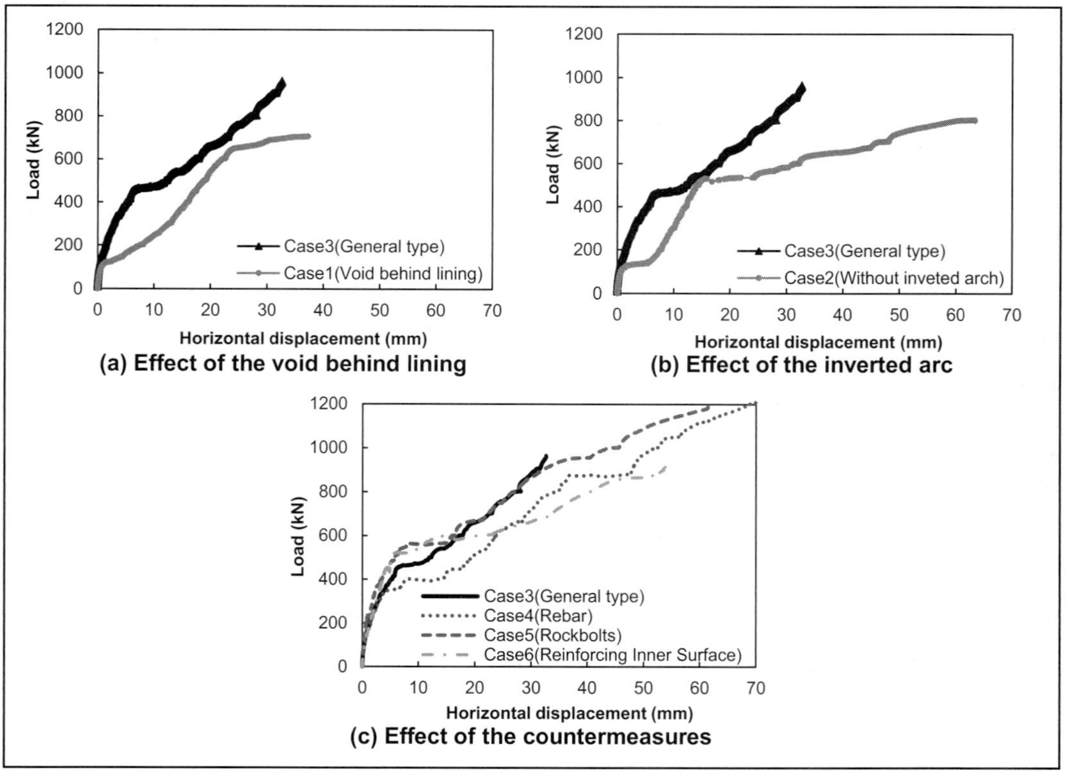


Figure 5. Load–displacement

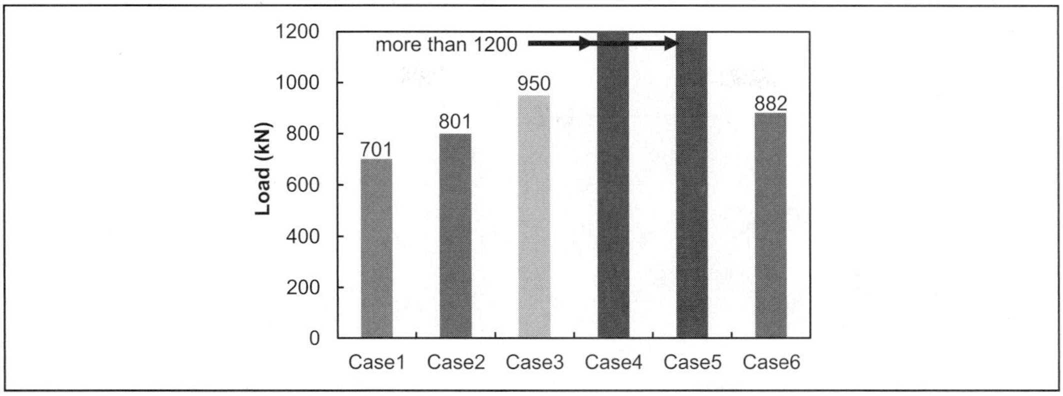


**Figure 6. Maximum load of all cases**

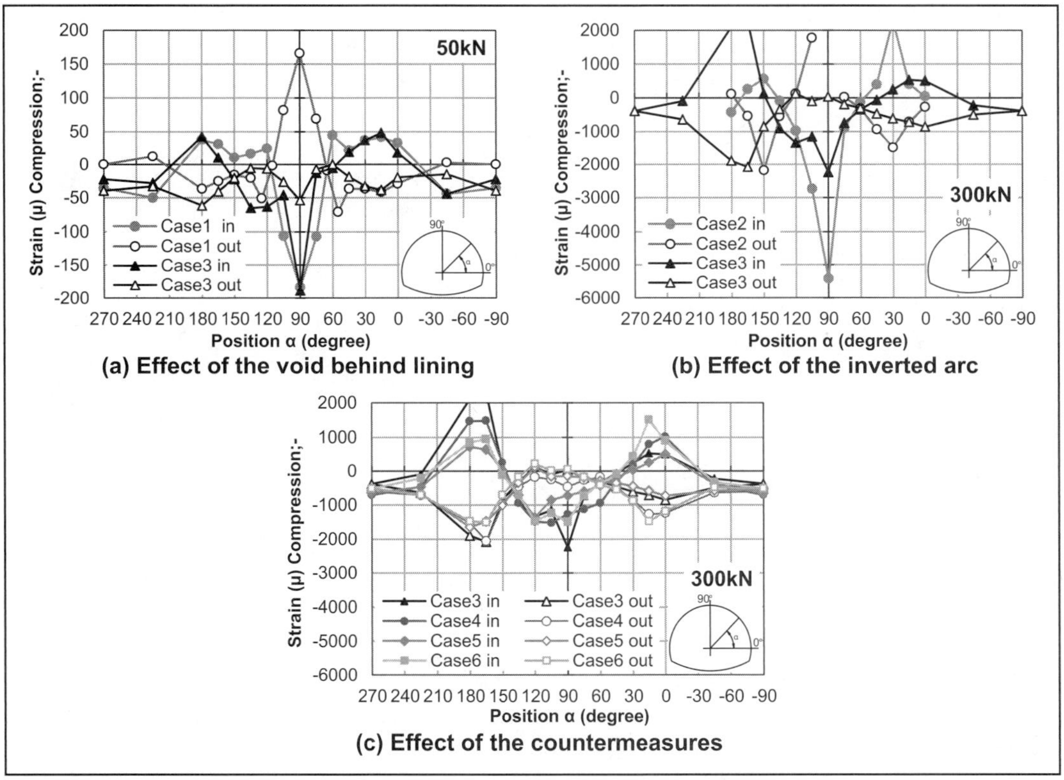


**Figure 7. Strain distribution**

load-displacement curves. Changes in linear gradient occurred at 170 and 650kN in Case 1, and 450kN in Case 3. These changes closely agree with the occurrence of cracks at the crown and shoulder in Case 1, and at the joint of the sidewall and invert in Case 3, as shown in Figure 8. The failure load in Case 3 was larger than that in Case 1, as illustrated in Figure 6.

Figure 7a shows the strain distribution on the lining at 50kN in Case 1 and Case 3. Large tensile strain occurred on the inner surface at the crown and outer surface at the shoulder in Case 1 compared with Case 3, even though the applied load was small. This means that the bending moment occurred at the crown and shoulder when energy from ground reaction forces was not obtained due to the presence of void spaces behind the lining.

The compressive strain in Case 3 occurred on the inner and outer surfaces at 45° to 150°. Tensile

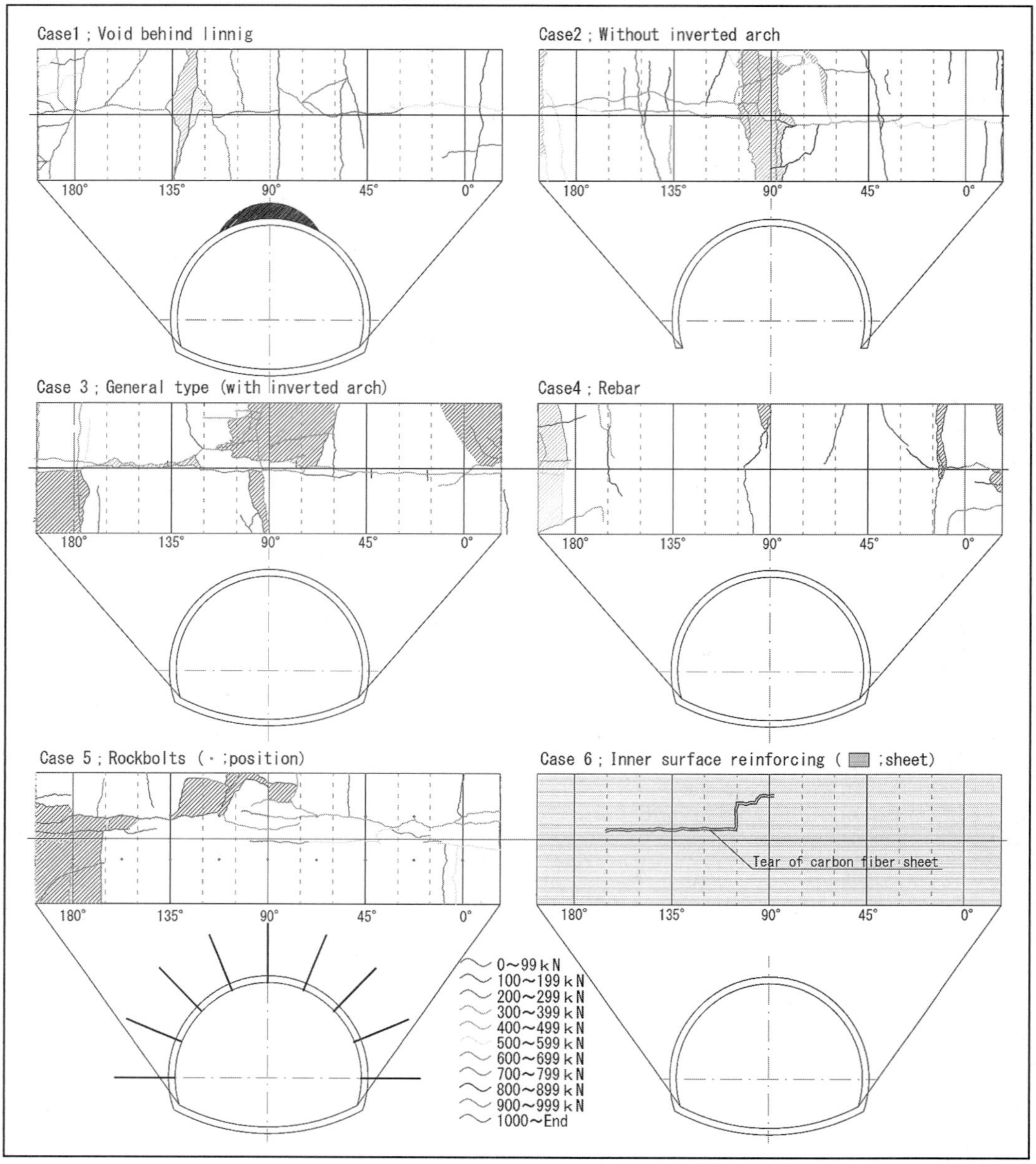


**Figure 8. The Development view of cracking of lining concrete models**

strain occurred on the inner surface and compressive strain occurred on the outer surface at 0° to 30° and 165° to 180° in the lining model.

These results show that compression failure occurs at the crown when horizontal compressive deformation is dominant during an earthquake, while flexural compression failure at the crown and flexural crack at the shoulder occurs when void spaces are present, especially in the case of construction by conventional timber lagging method. These phenomena are similar to the Type-II damage mode in Figure 3.

In Case 1 in Figure 8, three cracks occurred in the longitudinal direction at 60°, 90° and 120° at the initial stage of loading. Flexural compression failure eventually occurred at the crown and shoulder. The load at which longitudinal cracks occurred at the crown was larger in Case 3 than in Case 1. This kind of crack occurrence is due to axial compression force. However, the cracks in Case 3 are believed to be due to flexural compression force. Failure of the lining model in Case 3 occurs at the crown (shaded area in Figure 8).

Based on the above results, a tunnel with void spaces behind the lining at the crown is extremely vulnerable when load is applied horizontally caused by earthquake.

### Reinforcement with an Inverted Arch

Figure 5b compares Case 2 and Case 3 in terms of load and displacement. Changes in linear gradient occur at a load of about 150 and 500kN in Case 2, and 450kN in Case 3. As shown in Figure 8, these changes are likely due to crack occurrence, because the load at the time of the change in Figure 5 is in close agreement with the occurrence of cracks at the crown in Case 2, and at the joint of the sidewall and invert in Case 3. These phenomena are similar to the results stated above. As shown in Figure 6, the failure load in Case 3 was larger than that in Case 2.

Figure 7b shows the strain distribution in Case 2 (without inverted arch) and Case 3 (with one). Based on the results for Case 2, compressive strain occurred on the inner surface and tensile strain occurred on the outer surface at 75° to 120°. Compressive strain also occurred on the outer surface and tensile strain occurred on the inner surface at 0° to 45° and 150° to 180°. The value at the inner surface at 90° could not be measured since it exceeded the measurement limit of the strain gauge due to the large bending stress. The results for Case 3 are shown in the results stated above. The strain on the outer surface at 90° in Case 2 was larger than that in Case 3. These results show that the bending moment occurred at the crown in the case without an inverted arch; however, axial compression force occurred at the crown in the case with an inverted arch.

Based on the results for Case 2 shown in Figure 8, an initial crack occurred at the shoulder at 45° when the applied load was about 135kN. This crack is due to flexural compression failure. Subsequent cracks occurred at the crown and shoulder when the applied load was about 350kN. Flexural compression failure eventually occurred at the crown. In Case 3, shear failure occurred at the joint of the sidewall and inverted arch when the load was approximately 450kN. Shear failure of the lining model at the crown occurred as a result of axial compression force around a load of 950kN.

These results show that a tunnel with an invert has greater structural load-bearing capacity compared to a tunnel without the inverted arch.

### Effects of the Countermeasures

Figure 5c compares Case 3 (without any measures), Case 4 (lining reinforced with rebars), Case 5 (rock bolts) and Case 6 (inner surface reinforced with carbon fiber sheets). Change in linear gradient occurred at a load of 360kN in Case 4, 560kN in Case 5 and 650kN in Case 6. These phenomena of occurrence of the gradient change are similar to the results for Case 3 stated above. This change resulted in shear failure at the joint of the sidewall and inverted arch. It is considered that shear failure at the joint of the sidewall and inverted arch was due to stress concentration and controlled the load-bearing capacity. The failure load at the joint in Case 4 was about 20–30% smaller than that in the other cases, because the uniaxial compressive strength of the tunnel mortar in Case 4 was smaller than that in the other cases. As shown in Figure 6, the failure load in Case 3 was 880kN; however, the lining models in Case 4 and Case 5 did not collapse.

Based on the results shown in Figure 7c, these strain distribution modes are similar to Case 3.

As can be seen in Figure 8, crack occurrence in Case 4 was less than that in the other cases. The initial cracks occurred at 30° and 60° at an applied load of 95kN and subsequent cracks occurred at the crown at an applied load of 890kN. Failure eventually occurred at the joint of the sidewall and invert. However, failure did not occur at the crown.

In Case 5, several circumferential cracks occurred around the rock bolts. However, shear failure did not occur at the crown of the lining model, unlike Case 4. Also, longitudinal cracks did not occur at the crown in this case. Delamination at the crown of the lining occurred due to closure of these cracks.

In Case 6, cracks did not occur due to the carbon fiber sheets placed on the inner surface. Shear failure similar to that in Case 3 occurred at an applied load of 880kN. At the same time, the sheets were torn. The mortar of the lining model did not drop because the lining and sheets were bonded.

## CONCLUSION

Static loading experiments were carried out assuming a mountain tunnel affected by horizontal compressive deformation caused by an earthquake in order to clarify the mechanism of lining damage and the effect of countermeasures.

The results obtained from this study are as follows:

- The presence of the void space behind the lining at the crown reduces the load-bearing capacity of the tunnel structure because the lining is unable to obtain the ground reaction forces.
- The inverted arch strengthens the tunnel load-bearing capacity due to the formation of a ring structure and therefore the bending

moment that occurs in a tunnel without an invert changes to axial compressive force.

- The fracture of the joint between the inverted arch and sidewalls affects the load-bearing capacity of a tunnel with an inverted arch.
- Reinforcement of concrete lining by using rebars strengthens the tunnel load-bearing capacity because it constrains spalling and delamination of the tunnel lining concrete by preventing the development of cracks.
- Reinforcement of concrete lining by using rock bolts prevents collapse and strengthens the tunnel load-bearing capacity although crack generation is not prevented.
- Reinforcement of the inner surface by using carbon fiber sheets suppresses separation and exfoliation of tunnel lining concrete, although the load-bearing capacity is not increased.

In further research, we will examine realistic phenomena using numerical analysis, dynamic experiments and experiments on-site, because the conclusions presented in this paper are based on the results of laboratory-scale static loading experiments.

## REFERENCES

Kusaka A., Mashimo H., Mizukawa M., Morimoto S., "Fundamental Examination on Mechanism of Seismic Damage of Mountain Tunnel," ITA-AITES World Tunnel Congress 2009 Proceedings, O-11-18, pp. 1–7, 2009.

Kusaka A., Mashimo H., Isago N., Kadoyu K., "Seismic Behavior of Mountain Tunnel Affected by Difference of Lining Structure with Numerical Analysis, "ITA-AITES World Tunnel Congress 2011 Proceedings," pp. 450–458, 2011.

Mashimo H., "Damage of Roads Caused by 2004 Mid-Niigata Prefecture Earthquake," Tunnels and Underground, Vol. 36, No. 11, pp. 55–63, 2005 (in Japanese).

# 3D Finite Element Parametric Study of Large-Diameter, Underground, Segmentally Jointed Shafts

**Darin Kruse and Qifu Li**
Cobalt Construction Company

**ABSTRACT:** Use of large diameter segmentally lined shafts including top-down construction methods are explored for combined temporary and permanent earth retaining support. Due to the higher-than-typical diameter to segment thickness ratios (up to 300 to 1) in shafts of such large diameters, buckling becomes a critical issue. This paper presents a parametric feasibility analysis evaluating the global behavior of large diameter segmentally jointed shafts using 3D finite element analysis with special scrutiny given to the construction stages (critical state). Modeling of the soil-structural interaction and structural components including the precast wall segment joint interface are discussed in detail.

## INTRODUCTION

Worldwide, populations continue to "crowd" to the coasts and urban areas. These demographic changes are addressed, in part, by an increasing use of underground space. Underground structures are classically built rectangular in shape, which normally requires a two-step construction process beginning with installation of temporary shoring to support earth loads during the excavation, followed by the construction of the permanent structure. Attractive and practical alternatives are large diameter segmentally jointed shafts constructed using a top-down approach, referred to in this paper as segmentally lined shafts (SLS). As illustrated in Figure 1, the structure consists of a stacked series of rings formed from an assemblage of precast concrete segments, which, when fitted together, provide both temporary earth support during excavation and permanent structural support. A circular shaft sustains the lateral earth pressure mainly through hoop-stress mechanism. It is known that a cylinder shell structure with a large diameter to wall thickness ratio is prone to buckling under external pressure [1]. In typical designs of small diameter shafts and tunnel linings with small diameter to wall thickness ratios, nonlinear buckling analysis (i.e., nonlinear geometric analysis) is not considered, since the buckling capacities for those structures are much higher than typical pressures encountered at their design depths. With an economically viable shaft wall thickness, buckling becomes a critical issue for a circular shaft with a very large diameter.

This paper presents the feasibility study of constructing large diameter (up to 300 ft) circular SLS using 3D finite element analysis including soil-structural interaction (SSI). The focus of this paper is on the global structural stability issue only.

## FINITE ELEMENT MODELING AND ANALYSIS

### Linear Buckling Analysis

In the process of developing a finite element model to analyze the global behavior of SLS, linear buckling analysis is initially conducted for a prototype shaft model without considering any SSI. The model has the following geometric parameters: shaft radius R = 120 ft, depth h = 40 ft, and wall thickness t = 12 in. The monolithic shafts are assumed to be composed of concrete material with a Young's modulus of $E = 5\times10^6$ psi and a Poisson's ratio of $\upsilon = 0.25$. The shaft is assumed to be under pressure around its perimeter, varying linearly with depth, representing the soil pressure distribution. Figure 2a presents the result of the case where the base is free. It shows that for a perfect circular shaft with uniform circumferential pressure, the shaft will buckle under the triangular distribution pressure with only 1.6 psi at the base. Figure 2b shows the result of the case where the base is pinned, which simulates the connection of the base of the shaft to the mat slab foundation at the end of shaft construction. Once the base is pinned, the shaft buckles under the base pressure of 174 psi, which is much higher than a typical lateral earth pressure at the depth of 40 ft (around 15–30 psi). These results motivate the study of buckling behavior of such large diameter underground shafts before being connected to the mat slab foundation, since the surrounding soil not only imposes loads, but also provides support through SSI.

**Figure 1. Segmentally lined shaft illustration**

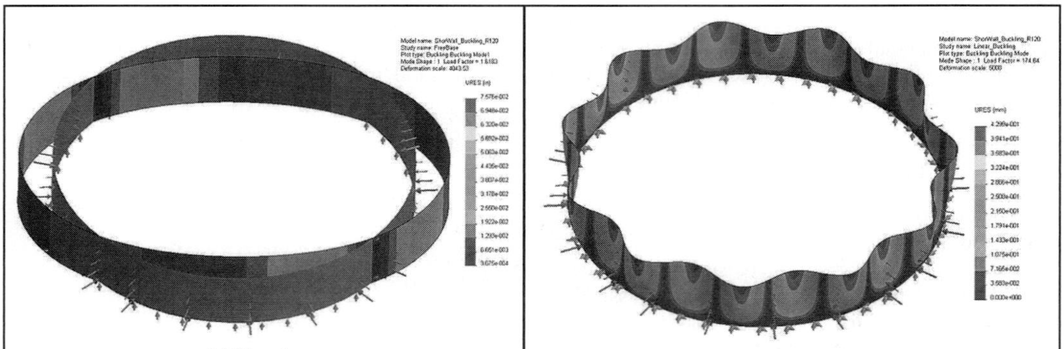

(a) Free base

(b) Pinned base

**Figure 2. Linear buckling analysis with a triangular pressure distribution**

### Nonlinear Buckling Analysis of Monolithic Concrete Shafts with SSI Using a Beam-Spring Model

It can be seen in Figure 2a that large diameter shafts without any constraints from the soil would buckle under very small pressures. In this section, a beam-spring type SSI model [2] is set up to study the buckling behavior of a shaft of similar geometry, taking SSI into account, as presented in Figure 3. The shaft geometry and parameters are the same as those studied in the linear buckling analysis section above. The base of the shaft is assumed to be unpinned. The soil surrounding the shaft is modeled using soil springs represented by surface interface elements in TNO DIANA [3]. The required inputs for surface interface elements are the normal and the shear interface stiffness moduli $K_n$ and $K_t$ in lb/in$^3$, corresponding to the normal and shear subgrade moduli, respectively. In the radial direction of the shaft, the soil is modeled as compression-only springs and in the circumferential direction, linear springs represent the soil. The shaft is modeled as a perfect cylinder using 20-node solid elements [3]. Due to the inherent nonlinearity of SSI, a nonlinear geometric analysis is required. To trigger the nonlinear buckling analysis, a non-uniform circumferential pressure distribution as shown in Figure 4 is imposed, where both lateral (vertical) and horizontal pressure distributions in each direction are included.

Figures 5 and 6 present the analysis results of models with $K_n$ = 50 lb/in$^3$ in compression and $K_t$ = 0.1 lb/in$^3$. The load deformation curve presented in Figure 5 shows that with SSI, the shaft deforms gradually as the load increases. Figure 6 illustrates the shape of deformation (buckling mode) under the given load pattern.

Figure 7 presents the effects of the normal subgrade modulus $K_n$ on the load deformation curves. It is observed that the shaft shows a slightly stiffer load deformation curve with increasing $K_n$ for medium to stiff soil. Figure 8 illustrates the shape deformation for a large normal subgrade modulus $K_n$ = 100 lb/in$^3$. Comparing the two curves, it can be observed that increasing the normal subgrade modulus does not change the mode of buckling of the shaft. The effects of the shear subgrade modulus $K_t$ on the load deformation curves are shown in Figure 9. It is evident that the shear stiffness modulus $K_t$ greatly affects the buckling behavior of the shaft. As shown in Figure 10, the buckling changes to a

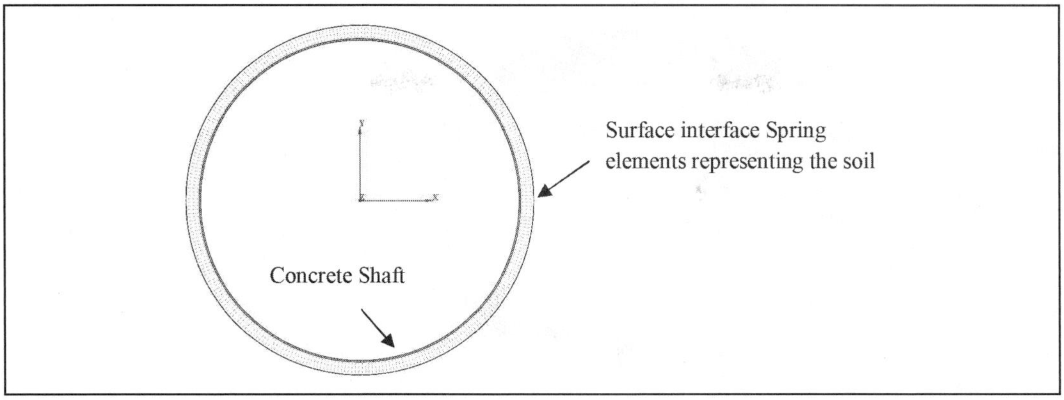


**Figure 3. SSI model with surface interface elements as soil springs**

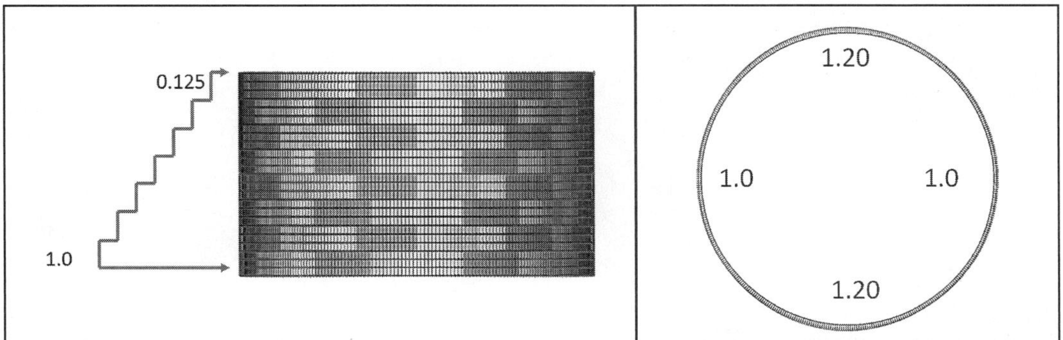


**(a) Lateral pressure distribution** **(b) Circumferential pressure distribution**

**Figure 4. Pressure distributions**

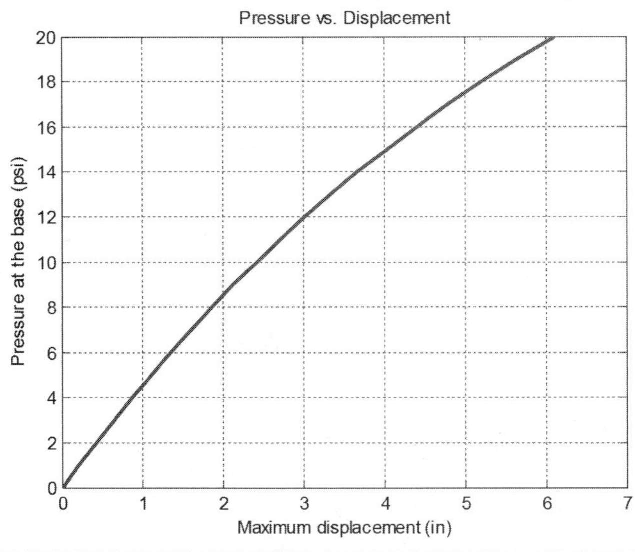


**Figure 5. Load deformation curve for $K_n$ = 50 lb/in$^3$ and $K_t$ = 0.1 lb/in$^3$**

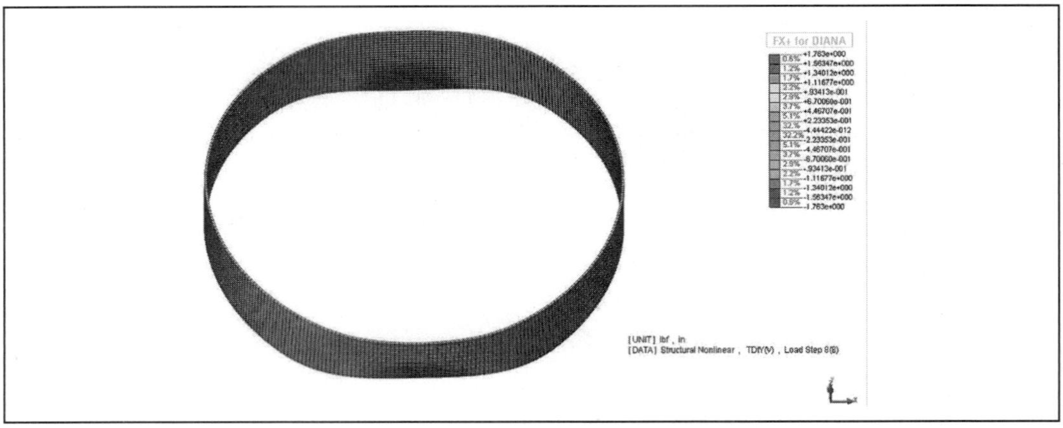

Figure 6. Shape of deformation

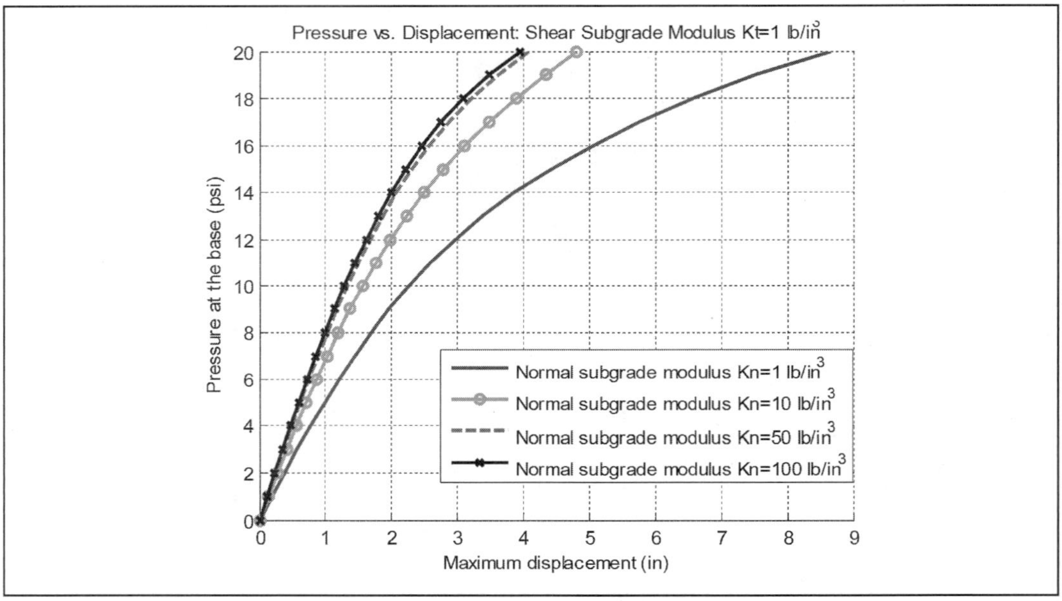


Figure 7. Load deformation curves for different normal subgrade moduli $K_n$

higher mode compared to that of Figure 6. It shows that there is a snap-through buckling from the load deformation curve in Figure 9 for the case of $K_t$ = 10 lb/in$^3$. Correspondingly, in typical small diameter tunnel lining designs only the normal subgrade modulus is considered while the shear subgrade modulus (tangential spring) is neglected [2]. To the authors' knowledge, there appears to be little available research on appropriate methods for determining shear subgrade modulus values.

Results presented in Figure 11 show the effects of different circumferential pressure eccentricities on the buckling capacity of the shaft. As expected, the buckling capacity of the shaft decreases with increasing pressure eccentricities.

### Nonlinear Buckling Analysis of a Segmentally Lined Shaft with SSI Using a Beam-Spring Model

In practice, the shaft consists of a series of stacked rings formed from an assemblage of precast concrete segments as illustrated in Figure 1. Nominal sizes of a precast segment are 5 ft high and 20 ft long. Typical segment joints are shown in Figure 12 and the corresponding finite element model is shown in Figure 13. The joints between precast wall segments

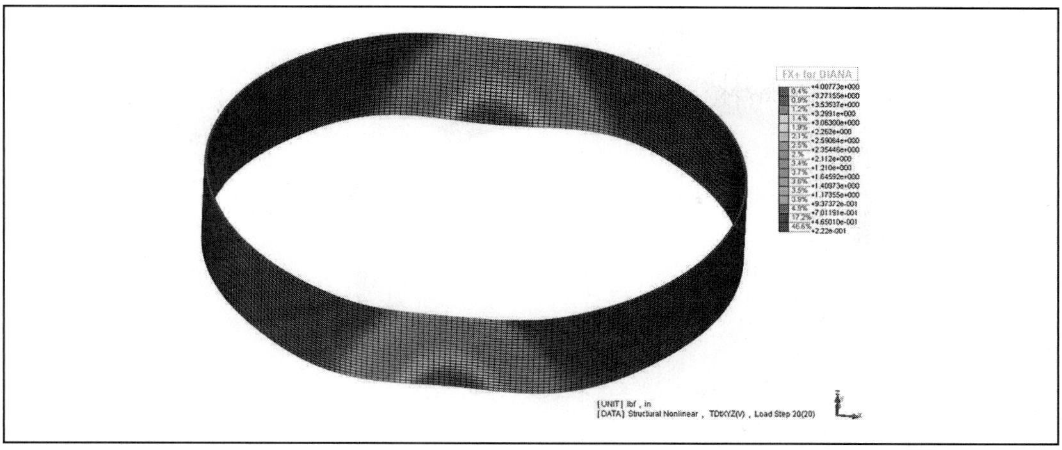


**Figure 8. Shape of deformation for $K_n$ = 100 lb/in$^3$**

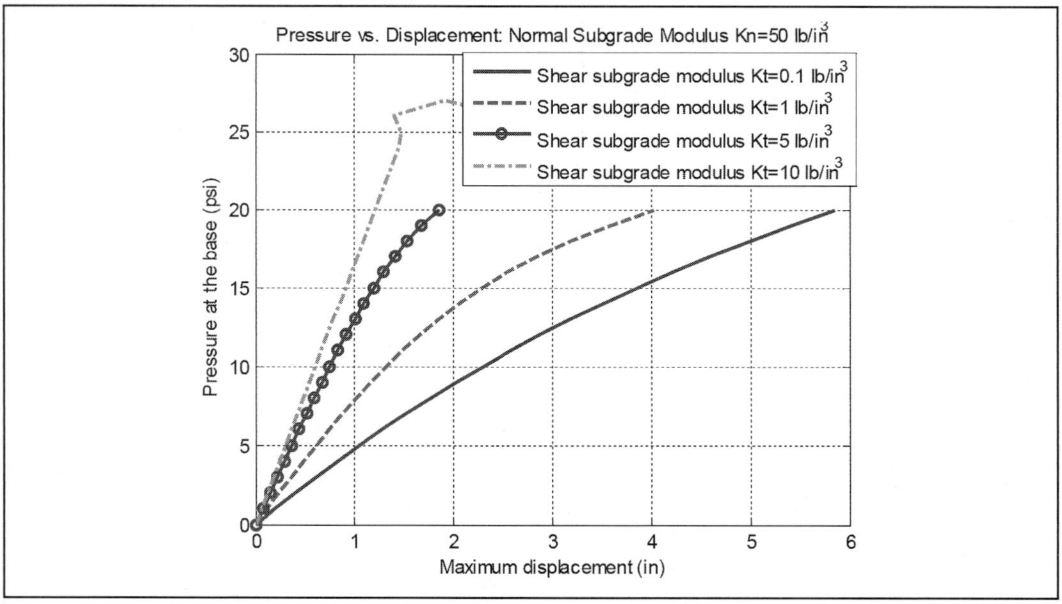


**Figure 9. Load deformation curves for different shear subgrade moduli $K_t$**

are modeled as nonlinear springs in normal and shear directions via surface interface elements [3]. The required material parameters for the interface elements are normal and shear interface stiffness moduli $K_n$ and $K_t$, which are input through traction (in the dimension of the stress) displacement relationships in normal and shear directions. An example of input data is shown in Figure 14 with normal traction versus displacement shown in Figure 14a and shear traction versus displacement shown in Figure 14b. Instead of modeling the joint components individually, the combination of all joint elements are composed into one equivalent joint interface modulus for each type of joint (vertical and horizontal) depending on the type and number of structural components.

Figure 15 shows the comparison between the monolithic model and the segmentally jointed model with soil moduli $K_n$ = 60 lb/in$^3$ and $K_t$ = 0.1 lb/ft$^3$. The segmentally jointed model has a lower buckling capacity than the monolithic model. These results highlight the importance of accurately estimating the segment joint stiffness.

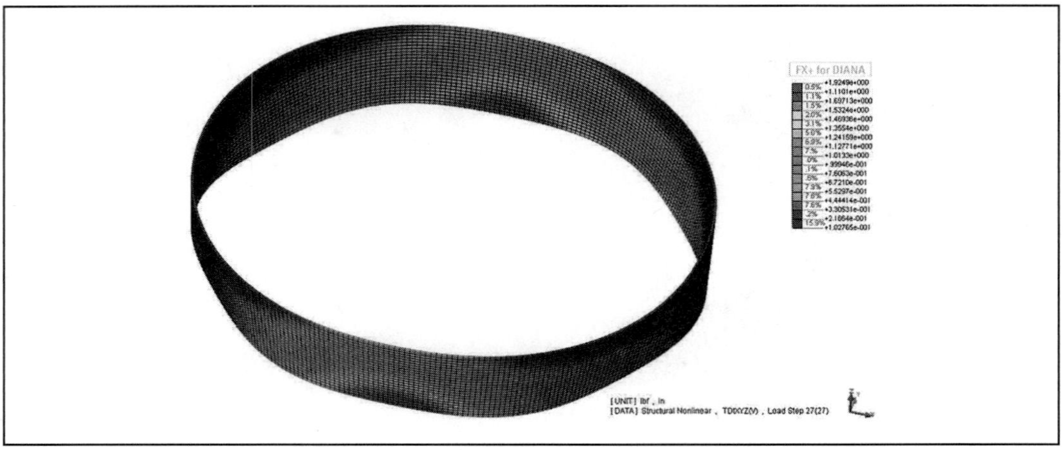


**Figure 10. Buckling mode with a relative higher shear subgrade modulus $K_t$ =10 lb/in$^3$**

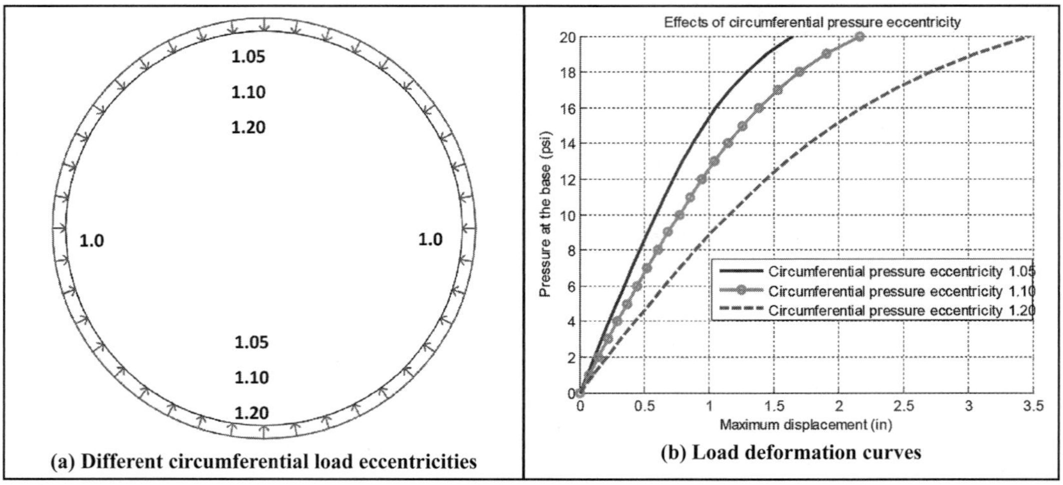


**Figure 11. Effects of different circumferential load eccentricities**

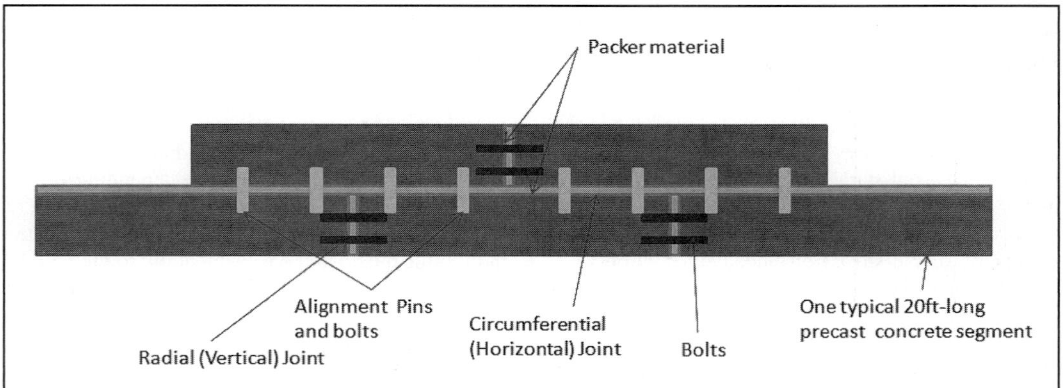


**Figure 12. Segment joints**

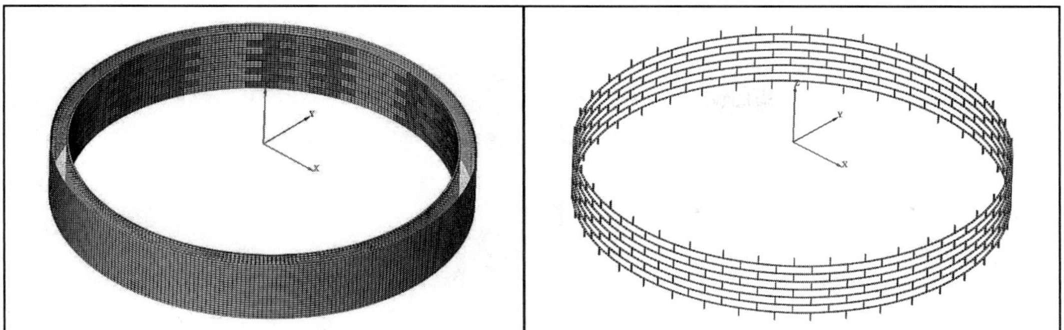

(a) Finite element model

(b) Vertical and horizontal segment joints modeled with surface interface elements

**Figure 13. SSI finite element model for the segmentally lined shaft**

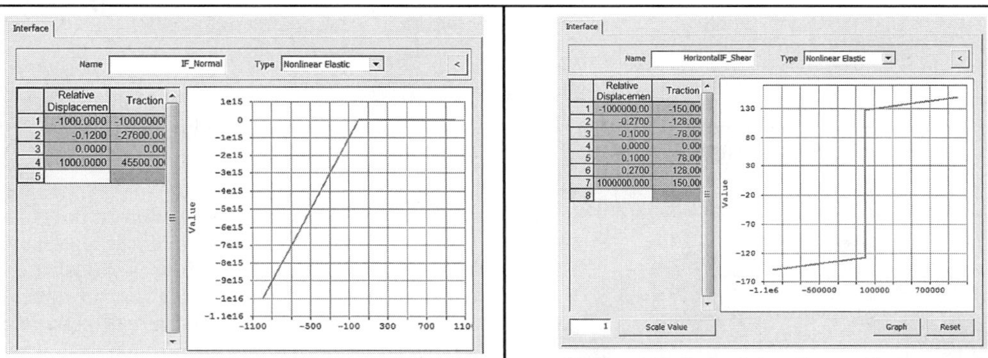


(a) Input for $K_n$ through traction vs. displacement

(b) Input $K_t$ through traction vs. displacement

**Figure 14. Inputs of interface stiffness moduli**

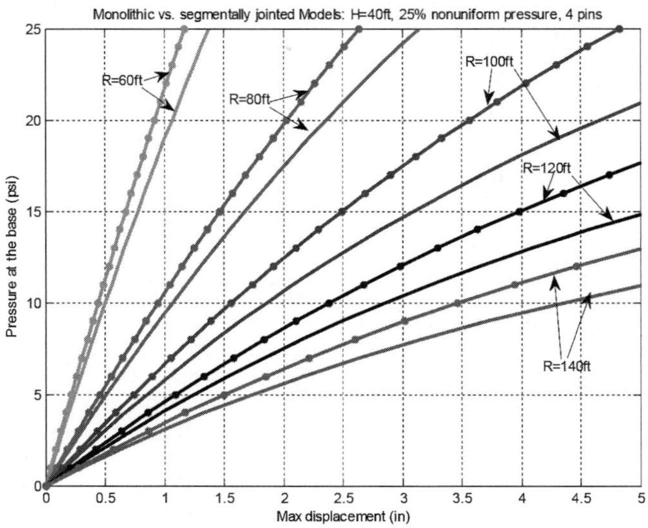


**Figure 15. Comparison between monolithic and jointed models (solid lines: segmental; solid lines with marks: monolithic)**

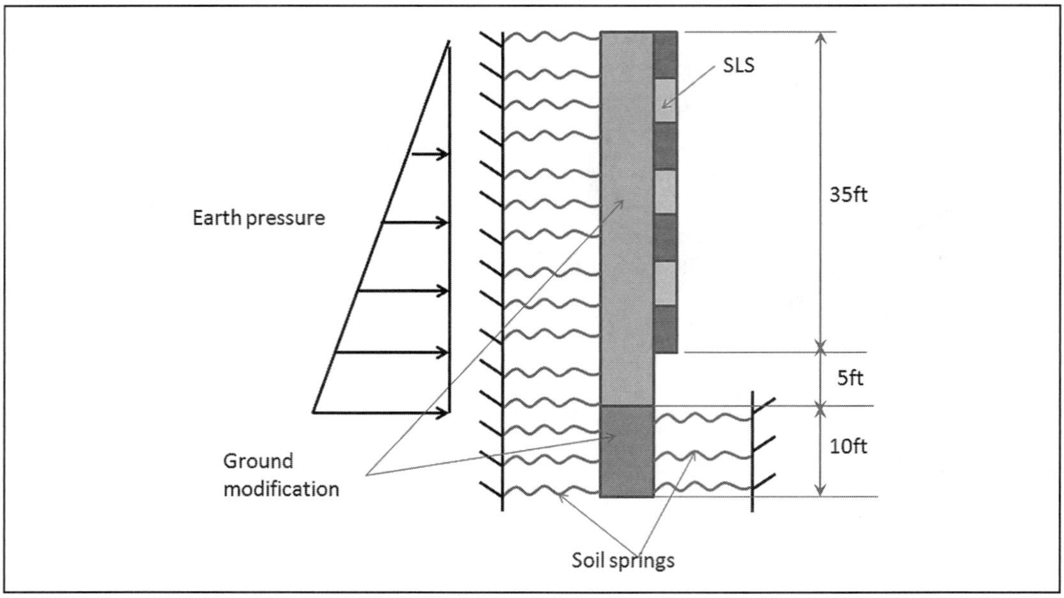


**Figure 16. Segmentally lined shaft with ground modification wall**

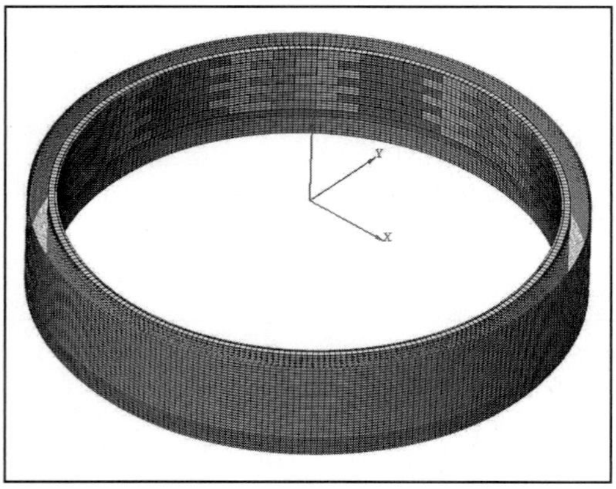

**Figure 17. Finite element model of a segmentally lined shaft plus a ground modification wall**

## Nonlinear Buckling Analysis of a Segmentally Lined Shaft plus a Ground Modification Wall

It is evident from the results of the analysis so far that for diameters greater than 200 ft (R > 100 ft), a 40-ft deep SLS with 12-in thick segments is not adequately strong to sustain the earth pressures without exceeding the allowable 3 in of deformation. In this model, a 3-ft thick cement mixed soil wall is added around the perimeter of the SLS structure to create a composite integrated structure, as illustrated in Figure 16. The corresponding finite element model is shown in Figure 17. In addition to becoming a structural component, the cement mixed soil wall also serves for other construction purposes such as ground control and ground water seepage cut off. The mixed soil to shaft interface is modeled as a compression only element in the normal direction with a very high normal interface stiffness value. Additionally, the ground modification soil-cement wall is assumed to be 10 ft deeper than the finished shaft structure. The Young's modulus used for the mixed soil corresponds to the unconfined compressive strength $q_u$ ranging from $100q_u$ to $1{,}000q_u$ psi [4]. The segment joints are modeled similar to the model presented in the previous section. The excavation of the last ring is considered to be the critical condition and is modeled as such. A relative low value of Young's modulus of mixed soil is used in the preliminary analysis to account for the cracking of cement mixed soil since this type of soil is brittle with very low tensile capacity. The model of a monolithic shaft plus a ground modification wall is used as a benchmark for comparison.

The results of adding a 3-ft thick cement mixed soil to the structure are presented in Figure 18. Radii of the shaft range from 100 ft to 150 ft. It shows that concrete shaft integrated with the mixed soil wall greatly increases the buckling capacity, which makes the construction of large diameter shafts with R > 100 ft and depths up to 40 ft feasible.

## Nonlinear Buckling Analysis of a Full-3D SSI Model

Nonlinear finite element analysis of the construction stage of an SLS through a full 3D SSI model

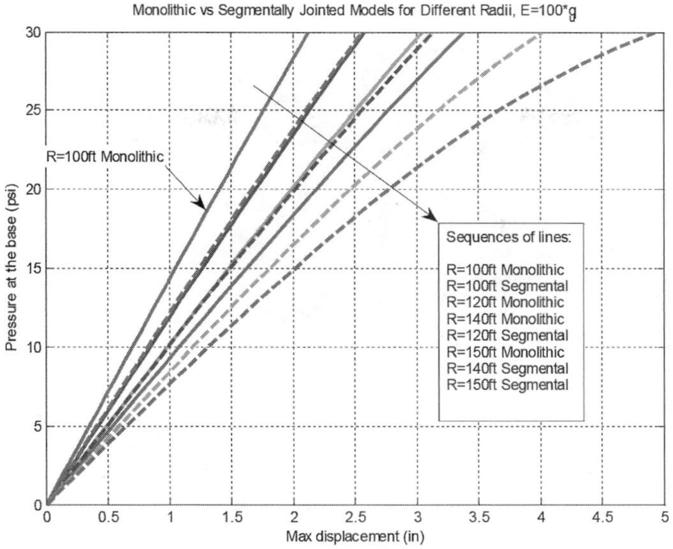


**Figure 18. Load deformation curves for SLS plus a ground modification wall**

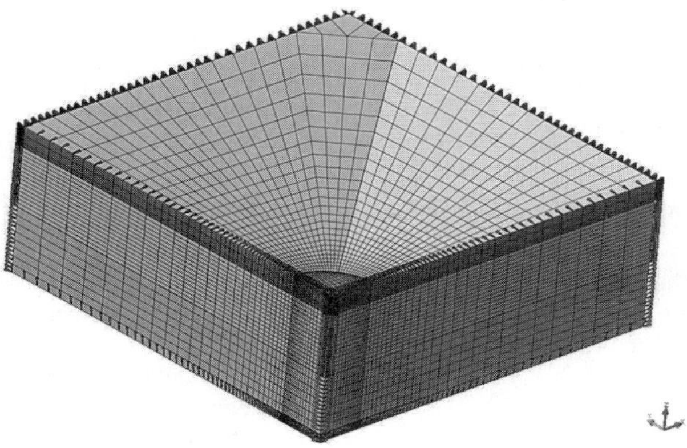

**Figure 19. Full 3D finite element model**

is explored in this section with the surrounding soil domain included. Figure 19 illustrates the corresponding finite element model. Utilizing symmetric conditions, only one quarter of the structure system is modeled to improve computational efficiency. The parameters of the concrete SLS and the ground modification wall are the same as those used in the previous section. Soil structural interface and segment joint interfaces are also modeled similar to the previous analysis. Gravity procedure is used to simulate the loading from soil domain during the excavation. The density of the soil in one section is set to be 1.2 times heavier than the rest to generate a non-uniform lateral pressure on the SLS for the nonlinear geometric analysis. The soil layers are modeled using linear elastic constitutive models with the following properties: layer 1 (0–20 ft), Young's modulus E = 1,000 psi, Poisson's ratio $\upsilon = 0.35$; layer 2 (20–40ft), Young's modulus E = 2,000 psi, Poisson's ratio $\upsilon = 0.35$; layer 3 (below 40 ft), Young's modulus E = 3,000 psi, Poisson's ratio $\upsilon = 0.35$; nominal unit weight of the soil $\gamma = 110$ lb/ft$^3$. In this preliminary analysis, the whole soil excavation from 0 to 40 ft is assumed to happen at once. Figure 20 shows the total deformation after the excavation. The soil rebounds at the center, and the surrounding soil pushes the concrete SLS moving upward. Figure 21 shows the

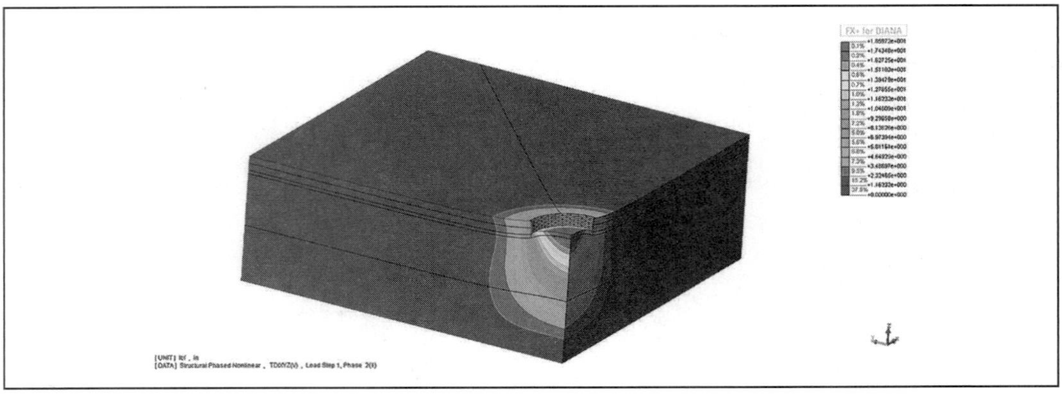

**Figure 20. Total deformation after the 40-ft deep excavation**

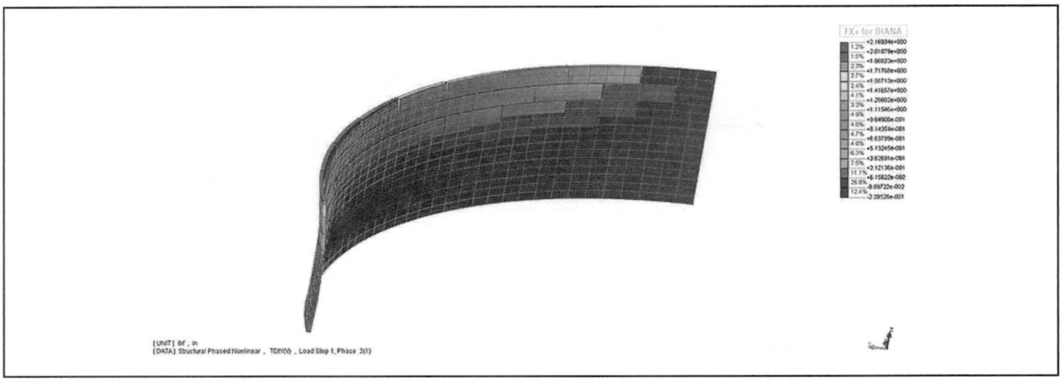

**Figure 21. Deformation of the SLS after the 40-ft deep excavation**

lateral deformation of the concrete SLS. The full 3D SSI SLS shows an improved buckling capacity with less lateral deformation than previous models.

## CONCLUSIONS

In this paper, different methods of analysis have been presented, which were used to study the global stability behavior of very large diameter circular underground SLSs (<300-ft diameter and <40-ft depth) during the construction stage. The results show that buckling is a critical issue. Ground surrounding the shaft not only imposes loading on the shaft but also provides support through SSI, which makes constructing such underground structures using the proposed top-down approach more feasible. In addition to geometric parameters such as the shaft diameter and the segment wall thickness, the segment joint stiffness and the shear interface stiffness of the SSI have significant influence on the buckling capacity of the shaft. The results presented in this paper represent an initial but significant step toward future large diameter SLS study, i.e., adopting site specific soil constitutive models and conducting detailed construction stage analyses and seismic analyses.

## REFERENCES


[1] S. P. Timoshenko and J. M. Gere, Theory of Elastic Stability, McGraw-Hill, 2nd ed., 1961.

[2] FHWA, Technical Manual for Design and Construction of Road Tunnels, 2009.

[3] *TNO DIANA User's Manual*, version 9.4.4.

[4] The Deep Mixing Method: Principle, Design and Construction, Edited by Coastal Development Institute Tokyo, 2002.

# Design and Construction Observations of the Neutrino Detection Chamber at Fermi National Accelerator Laboratory

**Mohamed Gamal, Joseph Wiedemann, and Arthur McGinn**
Brierley Associates

**Paul Beljan**
Kiewit Constructors

**ABSTRACT:** The design approach used to build the new Near Detector Hall (NDH) at Fermilab without jeopardizing the integrity of the existing MINOS Hall and MINOS Access Tunnel is discussed. The NDH is separated from the existing access tunnel by a relatively thin pillar that is 9.5 ft wide at its thinnest section. The relatively weak Scales Formation and the thin pillar necessitated some changes by the contractor's designer to the original design proposed by the owner's engineer. The alternative design implemented shorter excavation drifts and use of pre-tensioned rock bolts in the pillar and at selected locations in the existing access tunnel to minimize movements. The competent rock in the floor beneath the pillar played an important role in minimizing the deformation during construction. A comparison between predicted and observed responses for the underground openings is also presented.

## INTRODUCTION AND BACKGROUND

The Near Detector Hall (NDH) is a new underground facility that was built adjacent to the existing underground Main Injector Neutrino Oscillation Search (MINOS) Hall at Fermi National Accelerator Laboratory (Fermilab) Neutrinos Main Injector (NuMI) facility. The cavern is 350 feet underground, and is 20-ft wide, 22-ft high and 75-ft long. The new facility will house a 215 metric-ton Near Detector apparatus. A Far Detector facility was built before the NDH construction to house a 14,000-ton detector at Ash River in northern Minnesota. The neutrinos will travel the 500-mile distance from Fermilab to Ash River in less than three milliseconds, enough time for the neutrinos to transform into another type of neutrino.

The NDH was built by first excavating a main access passageway from the adjacent MINOS Access Tunnel as shown in Figure 1 which is a shotcrete face tunnel with rock bolts. The access passageway is approximately 14.5-ft wide by 19-ft high and about 10-ft long, and is located at the southern end of the NDH. Per Fermilab requirements, an Emergency Access Passageway with a minimum width of about 4 ft × 8-ft high, and about 19-ft long is located at the north end of the NDH. The emergency passageway connects to the MINOS Access Tunnel which has a finished cross sectional dimension of 28.5 ft in height and 27.5 ft in width. The crown elevations for the NDH and the MINOS Access Tunnel are 434 and 428.5 ft (amsl), respectively. Finally, the project also involved the construction of an equipment Alcove which is located along the west wall of the NDH and is approximately 10-ft wide, 21-ft long, and 19-ft high. A lightweight, moveable aluminum truss platform was also included in the NDH design package.

The most critical design issues for the NDH complex were preservation of the structural integrity of the pillar separating the two underground openings of the MINOS Access Tunnel and NDH cavern, and maintaining the overall integrity of the new and existing facilities by minimizing distress level. The pillar separating the two openings was about 9.5 ft at its thinnest section, increasing in width to about 19 ft. at the northern part of the NDH.

The contractor's designer (Brierley Associates) did a pre-bid analysis of ground conditions and anticipated behavior for the excavation of the NDH and ancillary passageways. 2D and 3D finite element modeling was done by Brierley to evaluate the stresses induced in the rock pillar under different construction and support scenarios. Based on these studies an alternative design and construction approach was proposed by Brierley Associates as discussed later.

## GEOLOGIC SETTING

The existing MINOS Access Tunnel and Hall are situated almost entirely within the shale/mudstone of the Scales Formation. The NDH is also

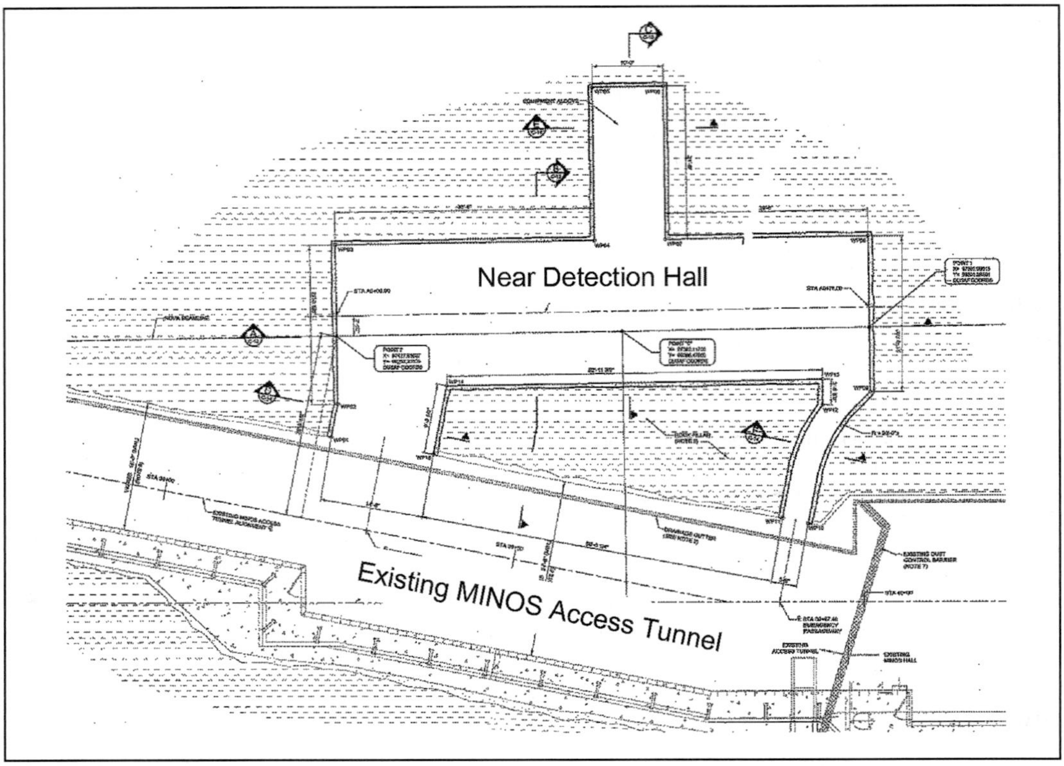


**Figure 1. Layout of the NDH and the adjacent access tunnel showing the pillar separating the two structures**

situated entirely within the same formation which is a relatively massive and uniform dark brown-gray to gray-black shale/mudstone. According to the Geotechnical Baseline Report, the formation is characterized as a relatively weak to moderately strong rock with average unconfined compressive strength of approximately 3,900 psi. The thickness of the Scales Formation is approximately 40-ft and is encountered between elev. 404 and elev. 446 ft. Field observations showed that at about El. 410, the shale became harder and more dolomitic, and this played a major role for the stability of the thin pillar between the NDH and the MINOS Access Tunnel.

The Scales Formation is overlain by the Ft. Atkinson and Brainard formations. The Ft. Atkinson is argillaceous, medium-gray limestone/dolomite, moderately porous and vuggy, and approximately 2-ft thick. The Brainard Formation is a thinly to thickly bedded, light to medium green-gray, calcareous siltstone with chert nodules and interbeds of dolomitic limestone, and is approximately 100-ft thick.

A massive dolomitic limestone of the Wise Lake Formation (Galena-Platteville Group) underlies the Scales Formation and is reported to be on the order of 120-ft to 150-ft thick. The top of the Wise Lake Formation is at about elev. 404 ft amsl, which is the invert elevation for the NDH.

## GROUND CONDITIONS

As described earlier the entire project was constructed within the Scales Formation, which is a massive shale unit. The material properties of the unit are listed:

- Unit weight, pcf: 162 ±5
- Uniaxial compressive strength, psi: 3,900 ±1,000
- Swelling potential,%: 2.7 ±0.1
- Slake durability: 89 (range 65 to 96)

A limited number of vertical joints were encountered, with two well defined sets striking as NE and NW trending joints. The NE set strikes 055° to 060° and the NW set strikes 310° to 315°. Only the NE set was encountered within the rock pillar separating the MINOS Access Tunnel from the NDH chamber, with joint spacing of about 10-ft apart. The length and spacing of less persistent joints ranged about 2 to 10 ft. No fracture zones were expected within the project excavations. Within the Scales Formation the joints were generally dry, planar, smooth, tight and fresh.

The Scales Formation is very thinly to thinly bedded and with nearly horizontal beds. Slaking of the Scales Formation was expected upon exposure to moist air or direct water for extended periods. Therefore, it was necessary to shotcrete the exposed rock surface as soon as practical after excavation.

The baseline permeability of the Scales Formation is $10^{-6}$ cm/sec, which indicates that steady-state infiltration should be less than 10 gallons per minute (gpm). Transient or flush flows may increase to 25 gpm. Four prominent seepage zones were observed between Elev. 612 and Elev. 512. (The top of the Scales Formation is at approximately Elev. 512). No seepage was reported from the rock units below the Ft. Atkinson formation.

Localized higher inflows occurred during the construction of the NDH through instrumentation bore holes in the existing MINOS Access Tunnel crown that penetrated the Ft. Atkinson and Brainard Formations. Permeation grouting was used to seal these holes and control the inflows to maintain the integrity of the Scales Formation.

The Rock Mass Rating (RMR) and the Rock Mass Quality (Q) for the Scales Formation were calculated to be 68 and 21, respectively. These values indicate the rock is generally of "good" quality. At intersections, the Q value decreases to a range of 5 to 10, which categorizes the rock at these locations as "fair."

## IN-SITU STRESS CONDITIONS

According to the GBR, previous construction experience in the Maquoketa Group had exhibited stabbing and overbreaks in the crown and invert of the TBM-mined MINOS tunnel. This behavior was assumed to be related to relatively high horizontal stresses, and suggests that the maximum horizontal stress to vertical stress ratio ($K_0$) within the Maquoketa Group, more specifically the Brainard formation, may be within the range of 3:1 to 5:1 as was indicated in GBR. Based on that, a $K_0$ value of 3 was indicated as the design value for the the Scales Formation.

The project site is in Batavia which is located 40 miles west of Chicago. Areas around Chicago and its suburbs are known to have high insitu stresses in the bedrock (Bauer et al., 1991). However, it is the authors' opinion that $K_0$ values between 1 and 3 in the Maquoketa Group, and between 1.0 to 1.5 range in the Scales Formation are more realistic for the following reasons:

1. It was found from the review of the original work by (Bauer et al., 1991). that the orientation of the major horizontal principal stress is well defined and is about N60E. The long axis of the proposed NDH is at about 33 degrees with the major principal stress direction. If the angle was 90 degrees, a $K_0$ value between 3 and 5 would have fully manifested itself, with the major horizontal stress acting perpendicular on the vertical walls.
2. The orientation of the intermediate stress is not as well defined (Hashash and Cording, 2002); however, Bauer (1991) indicated that it is oriented at about N30W. The orientation of the long axis of the NDH relative to the intermediate horizontal principal stress is about 57 degrees. (Bauer et al., 1991) reported $K_0$ between 1 to 3 for the intermediate stress while (Hashash and Cording, 2002) reported a value of 1.4 at a depth of 300 ft.
3. Bauer (1991) indicated that no insitu stress measurements could be made in the shale, and that the insitu stress measurements were obtained from testing the stiffer Wise Lake Formation and the deeper Platteville Group.
4. The shale in the Scales Formation is sandwiched between two significantly stiffer rock layers; the Brainard siltstone at the top and the Wise Lake Formation at the bottom. The two stiffer layers are expected to divert the high insitu stresses away from the shale especially if the higher creep potential of the shale is considered as this further helps to relieve locked-in stresses.
5. Lower $K_0$ is also supported by actual field observations during construction of the NDH, where no slabbing and only limited jointing was recorded in the shale of the Scales Formation.
6. The deformation patterns observed at the crown and at the side wall of the MINOS Hall, over the 4-month construction period, which are discussed later, also supported the lower $K_0$ value.

The impact of the $K_0$ value is better understood by studying the plots presented in Figure 2, which show the tangential stresses using closed form elastic solution for a circular opening (Poulos and Davis, 1974) for $K_0$ values ranging from ⅓ to 5. The plots show that for $K_0$ values ranging from ⅓ to 3 the stresses around the opening remain in compression at the crown, invert and springline. For $K_0$ values less than ⅓, the crown experiences tension while the springline remains in compression. At $K_0$ values greater than 3, the opposite occurs with the rock at springline experiencing tension at much higher magnitudes than those experienced in the crown for $K_0$ values less than ⅓. At $K_0$ of 5, the tensile stress at springline is 2 times the overburden pressure, which corresponds

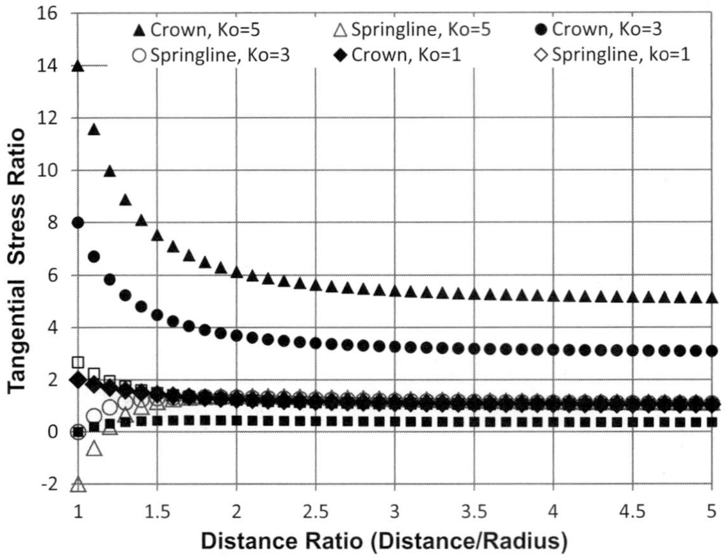


**Figure 2. Variation of tangential stress with distance at the crown and at the springline for different $K_0$ values (open symbols are for springline data and solid symbols for crown)**

to a tensile stress of about 667 psi. The compression stress at the crown for $K_0$ of 5 is 14 times the overburden pressure which corresponds a compressive stress of about 5,000 psi. The average unconfined compressive strength of the intact Brainard is 6,000 psi with an intact indirect tensile strength of 580 psi. For the Scales Formation, the corresponding values were 3,900 psi and 400 psi respectively. If the $K_0$ value was indeed 5, significant yielding and rock bursting and crushing in the crown in addition to the formation of tensile failure zones in the vertical walls should have been observed but none was observed during the construction of the access tunnel and the NDH.

The monitored post-excavation behavior of the MINOS Hall showed a total displacement of 0.7 inches in the crown and 0.2 inches in the side walls. This deformations pattern is not consistent with $K_0$ of 5 or even $K_0$ of 3, where the vertical wall deformations lateral deformation should have been 3 to 6 times higher than their counterparts at the crown.

From all of the above, the observed slabbing that occurred during the construction of the MINOS Access Tunnel was probably caused by a combination of moderately high insitu stresses and unconformity of the dolomite interbeds in the siltstone of the Brainard Formation.

## NUMERICAL MODELING RESULTS

As described earlier, several 2D and 3D finite element analyses (FEA) were executed to determine the most effective support system and construction sequence that could be implemented to maximize stability for the proposed NDH and the existing MINOS Access Tunnel. These initial FEA runs led to the recommendation of the following changes by Brierley Associates:

1. Use of 8 ft long drifts instead of the 10 ft drifts that were planned in the bid document. This should minimize exposure of the Scales Formation to moisture and should expedite installation of rock support.
2. Rather than using tie-rods to reinforce the pillars as was proposed by the owner's designer, grouted, prestressed bolts were first installed from the existing MINOS Access Tunnel before the start of the new NDH excavation. The advantage of using prestressed bolts, spaced at 5 to 6 feet, before the start of excavation was to provide some confinement to the pillar, thus reducing the potential of opening and loosening of small discontinuities under the loads induced by the excavation. For tie rods, the potential existed that these small discontinuity features may open and may well propagate before the tie rods are installed, thus weakening and jeopardizing the integrity of the pillar, in the absence of initial confinement.
3. To improve load arching above the relatively thin pillar, spot prestressed rock bolts were

**Table 1. Engineering and physical properties for different rock strata**

| Formation | GSI | $\sigma_{ci}$ (ksi) | $m_b$ | s | a | $E_{rm}$ (ksi) | $\gamma$ (pcf) | Material Type |
|---|---|---|---|---|---|---|---|---|
| Brainard | 65 | 6.0 | 2.006 | 0.0205 | 0.502 | 660 | 160 | Elastic-plastic |
| Scales | 70 | 3.9 | 2.005 | 0.036 | 0.501 | 187 | 162 | Elastic-plastic |
| Scales (D=0.2) | 60 | 3.5 | 1.5 | 0.015 | 0.503 | 147 | 162 | Elastic-plastic |
| Wise Lake | 65 | 12.0 | 2.865 | 0.0205 | 0.502 | 1,200 | 172 | Elastic-plastic |

**Table 2. Rock bolts layout and properties**

| Property | MINOS Access Tunnel | Near Detector Hall | Pillar |
|---|---|---|---|
| Bar type | Steel (fully bonded) | Steel (fully bonded) | Steel (fully bonded) |
| Diameter | 1 inch | 1 inch | 0.9 inch |
| Length | 8-ft | 8-ft | 8-FT |
| Spacing | 6-ft × 6-ft | 6-ft × 6-ft | 5.5-ft × 5.5-ft |
| Grade/yield strength | 60 ksi/47,400 lbs | 60 ksi/47,400 lbs | 60 ksi/69,200 lbs |
| Pre-tension force | 20,000 | 10,000 | 10,000 |
| Young's modulus | 29,000 ksi | 29,000 ksi | 29,000 ksi |

installed in the existing MINOS Access Tunnel in the area adjacent to the NDH.

The final FEA model simulated the excavation using a $K_o$ condition of 3 as requested by the owner. To simulate the effect of the drill-and-blast technique used in excavating the existing MINOS Access Tunnel, a 3-ft weaker zone with a disturbance factor (D) of 0.2 was used around the perimeter of the existing MINOS Access Tunnel. The disturbance factor was used to modify the strength of the Scales Formation within the blast disturbed zone. The $K_o$ value within the blast disturbed zone was taken as 1.0. Table 1 provides a summary of the strength parameters used for the Hoek and Brown non-linear material model and other physical properties such as the unit weight (g) for different strata. Table 2 summarizes the proposed support system for different project components, which was eventually implemented in the field. For modeling purposes and to be consistent with the overbreak observed for the existing access tunnel, the excavated MINOS Access Tunnel dimensions ranged from 32 ft wide at the bottom to about 26.5 ft at the top and the excavated height was assumed 29 ft. The excavated dimension of the NDH was modeled as a 23-ft high by 21-ft wide opening.

The excavation of the face of the proposed NDH was done using a CAT 314 excavator which can be equipped with a grinder attachment or hydraulic hammer. The excavation was done using a 10-ft bench and a top heading of 13 ft for the NDH. Field observations of good rock quality allowed the increase of the top heading to 18 ft leaving a 5-ft bench. For the existing tunnel, a top heading of 14 ft was used for modeling its excavation. Two to four inches thick fiber reinforced shotcrete was also used as part of the rock support system.

Figures 3 and 4 depict the magnitude of the resultant displacements before and after the construction of the NDH for $K_o = 3$ condition. Figure 3 shows that the displacement at the crown of the existing access as obtained from the 3D model was 0.5 in, which is in good agreement with the actual field measurement of about 0.7 in. The displacements of the sub-vertical wall as predicted by the model, were significantly higher than the measured 0.2 in. and averaged 0.5 in. This difference between the measured field displacements and those obtained from the 3D FE at access tunnel completion further supports the argument for a $K_o$ between 1.0 and 1.5. The use of $K_o$ between 1.0 and 1.5 with a top heading of 14 ft would have caused slight increase in the crown deflection, due to the reduction of lateral confinement at lower $K_o$ values, and would have caused the displacement of the sub-vertical wall to decrease to a value closer to 0.20 inches. It is important to realize that at $K_o$ of 1, the vertical and horizontal displacements are not expected to be the same because of the staged construction approach of a heading and a bench used in modeling. Note that the displacements were the smallest at the corners because of high confinement introduced by the high compressive stresses at these corners.

Figure 4 shows the effectiveness of using the prestressed bolts in minimizing the pillar displacement. The displacement values were smaller at and around the relatively thin pillar than the values obtained at other locations at the opposite vertical walls where no prestressed bolts were used. Another important factor that contributed to limited displacement at the pillar is the existence of competent rock in the floor beneath the pillar. The Wise Lake

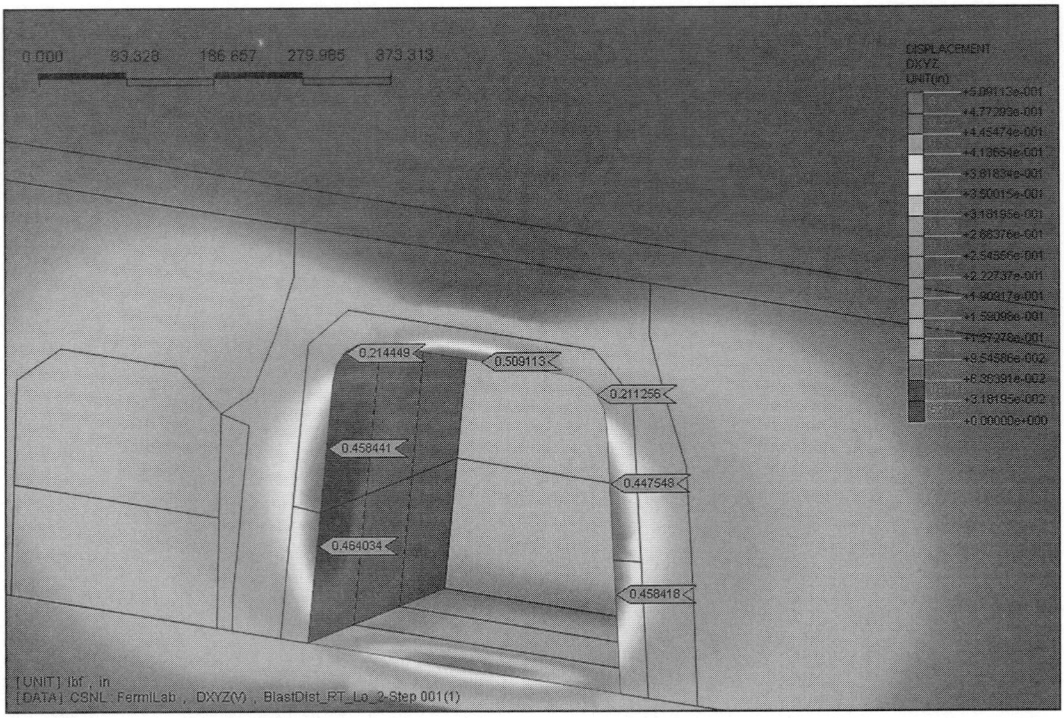


**Figure 3. Displacement magnitudes at different locations of the access tunnel before excavating the NDH**

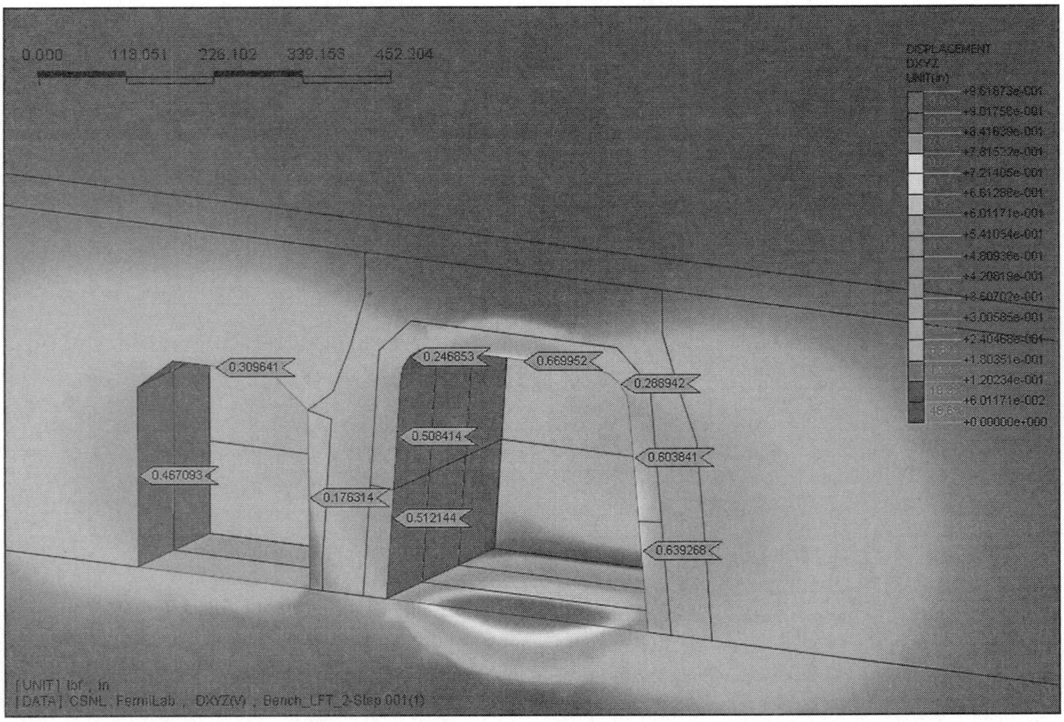


**Figure 4. Displacement magnitudes at different locations of the access tunnel and the NDH after excavating the NDH**

Formation is a massive dolomitic limestone that is up 150 ft thick. The average unconfined compressive strength of that formation is 12,000 psi with intact tensile strength of 1,000 psi. The competent limestone of the Wise Lake Formation minimized the vertical movement, maintained high factor of safety against bearing failure and provided good confinement at the bottom of the pillar, in addition to the confinement introduced by the prestressed rock bolts along the height of the pillar. The change in displacements in the existing access tunnel ranged from about 0.15 inches to 0.20 inches.

Table 3 compares the measured change in displacements in the existing MINOS Access Tunnel against the predicted displacement change from the 3D modeling at 3 different points, the locations of which are shown in Figure 5. The measured readings are the average of the survey readings at two different stations within the thinner section of the pillar. Table 3 shows good agreement between predicted and measured deformation except for the point in the pillar. However, both the model and the measurements show that the maximum deformation change did not occur at the reinforced pillar, but took place at the unreinforced vertical wall opposite to the reinforced pillar. The important point taken from the displacement measurements is not the degree of agreement between predicted and observed change in displacements, but that these observations confirmed successful arching of the load above the pillar, thus reducing the loading and deformation levels at the pillar.

**Table 3. Comparison of measured change in displacement using survey targets against predicted change in displacements in MINOS Access Tunnel**

| Point # | Measured Displacement (in) | Predicted Displacement (in) |
|---|---|---|
| 1 | 0.15 | 0.07 |
| 2 | 0.18 | 0.17 |
| 3 | 0.20 | 0.17 |

Figures 6 through 9 show the maximum and minimum principal stresses before and after the construction of the NDH for $K_0 = 3$ condition. The negative sign in these plots indicate compression stress and the positive indicates tensile stress. The fairly uniform stress reduction shown in Figure 6, denoted by the orange/yellow/red zones within the blast disturbed zone, is attributed to the stress arching within the less stiff blast disturbed zone to the stiffer more competent rock outside that zone. The contour lines concentration in areas just outside the disturbed zone indicates a rapid change of stress gradient as a result of the stiffer rock mass ability to bridge the weak

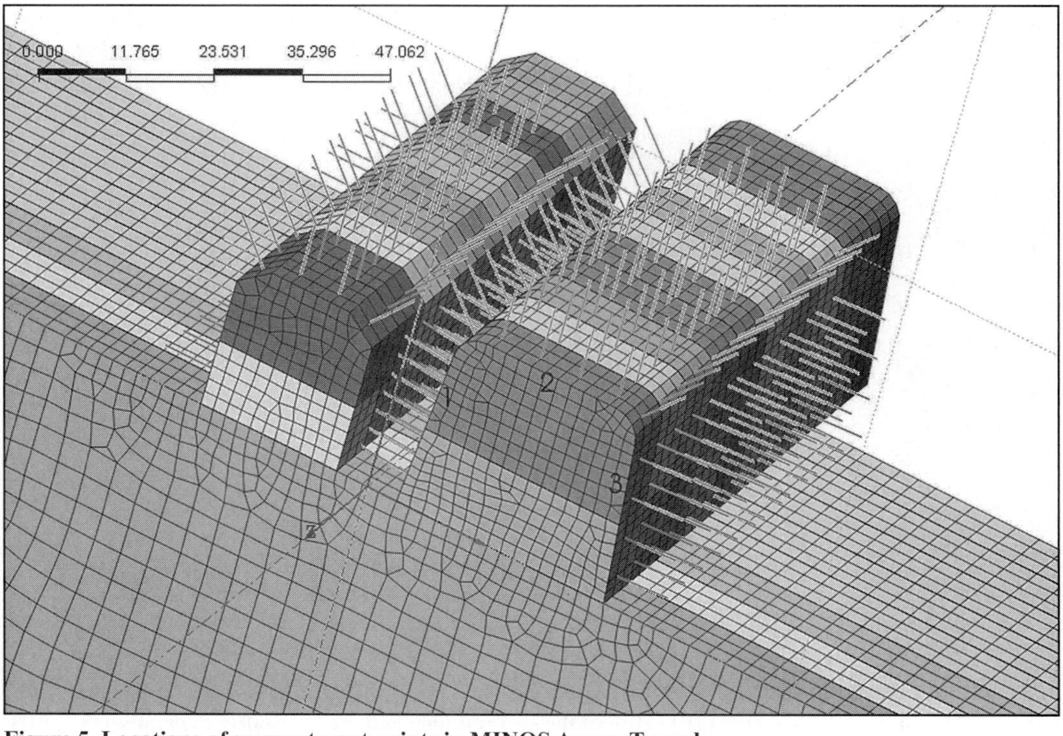


**Figure 5. Locations of survey target points in MINOS Access Tunnel**

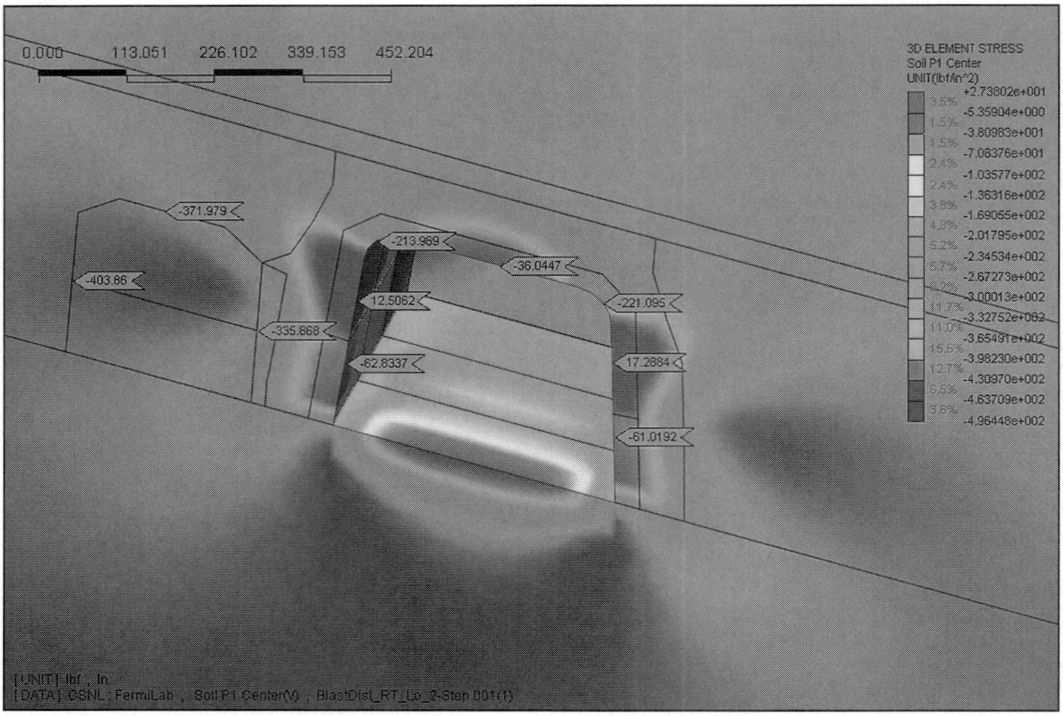


**Figure 6. Minimum principal stress (P1) around the access tunnel before excavating the NDH**

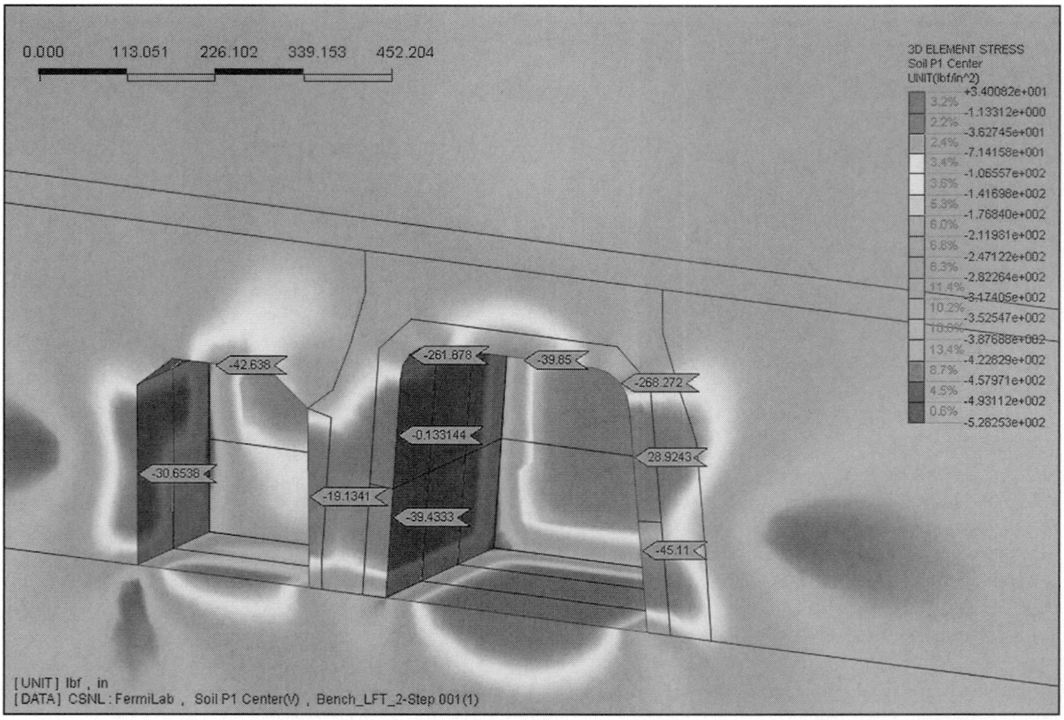


**Figure 7. Minimum principal stress (P1) around the access tunnel and the NDH after excavating the NDH**

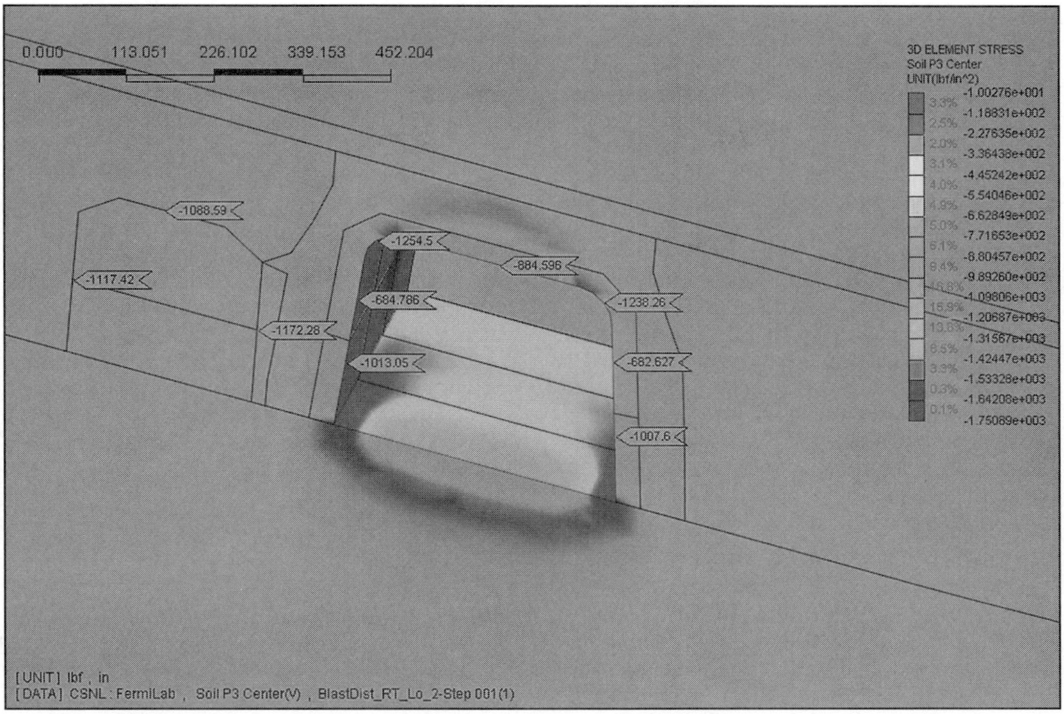


**Figure 8. Maximum principal stress (P3) around the access tunnel before excavating the NDH**

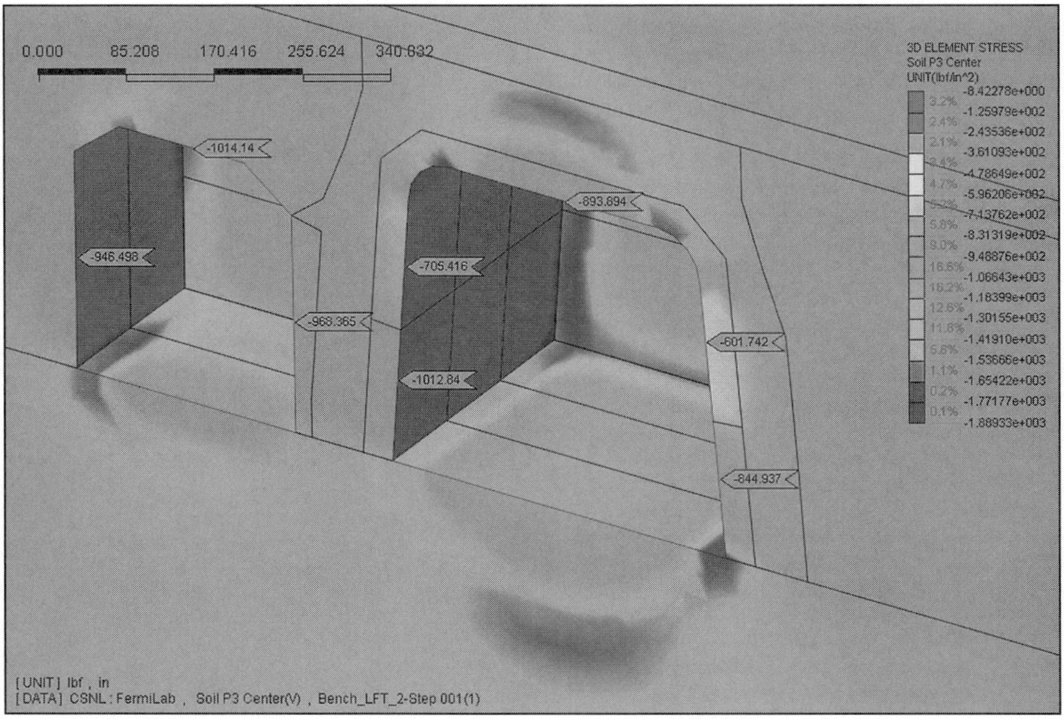


**Figure 9. Maximum principal stress (P3) around the access tunnel and the NDH after excavating the NDH**

disturbed zone. The stress concentrations denoted by the green and light blue are attributed to the change in stiffness between the Scales Formation and the stiffer underlying Wise Lake Formation. Figures 6 and 7 show that more than 91% of the pillar volume remained in compression after excavating the NDH, and that a very small pre-existing tensile stress zone that formed during the construction of Minos Access Tunnel remained almost unchanged after the construction of the NDH. This confirmed the effectiveness of the proposed support and construction sequence. The maximum tensile stress value was less than 30 psi near the Minos Access Tunnel. Another, much smaller, localized tensile zone near the NDH had tensile stress values less than 10 psi. These stress levels are reasonable when compared to the strength parameters of the rock.

## CONCLUSIONS

Observations made during the construction of the Near Detector Hall (NDH) confirmed the effectiveness of the rock support system and construction approach. Installing prestressed bolts in combination with competent rock in the floor beneath the relatively thin pillar minimized the resultant deformation, thus minimizing the load on the pillar. Reinforcing the pillar enhanced the rock ability to arch the load away from the thin pillar as was reflected by the increase in the deformation of the MINOS Access Tunnel's wall opposite to the pillar. The competent floor maintained a high factor of safety against pillar bearing failure and provided good confinement at the bottom of the pillar in addition to the confinement introduced by the prestressed rock bolts along the height of the pillar, and these factors further helped in keeping the displacement to a minimum.

The observations confirmed that lower $K_0$ between 1 and 1.5 are more realistic for the NDH built entirely in the shale of the Scales Formation. This was explained in terms of the sensitivity of the insitu stresses to the orientation of the NDH; the impact of the stiffer layers just above and below the shale, and finally the creep potential of the shale compared to that of the surrounding stiff layers.

## REFERENCES

Bauer, R.A., Curry, B.B., Graese, A.M., Vaiden, R.C., Su, W.J., and Hasek, M.J. (1991). Geotechnical Properties of Selected Pleistocene, Silurian, and Ordovician Deposits of Northeastern Illinois, *Illinois State Geological Survey*, (139), 1–64.

Hashash, Y.M.A., Cording, E.J. and Oh, J.M. (2002). Analysis of major shearing of a rock ridge. *International Journal of Rock Mechanics and Mining Sciences*, (39), 945–957.

Poulos, H.G., and Davis, E.H. (1974). *Elastic Solutions for Soil and Rock Mechanics*. New York, NY: John Wiley & Sons.

# Self-Bearing Shotcrete in Lieu of Self-Consolidating Concrete for Tunnel Rehabilitation

**Axel G. Nitschke, Vojtech Gall, and Sandeep Pyakurel**
Gall Zeidler Consultants

**Denis Mazzoni**
Swank Construction Company

**John Becker**
Coastal Gunite Construction Company

**ABSTRACT:** Pennsylvania Department of Transportation's almost 100 year old Liberty Tunnels underwent significant rehabilitation measures, including the replacement of aging ventilation arch walls. The arch walls act as tunnel shaped jet structures for fresh air supply. The original design foresaw the use of self-consolidating concrete and formwork for the replacement. The authors developed an alternative concept and subsequent design for a shotcrete arch wall solution in lieu of the self-consolidating concrete foreseen by the contract design. The alternate shotcrete concept provided the contractor with a schedule and cost saving solution under the given tight, two week long, construction window allowed during a complete shutdown of the tunnel.

## PROJECT OVERVIEW

The alignment of the Liberty Tunnels crosses Mount Washington in Pittsburgh, Pennsylvania. The twin tunnels provide a direct route from the South Hills suburbs to Pittsburgh and ease the commute from and to downtown Pittsburgh. The Liberty Tunnels are horseshoe shaped tubes. Each tube serves one direction and has an overall length of 1,795 m (5,888 ft).

The tunnels were opened to traffic in January 1924 and have gone through a series of upgrades and repairs during their service life. Originally the tunnels had no ventilation system, because the expected traffic volume through the tunnels was very limited. However, this was subject to change already shortly after the tunnels were opened to traffic and the traffic flow exceeded the predicted numbers. Only limited numbers of vehicles were permitted through the tunnels to keep the exhaust gases below dangerous levels. In 1928, the tunnels were upgraded and a ventilation system was designed to accommodate the increasing traffic flow. Two vertical vent shafts were constructed to draw exhaust from the midpoint of each tunnel and force a supply of fresh air into the tunnel through the so-called "arch walls." An arch wall is an arch structure, which is offset from the structural lining of the tunnel to provide for air-channels. The ventilation arch wall section acts like macroscopic air nozzle; fresh air is supplied from the ventilation shaft and pushed along the vent supply area on either side of the arch wall. At the end of the nozzle the arch walls are open, allowing the fresh air to enter into the tunnel, away from the exhaust point (see Figure 1, 2, 5, and 10).

Swank Construction Company (Swank) was awarded the Liberty Tunnels Rehabilitation project by the Pennsylvania Department of Transportation (PennDOT) in May 2013. The project included amongst other scope, the demolition and renewal of the ventilation arch walls inside the tunnels, close to the ventilation shaft. The subject section is located between STA 13+029 and STA 13+110 in the inbound tunnel and between STA 12+991 and STA 12+910 in the outbound tunnel. Each arch wall section is 24.7 m (81 ft) long. The existing ventilation arch wall has an intrados radius of 4.05 m (13 ft, 3¼ in.) and spans the entire arch with an opening angle of 180 degrees. In addition, two vertical walls divide the void space for the air supply along the left and right side of the arch wall (see Figure 3).

Structurally, the arch wall section can be divided into three sections from left to right in Figures 1 and 2: (1) merging area from the shaft, (2) full-arch area, where the arch wall is closed at the bottom, and (3) suspended arch area, where the arch wall is open at the bottom to provide an outlet for the fresh air (see also Figure 10). This paper focuses on the full-arch area (center) and does not address the merging area from the shaft (left) or the suspended arch area (right).

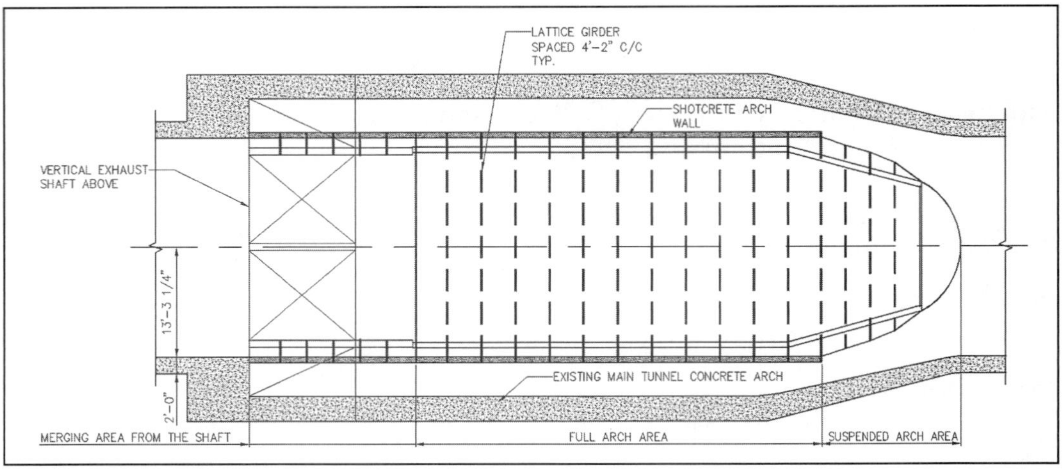


**Figure 1. tunnel ventilation arch wall section plan view**

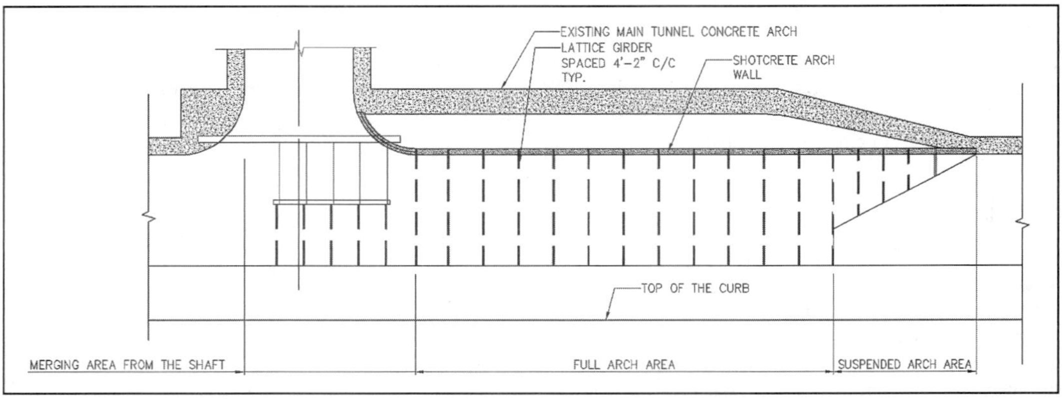


**Figure 2. Tunnel ventilation arch wall section—longitudinal section**

The original arch wall used u-shaped steel profiles as structural members, which were tied with radial hangers to the structural tunnel arch above. The fresh air travelled through the void space along the left and the right side wall. The center part was not utilized for ventilation. Vertical walls separated the center part from the sidewall areas, as shown in Figure 3.

All steel members, including the radial hangers (Figure 3), were later embedded in concrete to provide protection against corrosion as shown in the photograph in Figure 4.

The original rehabilitation design proposed the same structural approach with u-shaped beams and hangers embedded in reinforced, self-consolidating concrete.

Gall Zeidler Consultants (GZ), in cooperation with Swank Construction and Coastal Gunite, provided an alternate design and construction concept for the Liberty Tunnels Rehabilitation project. The proposed alternate concept used a self-bearing shotcrete arch wall, which allowed avoiding the utilization of the radial hangers as well as the cast-in-place, self-consolidating concrete.

## REHABILITATION DESIGN

### Original Design

The original design proposed demolishing and renewing the existing ventilation arch walls, following the original design approach with u-shaped steel beam and radial hangers embedded in concrete (Figure 3 and 4). The concrete arch was supposed to be reinforced with welded wire fabric. During the arch wall demolition it was intended to utilize the existing steel framing hangers, which are in good condition and replace hangers, which are deteriorated. A curved steel formwork, forming both sides of the free-standing arch wall was supposed to be used to form the cast-in-place arch. In addition, the

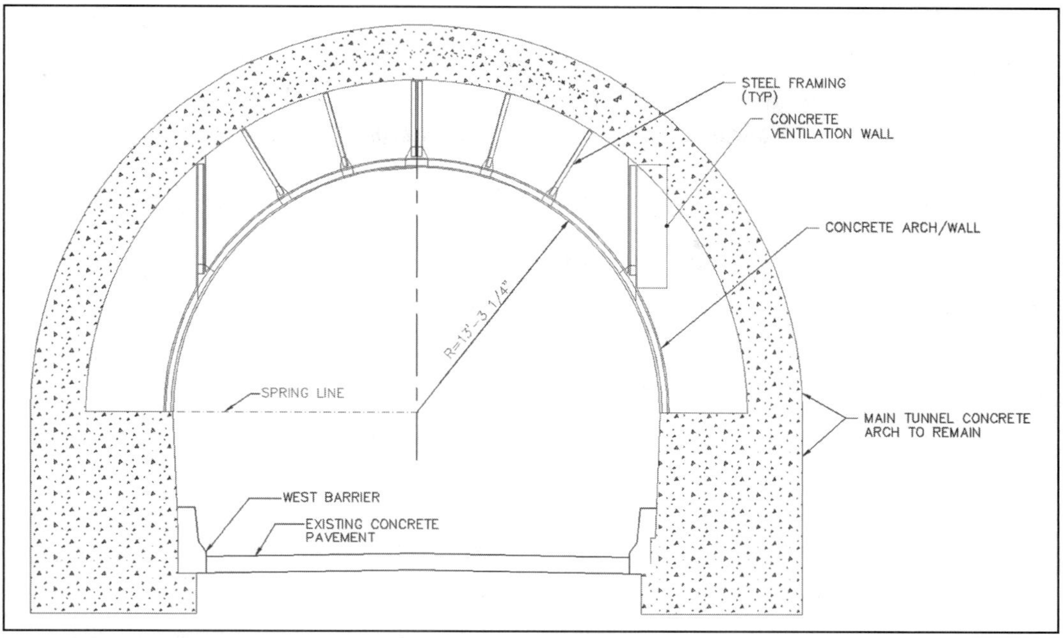


**Figure 3. Existing tunnel ventilation arch wall section—cross section**

**Figure 4. Embedded hangers in existing void space between main tunnel and ventilation arch wall**

two vertical walls and concrete embedment of the hangers on top of the arch were to be formed and poured as well. To account for the relatively thin, reinforced walls, limited accessibility and tight schedule during the given shutdown period the use of self-consolidating concrete was foreseen.

Self-consolidating concrete is a high-performance concrete that can flow easily into tight and constricted spaces without segregating and without requiring vibration. This was required due to the very limited accessibility. However, fresh self-consolidating concrete exerts high hydrostatic stresses, which have to be born by the formwork, and has a risk to rupture the formwork and create concrete blowouts. Therefore ordinary formwork could not be used for the envisioned application and required stronger formwork either made of steel or very strong timber formwork. The formwork also had to be embedded with studs and anchors of sufficient strength to prevent concrete blowouts or lifting from hydraulic

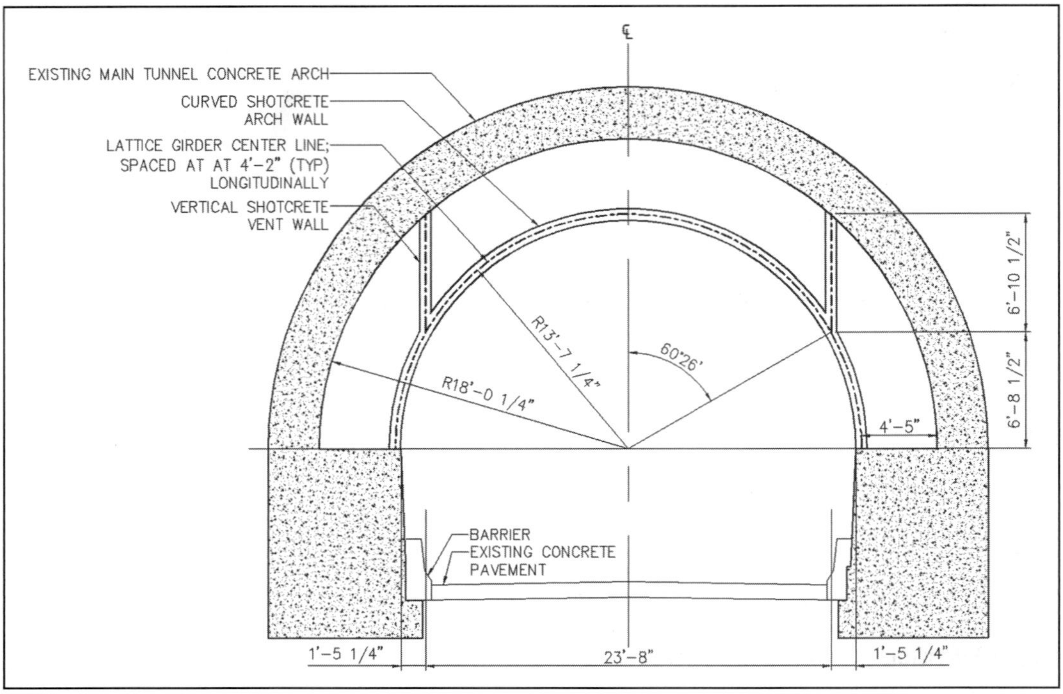


**Figure 5. Alternative design self-bearing shotcrete arch section**

stresses especially at the lower part of the formwork. Such custom-made formwork incurs high costs, especially due to its very limited reuse at the given application. In addition, the schedule impact by the risk of blowouts or deformation of the formwork was considered very high by the Contractor, because the limited shutdown period of the tunnel left no time for on site adjustments.

Another uncertainty was posed by the reuse of the existing hangers, which were embedded in concrete. To evaluate a potential reusability of the hangers the existing arch wall had to be demolished first, while the shutdown period had already started. The number of deteriorated hangers or hangers, which were damaged during the demolishing, was therefore unknown at the start of construction. Further, sorting out the hangers and replacing the deteriorated ones was considered a time consuming activity in itself. The hangers also posed an additional hindrance during formwork installation.

## Alternative Design

To reduce the schedule risk and provide cost savings an alternative design was developed in order to simplify the construction process. The alternative design focused on two critical aspects of the original design (1) the use of self-consolidating, cast-in-place concrete and (2) the structural utilization of hangers.

The cast-in-place concrete was avoided by the introduction of shotcrete, while the structural system was changed into a self-bearing arch, avoiding hangers as structural members. The latter allowed the complete removal of all hangers during the demolishing process, independently from their condition, without the need of replacement.

The alternative design utilized self-bearing shotcrete arch for the ventilation arch wall. The self-bearing shotcrete arch concept is often used to extend the underground section of a mined tunnel into the open portal area by providing a free-standing arch or so-called shotcrete canopy. Recent examples for the utilization of shotcrete canopies can be found at the Weehawken Tunnel, New Jersey and Devil's Slide Tunnel, California. Similarly, FHWA Technical Manual for Design and Construction of Road Tunnels (FHWA, 2009) also describes shotcrete canopy use and construction techniques.

Shotcrete canopies utilize the same materials as typically used for tunnel shotcrete linings for ground support. These materials are shotcrete, lattice girders, and reinforcement. While the initial lining during tunnel excavation and support is applied against the ground, an artificial surface on the backside has to be provided for a free-standing arch to allow for the built-up of the shotcrete lining. In case of the Liberty Tunnel relatively lightweight plywood was used, which could be easily removed at completion.

Alternatively expanded metal sheets may be used at the backside.

The cross section in Figure 5 provides a typical situation for the self-bearing shotcrete arch of the alternative design. It has to be noted that the two vertical walls shown in Figure 5 do not have any structural function and are for ventilation purposes, only. Structurally the arch wall supports itself as a free-standing, self-bearing arch, loaded by the weight of the two vertical overlying walls. Additional hangers such as in the original design are not necessary for the structural system.

The arch walls and the vertical walls have embedded lattice girders at a typical spacing of 1.27 m (4 ft 2 in.) center to center. The lattice girders were bolted on the abutment and anchored with undercut anchors at the tunnel main arch to provide stability during construction. The lattice girders were structurally not utilized in the design, despite the fact they provide additional reinforcement. The primary purpose of the lattice girders was the provision of a geometrical template for the sprayed shotcrete and temporary support for the reinforcement and shotcrete during the construction process. Structurally the arch was designed with a wall thickness of 6 inches. However, to account for construction and wall thickness tolerances in the design the weight of the wall was assumed for a wall thickness of 8 inches.

Linear elastic beam models were used to calculate structural forces acting on the ventilation arch walls. All ground and external loads are born by the main tunnel arch and do not affect the inner, self-bearing arch. The design loads considered self-weight of the structure, loads from overlying vertical walls, and earthquake load and air pressure of 2.39 kPa (50 psf) radially to the walls. The design resulted in minimum reinforcement in the arches as required per ACI 318 (ACI, 2005). The arches were reinforced with two layers of welded wire fabric, W9×W9 at 6 inch center to center spacing in both directions. The minimum reinforcement was important for the durability of the arch including controlling cracking from shrinkage and temperature changes. Following PennDOT's requirements all reinforcement as well as the lattice girders were galvanized.

## CONSTRUCTION SEQUENCE

Due to the two layers of reinforcement as well as the vertical walls on top of the arch walls, the design directed a mandatory construction sequence, which had to be followed by the Contractor during construction. The construction sequence is described below and shown in Figure 6.

The construction started with demolition of the existing ventilation arch wall (Step 1). In the second step, lattice girders were installed along the arch periphery as well as for the two vertical wall sections. The lattice girders were secured with undercut anchors at the top and dowels at the bottom of the arch of the main tunnel lining. The lattice girders were comprised of a three-piece arch plus one piece each for each vertical ventilation wall on either side. The lattice girder pieces were connected with bolts using butt plates welded at the end of each lattice girder section. In Step 3 a light plywood formwork was setup along with first layer of welded wire fabric at the extrados side of the lattice girder. To ensure sufficient concrete cover, spacers were used between the reinforcement and the plywood. In this step it is important to highlight that the center part of the arch had to be left open to provide access for the construction of the vertical walls. In Step 4 shotcrete was applied at the rounded as well at the vertical wall sections—excluding the center part. Only the vertical wall sections were completed to full thickness and with both layers of reinforcement, while the intrados layer of reinforcement at the arch wall sidewall was left out for later completion. During Step 5, the center arch section was closed by installation of the plywood and reinforcement at the extrados side of the arch. In the last step the center arch section was sprayed up to the intrados layer of reinforcement, followed by the installation of the intrados layer of reinforcement along the entire arch and completion of the shotcrete arch wall to full thickness, including a trowel finish. In a final step the plywood at the backside was removed, completing the arch wall construction.

## EXPERIENCE AND CHALLENGES DURING CONSTRUCTION

As mentioned before, the Liberty Tunnels are a primary commuting route to and from downtown Pittsburgh. Due to their importance the allowable shutdown period was very limited and demanded a very tight and compact construction schedule. In an early planning stage the ventilation arch walls were identified as a potential high impact risk for the schedule and were the controlling operation during the closure period for different rehabilitation measures in the tunnel.

The construction was split into two phases, phase 1 for the southbound tunnel and phase 2 for the northbound tunnel. As part of the bid documents PennDOT set forth 18 day closures per phase to run consecutively between the 4th of July and Labor Day. Failure to meet the 18-day closure would result in a penalty of $ 40,000 per day. During the planning phase it was apparent that meeting the 18-day restriction with the original design would be extremely challenging and alternatives were investigated. During development stages of the alternative design it was determined the arch walls could be completed in 16 days. This led to a schedule change that

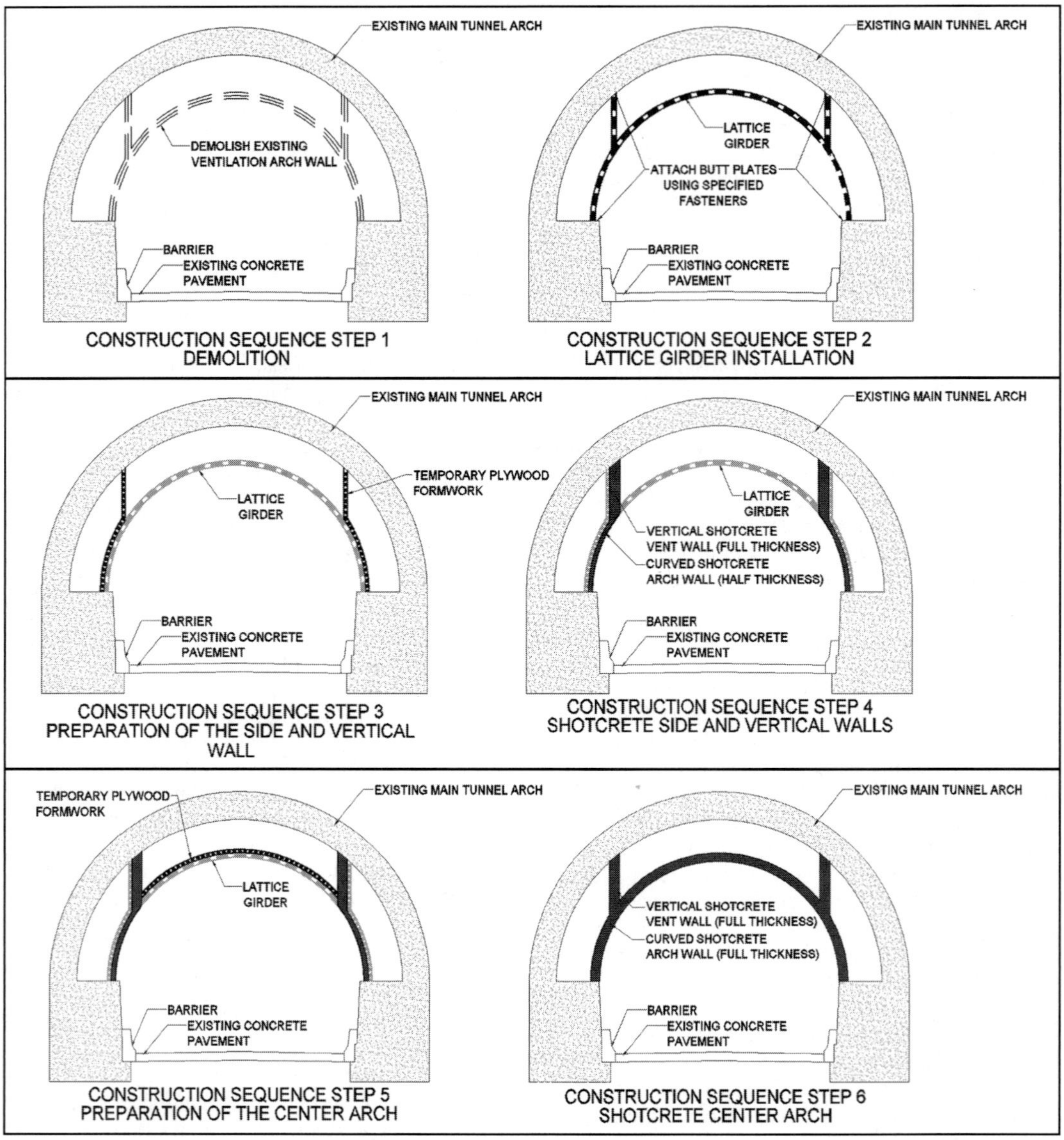


**Figure 6. Typical construction sequence**

reduced the allowable closure periods to be reduced to 16 day per phase in exchange for 2 weekend closures prior to each long-term closure. The two weekend closures were used on areas not associated with the ventilation arch walls. This allowed Swank to complete other activities prior to the scheduled long-term closure, which internally freed up resources and allowed to focus on the ventilation arch walls during the long-term closures.

Already one hour after the start of the closure the demolition of the existing arch walls began followed by the installation of the new shotcrete arch walls, following the construction sequence as described above. In the southbound tunnel (Phase 1) the work was completed with the second layer of shotcrete just hours before the opening of the tunnel for traffic. The delays were caused by logistical problems and a delayed delivery of lattice girders and undercut anchors. However, the phase 2 construction (northbound) was completed in around 14 days and much quicker than the southbound and 2 days under the maximum allowable 16 days, due higher efficiency in the sites' logistic and lessons learned as well as learning curve effects from the previous phase.

**Figure 7. Construction sequence Step 3—Lattice girder and extrados reinforcement sidewall sections**

**Figure 8. Construction sequence Step 4—Curved and vertical sidewall sections are shotcreted**

Crew assembly and size varied for each operation of the construction. Swank's portion of the installation was mainly demolition and installation of lattice girders, rebar and forming. In total, approximately 20 to 25 men per 12 hour shift worked on Swank's crew during the construction of the arch. Coastal Gunite was responsible for the installation of shotcrete and worked with a crew of 9 to 12 people per 12 hour shift.

Figure 7 depicts construction sequence Step 3. The lattice girders of the arch walls as well as the vertical walls sections have been erected. The extrados layer of reinforcement as well as plywood in the back has been installed. The center arch section is open to allow shotcreting of the vertical walls. In the back the suspended section of the arch wall, acting as a ventilation nozzle, which was not discussed in detail in this paper, can be seen.

Figure 8 shows the situation during construction sequence Step 4. The curved and vertical sidewall sections are already partially shotcreted. During this step the vertical sidewall sections will be completed with both layers of reinforcement and shotcreted to full thickness.

Figure 9 shows the construction sequence at Step 5. As soon as the vertical wall sections are completed, access through the center arch section is no longer needed. In this step the plywood and reinforcement in the center arch section can be installed and shotcreted. After this step the intrados level of reinforcement, covered by the final layer of trowel-finished shotctete can be installed.

Figure 10 shows the arch wall section after its rehabilitation, looking into the air nozzle opening. The smooth trowel-finish of the shotcrete makes it difficult to recognize that shotcrete in lieu of cast-in-place concrete was used.

The design specified stringent experience requirements for the shotcrete applicator to ensure the required high quality. Swank decided therefore

**Figure 9. Construction sequence Step 5—preparation of center arch section**

**Figure 10. Finished rehabilitation**

to subcontract the shotcrete work to the shotcrete specialists Coastal Gunite. The shotcrete was placed using an Allentown (Putzmeister) Elite 40 Shotcrete Pump. The concrete material was brought into the tunnel dry in bulk sacks and mixed inside a ready mix truck in order to both use the material specified and have enough available to place it in sufficient quantity given the tight construction schedule. The material itself included poly-fibers and a corrosion inhibitor. Excluding the finish coat, the shotcrete process involved the addition of a liquid accelerator at the shotcrete nozzle to reach the specified set times and meet the early strength requirements per design.

The Contractor was able to demonstrate proficiency encapsulating the lattice girders and both layers of mesh simultaneously by alternating the location of shotcrete and mesh installation. This approach allowed the project to move forward more quickly when needed. Overall the shotcrete was placed in three lifts per wall. The first layer of shotcrete was placed encapsulating the first layer of mesh and left enough of the lattice girder exposed such that the second layer could be installed. The second placement encapsulated all of the steel and was left rough so that a monolithic finish coat could be applied last as to be more appealing aesthetically.

The final layer was finished with a broom and was sprayed with a curing compound to attain proper cure and avoid surface cracking.

## CONCLUSION

For the Liberty Tunnel Rehabilitation project time was of the essence due to a short and limited closure of the tunnel. The alternative design of self-bearing shotcrete ventilation arch wall provided the contractor greater flexibility and reduced construction risk during the ventilation arch wall installation. The alternative design cut down the time required for the installation of an otherwise heavy formwork, which would have been required by the self-consolidating concrete. Further it was not required to retain the existing hangers supporting the ventilation arch walls since the shotcrete arch wall was self-supported.

The rehabilitation work conducted in the Liberty Tunnel showed a unique collaborative effort between the designer and contractors, which allowed the work to be completed on schedule, below the original cost, and with much less construction risk for the contractor as well as the owner.

The presented approach of the self-bearing shotcrete arch is a showcase for similar rehabilitation and repair works of aging tunnels, which have to be rehabilitated with a high quality under a limited time frame for tunnel closures and within the given budget.

## REFERENCES

American Concrete Institute (ACI)—Building Code Requirements for Structural Concrete, 2005 (ACI 318-05).

Federal Highway Administration (FHWA) 2009. "Technical Manual for Design and Construction of Road Tunnels—Civil Element," FHWA-NHI-09-010, Washington DC.

# Design and Analysis Techniques for High Head Tunnel Plugs

**Soumendra Sen**
New York City Department of Environmental Protection

**Z. Bade Sozer**
Jacobs Associates

**Daniel Ebin**
Jacobs Associates

**ABSTRACT:** The Rondout-West Branch Tunnel supplies approximately half of New York City's drinking water and is leaking. A bypass tunnel is being designed around the leakage area. When complete, the bypass tunnel project will have three permanent concrete plugs, two of which will be installed by removing the existing lining from the leaking tunnel. This paper discusses plug design, with an emphasis on leakage-free criteria, shear key design, and construction considerations. The design was optimized by computational modeling, which provides insight into the plug behavior. The plug design is unique because of the schedule challenges and extreme hydrostatic loading condition of 36 bar.

## INTRODUCTION

The Rondout-West Branch Tunnel (RWBT), a segment of the Delaware Aqueduct, was built from 1937 to 1944 and provides about 50% of New York City's total water supply. Monitoring and tunnel operations have shown that the RWBT is leaking up to 35 million gallons per day (MGD). In particular, two locations have been identified as areas of concern: Roseton and Wawarsing, as shown in Figure 1. The Wawarsing reach of the aqueduct will be repaired with an extensive grouting program when the RWBT is dewatered. However, because of the extensive nature of the leakage in the Roseton area, construction of the Rondout-West Bypass Tunnel has been proposed to divert the flow around this area. Two permanent concrete plugs will be installed in the existing RWBT at the intersections with the Bypass Tunnel to effectively seal off and isolate the leaking portion of the RWBT in the Roseton area. The construction of the plugs is on the critical path for the time-sensitive connection to RWBT. Therefore, a simple, practical, and effective design is required for the two RWBT plugs.

### Concrete Plug Use in Tunnels

Permanent plugs are commonly used for hydro facilities to block off construction access points. High head plugs have been successfully designed for deep mines of South Africa with reasonably hard rock and relatively high water pressures at deep ground levels (Garrett and Campbell Pitt, 1958, 1961; Lancaster, 1964). For the RWBT, the two plugs are not designed to support ground loads, but are built to resist the maximum internal pressure. Resistance to the applied hydraulic force is achieved by the mechanical interlock between the plug concrete and the rough excavation surface of the rock. In addition to satisfying the structural strength criteria, the plugs are designed to be leakage-free. The length of the plug is more often governed by leakage-free criteria rather than the structural strength criteria (Garrett and Campbell Pitt, 1958, 1961). This has been the case as well for the RWBT plugs.

### Purpose and Construction of Permanent Concrete Plugs in the RWBT

The permanent concrete plugs to seal off the abandoned section of the RWBT are designed as plain concrete parallel plugs with shear keys and minimal reinforcement at the face and will be installed as shown in Figures 2 and 3. The plugs are designed to be permanent and maintenance free for the 100-year design life of the Bypass Tunnel.

Bypass Tunnel excavation will be performed in phases. At the initial phase, a majority of the tunnel excavation and final lining installation will be completed up to a 100-foot (30.5 m) stand-off distance from the pressurized RWBT. To finalize the Bypass Tunnel connections to RWBT, the RWBT will be dewatered. However, because of the leaking characteristics of the existing RWBT lining, it is assumed

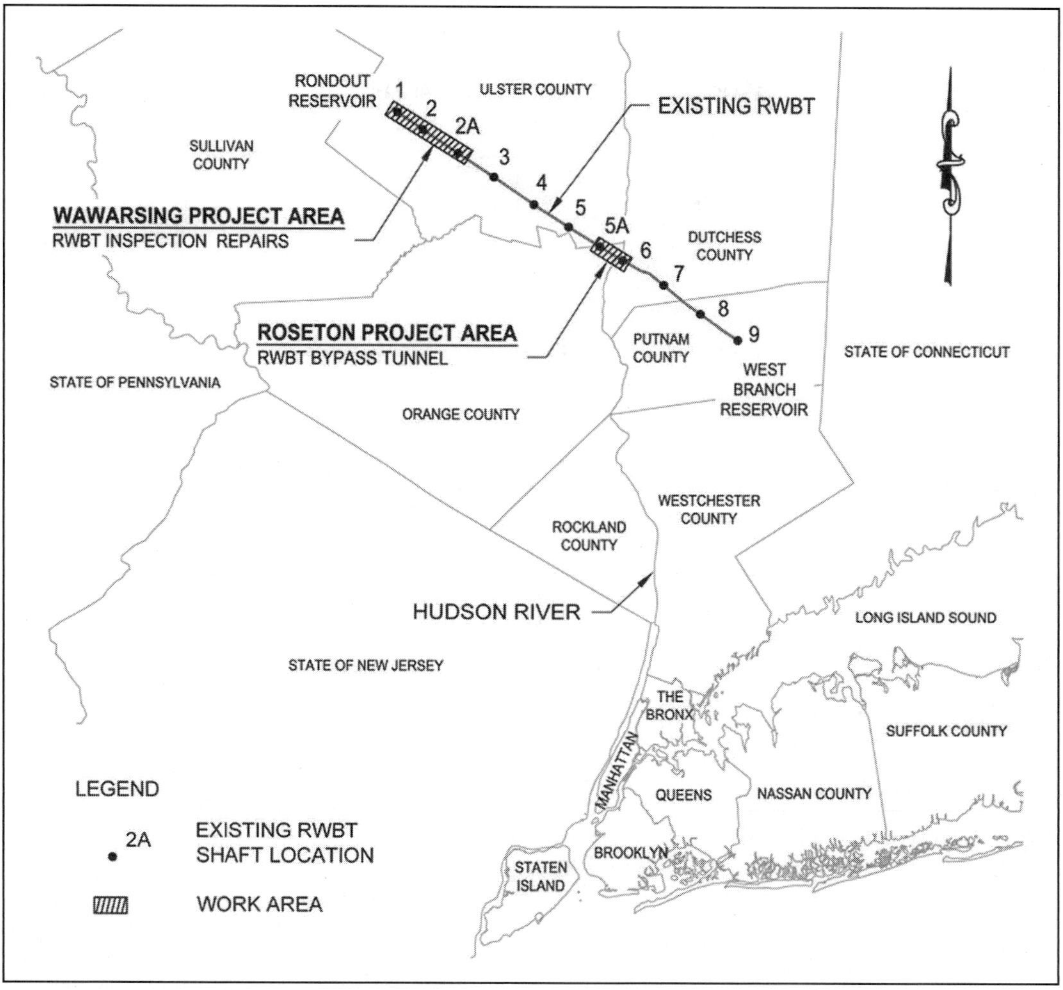


**Figure 1. Project overall plan**

that groundwater will infiltrate into the existing RWBT. Thus, when the Bypass Tunnel is excavated to break into the existing RWBT, the infiltrating water into the RWBT will have to be diverted via berms or dams in order to keep the intersections dry to minimize impact to plug construction.

For the installation of the plugs, existing RWBT lining will be removed to ensure contact with sound rock. Once the rock is exposed, any loose rock will be removed by scaling. Although the analysis does not require shear keys, to provide further conservatism in design, four shear keys will be installed in the middle portion of the plug. The shear keys installed below the springline are designed as 2-foot-rise (0.6 m), 8-foot-run (2.4 m) triangular shear keys.

The plugs are designed to resist the hydrostatic loading and to be leakage free. Contact grouting as well as quality control procedures, including concrete curing pipes during plug concrete pour, are required to prevent any water infiltration through the plug. After the plug concrete achieves initial set (typically 18–24 hours after placement), the contractor will perform pressure grouting from near the plug face to lower the hydraulic conductivity of the surrounding rock and seal any water pathways within the rock. The contractor is also required to perform contact grouting through preinstalled steel tube-a-manchette grout pipe at a minimum of 14 days after plug concrete is poured. The requirements for pressure and contact grouting are currently being developed to minimize the critical path schedule and will be prescribed by the contract documents to ensure that there is no leakage once the Bypass Tunnel goes into operation.

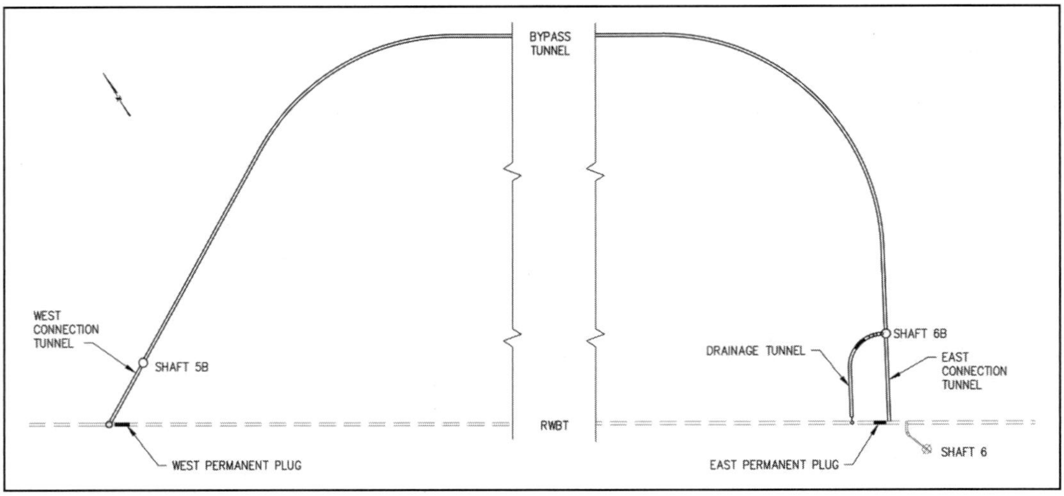


**Figure 2. Plan view of bypass tunnel**

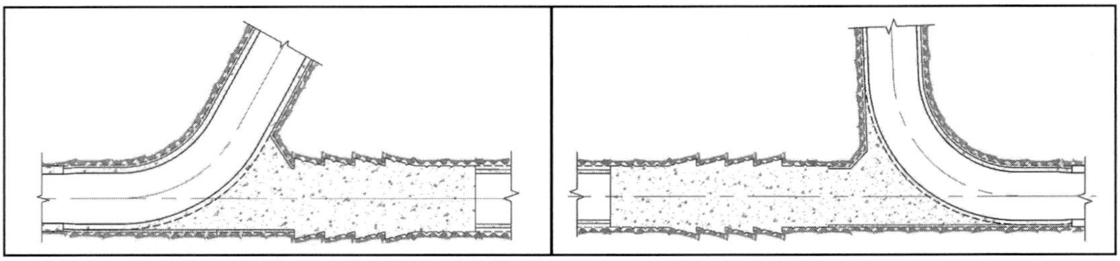

**Figure 3. West (left) and East (right) permanent plugs**

## ANALYSIS

The hydrostatic loading conditions will be significant for both plugs. For the maximum loading condition, right after Bypass Tunnel is put into operation, the plugs will be conservatively loaded with 1,200 feet (366 m) of internal hydraulic head on the operational side (Bypass Tunnel) of the intersections, with no pressure on the abandoned section. For equilibrium conditions, the 1,200 feet of internal hydraulic head on the operational side will be counteracted by 600 feet (183 m) of internal head on the abandoned side, assuming the abandoned Roseton area comes to equilibrium with the Hudson River head.

The successful sealing capacity of a plug is a function of the surrounding rock properties and is diminished by any discontinuities or fracture planes. Both plugs will be poured against shale units with hydraulic conductivities of less than $10^{-6}$ cm/sec. As discussed by Lang (1999), the allowable hydraulic gradient for the plug is calculated by dividing the design head of water by the plug length based on the surrounding rock characteristics. The allowable hydraulic gradient for "good rock"—which is defined as "hard to moderately hard, moderately jointed," with RMR values between 61 and 80—is 10 to 14 (Lang, 1999). In the East and West Intersections, where the plugs will be, the rock has RMR values of 66 and 68, respectively. The rock is modeled with a modulus of elasticity of 1,000 ksi and a Poisson's ratio of 0.3. Thus, for equilibrium loading conditions of a differential head of 600 feet, the minimum length required is 60 feet (18 m) based on a lower allowable hydraulic gradient of 10. The structural capacity of a 60-foot-long concrete plug is checked using STAAD Pro, as described below.

For loading conditions, a load factor of 1.0 is assumed for the conservative maximum hydrostatic head. The strength reduction factor $\phi$ for the concrete shear strength is assumed to be 0.6 (ACI 318-11, Section 9.3.5). A strength reduction factor of 0.5 is used for the rock shear strength to correspond to a factor of safety of 2.0.

As mentioned above, the construction of the plugs will be on the critical path, and the plug may have to be loaded before it attains full strength. Thus, the analysis was performed for a minimum 7-day strength of 2,250 psi concrete.

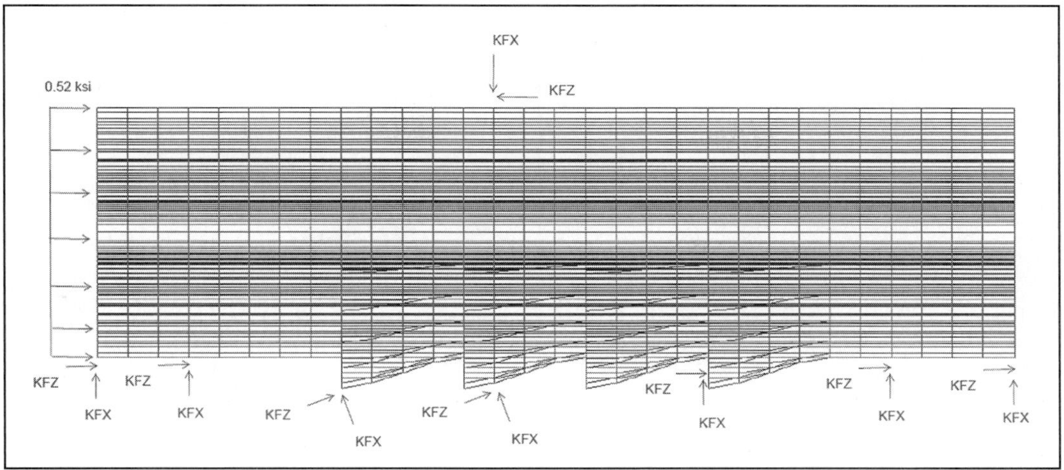


**Figure 4. Schematic elevation view of spring supports and loading**

### Finite Element Modeling

Eight-noded, solid isoparametric elements were used to build the three-dimensional concrete plug models in STAAD Pro. Each node has three translational degrees of freedom. The typical element size was 2 feet × 2 feet × 2 feet. Rock confinement was modeled with uniaxial or biaxial springs for each degree of freedom at the boundary nodes of the model. Uniaxial springs that release in tension are used to model the compression resistance provided by rock confinement. These springs are positioned perpendicular to bearing surfaces of the model. The spring constant is calculated based on the tributary area, the rock mass modulus, the Poisson's ratio of the rock, and the excavated radius of the plug. Biaxial springs are used to model the shear resistance provided by the interface between rock and concrete. These shear springs are applied parallel to bearing surfaces. Figure 4 shows, in an elevation view, how the spring supports are arranged on various parts of the model and the loading on the face.

Hydrostatic and rock loadings are not considered as these loads would increase the clamping action of the model and behave similar to the rock confinement provided by the axial springs.

The results of the model were checked for tension, compression, and punching shear. The allowable stresses were calculated per recommendations for plain concrete from ACI 318, Chapter 22 (ACI, 2011). The maximum tensile and compressive stresses were found to be in the longitudinal direction of the plug, as shown in Figure 5. The maximum tension occurs at the loading face and has a value of 96 psi (142 psi allowable). The maximum compression occurs about 4 feet (1.2 m) into the plug, located in the center of the light gray region, and has a value of 545 psi (810 psi allowable).

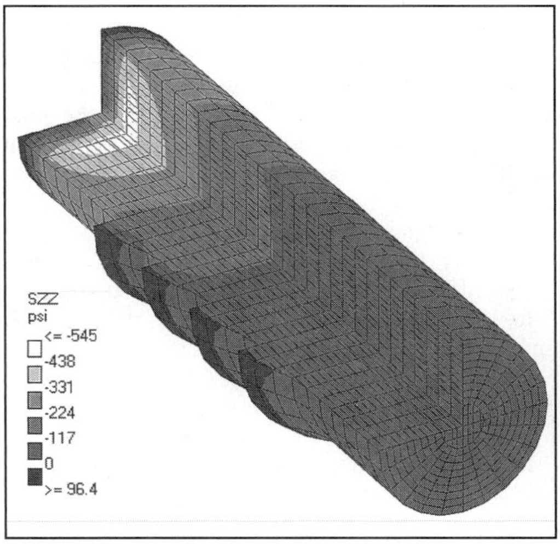


**Figure 5. Longitudinal compression and tension, three-quarter view**

Stress results from STAAD can only be displayed in the global coordinate system for solid elements. Therefore, two sets of stress contours are generated to check for punching shear, as shown in Figure 6. The allowable punching shear stress is 76 psi. The maximum shear stress on the horizontal plane occurs in the crown, approximately 4 feet from the face, as represented by the dark gray, and has a value of 89 psi. The maximum shear stress on the vertical plane, as represented by the dark gray, occurs in a similar location but at the springline, as well as where the shear key starts at the springline, and has a value of 89 psi. These values are greater than the allowable stress. However, the slightly overstressed zones are localized, and finite element

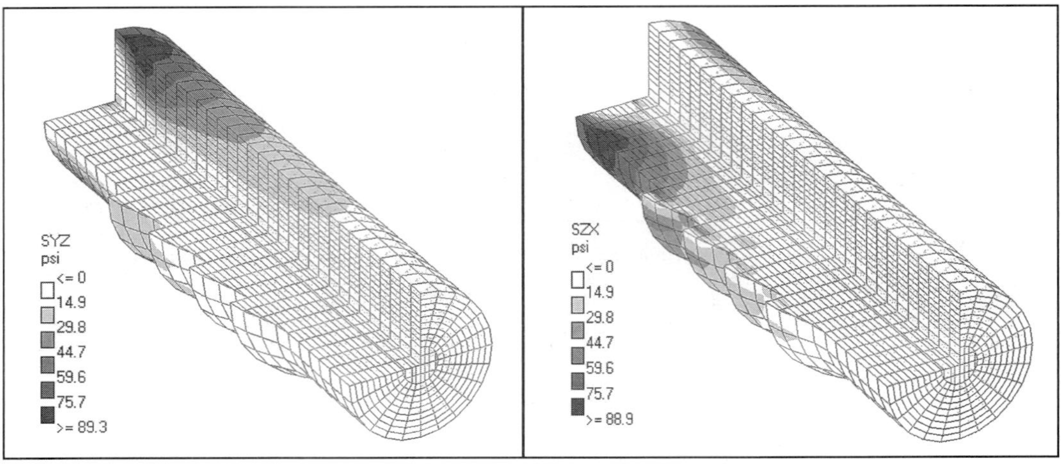


**Figure 6. Longitudinal shear stress on the horizontal (left) and vertical (right) planes, three-quarter view**

models supported with shear springs are known to overestimate stress concentrations. Thus, it is concluded that the stresses will get redistributed over a larger area, reducing to less than the allowable stress.

The results presented above are for the final configuration of the permanent plugs to be used in the RWBT. These four 8-foot-long, 2-foot-wide triangular shear keys from springline to invert act as wedges, driving the plug into the crown of the opening, almost like a modified tapered plug. The keys are sized such that they can be excavated with one 8-foot-long shot round and the effects of overbreak amplified, thereby increasing the mechanical bond between concrete and intact rock. The design of the plugs also includes a reinforcement cage at the loading face to mitigate any minor cracking.

Modeling results also showed that an insignificant amount of load was being transferred to the shear keys. A parametric analysis on the concrete compressive strength showed insignificant change in load transfer. The possibility of a straight plug with no shear keys was modeled to show that the plug can withstand the internal pressure with no shear keys. However, keys were included in the final configuration for additional conservatism and to increase the flow path along the plug.

## RECOMMENDATIONS AND CONCLUSIONS

The purpose of the Bypass Tunnel is to fix the leakage in the RWBT. The permanent concrete plugs are a key aspect in isolating the leaking portion of the existing RWBT from the Bypass Tunnel and the rest of the RWBT alignment once the existing tunnel is put back into service. The challenge in designing these plugs is to be able to provide enough capacity to resist high internal water pressures with concrete that has not reached its 28-day design strength and enough length to limit the infiltration of water through the rock when equilibrium has been reached. As such, the design assumed that the concrete gains only 50% of the design compressive strength in 7 days, after which the RWBT could be put back into service and the full 1,200 feet of internal head would be applied to the plugs.

The proposed plug design satisfies the requirements for a leakage-free and structurally engineered bulkhead as a solution to the problem. Furthermore, with only minimal face reinforcement and four shear keys, it is a constructible design with minimal impact to the project's critical path.

## REFERENCES

American Concrete Institute. 2011. ACI 318-11: Building Code Requirements for Structural Concrete and Commentary, Sections 9.3.5, 22–10.

Garrett, W.S., and L.T. Campbell Pitt. 1958. Tests on an experimental underground bulkhead for high pressures. *Journal of the South African Institute of Mining and Metallurgy*, October 1958, 123–143.

Garrett, W.S., and L.T. Campbell Pitt. 1961. Design and construction of underground bulkheads and water barriers. In *Trans. 7th Commonwealth Mining and Metallurgy Congress, South African Institute of Mining and Metallurgy, Johannesburg*. 1,283–1,299 (disc. 1,300–1,301).

Lancaster, F.P. August 1964. Report on the Research into Underground Plugs. Transvaal and Orange Free State Chamber of Mines. Research Report No. 27/64.

Lang, B. 1999. Permanent sealing of tunnels to retain tailings or acid rock drainage. In *Mine, Water & Environment*, 647–655.

## TRACK 2: DESIGN

## Session 3: Ground Movement and Structures Analysis

Michael Torsiello, Chair

# Impact of EPB Tunneling on Pile Foundations and Existing Tunnels

**Mahmoud Sepehrmanesh, Verya Nasri, and Navid Allahverdi**
AECOM

**ABSTRACT:** This paper discusses the challenges posed in designing an underground light rail transit project in downtown Los Angeles and the solutions proposed to economically meet these challenges. The project under study is one of the capital projects envisioned to expand the public transportation in the city of Los Angeles. The project proposes tunnel construction for approximately 1.6 kilometers long 6.7 m diameter twin TBM tunnels in soft ground with precast concrete segmental lining. Three dimensional finite element modeling was employed to model TBM operation and evaluate ground deformation occurring as a result of tunneling. Finite element analyses performed to ensure the load carrying and structural integrity of existing piles located in the vicinity of tunnels. The results showed that piles can sustain the tunneling induced forces with satisfactory margin of safety. Also, the impact of tunneling under existing subway tunnels was studied.

## INTRODUCTION

The focus of this paper is on adopting advanced three dimensional finite element modeling in tunneling and underground engineering applications. Finite element modeling is an indispensable tool in assessing the impact of underground excavation on existing structures and utilities. Finite element enables designers to virtually simulate different stages of the construction process with desired level of details. In practice, detailed finite element modeling is performed during final stages of design to further refine preliminary designs based on simplified or approximate methods.

The Regional Connector Transit Corridor (RCTC) is one of the capital projects envisioned to expand the public transportation in the city of Los Angeles. The alignment will connect two existing subway lines currently terminated in the downtown area. The preliminary design phase included approximately 1.6 kilometers long 6.7 m diameter twin TBM tunnels in soft ground. Earth pressure balance (EPB) shield tunneling was proposed for excavating the tunnels. A joint venture of AECOM and PB performed the preliminary design of the RCTC for the Los Angeles County Metropolitan Transportation Authority. The RCTC will be constructed by a design build contractor which will provide a more detailed final analysis and design of the project.

The geometrical constraints of connecting two existing stations and cost reduction considerations demanded setting up the alignment in close proximity to existing buildings, structures, and utilities. Designing a transit facility in such a congested downtown area presented a number of design challenges. For instance, (1) the proposed alignment passed under an existing bridge in proximity of its pile foundations which are supporting the bridge piers. The alignment of TBM-bored tunnels indicated a minimum of 0.75 m separation between the future tunnels and the existing piles; (2) the proposed alignment passed under an existing operational underground subway line with 1.5 m of vertical separation between tunnels. Raising the proposed alignment so close to the existing tunnel allowed reducing the depth of the adjacent cut and cover stations and hence reducing construction costs. In the following sections, detailed analysis methods and results are provided for mentioned tunnel crossing.

Three dimensional finite element models were developed to ensure the minimal impact of tunneling on the existing subway tunnels and existing bridge structure. The adequacy of load carrying capacity and structural integrity of piles were ensured during and after construction of tunnels. The finite element model accounted for staged construction and detailed shield-driven TBM processes including applying the balancing face pressure as well as injecting bentonite slurry through the TBM shield. The model took into account all relevant components of the construction process including the nonlinear soil behavior, shield tunneling, segmental lining installation and the tail void grouting.

## GEOLOGY AND SUBSURFACE INVESTIGATION

The proposed alignment is located in the northern portion of the Los Angeles Basin. This basin is a major elongated northwest-trending structural depression that has been filled with sediments up to 4,000 meters thick since middle Miocene time. On a local geologic scale, the alignment will traverse the southeastern end of the Elysian Park Hills and the ancient Los Angeles River floodplain. The Elysian Hills comprise the low-lying hills west of the Los Angeles River and southeast of the eastern end of the Santa Monica Mountains.

The alignment will encounter several geologic units that range in age from Pliocene to recent. The geologic units that will be encountered within the proposed tunnel alignment and station boxes are the Pliocene-age sedimentary strata of the Fernando Formation, Holocene to probable Late Pleistocene Alluvium, and historical/recent artificial fill. Artificial fill has been placed at various locations along the alignment such as utility trench backfills, structure backfills, roadway embankments, and areas overlying both existing and abandoned tunnels. Holocene to probable late Pleistocene-age alluvial deposits are present along the alignment beneath variably thick artificial fill. Overlying the Fernando Formation are alluvial deposits comprised primarily of inter-layered clays, silts, fine sands, and sand layers containing variable gravel and cobbles. Coarser grained alluvium comprised of poorly to well-graded sand with variable gravel and cobble content was reported in the lower portion of the alluvium above the Fernando Formation. Pliocene-age, sedimentary bedrock was mapped along portions of the alignment. The Fernando Formation is comprised predominantly of massive siltstone, with some interbeds of sandstone and conglomerate and well-cemented, fine-grained silty sandstone. Bedding dip inclinations range from approximately 70 to 75 degrees with dip vectors that range from N168 to N191.

The project alignment is located within the Los Angeles Forebay Area. Groundwater in the Los Angeles Forebay occurs primarily in the Quaternary-age sediments. This is due to the relatively low permeability of the underlying bedrock of the Fernando Formation. Aquifers in the Los Angeles Forebay area include the Semi-perched, the Gaspur, the Exposition, the Gardena, and the Gage. Because bedrock is relatively shallow and the water-bearing sediments are relatively thin along the majority of the alignment, only the Semi-perched aquifer is present in the project area. The Semi-perched Aquifer generally consists of the older sediments (Pleistocene-age) and locally the younger sediments (Pleistocene-age) overlying the bedrock, whereas the Gaspur Aquifer consists of the coarser-grained younger sediments in channel areas. A groundwater level contour map of the Los Angeles Quadrangle indicates groundwater depths ranged from historical highs of about 5 to 15 meters below ground surface east of the Bunker Hill area with a general southward gradient. It should be noted that shallow groundwater levels are typically influenced by seasonal rainfall and infiltration in addition to potential localized groundwater extraction.

## FINITE ELEMENT ANALYSIS

The complex and dynamic nature of shield-driven tunnel excavation, staged construction, segmental lining installation process, tail void grouting, and hydro-mechanical coupling in the surrounding ground preclude the use of traditional two-dimensional numerical analysis tools for modeling the ground behavior and structural response for this particular project. Therefore, three-dimensional, non-linear modeling approach, using state-of-the-art analysis program Midas Geotechnical & Tunneling Analysis System, MIDAS/GTS (2011) was adopted to evaluate the ground response and impact of tunneling on existing adjacent structures.

During the past three decades, a vast amount of effort has been expended to numerically simulate the shield-driven TBM tunneling processes and construction operation to accurately estimate the induced ground settlement. Among the latest attempts, (Kasper and Meschke 2004) developed a three-dimensional finite element model to study the influence of the soil and grout material properties and the cover on the surface settlements, loading and deformation of the tunnel lining and steering of the TBM. They modeled the TBM as a rigid movable body in frictional contact with soil. Their simulations employ a two-field finite element formulation to solve the strain field and pore-water pressure in soil and grout materials. Based on a number of parametric studies, (Kasper and Meschke 2006) concluded that: (1) strength characteristics and the over-consolidation ratio are major factors influencing the soil deformation in the vicinity of the shield machine and surface settlements, (2) for soils with a high permeability, larger final settlements observed only after full consolidation was observed, and (3) the cover of the tunnel is the most important factor in determining the forces developed in the lining.

### Finite Element Modeling Approach

The adopted 3D analysis approach allowed modeling the geometry of tunnel and excavation staging in order to evaluate the full impact of excavation progression on existing structures. The size of the model was determined in such a way to minimize the boundary effects on the analysis results while allowing the analysis to be performed efficiently. The finite element mesh was consisting of tetrahedron solid elements. A small element size of 0.6 m was used in the vicinity of the tunnels. In areas far away from the tunnels, the maximum element size was increased to 3.0 m.

### Modeling EPB TBM Process

Applying face pressure and shield bentonite slurry pressure, installing segmental rings, and tail void grouting are among features that were considered in the analysis in order to allow an accurate simulation of the EPB tunneling operations. The TBM excavation advances were modeled in 1.5 m intervals which is the length of one ring. The face of excavation was immediately pressurized after excavating each drift in order to reduce the settlement due to face loss. The face pressure was assumed to be constant for ease of application. The applied balancing face pressure was set equal to the horizontal in-situ stress at the centerline of the tunnel.

In order to model the conical shield support, compression-only gap elements were used to model the conical shield and the variable gap between the ground and the shield. The maximum gap was considered to be 7.5 cm at the tail of the shield. The length of the shield was assumed to be 4.5 m which is equal to three drifts with 1.5 m in length. Bentonite slurry pressure was applied through the length of shield, i.e., over 4.5 m behind the face. This slurry pressure prevents the soil from moving in and reduces the volume of shield ground loss and consequently reduces the ground deformation and settlement. The slurry pressure value was considered as the mean in-situ vertical and lateral stresses at the tunnels' springline elevation. By increasing bentonite slurry pressure, the crown deflection of tunnels as well as ground convergence will decrease. Theoretically, there is a pressure at which the settlement will completely diminish. Pressures in excess of this value will result in heaving of the ground surface.

Precast concrete segmental rings were installed behind the shield. The first 1.5 m behind the shield representing the ring under installation was assumed without any support; however, prior rings installed provided full support to the excavation. The thickness of the segments was assumed to be 25 cm with an additional 5 cm of hardened backfill grout injected behind the segments. A reduction factor of 0.80 was applied to the flexural stiffness of the rings to account for the effects of segment joints as suggested in (Lee and Ge 2001).

The in-situ stresses were initialized through prescribing at-rest lateral pressure coefficient. Surcharges due to the bridge service load and seismic load were applied on the pier columns during the initialization stage. All displacements were reset to zero in the initial stage. Mohr-Coulomb failure criterion was adopted for rock behavior. The displacement degrees of freedom at the bottom face of the model were fixed in all directions; however, only out-of-plane displacements were fixed on the four side faces of the model.

### Verification of Analysis Results

A check on the order of magnitude accuracy of three dimensional finite element results was made via comparing results with an approximate method described in (Chen et al. 1999). Also, convergence studies for different finite element meshes were performed to guarantee the convergence of results.

## TUNNEL CROSSING UNDER EXISTING BRIDGE

This section discusses the results of advanced 3D numerical studies which were conducted to assess the impact of tunneling-induced ground movements on the existing bridge piles. The proposed tunnel alignment runs between the axes 2, and 3 of the bridge piers and columns as shown in Figure 1. The piers and columns of the bridge are resting on deep foundations including piles and caissons. Soil movement as a result of tunnel excavation induces additional forces in the piles. The additional forces may potentially distress the structural integrity of the piles and the super-structure.

The profile of the TBM-bored tunnels indicated a minimum of 0.75 m separation between the future tunnels and the existing piles. It is evident that small separation between the bored tunnels and the existing piles will result in a reduction of skin resistance and tip bearing capacity of the piles depending on the relative location of tunnel with respect to the pile.

The tunneling excavation causes both axial and lateral deformations in piles located close to the tunnel. The maximum lateral deformation in the pile occurs about the depth of the tunnel's springline as the surrounding soil medium converges toward the center of the tunnel. As detailed in (Chen et al. 1999), the vertical soil movement above the tunnel's springline is generally downward and tends to impose negative skin friction on the pile, causing settlement and possible reduction in the pile load-carrying capacity; however, the vertical soil movement below the tunnel's springline is upward and will cause pile heave. As a result of pile deformation, additional axial force and bending moments will be induced in the piles. The key factor in pile's response and induced forces is the ratio of pile length to the tunnel cover. The pile behavior is rather different for long piles (piles whose tip are below the tunnel's springline) and short piles (piles whose tip are above the tunnel's springline) because maximum lateral soil movements occur about the tunnel springline.

The forces induced in piles as a result of the tunnel excavation were calculated and added to the existing service forces in the piles. Service forces are due to the dead load of the super-structure and traffic loads. Additionally, a lateral load equal to 10% of vertical load was considered at the bridge's deck

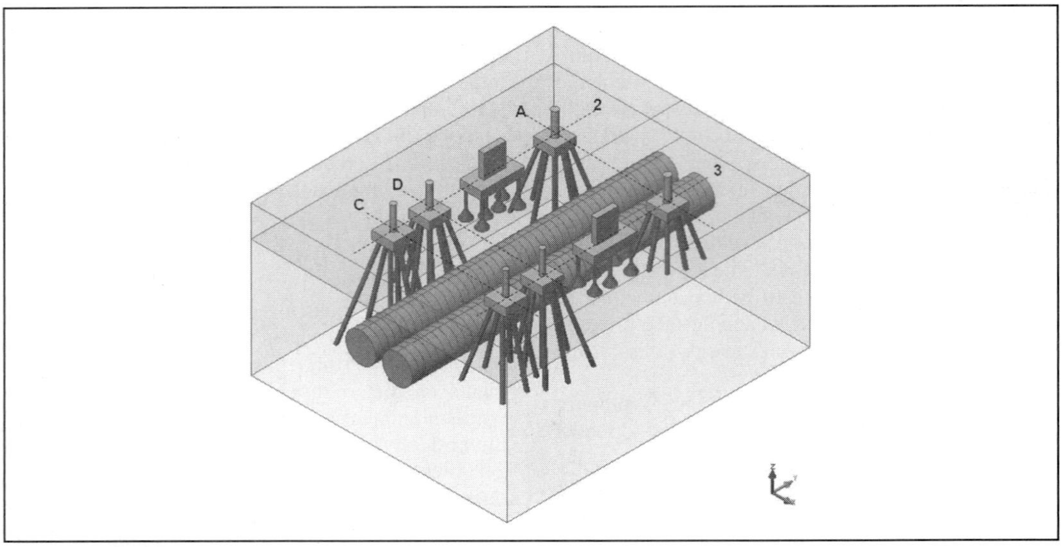


**Figure 1. Sketch of tunnels and piles of existing bridge**

**Table 1. Axial force and bending moment in piles**

| Pile Group | Axial Force Under Service Load (kN) | Axial Force After Tunneling (kN) | Bending Moment Under Service Load (kN-m) | Bending Moment After Tunneling (kN-m) |
|---|---|---|---|---|
| 2A | 480 | 915 | 4.6 | 28.9 |
| 3A | 512 | 488 | 5.2 | 34.0 |
| 2C | 612 | 863 | 8.7 | 24.7 |
| 3C | 603 | 885 | 12.2 | 21.5 |
| 2D | 477 | 986 | 7.4 | 23.3 |
| 3D | 512 | 959 | 9.1 | 34.2 |

level to account for lateral seismic loads. It is noteworthy to distinguish piles belonging to pile groups of 3A and 2D, 3D (shown in Figure 1) when interpreting the results. Piles in pile group 3A are relatively shorter than piles in the rest of pile groups. On the other hand, piles in pile groups 2D, and 3D have the minimum separation with the tunnel.

## Pile-cap Settlement and Pile Forces

The ground convergence, pile-cap settlements and induced forces in the piles can be controlled via applying pressurized bentonite slurry through the shield. By increasing the bentonite pressure, the tunnel convergence, pile disturbances, and ground settlement will decrease. The value of applied pressure was considered as the mean of in-situ vertical and horizontal stresses at the tunnel's springline elevation.

The induced axial force and bending moment when applying bentonite slurry pressure are presented in Table 1. The bending moment reported in Table 1 corresponds to the bending moment associated with pile deformation transverse to the tunneling direction. The final forces induced as a result of tunneling will remain in the piles permanently.

## Pile Strength

Structural integrity of piles was investigated for combined effects of axial force and bending moment via interaction diagram curves. As such, interaction diagrams were developed for 0.70 m circular section plain concrete piles according to the ACI-318 code provisions. Figure 2 shows the ultimate axial and bending moment pairs observed in each pile group. The ultimate factored forces were obtained by applying a uniform load factor of 1.5 to the results obtained from analysis. As observed, the order of axial force in piles is about the same except for piles in pile group 3A (piles with shortest length). Based on interaction diagram, the largest demand-to-capacity ratio belongs to pile groups 2D and 3D (piles with least separation with the tunnels). This demand-to-capacity ratio is around 0.5.

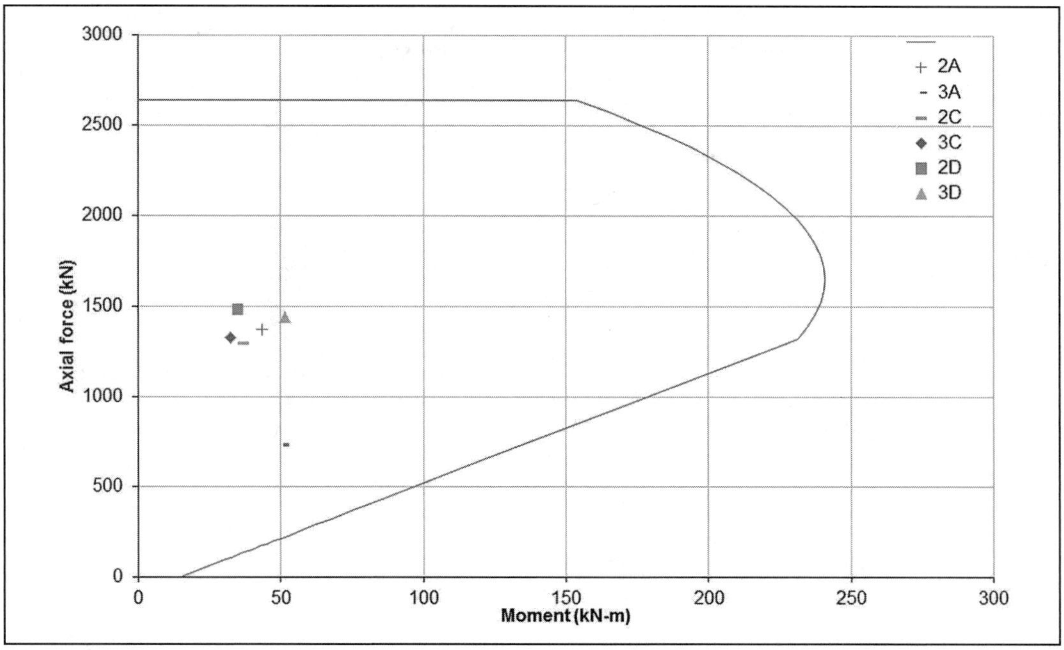


**Figure 2. Interaction diagram curve for piles in all pile groups**

**Pile Load Carrying Capacity**

The load carrying capacity of piles was evaluated considering pile tip bearing as well as frictional skin resistance contributions. The forces developed did not exceed the load carrying capacity of piles.

## TUNNEL CROSSING UNDER EXISTING SUBWAY TUNNELS

The proposed alignment passed under an existing operational underground subway line—Red Line—within a few feet of vertical separation between tunnels. Three dimensional finite element modeling employed to evaluate the settlement under the existing Red Line tunnels. The geometry of the tunnel crossing is shown in Figure 3. The vertical separation between the proposed tunnels and Red Line tunnels was set at 1.5 meters after raising the proposed tunnel vertical alignment to reduce excavation volume of neighboring stations.

The additional stresses and strains induced in the Red Line tunnel lining as a result of tunnel excavation, were calculated as the difference between the lining stresses/strains determined after the completion of Red Line tunnels and those obtained after completion of the new tunnel construction. The construction sequences of Red Line tunnels were modeled in order to obtain a realistic evaluation of existing stress in tunnel linings before commencing tunnel construction. As such, the Red Line tunnels were excavated one at a time in 5 meters drifts. CIP concrete linings were installed after finishing the excavations. After installing liners of Red Line tunnels, displacements were reset to zero.

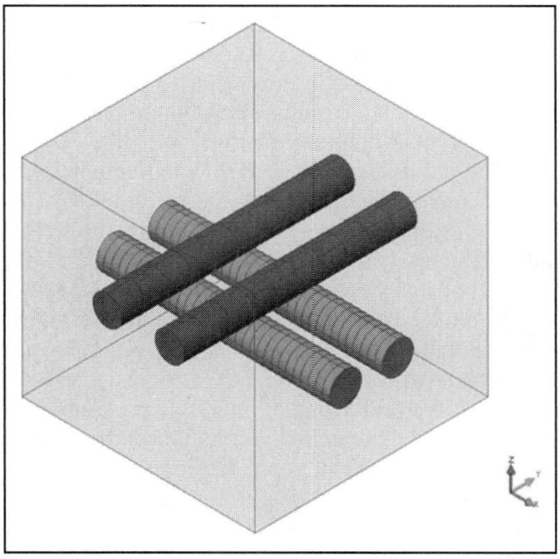

**Figure 3. Proposed tunnels shown beneath the Red Line tunnels**

In total, 98 construction stages were defined in the model to represent the construction processes of Red Line and proposed tunnels. Construction stages from 1 to 22 designated stages for constructing Red Line tunnels, while stages 23 to 98 represented

**Table 2. Tunneling-induced principal tensile stress at the invert of Red Line tunnel lining**

| Case No. | Slurry Pressure (kPa) | Tensile Stress Before Tunneling (kPa) | Tensile Stress After Tunneling (kPa) | Induced Tensile Stress in Existing Tunnels Lining (kPa) |
|---|---|---|---|---|
| 1 | 0 | 2,460 | 3,150 | 690 |
| 2 | 360 | 2,460 | 2,700 | 240 |
| 3 | 400 | 2,460 | 2,675 | 215 |
| 4 | 450 | 2,460 | 2,650 | 190 |
| 5 | 490 | 2,460 | 2,600 | 140 |

**Table 3. Tunneling-induced principal compressive stress at the crown of Red Line tunnel lining**

| Case No. | Slurry Pressure (kPa) | Compressive Stress Before Tunneling (kPa) | Compressive Stress After Tunneling (kPa) | Induced Compressive Stress in Existing Tunnels Lining (kPa) |
|---|---|---|---|---|
| 1 | 0 | 2,275 | 2,570 | 295 |
| 2 | 360 | 2,275 | 2,415 | 140 |
| 3 | 400 | 2,275 | 2,430 | 155 |
| 4 | 450 | 2,275 | 2,450 | 175 |
| 5 | 490 | 2,275 | 2,455 | 180 |

proposed tunnels construction. The construction operations for the existing Red Line tunnels are simulated by end of stage 22. In this stage the Red Line tunnels are bored and the cast in place concrete lining is installed. The displacements after implementing stage 22 are reset to zero and the principal tensile and compressive stresses are recorded in order to be compared with the results obtained from the final stage of proposed tunnel construction.

The induced principal tensile and compressive stresses are respectively presented in Table 2 and Table 3 for different bentonite slurry pressures. The induced stresses in the Red Line tunnel lining are a function of slurry pressure since all other parameters in the model remain unchanged. Different slurry pressure values correspond to 0, 80, 90, 100, and 110 percent of the mean in-situ vertical and lateral stresses at the center line of tunnel. For example, as shown in Table 2 and Table 3, the induced principal tensile stress is 240 kPa and the induced principal compressive stress is 140 kPa for Case 2 in which slurry pressure reach 360 kPa. It is noteworthy to mention that all tensile stress readings correspond to the invert of the Red Line lining located just above the proposed tunnels; while compressive stresses are measured at the crown of Red Line lining above the proposed tunnels.

In addition to induced stress/strains developed in the Red Line tunnel lining, the amount of maximum settlement/heave occurred at the invert of Red Line was critical for assessing the potential level of damage to the Red Line tunnels. The deformation readings at the final stage are exclusive to the future tunneling since all displacements prior to new tunnel construction were reset to zero. The maximum vertical settlements at the invert of the Red Line tunnels were compared for all bentonite slurry pressure cases. Table 4 summarizes the maximum deflection at the crown of the proposed tunnels and the maximum settlement at the invert of Red Line Tunnels. A positive value indicates upward movement (heave). Figure 4 shows the profile of settlement along the invert of the existing Red Line Tunnels for different values of bentonite pressure. As observed, tunnel excavation can be performed with negligible settlements developed under the Red Line provided proper amount of slurry pressure is applied. Cases 2 and 3 are representing bentonite pressures that resulted in very small settlements in the Red Line invert.

## CONCLUSION

The objective of this paper is to present the methodology and results of comprehensive three-dimensional finite element analyses which were performed to assess the potential impacts of tunneling under an existing subway tunnel as well as potential impact of tunneling under a bridge.

Advanced three dimensional finite element modeling was performed to assess the impact of tunneling on the pile foundations of an existing bridge. The results indicated that tunneling-induced forces in the piles can be mitigated via applying bentonite pressure throughout the shield. It was shown that piles can safely withstand the additional forces due to tunneling. Deformation of piles and settlement experienced under pile-caps were shown to be negligibly small.

**Table 4. Maximum vertical displacements**

| Case No. | Slurry Pressure (kPa) | Deflection at Crown of Proposed Tunnels (mm) | Settlement at Invert of Existing Tunnels (mm) |
|---|---|---|---|
| 1 | 0 | –8 | –5 |
| 2 | 360 | –2 | –1 |
| 3 | 400 | 0 | +1 |
| 4 | 450 | +1 | +2 |
| 5 | 490 | +2 | +3 |

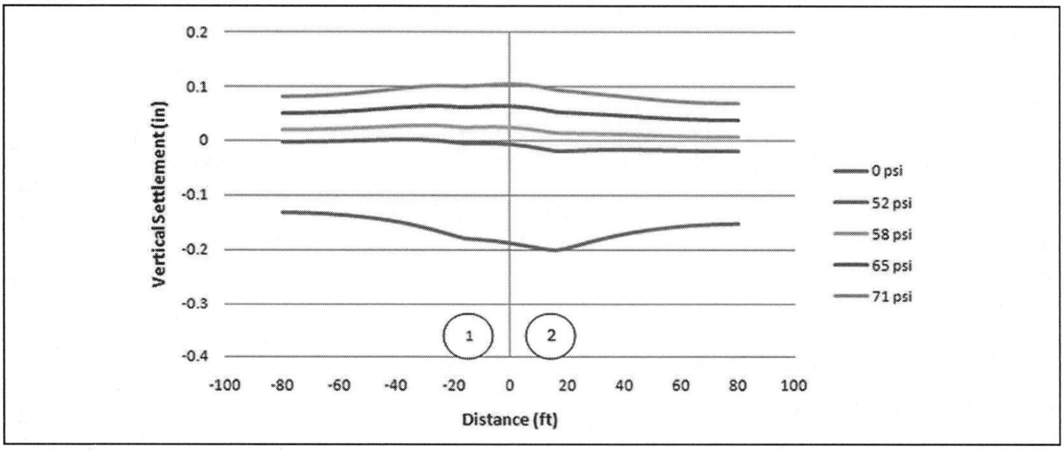


**Figure 4. Predicted Settlement/heave along the invert of Red Line tunnels for different slurry pressures**

In order to assess the impact of tunneling-induced ground movements on the existing Red Line Tunnels and to investigate the possibility of raising the proposed tunnel vertical profile, a comprehensive parametric study was conducted which utilized advanced three dimensional numerical modeling and analysis for Earth Pressure Balance (EPB) TBM driven tunneling. The parametric study of calibrated TBM bentonite pressure was conducted to demonstrate the effectiveness of this measure to mitigate the impacts of tunneling and control ground movements. This study has demonstrated that the predicted tunneling-induced ground settlements under the tunnel invert and the stresses/strains in the lining of the existing Red Line Tunnels can be effectively controlled by calibrated TBM tunneling. Therefore, it was recommended to raise the proposed tunnel vertical profile to within a quarter of the tunnel diameter (about 1.5 meters) separation from the existing Red Line Tunnels. This recommendation will result in a reduction of the depth of the cut-and-cover excavation for the proposed neighboring stations and consequently, will reduce the cost of construction for these stations.

## ACKNOWLEDGMENT

The authors would like to thank Girish Roy, the Deputy Executive Officer at the Los Angeles County Metropolitan Transportation Authority for his permission to publish this paper.

## REFERENCES

Chen, L.T., Poulos, H. G., Loganathan, N. 1999. Pile Response Caused by Tunneling, J. Geotech. Geoenviron. Eng., No. 125, pp. 207–215.

Kasper, T., Meschke, G. 2004. A 3D Finite Element Simulation Model for TBM Tunneling in Soft Ground, Int. J. Numer. Anal. Meth. Geomech., No. 28, pp. 1,441–1,460.

Kasper, T., Meschke, G. 2006. A Numerical Study of the Effect of Soil and Grout Material Properties and Cover Depth in Shield Tunneling, Computers and Geotechnics, No. 33, pp. 234–247.

Lee, K.M., Ge, X.W. 2001. The Equivalence of a Jointed Shield-driven Tunnel Lining to a Continuous Ring Structure, Can. Geotech J., No. 38, pp. 461–483.

MIDAS IT 2011. MIDAS/GTS User Manual.

# Site Characteristic Curve and Risk Assessment for Buildings Around Large, Deep Excavations

**S.H. Jason Choi, Xiaomin You, and Colin Barratt**
Parsons Brinckerhoff

**Edward J. Cording**
Department of Civil and Environmental Engineering, University of Illinois at Urbana-Champaign

**ABSTRACT:** During the design of large deep excavations on urban transit projects, both lateral and vertical ground movements and their impact on adjacent structures need to be addressed. An excavation support system should be selected to limit the movements to acceptable levels. This study utilizes a hypothetical subway station excavation in medium dense sand to demonstrate how to develop site characteristic curves, which can be incorporated into a risk analysis to determine the ground movement limits for controlling building damage and modifying shoring design.

## INTRODUCTION

Underground urban transit stations are often constructed within the public right of way using "cut-and-cover" construction, where the box is fully excavated; the final structure is built; the excavation is backfilled over; and the surface is restored. The excavation support system is a combination of support wall and lateral bracing. Support walls include soldier pile and lagging, tangent pile, secant pile and slurry walls amongst others. Lateral bracing elements include cross lot struts, rakers, corner bracing, and tiebacks. Figure 1 illustrates a typical cut-and-cover construction sequence.

The majority of ground movement associated with cut-and-cover construction occurs during the excavation progresses deeper and the temporary bracing is removed as the permanent structures are erected. During each stage of construction, the excavation support (shoring) deflects laterally together with the surrounding soils, causing vertical ground movements. Building response to lateral deflection is assessed based on the ground settlement at the building's foundation level and lateral ground movement beneath the foundation.

To understand the impacts of wall deflections on the degree of building damage, a parametric study can be performed and site characteristic curves can be developed. The variables that define the site characteristic curves are maximum shoring deflection versus number of buildings that exceed a level of damage. The site characteristic curves are used to assess the building risk rating of the overall project. Using the site characteristic curves and the overall building risk rating system, the maximum wall deflection target can be selected for shoring design.

In shoring design, the limiting equilibrium method is commonly used to configure the shoring system, as described in DM-7 (U.S. Navy Design Manual) or FHWA-RD-75 (Federal Highway Design and Construction Summary). This design methodology focuses on force and moment equilibrium and designs the shoring elements based on strength.

A semi-empirical approach combined with historical shoring deflection measurements and numerical analysis for stiff-dense drained soils, introduced by Cording (1984), can be used to define relationship of the maximum shoring deflection to the stiffness of excavation support system and soil stiffness such that the shoring design can be modified based on the maximum wall deflection limit. Further evaluations are made using numerical methods to model excavations and support systems.

## GROUND MOVEMENTS CAUSED BY DEEP EXCAVATIONS

The patterns of lateral wall deflection include "cantilever deflection" due to excavation prior to placing the deck beam, and "bulging deflection" that develops below brace levels as the excavation proceeds and bracing elements are installed. To predict horizontal and vertical ground movements, the wall deflection profile and limits of ground movement behind the wall (Figure 2) are determined based on soil parameters and excavation size described by Clough and O'Rourke (1990).

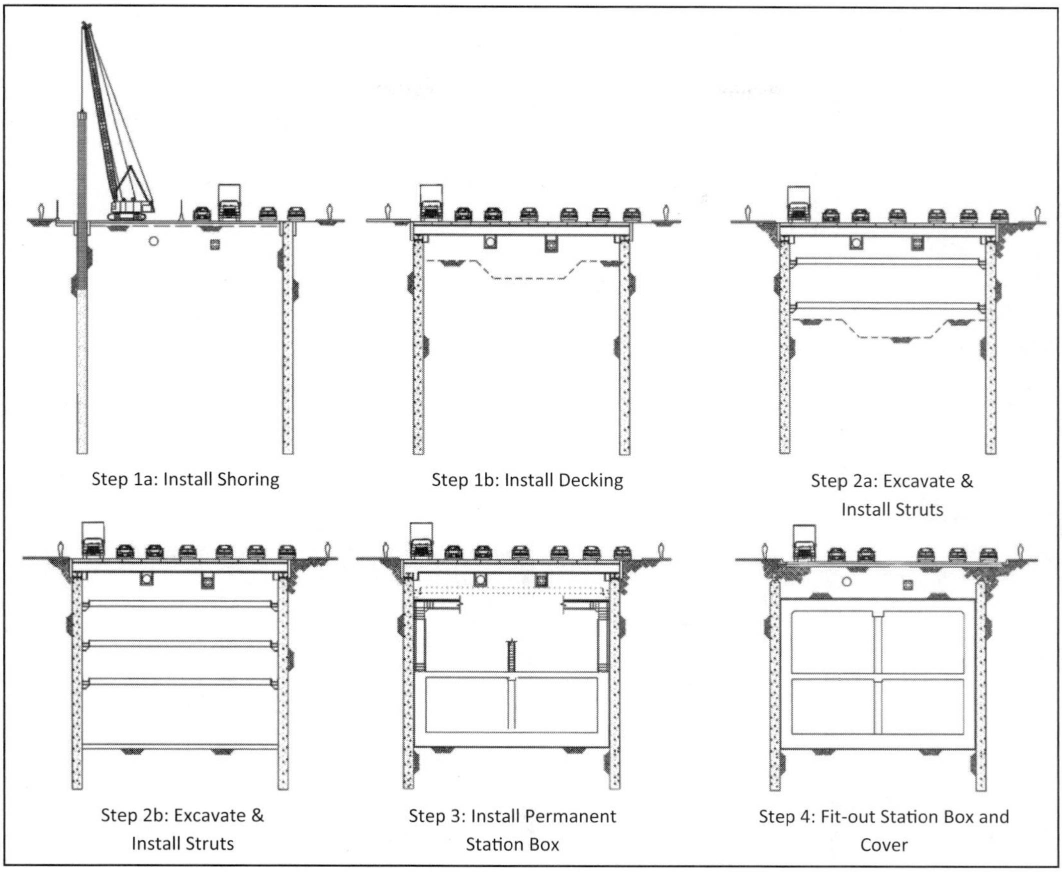


**Figure 1. Cut-and-cover construction sequence**

With the wall deflection profile defined, the surface settlement ($s_w$) at the face of the excavation is estimated assuming the volume of soil displaced by the wall ($V_s$) equals the settlement volume:

$$V_s = \frac{s_w D}{2} \rightarrow s_w = \frac{3V_s}{D}$$

(assuming parabolic distribution)

This is a conservative assumption for stiff soils. The ground surface settlement ($s$) at a distance $x$ from the excavation face is then determined as a parabolic distribution:

$$s = s_w \left( \frac{D-x}{D} \right)^2$$

The impact of ground movement at building basement or foundation level is predicted by determining the ground displacement at the foundation depth using the same method as for ground surface settlements. The vertical ground settlement at the face of the excavation at the foundation depth is estimated assuming the volume of soil displaced by the wall deflection below the foundation depth is equal to the settlement volume at the foundation depth.

For the subsurface lateral ground movement, it is assumed the greatest movement occurs at the face of the wall and equals the wall deflection, and decreases linearly to zero at a distance from the face of the wall at the limit of ground movement.

For buildings on deep foundations, the estimated ground movements are considered at the mid-depth of the foundation. The estimated ground movements are assumed to be equal to the building deformation for the building response evaluation discussed in the later section.

For the parametric study, a simplified wall deflection profile based on available historical data is developed (Figure 3). For simplicity, the wall deflection profile is described in terms of the maximum wall deflection ($d_2$), the excavation depth ($H_w$), and the depth to limit of ground movement ($H_p$).

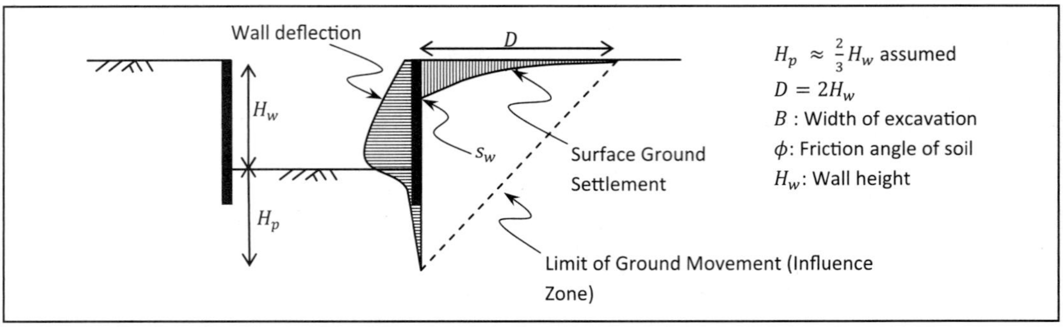


**Figure 2. Ground movement limit around deep excavation**

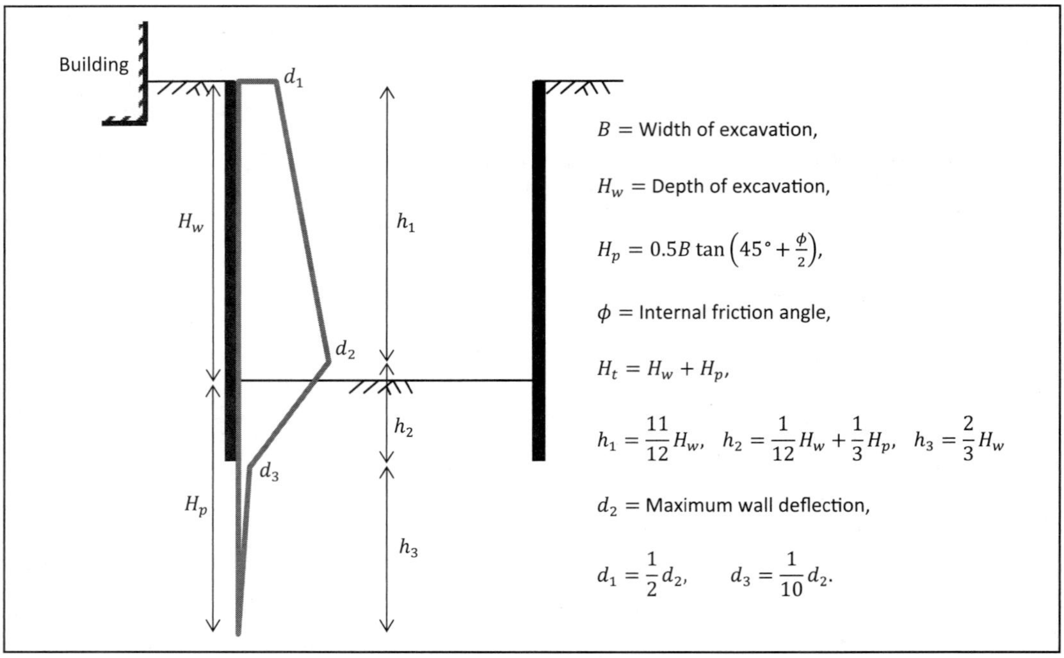


**Figure 3. Wall deflection shape**

## BUILDING RESPONSE TO GROUND MOVEMENT

Buildings with foundations situated within the influence zone of a deep excavation can settle differentially and distort as the underlying soils move. The degree of damage to a building can be evaluated based on the angular distortion and horizontal strain within the soil at the building foundation (basement) level. An initial assessment of the potential damage is performed assuming movements of the ground at the base of the building are equal to the building deformations. Movements of the ground at the base of the building are assumed equal to the building deformations. This is a conservative assumption for stiffer modern frame structures. For example, structures with grade beams will have horizontal strains that are significantly reduced from those determined from the ground displacements.

Burland and Wroth (1974) developed a framework for relating settlement of masonry buildings to their distortion and structural performance by analyzing the building as a beam or deep beam for a range of E/G. Boscardin and Cording (1989) added the effect of horizontal strains to the angular distortions for deep beams having a length/height (L/H) ratio of 1. This building damage criterion was modified (Cording et al. 2001; Cording et al. 2010) so that boundaries of the damage zone represent the average principal strain ($\varepsilon_P$), which is derived from the average angular distortion ($\beta$) and lateral

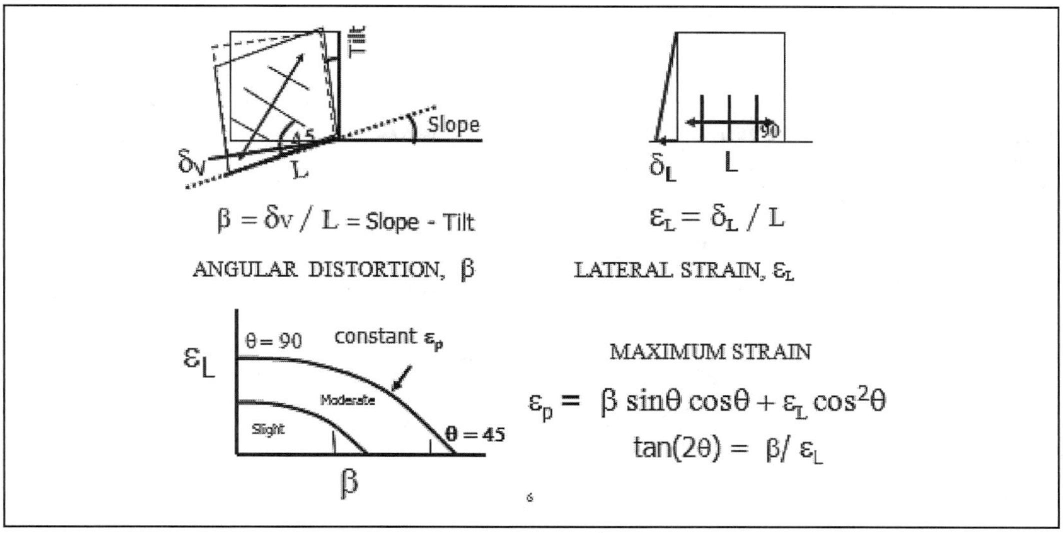


**Figure 4. Angular distortion and strains imposed on a structural element by ground movement (Cording et al., 2010)**

strain ($\varepsilon_L$) that develops within a structural bay or section of the building due to the vertical and horizontal movements of the building (Figure 4) that occur at the base.

The lateral strain is determined as the extension of the base divided by the base length. The angular distortion is determined as the differential settlement (settlement slope) minus the tilt of a bay or section. The average angular distortion when the entire building sits within the influence zone equals the tilt of the building subtracted from the average differential settlement. When the building sits partially within the influence zone, it is assumed that the building does not tilt and the differential settlement over the portion of the building within the settlement trough equals the average angular distortion of the building.

Table 1 categorizes the degree of damage based on the average principal strain of a building (Burland and Wroth 1974). The same categories are also presented in Figure 5. The boundaries between damage categories are set to represent a constant principal strain (Cording et al., 2001).

## BUILDING ASSESSMENT

The methodology described above was applied to 86 buildings that are fully or partially founded within the influence zone around seven excavation sites. For each building; the distance between building line and edge of excavation, the breadth of the building behind the excavation, and the building foundation type and depth below ground level were known. A Damage Category (DC) for each building was determined for different maximum wall deflections for the wall deflection profile described previously in Figure 3. The higher the maximum wall deflection the more buildings exceeded the selected DC limit. Table 2 summarizes the number and percentage of buildings with DC exceeding given values (0 through 4). Generally, damage caused to buildings with DC equal to 1 or less is considered acceptable but other project specific criteria can be applied. Figure 6 shows the building response plots for 86 buildings based on maximum wall deflections of 12.7-mm (0.5-in) and 38.1-mm (1.5-in).

## SITE CHARACTERISTIC CURVE

The building assessment results (Table 2) are presented as a plot of maximum wall deflection versus percentage of buildings with DC exceeding given values. Figure 7 presents the site characteristic curves for DC's exceeding 2, 3, and 4. The site characteristic curves provide a visual means of managing information for the buildings around excavation sites in the project and a way to set the general requirements to limit the DC to less than one (acceptable limit) for the majority of the buildings. Those buildings that have a DC greater than one can be identified and evaluated individually. The site characteristic curve for DC>1 shows the number of buildings at risk increased significantly when the maximum deflection increased from 25.4- to 38.1-mm (1.0- to 1.5-in). Therefore, a reasonable upper bound limit or target for maximum wall deflection would be 25.4-mm (1.0-in).

**Table 1. Building damage classification**

| Damage Category | Damage Classification | Description of Typical Damage and Likely Forms of Repair for Typical Masonry Buildings | Maximum Principal Extension Strain* (%) |
|---|---|---|---|
| 0 | Negligible (NGL) | Hairline cracks | < 0.05 |
| 1 | Very slight (VS) | Fine cracks easily treated during normal redecoration. Perhaps isolated slight fracture in building. Cracks in exterior brickwork visible upon close inspection. | 0.05 to 0.075 |
| 2 | Slight | Cracks easily filled. Redecoration probably required. Several slight fractures inside building. Exterior cracks visible: some repointing may be required for weather-tightness. Doors and windows may stick slightly. | 0.075 to 0.15 |
| 3 | Moderate | Cracks may require cutting out and patching. Recurrent cracks can be masked by suitable linings. Re-pointing and possibly replacement of a small amount of exterior brickwork may be required. Doors and windows sticking. Utility services may be interrupted. Weather-tightness often impaired. | 0.015 to 0.3 |
| 4 | Severe | Extensive repair involving removal and replacement of sections of walls, especially over doors and windows required. Windows and frames distorted. Floor slopes noticeably. Walls lean or bulge noticeably, some loss of bearing in beams. Utility services disrupted. | 0.3 to 0.6† |
| 5 | Very severe | Major repair required involving partial or complete reconstruction. Beams lose bearing, walls lean badly and require shoring. Windows broken by distortion. Danger of instability. | Greater than 0.6† |

* Introduced by Boscardin & Cording (1989).
† Modified from Boscardin & Cording (1989) for computation purpose.

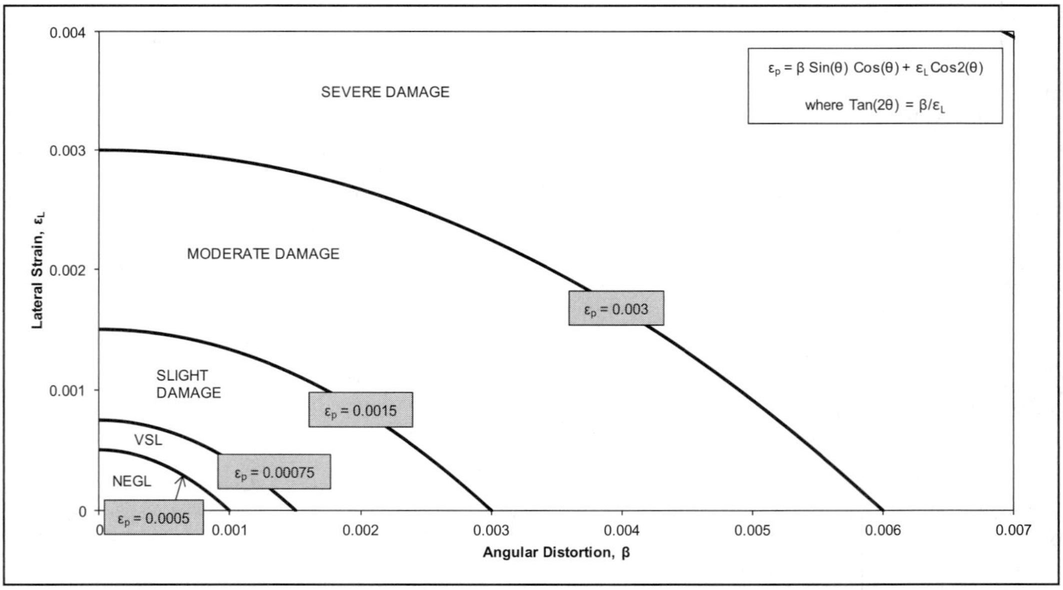


**Figure 5. Building damage classification chart (Cording et al. 2001; Cording et al. 2010, modified after Boscardin and Cording 1989)**

**Table 2. Number and percentage of buildings exceeding selected DC limits**

| Maximum Wall Deflection | Number of Buildings (% Buildings) | | | | |
|---|---|---|---|---|---|
| | DC >0 | DC >1 | DC >2 | DC >3 | DC >4 |
| 12.7-mm (0.5-in) | 2 (2%) | 0 (0%) | 0 (0%) | 0 (0%) | 0 (0%) |
| 25.4-mm (1.0-in) | 52 (60%) | 9 (10%) | 0 (0%) | 0 (0%) | 0 (0%) |
| 38.1-mm (1.5-in) | 83 (97%) | 52 (60%) | 2 (2%) | 0 (0%) | 0 (0%) |
| 50.8-mm (2.0-in) | 86 (100%) | 81 (94%) | 9 (10%) | 0 (0%) | 0 (0%) |
| 63.5-mm (2.5-in) | 86 (100%) | 86 (100%) | 23 (27%) | 0 (0%) | 0 (0%) |
| 76.2-mm (3.0-in) | 86 (100%) | 86 (100%) | 52 (60%) | 2 (2%) | 0 (0%) |

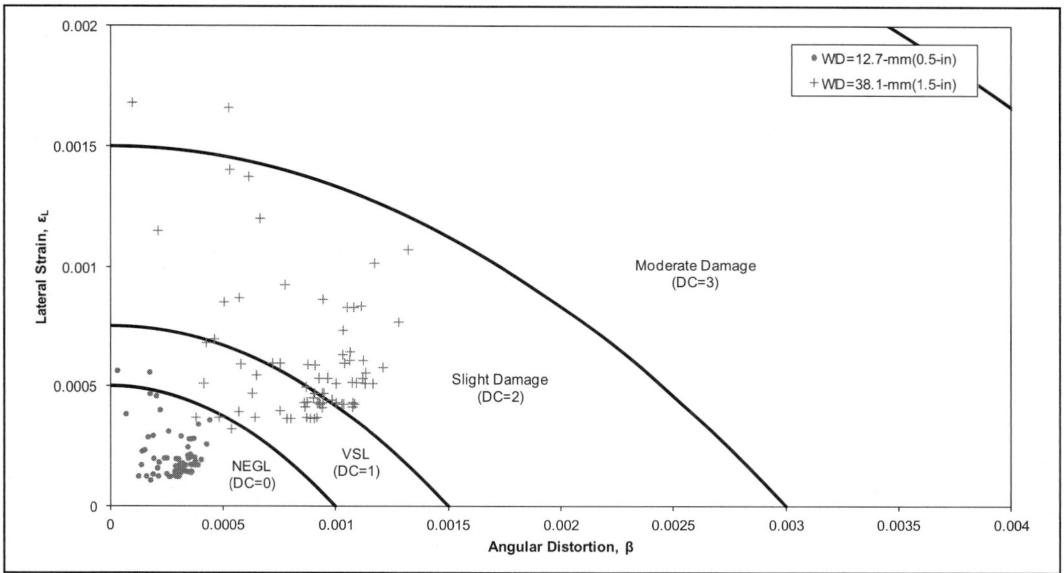


**Figure 6. Building response plot for maximum wall deflection of 12.7-mm (0.5-in) and 38.1-mm (1.5-in)**

## RISK ASSESSMENT

This section demonstrates that the overall project building risk rating due to the excavations can be evaluated based on the site characteristic curves. First, the percentage of buildings falling into each DC is obtained by the vertical distance between two neighboring curves. Varying the maximum wall deflection, a set of frequencies of buildings for each DC is obtained and summarized in Table 3.

Second, Frequency Category (FC) is assigned to each frequency of buildings per pre-defined matrix of FC (Table 4). Accordingly, FCs are determined for given DCs and maximum wall deflections as shown in Table 5.

Third, the Risk Score for the building damage due to excavation is obtained by:

Risk Score = Σ(DC×FC), for a given maximum wall deflection

DC = 0 or 1 are considered to be acceptable in terms of building protection, thus the Risk Score only takes DC's into account if it is higher than one. Risk Scores for various maximum wall deflections are calculated and summarized in Table 5.

Last, Risk Level is determined based on Risk Score per pre-defined rating system: Low when 0≤Risk Score<5, Medium when 5≤Risk Score<10, and High when Risk Score≥10. Risk Levels for various maximum wall deflections are summarized in Table 5. If the building risk for the project is 'High', maximum wall deflection shall be lowered requiring shoring design modification. In Table 5, the overall building risk is rated as 'High' when maximum wall deflection reaches 38.1-mm (1.5-in), and the overall building risk becomes 'Low' when maximum wall deflection is reduced to 25.4-mm (1.0-in). Also, on Figure 7, the percent buildings with DC exceeding one increases dramatically after the maximum wall deflection of 25.4-mm (1.0-in). Therefore, the maximum wall deflection of 25.4-mm (1.0-in) can be selected as the target wall deflection that would optimize the risk and the shoring cost.

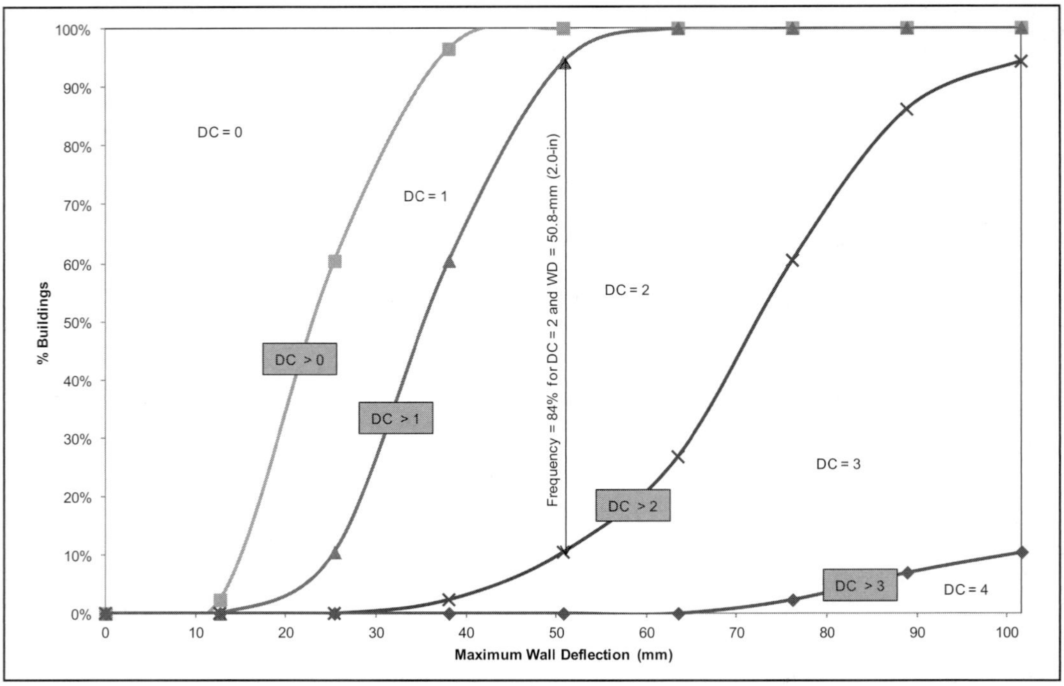


**Figure 7. Percentage of buildings per damage category**

**Table 3. Frequency of buildings per damage category**

| Maximum Wall Deflection | Frequency of Buildings DC=0 | DC=1 | DC=2 | DC=3 | DC=4 |
|---|---|---|---|---|---|
| 12.7-mm (0.5-in) | 98% | 2% | 0% | 0% | 0% |
| 25.4-mm (1.0-in) | 40% | 50% | 10% | 0% | 0% |
| 38.1-mm (1.5-in) | 3% | 36% | 58% | 2% | 0% |
| 50.8-mm (2.0-in) | 0% | 6% | 84% | 10% | 0% |
| 63.5-mm (2.5-in) | 0% | 0% | 73% | 27% | 0% |
| 76.2-mm (3.0-in) | 0% | 0% | 40% | 58% | 2% |

## SHORING DESIGN AND MODIFICATION

The limiting equilibrium method is commonly used to design a shoring system. However, this approach does not consider the anticipated ground movement which is an essential risk factor of the shoring system with respect to the building protection. Cording (1984) introduced a semi-empirical approach combined with historical shoring deflection measurements, in order to correlate the soil/shoring parameters to the shoring deflection. This approach considers the ground movement as a function of the stiffness of the shoring system and the stiffness of the soil. Figure 8 describes the semi-empirical approach. Using the equation on Figure 8, the average wall deflection (approximately a half of the maximum wall deflection) is estimated based on the soil parameters and the shoring configuration.

**Table 4. Matrix of frequency category (FC)**

| FC | Frequency |
|---|---|
| 0 | Frequency = 0% |
| 1 | 0% < Frequency ≤ 5% |
| 2 | 5% < Frequency ≤ 15% |
| 3 | 15% < Frequency ≤ 35% |
| 4 | 35% < Frequency ≤ 65% |
| 5 | 65% < Frequency ≤ 100% |

Then, the estimated maximum wall deflection retunes the Risk Level using Table 5. If the Risk Level is 'High' or the estimated maximum wall deflection is greater than the target selected, the shoring system configuration can be modified. This process may be iteratively performed until the Risk Level or the maximum wall deflection meets the target. In case of soldier pile and lagging system, the following modification can be considered to lower Risk Level, while minimizing the number of types of materials and equipment and limiting disruption to the pile and cross bracing installation process:

- Reduce the soldier pile spacing to stiffen the shoring wall. This has the added benefit of reducing the area of unsupported ground exposed between excavation advancing and lagging installation and associated loss of ground due to the installation process.

**Table 5. Risk rating summary**

| Maximum Wall Deflection | Frequency Category (FC) DC=0 | DC=1 | DC=2 | DC=3 | DC=4 | Risk Score | Risk Level |
|---|---|---|---|---|---|---|---|
| 12.7-mm (0.5-in) | 5 | 1 | 0 | 0 | 0 | 0 | Low |
| 25.4-mm (1.0-in) | 4 | 4 | 2 | 0 | 0 | 4 | Low |
| 38.1-mm (1.5-in) | 1 | 4 | 4 | 1 | 0 | 11 | High |
| 50.8-mm (2.0-in) | 0 | 2 | 5 | 2 | 0 | 16 | High |
| 63.5-mm (2.5-in) | 0 | 0 | 5 | 3 | 0 | 19 | High |
| 76.2-mm (3.0-in) | 0 | 0 | 4 | 4 | 1 | 24 | High |

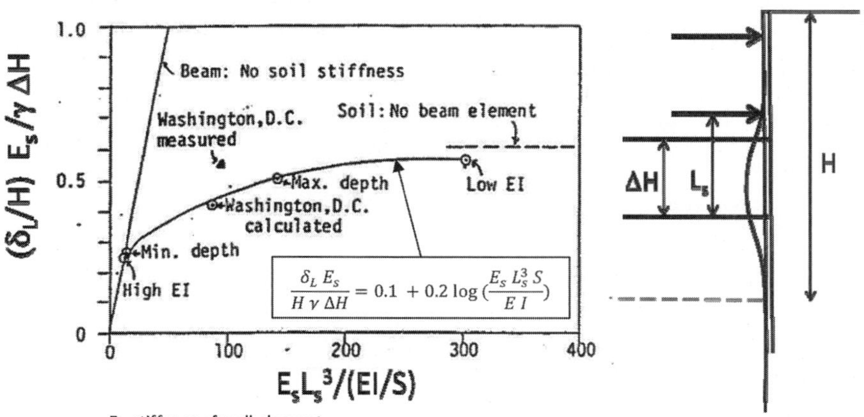


E: stiffness of wall element

I/S: moment of inertia, I, of wall element/horizontal spacing, S, of wall element

H: depth of cut

ΔH: vertical distance between struts or tiebacks

Ls: depth excavated below strut level before setting next strut

Es: soil stiffness (Young's Modulus), for strains on order of 0.1 %

γ: soil unit weight

$\delta_L$: average wall deflection

**Figure 8. Estimating lateral displacement for braced cut (Cording 1984)**

- Add strut/tieback level(s) and reduce vertical spaces of struts, thereby, reducing the depth of excavation between successive lifts and stiffening the shoring wall system.
- Increase strut tieback preload.
- Switch to semi-rigid wall system such as secant/tangent piles at the sensitive building locations without major equipment adjustment. Tangent piles generally can be installed with the same equipment. Equipment for secant piles where the secondary pile is installed by first cutting through the primary pile requires additional equipment with respect to the casing and therefore, thought in advance of mobilizing equipment to the work site.

## CONCLUSION

Assessment of building settlement due to deep station excavations is one of the most critical issues to be addressed in the preliminary engineering stage for an underground urban transit project. Once sufficient building, geotechnical, and design data have been obtained, the overall project behavior in terms of building damage can be evaluated and quantified by Site Characteristic Curves. Based on the Site Characteristic Curves, the Risk Level assessment can be performed and the maximum wall deflection limit can be determined. Then the shoring design is modified by the semi-empirical approach to reduce the Risk Level to an acceptable level to optimize the cost and the risk.

## REFERENCES

Boscardin, M.D. and Cording, E.J. 1989. "Building Response to Excavation-Induced Settlement," Journal of Geotechnical Engineering, American Society of Civil Engineers, Vol. 115, No.1, pp. 1–21.

Burland, J.B. and Wroth, C.P. 1974. "Settlement of buildings and associated damage," Proceedings of the Conference on Settlement of Structures, Pentech Press, London, England, pp. 611–654.

Clough, G.W. and O'Rourke, T.D. 1990. "Construction Induced Movement of In-Situ Walls," Proceedings, Design and Performance of Earth Retaining Structures, Cornell University, American Society of Civil Engineers, Geotechnical Special Publication No. 25, pp. 439–470.

Cording, E.J. 1984. "Use of Empirical Data for Braced Excavations and Tunnels in Soil," Chicago Lecture Series, Illinois Section ASCE, September.

Cording, E.J., Long, J.H., Son, M., and Laefer, D. F. 2001. "Modeling and analysis of excavation-induced building distortion and damage using a strain-based damage criterion," London Conference on Responses of Buildings to Excavation-Induced Ground Movements, Imperial College, London.

Cording, E.J., Long, J.H., Son, M., Laefer, D., and Ghahreman, B. 2010. "Assessment of Excavation-Induced Building Damage," Earth Retention Conference 3, Bellevue, WA. August.

# Ground Deformation from Multiple Tunnel Openings: Analysis of Queens Bored Tunnels

**Michael Mooney and Jacob Grasmick**
Colorado School of Mines

**April Clemmensen**
U.S. Air Force Academy

**Andrew Thompson and Eric Prantil**
Hatch Mott MacDonald

**Brett Robinson**
Traylor Bros., Inc.

**ABSTRACT:** The East Side Access Queens bored tunnels project involved the construction of four near surface, closely spaced metro transit tunnels beneath the rail yards and mainline railroad tracks.

The close proximity of the tunnels provides a unique opportunity to examine the influence of multiple closely spaced tunnel openings on ground deformation, particularly the accumulation of vertical surface deflection due to consecutive tunnels. This paper presents measured transverse vertical deflection profiles from the project, comparing observed with the established theory of Gaussian surface deflection response. Heave behavior prevailed at most cross sections; the transverse heave and settlement profiles caused by individual tunnel excavations exhibited Gaussian form; however, the width of the deflection profiles and the position with respect to tunnel centerline deviates from conventional theory.

## INTRODUCTION

The East Side Access Queens bored tunnels project involved the construction of four near surface, closely spaced metro transit tunnels beneath the rail yards and mainline railroad tracks in Sunnyside yards in Queens, New York (see Figure 1). The tunnels were excavated using two 6.9 m (22.5 ft) diameter Herrenknecht slurry shield TBMs primarily through highly variable glacial till soils and outwash deposits.

The close proximity of the four tunnels coupled with their shallow depth provides a unique opportunity to examine the influence of multiple closely spaced tunnel openings on ground deformation. Classical analysis of tunneling-induced ground deformation in a greenfield environment (no buildings) suggests that a Gaussian-shaped longitudinal and transverse deflection profile develops at the ground surface centered above a single tunnel (Peck 1969). With twin tunnels, the resulting surface deflection can be symmetric over the mid-point of two tunnels or asymmetric and shifted towards one tunnel (Cording and Hansmire 1975, Suwansawat and Einstein 2007). Suwansawat (2006) proposed a superposition technique to attribute surface settlement to individual tunnels in a twin tunnel environment is estimated via superposition of individual transverse Gaussian settlement profiles. Suwansawat and Einstein (2007) and Chapman et al. (2007) used this technique to show that the increment of transverse surface settlement caused by each tunnel in twin tunnel configurations (side by side and stacked) is Gaussian in shape and centered above the individual tunnel. This has not been explored extensively and has not been examined for more than two tunnels. The cross-section of four tunnels in the Queens bored tunnels project allows for a more detailed assessment.

This paper begins with a brief background to the project and to classical analysis of surface settlement followed by the layout of ground deformation instrumentation used and cross sections analyzed. Cumulative and incremental transverse surface deflection profiles are then presented at select locations along the alignment. The Gaussian shapes of the profiles are analyzed in terms of Gaussian

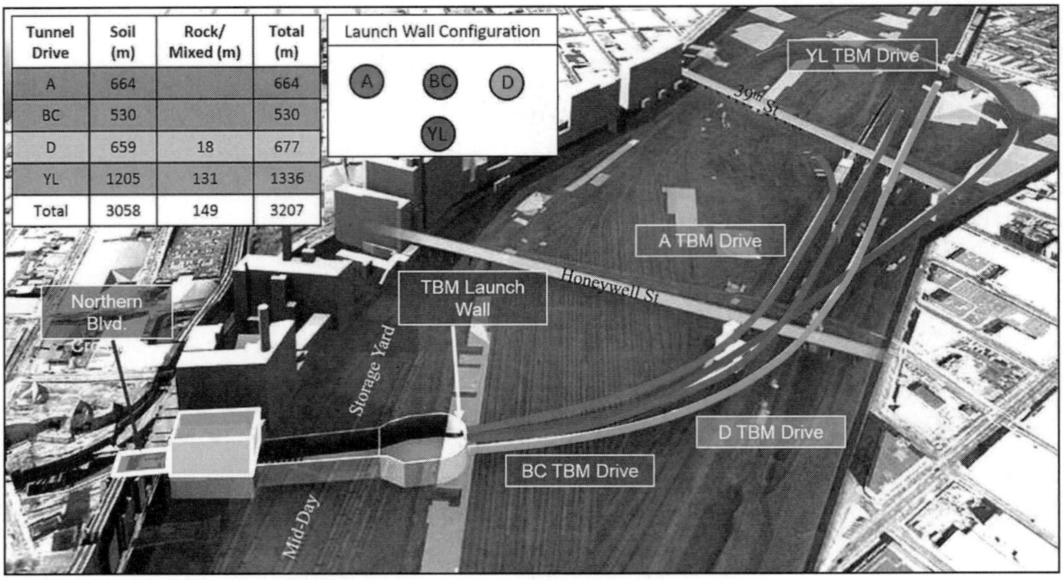


**Figure 1. Rendering of Queens bored tunnels (courtesy MTA)**

behavior and the applicability of the superposition technique is analyzed.

## BACKGROUND

The four tunnels totaling 3,207 m in length (refer to table in Figure 1 for individual tunnel lengths) were constructed by the joint venture of Granite Construction Northeast, Inc., Traylor Bros., Inc., and Frontier-Kemper Constructors, Inc. in 2011 and 2012. Two 6.9 m (22.5 ft) diameter Herrenknecht slurry shield TBMs were used. The cross-section at the launch wall (Figure 1) illustrates the four tunnel configuration. At the launch wall, excavation of tunnel YL (yard lead) began at a depth of 22.9 m below the existing ground surface. Tunnel A began 11.9 m deep and tunnels D and BC 11.7 m deep. Tunnel YL was driven first, followed by tunnels A, D and BC. The project is described in detail in Robinson & Wehrli (2013a,b).

### Geology

The ground conditions primarily consisted of highly variable glacial till soils and outwash deposits. 18 different strata ranging from gneiss bedrock to clay lenses were discovered along the alignment during the geotechnical site investigation (67 borings). These strata were highly variable and not uniform across the entire project site. The area mostly consists of well graded sand (SW), poorly graded sand (SP) and silty sand (SM), with some small lenses of clayey sand (SC), clay (C), silt (M) and gravel (G) present. There is a layer of decomposed gneiss resting atop the fractured gneiss bedrock. The first 130 m of tunnel YL was excavated in fractured bedrock while the other three tunnels were excavated in soil.

### Estimating Ground Deformation

Classical analysis of tunneling-induced ground settlement in a greenfield environment (no buildings) suggests that the settlement profile that develop at the ground surface is transversely Gaussian in shape and centered above a single tunnel (Figure 2). The transverse settlement basin or trough is characterized by a width parameter $i$ that is a function of soil type and depth to tunnel center $z_0$ (see Equation 1). The majority of reported $K$ values range from 0.4–0.6 in clay and 0.25–0.45 in sand (see Mair and Taylor 1997). If the strata is layered, New and O'Reilly (1991) proposed a revised relationship for $i$ that is based on a weighted average of layers. It is worth mentioning that these values of $K$ and the principle of Gaussian behavior in general has been used to explain settlement behavior. To the authors' knowledge, Gaussian behavior and these values has not been applied to tunnel-induced surface heave. The volume of the transverse deflection basin $V_s$ is determined from Equation 2 where $s_{max}$ is the maximum deflection. $V_s$ is related to the ground loss $V_L$ that occurs at tunnel depth. The settlement at any position $y$ is given by Equation 3.

$$i = Kz_0 \tag{1}$$

$$V_s = \sqrt{2\pi} i s_{max} \tag{2}$$

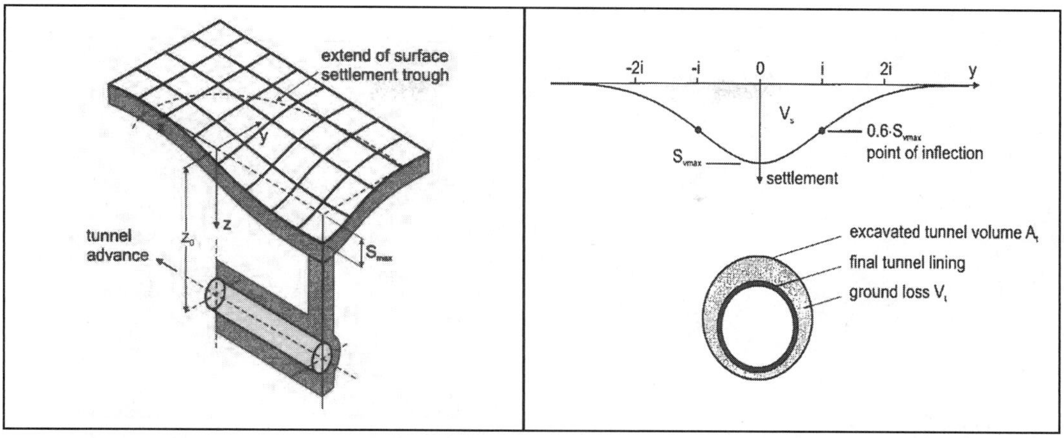


**Figure 2. Classical Gaussian shape settlement trough both longitudinally and transversely where the volume of the transverse settlement profile $V_s$ is typically equal to the ground loss $V_L$ at the tunnel depth (after Mair & Taylor 1997)**

$$s = s_{\max} \exp\left(\frac{-y^2}{2i^2}\right) \tag{3}$$

A fairly significant body of data indicates that $V_s$ manifested at the surface is approximately equal to $V_L$ in cohesive soils and that $V_s$ is less than $V_L$ in granular soils (due to dilation of the soil from the surface to the tunnel crown). Conservatively, $V_s$ is often assumed equal to $V_L$ in settlement prediction for granular soils. The four most significant contributors to ground loss during tunneling, assuming good practice, include (1) preconvergence and convergence due to stress relief at the face; (2) convergence around the forward shield due to the overcut annulus; (3) convergence at the tail shield due to the annulus outside the segmental concrete liners; and (4) convergence due to liner deformation after grouting (Mair & Taylor 2007). The increased use of pressurized face TBMs to minimize stress relief combined with shield annulus bentonite injection and two-part grouting around the segments from the tail shield has reduced volume loss on tunnel projects from a few percent (of the excavated volume) a decade ago to less than 0.5% on more recent projects.

It has been shown that when twin tunnels are excavated, the cumulative surface settlement can be estimated via superposition of individual transverse Gaussian settlement profiles (Suwansawat and Einstein 2007, Chapman et al. 2007). Each surface settlement profile can be related to $V_s$ and $V_L$. Suwansawat and Einstein (2007) showed that for side by side twin tunnels that the subtraction of the first tunnel's surface settlement profile from the cumulative surface settlement profile yielded a Gaussian profile centered above the second tunnel. They showed a similar outcome when analyzing surface deformation for two stacked tunnels. This is a helpful technique as it enables a reasonable estimate of ground deformation in the presence of two tunnels by superposition of the Gaussian profiles.

## Instrumentation

The layout of ground deformation instrumentation for the Queens bored tunnels is shown in Figure 3. Because rail traffic continued throughout construction, a sizable array of settlement monitoring points was established on the ground and on the train tracks throughout the Sunnyside yards. This included approximately 330 surface settlement monitoring points on the ground and over 1500 monitoring points on rail tracks. In addition, over 500 automated motorized total station (AMTS) survey prisms were deployed to monitor track movement and 15 multi-point borehole extensometers (MPBX) were installed, the majority with measurement points immediately above the tunnel crown, 1–2 m below the surface, and at the mid-depth to the crown. Ground and rail deflection data primarily at the northwest end (beginning) of the alignment were collected by manual survey (leveling staff or total station) with a frequency of once per day or less. Not all points were measured each day and the time of day was not recorded; a rolling approach was used depending on where the tunnel headings were. AMTS data were collected as frequently as once per hour.

The uncertainty/error in manual survey stems from both instrumentation uncertainty and systematic or operator error. Instrumentation uncertainty for total station surveys typically falls in the 2–3 mm range. Systematic and operator errors can include poor baseline readings, improper backsighting, incorrect use of survey equipment, etc. These

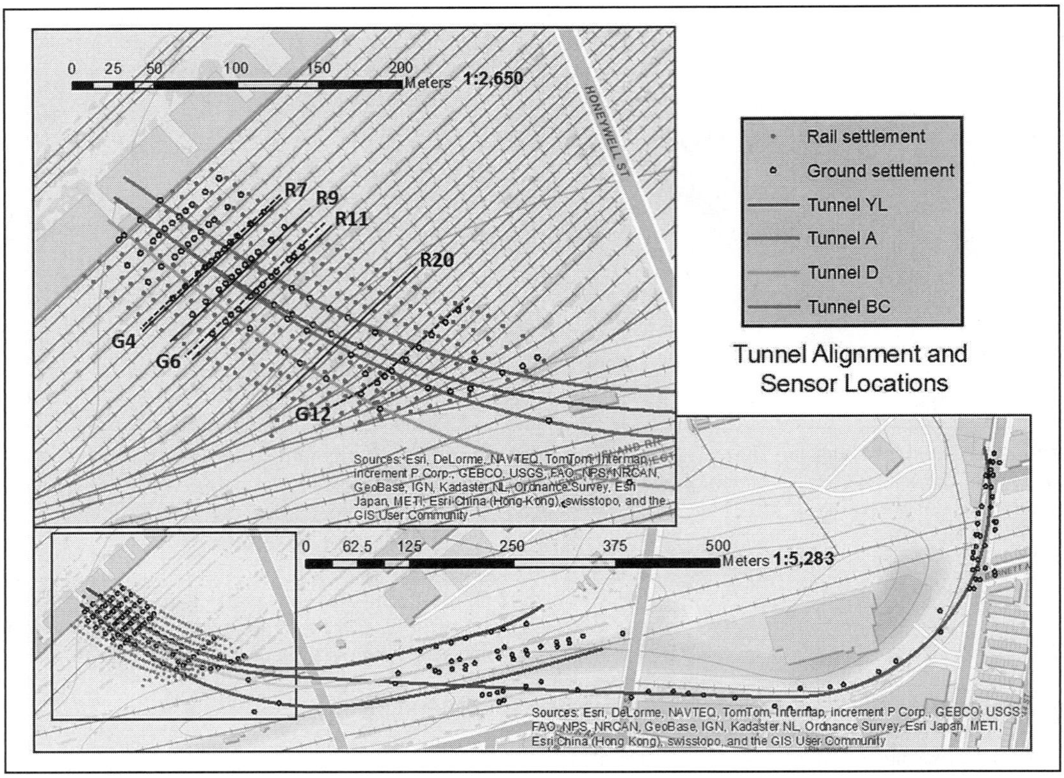


**Figure 3. Layout of ground deformation instrumentation (R=rail; G=ground)**

systematic and operator errors can usually be identified when analyzing day-to-day readings, and considerable effort was taken to remove these errors from the data presented herein.

## RESULTS

Both longitudinal and transverse deflection profiles were examined; only transverse profiles are presented here given paper length constraints. The day that baseline settlement readings were conducted (May 7, 2011) is established as day 0 for the purposes of data presentation. Tunnel YL excavation began on day 25 (ended day 284), tunnel A began on day 92 (ended day 229), tunnel D began on day 325 (ended day 389), and tunnel BC began on day 363 (ended day 443). To provide a sense for how the sequence of tunneling impacted ground settlement, Table 1 summarizes the time frame when each tunnel heading passed under cross section R7 (see Figure 3 for reference). For each tunnel, the days when the tunnel heading was 50 m behind (denoted –50 m), at the cross section (denoted 0 m) and 50 m ahead are shown. As is conveyed in Table 1, each tunnel passed cross section R7 (and all cross sections) quite independently.

**Table 1. Days when tunnels cross R7**

| Tunnel | –50 m | 0 m | 50 m |
|---|---|---|---|
| YL | 25 | 40 | 62 |
| A | 92 | 110 | 116 |
| D | 325 | 334 | 336 |
| BC | 363 | 369 | 374 |

Adjacent transverse vertical deflection profiles from both ground (G) and top of rail (R) measurements are presented for comparison in Figure 4. Deflection profiles are presented for each tunnel excavation after the respective tunnel is approximately 50m past the transverse profile. For all surface deflection data presented in this paper, geotechnical sign convention is used where positive corresponds to settlement and negative to heave.

In general, ground and rail transverse deflection profiles exhibit similar behavior both in magnitude and lateral ($y$) offset of peak deflection and shape of the deflection profile. There was some concern that top of rail deflection would not be representative of greenfield deflections due to possible bridging of the stiffer rail tracks. A comparison of ground and rail deflections illustrate that the rail track did

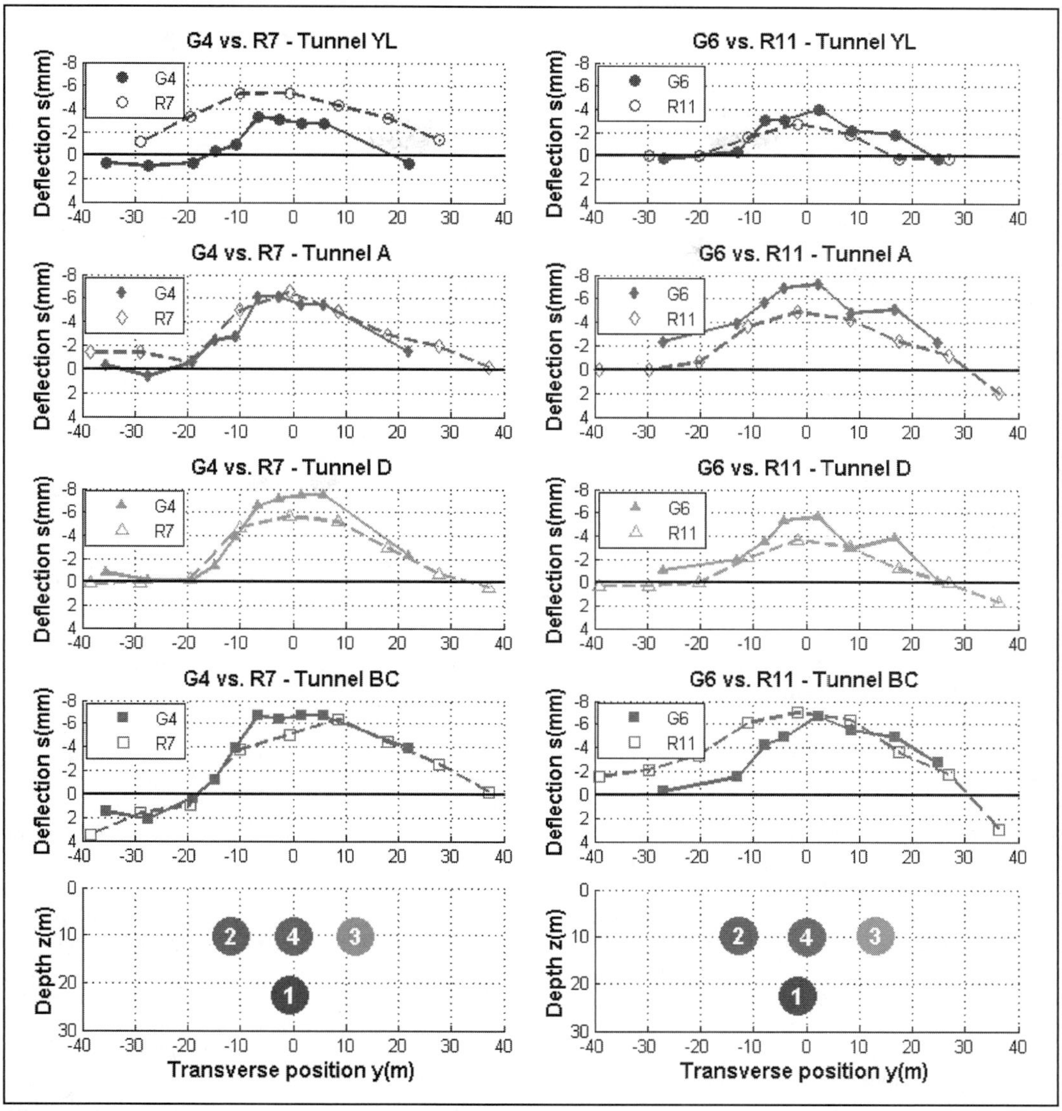


**Figure 4. Comparison of ground and rail deflection profiles**

not produce the bridging effect and, therefore, can be assumed to be representative of greenfield deflections. Top of rail deflection data yielded smoother transverse profiles than the ground deflection data. This suggests some uncertainty in analysis of individual ground deflection data, perhaps due to movements of the rebar stakes used to mark the ground deformation points. For these reasons, top of rail deflections will be presented hereafter.

The transverse surface deflection profiles measured at cross sections R9 and R11 (25 m apart) are shown in Figure 5. These data reflect the as-measured cumulative vertical deflections collected when each TBM heading was at least five diameters beyond each cross section to ensure that all effects on immediate settlement/heave had been captured. The R9 and R11 profiles primarily reflect ground heaving rather than settlement. Such a ground response is the result of elevated slurry face support pressure used to mitigate settlement during this project. By following the tunneling sequence, Figure 5a shows 2.5 mm maximum heave due to tunnel YL, followed by a negligible increase in heave due to tunnel A. A slight net settlement occurred during tunnel D construction. Finally, tunnel BC construction induced additional heave (7 mm cumulative). A similar sequence of transverse deflection response is reflected in cross section R11 that is 25 m beyond R9 (Figure 5b).

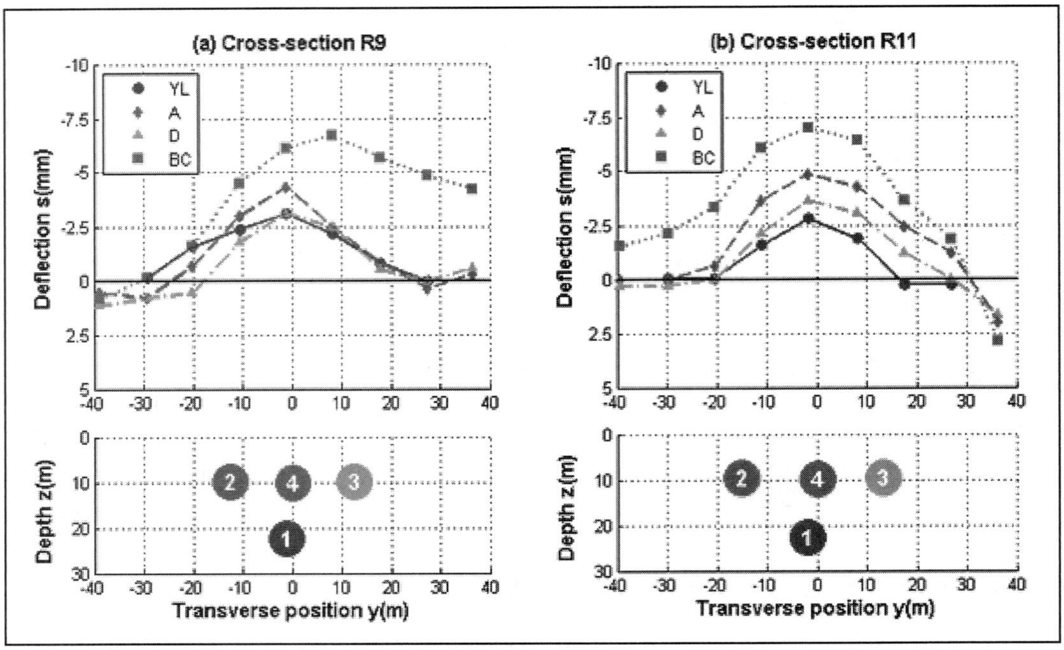


**Figure 5. Transverse surface deflection profiles (a) R9 and (b) R11 after passage of each of the four TBM headings (+ is settlement; – is heave)**

According to the Suwansawat (2006) superposition technique, the incremental surface settlement incurred from any single tunnel should be transversely Gaussian in shape and centered over the respective tunnel. Here we apply this technique to both heave and settlement behavior, recognizing that the application to heaving is beyond what Suwansawat (2006) intended. The R9 data from tunnel YL and the increments thereafter are presented in Figure 6. ΔA was determined by subtracting tunnel YL induced deflection from tunnel A deflection, ΔD was determined by subtracting the cumulative deflection after tunnel A from the cumulative deflection after tunnel D, etc. Each profile was then fit with a Gaussian curve (per Equation 3). Figure 6a shows that the heaving induced during tunnel YL construction exhibits Gaussian behavior, suggesting that the Gaussian response applies both to settlement (per the literature) and heaving as shown here. The transverse heave profile is centered above tunnel YL with a width parameter $i = 16$ m, and combined with a depth to tunnel center = 23 m, the resulting $K = 0.7$. This is slightly greater than the 0.25–0.6 range of $K$ values published in the literature for settlement behavior, but within reason.

The increment of transverse heave due to tunnel A construction was also found to be Gaussian in shape with a smaller width factor ($i = 9$ m). Greenfield settlement theory, however, would suggest a narrower deflection profile with $i$ closer to 4–5 m for a depth to 7 m. This difference may be due to the influence of tunnel YL and/or as a result of heave vs. settlement. Further, a more significant finding is that the symmetry axis of the settlement profile is offset approximately 10 m from the centerline of tunnel A (see Figure 6a). A number of factors likely contribute to the higher $i$ and the offset. The excavation of tunnel YL redistributes the ground stress in the vicinity of YL, e.g., arching sheds vertical stress above YL to the YL spring lines and influences spring line horizontal stresses. This creates a non-homogeneous stress field through which tunnel A is excavated. Strain follows stress, and therefore the deformation will be asymmetric. This effect is likely amplified by the non-uniform geology at this location (see Figure 7).

The increment of transverse settlement that occurred during tunnel D construction does exhibit Gaussian behavior but is not centered over tunnel D's center line. The fitted width of the Gaussian response ($i = 12$ m) is greater than what greenfield Gaussian theory predicts ($i = 4$–$5$ m assuming Equation 1 and granular soil). The increment of transverse heave due to tunnel BC construction also exhibits Gaussian response with a much greater width ($i = 27$ m) than greenfield theory would suggest ($i = 4$–$5$ m). This increment of transverse settlement is not aligned with the centerline of tunnel BC. Finally, when the fitted Gaussian distributions are added as shown in Figure 6b, the cumulative transverse settlement

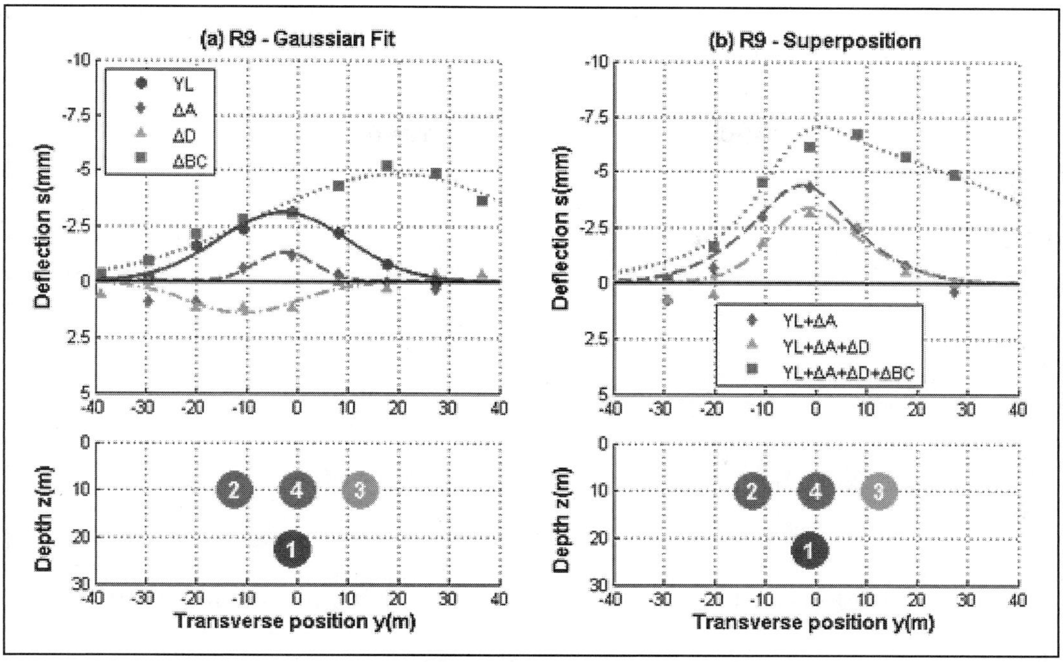


**Figure 6. (a) Incremental R9 surface deflections and their Gaussian fits; (b) a summation of the fitted Gaussian increments and their match with experimental data**

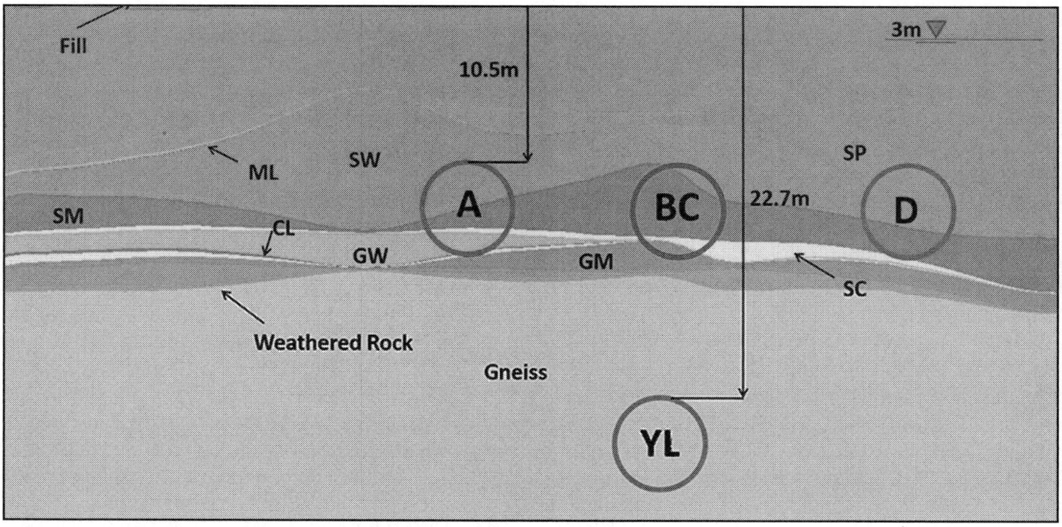


**Figure 7. Estimated geological profile at cross section R9**

profiles match well with the observed deformations in Figure 5 (RMSE < 1 mm), supporting the notion that transverse heave and settlement profiles due to individual tunnels are well represented by Gaussian curves.

Cross section R11 deflection response due to tunnel YL and the increments thereafter are presented in Figure 8 with their respective Gaussian curve fits. Figure 8a shows that all tunnels induce settlement or heave that clearly exhibits Gaussian behavior, further supporting the notion that Gaussian behavior applies for both settlement and heave. For tunnel YL, the width parameter $i$ = 8 m combined with a depth to tunnel = 23 m results in an average

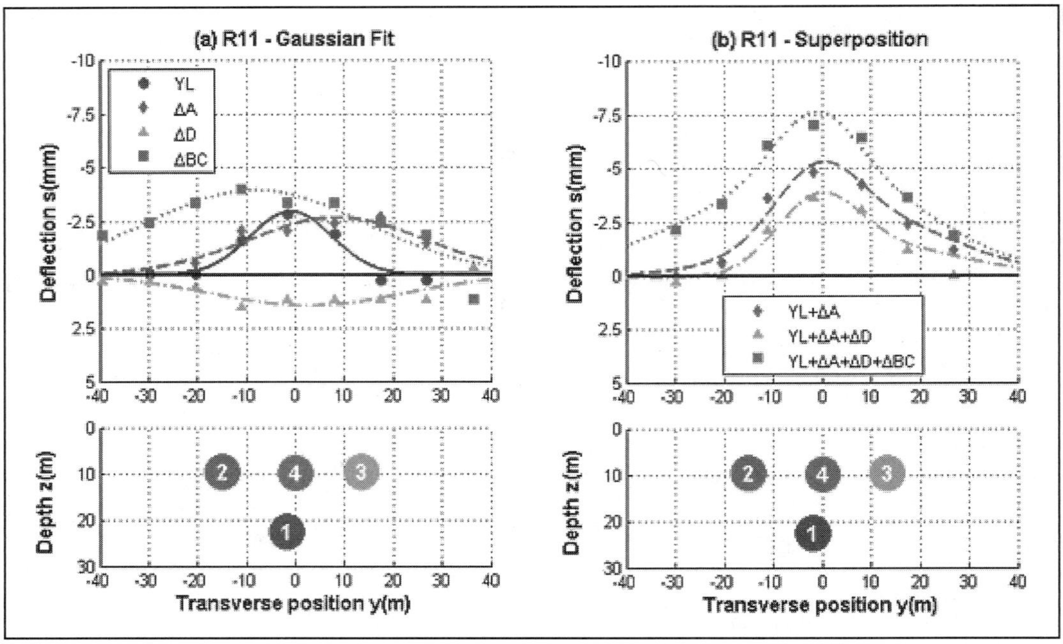


**Figure 8. (a) Incremental R11 surface deflections and their Gaussian fits; (b) a summation of the fitted Gaussian increments and their match with experimental data**

$K = 0.4$. This is a reasonable value within the range of published literature for settlement behavior. The increment of transverse heave due to tunnel A construction was similar to tunnel YL. Its response is Gaussian in shape but with a larger width factor ($i = 18$ m). Greenfield theory would suggest tunnel A would cause a narrower deflection profile (width factor closer to 4–5 m) because the depth to tunnel is much smaller (9.5 m).

The increment of transverse settlement that occurred during tunnel D also exhibits Gaussian behavior but is not centered over tunnel D's location. The fitted Gaussian width factor ($i = 20$ m) does not match greenfield theory using published data ($i = 4$–5 m). The increment of transverse heave due to tunnel BC construction also exhibits Gaussian response with an even greater width ($i = 23$ m) than classical approach would suggest ($i = 4$–5 m). This increment of transverse heave is more aligned with the centerline of tunnel BC than increments in settlement/heave due to tunnels A and D. When the fitted Gaussian distributions are added as shown in Figure 8b, the cumulative transverse deflection profiles match well with the observed deformations (RMSE < 1 mm). The geological cross section for R11 presented in Figure 9 once again reflects the non-homogeneous geology in which the TBMs were driven.

Figure 10a illustrates the measured deflection at transverse cross section R20 after passage of each TBM. R20 experienced nearly zero deflection due to tunnel YL, followed by subtle heaving due to tunnel A, a reduction in heave after tunnel D and a subtle increase in heave after tunnel BC. The incremental deflections attributed to each tunnel are shown in Figure 10b. Each incremental profile exhibits a Gaussian shape. The profiles induced by tunnels YL and D are offset from center while the profiles induced by tunnels A and BC are centered above their respective tunnels. Further, the profiles induced by tunnels A and BC exhibit width factors ($i = 5$ m and 8 m, respectively) that are consistent with and slightly greater than classic approach would predict. However, the settlement profile due to tunnel D exhibits a width factor $i = 10$ m that is greater by a factor of 2 than that predicted by greenfield theory. When the fitted Gaussian curves are added as shown in Figure 10c, the cumulative transverse deflection profiles do not match as well with the observed deflections as compared to sections R9 and R11 (RMSE < 2 mm).

A summary of the observed width parameters $i$ and $K$ estimated by fitting Gaussian responses to 112 transverse deflection profiles (28 cross sections × 4 tunnel advances at each) is presented in Table 2. The deflection profiles in both settlement and heave situations consistently exhibited Gaussian behavior. The measured width parameters $i$ and $K$ for tunnel YL were found to be greater than the range and mean reported in the literature for predominantly granular

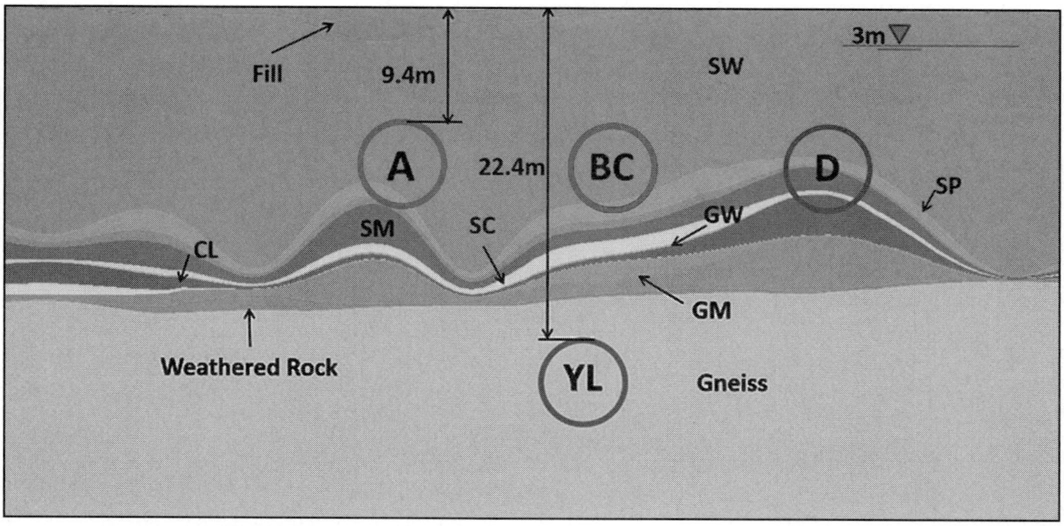


**Figure 9. Estimated geological profile at cross section R11**

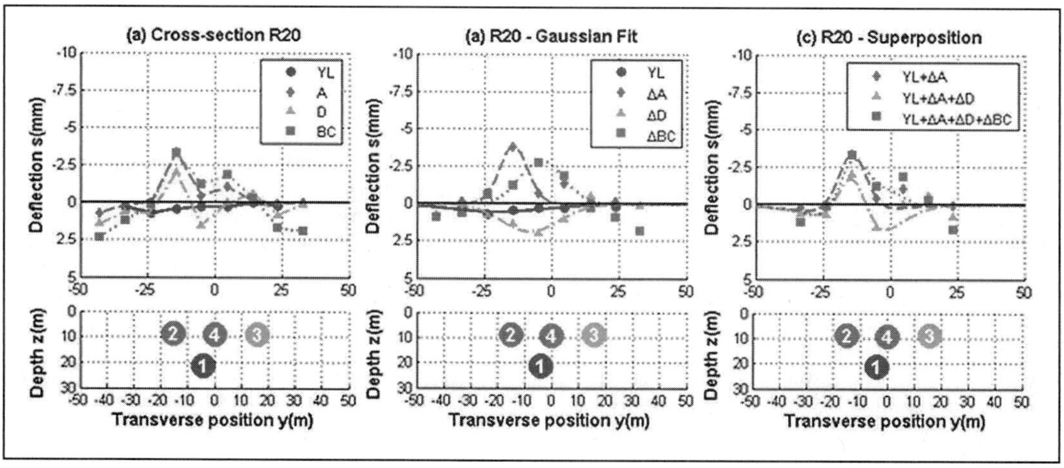


**Figure 10. (a) Incremental R20 surface deflections and their Gaussian fits; (b) a summation of the fitted Gaussian increments and their match with experimental data**

soils ($K$ is typically 0.35–0.45). Incremental heave and settlement profiles for tunnels A, BC and D exhibited considerably greater widths than reported in the literature for single tunnels (3×–4× greater) while tunnel YL yielded a slightly higher width parameter (1.5×–2.0× greater).

Furthermore, the peak of the incremental transverse deflection profiles did not coincide with centerlines of the tunnels, specifically for A, D and BC. A summary of the y-offset (– left, + right of tunnel centerline) from the tunnel center to the centerline of the Gaussian curves is presented in Table 3. The average y-offset for tunnel YL of 5.1 m is reasonable in comparison to average y-offsets of tunnels A, D and BC. A likely cause is a combination of the altered stress field created by each tunnel (leads to a heterogeneous stress field) combined with heterogeneity in the geological cross section. This phenomenon will be further explored through finite element modeling of the geological conditions combined with staged adjacent tunnel openings (beyond the scope of this paper).

## CONCLUSIONS

A progressive analysis of transverse surface deflection profiles was carried out at 28 cross sections as each of the four tunnels was constructed (112 profiles). While settlement did occur in some areas, the majority of deflection profiles observed revealed

**Table 2. Summary of width parameters for Gaussian curve**

| | YL | | ΔA | | ΔD | | ΔBC | |
|---|---|---|---|---|---|---|---|---|
| | *i* | *K* | *i* | *K* | *i* | *K* | *i* | *K* |
| Mean | 16.1 | 0.7 | 14.5 | 1.5 | 15.8 | 1.7 | 14.2 | 1.5 |
| Min | 3.9 | 0.2 | 3.0 | 0.3 | 3.3 | 0.4 | 3.1 | 0.3 |
| Max | 36.6 | 1.7 | 31.5 | 3.6 | 43.8 | 4.9 | 29.8 | 2.9 |
| Std dev | 8.3 | 0.4 | 8.7 | 1.0 | 10.7 | 1.2 | 8.6 | 0.8 |

**Table 3. Summary of y-offset (m) for fitted Gaussian curves**

| | YL | ΔA | ΔD | ΔBC |
|---|---|---|---|---|
| Avg | 5.1 | 16.3 | 14.8 | 12.2 |
| Max | 13.8 | 38.9 | 13.6 | 35.7 |
| Min | –15.9 | –13.3 | –47.0 | –31.4 |
| Std dev | 7.3 | 14.4 | 14.0 | 16.6 |

heaving behavior. An analysis of the incremental profiles, i.e., deflections due to individual tunnels, showed that both settlement and heave profiles exhibited Gaussian response. The majority of the fitted Gaussian surface deflection profiles induced by tunnel YL excavation were centered above YL and exhibited trough width parameters *i* and *K* that are consistent to slightly greater than those reported in the literature. The observed width parameters *i* and *K* for incremental deflection caused by tunnels A, D and BC deviated significantly from those reported in the literature for single tunnels in greenfield conditions. Specifically, trough widths were found to be 3–4 times greater than expected for single tunnels at these depths in granular soil. In addition, the incremental deflection profiles for the subsequent tunnels A, D and BC were not aligned directly above the respective tunnel. The results show that one cannot assume superposition of deflection profiles from individual tunnels driven in undisturbed ground as a method to estimate cumulative deflection profiles. These findings illustrate that the prediction of deflections due to the second, third and fourth tunnels in a closely-spaced multi-tunnel environment is not simple. The stress field in the ground is altered by each tunnel excavation, and excavation of a subsequent tunnel through an asymmetric stress field will yield asymmetric deformation. In this project, the heterogeneous geological conditions at many cross sections likely contributed to the results.

## ACKNOWLEDGMENTS

The authors thank the New York City MTA for the opportunity to work with the project data. The authors also thank a number of individuals for their help, namely Chris Fagan (Geocomp) and Robert Duran, Robert Godinez, and Richard Moore (all from Colorado School of Mines).

## REFERENCES

Chapman, D.N., Ahn, S.K. and Hunt, D.V.L. (2007) "Investigating ground movements caused by the construction of multiple tunnels in soft ground using laboratory model tests," *Canadian Geotechnical Journal*, 44, 631–643.

Cording, E.J. and Hansmire, W.H. (1975) "Deflections around soft ground tunnels," *General Rep. 5th Pan American Conf. Soil Mech. and Found. Engineering*, Session IV, 571–632.

Ercelebi, S.G., Copur, H. and Ocak, I. (2011) "Surface settlement predictions for Istanbul Metro tunnels excavated by EPB-TBM,' *Environ. Earth Sci.*, 62, 357–365.

Mair, R.J. and Taylor, R.N. (1997) "Bore tunneling in the urban environment," Proc. 14th Intl. Conf. Soil Mechanics and Foundation Engineering, 4, 2,353–2,385.

New, B.M. and O'Reilly, M.P. (1991) "Tunnelling induced ground movements; predicting their magnitude and effects," Proc. 4th Intl. Conf. on Ground Movements and Structures, Cardiff, invited review paper, Pentech Press, 671–697.

Peck, R.B. (1969) "Deep excavations and tunneling in soft ground," Proc. 7th Int. Conf. Soil Mechanics & Foundation Engineering, State of the Art Volume, Mexico City, 225–290.

Robinson, B. and Wehrli, J.M. (2013a) "East Side Access—Queens bored tunnels case study,' Proc. 21st Rapid Excavation and Tunneling Conference, Washington, DC, 1,014–1,041.

Robinson, B. and Wehrli, J.M. (2013b) "East Side Access—Queens bored tunnels engineering challenges,' Proc. 21st Rapid Excavation and Tunneling Conference, Washington, D.C., 1,086–1,118.

Suwansawat, S. (2006) "Superposition technique for mapping surface settlement troughs over twin tunnels," Proc. Intl. Symp. on Underground Excavation and Tunnelling, Bangkok, Thailand.

Suwansawat, S. and Einstein, H.H. (2007) "Describing settlement troughs over twin tunnels using a superposition technique," *J. Geotechnical & Geoenvironmental Engineering*, ASCE, 133(4), 445–468.

# Design Considerations and Enabling Works for the San Diego Central Courthouse Inmate Transfer Tunnel

**Albert Ruiz, Paul Gabryszak, Alan Haldin, Stuart Warren**
Hatch Mott MacDonald

**ABSTRACT**: The Inmate Transfer Tunnel may be constructed to provide for secure movements of inmates and officers between the new San Diego Central Courthouse and the existing County Jail, located 2 blocks east. The tunnel would be constructed immediately below the existing Courthouse, which must remain in service during tunneling, and will pass through an active fault. This paper summarizes challenges associated with tunneling in close proximity to the building foundations while controlling ground and building movements, seismic and fault displacement considerations, toolbox provisions, and restricted access for ground improvement and remedial works from within the existing Courthouse basement.

## PROJECT DESCRIPTION

The Judiciary Council Administrative Office of Courts of California will be constructing a new Central Courthouse in downtown San Diego. The new Central Courthouse will replace the existing County Courthouse that resides one block to the east of the proposed Central Courthouse. One block east of the existing County Courthouse is the existing County Jail. The Jail is connected to the existing Courthouse via an enclosed pedestrian bridge on the 2nd story. In order to facilitate secure future inmate movements from the existing County Jail to the new Central Courthouse, a new pedestrian tunnel, referred to as the Inmate Transfer Tunnel (Tunnel), has been designed to link the two structures beneath the existing courthouse, as shown in Figure 1.

A Tunnel Design Options Study was undertaken during the Schematic Design phase to determine the most appropriate means of constructing the Tunnel. Due to surface access restrictions, options were limited to only mined alternatives including Sequential Excavation Method (SEM), Digger Shield and Jacked Box. In developing the recommended tunneling method, the study considered various factors such as site constraints, ground conditions, ground risk factors, constructability and contracting risk, construction schedule and order-of-magnitude costs. The study concluded that SEM Tunneling was the preferred method, based primarily constructability considerations.

The Tunnel runs in a west-east direction from the new Central Courthouse to a new Transfer Structure, adjacent to the existing County Jail. The length of tunnel will be approximately 329'. The primary cross sectional requirement for the Tunnel is a 9' wide by 8'-7" high pedestrian clearance envelope. Additional space for utilities and drainage through the tunnel needs to be accommodated outside of this envelope. The most structurally efficient shape of a SEM Tunnel in these ground conditions is nominally "egg-shaped," which facilitates space for utilities and drainage in the annular space above and below the pedestrian clearance envelope. From a structural aspect, the final lining walls could have remained a constant thickness, however the inside tunnel walls were designed vertical for security purposes. Prior to completion of final design, a full size mock-up of the tunnel was constructed and it was determined that curved walls posed a potential risk to guards in a situation involving an inmate pushing a guard against the wall, making it difficult for the guard to maintain balance. Figure 2 shows the final geometry.

## GEOLOGICAL SETTING

### Soils

The tunnel would be constructed in the Bay Point Formation, which consists of silty sand; poorly graded sand with silt, locally with cobbles; and clayey sand. Layers and lenses of gravel and cobbles are contained within the formation locally and often encountered near the contact between the Bay Point Formation and the overlaying alluvium.

SPT blow counts recorded within the tunnel envelope (Elevation +9 ft 2 inch to –19 ft 1 inch) are typically ≥50 indicating very dense soils. When continuous coring was attempted, there was typically poor recovery and the samples were friable and lacked sufficient cohesion or cementation for pocket penetration testing in the field or unconfined compressive strength testing in the laboratory.

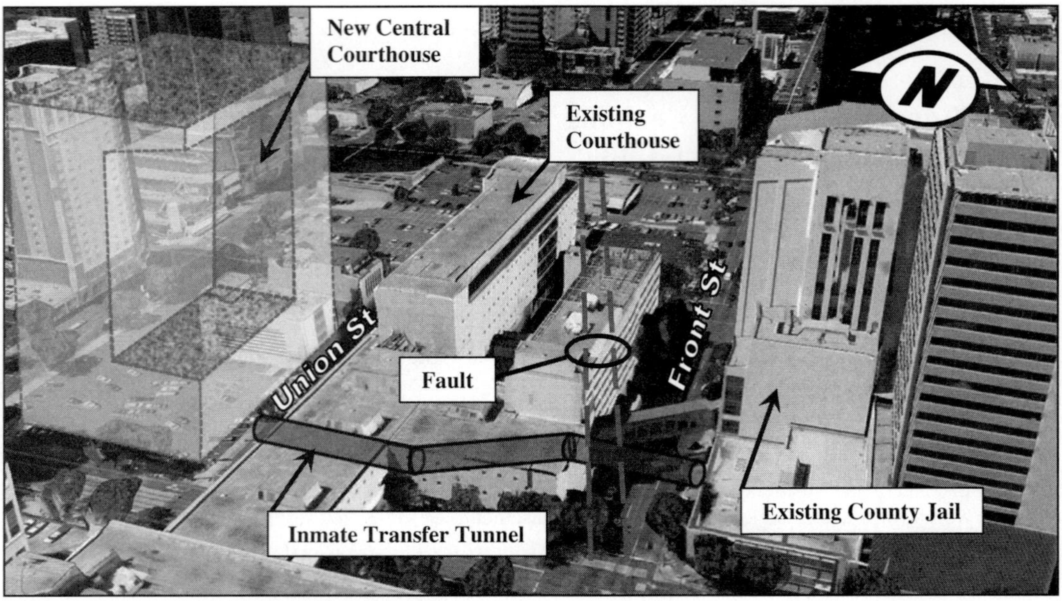


**Figure 1. Overview of tunnel and structures**

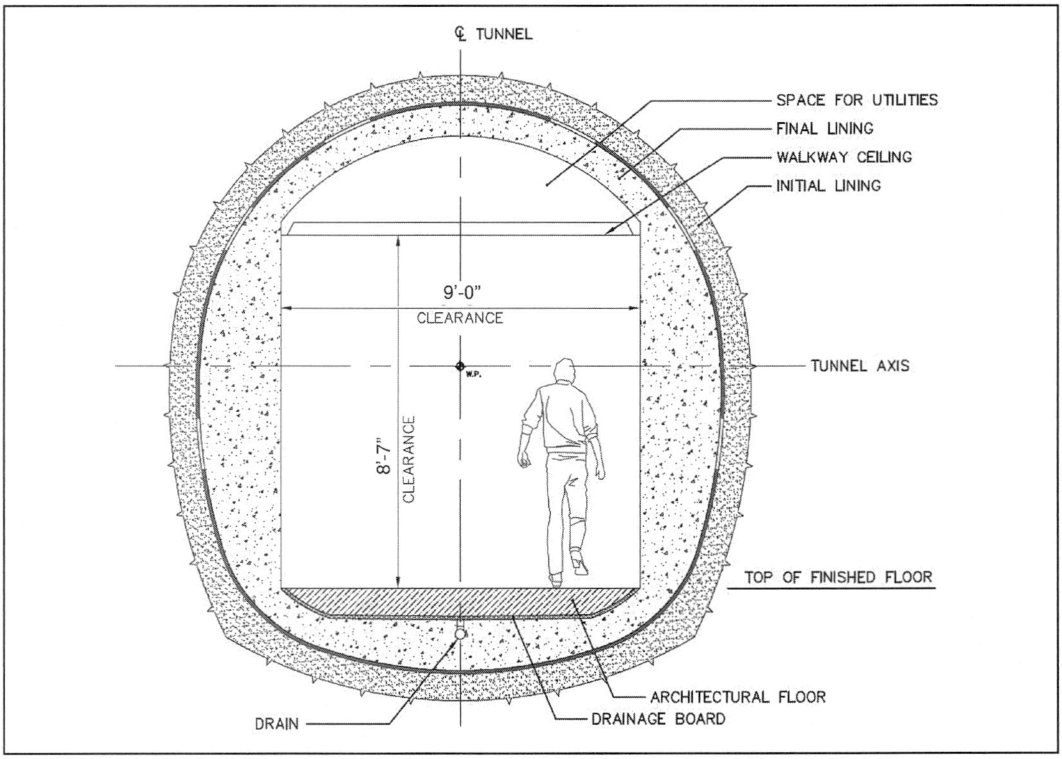


**Figure 2. Tunnel cross section**

### Groundwater

The groundwater table is typically at elevation +2 to –6.5 feet Mean Sea Level (ft MSL). Tunnel crown elevation varies nominally between +8 ft MSL and –1 ft MSL, and therefore part to all of the tunnel excavation will be in saturated soils below the water table.

The results of aquifer testing (groundwater slug tests) performed on two wells (B-102 and B-103) indicated an average hydraulic conductivity of 12.4 gallons per day per square foot (gpd/ft$^2$) or $5.8 \times 10^{-4}$ centimeters per second (cm/sec) and storativity of $8.3 \times 10^{-4}$. In addition, laboratory permeability testing of representative soil samples collected from three boreholes drilled along the tunnel alignment, showed a range of hydraulic conductivity from $6.2 \times 10^{-3}$ cm/sec at the west end of the tunnel alignment to $4.5 \times 10^{-5}$ cm/sec east of the fault.

### San Diego Fault

Downtown San Diego lies within the Rose Canyon Fault Zone. Onshore in downtown San Diego, there are two active fault zones that are designated as Alquist- Priolo Earthquake Fault Zones (EFZ) by the California State Geologist. Active faulting has been demonstrated in these zones and any development within the designated zones requires fault hazard investigations.

The San Diego Fault is considered active in the locations mapped in downtown San Diego. The fault has a limited length of approximately 1 km and probably does not act as an independent seismic source). The tunnel alignment crosses the San Diego Fault Zone at a location outside of the Alquist-Priolo EFZ. Fault hazard assessments were performed with a comprehensive program of Cone penetration Tests, borings and fault trenches. The fault was exposed in an investigation trench located in B Street. The fault consists of a primary zone of dislocation approximately 5 feet in true width, consisting of multiple, steeply dipping fault planes or shears, that have dislocated, tilted and warped the bedding in the Bay Point Formation. Occasional shears and minor tilting and warping of beds will be present in the Bay Point Formation up to 25 feet from the primary fault zone. The groundwater surface is an estimated 8 feet higher on the east side of the fault zone than on the west side of the fault zone.

A Deterministic Fault Hazard Assessment was performed during preliminary engineering. This indicated that a maximum 15-inch offset could occur within the primary fault zone during the magnitude 6.5 earthquake that the fault is considered capable of producing. Secondary fault zones adjacent to the primary rupture zone could have 2 inches of distributed offset, with up to 5 inches of tilt and warp.

A Probabilistic Fault Hazard Assessment was performed during final design. This concluded that the total mean primary fault displacement hazard is zero at an annual probability of exceedance of $7.5 \times 10^{-5}$ or a return period of about 15,000 years. Hence the probability of primary surface faulting at the tunnel that would be of engineering significance is very low. Given the low primary surface fault displacement hazard at the site, the secondary fault displacement hazard is even lower.

### Liquefaction Potential

The Bay Point Formation has very low liquefaction potential, and as such, no special provisions were required in the design.

## TUNNELING CHALLENGES AND CONSIDERATIONS

Although the Tunnel is relatively short at 329 ft, it encompasses a number of complex challenges that needed to be taken into consideration during design. The primary challenges include controlling ground movements while tunneling below the existing Courthouse foundations, crossing through an active fault-zone, tunneling in mixed face conditions, and overcoming restricted access and staging for tunnel enabling operations. These challenges are further summarized below.

### Tunneling Below the Existing Courthouse

In order to minimize concerns with vertical circulation as well as construction costs, the tunnel alignment was initially proposed at an elevation that would limit the depth of the portals at the New Central Courthouse and the Existing County Jail. As the tunnel will be used to transfer inmates, for security reasons it was desirable to have a straight tunnel to maintain a visual line-of-sight through the tunnel. The initial profile proposed to satisfy these objectives is shown in Figure 3.

The initial proposed alignment also resulted in a number of isolated footings being directly above the tunnel, including a deep foundation for an elevator pit, as shown in Figure 4.

A strict requirement was imposed on the project that full functionality of the existing Courthouse had to be maintained during construction. As such, the presence of concentrated loads directly above the tunnel posed a large challenge with regards to controlling building movements during tunneling. Sensitivity to ground settlement is high due in large part to the foundation depth and type. The building is primarily supported on isolated footings, which are prone to differential settlements, more so than continuous strip footings or raft foundations that have the ability to distribute bearing loads across a larger area. In additional to challenges associated with mitigating building movements, the concentrated loads directly above the tunnel crown also raised concerns regarding stability of the ground in the tunnel periphery and face exposed during excavation.

### Settlement Analysis

In general, the process of tunneling relieves existing stresses within the soil structure around the excavation allowing inward movement of the ground prior to the installation of the final tunnel support. This effect is referred to as ground loss and is quantified by comparing ground loss value as a percentage of the total excavation. Settlements and ground movements associated with tunneling manifest themselves at the ground surface in the form of a Gaussian Curve shaped trough, generally centered above the tunnel alignment. The extent of tunneling induced settlements is dependent on a number of factors including surrounding geotechnical parameters, amount and competency of ground cover, dewatering requirements, proximity to structures and foundations, and loads associated with those foundations.

Early in the design phase, a settlement study was performed to establish tolerable settlement levels of the existing Courthouse foundations. Several commonly accepted methods for predicting building damage were adopted that included assessments of angular distortions and associated tensile strains in the structure. Angular distortion is a measure of the shearing distortion of a structure and is often approximated as the rotation, due to settlement, of the straight line between two reference points on a structure, such as adjacent foundations.

A literature study of historical damage classifications of building structures similar in construction to the existing Courthouse indicated that angular distortions of less than 1/500 would typically yield cosmetic damage only. Cosmetic damage consists of non-structural cracks such as in internal wall finishes or mortar joints in brickwork. Angular distortions of greater than 1/150 are expected to result in damage that could reduce structural capacity of building members, thus limiting the functionality of the structure.

A settlement analysis concluded that for the initial proposed alignment, a 2% volume loss would create a maximum deflection ratio of 1/150

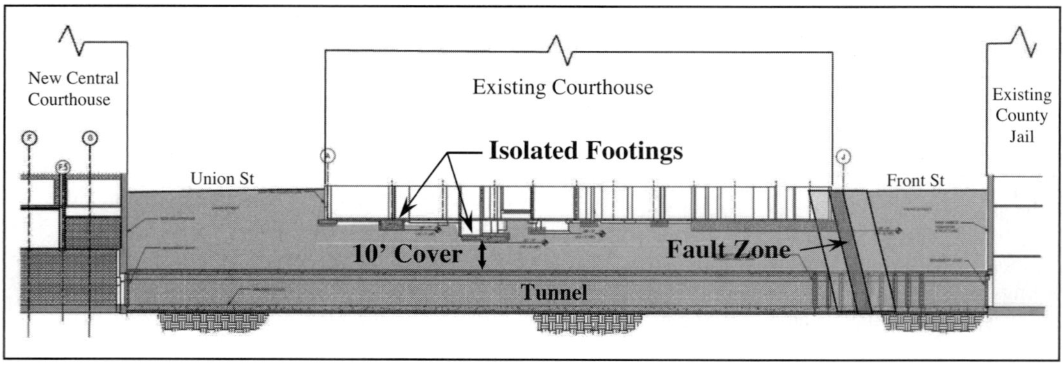


**Figure 3. Initial proposed tunnel profile**

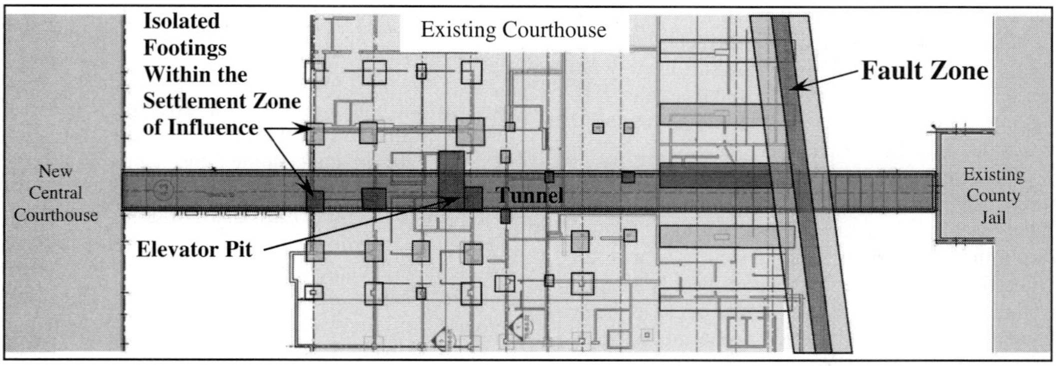


**Figure 4. Initial proposed tunnel plan**

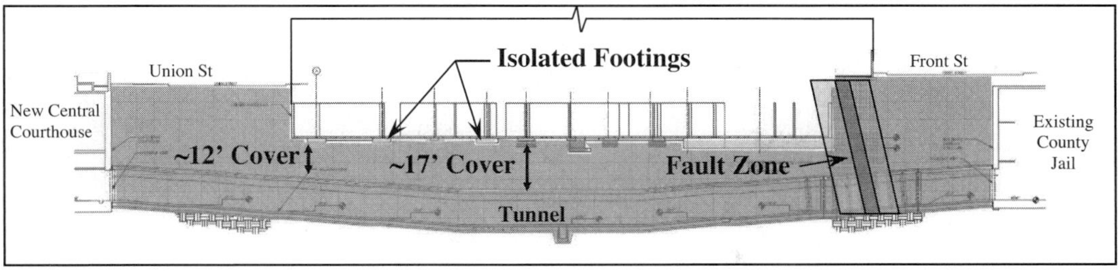


**Figure 5. Final tunnel profile**

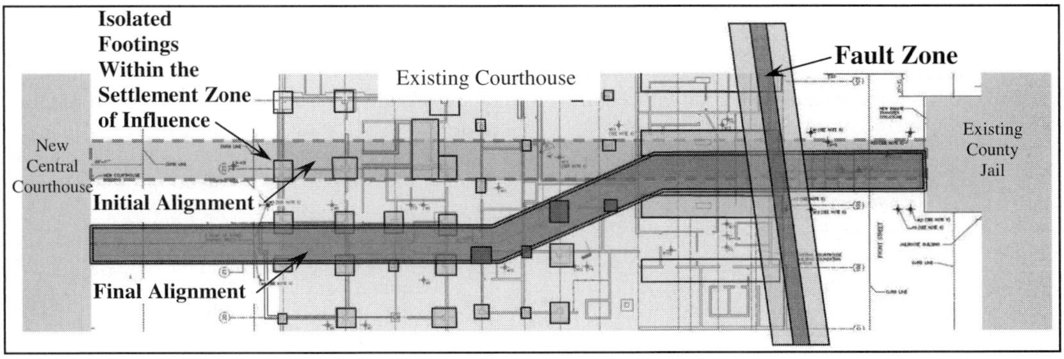


**Figure 6. Final tunnel plan**

between adjacent foundations, while a 0.5% volume loss would result in a maximum deflection ratio of 1/500. Achieving a minimum volume loss of 2% during tunneling was considered the lower achievable bound without undertaking extensive settlement mitigation measures; however a 2% volume loss is expected to yield results on the verge of structural damage. As a result, settlement mitigation measures would need to be introduced that could include widespread permeation grouting, compensation grouting, underpinning, and barrel-vaulting, among others to limit damage to tolerable levels.

As part of a value engineering exercise, an analysis was conducted to determine if potential settlement mitigation costs could be reduced by altering the tunnel alignment. By sloping the tunnel downwards from each portal, an additional 7' of cover could be realized below the most sensitive foundations, as shown in Figure 5.

Additionally, by introducing a horizontal "dogleg" into the alignment, and slightly offsetting the position on the tunnel, the elevator pit could be avoided and the number of building foundation directly above the tunnel could be significantly reduced, as shown in Figure 6.

A subsequent settlement analysis was performed based on the revised alignment. The results indicate that volume losses of 2% will yield deflection ratios less than 1/500 in most cases. At a few locations however, adjacent foundations are predicted to see deflection ratios as high as 1/300. At these levels, settlement damage typically includes doors and windows sticking and cracks in masonry joints requiring repointing.

Recognizing that the existing Courthouse will be decommissioned after opening of the new Central Courthouse, it was agreed upon that some damage could be tolerated, provided the damage does not impair the functionality of the facility. Therefore, a criterion for the maximum deflection ratio limit was set at 1/300.

Since actual ground movements that occur during tunneling can often vary from those predicted during design and analysis, a series of "toolbox items," commonly associated with SEM Tunneling, have been developed and included in the Contract Documents. Toolbox items are additional support elements or measures available for use during tunneling to supplement the standard ground support. Toolbox items deemed suitable for this tunnel are described further in this paper.

### SEM Standard Excavation and Ground Support

Based on modelling and structural analysis, a base design was developed that included a full width heading/bench/invert excavation sequence identified as System A, as shown in Figure 7.

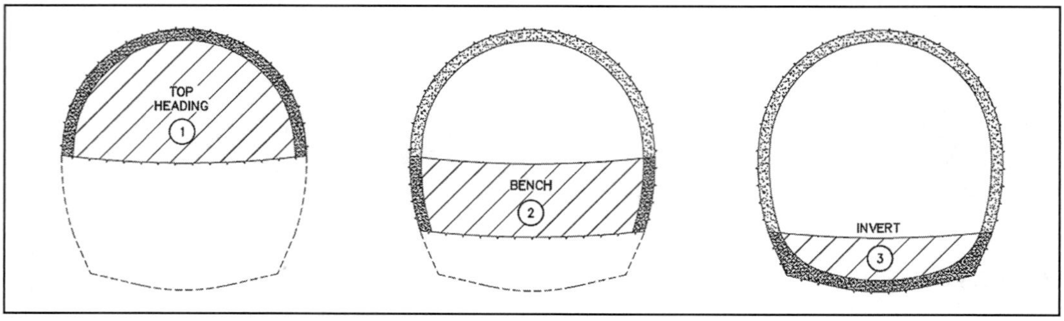


**Figure 7. System A excavation and initial support system**

Soil-structure modeling and analysis has indicated the following excavation and support requirements are considered reasonable for System A:

- Excavation sequence consisting of full width Heading/Bench/Invert (3 stage)
- Excavation round length ~2.5'
- Minimum distance between advancing Heading/Bench/Invert faces ~8.5'
- Minimum shotcrete thickness ~8"

In critical areas, such as the reaches below the exterior walls of the existing courthouse where cover to the building foundations is limited, through the fault zone and at the portals, grouted pipe spiles have been prescribed as a standard support elements. This is considered as a risk mitigation measure as much as a settlement mitigation measure.

### Toolbox Items

Toolbox items are commonly associated with SEM tunnel construction and are installed supplemental to the standard support measures. These items are primarily used to deal with localized pockets of lesser quality ground or other conditions that could otherwise result in excessive settlements or instabilities when relying upon standard support measures alone. The decision of when and which toolbox items should be implemented occurs during tunneling operations in reaction to the ground response.

Anticipated toolbox items for the construction of the Tunnel include:

- Grouted pipe spiling
- Permeation grouting
- Face wedge
- Split drift excavation and initial support

The split drift excavation sequence identified, as System B, has been developed to provide tighter control of ground movements in areas that may be deemed sensitive to settlements. This is accomplished by reducing the excavated volume prior to installation of ground support through the use of smaller drifts, as shown Figure 8.

The soil-structure modelling and analysis has indicated the following excavation and support requirements are considered reasonable:

- Excavation sequence consisting of split Heading/Bench/Invert (6 stage)
- Excavation round length ~2.5'
- Minimum shotcrete thickness ~8"

### Fault Provisions

Based on the results of the Probabilistic Fault Hazard Assessment, regulations and codes do not require any fault rupture in the fault zone to be accommodated in the design of the tunnel. However, the Administrative Office of Courts and the design team considered it prudent to design the fault crossing to allow longitudinal lining flexability in the area of the fault. The design accounts for this flexibility in the 40-foot length of tunnel crossed by the fault zone, by using a combination of movement joints and compressible joints. Each movement joint is designed to accommodate a maximum of 3 inches of movement while remaining water tight.

### Dewatering

Dewatering will be necessary to draw down the water table below the tunnel invert, to allow for a safe SEM Tunneling environment, and prevent hydrostatic build-up behind the initial shotcrete lining. Dewatering also has a potential to enhance ground design parameters by introducing an apparent cohesion into sandy material.

Dewatering will need to be accomplished from numerous locations to be effective and prevent a larger than necessary drawdown affect. Dewatering activities will occur from ground surface on Union and Front Streets, as well as from within the existing County Courthouse basement. A walkthrough of the existing Courthouse basement to determine possible

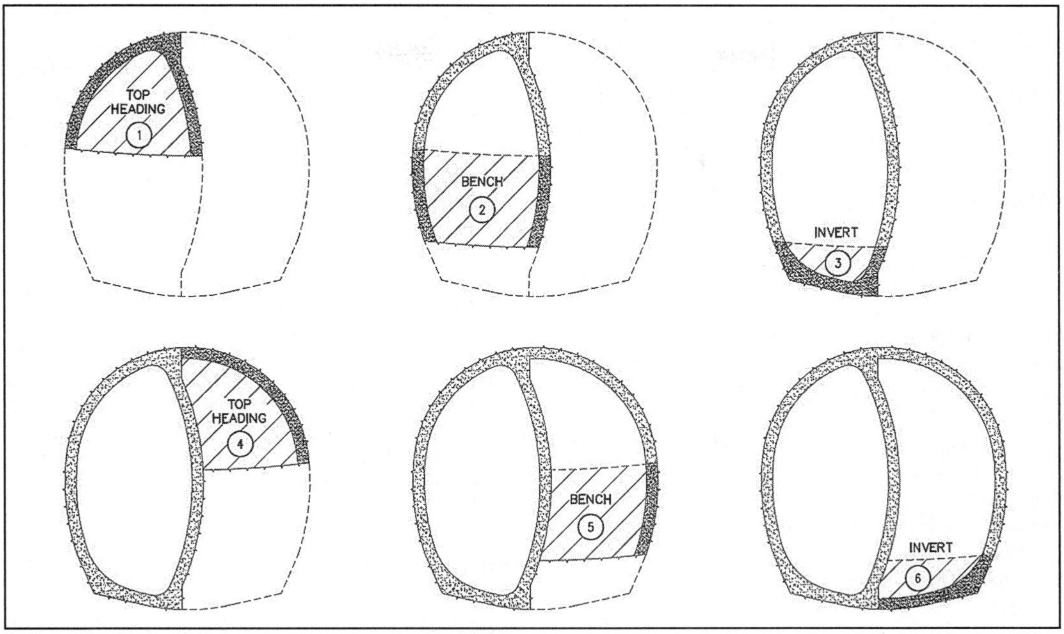


**Figure 8. System B excavation and initial support system**

locations for the wells was recently undertaken and possible well locations are indicated on the drawings. The basement of the existing Courthouse is separated into two independent jurisdictions, the courthouse and the county jail, and movements between the two jurisdictions are restricted. As such, two independent dewatering systems will be required.

### Instrumentation and Monitoring

An instrumentation and monitoring plan will be developed to ensure tunnel convergence and surface settlements are within the allowable movements to prevent damage to existing structures, utilities and other facilities. Anticipated instrumentation will include:

- Tunnel Convergence Monitoring Targets
- Multipoint Borehole Extensometers
- Ground Settlement Monitoring Points
- Utility Settlement Monitoring Points
- High Sensitivity Building Settlement Sensors
- High Sensitivity Settlement Sensor Reservoirs
- Strain gauges on lattice girders
- Inclinometers installed adjacent to existing elevator hydraulic cylinder

Of particular importance is the ability to monitor the existing County Courthouse for movements. A "high-sensitivity" monitoring system is anticipated to be established in the basement of the courthouse, consisting of settlement sensor reservoirs set up on columns and foundations within the settlement zone of influence. The system will provide a continuous monitoring readout, and instruments will provide real-time movement data. Provisions for access to the basement will need to be made throughout tunneling operations for instrumentation monitoring and maintenance purposes.

The hydraulic actuation cylinder shaft located underneath the elevator of the existing courthouse is especially susceptible to ground movements and is located within 6 feet of the tunnel lining. An inclinometer will be installed adjacent to the hydraulic cylinder to monitor ground for movements within the vicinity.

As noted previously, the basement of the existing Courthouse is separated into two independent jurisdictions, the courthouse and the county jail, and movements between the two jurisdictions are restricted. As such, two independent instrumentation and monitoring programs will be required; one for each jurisdiction.

Response Values, also known as trigger values, were established for each type of instrument. Response Values will consist of Threshold Values and Limiting Values.

Threshold Values are set lower than Limiting Values and provide advance notification of ground movements that are trending towards damaging levels, such that mitigative measures can be employed prior to movements reaching Limiting Values.

When a Limiting Values is approached or reached, typically an immediate suspension of excavation activities is implemented until movements can be controlled and corrective measures put into place to prevent further movement.

Response Values are shown in Table 1.

## CONSTRUCTION ACCESS AND STAGING

### Tunneling

The tunnel design had to take into account that all tunneling operations will be conducted from within the new Central Courthouse basement levels. The Central Courthouse will have three basement levels, B1, B2 and B3, with B3 being the lowest and at tunnel level.

Although the basement levels will be constructed prior to commencing tunneling operations, the Central Courthouse superstructure will be construction concurrently with tunnel construction. Access to and staging areas were developed and coordinated with the Courthouse architect and contractor to facilitate both operations.

All trucks delivering equipment and materials to the tunnel, as well as removing tunnel spoils, will utilize the B1 and B2 level access ramps to grade. The B2 level floor above the B3 level will not be cast until after tunneling operations are complete, allowing access for tunneling operations. However, the access ramp between B2 and B3 will pass over part of this area, which will locally restrict vertical clearance. Tunnel construction personnel will access the B3 level via a temporary steel staircase from B2.

Materials and equipment for tunnel construction will be lowered to the B3 level from B2 by a mobile crane. Due to the restricted vertical access at level B3, a bucket elevator could be used to remove muck from the tunnel to the B2 level. A dumper will be used to transport muck from the tunnel face to the bucket elevator.

The tunnel will be launched from the B3 level and breakout through a secant pile wall built against the walls of the basement. Tunneling operations will be on a continuous 24-hr working schedule, however, trucking operations above ground will be limited to standard daytime working hours. As such, tunnel mucking operations during daytime hours will consist of muck being loaded from the bucket conveyor into trucks for direct transportation from the site. During off hours, the muck from the bucket elevator will need to be temporarily stockpiled on the B2 level floor using a muck skip, and loaded onto trucks during daytime working hours.

**Table 1. Monitoring response values**

| | Instrument Response Values | |
|---|---|---|
| **Instrument Type** | **Threshold** | **Limiting** |
| Convergence monitoring targets | Absolute displacement 0.15% | Absolute displacement 0.20% |
| Roof leveling points | Vertical displacement 0.15% | Vertical displacement 0.20% |
| Lattice girder strain gauges | Strain 0.0010 | Strain 0.0014 |

### Existing Courthouse Basement

The basement of the existing Courthouse is separated into two independent jurisdictions, the Courthouse and the County Jail, and movements between the two jurisdictions are restricted. As such, independent programs for tunneling enabling works, such as dewatering and instrumentation and monitoring will be required.

### Adjacent and Future Construction Considerations

As noted previously, the existing Courthouse is expected to be demolished at some point in the future, and the land will be redeveloped. To account for these future works, the tunnel was analyzed to account for various potential loading conditions.

The "As is" scenario. This considers ground and ground water loadings, loads from the existing County Courthouse foundations, and Union and Front Street as is at present.

The "Demolition" Scenario. This considers ground and ground water loadings, but no loads from the existing County Courthouse, as this will be demolished to the level of the existing basement floor slab. The basement floor slab and walls will not be demolished in order to provide support to the basement opening, and will be supported as required. This scenario has been satisfactorily reviewed and concluded that the reduction in overburden loads does not induce tunnel floatation.

The "Park and/or Parking Lot" Scenario. This considers as with the previous scenario that the County Courthouse will be demolished and ground and ground water loadings will be contemplated, without loads from the existing County Courthouse, and with the basement floor slab and walls remaining to provide support to the basement opening. However, the basement will be partially or completely backfilled to form a park and/or parking lot.

## CONCLUSIONS

A number of difficult existing conditions had to be addressed in the design of this relatively short, but challenging project. The primary challenges include controlling ground movements while tunneling below the existing Courthouse foundations, designing a dewatering system to be installed and operated within the low headroom and limited access of the basements of the existing courthouse; crossing through an active fault-zone, and overcoming restricted access and staging for tunnel enabling operations. Although designed to 100% final design, at time of publication, the tunnel is not included in the initial development of the New Central Courthouse, but may be constructed at a later date.

## ACKNOWLEDGMENTS

The authors wish to thank Clifford Ham, Principal Architect with the Judicial Council of California Administrative Office of the Courts, for permission to publish this paper. We wish to thank Steven Sobel, Director, and Peter Lee and Frank Grima of Skidmore Owings & Merril LLP, Architects for the New Central Courthouse; Rob Stroop, Senior Project Geotechnical Engineer of URS, the geotechnical consultants for the project; and Trent Cohen at Hatch Mott MacDonald, for their the support and assistance during the design of the project. The photographs in Figure 3a and 3b are used with the permission of Rob Stroop of URS.

# Back Analysis of Observed Measurements for Optimised SCL Tunnel Design

**Angelos Gakis, Stephen Flynn, and Ali Nasekhian**
Dr Sauer & Partners Ltd

**ABSTRACT:** The Crossrail Farringdon Station is a deep level station with two platform tunnels, 300 metres long with the tunnel axis being at a depth of approximately 30 metres below street level, linked by means of eight cross passages and two concourse tunnels. The platform tunnels will be enlarged from the TBM bored Crossrail running tunnels using sprayed concrete lining (SCL) tunnelling method, after the removal of the TBM segments. The station is located in challenging ground conditions due to the presence of several faults and the heterogeneity of the Lambeth Group, which call for a variety of temporary works contingency measures and in-tunnel depressurisation scheme.

Back analyses for the already constructed cross passage CP1 were performed using finite element (FE) analysis in order to calibrate them against observed deformations, deriving valuable information for future design of similar SCL tunnels. The accuracy of assumptions made in the initial design stages had to be checked and such parameters as ground relaxation factor for construction advance, at rest earth pressure coefficient and soil stiffness were back-calculated and compared with the initial values.

The result of this work is refined FE modelling input parameters calibrated against measured deformation results, which demonstrated a greater accuracy for predicting both, in-tunnel and surface deformations and thus allowing for a better understanding of the soil-structure interaction to be anticipated in the remaining stages of SCL tunnel excavation for the completion of the project.

## CROSSRAIL FARRINGDON STATION

The Farringdon Station is one of the 8 stations of the Crossrail project in London that will provide an interchange between Crossrail and existing London Underground networks at the Eastern Ticket Hall/ Barbican Station and an interchange between Crossrail and Thameslink at the western ticket hall/ Farringdon side. This future station is a deep level station with two platform tunnels stretching between the existing London Underground Farringdon and Barbican Stations, with their tunnel axis being at a depth of approximately 30 metres below street level.

The contractor, BAM Ferrovial Kier Joint Venture (BFK), formed between BAM Nuttall Ltd, Ferrovial Agroman (UK) Ltd and Kier Construction Ltd, was awarded the construction contract (C435) for the main construction works at Farringdon station by Crossrail Limited (CRL) in November 2011. The Dr. Sauer & Partners Company Limited (DSP) has subsequently been contracted by BFK to provide sprayed concrete lining (SCL) design and site supervision.

The structures at Farringdon Station are:

- Escalator inclines ES1 and ES2,
- Eastbound platform tunnel (PTE) and westbound platform tunnel (PTW),
- Lower concourse tunnels CH1 and CH2,
- Cross passages CP1, CP2a, CP2b, CP3a, CP3b, CP4, CP6a, CP6b, CP7, CP8 and CP9,
- Ventilation adits VA1 and VA2,
- Platform extension tunnels PL1, PL2, PL3 and PL4,
- Stub tunnels STE1, STW1, and STW2,
- Lift shaft passage (LP1) connection to Thameslink lift shaft and
- Temporary tunnels—PL2RC wraparound, CP1-CH1 connection.

At the time that this article was written (October 2013), the completed tunnels in Farringdon station were the "Early Western Tunnels," STW2-PL1, CP1 and TBM reception soft-eyes in shafts SH-W1 and SH-W2, as well as CP1-CH1 connection adit, PL2RC Wraparound, CH1 pilot tunnel and westbound TBM pilot tunnel (shown in light blue colour in Figure 1). This paper focuses on CP1, a cross passage of approximately 35m long and cross sectional area of approximately 22m$^2$ (7.2m height × 6.2m width), connecting the Eastbound and Westbound platform tunnels (see Figure 1).

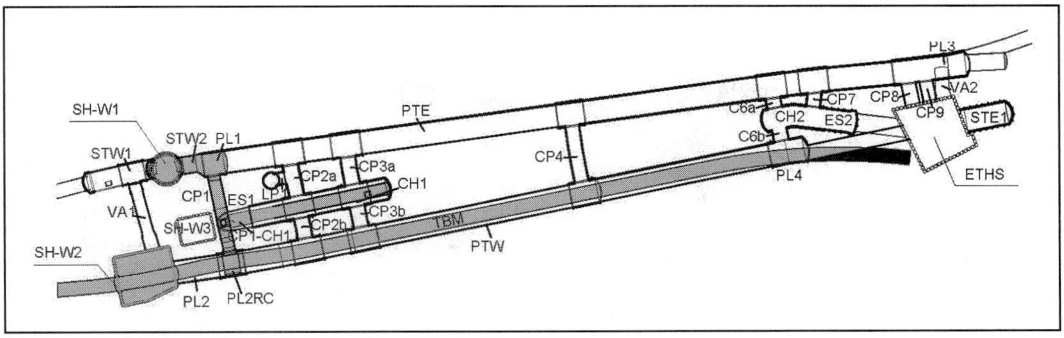


**Figure 1. Farringdon Station plan view**

**Table 1. Ground parameters under undrained conditions, except for Upper Strata and Thanet Sand which are drained**

| Soil Properties | | Upper Strata | LC (A2) Unfaulted | LC (A2) Faulted | Lambeth Group Unfaulted | Lambeth Group Faulted | Thanet Sand Unfaulted | Thanet Sand Faulted |
|---|---|---|---|---|---|---|---|---|
| Unit weight | [kN/m$^3$] | 17 | 20 | 20 | 21 | 21 | 21 | 21 |
| Young's modulus | [MPa] | 10 | 40+3.7z1 | 33+3z1 | 36+5.9z1 | 30+4.9z1 | 209+4.3z2 | 180+3.9z2 |
| Poisson's ratio | [–] | 0.2 | 0.495 | 0.495 | 0.495 | 0.495 | 0.2 | 0.2 |
| Shear strength | [kPa] | - | 85+6.5z1 | 40+6.7z1 | 95+10z1 | 55+8.8z1 | N/A | N/A |
| Friction angle | [°] | 31 | 0 | 0 | 0 | 0 | 39 | 34 |
| $K_0$ — value | [–] | 0.5 | 1.2 | 1.2 | 1.2 | 1.2 | 1.0 | 1.0 |

## DESIGN CONSIDERATIONS

Construction of urban tunnels calls for measurements of surface settlement and in-tunnel deformations. Factors such as nonlinear behaviour of soil, stress history, overburden and diameter of tunnel have a major influence on the soil-structure interaction.

The main purpose of this study is to investigate the performance of 2D and 3D finite element models under the prescribed design assumptions and to specify constraints for the analysis. An initial assessment of the accuracy of the preliminary analysis, were worst-case faulted ground material properties were used, was performed by comparison to the actual results.

2D & 3D finite element analyses were carried out to determine the in-tunnel deformations associated with the excavation and primary lining construction. This section describes the analysis methodology, soil and sprayed concrete properties, modelling stages and comparison of simulated results with actual deformations. In an effort to determine a more representative model, a parametric study and a back analysis was performed.

### 3D Finite Element Analysis

It is common practice to assume a so called relaxation factor in order to consider three dimensional effects in a 2D analysis. It is necessary to validate assumptions in 2D analysis to take into account 3D geometrical effects, construction sequences, and stress variations due to previous tunnelling activities. All these assumptions are highly dependent on the complexity of tunnel geometry and subsoil conditions and thus, estimating a reasonable and sound factor can be a difficult task. However, 3D FE modelling provides an efficient tool to investigate three-dimensional ground/structure interaction, face stability and gives information regarding volume loss and ground movements. In addition, since the plane strain conditions cannot apply at tunnel breakouts or tunnel junctions, 3D analysis is necessary in order to analyse tunnel lining and ground stability.

ABAQUS Version 6.12 released 2011 (Dassault Systemes Simulia Company) was employed to perform the numerical analyses. The ground was modelled using linear tetrahedral elements and sprayed concrete linings were modelled using linear triangular shell elements. Subsoil layers were modelled with the elastic-perfectly plastic Mohr-Coulomb model. The utilised properties of the soil are presented in Table 1.

The dimensions of the model boundaries were 145m × 120m × 44m (see Figure 2). The model was comprised of approximately 600,000 solid elements, 50,000 shell elements and 191 analysis steps. The construction shafts (SH-W1 and SH-W2), were modelled as a wished-in-place structure (i.e., the

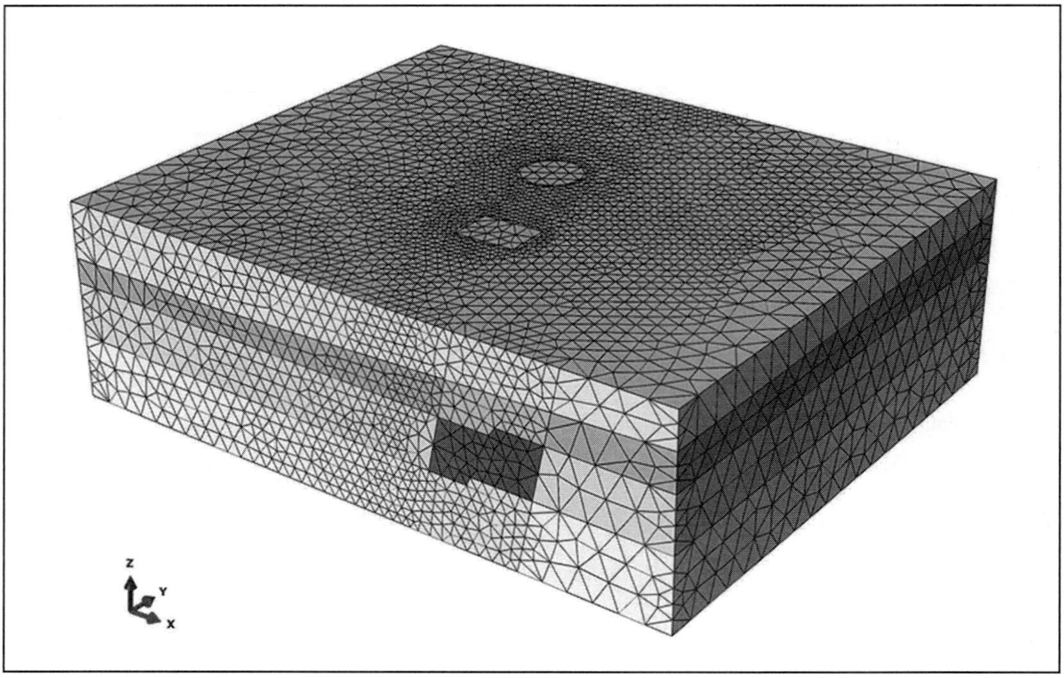


**Figure 2. Finite Element mesh and extent of the 3D model**

entire soil material was removed in one step and the ground support was simultaneously activated) following a relaxation step in which 50% of soil stiffness was reduced.

Undrained soil parameters were used in the analysis to account for the "fast" construction in comparison to the time of consolidation. The numerical analyses have been undertaken on the basis of a total stress analysis during which no pore water pressure was generated.

### *Ground Conditions and Model Geometry*

The geological formations encountered in the area of Farringdon Station, are the typical of the London basin, i.e., the upper strata comprising man-made deposits (Made Ground), Alluvium and River Terrace Deposits (mainly in the Eastern part) overlaying London Clay, Lambeth Group, Thanet Sand and Chalk. In the FE model Chalk layer was considered as the bed rock and therefore the bottom boundary of the model was the top of the Chalk layer. An important feature was that the thickness of the London Clay layer was significantly thinner in the western part relative to the eastern part, varying between 4 to 10m and contains London Clay A2 subdivision only. The layout of subsoil layers used in the FE model in the region around Western Ticket Hall is illustrated in Figure 3 and soil parameters used in the FE models are given in Table 1.

The main characteristics of the geology in Farringdon station in the area subject to this FE model are the following:

- The tunnel horizon was predominantly within the Lambeth Group, a heterogeneous formation comprising stiff to very stiff, over consolidated clays with inter-bedded sand lenses of unknown orientation, size and water pressure regime.
- The presence of Farringdon Station Fault which affects the thickness, the elevation and the continuity of the top/bottom of the soil layers. This Fault has been taken into account in the FE model by defining a zone with "faulted" material properties. Boundaries of the zone have been obtained by offsetting ±10 metres the fault plane identified in the geological section drawings in Geotechnical Interpretative Report. Figure 3 shows the geometry of the Fault zone in the FE model.
- London Clay contained only subdivision layer A2 with a slight inclination towards west. The thickness varied approximately between 4 to 10 metres.

### *Steel Fibre Reinforced Sprayed Concrete Model*

The primary SCL consisted of a fibre reinforced sprayed concrete lining. The material properties

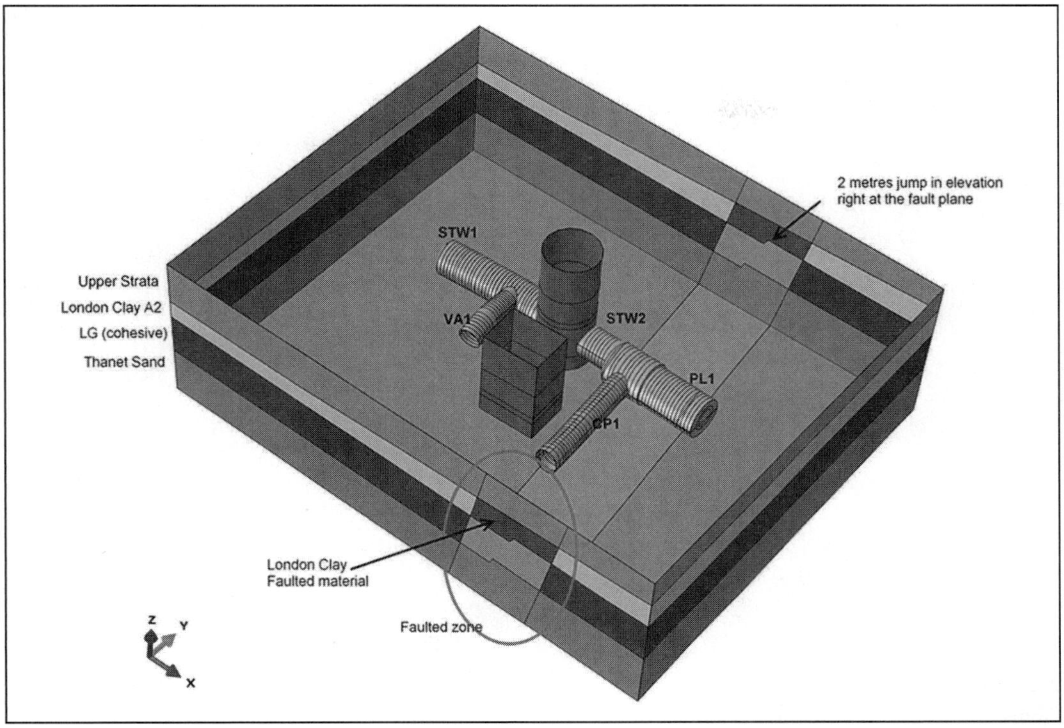


**Figure 3. Fault zone and subsoil layers in the 3D FE model**

utilised for the sprayed concrete tunnel linings are given in the Table 2. For the primary lining design, the 28-days compressive strength was considered.

In the 3D-FE analysis elasto-plastic material model was used to simulate the post failure behaviour of the SCL lining in tension, i.e., tension cut-off 0.4 MPa was considered for the tunnel linings based on the residual flexural tensile strength. However, no cap has been applied to the compressive strength of the concrete. For this purpose, the "concrete damaged plasticity model" which is a software-defined material model in ABAQUS has been employed.

## 2D Finite Element Analysis

Construction and installation of the primary lining was modelled following the prescribed excavation and support sequences of top heading, bench and invert (see Figure 4). The finite element software package Phase2 Version 8.0 (Rocscience) was used to carry out the numerical analysis. In the 2D model approximately eleven thousand 3-noded triangular solid material elements were used. The Mohr-Coulomb constitutive model was used to simulate the elasto-plastic behaviour for the elements forming the ground. The sprayed concrete lining in the model followed an elastic material behaviour (see Table 2).

**Table 2. Parameters for the steel fibre reinforced sprayed concrete used in the FE models**

| Parameter | Value |
|---|---|
| Characteristic cylindrical compressive strength of SFRC (28-days) | 28 MPa |
| Characteristic residual tensile strength for SFRC | 0.45 MPa |
| Elastic modulus used in FE analysis (Primary Lining only) | 13 GPa |
| Poisson's ratio | 0.2 |

For 2D analyses, a relaxation factor ($\lambda$) was applied in order to simulate the deformation that takes place prior to the installation of the support (Figure 5). The approach is referred to as the convergence-confinement method (Potts & Zdravkovic, 2001), to which the proportion of unloading before lining construction is prescribed and volume loss is a predicted value. Hence in the initial condition where no excavation has taken place, a value of $\lambda$=0 is applied whereas when the installation of the primary lining is complete, $\lambda$=1. The applied internal pressure reduction method was selected in order to simulate the ground relaxation. Hence, at all stages the internal pressure ($p$) applied, is a fraction of the

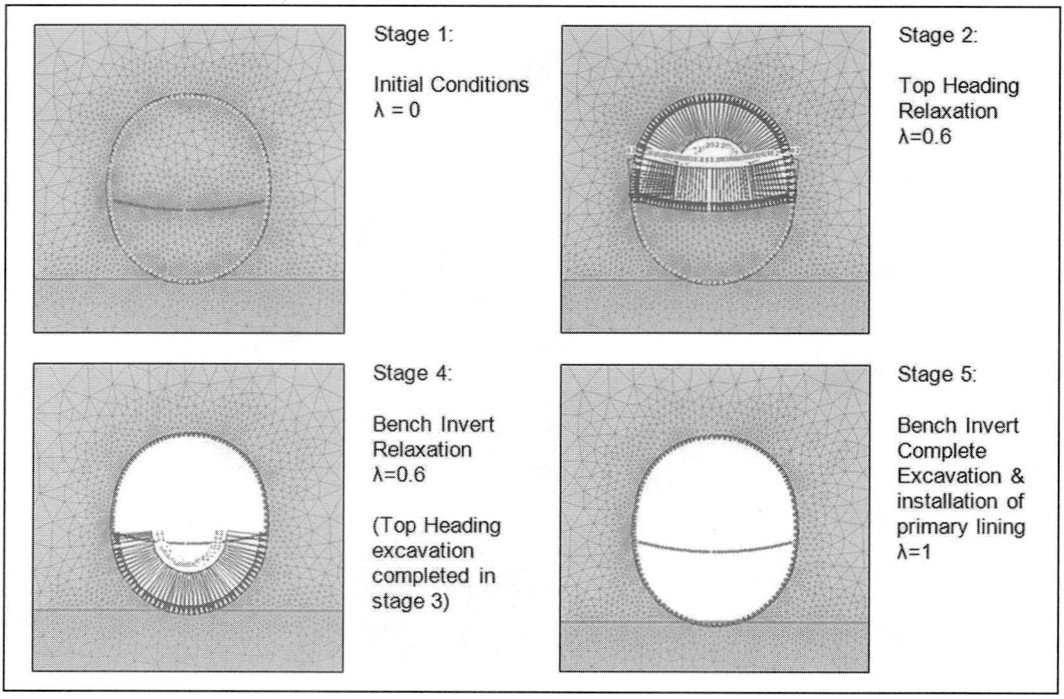


**Figure 4. Typical excavation and support stages in the 2D FE analysis**

in-situ pressure ($p_0$) that depends on the relaxation factor ($\lambda$) and is expressed by the following equation:

$$p = (1 - \lambda) \cdot p_0 \tag{1}$$

Prior to the installation of the primary lining, the pressure, $p$, is reduced from $p_0$ (in-situ state) to $(1 - \lambda) \cdot p_0$, with subsequent deformation taking place. When the excavation is completed and the support system is installed, the internal pressure is removed ($p = 0$) (see Figure 4).

A relaxation factor = 0.6 was used in the design of CP1, derived from an axisymmetric finite analysis model of equivalent excavation area, performed using Phase2 and later confirmed by the 3D finite element model.

During the parametric study presented in the next sections, it became apparent that the relaxation factor has a significant influence on the results, hence it has been refined for the optimised, final analyses.

## IN-TUNNEL MONITORING DATA

Monitoring cross sections were installed in several locations along CP1 tunnel, at a distance of 1.1, 6.1, 18.1, 22.0, 26.2 and 30.2. For the purpose of this report, the monitoring array at tunnel meter 18.100, installed approximately in the mid-length of the cross passage was selected. Being equally distant from the breakout area (junction with PL1 tunnel) and the headwall (end of CP1), this position was less affected by end effects, allowing for a comparison against the results of both 2D and 3D Finite Element analyses.

Monitoring of this cross section was carried out on a daily basis during the excavation of CP1. It has to be noted that due to the relatively small magnitude of the displacements, the results were slightly affected by the accuracy of the readings, but were in agreement with the results from the adjacent monitoring cross sections in CP1. The final monitoring data following the completion of the excavation of CP1 are presented in Figure 6. The crown monitoring point, M1, settled by approximately 7mm, as the upper side points M4 and M5. Smaller settlements were observed in the lower side points M6 and M7 (approximately 3 to 4mm). In terms of transverse displacements, points M4 and M5 exhibited approximately 5mm of inward movement, whereas points M6 and M7 moved by approximately 3mm. The 1.8mm transverse displacement at point M1 is attributed to the accuracy of the readings, as in the adjacent monitoring cross sections M1 also settled by approximately 7mm, but there was no transverse displacement.

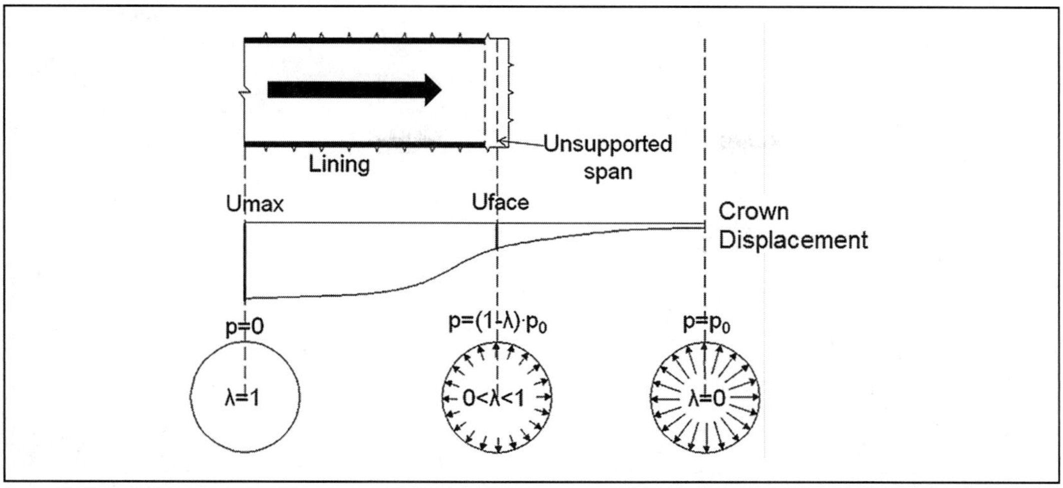


**Figure 5. Schematic of the convergence-confinement method (Spyridis, Gakis & Bedi 2012)**

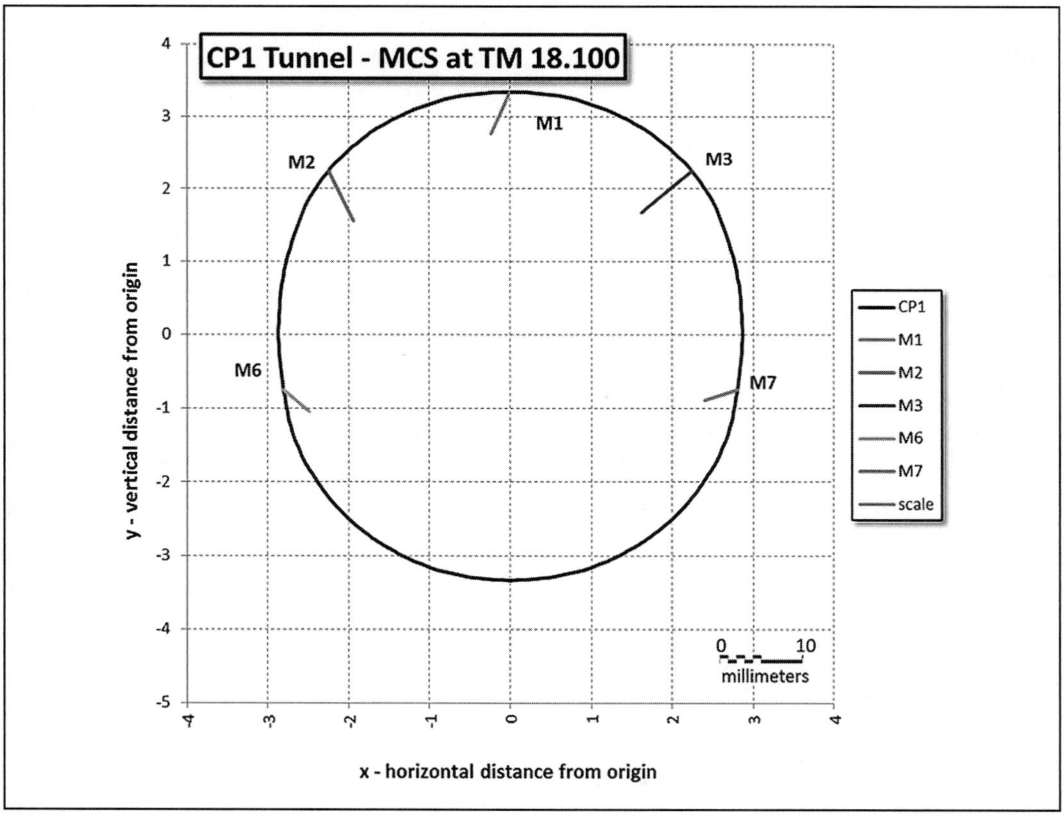


**Figure 6. In-tunnel monitoring data for CP1, monitoring cross section at 18.100 tunnel meters**

Table 3. List of the analyses performed during the parametric study and the corresponding results

| no. | | Analysis | Stratigraphy | λ | Soil paramters | $k_o$ | Unloading Wedge Stifness | In-Tunnel Deformation (mm) Settlements and Horizontal displacements to the right are positive | | | | | | | | | |
|---|---|---|---|---|---|---|---|---|---|---|---|---|---|---|---|---|---|
| | | | | | | | | M1 | | M4 | | M5 | | M6 | | M7 | |
| | | | | | | | | X | Y | X | Y | X | Y | X | Y | X | Y |
| 1 | 3D design analysis | | Design | - | Mixed | 1.2 | - | -1 | -3 | 2 | -2 | -4 | -3 | 4 | 0 | 3 | 0 |
| 2 | 2D | Design | Design | 0.6 | Unfaulted | 1.2 | - | -12 | -25 | 16 | -18 | -16 | -18 | 2 | 2 | -2 | 2 |
| 3 | 2D | Stratigraphy | Design | 0.6 | Faulted | 1.2 | - | -9 | -55 | 31 | -60 | -31 | -60 | 3 | 0 | -3 | 0 |
| 4 | 2D | Stratigraphy | As-Built | 0.6 | Faulted | 1.2 | - | -2 | -44 | 28 | -64 | -28 | -64 | 3 | 4 | -3 | 4 |
| 5 | 2D | $k_o$ | As-Built | 0.6 | Unfaulted | 1.2 | - | 0 | -12 | 12 | -18 | -12 | -18 | 2 | 2 | -2 | 2 |
| 6 | 2D | $k_o$ | As-Built | 0.6 | Unfaulted | 0.65 | - | 0 | -11 | 6 | -14 | -6 | -14 | 0 | 1 | 0 | 1 |
| 7 | 2D | $k_o$ | As-Built | 0.6 | Unfaulted | 1.35 | - | 0 | -16 | 14 | -24 | -14 | -24 | 2 | 2 | -2 | 2 |
| 8 | 2D | Unloading Wedge | As-Built | 0.6 | Unfaulted | 1.2 | E x 3 | 0 | -14 | 12 | -23 | -12 | -23 | 2 | 2 | -2 | 2 |
| 9 | 2D | Unloading Wedge | As-Built | 0.6 | Unfaulted | 1.2 | E x 4 | 0 | -14 | 13 | -22 | -13 | -22 | 2 | 2 | -2 | 2 |
| 10 | 2D | Unloading Wedge | As-Built | 0.6 | Unfaulted | 1.2 | E x 5 | 0 | -14 | 13 | -24 | -13 | -24 | 2 | 1 | -2 | 1 |
| 11 | 2D | Localized $k_o$ reduction | As-Built | 0.6 | Unfaulted | 0.65 | - | 0 | -11 | 1 | -8 | -1 | -8 | 1 | 2 | -1 | 2 |
| 12 | 2D | Relaxation factor, λ | As-Built | 0.3 | Unfaulted | 1.2 | - | -3 | -12 | 12 | -17 | -12 | -17 | 3 | 3 | -3 | 3 |
| 13 | 2D | Relaxation factor, λ | As-Built | 0.4 | Unfaulted | 1.2 | - | -3 | -12 | 12 | -18 | -12 | -18 | 3 | 1 | -3 | 1 |
| 14 | 2D | Relaxation factor, λ | As-Built | 0.5 | Unfaulted | 1.2 | - | -3 | -12 | 12 | -17 | -12 | -17 | 2 | 2 | -2 | 2 |
| 15 | 2D | Relaxation factor, λ | As-Built | 0.6 | Unfaulted | 1.2 | - | 0 | -12 | 12 | -18 | -12 | -18 | 2 | 2 | -2 | 2 |
| 16 | 2D | Relaxation factor, λ | As-Built | 0.7 | Unfaulted | 1.2 | - | 0 | -11 | 12 | -18 | -12 | -18 | 1 | 1 | -1 | 1 |
| 17 | 2D | Relaxation factor, λ | As-Built | 0.8 | Unfaulted | 1.2 | - | 0 | -10 | 10 | -17 | -10 | -17 | 1 | 1 | -1 | 1 |
| 19 | 3D Optimised | | Design | - | Mixed | 0.65 | - | 0 | -5 | 4 | -3 | -4 | -5 | 3 | 3 | -3 | 3 |
| 20 | 2D Optimised | | As-Built | 0.8 | Unfaulted | 0.65 | - | 0 | -6 | 5 | -8 | -5 | -8 | 0 | 0 | 0 | 0 |
| Actual displacements | | | | | | | | -1.8 | -7 | 5 | -7 | -5 | -7 | 3 | -4 | -3 | -3 |

## OPTIMISATION OF FE MODELS

Using the 2D and 3D finite element models of the design as a baseline, the influence of various parameters was investigated (parametric study). The purpose of this study was to assess the effect that these parameters had independently on the displacements of the tunnel section and to try and derive an optimised model that exhibited similar behaviour to the monitoring section at tunnel meter 18.1 of CP1.

### Parametric Study on the Influence of Various Parameters

The main parameters that were investigated are:

- The stratigraphy: (as assumed during the design stage and as observed through the excavation of CP1)—"design" was the stratigraphy used in the design models, based on the original site investigation data, "as-built" is the actual stratigraphy, as derived through the excavation of the tunnels.
- The relaxation factor, λ: a wide range of relaxation factors, varying between 0.3 and 0.8 was tested.
- The parameters of the soil: during the design phase, the "faulted" parameters were used, accounting for a potential degradation of the material properties in the proximity of faults. It was however proved that CP1 was not affected by faulting, as the fault was encountered in a different position than expected without intersecting CP1, hence "unfaulted" parameters were utilised in the majority of the parametric analyses.
- The at-rest earth pressure coefficient, $k_0$: the effect of $k_0$ was also investigated, carrying out analyses with values of 1.2 (design value), 1.35 and 0.65.
- The stiffness of the unloaded material: in most cases, stiff, over consolidated clays exhibit a stiffer response when unloaded. Using Mohr-Coulomb constitutive model, does not allow for this behaviour to be simulated. Therefore, an alternative method was used, applying an increased stiffness (3 to 5 times the original Young's modulus—"E ×3" to "E × 5") on a soil wedge, starting from the top-heading/bench interface, extending to the bottom of the model geometry with a 45 degrees angle.

Analyses 1 to 17 in Table 3, constitute the framework of this parametric study. These analyses were grouped according to the parameter that was investigated:

- 1–2: analyses carried out at the design state using 2D and 3D finite element modelling respectively.
- 3–4: effect of the stratigraphy, applying either the stratigraphy used in the design analyses, or the "as-built" stratigraphy.
- 5–7: effect of at-rest earth pressure coefficient, $k_0$.
- 8–10: effect of higher Young's modulus in the unloading wedge (as described above).
- 11: effect of reducing locally, assuming a zone of reduced $k_0 = 0.65$, around the tunnel excavation (3 radii from tunnel axis laterally) while the global $k_0$ remains 1.2, after Potts & Zdravkovic (2001).
- 12–17: effect of the relaxation factor.

### Back Analysed Model Results

The results of the analyses preformed in the context of the parametric study (analyses 3 to 17 of Table 3), were compared to the actual displacements measured in the monitoring cross section at chainage 18.1 (refer to Table 3 and Figure 6) and the impact of each parameter on the tunnel deformation was identified.

The outcome of this comparison is that $k_0$ the relaxation factor and the use of faulted/unfaulted parameters played a principal role. Thereafter, by varying the values of sensitive parameters identified in the parametric study both 3D and 2D models were optimised to fit the FE results to the actual measurements. The optimised model results are presented in Table 3 (analyses 19 and 20 respectively).

For the 3D model, a notable match was achieved by reducing $k_0$ from 1.2 to 0.65. For the 2D model, the optimised analysis used a relaxation factor 80%, $k_0$ of 0.65 and unfaulted soil parameters.

## CONCLUSIONS

The in-tunnel measurements of cross passage CP1, were evaluated utilising 2D and 3D finite element models, along with an extensive study on the influence of various parameters.

The main conclusions of this research suggest that:

- The soil in this geological environment exhibited a high ground relaxation prior to the installation of the primary lining. Additionally, the value of $k_0 = 1.2$ that was used in the design, overestimated slightly the magnitude of the displacements. $k_0 = 0.65$ gave results closer to the actual displacements.
- The use of faulted parameters for CP1 was slightly conservative. The application of unfaulted parameters proved to be more realistic in this study, as confirmed also by the actual excavation of the tunnel that did not encounter any fault or disturbed materials.

The 2D and 3D back-analysis showed good agreement with field data. The outcomes are very promising for future optimisation of two dimensional plain strain models. However, its application is problem specific. The relaxation factor estimated from an axisymmetric model requires further investigation. Lower $k_0$ analyses exhibited better match with in tunnel monitoring data. The results of this study may be adopted in future design for tunnelling in Lambeth Group formation.

## ACKNOWLEDGMENTS

The authors would like to acknowledge the contribution of BFK joint venture for the support and the guidance towards the completion of this study, Crossrail Ltd. for the motivation and for granting the permission for this paper to get published.

## REFERENCES

Potts, D.M., and Zdravković, L. (2001) Finite Element Analysis in Geotechnical Engineering: Theory and Application. London: Thomas Telford, 2001.

Simulia (1992) ABAQUS: User's Manual. [Providence, R.I.]: Hibbitt, Karlsson & Sorensen, 1992.

Spyridis, P., Gakis, A., Bedi, A. (2012). Numerical Modelling. Presentation to the BTS-YM, London, October 2012.

# Utility Protection While Tunneling in an Urban Environment for the DC Clean Rivers Project

**Peter Kottke and Joel Kantola**
Jacobs Associates

**Carlton M. Ray and William Levy**
DC Water

**ABSTRACT:** The District of Colombia Water and Sewer Authority is implementing the $2.6 billion Clean Rivers Project (DCCR) to control combined sewer overflows to the Anacostia and Potomac rivers. Currently, DC Water is designing and constructing the First Street Tunnel (FST), the third of four soft ground storage and conveyance tunnels being built as part of DCCR. The FST is located in an urban setting, beneath several of DC Water's water and sewer service lines. This paper addresses the technical and contractual measures taken by DC Water to protect its utilities that are within the influence zone of the FST tunneling operations.

## BACKGROUND

The DC Water and Sewer Authority (DC Water) collects and treats wastewater for the District of Columbia (DC) and the surrounding communities in Virginia and Maryland. The combined sewer collection system in DC carries both dry weather flow (sewage) and wet weather flow (storm water runoff combined with sewage). During heavy rain events, the sewer system is designed to discharge wet weather flows directly into Anacostia and Potomac rivers through combined sewer overflows (CSO). As part of a negotiated agreement with the U.S Environmental Protection Agency, DC Water is implementing a long term control plan, referred to as the DC Clean Rivers (DCCR) Project, to significantly reduce the amount of CSOs discharged into the rivers. The DCCR consists of four soft ground tunnel contracts totaling 12 miles (20 km). These tunnels will store and convey the wet weather flows to the Blue Plains Advanced Wastewater Treatment Plant for treatment prior to discharge into the Potomac River. The four tunnel contracts are shown in Figure 1, with pertinent information provided in Table 1.

The aforementioned tunnels are being mined through unlithified sediments from the Cretaceous Period, known locally as the Potomac Group (KP). In the tunneling locales, the KP consists of two formations: the overlying Patapsco/Arundel (P/A) and the underlying Patuxent (PTX). The KP (P/A) is predominantly stiff to hard silts and clays with occasional lenses of sand. The KP (PTX) is predominantly dense sands (silty/clayey) with occasional lenses of clean gravels and stiff clays. Above the KP, the stratigraphy is typically alluvial deposits from the Quaternary Period and/or fill to the ground surface. The FST tunnel horizon is entirely in the lower KP (PTX) formation, as seen in the soil profile provided in Figure 2.

DC Water is concerned about the impact of the soft ground tunneling operations beneath its existing water and sewer service lines. Several of these utilities are critical to the District's infrastructure, and in some cases an interruption of service cannot be tolerated. Identifying these critical utilities, and the measures required to protect them, is of paramount importance. The Design-Build process used for the tunnel contracts listed in Table 1, primarily the lengthy collaboration period with the shortlisted teams, helped form the contractual and technical approach to protecting utilities as this approach has evolved into what is currently being used on the FST.

## APPROACH TO UTILITY PROTECTION

The FST approach to utility protection is a multi-stepped process that includes the following:

- Identification of utilities potentially effected by tunneling
- As-built condition of existing utilities
- Ground loss assumptions
- Settlement and damage estimation
- Tier classification of utilities
- Construction Impact Assessment Report (CIAR)
- Payment allowance and incentive for best mining practice

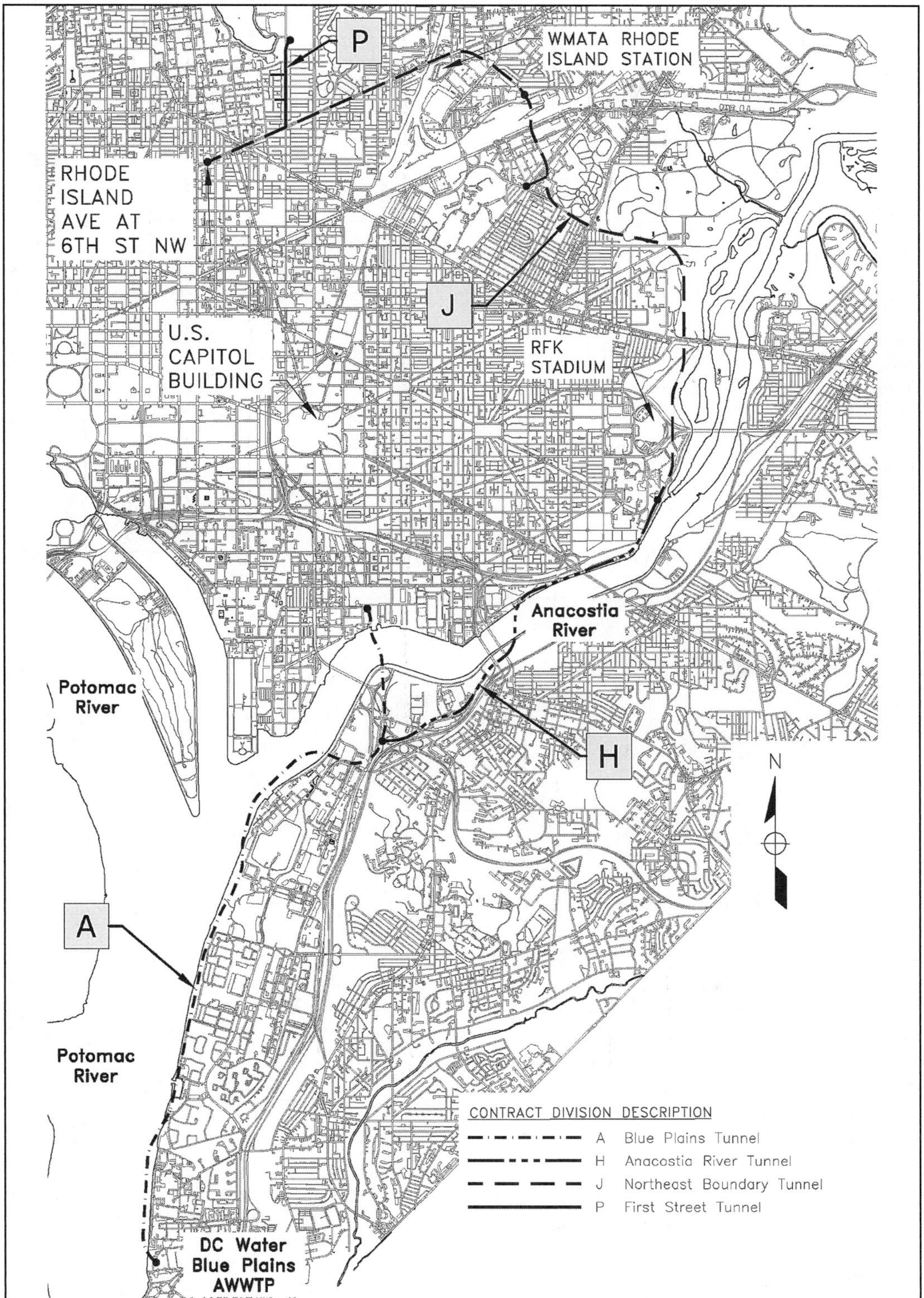


**Figure 1. Locations of DCCR tunnels within the District of Columbia**

**Table 1. DCCR tunnel contract summary**

| Contract Division and Name | Inside Diameter, ft (m) | Length, ft (m) | Number of Shafts (diversions) | Tunnel Inv. Depth, ft (m) | Ground Conditions | Procurement Method | Notice to Proceed | Contract Value ($ million) |
|---|---|---|---|---|---|---|---|---|
| Division A<br>Blue Plains Tunnel<br>(BPT) | 23<br>(7 m) | 24,225<br>(7,384 m) | 5(3) | 100–130<br>(30–40 m) | Clay & silty sand | Design-build | April 2010 | 330 |
| Division H<br>Anacostia River Tunnel<br>(ART) | 23<br>(7 m) | 12,485<br>(3,805 m) | 6(5) | 100–125<br>(30–38 m) | Clay & silty sand | Design-build & design-bid-build | June 2012 | 254 |
| Division P<br>First Street Tunnel<br>(FST) | 19<br>(5.8 m) | 2,700<br>(823 m) | 4(3) | 80–160<br>(24–49 m) | Silty sand | Design-build & design-bid-build | October 2013 | 158 |
| Division J<br>Northeast Boundary Tunnel<br>(NEBT) | 23<br>(7 m) | 26,700<br>(8,138 m) | 7(6) | 60–160<br>(18–49 m) | Clay & silty sand | Design-build | 2016 (plan) | Not bid yet |

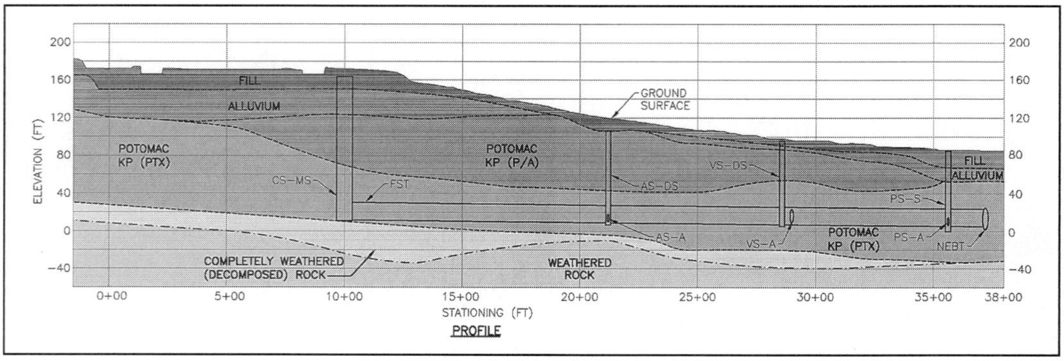


**Figure 2. Geologic stratigraphy of the First Street Tunnel**

## IDENTIFICATION OF UTILITIES POTENTIALLY AFFECTED BY TUNNELING

The first step when starting a utility protection program for a tunnel project is to estimate the generalized zone of influence (ZOI) of the tunnel and to begin identifying utilities that may be impacted by the tunneling. The initial ZOI estimate is used by the utility research group to identify utilities, and assign utility identification numbers (ID), discussed below. The initial ZOI is typically developed well in advance of detailed geotechnical knowledge about the tunneling environment and is based on generalized parameters such as material type, tunnel diameter, and tunnel depth. The ZOI can be estimated using methods developed by Peck (1969), O'Reilly (1982), or others.

The initial ZOI should be estimated such that the design ZOI will fall within it, but the estimate should not be overly conservative. Erring on the conservative side can result in significantly more time spent on identifying and researching utilities that will not be affected by the tunneling. For example, using an initial ZOI of 50 degrees rather than 45 degrees will tend to result in approximately 17 percent more utilities requiring identification and research efforts (Figure 3).

Several additional items require consideration during the utility identification phase, but are not the subject of this paper:

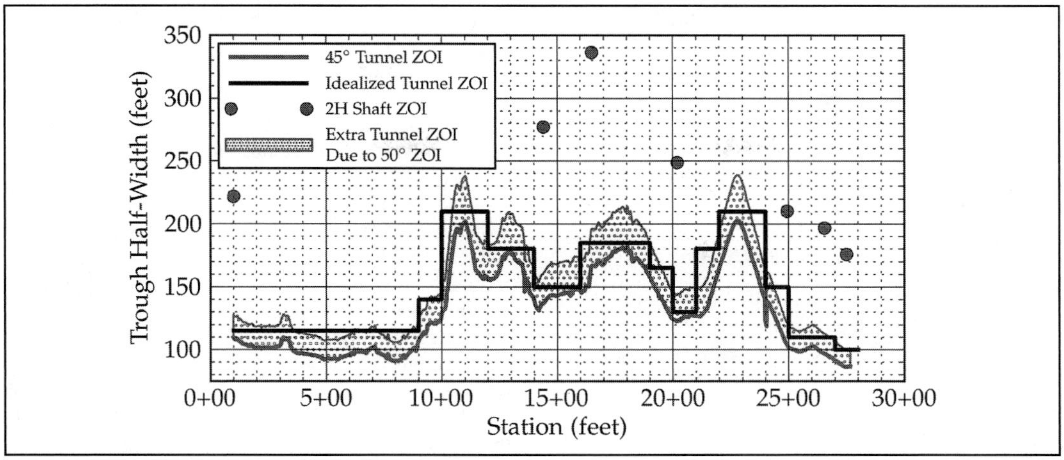


**Figure 3. Utility Research ZOIs for the NEBT**

- Deep shafts typically have a much wider ZOI than the tunnel.
- Construction staging areas, utility relocations, and near surface structures may be located outside the tunnel ZOI.

Additionally, in variable topographic or geologic environments, the ZOI estimate may vary widely, and creating a constant or stepped idealized ZOI is helpful to utility researchers for laying out their research zones (Figure 3).

## AS-BUILT CONDITION OF EXISTING UTILITIES

During the utility survey, the utility research group identifies utility locations, properties, and existing conditions through the use of:

- Counter maps (DC Water system maps)
- Geographical information system (GIS) databases
- Surveys
- Archival records including as-built drawings and specifications
- Potholing
- Archived inspection videos or reports

Each utility identified is assigned a unique ID based on the type of utility (storm, sewer, water, gas, electric, etc.). If useful, utility segments can be subdivided by appending a letter. For example, on the FST, the 90 in. (2,285 mm) portion of the Northeast Boundary Relief Sewer was identified as 010450A, and the 72 in. (1,830 mm) portion was identified as 010450B. The ID prefix (the first three numerals), "010," identifies the utility as a sewer and the last three numerals are the unique utility identifier. The full unique ID (010450A) is referenced throughout the impact analysis process and is also referenced in the Request for Proposal (RFP) documents and by the design-builder during design and construction. Separate ID prefixes are assigned depending on the structure type: sanitary sewer, storm sewer, water, electric, communication, buildings, bridges, walls, etc. The use of an unique ID provides an uncomplicated, unambiguous reference to a specific utility, which prevents confusion.

After assigning the utility ID the researchers review as-built information and populate a spreadsheet, which summarizes pertinent information such as size, depth, material type, etc. The utility IDs and spreadsheet will later be developed into the geotechnical instrumentation drawings included in the RFP documents. The as-built drawings, specifications, and inspection reports and videos are sorted by utility ID and are provided to the design-builder as non-contractual information.

After identifying water mains within the tunnel ZOI, DC Water performs water service redundancy studies along the tunnel alignment to ensure that temporary shutdown, or removal, of a service line does not cause a loss in water supply to customers. Based on this study, valves are identified that can isolate a service line within the influence zone of the tunnel, and then they are subsequently tested in the field. Based on the field testing, inoperable valves are replaced prior to tunneling beneath the utility.

Depending on the length and location of the tunnel, the utility research phase can be a long lead time item. Tunnels in urban environments, such as the FST, may have one major utility (wet utility, >24 in. [610 mm]) crossing every 160 ft (49 m) and parallel major utilities for the tunnel's entire length, whereas tunnels in suburban environments, such as the BPT and ART, may only have one major utility crossing every 330 ft (100 m) and no parallel utilities for any significant length.

After identification of utility locations and properties, the settlement characteristics of the tunnel need to be determined, based on detailed geotechnical investigation, and utilities need to be evaluated based on priority relating to material type and crossing orientation.

## GROUND LOSS ASSUMPTIONS

The preliminary assessment of the damage potential to utilities, used in the development of the RFP documents, is based on the amount of settlement expected during mining. Predicting settlement is a function of the assumed ground loss expected during tunneling. To minimize ground loss, the RFP documents require the tunnel to be constructed with a state-of-the-art earth pressure balance or slurry pressure balance tunnel boring machine (TBM) that is capable of, and is required to perform, the following functions:

- Apply and maintain adequate pressure to the freshly excavated heading
- Inject bentonite slurry, under pressure, into TBM overcut annulus
- Inject tail skin grouting into the annulus during advance
- Monitor, in real time, TBM performance (pressure, thrust, advance rate, etc.) and compare to ground loss indicators (settlement monitors, belt scales, etc.)

**Table 2. Ground loss assumptions**

| Contract | Notice to Proceed | Ground Loss Assumption |
|---|---|---|
| BPT | April 2010 | 1% |
| ART | June 2012 | 0.5% |
| FST | October 2013 | 0.5% |
| NEBT | December 2016 (planned) | ≤0.5% |

The assumed ground loss used for predicting utility damage for each DCCR contract is provided in Table 2. As shown, the ground loss requirement was reduced with each contract. Advances in controlling ground loss with pressurized face mining have rapidly accelerated over the last 10 years, and traditional assumptions on ground loss are constantly being revised based on field results from around the world. The DC Water Peer Review Board, based on its collective experience, advised the DC Water design team on the appropriate ground loss assumptions to use for each tunnel contract.

## SETTLEMENT AND DAMAGE ESTIMATION

Near surface settlements due to tunnel excavations have been studied by Peck (1969), Cording (1975), O'Reilly (1982), and others. Reported trough widths vary widely, potentially dependent on tunnel face volume loss and small strain behavior, but typically conform to a normal (Gaussian) probability density function (PDF) when viewed in the transverse (perpendicular to the tunnel axis) direction and a normal cumulative distribution function (CDF) when viewed parallel to the tunnel axis (Attewell and Woodman, 1982). Trough width parameter recommendations are typically provided for cohesive or cohesionless soils and do not consider the effect of a layered system.

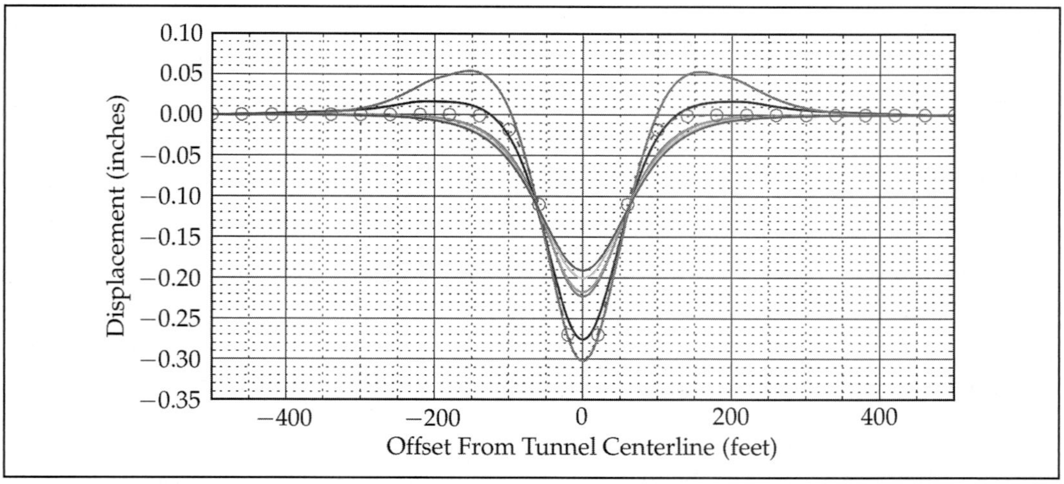


**Figure 4. Transverse vertical settlement profile for material property sensitivity analysis performed in FLAC2D during design of FST**

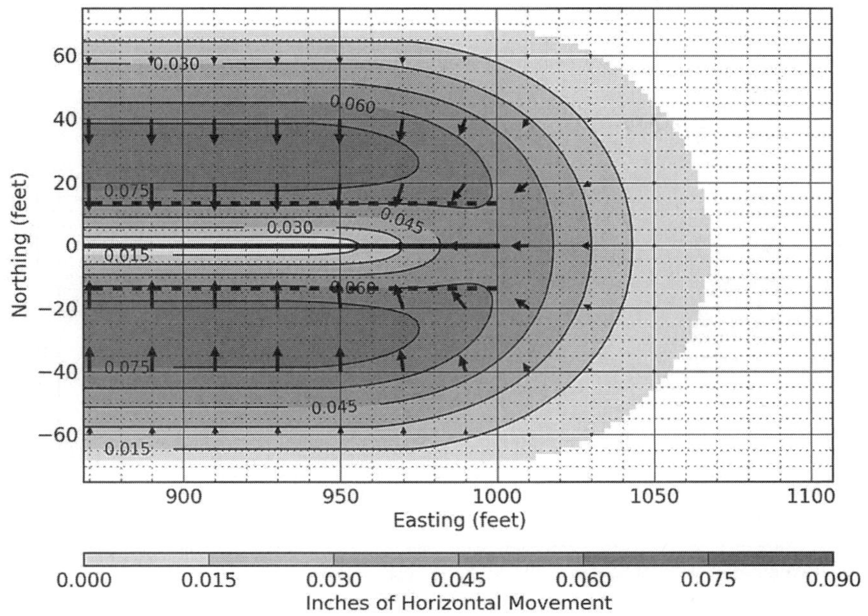


**Figure 5. Typical contours of horizontal movement (in.) for tunnel advancing left to right (solid black centerline, dashed excavation line; vectors show direction of horizontal movement)**

To better evaluate the effect of a layered geologic system on the width of the settlement trough, a FLAC2D model (Itasca, 2005) is developed to evaluate the small strain behavior of the soils. Sensitivity analyses are performed on material models (elastic and plastic), absolute stiffnesses, in situ stresses, and modulus reduction curves. In general, the settlement outputs from FLAC conform to the Gaussian distribution assumption, and the most conservative (narrowest) settlement trough is chosen for the design (see Figure 4).

When considering utilities that cross or are parallel to (and near) the tunnel centerline, the narrowest trough will tend to result in the most critical movements. Depending on the orientation of the utility with respect to the tunnel, either a PDF or CDF curve is used. If the utility is parallel to the tunnel and offset from the centerline, the CDF curve is scaled to account for reduced movements away from the centerline (Figures 5 and 6). The estimated settlements are used to evaluate the response of utilities based on utility material, joint type, and design.

Wet utilities in the DC area are typically constructed of one of five material types: brick and mortar, cast iron (CIP), ductile iron (DIP), prestressed concrete cylinder pipe (PCCP), or reinforced concrete pipe (RCP). Table 3 summarizes the evaluated failure mechanics for pipes of these material types.

An in-depth explanation of these mechanics is beyond the scope of this paper. Estimated strain in the brick and mortar utilities is compared to allowable strains published by Boscardin and Cording (1989) and Son and Cording (2005). Failure mode evaluation for the CIP and DIP pipes are the subject of Bracegirdle and Main (1996). RCP and PCCP are evaluated based on the as-built joint design and spacing. Additionally, PCCP joints are susceptible to joint lip shear and mortar cracking and need to be evaluated for longitudinal and bending strains to prevent joint lip shear and combined tensile strains to limit mortar cracking (Jeyapalan, 2014).

Pile-supported utilities require additional analysis steps because of variations in horizontal and vertical ground movements along the length of the pile. The pile type (frictional versus end bearing) as well as bearing elevation (above or below the ZOI) play a role in the response of utilities to settlements. Settlements at the utility elevation may be acceptable, but settlements at the bearing elevation of the pile may be significantly greater. Depending on the ground conditions and pile behavior, downdrag on the pile may result in additional utility displacement.

## TIER CLASSIFICATION OF UTILITIES

Based on the preliminary estimate of settlement and the allowable movement for a utility, a tier classification is assigned to each utility. The tier classification provides the mandatory required level of effort that the shortlisted teams are expected to carry in their

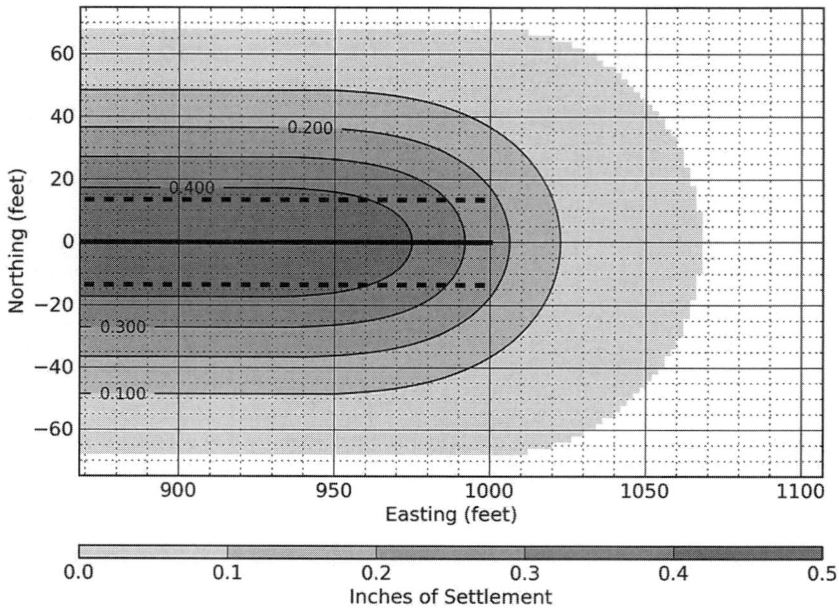


**Figure 6. Typical contours of vertical movement (in.) along and in front of tunnel advancing left to right (solid black centerline, dashed excavation line)**

**Table 3. Evaluated utility failure modes due to ground movement**

| | Material Type | | | | |
|---|---|---|---|---|---|
| **Failure Mode** | **Brick and Mortar** | **CIP** | **DIP** | **RCP** | **PCCP** |
| Joint Pull-Apart | × | ✓ | ✓ | ✓ | ✓ |
| Tensile Strain | ✓ | ✓ | ✓ | × | ✓ |
| Joint Rotation | × | ✓ | ✓ | ✓ | ✓ |
| Differential Movement | × | × | ✓ | × | × |
| Lip Shear | × | × | × | × | ✓ |

× = Failure mode typically not applicable to material type or utility design
✓ = Failure mode applicable to material type or utility design

cost proposal. The tiers are broken down into four levels:

- *Tier 1A*: Utilities likely to require protection measures prior to tunneling in order to prevent catastrophic conditions or collapse. Minimum protection measures are prescribed in the contract for Tier 1A utilities (e.g., replacement with new service line, strengthening, compensation grouting, ground improvement, etc.). A Detailed Assessment is required (described below).
- *Tier 1B*: Catastrophic conditions, or collapse, are unlikely to occur, but repairs are expected to be required after tunneling is completed (crack injection, mortar repair, etc.) A Detailed Assessment is required (described below).
- *Tier 2*: Utilities that are not likely to be damaged by tunneling. At a minimum, Tier 2 utilities require a Preliminary Assessment (described below) by the design-builder to ensure tunnel operation will not impact the utility.
- *Tier 2X*: Critical utilities to the District's infrastructure, but not likely to be damaged. Tier 2X utilities require a Detailed Assessment by the design-builder (described below) to ensure tunnel operation will not impact the utility.

The tier classification and allowable movements used in the RFP documents do not account for means and methods or the actual condition of the service lines, and are not applicable for more than a screening level of analysis. The RFP documents include provisions to require the winning team to revisit the tier classification and adjust allowable movements based on the mandatory Pre-Construction Condition Survey (PCCS) combined with the as-built records and a higher level of analysis. This process of revisiting the utility protection requirements in the RFP documents is referred to as the Construction Impact Assessment Report (CIAR).

## CONSTRUCTION IMPACT ASSESSMENT REPORT

The design-builder is responsible for generating a Construction Impact Assessment Report (CIAR) for all utilities within the influence zone. Prior to completing the CIAR, the design-builder is required to perform a PCCS of the utilities and incorporate the results into the CIAR. The RFP documents specify the type of PCCS required for each utility. Typically, for each utility these types are:

- Sewers: Manned entry and/or CCTV
- Critical watermains: Leak detection testing

The inspection results are used to account for the condition of the utility in the CIAR analysis. After construction, or after maximum settlement levels are exceeded, a Post-Construction Condition Survey, using the same inspection requirements as the PCCS, is required to determine if the utility is damaged.

The level of analysis performed in the CIAR depends on the tier rating of the utility. If the structure is a Tier 2 utility, the CIAR process requires a Preliminary Assessment, which includes:

- Use of empirical methods to:
  - Identify the influence zone of the tunnel
  - Estimate ground movements for the utility
  - Estimate damage to the utility using published allowable displacements
- Confirmation of the RFP tier classification
  - If the RFP tier classification cannot be confirmed and must be elevated to a Tier 1 utility, the design-builder is required to perform a Detailed Assessment.

A Detailed Assessment is required for fragile utilities identified during the PCCS and for any utility with a tier level of 1A, 1B, or 2X. The Detailed Assessment includes:

- Use of numerical modeling techniques (geomechanical and structural) and closed form solutions to:
  - Identify the influence zone of the tunnel
  - Estimate ground movements for the utility
  - Estimate damage to the utility using published allowable displacements
  - For Tier 1B structures, demonstrate the effectiveness of ground improvement, underpinning, strengthening, or other protection measures.
- Confirmation or revision of the RFP tier classification

## PAYMENT ALLOWANCE AND BEST MINING PRACTICES INCENTIVE

Allowing the design-builder to reassess the tier classification of utilities opens the door for a change in contract scope of work. To offset an increase in contract value, a payment item is added to the contract: "Tier 1A and 1B Allowance." The payment item covers the design-builder's unanticipated costs associated with implementing additional protective measures for utilities that were reclassified from Tier 2 or 2X to Tier 1A or 1B.

The payment item does not cover costs associated with the following:

- TBM configuration and the means and methods used to advance the tunnel heading that are capable of achieving a ground loss that is less than the assumed amount.
- Repair or replacement of Tier 1B, 2, and 2X utilities (as classified in the CIAR) when maximum settlement levels are reached or exceeded.
- Repair or replacement of Tier 1A utilities, above the costs of the original protective effort compensated by the Tier 1A and 1B Allowance Item. In other words, if the protective measure implemented by the design-builder fails to perform as designed, DC Water will not compensate the design-builder for the remedial work.

In order to give the design-builder an incentive to use best mining practices, such that protective measures are not necessary, the design-builder receives 30% of any unused portion of the Tier 1A and 1B Allowance upon Final Completion.

## CONCLUSIONS

Utility identification, settlement analysis, and utility protection are critical to the success and mitigation of tunneling impacts in urban environments. The utility impact analysis process benefits the owner, the design-builder and the public. By identifying and evaluating utilities during the RFP phase of design-build contracts, the owner shares risk with the design-builder and improves bid price by providing the bidder with accurate, realistic predictions of damage to existing utilities. The design-builder is able to use this information to accurately estimate the amount of resources required to analyze and mitigate damage to utilities. The risk is not wholly allocated to the owner's side because the design-builder is required to reevaluate damage estimations and utility categorizations prior to initiating tunneling. If recategorization is required, the bid price contains an allowance item to be used for contingency measures and recategorization. After final completion, if the design-builder has maintained best practices and settlements are within the RFP estimate, the design-builder can recoup a portion of the contingency pool. If the design-builder fails to maintain best practices and settlements are greater than allowed, the design-builder is responsible for utility repair.

While the public is typically not directly affected by tunnel construction activities, utility disruptions and street closures caused by settlements due to tunneling indirectly affect them and create a negative perception of tunneling. Properly identifying utility damage risks allows proactive mitigation to prevent damage and disruption.

## REFERENCES


Attewell, P.B. and Woodman, J.P. 1982. Predicting the dynamics of ground settlement and its derivatives caused by tunneling in soil. *Ground Engineering*, 15(8), 13–22.

Boscardin, M.D. and Cording, E.J. 1989. Building Response to Excavation-Induced Settlement, *Journal of Geotechnical Engineering, ASCE*, 115(1), 1–21.

Bracegirdle, A. and Main, R.J. 1996. A methodology for evaluation potential damage to cast iron pipes induced by tunnelling. *Geotechnical Aspects of Underground Construction in Soft Ground*, 659–664.

Cording, E.J. and Hansmire, W.H. 1975. Displacements around soft ground tunnels. In *Panamerican Conference on Soil Mechanics and Foundation Engineering*.

Itasca. 2005. FLAC Version 5.0371 software.

Jeyapalan, J.K. 2014. Tunnel and Shaft Impact Analyses for DC Clean Rivers Project to Protect Critical Pipelines Carrying Water and Sewage. NASTT, 2014, unpublished.

Mair, R.J., and Taylor, R.N. 1997. Bored Tunneling in the Urban Environment, Proceedings of the 14th International Conference on Soil Mechanics and Foundation Engineering, Hamburg, 2353–2385.

O'Reilly, M.P., and New, B.M. 1982. Settlements above tunnels in United Kingdom—Their magnitude and prediction. *Tunnelling 82, London, IMM*, 173–181.

Peck, R.B. 1969; Deep excavations and tunnels in soft ground. In *Proceedings of the 7th International Conference on Soil Mechanics and Foundation Engineering, Mexico City, State of the Art Volume*, 225–290.

Son, M. and Cording, E.J. 2005. Estimation of building damage due to excavation-induced ground movements. *Journal of Geotechnical and Geoenvironmental Engineering, ASCE*, 131(2), 162–177.

# Settlement Screening Analysis for the Baltimore Red Line

**James Parkes, Harald Cordes, Sotirios Vardakos, and Jason Kotova**
Parsons Brinckerhoff

**ABSTRACT:** The Baltimore Red Line is a proposed light rail transit line that includes a 3.4-mile Downtown Tunnel (DTT) segment through Baltimore City. The DTT consists of twin bored, 23-foot diameter tunnels, and five cut-and-cover underground stations. Construction of the DTT poses potential risk of excavation impacts to adjacent structures, which include masonry residential and commercial buildings, modern high rises, and historic structures. As part of a multi-stage evaluation, an initial screening assessment has been performed using analytical and empirical methods to estimate settlements and related impacts due to tunnel and station excavations. The screening assessment included detailed estimation of volume losses, comparison of the results of different methods, damage thresholds for buildings and utilities, and identification of shortcomings of the methods.

## INTRODUCTION

The Baltimore Red Line is a proposed light rail transit line for the Maryland Transit Administration (MTA) that will run from East to West through Baltimore City and County. The project includes the Downtown Tunnel (DTT) segment consisting of 3.4 miles of twin running tunnels, two portals, and five underground stations. The tunnels have an approximate outside diameter of 23 feet and will be mined using pressurized face tunnel boring machines (TBMs). The stations have two or three levels underground and have approximate plan dimensions of 66-feet wide by 285-feet long. Portals and stations will be built using cut-and-cover techniques with slurry walls utilized for both temporary support of excavation (SOE) and permanent structural walls. The DTT passes through an urban area including the Central Business District near the Inner Harbor as well as the Harbor East and Fell's Point neighborhoods. The DTT alignment is shown in Figure 1. Existing structures along the alignment vary considerably and include modern steel and concrete framed low- to high-rise structures to 19th century brick row homes, commercial buildings, and warehouses, as well as a dense network of utilities.

## GEOTECHNICAL CONDITIONS ALONG THE ALIGNMENT

The invert depth for the tunnels varies from 45 feet at the portals to up to 100 feet, with the majority of invert levels in the range of 65 to 85 feet in order to connect at the two- and three-level underground stations. Subsurface conditions along the DTT consist of variable fill and Coastal Plain sediments overlying crystalline rock of the Piedmont Plateau. Predominant rock types include igneous and metamorphic rocks consisting primarily of amphibolite and gneiss, with lesser amount of schist, marble, and other rock types.

Overlying the rock are Residual Soil and Transition Group materials, which consist of overburden derived from in-situ weathering of the parent bedrock. Residual Soil is completely decomposed and does not contain any relict rock characteristics. Residual Soils are only present in areas of limited extent. Transition Group materials are highly weathered and consist of soil-like materials that retain relict rock fabric and joints. These materials exhibit both soil and rock behavior. Residual Soil and Transition Group materials are very dense or hard based on SPT N-values.

Overlying the Residual Soil and Transition Group are Coastal Plain sediments consisting of Cretaceous and post-Cretaceous sediments. Cretaceous sediments are highly variable and include sand, gravel, and clay layers, although most of these sediments are granular with isolated interbedded clay layers. Cretaceous soils vary from clean to silty or clayey sands and gravels and are very dense.

The eastern portion of the DTT, from the Inner Harbor Station to the east portal, includes areas of in-filled marshes or reclaimed land. West of the Inner Harbor Station, the DTT runs through an upland area. Post-Cretaceous sediments are present in the eastern portion and are thicker in areas of reclaimed land. These sediments are similar to the Cretaceous sediments, although tend to be medium dense sands and gravels with occasional medium stiff to stiff interbedded clay. In areas of reclaimed land or former marsh, loose sands and soft organic silts are also present.

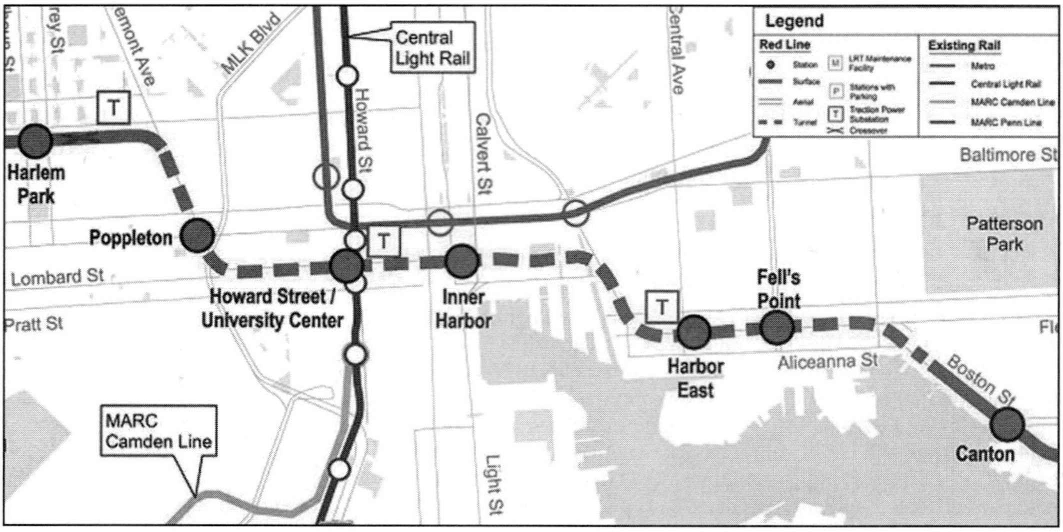


**Figure 1. DTT alignment location**

Fill overlies the post-Cretaceous sediments in the eastern section and the Cretaceous sediments in the western section of the DTT. Fill is highly variable in density and composition and includes brick, timbers, and other debris.

Groundwater levels are generally within 5 to 15 feet of the ground surface along the entire DTT. The Cretaceous soils are regional aquifers and a limited tidal influence is observed along the existing and former waterfront.

The DTT running tunnels encounter all of the materials described above except fill. The majority of the tunnels run through rock, Transition Group, and Cretaceous sediments. Mining conditions vary from full face conditions in rock, Transition Group, and Cretaceous soils, to mixed ground conditions consisting of a combination of these materials.

The station and portal excavations will encounter all of the materials described above. Some stations, such as Poppleton and Howard Street Stations, will encounter fill, Cretaceous soils, Transition Group, and rock. Other Stations, such as Inner Harbor and Fell's Point, will encounter post-Cretaceous sediments but not rock.

## EXCAVATION IMPACTS ON ADJACENT STRUCTURES STUDY

A settlement screening analysis has been performed for a Preliminary Engineering (PE) study on Excavation Impacts to Adjacent Structures (EIAS). The EIAS study was developed to assess ground deformations due to excavation of the tunnels and stations and the impact of those deformations on adjacent structures.

### EIAS Background and Assumptions

The EIAS study involved analytical, empirical, and numerical analyses. The EIAS study pertained only to movements caused by excavations. Groundwater drawdown and blasting impacts are not addressed. Catastrophic events, such as tunnel blow-ins or SOE failure, are not considered either.

### Staged Approach for Excavation Impact Assessments

The EIAS assessment follows a rational multi-stage process. This general process is applicable to both mined tunnel and cut-and-cover excavations. It consists of:

1. Perform a screening assessment using analytical or empirical methods to estimate the limits and magnitude of settlements. Damage assessment criteria are used to evaluate, whether adjacent structures may be at risk. The Stage 1 assessment is the focus of this paper.
2. Perform a detailed assessment of locations identified in Stage 1, including consideration for foundation loads. Numerical modeling is used for the Stage 2 analyses for the Red Line because it allows for a more detailed analysis and facilitates parametric studies. Results are used to further assess damage potential.
3. Perform detailed site specific analysis using soil-structure interaction and structural response. This considers the structure type, condition, and building stiffness to determine

how much movement the structure can tolerate.

4. Development of mitigation strategies based on Stages 1 through 3. These may include changing the location of an underground structure, ground improvements, structural improvements, and instrumentation and monitoring.

Stages 1 through 4 are generally performed in order, although in some cases a stage may be omitted (Stage 2 may progress to Stage 4 to assess mitigation measures) or an iterative approach may be used (Stages 2 or 3 may be revisited to assess the results of Stage 4).

### *Building Damage Criteria*

Stage 1 assessments are used as a screening tool to identify the zone of ground movements, estimate settlement contours, and classify structures into risk categories. Ground movements and deformation slopes using Stage 1 assessments are developed from analytical or empirical approaches, generally with spreadsheets, to allow quick assessment of a number of sections covering a large alignment area. Damage assessment criteria for Stage 1 assessments, presented in Figure 2 are simplified in order to assess a large volume of structures quickly. The damage assessment criteria are for masonry structures and are considered conservative for steel or concrete frames structures that can tolerate larger movements.

Generally, structures with "very slight" or "slight" risk did not warrant a Stage 2 analysis. Structures at "slight" damage risk will still require monitoring and may require minor repairs, such as crack filling and repointing, but are not anticipated to require significant mitigation measures. However, some judgment must be applied; for example, historic structures or structures in poor condition may be identified for subsequent Stage 2 analyses despite a low risk classification. All buildings located directly above the tunnels were considered to warrant a Stage 2 assessment because of the increased risk of mining directly underneath the structure. The results of Stage 2 assessments are evaluated using additional criteria, including horizontal strains and angular distortions.

### *Utility Damage Criteria*

Deformation thresholds for buried utilities have been developed based on references (Attewell et al. 1986; O'Rourke and Trautmann 1982), similar projects, and project-specific considerations. Deformation criteria are summarized in Table 1; values exceeding the thresholds indicate that a utility may be at-risk of damage. Both vertical and lateral ground movements are needed to assess utility damage potential.

Three modes of stress for buried pipelines are used to assess damage potential:

1. Straining of the pipe caused by flexural deformations that result in pipe rupture or intolerable deformation
2. Opening of joints due to rotation between pipe segments
3. Tensile pull-apart of joints caused by tensile axial movements along the pipeline

Deformation thresholds are reduced from published values for new pipelines to include consideration for pre-existing deformations/strains. Not all thresholds are applicable to all pipelines; joints opening and pull-apart are not considered for brick, cast-in-place concrete, or welded steel pipe. Joint pull apart is not considered for pipelines less than 8 inches in diameter because they behave like flexible pipelines (O'Rourke and Trautmann 1982). Cast iron pipes, which are generally water mains, are more critical because they provide water supply for fire fighters. Cast iron pipes are analyzed for joint deformation for diameters greater than 6 inches, and threshold values for cast iron pipes are very stringent because of the fire-life safety considerations. The threshold values in Table 1 are used for initial screening analyses; final design values may be revised based on evaluation by the utility owners.

## STAGE 1 SCREENING ASSESSMENTS FOR MINED TUNNELS

The Stage 1 assessment for mined tunnels used an analytical approach developed by Loganathan (2011) to estimate ground movements. Ground deformations and closed form equations are presented in Figure 3.

The analytical approach is used to develop settlements for "green field" conditions. Limits and contours of settlements are used to assess damage risk using the criteria in Figure 2 and Table 1. This approach assumes that settlements result from volume losses due to tunneling. Therefore, volume loss estimates are key to the assessment.

### Estimation of Volume Losses

Volume loss is the amount of soil excavated in addition to the theoretical tunnel volume, expressed as a percentage of the theoretical tunnel volume. Volume losses for TBM mined tunnels result from:

- Soils running, flowing, or collapsing into the face of an advancing tunnel.
- Closure of soils around the over-cut of the tunneling shield.

| Building Damage Classification After Berland (1995) and Mair et al. (1996) | | | | | Approximately Equivalent Ground Settlement and Slopes (after Rankin 1988) | |
|---|---|---|---|---|---|---|
| 1 | 2 | 3 | 4 | 5 | 6 | 7 |
| **Risk Category** | **Description of Degree of Damage** | **Description of Typical And Likely Forms of Repair for Typical Masonry Buildings** | **Approx. Crack Width (in)** | **Max. Tensile Strain (%)** | **Max Slope of Ground** | **Max Settlement of Building (in)** |
| 0 | Negligible | Hairline cracks | 0.004 | Less than 0.05 | | |
| 1 | Very slight | Fine cracks easily treated during normal redecoration. Perhaps isolated slight fracture in building. Cracks in exterior visible upon close inspection. | 0.004 to 0.04 | 0.05 to 0.075 | Less than 1:500 | Less than 0.4 |
| 2 | Slight | Cracks easily filled. Redecoration probably required. Several slight fractures inside building. Exterior cracks visible; some repainting may be required for weather-tightness. Doors and windows may stick slightly. | 0.04 to 0.2 | 0.075 to 0.15 | 1:500 to 1:200 | 0.4 to 2.0 |
| 3 | Moderate | Cracks may require cutting out and patching. Recurrent cracks can be masked by suitable linings. Brick pointing and possible replacement of a small amount of exterior brickwork may be required. Doors and windows sticking. Utility services may be interrupted. Weather tightness often impaired. | 0.2 to 0.6, or a number of cracks greater than 0.12 | 0.15 to 0.3 | 1:200 to 1:50 | 2.0 to 3.0 |
| 4 | Severe | Extensive repair involving removal and replacement of walls especially over door and windows required. Window and door frames distorted. Floor slopes noticeably. Walls lean or bulge noticeably. Some loss of bearing in beams. Utility services disrupted. | 0.6 to 1.0, but also depends on number of cracks | Greater than 0.3 | 1:200 to 1:50 | Greater than 3.0 |
| 5 | Very severe | Major repair required involving partial or complete reconstruction. Beams lose bearing, walls lean badly and required shoring. Windows broken by distortion. Danger of instability. | Usually greater than 1.0 but depends on number of cracks | | Greater than 1:50 | Greater than 3.0 |

Source: Adapted from Loganathan 2011.

**Figure 2. Analytical/empirical damage assessment criteria**

- Closure of the tail void space between the excavated diameter and the tunnel liner.

Volume losses vary by ground conditions and the alignment geometry. Higher losses occur when mining in multiple soil types and on curves, inclines, or areas with reduced tunnel separation (pillar). The DTT tunnel was subdivided into 60 sections, ranging in length from 75 to 1150 feet, based on mining conditions or alignment geometry.

Volume losses were estimated for "average" TBM control based on references, case histories, and engineering judgment. Volume loss estimates were intended to be reasonable yet conservative for the initial screening assessment. Estimates were initially developed for mining in full face conditions in one material type, then those values were adjusted for mining in mixed face conditions consisting of multiple material types. High and low values for "poor" and "good" TBM control were also developed for

**Table 1. Utility deformation thresholds for Stage 1 assessment**

| Utility Material | Dia. (in) | Allowable Joint Pull-Apart (in) | Allowable Joint Rotation (degrees) | Allowable Tensile Strain (microstrain) |
|---|---|---|---|---|
| Brick | All | N/A | N/A | 150 |
| Welded steel pipe (WSP) | All | N/A | N/A | 600 |
| Cast-in-place concrete (CIP) | All | N/A | N/A | 300 |
| Reinforced concrete pipe (RCP) | All | 0.40 | 0.25 | 300 |
| Terra cottta pipe (TCP) | All | 0.40 | 0.25 | 150 |
| Vitrified clay pipe (VCP) | All | 0.40 | 0.25 | 300 |
| Ductile iron pipe (DIP) | All | 0.40 | 0.25 | 600 |
| Cast Iron Pipe (CI) | 6 | 0.08 | 1.10 | 400 |
| | 8 | 0.08 | 0.90 | |
| | 10 | 0.08 | 0.80 | |
| | 12 | 0.08 | 0.70 | |
| | 16 | 0.08 | 0.50 | |
| | 20 | 0.07 | 0.40 | |
| | 24 | 0.06 | 0.30 | |
| | 30 | 0.06 | 0.20 | |
| | 40 | 0.05 | 0.15 | |

parametric study. Adjustment factors were developed for mining on curves, on inclines, and reduced pillar width. The resulting volume losses, summarized in Table 2, were used to estimate the amount of settlement for each section of the DTT.

The terms "good," "average," and "poor" are relative and subject to engineering judgment. "Average" workmanship relates to properly trained staff and equipment being properly used, such as timely adjustments to ground conditions, use of appropriate face pressures, monitoring of muck volumes versus theoretical excavation volume, proper tail void grouting, etc.

### Estimation of Settlements and Zone of Ground Movements

A settlement trough was calculated for each DTT section. A limiting value of 0.05 times the maximum settlement was used to define the limits of the trough since the Gaussian function approaches, but never equals zero. A trough was developed for each tunnel. Superposition was used to estimate overall ground displacement pattern, from which contours of settlement were developed to assess differential settlement and angular distortion at building locations. An example of ground displacement estimates is shown in Figure 4.

As a reasonability check, the maximum settlement for a single tunnel was calculated using the empirical method by Mair (1993). This method is more simplified but is based on observed ground movements and is accepted in industry practice. Results of the empirical method were generally found to be within 10% of the analytical results.

### Screening Assessment for Mined Tunnels

Settlement contours were developed and overlaid on the alignment plan as shown in Figure 5. Maximum and differential settlements were determined at each building within the zone of settlement and compared with the thresholds in Figure 2. This process was performed for the ground surface and one story below the surface to account for basements; "average" and "poor" TBM control levels were assessed. Structures at "moderate" risk or worse were identified for Stage 2 assessments. Results of the Stage 1 assessment are depicted in Figure 6.

Historic structures and structures that the tunnels pass directly underneath were automatically identified for Stage 2 assessments. Structures that differ from the assumed structure type inherent to the published damage criteria (Figure 2) were also identified for Stage 2 assessments. The published thresholds in Figure 2 were developed for low to midrise shallow bearing structures, generally with length to height ratios (L/H) of 1 or more. Two examples of structures identified for Stage 2 assessments are a tower with an L/H ratio of 0.12 and an existing underground tunnel. A total of eight structures were identified for Stage 2 assessments.

Utilities were evaluated in a similar fashion. Results indicated that approximately 44 percent of the utility crossings were considered at-risk. This does not mean that these utilities will fail; but rather additional analyses and/or mitigation measures, including monitoring, are warranted. Only highly critical utilities were identified for Stage 2 assessments, for a total of 8 additional Stage 2 assessments. Other utilities will experience similar levels of deformation, so

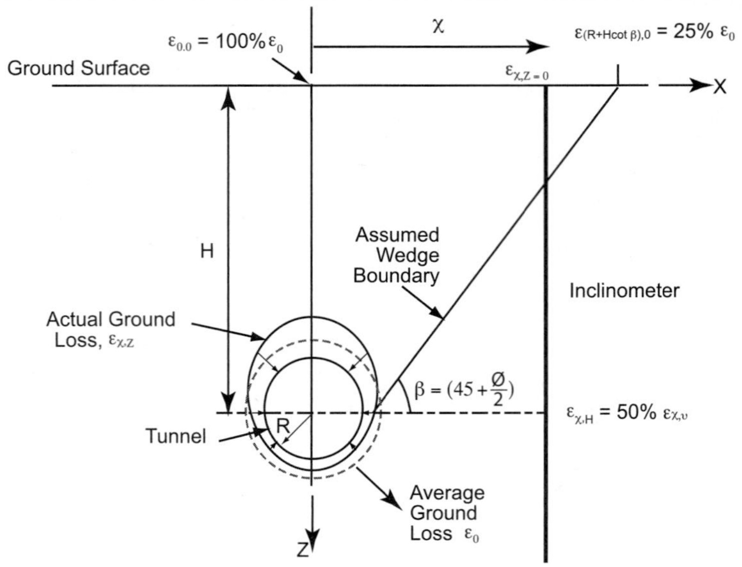


**Surface Settlement**

$$U_{z=0} = \varepsilon_0 R^2 . \frac{4H(1-\nu)}{H^2+x^2} . \exp\left\{-\frac{1.38x^2}{(H\cot\beta+R)^2}\right\}$$

**Subsurface Settlement**

$$U_z = \varepsilon_0 R^2 \left(-\frac{z-H}{x^2+(z-H)^2} + (3-4\nu)\frac{z+H}{x^2+(z+H)^2} - \frac{2z[x^2-(z+H)^2]}{[x^2+(z+H)^2]^2}\right)$$

$$\cdot \exp\left\{-\left[\frac{1.38x^2}{(H\cot\beta+R)^2} + \frac{0.69z^2}{H^2}\right]\right\}$$

**Lateral Deformation**

$$U_x = -\varepsilon_0 R^2 x \left[\frac{1}{x^2+(H-z)^2} + \frac{3-4\nu}{x^2+(H+z)^2} - \frac{4z(z+H)}{(x^2+(H+z)^2)^2}\right]$$

$$\cdot \exp\left\{-\left[\frac{1.38x^2}{(H\cot\beta+R)^2} + \frac{0.69z^2}{H^2}\right]\right\}$$

where:

$U_{z=0}$ = Ground surface settlement at transverse distance to centerline
$U_z$ = subsurface settlement at transverse distance to centerline
$U_x$ = Lateral deformations
$\varepsilon_0$ = average ground loss ratio
$R$= Radius of the tunnel
$H$ = depth of the tunnel below ground at springline
$z$ = depth below ground surface
$x$ = lateral distance from tunnel center line
$\beta$ = limit angle
$\nu$ = Poisson's ratio of soil

Source: Loganathan 2011.

**Figure 3. Ground deformation patterns and equations for closed form solution**

**Table 2. Summary of volume losses for Stage 1 assessment of Red Line DTT tunnels**

| Ground Conditions at Tunnel Face | Volume Losses, Straight Tunnel "Average" TBM Control | "Good" TBM Control | "Poor" TBM Control | Adjustment Factors for Alignment Conditions Vertical Curve | Inclined Alignment | Horizontal Curve (R > 900') | Horizontal Curve (R < 900') | Reduced Pillar Width |
|---|---|---|---|---|---|---|---|---|
| Cretaceous | 0.65% | 0.45% | 0.85% | +0.15% | +0.15% | +0.10% | +0.25% | +0.40% |
| Transition group | 0.75% | 0.50% | 1.00% | +0.15% | — | +0.10% | +0.25% | +0.40% |
| Post-Cretaceous + Cretaceous | 0.80% | 0.60% | 1.00% | +0.15% | +0.15% | +0.10% | +0.25% | +0.40% |
| Cretaceous + transition group | 0.90% | 0.65% | 1.15% | +0.15% | +0.15% | +0.10% | +0.25% | +0.40% |
| Cretaceous + transition group + rock | 1.00% | 0.75% | 1.50% | +0.20% | +0.20% | +0.15% | +0.30% | +0.50% |
| Transition group + rock | 1.00% | 0.75% | 1.50% | +0.20% | — | +0.15% | +0.30% | +0.50% |
| Rock | 0.20% | 0.10% | 0.30% | +0.05% | — | +0.05% | +0.10% | +0.10% |

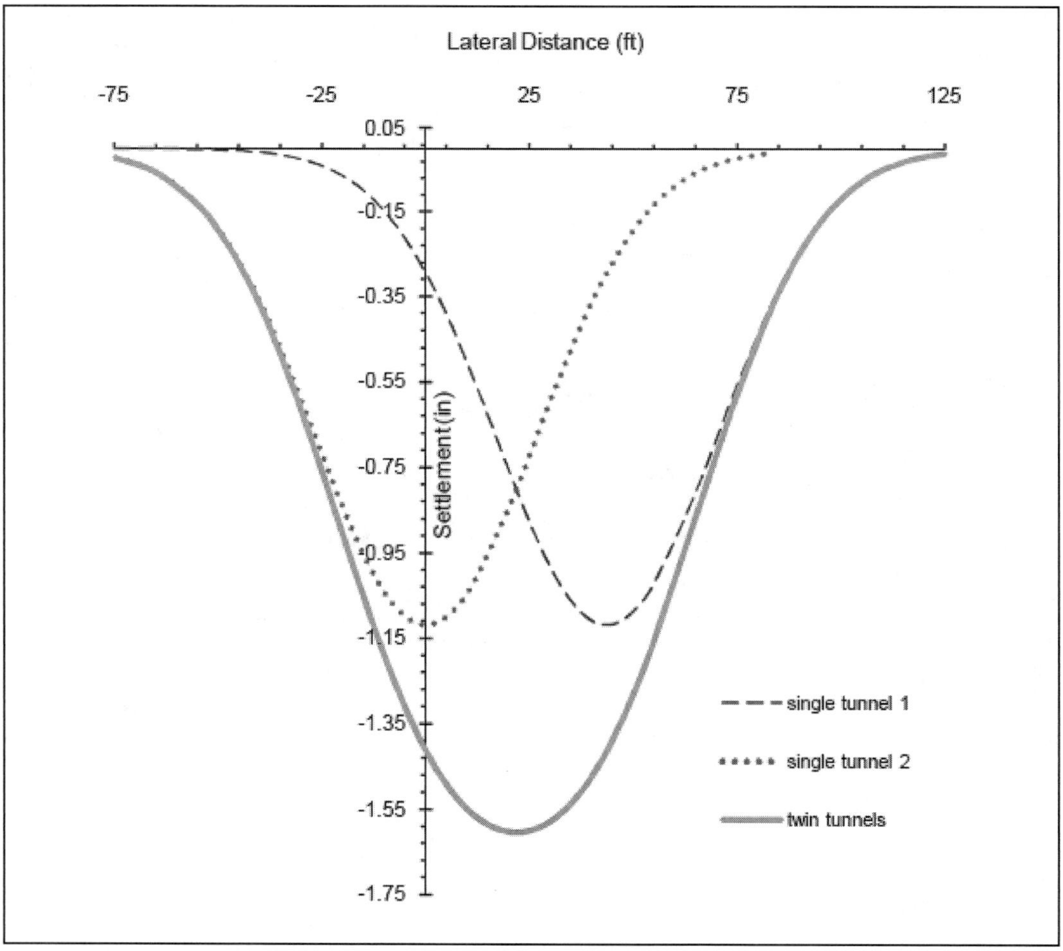


**Figure 4. Example of ground displacements estimated from analytical approach**

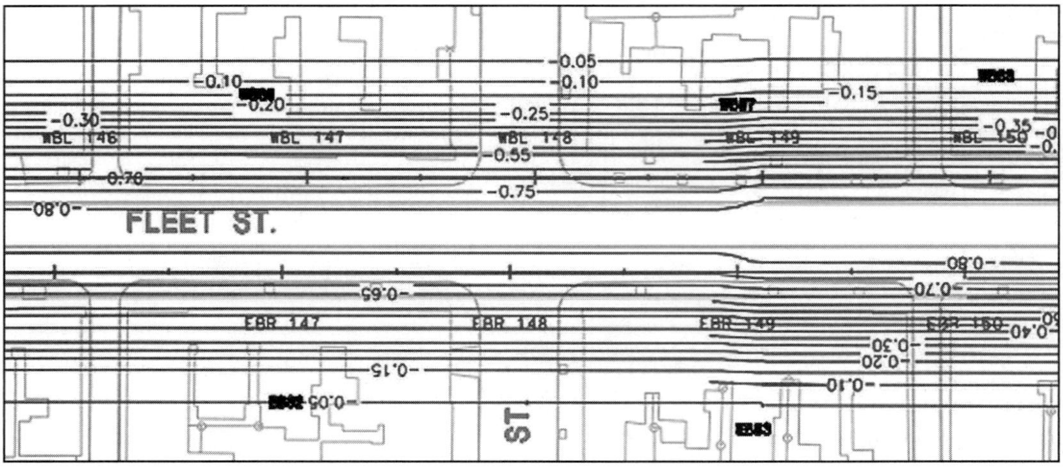

Figure 5. Example of settlement contours (inches) developed from analytical approach

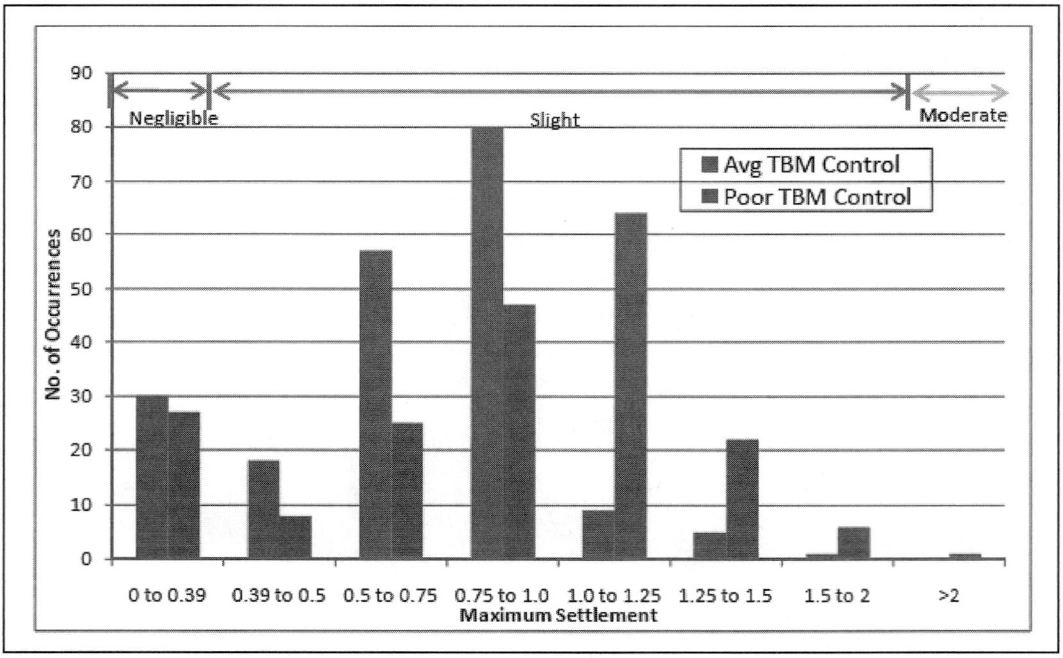

Figure 6. Histogram of maximum ground settlement at structures for the DTT Stage 1 assessment

those Stage 2 assessments will be applicable to other utilities in similar ground conditions.

### *Limitations of Screening Assessment*

The screening assessment relies on a number of simplifying assumptions. The following limitations are noted for the Stage 1 assessment:

- Applies to a "green field" condition, no external structure loads are accounted for.
- Applies to soft ground conditions; values for rock were included for consistency and completeness.
- Settlement trough volume is equal to the volume losses; no soil arching or volume expansion is considered.
- Movements caused by each tunnel are assumed to be equal.
- No effect from mining of the first tunnel is considered on the second tunnel.

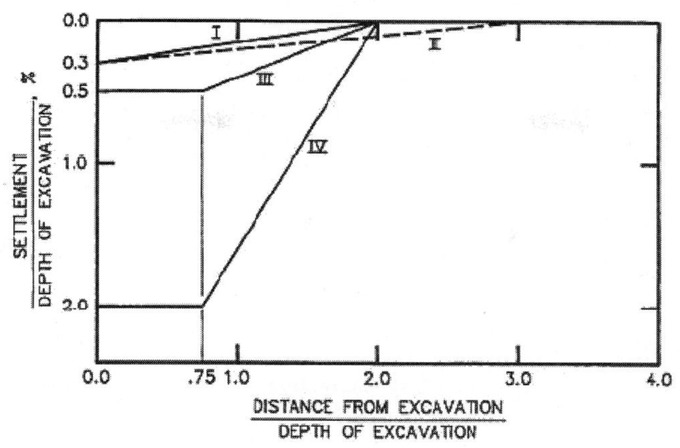


Source: AASHTO 2011.

**Figure 7. Empirical method for estimating settlements adjacent to braced walls**

- Damage criteria pertain to masonry structures on shallow foundations with L/H ratios of 1 or more.

## STAGE 1 SCREENING ASSESSMENTS FOR CUT-AND-COVER STRUCTURES

Similar to the mined tunnels, a Stage 1 screening assessment using empirical methods was performed to estimate ground movements due to excavation of the cut-and-cover stations and portals.

### Stage 1 Estimation of Settlements for Cut-and-Cover Structures

Cut-and-cover structures were designed using AASHTO (2011) specifications, which includes an empirical method for estimating movements caused by multi-level braced SOE systems. This empirical method was developed by Peck (1969) and expanded by Clough and O'Rourke (1990). Settlements are estimated as percentage of the excavation depth as shown in Figure 7.

Cut-and-cover structures will be built using slurry walls. Settlements will occur as a result of volume losses during slurry wall excavation. An empirical method, presented in Figure 8, was developed to estimate such settlements based on data presented in Clough and O'Rourke (1990). Total settlements at cut-and-cover excavations were estimated using superposition of the results from Figures 7 and 8.

### Results of Stage 1 Assessment for Cut-and-Cover Structures

Settlements adjacent to portals and stations were estimated to be as high as 2.1 to 4.6 inches, resulting in damage risk classifications of "moderate" to "very severe." Therefore, all cut-and-cover excavations require Stage 2 assessments. It is emphasized that these settlements were not considered highly accurate, but rather were taken as an indication that Stage 2 assessments were required. As of the date of this writing, limited Stage 2 cut-and-cover numerical modeling has been performed and results for settlements are in the range of 0.7 to 1.8 inches adjacent to station excavations. These limited Stage 2 results suggest that the Stage 1 results are not highly accurate.

### Limitations of Stage 1 Assessment for Cut-and-Cover Structures

The empirical screening for braced excavations is very simplified. The following limitations are noted:

- Settlements are based solely on depth of excavation, H. Deeper excavations result in greater settlements.
- Figure 7 was developed based on data for all wall types, including flexible walls such as sheetpiles and soldier piles and lagging. Slurry walls are stiffer and will experience less movement.

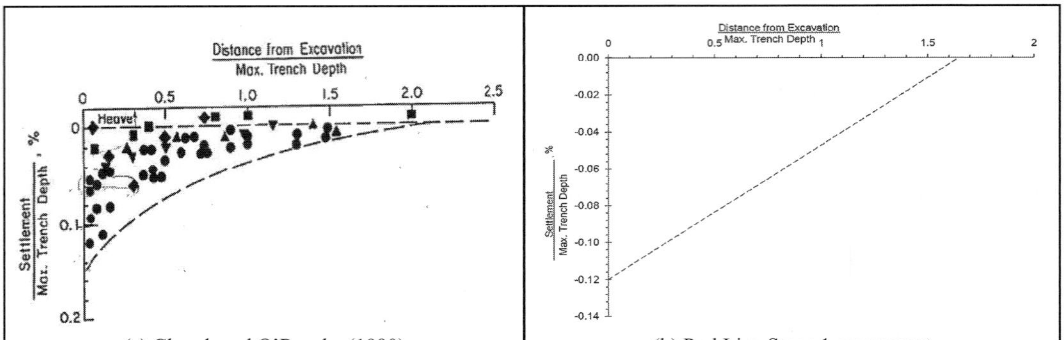


(a) Clough and O'Rourke (1990)

(b) Red Line Stage 1 assessment

**Figure 8. Diagrams for settlements induced by excavation of slurry wall elements**

- Figure 7 indicates maximum settlement for granular soils and stiff clays of 0.3% of H. Clough and O'Rourke (1990) indicate that for excavations in sands, stiff clays, and residual soils (comparable with DTT soils) tend to average about 0.15% of H, which is half of the Figure 7 maximum settlement.
- Figure 7 is based on uniform soil conditions in which the wall toe may deflect laterally. Slurry walls for DTT stations will toe into rock, so wall deflections may be less.
- As a comparison, the Charles Center Station excavation for the existing Baltimore Metro was 66 feet deep and used slurry walls in similar conditions. The Stage 1 approach would predict settlements up to 3.5 inches. Actual observed settlements at adjacent buildings were 0.3 to 1.0 inches (Zeigler et al. 1984).

## SUMMARY AND CONCLUSIONS

A rational multi-stage approach has been developed for evaluating excavation impacts on adjacent structures for the Baltimore Red Line. The approach involves a Stage 1 analytical or empirical screening assessment to identify at-risk structures, a Stage 2 detailed analysis to refine ground movement estimates, a Stage 3 assessment of soil-structure interaction and structural response, and a Stage 4 development of mitigation measures. A Stage 1 screening assessment has been developed based on published procedures and judgment for mined tunnels and cut-and-cover structures. The Stage 1 procedure, results, and limitations have been presented.

## ACKNOWLEDGMENTS

The authors would like to acknowledge Dick Flanagan of the General Engineering Consultant team who provided overall technical guidance and review of the analyses and reports, as well as the Program Management Team who provided review of the design memoranda for the MTA.

## REFERENCES

American Association of State Highway and Transportation Officials (AASHTO), 2011, *AASHTO LRFD Bridge Design Specifications*, Sixth Edition.

Attewell, P.B., Yeates, J., and Selby, A.R., 1986, "Soil Movements Induced by Tunneling and their Effects on Pipelines and Structures."

Clough, G. W., and O'Rourke, T. D., 1990, "Construction Induced Movements of In-Situ Walls," *Proceedings, ASCE Specialty Congress, Design and Performance of Earth Retaining Structures*, Cornell University, Ithaca, NY.

Loganathan, Nagan, 2011, *An Innovative Method For Assessing Tunneling-Induced Risks to Adjacent Structure,* PB 2009 William Barclay Parsons Fellowship, Monograph 25, Parsons Brinckerhoff, NY, NY.

Mair, R.J., 1993 "*Developments in Geotechnical Engineering Research: Application to Tunnels and Deep Excavations,*" Proceedings of Institution of Civil Engineers: Civil Engineering, 93. pp. 27–41.

O'Rourke, T.D. and Trautmann, C.H., 1982 "Buried Pipeline Response to Tunneling Ground Movements," Europipe '82 Conference, Basel, Switzerland.

Peck, R. B., 1969, "Deep Excavations and Tunneling in Soft Ground," Proceedings from the 7th International Conference on Soil Mechanics and Foundation Engineering, Mexico City.

Zeigler, E.J., Wirth, J. L., Miller, J. T., 1984, "Slurry Trench Wall Replaces Structure Underpinning," *International Conference on Case Histories in Geotechnical Engineering*, Vol. III, University of Missouri-Rolla, Rolla, Missouri.

# Protecting Existing Infrastructure Near Shafts

**Moussa Wone and Ronald Bizzarri**
DC Water

**Mohamed Younis**
Aldea Services LLC

**Lei (Kevin) Fu**
URS

**Brad Murray**
Jacobs Associates

**ABSTRACT:** Staging areas for shaft construction in urban and suburban localities can be congested with existing aboveground and belowground infrastructure, and often an assessment of the impacts from construction is required. Detailed engineering analyses of the magnitude of impacts from construction-related ground movements, surface surcharge loads, and vibrations can be used to identify structures that require special construction measures. This paper discusses the various means employed at the shaft sites for the DC Water Blue Plains Tunnel project for the protection of structures using diaphragm wall settlement cutoff, deep-soil mixing, and structure strengthening, and compares predicted versus actual performance.

## INTRODUCTION AND PROJECT BACKGROUND

The Blue Plains Tunnel (BPT) project is a component of a larger scheme to control combined sewer overflows (CSOs) to the District of Columbia's waterways, called the Long Term Control Plan (LTCP; See Figure 1). The LTCP is designed to meet the CSO control objectives of DC Water and to meet water quality standards in the District of Columbia. The BPT project consists of:

- Blue Plains Tunnel (BPT): An approximately 24,000-foot-long, 23-foot internal diameter (ID) tunnel.
- Blue Plains Tunnel Screening Shaft (BPT-SS): A 76-foot ID, 153.5-foot-deep screening shaft, also used to launch the TBM to mine the BPT, located on the Blue Plains Advanced Wastewater Treatment Plant (BPAWWTP) site.
- Blue Plains Tunnel Dewatering Shaft (BPT-DS): A 172-foot ID, 169 foot-deep dewatering pumping station shaft located on the BPAWWTP site.
- Bolling Air Force Base Drop Shaft (BAFB-DS): A 50-foot ID, 132 foot-deep overflow/drop shaft for connecting the Joint Base Anacostia-Bolling (JBAB) Overflow Structure to the BPT.
- Poplar Point Drop/Junction Shaft (PP-JS): A 55-foot ID, 124-foot-deep combined drop/junction shaft, located on District of Columbia government land.
- Surge Chamber and Approach Channel at PP-JS: The approach channel will connect the future Main Outfall Sewer Diversion Chamber (MOS-DC). The MOS-DC will be built over the existing modified twin sewers (under a different contract) for directing flow from the West Influent Sewer and East Influent Sewer to the PP-JS. A surge chamber will be constructed where the approach channel connects with the vortex generator in the PP-JS to facilitate handling the flow.
- Main Pumping Station Drop Shaft (MPS-DS): A 55-foot ID, 108.5-foot-deep drop shaft at DC Water's Main Pumping Station near 2nd Street and Tingey Street SE. This shaft will be used to convey diversions from CSO 9, 11A, 12, 13, and 14 diversion chambers.

To address concerns about the effects that BPT shaft construction would have existing infrastructure, a detailed assessment was performed for each shaft site to identify potential impacts and risks to existing structures and utilities. Site-specific plans were then developed to ensure that existing infrastructure was protected and that utility services would

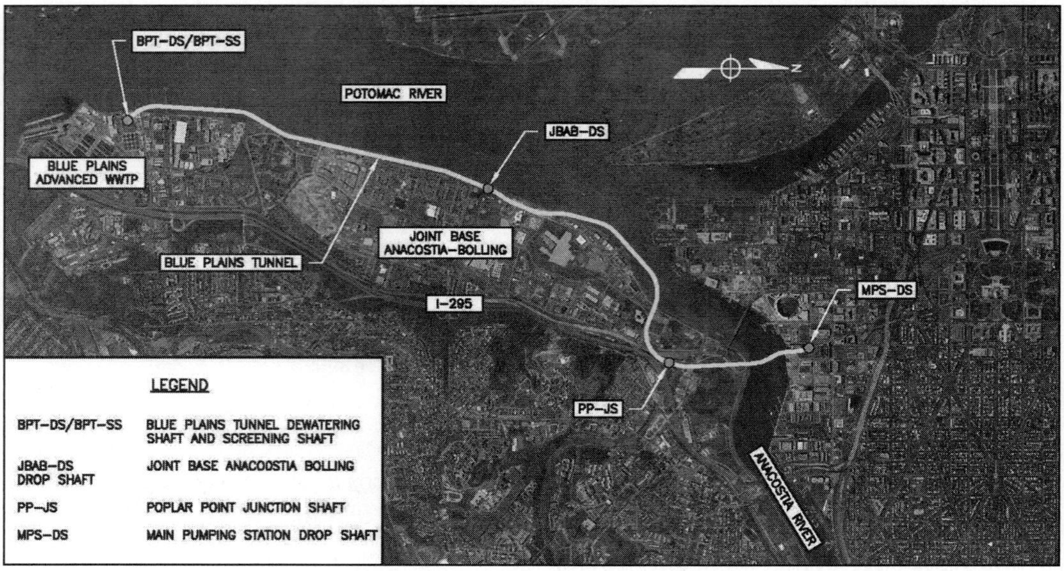


**Figure 1. Blue Plains Tunnel project site plan**

not be interrupted. The construction impact analysis approach and various means employed to protect these existing structures during shaft construction are described herein following a summary of the geologic conditions and shaft construction methods.

## GEOLOGICAL SETTING

The Blue Plains Tunnel Project site is located within the Atlantic Coastal Plain physiographic province. The Atlantic Coastal Plain comprises a wide belt of sedimentary deposits overlying crystalline bedrock. The natural deposits that underlie this region consist of Cretaceous-age formations, which are the oldest Coastal Plain sediments. Geologically recent alluvium is often present in the vicinity of historic and extant waterways. In some areas, the ground surface, as well as the course of the tributary stream, has been altered by placement of artificial fills.

The Cretaceous-age sediments, known collectively in this setting as the Potomac Group, consist of dense sands and gravels with variable fractions of fines, and very stiff to hard overconsolidated clays and silts. Although the clays and silts are typically very hard, the presence of slikensides (previous shear surfaces) often reduces the shear strength of soil mass. Man-made fills overlie the natural materials in many portions of the site. These fills were placed principally to develop various areas of the project site. The fills in such areas typically consist of soils that were locally available at the time of placement and, as such, are sometimes difficult to differentiate from undisturbed natural soils.

## SUBSURFACE CONDITIONS

The sites are covered with fill underlain by alluvium deposits. The alluvium is underlain by Potomac Group soils. Below is a summary of soil formation and groundwater conditions at the four shafts' construction sites.

### Strata

- **Fill deposits** include all types of locally derived soils and decomposed rock, and were placed by a range of methods including dumping and hydraulic filling. Fill contains fragments of construction debris, including wood, concrete, metal, cinders, and trash in varying amounts, and in some areas contains inclusions of organic materials. Fill is more frequently granular than fine-grained and will typically be saturated unless its position is above the normal range of groundwater levels.
- **Alluvium deposits** will be encountered below the fill at the shaft sites. Alluvium deposits, at many locations, consist of loose and soft silt, clay, and fine sand, with varying amounts of organic material. Sand and gravel deposits are also present at some locations, typically beneath the fine-grained material at the boundary with the underlying Potomac Group.
- **Potomac Group soils** underlie the alluvium deposits. These soils were previously overlain by several hundred feet of soils

deposits that were eroded away. The fine grained cohesive soils (Patapsco/Arundel (P/A) Formation) are usually hard and overconsolidated, and the coarse-grained cohesionless soils (Patuxent (PTX) Formation) are typically dense to very dense, owing to the prestress effect of the former overburden materials. The P/A Formation is not present at the MPS site. At this site The PTX Formation underlies the alluvium.

### Groundwater Levels

- **Groundwater levels** measured in the upper unconfined aquifer in the fill and alluvium ranged from approximately El. –9 to El. 3 feet. Piezometric levels measured in the semiconfined and confined water-bearing layers and lenses above El. –130 feet in the Potomac Group soils ranged from approximately El. –12 to El. –6. Piezometric levels measured in the confined water-bearing layers and lenses below El. –200 feet in the Potomac Group soils ranged from approximately El. –25 to El. –23 feet. Piezometric levels measured in fill and alluvium, as well as in Potomac Group soils, show clear tidal fluctuations of 0.5 to 1.0 foot in response to the approximately 3-foot tidal range in the Potomac and Anacostia rivers. Ground surface of the shaft sites varies from approximately El. 10 to El. 30 feet.

## SHAFT CONSTRUCTION METHODOLOGY

Support of the shafts consists of a circular configuration of slurry wall/diaphragm wall (D-Wall) and cast-in-place (CIP) concrete lining. The adopted methodology for the construction of the shafts is summarized below.

### BPT-DS and BPT-SS Shafts

- Install temporary guide walls.
- Excavate and install the diaphragm wall.
- Activate the depressurization well system to dewater granular soil layers below the shaft invert.
- Excavate BPT-DS/BPT-SS to the bottom of the temporary slabs.
- Construct the cast-in-place lining in the BPT-SS shaft up to a height of about 50 feet above the temporary slab level and provide an opening in the interconnection wall to connect the BPT-DS/BPT-SS shafts.
- Demolish the BPT-DS temporary slab after the tunnel boring machine (TBM) launch.
- Excavate to the bottom of the BPT-DS permanent slab.
- Construct BPT-DS slab and CIP inside liner.
- Demolish BPT-SS temporary slab and excavate below the bottom of the BPT-SS temporary slab and concrete collar.
- Excavate to the bottom of the BPT-SS permanent base slab.
- Construct BPT-SS base slab.
- Complete BPT-SS CIP lining to the finished elevation.

### BAFB-DS, PP-JS, and MPS-DS Shafts

- Physically locate all underground utilities and structures within the excavation area.
- Pre-excavate as necessary.
- Install temporary guide walls.
- Excavate and install the diaphragm wall.
- Excavate top portion of shaft.
- Flood the shaft and then excavate in the wet the balance to the following slab levels:
  - JBAB-DS: EL –123.00 feet
  - PP-DS: EL –115.25 feet
  - MPS-DS: EL –109.75 feet
- Pour the tremie slab.
- Construct CIP inside lining and CIP tunnel collar in the dry.
- Backfill shaft with flowable fill up to 8 feet above the tunnel crown and flood shaft to balance groundwater table prior to launch of the TBM.
- Complete TBM mining and dewatering of shaft.
- Remove flow fill and apply final treatment to exposed saw cut segmental lining along with annulus grout.

## NUMERICAL MODELING OF CONSTRUCTION IMPACTS TO ADJACENT STRUCTURES

### Key Structures Evaluated at Each Construction Site

Several key structures, classified as fragile or sensitive, exist along the BPT alignment. Of particular concern are some structures that required mitigation measures to maintain their stability and functionality, including the East Side Interceptor (ESI), Tiber Creek Sewer, and New Jersey Avenue Sewer. The analysis aimed at investigating the potential for loss of stability and/or functionality due to ground movement as a result of construction activities. The analysis also investigated potential mitigation measures to stabilize these structures, if required.

The numerical analysis consisted of 3-dimensional (3D) geomechanical modeling to estimate ground response to construction activities. The 3D geomechanical modeling was followed by structural modeling and analyses to estimate the impact

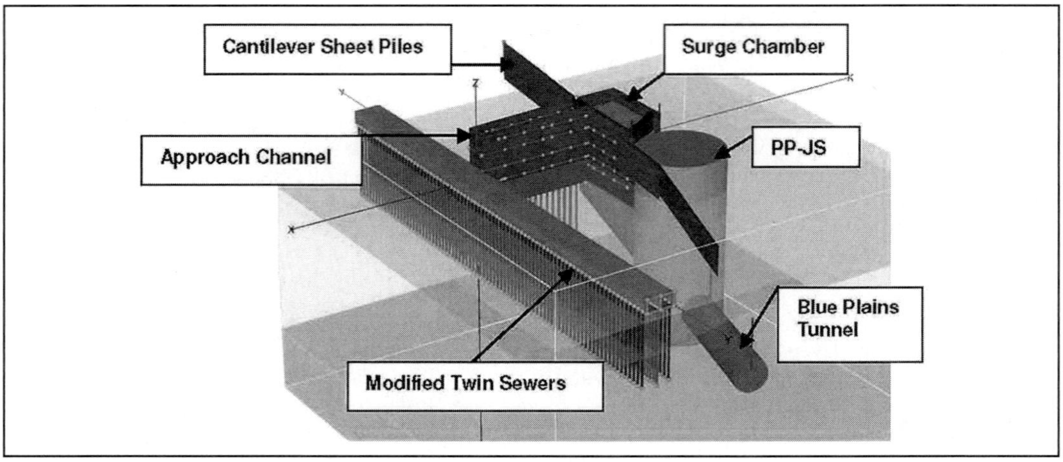


**Figure 2. 3D FE model at poplar point drop/junction shaft**

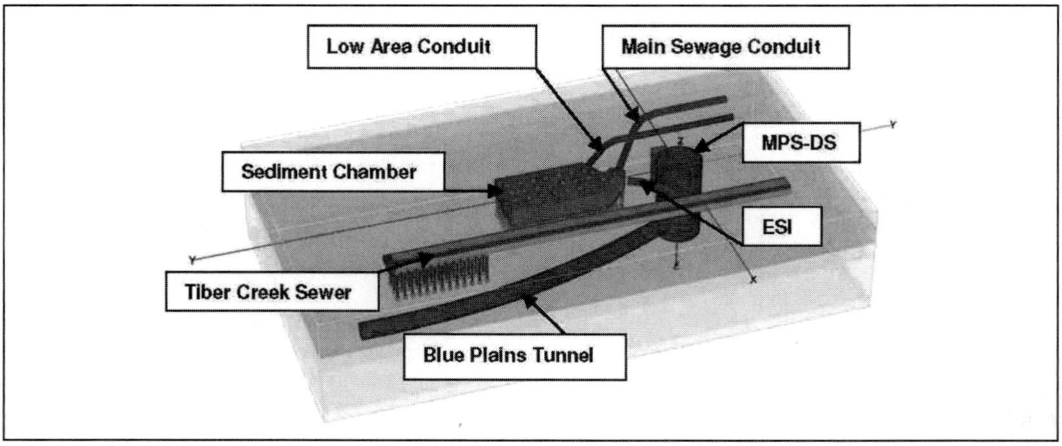


**Figure 3. 3D FE model at main pumping station drop shaft**

on existing structures based on the estimated ground responses predicted by the 3D geomechanical modeling.

The numerical models were developed using the Plaxis 3D software package. Plaxis 3D uses the finite element (FE) method to model ground deformation due to construction activities. It has some structural capabilities, which allow us to model the responses of underground structures and foundations to ground movements. See Figures 2 and 3 for FE models of the Poplar Point/Drop Shaft and Main Pumping Station Drop Shaft.

The lateral capacity of the pile foundation of the Tiber Creek Sewer was assessed using the L-Pile software package. The horizontal deformations at the pile foundations obtained from Plaxis 3D were used as input for the LPile analysis. Further to the analyses described above, structural checks and structural modeling using STAAD Pro software were carried out to evaluate the impact on the existing structures and pile foundation to determine whether protection/mitigation measures are required.

The Plaxis 3D model takes into consideration the existing conditions (including the structural conditions), construction activities, construction methodologies, and construction sequence. The next sections explain in detail our procedures for developing and executing the 3D model.

## DIAPHRAGM WALL FOR SETTLEMENT CUTOFF DURING SHAFT BREAKOUT

The BPAWWTP is among the largest wastewater treatment plants in the world (*Engineering News Record*, April 2, 2012), and the site of the initial launch of the Blue Plains Tunnel earth pressure balance tunnel boring machine (EPB TBM). The launch

**Table 1. Summary of model displacement results for the Blue Plains site**

| | Predicted Movements (in.) | | | | |
|---|---|---|---|---|---|
| **Construction Stage** | **Maximum Lateral Movement*** | **Tunnel Crown Settlement** | **Ground Surface Settlement** | **Tank Foundation Settlement** | **Tank Differential Settlement** |
| D-wall construction and shaft excavation | 1.0 | — | 0.23 | — | — |
| Tunnel excavation | 0.28 | 0.58 | 0.12 | — | — |
| At end of construction | 1.2 | 0.58 | 0.24 | 0.56 | 0.25 |

*The maximum movements from shaft and tunnel excavation stages do not occur at the same location.

was staged from the BPT-SS, which is in proximity to a number of critical treatment plant facilities, including several gravity thickener tanks and an operations control building. The tanks were constructed of reinforced concrete and are pile supported. The tips of the timber piles are shallower than the shaft and tunnel excavations and were predicted to be located within the zone of influence of the shaft tunnel excavations. The alignment of the tunnel is directly below one of the thickener tanks. DC Water had initial concerns that differential settlement could create a problem for the gravity thickener tank paddles.

Three-dimensional geomechanical/structural modeling of potential ground movement resulting from diaphragm wall panel construction, dewatering around the shaft, shaft excavation, and tunnel excavation (assuming 1% volume loss) was performed to assess the impact to the adjacent gravity thickener tanks and other structures. The results are summarized in Table 1. The most critical tank was predicted to experience a maximum of 0.25 inch of differential settlement and angular distortion of less than 1/2500. Structural calculations determined that the tank walls, floor slab, and piles would not be damaged by the predicted deformations.

The modeling results suggested that mitigation would not necessarily be required to protect the gravity thickener tanks, but risk assessment considerations had been given, even before the construction impact assessment modeling, that means were needed to ensure the stability of the ground during the shaft breakout. Ground improvement for shaft breakouts for soft ground tunnels is a common practice, and the initial plan for this project consisted of creating a block of overlapping jet grout columns immediately outside the Screening Shaft wall for this purpose. Further evaluations concluded that the Potomac Group clays would not be amenable to efficient jet grouting, and plans shifted toward creating a cutoff around the shaft breakout consisting of diaphragm walls. The diaphragm wall cutoff was approximately 60 feet long and 40 feet wide and was constructed using the same equipment used for the shaft walls. The wall thickness was 5 feet and had a compressive strength of at least 500 psi. Figure 4 shows the Dewatering and Screening Shaft site; Figure 5 illustrates the location of the cutoff wall at the site.

Monitoring of ground and structure movements was conducted using inclinometers, multipoint borehole extensometers, vibrating wire piezometers, and optically surveyed ground and structure monitoring points on the internal surfaces of the D-walls and adjacent structures. The instrumentation was laid out to enable monitoring of the shaft and tunnel excavations and permit a comparison between the predicted and actual movements. Actual movements are summarized in Table 2. It is noted that in-place inclinometer sensors installed in casings adjacent to and within diaphragm wall panels for monitoring horizontal movements during the panel construction and shaft excavations were found to be inaccurate and predicted erroneously high movements. Manual readings of the casing deflections using the traditional force-acceleration probe were generally consistent with the readings from the structure monitoring points on the diaphragm walls and confirmed the inaccuracy of the in-place sensors.

There were no incidents of significant ground loss during the diaphragm wall construction and shaft excavation, up to the temporary slab elevations, as well as the shaft breakout during tunnel construction. The monitoring results during shaft construction confirmed the prediction of the numerical modeling results. However, the actual settlements during the shaft excavation, including shaft breakout, show that a volume loss much lower than 1% was achieved by the EPB TBM. The resolution of the optical survey was not able to detect difference in settlements outside of the diaphragm wall cutoff during the breakout. At the time of this writing, approximately 1,120 feet of tunneling have been completed and the maximum settlement recorded by a MPBX, which is 6 feet above the tunnel crown, is about 0.2 inch.

## DEEP SOIL MIXING TO CONTROL D-WALL TRENCH INSTABILITY

Practical considerations about maintaining the stability of loose and soft fill and alluvial soils during the construction of MPS-DS diaphragm walls adjacent to an existing sewer were the primary motive

**Figure 4. Blue Plains dewatering and screening shaft site**

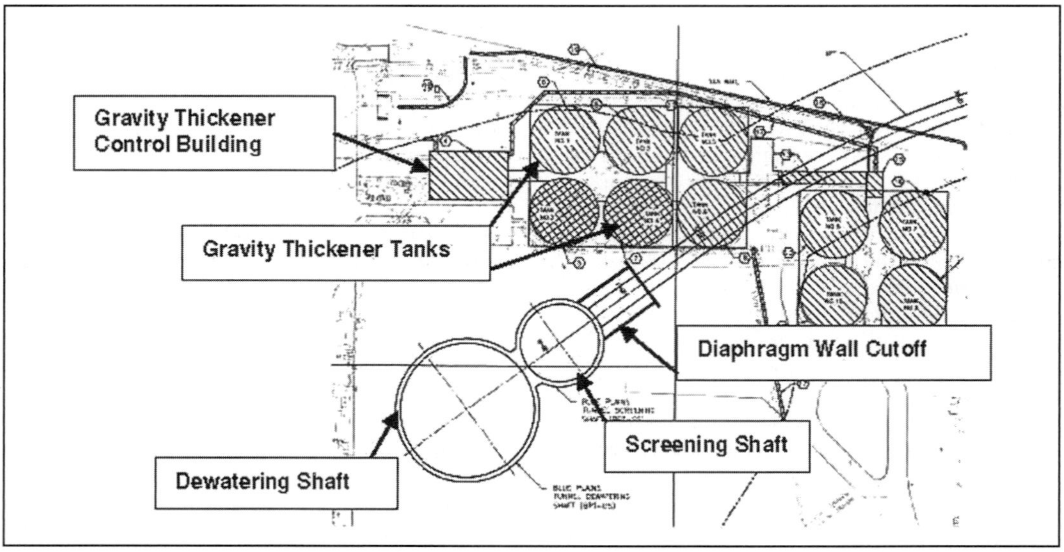


**Figure 5. Diaphragm wall cutoff at dewatering and screen shaft site**

in employing a ground improvement method at the Main Pumping Station site (Figure 6). The drop shaft is situated next to a number of wet and dry weather trunk and interceptor sewers that were constructed circa 1900. These are unreinforced concrete and masonry arch sewers with cross-sectional dimensions ranging from 4.5 feet to 26.5 feet. The East Side Interceptor (ESI), a 5-foot-diameter concrete-masonry sewer, runs tangentially to the drop shaft and is offset 7 feet to the outside of the closest diaphragm wall panel. Taking the ESI out of service during construction was not possible because of continuous high flows (which also made it impossible to inspect), and bypassing it was impracticable.

Numerical modeling of the shaft and tunnel construction–induced ground movements showed that the resulting sewer deformations should have little impact on the structures as long as trench stability was maintained. Deep soil mix columns were proposed between the ESI and MPS-DS, but the area of treatment was later expanded to form a closed ring ground improvement around the shaft to provide additional assurance of protection to the other sewers. Numerical modeling of the effect of ground improvement confirmed that the proposed ground

**Table 2. Summary of instrumentation monitoring results at the Blue Plains shaft site**

| Construction Stage | Maximum Lateral Movement (in.) | Tunnel Crown Settlement (in.) | Ground Surface Settlement (in.) | Tank Settlement (in.) |
|---|---|---|---|---|
| D-wall construction | 0.85* | — | 0.35‡ | <0.15‡ |
| Shaft excavation | 0.0.43* | — | 0.63† | <0.15‡ |
| D-wall construction and shaft excavation | 1.05* | — | — | <0.15‡ |
| Tunnel excavation | <0.15‡ | 0.13† | <0.15‡ | 0.1† |

* Measured by manual inclinometers.
† Measured by extensometers.
‡ Measured by structure monitoring points—reading noise is estimated to be ±0.15 in.

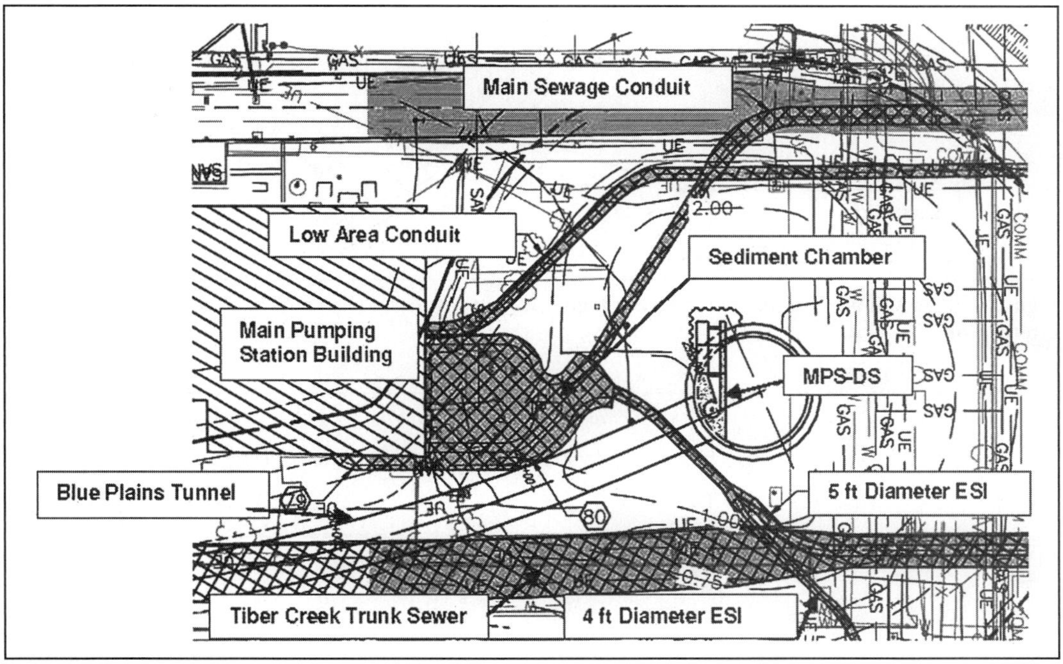


**Figure 6. Main pumping station site**

improvement would reduce overall ground movements affecting the other sewers.

The ground improvement design called for a minimum soilcrete compressive strength of 200 psi to a depth of 58 feet below ground surface, approximately 10 feet into the stiffer Potomac Group soils. Deep soil mixing was to be supplemented by jet grouting in areas where deep soil mixing was not feasible because of obstructions in the fill, although this did not become necessary during construction. Deep soil mixing was completed using a combination of 3-foot-diameter and 8-foot-diameter columns, as depicted in Figure 7. Some test samples of the in situ soilcrete adjacent to the ESI did not meet the required 200 psi strength, and those columns were reinforced with a #10 steel bar to compensate.

Horizontal movements of the ESI were monitored with inclinometers. The inclinometer casing deflection profile shown in Figure 8 confirms that stability of the fill and alluvial soils was achieved during shaft construction. The break in the deflection curve at El. 50 feet corresponds to the bottom of the ground improvement, indicating that the ground improvement had a mitigating effect in reducing the overall ground movements to less than 0.2 inch.

## STRENGTHENING EXISTING LARGE-DIAMETER SEWER

The Tiber Creek Trunk Sewer is a large-diameter, 100+ year old plain concrete horseshoe-shaped structure. It has a structural crack in the crown over a distance of less than 100 feet. The crack occurred

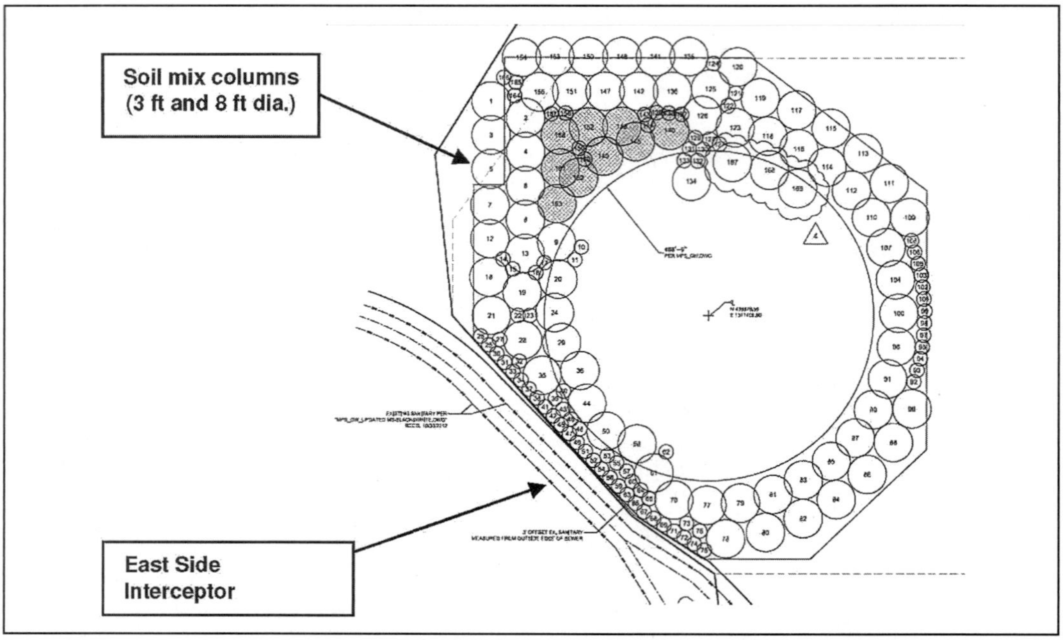


**Figure 7. Deep soil mix columns around the main pumping station drop shaft and between the east side interceptor**

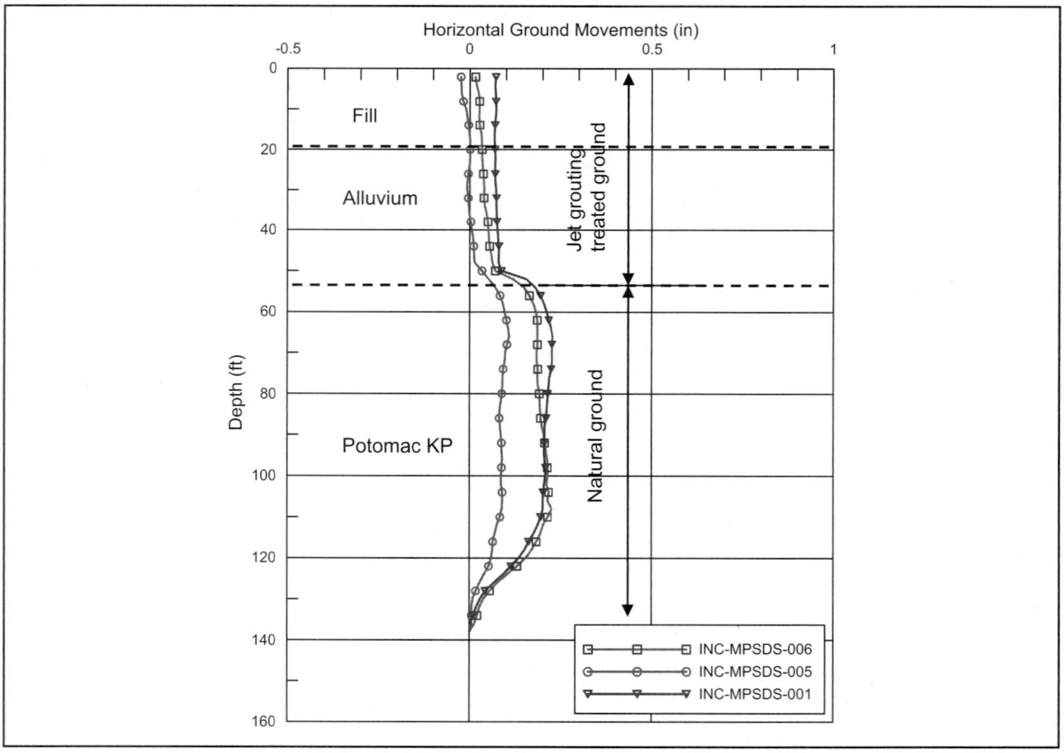


**Figure 8. Inclinometer casing deflection profiles adjacent to the main pumping station drop shaft**

**Figure 9. Steel ribs being installed in the Tiber Creek sewer**

in an area where the cross section transitions from a single arch with a 14-foot span to a double arch with a 12-foot span each. The maximum free span is 26.5 feet at the downstream end of the transition. The transition to the double arch is shown in Figure 9. The concrete is of variable quality. Cores taken from the walls supporting the arch show poor consolidation with strengths of just 800 to 1,500 psi. The base slab is also in poor condition in some locations. The arch concrete is of better quality, with strengths typically over 3,000 psi. It was anticipated that this sewer would be subjected to ground movements from the excavation of the Main Pumping Station Drop Shaft and the Blue Plains Tunnel. Furthermore, the sewer has just several feet of soil cover and would also be subjected to construction surcharge loads.

The following are factors considered in the analysis and design of protection/mitigation measures:

- Feasibility of reinforcing systems given the location and site restrictions
- Accessibility to the structures
- Minimizing alterations to the existing structures
- Maintenance of functionality during the implementation of the protection/mitigation measures (some sewer lines could not be taken off line at any given time)
- Schedule considerations and limitations (some of these protection measures were on the critical pass)

In addition to the factors described above, the following design criteria were considered critical during the design and analysis phase:

- Age and condition of existing condition: Modeling of existing concrete strength and cracks
- Deformation: Tolerable deformation for maintaining structure functionality

The existing Tiber Creek sewer at the MPS site was reinforced using W8×35 steel ribs. Figure 9 shows the steel ribs installed inside the Tiber Creek Sewer. Crack gauges were installed inside the sewers to monitor the existing cracks in the sewers. Strain gauges were attached to the steel ribs to monitoring the deformation of the steel ribs. The maximum increase in the crack width was about 0.03 inch.

## CONCLUSIONS

Instead of a one approach fits all solution for the protection of structures on the Blue Plains Tunnel Project, the combination of a staged assessment approach with customized, site-specific protection measures was used. This tailored approach helped to identify and develop protection measures for the structures at risk within the construction zone of influence. The measures taken include, among others:

- Designing and constructing cutoff diaphragm walls to permit a safe startup of the tunnel mining and reduce settlement risks for the gravity thickener tanks
- Implementing ground improvement (soil mixing) along with an instrumentation and monitoring system to maintain the stability of loose and soft fill and alluvial soils during diaphragm wall panel construction adjacent to a sewer that cannot be taken out service
- Physical reinforcement of the exiting Tiber Creek Sewer using steel ribs

The decision process for developing specific protection measures was based on numerical analyses and practical considerations. Construction monitoring data have, in many ways, confirmed the numerical analyses. Despite the ages and the conditions of the sewers and structures, the approach developed and implemented was very effective and so far no structural damage has been recorded.

# TRACK 2: DESIGN

## Session 4: Water and Wastewater Conveyance

John Bednarski, Chair

# Fixing the Leaks of the Rondout-West Branch Tunnel: Repairs and a Bypass Tunnel

**David Crouthamel and Dan Van Roosendaal**
Jacobs Associates

**Edward (Ted) Dowey and Ivan Zgaljic**
New York Department of Environmental Protection

**ABSTRACT:** The Rondout-West Branch Tunnel (RWBT) carries roughly half of New York City's drinking water from upstate watersheds to the Westchester County reservoirs. The tunnel is leaking up to 35 million gallons per day in two different locations. This paper details the design of the Bypass Tunnel that will eliminate the Roseton leak, and the design of the in-tunnel repairs for the Wawarsing leak. Major challenges for the construction of the new Bypass Tunnel include tunneling through fractured and faulted ground while managing up to 800 feet of head (24 bar) of external water pressure, constructing a tunnel lining to withstand 1,200 feet (36 bar) of internal pressure, and connecting the Bypass Tunnel to the RWBT in a minimum amount of time. Other challenges include the design of an inspection program for the entire 45-mile tunnel length and planning an in-tunnel grouting program to repair the Wawarsing leak.

## INTRODUCTION

New York City (the City) uses 1.1 billion gallons of water every day. The water comes from three upstate watersheds and travels by gravity to the City through three aqueducts operated by the New York City Department of Environmental Protection (DEP). The Delaware watershed is located west of the Hudson River and extends as far as 125 miles from the City. The Rondout-West Branch Tunnel (RWBT) brings water from the Delaware watershed to the West Branch Reservoir east of the Hudson. The RWBT is a 45-mile-long deep rock pressure tunnel, is 13.5 feet in diameter, has a nominal capacity of 890 million gallons per day (MGD), and was put into service in 1944. This tunnel carries about 50% of the City's water.

The RWBT is leaking on the order of 15 to 35 MGD in two different locations. The Roseton leak is at the low point of the tunnel under the Hudson River. The other leakage area is 21.5 miles west of Roseton under the Rondout Valley in the Town of Wawarsing. Because of the separate locations and different characteristics of the leaks, two unique repairs are being pursued.

## EXISTING CONDITIONS IN THE RWBT

The profile of the RWBT is shown in Figure 1. The RWBT is a gravity-driven pressure tunnel controlled by six valved, influent lines at the Rondout Effluent Chamber (REC), and includes 11 shafts. Shaft 6 is the low point of the tunnel on the east end of the Hudson River crossing and is equipped with dewatering facilities. The minimum ground cover is about 330 feet near Shaft 9, and the maximum is about 2,400 feet near the Shawangunk Mountains between Shafts 2A and 3. The profile also shows the hydraulic grade line of the tunnel under typical operating conditions. The internal operating head within the tunnel is approximately 600 feet greater than the external hydrostatic head at Roseton. At Wawarsing, the internal head exceeds the hydrostatic head by approximately 380 feet.

The 45-mile-long RWBT traverses four geologic provinces and includes several rock types—ranging from sandstone, shale, and limestone of the Rondout Valley Province (west end of the tunnel); to shale, conglomerates, limestone, dolomite, and quartzite in the Shawangunk-Range and Hudson-Mohawk Lowlands Provinces (west, central and Hudson River crossing); to gneiss of the Highlands Province (east end of the tunnel).

Most reaches of the RWBT were originally constructed in fair to good ground conditions with drill-and-blast methods utilizing several simultaneous contracts on different headings. The typical excavated tunnel diameter was about 16 feet. Temporary support, when required, consisted of steel ribs with partial steel channel lagging and a protective coating consisting of gunite applied to certain areas of slabby or deteriorated rock. For most of the length of the RWBT, the final lining consists of an unreinforced

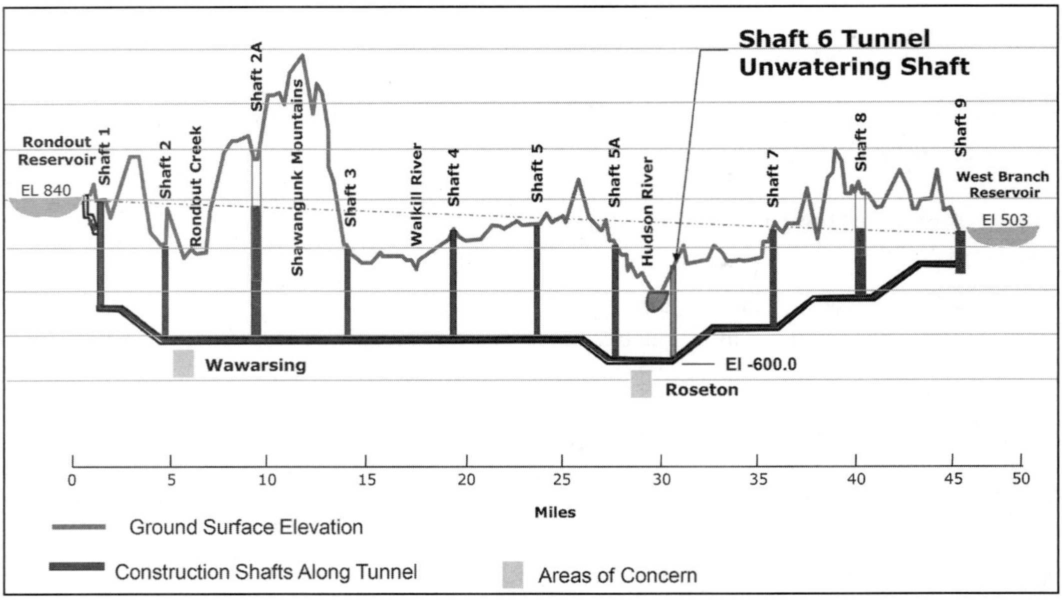


**Figure 1. RWBT profile**

concrete lining designed to take the full groundwater head upon dewatering (NYC-BWS, 1942). A typical section of the final lining is shown in Figure 2. The design of the tunnel relied on intimate contact between the lining and the surrounding rock to resist internal hydrostatic pressures. Accordingly, construction records indicate that systematic contact grouting was used behind the concrete lining to ensure that contact between the lining and the ground was achieved.

At Roseton, the tunnel was excavated through the Wappinger Group, which is a slightly metamorphosed dolomitic limestone or dolomite that has been faulted and gently folded. Faulted zones are highly fractured and weathered. In one reach, known as the "major fault zone," inflows of up to 1,800 gpm were encountered during construction.

At Wawarsing beneath the Rondout Creek Valley, the tunnel was excavated through a series of highly dipping sedimentary rocks composed of limestone, sandstone, and shale, which are faulted along the bedding and are directly connected to the overlying unconsolidated aquifer in the Rondout Creek Valley. Records show that inflows of up to 9,000 gpm were encountered in this faulted area during construction. To seal off these significant water inflows, in the Roseton and Wawarsing areas, a special steel-concrete composite "interliner" (Figure 3) was used in the faulted zones. These interliners are approximately 1,100 feet long and 440 feet long in Roseton and Wawarsing, respectively, and consist of a bolted, segmental steel lining approximately 16 feet in diameter surrounded by reinforced backfill concrete. An unreinforced concrete lining was cast inside of the steel to establish a uniform tunnel diameter of 13.5 feet.

### Leakage Conditions at Roseton

In the late 1990s, apparent leaks in the tunnel were noticed in the Roseton area. The surface expressions of the leaks formed a continuous stream on the west of the Hudson River, and water can now be seen bubbling out of the river in several locations during low tide. The leaks were surveyed and quantified to the extent possible, but a precise measure of the exfiltration has been elusive. The leaks are observed to vary with changing pressure conditions in the RWBT. Upon discovery of the leaks, the DEP began an investigation program that included a horizontal boring in close proximity to the tunnel in Roseton to assess the ground conditions near the leak. Rock conditions in the boring were similar to those recorded in the tunnel when excavated, but no major leakage pathways were found. The DEP also surveyed and photographed inside the full 45 miles of tunnel with an autonomous underwater vehicle (AUV). The results of these 2003 and 2009 surveys show no failure of the tunnel lining. In fact, it is believed that the tunnel is not in any danger of imminent collapse. Maps of the cracks in the tunnel lining were developed from the AUV photographs. Areas of cracking were identified, most of which were located near the surface leaks. While the leaks developed in the late 1990s, it appears that the rate of increase in the leaks has stabilized over the last five years.

### Leakage Conditions at Wawarsing

In the early 2000s, residents of Wawarsing complained of flooded basements. There was also evidence of fluctuations in the groundwater when the RWBT was shut down. The groundwater regime in the Wawarsing area, which includes the interconnected bedrock and unconsolidated aquifers, was studied by the U.S. Geological Survey (USGS). The conclusion of the USGS report (Stumm, et al., 2012) states: "Precipitation and other seasonal effects have the largest influence on water levels in the unconsolidated aquifer. Tunnel leakage from the bedrock was measurable in the unconsolidated aquifer. In the unconsolidated aquifer, elevated water levels due to tunnel-leakage influence, when combined with seasonally high water levels, can exacerbate or create localized basement flooding." Evidence of leakage into the bedrock—documented in the USGS report, the historical record of problematic water-bearing rock at tunnel level, and the recently developed crack mapping as a result of the AUV—pointed to a need for tunnel repair in the Wawarsing area.

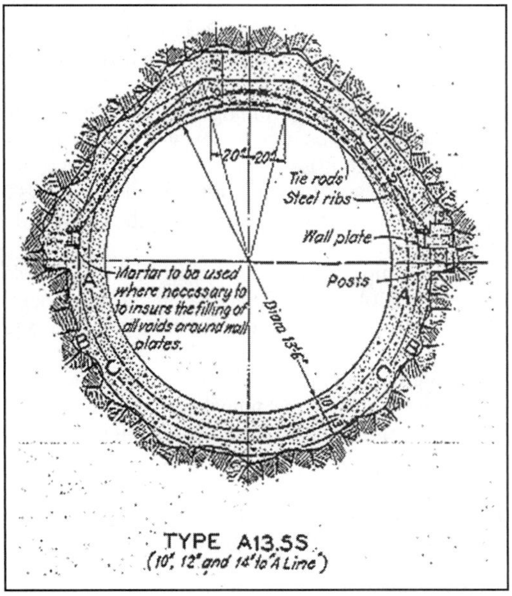


**Figure 2. RWBT typical final lining section**

## TUNNEL REPAIRS OVERVIEW

### Repair Method at Roseton: A Bypass Tunnel

Construction of a bypass tunnel at the Roseton leak site was chosen for the following reasons:

1. There are groundwater connections from the RWBT to the Hudson River. Groundwater inflows (into a dewatered tunnel), presumed to follow the paths of the exfiltration, could possibly provide an inexhaustible source of water, thereby making repairs very difficult and uncertain.
2. During shutdowns, inflows into the tunnel at Roseton occur at the low point of the tunnel,

**Figure 3. Steel-concrete interliner used in RWBT Roseton and Wawarsing areas**

and working effectively under shutdown conditions with substantial amounts of moving water, while maintaining pumping, would not be a preferred method for a planned repair. The safety ramifications of pump failure were determined to be not acceptable.
3. The logistics for accessing this area of the tunnel through existing shafts for major construction work were deemed not acceptable.
4. Combating water inflows while performing consolidation grouting did not lend itself to a defined timetable.
5. The duration of the outage to effect a repair from within the tunnel was determined to be too long and includes significant uncertainties, several of which could further prolong the shutdown.

Other remedies considered in the planning phase were: grouting from the surface, other internal repairs, and leaching lime into the cracks of the lining. All were rejected in favor of a bypass tunnel, which provided greater certainty of success.

### Repair Method at Wawarsing: Grouting/Internal Repairs

An in-tunnel repair for this section of the tunnel is being pursued for the following reasons:

1. The continuous slope of the RWBT in the Wawarsing area will allow for safe and efficient water handling.
2. Wawarsing is substantially closer to the upstream end of the tunnel, and therefore inundation is unlikely.
3. The extent of the area to be addressed and repaired is limited.
4. Access to the leaking area is relatively close to one of the shafts.
5. The internal repairs at Wawarsing can be conducted within the shutdown window needed for the Bypass connection.

Traditional methods of consolidation grouting and contact grouting will be employed to substantially reduce the leak in Wawarsing.

## BYPASS TUNNEL AT ROSETON

### General Overview of Alignment and Major Components

Figure 4 is a plan view of the Bypass Tunnel alignment. The completed tunnel will be 13,640 feet long, will traverse directly underneath the Hudson River, and will consist of the following main components:

1. Two access shafts located on the west (5B) and east (6B) sides of the Hudson River.
2. A 12,820-foot-long Main Drive Tunnel (Bypass Tunnel) under the Hudson River, driven from Shaft 5B to Shaft 6B using either conventional (drill-and-blast) methods or a tunnel boring machine (TBM).
3. Two Connection Tunnels, approximately 320 feet and 500 feet long, excavated between each shaft (5B and 6B) and the existing RWBT.
4. The existing pumping facility at Shaft 6, which will be used to dewater the RWBT.
5. A 450-foot-long Drainage Tunnel used to intercept water that infiltrates into the RWBT from the leaking reach beneath the Hudson River.
6. A sump and pumping facility at the bottom of Shaft 6B that will have a 20 MGD capacity to handle water generated from the Drainage Tunnel.
7. Permanent plugs that will isolate the leaking reach of the tunnel and the abandoned Drainage Tunnel from the completed Bypass Tunnel.

Each component of the Bypass Tunnel is discussed in more detail in the following sections.

### Geology

Figure 5 shows a plan view of the anticipated geologic conditions along the Bypass Tunnel alignment, as well as the locations and orientation of the exploratory borings used to characterize the geology of the Bypass alignment. The Bypass Tunnel will be excavated through Normanskill Shale, Wappinger Limestone, and Mount Merino Shale units.

### New Access Shafts for Bypass Construction

Two shafts are presently being constructed in order to build the Bypass Tunnel. The western shaft (Shaft 5B) is located on 30 acres directly over the RWBT and fronting New York State Route 9W. Shaft 5B is located in good rock, the Normanskill shale. It will serve as the primary construction shaft for the majority of the Bypass Tunnel as the proximity of the shaft to the state road eases many logistical challenges (muck removal, steel interlining transportation). The shaft has a depth of 880 feet and a 30-foot inside diameter, which was selected to accommodate

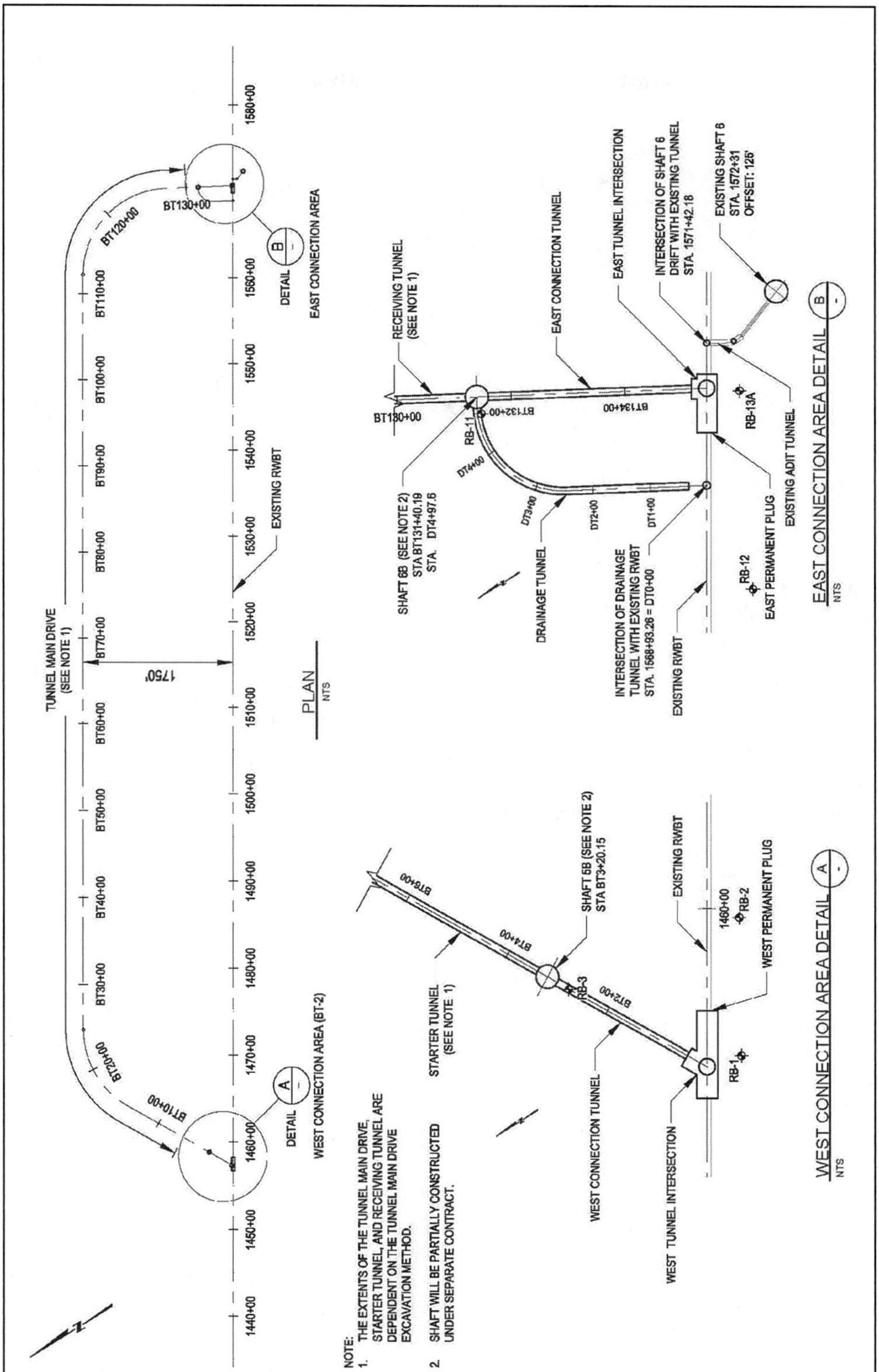


**Figure 4. Bypass tunnel alignment: Plan view**

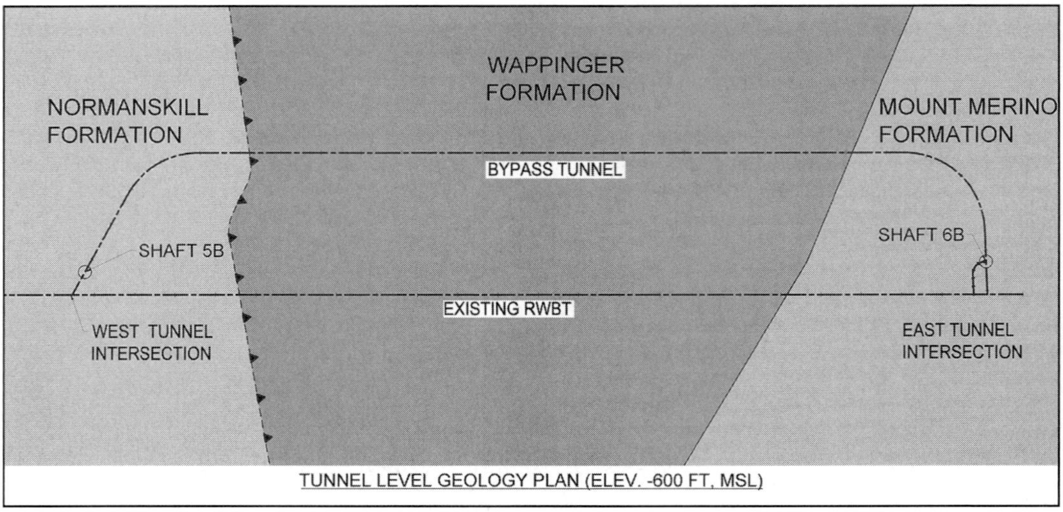


**Figure 5. Bypass Tunnel anticipated geologic conditions**

installation of a TBM and provide the tunnel with muck-handling equipment, access, ventilation, and utilities. After the tunnel is excavated and lined, this shaft will serve as the construction shaft during the connection to the RWBT.

Shaft 6B is located near existing Shaft 6 as the site has enough land to accommodate it, and the infrastructure of Shaft 6 complements the use of Shaft 6B. The Shaft 6B site is in favorable geology and is east of the Roseton leak. The total depth of Shaft 6B is 675 feet, reaching from the surface to the RWBT tunnel depth at an elevation of –600 feet. Shaft 6B will serve as the construction shaft for the Bypass Tunnel receiving chamber, the Connection Tunnel, the Drainage Tunnel, and the sump at the base of Shaft 6B. Shaft 6B includes five dewatering wells, which travel from the surface and terminate in the sump at the base of Shaft 6B. These dewatering pump wells offer redundancy, providing a location for future pumps to be installed should there be an issue with the Shaft 6 RWBT dewatering system. As a result of the additional dewatering pump wells, Shaft 6B is somewhat larger than Shaft 5B, at a 33-foot inside diameter.

The shaft final linings will be installed prior to making the connection to the RWBT. Also, the shaft caps will be fabricated on site so they can be quickly installed and pressurized to up to 600 feet of head after the connection. Therefore, a design challenge for the shafts was to provide a sufficient finished opening for the connection construction and still have the ability to close up the shafts quickly. This will be accomplished with an 18-foot-diameter access pipe embedded in refill concrete in the top section of the shaft. The access pipe will be capped with a transition piece and a 9-foot-diameter cap, all of which will be bolted together. A small subterranean access chamber will house the shaft caps.

### Drainage Tunnel and Pump Station

Because of the increased leakage of the RWBT, uncertainty exists whether the dewatering system for the existing tunnel will be sufficient to fully remove the additional inflow and maintain a safe working condition during tie-in. A new Drainage Tunnel, connected to a new pump station, at Shaft 6B, will augment the existing dewatering system. The Drainage Tunnel will be connected into the existing tunnel via drilled holes to maintain a safe and controlled condition at all times. The Drainage Tunnel and pump station will be equipped with control valves and monitoring systems to ensure that water flow through the Drainage Tunnel can be controlled and shut down in an emergency. The Drainage Tunnel and pump station will be designed for a maximum sustained dewatering rate of 20 MGD. Once the tie-in of the existing RWBT is completed, the Drainage Tunnel will be permanently sealed off. The remaining pump station, at Shaft 6B, will used as a supplemental dewatering facility tunnel access point.

### Bypass Tunnel Between Shafts

The Bypass Tunnel will be built between Shafts 5B and 6B and eventually connect to the RWBT. It is configured to be sufficiently separated from the RWBT to limit impacts from the existing tunnel and to maintain a safe working condition during construction.

### Excavation Methods and Initial Support Methods

The contractor will have the choice of selecting one of two excavation methods: conventional drill-and-blast or TBM. Drill-and-blast will use various conventional support methods, applicable to the anticipated ground and groundwater inflow conditions and to maintain a safe working environment during excavation. These support methods range from rock dowels to steel sets at variable spacings. It is anticipated that the TBM used would be a shielded hard-rock TBM, suitable for erection of bolted and gasketed segments for initial support. Power for the TBM will be provided by a pre-existing power source near Shaft 5B. The segments would be designed for the anticipated external head conditions of up to 775 feet.

### Final Lining

Since the tunnel will be carrying 1,200 feet of internal head, it will act as a pressure tunnel. The selection of final lining is based upon two criteria: (1) maintaining structural integrity, and (2) ensuring watertightness during operation of the pressure tunnel. Sufficient rock cover and confinement, as well as rock mass strength, must be considered to ensure the pressure tunnel can carry the high internal head without leakage or deterioration of the lining over its operational life (100 years). Analysis of the anticipated site conditions indicates that a majority of the Bypass Tunnel alignment will require a steel lining. The lining will have to be designed for the maximum internal head during operation and the external head during dewatered events or inspections. The remaining portion of the tunnel, which has sufficient rock cover and rock mass strength, will be lined with a reinforced cast-in-place concrete lining.

### Connection Tunnels from Shafts to RWBT

The remaining two Connection Tunnels from Shafts 5B and 6B to the RWBT will each be approximately 550 feet long. These tunnels will be excavated by conventional drill-and-blast means to within a 100-foot standoff or separation from the RWBT while it is operational. The anticipated ground cover and rock mass quality allow the two Connection Tunnels to be lined with a reinforced cast-in-place concrete lining.

### Connection Chambers

Once the RWBT is dewatered through the Drainage Tunnel and pump stations at Shaft 6 and Shaft 6B, the two Connection Tunnels will advance past the standoff position and excavate into the RWBT. The intercepts of the two Connection Tunnels with the RWBT will each result in an oversized chamber to accommodate a Connection Chamber. The Connection Chambers' final lining will be geometrically designed to accommodate hydraulic flow while minimizing head loss and the movement of future inspection equipment. The final lining of the Connection Chambers will be a reinforced cast-in-place concrete lining.

### Isolation of Abandoned RWBT and Permanent Plugs

To isolate the abandoned RWBT between the two connection points, permanent plugs will be installed within the existing RWBT adjacent to each Connection Chamber. The existing concrete lining of the RWBT will be removed, and 60-foot-long mass concrete plugs will be placed and keyed into the rock.

## CONNECTION CONSTRUCTION: SEQUENCING AND INTERRUPTIONS

The design of the Bypass Tunnel has been optimized through significant planning, analysis, and review to minimize the connection construction time in order to limit the interruption of water conveyance through the RWBT. The proposed sequence of the connection is as follows:

1. Dewater the RWBT using Shaft 6.
2. Upon realizing a steady-state level of inflow over a few days, start mining the Drainage Tunnel towards the RWBT.
3. Upon advancing the Drainage Tunnel to 25 feet of the RWBT, establish a work area and drill through blowout preventers into the RWBT.
4. Activate the Drainage Tunnel utilizing the drilled holes and the Drainage Tunnel piping, and start the Shaft 6B pumps.
5. Upon successful activation of the Drainage Tunnel, start mining the Connection Tunnels to the RWBT.
6. Hole through the Connection Tunnels into the RWBT.
7. Install a mandatory initial support and temporary lining in the Connection Tunnels and tunnel intersections. The temporary lining would have a design life of 15 years.
8. Prepare the existing RWBT for the permanent plugs.
9. Install three permanent plugs—two in the RWBT and one in the Drainage Tunnel.
10. Complete construction of the tunnel intersection and line the Connection Tunnels.
11. Demobilize and install the caps in Shafts 5B and 6B.
12. Restart the RWBT.

The goal is to complete the connection work in one 6- to 8-month shutdown period after completion of the Bypass Tunnel. Should the City not have enough water in the reservoirs, the connection can be interrupted at virtually any point and the connection work would resume the following year. Modeling of hydrology of the reservoirs using their 96-year history has determined that the connection will likely have a maximum of two interruptions and will be completed within a three-year period. The ability to interrupt the connection period also allows the RWBT to be shut down for smaller durations, which results in reducing the need to look for other water sources for use during the shutdown.

The time allotted for the connection will allow a time period to fully complete the Wawarsing repair.

## REPAIR WORK AT WAWARSING

The repair work at Wawarsing is anticipated to be conducted during the connection outage for the Bypass Tunnel. Conventional tunnel repair and treatment techniques will be used. The area of tunnel requiring repair is considered to be concentrated within three zones totaling less than 1,000 feet.

### Setting and Geologic Conditions

The Wawarsing area currently has a steel interlining. Leakage at the extremities of this lining through the existing cast-in-place lining is considered a possibility. As indicated earlier, the rock mass is made up of a complex sequence of limestone, sandstone, and shale that during the original excavation produced upward of 9,000 gpm of inflows. Inspection of the lining with AUV indicates that exfiltration is occurring through cracks in the cast-in-place lining.

### Repair and Grouting Program

During the connection outage, inspection, rock mass treatment, and repairs within the Wawarsing area will be undertaken. The first step is to inspect the existing lining conditions to secure a safe working condition. Structural repairs of the existing concrete lining will be made on an as-needed basis. Zones of high water inflow will be evaluated to determine a program to seal the rock mass outside of the lining with consolidation grouting. Consolidation grouting is considered necessary to reduce leakage through the highly variable ground mass. This would be followed by injection of cement immediately behind the concrete lining to ensure all voids behind the lining are sealed and to enhance performance of the concrete lining.

## INSPECTION PROCEDURES FOR REMAINDER OF TUNNEL

The RWBT will be inspected immediately following the tunnel dewatering for the Bypass Tunnel connections. Two inspections are planned for the RWBT: a safety assessment inspection and a condition assessment inspection.

The safety assessment inspection will allow experienced tunnel engineers to assess the condition of the dewatered tunnel and confirm that the tunnel integrity and environmental conditions are acceptable for personnel access. To perform the safety assessment, inspectors will rapidly move through the tunnel by vehicle and assess the condition of the tunnel lining and estimate the groundwater infiltration rate. Documentation will be limited to features that could possibly pose a safety hazard to entrants—such as tunnel lining leaks producing groundwater infiltration exceeding 50 gallons per minute, open tunnel lining cracks with an aperture exceeding 0.125 inch, and significant tunnel lining defects.

The objective of the condition assessment is to document the current condition of the tunnel to provide a baseline for future inspections. The inspection team will move through the tunnel at a slower pace than used for the safety assessment and document tunnel station, structural features (cracks and deformations), groundwater infiltration flow and size of orifice, and sediment intrusion. Visual condition assessment may include photo-documentation or LIDAR scanning and may be supplemented by non-destructive examination and destructive examination techniques.

## SCHEDULE

Prequalification of contractors will occur in mid-2014. The bid for the BT-2 Contract (all of the work after excavation and lining of Shaft 6B and Shaft 5B) is planned to be "on the street" in third to fourth quarter of 2014. The Notice to Proceed for the BT-2 Contract is planned for April 2015, but the construction site will not become available until April 2016 once the BT-1 shaft contracts are completed. A year is being allowed for TBM procurement and for development of groundwater treatment facilities and pipelines. The underground work is to be completed in 2021, with the connection period starting in October 2021. The construction would be completed by 2024.

## CONCLUSION

The Rondout-West Branch Tunnel, which is a vital part of the infrastructure providing water to the City of New York, is currently known to be leaking up to 35 MGD near the tunnel's Hudson River Crossing at Roseton, New York. The City is currently designing a new tunnel segment that will bypass this leakage zone. Dewatering of the tunnel to make the connection will also afford an opportunity to inspect the entire 45-mile length of the tunnel and make internal repairs to another leaking segment near Wawarsing, New York. This paper has described the key challenges facing the implementation of the Bypass Tunnel, the inspection and Wawarsing repairs, and the principal design components of the project.

## REFERENCES

City of New York, Board of Water Supply (NYC-BWS). 1942. The Delaware Water Supply for the City of New York—A Series of Fourteen Articles Reprinted from *Engineering News Record*, Article III: Designing the Aqueduct.

Frederick Stumm, Anthony Chu, Michael D. Como, and Michael L. Noll. 2012. Preliminary Analysis of the Hydrologic Effects of Temporary Shutdowns of the Rondout-West Branch Water Tunnel on the Groundwater-Flow System in Wawarsing, New York. Prepared in cooperation with the New York City Department of Environmental Protection. Reston, VA: USGS.

# Clean Design in Dirty Water: The Blue Plains Tunnel Segmental Lining

**Anthony Harding**
CH2M HILL

**Brett Zernich**
Traylor Skanska Jay Dee JV

**Moussa Wone**
DC Water

**ABSTRACT:** As a part of a much larger CSO program the design of the 23 ft (7m) ID segmental lining for the Blue Plains Tunnel offered significant benefits over traditional two pass systems. However, careful design was required to achieve the 100 year design life and provide structural resistance to an internal pressure that was higher than the external pressures on the lining. This paper describes how these and a number of other requirements were incorporated into the design to provide a ring that performs robustly for functional, cost and constructability criteria. Finally, comments are made on the performance of the ring in the tunnel and recommendations made for future designs.

## PROJECT BACKGROUND

The sewer system in the District of Columbia is comprised of both combined sewers and separate sanitary sewers. One third of the District is served by combined sewers. During dry weather, combined sewers carry both sewage and runoff from storms and convey them to the Blue Plains Advanced Waste Water Treatment Plant (BPAWWTP). Wastewater is then treated to remove pollutants before being discharged to the Potomac River. When the capacity of a sewer is exceeded during storms, the excess flow, called Combined Sewer Overflow (CSO), which is a mixture of sewage and storm water runoff, is discharged to the Anacostia and Potomac Rivers, Rock Creek and tributary waters, eventually discharging to the Chesapeake Bay. There are a total of 53 CSO outfalls in the District combined sewer system; fifteen of which discharge to the Anacostia River.

The Blue Plains Tunnel (BPT) project is a component of a larger scheme to control combined sewer overflows (CSOs) to the District of Columbia's waterways, called the Long Term Control Plan (LTCP). The LTCP is designed to meet the CSO control objectives of DC Water and to meet water quality standards in the District of Columbia.

The BPT project consists of:

- Blue Plains Tunnel (BPT): Approximately 7.3 km (24,000 ft) long, 7.01 m (23 ft) internal diameter (ID) tunnel.
- Blue Plains Tunnel Screening Shaft (BPT-SS): A screening shaft, also used to launch the TBM to mine the BPT, located on BPAWWTP site.
- Blue Plains Tunnel Dewatering Shaft (BPT-DS): A dewatering pumping station shaft, located on the BPAWWTP site.
- Bolling Air Force Base Drop Shaft (BAFB-DS): An overflow/drop shaft for connecting the Joint Base Anacostia-Bolling (JBAB) Overflow Structure to the BPT.
- Poplar Point Drop/Junction Shaft (PP-JS): A combination drop/junction shaft on District of Columbia government land.
- Surge Chamber and Approach Channel at PP-JS: The approach channel will connect the future Main Outfall Sewer Diversion Chamber (MOS-DC). The MOS-DC will be built over the existing modified twin sewers (under a different contract) for directing flow from the West Influent Sewer and East Influent Sewer to the PP-JS. A surge chamber will be constructed where the approach channel connects with the vortex generator in the PP-JS to facilitate handling the flow.
- Main Pumping Station Drop Shaft (MPS-DS): A drop shaft at DC Water's Main Pumping Station near 2nd Street and Tingey Street SE. This shaft will be used to convey diversions from CSO 9, 11A, 12, 13, and 14 diversion chambers.

This paper focusses on the design of the precast concrete segmental lining for the BPT.

## CLIENT REQUIREMENTS

The Clean Rivers program requires the construction of a significant length of tunnel in a short timeframe, at considerable expense. Therefore DC Water specified that the 100 year design life be met with a one-pass lining, as it would offer substantial cost and schedule savings. The winning tender addressed the problem of the aggressive environment on the two components of the traditional segmental lining: the concrete and the reinforcing steel. Firstly, the corrosive effects of the environment were modelled in a durability model, which showed that the area of compromised concrete strength could be limited to 1 inch, even after 100 years exposure, provided certain qualities of the mix were maintained.

The issue of corrosion of reinforcing steel is that as well as the water being aggressive to steel, the deterioration of the concrete reduces the effective cover to the steel, further increasing its vulnerability to corrosion. Furthermore, any corrosion of steel results in spalling of the concrete, which exposes more steel and accelerates the corrosion. Modelling of the conditions and increasing the cover can minimize the risk of corrosion, but there is a way of eliminating the risk altogether: to use steel fibers in lieu of conventional reinforcement. While surface corrosion of fibers does occur, the risk of fibers more than an inch from the surface of the concrete corroding is much less. Furthermore the consequences of corrosion are also much less as corrosion of fibers does not cause spalling of the concrete. While fiber corrosion can result in some loss of flexural strength, tunnel linings function predominantly in compression with low moments (typically within the middle third), so small losses in flexural strength do not reduce the robustness of the structural design. This makes tunnel linings an ideal structure for fiber reinforcement, as identified by a number of authors, including King (2005). Therefore replacing the conventional reinforcement with steel fibers offered a clear benefit to the project.

### Seismic Loads

The seismic requirements were to design for two events: a Maximum Design Earthquake (MDE) with a 2,475 year return period, a moment magnitude of 6.1 and distance of 108 kilometers; and an Ordinary Design Earthquake (ODE) with a 475 year return period, moment magnitude of 5.9, and an epicentral distance of 174 kilometers.

### Internal Pressure

The use of a one-pass lining system created another unusual requirement: on a two pass system the secondary lining is designed to resist the internal water pressures, including a transient surge pressure that was higher than the external water pressures. The absence of a secondary lining meant that the segmental lining would have to resist the load from the internal pressure.

## CONSTRUCTION REQUIREMENTS

The aim of any segmental lining design is to have a ring that is easy to manufacture transport and erect. This requires ongoing discussion between designer, manufacturer and constructor to refine the ring geometry, inserts, and other geometrical requirements. The first step of this process is to finalize the general arrangement of the ring. The adopted arrangement is shown in Figure 1.

The segmentation adopted is a 6+key arrangement, with three dowels in each segment and one in the key, giving a total of 19 dowels around the ring. The project adopted up and down rings (where the ring tapers vertically rather than horizontally) in preference to left and right rings. As the alignment follows a constant upward grade from the launch shaft at the Blue Plains Treatment plant, it has only horizontal curves with no vertical curves. Rotating the ring only two dowel positions provides around 60% of the effect of the taper in the horizontal axis, so only a few rotations from nominal position are required to follow all but the tightest curves on the alignment. Ideally these small rotations would meet the following criteria for easy builds:

- Counter key (first segment of build) as close as possible to the invert
- Key as close as possible to the crown

To achieve these objectives the nominal position of the key was provided one dowel position to the right and left of the crown respectively for the up and down rings. In these positions it was found that the segments 3 and 4 respectively were located in the invert, so these were selected as the counter keys. Thus a horizontal radius of more than half of the maximum could be achieved by simply rotating the left and right rings by two dowel positions, ensuring the counter key remained near the invert, and the key above axis.

Therefore after the segmentation definition, the key segment dimensions were set according to Table 1.

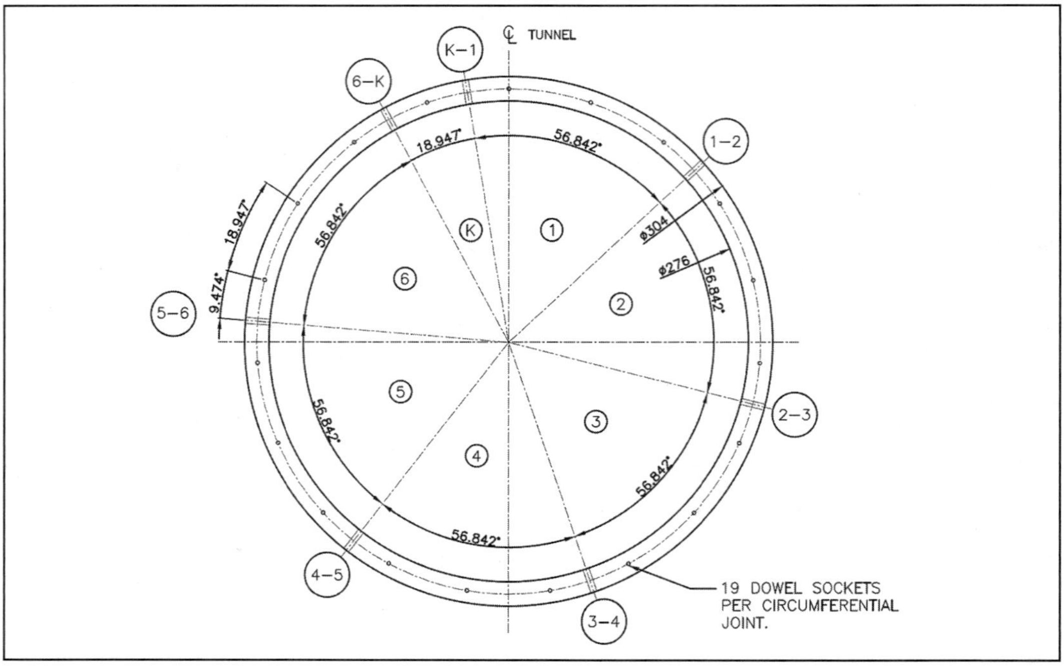


**Figure 1. General arrangement of down ring**

**Table 1. Lining key dimensions and features**

| Element | Value | Element | Value |
|---|---|---|---|
| Internal diameter | 7010 mm (276 inches) | Joint angle (from rectangular) | 9° |
| Ring length | 1829 mm (72 inches) | Lining design minimum radius | 580' |
| Thickness | 356 mm (14 inches) | Taper | Double taper, 80 mm (3.15 inches) total |
| Ring type | Up and Down tapered rings | Characteristic compressive strength | 48 MPa (7000 psi) |
| Segmentation | 6+1 Parallelogram | Longitudinal connectors | Spear bolt with plastic sockets |
| Segment type | Steel fiber reinforced concrete (SFRC) | Longitudinal joint alignment rods | 40 mm (1.57 inches) guide rods to longitudinal faces, 36" long |
| Key segment size | 18.947° nominal | Circumferential joint connectors | Push fit dowels |
| Other segment sizes | 56.842° nominal | Gasket | Dätwyler M 389 21 self-anchored gasket |

The minimum alignment radius for the project is 265 m (870 ft). The 3.15 inch taper on the ring provides a minimum theoretical radius of 176 m (580 ft), which provides sufficient taper to follow a recovery radius on the minimum alignment curve.

## Gaskets

Most of the embedded items for the segment were solutions quite commonly employed in the US. The exception was the use of a cast-in gasket, which had never been used in the US before but was preferred by the manufacturer and contractor. A thorough review by the designer was required to verify that it would meet the project requirements, which concluded that the risks of using a relatively new technology were outweighed by the following benefits of the cast-in system:

- Surety of anchoring the gasket could be verified visually from inspection of the segment.

**Table 2. Design sections adopted**

| Station | Depth to Tunnel Crown | Material | Reason for Selection |
|---|---|---|---|
| 10+00 | 35.0 m (115.0 ft) | Patapsco/Arundel fat clay | Deepest section and maximum $K_0$ |
| 85+00 | 29.8 m (97.9 ft) | Patapsco/Arundel clays overlying Patuxent sands | Mixed face |
| 135+00 | 33.5 m (110.0 ft) | Patapsco/Arundel clays overlying Patuxent sands | Mixed face (deep case) |
| 141+00 | 27.1 m (89.0 ft) | Alluvuim and Patapsco/Arundel | Shallow case with alluvium in the crown |
| 200+00 | 24.3 m (79.9 ft) | Patapsco/Arundel sands overlying Patuxent clays | Mixed face (shallow case) |
| 240+00 | 15.2 m (50.0 ft) | Patapsco/Arundel clays overlying Patuxent sands | Shallowest section |

- Health risks associated with the use of gasket glue were all but eliminated.
- Physical anchorage was likely to provide enhanced sealing performance over what the tests measure.

Initially the designer's main concern was whether the cast-in gaskets could be repaired effectively. The repair method provided relies on removing the gasket and filling the void to create a conventional gasket groove for a conventional gasket. Therefore upon review of the proposed method it was clear to the designer that a repaired cast-in gasket is no worse than the traditional gasket. In practice this was accomplished in the field and provided no downside to the manufacturer or constructor except for the time in removing the damaged gasket.

### Handling

As steel fiber reinforcement was to be employed, the flexural strength of the segments is significantly less than that of conventionally reinforced segments. This means that much more attention has to be paid to the various handling and storage arrangements adopted in the manufacturing plant, during transit to the job site, and at the tunneling site. Each of these stages was checked to ensure that excessive stresses would not be induced in the segments. Overall more than twenty separate lifting, transportation and stacking stages were identified, covering the various steps that the segments had to go through from the molds to the TBM erectors, resulting in more than ten separate load cases for analysis.

## GEOLOGY

The geological profile along the tunnel alignment consists of superficial deposits of fill and alluvium overlying Cretaceous age soils, the Patapsco/Arundel Formation (P/A) and the Patuxent Formation (PTX) of the Potomac Group. The Patapsco/Arundel Formation comprises sands and gravels with thick bands of silt and clay. The Patuxent Formation is similar, characterized predominantly by sand and gravel varying from silty and clayey to relatively clean, with silt and clay inter-beds.

The tunnel alignment is within the P/A and PTX formations along the alignment, apart from a very short stretch where an alluvium filled channel encroaches into the tunnel crown between Station 137+45 and Station 142+44.

As well as normal ground loads the soils of the Patapsco/Arundel Formation were somewhat challenging due to the high horizontal insitu stresses (design $K_0$ values of up to 1.4) and potential for swelling. The swelling was tackled through analysis demonstrating that the swelling pressures would never arise to more total pressure than the insitu stresses. This meant that a full overburden case had to be analyzed for the deepest section.

Groundwater along the alignment was determined by 3 different hydro-geologic zones: Fill and Alluvium (Upper Aquifer), Potomac Formation above elevation –130; and Potomac Formation below elevation –200. Design levels for these aquifers were from –3 m (–10 ft) to 4.4 m (14.3 ft).

## RING DESIGN

The first step of the ring design was to identify those sections along the alignment that would result in the worst combinations loads on the lining. This included sections with the potential to generate maximum axial force, maximum moment, and minimum axial force. The sections selected are as shown in Table 2.

### Ground Loads

All sections were initially checked using the closed form solutions from Muir Wood (1975) and Curtis (1976), and Duddeck and Erdmann (1985). This revealed the critical sections for further analysis.

Perhaps unsurprisingly, Station 10+00 was the critical case for hoop loads. Having the highest overburden in conjunction with the highest $K_0$ values,

**Table 3. Steel fiber reinforced concrete—key strength specification**

| Test | Value | Potential Issues and Limits of Specification |
|---|---|---|
| Compressive strength | 48 MPa (7000 psi) | Exceeding 50MPa can result in brittle behavior post-crack, and reduced post crack flexural strengths unless higher strength (and more costly) fibers are employed. |
| Equivalent flexural strength | 3.0 MPa (435 psi) | The authors' experience is that the practicalities of achieving a mix with the right ductility can make higher values difficult to achieve in the time usually allotted to mix design. |
| Tensile splitting strength | 4.3 MPa (624 psi) | While higher values can often be achieved, it is usually by increasing cement content, with consequent increase in compressive strength and potential reductions in equivalent flexural strengths. |

this section was also compromised by the fact that no relaxation of the ground could be allowed for due to the potential for swelling.

The closed form analyses of the mixed face conditions using the different stiffnesses of the materials revealed high levels of difference in bending moment at the section at Sta. 141+00. This meant that the influence of the different stiffness could create significant bending moments. Furthermore, the section was relatively shallow, with lower axial forces and hence less resistance to bending. Therefore this section was analyzed in a plane-strain continuum model using Phase2 2D finite element software. This analysis confirmed the section to be critical.

### Other Loads

Seismic loads were determined in accordance with Hashash (2001) and the effects of internal construction loads evaluated using a bedded beam model in Strand7 taken at the critical section 141+00. Seismic loads were combined with the results of the closed form analyses for all but this section to check for the worst effects during operation. For the construction cases the Strand7 model incorporated all loads (including ground loads) to complete the check of this section. As this section had the lowest hoop loads and bending moments closest to capacity, and also used the least stiff soil, it was logical to conclude that the other sections in stiffer soil with more residual capacity would remain within capacity under construction load cases.

### Durability

To demonstrate the durability of the lining for its design life, the corrosive effects of the environment were modelled in a durability model, which placed the following additional requirements on the lining design:

- Sacrificial layer of 25 mm (1 inch) for degradation of concrete due to hydrogen sulfide attack
- Additional allowance of 20 mm (0.79 inch) loss of flexural strength to allow for fiber loss due to carbonation and chloride-induced corrosion

### Steel Fiber Reinforced Concrete (SFRC) Design

Appropriate specification of SFRC is often crucial to its successful implementation. However, care was used throughout the design to ensure that concrete performance criteria that might prove difficult to achieve were not specified. It was not possible to verify what could be achieved with locally available aggregates, as the required test data simply wasn't available. Therefore moderately conservative values were specified based on project experience from other SFRC projects around the world, as described in Table 3.

The design for SFRC was generally in accordance with the intent of ACI-350, noting that SFRC is technically not covered by the ACI code. Flexural strengths were determined in accordance with King (2005), and were driven mainly by segment handling stresses. Bursting stresses for the TBM ram pressures and concentrated hoop loads on the longitudinal joint were used derive the maximum splitting strength.

## DESIGN FOR INTERNAL PRESSURE

There was a project requirement to design for an internal pressure that was 90 kPa (13 psi) higher than the external water pressure. This led to an overall requirement to design for a minimum pressure of:

$$P_{net} = u_{external} - u_{internal} + P'_{soil}$$

where $P_{net}$ is the net pressure on the lining; $u_{internal}$ is the internal pressure; $u_{external}$ is the external water pressure; and $P'_{soil}$ is the effective soil pressure. This equation shows that if the effective pressure from the soil is sufficiently low, the net pressure on the lining is negative, resulting in tension in the lining. Therefore the design needs to determine a *minimum* effective soil pressure to determine whether the lining could go into tension, and if so the maximum tension in the lining.

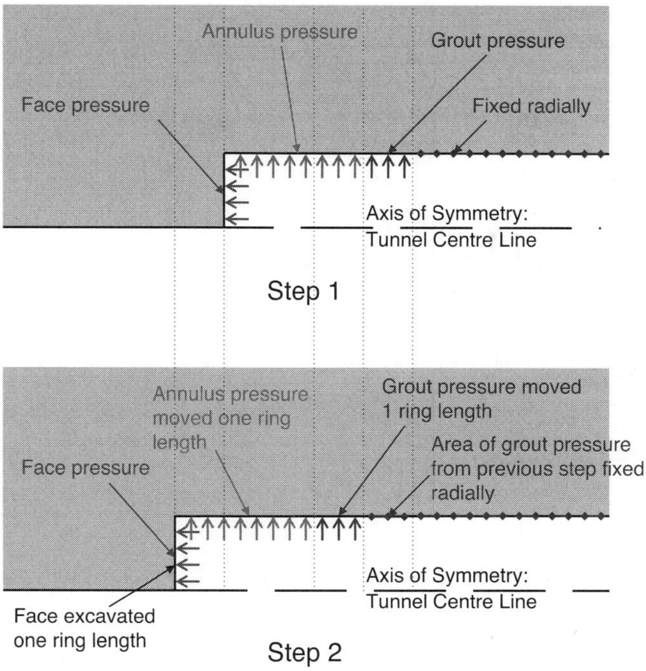


**Figure 2. Evaluation of pressures arising from TBM excavation**

This differs from conventional tunnel lining design, which aims to determine upperbound values: the *maximum* external loads that might be applied to the lining. In a normal tunnel lining design the designer would be aware that all the input parameters are not precisely known, but that there is a range into which they are likely to fall. The designer usually selects values for these inputs that will yield higher ground loads in the analysis than mean or 'best estimate' values. Furthermore, the analysis methods used implicitly make conservative assumptions about the behavior of the ground that result in increased loads being calculated. Even finite element models may use simplified stress-strain relationships, or make assumptions about the amount of relaxation that occurs along the length of the TBM prior to lining installation. Therefore the actual load from the soil is within a range, and in conventional tunnel design a value at the upper bound of this range is selected. However, for the internal pressure design a lower bound value for the effective soil pressure is required.

## Analysis of Minimum Pressures

For the sections in sands and gravels, it was judged that a soil load determined using Terzaghi's silo theory would provide a reasonable lower-bound answer. The resulting worst effective pressure was 65 kPa (9.4 psi), resulting in a maximum net internal pressure of 25 kPa (3.6 psi). While it is true that a finite element or finite difference analysis might provide a smaller number, this was judged to be too conservative given the restraint against inward movement provided by a TBM being managed to minimize settlement. Nevertheless, the results of this analysis were cross referenced against the more comprehensive analysis undertaken for the cohesive materials.

In the cohesive materials hand calculations using Terzaghi's equation with undrained parameters in a total stress analysis were used to determine the section that would provide the lowest load on the lining if sufficient relaxation were to occur. This was a section at Sta. 13+26, where the load provided was considerably less than the long term water pressure. Such a low pressure was considered to be excessively conservative so it was necessary to consider how much the operation of the TBM will restrain the inward movement of the ground to obtain more robust ground pressures on the lining post-construction. There is little guidance in the literature for this problem so a study was undertaken to establish a reasonable upperbound for the relaxation that occurs along the length of the TBM.

To this end a simple axisymmetric model of the TBM advance was created, as illustrated in Figure 2. This model sequentially excavated the tunnel one

**Table 4. Results of analysis**

| Input | | | | Results | | | | |
|---|---|---|---|---|---|---|---|---|
| | Lower-bound | Best Estimate (BE) | Upper-bound | Lower-bound | | BE | Upper-bound | |
| Input | | | | kPa (psi) | Change | kPa (psi) | kPa (psi) | Change |
| Face pressure | 193 kPa (28.0 psi) | 385 kPa (55.8 psi) | 578 kPa (83.8 psi) | 356 (51.6) | –0.3% | 357 (51.8) | 357 (51.8) | 0.0% |
| Annulus pressure | 100 kPa (14.5 psi) | 335 kPa (48.6 psi) | 620 kPa (89.9 psi) | 198 (28.7) | –44.5% | 357 (51.8) | 549 (79.6) | 53.8% |
| Grout pressure | 335 kPa (48.6 psi) | 435 kPa (63.1 psi) | 535 kPa (77.6 psi) | 326 (47.3) | –8.7% | 357 (51.8) | 387 (56.1) | 8.4% |
| Excavated length | 0.45 m (1.5 ft) ¼ ring length | 0.9 m (3 ft) ½ ring length | 1.8 m (6 ft) 1 ring length | 356 (51.6) | –0.3% | 357 (51.8) | 369 (53.5) | 3.4% |
| Soil stiffness parameters | 37.5 MPa (5437 psi) | 50 MPa (7249 psi) | 75 MPa (10874 psi) | 360 (52.2) | 0.8% | 357 (51.8) | 351 (50.9) | –1.7% |
| Soil strength parameters | $\phi$= 0.1° c=171 kPa (24.8 psi) | $\phi$= 0.1° c=228 kPa (33.1 psi) | $\phi$= 0.1° c=342 kPa (49.6 psi) | 357 (51.8) | 0.0% | 357 (51.8) | 357 (51.8) | 0.0% |
| Horizontal insitu stress | 0.58 MPa (84.1 psi) | 0.77 MPa (111.7 psi) | 1.16 MPa (168.2 psi) | 356 (51.6) | –0.3% | 357 (51.8) | 358 (51.9) | 0.3% |

ring length at a time. At each solution step a pressure was exerted on the face (representing the face pressure), the annulus (representing the fluid or material between the shield and the excavated ground), and the grout (representing 1 ring length of grout that has gelled but is still at or near the applied grout pressure). More than one ring length from the back of the shield the model was fixed in the radial direction to represent the comparatively stiff radial restraint of the fully grouted lining. The excavation and pressurization sequence was repeated some 24 times to yield a stable displacement profile 10 rings back from the back of the TBM.

The initial stresses in the model were as per the expected vertical stresses. The face pressure was set at 50 kPa above the long term water pressures, which was considered to be reasonable for excavating in a cohesive material. The annulus pressure was set to the long term hydrostatic pressure, and the grout pressure to 1 bar above this value. These values are all 'best estimate' values—essentially based on judgment. However, they are all subject to considerable variation during TBM operation, so the impacts of varying each of these variables was determined using a sensitivity analysis. The parameters varied and variation considered are shown in Table 4, with uncertain variables being varied by more than certain ones.

The potential for higher displacements resulting in the shield restraining inward movement was ignored because such restraint would result in a higher the final load in the lining, so the results are presented in Table 4 without considering the restraint of the tail shield.

The results clearly show that the strongest influence of the final pressure on the lining is the annulus pressure. The face pressure has a very small influence, probably because it is more than 1 diameter away from where the lining is 'locked in'. The grout pressure has a significant influence, but this is less apparent from the table because it has been varied by only ±25%, rather than the –70%/+90% of the annulus pressure. This result is of great concern if a designer wants a reasonably accurate prediction of loads, because the annulus pressure is probably the least commonly monitored pressure in the TBM, and while active pressurization systems, which inject slurry or other fluids into the annulus, are sometimes specified, they add cost and require additional controls to use effectively. Therefore the design sought to verify whether the design would work without the need to specify annulus pressures.

The first thing to note is that when the annulus pressure is at hydrostatic values (the best estimate analysis), the load locked in to the lining is 22 kPa (3 psi) higher than the annulus load, demonstrating the restraining effect of the TBM. This is a third of the Terzaghi load, providing confidence that the restraining effect of the TBM would likely result in loads in the granular material that were higher than the Terzaghi load. However, given the uncertainty in the annulus pressure in the cohesive units, the effects of long term changes in water pressure were considered in a consolidation model.

This was modelled in plane strain in Plaxis. The model was allowed to relax by the same amount as the worst case axisymmetrical model prior to lining installation (equivalent to an internal pressure of 198 kPa). This stage was modeled as an undrained material with pore pressure determination to simulate the installation process. The long term changes in pore pressures and resulting changes in load were

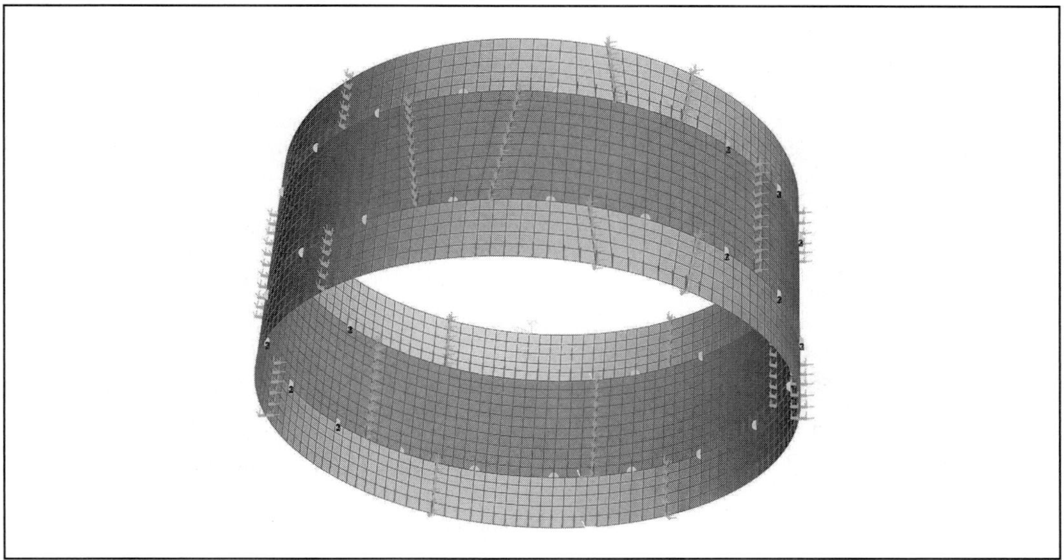

**Figure 3. Ring model for the check of the effects of internal pressure**

determined by a simulated consolidation process, providing an indication of the likely minimum long term load on the lining. The results showed that the long term consolidation effects result in a total external load some 18 kPa (2.6 psi) greater than the internal pressure, and 50 kPa (7 psi) greater than the load predicted by a Terzaghi analysis using long term parameters. Therefore the Terzaghi load was considered to offer a reasonable lower bound solution for the effective soil pressures in the cohesive units as well, resulting in a requirement to design the lining for a small net internal pressure of 25 kPa (3.6 psi).

### Check of Effects

To analyze the effects of the net pressure a 3D bedded beam model was used. This modelled one whole ring with two half rings either side. The segments are modelled as plates, and a symmetry boundary applied at the ends of the model. The ground was modelled as compression only springs, using the worst (lowest) soil stiffness for the alignment. Lateral springs were not used so the tendency of shear forces between lining and ground to restrain joint opening were conservatively ignored.

The dowels and bolts are included with appropriate stiffnesses in shear and tension. The model is illustrated in Figure 3.

The model was validated by a hand calculation that used the Kirsch equations for the expansion of a hole in an elastic continuum, in conjunction with the stiffness of the bolts in the longitudinal joints and the shear stiffness of the dowels in the radial joints.

The outward deflections of the model were low (maximum 1.23 mm or 0.05 in), and the gap opening at the longitudinal joints 1.2 mm (0.05 in). Therefore the gasket watertightness would not be compromised. Tension of the bolts was 65 kN (14 kips) per bolt in the direction of the ring, which equates to 75 kN (16.9 kips) in the direction of the bolt, while the contribution of the dowels was very small. Tension in the ring was 130 kN (29.2 kips) per segment, which equated to 200 kPa (29.0 psi), less than 10% of the tensile strength.

The bolt force resists 78% of the total load applied the ring, meaning that the ground load provides 20% of the resistance. While it might appear that the design relies on the bolts in the long term, it should be pointed out that if there were no bolts then the outward displacement would still only be in the order of 5mm. At this opening the guide rods would provide resistance to small forces that might cause the joints to displace, and once the transient internal pressure has passed the joints will close as the lining reverts to its normal state of acting in compression.

## CONSTRUCTION ISSUES

While the construction is ongoing, it is possible to note the following regarding the performance of the ring so far.

### Performance of Up/Down System

Adopting different counter keys for different ring types is an unusual approach, and had the potential to complicate the ring erection process because the operatives would have to get used to building the up and down rings in slightly different sequences. However, in practice the rings arrive at the TBM in stacks configured such that the segments unload into the correct sequence in the segments feeder, so this was not difficult to manage. At the time of writing the lining has been installed around the first 1,000 ft radius curve. The Up and Down rings have built well with the key near the crown of the tunnel in most cases.

### Guide Rods

The ring incorporates longitudinal guide rods on the radial joints. These rods enhanced the quality of the builds considerably and reduced the amount of steps and offsets normally seen with precast liners during start up while getting through the learning curve. The benefit of the guide rods was worth the extra cost in procurement and should be considered on future projects.

### Performance of Gaskets

The cast-in-place gaskets performed quite nicely in the field and were more robust during erection. "Squeezing gaskets" and bunching seemed to occur less during the build process. Traditional glued in gaskets can dislodge during erection, however this cast in gasket stayed locked into position. One disadvantage of cast-in-place gaskets is having to perform in-field repairs. This requires that the gasket be cut away from the segment and a traditional glued in place gasket repaired in its' place. This creates another step for repairing gaskets. However, the replacement of damaged or dislodged gaskets has been infrequent compared to conventional glued in gaskets so far.

At the time of writing, the gasket appears to exhibit good performance with respect to leakage criteria. Only minimal leakage through the gasket has been witnessed and could be considered essentially zero within the highest head ground water zone of the tunnel.

### Grouting Performance

While the analysis of the internal pressures robustly demonstrated that the lining could resist the internal pressures, the robustness of the solution also relies on good contact with the ground all the way around the ring. Therefore it was necessary to ensure that the annulus between TBM and ground was adequately grouted. To this end, the following requirements were incorporated into the specification:

- Minimum grout pressure of 1 bar above hydrostatic
- Maximum grout pressure of 2 bar above hydrostatic
- Volumes monitored to ensure that they are no less than 90% of the theoretical volume
- Systematic proof grouting 20% of the first 40 rings to verify the primary grouting procedure, ultimately reducing to once per 305 m (1000 ft) once the performance of the primary grouting system was verified

In practice, the grouting performance has been excellent. The associated maximum inward displacements have been cross referenced with the design relaxation to demonstrate that actual ground pressures are greater than the lower bound determined by the internal pressure analysis. Proof grouting records demonstrate the effectiveness of the grouting scheme, with an average of 0.7% of the primary grouting volume injected in the proof grout exercise, and a maximum of 1.3%.

### Diametrical Measuring Points

Diametrical measuring marks cast into the segment allowed survey crews to accurately measure diameter without having to guess at location. The marks could accurately be measured for true diameter measurements. This helped to quickly measure ring diameter in the field during survey campaigns.

Actual build performance to date has been very good. The survey data on the shape of the for the first 110 rings shows excellent circularity, with an average out of round of 0.13% and a maximum of 0.28%, compared to a 1% tolerance. Both squat (where tunnel sides move out and crown moves down) and ovalization (the opposite action to squat) have been observed in approximately equal measure, suggesting good build close to circular, and very little additional movement under grouting. The authors are of the view that this demonstrates the benefits of a trapezoidal/parallelogram arrangement, as the skewed joints assist in maintaining circularity of the ring.

### Key Movement

During the initial phases of mining, axial movement of up to 15 mm (half an inch) was witnessed on some of the keystones during the ring build process when the thrust jack was removed. This has been seen before on other projects and can be very alarming for the project team. Restraining devices were immediately installed to prevent the motion on all ring builds. A load cell was used to gather information on the amount of force the key was under to cause movement and tests were performed to determine if the ground or water pressure had anything to do with the movement. It was determined that this movement could occur directly after the ring build process within the tail shield indicating that the force from the gasket was driving the movement. At the time of writing a more sophisticated restraining device has been implemented to prevent motion, and on-going testing is being conducted for other potential solutions. It should be noted for future designs that the radial bolts could be angled more towards the axis of the tunnel, thus engaging the bolt into tension which would reduce the likely hood of this occurrence.

## CONCLUSIONS

The design of the segmental lining for the Blue Plains Tunnel offers a robust solution to the client's problem: a durable one-pass lining. In the process of designing the lining a few interesting problems and useful solutions have been developed as follows:

- Design using closed for solutions to start with has allowed the designer to get a good understanding of the ground structure interaction and allowed targeted use of finite element analysis tailored to the actual circumstances.
- A study of the internal pressure case has highlighted the difficulty of estimating lower bound ground loads, but also shown how the problem can be tackled in a robust manner.
- TBM operation, and specifically annulus and grouting pressures, have a significant effect on the load in the lining post construction. These are difficult to estimate and this should be considered before embarking on refined finite element analysis of this kind of problem.
- Use of up and down tapered rings instead of left and right offers good control of left and right curves while keeping the key in the crown most of the time.
- Use of cast-in gaskets is providing a more robust solution and better waterproofing.

Overall, client, designer, and contractor have worked together to produce a clean solution: one that meets the requirements in service, is buildable, and whose design is robust despite some unusual circumstances to design for.

## REFERENCES

Curtis, D.J., Discussion on reference 1, Geotechnique Vol 26 (1976), pp. 231–237.

Duddeck, D., Erdmann, J., On Structural Design Models for Tunnels in Soft Soil. Underground Space Vol 9, pp. 246–259, 1985.

Hashash Y., et al., Seismic design and analysis of underground structures, 2001, Tunneling And Underground Space Technology, v. 16, pp. 247–293.

King, M.R., The Design and Use of Steel Fiber Reinforced Concrete Segments, RETC, Seattle, June 2005.

Muir Wood, A.M., The circular tunnel in elastic ground. Géotechnique 25, No 1 (1975), pp. 115–127.

# Sanitation Districts of Los Angeles County Clearwater Program—Joint Water Pollution Control Plant Effluent Outfall Tunnel

**David Haug**
Sanitation Districts of Los Angeles County

**David Yankovich, Craig White, and Jon Kaneshiro**
Parsons Corporation

**Steve Dubnewych**
Jacobs Associates

**ABSTRACT:** In November 2012, the Sanitation Districts of Los Angeles County finalized a comprehensive Master Facilities Plan (MFP) for its Clearwater Program. One of the recommendations in the MFP was to construct a new Effluent Outfall Tunnel to provide sufficient capacity and redundancy for existing 8- and 12-foot diameter tunnels built in 1937 and 1958, respectively. The project will involve building a new 7-mile long, 18-foot internal diameter tunnel from the Districts' Joint Water Pollution Control Plant to a manifold structure at Royal Palms Beach. Connections to an existing, active 14-foot diameter force main and four existing ocean outfalls ranging from 5 to 10-foot diameter are required. This paper will discuss preliminary details of the tunnel final design phase and anticipated geological features for the tunnel alignment.

## INTRODUCTION

The Sanitation Districts of Los Angeles County (Sanitation Districts) are 23 independent special districts serving approximately 5.3 million residents in Los Angeles County. Seventeen of the districts that furnish sewerage services to the metropolitan Los Angeles area are signatory to a Joint Outfall Agreement that provides for a regional, interconnected system of facilities known as the Joint Outfall System (JOS). The JOS service area is shown in Figure 1. The JOS serves an area that encompasses 73 cities as well as unincorporated territory and parts of the City of Los Angeles. The JOS provides wastewater collection, treatment, reuse, and disposal for residential, commercial, and industrial users, and it includes seven treatment plants, the largest of which is the Joint Water Pollution Control Plant (JWPCP), located in the City of Carson. Currently, secondary treated effluent is conveyed through two 6-mile long parallel tunnels, 8- and 12-feet in diameter, from the JWPCP to a manifold structure located at Royal Palms Beach, near White Point on the Palos Verdes (PV) Peninsula, from which four seafloor outfalls extend offshore. The two main outfalls, 90- and 120-inches in diameter, extend approximately 1.5 miles offshore and discharge at a depth of approximately 200 feet below sea level. The other two outfalls, 60- and 72-inches in diameter, are used for additional capacity during heavy rain events and extend a shorter distance offshore and discharge at shallower depths. A schematic of the existing system is shown in Figure 2. Both tunnels and main outfalls are required to be in service at all times.

The Clearwater Program is a comprehensive planning effort undertaken by the Sanitation Districts to develop a Master Facilities Plan (MFP) and accompanying Environmental Impact Report/Environmental Impact Statement (EIR/EIS) documentation to guide the management and development of the JOS through the year 2050. One objective of the Clearwater Program is to provide overall system reliability by allowing for the inspection, maintenance, repair and replacement of aging infrastructure. The 8-foot tunnel was constructed in 1937 and the 12-foot tunnel in 1958. Because both tunnels are always in service and flow full every day, neither has been inspected for over 55 years. Another objective of the Clearwater Program is to ensure there is sufficient capacity within the JOS to meet the needs of future population growth. In January 1995, the JOS service area was inundated by two major back-to-back storm events. The resulting peak wastewater flows in the sewerage system from these storm events nearly exceeded the capacity of the JWPCP tunnel and ocean outfall system.

A multi-step, program-wide screening process was conducted and then followed by a project-specific

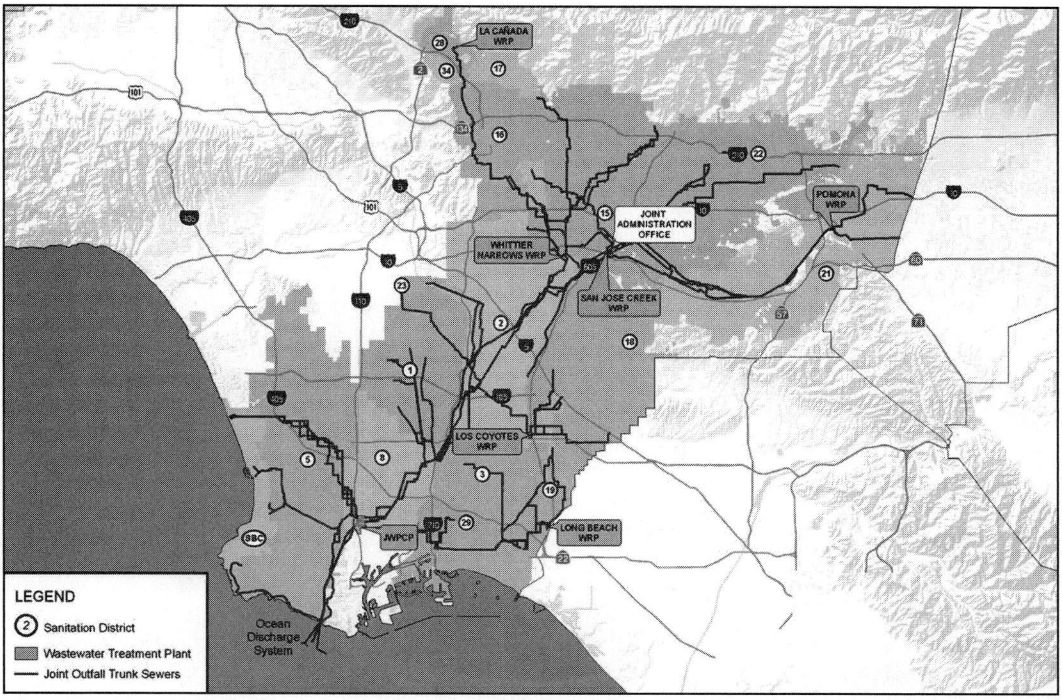


**Figure 1. Sanitation districts joint outfall system service area**

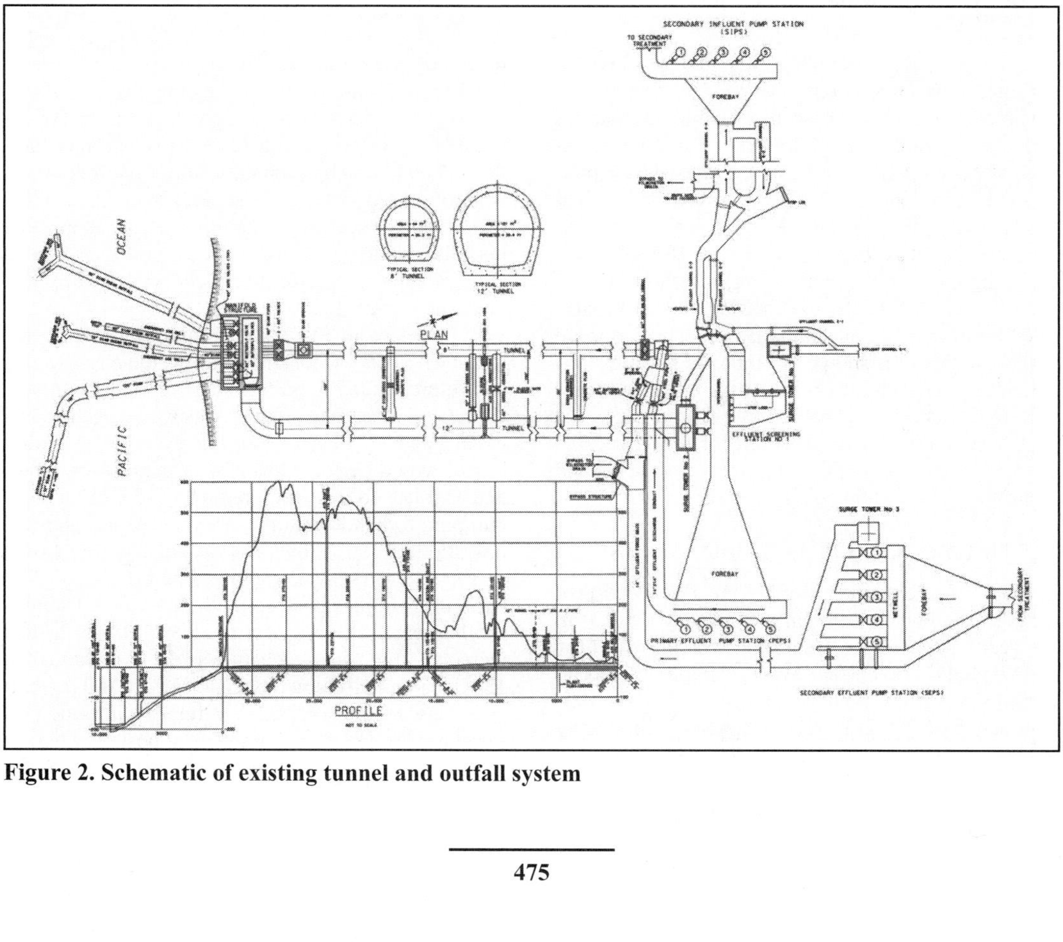


**Figure 2. Schematic of existing tunnel and outfall system**

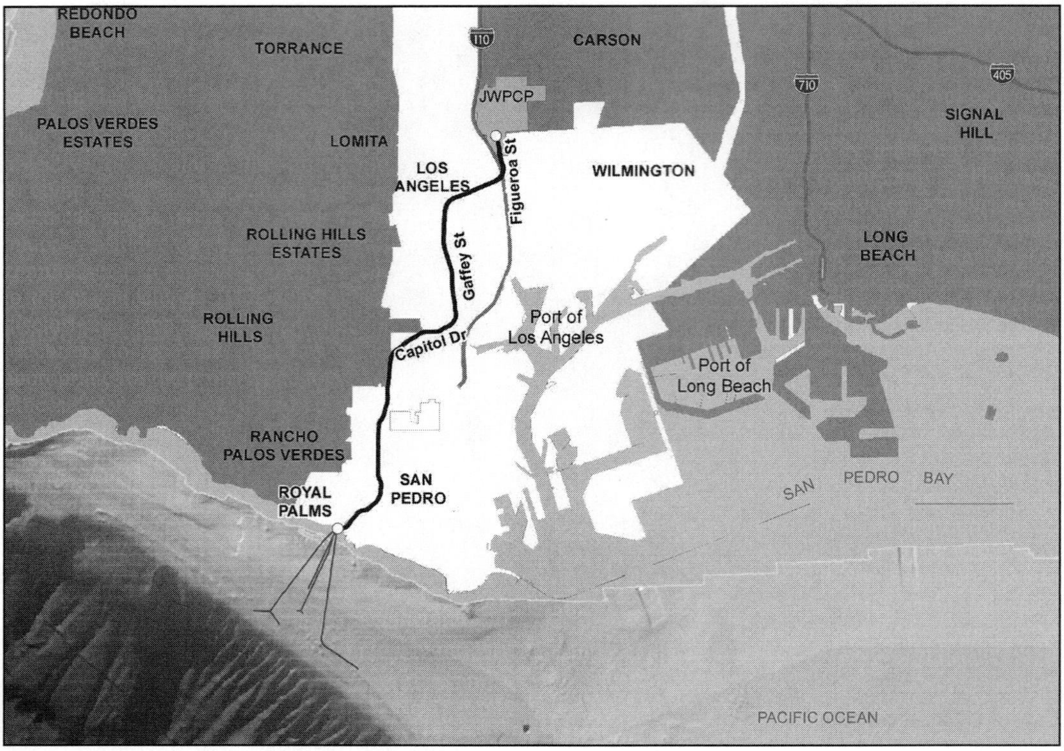


**Figure 3. Recommended tunnel alignment**

alternatives analysis, which evaluated over 50 potential alternatives and determined four highest ranked feasible alternatives. These four viable alternatives were then carried forward for detailed environmental analysis in the EIR. The EIR was released for public review in January 2012. Based on the detailed review of the four feasible alternatives, the recommended project is to construct a new 18-foot diameter, approximately 7-mile long, on-shore tunnel from the JWPCP to the existing White Point manifold structure at Royal Palms Beach. The proposed tunnel alignment is shown in Figure 3. The new tunnel, when connected to the existing ocean outfalls, will have adequate capacity to accommodate the peak wastewater flows projected for the year 2050 and will enable the Sanitation Districts to inspect and repair the existing tunnels, if necessary.

## TUNNEL DIAMETER DETERMINATION

In the MFP, the average flow at the JWPCP is projected to be 400 million gallons per day (MGD) in the year 2050 and the associated wet weather flow is 927 MGD. The current JWPCP tunnel and ocean outfall system has a maximum capacity of approximately 675 MGD. The system capacity is limited by the amount of internal pressure that can be exerted on the existing tunnels. The outfalls are capable of handling a greater internal pressure than the existing tunnels. The combined maximum capacity of the outfalls is greater than the 927 MGD projected storm flow. A condition assessment of the existing outfalls was conducted and found the pipes to be in excellent condition. If necessary, rehabilitation of the existing outfalls could be performed in the future to extend their remaining service life well beyond the 2050 planning horizon.

In selecting the diameter of the new tunnel, variables such as constructability, the present day and future hydraulic performance of the system, and operational and construction costs were analyzed and balanced against each other. For example, a smaller tunnel would perform better with present day flows and cost less to construct, but it would require more pumping to handle future flows and would increase operational cost. Conversely, a larger tunnel would convey the present day and future flows with less pumping, but the construction cost would be significantly greater. Diameters ranging from 14- to 22-feet were analyzed. Ultimately, it was determined that an 18-foot internal diameter tunnel gave the best balance between present day and future hydraulic performance while also being the most cost effective.

## GROUND CONDITIONS AND TUNNEL BORING MACHINE (TBM) SELECTION

The 18-foot internal diameter tunnel will be built using either an earth pressure balance (EPB), slurry pressure balance (Slurry), or a hybrid TBM. The outside diameter is expected to be approximately 21-feet in diameter, but will require refinement during the precast concrete segmental lining design. These pressurized-face machines are considered the only suitable means for underground excavation given the expected ground conditions along the alignment and for meeting other project requirements. The choice between the three types of TBM's is influenced by several factors, including grain size distribution; soil and rock strength; hazardous gases; and the feasibility of soil separation and muck disposal.

### Geologic Profile

The preliminary geological profile along the proposed tunnel alignment is shown in Figure 4. Along the alignment there are two distinct geological types of material the TBM will encounter, soil and rock. The northern part of the alignment will be located within Quaternary-aged deposits that include Holocene sediments consisting of fill, alluvium, and terrace deposits. These are underlain by Pleistocene sediments which include the Lakewood Formation and the San Pedro Formation. Both formations are primarily consolidated sediments and include aquifers which will have an impact on the selection of the machine type. The southern part of the alignment within the PV Hill will be located in rock-like material that includes Miocene-age Malaga Mudstone, Altamira Shale, and possible Miocene Volcanic rocks, San Onofre Breccia and possibly the Catalina Schist. The materials are anticipated to exhibit a range of ground behaviors, from soil-like or weak rock-like to raveling or squeezing ground conditions. Also, interbedded volcanic intrusive and extrusive beds as well as dolomite beds are expected which can exhibit strong rock properties. Hydrocarbons and hydrogen sulfide may also be encountered. With a single heading proposed, selecting a TBM that can accommodate both soil and rock will require additional geotechnical investigations to better define underground conditions.

### Groundwater Conditions

Along the alignment, four hydrogeological regimes were identified as shown in Figure 5. In the area surrounding the JWPCP (Regime 1), the groundwater level was measured at an elevation of approximately 15 feet below Mean Sea Level (MSL). The second regime is a band of injection wells called the Dominguez Gap Barrier. Potable water is injected into the ground to prevent saltwater from intruding into the Los Angeles Basin aquifers. The water level within the wells is usually kept at an elevation of +10 ft. MSL. Regime 3 is located south of the Dominguez Gap Barrier and reliable water level data is absent along the alignment. Given the proximity to the Pacific Ocean, the current assumption is that groundwater levels have generally equilibrated to sea level. The presence of groundwater in the bedrock formations (Regime 4) can be highly variable and vary greatly over short distances. Due to the limited data and the variability, no groundwater level is specified at this time for Regime 4. It is expected that the groundwater head along the alignment in the alluvium material will be less than 3.5 bars, while in the rock-like material of the PV Hill, should a zone of highly fractured rock filled with ground water extend from the ground surface to the tunnel, there is a possibility the hydrostatic pressure could reach as high as 11 or 12 bars.

### Cutterhead Selection

Regardless of which type of TBM is used, a bi-rotational cutterhead equipped with cutting tools to remove the ground will be utilized. A mixed ground cutter head will most likely be designed given the mixture of soft ground and weak rock. Back loading saddles or cutter boxes that allow the use of either disk cutters or rippers will most likely be incorporated into the cutterhead design. Based on preliminary data collected on the sedimentary rock along the alignment, the typical unconfined compressive strengths should be less than 150 lb/in$^2$, however, the lenses of higher strength material could be 5,000 lb/in$^2$ or greater.

### TBM Interventions

Throughout the entire tunnel length, TBM cutterhead interventions will be necessary. Ideally the interventions will be performed under free air, but access to the cutterhead while in soft ground or highly fractured rock beneath the water table may require the use of compressed air or possibly a mixed-gas environment. If longer interventions are required to perform repair work or change multiple cutters, the use of mixed gas under saturation conditions may be necessary. Working under saturation conditions will require the use of a saturation diving shuttle to transport the workers from the hyperbaric living quarters on the surface to the airlock on the TBM. While an 18-foot internal diameter tunnel will have enough space for ventilation, switches and pumps, there may not be enough space within the TBM for the shuttle to be connected directly to the bulkhead. To connect the shuttle directly to the bulkhead airlock, equipment

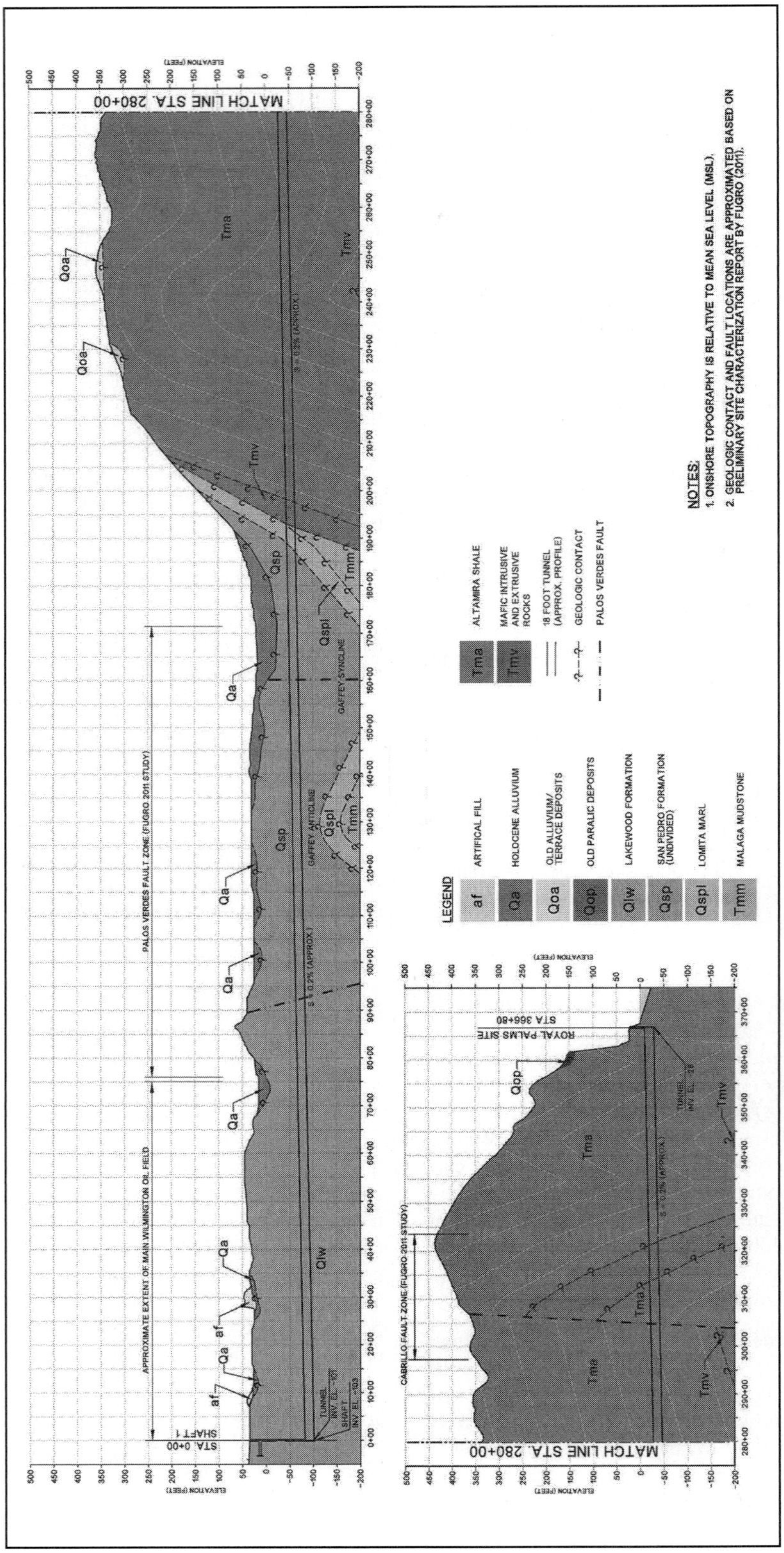


**Figure 4. Preliminary geological profile along the tunnel alignment**

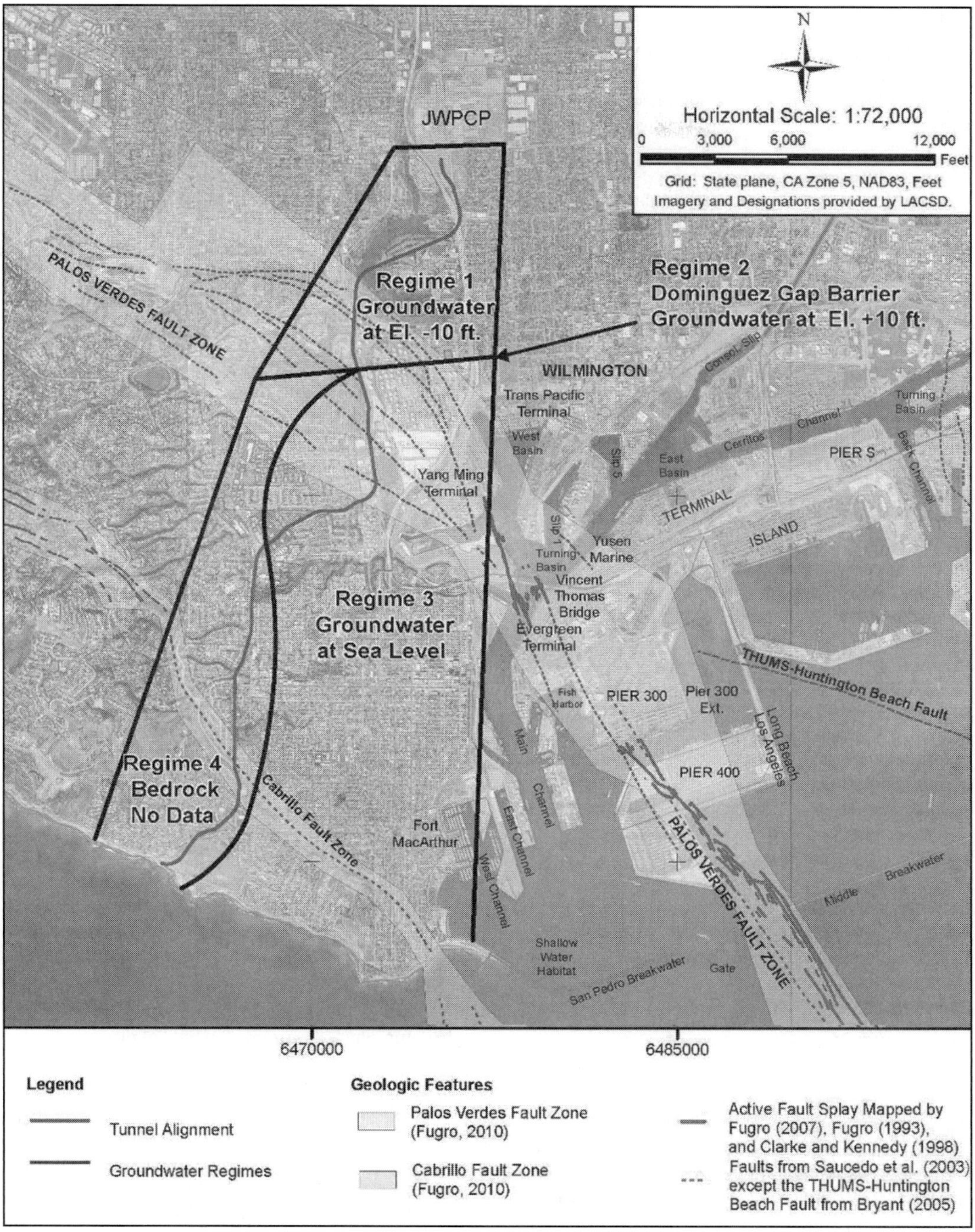


**Figure 5. Groundwater regimes along the tunnel alignment**

will need to be removed to create a large enough space for the shuttle. Connecting the shuttle to the TBM bulkhead airlock by a transfer tube appears to be more appropriate for the proposed tunnel. As the geotechnical conditions along the alignment are better defined, it might be possible to avoid saturation diving conditions by the use of ports in the TBM to provide for grouting, or ground freezing, to create a conditioned environment that has a reduced pressure at the cutter face.

### Muck Handling

A major area of analysis in the Clearwater Program EIR was the effect the tunneling operations will have on the air pollution and greenhouse gases in the surrounding environment while using diesel locomotives. The baseline for the analysis was an EPB TBM because larger horsepower locomotives will be required to transport the loaded muck cars. To mitigate the impact, different types of locomotives such as electric or natural gas were investigated. The analysis determined the batteries on electric locomotives would drain rapidly when the alignment becomes significantly long. The amount of time necessary to recharge the batteries was determined to have a significant impact on the production rate of the TBM. Either slower advance rates would be realized, or additional locomotives would be required to maintain an adequate amount of trains entering the tunnel with supplies. Neither option was deemed an acceptable alternative. The use of natural gas locomotives inside the confined space of the tunnel was determined impracticable due to safety concerns. Diesel locomotives were deemed the only power source capable of handing the transportation of the muck and supply cars over the entire length of the tunnel. Utilizing conveyor belts for muck disposal would reduce the amount of emissions in the tunnel by allowing small horsepower locomotives to be used. The reduction in emissions was not analyzed in the EIR because it was assumed the locomotives used with a Slurry TBM would be the same horsepower with an EPB TBM using a conveyor belt as the muck disposal method. To reduce the amount of diesel particulates entering the tunnel environment and being exhausted into the surrounding community, a Tier 4 engine on the locomotives was mandated as part of the mitigation measures.

## WILMINGTON OIL FIELD

The northern part of the alignment is located in the southwestern margin of the Wilmington Oil Field. The oil producing strata of the oil field is located at depths of approximately 2,500 to 4,000 feet below the ground surface. As a result, the probability of encountering natural oil deposits during tunneling is low. The oil field, however, contains numerous active, idle, and abandoned oil wells. Few wells are located in the vicinity of the alignment, but there is a possibility of encountering unknown abandoned wells. As part of the substructure utility research, methods such as magnetometer or ground penetrating radar may be utilized to identify any unmarked wells. The specifications for the project will also have provisions for removal of a well should one be encountered during tunneling operations. How the well will be removed, either through the TBM with an intervention, or from the surface, has not been determined at the time this paper was written.

## PRECAST CONCRETE SEGMENTAL TUNNEL LINING

Segmental precast concrete lining systems are typically used for tunnel excavation using EPB or Slurry TBM's. For the project, both one-pass and two-pass systems were considered. Although a one-pass lining system is more suitable, some type of special lining may be required to contend with the internal operating pressure to prevent leakage and deal with any offset created by movement along the PV Fault. For the projected wet weather flow of 927 MGD, the head on the system at the JWPCP could be approximately 100 to 115 feet. Although the effluent is treated to secondary levels, any leakage into the surrounding groundwater table is not acceptable. A possible design to handle the high internal pressure includes installation of tension reinforcement in the hoop direction with special connections carrying the tension across the radial and shear joints at the circumferential joints. To limit cracking and leakage, the quantity of reinforcement steel may be large resulting in a "waffle slab" segment where the segment is thicker around the joints and thinner in the middle. The design will allow adequate space for bolt connections while minimizing the volume of high-strength concrete. After installation, the waffle pockets could be filled in with a low-strength concrete to provide a smooth pipe-like finish. Another design to handle the high internal pressures includes installation of a post-tensioning strand inserted in a continuous circumferential duct embedded in the segments.

Within the PV Fault Zone, a two-pass system incorporating a 16-foot ID steel pipe inside the 18-foot ID precast concrete segmental liner is proposed to contend with any displacement of the fault and prevent the effluent from leaking out of the tunnel. The fault zone is shown in Figure 5. In advance of comprehensive geotechnical explorations, the assumption was made that the lining would extend the whole fault width of approximately 5,000 feet. An illustration of the crossing is shown in Figure 6. The annular space would be backfilled with low strength grout or crushable concrete. After the PV Fault characteristics are better defined, the length of two-pass lining may be reduced, or possibly eliminated entirely.

## SEISMIC SETTING

Preliminary seismic design criteria for the project was developed from seismic design criteria used for similar projects, the service life of the project, and the geologic conditions in the project area. Three levels

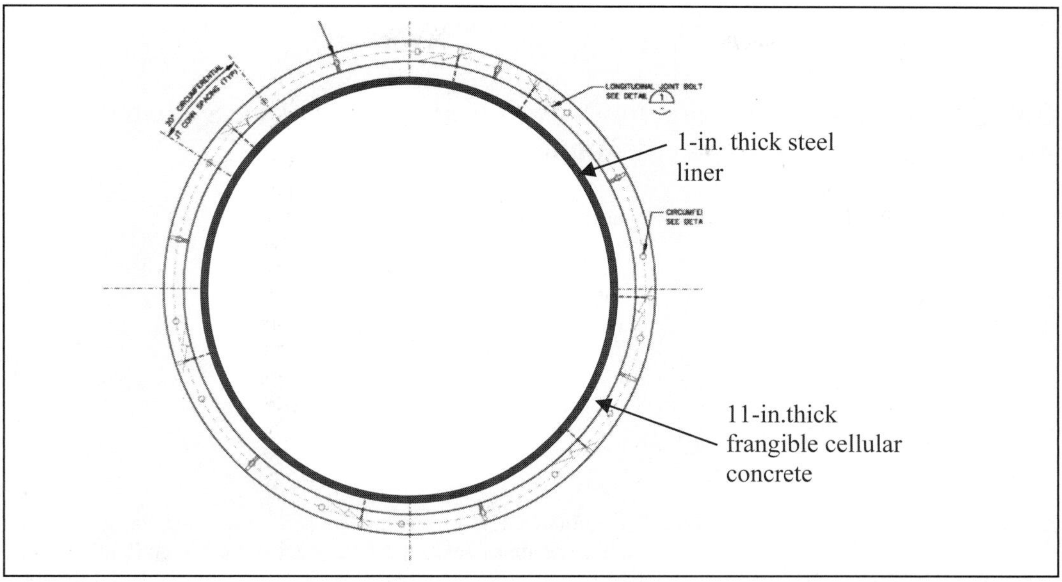


**Figure 6. Typical cross section of 16-ft ID liner at PV Fault crossing**

of seismic exposure were considered for the project which corresponded to a 475-, 975-, and 2475-year average return period. Evaluations were performed to identify potential geotechnical and seismic issues that could pose hazards to the structural integrity of the tunnel. The principal hazards were determined to result from wave propagation (WP) and permanent ground deformation (PGD). Seismically-induced WP will stress the tunnel liner and result in strains. WP may also produce hydrodynamic forces, which could produce a water hammer within the pipeline. In loose to soft-to-medium dense soils, PGD consists of liquefaction-induced settlement. Fault offsets of different magnitudes may result from PGD, causing shear, tension, and/or compression that may lead to failure or collapse of the tunnel. A return period of 975-yr was selected for the design of the tunnel, which corresponds to adisplacement on the PV Fault of 1.0 to 1.3 feet resulting from a magnitude 7.3 earthquake.

## CONCLUSION

Once the geotechnical investigation begins for final design, several of the issues related to completing the design of the tunnel and associated structures will be determined. Final design, which is currently underway, is expected to take approximately 2.5 to 3 years to complete. For the geotechnical program, 54 borings totaling approximately 14,000 feet are anticipated to be drilled. Construction of the tunnel and associated structures is envisioned to be packaged under a single contract. Advertising and bidding is tentatively set for late 2015 or early 2016. The construction duration is estimated to take approximately 7.5 years after notice to proceed is given. The 18-ft tunnel and associated structures, once built, will ensure the needs of the JOS are fulfilled for many decades to come.

## ACKNOWLEDGMENTS

The authors thank their respective employers for permission to publish this paper. Appreciation is also due to C. Jin, T. Sung, J. Ganz, S. Highter and A. Chang of the Districts; and Fugro Consultants, the geotechnical program consultant.

## REFERENCES

Dubnewych, S., M. Torsiello, J. Kaneshiro, and D. Haug, 2009, Design Considerations and Evaluation Process for a New Tunnel and Ocean Outfall Project, Rapid Excavation and Tunneling Conference 2009 Proceedings.

Fugro Consultants, Inc., 2011, Preliminary Site Characterization Report JWPCP Tunnel and Ocean Outfall Project.

Parsons, in association with Jacobs Associates and Blank & Veatch, 2011, Professional Design Services for the Preliminary Engineering of the Joint Water Pollution Control Plant Tunnel and Ocean Outfall, *Feasibility Study Report*.

Sanitation Districts of Los Angeles County, 2011, Clearwater Program Environmental Impact Report/Environmental Impact Statement, *Final*.

# Going Deep: Upgrading Honolulu's Wastewater Conveyance System with a Large-Diameter Gravity Tunnel

**Stephen J. Klein, Thomas W. Pennington, and Phaidra D. Campbell**
Jacobs Associates

**Richard Harada**
Wilson Okamoto Corporation

**Glenn Okita**
City and County of Honolulu

**ABSTRACT:** The City and County of Honolulu (CCH) is constructing a deep wastewater storage and conveyance tunnel on the windward side of Oahu. The approximately 3 mile (4.8 km) long, 10 ft (3 m) diameter tunnel will help bring Honolulu into compliance with a 2010 Consent Decree between CCH, EPA, and local stakeholders. The tunnel, combined with improvements to treatment facilities in Kaneohe and Kailua, replaces an existing pump station and force main system and consolidates flows from over 200 miles (320 km) of collection lines to a single storage and conveyance system. This paper discusses major planning and design considerations for the tunnel, and the role Oahu's sensitive environs played in its selection.

## INTRODUCTION

The City and County of Honolulu (CCH), Department of Environmental Services is undertaking several improvements to its existing wastewater collection, treatment, and disposal system in the Kaneohe-Kailua-Kahaluu service area in the Koolaupoko District on the windward side of Oahu, Hawaii. The largest of these improvements involves construction of a 10 ft (3 m) inner diameter storage and conveyance tunnel which is intended to replace an existing 42 in. (1,070 mm) inner diameter force main between the Kaneohe Wastewater Pretreatment Facility (KWPTF) and the Kailua Regional Wastewater Treatment Plant (KRWWTP). The new tunnel will provide storage for peak wet-weather flows, thereby preventing spills that would occur when wastewater flows exceed the capacity of the existing KRWWTP.

Several improvements are also planned at the KWPTF and KRWWTP, including a new 37 million gallon per day (mgd) (140 million liter per day) pump station and headworks facility at the downstream end of the tunnel, several diversion structures and pipelines to divert and consolidate flows away from the existing treatment facilities and existing pump stations, odor control facilities, and electrical and fiber optic upgrades. Once the tunnel is operational, it will allow the influent pump station and pretreatment equipment at the KWPTF to be decommissioned. Flows will then be pumped from the tunnel from the new pump station at the KRWWTP, and ultimately treated for release through the Mokapu Outfall, which extends approximately 5,000 ft (152 m) offshore.

## EXISTING FACILITIES

The existing Kailua-Kaneohe-Kahaluu wastewater service area is in the Koolaupoko District on the windward (northeast) side of the island of Oahu. The service area encompasses approximately 36,500 acres (14,770 hectares) and includes the suburban communities of Kailua and Kaneohe, and the rural-agricultural community of Kahaluu. The existing collection system consists of approximately 200 mile (320 km) of gravity lines and force mains ranging in diameter from 6 to 66 in. (150 to 1,675 mm), and 23 wastewater pump stations. An important element of the collection system is an existing 42 in. (1,070 mm) diameter force main, Force Main No. 1, which currently conveys pretreated flows collected at the KWPTF and conveys them to the treatment plant at the KRWWTP. As part of the stipulated 2010 Consent Decree between the U.S. Environmental Protection Agency (EPA), CCH, and local stakeholders, CCH is required to design and construct a gravity tunnel between the KWPTF and KRWWTP to handle foreseeable wet weather flows so that future wastewater overflows are prevented.

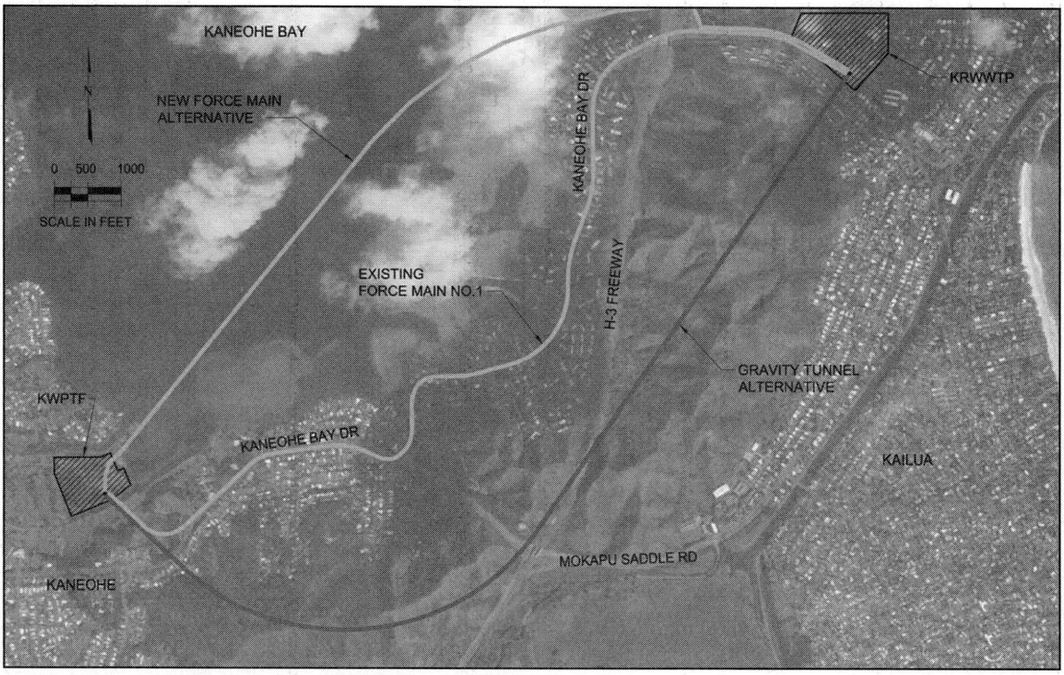


**Figure 1. Force main replacement alignment alternatives**

## ALIGNMENT SELECTION

In preparation for the 2010 Consent Decree, several alternatives were considered to supplement/replace the existing Force Main No. 1 (Figure 1). Two basic options were evaluated: (1) a new force main, and (2) a gravity conveyance/storage tunnel. The objective of each alternative is to convey wastewater from the KWPTF to the KRWWTP sites and to address the problem of peak infiltration and inflows that can occur during wet weather and result in overflow spills. The gravity tunnel alternative addresses this problem by providing sufficient storage capacity to hold peak flows until they can be treated after the storm passes, whereas the force main alternative would require construction of equalization/storage basins to accommodate peak flows.

For the force main alternative, two alternative alignments were considered. One alternative involved an overland route, essentially parallel to the existing force main along Kaneohe Bay Drive. This alternative would be primarily excavated by open cut trenching methods. Although relatively inexpensive, this alternative was eliminated during the planning stage because of the risk of damaging the existing force main during construction and the impacts to the community, including road closures and disruption to adjacent residences. The other alignment alternative involved construction of the force main beneath Kaneohe Bay and would consist of approximately 11,000 ft (3,350 m) of 42 in. (1,070 mm) diameter pipe constructed by trenchless methods, either horizontal directional drilling (HDD) or microtunneling. This alternative was preferred over the Kaneohe Bay Drive route because of its lower cost and reduced impact to the residential communities of Kaneohe and Kailua. However, there were substantial construction, permitting, and environmental risks due to the alignment extending beneath the protected waters of Kaneohe Bay where access during construction would be extremely limited. Other drawbacks of the two force main alternatives include requirements for a flow equalization (i.e., storage) basin to provide storage during peak wet weather flows at the KWPTF since the force main would not have the capacity to store these flows, and continued operation and maintenance of the existing Kaneohe Effluent Pump Station.

For the gravity tunnel alternative, two alignments were initially studied. The first followed the route of the existing force main alignment along Kaneohe Bay Drive, and the second followed an in-land route beneath an undeveloped ridge of the Oneawa Hills (interpreted to be a portion of the rim of the collapsed Ko'olau Volcano caldera). The gravity tunnel alignments would require a 10 to 13 ft (3 to 4 meter) inside diameter tunnel that begins at the existing KWPTF site in Kaneohe and terminates at the KRWWTP site in Kailua. The gravity tunnel would be much larger than the force main

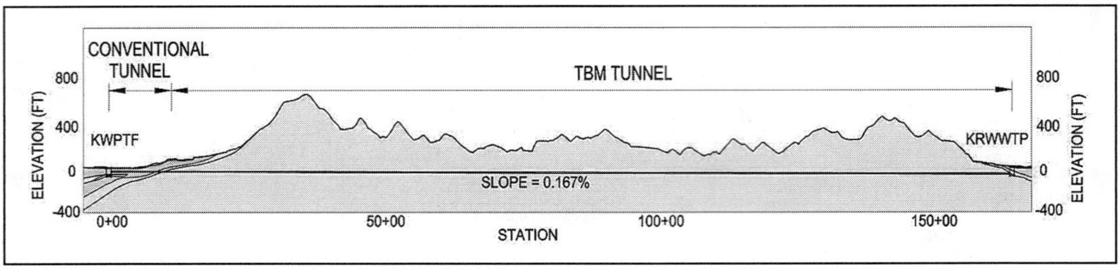


**Figure 2. Profile of Kaneohe/Kailua gravity sewer tunnel**

alternatives so it could be used to store peak flows, while also providing conveyance for daily flows between KWPTF and KRWWTP. Another advantage to the gravity tunnel alternative is that additional cost savings could be realized by the decommissioning of the pre-treatment facility and pump stations at the existing KWPTF, and decommissioning of Force Main No. 1. In comparison to the force main alternatives, the gravity tunnel alternative had an overall lower life cycle cost and also carried fewer construction risks than the force main alternatives.

The in-land tunnel alignment was the preferred gravity tunnel alignment since it was the most direct route between the KWPTF and KRWWTP, and it also had the most uniform geology. The alignment following Kaneohe Bay Drive was much more complex geologically, transitioning from saturated, soft ground to strong basalt four times, which would require a pressurized-face, soft ground/hard rock tunnel boring machine (TBM). Most of the in-land route could be constructed using a hard rock, main beam TBM, which would achieve higher advance rates, require less ground support, and consequently be much more economical. Also, the in-land alignment has a much lower potential for surface settlement and impacts to existing facilities overlying the tunnel since it would extend beneath the largely undeveloped Oneawa Hills.

After preparation of an Environmental Impact Statement, including consultation with the public and CCH, it was decided that the gravity tunnel following the in-land route was the preferred alternative for the project and that this design would be incorporated in the Consent Decree requirements. During final design, further refinements were made to the alignment to minimize the encroachment on privately owned properties and to provide a minimum radius of horizontal curve for excavation equipment. The tunnel profile was also optimized to minimize the tunnel depth. This would reduce construction cost of facilities, while maintaining a slope of 0.1% to maintain minimum flow velocity for conveyance and would allow for adequate sediment removal during tunnel operation. The finalized tunnel profile is shown in Figure 2.

## SUBSURFACE CONDITIONS

The project is located within the limits of the caldera of the former Ko'olau Volcano, which was one of the two shield volcanoes that formed the island of Oahu more than three million years ago. After the Ko'olau Volcano's shield-building stage, a period of quiescence began. During this time, the island's form was modified by erosional forces, including the occurrence of the massive Nuuanu Landslide that removed much of the eastern flank of the Ko'olau Volcano. The remaining western portion of the caldera wall now forms the summit of the Ko'olau Range to the west of the project site. It is believed the volcano underwent a rejuvenation stage less than 800,000 years ago when lava was confined by the walls of the collapsed volcano, resulting in thick, horizontally bedded lavas, and volcanic breccias. These deposits were also subjected to chemical and hydrothermal alteration from gasses and water within the caldera.

Volcanic dikes are common within the Ko'olau caldera and there can be as many as 1,000 dikes per mile (1,600 dikes per km) of horizontal distance, and average more than 100 per mile (160 per km) (Macdonald, 1956). Based on an examination of the frequency of dikes, lava flows, and cross-cutting dike relationships, it is believed that there may be an inner caldera and an outer caldera, which formed at different times. As a result of the project being within the collapsed caldera, the geology in this area is different and more complex than in areas outside of the caldera, where bedrock deposits consist largely of flow basalts that formed along flanks of the volcanic shield.

About 95% of the proposed tunnel alignment will be constructed through basalt rock, which is anticipated to consist primarily of dike complex rocks that formed from magma intruding into existing basalt breccia and basalt lava flows. Typically, the dikes are sheet-like structures, with steep dip angles and northwest trends (Walker, 1987). However, orientations vary and dikes often intersect one another, as shown in Figure 3. Dike complex, basalt breccia, and basalt lava flow rocks range from slightly to highly weathered, moderately hard to very hard, and medium strong to strong. Dike thicknesses are anticipated to vary, ranging from a few inches to tens of feet thick with an average thickness of 2.5 ft (0.75 m). Compressive strengths of the basalt rocks are as high as 36,000 psi (250 MPa) and average about 11,500 psi (79 MPa).

The remaining 5% of the proposed tunnel alignment will be constructed through Estuarine Deposits, which consist of organic silts and clays with varying amounts of sand and gravel, and organic material. These materials predominantly had blow counts of 4 blows per foot or less, which indicate very soft to soft soils. It is anticipated that this material will exhibit squeezing and fast raveling behavior when excavated; therefore, ground stabilization consisting of jet grouting will be required prior to excavating the tunnel in this section.

Groundwater along the project alignment occurs in two zones: a shallow perched aquifer in surficial soil deposits, and at depth in the bedrock. The primary sources of groundwater are from rainfall infiltration and from hydraulic connection with the adjacent Kaneohe Bay and Kailua Bay. At the KWPTF and KRWWTP sites, groundwater is brackish and groundwater levels are subject to seasonal and tidal fluctuations and vary from 7 to 21 ft (2 to 6.5 m) below the ground surface. The groundwater along the tunnel alignment generally varies from 80 to 190 ft (24 to 58 m) above the tunnel invert.

**Figure 3. Road cut near project site showing basalt dike intrusions**

## DESIGN AND CONSTRUCTION CONSIDERATIONS

The Kaneohe/Kailua Sewer Tunnel Project involves design and construction of 16,340 ft (4,980 m) of tunnel and excavation of three primary permanent shafts. Primary design and construction considerations for these structures are discussed in the following sections.

### Tunnel Lining

Because of the corrosive environment anticipated for the gravity tunnel, the tunnel lining is required to provide corrosion protection internally and externally throughout the minimum design life of 100 years. The design life is defined to be the expected life before replacement, major rehabilitation, or structural repairs are needed. There are various approaches to providing corrosion resistance to wastewater conveyance systems; however, the owner preferred the option of using a corrosion resistant lining (CRL) system. A CRL is a self-supporting structural system that is made of inert material, typically polymer-based, that is inherently resistant to hydrogen sulfide corrosion attack. Typical CRL systems used for this application include glass fiber-reinforced thermosetting resin pipe (GFRP), polymer concrete pipes, and PVC linings that are either cast into or chemically adhered to an existing concrete lining. The main factors that influence the selection of a particular CRL system are the required design life, tunnel diameter, external hydrostatic pressures, and cost.

After an analysis of each of the above tunnel lining options, the GFRP pipe was determined to be the most viable solution to use for the gravity tunnel. The structural design of the lining will also allow for backfill grouting of the annulus between the GFRP and the excavated tunnel wall. It is anticipated that the backfill material will be low-density cellular concrete backfill with low viscosity, permitting it to be pumped over the length of the tunnel. This GFRP pipe is generally impermeable and is connected to pipe segments using restrained pipe joints, typically gasketed bell-and-spigot or coupling joints.

Design considerations for the pipe included determining the pipe stiffness class and pressure class per the industry standard requirements set forth by the American Water Works Association (AWWA, 2005). Since the pipe is designed as a gravity pipe, little to no internal pressures will be introduced to the tunnel lining; therefore, a minimum pressure class of 25 psi (0.17 MPa) was considered acceptable, as confirmed by the hydraulic modeling conducted for the tunnel. The flexural requirements included buckling theory and minimum deformations induced from maximum external loading, which was a combination of the highest hydrostatic head and ground conditions. In addition, single-lobe buckling procedures were calculated using the Jacobsen Theory (1974), which estimates the potential gap between the tunnel and surrounding backfill, where buckling would potentially occur. A minimum stiffness class of 36 psi (0.25 MPa) was determined to be required to meet this condition with a factor of safety of 1.5.

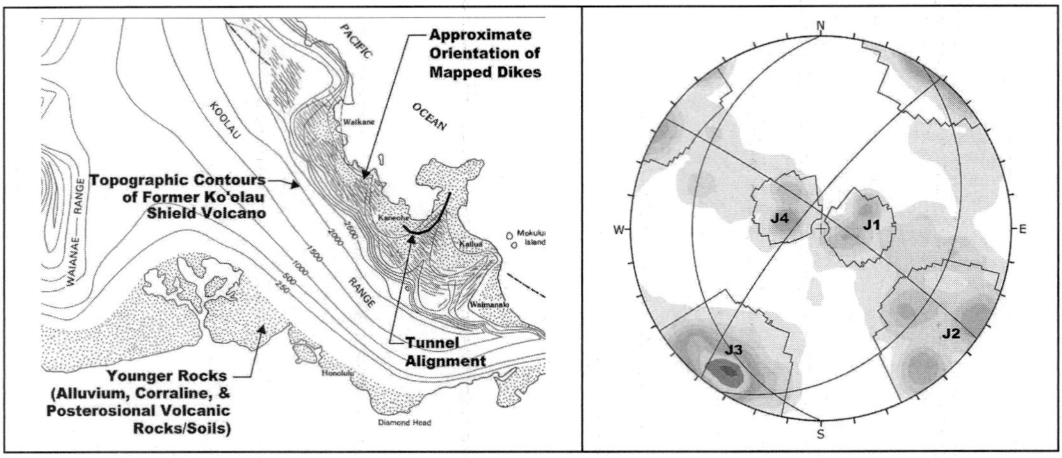


**Figure 4. Distribution of dike complex rock near project site (left) (after Walker, 1987) and mapped discontinuity orientations along tunnel alignment (right)**

Protection of the pipe during temporary loading conditions associated with transportation, handling, storage, and installation of the pipe, will be the responsibility of the contractor.

Seismic design loads on the tunnel were also characterized in terms of deformations and strains imposed on the pipe lining structure by the surrounding ground during earthquake shaking. Based on the design 2,500-year return period, the estimated short period peak ground acceleration (PGA) is 0.64g. The two modes of earthquake-induced deformation included ovaling and flexural/axial deformations induced by vertically propagating shear waves and horizontally propagating waves, respectively. The procedure followed the method endorsed by the International Tunnelling Association Working Group (Hashash et al., 2001).

## Tunnel Excavation by TBM

Because of the anticipated high quality of the rock mass, the majority of the tunnel (approximately 95% of tunnel alignment) will be excavated using a main beam tunnel boring machine (TBM) with a two-pass final lining approach. Based on the required finished diameter of the tunnel, the TBM tunnel will likely have an excavated diameter of about 13 to 15 ft (4 to 4.5 m). Rock cover above the tunnel will range from approximately 80 to 670 ft (24 to 204 m). TBM excavation is anticipated to be through unweathered to slightly weathered, hard, medium strong to extremely strong, and moderately fractured to massive basalt rock. However, at dike boundaries it is anticipated that the rock mass will be weakened locally and tend to have more fractures. Discontinuities were observed both in the basalt rock core and from road cuts near the tunnel alignment, which consisted primarily of variable orientation joints that are smooth to slightly rough with tight to narrow apertures. The most dominant discontinuity, Joint Set 3 (J3), as shown in Figure 4, is interpreted to correspond to the predominantly northwest-southeast strike of dikes in the vicinity of the project.

Discrete shear zones are anticipated within the rock mass, which will trend northwest or northeast at an inclination greater than 60 degrees from horizontal and will range in apparent width from a few inches to 10 ft (3 m). Because of the near-vertical orientation of the shear zones, these were difficult to detect during the subsurface investigation program from the predominantly vertical borings; however, mapping performed at a nearby quarry indicated these zones are present and likely persistent at tunnel depth. As part of the geotechnical baseline report established for the project, it was assumed that a total of 10 shear zones will be encountered, each ranging in width from 2 to 10 ft (0.6 to 3 m).

For the purposes of describing rock mass conditions in the excavated tunnel, three rock mass types (RMTs) were defined based on the physical characteristics of the rock and its anticipated behavior in the tunnel. The three RMTs are:

- B1: "Good rock," primarily characterized by unweathered to slightly weathered, massive/moderately jointed, and strong to very strong rock.
- B2: "Fair rock," primarily characterized by unweathered to moderately weathered, moderately blocky and seamy, and strong rock.
- B3: "Very poor to poor rock," primarily characterized by unweathered to highly weathered, very blocky and seamy rock with crushed zones, and variable strengths from very strong to very weak.

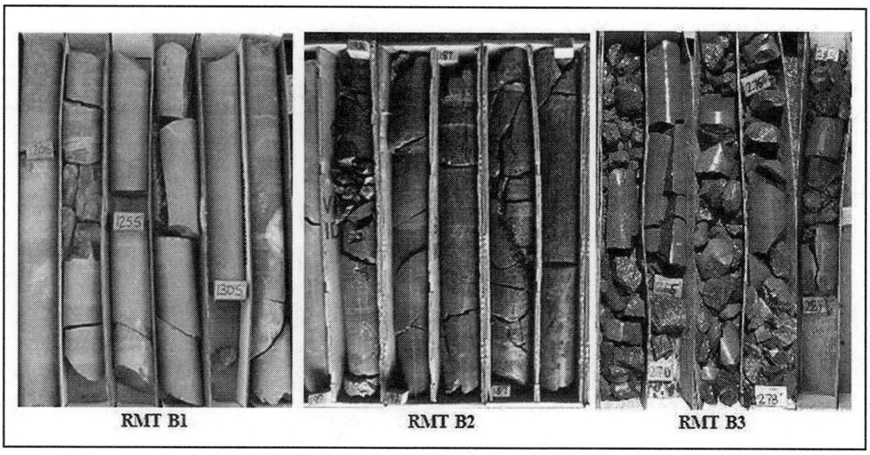


**Figure 5. Examples of RMTs anticipated along tunnel alignment**

The RMTs were developed based on a review of boring log data and were generally characterized using rock quality designation (RQD), rock mass rating (RMR) (Bieniawski, 1988), and Tunneling Quality Index (Q) (Barton, 1988). The RMTs also provide a basis for the ground characterization along the tunnel alignment and will be used during construction to confirm the rock mass conditions assumed during design. Figure 5 shows examples of the three RMT classifications from borings.

Initial support for the tunnel excavation will consist mainly of rock bolts and steel rib supports for the more fractured rock (RMT B3). Pre-excavation grouting will be required to reduce water inflows into the tunnel by establishing a zone of low permeability ground around the tunnel and will serve as a groundwater cutoff. Based on estimates of rock mass type along the tunnel, and in consideration of the presence of shear zones, it is expected that approximately 2,000 ft (610 m) of the TBM mined tunnel will require pre-excavation grouting.

## Tunnel Excavation by Conventionally Mined Methods

It is anticipated that conventionally mined tunnel excavation methods will be used at two locations along the tunnel alignment:

- Along the starter tunnel at the KRWWTP site, for a distance of approximately 150 ft (45 m)
- Through jet grout stabilized soils and mixed face conditions near the KWPTF site for a distance of about 1,000 ft (305 m).

The starter tunnel will extend from the mining shaft (also referred to as the Kailua TIPS Shaft) at the KRWWTP site and will be approximately 70 ft (21 m) below the ground surface. It will be sized to allow the TBM to launch and to facilitate construction of a future cast-in-place (CIP) transition structure. The 60 ft (18.3 m) long transition structure is designed to convey flows from the tunnel to the future influent pump station within the Kailua TIPS Shaft structure. It is anticipated that the starter tunnel will encounter a mixed face of weathered basalt and basalt and will require ground improvement using pre-excavation grouting. Because of the potential for unstable ground conditions, initial support through this tunnel section will consist of steel ribs or lattice girders with shotcrete lagging. Anticipated excavation methods include drill-and-blast operations; however, strict vibration criteria are in place to limit vibration impacts to nearby residents.

Excavation of the last 1,000 ft (305 m) of tunnel excavation will extend from the Kaneohe Shaft at the KWPTF site to the base of the Oneawa Hills ridge. In this area the tunnel will range from 40 to 120 ft (12 to 36.5 m) below the ground surface. The soils along this portion of the tunnel are not suitable for excavation by a main beam TBM. Therefore, the tunnel excavation will be sized to allow the TBM to be walked out of the tunnel to the Kaneohe Shaft for disassembly and demobilization. The first 500 ft (150 m) of this portion of the tunnel will encounter mixed face conditions of weathered basalt and older alluvium. The older alluvium generally consists of medium stiff to hard elastic silt and fat clays with varying amounts of sand, gravel, cobbles, and boulders. The ground conditions are expected to be firm to raveling ground; therefore, the initial support through this tunnel section will consist of steel ribs with shotcrete or timber lagging, and/or grouted spiles or forepoling presupport. The last 500 ft (150 m) will be located beneath the Bay View Golf Park and Kawa Stream to the Kaneohe shaft, and will encounter Estuarine Deposits. The ground conditions

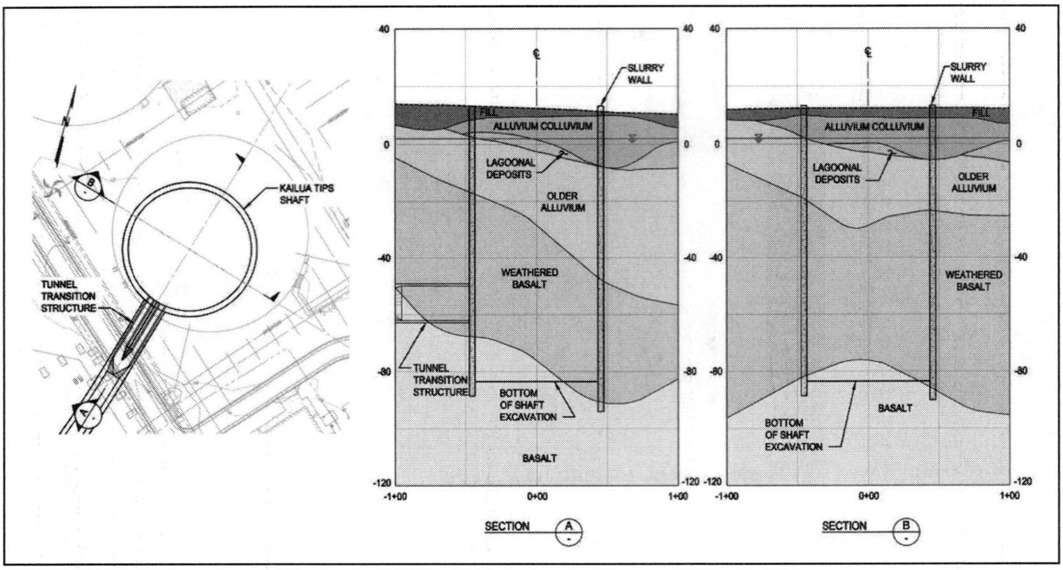


**Figure 6. Plan and geologic profile of the Kailua TIPS shaft**

for untreated Estuarine Deposits are expected to be squeezing or fast raveling; therefore, these soils will require ground stabilization by jet grouting methods prior to excavation. After proper treatment and curing, the jet grout stabilized soil should exhibit the characteristics of firm ground, allowing for installation of initial support of steel ribs with shotcrete lagging. Anticipated excavation methods along this portion of the tunnel include road header and hydraulic excavation. Drilling and blasting are not permitted because of the proximity of the tunnel to nearby residents and the risk of vibrations cracking and compromising the integrity of the jet grout stabilized soils.

## Kailua TIPS Shaft

The Kailua TIPS Shaft is located on the KRWWTP site and will be used to launch the TBM and to stage the tunnel excavation work. The future influent pump station will also be constructed inside the TIPS shaft excavation after tunnel construction is complete and the TIPS shaft requires a minimum clear shaft diameter of 87 ft (26.5 m). The subsurface materials at the shaft are complex and vary with depth and thickness below the ground surface. The materials generally consist of artificial fill, alluvium, lagoonal deposits, older alluvium, weathered basalt, and basalt. Because of the presence of the weak, compressible soil deposits above the weathered basalt and high groundwater table at the shaft, a rigid, watertight excavation support system is required to maintain excavation stability and to avoid surface subsidence due to lateral displacement of the support system and groundwater drawdowns. Therefore, slurry walls extending to the full depth of the shaft excavation and weathered basalt for water cutoff are required as part of the excavation support system. The circular slurry wall will be formed by constructing a series of rectangular panels excavated by a hydromill and rock chisels. Figure 6 includes cross sections through the shaft, parallel and perpendicular to the tunnel alignment, showing the complex geology and the required depths of the slurry walls.

Because of the potential for groundwater inflows and the presence of compressible soils within the KRWWTP, additional cutoff measures are required. These methods include staged cutoff grouting performed below the slurry walls prior to shaft excavation. Grouting will be performed through steel sleeve pipes affixed to the slurry wall panel reinforcement cages and cast in the slurry walls.

## Kaneohe Shaft

The Kaneohe Shaft is located at the KWPTF site and will be used to retrieve the TBM after tunnel excavation is completed. It will also serve as the primary access point to the tunnel for installation of the GFRP final lining system, including backfilling. Upon completion of the tunnel construction, the shaft will be built out with a final concrete lining. In addition, two vortex drops and a junction structure will be constructed adjacent to the shaft to separate flows into the tunnel from upstream diversion structures. The sizing of the shaft is controlled by construction requirements, which will require a 30 ft diameter (9 m) opening to allow for removal of the TBM and installation of the GFRP lining.

Subsurface conditions at the shaft are similar to those expected for the conventionally mined tunnel section and include artificial fill overlying estuarine deposits and recent alluvium. Similar to the Kailua TIPS Shaft, the presence of weak, compressible soils and a high groundwater table requires that a fully watertight excavation support system be used. Therefore, slurry walls extending to the full depth of the shaft combined with excavating the shaft in the wet and placing a watertight tremie concrete invert slab will be required to provide full groundwater cutoff.

Considerable coordination went into designing the Kaneohe Shaft to account for the future junction structure. The junction structure will be located immediately to the west of the shaft and will connect into the back of the slurry walls (i.e., connection wall) for its permanent condition. Prior to the junction structure excavation, the shaft will be subject to circumferential earth and groundwater pressures; however, during the excavation of the junction structure the shaft will experience an unbalance loading condition. This unbalanced loading condition causes an interruption of thrust that is typically carried by the circular shaft, which results in higher stress concentrations on the shaft at various locations within the shaft; in particular, at the tunnel opening and the connection wall. Therefore, additional structural elements and staging requirements are necessary during construction in order to maintain the structural integrity of the shaft. These requirements include: installing the shaft final lining and drop vortex prior to excavation of junction structure, increased reinforcement around the tunnel opening, and shear dowels at the connection wall to transfer loads to the final lining structure.

### Access Shaft

A vertical access shaft will be constructed at approximately the midpoint along the tunnel alignment between the KWPTF and KRWWTP. The shaft's function will be to allow for future inspections and maintenance activities for the tunnel. The shaft will be located at the Board of Water Supply Kapaa water tank site located along Mokapu Saddle Road near Interstate H-3. It will extend from the crown of the tunnel lining to the ground surface and will be approximately 276 ft (84 m) deep. To accommodate maintenance equipment anticipated to enter the access shaft, the internal diameter of the shaft was determined to be 8 ft (2.4 m). The shaft will also be fully lined with GFRP riser pipe and will tie into the GFRP tunnel lining through a Tee-section. This is similar to conventional manhole construction. A GFRP lid will be placed at the near surface to restrict any wastewater gases from exiting the tunnel and riser.

The subsurface conditions along the shaft include minor amounts (less than 10 ft [3 m]) of artificial fill and weathered basalt that overlie basalt breccia and dike complex deposits. The degree of fracturing within the basalt breccia generally ranges from very widely to closely fractured with localized zones of intensely fractured and crushed rock, typically located at dike margins. Because of the anticipated favorable quality of the rock, the access shaft will be excavated by raise bore methods. For this approach, the shaft will be excavated after the tunnel excavation is completed and prior to installation of the GFRP tunnel lining. Initial support, where necessary, will likely consist of shotcrete lining. In areas where raveling and sloughing of the shaft sidewalls are encountered, additional initial support in the form of steel ribs or rock bolts may be required.

## ENVIRONMENTAL CONSIDERATIONS

### Noise Control

Because of the proximity of the Kailua TIPS Shaft to nearby residences surrounding the KRWWTP, noise control measures will be implemented during construction. Construction noise criteria have been developed based on the noise control requirements in the Hawaii Department of Health (DOH) Title 11 Administrative Rules. Specific noise control measures that were developed include: construction of a temporary sound wall, sound-attenuating enclosures for all stationary equipment, noise-suppression devices for construction vehicles, and work hour limitations.

The temporary sound wall will encompass the KRWWTP staging area and was designed using acoustical modeling software to estimate sound levels at various locations within and around the construction site. Two acoustical phases were analyzed based on the construction activities that would generate the greatest noise potential: slurry wall construction and tunnel excavation construction. Typical noise levels for each type of equipment anticipated during these phases were used to complete the required height and noise canceling material to be used for the sound wall. Based on the acoustical model and the most strict noise requirements of Title 11 (nighttime noise limits of 45 dBA), the sound wall will need to be 10 to 27 ft (3 to 8.3 m) high and provide nearly full enclosure of the construction site, in order to meet state noise requirements at residential properties nearest to the KRWWTP (Figure 7). In addition, a sound absorbing material consisting of porous expanded polypropylene (PEPP) panels is required along the full height of the construction side of the wall.

## CONCLUSION

The improvements to the existing wastewater collection, treatment, and disposal system in the

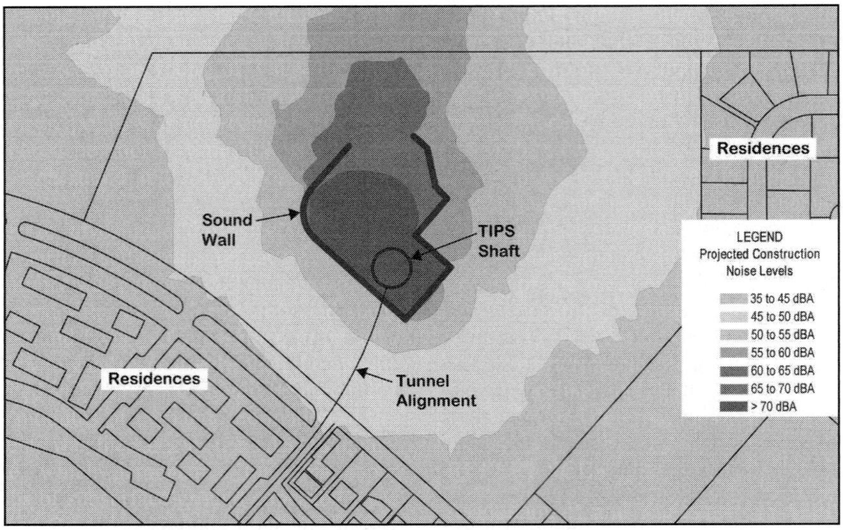


**Figure 7. Projected nighttime noise levels at KRWWTP**

Kaneohe-Kailua-Kahaluu service area on the island of Oahu, Hawaii included the detailed design and coordination for a deep wastewater storage and conveyance tunnel nearly 3 miles (4.8 km) long. As part of the preliminary engineering and design process, several tunnel alignments were evaluated and included a review of construction techniques and a preliminary assessment of risks associated with constructing a tunnel within the sensitive environs of Hawaii. The preferred tunnel alternative consists of a deep tunnel that optimizes the use available underground space, minimizes the impact to facilities near the ground surface, and provides a cost effective solution for CCH's wastewater storage and conveyance needs. In addition, the gravity tunnel alternative, including three shafts that will be used for construction, will also serve as permanent structures for future improvements. Construction of the tunnel is anticipated to begin in 2014, with completion expected in 2016 and completion of other follow-on improvements at the KRWWTP and KWPTF in 2018.

## REFERENCES

American Water Works Association (AWWA). 2005. *Fiberglass Pipe Design*, M45 Manual, 2nd Edition.

Barton, N. 1988. Rock mass classification and tunnel reinforcement selection using the Q-System, In *Rock Mass Classification Systems for Engineering Purposes*, L. Kirkaldie, ed. Philadelphia: ASTM Special Technical Publication 984.

Bieniawski, Z.T. 1988. The Rock Mass Rating (RMR) system (Geomechanics classification) in engineering practice. In *Rock Mass Classification Systems for Engineering Purposes*, L. Kirkaldie, ed. Philadelphia: ASTM Special Technical Publication 984.

Hashash, Y.M.A., Hook, J.J. Schmidt, B., and Yao, J.C. 2001. Seismic design and analysis of underground Structures. *Tunnelling and Underground Space Technology*, 16(4), 247–293.

Jacobsen, S. 1974. Buckling of Circular Rings and Cylindrical Tubes Under External Pressure. Water Power, Dec., 400–407.

Macdonald, G.A. 1956. The Structure of Hawaiian Volcanoes. *K. Ned. Geol. Mijnb. Genoot. Verh., Geol. Serv.,* 16, 274–295.

Walker, G.P.L. 1987. The dike complex of Koolau Volcano, Oahu: Internal structure of a Hawaiian rift zone. In *Volcanism in Hawaii, Vol. 2*. Edited by R.W. Decker, T.L. Wright, and P.H. Stauffer. US Geological Survey Professional Paper 1350, Washington, DC.

# Design of Steel Pipe Final Lining Subjected to Large, Seismically Induced Axial Strains

**Andrew McGlenn and Yiming Sun**
Jacobs Associates

**Allen Mitchell**
Greater Vancouver Water District

**ABSTRACT:** The Port Mann Main Water Supply Tunnel is a 1,000-m (3,280-ft) long two-pass tunnel. The tunnel is approximately 3.5-m (11.5-ft) in excavated diameter, and is sized to receive a 2.13-m (7.0-ft) OD pipeline as the final liner. The tunnel will be excavated by TBM between two shafts on either side of the Fraser River, British Columbia, Canada. The tunnel has precast segments as initial lining and steel pipe as final lining. It is located in an active seismic region, having a design earthquake with a 10,000-year return period and PGA of 0.7g. Axial strains of up to 0.4% are anticipated because of seismically induced lateral spreading. This paper discusses seismic design for the final lining to achieve the owner's seismic performance objectives.

## INTRODUCTION

The Greater Vancouver Water District (GVWD) is constructing the Port Mann Main Water Supply Tunnel under the Fraser River between Coquitlam and Surrey in British Columbia, Canada. The project serves as the long-term solution to replace the existing Fraser River water main crossing and accommodate future regional growth while also improving overall system reliability. The tunnel is 1,000-m (3,280-ft) long and approximately 3.5-m (11.5-ft) in excavated diameter, sized for a 2.13-m (7.0-ft) OD pipeline. It will be excavated in soil, below the depth of riverbed scour, at pressures of up to 6 bar using an earth pressure balanced tunnel boring machine (EPB TBM). Two approximately 65-m (210-ft) deep shafts will be constructed at the north and south sides of the river for the TBM launch and receiving purposes during construction. Figure 1 illustrates the tunnel vertical alignment.

The tunnel has a two-pass lining system: the watertight bolted and gasketed precast concrete segments as the initial lining, and the welded steel pipe as the final lining. The annulus space between the segmental lining and the steel pipe will be backfilled with lightweight concrete. A typical tunnel cross section is shown in Figure 2.

The tunnel is located in an active seismic region. The steel pipe final lining is required to remain functional following a design earthquake with a 10,000-year return period and a peak ground acceleration (PGA) of 0.7g. With this design earthquake, it is anticipated that large axial strains of potentially up to 0.4% will develop along the tunnel alignment due to seismically induced lateral spreading. This paper discusses seismic design for the steel pipe final lining and measures to minimize potential adverse effects of the design earthquake on the final lining to achieve the owner's seismic performance objectives.

## TUNNEL PROFILE AND GEOLOGIC CONDITIONS

The Port Mann Main Water Supply Tunnel has a slope of 1.47% from the South Shaft to the midpoint of the alignment and 0.26% for the rest of the alignment (see Figure 1). The high point of the tunnel is at the south end (at the South Shaft), where the tunnel springline is at about Elevation –49 m (–209 ft). The low point is at the north end (at the North Shaft), where the tunnel springline is at about Elevation –58 m (–190 ft). The ground cover above the tunnel varies along the alignment. The lowest cover over the tunnel crown (measured from the ground surface to the tunnel crown) of about 32 m (105 ft) occurs at the midpoint below the Fraser River. The highest cover over the tunnel crown is about 57 m (187 ft) and occurs at the low point of the tunnel alignment (at the North Shaft).

The soil deposits expected within the shaft and tunnel excavations can be classified into seven Tunnel Soil Groups (TSGs). These TSGs are summarized in Table 1. As indicated in Figure 1, the majority of the tunnel alignment is expected to be located within TSG1 and TSG2. Based on the geotechnical investigations completed for the project, rock consisting of sandstone, siltstone, and mudstone exists below the soil deposits at each shaft site.

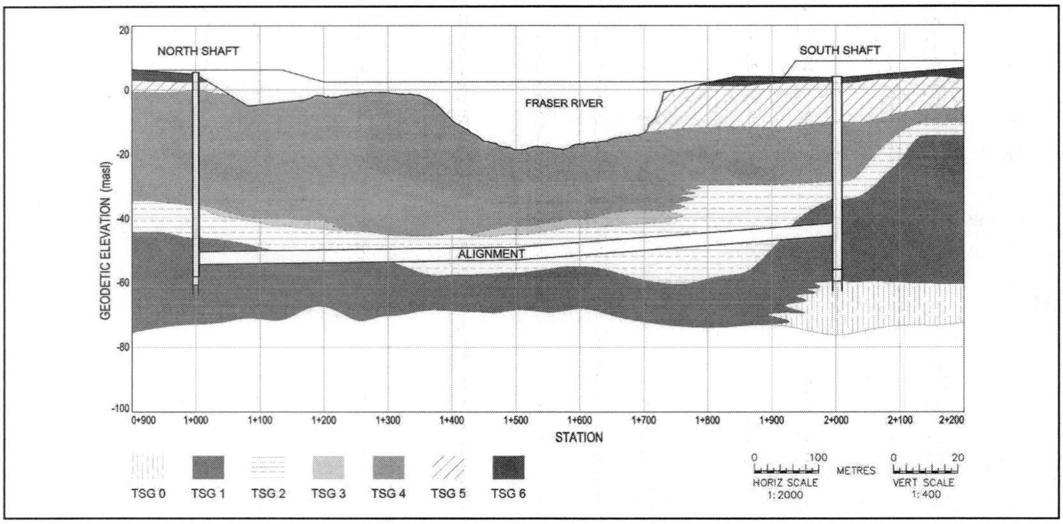


**Figure 1. Geologic profile and tunnel vertical alignment**

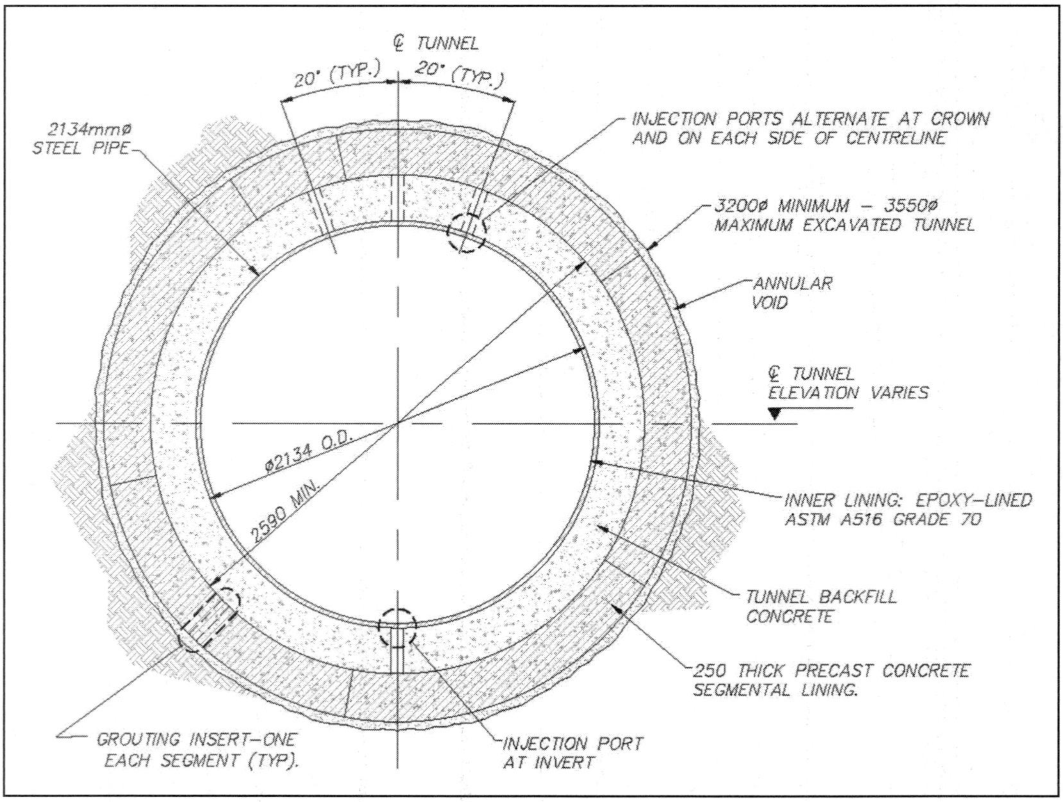


**Figure 2. Typical tunnel cross section**

**Table 1. Summary of characteristics of tunnel soil groups (TSGs)**

| ID of TSG | Characteristics |
|---|---|
| TSG0 | Silty clay to clayey silt, consisting of clays and silts deposited during glacial advance. |
| TSG1 | Silty sand, sand and gravel, and silty clay. TSG1 is a till-like, very dense or hard, poorly sorted heterogeneous mixture of clay, silt, sand, gravel, cobbles, and boulders. |
| TSG2 | Silty clay. TSG2 overlies TSG1 and typically consists of soft to very stiff silty clay to clayey silt with varying plasticity. Infrequent cobbles and scattered layers and/or lenses of coarse-grained soils are also part of this unit. |
| TSG3 | Gravel. The relatively thin (about 1.0 to 2.5 m [3 to 8 ft] thick), and flat-lying gravel layer is compact to dense, and contains variable amounts of sand. |
| TSG4 | Sand with gravel, and silt interlayers. The sands are very loose to very dense, although they are typically compact. Scattered within TSG4 are layers and lenses of silts and clays and cobbles. |
| TSG5 | Peat, sand, and silt, consisting of loose to compact (or firm to stiff) interbedded sand, silt, and amorphous and fibrous peat with wood fragments. |
| TSG6 | Fill, consisting of fill deposits that mantle alluvial deposits along both the north and south Fraser River shorelines. The fill is a heterogeneous mixture of compact to dense sand, gravel, cobbles, rubble, and organics. |

## DESIGN EARTHQUAKE AND SEISMIC PERFORMANCE CRITERIA

The project is located in a seismically active area. The site seismicity results from the thrusting (subducting) of the Juan de Fuca plate beneath the Continental North American Plate. The offshore plate tectonic setup has resulted in shallow crustal earthquakes occurring within the Continental plate, deep intraplate earthquakes occurring in the subducting plate, and interplate earthquakes occurring at the contact between the plates. Over the past several decades, intraplate earthquakes have occurred at regular intervals—Campbell River (M7.3, 1946), Olympia (M7.1, 1949), Seattle/Tacoma (M6.5, 1965), and Nisqually (M6.8, 2001). A site-specific seismic hazard study completed for the project area indicated intraplate and interplate earthquakes dominating the seismic risk at the site. The controlling earthquake scenarios included a M7.25 intraplate earthquake occurring at a distance of 55 km (35 mi) from the area and an M8.8 interplate subduction earthquake occurring at a distance of 160 km (100 mi) from the area.

The GVWD classified the new Port Mann Main Water Supply Tunnel crossing as a "Level 1" facility, which is required to withstand and remain functional following the Maximum Credible Earthquake (MCE) (Davidson et al., 2013). To meet this seismic design requirement, catastrophic damage to the facilities during the design earthquake must be avoided. The following design earthquake and seismic performance criteria were adopted for the project facilities:

- Maximum Credible Earthquake (MCE) that corresponds to a return period of 10,000 years with a PGA of 0.7g, as per a site-specific seismic hazard analysis completed by Abrahamson (2006). Under this level of shaking, the facilities may experience some distress, such as cracking and minor leakage, but are expected to remain operational at full capacity.

To evaluate the final lining performance quantitatively, a stress criterion specified in the project design criteria document (Jacobs Associates, 2008) was used. This criterion states that the maximum allowable tensile stress for the tunnel final lining and pipeline is 50 percent of the yield stress for normal operating conditions, and 75 percent of the yield stress for short-term transient conditions, such as an earthquake event. For the ASTM A516M Grade 60 or Grade 70 steel (with a minimum yield strength of 260 MPa [38 ksi]) proposed for the final lining, the allowable stress is 195 MPa (28 ksi) for short-term transient (combined static and seismic) conditions.

## KEY SEISMIC DESIGN CHALLENGES

Generally, a tunnel will respond in three ways to deformations imposed by the surrounding ground during seismically induced ground motions: axial compression and extension, longitudinal bending or "snaking," and ovaling or racking (Hashash, et al., 2001). Axial deformations in tunnels are generated by the components of seismic waves that produce motions parallel to the axis of the tunnel and cause alternating compression and tension. Bending deformations are caused by the components of seismic waves producing particle motions perpendicular to the longitudinal axis. Ovaling or racking deformations in a tunnel structure develop when shear waves propagate normal or nearly normal to the tunnel axis, resulting in a distortion of the cross-sectional shape of the tunnel lining.

In addition to these three modes of responses, the Port Mann Main Water Supply Tunnel will also be subject to the axial ground deformations due to lateral spreading (Sandwell, 2008). Site-specific ground response analyses were undertaken to quantify the magnitude and pattern of ground deformations resulting from the design earthquake scenarios (Abrahamson, 2006; Davidson et al., 2013). Of particular interest were the profile and magnitude of the permanent ground deformations caused by earthquake-induced lateral spreading occurring towards the Fraser River and the resulting interaction with the shafts and tunnel (Sandwell, 2008). The magnitude of the permanent ground deformations was estimated to be very high so that significant axial strains could be developed along the installed steel pipe final lining if the relative displacements between the ground and the final lining would not occur. Figure 3a shows the predicted axial displacement profiles along the tunnel caused by the lateral spreading following the design MCE event (X, Y, and Z components in the figure denote the axial, lateral, and vertical displacements, respectively). The estimated axial strains associated with the axial displacements are illustrated in Figure 3b. As indicated, axial strains up to approximately 0.4 percent in both tension and compression are predicted. This level of strain is beyond the yield limit of steel pipe. Therefore, the steel pipe final lining should be designed to accommodate this level of axial strains during its design life of 100 years.

To enhance the resistance of steel pipe final lining against the seismically induced ground deformations, feasible mitigation measures were evaluated during design to determine their effectiveness in resisting the ground motions. Two of the mitigation measures evaluated are:

- Specify the strength requirements for backfill concrete to provide stiffness to achieve composite action of combined segmental lining, backfill concrete, and steel pipe to minimize potential adverse effects of seismic ground deformations.
- Install circumferential anchor rings along the outside of steel pipe final lining to improve the composite action and resistance against axial deformations.

Evaluation of the interaction of ground-steel pipe deformations and steel pipe performance, and required mitigation measures to ensure pipeline functionality during and after the earthquake are the key seismic design challenges for this project.

## METHODS OF SEISMIC DESIGN ANALYSIS

Two different types of seismic analysis were performed to evaluate the effect of ground deformations caused by the design earthquakes. These two types of analyses can be summarized as follows:

- In **Analysis A** (Axial Effect), the effect of large axial tensile and compressive deformations of the tunnel caused by seismically induced lateral spreading of the ground was evaluated. This analysis focused on the quantification of potential plastic strains that would be developed in the steel pipe when it was subjected to large axial tensile and compressive forces. The analysis also evaluated the effect of the composite action of the tunnel lining system that consists of concrete segmental lining, backfill concrete, and steel pipe final lining.
- In **Analysis B** (Racking Effect), the effect of ovaling/racking deformations of the tunnel during strong earthquake ground motions was analyzed. This analysis concentrated on the magnitude of stresses developed in the steel pipe when it was subjected to racking deformations.

In this paper, only the axial effect based on Analysis A is discussed because this effect is unique to this tunnel project and more critical compared to the racking effect evaluated in Analysis B. In Analysis A, three-dimensional (3D) numerical analyses were employed. These analyses were performed using the FLAC3D program (Fast Lagrangian Analysis of Continua in 3 Dimensions) Version 3.0 (Itasca, 2005). For simplicity, the FLAC3D models developed for these analyses included only the tunnel lining system that consists of segmental lining, backfill concrete, and steel pipe final lining, while the ground adjacent to the tunnel was ignored (see Figures 4a and 5a). However, the effect of ground was indirectly accounted for by applying external pressure on the outside surface of the segmental lining. The external pressure represents the long-term ground loads and groundwater pressure on the tunnel lining system. The external pressure was assumed to be equal to 1.15 MPa (167 psi). In addition, an internal pressure equal to 2.5 MPa (360 psi), which represents the hydrostatic pressure on the inside surface of the steel pipe, was also considered in the analyses.

Key assumptions used in the FLAC3D analyses are as follows:

- The effect of lateral spreading on the tunnel can be represented by either axial displacements or forces experienced by the tunnel lining system. The effect of bending of the tunnel caused by lateral spreading is neglected.
- The annular space between the steel pipe final lining and the segmental lining is completely backfilled with concrete. Local buckling due to partial backfill confinement is neglected.
- The steel pipe final lining is not bonded to the backfill concrete, and will slip relative to the concrete under axial displacements or forces. The interaction between the steel pipe and the backfill concrete is controlled in the FLAC3D models by an interface placed between them.

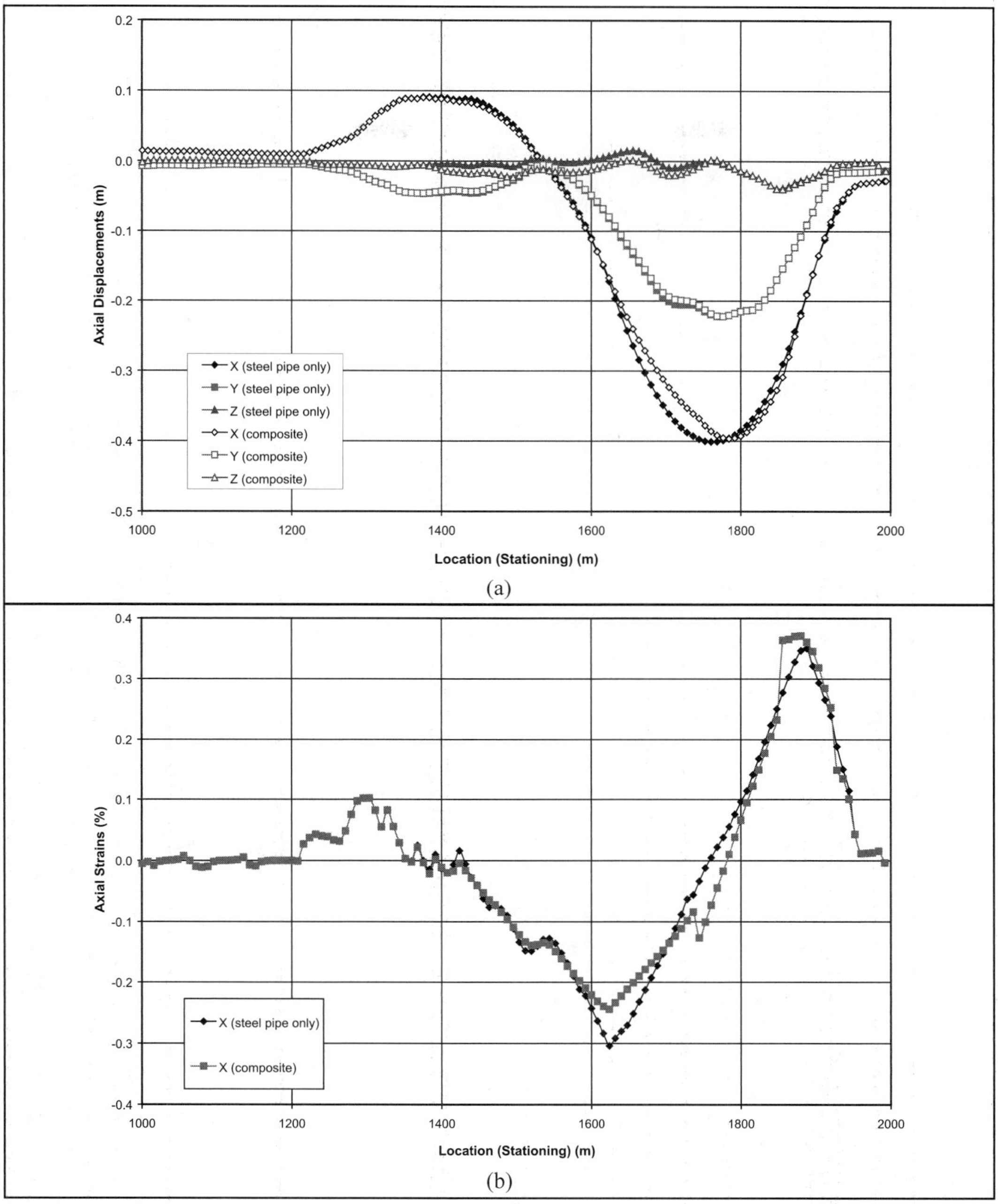


**Figure 3. (a) Displacement profiles and (b) axial strain profiles along tunnel due to lateral spreading**

- The external pressure (ground loads and groundwater pressure) is equal all around the tunnel.
- The behaviors of segmental lining and backfill concrete are governed by the Mohr-Coulomb failure criterion.
- The behavior of steel pipe final lining is governed by the von Mises yield criterion.

In a FLAC3D analysis, the axial displacements that result in either tensile or compressive strains in the steel pipe were prescribed as a boundary condition. The magnitude of the prescribed axial displacements was determined based on the axial displacement profiles generated from the analyses carried out by Sandwell (2008), as shown in Figure 3a (X component denotes the axial displacement in

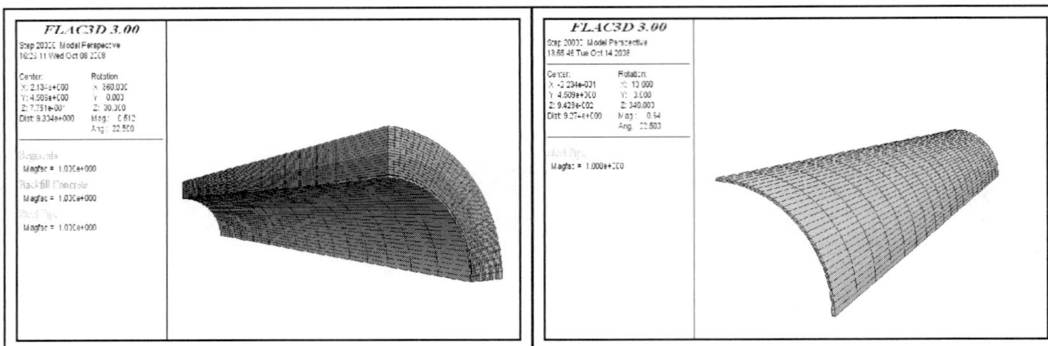


(a) Composite lining system | (b) Steel pipe

**Figure 4. Configuration of a FLAC3D model for steel pipe without anchor rings**

the figure). The corresponding axial strain profiles along the tunnel were then estimated based on the axial displacement profiles, and are presented in Figure 3b. As indicated, the maximum axial tensile and compressive strains are about 0.35 percent and –0.30 percent, respectively. These maximum axial strains were used as a basis in model calibrations of the FLAC3D analyses to determine the magnitude of prescribed axial displacements.

Since the displacement profiles shown in Figure 3a were generated from an elastic analysis (Sandwell, 2008), the axial strains are regarded as elastic strains and can only be used as such. Therefore, the following steps were taken in the analyses:

- **Step 1:** Establish a baseline condition for the steel pipe performance by conducting an elastic analysis. The model for this analysis contained the steel pipe only. In this analysis, the prescribed axial (either tensile or compressive) displacements were calibrated to achieve a desired axial elastic strain equal to about 0.35 percent in tension or about –0.30 percent in compression.
- **Step 2:** Perform a plastic analysis with the steel pipe only. In this analysis, the prescribed axial displacements were obtained from Step 1 for the corresponding tension or compression case. The calculated axial strains were considered as the anticipated maximum strains that would be experienced by the steel pipe when it would act alone under axial forces due to lateral spreading.
- **Step 3:** Perform a plastic analysis with a composite lining system. In this analysis, the prescribed axial displacements were obtained from Step 1 for the corresponding tension or compression case. The calculated axial strains were considered as the anticipated maximum strains. These maximum strains would be experienced by the steel pipe when a composite action of the lining system would occur under axial forces due to lateral spreading. A comparison of the calculated axial strains from Steps 2 and 3 could provide an understanding of the effect of the tunnel lining system on the potential steel pipe performance.

In addition, the effects of stiffness or strength of backfill concrete, use of composite anchor rings welded outside the steel pipe surface, and internal pressure were also evaluated in the FLAC3D analyses. The compressive strength values assumed for the backfill concrete ranged from 3.4 MPa (500 psi) to 13.8 MPa (2,000 psi). The spacings assumed for the composite anchor rings were 5 and 10 m (16.4 and 32.8 ft). Over twenty (20) FLAC3D runs were carried out.

Because of the axisymmetrical conditions assumed for the analyses, only a quarter of the lining system was included in the FLAC3D models. Figure 4 shows the configurations of a typical FLAC3D model and steel pipe without the anchor rings. Figure 5 presents the configurations of a typical FLAC3D model and steel pipe with the anchor rings.

Key inputs to the seismic analyses include material properties for concrete segmental lining, backfill concrete, and steel pipe, and dimensions of tunnel segmental lining and steel pipe final lining. These properties are presented in Tables 2 and 3.

## ESTIMATED SEISMIC PERFORMANCE OF TUNNEL

The seismic design evaluation was focused on the effect of backfill concrete strength/stiffness and anchor rings on the resistance of composite lining system against the axial deformations. The assumed ranges of backfill concrete strength and

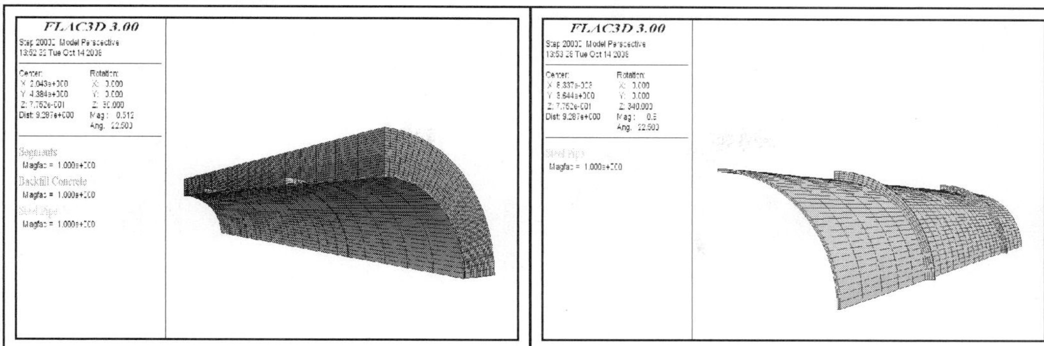


(a) Composite lining system | (b) Steel pipe with anchor rings

**Figure 5. Configuration of a FLAC3D model for steel pipe with anchor rings**

**Table 2. Mechanical properties of concrete and steel**

| Material Constant | Structural Concrete for Segments | Backfill Concrete* | Steel |
|---|---|---|---|
| Elastic Modulus, GPa | 30.4 (4,410 ksi) | 1.7–6.9 (250–1,000 ksi) | 200 (29,000 ksi) |
| Poisson's Ratio | 0.2 | 0.2 | 0.3 |
| Strength, MPa | 41.4 (6,000 psi) | 3.4–13.8 (500–2,000 psi) | 260 and 290 (38 and 42 ksi) |

*Elastic modulus estimated based on the correlation of $E=500f'_c$, where $E$ is the elastic modulus and $f'_c$ is the compressive strength.

**Table 3. Properties of segmental and steel pipe linings**

| Parameter | Segments | Steel Pipe | Steel Anchor Rings |
|---|---|---|---|
| Thickness, m | 0.25 (10 in.) | 0.025 (1 in.) | 0.025 (1 in.) |
| Area per meter, $m^2$/m | 0.25 (10 in.) | 0.025 (1 in.) | N/A |
| Width, m | N/A | N/A | 0.1 (4 in.) |
| Spacing, m | N/A | N/A | 5.0–10.0 (16.4–32.8 ft) |

corresponding stiffness and anchor ring spacing in the analyses are provided in Tables 2 and 3. The results of the FLAC3D in terms of the effect of large axial deformations are discussed below.

As discussed above, the FLAC3D analyses considered three different conditions: (1) steel pipe only assuming only elastic behavior, regarded as a baseline condition; (2) steel pipe only allowing plastic behavior; and (3) composite lining system allowing plastic behavior. Calculated axial stresses developed in the steel pipe final lining when subject to the axial tensile and compressive deformations are presented in Figures 6 to 8 for these conditions, respectively. The maximum axial strains and stresses calculated in the steel pipe are summarized in Table 4. The results from these analyses can be summarized as follows:

- When subjected to tension, the steel pipe is expected to experience large plastic strains ranging from about 0.4 to 1.8 percent. These strains are dependent on the magnitude of axial tensile forces and deformations of the ground, and appear to be less sensitive to the composite action of the tunnel lining system. Also, the steel pipe performance under tension is expected to be less sensitive to the changes in stiffness of backfill concrete.
- When subjected to compression, the steel pipe is expected to experience large plastic strains ranging from about –0.3 to –0.7 percent. Similarly, these strains are also dependent on the magnitude of axial compressive forces and deformations. With a backfill concrete strength greater than 6.9 MPa (1,000 psi), the composite action of the lining system is shown to be effective in minimizing the maximum strain developed in the steel pipe to about –0.4 percent. This composite action is expected to occur under compression even when relative displacements

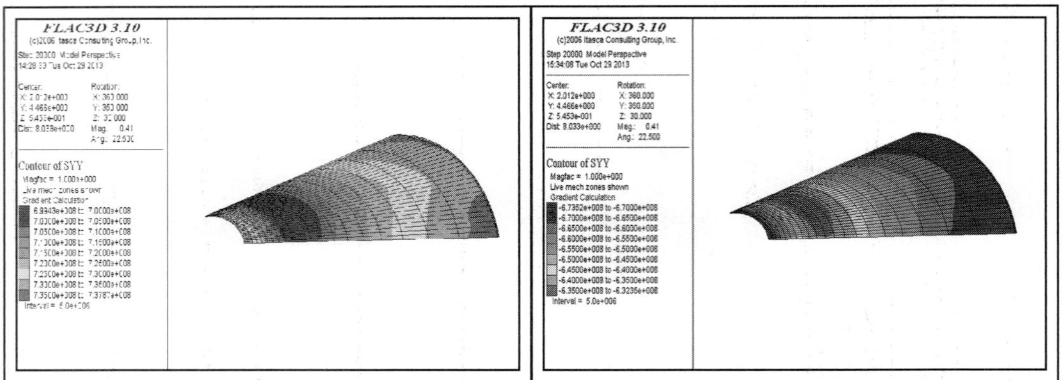


(a) In tension

(b) In compression

**Figure 6. Calculated axial stresses (in Pa) in steel pipe for steel pipe only model (baseline condition, elastic model)**

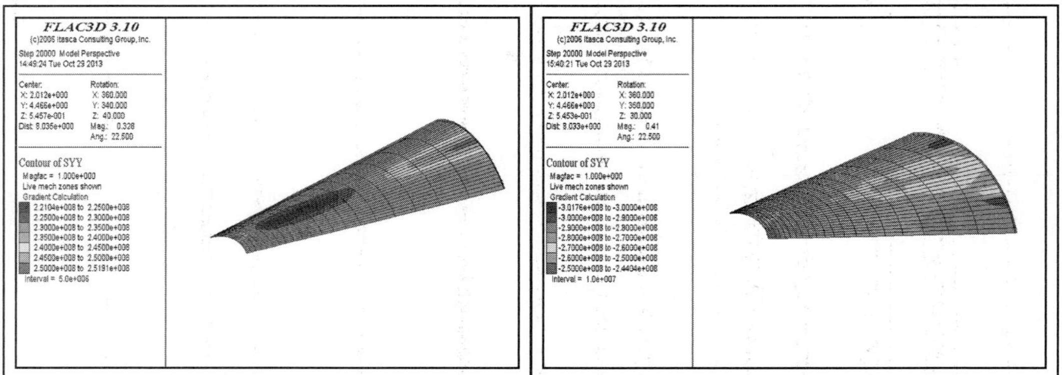


(a) In tension

(b) In compression

**Figure 7. Calculated axial stresses (in Pa) in steel pipe for steel pipe only model (plastic model)**

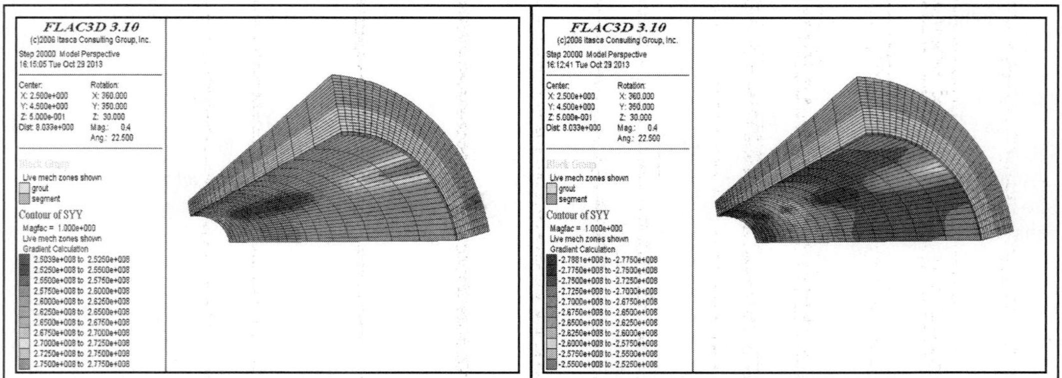


(a) In Tension

(b) In compression

**Figure 8. Calculated axial stresses (in Pa) in steel pipe for composite lining system model (plastic model)**

**Table 4. Results of FLAC3D analyses for cases with internal pressure**

| | Tension | | | | Compression | | | |
|---|---|---|---|---|---|---|---|---|
| **Parameter** | **Baseline (Pipe Only, Elastic Model)** | **Pipe Only (Plastic Model)** | **Composite (Plastic Model, Without Anchor Rings)** | **Composite (Plastic Model, With Anchor Rings)** | **Baseline (Pipe Only, Elastic Model)** | **Pipe Only (Plastic Model)** | **Composite (Plastic Model, Without Anchor Rings)** | **Composite (Plastic Model, With Anchor Rings)** |
| Axial strain (%) | 0.36 | 1.62 | 1.81 | 1.73 | –0.31 | –0.73 | –0.32 | –0.33 |
| Axial strain/ yield strain (±0.13%) | 2.8 | 12.5 | 13.9 | 13.3 | 2.4 | 5.6 | 2.5 | 2.5 |
| Max. axial stress (MPa) | 780 | 321 | 337 | 335 | –635 | –244 | –231 | –241 |
| Max axial stress/yield stress (±260 MPa) | 3.0 | 1.2 | 1.3 | 1.3 | 2.4 | 0.9 | 0.9 | 0.9 |

between the backfill concrete and the steel pipe final lining would happen. According to the *Seismic Guidelines for Water Pipelines* (American Lifelines Alliance, 2005), a conservative estimate of the onset of local buckling in a butt-welded, free standing pipe (without backfill) with a thickness of 25 mm (1 inch) and diameter of 2.13 m (7 ft) ranges from approximately –0.4 to –0.5 percent. Therefore, the estimated maximum strain developed in the steel pipe is expected to remain below the limit of the onset of local buckling if the backfill concrete is stiff enough to support the steel pipe. An annular gap/void may potentially exist between the steel pipe final lining and the backfill concrete. The effect of the annular gap/void on local buckling has been accounted for in the selected steel pipe thickness.

- Results from a few analyses completed indicate that use of anchor rings welded outside the steel pipe for increasing the composite effect does not demonstrate any significant benefit in reducing the plastic strains in both tensile and compressive conditions.

## DESIGN RECOMMENDATIONS AND SUMMARY

The following conclusions are drawn from the seismic design analyses of the steel pipe final lining for the Port Mann Main Water Supply Tunnel:

- The proposed steel pipe final lining is expected to experience large tensile strains when subjected to large axial displacements and forces caused by the seismically induced lateral spreading. However, because of the ductility of steel, these tensile strains are not expected to be great enough to cause a rupture of the steel pipe final lining.
- The estimated maximum compressive strain is below the limit of the onset of local buckling for a welded pipe. Potential for a buckling failure of the steel pipe due to lateral spreading is low. The worst case behavior at point of maximum compression is anticipated to be a localized wrinkling of the pipe due to the presence of the concrete backfill, thus only creating a small reduction in cross sectional area of pipe and maintaining flow capacity.
- A maximum weld hardness value of 250HV is recommended to maintain ductility at the welded pipe butt joints, accompanied by a rigorous QC program during construction.
- A minimum compressive strength of 6.9 MPa (1,000 psi) is recommended for the backfill concrete in order to provide adequate stiffness to support the steel pipe during the design MCE earthquakes.
- The lining system without anchor rings is expected to be able to act compositely under compression. Therefore, use of anchor rings to enhance the composite action is not recommended.
- The effect of longitudinal bending caused by seismic ground motions and lateral spreading was not analyzed in the FLAC3D analyses. This bending is expected to induce additional axial stresses in the steel pipe. The stress concentrations caused by the longitudinal bending may be an issue, especially in

areas near the intersections between the steel pipe and the shafts. These stress concentrations can be mitigated by increasing the stiffness of backfill concrete in these areas, based on the analyses completed for the Bay Tunnel Project (Jacobs Associates, 2010). Therefore, use of structural concrete with a minimum compressive strength of 27.6 MPa (4,000 psi) for backfill of the short tunnel sections (10 meters [33 ft]) adjacent to the shafts is recommended.

## ACKNOWLEDGMENTS

The authors would like to acknowledge the many contributions of Ausenco Sandwell, Jacobs Associates, Golder Associates, Hatch Mott MacDonald, and their subconsultants, as well as the GVWD staff in the design of this project. The contents of this paper reflect the views of the authors, who are responsible for the facts and accuracy of the data presented herein, and do not necessarily reflect the official views or policies of the GVWD. This paper does not constitute a standard, specification, or regulation.

## REFERENCES

Abrahamson, N.A. 2006. Seismic hazard assessment: Problems with current practice and future developments. In *Proceedings of the First European Conference on Earthquake Engineering and Seismology, Geneva*, Paper K2.

American Lifelines Alliance. 2005. *Seismic Guidelines for Water Pipelines*. R80.01.01 Rev. 0.

Gregg W. Davidson, F. Huber, and M.D. Gant. 2013. Port Mann Main Water Supply Tunnel: Improving system reliability. In *Proc. of 2013 Rapid Excavation and Tunneling Conference*, 160–170.

Hashash, Y.M.A., J.J. Hook, B. Schmidt, and J.L. Yao. 2001. Seismic design and analysis of underground structures. *Tunnelling and Underground Space Technology*, 16, 247–293.

Itasca Consulting Group. 2005. *Fast Lagrangian Analysis of Continua in 3 Dimensions* (FLAC3D), Version 3.0. Minneapolis, MN: Itasca Consulting Group.

Jacobs Associates. 2008. *Design Criteria for Shafts and Tunnels*. Technical Memo from G. Davidson, JA to J. Cross, Sandwell, Oct 6, 2008, Port Mann Water Supply Tunnel Project.

Jacobs Associates, 2010. *Technical Memorandum No. 7: Seismic Analysis*, Rev. 3, from J. Caulfield, JA to J. Wong, SFPUC, Dec. 1, 2010, Bay Tunnel Project.

Sandwell, 2008. *Structural Analysis of the Tunnel*. Draft Report (File No. 142978) prepared by S. Iranpour, Aug. 12, 2008, Port Mann Water Supply Tunnel Project.

# Cleaning Hartford Waterways Through Underground Storage

**Verya Nasri and William Bent**
AECOM

**William Hogan**
Hartford Metropolitan District Commission

**ABSTRACT:** The South Hartford Conveyance and Storage Tunnel (SHCST) is a major component of the Hartford Metropolitan District's Clean Water Project (CWP). This tunnel is intended to capture and store Combined Sewer Overflows (CSO) from the southern portion of Hartford, CT and Sanitary Sewer Overflows (SSO) from West Hartford and Newington, CT. The project is estimated to cost approximately $500 MM and will be constructed in multiple construction contracts. The project components include a deep rock tunnel 21,800 feet in length with a 25 feet excavated diameter, several miles of consolidation sewers, multiple hydraulic drop shafts with deaeration chambers and a 40 MGD tunnel dewatering pump station. Final design for this project is currently underway. This paper discusses the major aspects of the final design of the SHCST.

## INTRODUCTION

The South Hartford Conveyance and Storage Tunnel (SHCST) project is a significant component of the Hartford Metropolitan District's (MDC) Long Term Control Plan (LTCP) which is overseen by the Connecticut Department of Energy and Environmental Protection (CTDEEP). This project will address a portion of the MDC's Clean Water Project (CWP), which will reduce combined sewer overflows (CSOs); eliminate sanitary sewer overflows (SSOs); and reduce nitrogen released into the Connecticut River.

The purpose of the SHCST project is to eliminate West Hartford and Newington SSOs, eliminate Franklin Area CSOs discharging to Wethersfield Cove and to minimize CSO discharges to the South Branch Park River. The locations of each overflow are shown in Figure 1.

In 2010, the District prepared a Preliminary Design Report (PDR) for the SHCST project, which included relief of the Folly Brook Trunk Sewer and proposed to keep the TBM retrieval shaft within the City limits of Hartford. Figure 1 also shows the 2010 PDR recommended tunnel route. Subsequent to the PDR, the objectives of the SHCST have slightly shifted. In accordance with ongoing revisions to the LTCP, relief of the Folly Brook Trunk Sewer is no longer necessary. Additionally, the MDC has decided to perform less sewer separation in the Franklin Avenue drainage area. To replace the sewer separation, new relief points are proposed within the Franklin area and will be diverted to the SHCST. Figure 1 also shows the current recommended tunnel route (Alignment F).

During 2012, the MDC conducted an evaluation of the potential of connecting the proposed North Tunnel (originally proposed as an independent tunnel with its own pump station) into the South Hartford Conveyance and Storage Tunnel. This evaluation concluded that the two proposed tunnels could reasonably be connected together and operated as a single tunnel system utilizing the tunnel pump station at the eastern terminus of the South Hartford Tunnel (Figure 2). It also was concluded that this alternative was less costly than two independent tunnel systems.

During dry weather, the SHCST will not receive flow as the existing MDC collection system can adequately convey flow to the Hartford Water Pollution Control Facility (HWPCF). During wet weather, when the capacity of the existing collection system is exceeded, the SHCST will receive overflows that would have previously discharged directly to receiving waters.

New diversion structures will be constructed at each CSO/SSO relief point to divert overflows to new consolidation sewers (near surface). These, in-turn, will discharge flow to hydraulic drop shafts which will convey the flow in a controlled manner to the deep rock storage tunnel. Once in the tunnel, flow will be pumped to the new headworks at the HWPCF. The components of the SHCST project described in this paper are as follows:

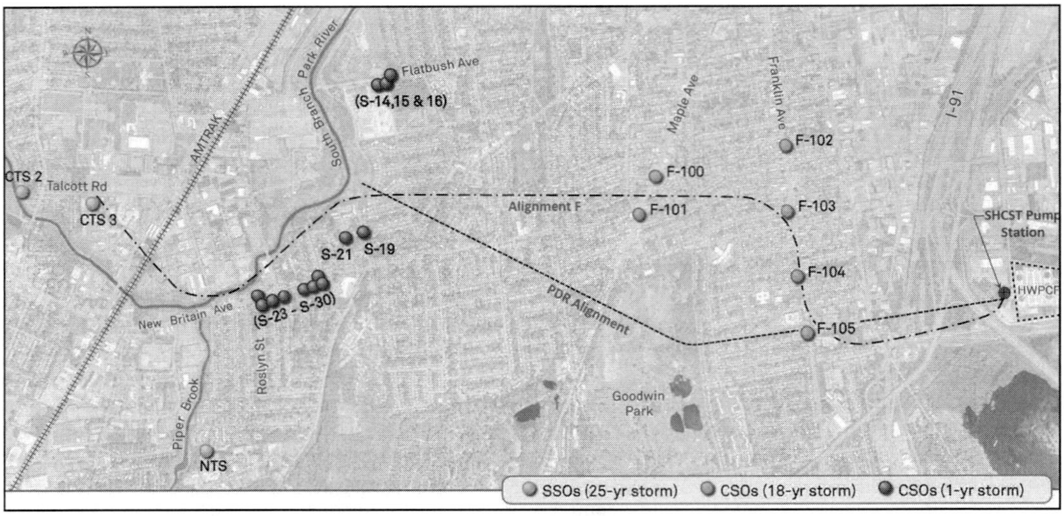


**Figure 1. Recommended SHCST tunnel alignment and location of CSOs and SSOs**

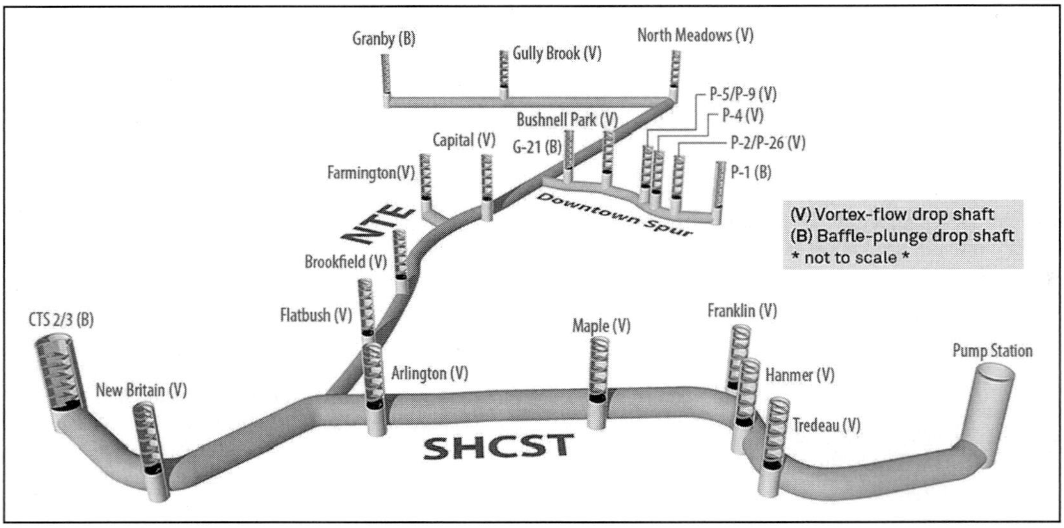


**Figure 2. SHCST and north tunnel integration**

- Deep rock tunnel (22 feet internal diameter and 21,800 feet long) with a TBM launch shaft near the HWPCF in Hartford and a TBM retrieval shaft in West Hartford
- 9,500 feet of near-surface consolidation sewers (24 to 66 inches in diameter)
- Six hydraulic drop shafts
- 40 MGD tunnel dewatering pump station
- Odor control at all potential air release points.

## BASIS OF DESIGN

The sizing of the tunnel was based on the volumes from the 1-year, 18-year and 25-year design storm models per the LTCP and updated collection system modeling from the MDC's Program Management Consultant. The LTCP specified a different level of control for each tributary area. Table 1 shows the peak flows and volumes to be stored in the SHCST for each major source and respective design storm.

Surge, air entrainment and pressure waves can occur in CSO tunnels filling rapidly, with detrimental results such as geysering, blowback and flow instabilities. Based on the hydraulic analysis, it appears that surge in the SHCST is unlikely, due to the relatively large tunnel diameter in comparison to the incoming peak flows.

**Table 1.Tributary overflows to the SHCST**

| Contribution | Design Storm | Peak Flow (MGD) | Volume (Mgal) |
|---|---|---|---|
| West Hartford/ Newington SSO | 25-yr | 27 | 17 |
| South Branch Park River CSO | 1-yr | 68 | 6 |
| Franklin Area Relief | 18-yr | 313 | 39 |
| Total | | | 62 |

Sediment deposition can present an ongoing maintenance burden if not controlled. Based upon the initial sediment deposition analysis and modeling, a slope of 0.1% appears adequate for the deep rock tunnel to cost-effectively minimize sediment deposition issues. This slope is consistent with the state of practice for other large diameter CSO tunnels as steeper slopes will increase project cost. The tunnel will still require periodic maintenance to remove sediment build-up over the life of the facility.

The capacity of the tunnel dewatering pump station has been established by the MDC as 40 MGD. A detailed analysis of the combined tunnel system is underway to determine final pump out times for the various wet weather events. A typical CSO tunnel is dewatered in 24 to 48 hours.

The operation of the tunnel must not result in odor complaints. As such, odor control has been assumed at each drop shaft location.

An alignment study was conducted to evaluate various configurations of rock tunnels and consolidation conduits. Seven (7) conceptual rock tunnel alignments and associated consolidation conduit options were developed and evaluated. The purpose of this alignment study was to identify a cost effective and acceptable tunnel alignment that balances the expectations of the many stakeholders impacted by the project.

All the alignments began in property owned by the District adjacent to the HWPCF. However, three different locations were identified as possible deep rock termination points. Two of these locations were located in space owned by various City of Hartford departments on the east side of the South Branch of the Park River and the third was in an unused parking lot on Talcott Road in a light industrial area on the west side of the river (in West Hartford). This third location significantly reduced the length of consolidation conduits and allowed the South Branch of the Park River to be crossed in rock using the deep rock tunnel instead of crossing the river with a shallower and more risky consolidation conduit.

A systematic approach was established to comparatively score each alternative. The cost estimate was used as the quantitative assessment for each alternative and was not included in the weighted ranking, which is the qualitative assessment. Three stakeholder impact categories were defined which consisted of High, Medium, and Low impact evaluation factors. Each evaluation factor was given a raw score and a weight which depended on its category. The score of each alignment alternative was then determined as the weighted sum of all evaluation factors for that alternative. Alignment F, shown in Figure 3, was identified as the preferred alignment and recommended to advance to final design. This alignment provides the maximum reduction in consolidation conduit length which reduces the associated cost, business impacts and construction risk.

## GEOTECHNICAL SETTINGS

The site area lies in the Central Lowlands physiographic province that extends in a north-south direction in the middle of the state. The central lowland area consists mainly of the sedimentary rocks and the associated igneous basalts of Triassic and Jurassic age. The Hartford Basin of Connecticut and southern Massachusetts is a half graben in structure, 90 miles long, and filled with approximately 13,000 feet of sedimentary rocks, and basaltic lavas and intrusions (Hubert et al., 1978). The source area for the sedimentary rocks was mainly the metamorphic rocks of the Eastern Highlands. Volcanic flows separated the deposition of the lacustrine and fluvial deposits, which were derived from the erosion of the highlands to the east. Displacements along the faults continued throughout the depositional period. The depositional sequence resulted in a series of features including the alluvial fan, lake, alluvial mudflats and floodplain deposits separated by basaltic flows.

Following the deposition of most of the sediments, the tectonic activity continued along the east edge of the basin. Displacements along the eastern border fault rotated the basin downward to the east that resulted in the easterly dipping beds. The Jurassic extensional tectonics is associated with the separation of the continents. That was the last major tectonic episode affecting the geology of the region. Age dating of the Triassic/Jurassic faulting in southern Connecticut has indicated that the last activity along the faults occurred approximately 175 million years ago (NNEC, 1975). All faults in the project area are therefore considered to be inactive.

The region has undergone a period of glaciations that has reshaped the terrain. Glaciers ground down the area's peaks, scraping away any weak or weathered rock and laying down a heterogeneous layer of ground-up rock. This till layer is present over much of the lower lying bedrock surfaces. The sediments of Glacial Lake Hitchcock filled in the deeply-incised Connecticut River Valley. The lake deposits are present in varying forms from Rocky Hill, Connecticut to Northern Vermont. Glaciers

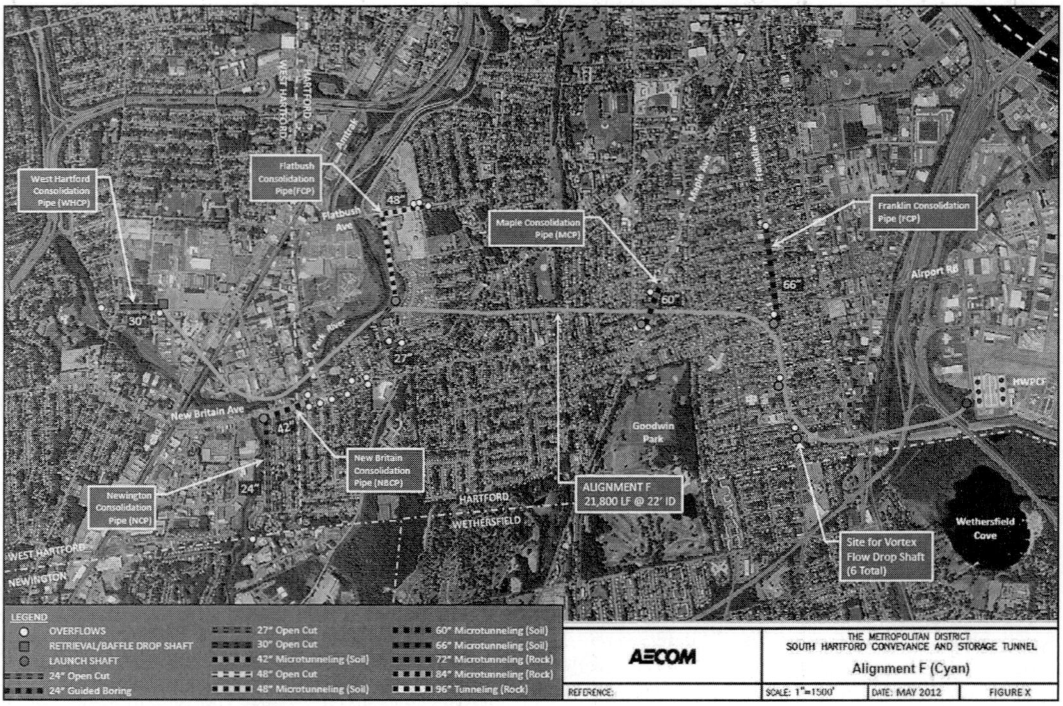


**Figure 3. Selected SHCST alignment (Alternative F)**

shaped the topography and left the area with much of the topographic relief present today. More recent alluvial deposits are common along the Connecticut and Park Rivers and their tributaries.

In the site area, the following soils are present overlying the bedrock, in general order of sequence from ground surface downwards: Artificial Fill, Alluvium, Beach Deposits of Lake Hitchcock, Glaciolacustrine Deposits, Glaciofluvial Deposits, and Glacial Till. Bedrock is not widely exposed in the project area. The formations that are in the general vicinity of the project and potentially could be encountered along the proposed tunnel alignment are the Portland Arkose, the Hampden Basalt, and the East Berlin Formation. These units consist of shale and basalt with fractured and fault zones (Figure 4).

The final geotechnical investigation program consists of 55 deep rock borings, 50 shallow borings, and 5 geophysical survey lines. The program includes geophysical logging (acoustic televiewer) performed in 21 of the deep borings and 5 of the shallow rock borings, water pressure (packer) testing performed in 30 of the deep borings and 8 of the shallow rock borings, 6 in-situ stress determinations in two deep boreholes, falling head tests completed in the soil profile in selected borings, observation wells installed in 13 of the deep borings as well as 13 of the shallow borings, and 22 vibrating wire piezometers installed in 16 borings. The program also included the monitoring of groundwater levels, and the completion of soil and rock laboratory testing.

## MAIN TUNNEL

The deep rock tunnel will be approximately 21,800 feet in length and have a finished internal diameter of 22 feet. The tunnel will be excavated by a hard rock Tunnel Boring Machine (TBM). The tunnel profile is entirely within bedrock. There are several different types of rock TBMs which are manufactured to operate in specific types of ground conditions. These include main beam, single shield, double shield, and convertible (hybrid) hard rock/earth pressure balance (EPB) machines. The selection of the appropriate type of the TBM is an important decision which will impact the type of final lining, construction safety, quality, cost and schedule. The final recommendation on the type of rock TBM will be based on several factors among which rock and groundwater conditions along the tunnel alignment represent very important considerations. This selection will be based directly on the borehole data obtained from the final design geotechnical investigation program.

It is anticipated that the rock mass along the tunnel alignment will primarily consist of competent shale, sandstone, and basalt bedding dipping 10 to 20 degrees with occasional known fault zones. It

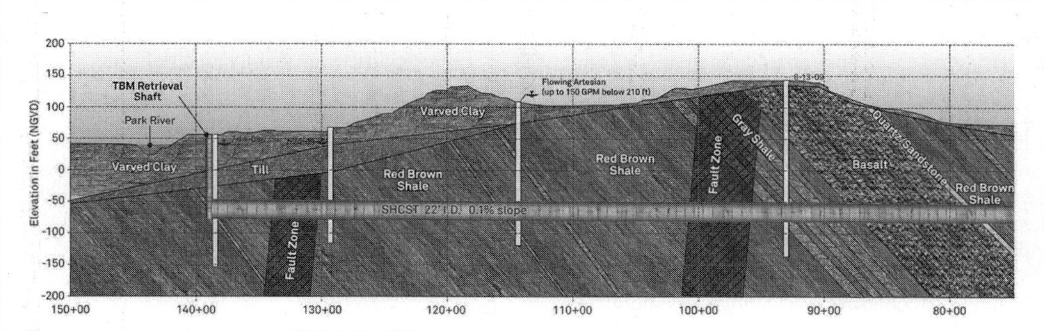


**Figure 4. Geologic profile: SHCST selected alignment (Alternative F)**

may also contain diabase dikes which, if encountered, may contain fractured rock and flowing water.

The size of the construction shafts will depend on the TBM diameter, TBM type, and the dimensions of the permanent structures that will be housed in each shaft. For a 25-ft diameter TBM (required to excavate the 22-ft ID tunnel), the minimum clear shaft diameters that are required to allow launching and retrieval of the TBM are 35 feet and 30 feet, respectively. Larger diameters may be required to accommodate the permanent structures or to suit the contractor's means and methods.

Key considerations in selecting the appropriate construction methods include preventing groundwater drawdown and providing support of excavation. The shafts will be excavated using two methods for ground support. Slurry wall panels, laid out to approximate a circular shape, will extend from top of grade through overburden and will anchor into top of competent rock. The slurry walls will act as temporary support walls during construction and as the permanent final liner.

Through the rock, the shaft will be excavated using drill-and-blast method and the rock face will be supported using rock dowels and sprayed shotcrete.

Starter and tail tunnels will be required to assemble the TBM and to store equipment and muck cars. The starter and tail tunnels will be excavated by drill-and-blast method with a horseshoe cross-section.

One and two-pass lining systems are both considered viable options for the SHCST. The final recommendation of the tunnel lining system will depend on ground and groundwater conditions along the tunnel alignment and the construction cost of each option. Both options should be carried forward through final design phase.

The anticipated ground conditions along the tunnel alignment necessitate the use of final lining for the tunnel to meet the design criteria and ensure long term stability, durability, and hydraulic performance. Viable lining options for SHCST are cast-in-place concrete (CIP) and precast concrete segments.

Important considerations in selecting the type of tunnel lining include the following:

- Durability and ability to withstand the service environment without significant degradation during the tunnel design life
- Constructability
- Life-cycle cost

A quantitative approach, adopted by EPA and ASCE, is used to assess the corrosion of the final lining. This approach estimates the loss of material as a function of time, concrete properties and CSO characteristics.

The recommendations for advancing the tunnel design are summarized below:

- Define geotechnical parameters for tunnel analysis and design.
- Perform groundwater infiltration and ground settlement analysis to quantify the risk of consolidation settlement due to dewatering.
- Analyze geotechnical data to support the selection of the tunnel lining system and type of TBM. Based on the available geotechnical information and construction cost estimate, both tunnel lining options, namely cast-in-place concrete and precast concrete segmental rings, should be carried forward during the final design.

Site plans were prepared to identify existing site conditions, areas for site access, staging and operations, work zone layouts and constraints, equipment and materials storage, utility protection and relocations, site drainage and grading, erosion and sedimentation controls, and electrical power requirements. A temporary site plan and a permanent site plan were developed at the TBM launch site and tunnel pump station. The temporary site plan designates specific areas during construction for the tunnel boring machine, the tunnel crane pad, the tunnel mucking operations, short and long term storage areas for tunnel segments, the pump station crane pad, contractor offices, workshops, storage areas and parking areas. The permanent site plan identifies the locations of the tunnel pump station, screening/ degritting building, HVAC and electric buildings, and odor control facilities.

A conceptual planning level cost estimate, schedule and contract packaging was performed. Costs from similar historical projects were obtained and utilized to develop unit costs and extrapolated for the SHCST project. A detailed cost estimate was performed to estimate the construction cost of the main deep rock tunnel, TBM launch shaft, and TBM retrieval shaft associated with the selected Alignment F. Two construction options were considered in the detailed cost estimate, namely tunnel excavation by a double shield TBM along with installation of precast concrete segmental rings and tunnel excavation by a main beam TBM followed by installation of initial rock support and cast-in-place concrete final lining. The cost estimate for the entire SHCST Project is approximately $500 MM. The project construction duration is estimated at approximately 72 months (6 years).

The recommended contract packaging is to release six construction contracts: (1) Preliminary Utility Relocation, (2) Tunnel, (3) Pump Station, (4) Franklin/ Maple Consolidation Conduits, (5) Flatbush/Arlington/Newington/ New Britain Consolidation Conduits, and (6) West Hartford Consolidation Conduit. The contracts were grouped to align construction skill sets but allow for the phased release of the bid packages. The overall construction schedule is to be coordinated such that the tunnel, pump station and consolidation conduit contracts are constructed independently but conclude coincidently.

MDC management has stated that a goal for the project is that odor complaints must not occur. Therefore, the odor control strategy for the SHSCT system is focused on minimizing odors from the two main shafts at the tunnel ends and at the six intermediate drop shaft sites. Ventilation rates of approximately 80,000 to 85,000 CFM have been estimated for both the upstream and downstream shaft. Ventilation rates ranging from 2,300 to 7,500 CFM have been estimated for the intermediate drop shafts.

Active fan driven odor control systems are recommended at the tunnel ends and passive systems are proposed for the six intermediate drop shafts. Activated carbon is recommended as the odor control treatment process. The odor control systems can either be located in buildings above grade and possibly even below grade in vaults, particularly for smaller systems. This is to address visual impacts in neighborhoods from these industrial type treatment systems. Early estimates of foot print size indicate the larger odor control facilities at the tunnel ends can be roughly 2,000 square feet in area and the smaller systems at the intermediate drop shafts can be roughly 300 square feet in overall size.

## DROP SHAFTS

Six hydraulic drop shafts are used to convey flow in a controlled manner from the shallower consolidation conduits to the deep rock tunnel. A two-level screening process was used to assess the characteristics of each site and to recommend either a baffle-plunge or tangential vortex based upon cost effectiveness, hydraulic performance, and operation and maintenance considerations (Figure 5).

The tangential vortex drop structure type was selected for all of the sites along the tunnel alignment (with the exception of the TBM retrieval site) due to its widely accepted use for deep rock CSO storage tunnels, history of acceptable performance, and cost effectiveness when compared to the baffle-plunge drop structure. The baffle-plunge drop structure type was selected for the deep rock tunnel TBM retrieval site because of the existence of the larger diameter shaft being constructed at this site for the TBM retrieval. Once such a large shaft is present, the baffle-plunge becomes ideally suited for such applications because of its compact surface area impact. Based on the drop shaft selections, potential operations criteria and maintenance requirements were developed for each of the proposed drop structure sites.

## CSO/SSO CONSOLIDATION CONDUITS

New diversion structures constructed near existing CSO/SSO locations will utilize transverse or side flow weirs to direct the design overflows from existing pipes into the consolidation conduits. These conduits then convey flows to the deep rock tunnel through either vortex or baffle drop shafts.

The consolidation conduits will be installed using a combination of microtunneling, guided boring, shallow rock tunneling, and open cut construction techniques. It is anticipated that three consolidation pipe branches along the selected alignment will be installed using microtunneling methods. This includes a 24-inch guided bore of the Newington Consolidation Pipe (NCP), a 42-inch microtunnel installation of the New Britain Consolidation Pipe (NBCP), and a 48-inch microtunnel installation of the Flatbush Consolidation Pipe (FCP). When considering microtunneling as the likely means of installation, effort has been made to locate conduits within

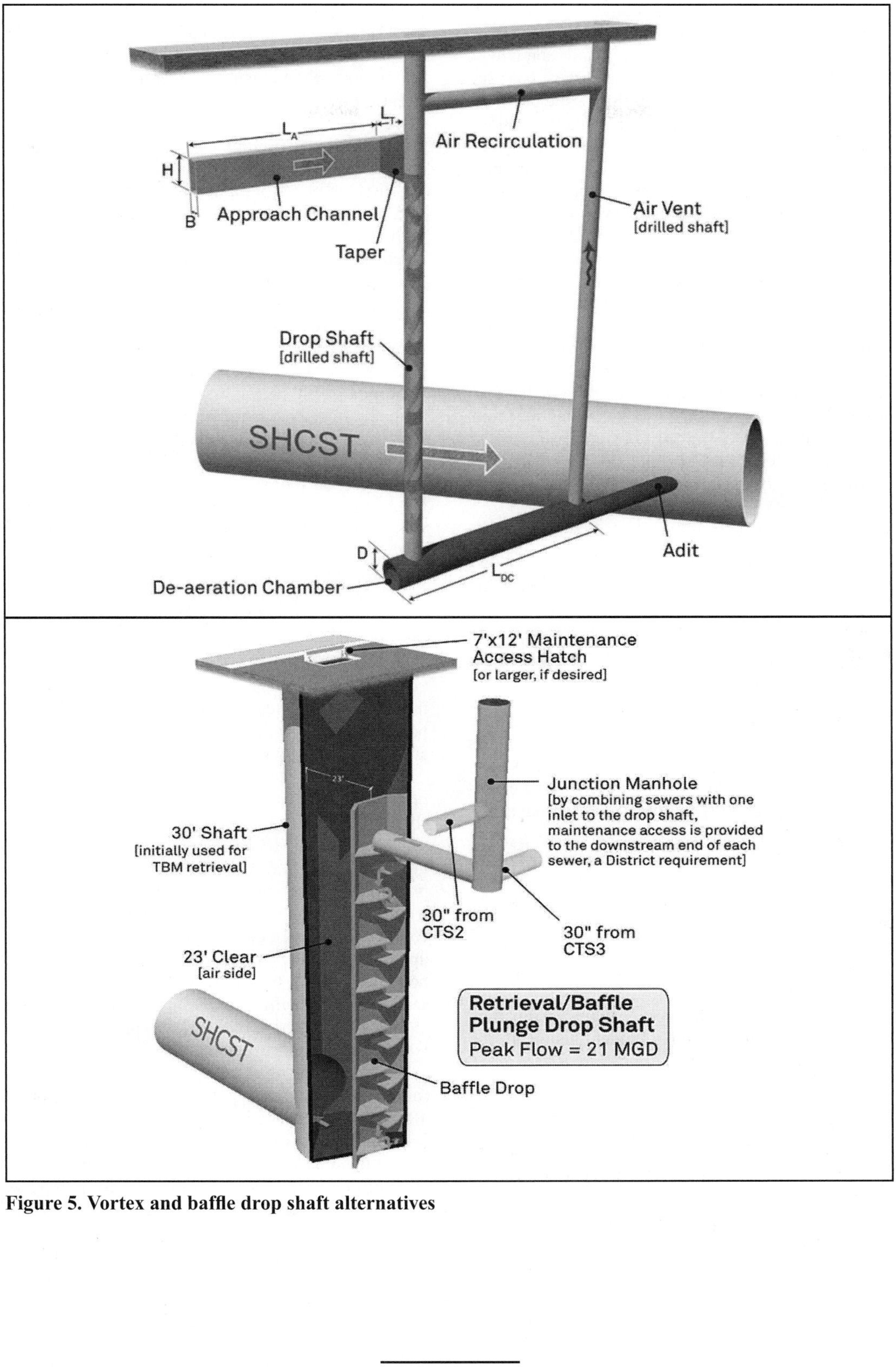


**Figure 5. Vortex and baffle drop shaft alternatives**

soil; however, there is the potential for mixed-face microtunneling in areas of till.

The open cut method of pipe installation will be utilized for installation of the 30-inch West Hartford Consolidation Pipe (WHCP), the southern section of the 24-inch NCP, and the 27-inch Arlington Consolidation Pipe (ACP). The open cut method creates more temporary disturbance to traffic, business and residences as this work is performed primarily within the roadways; however, it may be the preferred installation method due to the depth of the pipe, geotechnical conditions, and cost considerations. Open cut installations typically will be shallower than microtunneling installations.

Based on existing geotechnical information, it is anticipated that the 66-inch Franklin Avenue Consolidation Pipe (FACP) and the 60-inch Maple Avenue Consolidation Pipe (MACP) will be constructed using an open face rock tunneling machine. Consideration is given in final design to standardizing the diameters of these tunnel consolidation sewers to potentially reduce costs.

## PUMP STATION

The TPS is designed to pump out the SHCST following storm events so that the flow can be treated at the HWPCF. At this point, stored flows will receive adequate treatment prior to discharge to the Connecticut River. The proposed TPS will be located within the HWPCF complex.

The TPS will be designed to pump out at a maximum 40 MGD capacity. This rate will allow the 62 MG SHCST to be pumped out within 37 hours (1.55 days). The proposed tunnel invert elevation at the TPS site is –170 feet and the discharge elevation at the plant is +6. Therefore, the total maximum static head is 176 feet.

The recommended pump equipment consists of four 13.3 MGD vertical non-clog centrifugal pumps. This will provide a firm pumping capacity of 40 MGD with one unit out of service. Variable frequency drives (VFDs) are recommended for the pumps as turn-down capability to approximately 4 to 5 MGD can be achieved.

The TPS will discharge directly to the new Headworks facility currently under design at the HWPCF. The force main is currently sized to be 36-inches in diameter. The recommended connection point at the discharge end is at a new junction structure just upstream of the new influent pumping station. A surge tank will be provided on the discharge force main to minimize surges in the system. The surge tank will be situated at the TPS site.

Two pump station configurations are presented as the finalist options. One of these is a cavern pump station and the other is a circular pump station with a suction header pipe system (Figure 6). The two configurations are comparable in overall cost and the cavern pump station has some advantages in terms of non-cost criteria, mainly centered on maintenance attention associated with crane lifts. To allow for a more informed decision on pump station type, MDC personnel visited both type of facilities at other deep operating installations. Following those site visits, the circular pump station layout was fitted with a bridge crane at the lower level; this essentially made both configurations the same from a maintenance perspective. The City of Hartford building department office was also consulted at this time to assess requirements for this deep pump station to comply with the current 2005 Connecticut State Code governing this facility. A second means of egress (i.e., a second stair tower) and compartmentation of the floor area of the below grade levels were identified as the more extensive requirements of the code. The layouts for both finalist pump station configurations were then modified to address these more significant building code requirements. A comparative assessment of the capital costs of these two configurations was then prepared and it is concluded that the cost of the cavern pump station is less than that of the circular pump station. On this basis, the cavern pump station is recommended for the project.

A new 9,800 kW overhead electrical power service from CL&P will be required for the tunnel boring machine (TBM). This power feed will be converted to a permanent power feed for the TPS, once the TPS is completed and made operational. Current power requirements for the TPS and related facilities are on the order of 3,055 kW.

Screenings and grit capture will be accomplished in a separate 35-foot diameter dedicated shaft. The shaft which will be used as the launch shaft for the TBM tunnel will be converted to the grit/screenings shaft. Bar screens will be provided to protect the TPS pumps from solids and debris which would either clog or damage the pumps. A rake lowered by crane will either push or pull the screenings up from the shaft. Grit and other heavier debris will be removed from the pit by a clamshell bucket. The screenings shaft will be used for tunnel construction, allowing construction of the TPS to proceed in parallel with tunnel construction.

The TPS and the Grit/Screenings facility will be roughly 150 feet apart and will be connected with a 48-inch diameter suction header. An at-grade building will be constructed over the below grade pump station to house support facilities critical to the operation of the pump station and to allow for pump station access and egress. Personnel access/egress will be by elevator. A separate stair tower will be provided for emergency situations. The grit/screenings

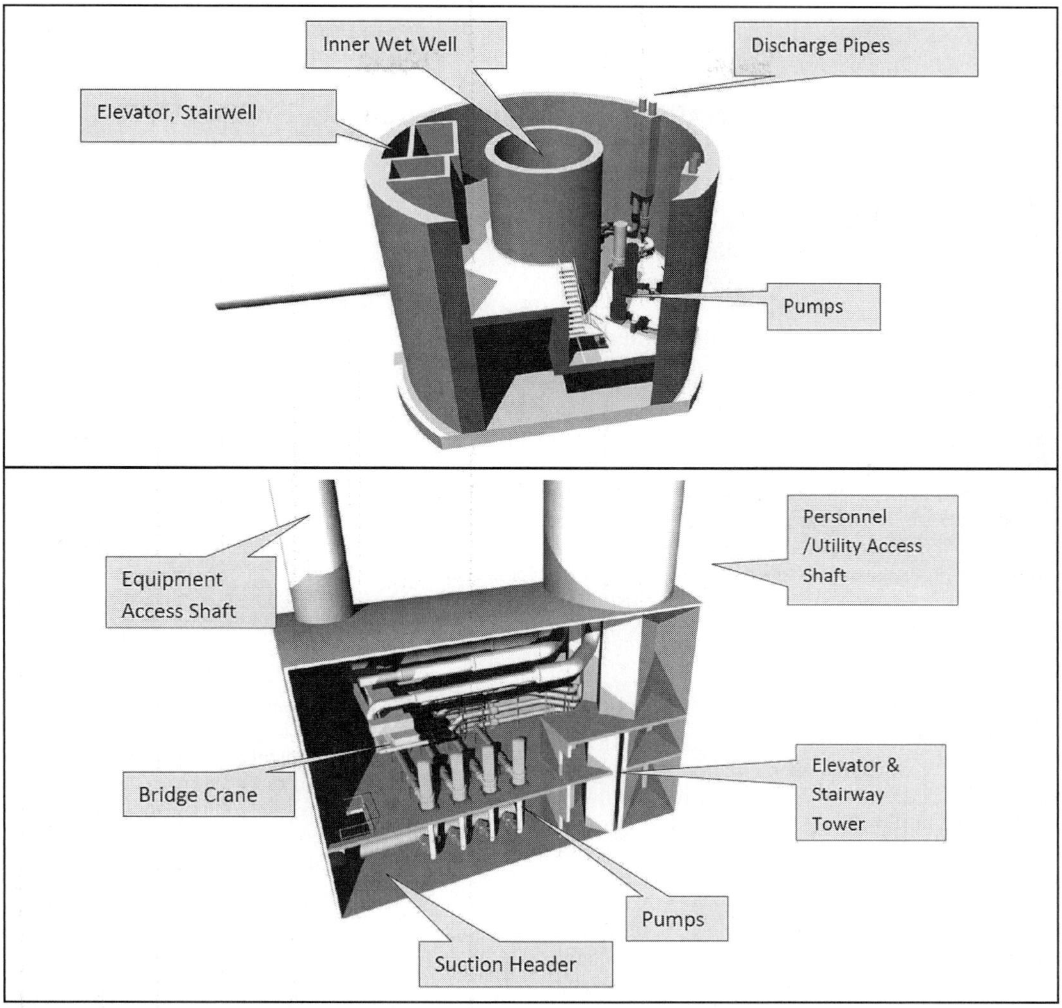


**Figure 6. Cavern and shaft pump station alternatives**

facility will also be enclosed in a building to better contain odors and to promote a more visually appealing facility to neighboring businesses.

## CONCLUSIONS

This paper presents the design of a deep rock conveyance and storage tunnel, drop shafts, consolidation conduits, and a pump station in Hartford, CT. The geological settings and subsurface investigation program are discussed and the general aspects of the preferred alignment selection are described. Relevant alternatives for the drop shafts and the pump station are explained and recommended options are presented.

## REFERENCES

Hubert, J.; Reed, F.; Dowdall, W. and Gilchrist, J. 1978. *Guide to the Mesozoic Redbeds of Central Connecticut*, Guidebook No. 4, Connecticut Geology and Natural History Survey, Hartford, CT.

Nasri, V.; Sullivan, J.; Pelletier, A. and McCarthy, B. 2013. *Design of South Hartford Conveyance and Storage Tunnel*, Rapid Excavation Tunneling Conference 2013, Washington, DC, June 23–26, 2013, pp. 546–556.

Northeast Nuclear Energy Company 1975. Geologic Mapping of the Bedrock Surface, Millstone Nuclear Power Station Unit 3, Docket No. 50-423.

# Zone 1 Interconnecting Watermain: Proudly Tunneling Under the Challenges Ahead

**Gianpiero Vancheri**
R.V. Anderson Associates Limited

**Marek Braczek**
Halton Region—Design and Construction

**ABSTRACT:** This paper will show how innovations in tunnel planning and design were developed to suit challenging conditions; including crossings of environmentally-regulated waterways, the avoidance of high pressure oil pipelines, hydrocarbon plumes from previous spill contamination, crossing existing rail infrastructure, a future grade separation, major highway, and other infrastructure.

With so many limiting criteria, hazards, and stakeholder requirements to be incorporated, examples of the simple but thorough risk assessment and management tools used are presented.

Constructability is a key factor of successful tunnel projects and examples of design to allow for flexibility in the Contractor's selection of means, methods, and phasing are presented.

## BACKGROUND

The Regional Municipality of Halton (Halton Region) is a regional municipality of Ontario, Canada. Located less than 40km to the southwest of Toronto in Southern Ontario, Halton Region comprises the City of Burlington and the towns of Oakville, Milton, and Halton Hills.

The Zone 1 Interconnecting Watermain project was identified in Halton *Region's South Halton Water and Wastewater Masterplan* as a large diameter watermain required for conveyance of treated water between the Burloak Water Purification Plant (BWPP) and the Kitchen Reservoir. The primary function of this watermain is a supplementary water supply to the Kitchen Reservoir and also to serve as the supply source to a future Zone 2 Booster Pumping Station (BPS) to be constructed at a site purchased by Halton Region for this purpose immediately south of the Queen Elizabeth Way (QEW), a 400 Series Highway (analogous to an Interstate Highway in the US).

The Environmental Study Report (ESR) was completed in October 2011 and recommended this large diameter watermain be located within a corridor following Burloak Drive from the BWPP at Rebecca Street to Upper Middle Road and along the unopened road allowance across Bronte Creek to the Kitchen Reservoir.

The final pipe sizes identified in the master planning process are noted as 1800mm (70") diameter from BWPP to just south of the Burloak Drive/ QEW interchange at a site proposed for the future Zone 2 BPS and 1500mm (60") diameter from the Zone 2 BPS site to the Kitchen Reservoir. The watermain and tunnel alignment is shown on Figure 1.

With the route of the watermain constrained as per the EA, the preferred solution incorporated the majority of the watermain installed as open cut with tunnel installations limited to crossings of environmental features, infrastructure obstructions, and protected watercourses.

Evaluation was carried out, and further assessed during Risk Management Workshops that followed, resulting in the design of the watermain to be installed by tunnelling for the entire length save for yard piping at the facilities sites.

## REGIONAL AND LOCAL GEOLOGIC CONDITIONS

The project area and surrounding region are characterized by shallow bedrock overlain by a thin glacial till deposit known as the Halton Till, which is a matrix of gravel, clay, and silt. The shale formations underlying the project area are of Upper Ordovician age. The uppermost of these is the Queenston Formation which overlies the Georgian Bay Formation. The entire tunneled length of the watermain is anticipated to be constructed entirely within the Queenston Formation.

The Queenston Formation is characterized by its color; and contains red siltstone, minor green shale and siltstone, with variable calcareous siltstone to sandstone and limestone interbeds.

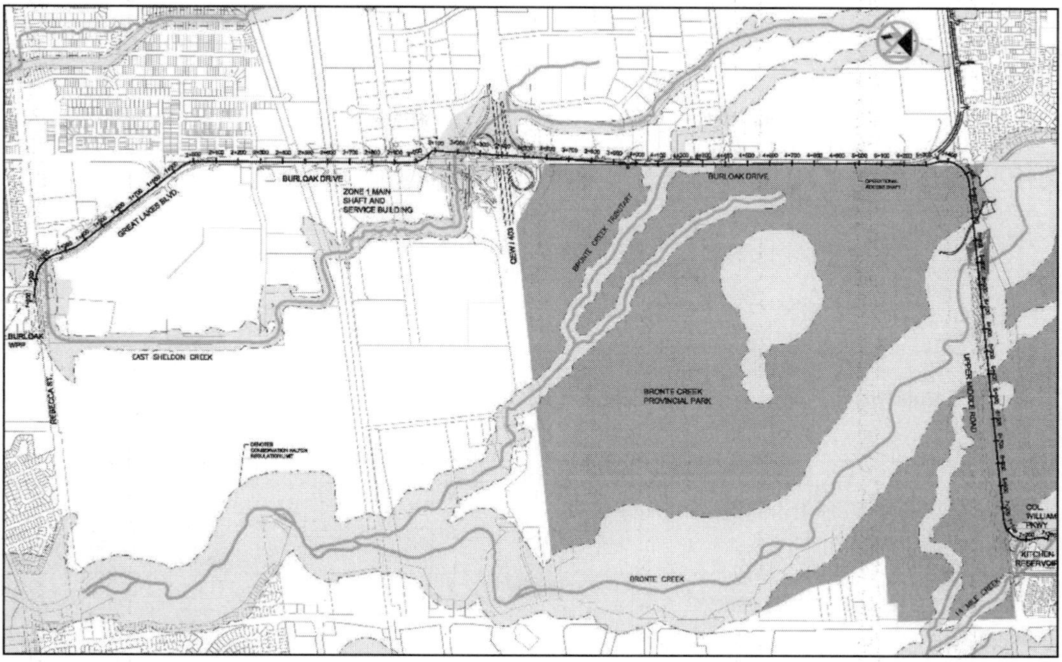


**Figure 1. Tunnel route**

The Queenston Shale possesses stress characteristics attributed to locked-in horizontal stress relief and stresses associated with tectonic movements which can result in crush zones and jointing. The major principal stress has been measured as 5MPa (725 psi).

Circular bored tunnels excavated by controlled mechanical methods, such as Tunnel Boring Machine (TBM), minimize the damage/fracture zone of the rock from the mined face which allows the tunnel face to achieve a state of equilibrium balanced around the circular form. The largest stresses act in the horizontal plane and are able to flow around the balanced circular excavation.

The intact rock properties of the Queenston shale sampled along the proposed tunnel alignment including Rock Quality Designation (RQD), compressive strength; along with the published in-situ stresses in the shale, lends the primary liner system design to a light support system typically comprising steel ribs fixed by rockbolts into the tunnel crown with elements of timber or steel wire mesh spanning longitudinally between the steel ribs (Figure 2). The main purpose of the primary liner is to apply positive support to the mined rock surface, prevent mobilization of the rock, and to stabilize the tunnel crown. The primary support to the tunnel roof is to

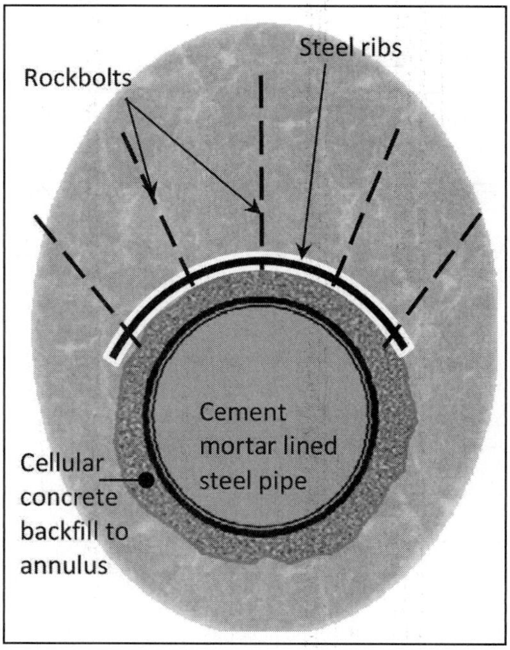


**Figure 2. Typical tunnel installed watermain section**

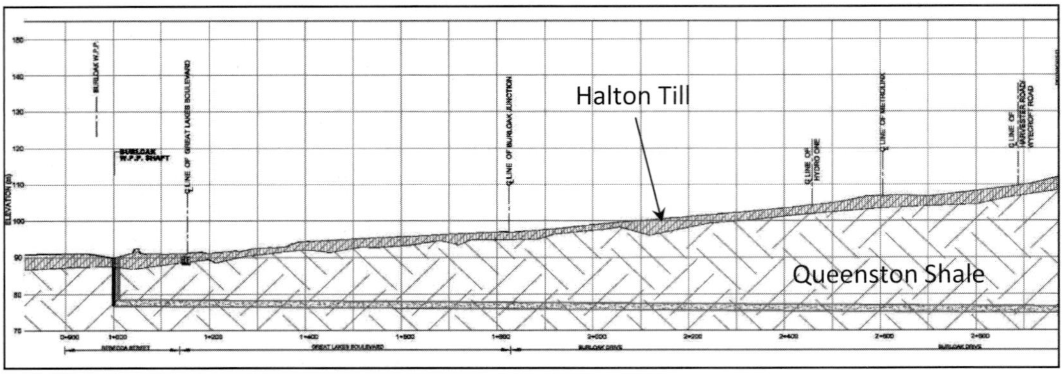


**Figure 3. Tunnel profile along chainage 1+000m to 3+000m (6,562' in length)**

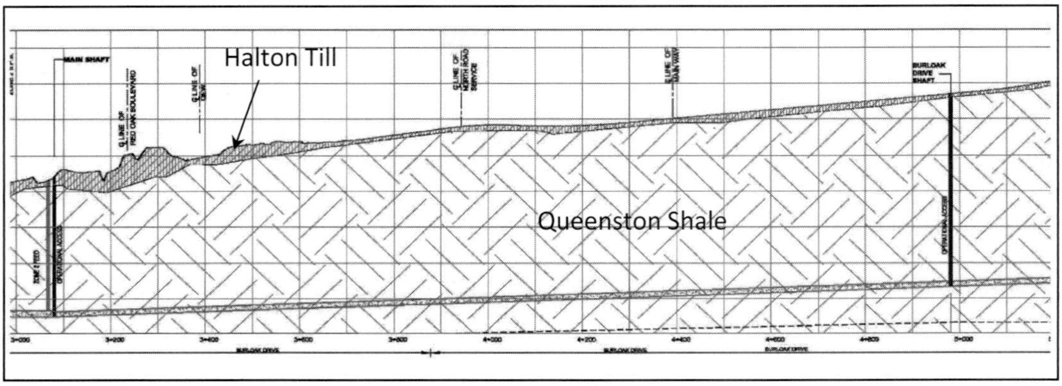


**Figure 4. Tunnel profile along chainage 3+000m to 5+200m (7,218' in length)**

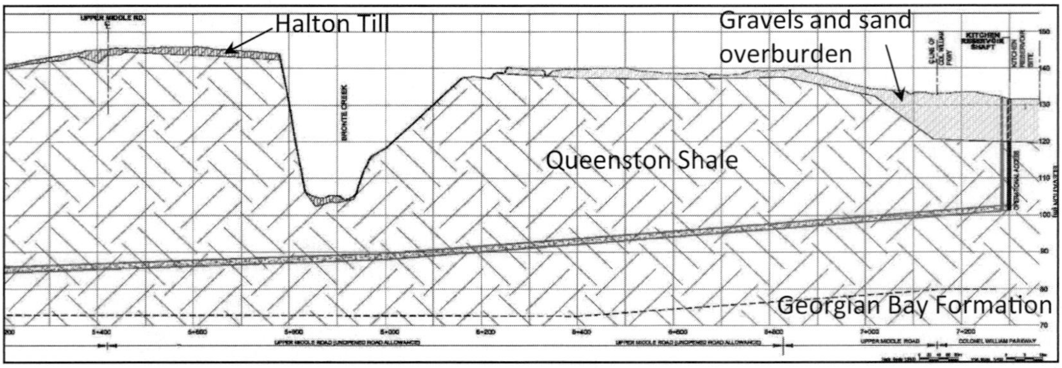


**Figure 5. Tunnel profile along chainage 5+200m to 7+300m (6,890' in length)**

be installed immediately behind the advance of the TBM shield. See Figures 3, 4, and 5.

The crossing of Bronte Creek, a 35m (115') deep valley within an environmentally sensitive Provincial Park, raised a concern because of the potential presence of a buried geological valley. Buried channels in the bedrock in-filled with water bearing glacial deposits and abrupt changes in rock quality are known to be present in the sedimentary bedrock in Southern Ontario and generally a geological feature visible from grade can be indicative of a feature buried below such as a fault zone.

Should a TBM intended for rock excavation mine into a water bearing soft ground valley, catastrophic inflows of water and soil could occur and halt progress.

Another hazard with the potential to be realized was the presence of large boulders having historically rolled into the valley after its glacial formation and subsequently overlain by weathered rock, till.

## ENVIRONMENTAL SENSITIVITY

Bronte Creek Park is an Ontario Provincial Park and a place of natural beauty that provides leisure facilities including hiking trails, swimming, fishing, biking trails, natural heritage education, birding, winter activities, and camping.

Bronte Creek which runs through the park is regulated by the local conservation authority, Conservation Halton (CH), and is home to several aquatic and terrestrial species at risk (SAR). To protect SAR, the Ontario Ministry of Natural Resources (MNR) has passed some of the toughest legislation in North America. Due to the environmental sensitivity of the Bronte Creek study area, initial geotechnical investigations carried out during the Environmental Assessment were limited to a geophysical study using seismic-refraction testing that identified two anomalous zones—one in the valley, the other on the east side of the valley

Not only is there a crossing of Bronte Creek within the park, but a tributary of Bronte Creek has to be crossed along the route of the tunnelled watermain which is subject to the same legislation as are two other crossing CH regulated areas and SAR protected areas local to the two crossings of East Sheldon Creek.

## METROLINX

The crossing of the Metrolinx rail lines not only has to satisfy current layout arrangements but also a future grade separation that lowers the existing road elevation below the rail lines.

At the proposed railway crossing, the tunnel depth is approximately 30m (98') and this cover incorporates in excess of 25m (82') of intact rock. This depth will avoid future conflicts and prevent the need for any increased work at grade during construction.

For the length of the proposed crossing, the watermain (carrier pipe) is held clear of and installed within a casing pipe designed in accordance with Transport Canada's *Standards Respecting Pipeline Crossing Under Railways* TCE-10, AREMA, and Metrolinx' Crossing requirements (Figure 6.).

To prevent formation of a watercourse below the railway right of way and mitigate the risk of washout or lifting of the ground cover to the watermain, a pressure relief arrangement was designed. This system incorporates a standpipe to provide a

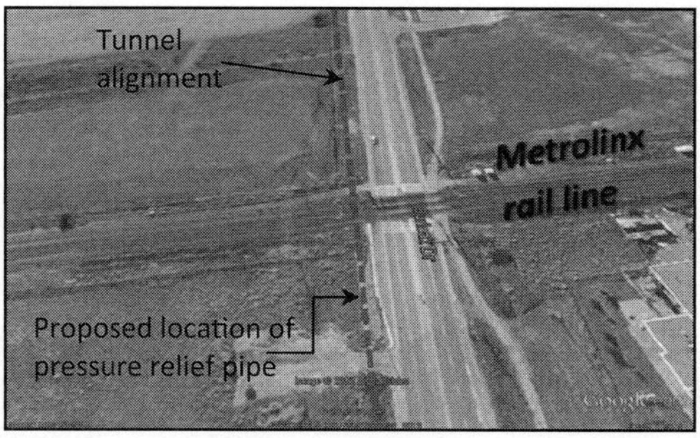


**Figure 6. Tunnnel alignment at Metrolinx crossing**

direct hydraulic link from inside the casing pipe for any leakage of the pressurized potable water from the carrier pipe to be discharged at grade into a ditch within the road allowance (Figure 7).

This arrangement provides a visible, live monitoring capability in order that the valves are closed and repair can be carried out from inside the carrier pipe.

## CROSSING OF THE QUEEN ELIZABETH WAY

The crossing of the Queen Elizabeth Way (QEW) includes crossing the eastbound on-ramp and westbound off-ramp, and the QEW itself (Figure 8).

Approvals constraints of the Ministry of Transport (MTO) associated with a 400 Series Highway require the tunnel to have a full circumference steel primary liner for the length of the crossing. For this requirement, a steel plate primary liner system was designed and detailed (Figure 9).

## HYDROCARBON PLUME

A previous crude oil spill occurred within the unopened road allowance along the route of the proposed tunneled watermain and where there is a change in heading direction by a full 90° (Figure 9). Existing remedial and spill containment measures are in place treating contaminated groundwater and any works local to this area must not affect the existing groundwater flow or gradient.

Property constraints in this area restricted the horizontal alignment of the tunnel to a narrow corridor. The route of the watermain changes in direction through a full right angle and with bends slowing down mining production and tunnel advance during construction, a more favorable arrangement would be to have a TBM turning shaft at the intersection of Burloak Drive and Upper Middle Road. However, any excavation through the horizontally bedded planes of rock for such a vertical shaft posed a risk of creating a direct connection between the contaminant and the tunnel.

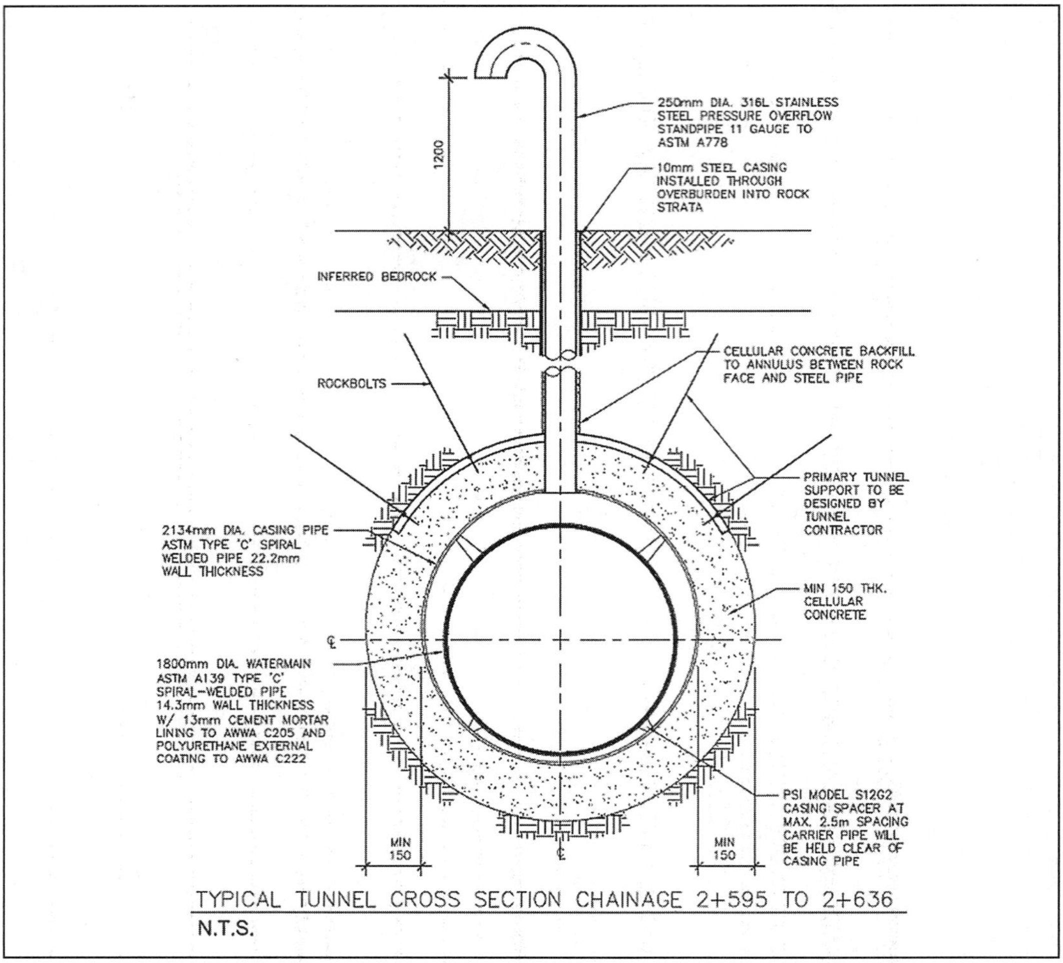


**Figure 7. Pressure relief arrangement**

**Figure 8. Crossing of the Queen Elizabeth Way**

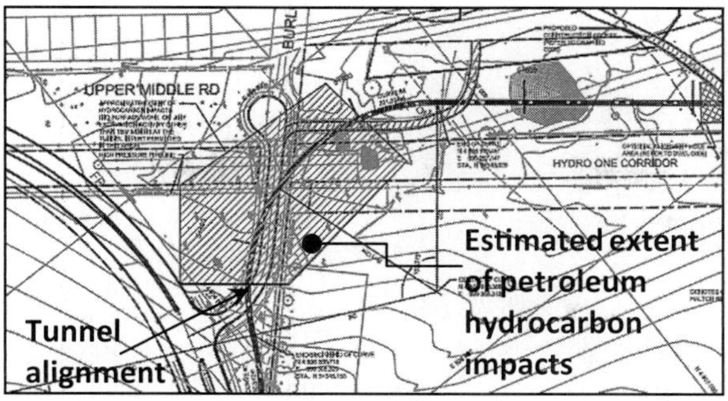


**Figure 9. Tunnel alignment at location of hydrocarbon plumet**

The drilling of alignment holes within this area could also pose a similar risk and is therefore not allowed within the contract. This restriction gave greater importance to being able to secure an area for an alignment hole in the Hydro One utility corridor which runs through the park.

## DESIGNER'S RISK ASSESSMENT

Recognizing that the opportunity to minimize risks is highest during the early feasibility stage of a project, a Designer's Risk Assessment was written upon commencing the Preliminary Design.

This dynamic document is continuously updated, identifying and mitigating risk items throughout the life of the project.

The Designer's Risk Assessment (DRA) used is a simple qualitative tool in evaluating known risks particular to this project using the definition of risk as:

$$\text{Risk} = \text{Likelihood of occurrence of the hazard} \times \text{Severity of the hazard}$$

Upon identifying each risk item, actions are carried out in the following order of priority:

1. *Eliminate the risk*—This is achieved by altering the design in such a manner that the potential for a hazard to be realized is removed by not carrying out a particular action that has been identified as carrying risk.
2. *Mitigate the risk*—This is achieved by reducing the likelihood of the hazard occurring or reducing the severity of the outcome of the hazard being realized.
3. *Inform all involved of the risk*—Once a risk has been identified, affected parties are clearly informed of known and identified risks for the purposes of Risk Sharing/Allocation and to also define the extent of known risks.

The DRA for this project incorporates experience of claims submitted on previous tunnelling projects and engineering opinion of risks associated with the construction of the Design details.

## RISK MANAGEMENT WORKSHOPS

The Risk Management Workshops were a series of workshops intended to supplement design and engage stakeholders. There were three workshops throughout the design process with focus shifting on topics as the design progressed from the preliminary phase through to the detailed design phase as did the involvement of stakeholders. A Risk Panel comprising specialists outside of the design team was assembled representing aspects including Tunnel Design, Tunnel Construction, Geotechnical and Subsurface Conditions, Watermain Design, and Project Management.

The workshops were able to identify key risk items and provided a forum for members of the design team, risk panel, and stakeholders to voice their respective concerns of hazards affecting them or under their control with discussion and understanding of acceptable mitigation measures presented, developed, and ultimately accepted.

One of the major technical issues focussed upon within the first workshop was the design and construction of the tunnel across the Bronte Creek valley. Conservation Halton (CH) and Ministry of Natural Resources (MNR) management policies associated with sensitive environmental areas, species at risk and cold water fishery designations had restricted access to the creek valley preventing the ability to collect the level of geotechnical information required for design and to verify or explain the anomalous zones identified in the seismic refraction study.

The design and construction risk concerns were well represented by the Design Team and the Risk Panel were able to effectively communicate the severity associated with the hazards of tunnelling beneath the creek with insufficient geotechnical information; the hazards associated with encountering a severe fault or hidden geological feature and

**Figure 10. Rock core sampling in the valley floor**

**Figure 11. Intact rock cores with breaks from sampling equipment**

the scale of potential impact, and effort required to address a hazard of this severity being realized.

Terraprobe, the geotechnical sub consultant member of the Design Team, devised the innovative solution of minimizing environmental impacts by utilizing modified concrete coring equipment to obtain rock core samples beneath the valley floor (Figure 10). The Environmental specialist member of the Design Team, LGL Limited, helped develop an access plan to the restricted areas to the acceptance of the stakeholders and we were able to obtain sufficient data to identify the magnitude of the risk associated with tunneling beneath the creek, reduce the likelihood of its occurrence by setting the vertical alignment of the tunnel using the field data retrieved and now having sufficient rock data for design (Figure 11).

Workshops dealt with subsequent elements of design such as operational features including access and maintenance of the watermain and shafts. A final workshop was carried out at the 90% stage of completion and provided one more opportunity to confirm the risk allocation decisions with the most information at hand for that final "second thought" on the proposed approach to managing the risks for the project. At this time some of the key decisions and information for consideration of baselines and statements within the GBR were uncovered.

## THE GBR: THE APPROACH TAKEN

The GBR does not eliminate risk, but rather is a method for sharing risks between the Owner and the Contractor. While contractually setting baselines for bidders to carry cost and select appropriate means and methods, the GBR defines the "line in the sand" by which a valid claim is measured.

The surrounding geologic conditions that we have designed the tunnel for has been well studied with a number of technical papers based on case studies written since the late 1970s. Many tunnelling projects have been carried out within this formation resulting in well-defined means and methods established and the properties of this formation have been found to be typically constant over large areas of southern Ontario. This understanding has been incorporated within setting the baselines in the GBR.

The GBR for this project was developed to portray conditions that are on the conservative side of realistic conditions expected to be encountered. The conditions reflect the findings of the geotechnical investigation, experience of the conditions within the rock formation that is anticipated to occur, and comments received during the Risk Workshops and GBR Review sessions involving members of the Risk Panel and the Design team. This approach was taken in order that the Owner does not pay for excessive adverse conditions that are not likely to occur and that only fairly adverse conditions encountered are valid as claims for a differing site condition.

## DESIGN FOR CONSTRUCTABILITY

> *"Contracting is about profit, nothing else matters. If a contractor doesn't try and do that, you don't want them as they're up to something else. It's done by reducing time, material costs, labour costs, complexity, and risk and increasing speed and safety by*

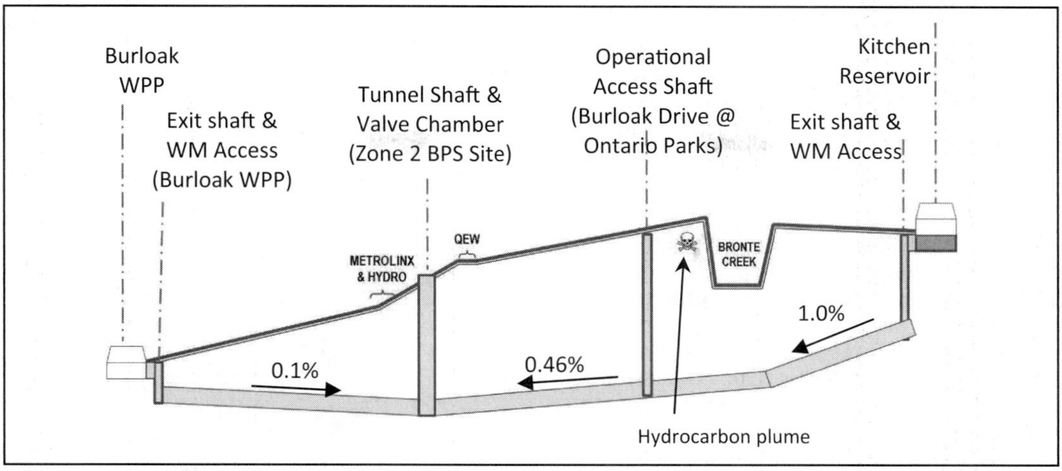


**Figure 12. Schematic of vertical alignment**

> *making the right choices from the many that are available."* (McNally 2013)

As well as designing for the physical conditions, strength requirements, and operational end user requirements, a major consideration in the design of the project is constructability. The intention was to allow an amount of flexibility for the tunnel construction specialist to determine the most appropriate means and methods to construct the tunnel as safely and efficiently as possible to meet the schedule.

The schematic in Figure 12 outlines the vertical alignment of the tunneled watermain in the final design. The constraining elevations occurred at the Burloak WPP and under the Bronte Creek. In both cases the elevations were set to obtain adequate competent rock cover for design without requiring onerous primary support systems restricting the size and type of TBM and increasing the complexity and cost of the entire tunnelling operation.

The design exploits the land area available at the site of the Zone 2 BPS site and allows for the majority of tunnel operations to be launched and serviced from there. In having one low-point at this shaft site there is the flexibility for a contractor to have two TBM headings in opposite directions mining uphill. During mining, muck is transported back to the servicing shaft along a rail system. Due to the associated spatial requirements this operation prohibits any other works being able to take place, a situation often referred to as "Tube Lock." The splitting of the tunnel into three lengths allows for phasing of the construction works to take place so that upon completion of mining within one particular length, the watermain can be laid and joints connected while another length of tunnel is excavated.

When considering the safety aspects during construction, the shaft layout allows for the maximum length for access to a shaft from the tunnel to be approximately 2km.

The environmental sensitivities associated with work in the Bronte Creek Park meant the provision for constructing alignment holes was restricted. The design acknowledged the need for tunnel servicing shafts and alignment holes. A permit was obtained on a minimally sized working area for the construction of an alignment hole within the park in an existing Hydro ONE corridor with construction access granted by the Park.

The horizontal alignment design also allows for practical layout of alignment holes to keep them out of the middle of the road.

The tunnel bore diameter is to the Contractor's design and as such allows the market to dictate the size of the excavation. A number of factors will be considered by the bidding contractors including TBM availability for the period of the Contract and material related costs. Increases in the diameter of the tunnel excavation impact the amount of mined excavation spoil (muck) requiring removal from site and the volume of cellular concrete backfill.

Based on the rock formation, availability of TBMs of known local Contractors, and previous projects similar in scope and size, the TBM bore diameter to mine the tunnel is anticipated to be a minimum of 2.44m (8') diameter and a maximum of 3.05m diameter (10'). It is anticipated that a *Main Beam* TBM will be used to construct the tunnel in this formation allowing for a minimal primary liner supporting only the crown or partial circumference of the tunnel.

## DESIGN FOR ROCK CONDITIONS

Current tunnel design practice for tunnels in southern Ontario mostly follows the results of study and recommendations made by Dr. Lo of the Geotechnical Research Centre at Western University and his coworkers in the 1980s and early 1990s(Lo and Yuen 1981; Lo, Cooke and Dunbar 1986, and Lo and Lee 1990).

*Time dependent deformation* (TDD) is manifested by swelling in the shale which can exert massive stresses on any element cast against the rock. The use of a weak, deformable liner around the watermain pipe acts to buffer the rock stress and protect the watermain. A low strength, cellular concrete is used to backfill the annulus between the pipe and the mined rock surface.

The cellular concrete is strong enough to resist the external hydrostatic pressure from the water table up to grade and can contribute to the composite strength of the pipe in steel pipe design. It is also designed to have a compressive strength low enough to allow the face of the shale to swell and compress the cellular concrete and reducing the load transmitted to the pipe. The cellular concrete is a closed cell specification to prevent the ingress of groundwater into the mined excavation that could cause the watermain to corrode.

## OPERATIONAL ACCESS SHAFTS

Due to the depth of the watermain ranging from 15m to 60m (49' to 197') below grade and the annulus around the pipe wall filled with cellular concrete, the only access to the watermain would be via entry to the inside at the tunnelled pipe elevation. The shaft design had to accommodate these access requirements.

Permanent access to the watermain at the tunnel elevation will be at four (4) shaft locations exploiting shafts used for construction.

The shafts will be used to launch and retrieve monitoring/inspection devices under live flow conditions. If repairs are required to be carried out to the watermain, each shaft has blind flanges for man entry inside the watermain under drained down, empty conditions.

The main tunnelling operations are to be carried out from the Zone 2 BPS site which will house the Zone 1 Main Shaft and a service building in the finished scheme, over the watermain and valve arrangements (Figure 13).

The original TBM launch shaft will be filled with mass fill concrete encasing the steel watermain rising mains, to form a 5m × 8m (16' × 26') finished access shaft.

Operational benefits of having the watermain valves at a higher elevation and within the building are realized with access to the bottom of the 35m (115') shaft only required for man entry inside the watermain. This shaft will be the point of entry during requirements for prolonged entry.

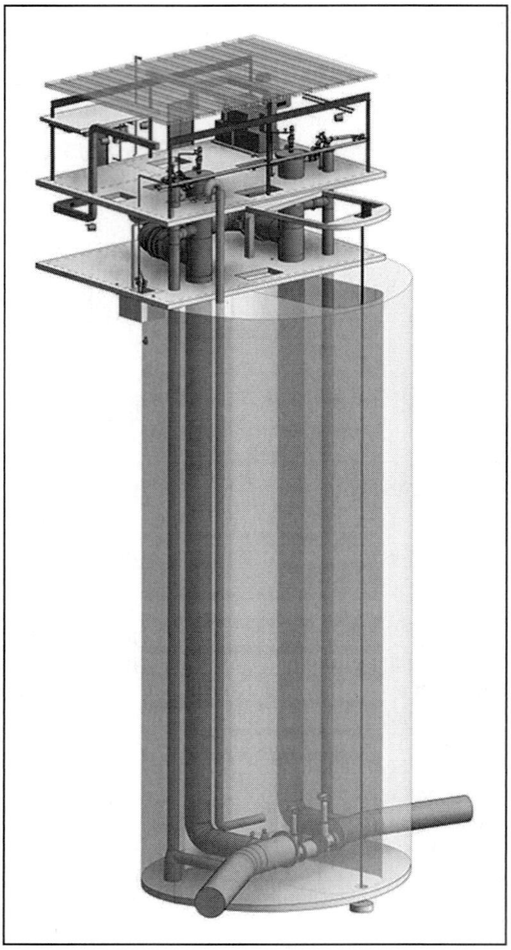

**Figure 13. Zone 1 main shaft**

At the southern end of the tunnel at the Burloak WPP and the northern end at the Kitchen reservoir, the shaft arrangements will comprise a watermain riser shaft and an operational access shaft (Figure 14).

To bring the watermain up to the open cut yard piping elevation from the tunnel elevation; the watermain riser shaft depth will be 15m (49') at the Burloak WPP and 30m (98') at Kitchen reservoir.

The operational access shaft at each site is a 3.6m (12') finished shaft large enough to lower a man basket for access to the watermain at the tunnel invert for launching or retrieving monitoring equipment and entry into the pipe for maintenance as required.

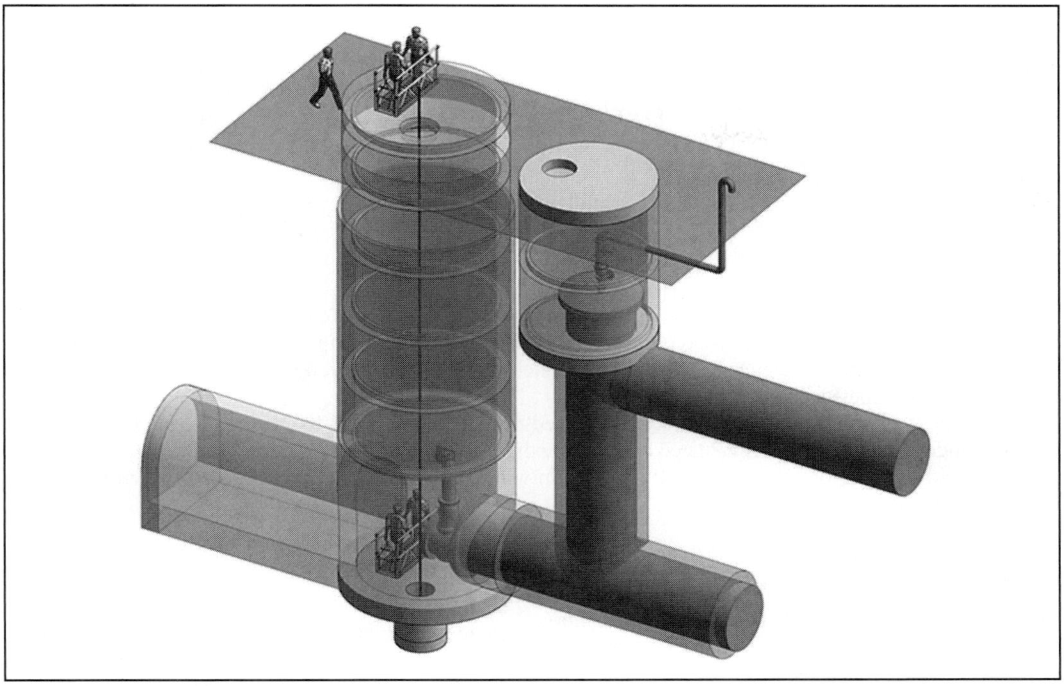

**Figure 14. Shaft arrangement at Burloak WPP**

The shaft at Burloak Drive and Ontario Parks is the deepest of the shafts at 60m (197') depth. A 3.6m (12') finished diameter shaft will provide access for man entry to the watermain under drained down conditions through a blind flange only. Unlike the other shafts, this shaft is not located within the owner's property and is likely to be only used in major repair events.

## CONCLUSIONS

- Design has to focus significant attention to construction methods more than they would on other construction projects.
- Flexibility in phasing construction and means and methods available will work to attract lower bids and optimize schedule.
- It is important to identify risk at the earliest stage of design and evaluate who is affected or can control this risk.
- The engagement of stakeholders in the risk assessment process can lead to more widely accepted mitigation measures.

## REFERENCES


Lo, K.Y., and Yuen, C.M.K. 1981. Design of tunnel lining in rock for long-term time effects. Canadian Geotechnical Journal, 18:24–39.

Lo, K.Y., Cooke, B.H., and Dunbar, D.D. 1986. Design of buried structures in squeezing rock in Toronto, Canada. In Proceedings of the 6th Canadian Tunnelling Conference, Niagara Falls, Ont. Tunnelling Association of Canada, Toronto, Ont. pp. 159–198.

Lo, K.Y., and Lee, Y.N. 1990. Time-dependent deformation behaviour of Queenston shale. Canadian Geotechnical Journal, 27:461–471.

McNally, M. 2013. Construction Issues relating to the use of Tunnel Boring Machines for underground construction. Lecture during Tunnels and Underground Structures in Rock short course, Geotechnical Research Centre, Western University, London, Ont.(unpublished).

# Mid-Halton Outfall Tunnel Overbreak Evaluation

**Donald C. Wotring and Gary Kramer**
Hatch Mott MacDonald

**ABSTRACT:** As part of recent wastewater facility upgrades for Halton Region, Ontario Canada, an outfall sewer tunnel will be constructed to convey treated effluent from the treatment plant to discharge into Lake Ontario. The 3.6-meter diameter, 6,255 m long tunnel will be mined through Georgian Bay shale that is interbedded with limestone and siltstone.

This paper presents the design methodology used to predict overbreak risk using two-dimensional finite element modeling. Overbreak is evaluated by coupling the material behavior with three Tunnel Boring Machine (TBM) shield and support configurations: main beam with actively-propped support, main beam with finger shield, and single shield.

## INTRODUCTION

The purpose of this paper is to present the design methodology and results of a tunnel overbreak evaluation, using two-dimensional finite element analyses. In addition to local experience and construction practice, the results were used to help in the development of baseline values to be included in the geotechnical baseline report and to develop specifications that adequately manage risk associated with various locally-preferred TBM shields and support types.

In particular, the proprietary McNally (2002, 2009) system has had great success locally and this tunnel support system is evaluated and compared with cantilever finger shield and single shield supports.

The Mid-Halton wastewater treatment plant (WWTP) is owned and operated by the Region of Halton. From its location in Oakville (west of Toronto), the WWTP services portions of the towns of Oakville, Milton, and Halton Hills, Ontario. As a part of the planned facility upgrades, an outfall sewer tunnel will be constructed to convey treated effluent from the WWTP eastward to discharge into Lake Ontario.

The site is approximately 40 km southwest of downtown Toronto and has total length of 6,255 m. The tunnel includes an onshore and offshore reach. The onshore reach is 4,085 m and extends from the mining shaft at the WWTP to an intermediate shaft at Coronation Park along the Lake Ontario shore. The offshore reach is 2,170 m and extends from the intermediate shaft to its terminus at an outfall array beneath Lake Ontario. The project alignment is shown in Figure 1.

The TBM will mine through Georgian Bay shale, which is interbedded with limestone and siltstone. The tunnel will be lined with cast-in-place, plain concrete to an internal diameter of 2.6 meters. The tunnel bore diameter is expected to be up to 3.6 meters.

The WWTP shaft will serve as the TBM launch shaft and will be completed as a reinforced concrete baffled drop structure. The Coronation Park Shaft will serve as a working shaft and will be completed as a reinforced concrete access structure.

The final 300 m in the offshore reach will contain an array of 18 diffusers. The 500 mm internal diameter risers will be capped with a diffuser port and rip-rap armor above the lake bottom.

## SUBSURFACE CONDITIONS

Southern Ontario is underlain by a thin mantle of glacial soils that overlie a thick sequence of Palaeozoic sedimentary rocks deposited directly on the Precambrian basement rock. The sedimentary rocks derived from marine sediments deposited approximately 325 to 570 million years ago. Within the region, the uppermost bedrock along the Lake Ontario shoreline is Queenston shale of the late Ordovician period. In the project vicinity, the Queenston shale has largely been eroded and is only approximately 25 meters thick beneath the WWTP and pinches out completely offshore. The Queenston shale conformably overlies the Georgian Bay shale. In the project vicinity, the Georgian Bay shale is located at the bedrock surface beneath Lake Ontario as well as onshore to the northeast starting from approximately Mississauga to Toronto.

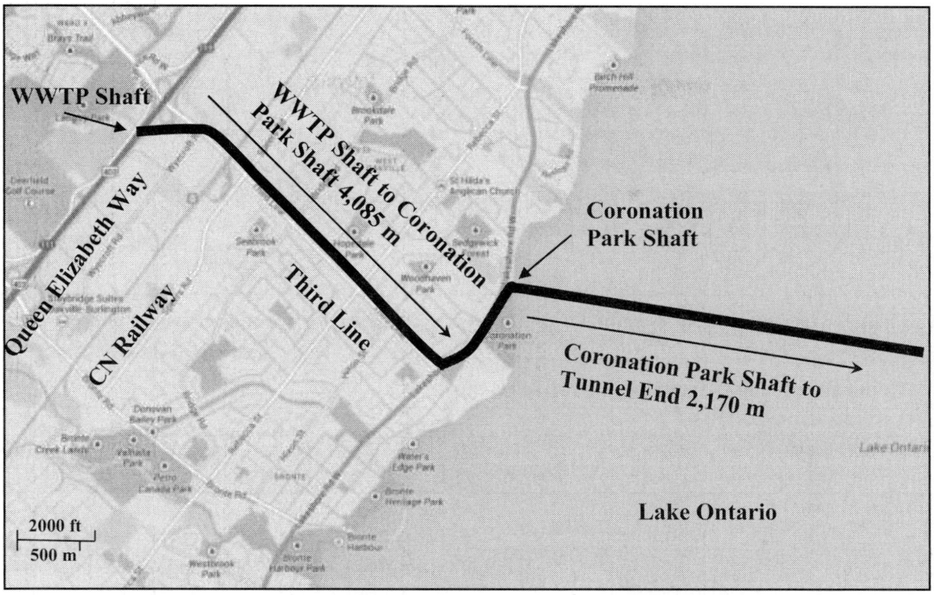


**Figure 1. Mid-Halton outfall tunnel project alignment (image after Google, 2014)**

**Table 1. Georgian Bay shale strength and stiffness parameters used in design**

| Parameter | Min. | Max. | Average | Design |
|---|---|---|---|---|
| Rock Quality Designation, RQD (%) | 0 | 100 | 87 | — |
| Unit Weight, $\gamma$ (kN/m$^3$) | 23.5 | 29.4 | 26.5 | 27.0 |
| UCS of intact rock, $\sigma_{ci}$ (MPa) | 7.1 | 35.0 | 19 | 18.5 |
| Ratio of Rock Modulus to UCS of intact rock, $E_i/\sigma_{ci}$ | 39 | 342 | 156 | 150 |

The Georgian Bay Shale can be generally characterized by the following:

- Massive appearing, highly fissile rock with widely spaced near-vertical joints and very closely spaced sub-horizontal bedding planes.
- High in-situ horizontal principal stresses of approximately 6.9 MPa and 2.5 MPa, where minor principal stress is the vertical stress.
- Cross-anisotropic deformation, swelling, and strength behavior.
- Highly susceptible to slaking upon exposure.
- Relatively soft, brittle failure in unconfined compression.
- Low tensile strength (i.e., approximately 5–10% of unconfined compression strength) across bedding planes and an order of magnitude less to virtually nil tensile strength normal to bedding planes.
- Time-dependent deformation behavior that is dependent on in-situ stress state and exposure to wetting or humidity.

The Georgian Bay shale parameters used for design are summarized in Table 1.

## NUMERICAL MODELING

### Constitutive Models

The Generalized Hoek-Brown constitutive model was used and an attempt was made to match modeled stress-strain behavior with expected rock mass behavior for various TBM shield/support configurations. Three different stress-strain scenarios were used:

- **Elastic-perfectly plastic (EPP)**—Used when modeling the McNally (2002, 2009) support system on a modified main beam TBM. With EPP behavior, the modeled rock mass will maintain peak strength upon failure This behavior is deemed reasonable since the McNally (2002, 2009) system maintains full roof support and limits deflection below other shield types, which helps prevent strength softening.
- **Elastic-softened plastic (ESP)**—Used when modeling a cantilever finger shield

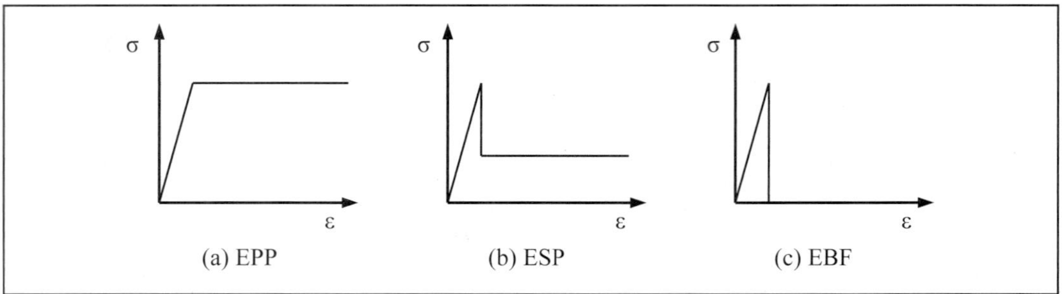


**Figure 2. Stress-strain behavior is uniaxial compression for constitutive model scenarios**

**Table 2. Generalized Hoek-Brown constitutive parameters**

| Constitutive Model | | GSI | $m_b$ | s | a |
|---|---|---|---|---|---|
| Elastic-Perfectly Plastic | Peak | 65 | 2.865 | 0.0205 | 0.502 |
| | Residual | 65 | 2.865 | 0.0205 | 0.502 |
| Elastic-Softened Plastic | Peak | 65 | 2.865 | 0.0205 | 0.502 |
| | Residual | 45 | 1.403 | 0.0022 | 0.508 |
| Elastic-Brittle Failure | Peak | 65 | 2.865 | 0.0205 | 0.502 |
| | Residual | 5 | ~0 | ~0 | ~0 |

main beam TBM or single shields. With ESP behavior, the modeled rock mass will experience immediate strength reduction upon reaching failure, but does maintain a non-zero residual strength. This behavior is deemed reasonable since the cantilever finger shield and single shield allow more crown displacement than the McNally (2002, 2009) system which increases the risk of slabbing, loosening, and overall strength reduction. However, with both of these support systems, massive rock fallout is prevented.

- **Elastic-brittle failure (EBF)**—Used where neither a shield nor initial support are provided. Rock fallout will occur until a stable crown develops. For this scenario, the rock mass is assumed to not mobilize residual strength.

The stress-strain behavior in uniaxial compression is shown schematically in Figure 2 for each scenario.

The above described stress-strain behavior may not model the expected rock behavior for each support system perfectly—in fact they probably do not. However, it is believed they are reasonable the evaluation intent and it was desired to use simple and commonly used stress-stain models for comparison. The results associated for any given modeled support system are deemed less important than the difference in the results between the methods.

The Hoek-Brown constant $m_i$ was assumed to be ten and the disturbance factor was assumed to be zero. The remaining Generalized Hoek-Brown parameters are summarized in Table 2.

**Finite Element Model**

Phase2, V8.0 (RocScience, 2012) two-dimensional finite element software was used for this evaluation. It was assumed that initial support is installed approximately one tunnel diameter behind the tunnel face. In order to reasonably estimate the amount of radial convergence that would occur before support installation, and at what modeled stage, the longitudinal deformation profile (LDP) and characteristic curve were developed.

The LDP relates radial deformation with distance from the tunnel face, x. Closed form solutions (Vlachopoulos and Diederichs, 2009) are available to estimate this profile. For this evaluation, the LDP was developed by modeling tunnel construction using a two-dimensional, axisymmetric finite element model, the details of which are not discussed in this paper. The LDP for EPP and ESP scenarios are shown in Figure 3.

The LDPs show that if initial support could be installed at the tunnel face (i.e., x = 0), approximately 14% of the maximum radial deformation would have already occurred. By the time the initial support is installed (say x/R = 2), over 90% of the radial displacement has already occurred. The initial support members will likely be lightly loaded. However, crown fallout may occur immediately behind the tunnel face. The TBM shield is critical in managing deformation and crown overbreak.

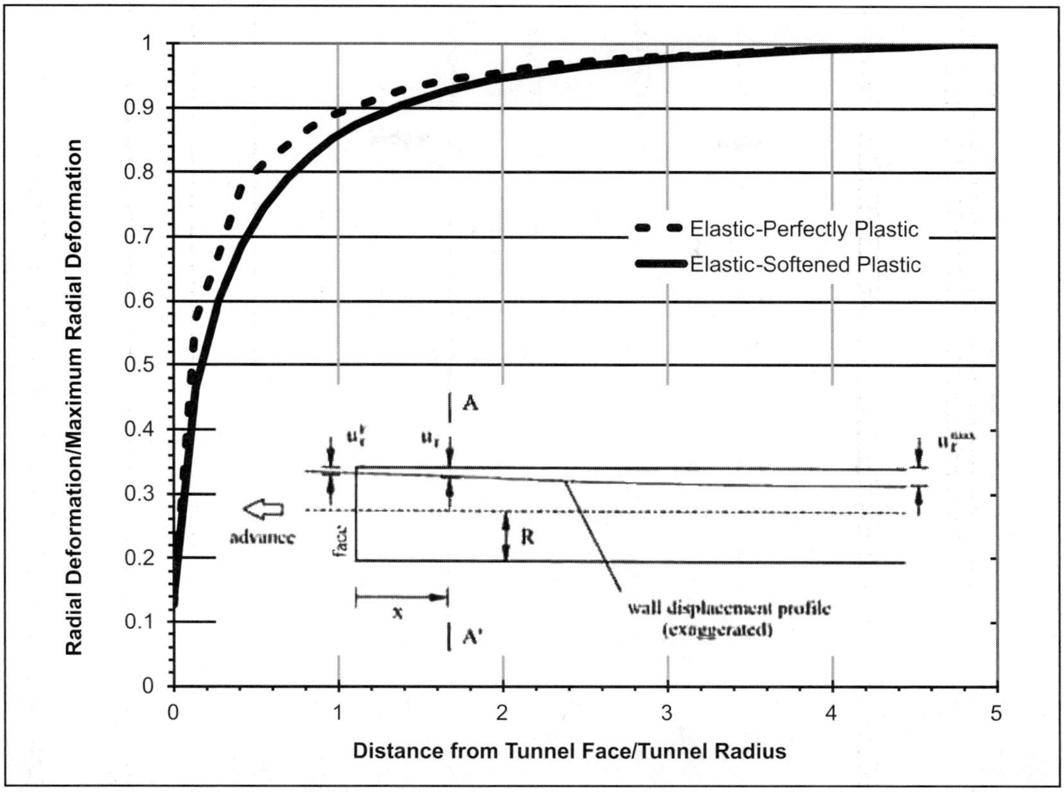


**Figure 3. Longitudinal deformation profiles used in design**

The core reduction method was used to develop characteristic curves. Therefore, the characteristic curves relate normalized core material stiffness ($E_i$/$E_o$) with normalized radial deformation ($u_r/u_{r,max}$). Figure 4 shows the model used to develop characteristic curves and to evaluate initial support.

The model boundaries are approximately eight tunnel diameters away from the tunnel excavation. Pinned connections were assigned to the boundaries. Six-node triangular elements are used and a gradually increasing mesh density was assigned near the excavation.

## OVERBREAK EVALUATION

### Main Beam TBM

Two different main beam TBM shields were evaluated. One shield, which was termed an actively-propped shield, is based on a patented system developed by McNally (2002, 2009). The other shield is a standard cantilever finger shield. A TBM shield modified for the McNally (2002, 2009) system is shown in Figure 5a, which is a photo taken at The Robbins Company Solon, OH plant.

The TBM shield in Figure 5a includes rectangular slots within the shield skin. This shield can be used as a traditional finger shield by securing metal finger plates (as shown in Figure 5a). Alternatively, the shield can be used as an actively-propped support system by inserting steel reinforcing bars or wooden slats (see Willis et al., 2012). These longitudinal roof support members are in constant contact with the tunnel roof and span between the shield and first initial support set behind the shield, as shown in Figure 5b.

The TBM shield is modeled as a pressure internal to the tunnel bore having a maximum value of 200 kPa. The shield pressure is applied to the upper 90 degrees of the tunnel bore and an equivalent reaction is modeled in the lower 90 degrees. The modeled initial support consists of four 1.5-meter long, No. 25 rock dowels ($F_y = 400$ MPa) installed through a C150×12 channel strap in the tunnel crown. The initial support sets were spaced every 1.2 meters along the tunnel.

The main beam TBM modeling sequence is shown in Table 3 for the actively-propped shield (i.e., EPP behavior). Only major stages are illustrated

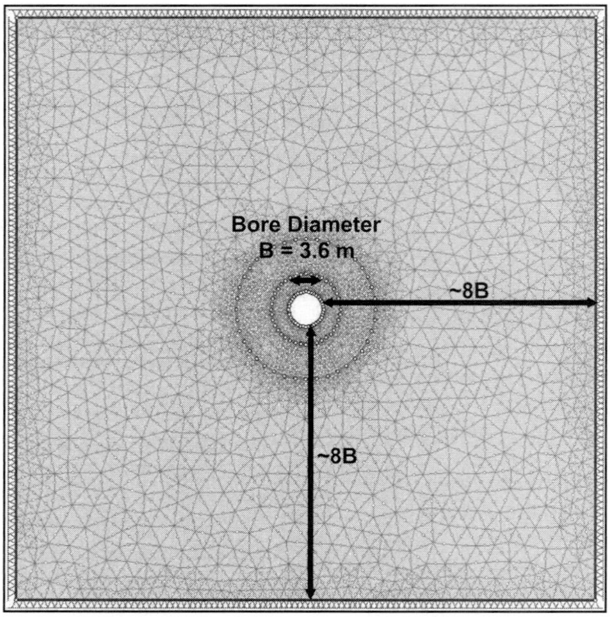


**Figure 4. Finite element model used for characteristic curves and initial support evaluation**

| (a) Actively-Propped shield<br>McNally (2002, 2009) and Robbins (2013) | (b) Actively-propped support<br>(Robbins, 2013 and Willis et al., 2012) |
|---|---|

**Figure 5. Main beam TBM shield for actively propped support**

in this sequence. A very similar sequence was used to model the cantilever finger shield support (i.e., ESP behavior). To prevent shocking the model, the shield pressure was decreased gradually, which gradually increased resistance in the initial support members.

### Single-Shield TBM

For a single-shield TBM, rock converges into the annular space and onto the shield. The single-shield was modeled as a full-circumference liner having the flexural rigidity equivalent to a 25 mm thick steel skin that is supported internally with W25×149 curved steel members spaced at 1.5 m. The modeled initial support consists of W100×19.3 steel ribs spaced at 1.2 m. The single-shield TBM modeling sequence is shown in Table 4. Only major stages are illustrated in this sequence. To prevent shocking the model, the shield flexural rigidity was decreased

**Table 3. Modeled construction sequence for actively-propped shield (EPP behavior)**

| Condition | Far Field | At Face | Shield | Support | Reduce Shield | Shield Removed |
|---|---|---|---|---|---|---|
| x/R | –5 | 0 | 0.6 | 2 | 2+ | 2+ |
| $u_r/u_{r,max}$ | 0 | 0.14 | 0.76 | ~0.95 | ~1 | ~1 |
| $E_i/E_o$ | 1.0 | 0.96 | 0.6 | 0 | 0 | 0 |
| TBM shield | — | — | 200 kPa | 200 kPa | 80 kPa | 0 |
| Initial support | — | — | — | Dowels | Dowels | Dowels |
| Modeled stage | 0 | 1 | 5 | 13 | 14–18 | 19 |
| Tunnel and support schematic | 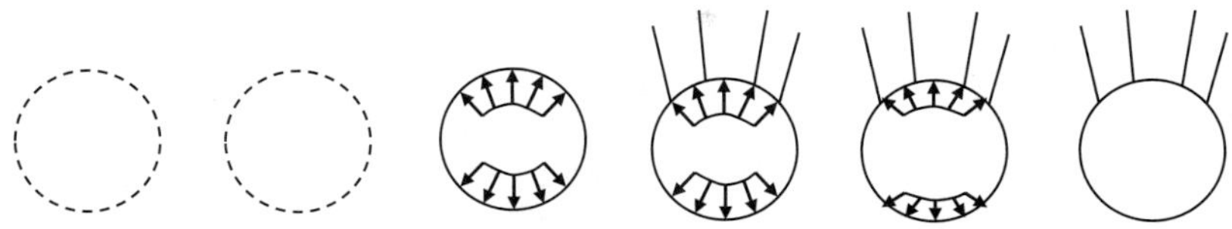 | | | | | |

**Table 4. Modeled construction sequence for single-shield (ESP behavior)**

| Condition | Far Field | At Face | Shield | Support | Reduce Shield | Shield Removed |
|---|---|---|---|---|---|---|
| x/R | –5 | 0 | 0.9 | 2+ | 2+ | 2+ |
| $u_r/u_{r,max}$ | 0 | 0.13 | 0.87 | 1.0 | 1.0 | 1.0 |
| $E_i/E_o$ | 1.0 | 0.92 | 0.06 | 0 | 0 | 0 |
| TBM shield | — | — | 1.0 | 1.0 | 0.5 | 0 |
| Initial support | — | — | — | W100×19.3 | W100×19.3 | W100×19.3 |
| Modeled stage | 0 | 2 | 10 | 14 | 15–16 | 17 |
| Tunnel and support schematic | 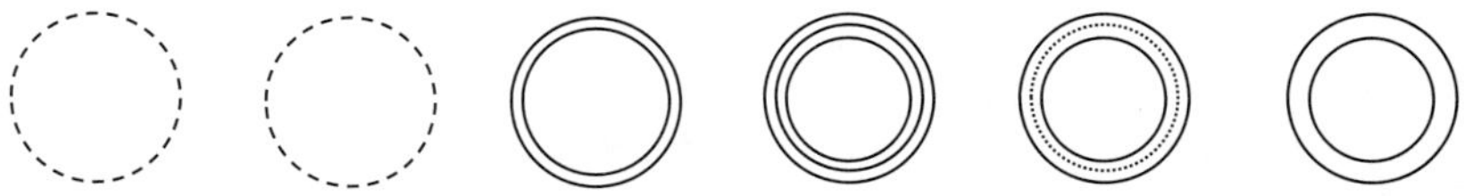 | | | | | |

gradually, which gradually increased resistance in the initial support member.

### No-Support Scenario

To create an upper bound to the expected overbreak potential, a scenario where neither a shield nor initial support is installed was also evaluated. If roof support was not provided, it is expected that slabbing and fallout would occur in the tunnel crown until the rock achieves a stable condition. The finite element method cannot accurately model fallout progression. However, to estimate possible fallout height, the EBF material constitutive behavior is used. This material behavior will predict the greatest plastic zone in the roof, and is expected to be conservative because the meta-stable rock adjacent to the roof fallout zone will actually mobilize residual strength.

## RESULTS AND SUMMARY

Figure 6 summarizes for each scenario, the resulting plastic zone height, $H_p$, above the tunnel crown. As expected, the no-support scenario provides an upper bound and has a plastic zone height equal to 1.4 times the tunnel bore diameter, D. The main beam TBM with actively-propped support (i.e., McNally, 2002, 2009) had the smallest plastic zone height ratio of approximately 0.2. The single shield and main beam with finger shield support had progressively greater plastic zone height ratios.

For comparison, superimposed on Figure 6 is the plastic zone height relationship developed by Martin et al. (1999), which is shown below.

$$\frac{H_p}{D} = 0.25 + 0.625\frac{(3\sigma_1 - \sigma_3)}{\sigma_{ci}}$$

The major and minor principal stresses are $\sigma_1$ and $\sigma_3$, respectively. All other parameters are as defined previously.

Also superimposed on Figure 6 is a line that represents the normalized rock dowel length, L. For finger shield evaluation, the dowels could potentially be installed entirely within a plastic zone, increasing risk of unacceptable crown fallout or overbreak.

Figure 7 shows tunnel crown maximum vertical displacement for each scenario at each modeled stage (also refer to Tables 3 and 4). As stated previously, the difference in behavior between systems was more important for the evaluation than the actual displacement magnitudes. The results indicate the actively-propped McNally (2002, 2009) support

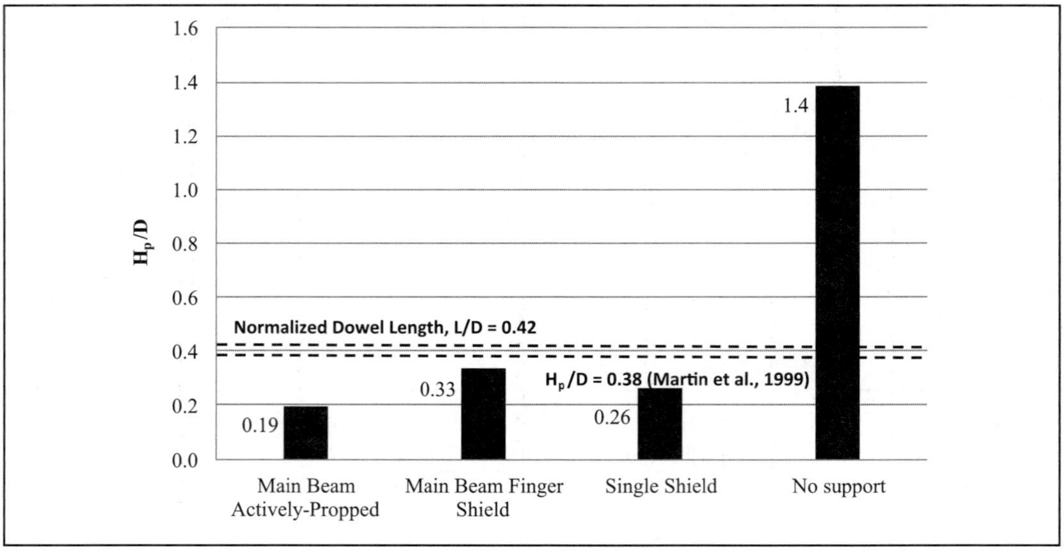


**Figure 6. Plastic zone height comparison**

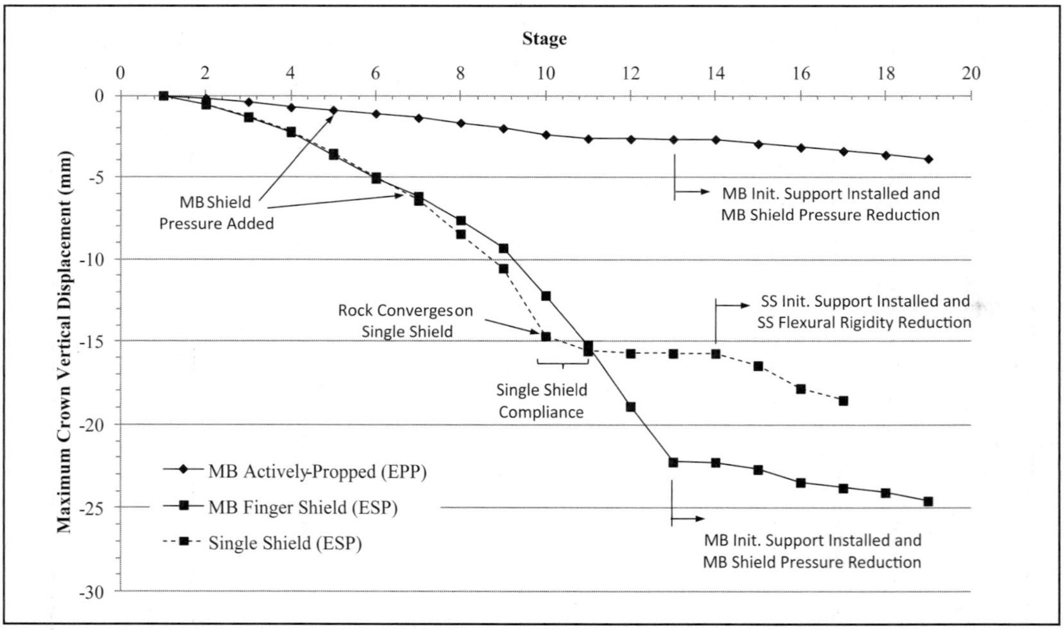


**Figure 7. Main beam and single-shield TBM evaluation results**

system may limit crown deformation to five times less the deformation expected for the cantilever finger shield and approximately 4 times less than for the single shield.

The modeling methodology presented herein is simply a tool that was used in combination with previous design and local-construction experience. At the time this paper was developed, construction had not started for this project, so the methodology and results presented herein cannot be validated with field measured data. Furthermore, it is difficult to reliably measure crown displacement during construction, particularly for the actively-propped system. By the time the instrumentation is in place, much of the displacement has already occurred. As a result, there is little reliable crown displacement data for similar sized tunnels mined locally. While these results cannot be compared with actual data measured during tunneling operations on this project or previous local projects, the results of this evaluation further illustrate what local experience has already shown—that a single-shield TBM with ribs and lagging support, or a main beam TBM with actively-propped support (McNally, 2002, 2009) are sufficient in preventing significant overbreak and rock mass fallout. In addition, the actively-propped support has the advantage that there is constant tunnel roof support that is tight up against the rock, which further reduces the risk of overbreak.

## REFERENCES

C&M McNally Engineering (McNally), 2002. Method and apparatus for feeding a tunnel roof support system from the roof shield of a TBM, US Patent 6468000 B2.

C&M McNally Engineering (McNally), 2009. Method and apparatus for feeding a tunnel roof support system from the roof shield of a TBM, Canadian Patent 2297456 C.

Google Maps, 2014. Oakville, Ontario map, https://maps.google.com/, copyright 2014.

Martin, C.D., Kaiser, P.K., and D.R. McCreath. 1999. Hoek-Brown parameters for predicting the depth of brittle failure around tunnels. Can. Geotech. J. 36. 136–151.

The Robbins Company, 2013. Olmos Trans-Andean Tunnel. www.therobbinscompany.com/case-study/olmos/. accessed 2013.

RocScience, Inc., Phase2, finite element software, V. 8.0, 2012.

Vlachopoulos, N. and Diederichs, M.S. 2009. Improved longitudinal displacement profiles for convergence confinement analysis of tunnels in squeezing ground. Journal of Rock Mechanics and Rock Engineering. 42 (2), 131–146.

Willis, D., Roby, J., and Askilsrud, O.G. 2012. Extreme machine modifications in difficult ground at the world's second deepest civil works tunnel. North American Tunneling Proceedings. Indianapolis, IN.

# TRACK 2: DESIGN

## Session 5: Design of Transit Tunnels

Axel Nitschke, Chair

# SEM Design Optimization for an Underground Crossover Cavern

**Pooyan Asadollahi and Bradford F. Townsend**
Parsons Corporation

**ABSTRACT:** A crossover cavern with a length of 354 feet is to be excavated using Sequential Excavation Method (SEM) through clayey siltstone of the Fernando formation in Los Angeles, California, as a part of Regional Connector Transit Corridor Project. This paper presents 2D and 3D finite element analyses performed to optimize the excavation sequences to ensure safe construction and minimize surface settlement. Sensitivity analyses were performed to investigate the effect of strength and deformability of the ground as well as the number, length, and sequence of drifts on the stability of the cavern and the surface settlement.

## INTRODUCTION

The Regional Connector Transit Corridor (RCTC) project provides a 1.8 mile-long connection between LA Metro LRT Gold Line, from Pasadena and East Los Angeles, to the Blue Line to Long Beach and new Expo Line to Culver City. The project includes 3,801 feet of cut and cover sections, 4,580 feet twin-bored tunnels, three stations, one crossover cavern, and four cross-passages. The RCTC project will be constructed through a Design-Build contract.

The crossover cavern with a length of 354 feet, located east of the 2nd/Broadway station, will be excavated conventionally using Sequential Excavation Method (SEM). This paper presents the analyses to design the cavern excavation sequences performed during the proposal stage by Parsons as the lead designer for the Shea-Walsh-Parsons (SWP) Joint Venture.

It should be acknowledged that at the time of preparation of this paper the Design-Build contract was not awarded. In addition, the geometry of crossover was changed by an addendum. This paper is based on the geometry given in the original Request for Proposal (RFP), not the current RFP as modified by the addendum. SWP final proposed sequences were adjusted accordingly but are not presented in this paper.

The crossover cavern geometry has been illustrated in Figure 1 together with the subsurface ground layers. The width and height of excavation is 57-ft 9-inch and 31-ft 10-inch, respectively. The cavern crown is on average 60 feet below the ground surface. The properties of ground layers were extracted from the geotechnical site investigation provided in the RFP documents or interpreted following empirical approaches presented in Kulhawy and Mayne (1990) and Hoek (2013).

The following supporting system was suggested based on previous experience:

- Initial support: 12-inch shotcrete with minimum 28-day compressive strength of 4,000 psi and reinforced by 70 pounds per cubic yard of steel fiber reinforcement; Lattice girder to be installed at 5 feet spacing for shaping;
- Pre-support: spilling of fully grouted #8 bolts with length of 20 feet inclined at 10 degrees with overlap of 5 feet; and
- SEM Toolbox Options (to be utilized as necessary): 40-foot-long fiberglass face dowels with diameter of ¾ inch to be installed with spacing of 5 feet center-to-center; and/or center core.

## DESIGN CRITERIA AND PHILOSOPHY

RFP documents defined the action level and the maximum acceptable angular distortion of existing buildings caused by RCTC construction to be 1/1000 and 1/750, respectively. The "Building and Adjacent Structure Protection Report," which was a part of the RFP documents, identified three existing structures that would be affected by the cavern excavation. By correlation analysis of the results presented in this report, it was interpreted that a maximum surface settlement of one-inch would cause distortion of 1/1500 on these three existing structures. Thus, it was determined that the maximum surface settlement had to be limited to less than 1.5 inches in order not to exceed the action level of impact on these existing structures.

To minimize induced ground deformation in SEM excavation, construction sequences should be optimized. Figure 2 illustrates the tunneling

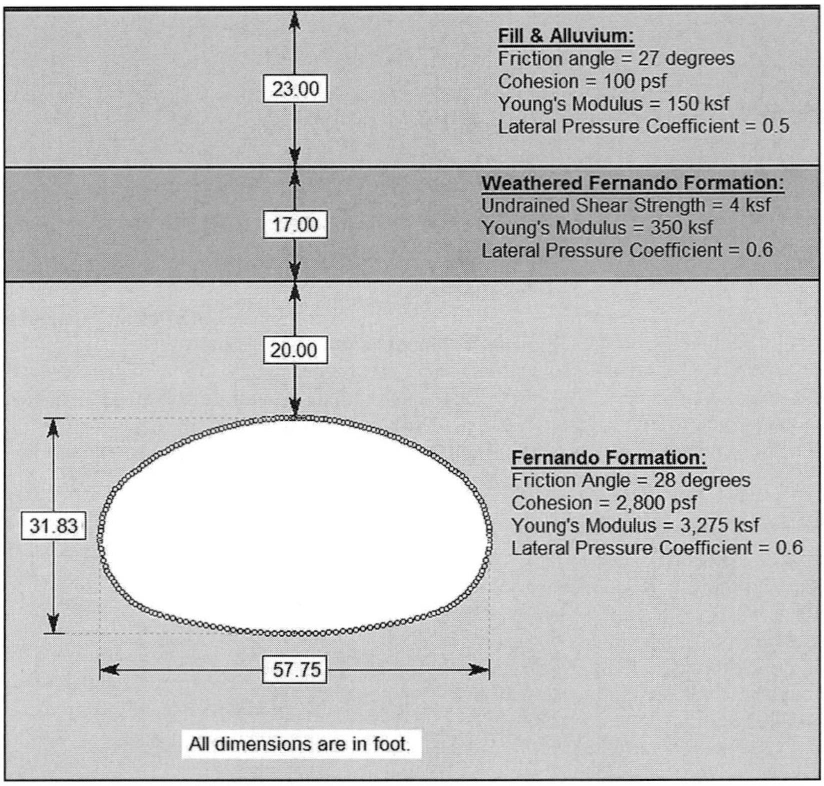


**Figure 1. Crossover cavern geometry and subsurface ground condition**

induced deformation. Considering the fact that pre-convergence and convergence are consequences of extrusion, controlling extrusion will automatically reduces pre-convergence and convergence. Control of extrusion is achieved by modifying rigidity and thus the strength and deformability of core-face.

In addition to minimizing ground deformations, for a safe underground construction, the stresses developed in the supporting system should not exceed the elements capacity with appropriate factor of safety.

Finally, it should be noted that, considering the construction means and methods and the conducted risk analysis, SWP wanted to limit the height of each excavation drift to 21 feet.

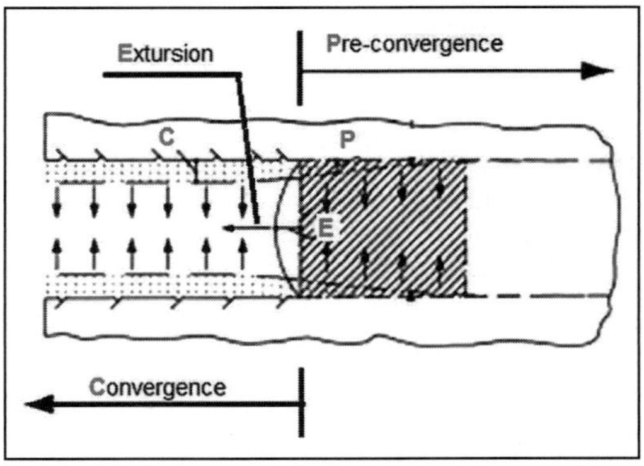


**Figure 2. Tunneling induced deformations (Asadollahi and Kaneshiro 2013)**

## ANALYSIS METHODOLOGY

Except for portals and break-ins/outs, underground constructions with length of more than 200 feet can be properly modeled in 2D using Plane Strain concepts with considerations for the 3D effect of distance from the excavation face.

Finite element 2D numerical analysis was performed using Phase 2 version 8. The distance from excavation face was simulated by softening the ground material inside the excavated area and based on graphs presented in Vlachopoulos and Diederichs (2009).

The stages of modeling were as follows:

- Simulating in situ condition, which represents ground behavior before any construction activity. A surcharge load of 1 kips per

**Table 1. Summary of 2D finite element analyses and results**

<table>
<tr><th>Alt. No.</th><th>Construction Sequences Description</th><th>Summary of Results</th></tr>
<tr><td>1</td><td>T all the way through followed by B.<br>T: 21 feet high; two left and right drifts; 5-foot excavation rounds.<br>B: 15-foot excavation rounds<br>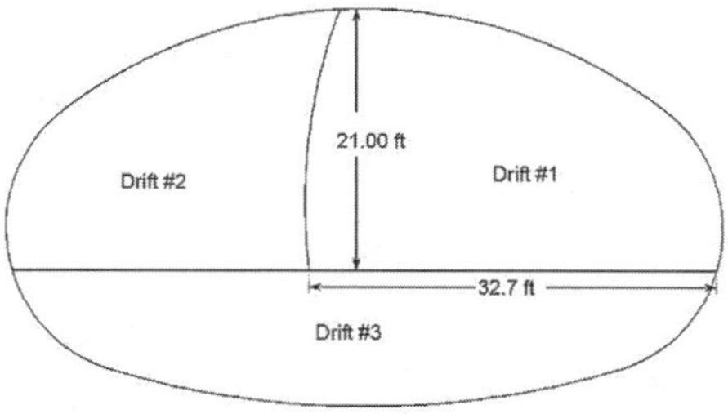
</td><td>The plastic points at sidewalls were observed to occur after excavation of T and before applying shotcrete at B.<br><br>Shotcrete yields at the crown and sidewalls/springline of the cavern because the sharp connection of center-wall and crown causes high concentration of stresses.</td></tr>
<tr><td>2</td><td>Alternative #1 plus "Enhanced Elephant Foot," which is defined as a triangular over-excavation to be filled with lattice girder and shotcrete before proceeding to the bench excavation</td><td>No shotcrete yielding occurs at the springlines. Shotcrete yields at the cavern (the same as Alternative # 1).<br>$S_{max}$ = 1.61 inches.</td></tr>
<tr><td>3</td><td>Alternative #2 with different center-wall location(s) and inclination(s)</td><td>The shotcrete yields at the crown the same as Alternatives #1 and #2.</td></tr>
<tr><td>4</td><td>Alternative #2 with thicker shotcrete (both 14-inch and 18-inch)</td><td>The stiffer support attracted more loads and the yielding points were still developed at the crown.</td></tr>
<tr><td>5</td><td>Alternative #2 with a pattern of 10-foot #8 fully grouted rock bolts spaced at 3 feet center-to-center</td><td>Reinforcing the ground with bolt did not prevent shotcrete from yielding because the yielding was not due to failure of rock mass.</td></tr>
<tr><td>6</td><td>• T all the way through followed by B.<br>• T plus Enhanced Elephant Foot: 21 feet high; 5-foot excavation rounds.<br>• B: 15-foot excavation rounds.<br>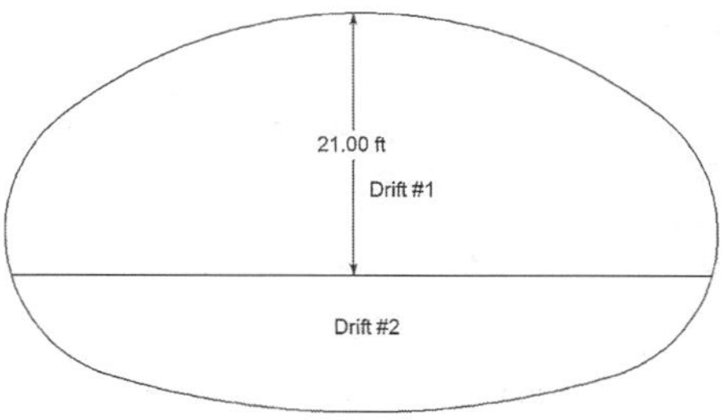
</td><td>Eliminating the center-wall provides better arch effect due to interaction of shotcrete and rock mass. No yielding occurs in shotcrete. $S_{max}$ = 1.56 inches; see distribution below:<br>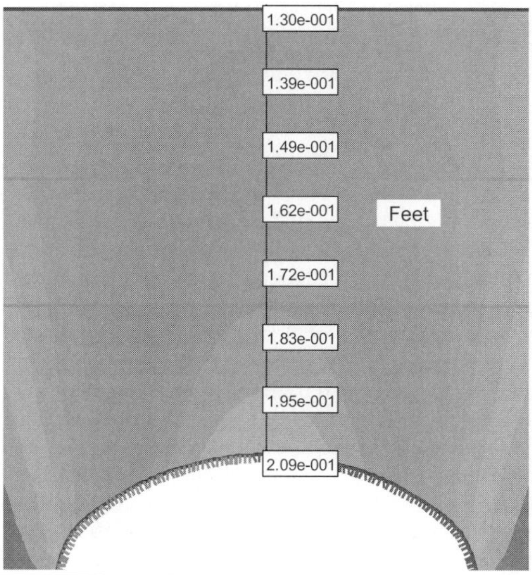
</td></tr>
</table>

Notes:

- **T** means Top Heading; **B** means Bench.
- $S_{max}$ is the calculated maximum surface settlement for the corresponding construction sequences.

square foot was applied to approximate existing/future structures.
- For each of the drifts:
  - Softening the ground inside the excavation area: the ratio of softening depends on the plastic zone around the excavation and was estimated using Vlachopoulos' and Diederichs' (2009) graphs. The softened ground is weightless and has Mohr-Coulomb constitutive model with the same shear strength properties as of the excavated ground. The modulus of elasticity of softened ground is a fraction of the Young's modulus of in situ material. This fraction was obtained with trial-and-error based on the ratio of softening determined earlier.
  - Removing the material inside the excavated area and applying green shotcrete, whose strength and deformability are ⅓ of 28-day properties of shotcrete.
  - Changing the properties of shotcrete to 28-days Young's modulus and strength.

In Phase 2 analysis, the contribution of spiling was ignored becauzse of the limitation of 2D modeling in simulating behavior of 3D problems.

3D finite element analysis was performed using Plaxis 3D version 2012.02. Typically, ground deformion extends up to a distance of twice of the tunnel diameter from the excavation face. Therefore, 100 feet of the cavern was simulated using Plaxis 3D. To ensure safe underground construction, the compressive strength of shotcrete was assumed to be 2000 psi, which is the minimum value that shall be reached before excavating the next round.

## CONSTRUCTION SEQUENCE OPTIMIZATION

Several construction sequence alternatives were analyzed using Phase 2 to find an alternative with which the stresses developed in the supporting system do not exceed the elements capacity. Table 1 summarizes the results of the analyses.

Alternative #6 of Table 1 was also analyzed using Plaxis 3D. The results of 2D and 3D analyses were aligned with each other with less than 7% difference in stresses and ground deformations. 2D finite element analysis underestimated the surface settlement while overestimated the stresses in shotcrete. The overestimation of stress in 2D analysis was intentionally done by adopting conservative relaxation factor in order to design safe underground construction.

Figure 3 presents the vertical deformation curve determined using Plaxis 3D analysis. The vertical settlement contours become horizontal at twice of tunnel diameter from the excavation face, which confirms the initial assumption of modeling 100 feet of tunnel.

In order to reduce induced settlement and impact on existing structures, excavation sequences can be optimized by having staggered excavation or considering effects of spiling, face dowel, and center core. These elements cannot be simulated in 2D plane strain numerical models. Therefore, Plaxis 3D analysis was performed to optimize the excavation sequences. The results are summarized in Table 2.

## SENSITIVITY ANALYSES

It can be seen in Table 2 that adding face dowels to Alternative #10 does not reduce the surface settlement. The reason behind this is either of the following:

1. The 3-round staggered approach is the best sequences to control deformation that can hardly be improved; or
2. The ground deformation is mainly due to convergence/pre-convergence rather than extrusion; or
3. The face dowels are not effective in stabilizing face and reducing extrusion.

To be able to better understand the reason, another analysis was performed in which face dowels were simulated but the top heading would be excavated all the way through followed by the bench (i.e., Alternative #7 plus face dowels). It was found that the simulated face dowels decrease the surface settlement by 3%.

Another model was built with center core, which reduced surface settlement by approximately 13%, roughly 10% more effectively compared to the face dowel. The reason is the center core not only stabilizes the face and reduces extrusion but also is a kind of wished-in-place support (i.e., lining before excavation).

Figure 4a presents plastic points around excavation for Alternative #10 plus center core. It can be seen that the number of plastic points is reduced by introducing center core support compared to those of Alternative #10 shown in Table 2. This is more evident at the failure plane perpendicular to the spiling.

Figure 4b illustrates plastic points around the excavation for Alternative #10 plus face dowel and center core. It shows that adding face dowels slightly reduced the number of plastic points which is not easily noticeable.

Center core and face dowels increase the cost and time required to finish each round. In addition, these measures are not significantly effective in stability of the cavern pre-supported by spiling and excavated following pattern of staggered top heading and

**Table 2. Reducing surface settlement by optimization of excavation sequences; 3D finite element analyses**

| No. | Construction Sequences Description | Summary of Results |
|---|---|---|
| 7 | Alternative #6 plus spiling:<br> | $S_{max}$ = 1.51 inches<br>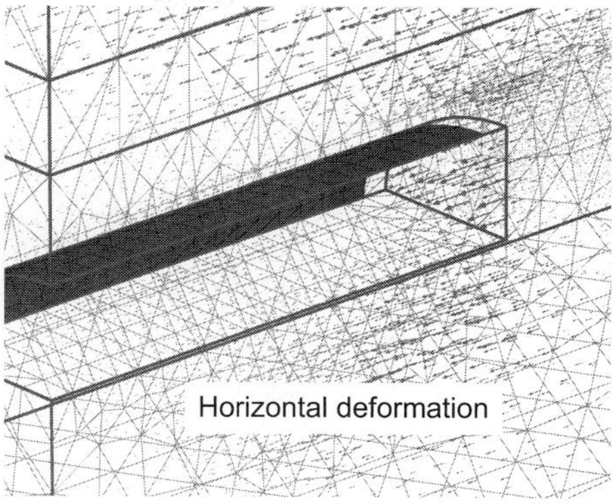 |
| 8 | Alternative #7 plus 4-inch face shotcrete | $S_{max}$ = 1.35 inches |
| 9 | Alternative #7 plus Center Core | $S_{max}$ = 1.3 inches |
| 10 | Staggered Excavation:<br>• **T** plus Enhanced Elephant Foot: 21 feet high; 5-foot excavation rounds.<br>• **B**: 15-foot excavation rounds.<br>• 3-round (15-foot) lag between **T** and **B** at all time. When the distance between the faces of top heading and bench become 30 feet, 15 feet of bench will be excavated. The lagging between top heading and bench is always between 15 and 30 ft.<br>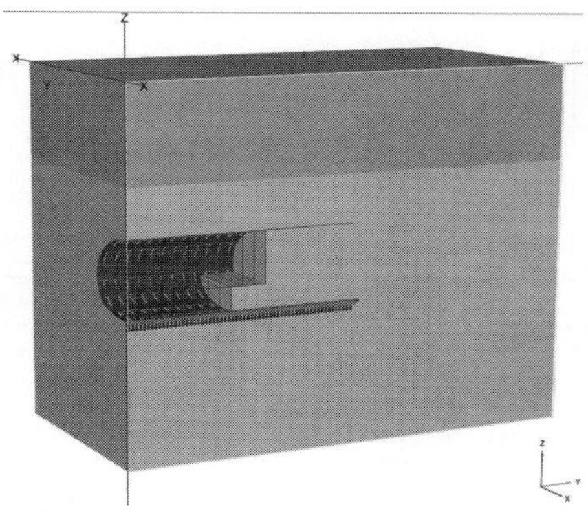 | $S_{max}$ = 1.15 inches<br>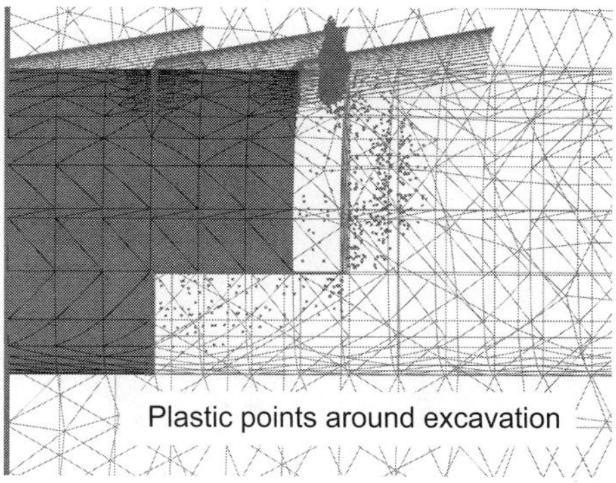 |
| 11 | Alternative #10 but with 6-round (30-foot) lag between **T** and **B** | $S_{max}$ = 1.29 inches |
| 12 | Alternative #10 but with 10-round (50-foot) lag between **T** and **B** | $S_{max}$ = 1.37 inches |
| 13 | Alternative #10 plus Face dowel | $S_{max}$ = 1.15 inches |

Notes:

- **T** means Top Heading; **B** means Bench.
- $S_{max}$ is the calculated maximum surface settlement for the corresponding construction sequences.

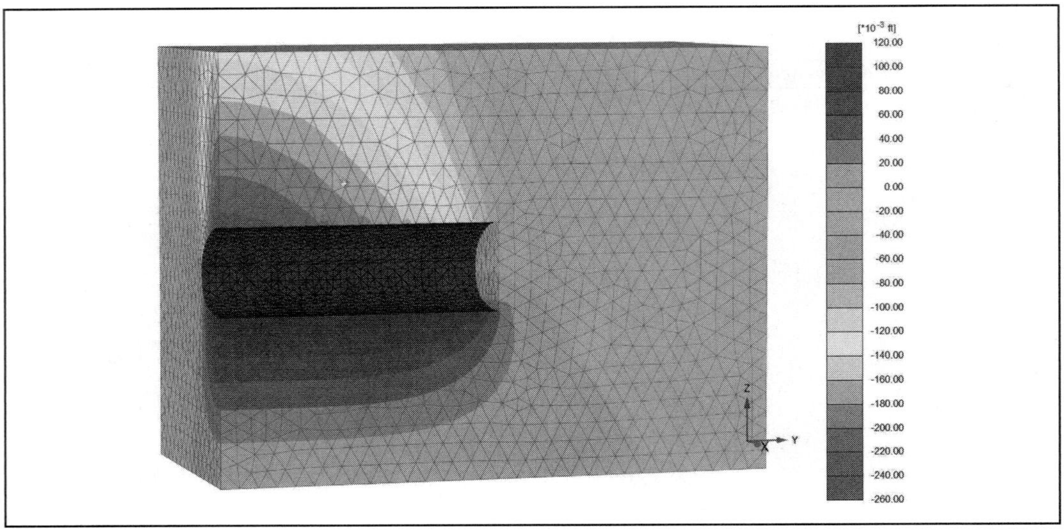


**Figure 3. Vertical ground deformation; Plaxis 3D analysis; Alternative 6**

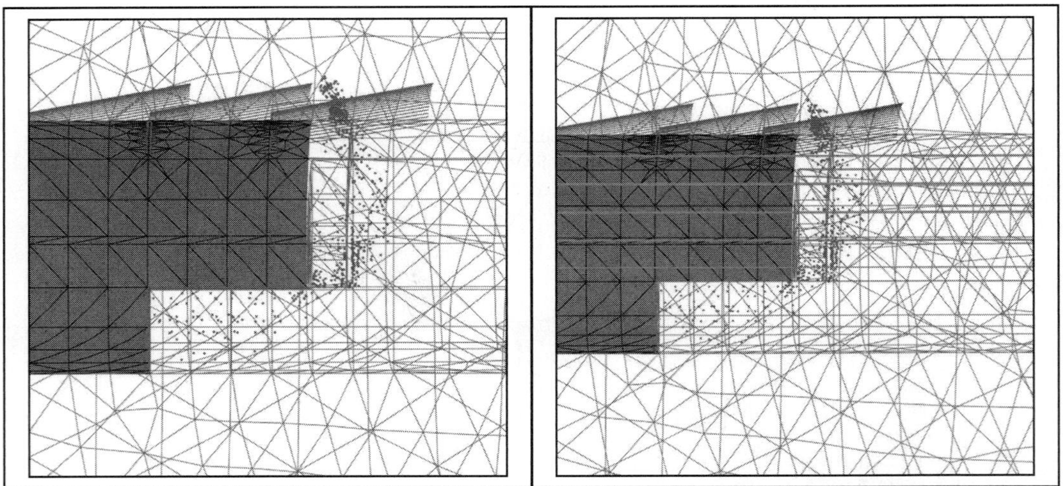

| (a) Spiling & center core | (b) Spiling & center core & face dowels |
|---|---|

**Figure 4. Plastic point around the excavation**

**Table 3. Impact of cavern excavation on the existing structures**

| Construction Sequences | Distortion |
|---|---|
| **T** then **B:**<br>**T** all the way through followed by **B**.<br>**T** plus Enhanced Elephant Foot: 21 feet high; 5-foot excavation rounds.<br>**B:** 15-foot excavation rounds. | 1/900 |
| **T** then **B** plus spiling | 1/1000 |
| **T** then **B** plus spiling and center core | 1/1160 |
| **T** then **B** plus spiling and face dowels | 1/1030 |
| **T** plus Enhanced Elephant Foot: 21 feet high; 5-foot excavation rounds.<br>**B:** 15-foot excavation rounds.<br>3-round (15-foot) lag between **T** and **B** at all time<br>Spiling | 1/1300 |

Note: Action level of distortion: 1/1000.

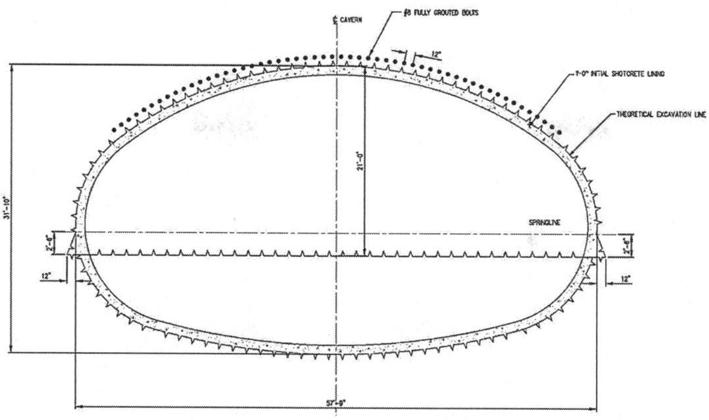


**Crossover cavern initial support cross-section**

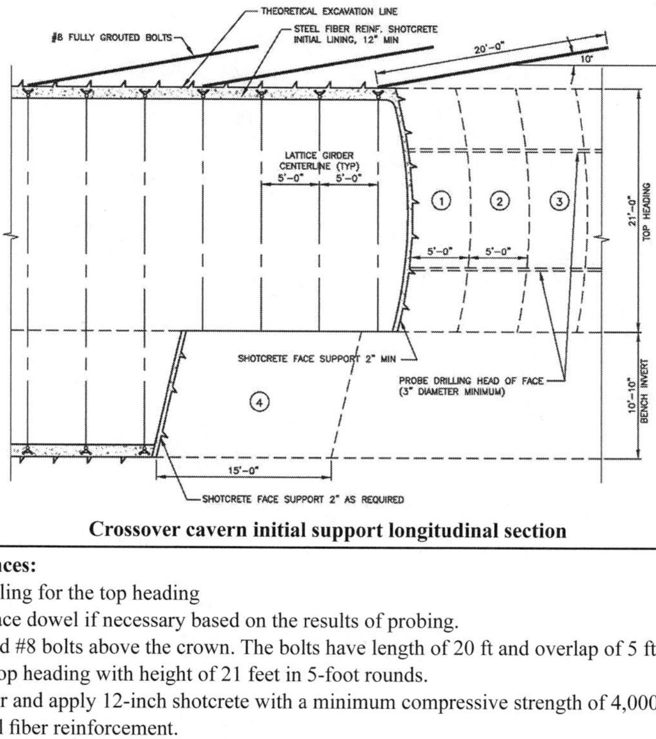


**Crossover cavern initial support longitudinal section**

**Construction Sequences:**

1. Perform probe drilling for the top heading
2. Install fiberglass face dowel if necessary based on the results of probing.
3. Install fully grouted #8 bolts above the crown. The bolts have length of 20 ft and overlap of 5 ft.
4. Excavate the full top heading with height of 21 feet in 5-foot rounds.
5. Install lattice girder and apply 12-inch shotcrete with a minimum compressive strength of 4,000 psi and reinforced by 70 lb/cy of steel fiber reinforcement.
6. 2-inch face sealing shotcrete as required based on ground condition.
7. Allow shotcrete to reach a minimum compressive strength of 2,000 psi before the next round.
8. Repeat stage 4 through 7 for the next round.
9. After each 3 rounds of top heading excavation:
   a. Install the next round of bolt canopy (Stage 3).
   b. Excavate the bench/invert in 15-foot round with lagging of 15 feet from top heading.
   c. Install lattice girder for the bench/invert and apply 12-inch shotcrete with a minimum compressive strength of 4,000 psi and reinforced by 70 lb/cy of steel fiber reinforcement.
   d. 2-inch face sealing shotcreted for the bench/invert as required. The face should have inclination of 1H:2V.
   e. Allow shotcrete to reach a minimum compressive strength of 2,000 psi before the next round.

**Figure 5. Recommended construction sequences for the crossover cavern**

bench with 3-round lagging. Thus, these measures will remain as toolbox and can be adopted as necessary.

A shorter excavation round of 3 feet together with spiling and center core was also analyzed. The number of plastic points was not reduced. The only advantage would be shorter time that is required to excavate each round and apply shotcrete initial support.

In order to investigate the effect of ground properties on surface settlement and cavern stability, a quick sensitivity was performed. The lower-bound and upper-bound values of deformability and shear strength of ground layers were assumed to be 50% lower and 50% higher, respectively, than the magnitudes presented in Figure 1. It should be noted that the Young's modulus of Fernando Formation in the RFP documents was the same as the selected upper-bound value, which demonstrate that this paper interpretations of subsurface ground condition were already conservative. Using the upper-bound properties, the max settlement would be 0.78 inch. The lower-bound properties gave the max settlement of 1.7 inch and some plastic points may be developed above the crown before applying the shotcrete, which may affect local stability of the underground construction.

## SUMMARY AND CONCLUSION

Table 3 summarizes the predicted impact of the cavern construction excavated following different sequences on the existing structures.

Figure 5 presents the recommended construction sequences concluded from the analyses described in this paper.

The general conclusions that can be made from this analysis are as follows:

- "Softening ground" approach together with Vlachopoulos' and Diederichs; (2009) graphs can properly simulate 3D effect of distance from the excavation face in 2D plane strain models.
- The center core is much more effective compared to face dowels in supporting the tunnel face and reducing the extrusion and ground surface settlement.
- Staggered heading and bench approach is one of the most effective measures in reducing construction impacts of conventional excavation on existing structures because it provides early ring closure thereby effectively stiffing the liner and is most effective in control of ground deformations.

## REFERENCES

Asadollahi, P, and Kaneshiro, JY (2013), "Tunnel and Ground Support Design," Tunneling Short Course, Colorado School of Mine, September 18–20, 2013.

Hoek, E (2013), "Practical Rock Mechanics," Accessed online through Rocscience website.

Kulhawy, FH, and Mayne, PW (1990), "Manual on Estimating Soil Properties for Foundation Design." Cornell University, Itasca, New York.

RCTC Geotechnical Baseline Report (2012), Task # PH2.7.9.14, by the Connector Partnership.3.

Vlachopoulos, N. and Diederichs, M.S. (2009). Improved longitudinal displacement profiles for convergence-confinement analysis of deep tunnels. Rock Mechanics and Rock Engineering, 42(2), 131–146.

# Optimized Modeling Approach to Tunnel Ventilation Analysis of Underground Subway Systems Equipped with Platform Edge Doors

**Nima Eslaminasab, Christopher Tattersall, and Stefania Jablonski**
Hatch Mott MacDonald

**Amir Golpaygan**
Hatch Associates

**Yousef Kimiagar**
Delcan Corporation

**ABSTRACT:** One-dimensional Tunnel Ventilation System (TVS) Analysis is a cost-effective and fast approach to analyze underground transit TVSs. In special cases, Computational Fluid Dynamics (CFD) analyses are customary to use. The complexity and associated cost of employing CFD precludes using it for modeling complete subway systems. One example of such a situation, where CFD is employed, is in transit stations equipped with Platform Edge Doors (PEDs).

In this paper, a one-dimensional TVS model of a simple real-life underground transit system equipped with PEDs is developed using Subway Environment Simulation (SES). At the same time a 3D CFD model is developed and analyzed. Comparing the results and performing a parametric study, a new modeling approach, to capture the effect of PEDs, using one-dimensional SES is introduced and discussed.

## INTRODUCTION

Tunnel Ventilation Systems (TVSs), typically comprised of series of shafts (relief and vent shafts), dampers, and fans (axial and jet fans), are a vital part of underground passenger rail systems. The main objectives of a TVS are to: (a) provide enough fresh air to the system and to maintain the tunnel/station temperatures during normal operation, and (b) maintain a tenable environment during emergency case (fire) in tunnels or stations. Figure 1 shows typical TVS facilities and equipment at a typical transit station under these two different modes of operation: Normal Mode (Left) where the train piston effect pushes the air in and out of the ventilation shafts, and Emergency Mode (Right) in which the axial ventilation fans move the air in and out of the shafts. There are commonly four vent shafts and four ventilation fans in a typical modern transit station.

NFPA 130 is the guiding code for the design of underground transit station in North America. It requires that the design of TVSs be supported by comprehensive engineering analyses. This is customarily accomplished using commercial software programs to analyze the ventilation system and to confirm its satisfactory performance. The analyses include dynamic, time-dependent modeling of the aerodynamics, thermodynamics and dynamic train movement phenomena to calculate the air-flows and temperatures at stations and tunnels. The two most commonly used methods to analyze TVSs are:

- One-dimensional analyses of the system using Subway Environment Simulation (SES) software package
- Three-dimensional computational fluid analysis (CFD), mainly using ANSYS, FDS, or similar proprietary software packages

The choice of methods is a function of cost and performance. Three-dimensional analyses using CFD can produce more realistic results compared to the one dimensional analysis. However, the overall cost of the CFD analysis and the time required to set up and run the models and to verify the results is immense, and investigating alternatives or what-if scenarios with CFD is thoroughly impractical at this time. Current industry practice is therefore to employ CFD only to investigate the air-flow patterns (and not addressing train dynamics or thermodynamic issues) in specific cases where complex air-flow patterns invalidate the basic one-dimensional air-flow premise, such as stations, cross-overs and similar structures.

On the other hand, over many years SES has proved to be a powerful and cost effective

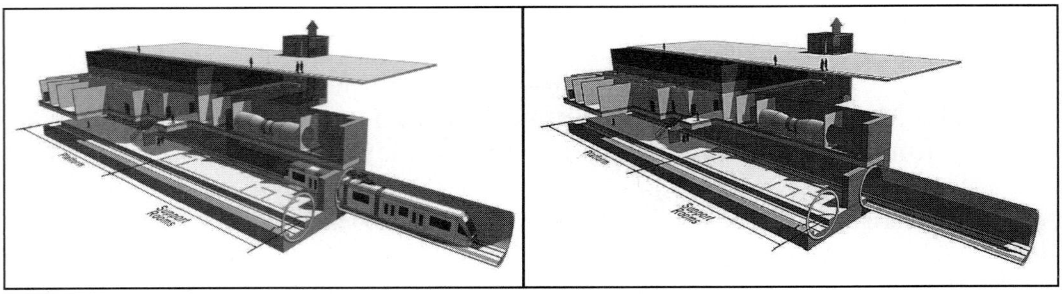

**Figure 1. Typical TVS facilities at a transit station**

one-dimensional tool to analyze TVS for underground transit systems. Results of SES analyses have been validated using CFD analysis and by field measurements and shown to provide acceptable levels of accuracy in cases where the effects of three-dimensional air-flow patterns are small.

## TOOLS FOR THE PROPER AIR-FLOW ANALYSIS OF PEDS

The use of platform edge doors (PEDs) is increasingly being specified on transit systems around the world (see Figure 2), due to the significant improvements in passenger and operation safety that they provide.

Implementation of PEDs has been made easier and more economical as a result of advanced train automatic control systems that are able to precisely align trains with the PEDs. Implementing PEDs generally follows three scenarios:

1. Installation of PEDs during the construction of new transit projects,
2. Including provisions for the future installation of PEDs, and
3. Retrofitting PEDs into existing underground transit stations.

During normal operation, temperature and fresh air circulation at underground transit stations are highly dependent on the train piston effects. Installation of PEDs may dramatically change air-flow patterns at stations. To gain a better understanding the air-flow patterns need to be closely analyzed and studied. CFD is the logical tool to perform these analyses due to the possible complex air-flow patterns, however PEDs really only affect the air-flow patterns during normal operation, when the trains are moving and the PEDs are closed (in Emergency Mode, airflow generally bypasses the platform area—see Figure 1). Using CFD to model the TVSs during normal operation will involve including the effect of train dynamics and train/surrounding thermodynamics in the model and over a large number of time steps. This would make CFD simulations very costly and time consuming.

If it were possible to account for the effects of PEDs in a conventional SES model, this problem could be overcome. The authors propose to do this by establishing a set of guidelines to account for the effects of PEDs in one-dimensional (SES) analyses. The methodology employed to develop these guidelines was to model a generic underground station/tunnels system using a conventional SES approach, then modify the parameters to account for effect of the PEDs as determined from separate CFD modeling results.

## TYPICAL UNDERGROUND STATION/ TUNNELS MODELING AND SIMULATION

### Typical Underground Station/Tunnels Configuration

For this study the authors used a simplified underground station with twin tunnels leading to portals in each direction. The twin tunnels are both 100 m long × 5.7 m diameter twin tunnels. The station consists of a 95 m long platform and two sets of stairs/escalators connecting the platform level to the concourse level. The concourse is connected to the ground level via two sets of stairs. There is one 14 $m^2$ ventilation shaft for each tunnel (four in total). The shafts are located at the tunnel/station junction and run from tunnel to ground level. Figure 3 depicts this configuration.

### System Modeling and Simulation

To correctly simulate the system in SES additional data are required, including climatic data, train physical information and train dynamic information, including speed and acceleration. For this exercise, data from a recent project was used. (Due to confidentiality agreements, it is not possible to provide this data in this paper.)

For this simulation, only one train is considered to be in the system. The train starts moving from Portal #2 (100 m away from the station), passes through the station without stopping, and stops at the other end of the tunnel, Portal #4. The maximum speed attained is 35.5 km/h along the route with the maximum coasting velocity of 25 km/h at the

**Figure 2. Platform edge doors in underground station**

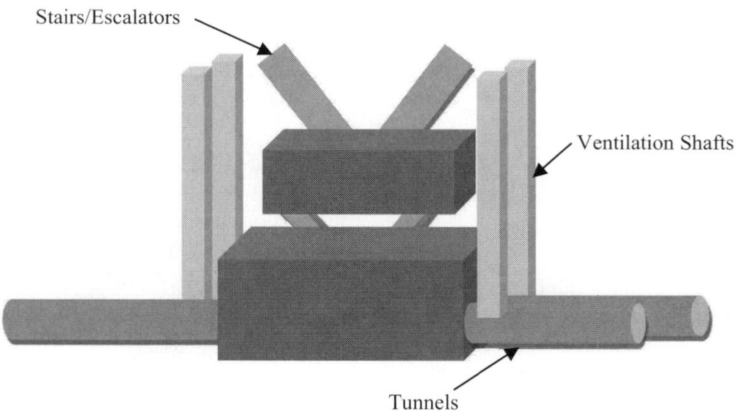


**Figure 3. Simplified underground station and twin tunnels configuration**

station. A schematic diagram of the modeled system is shown in Figure 4.

For clarity and ease of comparison the numbering system presented in this schematic is used for all simulation results presented hereafter.

In all the simulations that follow, the train moves from Portal #2 toward Portal #4 and the air-flow at Portal #1, Portal #3, Shafts #1, #2, #3 and #4, and Stair #1and #2 are recorded and compared.

## Conventional SES Model

Figure 5 shows a simplified node diagram for the conventional SES model. The model is comprised of linear segments, sub-segments and connecting nodes. The main elements of the model are the four tunnel segments, station platform and concourse segments, and tunnel ventilation shafts. The station platform line segment in the node diagram represents both the platform area and the track area within the station. The station platform is modeled as a single module, connected to the tunnels at both ends. There are no PEDs modeled in this system, as the conventional modeling approach does not allow adding any provision for it.

The system has been simulated to obtain the air-flow patterns in the tunnels due to the train piston effect. It takes approximately 50 seconds for a train to pass through the entire system. However, the results only present the first 22 seconds, which is the time taken by the train to approach the station. As depicted in Figure 6 to Figure 8, the amount of air-flow increases at all portals, shafts, and stair cases leading to grade as the train approaches the platform. The air-flow pattern at Portal #3 and #4 and Shafts #3

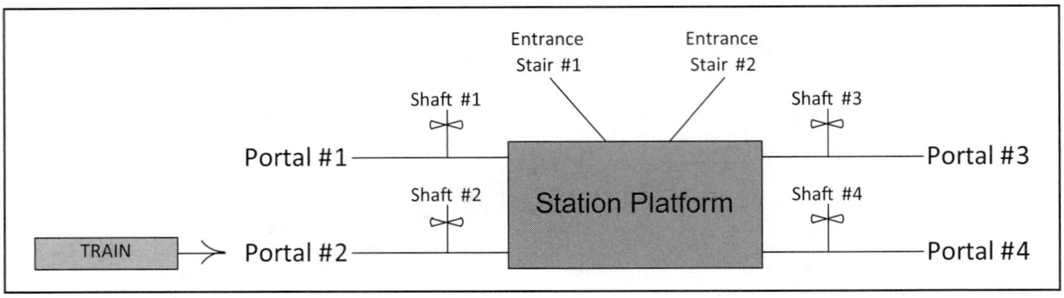


**Figure 4. System's schematic**

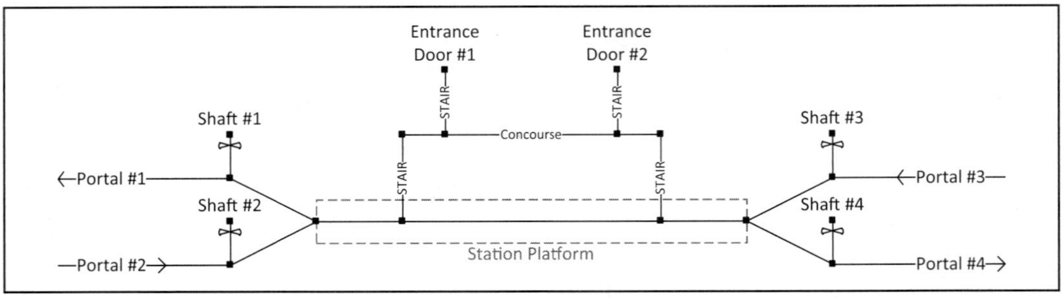


**Figure 5. Conventional SES model**

and #4 exhibit very similar patterns (as a result, the curves are plotted on top of each other).

### CFD Model

Computational Fluid Dynamics (CFD) was used to model the same system: to simulate the effects of train movement and the airflow resulting from piston effects in the system. Modeling was carried out using ANSYS Fluent®. This is a general purpose finite volume based commercial solver which allows for fully unstructured meshes.

Two different models were set-up and analyzed: one without PEDs and a second model with 2.2 m high PEDs. A linear approximation of the air-flow through Portal#2 obtained from the SES simulation (Figure 6) was used as the boundary condition for the CFD simulations. Figure 9 depicts the CFD model with PEDs, used for both simulations.

The results of the CFD model with no PEDs was obtained and compared with the original SES model, and good agreement was observed.

Figure 10 and Figure 11 depict the airflow patterns for the two CFD models. As expected, adding the PEDs into the system alters the air-flow patterns: for instance the air-flow at Portal #1 decreases from a maximum of 25 $m^3/s$ to around 21 $m^3/s$ as a result of adding PEDs. The air-flows at Portals #3 and #4 do not change significantly. We can also observe changes in the air-flow patterns in the ventilation shafts: for instance the maximum air-flow at ventilation Shaft #2 is reduced from 13 $m^3/s$ to around 11 $m^3/s$ as depicted in Figure 11. Although the air-flows at the stairs also change, these are minor and are not displayed.

The original SES modeling as described earlier does not provide an option to add PEDs into the model. This is due to a software limit on the number of branches connected to a single node (the maximum is three). The methodology presented in the next section will provide the required flexibility into the SES modelling process to incorporate PEDs into the SES model.

### Modified SES Model

Figure 12 depicts the modified SES model used for this exercise. In this model the platform and track areas are not combined into a single line segment, but rather are connected by line segments (to be referred to as "PED connections") that represent the air-flow constriction caused by the PEDs. By altering the properties of these connections, a system with no PEDs or a system with PEDs of specific heights can be modeled in the SES.

To validate the modified SES model, it was run with fully open PED connections (effectively representing the case of no PEDs installed) and the results were compared with the original SES model. Figure 13 shows the resulting airflows at the portals from the two models. The results are comparable: the air-flow patterns and the air-flow rates of the

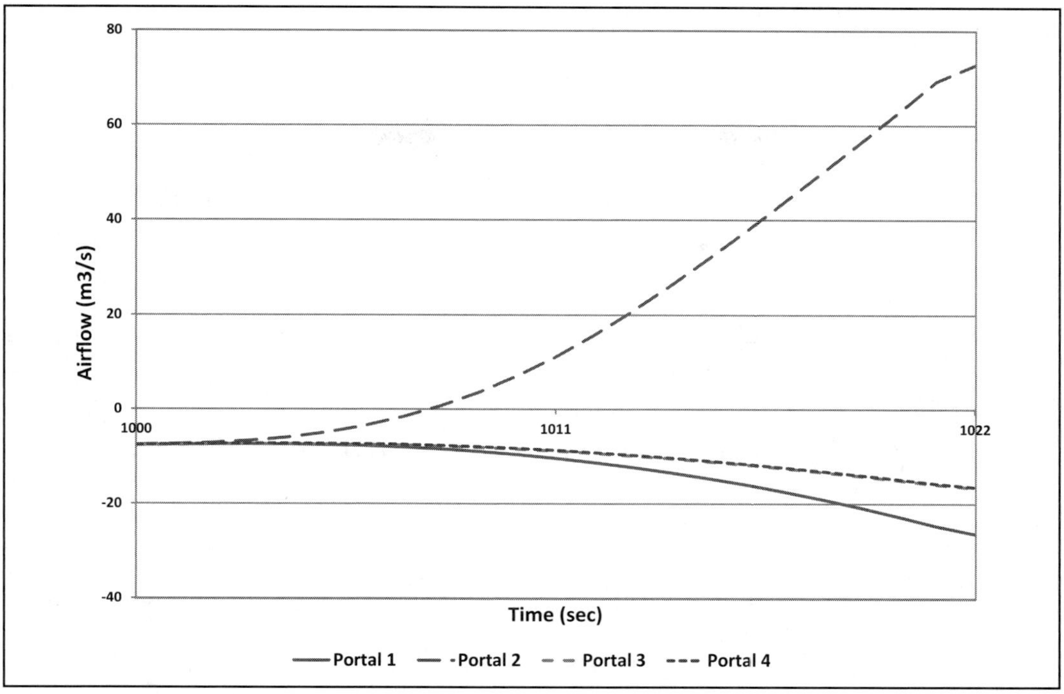


**Figure 6. Air-flow at portals—Original SES model**

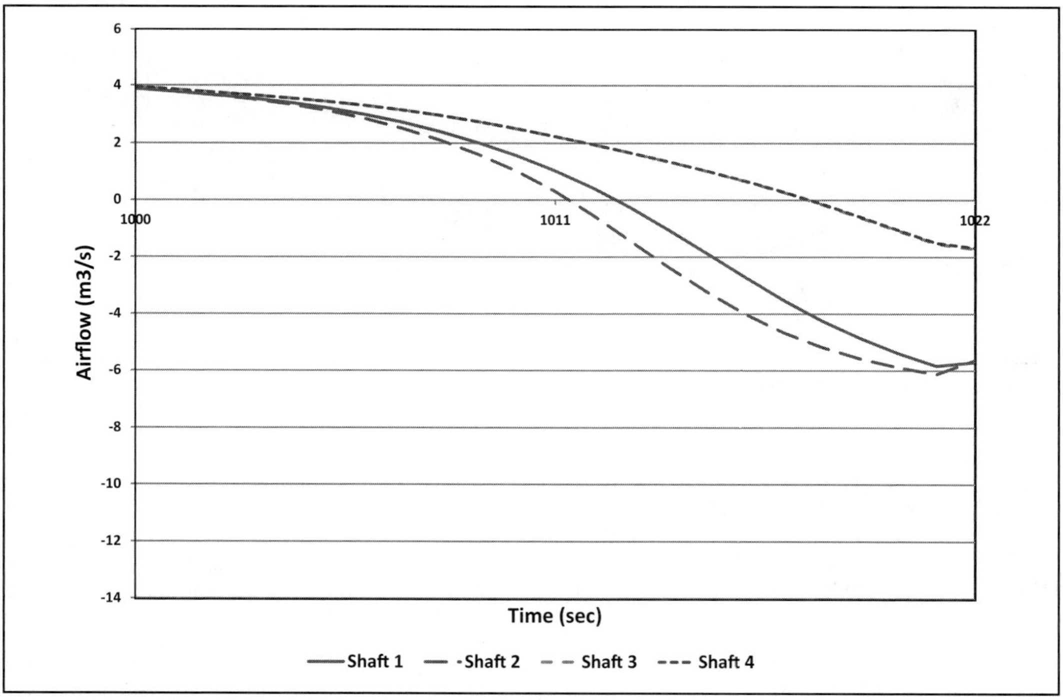


**Figure 7. Air-flow at ventilation shafts—Original SES model**

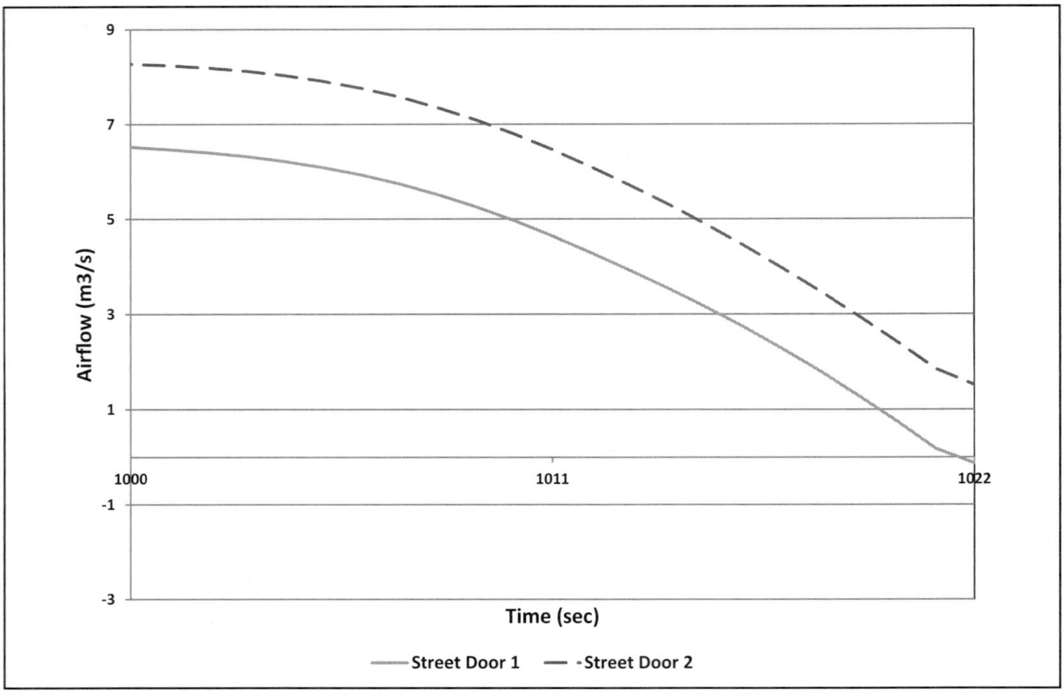


**Figure 8. Air-flow at stairs—Original SES model**

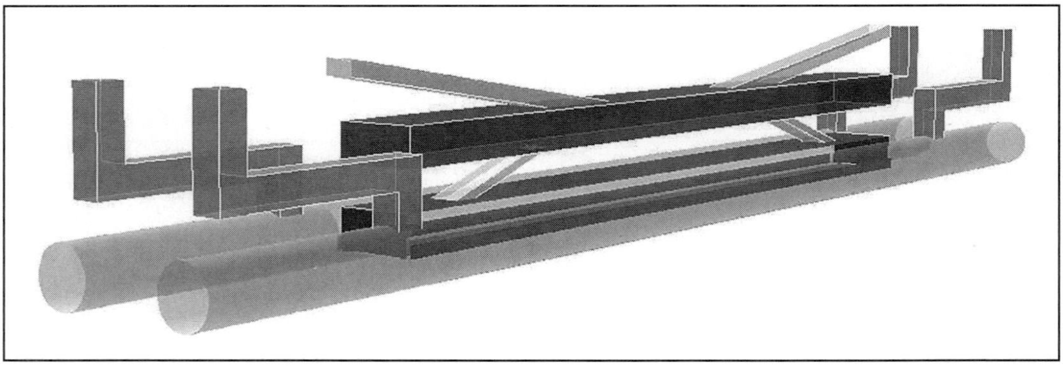

**Figure 9. CFD model complete with PEDs**

two models are very similar. However as the train approaches the station platform, the air-flow results begin to diverge. This is both expected (the models' structures are different) and not very significant.

## DISCUSSION AND CONCLUSION

Selected results from the CFD modeling, the original SES and the modified SES analyses for the case of the station equipped with PEDs are shown in Figure 14 and Figure 15.

The primary observation that can be made is that the modified SES model simulation is responsive to the presence of the PED connections in a way that is similar to the CFD results. The results suggest that predictions using a modified SES approach will be more accurate (i.e., closer to CFD results). For instance looking at the air-flows through ventilation shafts, it is clear that the modified SES results (small dashed line) show a modest overall improvement over the original SES model. Clearly, the data obtained using the modified SES model is showing trends similar to the ones exhibited by the CFD results; however they are not fully accurate. For such a complex problem, it was never expected

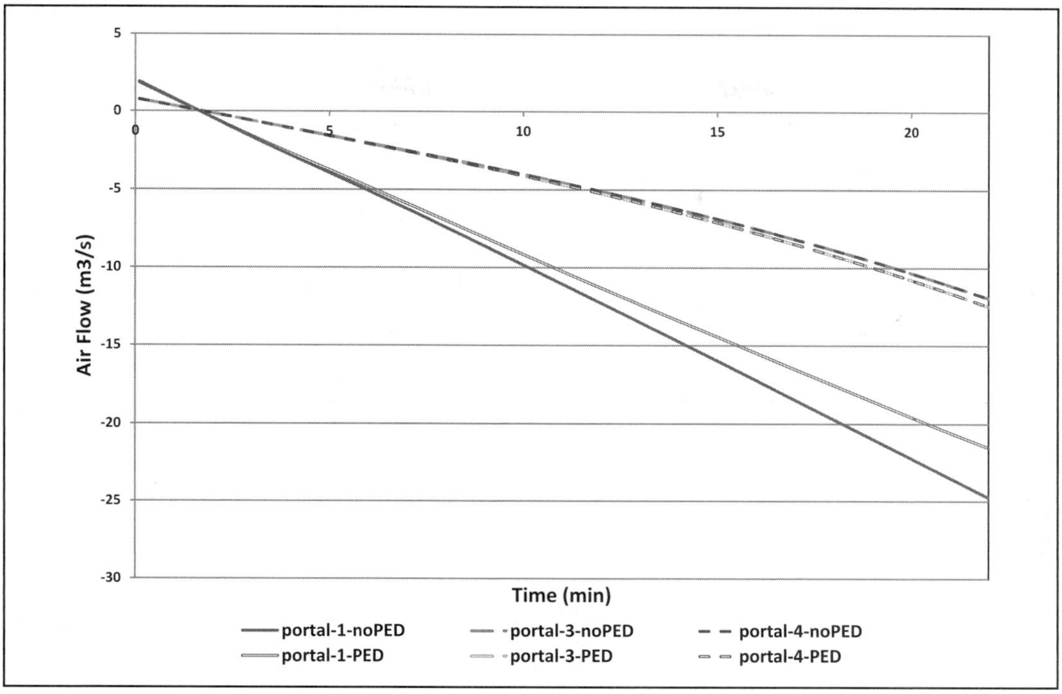


**Figure 10. Air-flow at Portals—CFD model with and without PED**

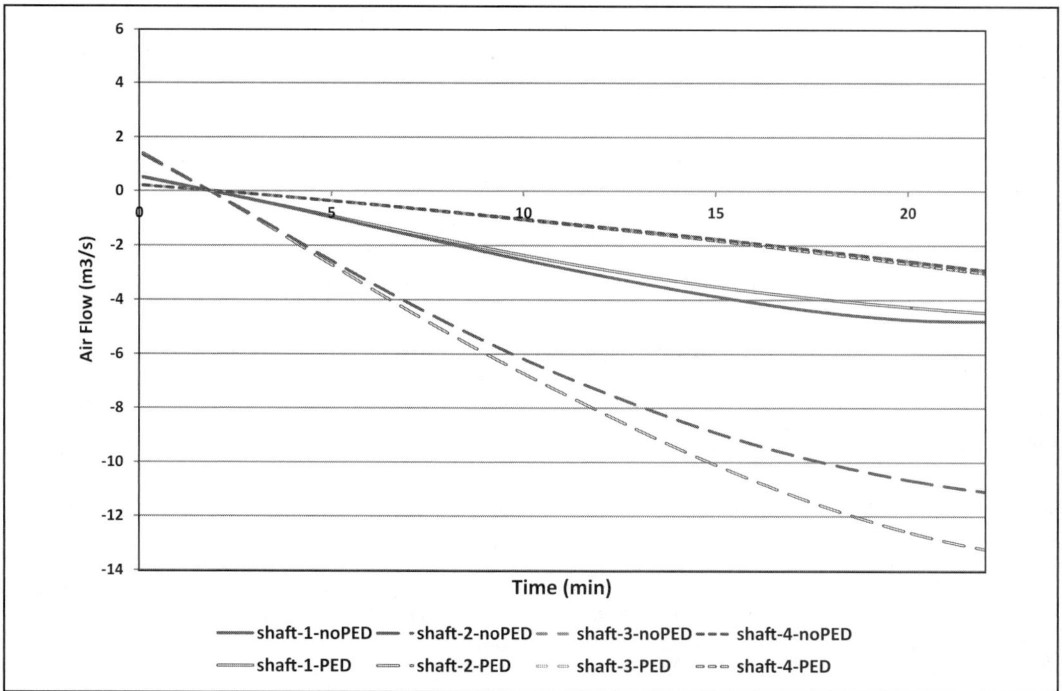


**Figure 11. Air-flow at ventilation shafts—CFD model with and without PED**

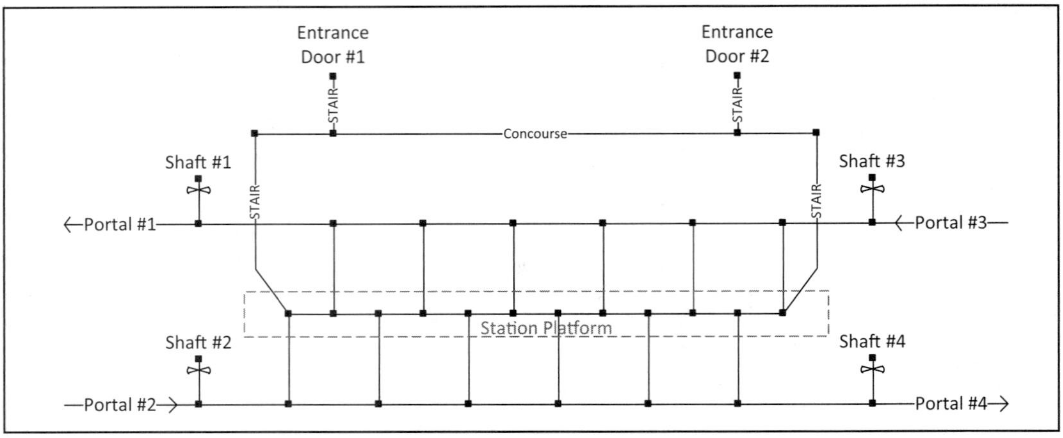


**Figure 12. Modified SES node diagram**

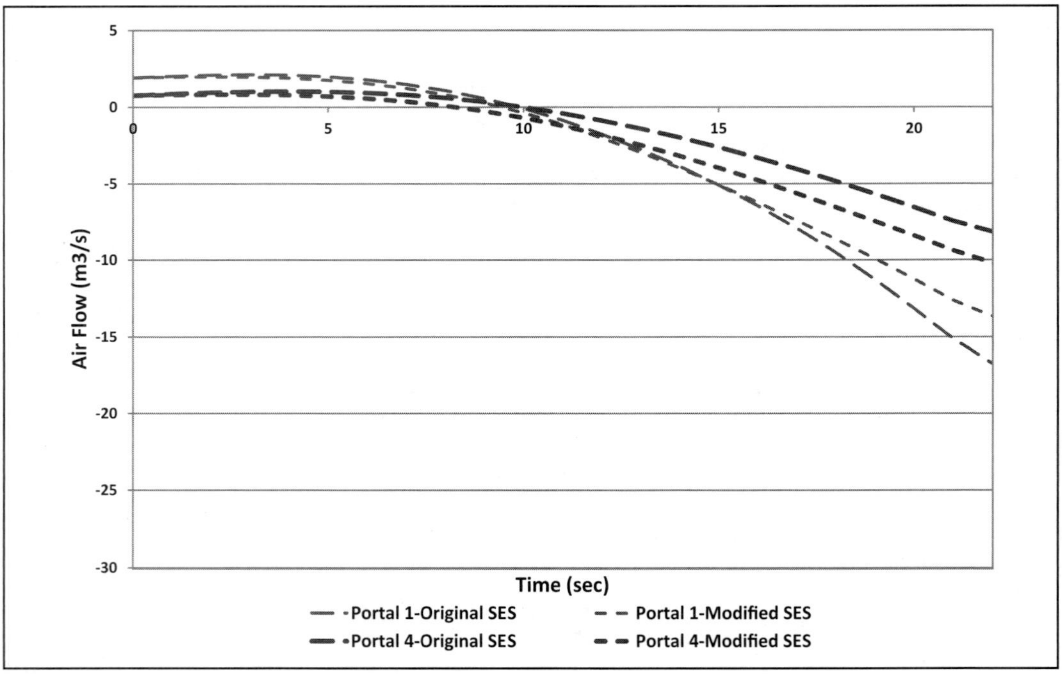


**Figure 13. Comparison of the original and modified SES results for portal airflows**

that a one-dimensional analysis would capture three dimensional air-flow patterns and completely match the CFD results.

The results, however, are very promising for employing this methodology. As well, there are other parameters that potentially affect the outcome of the SES simulation: for instance the ratio of PED height to tunnel diameter, as well as the assigned friction factors (K factors). These should be investigated in order to get a better understanding of the system and optimize the results of the modified SES model even further.

Based on the results, it is recommended to use a modified version of the SES modeling in conjunction with CFD analysis when dealing with PEDs. In an underground transit system consisting of tunnels and multiple stations, it is recommended to confirm and refine the modified SES results for one case with the CFD modeling and then implement the outcome into the remainder of the system.

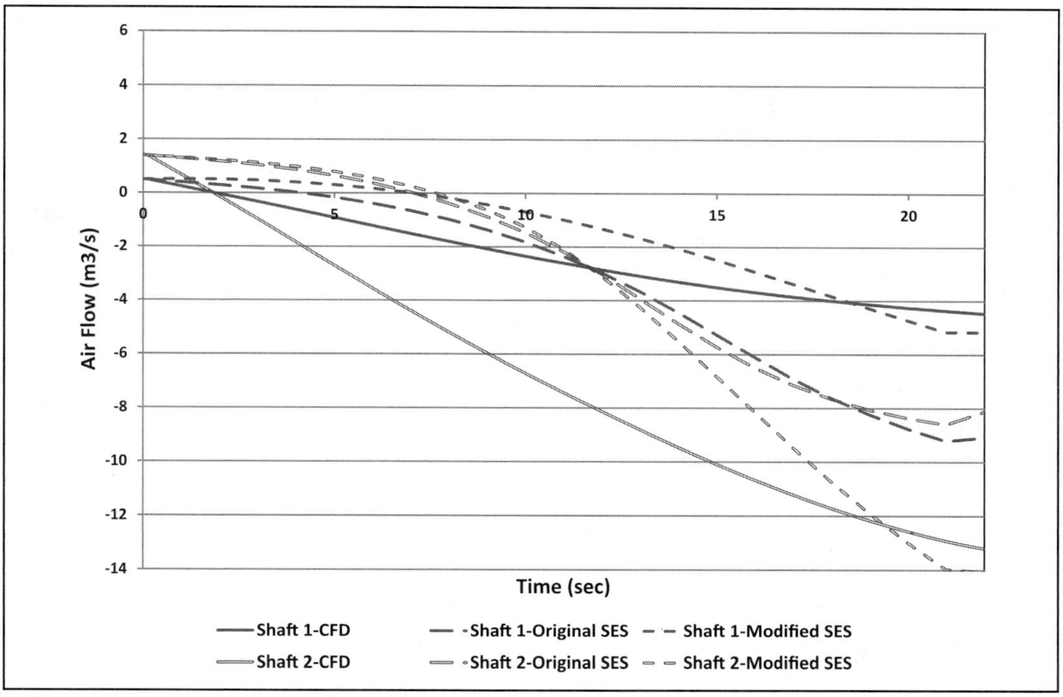


**Figure 14. Comparison of air-flow through ventilation shafts**

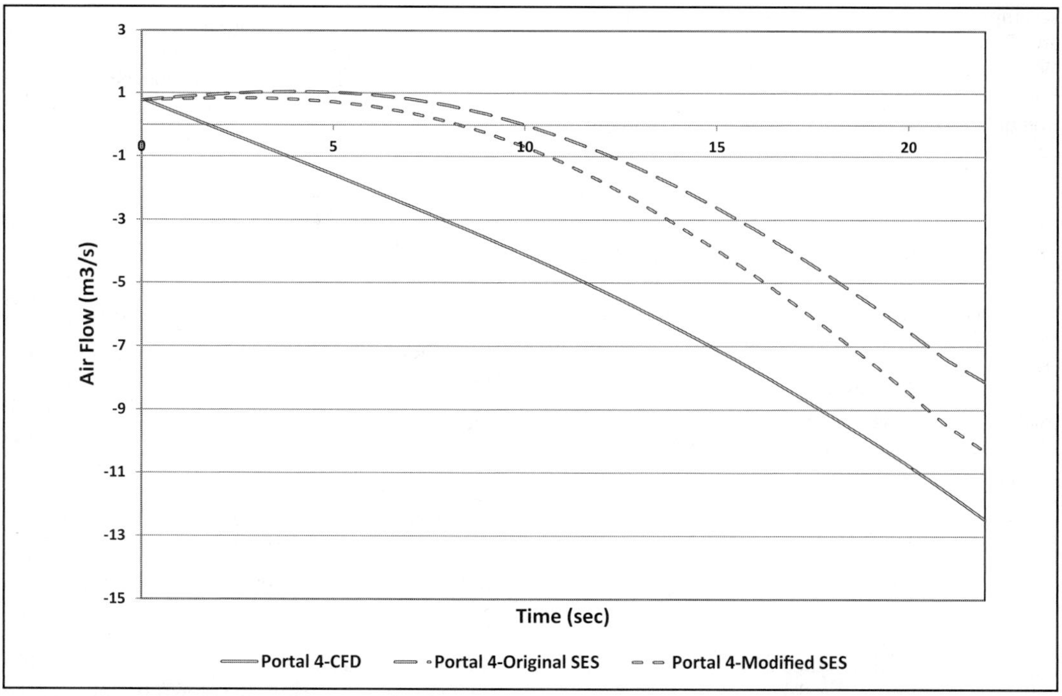


**Figure 15. Comparison of air-flow through Portal #4**

# East Link—Development of the Downtown Bellevue Tunnel

**Derek Penrice, Michael Lehnen, and Michael Wongkaew**
Hatch Mott MacDonald

**Brian Patton and Joel Theodore**
Sound Transit

**ABSTRACT:** Sound Transit's East Link Project, a $2.8 billion, 14-mile light rail (LRT) extension, will connect downtown Seattle with the rapidly developing Eastside cities of Bellevue and Redmond. Among the most technically and politically challenging aspects of the Project is the LRT segment through downtown Bellevue, which was conceptually engineered as a cut and cover tunnel. Committed to reducing public impacts during construction, Sound Transit studied the feasibility of constructing the tunnel using sequential excavation methods (SEM). The paper describes the risks and opportunities that led to the adoption of SEM as the preferred construction method for the tunnel, and the subsequent development of the tunnel design and configuration.

## INTRODUCTION

Sound Transit's (ST) East Link Project is a voter-approved $2.8 billion, 14-mile extension of ST's existing light rail transit (LRT) system from downtown Seattle, across Lake Washington via I-90, serving Mercer Island and the cities of Bellevue and Redmond on the east side of Lake Washington. With targeted completion in 2023, East Link will provide riders with an efficient and reliable connection between the largest population and employment centers on the Eastside and downtown Seattle. The project alignment is indicated in Figure 1.

ST completed the Preliminary Engineering (PE) phase of the East Link project in late 2011. Subsequently the final design contract for the easternmost seven miles of the project was awarded to the H-J-H Team, a joint venture comprised of HNTB, Jacobs Engineering and Hatch Mott MacDonald (HMM), in February 2012. H-J-H's final design scope includes approximately 7 miles of double-track LRT and stations between the I-90 Flyover (approximately the east end of the I-90 East Channel Bridge) and the Overlake Transit Center Station, adjacent to the Microsoft world headquarters. The detailed scope of the project includes the following:

- 8,900 feet of at-grade guideway, including storage track,
- 11,000 feet of retained cuts
- 2,100 feet of retained fill,
- 400 feet of trestle structure
- 16,200 feet of aerial guideway, including long span crossings over the I-90 and I-405 freeways
- 2,300 feet of tunnel within the City of Bellevue— the Downtown Bellevue Tunnel (DBT)
- Eight stations including Bellevue Transit Center Station (BTC) a hybrid underground station, four at-grade stations (East Main Station, 130th Station, Overlake Village Station and Overlake Transit Center Station), one retained cut station (120th Station), and two elevated stations (South Bellevue and Hospital)
- Parking facilities at three stations (parking structures at South Bellevue Station and Overlake Transit Center Station and a surface lot at 130th Station

The focus of this paper is the DBT, which begins immediately north of the East Main Station, and travels approximately one half mile under 110th Ave NE between Main Street and NE 6th Ave in downtown Bellevue; terminating at the Downtown Bellevue Transit Center (BTC) Station. The location and extent of the DBT is indicated in Figure 2.

### Background

Prior to the commencement of final design, ST and the City of Bellevue (COB) executed a Memorandum of Understanding (MOU) for funding and construction of the DBT. The MOU established a collaborative framework for ST and the COB to share the additional cost of a tunnel in downtown Bellevue to the East Link project. Due to tight budgetary constraints, there was a need to significantly reduce project costs after completion of the PE phase of the project.

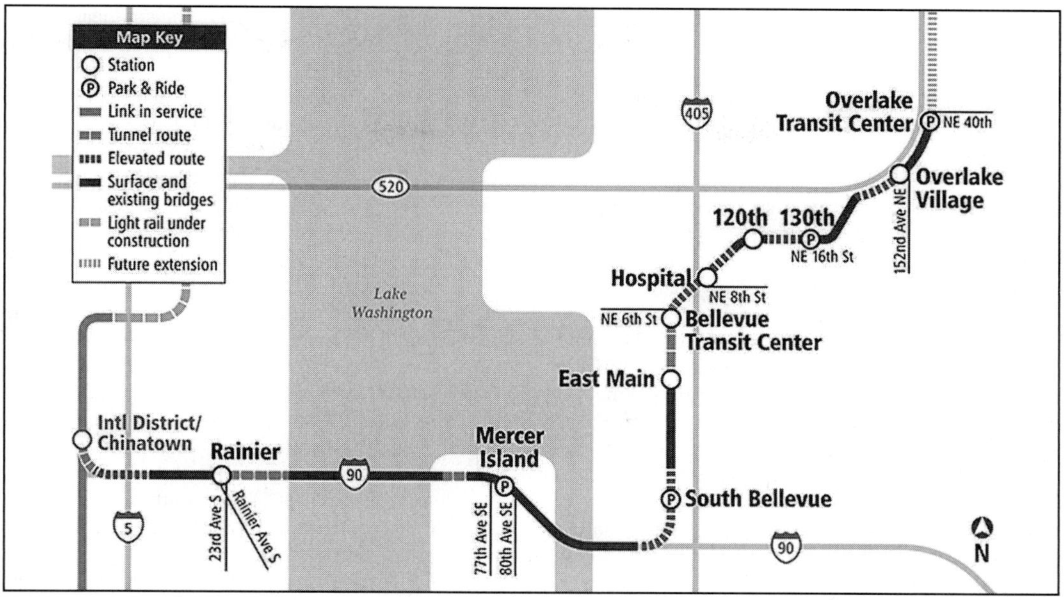


Figure 1. East Link project alignment

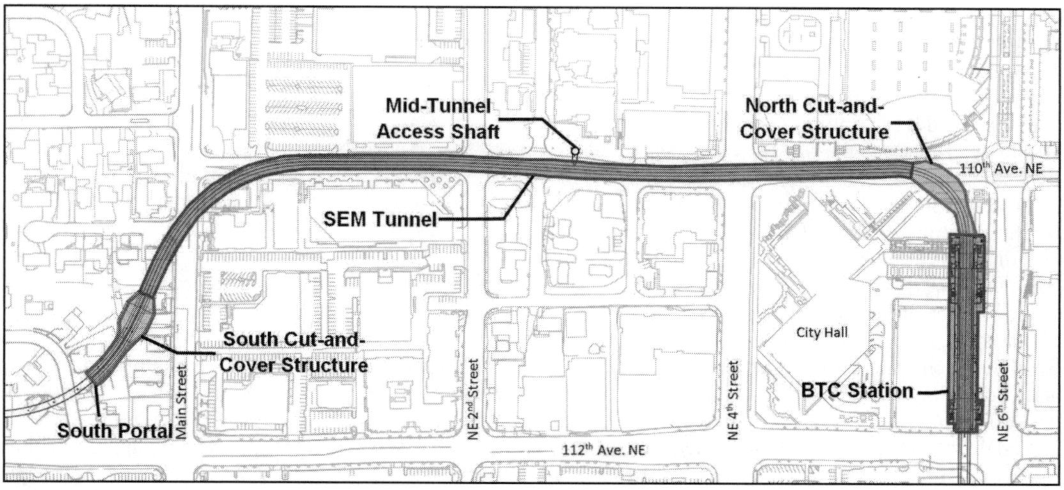


Figure 2. DBT Alignment, BTC Station relocated to NE 6th Street

The PE design for the DBT was a cut-and-cover tunnel, which included the BTC station in the northern half of the tunnel under 110th Ave NE. The short lengths of the tunnel to either side of the BTC station did not warrant use of mined or bored methods. The PE design comprised of a temporary soldier pile and lagging support of excavation (SOE) system, and a permanent cast-in-place (CIP) concrete twin cell box structure constructed inside the SOE walls. The PE design also included a temporary traffic decking system along 110th Ave NE to allow vehicular traffic to be maintained during construction of the DBT and BTC station.

Due to the aforementioned budgetary constraints and ST's Project Control Policy and Procedure ST undertook a formal Value Engineering (VE) exercise after completion of the PE design in an effort to identify potential cost saving ideas. The outcome of the VE effort included several ideas related to the DBT and BTC as listed below:

- Utilize a load bearing center wall in the cut-and-cover box structure
- Eliminate the tunnel waterproofing
- Replace a portion of the cut-and-cover box with a retained cut section

- Utilize reinforced concrete slurry diaphragm walls as both SOE and permanent structure walls
- Utilize a stacked tunnel configuration
- Relocate the BTC station from under 110th Ave NE, to NE 6th Street

Upon issuance of Notice to Proceed (NTP), ST requested that H-J-H evaluate these and other VE ideas. All VE ideas carried forward into the final design Early Work phase were re-named Cost Savings Ideas (CSIs). During evaluation of the CSIs, it became readily apparent that a significant cost reduction could be realized by relocating the BTC station. The PE station configuration, with separate underground mezzanine and platform levels resulted in deep excavations for the station itself and for the adjacent guideway tunnel, which in turn translated into significant construction costs. Moving the station to NE 6th Street resulted in a part cut and cover part at-grade station which could be constructed at far less cost.

The relocation of the BTC station resulted in a continuous tunnel segment along 110th Avenue NE between Main Street and NE 6th Street, which afforded the opportunity to re-evaluate the proposed tunnel construction method.

HMM, as tunnel designer for H-J-H, was requested by ST to perform an initial assessment of alternative tunneling methods for the newly reconfigured DBT and BTC Station. The purpose of the study was to determine the feasibility, cost/schedule effectiveness, and construction impacts and requirements associated with the use of mined tunneling methods for the construction of the DBT. The advancement of the study was predicated upon the outcome of an initial review workshop involving HMM and ST project and senior staff. The workshop included a review of the site geology; track alignment; adjacent and overlying constraints including buildings, utilities and excavation support elements; and potential tunnel configurations and construction methods. It was agreed that the short length of the tunnel, constrained right of way, tight radii, and presence of obstructions did not warrant use of Tunnel Boring Machine. It was also recognized that an SEM tunnel option faced multiple technical challenges, but it was agreed that there were no fatal flaws with the concept. The workshop concluded that SEM held promise as a potentially cost effective and less disruptive approach for construction of the DBT, and HMM was directed to proceed with the full study.

## THE SEM TUNNEL CONCEPT

### Alignment and Profile

The DBT is located within the central business district of the City of Bellevue. The south portal is located on land to be acquired to the southeast of the intersection of Main Street and 110th Place SE. From this location the alignment curves in a north-westerly direction onto 110th Avenue NE, and remains within the public right of way until immediately south of the intersection with NE 6th Street. A second horizontal curve, to the north-east allows the DBT to interface with the Bellevue Transit Center (BTC) Station, parallel to and immediately south of NE 6th Street. The alignment is indicated in Figure 2.

A goal of the PE design was to minimize the tunnel profile depth to minimize the costs of the cut and cover tunnel construction. This goal was constrained by the requirements for BTC Station. Conversely, there is no cost penalty associated with SEM for a deeper profile. As the SEM construction method relies on the ability of the surrounding soils to be self-supporting for a limited period of time, a deeper profile provides greater opportunity for the tunnel to be constructed within stronger, undisturbed soils. Unlike the horizontal alignment, which was essentially fixed to comply with the Final EIS/EIR, H-J-H had some flexibility to adjust the tunnel profile, while continuing to meet a number of identified profile constraints, including the following:

- Compliance with the ST Design Criteria Manual (DCM) for maximum grades for guideways and station platforms
- Complex soil and groundwater conditions between NE 4th Street and NE 6th Street
- Deep utility crossings at NE 4th Street
- Need for emergency access to NE 6th Street from the City Hall parking garage over the trackway
- Providing minimum clearances at the I-405 overcrossing, immediately east of BTC

The profile developed for the SEM Option is shown in Figure 3. The goal with the profile was to stay as deep as practicably possible, for as long as possible. This resulted in a profile grade of approximately 0.54% in the southern portion of the alignment. Beyond NE 2nd Street, the grade increases to a steep but operationally acceptable 5.7% to meet the BTC Station and I-405 location constraints.

### Existing Conditions

#### *Geology*

The prevailing ground conditions along the DBT alignment have been studied through two phases of investigation and testing. An initial phase of investigation and testing was completed during the project PE Phase. This assessment was supplemented by an additional program performed by H-J-H during Final Design, required to fill in gaps in the existing

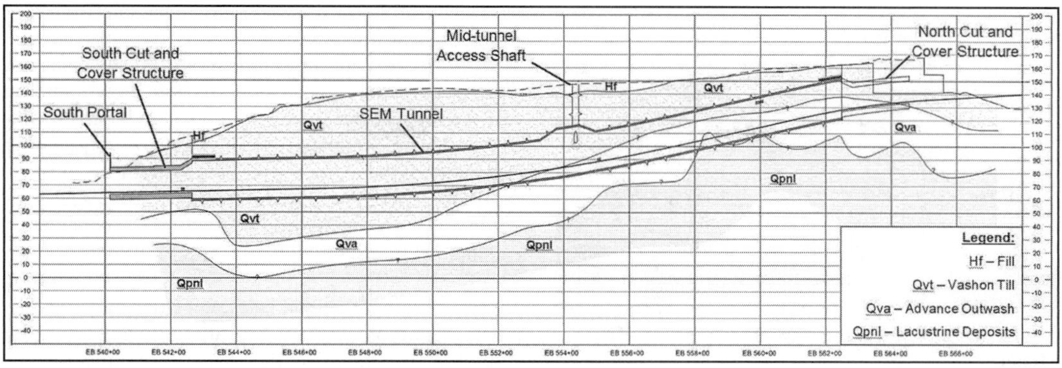


**Figure 3. Tunnel profile and anticipated soils conditions**

data, and to support the change in construction methods and the deeper tunnel profile. H-J-H initially performed nine additional borings totaling approximately 750 linear feet of drilling. Six of the borings were performed with sonic drilling techniques to better characterize the subsurface conditions. The remaining three borings were performed using mud rotary drilling methods to accommodate pressuremeter testing. A further four borings have since been added to the program to accommodate changes in scope and to provide further clarity on the anticipated conditions.

Based upon the investigations performed, the soils along the tunnel alignment have been grouped into four principal soil units as follows:

- Fill (Hf): predominantly loose to medium dense silty sand and sandy silt with varying amounts of gravel and organic matter. The fill layer is expected to vary in thickness from 0.5 feet to approximately 5 feet. The fill materials are expected to have little or no stand-up time and may flow if saturated.
- Vashon Till (Qvt): predominantly very dense silty sand with varying amounts of gravel and cobbles, some hard silt with varying amounts of sand and occasional very dense sand layers with varying amounts of silt. The till is expected to have good standup time from several minutes to over a year, though more limited stand up time will occur in the presence of lenses of sand or gravel.
- Advance Outwash Deposit (Qva): predominantly very dense silty sand and gravel with varying amounts of sand, silt, cobbles, and boulders. The sand and gravel layers may standup for a few minutes if moist, but will likely ravel if dry and flow if wet. Other projects in the vicinity of the DBT have experienced difficult and unstable ground conditions in this unit with less than a foot of groundwater pressure.
- Lacustrine Deposit (Qpnl): predominantly hard clay with hard silt and very dense sand layers. The silt and clay unit has variable excavation behavior. Silt with only a few percent clay content may have good standup time of hours to weeks. The more plastic clays are often fractured, which may result in slabbing and displacement of large fracture bounded wedges during and shortly after excavation unless immediately supported with shotcrete.

The subsurface soils along the tunnel alignment, as shown in Figure 3, generally consist of a thin layer of fill material overlying very dense overconsolidated glacial soils including till and advance outwash sand and gravel, which in turn overlie pre-vashon lacustrine deposits comprising silts and clays. A continuous outwash layer was encountered in borings north of NE 3rd Place with thickness ranging from 10 to 40 feet. Cobbles and boulders were inferred based on drilling action during field explorations. A continuous layer of lacustrine deposits that continued to boring termination depths was generally encountered in borings north of NE 2nd Street at depths ranging from 61 to 103 feet below ground surface.

Between the south portal and NE 2nd Street, along the southern 1,100 feet of the alignment, the excavation will be constructed entirely within the glacial till. North of NE 2nd Street, the excavation will still be predominantly within the till. However, for a length of approximately 700 feet, part of the bench and invert of the tunnel is anticipated to be within the outwash, and for a more limited extent of approximately 150 feet the lacustrine materials are anticipated to be encountered within the tunnel invert.

The majority of the completed borings have since been converted to piezometers, with readings

**Figure 4. Existing buildings, north of NE 4th Street**

taken at bi-monthly intervals. Based on work performed during the PE phase and the current field investigations and monitoring program, groundwater along the tunnel alignment is anticipated to grade upward from south to north from elevation 90 feet at Main Street to elevation 130 feet at NE 6th Street. With the exception of perched groundwater retained within sand lenses in the till, the excavations from the south portal to NE 3rd Street are expected to be substantially dry. North of NE 3rd Street, groundwater is anticipated within the advance outwash deposits. To improve the stability of the outwash materials, it is anticipated that dewatering of this layer will be necessary. However a pump test, proposed to determine the permeability of the outwash layer had to be abandoned when no groundwater was encountered in the well boring. This anomaly will be further investigated and the need for, and means and methods of dewatering this layer will continue to be studied over the course of Final Design.

### *Existing Buildings and Utilities*

The tunnel alignment is bounded by a number of commercial, residential and municipal buildings of varying height and construction type. South of NE 2nd Street the buildings are low rise, and pose relatively few concerns for construction. However, the buildings north of NE 2nd Street are larger, and of significant economic and cultural importance. These buildings include the 25 story Skyline Tower, the 26 story City Center Plaza Tower, leased by Microsoft Corporation, and Bellevue City Hall, shown in Figure 4. The City is justifiably proud of the City Hall Building, which is noted for its award-winning aesthetics and public art features.

In addition to the existing buildings a 16 story Marriott Courtyard Hotel is currently under construction immediately east of the tunnel alignment on 110th Avenue NE between NE 2nd Street and NE 4th Street. While the addition of another high rise building adjacent to the alignment presents additional challenge, the excavation of the tower basement has yielded useful information for the characterization of the geology for the DBT.

Existing buildings records indicate that all adjacent buildings are supported on a combination of perimeter strip footings and isolated spread footings. This is indicative of the strength and bearing capacity of the native soils. Almost all of the existing buildings have underground parking garages ranging from one to six underground levels. The deeper basements occur at the high-rise office buildings at the north end of the alignment at which point the tunnel invert is higher than the basement excavation depths. The separation between the building basements and the tunnel extrados as the tunnel turns towards NE 6th Street will be as low as 4 feet.

Composite utility plans illustrate that 110th Avenue NE reflects a typical urban setting with a relatively dense array of utilities including water supply, storm water, sanitary sewer, gas, electric and telecommunications lines both paralleling and traversing the tunnel alignment. Of particular note and concern for the SEM Tunnel are a cast iron water main, high pressure gas lines, and fiber optic lines serving a number of buildings along the alignment including the Skyline Tower and Bellevue City Hall.

### *Other Obstructions*

The majority of the existing buildings adjacent to the DBT include multi-level basements for resident and tenant parking. The basement excavations have been supported with either soldier piles and lagging with tie-backs or soil nail walls with shotcrete facing. In each case the tie-back or soil nail projects beyond the excavation into the public right of way. Based on typical excavation support construction practices, it is expected that the tiebacks and soil nails have been left in place.

The tunnel profile avoids any conflict with existing excavation support elements until the alignment is north of NE 2nd Place. Thereafter interference with soil nails and tie-backs is unavoidable. The City of Bellevue requires tie-backs extending into the public right of way to be de-tensioned after use. Since the majority of the buildings along 110th Street NE were constructed from 1980 onwards, this ordinance was in place and it is expected that it was followed; however, the potential of encountering a tensioned tieback remains. It is to be expected that the removal of tiebacks will require some additional care on the part of the contractor to prevent worker injury. The tunnel could potentially conflict with several hundred soil nails or tiebacks, and the implications for the project cost and schedule may be significant. The impact of removing these obstructions has been addressed within the project risk register and cost estimate as an allowance related to loss of productivity. The excavations for the Marriott

Courtyard Hotel basement encountered several very large boulders at an elevation consistent with the DBT horizon. Boulders of this size were not reported in other basement excavations. However, the potential for encountering a similar sized boulder within the tunnel excavation, a short distance away, cannot be ignored.

### SEM Tunnel Configuration

The DBT is approximately 2,490 feet in length. The DBT comprises of relatively short sections of cut and cover tunnel, of approximately 250 feet in length at its northern and southern extremes and a central SEM section of approximately 1,990 feet in length.

The extents of each of these construction methods were determined based upon requirements for the tunnel profile, the prevailing ground conditions, and also upon requirements for emergency ventilation.

To provide appropriate ventilation response in the event of a fire incident within the DBT, jet fans are required over the extent of the tunnel. The fans will be located in niches at the south and north portal areas, and at the midpoint of the tunnel, approximately. ST's preference is to have fans mounted in the sides of the tunnel as opposed to the ceiling, to minimize maintenance costs and avoid the need to de-energize the overhead contact system during fan maintenance. The resultant tunnel opening size at the fan niches at the south and north portals, in conjunction with relatively shallow cover dictated the limited use of cut and cover tunnel. The extent of each of the cut and cover sections was in turn dictated by requirements for fan performance.

The central fan niche will be accommodated by an enlargement of the typical SEM section. As using side mounted jet fans would have resulted in an overly large, unconstructable SEM cross-section, a saccardo nozzle system was developed, comprising a separate fan attic containing two fans with dampers to direct flow to either bore. This concept helped minimize the size of the SEM enlargement. The original expectation was that access to the attic and means to install and remove the fan equipment would be accommodated from the trackway. However, at the request of ST's Maintenance and Facilities Group, an access shaft and connecting adit will be provided to facilitate maintenance access and equipment exchange from the surface. The shaft and adit were added to the project scope after the conclusion of the SEM feasibility study. The capital costs for the shaft and adit are expected to be offset by reduced maintenance costs over the life of the facility and the avoidance of train service shutdown during maintenance operations.

The intent of the SEM concept was to maximize the length of SEM, taking advantage of the strength of the native soils, while still maintaining a practical minimum cover depth at the SEM to cut and cover interfaces. While a normal rule of thumb would be to provide at least one diameter of cover, there are several examples of large diameter SEM tunnels mined with relatively shallow cover, including the following:

- Fort Canning Tunnel, Singapore: Minimum cover of 9 feet in saturated soils comprising dense clayey silt, clay, sand lenses and boulders
- Tysons Corner Tunnels, Fairfax, VA: Minimum cover of 15 feet in soils comprising dense silts, low clay content, and sand. Variable groundwater from below invert up to spring line
- Dulles Airport Pedestrian Walkback Tunnel, VA: Minimum cover of 15 feet in soils comprising silt, clay and clayey sand in tunnel crown, and section generally within weathered siltstone. Groundwater table above rock surface.

In each of these cases the use of pre-support, comprising a grouted pipe canopy or other means, was introduced to compensate for the cover limitation.

The cover at the south SEM portal is expected to be approximately 15 feet. In this area, south of Main Street there are no direct building or utility impacts, the excavation is completely within the till material, and groundwater is not expected within the tunnel cross section, As the ground cover over the tunnel increases rapidly, the limited cover at the south portal is expected to be manageable with the installation of pre-support. At the north SEM portal, there is approximately 12 feet of cover. While borings have indicated that the material is almost entirely till, with limited to no fill material present, the continuity of the till material has been disrupted by excavations for utility trenches. However, with the use of appropriate pre-support, analysis conducted to date has demonstrated the feasibility of driving the SEM tunnel at this depth.

The tunnel cross-section must be sufficiently large to accommodate ST operating, maintenance and life safety requirements. A number of potential permutations exist for the configuration of the tunnel, which could be constructed with each track in a dedicated excavation or bore (twin-bore), or with both tracks within a single opening (single bore). The single bore cross-section was preferred for several reasons:

- The adjacent BTC and East Main Street Stations feature side platforms, which more readily accommodate the single bore track separation.

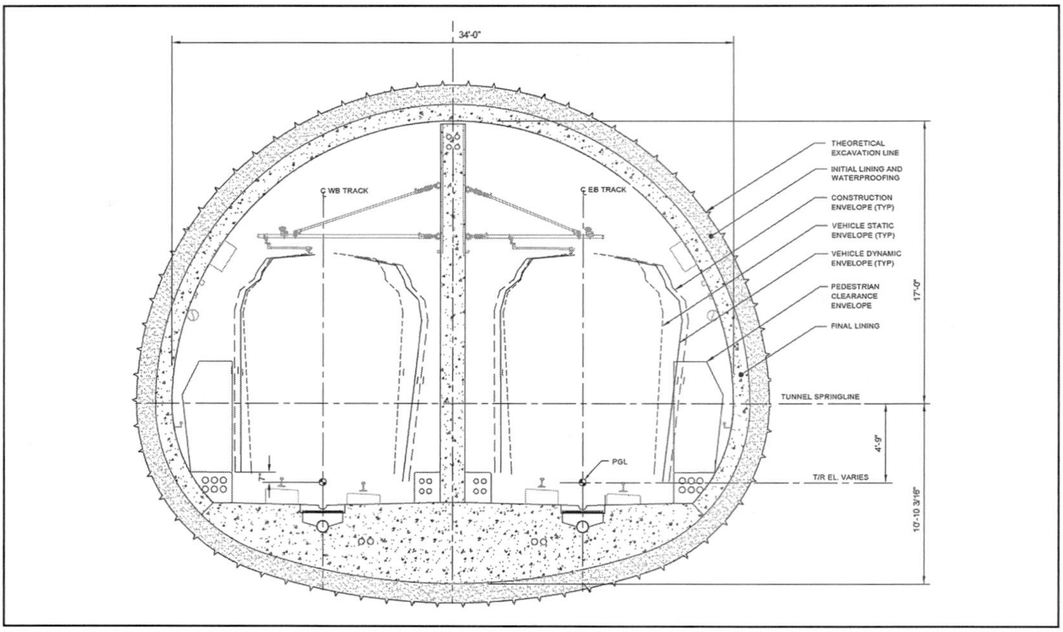


**Figure 5. Typical SEM tunnel cross-section**

- More readily accommodates tunnel ventilation requirements, particularly at the central fan niche
- Central street location minimizes the ROW needs and extent of interference with existing soil nails and tiebacks.
- Larger opening size improves contractor productivity–constructability, flexibility and ease of use of equipment etc.
- Though ultimately not necessary for this tunnel, cross passages or maintenance access can be easily provided in central dividing wall.

The resulting tunnel cross section based upon clearance and structure requirements has excavated dimensions of approximately 38'-0" in width and 30'-0" in height, as indicated in Figure 5. The center wall within the SEM tunnel is not a primary structural element. The wall provides fire and ventilation zone separation between the tracks and permits trains in opposite directions to be in the tunnel at the same time.

The mid tunnel access shaft will be approximately 20 feet in diameter and 50 feet in depth. The shaft diameter is sized to accommodate stair access to the mid-tunnel fan room and fan installation and removal. The shaft will be constructed almost entirely within the till, which is anticipated to have excellent stand up time. It is expected that the shaft excavation and initial support will also be by SEM, though contractors will have the option of using liner plates or soldier piles and lagging at their discretion. The adit connecting the shaft and the fan room will be approximately the same size as a typical transit cross passage excavation, with an excavated diameter of approximately 12 feet. The adit will be approximately 15 feet in length. Again it is anticipated that the adit will be excavated and supported using SEM. Due to the limited length and unusual cross section, the final shaft lining will also be shotcrete. The capital costs for the shaft and adit are expected to be offset by reduced maintenance costs over the service life of the project and avoidance of train service shut down with loss of revenue.

### Design Approach

A key aspect in determining the feasibility of the SEM concept was in performing sufficient appropriate analysis to:

- Determine appropriate excavation sequences and initial support requirements
- Demonstrate ground movements, and corresponding building and utility movements can be maintained within reasonable, acceptable limits
- Determine final lining requirements

The design of the SEM tunnel, the mid-tunnel access shaft, connecting adit, and the cut and cover structures is in accordance with the ST Design Criteria Manual (DCM). The DCM also adopts by reference other national design standards, including AASHTO LRFD Bridge Design Specifications, ACI Building

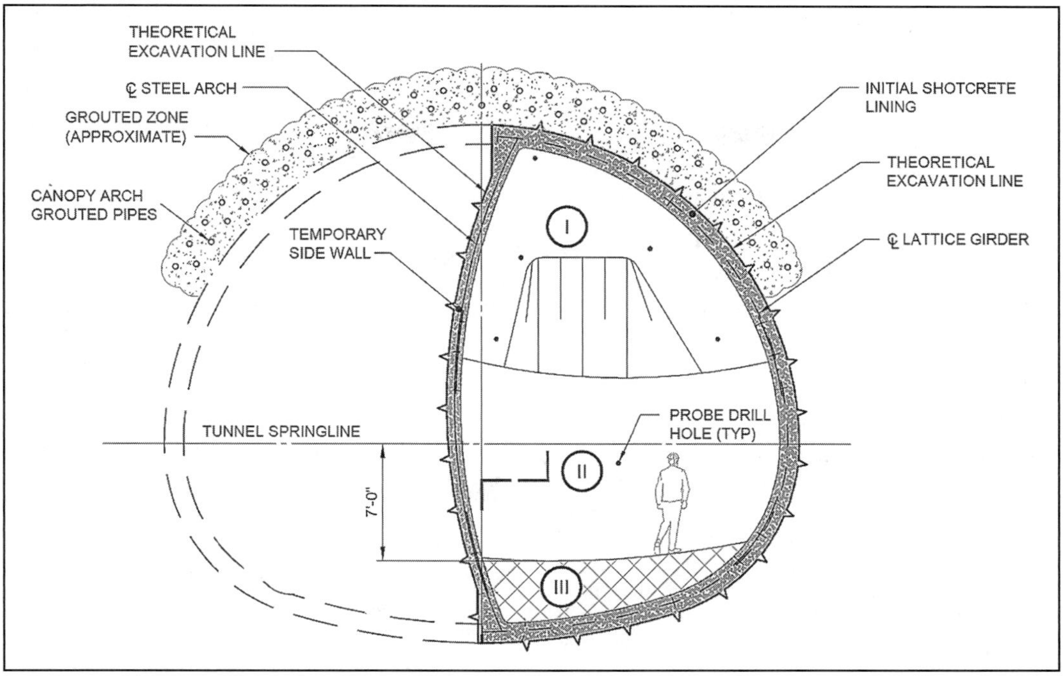


**Figure 6. Typical SEM tunnel excavation sequence**

Code Requirements for Structural Concrete (ACI 318) and FHWA Technical Manual for Design and Construction of Road Tunnels (FHWA-NHI-09-010). Notable ST design requirements applicable to the project include 100-year design life, mandatory use of numerical analysis methods to determine ground response and ground loads, seismic loading during construction condition, dual-level seismic design requirements and the corresponding allowable concrete and reinforcing steel strain limits for each seismic design level.

### *Excavation Sequence*

The design of the SEM tunnel is based on the principle that the disturbance of the surrounding ground must be minimized to mobilize the maximum self-supporting capacity of the soil and minimize risk of excessive ground movements. Thus, the excavation sequence and geometry must be optimized to promote smooth redistribution of in-situ stress field, allow timely support installation and ring closure while facilitating constructability.

With the benefit of learning from the successful SEM tunneling in similar ground for two platform tunnels of similar dimensions on Sound Transit's existing LRT system (Central Link Light Rail Contract C710–Beacon Hill Tunnel), the proposed excavation sequence comprises subdivision of the tunnel into two sidewall drifts, each with top heading, bench and invert. A round length of 4 feet is being considered at the current level design development. Figure 6 shows the proposed excavation sequence in cross section.

Initial support will principally comprise lattice girders and steel fiber reinforced shotcrete lining. Minimum pre-support, probing, and drainage measures will be prescribed, with allowance for supplemental toolbox items also included in the contract documents. This approach proved successful on ST's Beacon Hill Project.

### *Analysis Methods*

Multiple methods of analysis were used to evaluate the potential response of the ground to SEM tunneling and determine the load effects in the initial supports and final lining. The SEM tunnel is a linear structure of relatively constant cross section. Accordingly, two-dimensional numerical analyses performed in FLAC and Phase2 software were used to capture the plane-strain behaviors. Relaxation of the ground ahead of the analysis section is simulated by reduction of the ground moduli, as described in FHWA-NHI-09-010. Multiple analysis cases were performed to parametrically study the potential range of responses.

Three-dimensional numerical analyses in FLAC3D software are being used judiciously to confirm the effects of pre-support and the relaxation factors that are assumed in the two-dimensional analyses, and to study the complex three-dimensional response near the junction of the enlarged SEM tunnel, connecting adit and the mid-tunnel access shaft.

Beam-springs models were also used to determine the load effects in the lining. Because beam-spring models assume linearly elastic responses, they are well adapted to the Load and Resistance Factor Design (LRFD) approach for the lining which applies different load factors for different load types, such as dead loads, live loads, or ground loads.

### *Final Lining and Waterproofing*

The thickness and reinforcement of the final lining were designed based on the envelope of the load effects determined from the multiple methods of analysis. The final lining is currently detailed as a second-pass cast-in-place reinforced concrete with no composite action with the initial lining.

The final lining will be fully encapsulated with waterproofing membrane. The current design is based on PVC sheet waterproofing membrane that is compartmentalized with grids of water barriers. Provisions for injection sealing are specified using a pattern of grouting ports and re-injectable grouting houses.

An alternative method to construct and waterproof the final lining is being evaluated. This includes the use of steel fiber reinforced shotcrete final lining, in conjunction with spray applied waterproofing membrane.

### *Building and Utility Protection*

Twelve buildings and many utilities are located within the zone of influence of the DBT construction. An assessment program was conducted to identify the potential impacts from tunneling-induced ground movements on these buildings and utilities.

Ground movement contours were generated from the empirical Gaussian settlement trough that had been calibrated with the results from the 2D and 3D numerical analyses as well as case histories from similar SEM tunneling projects. The anticipated volume losses vary between 0.1 to 0.8 percent along the tunnel alignment due to multiple factors such as tunnel depth and soil units. At the portals, the anticipated ground movements from the excavation to construct the cut-and-cover structure were added to the tunneling induced ground movements.

For buildings, the screening criteria for potential ground movement impacts follow that of Boscardin and Cording (1998), Bjerrum (1963) and Wahls (1981). The utility tolerable deformation criteria were established from Attewell et al. (1986) and the criteria used on other tunneling projects in the Seattle area including the Sound Transit North Link Project and the WSDOT SR 99 Bored Tunnel Project.

Based upon empirical analyses conducted for the SEM feasibility study, preconstruction mitigation measures are anticipated for some high risk utilities and one small low-rise building, which the tunnel passes directly underneath and is scheduled to be purchased by ST for contractor staging. It is anticipated that the extent of the mitigation measures will be reduced through more refined analysis during final design. The building and utility protection work is in progress as of this writing. An instrumentation and monitoring program will be established to enable early detection of movements during construction so mitigation responses can be implemented in a timely manner.

The analyses provided the Project Team with confidence that a reasonable excavation sequence could be made to work, and that corresponding excavation induced impacts on adjacent and overlying infrastructure could be managed by appropriate excavation sequence, presupport, and initial support requirements.

## Cost Estimate and Risk Assessment

While the analyses demonstrated the feasibility of the SEM concept, there was no incentive for ST to adopt the mined tunnel unless it was demonstrated to offer substantial reduction in traffic disruption and general public disruption during construction, and, construction cost and/or schedule savings. An important part of the study was the comparative analysis of cost, schedule and risk for the SEM and Cut and Cover alternatives. For the SEM tunnel a number of key assumptions related to construction staging, operations and productivity were necessary.

- A single tunnel heading will be operated from the primary staging area at the south portal.
- Tunneling operations will continue 24 hours per day, 7 days per week.
- Typical tunnel advance rate was based upon the Sound Transit Beacon Hill Station Project platform tunnels, which are similar in cross section and excavation sequence. An estimated average full face advance of 3 feet per day was assumed.

As the SEM tunnel design was not sufficiently advanced, a number of allowances were included within the cost estimate for a number of cost significant items, including surface and tunnel-based instrumentation and monitoring, supplemental support or 'Toolbox' elements, removal of obstructions comprising an allowance for removal cost and a loss of productivity and ground improvement, comprising surface-based dewatering in the vicinity of NE 4th Street. Construction costs for both alternatives included excavation, initial support and construction of the tunnel structures, temporary provisions for dewatering, instrumentation and monitoring, traffic control, utility relocations and surface works and tunnel operating and fire life safety systems. Contractor indirect costs such as office and field staff, equipment, overhead and profit, general conditions, builders risk, bonding and taxes, etc. were also included.

In recognition of the fact that the risks associated with each construction method are significantly different, a quantitative risk assessment was performed to enable a risk-based contingency to be identified for each alternative. A team comprising ST and H-J-H personnel identified, ranked, and provided mitigations for significant risks associated with the design, construction, and procurement of each option. A primary risk for the cut and cover tunnel included utility relocations and impacts, and the extent of traffic decking and disruptions required on 110th Avenue NE. It was originally anticipated that a full closure concept with limited decking would be acceptable to the City of Bellevue. However, the City had more recently expressed severe concerns over traffic disruption with the concept. The cost implications associated with decking the entire excavation were significant. Primary risks for the SEM tunnel included geotechnical conditions, productivity, allowable working hours, and market conditions.

Based on the evaluations and analyses performed as part of the Feasibility Study, the construction cost of the SEM Tunnel was estimated at \$104.2 million, inclusive of risk based contingency, and was expected to take approximately 3 years to construct. The cut-and-cover alternative was estimated at \$117.9 million, also inclusive of risk based contingency and has an anticipated construction duration of 40 months.

### Study Conclusion

The SEM Feasibility study concluded that the use of SEM for the DBT was feasible. Furthermore, the results of the risk assessment and construction cost and schedule estimates suggested that the SEM alternative was advantageous to ST and to the City of Bellevue. H-J-H therefore recommended that the Project move forward with the SEM alternative. Correspondingly, ST and the City of Bellevue approved the SEM concept in April of 2013.

## PROJECT STATUS

Since the approval of the SEM concept, the design of the DBT has progressed to a 60% level of design completion. Beyond 60% the design and configuration will be refined as part of design progression, based upon the outcomes of ongoing geotechnical and material studies. Proposed design and construction enhancements will include the use of laser profiling to establish excavation lines/volumes and shotcrete thickness, use of spray applied waterproofing membrane, and a fiber reinforced shotcrete final lining. Final design completion is scheduled for late 2014, with bids solicited in 2015.

## ACKNOWLEDGMENTS

The authors would like to thank Sound Transit for the permission to publish this paper. In addition, the contributions of the H-J-H Team are gratefully acknowledged—HNTB Corporation, Jacobs Engineering, Golder Associates, and Shannon and Wilson.

## REFERENCES

AASHTO. 2012. *AASHTO LRFD Bridge Design Specifications, 6th Edition.* Washington, DC: American Association of State Highway and Transportation Officials.

Bjerrum, L. 1963. Discussion Session. In *Proceedings of European Conference on Soil Mechanics and Foundation Engineering,* Wiesbaden, Germany. 3:135–137.

Boscardin, M.D. and Cording, E.J. 1989. Building Response to Excavation-Induced Settlement. *Journal of Geotechnical Engineering.* 115(1):1–21.

CH2M Hill. 2011. *Sound Transit East Link Final Geotechnical Data and Considerations Report for Preliminary Engineering.*

FHWA. 2009. *Technical Manual for Design and Construction of Road Tunnels—Civil Elements.* Publication No. FHWA-NHI-09-010. Washington, DC: Federal Highway Administration.

H-J-H Final Design Partners. 2013. *East Link South Bellevue to Overlake Transit Center, Package E330 Draft Geotechnical Data Report, 60% Submittal.*

Sound Transit. 2012. *Link Design Criteria Manual, Revision 2.* Seattle, WA: Central Puget Sound Regional Transit Authority.

Wahls, H.E. 1981. Tolerable Settlement of Buildings. *Journal of the Geotechnical Engineering Division.* 107(11):1,489–1,504.

# Engineering of Cooks Lane Tunnel: An Overview of Challenges

**S. Rashidi, V. Nasri, and T. Kirby**
AECOM

**ABSTRACT:** The Cooks Lane Tunnel (CLT)—with an approximate length of 6,300 feet inclusive of approaches at both ends–is the shorter of the two tunnels envisioned as part of the Maryland Transit Administration's Baltimore Red Line. Variable geotechnical conditions, tunneling below groundwater table, mixed-face tunnel excavation, tunneling adjacent to the existing buildings and utilities, and open cut excavation in urban environment characterize some of the design challenges of the CLT. This paper describes the current proposed design and construction methodology for the CLT and presents the results of numerical modeling and analysis performed in order to support the Preliminary Engineering.

## INTRODUCTION

The Baltimore Red Line Project is a proposed 14.1-mile long east-west Light Rail Transit (LRT) line envisioned to connect the areas of Woodlawn, Edmondson Village, West Baltimore, downtown Baltimore, Inner Harbor East, Fells Point, Canton and the Johns Hopkins Bayview Medical Center Campus. The Red Line LRT System has two tunnel segments; the Cooks Lane Tunnel (CLT) and the Downtown Tunnel (DTT). The CLT segment–roughly 6,300 feet long -commences at the west portal located at the highway ramp for I-70 (to be removed) and terminates at the east portal which is at the intersection of Edmondson Avenue (US Route 40) and Glen Allen Drive. This segment of the project consists of the following construction components: approximately 4,786 feet of tunnels, 469 feet of cut and cover tunnel, and 1,045 feet of retained cut (U) section. The approximate horizontal alignment for the Red Line LRT Project is shown in Figure 1.

## GROUND AND GROUNDWATER CONDITIONS

The CLT will be excavated beneath the groundwater level, and in a range of ground conditions that are described as high strength and highly abrasive rock, in addition to mixed face of rock overlain by Transition Group material (TGM), and three fault zones, each with distinct properties. The ground geological condition is classified based on the International Society of Rock Mechanics (ISRM 1982) system of grading as shown in Table 1. Ground Classes I, II, and III represent rock and Ground Classes IV and V represent Transition Group materials.

A major portion of the tunnel profile will traverse through class IV and V material which is completely weathered and relatively permeable. This portion consists of the first 800-ft of the tunnel drive which starts at the west portal and the last 400-ft just before the east portal where the tunnel boring machine (TBM) will be extracted. The remaining tunnel path between these two zones is through competent rock as well as various combinations of ground types that create challenging mixed-face excavation conditions. Top of "competent" rock was defined as the level below which recovery with an NQ3 triple-tube core barrel is greater than 50 percent. In terms of the geotechnical ground class descriptions presented in Table 1, this definition is equivalent to a Ground Class III or better rock. Depth of this level ranges from about 16 feet along the central part of the alignment, where ground surface is highest, to greater than about 70 to 80 feet at inferred fault zones. Mixed-face conditions are concentrated near the two ends of the tunnel, adjacent to the cut and cover sections (Figure 2). Excavations for both the west cut-and-cover and retained cut section and the east cut-and-cover and retained cut section will be in all three types of earth materials:"competent" rock, the Transition Group material, and overburden.

Groundwater levels along the proposed CLT alignment are generally near the top of the Transition Group, within about 30 feet of the ground surface.

Overburden permeability is likely to be low ($10^{-7}$ to $10^{-5}$ cm/sec) in the clay-rich residual soils but higher in the localized sandy zones. Permeability in the Transition Group is expected to be generally low to moderate ($10^{-5}$ to $10^{-3}$ cm/sec) but much higher locally ($10^{-2}$ to $10^{-3}$ cm/sec) at open relict fractures, which could produce significant inflows.

Water-bearing properties of rock along the alignment are generally defined by fracture flow, with low permeability of intact rock. Rock mass permeability

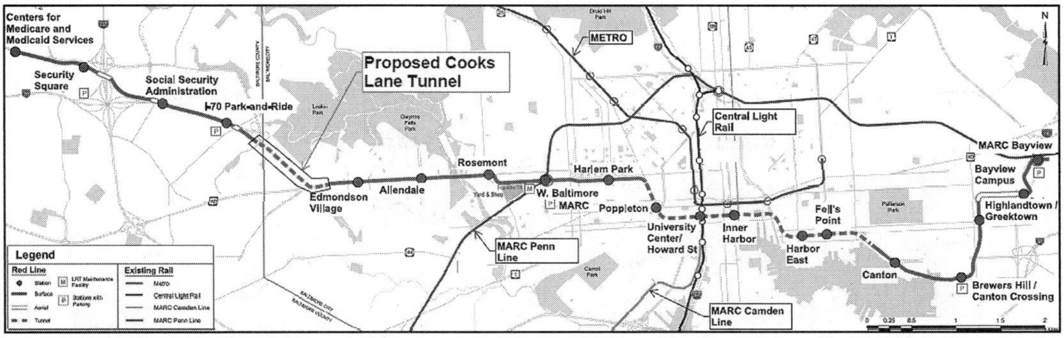


**Figure 1. Red Line LRT Project Plan**

**Table 1. Ground class descriptions**

| Ground Class | Description (ISRM Weathering Grades) |
|---|---|
| V | Completely weathered rock where all material is decomposed and disintegrated to soil but with original rock mass structure remains intact; disintegrates when agitated in water. |
| IV | Highly weathered rock where more than half is weathered to soil, does not disintegrate when agitated in water. |
| III | Fair to poor quality, closely to very closely fractured, slightly to moderately weathered rock |
| II | Good quality, moderately fractured, fresh to moderately weathered rock |
| I | Excellent quality, widely fractured, fresh to slightly weathered rock |

is expected to be highest in the fractured rock associated with fault zones. Results of packer permeability tests confirm that permeability in the rock mass is generally low ($10^{-7}$ to $10^{-5}$ cm/sec), with higher permeability ($10^{-4}$ to $10^{-3}$ cm/sec) in localized zones of closely spaced interconnected fractures or faulting. Preliminary information suggests that artesian conditions may have developed in deeper fractured rock at either end of the alignment.

Due to the high percentage of mafic minerals in much of the rock along the proposed Cooks Lane Tunnel alignment, groundwater is expected to be highly alkaline.

## TUNNEL CONSTRUCTION METHOD

The challenging geologic conditions along the proposed CLT alignment required a detailed study to determine the most appropriate and cost effective construction technique for the tunnel. The factors that will contribute to the preferred excavation method include: overall construction cost, construction duration, suitability of a particular method to the ground conditions, project site constraints, tunneling lengths, tunneling risks, and availability of appropriate expertise. Each of the construction methods offers advantages and disadvantages in their application for construction of CLT.

As stated earlier, the excavation adjacent to the tunnel portals has to take place in Transition Group material, which is soil-like material as well as mixed-face zones of Transition Group material overlying "competent" rock. It is also important to note that approximately 3,200 feet, or 67% of the tunnel drive is expected to be in competent rock with a rock cover thickness of at least 1 to 2 times tunnel diameter over the crown of the tunnel. The CLT excavation in rock is expected to encounter mostly mafic rocks, including amphibolite and diorite, amphibole (actinolite) schist, and amphibole gneiss. Mica schist is also present in the western portion of the alignment. The median unconfined compressive strength (UCS) for the amphibolite and amphibole gneiss rocks at CLT is about 32,000 pounds per square inch (psi). Excluding the relatively few tests for which failure occurred along an existing discontinuity, UCS typically ranged from about 15,000 to 54,000 psi for intact rock failures.

The results of SINTEF tests on samples of amphibolite and amphibole gneiss are summarized in Table 2.

The construction methods considered included cut and cover, NATM, and excavation by TBM. A discussion of applicability of each method is presented in the following subsections.

### Cut and Cover

The use of cut-and-cover construction method for the construction of the entire tunnel was not considered a viable option due to high cost and its disruptive impact on the surface roads and neighboring properties.

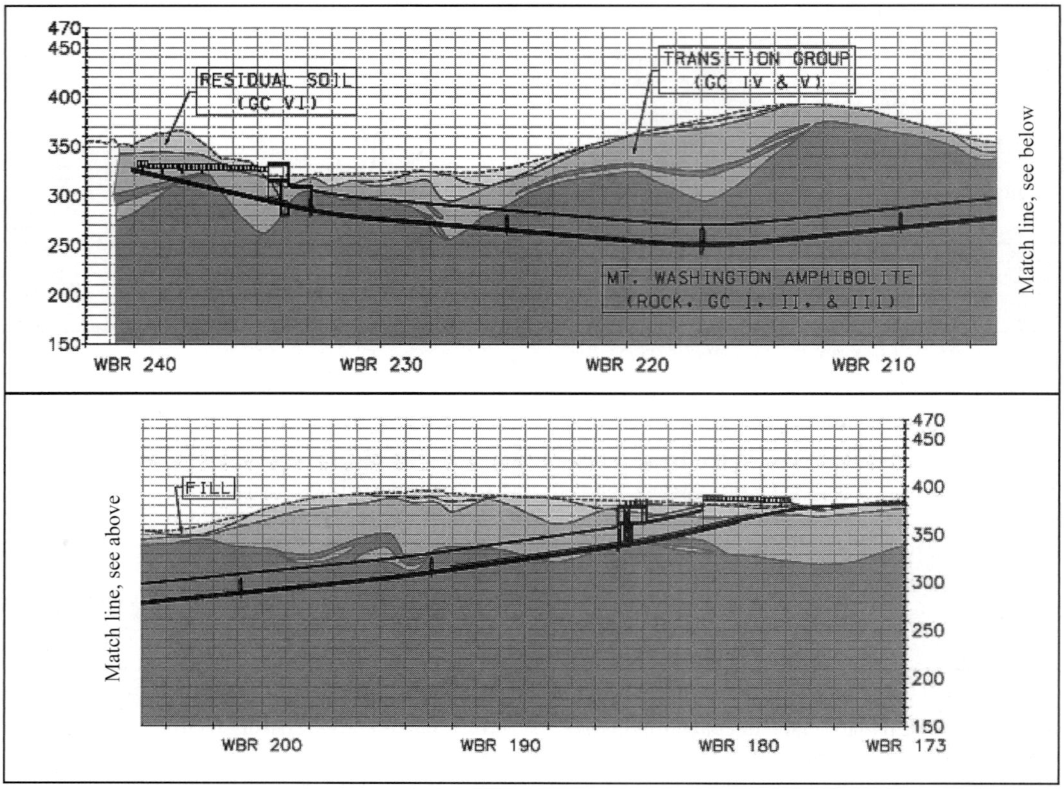


**Figure 2. Cooks Lane Tunnel geological profile**

**Table 2. Results of SINTEF tests**

| Drillability Index | Rating |
|---|---|
| Drilling rate index (DRI) | Extremely low |
| Bit wear index (BWI) | Medium to very high |
| Cutter life index (CLI) | Medium |

### New Austrian Tunneling Method (NATM)

Rock excavation by NATM, also known as Sequential Excavation Method (SEM), can be done using drilling-and-blasting method, road header, or a combination of the two. The advantage of this method is the adaptability, relatively quick commissioning and lower capital investment as compared to excavation using a Tunnel Boring Machine (TBM).

Prior excavation experience through Transition Group material near proposed CLT construction site has shown that this material is highly unstable once disturbed requiring extensive stabilization efforts. The other issue is the abrasiveness and high strength of the competent rock formation along a large portion of the CLT alignment that limits the excavation method to drilling-and-blasting. Lower excavation rate, increased construction risks, blasting-induced noise and vibration, and the length of the tunnel are among the factors that make this option less desirable than tunnel excavation by TBM.

### Tunnel Excavation by TBM

TBMs offer significant advantages with respect to excavation advance rates, reducing ground-borne vibrations, face stabilization and ground settlement control. The use of a TBM will allow for significantly higher production rates as compared with other methods of tunneling. It is anticipated that a TBM will be able to bore through the existing ground at the proposed CLT horizon at an average rate of 40 feet/day. However, the advance rate will be re-evaluated as laboratory and additional site data becomes available.

Use of a TBM also offers comparative benefits with respect to impacts on the adjacent properties. For most part, all of the construction activities will be focused around the launch pit, which is located away from most of the stakeholders. TBM extraction at the end of the drive on Edmondson Ave is a short duration activity.

## TBM SELECTION

A critical element of this project is control of groundwater inflow. Based on previous experience, with the anticipated poor ground behavior especially within the Transition Group material and mixed-face zones combined with the relatively large excavation (approximately 23 feet), use of compressed air TBM would be a risky endeavor.

In the past 20 years tunnel excavation in a challenging environment such as CLT has made spectacular improvements with excavation control by the application of pressurized-face shielded TBMs; such as Earth Pressure Balanced (EPB) or Slurry face (SF) TBMs. Regardless of the TBM type used, there are challenges when tunneling in mixed grounds such as uneven/unbalanced cutter force distribution at excavation face between the rock and soil. In such situation the cutters on rock attract more applied thrust than those on soil causing frequent impact loading and intense hammering effect on cutters and bearings resulting in high cutter wear and damage. The TBM operator will need to lower both thrust pressure and reduce advancing rate resulting in lower cutting efficiency. Other potential issues include excessive over-cutting of soil, leading to large ground settlement, high groundwater seepage at interfaces, jam of roller and cutter bearings, and difficulties in removal of mixed muck from the excavation chamber.

An alternative for circumventing the potential complications for excavating tunnels in mixed ground using TBM is to either modify the design of TBM to suit the ground conditions or conditioning the ground to suit the available TBMs.

Any TBM to be utilized will need to excavate tunnel sections at various locations along the alignment consisting of full-face competent rock (high strength and highly abrasive) and mixed-face conditions consisting of rock overlain by Transition Group (highly weathered or completely weathered). The TBM while excavating within the full-face competent rock sections will not need pressurized-face support to maintain face instability. It is also unlikely that while within competent rock, it would be required to operate with pressurized-face support due to ground permeability characteristics. However, in locations exhibiting a mixed-face condition of rock overlain by Transition Group as well as zones passing entirely through Transition Group material will require the TBM to operate in pressurized mode to maintain tunnel face stability. When properly configured, specifically EPB/hard rock hybrid machines, pressurized-face TBMs are capable of efficiently excavating a full-face of competent rock as well as mixed-face condition. For this reason, an EPB machine with hard rock cutting capability is being recommended for the Cooks Lane Tunnel.

## TUNNEL GEOMETRY

The two TBM-bored tunnel options include single-bore, dual-track (large diameter) tunnel and twin-bore, single-track (small diameter) tunnel.

### Single-Bore, Dual-Track

The single-bore option consists of dual tracks separated by a fire-rated wall to satisfy NFPA 130 requirements. Due to the need for dual tracks and a fire-rate wall, the preliminary inside tunnel diameter was established at approximately 34 feet. The advantages of single-bore, dual-track option include:

- Provides potential cost savings compared with twin-bore and SEM construction
- Requires single TBM extraction effort at the east portal
- Provides opportunity to reduce construction duration compared with twin bore option
- Eliminates the need for dedicated ventilation structures at portals
- Minimizes foot-print impact for construction staging
- Eliminates ROW impacts along Cooks Lane
- Eliminates the need for cross-passages and associated risks with penetrating lining for cross passages construction
- Provides ample systems space within the tunnel envelope

### Twin-Bore, Single-Track

The twin-bore option consists of driving parallel tunnels between portals; each carrying one track. This results in an inside tunnel diameter equal to 20 feet, approximately 14 feet smaller than that of the single-bore. The advantages of the twin-bore, single-track option include:

- It is easier to maintain tunnel face in competent rock and minimize mixed-face tunneling due to the smaller bore diameter
- The smaller cross-section results in less muck being generated
- Conforms to a more common size
- Provides additional cover (compared to single bore option) under properties at the intersection of Cooks Lane and Edmondson Avenue
- Provides additional cover (compared to single bore option) under utilities at west portal; alternatively, the profile gradient for the west approach can be reduced

The current design has adopted the twin-bore, single-track option.

## GROUNDWATER CONTROL

Based on the anticipated ground and groundwater conditions along the proposed CLT alignment, a convertible hard rock TBM capable of operating in both open and pressurized-face modes has been recommended. The machine will be operated in pressurized-face mode during tunneling through all soft ground, mixed-face conditions, Transition Group material, and short stretches of fractured rock with high groundwater inflow and should be converted to open mode (i.e., unpressurized) during tunneling through competent rock.

The TBM-bored tunnels require a gasketed precast concrete segmental lining to prevent the inflow of groundwater into the tunnel over the lined portion of the tunnel. The groundwater control measures should therefore, provide positive control of the inflow from the advancing tunnel face. Groundwater control during pressurized-face tunneling using a convertible Earth Pressure Balance TBM is achieved by the formation of a soil plug inside the face plenum (i.e., excavation chamber) to balance earth and hydrostatic pressures. The face pressure is primarily maintained by the screw conveyor operations and the presence of a soil plug.

Should the inflow of groundwater or loose materials during open mode tunneling in rock start to increase, it can be controlled by changing the operation mode from open face into a pressurized face mode to control the groundwater and material. For the pressurized face mode in rock, ground conditioning material will need to be added to facilitate the formation of the plug inside the screw conveyor since rock spoil typically has characteristics that are not conducive to plug formation.

When the TBM is operating in competent rock (open mode) the following groundwater control measures are anticipated depending on the expected amount of groundwater inflow:

- Dewatering at the tunnel face
- Drainage from probe holes
- Rock mass grouting

## MUCK HANDLING AND REMOVAL

More than 65% of the CLT excavation will be done in rock, where the TBM can operate in open face mode. Muck resulting from hard rock chipping is predominantly granular and includes a high percentage of gravel and possibly small cobble-size particles. As a result of chipping mechanism, the gravel and cobble-size particles within the muck are elongated in one direction. The process of chipping also generates sand and silt-size particles. Combined with infill in joints, fractures, and weathered seams, the resulting muck from hard rock TBM excavation is generally a coarse grained material consisting predominantly of gravel-size rock chips, but also includes significant percentages of sand and fines. This material generally classifies as a silty or clayey gravel depending on the nature and volume of fine-grained weathered rock or joint infill within the overall rock mass.

The open cut excavations will be performed using conventional earthwork equipment and will result in muck consistent with conventional bulk excavations. However, it is worth noting that the majority of the excavated materials will be below the groundwater table and may therefore, be in a saturated condition when excavated. The portal excavations will proceed from the ground surface to the tunnel depths, and will thus encounter all soil and rock strata above the base of the structure. As a result, muck from these excavations through the fill material that may include miscellaneous debris and obstructions.

### Hazardous Materials

Preliminary investigations indicated that naturally occurring asbestos minerals may be present in rock to be excavated for the CLT. These minerals pose a potential inhalation hazard if they are disturbed during excavation and allowed to become airborne, requiring worker protection and dust control. Specialized handling and disposal at an approved facility are also required for excavated asbestos-containing rock. Additional testing is required before naturally occurring asbestos can be ruled out along the CLT alignment.

Radon gas is another potential naturally occurring hazardous material. The Red Line project is in the US EPA Radon Zones 1 and 2 (high to moderate radon potential). The radon source is most likely the quartz-rich crystalline rock, but pockets of high radon can also occur in sediments. Radon gas would not pose a hazard for workers during excavation because the tunnels will be ventilated, and the workers will not have long-term exposure. Gabbroic rock types such as those at Cooks Lane often contain sulfide minerals, including pyrite, as observed in recovered CLT rock core samples. Sulfide minerals can potentially produce hydrogen sulfide gas as well as potentially corrosive groundwater, both of which will require consideration for construction and muck handling.

The CLT segment passes through an area that has experienced urban development, re-development, and industrial activity since Baltimore was founded in 1729. As with many industrial activities over the last few hundred years, industrial practices have changed and developed over time. Manufactured hazardous materials are likely to have been discharged, either intentionally or unintentionally, into the subsurface due to the various commercial and industrial operations throughout this area. Both solid and liquid hazardous materials of varying concentrations are likely to be present in isolated locations within the general area of the CLT. This is typical of many cities of comparable age and development history throughout the country and is a potential issue on any large underground project.

Contaminated soil and groundwater, if encountered, will require special handling and treatment for disposal. Tunnel construction may also affect the direction and transport rate of any existing contaminant plumes. A detailed assessment of the depths and strata that may include hazardous contaminants has not been performed yet. Further study is underway to assess whether the bored tunnels will encounter any hazardous contaminants.

### Traffic Impacts of Muck Disposal

The largest source of construction traffic will be the transport of excavated materials from the tunnel to various permanent disposal areas. Tunnel excavation will generate large volumes of muck. It is anticipated that tunnel construction will proceed as one heading at a time from the west portal. Muck will be hauled away using 3-axle dump trucks (maximum 20 cubic yard capacity), assuming a maximum allowable fully loaded truck weight of no more than 55 kips—based on the State of Maryland regulations. The daily truck traffic volume is proportional to the volume of the excavated material per day. For the bored tunnels, this will be directly proportional to the TBM advance rate. For the retained cut and the cut and cover segments, it will be proportional to the staged excavation progress. The total estimated muck volume for the CLT is presented in Table 3.

An average TBM advance rate of 40 feet per day is currently assumed based on the anticipated ground conditions along the CLT alignment. The estimated number of construction truck trips per day for one tunnel heading and excavation advancing rate of 40-ft/day is 62 truck loads per day.

**Table 3. Total bulk volume of muck for each ground class**

| Ground Class | Bulk Volume (cy) |
|---|---|
| Fill | 87,949 |
| Transition group material | 136,778 |
| Rock (I,II,III) | 289,188 |
| Total | 513,915 |

## TUNNEL NUMERICAL ANALYSIS

The literature suggests that for parallel tunnels of diameter D which are separated by a pillar of width W, the interaction effects are small at $W/D = 1$ and vanish at $W/D \geq 2$ (Ghaboussi and Ranken, 1977). The Cooks Lane twin tunnels are approximately 10-feet apart which is less than one tunnel diameter (22-ft). Excavating the tunnels in such close proximity will cause interaction between the two during the construction. The TBM excavating the first tunnel modifies the state of in-situ stresses and causes disturbance within the soil or rock surrounding that tunnel. The size of this affected zone depends on the TBM operation, ground type, in-situ stresses, tunnel depth, tunnel diameter, and characteristics of tunnel support system. Excavating the second tunnel will also create the same effect. If the distance between the two tunnels is small, these two zones will overlap. Such overlap, or interaction, manifests itself as changes in the stress field around the first tunnel resulting additional stresses in the concrete liner of this tunnel. These additional stresses must be taken into account in designing the pre-cast concrete liner of the tunnels and may also require pillar strengthening where deemed necessary.

Several analyses were performed to evaluate the zone of ground disturbance around each tunnel, degree of overlap between the disturbed zones, ground settlement, and the increase in liner forces and moments of the first tunnel due to the excavation of the second tunnel. These analyses were performed for different ground conditions surrounding the tunnels and by employing three levels of ground loss (0.5%, 1%, and 1.2%). Figure 3 shows the ground relative shear stress distribution after completion of the second tunnel for the case where tunnels are excavated in Transition Group material and the ground loss is 0.5%. The darker shaded zones around the tunnels in this figure are indicative of zones where shear strength of the soil has been exceeded. Therefore, when the tunnels are bored within the Transition Group material, regardless of percentage ground loss assumed in the analysis, the pillar experiences plastification. The zones of ground shear failure expand as the percentage of the ground loss increases from 0.5% to 1.2%.

The change in liner axial force and bending moment of the first tunnel due to excavation of second tunnel is shown in Figure 4 at four quarter points of the liner. The results shown in this figure are obtained for the case where the tunnels are excavated in Transition Group material.

## TUNNEL EXCAVATION IMPACT ON EXISTING STRUCTURES

The existing buildings alongside Cooks Lane are adjacent to the CLT. Just before Cooks Lane intersects with Edmondson Avenue, the tunnels turn towards the east passing directly beneath two 2-story homes. Approximately 2,800 feet of the tunnels will be bored in competent rock with at least one tunnel diameter of rock cover above tunnel crown.

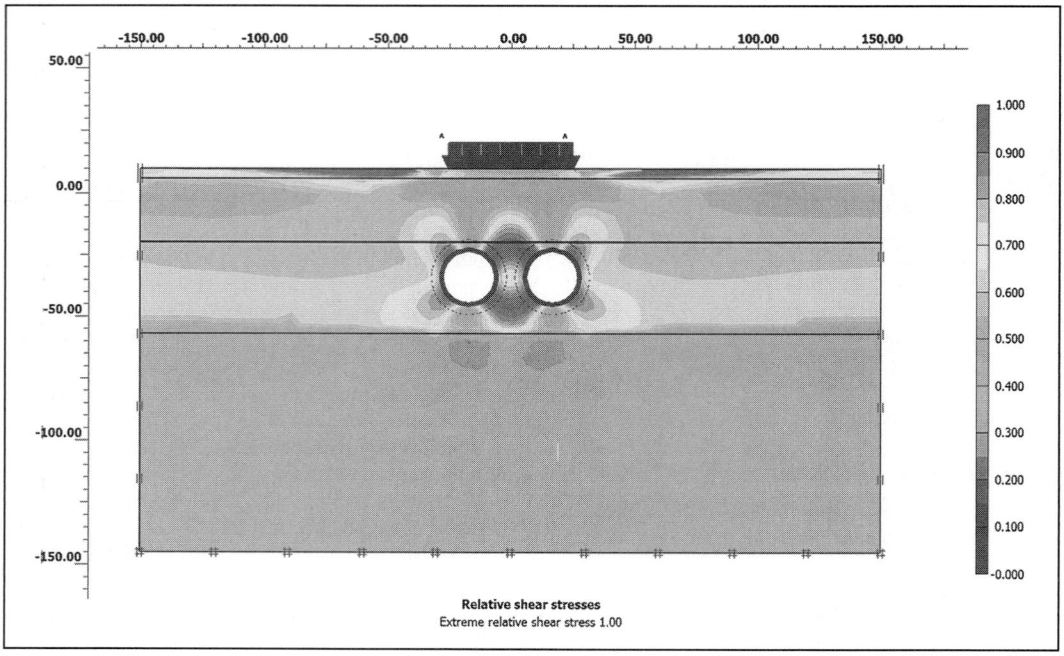


**Figure 3. Ground relative shear stress (TGM, 0.5% ground loss)**

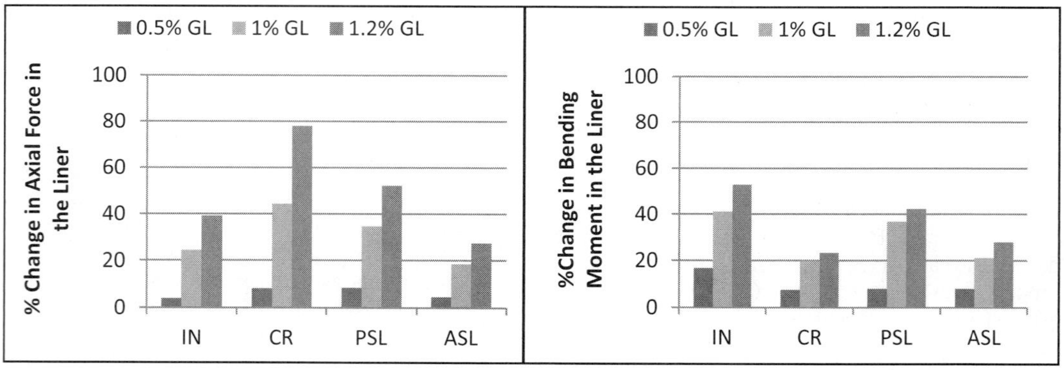


IN: Tunnel Invert, PSL: Tunnel Springline (Pillar-side), CR: Tunnel Crown, OSL: Tunnel Springline (Opposite to PSL)

**Figure 4. Increase in the liner forces of the first bored tunnel (TGM)**

The ground surface movement due to tunneling in this area will be negligible. The ground movements resulting from tunnel excavation in Transition Group materials and mixed-face were estimated during Preliminary Engineering using two-dimensional numerical analysis. The impact of tunneling on the existing buildings was then evaluated in accordance with the methodology proposed by Boscardin and Cording (1989). The results of this study have indicated that in general the building damage caused by tunneling would be negligible provided that the ground loss due to tunneling is properly controlled.

## CONCLUSIONS

This paper presents the major design and construction considerations for the Preliminary Engineering of the Cooks Lane Tunnel including construction methodology, viable tunnel configurations, groundwater control during construction, and tunnel muck removal. The paper also presents the results of tunnel numerical analyses performed to date. The engineering and geotechnical investigation for the Red Line LRT project are still ongoing and the design evolves as new information becomes available.

## REFERENCES

Boscardin, M.D. and Cording, E.J. 1989. Building response to excavation-induced settlement, *Journal of Geotechnical Engineering*, Volume 115, No.1, paper 23066, 1–21.

Ghaboussi, J., and Ranken, R.E. 1977. Interaction between two parallel tunnels, *International Journal for Numerical and Analytical Methods in Geomechanics*, 1: 75–103.

International Society for Rock Mechanics 1982. ISRM Suggested Methods: Rock Characterization, Testing and Monitoring (E.T. Brown, ed), Pergammon Press, Oxford.

# Innovations in Ventilation and Fire/Life Safety for an Urban Mega-Tunnel

**Michelle Smith**
Los Angeles County Metropolitan Transportation Authority

**Bernard Höpperger, Alexander Rudolf, Reinhard Gertl, and Marc Knickerbocker**
ILF

**C. Yoga Chandran and Tom Ionta**
CH2M HILL

**Derek Higa**
California Department of Transportation

**ABSTRACT:** A dual-bore freeway tunnel approximately 17.8 m (58.5 ft) in diameter and 7.9 km (4.9 mi) long is one of four multimodal alternatives under study in Southern California. Tunnel ventilation systems for tunnels of this magnitude would typically include intermediate shafts. In response to community concerns about preserving scenic and aesthetic resources, a new design that eliminates the need for intermediate ventilation shafts was developed for the State Route 710 (SR-710) freeway tunnel alternative. The related fire detection, fire suppression and other safety systems are also described. This paper highlights the innovative approach used for the ventilation scheme and presents the current ventilation and fire/life safety design for what would be the largest and longest freeway tunnel in the USA.

## INTRODUCTION

### Background

The SR-710 transportation corridor was originally envisioned to extend north from the City of Long Beach to the I-210/SR-134 and SR-710 interchange in the City of Pasadena. The segment between I-10 and the I-210/SR-134 and SR-710 interchange is the only uncompleted section.

For decades, planning efforts to improve mobility and relieve congestion on local arterials and nearby freeways were limited to a surface extension of the SR-710 freeway. Now, the California Department of Transportation (Caltrans) and Los Angeles County Metropolitan Transportation Authority (LA Metro) are considering a range of alternatives to address the problem.

In 2011, LA Metro contracted with the CH2M Hill team to conduct an environmental study to identify project alternatives to address the traffic congestion within and beyond the SR-710 corridor. LA Metro is the contracting agency for the environmental study, and Caltrans is the lead agency assigned authority to ensure the study is conducted in compliance with the National Environmental Policy Act (NEPA) and the California Environmental Quality Act (CEQA). Alternatives in the study include light rail transit (LRT), bus rapid transit (BRT), transportation system management/transportation demand management strategies (TSM/TDM), a freeway tunnel and No Build.

The freeway tunnel alternative is feasible due to advances in tunnel boring machine technology over the last 25 years, allowing greatly increased tunnel diameters. For this alternative LA Metro and Caltrans are considering dual-bore tunnels with 17.8 m (58.5 ft) diameters and approximately 7.9 km (4.9 mi) long between East Los Angeles and Pasadena. If this alternative is selected, it would rank as one of the largest urban mega-tunnels in the world.

The ventilation and related fire/life safety considerations for this mega-tunnel are the focus of this paper.

The SR-710 North Study area is situated in the east/northeast Los Angeles and west San Gabriel Valley area.

As depicted in Figure 1, the area is approximately 260 km$^2$ (100 mi$^2$) and generally bounded by

- I-210 on the north
- I-605 on the east

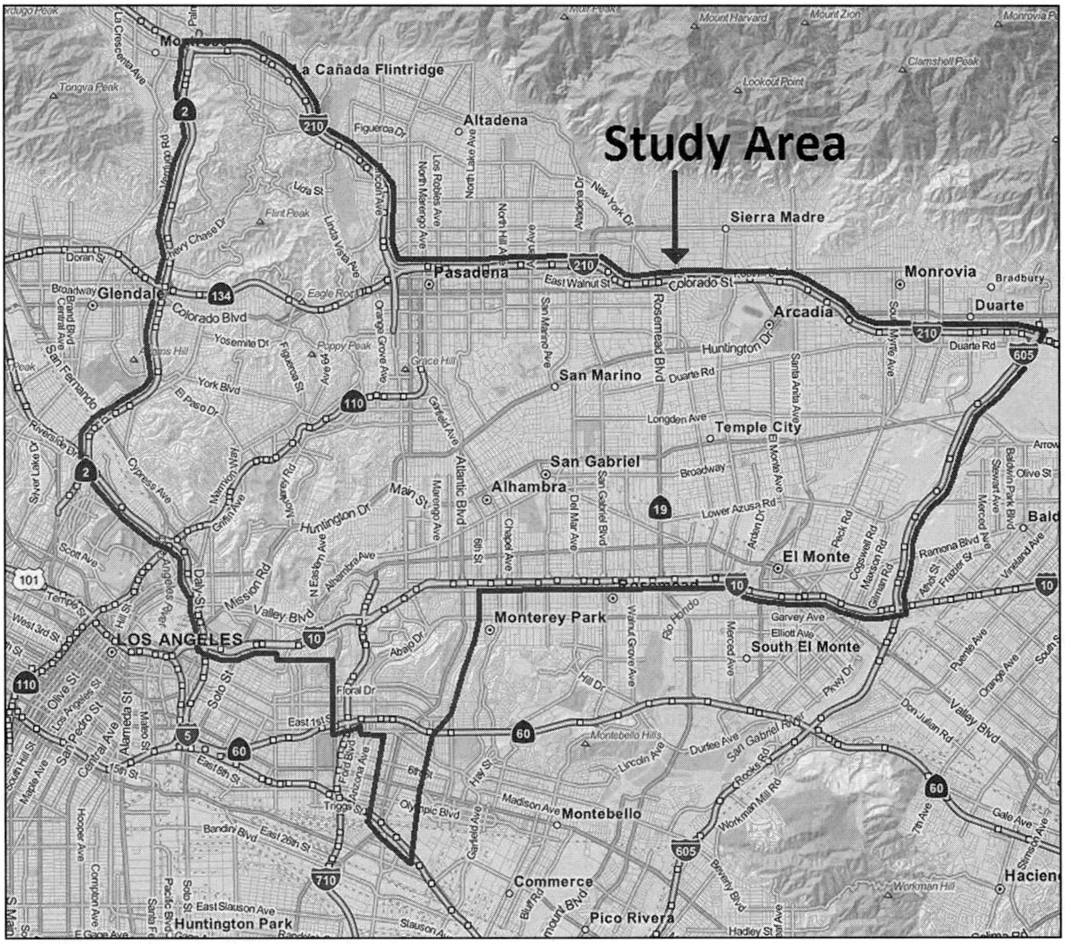


**Figure 1. SR-710 North study area**

- I-10 on the south
- I-5 /SR-2 on the west

The freeway tunnel alternative alignment starts at the existing southern stub of SR-710 in Alhambra, just north of I-10, and connects to the existing northern stub of SR-710, south of the I-210/SR-134 and SR-710 interchange in Pasadena (Figure 2).

The entire length of the alignment is 10.1 km (6.3 mi) with

- 6.8 km (4.2 mi) bored tunnel
- 1.1 km (0.7 mi) cut and cover
- 2.8 km (1.4 mi) at-grade

The bored tunnels would be located about 36.6–85.3 m (120–280 ft) below the surface.

Short cut and cover segments at the south and north termini provide access via portals to the bored tunnels. The southern terminus would be located south of Valley Boulevard. The portal at the northern terminus would be located north of Del Mar Boulevard. No intermediate interchanges are planned.

## Freeway Tunnel Alternative

The freeway tunnel alternative has two design variations: a dual-bore tunnel and a single-bore tunnel. Figure 3 illustrates the single-bore and dual-bore tunnel design variations.

The dual-bore variation has two side-by-side tunnels (one northbound and one southbound), with two levels and two lanes of unidirectional traffic and shoulder per level for a total of four lanes per tunnel. Vehicle cross-passages between the bores for emergency use would be provided, nominally spaced every 914 m (3,000 ft).

Both design variations include the following tunnel support systems.

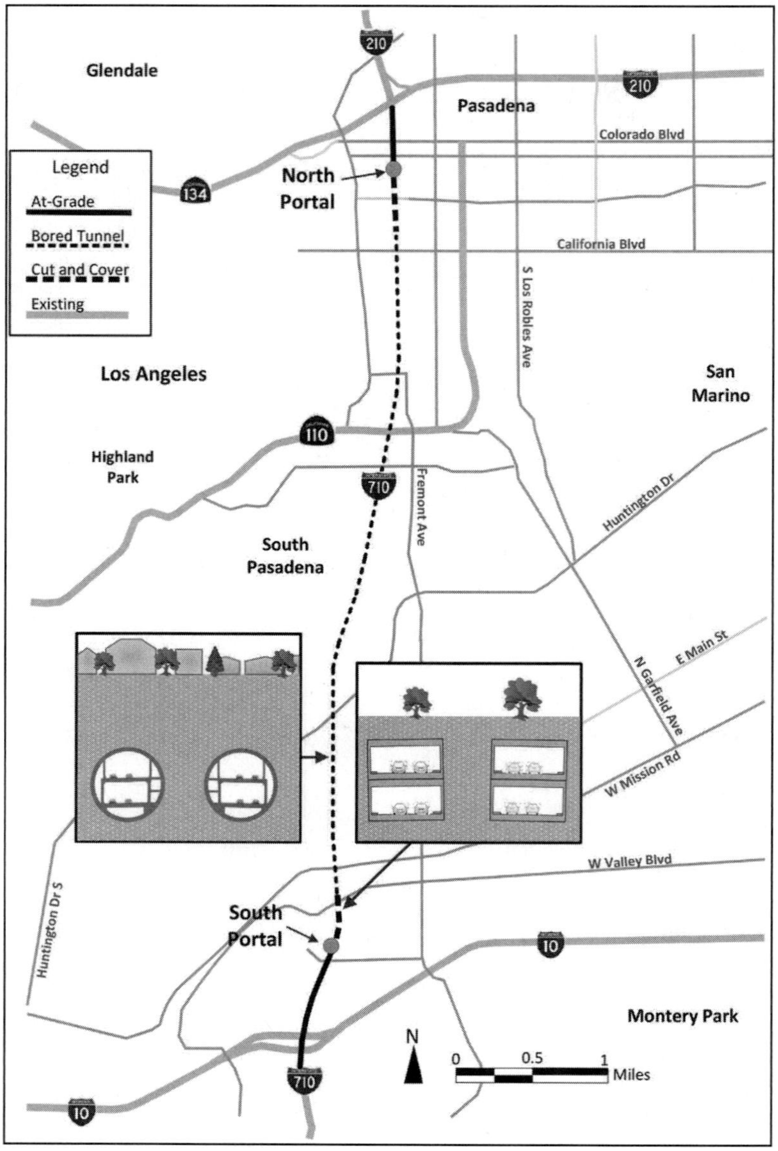


Figure 2. Freeway tunnel alternative alignment

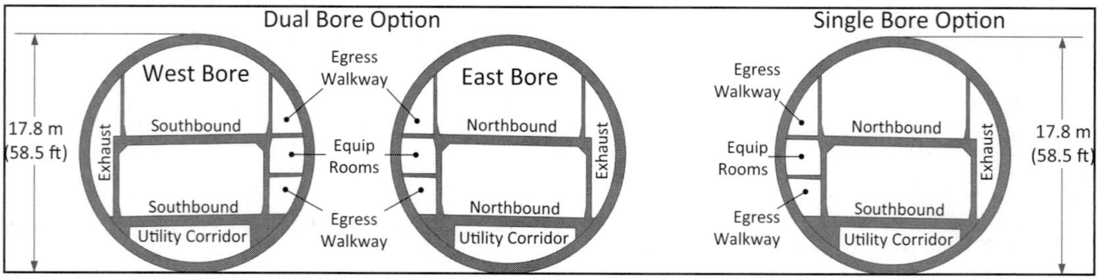


Figure 3. Dual- and single-bore tunnels

- Ventilation system
  - Exhaust fans at each portal
  - Single exhaust duct along the entire length of the tunnel
  - Jet fans within the traffic area of the tunnel
  - Air scrubbers at the portals
- Evacuation
  - Protected egress pedestrian walkways
  - Vehicles cross passages between bores
- Fire detection systems
  - Linear heat detection
  - Optical detectors
  - Video image detection
- Fire suppression
  - Deluge foam water system (fixed fire fighting system)
  - Standpipe and hose system
  - Fire extinguishers
- Communications
  - Variable message signs
  - Emergency telephones
  - Public safety radio
  - Wireless broadband network
- Operations and Maintenance Center (OMC) buildings at both the portals
  - 24-hour surveillance
  - Co-located first responders and fire fighting vehicles

There would be no operational restrictions for the tunnel, with the exception of vehicles carrying flammable or hazardous materials. However, it should be noted "Toll" and "No Truck" scenarios will be evaluated as well.

### History of the Ventilation System

In 2006, LA Metro commissioned a feasibility study to determine if tunneling would be a viable option. The conceptual evaluation of the tunnel ventilation system included intermediate ventilation shafts along the tunnel alignment, as shown in Figure 4 from the feasibility study.

Due to community and stakeholder concerns about intermediate shafts, alternative designs including a separate ventilation tunnel were designed. CH2M HILL's team is proposing an innovative ventilation concept and design approach that eliminates both the intermediate ventilation shafts and the need for a separate ventilation tunnel.

## PROPOSED VENTILATION CONCEPT

The ventilation concept for normal operation is longitudinal ventilation using jet fans. In the case of a fire, the concept allows smoke extraction locally with automated exhaust dampers. Ventilated emergency walkways are provided along the tunnel. There are air scrubbers at the exit portals of both tunnels.

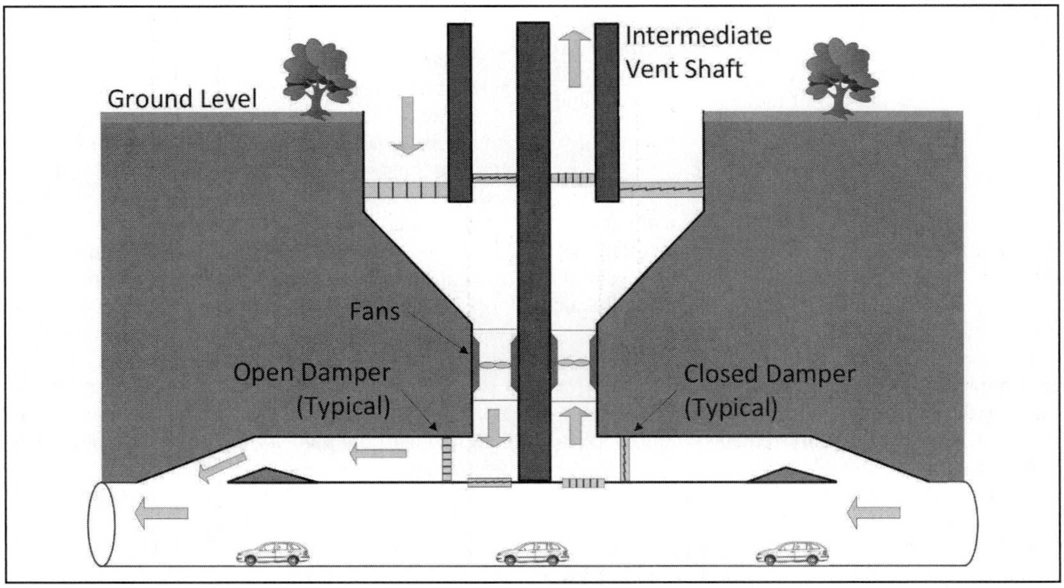


**Figure 4. Example of a mid-tunnel ventilation arrangement with Saccardo nozzles for longitudinal ventilation**

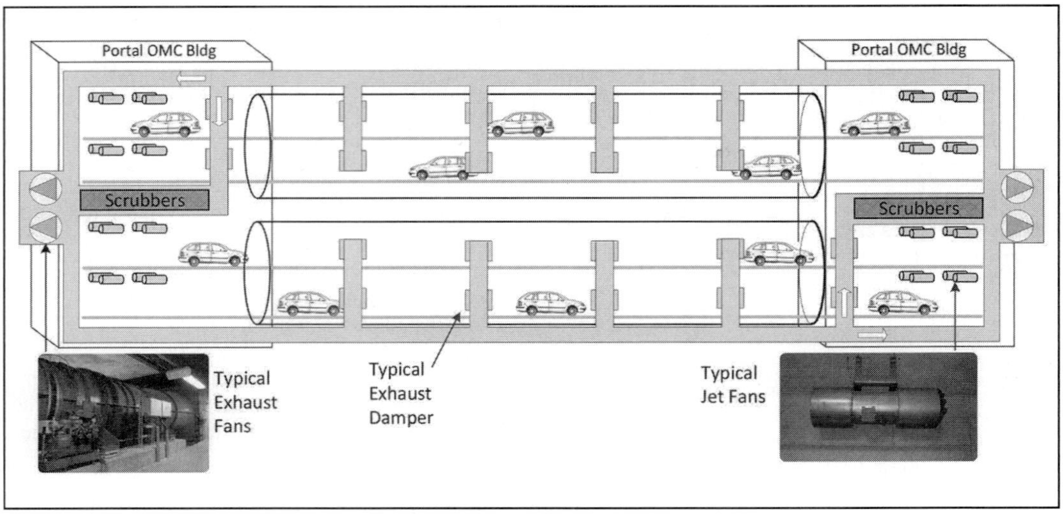


**Figure 5. Schematic representation of the ventilation system**

Figure 5 shows a schematic of the tunnel system. Both tunnels show a double deck and jet fans in the cut and cover section. Parallel to the traffic area, the exhaust duct is connected via dampers to the exhaust fans at the portals. The by-pass for air scrubbing is as indicated.

Besides eliminating the impact of intermediate vent shafts in the neighborhoods, this innovative ventilation concept also reduces the complexity and cost of the underground installation. The exhaust duct is efficiently incorporated in the lateral spaces of tunnel cross sections which are not useful for roadway area. Adequate space for the jet fans is accommodated by locating them outside of the bores in the cut and cover areas.

### Ventilation Objectives

During normal operation with free flowing traffic, the tunnel is self-ventilated by the "piston effect" of the traffic. Under heavier traffic conditions, ventilation would be mechanically assisted to keep the air quality in the tunnel within required limits for opacity, carbon monoxide and nitrogen oxides. A separate environmental investigation is being performed for the ventilation shaft exhaust and dispersion to confirm that outside air quality remains within acceptable limits. Used tunnel air is bypassed into the exhaust scrubbers near the portals.

In case of a fire, the ventilation system will provide safe egress in the enclosed and pressurized walkways with emergency exits spaced every 183 m (600 ft) or less. Local smoke extraction will happen via controllable dampers opened adjacent to the fire location. Longitudinal flow will be controlled with jet fans to maintain smoke-free zones upstream of the incident location.

### Elements of the Ventilation System

As shown in Figure 6, an exhaust duct is located adjacent to the traffic lanes over the entire tunnel length. The exhaust duct is connected to both decks via dampers every 91.4–106.7 m (300–350 ft). These dampers can be opened or closed as needed to extract smoke close to a fire location.

Under normal modes, ventilation will take advantage of the piston effect of traffic but boosted by fans as needed.

The OMC buildings are provided in the cut and cover sections near both portals located in-between the two tunnel tubes. They are equipped with two exhaust fans which allow extraction of smoke from both tunnels.

The exhaust fans are designed for a temperature resistance of 399°C (750°F) over two hours. Smoke will be extracted by operating the exhaust fans at both buildings in combination with two open dampers downstream of the incident location.

Jet fans are located in the cut and cover sections as shown in Figure 7. They allow control of the longitudinal flow velocity during normal and emergency operation.

Under normal operation, the tunnel air will be "scrubbed" for air quality reasons, thereby reducing emissions at the ventilation shafts (Figure 8). The current design considers filters capable of greatly reducing particulate matter. The filters are capable of removing 90–95% of $PM_{10}$ particles (10-µm-wide) and 80–85% of $PM_{2.5}$ particles (2.5-µm-wide). At the present time there are no technologies available,

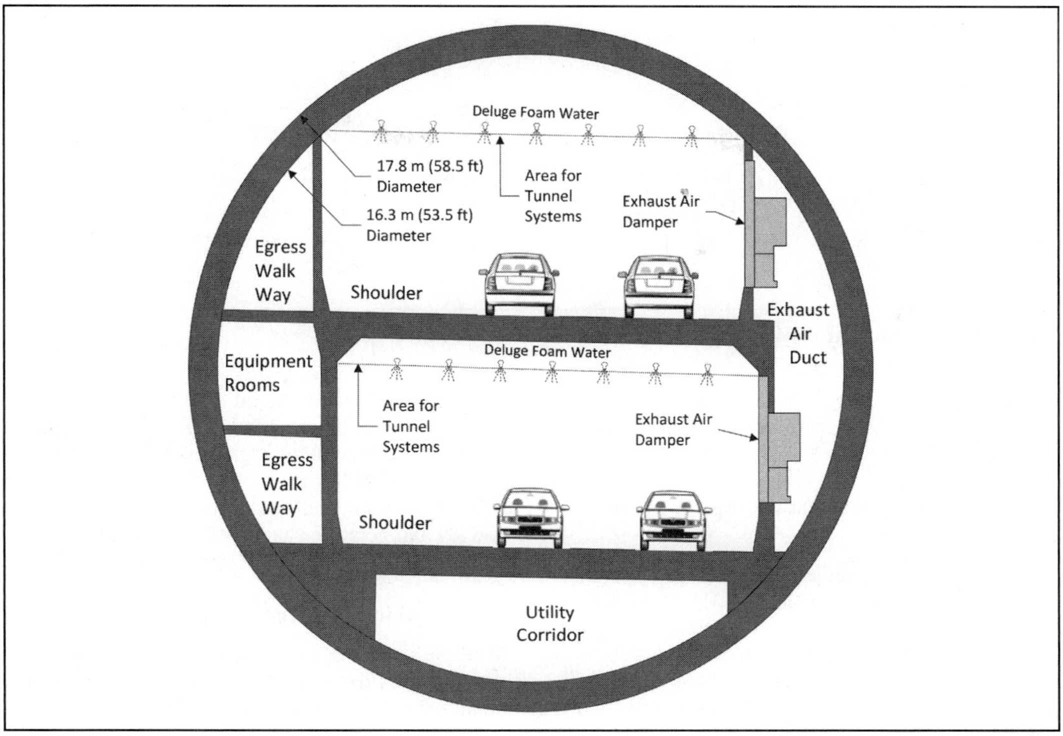


**Figure 6. Tunnel bore cross section**

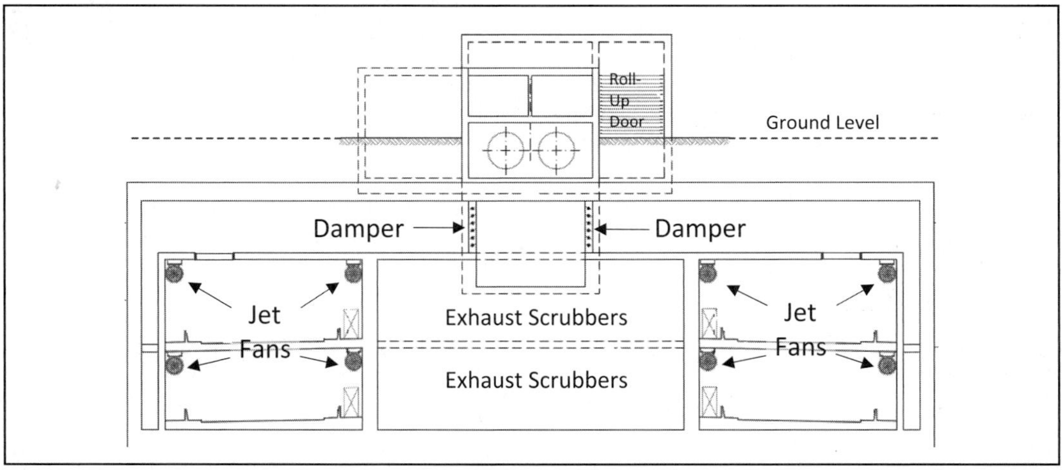


**Figure 7. Cross section of cut and cover of the North Portal OMC Building**

such as catalytic converters, able to filter gaseous emissions such as oxides of nitrogen and hydrocarbons from large volumes of exhaust.

Each OMC building extends from the tunnel level to above ground. The final exhaust air locations have not been determined yet. One possibility is to extend the exhaust duct from the south portal area to the I-10/SR-710 interchange. For the north portal one concept is to incorporate the vent shaft into the architecture of the OMC building.

During an emergency event, preventing smoke from entering the emergency walkway area is of paramount importance. To keep the walkway smoke-free, supply fans, located in the OMC buildings, will

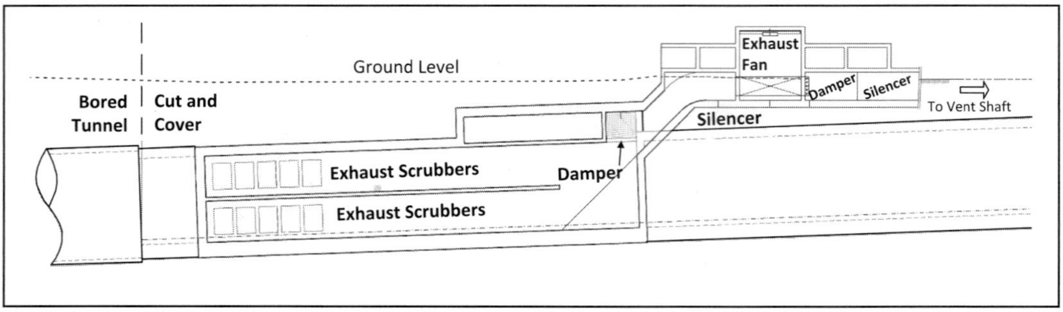


**Figure 8. Longitudinal section of the North Portal OMC Building**

create positive pressure in the walkways. This excess pressure causes fresh air to flow from the walkway into the traffic area.

### Preliminary Technical Data of the Ventilation System

The fresh air requirement for normal operation was determined using EMFAC2011, California's model for estimating on-road vehicle emissions. The required amount of fresh air is about 570 $m^3/s$ (1.2 million cfm) for the southbound (downhill) tunnel and about 650 $m^3/s$ (1.4 million cfm) for the northbound (uphill) tunnel.

For emergency operation the exhaust fans are designed with rotor diameters of approximately 3.7 m (12 ft) and approximately 1500 hp. The remotely controlled smoke exhaust dampers are located every 91.4–106.7 m (300–350 ft) and their cross section will be approximately 11.1 $m^2$ (120 $ft^2$). Each deck is equipped with approximately 20 jet fans in the cut and cover section with a motor power of approximately 120 hp each. The supply fans for ventilating the walkways are located in the portal OMC building. Their design flow rate is approximately 30.7 $m^3/s$ (65,000 cfm).

## MAJOR ELEMENTS OF THE FIRE/LIFE SAFETY SYSTEMS

### Fire Detection System

Fire detection systems are provided to detect a fire and its location. Several independent fire detection systems are provided.

- Linear heat detection
- Optical detectors
- Video image detection (CCTV Analytics)

Most tunnels around the world have just one or two of these systems. These three systems provide high redundancy thereby increasing safety. A study was performed to predict the expected performance for each of the detection systems using data from existing road tunnels. Based on this real-world data, it is estimated that the combined detection systems will be capable of detecting 80% of fires within 30 s and 100% of fires within 65 s. Small fires are more challenging to detect because of the limited smoke and low temperature change within the first few minutes. Larger fires will be detected within seconds.

### Positive Alarm Sequence

Multiple detection systems must be combined in a way that provides situational awareness clearly to the operators who will make the fire response decisions. To prevent false alarms from triggering a response, the system will be configured with a *positive alarm sequence* as prescribed by NFPA 72, 6.8.1.3. The operator will be able to quickly confirm actual incidents and precise roadway location using the CCTV system. If a false alarm is determined, the automated response can be suspended and the system reset by the operator.

### Fixed Fire Fighting System (FFFS)

Several coordination meetings with the LA Fire and Police Departments were conducted to determine the design relevant heat release rates and fire growth curves. Fire scenarios included the possibility of a fire spreading from one vehicle to another. Modeling shows that an FFFS would be able to contain the fire growth curve below 5 MW. If the FFFS was unavailable, the ventilation system is designed for a 100 MW dry fire.

Consequently, the modeling and analyses lead to an FFFS designed as a deluge foam water system per NFPA 16, the standard for foam-water sprinkler systems. Foam concentrate and water would be pumped separately from the north OMC building throughout the tunnels. The foam concentrate and water would be mixed proportionally in each of 127 zones per roadway. In addition, a standpipe and hose system will be provided for supplying fire water for up to one hour per the requirements of NFPA 14 and NFPA 502.

The combination of the FFFS, 3 independent fire detection systems and a ventilation system designed for 100 MW will allow compressed natural gas (CNG) buses in the tunnels.

### Co-Location of First Responders

In addition to the normal tunnel control operations centers at the OMC buildings, accommodations for 24-hour staffing of first responders at both portals are included. These crews with foam fire fighting vehicles ready at the portals will reduce response times to a minimum. Except for the Eisenhower Johnson Memorial Tunnel in Colorado, no other tunnels in North America have been identified with co-located fire response crews.

### Tunnel Broadband Wireless Network

Until recently, first responders at the scene have been limited to voice communications with the incident command center. With the advent of licensing in the 4.9 GHz spectrum for the exclusive use of public safety agencies, the design includes a broadband communications network in the tunnels. This will allow greater situational awareness for first responders with live video transmitted to and from the emergency vehicles. In the future this network could also provide the communications infrastructure necessary for remotely controlled robotic vehicles (i.e., firefighting drones).

## SUMMARY

The general design concept of the normal and emergency ventilation systems for the freeway tunnel alternative was described. The proposed ventilation system will avoid the need for intermediate exhaust shafts with a single stack at both portals. Air scrubbers will filter most of the emission particles. The feasibility of eliminating hydrocarbons and nitrogen oxides is under evaluation.

In the event of a fire, smoke will be pulled into an exhaust duct by dampers at the fire location and vented from either portal stack. Emergency walkways will be pressurized with fresh air for safe egress. The fire response and ventilation scenarios will be triggered by a triple redundant fire detection system as moderated by the operator through the positive alarm sequence. A fixed fire suppression system will contain the heat release rate below 5 MW. First responders and fire vehicles will be co-located at the portal OMC buildings.

## STUDY STATUS

The study team completed the Alternatives Analysis phase of the project in early 2013. The analysis recommended the four multi-modal alternatives mentioned herein as well as No-Build, which are currently being considered for the draft environmental document. Refinements as well as technical studies of all multimodal alternatives (LRT, BRT, TSM/TDM, freeway tunnel), with appropriate mitigation measures, continued into early 2014. It is expected that the draft environmental document will be circulated in 2014 and that the final environmental document and Record of Decision will be issued in 2015. The study team currently is in the process of developing details of the fire/life safety system and the ventilation design for the freeway tunnel alternative.

## ACKNOWLEDGMENTS

The authors would like to thank their respective employers for permission to publish this paper. Special thanks are also due to others on the team for their help with this paper including M. Crow (Metro), C. Govan (Metro) and F. Quon (Metro).

## REFERENCES

EMFAC2011, www.arb.ca.gov/msei/modeling.htm, January, 2014.

NFPA 14: Standard for the Installation of Standpipe and Hose Systems, 2013 Edition.

NFPA 16: Standard for the Installation of Foam-Water Sprinkler and Foam-Water Spray Systems, 2011 Edition.

NFPA 72: National Fire Alarm and Signaling Code, 2013 Edition.

NFPA 502: Standard for Road Tunnels, Bridges, and Other Limited Access Highways, 2014 Edition.

# Second Avenue Subway Project, New York: Design and Construction of the 96th Street Station

**Michael Trabold**
AECOM

**Richard Giffen**
Arup

**ABSTRACT:** Phase 1 of the Second Avenue Subway is a $4.4 billion project. The first phase will consist of 3 new stations, one refurbished station, and TBM tunnels connecting the stations. Of the three new stations, 2 are mined (72nd St. and 86th St.), and the 96th St. Station is a cut–and–cover station located on the upper east side of the borough of Manhattan in New York City.

This paper will discuss the design, planning, and construction of the 1500 foot long station at 96th Street. It will cover: slurry wall design and construction; temporary bracing and excavation, instrumentation and building monitoring; waterproofing; planning of entrance locations and ancillary buildings, and other aspects of building a cut-and-cover station in a densely populated urban environment.

## PROJECT OVERVIEW

The Second Avenue Subway will be the first major expansion of New York's subway system in over 50 years. To be constructed in four phases, it will run from 125th St. in Harlem to Hanover Square in the Financial District (see Figure 1). For most of its length, the alignment is directly under the Second Avenue roadway.

Phase 1 is under construction and consists of twin bored tunnels, 3 new stations at 96th St., 86th St. and 72nd St., and major modifications to an existing station at 63rd St. The cost of Phase 1 is $4.4 billion and it is scheduled to be completed by 2017.

The 96th St Station will be constructed under three separate contracts, for an approximate total cost of $750 million.

## SITE GEOLOGY

The 96th Street Station consists of open cut excavation from 92nd St. to 99th St. on the Upper East Side of Manhattan. See Figure 2. The geology to the south of 92nd Street allows for mining construction techniques, while the geology to the north leads to open cut construction along the alignment.

At 92nd Street bedrock is approximately 10 feet below street level and increases to 150 feet at 97th St. Groundwater is generally found 10 to 12 feet below ground level.

## DESIGN CONSIDERATIONS

### Geometry

The station geometry is defined by numerous codes and requirements with platform width and length being the driving factors. New York City Transit (NYCT), who maintains and operates the subway, requires 600 foot long platforms for all stations, and desired 30 foot wide platforms for the new stations. Each station requires additional space for electrical, power, signal, and communication rooms, as well as general station rooms including storage, station maintenance and office space. In addition, the 96th St. Station is a terminal station in Phase I, which requires a track crossover, additional rooms and facilities for crew quarters and maintenance shops. When all the requirements are met, the 96th St. Station is approximately 1500 foot long, 60 foot wide and 70 foot deep.

## TYPICAL STATION CONSTRUCTION

### Main Station

The typical construction consisted of installation of support of excavation (SOE) walls along 2nd Ave. by either the secant pile method (where rock level was within the excavation elevations) or by the slurry wall method when the entire depth was in soft ground (the majority of the box length). See Figure 3.

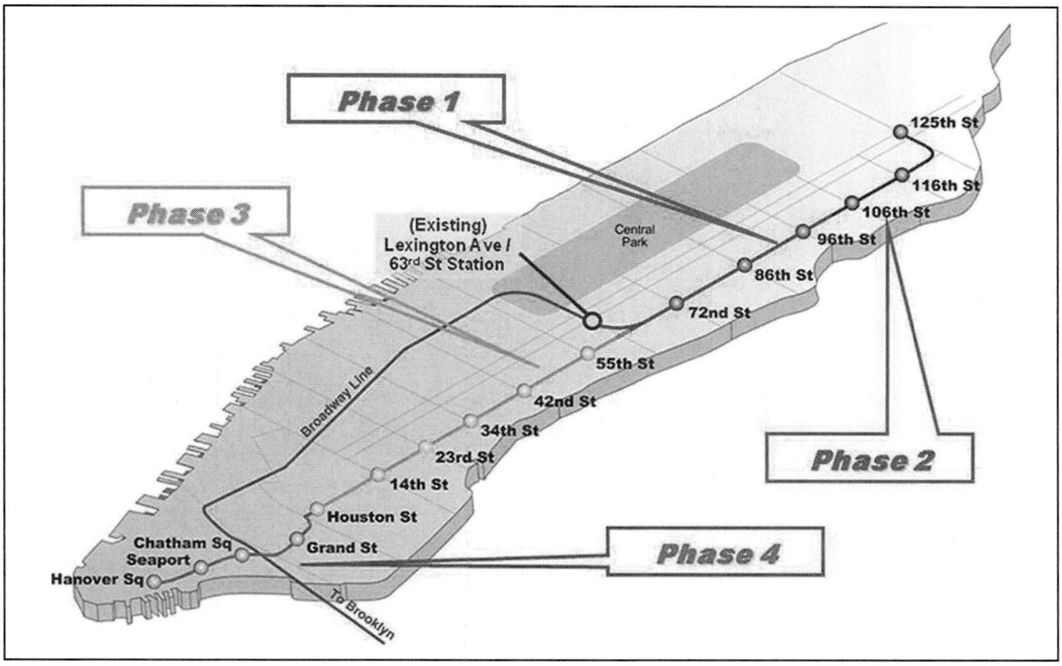


**Figure 1. Project map**

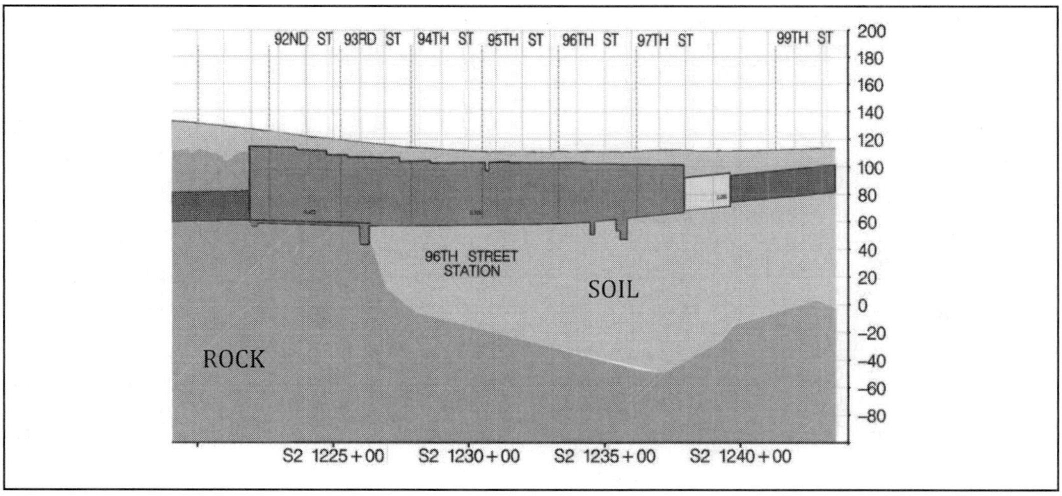


**Figure 2. 96th St. Station elevation and geological profile**

Prior to the installation of the SOE walls, existing utilities have to be relocated to avoid both the installation of the SOE walls, and the envelope of the new station construction. Most utilities are located beneath the roadway at varying depths between 6 to 10 feet. Typically the shallow utilities can remain in place while being temporarily supported during construction, but the remaining utilities have to be moved to outside the new station envelope into the space between the SOE walls and the adjacent buildings.

In the areas of high rock, the chosen method of SOE walls was temporary secant pile walls. In the areas of low rock level (typically north of 93rd St.) permanent slurry walls were used. Both polymer and bentonite were used as the slurry component, with polymer tried first and then replaced by bentonite with greater success.

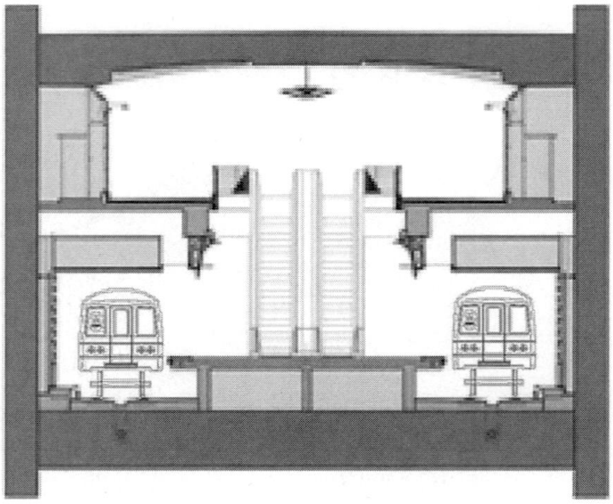

**Figure 3. Typical cross section—96th St. Station**

The use of temporary secant or permanent slurry SOE wall was primarily dependant on the depth of rock. The preferred form of final structure is a new permanent structure built inside a temporary SOE wall. This enables full continuity of waterproofing. However when the rock is significantly deeper than the invert slab a secant pile wall is not feasible and a thicker slurry wall would be required. Due to space constraints there is inadequate room for a temporary slurry wall and an internal permanent liner so these had to be combined, and the slurry wall was used as the permanent wall.

Once the SOE walls are completed, the excavation of soil and rock is undertaken in stages. The sequencing is critical and impacts the design of the SOE walls. The first 5 feet of soil is excavated, followed by the installation of the roadway and sidewalk decking beams. The decking beams are pre-loaded to act as the first level of struts that brace the SOE walls. As the excavation continues down towards the invert, wales and struts are installed as soil and rock is continually removed between the walls. Excavation is allowed to proceed below to the next level only after the level of bracing is installed and jacked into place. Once the excavation reaches its final elevation, typically 70 feet below street level, a concrete mud mat is placed as a working platform. The new station is then constructed 'bottom up' removing the temporary struts and wales as the new station elements are built as the construction proceeds towards the surface.

Longitudinally along 2nd Ave. there were also restrictions on the limits of excavation and unsupported SOE walls. The maximum length of open excavation was 100 feet. If the contractor desired to excavate at multiple locations, a minimum distance of 200 feet was required between unsupported areas. These restrictions were put in place to limit deflections of the SOE walls and minimize impact of adjacent properties.

In the areas of slurry walls, the invert, mezzanine beams and roof slab are connected to the slurry wall through couplers to provide continuity moment resistance.

In areas of secant piles, a cast in place structural box was constructed with keyed construction joints between the slabs and walls. See Figure 4.

### Entrances

To serve the predicted ridership, 96th Street will have 3 new entrances. Typical NYCT subway entrances are located on the sidewalk. These locations can lead to street congestion, as the entrances become 'choke' points for pedestrians at busy intersections. NYCT, in conjunction with New York City Department of Transportation, prefer to have the entrances located in adjacent buildings or plaza areas to ease street congestion.

At the corners of 94th St. and 96th St., two high rise buildings are set back from their property lines, leaving plaza areas that will be utilized for two of the three new entrances.

The third entrance will be located within a portion of an existing high rise building at the southwest corner of 94th St. and 2nd Ave. The construction of this entrance required substantial structural modifications to the building elements to accommodate

Figure 4. Waterproofing at south end of station

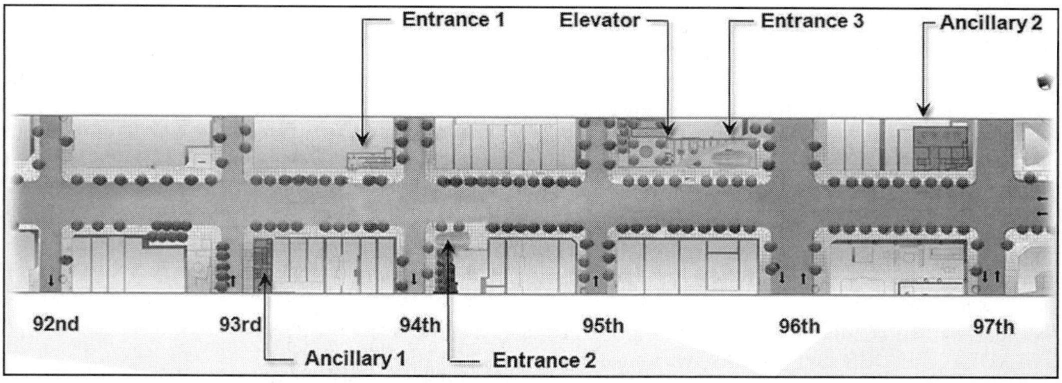


Figure 5. Plan view of 96th St. Station entrance and ancillary locations

the new entrance, which included new stairs and an escalator.

### Ancillary Buildings

The ancillary buildings house the ventilation equipment for the new station as well as associated electrical and mechanical equipment. They are located at both the southern and northern ends of the station. This is to avoid the public areas and to provide an efficient smoke management system as both ancillary buildings host fans to push or pull air during a smoke condition.

For 96th St Station two corner lots were purchased and demolished for the new ancillary buildings and they were both located adjacent to four story brick masonry buildings. See Figure 5.

The ancillary buildings are generally four story buildings, and extend below ground to the invert level of the station—approximately 60 feet below grade. The proposed SOE walls are secant pile walls and are designed to carry both temporary and permanent loads, unlike the secant pile walls for the main station, which are only designed for temporary loading. The secant pile wall was designed for the temporary loading conditions and for earth pressure in the permanent condition. A two foot reinforced concrete wall was constructed inside of the secant pile wall and designed to resist the hydrostatic pressures in the permanent condition and the vertical loading from the new ancillary building. The two walls are designed to act independently of each other.

### Load Cases

The 96th Street Station was designed for dead, superimposed dead and live loads including traffic and equipment loading, soil and rock loads, groundwater

**Figure 6. North end of station**

surcharge, building surcharge, temperature effects, earthquake and construction loading. A site specific earthquake analysis was undertaken for the station. Both balanced and unbalanced loading was analyzed.

The construction sequencing had an impact on the final design of the station. The 96th St station was designed as bottom up station, so the levels of bracing and sequencing of construction had to be taken into account in the final design of the slurry wall.

### Structural Analysis

The station was designed and analyzed for both the stresses locked in during construction as well as the final loading condition. The software programs STAADPro, PLAXIS and ETABS were predominantly used for the analysis. For the typical areas, 2-D analysis was used. At more complex areas, 3-D analysis was used. These areas included locations where the entrances and ancillary buildings junctioned with the main station.

The invert design was controlled by hydrostatic loads. The invert was approximately 70 feet below street level, and 60 feet below the water table, and was designed to span the 60 feet between the slurry walls. The typical slab was 6 feet deep.

To provide users with the best station experience it was decided to have both the platform and mezzanine areas column free. During the conceptual phase of the project, both steel and concrete forms of construction were investigated.

The large openings in the mezzanine for the new stair and escalators resulted in a complicated beam and slab arrangement with high loads at the connection to the slurry wall. To accurately model the mezzanine, it was analyzed in STAADPro with appropriate properties to represent the stiffness and flexibility of the mezzanine openings. See Figure 6.

The mezzanine beams that span around the openings were designed separately due to the large moments and axial loads experienced by these beam-columns. These beams had to also be designed for moment magnification due to the eccentricity of the loads.

The typical mezzanine beam was 30 inches wide and 42 inches deep, and the mezzanine slab was 8 inches thick.

The roof was designed to span the 60 feet between the slurry walls and to resist both the dead load from the soil above as well as the live traffic loading. The roof slab was typically 4 feet in depth.

The slurry walls were designed to resist both the temporary and permanent loading conditions. The assumed construction sequencing was indicated in the contract documents, as were the stiffness requirements of the temporary bracing. The design of the temporary bracing was by the Contractor.

The slurry walls were designed as vertically spanning elements, 4 feet thick, and to provide the necessary groundwater cut-off depths were at some locations 120 feet deep. The results of the analysis required a heavily reinforced wall. At some areas 3 layers of rebar were required, with #18 rebar used in some locations.

The typical width of a slurry wall panel was 20 feet, and the contractor installed the reinforcing steel in two 9 foot wide cages.

Horizontal reinforcing was primarily for shrinkage. The reinforcing for the wall—both flexural and shear—coupled with the need for tremie pipes to place the concrete and couplers to connect the wall reinforcing with the invert, mezzanine and roof slabs resulted in a crowded cages.

The static model contained the load combination requirements for maximum and minimum loading

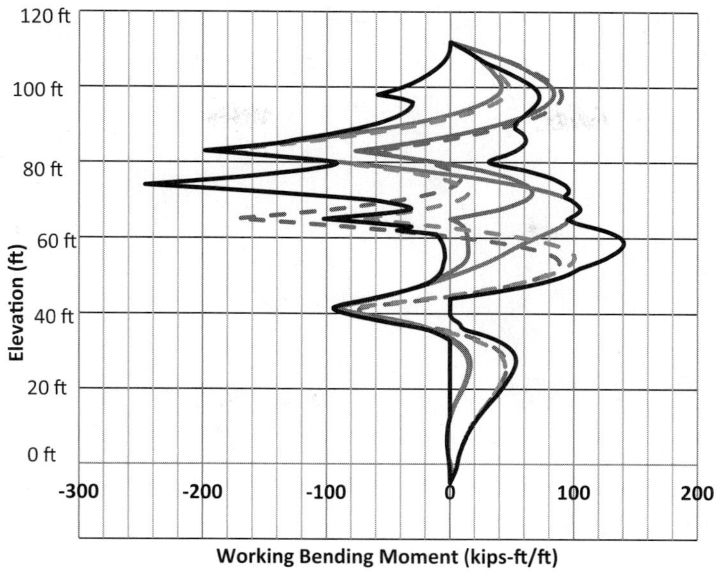


**Figure 7. Slurry wall moment envelopes**

and the unbalanced pressure models. Earthquake modeling was also undertaken on the static models.

Figure 7 shows the resultant moment force diagram for a typical slurry wall panel, containing both the static and soil-interaction results.

For complex areas, particularly at Entrances and Ancillary buildings, a 3D static model was used. At the northern ancillary, the columns were not founded on piles and so the roof, mezzanine and invert were affected by the loads on the roof and the uplift on the invert slab. These were analyzed in both 2D and 3D to assess the "pin cushion" effect of the columns on the slabs.

### Waterproofing

The decision to use slurry walls as permanent walls eliminated the ability to fully encapsulate the new station in waterproofing. In those areas, a waterproofing membrane was used for both the invert and roof slabs. To minimize the water leakage through the slurry walls, PVC waterstops were used between the panels and a low permeability concrete was specified. To further minimize water infiltration, post-grouting of the slurry wall joints with micro fine cement was performed at the areas of the roof and invert slabs.

To minimize the water through the intersection of the slurry wall and invert and roof slabs, hydrophilic water stops and re-injectable grout tubes were installed.

At the areas where the secant piles were used as the SOE, cast in place walls were constructed, the waterproofing membrane fully encapsulated the structure. In these areas a PVC membrane was used with water barriers to compartmentalize the areas and minimize the ability of any water leakages to migrate. The compartmentalized areas also had remedial grout tubes to allow for grouting in case any leaks arise.

## CONSTRUCTION CONSTRAINTS

Working in a dense urban environment makes a construction project of this magnitude difficult. With property at a premium, there is very little open land for laydown areas, work storage and other construction related activities needed to build a new station.

Since Second Avenue is a six lane main through fare for south-bound traffic in the borough of Manhattan, the NYC Department of Transportation mandated that four lanes of traffic remain open during both the morning and evening rush hours (Figure 8).

Since this was a residential area, NYC DOT also had restrictions on work hours during weekdays and weekends, and noise restrictions the contractor had to adhere to. The contractor was allowed to work week days for 7am until 10pm and from 10am to 7pm on weekends. The daytime weekday noise restrictions were 75dBA or Background (+5dBa) whichever is higher, and background +15dBA for impulsive sound, which is 2 seconds or less. Nighttime and weekends were lower and varied from 60dBA to 65dBA.

**Figure 8. Street view of work zone and travel lanes**

**Figure 9. Installation of slurry wall reinforcing cage**

## INSTRUMENTATION, BUILDING MONITORING, AND PROTECTION

The buildings along 2nd Ave between 93rd Street and 97th Street vary between 30 story modern high rise buildings built on pile or deep foundations and 4 to 5 story brick and timber buildings constructed on masonry rubble shallow foundation walls.

Extensive instrumentation was installed prior to any construction work commencing and continuous monitoring was undertaken during the work. To monitor the buildings, an Automated Motorized Total Station (AMTS) system was used for the project. An AMTS consists of fully automated monitoring units and optical survey prisms mounted on the buildings. The system allowed the contractor to remotely monitor the three components of movements of the buildings in real time during construction operations. Other instruments used to monitor the buildings included tilt sensors, inclinometers and seismographs.

Piezometers were used to monitor water levels both inside and outside of the mass excavation. Inclinometers were installed inside both the slurry and secant pile walls to monitor wall deflections. Ground deformation monitoring points were also installed to monitor any ground movements outside the SOE walls. See Figure 9.

Even with the building repairs, the existing buildings would undergo differential building settlements due to the construction activities. To limit these settlements, various protective measures were employed. Underpinning using mini piles installed from within the basement, and in areas where underpinning was not feasible jet grouting and compensation grouting was used.

### Building Movements

Building movements varied but typically did not exceed 1.5 inches during construction of the utility relocations, slurry walls and mass excavation for the new station. In the more critical areas adjacent to the ancillary structures, protection and strengthening measures limited the building movements to less than ½ inch.

## CONCLUSION

The design and construction of the 96th St. Station presented many challenges. Geological conditions, mainly a sharply drop off in the rock elevation at the south end of the station; space constraints, fitting the desired station layout in the area available between the adjacent building property lines; dense urban environment, working in close proximity to adjacent 100 year old buildings along the right-of-way; construction constraints, working in a residential neighborhood. These challenges led the design team to come up with viable solutions. Some of the solutions were two types of SOE walls, use of slurry wall as the permanent structure, protective measures for the older buildings and managing traffic and work procedures to be able to construct a new subway station in one of the busiest residential areas in New York City.

# Cost-Effectiveness Analysis for Transit Tunnel Security Based on Blast Analysis

**Xiaomin You and Joe O'Carroll**
Parsons Brinckerhoff

**ABSTRACT:** Because of the accessibility and potential impacts on human lives and economic activity, transit tunnels can be attractive terrorist targets. This paper discusses the necessity of blast protective design and threat level for a transit tunnel. 3-dimensional finite element models were developed based on typical transit tunnel geometry to investigate the behavior of structures under blast loading. Key design factors, such as the thickness of lining, additional steel bars, use of steel fibers, and other strengthening measures were considered and their effectiveness to minimize the damage was investigated. Conventional approaches to blast protection (i.e., increasing the lining thickness and additional rebar) were compared with some secondary protection measures. The paper concludes with a simple cost-effectiveness analysis.

## INTRODUCTION

According to The Blue Ribbon Panel on Bridge and Security (2003), there are more than 200 transit tunnels in the United States. Many of these tunnels are running through urban areas with high population densities carrying millions of passengers daily. Because of the easy accessibility from open environments and potential impacts on human lives and economic activity, transit tunnels can be attractive terrorist targets. Furthermore, a confined underground space can make a terrorist attack particularly catastrophic due to confined blast events and potential ground collapse. Table 1 summarizes various subway attacks in recent years, and shows the consequence measured in loss of lives and injuries. Most of the terrorist attacks hit the transport system during rush hour. Catastrophic damage to infrastructure is also a matter of concern, which could result in operation disruption and significant economic losses. Owners and operators should be aware of the benefit of a protective design in improving the safety and security of infrastructure facilities.

Transit tunnels are viewed as high-risk, high-damage potential targets. In order to assess the risk and vulnerability and to analyze the blast impacts on tunnel structures, Munfakh (2008) proposed an approach called TARIF, which has been applied to several tunnels. TARIF includes five steps: (1) identify threat; (2) evaluate assets; (3) calculate risk; (4) analyze impact and (5) provide fix. Choi (2009) developed a similar protective design steps as show in Figure 1. The proposed protection measures can be evaluated through a cost-benefit analysis to determine the preferred and effective solutions.

This paper discusses blast protective design and threat levels for a transit tunnel. 3-dimensional finite element models were developed based on typical transit tunnel geometry to investigate the behavior of structures under the assumed threat level. Key design factors, such as the thickness of lining, additional steel bars, use of steel fibers, and other strengthening measures were considered and their effectiveness to minimize the damage was investigated. Conventional approaches to blast protection (i.e., increasing the lining thickness and additional rebar) were compared with some secondary protection measures. It concludes with a simple cost-benefit analysis.

It should be noted that this paper did not discuss the progressive failure subsequent to the blast loading. According to Sung et al. (2010), progressive failure may occur due to weaken structure and ground and thus post-blast analysis is an important step to evaluate further damage post blast. The post-blast tunnel behavior shall be considered in the future.

## THREAT LEVELS

The explosive charge weight and the stand-off are two typical parameters used to define a blast threat. The charge weight is measured in equivalent pounds of TNT and the stand-off is the distance measured from the center of gravity of the charge to the bearing surface of the target structure.

According to the Federal Emergency Management Agency (FEMA) (2005), for design purposes, a briefcase, backpack or suitcase bomb can hold approximately 50 pounds, and a pipe bomb is generally in the range of 5 pounds of TNT or equivalent. The bombing incidents listed in Table 1

**Table 1. Summary of terrorist attacks on transit systems**

| Date | City | Attack | Consequence |
|---|---|---|---|
| December 29, 2013 | Volgograd, Russia | A suicide bomb detonated with a force equivalent of 22 lb. TNT at a train station. | 18 people were killed and over 44 injured. |
| March 29, 2010 | Moscow, Russia | A double suicide bombing at two stations of the Moscow Metro during the morning rush hour. | At least 40 people were killed, and over 100 injured. |
| July 7, 2005 | London, UK | Four explosions (3 on the London Underground and 1 on a bus) on the public transport network during the morning rush hour. | 52 civilians were killed and over 700 more were injured. |
| March 11, 2004 | Madrid, Spain | Ten near-simultaneous explosions occur on Madrid's commuter train system during the morning rush hour. | The explosions killed 191 people and wounded 1,800. |
| February 6, 2004 | Moscow, Russia | Suicide bomber detonates bomb near a subway station during the morning rush hour. | 41 people were killed and approximately 120 people were injured. |

Note: Data from Wikipedia, http://en.wikipedia.org/wiki/List_of_terrorist_incidents_involving_railway_systems.

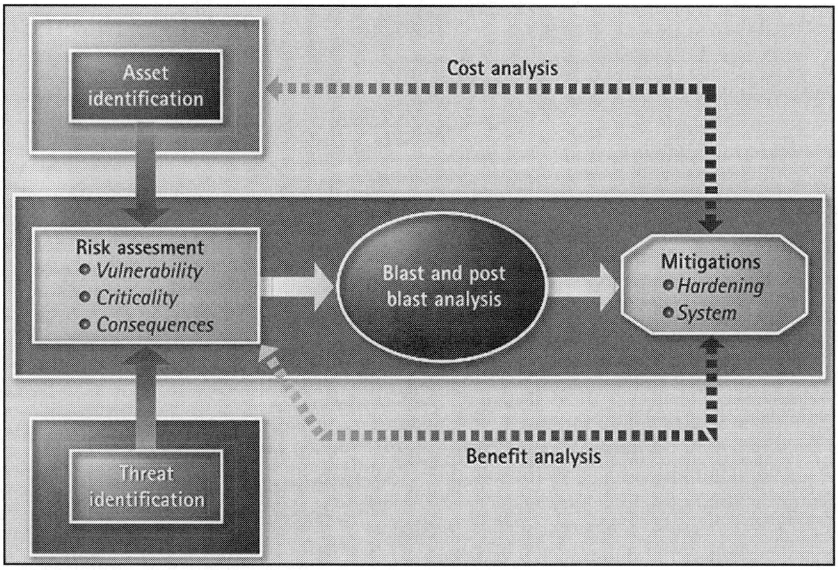


**Figure 1. Protective design steps for tunnel security (Choi, 2009)**

and recent terrorist attacks indicated an increasing trend of using backpack bombs. Considering access restrictions to a transit tunnel, a backpack bomb was selected as the predominant mode of explosive attack for this study. In order to evaluate the tunnel structures in a reasonable worse scenario, a single backpack bomb with a conservative estimate of charge weight of 100 pounds of TNT or equivalent is assumed.

This study did not consider multiple explosions of the same total charge weight, nor did it consider explosive confinement (casing, containers, etc. Sung (2009) recommended that, in spite of the fact that explosive confinement might have impacts on the consequence, the impacts of confinement are relatively less important in determining the structural response subject to the blast loadings. Therefore, for the purpose of this study, the impact of confinement was neglected and the charge was assumed to be spherical in shape.

## DAMAGE UNDER BLAST LOADING

The structural performance of tunnels under blast loading could be impacted by various factors. Some critical factors include construction type of tunnel, ground condition, tunnel depth, and properties of tunnel lining. TCRP/NCHRP (2006) developed relative severity of tunnel damage based on some of the critical factors, such as construction method, ground type, support type and tunnel depth, etc. The severity

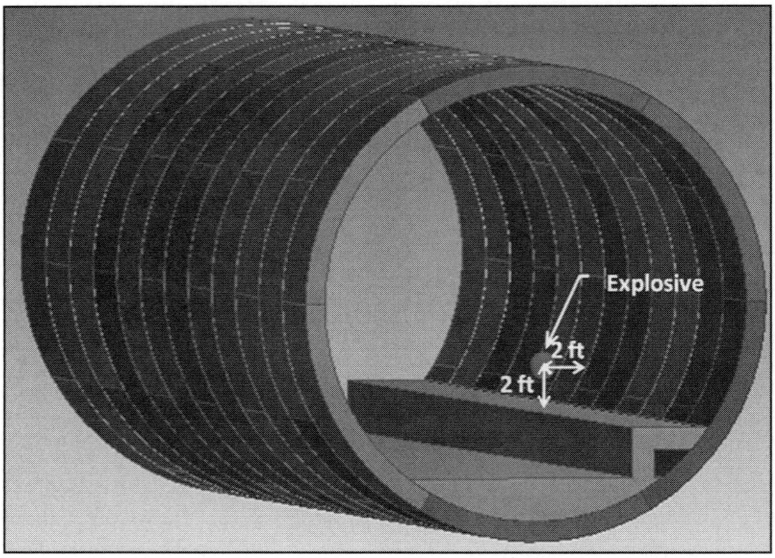


**Figure 2. Cross section in 3D model**

ratings were prepared in a qualitative manner based on recent tunnel security project experience and experts' opinions.

Proper quantitative assessment of the impact of explosions on underground facilities relies on sophisticated analytical simulations and the application of numerical analyses that take into account several factors representing the explosion, the structure and the ground. Sung (2009) pointed out that when an explosion occurs in a tunnel structure, the peak pressure and the impulse associated with the shock wave are high and amplified by the confining tunnel structure. Because of the combined effects of the explosion and the amplification of the blast pressures, the distribution of the shock loads on any one surface is non-uniform and complicated. This study utilized the quantitative analysis method, i.e., three-dimensional coupled Euler-Lagrange nonlinear transient dynamic analysis, to analyze the structural performance of tunnel lining under blast loadings.

## BLAST ANALYSIS

### Computer Program

The three-dimensional coupled Euler-Lagrange nonlinear finite element blast analysis was conducted using the commercial computer program ANSYS AUTODYN. It is an explicit numerical tool developed and verified for a wide range of impact penetration and blast problems. It simulates nonlinear dynamics, large strains and deformations, fluid-structure interactions, explosions, shock and blast waves, impact and penetration, contacts and interactions between structures.

### Geometry

In this study numerical analysis was performed based on an assumption that the tunnel was constructed with precast concrete segmental lining. The tunnel was excavated by a Tunnel Boring Machine (TBM) at shallow depth in soft ground. As most transit tunnels have a similar geometry, the following assumptions are made in the 3D numerical model shown in Figure 2 for the base case:

- Transit tunnel with an internal diameter of 20 feet.
- For purpose of simplicity, one ring is composed of 6 identical reinforced concrete segments with a width of 5 feet per ring.
- Precast segmental lining: a thick with of 11-inch with #5 longitudinal reinforcing bars at a spacing of 6-inch, i.e., 1 percent reinforcement ratio, and minimum transverse reinforcement.
- Concrete compressive strength of 6,000 psi for tunnel lining.
- Concrete compressive strength of 3,500 psi for interior structures.
- Yield strength of 60 ksi for steel rebar and 50 ksi for bolts and dowels.
- Two neighboring segments connected by two radial bolts.
- Two neighboring rings connected by 12 equally distributed steel dowels.
- Each segmental tunnel has a concrete invert and an emergency sidewalk resting on the segment panels.

The explosive was placed close to the tunnel wall. Generally, the blast pressure drops exponentially as stand-off distance increases. The greater the stand-off distance is, the less the explosive threat to the structure. Therefore, a small standoff distance, 2 ft., was chosen to simulate an explosion occurring close to the tunnel wall.

The base case was then served as the baseline to investigate the effect of protection measures. Some key design factors were varied as in protection measures, for example, increase of lining thickness, additional amount of steel rebar, use of steel fibers, and other strengthening measures. Details about the protection measures are presented in later sections.

## Material Models

### *Jones-Wilkins-Lee Equation of State (EOS) for High Explosives*

High explosives are chemical substances which, when subjected to suitable stimuli, react chemically to give a very rapid (order of microseconds) release of energy. A detonation wave is assumed to be a discontinuity that propagates through the unreacted material. Instantaneously, the detonation wave liberates energy and transforms the explosive into detonating products.

The expansion of high explosive products is modeled using the Jones-Wilkins -Lee (JWL) equation of state. The JWL equation of state describes the detonation product expansion down to a pressure of 1 kbar for high-energy explosive materials and has been proposed by JWL.

### *Concrete*

For the steel reinforced concrete, this analysis contained herein modeled concrete and steel rebar separately.

The extent of damage for concrete materials under explosion loads depends on the strain-rate effect (dynamic response). To incorporate the dynamic response of concrete materials, the current analysis uses a strain-rate-dependent concrete model—the RHT concrete model, developed by Riedal et al. (1999, 2009). It is particularly useful for modeling the dynamic loading of concrete.

### *Steel*

To model steel rebar, steel plate, steel bolts and dowels, the STEEL 4340 model from AUTODYN material library was used. This material model uses the Johnson-Cook Strength model, which is suited to representing the strength behavior of materials, typically metals, subjected to large strains, high strain rates, and high temperatures. Yield strength for steel rebar and steel plate was assumed as 60 ksi and steel bolts and dowels 50 ksi.

### *Steel Fiber Reinforced Concrete (SFRC)*

The above RHT concrete model is developed for plain concrete. For SFRC, It is reasonable to use RHT with some parameters revised based on its properties. To take into account the impact of reinforcement steel fiber, SFRC was modeled using a smeared model in which the steel fibers is assumed to be uniformly distributed over the concrete elements and the parameters in RHT concrete model were revised based on the selected steel fiber dosage.

Based on the study conducted by Smith (2011), the typical dosages for precast concrete lining are between 50 lb/cy and 100 lb/cy. This paper assumed a dosage of steel fiber 80 lb/cy for both tunnel lining and interior structures.

Smith (2011) found that the compressive strength of SFRC at a typical dosage can be up to 20% higher than for the same concrete mix without fibers. Consequently, the compressive strengths for SFRC used in this study were increased to 7.2 ksi for tunnel lining and 4.2 ksi for interior structures, respectively.

Regarding the tensile strength, Smith (2011) indicated a ratio between SFRC and plain concrete can be calculated by

$$\text{Ratio} = 0.002x + 1$$

where $x$ is the fiber dosage in lb/cy. For 80 lb/cy of steel fiber, the tensile strength of SFRC can be increased by 16% compared to plain concrete.

Steel fibers can significantly increase the shear strength of concrete members. Smith (2011) indicated that the shear capacity can be increased by 20% to 30% at a dosage of 66 lb/cy, and 35% to 65% with 132 lb/cy steel fibers. Within this paper, it was assumed the shear strength increased by 30%.

Another major benefit of adding steel fibers to concrete is that they can improve the ductility and therefore SFRC can reach a higher failure strain than plain concrete. Gebbeken and Greulich (2003) proposed following function to estimate the failure strain in correspondence to the fiber volume fraction, $V_f$:

$$\text{Failure strain} = 0.0091e^{0.4002V_f}$$

A fiber dosage of 80 lb/cy is equivalent to a fiber volume fraction of 0.6%. Thus the failure strain for the SFRC is obtained as 1.16%.

### *Geomaterial Models*

Within the ANSYS AUTODYN program, the Drucker-Prager Strength Linear model can be used to represent the behavior of soils and rocks. The

**Table 2. Soil parameters used in the blast analysis**

| Density (pcf) | Bulk Modulus (psi) | Shear Modulus (psi) | Cohesion (psi) | Friction Angle (degree) |
|---|---|---|---|---|
| 130 | 2.22E+04 | 7.41E+03 | 0 | 35 |

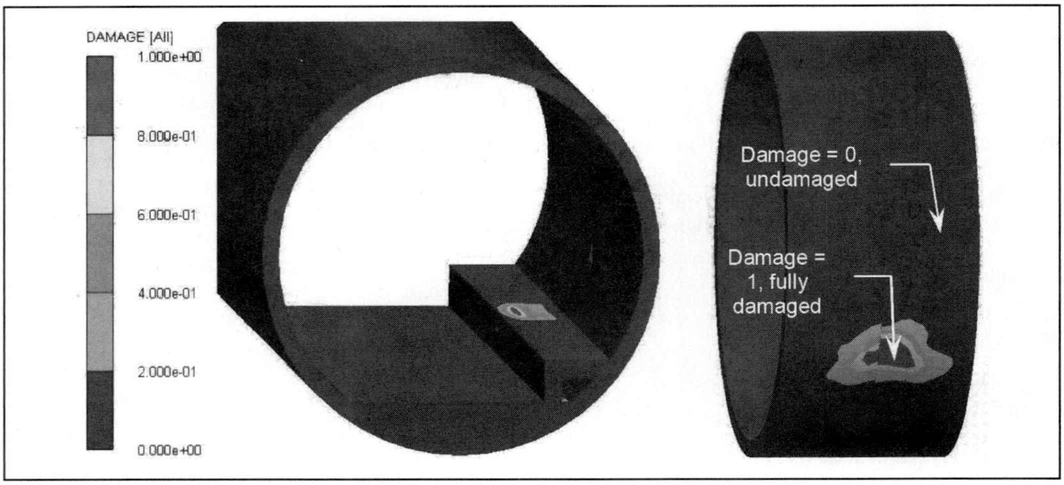


**Figure 3. Blast-induced structural damage in base case**

elastic properties for the soils are of paramount importance in blast analyses; specifically required were bulk modulus, K, and shear modulus G. Table 2 summarizes the assumed parameters for the soft soil used in this analysis.

### *Aluminum Foam Panel*

Usually, aluminum foam panel has a sandwich structure composed of two metallic cover sheets and a foamed aluminum core. Due to its advantages in high impact energy absorption, vibration reduction and light weight, aluminum foam panels are oftentimes applied to the bearing faces of protected structures to mitigate blast effects. As the foamed aluminum core is the major material to absorb impact energy, the metallic cover sheets are not included in the 3D model.

Boey (2009) investigated the characteristics of aluminum foam and developed a P-α compaction model to describe their dynamic compaction during an impact event. This study assumed the same aluminum foam and utilized the P-α compaction model and parameters as developed by Boey (2009).

### Blast-Induced Structural Damage

Sung and Munfakh (2009) described the process of tunnel failure under blast loadings. Tunnel failure is initiated from an overstress in the lining caused by an explosion. This may lead to failure of the lining if the strength of the lining material is less than the applied stress. The failure of the lining may be restricted to be a local failure such as spalling or local breach. When the tunnel lining is damaged locally or globally, failure of surrounding ground (collapse) and/or inundation with water (flooding) may follow.

Figure 3 presents the blast-induced structural damage obtained from the base case, in which the 11-inch thick steel reinforced concrete tunnel lining, 1% circumferential reinforcement ratio, was subject to blast loading due to explosion of 100 lb. TNT at a stand-off distance of 2 ft.

The analysis contained herein used the parameter Damage to measure the damage extent. Damage = 0 represents undamaged; while Damage = 1 means fully damaged states as shown in Figure 3. The analysis results indicated that the blast-induced damage in the base case was localized. The blast-induced damage area in the lining was about 2.6 square feet.

## PROTECTION MEASURES

In order to improve the structural and operational security and safety of transit tunnels, some protection measures can be considered if severe damage is expected.

A thicker lining or additional steel rebar are two traditional protection measures. They are relative easy for design and construction and thus being considered before addition of other protective materials. In addition, steel fiber reinforced concrete is considered as a protection measure in this study. Steel

**Table 3. Protection measures**

| | Lining Thickness | Steel Rebar | Steel Fiber | Steel Plate | Aluminum Foam Panel |
|---|---|---|---|---|---|
| Base Case | 11" | Rebar 1* | — | — | — |
| Measure 1: 12" thick lining | 12" | Rebar 1 | — | — | — |
| Measure 2: Additional rebar | 11" | Rebar 2† | — | — | — |
| Measure 3: Use of SFRC | 11" | — | 80 lb/cy | — | — |
| Measure 4: Interior steel plate | 11" | Rebar 1 | — | 1" | — |
| Measure 5: Interior aluminum foam panel | 11" | Rebar 1 | — | — | 4" |
| Measure 6: 15" thick lining+ additional rebar | 15" | Rebar 3‡ | — | — | — |

* In the cases with Rebar 1, the concrete lining was reinforced by #5 circumferential reinforcement at spacing of 6" and #3 transverse reinforcement at spacing of 12." The ratio for circumferential reinforcement is about 1%.
† In the cases with Rebar 2, the concrete lining was reinforced by #5 circumferential reinforcement at spacing of 3" and #3 transverse reinforcement at spacing of 6." The ratio for circumferential reinforcement is about 2%.
‡ In the case with Rebar 3, the 15" lining was reinforced by #6 circumferential reinforcement at a spacing of 3" and #3 transverse reinforcement at a spacing of 3." The ratio for circumferential reinforcement is about 2%.

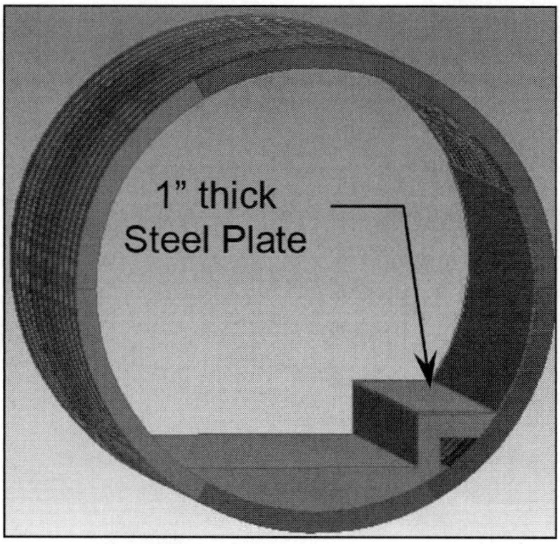


**Figure 4. Interior steel plate**

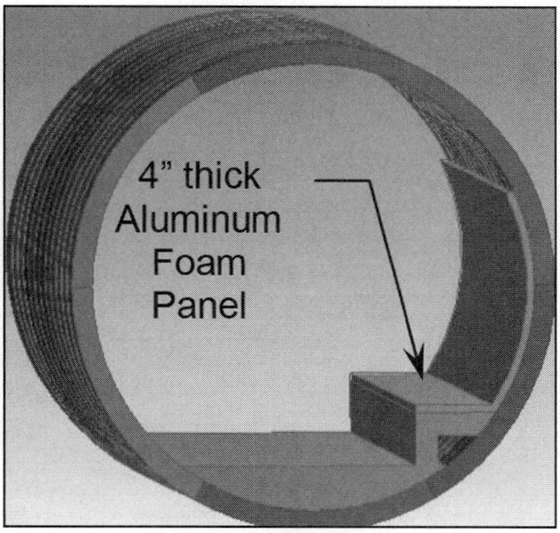


**Figure 5. Aluminum foam panel**

plate and aluminum foam panel are also alternate protection measures but they could be more costly and require extra labor and time for installation and maintenance. Table 3 summarized the parameters for the above measures. A series of numerical analysis were performed to quantify the effectiveness of the proposed measures.

Several other protection materials were introduced by Sung (2012), such as Micro-meshed reinforced concrete, Fiber reinforced polymers, Polyurea, and BlastWrap, which are not discussed in this paper and can be studied in a future cost-effectiveness analysis.

For the purpose of cost comparison, the steel reinforced concrete lining cost in the base case is assumed as $1600 per foot based on historic data from similar projects. This study only considered the initial costs for the proposed measures. The durability and life-cycle cost was not included in following cost-effective analysis. See Figures 4 and 5.

### Measure 1: 12-inch Thick Lining

In general, a thicker tunnel lining tends to perform better under extreme loading events. In the first scenario, the lining thickness was increased from 11 inch to 12 inch. By increasing the tunnel thickness by 1 inch, the volume of concrete lining increase by 10%. Therefore the cost for the precast lining could increase approximately by 5% to 10% compared to the base case.

Compared with the base case, the damage in this case (Figure 6) is 1.7 square feet, reduced by 35%.

### Measure 2: Additional Rebar

Another typical protection measure is to have additional steel reinforcing bars by reducing the spacing

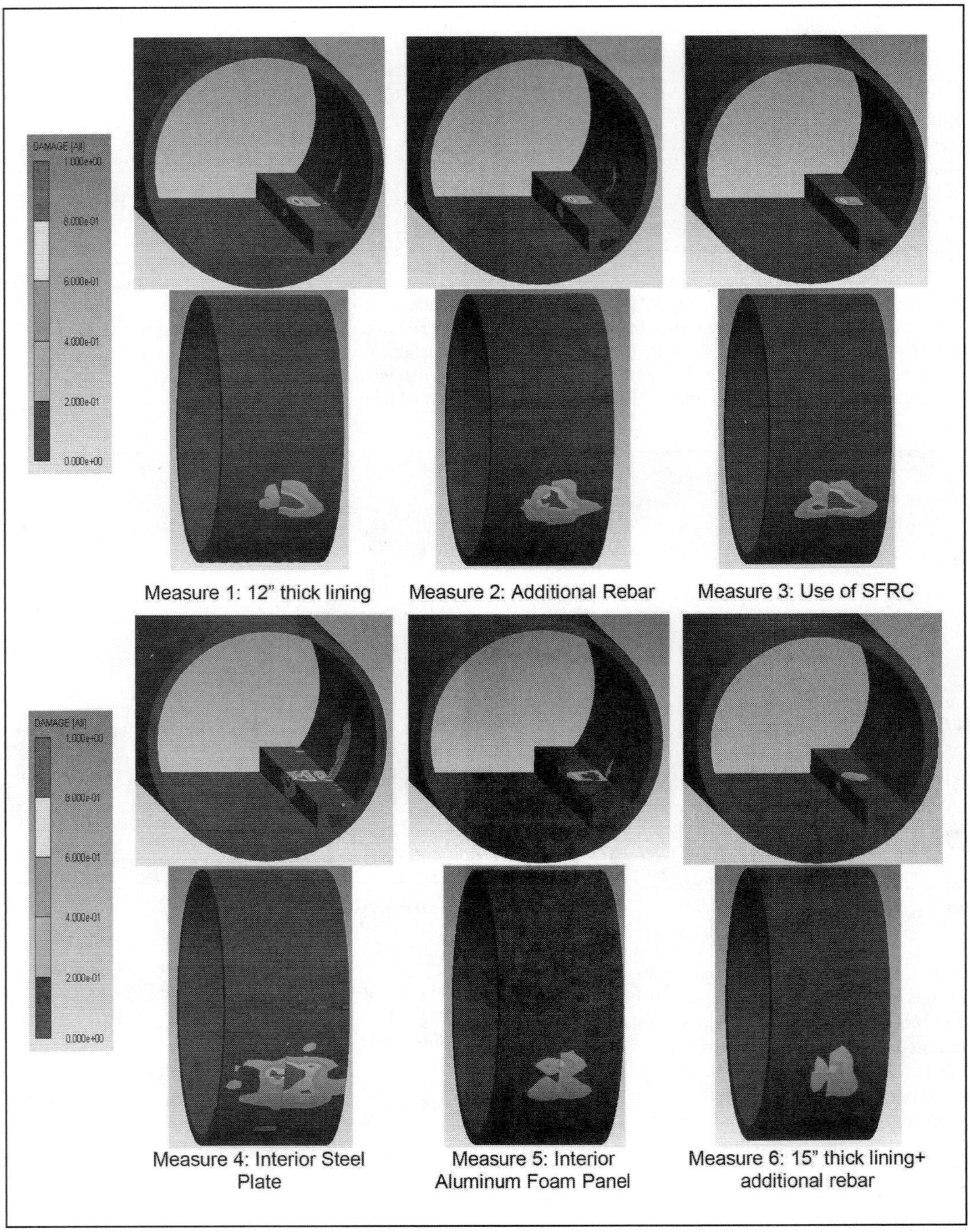


**Figure 6. Blast-induced structural damage**

in the concrete lining. In this analysis scenario, compared to the base case the spacing between bars was reduced by half and thus the total number of bars was doubled. By doing that, the cost for the precast lining could increase approximately by 20% to 40% based on experience from similar projects.

The damage at the tunnel lining in this scenario (Figure 6) is obtained at 1.4 square feet, reduced by 45% compared to the base case.

**Measure 3: Use of SFRC**

This scenario considered utilizing steel fiber reinforcement concrete with dosage of 80 lb/cy. Based on the cost data from tunnels using SFRC, there could be a saving around 15% to 20% compared to the steel reinforced concrete.

Figure 6 indicated that the steel fiber reinforced concrete could have higher blast resistant capacity when compared to steel reinforced concrete, as steel fibers could produce tougher concrete that hold together fragments after loading and thus the damage area was reduced to 2.3 square feet, reduced by 14% compared to the base case.

**Measure 4: Interior Steel Plate**

This analysis considers a worst case, in which an explosion is simulated occurring close to the tunnel wall and could lead to severe structural damage. Therefore, a 1-inch thick and 13.5-ft wide interior steel plate was bonded along the sidewalk and tunnel side wall. Additional material cost is estimated approximately at $ 500 to 550* per foot, which is about 30% to 35% more costly than the stee reinforced concrete lining in the base case.

In addition to higher cost, interior clearances for installation of steel plates could be a potential constructability issue due to Fire Life Safety requirements.

Figure 6 shows that the concrete lining and the sidewalk suffers more damage than the base case. It can be explained as a result of the sudden compaction between the steel plate and the concrete internal surface during the explosion and the 1-inch thick steel plate cannot provide sufficient impact energy absorption. Figure 7 also indicated the steel plate became plastic during due to the blast loading. Therefore, an interior 1-inch thick steel plate is not effective to mitigate the damage.

**Measure 5: Interior Aluminum Foam Panel**

Boey (2009) investigated the characteristics of porous materials such as Aluminum Foam. It was proved that the porous foam to be a good shock attenuator. The porous material efficiently delays the shock wave propagation and attenuates the amplitude by absorbing the kinetic energy through compaction of the material.

This protection measure considers a 4-inch thick and 13.5-ft wide interior aluminum foam panel (Figure 5). Additional material cost could be around $4,000† per foot, about 2.5 times of the cost for the steel reinforced lining in the base case. Similar constructabilty issue also exists in this measure—interior clearances for installation of the panel may not meet Fire Life Safety requirements.

Figure 8 shows that addition of an aluminum foam panel may be able to nearly fully mitigate the blast induced damage to tunnel lining, which may still have some plastic deformation; while the aluminum foam panel was severely damaged as the sacrifice structure under a blast loading.

**Measure 6: 15" Tunnel Lining + Additional Rebar**

This protection measure considers a 15-inch thick concrete lining reinforced by steel bars at a smaller spacing. The tunnel lining in this scenario (Figure 6) experienced some plastic deformation but the damage on the lining is nearly fully mitigated. Compared to the base case, the cost of tunnel lining, including additional excavation cost, could be doubled.

**CONCLUSIONS**

This paper discussed the necessity of blast protective design and threat levels for a transit tunnel. 3-dimensional finite element models were developed based on typical transit tunnel geometry to investigate the behavior of structures under blast loading from 100 lb. TNT. The analysis results indicated that local damage was observed in the tunnel lining.

Key design factors, such as the thickness of lining and reinforcement rebar and other strengthening measures were considered in this analysis. Their effectiveness to minimize the damage was investigated. Table 4 provides a summary of cost and effective analysis results for all proposed protection measures.

- Measure 1 (increasing the lining by 1-inch) and Measure 2 (double steel bars) can reduce but not fully mitigate the blast induced structural damage.
- Measure 3 (use of SFRC) does not show a significant increase in blast resistance capability with a dosage of 80 lb/cy.

*Cost was estimate based on $35 to $40 per square foot based on information from http://www.metalsdepot.com/index.php.

†Cost was estimated based on information from https://www.inventables.com/technologies/aluminum-foam.

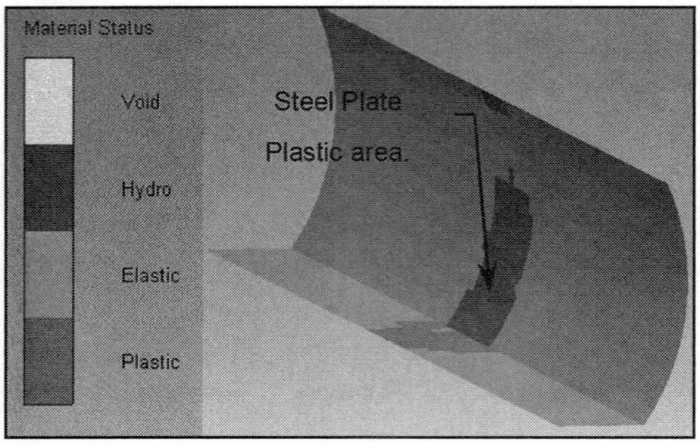


Figure 7. Steel plate under blast loading

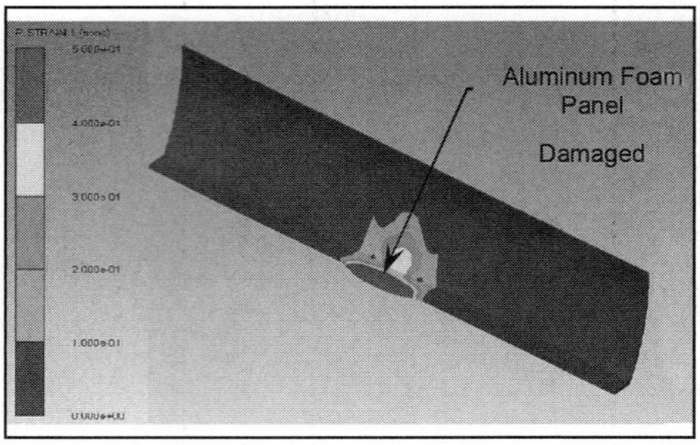


Figure 8. Aluminum foam panel

**Table 4. Summary of cost and effectiveness**

| | Protection Measures | Change in Damaged Lining Under an Explosion of 100 lb. TNT | Cost Increase* |
|---|---|---|---|
| Measure 1 | Increase concrete lining by 1" | –35% | +5% to +10% |
| Measure 2 | Double steel bars by reducing spacing | –45% | +20% to +40% |
| Measure 3 | Use of SFRC | –14% | –15% to –20% |
| Measure 4 | Installation of interior 1" steel plate | Not effective | +30% to +35% |
| Measure 5 | 4" Interior aluminum foam panel | –99% | About +250% |
| Measure 6 | 15" thick lining+ more steel bars | –99% | About +100% |

* For the purpose of cost comparison, the concrete lining cost in the base case is assumed as $1600 per foot per historic data from similar projects.

- Measure 4 (a 1-inch thick interior steel plate) is not effective to mitigate the damage.
- The analysis shows that Measure 5 (a 4-inch thick interior aluminum foam panel) is very effective to minimize the blast induced structural damage but it is also a costly protection material.
- Measure 6 considered a thicker lining with more steel bars. The blast analysis shows the tunnel lining may experience some plastic deformation but the damage is nearly fully mitigated. Compared to the base case, the cost of tunnel lining, including additional excavation cost, could be doubled.

Other products such as Micro-meshed reinforced concrete, Fiber reinforced polymers, Polyurea, and BlastWrap are available in the market for blast protection. Also, optimization of the design of anchored hooks and transverse reinforcement bars can be considered to further reduce the structural damage.

This study did not investigate the progressive failure subsequent to the blast loading, the durability and life-cycle cost for the proposed protective measures. The post-blast tunnel behavior and the durability of the proposed measure shall be considered in the future.

It should be noted that above results and discussion presented in this paper are not for any actual project and neither should be implemented in any design without proper case by case analysis being performed. The objective of this study is to provide a guideline to elicit industry discussion on the value to owners and operators of underground infrastructure blast protective design of tunnel linings when compared to current practice. Future studies will investigate the impact of explosions on cut and cover and sequentially excavated methods of building subway tunnels.

## ACKOWLEDGMENT

The authors would like to thank ANSYS, Inc for providing software tools ANSYS AUTODYN and ANSYS DesignModeler for this study.

## REFERENCES

ANSYS Mechanical User Guide R14.5. ANSYS, 2012.

Boey, C.W. "Investigation of Shock Wave Attenuation in Porous Materials," Master Thesis, Naval Postgraduate School, 2009.

Choi, S. "Tunnel Stability under Explosion," Parsons Brinckerhoff William Barclay Parsons Fellowship Monograph, 2009.

Choi, S. "Protective Design Guideline of Tunnels," Fifth International Symposium on Tunnel Safety & Security, New York, NY, 2012.

Choi, S., Chaney, I., and Moon, T. "Blast and Post Blast Behavior of Tunnels," North America Tunneling Conference, San Francisco, CA, 2010.

Choi, S., Munfakh, G. "Tunnel Design Under Explosion," Proceedings of Third International Conference on Protection of Structures Against Hazards, Padova, Italy, 2006.

David, S. "Fiber-Reinforced Concrete for Precast Tunnel Structures," Parsons Brinckerhoff William Barclay Parsons Fellowship Monograph, 2011.

FEMA. "Risk Assessment A How-To Guide to Mitigate Potential Terrorist Attacks Against Buildings," 2005.

Gebbeken, N. and Greulich, S. "A new material model for SFRC under high dynamic loadings," Proceedings of 11th International Symposium on Interaction of the Effects of Munitions with Structures, Mannheim, Germany, 2003.

Munfakh, G. "Fixing a TARIF for Security," World Tunnelling, 2008.

Quan, X., Birnbaum, N.K., Cowler, M.S., Berber, B.I., Clegg, R.A., and Hayhurst, C.J. "Numerical Simulation of Structural Deformation Under Shock and Impact Loads Using a Coupled Multi-Solver Approach," Structures Under Shock and Impact Loadings, Changsha, China, 2003.

Riedel, W., Thoma, K., Hiermaier, S., and Schmolinske, E., "Penetration of Reinforced Concrete by BETA-B-500, Numerical Analysis using a New Macroscopic Concrete Model for Hydrocodes," International Symposium, Interaction of the Effects of Munitions with Structures, Berlin Strausberg, 1999.

Riedel, W., Kawai, N., and Kondo, K., "Numerical assessment for impact strength measurements in concrete materials," International Journal of Impact Engineering, 36(2009) 283–293.

The Blue Ribbon Panel on Bridge and Security, "Recommendations for Bridge and Tunnel Security," FHWA, 2003.

TCRP/NCHRP. "Making Transportation Tunnels Safe and Secure," Transportation Research Board, Report 525, Volume 12, 2006.

# Istanbul Strait Road Tube Crossing Project: Independent Design Verification Engineer's Perspective

**Sanja Zlatanic, Yang Jiang, and Nasri Munfah**
HNTB Corporation

**Mehmet Ozturk**
YMSK Joint Venture

**Vojtech Gall**
Gall Zeidler Consultants

**ABSTRACT**: The iconic Istanbul Strait Road Tube Crossing Project (Eurasia Tunnel) will ease traffic congestion over the Bosphorus Strait separating Europe from Asia. Part 2 of the project is approximately 3.36 miles (5.40 km) long and includes, among other structures, 2.1 miles (3.4 km) of a bored tunnel under the Bosphorus Strait and 0.62 mile (1 km) of twin NATM tunnels on the Asian side. The region's variable geology, hydrology and propensity for seismic activity, combined with high water pressure and large-diameter/double-deck tunnel configuration, make the Eurasia Tunnel one of today's most challenging and complex projects under construction.

The paper focuses on the main features of the project's Part 2 and its significant challenges from the perspective of independent design verification (IDV) engineer. It also outlines the IDV review process and its benefits to the project.

## SEGMENTAL APPROACH TO A MEGA PROJECT

The Istanbul Strait Road Tunnel Crossing Project (Figure 1), initiated in 2011, is designed to ease traffic congestion over the Bosphorus Strait and provide an alternative road link approximately 0.62 mile (1.0 km) south of the Marmaray tunnel recently opened to rail traffic. Targeted for completion in 2017, this new 9.7 miles (14.6 km) long route will shorten the time it takes to cross the 2.30 miles (3.7 km) wide strait that separates the two continents, bringing important economic benefits to the entire region. The overall goal of this $1.25 billion project is to:

- Improve connections to a wide network of highways
- Increase capacity across the Bosphorus by 100,000 vehicles a day
- Save motorists up to 45 minutes of commute time in each direction
- Provide a blueprint for funding future infrastructure projects in the Republic of Turkey.

The Republic of Turkey, Ministry of Transport, Maritime Affairs and Communications, and Directorate General of Infrastructure Investments (Employer) have assigned ATAŞ (Avrasya Tüneli İşletme İnşaat ve Yatırım A.Ş.) to design, build and operate the Eurasia Tunnel for 26 years and to transfer the facility to the public authority at the end of this period. ATAŞ is a joint venture (YMSKJV) of a Turkish company, Yapı Merkezi, and a Korean company, SK Engineering and Construction Co. Ltd. Each firm is well-known for their successes in large-scale infrastructure and transportation projects. ATAŞ is tasked to put the tunnel into operation 55 months from the day the contract was executed in 2012.

The project consists of three parts. Parts 1 and 3 primarily contain road widening work on European and Asian sides, respectively. Part 2, a most complex part of the project, is 3.36 miles (5.4 km) long and includes 2.1 miles (3.4 km) of a large diameter bored tunnel under the Bosphorus Strait (Figure 2), 0.62 mile (1 km) of twin NATM tunnels on Asian side, approach roads, toll plazas, ventilation facilities, tunnel control facilities, U-sections, portal structures, cut and cover tunnels, and transition boxes on both European and Asian sides.

Parsons Brinckerhoff Inc. (Designer) is performing final design, and HNTB Corporation is the independent verification (IDV) engineer. HNTB is performing design review and independent check

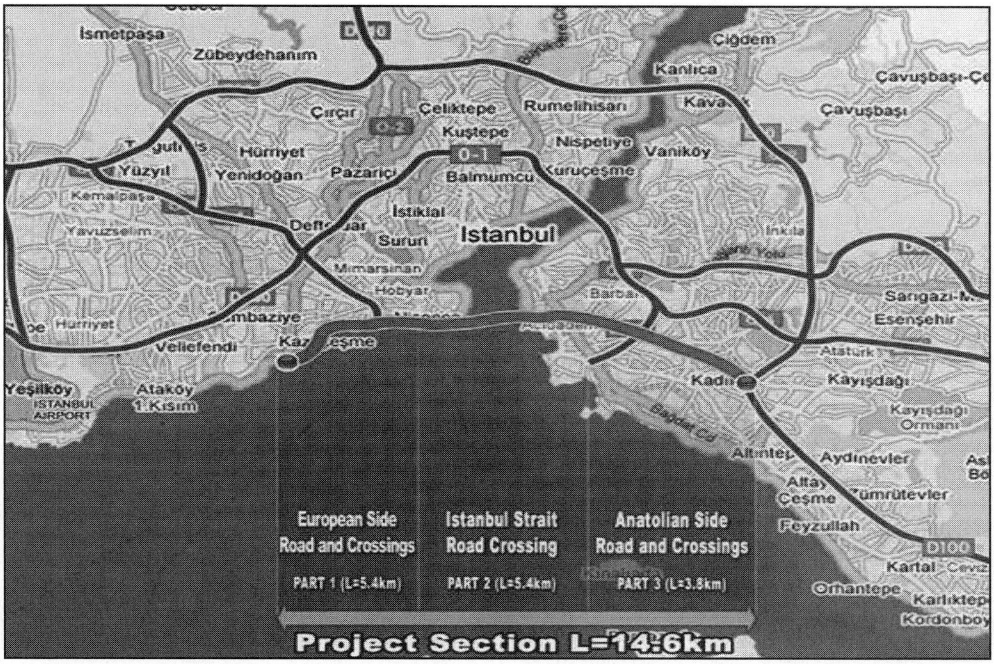


**Figure 1. Setting and location of Istanbul Strait Road Tube Crossing Project**

for category 3 structures of Part 2 of the project (per Eurocode), specifically pertaining to:

- Tunnel structures more than 10 feet (3 m) in equivalent diameter
- Structures with high structural redundancy
- Structures containing unconventional, innovative or complex design aspects
- Bridges with spans exceeding 164 feet (50 m) and/or with skews greater than 45°
- Difficult foundation problems
- Retaining walls with an effective retained height of 46 feet (14 m) or greater

## APPROACH TO RISK MITIGATION

The project is coupled with numerous technical challenges with potential for high construction and commercial risks. In order to minimize the project risk exposure and meet Employer's Requirements (ER), ATAS established a system to control the risks during all phases of the project execution, including conceptual planning, final design, construction, and facility operation. An important component of such system is to assign an Independent Design Verifier (IDV), who is to ensure that the project's major design risks are addressed by the Designer through design deliverables in accordance with planned construction Programme (Schedule). In light of the project's accelerated schedule, IDV role is critical in reviewing the design for correctness, identifying potential risk issues, verifying corrective actions, and minimizing potential that any critical issue may "fall through the cracks."

## PROJECT COMPLEX ISSUES

The following is a brief summary of the most critical technical issues that design and the design verification process need to address; these include the region's complex geology, hydrology, high seismic activity, and high water pressure imposed to a large-diameter bored tunnel.

### Geology and Hydrogeology

The project geology which the tunnel boring machine (TBM) will encounter generally consists of the Trakya bedrock underlying the alluvial sediments at the bottom of the channel. The bedrock, sedimentary in nature, is composed of inter-layered siltstones and sandstones, and it is systematically intruded by numerous volcanic/igneous dikes of diabase, andesite or dacite. The channel deposits, primarily alluvial in origin, vary in soil type and characteristics with depth, but are mainly an inter-layered mixture of sands, silts and clays. The dikes in the bedrock were reportedly encountered during Marmaray tunnels construction at a frequency of approximately 230 feet (70 m); their spacing, however, may vary and could extend up to 330 to 660 feet (100 to 200 m). The Istanbul Metro tunnels reported dike thicknesses varying from approximately 45 to 60 feet (14 to 18 m). The Trakya bedrock adjacent to the intrusive dikes is expected to be intensely fractured and weathered, more so than the host bedrock away from the intrusions.

Independent of these igneous intrusions in the Trakya bedrock, there is also a system of faults penetrating through the Trakya Formation. The adjacent Marmaray tunnels documented numerous smaller faults in the Trakya Formation; those were encountered by tunnel excavations approximately every 30 feet (9 m) and were reported as randomly oriented. In addition to mixed face conditions of rock and soil materials, highly variable rock strengths, abrasive mineralogy, and presence of stiff blocks embedded in soft matrix, the project alignment is subjected to a high water pressure. Maximum water pressure of 11 bars is anticipated at the mid-channel zone.

Tunneling in mixed face conditions will occur in three potential geological situations: alluvial overburden materials overlying weathered to moderately weathered Trakya Formation sedimentary bedrock; interface between Trakya sedimentary bedrock and volcanic dike intrusions; and fault zones passing through the Trakya Formation (Figure 3). Considering that the subsurface profile is usually represented by a linear interpolation between adjacent borings, the top of rock is expected to vary in elevation as an undulating and inclined surface; therefore, it is anticipated that the tunnel boring machine (TBM) will be in and out of mixed face conditions for some undetermined length at the interface between the Trakya bedrock and alluvial overburden.

**Figure 2. Double-deck tunnel (44.30 Ft dia.)**

## Region's High Seismicity

In the past two millennia the Marmara region has been the crossroads between the East and the West. Being a continuously populated region and having Istanbul as its center (and at times the capital of Byzantine, Roman and Ottoman Empires), the historical seismic records are relatively complete. The long-term seismicity of the Marmara region is illustrated in Figure 4.

The earthquake records indicate that, on average, at least one medium intensity (Io=VII–VIII) earthquake has affected Istanbul every 50 years. The proposed bored tunnel is passing under the Bosphorus channel in close proximity of the Main Marmara Fault, which makes a thorough assessment of the earthquake hazard important for the optimum design and safety of the tunnel.

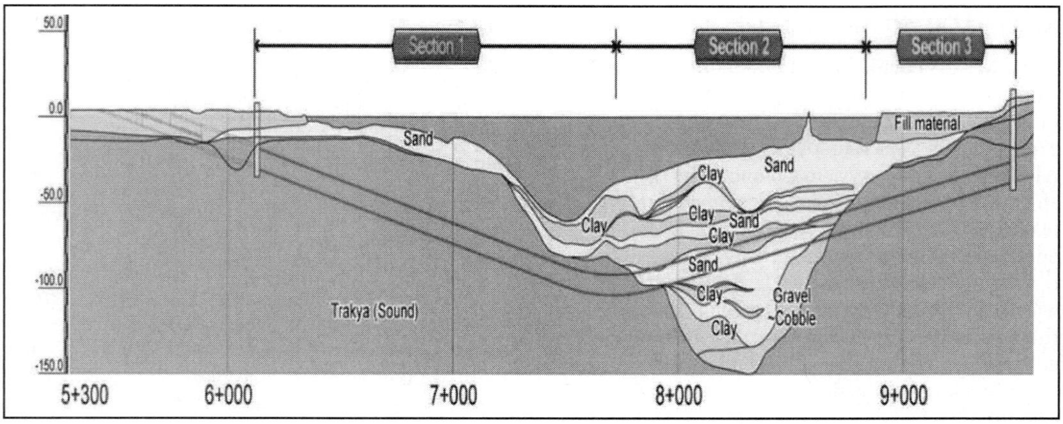


**Figure 3. Eurasia tunnel layout**

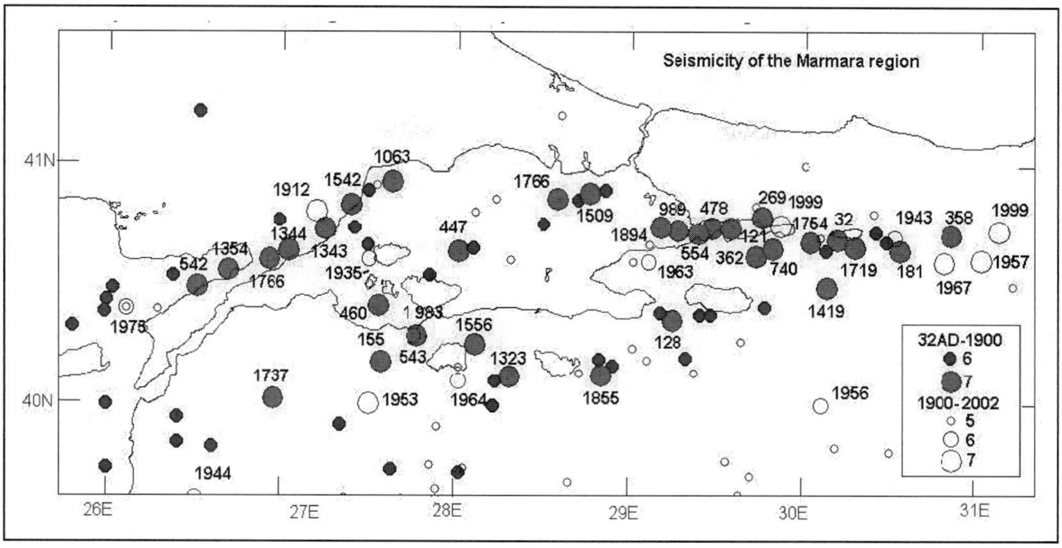


**Figure 4. Seismicity of the Marmara Region**

## TBM Technology Met Complex Requirements

The tunnel must bore through both Trakya bedrock (on both European and Asian sides) and the alluvial sediments at the bottom of the channel. It is likely the large diameter bore will encounter the interface between the relatively hard Trakya bedrock and the soft alluvial channel deposits at least at two locations. It is critical to the success of the tunneling that the excavation's face and its full perimeter are controlled at all times to minimize losses of ground and movements of the overlying ground, and to avoid potential loss of the face stability. The TBM is required to operate in closed face mode to minimize the possibility of large uncontrolled ground losses and resulting subsidence or loss of the tunnel face under water. Also, the TBM technical specifications need to address the geotechnical parameters established for soils and rock, the identified hydraulic conditions, intervention possibilities, segment installation requirements, and appropriate grouting and probing.

The TBM features met these advanced technological requirements. A mix-shield TBM with a diameter of 44.82 ft (13.66 m) was custom-made by Herrenknecht AG in Schwanau, Germany, to handle the project challenges (Figure 5). Due to the presence of both soft marine sediments and hard rock formations along the tunnel alignment, the TBM is designed to handle mixed-face conditions (Table 1). The machine's daily advancement is planned at 26–33 feet (8 to 10 m).

The TBM arrived at the project site in September 2013, and it will begin tunneling in April 2014. Even with the 11-bar high water pressure at the tunnel face, the TBM has been engineered so that the contractor can change the cutting tools quickly and safely under atmospheric pressure. This operation of complete cutting wheel replacement would occur from the rear of the machine, where all disc cutters and a large number of the cutting knives could be accessed and changed safely. A new type of cutting wheel is provided that reduces access time for maintenance work under pressurized air.

The mixshield is equipped with a special, newly developed lock system. It allows pressurized air access at well over 5 bars when necessary. To detect strong material wear early and to tackle necessary maintenance accesses in a targeted manner, wear detectors are integrated into the excavation tools as well as in the steel construction of the cutting wheel. Moreover, the disc cutters are equipped with the Disc Cutter Rotation Monitoring system (DCRM), developed by Herrenknecht. This system provides data about the rotational movement and temperature of the disc cutters to the machine operator in real time within the control container such that condition of the tools could be assessed and change intervals better planned.

The machine will be assembled and launched from the Asia transition box shown on Figure 6. The design aspect of the launch box presented a different set of challenges; some of those are addressed below

**Figure 5. The 44.82 ft dia. mixshield TBM by Herrenknecht**

**Table 1. Herrenknecht S-762 mixshield TBM features**

| | |
|---|---|
| Diameter | 44.82 ft (13.66 m) |
| Cutterhead power | 4,900 kW |
| Nominal torque | 17,200 kip-ft (23,290 kNm) |

through description of the IDV recommendations for this specific element of design.

## INDEPENDENT DESIGN VERIFICATION (IDV) PROCESS

The IDV review of Category 3 structures of Part 2 of the project, as defined previously, primarily focuses on review of critical design issues through design reviews and independent analyses. This is to verify compliance with Employer's Requirements and identified design codes and standards, primarily major American standards and Eurocode, as well as Turkish standards, whichever happened to be governing. Through IDV, major design components are evaluated, including design assumptions and methodology for both temporary and permanent structures, excavation methods (to suit the predominant ground conditions), sequencing and support of excavation for the tunnel and the approaches, technical specifications, and quality and durability of materials selected. In addition, the following elements have been reviewed and verified: the alignment; spaceproofing of all permanent structures including tunnel cross-sections to accommodate all required utilities and services and meet operational standards; geotechnical parameters including both static and dynamic/seismic parameters; structural systems and details; emergency ventilation and fire-life safety provisions; electrical and mechanical design parameters; tunnel lighting, and systems and operation components including facilities and buildings. The IDV reviews take place within two phases of the design, Basic Design and Detailed Design. Figure 7 illustrates the checklist of the IDV process as it pertains to design deliverables review and verification.

In Basic Design configuration of the structures is fixed, drawings are produced, and dimensions are given to the geometry of major design elements. Also, cross sections of members are fixed and dimensioned, general specifications for design and construction are established, and compliances to specific standards and codes are identified. The design is developed in detail for critical elements of the design, together with technical specifications. Verification of Basic Design assures that the YMSKJV and the Employer are enabled to evaluate the inherent risks. Verification process of Detailed Design commences with review of the Basic Design making sure it is fully developed into Detailed Design and Basic Design comments are closed.

In Detailed Design, final analyses of all aspects of the design are completed, technical specifications finalized, and focus of the Designer is on the mechanics of translating the dimensioned members into buildable and practical work elements. Detailed Design is presented through calculations, drawings and technical specifications. IDV included independent calculations for most critical design components: bored tunnel (including seismic checks), NATM tunnel, transition structures, cut and cover tunnel, retaining walls, U-sections, portals, temporary support of excavation, system buildings, ventilation analyses, and alignment and drainage calculations. After both Basic and Detailed Designs are verified IDV issues design verification certificate (DVE) allowing YMSKJV to proceed with construction of verified design packages and obtain the Employer's and Lender's (LTA) approvals. For schematic of IDV process refer to Figure 8.

**Figure 6. Asia transition box—TBM launch shaft**

**INDEPENDENT DESIGN REVIEW PROCESS FOR BASIC AND DETAILED DESIGN CATEGORY 3 STRUCTURES**

**Independent Analysis and early work activities**

- Independent analysis (as determined by IDV) will be performed in advance of the Basic Design and Detailed Design review. IDV will verify that independent analyses will support Designer's conclusions.
- A Value Engineering Workshop will be performed jointly with YSHJV for elements agreed to by YSHJV prior to the start of the review of the Basic Design. The VE will evaluate if design concepts can be improved to minimize risks, cost, and project duration. Identified improvements will be provided to YSHJV for implementation.

BASIC DESIGN

- Does design comply with ER and required codes and standards?
- Does Basic Design meet level of detail required?
- Are concepts correct and valid including meeting operational, maintenance and safety requirements?
- Is design constructible?
- Is Basic Design suitable for advancement to Detailed Design?
- Are specifications consistent with drawings and good practice?
- Are calculations verified and checked and are they supporting the Basic Design and meeting the level of details consistent with the Basic Design?
- Do independent analyses (where applicable) support Designer's conclusions?
- Are risks and hazards identified and addressed in the design?

DETAILED DESIGN

- Does Design comply with ER and required codes and standards?
- Are Detailed Design methodologies correct?
- Are agreed actions in Basic Design incorporated in the Detailed Design?
- Is Detailed Design complete?
- Are drawings acceptable and suitable for Detailed Design?
- Are specifications consistent with drawings and good practice? Are they completed to the Detailed Design level?
- Are calculations verified and checked by Designer, are they supporting the Detailed Design and meeting the level of detail required?
- Is the design ready for construction?

**Figure 7. Checklist of IDV for basic and detailed designs including value engineering**

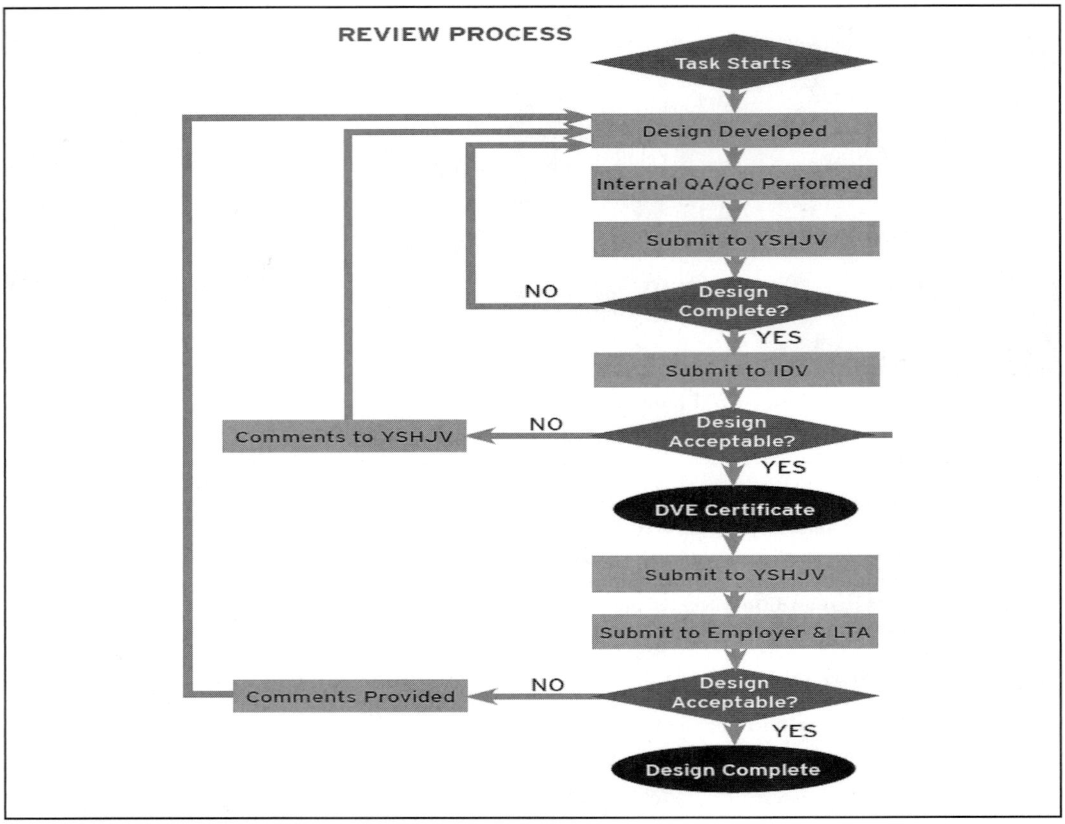


**Figure 8. Schematic of IDV review process**

## IDV: VALUE ADDED

The review of tunnel interior spaces has taken into consideration roadway configuration, including VMS, signage, lighting, barrier protection, and traffic clearance envelopes; provisions for emergency egress; vehicle breakdown considerations; structural member sizes; structural fire protection; tunnel lining segment type and configuration; tunnel ventilation and mechanical systems and equipment layout, and construction tolerances. Full inclusion of all required systems constituting the interior of the tunnel, their integration in accordance with Employer's Requirements (ER) and identification of potential interferences have been of primary importance while reviewing the tunnel design documents. The IDV review of structural design of the precast-concrete segmental tunnel lining is another key technical element. The lining design has been verified for the insitu and internal loadings including the earth and hydrostatic loads imposed by the surrounding ground and surcharges above, as well as loads imposed during the tunneling process, including jacking forces, handling and erection loads specified within the Design Manual prepared per Employer's Requirements. The IDV also verified the design of the precast concrete tunnel lining in accordance with AFTES Recommendations for the Design, Sizing and Construction of Precast Concrete Segments Installed at the Rear of a Tunnel Boring Machine (TBM), 1997. The IDV review verified that specifications are in place to ensure that the lining segments be produced using close tolerances in machined steel forms and cured until full strength is reached before use. In addition, the segments were verified to be of high quality, interchangeable, durable, extremely impermeable to salt water, and resistant to chloride and sulfate attack. The IDV review process included final bolting assembly, grout ducts, segment gaskets, fireproofing requirements, and the erection handling system. Inclusion of seismic joints had been verified at the appropriate locations through coordination with Designer. Structural independent verification was carried out for the tunnel interior structures including the bottom roadway slab, the upper roadway deck slab and corbel supports. Some of the IDV recommendations are as follows.

## Bored Tunnel Seismic Design

It was verified that the effects of soil structure interaction are taken into account in the analysis and design and that the tunnel is investigated for at least the three primary modes of deformation that occur during seismic shaking: ovaling/racking, axial and curvature deformations. The ovaling/racking deformation is caused primarily by seismic waves propagating perpendicular to the tunnel longitudinal axis (referred to as transverse ovaling/ racking analysis hereafter). Vertically propagating shear waves are generally considered the most critical type of waves for this mode of deformation. The axial and curvature deformations are induced by components of seismic waves that propagate along the longitudinal axis and/ or by spatially varying ground motions resulting from local soil/site effects. Since there are considerable structural discontinuities that exist at soil/rock interfaces, at changes in structure section, at changes in the type of construction, and at tunnel segmental joints, it was verified that these discontinuities were evaluated and that structures remain watertight during and after earthquake.

It was verified that during strong earthquake excitations the tunnel embedded within the soft alluvial channel deposit experiences large transient displacements in the transverse and longitudinal directions, in general conformance with the free-field ground shaking displacements. It was also verified that the tunnel embedded in the Trakya rock formation experienced much less shaking displacements, resulting in significant differential movements between tunnel in the soil and tunnel in the rock. It is generally difficult to reinforce the tunnel structure to resist such differential displacements due to high stresses usually developed where the tunnel intersects the soil/rock interface. Therefore, it was expected that flexible joints are to be introduced at transition zones, however, the location of such joints have been found to coincide with locations of largest axial forces; also, it was clear that Designer considered uncertainty to the exact location of the stiffness changes. IDV questioned the location of seismic joints and suggested a work plan that included further non-linear evaluations/numerical analyses and plasticity based techniques due to highly specialized nature of seismic joints design and manufacturing.

Through IDV process we recognized that the seismic design of tunnel structures is heavily dependent on the modeling of the soils and the structures through soil-structure interaction. The following betterment ideas are also introduced on the seismic analysis of the bored tunnel; considerations of these ideas resulted in more realistic seismic displacement and force demands predictions and in a more economical liner and joint design overall.

1. Numerical modeling of the ground (rock and soil):
   - The properties of the highly weathered Trakya Formation on the western flank are important for the tunnel seismic response as the material is at the transition of the bedrock and alluvial channel. Realistic material properties based on measured shear wave velocities for this region could potentially result in better seismic behaviors of the liner and the seismic joints and these were recommended. For example, the Poisson's ratios presented in the Seismic Design Report are consistent with values for saturated clays (0.45 or greater) and they are calculated from the measured shear wave and compression wave velocities. The measured compression wave velocities in the soil deposits are close to the compression wave velocity of salt-water (1500 m/sec). Since an in-situ testing will measure the compression velocity of the water and not that of the soil, it appeared that the reported compression wave velocities are not the actual compression wave velocity of the soil material. This also resulted in a higher value of the low-strain Young's modulus. The Young's modulus and Poisson's values impact the analyses and design, especially finite element or finite difference models.
   - The interface friction capacity for the soil/ rock could include a cohesion component. The impact is reduction of the axial force that has to be resisted by the tunnel section.
   - Evaluation of the impact of stiffening the alluvium by jet grouting beneath the tunnel at transition between the zones where the tunnel leaves the bedrock and penetrates into the alluvium was recommended. The concept is to provide a gradual change between the support conditions for the tunnel to modify the seismic response for the tunnel.
2. Structural modeling of the liner:
   - Tensile seismic behavior of the liner is important because water tightness and gasket design is more related to the liner tensile behavior. It was recommended that due consideration is given to the tensile behavior when modeling the liner during the time history analyses.
   - Seismic analysis and design of bored tunnel are interdependent because of nonlinear nature of the system. The structural model has to accurately reflect the design, i.e., realistic sections properties that

include the effect of uncracked or cracked concrete and reinforcement; type, size, and number of bolts at the circumferential joints; effect of cam/socket system, etc. It was recommended that strain values in concrete and reinforcement are demonstrated in accordance with the limits specified in the Design Manual for appropriate levels of earthquake. Also, the design should demonstrate that the gasket is water tight with appropriate factor of safety for appropriate levels of earthquake.

- Investigation of the possibility of eliminating seismic joints through use of refined non-linear analyses and plasticity based techniques was suggested.
- Analytical techniques to improve monitoring of circumferential joint performance and strain demands in the tunnel were recommended.

### Asia Transition Box and NATM Recommendations

The following recommendations were provided in regard to Asia Transition Box, serving as the TBM launch shaft, and the adjacent NATM tunnels:

- Refined the sequential excavation and support of the open-cut lifts in relation to the specific timing of the support installation for the purpose of risk mitigation.
- Suggested casting of the concrete box structure against the support of excavation to avoid large volume of excavation and subsequent backfill resulting in economy and schedule benefits.
- Discussed the use of drained vs. undrained design parameters in view of the fact that installation of a systematic dewatering system was not foreseen to obtain realistic load combinations for the support of excavation system.
- Suggested construction schedule-related evaluation of the temporary surcharge around the excavation perimeter to rationalize factors of safety required for support of excavation elements and obtain more practical design.
- Requested development of a toolbox of contingency measures and a systematic contingency plan in order to provide adjustment to conditions encountered and support optimization for the support of excavation system.
- Suggested installation of additional instruments for a more vigorous monitoring scheme at high-risk areas of the project.
- Conducted an independent numerical analysis for the NATM tunnels temporary and permanent supports (Figure 9), and provided comments to achieve design robustness.
- Suggested use of steel fiber reinforced shotcrete in lieu of welded-wire-fabric reinforced shotcrete for safety benefits and economy of construction.

**Figure 9. Finite Element analysis model of Asia NATM Tunnel**

### Miscellaneous Recommendations

- The tunnel segmental liner concrete mix is specified to resist the design loads and service life/durability requirements. The IDV verified that the concrete mix met durability and fire protection performance requirements and recommended inclusion of micro polypropylene fibers into the concrete mix to improve liner performance during design fire event. It was noted that the liner surface exposed to the design fire in the tunnel is to be fireproofed.
- Suggested careful integration of the traffic management with operations considering that the tunnel ventilation system, designed as a longitudinal system acting on a push-pull basis, is predicated on the fact that the minimum vehicle speed in the tunnel needs to be maintained at 20 km/hr. Considering that Istanbul is the second most congested city in the world, after Moscow, the traffic management system integration in the tunnel with that of the approach roads outside the tunnel during different times of city-wide congestion becomes a critical component of fire life safety. Tunnel roadway design and traffic controls must facilitate proper longitudinal ventilation of the tunnel.
- In terms of interior structures design, it was noted that the randomness of the precast tunnel liner end joints and their hardware would often prevent installation of post-installed anchors acting as a upper roadway deck supports.

- In order to reduce maintenance of the pumping system, cascade drainage pumping scheme was suggested; this system would require less maintenance due to more robust pump selections.
- Recommendations were included for the formation of Fire Life Safety Committee, which would include authority having jurisdiction, first responders, operators and owners, in project development process to assure tunnel ventilation system is capable of developing tenable environment for all expected activities during a design fire incident.
- Input was provided regarding exposed PVC cable and conduit installation in the tunnel utility corridor and requirement to comply with NFPA 502.
- It was suggested that all security system devices, rated for use on fire alarm equipment, are monitored through the security and SCADA system.
- A recommendation was made that the tunnel jet fans and damper control be provided through the SCADA system due to greater flexibility and control, rather than through the fire alarm system.
- Suggestions were made regarding assigning correct Safe Stopping Distance (SSD) for development of Threshold Zone length for lighting design.
- Recommendations in terms of Threshold Zone luminance level included use of polar diagrams to assess 'brightness of the sky' and its effect on required luminance levels at the tunnel portals. Adjustments to the length of Transition Zone were required based on CIE 88 eye adaptation curve.

## CONCLUSION

The IDV reviews provided multiple recommendations that were adopted by Designer. The recommendations have assured the design compliance with Employer's Requirements and design criteria, codes and standards, have reduced construction risks, improved constructability, increased safety and efficiency of the facilities, and decreased operational costs.

# TRACK 3: PLANNING

## Session 1: Project Delivery I—Design and Management

Ivan Hee, Chair

# TBM Procurement Aspects (for Owner's Eyes Only)

**Dan Ifrim and Derek Zoldy**
AECOM

**ABSTRACT:** Among tunneling projects challenges the delivery options and issues related to each of the delivery method elected is one of them. Mechanized tunneling is a safe and effective alternative to other methods in terms of schedule and cost. Its success involves a comprehensive and interdisciplinary consideration of all contributing factors such as technical specifications, design and procurement of equipment and tunnel linings.

The objective of this paper is to address the qualitative and quantitative approach together with the pros and cons of Owner procured TBMs and Concrete Segmental Linings.

The study includes the Owner, Contractor, and Consultant involvement in prescriptive and performance based specifications, TBM ownership and novation options, a summary of Owner and Contractor responsibilities, risk sharing as well as risk mitigation strategies.

## INTRODUCTION

In the past decade, the configuration and operating characteristics of Tunnel Boring Machines (TBMs) and the level to which the TBM characteristics should be specified and detailed have become a sensitive topic of discussion by Owners and their consultants.

Typically for the definition and discussions of these characteristics and specifications the Owners have used review processes such as Value Engineering and Constructability Reviews prior to tendering. Although that helped gaining knowledge from a wide range of industry experts and experienced contractors; the specifications could still become the subject of discussion and disputes during construction.

Usually the Owners decision is highly influenced by upfront visible cost and budget. Other factors of influence include the advisory team, Design Engineers, Geotechnical, and Tunneling experts along with Construction Managers, Contractors, Municipality Board members and Lobbyists.

Since the Owner is ultimately responsible for the conception, planning, financing, design, construction, and commissioning of all underground works, a huge burden on decision making lays on Owners Project Manager and its advisory group.

Only at the end of the job one could tell if the decision was right or wrong based on the success of the project.

## CONSTRUCTION PROCUREMENT CONSIDERATIONS

There are several tunnel construction procurement options available. The factors in making the right decisions are the objectives and constrains of the project as well as the resources, skills, and experience of the Owner organization. It was demonstrated that Value Engineering and Constructability Reviews have an important role in defining the balance between "Prescriptive" and "Performance" to set the path for a successful project.

TBM procurement process is strongly related with the project procurement type and the level of detail specified by the Owner's Engineer. The outcome of the procurement options analysis will point toward the level of prescriptivism for TBM characteristics.

Figure 1 illustrates the Owner's decision making process.

### The Prescriptive Option

The prescriptive option defines all TBM characteristics, sequence of assembly, launching, and operations as well as details of the ground support type and installation sequence and operations.

The prescriptive approach implies that the owner's engineers, advisors, and experts have the knowledge and capacity to determine the optimum TBM type, characteristics, methods, and sequence of operations for the specific project requirements.

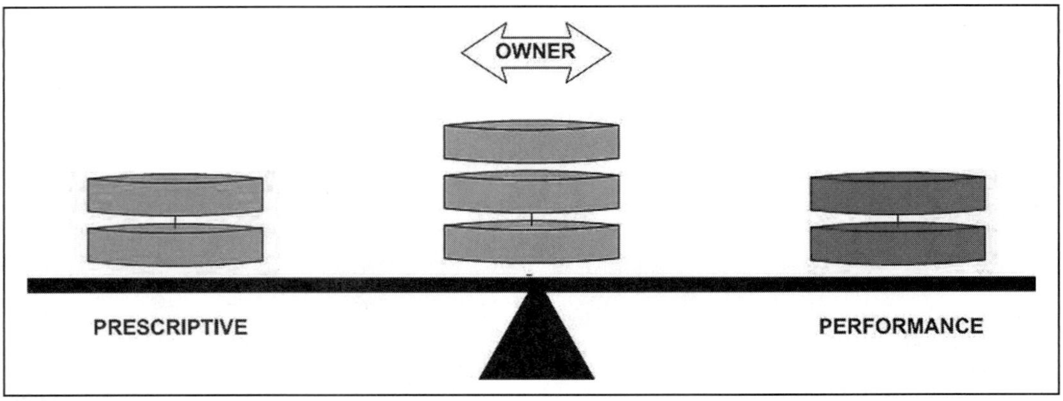


**Figure 1. Prescriptive vs. performance balance**

### The Performance Option

The performance option reduces considerably the role of the Design Engineer. Under this approach the contractor is required to meet key project performance requirements. The contractor has the liberty to select the TBM type, tunnelling methods, and sequence of operations to meet the contractual requirements.

The performance option implies a very rigorous selection of contractors at qualification phase.

The qualified contractors must have all knowledge and capacity to determine the optimum TBM type, characteristics, methods, and sequence of operations for the specific project requirements.

Past experience with similar projects in nature and complexity, possession of skilled engineers, construction management, TBM operators, mechanics, providing proof of using industry best practices and good references are key requirements for qualifications.

### The Design Build Option

The design build option is a form of Performance option that requires the Contractor to have a Design Engineering firm on board. The option will not be discussed in this paper since does not differ from the Performance option in terms of TBM procurement.

## TBM PROCUREMENT OPTIONS

The Owner faces several challenges is selecting one approach vs. another.

- Select a procurement model based on careful and informed deliberation of all aspects of the project, as well as a self-assessment of their own organization's capability to administer the selected approach [1];
- Qualify the Design Engineers based on Qualifications rather than Price[1];
- Verify that bidders are qualified by conducting prequalification program to ensure that all bidders meet the procurement approach criteria. (Qualification evaluation submittal should be separated by the bid price, as sometime the Owners may be tempted to oversee some of the missing skills when the price is low);
- Ensure safety and integrity of community and infrastructure is not compromised by the selected approach;
- Ensure the Project satisfy the needs in terms of Schedule and Budget without compromising quality;
- Ensure a fair amount of ground investigations to provide a confident level of information.

It is generally known that in a low bid tender practice contractors will not purchase sophisticated TBMs which may be necessary for the work; they will select a machine that can give the minimum requirements and the project will most likely suffer.

Reiterating the balance required between Prescriptive and Performance, the Owner will need to ensure that all TBM critical characteristics are specified. Beside TBM critical characteristics it is important that Contractors personnel is familiar with the machine and has the skills and competence to operate it as per the manufacturer specifications. TBM regular planned maintenance is another important factor that needs to be enforced through submittals, to ensure that the machine performs at all time as per manufacturer specifications.

There are several options for TBM procurement; Owner Procured, Owner procured and Novated to Contractor and Contractor Procured. These options will be discussed further highlighting the pros and

**Table 1. Owner-procured TBMs pros and cons**

| Advantages, Owner-Procured Machines | Disadvantages, Owner-Procured Machines |
|---|---|
| Schedule mitigation of minimum 12 months | Extensive Construction Management cost |
| Cost saving for multiple use of TBMs, in the same procurement type | Potential Contractor claims related to the difficulty in operating owner's TBM |
| Responsibilities for changed ground conditions are clearly identified | Contractor claims for TBM breakdowns and maintenance requirement becomes owner's fault |
| Cost savings from reduced contractor bid contingency -risks mitigation cost by the owner | Owner is directly liable for claims, disputes and litigation |
| Ground stabilization and mitigation requirements are specified by Owner's Engineers | Owner is directly responsible to community for problems related to settlements, delays and cost overruns |
| TBM reflects project characteristics as determined by owner/design and geotechnical engineers plus expert advisors | Contractor claims for different ground conditions deems owner's machine unsuitable |
| Ground stabilization and mitigation requirements are specified by Owner's Engineers | Detailed Contract Specifications (must deal with expected production rates, maintenance requirements, down-time and methodologies to deal with unexpected conditions) |
| Owner's Engineer requirements are satisfied | |

cons of each one of them. The options are discussed further below.

### Owner-Procured TBMs

Typically the Owner procures the Tunnel Boring Machine and tunnel liners (where applicable) and makes it available to the selected Contractor to build the tunnel. In this case the Owner is responsible for providing TBM consumables and spare parts, deal with machine breakdowns, etc.

This procurement option assumes that the Owner's Engineers are adequately knowledgeable of tunnelling construction requirements in the existing ground conditions and be able to pre-determine necessary tunneling requirements and correctly specify the TBM type and characteristics.

For this type of procurement the designer needs to develop a very detailed TBM technical specification. The designer needs to have the capability and possess the skills and knowledge in tunnel construction applications and more important in Tunnel Boring Machines.

The Design Engineer capabilities should include but not be limited to all aspects of TBM mining, TBM performance in different type of ground, ways to counter difficult ground, TBM logistics, asses the particularities of the job site and launching shaft to provide valuable specification requirements for the TBM manufacturing. The Design Engineer should be generally capable of defining procurement documents for the TBM and ancillary equipment including the segmental liner moulds.

Where the designer lacks any of these skills or knowledge shall be capable of identifying help from industry experts.

During prequalification and bidding process the Owner should impose criteria and carefully screen the proponents to ensure that the successful bidding Contractor possess the skills and capability to operate, troubleshoot, and provide adequate maintenance to the TBM to ensure efficient operation at a satisfactory production rate.

The Owner shall ensure that all scheduled maintenance is witnessed by its engineers and the TBM recommended operating parameters are not exceeded at any time.

The Contractor can be offered financial incentives to operate and maintain the TBM and ancillary equipment in good conditions while performing all ground support and tunneling requirements.

*The Owner needs to:*

- Have the Council backup and support for the procurement option
- Accept the work load and financial effort to support this procurement option
- Fairly recognize changed ground conditions when necessary
- Ensure clear and fair prequalification process

Table 1 lists a number of pros and cons for the Owner-procured TBM option [3].

### Owners Procured Tunnel Boring Machines and Novated to Contractor

**"Novation"** = *"The substitution of a new contract for an old one. The new agreement extinguishes the rights and obligations that were in effect under the old agreement."*

This option is a similar to the Owner's procured TBM option in which procurement process is handled by Owner with support from Design and Geotechnical Engineers, Consultants and participation from prequalified Contractors. Ultimately, prior to TBM delivery the ownership is transferred to Contractor.

**Table 2. Contractor-procured TBMs pros and cons**

| Advantages, Contractor-Procured Machines | Disadvantages, Contractor-Procured Machines |
|---|---|
| Owner has less responsibility for machine breakdowns or tunnelling difficulties | Owner has reduced input into tunnel boring machine characteristics |
| The Owner doesn't need to deal with a separate procurement process | Owner has reduced involvement in TBM characteristics and design input |
| Owner's Engineers and Consultants expertise in TBMs characteristics does not need to be extensive. | Tunnel boring machine will be "low-bid"—a basic and limited capabilities machine may result |
| Agency not concerned about disposal of tunnel boring machine at end of work | Possibility of unsuitable TBM for ground conditions causes delays and project costs overruns |
| TBM maintenance, consumables are contractor's responsibility | Contractor reluctance to new technology may result in a non productive machine |
| TBM transportation, launching and removal are contractor's responsibility | Longer construction schedule |
| Project Cash flow improvement by not paying upfront for TBM, segment forms and spare parts | A balanced prescriptive specification approach is needed to ensure TBM suitability |

All the benefits of the "Owner Procured" still apply; additionally the Contractor is direct involved in the specification of the machine, and becomes direct responsible for the performance of the machine.

This purchasing option was used in the St. Clair US-Canada[5] and Aguas Argentinas, Buenos Aires[6] projects. It is a method to achieve full tunnel machine design detailing and procurement by the owner, with formally structured owner, engineer and contractor involvement.

This purchasing option it is recommended for projects with schedule constrains and represents a better risk mitigation by detailing the TBM specifications and share the risk of eventual TBM performance issues in unexpected ground with the Contractor.

### Contractor-Procured Tunnel Boring Machines

For the Contractor Procured TBMs it is advisable that a high level of TBM characteristics are specified by the Owner to ensure the comfort level that the successful contractor will not utilize a non-compatible TBM.

Most of the contracts are written this way and it seems that the Owners have confidence that the Contractor will propose the right TBM. The recent innovations and developments in TBM manufacturing and owners educational process through conferences and seminaries are driving the tendency to increase the level of TBM specifications details.

The tendency pushes the Design Engineering and Consulting firms to keep up with the developments and refrain from using standard specifications and customize the TBM specification to suit the ground conditions and Owner preferences.

The specifications level and flexibility switches some of the responsibilities from Contractor to Owner and creates a higher standard for the bid where all contractors will have to comply with the minimum TBM requirements. The level of responsibility shifting from Contractor to Owner will still maintain a competitive environment at bidding and promote a new lever of low bid.

Table 2 lists a number of pros and cons of the Contractor procured option [3].

It is important that the Owner retains knowledgeable Design and Geotechnical Engineers that will have the capability to evaluate the soil conditions and produce the TBM technical specifications to ensure an adequate TBM match.

The detailed TBM specification could fire back to Owner if difficult ground conditions are encountered and the TBM does not perform per engineer's expectations. In such situations the contractor will not hesitate blaming the problem on the TBM.

Regardless the type of specification, detailed or not, the Owner is still substantially involved through the soil conditions and Geotechnical Baseline Report and has the obligation to review and approve the Contractor proposed equipment, means and method of construction.

Some Owners prefer not to get involved in such situations and deal with the risk associated with detailed specifications and have the tendency to move towards performance specifications to shift liabilities to Contractor.

## OWNER-PROCURED TBM EXAMPLES

### Introduction

All data presented in this sub chapter is public data available on the internet. For data source please refer to references 7 through 11.

### Toronto Transit Commission (TTC) Rapid Transit Expansion Program (RTEP) 1994

Sheppard, Eglinton, and Spadina rapid transit tunnel projects. Two owner procured Earth Pressure Balanced Machines (EPBM) TBMs. Contractor bids on installation and operation of TBMs and installation of pre-purchased pre-cast tunnel liner.

Two TBMs were procured by TTC from Lovat Equipment Inc. prior to commencement of the Twin Tunnels Contract [2].

The Contractor completed the construction 6.4 km of twin tunnel with five stations to connect Yonge Subway with Don Mills Road.

- **Owner:** Toronto transit Commission
- **TBM Supplier and Model:** LOVAT Model ME232SE
- **Contractor:** McNally-PCL-Foundation JV

### Saint Clair River Tunnel Between Sarnia, Ontario and Port Huron, Michigan, 1992

Negotiated compressed procurement process, design by consultants, owner procurement of tunnel boring machine and tunnel liners, contractor bids on installation and operation of TBMs and installation of pre-purchased pre-cast tunnel liner, construction management services by design consultants.

- **Owner:** St. Clair River Tunnel Company–Canadian National Railways
- **TBM Supplier and Model:** LOVAT Model ME375SE
- **Contractor:** JV led by Traylor Bros of the US with Frontier Kemper (US), Wayss & Freytag (Germany), and Foundation Company (Canada)

### South East Collector Trunk Sewer (SEC)—York Region, Ontario, 2009

The SEC project consists of a series of gravity sewers, pumping stations, forcemains and an 80-mgd wastewater treatment plant, in the Regions of York and Durham.

For the South East Collector Trunk Sewer tunnels York Region procurement included 4 Earth Pressure Balance TBMs from Caterpillar Tunnelling Canada Inc. (formerly Lovat Inc.)

- **Owner:** York Region
- **TBM Supplier and Model:** Caterpillar (Lovat)–RME142SE
- **Contractor:** Strabag

### Toronto York-Spadina Subway Extension—Toronto Transit Commission, Toronto, Ontario, 2009

The Toronto-York Spadina Subway Extension Project (TYSSE) included tunnelling an 8.6 kilometre of subway extension of the Yonge-University-Spadina subway line to Vaughan Corporate Centre in the northwest part of GTA. For the excavation of the TYSSE tunnel sections, Toronto Transit Commission (TTC) purchased four (4) Earth Pressure Balance TBMs from Caterpillar Tunnelling Canada Inc. (formerly Lovat Inc.).

- **Owner:** Toronto transit Commission
- **TBM Supplier and Model:** Caterpillar (Lovat)–RME142SE
- **Contractors:**
  Contract 1: JV of McNally, Kiewit and Aecon
  Contract 2: JV of OHL and FCC.

### Eglinton Crosstown LRT—Metrolinx, Toronto, Ontario, 2010

New LRT project, includes 25 stations and over 10km twin tunnels, The procurement included 4 Earth Pressure Balance TBMs from Caterpillar Tunnelling Canada Inc. (formerly Lovat Inc.).

The project was split in two contracts; first contract awarded TBM delivered in the spring and respectively summer of 2013. The second contract is not yet awarded.

- **Owner:** Metrolinx
- **TBM Supplier and Model:** Caterpillar (Lovat)–RME256SE
- **Contractors:**
  Contract 1: JV of Kenny Construction Co., Kenaidan Contracting Ltd., Obayashi Canada Ltd. and Technicore Underground Inc.
  Contract 2: JV of Aecon and Dragados

## OWNER PURCHASING OPTION DISCUSSION

This discussion is aiming to make the Owners aware of the pros and cons of the Owners Purchase Option for TBMs and ancillary equipment.

We will not discuss the pros and cons for Contractor Purchase Option since they are somewhat opposite of the Owners Purchase Option.

### Purchasing Option Assumed Advantages Effect on Tunnelling Project

- **Schedule mitigation of minimum 12 months**
  Early commitment to purchase the TBMs give the manufacturer the possibility to place

Purchase Orders for long lead components such as Main Drive Bearing, Gearboxes, Substations, VFD's etc. Schedule mitigation by advance TBM purchase is a smart approach; however in my opinion the purchase should be Novated to Contractor. Contractor TBM ownership will give the Owner the guarantee of Contractor motivation for TBM maintenance and performance.

- **Cost saving for multiple use of TBMs, in the same procurement type**
  The TBM cost can be amortized over several projects. TBM cost however should take in consideration the cost for conservation, storage and refurbishment cost for the future projects.
- **Responsibilities for changed ground conditions are clearly identified**
  The Owner is ensured that the TBM was built to counter the ground conditions as identified in the GBR.
- **Cost savings from reduced contractor bid contingency—risks mitigation cost by the owner**
  These savings are in terms of Contractor markup for purchasing process along with TBM transportation and commissioning.
- **Owner fully specified ground stabilization and mitigation requirements**
  The Owners have the guarantee that the TBMs are equipped with all systems required to provide ground stabilization in unforeseen situations that were identified in the risk register and there are no substitutes or omissions.
- **TBMs reflects the project characteristics as determined by Owner, Design and Geotechnical Engineers and Consultants**
  The Owners have the guarantee that the TBMs are equipped with all systems required to tunnel under the ground conditions identified in the GBR and there are no substitutes or omissions.
- **Ground stabilization and mitigation requirements are specified by Owner's Engineers**
  The Owners have the guarantee that the TBMs are equipped with all systems required to provide ground stabilization in unforeseen situations that were identified in the risk register and there are no substitutes or omissions.
- **Owner's Engineers requirements are satisfied**
  Owner Designers and Geotechnical Engineers satisfaction to have had identified the ground conditions and had the design, specifications and risk mitigation plan materialized.

**Purchasing Option Assumed Disadvantages Effect on Tunnelling Project**

- **Extensive Construction Management cost**
  The extensive work increases the project overall cost, however ensures that all prescriptive specifications are enforced.
- **Potential Contractor claims related to the difficulty in operating Owner's TBMs**
  Typically Contractors will complain if the TBM requires unusual skills such as PLC technicians, Compress Air Trained Workers, etc.
- **Owner is directly liable for claims, disputes and litigations**
  The Owner is responsible for all TBM related cost including transportation, storage, assembly and commissioning, spare parts and consumables. Additionally the Owners are paying for TBMs breakdowns and repairs and have the burden to deal with the manufacturers while not being aware of all technical issues that produced the breakdown. Again in my opinion purchase novation to Contractor is a better approach to alleviate all these problems.
- **Owner is directly responsible to community for problems related to settlements, delays and cost overruns**
  The Owner will have to deal with any settlements problems, if occurs from other cause than improper selection of TBMs. The owner is also accountable for project delays and cost overruns due to TBMs breakdowns.
- **Contractor claims for different ground conditions deems owner's machine unsuitable**
  This is a territory where the Owners Engineers needs to be very knowledgeable and be able to interpret the Geotechnical Data to avoid the situation.
- **Detailed Contract Specifications (must deal with expected production rates, maintenance requirements, down-time and methodologies to deal with unexpected conditions)**
  The Design Engineer and Consultants were experienced enough to appreciate the expected TBM production rates and have them specified.

## CONCLUSION

A tunneling project success depends on many factors, one of which is the subject of this paper. A careful balance between the cost and effort spent on excavation equipment/ technique versus cost of ground modification it is very important. If the balance leans one way or another that could lead to project delays and cost increase. A good example is in reference paper 4.

"The price to pay" for employing an open-face digger shield in coarse alluvium, is extensive ground modification including dewatering and/or grouting. Investing in a full-face Tunnel Boring Machine (TBM), on the other hand, reduces grouting costs; and employing Earth Pressure Balance (EPB) or Slurry Shields could even eliminate the need for dewatering"[4].

Therefore, tunneling option depends strongly on economics and upfront investment decisions. Unfortunately, performance specifications allow such decisions to be made by low-bid contractors willing to take the risk of "getting by" with less on tunneling equipment/technique, without having to pay the price for ground modification.

The solution to ensure risk mitigation is to shift the Owners focus from "Performance" to "Prescriptive" based specifications. With today rapid technology advance the Owners should demand their Consulting Engineers to present all options of tunnelling methods in the preliminary design phase to give them the option of taking an educated risk management decision.

It was noted that North American Contractors are more reluctant to apply new technologies on their own, without being imposed by prescriptive specifications. Prescriptive specifications are the way to implement technical progress and ensure risk mitigation.

Shifting the risk ownership from Contractor to Owner is another item for debate. The question is how many owners are or will be willing to accept it.

## REFERENCES


[1] UCA of SME. Recommended Contract Practices for Underground Construction. 2008.

[2] B.N. Pennington & J.R. Collins. Recent Contracting Practices Sheppard Twin Tunnels Project. 1998.

[3] John J. Reilly. Owner Responsibilities, Selection of Tunnel Boring Machines. 1996.

[4] Wolfgang Roth, Dames & Moore. The Case Against Performance Specs in Tunneling; NAT. 1996.

[5] Martin & MacLennan. The Negotiated Compressed Process Before and After. RETC Conference Proceedings. 1995.

[6] Martin & Leray. Negotiated Compressed Process for the World's Largest Privatized Water Supply System. ITA/NAT Proceedings. 1996.

[7] www.tunnelmachines.com/lovat/.

[8] www.tunneltalk.com.

[9] http://tbm.cat.com/.

[10] www.york.ca.

[11] www.mcnallycorp.com/

# Saving Money with a Coordinated Fire Life Safety Design

**Baljinder Bassi, Nishant Nayan, and Marc Morgan**
CH2M HILL

**ABSTRACT:** In the transportation industry, Fire Life Safety (FLS) and tunnel ventilation form an essential part of any tunneling design. The reasons that safety and comfort are so important are that an unsafe/uncomfortable system will not be successful commercially or politically. However, because the cost of FLS solutions represents only a small percentage of the total cost, the requirement for early coordination and design is often overlooked. This paper demonstrates the key aspects of a well-coordinated FLS design should be taken into account in order to minimize total project construction, operation and maintenance costs.

## INTRODUCTION

In the transportation industry, Fire Life Safety (FLS) and tunnel ventilation form an essential part of any tunneling design. The reasons why safety and comfort is so important is quite simple, an unsafe/uncomfortable system will not be successful commercially or politically. However, because the cost for design and construction of tunnel ventilation and FLS solutions themselves represent only a very small percentage of the total cost, the requirement for coordination and design at an early stage is often overlooked. But their impact on the civil and geotechnical design and costs is immense. This is because a coordinated design has a major impact/influence on fundamental design elements, including:

- Ventilation plant room sizes
- Jet/Booster fan niche dimensions
- Ventilation/draft relief shaft dimensions and locations
- The requirement of air cleaning equipment in road tunnels
- Electrical switch rooms, sub-stations, power distribution, cable ducts, and banks
- Tunnel geometry—TBM tunnel diameter or floor to ceiling height in the case of cut and cover tunnels
- General architectural design, including location of stair cores
- Egress stairs/escape stairs/cross passages design, spacing, and configuration
- The use of passive and active means of smoke control

The impact of bad design or not considering fire and life safety at an early stage often leads to inefficient FLS and pollution control systems and increased construction, operation, and maintenance costs. In the event of an emergency, these critical systems may not perform as intended if the design is not performed correctly, and hence the safety of the evacuating passengers may be compromised.

The cost of the remedial work because of undersized or ill-conceived fire life safety components after construction is completed obviously involves higher costs than would be required initially, and this does not even consider the impact to a company's reputation.

This paper demonstrates the key aspects of a well-coordinated tunnel ventilation design which should be taken into account early in the design process.

## BACKGROUND

A successful design will consider comfort and safety with paramount importance. The comfort and safety of tunnel occupants is primarily concerned with temperature, air velocity, air quality, pressure transients, and fires. The following describes how comfort and safety are applied to design principles.

### Comfort

#### *Temperature*

Temperature can adversely affect patron comfort. A system that is too warm or too cold may fail to attract patrons. Additionally, the movement of air increases patron heat loss. This tends to enhance patron comfort during warm weather. Temperature criteria are developed to provide a satisfactory level of patron comfort and safety and a suitable environment for the operation of mechanical and electrical (M&E) wayside equipment in the tunnels and on board the vehicles. Of primary concern is the maximum allowable air temperature in order to maintain efficient operation of the proposed air conditioning condensers. Modern trains typically have condensers located on the train rooftops, and in this case the air temperature of the stratified layer of air above the train should be considered. However, the location of the condensers on older rolling stock is often maybe beneath the passenger cab and so temperature at low level also needs to be considered. Information from the manufacturer of the train cars is used to increase design efficiency.

During congested periods, the temperature in a tunnels after a train is stopped in a for a prolonged period of time can increase significantly. If temperatures are found to be excessive, ventilation response modes are identified. A defined ventilation response during congested operations can keep the tunnel temperature within the operating range of the air conditioning equipment onboard trains, thereby keeping passengers comfortable on stationary trains. The comfort of passengers stranded on board congested trains is the priority over people in stations because station occupants have the ability to move to a more comfortable environment.

### *Air Velocity*

Excessive air velocities can be become a nuisance to both patrons and employees, and can result in moving dust and generating noise. The one-dimensional model of normal operations is used to analyze system characteristics with a varying number of trains per hour. Train movement can be adjusted to model two trains arriving at stations simultaneously, one train arriving at a station with a stationary train at the platform and two trains simultaneously leaving a station. Where peak velocities exceed safe limits or other design criteria are exceeded, the model ventilation shaft dimensions will be adjusted, to the extent possible, so that the requirements of passenger comfort and safety are achieved. Where possible, shaft and tunnel dimensions are optimized so that costs for civil and structural work can be minimized. Tunnel and shaft dimensions can affect portal pressure transients, so a pressure transient analysis is undertaken during design.

### *Pressure Transients*

Portal pressure transients are important because rapid fluctuations in pressure can occur onboard and around a train, especially as trains are passing each other. For lower speed trains like typical subways, the rapid pressure changes can be felt by passengers aurally and cause discomfort. It is important that these pressure changes are analyzed because typical subway cars are unsealed, and therefore immediately affected by the pressure conditions around the train. High speed trains are usually sealed, but still have to be concerned about pressure transients.

High speed trains create such extreme and rapid pressure transients that passengers can be uncomfortable even onboard sealed trains. High speed train sealing helps lessen the effect of pressure transients for passenger's onboard trains by making the inside of the train adjust to rapid changes in external pressure at a slower rate. Each train manufacturer can provide the rate at which the pressure onboard a train is affected by pressure around the train, and the rate provided includes all paths where air could leak such as windows, doors, joints, and ventilation intakes and exhausts. This rate of change is used to design the tunnel and its operating limits. Itt is important to remember that train sealing provides a subjective level of aural comfort, and limits and solutions can vary from one jurisdiction to another. A coordinated design will consider the pressure transients early in design to avoid expensive retrofits and operational limits for pressure mitigation. This is especially important when train headways may be reduced for future operations.

## Safety

### *Level of Protection*

The recommendations/requirements for almost all national and international guidelines for fire life safety in underground or enclosed tunnels and stations can be separated into two key components:

1. Maintaining a tenable environment in the direction of evacuation
2. Communications and the ability of patrons and staff to self-rescue

It is the designer's responsibility to provide a safe environment in which occupants can either self-rescue or safely remain in place until first responders arrive. The means by which this is achieved are numerous and can include points of egress, emergency ventilation, automatic fire detection and suppression systems amongst others. However, the design of the fire and life safety systems should not be undertaken as a secondary consideration that is designed to 'fit' into the structural envelope that has been or is being designed separately.

**Fire Size and Growth Rate.** The first step in efficiently designing a safe system is determining the proper design criteria including the fire size and growth rate. The fire size and growth rate need to be representative of a realistic emergency scenario. Designing for a fire that is too large or grows too fast can rapidly escalate the cost of a project. For example, an unrealistically large fire may require excessively high velocities, which may start a domino effect whereby more or larger fans are needed, requiring larger ventilation plant rooms for the fans, larger ducts, larger street gratings, larger fan niches, greater generator capacity, etc. the escalation in cost can get out of control quickly.

The fire growth rate can also greatly affect the civil design because of delays associated with detection and initiation. It takes a discrete amount of time to detect a fire, activate the fans, allow the fans to ramp up to full speed, and allow airflow to develop.

In addition the fire size and the growth rate also have an impact on the structural integrity of the tunnel.

**Smoke.** Hot smoke from a fire in a tunnel rises due to buoyancy forces. Due to a complex process of mass and heat exchange, the smoke is gradually cooled and mixes with the air. After a period of time, both upstream and downstream sections of the tunnel can be completely filled with smoke. Thus stratification is a temporary phenomenon, and experience shows that it may not last for more than about 15 minutes unless it is maintained by appropriate ventilation including extraction from the ceiling and/or control of longitudinal airflow. This time period is essential for tunnel users to egress from the emergency location.

**Pressure Transients.** Pressure transients are also a safety concern because they can cause aural harm to people in or around a high speed train. Although trains are sealed for comfort, the train and tunnel must be designed to meet medical safety criteria in the event the sealing is broken or malfunctioning. People such as maintenance personnel in the tunnel or adjacent rooms could also be aurally harmed by passing trains if the tunnel and train are not designed properly. Designing the cross section of the tunnel properly and efficiently will promote safety, avoid over design, and reduce capital cost.

Unfortunately, this is often the case because of the costs involved, the design of the tunnels, stations and other structural components is undertaken well in advance of the fire and life safety systems that are typically designed in accordance with NFPA standards.

### Performance vs. Prescriptive Design Approach

Two of the most widely adopted guides are NFPA 130* and NFPA 502†. In North America they are the most common reference standards adopted by the Authority Having Jurisdiction (AHJ). It is generally assumed that both of these documents are prescriptive in that the requirements are stated in terms of what must be provided. However, much of NFPA 130 actually requires extensive interpretation, and some requirements, such as smoke control in tunnels, are expressed in terms of objectives to be achieved as opposed to a prescriptive manner. This allows the designer much freedom to choose the approach to be adopted. The effectiveness of the approach must be demonstrated, usually by simulation. Further, the NFPA guides allow alternative solutions to all prescriptive requirements, providing an equal level of safety compared with a compliant solution can be demonstrated. Therefore, each system with its own constraints can balance how safety is achieved by evaluating costs of each safety measure versus how the measure fits into the safety framework, which will tend to select those measures that best fit the local conditions and avoid those that are difficult or costly to install. The key is that the overall level of safety must be equivalent to a compliant solution.

Designing to NFPA 130 or NFPA 502 generally requires experience and appropriate skills in analysis to support the design process. The ability to determine the most appropriate approaches to safety for a given system whilst maintaining the overall level of safety is an important requirement on the designer. As mentioned above, the protection of users can be achieved in many ways and these can split into two forms; active and passive measures.

## COORDINATING THE DESIGN

Design coordination can either involve passive measures, active measure, or a combination of both.

### Active Measures

#### *Ventilation*

Ventilation is the most common form of active measure. The emergency ventilation system generally consists of ventilation shafts and emergency fan equipment located strategically along the alignment. The ventilation system may incorporate equipment located within plant rooms or jet fans (often also called booster fans) distributed along the tunnel.

It should be remembered that the ventilation system may also be part of the tunnel operating strategies for normal and emergency conditions. Hence the systems are often sized and designed for pollutant control, tunnel congestion, and fire and smoke conditions.

The mechanical tunnel ventilation system should not be confused with any HVAC used for comfort control or air cleaning equipment such as electrostatic precipitators. The primary goal of the

* NFPA 130, 2014 Edition, Standard for Fixed Guideway Transit and Passenger Rail Systems.

† NFPA 502, 2011 Edition, Standard for Road Tunnels, Bridges, and Other Limited Access Highways.

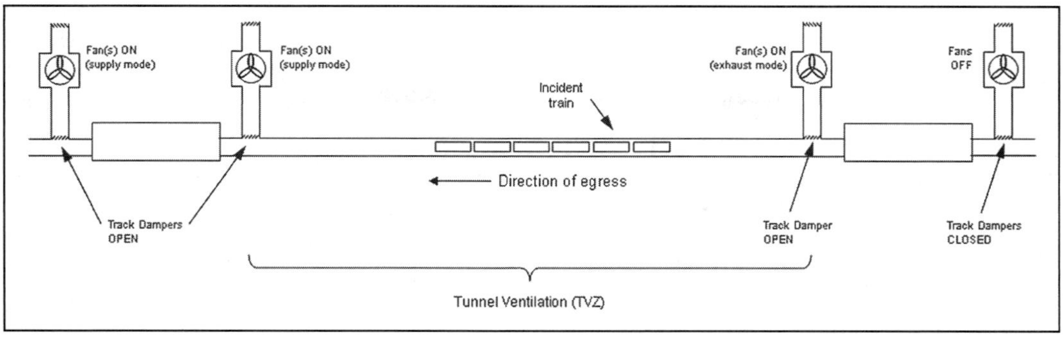


**Figure 1. Multiple fan response strategies**

mechanical tunnel ventilation system is to increase passenger safety and aid egress during an emergency, such as a fire within the tunnel network. Engineering analysis is required in order to size the mechanical ventilation plant, coordinate locations, and identify which fans should be operated for potential locations where a train may be on fire. Oversized fans or operating too many fans is undesirable as this may hinder passenger egress because air velocities are too high. Conversely, by under sizing the ventilation plant or not operating sufficient numbers of fans, it may not be possible to control the direction of smoke movement and prevent 'back layering'. This would again negatively impact safety. These calculations are normally undertaken using specialist software which include models for airflow within tunnels, stations and shafts, and the effects of fans, rolling stock operations, traffic density, ambient conditions, and so forth. Non-incident trains, stations or backed up traffic should be protected from the effects of smoke in a tunnel, so the ventilation system must be designed to avoid smoke being pushed towards egress routes or overshooting ventilation extract points and passing into the rest of the system.

Jet fans in tunnels are often considered the best option because the equipment is cheaper than the equivalent fan plant required in shafts or at stations and takes up far less room. However, this type of equipment is far better suited to road tunnels than mass transit or rail tunnels. Maintenance in rail tunnels is an ongoing issue because of the impact to operations. Whereas in a well-designed road tunnel with more than one traffic lane–in each direction, it may be possible to gain access but still keep some traffic lanes open, in a rail tunnel this is not an option.

In short tunnels it is often not possible to generate the necessary longitudinal flow using jet fans due to fan spacing constraints. Fan interaction and distances to portals means that you cannot place fans too close together. Therefore the maximum longitude flow that can be generated is limited and so it may not be possible to ventilate for the design fire size.

In very long tunnels, providing primary and backup power to remote fans that may be a significant distance from the power source should also be considered as this impacts cable sizes, duct banks, electrical plant room dimensions and so on.

In order to minimize costs, the mechanical ventilation system and draft relief shafts should be designed so that they share a common shaft. Airflow is directed in the desired direction using specially configured dampers. It is also possible and desirable to operate multiple fans so that fan equipment is sized economically, and response strategies utilize as much equipment as is required and available. This enables the designer to reduce fan sizes, physical shaft dimension, and power infrastructure costs and, potentially, incorporate redundancy for fan failure. A good example of this application is in the case of tunnel fire scenarios where a longitudinal or 'push-pull' ventilation strategy is desirable. The direction of air flow is intended to facilitate passenger egress and provide a tenable environment in which passengers can self-rescue. The number of fans supplying air to the tunnel can vary but in general; it is not common practice to exhaust from multiple sites as the possibility exists of 'pulling' smoke and toxic gases into the station environments. Figure 1 shows a fire response strategy for a hypothetical system.

The engineering analysis and simulations to be undertaken as part of the design process will determine the exact modes of operation for the mechanical ventilation and dampers at stations on either side of the incident tunnel ventilation zone. However, it may be necessary to promote air flow in the desired direction by closing track dampers at locations where the ventilation plant is not necessarily active. This is an effective way of reducing 'leakage' through to atmosphere and more efficiently promoting air flow in the desired tunnel sections.

Another important factor to consider is the cross sectional area of the tunnel. The means of determining if the tunnel ventilation system will be effective in providing a safe means of egress is by the calculation of the critical velocity.* The tunnel system is then modeled in one-dimensional simulation software, and analysis of the ventilation response strategies (modes of operations) is conducted. Physical tunnel parameters, such as the cross sectional area, tunnel perimeter, hydraulic diameter, lining, and grade all play a major part in the calculation of the critical velocity. One result of a larger cross sectional area is that more air is required in order to achieve critical velocity. Hence, larger ventilation plants will be required in order to ventilate tunnel ventilation zones where sections of twin track single bore tunnels exist between stations. Larger ventilation plants require larger mechanical plant rooms, larger primary electrical supply capacity, and if emergency power is in the form of on-site generators, larger emergency generators and diesel storage tanks. The need to provide a greater volume of air will also impact the size of ventilation shafts.

It is important to select the most appropriate and cost effective ventilation system early in design. Addressing ventilation later in design may limit the available options that do not involve a full redesign of the civil works.

### *Fixed Fire Fighting Systems (FFFS)*

FFFS in transit tunnels and public areas of stations are comprised of standpipes and sprinkler suppression systems. There are many different options when designing sprinklers including deluge, foam, mist, wet, dry, automatic, manual, etc.

The use of a sprinkler system can have a large influence on the size and growth rate of a fire. Sprinklers have been used as a justification to reduce the design fire size and fire growth rate. This in turn can help reduce the size and cost of the ventilation system and this will therefore impact the size of shafts, plant rooms and electrical loads.

### *Fire Detection*

The fire detection system determines how quickly a fire or emergency is detected and in turn how quickly an appropriate response can be initiated. Typical detection systems include manual pull stations, heat& smoke detectors, and manual or automated CCTV systems. Often a combination of these systems is used to ensure redundancy and limit false alarms. The key is that the detection system is able to quickly detect and identify the location of the fire. This will save lives for obvious reasons and could also be used to reduce costs because, for example, early detection initiates the FFFS, the FFFS ensures the fire does not grow too large and so reduce the size and capacity of the ventilation equipment and structural damage during a fire.

## Passive Measure

### *Ventilation*

Ventilation generally involves the design of ventilation shafts for normal operations and the design of aids for smoke control such as down-stands at stations.

In a rail system, the goal of a 'passive' ventilation systems is primarily to relieve pressure in the tunnels and exchange air with outside. The primary mechanism by which the tunnels are ventilated is through the "piston" action. As a train moves through the tunnels, the train 'pushes' an air column in front of itself causing a pressure differential along the length of the train. As the train accelerates, this pressure increases, and wherever changes in tunnel geometry or connections occur, pressure transients/ fluctuations occur. Very high pressures or pressure transients can give rise to severe discomfort for passengers on board the trains, as well as patrons within the stations. Pressure is relieved via draft (often also called piston) relief shafts, station exits and entrance ways, and other tunnels. However, with Platform Screen Doors (PSDs) installed, the relief of pressure via station ways and also the impact of pressure and the resulting high velocities within stations is reduced thereby increasing passenger comfort. In addition, high pressures can affect the function of wayside equipment within the tunnel.

However, the piston effect should not be treated as an undesirable nuisance. The aim of the passive ventilation system is, during normal operations, to make use of the piston effect in order to manage temperatures within the tunnel system. As trains move through the system, heat is generated by the on board air conditioning, braking systems, way side equipment, people, lighting, etc. This equipment can result in very high temperatures within the tunnel. This cooling load can best be controlled by removing warm air from the tunnels and introducing fresh air into the tunnels.

In order to take advantage of the piston action, air is exchanged with the outside environment via open draft relief dampers that connect the tunnels to the outside environment and, to a lesser extent, through the stations and portal. In general, draft relief shafts are located at or near all station approach tunnels. The location of the draft relief shafts and how they interconnect with the tunnel system must be carefully considered and engineered so that the passive systems are coordinated with the emergency

* NFPA 130 (2014) Section 3.3.13 The minimum steady-state velocity of the ventilation airflow moving toward the fire within a tunnel or passageway that is required to prevent backlayering at the fire site.

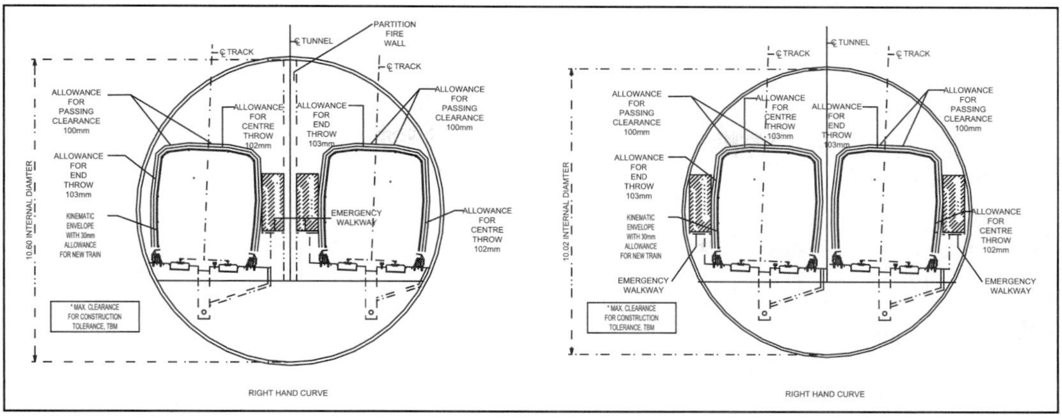


**Figure 2. Large bore diameter tunnel**

ventilation systems in order to maximize effectiveness and minimize costs.

Passive measures used for smoke control include down stands, enclosures around vertical transportation and provision of safe refuges. Smoke reservoirs can be used at a high level to collect buoyant smoke in the early stages of a fire before mechanical ventilation is activated. Platform screen doors are able to help keep smoke within the tunnel, thereby limiting the spread of smoke at the early stages of a fire. Effective use of passive ventilation measures can significantly reduce construction and running costs.

### *Platform Screen Doors (PSDs)*

PSDs are often not considered because of expensive train control systems required, PSDs can significantly reduce emergency fan plant requirements. They can also improve patron safety and comfort during normal operations. This is especially true in locations where the stations are air conditioned and they therefore help reduce the running costs associated with station HVAC plant. Hence pay back if calculated correctly could be very attractive. But the other advantage is with the emergency ventilation. The major points at which air is able to exit the system are at the portal and stations. By installing full height PSDs the efficiency of the emergency ventilation is increased for all tunnel based fire incidents. In addition, if the PSDs and any overhead track exhaust systems are designed correctly, station based fires may result in less smoke entering the station platform area and this improving tenability in these areas. Full height PSDs may be considered passive smoke control measures, as previously described because physical barrier helps contain smoke.

## EGRESS

Consider a large bore TBM tunnel in the 33 ft + diameter range. On the face of it, reducing the diameter of the TBM will save a lot of money, potentially reduce construction time and possibly help minimize issues such as ground settlement. The tunnel could be a common twin track metro tunnel with trains traveling in opposite direction located within the same bore. The contractor in this case was advised by the FLS design team to opt for a slightly larger TBM than the structural and geotechnical teams had identified as necessary, in order to allow for the construction of a center dividing wall and the use of cross passages. Figure 2 shows side by side the two tunnel cross sections that were considered. Unfortunately the contractor ordered the TBM before the complete FLS egress design requirements were issued and did not factor in egress stairs when they undertook their cost and schedule comparisons.

With the center dividing wall the single bore tunnel is effectively split into two. If the wall is designed as a fire barrier the non-incident track provides a safe route to egress with the use of sliding fire rated doors spaced every 800 ft. With no ability to link to a parallel tunnel and make use of cross passageways, egress stairs spaced every 2,500 ft must be constructed. The cost of providing the stairs to grade and the impact to schedule was not considered. In addition, the availability of land at grade, with some of the most expensive real estate in the world, and the disruption in very densely populated parts of the city has left the design build contractor political and cost headache. Furthermore, by splitting the tunnel into two smaller tunnels with smaller cross sections the air flow rates required are significantly reduced, and so continues the domino effect.

# Balancing Innovation, Risk, and Cost: Innovative Procurement Approach for Sewer Tunnel Rehabilitation

**F. Klingler**
FK Engineering Associates

**J. Edberg**
NTH Consultants, Ltd.

**M. McMahon, S. Lockhart, and K. Graboske**
Office of the Oakland County Water Resources Commissioner and Engineer for the Oakland Macomb Interceptor Drain Drainage District

**ABSTRACT:** In an industry where innovation, risk, and cost are intertwined, Contract 4 of the $160 million OMID sewer rehabilitation program presented an especially difficult problem: How to attain a competitive price for tunnel lining product, when the project conditions and technical requirements severely limit choices. While unique and complex underground challenges are often solved by including a sole-source requirement in the bidding documents, the design team for this project implemented a different approach. The final result was a winning bid that used fewer access points, included a very high quality lining system, and was 28% below the Engineer's cost estimate.

## INTRODUCTION

The 33 km (21 mi) long Oakland Macomb Interceptor Drain (OMID) sewer was built in the 1970s to serve about 830,000 users in the adjacent counties of Oakland and Macomb, Michigan. The sewer is owned and operated by the Oakland Macomb Interceptor Drain Drainage District (OMIDDD), which represents both counties in the operation and administration of the system. The sewer ranges from 0.9 m (3.0 ft) up to 3.9 m (12.75 ft) in diameter and is about 25 m to 32 m (80 to 100 ft) deep. It is non-redundant, meaning it cannot be diverted for repair purposes.

Several collapses or near collapses of the sewer have occurred since construction, each requiring an over-ground pumping bypass of up to 17.4 cms (520 cfs) of flow. After the most recent failure in 2004, an inspection revealed severe distress through much of the system. A global approach for accessing and repairing was subsequently developed, which involved an in-system flow storage concept, together with construction of access structures at various points in the system. The access and flow-storage system allowed for access at strategic points along the system, and provided for internal repairs in 7 to 15 hour periods, depending on the location within the system. During these repair periods, flow would be held behind flow control gates, and stored within upstream portions of the sewer. The flow would then be released over 9 to 17 hour periods to create available storage for the next work day.

The access and flow-storage system was built under Contract Nos. 1 and 2 of the $160,000,000 OMID rehabilitation program. Grouting and other repairs were performed under Contract No. 3 of the program. Contract 4 of the program involves installation of new lining within four sections of the OMID system, each ranging from about 260 m (800 feet) to about 4800 m (15,000 ft). The total amount of new structural lining under Contract 4 will be about 7,400 m (24,000 ft). The procurement of Contract 4 is the subject of this paper.

## EARLY RECOGNITION OF LINING CHALLENGES

During the initial planning and basis-of-design stages for Contract 4 of the project, it became clear that full scale relining of the pipe would be necessary for a major portion of the existing alignment. At the same time, a number of challenges were recognized, that together might severely limit the options for such relining. Major project challenges included:

- Sewer Depth and Access: The area of the sewer to be relined is 25 to 32 m (80 to 100 feet) below ground surface, making

entry and egress from existing shaft locations somewhat difficult. Further, the depth ensured that any new access shafts would be a major cost to the project.
- Ground Conditions: The sewer extends through several areas of predominantly silt and fine sands. In addition, the depth below groundwater table results in groundwater pressure head against the lining of 24 m to 32 m (75 ft to 100 ft)—more than enough to drive silt and fine sand particles through very small cracks in the existing lining or any new lining. Therefore, the new lining needed to be essentially impermeable.
- Sewer Access: While Contracts 1 and 2 involved installation of 5 access structures, it became evident that depending on the lining system employed, additional access structures would be needed. A separate issue involved the fact that a dewatered sewer could only be accomplished in 7 to 15 hour intervals, which would severely limit the access time for repairs.
- Initial Sewer Construction: Most of the sewer was constructed as a tunnel in deep soft ground with a steel rib primary liner and an unreinforced cast-in-place concrete secondary liner. The type of sewer wall combined with the evidenced deterioration and the ground conditions, resulted in a requirement for a structural lining.

Initial efforts to identify potential lining options that would adequately address the challenges revealed a very limited number of products that might be adequate. In fact, no product could be identified that possessed a proven track record under similar conditions and project requirements. Further, research identified a number of examples where one or more of the conditions of the OMID project had resulted in major problems (and in some cases catastrophic failures) with respect to lining installation and/or performance. In short, it became clear that given the depth, diameter, and system constraints involved in the proposed relining under Contract 4, the scope and scale of the proposed work was unprecedented.

As a result of research and outreach efforts by the engineer and owner, a number of lining manufacturers as well as contractors expressed interest in the project. Many of them indicated that their product could be designed to overcome the installation and performance challenges. Based on the initial research, it appeared that certain products might be able to meet the challenges, while other products likely would not. In any case, the available manufacturer information was inadequate to make a determination, and it became obvious that most if not all the lining manufacturers and contractors did not fully understand the challenges that the project presented. Further, a separate concern developed, that if a specific product or class of products were specified based on the initial information, the pool of prospective product providers would be very limited, and price competition would be minimal. In consideration of this, the engineer and owner determined that a major goal of the bidding process would be to allow bidding using as many types of lining systems as possible, without compromising the intended quality of finished product, and without increasing risk of successful installation. Further, it became apparent that because of the complexity of the project challenges, manufacturers and contractors would need to be fully aware of the depth of the challenges and risks. Such awareness is often very difficult to achieve in the relatively short time that most projects allow for bidding.

For all of the above reasons, a two-stage approach to procurement was developed. Stage 1 consisted of a "Request for Technical Submission" (RFTS) phase where lining manufacturers were invited to submit detailed product information. After detailed review and analysis of manufacturer's submittals, seven products were "pre-approved." Stage 2 consisted of a final design and bidding stage, during which contractors were invited to bid on a base and alternate scope of work, using one (or a combination) of the pre-approved lining systems.

## STAGE 1: REQUEST FOR TECHNICAL SUBMISSION

In addition to the lining products that were initially identified as having some track record with respect to at least some of the project challenges, a number of other products were identified that warranted further consideration. Some of these products could demonstrate the ability to overcome certain challenges, but were relatively unproven with respect to other challenges. Others that showed promise had only been used on a much smaller scale and/or had very little track record in the United Stated. None of the products identified could demonstrate experience overcoming the combined challenges of Contract 4.

It was determined that the initial stage of the procurement would need to be aimed at manufacturers of lining products. It would need to clearly define the existing conditions, clearly define the installation and performance challenges, and clearly define the installation and performance requirements. Since it would be a new and complex process, some of the information requested in the RFTS might require testing and/or design analysis by the manufacturer to prove their product capable of meeting the project requirements. To allow for development of such information, a three-month period was programmed into the RFTS stage of the procurement.

### Communication of Project Challenges

The RFTS document was developed with the objective of obtaining the information necessary to compare different products with respect to their technical merits, without consideration of price. A major consideration of the process was accurately communicating project conditions and challenges to the manufacturers. This communication and full disclosure at such an early stage allowed some manufacturers to recognize that their product was not appropriate for the project, and to withdraw from the process. For others, clear and comprehensive communication of the project challenges and conditions allowed manufacturers to better demonstrate the ability of their product to overcome such challenges.

The specific challenges and conditions that were conveyed in the RFTS document included summary of numerous issues, as well as providing backup and references wherever warranted. A number of issues were summarized for the lining manufacturers to consider. These included the following major issues, as well as a number of others:

- **Access**—Definition and successful communication of this issue to manufacturers was considered critical to their ability to evaluate the applicability of their product, and to communicate such applicability to the engineer and owner.
  - **Shafts:** The specifics of existing access structures were defined, and details were provided regarding allowable locations for additional access structures. On this basis, manufacturers could determine if their product was feasible given the existing and potential access.
  - **Flow Control:** The RFTS described the flow control system in detail, referencing the "Hydraulic Report for Flow Control" and providing a summary for the work proposed under Contract 4. Descriptions of the existing flow control system were provided, together with information on expected wet and dry weather flows, and how these would be controlled to facilitate the construction. On this basis, manufacturers could determine if the installation period available each day between flow release periods would be adequate for successful installation of their product.
- **As-Built Lining**—A summary of the condition of the as-constructed lining for the various reaches was provided, including lining thickness and type (primary and secondary), concrete strength, tunnel cross section, joint details, etc. These issues were defined for the four general reaches to be relined. The existing line and grade of the tunnel was provided, as the sudden line and grade changes in some areas would potentially limit slip-line pipe lengths for certain manufacturers. The locations of curves and deflected portions of the alignment were also identified as they would be critical for most lining technologies. Drastic grade changes, ramps, and areas with negative slope were identified, which generally would be an important consideration for any technology.
- **Existing Tunnel Conditions**—The condition of the underground system was defined:
  - **Tunnel Structure**—The average condition of the tunnel in the individual reaches to be relined were defined in terms of Pipeline Assessment Certification Program (PACP) grades, as well as general descriptions of the structural and serviceability conditions. Details of past inspections including laser profiling to determine section loss of the concrete lining were included.
  - **Subsurface Conditions**—The combined conditions that resulted in a requirement for an impermeable lining system were detailed. This included definition of the silt and fine sand condition, combined with the high exterior water head acting on the tunnel lining.
  - **Tunnel Atmospheric Conditions**—The corrosive environment of the interceptor was described and test information provided. Moderate to high concentrations of hydrogen sulfide, and smaller quantities of other potentially corrosive compounds (including petroleum products) were identified.
- **Definition of the Proposed Contract 4 Requirements**—The RFTS provided preliminary design for the proposed construction, including drawings, and minimum requirements. It also detailed structural design and performance criteria; and criteria for corrosion resistance, water tightness, and abrasion resistance.
- **Other Requirements**—For the purposes of the RFTS respondents' evaluation of necessary manufacturing rates, shipping requirements, and installation rates, a generalized Gantt chart schedule was prepared and included in the RFTS document. Various other requirements that were expected to potentially impact the product manufacturers were defined in the RFTS, including

insurance requirements, warranty requirements, and potential conflicts with other ongoing projects.

### Stage 1 Submission Requirements

An important goal of the RFTS was to adequately communicate the submission requirements, in order to get consistent submissions, and to better allow evaluation of the submitted information. All of the requirements were specified in such a way that Grout-in-Place-Lining (GIPL) systems, Slip-line systems, and hybrid systems could submit.

Manufacturers were required to complete a "Basic Information and Experience Documentation Questionnaire" which included requests for financial and business information, similar successful and unsuccessful projects, manufacturing plant locations and capacity, etc. Manufacturers were requested to provide general information on a quality control program for production and handling of their products; including details such as materials testing for the products used for the liner, grout (either structural or non-structural), bond testing between the grout and the liner (if applicable), installation tolerances, water-tightness of seams/joints/etc., surface preparation of the existing liner, pull out/bond testing (if applicable), testing methods for voids in the grout installation, testing to establish minimum requirements for protection of the product during transportation, handling and installation, etc.

Specific Structural Evaluations and Test Data were required including calculations, evaluations, boundary condition assumptions, material property assumptions, test data, etc. Documentation and test data were required to demonstrate water tightness and resistance to corrosion.

The RFTS submissions were required to include a detailed installation narrative for the proposed liner system, addressing various practical aspects of the installation, and describing how their system would be installed, while addressing all of the system constraints and performance requirements.

In order to develop further interest in the marketplace, advertisements for the RFTS were placed in various national tunneling and underground rehabilitation publications. The design team also reached out to manufacturers that they felt might be interested in the project. In addition to lining manufacturers, contractors were encouraged to procure and examine the RFTS documents. This brought contractors and contractor-supplier teams into the process early.

During development of the RFTS as well as during the 3-month response time, the engineer and owner encouraged manufacturers to invite Engineer and Owner personnel (at Owner expense) to visit manufacturing facilities and product installation sites to demonstrate the capabilities and applicability of their product to this project. About half of the manufacturers that ultimately submitted technical documents extended such invitations. In all cases, the visits proved critical to evaluating the merits and potential problems with the lining systems. In the writers' opinions, the visits to active lining installation sites proved to be the most beneficial.

### Evaluation of Manufacturer Submissions

Eleven submissions were received from manufacturers. These included submissions from three solid-wall glass fiber reinforced polymer mortar pipe slip-line systems, one composite-wall slip line system, three grout-in-place-liner (GIPL) systems, two hybrids, and two applied systems. Following a three-month evaluation period including three workshops and a number of interviews of respondents, the design team pre-approved seven of the submitted systems for the bidding phase of the project.

Pre-approved products included the slip line systems, and three GIPL systems. None of the hybrid systems and none of the applied systems were able to show in their technical submissions that their product would meet the minimum requirements of the RFTS.

Although the pre-approved product manufacturers were able to show that their products would meet the specified performance requirements, most of them were unable to convincingly demonstrate the product installation challenges could be adequately addressed. In most cases, it was left to the designer and future contractor to determine how best to overcome such challenges. For this reason, the final design and preparation of the contract bidding documents was especially challenging.

## FINAL DESIGN STAGE

The final design stage of the procurement effort overlapped the technical submission evaluation period by about three months. The design team found during this time, that questions arising during the submission evaluation stage were particularly useful in further developing the design. For example, certain challenges were identified that had not been considered before; including specific components of water-tightness, specifying the proper structural number for the finished product, and host-pipe preparation specification.

It also became clear that three of the slip line products were better capable of handling installation and long term performance challenges and risk than the other four pre-approved products. For this reason, an alternate bidding approach was adopted. The approach required submittal of a base bid that used any of three specified slip-line products and a voluntary alternate bid that could include any combination of the seven pre-approved products. There were a number of other bidding requirements including submission of a project schedule, installation narrative, and detailed qualifications worksheet. Bidders were encouraged to provide a price for an extended warranty as value added, for consideration by the Owner.

The bid evaluation approach was developed in advance, to consider the combined information submitted by the bidder (price, installation narrative, qualifications, schedule, etc.), as well as the differences between the alternate bids and the base bids. It was also determined in advance that all responsive and responsible base bids would be evaluated on the basis of price only.

The bid forms and the bidding process in general, were structured in such a way to promote efficiency and innovation by the bidders. For example:

- The project bidding documents allowed contractors to access the sewer through three existing access shafts and/or up to six additional shafts to be constructed at predetermined locations. The bidders were allowed to determine the number and locations of access shafts that would be necessary for installation of their chosen lining product.
- The project bidding documents included two shaft types/sizes that would be allowed at each of the six potential new shaft locations. Bidders were allowed to determine the type/size of access shafts that would be necessary for installation of their chosen lining product.
- Combinations of the three base-bid-approved products were allowed for the base bid, and combinations of any of the pre-approved products were allowed for the alternate bids.
- An owner-controlled insurance program was utilized to achieve a balance of insurance cost versus risk between bidders.
- Bidders were invited to witness the flow control operations ongoing for the previous contract, to allow for innovative scheduling and staging of the work.

The Engineer's opinion of probable cost was $64.1 million. A total of seven base bids and two alternate bids were received. Base bids ranged from $46.4 million to $68.9 million, and alternate bids ranged from $43.7 million to $60.7 million. After evaluation, the low base bid, submitted by Jay Dee Contractors, Inc., was determined to have the greatest value to the Owner, and Contract 4 was awarded to Jay Dee for the low base bid amount.

## CONCLUSIONS

The writers believe that the procurement approach used for Contract 4 of the $160 million OMID sewer rehabilitation program, allowed for healthy competition among manufacturers without compromising quality; and allowed contractors to be innovative while minimizing risk. To the writers' knowledge, the approach was unique for this size and type of project. Positive aspects of the process used for this project are summarized:

- Better Understanding of the Project by Manufacturers and Contractors—The lining of the OMID Sewer under Contract 4 is virtually unprecedented in scope, complexity, and size. The RFTS process provided a forum for manufacturers and contractors to fully understand the adverse conditions and constraints of the project well in advance of the bid period, so that manufacturers and contractors could form teams early and strategize regarding innovative approaches. This level of advance information is unusual in the underground industry.
- Increased Competition—This project was very complex, with very specific needs regarding the lining component of the work. Such specific needs are often answered by sole sourcing. The RFTS process allowed for contractors to choose from multiple different lining systems and therefore increased competition among lining manufacturers. In addition, the long period between distribution of the design information contained in the RFTS documents and the final bid date, allowed for increased interest among qualified contractors in the underground industry. This is probably best demonstrated by the fact that seven bids were received, which the writers consider very favorable for a project of this level of difficulty and specialization. The writers believe the number of lining systems allowed, together with the advance publicity of the project from the RFTS solicitation, created significant competition among both manufacturers and contractors, and likely contributed to the very favorable bid amounts.
- Better Understanding of Product Issues by the Design Team—The RFTS process provided the design team with a level of information and access to manufacturer data that is not typically available for most projects. This allowed for innovative final design, and the ability to design for specific issues

and challenges raised by the manufacturers. While the project is only in the early stages of construction at this writing, the writers believe that the level of understanding of issues raised by the manufacturers, will lead to less problems during construction.

- Increased Innovation by Manufacturers and Contractors—The bid documents allowed for Contractors to include the number and type/size of shafts, and shaft locations that accommodated their chosen lining product. With up to 6 months that manufacturers and contractors had to review the documents and develop strategies, this approach encouraged contractors to develop methods of installation that would minimize the number of shafts, each valued at about $2 million to $5 million. The winning bidder developed an approach that allowed installation of the lining with only 2 new shafts, which the writers believe contributed to the favorable low bid amount.

The process itself was not without challenges, and these should be considered by anyone considering the process for their project. Some of the larger challenges are summarized:

- The RFTS process added about 4 months onto the design schedule, as well as the associated cost for the design team to develop the documents, hold workshops, conduct an RFTS advertisement and distribution, perform manufacturer site visits, and evaluate the documents.
- The process added a level of complexity (and cost) to the design, because the design needed to be conducted to accommodate whichever product(s) would be part of the winning bid. For example, separate specifications and drawings needed to be developed for the various lining systems that were pre-approved and allowed for the bidding.

In the end, the procurement process used for Contract 4 of the OMID Sewer rehabilitation, had significantly positive results. The process provided for manufacturers of different lining systems to compete, and allowed bidders to be innovative in their installation approach. The result was a winning bid that used fewer shafts, included one of the Owner's preferred lining systems, and was 28% below the Engineer's opinion of probable construction cost.

# Deep Rock Tunnel Connector Project Construction Status Update

**John Morgan and Tim Shutters**
Citizens Energy Group

**ABSTRACT:** Citizens Energy Group issued Notice to Proceed for the construction of a 5.5 m (18 ft) diameter deep rock tunnel to S-K, JV (J.F. Shea & Kiewit Infrastructure) on December 16, 2011. The project includes three (3) tangential vortex drop shafts, three (3) corresponding vent shafts, three (3) utility shafts, one (1) launch, and one (1) retrieval shaft. The Contractor's plan is to construct all shafts as early in the project as possible. Currently, the launch shaft, retrieval shaft, the three utility shafts and all three drop shafts are in the construction phase. The tunnel boring machine (TBM) launch date was March 15, 2013. The following is an update on the construction progress of the project.

## PROJECT OVERVIEW

The project is part of the United States Environmental Protection Agency (EPA) approved Long Term Control program which was formally approved in 2006. The tunnel work is included in the approved Consent Decree amended in 2009 & 2010. A portion of the Amendments cover the construction of a tunnel storage system. The project referenced in this document, Deep Rock Tunnel Connector (DRTC) project, see Figure 1, comes as a result of the 2009 Amendment and is one of five (5) portions of an overall deep tunnel storage system.

This includes the initial portion of a system capable of storing 946 million liters (250 million gallons) of combined sewage that would overflow into local rivers and streams. Total distance for the 5.5 m (18 ft) diameter tunnel is 40.2 km (25 mi), see Figure 2.

The intent of the system is to clean up the local waterways. Reducing the overflow of raw sewage is the goal of the tunnel storage system. Capturing the raw sewage and storing it until the wastewater treatment plants have capacity to properly cleanse the liquid via treatment is the overall goal. The flow is rerouted via drop shafts to the deep tunnel storage system, from discharge into local waterways, to one of two wastewater treatment facilities in Indianapolis. At our Southport Advanced Wastewater Treatment Plant a deep tunnel pump station is currently under construction. The intent of the pump station is to lift the raw sewage and deliver the flow to the wastewater treatment plant.

The project includes approximately 12.87 km (8 mi) of 5.5 m (18 ft) finished tunnel and eleven (11) shafts. The shafts include collection of three (3) combined sewer overflows via tangential vortex drop structure and three (3) corresponding vent shafts. The intent of the drop shafts is to deliver the flow to the tunnel. The vent shafts are used to allow air to escape prior to the flow entering the tunnel. Also included are three (3) utility shafts. The intent of the utility shafts is multipurpose. The launch and retrieval shaft make up the remaining shafts for the project.

The design intent of the utility shafts was to assist in providing needed support for construction operations (air handling/concrete delivery for tunnel liner construction/potential power drop). For this project, the contractor chose not to use the structures for TBM power delivery. The contractor elected to install booster stations throughout the tunnel. As this was part of a "means and methods" decision the choice was entirely up to the contractor. Once the tunnel is complete, and becomes operational, the shafts will be used for maintenance access when needed.

The utility shafts are also meant to provide tunnel access for any potential maintenance issues that may arise once the system is put into operation. The need came as a result of nearly 9.66 km (6 miles) of this portion of the system with no overflows to intercept resulting in no shafts allowing access. Additionally, this is the lower portion of the gravity fed tunnel as it terminates at a pump station. Given the fact that the flow stored in the tunnel as a result of an overflow event will not be introduced into the wastewater treatment plant until capacity is available there to properly treat such, there is a chance that settlement of solids that would otherwise be suspended could create a need for additional maintenance. These shafts will allow for such maintenance to take place.

The two remaining shafts, launch and retrieval, will serve obvious purposes for construction machinery and mucking operations (the retrieval shaft will

Figure 1. Deep rock tunnel connector alignment

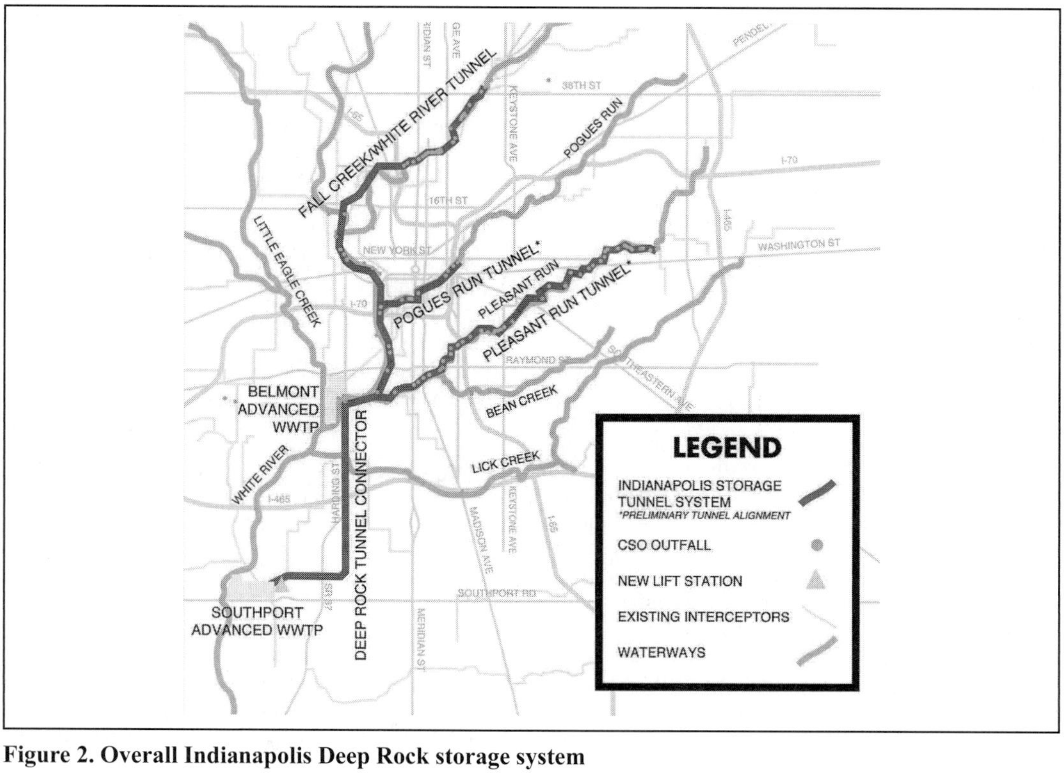


Figure 2. Overall Indianapolis Deep Rock storage system

serve as the launch shaft for the TBM on the next phase). They also serve as one of many delivery points needed to complete the installation of the concrete lining.

The Contractor's plan from the beginning was to construct all shafts early in the project. Once all shafts are constructed, the focus can be shifted to the TBM (assembly, launching, and tunneling).

## PROJECT COORDINATION

Since the project received Notice to Proceed weekly progress meetings have taken place. The original intent was to keep information flowing between Contractor, Designer, Construction Inspection (CI) Team and Owner. Also, given the fact that the Contractor had never worked in Indianapolis prior to this project the Owner found that helping the Contractor with points of contact to be a critical role in the success of the project.

At the first Dispute Resolution Board (DRB) meeting there was conversation that the weekly meetings may end and be scheduled for every other week or monthly. The DRB strongly suggested that the weekly meetings continue. This was to help keep communication open.

## LAUNCH SHAFT

The construction of the launch shaft began in mid-2012. The circular shaft is approximately 76.5 m (251 ft) in depth, see Figure 3. A concrete pad was installed to help keep the area clean and surface uniform throughout the slurry wall construction process. The initial 30.5 m (100 ft), 13.41 m (44 ft) finished inside diameter, was constructed via slurry wall method through alluvium. The first panel took longer to construct than the next two combined. The remaining 46.02 m (151 ft), 10.67 m (35 ft) finished inside diameter, was constructed via drill and shoot operation. Generally, two shots per week took place with mucking operations taking place between shots. Each shot resulted in approximately 3.66 m (12 ft) of disturbance.

The contact grouting phase at the alluvium/bedrock interface went well. After the soil was removed from the structure it was found that additional grout was needed behind the slurry wall panels in select areas. During the time period when this work was being completed, summer of 2012, Central Indiana suffered from a lack of rain. This was found to be beneficial to the project as it promoted better conditions for construction than previously anticipated.

At 30.5 m (100 ft) depth, bedrock was encountered. Following the excavation of the alluvium material, bedrock was removed via drill and blast method. The top 9.1 m (30 ft) of bedrock in this area is shale. The remaining bedrock 36.88 m (121 ft) consists of Jeffersonville Limestone. Specifically, the Vernon Fork Limestone Member deposited during the Devonian Era, see Figure 4.

While the design took place discussions were coordinated with local mining organizations. During these discussions information was shared regarding the strength of the limestone at various depths. This coordination helped to identify the overall depth of the tunnel for the deep rock system. Although the limestone is strong and holds it's shape with minimal to no support at the depth we are using, the material is not found to be highly marketable for an aggregate operation.

During design the geotechnical investigation identified a high level of methane existed. Although the level was not above safe limits the Contractor was required to monitor methane levels during excavation in this zone. Fortunately, no problems were encountered during construction.

The contractor's plan was to temporarily cease the drill and blast operation when the depth of the launch shaft reached approximately 68.6 m (225 ft) below grade. This was done in order to construct the concrete liner for the structure. This was the approximate depth of the tunnel crown. The shaft lining construction sequence chosen by the Contractor was logical.

## RETRIEVAL SHAFT

The retrieval shaft is very similar to the launch shaft in shape and diameter, see Figure 5. The slurry wall construction phase was moved to the retrieval shaft once the launch shaft was completed. The operation included installation of a concrete pad, similar to the launch shaft. In late 2012 this construction phase began. Several challenges were encountered at this location. This was due to the fact that up until approximately 80 years ago the river actually flowed through this corridor. Due to a flood event in Indianapolis in 1913, the river was re-routed. This area was filled in as part of the work related to the river re-routing. The fill used resulted in some construction challenges for this project.

The spoil in this area has been totally removed. Depth from ground surface to top of bedrock is similar to the launch shaft previously discussed. In this area the shale layer is absent. Grouting of the interface between the alluvium and bedrock has taken place.

## TAIL AND STARTER TUNNEL

The tail and starter tunnels were constructed during the Fall of 2012. Both the tail and starter tunnels were constructed via drill and shoot method exposing the entire circumference along the way. The tail tunnel was approximately 22.9 m (75 ft) in length. The

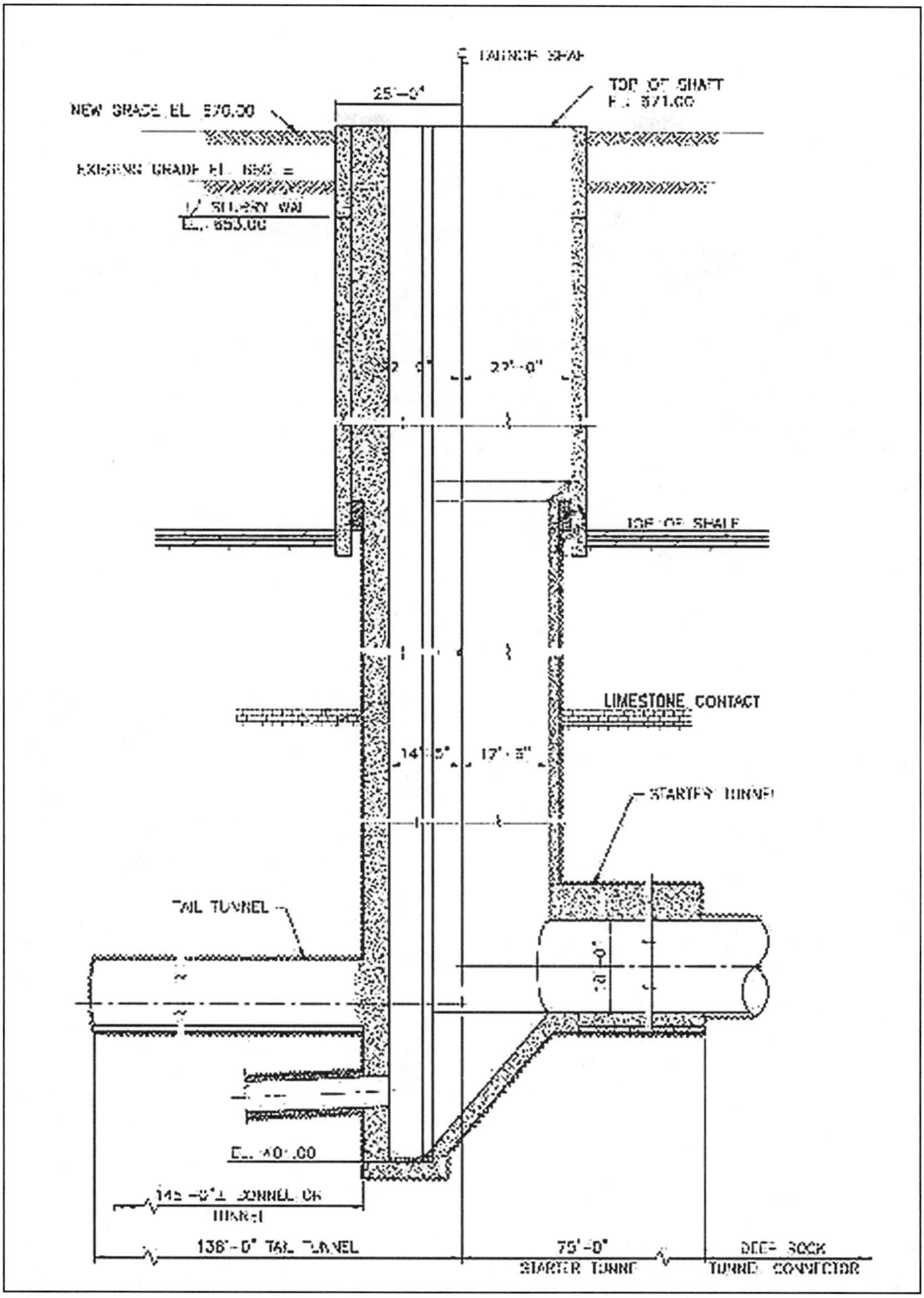


**Figure 3. Launch shaft configuration**

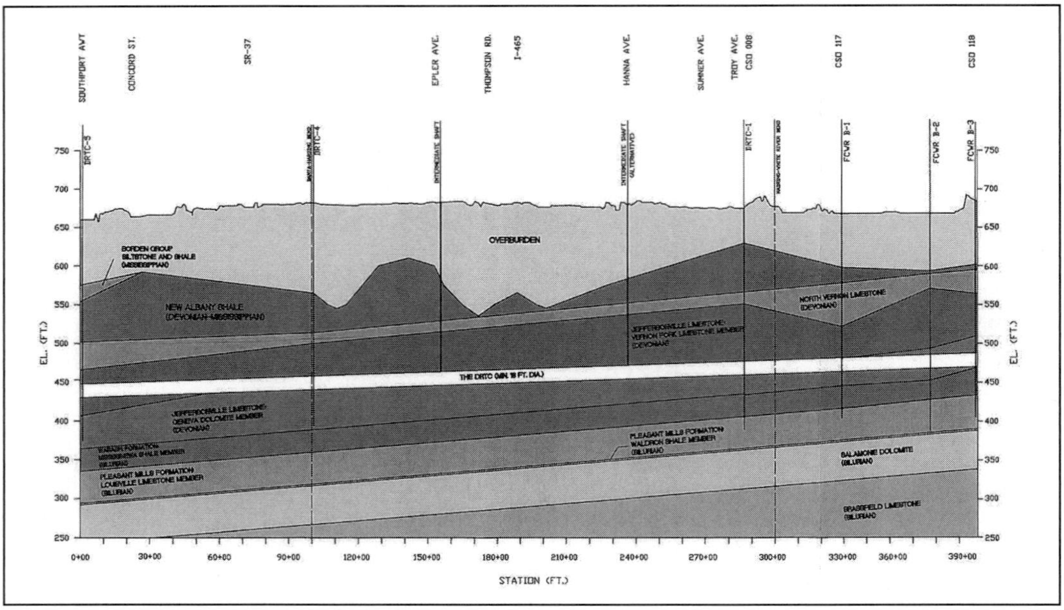


**Figure 4. Geological profile**

starter tunnel was 137.2 m (450 ft) long. Both starter and tail tunnel lengths were measured from center of launch shaft. The two tunnels were approximately 7 m (23 ft) diameter. The starter and tail tunnels are required in order to provide the space necessary to assemble and launch the TBM.

## TANGENTIAL VORTEX DROP SHAFTS

As previously reported, the project includes three (3) drop shafts. Each drop shaft location includes 2 shafts, see Figure 6. One for the tangential vortex drop installation. One for the air vent shaft slightly downstream. Thus far the Contractor has chosen to drill pilot holes through the bedrock at each drop shaft location. No bedrock has been removed yet as the TBM has not passed the corresponding location.

To date the majority of work has taken place on the drop shaft directly connecting CSO 117. The Contractor installed slurry wall panels at the CSO 117 location. Construction of the slurry panels was without interruption. As these facilities are constructed we have encountered problems below grade with existing infrastructure. Given the fact that the existing piping is approximately 100 years old challenges exist in what we have encountered. We have discovered several old abandoned structures that were not anticipated. The footprint of the area is relatively confined with many operational and abandoned pipes. To allow enough space for the new infrastructure much of the older, previously abandoned, had to be removed. There was simply not enough space for both to exist.

As with any project, we have learned many lessons on this project. We know we'll never have a design that lacks challenges. However, valuable lessons learned here include identifying the total area needed for the construction equipment. Accepting the fact that the existing infrastructure was not backfilled properly (to current standards) and accepting the fact that we simply need to remove and replace the piping and structures within the area to prevent crushing it during the new construction.

It is likely more cost effective to include such work in the original proposal than try to negotiate the pricing after the fact. It also helps prevent delays to the overall schedule due to unanticipated work.

The drop and vent shaft casings, down to bedrock elevation, at CSO 008 have been installed. This location is very close to the wastewater treatment plant. Similar challenges were found here also. Adding to the challenges was the proximity to the treatment plant and flow levels in the existing system as a result of wet weather events.

Much surface work remains in the area of this work. The approach channel has not been constructed. The work has been started recently and will not likely be completed for several months. The contractor plans to install the adit connecting the overflow to the tunnel then construct the shaft portion through bedrock via raised bore method.

The drop and vent shaft casings, down to bedrock elevation, at CSO 118 have been installed. All

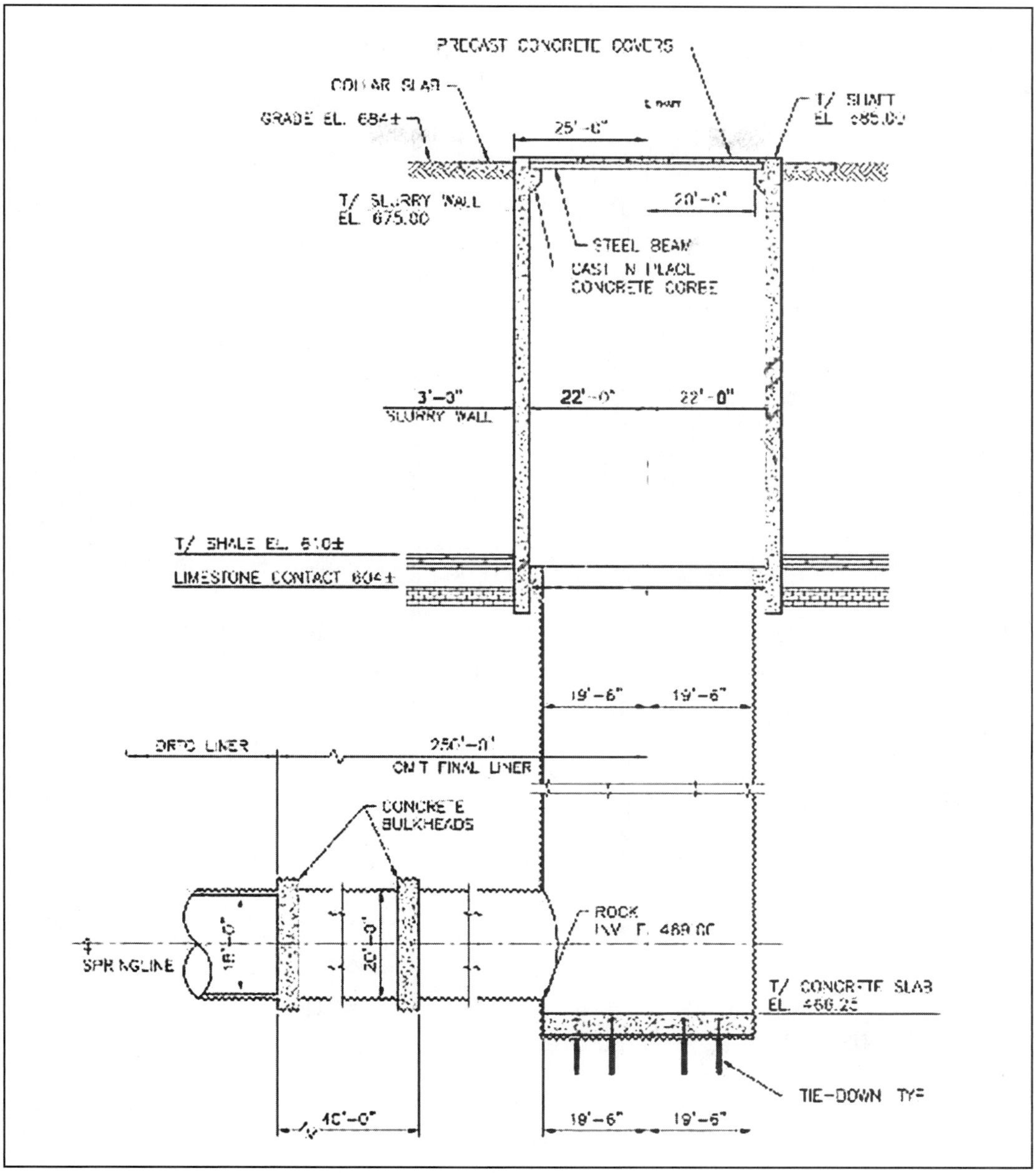


**Figure 5. Retrieval shaft configuration**

other work at this location is currently on hold due to a required power line relocation. Currently an aerial 138 kv crosses this project site. In working with the local electric utility the work will take place soon.

Two additional challenges existed for CSO 008 and CSO 118 locations as well. During the period between bid opening and contract award the ownership of the wastewater utility changed. Unfortunately, there were some errors in the documentation transferring the property. Some of those errors resulted in minor delays. Cooperation from the previous owner helped to allow property access and minimize delays for some of the earlier work under this contract.

## UTILITY SHAFTS

The project includes three (3) utility shafts. All are located on the north/south portion of the project alignment and are identical in shape and size, see Figure 7. The shafts are located approximately every 4.3 km (2.7 mi) thus splitting the total length of the

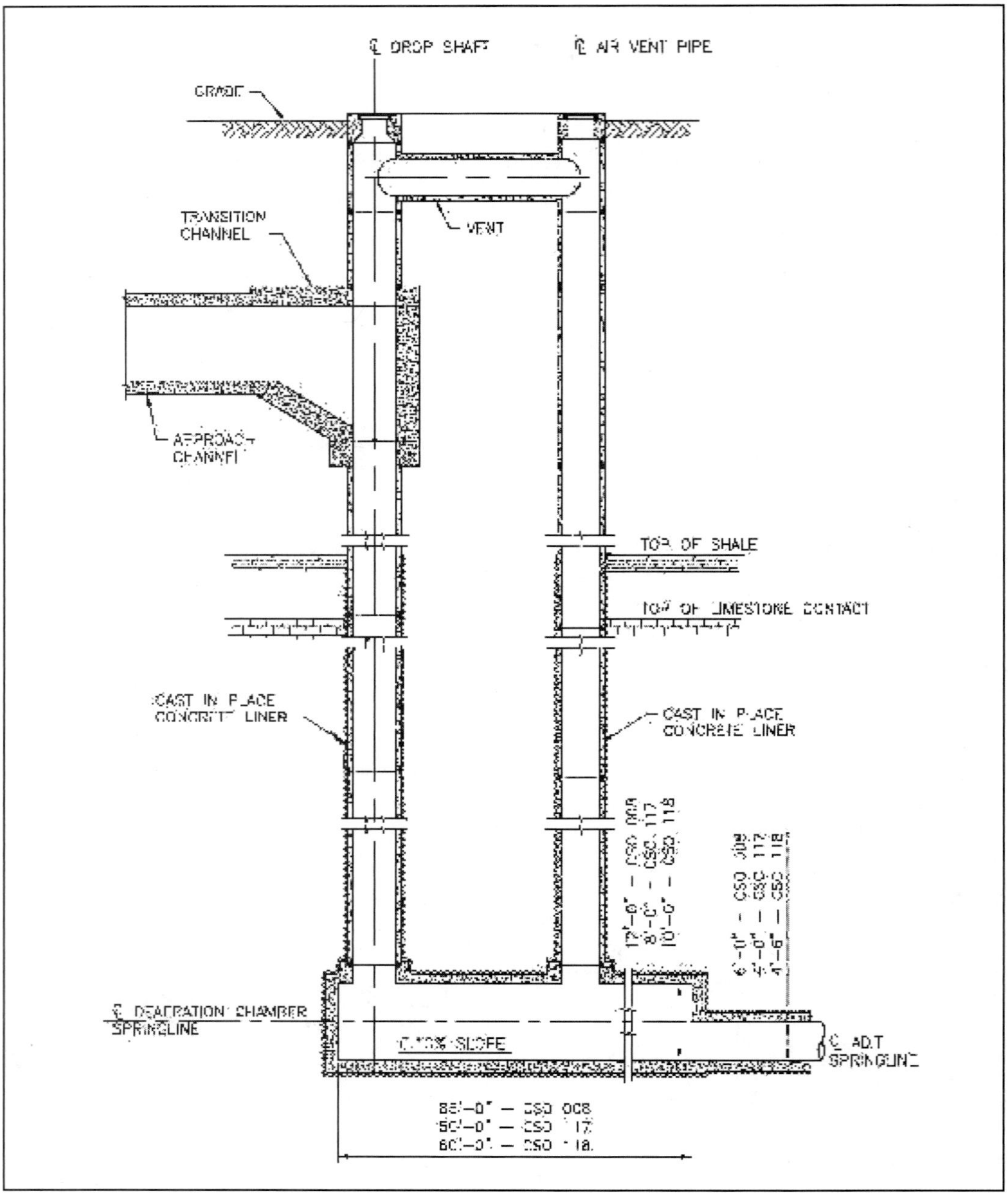


**Figure 6. Tangential vortex shaft configuration**

project alignment into thirds from a maintenance access standpoint.

In order to construct Utility Shaft #1, extensive utility relocation was necessary. The relocations took eight (8) months to be completed. The original intent was to introduce clean air at this location. However, due to the proximity of homes to the shaft location the contractor chose to extend the existing air line and utilize Utility Shaft #2 as the next location to add air to the tunnel. This was related to the location of the shaft in an industrial/commercial type area.

Utility Shaft #2 is the next shaft along the tunnel alignment. Its location is approximately halfway through the total tunnel length. After the Notice to Proceed, a privately owned fiber optic line serving a local college was discovered. Relocation of the fiber

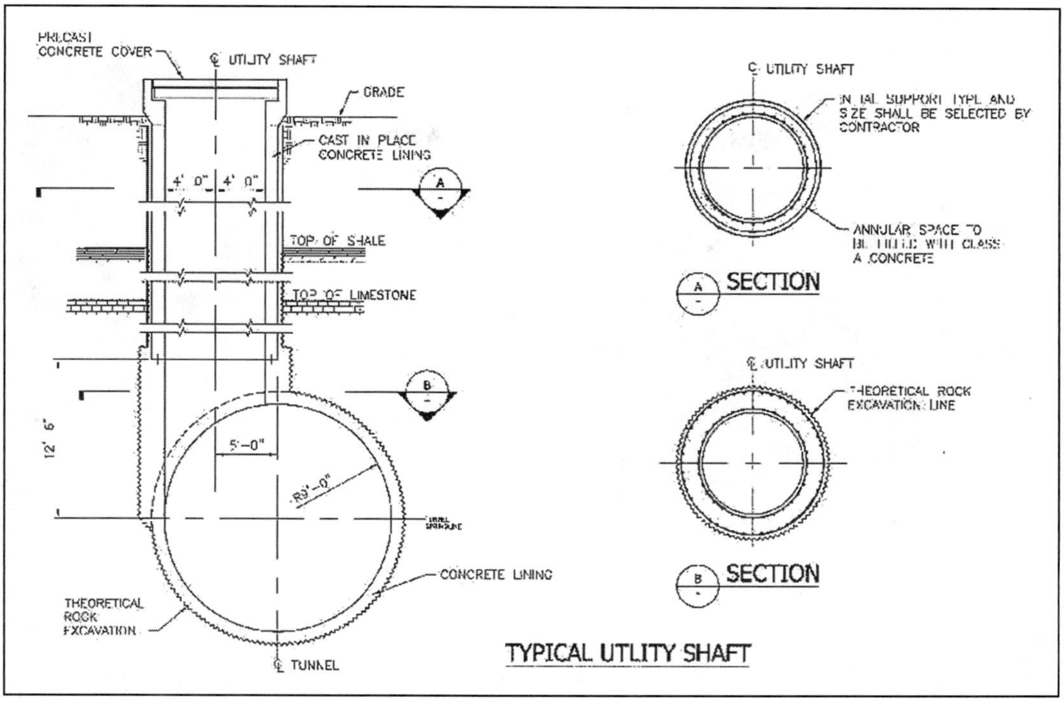


**Figure 7. Utility shaft configuration**

optic created minimal disruption to the related work in this location.

Utility Shaft #3 is located just south of the first drop shaft for an overflow. The utility shaft was needed due to the adit length for the drop shaft. From this point upstream, the tunnel is set to be located under, or directly adjacent to, the White River prior to arriving at the retrieval shaft.

## TUNNEL BORING

After receiving a significant overhaul at JF Shea's Mount Pleasant, Pennsylvania facility, the refurbished TBM, a Robbins Main Beam TBM Model #MB 203-205-4 (see Figure 8), was transported to the project site. This machine was originally commissioned in the mid 1970s. Testing of the machine at the facility in Pennsylvania was required. All testing proved satisfactory in October 2012. The machine was then disassembled and loaded for transport to Indianapolis. The final load arrived at the project site in late December.

The machine was reassembled on the surface at the project site in sections to ease in the ultimate assembly in the starter tunnel. Once the machine had been lowered and the pain staking operation of reassembly commenced, additional start up testing took place. This process took nearly two months to complete.

A special cutterhead was manufactured by Robbins for this specific project. The diameter of the new cutterhead was 6.15 m (20 ft 2 in). The slightly oversized unit allowed for the proper diameter volume of mining as well as the required 0.3048 m (12 in) thick clear concrete full diameter liner installation.

The TBM was launched March 15, 2013. As with any rock tunneling project, many challenges were found in making the machine fully operational. After a couple weeks they were overcome and daily production began to increase. The TBM currently holds 3 world records for rock tunneling production.

The production volume further supported the established depth of the deep rock tunnel. This depth was established via a detailed geotechnical investigation. Samples were taken, on average, every 304.8 m (1,000 ft) to help establish consistency in the strata.

On May 10, 2013 124.93 m (409.89 lf) of tunnel mining took place in a 24 hour period. This footage set a new world record for tunnel production in a single day. Two other records were broken on the project also. Those being the average weekly and monthly production records for the diameter range

**Figure 8. Tunnel boring machine**

of 6 to 7 meters. During the week of June 10, 2013 515.12 m (1,690.04 lf) of rock tunnel was bored. The average month production in May 2013 of 1,754.17 m (5,755.15 lf), was also achieved on the project.

The tunnel boring on the project is nearly complete and the concrete lining will soon be installed. The lining was designed at 0.3048 m (12 in) thickness. We anticipate the installation of the liner will take approximately 18 months.

## CONCLUSIONS

The project construction to date has not been without challenges. The contractor, construction inspection team, and owner have worked closely together to overcome these challenges. During this construction period, many lessons have been learned along the way. We continue to chart the lessons so that we can grow from them and become a more experienced owner in this type of construction.

# Best Practices for Utilizing Escrow Bid Documents as a Dispute Resolution Tool

**Valerie R. Wollet and Robert G. Scott**
H.R. Gray

## INTRODUCTION

Most tunnel construction projects are inherently risky, and not all risk can be mitigated or contracted away. Contractors may submit claims through the course of a project requesting additional compensation in the form of time and/or money.

Several dispute resolution approaches have been used in the construction industry. The use of escrow bid documents (EBD) has historically been a successful tool for preventing and quickly resolving disputes. Requirements for EBD, as part of the contract documents, specify that the Contractor compiles all information that was generated to prepare the bid price for the project. These documents are held in a secure, neutral location (in escrow) for the duration of construction. These documents are collectively referred to as EBD. The contents of the EBD can be referenced when negotiating prices adjustments, contract modifications, and claims and disputes settlements. This allows disputes to be resolved quickly and relatively inexpensively.

Use of EBDs together with other alternative dispute resolution methods, such as a properly written geotechnical baseline report (GBR) and Dispute Resolution Boards (DRB) can, in some cases, greatly minimize the impact of claims on underground construction projects.

Incorporation of EBDs should be considered on projects with high levels of inherent risk such as underground construction, projects using innovative construction means and methods, and other complex jobs. It is recommended that Owners incorporate EBD provisions as part of the contract documents for proposed tunnel construction projects. The intent of this paper is to provide a description of the aspects of EBDs as part of a dispute resolution method and to inform Owners and Contractors about why it is in their best interest to utilize EBDs.

## HISTORY

The first construction project to include provisions requiring the Contractor to submit EBDs was the Colorado Department of Highways Second Bore of the Eisenhower Memorial Tunnel which was constructed between 1975 and 1979. (UTRC 1991) This vehicular tunnel was the second of two tunnels excavated through the Rocky Mountains as part of Interstate 70, west of Denver, Colorado.

It is important to note that several of the innovative contracting methods that were developed on the Second Bore project directly resulted from the Colorado Department of Highways planning team's thorough analysis of the factors that contributed to multiple difficulties that occurred throughout construction of the First Bore, which was completed between 1968 and 1973. The concept of EBDs was one of many lessons learned from construction of the First Bore that contributed to the successful construction of the second bore of the Eisenhower Memorial Tunnel. The project was completed without major claims or delays and, "the adversary relationship common to many contracts did not develop" (McOllough 1981).

With success using EBD in conjunction with other dispute resolution methods on the Second Bore of the Eisenhower Memorial Tunnel, subsequent tunnel construction projects began to follow suit throughout the 1980s. (Johnson et al. 1983; UTRC 1991) Inclusion of EBD has since become an industry standard on tunnel construction projects.

Contract language for EBD provisions used in the tunnel industry today has remained generally unchanged since that developed for the Second Bore of the Eisenhower Memorial Tunnel (McOllough 1981).

## OVERVIEW

Provided that the minimum Contractor qualifications are met, cost ultimately plays into the consideration of how a tunnel construction contract is awarded, particularly when the Owner is a public entity. To protect the interests of both the Owner and the Contractor, EBDs are one of the many tools that can be included in the contract documents.

The information included in the EBD represents the Contractor's full understanding of the project and the intended means and methods of construction at the time of bid. This should include all backup information that the Contractor used to arrive at the bid price, such as the following:

- Assumptions about ground behavior
- Means and methods
- Sequencing
- Activity duration
- Production rates
- Resource management
- Quantity take-offs
- Calculations
- Subcontractor and supplier quotes

## BENEFITS

EBDS are a useful and cost effective tool for evaluating the cost differences between bidding assumptions and actual quantities and costs incurred. They can be used to determine the value of change orders for additional work or credits for deleted work.

Inclusion of EBD provisions can decrease the number and intensity of disputes. The openness of sharing information sets a tone of fairness and teamwork from the onset of the project. Knowing that the information can be accessed encourages good faith efforts from all parties to resolve disputes oftentimes without opening the EBD.

It should be noted that the Owner must be disciplined such that the EBD are not hastily opened for every little issue that arises, which could lead to adversarial relationships among team members. Good faith efforts should be expended to resolve disputes before referring to the EBD.

The attitude and relationships between the parties can have almost as much impact on a project's success as the technical components. Cooperative relationships on a project can make resolving disputes much easier for everyone involved.

## CONCERNS

The purpose of EBDs is to preserve the Contractor's assumptions at bid time. EBDs should not be used for pre-award evaluation of the proposed construction means and methods or to assess Contractor qualifications or cost estimating methods.

Contractors may have concerns about providing proprietary means and methods information to Owners that are oftentimes public entities. Should the Contractor's confidential information regarding how they perform their business become public, that Contractor could potentially lose a substantial amount of money to their competitors. Owners must recognize that the EBD is proprietary information and considered a trade secret as defined by the United States Freedom of Information Act (FOIA), Exemption 4 (FOIA, 5 U.S.C. § 552). As such, this information is considered privileged and is exempt from the Required Release of Public Records for work within the United States. Similar laws exist in other countries but it is imperative to verify the specific laws pertaining to the country where the work will occur.

As further protection from disclosure of information, many times the contract documents include language clearly stating that the EBD contents are at all times the sole property of the Contractor and are returned to the Contractor after the contract is closed out.

Contractors may also have apprehensions about extra time to prepare EBD. However, it is not the intent to cause more work for the Contractors when requiring EBD. Contractors should be permitted to submit documents in their standard formatting used for bid preparation. This information is being assembled for bid preparation anyway and should not require additional effort by the Contractors when assembling their bids.

It is to the Contractor's advantage to incorporate as much documentation of assumptions into the EBD as possible so that information is available to support a claim should it arise. If any information is not included, it is as though the Contractor did not look at or rely on that data at bid time. When all assumptions are documented, a claim's merit can quickly be agreed upon and everyone can move on to quantum negotiations and focus on completing the work.

## PROCESS

Within a specified time period after bid opening (generally within two to five working days) the three apparent lowest bidders submit their EBD to the Owner's Construction Management Team. If the Contractors do not submit their EBD in a timely manner, this should be considered sufficient cause to reject their bid. All EBD packages remain sealed. If the contract is not awarded to the apparent low bidder, only then will the next lowest bidder's EBD be opened and reviewed in the same manner. Once the contract is awarded, the unsuccessful bidders' EBDs are returned.

The successful bidder will then meet with the Construction Management Team to open their sealed EBD package and review its contents. This review generally takes two to three hours for major tunnel projects, depending on how well-organized the documents are. The purpose of this review is to verify that the contents are authentic, legible, and complete. This review is not intended to constitute approval of the Contractor's proposed construction methods, estimating assumptions, or interpretation of the contract documents. If all of the required documentation is not included in the EBD, it is up to the Construction Management Team's discretion as to whether to allow the bidder to submit additional information or to reject the bid.

After review, both the Contractor and the Construction Management Team representatives will take the EBD to a secure location, which is specified in the contract documents, and store the contents.

After the contract is awarded, the Contractor and the Construction Management Team will each designate representatives that are authorized to examine the EBD in writing to the other party. Either the Contractor or the Owner's Construction Management Team can request that the EBD be opened. The EBD will only be accessed in the presence of both parties. Many times, projects also allow members of the DRB to have access to the EBD in the presence of both parties.

Once construction is completed, all claims are resolved, final payment has been accepted by the Contractor, any warranty periods have expired, and has the Owner's Construction Management Team's permission, the Contractor then retains the contents of the box.

## CONTRACT DOCUMENTS

It is recommended that the General Conditions portion of the contract documents contain specifications including the requirement of EBD on tunnel construction projects.

The Contractor maintains ownership of the EBD at all times due to the proprietary and confidential nature. As discussed previously, the Owner should take care to safeguard this information while in their possession because it could be valuable to the Contractor's competitors. As discussed previously, all EBD information is proprietary material that is protected in the United States under the FOIA, Exemption 4 (FOIA, 5 U.S.C. § 552). Many Owners also include provisions in the contract documents conveying additional protection of the Contractor's proprietary information. The following is an example of language that could be included in contract documents to further safeguard the Contractor: "The EBDs are and shall always remain the property of the Contractor subject only to joint review by the Owner and the Contractor."

Requiring EBD is not meant to cause more work for the Contractors. Therefore, it is not recommended to specify items such as formatting. Contractors should be permitted to submit documents in their standard formatting used for bid preparation. However, it is typically required that the EBD are in the same language as the contract.

The Owner can specify that the EBD be stored in any secure location. For the convenience of the parties, a safety deposit box in a bank that is located fairly close to the project is typical. However, the Owner may choose an escrow agent in any location. The Owner typically pays for the escrow services, i.e., safety deposit box fees, which has a minimal cost.

## CONTENTS

The EBD should itemize the estimated cost of work for each line item contained in the bid schedule. Allowance line items or other line item costs that were provided by the Owner do not need to be included in the EBD. Total estimated item costs should indicate the allocation of typical cost categories, including:

- Direct labor
- Repair labor
- Equipment ownership
- Equipment operation
- Expendable materials
- Permanent materials
- Subcontracts
- Indirect costs
- Contingencies
- Profit

The EBD should contain all supporting information that the Contractor used to arrive at the bid price. This includes all calculations, quotes, add/deduct sheets, notes, sketches, reports, and other documentation related to the following:

- Quantity takeoffs
- Estimated crews
- Equipment
- Schedules
- Production and progress rates
- Close-out details and adjustments
- Subcontractor comparisons
- Supplier comparisons

Items directly related to tunneling should be specifically stated in the contract documents to be required in the EBD. This includes all assumptions (calculations and/or text) founded on the baseline subsurface conditions presented in the contract documents that form the basis for the Contractor's selected means, methods, and equipment. This will provide documentation of the Contractor's assumptions and intents at the time of bid, which would be extremely valuable if a differing site conditions claim should arise during construction. The following should be included at a minimum:

- Anticipated ground conditions for different reaches
- Tunnel Boring Machine (TBM) penetration rates for various ground conditions

- Estimated TBM utilization for various ground conditions
- Assumed support of excavation for various ground conditions
- Estimated TBM downtime (including scheduled maintenance, equipment breakdowns, downtime due to gas, etc.)
- Assessment of geotechnical, hydrogeological, TBM, and other construction variables affecting performance and production
- Cutter costs and anticipated number and frequency of cutter changes (including rings, hubs, bearings, mounting brackets, etc.)
- TBM manufacturer's proposal for equipment and performance

All costs should be included and identified in the EBD. Some Owners may choose to identify a minimum estimated unit cost (i.e., $10,000) requiring a detailed cost breakdown provided that all direct and indirect costs, as applicable, have been allocated appropriately. This prevents the Contractor from having to include all minor subcontractors and material supplier cost breakdowns; however it is recommended that the contract documents do not prohibit inclusion of this information.

## SUBCONTRACTORS

If any portion of the work is subcontracted in the bid price, the Owner should include contract provisions requiring that each Subcontractor also provide EBD to be included with the Contractor's in a sealed envelope. These documents will be treated in the same manner as the Contractor's EBD.

During the EBD review process, the Construction Management Team meets with the Contractor, who has their primary subcontractors wait outside the meeting room until the Contractor and Construction Management Team are ready to discuss the portion of the bid where a specific subcontractor was used. That subcontractor then comes into the room and opens their sealed documents. The group reviews their information for completeness, reseals the envelope while everyone is in the room, and subcontractor leaves. The review then proceeds with the next subcontractor.

If the Contractor decides to subcontract a portion of the work after the contract is awarded, it is recommended that the Owner require the Subcontractor submit their EBD to the Contractor before accepting the subcontract. This is in the best interest of the Contractor because they are ultimately responsible for the performance bond including subcontracted work.

## CONCLUSIONS

Underground construction inherently includes many risks that cannot be contracted away. Because of the risky nature of the business, it is prudent to have as many tools available that can be used to sort out any disputes that may arise during construction. Therefore, when utilized with other alternative dispute resolution tools such as a GBR and DRB, EBDs can, in some cases, greatly minimize the impact of claims, in the form of time and/or money, on underground construction projects.

It is recommended that Owners strongly consider utilizing EBDs on their proposed tunnel construction projects. Due to the inherent risk and complex nature, even the most well-planned tunnel construction projects will likely have claims of some sort. EBDs have historically been a successful tool, from both Owners' and Contractors' perspectives, for reducing animosity and quickly resolving disputes. EBDs enable the evaluation of cost differences between Contractors' bidding assumptions and actual quantities/costs incurred, which can be used to facilitate the determination of change order values.

Experience has shown that inclusion of EBD provisions can decrease the number and intensity of disputes. Additionally, the openness of sharing information sets a tone of fairness and teamwork from the onset of construction. It is a cost effective and sound risk management practice for tunnel construction projects. For these same reasons, inclusion of EBDs on tunnel construction projects has become an industry standard since the 1970s.

## REFERENCES

Freedom of Information Act (FOIA), 5 U.S.C. § 552.

Johnson, E.B., Holloway, L.J., and Kjerbol, G. 1983. Design of Mount Baker Ridge Freeway Tunnel in Seattle. Rapid Excavation and Tunneling Conference Proceedings—1983: SME.

McOllough, P. R. 1981. Eisenhower Memorial Tunnel—How Colorado Department of Highways Improved Contracting Practices and Management. Rapid Excavation and Tunneling Conference Proceedings—1981: SME.

Underground Technology Research Council (UTRC) Technical Committee on Contracting Practices. 1991. Avoiding and Resolving Disputes During Construction: Successful Practices and Guidelines. New York: American Society of Civil Engineers.

# TRACK 3: PLANNING

## Session 2: Risk and Cost Management

Brian Harris, Chair

# Active Risk Management—An Owner's Manual for Mega-Projects

**James J. Brady**
Geocomp Corporation

*Risk Management decisions that will empower Owners, save money, and reduce risks.*

**ABSTRACT:** This paper discusses an important topic where early participation by Owners in an Active Risk Management Process will both save money and mitigate risks while maximizing an Owner's ability to manage program funds. The area explored is the development and management of program-level and project-level Contingency budgets; including recommendations on how best to use project Allowances as well as some long-term strategies for managing Program-level contingencies from inception to commissioning. The ideas and recommendations presented in this paper are based upon first-hand observations of the inner workings (and areas for improvement) of many of the recent larger tunneling programs in the United States.

## INTRODUCTION

Risk management is one of the most important topics in the tunneling industry today. There are two compelling reasons for this; there has been a history of large cost overruns on several major tunneling projects and the trend over the last twenty years, which shows no signs of abating, is toward yet ever larger, ever more complex, projects. These projects are often combined into multi-billion dollar Programs (the Mega-Project/Mega-Program phenomenon). The timeline of many of these major tunneling programs, are measured in decades not years, and the numerous disciplines involved, often pushing the cutting edges of the current technology and research, mean that there is no longer any single person that is capable of possessing the entirety of skill sets, engineering expertise, cost and scheduling, or specification knowledge that is required to complete the work. In essence, the best program management team is akin to an organism that relies on the skills of the many in order to function at all. This is in sharp contrast to the tunneling industry of 30 years ago, when one small engineering firm was typically capable of designing the project in its entirety and most of the projects could be constructed with a seasoned Project Manager (PM), Project Engineer (PE), and Superintendent. Today, the complexities are higher, the dollars are bigger, the timelines are longer and the technologies are being pushed every step of the way; it is no wonder risk management is such a big topic. The price of failure has also never been higher because many of the projects are driven by Federal court orders and consent decrees with large fines and liquidated damages at stake. But that is only part of the story, the underlying story for tunneling Mega-Projects is the human story; in particular the transfer of knowledge between humans. It is counterintuitive to our modern concept of ever accelerating rates of change and had be experienced first-hand on these earlier projects, to be understood before it could be explained.

## THE DIFFERING PACES OF PROGRESS

We live in an age where technology advances at historically unprecedented speed; currently doubling at approximately every four years. This is astounding, and it affects in a fundamental way, how we think about the future. Mega-Project designs deliberately push the technological envelope because with the long delivery timelines, it is understood that to merely perform a design to the best of today's existing standards will in all probability be obsolete by the time it is completed. In fact, a common risk identified on Environmental Protection Agency (EPA)—mandated CSO reduction programs is worded to the effect that "the regulations will change before the project is completed rendering the completed program non-compliant." Running counter to this, however, is how the human element is incorporated into this brave new world.

Every Owner that has embarked on a major tunneling program or Mega-Project has endeavored to place around themselves the best team they could solicit to perform the work. Often the "key" members of these teams are required in the proposal stage to attest to their long-term commitment to that project in an effort by Owner's to keep those good managers and engineers from moving on to leading up charge for the next big project. In this and many other mitigation efforts, most Owners and their pre-design teams have wisely been very proactive on Risk Management, embarking on identification and mitigation strategies commendably early in the process. This has aided the team's ability to get their collective thoughts around the complications, challenges and deadlines they face. However, regardless of their risk management strategy, it is safe to say that every Owner and the team have gained new insights and understanding of the risks involved as they have moved through their programs. But, here is where the system runs contrary to our rapidly-changing times. While technological advance may double every four years, twenty years ago, the typical tunneling project lasted between 2.5 and 3 years releasing its human talent for other projects. Now these mega-projects are absorbing talent for the better part of decades at a time. The result of this change is that these often hard won "lessons learned" are slower to circulate into the industry at large. The consequence of this is that Owners are often repeating the mistakes of their predecessors as they learn the game anew and are thus unable to take advantage of the risk mitigation strategies learned by Owners that have already been down the same roads. This is further compounded by the litigious nature driving many of these projects which leaves most Owners reluctant to transmit their "lessons learned" for fear of the political repercussions.

## DIFFERENCES IN RISK MANAGEMENT PERSPECTIVE: OWNERS VS. CONTRACTORS

A common theme observed in all of these large programs is that the Owners end up reinventing the wheel in the critical area of risk allocation, risk management and contingency budgeting time and time again. While Owner's have, in the main, shown more foresight than the industry in adopting Active Risk Management practices, and generally work hard to do a good job of identifying risk within their selected project's pre-design team, they have repeatedly failed to find ways to motivate the Contractor/ Builder's team to rise up to a similar level of commitment. At the core of this phenomenon appears to be a misconception that these contractor "teams" are long-standing entities that are used to working together and have been preparing for their place in this particular project for as long as the Owner has been thinking about it. Nothing could be further from the truth. As an illustration, earlier this year, the author was asked to facilitate a Risk Management evaluation and provide Contingency budget recommendations for a contracting joint venture pursuing a $200 million-plus project. This "team" was one of three teams pre-selected due to their excellent qualifications at an RFQ stage to proceed on to the final bid stage and the results of this undertaking should make every Owner think critically about their own risk management programs.

First, the "team" could not attend a single risk management workshop, instead two risk workshops had to be set up. Why? Because aside from this joint-venture, they typically worked as competitors and each company was afraid of giving its estimating and other construction secrets away to the other. The two separate workshops proved to be real eye-openers. On this particular project there were liquidated damages for failure to complete on time as well as up to $5.75 million dollars in incentives for completion of the work at or before certain milestones. What made the separate workshops particularly interesting was that while one firm believed they had an 85% chance of making at least some of the $5.75 million in incentives, the other firm believed it had only a 14% chance of making any incentive at all! In an age when these large projects, by their very size and bonding requirements alone require joint ventures, this fundamental difference in outlook needs to be appreciated by every Owner.

The second result of this exercise was less obvious, but has implications for Owners who are trying to get their hands around what the "other" side is grappling with. On the surface, before the risk management workshops were performed, both Joint Venture (JV) partners unquestioningly believed, because they both nominally performed approximately half the dollar value of the work, that they faced "equal" risk on the project. However, the second part of this Active Risk Management exercise was to provide contingency recommendations (a quantitative analysis) of the risks facing the joint venture. Since they performed this exercise independently, the results provided the team with a late-developing, but important insight. It turned out that, due to the different risk exposures arising from the differing work activities performed by the team members, one partner carried six times the risk on the job when compared to their partner. Needless to say, the job took on a whole new persona when the JV team met after the Risk Analysis Workshop to wrap things up. For those Owners unfamiliar with the private sector, both of the above results were earth-shattering to this JV. They bore huge implications upon the project and even to the very fabric of the JV relationship going forward. This is information that Owner's want their JV Teams to work through and come to terms with before they get on their projects; not after.

The question is; what can Owner's do to facilitate this effort and make it part of the culture of chasing Mega-Project work? One recommendation is to require teams soliciting work to perform their own risk workshop analysis and submit them as part of their proposals. They can be performed either independently or as the JV. Potentially, how this information is gathered in and of itself, can be potentially telling to an Owner, indicating the degree of cohesiveness possessed by the JV-team. However, it is important to keep in mind that shared results of separate workshops may often be more candid, and therefore more informative, than a cursory "JV-Team" workshop. Either way, this is an important communication tool that will give the Owner insight as to how much their own identified project risk has been digested by each prospective team. In addition, it helps illuminate how insightful each team is in identifying new, and heretofore unidentified, project risks. By nature of their different role in the project, Owners should expect that contracting JV teams will bring an assortment of new risks to light that were not previously identified. This should not be considered a deficiency on the part of the Owner's pre-design team, who often has had several years to evaluate project risk, but rather fresh input from a different vantage point because each group's respective role sensitizes them to different project risks. Simply stated, these varying perspectives need to be reconciled to make any project, but especially Mega-Projects, function in a healthy manner in the coming years ahead. This is an important concept for Owner's embarking on mega-project work because the contractor-provided risk information is the prelude to a long-term relationship and it is critical to digest and understand where the newest member of the program (your Contractor JV Team) is coming from in terms of understanding both its threats and potential opportunities in the upcoming work.

One of the core benefits of an Active Risk Management Program is communication. Owner's must remember that the nature of this work typically requires joint ventures between contracting partners; these may be previously well-established, solid working relationships, but often these are marriages of convenience, and sometimes more like shotgun weddings when viable "courtships" fall through and leave firms settling for a partner that still "fits the bill." For example, as this paper is being written, Joint Venture teams are preparing bids for the LACMTA Westside Extension Project. This project has been heavily publicized, has been followed heavily by the tunneling contracting community and is estimated to cost in excess of $1.2 Billion dollars. Even on a project of this magnitude, at least one of the JV Teams in question did not begin getting together to estimate this project until less than 6 weeks before the original Bid Date. Owners and their pre-design Teams spend years on the run-up to a major program but the entities doing the work often don't start seriously examining the work in detail until about 40 days before the bid is due. This is a reality Owner's ignore at their peril. Requiring a Risk Workshop by perspective JV Teams can aid not only as vehicle for communicating their comparative understandings of project risk to the Owner but are also valuable for the JV Team's themselves as a powerful tool that improves the their understanding of the work.

## SHARING RISK INFORMATION IS A TWO-WAY STREET—A PRELUDE TO HONEST DIALOG

Owner's (or more accurately, their Legal Departments) still typically shy away from providing the results of their own internal Risk Workshops and Risk Registers to perspective contracting teams for fear that it will be used against them. There are pros and cons to the idea of sharing this information and no doubt the institutional inertia will not change the mind-set of legal departments any time soon. However, the future of effective project management on mega-projects will inevitably drive the Owner's toward sharing this information. If the improved risk communication that an Active Risk Management program engenders is to be properly harvested, Owners need to take the lead in sharing the results of their initiatives. With that said, there is nothing that requires the Owner to show the results of their risk analysis before their perspective construction teams have prepared and submitted their own. A recommended strategy would be to require the JV teams to prepare and submit their own independent risk registers (in a standardized format provided by the Owner) as part of the prequalification phase. During the subsequent bid phase, the Owner can share its own developed set of Construction Risks with each of the JV Teams so that these risks are disclosed and fairly evaluated by the teams prior to Bid. During, the final negotiations phase, the JV team would update its Risk Register and quantify it to provide both justification for the construction contingency carried by the team as well as give the Owner an additional tool to evaluate the project awareness, innovation, and quality of the various teams.

## QUANTIFYING AND MANAGING CONTINGENCY

Potentially the greatest benefit to Owners as a result of adopting an Active Risk Management program is the ability to quantitatively analyze their programs and projects in order to determine required contingency dollars. Most agencies are still slowly inching their way forward with this concept; each seemingly in the dark to the lessons learned by its numerous predecessor authorities that are further along on their programs (EPA Consent Decree CSO programs and FTA Subway Programs both have established track records). The FTA has probably done the most work along the lines of managing project contingency budgets, borrowing heavily from the concepts developed by the Department of Defense and later used by NASA.

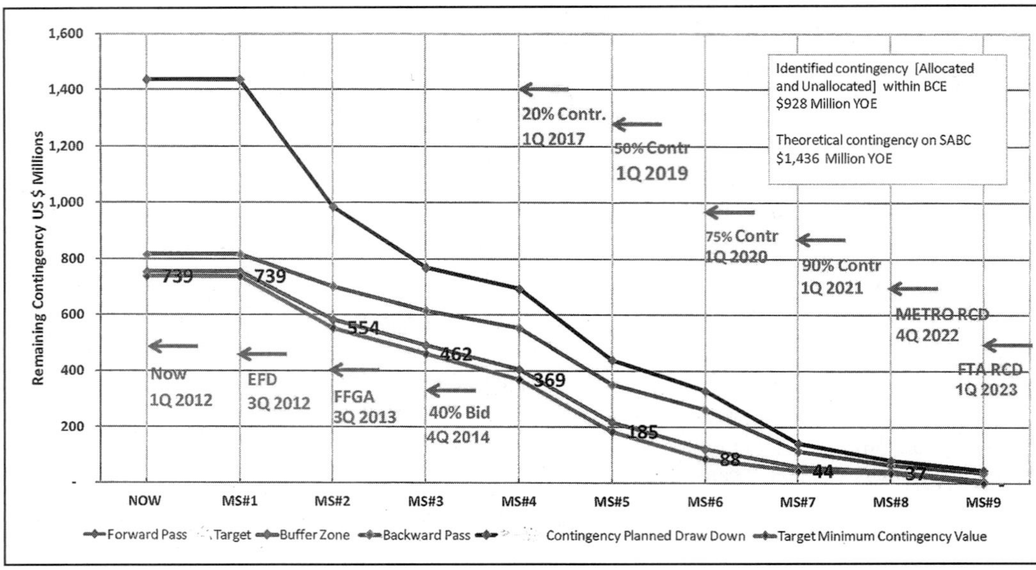


**Figure 1. Typical FTA cost contingency drawdown curve**

The FTA concept is straightforward enough, the general idea is that the contingency budget is continually and tracked and monitored against pre-established benchmarks as it is drawn down during the course of a project. However their presentation and calculations are opaque and difficult to disseminate. The other problem is that even when the concepts are understood, most Owners are employees of government agencies that treat the word "contingency" as an anathema. In government-speak, "contingency" is readily translated into "slush fund" and this stigma leads Owners working for government agencies to hide from confronting the fact that the contingency on these projects is necessary, real, and needs to be faced squarely and managed properly if they are ever going to recapture true restraint in public spending. Partly this is an education issue; politicians all too often live under the delusion that the Engineer's Estimate of the price of $2.2 Billion dollar CSO system upgrade spanning 20 years is akin to the price of a hamburger at McDonalds; that these two "prices" have the same level of exactitude and that any additional costs must be the result of either poor management or skullduggery. The Owners representatives at the project level realize that project contingency is required but make every effort to bury it in an effort to prevent second-guessing by the politicians and the public they represent. This draws them into the tactical error still made by almost every agency to this day which is that they try to manage contingency at too low a level resulting in incredible waste and inefficiency. In the case of the multi-year program, instead of managing contingency at the Program level, the contingency dollars get subdivided into "packets" that get added into the budget of the Program's constituent projects. The military analogy is that the army's "reserves" get ladled out to the front lines in advance of the battle. In this analogy, every Project Manager is a Division commander who then jealously keep their reserve allocation under lock and key, almost invariably guaranteeing that the Program is caught wrong-footed for the real cost risks that develop. On top of that, once these contingency dollars get allocated down to a specific project (say a 5 year Design/Build tunnel project), due to most governmental agency accounting rules they cannot be freed up until that project is completed. In the above example, let's say that amount is $46 million dollars; that's $46 million dollars that is locked down for *at least* 5 years and is unavailable for addressing contingencies on other concurrent projects or (more critically to the Owner) funding the follow-on projects in their Program. Most major programs find themselves cash starved after their initial slate of projects get underway and the lack of proper management of contingency reserves is a major contributor to the paucity of funds that seems to settle in around year four to five of these major programs.

In order to be effective, contingency needs to managed at the Program Level, using Quantitative Risk Analysis to establish the initial contingency budget for each constituent project. The sum of these contingency budgets would then be managed at the program level by the Owner's Risk Manager who would function as the Manager of the Contingency Fund. He would report Quarterly to the Owner's Program Manager and together both would "sign-off" on the allocation of contingency funds for actual risks encountered over the quarter. In addition, both signatures would be required to certify that nominal contingency funds allotted to a project had ceased to materialize and could therefore be removed from that projects' contingency reserve ledger. It is important to point out that Quantitative Risk Analysis, typically performed through a "Monte Carlo" simulation which "is a problem solving technique used to approximate the probability of certain outcomes by running multiple trial runs, called simulations, using random variables," does NOT provide line item level contingency dollars for each delineated risk.* Rather, the simulations provide an overall dollar value for *the pool* of identified risks at the percentage level of program risk-aversion established by the agency (typically described as a Confidence Interval (CI)). The way to properly draw down the contingency reserve for the project is to rerun the Quantitative Risk Model after removing and/or reducing the risks that have either passed or have been verifiably reduced in either their probability or consequences (typically due to successful risk mitigation efforts). The difference between the initial budget at the established CI and the current model output at the same CI is the amount that can be legitimately drawn down from that project's nominal contingency reserve and put to different uses. This is another key concept for Owner's to understand: a Quantitative Analysis provides you the answer to the overall question of contingency budgeting, for example. It does answer the question, "What contingency budget do I need on this tunnel project to assure me that 90% of the time I will have enough in reserve to pay for the project?. However, *it DOES NOT answer the previous question accurately at the individual risk level*; for example, "The Quantitative Analysis has calculated that a $33.7 Million overall Contingency reserve will see me through 90% of the time (90%CI). Well, Risk #143 is "Main Shaft Collapse during Construction." What is the 90% CI dollar figure associated with that particular risk? The probabilistic model does not directly answer this question (why? Because the outcome for low probability/high consequence events typically the 90th percentile run of a simulation that has a 1/1000th chance of occurrence is $0 (i.e., at least 99% of the time, this threat doesn't materialize at all). However, the model does capture the overall impact of these events by using large numbers of simulations which do capture the dollar impact of rare events in the "tails" of their distribution curves. That is, the overall contingency budget does capture these events with sufficient simulations. The proper way to assess the elimination of certain project risks is to re-run the model with these events zeroed out. The new contingency budget at the same confidence interval reflects the "real" dollar credit to the project of these events not transpiring.

*Definition by Investopedia.

## CONCLUSION

Risk management is and will continue to be one of the most important topics in the tunneling industry for the foreseeable future because of the need to implement methods to correct a track record of major cost overruns on large projects and because the needs for tunneling are driving the industry towards ever larger and more complex projects.

The current climate works against the dissemination of wisdom from lessons-learned on previous projects because the Mega-Projects are increasing the average time individuals spend on each project and because the litigious and political forces driving many of these projects, not to mention their high costs, greatly inhibit transparency and the willingness to acknowledge anything that might be perceived as failure or a mistake.

This lack of openness has led to a pattern of authorities repeating the faltering steps of their sister agencies who have preceded them down the Mega-Project path on similar efforts resulting in a repeated wasting of time and money.

Better communication is a major benefit from an Active Risk Management Program. Owners embarking on Mega-Projects need to understand that the Joint Ventures that form to build them often have little to no previous track record of collaboration, and typically have a very short time-frame in which serious examination of the Mega-Project takes place before bids are submitted. Further, the constituents of these teams are often unaware that they have widely varying appreciations of the risks involved in the project. All these issues underscore the need encourage that JV teams embark upon Active Risk Management practices from the outset of the project to increase their ability to both communicate internally as well as externally with the Owner.

Sharing risk information is ultimately a good thing. It should be a two-way street though, and a method is suggested above that would allow both the Owner and JV teams to share information without the discussion of risks becoming prejudicial to either party.

Contingency needs to be managed at the Program, and not the Project, level by a designated Risk Manager. Several Federal agencies have developed a system that would allow this effort to be planned and managed for Mega-Projects and long-term Programs (although not the position of Risk Manager at this level).

Contingency budgets are best developed through Quantitative Risk Analysis of a statistical risk model built up from the identified project risks. This process needs to be managed, maintained and periodically updated throughout the life of the project.

## PROJECTS CONSIDERED

Central Artery Tunnel
Port of Miami Tunnel
City of Portland's Willamette River CSO Program
DCWater's Clean Rivers Program (incl. Blue Plains Tunnel and Anacostia River Tunnel)
NYCMTA's East Side Access Project
NYCMTA's 2nd Avenue Subway Project
LA Metro's Regional Connector Project
LA Metro's Westside Subway Extension Project

# Risk and Contingency Management Planning on LACMTA's Purple Line Extension and Regional Connector Projects

**David Davies**
Parsons Brinckerhoff

**Rick Wilson**
Los Angeles Metropolitan Transit Authority

**ABSTRACT:** The Los Angeles Metropolitan Transit Authority (LACMTA) are adopting a risk-informed management process on their Regional Connector (RC) and Purple Line (Westside) Extension Projects (PLE) to review and validate project scopes, schedules, budgets and to analyze ongoing project development and management. This paper will present an overview of the development of the Risk and Contingency Management Plans that provide LACMTA and their funding partner, the Federal Transit Administration (FTA), greater confidence in safeguarding project budgets and schedules while at the same time streamlining the project's assessment process at required key project milestones.

## INTRODUCTION

The PLE and RC projects were conceived as part of LACMTA's 30/10 Initiative. This initiative utilizes the long-term revenue from the Measure R sales tax as collateral for long-term bonds and a federal loan, which will allow Metro to build 12 key mass transit projects in 10 years, rather than 30 years.

The PLE is a continuation of the existing Metro Purple Line and is located entirely within Los Angeles County, California. LACMTA ultimately proposes constructing an approximate 9-mile heavy rail transit (HRT) line, extending from the Wilshire/Western subway station to a new western terminus near the Veterans Affairs West Los Angeles Medical Center (VA Hospital), west of Interstate 405. The PLE will be a double-tracked, third-rail exclusive guideway heavy rail system built primarily under Wilshire Boulevard. The project scope includes the construction of nine fully underground stations, the procurement of up to 58 new heavy rail vehicles (HRVs), improvements to the existing Division 20 Rail Storage and Maintenance Yard in downtown Los Angeles, and improvements to the Rail Control Center supporting LACMTA rail operations.

The RC Project is a 1.9-mile, dual-track, fully underground light rail transit (LRT) service through downtown Los Angeles that will connect the existing Blue Line LRT service and the recently opened Exposition Line LRT service to the existing Gold Line LRT service at Little Tokyo. The Blue, Gold and Exposition LRT lines operate on a dual track with an overhead contact system (OCS) that provides power to the light rail vehicles (LRV). The RC will utilize the existing 7th Street/Metro Center Station and include three new underground LRT stations.

Both projects are being partly funded through the Federal Transit Authority New Starts program.

New Starts projects are defined as projects that are:

a. Projects with a capital cost of \$250m or greater, or
b. Seeking \$75m or more in funding.

The FTA New Starts and Small Starts Program (http://www.fta.dot.gov/12304.html) is the federal government's primary financial resource for supporting locally-planned, implemented, and operated transit "guideway" capital investments.

The Moving Ahead for Progress in the 21st Century Act (P.L. 112-141), MAP-21 directs FTA to evaluate and rate candidate New Starts projects as an input to federal funding decisions through a formal risk assessment process at specific milestones.

The FTA has developed a risk review process intended to:

- Inform the FTA about the projects risk
- Provide the Grantee (in this case LACMTA) with the recommendations to strengthen the project, and
- Provide FTA with highly confident project cost and schedule targets

**Table 1. Project risk scoring matrix**

| | Low (1) | Med (2) | High (3) | Very High (4) | Significant (5) | | Legend |
|---|---|---|---|---|---|---|---|
| Probability | <10% | 10–50% | 50–75% | 75–90% | >90% | Risk Score | Low ( <= 3) |
| Cost Impact | <$250K | $250K–$1M | $ 1–3 M | $3–10 M | >$10 M | | Medium( 3–10) |
| Time Impact | <1 Mth | 1–3 Mths | 3–6 Mths | 6–12 Mths | >12 Mths | | High ( >=10) |

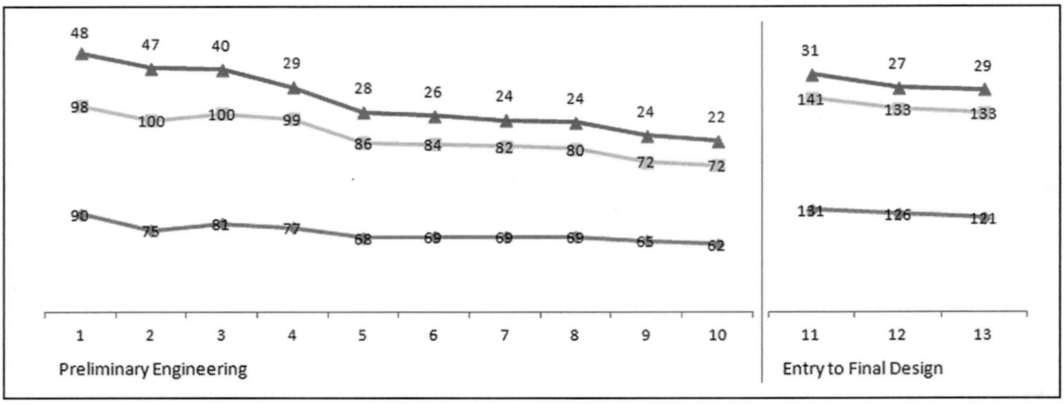


**Figure 1. Project risk trending**

## QUALITATIVE RISK ANALYSIS

On completion of the Projects Alternate Analysis a comprehensive risk register was developed which was used, by an a dedicated Risk Manager, to monitor and mitigate project risks through the preliminary engineering and final Design phases of the project (See Table 1).

Risks were identified through a series of risk identification workshops, at key project milestones, and through continuous monitoring of the project's risk profile.

Risks were assessed based on the following sequence:

- Likelihood of occurrence
- Estimated (and most likely) cost impact range
- Estimated (and most likely) schedule delay range

The risk score was determined by averaging the cost and schedule scores and multiplying the average by the likelihood score. For example:

- Likelihood of occurrence (Probability) = 3
- Probable Cost Impact Score = 4
- Probable Time Impact Score = 2
- Resulting Risk Score = 3 × (4 + 2)/2 = 9

Each risk was assigned to an appropriate risk level—low, medium, or high—according to the risk score. In the previous example the risk is at medium level as its risk score (9) falls in between 3 and 10.

Once the resulting risk score for each risk had been established the risks were sorted by "rank" score to determine those risks with the greatest potential severity impact. The "rank" scoring was then used to determine which risks required the greatest level of management and mitigation.

For the risks with the highest "rank" score, risk owners and mitigation strategies were assigned. During the risk management process, the assigned risks were reviewed on a monthly basis with the risk owners and appropriately updated. The updated risk register was then presented and discussed with the FTA at a monthly risk review meetings and a subsequent monthly risk report was issued to all project stakeholders (Figrue 1).

The Project Risk Trending chart demonstrates effective risk management of project risks during the Preliminary Engineering and Final Design phases. At the start of Preliminary Engineering 48 high, 98 medium and 90 low risks were identified that were managed down to 22 high, 72 medium and 62 low risks by the end of Preliminary Engineering. Similarly, for the Final Design phase there were

**Table 2. Extract from project risk register**

| ID | Description | FTA Milestone | C | T | P | Score | Review Comments |
|---|---|---|---|---|---|---|---|
| 26 | Local matching funding may be insufficient to meet cash flow for full build-out of 9 mile section. | Entry to FD | 5 | 5 | 3 | 15 | 9/28/2012 - Measure J, which will be on Los Angeles County's ballot this November, will extend the 30-year Measure R sales tax by another 30 years (to 2069) in order to accelerate the construction of 7 transit and up to 8 highway projects over the next decade. Risk #26 needs to be reassessed after Nov 7th. 11/28/2012 - Risk closed. Measure J failed and Section 1 will be built with available funding. |
| 27 | Federal funds may not be available to meet cash flow for full build out of 9 mile section. | Entry to FD | 5 | 5 | 3 | 15 | 1/12/12 - Reviewed at EFD Risk W/S # 1 - No change, Assigned as a Program Risk 11/28/2012 - Risk closed. Measure J failed and Section 1 will be built with available funding. |
| 583 | Interface risk between three system constractors. | 20% Construction | 3 | 2 | 3 | 7.5 | 9/28/2012 - Newly identified risk. 11/28/2012 - Risk closed. Not a risk to Section 1. |
| 571 | Continuing the development of two route options past PE stage | Entry to FD | 3 | 4 | 2 | 7 | 11/28/2012 - Risk closed. Not a risk to Section 1. |
| 403 | Based on ridership analysis, 104 vehicles have been reduced to 78 vehicles during FD. Ridership results may be too low. | Bid to DB | 5 | 1 | 2 | 6 | 5/31/2012 - RFMP completed. Metro is currently reviewing it. 11/28/2012 - Risk closed. The updated RFMP verified the number of vehicles (78). |
| 576 | Decreased headways in future may require special vehicle design featues (clarify) | Bid to DB | 3 | 2 | 2 | 5 | 11/28/2012 - Risk closed. Will be standard vehicles, no special design features. |

31 high, 141 medium and 131 low risks managed down to 29 high, 133 medium and 121 low risks by the end of Final Design (Table 2).

## QUANTITATIVE RISK ANALYSIS

A cost and schedule risk analysis was carried out in line with the risk assessment guidelines outlined in FTA's Oversight Procedure 40 and utilizes the FTA cost risk assessment workbook.

## COST RISK ANALYSIS

Based upon historical information, FTA has developed a model that takes the most optimistic cost estimate (free of contingency with a 10 percent likelihood of success) and the most pessimistic estimate (termed the 90th percentile) to which a LogNormal distribution curve is applied. This results in a cumulative density function (or "S" curve) of likely project cost ranges versus probability. The intention is to produce a more accurate and realistic end cost forecast based on past trends. The multiplication factors between the 10th percentile and the 90th percentile are known as the "Beta factors," now renamed in FTA's latest Oversight Procedure (OP40) as the Beta Risk Factor or "BRF." The modeling process has been called a "top-down" analysis in contrast with the traditional risk register-based Monte Carlo analysis that is referred to as the "bottom-up" approach.

The Top-Down Beta Risk Factor Analysis applies BRFs to a Base Cost Estimate (BCE). The BCE is conditioned by stripping out allocated and unallocated contingency and further reduced for embedded, latent, or patent buried contingency. FTA developed a profile representing progressive risk reduction across the delivery cycle based loosely around historic trends and adjusted for real-life experiences. Figure 2 shows the FTA's beta reduction factor allocation diagram.

## SCHEDULE RISK ANALYSIS

LACMTA's master schedule was used as the basis for the schedule risk model. A summarized schedule was developed and activity durations were collapsed and adjusted to reflect their most optimistic duration, free from any allowances for risk or "latent float." Latent float is float included in activity durations and is better pulled out and separated as specific float durations, termed "buffer" or intermediate float. The expression "buffer" describes float inserted within the critical or near critical path activities to absorb delays within the schedule directly related to perceived risks and uncertainty in those preceding activities.

The risk schedule is a simplified critical path network and, while trying to incorporate all sections of the project for completeness, only includes those activities believed to be key and critical to the project's completion and relevant to the risk assessment.

Three-point estimates for each activity in the risk schedule model (minimum, most likely, and maximum) have been developed.

Where activities have been determined to have discrete risks beyond those that could be captured

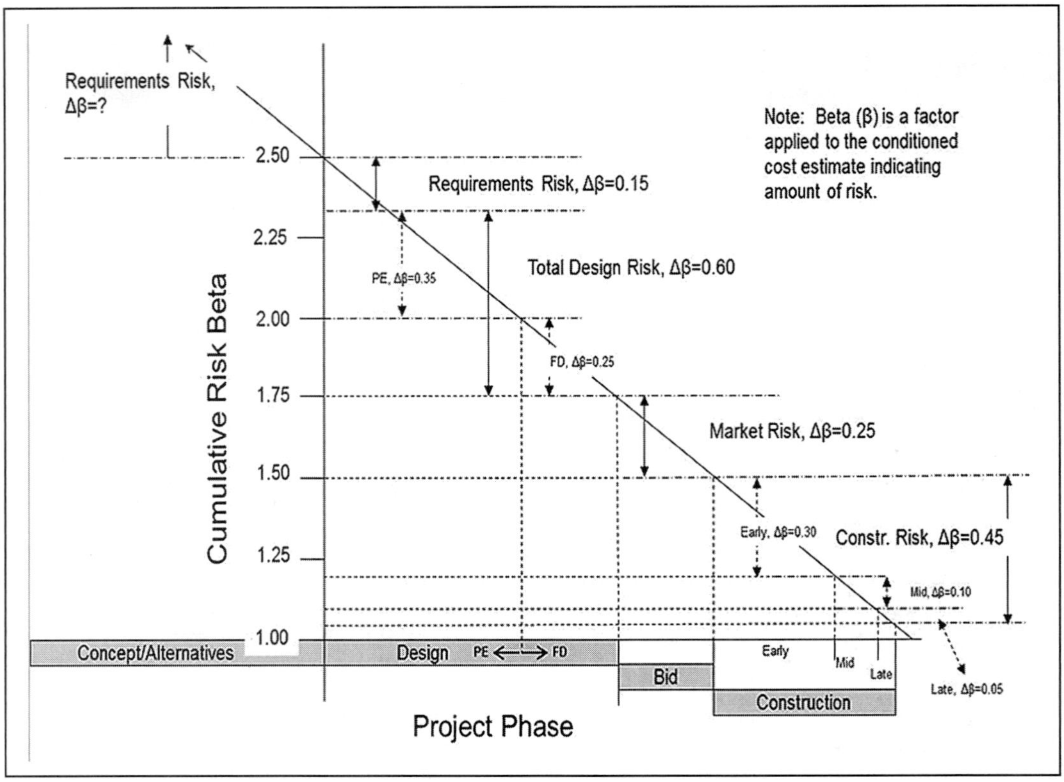


**Figure 2. FTA beta risk factor allocation diagram**

in range on the applicable activity, these have been modeled using the "task existence" function where unusual risk events associated with a given activity are considered through the incorporation of successor activities in the schedule model.

The PMOC shall "step back" sequentially through various completion milestones for the project and shall estimate the minimum amount of schedule contingency required to complete the project on schedule, in consideration of risks identified in this Oversight Procedure.

The schedule contingency recommendations were developed using these fundamental assumptions:

- At the Revenue Operations Date (ROD), schedule contingency requirements have been reduced to a minimum requirement or possibly eliminated,
- At the point of 100% complete with bid (for Design-Bid-Build) or 100% subcontracted (for Design-Build or CM-GC), the project should have sufficient schedule contingency available to absorb a schedule delay equivalent to 20% of the duration from Entry into Final Design through Revenue Operations.

In the Figure 3 example LACMTA project schedule had a target end date of 26-Dec-19. After collapsing and adjusting (C&A) to the optimized schedule duration the C&A end date was calculated as 9-April-19, 9 months in advance of the LACMTA target end date. The FTA float requirement on the C&A schedule was calculated as 17 months and this was added to the C&A schedule end date to return an FTA Target end date of 29-Jul-20, which is 8 months after the LACMTA target end date.

LACMTA are then required to manage the project schedule within the FTA Target end date.

## CONTINGENCY DRAWDOWN

At specific project milestones, the FTA requires the project to evaluate the project risk exposure at those points and evaluate the minimum contingency value that may be "drawn down" or expended at that point. As the project progresses through these project milestones the project cost and schedules are evaluated to ensure that they remain within the calculated minimum contingency values. If the project falls below or is trending to fall below the minimum contingency values then the grantee must implement mitigation measures to replenish the contingency values. The

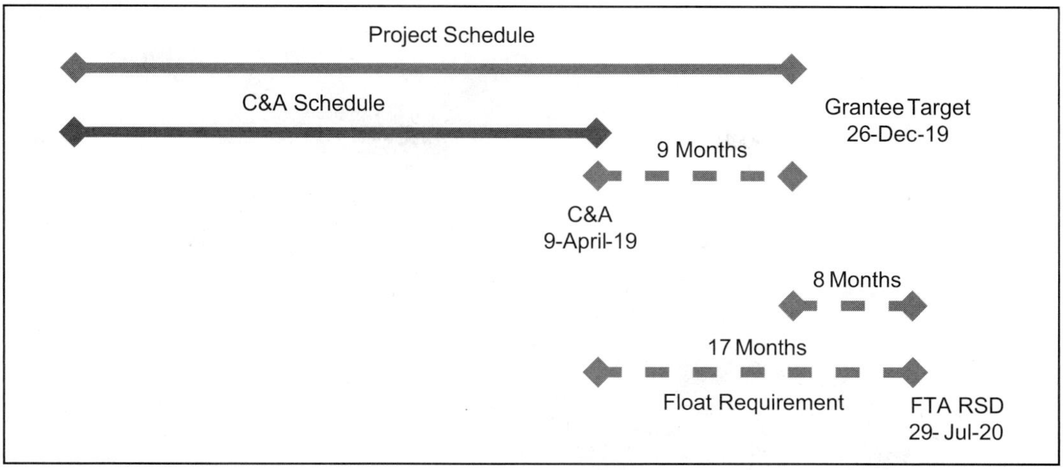


**Figure 3. Diagrammatic representation of FTA schedule buffer float requirement**

management of the minimum contingency values at project milestones is used to protect from inappropriately early drawdown of contingency values.

## RISK AND CONTINGENCY MANAGEMENT PLAN (RCMP)

The RCMP is a section of the LACMTA's Project Management Plan (PMP). The purpose of the RCMP is to highlight specific areas of management focus, as identified through the risk review process, and to provide a means for LACMTA to monitor progress as the project moves forward.

The LACMTA RCMP integrating the:

- Identification
- Assessment
- Response plans
- Management of the risk process, and
- Contingency management and drawdown control curves

## RISK MITIGATIONS

Throughout the project's design phases LACMTA adopted a risk informed approach to the design development.

Several design mitigations were adopted which reduced the overall risk profile of the projects.

Significant design mitigations were:

- Drafting of Request for Qualifications and Request for Proposals Documentation
- Early utility relocations
- Exploratory Shaft
- Use of EPBM's designed for excavating in gassy ground conditions

## DRAFTING OF RFQ/RFP DOCUMENTATION

When drafting the RFQ and RFP for the major Design-Build contracts LACMTA incorporated technical advancements and improved contract terms, based on "lessons learned" from successful management of LACMTA Gold Line Eastside Extension Project, Industry Constructability Review and uniformity in approach with other LACMTA rail projects.

## EARLY UTILITY RELOCATIONS

To reduce the potential schedule and cost risks to the critical path construction activities, three Design-Bid-Build contracts were issued for relocating water, power and sewer lines in advance of awarding the major Design-Build contract that included the tunneling and subway station excavation and construction.

The utility relocation risk was further mitigated by the use of extensive utility surveys incorporated into three dimensional Virtual Design and Construction VDC models to optimize the placement of station footprints to minimize the number of utilities required to be relocated or supported in place (Figure 4).

## EXPLORATORY SHAFT

The temporary exploratory shaft is being constructed to gather data related to soil conditions, gassy ground and ground water to assist in the geotechnical design of the Wilshire/Fairfax Station and tunnels. Risks associated with potential construction delays during the discovery and excavation of prehistoric fossils will be mitigated through planning of early construction activities.

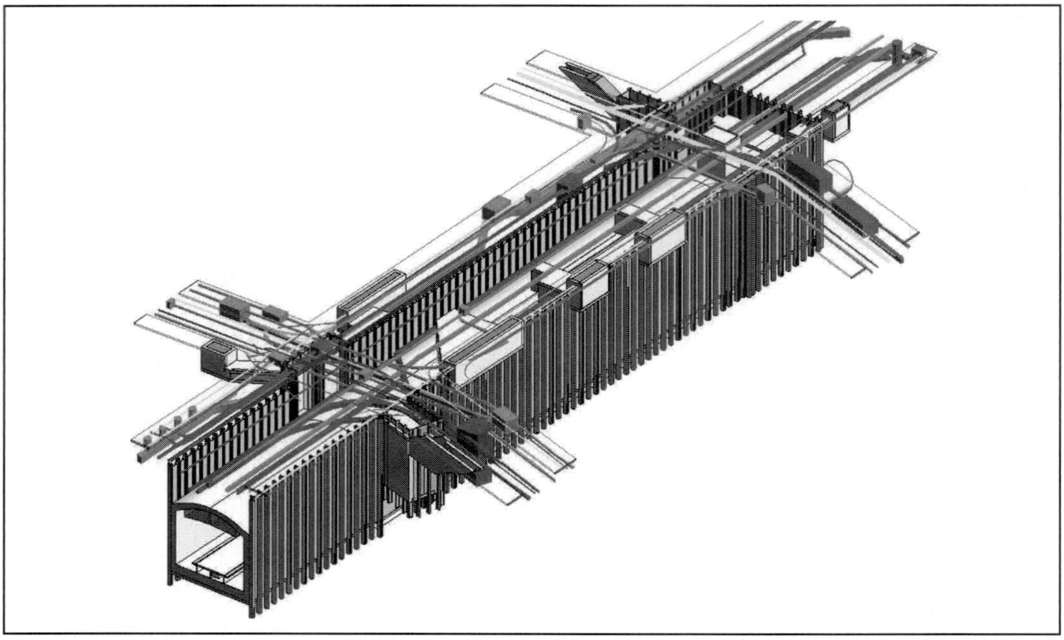

**Figure 4. Three dimensional (3D) station model illustrating utility interfaces**

## USE OF EARTH PRESSURE BALANCE TUNNELING MACHINES (EPBM) FOR LACMTA PROJECTS

In 1985, Congress enacted a bill to ban federal funding for the expansion of the Red Line Subway. This was in response to a methane gas explosion in the Fairfax District.

In October 2005, the American Public Transportation Association (APTA) conducted a review of Wilshire Corridor tunneling. The panel evaluated new advances in worldwide tunneling technology, as well as safety of building and operating transit tunnels in the identified hazard zone along Wilshire Boulevard. The panel concluded that such tunneling would be feasible and could be undertaken at no greater risk than other subway systems in the U.S.A. In December 2007, Congress repealed the federal prohibition on Subway construction along Wilshire Boulevard.

Use of EPBMs for LACMTA projects follows the recommendation in the 1995 report of a specially convened Tunneling Advisory Panel entitled "Report on Tunneling Feasibility and Performance." In acceptance of the report, LACMTA has instituted the policy, to reduce or avoid construction risk of excessive settlement with open face tunnel shields, by requiring pressurized-face tunneling.

Tunnels and stations will be built to provide a redundant protection system against gas intrusion. This might include:

- Physical barriers to keep gas out of the tunnels
- Addition of an inner CIP lining to sandwich an High Density Polyethylene (HDPE) barrier
- High volume ventilation systems
- Gas detection systems with alarms
- Emergency ventilation triggered by the gas detection systems.

During construction and, operations, safety codes require rigorous and continuous gas monitoring, alarms, automatic equipment shut-off and additional personnel training.

## CONCLUSION

The FTA New Starts and Small Starts Program (http://www.fta.dot.gov/12304.html) is the federal government's primary financial resource for supporting locally-planned, implemented, and operated transit "guideway" capital investments.

The FTA risk review process, within the New Starts and Small Starts Program, outlines the requirements for risk identification, assessment, mitigation, contingency policy and tracking, and organizational discipline and provided a robust set of tools for risk management of the LACMTA projects.

By adhering to the FTA risk review process LACMTA successfully inform the FTA about the projects risk, provide the LACMTA with the recommendations to strengthen the project, and provided highly confident, FTA, project cost and schedule targets.

Through this effective risk management program the LACMTA was able to identify and manage significant project risks through the application of a risk informed project design.

## REFERENCES

FTA White Paper (a.1 rev 08)—Risk-Informed Project Oversight at the FTA (Sillars, Ehrler, Reich).

ITA Guidelines for Tunneling Risk Management: International Tunnelling and Underground Space Association, Working Group No. 2, Tunnelling and Space Technology, Vol. 19, pp. 217–237, 2004.

ITIG A Code of Practice for Risk Management of Tunnelling Works, The International Tunnelling Insurance Group, 2006.

US DOT Federal Transit Administration, TPM-20 Office of Engineering, Project Management Oversight, OP-40 Risk and Contingency Review, Rev. 2, May 2010.

# Risk Analysis and Management on Toronto Light Rail Transit Program Projects—A Case Study

**Suresh Kataria**
Parsons Brinckerhoff

**Iqbal Hassan**
Metrolinx

**ABSTRACT:** Construction projects are typically exposed to socio-political, legal & environmental, operational and market related risks increasing project costs and causing delays. A robust risk management plan provides a project with a systematic process for identifying, assessing, evaluating, managing and documenting risks that could jeopardize the success of the project. This paper presents an application of a bottom-up risk management approach to the Metrolinx Toronto Light Rail Transit Program projects that identifies, quantifies and correlates risk and uncertainty, analyzing the collective impact of project risks using the Monte Carlo simulation technique to determine potential cost outcomes and associated confidence levels.

## INTRODUCTION

Program Managers and Directors of major transit programs in North America now say they have a better understanding of major risks facing their projects than they did a few years ago. This, in part, is as a result of the Federal Transit Administration (FTA) in the United States making risk assessments a mandatory requirement on transit projects in order for a transit agency to obtain federal funding and in part a growing awareness of how a formal project risk management process is increasingly becoming a culture in the delivery of major infrastructure projects (Federal Transit Administration, 2008). Yet many will tell you they need to have a better handle on strategic risk. There is a growing need for program managers to embrace a more enterprise wide view of risks related to strategy and delivery of a mega project, thus enabling the organization to be in a better position to achieve its strategic objectives.

Major public transportation projects are unique in nature. These present, at the early stages of the project, a high level of risk exposure whereas very limited information is available on the project risks (see Figure 1). Even though little is known at the early stages, the projects still need to go forward· and a risk management approach should be applied throughout the project's life cycle (Project Management Institute, 2009). The risk management process recognizes this situation and helps develop a broader range of estimates to account for more realistic project plans.

As the projects progresses, more information is available on the project risks. The individual risk components are decreased in both number and potential cumulative impact which reduces the range around specific costs and durations. More realistic project plans are achieved. As noted in FTA's *Risk Analysis Methodologies and Procedures,* the key benefit of a systematic evaluation of project risks is that it provides a project owner with increased confidence '*...that appropriate cost and schedule allowances have been established and, as a result, the project is more likely to be completed on time and within budget'* (Federal Transit Administration, 2004).

Although the impact of risks occurring on the transportation projects is multifold including cost and schedule, health & safety, reputation and legal impacts, the purpose of this risk assessment effort is to document mainly, the impact of risks to the project delivery in terms of project cost and delivery schedule. This is, by no means, indicative that the assessment of other risks types is not required. Other risks related directly to health, safety and security are, and should be, assessed using the other established processes such as preliminary hazard analysis, threat and vulnerability analysis, and other industry recognized processes.

This paper presents how a bottom-up risk management approach was used to identify, assess and manage risks on the Metrolinx Toronto Light Rail Transit (LRT) Program. A case study of the Eglinton Crosstown LRT project is used to discuss the overall risk assessment process. One section summarizes the risk management process as followed on the Program. Another section describes the details of the

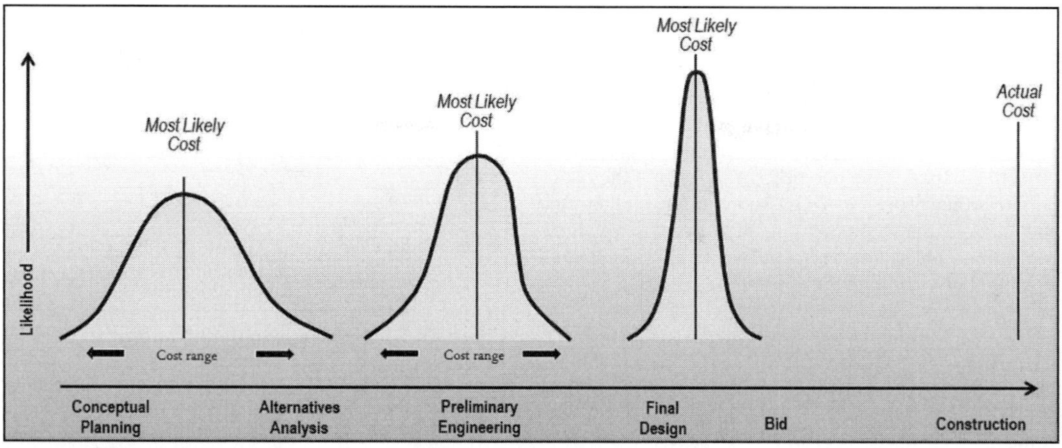


**Figure 1. Changing uncertainty with project progression**

risk management process as applied on the design and construction Eglinton Crosstown LRT project including the application of contractual risk allocation on the construction contract for a 5.5 km long bored twin tunnels. The conclusions are discussed at the end of the paper followed by a list of references.

## RISK ASSESSMENT ON CONSTRUCTION PROJECTS

All public transportation projects encounter a multitude of risks from conceptual design through to the operations that hamper the projects' primary goals of finishing the planned scope within planned budget and time. These risks range from environmental, social, political and funding risks in the conceptual stage, project execution risks during the design and construction stage, market risks during the procurement stage and the administration and operations risks during the revenue service stage.

From a project delivery perspective, most of these risks ultimately result in project cost overruns and schedule delays which then result in additional cost overruns when considering the loss of revenue due to delayed service operations. Where engineering issues have been encountered on large (>$100M) transportation projects, these risks have been largely controllable. By comparison, stakeholder, third-party and real-estate acquisition issues were less controllable and had larger impacts on the project definition. Some of the most significant risks to capital costs and schedule are time to achieve political consensus and acquisition of private property (Flyvbjerg, Holm, & Buhl, 2002).

### Risk Management Process

The goal of risk management process is to use a formal approach to improve project delivery (schedule and budgetary performance) through more proactive management of risks. The risk assessment process used on this program was a mixture of qualitative and quantitative analyses wherein the program level delivery risks were assessed on a qualitative level and the project specific risks were assessed for quantitative cost and schedule impacts due to individual risk events to the delivery of the projects.

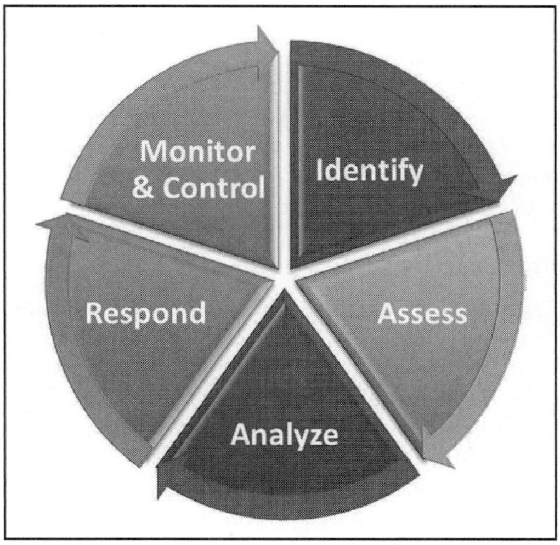


**Figure 2. Risk management is process driven**

Risk management provides a systematic approach to identify and prioritize risks as well as action-oriented information to program and project managers to assist in the mitigation or avoidance of undesirable project outcomes and the 'capture' or enhancement of opportunities. Risk management is not an optional activity, nor is it a substitute for other project management processes. The risk management process (see Figure 2), as defined by Project Management Institute (PMI), is process driven and aims to identify and assess risks and helps prioritizing efforts for effective management of identified risks (Project Management Institute, 2013). It is essential to successful management of program objectives. It adds the perspective of risks to the outputs of other processes (e.g., scheduling, budgeting and change management) and adds to their value by taking uncertainty into account to understand and manage challenges to the program's successful completion.

Effective management of project risks is necessary to significantly increase the chances of delivering a successful project and should be managed through appropriate management plans and procedures. Unlike the FTA's guidance on risk management of capital projects, FTA Oversight Procedure 40 (Federal Transit Administration, 2008), which requires development of a risk management plan for the capital projects, the equivalent governing agency does not exist in Canada and hence a risk management plan was not required. The Program Manager, realizing the benefits of a formal risk management program, adopted the FTA like approach and mandated to have established a formal risk management program for the LRT projects in Toronto.

### Project Contingency Assessment

Traditionally, on the major capital expansion projects, the cost and schedule contingencies are set based on the agencies' contingency management policies wherein, a defined percentage of the base cost estimate (or baseline schedule) is applied as contingency for all projects. Although this approach, based on past projects within the agency, is a good indicator of the required project contingencies, it has two potential shortcomings. Firstly, this policy relies on active implementation of the lessons learned from previous projects which are, generally, not implemented effectively due to lack of formal documentation of lessons learned on past projects. This leads to budgeting for high confidence targets, locks up capital, and discourages risk management and mitigation.

Secondly, as the major capital expansion projects are being built after many years and, as such, the historical project information does not accurately measure the impacts of the changing technical and socio-political factors. The FTA recognized these and other issues as weaknesses and developed guidelines to use a quantitative risk analysis approach to assess the project specific risks which ultimately then be used to establish project contingencies (Federal Transit Administration, 2008).

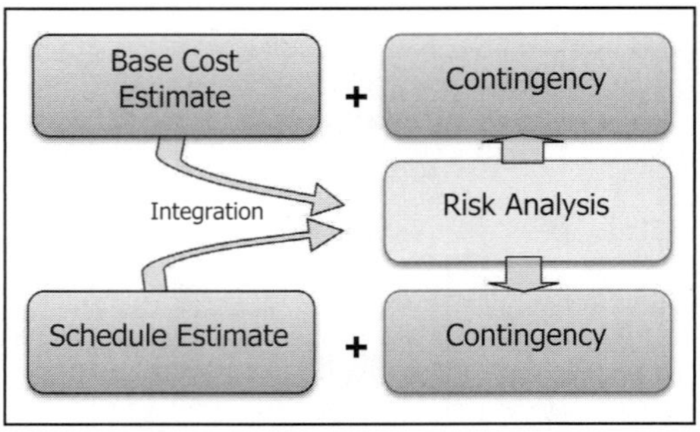


**Figure 3. Risk analysis to define and/or validate project contingencies**

Under the quantitative risk analysis approach, the cost estimate(s) and schedule(s) are stripped of all contingencies, allocated and unallocated, and are modified to reflect an optimistic baseline, a 'blue-sky' scenario. Risk analysis is then performed to estimate the variability in cost estimates and schedule, using Monte Carlo simulation, to gain a better understanding of the impacts due to the known-unknowns. The outcome of the risk analysis is used to, validate, or in some cases, define, the cost and schedule contingencies required for the projects (see Figure 3). Risk management adds value to the project by taking risks into account and provides a basis for estimating the amount of cost and schedule contingency reserves needed to cover for potential risks with a required level of confidence to meet project objectives.

## CASE STUDY—TRANSIT EXPANSION PROGRAM

### Program Description

The Government of Ontario, based on the recommendations from Metrolinx, committed $8.4 billion funding in 2008 to build four new Light Rail Transit (LRT) lines, a network totaling of 52 kilometers of light rail transit running underground and on the street in Toronto. These four LRT lines included Eglinton Crosstown LRT, Scarborough RT, Finch West LRT and Sheppard East LRT (see Figure 4). In summer 2013, this plan was revised and accordingly Scarborough RT planned LRT system was replaced to have heavy rail subway line instead. Metrolinx will own the LRT lines and the TTC will operate them.

### Program Organization Structure

The program organization structure, as shown in the Figure 5, comprised of the Rapid Transit Implementation Metrolinx (funding agency) which makes governance decisions on a monthly basis and

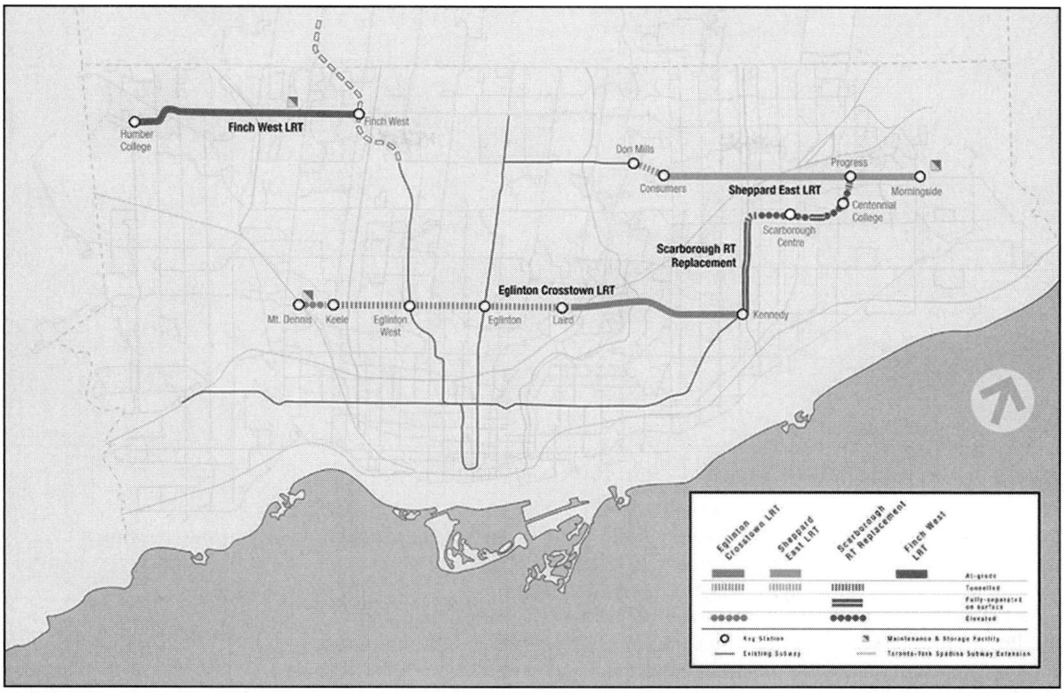


**Figure 4. Toronto light rail transit projects as part of *The Big Move***

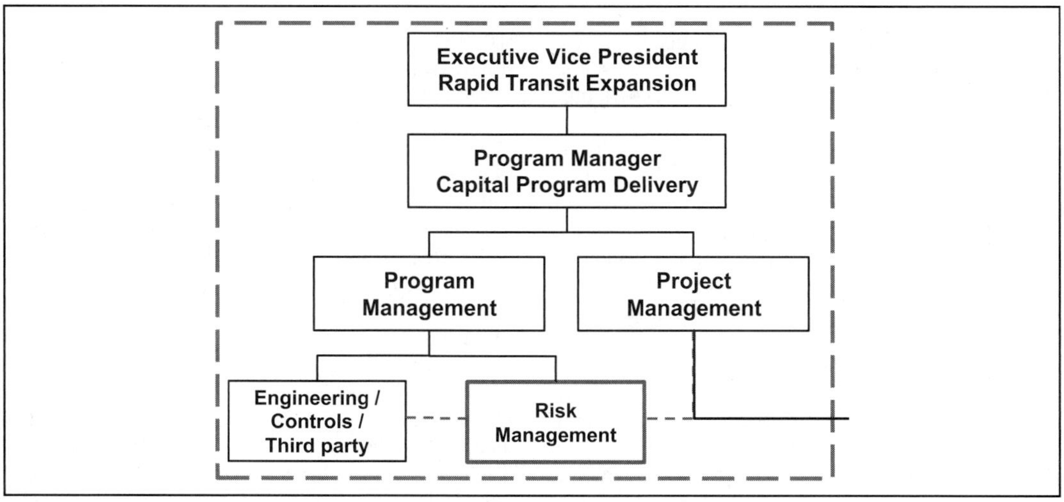


**Figure 5. Transit expansion program organization structure**

Capital Program Delivery team which deals with the day-to-day management of the program. The Program Management recognized the role for active risk management in decision making and retained Parsons Brinckerhoff to provide engineering and program management services of which program and project delivery risk assessments were a key part.

### Program Risk Management Approach

All capital transit projects are uncertain as they are unique, complex and involve multiple stakeholders. This Project uncertainty is inevitable and can be controlled through the use of a structured and disciplined process (Federal Transit Administration, 2008) (Project Management Institute, 2009). A Risk

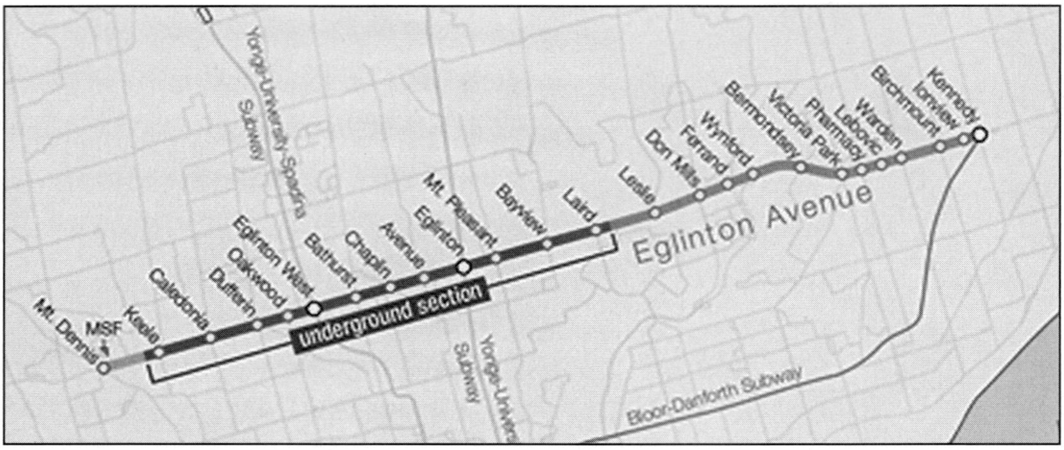


**Figure 6. Map of Eglinton Crosstown Light Rail Transit line**

Management Plan (RMP) was developed for the Program that outlined a systematic process for identifying, assessing, evaluating, managing, and documenting risks that could jeopardize the success of the Program. The RMP laid out roles and responsibilities to enable open risk awareness culture within the program delivery team and with stakeholders at the same time creating a sense of accountability and ownership so the people with corresponding responsibilities were aware of what was expected of them.

The RMP also described the process of risk identification, assessment and analysis being the evaluation and quantification of risk and uncertainty to the program schedule and budget. The results of the risk analysis were used to support the decision making process of the amount of schedule and cost contingency required at any given time given the available options to mitigate risk. This RMP and the policies and procedures contained within support the program goals to deliver a successful program of transit projects within the assigned budgets and agreed timeframes.

Risks to the Transit Expansion Program were managed both at the program level as well as at project level. The program level risk management effort attempts to identify "institutional" risks related to program scope, stakeholders' requirements, resources (internal and external), market and funding. The project level risk management dealt with identifying, assessing and managing location specific risks such as unknown site conditions, tunneling, excavation, right-of-way needs, etc. Although the risks were managed on all four LRT lines, this paper focuses on the application of risk management process on the Eglinton Crosstown project.

## Risk Management on Eglinton Crosstown Project

### *Project Description*

The Eglinton Crosstown LRT (ECLRT) is the cornerstone of the Toronto LRT projects with over 50% of the total program cost. It was originally planned as a 31 km long Light Rail Transit line from the Toronto Pearson International Airport to Kennedy Road. Of the proposed 31-km (19.4-mile) line, approximately 10 kilometers (6.3 miles) between Black Creek and Laird Avenue was planned to be constructed underground with the balance of the alignment located in dedicated tracks that were to be reserved within the existing Eglinton Avenue right-of-way.

This approach was revised to reflect the political nature of the City and as such the alignment was reduced to a total of approximately 20 km (12.5 miles) LRT line between Mt. Dennis to the west and Kennedy Station to the east. In this revised configuration, as shown in the Figure 6, the underground portion of the alignment remained unchanged and the project will have a total of 13 underground stations in a busy mid-town corridor along the Eglinton Avenue corridor which served residential/business needs. Construction on this corridor is challenging considering a large number of existing utilities within its right-of-way along with a high volume of existing vehicular traffic along the Eglinton Avenue.

In order to progress the project expeditiously, certain activities have commenced in advance which, as of November 2013, includes:

- Procurement of four (4) Tunnel Boring Machines (TBMs)

| | Low (1) | Medium (2) | High (3) | Very High (4) | Significant (5) |
|---|---|---|---|---|---|
| Likelihood | <10% | 10% to 50% | 50% to 75% | 75% to 90% | >90% |
| Cost impact ($) | <$250K | $250K to $1M | $1M to $5M | >$5M & <$15M | >$15M |
| Time impact | Less than 1 month | 1 month to 3 months | 3 months to 6 months | 6 months to 12 months | >12 months |

**Figure 7. Risk scoring matrix used to assess project risks**

- The tunnel pre-cast concrete lining manufacturing contract to provide the linings for approximately 10 km of the twin tunnels
- TBM Launch Shaft construction was completed in winter 2012 at Black Creek
- Twin tunnel contract from Black Creek to Yonge Station awarded in October 2012 and work of tunneling with planned completion in early 2017
- Remaining twin tunnel portion from Brentcliffe to Yonge Station awarded to in October 2013 for planned completion date of late 2016
- Concept design for all stations completed and Alternative Financing Procurement (AFP) was released in December 2013
- At-grade and grade separation options have been finalized and these concepts became part of the scope of AFP

In 2011, the ECLRT project management team prepared the Project Implementation Plan (PIP). The PIP provided an outline of the project implementation schedule (IS), cost estimate and the contract packaging plan and the guidelines for delivering the ECLRT project; both as an individual project and as an integral part of the overall Eglinton-Scarborough Crosstown transit line.

Recognizing the level of uncertainty at the preliminary engineering stage, assumptions were made for construction techniques, sequencing and productivity. The Assessment of the underlying and risks with the assumed implementation strategy was important to identify whether the underlying assumptions were overly optimistic or conservative.

A bottom-up risk identification and assessment of the PIP was carried out through a series of workshops which included individuals from project team, design consultants and individual from Engineering, Project Controls, Procurement and Property Acquisition. The risk workshops were organized to identify significant risks, uncertainties, opportunities, and assess the associated magnitude of occurrence, possible range of impacts to the contract schedule based on the best information available and to present the expected schedule durations for each of the packages before mitigation of the risks.

### *Risk Identification & Assessment*

The risk analysis effort was intended to stress test the implementation schedule, analyze schedule risk exposure and provide recommendations for the schedule contingency required at that stage. A schedule risk analysis model was developed based upon identified possible impacts of specific risks along with an assessment of the uncertainty surrounding major components of the activities in the implementation schedule.

Various project-related assumptions in the implementation schedule were reviewed and schedule risks were identified and assessed for potential severity. These include risks related to delay in design completion, procurement process, utility relocation, system integration, etc. Three-point estimates for duration were also made to account for the inherent uncertainty on assumed productivity for delivering certain tasks.

Risk assessment included determination of the importance of the risk along with the likelihood of its occurring. A 5-point risk assessment matrix, as shown in Figure 7, was used as a means to 'score' and 'rank' the identified risks. As the project was still at early design stage and only conceptual design was developed for various structures such as underground stations, the potential impacts and likelihoods assessed for risks were 'best estimates' of the team assessing it. The identified risks were then documented and a comprehensive risk register was developed with qualitative and, where possible, quantitative analyses. Each of these quantified schedule risks had a probability of occurrence and, typically, a range of possible impacts.

After identification of schedule risks to the critical path, or near critical paths, a high-level schedule, based on the implementation schedule, was developed to analyze impacts of identified risks and duration uncertainty on the project. This high-level schedule was called 'risk schedule' and also served as a management schedule to analyze major project risks.

### *Quantitative Risk Analysis*

The two principal components of schedule risk analysis are: a) the duration uncertainty associated

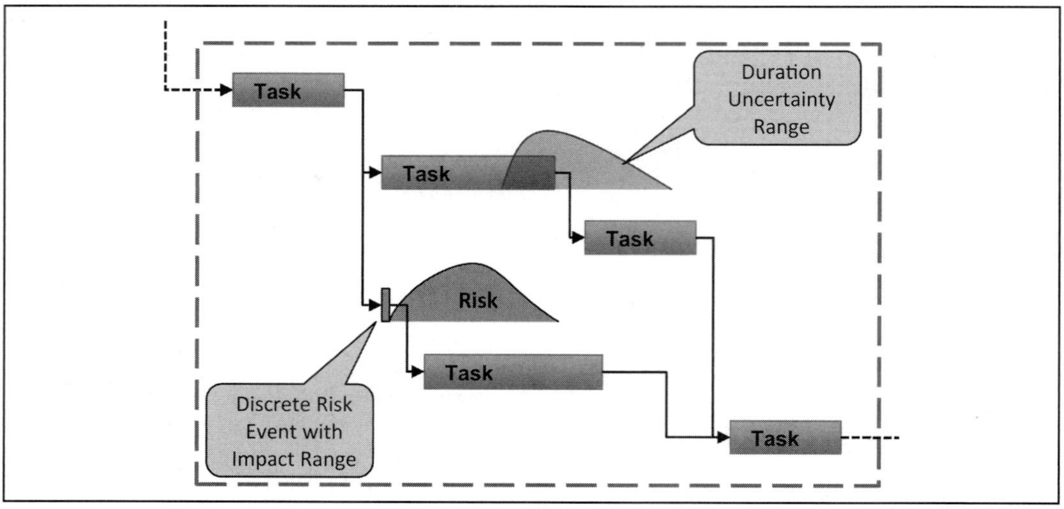


**Figure 8. Activities in schedule with duration uncertainty range and risk event**

with individual tasks or activities and, b) the risk events that may impact the schedule. It is premised on the underlying schedule's activities and logic. The implementation schedule, developed with planned durations and relationships between activities, is the 'deterministic' schedule. This deterministic schedule is imported to Oracle Primavera Risk Analysis and the deterministic (single point estimates) durations were replaced with duration ranges (Minimum and Maximum) or three-point estimates (Minimum, Most Likely, Maximum), wherein:

- *Minimum (Min):* the shortest duration that could reasonably be achieved given work assumptions; a 'blue sky' duration estimate
- *Most Likely (ML):* the amount of time assessors expect the activity to take
- *Maximum (Max):* the longest the activity is expected to take under the given assumptions; note that these were based only on the inherent nature of the activity in question, it did not take into account such discrete risks as tunnel collapse or failure to acquire necessary right-of-way

This operation essentially replaces the assumptions necessarily made in order to arrive at a single-point estimate of duration in the deterministic schedule with an explicit consideration of the duration uncertainty. Furthermore, identified discrete schedule risks were applied to this uncertainty impacted schedule where their potential impact has not already been accounted for by the applied uncertainty. These risk events were then associated with the appropriate activity within the schedule, essentially adding an activity to the schedule, though one which does not always occur (see Figure 8). Since planned activities are part of the original schedule, they 'happen' on every cycle and the variability are entirely due to the uncertainty of their duration, not their existence. This contrasts with risk events, which may or may not happen on any particular iteration, depending on the assessed probability of the schedule risk.

Monte Carlo simulations were then performed on this risk and uncertainty impacted schedule to generate probability distribution curves. Latin Hypercube sampling was employed to ensure that the full range of each distribution is sampled. For planned activities with duration uncertainty, on each iteration the program randomly selects a value (duration) within the distribution defined by Minimum/Most Likely/Maximum values associated with the activity. Values closer to 'most likely' be selected more often while values closer to the extreme ends, either very optimistic or very pessimistic, selected less often.

On the other hand, activities or risk events for which a most likely duration cannot be determined have only optimistic and pessimistic durations, forming a rectangular distribution, with any value from optimistic to pessimistic equally likely to be selected. The software program then summarizes the simulations in the form of a statistical report and cumulative 'S' curve (see Figure 9), providing confidence levels in achieving the whole or any part of the schedule.

***Risk Analysis***

During the risk assessment process, at the early stage of the project, it was recognized that major decisions were still yet to be made regarding the scope of the

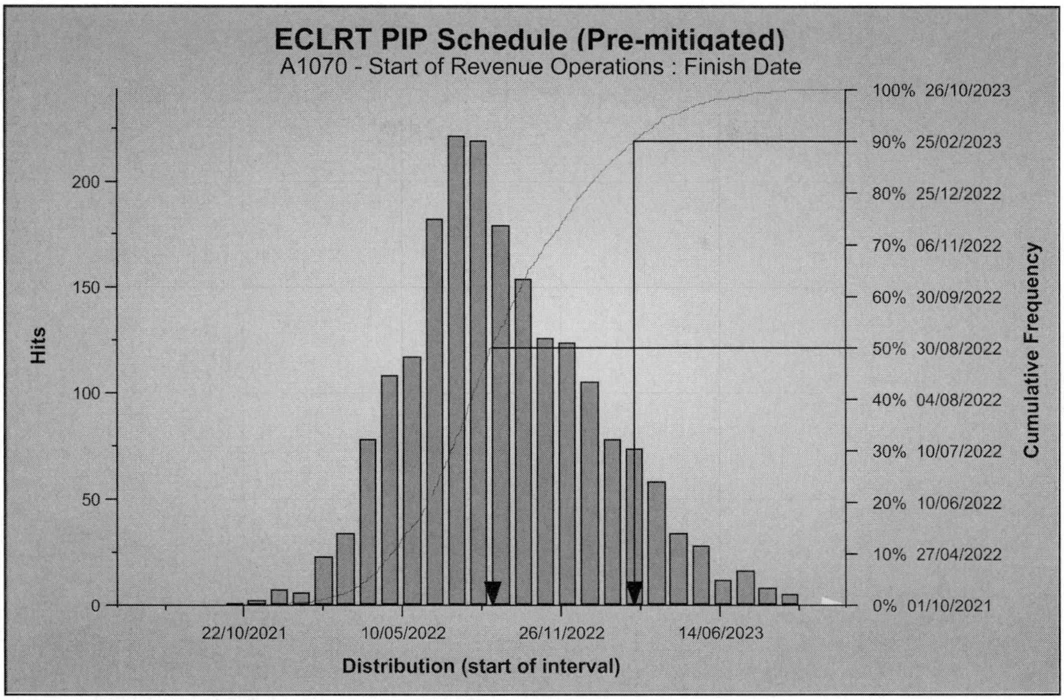


**Figure 9. Monte Carlo simulation generated 'S' curve showing finish date with associated confidence levels**

project, specific locations of individual stations and layouts, systems scope and technology, etc. all of which could influence the outcome of the risk analysis. Also, because the cost estimate was not developed to the level consistent with the PIP, the effect of risks on the project budget were not considered in schedule analysis and revisited once the cost estimate is further developed.

Schedule risk assessment was performed to validate the underlying project assumptions of the project implementation schedule from a risk perspective, gain an understanding of the potential schedule delays in delivering the Eglinton Crosstown LRT and the sensitivity of the schedule to risk, and to provide a platform for the project team to manage risk throughout each phase of the project.

The Monte Carlo analysis results based on the identified risks and duration uncertainty provided a probabilistic profile of the project finish date. The quantitative risk analysis explores the uncertainty in estimated durations and provided alternative dates and critical paths that were more realistic based on identified project risks. The schedule risk analysis determined a significant impact to the planned ECLRT revenue operations prior to mitigation from the disruption and lost productivity in station construction due to simultaneous construction of all subway stations along Eglinton Avenue and from the potential for delay in executing major utility relocations.

### *Risk Mitigation and Schedule Contingency*

The analysis indicates the level of risk exposure for various schedule activities which helped in assessing the potential schedule delays and the sensitivity of the individual 'discrete' risks to the overall project finish date. Risk owners were then identified for each risk and mitigation strategies were developed which included an option to transfer risks to the contractor if the team believed that the contractor would be in a best position to mitigate those risks. Risk levels were then adjusted to account based on planned and/or executed mitigation actions. Pre-mitigation and post mitigation scenarios were developed and assessed to compare the effectiveness of the developed mitigation plans.

The Monte Carlo analysis, performed again on the post-mitigated risk exposure, provided a probabilistic profile of the project finish date (see Figure 9). Recognizing that the project was still at the preliminary to mid-design stage with the underground stations still to be designed, a 50% confidence level was chosen as a lower limit to account for the possible risk mitigation and/or avoidance during the detail design stage. For the upper limit, a 90% confidence level was chosen to account for the fact that the project is located on a narrow right-of-way with the Eglinton Avenue being one of the busiest bus routes in Toronto and the opportunity to fully mitigate and/or avoid risks is limited. The 50% confidence and 90% confidence levels obtained from the Monte Carlo distribution were compared against the planned finish date and were used in setting the schedule contingency against the 'baseline' schedule.

### *Contractual Risk Allocation*

The owner, Metrolinx, understanding the complexity of the project, especially tunneling within the busiest neighborhood and recognizing the benefits from a formal risk assessment, decided to provide copies of the contract-specific risk registers available to potential bidders for the twin-bore tunnel construction contracts. On the major construction projects, risks never go away; they get priced either during the tender stage or later in terms of construction claims. The more information available on the risks, the more competitive a price of risks can be. By providing the risk information in the bid process, an opportunity was provided to the bidders to come up with mitigation for potential risks and price the scope of work competitively.

For the tunnel construction contract, the contractor was also required to meet or exceed the recommended industry tunneling risk management best practices established by the International Tunneling Insurance Group (International Tunnelling Insurance Group, 2012). The code sets out recommended practice for the identification of risks, their allocation between the parties in the contract, and the management and control of risks through the use of Risk Assessments and Risk Registers.

## CONCLUSIONS

An effective risk management process is essential for the successful management of a construction program. Also, the earlier that risk management can be used in the project life cycle, the more realistic project plans and expectation of results can be. This paper focuses on the application and details of a formal risk management process used to reduce and/or manage the risks on Eglinton Crosstown LRT project start from the early stages of project development. A positive risk identification, risk communication and mitigation process was followed and a risk-based decision making process was adopted and provided a platform for the project team to manage risk throughout each phase of the project.

Schedule risk assessment was performed to approach validation of the Eglinton Crosstown LRT project implementation schedule from a risk perspective, develop a better understanding of the potential schedule delays in delivering the project and the sensitivity of the schedule to risk. The goal was to identify significant risks, uncertainties, and opportunities and to assess the associated impact to the project schedule. The schedule risk assessment determined the significant risks to the planned project revenue operations and allowed project management to develop mitigation plans.

## REFERENCES


Federal Transit Administration. (2004). *Risk Analysis Methodologies and Procedures.* U.S. Department of Transportation, Washington, DC.

Federal Transit Administration. (2008). *Project Management Oversight Procedure 40—Risk Assessment and Mitigation Review.* U.S. Department of Transportation, Washington, DC.

Flyvbjerg, B., Holm, M. S., & Buhl, S. (2002). Underestimating Costs in Public Works Projects. *Journal of American Planning Association, 68*(3), 279–295.

International Tunnelling Insurance Group. (2012, May). A Code of Practice for Risk Management of Tunnel Works. 2nd *Edition*. The International Tunnelling Insurance Group.

Project Management Institute. (2009). *Practice Standard For Project Risk Management.* Newtown Square, PA, U.S.A.: Project Management Institute, Inc.

Project Management Institute. (2013). *A Guide to the Project Management Body of Knowledge (PMBOK)* (5th ed.). Project Management Institute, Inc.

Transportation Research Board. (2005). *TCRP Project G-07 Managing Capital Costs of Major Federally Funded Public Transportation Projects.* U.S. Department of Transportation.

Transportation Research Board. (2006). *TCRP Research Results Digest 78–Managing Capital Costs of Major Federally Funded Public Transportation Projects.* U.S. Department of Transportation.

# Risk Management Using the Decision Aids for Tunneling

**Sangyoon Min**
Parsons Corporation

**ABSTRACT:** DAT (Decision Aids for Tunneling) are a computer-based tool, which can be used to quantify risks and uncertainties associated with tunnel projects. The typical outcome from the DAT simulations can provide the probabilistic distributions of construction cost and time reflecting overall uncertainties and risks of the project. This paper presents an overall risk management process applied to a road tunnel project in Korea. This includes a typical semi-quantitative risk assessment and a quantitative risk assessment using DAT. Various applications of the DAT are presented in the paper including The various applications of the DAT are presented including determination of the rock classification/tunnel support patterns along the tunnel profile, and the verification of effectiveness of risk mitigation measures (e.g., schedule recovery methods). Most importantly, DAT were used to determine the best alternative design based on the probabilistic distribution of total construction time and cost considering overall risks associated with the project.

## INTRODUCTION

Tunneling and underground construction involve a high degree of risks and uncertainties, more so than other civil engineering projects. Before construction, geologic/geotechnical conditions are largely unknown. Even during construction, a variety of risks and uncertainties still remain due to, for example, varying effects of human and equipment performance, material properties and unforeseen construction events. Due to various types of potential risks associated with the construction of the tunnels and the inherent uncertainties, there may be significant cost overrun and schedule delay risks as well as safety and environmental risks. Therefore, a systematic risk management plan and its implementation need to be considered as an integral part of the project in order to control, mitigate, and manage potential risks to the project. All this effort in turn leads to minimization of potential physical loss or damage and associated delays, and achievement of the project ultimate goals.

This paper outlines the implementation of the risk assessment to one of the road tunnel projects in Korea using both qualitative (or semi-quantitative) risk assessment and quantitative risk assessment using the Decision Aids for Tunneling (DAT). The DAT are a computer based method with which probabilistic characteristics of the risks can be captured and their overall impacts on the project can be measured. The results of the DAT can be used for various decision making processes.

In the alternative design bid contract, the design-builder are supposed to propose an alternative design, which improves the owner's original design by eliminating major risks with effective mitigation measures and reducing construction cost and time.

For this reason, the DAT were implemented to verify the effectiveness of risk mitigation measures and to select the best design alternative considering overall risks and uncertainties.

## RISK ASSESSMENT

Risk assessment process generally consists of risk analysis, risk evaluation, risk mitigation, risk re-evaluation, and risk monitoring/control as shown in Figure 1. Each step shall be recorded in the risk register, which is a check list for all parties involved in the project (e.g., owner, contractor, and designer) during the design and construction phases to provide guidance for corrective measures to mitigate and reduce the unacceptable or significant level of potential risk scenarios to reasonable or acceptable risk level.

The established risk register shall be continually reviewed and revised through the life of the project to track and manage identified risks.

### Risk Analysis

Risk analysis includes categorization and identification of risks, which have the potential to negatively impact a project. A semi-quantitative (or qualitative) risk approach is the most common methodology to be used for the risk assessment. In this semi-quantitative risk approach, quantification of risks is a structured process identifying both the probability (i.e., frequency) and extent of consequences (i.e., severity) of the event. For risk scoring, the frequency

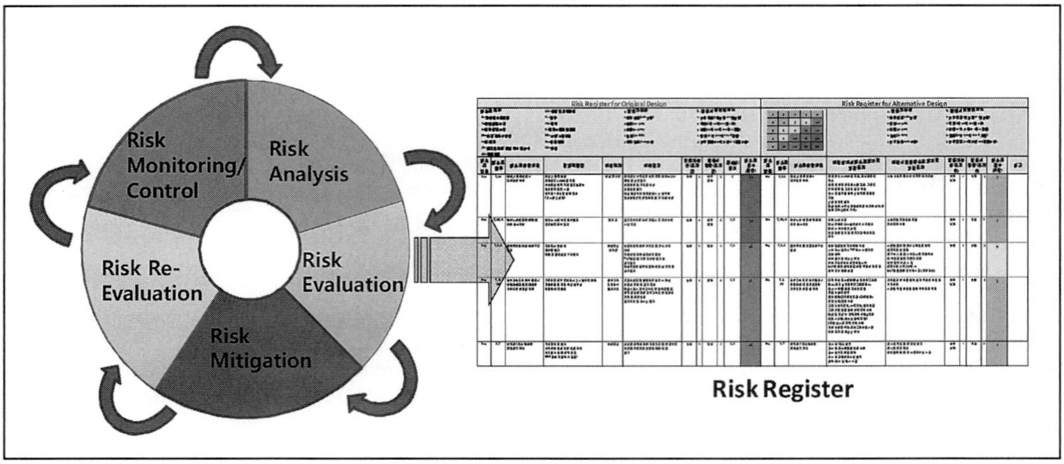


**Figure 1. Risk assessment process**

**Risk Scoring Matrix**

Frequency x Severity = Risk

| Frequency | | Very Low 1 | Low 2 | Medium 3 | High 4 | Very High 5 |
|---|---|---|---|---|---|---|
| Very Unlikely | 1 | 1 | 2 | 3 | 4 | 5 |
| Unlikely | 2 | 2 | 4 | 6 | 8 | 10 |
| Possible | 3 | 3 | 6 | 9 | 12 | 15 |
| Likely | 4 | 4 | 8 | 12 | 16 | 20 |
| Very Likely | 5 | 5 | 10 | 15 | 20 | 25 |

Severity → Increasing Risk; Frequency ↓ Increasing Risk

| Risk Class |
|---|
| Negligible 1-5 |
| Tolerable 6-9 |
| Undesirable 10-16 |
| Intolerable 17-25 |

**Figure 2. Risk scoring matrix and risk class**

and severity of each risk are classified into a number categories, and for each category, a numerical numbering system is assigned to each risk (e.g., numerical values from 1 to 5 are assigned to probabilities ranging from very unlikely, which corresponds to less 10% of chance to very likely which represents greater than 80% of risk occurrence). Typically, this scoring system is developed based on the guidelines from the British Tunneling Society (BTS) and International Tunneling Association (ITA).

Major causes of risks include, for instance unforeseen adverse conditions, significant deviations from assumed baseline conditions, equipment failure or malfunction, human error resulting in a construction accident, poor coordination, inadequate designs, specifications and program, and the occurrence of extreme events not anticipated. Potential locations of events shall also be identified since the frequency and potential impact of risks can vary even for the same type of risk scenario depending on the actual locations and areas affected by the risks occurring. See Figure 2.

In addition to the semi-quantitative risk approach described above, a quantitative risk approach was also applied to this study, which will be discussed later.

### Risk Evaluation

Risk evaluation is a process to evaluate each of the risks based on the results of the risk analysis, and classify them into different levels which represent the significance of the risks. This risk level will be compared with risk acceptance criteria or other decision criteria in order to establish the risk mitigation measures.

The overall risk evaluation process includes rating risks based on product of likelihood occurrence and potential impact. According to the risk scoring matrix shown below, the significance of risks is classified into four risk levels depending on risk score raging from negligible to intolerable.

### Risk Mitigation

Risk mitigation is a process to establish measures and actions to reduce/minimize the frequency of risks, alleviate the severity of consequences of risks, and/or transfer or allocate risks to the most appropriate party(ies). In order to manage and control the risks to an acceptance level, suitable control and mitigation measures need to be established. Timely consideration and actions are of the essence in risk mitigation measures. The aim is to anticipate and put in place effective proactive preventative measures. Contingency and emergency plans must be devised, implemented and maintained throughout the entire project period to address foreseeable accidents and emergencies.

### Risk Re-Evaluation

At this stage, all assessed risks should be revisited to establish any changes reflected by the mitigation measures. At this point, it should be able to establish whether implementation of a set of risk-mitigating actions will in fact reduce the risk to an acceptable level. Where risks cannot be managed to an acceptable level, they should be highlighted. For the identified risks with "Intolerable" level after mitigation, other approaches must be explored (e.g., an alternative construction methodologies) or risks may need to be transferred to external parties (e.g., contract insurance). In addition, the risks with "significant" level after mitigation also need to be re-assessed. The scope of the risk assessment at this stage of the project includes the identification of these residual risks after mitigation which have not been reduced to an acceptable risk level. Further investigation and additional steps to control and manage these "residual risks" shall be taken.

For this project, all assessed risks from the original design were re-evaluated considering proposed mitigation measures actions developed for the alternative design, and they were compared to each other. Details on this comparison and analysis will be discussed later.

## QUANTITATIVE RISK ASSESSMENT

The quantitative risk assessment shall be performed with the semi-quantitative in order to capture risks and uncertainties associated with the project and compare the original design with the alternative design in terms of probabilistic distribution of time and cost. The DAT were used for this purpose.

### Overview of Decision Aids for Tunneling

The DAT are a computer based tool used for risk analysis. The DAT can capture uncertainties and risks in tunnel projects. Figure 3 shows one of the typical results of the DAT simulation, which is called time-cost scattergram. A time-cost pair represents the result of one simulation and by conducting a large number of simulations; this scattergram reflects the overall risks and uncertainties associated with a particular project.

The DAT essentially consist of two major components; the geologic module and construction module. Geologic information such as areas, zones, ground parameters are defined for the geologic module, and the geologic module generates a possible profile of ground classes along the tunnel. For each zone, the ground parameters such as rock type and water conditions are defined. The states of each ground parameter are defined by their average lengths and transition probability which represents probability one state follows another state. A ground class profile can be obtained based on the combination of different ground parameter states (Figure 4). For construction simulation, construction methods are determined by the combination of the tunnel geometries and ground classes obtained from the geologic module. Each construction method is defined by a series of activities which is a basic unit of construction simulation. Each activity is then associated with time, cost and resource equations and their variables, and risk parameters, which can be defined probabilistically (Figure 5). As a result, by performing construction simulation, one can obtain the distribution of time and cost, and resource usages.

### Project Overview and DAT Input

The road tunnel project in this study includes construction of 4.5-mile long twin tunnels, with two undirectional traffic lanes in each bore, 18 pedestrian cross-passages, 10 vehicle cross-passages and two shafts. A traditional tunneling method with drill

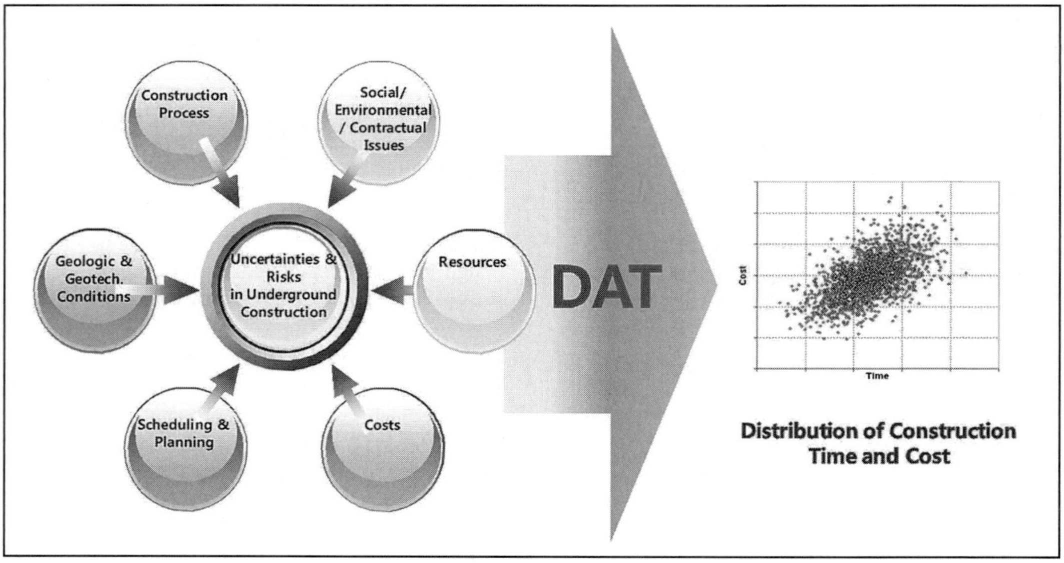


**Figure 3. Overview of decision aids for tunnel (DAT)**

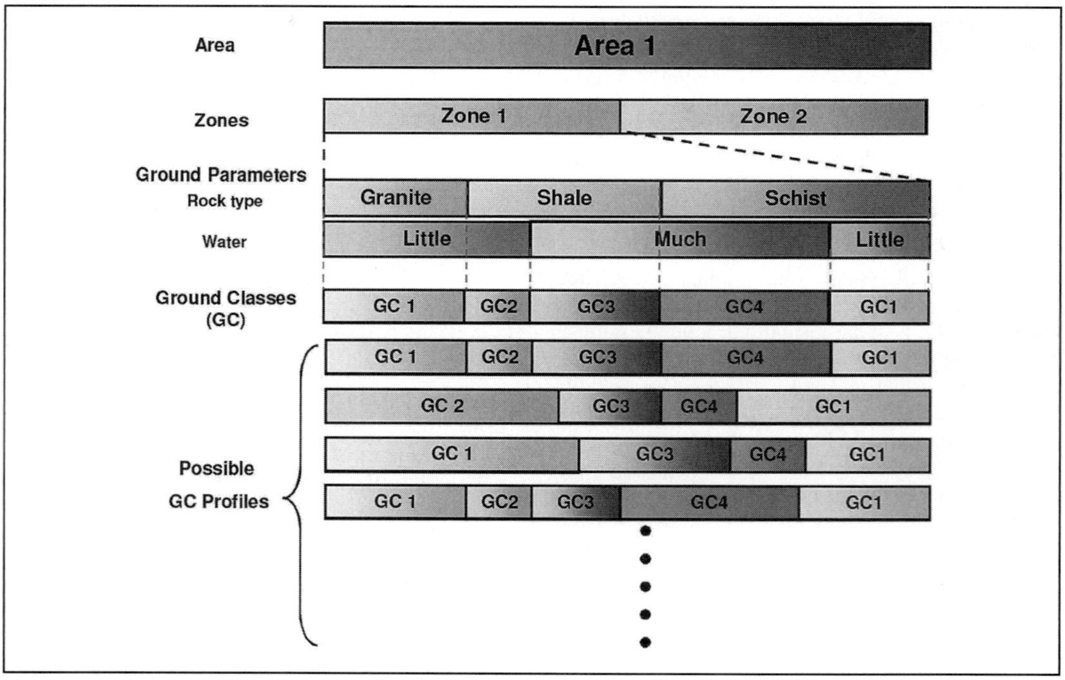


**Figure 4. Geologic module**

and blast was used for tunnel excavation under hard rock conditions. Figure 6 shows tunnel layouts of the original and alternative design, and geologic profile along the tunnel alignment.

For DAT input, the project area is divided into 5 areas, 9 zones and 11 ground parameter sets considering the ground parameter states such as RMR, electric resistivity, fault/fracture and etc. to generate the ground class profile. For construction simulation, 12 different tunnel support patters were defined with probabilistic ranges of the method variables such as the production rate and unit cost.

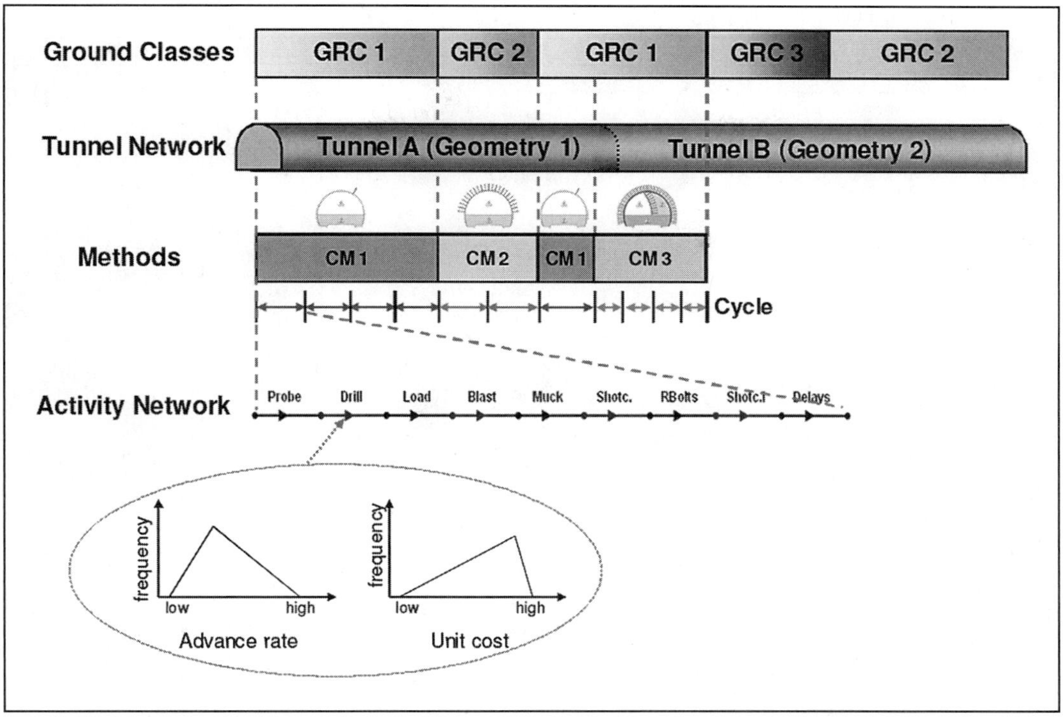


**Figure 5. Construction module**

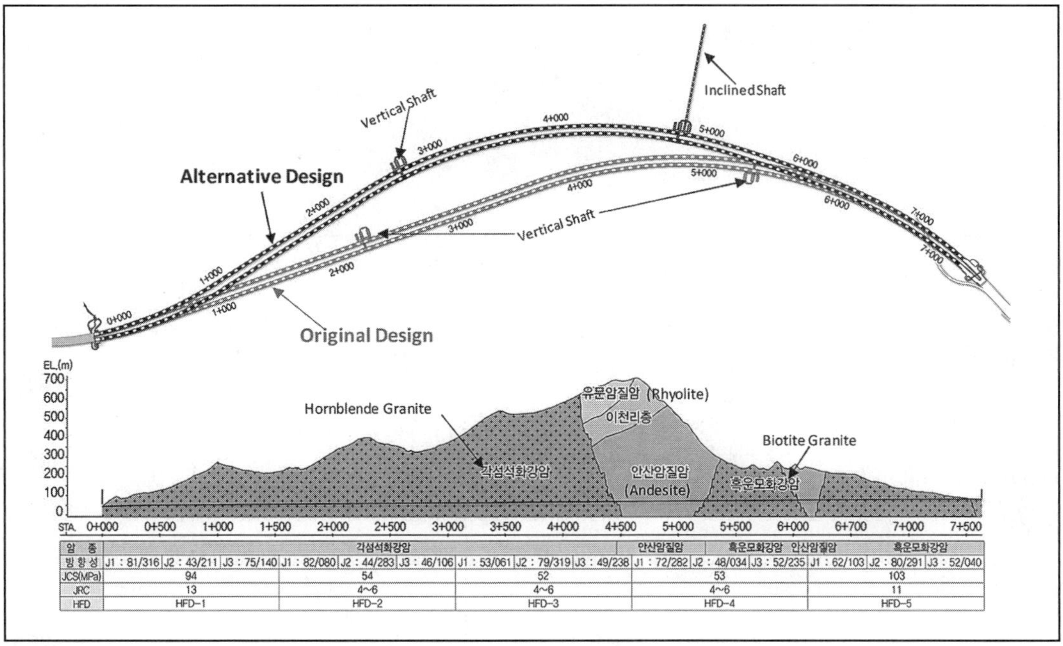


| 암 종 | 각섬석화강암 | | | | | | | | | 안산암질암 | 흑운모화강암 안산암질암 | | | 흑운모화강암 | |
|---|---|---|---|---|---|---|---|---|---|---|---|---|---|---|---|
| 방향성 | J1 : 81/316 | J2 : 43/211 | J3 : 75/140 | J1 : 82/080 | J2 : 44/283 | J3 : 46/106 | J1 : 53/061 | J2 : 79/319 | J3 : 49/238 | J1 : 72/282 | J2 : 48/034 | J3 : 52/235 | J1 : 62/103 | J2 : 80/291 | J3 : 52/040 |
| JCS(MPa) | 94 | | | 54 | | | 52 | | | 53 | | | 103 | | |
| JRC | 13 | | | 4~6 | | | 4~6 | | | 4~6 | | | 11 | | |
| HFD | HFD-1 | | | HFD-2 | | | HFD-3 | | | HFD-4 | | | HFD-5 | | |

**Figure 6. Tunnel layout (original design vs. alternative design) and geologic profile**

**Table 1. Risks for each risk category**

| Risk Categories | No. of Risks[1] | Before Mitigation (Original Design) | | | | No. of Risks[1] | After Mitigation (Alternative Design) | | | |
|---|---|---|---|---|---|---|---|---|---|---|
| | | Negligible | Tolerable | Significant | Intolerable | | Negligible | Tolerable | Significant | Intolerable |
| Geologic Conditions | 8 | 2 | 2 | 4 | 0 | 8 | 7 | 1 | 0 | 0 |
| Tunnel Design/Construction | 17 | 3 | 8 | 4 | 2 | 17 | 16 | 1 | 0 | 0 |
| Inclined/Vertical Shafts | 9 | 2 | 2 | 5 | 0 | 9 | 8 | 1 | 0 | 0 |
| Environmental/GW Impacts | 9 | 4 | 4 | 1 | 0 | 9 | 9 | 0 | 0 | 0 |
| Permitting | 1 | 0 | 0 | 1 | 0 | 1 | 0 | 1 | 0 | 0 |
| Spoils Disposal/Tunnel Watertreatment | 2 | 2 | 0 | 0 | 0 | 2 | 2 | 0 | 0 | 0 |
| Community/Neighbors | 7 | 3 | 2 | 1 | 1 | 7 | 7 | 0 | 0 | 0 |
| Operations and Maintenance | 6 | 3 | 1 | 2 | 0 | 6 | 6 | 0 | 0 | 0 |
| Portals | 2 | 1 | 0 | 1 | 0 | 2 | 2 | 0 | 0 | 0 |
| Power | 10 | 8 | 2 | 0 | 0 | 10 | 10 | 0 | 0 | 0 |
| Right of Way | 2 | 0 | 0 | 1 | 1 | 2 | 1 | 1 | 0 | 0 |
| Market Foprces/Bid Climate | 1 | 0 | 1 | 0 | 0 | 1 | 0 | 1 | 0 | 0 |
| Safety | 7 | 2 | 2 | 2 | 1 | 7 | 7 | 0 | 0 | 0 |
| Quality | 2 | 2 | 0 | 0 | 0 | 2 | 2 | 0 | 0 | 0 |
| **Total** | **83** | **32** | **24** | **22** | **5** | **83** | **77** | **6** | **0** | **0** |

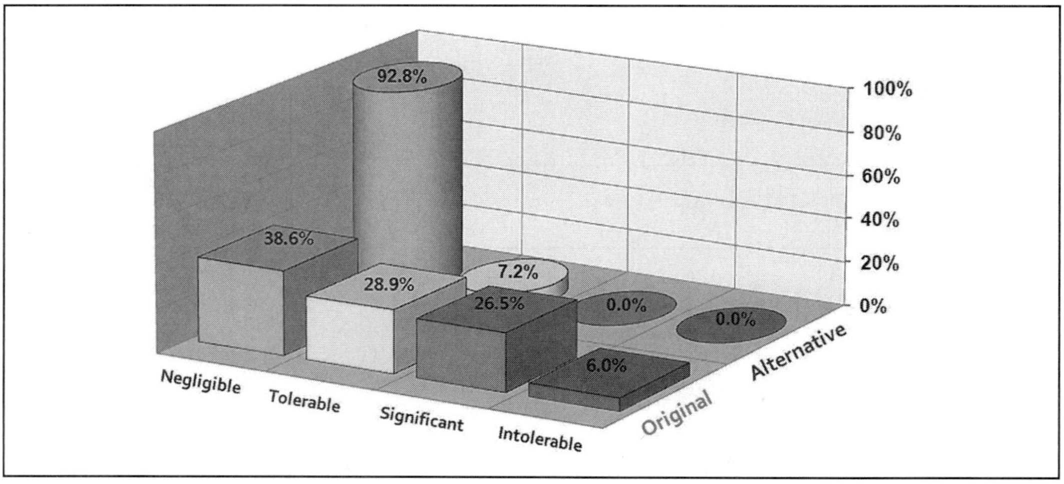


**Figure 7. Comparison between original and alternative designs**

## SEMI-QUANTITATIVE RISK ASSESSMENT RESULTS

All conceivable hazardous events threatening the project were identified. Various risk scenarios were grouped into 11 different risk categories as shown in Table 1. A distribution of the potential risks in each risk category (Table 1) shows tunnel design/construction, power, and environmental/groundwater impacts are the risk categories with a higher priority to be managed and controlled. The risk levels between the original design and the alternative design have been compared and analyzed based on the risk registers. Figure 7 shows that the risks with higher risk levels are dramatically reduced and all risks are well controlled and managed by effective mitigation measures proposed/implemented for the alternative design. All risks both with "tolerable" and "significant" risk levels in the original design have been eliminated in the alternative design while all the risks of the alternative design belong to either "negligible (92.8%)" or "tolerable (7.2%)" risk levels.

## QUANTITATIVE RISK ASSESSMENT RESULTS

Various applications of the DAT have been performed for the quantitative risk assessment.

DAT provide distribution of the ground classes (rock classification) as shown in Figure 8. This information was used to determine optimal tunnel support patterns considering uncertainties in geologic conditions.

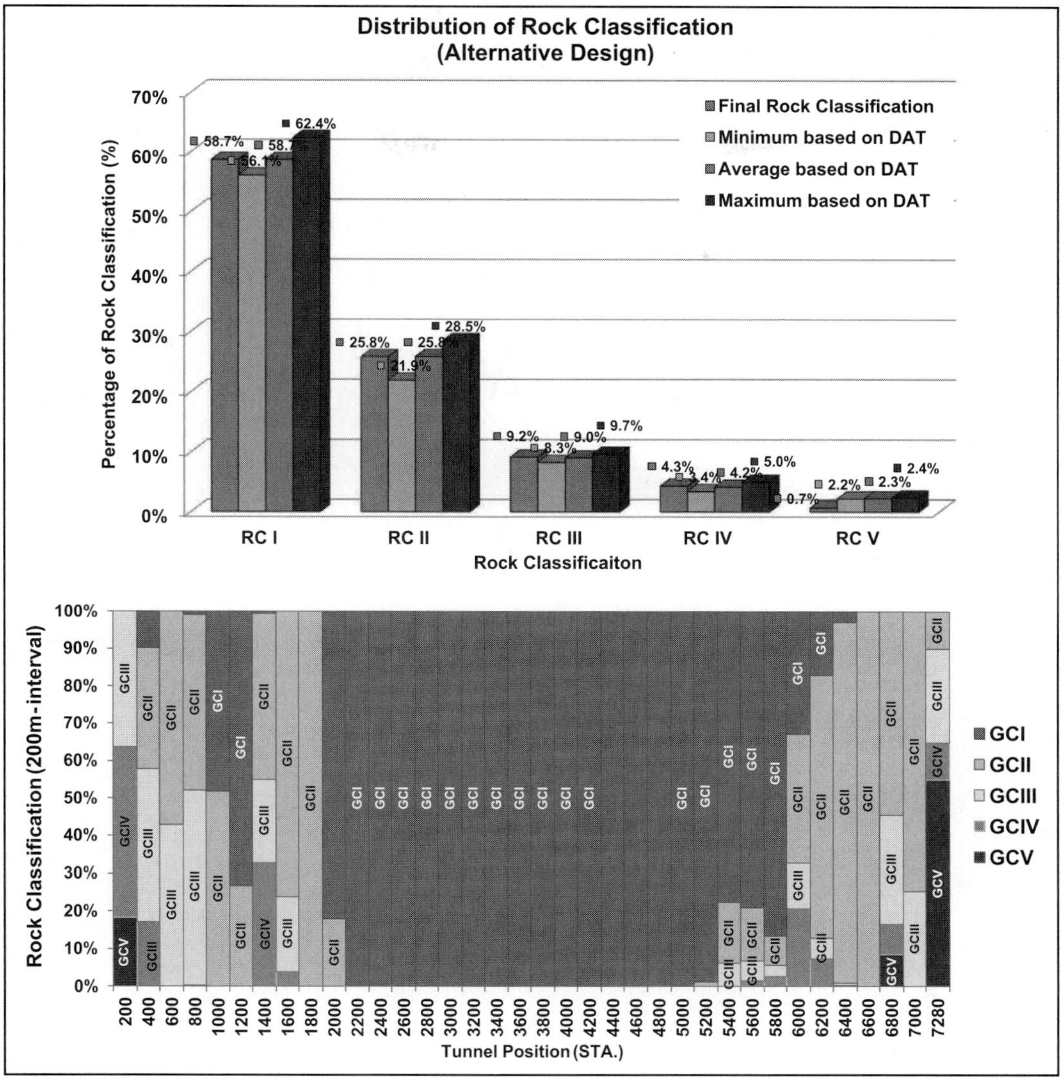


**Figure 8. Distribution of rock classification from DAT**

Throughout the semi-quantitative risk assessment, the potential risks causing any schedule delays have been identified (e.g., civil complaint, machine break-down, seasonal restriction of habitat, and breaking through fracture zones) and schedule recovery or improvement measures have been also developed including four additional sets of lining installation from the inclined shaft (the inclined shaft in alternative design replaces with the vertical shaft in original design as shown in Figure 6. DAT were used to verify the efficiency of risk mitigation measures by factoring the potential risks and their impacts on schedule into the model. Simulation result shown in Figure 9 illustrates the application of schedule recovery method can save approximately 1.8 months of schedule.

DAT simulation results also show that construction with the original design would exceed a planned construction completion schedule (60 months) and budget while construction with the alternative design could be within schedule and budget. In fact, the simulation results show that the construction with the alternative design gains 6-month floating time (Figure 10).

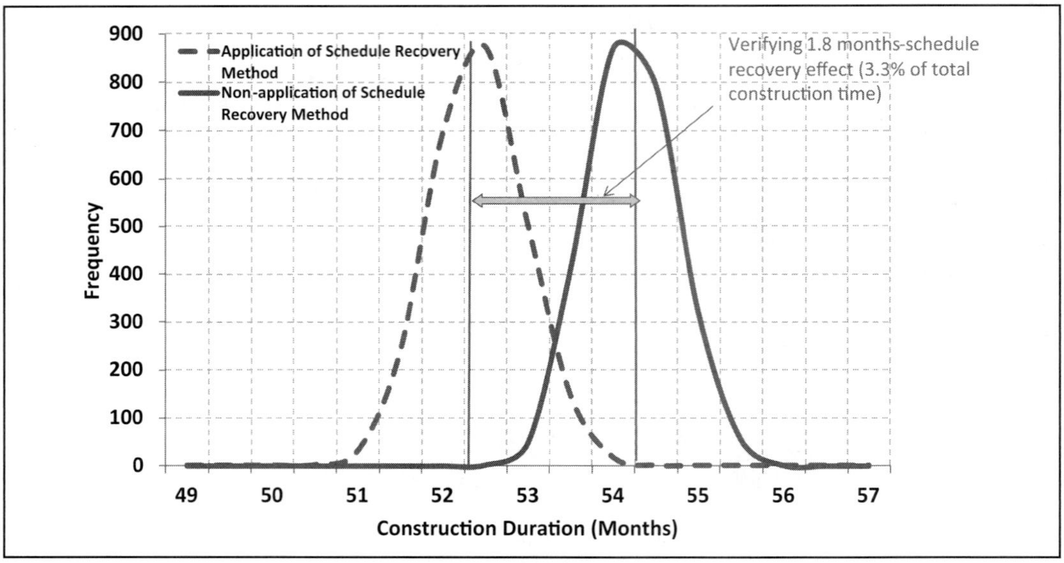


**Figure 9. Verification of efficiency of risk mitigation measures**

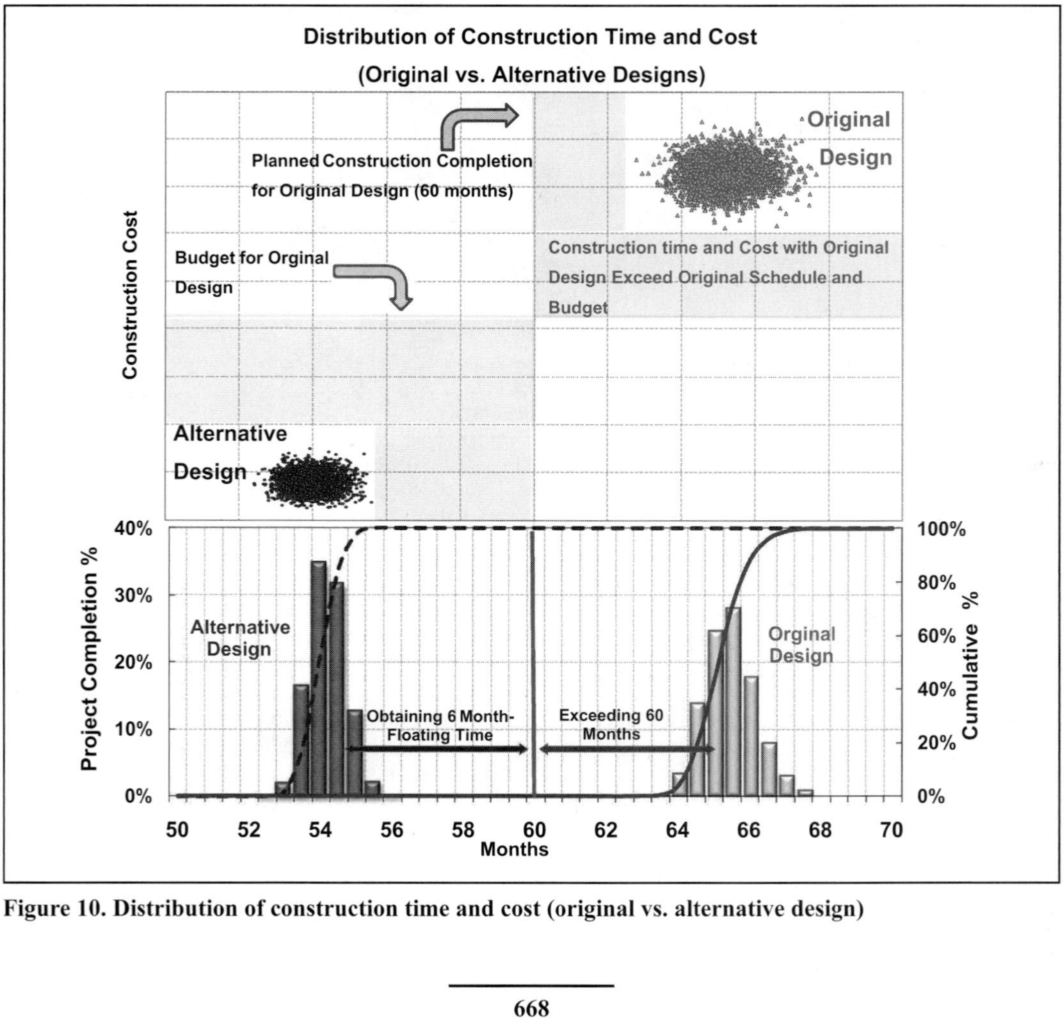


**Figure 10. Distribution of construction time and cost (original vs. alternative design)**

## CONCLUSIONS

Combination of semi-quantitative and quantitative using the DAT was successfully applied to the road tunnel project to control and manage risks. Throughout the systematic risk management approach, the optimal alternative design was developed, which not only reduces total construction cost and time but reduces risks from the original design significantly. The DAT was used to capture risks associated with the project and verify information identified and collected from the semi-quantitative risk assessment.

## REFERENCES

H.H. Einstein, "The Decision Aids for Tunnelling (DAT)—A Brief Review," Tunnelling Technology, Vol. 5, 37–49, 2002.

Min, S.Y. "Development of the resource model for the decision aids for tunneling (DAT)." Ph.D. dissertation, Dept. of Civil Eng., School of Eng., Massachusetts Institute of Technology, Cambridge, 2008.

Min, S.Y., Lee, J.S., Kim, T.K., Einstein, H.H., Design and construction of a road tunnel in Korea including application of the Decision Aids for Tunneling—A case study, Tunnelling and Underground Space Technology, 23. 91–102, 2008.

# Risk Registers and Their Use as a Contract Document

**Robert Goodfellow**
Aldea Services LLC

**Joe O'Carroll**
Parsons Brinckerhoff

**Spyridon Konstantis**
Qatar Rail

**ABSTRACT:** Since the international version of the Code of Practice for Risk Management of Tunnel Works was introduced by ITIG in 2006, the use of a risk register to identify and manage project risk is increasingly becoming standard practice. The risk register has generally not been used during procurement or presented as a contract document for design-bid-build contracts. Can a risk register become a contract document in fixed price contract procurement? This paper looks around the World for guidance on how to better manage risk by transferring risk registers through procurement as a contract document. Tools such as pre-award risk negotiation, early notice of risks, and risk-related release of contingency will be assessed.

## INTRODUCTION AND OBJECTIVES OF THE PAPER

Effective management of risk is the most significant single predictor of success or failure of underground projects. If risk can be managed clearly within the terms of the contract, projects are generally successful. Other projects on the other hand fail because risks were not properly identified or the allocation of known as well as unforeseen (and unforeseeable) risks was not clear. The tools and procedures in this paper seek to remedy this lack of clarity by increasing the bid-phase discussion of risk and risk allocation.

Current North American practice with respect to the use of risk registers on Design-Build (DB) and Design-Bid-Build (DBB) project procurement varies widely but can be characterized generally as follows:

Design-Build (DB):

- Projects use a risk register during the planning and reference design process.
- Risk Register is used in procurement as a discussion item in one-on-one meetings with the design-builder.
- Risk Register is not a contract document.
- Bid process and extended discussions makes risk-related requirements clearer to contractor.
- Risk Register is frequently used in the construction phase—particularly through detailed design

Design-Bid-Build (DBB):

- Projects use a risk register in planning and design stages.
- Risk register is sometimes used in construction phase.
- The risk register is not a contract document.
- The risk management process has no contractual standing and not discussed during the procurement phase.
- Contractor participation is inconsistent during construction even when owner and CM are promoting the use of a risk register—primarily due to the lack of contractual necessity to participate.

The conventional process of risk management is commonly applied by designers and can be seen in Figure 1.

### Development of the Code of Practice for Risk Management of Tunnel Works

The Code of Practice for Risk Management of Tunnel Works (hereafter referred to as the 'Code') was initially developed and published by the Association of British Insurers (ABI) and the British Tunnelling Society (BTS) in 2003 following an untenable loss history on tunnel and underground works projects. The international version of the Code was published in 2006 by the International Tunnelling Insurance Group (ITIG) and is endorsed by the International

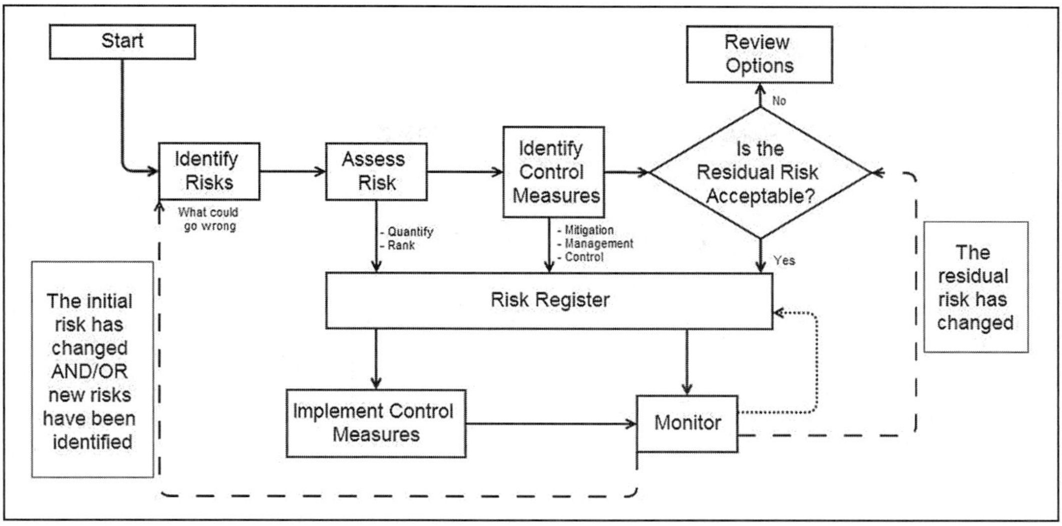


**Figure 1. Risk management process flow chart**

Tunnelling Association (ITA). While the nomenclature of Code has caused some consternation in the US, this document really amounts to a guideline to best management practices and recommendations. A new North American version of this Code (published in 2014) will carry the title of Best Management Practices (BMP) to make sure this is more clearly understood. The BMP will reflect conventional North American practice and provide guidance for risk management using DBB procurement.

### Objective of the Code

The Code illustrates the importance of risk management. Its main objective is to promote and secure best practice for the minimization and management of risks associated with the design and construction of 'Tunnel Works' (defined as tunnels, caverns, shafts and associated underground structures and including the renovation of existing underground structures) in order to:

- Reduce the probability of occurrence and size of claims
- Increase transparency and certainty on financial exposure and risk transfer
- Provide insurers with a better understanding of the project-specific risks during the underwriting process and an 'auditable' trail of the project's risk management during design and construction

The Code sets out project management procedures and systems for contract procurement, design and construction activities, practice for the identification of risks, their allocation between the parties to a contract and Contract Insurers, and the management and control of risks through the use of risk assessments and risk registers.

### Fundamental Principles of the Code

The code was developed to operate in parallel with existing local standards, statutory and legislative duties and responsibilities. Project specific hazard identification is to be carried out during each of the four identified project stages, i.e., project development; construction contract procurement; design and construction stage. In each stage risks must be assessed and recorded in a risk register including risk, mitigation and responsibility allocation. The risk registers are to be a 'live' document, continuously updated and reviewed.

Identified risks are to be appropriately managed to ensure reduction to a level 'As Low As Reasonably Practicable' (ALARP principle). Risk assessments and registers are to be "cascaded" throughout the project phases to ensure that all key project parties are aware of previously identified hazards and associated assessed risks. In the framework of the Code, insurance should not be considered as a contingency or mitigation measure in risk assessments for identified risks.

With particular reference to risk registers, the Code states that risk registers shall clearly identify and clarify ownership of risks and shall detail clearly and concisely how the risks are to be allocated, controlled, mitigated and managed. The systems used to track risks shall enable the management and mitigation of risks through contingency measures and controls to be monitored through all stages of a project.

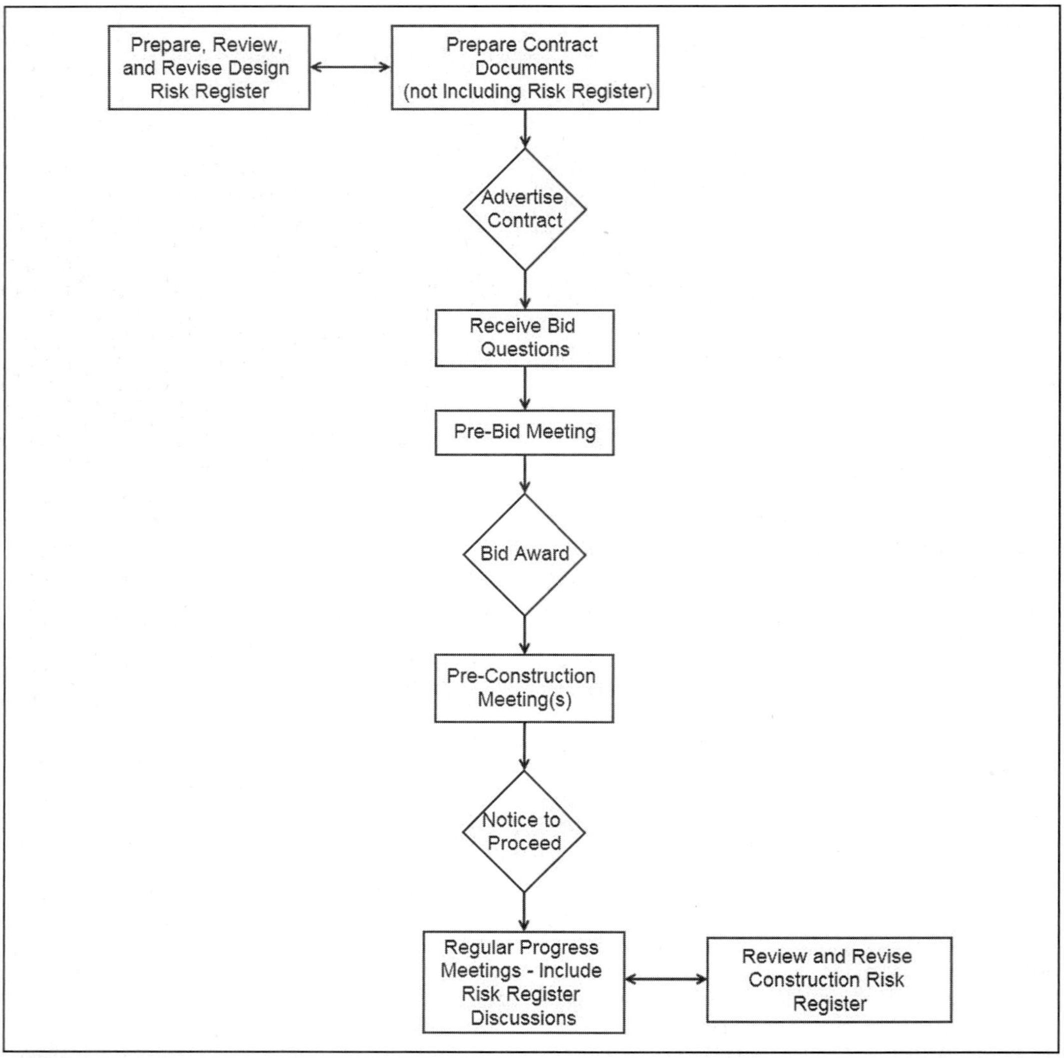


**Figure 2. Workflow for standard design-bid-build procurement without consideration of risk register**

Several Authors have written in detail on the application of the code to projects including Goodfellow and Mellors (2007), Spencer (2008), and Reiner (2011).

## CURRENT PROCUREMENT OF TUNNEL PROJECTS

Figure 2 shows a conventional process of design-bid-build procurement. With regard to procurement the Code provides the following guidance:

- Contract Documentation (as well as sub-contact documentation for Tunnel Works as appropriate) shall include full disclosure of those hazards and associated risks identified at the Project Development Stage for the preferred project option (or options) in the form of a project Risk Assessment.
- For all contracts, the tenderer shall be required through the Contract Documentation, to provide the Tender Risk Register for the benefit of the Contract Insurers. This Tender Risk Register should demonstrate how the tender submission adequately and appropriately caters for risks identified and to be allocated to the Contractor including their management and control procedures, proposed contingency measures and the cost and [schedule] implications of the implementation of contingency measures.

From the Authors' experience, North American practice has more recently including similar practices as laid out in the Code in so much as risk registers can be (and have been) prepared during the project development stage. However in the contract procurement stage, these registers (as registers) may or may not have been part of the bid documents. It is more likely that risks being carried through to the construction stage have been described in contract documents such as the geotechnical baseline report (GBR)) in a narrative form.

Fundamentally the Code recommends clear identification and notification of project risks through the procurement process and recommends the tool for this transfer to be the risk register. Outside of North America the Code has been used most frequently for design-build procurement where there is the opportunity to discuss with each bidder, one-on-one, their perceived risk and how these are handled in the bid documents. To some extent this has occurred on major DB projects in North America such as the Alaskan Way Viaduct SR 99 Tunnel project and the Port of Miami Tunnel project. The challenge has always been how the same process can be represented fairly and openly through a design-bid-build hard money bidding process?

The conventional DBB approach to procurement of tunnel projects has frequently led to misunderstandings of the intent of contract documents by contractors that result in claims and disputes. The intent of the documents and which party bears the risk is frequently discussed during a dispute process but there is no way to convey this intent outside the contract documents. One of the primary drivers of this paper is to offer recommendations to improve the presentation of project risk such that the intent of risk allocation and how risk is to be managed during construction is clearly defined.

## RISK ALLOCATION REPORT

The North American BMP specifically addresses the issue of transfer of carefully thought out risk documentation through the procurement phase and into construction of projects. The BMP specifically addresses the more conventional North American practice of design-bid-build procurement. A new tool that is being recommended is a risk allocation report (RAR). The RAR is a narrative of the risk register at the conclusion of design (equally applicable to the final reference design as it is to 100% complete design). The aim of the RAR is to provide clear explanatory language of how each risk in the risk register is allocated in the contract documents.

The RAR would then accompany the risk register and also be a contract document that takes its place alongside the drawings, specifications, GBR, GDR and the contract agreement. The RAR would provide direct reference to other parts of the contract documents pointing out clearly and providing whatever support narrative that is necessary to clarify and explain the intent of the risk allocation. It is anticipated that there would be places where reference and explanation of payment terms would be part of this narrative.

It is important that the narrative be clear and definitive in a similar spirit to the wording of a GBR and that any and all exculpatory language be eliminated from the RAR. It should be expected that a simple reference to a clear contract clause would be sufficient in most cases. However, risks that need additional narrative for more clear explanation are the most important, obviously the most difficult to allocate clearly with a simple contractual clause, and therefore these are the risks most liable to be the subject of disputes and claims. The primary purpose of the Risk Allocation Report is to clarify the intent and letter of the risk allocation in the contract documents.

It is clear that the use of a risk register through the procurement process would require that the risk register be modified at distinct stages of its development. The procurement risk register would not contain the funding risks and many other retired risks regarding planning, alignment selection or other permit activities contained on the planning and design risk register. Only those risks allocated in the contract and specific risks completely mitigated that, in the judgment of the owner, it would help the project for the contractor to see the "what" and the "how" of the design risk process would appear on the procurement risk register. Similarly, the construction risk register would include risks added by the contractor related to their means and methods and then be monitored jointly by the project team throughout construction and commissioning of the facility.

Risk is managed more consistently and collaboratively through procurement and construction when a more detailed discussion of risk in the contract documents is undertaken. Using the risk register in concert with a RAR as a contract document improves the clarity of this discussion.

## CONTRACTUAL RISK SHARING PRINCIPLES

Common wisdom in construction has been that "the party that can best manage the risk should bear the risk." As a result, traditional construction contracts shift risk among the various participants, and sometimes, despite the common wisdom, the party who bears the risk is the one with the least bargaining power rather than the one best able to manage the risk. This assumes also that there is one, and only one, party that can effectively manage the risk. In reality the actions of various project participants, external stakeholder or other events outside of the owner's or contractors control, can influence the occurrence or magnitude of a risk.

When a party is allocated a risk that it cannot adequately control, it will seek to protect itself against that risk in one or both of two ways: either by increasing its contract price in order to build in additional contingency to monetize the risk; or later in the project, by engaging in adversarial behavior, such as bringing claims or demanding change orders so as to recoup damages resulting from the risk.

Inflated contingencies to take account for this inappropriate allocation then get multiplied throughout the supply chain, as risk contingency gets stacked on top of contingency. The result is that an owner may either abandon the project as unaffordable or tie up a larger proportion of its funds than is actually necessary to address the project risks. This also results in a lost opportunity, since the owner is unable to use those funds for other important goals.

Owners are often accused of risk shedding rather than risk sharing and it is worthwhile to state that it is rarely possible to completely shed risk on their projects. Attempting to shed risk frequently results in higher initial bids without eliminating the specter of claims on the project. The potential for a failed project will always reflect badly on the owner, impact negatively his appetite for future underground work and form a negative public opinion and attitude.

It is important for the owner to recognize that they set the tone for projects and it is the authors' opinion that owners must set in place contract terms that build trust with the contractor community and endorse and promote a collaborative means to solve the problems that will inevitably occur on site. This approach will lead to successful and lower cost projects that will build the trust of the public and lead to more underground infrastructure investment.

The contract, as ever, must set up the basic boundaries of risk allocation and risk sharing. This is much more easily done through the interaction of DB procurement than it is in more traditional hard money and low bid environment usually seen in DBB.

The traditional risk-shifting approach of DBB contracts provides no commercial incentive to the parties "not at risk" to offer help to the risk-bearing party. Instead, project participants have economic motives to view those problems as "someone else's" rather than "ours." This traditional approach results in each party trying to optimize its own part of the project rather than optimizing the project in its entirety.

Rather than simply shifting risk the authors propose the idea of a contract having a risk-sharing incentive, in which the contractor and the owner share in the savings to the project budget if the risks are adequately managed or not realized such that the project is on or under budget. The risk sharing incentive can be based on a pre-negotiated ratio (e.g., 70/30) where the Owner retains 70% of the assigned risk value and the contractor is rewarded 30% of the assigned risk value when both parties agree the project has reached a point at which that particular risk can be confidently retired. These could be key milestones in a project and be referred to as Contingency Release Points (CRP). Figure 3 shows a typical distribution of contingency and the project milestones that can be used to release the risk contingency to the contractor.

There will be a range of contingency distribution ratios for different risks but this must be thought through carefully by the Designer and owner after due consideration of the project risks. At final completion, the remaining contingency can be distributed to the parties and the incentive payment for strong risk collaboration can be realized.

## CONTINGENCY ADMINISTRATION AND DRAWDOWN ACCOUNTS DURING CONSTRUCTION

How do you establish what risk contingency values should be? The first step is the owner making a quantitative assessment of each risk being carried over into the construction phase and a collective assessment of the all identified risks to establish an overall contingency for the project. These risks will have been captured in the risk register and the RAR, without any monetary value. The contractor can be asked to "price" these risks separate from his contract price for performing the works as described in the contract documents. The bids can still be considered competitive in terms of the owner evaluating the bidder's 'base price plus risk contingency'. The owner can also compare each bidder's evaluation of the risk cost compared to its own assessment. This will provide a check on its reasonableness and an indication of the contractor's commitment to an equitable risk sharing strategy.

A contingency drawdown plan is a key component of risk management that should be established by an owner early in a project aligning closely with the level of risk mitigation required at each stage of the project. Hold points to check the adequacy of contingency throughout the lifecycle of the project (e.g., environmental clearance; end of preliminary engineering; completion of geotechnical investigations; final design and award of contracts). The contingency in the project budget when a contract is awarded must be consistent with the residual risk carried over from the design phase into the construction phase. A distribution of the contingency drawdown throughout the project that reduces as project risks are retired is shown in Figure 3.

Arbitrarily including a percentage (e.g., 5% or 10% over the lowest bid) as the contingent amount

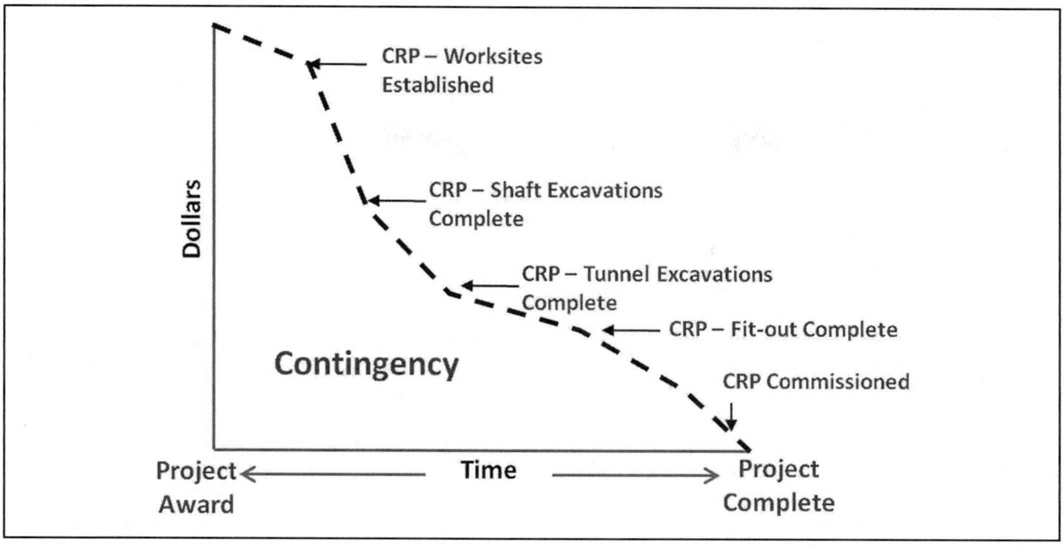


**Figure 3. Example contingency drawdown with possible contingency release points (CRP)**

is not recommended. The contingency should be established through a thorough analysis of the risks being carried over into construction and how these risks have been allocated. The owner may choose to retain a large amount of risk, develop mechanisms for sharing risk with the contractor, or transfer risk to the contractor or another third party. Whichever management approach is adopted it is important that every contract be considered unique and analyzed accordingly.

A point will exist where mitigation becomes increasingly difficult to implement and beyond which risk acceptance through the application of project contingency is the only effective means to treat project risk. This "break point" between risk reduction and risk acceptance typically occurs at the point where all market risk has been mitigated and the contract has been awarded. The procurement strategy and the procurement phase becomes a critical element in management of risk and control of the budget through the remainder of the project. How risks are allocated and how risks are shared will ultimately determine the success of a tunnel and or underground construction project.

The idea is to establish trusting relationships in which the contractor and owner focus on overall project outcomes rather than individual responsibilities, as well as work collaboratively to find solutions rather than shift blame. Once procurement is complete and construction begins, all project risks become a commercial issue and a fair and equitable incentive distribution is aimed at helping unify the parties' goals in terms of managing risk for successful completion of the project. By sharing risk, all project participants have a financial stake in effectively identifying and mitigating risks that in traditional projects would be "someone else's problem." This leads to a reduced overall project risk profile as well as a more equitable approach to risk management. When another's problem will have a direct impact on your bottom line, you are more likely to offer help in solving the problem—promoting an "all for one, one for all" culture with everyone trying to reduce risk in their own way.

## GLOBAL DEVELOPMENTS IN RISK MANAGEMENT PRACTICE

### New Engineering Contract

The New Engineering Contract (NEC) created by the UK Institution of Civil Engineers and published in 1993, guides the drafting of documents on civil engineering and construction projects for the purpose of obtaining bid, awarding and administering contracts. The NEC contracts form a suite of contracts, with NEC being the brand name for the "umbrella" of contracts. When it was first launched in 1993, it was simply the "New Engineering Contract." This specific contract has been renamed the "Engineering and Construction Contract"(ECC) which is the main contract used for any construction based project (available at www.NECcontract.com with additional commentary provided at: http://www.neccontract.com/documents/contracts/Guidance%20Notes/NEC3_EEC.pdf). Numerous changes have been made to improve and enhance the ECC document including the addition of a Risk Register as a contractual item.

The NEC allocates the risks between the parties to the contract clearly and simply. But it also helps to reduce the likelihood of those risks occurring and their subsequent impact, if they do occur, by the application of collaborative foresight and risk reduction procedures. In this way, it aims to improve the outcome of projects generally for parties whose interests might seem to be opposed. To reinforce the pro-active approach to dealing with risk, the ECC now includes provisions for managing project risk through a 'risk register'. The risk register is not a compilation of contractual risks between the parties—these are reflected through "compensation events"- it is a complete list of all project risks.

A pre-contract project risk register is not required by the NEC but is considered good practice. However, ECC does include a requirement for a risk register post contract: it is a post-contract risk-management tool and is not the place for risk allocation. Initially, the Employer shares with the contractor risks they are aware of at tender (bid) stage (provided as part of Contract Data Part 1) and the contractor, in their tender return, adds to this list any risks that they may be aware of in addition to those identified by the owner (employer) (provided as part of Contract Data Part 2). The consolidated list then forms the initial risk register for the construction phase of the project. Once the contract commences this risk register is used as a live document and updated to capture any further risks that may be identified through an early warning process.

### *Risk Reduction Meetings*

The early warning process is designed to ensure that the parties to the contract are made aware as soon as possible of any event which may increase the amount that the Employer has to pay; delay completion of the works; impair the performance of the works once completed, or affect others working on the project. The contract requires the parties to meet, to seek mutually beneficial solutions to overcome these problems, and to operate a formal risk register of notified events. At the risk-reduction meeting (formerly known as an 'early warning meeting' but changed to convey what was always the purpose of the meeting) the risk(s) is discussed, the register revised to record decisions made or amend the potential impact or probability of occurrence and where appropriate if the risk has passed for it to be formally retired.

Compensation events on the other hand are pre-defined events in the contract which are at the risk of the employer, and which may lead to the payment to the contractor changing or the completion date, i.e., the date by which the contractor is required to complete the works, being extended. A principle of the ECC is that, when such an event occurs, the project manager, acting on behalf of the employer and in communication with him, should, whenever possible, be presented with options for dealing with the problem from which he can choose, directed by the interests of the owner.

The ECC is not without its problems and has attracted considerable legal scrutiny, particularly in the execution of listed compensation events, the obligations of the owner and the contractor in terms of timely notification of a compensation event occurring and the ambiguity that can arise as to whether an event is dealt with as a compensation event or, if the contractor fails to notify the owner of such an event, whether he can then raise it under the early warning process and get compensation for the agreed solution. The early warning and notification process is described in an article on the website of the international law office (http://www.internationallawoffice.com/newsletters/detail.aspx?g=f0bb6d79-d4c2-47bf-9e26-29b7f7f5ab9f).

It is not the intent of this paper to analyze the details of how the EEC3 is used as a contract form, or imply that similar contract forms would or could be applicable in North America but merely to make observations on the intent of current contracting practices outside of North America to include a risk register and a formal risk reduction process as a contractual requirement on both the owner and the contractor and to solicit discussion on this important topic.

### *Crossrail*

The Crossrail program in London is an example where an NEC Target Cost form of contract has been used and 27 different risks were explicitly described in contract procurement documents provided to teams by the owner. The tenderers were required to respond with their own risks and then risk reduction meetings are held regularly throughout construction using the risk register as the basis for these discussions.

## Alaskan Way Viaduct Replacement Program

The Alaskan Way Viaduct Replacement Program (AWVP) has used the Washington DOT standard approach to cost and risk evaluation known as CEVP (Cost Estimate Validation Process). During the preliminary and final design phases, this process involves expert solicitation of quantitative risk assessments and collation of this work into a base and risk cost and schedule evaluation for more robust estimation of final project costs. This method has proven to be an excellent way of budgeting major infrastructure work (Reilly, 2008). The AWVP used a series of contingency funds during procurement of this $1.9Bn megaproject in order to clarify and quantify the risks associated with a wide range of specific activities, including insurance costs; deformation and repair of

| Delay: | | | | | |
|---|---|---|---|---|---|
| | **Disastrous** | **Severe** | **Serious** | **Considerable** | **Insignificant** |
| Delay (months per Hazard) | >12 | 6–12 | 3–6 | 1–3 | <1 |
| | | | | | |
| Economic loss to Metrolinx: | | | | | |
| | **Disastrous** | **Severe** | **Serious** | **Considerable** | **Insignificant** |
| Loss in Millions (CND) | >$15M | $5M–$15M | $1M–$5M | $250K–$1M | <$250K |

**Figure 4. Extract from the ECLRT Specifications Section 01 35 30 Risk Management—consequence classification for assessment of hazards**

third party structures and utilities; and differing site conditions and hyperbaric interventions.

These funds are accessed through specific contract clauses and are drawn down during the project once the relevant risks are either realized or retired. Funds not used during the project are distributed between the owner and contractor in varying proportions, from 75% given to the contractor in certain funds to 100% retained by the owner in other funds. The owner and contractor jointly track the use of these funds and the exposure to relevant risks to make sure that these funds continue to be sufficient if risks are realized over the remainder of the project. This system of cost and risk management has worked well through procurement and in the early stages of construction of the tunnel project.

### Eglinton Crosstown Light Rail Transit Project

On the recently awarded Eglinton Crosstown Light Rail Transit (ECLRT) tunnel contracts in Toronto Canada, Metrolinx dedicated a section of the project Specifications to Risk Management with an overall requirement for the Contractor to "Meet or exceed the recommended industry tunnelling risk management best practices established by the International Tunnelling Insurance Group." The specifications require the Contractor to confirm the owners of risks, actions and measures to mitigate the impact of the risks during the construction phase including risks identified by the contractor as well as contract-related brought forward from Metrolinx's risk register. The risk register was to be freely available in a format that can be easily shared with all parties. In addition a joint risk management team made up of Metrolinx's Representative and the Contractor were required to meet every two weeks to review the project risk register and practices on site that are at variance with the Contractor's risk mitigation measures, review the results of event investigations and to identify and review new risks, together with corresponding prevention and mitigation recommendations and action plans (Figure 4).

Metrolinx provided a risk event list in register format for the Contractor to assess risks to safety and security, delays to permitting, financial/commercial/contractual events, logistics, site access, construction and the environment. In the specifications qualitative criteria for the assessments (severity) of risk to health and safety, third party property, the environment, delays in contract completions, economic loss to Metrolinx and loss of goodwill.

## INSURANCE COSTS AND RISK REGISTERS

The existence of a consistent risk register from planning, design, through procurement and its use as a contract document could also contribute towards a more realistic and informed decision making process and potentially for significant cost savings in relation to the insurance costs of a tunnel project.

As mentioned in an example above, the AWVP used a series of contingency funds during procurement in order to clarify and quantify the risks associated with a wide range of specific activities. One of these funds was set up for insurance cost.

Insurance costs are associated with insurance premiums which in turn depend among other items on the insurance market capacity and availability, the insurer's risk appetite and the Probable Maximum Loss (PML) study (Denney, Konstantis and Tillie, 2014).

The PML study is a key element of the underwriting process and the insurer's evaluation and consideration of the risk as it helps them to decide on the proportion of the risk that they are willing to retain and the necessity to arrange for reinsurance for their risk share. The above considerations have a direct impact on the insurance premium and costs.

For a construction project the PML can be defined as follows: *"The Probable Maximum Loss is an estimate of the maximum loss which could be sustained by the insurers as a result of any one occurrence considered by the underwriter to be within the realms of probability. This ignores such coincidences and catastrophes which are remote possibilities, but which remain highly improbable."*

In order therefore to assess the PML on a realistic and substantiated basis, a basic process must be followed with the core elements being among others (Heller, 2002):

- What could cause damage to the works (Identify Risks box in Figure 1)
- What is the maximum loss scenario, i.e., what is the most catastrophically possible event within the bounds of reason, which could give rise to the maximum physical loss or damage to the works (Assess Risks box in Figure 1)
- What is the PML value associated with the loss scenario, i.e., taking all things into consideration what is the cost of reinstating the portion of works lost or damaged in the PML scenario (note that this may be considerably more than the original cost of the damaged element of the works and where this is the case advice should be sought from the underwriters as there may be limitations in the insurance policy).

The latter point is related to the risk/cost contingencies that are allocated during project procurement and the total cost of risk per activity/identified risk, including the retained cost of risk, the insurance premiums for the insured part of the risk and the contingency fund for uninsured events (these can be remote possibilities with very high consequences which still remain highly improbable, also known as 'black swan' scenarios).

It can therefore be argued that reasonable estimations of the insurance costs are feasible when the PML assessment is based not only on sound engineering principles but most importantly on a realistic framework within the realms of probability. This framework can be provided by the project specific risk registers and risk documentation developed and 'cascaded' throughout the different project stages, as highlighted earlier in previous sections of this paper.

## SUMMARY OF RECOMMENDED PROPOSED PROCESS

The process proposed as a new standard of practice for risk management on DBB projects is shown in Figure 5. The addition of an RAR alongside a risk register as one of the contract documents is a significant change from current practice.

Reinforcing this change is a mandatory pre-bid risk meeting with prospective bidders—carried out with or at a separate time to the conventional pre-bid meeting. This meeting goes through the risk register and RAR to fully explain the intent of the contract as well as setting forth how known and unknown risks will be identified and paid for during the contract. This meeting also describes how new risks will be identified and the working relationship that is sought between parties to the contract for the most successful prosecution of the project.

## CONCLUSIONS AND RECOMMENDATIONS

There is a need to continuously strive for better ways to manage the risk on underground projects. The following activities are proposed to promote the improved management of risk on tunnel projects:

- A risk register should be included as a contract document for major tunnel and underground construction projects.
- The risk register should be supported by a risk allocation report to clarify how individual project risks have been allocated in the contract.
- The design risk register should have non-relevant finance and design risks removed on conversion into the procurement risk register.
- The procurement risk register should have contractor's identified risks included upon conversion into the construction risks register.
- Risk contingency should be held by the owner or applied in allowance items as part of the contract. The funds should be used to pay for risks that manifest during construction and the funds should then be released in accordance with prescribed contract-defined proportions if risks are not realized.
- Project owners must recognize that they set the tone for a collaborative or adversarial relationship with the contractor based on the contract terms that are put together.
- Contractors must understand and recognize when the owner puts forward a progressive and collaborative risk sharing contract and the contractor must then trust that the owner will follow through with the contractual promises.
- Designers must recognize that in this form of contract their design must be well thought out and consistent and that both what is known and what is unknown must be paid for, either within the base bid or in risk-based contingencies.
- Risk consultants and advisors carrying out or reviewing Probable Maximum Loss studies must base their assumptions and estimations on sound engineering principles and most importantly on a realistic framework which can be provided by project specific

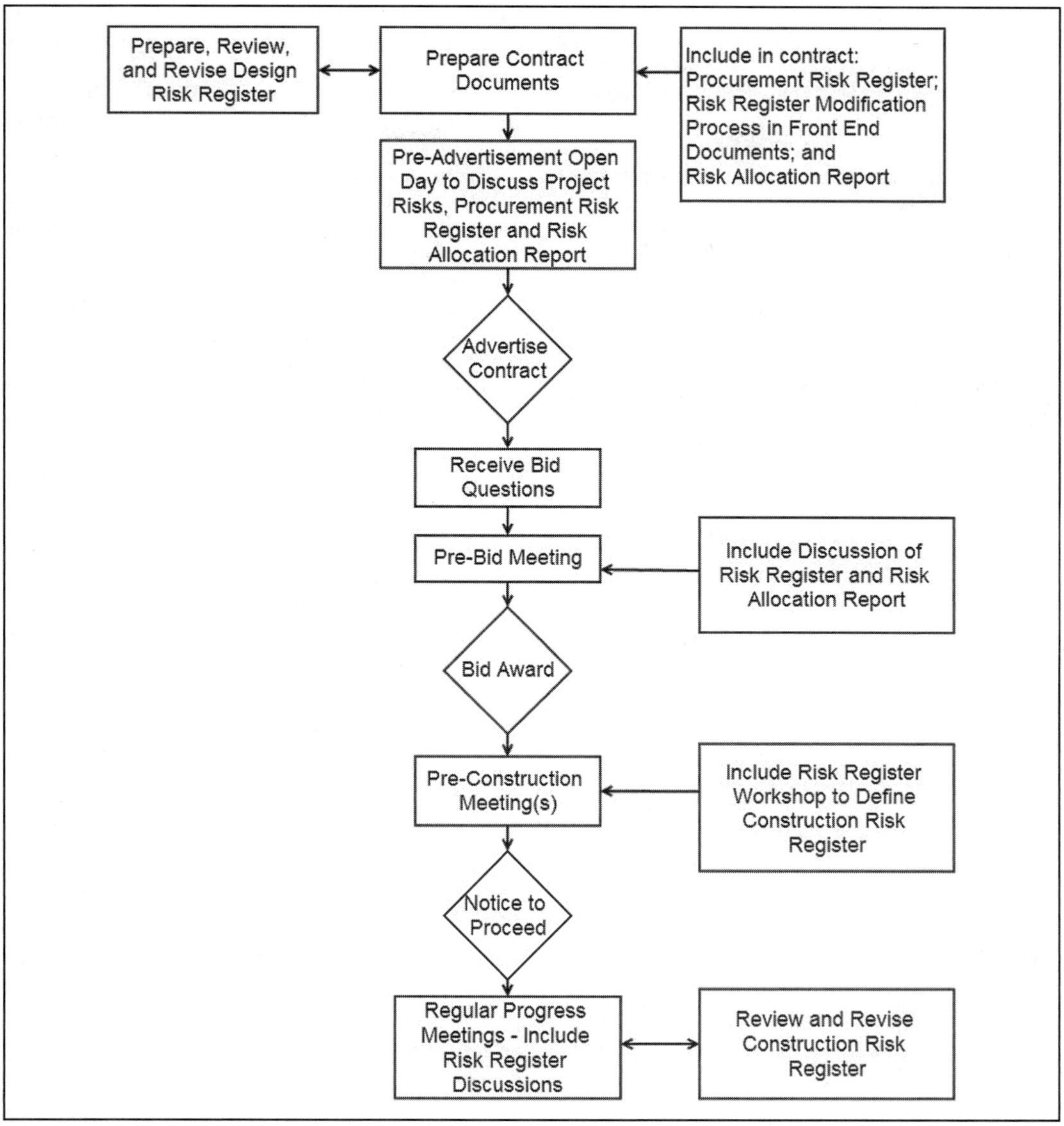


**Figure 5. Proposed process for design-bid-build procurement using the risk register as a contract document**

risk registers and risk documentation developed and 'cascaded' throughout the different project stages.
- The construction manager must understand the form of contract as well as the design and must also administer the project fairly in accordance with the risk terms defined in the contract.

## REFERENCES

Denney, B., Konstantis, S. and Tillie, S. 2014 *Proactive risk management–Helping to drive down insurance premiums and project costs.* Proceedings of the World Tunnel Congress 2014—Tunnels for a better Life. Foz do Iguaçu, Brazil. (accepted for publication).

Goodfellow, R.J.F. and Mellors, T.W., 2007. *Cracking the Code—Assessing Implementation in the United States of the Codes of Practice for Risk Management of Tunnel Works*. Proceedings RETC 2007, Toronto.

Heller, H. 2002. *Possible Maximum Loss Assessment of Civil Engineering Projects*. Working Group Paper, IMIA.

http://www.echarris.com/pdf/7531_NEC_THE%20RISK%20OF%20THE%20RISK%20REGISTER%20UNDER%20THE%20NEC_UPDATED%20SEPT%202011.pdf.

http://www.internationallawoffice.com/newsletters/detail.aspx?g=f0bb6d79-d4c2-47bf-9e26-29b7f7f5ab9f.

http://www.neccontract.com/documents/contracts/Guidance%20Notes/NEC3_EEC.pdf.

ITIG, 2006. *A Code of Practice for Risk Management of Tunnel Works*. International Tunnel Insurers Group.

Mellors, T.W., and Southcott, D., eds., 2003. *The Joint Code of Practice for Risk Management of Tunnel Works in the UK*. British Tunnelling Society and Association of British Insurers.

Reilly, J.J., 2008. Probable Cost Estimating and Risk Management Proc. North American Tunneling Conference 2008, San Francisco.

Reiner, H., 2011: Developments in the Tunnelling Industry following introduction of the Tunnelling Code of Practice. Amsterdam. IMIA Annual Conference.

Spencer, M., 2008: The Tunnel Code of Practice for Risk Management—5 Years On, IMIA Short Paper.

# Planning for What Could Go Wrong, When in Fact It Could Go Right—The Importance of Risk Management on Tunneling and Underground Construction Projects

**Joe O'Carroll**
Parsons Brinckerhoff

**ABSTRACT:** Regardless of whether a tunneling program or project is in the planning, design, or execution phase tackling risks early on is more time and cost-effective than responding to problems. From inception, projects carry more risk than anticipated resulting in missing target dates and experiencing cost increases. As projects unfold, new risks arise, and they become problems if not identified and acted on immediately. This paper outlines key risk management strategies for owners, tunnel designers and constructors that eliminate, or lessen the severity of, as much risk as possible from the earliest project planning and design stages.

## INTRODUCTION

Several major impediments threaten the successful delivery of tunnel and underground construction projects. In addition to technical design and construction risks associated with tunnel projects many of which are well understood—budgetary pressures, funding constraints and the complexity of procuring, administering and executing multiple concurrent projects in a large program could also have a catastrophic impact on the program's objectives. Contributing factors often include poorly managed design, procurement, construction, managerial, organizational, market and stakeholder risks, as well as non-conforming inputs or outputs to the costs and schedule, or any other potential inabilities to deliver the required or desired results.

A strategy to prevent 'failure' of a project requires a proactive and holistic approach to risk management. Through planning, identification, analysis, management, monitoring and control of the project's risks (both threats and opportunities), attention to management of risk will increase the probability and impact of positive events and decrease the probability and impact of events adverse to a project's objectives.

## START BY PLANNING FOR SUCCESS

Managing risks, and their potential impacts, is critical to achieving a project's overall cost, schedule, safety and performance objectives. As a process and management approach, risk management is an integral part of a successful project delivery. By promoting proactive risk management strategies that eliminate, or lessen the severity, of as much risk as possible from the earliest project planning and design stages, tunneling and underground construction projects can be successfully delivered, with a higher degree of certainty in terms of cost and time, despite many political, financial or regulatory obstacles and market conditions.

A successful project is one that will have met all of the following, as a minimum: be deemed to have realized the opportunities (goals and objectives) identified for the project; completed within cost and schedule goals; achieved the quality and performance expected by owner and stakeholders; and engendered no adverse political or stakeholder reaction throughout its design, construction and commissioning. Risk management is an explicit, structured process and a tool to help manage uncertain events, so as to maximize chances of achieving a successful project. As such it will help all stakeholders, owner, designer, contractor and other third parties understand and manage the relationships between their business environment, their strategic objectives, the risk to achieving these objectives and their actual performance.

Often, by the time a project is under way, managers are forced into a reactive stance, responding to each threat as it emerges. This reactive stance results in perpetual firefighting with little chance of regaining control. To be effective, project managers and project engineers from the earliest planning stage must identify, evaluate, communicate, and prioritize risks on the basis of likelihood and severity of effect on the project.

The projects we, as planners, engineers and constructors, undertake in today's business environment continue to push the envelope in terms of technical difficulty, size, overall complexity, financing challenges and project delivery. Risk management

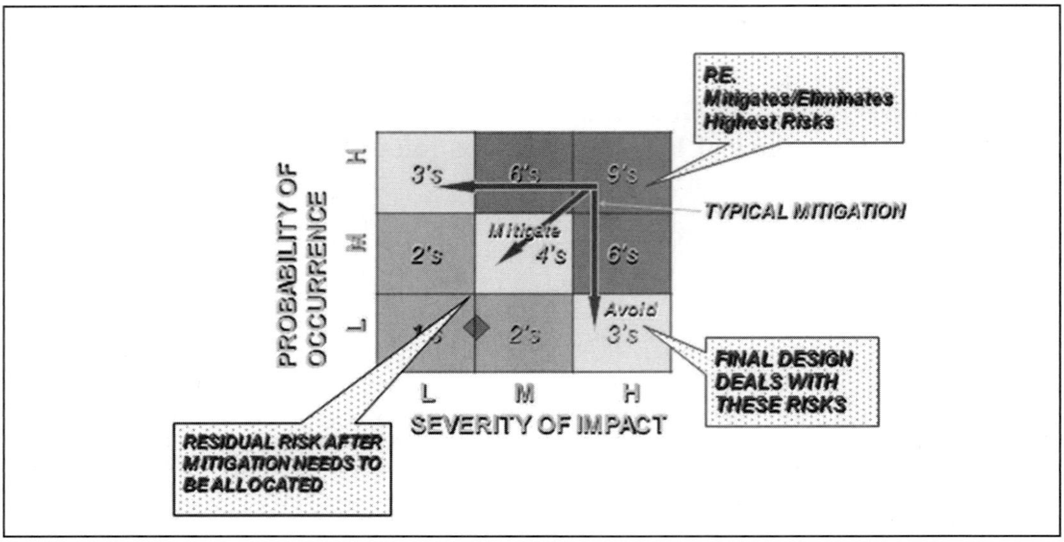


**Figure 1. Management philosophy**

**PROJECT RISK REGISTER *(Extract)***
Project X, Toronto Canada
REV : 0
DATE PREPARED: July 28, 2010

| Category | Risk Ref. I.D. | Risk Description |
|---|---|---|
| **6.0 PLANNING** | | |
| | 6.0.1 | Delivery of design and construction does not align with the requirements of the EA documents that leads to undertakings not as described (scope changes), concepts oversimplified, design poorly translated & construction poorly implemented. |
| | 6.0.2 | Challenges to the EA documents with potential for court protests result in costly legal processes and delays |
| | 6.0.3 | Public feel they have not be given adequate "voice" at the public meeting and open houses leads potential for increased protests. |
| | 6.0.4 | Concept plans are hurried and budgets inadequately reflect the true costs for delivery the program. |
| | 6.0.5 | Ridership forecasts overestimates/underestimates ridership impacting vehicle requirements impacting costs and operational capacity. |

**Figure 2. Project risk register**

is not a new concept to the world of tunneling and underground construction. We are acutely aware of what could go wrong on a project and do our best to plan accordingly.

In recent years, we have put tools in place such as the Geotechnical Baseline Report that communicate risk and a changed condition clause in the contract that allow a mechanism for risk allocation and sharing. While there is no arguing that these two critical components of managing risk on an underground project have contributed to increase the opportunity for projects to be more successful there are so many more 'softer' factors that force a project, from inception through to completion, to go 'wrong'. There are many simple steps that can be taken to provide a higher level of confidence that a project can go 'right'. Managing risk has always been at the heart of the underground construction industry however in recent years the benefits of having a formal, systematic and comprehensive risk management program in place that can be carried through all phases of a project is being realized by all stakeholders in a project be they owners, designers or contractors, financers or insurers. A risk-based approach relies on teams understanding their missions and a relentless focus on early identification and prevention of problems thereby increasing the confidence in a within time and on budget delivery.

## UNDERSTANDING "SOFT" RISKS AND HOW TO MANAGE THEM

Below are examples of 'softer' risks that are equally important in managing and delivering a tunnel and underground construction project and suggested steps to mitigate and manage them.

### Not Properly Identifying (and Articulating) the Risks Beforehand, or Not Allocating These Risks in a Clear (Unambiguous) and Balanced Manner

This is best managed by a clear articulation and understanding by all stakeholders on the project of strategic, technical, environmental, financial, economic, political, operational, schedule & resource risks. In addition to the common documents an owner procures during the early phases of a project such as Conceptual Engineering Report (CER); Basis of Design Report (BOD); Construction Phasing and Contract Strategy Report it is recommended that the owner also develop a Risk Register, a Risk and Contingency Management Plan (RCMP) and a Risk Allocation Report (RAR).

Also important is letting a party bear the risk that it can control i.e., the Owner should take on the risk of geological conditions, environmental and other types of permits necessary to implement the project, authorizations, land use, advance long lead utility relocations; the Contractor should take on the risk of means and methods, productivity, suppliers, sub-contractors, workforce, materials, equipment, and others; and the design Consultant should bear the risk of design performance.

### Tunneling Projects Face a Multitude of Risk to Health, Safety, Third Party Property, the Environment, and Community Infrastructure

Risk-based approaches designed to manage known environmental risks, anticipate unforeseen risks, and pro-actively intervene to streamline project delivery are highly recommended. Early project definition that communicates environmental, utility and right of way impacts to all stakeholders and the community in which the project is being constructed will pay dividends in the long run. When it comes to interacting with the public communication is "king." A well-developed risk management plan should address and help manage risk to:

- Health and safety of third parties;
- Third party property, including utilities, existing buildings and structures, cultural heritage buildings and above and below ground infrastructure;
- The environment including possible land, water, air and noise pollution;
- Community infrastructure, i.e., extent of traffic lane and sidewalk closures, barricades, in excess of what the community expects
- Local businesses and residents from more noise, dust, dirt and vibration than the city regulations permit or the community expects; and
- Health and safety of workers, many of whom will be hoping for employment or other business opportunities as a result of the project being in their neighborhood.

## PREVENTING PROCUREMENT RISK

A major risk on tunneling projects is that the owner enters a contract based on expectations of co-operation, not conflict, and assumes that most of the objectives of the owner and the contractor are the same. Several risks result from this risk not least of which is:

- Lack of clarity and direction when disputes arise
- Not finishing the project on time
- Not finishing within budget and at the agreed-upon level of quality
- Misunderstanding by contractors of the objectives of the owner

A well-developed comprehensive risk management plan will provide:

- A clear definition of the scope of works and risk allocation
- Well prepared bid documents that give an unambiguous set of conditions as well as clear requirements on risk allocation
- Well-defined rules for acceptance and take-over by the Owner
- A balanced sharing of risk and conflict resolution schemes that secure a quick resolution of conflicts

## AVOIDING COST OVERRUNS AND SCHEDULE DELAYS FROM THE START

Major tunneling projects are at high risk to cost overruns and construction delays because of the linear nature of construction. It is unlikely that a project will not be completed due to technical reasons however cost overruns and construction delays are probable and an important source of risk. The first step in preventing cost overruns and schedule delays in through development of robust cost estimates and construction schedules by estimators experienced in preparing bid-like cost estimates and schedules for contractors on major tunnel projects followed by the identification and management of risks, and their potential impacts, that could adversely impact the out completion cost of the project.

Risk identification can be conducted through a series of risk interviews and workshops engaging all key project participants and stakeholders. The objective is to examine baseline assumptions and assess the impact of risk to completion within the desired cost and schedule objective. The process is supported by a Monte Carlo analysis on the both cost and schedule taking into account all identified and quantified risks producing a probability distribution of the project's possible complete dates and costs. This level of risk analysis provides a quantitative basis for levels of confidence; serve's to prioritize attention on risk most likely to have a significant impact and establishes required cost and schedule contingency levels.

Major tunnel and underground construction projects run the risk of cost overruns due to:

- Underestimation of construction costs often as a result of needing to stay within a politically acceptable cost estimate
- Underestimation of risk and its financial consequences
- Inadequate contingencies to cover risk
- Lack of understanding as to the purpose and management of contingencies
- Poor cost control
- Additional financing expenses
- Underestimated or increased administrative costs
- Design and performance requirements changes
- Ambiguity in the contract documents leading to low bids followed by claims and contractual disputes

Several risks that contribute to delays on major tunneling and underground projects include:

- Underestimation of construction schedule often as a result of needing to stay within a politically acceptable timeframe for completion
- Problems with a project's organizational structure resulting in lack of decision making
- Environmental mandates not being met or being challenged
- Procurement delays due to stakeholders in a project having different interests
- Unclear contractual responsibilities
- Unforeseen geological difficulties
- Mechanical problems with selected excavation equipment
- Lower than expected tunneling progress
- Late changes in the design including technology changes

Steps that can be taken to more effectively manage these risks and keep your project on the right track include:

- Establishment of agreements with the regulatory agencies to guide the review of the project; by defining clear roles and responsibilities, review and approval timeframes, agreed upon methodologies, funding of dedicated agency staff, and a issue resolution process.
- Establishment of program level agreements for threatened and endangered species, stream and wetland mitigation, or other environmental impacts that may affect procurement schedules.
- Designing to budget
- Creating geotechnical baselines as a basis for the contract
- Risk analysis to determine construction contingency in cost estimating
- Communicate the concept of risks early in the process to help all stakeholders understand tradeoffs.
- Provide transparency in financial reporting system.

## Designing to Budget

Scope and budget creep during the design phase of a program places major pressures on those responsible for delivery within strict financial and funding constraints. In the planning phase developing project level budget ranges in association with a risk based contingency will provide decision makers and their partners with a more accurate assessment of ultimate project and program costs and a realistic confidence level in achieving their target.

Once a robust and realistic budget is established it is not unreasonable to then expect designers to 'design to budget' in order to more effectively reduce the potential for scope and budget creep prior to award of contracts. Early definition and mitigation of risks associated with environmental constraints, right of way acquisition requirements, utility relocations, construction costs and schedule expectations are key to successful delivery of procurement documents that will result in construction bids being received well within established budgets.

### Market Forces Negatively Impact Project Costs

One of the most influential impacts to cost certainty is the prevailing local, national and world economic climate. As was seen in 2004 through 2007, significant over heating prevailed driving up actual bid construction costs well over reported building cost inflation indices. Over the past few years the depressed economic conditions has, and continues to have, an over-riding influence on actual bid costs. It is uncertain how the economic climate over the next few years will influence the bidding market although one could speculate it could continue to be significant. Traditional trends show sharp depression followed by sharp return to overheating in the construction industry and, the deeper the recession, or in this case the depression, the potentially greater the upswing driven by diminished capacity in material, plant and labor supply chains unable to ramp up to meet the sudden up-turn in demand.

A strategy to manage market force surprises could include an analysis of the following market related scenarios: recession ending with slow recovery, recession ends with fast recovery, prolonged recession with a slow recovery and prolonged recession with fast recovery. Controlling market forces on a project is virtually impossible but understanding their impact and preparing accordingly is critical to achieving cost certainty. Depending upon the prevailing market conditions progressing quickly with a project may capture the benefit of continuing depressed bids to the advantage of the project.

## PREVENTING GEOTECHNICAL RISK BECOMING TOO ONEROUS

Underground projects present unique challenges that increase risk more than any other type of new infrastructure. Risk is increased significantly when the ground conditions are poorly understood and misrepresented in the contract. Irrespective of location and type the ground conditions will always present an element of risk in the excavation of a tunnel or underground structure.

This is usually caused by:

- Inappropriate site investigation techniques, drilling, sampling and lab testing resulting in poor geotechnical interpretations and baselines
- Unknown buried obstructions (piling, logs, concrete, sheeting, harbor remnants, boulders) along alignment
- Insufficient geotechnical data provided to the contractor at time of bid

And results in:

- Unforeseen and adverse ground conditions resulting in a delay or stoppage to tunneling
- Settlement damage or distortion to utilities or other structures due to ground loss going undetected until it effects roadways, utilities, nearby buildings resulting in loss of services, cracked sidewalks, windows and doors of surrounding buildings being out of plumb
- Excessive settlement of existing structures leads to unacceptable structural damage of masonry buildings
- Uncontrolled loss of pressure at the tunnel face during excavation causes settlement, sinkholes or in extreme cases complete collapse of the ground above the tunnel

This risk need not become unnecessarily onerous if managed properly from the early phases of a project. Steps in which this risk can be minimized include:

- Carry out a comprehensive desk top study on the historical usage of the project area to get a greater understanding of what might be below the surface in terms of buried structures, industrial debris, disused basements, pipeline building foundations etc.
- Plan a comprehensive phased geotechnical program that includes both physical and geophysical investigations where appropriate.
- Identify and prioritize "hot spots" which would be significant in determining both the feasibility and methodology needed for tunnel construction.
- Depending on results of pre-design investigations, define scope for geotechnical investigations for final design including investigating the presence of methane gas, contaminants, fault zones, mixed face conditions as these are likely to significantly increase tunneling and shaft construction costs.

## CONCLUSIONS

Risk is inherent in most human endeavors. Controlling that risk is often the difference between success and failure. To be successful on today's tunnel and underground construction projects, risk management must be viewed as an explicit, structured process and an integral part of project development—a way of thinking as well as a tool to help manage uncertain events.

Top Left: Seattle Seawall Construction (circa 1930) Top Right : Alaskan Way Viaduct Construction (circa1951). Bottom Left: Modeling the subsurface environment during preliminary design for the SR99 Tunnel (PB. 2010)

**Figure 3.**

To achieve success one must start by planning for success. Risk management strategies that eliminate, or lessen the severity of, as much risk as possible from the earliest project planning and design stages is an important step in achieving this goal. A risk-based approach relies on all participating parties in a project understanding their missions and a relentless focus on early identification and prevention of problems thereby increasing the confidence in a within time and on budget delivery. A well-developed comprehensive risk management plan will provide a clear definition of the scope of works and risk allocation, well prepared bid documents that give an unambiguous set of conditions as well as clear requirements on risk allocation, well-defined rules for acceptance and take-over by the owner and a balanced sharing of risk and conflict resolution schemes that secure a quick resolution of conflicts.

Key to avoiding cost overruns and schedule delays are designing to budget; creating geotechnical baselines as a basis for the contract, and adopting risk analysis techniques to ensure reasonable contingencies are established in the cost estimate and schedule, supported by a contingency management plan that preserves appropriate contingencies through all phases of a project from concept to commissioning.

When owners, designers, constructors and other stakeholders commit to jointly identifying and mitigating risks through the comprehensive assessment of risk value, use of risk workshops, development of an "actionable" risk registers, risk analysis and the development of risk and contingency management plans our industry will benefit from having more projects with fewer delays, less cost increases, reduced environmental or other third party impacts, and ultimately decreased risk to operational safety and reliability of our tunnel or underground construction projects.

# TRACK 3: PLANNING

## Session 3: Geotechnical and Third-Party Planning

Sean Harvey, Chair

# Gas Studies for the Westside Subway Extension, Section 1

**Roy F. Cook and Shankam Jagannath**
Parsons Brinckerhoff

**Francis Fong**
FPA Underground Inc.

**ABSTRACT:** Methane and hydrogen sulfide gases are present in the ground beneath Los Angeles, California. These gases present a challenge to tunnel construction for Metro's Westside Subway Extension (WSE) being built to extend its Heavy Rail Transit System. For Section 1 of the project, an extensive investigation including more than fifty gas monitoring wells has been carried out to characterize the gases both in the groundwater and in the vadose zone. The paper describes the investigations, gas characterizations, and the uncertainties associated with gas data collection.

## INTRODUCTION

Methane and hydrogen sulfide are found in the near surface soil deposits in which the tunnels and subway stations for the Westside Subway Extension (WSE) will be built. They originate from hydrocarbon deposits found in the deep sedimentary rocks of the Los Angeles basin. The gases are found in the vadose zone, and below the water table dissolved in the groundwater and as bubbles. The potential explosive nature of methane and the toxicity of hydrogen sulfide present challenges to working underground that must be dealt with when planning, building and operating the tunnels and subway stations.

For the WSE Project, an investigation program has sought to characterize the gases along an alignment following Wilshire Boulevard from the existing Wilshire/Western Subway Station to just beyond the intersection of Wilshire Boulevard with La Cienega Boulevard (see Figure 1). The data obtained will provide a basis for estimating the quantity of gas that would off gas from the ground and groundwater during excavation. The investigations have included:

- Sampling gas concentrations during drilling
- Periodic measurements of gases in the head space of groundwater monitoring wells
- Periodic monitoring of gas concentrations and pressures in multi-stage gas probes
- Periodic collection and analysis of gas and water samples collected from multi-stage gas probes
- Measurement of gas flux (flow from ground)

## GEOLOGY

The Los Angeles basin is a major elongated northwest-trending deep structural depression. The subway will be built within the following geologic formations within the basin:

- Alluvium—loose to dense sands and gravels (stream channel deposits) and interbedded medium stiff to hard silts and clays, and gravelly silts and clays (fan and overbank deposits).
- Lakewood Formation—interbedded dense clayey sands, silty sands and poorly graded sands, and very stiff to hard silts and clays.
- San Pedro Formation—interbedded dense sand and silty sand and stiff to hard silt layers.
- Fernando Formation—massive, stiff to hard, friable, extremely weak to weak siltstone and claystone with thin sandstone interbeds and calcium carbonate cemented concretions.

The Los Angeles basin is a major source of hydrocarbons. Methane, hydrogen sulfide and asphlatum associated with these deposits migrate through fractures to the ground surface and reside in the near surface soil deposits. Gas can be found throughout the alignment but are particularly associated with the asphalt impacted soils found along Wilshire Boulevard within approximately 750 meters east and west of Hancock Park where the La Brea Tar Pits are located (Metro, 2012).

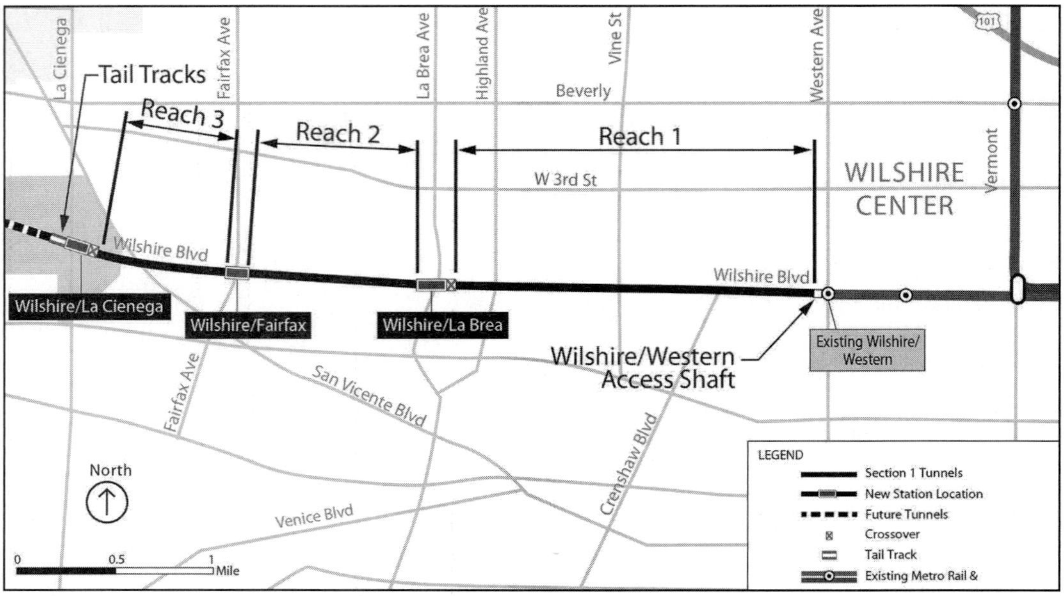


**Figure 1. WSE Section 1 alignment**

## CHARACTERIZATION OF GASES

***Methane*** is a naturally occurring gas associated with the decomposition of organic materials. Methane is common in oil and gas fields. Methane gas is not highly toxic, rather it asphyxiates as it displaces oxygen; but it is explosive when its concentration is between five and fifteen percent in air at atmospheric pressure. Methane (density ~0.72 g/l) is lighter than air and it tends to rise through the ground and dissipate. It is moderately soluble in water with approximately 40 to 50 cubic centimeters dissolving in a liter of water at atmospheric pressure.

Isotopic analysis of methane samples collected from gas probes along the alignment suggest the gas is predominantly of thermogenic origin consistent with an oil field source and seeps up from the hydrocarbon deposits below. However, bacterial breakdown of hydrocarbons near surface has also been put forward as a source for methane in the vadose zone (Kim and Crowley, 2007).

***Hydrogen Sulfide*** is produced by the anaerobic decomposition of organic and inorganic matter containing sulfur. It is highly toxic when inhaled. Its flammability range is from four to forty six percent and it is corrosive. Hydrogen sulfide (density ~1.54 g/l at atmospheric pressure) is heavier than air and within the ground tends to accumulate above the groundwater table and within depressions underlain by more impermeable material. It is highly soluble in water with approximately 2,800 cubic centimeters dissolving in a liter of water at atmospheric pressure.

Anaerobic conditions typically prevail in the subsurface when significant levels of biodegradable organic compounds such as asphalt, oil, or plant materials are present. Under these conditions, methane and carbon dioxide are typically present at approximately equal concentrations and account for nearly all of the gas in the soil.

## EXPERIENCE WITH SUBSURFACE GASES IN LOS ANGELES

Following the "Ross Dress-For-Less Fire" (Hamilton D.H. and R.L. Meehan, 1992), the City of Los Angeles created a Methane Risk Zone Map in 1985 (subsequently updated) and adopted regulations for construction of buildings and other structures within the methane zone (Figure 2).

As a result of the scrutiny given to methane in the area, Metro has studied subsurface gas conditions for projects in the Mid Cities area of Los Angeles from the mid 1980s (Elioff et al., 1995). As well as finding methane, the studies found high levels of hydrogen sulfide south of Wilshire Boulevard along Crenshaw and Pico Boulevards. The highest gas readings occurred in the San Pedro Formation, generally in unsaturated zones capped by the less permeable Lakewood Formation and/or zones of groundwater. Hydrogen sulfide concentrations up to approximately 20,000 parts per million (ppm) were measured.

With the renewed interest in a subway to the west, an alignment along Wilshire Boulevard was re-investigated. This alignment with the exception

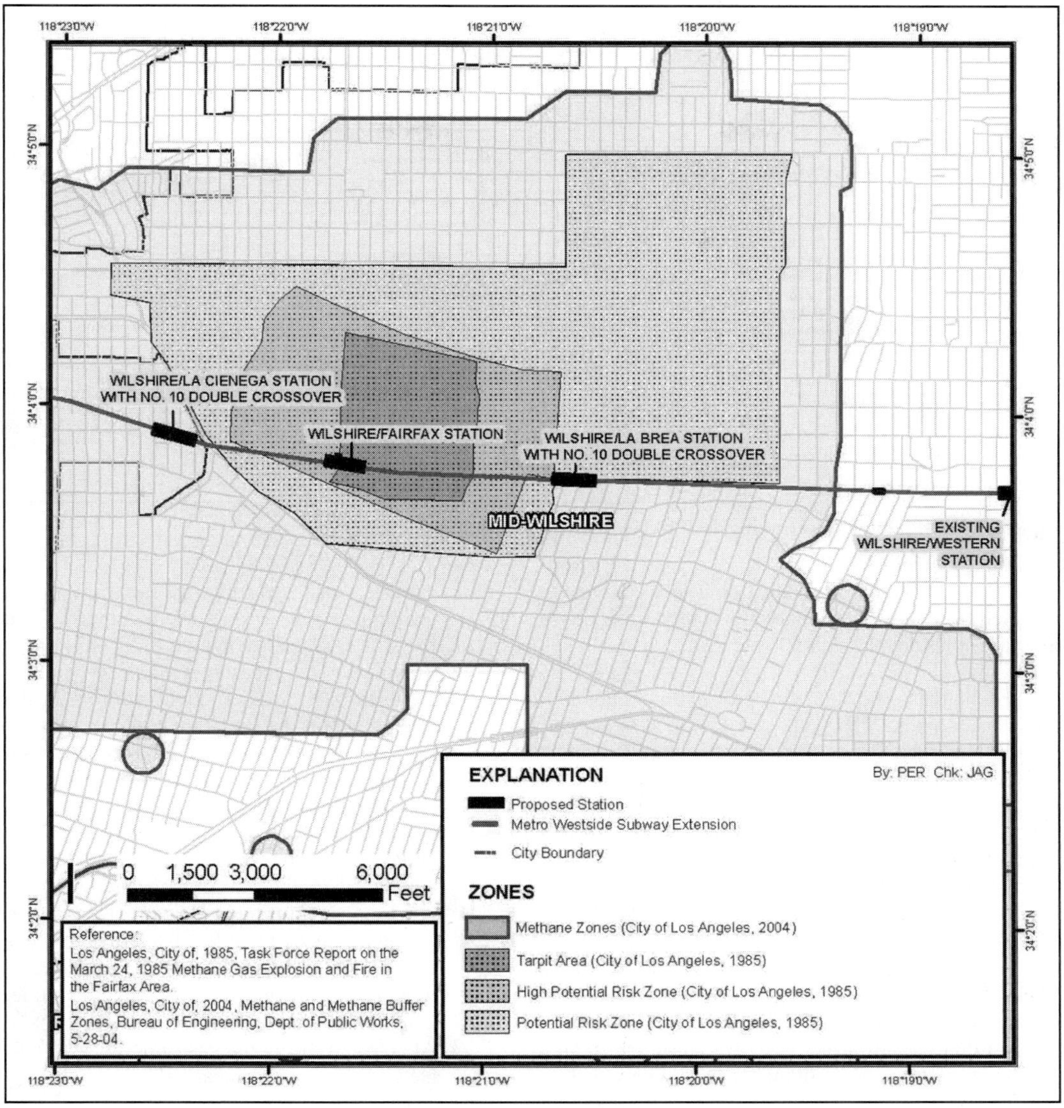


**Figure 2. City of Los Angeles methane and methane buffer zone map showing Section 1 of WSE**

of approximately 1,000 meters to the east passes through an area designated as a "Methane Zone" on the City's "Methane and Methane Buffer Zone" map (City of Los Angeles, 2004) and areas designated as the "Potential Risk Zone," "High Potential Risk Zone" and "Tar Pit Area." In these zones, risks of encountering methane are considered elevated.

## SUBSURFACE INVESTIGATION OF GASES

The current investigations focused on characterizing gas conditions with a program of monitoring wells containing multi-stage probes where gas measurements were taken and from which gas and groundwater samples were taken and analyzed for their gas compositions. The results from previous investigations were combined with the data from the fifty four wells installed for the WSE project.

### Gas Monitoring Well Installation

Gas monitoring wells typically consisted of one to three nested probes monitoring gases and one to two PVC standpipes (see Figure 3). The configuration provides for measurement of gas concentrations and pressures within the vadose zone, and concentrations

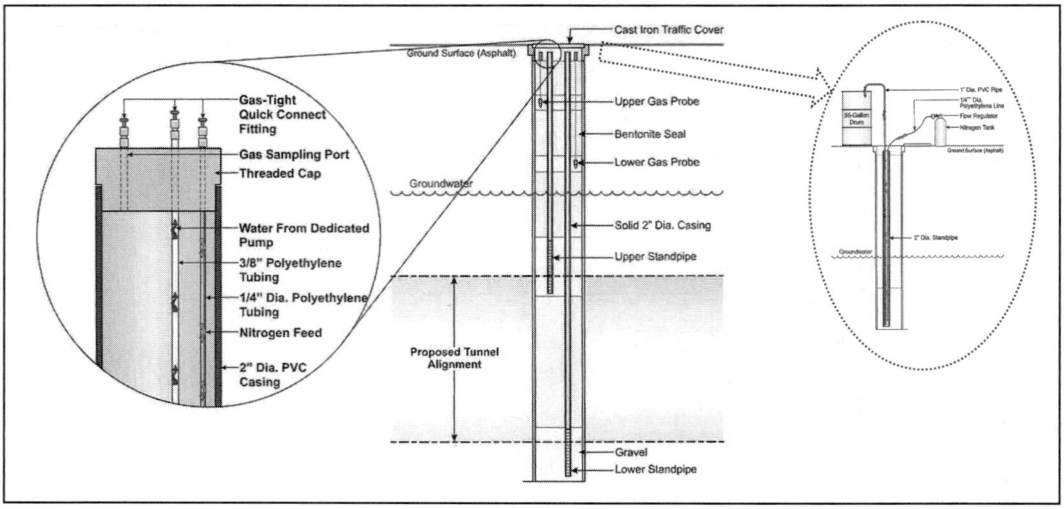


**Figure 3. Typical gas monitoring well development (GeoKinetics, 2012)**

of gases dissolved in groundwater. The wells were installed in hollow-stem auger and rotary-wash borings and holes developed for Cone Penetrometer Test soundings.

Each standpipe and gas probe was immediately sealed with a PVC cap, and equipped with gas tight fittings such that connection of a purge line and sample collection line could be made without unsealing the in-hole installation. The monitoring well was developed using nitrogen air lift methods to reduce introducing air as a contaminant into the ground. Each standpipe was purged until the water removed from the hole was without observable suspended sediment (see Figure 3).

Table 1 presents the number of probes and standpipes (including historic wells) installed within tunnel reaches and at station sites. In most cases, a dual casing nested well was installed in an 28.5 centimeters outside diameter hole with 1.5- to 3-meter long screens. Some wells consisted of a single 5 centimeter diameter PVC casing placed in a 18.5 centimeter diameter hole (with 3- to 6-meter) long screened intervals in the casing). The specific well configuration (i.e., number of gas probes, depth of gas probes, number of standpipes, and depth of screened intervals) was determined in the field based on the lithologic conditions. Screened intervals were adjusted where necessary to avoid oil or tar-bearing sands that would clog the screens. Soil gas probes were typically installed above the groundwater level encountered during drilling.

## Gas Measurement and Sampling

Gas measurement was performed periodically in wells using the following approaches:

1. Pressure was measured in each probe and standpipe using a Magnehelic gauge with a resolution of approximately 1.25 millimeter of water. The measurement was typically made through a quick-connect fitting in the sealing cap of the standpipe or gas probe.
2. Gas concentrations were measured in each standpipe (head-space measurement) and gas probe using multi-gas infrared gas analyzers (Landtec GEM-200 Plus or GA-90 and/or Qrae Plus Models). The analyzer was connected to the gas probe/standpipe by a quick-connect fitting in the cap. The analyzer extracted gas by pumping at a rate of approximately 500 cubic centimeters per minute and monitored for methane, hydrogen sulfide, oxygen, and carbon dioxide.
3. At selected locations, gas samples were collected for laboratory analysis.

## Groundwater Sampling for Analysis of Dissolved Gases

After the standpipes were developed and purged, groundwater samples were collected periodically for analyses of dissolved hydrogen sulfide, methane, and other gases (carbon dioxide, ethane) as well as the presence of volatile organic compounds and metals.

Sample collection used a pneumatic pump driven by compressed nitrogen (to prevent the introduction of air as a contaminant) to force groundwater samples into sealed, clear, Schedule 40 PVC sampling containers. A gas-tight quick-connect fitting on one end of the container connected the pump discharge line to the well cap. Another gas-tight quick-connect fitting on the other end of the container was

**Table 1. Details pertaining to gas probes and standpipes in gas monitoring wells**

| Location | # of Gas Monitoring Wells | # of Standpipes | # of Gas Probes | Total # of Sampling Points per Location |
|---|---|---|---|---|
| Tunnel Reach 1 | 11 | 22 | 12 | 34 |
| Wilshire/La Brea Station | 2 | 2 | 6 | 8 |
| Tunnel Reach 2 | 16 | 29 | 27 | 56 |
| Wilshire/Fairfax Station | 14 | 21 | 38 | 59 |
| Tunnel Reach 3 | 5 | 7 | 11 | 18 |
| Wilshire/La Cienega Station & Tail Tracks | 6 | 5 | 13 | 18 |
| **Totals** | **54** | **86** | **107** | **193** |

connected to an adjustable back-pressure valve. Prior to sampling, a valve was adjusted to maintain a back-pressure equivalent to the hydrostatic pressure at the bottom of the standpipe. Several volumes of groundwater were then purged through the container using the nitrogen driven pneumatic pump. After purging, the sampling container was filled, the quick-connect fittings were detached and the container was transported to a laboratory for analysis of dissolved gases, hydrocarbons, and metals.

### Dissolved Gas Extraction

Groundwater samples were collected from the standpipes in 5 to 10 liter Tedlar bags. The bags were evacuated and sealed prior to sample collection. The groundwater was purged from the standpipes using nitrogen-driven pneumatic pumps. The groundwater was maintained at, or above, its in-situ hydrostatic pressure until it entered the Tedlar bag. Once filled, the sealed bags were transported to a laboratory and placed in a vacuum chamber. The pressure in the chamber was reduced to less than one percent of atmospheric pressure so that the dissolved gases in the sample could exsolve. At that point, atmospheric pressure was restored and the volume of accumulated gas was measured. The evolved gas was then extracted from the Tedlar bag using a syringe and injected into a train of infrared gas analyzers to quantify methane, hydrogen sulfide, oxygen, and carbon dioxide.

### BAT® Groundwater/Gas Sampling in CPTs

BAT® groundwater sampling was performed in association with Cone Penetrometer Testing (CPT). The BAT® procedure takes groundwater samples at depth while maintaining the in-situ pressure (Blegen et al., 1988). This ensures that the dissolved gases will not evolve from the groundwater prior to the laboratory testing.

The CPT was advanced to the desired depth and then an evacuated BAT® sampler was lowered down inside the CPT drill rods onto a BAT® filter tip using an extension cable. By gravity, a double-ended needle on the sampler penetrates the septum in the filter tip and the septum of the sampler itself and collects both water and gas samples. The groundwater pressure and suction from the sampler draws groundwater and gas into the sampler. Upon lifting the BAT® sampler, the flexible septa in both the filter tip and the sample tube automatically reseal. The liquid and gas sample is thereby kept hermetically sealed. Once removed from the CPT drill rods, the sample can be sent to a laboratory for analysis.

### Impact of Air Contamination on Gas Measurements

Samples representative of in situ gas conditions underground are difficult to make because of sample contamination with air. Air is introduced during drilling, installation, development and monitoring of standpipes and gas probes. It can dilute gas concentrations but also alters the chemical composition of the gas mix as the components of air, most notably oxygen reacting with the underground gases. When air is introduced into an anaerobic environment, oxygen can be consumed by the aerobic biodegradation of organic compounds that are present with carbon dioxide being the primary by-product. In addition, nitrogen can be converted into ammonium and various nitrites and nitrates by nitrogen fixing bacteria. Hydrogen generated by these processes is typically converted into other compounds and does not persist.

Field procedures can minimize contamination by air but cannot fully eliminate it. The impact can be further aggravated by barometric pressure variations. They can drive air through leakage paths thereby changing gas concentrations and chemical compositions. Daily fluctuations of about one to two inches of water are normal in the Los Angeles Basin with pressures typically falling during the day as the atmosphere warms and rises at night as temperatures cool. The amount of oxygen and nitrogen that may be introduced to the underground environment by the drilling, well installation and sampling activities is difficult to assess as these gases have different degrees of activity and their relative concentrations change over time when in contact with other gases.

Oxygen is generally consumed preferentially before the nitrogen. The time required for these processes to run to completion is dependent on the amount of air introduced to the subsurface, the amount of organic matter present, soil chemistry and the bacteriological environment. In a typical anaerobic setting, the amount of time required for a system to recover could range from a few weeks to several months.

For gas probe readings, when oxygen is not present or present in only low concentrations, the sum of concentrations of oxygen, carbon dioxide and methane typically approaches 100 percent. Where they do not, it is likely that nitrogen and argon account for the balance of the gas that is "missing."

### Impact of Asphalt on Gas Measurements

Gas monitoring in wells and probes was adversely impacted in the asphalt impacted soils. The impacts have included:

*Clogging of Well screens*—In monitoring wells installed in the early phase of the investigation, screens placed in the asphalt impacted soils clogged with oil and asphalt. To avoid this, screens installed in monitoring wells placed later were located in soils either above or below the asphalt impacted soils.

*Unsuccessful sampling of groundwater in CPTs*—BAT® sampling of groundwater (for analysis of dissolved gases) proved unsuccessful in three out of four holes in which the system was tried. The failures were apparently caused by asphalt from the asphalt-impacted soils smearing the porous filter membranes of the CPTs as they were driven into the ground. The smear impeded flow through the filter membranes into the BAT® samplers to such an extent that water samples could not be recovered.

*Seeps of Asphalt-Water at Ground Surface from Standpipes*—At five locations where standpipes were installed in asphalt impacted soils, the standpipes were completely filled and asphalt would seep out of standpipes when they were uncapped. At other installations, asphalt forced its way up the borehole past the well seal, emerging at the ground surface. Figure 4 shows the condition at top of asphalt filled well with Figure 5 showing the condition following clean up. A possible explanation for this phenomenon is that the screened section of the standpipe intercepts a zone of asphalt impacted soil at depth that is saturated with fluid (water-asphalt mix).The water-asphalt mix enters the standpipe under hydrostatic pressure. As the fluid enters and rises up the standpipe, the confining pressure drops and gases exsolve. The exsolving gas reduces the effective density of the fluid within the standpipe and induces the flow of the water-asphalt-gas mix up the standpipe resulting in overflowing of the standpipe.

**Figure 4. Top of monitoring well–asphalt filled**

**Figure 5. Cleaned-up monitoring well**

### Results and Discussions

Methane, hydrogen sulfide and gas pressures and extensive tar-impacted soils were routinely monitored along the tunnel alignment (see Table 2). High gas concentrations and high gas pressures were measured In the Fairfax area, which is the vicinity of the La Brea Tar Pits and the area where the asphalt impacted soils are found.

The alignment west of South Dunsmuir Avenue (about 500 meters west of Wilshire/La Brea Station) in Reach 2 has consistently shown elevated gas conditions, where, "elevated" for the purposes of this discussion has been defined as locations where:

**Table 2. Highest recorded gas concentrations/gas pressures from gas probes and standpipes in gas monitoring wells within Section 1 of WSE (monitoring period 2009–2013)**

| Location | Methane (%) | Hydrogen Sulfide (ppm) | Gas Pressure (cm of water) |
|---|---|---|---|
| Tunnel Reach 1 | 1.2 | 1.0 | 0 |
| Wilshire/La Brea Station | 0.7 | 0.1 | 2 |
| Tunnel Reach 2 | 91.5 | 460 | 2,144 |
| Wilshire/Fairfax Station | 100 | 6,500 | 704 |
| Tunnel Reach 3 | 99 | 415 | 41 |
| Wilshire/La Cienega Station & Tail Tracks | 6.3 | 4.0 | 4 |

- Methane levels are above five percent or
- Hydrogen Sulfide levels are above five ppm

The subsurface gas pressures measured in the gas probes and standpipes are generally consistent with the hydrostatic head. However, exceptions were measured in some locations where persistent gas pressures up to 250 centimeters of water were measured above the hydrostatic head. Artesian gas pressures are likely the result of gas and groundwater trapped beneath more impermeable layers.

When totaled, the gas concentrations do not generally total 100 percent as all gases are not accounted for. Since contamination by air occurs, nitrogen, carbon dioxide and argon likely account for the balance of gas that is "missing" in the monitoring result totals.

## GAS FLUX TESTING

As part of the investigation of subsurface gas conditions, a testing and sampling program using a downhole flux chamber was performed in two borings in the vicinity of La Brea Tar pits (Schmidt and Fong, 2013). The testing followed procedures developed by the Environmental Protection Agency (EPA) for obtaining gas data from borings drilled into waste materials (US EPA, 1986). The chamber is an acrylic cylinder with a 0.0032 $m^2$ exposed surface area and a volume of approximately 0.0064 $m^3$.

The flux chamber fits inside the hollow-stem of a typical auger used for site investigations. To take measurements, auger drilling is stopped and the chamber is lowered down the inside of the hollow stem auger onto the bottom surface of the boring. Clean, dry sweep air (zero grade air) is introduced into the chamber at a controlled rate of about 1 liter per minute and the concentration of the species of interest in the sweep air is measured at the exit of the chamber. Gas measurements were made using a TVA-1000 analyzer, a portable unit that provides continuous real time data monitoring of total hydrocarbons and a Jerome 631X hydrogen sulfide analyzer. At each sampling location, measurements were at different times to establish changes in gas concentrations over time.

The EPA sampling procedure was modified to account for groundwater and asphalt/sand/water slurry mix filling the boring and rising into the chamber. When this happened, either the chamber was suspended above the water or slurry surface in the boring, or cuttings and groundwater bailed from the boring were tested at ground surface. The subsurface flux testing and sample collection was carried out at multiple depths in two borings—Boring M-351 and Boring M-352. Table 3 provides the qualitative metric used for field data characterization.

Observations from the field testing indicated:

- *Gas in the Vadose zone*–Low to moderate gas concentrations
- *Gas in the Asphalt Impacted Soils*—High to very high concentrations (including flameout) of methane equivalent (below a depth of 7 meters in M-351 and 12 meters in M-352) and low concentrations of hydrogen sulfide throughout both borings with occasional high concentrations associated with groundwater in flux chamber.
- *Gas in the spoil materials*—Low to moderate concentrations in spoil tested at the surface.
- *Gas in groundwater bailed from the borings*—Low methane concentrations and low to moderate hydrogen sulfide exsolving from groundwater when tested after a time period of 15 to 20 minutes.

Gas samples analyzed in the laboratory provide a quantitative estimate of gas emissions from the formation at the time the samples were collected. The results do not give peak emission rates that may occur immediately the ground is disturbed and before sampling. They also do not represent long-term offgassing results. They are a product of the boring and sampling techniques used, they reflect conditions within the boring at the time of sampling (i.e., presence of groundwater or asphalt), the disturbance of the ground, and the specific gases analyzed.

**Table 3. Qualitative designation of gas concentrations for flux chamber testing**

| Gas Concentration | | Designation |
|---|---|---|
| **For LEL based on Methane Equivalent (LEL=10,000 ppm, UEL=150,000 ppm)** | | |
| 0 to 2,500 ppm | 0 to 5% LEL | Low |
| 2,500 ppm to 10,000 ppm | 5% LEL to 20% LEL | Mod |
| 10,000 ppm to 50,000 ppm | 20% LEL to 100% LEL | High |
| 50,000 ppm to 100,000 ppm | 100% LEL to 10% methane | Very high |
| Greater than 100,000 ppm (10% methane) | >10% methane | Flame-out |
| **For Hydrogen Sulfide** | | |
| 0 to 5 ppm | | Low |
| 5 to 10 ppm | | Mod |
| 10 to 20 ppm | | High |
| Greater than 20 ppm | | Very high |

Generally, methane flux increased with depth but there was no apparent correlation between ground conditions and gas flux levels. Maximum methane flux was 100,850 mg/m$^2$/min at a depth of 25 meters in Boring M-351 and 97,490 mg/m$^2$/min at a depth of 17 meters in Boring M-352. The dominant hydrocarbon range measured in samples in Boring M-351 was C8 (e.g., octane), which is typically associated with gasoline or refined petroleum product. Samples from Boring M-352 were primarily in the hydrocarbon range of C11+ (e.g., heavy organic petroleum compounds, waxes, asphalt and tar) and were aromatic and paraffin in character (Boduszynski et al., 1998). It is hypothesized that the compounds in Boring M-351 are associated with a gas station from 1950s that occupied the site; whereas the gas emissions from Boring M-352 are more representative of the flux from natural asphalt impacted soils.

Reduced sulfur compounds were measured intermittently (present in three out of nine samples tested from the two borings) at shallow depths with hydrogen sulfide being the dominant reduced sulfur compound. The maximum hydrogen sulfide flux was 18 mg/m$^2$/min and was measured in the sample taken in Boring M-352 at a depth of 10 meters. This sample was taken with the flux chamber flooded with groundwater and is likely to represent off-gassing of the groundwater (from a shallower depth) rather than being representative of the gas level in the ground.

## ESTIMATION OF GAS RELEASES FROM GROUND

The primary source of gas appears to be release from the groundwater. Excavation for the underground works will release gas since groundwater pressure will decrease and gas wills exsolve from it. Estimating the volume of gases off-gassing from the excavated material can be made based upon estimating:

- Amount of groundwater in excavated material from soil porosities
- Groundwater pressure based on depth
- Dissolved gas concentrations in equilibrium in groundwater using Henry's Law
- Off-gassing based on the reduction in groundwater pressure due to excavation

## CONCLUSIONS

Gas conditions have been characterized along the WSE alignment in Los Angeles by an extensive gas monitoring program. The program has identified the many issues, particularly associated with sampling and sample contamination by air and has introduced sampling techniques to overcome them. The program has taken measurements that confirm the presence of methane and hydrogen sulfide over much of the alignment and the correlation of these areas with the methane zones identified by the City of Los Angeles.

## ACKNOWLEDGMENTS

The Authors acknowledge the technical input provided by AMEC, GeoKinetics and C.E. Schmidt. They thank the Los Angeles Metropolitan Transit Authority for permission to present this work.

## REFERENCES

Blegen R.P., J.W. Hess and J.E. Denne, 1988, Field comparison of Ground-water sampling Devices, NWWA Second Annual Outdoor Action Conference, Las Vegas, Nevada.

Boduszynskit, M.M., C.E. Rechsteiner, A.S.G Shafizadeh and R.M.K. Carlson, 1998, Composition and Properties of Heavy Crudes, UNITAR Center for Heavy Crude and Tar Sands.

City of Los Angeles, 2004, Bureau of Engineering, Department of Public Works, Methane and Methane Buffer Zones.

Elioff M.A., T.P. Smirnoff, P.F. Ryan, J.B. Putnam and B.M.Ghadiali, 1995 Geotchnical Investigations and Design Alternatives for Tunneling in the Presence of Hydrogen Sulfide—Los Angeles, RETC.

GeoKinetics, 2012, Internal Communication.

Hamilton D.H and R.L. Meehan, 1992, Cause of the 1985 Ross Store Explosion and Other Gas Ventings, Fairfax District, Los Angeles, Engineering Geology Practice in Southern California, Association of Engineering Geologists, Special Publication No. 4, 1992.

Kim J.S., and D.E. Crowley, 2007, Microbial diversity in natural asphalts of the Rancho La Brea Tar Pits, Applied and Environmental Microbiology, pp. 4,579–91.

Metro 2012, Los Angeles County Transportation Authority, Geotechnical Baseline Report for Exploratory Shaft.

Schmidt C.E. and F. Fong, 2013, Internal Technical Memorandum, Summary of Testing Results for Air Pathway Analysis Conducted at the Exploratory Shaft.

US EPA, 1986, User's Guide for the Measurement of Gaseous Emissions from Subsurface Wastes using a Downhole Flux Chamber, Office of Research and Development, Cincinnati, Ohio.

# Development and Maintenance of a High Level Safety Program and Culture in the Highly Challenging New York City Labor Market

**Mike Ceglio**
Skanska USA Civil Northeast

**Mike Vigna**
Hirani Engineering/Consultant Construction Management Team

**ABSTRACT:** Stakeholder's expectations were high for a safe and successful outcome going into to the construction phase for a project involving the mining of a station cavern in a densely populated neighborhood in New York City. Initial preparations from the owner MTA, required an assessment of the contractors safety personnel qualifications and overall safety program prior to the commencement of work. In addition, the contractor, Skanska USA Civil, in joint venture with Traylor Brothers, performed a complete hazard analysis of all risks involved using nationally recognized certification programs in safety (OHSAS 18001) and environmental ISO 14001. A key component to the success throughout the project was the open relationship that was maintained between the NYC unions, the client, and the contractor. In addition there were many elements that influenced the success of safety performance on the job. The client would hold daily site inspections with the contractor as well as hold monthly meetings with all contracts that made up the 2nd Avenue subway project to discuss project compliance and lessons learned. The contractor had a safety program that was driven from the top leadership down to craft and had an extensive training program for all employees. Another factor that played an important role was the buy in towards safety from the craft employees. Lastly the project had a strong relationship with the FDNY (Fire Department of New York) for preparations in fire and life safety.

## INTRODUCTION

New York City Transit (NYCT), for the first time in over sixty years, is expanding their subway system with the three-phase Second Avenue Subway (SAS) Project. The first phase of the project, includes new tunnels from 105th Street to 63rd Street, with new stations at 96th, 86th, and 72nd Streets, and new entrances to the existing Lexington Avenue/63rd Street Station at 63rd Street and Third Avenue. The 86th Street Station Cavern Mining and Heavy Civil / Structural Contract now under construction includes the removal of approximately 155,000 bank cubic yards (BCY) of rock, and the placement of the permanent Station concrete lining. See Figure 1.

The Metropolitan Transportation Authority Capital Construction (MTACC) is the owner and the design engineer is the joint venture AECOM/Arup. The consultant construction manager is Parsons Brinckerhoff (PB/CCM), and the contractor is Skanska/Traylor joint venture (STJV). The estimated daily ridership for Phase 1 is expected to be 213,000 with a target completion date of December 2016, at a cost of $4.451 Billion.

### Phase 1 Overview

During Phase 1, there are four concurrent station cavern construction contracts in progress at 96th, 86th, 72nd, and 63rd Streets. However, the first contract consisted of two, 6.7m diameter parallel tunnels located along Second Avenue which were mined by tunnel boring machines (TBM's). The total mined length was 3,901m; the S1 (West) tunnel was 2,377m, and the S2 (East) tunnel was 1,524m.

The 86th Street Station extends between 83rd and 87th Streets, and includes north and south entrances, and two ancillary buildings housing station ventilation equipment. Contract C-26008 (5B), the subjuect of this paper, covers the mining of a 286.5m long rock cavern for the station at 86th Street and heavy civil structural work. The arrangment of the underground spaces for station, entrance structures and ancillary buildings is shown above in Figure 2. The Contract was awarded to a Joint Venture of Skankska USA Civil Northeast and Traylor Brothers (STJV) in August, 2011 at a cost of $302 Milion. The contract duration is 37 months and consists of excavating 122,300 cubic meters of rock, spraying 15,290 cubic meters of shotcrete, installing 3,175,000 kg of reinforcing steel and 53,520 cubic meters of structural concrete.

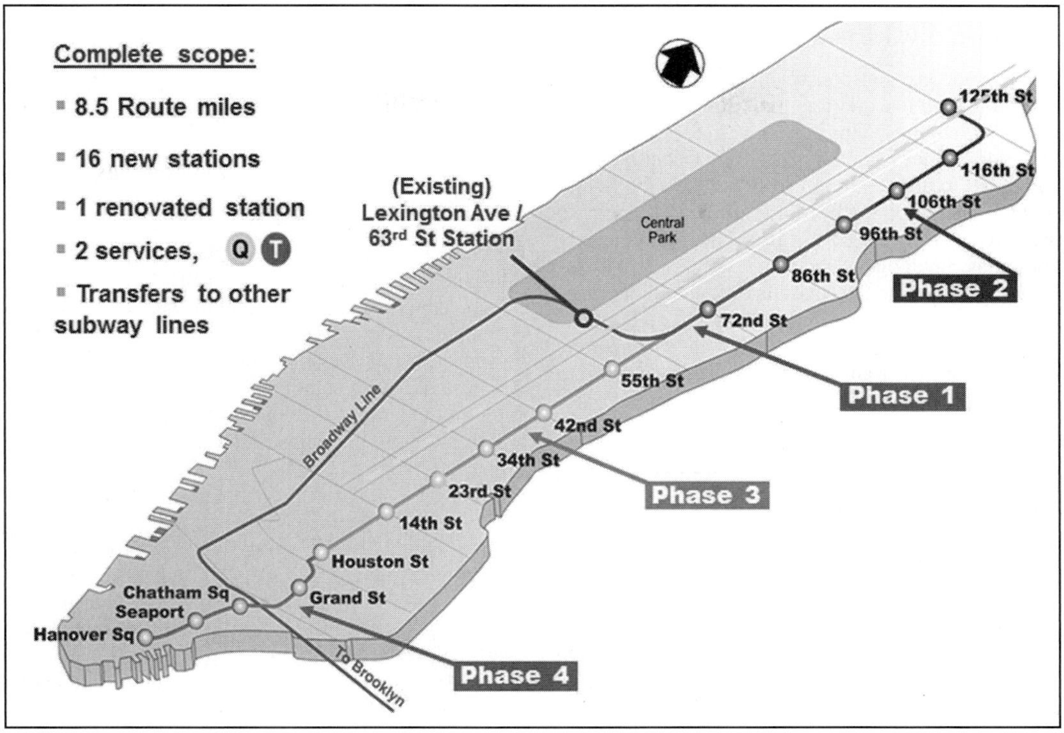


Figure 1. Second Avenue Subway—project plan

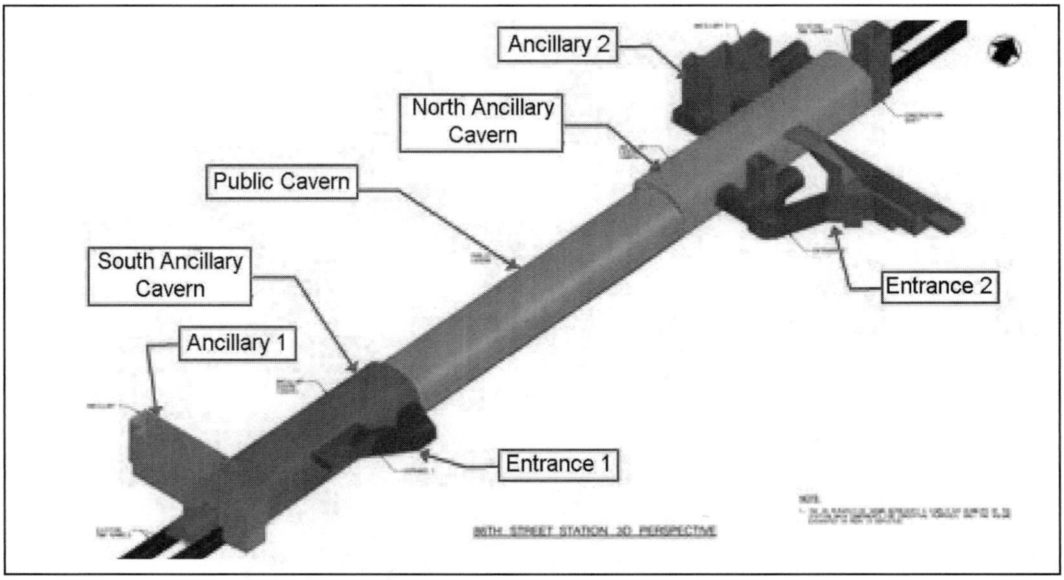


Figure 2. 86th Street Station layout

### Historical Background of the NYC Underground Construction Work Force

Early Underground construction for tunnels and deep foundation work was, and still is considered, very challenging to perform even under the best of circumstances and conditions, especially when compared to above ground work. In New York City (NYC), this work has been traditionally assigned to the Tunnel Workers of Local 147 who are widely known as Sandhogs." The sandhogs were responsible for constructing the caissons for the foundation of the Brooklyn Bridge as well as the hand dug tunnels for early sections of the NYC subway system. The work was difficult and extremely hazardous due to the means and methods available at that time. Worker injuries and deaths were common and, unfortunately, an expected occurrence (See Figure 3). Today, Local 147 sandhogs are still the primary labor force used to mine and excavate the shafts, tunnels and caverns in NYC including the 2nd Ave Subway project.

DARING SAND HOG FALLS TO HIS DEATH

Apert's Body Lay for Hours at the Bottom of a Caisson in the McAdoo Tunnel.

WORKERS LAUGH AT MISHAP

Dangers of Their Calling So Great and Accidents So Common That They Are Indifferent.

**Figure 3. NY Times, May 20, 1907**

### Cavern Excavation Means and Methods

#### *Sequential Excavation Method (SEM)*

The cavern was mined using the sequential excavation method (SEM). The main SEM characteristics include a defined round length, support measures (including shotcrete), multiple drifts/headings, with support installed every round, pre or localized support, and instrumentation.

After sinking the construction shafts at 83rd and 87th Streets, the top heading was drilled and blasted. The top heading was divided into three sections the center pilot, and east and west slashes Numbers 1–3, followed by intermediate bench No. 4, and bottom bench No. 5. See Figure 4. Two main top headings were excavated from the north and south towards one another. During excavation of the intermediate bench, shot rock fell into the previously mined tunnels below.

At Skanska USA Civil safety comes first above all else. Safety is driven by top management to help achieve the company's ultimate goal of zero accidents. "I understand the level of commitment required to achieve zero, and I know it's not easy. That being said, in my time at Skanska no matter what challenges we have faced, we've always come through on top because our people are dedicated to strengthening this company and its value. So I challenge all of you and myself to "Lead in Safety." Leading in Safety means never walking by an unsafe condition without having it corrected;asking our craft if the operation they are working on has any aspects that they feel are unsafe; not tolerating behavior that doesn't fit our safety mission; and remaining firm in

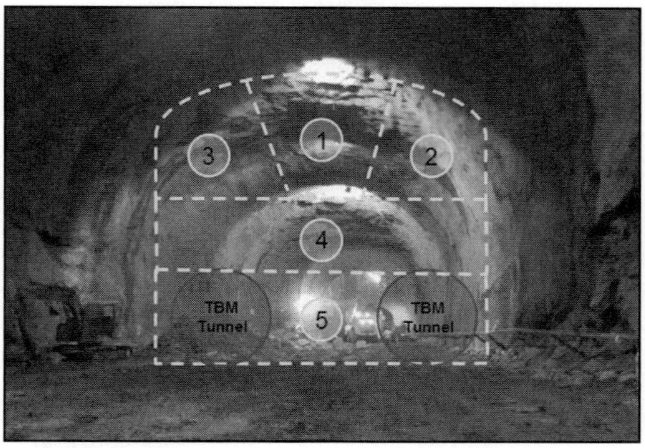


**Figure 4. Sequential excavation sequence**

our commitment to an injury–ree environment. We will never achieve zero if we have just one safety leader. Achieving zero can only happen if we are all safety leaders. Be leaders in safety not just at work, but at home. Safety should be a way of life for all of us, every day. If we can deliver on that aspiration, the generations to follow will only know the safe way of working and we will all have achieved something we can all be proud of. You have my unwaivering commitment and support in our quest to achieve ZERO." Richard Cavallaro, President and CEO, Skanska USA Civil.

Skanska/Traylor's Safety program is a nationally recognized program that makes environmental health and safety an integral part of everyone's work day. Before the 86th Street Cavern project started, the contractor joint venture partners sat down to develop a site specific safety plan. The team identified all environmental aspects and hazards the project would face during the construction phase and controls were put in place to mitigate the risks involved. Four aspects/hazards are picked for review on a monthly basis with the management team to gauge how well the program is being implemented on the job. This is an opportunity to make any changes or additions to the program and to discuss what is working and what is not.

In addition to the monthly management review meeting, a monthly safety committee meeting is conducted with all union representatives on the job site and STJV management. This is an opportunity to get feedback from the craft and again to find out what is working for the team and to identify areas that we need to improve on. It also gives an opportunity to review any incidents from the month and to communicate any corrective actions.

In order to monitor and ensure compliance on the job, daily site walkthroughs with the owner representatives,Owner Controlled Insurance Program (OCIP) representatives, and construction managers are conducted. Deficiencies are either corrected and abated immediately or the issue is brought to the team so a plan can be put in place to eliminate the issue.

Communication to the craft is the most important way to promote safety on the job. STJV has daily safety meetings before every shift to discuss any compliance issues from the prior day and to communicate safety awareness on the job. All foremen, walking bosses, supervisors, safety managers, and project managers attend these meetings. Immediately following the meetings, all foremen conduct a documented daily job briefing with their crews on what was discussed in the meeting and any hazards that are identified on their task specific safety plan. If the foremen or crew identifies any additional risks before starting work, this will also be a part of the daily job briefing.

On top of the daily meeting, every Monday morning all craft employees are addressed by the safety department with a weekly "toolbox talk" or "in-field training session" that is most relevant to the work being performed at the time. Topic's can include but are not limited to, skil saw safety, power tool safety, fall protection, ladders and compressed gas safety. During these training session the entire underground work force is addressed with a mega phone and visual aids and or physical demonstrations are peformed so the craft could witness safe use and practices in action. Through trial and error, this method of safety training proved to be more effective then classroom training. Immediately following, the project personnel break out into a group "Stretch and Flex" sessions to warm up the body before the day's work. The "Stretch and Flex" program is encouraged but not mandatory.

Skanska's safety vision is "To be the world leader in construction-related safety performance with consistent improvements towards an injury-free workplace." The company's ultimate goal is "zero accidents" and before starting the 86th Street project the company was well aware of the reputation of the underground safety culture and knew that in order to improve safety, a tremendous change would have to be made. The most obvious change was a complete transformation of the underground construction culture.

As you've read in the paper above, safety has been a top priority for the underground construction workers in New York City. STJV communicated this message to the Local 147 union delegates who agreed that safety on the job needed to be improved and they were willing to cooperate and work with the joint venture to make this possible. The main focus area with this initiative was "training." Before the first underground operation started, the crews were brought to the job early to learn about the STJV's safety culture and to undergo a series of mandatory training sessions to work on the project. Training sessions consisted of jobsite orientations, general awareness and competent person training in numerous OSHA subparts, aerial lift operator training, certificates of fitness from the Fire Department of New York, and crane signal and rigging training. In addition, all workers were given a 2-hour "Injury Free Environment" training conducted by the project executives on the project. "Injury Free Environment" training consists of delivering a message about bringing safety to a personal level and getting to know the people around you. "IFE" is about making safety a

**Figure 5. Wet scrubber unit**

part of your daily routine, not only at work but at home as well. It is about letting your co-worker know that you care for them and you care about their safety. It's about looking out for one another and believing that zero injuries is an attainable goal. This is a type of training that brings safety to another level and sets STJV apart from other companies in the industry.

In the early stages of the job there was definitely some apprehension by the Sandhogs because it was the first time they have ever worked for Skanska/Traylor JV, and to some it was unclear as to why the company cared so much about safety. Most were not used to safety being such a big part of their daily routine. Others assumed the company preached safety because it brought down their insurance premiums. Over time, through building relationships and being consistent with the message that safety is our "number one priority," the Sandhogs started to believe that the company cared about the well-being of their employees and truly believed that "zero accidents" was achievable.

In order to ensure worker safety during the blasting, rock excavation, and shotcreting stages of the project, consistent communication, training, and innovative equipment and technology were implemented. During the blasting phase, a wet scrubber unit was installed and used to vent the dust and fumes and clear the air. This was beneficial to both the worker's and the community. Figure 5 shows a picture of the unit used to pull contaminated air in, filter and wet down airborne particulates and exhaust clean scrubbed air into the atmosphere. Figure 6 shows the entire system with the fresh air intake in place.

The project also created "The Green Light Procedure" for preventing project staff and workers from re-entering the blast location before safe air levels were achieved. This was a culture shock to most underground workers that were accustomed to re-entering the space immediately following a blast or staying underground during a blast. Workers were prohibited from re-entering the blast location until carbon monoxide levels were at or below the OSHA Permissible Exposure Limit PEL. Safety personnel would send the gas meter down the shaft on a rope and then pull it up to determine the air quality underground. This process would continue until a safe air quality was achieved and the "Green Light" was given to underground crews to return back to work.

During the rock excavation stage, rock support was essential in ensuring worker safety. Because there were three shifts operating during this phase of the project, communication from one shift to the next played a crucial role in keeping everyone abreast of bad rock conditions and unprotected areas. The project safety protocol also mandated that scaling would take place on every shift to make certain all workers were protected from falling rock. During the drilling stages it was imperative that crews created an exclusion zone around the boom of the drill rigs to prevent workers from being struck by falling debris since the vibration often loosened the surrounding rock mass.

The shotcrete phase introduced a new set of overhead hazards. The biggest one being falling shotcrete during and after application. To mitigate

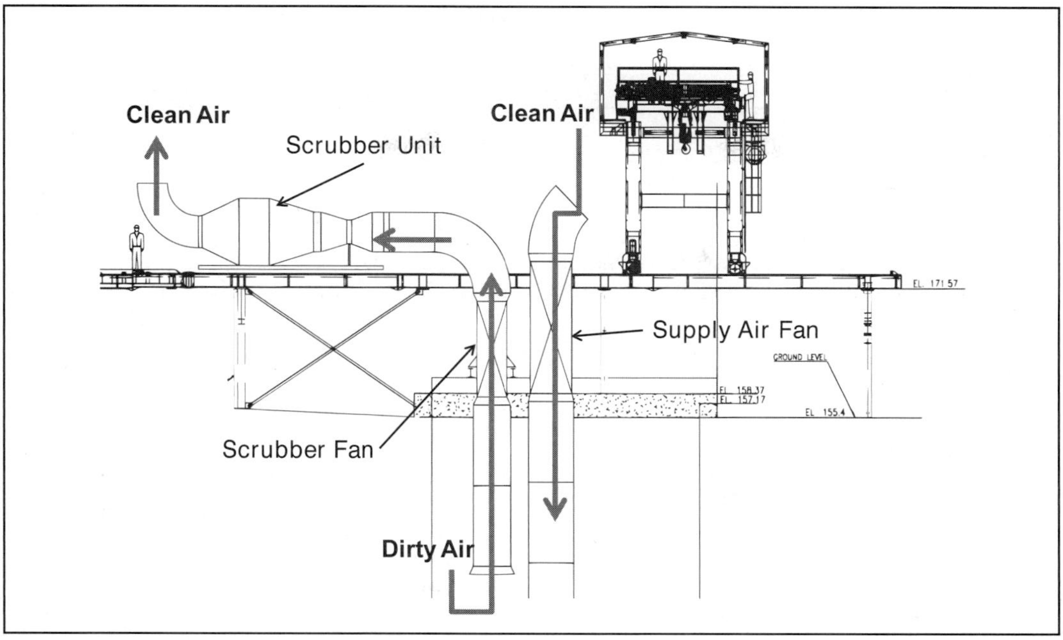


**Figure 6. Wet scrubber system with clean air intake**

the risk a procedure was put in place to barricade off the areas around overhead spraying and for the crew to work themselves away from the area of operation to ensure they were not exposed. The MTA also mandated a 90 minute waiting period for all overhead sprayed shotcrete. This meant that no one was allowed to enter under a freshly sprayed area for 90 minutes after application. Since this was a new set of guidelines for underground construction the team came up with a plan to cut the cavern down the middle and spray on one side the other side could act as safe access.

Silica exposures were another type of hazard that had to be dealt with during the shotcrete phase. All employees received medical evaluation questionnaires which needed to be approved by a physician before anyone was fitted in a respirator. Dust boss's (Large dust suppression fans powered by electric and water) were strategically placed in relation to where the application was for a given day. These units drastically reduced the airborne particulates generated during this type of operation. Continuous monitoring for silica exposures were taken to determine areas that required respiratory protection and areas that did not. Zones were set up using traffic barrels and marking tape to delineate the areas. As discussed, the wet scrubber system played an important role in keeping dust levels down.

## Overview of the MTACC Contractor Safety Requirements and Policy and the Oversight Function by the CCM

MTACC has strict safety requirements for all contract work on the SAS project. It mandates the safety and security of the public, private property, MTA employees, and Contractor employees as well as their subcontractors operate in compliance with federal OSHA regulations, New York State Uniform Fire Prevention and Building code, federal Environmental Protection Administration (EPA), National Fire Protection Association (NFPA) as well as other state and local regulations. MTACC has tasked the oversight function to the Consultant Construction Management (CCM) team which is headed by Parsons Brinkerhoff and selected subconsultants. In addition to monitoring compliance with the above stated regulations, the CCM Safety Manager has the responsibility of providing basic oversight tasks that monitor the safety performance of the Contractor through the duration of the work. The following are examples of the tasks that are performed in this function:

- Review of Contractor submitted Safe Work Plans (SWP). A safe work plan is required for individual and significant construction activities that will be implemented throughout the course of the work. The SWP encapsulates

**Table 1. MTACC quarterly safety audit categories—Figure 7**

| CATEGORIES | | | |
|---|---|---|---|
| **MANAGEMENT** | | | **Percentage** |
| | Points | | |
| Category 1 | 865 | Supervision / Organization | 100.00% |
| Category 2 | 455 | Programs: Accident Prevention & HAZCOM | 100.00% |
| | | | |
| **MEANS & METHODS** | | | |
| Category 3 | 990 | General Safety / Housekeeping | 100.00% |
| Category 4 | 485 | Motor Vehicles / Heavy Equipment | 100.00% |
| Category 5 | 395 | Barricades | 100.00% |
| Category 6 | 1280 | Cranes / Derricks / Hoists / Conveyors | 100.00% |
| Category 7 | 820 | Scaffolds & Man Lifts | 100.00% |
| Category 8 | 670 | Ladders / Stairways | 100.00% |
| Category 9 | 630 | Fall Protection | 100.00% |
| Category 10 | 595 | Tools (Hand and Power) | 100.00% |
| Category 11 | 690 | Fire Protection / Prevention | 100.00% |
| Category 12 | 240 | Lockout / Tag Out | 100.00% |
| | | | |
| **OPERATIONS** | | | |
| Category 13 | 445 | Demolition | 100.00% |
| Category 14 | 415 | Excavations | 100.00% |
| Category 15 | 160 | Concrete and Masonry Construction | 100.00% |
| Category 16 | 700 | Steel Erection | 100.00% |
| Category 17 | 740 | Welding and Cutting | 100.00% |
| Category 18 | 780 | Electrical | 100.00% |
| Category 19 | 1010 | Track Safety | 100.00% |
| | | | |
| **SPECIALS** | | | |
| Category 20 | 370 | Slings and Rigging Hardware | 100.00% |
| Category 21 | 350 | Confined Space Operations | 100.00% |
| Category 22 | 1405 | Underground Construction: Caissons, Cofferdams & Compressed Air | 100.00% |
| Category 23 | 605 | Blasting and Use of Explosives | 100.00% |
| Category 24 | 550 | Diving Operations & Marine Work | 100.00% |
| Category 25 | 385 | Power Transmission and Distribution | 100.00% |
| Category 26 | 300 | Recordable Injuries | 100.00% |

the basic procedures that are required for workers to follow in order to perform the activity safely. A review by the CCM prior to commence of work is performed to determine if all elements are adequately addressed for worker protection. Once the SWP has been approved by the CCM, it is subject to a joint assessment in the field by members of the Contractor, CCM Safety and sometimes the MTACC depending on the scope of the activity to be performed. This step is taken to ensure that the safety elements implemented by the Contractor are viable and effective. Any findings during obtained during the field assessment can be applied to the plan to improve its effectiveness.

- Review of work related injuries and near miss reporting are critical in order to prevent similar events from reoccurring. The Contractor submitted injury and near miss reports are reviewed for root cause analysis and corrective measures that will be implemented to avoid similar events. A post-accident review meeting for serious especially those injuries resulting in a worker's time away from the job due to injury are conducted with the Contractor's Project Manager and CCM Safety Manager. The result of the meeting is that all parties come away with a clear expectation of how an operation will proceed going forward.
- The Contractor's overall safety program is subject to quarterly safety audits that measure adherence to the Projects safety requirements as well as measuring the Contractor management's participation within their own

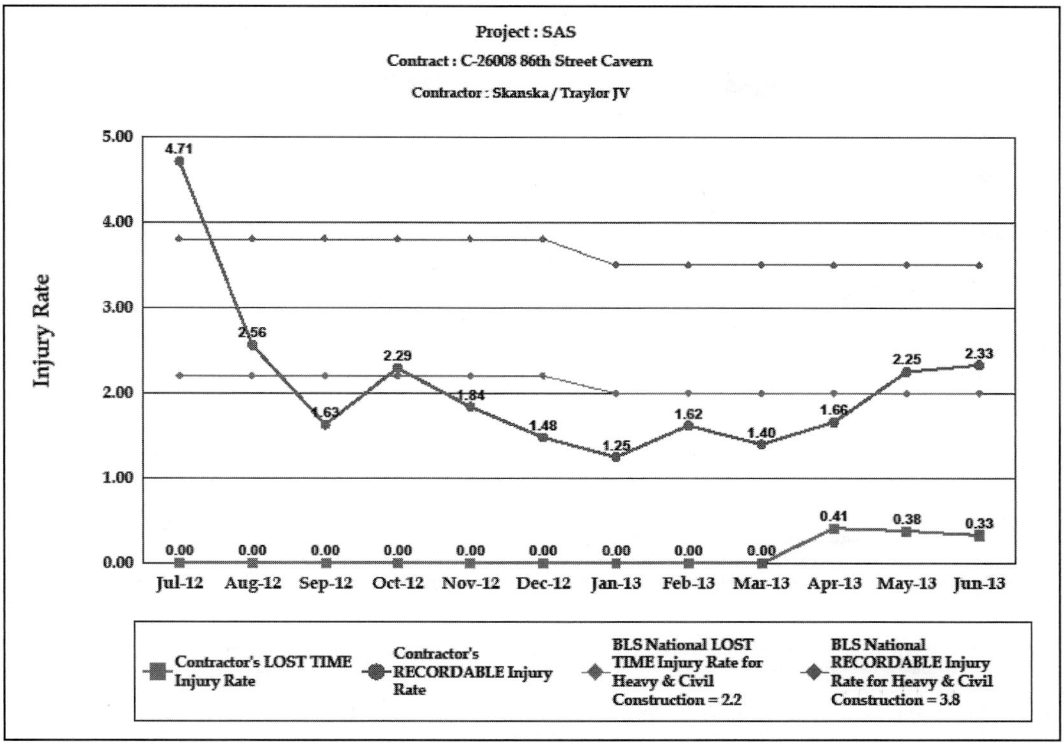


**Figure 7. Cumulative profiles of lost time and recordable injury rates**

program. The audit is broken down into relevant categories in order to accurately assess the work performed during the previous quarter. Table 1 depicts the parameters that are reviewed during the audit process.

- The CCM conducts a monthly safety meeting with all construction contractors performing work on the Project. The meeting is an opportunity for contractors to share experiences, both positive and negative, with others and serves as a lessons learned forum. Also, monthly performance data for work related injuries is shared and further discussion may take place dependent upon any upward trending in injuries occurring.

Both the contractor and owner's representatives had a strong working relationship with the Fire Department of New York which helped address major life safety concerns on the project. As part of the contract, the contractor was required to install and maintain a functional fire standpipe to the underground operations. Throughout the blasting phase this was not feasible because of the scope of work, the FDNY gave variances which allowed the contractor to keep 125 LB. portable fire extinguisher units spread out throughout the jobsite. All mining equipment was also equipped with ANSUL systems (Onboard fire suppression system.) A benefit to the blasting phase from a fire protection standpoint was there was very little fire load underground. Heavy equipment and electrical panels were the only things with the potential to burn, the remaining material was bedrock and blasted rock. Once the construction phase allowed, the contractor installed 10.16 CM steel and pvc pipe from street level to the underground operations. A FDNY connection was required every 91.44 M in order to comply with code compliance. The system had to be tested by a licensed plumber and had to hold 200 psi for 2hrs. The FDNY also performed monthly site inspections to keep up to date with any changes in access and to ensure they stayed familiar with site conditions in case a rescue was needed. The FDNY also attended project wide safety committee meetings on a monthly basis to communicate compliance issues and to keep everyone up to date with code.

Contractor safety performance is monitored on a monthly basis. Resulting trends in worker injury rates are tracked for both OSHA classified recordable injuries and recordable injuries resulting in days away from work (lost time). Customized

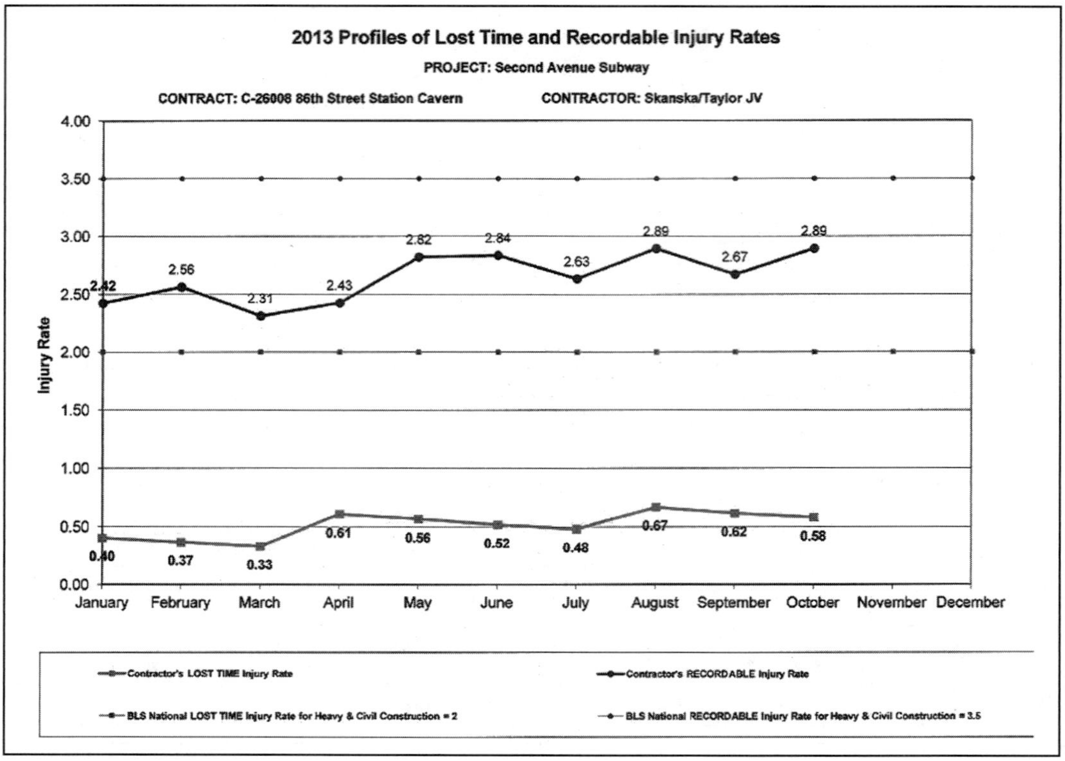


**Figure 8. Contract injury rates**

spreadsheets and graphs illustrating injury trends are included in monthly reports generated by MTACC. This provides thecontractor Project Management team with an illustration of the project safety performance for a period of time (see Figure 7).

## CLOSING REMARKS

All in all, the approach to safety on this project was a team based effort. Through collaboration by contractor, owner, OCIP and CCM the team was able to solve difficult challenges and deliver a safe productive project to date. The project set off over 600 blast attempts without incident and has worked over 1.2 million man-hours keeping injury rates far below industry standards (see Figure 8). The team understood the challenges that we would face on a project of this caliber and have taken a proactive approach to ensure a safe and successful completion. The local 147 sandhog union embraced the safety culture promoted on the job and the team looks forward to heading up additional contracts related to this mega project.

# Geotechnical Planning Advances for TBM Projects in Glacial Geology

**Ulf Gwildis**
CDM Smith, Inc.

**Robert A. Robinson**
Shannon & Wilson, Inc.

**Richard Sage**
Sound Transit

**David Sowers**
Washington State Department of Transportation

**ABSTRACT:** Over the past three decades, a dozen TBM-driven tunnels have been constructed in the Seattle metropolitan area with its unique geological setting of interlayered glacial, glaciomarine, and non-glacial deposits. Several new tunnels—among them the Alaskan Way Viaduct Replacement Tunnel utilizing the world's largest-diameter TBM manufactured to date—are under design or construction. Experience gained from each of these projects resulted in new and innovative approaches in planning, design, contractual description of subsoil conditions, and formulation of contractual requirements for future tunnel drives. This paper analyzes the different approaches and outlines trends in dealing with the specific challenges of tunneling in complex glacial geology.

## TBM PROJECTS IN SEATTLE

The pronounced topography of the Seattle metropolitan area is one of the reasons for its rich history of tunneling projects over the past 130 years (ROBINSON et al. 2002, 2013). The past decades saw the use of pressurized face tunnel boring machines (TBM)—both earth pressure balance machines (EPBM) and slurry machines (STBM)—leading to the planning and construction of increasingly complex projects in terms of tunnel length, diameter, depth, interfacing with other system components such as underground rail stations and pump shafts, and impact potential of existing structures and groundwater resources. These projects include the tunnels of Sound Transit's light rail system, King County's Brightwater sewer conveyance system, and the Alaskan Way Viaduct Replacement Tunnel, a two-deck highway tunnel underneath Seattle's downtown area by the Washington Department of Transportation (Figures 1 and 2, Table 1). As the common denominator of these tunnels, glacially overconsolidated glacial and interglacial deposits and their specific characteristics—variability of soil types with different geomechanical tunneling behavior, frequent occurrence of boulders, in part high soil abrasiveness, complex hydrogeologic conditions to name a few—constitute unique challenges to TBM operations. This makes the Seattle area a focal point of new approaches in exploring and characterizing geotechnical conditions and specifying construction for mechanized tunneling contracts. Each project has contributed innovative solutions to unique challenges, which—combined with technical advances in TBM, instrumentation, and data processing technology—increased the knowledge base benefitting future tunnel projects.

## GLACIAL GEOLOGY

Glacial geology studies geologic features formed in conjunction with the movement of glaciers and related processes. These processes include abrasion, erosion, material transport and deposition by ice, meltwater, and wind, taking place in different environments ranging from terrestrial, fluvial, and lacustrine to marine. Material transport by glaciers and deposition as till leaves the soil components poorly sorted and little rounded, resulting in a material mix ranging from silt to boulder fraction, the coarse components often embedded in a finer grained matrix. Boulders being rock masses exhibiting the greatest resistance against the destructive forces during glacial transport typically have high strength. Boulders

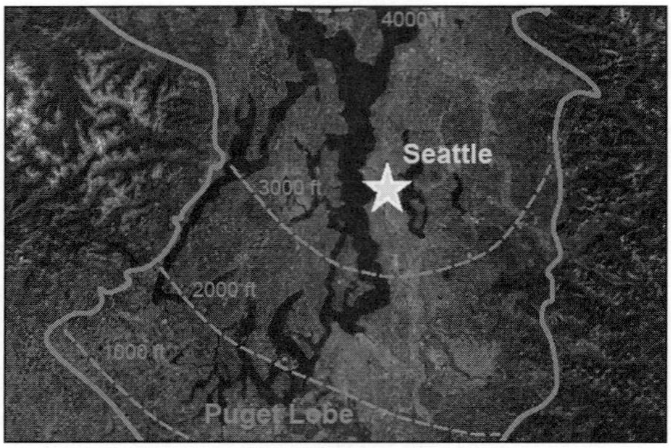


Figure 1. Puget Sound region with extent and thickness of glacial lobe (Vashon glaciation)

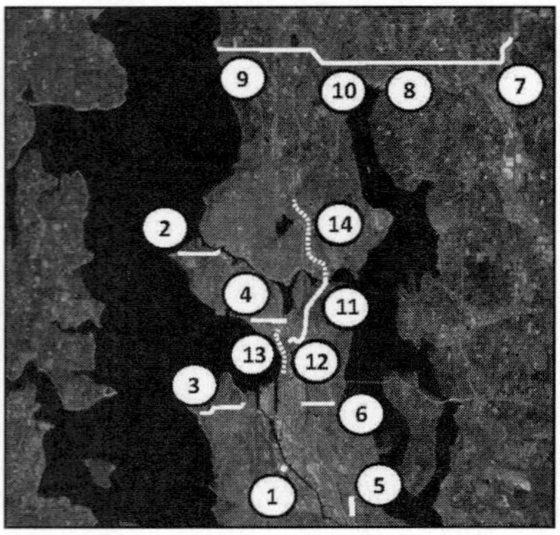


Figure 2. Seattle metropolitan area with TBM tunnel alignments

**Table 1. Seattle area TBM tunnel construction contracts**

| | Project | Function | Year of Completion | TBM Type | Number of Drives | Length (m) | Diameter (m) |
|---|---|---|---|---|---|---|---|
| 1 | Renton Sewer Tunnel ETS-6 | Wastewater | 1986 | EPBM* | 1 | 322 | 3.7 |
| 2 | Fort Lawton Parallel Tunnel | Wastewater | 1991 | EPBM* | 1 | 2,542 | 4.7 |
| 3 | West Seattle Tunnel | Wastewater | 1997 | EPBM* | 1 | 3,128 | 4 |
| 4 | Denny Way CSO | Storage | 2002 | EPBM | 1 | 1,890 | 5 |
| 5 | Henderson CSO | Storage | 2002 | EPBM | 1 | 946 | 5.1 |
| 6 | Beacon Hill Transit Tunnel | Lightrail | 2008 | EPBM | 2 | 2 × 1,310 | 6.5 |
| 7 | Brightwater East Contract | Wastewater | 2008 | EPBM | 1 | 4,231 | 5.9 |
| 8 | Brightwater Central Contract | Wastewater | 2011 | STBM | 2 | 6,651 | 5.4 |
| 9 | Brightwater West Contract | Wastewater | 2010 | EPBM | 1 | 6,424 | 4.7 |
| 10 | Brightwater BT3C Contract | Wastewater | 2011 | EPBM | 1 | 3,018 | 4.9 |
| 11 | U-Link U220 | Lightrail | 2013 | EPBM | 2 | 2 × 3,475 | 6.6 |
| 12 | U-Link U230 | Lightrail | 2013 | EPBM | 2 | 2 × 1,183 | 6.6 |
| 13 | Alaskan Way Viaduct Tunnel | Highway | On-going | EPBM | 1 | 2,825 | 17.5 |
| 14 | Northgate Link | Lightrail | On-going | EPBM | 6 | 5,617 | 6.5–6.6 |

* Partial EPBM with pressure relieving gate.

and other coarse components can accumulate at geologic contacts where deposition is succeeded by erosional processes. In comparison, material transported by meltwater and deposited in a fluvial environment as coarse grained or in a lacustrine environment as fine grained sediment shows a higher degree of sorting and rounding of its coarser components. Sediments formed in a marine depositional environment may be influenced by glaciers as soil material unloads off floating ice sheets (dropstones). The different stages and depositional environments of the glacial cycle generate various characteristic deposits. Their close proximity, space and time wise, results in a complex distribution and variation of the different soil types. Subsequent glaciation cycles resulting in partial erosion and new deposition add to the geological complexity. Depending on the ice loads, the sediments deposited before the retreat of the latest glacier are often overconsolidated and either very dense or hard.

The Puget Sound area has been shaped by glacial advances from the North and its geologic inventory includes the characteristic deposits of several glacial cycles (Table 2). The steep bluffs along the shoreline attest to the high material strength due to glacial overconsolidation. Their natural outcrops (Figure 3) as well as man-made excavations (Figure 4) reveal the complexity of the regional glacial geology.

## GEOTECHNICAL PLANNING

The process starts at the investigation phase with geological and geotechnical data collection from relevant prior projects and using existing outcrops, remote sensing, exploratory borings, groundwater

**Table 2. Characteristic Pleistocene deposits in the Seattle area**

| Sediment Type | Description |
|---|---|
| Glaciolacustrine deposit | Laminated to massive silt, clayey silt, and silty clay containing scattered lenses of coarser components; deposited in lowland or proglacial lakes; marks transition from glacial to non-glacial |
| Glaciomarine drift | Diamict of glacially derived debris of highly variable composition, often a clay and silt matrix with variable amounts of sand, gravel, cobbles, boulders; deposited in a marine environment |
| Advance outwash | Stratified sand and gravel with cobbles; deposited by streams and rivers flowing out from the front of the advancing ice sheet |
| Till | Compact diamict with a fine-grained to sandy matrix containing subrounded coarser components of gravel, cobble, and boulder fraction; glacially transported and deposited |
| Recessional outwash | Stratified sand and gravel with cobbles, moderately to well sorted, less common silty sand and silt; deposited in broad outwash channels carrying glacial meltwater (fluvial) and in slackwater environments (lacustrine) |
| Interglacial deposits | Clay, silt, sand, gravel, peat, and tephra layers deposited in lacustrine and fluvial environments during interglacial periods |

**Figure 3. Glacial deposits exposed at a Whidbey Island Bluff**

**Figure 4. Glacial deposits exposed in a seismic trench (Insert shows an erosional contact)**

monitoring over several annual cycles, field and laboratory testing, etc. The geotechnical design then provides input for the contract drawings and specifications as well as the geotechnical characterization, which becomes part of the contract documents. The publication 'Geotechnical Baseline Reports for Construction: Suggested Guidelines' (ASCE, 2007), commonly referred to as the 'Gold Book', provides a standard for geotechnical reporting for tunnel projects, expanding the scope of its 1997 predecessor to other underground construction work and by also considering the design-build delivery method. Geotechnical planning includes determining baseline elevations for groundwater, ground surface and existing structures along the tunnel alignment. During construction, data collection continues for quantifying construction impact. Geotechnical construction monitoring usually encompasses tracking the TBM operation and geotechnical conditions encountered. This serves the purpose of verifying design assumptions, exerting control for mitigating the risk of damages, and generating the data base necessary for evaluating unexpected system behavior or claims.

Table 3 lists new approaches and ideas introduced to TBM projects in the Seattle area over the past three decades. These must be viewed within the framework of geotechnical conditions and project planning objectives driving innovation. In the following, the specific challenges of glacial geology tunneling are discussed.

### Face Stability/Overexcavation

Glacially overconsolidated deposits with a significant percentage of fine-grained components such as some tills and lacustrine clays often allow for steep slopes and may exhibit a stand-up time of the unsupported tunnel face under certain conditions. This, however, is not the case for coarse-grained glacial deposits such as outwash, especially beneath the groundwater table, where the tunnel face is likely to

**Table 3. New approaches for Seattle area TBM projects**

| | Project | New Approaches/Innovations | Year |
|---|---|---|---|
| 1 | Renton Sewer Tunnel ETS-6 | Precursor EPBM with pressure relieving gate (no auger) | 1985 |
| 2 | Fort Lawton Parallel Tunnel | EPBM with capacity of mining in full EPB mode (not utilized) | 1989 |
| 3 | West Seattle Tunnel | Use of soil conditioning agents for wear mitigation (and finishing drive) | 1995 |
| | | Use of gasketed concrete liner segments | 1995 |
| 4 | Denny Way CSO | Use of EPBM in full EPB mode | 2001 |
| 6 | Beacon Hill Transit Tunnel | Use of sonic drilling for exploration | 1999 |
| | | Use of belt conveyor scales for tunnel muck | 2006 |
| | | Post-mining evaluation of operational parameters for void mitigation planning | 2009 |
| 7–10 | Brightwater Conveyance System | Exploration included four different drilling methods, soil abrasiveness testing (AVS/SAT), geologic age determination, lab testing for ground freeze design | 2003 |
| | | Geotech baseline using categorized face conditions as percentage of drive length | 2005 |
| | | Use of STBMs (Central Contract) | 2007 |
| | | Development of tracking procedure for face conditions | 2007 |
| | | Use of TBM operational data visualization software by owner | 2007 |
| | | Construction of underground connection of two TBM tunnels (BT3C Contract) | 2011 |
| 11 | U-Link U220 | Continuous face condition tracking and multi-stage TBM data checking by CM | 2011 |
| 12 | U-Link U230 | TBM equipped with gauge cutter wear (overcut) measuring device | 2011 |
| 13 | Alaskan Way Viaduct Replacement Tunnel | Twin-deck highway design resulting in world's largest TBM diameter to date | 2013 |
| | | Design-build TBM project | 2013 |
| | | Test section at start of TBM drive for verifying tunneling procedure | 2013 |
| | | Monitoring of tunnel muck and integrated Bentonite injection for mass balance | 2013 |
| | | TBM equipped with cutters that can be changed in atmospheric conditions | 2013 |

collapse without proper support. Raveling or flowing soil conditions quickly lead to significant overexcavation, resulting in voids above the tunnel and chimneys to the ground surface.

Early Seattle TBM projects (Table 1) specified the ability to operate in closed mode albeit limited to sections where face support would be needed (Fort Lawton, West Seattle); however, contractors preferred the partial EPBM technology with a pressure relieving gate rather than a cased screw auger. This partial EPBM proved ineffective in controlling flowing granular soils and limiting ground loss. Efforts to reduce the impact risk led to the specified requirement of TBM operation in closed mode (Denny Way, Brightwater). Experience of sinkholes and voids detected after completion of the Beacon Hill tunnel drives (Robinson et al., 2012; Figure 5) led to increased scrutiny of real-time monitoring of TBM operational parameters for indication of overexcavation (Brightwater, U-Link). The use of a refurbished EPBM at the BT3C section with high hydrostatic head also required addressing the risk of flowing soils entering in an uncontrolled manner through the conveyor screw, which resulted in modifications of the conveyor screw (added length, added guillotine doors, option of generating a grease plug) and monitoring procedures.

With larger excavation diameters the probability of encountering different soil types in the tunnel

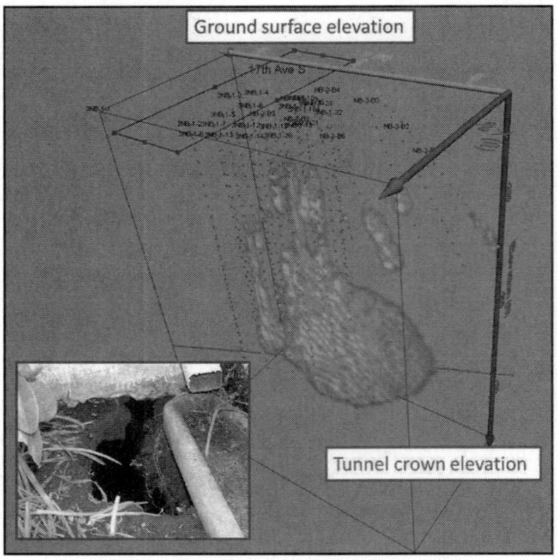


**Figure 5. TBM overexcavation effect at Beacon Hill drive (Insert shows void at ground surface)**

face and thereby the challenge of maintaining a stable face increases. For the Alaskan Way Viaduct Replacement Tunnel, an innovative solution of excavation volume monitoring combined with integrated grout injection was introduced for mitigating the risk of overexcavation (Figure 6). A test section at

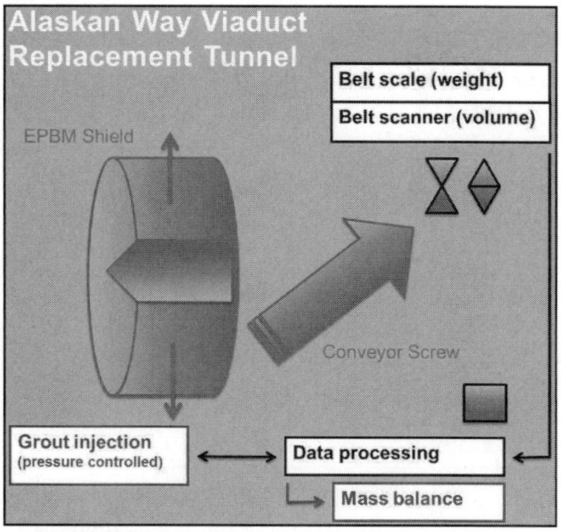


**Figure 6. Excavation volume control and automated Bentonite injection system (Alaskan Way)**

**Figure 7. Tunnel face in soil with stand-up time (Insert shows fast raveling ground)**

the beginning of the drive served for performance verification.

## Variability of Face Conditions

The deposition and erosion processes of subsequent glaciation cycles provide for frequent transitions of soil types, laterally and vertically. This in turn provides for high variability of tunnel face conditions, especially for alignments perpendicular to the direction of glacial advances. An increase of excavation diameters increases the likeliness of so-called mixed face conditions made up of two or more soil types of different tunneling behavior (Figure 7).

While most TBM projects provided geologic profiles with a detailed description of soil conditions (Fort Lawton, Denny Way, Beacon Hill, Alaskan Way Viaduct), more detailed qualitative and quantitative geotechnical baselines were developed where ground conditions were too complex to allow drawing geologic sections. For the Brightwater tunnels, the different soil types were grouped regarding their relevant tunneling behavior characteristics and the resulting groups were used for defining a set of typical tunnel face conditions. Quantification of the tunnel face conditions as percentages of the tunnel drive lengths then provided a baseline (Newby et al. 2008) (Figure 8). This approach was later adopted for other tunnel projects (U-Link).

The variability of face conditions is to be considered for planning the TBM advance and for selecting suitable locations for interventions. Intervention planning may vary depending on TBM and cutterhead type used but should always consider all available geotechnical data including tracking results. Large diameter TBMs such as the one used for the Alaskan Way Viaduct Replacement Tunnel can be designed for allowing inspection and replacement of cutting tools by backloading under atmospheric conditions.

## Face Condition Determination

Pressurized face tunneling allows access to the face only during interventions. However, for comparison with contractual geotechnical baselines continuous tracking may be necessary. A tracking procedure was developed for the Brightwater tunnels (Gwildis et al. 2009) (Figures 9 and 10) and has been adopted at subsequent projects such as the U-Link tunnels. The data base includes spoil samples, TBM operational parameters, inspection reports, groundwater monitoring measurements, etc. Critical elements of the procedure are verifiability of the data sources and reproducibility of the evaluation. Limitations include difficulties differentiating between mixed face conditions during EPBM tunneling, where the homogenization of the muck often requires laboratory index testing and comparison with according baselines for determining the presence or absence of a specific condition.

Irrespective of the resolution that tracking can achieve for a specific TBM project, the tracking results constitute a valuable basis for tasks ranging from planning inspection stops to evaluating the risk of third-party impact and claims. Even when considering a non-geotechnical baseline for tunneling performance such as the measured-mile method, the face conditions encountered need to be taken into account as a significant factor for most mining related processes.

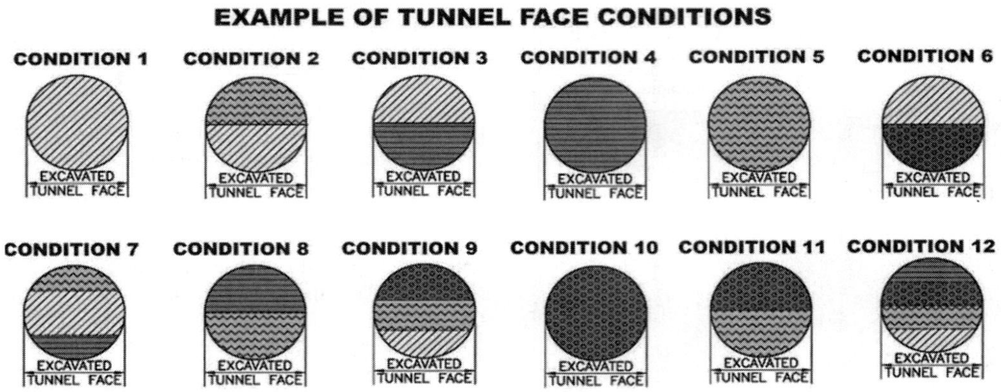


| CONDITION | % OF SECTION | | GRADATION mm | | | | | | | | | |
|---|---|---|---|---|---|---|---|---|---|---|---|---|
| | | | $D_{85}$ | | $D_{60}$ | | $D_{50}$ | | $D_{15}$ | | $D_{10}$ | |
| | BT-2 | BT-3 | MAX | MIN | MAX | MIN | MAX | MIN | MAX | MIN | MAX | MIN |
| 1 | 16-23% | 9-13% | 5 | 0.25 | 0.55 | 0.15 | 0.4 | 0.05 | 0.15 | 0.04 | 0.1 | 0.01 |
| 2 | 23-34% | 3-5% | 2.5 | 0.1 | 0.2 | 0.03 | 0.15 | 0.02 | 0.02 | 0.002 | 0.01 | 0.001 |
| 3 | 6-10% | 3-5% | 2 | 0.1 | 0.25 | 0.04 | 0.15 | 0.03 | 0.02 | 0.005 | 0.007 | 0.002 |
| 4 | 3-5% | 1-2% | 0.3 | 0.05 | 0.08 | 0.02 | 0.05 | 0.015 | 0.007 | 0.002 | 0.003 | 0.001 |
| 5 | 12-18% | 35-50% | 0.6 | 0.02 | 0.05 | 0.003 | 0.03 | 0.0015 | 0.001 | 0.0005 | 0.001 | 0.0001 |
| 6 | 0 | 14-22% | 4 | 0.25 | 0.7 | 0.14 | 0.5 | 0.1 | 0.2 | 0.04 | 0.15 | 0.02 |
| 7 | 3-5% | 4-6% | 2 | 0.1 | 0.25 | 0.04 | 0.15 | 0.03 | 0.02 | 0.002 | 0.005 | 0.0007 |
| 8 | 6-10% | 3-5% | 0.15 | 0.04 | 0.05 | 0.004 | 0.03 | 0.002 | 0.003 | 0.0007 | 0.002 | 0.0005 |
| 9 | 2-4% | 2-4% | 5 | 0.2 | 1 | 0.04 | 0.7 | 0.02 | 0.06 | 0.002 | 0.01 | 0.001 |
| 10 | 5-8% | 5-8% | 40 | 15 | 35 | 6 | 25 | 3 | 6 | 0.03 | 4 | 0.01 |
| 11 | 0 | 2-4% | 40 | 6 | 15 | 0.2 | 3 | 0.03 | 0.008 | 0.001 | 0.004 | 0.0005 |
| 12 | 3-5% | 0 | 25 | 2 | 0.6 | 0.08 | 0.25 | 0.03 | 0.01 | 0.002 | 0.005 | 0.001 |

**Figure 8. Geotechnical baseline approach using typical tunnel face conditions**

**Figure 9. Example of STBM spoil sample showing different phases**

## Boulders

Per the geologic processes described earlier, boulders typically have high strength and mineral hardness, and their occurrence depends on the type of glacial deposit or geologic contact. When encountering boulders, a TBM operation may be impacted, although a cutterhead equipped with disc cutters may grind through a boulder if it is held in place by a strong soil matrix. Up to a certain size, boulders may be broken up by ripper teeth due to impact forces. Manual removal of boulders with hydraulic drills, hydraulic splitting devices, or expanding grout will likely have a significant schedule impact (Figure 11).

During design, a main question is how to determine boulder baselines, which usually include size, number, strength and information on distribution. Due to scale issues, relying solely on data collected with exploratory borings may be insufficient. The results of boulder counts from local case history data, natural outcrops, cuts, or test pits need also be considered within the context of the geologic setting.

During construction, a main question is how to track boulders for comparison with contractual baselines. Other than for isolated events, efforts of reconstructing boulders broken up during the mining process may not be practical. Tracking of coarse components in the tunnel muck, however, will provide information regarding distribution of glacial and non-glacial deposits (Figure 12) assisting in quantifying impact. Other approaches such as the measured-mile method (Edgerton et al. 2012) may have some value in cases where boulder-bearing tunnel sections can be clearly distinguished, boulder impact on the progress rate is significant, and other factors can be quantified.

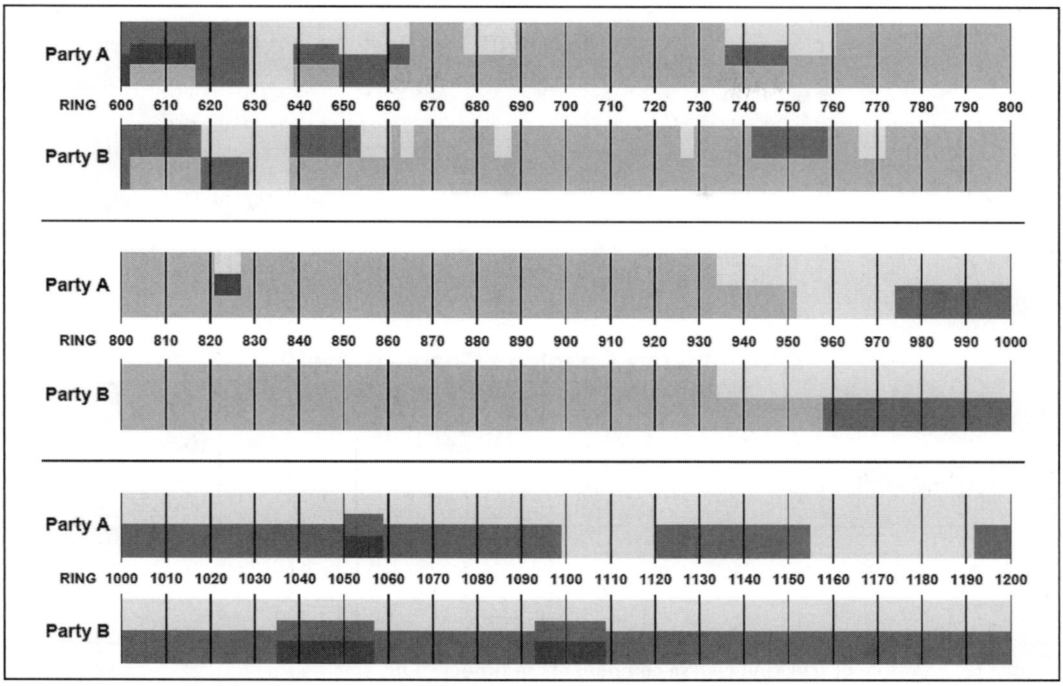


**Figure 10. Example of color-coded face condition tracking by two contract parties**

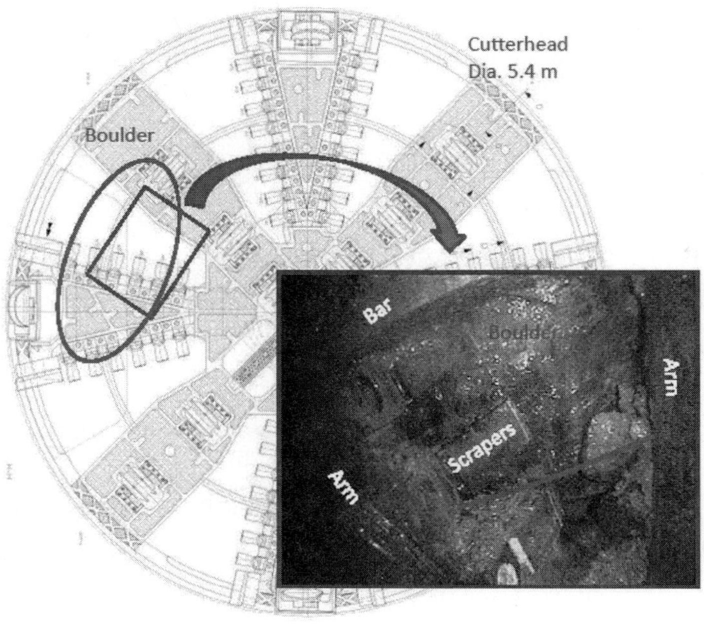


**Figure 11. Example of cutterhead blockage by boulder**

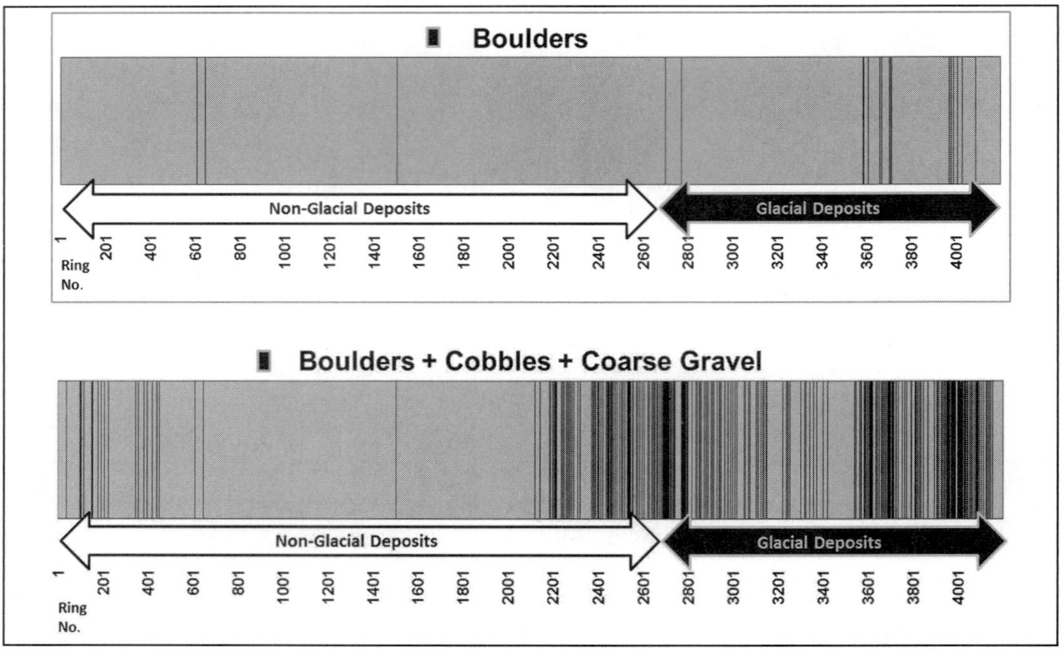


**Figure 12. Example of tracking coarse components in tunnel spoils**

### Soil Abrasiveness/Wear

Some glacial deposits can be highly abrasive due to factors such as the presence of angular particles of high mineral hardness and great bonding strength in a fine-grained matrix. In addition, cobbles and boulders may cause significant wear and damage due to impact forces.

Contract documents of early projects included qualitative warnings of abrasive soils and quantitative information on quartz and feldspars content. Brightwater was the first project using the SAT test, specifically developed for mechanized tunneling, as well as Miller Number testing for providing ranges and averages of values for soil abrasiveness.

Early projects experienced in part heavy TBM wear damage (West Seattle, Denny Way) when not using soil conditioners. Ever since, the use of soil conditioners is seen as an important means of mitigating wear. Regular tool inspections are required in order to prevent wear damage to the TBM (Figure 13). At Brightwater, regular wear measurements and tracking of face conditions enabled correlations between soil types and tool wear, albeit with different results for different tunnel drives and TBM types (Gwildis et al. 2010, Shinouda et al. 2011) (Figure 14).

**Figure 13. Severe wear of cutterhead and cutting tools**

### Soil Stickiness

Fine-grained glacial deposits such as lacustrine clays with high adhesive forces to metal surfaces pose a risk of less efficient mining, slow progress, additional expenses for conditioners, system interruptions due to clogging, etc. While the contract documents of early projects included categorization of fine-grained soils in terms of Atterberg Limits and consistency, more recent projects have provided geotechnical baselines with quantification of the stickiness potential based on Plasticity Index and Liquidity Index.

### Disturbed Zones

Zones of tectonic faulting and a-tectonic movement (e.g., buried landslides) have the potential of impacting TBM operations due to characteristics

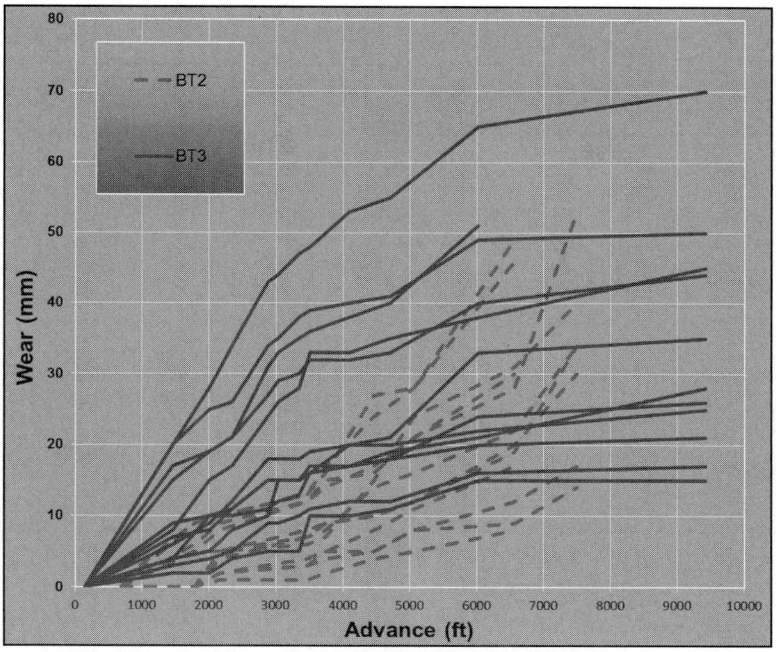


**Figure 14. Wear curves for tool positions of two drives with identical cutterhead assemblies**

such as decreased material strength, altered hydraulic conductivity, offset of geologic units, etc. In addition to logging of indicators of relative movement at the scale of drill cores such as slickensides in clay, the use of remote sensing methods such as LIDAR has become common. Baseline approaches may vary depending on how clearly and accurately disturbed and undisturbed alignment sections can be distinguished.

### Groundwater

Variability of glacial geology means variability of aquifer distribution, laterally and vertically. Tectonic features such as faults and related offsets can provide additional pathways or cut-offs for groundwater flow. Geologic complexity and alignment orientation relative to the direction of glacial advances may determine a suitable spacing of groundwater monitoring points, vertically and along the alignment. The objective of data collection over several annual cycles provides the data base for a continuous hydrostatic head baseline referenced to the tunnel elevation, e.g., the tunnel invert. During tunneling, the baseline can be verified to a certain degree by observing the impact of the TBM operation to nearby groundwater monitoring points. The impact pattern reflects the mining and ring-building cycle and is specific to the type of TBM used, as illustrated by Figures 15 ('bow wave' for an STBM) and 16 (drawdown for an EPBM).

### Other Subsoil Characteristics

Other subsoil characteristics to consider during planning and execution of TBM operations in glacial geology include the accumulation of methane gas from organic non-glacial deposits trapped by low-permeability units, the potential for squeezing ground conditions in deep tunnels, and high horizontal stresses from glacial loading and retreat. Soil pH may need to be considered for muck disposal.

## TRENDS

Technological advances have led to increasingly integrated data bases available for real-time evaluation of the interaction of TBM operation, subsoil conditions, and existing structures. New approaches to contractual description of geotechnical conditions and project delivery are being considered, the Alaskan Way Viaduct Replacement Tunnel being the first design-build TBM project in the Seattle area. This has opened the door for innovative solutions offered by competing contractors looking for a competitive advantage. At the same time, the complexity of glacial geology has not changed as a significant risk factor for unforeseen project impacts. A statute in the State of Washington excludes no-damage-for-delay clauses in construction contracts, thereby emphasizing the need for verifiable data collection and reproducible evaluation processes of cause-effect relationships and impact quantification. All these factors point to past, present, and future Seattle area TBM projects being valuable case studies.

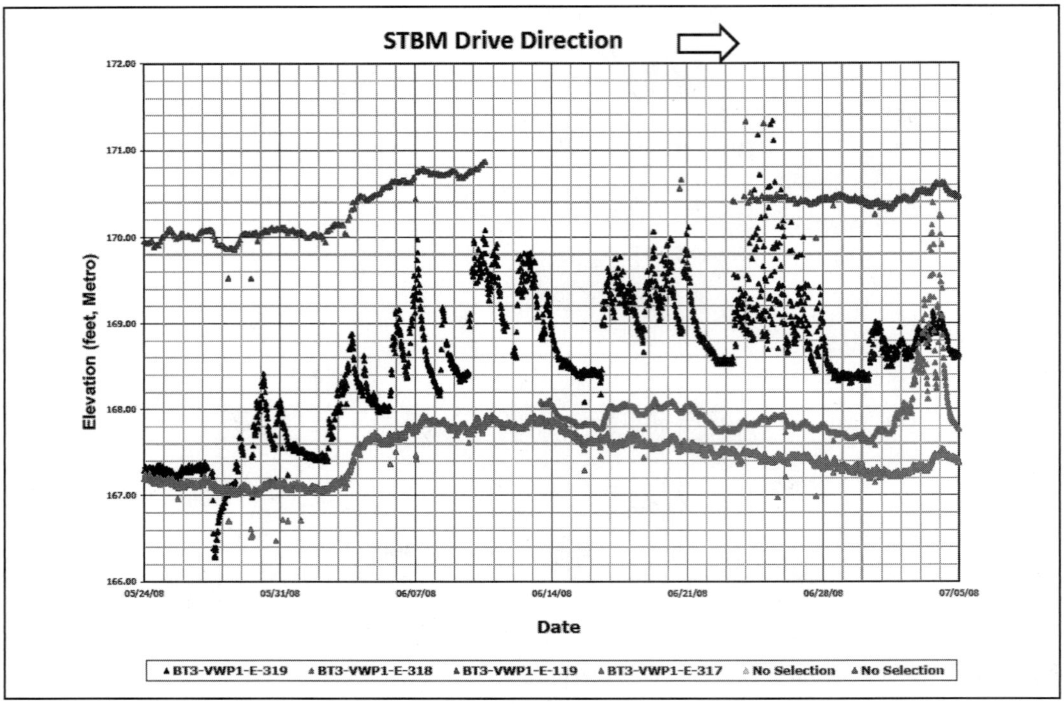


**Figure 15. Groundwater response to STBM operation**

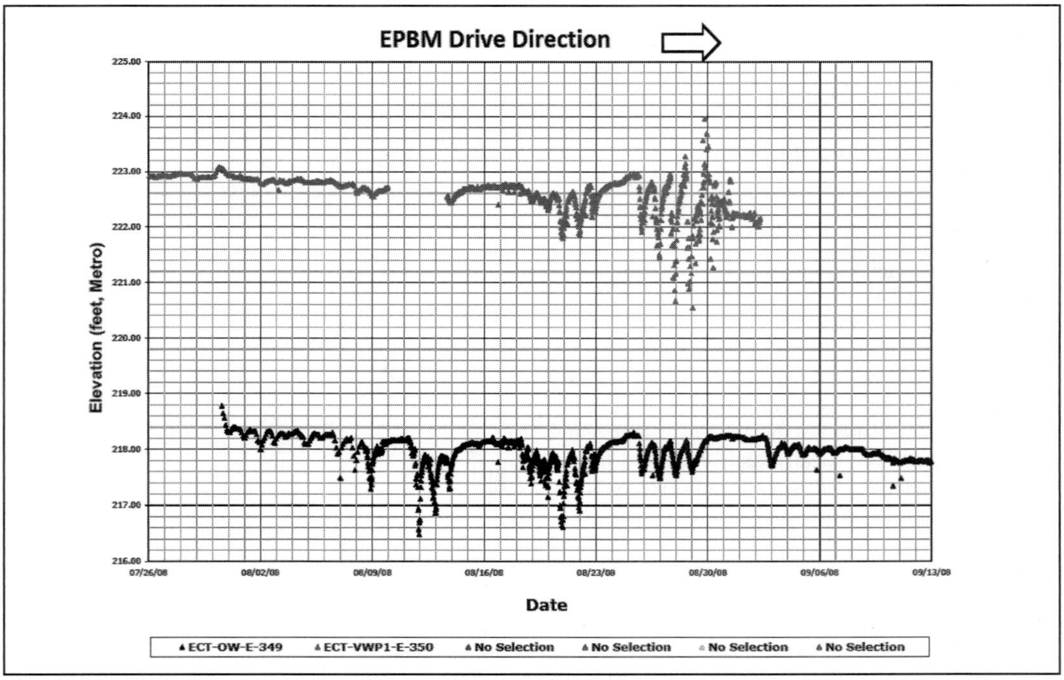


**Figure 16. Groundwater response to EPBM operation**

## REFERENCES

ASCE, 2007. Geotechnical Baseline Reports for Construction: Suggested Guidelines. Prepared by the Technical Committee on Geotechnical Reports of the Underground Technology Research Council, Randall J. Essex, Editor, ASCE, 2007.

Edgerton, W.W., Morgan, A.M., Bizzarri, R.E., 2012. Developing Functional Baselines. *NAT Proceedings*. Indianapolis, Indiana, June 2012.

Gwildis, U.G., Maday, L.E., Newby, J.E., 2009. Actual vs. Baseline Tracking during TBM Tunneling in highly variable Glacial Geology. *RETC Proceedings*. Las Vegas, Nevada, June 2009.

Gwildis, U.G., Sass, I., Rostami, J., Gilbert, M.B., 2010. Soil Abrasion Effects on TBM Tunneling. *ITA-AITES World Tunnel Congress Proceedings*. Vancouver, B.C., June 2010.

Newby, J.E., Gilbert, M.B., Maday, L.E., 2008. Establishing Geotechnical Baseline Values for Deep Soft Ground Tunnels. *NAT Proceedings*. San Francisco, California, June 2008.

Robinson, R.A., 2013. Lessons Learned from 130 Years of Tunneling in Seattle's Complex Soils. *RETC Proceedings*. Washington D.C., June 2013.

Robinson, R.A., Cox, E., Dirks, M., 2002. Tunneling in Seattle—A History of Innovation. *NAT Proceedings*. Seattle, Washington, June 2002.

Robinson, R., Sage, R., Clark, R., Cording, E., Raleigh, P., Wiggins, C., 2012. Conveyor Belt Weigh Scale Measurements, Face Pressures, and Related Ground Losses in EPBM Tunneling. *NAT Proceedings*. Indianapolis, Indiana, June 2012.

Shinouda, M.M., Gwildis, U.G., Wang, P., Hodder, W., 2011. Cutterhead Maintenance for EPB Tunnel Boring Machines. *RETC Proceedings*. San Francisco, California, June 2011.

# Finding a Balanced Contracting Approach to Pre-Excavation Grouting

**Ryan McCarter, John Stolz, and David Crouthamel**
Jacobs Associates

**David F. Tsztoo**
San Francisco Public Utilities Commission

**ABSTRACT:** Tunneling often requires pre-excavation grouting to reduce groundwater inflows and improve ground conditions. However, the work is expensive, the amount of grouting is a function of the contractor's means and methods, and its overall efficiency is debatable. Over the years, the industry has struggled with different payment mechanisms to administer this work, but typically few seem to assure the contractor of getting through the ground while also allowing the owner to control cost and schedule overruns. This paper examines a payment mechanism and contractual considerations used on the SFPUC's New Irvington Tunnel that may help meet these goals of cost and schedule control.

## PURPOSE OF PRE-EXCAVATION GROUTING

Underground construction methods and geotechnical risks dictate project-specific pre-excavation grouting programs. For the purpose of this paper, pre-excavation grouting comprises grout injection into the ground mass ahead of the advancing excavation, from within the tunnel, to reduce groundwater inflow and improve ground conditions. This practice applies to any excavation project that may benefit from pre-treating the ground, such as sinking a shaft or advancing a tunnel. Pre-excavation grouting almost always interrupts excavation advancement, so its application will impact the critical path of the project schedule, causing a costly burden to the project.

This cost premium must be weighed against the benefits of reducing groundwater inflows by pre-excavation grouting to limit impacts to tunnel excavation operations, improve the final lining conditions, and reduce impacts to groundwater resources. The act of pre-excavation grouting can also have a limited and secondary effect on improving ground behavior, such as standup time and overall strength of the ground mass even if its efficacy is difficult to predict. Therefore, excavation and support methods should be selected based on the anticipated ground conditions, with pre-excavation grouting reserved only as a last effort in ground improvement.

The unique elements of risk associated with each project define the purpose for performing pre-excavation grouting. All project stakeholders must be considered to determine the magnitude and ownership of risk each party will assume. The method of contracting the performance of pre-excavation grouting must take into account these various elements of risk and their impact on the project outcome, associated stakeholders, natural resources, existing structures, and their owners.

## STATEMENT OF PROBLEM

The primary problem in developing the contracting approach for pre-excavation grouting is determining the appropriate level of detail in the scope to address the risks and compensate the contractor fairly. Technically deficient scope, overly simplified or complex measurement units, and lack of identified elements of risks can individually or collectively derail the ability to meet the project goals. Pre-excavation grouting is performed in conjunction with the excavation activities and should be addressed with the same level of design importance and attention. The contracting approach must be compatible with the anticipated risks.

## OWNERSHIP OF RISK

There are many sources of risk and liability in underground work that involve complex construction sequences, linear construction paths, selection of means and methods, and subsurface conditions that can never be fully known. Understanding the nature and allocation of the risk is essential to the success of a project. Based on the premise that risk should be allocated to the party that has the most control over

**Table 1. Typical tunneling risks associated with pre-excavation grouting**

| Risk | Primary Risk Allocation | Management of Risk |
|---|---|---|
| Excessive inflow directly impacting tunnel work | Contractor | Pretreatment/selection of means and methods |
| Sustained inflows impacting final lining work | Contractor | Selection of means and methods |
| Settlement of adjacent structures | Contractor | Contractor determines means and methods based on specified protection requirements |
| Negative impacts on groundwater resources | Owner | Design and specify specific protection measures |
| Encountering contaminated zones | Owner | Design and specify specific protection measures |

it, Table 1 exhibits simplified examples of typical risks and their primary allocation and management.

To ensure the contractor is properly prepared, each risk must be identified and assessed individually to determine how it should be addressed and to whom it should be allocated. The obvious method to deliver these requirements is through the development of proper bid documents. Allocation of the risk, without proper bid documentation or lack of enforcement of the contract terms, will typically lead to increased costs and delays developed through disputes and claims. As long as the bid documents are clear and followed by both parties, the ability to navigate anticipated risk should be without incident. Under most cases, the primary allocation of risk falls on the owner to properly prepare adequate bid documents. The burden then falls on the contractor to successfully manage the risk in accordance with the contract.

## SCOPE OF WORK

The contract documents should clearly describe the factors that influence the scope for pre-excavation grouting and the responsibilities of each party. The lack of clear scope and understanding of who takes responsibility has historically corrupted the contracting approach for pre-excavation grouting and the division of responsibility.

Developing a clear scope of work requires a clear description of the anticipated site conditions, the establishment of mechanisms to measure the conditions in the field, and a fixed understanding of the means and methods. This creates a suitable baseline for the contractor to develop a basis of bid. If, on the other hand, there is uncertainty in the scope by the description of the site conditions or the selection of means and methods that influence the need for grouting, or both, then a complete scope cannot be developed. When there is uncertainty in the site conditions, the risks shift towards the owner. If the means and methods are uncertain, then the risks shift toward the contractor. Whoever owns the uncertainty, owns the risk.

Understanding the risk factors that will affect the scope of work dictates who should be in control of the scope of work. The control of the work can be described as either contractor-directed work or owner-directed work.

## COMPENSATION METHODS

Devising a fair compensation structure for pre-excavation grouting is challenging because it needs to balance the control of the work while meeting the project needs. The objectives of the owner typically do not align with the goals of the contractor. To share control over pre-excavation grouting, contract guidelines must be developed that establish minimum performance criteria and thresholds for compensation. The payment structure must complement the project risks and include all of the cost impacts associated with pre-excavation grouting. This allows the owner to meet global objectives without restricting the contractor to means and methods that it may not consider appropriate for the work.

These considerations make it clear that administration of pre-excavation grouting requires flexibility and therefore some kind of unit price payment mechanism since the amount of grouting is never known in advance. However, by properly accepting the risk for groundwater conditions, owners are exposed to the associated fallout risks: threshold inflows that are too high or too low; and inflow criteria that may be adequate for control of water inflows, but inadequate for stabilizing the ground. Depending on the bidding contractor's perception of where or how these thresholds are established, the bids for grouting may be unbalanced one way or the other. That would signal an intent to either use them beyond the project's need by overrunning the estimated quantities, or not use the bid items at all by bidding everything well below cost.

Payment mechanisms cannot be used to address technical issues associated with establishing the criteria for grouting. Their main goal is to provide flexibility by not only encouraging the selection of the most appropriate means and methods for performing the work, but also by devising the means to segregate the grouting effort from the excavation and ground support requirements, all while considering how this may impact the contractor's ability to construct the final lining.

The best way to accomplish this is through a combination of pay items that address risks in conducting the work by compensating for the amount of material used and the effort it takes to perform the work. Accordingly, the following payment mechanism guidelines were developed:

**Time + Material + Delay**

| | |
|---|---|
| Time | The time required to drill holes or pump grout. This would involve the heading crew and the outside plant support or a fraction thereof if the support services another operation, or a 24-hour operation on a single shift basis. |
| Material | The weight or volume of materials—cement, additives, and so on. |
| Delay | The time lost if the work critically delays project completion. The costs here include supervisory, home office, and other related costs. |

This payment structure requires strict guidelines for its administration; in particular, items measured by time must define the compensable components while establishing performance criteria. For example, to stimulate efficiency, time needed to mobilize/demobilize drilling or grouting equipment should not be measured. Another way to stimulate efficiency might be to require minimum performance criteria—things like grout pump and hole drilling capacity, and other factors affecting the time spent performing the work.

Another consideration is whether to exclude the grouting items from what is usually called the "Variations in Quantity Clause" commonly found in the contract's general conditions. Excluding such restrictions is preferable since there is no significant fixed cost element for the provision of materials, and that the selection of units of time itself eliminates the risk of cost variances.

The Time + Material + Delay payment mechanism was developed for the pre-excavation drilling and grouting of the New Irvington Tunnel project. This project exemplifies the success of this payment mechanism.

## NEW IRVINGTON TUNNEL PROJECT: OVERVIEW AND DESCRIPTION

### Background

The New Irvington Tunnel (NIT) is an integral project of the San Francisco Public Utilities Commission's (SFPUC) Water System Improvement Program to upgrade and augment the existing parallel tunnel with respect to seismic reliability and overall capacity. The new 3.5-mile-long tunnel will carry water through an 8.5-foot-diameter pipeline from the Sunol Valley to Fremont in Alameda County, California. Conventional tunneling methods included the use of roadheaders and blasting to excavate with steel arch sets to support the ground. Groundwater posed a significant challenge, which required great attention to pre-excavation drilling and grouting.

### Risk and Ownership of Risk

Risk was identified in the contract by determining risk posed by groundwater and ground conditions with allowable contractor means and methods. The pre-excavation grouting payment provisions were developed to offset risk with respect to time to complete the task of the pre-excavation grouting procedures. The risks included drilling time in highly variable and poor ground, which carried an uncertainty in time to advance holes ahead of the face. Likewise, with a large amount of grouting necessary to treat high volume, high head conditions, the time needed to inject grout into the rock mass was an uncertainty and therefore a risk. Payment provisions were developed to drill holes and place grout into the holes based upon minimum performance criteria, which the contractor was required to meet as a basis for bid. This was the threshold of risk intended to be owned and paid for by the owner.

The pre-excavation grouting program was configured to be directed by the contractor. As such, variable inflow criteria and incentivizing of the grouting program allowed the contractor to make the decision and limit the consumption of the grouting quantities. By performing the pre-excavation grouting within the performance requirements set forth in the contract and varying the inflow criteria to tolerable levels of risk, the contractor could manage the grouting program and potentially gain a tangible incentive.

### Drilling and Grouting Estimates

Pre-excavation drilling and grouting estimates were based on computational modeling that predicted inflow magnitude by location. This model relied heavily on records of the occurrence and expression of groundwater conditions from the existing tunnel records. Attendant impacts on tunnel advance were made available. Pre-excavation grouting estimates based on historical performance records of tunnels of similar size, groundwater pressure, and ground conditions were used to check and refine the model. Appropriate probe hole inflow trigger criteria for performing pre-excavation grouting were selected based on the head conditions and inflow magnitudes.

Table 2 summarizes the contract-specified pre-excavation grouting guidelines developed for NIT.

The Time + Material + Delay payment mechanism was implemented in the contract as follows in Table 3.

In order to stimulate efficiency of the contractor's pre-excavation grouting program, an incentive bonus was added to the contract provisions. The incentive was to be calculated at the end of the project and was equal to the unexpended sum of the sub-bid items, a through i, listed above. It guaranteed that the contractor would receive the entire bid amount for pre-excavation grouting, whether the quantities were consumed or not. The incentive promoted the contractor to only use pre-excavation grouting when necessary and avoided exploitation of the Time + Material + Delay payment mechanism. While pre-excavation drilling and grouting work was performed, the tunneling would be stopped; therefore, this incentive also would benefit the owner by conserving the project schedule.

## NEW IRVINGTON TUNNEL PROJECT: CONSTRUCTION

Notice to proceed on construction of the New Irvington Tunnel was issued in August 2010. The NIT contract allowed tunnel construction with up to four excavation headings: one tunnel heading from each of the two end portals and two more tunnel headings from an intermediate shaft. The first roadheader commenced tunnel excavation in March 2011 and two more roadheaders were mobilized to begin advancing the other headings in the following months. In October 2013, the last two headings met, completing the 18,660-foot-long tunnel.

All four headings faced significant challenges because of the rapidly changing ground conditions through several formations of sedimentary rock and fault zone crossings deep below the water table. Groundwater inflows greatly contributed to the difficulty tunneling through poor ground conditions.

Implementation of pre-excavation drilling and grouting was crucial to manage the risks of encountering high groundwater inflows at the tunnel face. A drill boom was mounted to each of the three roadheaders to perform pre-excavation drilling. Probe holes were typically drilled to a depth of 100 feet beyond the face at slight angles to terminate just outside the tunnel envelope. Each 2-inch-diameter hole was tested for gas, and the groundwater inflow rate was measured. The contractor devised a grouting program based on the contract criteria with various grout mix designs, hole patterns, and grouting phases. The contractor elected to attempt pumping grout into the face if any hole produced greater than 0.2 gallon per minute per foot (gpm/ft) of drilled hole length (e.g., 100 ft hole = 20 gpm). Cement grout was mixed and pumped from inside each tunnel heading.

**Table 2. NIT contract inflow thresholds for the New Irvington Tunnel**

| Inflow Rate* | Contract Directed Grouting | Contract Compensable Grouting |
|---|---|---|
| <0.2 gpm/ft | Not required | Not compensable |
| 0.2–0.5 gpm/ft | Not required | Compensable |
| >0.5 gpm/ft | Required | Compensable |

* Gallons per minute per foot of drilled hole length.

## NEW IRVINGTON TUNNEL PROJECT: CONDITIONS ENCOUNTERED

The nature of the anticipated tunneling challenges required constant observations of the field conditions and careful decision making. The construction management inspectors recorded crucial data necessary to ensure the project team was aware of the challenges faced each day. Data included detailed geologic descriptions, face map sketches, as-built details of the initial support steel sets, descriptions of pre-support methods, pre-excavation drilling and grouting tracking, and daily narratives of construction activities. Much of the data were converted into a database in order to develop construction summaries, graphs, and figures. In general, the ground conditions were more favorable than expected, with more competent rock and long reaches with negligible groundwater inflows. However, the most difficult areas were encountered in the last four months of tunneling with about 1,000 feet remaining between the final two headings.

Groundwater surges were encountered ahead of the face by drilling pre-excavation probe holes.

**Table 3. NIT sub-bid items and estimates for pre-excavation drilling and grouting**

| Bid Item | Bid Description | Type of Measurement | Estimate |
|---|---|---|---|
| a | Drilling for drainage or grout holes | Time | 2,850 hr |
| b | Grout Injection, Type III portland cement | Time | 1,660 hr |
| c | Grout injection, ultra-fine cement | Time | 890 hr |
| d | Grout injection, polyurethane grout | Time | 300 hr |
| e | Indirect cost hole drilling / grout injection | Delay | 900 hr |
| f | Type III portland cement | Material | 6,800 t |
| g | Ultra-fine cement | Material | 1,200 t |
| h | Polyurethane grout | Material | 1,200 gal |
| i | Sodium silicate grout | Material | 2,100 gal |

**Table 4. NIT sub-bid items: estimate versus actual**

| Item | Description | Estimate | Actual* | % |
|---|---|---|---|---|
| a | Drilling for drainage or grout holes | 2,850 hr | 2,867 hr | 100.6% |
| b | Grout Injection, Type III portland cement | 1,660 hr | 1,237 hr | 74.5% |
| c | Grout injection, ultra-fine cement | 890 hr | 125 hr | 14.0% |
| d | Grout injection, polyurethane grout | 300 hr | 12 hr | 4.0% |
| e | Indirect cost hole drilling / grout injection | 900 hr | 2,190 hr | 243.3% |
| f | Type III portland cement | 6,800 tons | 3,755 tons | 55.2% |
| g | Ultra-fine cement | 1,200 tons | 268 tons | 22.3% |
| h | Polyurethane grout | 1,200 gal | 117 gal | 9.8% |
| i | Sodium silicate grout | 2,100 gal | 273 gal | 13.0% |

*Actual values reflect the quantities at the time of this paper and are still subject to review and acceptance by the owner and contractor.

Pre-excavation grouting typically succeeded in restricting inflows down to a reasonable level to allow the tunneling operation to proceed with limited groundwater impact on the ground stability. It was also in the best interest of the contractor to complete the tunnel with the least amount of groundwater inflows possible to benefit the final lining installation operation. Long stretches of the tunnel encountered almost no groundwater at all, while other areas required several cycles of pre-excavation drilling and grouting to reduce the inflows to a manageable level. Inflow rates exceeded 300 gpm in some cases and achieved back pressures of up to 175 psi, equivalent to about 400 feet of head pressure. Post excavation residual groundwater inflows were pumped into a PVC pipeline and discharged into a water treatment plant at each heading portal. A limited number of distinct sources contributed to a majority of the 1,000 gpm average sustained total inflow at the completion of excavation. The pre-excavation grouting program was considered to be highly effective and vital to the success of the tunneling at NIT.

## NEW IRVINGTON TUNNEL: OUTCOMES OF PRE-EXCAVATION GROUTING

The effectiveness of the pre-excavation drilling and grouting program at NIT was dependent on the provisions outlined in the contract and the contractor's selected means and methods. Consistent probe drilling routines and verification drilling after grouting were crucial to assess the groundwater conditions ahead of the face. Recording the drilling distance where groundwater was encountered behind the tunnel face allowed the contractor to advance excavation in stages between grouting and effectively push the groundwater further back. Careful attention to the conditions and open communication with the project team contributed to the successful performance of the pre-excavation grouting program.

The tunnel headings encountered less groundwater overall than expected, thus requiring less pre-excavation grouting than anticipated in the bid quantity estimate. Table 4 summarizes the actual consumption of the sub-bid items. The time and material quantities for grouting were well below the estimated quantities, particularly for the special grout materials (c,d,g,h,i). However, contract administration decisions based on unforeseen conditions contributed to more drilling time compensated than expected in the bid quantity estimate resulting in the drilling sub-bid item (a) becoming completely consumed. Similarly, the indirect cost sub-bid item (e) also exceeded the estimate because of contract administration and payment decisions. The sum of the remaining sub-bid items left the contractor with an incentive of about 10% of the aggregate sum of the pre-excavation grouting bid items.

## CONTRACTUAL CONSIDERATIONS

The Time + Material + Delay payment mechanism successfully controlled project costs and schedule overruns at the New Irvington Tunnel while compensating the contractor fairly. While many considerations were built into the contract documents to support the payment mechanism developed, the following highlights some of the key topics that were examined when developing this pre-excavation grouting payment mechanism.

- *Tracking Quantities:* The unit price–based payment structure requires close attention and monitoring in the field by contractor and owner field representatives. The quantities must be tracked daily and summarized each month for accurate payment. The contract provisions must emphasize the importance of the field quantity tracking.
- *Performance Requirements:* The minimum requirements of the equipment, materials, and methods to be used must be clearly described and properly administered.
- *Compensable Criteria:* The inflow thresholds for pre-excavation drilling and grouting (as in Table 1) must be clearly described to

allow shared control of the risks between the contractor and owner. Distinct definitions of compensable work and non-compensable work must be included to allow accurate tracking and determination of compensable payment.

- *Contract Provisions:* The direct time and material are straightforward payment units. However, the calculation of delay or indirect time can become cumbersome. At NIT, multiple concurrent headings created difficulty in determining actual schedule impacts during pre-excavation drilling and grouting. Contract Provisions should be written carefully to describe time impact analysis methods to accurately track and calculate critical path delay by field personnel.
- *Time Extensions:* Delays caused by pre-excavation grouting are compensated under the indirect time sub-bid item in this payment mechanism. The total indirect time can be estimated for bid purposes. However, the actual critical delay is unknown before construction. This quantity of time should be excluded from the total contract duration and baseline schedule. Related time extensions should be carefully examined and administered through contract time extension change orders with zero cost associated.
- *Related Work:* Pre-excavation grouting is performed in conjunction with the excavation activities and is closely related to other activities such as dewatering, groundwater handling and treatment, and final lining installation. Provisions for related work must complement the pre-excavation grouting criteria and provisions.
- *Other Contract Documents:* Environmental, geotechnical, and hydrologic reports all influence the project pre-excavation grouting expectations. All contract documents should be compatible to ensure project needs are met.

## CONCLUSION

Many different factors influence the amount of pre-excavation grouting performed on any given project. By establishing a set price and reasonable quantity for each component, the Time + Material + Delay payment mechanism can provide equitable compensation under a variety of changing circumstances. However, this payment mechanism alone cannot assure the amount of grouting performed will achieve the owner's objectives. Setting the minimum inflow threshold for compensable and required grouting can ensure the owner's risks are addressed. Offering a tangible incentive tied to this payment mechanism encourages the contractor to employ means and methods that recognize the need for efficient performance of the pre-excavation grouting, but also integrates the operation into a systematic plan for execution.

Pre-excavation grouting often drives the success of tunneling projects and should be considered to be as important as the tunneling itself. Developing a comprehensive payment mechanism, such as the Time + Material + Delay approach described in this paper, benefits all parties and can collectively meet the goals of the project.

# Writing Better Geotechnical Baseline Reports

**Jack Raymer**
Jacobs Engineering

**ABSTRACT:** Geotechnical Baseline Reports (GBRs) are central to modern tunnel design but they are not easy to write and sometimes create more problems than they solve. This paper presents straightforward solutions to some of the more common pitfalls. Topics include the right way to manage risk, GBR language, what belongs in a GBR and what does not, and the reason why GBRs are difficult in the first place. This paper both builds upon and challenges the guidelines in the "Gold Book."

## INTRODUCTION

Geotechnical Baseline Reports (GBRs) have been in use for at least 20 years in their present form, with roots going back more than 40 years. Suggested guidelines for writing and using GBRs are described in the "Gold Book," by Essex (2007) and published by the American Society of Civil Engineers. Nevertheless, GBRs are not easy to write. Research by Heslop and Caruso (2013) has shown that GBRs have not improved significantly over the past 20 years, even as some of the industry's leaders have been expressing growing frustration, even proposing to take the "Geotechnical" out of GBR. There must be a problem somewhere.

The purpose of this paper is to provide some personal opinions of how to write better GBRs. These opinions were developed over 16 years of hands-on experience writing and then dealing with the good and not-so-good consequences of those GBRs during construction. This experience includes 10 major GBRs for over 40 miles of mostly hard-rock tunnel plus 22 larger-diameter construction or pump-station shafts. This experience was guided most notably by Ron Heuer, Gregg Korbin, Gary Brierley and Mike Robison, along with a number of other owners, contractors and consultants. Most of these opinions build on what is in the "Gold Book," whereas others challenge some of its points. Readers are encouraged to compare this paper with the "Gold Book" and then decide for themselves how to improve their own GBRs.

Heuer has said repeatedly "Don't play games." This has been a key piece of advice over the years and is a central concept of this paper. While it is always tempting to try to outsmart or manipulate the contractors, it seldom works and often backfires. A much better way is to tell the truth about the ground as best we are able so that the contractor can bid and execute the project as efficiently as possible, knowing that sometimes we will be right, sometimes we will be wrong, and many times the difference will be too small to matter.

## THE PURPOSE OF A GBR

The main purpose of a GBR is to tell bidding contractors what to expect about the ground with respect to the anticipated construction. GBRs give the bidders contractual certainty regarding ground conditions so that the bidders do not have to include large amounts of contingency in their bids or take undue risks regarding ground conditions. Ideally, the GBR creates a "level playing field," so all contractors bid the same ground conditions. GBRs are also used during construction to resolve disputes regarding the conditions and behaviors of the ground, but this role is really a consequence of the main purpose.

A GBR must be both unequivocal and trustworthy if it is to give the bidding contractors the confidence they need to bid accurately. These two concepts are central to this paper and will be developed in detail throughout. Another concept that runs through this paper is the position of contractor as the low-bidder. In most projects in the United States, the selected contractor is the low bidder by law. In order to be the low bidder, the contractor is forced by the process to be as optimistic as reasonably possible about the conditions and behavior of the ground, regardless of what might be said in the GBR. If the GBR is unclear or unreasonably conservative, then bidders must make their own interpretation, otherwise they will not be low and will never be the contractor.

## A RISK MANAGEMENT TOOL

GBRs are often described as a "risk management tool," but there are different interpretations as to what that means. The true risk-management value of a GBR is that it is unequivocal. A proper GBR takes a clear stand regarding ground conditions and behavior for every foot of the alignment so that the contractor knows what to bid on. By taking a clear stand, without hedging, the GBR takes away the very great risk that the low-bid contractor will take an optimistic view of the ground and then try to recover the losses from the owner if the interpretation was too optimistic.

The wrong way to manage risk is to think of the GBR as an insurance policy. In the insurance policy mindset, the engineer describes the ground "conservatively," meaning worse than he thinks it really is, so that the contractor puts more money in the bid, thereby reducing the potential for cost overruns. There are two fallacies with this thinking. The first is that the low bidder will have trusted the GBR enough to bid the falsely described conditions rather than the conditions indicated by their own analysis of the data. The second fallacy is that it is even possible to be "conservative" in describing the ground for something as complicated as mechanical tunneling, without creating additional, and sometimes even worse problems.

Being "conservative" causes the bidders to put money in the wrong places. If the GBR is conservative with respect to blocky and seamy ground, then the contractor is likely to put too much money in ground support and not enough money in penetration rate and cutter changes. While these misassignments might balance out, sometimes they do not, and if they cause the contractor to lose money, the contractor might have a valid claim. After all, if the contractor overbid on the ground support he must have underbid somewhere else otherwise he would not be the low bidder. Even worse, a "conservative" GBR might lead the contractor to choose the wrong TBM for the project. If the rock is described as much harder than it really is, then the low-bid contractor might opt for power and then be stuck with a machine that is not so good at negotiating difficult ground. Room in the heading is at a premium, especially in small-diameter tunnels. Larger motors and thrust cylinders means that there is less room for installing support or drilling grout holes.

## THE LANGUAGE OF EXPECTATIONS

The purpose of a GBR is to tell contractors what they should expect, not what they are expected to do—that latter role belongs to the Specifications and Drawings. GBRs should not contain phrases like "The Contractor shall..." or "...is required." For example, saying "The contractor shall install rock bolts," or "rock bolts are required to support the ground" places a requirement on the contractor, which is beyond the authority of a GBR. Furthermore, these kinds of statements tend to be redundant with the drawings and specifications, which is not good engineering practice in general. Instead, the GBR needs to stick with the language of expectations regarding the conditions and behaviors of the ground:

- "...[condition or behavior] is expected."
- "The ground is expected to...[behavior] if the Contractor does...[action]."
- "The ground is expected to consist of... [condition]."

For example, the GBR can say "The ground is expected to become unstable if the Contractor does not install rock bolts as shown on the drawings," or "Rock bolts are necessary to keep the ground stable." In these sentences, the GBR places no requirement on the contractor but does indicate expected ground behavior as the consequence if some action is not taken. The emphasis here is on the ground and ground behavior, which is the proper role of the GBR.

Sometimes it is appropriate in the GBR to emphasize or explain something that the Contractor must do. In these cases, the GBR should refer to the Specifications or Drawings. For example, if the GBR indicates that the tunnel is expected to make up to 500 gallons per minute of groundwater inflows, but the Specifications require a pumping capacity of 1000 gpm, the GBR could say "The tunnel is expected to produce up to 500 gpm but the Specifications require a pumping capacity of 1000 gpm for safety." In that sentence, the GBR is fulfilling its primary purpose by stating the expected condition. The GBR is also reporting on the fact that the Specifications say something different from the expected condition, without itself placing a requirement on the contractor. If the GBR did not report and explain this apparent inconsistency, then there would probably be a bid question asking about it.

The GBR should really not refer to the "expected construction," because the GBR does not have the authority to expect anything from the contractor. Instead, the GBR should refer to the "anticipated construction" or the "presumed construction." The word "expect" indicates obligation on the part of the person or thing that is expected to perform. The words "anticipate" and "presume" emphasize the preparations and thinking of the one waiting to see what will happen.

Some people like to use the word "will," as in the "ground will fail if the contractor does not install the support shown on the drawings." While certainly better than "the ground might fail...," it is also not exactly correct and goes against the principle of being trustworthy. The Engineer does not really know if the ground will or will not fail. It would be more truthful and more accurate for the engineer to say that the ground "could" fail. Then if the ground fails the contractor will have been sufficiently warned. A stronger statement, however, would be to say "the ground is expected to fail...," since the expectation belongs to the author of the GBR and not the ground. As a small nuance, the words "will" and "would" are about intention, "may" and "might" are about permission, "shall" and "should" are about obligation, and "can" and "could" are about potential. The ground has no intentions, and cares nothing about permissions or obligations. The ground does have potential, however.

## SOME GOOD PRACTICES

### Definitions

Terminology in the GBR should be defined in the context of the project. The best way to do this is at the point where the terms are used in the text. Long lists of definitions at the beginning or end of the GBR are generally less helpful because the definitions tend to be too generic and are out of context. Well-known, rigorous terms, like RQD or RMR should be defined by reference rather than redefined in the GBR. Terms that are defined sufficiently in a standard dictionary or standard undergraduate textbook generally do not need definitions unless they are used in a particular way or with some particular nuance. For example, if the GBR makes an important distinction between running and flowing sand, then differences in those terms should be explained, but if the GBR simply describes "running and flowing sand" without distinguishing them, then not much additional explanation should be needed. Long, belabored definitions and textbook explanations make the GBR confusing to read and detract from its usefulness.

### Explaining the Logic

GBRs are more believable, and therefore more trustworthy, when the authors explain the logic behind their baselines. For example, if the tunnel is expected to encounter a certain number of boulders that warrant interventions, then the authors should explain how they came to their conclusion. The basis may be from data in the GDR, from experience with similar jobs in similar ground, or other sources. The analysis can be based on statistical modeling or geological modeling. The explanation does not have to be lengthy, but does have to be clear enough so that the baseline does not seem arbitrary. If the explanation is longer than one paragraph, it should probably be put in an appendix or figure so as not to impede the flow of the GBR. Explanations are especially important if the authors of the GBR have developed a reputation for being overly conservative in the past.

### Quantitative Qualifiers

Variability is a natural condition of the ground; uncertainty is a condition of our knowledge about the ground (Raymer, 2010). A GBR must account for both to be trustworthy and unequivocal. The ground is a natural system that typically varies from place to place along the alignment. Sometimes this variability can be mapped, as on a profile. In other cases, the variability can be described statistically. Likewise, our knowledge is uncertain because the test borings are a very small sample of the rock to be encountered in the tunnel. The temptation is to draw fuzzy or dashed lines on profiles and contour maps, to use vague terms like "approximately" or "relatively," or to give wide, unqualified ranges, such as "The rock has RQD values that range from 0 to 100."

A better approach is to use clear values and sharp lines with quantitative qualifiers. Consider the example of the top of rock at a shaft. Saying "The top of rock occurs at El. 250 [feet]" is too simplistic, and therefore not trustworthy, because the top of rock is not going to be flat like a concrete slab would be flat. Saying "The top of rock occurs at approximately El. 250 and that it is relatively flat" is too vague, and therefore not unequivocal. A better approach would be to say "The top of rock occurs at El. 250 ±1," followed by a short explanation of why the author chose ±1. Does ±1 reflect the natural variability of the ground around the perimeter of the shaft? Does it reflect the uncertainty in the depth at which the borings encountered rock? Explanations are especially important where the expected range is large, such as El. 250 ±15. Does the author of the GBR have real reason to expect that the top of rock would vary over a range of 30 feet around the perimeter or over the area of the shaft? Or is the author just trying to give his own uncertainty broad cover? If the top of rock actually varies over a range of 30 feet, then that could preclude a number of construction methods that might actually be safer, less costly or more appropriate. If the top of rock is not known because there is no boring or not enough borings, then the GBR should explain the situation as truthfully as possible and the contract should provide a means of handling the uncertainty using unit rates or some other equitable approach.

Likewise, profiles and contour maps should be drawn with clear, sharp lines, not dashed lines. A range of allowable variability can be given as a note on the drawing: "Contours are accurate to ±2 feet," or "Contacts between rock layers on the profile are accurate to ±2 feet in the vertical dimension." As long as the range is small, then little explanation is needed, but if the range becomes large, then significant explanation will be needed to give the reader confidence that it is not simply arbitrary conservatism to avoid taking a clear stand.

Raymer (2010) showed how to calculate statistical variability and uncertainty for quantitative test results. Raymer showed that if sufficient data has been collected, the statistical uncertainty tends to be much smaller than the natural variability of the ground. Therefore, the GBR should focus on describing the variability as a statistical distribution. The best way to represent the expected values for the tunnel (or some defined part the tunnel) is often with a table. Figure 1 shows an example table based on 67 tests from an actual project. For the sake of clarity, the table should be explained in detail, so no one can claim they misunderstood the baseline. The percentiles in the table represent the natural variability within rock mass to be tunneled: 10 percent of the rock mass is expected to be weaker than 15.8 ±0.7 ksi. The plus or minus numbers (which may or may not be included) represent the statistical uncertainty. Note how the statisical uncertainty is actually quite low compared to the variability. No Disputes Review Board is going to listen to a contractor quibble over ±0.7 ksi if the average is 24 ksi. But if the engineer baselines 40 ksi to be "conservative," then the contractor will have had to ignore the baseline in order to be low bidder. This defeats the purpose of a GBR.

Giving the bidders the best estimate of ground conditions allows them to use the data in the way they see most appropriate. It also reduces design risk for the engineer because the best estimate does not pre-suppose what is "conservative."

| Unconfined Compressive Strength of Tunnel Rock | | |
|---|---|---|
| **Range of Rock in Tunnel** | **Percentile** | **UCS (ksi)** |
| 10 percent is weaker | $10^{th}$ | 15.8 ±0.7 |
| Average | $50^{th}$ | 24.1 ±0.6 |
| 10 percent is stronger | $90^{th}$ | 32.4 ±0.7 |
| Practical maximum | $99^{th}$ | 39.2 ±0.7 |
| *Values in this table are based on UCS tests from vertical NQ rock core and are valid at that scale and orientation* | | |

**Figure 1. Example of a table for baselining a statistical distribution of variable ground conditions**

## GBRs AND GDRs

A GBR is typically based on a Geotechnical Data Report (GDR). The purpose of a GDR is to present data about the ground rather than establish expectations. As a matter of good engineering practice, it is important that GBRs and GDRs stick to their respective roles and not overlap one another. If the two documents overlap, then the result can be ambiguity and conflicting statements. Ambiguity and conflicting statements give the low-bid contractor the opportunity to choose the more optimistic interpretation and then make a claim against the owner if the outcome was unfavorable.

For a GDR, "data" is factual information that is gathered, reduced, organized and presented according to scientifically repeatable procedures. The data is considered "factual" because the way it is obtained is scientifically repeatable—in other words, any appropriately skilled person could read the procedure and then perform the same tests or log the same core and obtain the same results. This is very different from predicting the ground conditions between the borings, which is the purview of the GBR.

The text of a GDR should focus on the procedures, rather than the findings, so that another person could repeat the work and obtain the same result. If the procedures followed published standards, then citing the standard is adequate description. If the procedures deviated significantly from the standard, then the deviation needs to be described in the GDR. If there is no suitable published standard for the procedure, then a full description of the procedure should be included in the GDR. There is nothing wrong with not following a published standard, as long as the procedure that was used is fully described in the GDR so that another person could repeat it.

Data includes the identification, description and stratification of rock and soil in a boring. Of anything that normally goes in a GDR, boring logs are the most subjective and the most likely to vary substantially from project to project. This high degree of variability from project to project is appropriate, because factors that are critical in one geologic setting are often trivial in another. The GDR must go into great detail as to how rocks and soils are named, identified and described on the boring logs. This is especially critical in rock, where there are often no universally recognized naming conventions—where one person's granite is another's nepheline syenite. A good way for a GDR to handle lithologic names is to define explicitly a limited number of names that are appropriate for the project and then use only those names on the boring logs. Names that do not appear on a boring log should not be defined; generic textbook definitions should not be used but only definitions that are explicitly adapted for the project. Furthermore, for lithologic names to be useful, they must be clearly recognizable using the tools available in the field. It does no good to make subtle distinctions that can only be made using thin sections or quantitative laboratory analysis, especially if those distinctions do not significantly affect ground behavior during tunneling.

Table 1 provides some guidelines for some items that belong and do not belong in GBRs and GDRs. To avoid repetition, things that belong in a GBR do not belong in a GDR and vice versa. Some of the items in the table need some explanation.

**Table 1. Guidelines for GDRs and GBRs**

| Things that should be in a GDR | Things that should be in a GBR |
|---|---|
| • Information on the regional geologic setting based on what is available from the literature or general experience in the area. *This includes stratigraphy, structure, oil and gas fields, underground mines, groundwater resources, etc.*<br>• Presentation of the data in clear format—boring logs, laboratory reports, survey data, core photos, field data.<br>• Maps showing locations of the borings and other field tests in relationship to the tunnel alignment.<br>• A summary table listing all of the borings and all of the tests or suites of tests that were done in each, including survey data.<br>• Detailed description of all field and laboratory procedures. *Standard procedures can be referenced, special procedures need to discussed so someone can reproduce them.*<br>• Summary tables listing the final results of each test based on standard or explained methods of data reduction for those tests. | • Brief description of the work in sufficient detail for the reader to understand the GBR.<br>• Geotechnical stratification of the ground, along with a geotechnical profile of the alignment. *The strata must be clearly defined and mappable at tunnel scale.*<br>• Summarized properties for each stratum that the tunnel or shafts will encounter. *For shafts, it may be better do this one shaft at a time since shafts are typically not close to each other.*<br>• Estimates of groundwater inflow for the tunnel and shafts.<br>• Description of ground behavior in response to excavation and how the support will interact with the ground.<br>• Identification of bad zones and how those zones can be negotiated using the anticipated tunneling methods.<br>• Explanation of other special geotechnical design issues, such as contamination, gas management, grouting, final linings, tunnel drainage, etc.<br>• Interpretive summaries of other similar projects in similar ground where the similarities are used as a basis of expectations. |
| **Things that should not be in a GDR** | **Things that should not be in a GBR** |
| • Statistical summaries of the data, such as histograms or averages. *Each data point needs to stand on its own.*<br>• Geotechnical profiles or interpretations.<br>• Things that are repeated or stated differently in the GBR. | • General textbook discussions about tunneling or geology.<br>• Scientific narratives on the geologic setting.<br>• Requirements on the Contractor. *Requirements of the design belong in the Specifications or Drawings.*<br>• Vague ranges that are not supported by the data or tightly quantifiable.<br>• Design recommendations or parameters for use by the Engineer as opposed to the Contractor.<br>• Unrealistic "baselines" designed to shift the risk from the Owner onto the Contractor. |

- *Literature Reviews.* Literature reviews on the regional geology are more appropriate in a GDR than a GBR because published papers can be regarded as "facts." Regardless of whether the information in the paper is right or wrong, any appropriately skilled reader should be able to read the paper and come to the same understanding of what was being said. Furthermore, if literature reviews are in a GBR, then it could be interpreted that the conditions described in some paper constitute the expected conditions for the project.
- *Statistical Data Summaries:* The GDR should include complete summary tables of reduced data, such as final results from packer test or uniaxial compressive strength tests, but should not include statistics on the data. The GDR should not provide averages, maxima, minima or standard deviations. The GDR should not provide histograms of the data. The reason is that statistics and histograms carry statistical inference, and statistical inference is a form of interpretation that belongs in the GBR. For example, if the GDR provides a histogram of results from strength testing, then a contractor could interpret that histogram as being representative of conditions to be encountered in the tunnel. Not only could this create a conflict with the GBR, it could also be incorrect. Statistics are only valid if certain conditions are met: the sampled domain must be representative of the ground in question, and within that domain, the samples must have been collected randomly. These issues can be overcome in a GBR because the GBR has the authority to state expectations based on interpretations and inference.
- *Profiles and Stratification:* The most essential role of a GBR is to predict the ground between the borings. The best way to do this, in most cases, is for the GBR to include a completely stratified profile along the tunnel alignment, with no blank spaces between

the borings. In some cases it is helpful to show the borings on the profile but in other cases it is not. For example, if the strata dips obliquely across the face of the tunnel, then borings that are not exactly on the centerline of the tunnel cannot be projected orthogonally onto the profile; if such a boring were shown at its true elevation and perpendicular offset to the centerline, the result would be a wrong profile. Likewise, Rock Quality Designation (RQD) measurements and other test results should not be shown on a profile but should assigned as statistically distributed properties to the geologic units shown on the profile. For example, in many cases, RQD is a function of the steeply dippling fractures encountered in a boring. Steeply dipping fractures seldom correlate horizontally along a profile, so plotting them on a profile is misleading with respect to the ground between the borings. A contractor could look at RQD measurements plotted on a profile and justifiably infer that the values shown would be representative of the ground halfway to the next boring. If the RQD values were found to be lower than shown, the Contractor could claim for unexpected support problems. If the RQD values were higher than expected, the Contractor could claim for reduced TBM production (Barton, 2000).

- *Design Recommendations:* Geotechnical design recommendations for the design engineer belong in neither the GBR nor GDR. GBRs and GDRs are written for the contractor, not the design engineer. Design recommendation for the engineer should go in a separate document that may or may not be shared with the contractor, such as a geotechnical design memorandum. The only exception is if some element of the design is left to the contractor. In that case, design recommendations should be rephrased into expected conditions and expected design parameters, and put in the GBR. In no cases should any design recommendations go in the GDR.

## SOME QUESTIONABLE PRACTICES

### Bolded Statements

Some people advocate putting baseline statements in bold, so as to distinguish them from things that are not baseline. I think this is both unnecessary and detrimental. First, the purpose of a GBR is to describe expected ground conditions. Therefore, everything that is in a GBR should be regarded as expected, and the GBR should not contain information about the ground that is not expected. This is part of the reason literature reviews are better placed in the GDR. In the GDR, a literature review is a neutral, journalistic report on what others have published about the ground, without explicitly accepting or rejecting how those reports might apply to the project.

Second, putting some statements about the ground in bold and others not in bold is playing games with the baseline. Why are some properties or characteristics of the ground baselined and others not? For example, why baseline unconfined compressive strength but not point-load strength? Both tests are repeatable and both tests provide useful, albeit different, information about the ground. If the author of a GBR is not confident that certain data in the GDR is useful for representing expected conditions, then it is better to simply leave it out of the GBR altogether or to explain in some detail why it is not useful. There are very many geologic factors that contribute to a successful or unsuccessful project. It is better that the author of the GBR simply provide good information unequivocally rather than to say one property is specifically baselined but another is not.

### Challenging Baselines

Some people say that the GBR should contain guidelines as to how the Contractor can challenge a particular baseline. I think this is inappropriate and detrimental because it is playing games. The purpose of the GBR is to describe the expected conditions, not set up a game with the contractor. If the GBR focuses on describing the expected conditions, and those descriptions are scientifically based rather than arbitrary, then good science contains all the rules necessary for the contractor to challenge a baseline that was both wrong and caused him to lose money. If the baseline was clearly wrong and caused damage, then the engineer and owner should admit it and come to an agreement quickly before damages and hard feelings multiply. If the contractor's claim is questionable, then the onus falls on the contractor to develop a proper scientific challenge to the baseline. In cases where more data has to be collected to challenge the baseline, then it is advisable for the contractor and the owner's engineer to work together in good faith to develop a mutually agreeable, scientifically based program to gather and analyze the data, so that the new data itself does not come into question. Raymer (2010) described how statistically based baselines can be challenged.

Rules for challenging a baseline should not be in a GBR for two reasons. First, only the Specifications and the Drawings have the authority to tell the contractor what he is required to do. The Specifications normally provide general rules for handling differing site conditions, including the role of the GBR.

Second, if specific rules are provided for challenging a specific baseline statement, then it destroys the scientific credibility of the statement, since the statement is subject to a contractual authority that supersedes science. Now the baseline statement may be seen as arbitrary and no longer trustworthy, which means the low-bid contractor might choose to reject it as not representative of the ground conditions.

## WHY ARE GBRs DIFFICULT?

GBRs are a great idea that has caused a lot of frustration in the industry. GBRs seem like they should be easy to write, but they are not. The concept of a GBR is simple—tell the contractor what he is entitled to expect about the ground. The problem is that the science behind GBRs is really quite difficult. If the science is not right, then the GBR will probably be wrong, and wrong GBRs lead to lost money, lots of claims and bad projects. It only takes one bad project to create a frustrated owner and a frustrated engineer. Once-burned engineers have tried to solve these problems by wordsmithing GBRs and playing all sorts of games. These efforts often backfire because the underlying problem was not the nuances of the wording or failure to define something rigorously enough, but that the engineer simply did not have an accurate understanding of the ground in the first place.

The science behind a GBR is predictive geology, as distinguished from descriptive geology. The goal of predictive geology is to understand what is happening between the borings, rather than describing what was encountered in the borings. Predictive geology is based on geologic models of how the earth fits together. Some of these models are conceptual, some are mathematical and some are statistical (Raymer, 2010); some are simple and some are quite sophisticated. Predictive geology is of minor importance in much of civil engineering, because borings are typically close together and the stress envelopes under structures dissipate rapidly with depth. Predictive geology is much more important in long, deep tunnels where borings are far apart, stress is concentrated in the heading, and the ground at tunnel depth is very different from the ground at the surface. The greatest applications for predictive geology, however, is with exploration for rare minerals and petroleum; this last context is where most research is performed.

Figure 2 shows the flow of thinking necessary to develop scientifically reasonable baselines. The first step is the subject of the GDR. The last three steps are the subject of the GBR. The difficult step is developing an accurate predictive geological model of the alignment. Once that is developed, then engineering models can be applied, such as

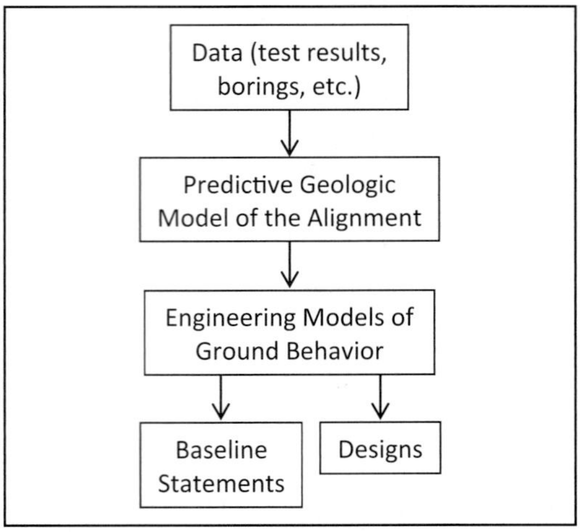


**Figure 2. Logic flow for developing geotechnical baselines**

wedge analysis, RMR and Q, finite element modeling, groundwater inflow calculations, and boreability models. Unfortunately, people like to jump into what they are good at. If tunnel engineers skim over the predictive geologic model in order to get into the engineering models, then the geologic model is likely to be simplistic. If the geologic model is simplistic, then the engineering models, no matter how sophisticated, cannot reflect adequately the ground behavior along the alignment. This leads to wrong baselines, poor design and sometimes bad projects.

## CONCLUSIONS

Projects generally go well for the owner and engineer when the contractor makes money by doing the work efficiently. When the contractor starts losing money, then all sorts of trouble can arise, including claims, cut corners, and undue risk taking. These troubles can quickly turn a promising project into a bad project where everyone loses—including the owner and engineer as well as the contractor. Therefore, the main goal of the engineer should be to help the contractor fulfill the contract as efficiently as possible. For tunnels, this starts with giving the contractor an accurate and well-organized GBR that is both trustworthy and unequivocal.

The engineer must never forget that the low bidder wins the contract. Any attempt to manipulate contractors into bidding higher by inflating baseline conditions is likely to backfire. Either the low bidder will ignore the baseline as arbitrary, or the low bidder will put money in the wrong places. In both cases, the engineer is will have contributed to chaos on the project that could result in significant losses for both the contractor and the owner.

## REFERENCES

Essex, Randall J., 2007, *Geotechnical Baseline Reports for Construction, Suggested Guidelines,* American Society of Civil Engineers, Reston, Virginia.

Heslop, Paul and Caruso, Christopher, 2013, "Recommendations on How Geotechnical Baseline Reports Can Be Prepared for Tunnel Projects, in *Rapid Excavation and Tunneling Conference, 2013 Proceedings,* edited by Michael J. DiPonio and Chris Dixon, Society of Mining Metallurgy and Exporation, Englewood, Colorado, pp. 343–357.

Raymer, Jack, 2010, "Geotechnical Variability and Uncertainty in Long Tunnels," in *North American Tunneling 2010*, edited by Eckert and others, Society of Mining Metallurgy and Exploration, Inc., Englewood, Colorado, pp. 316–322.

## ARTICLE DISCLAIMER

# A Systematic Approach for the Protection of Structures Adjacent to Bored and Cut-and-Cover Tunnels for the Regional Connector Transit Project

**Tung Vu and Paul Roy**
AECOM

**ABSTRACT:** The Regional Connector Transit Corridor Project will connect the existing Blue, Gold and Expo Lines which serve Santa Monica, Pasadena, Long Beach and the Eastside to downtown, but do not currently interconnect. Along the alignment three underground stations, one crossover mined cavern, and four cross-passages will be constructed. As the alignment runs through the heart of downtown Los Angeles, a number of existing buildings, structures, and utilities are located within the potential influence zone caused by the project's underground construction. During the advanced preliminary engineering phase, a building protection program was carried out to identify potential risks to the adjacent structures and suitable measures to mitigate the anticipated impacts. The paper discusses a systematic approach during advanced preliminary design to ensure protection of adjacent and buildings and structures along the alignment and a summary of the results and recommended mitigation measures.

## INTRODUCTION

The Regional Connector Transit Corridor (Regional Connector) project consists of approximately 580 m (1,900 ft) of cut-and-cover tunnel; two sections of twin-bored tunnels totaling approximately 1,460 m (4,800 ft) in length; three underground stations near the intersections of 2nd and Hope Streets, 2nd and Broadway, and 1st and Central Streets; one mined crossover cavern of approximately 90 m (300 ft) in length; and four cross-passages. Figure 1 schematically shows the project alignment and major components. The preliminary engineering assessment of potential impacts on adjacent buildings and structures was performed by the Connector Partnership Joint Venture that includes AECOM and Parsons Brinckerhoff, and their subconsultants.

## GEOLOGIC CONDITIONS

The main geological formations along the tunnel alignment consist of Younger Alluvium, Older Alluvium, and Fernando Formation. The Younger Alluvium consists primarily of medium dense silty, fine- to medium-grained, poorly graded to well-graded sand with some gravels and medium stiff to stiff silts and clays. The Older Alluvium consists of dense to very dense, poorly to well-graded sand with variable gravel and cobble contents. The Fernando Formation consists predominantly of extremely weak to very weak, massive, clayey siltstone with rare interbeds of well cemented, medium strong to strong, fined-grained sandstone. The clayey siltstone is generally moderately to highly weathered at shallow depths and slightly weathered to fresh at greater depths below the contact with the overlying alluviums.

The majority of bored tunnels will be excavated completely within the Fernando Formation, except for a stretch of approximately 300 m (1,000 ft) on the eastern end where they will be excavated in a full face of Older Alluvium or a mixed face of Fernando Formation and Older Alluvium. Cut-and-cover excavations will encounter these soil and rock strata at variable depths. The groundwater level varies from one meter (a couple of feet) below to about 18 m (60 ft) above the tunnel crown.

## EXISTING STRUCTURES AND UTILITIES ALONG TUNNEL ALIGNMENT

A total of 53 buildings along the entire tunnel alignment were determined to be in close proximity to the alignment and could be affected by the tunnel and station construction. These include 30 buildings adjacent to the bored tunnels, 3 buildings adjacent to the cavern, and 20 buildings adjacent to the cut-and-cover excavations. The adjacent buildings consist primarily of high-rise office buildings with underground basements along Flower Street and part of 2nd Street west of Main Street; and of one- to six-story retail and office buildings, parking structures, and one ten-story building east of Main Street.

The main adjacent underground structures and utilities include the 2nd Street Tunnel running along

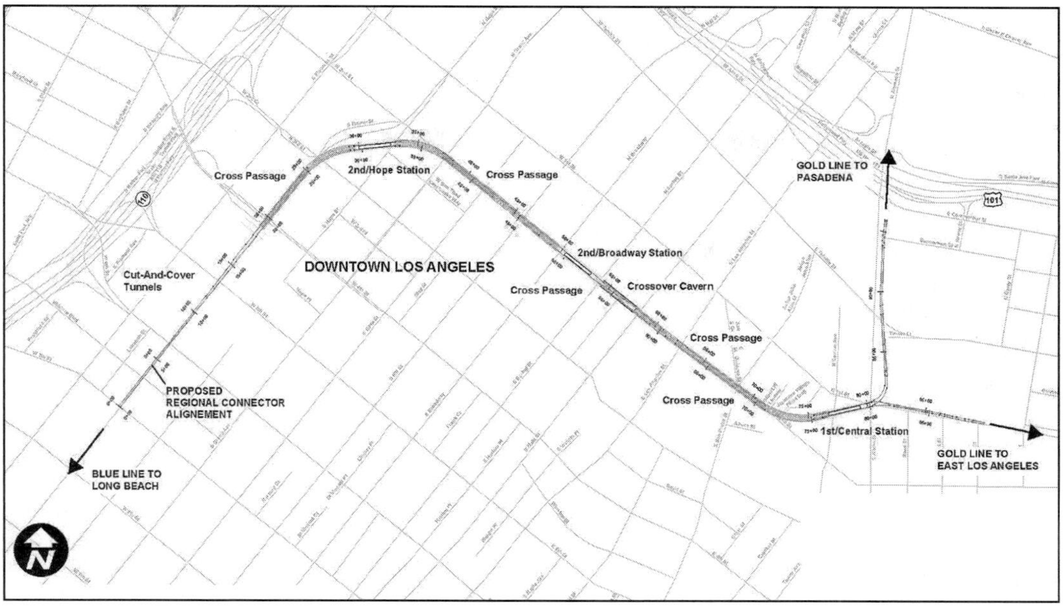


**Figure 1. Regional connector project alignment**

the Regional Connector alignment between Hill and Hope Streets; the Red Line Tunnels cross over the proposed alignment at Hill Street; foundations of the 4th Street Bridge and adjacent ramp; and the Grand Avenue Bridge piers located on both sides of the bored tunnels at Grand Avenue. In addition, there are a number of existing underground utilities of variable size and age located within the potential influence zone, including storm drains, sewer lines, water lines, gas lines, and telecommunication lines, and especially four major utilities including the Flower Street Storm Drain, Bunker Hill Central Plant Piping, Los Angeles County Storm Drain, and Alameda Storm Drain.

The information on existing conditions of the adjacent buildings, structures and utilities was obtained from various sources, including the prior studies, records available at the City of Los Angeles Department of Building and Safety, building walk-throughs (for selected buildings), and especially records from the property owners where the majority of building information was collected. During the building walk-throughs, photos were taken (where practical) to document the building existing conditions.

## IMPACTS ON ADJACENT BUILDINGS AND STRUCTURES

### Assessment Methodology

In an attempt to better identify and characterize potential risks associated with the critical adjacent buildings and structures, the potential impacts were evaluated in two stages: preliminary assessment and second stage assessment.

The preliminary stage involves the estimation of free-field settlements induced by the underground construction without considering the presence of the existing structures, and the screening of the existing buildings to be evaluated in the subsequent stage. The preliminary stage uses conservative screening criteria to assure that the buildings that are not considered in the second stage assessment will not be subject to damage levels more severe than "Negligible." A maximum absolute settlement of 6 mm (0.25 inch) and maximum settlement trough slope of 1/600 were adopted as screening thresholds for this project. All buildings and structures that have either absolute settlement or slope exceeding the above thresholds are evaluated in the second assessment stage.

The second stage focuses on evaluation of structural response to the estimated ground movements and severity of the possible damage, and to determine which buildings or structures are potentially at risk of being damaged, requiring mitigation or repair. The second stage assessment is more rigorous as the buildings' properties and structural behaviors are taken into account. The buildings adjacent to bored tunnels were evaluated using the Boscardin & Cording (1989) method while those adjacent to cut-and-cover excavations were evaluated using the Son & Cording (2005) method.

The Boscardin and Cording method is an empirical method that predicts potential damage to existing buildings and structures based on the critical tensile strains estimated using a deep beam model, which are a function of the building angular distortion and horizontal tensile strain. Depending on the location of the building relative to the tunnel excavations, different portions of the building can lie in a hogging or sagging zone which is separated from each other by the point of inflection of the settlement trough. Since the building portions in each zone experience different structural responses to the settlement and ground horizontal strains, they are considered separately, as recommended by Mair et al. (1996) and illustrated in Figure 2.

For the building portion located in a hogging zone, the neutral axis of the beam is assumed to be at the lower edge of the beam, and the maximum angular distortion is calculated using the equation recommended by Boscardin and Cording (1989); while in the sagging zone, the beam neutral axis is assumed to be at mid-height, and the angular distortion is calculated using the equation recommended by Walhs (1981).

Settlement calculations following the Boscardin and Cording method are performed using MathCAD software. In order to estimate the magnitude of expected damage to the structure, the calculated maximum values of angular distortion ($\beta_{max}$) and horizontal strain ($\varepsilon_{h,max}$) for each building are compared to limiting strain values by plotting in the chart illustrated in Figure 3 and correlating with the visual building damage classification presented in Table 1.

## Ground Movements Caused by Bored Tunnels

Ground movements induced by tunneling consist of both vertical and lateral movements in directions transverse and parallel to tunnel alignment. Ground movements transverse to the tunnel centerline are more critical to the adjacent buildings and utilities. The ground movements parallel to the tunnel excavation are considered less critical to the buildings and structures because the impact of longitudinal settlement is typically transitory, leveling off as the tunnel passes.

The induced settlements transverse to the proposed tunnels are estimated using the semi-empirical method that was originally proposed by Peck (1969), and subsequently updated by O'Reilly and New (1982 and 1992) and others. This method assumes that the shape of the settlement trough above a single tunnel follows a Gaussian distribution and that the volume of the settlement trough is equal to the total volume of lost ground during tunneling. The total settlements caused by two tunnels are the sum of the settlements caused by each individual tunnel, assuming superposition.

The shape of the settlement trough over a single tunnel is characterized by three main parameters: depth to the tunnel springline (z), the ground loss ($V_l$), and horizontal distance from the tunnel centerline to the point of inflection of the settlement profile curve ($i$). In this study, the depth $z$ is the vertical distance from the building or structure's foundation

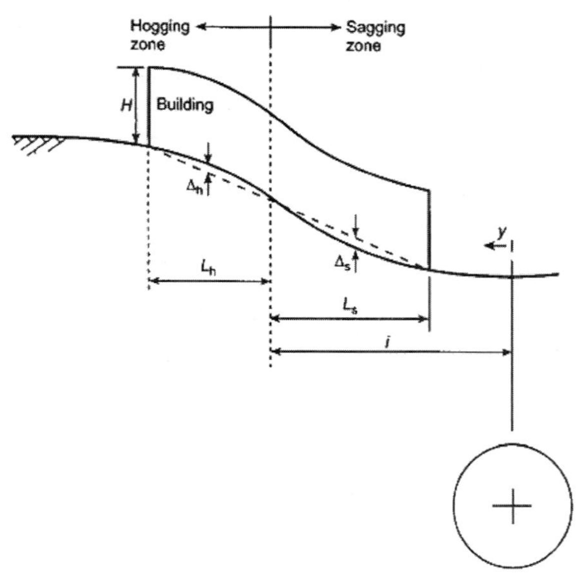


**Figure 2. Building deflection in hogging and sagging zones (Mair et al., 1996)**

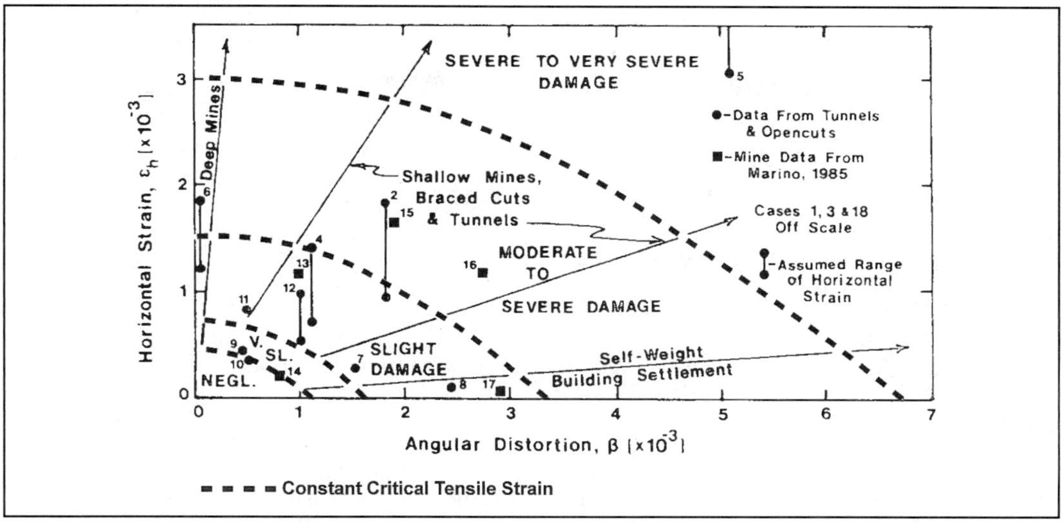


**Figure 3. Relationship of damage to angular distortion and horizontal strain (Boscardin and Cording, 1989)**

**Table 1. Classification of visible damage (Boscardin and Cording, 1989)**

| Damage Level | Description of Damage* | Approximate Width of Cracks,† mm |
|---|---|---|
| Negligible | Hairline cracks | <0.1 |
| Very slight | Fine cracks easily treated during normal redecoration. Perhaps isolated slight fracture in building. Cracks in exterior brickwork visible upon close inspection. | <1 |
| Slight | Cracks easily filled. Redecoration probably required. Several slight fractures inside building. Exterior cracks visible, some re-pointing may be required for weather tightness. Doors and windows may stick slightly. | <5 |
| Moderate | Cracks may require cutting out and patching. Recurrent cracks can be masked by suitable linings. Tuck-pointing and possibly replacement of a small amount of exterior brickwork may be required. Doors and windows sticking. Utility service may be interrupted. Weather tightness often impaired. | 5 to 15, or several cracks > 3 mm |
| Severe | Extensive repair involving removal and replacement of sections of walls, especially over doors and windows required. Windows and door frames distorted, floor slopes noticeably. Walls lean or bulge noticeably, some loss of bearing in beams. Utility service disrupted. | 15 to 25, also depends on number of cracks |
| Very severe | Major repair required involving partial or complete reconstruction. Beams lose bearing, walls lean badly and require shoring. Windows broken by distortion. Danger of instability. | Usually >25, depends on number of cracks |

* Location of damage in the building or structure must be considered when classifying degree of damage.
† Crack width is only one aspect of damage and should not be used alone as a direct measure of it.

bottom (or pile tip if buildings are founded on piles or caissons), or utility springline, to the proposed tunnel springline at the location of the structure under consideration. The settlements caused by a single tunnel excavation are predicted using the following equations:

$$S_{z(x)} = S_{z,\max} * e^{\left(-\frac{x^2}{2i^2}\right)}$$

$$S_{z,\max} = 0.313 * V_l * \frac{D^2}{i}$$

$$i = K * z$$

where:

$S_{z(x)}$ = settlement at location $x$ from tunnel centerline
$x$ = horizontal distance from tunnel centerline

$z$ = vertical distance from tunnel springline to point of analysis
$i$ = distance from tunnel centerline to point of inflection on settlement profile curve
$D$ = excavated tunnel diameter
$V_1$ = average ground loss
$K$ = trough width factor

The ground loss, $V_1$, and settlement trough width factor, $K$, are two important input parameters that need careful evaluations. Ground loss is the factor that has the most significant effects on the tunneling-induced ground movements. Limiting ground losses into tunnel excavations is the primary method to limit ground movements. Table 2 summarizes monitored ground losses associated with the use of EPBMs from recently completed tunnel projects worldwide. In all of these tunnel projects, the reported ground losses were typically achieved with a good control of face pressure, bentonite slurry injection in the annular gap around the TBMs, and tail-skin grouting to limit ground movement into the tunnel excavation. Higher ground losses were reported along a learning curve, or in tunnel sections where tail-skin grouting was not implemented or inadequate face pressures or slurry injection pressures were applied.

Based on the above reported ground loss and taking into account the mixed face conditions of alluvium and Fernando Formation on the eastern end of the alignment, a typical ground loss of 1.0% was assumed for the tunnel excavation in alluvium or mixed face conditions. For the tunnel excavation in Fernando Formation, which consists primarily of massive, weakly cemented, very weak to weak clayey siltstone, a typical ground loss of 0.5% is expected without bentonite injection around the shield.

The transverse distance from the tunnel centerline to the inflection point, ($i = K^*z$), is characterized by the depth to the tunnel springline, $z$, and a trough width factor $K$, which is a function of ground type. Table 3 shows the $K$ values of different soil types

**Table 2. Monitored ground losses of recent tunnel projects**

| Project Names | | Exc. Dia., m (ft) | Year | TBM | Geologic Conditions | Typical Ground Losses, % | References |
|---|---|---|---|---|---|---|---|
| Sao Paulo Metro Line 4—Lot 1, Sao Paulo, Brazil | | 9.50 (31.2) | 2009 | EPBM | Three formations: soil derived from the alteration of gneiss; interbedded high to medium plasticity clay and sandy clay with gravel; and interbedded medium stiff to hard clay with fine to coarse sands | < 0.4 | Pellegrini and Perruzza, 2009 |
| Barcelona Metro Line 9, Barcelona, Spain | Mas Blau to San Cosme Segment | 9.40 (30.8) | 2008 | EBPM | Submerged fine silty sands and clayey silts | 0.4 to 0.8 | Mignini et al., 2008 |
| | Segment IV-B (San Adria) | 11.95 (39.2) | 2007 | EPBM | Sands, clay, and silts overlying gravels with sands | 0.7 to 1.0 | Della Valle, 2007 |
| | Segment IV-C (Trajana) | 11.95 (39.2) | 2007 | EPBM | Mixed face of silts and sands or gravels in a sandy clay matrix overlying highly to completely weathered granodiorite | 0.2 to 0.6 | Della Valle, 2007 |
| | Fira to Park Logistic Segment | 9.4 (30.8) | 2006 | EPBM | Silty sands with sandy silts, silts and silty clays | 0.3 to 0.4 | Orfila et al., 2007, Della Valle, 2007 |
| Madrid South Bypass M-30 Tunnels, Madrid, Spain | | 15.0 (50.0) | 2007 | EPBM | Mixed face of sandy clay overlying hard clay with gypsum | 0.1 to 0.4 | Universidade da Coruña, 2008. |
| MTA Gold Line Eastside Extension, LA, USA | | 6.52 (21.4) | 2007 | EPBM | Mix of stiff to hard silt, lean clay, sandy clay, and loose to very dense sand and gravel | < 0.3 | Choueiry et al., 2007 |
| Channel Tunnel Rail Link, London, UK[(1)] | | 8.14 (36.7) | 2004 | EPBM | London clay: Stiff to hard clay | 0.3 to 0.8 | Bowers et al., 2005; Mair and Borghi, 2008. |
| | | | | | Fine and medium silty sand | 0.3 to 0 8 | |

**Table 3. Settlement trough width factors, *K***

| Soil Types | *K* Value |
|---|---|
| Artificial fill | 0.3 |
| Younger alluvium | 0.3 |
| Older alluvium—above groundwater table | 0.2 |
| Older alluvium—below groundwater table | 0.6 |
| Fernando formation | 0.4 |

selected for this preliminary engineering study as recommended by O'Reilly and New (1982) and Peck (1969). The composite trough width parameter $i$ of $N$ soil layers above the tunnel springline, each of thickness $z_N$, is calculated using the following equation recommended by O'Reilly and New (1992).

$$i = K_1 z_1 + K_2 z_2 + \ldots + K_N z_N$$

### Ground Movements Caused by Cut-and-Cover Excavations

The first practical approach for estimating ground movements caused by deep excavations was proposed by Peck (1969). Peck compiled data on ground settlement adjacent to temporary braced sheet pile and soldier pile walls and developed a chart that gave the ground settlement as a function of distance from excavation and type of soil. Since the publication of Peck's paper, other empirical and semi-empirical methods have been proposed to estimate ground movements caused by deep excavations. In this study, the vertical ground movement (settlement), horizontal ground movement, and settlement envelope are estimated following the empirical method proposed by Clough and O'Rourke (1990). Knowing the maximum settlement, the surface settlement envelope can be estimated using the dimensionless diagrams shown in Figure 4. In stiff clays, residual soils, and sands, maximum lateral wall movements and settlements of the retained soil average about 0.2% to 0.3% H, with a scattering of case history data up to 0.5% H.

### Unmitigated Effects of Ground Movements to Buildings and Structures

A total of 33 buildings adjacent to the bored tunnels and caverns are assessed with the preliminary analysis. Sixteen buildings are screened out of the second stage assessment and these buildings are considered as not being affected or negligibly affected by the tunneling-induced ground movements.

Seventeen buildings having maximum total settlement and slope that exceeded the above criteria were evaluated in the second stage assessment using the Boscardin and Cording method. Of these, nine buildings have a maximum anticipated damage level ranging from "Negligible" to "Very Slight"; one

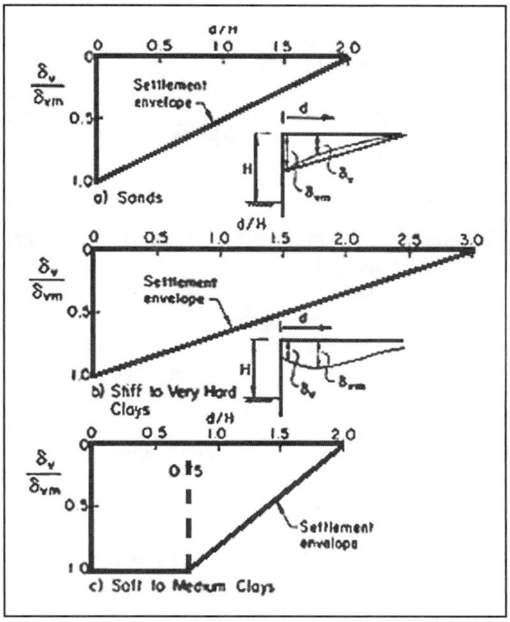


**Figure 4. Recommended dimensionless settlement profiles adjacent to excavations (after Clough and O'Rourke, 1990)**

building has a maximum anticipated damage level as "Slight"; four buildings have a maximum anticipated damage level as "Moderate"; and three buildings have a maximum anticipated damage level ranging from "Severe" to "Very Severe."

Additional analyses were performed for pile foundations located in the pile influence zone to evaluate the additional pile loads caused by tunneling-induced ground movements. This zone is defined by Jacobz et al. (2001) as a soil prism above the tunnel spring line that is limited by a 1:1 (45 degrees) upslope line on each side of the tunnel. A review of the buildings on piles or caissons along the proposed alignment indicates that the piles of the Angelus Plaza Parking Structure are more critical than the remaining piles and caissons. These piles were evaluated for the anticipated additional load caused by the lateral ground movements using the LPILE V5.0 program (Ensoft, Inc., 2005). The results from this pile analysis indicated that the internal forces in the pile caused by the lateral ground movements and vertical loads from the above structure are well below the pile capacity.

Twenty buildings adjacent to the cut-and-cover excavations were identified during the preliminary assessment stage. Five of these buildings were determined to be outside and fifteen buildings were within the approximate limits of the settlement trough. Of these fifteen buildings, nine buildings have maximum estimated settlements below the thresholds and were not analyzed further. The remaining six buildings having the maximum estimated settlements above the thresholds were subsequently analyzed in the second stage using numerical modeling (as discussed subsequently) to determine the potential damage levels.

### Numerical Modeling

Subsequent to the analyses using empirical methods, numerical modeling using PLAXIS computer program was performed for several buildings and structures that are either structurally critical or located adjacent to complex excavations. These include the Higgins Buildings, 2nd Street Tunnel, Redline Tunnels, 4th Street Bridge and Ramps, buildings adjacent to crossover cavern, Bunker Hill Central Plant piping, and the six buildings adjacent to cut-and-cover excavations that have maximum estimated settlements above the specified thresholds.

Numerical analysis allows different excavation sequences and initial ground support schemes to be modeled and the ground movements in each case to be determined. Since the numerical modeling procedures are based on a case-by-case basis, these numerical analyses are not presented in detail in this paper. Some details can be found in the papers previously published by Navid et al. (2012) and Bergeson et al. (2012).

The settlement, angular distortion and horizontal strain are calculated at the foundation level of the buildings and structures and used to assess the level of possible damage according to Boscardin and Cording's method. Results from these numerical analyses indicated that the maximum anticipated damage levels of the Higgins Building, 2nd Street Tunnel, Redline Tunnels and 4th Street Bridge and Ramps are "Negligible." Among the six buildings adjacent to cut-and-cover that were analyzed with numerical modeling, five buildings have the expected damage level of "Negligible" and one building has "Very Slight." Based on the results of the above analyses, the buildings and structures that are anticipated at higher risks and require mitigation measures are flagged for protection as illustrated in Figure 5.

## IMPACTS ON UTILITIES

Settlement impacts on buried pipeline utilities are typically caused by one or more of the following effects, as summarized in O'Rourke and Trautman (1982): (1) tensile pull-apart at joints; (2) opening of joints between pipe segments, θ, due to relative rotation between two pipe segments; and (3) straining of pipe caused by flexural deformations, $\varepsilon_b$, and lateral deformations, $\varepsilon_h$, that lead to rupture or intolerable deformation.

The first two effects primarily occur at well-defined joints and would be more likely to occur for fairly rigid, jointed pipes, such as concrete pipes or vitrified clay pipes (VCP). The third type of effect is caused by differential settlements and lateral ground movements, and is most likely to occur in flexible pipelines with well-designed rigid joints that can take significant rotation, such as welded steel pipelines or small (less than 20 cm in diameter) cast iron pipes (CI) and ductile iron pipes (DIP) (O'Rourke and Trautman, 1982). Schematics of each of these three modes of failure are shown in Figure 6.

The tensile pull-apart at the joints is typically only a factor if one end of the utility is fixed to some rigid object (i.e., building, manhole, etc.), or if joints are particularly sensitive, such as in cast iron pipes. O'Rourke and Trautmann (1982) reported an allowable axial joint slip of 25 mm (1.0 inch) for buried CI pipe. Attewell et al. (1986) also reported a range of allowable axial joint slip of 10 to 25 mm (0.4 to 1.0 inch) for sound CI water and gas mains with different joint packing materials (cited by Bracegirdle et al. 1996). Considering the potential existing joint deformations, a reasonable lower allowable limit of 10 mm (0.4 inch) was used for CI and other types of jointed pipes, such as concrete, VCP, or DIP, in this study.

Joint rotation failure will occur for rigid utilities with joints, or for any utility that has joints that allow rotation. For utilities transverse to a single tunnel excavation, the critical joint rotation point is directly above the sagging point of the settlement trough.

Tensile strains in utilities are estimated following procedures proposed by Attewell et al. (1986) that are summarized in the paper by Bracegirdle et al. (1996). In this study, the smaller utilities were approximately assumed to follow closely the ground settlement trough; thus, the utility bending tensile strain is calculated directly from the settlement trough hogging curvature. However, larger utilities and structures such as the Los Angeles County Storm Drain and Red Line Tunnels have a significant relative stiffness compared to the surrounding ground, hence modifying the settlement trough curvature. Consequently, the utility-ground interaction was considered by following the procedure outlined by Yeates (1985).

The utility additional strains caused by ground movements should be limited so that the utility total tensile strains are kept below the limiting strains at cracking. For cast iron pipes, an elastic tensile strain of 400 micro-strain can be derived from the design stress specified by codes (Attewell et al., 1986). For brick and ductile iron/steel utilities, limiting additional tensile strain of 150 and 600 micro-strain were assumed, respectively.

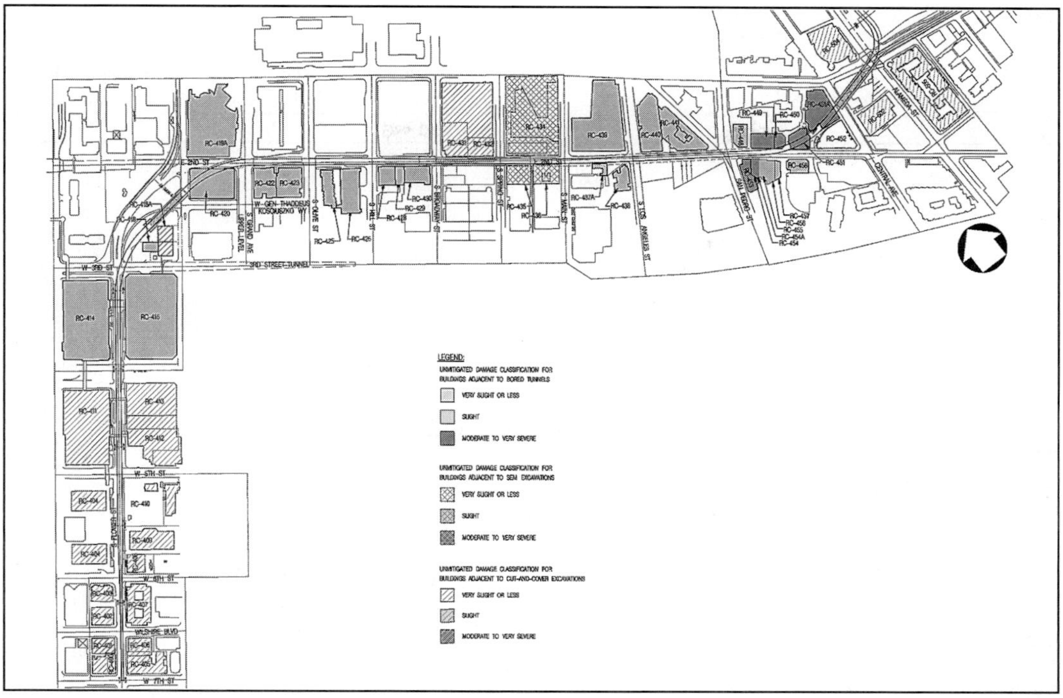

Figure 5. Adjacent buildings impacted by bored tunnels and cut-and-cover excavations

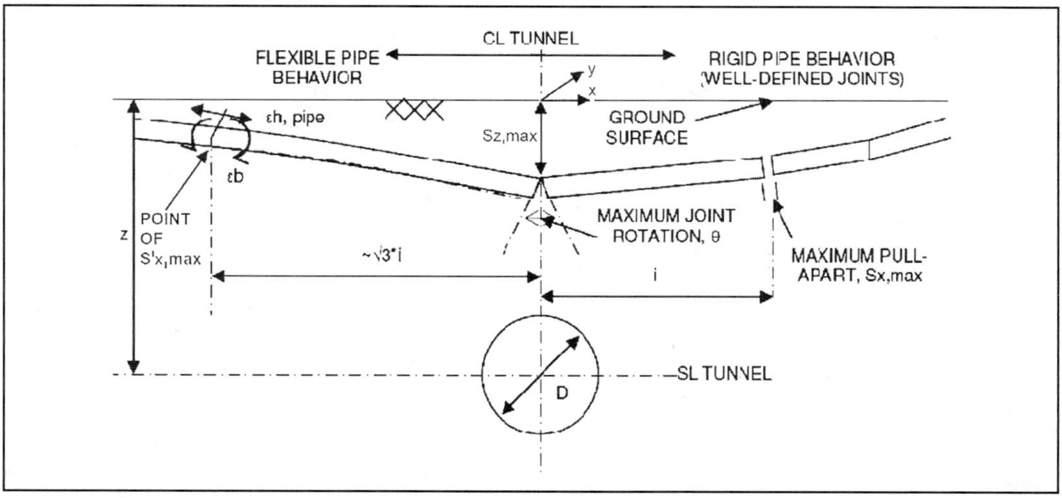


Figure 6. Utility impacts from tunneling induced ground movements

For each failure mode, a ratio that compares capacity to demand is calculated to estimate the impact of anticipated settlements on the utilities. This ratio can be interpreted as follows:

- Ratios $< 1.0$: The utility would likely be adversely impacted by construction and pro-active measures should be taken to prevent damage.
- $1.0 \leq$ Ratios $< 1.5$: Significant impacts are not expected but the utility may be affected by construction. Specific geotechnical instrumentation and surveys may be warranted to monitor soil and utility deformations.

**Table 4. Analytical results for utilities**

| No. | Utilities | Depth to Centerline (m) | Dimension (cm) | Orientation Relative to Tunnels | Minimum Capacity-to-Demand Ratio |
|---|---|---|---|---|---|
| | | **2nd and LA Street Intersection** | | | |
| 1 | Storm drain, RCP | 2.7 (9 ft) | 46 ID (18 in) | Transverse | 3.3 |
| 2 | Sewer, VCP | 4.5 (15 ft) | 40 ID (16 in) | Transverse | 2.9 |
| 3 | Sewer, RCP | 9 (31 ft) | 76 ID (30 in) | Transverse | 1.5 |
| 4 | Gas, CI | 1.2 (4 ft) | 15 ID (6 in) | Transverse | 1.9 |
| 5 | Water, CI | 1.2 (4 ft) | 30 ID (12 in) | Transverse | 1.8 |
| 6 | BP&L, URC | 1.2 (4 ft) | 100 by 64 (40 by 25-in) | Transverse | 1.6 |
| | | **2nd Between LA St and San Pedro St** | | | |
| 7 | Storm drain, Brick (R:67+00) | 1.5 (5 ft) | 78 ID (31 in) | Parallel | 1.1 |
| | | **2nd and San Pedro Street Intersection** | | | |
| 8 | Storm drain, RCP | 2.4 (8 ft) | 46 ID (18 in) | Transverse | 1.0/2.0* |
| 9 | Gas, CI | 0.6 (2 ft) | 7.5 ID (3 in) | Transverse | 0.9/1.8* |
| 10 | Water, CI | 1.8 (6 ft) | 30 ID (12 in) | Transverse | 0.6/1.2* |
| 11 | Water, CI | 1.5 (5 ft) | 20 ID (8 in) | Parallel | 0.6/1.2* |
| 12 | Sewer, VCP | 3 (10 ft) | 20 ID (8 in) | Parallel | 1.9/3.8* |
| 13 | Electric ducts, URC | 0.9 (3 ft) | 53 by 53 (21 by 21-in) | Parallel | 0.8/1.6* |

* First and second values based on values of ground volume loss of 1.0% and 0.5% respectively.

- Ratios ≥ 1.5: No adverse impacts are expected and no specific geotechnical instrumentation or monitoring will be required.

The results from analyses performed for the typical utilities located at the intersections of 2nd Street with Los Angeles Street and San Pedro Street are presented in Table 4.

## MITIGATION MEASURES AND ANTICIPATED EFFECTS

The analytical results indicate that the majority of the adjacent buildings, structures, and utilities have the anticipated damage levels of "Very Slight" or less severe, which require continuous monitoring only. The buildings and structures that have anticipated damage levels of "Moderate" or more severe require mitigation measures in advance. These include five buildings in the Little Tokyo area, the Bunker Hill Central Plant pipes crossing Flower Street, the Los Angeles County Storm Drain and some small utilities in the mixed-face tunneling zone on the eastern end of the alignment.

Mitigation measures recommended for this project consist of (1) controlling TBM ground loss with the advanced TBM technology and (2) grouting technology including permeation grouting, jet grouting, compaction grouting, and compensation grouting.

Ground loss into the tunnel excavation is the most important factor contributing to ground movements around tunnels. Ground loss is generally caused by a combination of three sources: over-excavation of unsupported, unstable ground at the face; intrusion of surrounding material into the annular space caused by the cutterhead overcut and shield conicity; and intrusion of surrounding material into the annular space between the outside skin of the shield and the outside surface of the primary support. The advanced TBM technology allows effective control of these three sources of ground loss through applying positive face pressures, shield bentonite injection, and tail-skin grouting.

Pressurized closed-face TBMs apply a positive pressure to the tunnel face, counterbalancing external earth and hydrostatic pressures; hence being able to limit ground loss at tunnel face to minimal amounts.

Shield Bentonite Injection: Monitored settlement data indicate that 40–50% of total volume loss occurs along the shield (Leca et al., 2006). A system of injection lines is incorporated in new EPBMs to allow a controllable slurry injection, leading to an immediate support of the surrounding ground.

Tail-skin Grouting: The annular gap between the excavated face and the extrados of the lining contributes 30–40% to total volume loss around a tunnel excavation (Leca et al., 2006). This annulus can be effectively filled with grout as the shield advances. In current TBM design, a system of grouting pipes is incorporated that allows continuous grout injection through the tail shield, providing immediate support.

The results from the analyses performed indicate that successfully controlling the ground loss below 0.5% will protect the Redline Tunnels, Broad Museum, and the utilities in the mixed-face zone from considerable damages.

Due to the sensitivity of the Redline Tunnels and the Broad Museum, mitigation measure in form of controlling TBM ground loss with the advanced TBM technology is required even though the anticipated damage levels from the analyses are "Very Slight."

Grouting was recommended as mitigation measures for the Little Tokyo buildings, the Bunker Hill Central Plant piping, and the utilities in the mixed-face zone. The grouting program includes performing permeation grouting or jet grouting prior to tunneling to create a supported zone around the tunnels, hence reducing ground loss due to tunneling. In addition, compensation grout pipes are also installed in advance underneath the building foundation in order to correct the buildings' excessive settlement when detected.

## GEOTECHNICAL INSTRUMENTATION

A geotechnical instrumentation and monitoring program is required to provide warning of potentially damaging settlements to existing buildings, structures, and utilities along the proposed alignment. Recommended geotechnical instrumentation for this project consists of the following.

Multiple position borehole extensometers (MPBXs): Each MPBX would be installed with at least 3 anchors; the deepest to be located about 1.5 m (5 ft) above the tunnel excavation and the other anchors would be located approximately at 3 to 4.5 m (10- to 15-ft) intervals above the lowest anchor.

Deep Benchmarks: Deep benchmarks are installed to provide a reference elevation for comparison of potential elevation changes measured by the MPBX, ground surface points, and building points. They must be installed with the tip at an elevation below the tunnel elevation in order to provide a stable reference that is not affected by tunneling or other near surface influences, such as temperature or moisture changes.

Groundwater Monitoring Wells or Piezometers: Including a standpipe piezometer or pressure transducer for monitoring groundwater elevations. These are required at certain locations along the alignment corridor to track the extent to which groundwater level lowering may occur.

Inclinometers: Inclinometers are required to monitor lateral ground movements due to station excavations. An inclinometer consists of a casing, probe, and readout indicator. The casing is installed within about 1.5 m (3 ft) of the excavation walls, extends below the excavation bottom level, and is grouted to allow the same lateral movements as the surrounding ground.

Ground Surface Settlement Points: These reference points may be installed in arrays that are perpendicular to the tunnel so as to help evaluate the extent of settlement associated with tunneling activities.

Building Monitoring Points: These are survey points usually installed on faces of critical buildings and structures or structures where damaging settlement are anticipated. The monitoring data can be recorded with conventional optical survey equipment or with real-time automated motorized total stations (AMTS).

Crackmeters: In certain cases, crackmeters can be used to monitor construction-related changes to existing cracks. These sensors may be manual or electronic (i.e., vibrating wire crack gauges).

Tiltmeters: In certain cases, tiltmeters can be installed on main structural components of buildings and structures to monitor structure tilt due to ground movements.

Convergence Monitoring Points: Convergence monitoring points are required at key locations to monitor for possible convergence of the bored tunnels, mined cross passages and the mined cavern for the cross-over structure.

Instrumentation Zone: An instrumentation zone is defined as a portion of a cut-and-cover excavation that contains equipment to monitor loads in support elements. For braced excavations, an instrumentation zone includes strain gauges installed on a specified number of struts. For a tie-back excavation, load cells would typically be installed on a group of tie-back anchors on opposing sides of the excavation. In either case, the load monitoring sensors would be read electronically using switch boxes, data loggers, and other associated equipment so that readings can be obtained in near "real time."

CCTV: Preconstruction survey of selected utilities can be performed by closed-circuit TV (CCTV) to examine the existing internal conditions of the utilities. Based on the results of the surveillance, additional mitigation or protection measures could be considered.

## INSTRUMENTATION MONITORING

Two phases of instrument readings are needed as described below.

Pre-construction readings: These readings are conducted to document proper operation of the instruments and document baseline readings for critical buildings and utilities.

Readings during construction: These readings are conducted to confirm proper operation of the instruments and to establish baseline (i.e., pre-construction) conditions. Sufficient measurements are needed to document stable and reliable readings.

Instrumentation monitoring requirements during construction will depend on the progress of the excavation work, the type of instrument, and the characteristics of the structures located nearby. Typically, load monitoring instruments, settlement monitoring instruments, and convergence monitoring instruments are monitored at least daily when excavation is occurring. As noted above, some of these instruments will be required to be monitored automatically, using data loggers.

Other instruments, such as inclinometers and piezometers typically monitor conditions that change less rapidly than the load and settlement instruments, and therefore are typically monitored less frequently. Typical monitoring frequency for these instruments is weekly. However, more frequent monitoring may be required in certain circumstances.

## CONCLUSION

A systematic approach was employed for the evaluation of potential risks associated with the ground movements caused by tunneling and cut-and-cover excavations of the Regional Connector project. The two stage evaluation proves to be an effective approach that allows elimination of non-critical structures and allows in depth assessment of critical structures. A careful review of settlement data of the previous tunneling projects in the similar geological conditions allows a more reasonable estimation of ground loss due to tunneling. Numerical modeling is necessary to estimate the ground movements caused by complex excavations or to assess the potential risks of the structures that are more sensitive to ground movements. This paper presents the work performed during the preliminary engineering phase. As the design-build delivery format is used for this project, the wining team will be responsible for the final assessment of potential damages to adjacent buildings and structures and the corresponding building protection program. It is believed that a combination of proper mitigation measures and a complete geotechnical instrumentation and monitoring program would effectively mitigate the potential risks due to tunneling-induced ground movements.

## REFERENCES


Attewell, P.B., Yeates, J., and Selby, A.R. (1986) Soil movements induced by tunnelling and their effects on pipelines and structures. Blackie, Glasgow.

Boscardin, M.D., and E.J. Cording. 1989. Building response to excavation-induced settlement, *Journal of Geotechnical Engineering Division*, American Society of Civil Engineers, 115(1), 1–21.

Jacobz, S.W., J.R. Standing, R.J. Mair, K. Soga, T. Hagiwara, and T. Sugiyama, (2001). Tunneling effects on driven piles. *Proceeding of International conference on response of buildings to excavation-induced ground movements*, CIRIA publication.

Leca, E., Y. Leblais, and K. Kuhnhenn. 2000. Underground works in soils and soft rock tunneling. Proceedings of International Conference on Geotechnical and Geological Engineering, 1, 220–268.

Mair, R., and F. Borghi. 2008. Influencia del acondicionamiento del material excavado en la operación y control de las tuneladoras a presión de tierra. *Jornada Técnica de Túneles con EPB. Simulación y Control de la Tuneladora.* Universitat Politècnica de Catalunya, Barcelona, Spain.

Mair, R.J., R.N. Taylor, and J.B. Burland. 1996. Prediction of ground movements and assessment of risk of building damage due to bored tunneling.*Proceedings of International Conference of Geotechnical Aspects of on Underground Construction in Soft Ground.* April 1996, London, UK, 713–718.

O'Reilly, M.P., and B.M. New. 1982. Settlements above tunnels in the United Kingdom, their magnitude and prediction, *Proceedings, Tunnelling '82*, 173–181.

O'Reilly, M.P., and B.M. New. 1991. Tunneling induced ground movement; predicting their magnitude and effects. *Proceedings of the 4th International conference on ground movements and structures*. London, Pentech Press, 671–697.

O'Rourke, T.D., and C.H. Trautman. 1982. Buried pipeline response to tunneling ground movements, *Proceedings Europipe '82 Conference, Switzerland*, Paper 1, 9–15.

Peck, R.B. 1969. Deep excavations and tunneling in soft ground, *Proceedings of the 7th International Conference on Soil Mechanics and Foundation Engineering*, Mexico City, State of the Art Volume, 225–290.

Yeates, J. 1985. The response of buried pipelines to ground movements caused by tunnelling in soil. *Proceedings 3rd International Conference Ground Movements and Structures*. Published as Ground Movements and Structures. Edited by J.D. Geddes. Pentech Press, Plymouth, 1985, 145–60.

# TRACK 3: PLANNING

## Session 4: Project Delivery II—Design/Build

Mike McKenna, Chair

# Use of Innovative Project Procurement as a Tool in Shaping the Design of Underground Megaprojects for the Crenshaw LAX Light Rail Transit Corridor Project in Los Angeles, California

**Kimberly Ong and Matthew Crow**
Los Angeles County Metropolitan Transportation Authority

**ABSTRACT:** The Crenshaw/LAX Transit Corridor Project is a $2.058 Bn. transportation infrastructure project that used a single Design/Build (DB) contract to procure a 13.5-km (8.5-mi) light rail transit line including twin bored tunnels, cut and cover tunnels, three underground stations, one elevated station, four at-grade stations as well as aerial guideway and bridge structures over major thoroughfares. The paper provides a summary of the procurement approach, from development of bridging documents, leveraging the proposers' alternative technical concepts, and the resultant design approach.

## INTRODUCTION

The Los Angeles County Metropolitan Transportation Authority (Metro) provides local and regional public transit within Los Angeles County that includes conventional busses, bus rapid transit, light rail transit (LRT) and heavy rail transit (HRT) transporting about 1.5 million passengers each weekday. The existing 142-km (88-mi) LRT and HRT rail network includes 32-km (20-mi) of tunnels. Metro has embarked on a major capital expansion program, as detailed elsewhere (Murthy 2014), to improve public transit service and mobility in Los Angeles County that includes expansion of rail transit lines and tunnels including the Crenshaw/LAX Transit Corridor Project (C/LAX).

## PROJECT DESCRIPTION

The C/LAX project is a north-south light rail transit line that will serve the cities of Los Angeles, Inglewood and El Segundo, and portions of unincorporated Los Angeles County. The guideway, presented on Figure 1, extends 13.5-km (8.5-mi) from the existing Metro Exposition Line, at Crenshaw and Exposition Boulevards, which opened in 2012, to a connection with the existing Metro Green Line at the Aviation/LAX Station, which opened in 1995. The new rail extension will offer an alternative transportation option to congested roadways and provide significant environmental benefits, economic development and employment opportunities throughout Los Angeles County. Riders will be able to make more efficient connections within the entire Metro rail system, municipal bus lines and other regional transportation services.

The line includes eight new stations, three of which are underground at Crenshaw/Exposition, Crenshaw/Martin Luther King, and Crenshaw/Vernon (Leimert Park). There is one aerial station at Aviation/Century with the other stations, Crenshaw/Slauson, Florence/West, Florence/La Brea, and Florence/Hindry being at-grade. All stations will provide safe convenient customer interfaces and site connectivity and will reflect Metro's system wide state-of-the art design standards.

New park and ride lots will be provided at Crenshaw/Exposition, Florence/West and Florence/La Brea. Additionally a new maintenance facility, the Southwestern Yard, will be constructed as a separate DB contract at Arbor Vitae/Bellanca.

The 13.5-km (8.5-mi) alignment provides grade separations with twin 5.8-m (19-ft) diameter bored 2.3-km (1.5-mi) mile long tunnels from Exposition Boulevard to Brynhurst Avenue and a 0.9-km (0.6-mi) long cut and cover tunnel from 59th Place to 67th Street. Aerial structures provide grade separations across La Brea Avenue, La Cienega Boulevard, I-405 Freeway, Manchester Avenue, and Century Boulevard. A below grade partially covered trench is planned near the ends of the Los Angeles International Airport runways, with the remainder of the guideways at-grade. The base project included only six stations, however the stations at Crenshaw/Vernon and Florence/Hindry were bid options that have been exercised.

## FUNDING

The project budget is $2.058 Bn. with a complex mix of funding sources, of which the majority is coming from Measure R, the half-cent sales tax

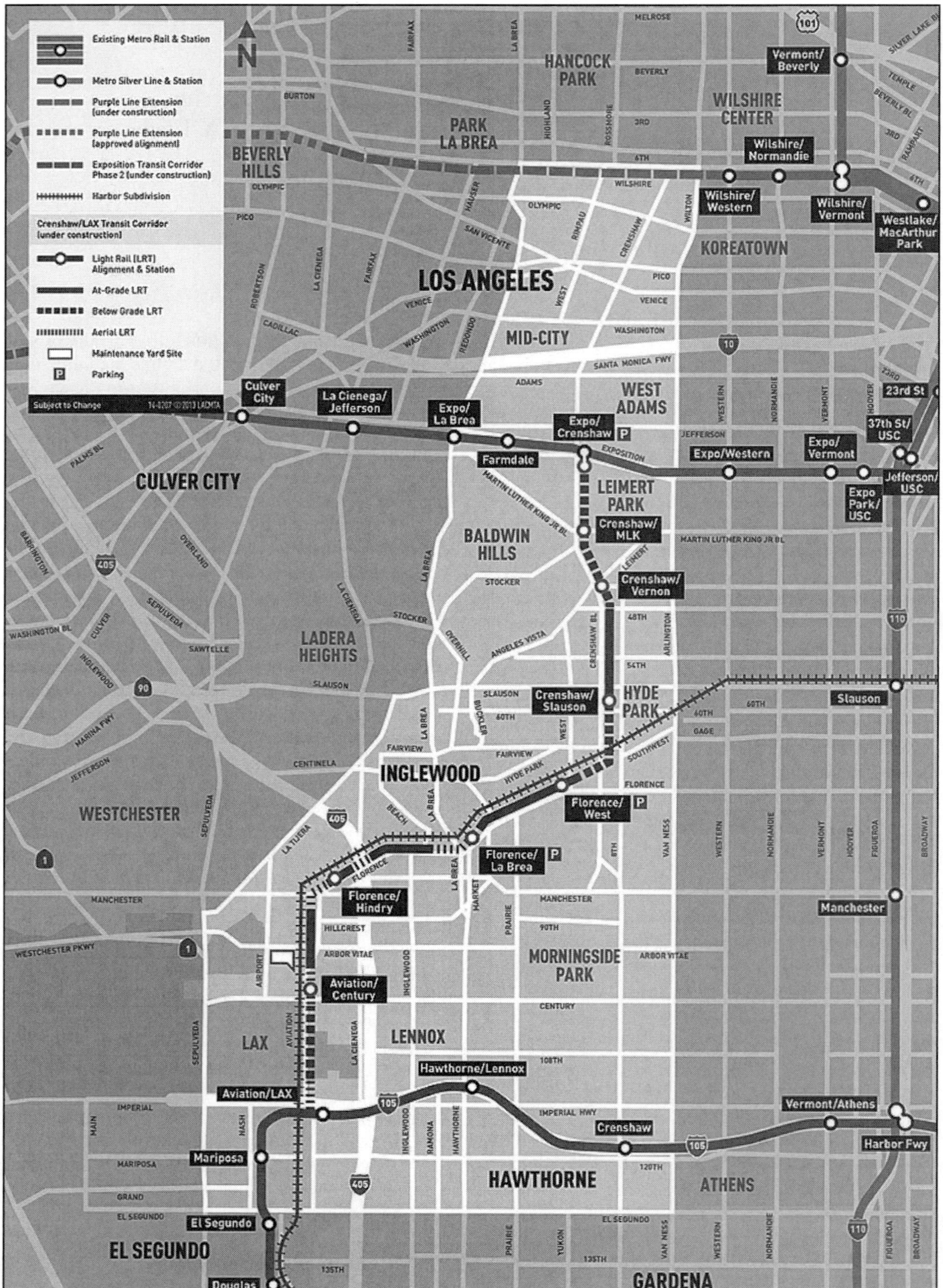


Figure 1. Crenshaw/LAX Transit Corridor Project alignment

initiative approved by Los Angeles County voters in November 2008 and a $546 M loan provided by the United States Department of Transportation under the Transportation Finance and Innovation Act.

## SCHEDULE

The C/LAX transit corridor was included in the first Los Angeles rail system plan in 1967 and has over the past four decades been the focus of numerous plans and studies. More recently the draft Environmental Impact Statement (DEIS) and Environmental Impact Report (DEIR) were completed in 2009, with the final environmental documents (FEIS/FEIR) being approved by the Metro Board in September 2011. A Record of Decision (ROD) was granted by the Federal Transportation Authority (FTA) in December 2011. Preliminary Engineering was initiated during the environmental phase and refined for the procurement documents. The anticipated planned opening of the project is expected in 2019, which will restore rail transit operations in the area that ceased in the 1950s with the demise of the Los Angeles Railway Yellow Cars that operated along Crenshaw Boulevard.

## PROCUREMENT PROCESS

On March 24, 2011, the Metro Board authorized the use of a DB project delivery method for the project using either the Public Utilities Code (PUC) Section 130242, for low bid or Public Contract Code (PCC) Sections 20209.5—20209.14, for negotiated procurement. A project strategy meeting was held in June 2011 and after discernment of the complex nature of the project and the need to hire the best contractor team to work with a diverse and active community, it was determined to use a two stage negotiated "best value" procurement as the preferred procurement strategy. This project was an excellent candidate for DB, as was best value, as it provides the maximum flexibility to proposers without violating basic technical requirements included in the established Metro rail design criteria and standards, as well as other contractual requirements. The use of this method was a departure from the sealed low bid process that Metro has generally used in the past. The method involves a two stage process defined by PCC Section 20209.7 and Metro's Procurement and Procedures. First a Request for Qualifications (RFQ) required a Statement of Qualifications (SOQ) to be submitted by interested responders, which was evaluated on pass/fail criteria. Then a Request for Proposal (RFP) was issued to qualified proposers, who were requested to submit separate technical and price proposals, which were then evaluated by a proposal evaluation team (PET) to make a best value determination. Following the completion of technical scoring price was added to the score to make a recommendation for award.

### Request for Qualifications

California PCC 20209.5 et seq. describes requirements that must be met when using DB delivery methodology in a best value procurement. Metro followed those requirements to pre-qualify DB entities using a standardized questionnaire developed by the State of California Director of the Department of Industrial Relations (DIR). The objective was to select respondents having a particular level of experience and past performance to match the level of complexity of this project, which was paramount to Metro. Unfortunately, the law does not have discretion to delete questions from the questionnaire; however Metro is permitted to supplement the questions with additional requirements to satisfy these needs. Metro also included the Metro Standard pre-qualification application. Metro adopted the scoring system suggested by the DIR to score the SOQs and the pre-qualification applications were evaluated in accordance with the Metro administrative code. All this necessitated a lengthy document. The RFQ was released in December 23, 2011, and SOQ were received on March 12, 2012. Four design build teams were found to be qualified: Crenshaw Transit Partners, (Fluor, Balfour Beatty, SA Healy J.V.); Skanska, Traylor, Kiewit, J.V.; URS, Dragados, Flatiron, J.V.; and Walsh Shea Corridor Constructors. Qualification was based on the scoring of the submitted SOQs, evaluation of pre-qualification applications and enabled participation in the second stage of the Process.

### Request for Proposals

The RFP was issued on June 22, 2012, which required formal technical and price proposals in accordance with the requirements defined in the RFP. A pre-proposal conference was held on July 10, 2012 attended by nearly 300 people. Proposals were received from the four qualified DB entities on December 6, 2012.

### Review of Proposals

The major activities in order to make a best value determination included two distinct phases, a pre-submittal of Alternative Technical Proposals (ATC) s during the RFP stage, which is described later, and a second phase comprising four discrete stages after receipt of submittal of proposals.

1. First evaluation of technical proposals by the PET supported by various subject matter experts. The PET provided initial technical scoring and ranking of proposers. Price proposals were evaluated by a separate cost

team and scores were combined after initial technical scores were determined. The overall scoring weighting was Price 45%, Project Management 30% and Technical Approach 25%.

2. Oral presentations were then requested by Metro with each of the Proposers. The PET evaluated if a contract award could be made or if a final amendment requesting a best and final offer (BAFO) should be exercised.
3. Since an award decision was not made in Stage 2, a competitive range was established and discussions were held with proposers to better understand proposals and opportunities for savings in preparation for release of a BAFO.
4. Finally the BAFO was received from interested proposers, and the PET re-evaluated to determine final scores and recommend a contract award to the Metro Board.

## ALTERNATIVE TECHNICAL CONCEPTS

### Alternative Technical Concept Submittal Requirements

The ATCs were required to be equal or better than the original requirements of the contract and each ATC submittal required specific elements to be addressed as follows:

1. Description of the proposed ATC, including drawings, product data, and other technical information, and a discussion of how the ATC will be used on the Project.
2. Justification for the use of the ATC, including a description of the objectives of the proposed ATC and a discussion of the reasons why ATC would be advantageous to Metro.
3. Technical requirements or other contractual documents, including the RFP that are inconsistent with the proposed ATC, and a description of deviations or modifications to these requirements.
4. Description of other projects or cases in which the proposed ATC has been used, and the results of such usage in achieving stated ATC objectives.
5. Anticipated impact on cost, including construction cost, project management cost, operations and maintenance cost, etc., or anticipated impacts on schedule resulting from its implementation.
6. Operational or maintenance impacts, including any life cycle impacts, or any change in operation or maintenance requirements, anticipated to result from the use of the proposed ATC.
7. Construction, environmental, or safety impacts that could be anticipated from the use of the proposed ATC, including any inconsistencies with or impacts on the Final EIS/EIR or the ROD or any mitigation measure required by or adopted in those documents.
8. Additional right of way or other property interests that would need to be acquired in connection with the implementation of the proposed ATC.
9. Risks to Metro, third parties, or the project resulting from the implementation of the ATC.

Shortly after issue of the RFP, initial ATCs were developed by the proposers and issued to Metro for review. Extensive engineering and cost information was not required for the initial ATCs. There was no discussion with proposers at this initial stage. With hindsight, early confidential one-on-one meetings would have benefited the initial ATC process by providing clarification to the proposers and providing needed Metro input on what constitutes an acceptable ATC. Of the 119 initial ATCs received, 88 were accepted, 27 were rejected and 4 were not considered ATCs.

Metro evaluated the ATCs and identified which ATCs had sufficient merit to be developed further as detailed ATCs. Detailed ATCs were then developed by the Proposers and issued to Metro for review. Confidential one-on-one meetings were then held between Metro and each of the Proposers to clarify the detailed ATCs. The detailed ATCs that were accepted by Metro could then be incorporated into both technical and price proposals. Metro chose not to approve ATCs conditionally in this process. A total of 47 detailed ATCs were submitted by the proposers, which were reviewed by Metro, of which 37 were accepted, 8 rejected and 2 not considered ATCs. The resulting range of cost savings was $50 to $90 million.

### Examples of ATCs

The ATCs received covered many aspects of the project, however due to the inherent prescriptive nature of underground construction for Metro; this limited the innovation related to the implementation of the underground works. This prescription based upon lessons learned in construction of previous transit tunnels in Los Angeles included the use of pressurized face tunnel boring machines to construct a single pass precast concrete lining with double gaskets and cut and cover stations. A few representative examples of ATCs are presented below.

**Figure 2. Alternative technical concept of clear span bridge across I-405 freeway**

### *Station Kit of Parts*

The base design in the procurement documents established a modular approach for station design and construction. Several months before the RFP was released, Metro completed a report that established a system wide concept for station design at both ground and above ground level, the station "Kit of Parts." The Station Kit of Parts was included in the RFP, and application of the Kit of Parts would be determined by the Proposers. The Proposers identified alternative concepts incorporating the Kit of Parts that reduced the construction cost while maintaining the intent to both standardize and provide station branding.

### *Grade Separation and Station at La Brea*

The base design at the La Brea intersection grade separated the LRT below a major roadway, La Brea Avenue, and located the La Brea station at-grade, but below the adjacent major street, Florence Avenue. ATCs were proposed for the guideway to bridge over La Brea Avenue and raise the station to be level with the adjacent street, Florence Avenue. The concept of grade separating over La Brea Avenue was initially evaluated in an earlier phase of the project and approved as part of the Locally Preferred Alternative. However, this was not incorporated into the final environmental approval due to factors including the uncertainty of abandonment of an adjacent Burlington Northern Santa Fe (BNSF) freight railroad track.

Allowing this alternative concept in the RFP process after a formal abandonment order was issued for BNSF did not require a supplemental environmental assessment or opening of the ROD, which was not allowed according to the procurement instructions. Subsequent discussions with the FTA allowed Metro to clear this change through a categorical exclusion process under National Environmental Policy Act (NEPA). By implementing this alternative concept, the construction cost was reduced, the construction impact to the community was greatly lessened, and the operations and visibility of the station was significantly improved.

### *Bridge Over I-405 Freeway*

The guideway crosses over the I-405, one of the busiest freeways in North America. The reference design required demolition of the existing freight railroad bridge and construction of foundations and bents for a multi-span bridge within the freeway, as it was considered more cost effective than a clear span bridge. An alternative concept proposed the guideway bridge as a clear span over the freeway (Figure 2.), which at first sight would appear more expensive; however, the existing freight bridge would be used to support the falsework for the cast in place structure and then the existing freight railroad bridge left in place for alternative uses. This reduces impact on freeway traffic significantly, which not only reduces costs and has potential schedule benefits, but also avoids unnecessary disruption to regional mobility during construction.

### *Optimizing Horizontal and Vertical Alignment Profiles*

Proposers were encouraged to optimize the alignment, particularly underground segments, where raising a station box by several feet could generate significant cost savings. One ATC went as far as to completely shift one of the underground stations allowing the elevation of the station to be raised by over ten feet. This design innovation required discussion with a third party who was brought in under a confidentiality agreement to solicit their comments and support before accepting this ATC. In the majority of ATCs that took advantage to optimize the alignment, the only restrictions placed on proposers were compliance with the approved environmental documents, Metro's design criteria, and no additional property acquisition.

## Benefits of ATCs

Including the ATCs as part of the procurement process was considered beneficial for both the proposers and for Metro. Benefits included improved contract documents, better understanding of the contract documents by the proposers, improved design solutions, and competitive pricing. In certain cases, improvement to the contract documents was made as amendments to the RFP for benefit of all proposers. The ATC process clarified the type of design innovation that is permissible and enabled critical review of contract technical requirements. Improved design solutions were facilitated as both designer and contractor worked collaboratively in developing solutions that can enhance or streamline project features while providing the team a competitive edge and lower costs to Metro. The ATCs resulted in "out of the box" solutions that were incorporated into the bid price and allowed execution upon contract award. Alternatively, if the ATC process had not been used, the alternative concepts may not have been identified to Metro until after the Notice to Proceed (NTP) as value engineering, which would reduce potential schedule benefit and only allowed Metro cost savings of 50% that are likely understated without competition. Upon contract award and payment of stipends to unsuccessful proposers, all ATCs became the intellectual property of Metro with freedom to direct implementation to the successful proposer. The confidential ATCs and clarification one-on-one meetings, within the extent allowed, enabled Metro to have technically and price competitive proposals.

## BEST AND FINAL OFFER PROCESS

### Discussions

Metro advised proposers that a recommendation for award was not made after initial evaluations and scoring. Questions were developed and discussions were coordinated over a one month period with proposers on both technical and price proposals. Discussions included additional ATCs and cost saving measures that were encouraged on an individual proposer basis and some instances reflected in changes to technical requirements that were put out to all proposers. These BAFO discussions resulted in a number of cost savings that were evident in all proposers' final offers. These included changes in requirements for deflection criteria for support of excavation walls for station boxes and cut and cover segments, which allowed proposers more flexibility in means and methods. Increased flexibility in hardscape and landscape treatments in station and plaza areas also realized savings. Additional laydown areas and an existing office facility on Metro-owned property for use as a co-located project office were provided at a new planned 18 acre maintenance facility adjacent to the C/LAX alignment. A Reduction in the number of key personnel that proposers were required to commit to the project full time allowed proposers more flexibility in staffing. Cost savings were also realized by a reduction in daily liquidated damages, a cap on maximum exposure on liquidated damages, a reduction in the cap on limit of liability on Builder's Risk insurance and limit on damages to correct non performing work.

## CONTRACT AWARD

On June 27, 2013, the Metro Board approved a base award of a 57 month, firm fixed price contract to Walsh/Shea Corridor Constructors (WSCC) for the final design and construction of the C/LAX project in the amount of $1,177,032,356 for the base project and $95,600,000 and an additional 3 months for station options at Vernon/Crenshaw (Leimert Park) and Florence/Hindry.

## CONCLUSIONS

This large and complex procurement of light rail infrastructure was clearly successful, however a number of issues deserve further discernment. Allowing confidential one-on-one discussions with proposers from initiation of the RFP, instead of after submittal of detailed ATCs could be considered, as a number of proposers commented that the ATC process was the only means to have dialogue with Metro and encouraged dialogue regardless if there is a formal ATC process or not. These considerations should include addressing concern of avoiding entering into negotiations with prospective proposers ahead of receiving formal proposals. The two-stage ATC provided a valuable screening process to eliminate concepts that were not deemed feasible for incorporation during the procurement process, which allowed the Proposers and Metro to concentrate resources on only those potentially viable concepts. Allowing the ATC process without any initial commitment to a BAFO process allowed discussion of alternative concepts early in the procurement process that could have been held until the BAFO phase, thus allowing a shortened duration for the BAFO phase.

## REFERENCE

Murthy, K. and Crow, M., 2014. Project Delivery Strategy for Major Transportation Projects in Los Angeles. North American Tunneling 2014 Conference Proceedings.

# Tunnel Construction by a Mixed-Use Real Estate Developer

**John F. Beesley and Christopher Allen**
DC Water and Sewer Authority

**Lou Williams**
Forest City Washington

**ABSTRACT:** The Washington Navy Yard was, for over 200 years since 1799, a center for naval shipbuilding and outfitting. Today, a portion of the site is being redeveloped as a vibrant waterfront neighborhood by Forest City Washington. DC Water's Clean Rivers project required a 340 m (1,120 ft), 1.8 m (6 ft) inside diameter tunnel beneath Tingey Street for control of combined sewer overflows to the Anacostia River. Forest City agreed to facilitate the Tingey Street construction since it has several projects being built concurrently in the immediate area. The Tingey Street diversion sewer project put the developer, Forest City, in a significant project coordination role for the publicly-funded USD $16 million tunnel. This paper describes how DC Water and Forest City executed this fast-tracked, design-build project.

## INTRODUCTION

The DC Clean Rivers Project is a $2.6 billion USD program that will reduce combined sewer overflows (CSO) to the Anacostia and Potomac rivers in the Washington, DC area. The project has been segmented into distinct construction packages that have been termed "Divisions." The DC Clean Rivers project is subject to a Consent Decree that was filed in the District of Columbia district court on March 25, 2005. The Consent Decree requires that Phase 1 of the program be completed by March 25, 2018; therefore, the Tingey Street diversion sewer needs to be placed in service by that date.

DC Water has significant schedule float for the Tingey Street diversion sewer, but Forest City Washington, the local real estate developer, is also in the midst of a major redevelopment program in the exact same area. It made a lot of sense to work with Forest City so that the construction of the adjacent projects could be completed simultaneously.

As seen in Figure 1, which shows the relatively new US Department of Transportation headquarters building, some of the area has already been redeveloped. The laydown yard for the launch shaft of the Tingey Street tunnel is in the foreground of Figure 1. This parcel of land, where the launch shaft laydown yard is located, is also slated for redevelopment by Forest City Washington in the near future.

Heavy construction, by its nature, is always a little messy and by entering into an agreement with Forest City Washington to construct the Tingey Street diversion sewer, DC Water could greatly reduce the overall negative impacts to the local community from the project. The negative impacts consisting of construction traffic, noise, detours, and torn-up streets are still unavoidable, but this unique agreement between DC Water and the real estate developer allowed Forest City Washington to control those impacts in conjunction with the other projects in the immediate area. As seen in Figure 2, construction of the manhole shaft at CSO-013 and construction of Forest City Washington's Twelve12, a mixed-use residential and commercial building, can easily appear to be the same project.

## DESIGN-BUILD PROCUREMENT

In the fall of 2011, DC Water and Forest City Washington concluded negotiations concerning the path forward for the Tingey Street diversion sewer. A contract was signed on January 27, 2012 and Forest City Washington, in consultation with DC Water, started the process to obtain a subcontractor for the Design-Build work. Forest City was paid a lump sum for Preconstruction Services that Forest City Washington used to fund final preparation of a Request for Proposal (RFP) and procure a design-build entity to construct the project.

The contract required Forest City Washington to evaluate the proposals, in conjunction with DC Water, and to select a design-build contractor for the tunnel and concrete structures. DC Water provided an initial set of drawings and specifications to establish the minimum design requirements in the RFP. The selected design-builder used the initial set of design documents to guide them in preparing their proposals, and later as a starting point for the final design.

**Figure 1. The Division B launch shaft laydown yard is across Tingey Street from the new US Department of Transportation headquarters building**

**Figure 2. The CSO-013 manhole shaft (cat excavator) sits directly in front of the Twelve12 building construction project. Casual observers would think they are the same project, but they are unrelated.**

During the spring of 2012, Forest City received proposals from four design-build teams for the Tingey Street diversion sewer. A fifth team had initially expressed interest in submitting a design-build proposal, but that team subsequently withdrew. The teams submitting proposals were:

- Corman/Bradshaw/Jenny
- Northeast Remsco/CDM Smith
- Southland Contracting/Brierley
- Ulliman Schutte/Aldea

After evaluating the proposals from both a technical and price standpoint, Forest City selected the design-build team of Northeast Remsco/CDM Smith. The technical aspects of the proposals and each proposer's approach to the project were carefully considered during the evaluation period by DC Water and Forest City Washington. Ultimately, the technical considerations were not significantly different from one another and the approaches did not vary substantially. The project ended up being awarded to the proposer with the lowest cost, as would be the case with a typical public works design-bid-build procurement.

The design-build team of Northeast Remsco/CDM Smith was given the green light to proceed and a Contract between Forest City and Northeast Remsco was finalized in late June 2012. This action kicked off the final design effort and CDM Smith began preparing the final design documents.

## FINAL DESIGN OF THE TINGEY STREET DIVERSION SEWER

The Request for Proposal that Forest City Washington issued included a design for the Tingey Street diversion sewer that was not 100 percent complete. Mandatory requirements, such as the diversion sewer pipe size and basic diversion structure features had been established. The design-build team set to work completing the final design configuration in July 2012. The inside diameter of the tunnel was changed from 1.68 m (66 in.) to 1.83 m (72 in.), but initially that was the only major design change that varied from the original RFP. The up-sizing had been presented in the original proposal that DC Water reviewed from Northeast Remsco, and DC Water had approved that change during the proposal preparation phase of the project.

In addition to the primary tunnel, a relatively short, pilot tube, guided auger-bore tunnel was also needed. The pilot tube guided auger-bore tunnel was 0.91 m (36 in.) diameter and approximately 33.5 m (110 ft) long. The design also required two cast-in-place diversion chambers at CSO-013 and CSO-014. Each structure included trash racks, stop logs, tide gates, access covers and other features to divert and control flows to the new Clean Rivers tunnel system.

Design packages were developed and submitted in tandem to both DC Water and Forest City Washington for review and concurrence as the final design proceeded during the fall of 2012. Ultimately DC Water approved the final design drawings and the final design specifications. Forest City Washington coordinated the process and reviewed the drawings and specifications to ensure that they were compatible with their other work that was occurring simultaneously in the same area.

Three major Forest City Washington projects were directly adjacent to Tingey Street diversion sewer work areas. The Forest City Washington properties are The Boilermaker Shops, a retail/commercial property; The Foundry Lofts, a mixed use property of apartments and ground level retail establishments; and Twelve12, a mixed-use residential and commercial property. Construction of Twelve12 and the Tingey Street diversion sewer occurred simultaneously and Forest City was in direct control of both construction operations. The residential units in The Foundry Lofts, shown in Figure 3, had been first occupied in the fall of 2011. Naturally, Forest City wanted the new residents to be impacted as little as possible by the project. Forest City reviewed the design so that conflicts and interfaces among these areas could be worked out during the construction phase of the project.

Jet grout columns covering a distance of about 107 m (350 ft), a cast-in-place diversion structure at CSO014, and the CSO-013 manhole shaft each were directly adjacent to the Twelve12 construction site and contractor staging areas. Forest City Washington was able to address pending issues during the design process with both contractors and was generally successful in maintaining harmony between both projects.

Forest City also coordinated design issues with other stakeholders along Tingey Street. After the design had been completed, another significant issue arose when the US Department of Transportation, a tenant in one of the adjacent buildings, objected to reducing Tingey Street to only a single lane during construction. Forest City was able to work with their design-build team and they adjusted the tunnel alignment so that the launch shaft could be moved farther from the street. The launch shaft, while remaining in the public right-of-way, was simply placed closer to the Forest City property line.

The construction staging area at the launch shaft eventually took up more of a parking lot, but since Forest City owns the parking lot, the design change was significantly easier to accomplish. In the end, the required design change mandated by the US

**Figure 3. The CSO-013 diversion chamber excavation, with Forest City's Foundry Lofts apartments in the background. Forest City worked directly with their residents and their contractor to minimize construction impacts.**

**Figure 4. A rendering of Forest City's Twelve12 property. The CSO-013 manhole shaft is approximately where the white car is located in the bottom-right of the figure.**

Department of Transportation was greatly facilitated since Forest City controlled the parcel of land adjacent to the launch shaft.

## CONSTRUCTION OF THE TINGEY STREET DIVERSION SEWER

Construction of the Tingey Street diversion sewer project moved into full swing in early 2013. The jacking pit for the primary 1.83 m (72 in.) tunnel was adjacent to DC Water's main pump station at the end of New Jersey Avenue southeast. At that location a separate shaft construction project, Division A, was also underway for the Clean Rivers program, but the Division A work did not impact the Tingey Street project. Other jobsites, however, were in the same project area and a large building under construction by Forest City, the Twelve12 development, was the principle "other" project that Forest City needed to coordinate.

The CSO-013 manhole shaft was in the immediate area where Forest City's Twelve12 mixed-use property was being constructed. As seen in Figure 4, if the manhole construction had been constructed at a later date, new tenants at the Twelve12 property would have been directly impacted by its construction. Though the shaft was constructed entirely in the public right-of-way, excavating the shaft during the same time period as the building construction was a natural benefit to the future property occupants. On numerous occasions both contractors working for Forest City Washington, Walsh Construction at Twelve12 and Northeast Remsco doing the tunnel, had equipment and men working immediately adjacent to one another.

The intersection of 4th Street Southeast and Tingey Street became very congested during some phases of construction. A large gravity sewer, the Eastside Interceptor, ran beneath the intersection and plans required that the sewer be supported in situ by a chemical grouting program. The drilling and chemical grouting contractor needed to occupy most of the intersection for the grouting program. Building construction also needed access through portions of the intersection that were destined for closure by the grouting contractor. However, since both contractors ultimately reported to Forest City Washington; coordination problems were resolved with Forest City prioritizing the work.

The intersection of 4th Street Southeast and Tingey Street is also situated directly over the Washington Metropolitan Area Transit Authority's (WMATA) Green Line. This heavy-rail, dual-tunnel subway is approximately 46 ft (14 m) below the Tingey Street tunnel. Monitoring in the subway tunnel was done remotely, but the intersection surface was also actively surveyed on a daily basis during the tunneling phase of construction. The intersection, as seen in Figures 2 and 4, was, therefore, the scene of numerous concurrent activities and Forest City was able to prioritize and coordinate the various parties that needed to work in that location.

The launch shaft for the diversion sewer tunnel was completed on schedule in July 2013. The receipt shaft took a little longer than planned, but it was successfully completed by the end of August. The 2.26 m (7 ft 5 in.) Herrenknecht AVND 1800 microtunnel boring machine (MTBM) "Go Yard" was assembled and placed in position the first week of September, see Figure 5. "Go Yard" was launched into full production on September 17 and completed a successful tunnel drive about 20 days later on October 7. Though the MTBM tunnel drive encountered some wood debris during the mining cycle, the drive was completed without significant delay.

The tunnel alignment consisted of alluvium, gravel, sand, and clay. The tunnel also passed through "fill" materials from past Washington Navy Yard activities; but fortunately "Go Yard" mined

**Figure 5. Northeast Remsco's microtunnel boring machine "Go Yard" being readied for launch in the shadow of the Washington National's baseball stadium**

through, or pushed aside, everything that the cutterhead encountered in the tunnel alignment. Due to the knowledge that some of the tunnel alignment was in material that had been filled in by the US Navy in the 1800s, the potential for encountering a significant underground obstruction was a constant concern while mining the last half of the tunnel.

Figure 6 is a picture of "Go Yard" after the drive was completed. Proper planning by the contractor, good coordination by Forest City Washington, and some luck that nothing significant was left behind by the US Navy in the alignment; resulted in a successful tunnel drive by Northeast Remsco Construction. Removal of the tunnel support equipment was uneventful and the annulus was grouted with a simple grout mix of cement, water and a small amount of bentonite.

In addition to the main tunnel, a short 0.91 m (36 in.) pilot tube guided auger-bore tunnel was also constructed at the intersection of 4th Street Southeast and Tingey Street. The contractor initially installed a 10 cm (4 in.) pilot tube between the CSO-013 diversion chamber shaft and the CSO-013 manhole shaft. The bore length was approximately 33.5 m (110 ft). After installing the pilot tube, a 0.91 m (36 in.) fiberglass reinforced pipe was jacked into position by following the pilot tube. The cuttings were removed using an auger and casing system that was nested inside the 0.91 m (36 in.) fiberglass pipe.

Concrete diversion structures were constructed on two active CSOs that drained areas to the north of the Washington Navy Yard in southeast Washington DC. Temporary flumes built from steel pipe were installed in both sewers for most of the work, but on a few occasions, before the flumes were installed, the contractor had to step aside when rains inundated the system. Fortunately, the months of August and September 2013 were drier than normal for the Washington DC area and the contractor was not significantly affected by combined-sewer overflows. The contractor was prepared to handle overflows and dealt with them as expected, but overflows basically did not materialize as often as expected.

The diversion structure at CSO-014 was located between two historic Washington Navy Yard buildings, as seen in Figure 7. This limited the construction work and staging area. Construction impacts to Building 74, as seen on the right in Figure 7, needed to be closely monitored since the building was only 3 feet from the excavation. Since Forest City occupied Building 74, they could easily monitor construction impacts to themselves and the other tenants in this 1930s era building.

Work on the cast-in-place diversion structures was completed in the latter part of 2013 by a local cast-in-place concrete subcontractor. Street and landscaping restoration was also performed in late 2013 and early 2014. The contract completion date between DC Water and Forest City was April 9, 2014. However, since the contract between DC Water and Forest City Washington included liquidated damages, Forest City required their contractor to reach substantial completion well before the April deadline. The contractor successfully met all of the completion milestones established by DC Water and Forest City Washington for the Tingey Street Diversion Sewer project.

## BENEFITS TO USING A REAL ESTATE DEVELOPER

On numerous occasions during this project it was evident that using Forest City Washington to build the tunnel and sewer diversion structures was worth the extra cost in fee and general conditions that DC Water paid to Forest City Washington. At times Forest City's contractors had to work side-by-side as they built totally separate projects. For example, in Figure 8 crews are seen installing jet grout columns for the tunnel project, and they are working directly

**Figure 6. "Go Yard" after completing the Tingey Street Diversion Sewer tunnel drive**

**Figure 7. The CSO-014 excavation and receipt shaft were immediately adjacent to two historic Washington Navy Yard buildings that Forest City is planning to renovate as part of the overall area redevelopment program**

adjacent to a transit mixer that is placing concrete for the Twelve12 building construction. This type of work occurred often. However, both contractors worked for Forest City Washington, so Forest City could quickly resolve any issues or interferences that might arise between contractors.

Table 1 is a brief summary of a few issues that were resolved relatively quickly by Forest City. Since Forest City was the controlling entity over much of the land and development in the area, their involvement in resolving these issues was crucial to solving the problems.

## CONCLUSION

DC Water entered into a unique arrangement with real estate developer Forest City Washington so that the Tingey Street diversion sewer project could be built quickly, and be less disruptive, to an emerging neighborhood. As the primary real estate developer for the area, Forest City was in a unique position to coordinate multiple construction projects in the same general area. By coordinating this Clean Rivers tunnel project, Forest City facilitated construction by managing job sites that overlapped and directly influenced each other. This arrangement placed a single decision-maker, Forest City Washington, in the unique position to facilitate the mitigation of negative impacts from numerous heavy construction areas. Subsequently, two separate owners did not have to manage their individual contractors around each other since DC Water had assigned the coordination responsibility of their project to Forest City Washington.

**Figure 8. Tunnel crews working adjacent to crews working on Forest City's Twelve12 mixed-use development. The transit mixer belonged to the building contractor, but the drill rig is part of the tunnel project. Both projects reported to the Forest City Washington.**

**Table 1. A few issues solved by The Yards real estate developer, Forest City Washington**

| Issue | Problem | Solution |
|---|---|---|
| Construction Staging–CSO-013 Manhole Shaft | Two contractors needed to work in the exact same location | Since DC Water used Forest City Washington as the Division B contractor, Forest City was in direct control of both builders. Forest City reviewed competing problems and could balance the needs of each of their subcontractors. |
| Relocate Launch Shaft | The US Department of Transportation objected to Tingey Street being reduced to only one lane. | Forest City Washington controlled the adjacent land (a paved parking lot) that would be impacted by the change. Forest City coordinated the encroachment of construction equipment onto their parking property without a lengthy delay. |
| Spoils Disposal for Jet Grouting and Tunneling | A suitable area for tunnel and jet grout spoils was not available. | Forest City Washington owns other parcels in the immediate area that DC Water could only have used through lengthy negotiations. Forest City allowed their subcontractors to access areas that otherwise would not have been available. |
| Equipment Staging and Pipe Storage | Contractor staging areas were very small | Forest City Washington permitted the tunnel contractor to use a vacant lot, as seen on Figure 9, on short notice and a temporary basis. |

**Figure 9. During one phase on the project, Forest City allowed storage of sewer pipe on a vacant lot for about 3 weeks. Small accommodations such as this helped the project succeed.**

# Specifying Design-Builder's Pressurized-Face Soft Ground Tunnel Boring Machines—Learning from the Past and Moving Forward, from Designer's Perspectives

**Peter Chou, Joseph Rigney, Bradford F. Townsend, and Jonathan Taylor**
Parsons Corporation

**ABSTRACT:** Pressurized-face Tunnel Boring Machines (TBM) are in high demand in urban areas in North America and internationally. Recent technology innovations, such as new conditioning admixtures, remote sensing/monitoring, advanced probing and grouting techniques, have enabled the construction of large diameter tunnels in shallow ground cover and difficult subsurface conditions with minimal ground loss. This is evident from a wide range of recent literature on pressurized-face tunneling projects. This paper provides a comparison of over 40 such tunnels based on their dimension, machine features, geological setting, ground loss, and ground improvement performed. Based on recent design-build tunnel experience, recommendations are presented for best practices in specifying soft ground TBMs to maximize performance and reduce risk.

## INTRODUCTION

Pressurized-face Tunnel Boring Machines (TBM) are in high demand; thanks to rapid population growth in the major cities. Aging transportation and utility infrastructures are in urgent needs of either an expansion or an upgrade. However, bored tunnel construction can cause major concerns such as adverse impacts to public safety, third party property damages, and construction delays or cost overrun due to inappropriate construction methods or insufficient design specifications. Contributing to technology innovations, the ability to control ground deformation when constructing large urban tunnels using pressurized-face TBMs in shallow, weak, saturated, complex mixed-ground conditions has greatly improved. Modern TBMs with advanced grouting techniques alongside conditioning admixtures and precise real-time monitoring can maintain face stability, and in return, reduce ground loss. This is evident from a wide range of recent pressurized-face tunneling project literatures.

There are many unique features available to TBMs, and the suitability of a TBM to the ground directly governs the cost and schedule of a project. It is recognized that some differing site condition (DSC) claims may have been submitted as a result of improper TBM selection or lack of special features on TBMs, rather than actual unexpected ground conditions. In addition to accelerating overall project schedule, one reason that alternative contracting methods such as design-build and Public-Private-Partnership (P3) have been more widely used in recent years is in attempts to allocate, minimize and mitigate risks associated with construction methods. So, how does one specify a Design-builder's TBM? There are perceptions in the industry that in weak, saturated ground or mixed face conditions, specifying either a slurry or earth pressure balance (EPB) TBM would be almost risk free. Although there have been plenty of successful projects in recent years, it should be noted that there were still projects resulting in significant DSC claims even when closed face TBMs were specified and used. Tunnel construction continues to expose all contractual parties to risks associated with unknown or unpredicted ground conditions in response to selected tunneling methods.

Prior to tunnel construction, the owner typically carries out sufficient preliminary investigations, both on existing facilities and future structures, evaluates tolerance of existing structures, and assesses various prevention and mitigation measures that may be required. The project-specific design criteria and technical specifications will then be issued, typically on a performance basis for design-build projects. The specifications focuses on typical TBM features related to a pressurized face, either EPB or slurry (or a slurry/EPB hybrid), probing and grouting ability in front of TBM, and tail void grouting, alongside some performance basis criteria monitored by instrumentation. However, following this process, there had been projects that experienced ground problems using the specified TBM. An example is the Port of Miami Tunnel (POMT), built in 2011–2013, where the ground was too porous for a conventional EPB machine. Fortunately, the contractor identified the concern at a very early stage, and the solution for POMT's 12.8 meter diameter machine was to modify the EPB to a hybrid system and perform additional ground treatment in the form of grouting and soil mixing.

A comparison of over 40 tunnels based on their dimensions, machine features, geological setting, ground losses and ground improvement applications have been analyzed in this paper. Based on data collected and review of recent design-build TBM specifications for tunnels in Los Angeles, Seattle, Miami and Washington DC, recommendations with regards to best practices in specifying soft ground TBMs from performance basis and risk reduction viewpoints are discussed.

## KEY COMPONENTS OF OWNER'S TBM SPECIFICATIONS

Designers recognize that most TBM equipment-related decisions are better left to the contractor. Therefore, for design-build projects, the Owners typically focus on principle TBM performance-based requirements in contract documents. As long as the contractor performs tunnel construction in such manner to minimize ground movement and control water inflow at all times under all conditions, the Owner can still benefit from the advantages of design-build projects: integral design and construction, simplified accountability and responsibilities, and risk sharing. However, is this the best strategy if a tunnel construction involves difficult ground conditions, high risk environments (third party and/or environmental impacts), and demanding design criteria? Below are some of the key components related to soft ground TBM (and tunneling operations) in the Owner's design-build project specifications.

- TBM Geometry
- TBM Selection (see Table 1 machine comparison)
- Lining and Grouting Procedure
- TBM Performance and Drive Tolerance
- Use of Foam and Additives (EPB)
- Special Feature, such as probing, air lock, retractable cutter
- Monitoring Bentonite Suspension (Slurry)
- Positive Closure and Safety

Also, other project-specific TBM performance (such as ground movement, or sometimes ground loss) and operational requirements are described in the most recent tunnel projects, and are listed below.

### TBM Ground Loss & Mitigation Methods

It is critical to the success of tunneling that the TBM face and its full perimeter be controlled to minimize ground loss. A pressurized TBM with good tail grouting technique is a must for soft ground tunnels to minimize uncontrolled ground losses, and resulting subsidence. A good quality grouting around the tunnels could even reduce ground loss associated

**Table 1. Comparison of EPB and slurry TBMs' Applications (modified from Thewes, 2008)**

| TBM Types | EPB | Slurry (Mixshield) |
|---|---|---|
| Suitable ground type | General fine grained soils; require significant conditioning efforts in granular and coarse grained soils | Best for open grade, and granular soils; require high slurry separation and danger of clogging in fine grained soils |
| Groundwater and compressed air intervention | Slower emptying head chamber and preparing compressed air intervention; use planned ground improvement intervention | Easier control and faster compressed air intervention, lower temperature in cutterhead chamber |
| Boulders | Higher wear on disc cutter, blocking screw conveyor possible | Less wear on disc cutter, stone crusher possible |
| Mix-face ground | High risks of instability and wear | Better control of face support and wear |
| Machine cost | Typically lower (therefore, more pressurized face TBMs on projects are EPB) | More expensive, less used machine available |
| Face control during stop | Desegregation of foam and soils in the chamber | Continuous uninterrupted face support |
| Settlement (face pressure control) | Less precise, fluctuation and non-linear distribution of face pressure support | Very precise (0.05 bar) and sensitive, linear distribution of face pressure support |
| Gassy ground | Possible exposure in tunnel | No exposure in tunnel |
| Contaminated soils | Possible exposition in tunnel, less muck for special treatment | No exposition in tunnel, more muck for special treatment |
| Muck disposal | Direct disposal possible | No direct disposal, higher cost |
| Operation | Complex for conditioning and face support in mix face or transitions; more interior space | More complex for slurry pipes and separation plant; less power consumption |
| Jobsite setup | Smaller | Larger form separation plant |

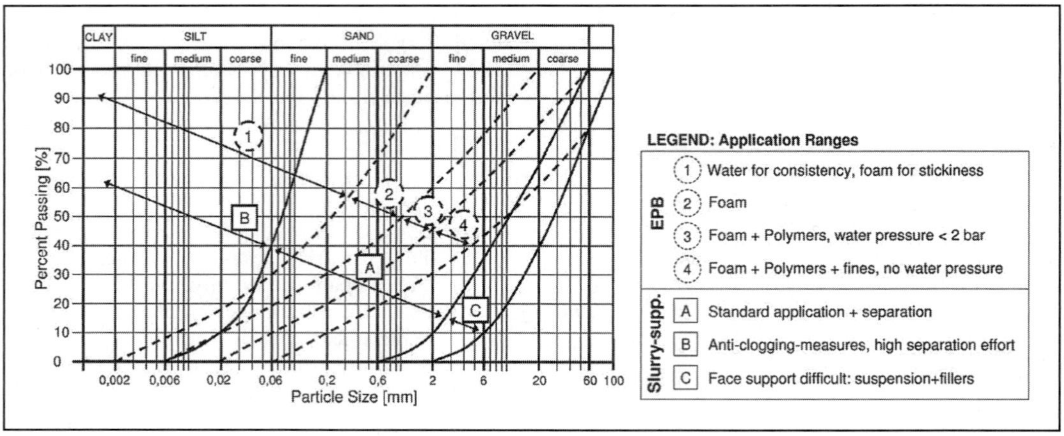


**Figure 1. Application ranges of pressurized face TBM with additives (Thewes, 2008)**

with soil consolidation by preventing leakage into tunnels.

Grouting voids as a two-stage grouting process can be considered to ensure complete ground to liner contact. After tail grouting (first stage), sometimes the second stage grouting takes place behind the trailing gear of the machine where grout holes are drilled through the grout ports in the lining, through the first stage grout. The design of the grout ports is a critical detail in the segment design in order to minimize the potential for these ports to be a source of leakage. The use of foam and additives is another popular solution to controlling ground loss. Using an automated and computer controlled injection system will condition the muck, and reduce both torque requirements and cutter wear. The new additives developed really push conventional EPB operation range to a new limit (see Figure 1). Because of this, owners typically only provide prescriptive requirements when a slurry TBM is the only choice.

### Mandatory Launching Method and Site

Larger tunnels require much greater quantity muck handling and removal out of tunnel, entry of precast concrete segments into the tunnel, and TBM power supply and ventilation at the launching site. Therefore, the location to handle spoils and direction of TBM launching could be prescriptive based on the Owner's environmental assessment and impact study (e.g., LA Metro's Regional Connector Transit Corridor). In addition to material handling, due to potential third party coordination or schedule concern (i.e., permitting), the Owner could attain the required TBM power supply in advance, which would result in a designated launching site (e.g., LA Metro's Purple Line Extension, DC Water First Street Tunnel).

### Minimum Number of TBMs Required, Condition of TBM

Based on the tunnel length and anticipated penetration rate, the Owner may want to specify the minimum number of TBMs to meet project schedule and reduce equipment breakdown risks (e.g., LA Metro's Purple Line Extension: minimum two TBMs). Sometimes, reconditioned/rebuilt TBM(s) are permitted subject to meeting the project specifications and the Owner's approval (e.g., LA Metro's Regional Connector Transit Corridor).

### Mandatory Protection of Existing Structure

Each project has specific areas of known risk of damage by construction. There are also community concerns for impacts of construction, and third party concerns for the impacts of construction on underground utilities and other underground infrastructure. In such cases, the Owner should establish mandatory requirements that the contractor must implement as an integral part of the construction. The specifications for mitigation usually have the following major elements:

- **Pre- and Post-Construction Survey:** Documentation of condition of existing structures.
- **Settlement Analysis:** An assessment of expected settlements and impact on structures, resulting from construction. The need for protection of the existing structures is a function of both settlement and the characteristics of structures.
- **Comprehensive Instrumentation and Monitoring:** Monitor ground and structure movements, and provide a maximum allowable settlement and deflection slope to mitigate risks.

- **Mitigation Actions:** Minimize, limit and avoid damage to structures due to settlement caused by tunneling. Mitigation techniques include mandatory pre-tunneling ground improvement underpinning, and compensation grouting during tunneling.

## LITERATURE REVIEW OF RECENT AND PAST TUNNEL PROJECTS

A case history review of over 40 pressurized face TBMs tunnels has been performed, mainly focusing on ground losses (and surface settlement) versus tunnel depth, cover-diameter (C/D) ratio, ground type, machine type and year built. The tunnel penetration rate is also compared. The reviewed tunnels are divided into two groups (built prior to and after 2000). Machine features and ground improvement methods, if performed and literature available, are also discussed and reviewed with TBM performance. The reviewed tunnels built prior to and after 2000 are summarized in Tables 2 and 3, respectively.

### Ground Loss and Settlement

Based on our literature review, the tunnel projects built in late 1990s which introduced ground improvement applications (e.g., Docklands, Madrid Metro Extension and Bangkok Metro) generally reduced ground loss to less than 1%. It is interesting to point out that in early 1990s, greater settlements seemed to occur on the projects using closed-face TBMs in cohesionless soils or alluvial soft clay without ground treatment (e.g., Taipei RTS). For Edmonton SLRT (Tweedie et al., 1989), up to 200 mm of settlement was reported in sand. After 1994 or so, ground improvements and tail grouting became a popular choice for soft ground tunneling; therefore, the magnitude of ground settlements began to reduce. However, in the late 1990s, the feasibility of closed-mode operation in the stiff to hard plastic clays became a concern. For Jubilee Line Extension (Jardine, 2001) as an example, greater ground loss (1.2%–1.7%) was reported for tunneling in stiff clays, compared to a ground loss of 0.3%–0.7% in granular soils when 1st generation computerized soil conditioning injection system and automated grout pump system became available. The tunnel projects built after 2000 that had employed advanced TBM performance monitoring technology, more precise and automated grouting and mucking volume monitoring, and performed ground improvements to critical sections had generally controlled ground loss to less than 0.5%. Early examples are Heathrow Airside Road Tunnel (Darby, 2003) where compressed air and interlock was introduced to assist face support (reduce foam use in stiff London Clay) for EPB.

For Channel Tunnel Rail Link (Bower et al., 2005), the use of additives resulted in a good control of the chamber pressure, and in return, the volume losses were generally on the order of 0.25% (settlement less than 15 mm) in Thanet Sand, and 0.25%–0.75% in London Clay. Trial runs were performed on the project, and the results indicated that its reduction in face pressure and use of soil conditioning increased ground loss to up to 3%. Reported ground loss and maximum settlement versus year built are shown in Figures 2 and 3, respectively.

"Early learning curve" issue still existed at the TBM start in more recent tunnels. As an example, overall settlement was about 2 mm (ground loss of 0.5%) for Denny Way CSO (Cochran, 2003), which utilized an EPB TBM with computerized soil conditioning, pumping and air pressure monitor; however, up to 90 mm (ground loss of 6%) was still recorded at the start (first 75m) of TBM penetration in sands with a shallow ground cover (10 m).

### Tunnel Depth and Diameter

In early 1990s, several empirical approaches (such as Clough and Schmidt, 1981; Mair, 1993) had provided the correlations among tunnel depth and diameter (or its ratio), ground loss, surface settlement and face stability number (Lee et al., 1992), based on past instrumentation data. From this review, it does appear that such correlation exists for tunnels constructed prior to 2000. However, through proper ground improvements and innovative TBM technology, surface settlements and ground losses become small, regardless of tunnel depth and diameter. The 15 m diameter Madrid Line 10 (M-30) tunnel (Fernandez, 2005) bored at a depth as shallow as 20 m only resulted in a reported 12 mm maximum settlement. Tunnel depth versus maximum reported surface settlements, and ratio of tunnel depth to diameter versus reported ground losses are shown in Figures 4 and 5, respectively. Figure 6 shows tunnel depth versus ratio of maximum settlement to tunnel diameter.

From literature review, if a bored tunnel is to be constructed in flowing ground (silty and sandy soils), there are important parameters to consider, including operating only in closed-mode, maintaining minimum face pressure, and preventing TBM stoppage outside of planned intervention zones. For example, one lesson learned is that stopping/slowing TBM without properly maintaining face pressure could result in higher ground loss (if without ground treatment in advance). At intervention, inspection or un-intended TBM stoppages, unfavorable settlements occurred from 100 mm (Bangkok Metro; Maconochie, et al., 2001) to up to 500 mm (at Sydney Airport; Nye, 1999) were observed in previous construction.

**Table 2. Case histories of tunnels built prior to 2000**

| Tunnel Project, before 2000 | TBM Type | Year Built | Reference | Depth (m) | Diameter (m) | Soil Type | E (ave) Soil (MPa) | Su (kN/m2) | Ko | Soil Unit Weight (kN/m3) | Tail Gap (mm) | Maximum Settlement (mm) | Reported Volume Loss (%) |
|---|---|---|---|---|---|---|---|---|---|---|---|---|---|
| San Francisco N-2, Line #4 | EPB | before 1983 | Chou and Bobet, 2002 | 9.7 | 3.7 | Soft Clay (Bay Mud) | - | 24 to 28 | - | - | 140 | 31 | 2 |
| San Francisco N-2, Line #2 | EPB | | | 9.7 | 3.7 | Soft Clay (Bay Mud) | - | | - | - | | 46 | 6.3 |
| Milwaukee CT-8-1 | EPB | before 1989 | Chou and Bobet, 2002 | 16 | 3.2 | Stiff clayey silt | 20 | 35 | 0.5 | 19 | 320 | 114 | 6.1 |
| Milwaukee CT-8-2 | | | | 12.8 | 3 | Medium stiff organic silt | 20 | 35 | 0.5 | 19 | 196 | 70 | 6.1 |
| Milwaukee NS-10-U | Slurry, jacked pipe | | | 7.9 | 2.25 | Medium stiff organic silt | 20 | 35 | 0.5 | 19 | 324 | 84 | - |
| Milwaukee NS-10-R | | | | 7.6 | 2.25 | Medium stiff organic silt | 20 | 35 | 0.5 | 19 | 78 | 55 | - |
| Milwaukee CT-5/6 | EPB | | | 10.7 | 3.23 | Soft to medium stiff organic silt | 20 | 35 | 0.5 | 19 | 98 | 40 | 6.8 |
| Milwaukee CT-7 | EPB | | | 7.4 | 3.57 | Soft to medium stiff organic silt | 20 | 35 | 0.5 | 19 | 103 | 56 | 9.7 |
| Edmonton SLRT, Phase 2, Canada | Slurry | 1986-1989 | Tweedie et al, 1989 | 6-16 | 6.3 | Silty sand and till | 25 to 50 | 25 | 0.85 to 1 | 20 | 75 | less than 15 | - |
| Taipei RTS, Hsintien Line 218 B1 | EPB | before 1993 | Moh, 1993 | 18.5 | 6.05 | Loose silty sand and soft to stiff silty clay | - | - | - | - | - | 20 to 30 | 1.3 to 2 |
| Villejust High Speed Rail, France | Slurry | 1988-1994 | Mair, 1996 | varies | 9.25 | Fine sands | - | - | - | - | - | - | 0.7 to 1.3 |
| Bangkok Water Transmission Tunnel (MWA) | EPB | 1994-1997 | Chou and Bobet, 2002 | 18 | 2.67 / 2.8 | Soft to stiff clay | 20 | 15 | 1 | 17 | 37 | 12 | 3 to 4 |
| Cairo Metro Line 2 | Slurry | 1994-1996 | Mair, 1996 | 18-28 | 9.48 | Nile alluvium | 4-13 | 75 | 0.5-0.7 | 19 | - | 5-25 (up to 73) | 0.2 to 1 |
| Mexico City Sewer | Slurry | before 1997 | Romo, 1997 | 12.75 | 4 | Soft Clay interbedded with silts | - | - | - | - | - | 25 | 2.2 |
| Docklands Light Railway, Lewisham Extension, London, UK | Slurry | 1996-1998 | Sugiyama et al., 1999 | 12-18 | 5.9 | Stiff to hard clays with gravels | - | 200 to 250 | - | - | - | 4 to 8 | 0.7 to 1.0 |
| | | | | 13-18 | | Stiff to hard clays and silty fine dense sand | - | | - | - | - | 5 to 12 | 0.5 to 0.9 |
| Second Heinenoord | Slurry | 1997-1999 | E.P. Van Jaarsveld et al., 1999 | 16.7 | 8.5 | Holocene, Dense Sand | 24 | - | - | 24 | - | 22 | 1.3 |
| Taipei RTS, Hsintien Line CH221 | Slurry | 1998 | Yang et al., 1997 | 8-16 | 6.25 | alluvial silty sand to soft to stiff clay | 22 to 30 | - | 0.45 | 20 | - | 27 to 55 | 0.6 to 1.1 |
| | Slurry | 1998 | | 25-30 | 6.25 | | 23 to 30 | - | 0.45 | 20 | - | - | 0.25 to 1.5 |
| New Southern Railway at Sydney Airport | Slurry | 1999 | Nye, 1999 | 17-22 | 10.7 | clays and sands | 60 | - | 0.6 | 20 | 150 | 6 | 0.2 |
| Madrid Metro Extension, Lines 4 and 9 | EPB | 1995 to 1999, and 1999 to 2003 | Melis et al., 2002 | 9.5-11.9 | 9.38 | Loamy sand, clayey sand, and clays | 80 to 170 | 100 to 150 | 1+ | 20 | - | 15-16 | - |
| Alexandria Wastewater Tunnel, Egypt | EPB, jacked pipe | before 2000 | Abdrabbo et al, 1999 | 13 | 2.86 | Silty clay and silty sand | - | 10 | - | 18.5 | - | 25 | 4 to 6 |
| Bangkok Metro (MRTA) - Blue Line | EPB | 1996-2001 | Maconochie, et al., 2001 | 12-28 | 6.47 | Stiff Clay & Dense Sand | 66 | 80 | - | 20 | 80 | 9 (initial)<br>15 (two weeks) | 0.5 to 1 |
| Jubilee Line Extension (Easten) Keetons Estate | EPB | 1999 | Burland, 2001; Withers, 2001 | 20 | 5 | Stiff clays | - | - | - | - | - | - | 1.2 to 1.7 |
| Jubilee Line Extension (Easten) Southwark Park | EPB | 2000 | Jardine, 2001; Withers, 2001 | 15-20 | 4.5 / 5 | Lambeth Group (i.e. sands, clays and gravels) | - | - | - | - | - | 8 to 25 | 0.3 to 0.7 |

Notes:

1. Up to 8 TBMs were used for one project (i.e., Taipei RTS, Madrid Metro Extension and Bangkok Metro); the table only reports the TBM type that used for the alignment/ contract which instrumentation data was reported. The maximum settlement reported in the table does not include anomaly, or incidents such as TBM break down.
2. Year built was the time that actual tunneling was conducted.
3. Depths reported were the depths at the instrumentation sections where surface settlement was reported; Tunnel depths vary along its alignments.
4. Soil parameters reported were from available back analyses of the reported maximum settlement and/or ground loss.

**Table 3. Case histories of tunnels built after 2000**

| Tunnel Project, After 2000 | Construction Method | Year Built | Reference | Depth (m) | Diameter (m) | Soil Type | E (ave) Soil (MPa) | Su (kN/m2) | Ko | Soil Unit Weight (kN/m3) | Tail Gap (mm) | Maximum Settlement (mm) | Reported Volume Loss (%) |
|---|---|---|---|---|---|---|---|---|---|---|---|---|---|
| Denny Way CSO, Seattle, WA | EPB | 2001-2004 | Cochran, 2003 | 10-30 | 4.5 / 5 | Saturated hard clay, abrasive sand and gravel | - | - | - | - | - | 2 (up to 90) | 0.5 (up to 6) |
| Airside Road Tunnel (Heathrow Airport) | EPB, compressed air | 2002 to 2003 | Darby, 2003 | 20 | 8.8 / 9.2 | Thames Gravel, London Clay; Stiff to hard clay | - | 150 to 200 | 1.5 | 20 | 200 | less than 15 | 0.2 to 0.35 |
| East Central Interceptor Sewer, Los Angeles | EPB | 2002-2003 | Crow and Holzhauser, 2003 | 9-105 | 4.7 | alluvial sand, gravel, silt and clay, thru active fault zone | - | - | - | - | - | generally < 19 (up to 39) | - |
| Channel Tunnel Rail Link, London, UK (Contract 220, 240 and 250) | EPB, compressed air | 2002-2004 | Bowers et al., 2005; | 12-25 | 8.2 | London Clay; Stiff to hard clay (Cont. 250) | - | 250 | - | - | - | 5 to 8 | 0.5 |
| | | | Mair and Borghi, 2008 | 25-35 | | Fine and medium silty sand (Thanet sands) | - | - | - | - | - | 5 to 15 | 0.2 to 0.8 |
| Toulouse Line B (France) Contracts 2 & 5 | EPB | 2002-2005 | Emeriault, Kastner, 2008 | 13-23 | 7.7 | Hard sandy clay with pockets dense sands | 165 | 300 | 1.7 | 22 | 200 | 2 | - |
| West Side CSO Tunnel, Portland, OR | Slurry | 2002-2006 | McDonald, 2007 | 15-20 | 5.1 | Alluvail sand/silt, gravel and Troutdale | - | - | 0.2-0.5 | - | - | - | - |
| MTA Gold Line Eastside Extension, L.A., CA | EPB | 2004 to 2007 | Choueiry et al., 2007 | 15 | 6.5 | Alluvial stiff clay, lean clay, sandy and sand and gravel | 48 | 24 | 0.8 | 21 | - | less than 6 (12 max) | < 0.3 |
| Barcelona Metro Line 9, Barcelona, Spain - Mas Blau to San Cosme Segment | EPB | 2006 | Mignini et al., 2008 | 14-20 | 9.4 | Submerged fine silty sands and clayey silts | 15 to 35 | 7 to 15 | 0.4 to 0.5 | 19 | - | 15 to 35 | 0.4 to 0.8 |
| Barcelona Metro Line 9, Barcelona, Spain - Fira to Park | EPB | 2006 | Orfila et. al, 2007 Della Valle, 2007 | varies | 9.4 | Silty sands with sandy silts, silts and silty clays | - | - | - | - | - | 10 | 0.3 to 0.4 |
| C-710 Beacon Hill Tunnel (Sound Transit) | EPB | 2005 to 2007 | Redmond et al, 2007 | 47 | 6.4 | Glacial soils, silty sands and gravel | 300 to 400 | 200 to 500 | 0.7 to 1 | - | - | - | - |
| Barcelona Metro Line 9, Barcelona, Spain - Segment IV-B | EPB | 2007 | Della Valle, 2007 | varies | 11.9 | Sands, clay, and silts overlying gravels with sands | - | - | - | - | - | - | 0.7 to 1.0 |
| Barcelona Metro Line 9, Barcelona, Spain - Segment IV-C | EPB | 2007 | Della Valle, 2007 | varies | 11.9 | Silts and sands and gravels in a sandy clay matrix overlying weathered rock | - | - | - | - | - | - | 0.2 to 0.6 |
| Madrid South Bypass M-30 Tunnels, Madrid, Spain | EPB | 2006-2008 | Universidade da Coruña. 2008 Fernandez, 2005 | 20-40 | 15.2 | Mixed face of sandy clay overlying hard clay with weathered rock | 220 | 60 | - | 20 | - | 12 | 0.1 to 0.4 |
| Italy High Speed Rail Bologna-Florence Line | EPB | 2008 | Do et al, 2013 | 15-25 | 8.3 / 9.4 | Alluvial clay and sandy soils | 150 | - | 0.5 | 20 | 150 | 30 | - |
| Seattle University Link Light Rail U220 | EPB | 2008 | Burdick et al, 2012 | 36 to 49 | 6.5 | Clay, silt, sand and gravel | - | - | - | - | - | less than 13 mm | - |
| Shanghai River Crossing - Chong Ming Tunnel | Slurry | 2006-2007 | Ng, et al , 2008 | 60 max | 14.53 / 15.43 | Soft clay and loose silty sand | - | - | - | - | - | - | - |
| Sao Paulo Metro Line 4 - Lot 1, San Paulo, Brazil | EPB | 2009 | Pellegrini and Perruzza. 2009 | varies | 9.5 | Clay and sandy clay with gravel, and stiff clay with coarse sands | - | - | - | - | - | - | < 0.4 |
| Crossrail, UK | EPB (Slurry for chalk) | 2011-2013 | Crossrail Website, 2013 | 25 and 40 max | 6.2 / 7.1 | London clay, sand and gravel | - | - | - | - | - | 4 to 12 | 0.4 to 0.9 |
| Italy Rome Metro Line B1 Tunnel | EPB | 2012 | Castellanza et al, 2008 | 15-40 | 10 | Alluvial loose silts and clay soils | - | - | - | - | - | 9 | 0.2 to 0.4 |
| Italy Milano Metro Line 5 Tunnel | EPB | 2013 | SELI website | varies | 6.8 | Silt, clay and sand | - | - | - | - | - | - | - |

Notes:

1. Up to 8 TBMs were used for one project (i.e., Channel Tunnel Rail Link and Crossrail); the table only reports the TBM type that used for the alignment/contract which instrumentation data was used.
2. Same as Table 2.

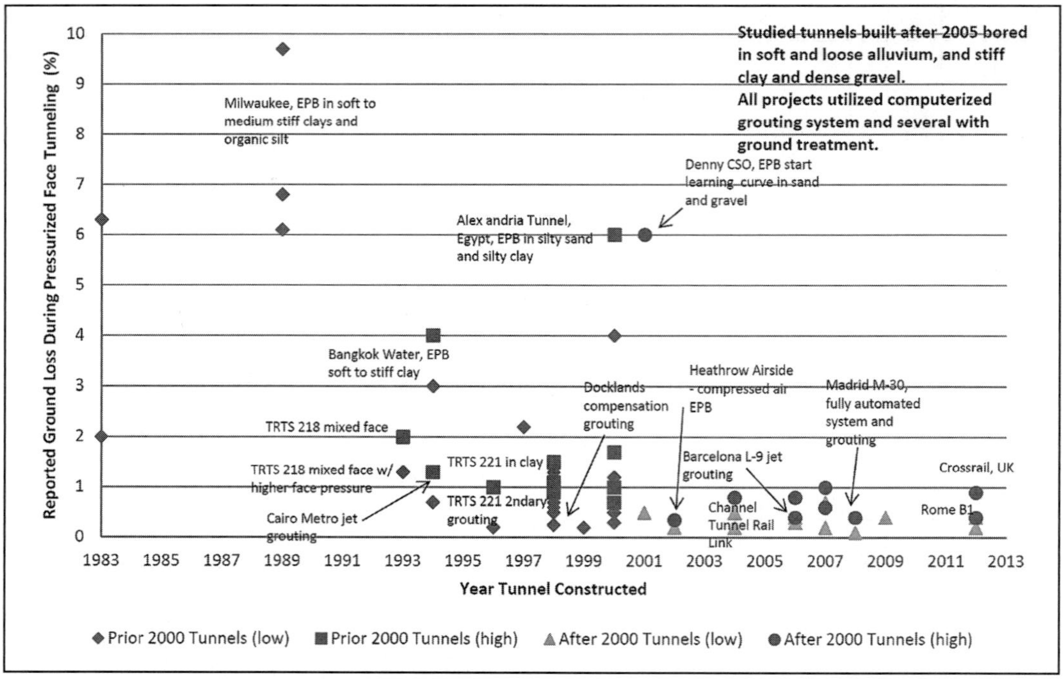


**Figure 2. Reported ground loss versus year tunnel built**

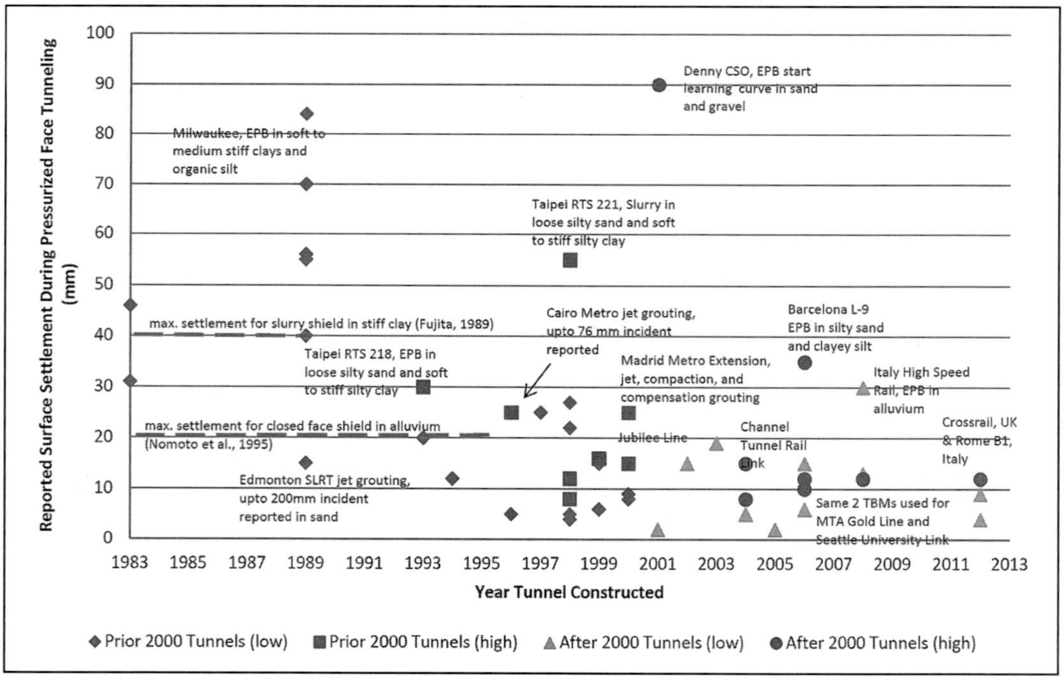


**Figure 3. Reported maximum surface settlement versus year tunnel built**

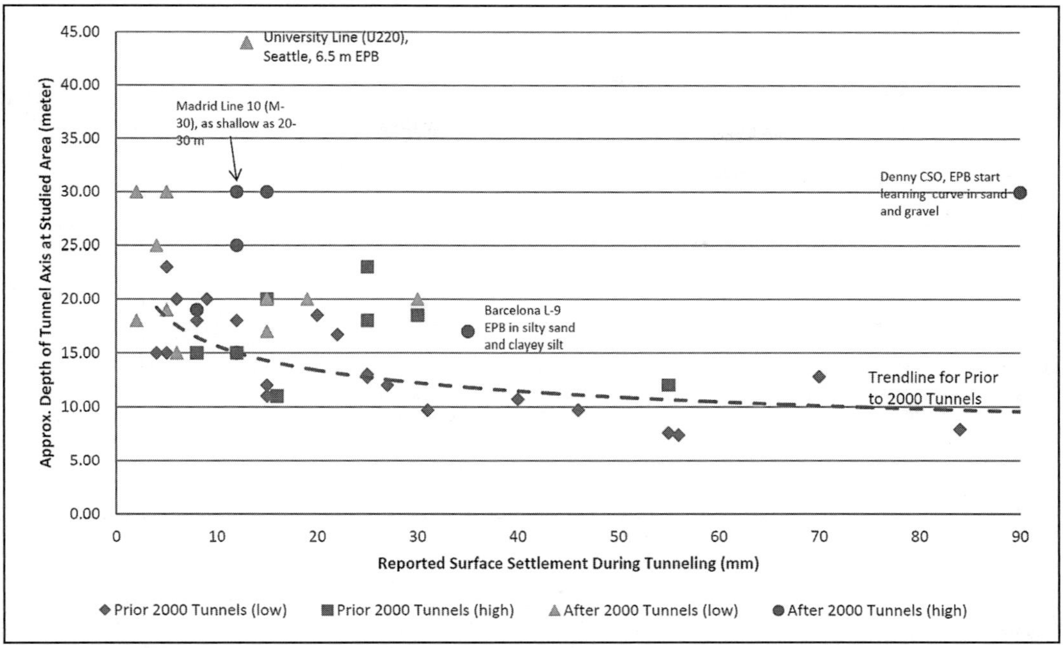


**Figure 4. Tunnel axis depth versus reported maximum surface settlement**

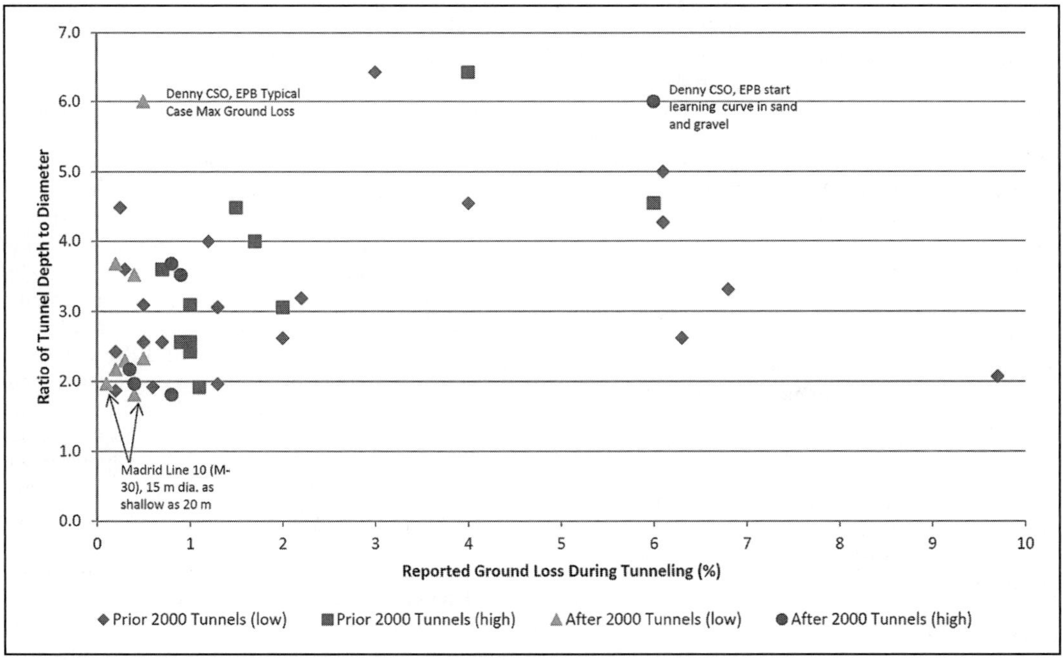


**Figure 5. Ratio of tunnel depth to diameter versus reported maximum ground loss**

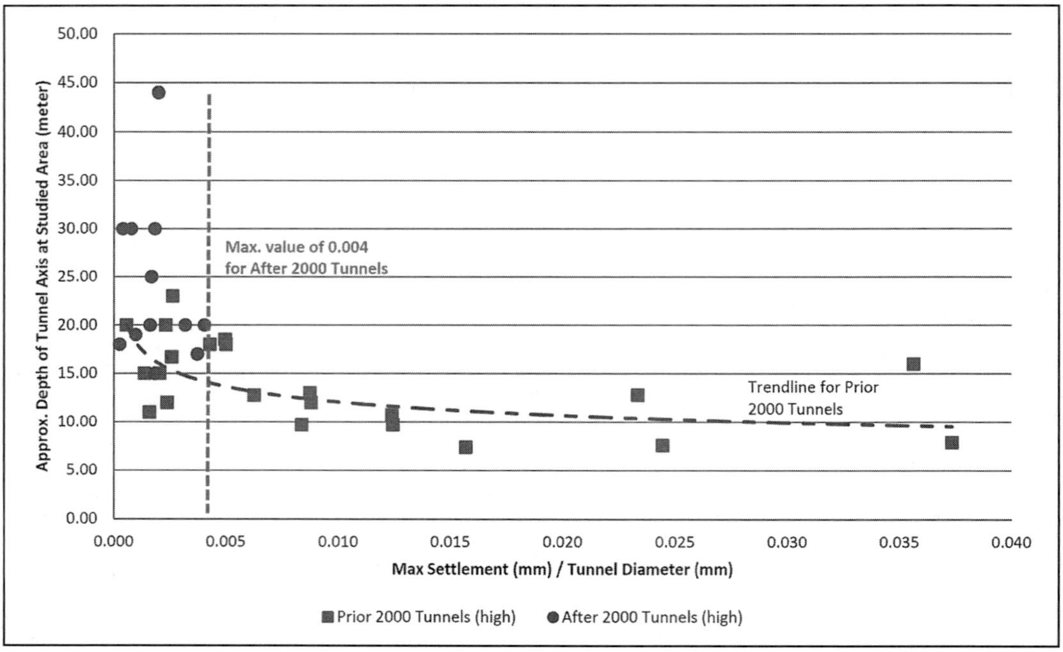


**Figure 6. Tunnel depth versus ratio of maximum settlement to tunnel diameter**

Literature also indicates that conventionally, slurry TBMs are the preferred choice for saturated gravelly and cobble ground, as crusher is generally efficient, and if there is a need to enter the cutterhead chamber, the slurry fluid can be readily evacuated and replaced with compressed air. However, after 2000, EPB machines had been successfully used in glacial cobble ground, such as Beacon Hill tunnel (Redmond et al., 2007).

## TBM Production and Performance

As tunneling performance has improved, ground deformation has lessened. With all the automated grouting and lining installation, and real-time performance monitoring techniques available, TBM production rates have also increased in time, regardless of excavated diameter, as shown in Figure 7. TBMs with over 100 MN thrust force and a torque of over 100 MNm are not uncommon, such as Madrid Line 10 (M-30) machine. The difference in production rate of EPB and Slurry TBMs appear to be less significant in recent tunnel projects.

## CONCLUSION

Tunneling methods must be selected for suitability considering the geologic and hydrologic conditions, possible impacts on the adjacent structures, compatibility with final ground support, safety and economy. The TBM's suitability to the ground conditions poses significant uncertainty to the project costs and schedule. From lessons learned, once the design-build team's means and methods, equipment, procedures, and sequences have been chosen and the tunnel is underway, changes can be extremely costly and difficult, and may not be even possible. Therefore, specification of all construction means and methods, sequences, equipment, and systems should consider the full range of ground conditions. Primary objectives of TBM specifications are to provide key mandatory TBM features or operation requirements which have demonstrated successful applications in promoting construction and public safety, and minimizing ground loss and third party impacts. Based on the review, both EPB and slurry machines performed well with any size and depth. Unless ground and environmental conditions forbid the use of one type of TBM over the other, the owner will permit both. Exceptions include Sydney Airport (Nye, 1999), where it was believed that a slurry TBM could control both the face pressure and the tail grout pressure best, and Heathrow Airside Road Tunnel (Darby, 2003), where an EPB machine delivered better production in stiff to hard London Clay.

Prescribing TBM type defeats the risk sharing goal for design-build projects. The specifications should mitigate unnecessary risks from the Owners by sharing responsibility of TBM selection and some of its secondary operation parameters with the contractor. Based on the literature review and our knowledge of recent TBM innovations, the following key

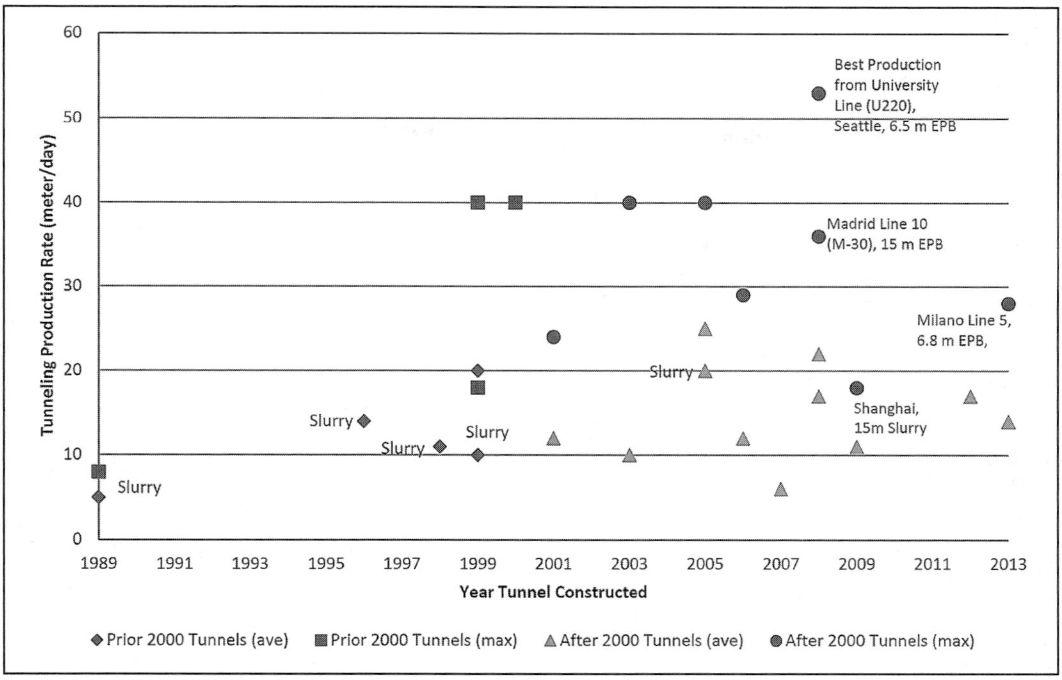


**Figure 7. Tunnel production rate versus year tunnel built**

TBM features are recommended to address in the design build tunnel specifications.

### Face Support

TBM shall always operate under positive pressure (less than at-rest pressure), sometimes no stoppage can be allowed (or request mandatory intervention zone), and always monitor and control pressure in the chamber (soil conditioning, spoil removal and advance rate). Utilization of air to aide in maintaining EPB pressure was quite successful at Heathrow Airside Road Tunnel (Darby, 2003), and mandatory compressed air locks (to access face when required) shall be considered.

### Probe Hole and Grouting Through TBM Face

Specifying minimum open ratio on cutterhead to deal with potential cobble/boulder and obstruction, and requesting extra open area for probing could be considered.

### Grouting Technology and Its Monitoring System

In soft ground, grouting through segments shall become secondary and supplemental. Tail grouting shall be mandatory, aiming at increasing strength and stiffness and/or reducing ground permeability, and shall consist of interlocking of annular grouting to minimize potential voids when the TBM advances.

- **Tail Grout** provided with a computer operated integrated grouting system which considers rate of advance, grout quantities, prevailing external pressure and related variables (grout setup). The grouting system must be able to handle very large and variable grout quantities at very low pressure, using a hydrophilic grout which solidifies quickly to avoid grout migration and free flow into the solution channels or voids. A two-component annular backfill grout that is sufficiently fluid to flow freely under pressure through the system and completely fills the annular space between the lining and the excavated ground as the TBM advances, prior to its initial set.
- **Compensation grouting** has resulted in significant reductions in ultimate ground losses and building settlements for soft ground tunneling. During the construction of the Madrid Metro Extension (Melis et al., 2002), jet grouting, compensation grouting and compaction grouting were used to improve the ground, and provided building protection with successful results. With compensation grouting, less than 8 mm of settlement was reported underneath buildings in Jubilee Line Extension (Jardine, 2001). Compensation grouting (ground treatment) shall be considered in urban soft ground tunneling, in addition to new TBM features.

### Additive Injection and Automatic Pressure Recover System

TBMs with automated soil conditioning and fine compensation injection system are quite common now. In order to guarantee a continuous confinement pressure at the face, TBMs employed in Barcelona's Metro Line 9 (Mignini et al., 2008) were equipped with an automatic pressure recovery system. These modernized EPB shields have been equipped with a grouting system to execute volume-controlled slurry bentonite injection in annular gap around the shield. Even in normally consolidated silts with ground covers ranging from 1.5 to 2 diameters, almost negligible settlement with ground losses well below 0.5% can be maintained. In Channel Tunnel Rail Link (Bowers et al., 2005), through continuous injection of bentonite around TBM prevent additional ground movements due to TBM steering overcut, and reduce ground loss to as low as 0.25%.

### Tunnel Performance Demonstration and Instrumentation Sections

Geotechnical instrumentation shall concentrate at specific locations to measure detailed subsurface ground deformation to provide a basis for evaluating tunneling performance, before tunneling under sensitive structure. To mitigate TBM "learning curve" concern (high ground loss at the TBM start, i.e., Jubilee Line Extension; Jardine, 2001), performance demonstration section is highly recommended to demonstrate the contractor has the ability to conduct tunneling to meet settlement performance requirements, protect the utilities in the street, and demonstrate that pressured-face tunneling minimizes ground movement.

### Real-time Integral Instrumentation and TBM Monitoring

The 'look ahead' techniques that are integral to the TBM using seismic, acoustic and electrical methods have experienced considerable growth in their development. Cutterhead monitoring has also experienced growth, with the recent advent of wireless data collection and transfer from the rotating cutterhead to the project control center. Real-time data of machine functions shall be transmitted via integrated system that provides remote access through internet or intranet to data and data reductions stored. Other areas of TBM monitoring such as grouting and muck measurement have also advanced; continuous time plots of face pressures (individual measurement and average), overcut annulus pressures outside the shield (individual measurement and average), volume and weight of material excavated and removed from the face, tail grouting and bentonite injection, extensometer and vibrating wire piezometer readings at the nearest locations to the tunnel face. As TBMs grow in size and are employed in challenging ground conditions, advances in monitoring will likely continue to help ensure efficient and productive tunneling operations.

### Safety Features

TBM specifications shall also address the safety features of the TBM, such as gas monitor with automated power shutoff, automatic fire suppression systems, and the ability to create a smoke cut-off "curtain" at the rear of the TBM.

## REFERENCES


Abdrabbo F.M., El-Nahhas F.M., El-Lateef H.A. 1999. Field measurements during construction of a Sewer tunnel. Geo-engineering for Underground Facilities. Proceedings of the Third National Conference, June 13–17, 1999. pp. 879–887.

Bowers, K., E. Woods, and F. Mimnagh. 2005. Urban tunneling experience on the UK's Channel Tunnel Rail Link. Rapid Excavation and Tunneling Conference Proceedings, Seattle, WA. Littleton, CO: Society of Mining Engineers. pp. 769–782.

Burdick M., Krulc M., McLane R., Brandt J. 2013. University Link Light Rail TBM Tunnel UWS to CHS Contract U220: Case History. Rapid Excavation and Tunneling Conference Proceedings, Washington DC. Littleton, CO: Society of Mining Engineers. pp. 987–1013.

Burland J.B. 2001. Assessment methods used in design building response to tunneling–Case studies from construction of the Jubilee Line Extension. London, Vol. I, Projects and methods, pp. 23–43.

Castellanza R., Bett A., Lambrughi A. 2008. Three-dimensional Numerical Models for Mechanized Excavations in Urban Areas. Jornada Técnica de Túneles con EPB. Simulación y Control de la Tuneladora. Universitat Politècnica de Catalunya, Barcelona, Spain. pp. 105–128.

Chou W., Bobet, A. 2002. Predictions of Ground Deformations in Shallow Tunnels in Clay. Tunneling and Underground Space Technology 17 (2002). pp. 3–19.

Choueiry, E., A. Elioff, J. Richards, and B. Robinson. 2007. Planning and construction of the Metro Gold Line Eastside Extension Tunnels, Los Angeles, California. Rapid Excavation and Tunneling Conference Proceedings, Toronto, Canada. Littleton, CO: Society of Mining Engineers. pp. 808–828.

Clough G.W. and Schmidt B. 1981. Excavations and Tunnelling, Soft Clay Engineering, Chapter 8, Edited by E W Brand and R P Brenner, Elsevier.

Cochran J., Pearia T., Robinson, R., Haynes C. 2003. Successful Application of EPB Tunneling for the Denny CSO Project, Seattle, Washington. Rapid Excavation and Tunneling Conference Proceedings, New Orleans, LA. Littleton, CO: Society of Mining Engineers. pp. 1104–1116.

Crow M., Holzhauser J. 2003. Performance of Four EPB-TBMs Above and Below the Groundwater Table on the ECIS Project, Los Angeles, CA, USA. Rapid Excavation and Tunneling Conference Proceedings, New Orleans, LA. Littleton, CO: Society of Mining Engineers. pp. 905–926.

Darby A.W. 2003. The Airside Road Tunnel Heathrow Airport, England. Rapid Excavation and Tunneling Conference Proceedings, New Orleans, LA. Littleton, CO: Society of Mining Engineers. pp. 638–647.

Della Valle, N. 2007. Influencia de los parámetros de excavación de las tuneladoras en la generación de asientos en superficie. Lectures of the workshop on Tunnels in the city. Universitat Politècnica de Catalunya, Barcelona, Spain.

Do N., Dias D., Oreste P., Djeran-Maigre I. 2013. 3D Modeling for Mechanized Tunneling in Soft Ground-influence of the Constitutive Model. American Journal of Applied Sciences 10 (8): pp. 863–875.

Emeriault F., Kastner R. 2008. Tunneling with Earth Pressure Balance Tunnel Boring Machines: Observations and Modeling. Jornada Técnica de Túneles con EPB. Simulación y Control de la Tuneladora. Universitat Politècnica de Catalunya, Barcelona, Spain. pp. 4–28.

Fernandez, E. 2005. The Madrid renewal inner ring Calle 30 with the largest EPB machines—Planning and results. Rapid Excavation and Tunneling Conference Proceedings, Seattle, WA. Littleton, CO: Society of Mining Engineers. pp. 769–782.

Fujita, K. 1989. Special lecture B: Underground construction, tunnel, underground transportation. Proceedings of 12th International Conference on SMFE, Rio de Janeiro, Vol. 4. pp. 2,159–2,176.

Jardine F.M. 2001. Response of Building to excavation-induced ground movements. Proceedings of the international conference held at Imperial College, London, UK, July 17–18, 2001.

Lee K.M., Rowe R.K., Lo K.Y. 1992. Subsidence owing to tunnelling. I. Estimating the gap parameter, Canadian Geotechnical Journal 29, pp. 929–940.

Maconochie D.J., Suwansawat S., Chang C.C. 2001. Rapid Excavation and Tunneling Conference Proceedings, San Diego, CA. Littleton, CO: Society of Mining Engineers. pp. 113–128.

Mair, R.J. 1993. Developments in geotechnical engineering research: application to tunnels and deep excavations, Unwin Memorial Lecture 1992, Proceedings Institution of Civil Engineers. Civil Engineering, Vol. 93, pp. 27–41.

Mair, R., and F. Borghi. 2008. Influencia del acondicionamiento del material excavado en la operación y control de las tuneladoras a presión de tierra. Jornada Técnica de Túneles con EPB. Simulación y Control de la Tuneladora. Universitat Politècnica de Catalunya, Barcelona, Spain.

Mair, R.J., Taylor, R and Burland, J. 1996. Prediction of ground movements and assessment of risk of building damage due to bored tunneling, Conference on Geotechnical Aspects of Underground Construction in Soft Ground, London.

McDonald J., Kabat J. 2007. Slurry Shield Tunneling in Portland-West Side CSO Tunnel Project. Rapid Excavation and Tunneling Conference Proceedings, Toronto, Canada. Littleton, CO: Society of Mining Engineers. pp. 830–842.

Melis, M., Medina, L. and Rodriguez. J.M. 2002. Prediction and analysis of subsidence induced by shield tunneling in the Madrid Metro extension. Canadian Geotec. J., 39: pp. 1,273–1,287.

Mignini, A., T. Orfila, M. Colomer, F. Bertagnolia. 2008. Surface settlement minimization in soft soil when excavating with an earth pressure balance shield, Barcelona Metro Line 9 Mas Blau–San Cosme station, tunneling work procedure. Jornada Técnica de Túneles con EPB. Simulación y Control de la Tuneladora. Universitat Politècnica de Catalunya, Barcelona, Spain.

Moh, I.C. and Hwang, R.N. 1993. Underground Construction of Taipei Transit Systems, Keynote Speech, 11th Southeast Asian Geotechnical Conference, May 5–8, Singapore.

Ng F., Frew R., Wong E., Ferguson G. 2008. Aspects of the Design and Construction of the Chong Ming Tunnel Shanghai. 13th Australian Tunneling Conference. Melbourne, Vic, 4–7 May 2008. pp. 253–263.

Nomoto, T., Mori, H. and Matsumoto, M. 1995. Overview of ground movement during shield tunneling: A survey on Japanese shield tunneling. Proceedings of Underground Construction in Soft Ground, Fujita and Kusakabe, (eds), Balkema. pp. 345–351.

Nye, E.J. 1999. The Soft Ground Bored Tunnel Under Sydney Airport. Tenth Australian Tunneling Conference. Melbourne, Vic, 21–24 March 1999. pp. 75–83.

Orfila, T., N. Moya, and N. Della Valle. 2007. Optimizing line 9's EPBM parameters. Tunnel and Tunneling International Magazine, April 2007.

Pellegrini, L., and P. Perruzza. 2009. Sao Paulo Metro Project—Control of settlements in variable soil conditions through EPB pressure and bicomponent backfill grout. Rapid Excavation and Tunneling Conference Proceedings, Las Vegas, NV. Littleton, CO: Society of Mining Engineers. pp. 1137–1153.

Redmond S., Tattersall C., Garavelli N., Kudo T., Raleigh P., Lehnen M. 2007. Driving the Twin Bore Running Tunnels by Earth Pressure Balance Machine on the Sound Transit C-710 Beacon Hill Station and Tunnels Project. Rapid Excavation and Tunneling Conference Proceedings, Toronto, Canada. Littleton, CO: Society of Mining Engineers. pp. 793–807.

Romo, M.P. 1997. Soil Movements Induced by Slurry Shield Tunneling. Proceedings of the 14th International Conference on Soil Mechanics and Foundation Engineering, Hamburg, 3, pp. 1,473–1,481.

Sugiyama et al. 1999. Observations of ground movements during tunnel construction by slurry shield method at the Dockland Light Railway Lewisham Extension—East London. Soils and Foundations, 39(3), pp. 99–112, June 1999. Japanese Geotechnical Society.

Thewes, M. 2008. Shield Tunneling Technology for the Mitigation of Geotechnical Risks. Invited Lecture + Publication. Proceedings of "Tunneles en zonas urbanas densamente pobladas -experiences internationals," Col.legi d'Enginyers de Camins, Canals i Ports, Barcelona.

Tweedie, R.W., Harris, M.C., Gerber, G.E., Eisenstein, Z. 1989. Ground and Structure Monitoring for the Edmonton SLRT–Phase II Tunnels. 1989 Tunnelling Association of Canada Annual Publication. pp. 61–72.

Universidade da Coruña. 2008. Construcción By-pass Sur de la M-30. http://caminos.udc.es/novedades/archivos/7ByPassM30.pdf.

Van Jaarsveld, E.P., Plekkenpol, J.W., and Messemaeckers van de Graaf, C.A. 1999. Ground deformations due to the boring of the Second Heinenoord Tunnel, Geotechnical Engineering for Transportation Infrastructure, Barends et al. (eds), Balkema, Rotterdam.

Withers A.D. 2001. Murdoch, Neptune and Clegg Houses in Moodkey Street, Rotherhithed. Building Response to Tunneling—Case studies from construction of the Jubilee Line Extension. London, Thomas Telford, pp. 811–828.

Yang, G.R., Yang, P.E., Fan, C.B., 1997. Ground settlement troughs induced by multiple shield tunneling. In: Proceedings of 2nd International Symposium on Structures and Foundations in Civil Engineering, Hong Kong, China, pp. 744–751.

# Design-Build Tunnel Contract for Los Angeles Regional Connector Project

**William H. Hansmire**
Parsons Brinckerhoff

**Girish Roy**
Los Angeles County Metropolitan Transportation Authority

**ABSTRACT:** The Regional Connector project is a $1.4 (US) billion, fast-tracked Los Angeles transit tunnel project. Preliminary engineering, design-build contract development, and Request for Proposal took place over a period of three years and Notice to Proceed for construction is expected in early 2014. The project includes 1.4 km (0.9 mi.) of pressurized-face tunneling, three cut-and-cover underground stations, an SEM-excavated crossover cavern, and approximately 400 m (1,300 ft) of cut-and-cover guide way. Risk management was an integral part of the project. Substantial effort was made to create Contract Documents with the right balance of prescriptive vs performance requirements and included extensive geotechnical investigations and a Geotechnical Baseline Report.

## INTRODUCTION

The project experience for this $1.4 (US) billion, fast-tracked Los Angeles transit tunnel project is presented for the 3+ years encompassing final environmental approval, preliminary engineering, design-build contract development, and Request for Proposal (RFP). The project will be constructed by cut and cover, pressurized-face tunneling, and sequential excavation method (SEM) tunneling. Award and Notice to Proceed (NTP) are expected in early 2014. In order to expedite the work, the Los Angeles County Metropolitan Transportation Authority (Metro) selected project delivery by a single design-build contract. The Regional Connector Transit Corridor Project (Regional Connector) will link three existing Metro light rail transit lines and create an entirely new operating system for light rail transit service crisscrossing Los Angeles. Constructed in downtown Los Angeles, the Regional Connector alignment is under major city streets, passes through the Little Tokyo district and near world famous cultural institutions.

## WHAT DOES REGIONAL CONNECTOR MEAN?

The Regional Connector is a critically important rail connection project overwhelmingly approved in 2008 by voters of the Measure R sales tax ordinance for Los Angeles County transportation improvements. It improves mobility through downtown Los Angeles, but its benefits also bring significant mobility improvements for transit commuters throughout Los Angeles County. The project will provide a one-seat, one-fare ride for commuters from Azusa to Long Beach and from East Los Angeles to Santa Monica without the need to transfer between rail lines for major east/west and north/south trips. The Regional Connector will form the link to create these north/south and east/west lines that will operate on the new trunk section in tunnel. See Figure 1.

The Regional Connector completes a 3.1 km-gap (1.9 mi.) between the Metro Blue Line/Metro Expo Line and the Metro Gold Line by providing a direct connection with three new stations planned for 1st Street/Central Avenue, 2nd Street/Broadway and 2nd Place/Hope Street. Once built, the Regional Connector Project will attract 17,000 new daily riders and provide access to more than 90,000 passengers saving commuters up to 20 minutes off their daily commutes.

As shown in Figure 2 to achieve the physical "connection" at the southern end of the project, the Regional Connector starts as a continuation of the light rail Blue Line, which since opening nearly 20 years ago, has been a terminal station and transfer connection to the heavy rail Red Line. A "knock-out" panel in the existing Blue Line tail track structure will be removed and the transit guideway will be extended by cut and cover for two blocks along Flower Street where the construction method changes to tunneling.

At the north end of the project, an underground junction ("wye") splitting the rail line into north and

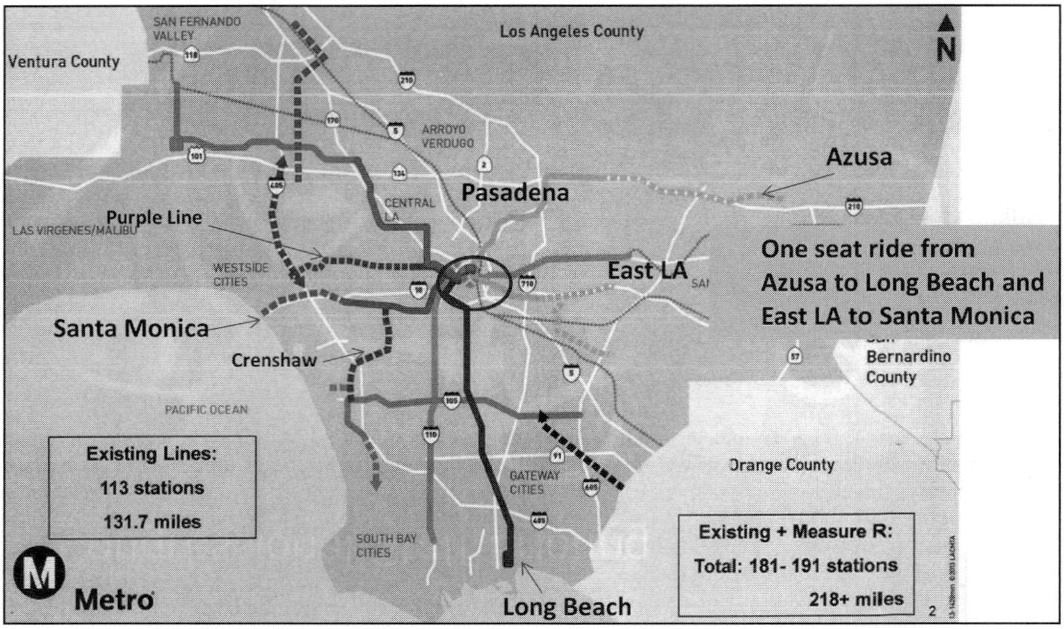


**Figure 1. Los Angeles Metro illustrating relationship of regional connector to all transit lines**

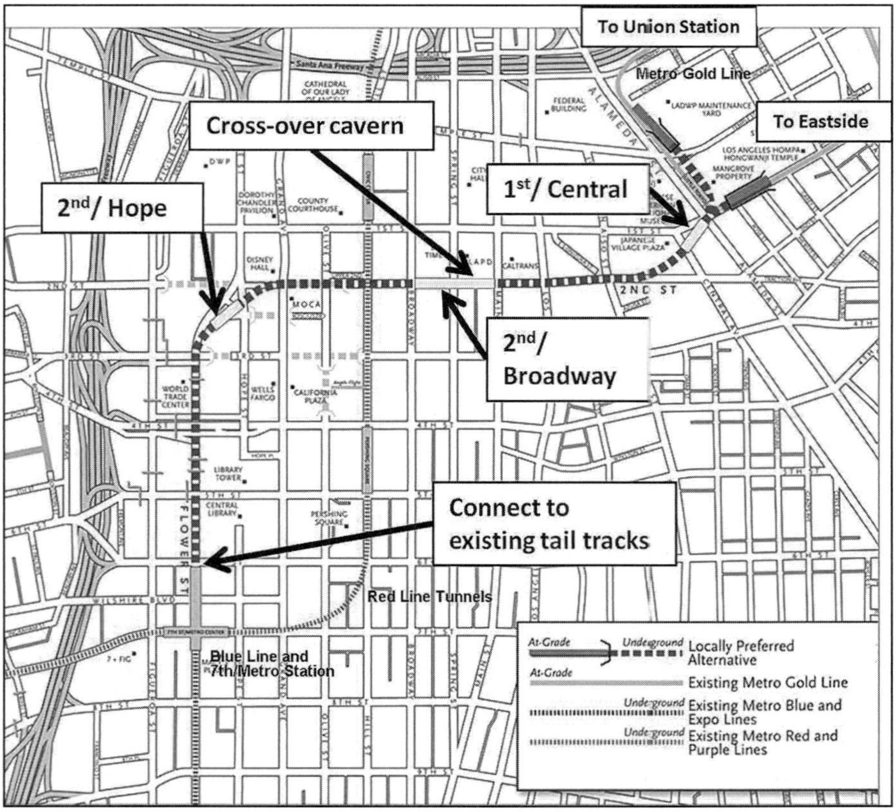


**Figure 2. Key features of LA Metro regional connector, downtown Los Angeles**

east directions will be constructed beneath the intersection of 1st and Alameda Streets. Two portals are required to accommodate the split, one for the north alignment toward Union Station (Azusa) and another for the east alignment to East Los Angeles.

## SCOPE AND SCHEDULE FOR DESIGN-BUILD CONTRACT

The contract is all-encompassing including construction of the tunnels, stations, and all mechanical, electrical, and systems work. Major features are summarized below.

- Fully underground (excluding portals to existing at-grade transit line)
- 3.1 km (1.9 mi) overall length
- 1.4 km (0.9 mi.) twin bore tunnels
- One-pass precast concrete segmental lining, 5.7-m (18 ft-10 in.) inside diameter
- 400 m (1,300 ft) cut-and-cover guideway in Flower Street
- 88 m (290 ft) crossover cavern constructed by SEM
- Three cut-and-cover underground stations
- Civil, structural, and architectural design
- Mechanical and electrical design
- Emergency tunnel ventilation design
- Systems design: train control, communications, traction power
- Traction power including overhead contact system (OCS)

Project schedule is as follows:

- March 31, 2012: Completion of Preliminary Engineering (PE)
- August 24, 2012: Request for Qualification (RFQ)
- July 29, 2012: Record of Decision (ROD)
- January 7, 2013: Request for Proposals made to 4 pre-qualified teams
- September 9, 2013: Design-Build Proposals submitted to Metro
- 2014: Notice to Proceed (NTP)
- 2020: Open to Revenue Service

Two years time was required to advance the project from the end of Preliminary Engineering in 2012, through completion of the environmental process and the Federal Transportation Administration (FTA) issuing the Record of Decision (ROD), procurement (RFQ, RFP), and start of the project in 2014. Opening to revenue service will be in two phases: Phase I, open to East Los Angeles (2,170 calendar days from NTP), and Phase II, open to Union Station/Azusa (2,300 days from NTP) and constitutes Substantial Completion. In round numbers, the project has a six-year duration.

The design/build contract has few intermediate milestones between start of the work and opening to revenue service. By intent, the design-build Contractor will have the freedom to plan, design, and construct the work in the sequence it chooses.

## GEOLOGIC CONDITIONS

Tunneling will be principally in weak rock, a poorly bedded to massive clayey siltstone to silty claystone, the Pliocene-age Fernando formation. Tunneling will start in Little Tokyo in coarse-grained alluvial soils, and within about 65 m (200 ft), tunneling transitions over about 135 m (400 ft) into a full face of weak rock (Fernando) for the remainder of the tunnel drives. Perched groundwater generally exists within the lower portion of the alluvial deposits, due to the relatively low permeability of the underlying Fernando formation. Due to its fine grained texture, typical massive condition (i.e., lack of well developed bedding planes), and lack of open fractures, the Fernando formation has a relatively low hydraulic conductivity.

The Fernando formation is poorly cemented and extremely weak to very weak rock (per ISRM, 1978). In the tunnel horizon it is typically slightly weathered to fresh bedrock. The unconfined compressive strength ranges from approximately 0.2 to 2.0 Mpa (25 psi to 300 psi). The Fernando formation contains cemented materials that are notably stronger than the typical siltstone/claystone material and are considered to be very strong rock. With regard to stickiness and clogging of an earth pressure balance machine (EPBM), the anticipated tendency for clogging is low to medium (Thewes and Burger, 2005). Miller abrasion testing indicated “low” to “moderate” abrasivity.

Excavations for 2nd/Hope and 2nd/Broadway Stations will be constructed principally in Fernando. Along Flower Street, the lower portion of the excavation will be in the Fernando. The cut and cover excavation for 1st/Central Station and the 1st Street and Alameda Street guideways will encounter fill and older and younger coarse-grained alluvium, Pleistocene-age and Holocene-age, respectively.

Most tunnels in the Los Angeles basin have hazardous gasses, primarily methane, but also hydrogen sulfide, that are associated with the oil producing formations underlying the area. The Regional Connector tunnels have been classified by CalOSHA as “Potentially Gassy.”

Geologic site characterization included project-specific borings undertaken during both conceptual engineering and later in Preliminary Engineering. The results of borings done for the many high-rise

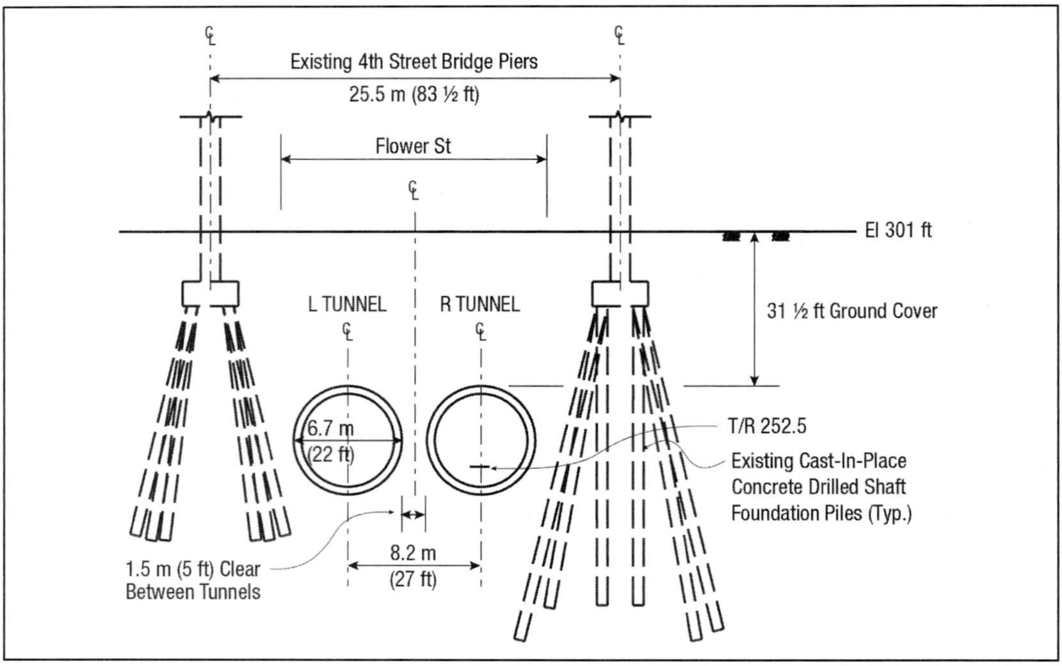


**Figure 3. Tunnel alignment constrained between 4th Street Bridge piers**

buildings along the alignment were used where relevant. A comprehensive Geotechnial Data Report (GDR) and Geotechnical Baseline Report (GBR) were prepared and included as contract documents.

## KEY PROJECT ELEMENTS AND ISSUES

### Existing Structures Set Tunnel Profile

Existing structures strongly affect the vertical profile. At each end of the project, the horizontal alignment and profile must meet top of rail for connection to the existing lines. The track alignment must conform to Metro Rail Design Criteria (MRDC) for horizontal and vertical curves and achieve operating speeds supporting train operation on 2½ min headways. The alignment has to accommodate the following existing structures and other conditions:

- *7th/Metro Center Tail Tracks:* at the south end, the project must connect to the end of existing Blue Line tail tracks, which will become running line when the Regional Connector opens.
- *4th Street Bridge Foundations:* As shown in Figure 3, the tunnels have to pass between foundations of this bridge. Protection of the bridge from the effects of tunneling is to be determined in Final Design and the Contractor must obtain City of Los Angeles agency approval. (See also Sepehrmanesh, et al. 2014.)
- *Red Line Tunnels:* The Regional Connector crosses the existing Metro Red Line tunnels (see Figure 2), which imposed a significant constraint on setting the tunnel profile. Shallow tunnels over the existing Red Line were not practical primarily due to the presence of utilities. A conservative 6 m (20 ft) of clearance below the existing tunnels was initially assumed. This resulted in additional cost for greater depth of the adjacent station and much steeper track grades for operations. After extensive analysis during Preliminary Engineering, the tunnel profile was raised to have about 1.5 m (5 ft) clearance between the extrados of existing and new tunnels (see Roy et al., 2012).
- *Little Tokyo Buildings:* at the north end of the project, the tunnels pass under several existing buildings with shallow cover. Building protection by compensation grouting is required. During the Draft EIS, the alignment had tunnel construction starting from a shaft on 2nd Street and the alignment had an operationally undesirable, very tight turn from 2nd Street to Alameda Street. At the start of Preliminary Engineering, a major alignment improvement was made to go under the Japanese Village Plaza buildings, which

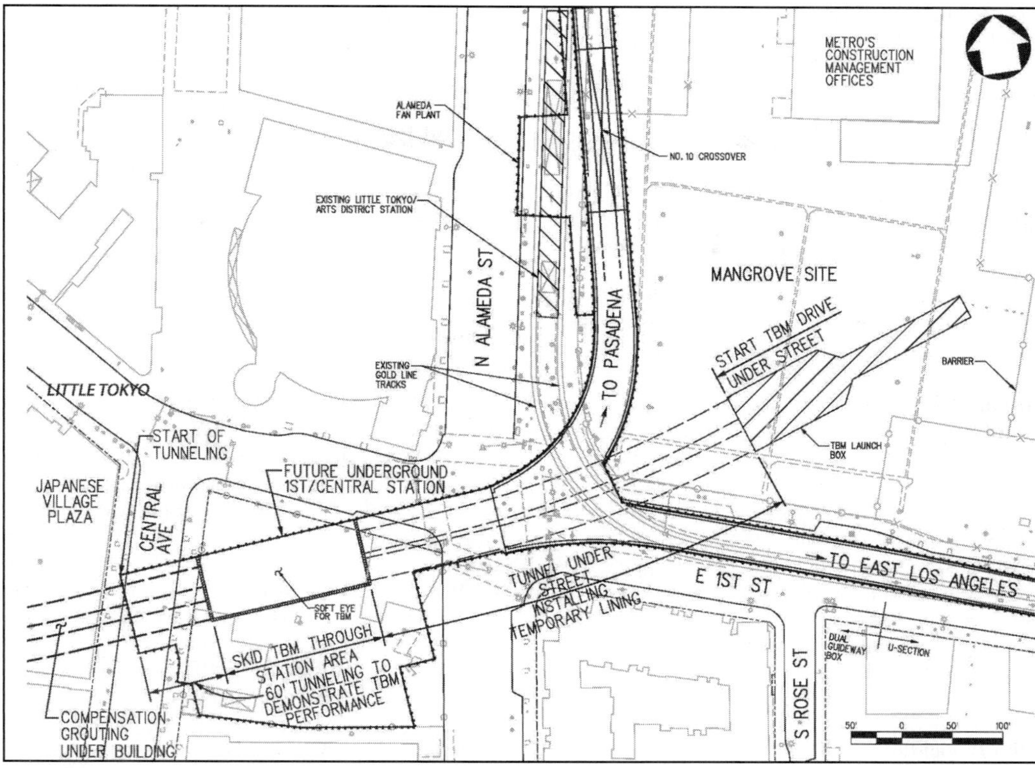


**Figure 4. Environmental mitigation requirement for start of tunneling outside of track alignment**

eliminated the tight curve and shaft construction on 2nd Street in Little Tokyo.

- *Buddhist Temple:* at the eastern end of the project at the tie-in to the existing Gold line, which at this location is at-grade (street-running), an environmental commitment was make to avoid visual impact to the Temple. The limits of work are set by the property boundary of the Buddhist Temple, which was found to be a major profile alignment constraint. From west to east after the alignment passes under the Little Tokyo buildings, thru the 1st/Central Station, then the Wye, a steep grade is required to get the rail profile to existing street grade at the Temple property line. The steep grade also prohibits installing a crossover on this leg of the project.

## Environmental Commitments Require Special Start for Tunneling

Tunnel worksites and the direction of tunneling had been identified as major concerns of the community. It was ultimately judged that insufficient space is available to permit efficient tunneling from west to east. This meant tunneling would have to be done from east to west, and start in the Little Tokyo community. Los Angeles' Little Tokyo is one of a few such communities remaining in the United States. The strong community concern was that transit tunnel construction could lead to the demise of Little Tokyo as a treasured cultural resource.

As a significant environmental mitigation, the commitment was made to require the start of tunneling not within the new transit station box structure in Little Tokyo, but to require launching the tunnel boring machine and service tunneling across the street in a City of Los Angeles owned lot known as the "Mangrove site." See Figure 4. The approximate 5 acres (2 ha) Mangrove site was not available as a construction work site during the early part of the environmental work but became available as the Preliminary Engineering was in progress. In the RFP, the Contractor was given the choice of either tunneling under Alameda Street and installing a temporary tunnel lining, or first excavating fully to the invert under a decked 1st and Alameda intersection, then skid the TBM to the station box to launch the TBM. This environmental commitment to launch the TBM from the Mangrove site became a winning solution to substantially mitigate construction impacts in Little Tokyo.

### Utilities and Decking on Flower Street

Approximately 400 m (1,300 ft) of the transit line will be cut and cover construction along Flower Street, a major downtown Los Angeles roadway. (TBM tunneling was considered not to be not feasible because of the presence of tiebacks from prior building construction. See below for details.) Flower Street has major and minor utilities throughout. Advanced relocation of these utilities will remove them from the design-build Contractor scope, and in turn, avoid the risk of additional cost and potential schedule delay with utility relocation. (See below for further utility details.) To further reduce the risk, Metro investigated the use of a "raised deck" on Flower Street that would elevate the existing roadway about 0.6 m (2 ft). This would reduce the need for most utility relocation and permit more utilities to be protected in place. The value of this approach is shorter construction duration, and less risk and less cost. After substantial community input, the raised decking will be limited to 250 mm (10 in.) above the existing street surface.

### Risk of Utility Delay Mitigated by Advance Utility Relocation

The numerous existing utilities pose a major risk of construction delay throughout the project wherever cut and cover construction is required for stations and non-tunneled line sections of guideway. Every utility was evaluated early in Preliminary Engineering to determine if it was reasonable for the design-build Contractor to relocate or maintain in place the utility with an acceptable amount of risk of delay and be able to meet the contract completion milestones. Where specific conditions exist (like a very old, fragile water line) or a clear conflict is indicated with cut and cover construction (such as utilities directly in conflict with future soldier piles for support of station excavations), advance relocation of the utility was put in to an Advance Utility Contract by Metro.

In summary, accommodation of utilities for the entire project was categorized as follows:

- *AR-3: Advanced utility relocation by Third Party Owner.* Typically "dry utilities" for communications (fiber optic) and gas. Starting in 2013, these are being relocated by each utility with costs paid by Metro.
- *AR-M: Advanced utility relocation by Metro Contract No. C0981R.* This Advanced Utility Contract is expected to start in early 2014 and be completed within approximately one year. It eliminates the risk of delay for specific major utilities for the main Regional Connector contract (Metro Contract No. C0980). Relocated utilities include water, high voltage power, and in some cases sewers.
- *DB-R: Utilities that are the responsibility of Metro Contract No. C0980.* Temporary or permanent utility relocation, protect in place, and restore as required by the specific situation in the Contractor's Final Design.

### Tunneling and Tunnel Lining

Tunnels are required to be constructed using a "pressurized face" tunnel boring machine (TBM). Slurry or earth pressure balance machine (EPBM) is the Contractor's choice. Based on the geologic conditions, an EPBM is most likely to be used.

The mandatory contract requirement for pressurized-face tunneling resulted from Metro's past tunneling experience. Metro tunnels completed in the 1990s had a "two-pass" lining: an initial lining of expanded pre-cast concrete segments, and final lining consisting of a water and hydrocarbon gas-impermeable membrane (HDPE) and cast-in-place concrete. Practical elimination of the explosion hazard of naturally occurring gasses for rail operations was achieved with this lining and ventilation. If minor volumes of gas enter the operating tunnel, they are purged by the operating train action (piston effect) or, if detected by instrumentation, by activation of the ventilation system.

These previous LA Metro tunnels were constructed using open (digger) shields, the traditional method at the time in Los Angeles. Tunneling with an open-face shield was possible as much of the ground is above groundwater level or in weak/weathered sedimentary rock. Although pressurized-face tunneling was an emerging technology in tunneling at that time in the 1990s, it was not deemed suitable for the relatively dry tunnel ground conditions in Los Angeles. Unacceptable ground surface settlement occurred in some areas along the Hollywood and North Hollywood Red Line tunnels using open shields. This experience led Metro to specify only pressurized-face tunnel boring machines (TBMs) on subsequent soft-ground tunnel contracts (Eisenstein et al. 1995). The Eastside Extension tunneling was completed in 2006 using an EPBM and a single-pass lining with double gaskets to provide redundancy against gas leakage. Following the Eastside Extension precedent, the precast segments for the Regional Connector will have a double gasket system, which based on the Eastside tunnel experience, has achieved gas tightness comparable to that with the two-pass tunnel linings and HDPE membrane.

Twin-bore tunneling will be in three sections between stations and launch sites for tunnel drives of approximately 600 m, 500 m , and 300 m (2,000 ft, 1,600 ft, and 1,000 ft) for a route length of twin bore tunneling totaling approximately 1,400 m (4,600 ft).

## Settlement and Building Protection

The design-build contract has a combination of mandatory and performance requirements to protect buildings and utilities. Action and Maximum Levels for angular distortion and total settlement of structures, facilities, and utilities adjacent to excavations are shown in Table 1.

At the start of tunneling in Little Tokyo, compensation grouting is a mandatory method required to protect the Japanese Village Plaza buildings. Where the tunnel passes close under the large 2nd Street storm drain and tunneling is in alluvium, mandatory ground improvement by permeation grouting is required as shown in Figure 5. For the balance of the project, the requirement is performance based without a mandatory method indicated in the Contract and will be determined in Final Design.

## Stations

Three cut and cover underground stations are planned. Center platforms are 152 m (270 ft) long to

**Table 1. Action and maximum levels angular distortion and settlement**

| | Action Level | Maximum Level |
|---|---|---|
| Angular distortion* | 1/1000 | 1/600 |
| Total settlement/heave | 9 mm (0.35 in.) | 13 mm (0.5 in.) |

*Average settlement slope or slope between building walls or columns whichever less.

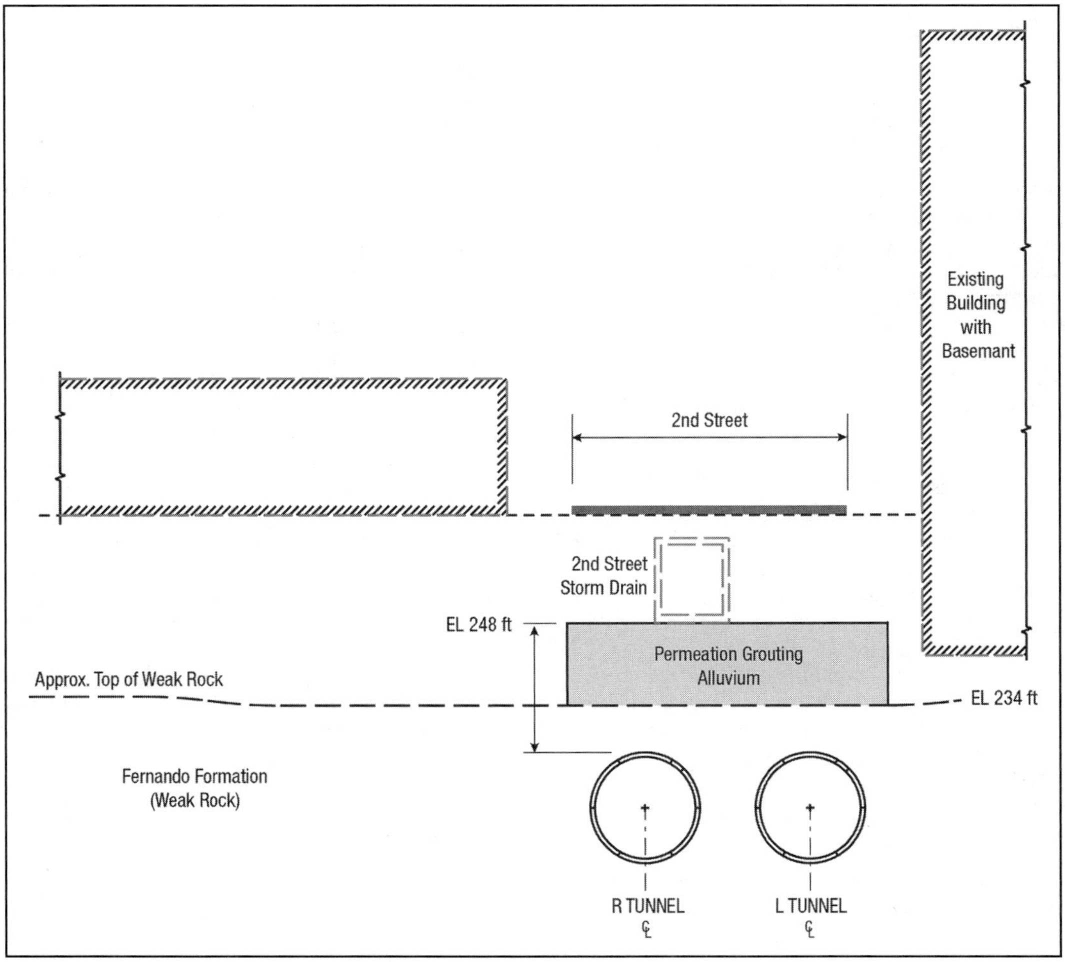


**Figure 5. Ground improvement required at specific locations**

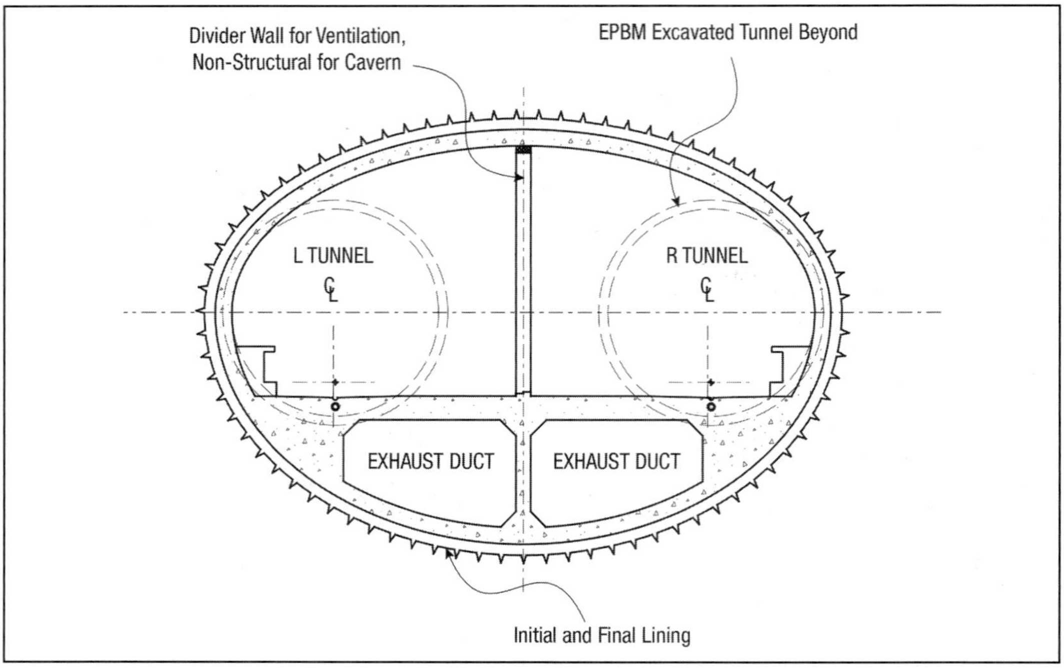


**Figure 6. Rail cross-over cavern with under duct for ventilation**

accommodate a three-car train, and have a minimum width of 7.6 m (25 ft -1½ in.). Emergency ventilation systems for tunnels and stations are to be designed for a rapid growth rate (arson) fire. Stations are described below and see Figure 2 for locations.

- *2nd/Hope Street Station.* Serves the Bunker Hill area and is adjacent to the iconic Walt Disney Concert Hall. This is the deepest station at about 34 m (112 ft) maximum depth and will be served by high-speed elevators, not escalators, a first for Metro.
- *2nd/Broadway Station*: Located within the heart of historic downtown Los Angeles and serves Los Angeles City Hall and Police Department Headquarters. It will be designed for a future overbuild of a high-rise building and has a depth of about 25 m (82 ft).
- *1st/Central Station*: Located in the heart of Little Tokyo. Shallowest station at about 11 m (37 ft) depth and has direct access to the station platform by escalators without a mezzanine level.

### Cross-Over Structure

Efficient rail operations require crossovers to be able to single-track trains to deal with equipment failure, emergency conditions, and maintenance. Without crossovers, the tunnel would have to completely close as the single-track train operating headways would become unacceptably long. It was important to have a crossover within the Regional Connector. To meet this need, one double crossover with No. 8 track turnouts is located just east of 2nd/Broadway Station. This 17 m wide (57 ft) finished, 88 m long (290 ft) structure is to be constructed by sequential excavation method (SEM) tunneling in the weak rock Fernando formation. See Figure 6. To achieve extraction of smoke during a fire emergency from both ends of the crossover, the large ventilation duct with a center support shown in Figure 6 the crossover tracks was added during the RFP period. The ventilation under duct connects to the emergency ventilation fans located at the adjacent end of the 2nd/Broadway Station platform.

Early in conceptual design during the DEIS, construction of this crossover was assumed to be as a cut and cover excavation, like for the immediately adjacent station. However upon further investigation during Preliminary Engineering, legal encroachments by an underground garage and sidewalk vaults associated with the adjacent structures meant that a full-width excavation could not be constructed without costly underpinning. See Figure 7. After study, the SEM-excavated cavern was determined

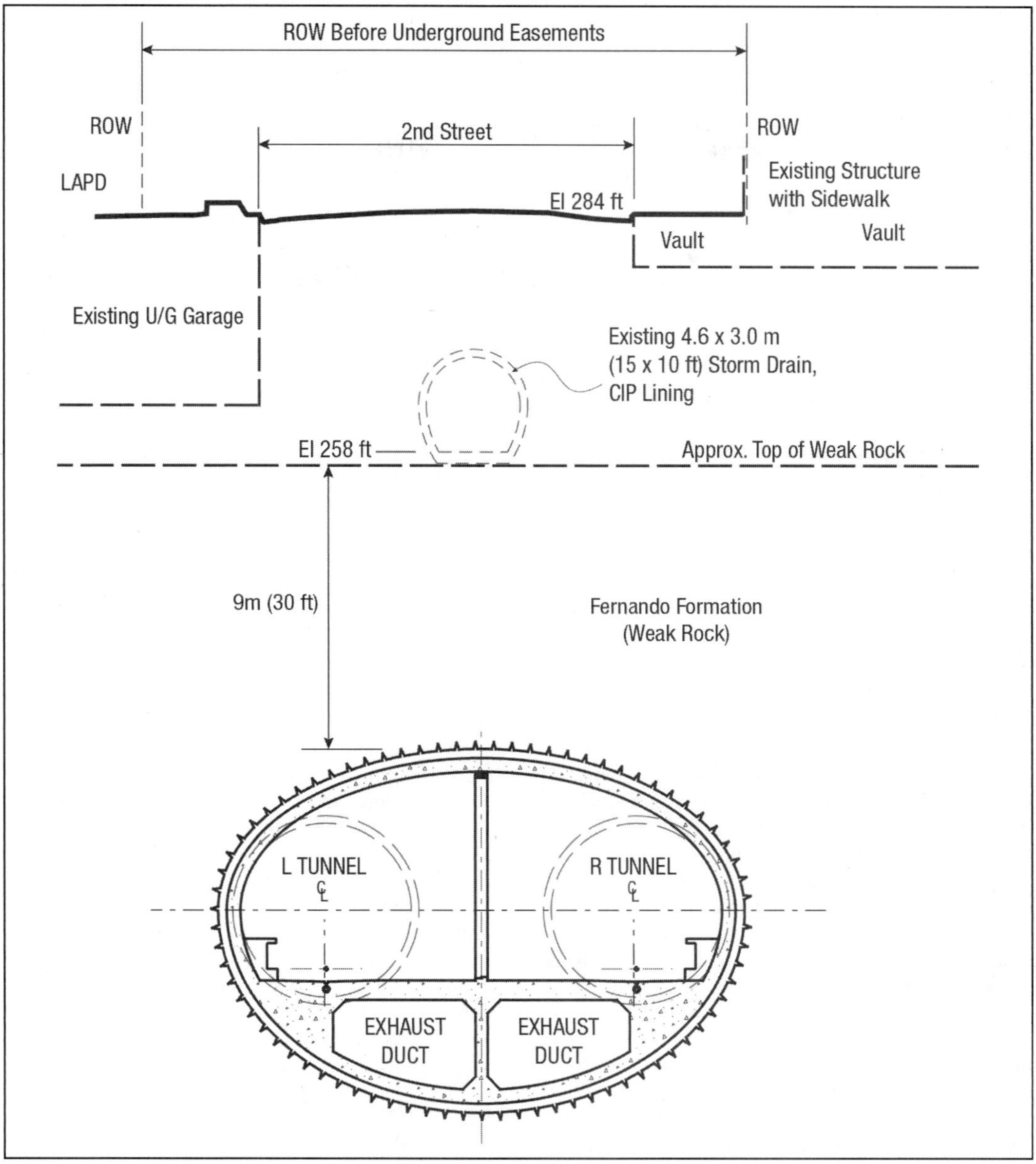


**Figure 7. Cross section of cross-over cavern constructed by SEM showing storm drain and existing buildings hanging over alignment**

to be constructible and was adopted in the project configuration.

### Cut and Cover Construction on Flower Street Required to Avoid Tie-Back Hazards to Tunneling

The existing deep basements and parking garages along Flower Street used tie-backs (steel bars or cables grouted in the ground) to laterally support the original excavations during construction. The steel tie-backs extend deep below ground across the width of Flower Street from both sides in varying number along the alignment and have been abandoned in place. Tie-backs exist every six to eight feet in this reach of the project and total in the hundreds. See the cross-section in Figure 8.

Tie-backs are a major hazard to tunneling with a closed-face (pressurized-face) TBM, whether EPBM

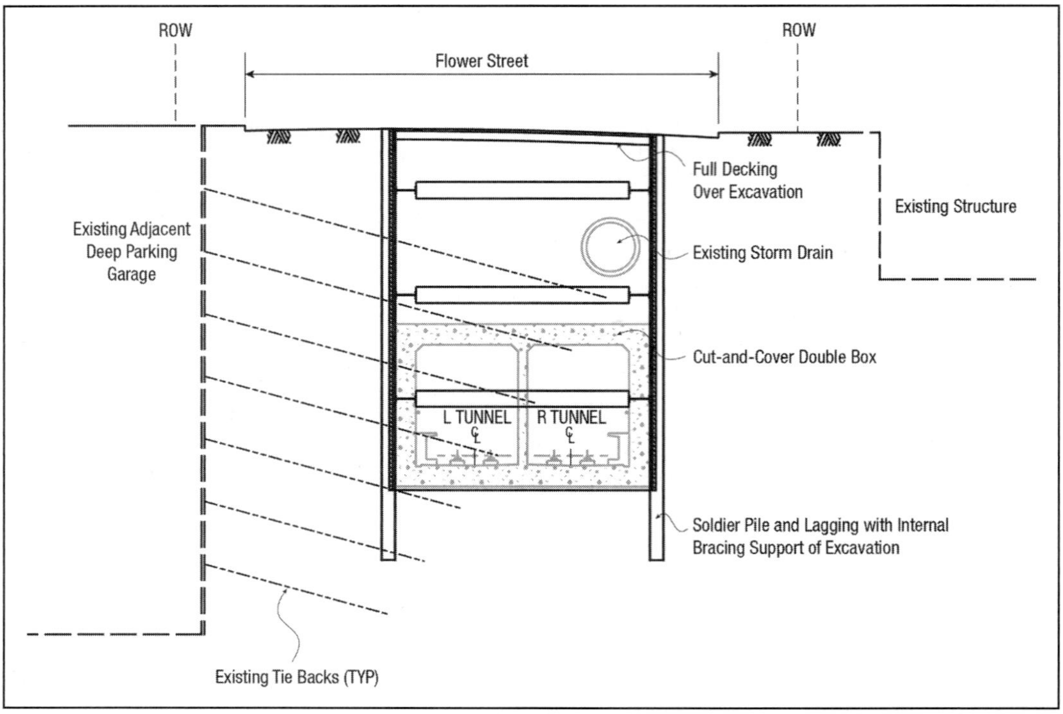


**Figure 8. Cross section on flower street showing existing tie-backs**

or slurry. The TBM cutterhead is not capable of cutting through or otherwise processing a steel tie-back without damage to the cutterhead. The TBM would need to stop advancing and substantial down time would be required to work within or ahead (in front) of the TBM cutterhead to manually remove a tie-back. The cutterhead is a huge barrier between tunnel workers and a tie-back that would have to be removed. The pressurized-face machine is designed to control excavation of the soils, which in reverse, practically prohibits tunnel worker access ahead of the machine. The machines are designed with some access ahead of the cutterhead, which makes access possible, but does not make the process of working ahead of the cutterhead easy or automatically safe.

Working through the spokes of the cutterhead or ahead of the cutterhead would add significant time to the construction schedule even if firm ground conditions are present. If ground water is present and soils are unstable, grouting would be required to create firm ground conditions or the work would have to be done under compressed air (hyperbaric conditions) with appropriate safety precautions instituted. Removal of one tie-back would likely have to be done in several sections to free the steel tendon from the ground and cutterhead. Dealing with one or two tie-backs in this manner might be practical. The result would still be a substantial delay and significant cost increase. Encountering hundreds of tie-backs, which is the case in this section of Flower Street, rendered the use of an EPBM not viable. For two blocks along Flower Street, construction therefore will be by cut and cover.

Removal of tie-backs in advance of tunneling is theoretically possible, but in practice, difficult to accomplish reliably. Where their location is fairly well known, a few tie-backs in the TBM path can be removed in advance. This situation exists in one location where up to twenty tie-backs will be removed directly from a deep trench at 3rd and Flower Streets, which allows tunneling a block farther to 4th Street.

## RELATIONSHIP OF RFP DOCUMENTS TO FINAL DESIGN

Metro carried the design to a Preliminary Engineering stage, which serves as the basis for the Request for Proposal (RFP) that was provided to design-build contactor teams. The Contractor's Engineer (or Architect) of Record will prepare Final Design Documents to adequately and completely depict and record the Contractor's detailed design. At the highest level, the Final Design must conform to the Metro Rail Design Criteria (MRDC) including Metro Rail Fire/Life-Safety Criteria. The MRDC provide the minimum criteria and adherence is required during the design, construction, testing, and commissioning of the Project.

In addition to criteria, the project-specific requirements are in the form of Project Definition Documents and are comprised of the following.

*Scope of Work*: Extensive statement of the project-specific requirements and has contractual precedence above that of General and Technical Requirement and Project Definition Drawings.

*General Requirements*: Largely non-technical requirements in the form of approximately 60 individual sections prescribing such matters relating to administration of the contract, but also includes technical scope related to environmental mitigation.

*Technical Requirements*: Approximately 280 section comprise what is referred to as the project-specific "TRs." TRs range from being prescriptive to performance-based. Initially, in the preparation of the RFP, it was thought that all TRs could be completely "performance based." However, it was found that this was not possible to go back to such basics given the long precedent of Los Angeles Metro projects and the Metro Rail Design Criteria. The lesson was that design-build does not mean exclusively performance-based requirements. Thus many TRs are prescriptive regarding technical details but generally have no contractor means and methods prescribed. That is, the TR states prescriptively what is required, but not necessarily how the work is to be planned, constructed, or installed. The exception is for the systems work (train control, communications, and traction power) where the TRs by their nature are typically performance based. However, the need to fit to existing systems, to achieve uniformity among projects, and to satisfy owner preferences meant many systems TRs are prescriptive.

*Project Definition Drawings*: These encompass all disciplines and have been developed to a sufficient level to define real estate requirements, to provide a basis for estimation of cost, and to establish the configuration upon with the FEIS/FEIR was prepared and approved. Approximately 900 drawings were prepared. For the Regional Connector, it was necessary to carry design development of stations further than might typically be required for design-build procurement.

## CONCLUSION

The Regional Connector is a comprehensive one-contract design-build tunnel project in downtown Los Angeles. This will not be just an extension of an existing transit line, but will create entirely new lines of transportation when it connects Metro's existing Blue, Expo, and Gold light rail transit lines. The significance and impact of this project to transportation in Los Angeles will be many times that of is modest 3.1 km (1.9 mi) length as it will make efficient public transpiration possible long distances across Los Angeles County.

## ACKNOWLEDGMENTS

Metro's Project Manager for the Regional Connector Transit Corridor Project is Girish Roy. Preliminary Engineering is by The Connector Partnership JV (CPJV), a joint venture of AECOM and Parsons Brinckerhoff. Bill Hansmire is the Design Manger for CPJV.

## REFERENCES

Eisenstein, Dan, et al., 1995, Los Angeles County Metro Rail Project: Report on Tunneling Feasibility and Performance.

Elioff, M.A., Hansmire, W.H., Abramson, L.W, 2012, 20+ km of Tunnel Design Underway for Extending Metros in Los Angeles, ITA—AITES World Tunnel Congress Proceedings, 2012, Bangkok, Thailand.

ISRM (International Society for Rock Mechanics), 1978. Rock Characterization Testing and Monitoring (ISRM Suggested Methods). Brown, E.T., ed., Pergamon Press, London, 211 pp.

Roy, P. Nasri, V., Roy, G., 2012, Los Angeles Regional Connector Project—Urban Downtown Tunnelling Challenges, ITA—AITES World Tunnel Congress Proceedings, 2012, Bangkok, Thailand.

Sepehrmanesh, M., Nasri, V., Allahverdi, N., 2014, Risk Assessment of Pile Foundations to Shield Tunneling, 2014 North American Tunneling Conference, Los Angeles (In Press).

Thewes, M. and Burger, W., 2005, Clogging of TBM Drives in Clay—Identification and Mitigation of Risks. Underground Space Use: Analysis of the Past and Lessons for the Future—Erdem & Solak (eds), pp. 737–742.

# Tunnel Construction Management for Design-Build Delivery

**James Wonneberg**
EPC Consultants Inc.

**Rafael Castro**
Jacobs Associates

**Christopher Allen**
DC Water and Sewer Authority

## INTRODUCTION

In recent years project owners have increasingly turned to Design-Build (DB) procurement to deliver large-diameter tunnel construction projects within compressed schedules. To capitalize on the benefits of DB delivery, the entire project team—including the construction manager—must achieve a fundamental shift in mindset and project management approach, as compared to the more traditional Design-Bid-Build (DBB) delivery. This paper provides a case history of construction management on the Blue Plains Tunnel as part of the DC Clean Rivers Project in Washington, DC and also provides key takeaways and recommendations for owners and construction managers that can be applied to other DB tunnel projects.

## DC CLEAN RIVERS PROJECT OVERVIEW

The District of Columbia Water and Sewer Authority (DC Water) provides water and wastewater service to the District of Columbia, as well as wastewater treatment for Montgomery and Prince George's counties in Maryland and Fairfax and Loudoun counties in Virginia.

Like many older cities in the United States, portions of the District are served by combined sewers dating back to the late 19th century, carrying both sewage and runoff from storms. When these sewers reach capacity during large rain events, a mixture of sewage and storm water runoff must be discharged directly into the District's waterways. These discharges are called Combined Sewer Overflows or CSOs and occur at a rate of 2.5 billion gallons per year in the District alone (DC Water 2002).

In 2005 DC Water entered into a Federal Consent Decree with the United States and District Government to implement DC Water's Long Term

**Figure 1. Anacostia River Tunnels**

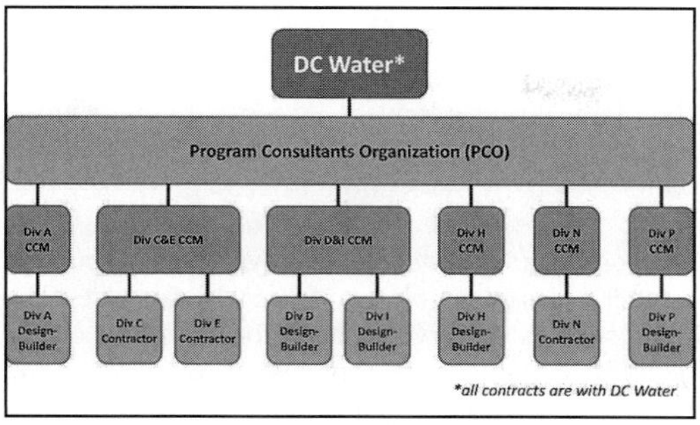


**Figure 2. Simplified organizational chart**

Control Plan (LTCP) to construct a system of tunnels and diversion sewers in order to reduce CSOs by 96 percent overall and 98 percent for the Anacostia River alone. This Consent Decree includes specific design, construction and operational deadlines and requires DC Water to pay damages if the deadlines are not met (DC Water 2002). Figure 1 illustrates the Anacostia River Tunnels portion of the Clean Rivers Project, scheduled to be completed in 2022.

In order to deliver this massive undertaking, valued at $2.5 Billion over a 20-year timeframe, DC Water established a new department within its engineering branch and procured a Program Consultants Organization (PCO) to provide preliminary design services and oversight of all final design, construction, design-build and construction management contracts across the program. Figure 2 presents a simplified organizational chart for construction management on the Clean Rivers Project. DC Water chose the DB delivery method for certain contracts in order to meet program schedule constraints driven by the Consent Decree, and also to capitalize on the added value of early contractor involvement in the final design process.

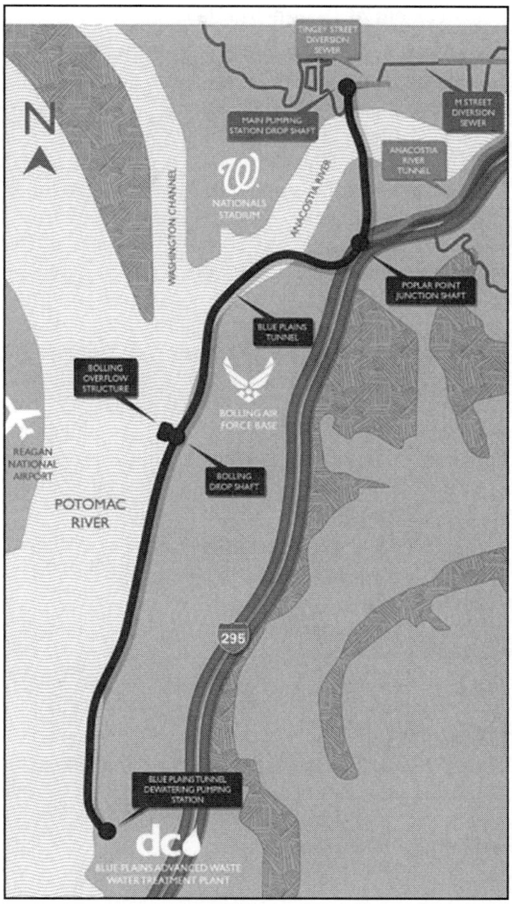


**Figure 3. Blue Plains Tunnel**

## BLUE PLAINS TUNNEL PROJECT OVERVIEW

### Scope of Work

Known as "Division A," the Blue Plains Tunnel Project scope of work is shown in Figure 3 and consists of the following major elements:

- A 72' I.D. × 150' deep screening shaft (BPT-SS) for use in mining the Blue Plains Tunnel and an adjoined 132' I.D. × 166' deep dewatering shaft (BPT-DS) for a future pumping station (by others), both located at DC Water's Blue Plains Advanced Wastewater Treatment Plant.
- A 50' I.D. × 132' deep drop/overflow shaft (BAFB-DS) with a vortex generator structure and approach channel, located within Joint Base Anacostia-Bolling.
- A 55' I.D. × 124' deep junction/drop shaft (PP-JS) with a vortex generator structure, a surge chamber and an approach channel passing beneath a roadway, located on District right-of-way at Poplar Point.
- A 55' I.D. × 109' deep drop shaft (MPS-DS) with a vortex generator structure and approach channel, located at DC Water's Main Pumping Station.
- A 23' I.D. × 24,200' long Blue Plains Tunnel (BPT), a soft-ground tunnel excavated from BPT-SS to MPS-DS using a pressurized face tunnel boring machine (TBM) and lined with bolted gasketed precast segments. Depth of cover above the tunnel varies along the alignment from 77' to 115'.

## Procurement of Design-Builder

As soon as the PCO completed the RFP Documents for the Blue Plains Tunnel project (equivalent to a 30% preliminary design), DC Water initiated a two-phase qualifications-based selection process for the DB contract. The first phase was issuance of a Request for Qualifications (RFQ). DC Water then announced a shortlist of the three most qualified DB proposers. To those shortlisted proposers, DC Water issued a Request for Proposal (RFP). The RFP established the scope of work for the DB contract, including specifications, drawings, GBR, GDR and reference documents, and required each proposer to submit a Technical Proposal and a Price Proposal. DC Water required each shortlisted proposer to participate in presentations, interviews and collaboration sessions before scoring the Technical Proposals. Price Proposals, using the bid form issued with the RFP, were submitted in sealed envelopes with the Technical Proposals, but were not opened until after evaluation and scoring of the Technical Proposals. The winning proposer was ultimately determined using a weighted score of 35% technical score and 65% price.

## Key Provisions of the Design-Build RFP Documents

The RFP Documents, issued during procurement as described above, included the following:

### *Mandatory Requirements*

The design-builder is required to comply with mandatory requirements for design and construction of both temporary and permanent work, including (a) all Division 1 Specifications, (b) technical specifications for certain aspects of the work to be further developed by the design-builder, (c) drawings establishing minimum requirements for shafts, near-surface structures, the tunnel, construction staging areas, and geotechnical instrumentation, (d) 100-year design life for all permanent structures, and (e) shaft and tunnel design criteria and load combinations.

### *Final Design Requirements*

The design-builder is required to retain qualified and licensed design professionals ("Engineer of Record" or "EoR") to prepare deliverables for owner review, including (a) Basic Design Report, (b) various design studies, and (c) advanced design submittals (60%, 100%, and Released for Construction or "RFC" sets) which include drawings, specifications and design calculations. The design-builder is permitted to separate advanced design submissions into "packages" by site or work element, however, the submittals listed are required for each package.

### *Protection of Structures*

The RFP Documents included a list of structures (buildings, bridges, sewers, levees, etc.) in the vicinity of the shafts and along the tunnel alignment that were potentially susceptible to damage from settlement, and required the design-builder' to prepare pre-construction condition surveys and construction impact assessment reports for each structure. The purpose of these surveys and reports is to; (a) obtain detailed condition assessments of each structure, (b) calculate expected ground movements associated with the work, using detailed numerical modeling for certain structures, and (c) determine whether protection schemes (such as grouting, etc.) are necessary. The design-builder must then implement the protection scheme(s) prior to starting work in the affected area(s). Costs for such measures are covered by an allowance included in the contract price.

### *Permits and Agreements*

The RFP Documents identified certain easements, agreements and permits to be obtained by the owner, then required the design-builder to obtain all other permits required to perform the work.

### *Construction-Phase Submittals*

The RFP Documents require construction-phase submittals for all work, including qualifications, product data, material tests and certifications, mix designs, shop/working drawings, method statements and contingency plans. The RFP Documents provided this list in boilerplate format and required the design-builder's EoR to further develop the submittal requirements as part of the Final Design Specifications for each work element. The RFP Documents also stipulated that all Construction-Phase Submittals must first be approved by the design-builder's EoR and QAQC Manager prior to submission to the owner.

### *Health, Safety and Environmental Controls*

The RFP Documents require the design-builder to provide a safety manager and safety representatives, develop a site-specific health and safety plan with sub-plans and task-specific plans, comply with OSHA and DC Water's Rolling Owner-Controlled Insurance Program (ROCIP) Construction Safety Manual, and coordinate with District Fire and Emergency Medical Services for joint training exercises and incident response procedures. The RFP Documents also require the design-builder to provide an Environmental Manager and develop an Environmental Protection Plan, with sub-plans, to control pollution and construction-generated wastes, and comply with all Federal and District regulations and permit requirements.

### *QAQC Requirements*

The RFP Documents require the design-builder to; (a) develop comprehensive design and construction QAQC plans, (b) provide a sufficient staff of qualified QC Representatives, distinct and separate from the design and production staff, (c) provide a qualified independent testing laboratory, (d) inspect and test all work using written procedures and checklists to ensure conformance with RFC Documents, and (e) submit all QAQC records to the owner on a monthly basis. The RFP Documents also state that the owner will establish an Independent Verification and Assurance (IVA) program for the sole benefit of the owner. The design-builder is required to provide notice of inspections and tests and furnish access to the work to assist the owner with implementation of the IVA program.

### *Contract Price and Time*

The Contract Price for all work is $330 Million, which includes allowances for certain types of shaft and tunnel delays and protection of structures measures, as well a contingency allowance used to fund change orders. The RFP Documents required the design-builder to substantially complete all work no later than 1,560 days (4+ years) after Notice to Proceed, and established five other Interim Milestones to facilitate turnover of certain construction staging areas for use by other Clean Rivers Project divisions.

### *Geotechnical Baseline Report (GBR) and Differing Site Condition (DSC) Clause*

The RFP Documents included a GBR to establish contractual baselines describing the anticipated ground conditions that would be encountered during construction, based on extensive geotechnical investigations conducted by the owner prior to issuing the RFP. The GBR discussed geologic settings, previous local tunnel construction experience, characteristics and baseline engineering properties for anticipated soil groups, and design and construction considerations for the shafts and tunnel. The DSC clause includes provisions for (a) timely notice of potential DSCs, (b) owner investigation, and (c) comparison to the contractual baselines in the GBR. If the actual conditions encountered meet the criteria of a DSC and the design-builder has followed the procedural requirements, the design-builder is entitled to request an appropriate change order.

### *Changes and Disputes*

The owner may request changes in the work and the design-builder may request relief due to any event or situation arising out of the work. Price adjustments are based on mutually-accepted lump sum pricing, unit prices, or are cost reimbursable following specified markups. Price and time adjustments are executed via change order and funded from the contingency allowance included in the contract price. The RFP Documents also establish a Dispute Review Board (DRB), arbitration, and mediation for resolution of contractual disputes between the parties.

### *Formal Partnering Process*

Formal partnering consists of selecting a mutually-agreed facilitator and establishing workshops to encourage resolution of conflicts at the lowest responsible management level. Formal partnering does not have any legal significance and does not supersede the procedural requirements for resolution of disputes.

## Procurement of Consultant Construction Manager (CCM)

DC Water initially considered ramping up in-house staff to provide construction management for the program. But the prospect of hiring over 100 soft-ground tunnel specialists in a matter of a few years convinced the organization to employ outside consultants. DC Water's RFP for CCM services required proposers to submit qualifications statements and technical proposals. DC Water then announced a short list of firms, conducted interviews with each, and made a final selection. The selected firm was then required to submit a price proposal which was used by DC Water to establish a not-to-exceed amount for the CCM agreement.

## Key Provisions of the CCM Agreement

The CCM is responsible for directing, coordinating and monitoring all aspects of the project, as approved by DC Water, except that the CCM shall not assume any authority of the Contracting Officer with regards to altering the terms, conditions or cost of the DB contract. The CCM is required to provide a full-time Resident Engineer, administrative support personnel and qualified inspection staff for surveillance of materials during construction, evaluating and inspecting workmanship, and monitoring daily for compliance with the construction contract documents. At least one CCM representative must be on site at all times that the contractor is working.

The CCM is required to implement construction management processes consistent with the PCO's Clean Rivers Project CM Plan, which was developed to ensure consistency in approach across the entire DC Clean Rivers Project.

The Blue Plains Tunnel CCM agreement included a total not-to-exceed amount of $22.7 Million over a six-year timeframe, with services

**Table 1. Scope of work comparison**

| Primary Responsibilities | Design-Bid-Build Typical Tunnel Project | | Design-Build Blue Plains Tunnel | |
|---|---|---|---|---|
| | Owner | Contractor | Owner | Design-Builder |
| Preliminary Design | X | | X | |
| Final Design | X | | | X |
| Construction Impact Assessments | X | | | X |
| Geotechnical Instrumentation Design | X | | | X |
| Protection of Structures Design | X | | | X |
| Permits | X | | | X |
| Construction Submittals | | X | | X |
| Install & Baseline Geotechnical Instrumentation | | X | | X |
| Construction | | X | | X |
| Construction Safety | | X | | X |
| Environmental Compliance | | X | | X |
| Quality Assurance / Quality Control | QA | QC | IVA | QA/QC |
| Monitor Geotechnical Instrumentation | | X | | X |
| Design Support During Construction | X | | | X |

(Rounded box) Indicates "Contract Work" for the contracting entity

authorized annually via work plans. This amounts to 6.8% of the Blue Plains Tunnel DB contract price. At the time of writing the CCM team consists of twenty-one full-time employees, including twelve inspectors covering round-the-clock construction at multiple sites.

## DESIGN-BUILD VS. DESIGN-BID-BUILD DELIVERY

The primary difference between DBB and DB delivery is that DB transfers final design and other engineering services into the contracting entity (design-builder's) scope of work. As shown in Table 1, the Blue Plains Tunnel RFP Documents transferred a significant amount of workload to the design-builder.

DB delivery requires an understanding of design-build methodology for all parties involved as follows:

- **Owners** establish project objectives and minimum requirements via the RFP documents, and then review the design-builder's design submittals to ensure compliance with RFP requirements.
- **Design-Builders** must now deliver a significant amount of final design and other engineering services "on the clock and within budget" that was previously done by others prior to NTP.
- **Construction Managers** must become intimately involved with a much broader scope of "Contract Work" requiring administration, coordination, project controls and reporting.

Many project owners select DB delivery because it allows the contracting entity to concurrently perform tasks that were traditionally non-concurrent, thereby compressing the overall project delivery schedule (Koch et al. 2010). By splitting final design into packages, a design-builder can quickly advance from final design to construction for certain work elements on the critical path, while other packages are still being designed by the EoR.

## KEY TAKEAWAYS AND RECOMMENDATIONS FROM THE CONSTRUCTION MANAGER'S PERSPECTIVE

The following points are based on lessons learned from the Blue Plains Tunnel project, currently underway, which has been the lead author's first experience with DB delivery. These recommendations are intended to promote thoughtful discussions amongst other project owners and construction managers as they develop strategies for delivery of DB tunnel projects of their own.

### Design-Build 101

On a DB project, the construction manager must embrace the fact that the project critical path now includes design and construction and all steps in between. The construction manager is now responsible for overseeing design, permitting and many other preparatory tasks that would have traditionally been managed by others *prior to NTP* on a DBB project.

As illustrated in Figure 4, construction is the *last step* in a lengthy and complex sequence of contract work. One of the construction manager's key roles is to ensure that the design-builder has a viable schedule to complete all of the pre-requisites for construction. To put it bluntly, "construction people" (CMs and contractors) cannot afford to ignore critical work in "design land."

## Final Design

One surprising and often overlooked aspect of DB delivery is that the owner's design review activities may be on the contracting entity's critical path! (Koch et al. 2010) Before construction can begin, the design-builder and owner must engage in a well-coordinated "dance," exchanging several design submissions and reviews for each package. Figure 5 illustrates the Blue Plains Tunnel design review process.

Due to this shared responsibility for a successful outcome, the construction manager is advised to play a central role in (a) tracking the design-builder's progress on design submissions, (b) ensuring timely owner reviews and resolution of comments, and (c) identifying opportunities to compress or simplify the Final Design process if needed for work elements that are particularly critical.

To further improve coordination, project teams are advised to designate one individual from each entity (design-builder, construction manager and owner's Design Reviewer) as "design coordinators." These three individuals should meet regularly to track progress, resolve issues, and seek management input if needed.

## Protection of Structures

On the Blue Plains Tunnel Project the decision of whether or not to implement mitigation measures to mitigate the impacts of settlement for nearby structures (such as grouting or other means) was left up to the design-builder's EoR. To account for this variable scope the owner included an allowance in the Contract Price; however, no such "allowance" was included in the Contract Time.

Under these circumstances, the project team's mission is therefore to (a) determine as early as possible whether actual mitigation measures are necessary, (b) design the mitigation, and (c) implement the mitigation prior to the start of the construction activity in question. In many cases, the inputs to this process are not readily available at NTP; for example the design-builder must advance Final Design packages, condition assessments and numerical modeling before a go/no-go decision can be made. Here again,

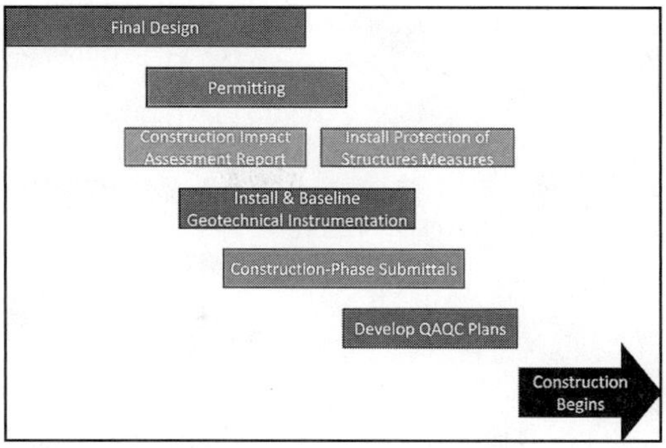


**Figure 4. Design-build sequence**

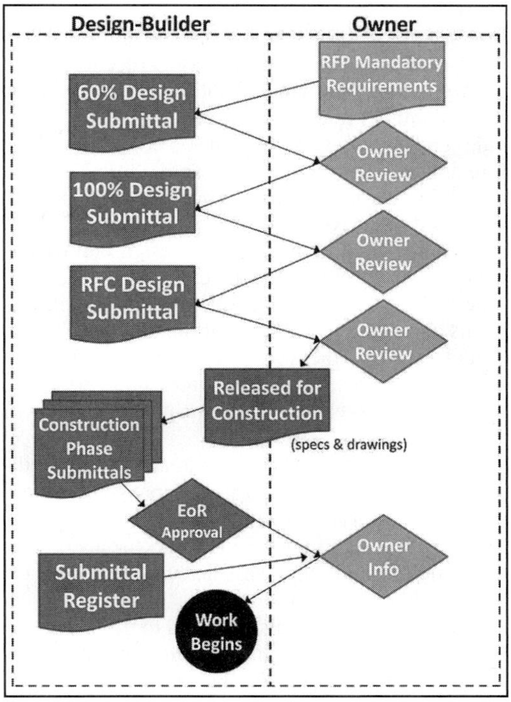


**Figure 5. Design review process**

the construction manager is advised to play a central role in tracking and facilitating this effort in order to keep the project on schedule.

The Blue Plains Tunnel project team implemented extensive mitigation measures at the Main Pumping Station site in order to protect sensitive below-grade structures, including soil mixing, steel sets inside a brick arch sewer, and concrete slabs at grade to withstand construction loads.

## Geotechnical Instrumentation

Today's tunnel projects rely on extensive geotechnical instrumentation systems to measure groundwater

**Figure 6. Instruments at BPAWWTP site**

levels, surface and subsurface movement, and movement of existing infrastructure, in order to protect the well-being of personnel and property. The Blue Plains Tunnel Project is no exception. Figure 6 provides an aerial view of instruments at just one of the four construction staging areas, and Figure 7 lists the design-builder's instrumentation responsibilities as stipulated in the RFP Documents.

At the time of writing the design-builder has installed a total of 600 instruments (many with multiple sensors), and installation of additional instruments ahead of the TBM is still underway. It's easy to underestimate the level of effort needed from the entire project team to successfully implement and manage a geotechnical instrumentation system of this magnitude.

The construction manager is advised to facilitate development of clear roles, responsibilities and processes for prompt resolution of all instrumentation-related issues. Figure 8 illustrates the instrumentation alert resolution process that was developed for the Blue Plains Tunnel Project.

Project teams are also advised to designate one individual from each entity (EoR, design-builder construction team, instrumentation subcontractor, and construction manager) as "instrumentation gurus." These individuals should meet regularly to track progress, resolve issues, and seek management input if needed.

## Construction Submittals

For many DBB projects, the construction manager's first assignment after NTP is to comb through the specifications and develop a list of *all* required submittals. For DB projects procured using preliminary design documents, this exercise is practically impossible because final design specifications do not exist at the time of NTP. To address this challenge, the RFP Documents for the Blue Plains Tunnel required the design-builder to provide a submittal register with each design package.

Design-Builder Responsibilities for Instrumentation:

- Develop the instrumentation design (RFC drawings & specs)
- Submit the instrumentation equipment and methods (construction submittals)
- Prepare QAQC procedures for installation
- Install instruments / perform QAQC
- Activate data monitoring system / train users
- Baseline instruments
- Survey / monitor instruments
- Respond to & resolve all alerts

**Figure 7. Design-builder's responsibilities**

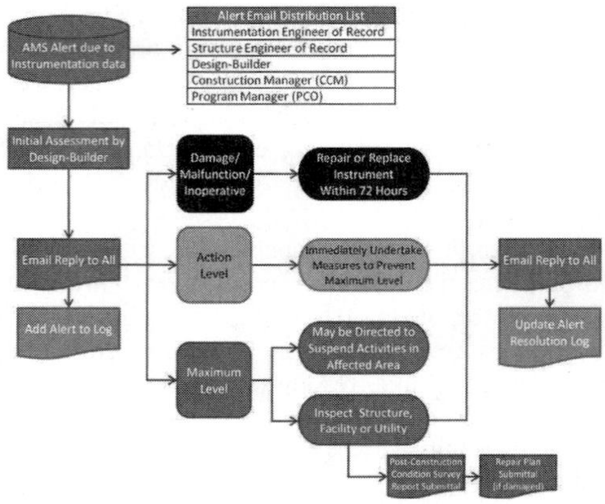


**Figure 8. Instrumentation alert resolution process**

During preparation of RFP Documents, project owners and construction managers should ask themselves what level of owner review is desired for the design-builder's construction submittals. Because the design-builder's EoR is now the key reviewer, the owner may elect to receive some or all of the construction submittals "for information" rather than "for review." If owner reviews are desired, will they occur before, after, or concurrent with EoR reviews? If the EoR initially approves a submittal, but the owner marks it "Revise & Resubmit," does the EoR have to re-review the resubmittal? For a project with thousands of submittals which are pre-requisites for construction, the answers to these procedural questions have significant implications for the design-builder's workload and schedule.

## QAQC

For the Blue Plains Tunnel Project the owner elected to transfer all QAQC responsibilities to the contracting entity (design-builder), and establish an owner Independent Verification and Assurance (IVA) program, as is now customary for FHWA, VDOT and other transportation projects. This transfer of responsibility requires a paradigm shift for the entire project team, many of whom have traditionally focused on satisfying the owner's inspector. Although contractor-controlled QAQC is different, QAQC is still QAQC. Construction managers are advised to support the design-builder in establishing a successful QAQC program featuring (a) a strong QAQC Manager, (b) clear, concise and useable QAQC Plan with standardized templates and workflows, (c) experienced QC Representatives with authority, and (d) diligent and organized record keeping.

In addition to a project-wide QAQC Plan, the RFP Documents require the design-builder to submit task-specific QC plans for owner review prior to the start of work on each element. Figure 9 illustrates this process.

When setting up a new project, owners must ask themselves what level of involvement is desired for the owner's IVA program. For instance, an owner may require the construction manager to verify anywhere from 0% to 100% of the design-builder's inspection, sampling and testing. This decision is the owner's prerogative and may be influenced by any number of factors such as the owner's budget, available expertise, prior experiences, or the expectations of the owner's governance structure. For the Blue Plains Tunnel Project the construction manager is required to provide full-time inspection services on all shifts, to "spot check" all work for conformance with RFC requirements. Figure 10 illustrates a chain of communication implemented by the project team early on in order to encourage resolution of QAQC-related issues at the lowest possible level.

## Design Changes & RFIs During Construction

Owners and construction managers must be prepared for the reality of further design changes after the final design is approved and Released for Construction. Such changes may arise from construction inquiries or issues, third party issues, revisions from the EoR, or owner changes. It is important to recognize that design changes arising during construction are always time-sensitive. When setting up a new project, the owner's team must ask themselves what level of involvement is desired. Will the owner review design changes after RFC? If so, will EoR and owner reviews be concurrent or sequential? What exactly will the owner review consist of? What happens if the EoR approves a change but the owner does not? Is owner approval required before construction can continue? Conducting this thought experiment in advance will also provide the owner's team with valuable perspective on overall roles and responsibilities of the Engineer of Record and the owner's design reviewer.

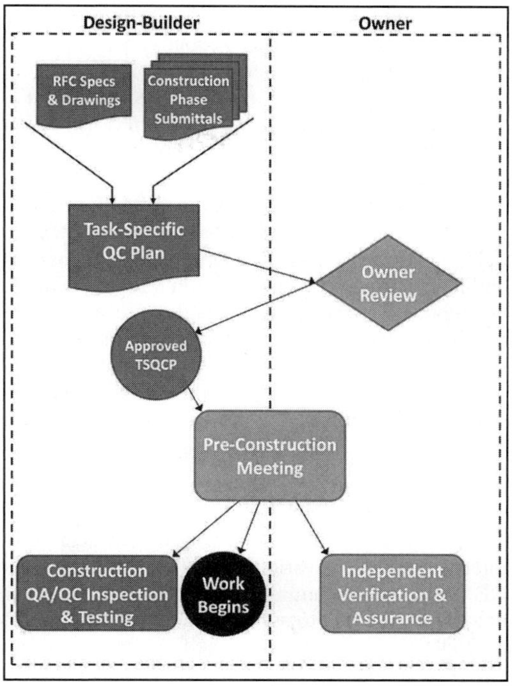


**Figure 9. Preparing for construction QAQC**

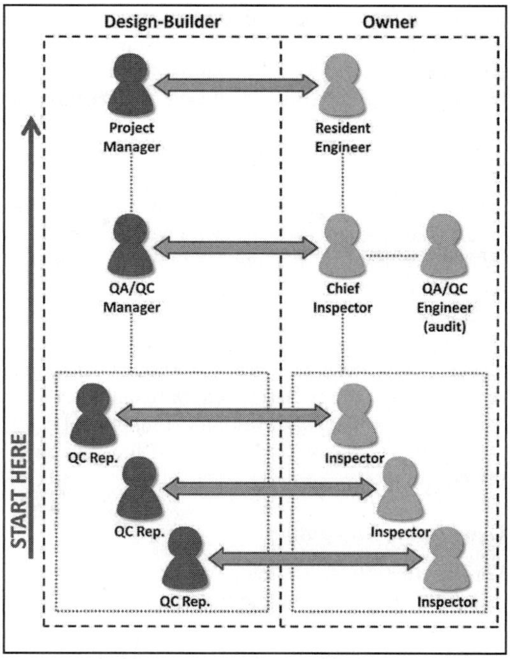


**Figure 10. QAQC chain of communication**

In addition to conventional Requests for Information (RFIs) from the contracting entity to the owner, DB delivery adds "internal" RFIs from the construction team to their EoR, as well as RFIs from the owner to the EoR. Owners and construction managers are advised to develop processes and procedures for *all three types* of RFIs.

Due to the fast-paced nature of DB delivery, it is critical that the design-builder's and owner's field personnel receive all approved design changes and RFIs pertinent to their work in a timely manner. Owners and construction managers are advised to require the design-builder to furnish approved copies of all such documentation for inspection at QAQC witness and hold points, since this generally constitutes the "last chance" to incorporate changes into the construction of the work.

### Document Control System

Developing a document control system that can fully integrate all of the various workflows needed for a DB project is a daunting task and warrants a paper unto itself. It's safe to assume that for any given DB project there is no "off the shelf" software system that can immediately handle all necessary workflows stipulated in that project's contract documents. Owners and construction managers are advised to carefully develop the contract procedural requirements *and* the owner's document control system in coordinated fashion, prior to issuance of the DB RFP Documents. Special attention should be given to tracking design changes during construction, as discussed above. Construction managers can provide valuable input by combing through pre-RFP documents to identify all of the various types of documents that will be generated, and asking themselves a series of questions for each type. Who generates it? Who reviews it? If there are multiple reviewers, what is the process? How are the documents and review comments exchanged? Are automated workflows needed? Are system-generated tracking reports needed? What format is needed for the final project record? Resolving these procedural challenges prior to NTP is essential for developing a functional document control system, and critical to the success of the DB project.

### Geographic Scheduling

For the Blue Plains Tunnel project, the authors have relied heavily on a Geographic Schedule (Figure 11) as an essential tool in the construction manager's toolbox. This one-page graphic is updated monthly using links to the design-builder's P6 schedule, and has become the project team's go-to document for explaining the scope of work, communicating overall progress and schedule, evaluating what-if scenarios,

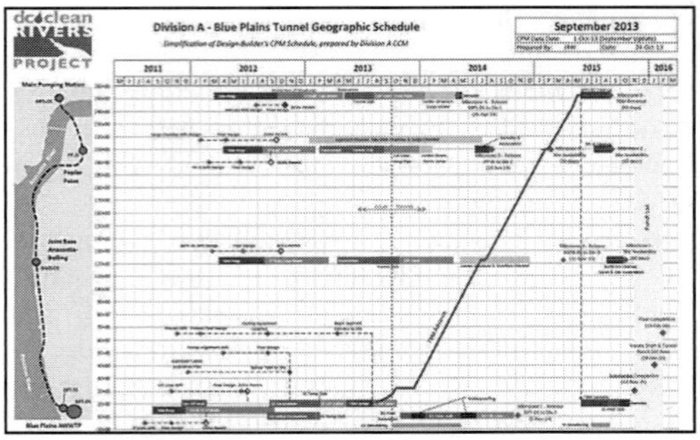

**Figure 11. Blue Plains Tunnel geographic schedule**

and making key project decisions. For more information on Geographic Scheduling, refer to the 2012 SME paper titled *Upgrade to a Geographic Schedule* (Wonneberg and Drake 2012).

### Integrated Data Monitoring System

Construction of shafts and tunnels introduces the need for sophisticated data monitoring software, both for geotechnical instrumentation and the tunnel boring machine. Owners and construction managers are advised to carefully evaluate their data monitoring and reporting needs, then solicit input from software vendors prior to issuance of the RFP Documents to ensure that the RFP requirements are consistent with (a) the project team's vision and (b) currently available software. What data will be collected and monitored? What types of reports are needed? Who will own and manage the monitoring system? What are the design-builder's responsibilities? What are the owner's responsibilities? What is the process for responding to system-generated alerts?

The owner's team has elected to procure Tunnel Process Control (TPC) software, by Tunnelsoft, Inc. (Kent, WA), for use across all of the tunnel contracts on the DC Clean Rivers Project. In addition to integrating TBM *and* geotechnical instrumentation data with fully customizable monitoring, reporting and visualization, this software enables the construction management team to generate daily tunnel inspection reports and carry out the owner's Independent Verification & Assurance of the design-builder's TBM muck and grout volumes and precast tunnel liner QAQC. Nearly all of the construction manager's daily workflows pertaining to the tunnel are completed within the TPC system. Figure 12 illustrates the system's capabilities.

### Change Management

DB delivery introduces the possibility of at least three different types of change orders: (a) changes affecting only design services, (b) changes affecting

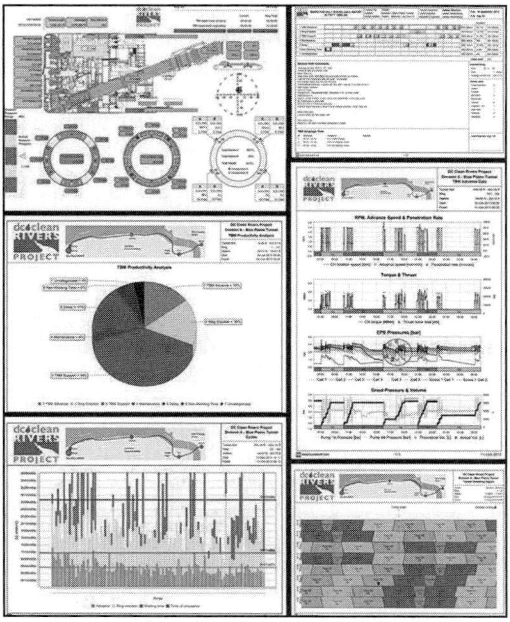

Figure 12. TPC system capabilities

design services and construction, and (c) changes affecting only construction. Construction managers coming from DBB projects may not be accustomed to negotiating design fees, therefore, owners and construction managers are advised to develop a strategy for dealing with design changes and then ensure that the contract's general conditions provide a framework for administering such changes.

For owner-requested changes requiring additional design services, construction managers may be able to implement a phased approach by authorizing the design-builder to design the change work, then negotiating construction cost and time impacts based on the completed design. This approach can benefit project owners by reducing contingency costs associated with uncertainties in the scope of work. The drawback is that the owner will need to incur the cost of the design before the full cost and time impact of the requested change is known.

### Formal Partnering Process

DB delivery requires a higher level of trust and partnering than DBB delivery, in order to achieve *design and construction* within the fixed budget and schedule (Koch et al. 2010). As noted above, DB delivery also puts the owner's design reviews on the design-builder's critical path. To that end, project executives are advised to take advantage of periodic partnering sessions and use them to overcome design and construction roadblocks as a team, in order to keep the project on schedule.

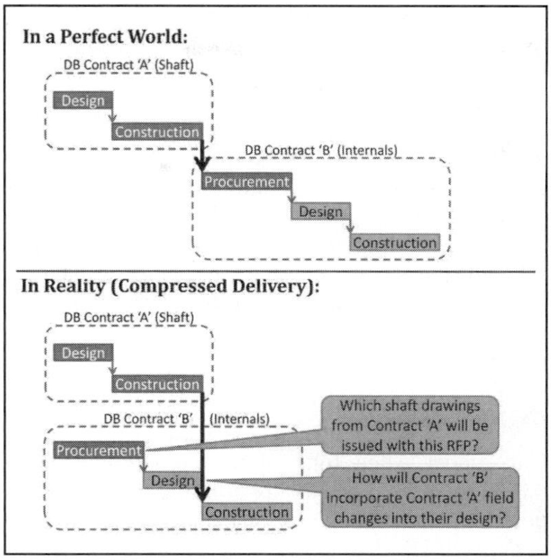


Figure 13. DB contract interface scenario

### Contract Interface (Program Management)

Although many of today's tunnel contracts are mega projects in their own right, they are often just one piece of a larger "program" to deliver new infrastructure for the project owner, such as a CSO system or a subway. Program management on this scale demands specialized expertise, most notably the ability to plan and manage all of the interfaces between the individual contracts in order to achieve the program-level objectives. Interfaces between tunnel contracts can be particularly challenging because each project's schedule is highly dependent on actual vs. planned TBM advance rates. Program managers are advised to develop and maintain a one-page program-wide geographic schedule, with linear schedules for each of the tunnel contracts, in order to visualize how all of the pieces of the puzzle fit together.

In addition to the *construction* interfaces described above, program managers in DB delivery must also carefully and actively coordinate *design* interfaces. Consider the following scenario: Design-Builder 'A' designs and constructs a deep shaft, then turns over the site to Design-Builder 'B' who will design and construct internal structures and facilities within the shaft. Although this may seem straightforward, several complexities lie beneath the surface. In a perfect world the RFP Documents for Design-Builder 'B' *should* provide a complete set of final design and as-built documents for the shaft. But in reality, a compressed program schedule will dictate procurement of Design-Builder 'B' during shaft construction, before the as-built details of the shaft are known. Figure 13 illustrates this challenge.

Project owners should be aware that the owner nearly always carries the risk (e.g., cost and time) if an interface between two or more contracts goes awry. Oftentimes when working within compressed delivery schedules there is no perfect solution to contract interfaces. But program managers can manage and reduce the Owner's risks by (a) developing a cohesive and centralized strategy for managing contract interfaces, (b) committing expert resources to this task, and (c) developing robust tracking tools capable of pulling data from various monthly CPM schedule updates in order to monitor *design* and *construction* interfaces in real-time across the entire program. Such measures will pay great dividends in the overall program outcome.

## CONCLUSION

Christopher Allen (one of the co-authors of this paper) is known for emphasizing the following two points which in many ways encapsulate the risks and rewards of DB delivery:

> *"Schedule slippage during the design phase is lost forever."*

> *"Project teams that understand how design-build works can achieve beautiful things."*

It is the authors' hope that readers will build upon the key takeaways and recommendations included herein, to plan and execute successful DB projects of their own. We also encourage readers with DB experience to share their lessons learned in similar fashion so that the tunnel industry as a whole may continuously improve the DB delivery method, which is becoming more and more prevalent throughout the industry.

## REFERENCES

DC Water, 2002, *WASA's Recommended Combined Sewer System Long Term Control Plan, Executive Summary,* Washington, DC (http://www.dcwater.com/workzones/projects/pdfs/ltcp/Executive_Summary.pdf).

Koch, J.E., Gransberg, D.D., Molenaar, K.R. 2010. *Project Administration for Design-Build Contracts—A Primer for Owners, Engineers, and Contractors.* Reston, VA: ASCE Press.

Wonneberg, J.R., Drake, R.D. 2012. Upgrade to a Geographic Schedule. In *2012 Proceedings of North American Tunneling*, Indianapolis, IN, June 24–27, 2012, Englewood, CO: Society for Mining, Metallurgy, and Exploration, Inc. (SME).

# Project Delivery Strategy for Major Transportation Projects in Los Angeles

**K.N. Murthy and Matthew Crow**
Los Angeles County Metropolitan Transportation Authority

**ABSTRACT:** Measure R, which was passed on November 05, 2008 provides $40 billion over the next 30 years with funds being used for both direct funding and as collateral for long term bonds to be spent on projects in Los Angeles County, including major new rail transit and highway tunnels. Metro has recently used a successful combination of traditional and alternative delivery methods on major transportation projects, including the I-405 freeway widening, Metro Gold Line Eastside Light Rail Extension and Orange Line Bus Rapid Transit projects. The paper reviews the project delivery methods used to date on projects and shares the rationale for the selection of the design build (DB) procurement strategy being used for the delivery of the Crenshaw LAX Transit Corridor, Regional Connector Transit Corridor and Purple Line Extension projects. This delivery method was influenced by a commitment to accelerate the overall delivery schedule, facilitate local job creation, improve the quality of life for the region, and was based upon consideration of methods used by other transit agencies and the regulatory framework of options available to the Agency for these three underground construction projects. The paper also explains how the approach has maintained control of key underground construction requirements for the safe and efficient construction and operation, adoption of appropriate standardized approaches, while allowing freedom for the design builder to provide market based integrated solutions.

## EXISTING AND PLANNED L.A. METRO RAIL SYSTEM

The existing rail transit system in Los Angeles (L.A.) comprises a 140-km (88-mi) long network of both Heavy Rail Transit (HRT) running in tunnel with 750-V DC third rail traction power supply and Light Rail Transit (LRT) running generally at grade with sections of tunnel and aerial structures with 750-V DC overhead catenary traction power supply. The existing and planned Metro LRT and HRT rail lines are presented on Figure 1.

In total about 30km (18-mi), of the 140 (88-mi) long system is underground. The first line to enter service in 1990 was the 35-km (22-mi) Blue line Light Rail Transit (LRT), which is generally at grade with a 0.4-km (0.25-mi) length of cut and cover tunnel in Downtown Los Angeles. This was followed in 1993 with a 7.5-km (4.8-mi) long operating segment of the Red Line HRT subway, which came of age this year, having opened in 1993. A summary of the other completed Metro rail lines presented in Table 1 includes the Green Line, Red Line to North Hollywood, Purple Line to Wilshire / Western, Gold Line to Sierra Madre and Expo Line to Culver City.

There are two projects presented in Table 2, the Gold Line Foothills Extension and Expo Line Phase 2 that are currently being constructed by Joint Powers Authorities, which on completion will be owned and operated by Metro.

The three principal transit projects being constructed or in procurement by Metro are the Crenshaw/LAX Transit Corridor (C/LAX), Regional Connector Transit Corridor (RC), and the Purple Line Extension (PLE), which was formerly called the Westside Subway Extension or Subway to the Sea. The RC and C/LAX are LRT and the PLE is HRT, with the PLE and RC guideways constructed entirely within tunnel and the C/LAX having 3.2-km (2.0-mi) of tunneled guideway and three underground stations. A brief summary of these three projects is presented below, as detailed descriptions of these projects have been presented elsewhere (Wallis 2014).

### Crenshaw LAX Transit Corridor

This twin track 13.7 km (8.5-mi) long LRT line follows a north/south corridor that serves the cities of Los Angeles, Inglewood, Hawthorne and El Segundo, as well as unincorporated L.A. County, with 8 stations, aerial grade separations, below grade, at-grade construction, and a maintenance facility yard. The project alignment presented in Figure 2 provides for connections with the L.A. International Airport (LAX), as well as links to the Metro Green Line, the Exposition Line and countywide bus network. It is planned to open to passengers is in 2019.

**Figure 1. Metro rail transit lines in operation, construction or procurement**

**Table 1. Existing LACMTA rail transit lines**

| Line | Opening Year | Length [mi] | Type | Construction | Destinations |
|---|---|---|---|---|---|
| Blue Line | 1990 | 22 | LRT | At grade with short cut & cover tunnel and one underground station. | Downtown L.A. Long Beach. |
| Red Line | 1993 | 17.4 | HRT | Underground stations, tunneled guideway. | Downtown L.A. N. Hollywood. |
| Purple Line | 1993 | [Note*] | HRT | Underground stations and tunneled guideway. | Downtown L.A. Mid-Wilshire |
| Green Line | 1995 | 20 | LRT | At grade and elevated. | Redondo Beach Norwalk. |
| Gold Line | 2002 | 19.7 | LRT | At grade, elevated and bored / cut & cover tunnels with two underground stations. | East L.A. Pasadena Downtown L.A. |
| Expo Line Ph. 1 | 2012 | 8.6 | LRT | At grade and elevated. | Downtown L.A. Culver City |

* Included with Red Line.

**Table 2. Metro rail transit lines under construction by Joint Powers Authorities**

| Line | Opening Year | Length [mi] | Type | Construction | Destination |
|---|---|---|---|---|---|
| Gold Line Foothills Extension | 2016 | 11.5 | LRT | At grade | Pasadena Montclair |
| Expo Line Ph.2 | 2016 | 6.6 | LRT | At grade | Culver City Santa Monica |

*Anticipated opening for revenue operations.

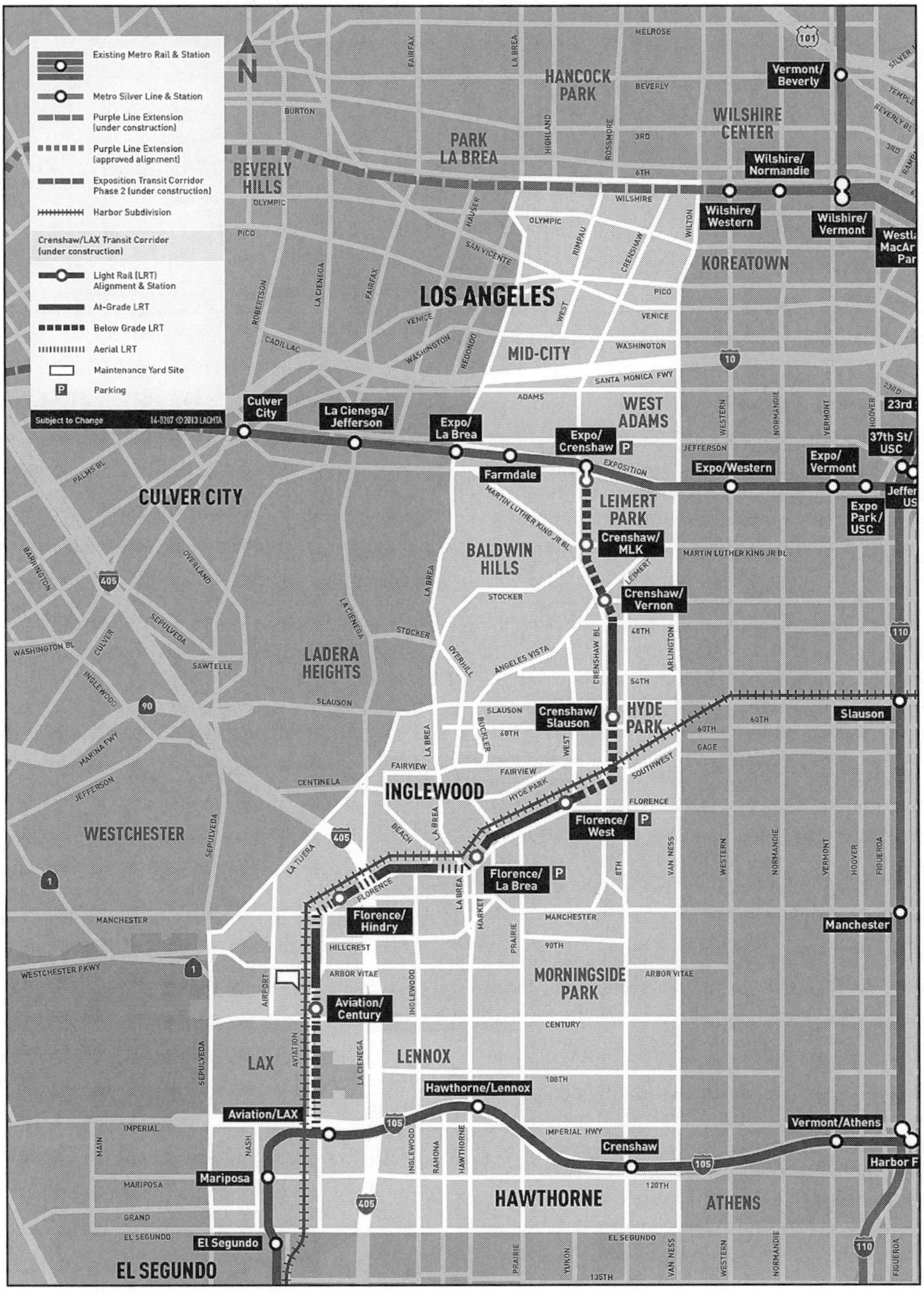


**Figure 2. Crenshaw/LAX Transit Corridor alignment**

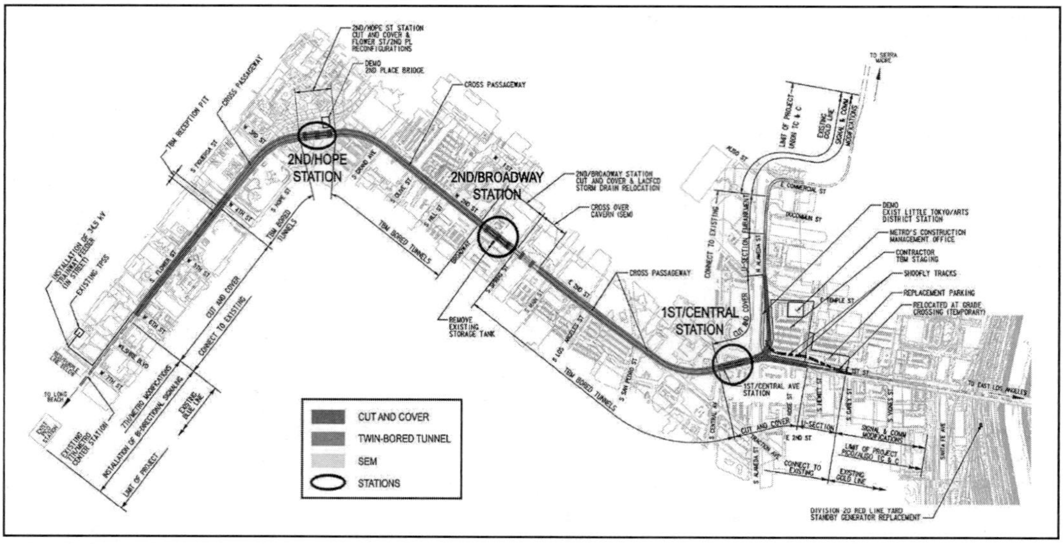


**Figure 3. Regional connector transit corridor alignment**

### Regional Connector Transit Corridor

This twin track 3-km (1.9-mi) underground LRT line in Downtown L.A. connects the Metro Gold Line, Metro Blue Line, and Expo Line, as shown in Figure 3. This serves additional downtown destinations, as well as providing a 20 minute commute reduction by offering a one seat ride from Azusa to Long Beach and East LA to Santa Monica or a single same platform change to reach all destinations on these lines. It will have three cut and cover underground stations at 2nd / Hope, 2nd / Broadway and 1st / Central. The works include a cut and cover tunnel guideway along Flower Street connecting into the existing tail tracks of the 7th and Metro Station, twin TBM bored underground guideway tunnels along 2nd Street with a crossover cavern east of 2nd and Broadway Station. It is planned to be open in 2020.

### Purple Line Extension

The alignment of the PLE is an extension of the existing Purple Line at Western/Wilshire along the Wilshire Corridor to near the Veterans Affairs Hospital in Brentwood and is in three sections as shown on Figure 4.

The final environmental documents, Environmental Impact Statement and Report (EIS/EIR) for all three sections were approved in May 2012 and includes 14 km (8.7-mis) of twin bore mined tunnel guideways with 7 cut and cover underground stations. The first phase, Segment 1, to La Cienega is planned to be open in 2023.

## CHARACTERISTICS OF PRINCIPAL METRO RAIL TRANSIT PROJECTS

Clearly these three projects have differences, the most notable being the type of train, LRT or HRT, however there are many common characteristics including; construction costs over $1B, urban environments, connectivity with existing rail transit systems, extensive utility diversions, entirely or significant sections of tunneled guideway, community expectations, seismic engineering requirements, and the potential for explosive/toxic gasses in ground. The characteristics of the projects help determine the procurement method as well as project requirements.

## FINANCIAL CONSIDERATIONS

### Funding

Funding for Metro comes from a complex mix of funds including, fare revenue, Proposition A and C sales taxes, Federal and State grants, interest income/bonds, and other local revenue. Measure R, another sales tax, which was approved by a two-thirds majority of LA county voters in November 2008 clearly showed that the local population wanted more improvements to aid mobility. This alone commits a projected $40 Bn. to traffic relief and transportation upgrades over the next 30 years. It is estimated to create over 200,000 construction jobs and infuse $32 Bn. into the local economy. When this went into effect in 1 July 2009, it became the primary driver to hasten the delivery of these three projects.

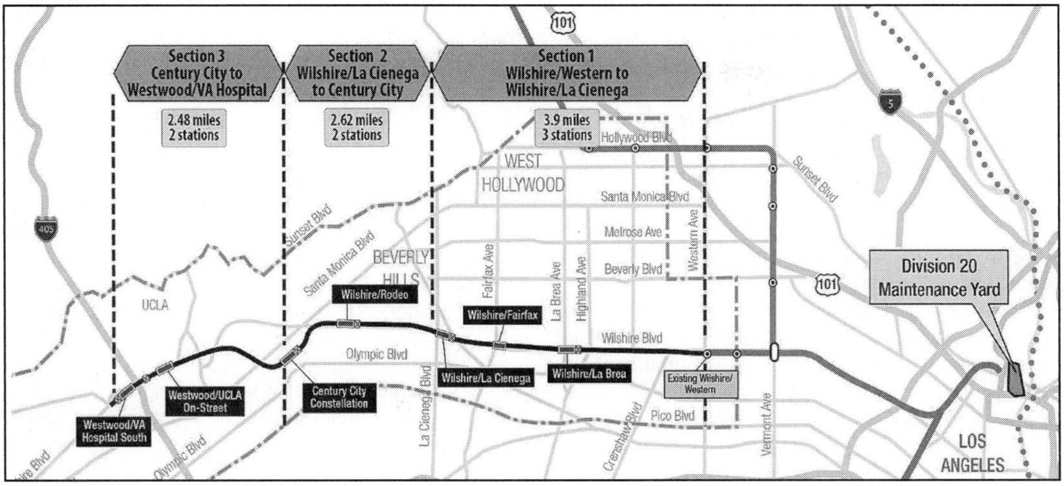


**Figure 4. Purple Line extension alignment**

### State of Market

A significant upward trend in project bid costs was observed starting in the mid 2012 timeframe, which Metro considered could impact the ability to deliver the Measure R and other projects within existing budgets. To consider and mitigate this impact, a market conditions analysis was performed. The report confirmed that increased bid prices were not only a function of the recovering Los Angeles construction market, but also the size of the Metro program, specific Metro requirements, processes and procedures, and the inherent risk of mega projects. The report found no evidence that choice of delivery method design-bid-build (DBB) or DB impacted bid prices. Although many of these impacts are outside the control of Metro, key findings that are being addressed include a review of estimating methodology and procedures, formal risk analysis procedures with Monte Carlo simulation, review of contract documents for fair allocation of risk, provision of appropriate liability requirements and industry standard insurance requirements, review of change order procedures, continued use of advance utility relocation packages, establishment of robust interagency co-operative agreements, realistic Disadvantaged Business Enterprise (DBE) goals, assessment of labor demand, appropriate Metro staffing levels, and embracing partnering on all levels.

## KEY ISSUES AND OBJECTIVES

In general the key issues being addressed are adherence to cost and schedule, avoid tunneling mishaps, connecting to existing operational rail transit lines, significant tunneling and underground stations, and design aligned with chosen construction methodology. These can be achieved by overlap or zippering of design and construction, prescription of robust underground construction methodology, seamless connection to existing transit system, allowing innovation within the framework of established Metro standards, and early contractor involvement.

The goal of meeting project budget and schedule is achieved by allowing savings from efficient contractor means and methods, incentivizing the contractor to be efficient, managing of key interfaces, reducing potential for disputes/claims/litigation, and adequate levels of MTA resources and consultant expenditures in preliminary engineering and final design phases.

Reducing cost, schedule and technical risk to as low as reasonably practical (ALARP) is achieved by requiring equitable sharing and allocation of risk to the party best able to manage, by choosing a contractual mechanism to do this, adopting a collaborative/partnering approach, and use of incentives/liquidated damages (LD)s and appropriate risk sharing.

Other requirements include TBM minimum requirements (pressurized tunneling system), management of TBM operations and grouting to minimize groundloss, segment design, settlement criteria, risk allocation, risk sharing, allowances and provisions.

## PRESCRIPTIVE CONSTRUCTION ELEMENTS

Certain design elements remain prescriptive in order to meet Metro needs based on experience and operational requirements as presented below:

***Station Design***—3rd party agreements with respect to entrances, ventilation stacks and construction impacts, Metro operations, patron access, station footprint, relocations on utility contract, systems and operations requirements.

***Station and Cut & Cover Support of Excavation***—Metro standard structural drawings provide mandatory requirements for design including loading diagrams.

***Station and Cut & Cover Water / Gas Proofing***—Metro standard drawings that present complete encasement in HDPE membrane with compartmentalization with water bar and cast in grout lines to enable effective defect location and repair, details of joints, corners, testing, surface treatment, protection and treatment of penetrations.

***Tunnel Lining***—Utilize a one-pass precast concrete segmental lining with double gaskets and cross gasket with dowels and bolts to maintain gasket seal, mandatory leakage requirements, which provide a robust system with the ability to detect and repair any leaks.

***Tunneling***—Pressurized tunneling system including pressurizing the TBM face to prevent water/soil inflow and ground loss, pressurized bentonite injection around perimeter of shield, pressurized annular grout through TBM tailshield, monitoring of pressures, procedure for interventions, filling of plenum after interventions, adding material when plenum pressure drops, ground monitoring instrumentation, integrated monitoring of machine functions, surface and building monitoring instrumentation and data management, and specialists coordination of instrumentation results with monitoring of and readout of machine functions.

***Stations and Finishes***—Aspects of design, integration, quality of materials, and rider flow patterns, waterproofing of station box, proactive systems for interface elimination, consideration of options and maintenance requirements, and design of interfaces with existing systems.

***External Stakeholder and Environmental Requirements***—including environmental commitments (EIS/EIR), Federal Transportation Administration (FTA) provisions (Record of Decision), political support and positive community communications.

Other critical success factors identified include prescription of key design builder staff, following best practice for management by Metro and DB teams, and use of Metro standards and key requirements, as these are established transit systems.

## RISK MANAGEMENT

Management of risk takes many forms at different phases on projects of this complexity. First, close coordination with the FTA was maintained in providing a "roadmap" for key project deliverables, which define scope, costs, schedule and a risk register/risk and contingency management plan. Monthly full funding grant agreement (FFGA) checklist meetings are being held between Metro and the FTA to complete the requirements towards receiving an FFGA. Throughout the process the rigorous FTA procedures and risk register were maintained.

There are many risks that were addressed, however two of the principal risks discussed here are relate to utility relocations and the ground conditions.

Relocation of utilities risks are managed in the following ways; master cooperative agreements with utilities and cities and early utility relocation contracts using the DBB process; defined Metro / third party submittal review turnaround; three dimensional modeling of station locations to minimize impacts to utilities; and third party coordination with various city agencies, utility companies, major stakeholders and property owners.

Underground construction risk mitigation is managed in tunneling by use of pressurized closed face TBMs with only cross passages constructed using hand mining/sequential excavation methods and excavation support using one pass pre-cast concrete segments with cross passages using cast-in-place concrete. For underground station excavation a cut and cover system with a temporary road deck is to be used. Tunnel liner segment gaskets and an impermeable hydrocarbon resistant membrane provide water/gas barrier in tunnels and cut and cover structures respectively. Complete geotechnical studies for underground work were completed for all three projects including 100% design level geotechnical data reports (GDR) and a geotechnical baseline reports (GBR) that clearly apportion underground conditions risk between Metro and the constructor. Retaining a small focused and independent tunnel advisory panel (TAP) ensures best practice is maintained. Seismic engineering considerations are addressed by adherence to the established Metro seismic design criteria developed with best-in-class seismic engineers.

Risk will also be managed by technical advancements and improved contract terms being based on "Lessons Learned" from successful management of Metro Gold Line Eastside Extension Project, industry constructability reviews and uniformity in approach with other Metro rail projects.

## DEVELOPMENT OF CONTRACT STRATEGY

### Contract Packaging Options

Numerous general contracting delivery methods could be considered and are described in detail in other publications and include, Design-Bid-Build (DBB), DB, DB with best value (DB/BV), Enhanced DB with early contractor involvement (EDB/ECI), CMGC/CM at Risk, Modified CMGC (Boston Green Line/UDOT Mountain View Corridor), Cost reimbursable with target cost (Portland CSO, Expo Phase 2), Cost reimbursable with target cost and pain-gain provisions (UK CTRL), Consensus documents / relationship contracting, Alliancing (Full relationship contracting—(UK, Australian & NZ projects), Alternative financing & procurement (Ontario & Alberta Canada). These can be simply categorized as DBB, DB, or alternative project delivery methods.

Contract packaging options also considered the choice of multiple contracts for individual civil and systems components or a single contract for civil and systems.

Early LRT and HRT tunneling projects were procured as DBB contracts with the lowest competitive bidder building the works to final designs prepared by Metro's designers, however as can be seen in Table 3, more recently other procurement methods have been used by either Metro or by Joint Powers Authorities for Metro. The most recent Gold Line Eastside Extension was successfully procured as a hybrid contract, with DBB for the 3.2-km (2-mi) length of bored tunnel guideway and initial support of underground stations, and DB contracts for the underground station structures and at grade sections of the alignment.

### Legislation

Metro is governed by two laws that allow it to utilize a DB delivery process: California Public Utilities Code (PUC) 130242, a low bid selection and California Public Contract Code (PCC) 20209, a best value selection.

### Rationale for Metro Design Build Procurement

Although considered of interest in meeting the program goals, Metro procurement and County Counsel advised that early contractor involvement similar to the Expo Phase 2 DB contract, in essence involving "Dueling Designers" or other alternative project delivery methods cannot be used at this time. Thus the choice is either DBB or DB with either a low bid selection or best value selection.

For the current projects, Metro is making a departure from its previous procurement strategies with the three major LRT/HRT projects, which are all being procured as DB contracts integrating all civil and systems work using PCC 20209.5-2029.14, a sealed bid or a negotiated "full trade off best value procurement" process based on preliminary engineering designs. This procurement includes qualification of design builders using a State Department of Industrial Relations questionnaire, request for proposals, and evaluation scoring based on price, project management, and technical approach.

This decision by the Metro Board, like all decisions, has its advantages and disadvantages. To be efficient and cost conscious Metro has encouraged the use of new and innovative construction methods and are also focused on compatible technology with the existing ones for operation and maintenance purposes. Metro has the advantage of being a mature rail transit agency, having over 20 years' experience as both an operator and developer of similar LRT and HRT systems. Metro has delivered these type of projects before, having in the past performed all the individual technical components that are now required. The design requirements and standards have been established over many years for both LRT and HRT systems. This past experience and lessons learned has provided Metro with a solid understanding of design requirements. As these new projects are comparable to these earlier works, there is no need to expend time and public funds completing a 100% detailed design ahead of engaging a DB contractor for the new works. Systems requirements in the DB proposals are included as performance specifications and the new works are extensions of existing lines and require continuity and compatibility with those earlier installations. In general, Metro gains 12–18 months on the overall procurement by overlapping detailed design with construction. Risk in DB procurement is more equitably shared between the owner and the constructor with the final design having to comply with strict requirements. Although there is some room for innovation on means and methods, aspects of the performance requirements are prescriptive, requiring strict compliance on technical and performance criteria. In particular there is little tolerance for experimentation in underground construction that might lead to surface settlement.

### Early Contractor Involvement

Early contractor involvement (ECI) was considered essential in the successful implementation of these projects. Although ECI could not be included as a formal part of the contract procurement, the process was included as a step leading to issue of the procurement documents. For example, the PLE project included a comprehensive industry constructability review comprising the following process. Metro compiled selected draft technical documents, drawings, specifications, and technical reports and issued them to potential prime proposers along with a list of questions for the proposers to consider during the industry constructability review period. This industry constructability review focused on constructability and tunneling issues, alternative means and methods and potential innovative cost saving ideas. Metro then conducted "One-on-One" confidential meetings with prime proposers to discuss specific technical comments related to written responses to Metro questions prior to the issue of the RFQ.

### Alternative Technical Concepts

PUC 20209.5-2029.14 allows Metro to accept alternative technical concepts (ATC)s after issue of the RFP during the development of the DBs proposals. This was used most effectively on the C/LAX project, as the project constraints provided sufficient

**Table 3. Previous contracting and delivery methods used for metro rail transit projects**

| Line | Construction Agency | Procurement |
|---|---|---|
| Metro Blue line | LACMTA | DBB |
| Metro Green Line | LACMTA/ Caltrans | DBB |
| Metro Red Line Segments 1,2, & 3 | LACMTA | DBB for stations, tunnels and Civil Works. Systems DB equivalent (performance based) |
| Metro Gold line Eastside Extension | LACMTA | DBB for tunnels and Station SOE. DB for stations, civil, trackwork, and systems. |
| Metro Gold Line Union Station to Pasadena | Joint Powers Authority | DB |
| Metro Gold line Pasadena to Azuza | Joint Powers Authority | DB |
| Exposition Line Phase 1 | Joint Powers Authority | Modified DB Approach as using upset (target) prices and negotiated construction packages. Reimbursable Target Cost. |
| Exposition Line Phase 2 | Joint Powers Authority | Two Stage Procurement, selection of two DB teams for finalization of PE to approx. 30% and construction input. RFP issued to the selected DB teams with competitive selection based on PE performance, price and qualifications. |

ability for the DB to identify and develop other design concepts, as the project comprised a variable mix of underground, at grade and aerial guideways and stations. ATCs for the C/LAX allowed the proposers to provide Metro with innovative and cost savings ideas. The ATCs were submitted in two separate steps with an initial outline for compliance with contract requirements, followed by a detailed submittal and finally accepted ATCs were included in the individual proposals. This process is explained further elsewhere (Ong 2014). Although the ATC process provided good value for the aerial and at grade segments, their use was found to be limited somewhat for the underground stations and guideways because of the inherent prescribed nature of the associated contract performance requirements. The RC and PLE projects, which have guideways and stations entirely underground did not include the ATC process in development of the constructor's proposals.

### Advanced Utility Contracts

Advance utility contracts are being used to avoid delays to the main contracts by relocation of the complex network of existing utilities which lie beneath the busy urban streets. In order to ensure compatibility with the future DB main construction contracts and discernment of risk, the DBB delivery method was selected for the advanced Utility Relocation Packages for each of these three projects.

## CURRENT STATUS OF PROCUREMENT

A total of four proposers responded to the solicitation for the C/LAX project on December 6, 2012; Crenshaw Transit Partners (Fluor/Balfour Beatty/ SA Healy J.V.); Skanska/Traylor/Keiwit J.V.; URS/ Dragados/Flatiron J.V.; and Walsh Shea Corridor Constructors. The Contract was awarded to Walsh Shea Corridor Constructors in May 2013, with a proposal following the clearly defined contract rules that provided the best value to Metro and not only had the highest technical score, but was the lowest price of $1.272 Bn. The RC and PLE DB contracts are currently in procurement.

## CONCLUDING REMARKS

- Metro has a large program of expansion of their existing rail transit network.
- DB was selected for the procurement of the C/LAX, RC and PLE rail transit projects.
- DBB was selected for associated advance utility contracts.
- Careful use of early contractor involvement is an important tool in procurement of major transit projects.
- Market trends and changes can and must be addressed by mitigation measures.
- ATC can be used effectively on appropriate projects.
- Prescription of some technical requirements is appropriate in DB procurement.

## REFERENCES

Ong, K. and Crow, M., 2014. Use of innovative project procurement as a tool in shaping the design of underground megaprojects for the Crenshaw LAX Light Rail Transit Corridor Project in Los Angeles. North American Tunneling 2014 Conference Proceedings.

Wallis, S., 2013. Los Angeles Scopes Mega Metro Expansions. Article from Tunnel Talk Magazine, November 2013.

# Risk Management for Design Build Projects

**Subal Sarkar**
URS Corp

**ABSTRACT:** On DB projects, risks of project design and development is transferred to a contractor at an early stage and they are exposed to additional risks. The contractor takes risks of unknown and inadequate project and site conditions that have not been fully developed at the time of bidding. Although risk management and risk sharing are accepted in underground construction projects and mega projects, not all owners see the benefits and some use inequitable risk sharing that contractor must cover in his pricing. This paper discusses risk management issues and suggests partnering with the contractor in risk management.

## INTRODUCTION

Design Build (DB) project delivery system is gaining more and more popularity in the US in the infrastructure and transportation projects including projects involving high risk underground construction. Major appeals being accelerated project delivery and that it allows owners to deal with a single point of contact for both final design and construction. For mega projects, variation of the DB delivery system includes design, build, operate and maintain (DBOM); design, build, finance, operate and maintain (DFBOM) and public private partnership (PPP). While accelerated delivery on many projects have been reported by Federal Highway Administration and other agencies for DB process, claims for reduced costs and reduced risks are yet to be fully validated. DB projects expose contractor risks that they do not have under traditional delivery method. Under conventional delivery method, project owners generally provide detailed plans and specifications that have implied warranty. The contractor may not be liable for loss or damage from insufficient or defective information in plans and contract documents. In DB delivery, contractor takes risks of unknown and inadequate project and site conditions that have not been fully developed at the time of bidding as well as the risk of final design. In addition, megaprojects carry risks of financing, operations and maintenance. Risk management and risk sharing is therefore relevant for DB delivery projects and in particular very pertinent for megaprojects. Although risk management and risk sharing principles are accepted in underground construction projects and megaprojects, not all owners see the benefits in its implementation or owners use inequitable risk sharing such that contractors must cover additional risks in their pricing.

## RISK MANAGEMENT APPROACH

Risk management process involves the identification, assessment and prioritization of risks followed by coordinated efforts to minimize, monitor and control the probability and/or impact of risk events. For DB and megaprojects, proactive risk management which is an ongoing and iterative process that should be initiated by the owner at the planning stage and conducted throughout the life cycle of the project. The risk management should be an integral part of the organization and decision making processes, be systematic and structured, explicitly address uncertainty and assumptions, be iterative and responsive to changes, and be capable of continual improvement and enhancement.

The process should involve defining the scope and objectives (including that of various stakeholders), framework for the activity and agenda for identification, developing an analysis method involved in the process and mitigation and solution using technological, staff and project organizational resources.

For DB project delivery, the owners have the opportunity to reduce the risk exposure through cooperation and collaboration with the contractor in jointly developing risk management plan and implementing via partnering and maintaining open communications with the contractor during design, construction, and operation and maintenance phases. This will allow innovation and efficiency in technology selection and dealing with unforeseen risks.

The benefits of proactive risk management are:

- Better and definitive understanding of risks and their effects on the project
- Better contingency planning and selection of appropriate mitigation measures when they occur

- Feed back into planning and design process in terms of ways of avoiding, preventing and minimizing risks
- Feed back into construction and operation of the project for risk scenarios, response and contingency planning
- Reduced risk exposure

**Identification of Risk Elements**

Significant risk elements and project cost drivers can be identified from the past experience at the planning phase. Risk charting or a risk register should be initiated at the planning stage. Further, detailed risk elements related to selected design and construction method may be developed at subsequent project phases and perhaps in collaboration with the contractor and other stakeholders.

**Assessment of Risks**

Once risks have been identified, they must then be assessed as to their potential severity of impact and to the probability of occurrence. The fundamental difficulty in risk assessment is determining the probability of occurrence since accurate statistical information is not available on past incidents. Furthermore, severity of the consequences is often difficult to predict. Nonetheless best educated opinions of experts in the field and available statistical data from case histories should be used for primary sources of information. Nevertheless, the information generated would be adequate to understand primary risks and that the risk management decisions and risk prioritization can be made. For the most part magnitude of risk can be quantified as probability of occurrence multiplied by severity of consequence. Both the probability of occurrence and consequence of impact may be assessed on a scale of 1 to 5. The magnitude of the risk thus can be 1 through 25, one being the lowest and 25 being the highest magnitude. A risk matrix can be developed with one axis being probability of risk and the other axis being severity of consequence. The risks can be qualitatively categorized as low, medium and high. Strategies should be developed as to how each category should be treated for mitigation.

In some sensitive cases where a high risk element requires a relatively accurate projection of high and low that can be used for reducing project contingencies such as a fixed cost, Monte Carlo Simulation may be used. In this simulation, approximate probability of certain outcome is solved by running multiple trial runs using random variables to produce and bracket a risk profile.

**Risk Mitigation**

Once risks have been identified and assessed, the risks can be managed/mitigated using the following options:

- Avoidance (eliminate, find alternative solutions)
- Reduction (optimize the risk by deliberate action)
- Transfer (insure or transfer to other stakeholder)
- Sharing (share with contractor, insurer or third party)
- Retention (owner accepts and budget)

**Risk Management for DB and Megaprojects**

Generally accepted risk elements and risk issues for DB and megaprojects and various risk mitigation options during various project phases are discussed in the following sections. In this paper the consortium undertaking the project including the construction contractor is called as DB contractor or contractor.

***Planning Phase***

Risk management starts with the planning phase when a risk register should be prepared to include generally accepted and known significant cost drivers and risk elements. The risk elements that may be dealt at this stage are: financing, revenue forecasting, maintenance and operating costs, alignment selection, environmental permits, rights-of-way (ROW) and property acquisition, and project schedule. The schedule risk associated with obtaining state legislative approval, if necessary, for alternative delivery and procurement process is not discussed in this paper. The process must include stakeholders from this stage and involve them as appropriate in the risk management process.

- **Financing.** Financing for megaprojects are very complex that involves funding capital cost of construction from various sources including grants from federal, state, county and municipal funds; floating project bonds; local tax referendum; contractor financing; and other sources. In addition, financing or subsidy for operation and maintenance costs may be obtained from outside sources over the length of the DFBOM or PPP contract. Some of these factors are influenced by political and economical environment at the time. Unless full financing is obtained from the contractor, the owner takes the risk of shortfall. Strategies may be developed for sharing the financing and associated risks with the contractor and insurance company. Cost of borrowing from the contractor may be higher than if the agency can float bond to cover the costs. Opportunities exist to transfer or sharing this risk with the DB contractor. In certain projects, particularly in the overseas projects, various concessions are included to generate additional revenue.

- **Revenue Forecasting.** Revenue forecasting is based on numerous assumptions and complex modeling from collected data that must be materialized over the life of the contract. Optimistic revenue forecasts and lack of subsidy or strategy for additional revenue generation have resulted in significant revenue shortfall and financial hardship and/or bankruptcy in several of the world's transportation megaprojects including US toll road projects. This risk is too high to be borne by the contractor alone and perhaps could be considered for sharing with the project owner. In 2009 Florida agreed to pay the contractor (private investors) a fixed sum annually over 30 years to renovate and add lanes to an existing interstate highway near Fort Lauderdale. Illinois and Indiana are offering contractors set payments instead of toll revenue.
- **Maintenance and Operating Costs.** Maintenance is contractor's responsibility in DBOM and PPP contracts and the risk is assigned to him. However, operating costs in the deficit revenue situation as mentioned above may be paid or subsidized. In the transit industry, operating subsidy is often provided by the Federal and state government agencies to maintain transit service to certain corridors. On the New Jersey Transit's Hudson Bergen Light Rail System (HBLRTS), an early FTA demonstration project using DBOM, a fixed sum of the annual subsidy received was passed through to the contractor.
- **Alignment Selection.** Locally preferred alignment (LPA) is generally selected during the environmental impact statement (EIS) or environmental assessment (EA) process and is often based on ridership corridors, available ROW and minimum impact to the community and environment. Further refinement is made during design phase based on geologic risks, ground contamination risks, utility conflict, risk of certain property acquisition, and high risk of construction through difficult geology. Opportunity exists for risk avoidance during planning and design process. Probability of risk of major alignment change after the EIS process is small but its consequence may be very high impacting project schedule and cost. The risk is generally retained by the owner. Alignment change may require a supplemental EIS or EA for federally funded projects and renegotiation of cost and schedule with the contractor. On the HBLRTS project, a large segment of waterfront alignment had to be relocated inland after the selection of the DBOM contractor. This was required because of change in local political support during a mayoral election. In two other instances involving conventional design bid build delivery, New York's East Side Access Project and Singapore Underground Roads project, major alignment relocations were necessary during the design phase to avoid high risk underground construction below sensitive and hi rise buildings that otherwise required risks associated with underpinning of numerous buildings.
- **Environmental Permits.** Environmental permits require significant lead time and generally obtained by the owner who retains the risks unless contractor's means and methods change the requirements of the permit conditions. Since the permit process is long, often a contractor is selected before the final permits are obtained. In this process the contractor is given the copies of permit applications and it is stipulated that the contractor will be bound by the permit conditions as provided by the permitting agency. Risk of unanticipated and costly conditions not stipulated in the contract may remain with the owner.
- **R.O.W and Property Acquisition.** For a linear transportation project, numerous properties must be acquired along the ROW before commencement of construction. Acquisition and condemnation, if needed, is a long process. Often a DB contractor is on board as the process continues through the design period. Because it is a long process, in some instances, as experienced on the HBLRTS project, certain properties change hands in more than one time during the negotiation process making it extremely difficult to close on some properties. The risk of acquiring the permanent ROW is retained by the owner, however, risk of acquiring the temporary ROW during construction is generally transferred to the contractor.
- **Project Schedule.** Project completion schedule is driven by many factors including political and public pressure to open service

on given date, early revenue generation and cost of borrowing money. This risk is generally transferred to the contractor and various penalty and incentive packages are included in the contract. However, risks associated with delayed environmental permits, ROW acquisition and alignment changes, generally retained by the owner but may be shared between the owner and the contractor for DBOM and PPP projects.

### *Design Phase*

Most of the major risk elements associated with the construction are entered in the register by the owner's engineer during the preliminary engineering design. At this time mitigation strategies are developed for addressing the risks in the process of design, preparation of bid documents and during construction. Since DB contractor will be on board for the final design, he should be a part of the risk management process. Appropriate and accepted risk management practices such as dispute resolution board, escrow bid documents and such accepted tools should be incorporated in the contract. Generally accepted and known risk elements, issues related to them and mitigation strategy are discussed below.

- **Contract Packaging and Bidding Strategy.** Risks associated with contract packaging and bidding can be brainstormed internally and possibly shared with and input obtained from the industry outreach meetings. If there is a segment of high risk underground construction involved in an otherwise straight forward linear project, the risk can be dealt during construction packaging and during contract negotiation. Such examples are, a tunnel and underground station construction was included in HBLRTS project and an intake tunnel in the Elm Ridge Generating Station,a DB power plant project in Wisconsin. Although costs of the tunnel in both cases were 10–15 percent of the total construction, it was a high risk element that in both cases owner and the contractor negotiated separately as a turnkey element with a set aside cost within the overall contract. Because of stringent budget restrictions on the power plant project, the owner engaged an independent engineering firm to perform cost risk analysis using Monte Carlo simulation and retained the firm to monitor the tunnel procurement process.. On the HBLRTS project the owner's engineer developed the final design and construction plans of the tunnel and station cavern.
- **Geotechnical/Subsurface Conditions.** Risk associated with subsurface conditions which deviates from what could be reasonably anticipated during bidding stage, otherwise known as "differing site conditions," impact the construction performance and constitute major changed condition claims particularly in underground construction. The result could be increased cost, delay and may even require change in means and methods of construction. An extreme example is encountering gaseous grounds that have had major impact on tunnel projects in Los Angeles, Detroit and elsewhere. The risks associated with changed conditions are generally retained by the owner. However, because substantial financial arrangement is involved in DFBOM or PPP contract, the initial and a limited amount of cost resulting from differing site conditions may be shared with the contractor as was done recently on the recently constructed Port of Miami Tunnel project.

  Paucity of geotechnical data in the contract are much too common in the DB delivery because it is assumed that the level of geotechnical data should be at the rate of design development which is typically 15 to 30% of completion. However, it is proven that comprehensive geotechnical investigation and quality data reduces risk, contract contingency and claims. In the HBLRTS, one of the earlier FTA demonstration project involving DBOM, the owner equipped with comments received during industry outreach and data presented by its engineer was able to obtain FTA approval to collect and present comprehensive geotechnical data in the DBOM contract. In most underground construction projects involving design-bid-build delivery, comprehensive geotechnical data presented in the geotechnical data report and contract baseline conditions defined in the geotechnical baseline report, provide the basis for triggering differing or changed conditions and basis for price adjustment. Experience and construction case histories of tunnels demonstrate that comprehensive geotechnical data reduce price contingency and risk. This principle should also apply to any major linear construction involving viaducts, bridges and embankments through complex geologic settings and also to DB project delivery.
- **Utilities and Buried Structures.** Unidentified and significant utilities and

obstructions that could not be anticipated at the bidding stage could be treated as differing ground conditions and the risk generally retained by the owner. Utility investigation is particularly important for construction in major urban areas and obstruction investigation is important for waterway crossings. One major underground construction contract of the 2nd Avenue subway construction in New York City suffered significant delays and incurred costs during relocation of numerous unidentified utilities and house connections at greater depths than shown on plans. In addition, utility relocation caused significant foundation movements under older residential buildings that some required unanticipated underpinning and tenant relocation. On the same project, properly designed and carefully constructed deep underground station did have little further effect on adjacent buildings. Unknown buried seawalls and old waterfront structures are not uncommon along waterways. A buried ship wreck across the Fort McHenry Tunnel alignment in Baltimore Harbor was identified from extensive search of historic navigational records and the conflict was avoided during design. In another project under the Elizabeth River in Portsmouth, Virginia, a sunken steel barge was encountered that stopped a horizontal directional drilling construction across the river. The barge had to be removed before resuming construction that resulted in significant delay and additional cost.

- **Contaminated Ground and Groundwater.** The risk of unanticipated contaminated ground and groundwater encountered during construction that requires special handling and disposal and that materially affect the contract generally retained by the owner. However, as discussed above in the geotechnical and subsurface section, the risk perhaps may be negotiated and shared with the DBOM/PPP contractor.
- **Technology and Equipment Selection.** This is mainly applicable to Tunneling and underground projects. Selection of tunneling technology, tunnel boring machine (TBM), construction means and methods and prosecution of work for which the contractor can exercise reasonable control should be the contractor's risk. Where such control is impaired by the owner or third party action, the risk may be transferred to the owner. In certain megaprojects a TBM may cost over $40 million and most likely be manufactured in Japan or other overseas locations. TBM transportation and delivery to the site and risk of losing the TBM during transportation pose high cost risk. Although the risk can be partially offset with insurance, residual risk remains that may be transferred to the contractor or shared. On the Port of Miami Tunnel project, this risk was shared between the owner and the contractor.
- **Impact on Third Party Properties and Utilities.** Ground deformation resulting from heavy construction and underground construction impact adjacent structures, even if normal standard of care is maintained in design and construction. Some affected structures may require repair and rehabilitation. This risk may be shared by the insurer, owner and the contractor. In certain underground construction projects, a project contingency fund may be created to cover this risk. On the Alaskan Way Viaduct project such a fund has been created. Similar contingency fund was used successfully during Lexington Market subway station construction project in Baltimore, Maryland in 1970s built by conventional design-bid -build contract.
- **Third Party Approvals and Permits.** Risk of third party approval required for access to buildings, utilities and other structures for protection and underpinning generally are allocated to the contractor.
- **Temporary and Permanent Ground Support for Underground Construction.** All risks relating to the temporary ground support for underground construction should be allocated to the contractor. On the recent #7 Subway Extension, aDB project in New York City, the design required a minimum temporary ground support in tunnels and caverns. However, the contractor requested the owner to remove all temporary support requirements from the plans and agreed to take the associated risk and reduce the contract price. The contract documents were modified with ground support performance requirements and set deformation limits that proved to be beneficial to both parties. This is an example of good risk management through collaboration.
- **Handling of Excavated Material, Transport and Disposal.** This is primarily applicable to underground construction. Various risks associated with handling, stock piling, processing, transport and disposal may

be allocated to the contractor. If excavated muck cannot be disposed at the rate required to maintain tunneling progress, delays can be expected. All risks associated with muck handling and disposal may be allocated to the contractor. However, risks associated with property acquisition for staging areas, trucking restrictions imposed by municipal and state authorities over and beyond that has been stipulated in the contract, and obtaining permit for a designated disposal facility generally be retained by the owner. However, under a DBOM/PPP contract opportunity may exist to share the risk.

- **Natural Disasters and Force Majeure.** Truly unpredictable risk of natural disaster or force majeure should be allocated to the insurer unless the owner wants to be self insured and retain the risk.
- **Economic Factors and Labor Contracts.** Although on megaprojects DB contractor will negotiate a long term labor agreement, risk may not be eliminated for long term projects. Risk associated with unexpected economic factors, significant labor cost escalation and general strikes are not easily allocated, however, may be negotiated to allow for some relief to the contractor.

### *Construction Phase*

Risk register should be updated, resolutions noted and new risk elements are added as discovered during construction and clearly communicated with parties involved. The risk of construction safety is transferred to the contractor. Through partnering with the contractor during construction, prompt resolution of unforeseen risks is possible. A proactive risk management effort should reduce the impact on project schedule and cost.

## CONCLUSION

A proactive risk management should be an integral part of all underground construction and mega projects. Each project has its own set of risks and how one uses the risk management process may depend on project complexities; constituent involvement, organizational structure, resources and approach; risk tolerance and such factors. However, establishing an effective and efficient risk management process and equitable risk sharing within the project framework is essential. For the DB and megaprojects, the owner should consider actively involving the constituents and the DB contractor for their collective contribution and sharing of lessons learned from previous experiences to contribute and bear positively on the success of the risk management plan and that of the project.

# Design/Build a Panacea?—No

**Timothy P. Smirnoff**
Parsons Brinckerhoff

**ABSTRACT:** The current trend for large and small public properties to build major infrastructure is to go Design/Build. Owners believe the Design/Build approach will shorten the time from preliminary engineering and environmental clearance to delivery while saving money. For Design/Builders and their engineers this seems to be a false economy and saves neither money nor time and in many cases leaves the Owner's lacking. This paper will examine some of the myths in Design/Build and shifts of risks to the Design/Builder as they go through the protracted and expensive process of developing design to meet project requirements and assume contractual risks that are often either attainable or economical. For the responsible final design engineer it forces vast compromises in design to attain the sometimes unattainable low bid.

## INTRODUCTION

Until the later part of the 20th century, contracting for most public infrastructure projects had relied on the traditional Design-Bid-Build (DBB) procurement method. This two step process relied on the owner with his staff and in many instances his engineering consultants preparing the design with associated contract documents and subsequently issuing a construction contract to be performed by a Construction Contractor for a firm fixed price. Almost always, as prescribed by state or federal law, the award was made to the "lowest responsible" bidder. This process was instituted as a contracting approach to allow fixed points of responsibility for both the design and construction, and allay fears and occurrences of fraud, collusion, and other shortcomings that had been apparent in times past, and was thought to give the "people," whose monies were to be spent, the fairest and fixed price for the specified work. This process has worked well for most highways, at grade and elevated rail/transit and similar projects where public agencies maintained a high control of the design and construction, and were there was relatively low risks in the work or geotechnical conditions. These risks were understood by both the Owners and the contracting community.

For complex project, especially underground projects which were becoming larger and more costly, by the end of the century the track record for DBB had dampened most owners and the public enthusiasm, and become stellar examples for their claims, cost and schedule overruns. Moreover these increases were non-productive, and resulted in protracted disputes and costly litigation which results rarely satisfied either party. By the mid 1970 the U.S. National Committed on Tunneling Technology (1974) acknowledged that:

> "...contracting practices in the United States are inadequate for even past methods and constitute a serious barrier to...new technology and to the most economical... tunneling."

Part of this report proposed a number of contracting processes to help modify and improve the DBB contracting model, many of which are now standard within the underground construction industry such as:

- Disclose All Information
- Eliminate Disclaimers
- Recognize Differing Site Conditions
- Distinguish Responsibilities for Tunnel Ground Support
  - Initial—Contractor
  - Final/Permanent—Engineer/Owner
- Contingent or supplemental prices of unknown or unquantifiable quantities and priced contacts with a bill of quantities

With these recommendations and some variations of the DBB contract, e.g., incentive/disincentives, and some assignment of risks, DBB contracting remains the most prominent public agency procurement means.

## PROJECT DELIVERY MODELS

In the mid 1980s both in the United Kingdom and in the US, the first steps were made to move away from conventional contracting to a number of variants.

**Table 1. Project delivery models**

| Acronym | Model |
|---|---|
| DBB | Design-Bid-Build—Traditional method with design by owners engineer put out to bid |
| DB or D+C | Design-Build (sometimes called Design and Construct)—integrates the two phases to produce savings in time and cost, by incorporating contractors' means and methods into the design phase |
| DBM or DCM | Design Build (or Construct) Maintain—provides a fuller life cycle review (shadow too is one method of payment) with contractor retained for maintenance to encourage a more durable design |
| BOT | Build Operate Transfer—Allows builder concession to operate the facility and derive revenue over the concession period before transferring it back to government authority |
| BOOT | Build Own Operate Transfer—similar to BOT, but usually incorporates the property development rights linked with the infrastructure (e.g., air rights over rail stations). |
| BLT | Build Lease Transfer—similar to BOT except the builder gets a concession to lease the use of the facility before transferring it back to the government authority. |
| DBFO | Design Build Finance Operate—the private sector provides a facility and a service. The service is sold to the public sector at a fee. |
| PPP | Public Private Partnership—an agreement, usually based on one of the above models, where the government takes on a role of actively controlling some of the risks with involvement of statutory authorities who eventually take operational control. |

Table 1, taken from Caiden presents of summary of the now common forms of Project Delivery models in use and the alphabet soup of initials.

The UK's institution of Civil Engineers (ICE) in 1985, published the New Engineering Contract (NEC), which was later published in 1993 and more recently in 1995 as the Engineering and Construction Contract (ECC) 3.

The evolution of new processes for contracting practices appears to take decades to advance, continues well into the 21st Century. From 1987, the Transportation Research Board (TRB) began the process to identify innovative contracting practices which were later developed in 1990s by the Federal Highways Administration to established a means implementing alternative contracting models ((TRB Circular 386:" Innovative Contracting Practices) and currently and have been adopted by the FHWA and AASHTO (2005) are no longer considered to be experimental.

## CURRENT ALTERNATIVE DELIVERY METHODS

The current trend for many transit properties and transportation agencies is to turn to alternative contracting methods to continue providing the needed improvements. Increasingly, owners seek to minimize dollar outlay and long term dollar commitments at little or no cost to the agency or tax payers, the philosophy being "Let the User Pay for it." The paramount objective for the use of the DB approach and its many derivatives as espoused by many owners is to get the project built within the quickest schedule, at a reasonable cost, and with a single point-of-responsibility. Generally, the sooner the project is completed, the sooner investment income is realized (e.g., space lease income, highway toll income, utility generation revenues, etc.). A side benefit, some believe that can result from a DB approach, is enhanced technical integration and constructability. The DB contractor remains responsible to construct a project that is acceptable to the owner, assumes all risks allocated by the contract, and guarantees that the project will perform to the standards set by the specifications. Of the many derivative approaches for Design/Build several have gained favor for the perceived advantages of time of delivery and lower costs. These design/build methods may take several forms but the most common in the United States are:

- Design/Build (DB)-traditional approach
- Design/Build/Operate/Maintain (DBOM or BOM)
- Design/Build/Own/Operate/Transfer—BOT or BOOT

In most instances the theme is similar, the owner provides the motive, and the project preliminary design and usually environmental consents, and the Design/Builder or concessionaire provides financing, design, construction, usually quality control etc. and recovers the cost with a toll or use fee. In some instances the agency may chose to provide an availability fee or similar mechanism to pay for the project where costs can be covered by the agency.

To often latterly, increasingly toll revenue projections/patronage estimates have been notoriously inflated and unattainable to ensure the solvency of the project which has increasingly lead to the financial failure of a number of tunnel projects. Notably in Australia where it became clear the toll revenue for usage for these tunnel projects was well short of that needed to cover the construction cost, over the last 15 years, most of the privately owned toll road project have fallen into receivership or administration within a short time of opening. Similarly, transportation projects in the US such as the South Beach Expressway in San Diego, CA, or Beach Express Bridge in Orange Beach, Ala have had similar fates. To overcome this some owners have turned to the use of availability payments, according to the tunnel's performance rather than on actual traffic volumes. (East-West Link Melbourne, Australia, and Port of Miami Tunnel, Miami, and Florida are two recent examples).

In most instances the owner retains an engineer to prepare the required environmental and preliminary engineering necessary to define the project, the needed mitigations, and the develop project specific performance specifications and related requirements for the major items of work, and in some cases reference specifications for a tender brief or request for proposal.

A request for proposal is then issued. In many instances a precursor is the issuance of a RFQ and prequalification process is instituted to ensure selection of financially and technically qualified firms to provide the work. Once the RFP are submitted and a selection made the DB proposing entity has been selected, the Designer-of-Record then is tasked with preparation of the final design, including Specifications that will be used by the DB contractor to construct the project and to provide quality control (QC).

## PROMINENT ADVANTAGES

Among the most prominent obstacles or short comings seen in this "traditional DBB" are:

- All the design and planning work must be completed before the construction contracts are let this sequential process extends the time for completion.
- The track record of substantial cost and schedule overruns—the record for most large public infrastructure projects is dismal for meeting original budgeted cost and schedule.
- Shedding of risk to the Contractor—promise to deliver the specified project for the fixed cost of the contract. Most public agencies are risk adverse and the record of contentious and protracted litigation is all to familiar to these agencies. Many are bound by law not to pay claims without a court order to do so.
- Lack of Contractor/Construction input at the early stages of design loses the opportunity for practical suggestions and methods with potential cost and schedule savings. Where a Value Engineering Cost Proposal provision exists in a contract with cost savings sharing, many owners feel they do not get the full value in monetary savings.
- Disincentive to innovation and prominently. Build it like the plans say.
- The inefficient, adversarial relationship leading to claims, disputes and litigation.
- With the dwindling of public funds for infrastructure, alternative contracting becomes attractive to finance the work.
- Both Political and Public perceptions of agency inability to deliver projects and the frustration at delays, and time scale for completing.

In the last few years major transportation/transit agencies—New York City's MTA East Side Access Project, Second Avenue Subway), Los Angeles Country MTA (East-Side Extension, Crenshaw LRT), Florida State DOT (Port of Miami Tunnel), Washington State DOT (Alaska Way Viaduct Replacement Project), and many others have used DB as a procurement method for heavy underground works. United States contracting practice is much further behind Europe, Asia, and Australia where DB, and variations like BOOT, Public Private Partnerships have long been used.

Hence the current lunge to Design/Build by many owners. DB is proposed to be better for all the parties, Owner and Contractor as it inherently provides the integration of the design and construction processes and places the foundation of risk of design, construction and largely subsurface conditions in a single point.

## CURRENT COMPLAINTS

But already there are some "chinks in the amour." In addition to the financial failures described above, following are other DB issues from various perspectives.

### Chief Owner Complaints

- We don't get what we wanted:
  - The product does not conform or provide the system we wanted;
  - Everything from the lowest cost provider, without regard to operations and maintenance.

  Corollary to that is Owners often feel they have lost control of the delivered project and lost control of the design and quality.
- No standardization of equipment and systems with multiple contracts.

  With multiple DB contracts in a large scale project, motors, fans, escalators, elevators, control systems, etc are likely to vary requiring storage of many multiple variations of the components and each having different installation, maintenance and operations constraints and requirements. These duplications impose burdensome requirements for owner staff training, retaining specialized equipment, and confusion.

- The contractor builds what he wants.

  They a have little regard for the contract provisions if they want or have a money saving idea.
- The Tender/Bid and often ensuing negotiations for a Best and Final offer (BAFO) process takes as much time as final design and the standard time for bidding of a conventional contract.

  The times for the pre-qualification, tender, following negotiation including BAFO extended periods and rarely involve maintenance and operations personal who will eventually operate the systems. Many bids contain bid alternatives and innovations that require redesign and negotiations as well as revisions to tender terms and conditions that must be distributed to the other potential proposers.
- Still requires Contract oversight and Q/C and Contract administration.

  The time for administration and oversight including QC/QA does not reduce substantially.
- The quality of the delivered project requires oversight during design and construction of experience and technically qualified staff which most authorities lack.

  Through attrition and in many cases no recent history in with similar work, agency staffs no longer have the cadre of experienced professionals to do this.
- We lack the experience to adequately qualify and administer such contracts.

  Cost Estimating, program scheduling, administration and contract changes and claims requires trained experienced Staff which most agencies no longer have, requiring a program manger and construction consultant at added cost.
- Depending on allocation of underground risk, and extent of unknown site conditions, many of the same disputes and claims still occur.

  Owners feel they get as many claims and disputes as before with traditional contracting and are in much the same position as before. Changes either in the design or instigated by the Contractor in the bidding phase require contract modifications and review much as before.

  The DB process has not eliminated differing site conditions claims and seems to have enabled Contractors instances to create a claim.
- We tried to provide for contingency items and overruns in the bid price, but the Contractor only thought it was their money to spend.
- In the end we saved neither time not total cost.

## Chief Contractor Complaints

- Owners don't know the cost of the project.

  Owner's budget estimates are often years old, not escalated nor compatible with the project tender documents and the tender price often comes as a big surprise.
- The Owner's don't know what they want.

  The Owner still wants to review, modify and change designs, review all the submittals, etc. Interminable meetings to review discuss project requirements, and all to often amend, alter, modify and/or change the project requirements….
- Owners don't know what is included in the work or more importantly NOT included in the work.

  The classic question from the Owner—where is the ########? Response—Where do I get paid for that and where is it specified?
- The Owner's does not have the permits, rights of way, public or environmental clearances, and most importantly utility relocations and protection of adjacent infrastructure needed to start or perform needed work, or provide sufficient time in the project schedule for these necessary items.
- Often no recovery for prequalification proposals and tender package design, and related efforts/costs.

  The cost of such proposals for major works is very high and requires a design subcontractor/partner who needs to be paid as the design progresses. With many proposals this work is not compensable through the contract terms.

  We generally do not have the experienced staff to supervise design staff and coordinate the design effort for large complicated projects including track work, ventilation, fire/life/Safety, etc.
- Owners still feel they can control my designer.
- Owners lack the political support or governmental agency control and approvals.

- Design teams—to conservative, not outcome oriented, mostly not work-wise. Still think their client is the Owner.
- Owner Design Reviews and Design control is time consuming and restrictive.
- Owners Construction Engineers and Design Reviewers lack credible construction experience.

**Final Design Engineers Complaints**

- We are just another sub to the prime contractor

  Contractor shop us just like a ordinary subcontractor fully studied and developed within the entire system.
- We are not finished when the design drawings are stamped.

  Contractors do not look at the drawings during development and design and once stamped and issued for construction many issue become evident and require design changes.
- Often once we stamp the drawings we never hear another word—even when they are changed in the field.

  The construction changes are not run through a change process or reviewed.
- The contractor builds what he wants.
- Lack field construction input during construction—especially in underground work with initial support.

  As conditions are uncovered underground, we are not notified or asked to check or modify support recommendations.
- We don't get any respect.
- The Contractor Tender Team and the Construction Team in the field at have entirely different ideas of the project/means and method/design requirements.

  The bid team is often replaced once the job is won with an entire new set of Contractor personnel, who have entirely different ideas for the design, its construction and the sequence.

## WHERE DO WE STAND

So where do we stand today:

| What we thought of as *Advantages* | *Current Perception* |
|---|---|
| • All the design and planning work must be completed before the construction contracts are let this sequential process extends the time for completion | • Design issued piecemeal and often incomplete to "get things going" |
| • The track record of substantial cost and schedule overruns—the record for most large public infrastructure projects is dismal for meeting original budgeted cost and schedule. | • Lack of good Owner Budget leads to contractors taking the blame for cost and schedule overruns |
| • Shedding of risk to the Contractor—promise to deliver the specified project for the fixed cost of the contract. Most public agencies are risk adverse and the record of contentious and protracted litigation is all to familiar to these agencies. Many are bound by law not to pay claims without a court order to do so.<br>• Lack of Contractor/Construction input at the early stages of design loses the opportunity for practical suggestions and methods with potential cost and schedule savings. | • Claims and litigation still result<br>• Increased costs for owner and no less litigation |
| • Where a Value Engineering Cost Proposal provision exists in a contract with cost savings sharing, many owners feel they do not get the full value in monetary savings | • Contractor input has often lead to cost savings and schedule savings |
| • Disincentive to innovation and prominently. Build it like the plans say | • Cost sharing is eliminated with the DB model if it is in the original proposal |
| • The inefficient, adversarial relationship leading to claims, disputes and litigation | • Innovation is key to a successful bid |
| • With the dwindling of public funds for infrastructure, alternative contracting becomes attractive | • Much the same |
| • Both Political and Public perceptions of agency inability to deliver projects and the frustration at delays, and time scale for completing | • Key Driver for Alternative Contracting<br>• Unless done professionally on both sides the same perceptions are possible. |
| | • No recovery for prequalification proposals and tender package design, and related efforts/costs |

## CONCLUSIONS

Design/Build was a much sought after prescription for the aches and pains of contracting and often when used with alternative contracting means to provide a means of financing design and construction with limited public funding. The experience of the last few years has shown that with every pill there are side effects and the outcome may not be as it was thought. Many of the same problems and complaints still exist and do not generally have easy or rapidly evolving solutions within the industry. Recent experience in the United States and Australia and other countries indicates that costly overruns, and schedule delays are still prevalent. In several notable examples in Australia and in the United States individual traffic or patronage demands have been vastly overstated so that upon completion the projects no longer have economic viability.

The jury is still out and with more projects slated to use DB and its alternatives, the processes and the records of performance and successes will be better defined and will certainly evolve to clarify and improve DB, as have most contracting practices in the past.

Owners, Engineers and Contractors must work together to develop a more efficient, effective and more equitable form of DB contracting and developing better models for both conceiving and delivering the infrastructure needed.

## REFERENCES

"Australia suffers toll concession failures," (2013) Nov., *Tunnel/Talk*, see http://www.tunneltalk.com/2013-Nov2013.

"Avoiding and Resolving Disputes in Underground Construction" (1989), Technical Committee on Contracting Practices of the Underground Technology Research Council, American Society of Civil Engineers.

"Better Contracting For Underground Construction" (1974), Standing Committee No. 4 Contracting Practices, United States National Committee on Tunneling Technology, National Academy of Science.

D. Caiden, P.J. Chamely and C .S. Covi, "Project Delivery Models and Risk Allocation for Whole life tunnel Management" (2005), RETC Proceedings (pp. 71–83).

"Improved Contracting Methods for Highway Construction Projects," (1999), NCHRP 10-49 Final Report, December.

"Innovative Contracting Practices, Transportation Research Board (1991),Circular 386, National Research Council, Washington D.C.

"Private Turnpikes' Investors Pay Toll as Traffic declines," *Wall Street Journal*, Nov, 21, 2013 (pp. 1 and 16).

"The Engineering and Construction Contract," (1995), The Institution of Civil Engineer, Thomas Telfor Ltd., London.

# TRACK 3: PLANNING

## Session 5: Future Projects and Industry Trends

Dave Haug, Chair

# Planning and Preliminary Design of the Ohio Canal Interceptor CSO Tunnel Project in Akron, Ohio

**David Mast**
AECOM Technologies, Inc.

**Michelle DiFiore**
City of Akron–Engineering Bureau

**ABSTRACT:** The Ohio Canal Interceptor Combined Sewage Storage and Conveyance Tunnel project is one of over 20 control measures described in the City of Akron's Combined Sewer Overflow (CSO) Plan. Project requirements include a minimum storage volume, maximum typical year overflow frequencies for multiple existing and new overflow structures, gravity dewatering of the conveyance system, and ability to convey both sanitary and 10-year, 1-hour (hr) storm flows. The preliminary engineering concept included 6,136 lineal feet of 27-foot inside diameter (ID) tunnel, mined through rock, soil, and mixed-face conditions, and the world's largest baffle drop structures. The paper discusses the City's CSO Program as well as Planning Studies, Preliminary Engineering and Design, and Value Engineering for the tunnel project.

## SUMMARY OF AKRON'S LONG TERM CONTROL PLAN

The City of Akron (City) is the fifth-largest city in the state of Ohio (see Figure 1), with a population of approximately 200,000 people and an area of approximately 62 square miles (City of Akron Website 2013). The City also provides sewer services to the Akron Metropolitan Area, approximately 183 square miles in area a population of over 356,000 people (Akron Oct. 2010 CSO LTCP 2010). A portion of the sewer system, generally centered on downtown Akron, consists of combined sewers conveying both sanitary and storm flow (gray areas on Figure 1). During large rain events, excess combined sewage must be relieved from the combined sewer. This relief occurs in underground regulators locally termed "Racks." At the Racks, dry weather flow drops through metal bar racks to an underflow pipe to be conveyed via interceptors to the City's Water Reclamation Facility (WRF). During wet weather flows, the underflow pipe acts as a regulator and excess flow continues over the rack and underflow pipe to an overflow pipe for conveyance to a nearby body of water. The overflow events are termed Combined Sewer Overflows (CSOs).

**Figure 1. Project location map**

Akron began a detailed assessment of water quality impacts from CSOs in the "Ohio Canal Combined Sewer Overflow Study" in 1991. This Study was completed in 1993. In anticipation of proposed CSO guidance, the City completed an additional 33 studies leading up to the submittal of an updated Facilities Plan and CSO Alternatives in 1999 in full compliance with its NPDES Permit. This plan was approved by Ohio EPA in 2002. At that time US EPA assumed the lead role in the EPA negotiations. Akron completed a $23 million storage basin in 2004 that eliminated 33% of the total CSO overflow volume. The City of Akron, Ohio EPA and US EPA reached an agreement on a Consent Decree (including the CSO LTCP) in November 2009. The proposed decree was then filed in the United States District Court for Northern District of Ohio, Eastern Division. As of the date of this paper, it has not been lodged.

## OHIO CANAL INTERCEPTOR TUNNEL CSO PLAN REQUIREMENTS

As part of the CSO LTCP, the City would design and construct the Ohio Canal Interceptor (OCI) Combined Sewage Storage and Conveyance Tunnel project (hereafter referred to as "OCIT project" or "OCIT system"). The general OCIT project area is within the heavy dashed oval shown on Figure 1. The OCIT system as currently envisioned includes a large diameter storage tunnel as well as related consolidation sewers and drop shafts to control CSOs from Racks that overflow to the Little Cuyahoga River and the Ohio & Erie Canal, (see Figure 2). The OCIT system must meet minimum design and performance criteria listed in Table 1, as well as additional criteria established by the City. A future Enhanced High Rate Treatment (EHRT) system, listed as an ACTIFLO™ facility, is also required to treat overflows from the OCI Tunnel.

## CITY REQUIREMENTS FOR OHIO CANAL INTERCEPTOR TUNNEL

In addition to Table 1 requirements, the City is also requiring the OCI Tunnel system convey dry weather sanitary flow and up to the 10-year, 1 hour storm flows from the majority of Racks in the project area. Table 2 lists the design flows for each Rack. Conveying the 10-year flow through the tunnel will prevent a large volume of combined sewage from entering the Ohio & Erie Canal.

Flows larger than the design storm will be conveyed to receiving waters, utilizing mostly existing overflow pipes. The City also expressed preference that the OCI Tunnel be designed to dewater by gravity to existing interceptors, and overflow by gravity. Due to local topography, this requirement appeared achievable, and an existing interceptor and the Little Cuyahoga River are present at the north end of the project area to receive flows (see Figure 1 and Figure 2).

## ADVANCED PLANNING STUDY

Horizontal tunnel alignment planning began as part of Facility Planning and continued through an OCI Storage Tunnel Advanced Planning Study (APS) in 2006 (OCI CSO APS 2006). The heavy dashed line in Figure 2 illustrates a straight line route from the southernmost Rack (#16) to the northernmost Rack (#24), and the heavy solid black line illustrates a route directly connecting each Rack in the control measure. In 2012, the preliminary design team and City used the criteria listed in Table 3 to identify 28 feasible horizontal tunnel alignment alternatives.

Widely spaced geotechnical borings were performed during the 2006 APS to assess conditions in the project corridor, particularly depth to bedrock. Geologic references suggest the project corridor is located on the east bank of an ancient bedrock valley likely created during a glacial melt period, with several side stream valleys merging from the east (see Figure 3). The combined APS and background research indicated a vertical tunnel profile mostly in bedrock might be achievable along at least one of the 28 candidate alignments while still meeting the City's requirement of dewatering by gravity.

## PRELIMINARY ENGINEERING REPORT (P.E.R.)—EHRT SITE EVALUATION

Between January and November 2012, the City and Preliminary Engineering Team (led by AECOM) progressed the OCIT project from advanced planning to preliminary design. The final preferred tunnel alignment selection process is illustrated in Figure 4.

As indicated in Table 1, the OCIT system has a performance criteria of no more than seven (7) CSOs in the typical year, and an estimated 350 MGD capacity EHRT must be built in the future to treat all of those seven (7) events. Therefore, there were advantages to tunnel alignments which terminated at a location with sufficient contiguous City-owned land necessary to build the EHRT. In addition, the north end of the project area was preferred for the mining laydown area due to land constraints at the urbanized south end.

The City and P.E.R. team identified four (4) candidate sites for the future EHRT facility. They used a modified paired analysis process to identify the most preferred EHRT site. During half-day workshops, the P.E.R. and City team compared each site one-to-one against each of the other three sites according to each of the nine (9) criteria shown in Figure 5. The preferred option of the two was assigned a point value to emphasize the relative difference. For example, for the criterion "Maintenance Complexity," one point

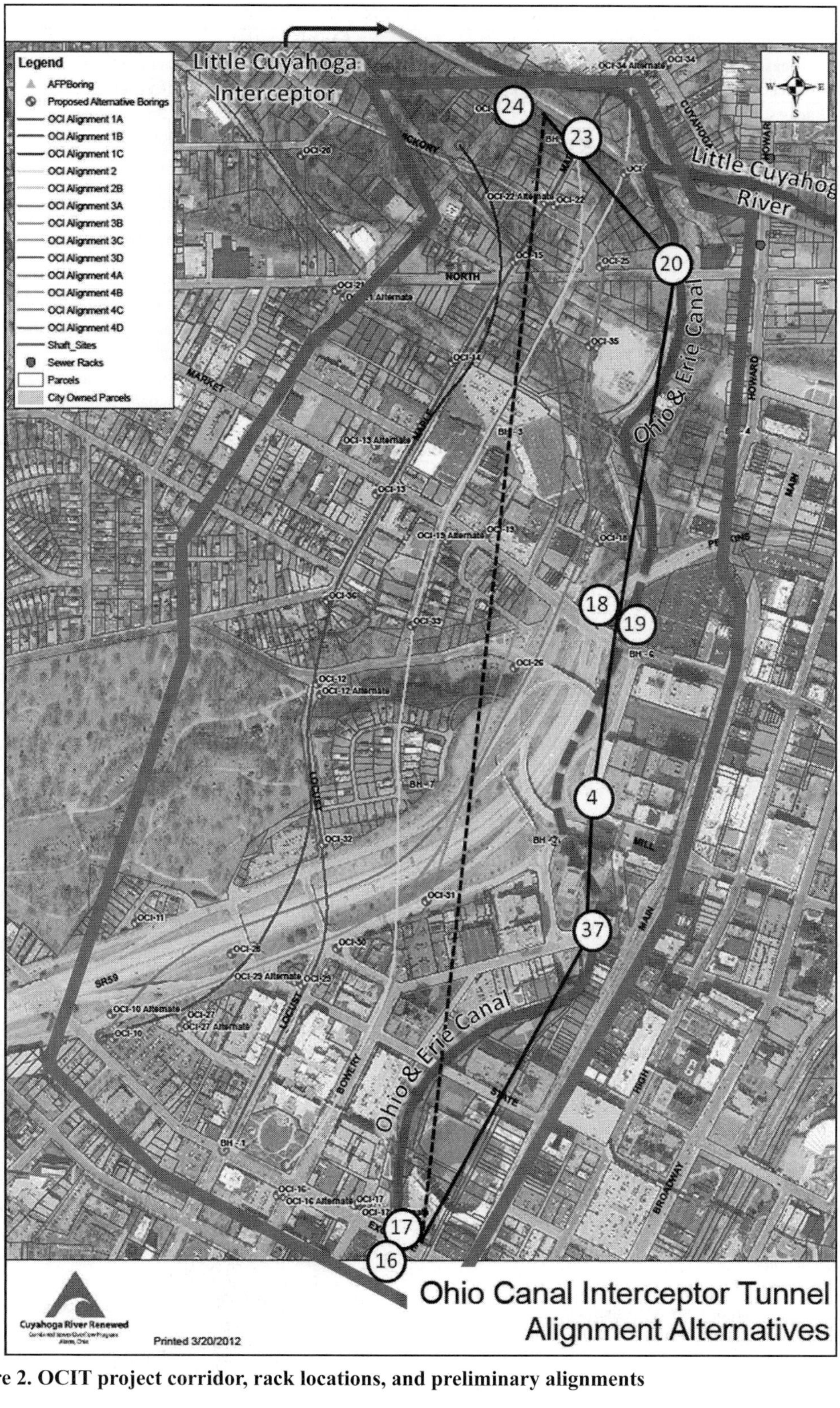


Figure 2. OCIT project corridor, rack locations, and preliminary alignments

**Table 1. CSO LTCP design and performance requirements for OCI tunnel and EHRT**

| Row # | Control Measure Location | Description | Design Criteria | Performance Criteria (Typical Year) |
|---|---|---|---|---|
| 11 | Racks 4, 16, 17, 18, 19, 20, 23, 24, 37 | Ohio Canal Tunnel—Construct a 28-foot internal diameter tunnel, 5,550 feet in length, or any other combination of diameter and length that achieves the design criteria. | Minimum storage volume of 25,600,000 gallons. This volume excludes conveyance tunnels, dewatering tunnels/sewers, adits, and drop shafts. | No more than 7 CSOs<br>Racks 4 and 37: 0 CSOs |
| 11a | Racks 16, 17, 28, 19, 20, 23, 24 | Ohio Canal Tunnel—ACTIFLO™ Ballasted Flocculation Unit or equivalent technology that meets the Design and Performance Criteria and Disinfection | An ACTIFLO™ ballasted flocculation unit (sand ballast technology) or an EPA and OEPA-approved equivalent technology that meets the Performance Criteria, including disinfection, with a minimum sustained design capacity of 300 million gallons per day (208,000 gallons per minute). The hydraulic loading rate shall not exceed the manufacturer's recommendation. | Treated discharges must not exceed the following discharge limitations:<br>1) 30 mg/l TSS<br>2) In the recreational season, no more than 10% of the samples taken during any 30 day period shall exceed 298/100 ml of *E. coli.*<br>3) The geometric mean of all the samples collected during the recreational season shall not exceed 126/100 ml of *E. coli.*<br>4) 0.024 mg/l residual chlorine |

**Table 2. Design flowrates for OCIT racks (Akron Oct. 2012 OCIT P.E.R. , 2012)**

| Rack | Design Flow Rate (MGD) | Design Storm |
|---|---|---|
| Rack 4 | 34 | Typical year |
| Rack 16 | 285 | 10-Year, 1-Hour |
| Rack 17 | 449 | 10-Year, 1-Hour |
| Rack 18 | 477 | 10-Year, 1-Hour |
| Rack 19 | 74 | 10-Year, 1-Hour |
| Rack 20 | 40 | 10-Year, 1-Hour |
| Rack 23 | 41 | 10-Year, 1-Hour |
| Rack 24 | 325 | 10-Year, 1-Hour |
| Rack 37 | 12 | Typical year |

**Table 3. OCIT APS tunnel alignment selection criteria**

- Minimum 1,000-foot tunnel boring machine turning radius,
- Maximize use of City of Akron Right-of-Way,
- Maximize use of City of Akron owned property,
- Minimize impact to Ohio Department of Transportation (ODOT) bridges,,
- Avoid cemeteries,
- Avoid hospital buildings,
- Avoid downtown congestion.
- Minimize impact of large drop shafts to central business district at the east edge of the project corridor.

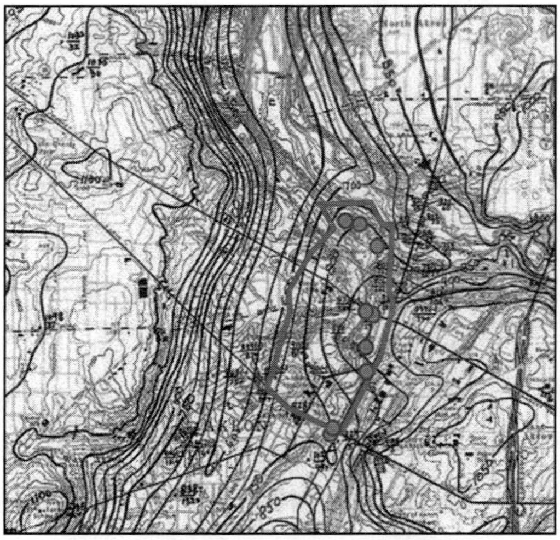

**Figure 3. OCIT project corridor on bedrock topography map (Vormelker 1996)**

was assigned to the preferred site if the two compared sites were "Nearly the Same," five points if the preferred site was "Moderately Better," and nine points for "Significantly Better." Maintenance Complexity was designed to consider equipment needs (such as pumps), necessary staff training and availability of qualified staff, monitoring and operating controls necessary, and changes in staffing levels for the City.

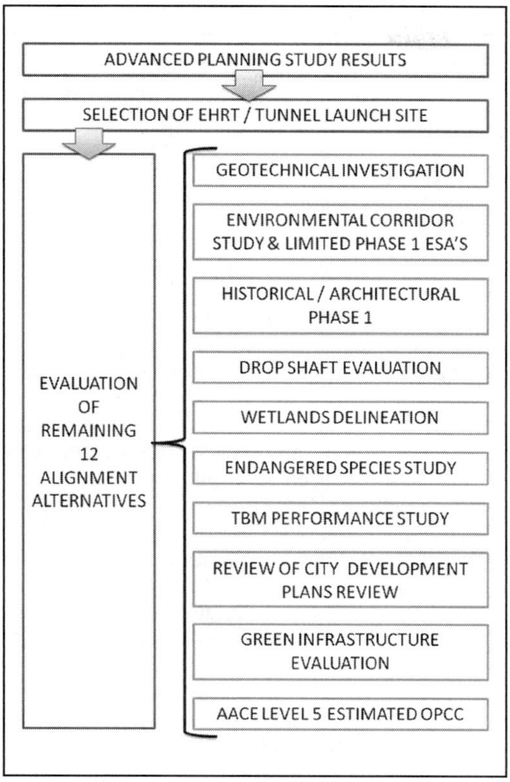


**Figure 4. Generalized OCI tunnel alignment selection process**

Table 4 illustrates the completed scoring matrix for the criterion Maintenance Complexity. Scores were input into the software program Criterium Decision Plus to confirm manual calculations and produce graphical result graphs (also on Figure 5).

One of the key differentiators for the selected site (Site 2) was its potential to accommodate a gravity-driven EHRT. Due to the existing topography, existing residential homes, and the alignment of the Little Cuyahoga River at the north end of the project corridor, terminating the tunnel at the other three (3) sites would have necessitated an approximately 350 MGD pumping station, residential home demolition, and/or a long 1,300 MGD capacity tunnel overflow discharge pipe. The relative lifecycle cost increase necessary for the pump station and the need to purchase and remove existing homes were significant contributors to the final selection.

## P.E.R.—FINAL TUNNEL ALIGNMENT EVALUATIONS

Once the EHRT site was chosen and number of feasible alignments narrowed, the City and P.E.R. team completed field investigations, historical research, modeling, and risk analyses. A new set of detailed selection criteria were established (see Table 5) and new relative weightings chosen. A two-day evaluation workshop was initially held to choose four (4) potential tunnel alignments, and then a second workshop resulted in selection of a final preferred alignment (alignment #4B). Due to the large number of

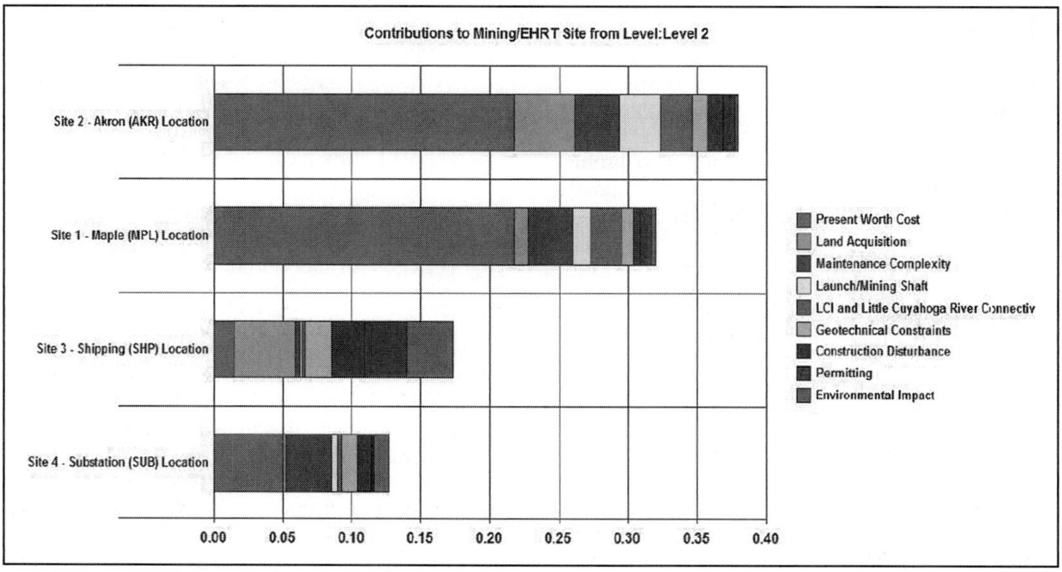


**Figure 5. Weighted criteria scores for EHRT/mining site selection workshops**

**Table 4. Scoring table for EHRT site selection criteria "Maintenance Complexity"**

| Criteria Category | Maintenance Complexity | | | | | |
|---|---|---|---|---|---|---|
| | Site 2—AKR | | Site 3—SHP | | Site 4 - SUB | |
| **Alternative** | **Preferred Alternative** | **Score** | **Alt.** | **Score** | **Alt** | **Score** |
| Site 1—MPL | 2 | 1 | 1 | 9 | 4 | 1 |
| Site 2—AKR | | | 2 | 9 | 2 | 1 |
| Site 3—SHP | | | | | 4 | 9 |
| Site 4—SUB | | | | | | |

**Table 5. Evaluation criteria for OCI storage and conveyance tunnel alignment selection**

**Feasibility of Implementation**
- Land acquisition
- Construction disturbance
- Geotechnical constraints
- Construction shaft locations
- Permitting
- Economic development
- Consolidation sewer alignments

**Economic Impact**
- Capital cost
- O&M cost

**Environmental Impact**
- Odor control
- Impact on habitat, erosion, etc.
- Historical/architectural impact

**Operations and Maintenance**
- Maintenance complexity

**Emergency Rescue Shafts**

alignment alternatives being considered, a simple weighted scoring process was used. Alignment #4B appeared to maximize the amount of rock cover over the tunnel, minimized necessary land purchases, and allowed more access to drop shaft locations. Figure 6 and Figure 7 illustrate plan and profiles for the selected tunnel alignment, including updated bedrock contours based on PER borings.

## PRELIMINARY DESIGN OF OCIT PROJECT

Once a preferred alignment was selected, AECOM prepared the following preliminary design documents:

- Preliminary Engineering Report containing historical data, design criteria and performance requirements, and detailing the process and decisions to date.
- Preliminary design drawings and list of required technical specifications.
- Geotechnical Data and Evaluation Reports, including boreability and TBM performance evaluations.
- List of properties within the permanent and temporary project limits.
- Preliminary hydraulic analyses and surge analyses.
- Condition assessments of existing infrastructure (sewers and Racks) to which the new facility will connect, and recommendations for repairs.
- List of required permits and regulatory / utility coordination necessary.
- List of project stakeholders who should be considered during design and construction.
- Preliminary due diligence studies to assess potential environmental contamination, wetlands, endangered species, historical / architectural structures, noise and vibration.
- Evaluation of potential for corrosion issues in the OCIT and consolidation sewer system.
- Preliminary list of possible project risks, later integrated into a full risk register.
- Considerations for control strategies as well as O&M strategies.
- Evaluation of possible green infrastructure opportunities.
- Opinion of probable construction cost (accuracy between AACE Class 4 and Class 5).
- Contracting considerations, such as options for local, disadvantaged, and small businesses.
- Evaluations of alternative technologies as they were presented to the City.
- Recommended future analyses for the Final Designer to perform.

A few of the more significant challenges encountered during the Preliminary Design are discussed below.

**Mining Through Mixed Reaches.** On the selected alignment, the tunnel zone necessary to achieve gravity dewatering at the downstream end meant the tunnel would have to be mined through outwash sands and gravels, transition to soft to moderately hard shale and siltstone bedrock, and then transition again back into soft ground. Risk was partially mitigated by purchasing land and buildings near the alignment as well as careful alignment "tweaks" to avoid structures. TBM production rates

**Figure 6. Selected OCI storage and conveyance tunnel P.E.R. alignment with estimated bedrock contours from P.E.R. investigations**

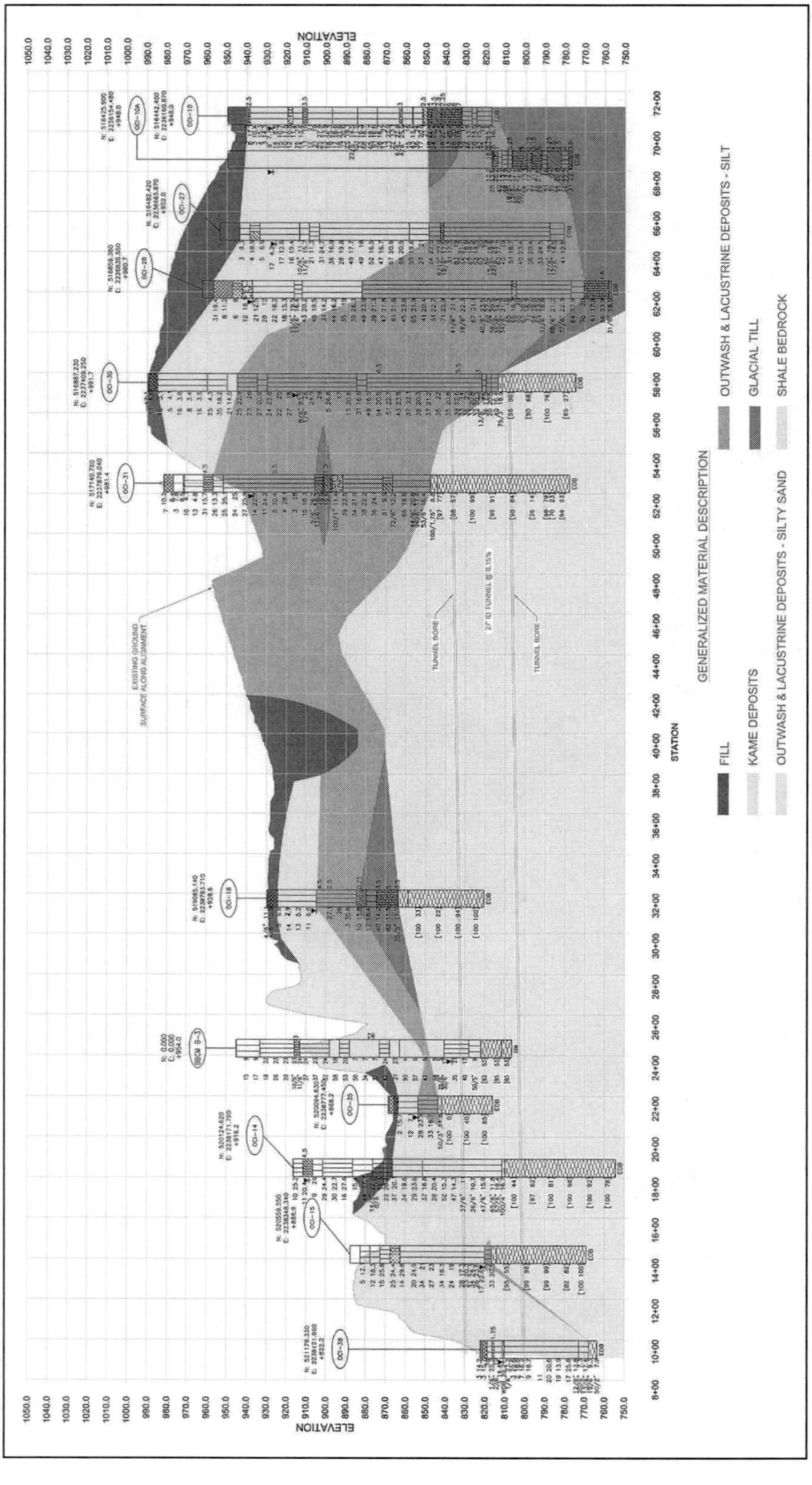


**Figure 7. Subsurface profile for selected OCI storage and conveyance tunnel preliminary design alignment (Alignment #4B)**

were estimated based on empirical analyses as well as historical data for similar tunnel projects. The approximately 30 ft OCI Tunnel TBM is anticipated to achieve an overall production rate of 6.5 feet per hour (ft/hr) in soft ground and mixed ground, and as much as 9 ft/hr in rock reaches. The average daily advance rate for the entire tunnel is anticipated to be around 45 feet per day (ft/day) (~16.6 m/day) and overall utilization of approximately 28% for an EPB type TBM. (OCIT P.E.R. 2012)

**St. Vincent–St. Mary Dump.** AECOM historical research and discussions with City engineering staff revealed that the preferred alignment will cross a former debris dump site located under the existing St. Vincent / St. Mary High School football field (north of Market Street). The dump area was reportedly a former ravine or low area next to the Ohio & Erie Canal filled in the late 19th and early 20th centuries. During football field construction, the school obtained a Rule 13 agreement with Ohio EPA to allow work on the landfill, and installed a membrane cap between the dump and football field. The anticipated soft ground to rock transition for the tunnel zone could occur a few hundred feet north of the football field. However, there was concern for dips and valleys in the bedrock surface under the dump limits, which could create a conduit for landfill leachate to reach the tunnel. AECOM recommended the Final Designer consider inclined borings as well as non-intrusive investigations, such as geophysical studies, be performed across the football field to better quantify risk exposure and provide evidence to Ohio EPA of an impermeable rock barrier between the tunnel and landfill. The goal would be for the agency to agree the tunnel muck and any groundwater inflow would not require special handling (i.e., non-applicability of Ohio EPA Rule 13, which governs work at known landfill sites).

**Drop Shaft Evaluations and Selection.** OCI Storage and Conveyance Tunnel drop shafts had to be designed to convey an extreme range of flows. To accomplish this, the AECOM preliminary design team evaluated a variety of drop shaft types, including baffle drops, tangential vortex drops, helicoidal vortex drop shafts, plunge drops, and combinations. While vortex drops seemed to be ideal for handling dry weather flows, baffle drops stood out for their ability to receive multiple inlet pipes at varying elevations. In addition, the vortex approach chamber and throat at the southern drop shaft would need to be built well below grade, in saturated cohesionless ground. The City also expressed a desire to perform all solids and debris handling at the downstream end of the OCI tunnel. This requirement means there is a high potential of large debris reaching the drop shafts. Based on the requirements described above, the P.E.R team recommended baffle drop shafts.

**Consolidation Sewer Alignments.** In order to convey flow from the City's existing sewer system to the tunnel, consolidation sewers from the Rack structures to the tunnel were considered. The majority of the Racks are located in urbanized, difficult to access, areas. The existing sewers are generally large diameters with significant dry weather flow. At the southern (upstream) terminus of the tunnel, two Racks are located 2,000 feet away from the planned retrieval shaft for the tunnel. This consolidation sewer was sized to be 12 feet in diameter, possibly requiring its own tunnel boring machine, and anticipated to be through mixed face conditions. This sewer extended past the Akron Children's Hospital and was located under a major "one way" thoroughfare in the City. Near the middle of the OCI tunnel, one Rack structure was in the central downtown business district and is located inside of a multilevel underground parking deck. Other mid-tunnel Rack structures are located on opposite sides of an innerbelt highway system and adjacent to a Federal courthouse, again requiring another tunneled sewer under the highway system. Near the downstream end of the tunnel (north end of project limits) Rack structures are located nearer to residential areas and a regional hike and bike trail, but are anticipated to be connected through microtunneled or other trenchless sewer construction. The main goals of the consolidation sewer alignment selection were to minimize costs and disturbance to the public.

## VALUE ENGINEERING PROCESS

Upon completion of the Preliminary Engineering Report, AECOM engaged the consulting firm Robinson, Stafford, and Rude, Inc. of Gulfport, Florida to lead a Value Engineering (VE) Study. VE Team members consisted of experts in various components of the OCIT project. Although the City and P.E.R. Team provided the VE Team with project constraints, the VE Team was asked to consider alternate tunnel alignments and alternate sewage storage options which might reduce risk and cost of the project. Thirty-five individual recommendations were presented for consideration, and nine (9) cost-savings ideas and another three (3) design suggestions were planned to be implemented as of late 2012. Total cost savings were estimated by the VE team to be approximately $64 Million.

The VE report did recommend an option with a reduced tunnel diameter in combination with a large storage tank. However, after the VE session, complete vetting of the VE "storage facility/reduced tunnel diameter" recommendation identified a significantly higher cost along with increased and unacceptable settlement risks for the storage facility. This recommendation was eliminated, greatly reducing the potential savings.

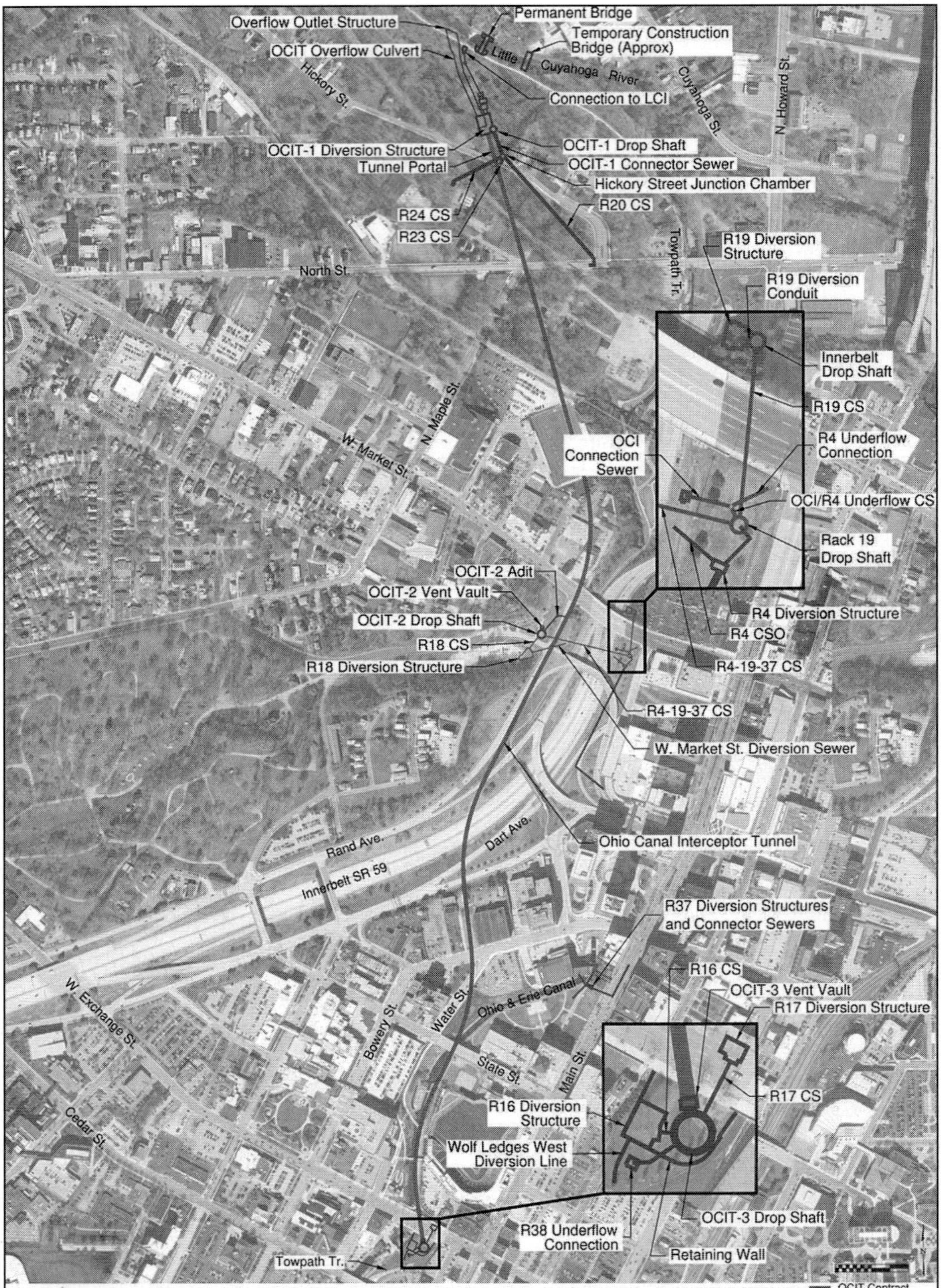


**Figure 8. OCIT Water Street Alignment and project structures**

Another significant risk reduction and potential cost savings was the VE team's recommendation to move the OCI Storage and Conveyance Tunnel alignment over approximately the southern half of the project. The new alignment, referred to as the "Water Street Alignment," is illustrated in Figure 8. The VE Team suggested this alignment outside of City ROW, because the newest geotechnical information strongly suggested that bedrock is present in the tunnel profile for the southern half of the new alignment. This alignment also eliminated a long consolidation sewer in difficult ground conditions.

## REFLECTIONS AND CONCLUSIONS

- In nearly all analyses, the lifecycle costs of a full time dry and wet weather pumping station exceeded the projected costs necessary to construct a gravity-drained tunnel, even when considering tunnel alignments passing through multiple ground conditions (soft ground and rock) as well as mixed face tunneling conditions (transition from soil to rock and vice versa).
- Value engineering process combined with increased amount of geotechnical investigation data was successful in realizing an OCI Storage and Conveyance Tunnel alignment with more favorable geologic conditions for tunneling and decreased consolidation sewer infrastructure. By giving VE team very few constraints, project value was improved despite the need to release constraints that were in place earlier in the tunnel alignment evaluations.
- There appears to be little experience in the U.S. designing large diameter, deep CSO storage and conveyance tunnels to also convey dry weather sanitary flows. Adding dry weather sanitary flows may require specially-designed dry flow drop structures within much larger wet weather drop structures. Important considerations include the following: how will debris and solids be handled at multiple upstream diversion structures or a single downstream end; odor control for systems conveying sanitary flows; potential for air entrainment due to dry weather drop structures inside wet weather structures.

## REFERENCES

"Akron Consent Decree 2010, Final CSO Long Term Control Plan Update Report." Akron Public Utilities Bureau, City of Akron, Ohio, 2010.

"City of Akron Consent Decree Ohio Canal Interceptor Tunnel Preliminary Engineering Report." 2012.

*City of Akron Website.* 2013. www.akronohio.gov (accessed January 13, 2014).

*City of Akron, Ohio CSO Program Website.* http://www.akroncso.com (accessed January 14, 2014).

"Ohio Canal CSO Alternatives Advanced Planning Study." 2006.

"Ohio Canal Interceptor Tunnel Preliminary Engineering Report." 2012.

Vormelker, J.D. and Swinford, E.M. "Bedrock Topography of the Akron West, Ohio, quadrangle." *National Geologic Data Map Database.* Ohio Division of Geological Survey. 1996. http://ngmdb.usgs.gov (accessed January 13, 2014).

# Geotechnical Investigation of the Northeast Interceptor Sewer Phase 2A Hollywood Fault Crossing

**Patrick Schmidt and Fred Burnett**
City of Los Angeles Department of Public Works, Bureau of Engineering

**ABSTRACT:** This paper presents a summary of the geotechnical investigations performed to evaluate the geotechnical conditions along the proposed Northeast Interceptor Sewer Phase 2A (NEIS 2A) tunnel crossing of the Hollywood Fault. The geotechnical studies for the NEIS 2A project, including the Hollywood Fault crossing, initiated with a planning study of proposed alternative alignments, followed with a pre-design study of a preferred alignment and then a design level study of the selected alignment. The design level study was performed in several phases. Exploratory borings with down-hole geophysics, packer and pressuremeter testing, Cone Penetrometer Testing, Remi Surveys and surface geophysics lines were used during the study. At the fault crossing, the NEIS 2A tunnel will have an invert depth of approximately 140 feet below the ground surface and will transition from Modelo Formation to Old Alluvium to Feliz Granodiorite and back into Modelo Formation. Along the fault zone, the tunneling machine will encounter squeezing and running ground conditions, methane gas and up to five bars of water pressure.

## INTRODUCTION

The Northeast Interceptor Sewer Phase 2A (NEIS 2A) project consists of two lines with a total length of approximately 3.8 miles. A regional map showing the location of the NEIS 2A alignment, along with some other major sewers, is shown on Figure 1.

Line 1, which crosses the Hollywood Fault, will be an eight-foot diameter sewer with a length of about 3 miles. Line 2 will be a seven-foot diameter sewer with a length of about 0.8 miles. One or more tunneling machines, with anticipated diameters on the order of 15 feet, will be used to tunnel the alignments. Earth Pressure Balance (EPB) or slurry Tunnel Boring Machines (TBM) will be used for the tunneling. Three shaft sites will be utilized to facilitate the tunneling.

In the area of the Hollywood Fault crossing, the project alignment extends from Griffith Park (east of Crystal Springs Drive), across Los Feliz Boulevard and south along Riverside Drive (west of the Golden State Freeway (I-5)) as shown on Figure 2. Through the fault zone, the sewer will have an invert depth of approximately 140 feet below the ground surface.

The active Hollywood fault trends east northeast along the base of the Santa Monica Mountains from the West Hollywood-Beverly Hills area to the Los Angeles River, a distance of approximately 8.7 miles. At the location of the NEIS 2A alignment, geologic maps (Lamar 1961, 1970; Webber et al., 1980 and Dibblee, 1991) show an approximate North 50° East trending fault trace located approximately 1,000 feet south of Los Feliz Boulevard (Figure 3).

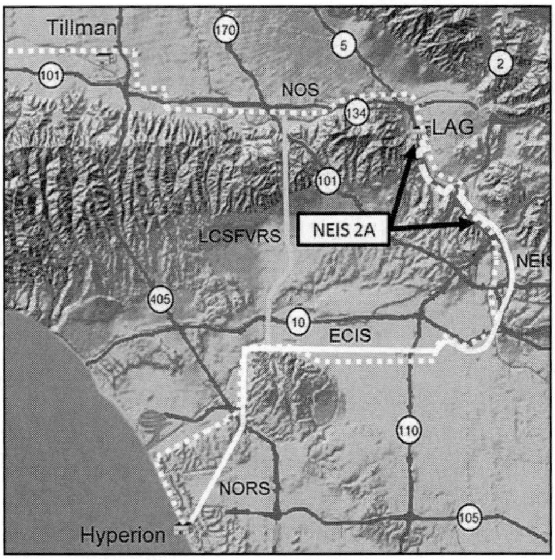


**Figure 1. Regional map**

The fault juxtaposes Feliz Granodiorite on the north against sedimentary bedrock on the south (Lamar, 1970; Dibblee, 1991). A second fault trace, inferred on the basis of topographic lineaments, is shown on several geologic maps (Weber et al., 1980; Dibblee, 1991; and Dolan et al., 1997, 2000) north of Los Feliz Boulevard west of the NEIS 2A alignment (Figure 3). Continuation of this fault trace eastward appears to cross the NEIS 2A alignment south of Los Feliz Boulevard and north of the other fault trace.

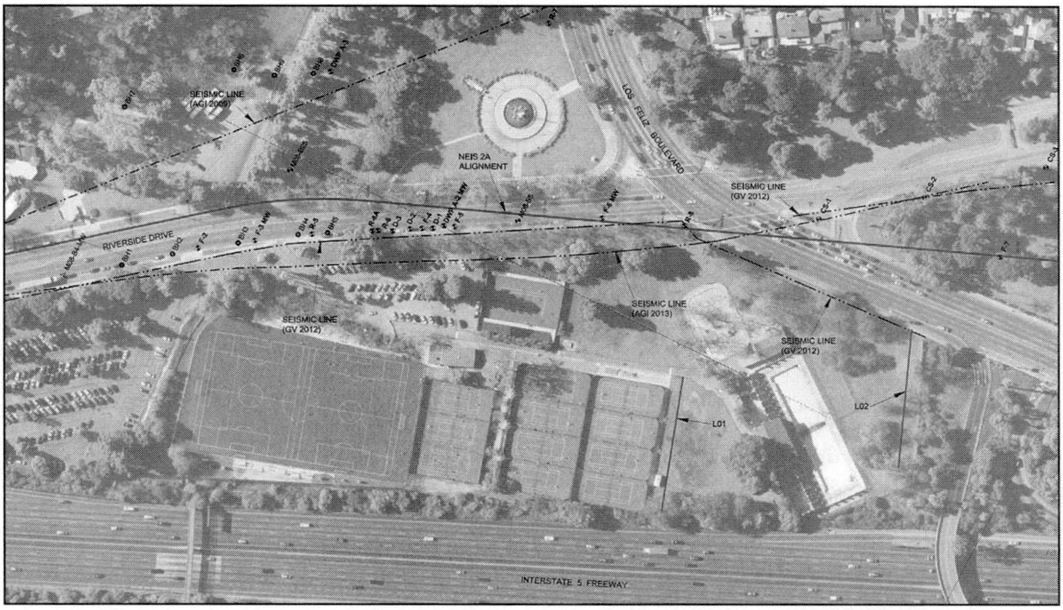


**Figure 2. NEIS 2A Project alignment through fault area showing field investigation locations**

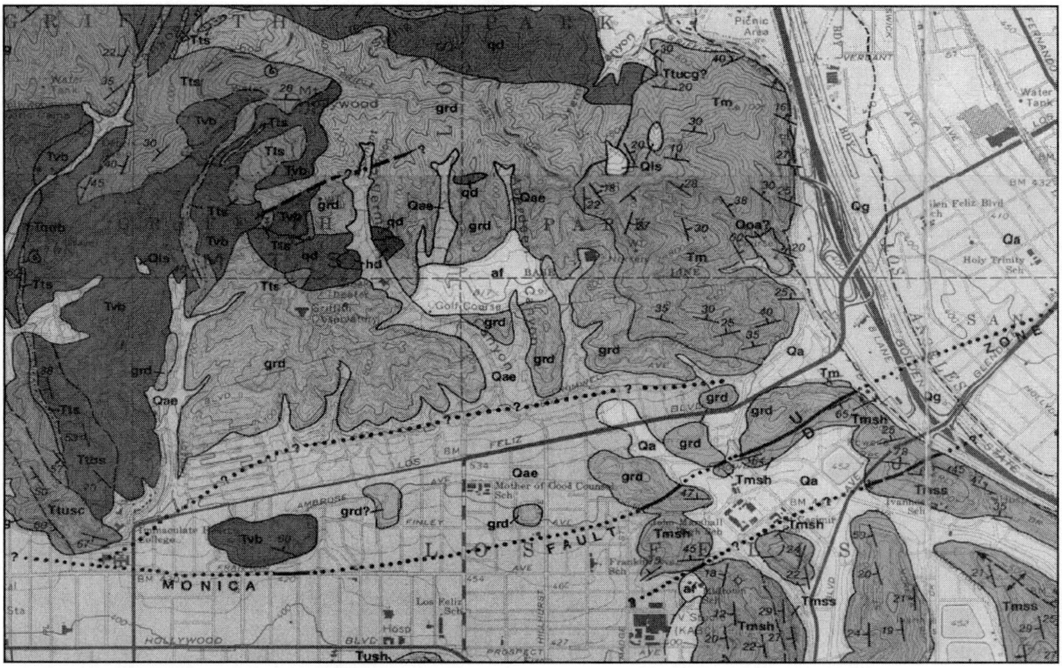


**Figure 3. Geologic map (Dibblee 1991)**

The geotechnical studies for the NEIS 2A project, including the Hollywood Fault crossing, initiated with a planning study of proposed alternative alignments, followed with a pre-design study of a preferred alignment and then a design level study of a modified selected alignment.

## PLANNING STUDY

A planning study was performed for the NEIS 2 project in 2004/2005 to evaluate several alignments being considered for the project. As part of this work, Dr. James Dolan, Professor at the University of Southern California, was retained to prepare a letter report regarding the current state of knowledge of the location, degree of activity and seismic potential of the Raymond, Hollywood, Santa Monica Fault system. Dr. Dolan concluded that the NEIS 2A alignment crosses the easternmost portion of the Hollywood Fault. Dr. Dolan's report includes a figure illustrating two fault traces projecting towards the NEIS 2A alignment as previously described.

## PRE-DESIGN LEVEL GEOTECHNICAL INVESTIGATION

The pre-design level geotechnical investigation kicked off in the fall of 2008 and extended for approximately two years. The project scope was much larger during the pre-design phase because it included two pieces of NEIS (2A and 2B) and the Glendale Burbank Interceptor Sewer (GBIS). The scope of work consisted of document review, field investigations, laboratory testing and preparation of a Pre-Design Geotechnical Data Report.

### Document Review

Documents reviewed in support of the Hollywood fault crossing included; (1) published geologic maps prepared by Hoots, Dibble, Lamar and Neuerburg; (2) a Geotechnical Data Report for a 96-inch water pipeline; (3) Caltrans and City of Los Angeles boring logs for nearby bridges and freeway ramps; (4) geotechnical reports for improvements to private properties near the alignment; and (5) fault data available on-line at the USGS website.

Several items of interest were identified during this review. Two of these were on properties west of the NEIS 2A alignment and south of Los Feliz Boulevard. One was that granitic rock was logged below alluvium in two of the borings drilled for the 96-inch water pipeline project. One of these borings, DWP A-3 is shown on Figure 2. A second item of interest was that a test pit log associated with a private property geotechnical study reportedly encountered faulting with granitic rock on the north and siltstone on the south.

In addition, Peterson (2008) presents projections of the Hollywood and Santa Monica alt 2 fault alignments as shown on Figure 4. These alignments

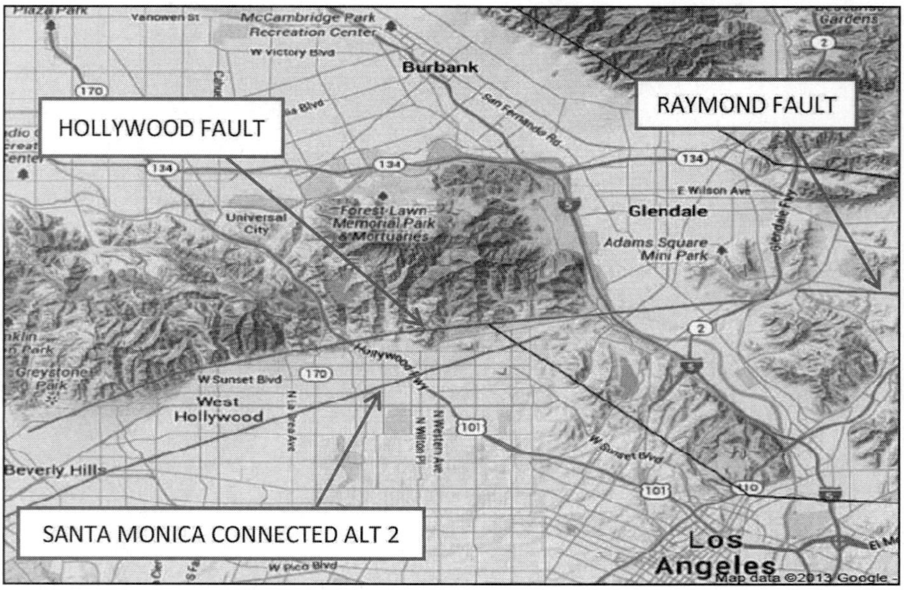


**Figure 4. Fault orientations from USGS (2008)**

may coincide with the strands of faulting presented in Dibblee 1991 (Figure 3) and Dolans 2004 letter report. Thus, it's possible that the Hollywood and Santa Monica fault systems merge within the fault zone being studied for the NEIS 2A project. Regardless, for the purpose of this paper, the fault crossing under study will be referred to as the Hollywood Fault Crossing.

### Field Investigation

The pre-design field investigation included field reconnaissance, geologic field mapping, drilling of exploratory borings, and a surface geophysical survey in the area of the fault. In addition, the pre-design phase was used to evaluate the usefulness of down-hole acoustic televiewer testing and refraction microtremor (ReMi) surveys for the project.

Two exploratory borings (M08-B4 and M08-B5) were selected in an effort to confine the limits of the fault zone. A third boring (M08-B25) was added in an effort to encounter granitic rock (mapped to the west and encountered in previous studies). These borings were advanced to depths of 150 to 170 feet below the ground surface. M08-B4 encountered Puente Formation (Lamar) at depth and M08-B5 encountered Modelo Formation (Hoots) at depth. Feliz Granodiorite was encountered within boring M08-B25 below bedrock logged as Modelo Formation.

Down-hole acoustic televiewer testing was performed in exploratory borings M08-B4 and M08-B25 to obtain data on discontinuity orientation (joint fractures, bedding, and shears) and frequency. The acoustic televiewer is a rotating ultrasonic device that electronically captures an orientated 360 degree view of the boring wall.

Exploratory boring M08-B4 was converted into a 2-inch groundwater monitoring well screened within the bedrock of the Puente Formation. Based on previous experience, mud cake development on the walls of the boring could prevent accurate measurement of methane within the formation. Therefore, a gas valve fitting was installed at the top of the PVC riser for later collection of air samples for gas analysis. A groundwater sample collected from the well was analyzed for several parameters including hydrogen sulfide. Hydrogen sulfide was not detected in the sample. Methane was not detected in the headspace air.

A 1,000-foot long geophysics surface survey (AGI 2009) was performed to generate a geophysical profile of subsurface layering to supplement the geotechnical exploration data in evaluating the location of possible fault traces associated with the Hollywood fault zone. The survey line was located south of Los Feliz Boulevard, about 100 to 300 feet west of the proposed NEIS 2A alignment.

Two ReMi surveys were conducted east of the NEIS 2A alignment and south of Los Feliz Boulevard. The purpose of the ReMi surveys was to evaluate differences in shear wave velocities of underlying geologic units in an effort to identify subsurface units including the alluvium/bedrock contact.

### Findings and Interim Conclusions

Based on the materials encountered in borings M08-B4, M08-B5 and M08-B25, a main fault was located between exploratory borings M08-B4 and M08-B5. Interpolation of the geophysical surface survey data resulted in the identification of two north dipping fault planes with three south-dipping secondary faults in the hanging wall merging at depth.

Following the pre-design phase it was also decided that Remi surveys would not be included in future phases of work. Although a valuable tool, it was felt that traditional geophysical surface surveys were more beneficial and cost effective in studying the fault crossing in this area.

## DESIGN LEVEL GEOTECHNICAL INVESTIGATION

The design level geotechnical study for the NEIS 2A project, including the fault crossing, was performed in several phases over a two-year period beginning in the summer of 2011. Some of the phasing was due, in part, to the way in which the alignment was subdivided into segments for the purpose of performing the geotechnical investigation.

During each phase of investigation, core barrel and core box photos were taken of all recovered core as was done during pre-design. During the design level investigations core logging also included 4-inch notations of the relative strength of the recovered core. The relative strength classifications were determined using hammers and knives to scrape and otherwise impact the rock. This data was later used to roughly correlate with Uniaxial Compressive Strengths and corrected Point Load Index strengths and to estimate the percentages of relative strength for each formation.

Water pressure (packer), pressuremeter, acoustic televiewer, and in-situ compression and shear wave velocity testing was performed in selected borings during the design level investigations. Borehole stability concerns prevented down-hole testing of some borings.

### Field Investigations

**Initial Design Level Exploratory Borings.** Following the pre-design phase, the project scope and alignment was refined. During this time, GEO utilized the City of Los Angeles Department of General Services drill crew to drill nine borings in the area of the fault crossing. The City drill crew utilizes truck mounted drill rigs equipped with continuous flight and hollow stem augers. Understanding the limitations of this equipment with regards to the subsurface conditions, these borings were drilled with the intent of determining alluvium thickness and with the hope of penetrating some depth into the underlying rock to identify the rock type. Four boring locations were selected west of the project alignment with consideration to information obtained during research of geotechnical reports for nearby private properties in addition to the previous borings drilled for the water line study and the pre-design borings. The other borings were selected between pre-design borings M08-B4 and M08-B5 within Riverside Drive. The borings were completed to depths of 26 to 77 feet below the ground surface with the two northern most borings west of the alignment (BH8 and BH9) encountering granitic bedrock beneath siltstone/sandstone. The other borings encountered alluvium overlying siltstone/sandstone bedrock and did not encounter granitic rock within the explored depths.

**Griffith Park; Los Feliz Boulevard to I-5 Shaft Site.** The field investigation for this segment included four exploratory borings (R-8 and CS-1 through CS-3) and a geophysics surface survey (GV 2012) in support of the fault crossing study. Although this portion of the project alignment is north of projected fault traces, we felt it was important to have closely spaced exploratory borings and a geophysical surface survey north of Los Feliz Boulevard. This data was considered necessary to evaluate the potential for fault traces projecting north of Los Feliz Boulevard and to compliment the geophysical surface survey performed during the pre-design phase and exploratory borings and geophysical surface surveys planned south of Los Feliz Boulevard in the next phase of study.

Exploratory borings R-8 and CS-1 were located at the southeast corner and northeast corners of the intersection of Riverside Drive and Los Feliz Boulevard, respectively, and borings CS-2 and CS-3 along Crystal Springs Drive north of the intersection. These borings were advanced to depths of 190 to 200 feet below the ground surface. Each of these borings encountered alluvium overlying Modelo Formation bedrock.

Borings R-8 and CS-1 were converted into 2-inch diameter groundwater monitoring wells screened within the bedrock of the Modelo Formation near the tunnel zone. Each well was fitted with a gas valve. During development, excessive gas bubbles were noticed in the pumped (submersible) and bailed water of wells CS-1 and R-8. At CS-1, pump cavitations and bubbles in the ejection line were noticed, but gradually lessened as well development progressed. Pump cavitations due to gas bubbles at R-8 rendered use of the pump impractical. Following surging, and after multiple attempts to use a submersible pump, a bailer was ultimately utilized to develop the R-8 well. Water bailed from R-8 appeared to be "frothing" in the top portion of the bailer and continued to create bubbles in the disposal drums. A gas meter was placed near the top of the boring during bailing and detected explosive gas above 20% of the Lower Explosive Limit. As a result, development of R-8 was terminated.

Hydrogen sulfide was not detected in groundwater samples collected from these wells. However, relatively high dissolved methane concentrations of 21,000 ug/l and 20,000 ug/l were measured in ground water samples collected from wells R-8 and CS-1, respectively. In addition, high methane gas concentrations of 160,000 and 93,000 parts per million (ppm) were measured in the headspace air of wells R-8 and CS-1, respectively.

The geophysical survey line was located along the east side of Crystal Springs Drive and extended 700 feet north of Los Feliz Boulevard. The line was offset approximately 40 feet (at Los Feliz Boulevard) to 150 feet (at CS-3) west of the proposed NEIS 2A alignment.

**Riverside Drive; Fletcher Drive to Los Feliz Boulevard.** The field investigation for this segment included three exploratory borings (R-5 through R-7), Cone Penetrometer Testing (CPT), and two geophysical survey lines in support of the fault crossing study.

CPT testing was included in the scope of work to evaluate whether this method of investigation was beneficial to the fault study. One geophysics line was located along the east side of Riverside Drive, extending south from Loz Feliz Boulevard. The other was along the south side of Los Feliz Boulevard east of Riverside Drive. Since the alignment of Los Feliz Boulevard is towards the northeast, the geophysics line along the south side had a northeast-southwest orientation that provided overlap with the geophysics survey line performed during the previous phase of investigation.

Exploratory borings R-5 and R-6 were located between pre-design borings M08-B4 and M08-B5 to fill in the data gap and boring R-7 was located west of the alignment near the end of the pre-design geophysics survey line. R-7 was drilled primarily for the purpose of performing a down-hole velocity survey. This data was needed to facilitate development of a scaled (horizontal and vertical scales in feet) section depicting the subsurface profile beneath the geophysics line conducted during the pre-design phase of investigation. Borehole stability problems led to the abandonment of exploratory boring R-6 at a depth of 105 feet below the ground surface (Feliz Granodiorite was encountered at 103 feet).

An additional boring, R-6A, was drilled about 10 feet south of R-6 and advanced to the target depth of 180 feet bgs. Borings R-5 and R-7 were advanced to depths of 180 feet below the ground surface. Boring R-5 encountered alluvium over Modelo Formation over Feliz Granodiorite. Boring R-6 encountered alluvium over old alluvium over Modelo Formation over Feliz Granodiorite. Boring R-7 encountered alluvium overlying Modelo Formation bedrock.

CPT was performed at five locations along the east side of Riverside Drive in the area of borings R-5 and R-6. Each of the CPT's was advanced to refusal, depths of approximately 31 feet to 75 feet below the ground surface. Shear wave testing was performed in four of the five CPTs.

Boring R-6A was converted into a 2-inch diameter groundwater monitoring well screened into the Feliz Granodiorite bedrock. A hydrogen sulfide concentration of 0.8 mg/l was measured for a groundwater sample collected from this well. Artesian groundwater conditions precluded measurement of methane in the headspace of this monitoring well. A pressure gauge was installed at the top of the R-6A well casing to measure water pressure at the ground surface.

One geophysics survey line (GV 2012) extended a distance of approximately 1,670 feet along the east side of Riverside Drive south of Los Feliz Boulevard and the other approximately 420 feet along the south side of Los Feliz Boulevard from Riverside Drive. The 1,670 foot survey line was sub-parallel to the NEIS 2A alignment with an offset of 0 feet to 80 feet. The 420-foot line was skewed approximately 20 degrees to the NEIS 2A alignment.

## Findings and Interim Conclusions Following the Griffith Park and Riverside Drive Phases of Study

Data from the Griffith Park and Riverside Drive phases of study were reviewed and evaluated before proceeding with the next phase of study. Some of the noticeable findings identified during these phases of study include; (1) High methane concentrations on the north and south sides of Los Feliz Boulevard; (2) Feliz Granodiorite bedrock at approximately the same depths in borings R-5, R-6(6A) and M08-B25 (Pre-Design), and (3) artesian groundwater within the Feliz Granodiorite bedrock. The pressure gauge installed on well R-6A was monitored during the course of this study and stabilized around 9.6 pounds per square inch.

No fault-like anomalies were interpreted for the geophysics line north of Los Feliz Boulevard. However, this line contained approximately 300 feet, between exploratory boring CS-1 and CS-3, of reduced data quality. Interpolation of the geophysical data south of Los Feliz Boulevard, along the east side of Riverside Drive, was also difficult. However, the geophysicists were able to identify an approximately 470-foot wide zone that lacked continuous, horizontal structure. This zone, labeled "Apparent Main Zone of Offset Reflectors" in the geophysics report contained six fault like anomalies. The report also noted additional disruptions in the reflectors outside the designated main fault-like zone that could be related to minor faulting. No fault-like anomalies were interpreted along the geophysics line on the south side of Los Feliz Boulevard, east of Riverside Drive.

## Further Field Investigations

**Fault Zone.** The next phase of investigation included six exploratory borings (F-2 through F-7) and an additional surface geophysical survey. Exploratory borings F-2 through F-6 were located along the east side of Riverside Drive, north and south of borings R-5 and R-6. Boring F-7 was located at a planned maintenance hole north of Los Feliz Boulevard. The purpose of borings F-2 through F-6 was to fill in the data gaps between the previous borings in order to develop a clearer picture of the subsurface conditions through the fault zone. Borehole stability problems led to the abandonment of exploratory boring F-5 at a depth of 140 feet below the ground surface. The other borings were advanced to depths of 160 feet below the ground surface.

Borings F-2, F-5, F-6 and F-7 encountered alluvium overlying Modelo Formation bedrock. Boring F-3 encountered alluvium over Modelo Formation over Feliz Granodiorite. Boring F-4 encountered alluvium over old alluvium over Feliz Granodiorite.

Borings F-2, F-6 and F-7 were converted into 2-inch diameter groundwater monitoring wells screened in the bedrock formations near the tunnel zone upon completion of drilling. Hydrogen sulfide concentrations of 0.036 mg/l and 0.061 mg/l were measured in samples collected from wells F-2 and F-7. Hydrogen sulfide was not detected in a groundwater sample collected from well F-6.

The headspace air in the wells was tested for methane. Methane gas concentrations of 540, 41,000 and 18,000 ppm were measured in wells F-2, F-6 and F-7, respectively. In addition, boring M08-B4 from the pre-design phase was sampled and a methane gas concentration of 17 ppm was measured.

A second 1,670-foot geophysics survey line (AGI 2013) was located approximately 10 to 30 feet east and about parallel to Crystal Springs Drive and Riverside Drive. The line extended from approximately 350 feet north to 1,200 feet south of Los Feliz Boulevard.

**Fault Zone Supplemental.** Exploratory boring F-4 of the previous phase of study encountered an anomaly of deep old alluvium, extending through the tunnel zone, overlying granitic rock. As a result, GEO decided to have the City drill crew drill three additional hollow stem auger borings in the area of boring F-4. These borings (D-1 through D-3) were drilled with the intent of further defining the limits of the deep old alluvium and to hopefully penetrate some depth into the underlying bedrock. These borings were completed to approximate depths of 97 and 110 feet below the ground surface. These borings encountered between 36 and 41 feet of alluvium underlain by older alluvium to 70 to 81 feet underlain by sedimentary bedrock. Feliz Granodiorite was encountered in borings D-2 and D-3 at 110 feet below the ground surface.

Figure 2 shows the locations of all the field explorations (CPT locations not shown for clarity) and geophysical survey lines performed for the geotechnical investigation of the portion of the NEIS 2A alignment that crosses the Hollywood fault zone.

## FINDINGS, INTERPRETATION, AND FINAL CONCLUSIONS

Some of the important items learned for the portion of the NEIS 2A alignment that extends through the Hollywood fault zone, as a result of the field studies, are summarized in the following paragraphs.

Several faults will be crossed during tunneling. Based on the data from the exploratory borings and the geophysics lines, eight or more faults will be crossed during tunneling. Two of these (labeled Fault A and Fault B1) are assumed to be main strands of the Hollywood Fault and are considered more significant for the planned tunneling. These faults sandwich granitic rock between old alluvium on the north and Modelo formation on the south and will be potentially problematic due to the nature of the altered, pulverized granitic rock and anticipated change in water pressure during the crossing of these faults. The other faults have created relatively narrow fault gouge zones within either the Puente or Modelo Formations. Due to the difficulties interpolating the geophysical data, it is possible additional faults will be encountered during tunneling through the area.

The first geophysics line along the east side of Riverside Drive interpreted a fault like anomalies zone, with six fault like anomalies within a width of approximately 470 feet. Additional disruptions were noted outside of this designated zone that could be related to minor faulting. The second geophysics line along the east side of Riverside Drive, in combination with the geophysics line performed during pre-design, interpreted eight faults over a width of approximately 1,100 feet. Based on this data and data from the exploratory borings, GEO extended the fault zone further to the north to create a fault zone width of 1,300 feet for the project.

A very narrow zone of old alluvium will be encountered during tunneling through the fault zone. The existence of this material within the tunnel zone was identified during the final phase of study. It would have been missed if it wasn't for one exploratory boring, F-4. Borings D-1 and D-2, located approximately 20 feet north and south of F-4, encountered bedrock well above the tunnel zone. Thus, the width of the old alluvium at tunnel depth is less than 40 feet along the boring transect.

High concentrations of methane gas will be encountered during tunneling within the fault zone. Along the project alignment, the highest levels of methane were measured north of the main fault traces in the area of Los Feliz Boulevard.

Within the Fault zone, water elevations in wells north of the artesian area are shallower than those to the south. Artesian groundwater pressure, up to five bars, will be encountered within the Feliz Granodiorite of the fault crossing zone. Maximum water pressure along the rest of the project alignment, at tunnel depths, is typically on the order of three to four bars or less. Thus, water pressure within the fault zone is the controlling water pressure for the project.

The condition of the granitic rock within the tunnel zone was found to be highly variable. Varying from intact, highly friable to completely pulverized. Some of the material is broken down to primarily sand size that will create a flowing ground condition under the water pressure.

Zones of relatively soft, clayey fault gouge material will be encountered during tunneling. The tunneling machine and constructed tunnel liner will be subject to squeezing ground conditions.

Additional work is needed in the fault zone to further assess the environmental conditions. Additional work, which is now underway, includes measurement of the soil gas pressure in selected monitoring wells and analyses of water samples for dissolved gases (hydrogen sulfide and methane concentrations). Isotopic fingerprinting is also being performed to determine whether the source of the methane is biogenic or thermogenic (petrogenic).

Interpretations of the data from the various phases of investigation lead to the development of the profile shown on Figure 5.

## ACKNOWLEDGMENTS

The authors would like to thank Mr. Ken Redd of the Bureau of Engineering Executive Division and Mr. Carlos Amaya of the Bureau of Engineering Wastewater Conveyance Engineering Division for their support of this paper. We also thank Mr. Pedro Garcia of the Geotechnical Engineering Group for his contribution to this paper and Mr. Eric Ramos and his crew for the numerous trips made to the field

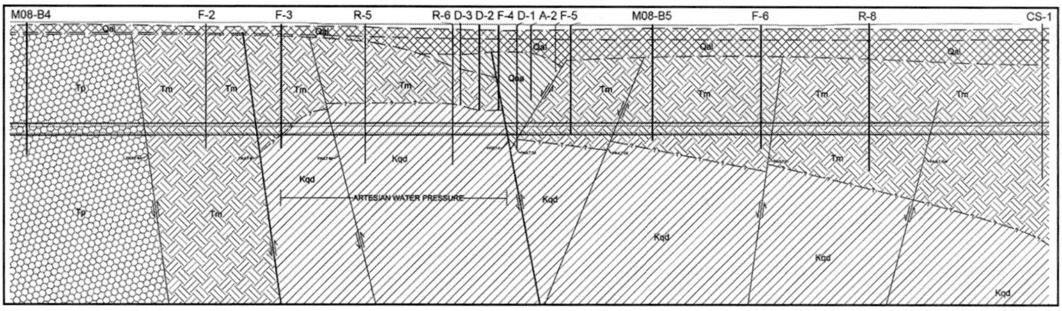


**Figure 5. Design profile of hollywood fault crossing**

to survey exploration locations. Finally, we would like to thank Amec Environmental & Infrastructure, Inc., Kleinfelder West Inc., the URS Corporation, and the City of Los Angeles Department of General Services Standards Division for their support of the field and laboratory investigations conducted for this project.

## REFERENCES

AMEC Environment & Infrastructure, Inc (2012), Geotechnical / Environmental Data Report Proposed Northeast Interceptor Sewer Phase 2A (NEIS 2A) Riverside Drive Segment.

AMEC Environment & Infrastructure, Inc (2013), Geotechnical / Environmental Data Report Northeast Interceptor Sewer Phase 2A (NEIS 2A) in the Area of the Hollywood Fault Crossing along Riverside Drive.

City of Los Angeles Department of Public Works (2014), Geotechnical Data Report: Northeast Interceptor Sewer Phase 2A (NEIS 2A), City of Los Angeles Department of Public Works, Bureau of Engineering, Geotechnical Engineering Group.

PB Americas, Inc. and Mactec (2010), Final Pre-Design Phase Geotechnical / Environmental Report, Proposed Northeast Interceptor Sewer (Phase 2) / Glendale Burbank Interceptor Sewer (NEIS Ph 2/GBIS), City of Los Angeles, California.

URS Corporation (2012), Geotechnical and Environmental Data Report for Northeast Interceptor Sewer Phase 2A (NEIS 2A), Griffith Park (Southern Portion).

# White River Collection Consolidation Sewer Phase 1 Project

**John Morgan**
Citizens Energy Group

**Jeremy Morris**
Christopher B. Burke Engineering, LLC

## PROJECT BACKGROUND

Citizens Energy Group (CEG) is implementing a federally-mandated Long Term Control Plan (LTCP) Consent Decree (CD) to abate combined sewer overflows that will dramatically improve Indianapolis' waterways. Included in this plan are provisions for a series of capital project control measures to be implemented. One of these capital projects is the White River Collection Consolidation Sewer (WR CCS) Phase 1 project.

The WR06, drop shaft for White River Deep Tunnel, portion of the WR CCS Phase 1 project is located immediately adjacent to the Bush Stadium redevelopment project currently being constructed by Core Redevelopment, as part of the 16 tech urban technology park, a City commissioned master plan. To avoid significant future surface disruptions, CEG worked with Core Redevelopment and the City of Indianapolis to fast track the project in order to complete construction in 2013. WR06 construction started in June and is currently scheduled to be substantially complete on December 17, 2013.

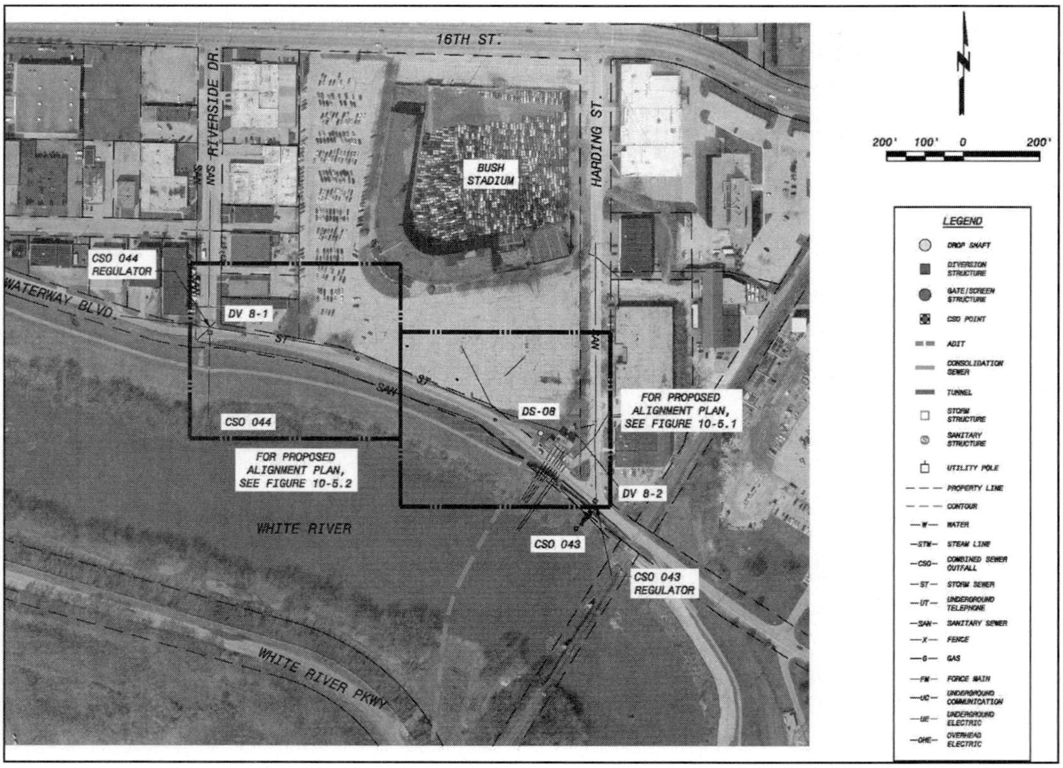


**Figure 1. Overall WR06 site**

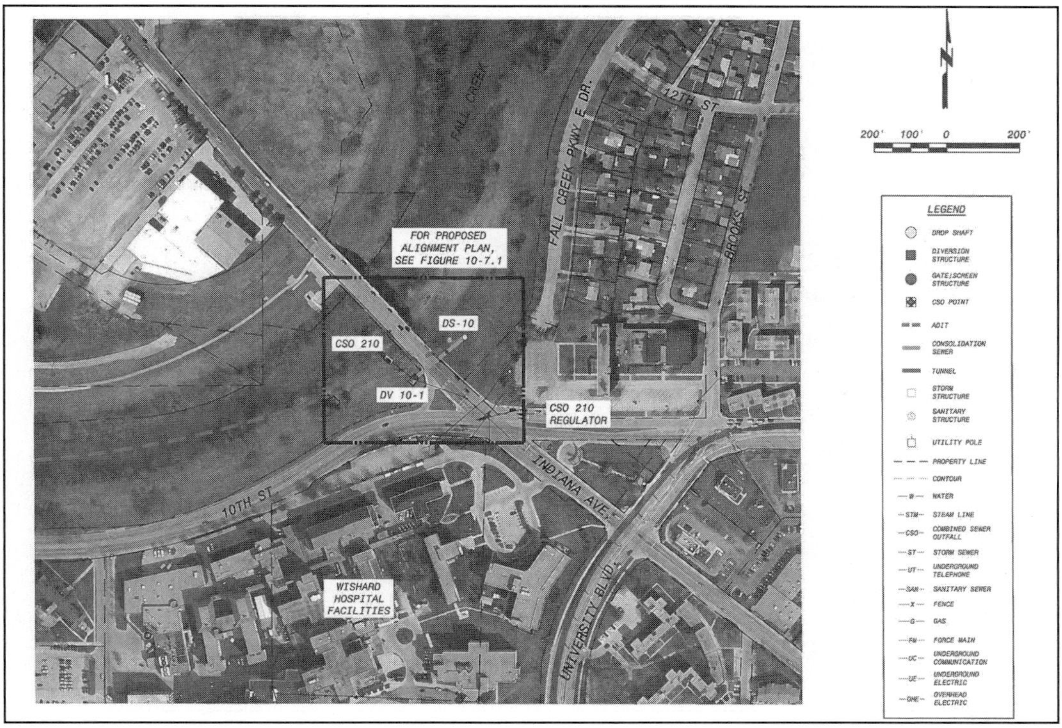


**Figure 2. Overall FC01 site**

**Figure 3. Bush Stadium during Cash for Clunkers**

**Figure 4. Bush Stadium during Cash for Clunkers**

The redevelopment of Bush Stadium was needed for Indianapolis. The stadium served as home of the Indianapolis Indians, AAA Baseball Team, from 1931 until 1996. At that time a new stadium was constructed in downtown Indianapolis (Victory Field) and the Indian's relocated. The Stadium sat idle until 1999. At that point it was turned into a race track until 2002. The .40 kilometer (.25 mile) dirt track did not result in the volume draw that some thought it would become. The Stadium was even rented out during the "Cash for Clunkers" program as a storage area by a local salvage yard for their surplus vehicles.

## DESIGN CHALLENGES, COORDINATION, AND COLLABORATION

During the early stage of design, the WR CCS Phase 1 project was to be constructed in the parking lot of Bush Stadium. During the design of the project, CEG had acquired the wastewater assets

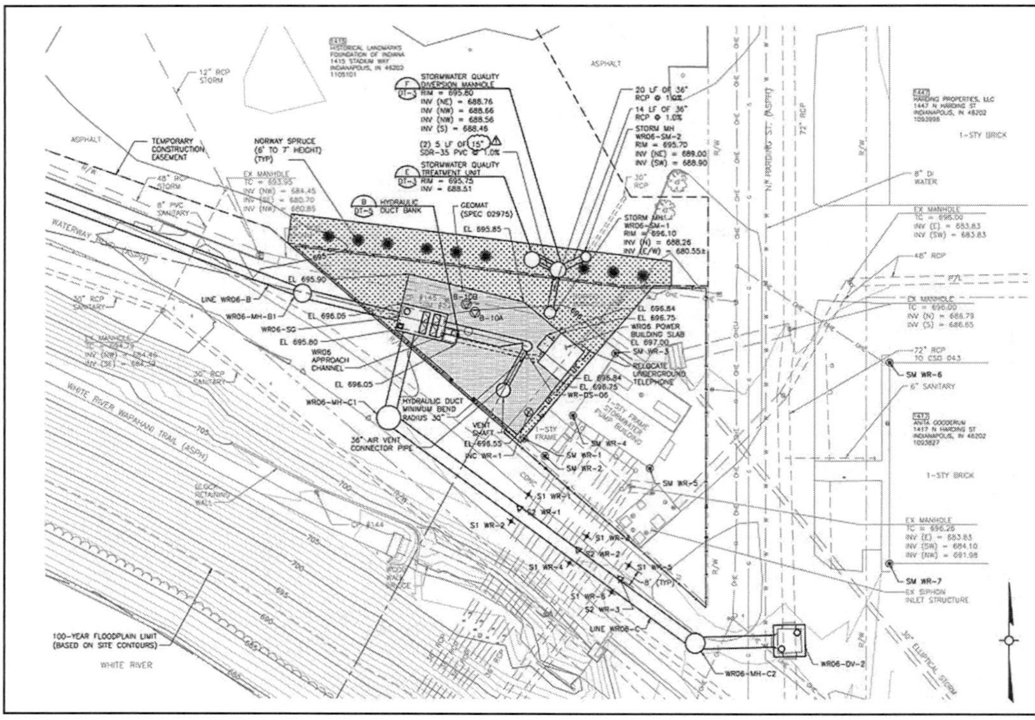


**Figure 5. WR06 site plan**

as well as operation, maintenance and all required Capital Improvements from the City of Indianapolis. Several months after the ownership transition to CEG, it was quickly discovered that parcel ownership, which was identified for the WR CCS Phase 1 project site, was transferred to Core Redevelopment. Once we were aware of this error, we discussed our issues with Core Redevelopment and the City of Indianapolis. After the initial discussion with the three parties, it was identified that the City has another project in the area. The three parties entered into an agreement. This agreement stated; CEG had to construct the project in a timely manner, Core had to transfer the land back to CEG, and the City had to pay for the Storm Water facilities, curb, gutters, and pavement restoration. The CEG Special Project Group strove to find many synergies and cost saving solutions amongst the three projects. Below is a list of many synergies and cost saving solutions.

### Synergies Between CEG and Core Redevelopment

- CEG acquired the land for free from Core Redevelopment in exchange to construct WR CCS Phase 1 in FY 2013 instead of FY 2015.
- Cost saving due to inflation
  - The average inflation rate from Jan. 2012–May 2013 is 1.9%. (1.9% × 3 years = 5.7%) The cost of the White River CCS Phase I in 2013 dollars is projected to be $15,800,000. The estimated savings from inflation is $900,600. The project would cost approximately $16,700,600 if the project was not accelerated.
- CEG will have the project substantially completed by the time Core is ready to start Phase II by September 1st, 2013.
- CEG and Core Redevelopment have maintained a good working relationship.

### Synergies Between CEG and the City of Indianapolis

- The City is paying for Storm Water facilities related to the Bush Stadium re-development.
- The City will restore the curb, gutters, and the pavement as part of their project.
- CEG removed guardrails, which were recycled to the City for future use.

### Synergies Between CEG and the Community

- CEG has utilized local vendors;
  - Bowen Engineering Corp (Contractor)
  - Christopher B. Burke (Construction Inspectors)
  - All M/W/VBE goals will be met on the project

- CEG had no additional cost for redesign since the initial design did not change.
- CEG is maintaining access for neighbors and students to the pedestrian bridge for the White River Wapahani Trail.

## FINAL PROJECT DESIGN AND SCHEDULE

The final design of this first phase of the Fall Creek/ White River Tunnel System includes construction within two project areas, WR06 and FC01 (drop shaft for Fall Creek Deep Tunnel included in project as bid alternate and was awarded with the project).

The WR06 project area along Waterway Boulevard between Harding Street and Riverside Drive includes the following:

- Three diversion structures to capture two combined sewer overflows, CSO 043 and 044, to the White River
- Seven large manholes with diameters varying from 2438 mm (96 in) to 3048 mm (120 in)
- Approximately 27 meters (90 feet) of 914 mm (36 in) diameter sewer
- Approximately 244 meters (800 feet) of 1,372 mm (54 in) diameter sewer
- Approximately 91 meters (300 feet) of 1,829 mm (72 in) diameter sewer
- One drop shaft and one vent shaft, both approximately 61 meters (200 feet) deep
- One screen and gate structure

The FC01 project area at the intersection of 10th Street and Indiana Avenue includes the following:

- One diversion structure to capture one combined sewer overflow, CSO 210, on Fall Creek
- Approximately 27 meters (90 feet) of 2,743 mm (108 in) diameter sewer
- One drop shaft and one vent shaft, both approximately 61 m (200 ft) deep
- One screen and gate structure

**Figure 6. Typical large diversion structure base slab**

**Figure 7. Typical drop shaft final liner**

**Figure 8. Typical large diameter manhole installation**

**Figure 9. 1372 mm (54 in) diameter connection to WR06-DV-1**

Figure 10. Typical 1829 mm (72 in) jack and bore

Figure 11. Typical 914 mm (36 in) sewer

Figure 12. Typical screen and gate structure

Figure 13. Typical screen and gate structure with approach channel

The WR06 project area will disrupt traffic along Waterway Boulevard between Harding Street and Riverside Drive. The FC01 project area will involve periodic traffic disruptions at 10th Street and Indiana Avenue.

### Overall Preliminary Project Schedule

- WR06 Construction Period: April–September 2013
- FC01 Construction Period: April–November 2013

## CONSTRUCTION CHALLENGES

### Indianapolis Power and Light (IPL) Utility Challenges at WR06 Project Site

One of the early challenges on this fast tracked project has been utility coordination. Multiple utilities require relocation in the vicinity of the Harding Street and Waterway Boulevard intersection in order to accommodate construction of the proposed diversion structure and associated excavation support system. Indianapolis Power and Light (IPL) has multiple overhead power lines at this intersection. The project contractor has had extensive coordination with IPL since the project started. The contract documents required the Contractor to take responsibility for all utility relocations related to the construction of the project.

IPL has relocated some of their overhead facilities in the area and intended to de-energize higher voltage power lines that could not be easily relocated in order to comply with OSHA regulations while Contractor performs the proposed work. Through Contractor's coordination efforts with IPL, it was recently realized the power lines to be de-energized for the WR06 project are one of two sources, this one being secondary power source, of power to CEG's Riverside Pump Station and well field. De-energizing these lines would impact the secondary power source to the Riverside Pump Station. If the Riverside Pump Station were to experience

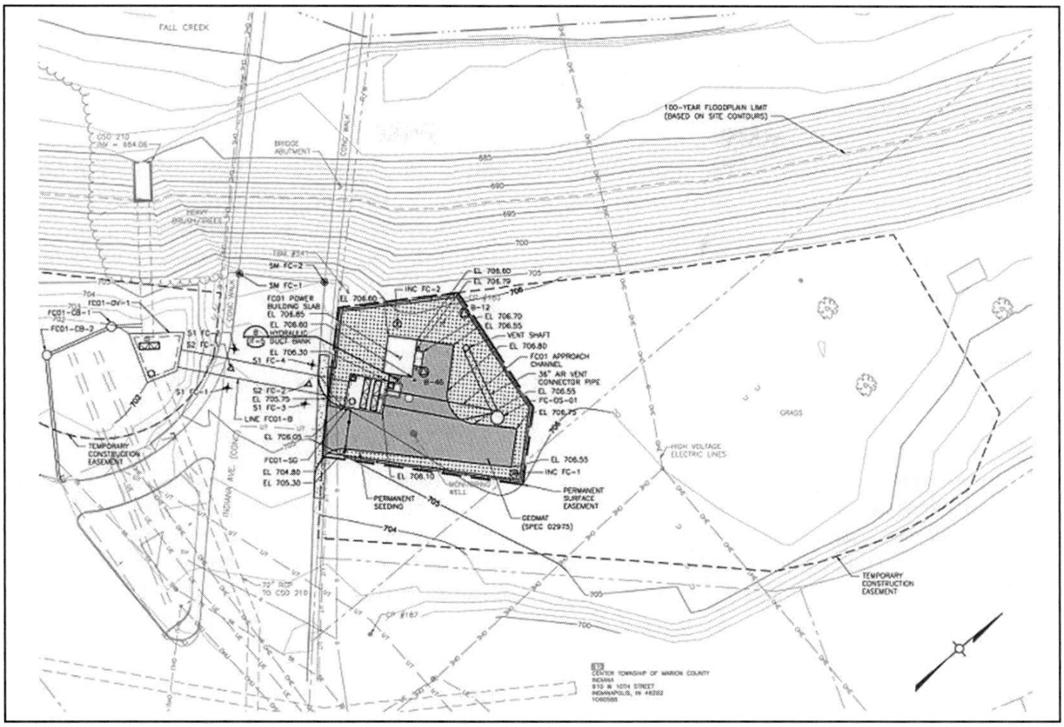


**Figure 14. FC01 site plan**

**Figure 15. Typical shaft installation**

**Figure 16. Typical screen and gate structure sheet pile support and excavation**

a power failure without a secondary power source, downtown Indianapolis could experience low water pressures and have limited fire protection until power is restored.

Based on Contractor's schedule, the lines would need to be de-energized for 6 to 8 weeks to complete the necessary sewer construction. The work would be accomplished by implementing three separate two-week power shut down periods with the lines re-energized for approximately one week between each shut down. This would reduce the time period that the secondary feed is out of service. Additionally, due to delays in the initial project notice to proceed, the outages were scheduled during a peak usage period (for both water and power consumption) creating additional project difficulties and risk.

### Evaluated Alternatives

The WR CCS project team had several meetings to discuss the alternatives to facilitate constructing the sewer improvements while still providing a reliable power source to the Riverside pump station facility. The following is a list of alternates considered and evaluated:

- Relocation of the diversion structure (WR06-DV-2).
- Reconfiguration of the Riverside substation to provide a redundant power source when lines are de-energized.
- Connect two temporary back-up generators for the secondary power source.
- Re-locate the power lines in conflict with sewer construction.
- Raise the power lines in conflict with sewer construction.
- Install new power lines from another source to the Riverside Pump Station.
- Allow scheduled secondary power feed outages with extensive coordination.
- Modify the WR CCS project scope of work/ schedule to allow time for IPL to provide an adequate redundant power source while west bank lines are de-energized.
- Allow a reduced duration of scheduled secondary power outages with extensive coordination due to increased BEC work hours/ shifts.

Ultimately, Citizens elected to proceed with the bottom alternative listed above. This alternative provided reduced risk compared to first alternative listed due to the reduced duration of scheduled secondary power outages. This alternative also limited impacts to the project schedule and associated surface restoration in the vicinity of the Harding St. and Waterway Blvd. intersection.

### Citizens Water Utility Challenges at FC01 Project Site

As construction began on the FC01 project site it was quickly realized during utility coordination efforts multiple 914.40 mm (36 in) distribution water mains were located within, or close to, the contractors propose support of excavation. These water mains were construction in the early 1960s of reinforced concrete pipe material. These mains provide the primary water supply to downtown Indianapolis. If interrupted, the downtown area water pressures would drop to a dangerously low level which would be insufficient for fire protection. In addition, if these lines were to rupture in the vicinity of this project, several important medical facilities and a hospital could experience flooding. As a result of the critical water mains proximity to the proposed diversion structure and associated support of excavation, the project team immediately began re-evaluation and re-design of the proposed diversion structure to eliminate the conflict and accomplish the project goals. The exhibits below detail the original design and final redesign of proposed infrastructure at the FC01 site. Currently, the project team expects a minimum overall cost increase associated with these fast tracked changes during construction.

### General Accessibility Challenges

The WRCCS Phase 1 project also has many challenges due to the locations where work is to be performed. Most combined sewer overflow (CSO) outfalls are very close to or in developed urban areas which typically result in open space constraints for construction. Local businesses, developments, and neighborhoods are often nearby during this work. Extensive coordination is needed to allow access for construction traffic, staging areas and the permanent space to install the components to consolidate and collect sewer overflow while working adjacent to others throughout construction. The WR06 and FC01 sites of the WRCCS Phase 1 project are both challenging site to access and stage from during this type of linear underground heavy civil infrastructure construction.

### WR06 Site Accessibility Challenges

Coordinating multiple simultaneous construction activities is a major role due to the fast track nature of this project at the WR06 site. Simultaneous phases of work are required to be performed to meet the timeframe while working immediately adjacent to the redevelopment of the existing Bush Stadium project.

The WR06 site is bordered by the White River and associated levee system and Riverside Boulevard to the south, the on-going Bush Stadium redevelopment to the north, Riverside Drive to the west, and Harding Street to the east. Construction traffic access to the WR06 site is contractually restricted to Riverside Drive only. The ongoing Bush Stadium redevelopment project limits extend to the right-of-way limits of Waterway Boulevard constraining the area for staging and further restricting access during construction. Recent road reconstruction of Harding St. and the pedestrian traffic to the Bush Stadium housing constrained access from the east. Continuous project coordination is required to successfully orchestrate the traffic for both contractors as the simultaneous project are completed.

Access is further constrained during the open cut sewer is installed from the WR06-SG structure

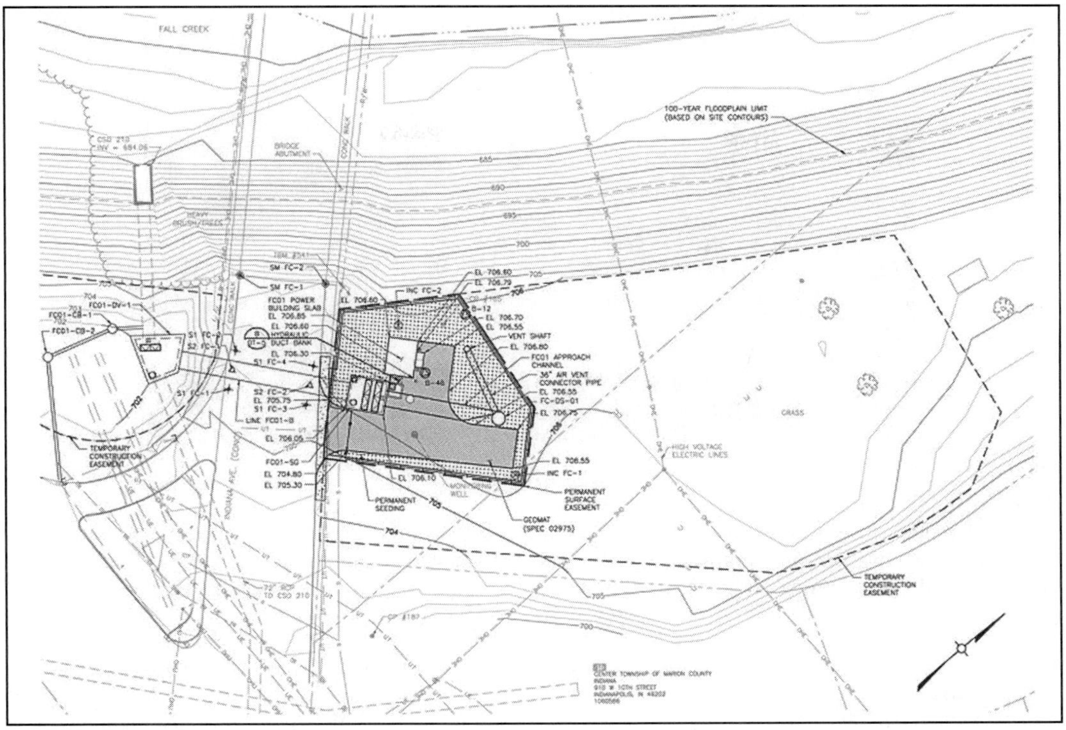

Figure 17. Original FC01 site plan

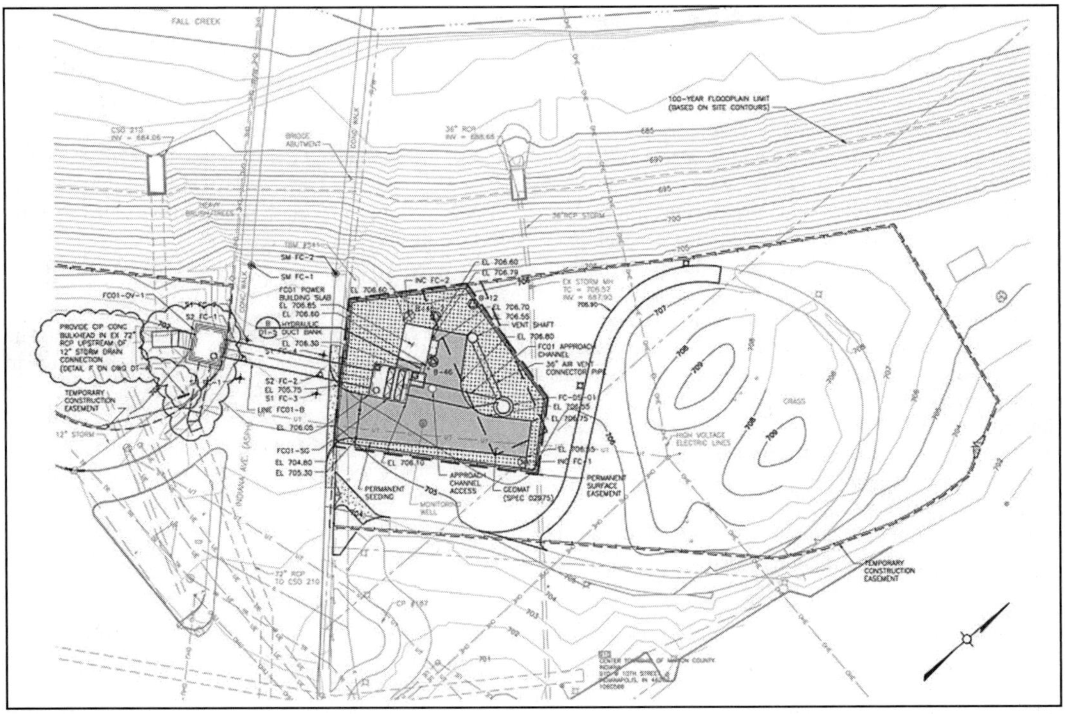

Figure 18. FC01 site plan changes

**Figure 19. Proximity of residents to the FC01 site**

on the east side of the project to the WR06-DV-1 structure on the west side of the project. This sewer line will divert the flow from CSO 044 located at Riverside Drive and Waterway Boulevard. WR06-DV-1 structure construction and the associated sewer lines further restrict the only contractually allowed access to the site.

The staging of the 54 inch RCP along Waterway further limits construction traffic access for the Bush Stadium Redevelopment project as well as the simultaneous WR06-SG structure work. Access for local businesses is also a challenge during these construction activities. The overall goal is not to negatively impact customer access to any business of residence along the project alignment.

Access to the tool supply company located at Waterway Boulevard and Riverside Drive must be maintained during all phases of work at the WR06 site. Multiple deliveries and on site customers are required to be allowed access during the construction of the WRCCS Phase 1 project. On the east end of the project site, adjacent to the WR06-DV-2 structure at the intersection of Waterway Boulevard and Harding Street, a local HVAC manufacturer also requires multiple daily deliveries. In addition to vehicular traffic, pedestrian traffic had to be maintained around the project site utilizing necessary barricades, pavement markings and safety fences. These boundaries guide the pedestrians safely through and/or around the projects limits. All these challenges needed to be addressed to achieve adequate access for all involved in active construction or being impacted by the construction during all phases of the work at the WR06 site.

### FC01 Site Accessibility Challenges

Site access for the installation of the work at the FC01 was impacted by two main constraints. The site is located immediately adjacent to the recently completed apartments (known as The Avenue), Fall Creek Parkway to the south and Indiana Avenue to the west. Indiana Avenue is not allowed contractually to be restricted or closed due to the road being a major thoroughfare for emergency traffic to the nearby hospitals. Located in the center of the FC01 site is a large utility pole carrying power lines for the primary feed of the Riverside water pump station, power sub-station, and nearby hospitals as well as cable and communication cables. Clearance under the lines did not allow large vehicle or crane access and assembly within the proposed working area. A significant amount of utility coordination was required to reroute the lines underground temporarily to allow passage of construction equipment across the site during construction.

### Challenges with Recent Development and Public Relations

The recent development immediately adjacent to the FC01 work site added challenges to the construction at the FC01 site. During the construction of The Avenue apartments, and business park adjacent to the FC01 site, an additional .46 m–.61 m (1.5 ft–2 ft) of fill was placed over the work site. The additional fill was not included in the original design drawings as the survey for design was performed prior to this work. The project team worked to cost effectively remove the additional material and prepared the site for construction to the elevations indicated in the drawings.

As with the WR06 site, the FC01 site is challenged by the close proximity of residential occupants. Unlike the WR06 site, the FC01 site had limited working hours identified in the Contract Documents (7:00 AM to 7:00 PM). The project team continually coordinates with the management of The

Figure 20. Proximity of residents at the WR06 site

Figure 21. Overburden casing with carbide steel cutting teeth

Avenue apartments to address concerns during all construction activities. The contractor performed the drilling of the drop shafts utilizing a night shift until drilling was complete without adversely affecting the adjacent occupants of The Avenue apartments. The contractor distributed flyers with information and advantages for working night shifts before these operations began. Some minor complaints were received during night shift for noise and lights. All complaints were immediately addressed and the night shift was discontinued as soon as the drilling work was completed. The project team's continued coordination efforts and personal attention have allowed construction to progress on schedule in close proximity of the residential development.

## UNCONVENTIONAL VENT SHAFT CONSTRUCTION AT THE WR06 SITE

Once the overburden drilling was complete for the vent shaft at the WR06 site a 3960 mm (156 in) diameter steel overburden casing was installed from

Figure 22. Float can top view

the surface to the bedrock to support the overburden excavation during the rock drilling operation. A series of cutting teeth were welded to the bottom of the overburden steel casing and the contractor rotated and drilled the casing into the bedrock in order to provide adequate embedment and a solid connection and seal at the bedrock interface.

The Manitowoc 4100 series II crane used to drill the overburden was utilized for the casing installation. The crane could not provide enough power to initiate rotation of casing while the casing was in contact with the bedrock. The crane was rated for a maximum load of 71,214 kg (157,000 lbs) and the combined weight of the drilling shaft and casing alone totaled 59,420 kg (131,000 lbs). Since the total load of the casing and drill rigging was too large to initiate rotation, the contractor had to improvise to provide the additional lift necessary to allow for rotation and drilling of the large casing into the bedrock. The contractor utilized an approximately 17 m long (55 ft), 2135 mm (7 ft) diameter float can placed inside of the overburden casing to provide the additional lift.

The float can was manufactured of a steel cylinder, open on the bottom, and sealed on the top with a welded steel plate. An air intake valve and two air release valves were installed through the top plate. Once the float can was placed inside the overburden casing and attached to the casing below the rotating arm of the crane, air was pumped into float can replacing the bentonite slurry. The float can, filled with air, provided the necessary lift on the overburden casing to assist the crane in suspending the casing above the bedrock and allow for drilling to begin. Once the overburden casing was in rotation, air was released and the casing advanced into the bedrock. Float can techniques such as these employed on this project are typically used in marine applications, however, this technique was proven to be very effective in this land based application.

Figure 23. Float can installation

## CONCLUSION

Fast tracked projects such as this one can be constructed with success. An important key to the success is working with all residents, business owners and corresponding utility owners during the design stages. Face to face meetings are an important key to the overall project success. Also, stressing the reality of the project is very important. Many conceptual projects are proposed. Sharing the fact that the project is not simply conceptual and will be constructed is an important key to the overall success.

## REFERENCES

Black & Veatch, September 2010. Prepared for Indianapolis Department of Public Works (DPW). *Fall Creek/White River Tunnel System Technical Memo for Project No. CS-18-072B.*

Black & Veatch, March 2013. Prepared for Citizens Energy Group. *Conformed Plans for Project No. 92TU00230 Fall Creek/White River Tunnel White River CCS Phase 1.*

Indiana Historical Landmark Foundation, 2010. *Historic Stadium Becomes Clunker Graveyard* http://www.wired.com/autopia/2010/12/historic-stadium-becomes-clunker-graveyard/).

# Design Considerations for Mega Tunnels

**Steve Dubnewych and Michael Torsiello**
Jacobs Associates

**C. Yoga Chandran**
CH2M HILL

**Michelle Smith**
Los Angeles County Metropolitan Transportation Authority

**Derek Higa**
California Department of Transportation

**ABSTRACT:** At approximately 60 feet (18.2 m) in diameter and about 4.5 miles long (7.2 km), the State Route 710 (SR 710) Freeway Tunnel alternative, one of four multimodal alternatives under study in Southern California, would be one of the largest and longest freeway tunnels in the world. Both single- and twin-bore tunnel configurations are being considered. Challenging geotechnical conditions identified along the tunnel alignments include mixed face conditions, high groundwater pressures, active earthquake faults, and formations with a potential for methane and hydrogen gas. This paper focuses on the design and construction challenges associated with the Freeway Tunnel alternative that is being analyzed to address mobility constraints in the study area.

## INTRODUCTION

### Background

The State Route 710 (SR 710) transportation corridor was originally envisioned to extend north from the City of Long Beach to the Interstate 210/SR 134 and SR 710 interchange in the City of Pasadena. A nearly 4.5-mile segment between Interstate 10 and the Interstate 210/SR 134 and SR 710 and interchange is the only uncompleted portion of the facility.

For decades, planning efforts to improve mobility and relieve congestion on local arterials and nearby freeways, resulting in part from the uncompleted portion of the SR 710 corridor, were limited to a surface extension of the SR 710. Today, the Los Angeles County Metropolitan Transportation Authority (LA Metro), in cooperation with the California Department of Transportation (Caltrans), is considering the design of one of the largest freeway tunnels in the world, along with four other alternatives that include a light rail transit (LRT), a bus rapid transit system (BRT), and transportation system management/transportation demand management strategies) (TSM/TDM) and No Build as potential transportation solutions. LA Metro is the contracting agency for the environmental document, and Caltrans is the lead agency with authority to ensure the study is being conducted in compliance with the National Environmental Policy Act (NEPA) and the California Environmental Quality Act (CEQA).

Due to advances in tunnel boring machines (TBMs) over the past 25 years that allow for much larger tunnel diameters, tunneling has been considered for all appropriate alternatives under consideration. Tunnel boring was found to be applicable for the LRT and Freeway Tunnel alternatives. Design considerations for the Freeway Tunnel alternative are the focus of this paper because of the unique geotechnical challenges within study area.

In 2011, LA Metro contracted with the CH2M Hill team to conduct an environmental study to identify project alternatives to address the traffic congestion within and beyond the SR 710 corridor. The study area for this SR 710 North Study, as depicted in Figure 1, is approximately 100 square miles (260 km$^2$) and is generally bounded by Interstate 210 (I-210) on the north, I-605 on the east, I-10 on the south, and I-5 and SR 2 on the west.

### Freeway Tunnel Alternative Description

As shown in Figure 2, the alignment for the Freeway Tunnel alternative starts at the existing southern stub of SR 710 in Alhambra, north of I-10, and connects

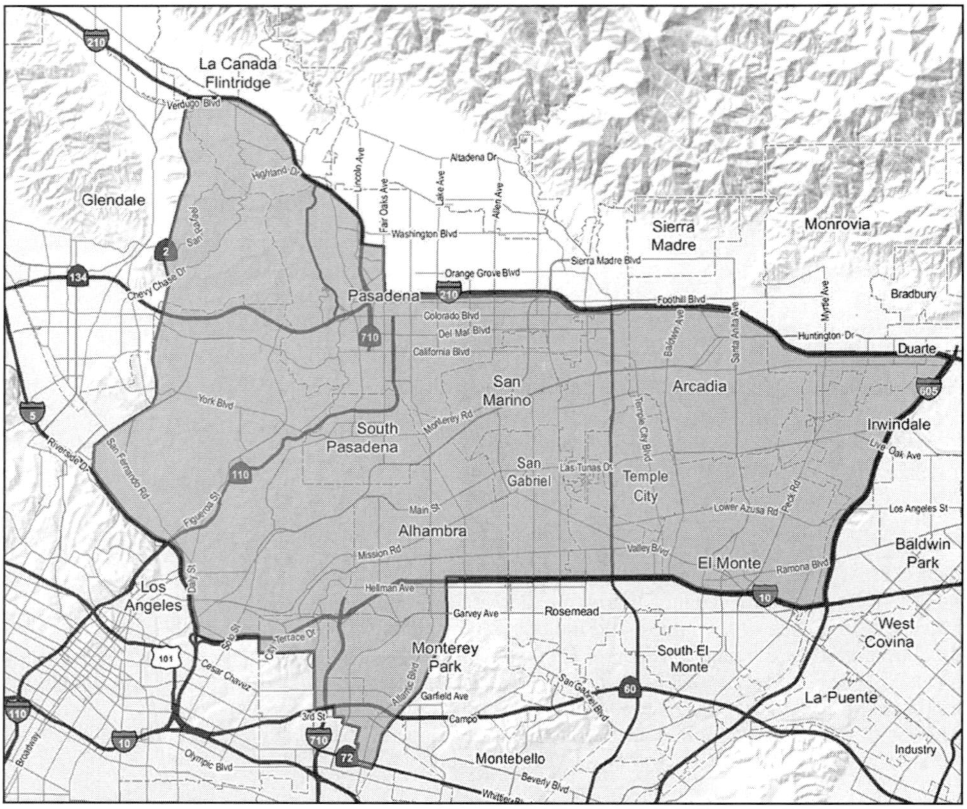


**Figure 1. Study area**

to the existing northern stub of SR 710, south of the I-210/SR 134 and SR 710 interchange in Pasadena. The Freeway Tunnel alternative has two design variations: a twin-bore tunnel and a single-bore tunnel. Both tunnel design variations would include the following tunnel support systems: emergency evacuation for pedestrians and vehicles, air scrubbers, a ventilation system consisting of exhaust fans at each portal, an exhaust duct along the entire length of the tunnel, fire detection and suppression systems, communications and surveillance systems, and 24-hour monitoring. An operations and maintenance building would be constructed at the north and south ends of the tunnel. There would be no operational restrictions for the tunnel, with the exception of vehicles carrying flammable or hazardous materials; however, it should be noted "Toll" and "No Truck" scenarios will be evaluated as well.

The twin-bore design variation includes two tunnels that independently convey northbound and southbound vehicles. The overall length of the improvements with this alternative is approximately 6.3 mi long (10.1 km), with two 4.2 mi (6.8 km) bored tunnels, each with 0.7 mi (1.1 km) of cut-and-cover tunnel, and 1.4 mi (2.3 km) of at-grade segments. This tunnel variation would consist of twin two-level bored tunnels with two lanes on each level and in each direction (see Figure 3). Each bored tunnel would have an excavated diameter of approximately 60 ft (18.2 m) and would be located approximately 120 to 280 ft (36.6 to 85.3 m) below the ground surface, except near the portal areas where the cover is less. Vehicle cross passages would be provided, connecting one tunnel to the other tunnel, for use in an emergency situation. It is assumed that the tunnels will be excavated with a pressurized-face TBM supported with a precast concrete segmental lining.

The single-bore design variation includes one tunnel having an excavated diameter of approximately 60 feet (18.2m) that carries both northbound and southbound vehicles. The single-bore tunnel would be in the same location as the northbound tunnel in the twin-bore tunnel design, but it would have two northbound lanes on one level and two southbound lanes on the other.

This Freeway Tunnel alternative is considered a mega project due to the fact that it would be the largest diameter tunnel in the world as well as its length and the timeframe needed to complete the project from its inception to completion. As planned,

**Figure 2. Freeway tunnel alternative plan**

it would have an excavated diameter larger than the SR-99 Tunnel currently under construction in Seattle, Washington.

A cross section of the twin-bore variation is shown in Figure 3; the single-bore variation would have a similar cross section, with only one tunnel bore.

## ANTICIPATED GEOLOGIC CONDITIONS

The project area encompasses portions of the San Gabriel Valley, San Rafael Hills, and the Repetto Hills, and is underlain by Quaternary-age alluvium, Tertiary-age sedimentary rocks, and Mesozoic-age crystalline igneous and metamorphic rocks.

The geologic units along the proposed Freeway Tunnel alignment include Artificial Fill, Alluvium, Fernando Formation, Puente Formation, Topanga Formation, and basement complex rocks (Wilson Quartz Diorite). Table 1 summarizes these primary geologic units, which are relevant to this study.

The geologic strata are deformed into a series of folds and faults. Frequent changes in bedding orientation due to folding and faulting are expected at the depths of the Freeway Tunnel. The geologic conditions in and around the fault zones are expected to vary widely, including narrow to wide zones of highly fractured rock and/or clayey gouge. While there are several faults that are located in the project area, the faults of most interest are the Raymond fault, the San Rafael fault, and the Eagle Rock fault; many of the faults along the alignment (including the Highland Park fault) are inactive. The Raymond fault is considered to be an active fault, and the San Rafael and Eagle Rock faults are considered

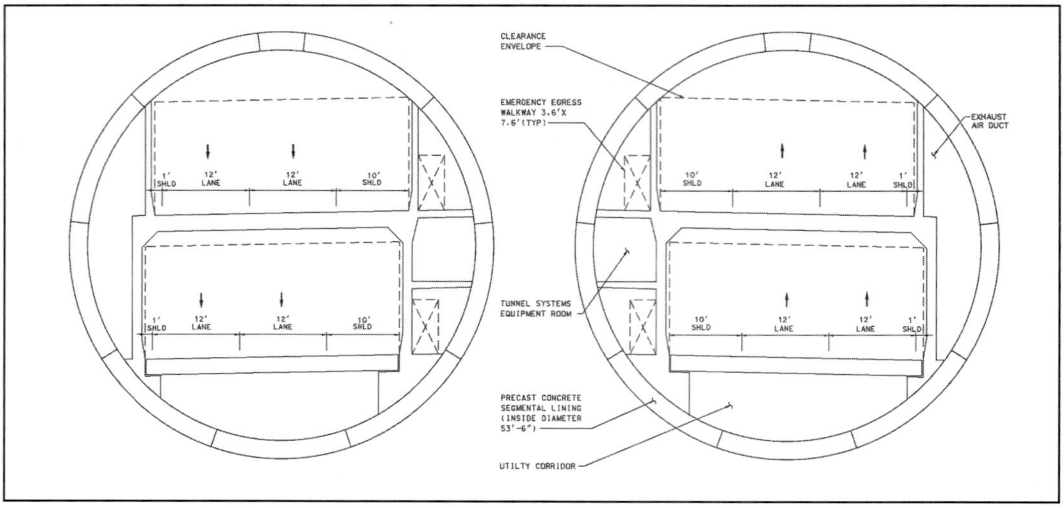


**Figure 3. Freeway twin-bore tunnel cross section**

**Table 1. Geologic units along freeway alignment with general descriptions**

| Geologic Unit | Approximate Length, ft (m) | General Description |
|---|---|---|
| Artificial Fill (Af) | N/A | Varies. Where encountered, observed to be fine-grained with some coarse-grained constituents. |
| Alluvium (Qal) | 4,600 (1,402 m) | Sand and gravel with scattered cobbles and boulders with layers/lenses of silt and clay. |
| Fernando (Tf) | 3,700 (1,128 m) | Predominantly claystone, mudstone, and siltstone with some sandstone and conglomerate. |
| Puente (Tp) | 3,600 (1,097 m) | Predominantly siltstone with interbedded sandstone. Potential for cemented layers and/or concretions. |
| Topanga (Tt) | 8,900 (2,713 m) | Predominantly siltstone with interbedded sandstone with some conglomerate. Potential for cemented layers and/or concretions. |
| Basement Complex Rocks (Wqd) | 1,600 (488 m) | Suite of lithologies, including diorite, monzonite, quartz diorite, quartz monzonite, and gneissic diorite. |

potentially active. These faults all cross the Freeway Tunnel alignment and could be capable of producing ground movements.

The depth to groundwater ranges from less than 10 feet (3 m) to approximately 175 feet (53.3 m) below ground surface along the tunnel alignment, resulting in groundwater levels of up to approximately 150 feet (45 m) above the Freeway Tunnel crown. The shallowest groundwater is expected at the south portal and along the southern portion of the alignment. A generalized geologic profile along the Freeway Tunnel alternative is shown in Figure 4.

## GEOTECHNICAL CONSIDERATIONS

Some key geotechnical considerations that would affect the tunnel design include high groundwater pressures, variable ground conditions, naturally occurring gas, and fault crossings. These conditions, as well as the design approaches being considered to handle the challenges, are discussed below.

### High Groundwater Pressures

Tunneling under high groundwater pressures may involve significant risk, including but not limited to the potential for groundwater inflows at the face of the excavation, high pressures acting on the tunnel lining, and negative impacts on the ability to perform interventions under free air for inspection and maintenance of the TBM cutterhead. Based on the data available, groundwater pressures could be as high as 6 bar (600 kPa) at the face of the Freeway Tunnel where the groundwater cover above the crown is at its maximum level.

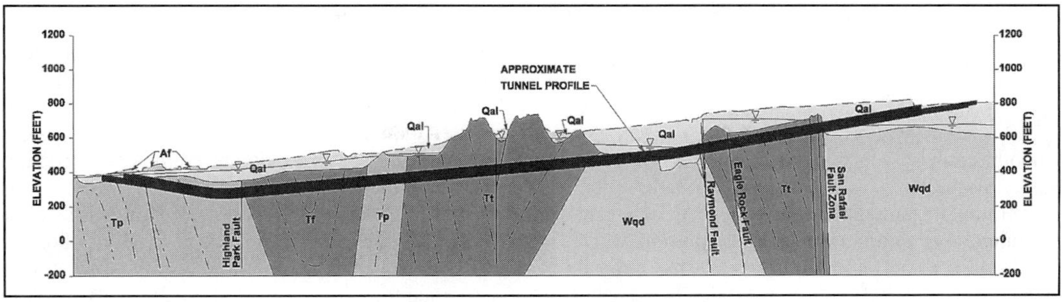


**Figure 4. Generalized geologic profile along freeway tunnel alternative**

### *Design Approach*

To overcome this challenge, the excavation of the tunnel will likely require a pressurized-face TBM (such as an earth pressure balance or slurry TBM), which is ideal for providing face control and mitigating the risk of high groundwater inflows. Additionally, a gasketed precast concrete segmental lining will be used to satisfy the long-term operational needs of the tunnel. These linings are designed to be essentially watertight.

Typically, when the cutting tools at the face of the TBM need to be replaced under full hydrostatic pressure, compressed air or mixed-gas hyperbaric interventions need to be performed. This is because of the need to maintain face stability to prevent ground from invading the tunnel and groundwater from flooding the tunnel. While in stable rock conditions, the face could be accessed under no or low pressures; however, the full hydrostatic pressure would need to be balanced in saturated alluvium and possibly the weak rock.

One new development that will make cutter changes much easier is the hollow spoke cutterhead. In machines of the size expected to be needed for the Freeway Tunnel alternative, it is now possible to design a cutterhead that allows cutting tool changes from within the spokes of the TBM cutterhead under free air. Changing the cutting equipment under free air is generally more time efficient and less risky than having to perform hyperbaric interventions. Bertha, the earth pressure balance (EPB) TBM currently mining the SR-99 Tunnel in Seattle, Washington, has been designed so that the majority of the cutting equipment can be changed in this fashion.

## Variable Ground Conditions

As indicated in Table 1 and Figure 4, there are several different geologic units anticipated along the tunnel alternatives. Variable ground conditions are more challenging for tunnel excavation than uniform conditions, especially where there are significant variations in strength of the ground. In transitions between soil and rock and from one formation to another, mixed-face conditions will be encountered; examples of the transitions include alluvium to soft rock, soft rock to hard rock, or hard rock to alluvium. Each transition would be different depending on the two different types of materials and the angle of the contact which impacts the amount of mixed face. Impacts include the possibility of slower TBM advance rates and the potential for stability issues, loss of ground, and surface settlement where mixed-face conditions are encountered.

### *Design Approach*

For the Freeway Tunnel alternative, there are few places where mixed-face conditions are expected; however, there are several transitions between geologic units. Because of the large tunnel diameter and the fact that the contacts between units are not vertical, the transition zones between geologic units could be long, resulting in significant lengths of mixed-face conditions. Additionally, the weak sedimentary rock formations are expected to have some inherent variability (such as the presence of cobbles or boulders, or cemented zones). To overcome this, a TBM designed to excavate all expected ground conditions should be specified and it should have the capability of controlling variable and unstable ground, especially at soil/rock interfaces. For EPB machines, the use of effective ground conditioning agents will be extremely important.

## Naturally Occurring Gas

The potential for naturally occurring gas, such as methane or hydrogen sulfide, is a significant design and construction issue. Although naturally occurring gas was not encountered in any of the borings drilled for the underground exploration program for this study, it is anticipated that gas could be encountered in several of the formations expected along these alignments based on experience tunneling in Los Angeles. Based on the previous experience in the Los Angeles basin, it is most likely expected in the

Puente Formation. As seen in Table 1, the Freeway Tunnel will encounter the Puente Formation for portions of the alignment (approximately 15 to 20%).

### *Design Approach*

Encountering gas during construction is primarily a safety issue. The atmosphere can be made safer by preventing hazardous concentrations of the gas in the tunnel and eliminating potential ignition sources. While this is an issue during construction, this issue can be mitigated during the design. Recently, the Sparvo Tunnel in Italy was successfully mined in formations with high concentrations of methane gas. The EPB TBM was designed with a complex safety system, including explosion-protected equipment, a fully enclosed conveyor belt for the excavated materials, and a permanent fresh air supply for all the workers in the tunnel (TunnelTalk, 2012). Additionally, the machine was outfitted with a permanent monitoring system to measure and record the concentrations of methane. A TBM with similar systems could be specified for this project. Also, in California this hazard is well recognized and the California Division of Occupational Safety and Health (Cal/OSHA) regulates tunnel construction to ensure that safe working conditions are maintained. The contractor must comply with the Cal/OSHA tunnel safety orders for dealing with naturally occurring gases during tunnel excavation.

## Fault Crossings

Tunnels and underground structures generally perform well in earthquakes, except where the tunnel crosses active faults or where there is other seismically induced ground failure such as slope failure or liquefaction. The displacements associated with these ground movements have the potential for shearing the tunnel structure, resulting in significant damage. The Freeway Tunnel alternative crosses several faults that have the potential for generating ground movements (offsets) if a seismic event occurred. The active and potentially active faults include the Raymond, Eagle Rock, and San Rafael faults (refer to Figure 4).

Caltrans uses a Safety Evaluation Earthquake and a Functional Evaluation Earthquake with return periods of 1,000 and 100 years, respectively. The anticipated horizontal displacement for the Freeway Tunnel could be up to approximately 1.6 feet (0.5 m), with a smaller vertical displacement at each of the fault crossings.

In addition to the potential for fault offset at the fault crossings, squeezing ground conditions are possible in and around the fault zones based on the geotechnical information available to date. Squeezing ground is most likely to be encountered in weak, sheared rock, fault gouge, and overstressed cohesive soils.

### *Design Approach*

As part of the conceptual design for each alternative, a fault crossing concept must be considered. The objective would be to design the structure to avoid collapse in an earthquake and at the same time have a system that could be repaired without major reconstruction to restore functionality after a design seismic event.

To accommodate the expected fault offset, an enlarged tunnel vault reach for each tunnel bore is being considered. This concept is similar to what was performed for LA Metro's Red Line tunnels crossing the Hollywood fault (Albino et al., 1999). The oversized vault excavation would be designed to accommodate the movement/offset from a seismic event. A conceptual sketch of this seismic vault is shown in plan in Figure 5. Construction of the tunnel vault reach at the fault zones poses a challenge not only in terms of constructability, but also because of its impact on the overall construction schedule. While there could be several ways to construct such a vault, the methods are limited by the lack of surface access to the tunnels for the Freeway Tunnel alternative.

Because of access limitations, one of the options could be to excavate the oversized vault from within the TBM-excavated tunnels. A simplified construction sequence would be as follows:

- Perform ground improvement and install rock dowels from within the already-excavated, segmentally-lined tunnel.
- Remove only one segmental lining ring at a time; excavate ground to achieve the final lining profile of the vault that is desired.
- Install initial lining (such as fiber-reinforced shotcrete) where segmental lining was removed and repeat this process for each ring until entire length of the vault is excavated.
- After vault excavation has been completed, install waterproofing and a cast-in-place concrete final lining over the initial lining.

This operation would have an impact on the TBM trailing gear, mucking operations, and installation of the roadway deck or rail, and would require specialized equipment to disassemble the segmental lining.

To address the squeezing conditions anticipated in the fault zones during TBM excavation, methods to reduce the ground pressure acting on the TBM should be taken. This is especially important along the body of the TBM, where high friction forces caused by the convergence of the ground can trap the machine. To overcome these forces, the TBM can be designed with:

- The capability of injecting bentonite into the annulus along the shield (which acts as a lubricant),

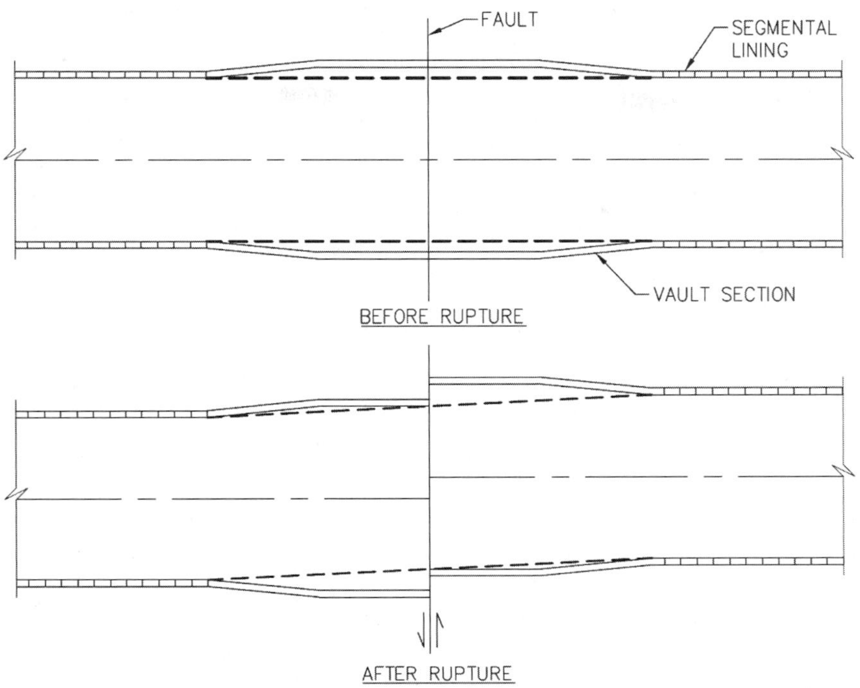


**Figure 5. Plan of seismic vault section before and after rupture**

- Adjustable gauge cutters to increase the overcut to accommodate this convergence, and
- A tapered shield so that the diameter is reduced from the cutterhead to the tail of the shield.

Additionally, continuous mining should occur in areas identified as having a risk for squeezing so that the ground does not have time to converge and trap the TBM.

## MEGA TUNNEL DESIGN CONSIDERATIONS

Several aspects of tunnel excavation that may be considered routine for average-sized tunnels can become complex for tunnels the size of the proposed Freeway Tunnel. There are several considerations for this large-diameter tunnel that have been evaluated as part of the preliminary phases of this study. These include construction power requirements, transportation of the TBMs to the portals, the handling of excavated material, right-of-way requirements, and the potential for surface settlement.

### Construction Power Requirements

TBMs and the other supporting equipment necessary to excavate the tunnel require a significant amount of power. Primary power is usually supplied by utilities via high voltage transmission to a substation at the tunnel's construction portal area, and the backup emergency power is supplied by generators. The construction power needed for the excavation of the Freeway Tunnel alternative could be more significant if the twin-bore variation is selected, as it is possible that up to four TBMs could be used to excavate the twin bores—two from each portal. It is anticipated that approximately 60 MW of power could be needed at each portal site to support the construction activities if the TBMs at each portal are mining simultaneously. It is expected that each mega TBM could require approximately 25 MW of power, with ancillary construction equipment at the portal areas making up the rest of the total construction power need. This amount of power will be required throughout the excavation process, and a redundant power supply (possibly provided by generators) would be needed for emergency use in the event of a power failure to provide power to the equipment necessary to maintain a safe environment for the workers in the tunnel. Additionally, the tunnel will require power for normal operations, once it is opened for traffic, but those needs are expected to be significantly less than the construction phase needs.

The power needs of 60 MW at each portal are significant—60 MW represents the amount of power necessary to supply approximately 50,000 households with power. While the power needs for the Freeway Tunnel's construction are significant, the design team is working with local power suppliers early on. This amount of power is typically not readily available at portal locations and will have to be routed to each of the construction portals which will take time. Because of the long lead time typically required to bring the power to each of the portal areas, early involvement from the power suppliers for a project of this size is essential.

### TBM Transportation

A TBM on the order of 60 feet in diameter will likely have a shield length of at least 60 feet as well as several backup gantries. While the TBM can be transported to the portal areas in parts and assembled on site, several of the individual parts are still very large and extremely heavy. Assuming the TBM will be transported to Los Angeles by sea from the TBM Manufacturer, upon reaching the port, the parts of the TBM will need to be transported to the two portal areas by road. Size and weight limitations could potentially be an issue for some of the larger and heavier components of a TBM such as the main bearing, the shield, and the cutterhead. It is a benefit to this project that both construction portals are adjacent to freeways.

### Right-of-Way Requirements

While freeway tunnels do not need the significant amounts of surface right-of-way (ROW) that surface freeways do, tunnels require permanent underground easements. The twin-bore Freeway Tunnel alternative consists of two tunnel bores over 60 feet in diameter (18.3 m), spaced approximately 60 feet apart. This requires a permanent underground tunnel easement approximately 200 feet wide (61 m) along its entire alignment. This is a significant consideration for mega tunnels.

Twin-bore light-rail tunnels can often be designed under city streets so that the tunnels fall within the public ROW; however, for mega tunnels, that is generally not possible, and ROW impacts must be considered. ROW impacts for this study are being performed by overlaying the design footprint of each alternative onto an assessor's parcel map to determine the number of parcels requiring permanent surface easements and permanent underground easements.

### Handling of Excavated Material

The excavated material from Freeway tunneling operations will be removed from the tunnels at the construction portals. For the twin-bore alternatives, assuming that mining of each bore is occurring simultaneously, a significant amount of excavated material could be generated daily at both the portal at the north and south ends of the bored tunnel alignment. It is expected that for schedule considerations, the Freeway Tunnel twin-bore alternative would be mined using two TBMs each from both the north and south portals simultaneously, for a total of four TBMs.

It is expected that approximately 4 million cubic yards of excavated material could be generated at each portal over the course of the bored tunnel excavation. In addition to the significant volume generated over the life of the project, the rate at which the TBMs excavate through the ground could impact the handling of the excavated materials. On an average day of mining, approximate 9,000 cubic yards of excavated material could be generated at each of the portals for the twin-bore alternative. This is a significant amount of material, which, if stockpiled 5 feet (1.5 m) high, would require approximately an acre of land at each portal's construction staging area. The material may need to be stockpiled at the site prior to transportation in order to dry, depending on its water content as it comes out of the TBM. Additionally, it may need to be stockpiled if the excavation rate exceeds the rate at which the muck can be hauled away. Hauling this amount of material per day would require hundreds of truck trips per portal, if trucks are to be used to haul the material. Coordinating all of the traffic in and out of the construction staging areas at the portals will be challenging. Potential disposal sites accessible by heavy rail are also under consideration.

However, both the north and south portal areas have direct freeway access, so this will mitigate the impacts of trucks on the surrounding communities and allow the contractor flexibility in its hauling operations. Additionally, at the south portal, there are existing heavy rail tracks adjacent to the land available for the construction staging area; the design team is researching the feasibility of using this rail for the hauling of excavated material.

### Potential Surface Settlement Trough

One of the variables that control the amount of expected surface settlement that occurs during a tunnel excavation project is the amount of ground loss while tunneling. The ground loss that occurs in soft ground formations is a function of several factors, including expected ground conditions, presence of groundwater, construction means and methods, and overall workmanship. Ground loss is often reported as the percentage of ground lost in the excavation, or the volume loss. Ground loss during excavation is typically caused by a combination of three general sources: face losses, shield losses, and tail losses. Of particular concern for the mega TBMs is the large annular gap between the diameter excavated by the TBM and the segmental lining extrados. This gap is expected to be on the order of 8 to 10 inches (203 to

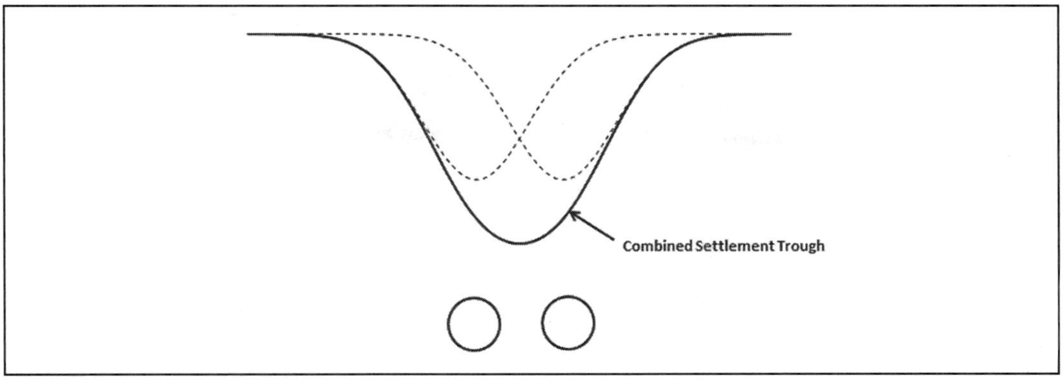


**Figure 6. Typical surface settlement above two tunnels**

254 mm) for a tunnel as large as the expected Freeway Tunnel. This gap occurs because of the shield geometry (assuming a tapered shield is used), large tail shield thickness, and the gap between the segment extrados and the inside diameter of the shield, and is typical in a large-diameter TBM. Shield and tail losses could therefore be the most significant source of volume losses for the Freeway Tunnel and must be accounted for.

Another consideration for this mega tunnel project is that the settlement troughs for each of the twin-bore tunnels will be additive. The total vertical ground movement caused by two tunnels is the sum of the ground movements caused by each individual tunnel, assuming the ground movement associated with each bore is independent of the other and can be superposed to estimate the combined trough due to both tunnels. Figure 6 shows a typical surface settlement trough above two tunnels. Depending on the shape of the troughs and the distance separating the two tunnels (pillar width), the superposition can amplify the maximum settlement seen. Additionally, for twin-bore mega tunnels, the zone of influence for the combined tunnels can be quite wide because of the geometry of the combined trough; however the width over which settlement is greater than 0.25 inches would be much less.

In subsequent phases of the study, more detailed stages of evaluation will be performed, which will include a structure-specific analysis to better understand each structure's response to the excavation-induced ground movements. If deemed necessary, the anticipated methodology for building protection will be to use compensation grouting, which involves carefully controlled injection of grout between underground excavations and structures requiring protection from settlement. With active monitoring, proper TBM control, and appropriate mitigation measures where necessary, surface settlement impacts to existing structures and utilities can be controlled.

## STUDY STATUS

The study team completed the Alternatives Analysis phase of the project in early 2013. The analysis recommended the four multi-modal alternatives mentioned herein as well as a No-Build alternative, which are currently being considered for the draft environmental document. Refinement as well as technical studies of all multimodal alternatives (LRT, BRT, TSM/TDM, Freeway Tunnel), with appropriate mitigation measures, continued into early 2014. It is expected that the draft environmental document will be circulated in 2014 and that the final environmental document and Record of Decision will be issued in 2015.

## ACKNOWLEDGMENTS

The authors would like to thank their respective employers for permission to publish this paper. Special thanks are also due to others on the team for their help with this paper including M. Crow, F. Quon and C. Govan (Metro); R. Raveendra (CH2M HILL), and S. Klein and K. Quinn (Jacobs Associates).

## REFERENCES

Albino, J.H., G. Roy, and J.E. Monsees. 1999. Crossing the Hollywood Fault using special seismic section. *In Rapid Excavation and Tunneling Conference Proceedings*. Littleton, CO: Society for Mining, Metallurgy, and Exploration, Inc., ASCE.

TunnelTalk. 2012. Herrenknecht News Release: Giant Sparvo EPBM Completes First Drive. August.

# Taking Safety to a Hole New Level—Making of the Ship Grant Video on Tunnel Training

**Anita Johnson**
Sound Transit

**Mandi Kime**
Associated General Contractors of Washington

With tunneling on the rise in the US, the inherent uncertainties and dangers of underground construction makes it imperative to have a comprehensive applicable training program for the workers and the visitors at a tunnel construction site. Available resources for awareness-level training in underground construction work are practically nonexistent or outdated. The last video addressing underground safety was produced 20 years ago for the mining industry and is not freely available to the public. The authors, being involved in the construction and tunneling industry for almost two decades, realized this lack of training resources and went through the Safety & Health Investment Projects grant process to prepare the training video.

This paper describes the author's goals for the training video, their approach to making the video and the challenges they faced while preparing the video. The paper also discusses the authors future outreach plans to spread their message of safety in underground construction through continuous interaction with the tunneling community.

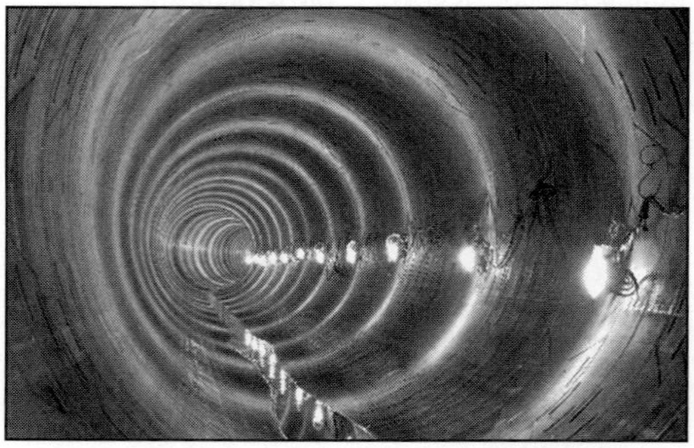

**Figure 1. Lewis and Clark, TBMs began excavation from the Nicolai Shaft. Lewis completed the journey north, under the river to the Confluent Shaft and then to the Swan Island Pump Station Shaft. Clark completed its journey and broke through on July 29th, 2005 at the Clay Street Shaft. This work is complete and the tunnel became 100% operational in December 2006.**

## INTRODUCTION

It's hard to write a technical piece about a topic that has so many other factors aside from the technical ones. This dynamic project involved teamwork, scheduling, overcoming challenges, and most of all a passion for people, and making sure they go home safety every night.

The vision for our H.O.L.E. (Hazard Observation & Labor Education) video project started over 10 years ago, inspired by the Portland, Oregon, West Side Combined Sewer Overflow (CSO) tunnel project. It was rapidly identified as a need after conducting multiple project orientations, creating several tiresome PowerPoint presentations, and after adlibbing to hundreds of visitors. It was determined that there was a need for a relevant and updated method for training tunnel workers. This training video was created as a result of playing 20 year old VHS tapes to hundreds of workers and visitors over and over again. It's available today because in 2003 a 20 year old VHS tape was the only means of audio/visual tunnel training available. At that time and for many years after, every contractor had to create their own program or system to train their workforce because there was nothing available to buy or even use as a template. It was a long time problem that needs an answer and with new tunneling technology, new safety standards, and a rise in tunnel projects across the U.S. made it was apparent that the industry needed a tool to implement and expand their tunnel safety program and a resource for training both workers and visitors alike. We needed a product that everyone could use and tailor to their specific operation and site considerations.

## IN THE BEGINNING

Fast forward to 2011, the State of Washington was in the middle of its third soft ground TBM tunnel

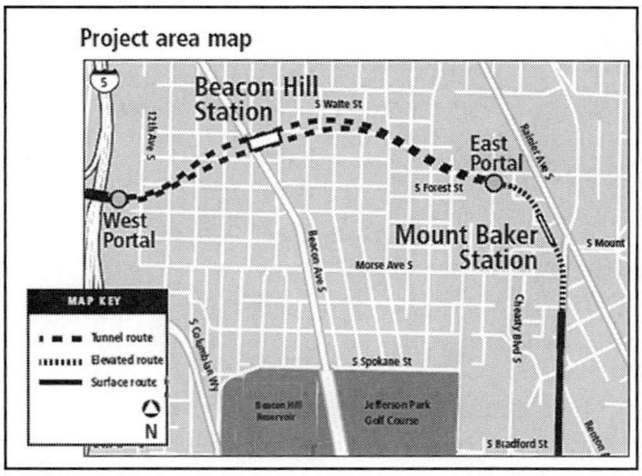


**Figure 2. Construction of the station began in March 2005 and was completed in July 2009. Obayashi Corporation was the general contractor and twin tunnels were excavated by a TBM named the "Emerald Mole."**

project (as of late) with the fourth and biggest on its way. Washington State started in 2005 with Sound Transit's Central Link (Beacon Hill Tunnel), then moved into the King County's Bright Water Treatment System in 2006, and was currently working with Sound Transit's University Link Light Rail Project. The U-Link project was where the majority of the video was filmed with the primary goal of being able to impact the next local project, The Alaskan Way Viaduct Replacement Tunnel (currently the largest TBM in North America).

In October 2011 funds were made available through the State of Washington Safety and Health Investment Projects (SHIP) grant for new and innovative safety projects, so we knew right away that we had an answer.

With the gap in education resources targeted towards tunnel training and with the incident and fatality rates on the rise, we now had a way to both solve our problem and educate workers. At that time, awareness level training in underground construction work were nonexistent and Private-sector mining fatalities were up by 74 percent in 2010, with an increase from 99 cases in 2009 to 172 cases in 2010.

## THE SHIP PROGRAM

The SHIP grant is Washington State's Labor and Industries ***S****afety and* ***H****ealth* ***I****nvestment* ***P****rojects* grant program. The program funds two types of projects.

- Safety and Health, which promotes safe and healthy work practices; and
- Return to Work, which supports projects that assists injured to return to work sooner.

**Figure 3. University Link Light Rail Tunnel, (UW to Capitol Hill)—NB TBM "Balto"—named after a Siberian Husky sled dog who led his team on the final leg of the 1925 serum run to Nome, Traylor Brothers-Frontier Kemper was the General Contractor.**

The SHIP grant program funds innovation and research projects.

All products developed as a result of a SHIP grant are available to the public and will be accessible for download from the SHIP website once the projects are completed. The SHIP grant program gets funded through the Washington State Department of Labor & Industries' Workers Compensation Medical Aid Fund. Of the 31 applicants who submitted project proposals in that SHIP Grant funding cycle, only 9 were selected for funding, including our project, subsequently the largest project funded.

## PROJECT ROLES

The Associated General Contractors (AGC) of Washington, Northwest Laborers-Employers Training Trust Fund (NWLETT), Integrity Safety Services (ISS) along with Anita Johnson came together to create the team of professionals that put the entire package together to help bridge the gap for training in a much needed area of construction.

Mrs. Kime and her team come with 31 years of construction experience. The mission of the AGC Safety Services Department is to help contractors take the work out of safety and regulatory compliance. We work to collaborate and share best practices within the Construction Safety community. The motto of the AGC is: Skill, Integrity & Responsibility. With that we aim to make having a safer work place easier for contractors to accomplish, while supporting employee empowerment, enhancing our environments, and community involvement. Mandi has served as the Director of Safety at AGC for over 6 years and worked with AGC for a combined 10 years. She has been a

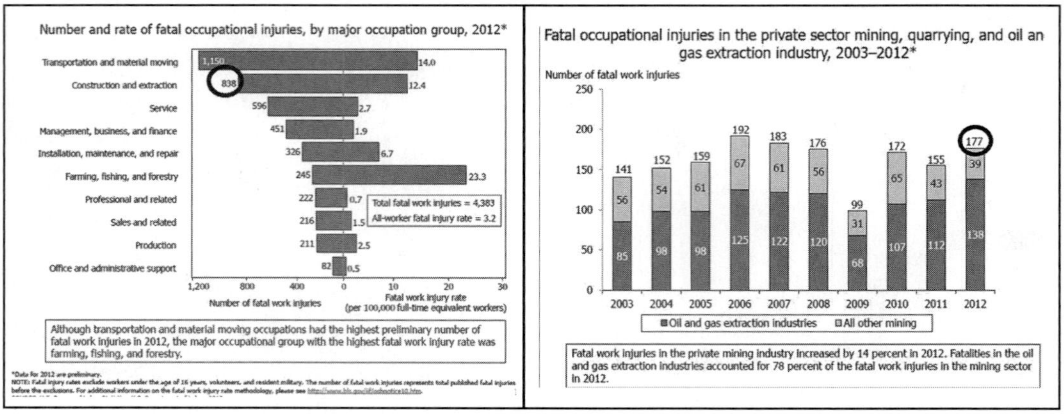


**Figures 4 and 5. U.S. Bureau of Labor Statistics, U.S. Department of Labor, 2013**

presenter at AGG of America Safety Committee Events, American Society of Safety Engineers Professional Development Meetings, multiple Washington Governor's Industrial Safety and Health Conferences, and every Washington Construction Safety Day events. She has been featured in 3 national publications ( Safety +Health, Engineering News Record, and Constructor Magazine) and several local trade publications as well as TV and Radio. She serves on various industry collaboration groups and regulatory committees in Washington, her connection to the Construction Safety Community was key to bringing collaborators to the table and implementation of the final products.

Mr. Warren (along with the training trust) brings an immense amount of practicality and applicability with 37 years in the construction industry: 12 years in the field and 26 years with the Training Program (14 of those as the director. The NWLETT's mission is to provide state-of-the-art construction training to the men and women in our region.

Integrity Safety Services (ISS) is a team of dedicated safety professionals with a passion to develop and implement workplace safety, corporate loss control, and regulatory compliance solutions. They bring 65 years of industry knowledge and 3 years' experience as a grant partner or grant manager. ISS excels in worker training, jobsite safety inspections, jobsite safety staffing, environmental testing, and recently added drug screening to their services. In addition, they can provide videography services through Integrity Productions, a division of Integrity Safety Services. Their primary contribution was videography as well as insight from previous grant projects to streamline our process and procedure.

Ms. Johnson has 22 years experiences in Health & Safety with 11 years specific to tunneling and worked on the 10 slurry machine used in the US (West Side CSO Tunnel- Portland, OR). She also coordinated the first hyperbaric team on a slurry machine in the US (144 interventions w/o incident) and was a 2010 ACCSH Meeting Presenter on Compressed Air. With her experience as a dedicated safety professional and her vision to implement a change in tunnel safety it was a natural fit for her to be a part of our groundbreaking project.

## UNDERGROUND STATISTICS

With tunneling on the rise in the U.S., the inherent uncertainties and dangers of underground construction makes it's imperative to have a comprehensive applicable training program for the workers and visitors at a tunnel construction site. Our vision was to create a video based Tunnel Safety Training & Hazard Awareness Program to reach all trades involved in underground construction. The need was there and the projects were all using training products that didn't address the current state of the industry. This project was created to help increase awareness and reduce incidents in underground construction by giving the industry a tool for implementing a critical component of their tunnel safety program.

In 2012, there were 838 fatal on-the-job injuries to construction workers—more than in any other single industry sector and nearly one out of every five work-related deaths in the U.S. that year. And Private sector mining fatalities were up 79% percent (2012) from 99 cases (2009).

We saw the need to raise awareness and create training consistency in the underground industry and be the catalyst to change one of the oldest occupations. The probability of injuries from an accident in tunnels is about the same as any other constructions site. However, if an accident does happen in a tunnel, the severity of injuries sustained can be significantly higher than that of other projects, and the rescue of injured workers can be vastly more complex. In some cases you could be two to three miles down the tunnel and *utilizing resources from several miles away*. It is imperative that your immediate workforce and emergency services (for example) are trained and communicating at the same level to best service workers. It is also essential to reach out to all trades in the industry to establish a consistent baseline for safety and hazard recognition. Before you can control hazards you need to know what the hazards are. The real time approach depicted in the H.O.L.E. video is intended to help repeat the same message needed to assist in those circumstances. This type of training is imperative in the underground industry because workers can be afraid of unfamiliar settings.

Since underground work is so unique we knew we needed to bridge the gap for workers between the dated awareness material being used in the industry and the current technology and practices in place today. What workers were being shown as examples of the underground environment was vastly different from today's tunnels, lighting, air quality, access, and general working conditions have improved thanks to technology. But today's technology was not reflected in training materials being used by our industry.

Awareness and education can help decrease that fear and reduce those statistics. Overall, efforts in occupational health and safety must aim to prevent industrial incidents and recognize the connection between worker health and safety, and the workplace. In other words, occupational health and safety encompasses the social, mental and physical well-being of workers that is the "whole person."

Successful occupational health and safety programs require the collaboration and participation of both employers and workers together to create a synergistic approach in keeping our workers safe. This training program used interviews and input from people at all levels of project participation, from visitor to project management to project ownership.

## COMMITTEE REPRESENTATION

At the start of the project we gathered all interested parties in an effort to help establish a baseline for an overall goal of a first of its kind, tunnel safety video. We aimed to create an extraordinary, interactive, and practical product that we all personally identified with as our grant committee had a collective 90years of construction experience, but the real key was to involve those who we were targeting with this innovative training.

This video project was a collaborative effort of multiple individuals using our collective knowledge to impact all workers in the field of underground construction and in an effort to pull from a vast majority of the industry to adequately represent end users,

**Figure 6. Every person in the tunnel leaves a numbered brass at the entrance and carries a matching brass on their person. A quick glance at the board can tell you who and how many people are underground at any given time.**

we were fortunate enough to pair with the following groups, positions, and/or areas of influence:

- Owners (project)
- Construction Management
- Project Management
- Contractors
- Unions
- Management
- Safety Personnel
- Workers
- Emergency Management Personnel
- Regulatory Agencies

As a result we identified multiple areas of importance, but we tried to cover those commonly encountered. In addition we wanted to make sure and portray the human factor into our training and not approach it from a compliance aspect. We ultimately came up with 15 topics (see Outputs/ products) that we felt would address our needs and the needs of the workers.

One common goal of the planning committee was to build ownership in safety amongst the workforce as that is key to culture shift. It was best summarized by Anita, "People learn when they can relate, if they relate they'll take ownership, the more people who have "ownership," the quicker it spreads and before you know it the industry has changed for the better."

## CONTENT AND FORMAT

We documented the primary hazards and safety solutions for the tunneling industry to help provide workers and visitors with a clear overview that will apply to all underground construction projects. By definition, "a 'hazard' is something that in itself may cause harm or injury" (Rajagopal, 2009).

Our "HOLE" approach was to increase the awareness of underground hazards and to decrease injuries or incidents in the underground industry. We looked at statistics, past incidents, and current trends. Some of the best information we collected was the interview snippets from our filming days. While many of these snippets did not make it into the final DVD, the content or message of them had impacts on the content and layout of our final product. Those were the factors that led us to the final list of sections depicted on the video. The list below is an overview of what you'll expect to see after watching our H.O.L.E. Tunnel Training Video.

- Access & Accountability
  - Signing in
  - Brass in/out
- PPE & Lighting
  - Personal
  - Task
  - Temporary
- Communications
  - Mine Phone
  - Leaky Feeder
  - Hard Line
- Air, Gas, & Ventilation
  - Hand held
  - Machine mounted
- Fire Prevention & Protection
  - Fire Extinguishers
  - Water deluge
  - Fire department hose connections
- Housekeeping
  - Walkways
  - Trailing gear
  - Ramps
  - Ladders
  - Rail
- Electrical Safety
- Flood Control
  - Pumps
  - Monitors
- Tunnel Rescue
  - Fire Department Coordination
  - First Responder Training
- Mechanical
  - Segment Feeder
  - Loci
  - Conveyor
  - Machines
- Hyperbaric
  - Basic Knowledge
  - History
  - Application
- Segment Erector Operations
- Stored Energy/ Lock out
  - Pressure
  - Hazard Recognition

## HOW TO USE PRODUCTS

The video was intended to target a variety of users to augment their already established tunnel safety training program and was created for workers and visitors alike. The total video is 60 minutes in length with the intent of being used in a variety of ways. It can be used as a standalone session for worker orientation, a standalone session for visitor orientation, or as multiple mini training (retraining) sessions. We wanted it to have availability to be used by all areas of the underground industry and to be generally specific. You might wonder what generally specific" means? It was our means of covering all tunnel related items with just enough detail to be specific for the industry, but general enough to be used worldwide. Any company, regulatory agency, owner, worker, or even physician (hyperbaric) could utilize the video and incorporate it into their corporate policy or project specific rules and regulations.

The video has a 50 minute worker portion, a 10-minute visitor portion, and 13 separate sections with the ability to be played individually for worker training or retraining. Along with the video itself, the full package includes:

- Orientation checklist (fillable template) to show which sections of the video were reviewed and any added site specific considerations covered during the orientation
- Worker acknowledgment form to document a worker's commitment to adhere to the video guidelines and site specific considerations
- Wallet reminder card to share the most common tunneling hazards and protective measures
- Hard hat sticker for the contractor to use in verification "at a glance " of training for those onsite
- Safety Poster to be placed on the jobsite in various locations as a reminder of the primary hazards and expected protective measures

## PROJECT CHALLENGES

This project was quite a learning process for the whole team. Our team had to overcome staffing changes, site clearance and access issues and schedule impacts. Some lessons learned include the following.

Administration. While the SHIP office worked hard to streamline the reporting process, it is still a

very large and time consuming part of the project and in the future, we will likely assign more hours to clerical assignments. It takes effort and organizational ability to manage the contractors working on the project, and to keep track of all the activities for reporting purposes, especially when you have so many moving parts to a project.

Schedule. We experienced several types of unavoidable delays on our project. The original funding date that the project was based on was 3 months before the funding was actually released. Further complicating the schedule, our project manager had to take leave during a portion of the grant, and we had to transition to a temporary project manager causing our grant project to start later than we anticipated. This impacted our ability to get footage on local projects in the order and manner in which we had planned. Also, our final products were held up in production at the replication facility. We learned you have to be able to adapt when the schedule is impacted. To overcome the project manager situation, we had to do more phone conferencing/email communicating. To overcome the late project start, we ended up needing to travel to out-of-state projects to procure our remaining shot list and interviews. While, it wasn't an original part of our proposal, our team acted quickly to assimilate a "plan B" and act on that plan before further delays could derail the project. While it was a lesson in procedure, and ultimately not included in our grant budget, our grant team felt passionately that this travel was crucial to the ultimate success of the project, our schedule and our goal to have the products ready before the Viaduct Tunnel-Seattle project was too far underway to use our products. As such the grant partners agreed to donate these expenses to the project. This was difficult because we needed to obtain site access from companies and project owners who were not originally on our project committee and then had to educate them on our project, how it would benefit them, and what we needed to film. We then had to travel and access the sites without disturbing the work going on or creating hazards/ distractions with our film crew. While the film crew was in transit to the first out-of-state project, a fatal accident occurred on that site. We were unable to access the project moving forward and thus the need for the second out-of-state site was created. The tragic loss of a worker during the course of creating this video project is a testament to the fact that more work needs to be done in worker safety and health in this industry. We dedicate our efforts on this project to that worker specifically as well as the many who lost their lives and been injured in this line of work. While we did not intend to have a multi-state project, the end result is a more thorough product because it features various sites and practices. During the development of this project we learned how excited the tunneling/ underground construction community was for our final product. To overcome issues we had getting the products reproduced fast enough for the demand, we posted all materials online and gave interested parties a way to access our products until we could get the individual copies sent directly.

Teamwork. We had many people coming together to make this a success. We found the process went best when we were all communicating in person, but that was a challenge with the complex schedules everyone on our team had. A future consideration would be to have monthly meetings in person, and all other meetings on the phone or email in order to maintain cohesiveness of the team.

## CONCLUSION

Our goals for this project was to have a clear, current and impactful training program available for use locally on the Seattle Viaduct Tunnel Project and to impact the tunneling industry as a "HOLE" in the future. We have achieved this as well as a commitment for our program to be used for all who access the tunnel. This program will impact hundreds of workers here in states, in Canada, and internationally as well. We have received requests from employees, employers, regulatory agencies, labor organizations, and associations/conferences. It's apparent that the industry is eager to use our program in all aspects of tunneling and that this proves the broad appeal for a cutting edge training program on a growing industry.

We want to thank the many individuals and organizations who contributed valuable expertise to this project including:

- Seattle Fire Department
- NW Laborer's-Employees Training Trust
- Sound Transit
- HealthForce
- Seattle Tunnel Partners
- Frank Coluccio Construction
- Frontier-Kemper Constructors
- JayDee Construction
- Seattle Tunnel Partners
- Traylor Brothers
- Barnard Construction Company
- James W Fowler
- Washington State Department of Transportation
- Seattle Tunnel and Rail Team

# University-Industry Collaboration in the Underground Construction and Tunneling Field

**Bryan Walter**
Colorado School of Mines

**ABSTRACT:** The underground construction and tunneling (UC&T) industry is rapidly expanding to meet global needs.[1] This expansion is a direct response for the pursuit of major infrastructure improvements in areas such as public transportation, water supply and storage, wastewater transmission and mining. Examples of this growth include the East Side Access project in NYC[2], the Chicago Deep Tunnel project[3], and the Alaskan Way Viaduct Replacement project in Seattle, Washington.[4] As with any rapidly expanding industrial field, a primary concern is meeting the demand for qualified and trained personnel. Any technical industry looks to the academic community to fulfill this need, and the UC&T industry is no exception. Consistent with the literature [5], [6], [7], [8], this paper presents findings from a series of interviews with both industry and academic players suggests that academia is not meeting this demand. Therefore, the focus of this paper is to explore the barriers and opportunities that affect prospects for increasing collaboration between academia and industry in the field of UC&T.

Investigation of the relationship between universities and the UC&T industry indicates potential problems. Analysis of how new graduates are hired, how the industry fosters academic contributions and how there are a limited number of venues that encourage student participation suggests that the lack of collaboration between academia and industry is failing to create adequate pathways to provide trained personnel and is potentially hindering innovation in the United States UC&T industry. As such, the goal of this paper is to investigate the barriers and opportunities that drive potential collaboration between academia and industry in the field of UC&T.

## UNIVERSITIES OFFERING COURSES IN UC&T

Because of the need for specialized training, the UC&T industry relies on universities to train incoming engineers, but few programs exist to fill this need. Currently there are 19 U.S. universities that offer courses in UC&T.[9] In addition, of the 19 schools, only one offers a graduate degree in UC&T, the Colorado School of Mines.

To help gain perspective as to why the number of schools that offer academic coursework on the subject is significant, a similar comparison can be found in the specialty field of explosive engineering. Explosive engineering is commonly practiced in the UC&T field, and therefore a representative analogy. To investigate this comparison, we can look to the number of ABET accredited programs within these two specialties. The explosive specialty is typically offered within a mining engineering program, and the UC&T specialty is generally offered within a civil and/or mining engineering program. For the explosive specialty, 64% of the accredited mining programs (14 in total) offer either courses or a minor related to explosive engineering. A similar comparison of the UC&T specialty reveals that less that 8% of the ABET accredited civil (231 in total) and mining (14 in total) programs offer courses related to this field. This shortage of academic opportunities, paralleled in the mining sector and discussed by McCarter [8], could also lead to an insufficient number of advanced degrees and ultimately a deficit in the number of faculty available to teach courses related to UC&T.

## LESSONS FROM INDUSTRY-UNIVERSITY COLLABORATIONS

Cases of successful and unsuccessful collaborations between academia and industry provide insight into the nature of constructive collaboration in the UC&T field. For example, a study by Gray et al[10] explored the sustainability of industry/university cooperative research centers (I/UCRCs). The I/UCRC was founded in the 1980s by the National Science Foundation (NSF) with the express purpose of fostering industry-academic relationships.[11] Under the initial program there were over fifty research centers created,with roughly two-thirds still in operation today. Of these, two notably successful programs are the Advanced Steel Processing and Products Research Center (ASPPRC) at the Colorado School of Mines (CSM) and the Center for Advanced Communication (CAC)at Villanova University. Both institutions have developed world-class research centers that have demonstrated the value of industry-academic consortiums, and their industry members have reaped the benefits. As stated in the report by Gray et al. [10], three areas that industry members benefit from are: (1) direct access to new knowledge[12], often in the form of a publication created by the center, (2) intimate access to the pool of students who helped create these advancements (typically as part of their thesis or dissertations), and (3) access to expensive laboratory equipment procured by the center.

In contrast, there are a number of I/UCRCs that have not succeeded in sustaining an industry/academic consortium. Several reasons, given by Gray et al. [13], for these failures are the following: (1) the lack of a PhD track program to sustain research, (2) non-tenured faculty in the center's leadership positions, (3) ineffective succession planning for the loss of key center personnel, (4) industry's reluctance to share or disseminate proprietary research, and (5) a lack of institution support from the hosting university.

To illustrate the importance of these condemning mechanisms, a closer look shows how they are used favorably by the two successful programs. First, Villanova University was able to expand their research capacity and visibility by adding a PhD program in electrical engineering. This increased the center's manpower and research sustainability, but came at a cost that the center could not initially afford. The institution, in support of the program, covered the startup costs, which in turn created synergy between the center and the university. Second, the ASPPRC at CSM has successfully transitioned through three generations of faculty leadership and "…is a model of continuity of leadership."[10] This foresight has played a large role in the center's success and its significant contribution to the nearly $60 million in CSM's 2012 research expenditures. In addition, ASPPRC produces 4–5 Masters degrees and, 2–3 PhD's every year, and has been responsible for over 385 technical publications.[10]

In contrast, there is an extensive body of literature that explores the capitalization of university research. Berman [14] and Washburn [15] explore this form of university-industry collaboration in their two separate texts; however, neither discuss nor offer a framework that fits the UC&T field. The UC&T industry is different than other technology sectors, such as nanotechnology, biotechnology or material science, where the university research is driving the innovation and creating market places. In the UC&T sector, the societal need to construct a tunnel often is created because of a challenging physical environment or social need.[16] The specific attributes of that environment are what drive the innovation: for example, constructing an underground subway to improve transportation in a congested city or solving a city's combined sewer overflow (CSO) to reduce contaminating water supplies.

## VENUES FOR COLLABORATION IN THE UC&T FIELD

To better understand the barriers and opportunities of collaboration the venues in which the industry participates in, which are conferences and short courses, will be explored.

### Conferences

In the UC&T field, conferences are highly attended venues; accordingly their content and attendance are representative metrics of the class of participants. The two prominent conferences held in the Unites States are the North American Tunneling Conference (NAT) and the Rapid Excavation and Tunneling Conference (RETC). Two areas that are discussed are the number of academic attendees and the number of academic publications at both NAT and RETC.

One indication of the disproportionate relationship between industry and academia is the limited number of academic participants, in comparison to industry participants, at the two major U.S. conferences. For example, at RETC 2013 the percentage of academic participants was 2.7% of 885 attendees and at NAT 2012 the percentage was 1.9% of 918 attendees. It is suspect that this is not the desired ratio for two reasons. First, the industry is supporting student participation and attendance via scholarships, [17] and second, at many of the conferences there are special committee meetings aimed at increasing student involvement. Despite these efforts, if a comparison between the 19 schools and the attendees is made, only 42% of the 19 universities had participants at RETC 2013 and 32% of the 19 universities had participants at NAT 2012.

These attendance percentages differ greatly to that of conference venues outside the UC&T community. By comparison the American Rock Mechanics Association (ARMA), an organization that hosts a technical symposium that focuses on rock mechanics, rock engineering and geomechanics, had 38% of their 632 attendees have an academic affiliation in 2013. Yet another example is the American Institute of Aeronautics (AIAA) that hosts the Structures, Structural Dynamics, and Materials (SDM) conference, which in 2013 had 43% of its 530 attendees have an academic affiliation.

Another indication of the poor university-industry relationship is the low number of academic publications at NAT and RETC. The conferences alternate years and typically have an attendance of over 1000 persons.[18] Over the last thirteen years there have been 1228 papers written and less than 6% (69 in total) of them included an author from an

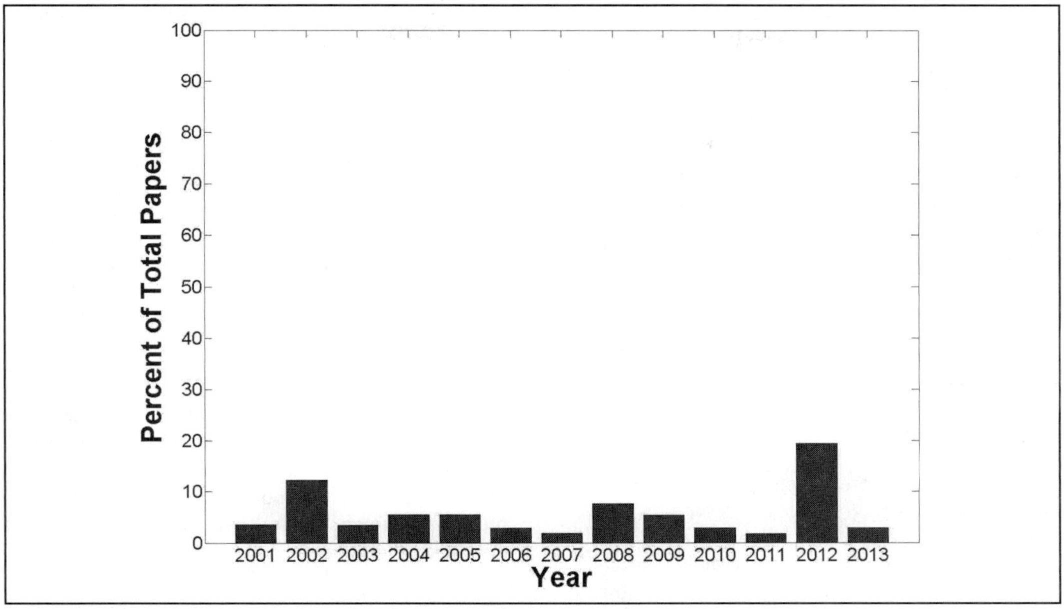


**Figure 1. Histogram of the percent of papers that had an academic contributor, where the data was collected from the North American Tunneling Conference (NAT) for the year: 2002, 2004, 2006, 2008, 2010, 2012 and from the Rapid Excavation and Tunneling Conference (RETC) for the year: 2001, 2003, 2005, 2007, 2009, 2011, 2013**

academic institution. A histogram of the collected results can be seen in Figure 1, where approx. 100 papers were submitted per conference.

As shown by Figure 1, the general trend was that less than 10% of the papers contained an academic author except for 2002 and 2012. In 2012, twenty academic papers were published, over twice the average of the other twelve years. This is significant because eight of the twenty were submitted by two graduate students pursuing research in the UC&T arena. Additionally, six of these papers were directly related to work completed during UC&T internships. This dramatic increase in publication shows the impact that academic inclusion had on the industry by nearly doubling the number of submitted papers and posing answers to challenging questions within the industry. Conversely in the 2002 year, it was also the lowest total number of papers of the thirteen years in the study.

As learned from attending these conferences and reviewing their mission statements, the purpose of the gatherings is for industry to get together and "learn about the most recent advances and breakthroughs in this unique field."[19] As this statement suggests, these conferences provide a venue where the industry gets together, shares knowledge, and furthers the industry, albeit "with the same old faces," as stated by a conference attendee from industry. So, for an industry seeking to encourage new recruits, it appears to be failing to attract the richest pools of candidates.

### Short Courses

Another forum that the UC&T industry uses to disseminate emerging technologies is short courses. These multi-day training sessions offer an "intensive course presented by a panel of...experts." [20] Often, these courses are hosted in conjunction with a major conference, but several are sole venues, such as the tunneling courses offered through CSM's Special Programs and Continuing Education (SPACE) program. These settings offer great opportunities to network, because representatives from all of the major UC&T organizations (owners, engineers, contractors, equipment manufacturers and consultants) typically participate. However, academics are almost completely devoid of attendance. For example, at CSM's 2011 and 2012 tunneling short courses, there were only 2 of 136 and 2 of 125 registered participants, in respective years, that possessed an academic affiliation.

The lack of participation by academians in these short courses is a missed opportunity. This is misfortunate for two reasons. The first is that these courses offer condensed training in relevant UC&T topics that would be difficult to reproduce in an academic course, even over multiple semesters. The primary reason it would be difficult to reproduce a similar academic course is the sheer number of topics, over thirty at CSM's 2013 tunneling short course, and the diverse backgrounds of each of the speakers. Secondly, the setting is ripe for learning about opportunities to collaborate, because the attendees are there to learn about UC&T methodologies, many of which potentially posses short coming. These deficiencies and the discussion of their resolutions present a pathway for collaborative research between industry and academia.

## METHODOLOGY

To further explore the barriers and opportunities that influence collaboration between industry and academia, interviews were conducted that targeted persons in these communities to gain their alternative perspectives on collaboration in the UC&T field. The interviewees were selected from both academia and industry in order to gain alternative perspectives. Well-known candidates in academia and industry were contacted. Then, at the conclusion of their interviews, the candidates were asked to recommend other prospects, and as expected, the list of potential interviewees snowballs from there. The interviews were conducted either face-to-face or via the telephone, vary in length up to nearly one hour. The interviews were initially conducted by asking several common questions and then moving to a more open discussion. For the remainder of this paper, the different interviewees will be coded as either industry subject X or academia subject Y. This designation is important in safeguarding the privacy of the subjects, as required by the Institutional Review Board exemption.

The first few interviews made it clear that the industry is "fragmented," as stated by academic subject #4, into the following five groups: (1) owners, who take ownership of the finished product(e.g., Denver RTD, Indianapolis Public Works), (2) consultants, who offer niche services like instrumentation or equipment audits (e.g., GEO-Instruments, Snyder Engineering) (3) designers, such as Jacobs Associates, MWH, etc., who might design the tunnel or provide project oversight, (4) contractors, such as Kiewit, Jay Dee, etc., who are responsible for physically building the underground structure or its excavation and (5) manufacturers, such as Robbins, AtlasCopco, etc., who are the entities building the equipment or machines that serve the industry. In light of this fragmentation, the scope of questions were broadened to address each of these groups in this interview process in order to gain a more representative understanding of the industry. Interview requests were sent to 28 persons, and the final pool consisted of 11 interviewees, 6 from academia and 5 from industry. The remainder of this paper draws upon these interviews to explore the barriers and opportunities to expanding university-industry collaboration.

## BARRIERS TO COLLABORATION

There are a number of roadblocks that inhibit collaboration in the UC&T field. Two of the more prominent obstructions are the contrasting incentives between industry and academia and the decline in government funding to support public infrastructure improvements, which is the bulk of UC&T work.

### Contrasting Incentives

A general sentiment from industry experts was that academia wants to produce publications and industry wants to increase profits, which are clearly different goals and therefore a barrier to collaboration. An interesting comment offered by industry subject #1 was that the industry needs solutions that are "immediately deployable" and suggested that results shouldn't coincide with the end of a semester. Another perceived distinction, shared by multiple interviewees, was that academic research needs to be more applied. In addition, during these interviews, there was often a strong biased overtone between the two groups. This implication is best summarized by two quotes, one from an academic and the other from an industry representative. Academic subject #6 referred to persons in industry as a bunch of "grey beards," and an industry subject #5 referred to academic professors as "modelers." These are both strong statements; nevertheless, they illustrate contrasting perceptions, real or perceived, that should be explored while investigating the university/industry relationship in the UC&T industry.

Additionally, several interviewees suggested that the research being done by industry is proprietary and being done primarily by equipment manufacturers, often being of a more applied nature. Industry subject #1 offered that industry research consisted of technologies that were "90% proven and 10% un-proven," indicating that innovation in the industry is conservative and being developed through small, incremental changes. This conservative and proprietary nature suggests that industry is supporting the majority of the research in the UC&T field, but not including academia in the process.

Another barrier to industry-academic partnerships is the speed at which construction projects need to be completed. This hurdle is exacerbated by the ever changing conditions after construction commences. Generally, UC&T projects are designed from geotechnical data reports (GDRs), which typically measure significantly less than 1% of the geology being impacted. Therefore, during construction, the design is often modified to meet the true geological condition. As can be imagined, this requires the industry to be both dynamic and flexible, two words that wreak havoc on a detailed and structured research plan. All is not lost, however, and much can be learned from these dynamic site conditions. In fact, this topic is the most common thesis of papers presented at tunneling conferences, where titles like "Lessons Learned from 130 Years of Tunneling in Seattle's Complex Soil" and "Innovative Approach to Muck Disposal and Ventilation During Drill-and-Blast Operations in a Densely Populated Urban Environment" are two examples of papers that were presented at RETC 2013. If these challenges were presented retrospectively, as academic research opportunities, then students (future employees) would get exposed to the industry and potential scientific discoveries could be realized to solve common industry problems.

An important industry example and focus of RETC 2013 [21] was a number of papers discussing soil conditioning. [22],[23], [24] As discussed in these papers, soil conditioning offers great benefits to the industry, but there is room for improvement. In this example, industry does not have the time to perform extensive studies; however, retrospectively allowing academia to mull over similar case studies (data sets) would help familiarize the students with the industry and produce more complete answers to the posed problems.

### Funding

Due to shrinking federal and municipal budgets, the UC&T industry has become fiercely competitive and introverted to minimize expenditures. The general opinion of multiple interviewees was that very little research is currently being done in the industry. Academic Subject #3 answered that "the margins are tight these days," and government funding has dried up. Multiple candidates described the 1970s and 1980s as a period when much of the industry's research was solidified. The US government was funding research through the Bureau of Mines for projects like the Yucca Mountain Project and using TBMs to create egress tunnels for the rapid deployment of missiles.[25] The Bureau of Mines has since been closed, and alternative sources of research funding have not yet to be forthcoming.

Alternatively, a \$5 million contribution was made to CSM's UC&T program. [26] One of the objectives of this donation was to develop an industry consortium and foster industry/academic collaboration in the UC&T industry. To date, this gift has revitalized CSMs semi-monthly UC&T seminars, facilitated multiple UC&T site visits, and has initiated the development of university/industry related research. While only time will tell, it appears that this funding, in concert with CSM's UC&T center, has made a significant step towards facilitating collaboration in the UC&T industry.

## PATHWAYS TO COLLABORATION

There are a number of pathways that foster collaboration between academia and industry in the UC&T field. Two of the more notable are internships and collaborative research projects that involve specialized testing equipment, often located at an academic institution.

### Internships

The majority of industry's interviewees stated that internships were the only way to obtain qualified persons, suggesting that internships are a primary vehicle for technical training, serving to enhance pathways of collaboration. The UC&T environment is typically very demanding, and requires personnel that can react quickly and improvise. Academic Subject #3 suggested, however, that this sort of dynamism is something that is difficult to teach in a classroom. Therefore, as a starting point, internships offer a gateway to train future UC&T personnel to its unique set of requirements, illuminating a pathway for future collaboration.

The original scope of this research did not include the collection of data on internship experiences from other students. However, having participated in two internships, one as an undergraduate and another as a graduate student, both experiences were valuable, but for different reasons. During the undergraduate internship there was a stronger focus on the "labor" component and a clear understanding that my purpose was to assist more senior employees. As a graduate student, I had more skills to offer and was able to contribute and even advise on several research projects. In fact, one of these investigations matured into my thesis topic, and as discussed previously, several others resulted in both conference and peer-reviewed academic journal papers.

An important aspect of internships, especially at the graduate level, is the dissemination of information. In my case, the employer (Jay Dee Contractors) has been very supportive in the publication of results obtained during and after the internship. This diffusion has afforded the inclusion of additional undergraduate and graduate students, consequentially diversifying the research and drawing more personnel to the UC&T industry. Yet, a fellow graduate student has been less fortunate in his internship endeavors. His work has remained proprietary, therefore limiting the bounds of collaboration. This is precisely one of the conclusions that was realized in the previously discussed findings, by Gray et al. in their sustainability study on I/UCRCs.

Nevertheless, the most beneficial aspects of my internships were establishing contact with industry members and carrying on the relationships after returning to school. Specifically, the associations developed during my graduate internship fostered pathways for collaboration in three ways: (1) it allowed me to grow my dissertation work, with industry support, (2) it created a network of experienced professional to query, and (3) it yielded a wealth of data and information, supplemental to my own research, that provided motivation for fellow students and faculty to develop additional research avenues.

### Collaborative Research

While there is very little research being done in the UC&T industry that is collaborative in nature, there are multiple research opportunities that could benefit from the synergy of academic-industry collaboration. Some of these opportunities arise from challenges facing the UC&T industry. Several examples are complex geology, surface settlement, and the requirements to tunnel under existing infrastructure. These obstacles provide opportunities for extensive industry-academic collaboration.

In particular, one historical example suggests a path for future collaboration. During the 1970s the Earth Mechanics Institute (EMI) was established at CSM. A major contribution of EMI was the construction of the Linear Cutting Machine (LCM), which was capable of performing full scale rock-cutting tests. The LCM was funded by both the NSF and industry. The results developed with this machine have both aided industry projects and facilitated academic research. Specifically, the LCM has been used to develop performance prediction models for many tunneling projects and has been the experimental source for multiple MSc and PhD degrees. This piece of equipment still provides valuable services to both the hard rock industry and academia today. However, there has been an increasing demand for underground construction to be performed in mixed soils, and a full scale test platform, like the LCM, does not exist for this medium. Developing a machine like this takes significant resources and should be developed jointly between industry and academia to ensure that both parties' objectives are met. At this juncture a university/industry common research goal, similar to the LCM but for mixed soil, has not been realized. However, such a venture might be beneficial to the university/industry relationship and contribute to advances in the industry.

## CONCLUSION

There are several barriers and opportunities affecting prospects for increasing collaboration between academia and industry in the field of UC&T. Several of these barriers are institutional, such as the low number of universities offering programs that support the UC&T field or the diminishing federal and municipal budgets. However, the majority of the hurdles are a result of dissimilar short term goals. In many cases, the long term goals are well aligned, but the frameworks are not in place to support collaboration. As this research has shown, there are several opportunities to bridge this gap through internships and collaborative research projects. However, the most obvious is for "U.S. companies…to be more supportive of their excellent university R&D systems," [27] and make an effort to include them in UC&T projects. This in turn will boost the number of individuals involved in the UC&T field and begin to close the loop on dissimilar goals by including academia in the process.

There are several directions that could be revised to further explore this research. The first would be to include more interviewees in the study. Specifically, obtaining more candidates from each of the identified industry segments (owners, consultants, designers, contractors, and manufacturers), twould offer a more holistic perspective. Second, the interview questions could be revised to probe the results obtained in this study. These revised questions, in conjunction with a larger interview pool, could be used to more specifically ascertain how the industry's segmentation either supports or blocks collaboration. Third, more statistical data could be collected, especially in regards to alternative UC&T venues, for both their attendance and publication ratios. Finally, a study of the affiliation of authors publishing in peer-reviewed journals, connected with UC&T research, could be conducted.

Additionally, there are other directions that could be exploited to further this research. The first would to become more active in the special committee meetings, typically held at conferences that are dedicated to promoting UC&T in academia. These sessions are generally not advertised, and their existence was learned through interviewees. Nevertheless, participating in these meeting would be beneficial. Additionally, the scope of this research was limited to activities in the United States, but the UC&T field is expanding globally. Further research could bolster this work to gain an international perspective on collaboration. Finally, interviews of recent interns, conducted on both the students and their employers, would offer tremendous insight into current collaboration pathways.

## REFERENCES


[1] "Tunnel Education at the Mile High Level," *Trenchless Technology*, pp. 11–12, Nov-2010.

[2] "Long Island Rail Road East Side Access," *Metropolitan Transportation Authority*, 2013. [Online]. Available: http://www.mta.info/capconstr/esas/. [Accessed: 07-Sep-2013].

[3] "Tunnel and Reservoir Plan (TARP)," *Metropolitan Water Reclamation District of Greater Chicago*, 2013. [Online]. Available: http://www.mwrd.org/irj/portal/anonymous/tarp. [Accessed: 07-Sep-2013].

[4] "Alaskan Way Viaduct Replacement Program," *Transportation, Washington State Department of*, 2013. [Online]. Available: http://www.wsdot.wa.gov/projects/Viaduct/.

[5] G. Brierley and A. Hawks, "Tunnel education in the United States," *Tunneling & Underground Construction*, no. June, pp. 18–21, 2010.

[6] P. Asadollahi and A. Zeytinci, "Sustainable development of infrastructures using underground spaces: role of academia," *ASEE Middle Atlantic Regional Conference*, no. Figure 1, 2011.

[7] J.K. Lemley, "Towards New Worlds in Tunneling: Advancing the 'Art of the Possible'," *Tunnelling and Underground Space Technology*, vol. 8, no. 1, pp. 3–6, Jan. 1993.

[8] M.K. McCarter, "Mining faculty in the United States: current status and sustainability," *Mining Engineering*, no. September, pp. 28–33, 2007.

[9] B. Walter, "Detecting Changing Geologic Conditions with Tunnel Boring Machines by Using Passive Vibration Measurements," Colorado School of Mines, 2013.

[10] D.O. Gray, L. Tornatzky, L. McGowen, and E. Sundrome, "Research Center Sustainability and Survival: Case Studies of Fidelity, Reinvention and Leadership of Industry/University Cooperative Research Centers," 2012.

[11] NSF, "Industry & University Cooperative Research Program (I/UCRC)," 2013. [Online]. Available: http://www.nsf.gov/eng/iip/iucrc/.

[12] D.O. Gray, "Making Team Science Better: Applying Improvement-Oriented Evaluation Principles to Evaluation of Cooperative Research Centers," *New Directions for Evaluation*, no. 118, pp. 73–87, 2008.

[13] D. Gray, E. Sundstrom, L.G. Tornatzky, and L. McGowen, "When Triple Helix unravels: A multi-case analysis of failures in industry-university cooperative research centres," *Industry and Higher Education*, vol. 25, no. 5, pp. 333–345, Oct. 2011.

[14] E.P. Berman, *Creating the Market University:How Academic Science Became an Economic Engine*. Princeton University Press, 2012, p. 265.

[15] J. Washburn, *University Inc.:The Corporate Corruption of Higher Education*. Basic Books, 2005, p. 328.

[16] G. West, *Innovation and the rise of the tunnelling industry*. Press Syndicate of the University of Cambridge, 1988, p. 355.

[17] "SME Student Scholarships," *Society for Mining, Metallurgy and Exploration (SME)*, 2013. [Online]. Available: http://www.smenet.org/scholarships/.

[18] SME, "RETC History: Attendance & Exhibitor Stats," 2013. [Online]. Available: http://www.retc.org/stats.cfm.

[19] M.A. Diponio and C. Dixon, "2013 Proceedings," in *RETC Proceededings*, 2013, p. 1350.

[20] "Pilot Tube Microtunneling/Microtunneling Short Course," *Colorado School of Mines*, 2013. [Online]. Available: http://csmspace.com/events/microtunnel/.

[21] S. Wallis and J. Green, "Technical delivery at RETC," *Tunnel Talk*, 2013. [Online]. Available: http://tunneltalk.com/RETC2013-roundup-27June2013-A-focus-on-technology-and-development.php.

[22] E.A. Gharahbagh, J. Rostami, K. Talebi, and J. Ibarra, "experimental study of impact of soil conditioning on soil abrasion and cutter," in *Rapid Excavation and Tunneling Conference 2013 Proceedings*, 2013, pp. 1,052–1,064.

[23] M.A. Diponio, J. Dee, B. Hagan, P. Raleigh, and J. Associates, "Keeping the chamber full: managing the 'air bubble' in EPB tunneling," in *Rapid Excavation and Tunneling Conference 2013 Proceedings*, 2013, pp. 1,065–1,073.

[24] M.M. Shinouda and J. Dee, "untangling the mystery of soil conditioning," in *Rapid Excavation and Tunneling Conference 2013 Proceedings*, 2013, pp. 1,074–1,085.

[25] Anonymous, "Schooling in earth systems," *T&T North America*, p. 14, 2000.

[26] Colorado School of Mines, "Alumnus Gives $5 million to Underground Construction and Tunneling Program, Scholarships," *Giving News*, 2012. [Online]. Available: http://giving.mines.edu/s/840/giveindex.aspx?sid=840&gid=1&pgid=2271.

[27] "2013 Tunneling Industry Forecast," *Tunnel Business Magazine*, 2013. [Online]. Available: http://www.tunnelingonline.com/2013-tunneling-industry-orecast/. [Accessed: 10-Jan-2013].

[28] "Links to Mining Schools," *Mine Safety and Health Administration*, 2013. [Online]. Available: http://www.msha.gov/SITEINF2.HTM. [Accessed: 07-Dec-2013].

[29] "List of Mining Schools," *Everything Mining*, 2013. [Online]. Available: http://www.everythingmining.com/list_of_mining_schools.htm. [Accessed: 07-Dec-2013].

# TRACK 4: CASE STUDIES

## Session 1: Ground Treatment/Control

William Dean, Chair

# Construction of the San Francisco Bay Tunnel

**R. John Caulfield**
Jacobs Engineering

**Ed Whitman, Jim Stevens, and Kit Fleming**
Michels/Jay Dee/Coluccio Joint Venture

**Johanna I. Wong**
San Francisco Public Utilities Commission

**ABSTRACT:** The Hetch Hetchy water system, owned and operated by the San Francisco Public Utilities Commission (SFPUC), serves 2.4 million people within the greater Bay Area. The system crosses under the San Francisco Bay and required replacement of 1920s-era seismically vulnerable pipelines with a tunneled section.

The Bay Tunnel was the first TBM-driven tunnel underneath the San Francisco Bay. It was constructed using an Earth Pressure Balance machine with specific ground conditioning capabilities. The tunnel is approximately eight kilometers (five miles) long with an excavated diameter of 4.6 meters (15 feet). The project encountered ground that varied from extremely hard, stiff clays with interbedded sands and silts, to a section of hard rock near the end of the tunnel drive.

This paper focuses on the construction means and methods used to facilitate the typical progress of approximately 38m (125 feet) per day over two ten-hour shifts.

## PROJECT BACKGROUND

The San Francisco Public Utilities Commission (SFPUC) manages a large water supply system of reservoirs, tunnels, pipelines, and treatment systems that stretches about 320km (200 mi.) from the Hetch Hetchy Reservoir in Yosemite National Park to the San Francisco Bay Area (Figure 1).

The SFPUC serves 2.4 million residential, commercial, and industrial customers. In November 2002, the SFPUC launched a $4.6 billion Water System Improvement Program (WSIP) to repair, replace, and seismically upgrade the aging 1920s era infrastructure.

During the WSIP engineering design studies, the SFPUC found two major water supply pipeline arteries in the system to be particularly vulnerable to seismic events. These pipelines travel from the City of Newark, above ground on wooden trestles, under the San Francisco Bay, on a 1920s era bridge, and then across marshy wetlands on a pile-supported trestle into the City of Menlo Park. The replacement of these pipelines, Bay Division Pipelines 1 and 2 (BDPL 1 & 2), with a more seismically robust tunnel (Bay Tunnel) was identified as a key element of the WSIP program.

The Bay Tunnel alignment is located between two shafts. The presence of environmentally-sensitive habitats on the Bay margins precluded using cut and cover pipelines which resulted in the need for an 8km (5 mi)-long tunnel with only launching and receiving shafts and no intermediate construction shafts. These two shafts are located on properties owned by SFPUC in the City of Newark (Newark Site) and the City of Menlo Park (Ravenswood Site). Figure 2 provides an aerial view of the Bay Tunnel alignment.

The project bids were opened on November 11, 2009. The lowest bid was submitted by a joint venture of the Michels Corporation of Brownsville, Wisconsin; Jay Dee Contractors of Livonia, Michigan and Frank Coluccio Construction Co. of Seattle, Washington (MJC). MJC's bid for the project was $215.3 million which was only $97,000 below the second bidder and $35 million under the Engineer's Estimate of $250 million for the project.

The construction Notice-to-Proceed was issued in April 2010 with a specified Final Completion date in May 2015.

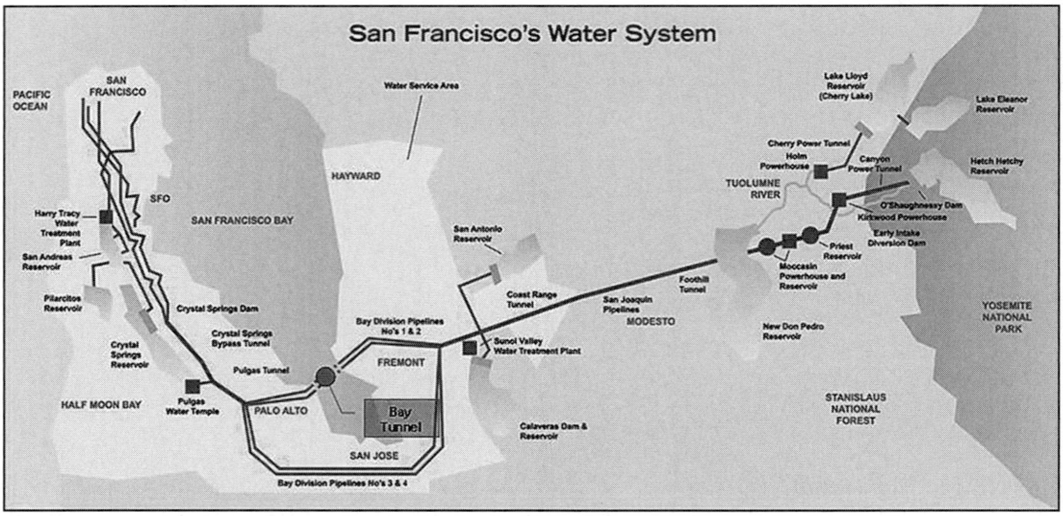


**Figure 1. Schematic diagram of the SFPUC's water distribution system**

**Figure 2. Aerial view of the Bay Tunnel alignment**

## RAVENSWOOD SITE LAYOUT AND LAUNCH SHAFT

The main construction staging area and location of the tunneling support facilities was the Ravenswood site located in the City of Menlo Park on SFPUC property. It encompasses about 56,250 square meters (13.9 acres) of level land and was readily accessible to main arterial highways which were more than sufficient to support the construction operations. The Ravenswood site layout and construction staging are depicted in Figure 3.

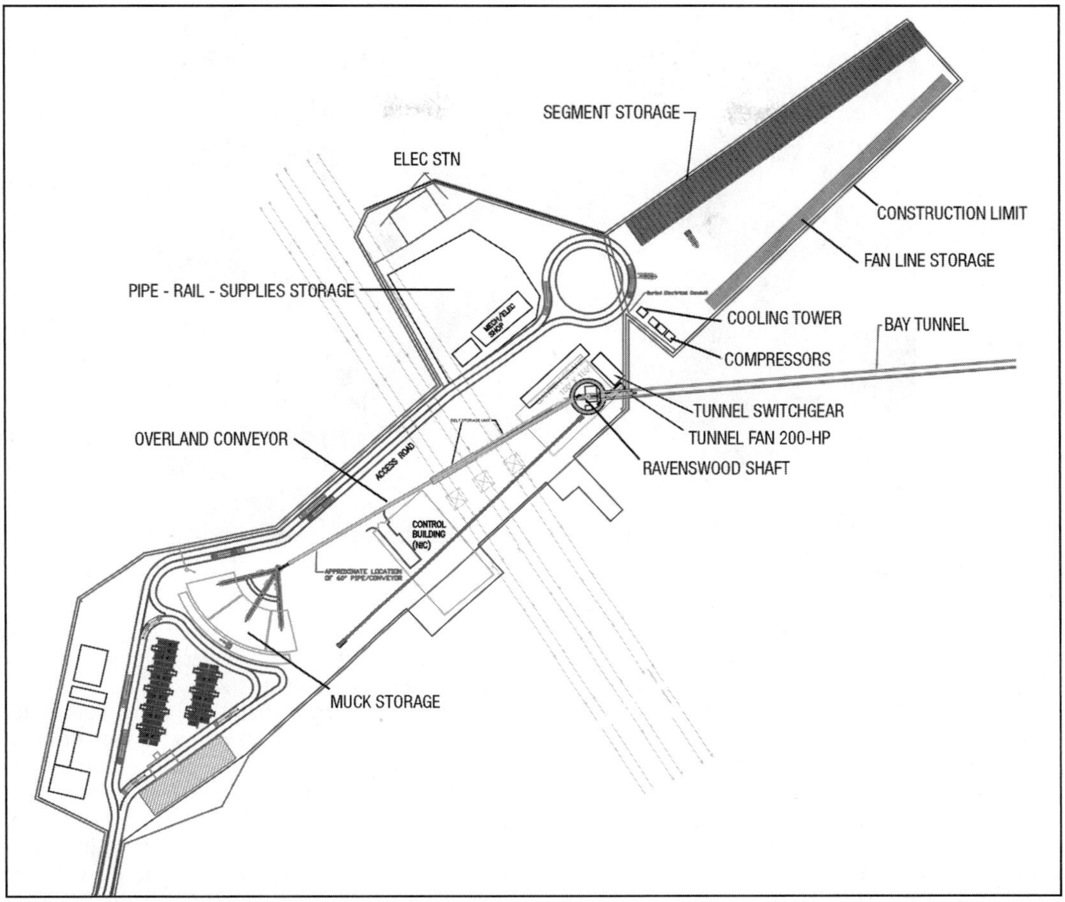


**Figure 3. Ravenswood site layout and construction staging**

The launch shaft for the Tunnel Boring Machine (TBM) was constructed at the Ravenswood site in the vicinity of the existing BDPL pipelines. Due to constraints on the release of groundwater into the adjacent wetlands and concerns over settlement, the shaft was required to be essentially watertight. A diaphragm slurry wall shaft was selected as the preferred method of construction. The shaft is 37.8m (124 feet) deep to the invert, with 43m (141 feet) deep slurry walls panels that extend below the shaft bottom. It has an outer diameter of 19.5m (64 feet) with 0.9m (3-foot) thick walls, resulting in an internal diameter of 17.7m (58 feet). The shaft was excavated in the wet, which required the placement of a 3.7m (12 foot) thick reinforced concrete tremie slab that was keyed and dowelled in to the shaft walls at the bottom (Wong, et al. 2011).

The geology in the shaft location generally consisted of 3.7m (12 feet) of fill and alluvium of soft silty clays and loose silty sand, overlying 5.2m (17 feet) of very soft to very stiff Bay Mud, 30.5m (100 feet) of San Antonio Formation sandy and silty clays with some loose sand layers, and then stiff to very stiff, high plasticity Old Bay Clay to a depth of 54.9m (180 feet). The geotechnical investigation included the installation of multilevel, vibrating-wire piezometers to monitor groundwater pressures near the two shaft sites. Piezometer readings indicate that the groundwater pressures were generally consistent at 1 to 3.4m (3 to 11 ft) below the ground surface and are influenced by tidal variations. Readings between piezometer levels indicated that there was some hydraulic conductivity between the individual geologic units. The interconnection of aquifers below the site led the designer to specify that they be designed to full hydrostatic loading.

The shaft panels were excavated using a clamshell bucket with heavy chisel teeth suspended from a 100 ton crawler crane. The six primary panels were 2.7m (9-feet) long by 0.9m (3-feet) wide. Each primary panel was reinforced with two W33×130 Grade 50 H-pile soldier beams as end stops, with the

**Figure 4. Compound mechanical seal at the EPBM break-out**

interior of the panel provided with a steel reinforcing cage. The six secondary panels were 8.2m (27-feet) long (three 2.7m bites) by 0.9m (3-feet) wide.

The secondary panel reinforcing was with steel cages that were placed in two top and two bottom sections that were clamped together during installation. Fiberglass reinforcement was used with one of the secondary panels in the area of the soft eye for the TBM break-out. A 1.5m (5-foot) by 0.3m (1 foot) deep shear key was blocked out along the slab/shaft interface and was used in conjunction with formsavers to allow for connection of the slab reinforcement. The shear keys were outfitted with jackpacks and plywood filler to facilitate clearing them prior to concrete placement.

The top portion of the shaft was excavated in the dry to the greatest extent possible and then subsequently in the wet down to the total shaft depth. The shaft was excavated using a Manitowoc 3900 crane outfitted with a 3 cubic yard clamshell bucket. Geotechnical instrumentation to detect any excessive shaft movements and/or deflections was monitored daily during the excavation.

Following the shaft excavation the shear keys were cleared by inflating the jackpacks and divers were used to make the dowel connections to the formsavers and the tremie slab reinforcement. The reinforcing of the tremie slab consisted of two mats with a total weight of approximately 45 tonnes (50 tons). The mats were lowered with their support structure using a 265 ton Liebherr hydraulic crane through the water in the shaft and tied-in to the dowels. Approximately 1,200 cubic yards of 41Mpa (6,000 psi) concrete was then placed in a mass pour using two 100 cubic yard per hour concrete pumps fed by 42 transit mixers delivering concrete from three separate batch plants.

Once the tremie slab had cured the water within the shaft was pumped out. Water disposal was an important aspect of the construction. At the Ravenswood site, the only economical discharge point for collected groundwater and construction water was the surrounding environmentally sensitive marsh and tidal flats that ultimately discharged into San Francisco Bay. Because of this, stringent water disposal standards were enforced on the project. The project contract documents required a water treatment facility that could sufficiently treat up to 125 litres per second (2000 gpm) in order to accommodate any uncontrolled large inflows into the launch shaft, particularly during the TBM break-out period.

Ground improvement by jet grouting had been specified for a minimum of 12.2m (40 feet) outside of the shaft and in the direction of the tunnel, within 2 tunnel diameters from the centerline of the tunnel, to create a seal outside the shaft to mitigate the inflow of water and soil upon break-out of the TBM through the soft eye. The area was of particular concern because of the close proximity to the existing in-service BDPL's. Through field observation of the ground conditions during the shaft excavation as well as additional subsurface exploration, MJC determined that the ground improvement zone could be exchanged for an alternate construction method. MJC proposed to substitute the jet grout block with a custom-built compound mechanical breakout seal, designed and manufactured by the Mutsubishi Rubber Company in Japan (see Figure 4). This seal was installed over the fiberglass rebar tunnel eye, at the bottom of the shaft. This seal was further supplemented by additional geotechnical monitoring together with a contingency compensation grouting plan to mitigate any settlement risk under the critical pipelines at a crucial stage of the project where volume loss often occurs.

## TUNNEL BORING MACHINE AND TUNNELING CONDITIONS

The Tunnel Boring Machine (TBM) for the Bay Tunnel Project was manufactured by Hitachi-Zosen of Japan. The TBM design was an Earth Pressure Balance Machine (EPBM) with an excavated diameter of 4.56m (15 feet) which incorporated features to facilitate excavation through the anticipated ground conditions along the alignment (see Figure 5).

The EPBM was capable of delivering 1,225kN/m$^2$ of propulsion force and an advance rate of 25.4cm (10 inches) per minute. The muck extraction was handled by a set of 624mm (24 inch) ribbon type screws with a capacity of 370m$^3$/hr (484 cy/hr). The cutterhead could deliver a torque of 2,387kN·m at 4.0 RPMs and was able to be dressed either with disc cutters or scrapers. However, through the entire drive the EPBM utilized scrapers bits only and managed to complete the drive without changing any of the cutterhead tools. Advance rates were exceptional for the drive with more than 61m (200ft) per day achieved on numerous occasions. The EPBM was outfitted with independent soil conditioning ports both on the cutterhead and in the mixing chamber which enabled MJC to implement different soil conditioning techniques simultaneously and maintain a better control over EPBM throughout the drive. The electrical cabinets were all outfitted with a purge and pressurized system and all the electrical controls and sensors were permissible in anticipation of a potential Cal/OSHA reclassification of the tunnel from "Potentially Gassy with Special Conditions" to "Gassy."

The EPBM was outfitted with 12 hydraulic thrust cylinders for propulsion and steering of the machine. The EPBM main and tail shields also incorporated articulation to help facilitate line and grade adjustments, as well as negotiating curves.

The launch of the EPBM started in August 2011. The EPBM shields and 25 ancillary gantries of the 230m (754 foot) long machine was launched from the shaft and, after 4 setup changes, completely assembled by the end of December 2011, two months ahead of schedule.

The geotechnical evaluations during the design phase resulted in the tunnel being situated primarily in the San Antonio Formation to optimize tunneling conditions, depth, and seismic performance. Figure 6 provides a generalized geologic profile along the tunnel alignment.

The San Antonio Formation consists of interbedded medium stiff to hard clays, silts and sands, with random perched brackish water pockets in confined lenses of silts, sands, and some gravel. In addition to the San Antonio Formation materials, at the end of the drive, it was necessary to tunnel through a 226m (740 foot) long section of Franciscan Formation bedrock that consisted mainly of highly fractured basalt and serpentine rock. The entire tunnel alignment is under the water table, potentially subject to approximately 3.2 bars (46 psi) of hydrostatic pressure.

**Figure 5. Assembly of the earth pressure balance machine**

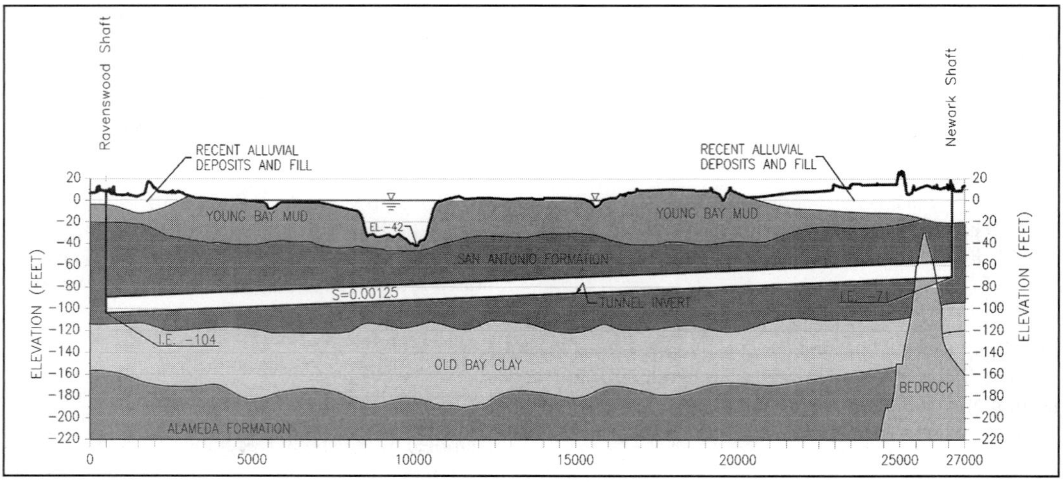


**Figure 6. Generalized geologic profile along the tunnel alignment**

The stiffness of the clays encountered was higher than anticipated and the soil conditioners originally intended for the drive were unable to penetrate the ground and proved, initially, to be ineffective at conditioning the muck. The "extruded bands" of clay discharged by the screw conveyor were so compact and dense they posed significant risks to the tunnel conveyor system, but also precluded the spoils from being extracted out of the shaft using the EPBM vertical hold conveyor, and also compromised the control of the EPBM while it was in close proximity to existing critical SFPUC infrastructure at the surface. Additionally, the risk of either plugging the machine's mixing chamber or spoils bridging over the screw intake was significant.

Using laboratory analysis of the clays, MJC first selected viable additives and associated proportioning capable of breaking down those clays. But the composition and proportions of the soil conditioners, however instrumental, were only part of the solution to the muck conditioning issues. To overcome the imperviousness of the stiff clays, MJC resorted to carefully adjusting the injection of raw soil conditioners (no air) at the face of the excavation, while foamed conditioners were simultaneously being injected into the mixing chamber through separate ports. As tunneling progressed, the proportions of additives were further adapted to optimize the muck conditioning and extraction in varying geologies, and better control the EPBM.

While lenses of coarse grained materials within the clays were predicted, the EPBM encountered numerous, unexpected, and rapidly changing face conditions of a much greater magnitude than previously considered by MJC. In many instances the EPBM encountered up to a full face of sand. Those extreme variations in ground conditions required modifications to MJC's original ground conditioning plan to accommodate such broad disparities. Furthermore, the early signs of those changing ground conditions and the great differences in the length of those lenses made it very difficult to repeatedly and diligently alternate the ground conditioning parameters and, by extension, control the EPBM effectively in those isolated zones at first. The lag between the face changing quasi-instantaneously from a full face of clay to a full face of sand and the blowing out of loose sand and water through the screw conveyor once the last "plug" of clay had exited, momentarily compromised the EPBM control until soil conditioning changes could be implemented. The lack of cohesiveness of those sands and the large amounts of ground water that they contained required the use of much dryer foams, with minimal amounts of added water; a drastic contrast to what had just been injected for most of that shove. Additionally, having no ability to quantify or relieve the mixing chamber's "air bubble," most of the wet material that had not fallen off the tunnel conveyor advancing tail piece, would uncontrollably blow out of the screw as compressed air bubbles trapped in the muck would violently exit the screw conveyors. In such conditions, soil conditioners, air ratios, screws and screw gates had to constantly be adjusted to respond to those unsteady ground conditions. Despite these conditions, for most of the drive, the geotechnical instrumentation indicated volume losses under 1% with most readings within the 0.2% to 0.6% range.

During the design phase, Soil Abrasion Tests (SAT) were conducted on the range of soil types expected to be encountered during tunnel excavation to provide a general indication of abrasion of the excavation tools. The SAT is a relatively new test procedure and is modeled on the abrasion value

(AVS) test originally developed for rock (Nilsen et al., 2006). The SAT results are compared to the standard AVS test results to determine the relative abrasivity of soils. The Soil Abrasion Test (SAT) mean values from ten samples selected from the tunnel envelope ranged from 3 to 23, which indicated "very low" to "medium" abrasivity. The actual ground conditions encountered during construction together with appropriate ground conditioning provided for excellent results in terms of abrasive wear. In fact, only marginal primary wear was observed on the excavation tools and the original dressing of the cutterhead was never changed over the entire five-mile tunnel drive.

Considering the exceptional longevity of the cutting tools selected and the condition of the cutterhead following an inspection stop, as well as the performance of the EPBM through the 72MPa concrete of the launch shaft, MJC decided, after tunneling 7.5km (4.7 miles) of the soft ground reach, to attempt to complete the Franciscan Formation bedrock reach with the original cutterhead dressing. One of the main difficulties to surmount through this rock reach was preventing the angular rock cuttings from locking in the screw conveyors. MJC was not able to batch sufficiently dense bentonite slurry to "carry" the cuttings through the screw conveyors. Ultimately, the bentonite was supplemented with stabilized tunnel backfill grout which proved very effective in increasing the density of the slurry and facilitating the extraction of the rock spoils.

The EPBM performed extremely well and the estimated production rates were exceeded by 50% within two months of the machine being completely assembled. It also provided excellent availability which allowed those production rates to be sustained throughout the drive with an average advance rate of 38m (125 feet) per day (using two 10-hour shifts), a peak of 68.6m (225 feet) per day and a record month of 850m (2788 feet) advanced. After a 17 month drive the machine finally holed through into the Newark receiving shaft in January 2013.

Due to the nature of the ground along the tunnel drive and an anticipated hydrostatic head of up to 3.2 bar (46 psi), MJC also prepared for hyperbaric interventions by screening and training personnel and assembling on site a complete hyperbaric emergency facility, capable of extracting, transporting, and dispensing emergency medical treatment to personnel under hyperbaric conditions. However, the tunnel work plan indicated that compressed air interventions would only be used when the ground conditions warranted them. Through careful selection of the EPBM inspection points, MJC was able to perform all of its nine scheduled cutterhead inspections under atmospheric conditions without incidents.

During construction planning, MJC had evaluated several innovative muck removal systems including high capacity concrete pumps, a 45m (147.6 foot) continuous vertical ribbon screw and a dedicated incline tunnel. The JV eventually chose to use a variable frequency drive (VFD) operated, composite conveyor system consisting of a 7km (4.3 mile) continuous tunnel conveyor, tripping at the shaft into a vertical hold conveyor. This vertical conveyor system was manufactured by Hirosawa Corporation in Japan, and included a set of overland and stacker conveyors on the surface. Despite requiring frequent adjustments, this system ultimately proved to be very effective and a much safer alternative to the traditional muck box approach.

The stockpiled tunnel muck was screened for hazardous materials to identify the appropriate disposal location and it was loaded into trucks with a Fuchs material handler equipped with a clam bucket. This allowed more than 160 trucks to be loaded in a 10 hour shift.

The majority of the tunnel muck was ultimately dispositioned for beneficial reuse in nearby quarry reclamation and levee restoration projects. Some elevated levels of chrysotile asbestos and naturally occurring heavy metals were found within the muck that was generated within the Franciscan Complex bedrock materials near the end of the tunnel drive. This required disposal as a classified hazardous waste. This also required enhanced Personal Protective Equipment (PPE) for underground personnel during excavation, as well as increased air quality monitoring.

The tunnel was constructed as a two-pass system. The first pass of initial ground support consisted of a bolted and gasketed, precast concrete segmental lining erected immediately behind the EPBM. The contract plans and specifications included a preliminary segmental lining design detailed enough for bidding. However, the final segmental lining design was modified by MJC for their means and methods of tunnel excavation. The arrangement of the segments is shown in Figure 7. Each 3.9m (12 foot-10 inch) I.D segmental ring consisted of six trapezoidal pieces. The segments were 254mm (10 inches) thick and 1.5m (5 feet) in length with a taper of 19mm (0.75 inches), which could facilitate a minimum curve radius of 177m (580 feet). The segment materials were comprised of 41 Mpa (6,000psi) concrete and reinforced with a dosage of 35.5kg/cubic meter (60 lbs/cu yd) of Maccoferri Wirand FF3 steel fiber. The segment joints were fitted with EDPM gaskets to minimize water inflows to the contract specified tolerances. The segment radial joints were provided with bolt connections, while the circumferential joints were provided with Sofrasar

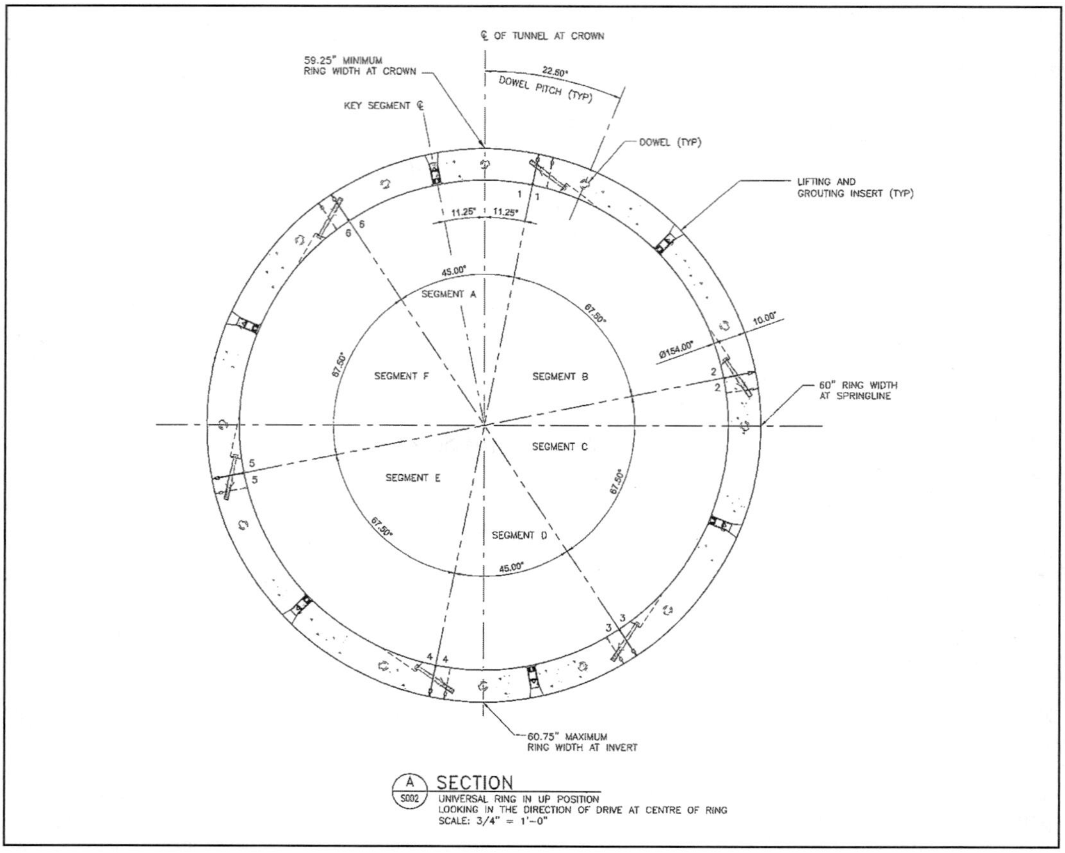


**Figure 7. Precast concrete segment arrangement**

self-locking dowels. Production builds of the rings in the tunnel had an average cycle time of about 10 to 11 minutes.

Following segment ring installation backfill grouting of the void between the segments and the surrounding ground was performed. MJC selected a Sagami-Servo RS-20LS-2 Automatic Mixing Plant as the surface backfill grout plant for mixing the A component of the two part backfill grouting to fill the annular space outside of the installed segmental lining transported via truck to the site. The backfill grouting was performed through grout ports in the segments.

The segments were manufactured at the Traylor Shea plant located approximately 135km (84 miles) away from the site near Stockton, California. Segments were transported to the Ravenswood site two rings at a time via truck and on-site storage.

## RECEIVING SHAFT AND DISPOSITION OF THE EPBM

The receiving shaft at the Newark site also provided its specific challenges. The design originally called for an 8.5m (27.9 foot) I.D. by 30m (98.5 ft) deep slurry wall or caisson shaft installed in the wet, located at the end of a narrow site, with environmentally protected wetlands, endangered species, and with the existing BDPL pipelines less than a meter from the shaft. The Newark shaft site was also located within plumes of contaminated groundwater with elevated levels of chlorinated volatile organic compounds so it was important to avoid cross contamination of the site aquifers. The groundwater required pretreatment to meet the discharge standards specified by the local sanitary sewer agency.

MJC proposed altering the specified excavation method as well as the size and location of the shaft. The shaft internal diameter was reduced to 6.4m (21 feet), and moved 6.7m (22 feet) away from the BDPLs. To prevent accidental spills of drilling fluids, reduce the footprint and ground loads, avoid employee exposure to the hazardous contaminants, prevent the spreading or cross-contamination of the aquifers and spoils, limit vibrations close to the BDPLs, and provide better control over the quality of the work, MJC opted for excavating the shaft in the dry using ground freezing techniques with zone freezing and bore-through freeze pipes as the initial shaft support method.

The jet grout ground improvement zone was also eliminated and replaced by a mechanical exit seal in tandem with a top hat. The hole-through procedure was similarly modified to abandon the EPBM by converting the shields into an extension of the mechanical seal system, acting as a 44mm (1.75 in) thick steel collar, extending through the tunnel eye, 12m (39.4 feet) into the ground.

## FINAL LINING

The final lining will consist of welded steel pipe 2.74 meter (108 inch) in finished diameter including a 16mm (⅝ inch) thick mortar lining. The annular space between the outside of the pipe and the initial support will be backfilled with cellular concrete.

Once tunnel excavation and final lining installation are complete, a steel riser pipe will be installed in the shafts and the annular space backfilled with a combination of concrete and controlled low strength material.

## SUMMARY AND CONCLUSIONS

The SFPUC's 8 kilometer (5 mile) long Bay Tunnel is a critical lifeline water supply facility for the greater San Francisco Bay Area communities. Replacement of the existing antiquated 1920s era pipeline system with the seismically robust Bay Tunnel was necessary to adequately address all of the project service requirements.

The ground conditions within the underlying San Antonio Formation were very well suited for Earth Pressure Balance tunneling technology and excellent advance rates of 38.1m/day (125 feet/day) average and 68.6m/day (225 feet/day) peak were achieved during construction. The tunnel excavation commenced in August 2011 and the hole-through occurred 17 months later at the receiving shaft in January 2013, approximately 8 months ahead of schedule.

## REFERENCES

Caulfield, R.J., Romero V., & Wong J. (2007). "Planning and Design of the Bay Tunnel," Proceedings, 2007 Rapid Excavation and Tunneling Conference, Society for Mining, Metallurgy, and Exploration, Inc./American Society of Civil Engineers, Toronto, Ontario, Canada, pp. 685–693.

Caulfield, R.J., Pawlik I., & Wong J. (2009). "Bay Tunnel—Design Challenges," Proceedings, 2009 Rapid Excavation and Tunneling Conference, Society for Mining, Metallurgy, and Exploration, Inc./American Society of Civil Engineers, Las Vegas, Nevada, pp. 60–71.

MJC Bay Tunnel JV (2010), TLPCS Design Package, Rev4.

Nilsen, B., Dahl, F., Holzhaeuser J., and Raleigh P. (2006). Abrasivity of Soils in TBM. Tunnelling, Tunnels and Tunnelling International.

Wong J., Mues R., & McVicker L. (2011), Shaft Design and Construction Considerations for the Bay Tunnel Project, Proceedings, 2011 Rapid Excavation and Tunneling Conference, Society for Mining, Metallurgy, and Exploration, Inc./ American Society of Civil Engineers, San Francisco, California, pp. 1,218–1,232.

# Design and Construction of the Seekonk Combined Sewer Overflow Interceptor Tunnel

**Mohammad Reza Jafari, William J. Cotter, and Roger M. Norton**
CDM Smith Inc.

**John A. Purciello**
Northeast Remsco Construction, Inc.

**ABSTRACT:** The Seekonk Combined Sewer Overflow Interceptor (SCSOI) project in Providence, Rhode Island is part of the Narragansett Bay Commission's (NBC) Combined Sewer Overflow (CSO) abatement program. The project consists of 7,210-ft of pipeline with diversion, regulator, and interceptor relief structures. This paper presents the technical challenges from the original consideration of open cut to the final design as microtunneling and through the evaluation of alternatives proposed by the contractor during the construction phase of the work. The alignment crossed below or within the zone of influence of structures, utility pipelines, bridge foundations, railroad tracks, and known obstructions in challenging ground that has potential for liquefaction.

## INTRODUCTION AND BACKGROUND

The Seekonk CSO Interceptor is a critical aspect of the three-phase CSO Abatement Program, being implemented in Providence, Rhode Island by the Narragansett Bay Commission. The interceptor, included in Phase II of the program, will capture in excess of 60 cubic feet of combined sewer overflow per second from the existing parallel Seekonk River Interceptor (SRI). To date, Phase 1 of the program, consisting of the Main Spine Tunnel and a major interceptor, has eliminated over 60 million gallons of combined sewer overflow from discharging to the bay.

The alignment of the SCSOI was initially considered to be constructed as an open cut project. However, political and regulatory requirements necessitated the use of open faced tunneling over several segments. These segments included the Historic India Point Park area, the section under an interstate highway, and beneath existing railroad tracks. As the design progressed, additional segments of trenchless construction were added until the majority of the alignment was to be completed by some form of tunneling. During the final refinement of the design, it was determined that the use of microtunnel technology would likely result in the most cost-effective solution. Microtunneling challenges associated with the difficult ground conditions, including non-plastic silts and sands, gravelly material, and sections of fill, along with tight utility crossings would be offset by traffic impact reduction, elimination of mammoth trench excavations with support of excavation systems, and surface restoration associated with open cut construction.

## SUBSURFACE EXPLORATION PROGRAM

Thirty-two borings spaced approximately 250 to 300 feet apart were drilled as part of the geotechnical investigation phase in the fall of 2007. Figure 1, on the following page, provides an overview of the tunnel alignment and the test boring locations. Generally, the borings were drilled to 1.5 times the expected depth of the excavation. The borings ranged in depth from 33 to 58 feet.

For each test boring, Standard Penetration Tests (SPT) was performed at approximately 5-ft intervals. Split spoon samples were collected in accordance with ASTM D1586. Representative soil samples were taken from each split spoon and stored in jars for later review and laboratory testing. Shelby tube samples were collected in accordance with ASTM D1587. In addition, nine field vane shear tests were performed in accordance with ASTM D2573.

This project did not have the typical Geotechnical Baseline Report (GBR). In the absence of a GBR, contract language within the project specifications was provided to help resolve any disputes regarding boulders, obstructions, payment, etc. that might occur during construction. A Geotechnical Data Report was included in the contract documents.

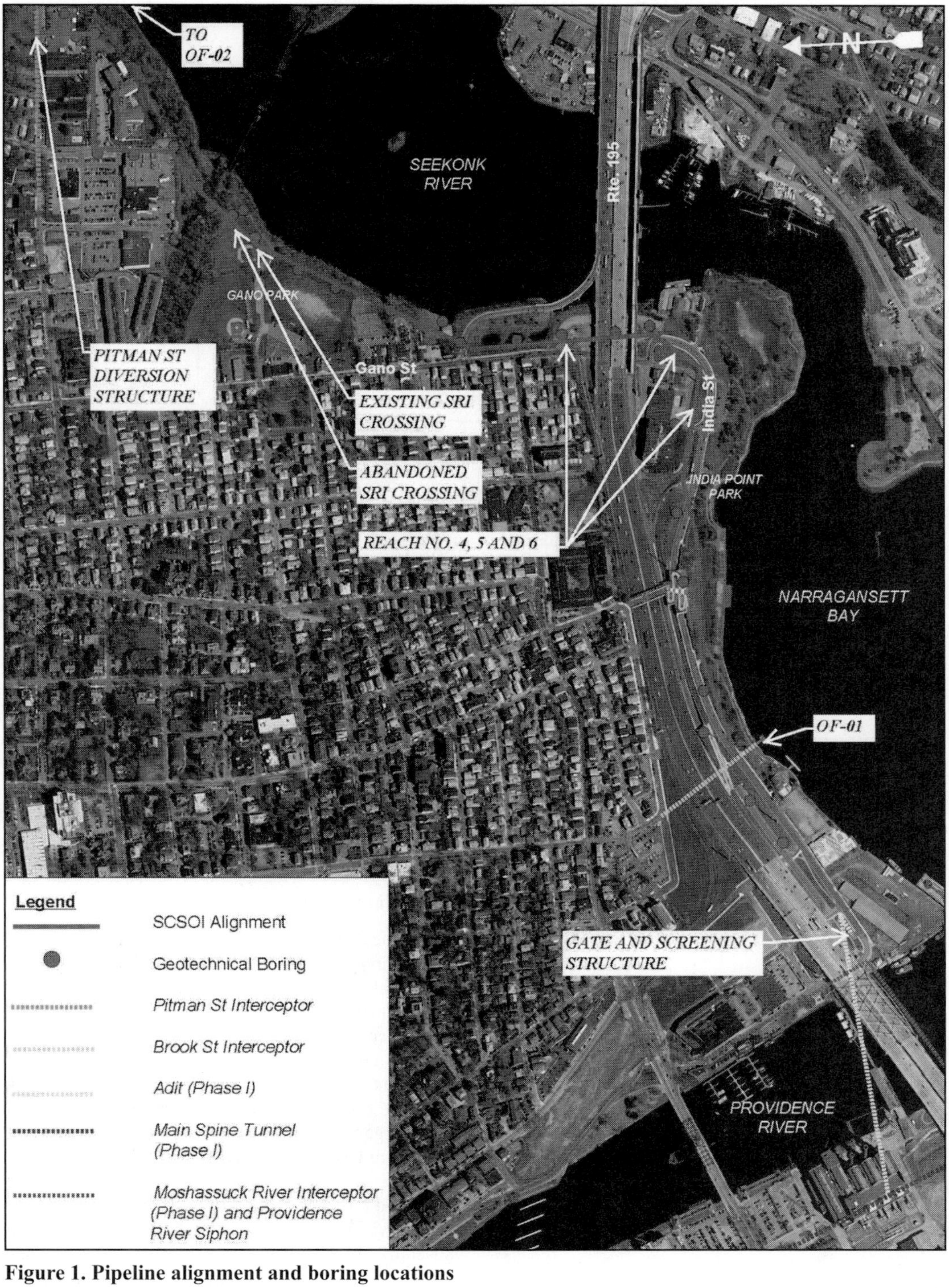


**Figure 1. Pipeline alignment and boring locations**

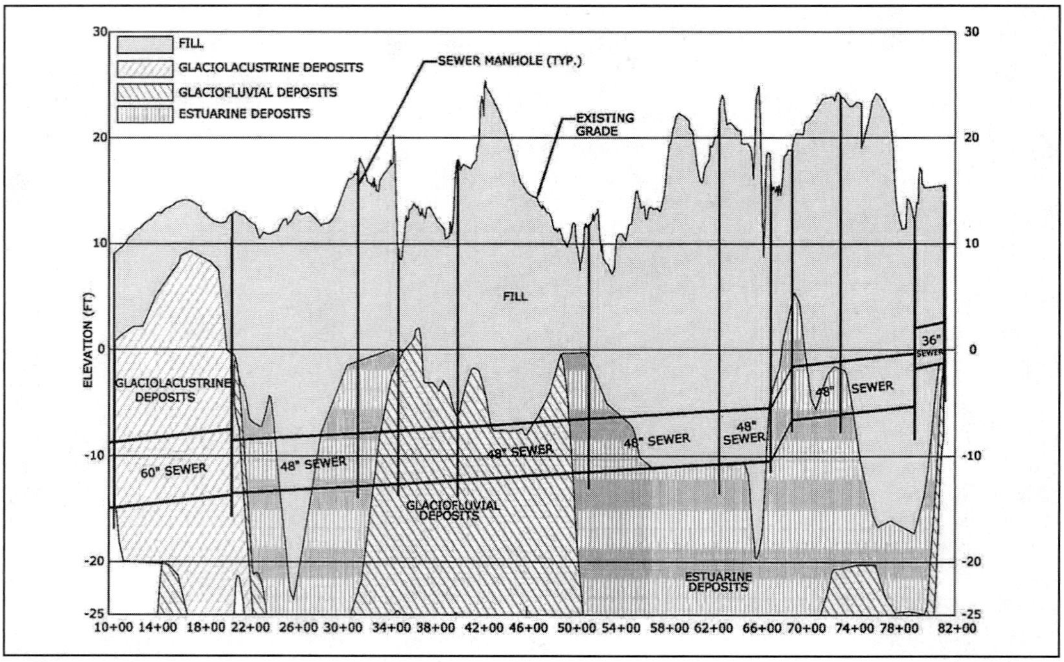


**Figure 2. Subsurface soil profile along the tunnel alignment**

## GENERAL GROUND BEHAVIOR AND SUBSURFACE CHARACTERIZATION

The test borings conducted along the SCSOI proposed alignment encountered varying thickness of Fill, Estuarine/Bottom Bay, Glaciofluvial, and Glaiolacustrine deposits as shown in Figure 2. Not all of these deposits were encountered at each boring location. The following are general descriptions of the deposits and their behavior during open face tunneling according to Tunnelman's Ground Classification as modified by Heuer (1974) from Terzaghi (1950).

### Fill

A layer of fill was encountered at ground surface at all 32 borings. The fill layers ranged in thickness from a minimum of 3.5-ft to a maximum of 37.5-ft, with an average thickness of 18-ft. The fill typically consisted of yellowish brown to olive brown to black, coarse to fine sand with varying amounts of silt and gravel. At some locations the fill also included various amounts of other materials such as brick, glass, wood, shells, and metal fragments. Obstructions were also encountered at some boring locations. The depth of the obstructions ranged from near the surface to approximately 30-ft below ground surface.

The fill soil group is highly variable in nature, ranging from soft to stiff cohesive soils to loose to dense granular soils. Therefore the fill soil group exhibits fast raveling to running behavior above the groundwater table and fast raveling to flowing behavior below the groundwater table. The SPT N-Values recorded in the fill layer varied considerably with a minimum of 1 blow per foot (bl/ft) and a maximum of 115 bl/ft. The average N-value for fill was 23 bl/ft.

### Estuarine/Bottom Bay Deposits

Estuarine or Bottom Bay Deposits were encountered underlying the fill layer at 15 test boring locations. The thickness of the estuarine/bottom bay deposits ranged from 13.5-ft to 35-ft with an average thickness encountered of 24-ft. Generally, the estuarine deposit consisted of either gray, medium dense, medium to fine sand or a gray, loose to very loose silt. At some test boring locations, peat, roots, shells, or other organic material were encountered. Estuarine or bottom bay deposits is anticipated to exhibit slow raveling to running behavior above groundwater table and fast raveling to flowing behavior below the groundwater table. Typically, where the stratum consisted of silt, the N-value ranged from 2 to 8 bl/ft and where the stratum consisted of sand, the N-values were more variable, ranging from 7 to 52 bl/ft.

Overall, the average N-value for Estuarine/Bottom Bay deposit was 8 bl/ft.

### Glaciolacustrine Deposits

Glaciolacustrine deposits were encountered underlying either fill or estuarine deposits at five of the test boring locations. At two of the locations, the borings were terminated prior to reaching the bottom of the glaciolacustrine deposit. Of the locations where borings were advanced through the deposit, the thickness ranged from 12.5-ft to 34-ft. Generally, the glaciolacustrine deposits were described as medium to fine sand with varying amounts of silt, coarse sand, coarse gravel, cobbles and clay; to fine sandy silt; to laminated to varved silt with varying amounts of clay and fine sand; to silt with varying amounts of clay and fine sand; to a lean clay with varying amounts of silt. Glaciolacustrine deposits exhibited firm to slow raveling behavior above the groundwater table and fast raveling to flowing behavior below the groundwater table. At some locations varves of silt and fine sand were encountered. The N-values ranged from 5 to 35 bl/ft with the average N-value of 21 bl/ft.

### Glaciofluvial Deposits

Glaciofluvial deposits were encountered at 24 test boring locations. Glaciofluvial deposits were typically the deepest deposit encountered. In some cases the deposits were encountered underlying a fill layer and continued until the bottom of the test boring. In many cases the borings were terminated prior to reaching the bottom of the glaciofluvial deposit. Therefore it was difficult to estimate the thickness of this deposit. Where encountered, the thickness of glaciofluvial deposits ranged from 3.5-ft to 49.5-ft with an average thickness of 19-ft.

Generally, this deposit ranged from brown to dark gray, dense, fine to coarse sand with varying amounts of gravel, silt and clay and numerous cobbles and boulders; to fine to coarse gravel with various amount of fine to coarse sand, silt, clay and numerous cobbles and boulders. Glaciofluvial deposit is anticipated to exhibit firm to slow raveling behavior above groundwater table and fast raveling to flowing behavior below the groundwater table. The N-values for Glaciofluvial Deposits ranged from 4 to 82 bl/ft with an average of 31 bl/ft.

### Groundwater Levels

Temporary monitoring wells were installed in 16 of the 32 borings. Monthly water levels were recorded since December 2007. Generally, the measured groundwater level in wells located within 200-ft of the Seekonk River ranged from El. 0.5 to 1.8 (ground elevations varying from 10.0 to 18.0). Wells located further than 200-ft from the Seekonk River indicated a wider range of observed groundwater levels ranging from El. 3.5 to 11.0 (ground surface elevations varying from 15.0 to 25.0). Generally the microtunneling operation was conducted below the groundwater table with the water table at 8.0-ft to 10.0-ft above the tunnel crown.

## PROJECT COMPONENTS, FEATURES AND RESTRICTIONS AFFECTING DESIGN, AND CONSTRUCTION

Key components of the SCSOI microtunneling project include:

- 1,020 LF 60-in RCP constructed by MTBM
- 5,940 LF 48-in RCP constructed by MTBM
- 250 LF 36-in RCP constructed open cut
- Microtunneled utility crossing with 1 foot of clearance
- 13 manholes, 2 diversion structures, 1 interceptor relief structure, and modifications to 1 regulator structure
- Replacement of a segment of 44-in × 64-in vertical, elliptical, brick pipe

In total there are 12 reaches (segments) of pipe installation. The main characteristics of each reach are shown in Table 1.

All the reaches are straight except Reach No. 2 and Reach No. 3 which are curved alignments. The construction method used was microtunneling using a Herrenknekht AVN 1200 slurry MTBM for the 48-in diameter pipe and an AVN 1500 for the 60-in diameter pipe. The launch pit for the MTBM was a circular shaped shaft constructed of steel sheet piles with steel ring beams for walers. Extensive geotechnical monitoring instrumentation consisting of observation wells, structure settlement points, inclinometers, surface monitoring points, utility settlement points, and seismographs were designed and installed along the tunnel alignment.

Along the project alignment, several above and below ground structures/features existed which impacted the design and construction of the project. Additionally, various design objectives had to be considered for design and construction of the SCSOI project.

## DIFFICULT SUBSURFACE CONDITIONS

Microtunneling operations in Reach Nos. 4, 5 and 6 were conducted mainly in glaciofluvial deposits. Numerous cobbles and boulders were encountered, causing a slowdown in MTBM forward progress, creating a sinkhole, and damaging the MTBM. Figure 3 shows the cobbles and boulders which the machine encountered during the construction. Figure 4 presents one of the sinkholes that

**Table 1. Main characteristics of pipeline reaches**

| Reach No. (segment) | Pipe Inside Diameter (inch) | Geometry of the Reach: Straight Alignment (length, ft) | Geometry of the Reach: Curved Alignment (radius/length, ft) | Method of Construction |
|---|---|---|---|---|
| 1 | 60 | 980 | | MTBM |
| 2 | 48 | | 2,864/1,054 | MTBM |
| 3 | 48 | | 1,528/306 | MTBM |
| 4 | 48 | 473 | | MTBM |
| 5 | 48 | 264 | | MTBM |
| 6 | 48 | 806 | | MTBM |
| 7 | 48 | 1,084 | | MTBM |
| 8 | 48 | 393 | | MTBM |
| 9 | 48 | 141 | | MTBM |
| 10 | 48 | 395 | | MTBM |
| 11 | 48 | 563 | | MTBM |
| 12 | 36 | 251 | | Open Cut |

**Figure 3. Cobbles and boulders in glaciofluvial deposits**

**Figure 4. Sinkhole due to microtunneling**

developed during microtunneling operation through Glaciofluvial deposits.

Figure 5 presents the MTBM condition before and after completing the microtunneling operation for the above mentioned reaches through glaciofluvial deposits. The machine suffered damage including broken teeth, partial wear to the diamond shaped surfacing on the face of the cutter head, near complete wear of diamond shaped surfacing on the sides of the cutter head, wear of the disc cutters, wear of bars in the crusher, and damage to the pumps.

The highly skilled MTBM machine operator was able to use the maximum capacity of the machine while sustaining minimum damage to machine. The operator's skill was also effective for keeping the machine on the designed alignment, both horizontally and vertically.

Microtunneling operations in Reach No. 7 were performed mainly in Fill Deposits. The MTBM traversed approximately 1,000-ft through the fill deposits. Due to its heterogeneous nature, the MTBM encountered various materials including timber, significant amounts of metals, clam and oyster shells that were deposited from an old seafood processing plant, and railroad spikes. Again, MTBM progress was impeded and some damage to the MTBM, pumps, and lines occurred, but there was no stoppage of the work. Figure 6 shows samples of metals which were collected from the on-site separation plant.

The shredded timber caused blockages at the slurry return lines and in the ports on the face of the machine. The blockages required the operator to resort to the use of radical backflushing. This technique reverses the flow of the slurry and allows a large volume of slurry to clear the blocked ports, a process similar to backwashing a filter. The large volumes of slurry being backflushed caused an unstable tunnel face that caused the MTBM to want to settle. Fortunately, an experienced soft ground MTBM operator was able to overcome this problem

Figure 5. MTBM condition, (a) before construction and (b) after completion of Reach Nos. 4 and 5

Figure 6. Metals extracted by MTBM collected at separation plant

and bring the tunnel home on line and grade. Figure 7 shows the effects of the large volumes of slurry used to backflush in the fill areas that created limited areas of slurry breakout, where the slurry breached the ground surface.

Figure 7. Slurry break out

## PASSING BELOW 42-IN PRESTRESSED CONCRETE CYLINDER PIPE (PCCP)

The tunnel alignment passes under the existing 42-in PCCP SRI in Reach No. 7. The PCCP was installed by open cut excavation using sheet piles for the support of excavation. The tip elevations of the sheet piles were below the invert of the new tunnel. In order to pass under the PCCP, the contractor had to support the existing PCCP and remove the sheet piles. A support system consisting of soldier piles,

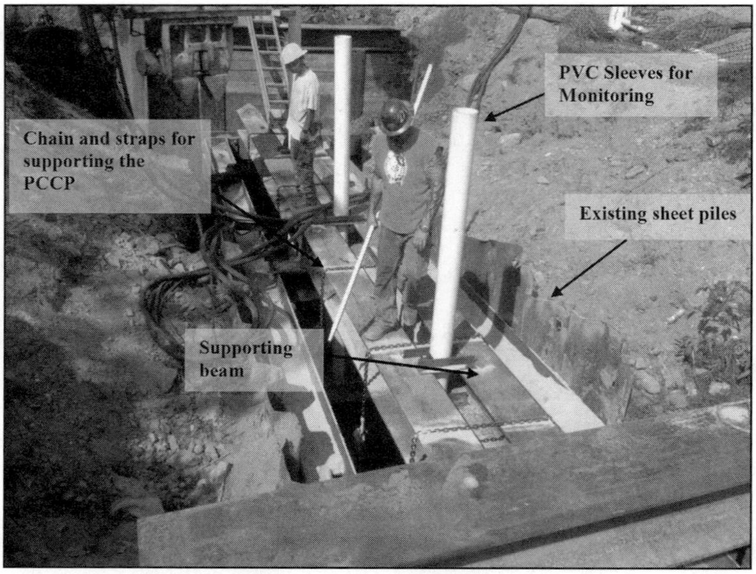


**Figure 8. Support system and utility monitoring for 42-in PCCP**

wales, and beams were installed and the PCCP was supported by the main beams with straps. Three PVC sleeves were installed along the PCCP to monitor the pipe movement. After supporting the pipe, the sheet piles were extracted and the MTBM passed under the existing 42-in PCCP. Figure 8 presents the support system and the PVC sleeves for monitoring purposes.

## CROSSING THROUGH TIMBER PILES SUPPORTING THE ABANDONED 42-IN CAST IRON (CI) SEEKONK RIVER INTERCEPTOR

Record drawings, dated January 1933 indicated that the timber pile foundation supporting an abandoned section of the 42-in Seekonk River Interceptor (SRI) might interfere with the microtunneling operation in Reach No. 8. Since the exact elevation and location of the pipe and the pile foundation were unknown, various investigative techniques were considered, such as test pits or Ground Penetration Radar (GPR), to identify the unknown parameters. A test pit 20-ft long and 12-ft wide and approximately 28-ft deep supported by a slide rail excavation support system was excavated. The CI pipe and the timber pile foundation were located and removed. Excavation of the test pit indicated that the elevation of the pipe and the pile foundation were lower than the expected elevation and would have been in conflict with the tunnel. Figures 9 and 10 show the excavation support and the exposed pile.

**Figure 9. Excavation support system**

**Figure 10. Exposed timber pile and cap**

## PROJECT STATUS

As of mid-October 2013, construction is approximately half complete with no major setbacks and vary few claims. The contractor has overcome all technical adversities and has effectively worked together with the Resident Engineer, Program Manager, and Design Engineer to develop innovative and cost saving modifications.

## CONCLUSIONS

There are many challenges in the use of microtunneling technology, particularly true in urban environments. Relics from previous construction efforts, areas of fill, existing voids, low strength materials, cobbles, boulders, abrasive materials, and existing utilities can all impede the progress of the MTBM. In old waterfront areas, like Providence, abandoned and filled piers and docks exist. All of these challenges have been encountered so far during construction of the SCSOI.

Research of historical archives and local geology and a robust subsurface investigation program have helped move this project towards success, but most critical to success are three factors: (1) capabilities of the MTBM operator, (2) Owner and public expectations, and (3) an extensive geotechnical program during design with the design engineer's support during construction. A highly skilled and capable specialty contractor will make the necessary adjustments during microtunneling operations to keep the momentum of forward progress. The Owner's realistic and flexible expectations, while never compromising the public welfare or the integrity of the project, fosters an environment where technical capabilities are leveraged and difficulties are recognized and addressed.

Looking back through the design phase, as the shift was made from open cut construction to microtunneling, and forward towards the successful completion of this project, all microtunneling issues encountered during construction have been able to be addressed. Microtunneling aspects requiring monitoring during construction include potential for sinkholes, additional investigations needed to identify anticipated obstructions or unknowns, and implementation and adherence to the geotechnical monitoring program. Also, of considerable note, are the benefits and cost savings realized by use of trenchless technologies on this project. The dramatic reduction in the amount of steel sheeting left-in-place, the lessened impacts to traffic, reduced disruption to the public and to the ground surface, and the added environmental benefit of reducing steel to be left-in-place have all contributed to a cost-effective project.

## REFERENCES

Hauer, R. 1974. Important ground parameters in soft ground tunneling. In *Proceedings of Specialty Conference on Subsurface Exploration for Underground Excavation and Heavy Construction*, ASCE, pp. 41–55.

Terzaghi, K. 1946. Rock Defects on tunnel support. In Rock Tunneling with Steel Support, ed. R.V. Proctor and T. White, pp. 15–99. Youngstown, Ohio: Commercial Shearing Co.

# Settlement Monitoring for the University Link Light Rail in Seattle, Washington

**Kimberly Mugg**
Parsons Brinckerhoff

**Richard Sage**
Sound Transit

**ABSTRACT:** An extensive instrumentation program for the University Link Light Rail, Link Contract U220 in Seattle, Washington, was successfully implemented to monitor both surface and subsurface movement due to mining of twin bore tunnels using Earth Pressure Balance Tunnel Boring Machines. This paper describes the challenges of instrument installation along the alignment and also highlights the most meaningful monitoring results, including measured ground movements from extensometers installed to depths immediately above tunnels and a comparison of measured surface and subsurface movements in two distinct geologic conditions. In both cases, tunneling was in glacially consolidated course-grained non-glacial deposits; however, one location was capped by non-glacially consolidated deposits to the surface, while the other location was capped by glacially consolidated soils to the surface. Settlement measurements and rates of settlement were different in each area, regardless of the tunnel face encountering similar soil types.

## INTRODUCTION

The University Link Light Rail is a $1.948 billion project for the Central Puget Sound Regional Transit Authority (Sound Transit) that will extend the existing light rail system from downtown Seattle to the University of Washington campus. The project consists of two cut-and-cover stations and 3.2 miles of twin-bore tunnels. The project is currently scheduled for completion in the 1st Quarter of 2016, at least 6 months ahead of schedule and $107 million below budget.

The tunnels were constructed under two separate contracts. The U220 contract included the installation of the excavation support for the University of Washington Station (UWS) and cross-over, excavation of the cross-over structure, 2.2 miles of twin-bore tunnels between the UWS and the Capitol Hill Station (CHS), and 16 cross passages constructed by sequential excavation mining method (SEM). The U230 contract included the installation of the excavation support for and excavation of the CHS, 0.8 mile of twin-bore tunnels between the CHS and downtown Seattle, and 5 cross passages using SEM.

The focus of this paper is on the geotechnical instrumentation and monitoring program for the U220 contract, in particular the tunneling portion of the contract. The U220 tunnels were constructed by Traylor Frontier-Kemper Joint Venture (TFKJV). TFKJV subcontracted the subsurface instrumentation installation, maintenance, and monitoring to Group Delta, a specialist in subsurface instrumentation. Surface and near-surface monitoring points were installed, maintained, and monitored by TFKJV.

The first part of this paper discusses the instrumentation and monitoring program and the implementation of the program during construction, including challenges with installation and roles during construction. The second part of this paper presents the results of a back-analysis of ground loss around the two tunnel boring machines (TBMs) based on vertical movements measured near the crown of the tunnels and provides a comparison of settlement data between two distinct geologic profiles.

The University Link Light Rail project was designed by a joint venture partnership of Jacobs Associates, HNTB, and AECOM. Construction management of both tunneling contracts was performed by a joint venture partnership of CH2M HILL and Jacobs Engineering.

## TUNNEL ALIGNMENT

The tunnel line and grade for the U220 contract did not present any particular challenges or problems during construction. Large-radius horizontal curves ranging from 2,552 feet (tightest) to 7,240 feet (widest) are present as shown in Figure 1. A vertical curve is present just past the Montlake Cut, as shown in Figure 2. Both tunnels were mined uphill at a 4.1 percent grade

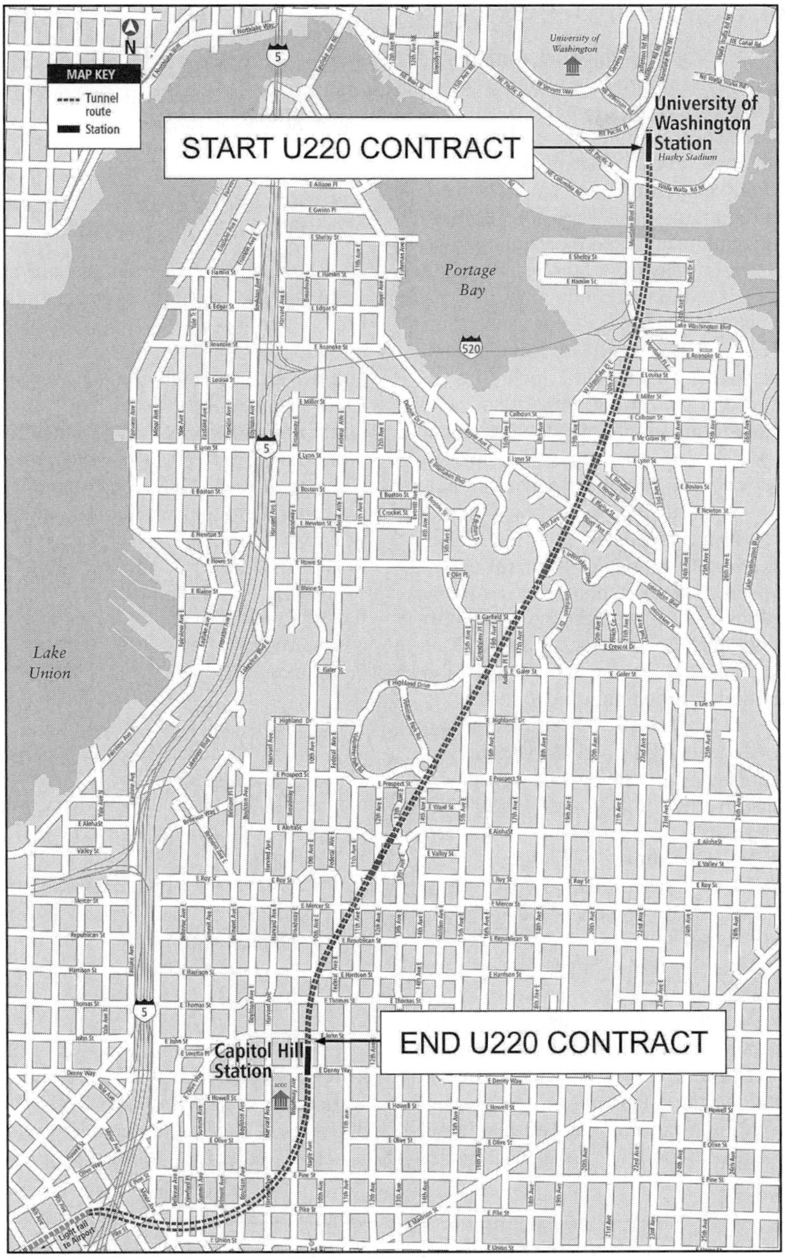


**Figure 1. University Link Light Rail, U220 tunnel contract limits**

from the UWS to the CHS, except for the downward 4.5 percent grade on the north side of the vertical curve in the vicinity of the Montlake Cut.

The tunnel profile in Figure 2 shows a relatively deep alignment, except at the Montlake Cut and near the CHS. The two tunnels are deeper than 75 feet for most of the alignment and have a maximum depth of approximately 315 feet.

## TBM AND TUNNEL GEOMETRY

The outside diameter of the segmental concrete lining was 6.248 meters (20.5 feet). Each of the two TBMs had a 6.560-meter shield diameter and a 6.570-meter cut diameter. The soil pillar width (springline to springline dimension) between the two tunnels ranged from 15 to 20 feet.

## SITE GEOLOGY AND GROUNDWATER

The project site is located within the central portion of the Puget Lowland physiographic province, a broad low-lying region in western Washington between the Cascade Mountains to the east, Olympic Mountains to the west, and the San Juan Islands to the north. The Puget Lowland is a seismically active region and the predominant soil deposits within the region are of glacial origin and/or glacially consolidated. The present-day topography shown in Figure 2 is the result of the last continental glacier.

Soil deposits encountered during subsurface exploration are of Quaternary and Holocene-age and consist of a wide range of soil types, including silt, clay, sand, and gravel. The stratigraphy of soil units is complicated due to at least six successive glaciations, in which the processes of glacial erosion and deposition have changed the landscape that was formed by prior glaciations. Because of the complex stratigraphy, the ground through which the tunnels were excavated is highly variable in lateral extent. There were portions of the southbound (SB) tunnel that encountered different conditions than the northbound (NB) tunnel, despite being at the same elevation and station and only having a 15- to 20-foot horizontal separation between the two tunnels.

The majority of the tunnels were mined through Glaciolacustrine deposits (Qpgl) consisting of very dense silt, clay, silty clay, or clayey silt. Non-glacial fluvial and lacustrine deposits (Qpnf, Qpnl) were encountered to a lesser extent and are typically represented by the "valleys" shown in Figure 2. These soils have been glacially consolidated by the weight of thousands of feet of ice above them and are generally dense to very dense.

Above the tunnels, the stratigraphy is comprised almost entirely of glacially consolidated soils. Non-glacially consolidated soils, consisting of artificial fill, recent alluvium, recent lacustrine deposits, and some landslide debris, were encountered in several borings along the alignment at depths ranging from very shallow to more than 100 feet. Non-glacially consolidated soils have not been exposed to the weight of ice and are less dense than glacially consolidated soils.

Groundwater pressures at the tunnel invert were in the range of 3.5 bar to 5.0 bar (117 to 167 feet of water head) for most of the alignment (Burdick et al. 2013).

## TUNNEL INSTRUMENTATION AND MONITORING PROGRAM

An extensive instrumentation and monitoring program for the project was specified to monitor and protect buildings and structures. The program included both a large number of monitoring points and a relatively high frequency of data collection. The robustness of the instrumentation program was partly due to the recent experience from the Beacon Hill tunnel project in Seattle, where voids were detected after completion of tunneling. A significant contribution to the design of the monitoring program was also provided by Seattle Public Utilities (SPU) to address concerns with the potential for settlement of utilities. In addition, there were several key surface features above the tunnels that were taken into consideration when developing this monitoring program, including:

- Montlake Cut Ship Canal—a narrow canal connection between Lake Washington and Lake Union
- State Route (SR) 520 overpass structure
- Large-diameter aging 54-inch diameter water main
- Historical structures located on property owned by the Seattle Parks and Recreation Department
- Several high-density residential neighborhoods

The Montlake Cut Ship Canal crossing had the shallowest cover over the tunnels and was monitored with inclinometers drilled from barges on either side of the cut. The SR 520 overpass was monitored with structure settlement points. Utility pipes, including

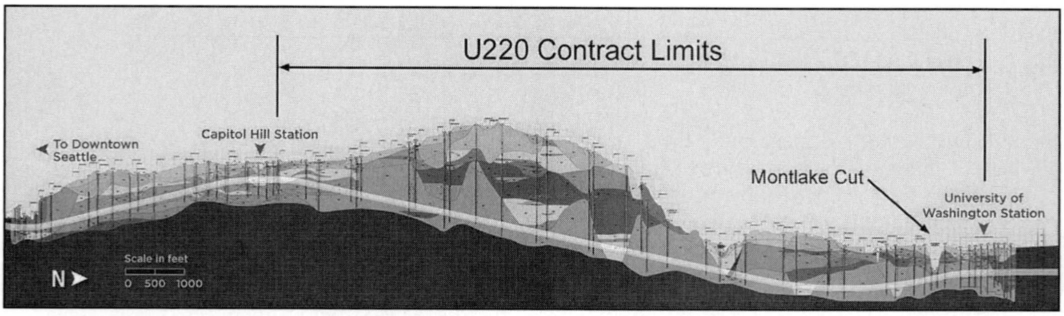


**Figure 2. Geologic profile and tunnel profile**

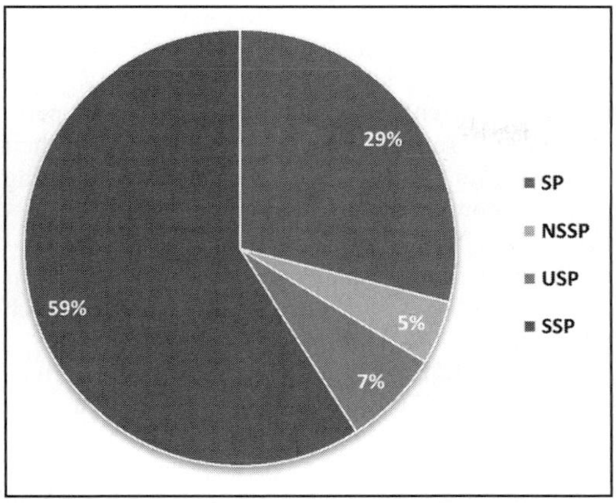


**Figure 3. Distribution of surface and near-surface instrumentation**

the large-diameter water main, were monitored with utility settlement points. Historical structures were monitored with extensometers and structure settlement points. Structure settlement points were also installed on all four corners of most residential dwellings within the subsurface tunnel easement zone. A total of 70 residential homes were monitored by up to four structure settlement points per structure.

### Instrumentation Types and Quantities

The tunnel and cross passage monitoring program included 549 surface or near-surface monitoring locations and 46 subsurface monitoring locations. Additionally, there were several vibrating wire and standpipe piezometers along the alignment that were monitored. Surface or near-surface monitoring points included:

- Surface Settlement Points (SP)—Survey nails installed in concrete or pavement
- Near-surface Settlement Points (NSSP)—Rebar grouted in place approximately 2 feet below grade
- Utility Settlement Points (USP)—Bars attached with epoxy mortar to the top of large utility pipes
- Structure Settlement Points (SSP)—Survey targets, adhesive-type or screw-type, attached to building foundations

The distribution of instruments installed at the surface or near-surface is shown in Figure 3. More than half of these settlement points were attached to buildings foundations.

Subsurface instrumentation included 44 boreholes with multi-point extensometers (MPBX) and 2 boreholes with deep inclinometers. MPBXs consisted of a recessed head inside a protective

**Figure 4. MPBX instrument head**

**Figure 5. MPBX fiberglass rod protective tubing**

enclosure, 6-millimeter diameter fiberglass rods in protective tubing, and up to five hydraulic borros anchors per instrument (Figures 4 and 5). The communications equipment for automated readings consisted of a data logger and wireless radio modem installed below grade at each instrument head and a number of cell modems distributed at the surface along the alignment to collect the data from several nearby instruments. Data was then uploaded to a server. ATLAS monitoring software, which is an internet-based data management system, was used to post-process the data. Post-processed data was then made available for review on ATLAS.

### Installation Challenges

In general, the tunnel alignment was advantageous from the standpoint of instrumentation installation and monitoring because the monitoring locations were typically on side roadways where traffic control was not required. This also allowed for several MPBXs to be installed directly above the tunnel crown with easy access for installation

**Figure 6. Crane lifting drill rig into place**

and easy access for maintenance or manual readings. However, there were several other challenges that were encountered during the installation phase, including utility interferences, difficult property owners, and access issues. Some of the more interesting challenges are described below.

**Installation of Deep MPBXs**—There were two 300-foot MPBXs installed in Volunteer Park for monitoring the tunnel excavations and the excavation of a cross passage located approximately 300 feet below a historic brick water tower. The water tower houses a steel water storage tank that is in active use by SPU. MPBX installations were particularly challenging in this area due to the depth required and because of difficult site access. The MPBXs installed at these two locations consisted of 6-millimeter (0.24-inch) diameter fiberglass rods, which at the required 300-foot depths, push the limits for being able to overcome buoyancy while keeping the rods in tension. Experience with installation of fiberglass rods at these depths is paramount to a successful installation. Other challenges involved the drilling of deep borings on property owned by the Seattle Parks and Recreation Department. The boreholes could not be partially completed and left overnight and the work had to be scheduled around community events taking place at the park. Access to the boring location for the instrument above the cross passage was limited due to trees and a steep slope leading up to the location. The drill rig needed to be crane-lifted into place as shown in Figure 6. Despite the challenges, drilling and installation for both of these instruments was successful.

**Installation of Inclinometers**—Borings for installation of inclinometers on each side of the Montlake Cut needed to be drilled from barges and required permits from the U.S. Army Corps of Engineers. Because of the lead time needed to acquire permits and the close proximity to the launch shaft, these were the first two instruments that were installed for the tunnels. Drilling and installation of these instruments was successfully completed within schedule and without incident.

**Right-of-entry Agreements**—Some property owners were sensitive about allowing access to their properties for monitoring, with one owner agreeing to allow monitoring to take place days after the passage of the TBM near the property and other owners allowing communication only between their attorneys and Sound Transit Community Outreach personnel.

**Utility Locates**—Prior to drilling two borings adjacent to SR 520 for installation of extensometers, a large water main crossing below SR 520 needed to be located to avoid conflicts. There was great difficulty locating the 54-inch diameter water main because there was limited as-built information and it is buried below an embankment near the off-ramp. SPU was unable to locate the water main to within an acceptable accuracy. Ground penetrating radar was ultimately used to locate the pipe to within a horizontal accuracy of ±3 feet and the boring locations were relocated outside the pipe zone.

### Monitoring

The contract frequency of monitoring was specified by assigning one of seven reading schedules to each instrument. Monitoring points specifically for monitoring tunneling-related deformation were assigned schedules based on horizontal proximity of the monitoring point to the TBM face, with frequencies up to twice daily when the TBM was within 200 feet of the instrument.

Monitoring of the instruments during construction was the responsibility of the contractor and the contractor's instrumentation specialist. The contractor was responsible for the timely reporting of data to Sound Transit's Construction Management (CM) team. The CM team's geotechnical engineer reviewed and evaluated instrumentation data on a daily basis during tunneling and periodically as part of the overall CM team's multi-tiered approach to evaluating TBM performance (Banerjee and Shorey 2012). Extensometer data was typically made available within 6 hours of data collection. Survey data was typically made available within 24 hours of data collection.

## MEASURED DEFORMATION AND GROUND LOSS DURING CONSTRUCTION

The trigger and maximum action levels of settlement for the surface instrumentation were set at 0.50 inch and 0.75 inch, respectively, for most surface points. In general, very little settlement was detected in surface instrumentation along the alignment and, when discounting "noise" inherent in the measurements,

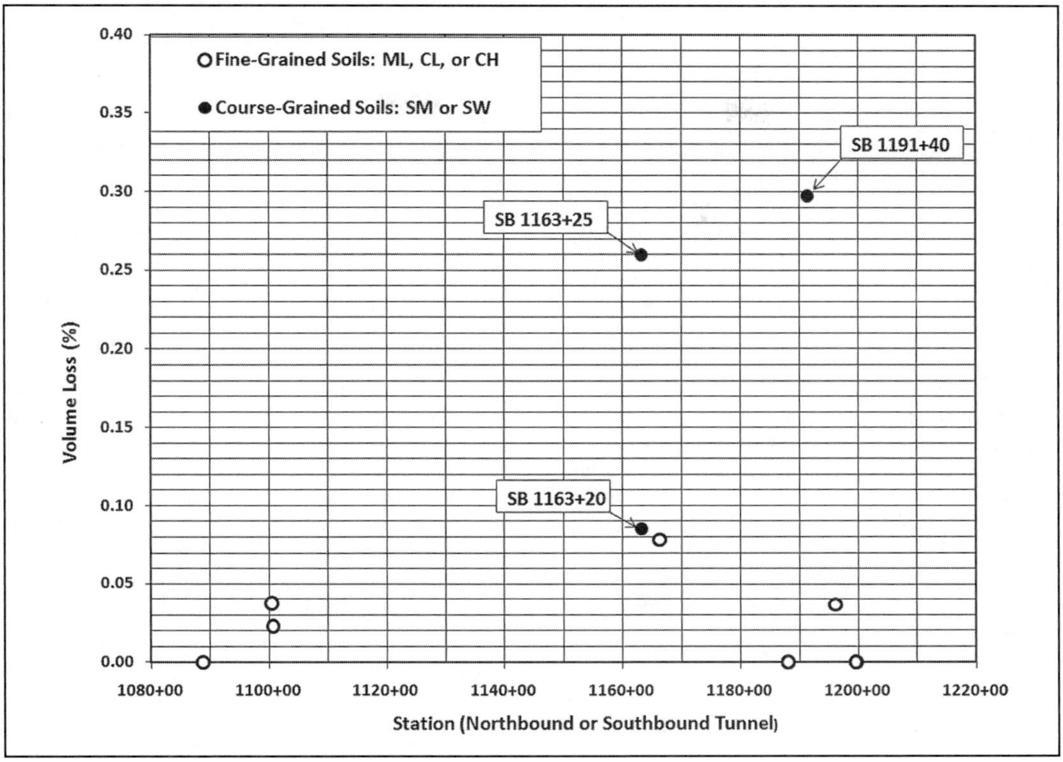


**Figure 7. Measured volume loss around earth pressure TBMs by soil type**

the values were typically below the specified action levels.

The trigger and maximum action levels of settlement for the extensometers varied by depth of instrument as well as excavation type being monitored. Lower thresholds were set for instruments over cross passage excavations because a quicker reaction time would be needed to control the excavation if it were to become unstable. Action levels for all instruments increased with the depth of the instrument because the largest movements were expected to occur closest to the excavation and the smallest movements were expected to occur closest to the ground surface. Trigger levels ranged from 0.3 inch to 1.8 inches. Maximum levels ranged from 0.5 inch to 3 inches. Subsurface movement was detected in several extensometer anchors. All movements were typically below the specified action levels with the maximum measured movements in the deepest anchors ranging from 0.0 inch to about 0.4 inch.

### Volume Loss Estimates

Several of the extensometer anchors were installed close enough to the tunnel crown so that representative ground loss percentages could be calculated from the measured movements using the equation:

$$V_L = \delta v * 2(r + y)$$

(Cording and Hansmire 1975)

where

$V_L$ = volume lost into the tunnel
$\delta v$ = deep vertical displacement
$r$ = tunnel radius
$y$ = distance of the settlement point above the tunnel crown

Volume loss was calculated from the maximum measured movement at anchors that were located less than 7 feet from the tunnel crown. Twelve out of 44 monitoring locations met this criterion. Values are plotted in Figure 7 by tunnel station location and soil type. The three highest volume losses occurred at approximately Stations SB 1163+20, NB 1163+25, and SB 1191+40 where the TBM mined through non-glacial fluvial deposits (Qpnf). Qpnf was encountered at other locations; however, only two of these areas were instrumented.

### Evaluation of Ground Loss Components

Ground loss around a closed-face, pressurized TBM occurs due to face losses, shield losses, and tail losses as shown in Figure 8. Since ground loss is associated with ground movement, deep settlement points or extensometers installed near the tunnel crown are indicative of where and how much ground loss occurs during tunneling.

The MPBX data at Stations SB 1191+40 and SB 1163+25, which correspond to the two highest volume loss percentages in Figure 7, are evaluated in greater detail to determine source of ground loss around the TBM. Plots of the data from MPBX E-391 at Station SB 1191+40 and MPBX E-361 at Station SB 1163+25 are shown in Figures 9 and 10, respectively. The vertical axis represents the downward movement and the horizontal axis represents the distance from the monitoring location to the TBM cutterhead.

In evaluating the field measurements, the TBM operating parameters were reviewed. There were no anomalies noted, such as excessive muck weight, low grout volumes, or irregular earth pressures, while mining at the instrumented locations.

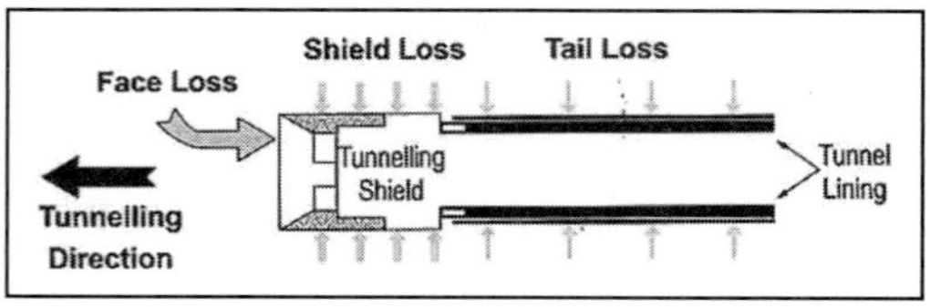


**Figure 8. Ground loss components around TBMs (Source: Loganathan 2011)**

E-391 was installed 4 feet from the SB tunnel centerline and the lowest anchor is installed approximately 5.5 feet above the tunnel crown. Measurements were recorded every 6 hours. E-361 was installed 2 feet from the SB tunnel centerline and the lowest anchor is installed approximately 4 feet above the tunnel crown. Measurements were recorded every 12 hours.

As shown in Figures 9 and 10, total movements were similar at both locations and the ground was effectively stabilized after grouting of the lining was completed. The measurements also give an indication of the source of ground loss being principally over the shield and some at the tail of the shield. However, because measurements were recorded not less than 6 hours apart, the sources of ground loss are open to interpretation. Table 1 is one interpretation of the components of ground loss based on data plotted in Figures 9 and 10.

## SURFACE SETTLEMENT COMPARISON

In the previous section, measured ground deformations above the TBM were evaluated at two tunnel locations, both with similar ground conditions encountered in the face during tunneling. Ground loss values were similar for both locations. Despite

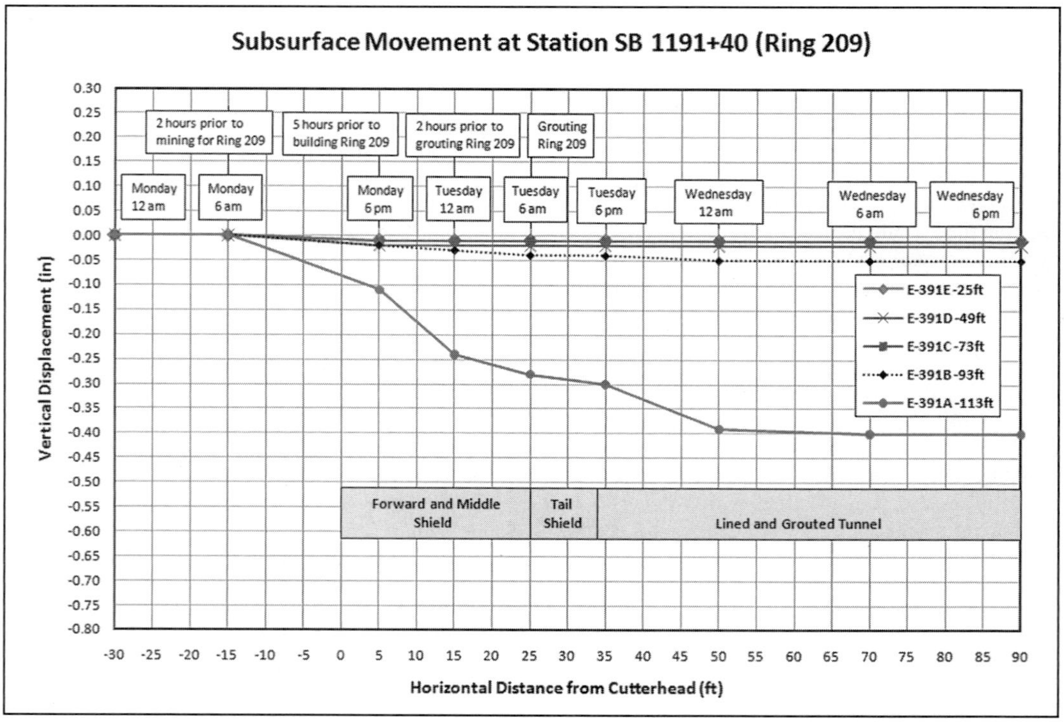


**Figure 9. Vertical displacements at SB 1191+40 (Ring 209) vs. TBM cutterhead position**

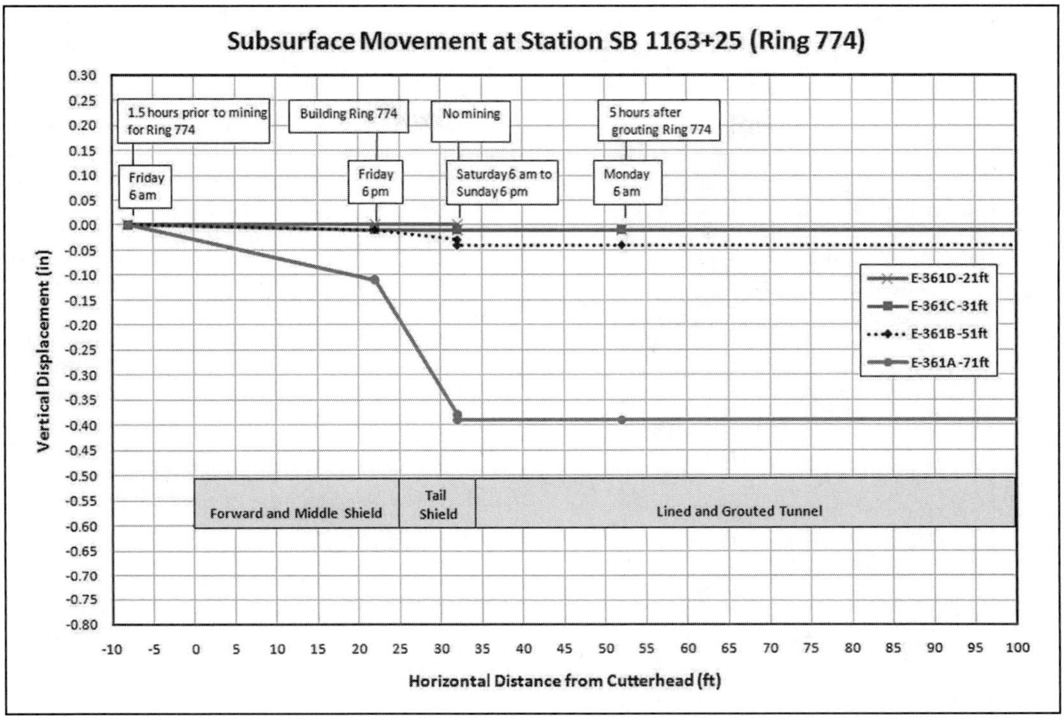


**Figure 10. Vertical displacements at SB 1163+25 (Ring 774) vs. TBM cutterhead position**

**Table 1. Measured volume of lost ground**

| | Volume ($ft^3/ft$) | | | |
|---|---|---|---|---|
| **Station** | **Face Loss** | **Shield Loss** | **Tail Loss** | **Total ($ft^3/ft$) (%)** |
| SB 1191+40 | 0.0 | 0.8 | 0.3 | 1.1 (0.30%) |
| SB 1163+25 | 0.0 | 1.0 | 0.0 | 1.0 (0.26%) |

similarities in ground type at the tunnel face, the subsurface profiles above the tunnels are distinctly different. In this section, an attempt is made to understand differences in the rate of settlement at the ground surface.

## Geologic Profile Comparison

The subsurface profiles at Stations SB 1191+40 and SB 1163+25 are shown in Figures 11 and 12, respectively. At SB 1191+40, the soil column above the tunnel consists of about 118 feet of glacially consolidated deposits. At SB 1163+25, the soil column above the tunnel consists of about 35 feet of non-glacially consolidated (referred to herein as "normally consolidated") deposits overlying 40 feet of glacially consolidated course-grained non-glacial deposits (Qpnf).

## Measured Movements

An array of surface settlement points installed perpendicular to the longitudinal axis of the tunnel was surveyed at each of the two locations. Surveying was performed by optical survey methods and accurate to within ±3 millimeters (0.12 inch). Figure 13 is a plot of the settlement vs. number of days since mining at Station SB 1191+40 and at Station SB 1163+25.

Solid lines in Figure 13 represent movement above the glacially consolidated soil cap and dashed lines represent movement above the normally consolidated soil cap. Average initial reaction times at each instrument location in Figure 13 are summarized in Table 2. The data confirms that settlement takes less time to reach the ground surface in a looser soil than in a very dense soil and, for the same ground loss around the TBM, the magnitude of settlement that reaches the ground surface will be less in a looser soil than in a very dense soil.

## CONCLUSION

The instrumentation program for the U220 tunnel contract was a significant undertaking that was successful in spite of some of the installation challenges. The instrumentation program was effective

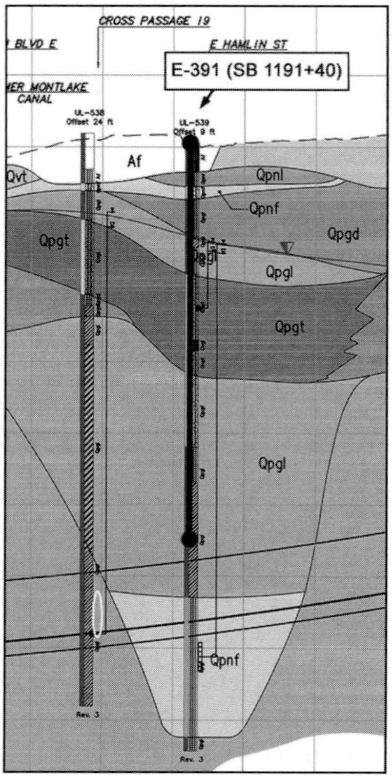


**Figure 11. Subsurface profile and MPBX at Station SB 1191+40 (courtesy of NTP)**

in monitoring the effects of tunneling, and many of the instruments allowed for an in-depth analysis of TBM performance. Volume loss on the order of 0.5 percent is considered acceptable performance for closed-face, pressurized TBM tunneling. The calculated average volume loss percentage on this project was 0.07 percent, demonstrating that very low volume losses can be achieved.

Extensometers provide a good indication of ground behavior around TBMs and should be utilized as much as possible on tunnel projects to indicate the effectiveness of the operation of the TBM and the control of the ground. Efforts should be made during construction to understand the relationship of construction activities at locations instrumented with extensometers to identify where changes in construction may need to take place. Optical survey can be an important measure of the general effects of tunneling after the tunneling has occurred but do not give a direct measurement of TBM performance.

Geology is an important parameter in understanding the rate of settlement propagation to the surface. Deep tunnels mined through glacially consolidated soils may result in deep settlement that may take longer to reach the surface or may not reach the surface at all; whereas, shallower tunnels mined through normally consolidated soils may see an almost immediate reaction to tunneling at the ground surface. The design of instrumentation programs on future tunnel projects may be able to take advantage

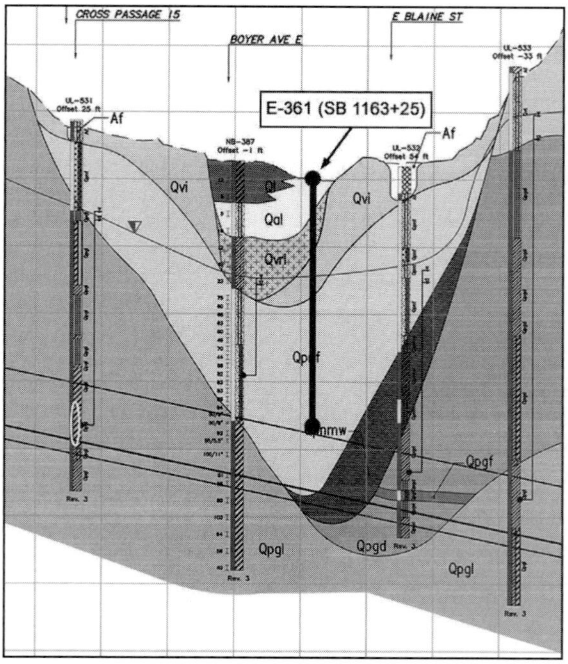


**Figure 12. Subsurface profile and MPBX at Station SB 1163+25 (courtesy of NTP)**

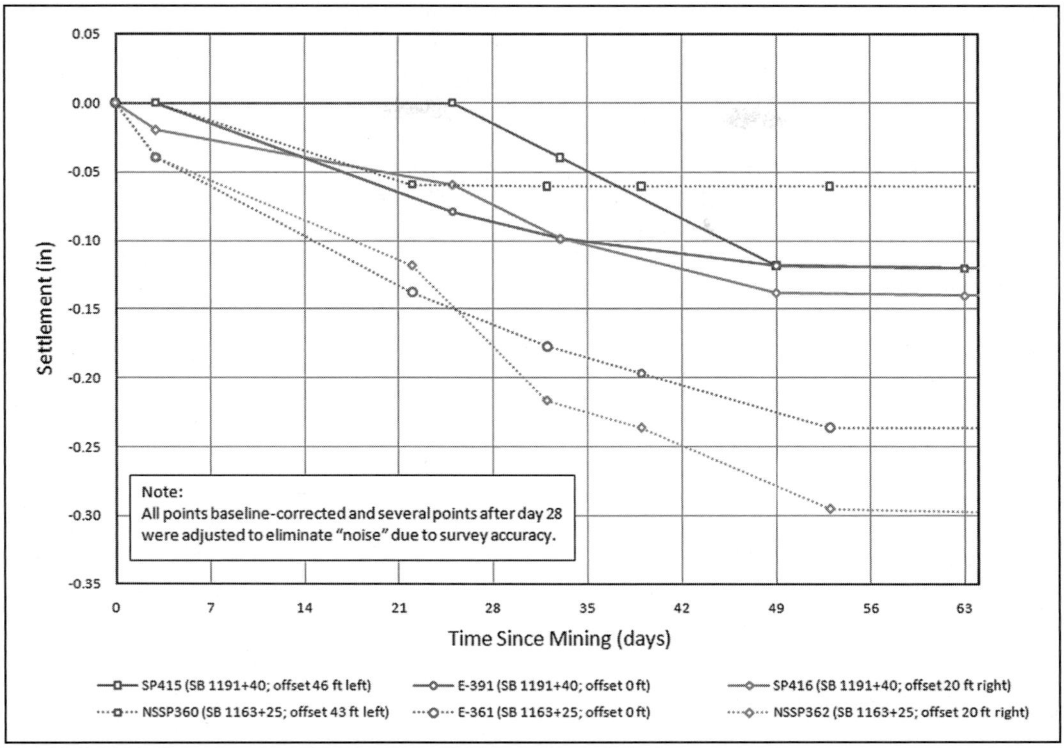


**Figure 13. Measured surface settlement vs. time since active mining**

**Table 2. Average initial reaction time at the ground surface since mining**

| Settlement Point | Station (offset) | Soil Cap* | Reaction Time† | Distance from TBM to Settlement Point‡ |
|---|---|---|---|---|
| SP415 | SB 1191+40 (46 ft left) | NC | 13 days | >> 300 ft |
| NSSP360 | SB 1163+25 (43 ft left) | GC | 29 days | >> 300 ft |
| E-391 | SB 1191+40 (0 ft) | NC | 1 day | 62 ft |
| E-361 | SB 1163+25 (0 ft) | GC | 18 days | >> 300 ft |
| SP416 | SB 1191+40 (20 ft right) | NC | 1 day | 62 ft |
| NSSP362 | SB 1163+25 ( 20 ft right) | GC | 1 day | 62 ft |

* NC= Soil column with normally consolidated soil cap; GC=Soil column with glacially consolidated soil cap.
† Average between last zero reading and first non-zero reading.
‡ Based on average advance rate of 62 feet/day.

of this knowledge and establish the monitoring frequency accordingly, specifying intense short periods of monitoring in some areas and less frequent, longer-term monitoring in others.

## ACKNOWLEDGMENTS

The authors are grateful for the permission granted by Sound Transit to publish this paper and for their cooperation. The authors would also like to acknowledge the contributions made by Ed Shorey of CH2M Hill, Richard McLane of Traylor Brothers, Inc., and Bill Hansmire of Parsons Brinckerhoff.

## REFERENCES

Banerjee, I., and Shorey, E. 2012. EPBM Tunnel Excavation Assessment for the University Link Project, Seattle. In *North American Tunneling; 2012 Proceedings*. Ed. M. Fowler, Palermo, R., Pintabona, R., Smithson Jr., M.; p. 140–145.

Burdick, M., Krulc, M., McLane, R., and Brandt, J. 2013. University Link Light Rail TBM Tunnel UWS to CHS Contract U220: Case History. In *Rapid Excavation and Tunneling Conference; 2013 Proceedings*. Ed. DiPonio, M.A., and Dixon, C.; p. 987–1,013.

Cording, E.J., and Hansmire, W.H. 1975. Displacements around Soft Ground Tunnels. *General Report: Session IV, Tunnels in Soil,* 5th *Panamerican Congress on Soil Mechanics and Foundation Engineering,* Buenos Aires, November, 1975.

Loganathan, N. 2009. An Innovative Method to Assess the Risk to Adjacent Structures Associated with Urban Tunneling. *PB 2009 William Barclay Parsons Fellowship Monograph 25;* Parsons Brinckerhoff.

Northlink Transit Partners. 2009. *Geotechnical Baseline Report, University Link Light Rail TBM Tunnels (UWS to CHS) Link Contract U220.*

# TBM/MTBM/HDD Rescues Using Ground Freezing

**Daniel Mageau, Larry Applegate, and Aaron McCain**
SoilFreeze

**ABSTRACT:** This paper presents case histories of five below-ground rescues of tunnel boring machines (TBM), microtunnel boring machines (MTBM) and horizontal directional drilling (HDD) using ground freezing as the primary technology. In all five cases, the tunnel machines/casing became stuck and the tunneling contractors had been unable to fix or retrieve the machines/casing. The case histories include:

- A 5' diameter MTBM stuck 10 feet from the receiving shaft in Los Angeles, CA
- A 3' diameter MTBM stuck 30 feet deep between a river and an airport runway in Renton, WA
- A 17' diameter TBM stuck 300 feet below a residential area in Lake Forest Park, WA
- A 3' diameter HDD pull-head stuck 65 feet beneath an interstate off-ramp in Marysville, WA
- An 18' diameter TBM stuck inside a broken, leaking portal 90 feet deep in Newark, CA

Ground freezing was used on all five projects to create a stable access to either repair or replace the tunneling equipment. Most of the freezing was accomplished with small, cost-effective freeze equipment using chilled calcium chloride brine pumped through either vertical, horizontal or angled steel freeze pipes. Two of the projects also involved the use of liquid nitrogen in addition to calcium chloride brine. Freeze pipes were installed from ground surface, from inside the access shaft, and from inside the tunnel. In all cases, a solid mass of frozen soil was created around, above and below the TBM/MTBM/HDD in sandy and silty, unstable soils with high groundwater conditions. In one case, the pressure on the frozen soil shoring was well over 5 bars (70 psi). The five rescue projects were all successful—the contractors were able to complete the tunnels and the project. An overview regarding the basic engineering design, construction difficulties and the final ground freezing solution is presented for each case history.

## INTRODUCTION

Tunneling is always a risky business and most of the risks are associated with unknown and variable soil and groundwater conditions. The more advanced tunneling technology becomes, the more contractors are willing to tunnel in difficult ground conditions. No matter how much planning or money goes into a project, risks that can stop a tunneling project still remain. When this happens, it is important for a tunneling contractor to know that there is a way out, that there is an affordable and effective technology available—ground freezing—to rescue the TBM, MTBM or HDD, if and when it gets stuck. The most important part of any construction project is to finish.

Ground freezing has been used for more than a century to stabilize loose, wet ground for temporary shoring during construction. Until recently, it was a very expensive stabilization method reserved for only large complex projects. With recent advances in freeze equipment (small, more efficient chillers) and engineering expertise (e.g., finite element modeling), ground freezing can now be employed quickly and cost effectively for a wide range of project sizes in nearly any type of soil, including running sand and silt, soft clay and even peat. Frozen soil is very strong (compressive strengths in sand are typically more than 80 ksf) and essentially impermeable ($>10^{-10}$ cm/sec; Mageau and Morgenstern, 1980). Moreover, freeze pipes can be installed at any angle, around and below obstacles and to great depths (some mining access shafts have been over 1,000 feet deep). Freezing involves a natural hardening of in-situ soil through removal of heat from the ground without mechanical modifications to the ground. Proof that the ground is fully frozen and ready for excavation is obtained from a series of temperatures sensors installed at key locations in the freeze zone. When the design target temperature is achieved we know the ground is frozen, strong and waterproof prior to the start of excavation—as opposed to grouting where grout penetration is not evident until after excavation. These qualities make ground freezing ideal for soil stabilization on most tunneling related projects, including machine rescues.

Frozen soil stabilization is most commonly achieved by circulating calcium chloride brine

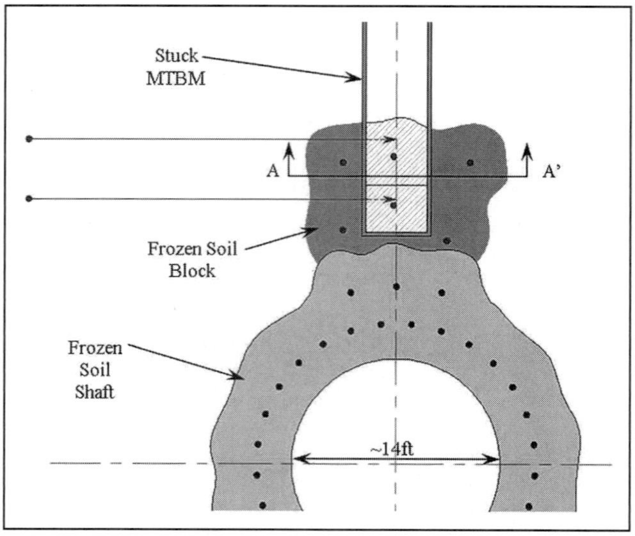


Figure 1. Plan view of frozen soil stabilization

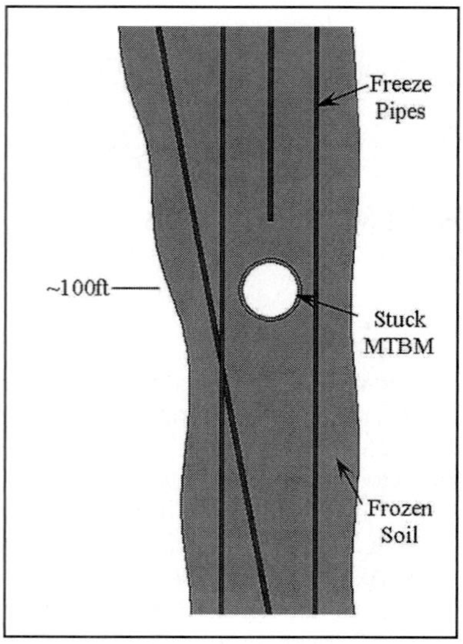


Figure 2. Elevation view—Section A-A'

(non-toxic salt water) through a series of 3 to 4 inch diameter steel freeze pipes spaced at 2 to 6 feet apart within the stabilization site. The freeze pipes have a cap welded to the bottom so the brine is completely contained within the freeze system—no brine goes into the ground. The brine is chilled to 20°F or colder as it is pumped though chillers. As the freeze pipes temperatures decrease, heat is extracted from the ground and the soil around the pipes freezes into a hard, waterproof condition in typically 3 to 6 weeks, depending on the pipe spacing and soil type (Mageau and Nixon, 2004). Once the ground in the targeted area is completely frozen, the required excavation can begin. After the machine rescue is completed, the freeze system is turned off and the ground thaws out back to its natural condition over a few months.

## ENGINEERING REQUIREMENTS

As with any shoring project, ground freezing soil stabilization requires appropriate engineering to achieve a stable, safe system. Because the shoring is made entirely of soil and groundwater that is frozen, a good understanding of the subsurface soil and groundwater conditions is needed at the start of the design. Usually, the project geotechnical report is sufficient for this, although addition borings and laboratory tests are sometimes done to augment the original report. A unique aspect of frozen soil shoring is the need for thermal analyses to calculate (1) rate of frost soil formation, (2) distribution of ground temperatures over time within the target region, (3) optimal spacing of the freeze pipes and (4) chiller requirements. Thermal analyses are accomplished using a finite element 2D computer program, such as TEMP/W. Once the thermal analyses are completed, a structural evaluation using finite element programs such as PLAXIS 2D and 3D are used to evaluate the expected deformation of the frozen soil and the factor of safety after excavation is completed. Frozen soil strength is a direct function of soil temperature (colder temperatures = stronger frozen soil), so it is important to couple thermal and structural analyses for frozen soil designs.

## CASE HISTORIES OF TUNNEL RESCUE PROJECTS

### Force Main Receiving Shaft MTBM Rescue—Los Angeles, CA

#### *Project Overview*

The project involved the installation of two new water lines approximately 1,500 feet long beneath Los Angeles Harbor. The 5-foot diameter casings for each line were installed using MTBM methods. A total of four MTBM access shafts (two 20-foot diameter jacking and two 13-foot diameter receiving) to depths of about 100 feet were constructed using frozen soil technology. Soil conditions consisted of loose to medium dense sand to depths of 120 feet over layers of stiff clay and dense sand. Saline groundwater was about 10 feet below ground surface (bgs). During the final stages of tunneling for the Force Main line, the MTBM had mechanical difficulties and eventually became stuck—just a few feet away from the outside edge of the frozen soil shoring wall. The contractor was not sure precisely where the cutter face was relative to the portal because of difficulties with tunneling in the final

40 feet. As workers inside the tunnel casing were attempting to fix the problems with the pumps near the cutter head, a seal at the machine head broke and the tunnel flooded rapidly. At this point, the MTBM was inoperable.

### *Ground Freezing Solution*

The contractor evaluated a number of options before selecting ground freezing to stabilize the ground around the stuck MTBM. The solution for completing the tunnel involved freezing a block of ground over, around, beneath and in front of the head of the machine. This frozen block would be connected thermally to the frozen soil shaft already in place to create a continuous zone of hard-frozen ground. An illustration of this frozen soil stabilization is shown in plan in Figure 1. The additional freeze pipes used for the stabilization block are shown as red circles (the black circles represent freeze pipes already in place for the frozen soil access shaft). Several batter freeze pipes (shown as red arrows) were installed to freeze beneath the MTBM. Figure 2 shows a cross section A-A' through the MTBM head encased in frozen soil. Figure 3 is a photo of the installed extra freeze pipes needed for the frozen block stabilization. For this project, which was completed in 2002, freezing for the stabilization extended from the ground surface to below the MTBM (about 100 feet). The frozen block at the force main receiving shaft site took about 4 weeks to form.

Once the stabilizing block was fully formed and hard-frozen, the contractor began to hand mine starting from the inside face of the frozen soil receiving shaft to the face of the stuck MTBM. Since neither the precise elevation nor the lateral position of the machine was known, the hand mine zone was enlarged to make sure the machine would be encountered. Eventually, the face of the machine was encountered—about 10 feet from the inside shaft wall face. Figure 4 shows the exposed face of the MTBM encased in hard-frozen sand after hand excavation. The contractor then constructed a permanent liner from the end of the MTBM to the receiving shaft to complete the tunnel.

**Figure 3. Extra freeze pipes in operation**

**Figure 4. MTBM face encased in frozen soil**

## Bryn Mawr MTBM Retrieval—Renton, WA

### *Project Overview*

A new 3-foot diameter storm water line was being installed under the Cedar River in Renton, Washington using MTBM methods. The river was only about 50 feet wide at this location. The tunneling contractor encountered unanticipated buried steel cables with his machine near the midpoint of the river. He was able to mine the machine to the other shore and eventually stopped between the river and the Boeing Renton airstrip at a depth of 30 feet bgs. Because of severe height restrictions from the airstrip and environmental issues related to the river, there were very few viable options available to the contractor to safely and quickly retrieve the machine so it could be repaired for tunneling the remainder of the line. Soils at this location consisted of loose sand and silt with extensive silts layers below the MTBM level, which provided excellent groundwater cutoff from below. The water level was about 5 feet bgs and was tied to the nearby river level.

### *Ground Freezing Solution*

The general contractor decided the best way to complete the water line was to build an interim (unplanned) access shaft over the stuck MTBM

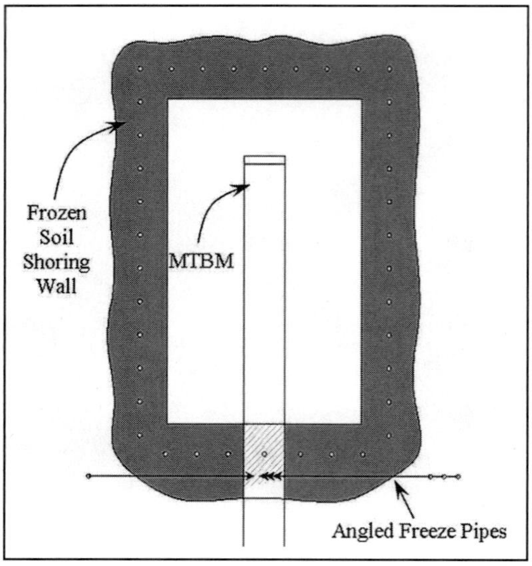


**Figure 5. Plan view of frozen soil shoring**

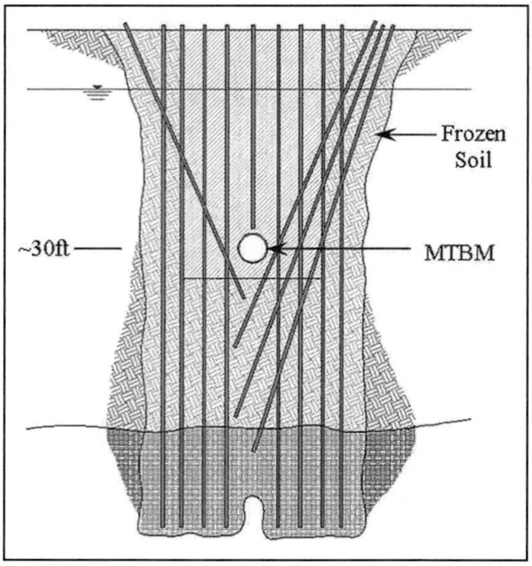


**Figure 6. Elevation view of frozen soil shoring**

between the river and the air strip. This would allow him to remove the machine, repair it and then mine out of the shaft with the repaired machine to complete the tunneling. Ground freezing was selected as the shoring method for this shaft because of its ability to create a structural wall in the loose sands and to cut off groundwater from the river and as well as below the shaft. It was acceptable from the environmental agencies because there was negligible risk of polluting the river, whereas grouting or concrete shoring posed a significant risk. Sheet piles were not feasible because of the height restrictions imposed by the airstrip and because sheets could not shore directly beneath the MTBM.

The 3-inch steel freeze pipes were installed around the rectangular-shape excavation zone by driving with a small hammer to about 60 feet bgs (see Figures 5 and 6). They were located within a trench around the site that would be covered with steel plates after installation to satisfy requirements of the airstrip to have limited aboveground obstructions. To freeze beneath the MTBM it was necessary to install a number of angled freeze pipes as shown in Figure 6. The freeze pipes were spaced very close—2 feet on center—to facilitate freezedown within 2 weeks in order to minimize delays in the project. After freezedown, the contractor excavated the unfrozen soil inside the frozen soil shoring with virtually no dewatering. The MTBM was removed from the ground as shown in Figure 7 and a large diameter pre-cast concrete manhole was installed to provide structural supported during tunneling operations (Figure 8). The contractor then repaired the MTBM, placed it back in the newly installed

**Figure 7. Removal of MTBM from shaft with no dewatering**

**Figure 8. Installing concrete manhole inside frozen shoring**

manhole and then mined out of this manhole to complete the tunnel.

### Brightwater BT3 Rescue of TBM—Lake Forest Park, WA

#### *Project Overview*

The 1.8 billion dollar Brightwater Stormwater Conveyance project involved a sewage treatment plant and 4 separate tunneling contracts spanning 12 miles. During mining of the third section (BT3) the 18-foot diameter TBM face became excessively worn 300 feet bgs in dense glacial sand and clay soils with over 5 bar of groundwater pressure. After several attempts to dewater and repair the face in-situ failed, the machine was deemed inoperable with 2 miles of tunneling remaining in this section. At this time the TBM was located directly below a quiet residential street with insufficient access for a rescue shaft. During the months of repair and mining attempts, significant quantities of soil in front of and surrounding the TBM were removed resulting in voids and highly disturbed, softened soil at the face.

#### *Ground Freezing Solution*

The owners and their engineers elected to use ground freezing to freeze a stabilized block of frozen soil in front of, over, below and around the stuck TBM (Mageau, et al., 2012). As this was being done, another tunneling contractor from the BT4 section was retained to complete the final 2 miles of the BT3 section. This was done by mining from an existing access shaft (where BT3 meets BT4) to the stuck BT3 machine using a smaller diameter TBM (16 foot) that had just successfully completed the BT4 section. The BT4 machine required significant refurbishing before it could start this unplanned new mining. The smaller BT4 machine would then mine directing into the empty shell of the larger BT3 to complete the tunnel.

In order for this innovative procedure to work, the stuck BT3 machine had to be completely gutted. All of the inside elements of the machine (conveyors, bulkhead walls, piping, cutter face, etc.) had to be removed by hand by torching small sections at a time, leaving only the 2-inch thick steel shell. To provide a safe environment for workers, a large frozen soil block was first created around BT3. This was done by installing 40 vertical freeze pipes from the ground surface to a level of 16 feet below the machine invert (depth of 330 feet). Because the site above the TBM was in a residential area with very limited electrical capacity, freezing was limited only to a 50-foot zone (16 feet above and 16 feet below the TBM) using zone freeze pipes that freeze only a targeted lower section of ground. Brine was circulated only in this lower section while the pipes above the 280 foot level were insulated to keep this zone from freezing. This resulted in significantly lower chiller (and electrical) requirements. Figures 9 and 10 show the side view and end view, respectively, of the frozen soil block around the BT3 machine.

Four additional angled freeze pipes were required to freeze soil beneath the TBM where the vertical freeze pipes could not reach. These pipes were installed in holes drilled at an angle from inside the machine. An illustration of these angle freeze pipes is shown in Figures 9 and 10. Brine was delivered to these four pipes via an extra vertical pipe that extended from ground surface to the TBM shield (blue line in Figure 10). Liquid nitrogen was used

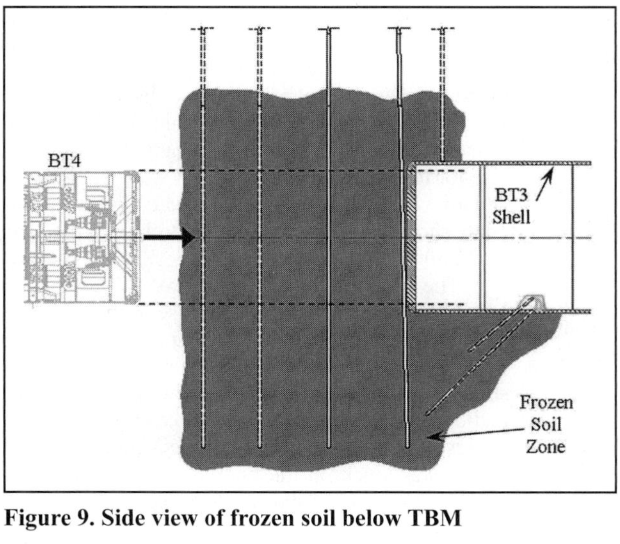


**Figure 9. Side view of frozen soil below TBM**

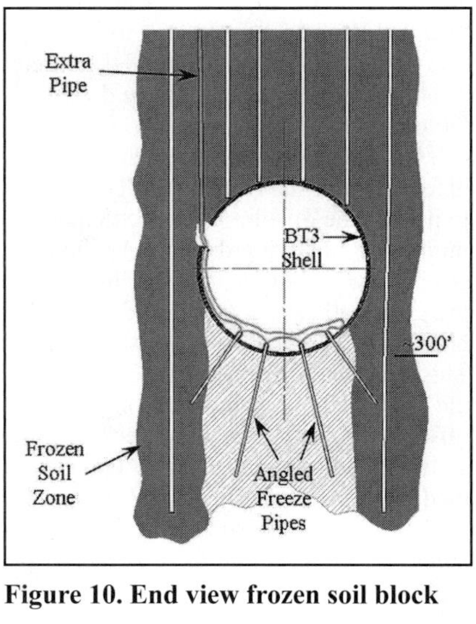


**Figure 10. End view frozen soil block**

**Figure 11. Removing cutter head near frozen soil**

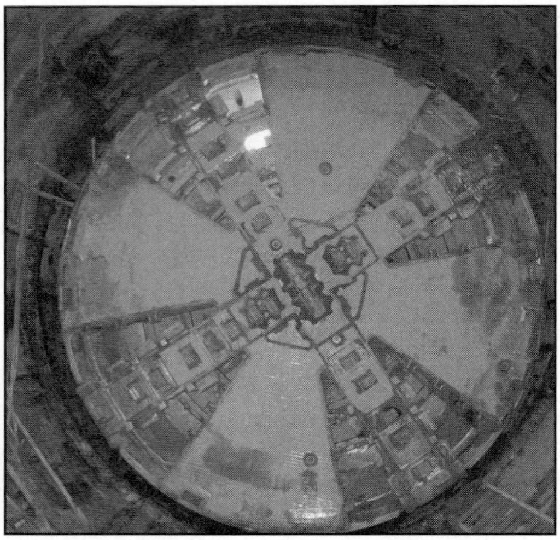

**Figure 12. BT4 successfully mined inside BT4**

to accelerate frost growth at the bottom of this extra pipe to create a stable frozen zone near the shield. A hole was then torched in the TBM shield and the bottom section cut out of the extra pipe to allow brine lines to be installed from the chillers at the surface to the four angled pipes inside the TBM. After the frozen block was fully formed (about 12 weeks) workers safely removed all inside elements of BT3 (Figure 11). The BT4 machine then mined through the frozen soil block and into the then empty BT3 shell to complete section three of the Brightwater tunneling project (Figure 12).

### Tulalip Water Line HDD Rescue—Marysville, WA

#### *Project Overview*

A new 3-foot diameter water line was being installed beneath an off-ramp from Interstate 5 near Marysville, Washington using HDD methods as part of an 8 mile line. The highway department required the contractor to first install a 5-foot diameter steel conductor casing beneath the off-ramp section before starting the HDD work. This was a precaution to protect the off-ramp from settlement should over-mining occur during drilling. As the contractor was pulling the 3-foot steel water line back, the end of the pull-head caught on the end of the 5-foot casing, preventing completion of the water line installation. This incident occurred 65 feet deep directly beneath the I-5 off-ramp. The area along the bottom of the off-ramp embankment was a sensitive wetlands. The soils beneath the embankment consisted of loose to medium dense, wet silty sand with water levels at the base of the embankment. The highway department would not allow the contractor to shut down the off-ramp even for short time periods, ruling out a vertical rescue shaft in the off-ramp lane.

#### *Ground Freezing Solution*

An inclined rescue shaft was design and installed from the bottom of the embankment to the intersection of the 3-foot pipe and 5-foot casing (rescue zone). This would allow workers to safely access the rescue zone to evaluate the nature of the problem and then to repair the problem so they could complete the water line. Figure 13 is a plan view of the freeze shaft site beneath the off-ramp. Figure 14 is an elevation view of the freeze pipes used to create the inclined frozen soil shaft (blue lines). Note that additional freeze pipes (brown lines) were installed from the median between I-5 and the off-ramp to the rescue zone in order to create a frozen soil groundwater cutoff plug at the bottom of the frozen soil shaft. A cylindrically frozen soil wall about 6 feet in thickness was created by the 22 angled freeze pipes after 5 weeks of freezing (McCain, et al., 2013). Because the contractor was uncertain about the precise location of the rescue zone, it was necessary to add a few extra freeze pipes to create a wider freeze zone, should the excavation need to be expanded at the bottom.

The photo in Figure 15 shows the completed freeze pipe installation near the base of the off-ramp embankment. The contractor hand-excavated the unfrozen soil inside the 10-foot diameter frozen soil shoring using high pressure water wands and pneumatic chipping guns. When the excavation reached the rescue zone, they found that the 3-foot water line casing was stuck on the bottom of the 5-foot conductor casing. They did need to expand the cavity

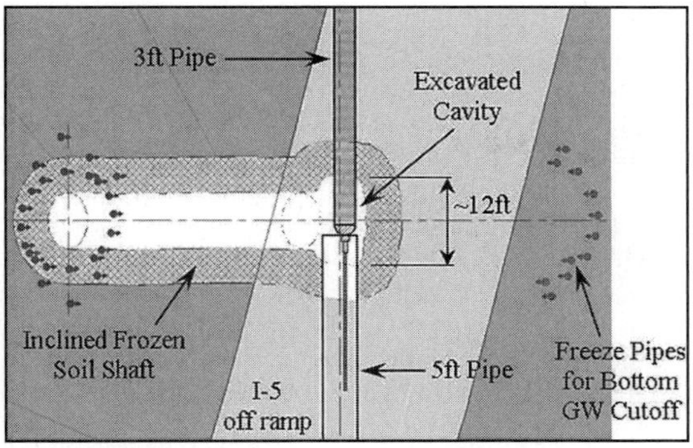


**Figure 13. Plan view of freeze pipes for rescue shaft**

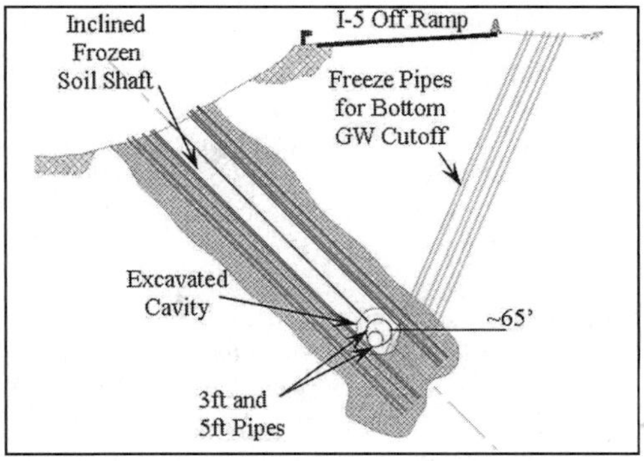


**Figure 14. Elevation view angled freeze pipes**

of the excavation in the rescue zone to complete the repair. The cavity in the frozen soil shoring was thereby increased to about 12 feet in width. The frozen soil shoring worked very well. Measured ground movements at the surface were less than ⅛ inch and no groundwater seepage into the rescue shaft was observed. The repair consisted of inserting a 240-foot long water line extension pipe through the conductor casing and connecting the end of this extension to the end of the stuck 3-foot water line. A photo of the completed water line section with the special beveled fittings (green pipe) is presented in Figure 16.

**Figure 15. Entrance to frozen soil rescue shaft**

**Figure 16. Completed water line section in frozen cave**

## Water Line Receiving Shaft Portal Repair—Newark, CA

### *Project Overview*

The City of San Francisco hired a tunneling contractor to install a pair of new water supply lines to the city beneath San Francisco Bay. The contractor used a 15-foot diameter TBM to install a steel casing into which the new water lines would be inserted. The 5-mile long tunnel project ended at a 21-foot diameter receiving shaft in Newark, CA. This receiving shaft was constructed using frozen soil shoring that extended from ground surface to 110 feet bgs. Patented zone freeze pipes were used in the central areas of the shaft to create a watertight frozen plug for cutting off groundwater inflow at the base and resisting hydrostatic pressures. A permanent concrete liner was installed over the face of the frozen soil shoring. Soils at this site consisted of loose to medium dense silty sand with some clay and gravel seams. Groundwater was about 5 feet bgs.

The contractor's TBM missed the center of the portal by about 10 inches on its way into the "top hat." The TBM ripped the seal and created a crack in the concrete liner that caused the shaft to flood with water and sand (the frozen soil shoring outside the portal had been decommissioned months prior). Several attempts were made to repair the crack by grouting outside the shaft but with no success. Water and sand flowed into the shaft after each unsuccessful attempt, eventually creating voids and highly disturbed loose soil conditions around and above the TBM.

### *Ground Freezing Solution*

The contractor ultimately elected to stabilize the loose sands around the TBM by freezing. This allowed him to open the "top hat" seal in the shaft and build a new concrete wall around the end of the TBM shield to cover the cracked zone without water or sand intrusion. Because the excessive disturbance and loss of soil around the portal area and the failed grouting procedures, the precise nature of the materials to be frozen was unknown. The materials surrounding the TBM were believed to be a mix of very loose sand and silt, grout, and water-filled voids. Also, the pathway for water into the shaft was not clear—it was possible that there were multiple fractures around the portal not visually identifiable. Therefore, special freezing techniques were required for this stabilization.

Freezing of the zone around the portal was accomplished using liquid nitrogen through ¾ inch diameter braded metal lines installed on the inside face of the TBM shield and in vertical pipes located just outside of the concrete shaft. Figure 17 shows a plan view of the frozen soil stabilization zone along with the vertical freeze pipes. Figure 18 shows an elevation view of the portal area with the TBM in place along with the circular freeze line lines through which liquid nitrogen flows inside the TBM. The braded metal lines were placed in four loops around the TBM and attached to the shield by hand. Then grout was hand-pasted over these lines to more evenly distribute the heat extraction along the shield. Finally, the braded metal line/concrete assembly was insulated to the extent practical using concrete insulation blankets. Figure 19 is a photo taken inside the TBM that shows freezing of the shield ceiling. Liquid nitrogen flowed to manifold lines that connected from liquid nitrogen tankers at the surface down to the TBM. These manifold lines were installed in 4-inch steel casing that pre-installed in a holed drilled specially for this stabilization procedure.

Three of the vertical pipes previously installed for shaft construction and then emptied of calcium chloride brine and abandoned were re-used as casing for the 1 inch liquid nitrogen lines. In addition to these, four new vertical pipes were installed in the ground for liquid nitrogen lines to augment the freezing and widen the freeze zone. Liquid nitrogen removes heat as it goes from liquid to vapor, which occurs at approximately –320°F. The vapor (nitrogen gas) must then be allow to escape into the atmosphere (Figure 20). Therefore, an open pipe system is used for this method of freezing.

Freezing with liquid nitrogen continued for several weeks, which is longer than the thermal analysis indicated. This was because there were more water-filled voids than anticipated and because ground water was moving into the shaft through the crack. After about a week, the contractor flooded the shaft to equalize water pressures and to stop the groundwater movement through the portal area. This accelerated the freezing and a solid frozen soil zone was quickly created around the portal. The contractor then removed the 'top hat' and was able to complete the connection of the TBM shield to the concrete shaft liner with no further inflow of groundwater or sand.

## CONCLUSIONS

Ground freezing was successfully used on five different tunnel rescue projects involving TBM,

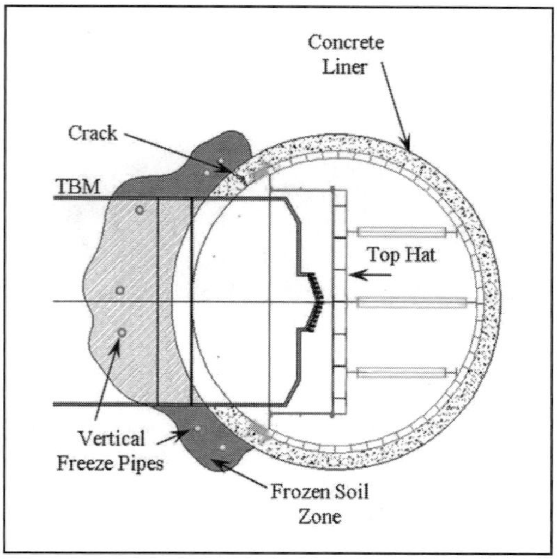


**Figure 17. Plan view of frozen stabilization zone**

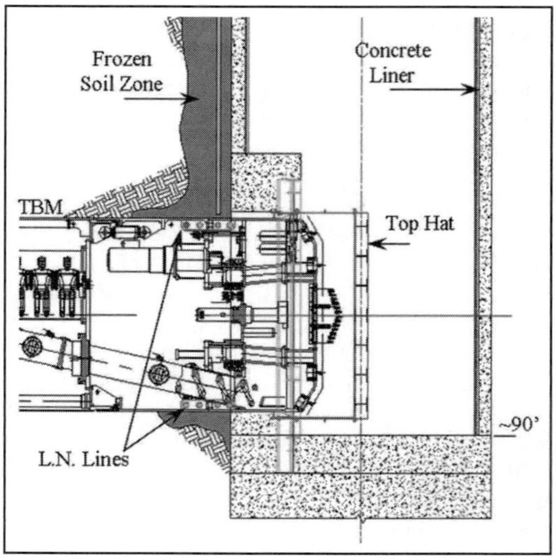


**Figure 18. Elevation view portal area freezing**

**Figure 19. LN lines freeing ceiling of TBM**

**Figure 20. LN vapor escaping from vertical freeze pipes**

MTBM and HDD technology. Each application illustrates a creative and unique way to use ground freezing for stabilizing complex ground conditions around non-uniform structures. Freeze pipes were installed vertically and at various angles to achieve a uniform and continuous frozen soil zone in front of, over, below and around tunnel casing. No matter how complex the conditions, properly designed and installed freeze systems can stabilize nearly any type of soil to allow a contractor safe access to a problem area for evaluation, repair, refurbishing, replacement of problem area. This allows tunneling contractors and owners to continue to push the technical limits of the their equipment with the knowledge that there is a proven and cost-effective technology they can use if problems underground do occur so they can complete the project.

## REFERENCES

Mageau, D.W. and Morgenstern, N.R., 1980. Observation on moisture migration in frozen soil. Canadian Geotechnical Journal, Vol. 17, No. 1, pp. 54–60.

Mageau, D.W. and Nixon, J.F. 2004. Chapter 6—Active Freeze Systems, ASCE Technical Council on Cold Regions Engineering Monograph, “Thermal Analysis, Construction, and Monitoring Methods for Frozen Ground,” Ed. David C. Esch. pp. 193–237.

Mageau, D.W., Applegate, L., and McCain, A. 2012. Frozen Soil Shoring to Dismantle a Disabled TBM 315 Feet Below Ground Surface. NASTT Pacific Northwest Trenchless Review, NASTT Pacific NW Chapter, pp 27–30.

McCain, A., Mageau, D., Shah. K., and Olson, M. 2013. Design and Construction of an Inclined Frozen Soil HDD Rescue Shaft. NASTT 2013 No-Dig Conference, Sacramento, California, March 3–7, 2013.

# Time-Dependent Movements on the Billy Bishop Toronto City Airport Pedestrian Tunnel, Ontario, Canada

**Jon Hurt, Sean Lee, Amirreza Ghasemi, Seth Pollak, and Andrew Cushing**
Arup

**ABSTRACT:** Billy Bishop Toronto City Airport Pedestrian Tunnel is a passenger access tunnel under the Western Channel of Lake Ontario. Southern Ontario is known for very high horizontal in-situ stresses and time dependent behavior of shale formations, with continued rock deformation and final linings experiencing distress in several projects, sometimes years after installation. This paper discusses construction sequencing, methods used for predicting time dependent deformation, instrumentation and monitoring, and comparison of the predicted against the actual movements during construction. Results from a custom-made FLAC subroutine along with a closed form solution are provided and the differences are discussed.

## INTRODUCTION

The Billy Bishop Pedestrian Tunnel is an underground pedestrian tunnel under construction for Billy Bishop Toronto City Airport (BBTCA), which is located on the Toronto Islands in Lake Ontario. The location of the BBTCA relative to downtown Toronto and the other major Toronto airports is shown in Figure 1.

## PROJECT HISTORY

In early 2010, the Toronto Port Authority (TPA) announced that it was seeking a private partner to construct the pedestrian tunnel. In July 2011, an agreement involving an exchange of land between the TPA and the City of Toronto permitted the tunnel project to go forward. Three consortia were invited to respond to the Request for Proposal (RFP) for the project, with bids being submitted in October 2011. In January 2012, a Public Private Partnership (P3) was formed between the TPA and Forum Infrastructure Partners, a consortium consisting of Forum Equity Partners (Developer and Equity Partner), PCL Constructors, Inc. (General Contractor), Technicore Underground (Shaft and Tunneling Contractor), Johnson Controls (Facilities Manager), Arup Canada, Inc. (Lead Designer—Structures and Tunneling), ZAS (Architect), and EXP (Geotechnical Engineer of Record), with groundbreaking for the project taking place in March 2012.

## PROJECT OVERVIEW

The island airport is separated from the mainland by the 120m wide Western Gap Channel, and passengers currently access the airport by Ferry. The project consists of the construction of shafts on the mainland and island sides of the channel and a tunnel between. On the mainland side, the shaft will accommodate six elevators. The island shaft contains two elevators and two banks of three escalators running up from the tunnel to the airport lobby, as shown in Figure 2. The internal dimensions of the tunnel are 9.3m wide and 6m high, to provide a spacious environment and accommodate two moving walkways. In addition, three utility conduits—water and sanitary force mains—will be run through the temporary construction works above the tunnel crown permanent lining. Each of the conduits consists of a thermally welded HDPE pipe installed within a temporary steel sleeve pipe. Including the mains within the tunnel project saved the City of Toronto approximately $10m.

The tunnel will be constructed within the horizontally bedded Georgian Bay Shale. This shale unit consists of 'typically moderately weathered to fresh, grey to dark grey, fine to very fine grained fissile shale interbedded with slightly weathered to fresh grey, fine grained calcareous siltstone and limestone Interbeds' [Project GBR]. There are two distinctive features of the shale in the Toronto region. One is a high horizontal stress regime, and the second is long-term time dependent swelling behavior which occurs when the following factors occur:

- Stress relief of the rock mass
- Availability of fresh water

The swelling is a consequence of the reduction in confined stress in the rock which occurs upon excavation in combination with a differential gradient in salinity between the saline rock porewater and freshwater from Lake Ontario or even humid air. Osmotic and diffusive processes result in a decrease in the

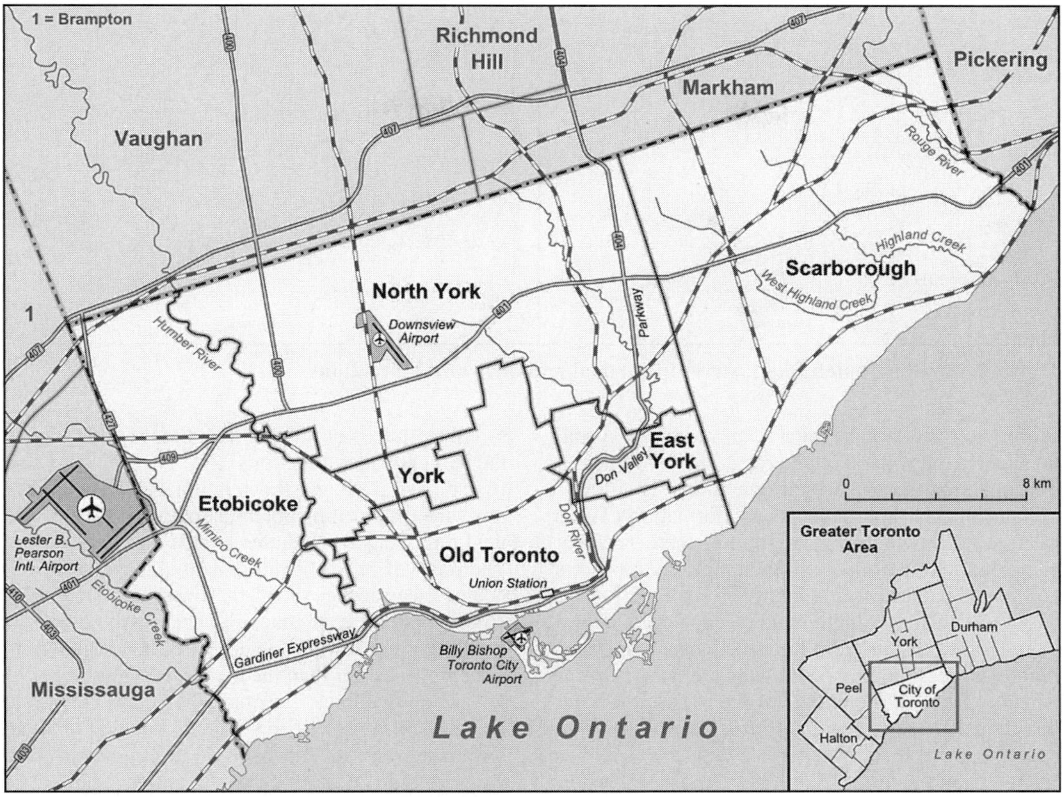


Figure 1. Location of the Billy Bishop Toronto City Airport

Figure 2. Proposed Billy Bishop Airport pedestrian tunnel (BBAPT) [Image by ZAS Architects]

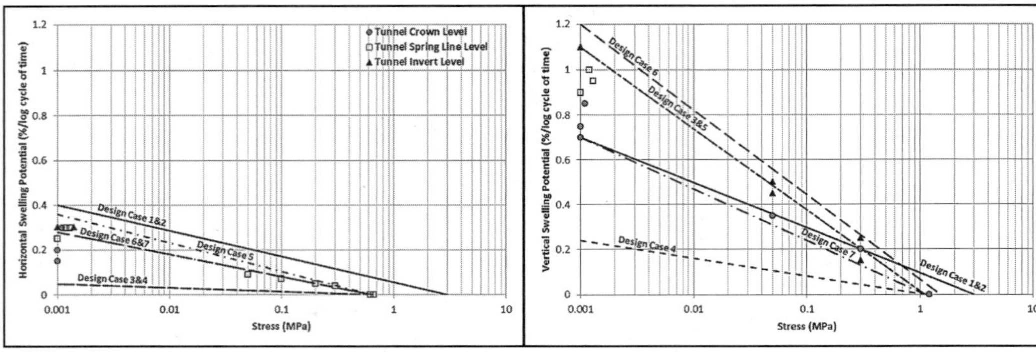


**Figure 3. Swelling potential vs. stress in vertical and horizontal directions**

salinity of the rock porewater achieved by an overall increase in the water content, resulting in volumetric expansion of the shale rock over time. The development of this time dependent deformation (TDD) relative to the time of installation of the permanent lining has a direct impact on the long-term moments and forces induced on the lining.

Following the methodology developed by Lo et al. (1978), results from free swell tests, semi-confined well tests and no-swell tests are used to identify the "Swelling Potential" of the rock in different directions. These tests are performed as follows:

- ***Free swell test:*** Sample is exposed to water; and vertical and horizontal deformation of sample in time is recorded. Eight of these tests were performed for the Billy Bishop project.
- ***Semi-confined swell test:*** Sample is exposed to water and a constant load is applied to the specimen. The deformation in the direction of the applied load is recorded in time. Twelve SCST tests were performed for the project.
- ***Null swell test:*** Sample is exposed to water and variable load is applied. No deformation is allowed in the direction of the applied load. Change of load in time is recorded. Three samples were tested for the project.

Swelling potential is defined as the average slope of the swelling strain versus the logarithm of time and is defined for a specific direction, since behavior in the vertical and horizontal directions is typically noticeably different. The swelling potential decreases as the applied pressure is increased. The pressure where swelling potential is zero and no swell occurs, is called the "Critical Stress" and is defined with the result of the no-swell test.

Hawlader, Lee, and Lo (2003) studied the impact of applied load on the swelling potential of different samples. They concluded that the applied stress in one principal stress direction reduces swelling strain not only in that direction but also in the perpendicular directions.

Figure 3 shows the relation between applied pressure and swelling potential in horizontal and vertical directions for samples of Billy Bishop project. The point of zero swelling potential (Critical stress) is also clear at the end of the lines. Similar to previous experience of other projects in the area, the swelling potential in the vertical direction is two to three times higher than the horizontal value.

The swelling potential of shales tends to increase with decreasing calcite content, and an increasing outward salt concentration gradient from the pore fluid of the rock to the ambient fluid (Lee and Lo, 1993). Therefore, calcite content and salt concentrations (salinity) of pore water in the rock samples were also considered in the tests.

## TUNNEL CONSTRUCTION

The construction sequence is shown in Figure 4. Once the seven TBM drift tunnels had been constructed and filled with concrete, two further TBM tunnels were driven (#8 and #9). These provided an opening that simplified the main excavation, Cut 1, which was excavated by breaking the rock around these pilot tunnels (Figure 5). With the arched roof and inclined side walls, Cut 1 was performed with the need for any additional rock support. Figure 4 also shows the location of the three force main utility conduits, one installed in each of three of the western drift bores (numbers 2, 4, and 7). The steel pipe sleeves were hung from the crown of each drift bore with steel cables, with a final HDPE pipe being installed incrementally within each steel sleeve.

The next stage in the sequence, Cut 2, involved excavating the final section of either side of the first cut. The sidewalls were bolted, with particular attention paid to ensuring the rock under the arch was fully supported. The final excavation stage, Cut 3, was the removal of the invert, and this was carried out in two stages—a rough cut for the bulk of the

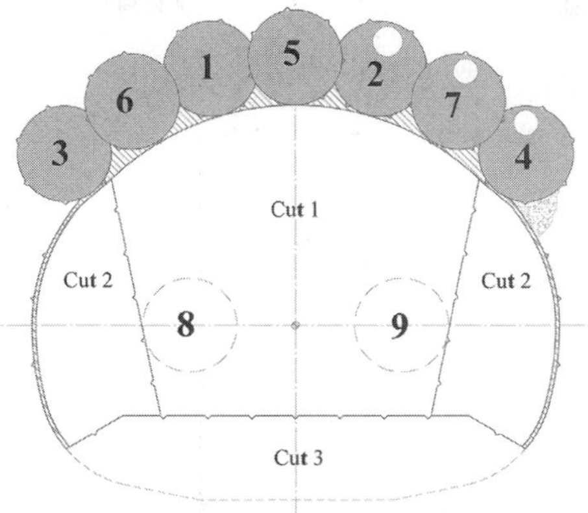


**Figure 4. Excavation sequence of proposed TBM drifts drilling and backfilling**

**Figure 5. Tunnel excavation—Cut 1**

excavation following by trimming to the required profile. Bulk excavation was typically performed by hoe-ramming and Dosco roadheader, and trimming by roadheader mounted on an excavator.

The presence of design engineers on the site allowed several modifications to the excavation sequence to allow optimization of the schedule, based on observations of the rock and monitoring results. This included allowing the Cut 1 excavation to proceed for the whole length of the tunnel before the Cut 2 commenced, and delaying the application of shotcrete on the sidewalls, which removed this operation from the critical path.

Following the excavation, the lining was placed. A fully tanked PVC compartmentalized membrane system was used. The reinforcement for the lining was prefabricated to allow for rapid assembly in the tunnel. After the invert concrete was poured, the arch concrete was placed using a 12m long arch form. The arch lining was designed with steel fiber reinforced concrete, with reinforcement only provided up to the shoulders, to maximize the speed of construction.

## SHORT-TERM TUNNEL SUPPORT DESIGN

The tunnel has a relatively shallow rock cover of eight meters for the 10.5m-wide excavation span. Given that the excavation is tunder the lake and there was not a comprehensive knowledge of the condition of the rock, a rather novel temporary lining ("pre-support") system was selected that involved a series of interlocking, horizontal TBM-driven secant drift bores which are backfilled with 15 MPa strength concrete to form an integral arched roof under which mass excavation can occur (see Figures 3 and 4). This mitigated the risk from encountering a water bearing feature under the channel, as the small 1.8m diameter TBMs would provide a much greater means of face stability than an open face SEM excavation.

**Table 1. Summary of key results for short-term (temporary) stability of backfilled arch and tunnel**

| $\sigma_H$ (MPa) | Max. Dowel Axial Force (kN) | Vertical Displacement Under Arch (mm) | Max. Displacement Under Arch Abutment (mm) | Max. Horizontal Tunnel Convergence (mm) | Depth of Plastic Zone Under Invert (m) | Max. Arch Compressive Stress (MPa) (without swell) |
|---|---|---|---|---|---|---|
| 2.5 | 176 | 11 | 13 | 5 | 0 | 4 |
| 6.9 | 395 | 27 | 40 | 7 | 4 | 8.5 |

The temporary support of the TBM bores was provided by steel ring beams at a nominal 1.2m spacing supplemented with plywood sheet lagging. The design was based on short term convergence confinement calculation (which included the effect of swelling for the time the bores were expected to remain open) and validated by the discrete element program UDEC (v.4.0) by Itasca Consulting Group, Inc. The UDEC model was used to model the entire construction sequence of excavating and backfilling each individual drift bore followed by the main tunnel excavation. Of particular interest was the horizontal shear displacement of the bedded shale rock and how this affected stress redistribution around and through the backfilled concrete arch. Additionally, the joints around and within the arch were assigned properties consistent with a concrete-concrete or concrete-rock interface created by the backfill pours.

Two cases of horizontal in situ stress were considered to bound the expected short term behavior. Key results are summarized in Table 1.

A small percentage of boundary joint length along the arch perimeter indicated debonding (Figure 6). However, the remaining bonded length was more than enough to prevent a downward translation of the structure. The internal concrete-concrete joints did not show any loss of contact or debonding (i.e., both joint normal and shear stiffnesses > 0) which indicated that no tensile forces exist within the arch, even in the critical case of a potential wedge failure in one of the side abutments. The results confirmed the assumption of the arch acting as a single, contiguous member in response to the stress distribution resulting from the main tunnel excavation (Figure 7).

The temporary main tunnel support design consisted of installing 32mm diameter rock dowels in the sidewall abutments under the backfilled arch once the final excavation profile was cut. The aim of the dowels was to stabilize any overbreak or wedge failures resulting from the occasional sub vertical joint dipping into the tunnel.

The final lining of the tunnel was design to maximize productivity during construction, and utilizes a conventional reinforced concrete invert slab with prefabricated reinforcement panels, and a concrete arch with a minimum 400mm thick steel fiber reinforced concrete (SFRC) lining with partial steel rebar reinforcement in the sidewalls to counter long term swelling pressures.

## DESIGN PROCEDURE FOR TIME-DEPENDENT DEFORMATION

There are two common design methods available to assess the impacts of TDD on the tunnel lining. Lo and Yuen (1981) developed a closed form solution method to predict the long term loads and displacement at any point in time in lining and rock. However, the closed-form solution method does not consider the effect of time-dependent swelling-induced rock stress on the swelling potential of the shale rock. As a result, the closed-form solutions for the final unlined rock swelling displacement and lining moments and forces are conservatively over-estimated. To obtain a more realistic estimate of the lining loads, a numerical model to account for swelling was developed by Itasca on behalf of Arup and implemented in the FLAC 2D finite difference program. The swelling rock constitutive model in Hawlader et al. (2003; 2005) was developed based on the Mohr-Coulomb elastic/perfectly plastic material model. This is based on the observations in the laboratory experiments that the swelling strains in the principal swelling directions of a Shale rock specimen increase linearly with the logarithm of time, and the swelling strains are reduced in both parallel and perpendicular directions by the application of stress on the rock specimen. In this project, the model formulation was implemented for use with the two-dimensional code FLAC in plane strain mode.

A "virtual critical stress" concept was introduced for vertical swelling in the zone between the rock surface and the depth where each of in-situ stress is equal to the critical stress ($\sigma_c$). This zone would have swelled before any construction activity started, and was assumed to be in a stable condition with the in-situ stress level. To avoid modeling vertical swelling in this zone, the critical stress zone in the vertical direction was set equal to the in-situ stress (a virtual critical stress, which is reduced from the original critical stress). The critical stress below this zone was maintained at the actual level.

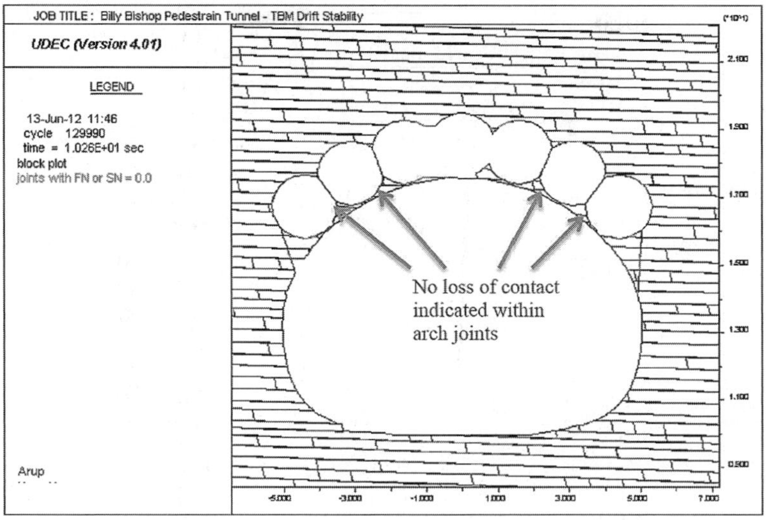


**Figure 6. Backfilled concrete arch stability modeling in UDEC showing zones of joint debonding upon excavation of main tunnel**

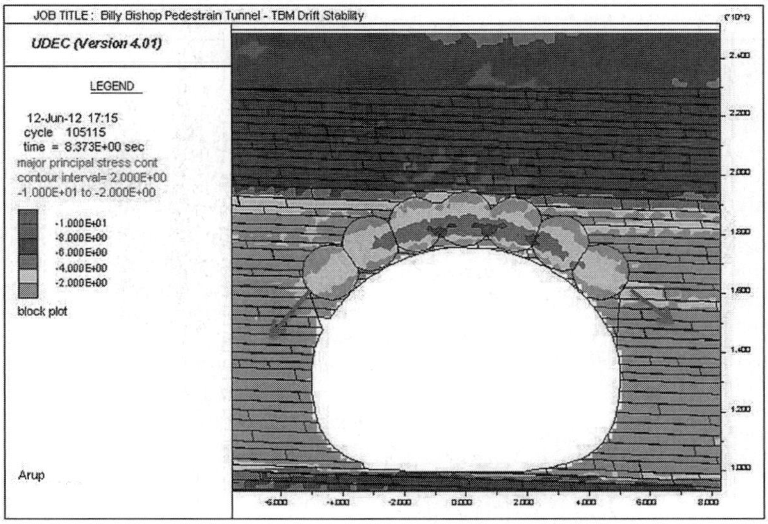


**Figure 7. Backfilled concrete arch stability modeling in UDEC showing maximum principal stress contour upon excavation of main tunnel**

### Derivation of Swelling Parameters

The time-dependent model requires eight input parameters/properties, i.e., the slope of the bedding plane ($\alpha$), the time ($t_0$) for initiation of swelling, Young's modulus ($E$), Poisson's ratio ($v$), the three free swell potentials in principal swelling directions ($m_{x(0)}$, $m_{y(0)}$ and $m_{z(0)}$), one pseudo-Poisson's ratio ($\mu$), a threshold stress ($\sigma_{th}$) (below which no swelling strain reduction occurs) and the critical stress ($\sigma_c$) (above which the swelling is suppressed completely).

A series of laboratory swelling tests on shale samples, along with in-situ rock stress measurements, were performed, obtained and used in the engineering analysis of the tunnel lining.

Table 2 lists the values that were interpreted from the tests. Values of other projects and the back

**Table 2. TDD parameters of the Georgian Bay shale at different projects in Southern Ontario**

| Project | | In-situ Stress (MPa) | Swell Potential | | Critical Stress (MPa) | Elastic Modulus (Mpa) | Time-Dependent Elastic Modulus (MPa) | Viscosity (MPa-day) |
|---|---|---|---|---|---|---|---|---|
| | | | Vertical | Horizontal | | | | |
| Billy Bishop Toronto City Airport Pedestrian Tunnel | GBR | 2.5–6.9 | 0.04 to 0.26% | 0.04 to 0.26% | 2 to 5 | 5,875 | 320 to 1,100 | 38,000 to 73,800 |
| | Project Specific Testing (Exp) | 8.2* | 0.7 to 1.1% | 0.15 to 0.3% | 0.64 | 4,000 | 440 | 29,500 |
| | Design | 2.5–8.2 | 0.7% (Base) 0.24 to 1.2% (Sensitivity Analysis) | 0.4% (Base) 0.05 to 0.45 (Sensitivity Analysis) | 3 (Base) 0.64 to 3 (Sensitivity Analysis) | | N/A | N/A |
| | Shaft Backanalysis | 4.7 (N-S), 5.1 (E-W) | 0.7 to 1.1% | 0.3 to 0.4% | 3 | 6,800 | N/A | N/A |
| Heart Lake Tunnel, Mississauga (Lo et al. 1979) | | Vert. 0.435 Hor. 2 to 8 (Sensitivity Analysis) | 0.42% | 0.13 to 0.17% | 1.9 to 2.6 | 8,300 to 12,000 | 250<br>490<br>140 | 41,200<br>52,600<br>20,800 |
| Deep Lake Water Cooling Centre Island Tunnels, Toronto (Lo and Micic 2010) | | — | — | 0.12% | — | 1,500 to 4,700 | 1100 | 73,600 |
| Skydome and John Street Tunnel, Toronto (Lo et al. 1987) | | Vert. 0.2 Hor. 10 | 0.2 to 0.62% | 0.04 to 0.26% | — | 2,200 to 2,300 | 380 | 38,000 |

*Single test, not considered representative.

analysis described below are also presented for comparison.

## Numerical Analysis

Continuum modeling using the finite difference code FLAC (v.7.0 with the Swello module) was used to design the final tunnel lining including the effect of the TDD behavior of the shale rock mass. As the temporary condition (prior to permanent lining installation) is expected persist for approximately 75 days, the effects of TDD deformation must be considered.

The numerical modeling included all stages of the construction sequence, including the boring and backfilling the seven interlocking TBM drift bores (as well as two lower pilot tunnels to facilitate conventional mass excavation underneath the resulting arch), the three cuts for the main tunnel excavation and the installation of the lining. Figure 8 shows the FLAC 2D model.

The initial designs were based on the permanent invert slab for the main tunnel being placed between 50 days after the bench excavation, with the permanent lining for the tunnel crown and sidewalls cast between 125 days after the main tunnel excavation. However, as excavation proceeded and monitoring data was collected from the shaft, the model was repeated with 30 days to invert placement and 75 days before placing the arch.

A series of seven design cases were considered using a selected range of parameters for in-situ stresses, swelling potential values, critical stresses and boundary conditions. The FLAC 2D runs were carried out to determine the long-term axial thrust and bending moment developed within the lining caused by the swelling rock mass. These analyses were carried out to a design life of 100 years, and were considered both with and without the additional loads caused by a full hydrostatic water pressure build up around the tunnel.

Figure 9 shows a comparison between the predictions of the FLAC model with the movements predicted by the closed form solution from Lo and Yuen. It can be seen that for this size excavation, the movements predicted by numerical analysis are considerably lower.

## MONITORING DURING CONSTRUCTION TO VALIDATE DESIGN APPROACH

To verify the design assumptions for the long-term swelling behavior of the shale rock during construction of the BBTCA Pedestrian Tunnel, a number of instrumentation and monitoring programs were implemented.

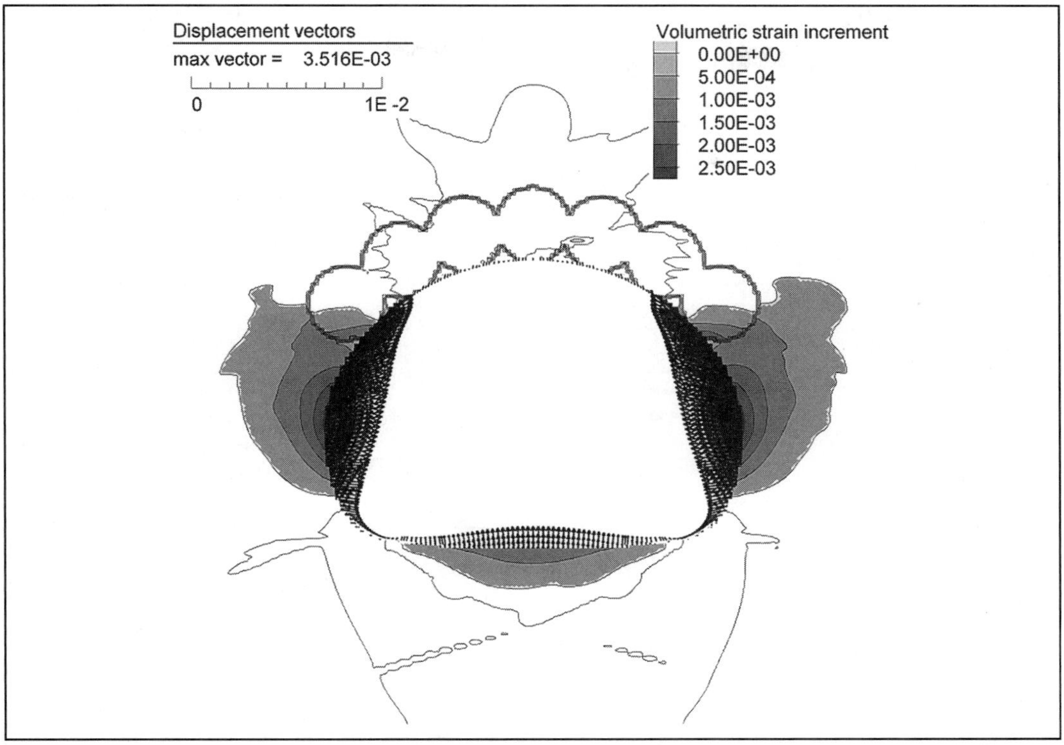


**Figure 8. FLAC 2D model of TBM drifts and main tunnel showing strain due to TDD in the rock and displacement of the sidewalls 150 days after excavation**

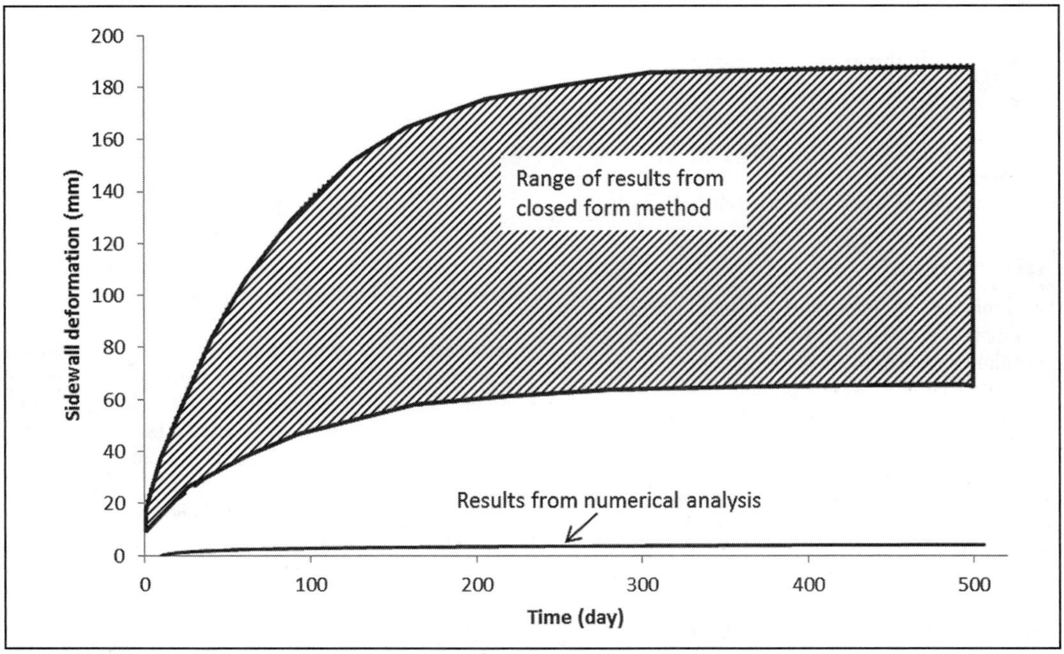


**Figure 9. Comparison of results from closed form solution and numerical analysis**

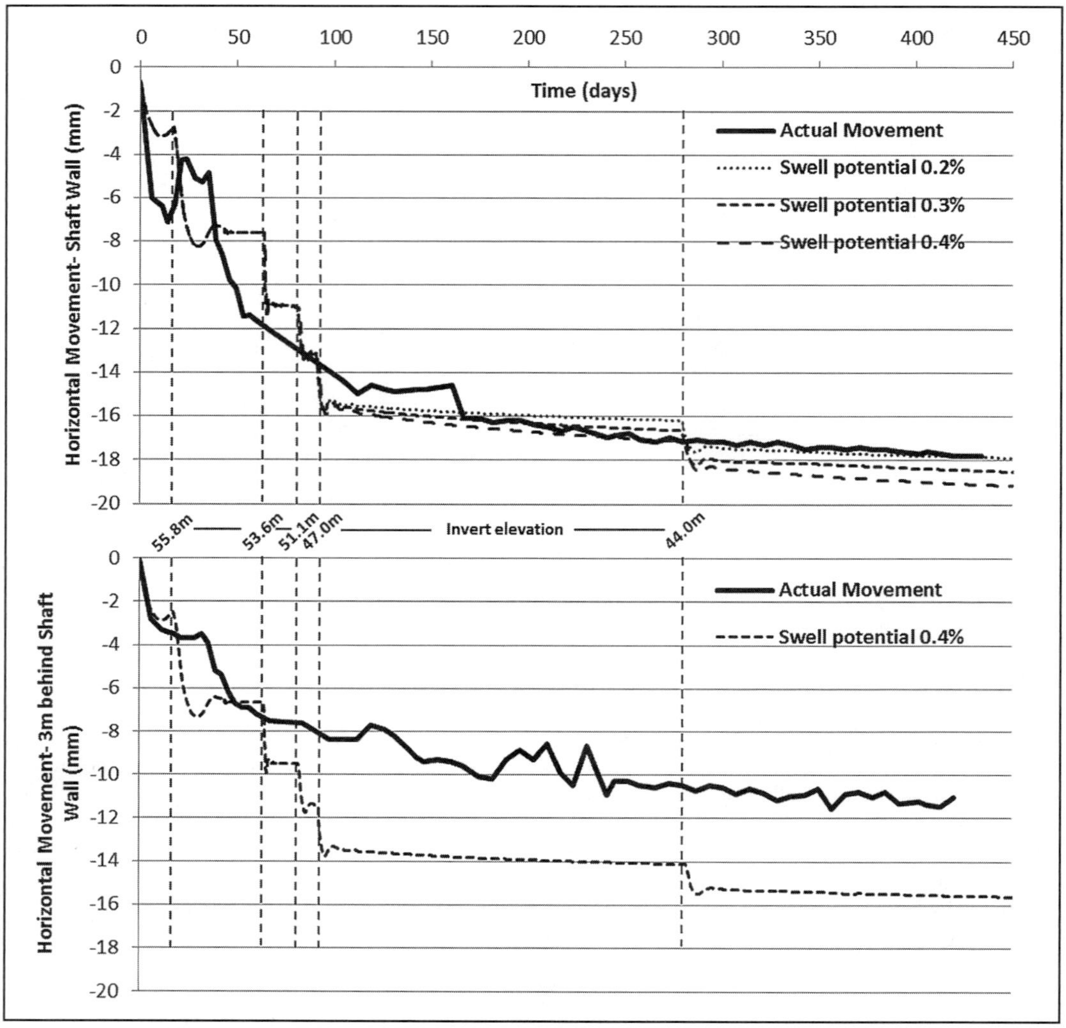


**Figure 10. Shaft back-analysis (elev. +57.4 m)**

### Mainland Shaft

The mainland shaft provided valuable information since it was the first excavation of the project in the Geotgian Bay Shale and remained open the longest. The initial support requirements consisted of bolts and mesh on the north and south walls to protect against loose blocks formed due to clusters of vertical joints. Three inclinometers were installed—two on the north side (0.5m and 3m from the excavation face) and one on the west side (0.5m from the face). The results showed elastic movements during each successive shaft excavation was performed, followed by very small TDD movements. It should be noted that the shaft walls were very wet, providing the ideal conditions for TDD to occur.

Back-analysis was complicated by the degree of restraint provided by the shaft invert. To model this, the elastic response to the shaft excavation was first modeled in 3D using Midas GTS. This allowed the appropriate movement at each inclinometer location to be obtained for each stage of the shaft excavation, as the shaft invert was gradually lowered and the restraining effect on the shaft movement reduced. This data was then used in a 2D FLAC model of a horizontal slice of the shaft, which used a support pressure on the inside of the excavation boundary to model the staged excavation of the shaft. The TDD routine was used to assess the TDD that occurred at each stage.

The back-analysis provided results that gave relatively good agreement with the recorded movements in the inclinometers, as shown in Figure 10.

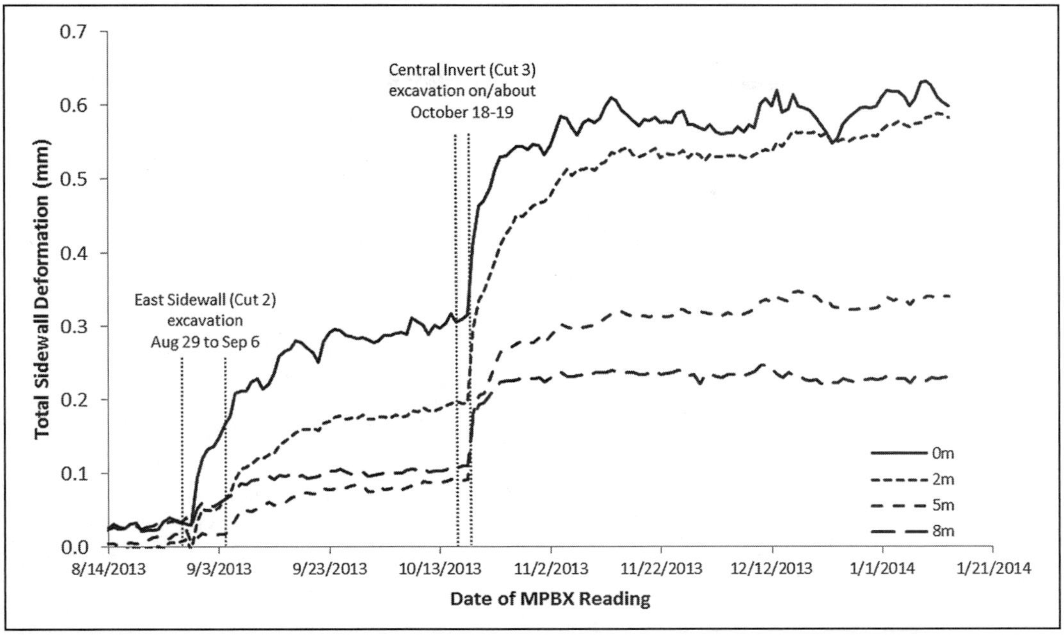


**Figure 11. Measured east wall (MPBX) movements from August 14, 2013, to January 21, 2014**

The movement of the shaft walls during excavation was used to investigate in-situ stress and elastic modulus. The ongoing TDD was used to investigate parameters for critical stress and the swell potentials. While there are a number of inter-related parameters for the swelling, meaning it was not possible with the data available to isolate each individual effect, the data indicated that the critical swell potential was higher than predicted by testing.

Within the limits of the accuracy of the recorded data and the analysis, the back-analysis provided justification for the following key parameters:

- Horizontal Stress: ~4.7 MPa (N-S), ~5.1 MPa (E-W)
- Horizontal Swell potential: In the range between 0.03 to 0.04%
- Critical Stress: ~3 MPa

## TBM-Driven Secant Bore

The radial deformation of the first 1.8m diameter TBM-driven secant drift bore were measured with time using tape extensometers at two separate locations. The installed support was light steel ribs, and these were not observed as being under load at the monitoirng locations. The drift tunnels were generally dry with the occassional water seep, particularly above a 200mm thick band of limestone that was located at tunnel springline. Within the limits of accuracy of the tape extensometeer (±0.5mm) a general convergance trend was detected which was open for approximately 10 weeks. The monitoring showed convergence on both horizontal and vertical chords, with the horizontal convergence of around 1mm, approximately double the vertical. The prediction model showed that TDD movement with a similar horizontal convergence, but slight divergence on the vertical chord.

## Tunnel Monitoring

Construction monitoring consisted of tunnel movement measurements obtained from optical survey of prism points, multi-point borehole extensometers (MPBXs) installed horizontally in the tunnel sidewalls, and measurements of horizontal convergence using a mechanical tape extensometer.

The MPBXs were installed horizontally within the final tunnel sidewalls to measure the tunnel wall movements during and after the tunnel excavation. They were installed 20m from the mainland shaft to allow installation as early as possible without the presence of the shaft excavation impacting the results. On the east wall, the base point of the MPBX was 10m into the rock, and 8m on the west side. The MPBXs were installed during the cut 1 excavation, with localized niches cut in the tunnel sidewalls. This allowed them to be in place when the cut 2 excavation took place, recording both the elastic ground movement and the TDD that occurred. Figure 11 shows a typical plot of the MPBX data.

The tape extensometer readings for the tunnel were recorded on a regular basis at twenty meter

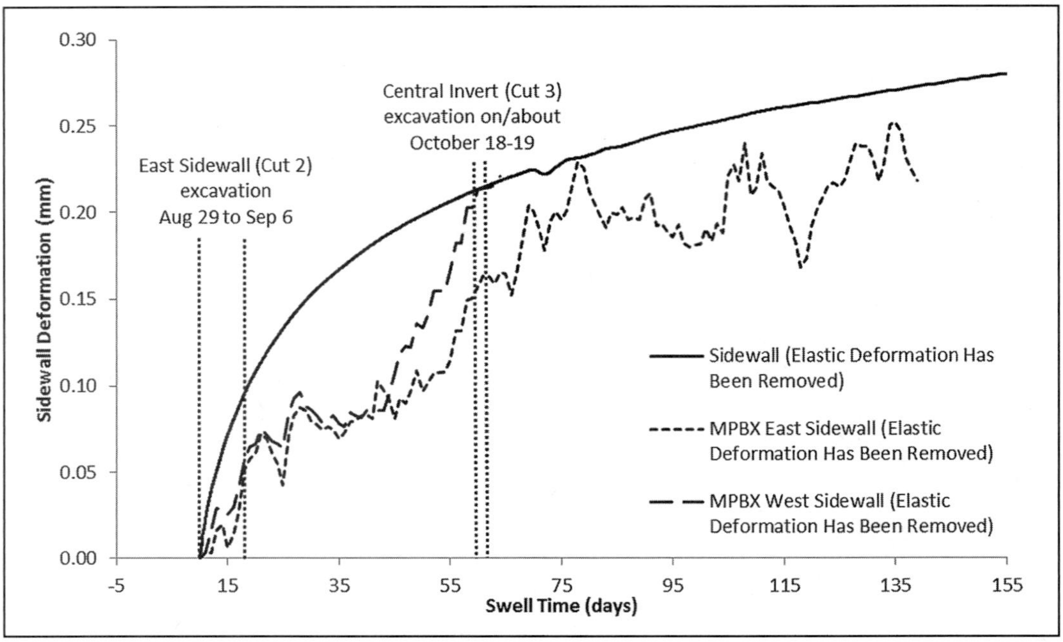


**Figure 12. Measured versus predicted TDD in the tunnel**

intervals along the tunnel, and displayed a similar pattern of movement, although with a slightly higher magnitude than the MPBX data. The range of movements recorded with the tape extensometer is shown on Figure 12, along with the MPBX data, and the predicted movements from the FLAC models used in the design. The survey targets were primarily intended to monitor the overall stability of the tunnel and the accuracy range was not high enough to track sidewall TDD movement in detail, but overall the survey measurements showed general agreement with the tape extensometer readings.

Figure 12 demonstrates that the measured TDD was less that the prediction. It was believed that this may have been due to a lack of water to support the swelling, since the tunnel was relatively dry. As a result, the design was reviewed to ensure that the design could accommodate a greater proportion of the TDD movements after the lining was installed.

## CONCLUSION

The project has used a customized routine within a FLAC2D model to design the tunnel lining to accommodate the loading from Time Dependent Deformation in the Georgian Bay Shale. Monitoring during construction has been performed to validate the parameters used and to justify the design approach.

## REFERENCES


Hawlader, B.C, Y.N. Lee, and K.Y. Lo. (2003) “Three-Dimensional Stress Effects on Time-Dependent Swelling Behavior of Shaly Rock,” *Canadian Geotechnical Journal*, Vol. 40, pp. 501–511.

Hawlader, B.C., K.Y. Lo, and I.D. Moore. (2005) “Analysis of Tunnels in Shaly Rock Considering Three-Dimensional Stress Effects on Swelling,” *Canadian Geotechnical Journal*, Vol. 42, pp. 1–12.

Lo, K.Y. and S. Micic. (2010) “Evaluation of swelling properties of shales for the design of underground structures,” ITA-AITES 2010 World Tunnel Congress, Vancouver, B.C. Canada.

Lo, K.Y. and C.M.K. Yuen. (1981) “Design of Tunnel Lining in Rock for Long Term Time Effects,” *Canadian Geotechnical Journal*, Vol. 18, pp. 24–39.

Lo, K.Y., R.S.C. Wai, J.H.L. Palmer and R.M. Quigley. (1978) “Time-dependent deformation of shaley rocks in Southern Ontario,” *Canadian Geotechnical Journal*, Volume 15, pp. 537–547.

# TRACK 4: CASE STUDIES

## Session 2: Water Control/Grouting

Derek Penrice, Chair

# New Irvington Tunnel Meets the Challenges

**Theodore Feldsher**
URS Corporation

**Glenn Boyce**
Jacobs Associates

**David F. Tsztoo**
San Francisco Public Utilities Commission

**Michael Cash**
Southland Tutor Perini

**Daniel McMaster**
Hatch Mott MacDonald

**ABSTRACT:** The New Irvington Tunnel, located in Alameda County, California, is a key component of San Francisco's Water System Improvement Program. Mining of the 3.5-mile tunnel started in March 2011 and was completed in October 2013. Installation of the final lining is projected for completion in Summer 2014 and the tunnel commissioning for service is planned in Fall 2014. This paper will describe some of the key challenges encountered during the project construction phase and how they were successfully addressed and managed by the project team. Some of the most significant challenges included a scarcity of experienced tunnel labor for conventional mining; very heavy groundwater inflows; difficult tunnel excavation including squeezing ground conditions; and gassy tunneling conditions which led to reclassification by CalOSHA during construction.

## PROJECT DESCRIPTION

The New Irvington Tunnel is a critical link in the regional water system operated by the San Francisco Public Utilities Commission (SFPUC). The tunnel is located in Alameda County, California and extends approximately 5.6 km (3.5 mi) from the Sunol Valley to the hills above the City of Fremont. The tunnel has a finished diameter of 2.6 m (8.5 ft.) and delivers water to over 2.5 million customers in the San Francisco Bay Area. The new tunnel is aligned parallel to the existing Irvington Tunnel, which was built between 1928 and 1931 as part of the original Hetch Hetchy Aqueduct System.

The tunnel profile is shown in Figure 1. The tunnel alignment crosses several secondary fault traces and is located between two major active fault systems. The Calaveras Fault is located about a quarter mile east of the portal in the Sunol Valley, and the Hayward Fault System is located about a mile west of the portal in Fremont. Either of these active faults could cause strong shaking and secondary fault offset which would seriously damage the existing tunnel and disrupt flow to the City's customers.

Prior to construction of the New Irvington Tunnel, high water demands prevented the City from taking the existing tunnel out of service for inspection or maintenance for many years. To provide seismic reliability and ensure continued reliable delivery of high quality water to all of its customers, the SFPUC determined that construction of a new parallel tunnel was necessary. The new and existing tunnels connect at each portal to complex large diameter pipeline systems. At the east (upstream) portal, the tunnels connect to four Alameda Creek Siphons, the fourth of which was recently built to survive seismic offset on the Calaveras Fault. At the west end, the tunnels connect through complex manifold systems with five Bay Division Pipelines, the fifth of which was recently added. Together, the New Irvington Tunnel and the upgraded large-diameter pipeline connections at each end represent critical lifeline components of San Francisco's water delivery system. These and over 80 other projects were built as part of San Francisco's $4.6 Billion Water System Improvement Program, which was initiated in 2002 to improve the overall reliability of the existing water system.

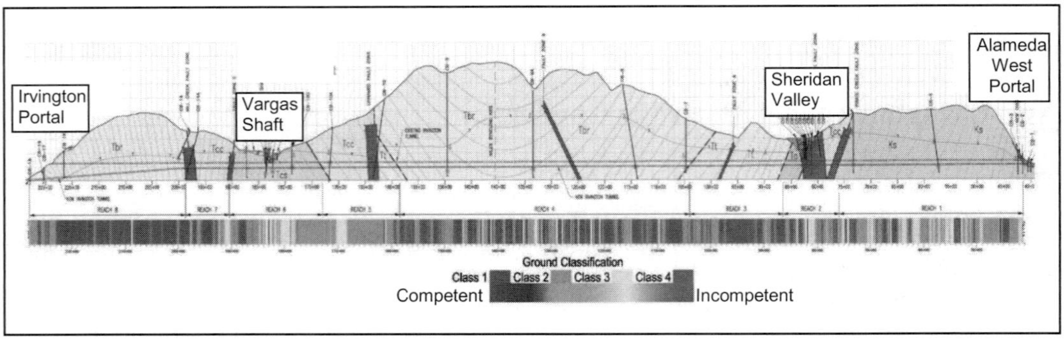


**Figure 1. Baseline geologic profile and actual ground classifications**

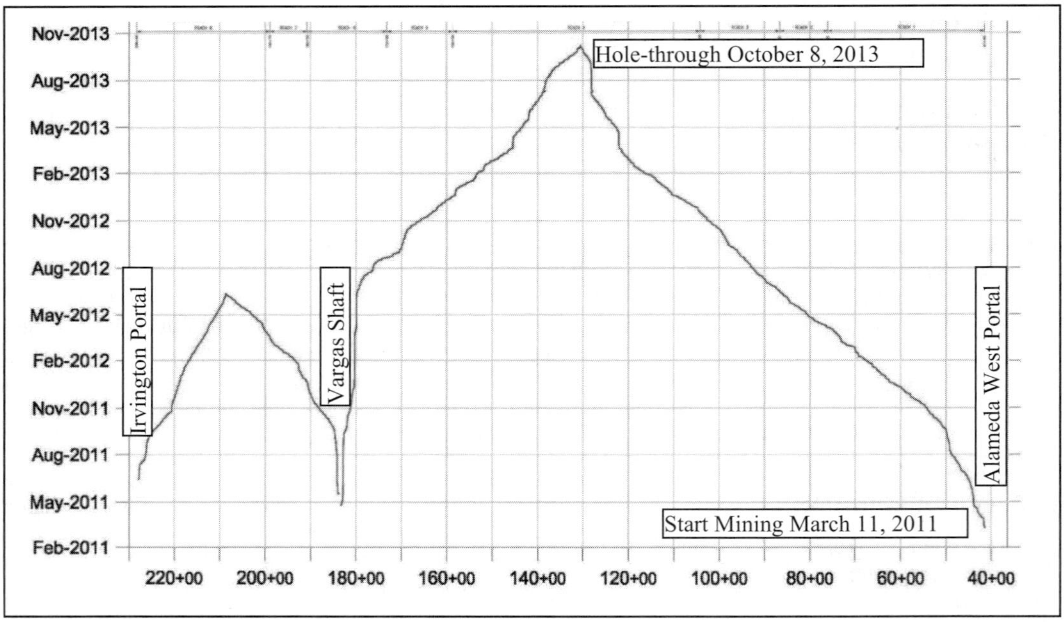


**Figure 2. Tunnel excavation progress by station**

The project final design phase was initiated in 2006, and included an extensive program of geotechnical investigations. More details on the geotechnical investigations and project design are presented in Boyce, et al., (2010a) and Boyce, et al., (2010b). The design was completed and the project went to bid in late 2009. Four bids were received, ranging from $237M to $293M. For comparison, the Engineer's Estimate was $253M. The Contract was awarded to the low bidder, a Joint Venture of Southland Tunneling and Tutor Perini (STP-JV), and construction started in August 2010. Tunnel mining started in March 2011 and was completed in October 2013. Figure 2 shows a plot of the tunnel excavation progress by station. Figure 3 shows a plot of total weekly tunneling production. At the time this paper was drafted, installation of the final liner was in progress, and the project construction was about 90 percent complete overall. The projected total construction cost was $256M, including $29M in contract change orders. The causes for some of the change orders are described in the following sections of this paper.

During construction, the new tunnel encountered difficult and highly variable ground conditions similar to those reported for the existing tunnel. The ground conditions included zones of weak, fractured, and sheared sandstone/siltstone/shale combined with high groundwater heads and high inflows. Extensive probe drilling, grouting, and drainage were carried out. Surface dewatering wells were installed to reduce the impact of inflows on tunnel excavation in one reach. The tunnel excavation was completed

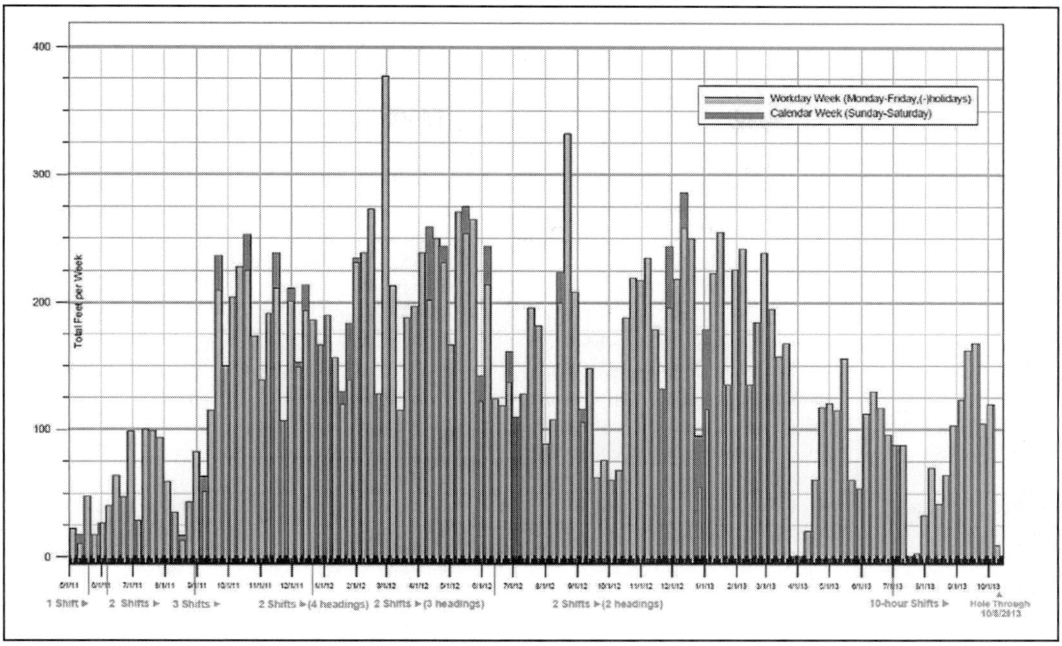


**Figure 3. Total weekly tunneling production (all headings combined)**

using conventional mining methods, including the use of roadheaders in combination with drill & blast methods. Steel sets with timber blocking and cribbing were the primary initial support measures, combined with channel spiling, invert struts, and other measures as needed to cope with the ground conditions encountered. The final tunnel liner consists of a continuous 8.5-diameter steel pipe backfilled with low density cellular concrete. The steel liner thicknesses and joint types were designed to resist external hydrostatic pressures and to accommodate up to 6 inches of secondary fault offset in four specific zones. The limits of these secondary fault zones were confirmed during construction based on geologic mapping performed in the tunnel headings. More details on the conditions encountered and the challenges to tunnel excavation and support are presented in the following sections.

## GEOLOGIC CONDITIONS AND GROUND CHARACTERIZATION

The New Irvington Tunnel is located in the Coast Range geologic region of the San Francisco Bay Area. The tunnel passes through an uplifted and deformed block of sedimentary rock formations bounded by the Hayward Fault on the west and the Calaveras Fault on the east. Activity on these fault systems has led to significant folding, fracturing, and shearing in the rocks along the tunnel alignment. Regional tectonic compression has uplifted the range and created folds that form at least one anticline and one large syncline in the site area. Although not crossed by any seismically active faults, the alignment contains four mapped secondary faults which can undergo sympathetic offset during a major event on the nearby active fault systems. The rock mass is generally composed of weak, intensely fractured and sheared sedimentary rocks (mainly sandstone, siltstone, interbedded siltstone/sandstone, and shale), and also includes some sections of stronger and more massive rock. The tunnel also intercepted a number of faults and shear zones with abundant clay gouge. A geologic profile along the alignment is shown in Figure 1.

Ground conditions along the tunnel alignment were divided into four ground classes to aid in the selection of tunnel excavation and support methods. The ground classes were defined based on the physical characteristics of the ground and its anticipated behavior during tunnel excavation. The ground class definitions, predominant ground behaviors, and key characteristics associated with each ground class are described in Table 1. The distribution of ground classes actually encountered during tunnel mining is illustrated in Figure 1. Overall, the actual distribution of ground classes was more favorable to tunnel mining than originally anticipated. For example, the Geotechnical Baseline Report stated that the combined footage of ground classes III and IV was expected to be 65% of the total alignment length. However, during mining the actual total encountered was only about 24%.

**Table 1. Definitions of ground classes**

| Ground Class Definitions | Typical Rock Characteristics | Typical Discontinuity Characteristics | Ground Behavior |
|---|---|---|---|
| I: Massive to Moderately Fractured Rock | Sandstone, siltstone, and interbedded siltstone/sandstone; weak to strong rock; slightly weathered to fresh | Very rough to rough; fresh to slightly weathered surfaces | Structurally controlled block instability; spalling |
| II: Highly Fractured Rock | Sandstone, siltstone, interbedded siltstone/sandstone, and shale; weak to moderately strong rock; highly to slightly weathered | Rough, smooth, or slickensided surfaces or bedding planes; moderately to highly weathered/altered surfaces with infillings of clay and/or sand | Slow raveling; fast raveling where flowing groundwater is encountered |
| III: Intensely Fractured Rock | Sandstone, siltstone, interbedded siltstone/sandstone, and shale; thinly bedded to laminated rock structure; very weak to moderately strong rock, may be friable, poorly cemented; highly to slightly weathered/altered | Smooth, slickensided surfaces; highly weathered/altered with occasional moderately wide clay/sand-filled joints, shears, and shear zones | Fast raveling, caving; potentially flowing ground |
| IV: Heavily Sheared/Faulted Rock with Clay Gouge/Infilling Materials | Heavily sheared rock including fault gouge, shattered rock, all with abundant clay; extremely weak to very weak rock; moderately to completely weathered/altered | Slickensided surfaces; highly weathered/altered with wide clay-filled joints, shears, and fault/shear zones | Squeezing; swelling; caving; fast raveling |

## TUNNEL EXCAVATION AND SUPPORT CHALLENGES

The tunnel excavation encountered difficult and highly variable ground conditions. As anticipated, unstable ground conditions were encountered throughout the tunnel, including but not limited to, raveling/caving, squeezing, running, and flowing conditions. The sheared nature of the rock strata, weak rock strengths, abundant clay infilling materials, intensely fractured rock mass, and high groundwater levels all required mitigation to maintain stability of the tunnel excavation.

Pre-support using channel and bar spiling was necessary in many areas to control raveling, caving, and crown instability, primarily in tunnel reaches with Ground Class III and IV conditions. Face support in conjunction with pre-support was required in some areas to control overbreak, raveling, running/flowing, and caving behaviors at the tunnel face. Face support measures included breastboards, mining with a top heading and bench, and numerous bulkheads to control fast raveling/flowing ground conditions. The bulkheads were also used to facilitate drilling and grouting for water inflow control and ground stabilization.

One of the more significant challenges associated with the tunnel mining was the high frequency of changes in "ground class" and hence ground characteristics in terms of rock soundness and strength. This high frequency of changes in ground class meant that the tunnel crews needed to be able to readily adapt to the variable conditions and be prepared for the use of different excavation techniques, requiring either drill and blast, roadheader excavation or hand excavation. The crews also needed to be able to rapidly deploy the use of different ground support systems. Eventually, the crews developed a hybrid system for improved efficiency. This system consisted of drill and blast excavation with a relatively light shot to loosen the rock mass and partially pull the round, followed by mining out with the roadheader to muck out the shot rock and trim the opening out to the full excavated dimensions. This approach helped minimize overbreak and maximize the utilization of the roadheaders.

Early in the job, it took time for the construction crews to develop and refine procedures for integrating drill and blast operations with roadheader excavation. Determining when each was most appropriate was often by trial and error. However, switching back and forth between the two methods was initially quite time consuming, so the mining efficiency had to be balanced with the time necessary to switch methods. In some areas the roadheaders were used to mine through relatively massive and strong sandstone (with unconfined compressive strengths approaching or exceeding 10,000 psi), resulting in relatively slow advance rates and high wear and tear on the equipment. Figure 4 shows one of the three roadheaders used on the project.

To enable the crews to rapidly deploy and adapt support types to the ground class classifications, a standardized support system was selected. It was decided that steel sets with timber lagging would be

**Figure 4. Antraquip roadheader being lowered into Vargas shaft**

used as the primary support type, changing the spacing of the steel sets depending on the ground classification. This enabled the crews to become very proficient with the installation of the initial support systems. Also, the materials required for ground support were stored in the tunnel heading, enabling the support to be installed quickly when poor ground conditions were encountered.

## SQUEEZING GROUND CHALLENGES

Squeezing conditions in underground construction can occur when the in-situ stresses around an excavation exceed the strength of the rock mass. Squeezing behavior usually requires a combination of high in-situ stresses (due to tunnel depth and/or regional geologic stress regimes) and low rock mass strengths. In shear zones or zones with many clay filled joints, the rock mass shear strength can be controlled by the properties of the infilling materials, which are often weak and clayey, and can lead to time dependent behavior under stress.

The New Irvington Tunnel encountered squeezing ground conditions in numerous areas during mining. The conditions were usually manifested by crushing and splitting of timber blocking and lagging, deformation of the steel sets, and/or heaving and cracking of the concrete invert. Some steel sets experienced inward convergence of the ribs, plunging of the foot blocks, and in a few cases buckling of the flanges. The effects typically occurred over a period of days to weeks following initial excavation and support. In a number of the squeezing areas, the rock conditions appeared relatively benign during initial excavation, and so appropriate initial support measures to resist squeezing were not initially considered necessary. However, when continued convergence was verified, remining was sometimes necessary to reestablish the minimum required clearance envelope and resupport the ground. This remining work was done hundreds of feet behind the face, which limited access to the face and as a result significantly delayed the overall tunneling advance. Typical resupport measures included jump sets, doubled up sets, and invert struts. In some areas, the squeezing first manifested as heave and cracking of the concrete invert slab, requiring removal and replacement of the slab, often in combination with installation of invert struts. Convergence monitoring was utilized in repaired areas to confirm full stabilization of the ground prior to installation of the final steel lining. Unfortunately, baseline convergence monitoring immediately following initial excavation was usually hampered by the presence of the roadheader and the muck removal gantry conveyor system. This delayed the convergence readings until after the roadheader had passed. Where monitoring could be conducted, convergence amounts of 6 to 8 inches or more were recorded prior to stabilization of the most actively squeezing areas.

## LABOR AVAILABILITY CHALLENGES

At the start of the project, the available labor force was relatively inexperienced with conventional tunnel mining under highly variable and very tough conditions. Most of the laborers who had any tunneling experience were used to tunnel boring machines, not conventional mining. In addition to skilled labor, the Contractor also had the challenge of locating and hiring experienced superintendents to train the crews. Eventually, a group of skilled superintendents was assembled, and they were able to train up multiple crews of tunnel miners.

Even with experienced superintendents, many of the general construction laborers available in

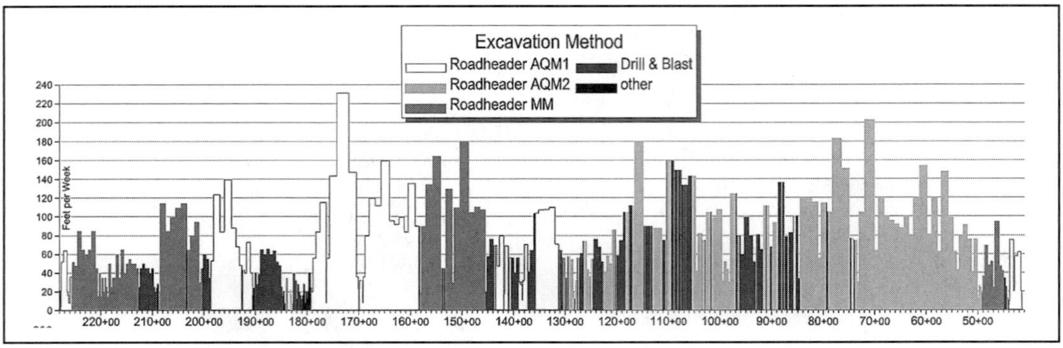


**Figure 5. Weekly advance rate per heading by station and mining method**

the project area were found to be poorly suited for the rigors and challenges of underground tunneling work. About 180 candidates were tried and less than a third were found satisfactory. Even the best candidates took 6 to 9 months of on the job training to learn how to deal with the difficult and variable ground conditions, support system installation procedures, drilling and grouting procedures, and different mining methods. The learning curve included time to develop basic mining skills, including safely reading the ground conditions, avoiding rock falls, installing initial support systems, and reacting quickly to groundwater inflows.

Getting the crews up the learning curve took a substantial amount of time and effort during the early portion of the project, and this appears to have impacted mining production. Figure 3 shows the overall tunnel mining production rate for the project. As indicated, the production rate increased markedly over the first year of construction. This illustrates the improved overall productivity as the crews developed their skills and mining techniques. Detailed data on weekly tunnel advance rate by station is summarized in Figure 5 along with the excavation methodologies used. Once the tunneling crews were established and advanced up the learning curve, some exceptional mining progress was achieved through very difficult conditions.

The low advance rates from station 180+00 to station 180+50 (as shown in Figure 5) resulted from both the difficult ground conditions and the difficulty in establishing the skilled labor force required for the project. While it would have been desirable to mine all four tunnel headings at the same time, this was found impractical due to staffing limitations. Meeting the required hiring targets from the local community was a priority, but it was decided that one of the tunnel headings should be placed on hold to ensure that the remaining three headings were staffed with skilled and competent personnel and the tunnel was advanced proficiently and safely. As a result, while tunneling was completed westward from the Vargas Shaft towards the Irvington Portal, tunneling eastward out of Vargas Shaft was temporarily put on hold (as shown in Figure 2). Any schedule delays incurred by concentrating the work in three headings were later mitigated by installing the final lining between Irvington Portal and Vargas Shaft concurrently with the tunnel excavation from Vargas Shaft to Alameda West Portal.

## GROUNDWATER INFLOW CHALLENGES

Pre-construction groundwater levels were as much as 370 ft. above the tunnel crown. Three-dimensional transient groundwater modeling during design predicted heavy groundwater inflows and extensive drawdowns resulting from tunnel mining (see Zhang et al. 2008 for details on the modeling). Based on the modeling results, the Geotechnical Baseline Report defined the maximum sustained portal flows at 850 gallons per minute (gpm) (URS and Jacobs Associates, 2009). Consistent with the modeling predictions, high inflows were encountered during tunnel mining in most of the reaches. The actual portal discharge flows measured at the on-site water treatment plants are shown in Figure 6. Figure 7 shows a photograph of one very heavy inflow area. Concentrated inflows often continued even after large amounts of pre-excavation grouting. Extensive panning and collection measures were later installed in such areas to allow placement of the final lining and cellular backfill. Selection and implementation of specific groundwater control measures during tunnel construction were the Contractor's responsibility. The primary groundwater control measures utilized on the project included pre-drainage and pre-excavation grouting.

The general performance objectives for pre-excavation grouting and pre-drainage included the following: (1) limiting the groundwater inflows at the tunnel face to a rate compatible with the selected tunnel construction means and methods, and (2) mitigating adverse ground behavior caused by heavy

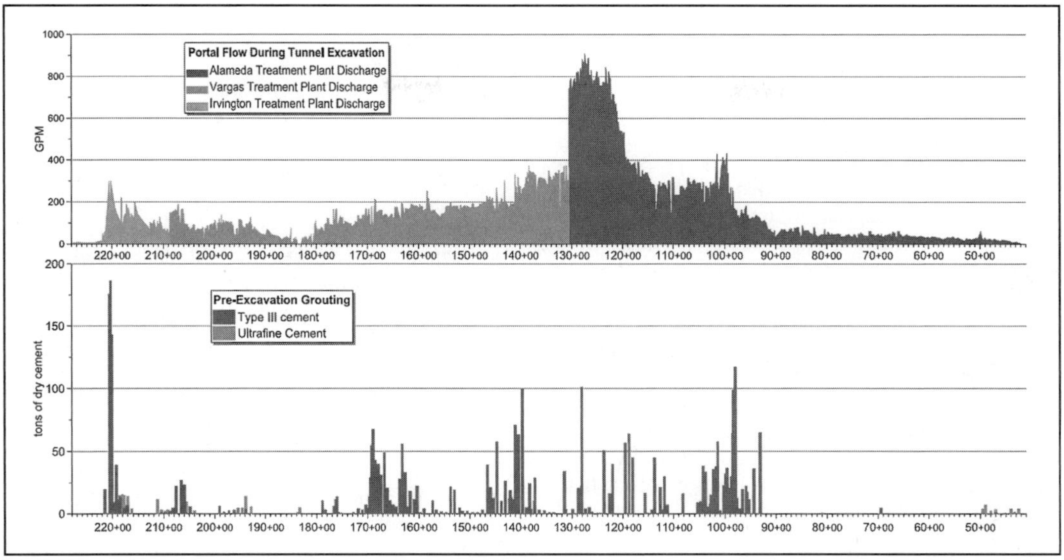


**Figure 6. Portal discharge flows and pre-excavation grouting**

**Figure 7. Tunnel heading in zone of high water inflow**

groundwater inflows as necessary to allow adequate installation of steel sets and initial support measures. During construction, the contractor frequently had to adjust and adapt the pre-excavation grouting and drainage techniques and procedures to accommodate very difficult ground and groundwater conditions.

### Probe Hole Drilling

Continuous probe hole drilling was required for the entire length of the tunnel. Two probe holes were required to be drilled out in front of each heading face at all times with a minimum 20 foot overlap. The intent was to provide a minimum 20-foot warning before the tunnel face encountered any adverse ground condition or zone of high water inflow. During construction, the Contractor opted to drill the 2-inch diameter probe holes out 100 to 120 feet ahead of the face to get more advance information on upcoming conditions. The probe holes were successfully able to identify problematic zones in front of the heading faces, including zones of soft ground as well as zones of high inflow. The Contractor worked with both Antriquip and Mitsu Miike to design and mount suitable probe drills on each roadheader. In addition to the continuous probe holes and the frequent grout holes, the drills were also used on several occasions to drill blast holes. Figure 4 shows the drill attachment mounted on one of the roadheaders.

### Pre-Drainage

Pre-drainage from within the tunnel was accomplished using probe holes, drilled around the perimeter of the face. Unless grouting was performed as described below, the holes were left open to drain and relieve hydrostatic pressures in front of the face as the headings were advanced. Pre-drainage generally supplemented pre-support measures such as breast boarding the face and driven channel spiling. Pre-drainage was implemented in some weak zones to improve ground behavior and prevent or mitigate the potential for flowing ground conditions. Pre-drainage was sometimes the only option in zones with clay-filled shears and clay fault gouge, where cement grout penetration was limited by the overall low hydraulic conductivity of the rock mass. Pre-drainage to lower groundwater pressures ahead of the face was generally not attempted in zones of high conductivity, because the high inflow volumes and high pressures were typically very slow to dissipate. Such areas were generally grouted instead.

### Pre-Excavation Grouting

Pre-excavation grouting was most often employed in areas of the tunnel where the rock mass was relatively strong and the hydraulic conductivity was controlled by open fracture networks. The decision of when to grout was made by the contractor, based in part on a specified probe hole inflow threshold. By contract, pre-excavation grouting was deemed compensable if the inflow threshold was met or exceeded by any probe hole in the tunnel face. The threshold was defined as 20 gpm or more flowing from any discrete water bearing feature, or 0.2 gpm per foot of probe hole (i.e., 20 gpm for a 100 foot hole). During construction, the Contractor opted to carry out pre-excavation grouting at every location where inflows exceeded the specified threshold. At numerous locations requiring grouting, probe hole inflows reached 200 to 300 gpm. After mining through these areas, grouted joints with apertures typically less than 1 inch wide were often observed. Figure 7 shows the total quantities of grout injection along the alignment.

The Contractor initially developed a pre-excavation grouting plan that included grout mixes for both Ultra-fine and Type III cement. However, after some trials early on in the project, it was concluded that the Type III cement grout achieved sufficient penetration of the joints and fractures to adequately control water inflows. Grout holes were typically angled out around the perimeter of the face from 5 to 30 degrees depending on the locations of the water bearing features and the type of ground encountered in the probe holes. The Contractor used a systematic approach of down staged drilling and grouting in lieu of the specified casing advancement system. Inflatable PVC packers 15 to 20 feet long were installed into each grout hole with rapid setting cement before grouting commenced. Up to two holes were grouted simultaneously. Grouting was performed using a portable grout plant mobilized on a flat car and set up just behind the roadheader, using dry bags of cement (Figure 8). The Contractor opted for use of portable plants in lieu of the specified remote grout batching system, which would have required plants located outside at each portal. Grout injection pressures varied depending on back pressures of the groundwater but typically did not exceed 300 psi. In all, about 8 million pounds (4,000 tons) of cement grout was injected throughout the length of the 3.5 mile tunnel.

### Grouting Payment Provisions and Contract Incentive

Pre-excavation Grouting measurement and payment provisions were developed to compensate the Contractor on a unit price basis for the following items:

- Drilling of holes for drainage and grouting (hourly)
- Grout injection of Portland Cement and Ultrafine Cement (hourly)
- Material cost of Portland Cement and Ultrafine Cement (ton)
- Material cost of polyurethane and sodium silicate grout (gallon)
- Indirect Cost of hole drilling and grout injection (hourly)

In addition to these unit price items, an incentive clause was established to encourage the Contractor to grout only when necessary for efficient tunnel advance. The Contractor was responsible for selection of tunnel construction means and methods, and for determining when to perform pre-excavation grouting for managing ground behavior and groundwater inflows. The incentive clause was based on the sum of the original bid prices for the grouting bid items, which totaled about $8.5M. The Contractor is entitled to any unspent balance remaining from this sum as a one-time incentive payment at the end of the contract, as long as the substantial completion schedule is met. At the time this paper was drafted, the unspent balance potentially available for the incentive was about $915k. More details in the contract payment provisions for pre-excavation grouting are presented in McCarter, et al. (2014).

**Figure 8. Pre-excavation grout plant in heading**

### Leakage from Existing Tunnel

The existing Irvington tunnel was in continuous operation during construction of the new tunnel, and this led to one of the most significant water inflow challenges. For most of the alignment, the new tunnel parallels the existing tunnel with a separation distance of less than 190 feet horizontally and 30 feet vertically. Also, for most of the alignment, the existing groundwater head is higher than or similar to the normal operating head in the existing tunnel. In one area near the Irvington Portal, where the separation between the tunnels was only about 75 feet, the Contractor encountered probe hole inflows consistently in the range of 200 to 300 gpm under substantial pressure. The flows did not dissipate appreciably with time, so drainage was not considered a viable mitigation option. The source of the flows was later found to be a series of open joints and fractures, roughly perpendicular to the tunnel alignment, with apertures ranging up to about 3 inches. These open joints were encountered over about 400 feet of the tunnel length. In addition to containing large amounts of stored water, the joints probably also had some degree of hydraulic connection to the existing tunnel. Water chemistry tests could not conclusively determine that the water source was leakage from the existing tunnel, but this is the most likely explanation since the existing tunnel reports indicated encountering little or no groundwater in this area. Treatment of the inflows included injection of large quantities of cement grout and additives including vermiculite, bentonite, and sawdust in effort to fill the large water bearing joints and fractures. In total, the Contractor injected over 1.4 million lbs of cement grout during the course of mining through this zone. Although many grouting cycles were required, the grouting ultimately proved effective at reducing the inflows and allowing the heading to advance safely. The water quality flowing through the existing tunnel was continuously monitored during this period of time, to make sure the injected grout did not infiltrate into the existing tunnel.

### Sheridan Valley Dewatering

Another section of tunnel where high inflows were expected was the zone through Sheridan Valley (designated as Reach 2, as shown in Figure 1). This zone was carefully evaluated during the geotechnical investigations and design phase, because it presented very tough mining conditions and heavy water inflows during the existing tunnel construction. Based on the geologic and pump test data from the fractured and highly permeable Oursan Sandstone Formation, this short reach was conclude to be capable of producing up to 800 gpm of groundwater inflow into the tunnel unless mitigated. Additionally, the tunnel crosses through the Sheridan fault zone in this reach, with expected ground conditions including intensely sheared and fractured rock and squeezing ground. Together with the high water table, these conditions presented a risk of running ground or other adverse behavior during tunnel mining.

To mitigate the risk of adverse ground behavior, the designers conclude that pre-drainage in the Sheridan Valley area was the best solution. Because of the relatively low cover in this reach, a surface dewatering well field was designed as the most efficient pre-drainage method. A temporary surface construction easement was acquired from the property owner, and the Contractor installed a total of 23 dewatering wells, each 270 feet deep, extending 30 feet below the tunnel invert. The wells were installed between stations 79+80 and 87+50. The wells started pumping in January 2012, several months prior to the date when the tunnel heading from Alameda West Portal (AWP) entered the reach. The initial discharge rate from the well field averaged about 900 gpm over the first several months and declined over the next 12 months to a steady state discharge of about 300 gpm. By the time the Alameda West heading advanced through the dewatered zone, the tunnel face and probe holes were effectively dry. The wells successfully lowered the groundwater below the tunnel invert and as a result the tunnel heading advanced through the reach without incident or delay. The dewatering wells are expected to operate for a period of about 30 months. The wells will not be turned off until after the final liner has been installed, the annulus has been backfilled with low density cellular grout, and contact grouting is completed. This is expected to occur sometime in the fall of 2014.

In early 2012, a number of private wells of the local property owners and ranchers along the new tunnel alignment began to show impacts as a result of the project dewatering. In anticipation of this event, the SFPUC had previously developed a Groundwater Management Plan outlining mitigation measures that would be undertaken to restore impacted water supplies. As part of the mitigation, SFPUC designed

**Figure 9. Refuge chamber used in tunnel heading**

and constructed a 4-inch HDPE surface distribution pipeline for discharge flows from the surface well dewatering field in the Sheridan Valley. The pipeline provided water primarily for irrigation and livestock, and was eventually connected to more than 12 different properties within the impacted area. The pipeline saved the SFPUC hundreds of thousands of dollars on the cost of delivering water by truck to the impacted properties. More details on the extensive and successful groundwater impact mitigation efforts carried out for the project are presented in Tsztoo, et al. (2014).

## CHALLENGES ASSOCIATED WITH TUNNEL RECLASSIFICATION

The initial tunnel classification of Potentially Gassy with Special Conditions was obtained during the design phase. The classification was based on available reports from the original tunnel construction, data from subsequent inspections, and data from the extensive field investigations conducted during design. CalOSHA subsequently reclassified the tunnel as Gassy less than four months after the start of mining. This presented a number of additional challenges to the construction team.

The original Irvington Tunnel construction reports from SFPUC's archives indicated a number of localized and relatively minor gas detections, as did subsequent tunnel inspections*. However, the investigations conducted during final design for the new tunnel reported no measurable flammable gas detections. Based on review of the available data, the Design Team concluded that limited occurrences of gas should be anticipated during mining, but expected that these could be successfully mitigated with adequate ventilation, monitoring, and implementation of basic safety precautions. CalOSHA agreed with this logic and supported the proposed tunnel classification as potentially gassy.

Even with the potentially gassy classification, the Design Team recognized the potential risk of gas detections and subsequent reclassification during construction. Therefore, the Contract required that all underground mining equipment must be permissible. Even though not mandated by the initial CalOSHA classification, this specification requirement substantially mitigated the cost and schedule risks of possible reclassification. Including this clause in the Contract mitigated the budget and schedule risk associated with procurement and mobilization of replacement equipment should reclassification occur.

The specified requirement for all permissible mining equipment proved to be prudent. In June 2011, a welder inadvertently ignited a pocket of methane gas that had collected in the crown of the Vargas West tunnel heading. Fortunately, no injuries occurred, but all mining operations were stopped and the tunnel was temporarily evacuated as a precaution. Following an investigation of the incident, CalOSHA reevaluated the safety risks in the tunnel. As a result of the reevaluation, CalOSHA subsequently re-classified the entire tunnel as Gassy with Special Conditions. Part of their rationale was that the mining operation still had more than two years to go, and mining had yet to be carried out in several locations where gas was anticipated based on detections in parallel locations in the existing tunnel. Based on the serious incident that occurred early in construction, the risk of additional problematic gas occurrences in the new tunnel was concluded by CalOSHA to be high enough to merit the need for heightened precautions and safety measures.

Following the reclassification, the project was subject to ramped up regulatory scrutiny including more frequent inspection visits by CalOSHA. The major mining equipment (including the roadheaders and locomotives) did not have to be replaced because it was originally specified to be permissible. However, the ventilation system had to be upgraded including installation of blast relief hatches at the portals, additional fans, and other features to ensure ample airflow at the tunnel face during excavation. Hot work permits were required for all welding operations and work with non-permissible equipment in the tunnel. Dedicated full time gas testers were required for each shift at each tunnel heading and tunnel crews had to complete additional safety training for working in gassy conditions. CalOSHA also required refuge chambers to be procured and installed in each tunnel heading that extended more than 5,000 feet from the portals (Figure 9). This requirement was strictly enforced, requiring a refuge

* Available records for the 25-mile long Coast Range Tunnel (constructed at the same time as the nearby Irvington Tunnel) reported very gassy conditions during construction, including at least one fatal explosion.

chamber in a heading that extended only about 5,400 feet before hole-through.

Contract change orders were subsequently issued for the ventilation system upgrades, the delays associated with hot work permits and working under gassy conditions, and the hiring of the dedicated gas monitoring staff. Other change orders were issued for the refuge chambers and for continued operations under the more rigorous regulatory permit requirements of the Gassy Tunnel Classification. The resulting total budget increase due to the tunnel reclassification change orders was about $18 million. The construction schedule was also extended by 97 days to account for the additional unanticipated inefficiency.

## LESSONS LEARNED AND CONCLUSIONS

The implementation of surface dewatering wells in the Sheridan Valley tunnel reach proved to be very beneficial. The wells lowered the groundwater in an area with two secondary fault zones. The tunnel excavation advance rates through the dewatered area were much better than if dewatering had not been conducted. As a secondary benefit, the dewatering well water was used to provide supplemental irrigation supplies to impacted local landowners during tunnel construction. The effectiveness of the dewatering program was highlighted during tunnel cleanup before the installation of the steel lining. Some of the wells were temporarily taken off line for replacement and repositioning. During that time, the completed tunnel saw significant increases in groundwater inflows, which disappeared after the dewatering wells were turned back on.

The pre-excavation grouting program also proved to be an effective means of controlling and reducing groundwater inflows into the tunnel. With conventional mining providing full access to the face, the Contractor was able to adapt to variable and difficult ground conditions for both drilling and grouting, changing drilling angles and locations to intercept discrete features as needed. In addition, the use of unit price bid items to pay for the drilling and grouting activities and the delay time associated with critical path grouting proved effective for achieving the desired technical results and beneficial for contract administration.

The Contract requirement for all underground mining equipment to be permissible proved to be a very prudent precaution, which paid off when CalOSHA reclassified the entire tunnel as Gassy with Special Conditions. Significant time and cost was saved that would otherwise have been incurred to switch the mining equipment out and replace it with permissible equipment.

Overall, the New Irvington Tunnel Project successfully overcame many difficult challenges, through the combined efforts of an integrated team including the Owner, Contractor, Designer, and Construction Manager. The completed tunnel project will serve as a critical component of SFPUC's upgraded water delivery lifeline system for the San Francisco Bay Area.

## REFERENCES


Boyce, G., T. Feldsher, S. Gambino, C. Jaramillo, and D. Tsztoo (2010a). "Geotechnical Investigations and Ground Characterization—The New Irvington Tunnel Project," *Proceedings of World Tunneling Conference 2010,* May, Paper 542.

Boyce, G., S. Klein, Y. Sun, T. Feldsher, and D. Tsztoo (2010b). "New Irvington Tunnel Design Challenges," *Proceedings of North American Tunneling 2010,* June, pp. 325–336.

McCarter, R., J. Stolz, D. Crouthamel, and D. Tsztoo (2014). "Finding a Balanced Contracting Approach to Pre-Excavation Grouting," *Proceedings of North American Tunneling Conference 2014*, June, Paper 104.

Tsztoo, D., B. Abelli-Amen, and B. Lauppe Rhodes (2014). "New Irvington Tunnel Groundwater Management Plan," *Proceedings of North American Tunneling Conference 2014*, June, Paper 163.

URS Corporation and Jacobs Associates (2009). *Geotechnical Baseline Report for the New Irvington Tunnel Project*, prepared for the San Francisco Public Utilities Commission, December.

Zhang, J., C. Jaramillo, and T. Feldsher (2008). "Transient Simulation of Groundwater Flow for Tunnel Construction Using Time-Variable Boundary Condition," *MODFLOW and More 2008: Ground Water and Public Policy—Conference Proceedings*, pages 405–409.

# Port of Miami Tunnel Formation Layer 7 Grouting: Off Shore Rock Grouting, Tunnel Monitoring, and Ground Freezing

**Luca Barison**
Nicholson Construction Company

## PROJECT SETUP

The Port of Miami Tunnel Project has been part of the City of Miami's long-term planning for the last 30 years. The port, which sits on Dodge Island within Biscayne Bay, is one of the busiest in the country, for both cargo and cruise ship traffic.

The tunnel was designed to increase access to the port by providing alternate routes for inbound and outbound cargo truck and cruise-related tour bus traffic. Thousands of vehicles, including cruise and cargo ship-related vehicles, make their way to and from the Port of Miami on a daily basis, causing congestion, damage to the existing roads and potential safety hazards for pedestrians. The traffic in and out of downtown Miami and the port is expected to reach 70,000 vehicles a day by 2033. It will connect Interstate 395 (I-395) and Florida State Road 836, which merge into State Route A1A at the city limit of Miami on MacArthur Causeway, as well as Interstate 95 directly to the Port of Miami (Figure 1).

Currently, there is a single main road that connects the port to the mainland, which is Port Boulevard.

The 3,900-foot structure will include twin tunnels, parts of which will be constructed beneath the main shipping channel within Biscayne Bay. Each tunnel will be 41 feet in diameter and will hold two lanes of traffic. At their lowest points, the tunnels will be 120-feet below sea level, a depth that will allow for cruise-ship traffic while maintaining some rock and limestone formation above the tunnels. The new tunnels are expected to handle up to 1.5 million trucks per year.

When finished, the tunnel will provide direct access between Miami's seaport and the two most heavily-traveled roadways, Interstates 395 and 95.

## UNIQUE SOIL CONDITIONS

At the project's inception, it was already known that there was a high degree of uncertainty as to what the geotechnical parameters would need to be based on the site's varying geology. Extensive sonic coring and cone penetration testing was performed to fully understand the ground conditions at the site. The testing effort lasted 22 months, making it one of the most extensive ground investigations ever performed in South Florida.

The GBR indicated that the site's strata was comprised, mainly, of fill, lagoon silts, siliceous sands and variably indurated limestones.

The site strata consisted of eight layers:

- Layer 1—fill from dredging
- Layer 2—silt and coastal sediments
- Layer 3—Miami Limestone (less than 5 MPa in strength)
- Layer 4—Transition zone between Miami formation and Fort Thompson formation
- Layer 5 and Layer 6—Fort Thompson and Anastasia Formation (respectively, with indurate rock layer strength of 35 MPa)
- Layer 7—Key Largo Formation, unusually thick at the project site
- Layer 8—Competent Tamiami Limestone

The soil conditions at the site were primarily made up of the Key Largo Formation, a porous coralline limestone that has formed as a result of lowering sea levels and the continued exposure and erosion of the coral reefs that existed beneath the Florida Keys and Miami. The Key Largo Formation is an unusually soft and porous limestone, making it a somewhat unstable geology. The formation is a highly dissolved, highly porous, coralline limestone, consisting of coral and limestone fragments. Some test samples returned a porosity of up to 85% and permeability of $10^{-2}$ m/s. The Key Largo limestone interfingers with the Miami Oolite, Anastasia and Fort Thompson Formations in the eastern areas of Miami-Dade County. The Key Largo Formation is a coralline limestone composed of coral heads encased in a matrix of calcarenite. Typically, the Key Largo Formation consists of a highly crystalline, very porous reef deposit containing corals, bryozoans and mollusks. It contains an organic framework of coral colonies and interstitial skeletal calcarenite. Based on the interpretation of the borings, sonic cores, CPT

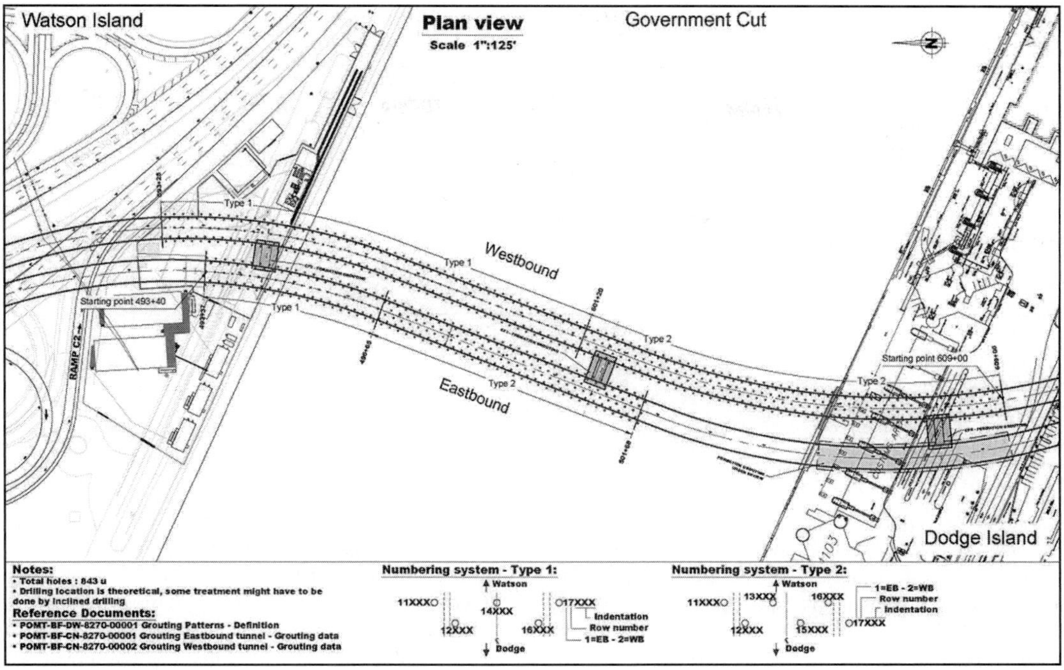


**Figure 1. Plan view**

soundings and large diameter test shaft, the degree of cementation within this stratum is erratic, and ranges from poorly to weakly cemented. (See Table 1 from Geotechnical Interpretive Report, dated February 4, 2011.)

German firm Herrenknecht, a world-renowned TBM builder, designed the $45 million TBM that is being used at the Port of Miami to effectively work in the unique geological conditions of the Government Cut. The machine, named Harriet, is 43 feet in diameter and 457-feet long. Harriet is the longest soft-diameter TBM that has been used to-date in the United States. The Port of Miami tunnel would be the first ever to be constructed in this type of soil condition.

## GROUTING PROGRAM

Tunnel excavation required the utilization of a pressurized face TBM. The high permeability of the formation and the risk of slurry loss led to the selection of an EPB TBM. However this was feasible only in the upper soils where the rock layers are better cemented. To maintain face and ground stability in the sections of the tunnel where the Key Largo formation was predominant a grouting campaign was implemented together with an innovative water controlled pressure mode at the TBM.

Variables for the grouting program were successively investigated and an extensive ground investigation campaign, including eight consecutive grout tests programs, was implemented. The investigation campaign included the following:

1. Concept of grouting (Compaction vs. Permeation)
2. Over 12 mix designs (Paris and Miami)
3. Delivery method (Upstage, Downstage, CFA)
4. Refusal Criteria (pressures, volumes, flow rates)
5. Hole spacing
6. Stage length
7. Methods for combating grout hole collapse
8. Sequence of grouting
9. Dodge Island vs. Watson Island

The results of the test programs led to a pumpable, stable mix with low strength, high penetrability but high thixotropy and excellent filtrate resistance which was a very difficult combination to achieve.

Nicholson was contracted to perform a one-of-a-kind grouting program to optimize the ground conditions ahead of the tunnel boring operation. The grouting program contained both an onshore and offshore component, of which the offshore component introduced several challenges.

Down-stage grouting was the technique chosen because of the unique soil conditions. Down-stage grouting would be performed by drilling down to the

**Table 1. Summary of results of field subsurface investigation (complementary & GBR/GDR) bored tunnel section—channel**

| Stratum | Approx. Top Elevation (ft, NGVD) | Stratum Thickness (feet) | SPT N-Value (blows/ft)[1] | Conventional Rotary Coring Data | | Sonic Core Data | CPT Data |
|---|---|---|---|---|---|---|---|
| | | | | Recovery (%) | RQD (%) | Recovery (%)[6] | $q_c$ (tons/ft$^2$)[2] |
| 5 | el -30 to el -46 | 5 to 26 (14) | 10 to Refusal (35) | 6 to 100 (59) | 0 to 82 (28) | 66 to 100 (90) | - |
| 6 | el -42 to el -65 | 17 to 36 (25) | 15 to Refusal (43) | 30 to 100 (86) | 0 to 100 (51) | 57 to 100 (92) | - |
| 7 | el -73 to el -83 | 21 to 55 (30) | 5 to Refusal (28) | 0 to 57 (28) | 0 to 24 (1) | 16 to 93 (63) | 0 to 411 (83) |
| 8 | el -108 to el -117 | -- | 6 to Refusal (38) | 0 to 100 (65) | 0 to 75 (19) | 100 | Refusal (typ.) |

tunnel invert (approximately 126 feet below grade of vertical depth at its lowest point with inclined holes up to 146 feet deep) and grouting approximately 40 feet up to the crown. Specified volumes of grout were pumped at specific pressures to reduce voids in the soil and to ensure that the grout would stay contained within the alignment.

Unlike typical grouting jobs, in which grout is pumped until refusal, Nicholson pumped specific volumes of grout at specific pressures to reduce voids and keep the grout contained within the tunnel alignment. Crews used low-mobility grout consisting of processed lake fill sand, bentonite, cement and chemical filtrate reducer, as specified by the general contractor.

The offshore program was performed under strict environmental restrictions designed to protect the port's shoreline. The 70-foot buffer required to maintain the integrity of the shoreline gave Nicholson the opportunity to design an innovative solution. This required finding a way to get the grout from the onshore plant onto the barge without touching the protected area. To accomplish this, a pipeline bridge was designed that would not only meet the environmental restrictions but also, not be affected by the changing tides. The pipeline bridge carried the grout from a fixed point on land to a barge. When the grout reached the first barge, it was routed to an agitation tank at each of the four drilling stations set up offshore. The plant capacity was sized to provide 1,000 CY/day. The grout was batched at the plant located on Watson Island, transferred to concrete trucks and delivered to the pipeline bridge.

Performing a massive grouting operation within one of the busiest cruise and cargo ship ports in the world requires a dynamic mobilization/demobilization plan that can be executed around the port's constant in and outbound traffic. Nicholson's grouting operations had to be completely demobilized when there were any cruise ships in port at Dodge Island, which meant that the company was mobilizing and demobilizing up to 10 barges per week. Barges were typically mobilized on Monday afternoons and completely demobilized by either Thursday or Friday at 2 AM, depending on the port's in and outbound cruise schedule. The busy cruise and cargo ship schedule left crews with small windows of working time on the job site ranging from 24 to 72 hours per window. This created the need to rethink the standard methodologies of drilling and grouting, which would have required several steps to complete a single hole. Nicholson's team had to maximize its time in the water once the barges were positioned, which meant developing new systems that included tooling, cranes using long leads and a way to expedite the transportation of grout from the land to the barges.

With the goal of optimizing work time production rates as much as possible, Nicholson called upon sister company, Bermingham Foundation Solutions, to assist in the design of a custom drilling and grouting system. After extensive industry research, brainstorming and collaboration between Nicholson and Bermingham, a unique methodology using innovative equipment was conceived so that the drilling and grouting process could be completed in a single pass, meaning, one insertion of the drill bit and drill string.

Bermingham designed all components with 3D CAD software and analyzed stresses with Finite Element Analysis (FEA.) Computational Fluid Dynamics (CFD) was also performed to calculate flow velocities, flow paths and pressure drops. All components of the innovative drilling system were designed, machine and assembled by Bermingham.

The system was designed to accommodate approximately 40 feet of water depth, 40 feet of competent limestone and 40 feet of grouting zone. The site's soil conditions dictated down-stage grouting, which then required that the grouting zone be

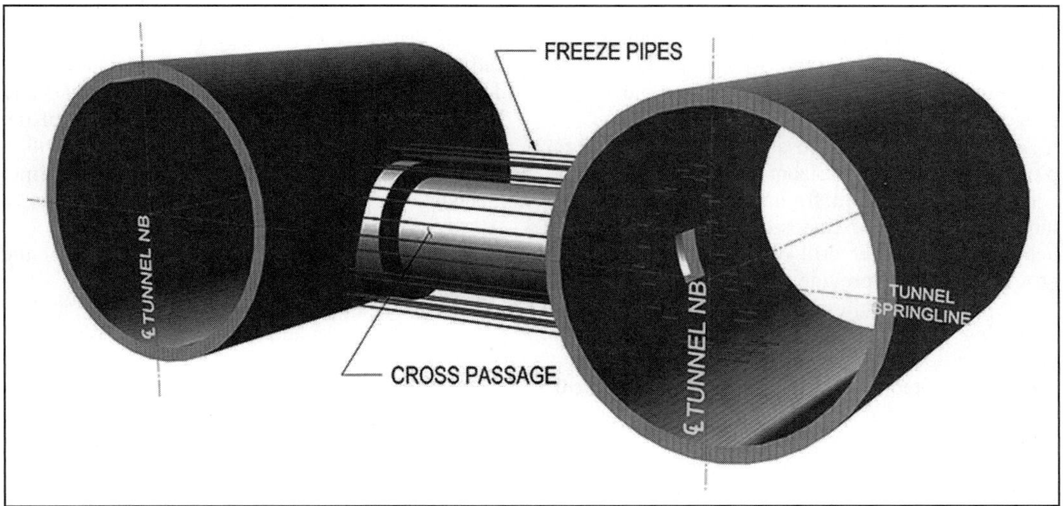


**Figure 2. Freeze pipes and cross passage**

filled with grout in four, 10-foot stages, starting with the upper stage and ending with the lower stage, but within each stage the grout needed to be pumped from the bottom of the stage to the top of the stage.

A 130-foot long, dual-walled drill string was utilized on a crane-mounted 140-foot Bermingham Vertical Travel Lead system for reverse circulation drilling. A seven-inch diameter crossover was designed to adapt between the custom drill string and the custom-designed drill-bit. The system was unique in that it was able to convert from drilling to grouting without removing any part of the bit or drill string from the hole. To keep compliant with the site's strict environmental restrictions, all cuttings from the holes were collected in tanks to avoid contact with the protected waters.

This process allowed Nicholson to perform the work on schedule and with actual production rates exceeding that of conventional, land-based drilling equipment. In addition, four rigs were used to maximize production rates in the short work windows.

Crews pumped low-mobility grout made up of processed lake fill sand, bentonite, cement and chemical filtrate reducer, as specified by the general contractor.

Nicholson drilled more than 1,000 grout holes and placed approximately 65,000 cubic yards of grout to support the port's tunneling operation. The company used GROUT I.T., a proprietary, automated computerized instrumentation system to monitor and record all grouting parameters, including pressure, volume, apparent lugeon and flow, in real-time. All drilling parameters were monitored and recorded using a separate drilling parameters recording system.

## STATE-OF-THE-ART MONITORING

Though most of the tunnel drives were subaqueous and would, therefore, cause only minor ground movements, the Customs Building, which sits directly above the tunnel alignment, required monitoring. It was determined during construction that the settlement of the Customs Building had to be monitored every four hours with an accuracy of 0.04 inches, which would be extremely costly using traditional surveying methods.

Nicholson's sister company, Soldata, Inc., was contracted to install a state-of-the-art, automated monitoring system to monitor the buildings, including the Customs Building, and surrounding structures on Dodge Island during tunnel construction.

Soldata also provided its reflectorless automatic surface settlement monitoring system, Centaur, for its first application in America.

## DRILLING AND INSTALLATION OF FREEZE PIPES FOR FREEZING OF THE CROSS PASSAGES

Five cross passages are required to connect the two tunnels of the POMT project. The geology of the site presented challenges for this portion of the project as well, and required specific ground treatment solutions. Ground freezing was selected by the client as the soil stabilization method for the construction and temporary support of excavation for Cross Passages 2 and 3 between the eastbound and westbound sections of the tunnel (Figure 2).

Cross Passages 2 and 3 are located approximately 100 feet (30m) below the watertable and needed to be constructed partly or wholly within the Key Largo limestone formation. The upper portion of the excavation for Cross Passage 2 is located in the Anastasia formation, which is also highly permeable. The high level of porosity of the Key Largo formation dictated the design of the ground freeze program.

At each of the two cross passages, 44 ea. freeze pipes were installed horizontally in a double concentric circular pattern from within the eastbound tunnel to touch the extrados of the westbound tunnel. Permanent outer drill casing were drilled and grouted in place to provide stable holes to install the freeze pipes. The drilling procedure for the horizontal casings required the installation of a blow-out prevention device to withstand the hydrostatic pressure from the tunnel exterior. Alignment of the drill casing was essential to prevent deviation of the casings during drilling. Once the casings were drilled and installed a gyroscopic survey was conducted to ensure that the design drilling tolerances were met. Verification of this tolerance was critical as deviation of the freeze pipes could potentially create a gap in the frozen earth mass. The drill casings were grouted in place using a cement-bentonite grout mix injected to the rock mass previously grouted during the tunnel grouting program. An outer and inner freeze pipe was installed in each casing. Piezometers and temperature thermistors were also installed to monitor the freezing progress. Freeze heads were then connected to the pipes to inject and circulate the brine to freeze the ground.

This is the first application of this technique in the state of Florida.

## SUMMARY AND CONCLUSION

A tunneling operation in challenging ground coupled with unique logistics required special methodology, design ingenuity and state of the art operational solutions to meet the schedule and technical challenges.

The project's innovations included using a proven drilling methodology (reverse circulation) and applying it to grout injection. The collaborative design of the single-pass system by Bermingham and Nicholson allowed the drilling and grouting of the holes in down stage without removing the tooling from the hole. The challenges that this system met included:

- Fulfilling the technical requirements of the client
- Being able to penetrate the rock in an efficient manner and drill holes in a variable and unstable formation
- Removing drill cuttings from the hole
- Injecting the low-mobility grout through the same system without removing and/or disconnecting the tooling from the hole
- Maintaining the drilling and grouting operation in both an environmentally and operationally safe manner

From the extensive investigation campaign, to the selection and injection of grout mixes and tunneling methods the project provided a series of firsts in the geotechnical foundation industry.

The project was successfully completed within the original ten-month schedule but with an additional 50% of drill footage required to complete the design. The changes to the equipment selection for the drilling and supply of grout enabled the Nicholson team to utilize fewer drills to drill the holes in a continuous manner without having to add or remove tooling from the hole after initial insertion. This helped to maintain a more efficient operation and allowed a productivity per rig of up to 40% higher than what was originally planned.

# Making the Tap: A New Deep Water Intake for the City of Austin's Water Treatment Plant No. 4

**Kevin Koeller**
AECOM

**Gregg Sherry**
Brierley Associates

**Satoshi Akai**
Obayashi

**ABSTRACT:** The City of Austin's Water Treatment Plant No. 4 (WTP4) is one of the few large scale deep water intakes in the nation. The tap into the lake was completed in 120 feet of water with a 12.5 diameter intake shaft 85 feet deep and a fabricated steel riser pipe, which was subsequently connected to a 9 foot diameter tunnel. Prior to completing the tap, a series of probing and pressure grouting operations were performed inside the tunnel. This tap was recently completed as planned and this paper will present a case history of this challenging and very successful project.

## BACKGROUND

Scheduled decommissioning of the City of Austin's (COA) original water treatment plant and Austin's growth were key factors in the construction of the Water Treatment Plant No. 4 (WTP4), the first new water treatment plant constructed by the COA in nearly half a century. In 2002, the COA hired a Carollo Engineers' team to perform preliminary site assessment engineering, environmental studies, design, and construction phase services for a new water treatment plant to be located in west Austin near Lake Travis. AECOM, a subcontractor for Carollo, was responsible for the raw water portion of the project, with Brierley Associates on the tunneling portion of the project. During design, it was determined that the new WTP4 facility would be planned for an initial treatment capacity of 50 MGD, with plans to expand to ultimate treatment capacity of 300 MGD. Due to the complexity of constructing deep water intakes within large water impoundments like Lake Travis, the raw water intake portion of the project was sized for the ultimate treatment capacity of 300 MGD.

The design of the WTP4 Raw Water System included a deep water intake at Lake Travis, a tunnel from the intake to the pump station, a new pump station, and the transmission main tunnel from the pump station to the water treatment plant site.

As shown in Figure 1 Hydraulic Profile, key components of the raw water intake included:

- Access Shaft: A 425 vertical foot 25.5-ft inside diameter shaft from ground surface at elevation 870.0 feet to elevation 445.0 feet.
- Raw Water Tunnel: A 9-ft diameter tunnel lined with a 12inch cast in place concrete liner approximately 4,386 ft long from Station 10+00 to 53+86, to intersect the Access Shaft at elevation 445.7 feet.
- Raw Water Intake Shaft: A 9-ft diameter steel lined shaft 82.75 feet long that was installed under water approximately 400 feet off shore in a depth of approximately 120 feet of water. The shaft extends vertically from the lake bottom elevation of 526 feet to the Raw Water Intake Tunnel at invert elevation 450 feet, to begin the 90 degree Tee connection to the Raw Water Intake Tunnel.
- Raw Water Intake: A 123.75 vertical foot high structure that comprised three intake screen structures and associated 9-ft diameter piping. The intake connected to the top of the raw water intake shaft and sloped upward toward the shore from the bottom of the lake.

## GEOLOGY

Results of the geotechnical investigations indicated that the access shaft, raw water tunnel, and intake were all located entirely in the Glen Rose Limestone Formation. The Glen Rose (Kgr) Formation is a relatively competent limestone formation that includes

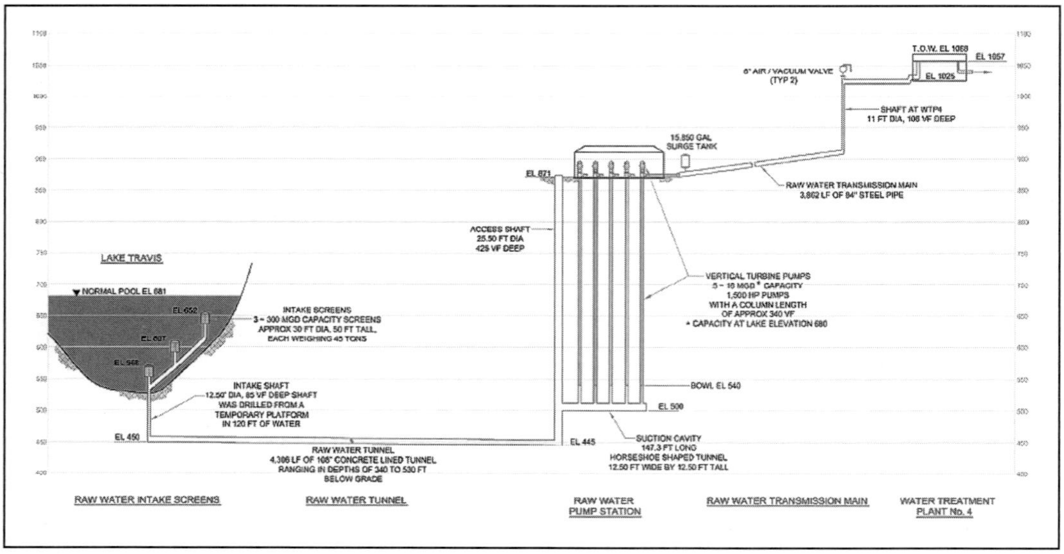


**Figure 1. Hydraulic profile**

alternating hard and soft beds of limestone, dolomitic limestone, and marl found to vary in thickness and hardness. Test results identified an RQD ranging from 100 to 42 with an average of 85.

## PROJECT RISKS

The geotechnical investigations suggested that the rock encountered during construction of the raw water tunnel would vary from very good to excellent. It was anticipated that less than 5% of the raw water tunnel would have regions with a poor-to-fair rock quality. The main concern in the raw water tunnel was flooding. Based upon the geotechnical information, the phreatic water level was located at elevation 720 feet, approximately 150 vertical feet below the top of the access shaft. Thus, a large portion of the access shaft and raw water tunnel would be constructed below the water table. The Glen Rose formation is generally considered tight, but discontinuities in the rock could be water bearing and a source of water infiltration. With the depth of the tunnel and ground water pressure, any significant water infiltration could have the potential for flooding the tunnel. To insure any discontinuities in the rock would be identified and mitigated, the lower tunnel was required to be probed prior to all tunneling operations. In addition, an extensive probing and grouting operation was established prior to making the lake tap connection.

## CONSTRUCTION

As the project started into the detailed design phase, the COA decided to use the Construction Manager at Risk (CMAR) method of project delivery for construction of the water treatment plant and corresponding raw water intake system. MWH Constructors was selected as the CMAR firm to manage and construct the project. A total of five subcontractors prepared proposals on the raw water intake system. After review, the project was awarded to Austin Hill Country Constructors, a joint venture of Obayashi USA and Manson Construction Co, with Obayashi performing all work on the land, including the tunnels, and Manson completing all marine work in Lake Travis.

## SHAFT CONSTRUCTION

The access shaft was excavated using an Antraquip AQM 100 shaft sinker specifically designed for vertical rock excavation. Designed as a 30-ft maximum 20-ft minimum inside diameter shaft from elevation 870.0 feet to 445.7 feet, the access shaft was constructed with a cast in place concrete liner that was a minimum of 12 inches thick and with a 28-day compressive design strength of 5,000 psi. The access shaft was excavated to an outside diameter of 28.5 feet with the final shaft lining placed using a 25.5 foot outside diameter circular steel form designed for use in a top-down concrete placement operation. Excavation and placement of the concrete shaft liner took approximately nine months to complete, with a pause in the middle of the shaft construction to work on other components of the project associated with the pump station (Figure 2).

## LOWER TUNNEL CONSTRUCTION

This project involved two main tunnels from the pump station site and the lake tap occurred in the deeper tunnel on the project. Termed the "lower tunnel," this raw water tunnel started at the bottom of the access shaft at a depth of 435.0 feet, traveled at an approximate 0.1% upward a distance of 4,386 linear feet, and terminated at a lake tap in Lake Travis. Similar to the construction of the access shaft, Obayashi selected a roadheader to construct the tunnels on the project. The machine used was an Antraquip AQM 150 roadheader with a 150 kW 480 V powered cutter head that utilized carbide bullet style picks. Gathering arms located on the front apron guide the excavated material to the centrally located conveyor systems that extend from the rear of the machine. The excavated material was transferred from the heading via the gantry conveyor to the waiting muck trains. The muck trains consisted of a diesel locomotive and five muck boxes.

Due to the Glen Rose Formation and overall generally competent material, the initial support was anticipated to be two pattern bolts in the roof at five foot spacing. It was estimated that approximately five cubic yards of material were generated for every foot of excavated tunnel so when all five muck boxes were full, approximately five feet of tunnel was excavated. This coincided well with the ground support spacing and Obayashi installed the ground support during the excavation downtime when the full muck train was swapped with the empty train.

The main concern constructing this tunnel below the water table was the possible presence of perched water or fractures that could allow water into the tunnel. Probing, which occurred before all mining operations in the raw water tunnel, consisted of a series of probe holes in the center of the tunnel two feet above the spring line. The first probe hole was one and three eighths of an inch in diameter and sixty feet in length. After excavating the first fifty feet of the tunnel, a second probe hole was drilled on center two feet below spring line, sixty feet in length. A third probe hole was then drilled after the next fifty feet of tunnel was excavated at the same elevation as the first probe hole. This sequence was repeated throughout the entire tunnel excavation, maintaining a minimum ten foot overlap between the probe holes. Excavation of the lower tunnel took approximately six months to complete.

## MARINE WORK

While the shaft was being excavated, marine work began in Lake Travis by Manson Construction. The raw water intake was a 123.75 vertical foot structure as shown in Figure 3 which was constructed in one of the deepest parts of the lake. Although the shoreline was in fairly close proximity, the steep face of the shoreline provided no immediate access or staging area. Accordingly, the marine work required a large amount of preparation and planning to install the intake structure a considerable distant from the temporary staging area. This required that an agreement be reached with a local property owner to rent a section of a local marina to stage and construct the temporary platforms required to construct the intake. Different from other large scale marine projects which were located along a coastline, Lake Travis was located inland and all marine equipment was limited to what could be hauled over road.

**Figure 2. Access shaft**

The size of the 9-ft diameter intake shaft and intake structure required significant offshore operations supported by floating structures and semi-permanent platforms. Accordingly, Manson planned and designed temporary work platforms that would be used to both drill the shaft for the riser pipe, set the intake riser, and construct the intake structure. The main temporary drilling platform consisted of a structure supported by four 48 inch diameter pipe piles that were 215 feet long to account for the depth of water in the lake. The majority of this work platform was constructed "in the dry," at a temporary staging area a distance from the intake site. Once this

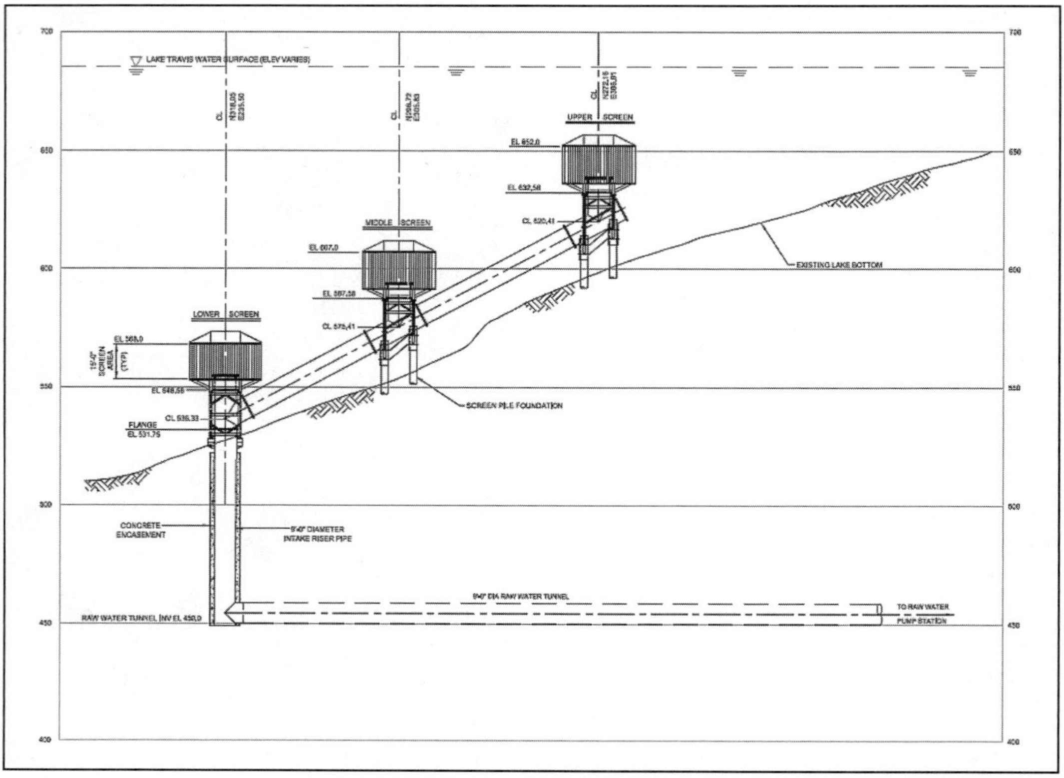


**Figure 3. Intake profile**

**Figure 4. Overview of Lake Travis**

temporary platform was constructed, it was lowered into the water and suspended on all sides by a square "moon pool" constructed of Flex-i-float barges. The moon pool was then floated to the intake site and positioned by a surveyor. This temporary platform served as a template to drive the 48 inch diameter pipe piles into suitable bedrock, to roughly elevation 504 feet for the east piles and 490 feet for the west piles. Once the piles were lofted, they were set by a barge mounted Manitowoc crane and driven using an IHC-SC200 impact hammer. Once the piles were seated to required tip elevation, the drive template was raised and secured at elevation 690.00 feet. The template was then decked with 12×12 crane mats for supporting the drill rig. See Figure 4.

The intake riser drill casing consisted of a 13 foot diameter by 210 foot long steel casing that was installed from the temporary work platform to the lake bottom. The casing was installed in three pieces using the top down construction. The first pipe segment was hung in a temporary support frame allowing the crane to unhook from the first suspended pipe segment and rig to the second pipe segment. The second pipe segment was set above the first pipe segment and suspended by the crane until the joint was aligned and welded. Once the casing reached the overall length of 210 feet, it was seated to approximate elevation 496 feet using an APE 600 vibratory hammer. The casing was surveyed during the driving to ensure a correct position within construction tolerances. Once the casing was in place, Manson's subcontractor Case Foundations, Inc. arrived on site and drilled the riser shaft using a Wirth PBA 933 Drill with 12.5 foot diameter drill

**Figure 5. Temporary platform**

**Figure 7. Intake riser**

**Figure 6. Intake drilling operations**

**Figure 8. Assembly of intake piping**

head. Drill platform was at elevation 690 feet and drilling started at approximate elevation 520 feet. The shaft was drilled to approximate elevation 440 feet. During these drilling operations, elevation in Lake Travis was approximately 640 feet. In order to ensure that any loose material did not interfere with the setting of the riser, an additional five feet of drilling depth was performed. Once the shaft was in place, the area immediately around the casing was excavated using divers to approximate elevation 515 feet to ensure that no run off material entered the shaft. Once this material was removed, the 13 foot casing was cut off below elevation 523.5 feet to ensure that it did not interfere with the riser installation (Figures 5 and 6).

The marine work on the WTP4 project was some of the most visible work to the public. Because all marine work was required to have minimal impacts to the surrounding area, the drilling operations in Lake Travis required stringent environmental attention. The intake riser drilling was performed using reverse circulation drilling and eight Baker sediment tanks and all drilling operations were closely monitored to insure that there was no discharge of any material into the lake (Figures 7 and 8).

Manson Construction planned to construct and install the intake riser pipe in one completely assembled piece. This required specialized construction and trucking to get to the project site. Fabricated in Louisiana, the 82.75 foot long piece was trucked to Lake Travis and off loaded directly onto a Flex-i-Float material barge. The intake pipe was later trucked separately and assembled at Manson's temporary staging area. Once the riser was staged on the material barge, remaining piping including the lower wye and lower screen support structure were attached. Once these pieces were bolted up, Manson floated the two material barges to the jobsite. The riser and lower wye were then set into the water, ballasted appropriately, and suspended over the drilled shaft by a winch mounted on the drill platform. The intake pipe was then lifted by the 4100 ringer barge crane and positioned at the appropriate angle for bolt up to be connected to the riser and lower wye. Once the piping was positioned in place, the crew bolted the two sections together so that the entire riser and

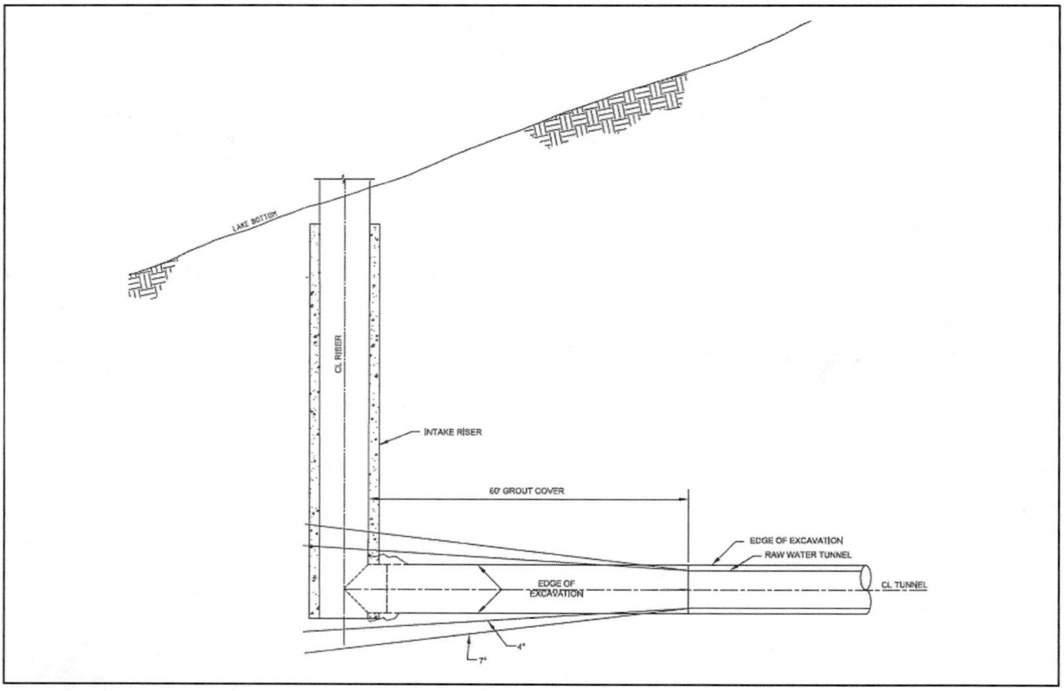


**Figure 9. Intake and tunnel section**

intake structure were assembled "in the dry." The intake riser and assembly were then slowly lowered in the water to design elevation. Progress and position were monitored by the use of two Mesotech side scanning sonar underwater survey installations. Plumb and location were monitored from the shoreline by means of a custom blind flange attached to the upper wye that had a vertical HSS section welded to the top which stuck out above the water (Figure 9).

Once the intake riser and associated pipe were set in place and everything was verified, crews made preparations to grout the annular space between the riser and the drilled shaft with cementitious grout. This was a critical step and key measurements were made to insure the structure was located in the right position. The grouting of the intake riser was completed in two stages, using tremie pipes and divers who carefully monitored all grouting operations.

## PRESSURE GROUTING

Marine construction was completed in July 2012 at the time when only approximately twenty percent of the raw tunnel was mined. After several additional months of mining, the lower tunnel reached close to Lake Travis in late November 2012. At this time, at a distance of approximately 60 feet from the intake riser, excavation of the tunnel halted and the lake tap pressure grouting operation started. In order to stage the drill for the grout curtain, a small niche was excavated in the rib of the tunnel at STA 51+21. The road header and gantry were then backed up past the niche and the drill was then maneuvered to the face and set up for drilling and pressure grouting. In accordance with specifications, a total of 30 holes were drilled, 12 in the inner ring looking out at four degrees and 18 drilled in the outer ring looking out at seven degrees.

Prior to and during the probing and pressure grouting operations, a significant amount of coordination and team work occurred. This was a critical stage of the project and any weathered rock or fissured rock could have continuity with the lake above, causing significant water pressure that could lead to injury or flooding. To manage this risk, the project team developed a communication plan and took all of the steps necessary to make any required adjustments to the drilling and pressure grouting operations to match any possible ground conditions encountered during the probing and grouting operations.

The initial grout mix was designed to have a water cementitious material ratio of 1:1 by weight. This grout mix was adjustable depending on ground conditions. If, after fifteen minutes of steady pumping, the pressure did not increase, the water cement ratio was reduced. Refusal was defined as less than one half of a gallon per minute measured over a consecutive five minute period.

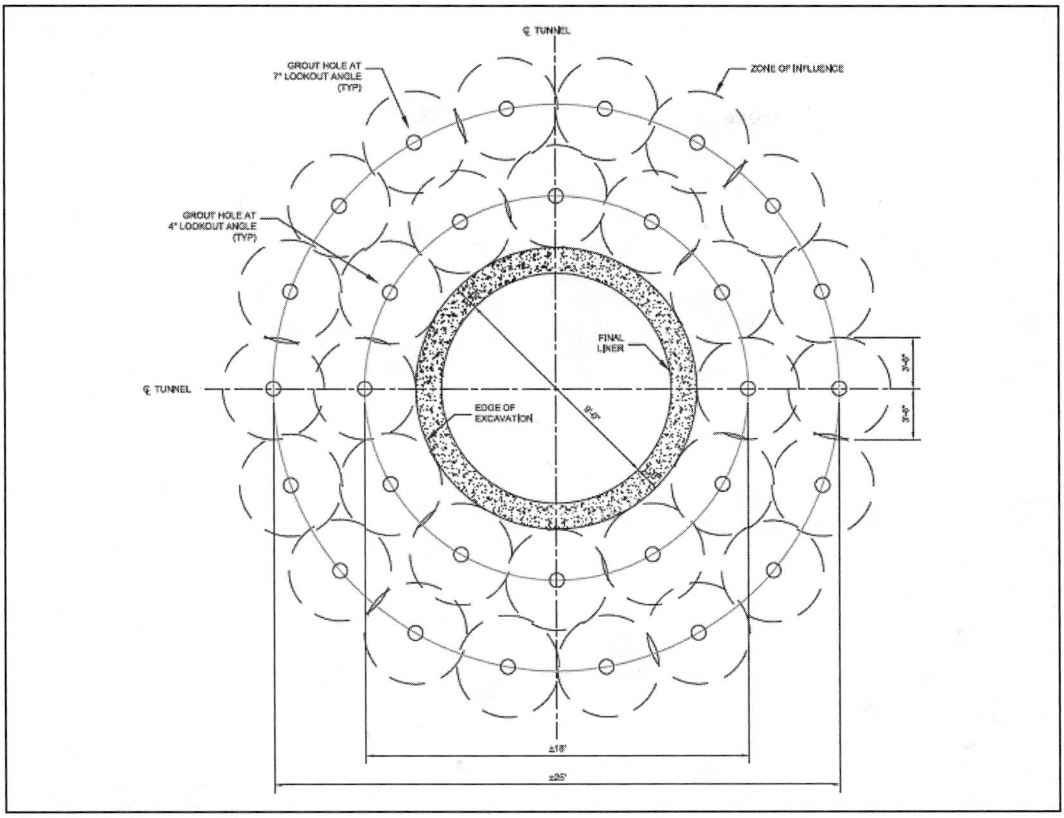


**Figure 10. Grouting section**

A number of trial batches of grout were prepared using two different types of cement: ultra fine and Type I/2 Portland. The ultrafine cement grout with the accelerator was developed to achieve the early strength that would be used to stop the flash water during boring. However, it was understood that due to the quicker set time, this grout would have limitations in grouting long cracks in the rock or grouting relatively large voids. Therefore, two more grout mixes, the ultra fine cement grout without accelerator and Type 1/2 Portland cement grout, were prepared in anticipation of needing a grout with a longer set time. Trial batches of these grouts were prepared and their properties, including set time and strength, were closely monitored. This provided several options for grouting depending on ground conditions encountered.

The need to isolate the probe drilling required that all holes be drilled through a packer fitted with a gate valve. To maintain the two inch diameter bore, a starter hole was drilled at four inch diameter approximately 10 feet long. A mechanical packer was utilized to seal the bore, with an additional steel plate fixed to the end of the packer and bolted to the tunnel face to provide both additional support and a secondary redundant means to seal the hole during grouting.

The pressure grouting operations involved a two step process of a series of outer ring holes and a series of inner ring holes as shown in Figure 10. In anticipation of encountering a larger number of voids, the outer ring holes were deemed the primary grout holes, 18 total. The inner ring holes were the secondary holes, 12 total, would fill any voids left by the primary grout holes. Grouting both the outer and inner rings would be a good indication that area inside the curtain was sealed against water ingress from the surrounding rock mass. As a double check, it was proposed to drill an additional probe hole after grouting operations to verify and monitor any water ingress. In anticipation that water injection or discharge testing would be used to determine the permeability of the holes, it was planned to grout the holes from the most permeable to the least. The maximum pressure for grouting operations was determined to be 120 psi, which was slightly higher than the water head of the Lake Travis. See Figures 11 and 12.

The grout plant was brought into the heading and set up approximately 100 feet from the heading

Figure 11. Probing operations

Figure 12. Completed pressure grouting

Figure 13. Crew at riser pipe

Figure 14. Crew at riser pipe

at the staging area and hoses were installed from the staging area to the heading. The pump lines were connected to the grout monitors, then to the packers located at each grout hole. The return line was connected down station of the grout monitor and ran back to the agitator tanks located in the staging area providing the option to circulate the grout material during any setup and downtime. Water was added to the mixing tank via the water meter. The meter only allowed the preset volume of water to be discharged at one time; this amount of water matched the water for the pre-determined water cement ratio. The meter was fitted with a totalizer to record the total water used during a grouting session.

The extensive up front preparations and good ground conditions made pressure grouting operations uneventful and as planned. Probing operations identified only a few holes with groundwater, all of which were manageable. Only a few of the probe holes took any grout, which significantly reduced the small amount of groundwater encountered. After completing the pressure grouting and sealing of grout holes, two verification holes were drilled on center line two foot above and below spring line of the tunnel. Fortunately, these holes did not make any water or show signs of continue water ingress and the grout curtain was deemed a success. The verification holes were left open and continually monitored as the excavation progressed.

After completing all pressure grouting operations, excavation of the remaining 60 LF to the riser resumed. The majority of this excavation was carried out with the AQM 150; the remaining material was removed using a hydraulic breaker and pneumatic hand tools.

At a distance of approximately 20 LF from the riser, crews stopped excavation and performed one remaining probe hole. This was done as a precautionary measure, just as a final check before the last section of excavation. This probe hole identified no water and crews proceeded, reaching the riser pipe on December 19, 2013. See Figures 13 and 14.

## LESSONS LEARNED

The following lessons were learned on the project.

- Tunneling operations involve, to a large extent, managing risk. On this project, the potential for flooding the tunnel was high and the team took the steps necessary to insure that the any potential problems were resolved before they occurred.
- A worst case scenario was anticipated and a solid plan prepared to overcome potential difficult situations. Fortunately, the project team encountered favorable ground conditions and no unanticipated situations were encountered during the lake tap. The detailed planning, completed prior to performing the work, inspired the crew and increased the safety culture on site.
- Careful and detailed survey efforts and procedures were implemented between the tunneling contractor, Obayashi, and marine contractor, Manson, to insure that the intake structure and tunnel were constructed in the correct locations. This was a challenging undertaking and completed successfully within ¼ of an inch.

## CONCLUSION

Challenging deep water intake projects require a significant amount of detailed planning, design, and execution. On the WTP4 raw water intake, the engineer and contractors prepared these documents and in addition formed a strong team, which combined with favorable ground conditions led to a project that was completed to specification, on scheduled and within budget.

# Application of Vacuum Dewatering Systems for SEM Cross Passage Constructions in Difficult Ground Conditions

**Nate Long, Brian Hagan, and Ehsan Alavi Gharahbagh**
Jay Dee Contractors, Inc.

**ABSTRACT:** Fine soils with low permeability can be challenging to dewater, subsequently making excavation by SEM difficult and hazardous. Vacuum dewatering can increase soil stability, reducing the chance of lost ground and surface settlement. However, this technique can have significant upfront costs, in time and resources, and ongoing daily tuning and maintenance requirements. Piezometers installed in the formations being dewatered and the vacuum system itself, can provide invaluable information if they can be read and interpreted in a near real-time framework. This system, although difficult can lead to excellent results in ground conditions that would otherwise require more costly ground improvement measures.

## INTRODUCTION

Considering that Sequential Excavation Method (SEM) Tunneling can be inherently difficult, dangerous, and costly, it is very important that the ground that is intended to be excavated via SEM is prepared as thoroughly, timely, and cost effectively as possible. Sometimes the ground conditions are such that no additional work is needed to prepare the surrounding ground. However, when surrounding ground water is present, the surrounding ground must be properly prepared in advance of the SEM tunneling. Some common methods to handle surrounding ground water in ground that is intended to be excavated via SEM tunneling include ground improvement via soil freezing, jet grouting, permeation grouting, and/or ground dewatering via surface wells, gravity drainage or vacuum drainage. Most of these methods can be done from the surface, while some of these methods can be done from within the tunnel.

The specific ground conditions and access to the ground from the surface are the main factors that affect the decisions of which method to use to treat problematic ground and whether to use the method from the surface or from within the tunnel. When the ground conditions involve significant ground water in layered soil types with some of those soil types having very fine soils and low permeability, then ground dewatering can be quite challenging. In these conditions, the dewatering well must be placed into the correct ground layer in order to properly extract the water, and the correct layers can be hard to target from the surface. Even though it is often a more expensive option, improving the ground via soil freezing, jet grouting or permeation grouting is often considered in these situations. If there is sufficient surface access, the methods to treat problematic ground are more commonly done from the surface because it is generally easier and safer to address problematic ground from the surface rather than from within the confine space of the tunnel. Furthermore, addressing problematic ground from the surface limits the amount of equipment and manpower that is needed inside the tunnel, which therefore limits the impact to other tunneling activities. Providing sufficient surface access can be very difficult though, especially if the tunnel is under a highly urban area.

The University Link Light Rail TBM Tunnels–CHS to PSST (U230) project in downtown, Seattle, Washington was a project that involved SEM tunneling in a highly urban environment with significant ground water, in highly layered soil conditions with low permeability. The U230 project consisted of twin bored 18'-10" inner diameter transit tunnels with cross passages connecting the tunnels at various locations. SEM was required for the cross passage excavation. On this project, it was determined that there were two viable methods to prepare the ground prior to SEM tunneling. One method was ground freezing from the surface and/or the tunnel, which is a common and well understood process. Many papers have been written with regards to the use of ground freezing to handle ground water in highly layered soil conditions with low permeability. The other method, which was a much less common method, was to use a vacuum drainage system that would be installed, operated, and maintained completely from within the tunnel. After weighing the advantages and disadvantages between the two methods, it was eventually decided that using a vacuum dewatering system from within the tunnel was the best approach. The following section of this paper is devoted to compare the ground freezing with vacuum dewatering system.

## GROUND FREEZING VS. VACUUM DRAINAGE SYSTEM

As previously mentioned, the U230 tunnel project Contractor and Owner had to weigh the advantages and disadvantages of ground freezing and a vacuum drainage system for addressing the ground to be excavated via SEM in the cross passage locations connecting the tunnels. The following narrative will explain the two methods (ground freezing and vacuum drainage system) and highlight the advantages and disadvantages:

Ground freezing is the process by which the ground water is frozen in situ to create an engineered composite material. This material is essentially impermeable and provides pre-support strength to excavations. Ground freezing has been used in shaft excavations for approximately 25 years. Ground freezing is a relatively new system for tunnel cross passages, but it has been used successfully in several projects such as the Port of Miami Tunnel Project. In general, ground freezing is an expensive and slow process. Brine or expendable agents can be used to induce the initial freeze.

Ground freezing from the surface is preferred compared to from inside the tunnel. Ground freezing from the surface allows work to continue in the tunnel, which is generally the critical path of any tunneling project. Thus, the ground freezing (which is a slow operation overall) would not interrupt any other activities in the tunnel. One of the main disadvantages of ground freezing from the surface involve impacting utilities and the surrounding community. Furthermore, as the depth of the required freeze increases, it becomes more difficult and sometimes impossible in some locations.

To prepare the site for ground freezing, the work area needs to be leveled to provide the drill rigs a stable work platform. As many tunnel projects are being constructed in urban environments, protection of utilities becomes a major concern in a ground freezing operation. The protection of utilities is not only a consideration during the drilling of the pipes but also during the establishment of the frozen block. If the utilities are located in close proximity to the freeze zone, they may require thermal protection in addition to physical protection.

Once the utilities in the area have been protected and the work site is sufficiently prepared, the installation of the freeze piping can begin. Rotary sonic is the preferred method for freeze pipe installation. Freeze pipes are usually made of steel. All joints should be welded and tested prior to being filled with brine. Couplers can leak and are not recommended. Once the freeze pipes are installed and tested the manifold, chillers and ancillary piping can be installed. Once this process is complete, freeze down of the block can begin.

The freezing process can take 6–12 weeks depending on the size of block. Thermal couples and piezometers are used to monitor the block. An array of thermocouples is installed in the brine system, the freeze pipes and the soil, both inside and outside of the freeze zone. Using a piezometer installed in the interior of the freeze zone the growth of block can be monitored, giving the well known pressure spike once the block is closed. The freezing process can be done with liquid nitrogen, which would shorten the required time for freezing, but it is significantly expensive compared to using brine.

The freeze must be maintained throughout the progression of the work with the same equipment that was used for initial freeze. Once the final lining of the excavation is completed and comes to strength, the freezing operation can be stopped. The chiller and manifolds are removed and the freeze pipes are drained of the brine. This allows the frozen zone to thaw naturally. After the brine has been removed the pipes can be abandoned by being backfilled with grout.

It is also possible to perform ground freezing from within the tunnel. The operations completed in the tunnel are similar to the work being done from the surface. One benefit of performing ground freezing from within the tunnel is that the utilities are usually no longer in the way of the drilling that would have taken place from the surface, and thus, the utilities do not have to be protected. Heave caused by freeze from within tunnels has not been observed to be a significant issue in the past. Freeze pipes installed in a tunnel are generally smaller in diameter than those installed on the surface, because smaller drills to install the pipes have to be used to fit within the confines of the tunnel. Since the pipes are smaller diameter, a tighter pattern of holes is used to ensure that a complete frozen formation is achieved. Chillers installed in tunnels have to be smaller than the chillers that could be installed on the surface, due to the constraints of the underground work zone. This compounds the need for smaller freeze pipes at closer spacing. Piezometers and thermal couples are also used in underground installations. The major disadvantage of performing the freeze from underground is that the freeze installation operation occupies a large percentage of the tunnel.

Installation of the vacuum system from within the tunnel is similar to the installation of the freezing system in the tunnel. The vacuum dewatering lances should be installed in a pattern resembling that of the freeze pipes. The goal is to create a zone of dewatered soil around the entire circumference of the excavation. The target thickness of this zone is

**Figure 1. Installed guillotine valve to control the inflow of loose ground into the tunnel**

a minimum of 10 feet. By applying vacuum to this soil, the internal friction of the soil can be increased. This increase in internal friction leads to an increase in shear strength, which in turn leads to an increase of the overall stability of the soil.

One major item of concern with the vacuum system is the inflow of loose ground into the tunnel. In some instances this ground will flow with no external factors. Without a predesigned system to stop this flow it can become problematic and potentially dangerous. Valves need to be installed on the tunnel lining to ensure that flowing ground will not enter the tunnel environment without controlling the rate (if necessary), Figure 1. Both self-drilled well points and pre-drilled and installed well points can be used for this application.

Plumbing of the vacuum system is functionally similar to that of the freeze system. Self-supporting hose must be used to keep the walls from collapsing upon themselves when the negative pressure is applied. Clear hoses are advantageous, since the process can be monitored visually and the flow rate can be estimated. The hoses should be run to a manifold that includes individual valves for each vacuum point. The valves need to be capable of adjusting the volume easily and accurately; this can be achieved with ball valves, screw valves, etc. This manifold can then be run to the vacuum pump. The closer the vacuum pump is to the dewatering area, the smaller amount of line loss that will influence the system. Placing the vacuum pump downhill of the dewatering area has the chance to increase the effectiveness, due to increased suction.

Piezometers in the well system should be installed in similar pattern to the thermal couples that are used in the freeze piping. The piezometers that are installed within a few feet of the vacuum lances are used to measure the status of the system. The piezometers farther from the suction locations help to establish the overall effectiveness of the entire system. The piezometers react almost identically to the thermal couples used in ground freezing; however they react significantly faster. Real time monitoring and data logging greatly enhance the efficiency of the vacuum system. In addition to the data monitoring, the plumbing system should be checked at least daily for damage, leaks or issues that could cause potential problems.

After the well system has reached the designed pressure (in the ground), it needs to be maintained throughout the excavation. The system requires extensive tuning during the entire excavation process. This is due to the fact that the pressure equilibrium changes constantly as the ground is removed and ground support (primarily shotcrete) is installed. Once the final lining has reached the required strength, the vacuum system can be turned off, the manifolds can be disassembled, and the vacuum lances can be backfilled with neat cement grout.

One of the main differences between ground freezing and the well system relates to the ground response when the system is inactive (turned off or momentarily disabled). Ground freezing creates a frozen block that is stable and can take many weeks to thaw. In contrast, ground modified by vacuum dewatering can return to a natural state in a matter of hours. This fast reaction time makes is imperative that continuous monitoring is undertaken. Continuous monitoring will allow for early detection of system problems, including malfunction and loss of suction due to excavation progression. System malfunctions need to be diagnosed and repaired immediately. Having a backup vacuum pump with automatic switching capabilities is a wise precaution.

A properly maintained well system will allow for safe excavation of the cross passage openings. Before every round of excavation the piezometer readings need to be analyzed. The excavation should not begin if the readings are not currently stable or slowly falling. Generally, after excavation has started the vacuum system will begin to lose suction and the pressures will rise. This is normal and not major cause for concern, however the support of excavation (shotcrete) should be installed promptly. The system should be checked for obvious defects;

any vacuum lances that were damaged during the excavation should be shut off, and damaged hoses should be repaired if possible or disconnected from the system.

After all repairs have been completed, the shotcrete will allow for the seal to re-establish and the pressure readings on the piezometers should begin to decrease almost immediately. Measurements taken 30 minutes after the shotcrete has been applied should show obvious downward trending data. If the pressure has not fallen within 30 minutes, then there are still issues with the system that need to be addressed. This process will be repeated throughout the excavation and support sequence.

Some basic cost and schedule impacts between a freeze system and a vacuum well system from within the tunnel include, Installing the freeze system from the surface has the advantage of keeping equipment out of the tunnel, but it greatly increases the amount of drilling that needs to be undertaken. Every foot of overburden adds 30–40 feet of vertical drilling for a typical cross passage. This also means an additional 30–40 feet of piping material and backfilling that must be completed.

Vacuum well systems and freezing systems that are installed from within the tunnel create impediments to traffic through the tunnel. All work on the opposite side from the portal is essentially impossible. The vacuum system can be installed with the same drill that is used for spiling along the excavation. The freeze system has to be installed with a cased hole, and generally requires specialized drilling equipment. The chillers are also customized equipment that generally have to be purchased for the job and have low resale value after the project is completed. Vacuum pumps, on the other hand, are readily available for rent or purchase and can be used for other dewatering applications, which increases the resale value. Furthermore, a freeze system is generally installed and operated by a subcontractor, which can lead to increased costs, whereas a vacuum well system can be installed and maintained by the Contractor.

## VACUUM DEWATERING SYSTEM APPLICATION IN U230 PROJECT

### University Link Light Rail Project (U230)

The Sound Transit University Link Extension is a 3.15 mile light rail extension that will run in twin bored tunnels from downtown Seattle to the University of Washington, with stations at Capitol Hill and on the University of Washington campus near Husky Stadium. The project is broken up into several parts, where the U230 portion dealt with the excavation of Capitol Hill Station and installation of 0.73 miles of twin-bored tunnels from the Capitol Hill Station (CHS) to the Pine Street stub tunnel (PSST) in downtown Seattle. The joint venture of Jay Dee Contractors, Inc., Frank Coluccio Construction Company, and Michels Corporation was awarded the U230 Project. The tunnels were approximately 21'-2" in diameter and were to be furnished with concrete segmental tunnel lining having an outside diameter of 20'-7"and an inside diameter of 18'-10". There were five (5) cross passages that needed to be excavated by using the Sequential Excavation Method between the twin tunnels.

### Geological Condition of the U230 Project

The Seattle area has a complicated geologic history. Six major glacial events have happened during last 2 million years with intervening non-glacial erosional and depositional periods. The geologic description of this project can be divided into fluvial deposits, glacial deposits, lacustrine and glacio-lacustrine deposits, all of which have been glacially over-ridden and are therefore highly over-consolidated ($1 < OCR < 4.3$ and OCR average of 2.5). Around 70% of the alignment is located in over-consolidated fine grain soils. The Soil Groups (SGs) defined for this project are Blue SG that represents over-consolidated fine-grained, plastic soils, Turquoise SG that represents over-consolidated fine-grained, non-plastic soils, Yellow SG that represents over-consolidated fine to coarse sand, with varying amounts of gravel, silt, and clay and Purple SG that represents normally consolidated fine to coarse sand, with varying amounts of gravel, silt, and clay (Purple SG is not present on the tunnel alignment and only encountered at PSST and CHS). Table 1 shows the permeability of the soil groups along the alignment.

### Cross Passages in U230

The cross passages were approximately 800 feet apart and 15 to 25 feet in length. The cross passages were broken into categories based upon the ground modification that would be required at each location. By design, there were three category one (1) cross passages that would not require ground modification and two category two (2) cross passages that would require dewatering. After the tunnels had been driven it was determined that one of the category one (1) cross passages should be a category two (2). Additionally one of the category two (2) cross passages contained distinctly difficult ground and perhaps should have been originally designed as a category three (3). Ground freezing was not an option due to schedule and access from the surface to this cross passage. This cross passage was dewatered using the vacuum system described herein.

### Geological Condition of Cross Passage No. 3

Figure 2 shows the geological condition of the cross passage three (CP3) compared to the actual encountered geology. As you can see in this figure, the blue soil group consist of over-consolidated fine-grained, plastic soils were originally interpreted for the entire excavation area of this cross passage. This soil group is defined as firm to slow raveling when unsupported in cross passage locations. On the other hand, the encountered geology at the right top part of the excavation area was Turquoise soil group which is defined as over-consolidated fine-grained, non-plastic soils. This soil group will flow when exposed in an unsupported face.

### Application of the Vacuum Dewatering System in CP3

The planned probe drilling design, prepared by Gall Zeidler Consultants, LLC, consisted of 18 probe holes with lengths varying from 10 feet to 30 feet.

**Table 1. Permeability of soil groups along the alignment**

| | Permeability (cm/sec) | |
|---|---|---|
| **Soil Group** | **Horizontal** | **Vertical** |
| Blue | $2\times10^{-6}$–$3\times10^{-5}$ | $1\times10^{-8}$–$1\times10^{-6}$ |
| Turquoise | $4\times10^{-4}$–$1\times10^{-3}$ | $1\times10^{-7}$–$1\times10^{-5}$ |
| Yellow | $1\times10^{-4}$–$7\times10^{-3}$ | $7\times10^{-5}$–$3\times10^{-4}$ |
| Yellow2* | $1\times10^{-6}$–$5\times10^{-5}$ | $1\times10^{-7}$–$1\times10^{-6}$ |
| Red | $1\times10^{-4}$–$5\times10^{-2}$ | $5\times10^{-5}$–$3\times10^{-2}$ |
| Purple | $1\times10^{-5}$–$1\times10^{-2}$ | $1\times10^{-7}$–$1\times10^{-4}$ |

*Till material that are part of the yellow soil group are called yellow2 in the GBR. Yellow2 SG has higher plasticity and fine content than yellow SG.

Drilling angles were planned from –35 degrees to 40 degrees. Drainage pipe would be installed, similar to Figure 3, allowing gravity to drain the surrounding soils. Before drilling began, using additional information obtained from other works, it was determined that a vacuum dewatering system would be required. The probe drilling for CP3 began on May 1, 2012 and was completed on May 12, 2012. At completion of the designed probing, 390 feet of self drilling drainage pipe had been installed in 14 holes and 47 feet of screened and driven well points installed in four (4) holes. In addition to the probe drilling and pipe installation, a piezometer (Figure 4) was installed 6 feet above springline and at a depth of approximately 6 feet.

This piezometer was monitored during the following weeks to confirm the depressurization of the soils surrounding the cross passage excavation. However, after nearly a month of dewatering it was determined that the pressure had not been reduced to an acceptable level for excavation. Figure 5 displays the monitoring of this piezometer. At this point, it was determined that additional dewatering drilling would be required.

Four (4) drainage pipes were drilled and installed, with a combined wet length of 78 feet on June 7, 2012. A few days prior a piezometer was installed in the opposite tunnel, to monitor the dewatering at the far end of the cross passage excavation. Continuing to prove insufficient, the system was expanded with five (5) more drainage pipes (102 feet) installed on June 14, 2012. Positive results were recorded over the next two weeks, but three (3) additional drainage pipes (44 feet) were needed to deal with a particularly difficult area just outside the breakout zone.

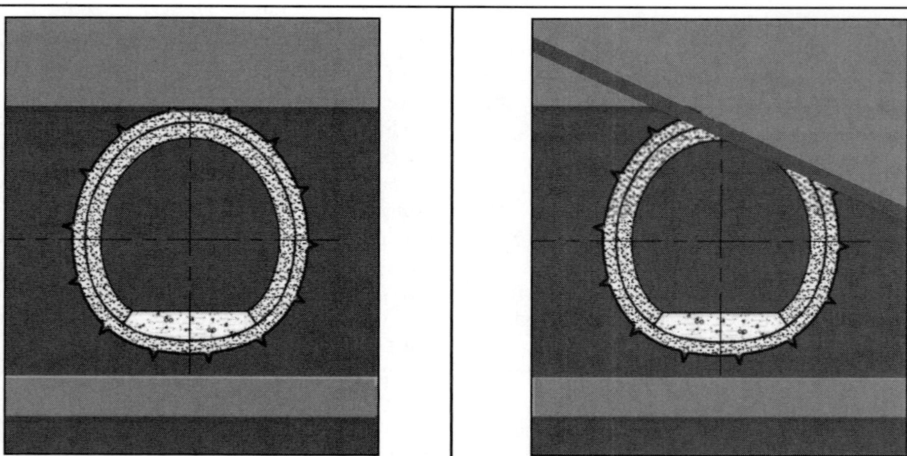

(a) (b)

**Figure 2. Geological condition of the CP3 (a) as described in the GBR and (b) encountered in the field**

Breakout began immediately following these last three (3) drainage holes on July 2, 2012. Promptly after breakout fast raveling and flowing ground behavior were encountered. The ground conditions consisted of very moist to wet, sandy silt and disturbed and slickenslided, silty clay. Before the first top heading could be completed a chimney formed in the sandy silt and broke out above the spiles. This void was grouted shortly after its formation and was estimated to be only 6.5 cubic feet. Water inflow was experienced at the segment breakout, maintaining a discharge of approximately five (5) gpm throughout. Excavation continued, utilizing small pocket excavations for the breakout and first three rounds. Especially during the first three rounds of excavation the vacuum system was critical to preserving the stability and safety of the excavation face. Consistent maintenance of the system was an ongoing challenge. Loss of pressure due to malfunction was a constant concern and was monitoring with a mechanical gauge attached to the vacuum manifold. Loss of pressure due to malfunction can be seen in Figure 6, including the rebounding of vacuum pressures. Eventually a piezometer was installed in the vacuum manifold, to allow for better tracking of the system. The loss of vacuum during excavation was also experienced, as seen in Figure 7. Note how the piezometer in the soil shows almost no delay compared to the one in the vacuum manifold. This loss of pressure increases the mobility of the surrounding

**Figure 3. Drilling and installation of drain pipes at cross passage**

**Figure 4. Installation of piezometer with electric rotary hammer drill**

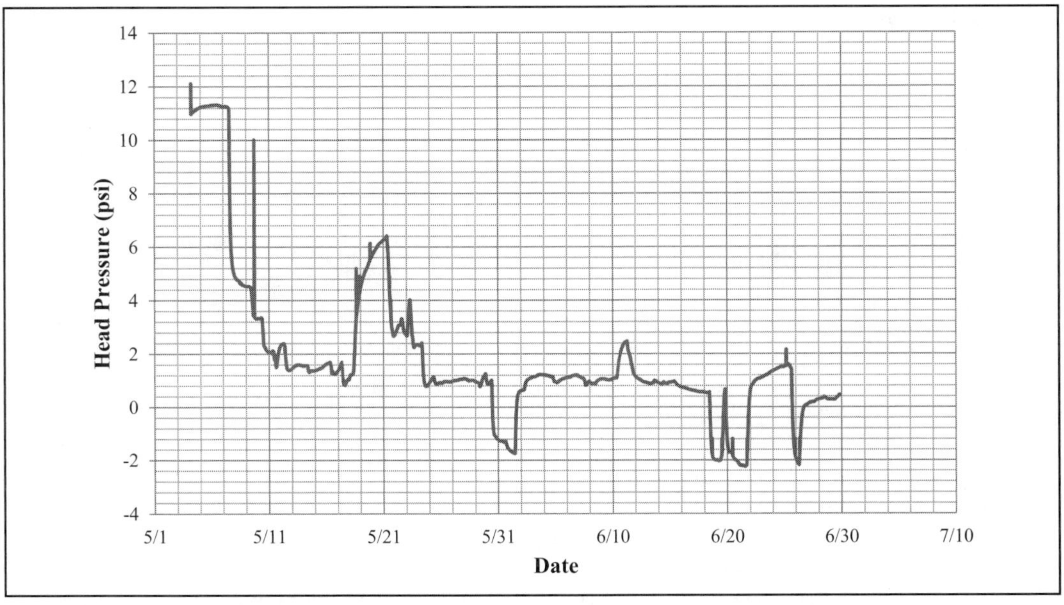


**Figure 5. Pre-excavation pressures at CP3**

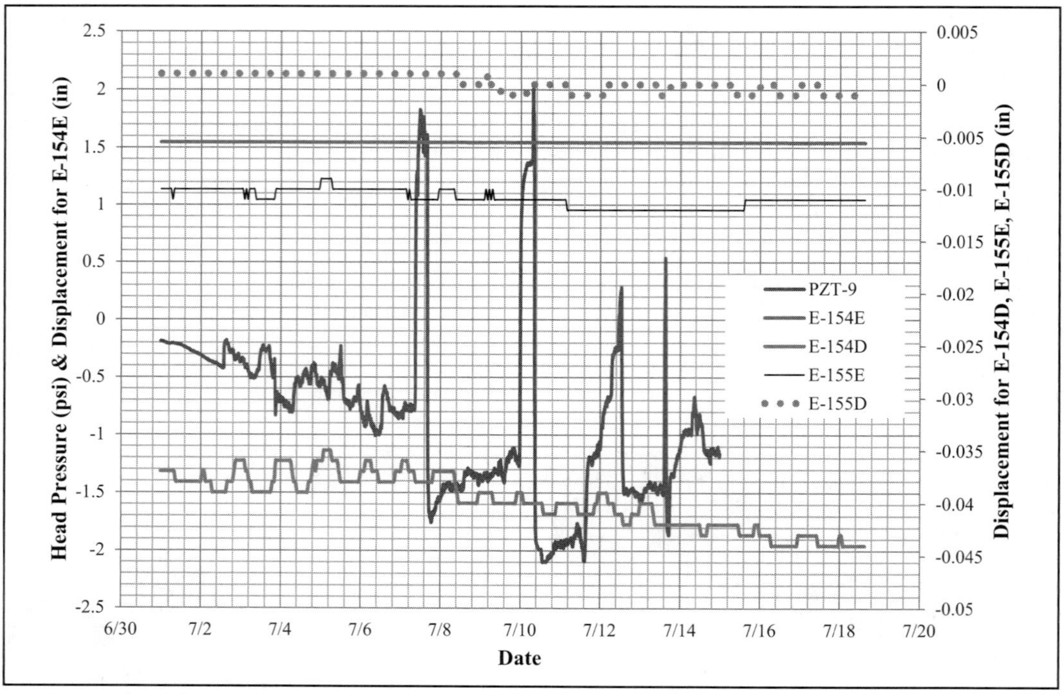


**Figure 6. Comparison between pressure and displacement at CP3**

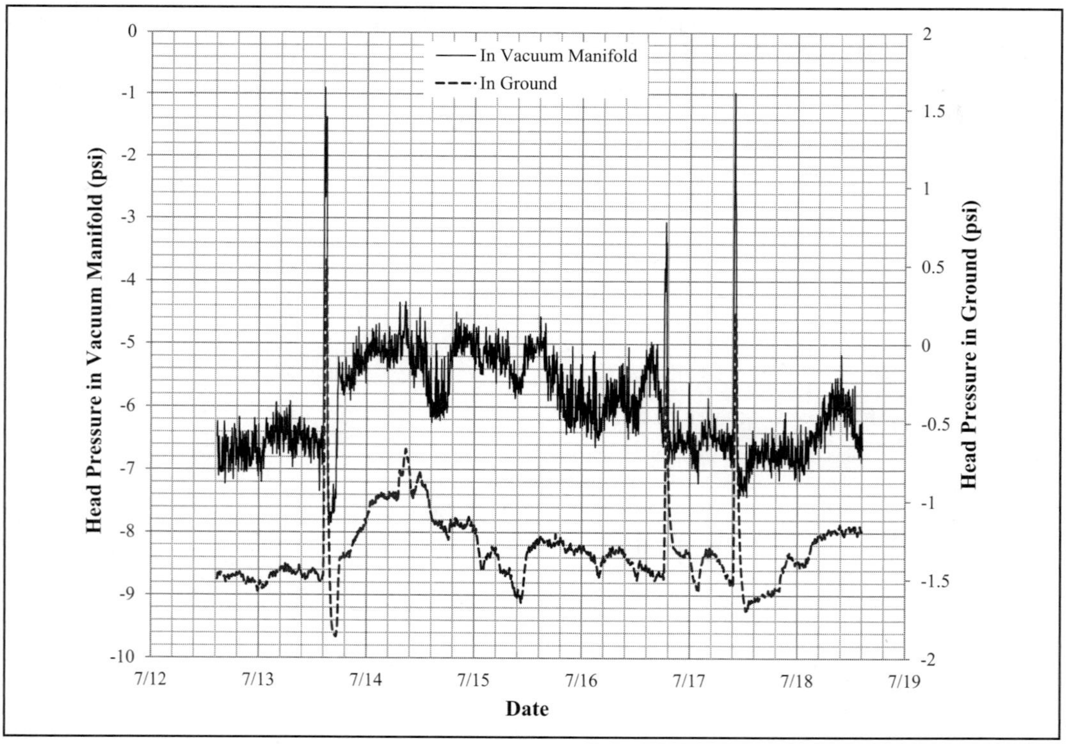


**Figure 7. Comparison between pressure in vacuum manifold and in ground during the excavation**

**Table 2. Excavation pre-support and support summary for CP3**

| | |
|---|---|
| Drainage holes installed | 32 |
| Length of drainage pipe installed (feet) | 62 |
| Piezometers | 7 |
| Pocket excavations | 65 |
| Lattice girders | 3 |
| Grouted IBO spiles | 44 |

soils, thus increasing the need for rapid support of excavation (shotcrete) over traditional SEM.

The rapid support of excavation, in keeping with the SEM principles, resulted in settlement readings that were excellent for the difficult ground conditions as reported in Figure 6. Extensometer data, that showed settlement during the TBM excavations in the area, do not show measureable movement throughout the entirety of the cross passage excavation. Figure 6 shows the bottom two points of the multipoint borehole extensometers above the CP3 and running tunnel connections. Additionally, the surfaced and building settlements points in the area remained stable and no movements were reported.

After round three, the ground conditions changed to steep faced clayey silt, which exhibited a blocky behavior but displayed increased stand-up time due to reduced moisture content. Rounds four through seven were completed with pocket excavation but were relatively uneventful. An igneous boulder, 4'7" × 2'6" × 2'5", was found and removed as part of round four excavation. Excavation was completed on July 25, 2012. The vacuum system was kept in place during the installation of the waterproofing membrane and final concrete. Leaving the vacuum system active, allowed the following works to progress unimpeded by groundwater in flows. The vacuum system was subsequently discontinued on September 13, 2012 and the drainage holes backfill grouted. Table 2 summarizes the CP3 excavation pre-support and support summary.

### Comparison Between Vacuum Dewatering and Ground Freezing in Case of Time and Cost

By the time the holes were backfilled, the vacuum system had been active for nearly 11 weeks and took nearly eight weeks to install. This installation time was extended by 4 weeks due to the inability of the original installation to effectively dewater the ground. If the system had been sufficiently designed from the outset, it would have effectively saved a month of schedule. Using this information, in hindsight, it leads to the opinion that the system should be over designed if schedule is more of a concern than direct cost.

The majority of the direct cost of the vacuum system is in the installation of the drainage pipes. The ongoing costs of the maintenance are not insignificant but they are significantly less than a ground freezing system. Considerations were made to install a ground freezing system from the surface or from inside the tunnel. Traditionally a surface installation would be the preferred choice, however, not having to install the system from the surface, in a dense, urban environment, is a cost savings, in addition to decreasing environmental and community impacts. The major burden to installing a ground freezing systems from inside the tunnel was amount of space that it would take up. This space was considered too valuable to other operations and this option was ruled out.

## CONCLUSIONS

SEM tunneling is difficult and expensive. The cost and adversity is compounded with troublesome ground conditions. Ground that is difficult to dewater can generally be frozen, either from the surface or from the tunnel. However urban environments and tunnels with multiple concurrent activities make these options at least undesirable and at worst impossible. This was the scenario experienced on the U230 project at CP3.

The most difficult areas of CP3 were made up of very moist to wet, sandy, silt with low permeability. Although unplanned at the beginning, the vacuum system proved to be very effective at reducing the mobilization of these soils during the excavation. U230 did not realize the substantial time savings that would have been available if the entire system was installed from the beginning. Overall cost and community impact were significantly reduced compared to a ground freezing system installed from the surface. CP3 of the U230 project was completed safely and with zero recorded ground movement and the effectiveness of the vacuum system played a significant role in that success.

## REFERENCES

University Link 230 Geotechnical Baseline Report (volume 6).

University Link 230 Geotechnical Data Report (volume 7).

# Compensation Grout Design for the San Francisco Central Subway Project

**M. Luis Piek**
Arup

**Robert Essler**
R.D. Geotech

**Alessandro Tricamo**
Barnard-Impregilo-Healy Joint Venture

**Mikael Calando**
Condon-Johnson/Nicholson Joint Venture

**ABSTRACT:** The use of compensation grouting in urban areas on tunnel projects has been widely used in Europe, with the list of projects successfully employing its use in the US steadily growing. While references exist for engineers to develop a compensation grouting design, most designs are site specific and require a fair amount of practical experience to properly adjust the program and advise during construction. This paper discusses the specific design approach implemented for the San Francisco Central Subway Project and may serve as reference on future projects in similar conditions. The design approach, preparatory works, pipe installation and ground preconditioning, active compensation grouting during TBM passage, and results are covered.

## INTRODUCTION

The SFMTA C1252 Central Subway Project consists of approximately two miles of surface and subsurface rail that will extend the Third Street Light Rail in a Northwest-Southeast link through the center of San Francisco. The Project will link Chinatown to the Downtown and South of Market Street (SOMA) areas of San Francisco, including AT&T Park. Contract 1252 includes construction of the 500ft launch box, retrieval shaft, approximately 1.7 miles of twin tunnels, and 5 cross passages. Within the Contract Drawings, areas were identified requiring compensation grouting, including particularly sensitive buildings and structures, which includes 100 year old buildings the BART tunnels. In June of 2012 Arup was contracted to BIHJV to complete the compensation grouting at those locations. The construction and execution of the compensation grouting program was subcontracted to Condon-Johnson/Nicholson Joint Venture (CJN).

The tunnel alignment begins at the launch box at Fourth Street between Bryant and Harrison Streets and extends northwest along Fourth Street, crossing Market Street at Stockton Street. The tunnel extends to Washington Square at Columbus Avenue and Union Street, with the retrieval shaft located at the former Pagoda Theatre site. To minimize disturbance to the downtown area of San Francisco, the tunnels are constructed using Earth Pressure Balance (EPB) TBMs. Compensation grout tubes were designed and installed at strategic locations to reduce or balance the ground loss due to passage of TBMs.

## ASSESSMENT OF GROUND MOVEMENTS

In order to establish where compensation grouting is necessary, the first step is to establish the potential ground moment and settlement related directly to the volume loss due to TBM passage. For design, two conditions are assumed:

- **Condition 1:** The worst credible settlement and effects on the buildings at a volume loss ($V_L$) of 1%. This condition is conservatively used for the design of the compensation grouting program
- **Condition 2:** The settlement and effects on the buildings due to the expected (probable) maximum volume loss as set by the contractor to calculate the limits of the operation parameters for the TBM during tunneling.

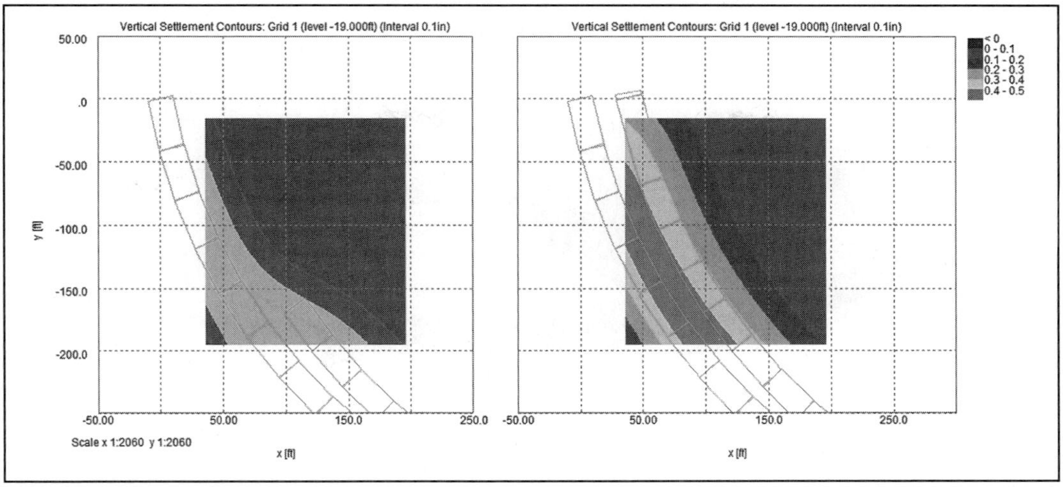


**Figure 1. Settlement contour plot of with TBM 1 passed and TBM 2 approaching and TBM2 passed**

For the subject project, the $V_L$ was set at 0.5%. It should be emphasized that the performance of the TBMs and the parameters to drive the TBMs selected by the contractor are critical to ensure the volume loss remains at or below their selected target volume loss.

The analyses were carried out using the software XDisp, developed in-house by Arup, which calculates the ground movements induced by tunneling, embedded wall excavations or mining works producing three dimensional displacements and strains widely using the accepted empirical approach as presented by O'Reilly & New 1982 and Attewell & Woodman 1982. Tunnels are modeled as excavations of circular cross-sections in soil and the main parameters in the analyses are the volume loss ($V_L$) and the shape of the settlement distribution curve, defined by the trough width parameter k.

Since variable soil conditions were expected along the alignment, two different k-values were considered to account for the variability of the bored profile and the overlaying soils. Since the thickness and the cohesion of the Colma Sand, Undifferentiated and Bay Mud layers vary along the alignment, an upper bound and a lower bound set of values for the trough width parameter were considered.

- k=0.45 for the first tunnel and k=0.55 for the second tunnel
- k=0.50 for the first tunnel and k=0.75 for the second tunnel

The settlement trough due to the excavation of the first tunnel is then superimposed to the settlement trough calculated for the second tunnel. A higher trough width parameter is assumed for the second trough due to the softening effect caused by the excavation of the first tunnel, which results in a different soil behavior when the second tunnel is bored. That is, the boring of the first tunnel induces strains in the soil which causes a change in the stiffness, and therefore in the ground response when the second tunnel is bored. (D.N. Chapman et al. 2004).

In the first pass assessment, the stiffness of the buildings is not considered and the displacements calculated assuming "greenfield" conditions to provide conservative maximum and differential settlements. Additionally, any benefit from the pre-conditioning of the soil is also not accounted for in the analyses. The pre-conditioning of the ground will stiffen the soil and facilitate some degree of arching within the soil. This is expected to redistribute the ground movements to a flatter settlement ground curve. If the settlement is found to be in excess of the allowable, additional calculations incorporating the building stiffness and/or the stiffening of the soil due to preconditioning can then be performed.

In conjunction with the ground movements, building surveys are conducted to gather information on the building type (e.g., masonry, wood framed, steel moment frame, braced frame) as well as the foundation type (e.g., shallow or deep, strip footings, piled, isolated columns) to assess the settlement levels that could cause structural damage to the buildings. Strain calculations based on the ground movements can be completed to determine the expected level of damage to the buildings. Examples of the settlement contours for the TBMs passing under Forever 21 are shown in Figure 1 and Figure 2.

The conclusion of the Ground Movement Assessment report was that at $V_L = 0.5\%$, the need for compensation grouting was minimal, however the risk profile of the project necessitated compensation

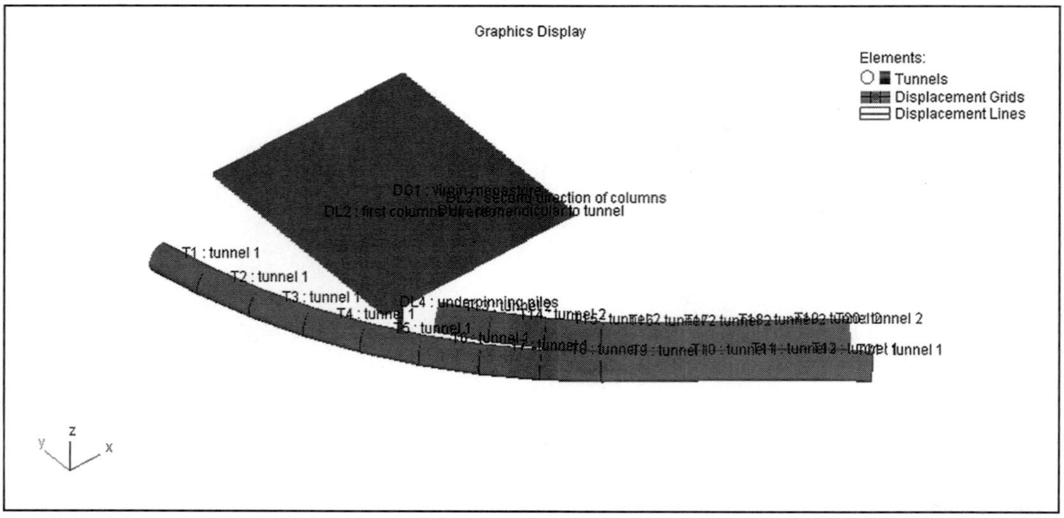


**Figure 2. 3D view with TBM2 in the section closest to the center of the building**

grouting at the $V_L = 1\%$ levels at the locations indicated in the Contract.

If building damage were to be expected by the greenfield approach, an increased level of rigor can be applied, as described by Mair 1996 for the surface buildings on the Jubilee Line Extension.

For the purposes of this paper, the compensation grout works are divided into three major areas, with the structures in the order encountered during TBM tunneling as follows:

1. TBM Launch and First 1500ft
   - Whole Foods, Building Block 3751-411 to 415, 367–399 Fourth Street
   - AT&T, Building Block 3751-105,112 and 155, 795 Folsom Street
2. 4th Street Crossing / BART
   - Old Navy, Building Block 3705-048A, 801–803 Market Street
   - Market Street Subway Tunnels (BART)
   - Forever 21, Building Block 0328-002, 2 Stockton Street
3. Green Street Shaft
   - Bank of America, Building Block 0130-001, 1435 Stockton Street

## COMPENSATION GROUTING DESIGN

### Design Approach

The area requiring compensation grout is first set out considering the influence area due to the passage of the TBM and further adjusted considering any minor ground loss due to installation of the grout tubes, the type of ground, the spacing of the injection ports, the mix design for pre-conditioning and during TBM passage, and the sequence under which pre-conditioning and compensation grouting passes are completed.

A certain level of redundancy should be included in any compensation grout design to accommodate any deviations of the Sleeve Port Pipes (SPP) from their target distal point, blocked ports, pipe breakages, or blockages. For instance, the loss of one pipe in an array should not jeopardize the grouting program; however, losing 2–3 adjacent pipes could be an issue as this might leave an area without any settlement mitigation. Likewise it is not necessary to have every port on every pipe working in order for the program to be effective. A robust level of redundancy for this compensation grouting program was included and considered appropriate for an urban environment with sensitive buildings and economically vital infrastructure.

Within the first 150ft of the tunnel drive, the northbound TBM passes directly in front the first of two large multistoried buildings founded on piles: the Whole Foods building, and approximately 600ft thereafter, the AT&T building. To protect the loss of skin-friction along the pile length, sub-vertical compensation grout tubes were planned along the building length between the TBM and the piles.

Selemetas et al. investigated the full scale performance of both end bearing and friction piles during a TBM transit. They concluded that there was evidence to suggest the existence of three zones around an EPB tunnel in London Clay, similar to those presented in previous studies (Kaalberg et al. 1999), in which pile head and ground surface settlements can be correlated.

Referring to Figure 3, piles with their bases located in Zone A were shown to settle 2–4mm more than the ground surface (Ratio of pile to ground

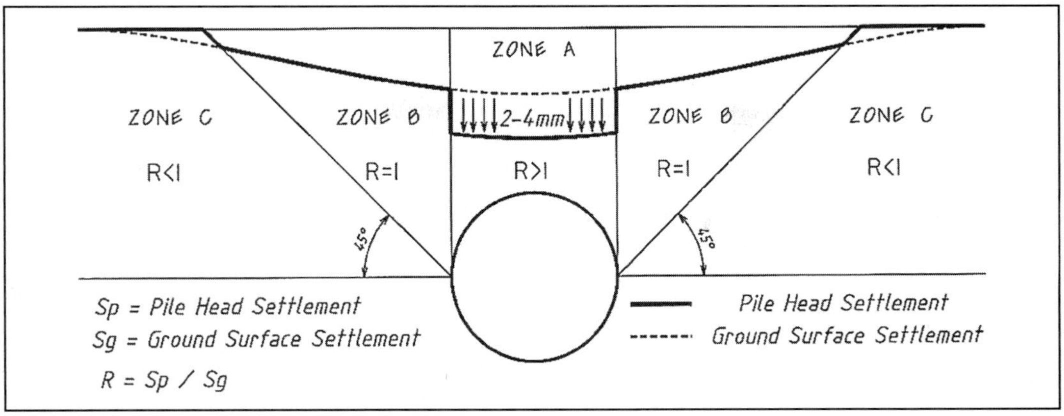


**Figure 3. Zones of settlement according to Selemetas**

settlement R>1). Piles with their bases in Zones B (defined by an angle of 45 between Zones A and C) settled by the same amount as the surface (R=1). Finally, piles with their bases in Zones C were found to settle less than the surface (R<1). Therefore for most practical applications reasonable predictions of pile settlement could be made by using the Gaussian curve as a reference frame. The critical zone is Zone A in which piles are likely to settle more than the ground due to the reduction in their base load. To counterbalance this loss in the base load the piles settled in order to mobilize the required shaft friction.

It is important to appreciate that the boundaries of Zone B are simplified by simply drawing two straight lines from the spring-line of the tunnel. The angle between Zones A and C is probably a function of the shearing resistance of the soil and the tunnelling volume loss and therefore is not likely to be constant. Based on the review of the Selemetas approach along with additional information from the Heinenoord study reported by Kaalberg, it seems probable that the zoned approach could be applied to the AT&T and WF buildings.

For the Whole Foods buildings the pile toes are beyond the zone of influence of the tunnel and within Zone C as defined above and were predicted to undergo only lateral movement with minimal settlement. The compensation grouting could only therefore be used to maintain or reapply horizontal stresses within the ground in an attempt to limit pile lateral movement.

For the AT&T building the pile toes are within Zone B, and could suffer both vertical and horizontal movement. It was concluded that the AT&T building would likely experience settlement by an amount similar to the surface settlement. Thus compensation grouting applied towards the base of the piles rather than along the length of the pile was important to provide adequate settlement control.

At Old Navy the southbound tunnel passes under the building and its foundations. An initial conceptual design included sub vertical pipes installed from sidewalk level of the east side of Fourth Street, terminating within a few feet of the west side of the tunnels. Arup improved the arrangement by moving the pipes away from the tunnel and closer to the level of the existing pile toes. When deciding the level of grout tubes the designer should consider the risk of imposing high grouting pressures on the tunnel during construction, but also placement of the pipes too close to individual foundations then the local stress distribution can affect the grouting efficiency. Movement of the pipes during the design as described decreased the risk to the tunnel during construction in accordance with best practices for compensation grouting and maintained the ports in location under the foundations to achieve the target pre-conditioning heave. This is explained in more detail in Essler et al. 2000.

Due to the very tight clearance between the new and existing BART tunnels, the compensation grout pipe configuration at BART included sub horizontal pipes to be installed very close to the tunnel crown, approximately 6ft above the tunnel crown. The pipe configuration at Forever 21 was also placed in very close vicinity to the tunnels due to the presence of pin piles installed as part of the early works utility relocation contract. Due to the vicinity of the pipes to the TBM operations, during tunnel activities a TBM exclusion zone was incorporated into the compensation grouting design to prevent face instability and grout ingress into the TBM mucking system. Typical exclusion zones are centered around the crown of the TBM with a radius of approximately half of the TBM diameter (Figure 4).

At the Bank of America Buildings a purpose-built shaft on the opposite side of the building from where the TBM will pass was used for compensation

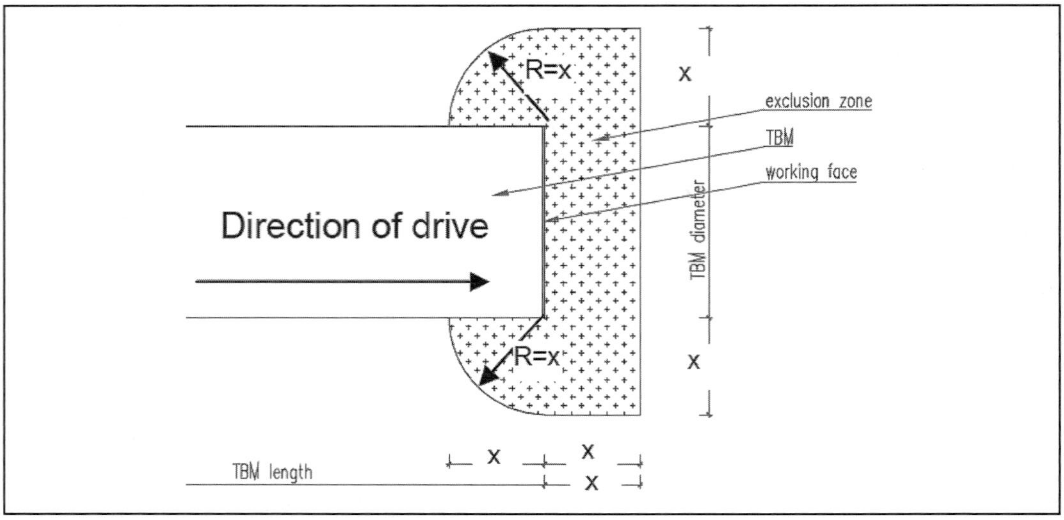


**Figure 4. Typical configuration of the TBM exclusion zone**

grout pipe installation. Locating the shaft so remote from the tunnel alignment means the compensation pipes will pass completely underneath the building, providing full access to the building foundation. The pipes are located approximately mid-way between the tunnel and foundations.

### Surface Array Design and Installation

Grouting arrays can be installed from the surface where TBMs pass adjacent to or partially beneath buildings. They are relatively simple to install, do not require extensive access, and can be installed using relatively standard drilling equipment. When installing surface grouting arrays in existing right of ways, especially in urban centers, existing utilities both known and unknown should be considered. Because the angle of installation is important, often the arrays are pushed to one side of the street, near the sidewalks. The boom height of the drill can overhang the sidewalk, necessitating closure, or in more extreme cases, the boom can come very close or even conflict with existing buildings. In addition, overhead power poles and telecommunication infrastructure should also be considered, as restrictions for operating equipment within a specified safe working distance are often imposed.

For the subject project, the use of a high-boom Klemm 803 was used to install the surface arrays at Whole Foods and AT&T as originally foreseen, however at Old Navy the presence of unknown utilities in the street required an adjustment to the design installation angle of compensation grouting tubes. This, in turn, caused the lowering of the drill boom, and resulted in a construction conflict with the Ross Dress for Less building across the street, requiring the mobilization of a shorter boom drill Comacchio MC-602.

Surface arrays should be designed so they are fairly flexible, meaning that if conflicts are discovered on the surface locally increasing or decreasing the spacing of an array by minor amounts (2 to 3 feet) is readily feasible. A designer should anticipate these changes and allow room in their design to for such accommodation. Further, it is helpful to the contractor if the compensation grout design includes a range of allowable installation angles, accompanied by an allowable change of installation angle between adjacent grout pipes. Indeed, at Whole Foods, AT&T, and Old Navy locations such adjustments were necessary.

When locating the grouting arrays, the intent should be to arrange the pattern in such a way that the injections occur close to and mostly underneath the building foundations and do not cause uneven lifting of the façade. The pattern should not be located too close to the finished TBM tunnels, which could compromise the effectiveness and ability to inject grout during the active compensation grouting phase.

The compensation grout design for the Whole Foods and AT&T buildings utilized surface arrays installed between the northbound TBM tunnel and the pile foundations to preserve skin friction along the pile lengths and offset settlements. An example cross section of the alignment of the surface array at Whole Foods and Old Navy are shown in Figure 5 and Figure 6, respectively.

SPPs were constructed of either Schedule 80 PVC or Schedule 40 steel with injection port spacings at 14" (40cm) intervals. Standard SPPs have injection ports drilled into the tubing at four locations spaced 90 degrees around the circumference of the pipe with the holes covered by a flexible rubber

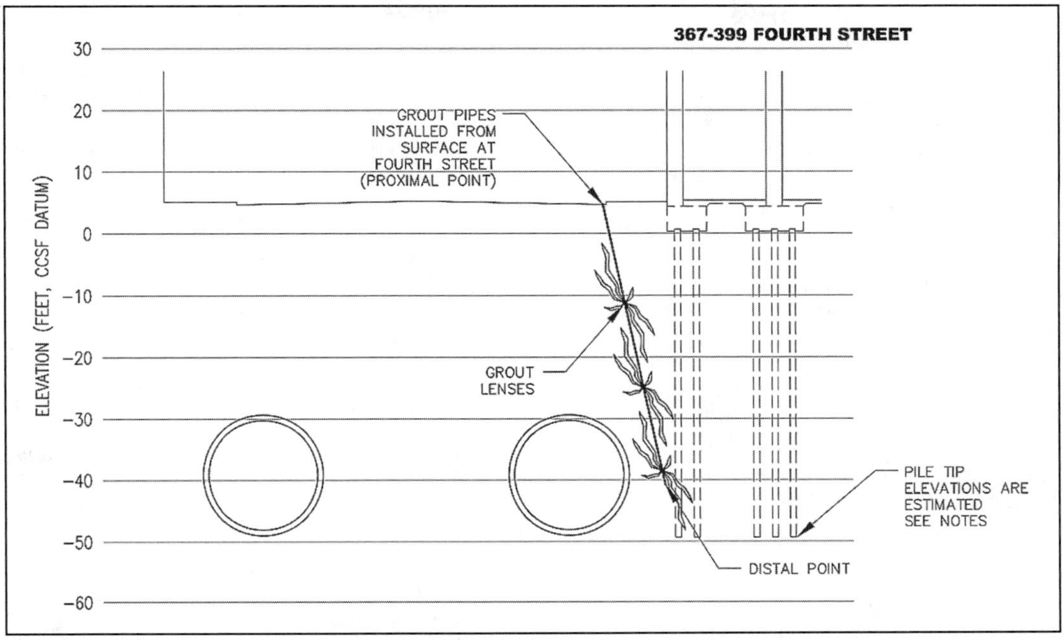


Figure 5. Compensation grout array at Whole Foods building

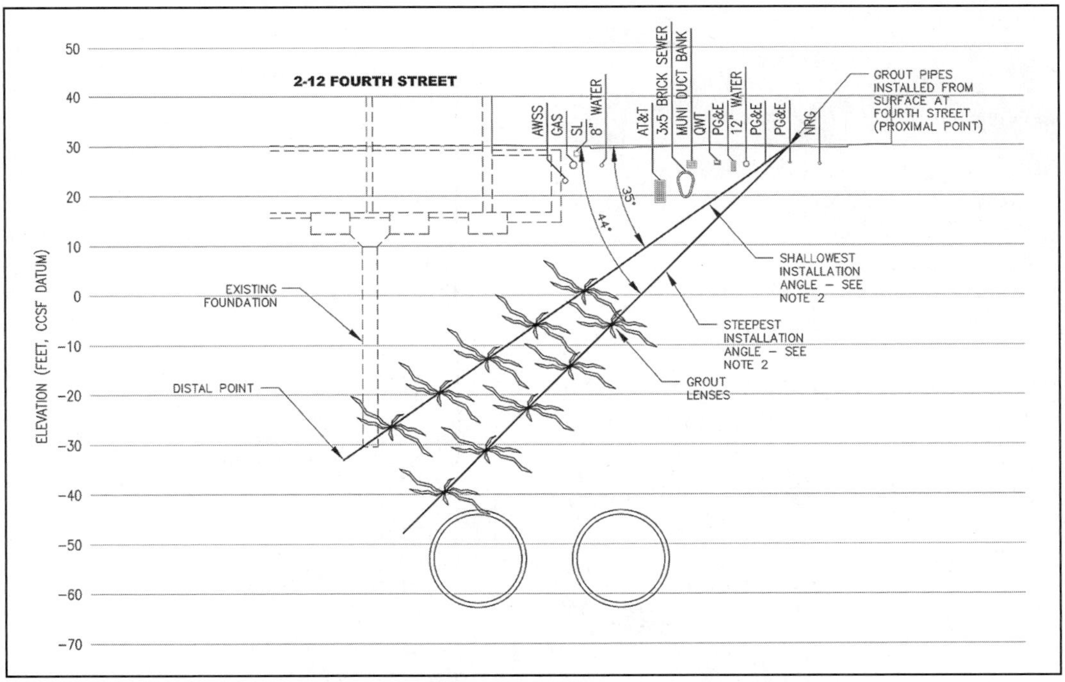


Figure 6. Compensation grout array at Old Navy building

sleeve which allows the grout to flow out into the surrounding environment and prevents it from flowing back into the tube once pumping has stopped. The type of rubber is important: the rubber should be made from a special compound which resists changes in durometer and extensibility due to UV and temperature. All rubber sleeves should be held in place with a light adhesive and taped down at both ends. While both recessed and non-recessed sleeve options exist, it is the recommendation of the authors that if possible the ports always be recessed such that the pipe is flush with no protrusions along the pipe barrel to allow for minimal resistance while advancing the pipe. Also, the connecting ends of the pipe should be machined to create a flush-bell connection system with the effect that the SPP diameters are constant on both the inside and outside.

At Whole Foods and AT&T 2-inch diameter PVC SPPs from C&M manufacturing were used. PVC pipes from another supplier were installed at Old Navy, however those pipes failed in splitting during preconditioning. The cause of the splitting was never definitely found, however higher pressures were used for preconditioning than at the

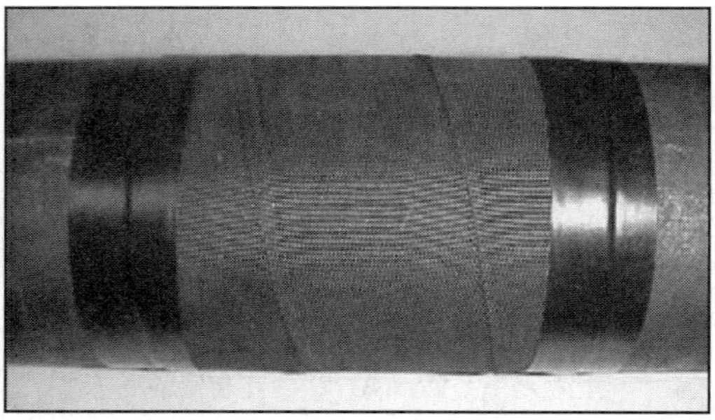

**Figure 7. 2-inch diameter steel SPP with protection over the grout port**

Whole Foods and AT&T buildings. Remedial work included overcoring the installed pipes with a 6" casing, injecting cement/bentonite slurry into the hole as the casing was pulled back, and installation of new 2-inch steel SPPs (supplied by Strata-Tech) grouted into place. An example of a 2-inch steel SPP is shown in Figure 7.

**Subsurface Grouting Array Design and Installation**

When TBMs pass directly underneath buildings, or when the extent of a surface array could be too disruptive to the surrounding community, subsurface arrays can be used. These are normally configured in access shafts, either single or multiple, along the alignment, however more creative arrangement configurations have been used on other projects (Essler et. al 2000). In circular shaft configurations, it is common to design the array as a fan, with the pipes extending outward from the center point of the shaft. At the proximal end (the end of the pipe protruding into the shaft) the pipes in the array cannot be located on the same level, as it would be physically impossible to install all the pipes, thus several rows of pipes are required, usually on the order of 3 or 4 depending on the spacing and density. With the azimuth passing through the shaft center, the equipment to drill the pipes can be mounted on a shaft located at the shaft center, allowing the drill equipment to be rotated and raised and lowered to drill the multitude of pipes, creating an efficient method to install the grout pipes. A photo of the pipe arrangement in the shaft at Ellis Street for the Central Subway Project clearly showing four rows of pipes installed is shown in Figure 8.

The design and construction of the compensation grout design in shafts should normally meet the following general requirements:

**Figure 8. Compensation grout array at Ellis Street shaft**

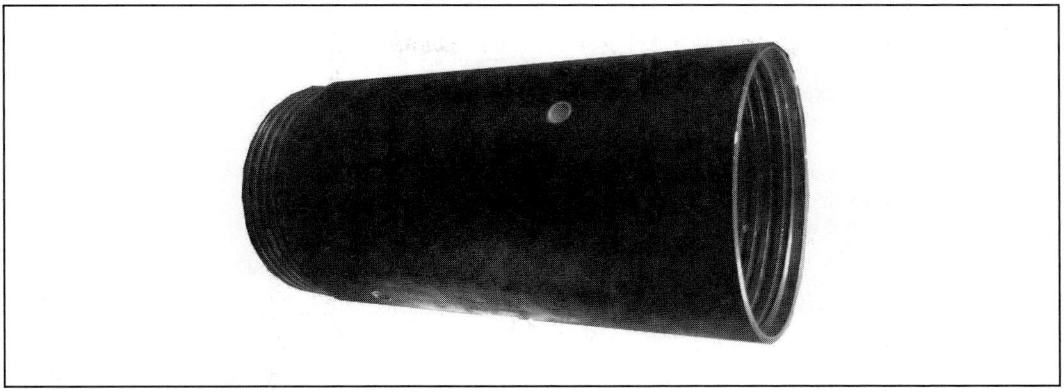

**Figure 9. Sample length of 6-inch diameter steel SPP**

1. Shafts should be deep enough to install the pipes to the levels and directions indicated. Normally pipe installations should be level or downwards. Upward drilling from within a shaft is difficult;
2. Shaft structural integrity should not be compromised by the penetration of the shaft wall of the grout pipes, exercise proper pipe spacing and level separation;
3. Shafts and penetrations by the SPPs must be watertight, lowering of the water table or flushing of fines from around the pipes and into the shaft can result in additional subsistence;
4. Design the pipe array such that the method of construction does not limit the construction methods of the contractor and thus necessitating a pipe layout redesign; and
5. Layout of the grout tubes should leave open locations for subsequent pipe installations if obstructions are encountered when installing pipes.

At Ellis Street Shaft, due to the required installation length (up to 205ft) and the requirement to maintain the pipes within a 2% tolerance, CJN elected to install single-use 6-inch diameter SPP pipes tipped with sacrificial full-face drag bits, which includes a pilot bit that drills the center part of the hole with a casing shoe welded to the pipe and a symmetrical ring bit. The result is a drillhead which reduces air leakage and prevents overdrilling.

Normally these large flush-bell pipes are used in pipe canopy applications when doing SEM-type tunnel excavations, and due to their higher moment of inertia deflect less over a longer distance allowing tighter installation tolerances. Two ports were installed per location at 180 degree separation, and clocked 90 degrees each 14" spacing. An example of the 6" steel SPP is shown in Figure 9.

Alternatives to install the pipes in the access shaft as used on the Gold Line (Robinson and Bragard, 2007) using HDD were investigated, however there was concern that volume loss could occur during the process of injecting cement/bentonite slurry in the hole as the drill string is pulled back if a leak in the system was developed.

The drill used to install the pipes was a modified Klemm 806 mast with double head. The mast is mounted on a platform to adjust height, inclination and azimuth. When drilling from shafts under high water pressure, a system was required that could deal with the high ground water pressures and prevent soil and groundwater ingress at all stages of the drilling operation. For this application the pipe was fitted with a check valve with a bentonite flush to advance the hole by injection through the bit. The use of a flush pipe system greatly helped accomplish the installation of the pipe using the stuffing box.

Care must be taken to prevent the pipes from becoming blocked with grout or soil, thus the protective wrapping of the ports prior to installation. However if it does occur, pipes can be cleaned using water injected under *low pressure* or by mechanical means; high pressure water injection (greater than 500psi) should be avoided because in granular soils water can enter the ground and cause injection ports to be wedged open with sand/gravel particles. This renders the pipe useless for grouting unless the leakage can be sealed which can be very time consuming.

It was not foreseen that the installation of the SPPs would result in any significant settlement, however if it did occur, sequenced drilling and pre-conditioning of the ground could have been used as one method to control any inadvertent ground movement due to SPP installation.

## Pipe Surveys

Often, surveys of pipes are required following installation to confirm that they have been installed to the plan position and depth required. For surface arrays where the pipes are at an angle where gravity assists with the installation, Shape Accel Arrays can be used, however our experience indicated that the Gyrosmart system provided better information for longer horizontal holes.

The Gyrosmart system is a MEMS memory gyro system for borehole surveying. The GyroSmart is built around a digital micro-gyro, which consists of a silicon sensor chip and an integrated circuit assembled in a ceramic (non-magnetic) package. The GyroSmart is not affected by magnetics and its small size makes it easy for surveying inside rods or a casing string. The miniature sensors detect changes in direction, azimuth and dip from a known starting point. The starting point was calculated based on survey marks laid out on site from the initial line of the instrument. With some post-processing, the system shows the relative dip and direction of the SPPs. In general the distal position should be within 3% of the hole length in any direction, however where distal tolerance is critical, i.e., beneath the BART tunnels a 2% tolerance should be followed.

In addition to standard pipe surveys, all compensation grout tubes installed below the BART tunnels required hold points for an additional survey: once they reached a minimum encroachment of 20 feet of the BART tunnels, and once after final installation depth. The goal was to ensure grout pipe installation was a minimum of 2 feet beyond the exterior of the BART tunnels, including all drill tolerances and allowable deviations.

While this requirement may be considered onerous, in general, where grout pipes are installed within 5ft of any underground structure (e.g., existing tunnels, basements, caverns) then a pipe installation survey should be carried out before the pipes are grouted in to ensure that they are within design tolerance. All surveys should report the Northing, Easting, and Elevation of the proximal and distal ends, and the contractor should provide an as-built drawing showing the plan position and levels of all pipes installed after installation to the Engineer. Although large pipes were used at Ellis Street, horizontal and vertical deviations of over 5ft and up to 2ft were measured, respectfully. The use of smaller diameter pipe would have likely resulted in higher deviations.

Once installed, the SPPs must be locked into the ground by placement of an annular grout. In general the same low strength, low bleed, stable grouts used pre-conditioning and active compensation grouting can be used for the annular grout.

## Grout Mixes

Typically, grout mixes for compensation grouting projects should be low bleed, stable grouts with a low pressure filtration coefficient. Cement/Bentonite grouts would be appropriate in the more clayey strata while granular deposits should utilize grouts that minimize penetration into the strata. Excessive permeation can cause problems with the subsequent use of the sleeves due to the high pressures required to penetrate through the grouted or partially grouted materials. For this reason a grout with a low strength needs to be formulated with a 28-day break strength of less than 500 psi recommended. For the project, CJN and Arup worked in conjunction to carry out trial mixes to carry out tests to verify the grout mix formulation prior to construction. Table 1 shows some of the grout mixes used.

**Table 1. Grout mixes**

| Component | Basic | Low W/C | Ultrafine |
|---|---|---|---|
| Cement (lb) | 94 | 94 | 44 |
| Bentonite (lb) | 5.3 | 4.0 | 3.3 |
| Water (lb) | 175 | 132 | 82 |
| Water (gal) | 21 | 15.8 | 9.8 |
| Silicate N38 (gal) | — | — | — |
| Super P: Mighty 150 by Kao (lb) | — | — | 1.3 |
| TOTAL (lb) | 274.7 | 229.9 | 130.7 |
| Target @ 28 days (psi) | 300 | 425 | 300 |

## Ground Preconditioning

The preconditioning of the ground is an important aspect of compensation grouting. It has the following benefits:

- It conditions the ground such that existing voids are filled and further injections of grout will cause predictable ground movements;
- It allows the designer to carry out an assessment of grouting efficiency (the ratio of change in ground volume or uplift to the volume of grout injected);
- It stiffens the ground thus reducing future movements;
- It creates a nominal upward movement that in itself will compensate for some movement during tunneling

In general the contractor should inject grout using a number of sequences or "passes." Each grouting area should be partitioned into plan sub-areas of approximately $10ft^2$. The total estimated volume of preconditioning grout to be injected per square foot, as well as the target refusal pressure and/or heave criteria is provided by the designer to the contractor. Providing these clear termination targets at the onset is important to give the contractor a clear goal.

Means of determining if grout is entering into pipes not being injected should be included in the means and methods, as well as determination of line-loss pressure in the pipes to achieve the target pressures in the ground. The volume and flow rate should be approximately be 2.5 to 3 gallons/minute and the injection process flow controlled with the pressure varied as required to maintain the required design flow rate. During preconditioning of the ground, record all the injection volumes, times, and pressures on a per-injection basis. Also note the mix design, time start, time stop, total injection time, injection rate, grout volumes, grouting pressures at point of injection, volume, and max pressure for each port injected. As a large amount of data is produced, the designer and contractor need to agree on how this data will be processed and presented for review and interpretation, and by whom.

For the Central Subway project, the total preconditioning volume was injected using a first and second pass approach in order to review the monitoring information, assess how the ground is reacting to the injections, allow any pore pressure reduction, and to assess if there is any movement detected. Normally a 50-50 split is used however other arrangements such as 70-30 can be used in certain circumstances, for instance if secondary grout takes are lower than expected. If possible, stagger injections with at least one shift between first and second pass injections to allow for pore pressure dissipation and gel of the grout.

For the subject project, the preconditioning of the ground was very successful, with filling of the void space in the ground and a slight heave (~⅛") of the buildings achieved.

In the Ellis Street shaft, where the large 6" steel pipes were used, very high pressures (+90 bars) were necessary to inject the preconditioning grout. All these pipes were located in the over-consolidated Colma formation, and care should be taken in performing injections into this type of dense material. In hindsight the presence of only two 180 degree opposed ports at 14" was not considered sufficient and at a minimum 4 ports 90 degrees separated every 14" would have been a great improvement. This is because the area where grout must pass is small, and injections can be hindered by blockages, or failed or partial opening of the ports. A sleeved arrangement in comparison could be considered superior, as even though it has the same 4 openings at each sleeve port the rubber sleeve separates the openings in the pipe from the ground, allowing the grout several flow paths if one is blocked or restricted.

## COMPENSATION GROUTING DURING TBM PASSAGE

Compensation grouting during TBM passage consists of the following steps:

1. The trigger levels for the start of compensation grouting are agreed;
2. The contractor should arrange for 24 hour manning of the grouting plant when the TBM is expected near the grouting zone. Good practice states that this should be when the TBM is 100 ft from the closest point to the grouting zone and continue until the TBM is 100ft from the last closest point of the grouting zone. As it may take time to arrange resources, it is good to find an agreed time period between the grouting subcontractor (if separate) with the tunneling contractor in advance of the TBM arrival based on a "best guess" rate of TBM advance;
3. The grouting contractor maintains their resources and reviews the monitoring data. When grouting is required, they will either manually or computer generate the required grouting passes based on advice from the grouting designer;
4. Grouting passes typically are centered over the zone of maximum settlement and comprise injections at approximately 3ft intervals in a number of pipes, with volumes of 15 to 25 gallons injected at a flow rate of 2.5 to 3 gallons per minute;
5. The rate of grout injection will most probably require 3 to 4 packers to be used at the same time. This is an extremely important point in the execution of any compensation grout design. During TBM passage there is a finite level of resources that can be applied to perform compensation grouting. This relates to the limited number of packers that can be used, the grouting efficiency and the grout flow rate. The emphasis needs to be to protect the existing structures from the damaged caused by potential settlement during the tunneling operations; however consideration has also been given to TBM progress, while providing the appropriate level of settlement mitigation. The grouting contractor needs to demonstrate in a developed method statement that they can provide the necessary equipment and staff to adequately react without limiting the advance of the TBMs.
6. The above process continues until the TBM has passed out of the grouting zone.

In general the same grout mixes to those used for the preconditioning can be employed for active compensation grouting. The process should be flow controlled with the pressures used during preconditioning used as a guide. No pressure limit is necessary unless the grouting is close enough to a structure or the tunnel to give cause for concern;

however a pressure limit may be defined and implemented for practical reasons during the performance of the injection. For the subject project a mandatory pressure limit was imposed only where the tunnels passed beneath BART where the compensation pipes were located relatively close to both the BART and the new tunnels.

### Compensation Grouting Pass 1

Compensation grouting is carried out by assessing the settlement that has occurred since the previous injections. For the first pass this will be the settlement since completion of pre-conditioning. The actual final grouting efficiencies measured during the latter stages of pre-conditioning are used as the starting point in order to calculate injection volumes. In general the compensation grouting area should be portioned into sub zones in the range say 10ft by 10ft or 15ft by 15ft and the settlement in each of these sub zones observed. Based on the individual settlements in each sub zone, the volume of grout is then calculated utilizing the latest efficiency factor. These grout volumes are then injected within each sub zone to compensate for the settlements at the start of grouting.

On completion of the grouting pass the settlements are observed again and the process repeated. The size of the sub zones will depend on the local geometry. It is common to assign sub zoning based on settlement monitoring points so that the measurement system is integrated with the settlement control. Most specialist compensation grouting contractors provide software that can assess the settlement and target the grout injection.

### Compensation Grouting Pass 2 and Subsequent Passes

Repeat of Pass 1 initially followed by specific subarea of injection based on the results of the monitoring. Grout volumes should be selected based on the results of previous injections.

For locations where the implementation of an exclusion zone around the face of the TBM is used, such as at the Market Street subway tunnels, grouting would take place immediately after the TBM has passed. In a case such as this, the most efficient and safest methodology to adopt is as follows:

1. Ensure good communication between the grouting contractor and the tunnel engineers;
2. The tunnel engineers reports the face position at intervals no greater than 30 minutes (15 minutes preferred);
3. The tunnel engineer must advise when the ring segments are being built and completed. Compensation grouting is to be *only performed once ring build is complete* and the TBM has recommenced excavation;
4. Following completion of the ring construction , the grouting pass is initiated while the TBM is advancing;
5. The grouting pass is either generated from the settlement monitoring system within the tunnels or commonly a grouting pass is agreed based on an agreed volume loss. The rate of advance is usually the length of one ring and thus the total volume loss volume can be readily calculated;
6. Grouting should be carried out considering any calculated pressure limits on the tunnel linings. The actual grouting pressures used during the preconditioning phases should be used as a guide to determine the magnitude of grouting pressures during compensation grouting;
7. The above process is repeated until the TBM has passed out of influence of the rings of the segmental tunnel lining;
8. Following TBM passage, secondary grouting to regain tunnel levels can be carried out.

It is worthwhile to again mention the importance for the grouting contractor to allow for redundancy in their grouting resources both in terms of equipment and labor.

## INSTRUMENTATION AND MONITORING

Volume loss is one of the key assumptions that are used in determining the extent and usage of a compensation grouting program. While the systems installed in TBMs to calculate volume loss have advanced, in the opinion of the authors the use of multi-point borehole extensometers (MPBX) continue to be the most straightforward way of calculating the actual movement occurring in the ground and the behavior of the soil as the TBM approaches and passes. Thus placing MPBX instruments ahead of critical zones will give the contractor an idea as to the level of effort they will need to exert when the TBM is passing. While our target volume loss was 0.5%, back calculations from the MPBXs showed approximately 0.35%.

In terms of reporting the results of a monitoring, a rapid response system is essential when carrying out both preconditioning of the ground and for compensation grouting during TBM passage. For example for the tunnel passing under the Market Street subway tunnels, if face pressures were difficult to control, it was estimated that the rate of settlement could approach 0.5 to 0.75 inch per hour. Thus the monitoring system needs to be able to generate the overall movement of the tunnels within short periods of time otherwise the information will become out of date very quickly. Generally a complete update of all monitoring points affected by the tunnel within 15 to 20 minutes is sufficient to allow the grouting engineer an overview of the settlement generation.

Most specialist grouting contractors have software that links the monitoring of the structure to the grouting and automatically generates the grouting passes. This could be satisfactory for compensation grouting when slower forms of tunnel construction such as SEM tunneling are used, however caution should be exercised to ensure the system does not suffer due to any time lag if the TBM is moving quickly.

## RESULTS

While a compensation grouting program could be considered by some as redundant and unnecessary if proper face pressure in the TBM is maintained, it should be emphasized that compensation grouting programs are not just a "safety net" in case face pressure is lost or the volume loss exceeds the target along one particular reach. The considerable work that takes place in preconditioning the ground in order to fill the void space, stiffen the ground, and induce heave into the surroundings gives the TBM operators that much more room for margin. No project is perfect and unknowns often arise, seemingly at the most inopportune of times.

When passing under BART, which is arguably one of the most critical pieces of transit infrastructure to the Bay Area, the settlement recorded by the instruments was almost negligible (~0.5mm). Using the average volume loss back-calculated from the MPBXs, this value would not have been achieved solely by the operation of the TBM. Indeed, preconditioning on the Central Subway project has played a definite role and thus far allowed the TBMs to pass without delay. Further, by having a system installed that can "zero" the ground settlement after the passage of one machine and prepare the ground for the passage for the second can quickly pay for itself if, or when, those unknowns are encountered.

Briefly in closing, the authors would like to call attention to those that assisted in making the compensation grouting program successful, namely: Matt Fowler, Pietro Fioravanti, Evelyn Sanchez, Sia Motlagh, Nick O'Riordan, Tom Richards, and the BIH and CJ-N crews. Your efforts made this all possible. We thank you.

## REFERENCES

[1] Compensation Grouting: Concept, Theory and Practice, R.D. Essler, R. Drooff and E. Falk, Advances in Grouting and Ground Modification (Geotechnical Special Publication No. 104), Geo-Denver 2000, Denver, Colorado, August 5–8, 2000.

[2] The Response of Full-Scale Piles to Tunnelling, D. Selemetas, R.J. Mair, and J.R. Standing, Proceedings of the 5th International Symposium TC28. Amsterdam, Netherlands, 15–17 June 2005.

[3] Evaluatie van de meetresultaten van het proefpalenprojek ter plaatse van de tweede Heinenoordtunnel (in Dutch). Kaalberg, F.J., Lengkeek, H.J., and Teunissen, E.A.H. (1999). Adviesbureau Noord/Zuidlijn Report no. R981382, Amsterdam.

[4] Assessment of risk of damage to buildings due to tunneling and excavation, Burland, J.B. (1995). Invited Special Lecture, 1st International Conference on Earthquake Geotechnical Engineering, IS Tokyo '95.

[5] Compaction Grouting to Control Movements During Tunnelling, Baker, W.H., Cording, E.J. and MacPherson, H.H.(1983) Underground Space, Vol. 7, Pergamon Press Ltd, pp. 205–212.

[6] Predicting the Settlements above Twin Tunnels Constructed in Soft Ground, D.N. Chapman, C.D.F. Rogers, D.V.L. Hunt, 2004.

[7] Settlements above tunnels in the United Kingdom—their magnitude and prediction, O'Reilly, M.P. and New, B.M. (1982). Tunnelling '82. pp. 173–181.

[8] Prediction of ground movements and assessment of risk of building damage due to bored tunneling, Mair, R.J., Taylor, R.N. and Burland, J.B. (1996) Proceedings of the International Symposium on Geotechnical Aspects of Underground Construction in Soft Ground, 713–718, Balkema, Rotterdam.

[9] Predicting the Dynamics of Ground Settlement and its Derivatives caused by Tunnelling in Soil, Attewell P.B., and Woodman, J.P. (1982) Ground Engineering, Vol. 15, No 8, pp. 13–22 and 36.

[10] Los Angeles Metro Gold Line Eastside Extension—Tunnel Construction Case History, Robinson B. and Bragard C., Proceedings of the RETC 2007 pp. 472–494.

# New Irvington Tunnel Groundwater Management Plan

**David F. Tsztoo and Betsy Lauppe Rhodes**
San Francisco Public Utilities Commission

**Bruce Abelli-Amen**
Baseline Environmental Consulting

**ABSTRACT:** The San Francisco Public Utilities Commission (SFPUC) is in the process of constructing a new 3.5-mile water transmission tunnel, the New Irvington Tunnel, in the San Francisco Bay Area. One of the challenges was protecting residential groundwater supply along the tunnel alignment during tunnel construction.

Construction of the Existing Irvington Tunnel in the late 1920s caused a substantial decline of groundwater levels and adversely affected surrounding wells, ponds, and springs. The legacy of this historic impact lived on with current residents. This paper discusses how the SFPUC is working with neighbors and ensuring that the local water supply is maintained during project construction.

## INTRODUCTION

### Hetch Hetchy Regional Water System

The water system that serves San Francisco and four surrounding counties is the Hetch Hetchy Regional Water System. It transports by gravity pristine snow melt from the Sierra Nevada Mountains through a series of pipelines and tunnels over 180 miles to 2.6 million customers in the San Francisco Bay Area (Figure 1).

The system crosses three major, active earthquake faults: the San Andreas Fault on the Peninsula, and the Hayward and Calaveras Faults in the East Bay. Each fault is capable of high magnitude shaking and offset displacements, which can inflict significant damage on the water system.

### Water System Improvement Program

In response to this long term threat to the integrity of the water system, the San Francisco Public Utilities Commission (SFPUC) initiated in 2002 the $4.6 billion Water System Improvement Program (WSIP) to protect its many water facilities from the risk of earthquake damage. The WSIP consist of 81 projects to reinforce, rehabilitate, or build anew to current seismic standards the major pipelines, tunnels, treatment plants, pump stations, and related facilities of the water system.

The existing Irvington Tunnel is located in the center of the water system, less than one mile east of the Hayward Fault and less than ½ mile west of the Calaveras Fault. The New Irvington Tunnel (NIT) is a key WSIP project and will build a redundant tunnel parallel to the existing tunnel within the same SFPUC right-of-way.

### Why Build the NIT?

The existing Irvington Tunnel is reaching the end of its useful life. It was built between 1927 and 1932 by drill and shoot, or blasting excavation. Heavy timbers and wood lagging were used for the initial tunnel support. It was finished with a cast in place concrete lining, but no rebar or reinforcing steel was used strengthen the lining (Figure 2).

The existing tunnel also has no current capacity for maintenance shutdowns. Water demands on the tunnel are such that it cannot be taken out of service for even one day. The tunnel must deliver approximately 200 million gallons per day (mgd) during the winter months, and up to 300 mgd during the summer months. The last time the tunnel was taken out of service for inspection was in 1966—some 47 years ago. It has been at least that long since any maintenance has been performed.

So, due to its age and substantially unreinforced and likely disrepair condition, the existing Irvington Tunnel is very vulnerable to major earthquake damage from either the Hayward Fault or the Calaveras Fault. The completed NIT will address these issues.

### Project Overview

The NIT project will provide a seismically-designed tunnel that can withstand earthquake damage, and allow the SFPUC to take the existing tunnel out of service for future maintenance and repairs. The

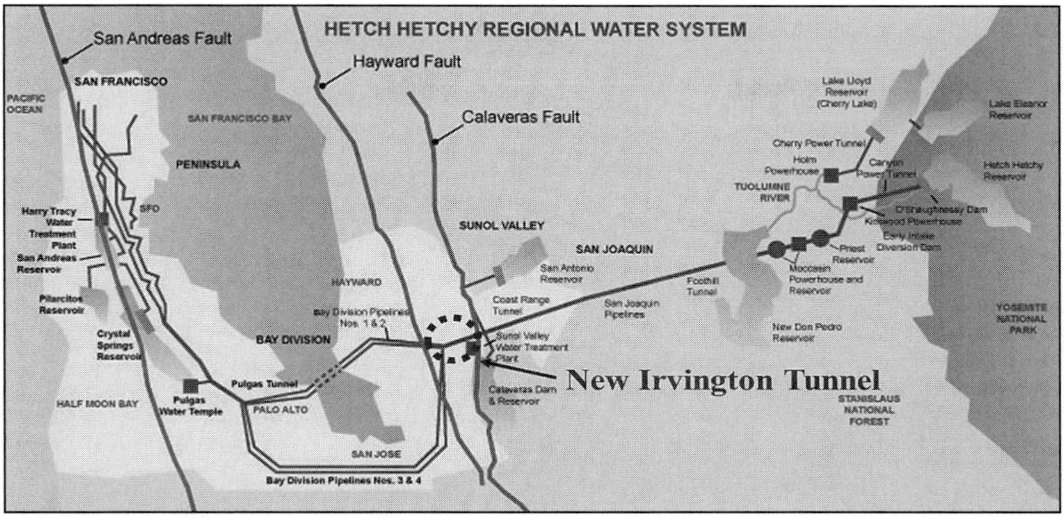


**Figure 1. Hetch Hetchy Regional Water System**

**Figure 2. Existing Irvington Tunnel**

design criteria required the new tunnel to able to withstand a Magnitude 6.7 earthquake and deliver 120 mgd of water within 24 hours after a major earthquake on the Calaveras Fault, and up to 229 mgd of water within 24 hours after a major earthquake on the San Andreas or Hayward Fault.

The project final design and environmental review were carried out from 2006 to 2010. In June 2010, the SFPUC awarded a $226.6 million construction contract to Southland/Tutor Perini Joint Venture for the project. It will be 18,660 feet in length and have a finished diameter of 8.5 feet. The final liner will be welded steel pipe for the entire length of the tunnel. Construction started in August 2010 and tunnel excavation began in March 2011.

The project is located within the existing SFPUC Irvington Tunnel right-of-way, between Calaveras Road near Sunol, California and Mission Boulevard in Fremont, California. The east end of the NIT is at Alameda West Portal (AWP) west of Calaveras Road, where the tunnel connects to water delivering pipelines. The west end of the NIT is downstream at Irvington Portal near Mission Boulevard, where the tunnel connects to pipelines,

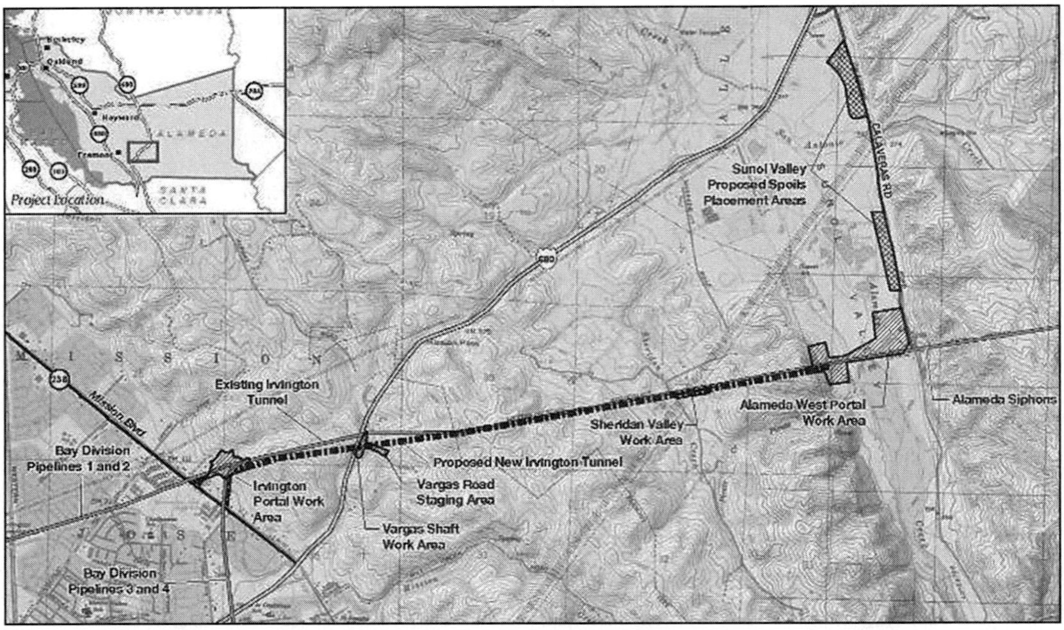


**Figure 3. New Irvington Tunnel project site map**

which carry the water across San Francisco Bay and around the southern end of the Bay through the city of San Jose (Figure 3).

About ¾-mile west of the AWP is the Sheridan Road Dewatering Site. Here the Contractor has drilled 23 surface wells 270 feet deep to intercept the tunnel alignment and remove groundwater from the tunnel excavation below and make the excavation process safer. About two miles to the west is the Vargas Shaft Site, adjacent to Vargas Road and Highway 680. From this 115-foot deep shaft are two additional tunnel headings west and east along the NIT alignment to allow the Contractor to expedite the excavation.

### Geology and Groundwater

The tunnel geology varies from hard, massive Briones Formation sandstone, to softer Claremont and Tice Shales, and to alluvium and highly fractured rock and clay gouge in the seven fault zones along the alignment. The design team considered the Sheridan Fault Zone in the Sheridan Valley to be the most challenging ground (Figure 4). The Sheridan Fault Zone is a 400 foot broad secondary fault zone with soft clays, fractured rock and relatively unstable ground for tunneling. The ground stability in this zone was expected to be further exacerbated by a high water table, perhaps 200 feet or more above the tunnel elevation. This presented a risk of flowing or running ground—dangerous conditions for tunneling.

It was above at the surface, at the Sheridan Valley Site, that the tunnel design engineers decided to specify surface dewatering wells to remove as much groundwater as possible from the tunnel alignment area, to improve the tunneling conditions. A series of 23 dewatering wells was drilled in this area, extending about 270 feet deep. The dewatering well field initially produced up to 900 gpm and then later tapered off to steady state rate of less than 300 gpm. The groundwater was discharged to the nearby Sheridan Creek.

The dewatering program was very successful at lowering the groundwater in this critical area for tunnel excavation and allowing the excavation to proceed more safely through what otherwise would have been very problematic ground conditions. But it created a problem for local Sheridan Valley residents who depended on well water for domestic consumption and irrigation for their farms and ranches.

## HISTORY OF MISTRUST

### Past Experience

The predecessors who constructed the first tunnel kept meticulous construction records. Those records showed losses of groundwater supply in some of the wells and springs above the alignment. There were no environmental measures undertaken to mitigate the losses. The issue of groundwater losses made for very adversarial relationships and lawsuits between the SFPUC and the Sheridan Valley residents.

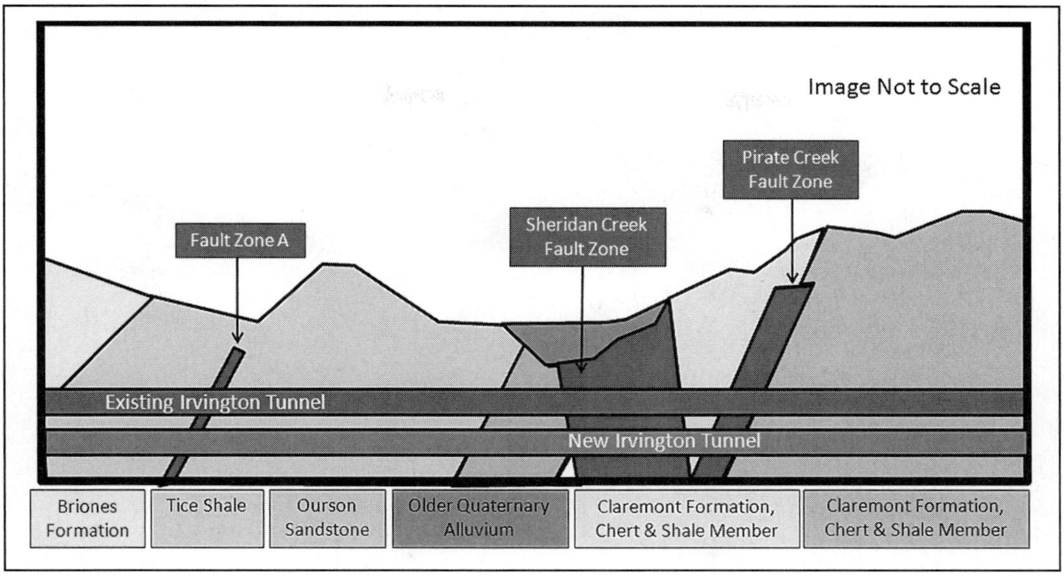


**Figure 4. Sheridan fault zone geology**

Most of the same families who owned property in the area back then still own them today. The families have an inherent distrust of the SFPUC for damage, or perceived damage, done to their wells that were never mitigated.

### Where It Started

It was no surprise that the neighbors in this area were extremely distrustful of the SFPUC when the NIT was being developed in the Planning Phase. However, the level of hostility was underestimated. The first meeting with the neighbors resulted in neighbors yelling, and threatening lawsuits. At that meeting, the neighbors were told that the SFPUC would make them whole if the tunnel affected their water supply, but they did not believe this.

### Bridging the Gap

A classic technique for bridging to hostile audiences is to reach out to a community leader who will talk and build a relationship with them, first. We found an individual who spent a lot of time with team members on the phone to explain the neighbors' concerns about our program and project. We, in turn, relayed information through him to the group and tested message points and program elements with him. Nine months later we were offered a second chance when the group met in the barn of a neighbor and asked us to attend to answer questions.

This time we were ready with a clear, consistent message and an approved plan for what we were prepared to do in the case of tunnel related water losses. If the tunnel affects your wells, we will lower your pump, we will provide you supplemental water, and if in the end your water does not recover, we will drill you a new well. The community's response was much more positive. They started responding to phone calls, and allowed us to launch our groundwater management program.

### Community Engagement

Be patient. We started outreach to the community a full four years before construction. We needed time to develop relationships with hostile communities. We met with property owners in their kitchens, on their driveways, or wherever and whenever they want to engage. At each meeting, we repeated the consistent message that if they suffer groundwater lost as a result of the NIT, that the SFPUC will make them whole by supplementing their water supply.

Should an organization not have the luxury of such a long lead time, the recommended course of action would not differ greatly. Continue to try to reach out directly to the community during environmental review and construction. If local ambassadors are not readily apparent, it might be appropriate to reach out to local elected officials or other agency's offices to inquire if they could assume that role, perhaps convene a meeting for you, or identify someone who could. Lastly, just because your project moves forward, do not stop trying to reach out to people in as many ways as possible.

Today, we hold quarterly coffees at a local coffee shop, we attend local events, staff booths at local fairs and in other ways be present in the community

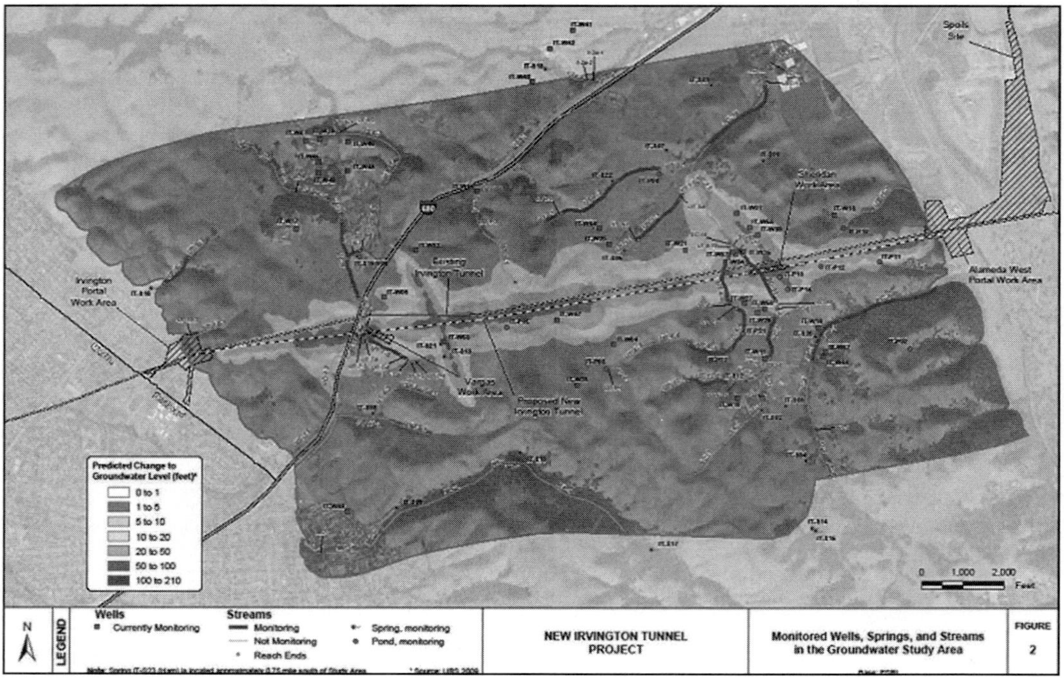


**Figure 5. Groundwater study area**

to show that we are not going to leave them without water. We are committed to them and to this process. We engaged a call center to take calls 24/7 from residents concerned about their water supply. And, the call center will contact a list of project team members until someone can respond to the resident. All of this was key to the planning and implementation of the Groundwater Management Program.

## PROPOSED MITIGATION APPROACH

### Groundwater Study

During the project planning and design, the SFPUC hired groundwater specialty consultants to study the potential impacts that tunnel dewatering would have on local wells and springs. The consultants relied on a groundwater model developed by the tunnel design engineers, which predicted the potential groundwater impacts due to excavation of the new tunnel. The model was based upon geologic and groundwater information obtained from an extensive subsurface investigation program as well as limited information available from the original Irvington Tunnel construction period. The study area was a mile north and a mile south of the NIT alignment as shown in Figure 5. The figure shows well and spring locations and predicted groundwater impacts due to tunneling.

Some 33 well water users were identified in this area. The model indicated about half of the wells would most likely be impacted and require mitigation to make up water losses from tunnel dewatering. This was borne out during NIT construction. Most of these affected well users reside in the Sheridan Valley along Sheridan Road.

### Data Collection

The main goal of the Groundwater Management Program was to maintain water supply for overlying property owners by providing another source of water to those who had their wells or springs dry up due to tunneling. The data collection approach was designed to help the team meet this goal.

It was recognized early in the planning and pre-construction period that the complex hydrogeology and the variability in the types of existing small water systems in the study area would preclude a "one size fits all" solution to monitoring and provision of supplemental water. The groundwater specialty consultants initiated collection of data about the individual water systems more than two years prior to the initiation of tunnel construction. During this pre-construction monitoring period, the field staff also began developing relationships with the overlying property owners and started to understand how each small water system operated and its water use patterns.

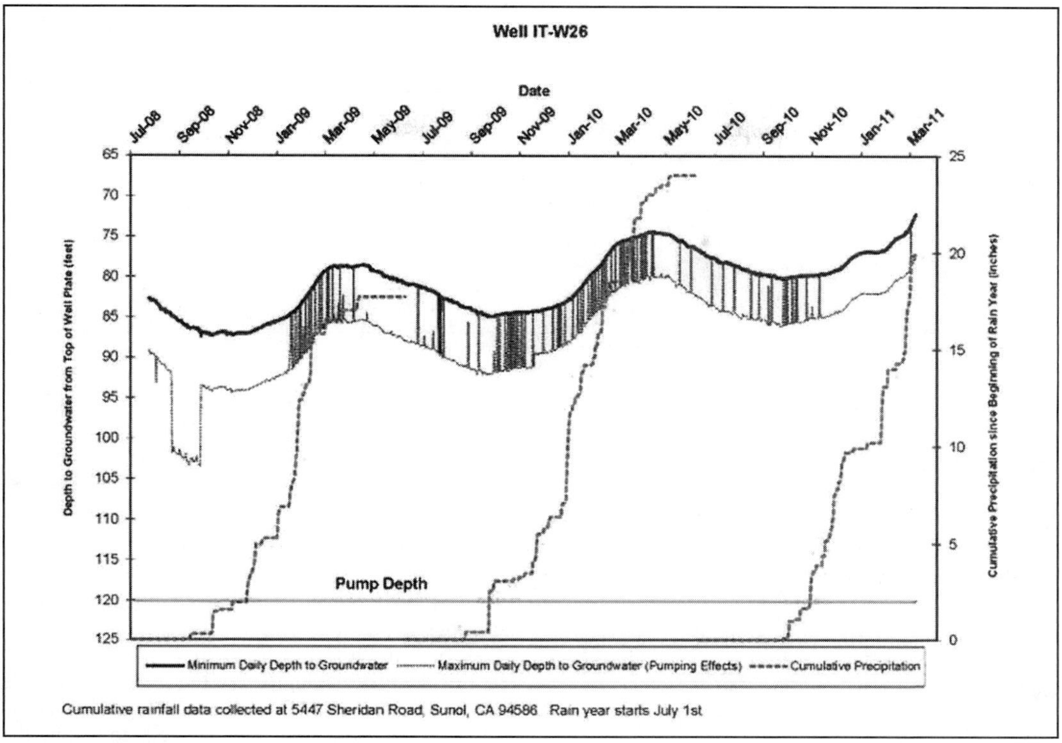


**Figure 6. Hydrograph for groundwater**

### What Did We Find?

The well and springs systems that we found at the individual properties ranged from shallow, small-diameter wells that were not properly sealed or up to code to modern large-diameter high-producing wells with integrated water treatment systems. Many naturally occurring springs are located in the Groundwater Study Area. Many of the property owners use spring water for their primary domestic supply. In some cases, these springs had been captured in cinder block or concrete vaults with a submersible pump lowered in and a piece of plywood placed over the top. Many of these systems were exposed to surface environmental contaminants.

Based on the observed land use, which in some cases included extensive landscaping, it was apparent that many of the property owners were high-volume water users.

### Determining Baseline Conditions

After locating, inspecting, and gaining an understanding of the how each water system operated, the specialty groundwater consulting team installed water level pressure transducers in each well so that groundwater levels could be continuously monitored. In addition, these transducers allowed the team to understand the response of the water-bearing zone to pumping by monitoring how far water levels dropped when water was being pumped. The team also installed water use totalizers so that the volume of water used by each property owner could be monitored over time. These data would help the team to design appropriate supplemental water supply strategies.

During the pre-construction period, the team made quarterly visits to each property to download the data. These visits also provided opportunities to continue to develop the relationships between field staff and the property owners.

### Hydrograph for Groundwater

Data downloaded from the water level pressure transducers allowed the specialty groundwater team to prepare groundwater level hydrographs (Figure 6). The typical pre-construction hydrograph shows a seasonal fluctuation of the groundwater level in response to precipitation and aquifer recharge. These hydrographs also show how much the water levels drop during active pumping. To assist in well vulnerability assessments and decision-making, additional information like well depth, pump depth, and cumulative precipitation data were also placed on the hydrographs.

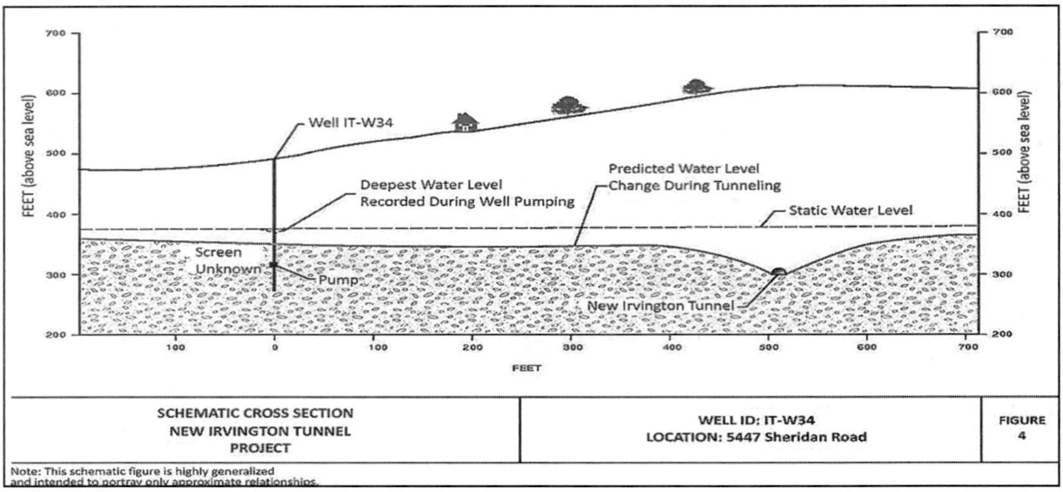


**Figure 7. Groundwater management plan schematic cross section**

### Site-Specific Groundwater Management Plan

Based on the baseline data collected during the pre-construction monitoring period, the team developed a specific groundwater management plan for each and every property owner's water system. These groundwater management plans included all the baseline data that had been collected, including hydrographs showing fluctuations in groundwater levels over time. The groundwater management plans also included a vulnerability assessment for each well system. This vulnerability assessment was, in part, illustrated using schematic cross-sections that showed the specific information related to that particular well (e.g., well depth, pump depth) and distance from the tunnel alignment (Figure 7). The groundwater management plans also included the specific actions that would be taken to ensure that water supply would be maintained. Depending on the specifics of the water system, these actions included installation of new storage tanks, upgrades to pumping and conveyance systems, and/or lowering of existing pumps in the well. In some cases, water systems were modified before tunneling began so that when a water loss occurred, supply of an alternate source of water could be completed within just a few hours.

The team met with the property owners in the study area and reviewed their specific groundwater management plan with them to be sure that they understood the data that had been collected and what actions would be taken if their water supply was interrupted.

### Contract Allowance

The groundwater management plans for so many residents represented a potentially large financial commitment. The SFPUC did not want to risk a shortfall of resources to properly mitigate the losses of groundwater through its low bid contracting process. So, an allowance of $5 million was set up as part of the contract and every bidder had to include this item in their bid. This allowance was derived by a worst case estimate of 33 new wells to restore the wells lost due to tunnel dewatering. The allowance became the funding source for the implementation of the groundwater management plan. As the need for mitigation work arises, the Contractor submits quotations for SFPUC review/approval and the work is paid out of the allowance. To date, approximately half of the allowance has been expended.

## GROUNDWATER PLAN IN ACTION

### The Effects Are Real

The acquisition and availability of more than two years of baseline data allowed the groundwater team to quickly evaluate whether changes in water system performance was likely a natural fluctuation or resulting from a tunnel-caused groundwater decline. In most cases, the groundwater level hydrographs for the baseline data showed regular and predictable seasonal fluctuations in groundwater levels, with higher groundwater levels occurring during the rainy season and lower groundwater levels during dry months. As shown in Figure 8, it is clear that a groundwater decline of far greater magnitude than one that could be attributed to seasonal fluctuation began in about January 2012. On many occasions, the baseline data collected allowed the SFPUC to quickly acknowledge responsibility and initiate the provision of alternate water supply, in most cases trucking potable water to the property to fill the storage tanks. Conversely, on many occasions the baseline clearly

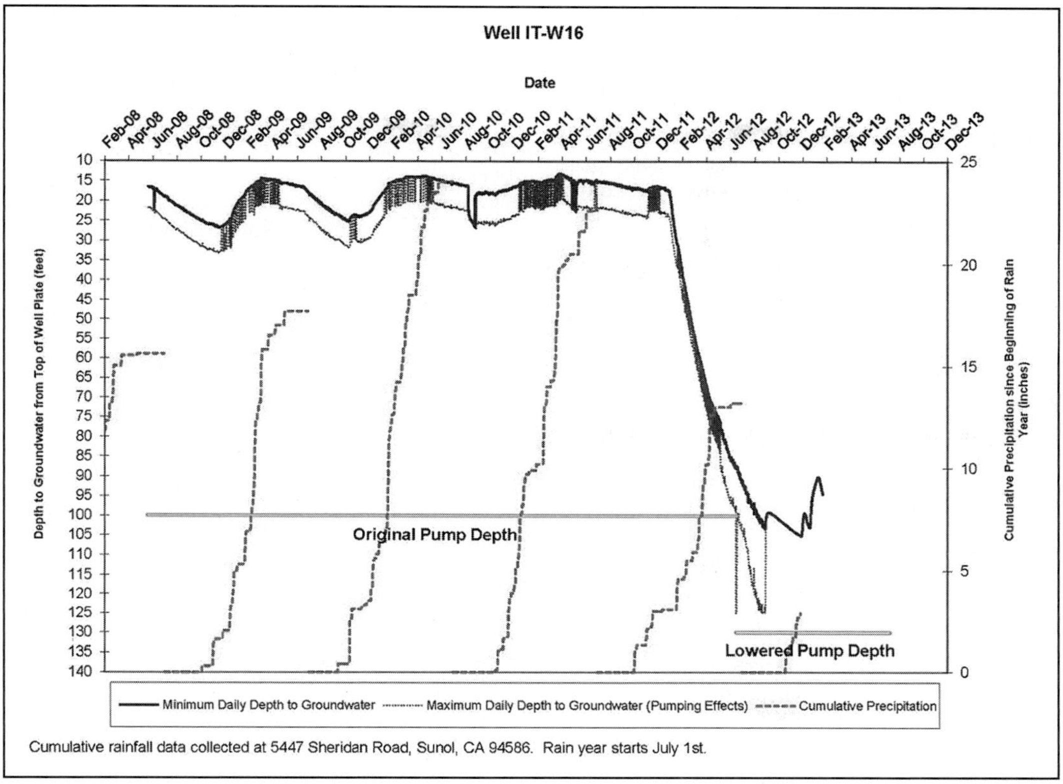


**Figure 8. The effects are real**

indicated that the tunneling was not the cause of the water system malfunction and the property owner was able to promptly contact their own water well contractor to correct the problem.

### Sheridan Dewatering Well Field and Pipeline

At the Sheridan Dewatering Field, SFPUC-installed wells were used to dewater the tunnel alignment in an area of saturated poor ground conditions to improve constructability and safety conditions in the tunnel. Groundwater extracted from the dewatering field was discharged at a rate ranging from 300 to 900 gpm to the adjacent Sheridan Creek. As their wells went dry, the property owners could see what they felt was their groundwater flowing out of the valley in Sheridan Creek. After several weeks of this situation, the field staff started hearing grumblings of resentment from the property owners.

With some input and cooperation from the property owners, the groundwater team designed and constructed a temporary pipeline system that captured a portion of this discharge from the dewatering field and distributed it back to the affected property owners for non-potable use (e.g., irrigation, livestock). This solution was very popular with the property owners and was a substantial budget saver for the SFPUC because the millions of gallons of water distributed through the pipeline did not have to be trucked in.

## LESSONS LEARNED

### Outreach Helps

Be patient. Early, prolonged and proactive outreach helps. Fully integrate your public outreach function into your planning and design process. Recognize issues early and acknowledge them. Be present in the community. Partner with stakeholders in planning so they take ownership. Follow through on commitments. Mistakes, delays and the unexpected happen. Own up to it, and make amends. You must be able to respond quickly.

### Walk The Talk

Once the problem has been identified, financially back up the needed mitigation measures and set up a way to address the problem during construction. The NIT did this by creating a $5 million allowance in the construction contract so that the Contractor can be directed to respond quickly with the appropriate water well or supplemental water supply subcontractor hired.

### Measures of Success

The New Irvington Tunnel team was able to develop and implement groundwater management plans with approximately 33 property owners. By word of mouth, some residents who were not part of the original phase of study heard about the benefits of having groundwater management plan protection and have joined the program. Since the contract allowance is far from expended, it was decided to accommodate the additional residents.

We were able to complete the environmental review process and design without a single protest. So far, through almost 3 years of construction, community relations have been inconsequential or cordial. As a matter of fact, one of the neighbors wrote to compliment the team on its efforts through the groundwater program.

We still allocate a tremendous number of resources and time at this issue alone. We meet weekly to discuss the program and the ongoing issues. We certainly have not crossed the goal line and are not spiking the proverbial ball, yet.

However, just to provide a snapshot of how far we've come, one of the property owners who stormed out of that first meeting shouting about his lawyer attends every coffee we host. He comes, sits down, asks questions, and before he leaves, he says the same thing. "I don't know what you're doing, guys, but keep doing it."

To that we say, we absolutely will!

# Cross-Town Tunnel Water Main Rehabilitation Design and Construction Challenges

**John I. Yao**
Jacobs Associates

**Duncan K. Mukira**
District of Columbia Water and Sewer Authority

**ABSTRACT:** The Cross-Town Tunnel is a 2.1-m (7-ft) diameter, 4.0-km (2.5-mile) long pressure tunnel that supplies water to Downtown Washington D.C. Leakage from the tunnel was detected on a major parkway located above the tunnel. Based on detailed studies of the tunnel, impermeable steel lining was installed over a 240.5-m (789-ft) long reach of the tunnel. An existing 7.9-m (26-ft) diameter, 36.6-m (120-ft) deep shaft was demolished for construction access, and was reconstructed after the steel lining installation. This paper discusses studies performed to identify the critical zone of leakage, as well as the design and construction challenges of the project.

## INTRODUCTION

The District of Columbia Water and Sewer Authority (DC Water) Cross-Town Tunnel was completed in 1984. This treated-water transmission tunnel is a portion of the west-to-east-aligned Cross-Town Transmission Main, which conveys water from the Washington Aqueduct Division's Dalecarlia Water Treatment Plant easterly to a northwest urban part of Washington D.C. The Cross-Town Tunnel facility is approximately 3,993-m (13,100-ft)-long, 2.1-m (7-ft)-inside-diameter (ID) tunnel, lined with 46-cm (18-in)-thick, nominally reinforced cast-in-place concrete. The tunnel was excavated in bedrock consisting primarily of gneiss by a tunnel boring machine (TBM). The tunnel facility also includes three shafts: the Foxhall Shaft at the west end of the tunnel, the 25th Street Shaft, and the Scott Circle Shaft at the east end of the tunnel. Both end shafts have a 1.8-m (6-ft)-ID conduit and the 25th Street Shaft conduit has a 1.2-m (4-ft) ID. Figure 1 shows the plan and profile of the Cross-Town Tunnel. The ground cover above the tunnel generally varies from approximately 37 m (120 ft) to 61 m (200 ft), except at a localized topographic depression at the crossing of Rock Creek, where the ground cover is as little as 18 m (60 ft) over the tunnel. The width of the topographic depression includes two steep side slopes and the four-lane Rock Creek Parkway for a total width of about 91 m (300 ft).

The tunnel was in operation for about 24 years without detectable leakage until December 2008, when leakage from the tunnel was identified as several wet spots on Rock Creek Parkway and adjacent slopes in an area where the tunnel crosses under Rock Creek Park. Testing showed that the water had residue chlorine similar to the water conveyed by the tunnel. In order to maintain safe conditions along Rock Creek Parkway and to continue providing District of Columbia residents with a reliable supply of drinking water, a study was performed to investigate the cause of leakage and to develop the appropriate rehabilitation scheme.

## LEAKAGE INVESTIGATION

For the leakage investigation, the Cross-Town Tunnel was isolated and taken out of service, and water transmission was maintained by redundant pipeline mains.

### ROV Inspection

A remotely operated vehicle (ROV) was used to perform preliminary reconnaissance of the tunnel. The ROV equipment was lowered through the water-filled 25th Street shaft conduit to the bottom, where the shaft formed a tee intersection with the tunnel. No specific large-sized structural feature that could have resulted in sudden appearance of substantial localized leakage was observed in the tunnel reaches traversed by the ROV. The more prominent crack features observed were circumferential shrinkage cracks in the lining (see Figure 2). Only a few longitudinal cracks were observed, and none showed any apparent significance. Likewise, only a few sloping discontinuities, likely cold joints from concrete

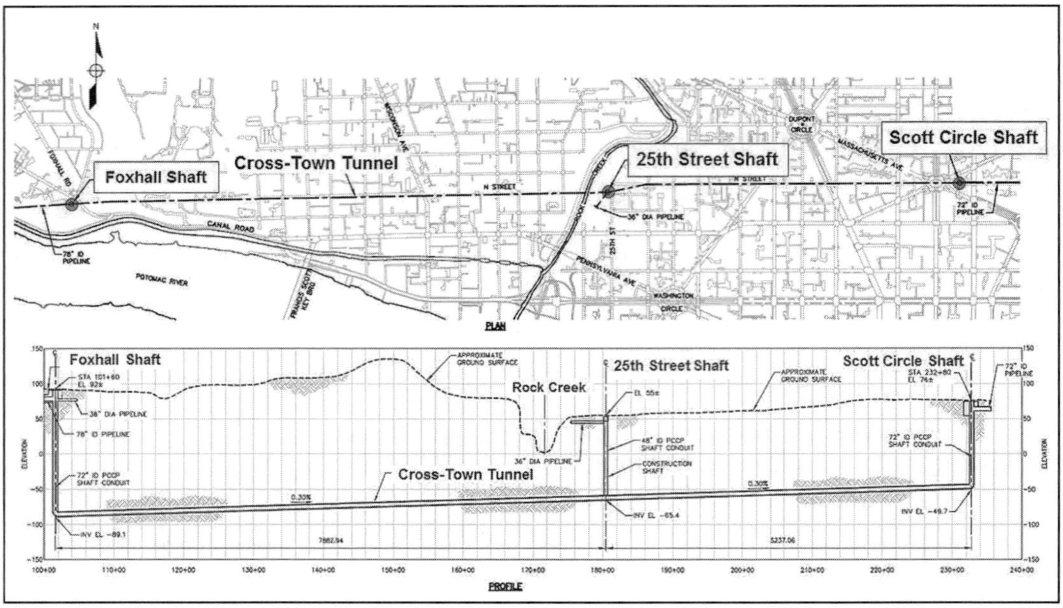


**Figure 1. Cross-Town Tunnel plan and profile**

placement operations, were seen. Thus, it appeared that leakage resulting in the observed seepage at surface level is due to permeable circumferential cracks in the lining. However, assessment of overall permeability is not possible from the ROV inspection because limited evaluation of crack characteristics such as width, cleanliness, and amount of precipitates, could be performed from the video images.

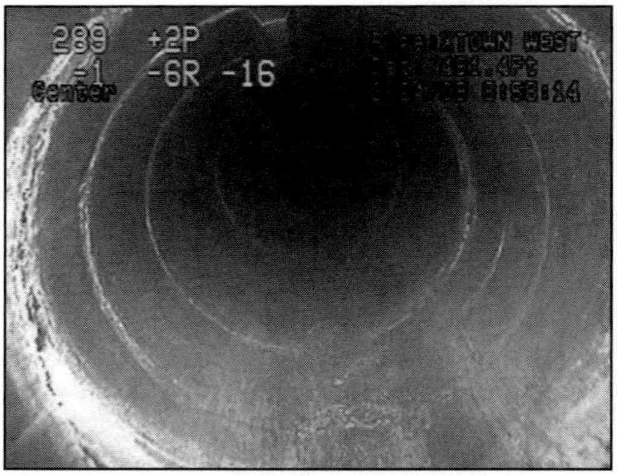


**Figure 2. Circumferential cracks with precipitates (white "rings")**

### Physical Tunnel Inspection

A physical inspection of the Cross-Town Tunnel lining was necessary for identifying the section of tunnel requiring rehabilitation to prevent exfiltration during future operation of the tunnel. A standard tunnel mapping form based on the peripheral geologic mapping method developed by the U.S. Army Corps of Engineers (EM 1110-1-1804) was used for recording significant observations along the tunnel alignment. The location of each crack and characteristics important for evaluating the permeability of the crack were recorded on the mapping form. Descriptors used for mapping are summarized in Table 1.

Circumferential cracks are the most prevalent type of cracks found in the tunnel. Spacing of the cracks along the tunnel is typically less than 3.0 m (10 ft). The width of the cracks typically ranges from hairline to 1.5 mm with most widths being hairline or 0.5 mm. Most of the cracks are partially healed. White precipitates, the most prevalent type of precipitates were observed on nearly all cracks. Dark brown, pink and yellow colored precipitates were

**Table 1. Descriptors used to describe crack characteristics**

| Characteristic | Descriptors Used |
|---|---|
| Aperture | Measured or estimated width in millimeters, or hairline |
| Degree of healing | Fully, partially, trace, or clean |
| Precipitates | Heavy, medium, low, or trace; and color |
| Water inflow | Flowing, dripping, wet, moist, or dry; and quantity in gallons per minute (gpm) when flow can be estimated or measured |

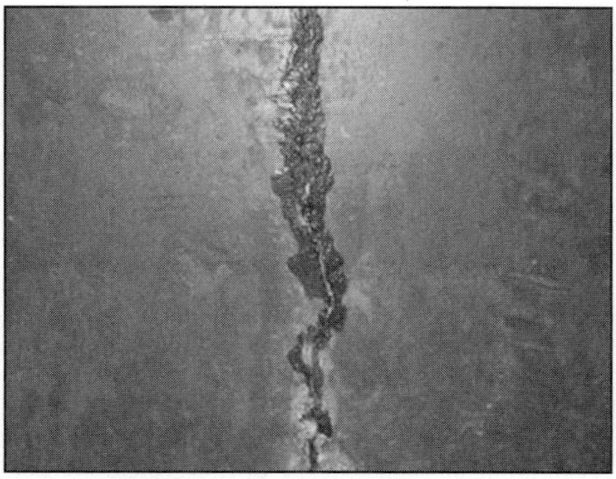

**Figure 3. Dark brown precipitates**

**Figure 4. Pink precipitates at cracks with water inflows**

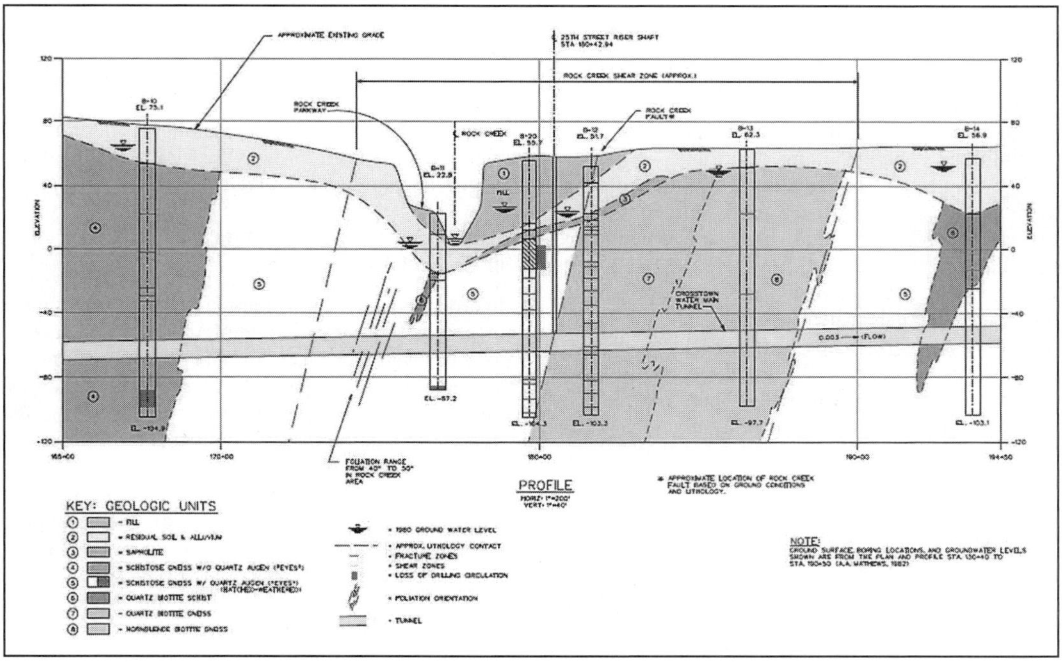


**Figure 5. Geologic profile in the area of Rock Creek**

also found (see Figures 3 and 4). The cracks with relatively high water inflow (i.e., flowing or dripping) appeared to have greater amount of colored precipitates, particularly the dark brown precipitates. Several samples of precipitates were taken for laboratory testing, but results do not reveal any significant characteristics that would provide additional information on factors relating to the tunnel leakage.

Water inflow ranges from dry to flowing with most categorized as dry, moist, or wet. Many cracks that were flowing or dripping concentrated between approximately Sta. 175+00 and Sta. 179+00 (see Figure 5 for stationing). The estimated or measured flows from the cracks range from much less than 0.1 gpm to less than 0.5 gpm. Figure 4 shows water flowing from a crack at the tunnel crown; however, water also flow in through other locations of the cracks, such as below the tunnel springline.

Longitudinal and inclined cracks were occasionally found in the tunnel. These cracks may be as short as one foot or longer than 6.1 m (20 ft) in length. The longer cracks typically follow along inclined construction joints. Most cracks are hairlines and fully healed, with no or trace amounts of precipitates. Water inflow typically ranges from dry to wet indicating that most of these cracks have little or no permeability.

## EVALUATION AND IDENTIFICATION OF CRITICAL TUNNEL REACH

The identification of the critical tunnel reach with respect to leakage involved the study of geologic conditions, depth of ground cover, and the tunnel lining based on mapping data from the physical inspection.

## CHARACTERISTICS OF CAST-IN-PLACE CONCRETE TUNNEL LININGS

Pressure tunnels lined with reinforced concrete lining leak to various degrees, mostly due to inherent circumferential shrinkage cracking, and occasionally longitudinal cracking associated with lining/ground interaction during operation as well as other design and construction issues. This type of lining is classified as semipermeable lining which exposes the rock mass to the water pressure present during operation of the tunnel. The combined effect of all cracks in a reinforced concrete lining determines its permeability characteristics. For a lining to be effective as a containment element to keep pressurized water in the tunnel conduit, the lining must be less permeable than the surrounding ground. Where the lining does act as a containment element to reduce leakage from the tunnel, there is a net pressure reduction across the lining from the inside to the outside of the lining and consequently some leakage control. Where the ground has low permeability, there is little benefit from the reinforcement in the concrete lining, because the surrounding ground is the element controlling leakage from the tunnel.

### Geologic Conditions and Depth of Ground Cover

The most significant topographic feature in the project area is the approximately 18-m (60-ft)-deep, north-south trending valley cut by Rock Creek that crosses the tunnel alignment near Sta. 177+00 (see Figure 5). The geologic profile in Figure 5 shows that the Cross-Town Tunnel has about 12 m (40 ft) of gneiss bedrock cover beneath about 6.1 m (20 ft) of alluvium under the Rock Creek channel. Rock Creek is associated with the Rock Creek Shear Zone. However, there is no strong indication of encountering significant ground instability or groundwater during the tunnel excavation, suggesting that the in-situ permeability under the creek was low when the tunnel was constructed.

The ratio of depth of cover, measured as feet of rock cover above the tunnel, to total hydraulic head from the static hydraulic gradient line (HGL) at the same location, called the Cover Ratio, is commonly required to be about 0.4 for an approximate safety factor of unity (Eskilsson, 1999). This rule of thumb was verified for the region in extensive hydraulic jacking testing for the Bi-County Tunnel (EBASCO, 1989) and by additional testing for an extension of the Bi-County Tunnel in 1996. The Cover Ratio for the tunnel crossing under Rock Creek is approximately 0.25, while the ratio is well above 0.5 along most of the tunnel alignment. Where the minimum cover required has not been provided for, there is risk for opening up of existing cracks from hydraulic jacking. This would result in increased leakage and progressive migration of water with high pressure throughout the rock in the area, and eventual recognition of such leakage at the ground surface. In spite of the presence of insufficient cover at the Rock Creek crossing, the tunnel has apparently operated without detectable leakage for more than two decades. The likely explanation may be that few naturally permeable fractures are present in the rock mass, or that the topographic depression is relatively narrow. With time, existing joints in the rock may have increased permeability due to erosion or dissolution of joint infillings resulting in migration of high pressure water through the rock mass, local rise of groundwater table, and detectable seepage.

### Evaluation of Mapping Data and Identification of Critical Tunnel Reach

In order to evaluate information gathered from tunnel mapping, locations of circumferential and diagonal/longitudinal cracks were plotted with tunnel station, along with key characteristics of each crack (see Figure 6). The degree of permeability shown in the figure is an interpreted parameter that is a function of aperture, degree of healing, precipitates, and water inflow observed. Permeability of a crack increases with wider aperture, lower degree of healing, higher amount of precipitates, and higher quantity of water inflow.

A study of the mapping data found that cracks with relatively high water inflow (i.e., flowing or dripping) appear to concentrate approximately between Sta. 175+00 and Sta. 179+00, but cracks rated with a relatively high degree of permeability (i.e., very or medium) appear to concentrate approximately between Sta. 173+00 and Sta. 179+00. The reason for the relatively high water inflow observed between Sta. 175+00 and Sta. 179+00 is likely related to the fact that this reach of the tunnel is under the general area of Rock Creek, which in theory, acts as an infinite source of water for the inflows. However, the absence of water inflow from a crack does not automatically indicate that the crack is impermeable, because tunnel reaches further away from Rock Creek may not have an infinite source of water, and may be dry due to drainage into the tunnel before

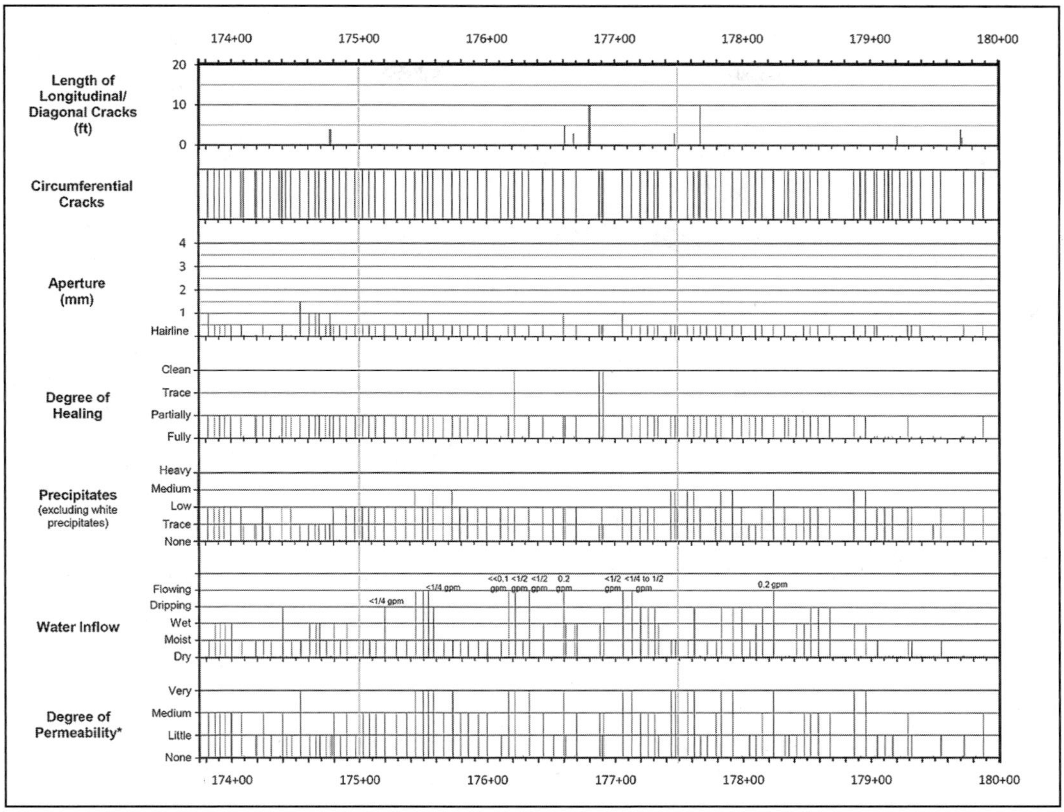


**Figure 6. Example of summary of tunnel mapping**

the inspection. Statistical analyses of the mapping data show that cracks with higher inflows generally have greater amounts of precipitates than cracks with lower inflows. Therefore, an indication that a crack without observed inflow is likely to be permeable is the presence of dark brown precipitates. These precipitates appeared to have been brought into the tunnel shortly following unwatering by water inflow.

## REHABILITATION DESIGN

Based on the results of the studies conducted, it was recommended that an impermeable lining be installed in the Cross-Town Tunnel between Sta. 173+00 and Sta. 180+25 for a total length of 221 m (725 ft). A total length of 38 m (125 ft) was added to the east end of the critical reach in order to: Address the presence of cracks and geological conditions within that section of the tunnel; and gain some distance away from the Rock Creek topographic depression area.

### Tunnel Steel Lining

The Cross-Town Tunnel was to be rehabilitated by installing steel lining and backfilling the annular space between the existing concrete lining and the new steel lining with low-shrink cement grout (see Figure 7). Main aspects of the steel lining design are summarized below:

- The lining has a finished inside diameter of 1.9 m (6 ft 3 in), with 9.5-mm (⅜-in) thick steel wall that is lined internally with 12.7-mm (½-in)-thick cement mortar lining. The diameter considers the required annular space for grout backfill and local variations in the existing lining.
- A pipe segment length of 7.6 m (25 ft) was selected. This length is identical to the form segment length used for placement of the Cross-Town Tunnel concrete lining and allows the field joints of the steel pipe segments to coincide with potential tunnel alignment deviations.
- Grout ports for annular space grouting are located in the crown, springline, and invert of each lining segment.
- The steel lining segments are joined by welding using an external backing bar ring with

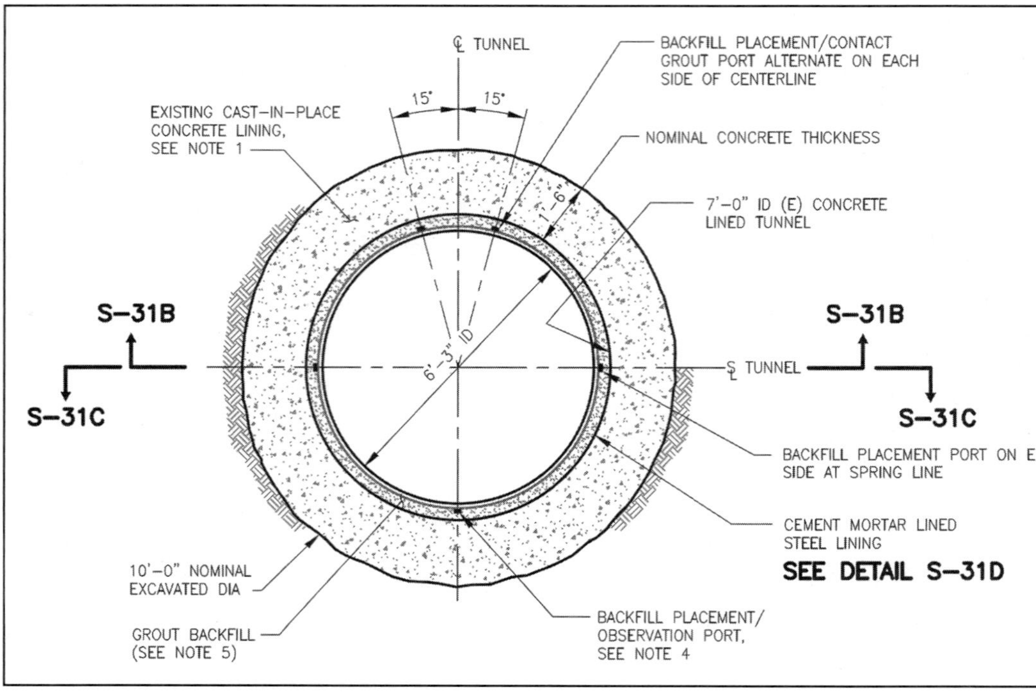


**Figure 7. Tunnel steel lining with annular space backfilled with low-shrink cement grout**

full penetration butt joints welded from inside the pipe. The holdback area at the joint (i.e., area without the shop-applied cement mortar lining) is repaired with field-applied cement mortar following the completion of welding.

Prior to steel lining installation, a grout seal consisting of two fans of consolidation grout holes would be installed at each end of the tunnel rehabilitation section (see Figure 8). The purpose of the grout fans is to prevent shunt flow of pressurized water behind the concrete and steel lining.

### 25th Street Shaft Demolition and Reconstruction

In order to access the Cross-Town Tunnel for the required rehabilitation work, it was necessary to demolish the 25th Street Shaft structure from the surface level down to the tunnel invert. The 25th Street Shaft was originally used as a work shaft for the construction of the Cross-Town Tunnel. Upon completion of tunnel construction, a 122-cm (48-in)-diameter shaft conduit, which provides a connection to a near surface 91-cm (36-in) water main, was installed, and the 7.9-m (26-ft)-diameter, approximately 36.6-m (120-ft)-deep shaft was backfilled with concrete. Although the original design specified 13.8 MPa (2,000 psi) low-shrink concrete backfill, construction records indicate 27.6 MPa (4,000 psi) and higher strength concrete backfill was place.

It was anticipated that the shaft would be demolished and opened to its original diameter for the tunnel rehabilitation. An initial support system consisting of ring beams and timber lagging was to be installed, unless an alternate system is proposed by the Contractor and approved by the Engineer. Blasting as a method for shaft demolition was not permitted due to some concerns that the shaft is located near a school, and high-rise office and apartment buildings. It was anticipated that special methods, such as the use of expansive mortar, in conjunction with mechanical methods (e.g., excavator with hydraulic hammer attachment) may be needed to meet project schedule. However, means and methods for demolition were ultimately left up to the contractor.

Upon completion of the tunnel steel lining, a 1.8-m (6-ft) ID shaft conduit would connect the tunnel to the 91-cm (36-in)-diameter water main and the annular space backfilled with concrete. At the bottom of the shaft, a steel tee would be welded on both ends to the steel tunnel lining and at the top to the shaft conduit. A new vault would be constructed at the top of the shaft for future access to the tunnel. Figure 9 shows the design for the reconstructed 25th Street Shaft.

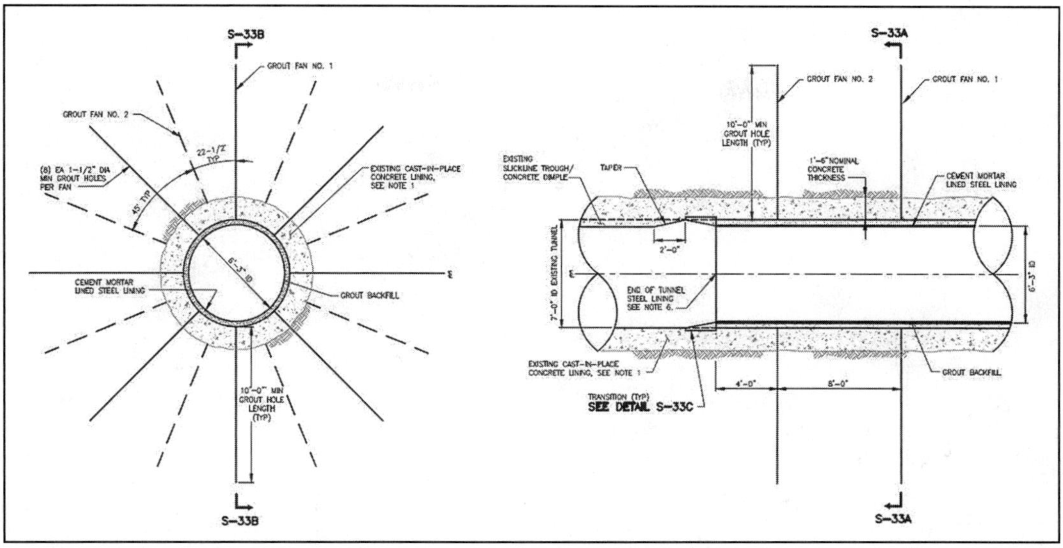


**Figure 8. Grout fans at each end of the tunnel rehabilitation section**

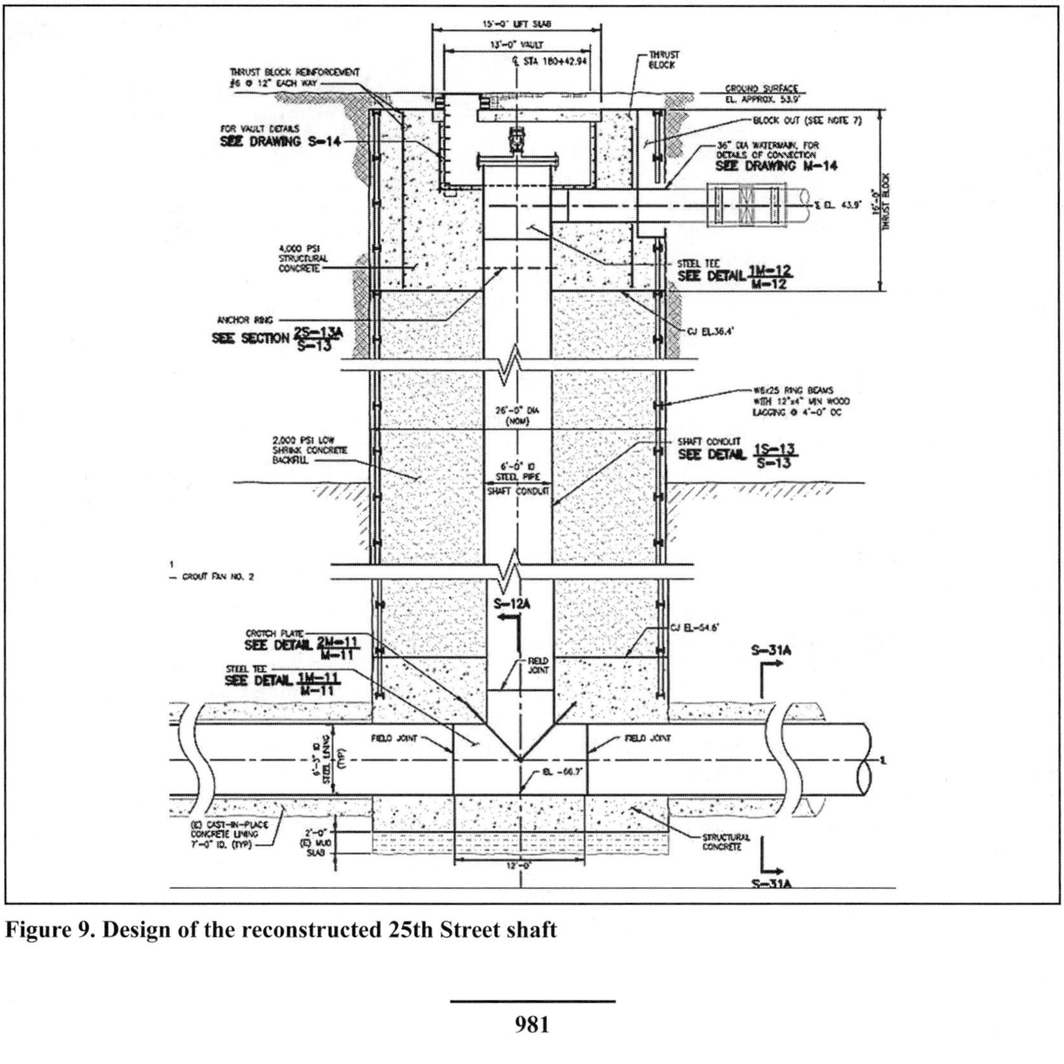


**Figure 9. Design of the reconstructed 25th Street shaft**

Figure 10. 25th Street shaft demolition

Figure 11. Lowering steel pipe into 25th Street shaft

Figure 12. Temporary steel ribs used to stiffen steel pipes during grout placement

## CONSTRUCTION

The construction notice to proceed for the Cross-Town Tunnel was issued in mid-August 2011. After some initial preparatory work including isolation of the tunnel from the transmission system and the construction of a temporary noise barrier wall around the shaft area, the 25th Street Shaft demolition began in mid-November 2011. The contractor selected to use mini-excavator with hoe-ram to chip out the shaft concrete. The concrete debris was then loaded into muck buckets, hoisted to the surface, and removed from the site (see Figure 10). The demolition rate at approximately 1/10-m (⅓-ft) per day was significantly less than planned. In February 2012, the contractor drilled multiple 6.1-m (20-ft)-deep relief holes in the shaft concrete in hope that these holes would increase the demolition rate by creating free face for the concrete to break into, but they had essentially no effect on the demolition rate. Since the Cross-Town Tunnel was required to be back in service by the end of May 2012 before the summer when water demand is typically at its annual peak, DC Water decided to temporarily suspend construction when the shaft demolition reached a depth of approximately 16.5 m (54 ft). A dished spigot bulkhead with at least 1.7 m (5.5 ft) thick of fiber-reinforced concrete backfill above the bulkhead were used to cap the PCCP shaft conduit at the center of the shaft. The spigot bulkhead, which fitted into an existing PCCP joint, provided a watertight seal, while the concrete backfill provided the weight to resist the upward water pressure when the tunnel in operation. The tunnel was put back into service for the summer months. The contractor would return to the site in the fall season when construction resumes.

The shaft demolition resumed in October 2012. The contractor added a work shift and the demolition rate increased to about 0.24 m (0.8 ft) per day. The shaft demolition was completed in January 2013 and the tunnel steel lining installation began shortly after (see Figure 11). Each steel pipe segment was carried into the tunnel using steerable wheeled dolly that attached to both ends of the pipe. Backfill grouting was performed after every 2 to 4 pipe segments, or generally 15 to 30 m (50 to 100 ft), were welded in place. Bulkhead at each end of section to be grouted consisted of brick and mortar with 4 valved weep holes. Contact grouting was performed from the grout ports in the crown. Temporary steel ribs were used to stiffen the steel pipes during grout placement (see Figure 12). Tunnel steel lining and the reconstruction of the 25th Street Shaft were completed in March and May 2013, respectively. After disinfection and pressure testing, the Cross-Town Tunnel was successfully put back into service. To date no leakage from the tunnel has been reported after the rehabilitation was completed.

## REFERENCES

EBASCO, 1989, Bi-county Water tunnel Evaluation of Repair Alternatives Final Report: Prepared for the Washington Suburban Sanitary Commission, April 1989.

Engineer Manual (EM 1110-1-1804), 2001. Engineering and Design Geotechnical Investigations, US Army Corps of Engineers. Washington.

Eskilsson, J.N., 1999, Design of Pressure Tunnels, Proceedings of the Geo-Engineering for Underground Facilities Conference, ASCE, University of Illinois, Urbana.

# TRACK 4: CASE STUDIES

## Session 3: NATM/SEM/Contracting Methods

Scott Rand, Chair

# University Link Project Cross Passage 17: Collaborative Approach Is the Key to Success

**Indra Banerjee, Ed Shorey, Paul Gasson, and Jim Salley**
CH2M HILL

**Matthew Burdick**
Traylor Bros., Inc.

**Sam Swartz**
Jacobs Associates

**Ginger Ferguson**
Sound Transit

**ABSTRACT:** Transit tunnels require cross passages between adjacent main tunnels to provide a safe means of egress or a safe haven in case of emergency. Cross passages are generally excavated under free air using mechanical (sometimes manual) means and present one of the major technical challenges of transit tunnel projects.

Cross Passage 17 is one of 16 cross passages constructed under Contract U220 of Central Puget Sound Regional Transit Authority's University Link Project in Seattle, WA. The cross passage was moved 35 feet south from its design location after review of data developed during TBM inspection stops indicated ground conditions in the originally planned location were poorly suited for excavation. Pre-excavation probe holes drilled at the revised location showed no signs for concern, allowing excavation to proceed using the Sequential Excavation Method beginning with excavation of the top heading. After completion of the top heading, ground water seepage into the excavation steadily increased reaching a maximum flow rate of more than 200 gallons per minute, along with about 20 to 40 cubic yards of sand adding to the difficulties.

This paper presents the cooperative process undertaken by the owner, designer, construction manager and contractor which resulted in development of the path forward to successfully complete the cross passage excavation, install the final lining and ensure long term stability of the area. Issues and challenges that were encountered during development and execution of the cross passage work are identified along with steps taken to mitigate them.

## INTRODUCTION

The excavation and initial ground support for cross passages on the University Link Project was performed utilizing Sequential Excavation Method (SEM) under free air conditions. The Geotechnical Baseline Report classified Cross Passage 17 (in its original location) as Category 1, where SEM techniques without additional stability measures could be utilized as opposed to Category 2, which would have required specialized SEM tool box items including pocket excavation and active groundwater management. Category 1 support was also specified for Cross Passage 17 in its revised location.

The greatest challenge for SEM construction under free air is management and mitigation of the impacts of any unforeseen events or ground conditions, so that their effects on project cost and schedule can be controlled. When these situations arise, the best approach for the project is to assemble a pool of ideas from all stakeholders to develop a workable solution and successful strategy. The difficulties encountered during construction of Cross Passage 17 on the U220 project provided an outstanding opportunity to fully utilize this procedure.

## PROJECT DESCRIPTION

University Link (U-Link) is the $1.95 billion, 3.15 mile extension of the Central Puget Sound Regional Transit Authority's (Sound Transit's) light rail system. It will run in twin-bored tunnels from Downtown Seattle north to the University of Washington, with stations at Capitol Hill and on

**Figure 1. University Link Project—Contract U220**

the University of Washington campus near Husky Stadium. The U-Link project is divided into multiple contract packages with the two tunneling contracts being Contract U220 and Contract U230.

A joint venture of Traylor Bros., Inc. and Frontier Kemper (TFK) was awarded the U220 contract for construction of the northern tunnel section. The major work on contract U220 includes construction of 11,400 foot long segmentally lined twin-bored tunnels using pressurized face techniques between the University of Washington Station (UWS) and the Capitol Hill Station, construction of 16 cross passages between the bored tunnels using sequential excavation methods, and civil and structural work for the UWS crossover. The tunnel alignment passes beneath dense residential and commercial neighborhoods of Seattle and includes the critical crossing of the Lake Washington Ship Canal at Montlake Cut, a man-made canal connecting Lake Washington to Lake Union. See Figure 1.

A joint venture of CH2M HILL Inc. and Jacobs Engineering provides Construction Management Services for the U-Link Project. Northlink Transit Partners JV (NTP), a joint venture of Jacobs Associates, AECOM and HNTB provided design services for the University Link System, including Contract U220.

The entire U220 tunnel alignment lies below the groundwater table with ground cover ranging from 12 feet to 300 feet above the tunnel. The soils encountered in the tunnel route consist of highly overconsolidated clays, silts, sands and gravels, of both glacial and non-glacial origin, in varying proportions. Soil conditions are often variable over short distances, both laterally and vertically.

## CROSS PASSAGE 17

Cross Passage 17, the fifth cross passage from the north end of the U220 tunnels, is situated beneath the Montlake neighborhood of Seattle, a residential area in close proximity to an elementary school; just south of a buried alluvial "valley" consisting primarily of granular soils. The ground condition was baselined to be glacially overridden silty clay and therefore the cross passage was designated as requiring Category 1 support in the GBR.

During the design phase soil investigation, boring UL-534 primarily defined the subsurface conditions for this cross passage, however the boring did not extend below the cross passage invert. The cross passage was under 150 feet of groundwater with ground cover of 170 feet. See Figure 2.

## PHASE 1—PRE-CONSTRUCTION OBSERVATION

During tunneling by the Tunnel Boring Machine (TBM), the Contract Specifications required that TFK perform an inspection stop at each planned cross passage location. The intent of the stop was to perform a visual inspection of the excavation face, assess ground conditions via a remote camera in the excavation chamber and verify ground stability such that the excavation and support requirements for each cross passage could be determined.

In September 2011, the TBM mining the Southbound tunnel reached the design location for Cross Passage 17. After following the cutter head inspection procedures, the ground was determined to be unsuitable for free air cross passage excavation as the monitored pressure within the excavation chamber never stabilized and the video feed indicated

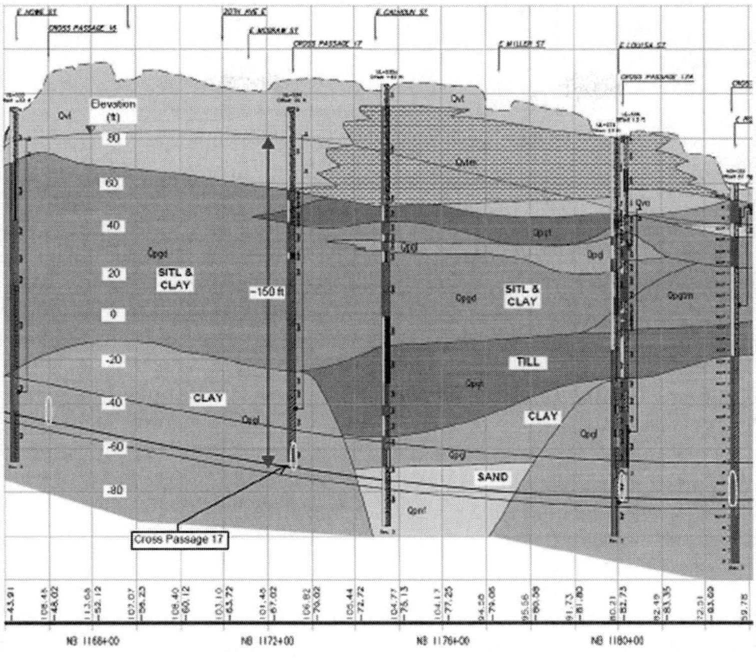


**Figure 2. Anticipated ground conditions—Cross Passage 17**

inflowing water. Subsequently, a second inspection stop was performed forty feet to the south. Better ground conditions were encountered at this inspection stop; the material removed from the conveyor belt in this area was stiff clay. Some groundwater inflow was observed in the second cutter head inspection stop.

In October 2011, the TBM mining the Northbound tunnel performed an inspection stop near the design location for Cross Passage 17 and confirmed that the ground conditions were not suitable for Category 1 construction.

The cross passage was moved 35 feet from its design location after joint review of the inspection stop information by TFK, Sound Transit, the CM team and the Designer.

Preconstruction probe drilling through the segmental lining at the revised location confirmed the presence of clayey soil and no groundwater inflow: although, the probes focused primarily on the top heading of the cross passages.

## PHASE 2—CROSS-PASSAGE CONSTRUCTION

Typical excavation sequence through a cross passage was to pre-support the two main tunnels around the openings, break out the lining in the Southbound tunnel, excavate the top heading clear across from one TBM tunnel (Southbound) to the other (Northbound) and repeat the process for the bench. After bench excavation was completed, the segmental lining intersected on the second TBM tunnel would be sawcut and removed. Pre-support in the main tunnels was selected and designed by TFK and consisted of a shotcrete shell. Ground conditions and water inflow at Cross Passage 17 required several modifications to this typical sequence.

### Excavation and Problem Discovery

The initial work at Cross Passage 17 consisting of saw cutting and removal of the segmental lining, excavation and support of the first top heading round, proceeded without event. Unfortunately, these conditions did not persist as the signs of trouble were first encountered during excavation of the second top heading round, with a small amount of water entering through the temporary invert of the top heading excavation directly adjacent to the segmental lining of the Southbound tunnel. This water inflow did not represent a serious complication or present a problem for overall stability of the excavation so the operation continued. Over the several days it took to complete excavating the top heading, the rate of water inflow continuously increased up to approximately 50 GPM.

After the completion of top heading excavation it was apparent that additional measures would have to be implemented prior to the start of bench

**Figure 3. Groundwater encountered at Cross Passage 17 channelized through probe hole**

excavation. The first measure undertaken was to drill a series of probe holes down through the temporary invert of the top heading through the future bench excavation and into the ground below. The first two holes were drilled adjacent to the southbound tunnel to approximately ten feet deep (one foot past the limit of the bench excavation) without encountering any additional water. The third probe hole was advanced to approximately the same elevation as the first two and encountered a pressurized sand layer. A geyser of sand and water, the full diameter of the probe hole, sprayed into and across the inside of the cross passage. See Figure 3. By the time an appropriate packer was tracked down and installed, nearly five cubic yards of sand had been deposited in the cross passage and in the Southbound tunnel. Sand was deposited in the Southbound tunnel to the extent that it blocked rail traffic from passing through the area. Elevated concentrations of methane and hydrogen sulfide were also encountered during the unintended sand and water discharge event. Subsequent to this initial event, a gas meter was permanently positioned inside the cross passage but fortunately this was the only time anything registered on the meter.

Based on the dramatic and erratic results obtained during the initial ad hoc probe drilling, the Designer devised a more systematic probe drilling program to more fully ascertain conditions beneath the cross passage. The second round of probe hole drilling encountered problems related to tight access and the subsequent limitations on the size of equipment that could be utilized to install the probe holes. Ultimately, the probe holes consisted of partially screened, three inch steel casings that were driven into the ground with a hand held pneumatic fence post driver. Air lances were used to remove material from within the casing when the penetration rate became unacceptably slow. Water was observed in a few of the second round probe holes but was not pressurized to the same extent as the initial probe hole.

### Remediation Issues

Once the exploratory phase was completed, the Team gathered to develop plans that would permit the bench to be safely excavated. All parties agreed that some form of dewatering would be required but there was vigorous discussion as to which methodology (in-tunnel or surface) would provide the best solution. The team decided to proceed with installation of an in-tunnel system first while preparations were made for a surface dewatering system in the event it proved necessary.

### In-Tunnel Dewatering

The in-tunnel dewatering system was to rely upon installation of a series of ten inclined small diameter, vacuum wells underneath and around the cross passage. In and of itself, installing this series of wells would normally not be difficult but this case had extenuating circumstances that complicated installation. Specifically, temporary support of the segmental lining around the breakout area was provided by a reinforced shotcrete shell (termed "propping") placed against the intrados of the segmental lining, over a width of twenty feet. Unfortunately, the most heavily loaded portion of the propping shell was the thickened beam located directly below the cross passage opening. Of course, this is precisely where the vacuum wells had to be installed. This conflict required close coordination and additional design review by the propping designer (Halcrow, a CH2M Hill Company) whereby the impact to the propping reinforcing steel of the six inch diameter core required at each well location was analyzed. Well locations were adjusted such that demand ratios of the propping remained at acceptable levels while the wells could still perform adequately. Observations during drilling of these wells also aided to further delineate the clay to sand contact, which appeared to dip to the west (towards the Southbound tunnel). See Figure 4.

Installation of the 10 wells occurred without complications. The hydrostatic pressure in the underlying sandy layer decreased under just gravity flow. However, the greatest difficulty encountered during well installation and operation was maintaining overall efficiency of the vacuum system. Despite several attempts at sealing all the possible interfaces where vacuum pressure could be lost, the entire system never achieved the desired efficiency. A significant effort was also expended on fine tuning the system by adjusting valves for each vacuum well. Multiple attempts were made to improve the

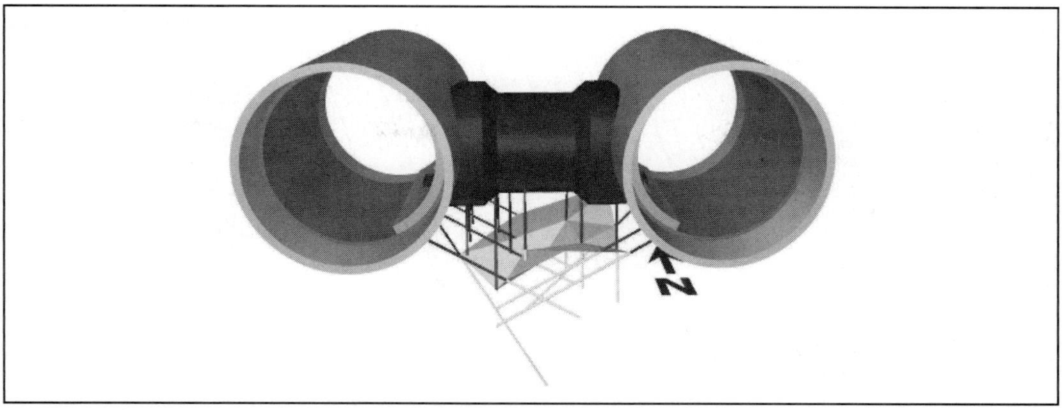


**Figure 4. 3D View of probe holes and well points at Cross Passage 17 with encountered sandy layer**

mechanical efficiency of the system (e.g., increased vacuum line size, increased header size, additional pumping capacity, etc.) with positive but not overwhelming improvement.

## Surface Dewatering

The surface dewatering component was to include up to four wells, incrementally drilled as their need arose, each with a six inch casing per the layout shown in Figure 5. Difficulties developed during planning for this work as the wells would be located in a residential area of Seattle with very narrow streets and an adjacent elementary school. Additional delays were encountered due to the permitting approval process not only for well drilling but also dewatering discharge to the local combined sewer system.

Surface dewatering well startup was plagued with numerous seemingly minor yet important issues—many of which were related to the electrical system provided to power the dewatering system. Early investigation into the electrical service in the area of the Cross Passage 17 surface dewatering system determined that what was available (single phase) was not capable of powering the three phase dewatering system. This meant a mobile generating plant with complete redundancy and automatic transfer switch would be required to operate the dewatering system. After struggling though the first few days of the dewatering operation, it was determined that the initial generators rented for the task were too small and that the automatic transfer switch did not function properly. Larger generators and a new transfer switch solved these problems. However, the larger generators were louder and burned more fuel—both conditions resulted in more complaints from the local residents. Improved noise mitigation efforts in the form of insulated dog houses around the generators helped quell the issues related to noise.

Of the three wells that were drilled, well number one had the most significant impact on the water level around the cross passage and created the most difficulty, both for well installation and within the cross passage below. This well suffered significant bottom heave that required extensive bailing, ultimately leading to the placement of far more sand pack material than originally intended. Additionally, the screen for well one broke during installation, reducing the well's efficiency to the point it had to be replaced. The redrilling of well number one had a tremendous impact on the formation around the cross passage, creating a direct conduit between the cross passage and an underlying but geologically separate pressurized sand aquifer, as inflow of water and sand into the tunnels picked up dramatically to nearly 500 GPM. In addition to concerns of overall cross passage stability, the inflows represented a severe problem for the rest of the project as it temporarily exceeded the pump capacity within the Southbound tunnel, resulting in nearly two feet of water collecting at the low point of the tunnel. The allowable discharge rate to the sewer for the main tunneling site was temporarily exceeded as recovery from this event proceeded. After additional pumps were mobilized, the tunnel was pumped dry revealing surprisingly significant amount of sand (20–30 cubic yards) had been deposited along the invert of the tunnel. In addition to abating inflows, the material that had been lost from the formation would have to be replaced at some point.

Overall capacity of well number one was an issue that needed to be overcome. Initially, each well was fitted with a 15 horsepower submersible pump capable of producing approximately 150 GPM. This proved insufficient in well number

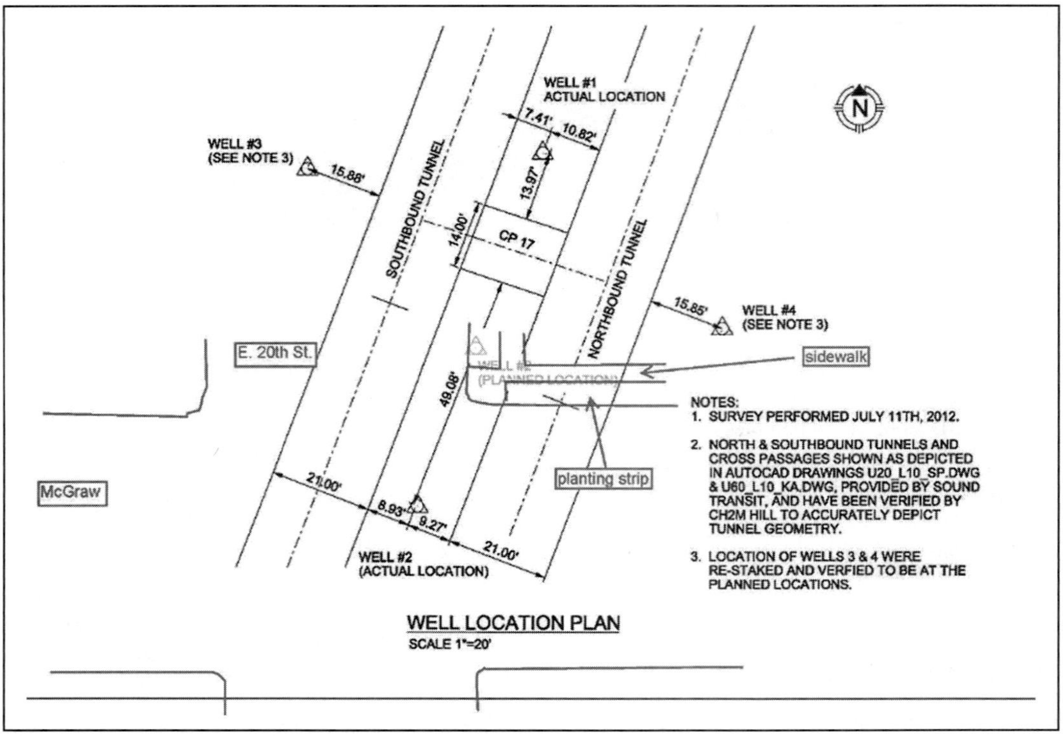


**Figure 5. Plan view of the surface dewatering setup**

one. Two increases in pump size were made—first to a 30 horsepower submersible pump and finally to a 60 horsepower in line shaft irrigation pump capable of removing nearly 500 GPM. The final pump change provided the necessary dewatering flow that lowered groundwater levels in the underlying sand layer to acceptable levels and permitted completion of the excavation of the cross passage.

### Bench Excavation

Actual excavation of the bench went very smoothly with only one significant issue encountered. During excavation of the third bench round, a "rabbit hole" was found on the north side of the cross passage near spring line. This hole was pointed directly toward well number one and was likely the source of a portion of the material heaved and bailed during installation of well number one. A layer of shotcrete (with two pipe ports penetrating it) was placed over the hole where it intersected the side wall of the cross passage. These two ports were later used to deliver a backfill material to the rabbit hole with the lower hole a delivery port and the upper hole serving as a vent. Despite the proximity of the underlying sand layer, no sand was actually encountered during bench excavation, with the exception of the rabbit hole. See Figure 6.

## PHASE 3—POST-CONSTRUCTION CHALLENGES

After completion of excavation and installation of waterproofing and the final lining, additional issues arose during removal of the temporary shotcrete support for the segmental lining and decommissioning of the in-tunnel dewatering system.

### Well Abandonment and Void Filling Issues

Subsequent to decommissioning the surface dewatering system, a preliminary attempt was made to grout the in-tunnel well points. However, while demolishing shotcrete in the southbound tunnel in January 2013 groundwater was seen flowing into the tunnel between the segmental lining and the propping. As a section of the propping was removed around one of the in-tunnel wells, groundwater inflows in excess of 50 GPM were observed and approximately 15 to 20 cubic yards of sand was transported into the tunnel from the annular space between the tunnel liner and the well point. A temporary steel cover was fabricated and bolted in place to contain and staunch the inflow until a permanent solution could be developed and enacted. The loss of sand was abated, while groundwater inflows continued. See Figure 7.

**Figure 6. Completion of excavation and initial ground support for Cross Passage 17**

As five of the nine remaining well points also penetrated the shotcrete, there were significant concerns about removing the shotcrete in proximity to the well points without additional episodes of ground loss and groundwater inflows. The other four well points also required verification of grouting and sealing prior to removal of temporary plates and final decommissioning.

### Remediation Approach

To successfully decommission the well points, the project team developed a detailed plan for approaching the problem at hand. This plan was developed with input from all parties, with key issues being safety of workers throughout the process, meeting schedule constraints to avoid impacting critical path activities, and technical resolution of the well point decommissioning. The overall goals of the plan were to be able to significantly reduce risk of any further ground loss and large inflows; using grouting to replace ground loss to the extent possible; and restore the segmental lining to conditions consistent with the 100-year design life. This plan consisted of numerous detailed steps, in general as follows:

- Assessing current conditions and risks, by carefully opening available ports, and performing dye tests to determine preferential flow paths;
- Relieving pressures around the tunnel lining through existing grout holes, in order to limit back-pressures during subsequent grouting, and to allow water into the tunnel in a more controlled fashion;
- Securing the temporary shotcrete support to the segmental lining with anchors around well point locations, to limit the risks from back-pressures, and to allow partial sealing around the perimeter of the shotcrete;

**Figure 7. Groundwater inflow encountered during well abandonment**

- Grouting well point holes through the segmental lining around the existing collars to reduce or even eliminate potential for additional material loss and inflows; Grouting was done through small diameter core holes, pre-existing packers, or adjacent grout ports (Figure 8);
- Verification of sealing of well points with small diameter core holes;
- Removal of temporary plates, packers, and shotcrete propping in and around the well points;
- Installation of temporary grout ports at each well point and subsequent replacement grouting around the lower portion of the segmental lining;
- And removal of temporary grout ports and final restoration of the tunnel lining; Final restoration included installation of stainless steel plates with gaskets, to provide

Figure 8. View of bolted shotcrete and grout ports

Figure 9. Well point 1 with final stainless-steel plates

redundancy to the grouting and patching of the well-points (Figure 9).

At various steps, hold points and go/no-go points were set up to verify that goals were being achieved prior to progressing to subsequent steps. The plan was also set up with the ability to assess conditions as the plan progressed, and to make adjustments as needed. Contingency plans and measures were developed and materials were on standby in the tunnel in case inflows and flowing ground was encountered. One of the well points did encounter a small amount of flowing ground and inflows during removal of a temporary cover plate, but ground loss was limited to a quarter cubic yard, and no additional issues were encountered.

Implementation of the plan took cooperation from all parties involved, and careful planning and execution. Work on the remediation steps began in earnest in late February 2013, and was completed by late March 2013. Frequent team meetings were held to assess progress at critical go/no-go points, and to determine if any adjustments to the plan were necessary. Daily updates were provided on construction progress. TFK and their tunnel grouting subcontractor, Condon-Johnson (CJA) worked closely together, sharing resources and expertise, and no significant issues were encountered. TFK was able to adjust their overall schedule and work plan to accommodate the work at Cross Passage 17, and no overall impact on the critical path occurred.

Overall, the plan to seal and decommission the well points was successfully implemented. Only minor issues were encountered during implementation, which the team was well prepared to deal with. A total of over nine cubic yards of grout was injected to seal up the well points, and provide replacement for the lost ground. While grout quantities injected did not equal the amount of ground lost, the amount of grout injected was estimated to be enough to consolidate ground below the cross passage and tunnels, and limit any long-term deformations. A deformation monitoring program has been recommended for implementation during early system operations, to verify that no further deformations of the lining occur.

## SUMMARY AND CONCLUSION

At the time of writing this paper, the project has achieved acceptance with all issues related to Cross Passage 17 being resolved.

In conclusion:

- Implementation of this process has been possible mainly because of the cooperative relationships between the Owner, CM team, Designer and Contractor. In projects where stakeholders do not share a comfortable working relationship, successful implementation of this approach would be a challenging task.
- Effective management of the cost and time to mitigate the unforeseen condition was also achieved by all parties.
- Investigative efforts during design and requirements for probing during construction were bolstered for subsequent Sound Transit contracts, to better identify ground conditions.
- We acknowledge the contribution of all members of TFK, CJA, Sound Transit, the CM team and NTP who participated in the difficult construction of Cross Passage 17.

# Unconventional Approach During Design and Construction of a 3-Cell-Cross Passage on Contract CQ-031 Queens Bored Tunnels in Queens, New York

**Jean-Marc Wehrli**
Traylor Bros., Inc.

**Norbert Fuegenschuh and Johannes Jaeger**
BeMo USA Inc.

**ABSTRACT:** Construction of cross passages on TBM projects proves to be a challenging task. Construction methods are different from the usual TBM approach, and require flexibility and often unconventional thinking by owners and contractors.

Contract CQ031 of the East Side Access Project in Queens, New York required the construction of a 3-cell-cross passage between the Yard Lead Tunnel and an emergency exit / ventilation structure. The Granite/Traylor/Frontier-Kemper Joint Venture created an innovative solution by applying NATM principles. The paper describes the design and execution of this sophisticated concept, including the challenges regarding ground improvement, construction phases, selection of equipment, permanent lining works and schedule issues.

## INTRODUCTION

The East Side Access project in New York City is sponsored by the Metropolitan Transportation Authority (MTA) and its purpose is to bring the Long Island Railroad from Queens into a new station built below Grand Central Terminal in Manhattan. Contract CQ031, a design-bid-build scheme that was awarded to the Joint-Venture of Granite Construction, Traylor Bros., Inc. and Frontier-Kemper Constructors, Inc. (GTF JV) in 2009, involved the construction of four 19.5 ft finished diameter bored tunnels with a total length of 10,620 lf. It included approach structures, one emergency exit and ventilation structure and miscellaneous appurtenant structures above ground in Queens' Sunnyside Yard.

## GEOLOGY AND GROUND TREATMENT

The Yard Lead Tunnel Emergency Exit (YLEE) Structure serves as a combined emergency exit and ventilation structure for the Yard Lead Tunnel. It is located adjacent to active loop tracks and the south abutment of the 39th Street Bridge. Construction of the YLEE comprised installation of permanent cast-in-place structures within a temporary watertight shaft, with a cross passage connection to the bored tunnel. The profile of the cross passage lies fully within Glacial Till, a highly variable and medium dense to very dense sand deposit with a fines content of 5 to 30%, gravel, and occasionally boulders. Overlaying the till is a 5 ft. thick layer of fill material. The overburden is approximately 24 ft, and the ground water table around 10 to 12 ft. above tunnel crown. Due to restrictions on groundwater dewatering rates in the work area, and for the purpose of improving ground stability, ground treatment around the cross passage was mandated by the contract and at a minimum had to extend from the shaft to the centerline of the bored tunnel. After consulting with geotechnical experts, the JV selected jet-grout, similar to the method included as a suggested method by the Owner. It was also agreed to extend the ground treatment across the full perimeter of the bored tunnel so as to enable use of the jet-grout zone for cutter changes on the TBM under atmospheric pressure. In addition, the design required the grout treatment to extend at least 10 ft. around the excavation profile of the cross passage as well as the TBM in any direction. Because the footprint of the jet-grout zone infringed on the existing loop tracks and signal cable troughs on the surface, the JV decided to reduce its size by encapsulating the area of treated ground with secant piles. This change provided the additional benefit of having the perimeter of the grout treatment secured from unexpected ground loss by walls of rigid piles.

## DESIGN APPROACH

The Yard Lead cross passage had to accommodate a 5 ft. wide emergency exit and two 10 ft. wide ventilation openings, each with a length of approximately 12 ft., requiring an excavation profile of around

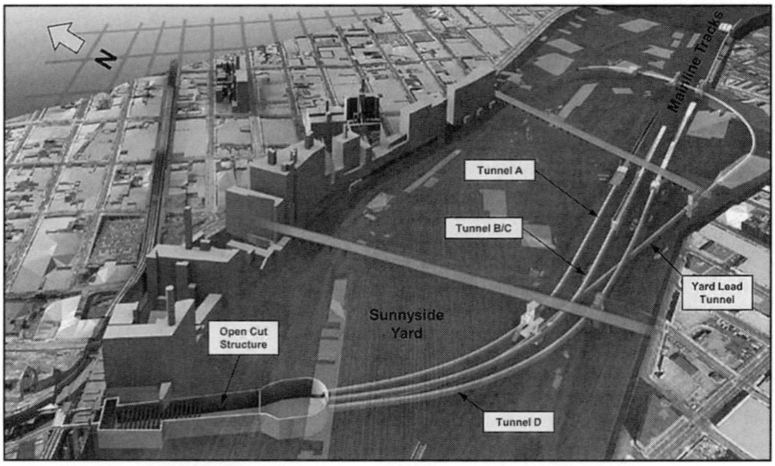


**Figure 1. Site layout for Contract CQ031**

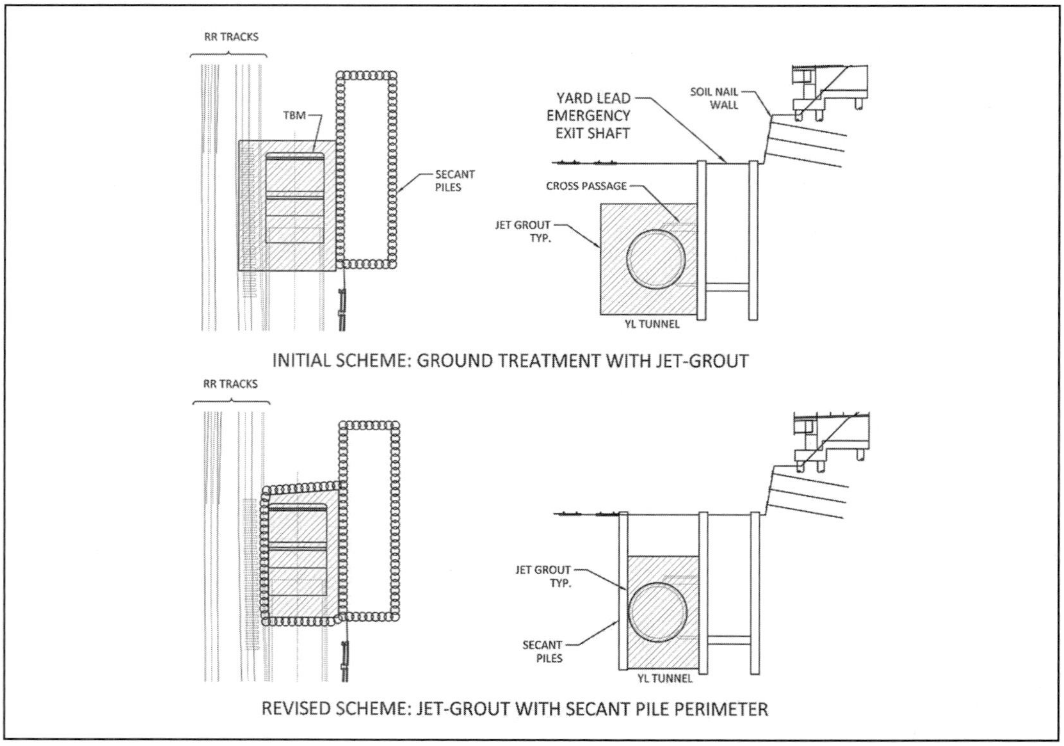


**Figure 2. Ground treatment scheme**

40 ft. width and 20 ft. height. The Owner's design was based on a rectangular excavation profile with steel girders and struts lagged with timber as temporary support in combination with a pre-drilled pipe canopy. The temporary support system for the precast concrete segmental liner of the bored tunnel at the cross passage opening, consisted of vertical steel braces spaced at 5 ft centers. At bid time, the JV realized that by applying an SEM based approach, the cross passage could be built more economically and in a shorter period of time, so a decision was made to hire the Austrian company Beton- und Monierbau USA Inc. (BeMo) to re-design the cross passage. The new design was based on excavating three drifts

using SEM, one after another with the center drift excavated last. Each drift would accommodate one of the three openings in the permanent concrete structure. Excavation of the center drift would require the partial removal of the shotcrete liner of the outer two drifts for tying together the three openings of the permanent structure.

## Cross-Section Geometry

Originally near rectangular, cellular cross sections were foreseen for the emergency exits. As the decision was made to use a shotcrete shell for temporary support, some changes in the excavation profile were required to optimize the profile structurally for shotcrete applications. The main goal was to avoid excessive bending moments requiring significant amounts of bar reinforcement. In shotcrete applications, high degrees of reinforcement increase the risk for poor encapsulation and shadowing. Therefore, a more rounded cross section, which carries loads more in axial compression than in bending, is favored.

## Numerical Modelling

A combination of 2D and 3D finite element models were used for designing the temporary support such as props for the running tunnel, for the temporary shotcrete lining, and for the permanent lining. The starting point was the analysis of the geostatic in-situ stresses, followed by pile installation and ground improvement. Then, modelling of the complete excavation/construction history was undertaken in order to assess the loads on props, the shotcrete structure and the permanent in-situ concrete lining as a result of the excavation and construction activities. In addition to ground pressure coming from construction, live loads, water pressure and seismic loads were applied to the final cast in place liner.

The initial support system for the adits is made up of pipe arches, a shotcrete shell with lattice girders and bar reinforcement. Additionally, elements of the cast in place concrete structure are part of the initial support system. The early loading of immature shotcrete, which is typical for shotcrete applications

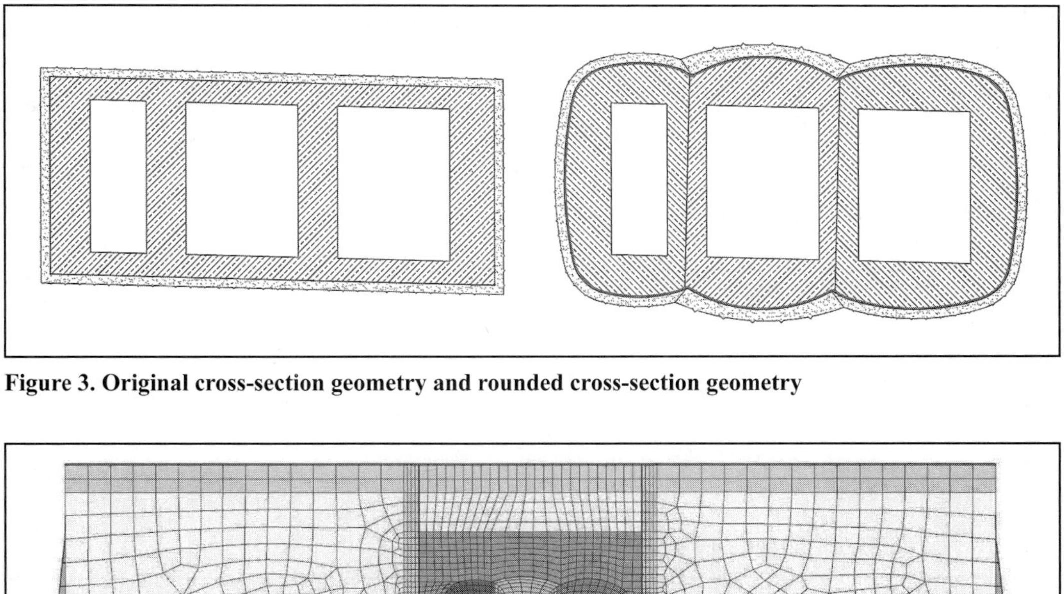

**Figure 3. Original cross-section geometry and rounded cross-section geometry**

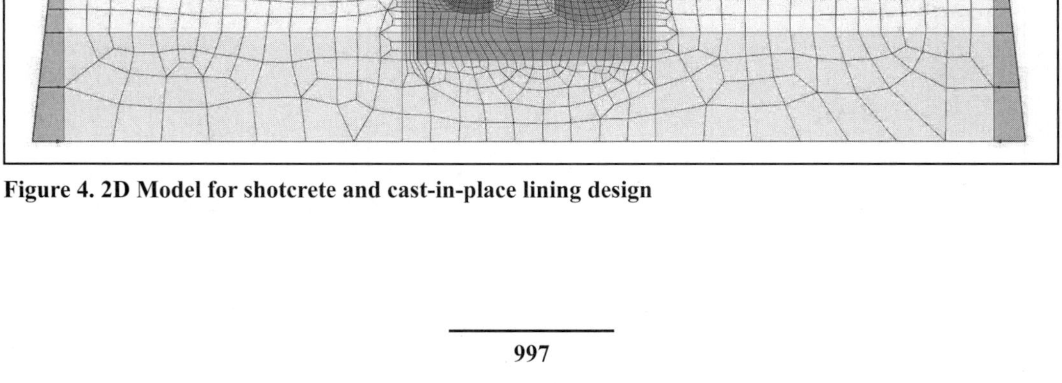

**Figure 4. 2D Model for shotcrete and cast-in-place lining design**

in tunneling, requires some special attention in the structural analysis. The problem is to realistically assess the shotcrete stiffness. To account for the ability of stress relaxation and creep, the hypothetical modulus of elasticity (HME) approach was chosen, which allows for modelling of sprayed concrete by means of Hooke's law. In general, values for HME are significantly lower than the elastic moduli suggested in the design codes for concrete [3]. Based on the loading condition, an HME of 1,088 ksi (7.5 GPa) was considered to be appropriate. In the final stage, a shotcrete strength of 5,000 psi was required. The compressive strength of the cast in place final concrete was 4,000 psi.

All adits were located within jet-grout material. The UCS design strength for jet-grout material was chosen to be 190 psi. Although averaged test data would suggest a significantly higher strength, this comparatively low value was chosen to account for possible non-homogenous zones, and for scatter in the test data results. The geotechnical parameters for glacial till, fill material and Gardiners clay are defined in [4] and show typical soft ground strength with friction angles between 32° and 38° with no cohesion.

## MINING

As a result of the redesign completed by BeMo, the New Austrian Tunneling Method (NATM) was selected to construct the 3-cell cross passage between the Yard Lead Tunnel and the combined emergency exit/ventilation shaft structure. The excavation sequence required that drifts 1 and 3 on the outside had to be constructed first, including waterproofing and cast-in-place lining installation before excavating the center drift (drift 2). This last step included demolition of the inner shotcrete walls and tedious construction of connection details.

### Construction Sequence in Detail

1. Pipe umbrella installation over drifts 1, 2 and 3 (drilled and installed from shaft)
2. Propping installation in main tunnel at CP tunnel 1 and 3
3. Excavation and support of drift 1
4. Excavation and support of drift 3
5. Waterproofing of drifts 1 & 3
6. CIP Inner lining for drifts 1 & 3
7. Relocation of propping to enable excavation of the center drift (start-up depending on development of concrete compressive strength in adjacent drifts)
8. Excavation and support of center drift
9. Waterproofing of center drift
10. CIP Inner lining for center drift

This construction sequence is basically an adaptation of a tunneling method that has been previously applied on large 3-cell subway station projects (examples: Bochum, Germany/ Dortmund, Germany/ Toronto, Canada).

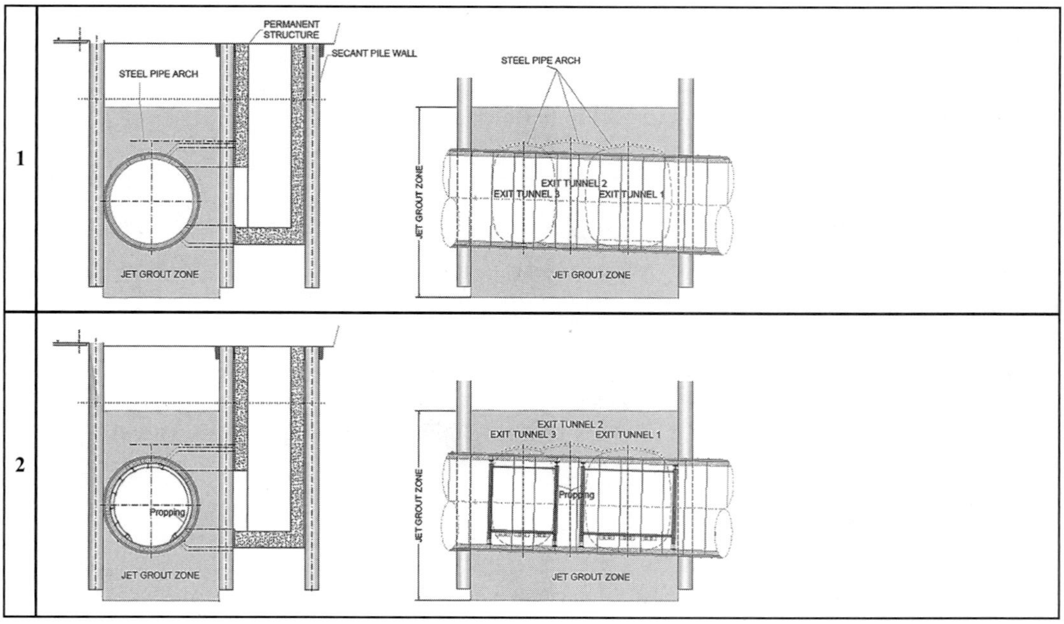


**Figure 5. Construction sequence** **(figure continues)**

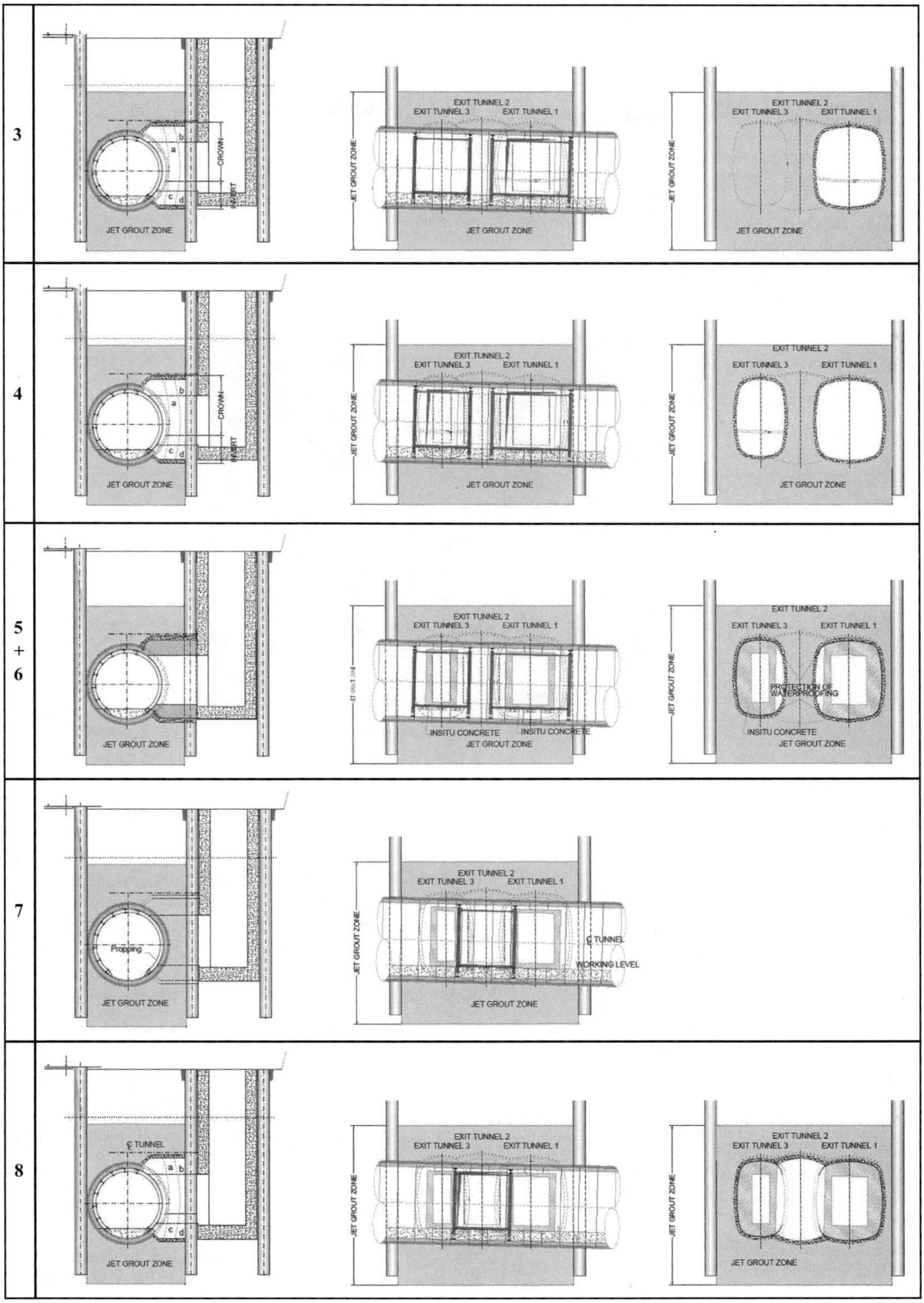


Figure 5. Construction sequence (continued) (figure continues)

9 + 10

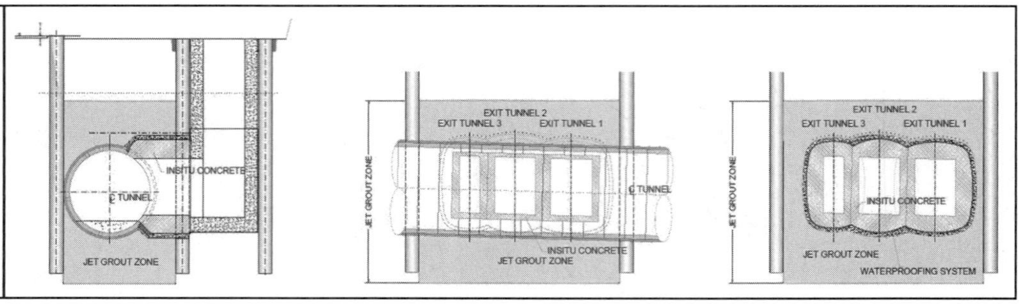


**Figure 5. Construction sequence (continued)**

**Figure 6. Foldable frame for bracing of the tunnel segmental lining**

## Pipe Umbrella

Since the combined emergency exit/ventilation shaft was located in close proximity to the Yard Lead Tunnel, the JV was able to install the pipe umbrella from within the shaft, basically covering the total length and width of the 3-cell-cross passage structure.

The umbrella consisting of 40 pieces of pipe in total, with a diameter 3.5 in. and a length of 24.5 ft., was drilled, put in place and grouted from within the shaft. The 3 ft. thick temporary shaft support, consisting of secant piles, had to be penetrated with 5 in diameter core drills at each pipe location.

Drilling of the pipes also provided an indication with respect to the ground conditions that would be encountered during the excavation of the cross passages. The jet-grouted ground behind the secant piles proved to be watertight and of very good consistency. The design specified a strength for the jet-grouting material of 500 psi and a permeability of not more than $10^{-5}$ cm/sec. In reality, the compressive strength tested on post-jet-grout core samples reached more than 3,000 psi, and exceeded the required design strength significantly. Packer tests confirmed that the permeability criteria were met. [2]

The propping for the TBM tunnel was designed as foldable steel bracings ("hamster cages"), which could be brought into the tunnel on a flatbed rail car [2]. Two circular bracings were put in place for drifts 1 and 3. Per the initial design, the final liner in drift 1 had to be in place before the excavation of drift 3 could begin. However, the JV was able to later persuade the designer to allow removal of the segments, and excavation of drift 3 concurrent with waterproofing and final liner installation in drift 1. This was based on ground conditions that turned out to be better than anticipated at the design stage and higher than expected in-situ strength of the jet-grout. This allowed the JV to cut a couple of weeks from the schedule.

The initial plan was to support the openings for drift 1 and 3 with one bracing each, and to move the one used for drift 3 to the center for excavation of drift 2, once the cast-in-place liner in drifts 1 and 3 had been poured and had attained 70% of the design strength. At that point, the bracing used for supporting the excavation of drift 1 was supposed to be

**Figure 7. Tunnel liner bracing and start-up of excavation for drift 1**

**Figure 8. Temporary working deck at the cross-passage location**

dismantled and taken out of the tunnel. Since the bracings also served as support for the work deck, their relocation to enable excavation of drift 2 turned out to become too cumbersome and time-consuming in the field. Therefore, it was decided to leave them both in place and add a cross beam between the two, and a vertical prop on either side of the future opening for drift 2.

To create a level surface for the tunneling equipment and material storage, a temporary wooden deck was installed in the area of the 3-cell structure. It consisted of crane mats with a thickness of 12 in supported by steel beams. An alternative would have been to fill the TBM tunnel with backfill material to a convenient working level.

### Excavation

Saw cutting of the pre-cast tunnel liner segments for future openings of the cross-passages was performed by a subcontractor, working off of a scaffold and utilizing a hydraulic driven circular concrete saw.

Prior to removal of the segments, the JV was required to drill at least eight probe holes per drift. Given the short length of the drifts and the ease in mobilization, it was decided to drill the probe holes manually using air-powered jack-legs.

Demolition and removal of the segments was accomplished using a mini excavator with a hydraulic breaker (chisel) attachment. It was decided to utilize the same equipment set-up for mining through the jet-grout zone due to its relatively high average strength of around 3,000 psi. Work started out at drift 1 using a CAT 305 mini excavator with a hydraulic breaker attachment. Soon, it was realized that this choice of equipment was not ideal, since the excavation rates achieved were considerably less than anticipated putting the accomplishment of a critical milestone in the schedule at risk. The use

**Figure 9. Brokk 330 at work**

of a roadheader attachment instead, in combination with the manual pre-drilling of relief holes through the excavation face using jack-legs did not yield better results so, after finishing the excavation of drift 1 two weeks behind schedule, the decision was made to change the type of excavation equipment. Drifts 3 and 2 were then excavated using a remote controlled electric-powered Brokk 330 demolition breaker. A series of attachments (roadheader, chisel, buckets) were available on site for the different tasks. The new equipment proved to be much more capable of coping with the circumstances, and excavation rates increased significantly allowing the JV to make up for lost time. The next two drifts were excavated in half the time compared to the first one. The Brokk excavator also proved to be the right choice for the break-out into the shaft through the 3-foot thick secant pile temporary support of the combined emergency exit/ventilation shaft structure.

Mucking was accomplished with the Brokk by replacing the breaker with a bucket. Using the Brokk, the muck was transferred from inside the cross passage into a muck box sitting on a flatcar on rails at the end of the temporary work deck. The boom of the Brokk turned out to be long enough to accomplish this task without tramming. Once full, the muck box was pulled back through the tunnel to the launch shaft where it was hoisted to the surface and dumped into a muck bin. After completion of drift 1, access from the launch shaft became unavailable and mucking had to be carried out via the approach structure at the other end of the TBM tunnel. Because the rails between the cross passage and the approach structure had already been removed, the JV decided to use a skid-steer for hauling the muck out of the tunnel. The skid-steer was able to drive up the approach ramp to the surface where the muck was discharged into roll-off bins.

The design prepared by the JV called for excavation lengths of 3 feet per round. The tunnel face was subdivided into a crown and bench/ invert section. Lattice girders, type Pantex 95/6/10 were installed in the crown and invert after each round of excavation. The subsequently installed shotcrete shell was 12 inches thick and had to be reinforced with 2 layers of wire mesh, type 4×4-4/4 per ASTM A-497. The tight radii in the cross section of the small tunnels required the wire mesh to be pre-bent.

The lattice girders were set to line and grade with the aid of a crosshair laser, mounted on a bracket installed to the liner of the TBM tunnel on the opposite side of the opening. Verticality of the girders was checked using a plumb-bob and level.

### Shotcrete

Shotcrete in the tunnel was applied using the dry-mix method. Shotcrete pot, material silo and air compressors, i.e., the whole supply line for shotcrete, were positioned on the surface beside the shaft. The 2.5 inch diameter shotcrete slick line was routed down the shaft through a 10 inch pre-drilled hole into the Yard Lead Tunnel. This hole was drilled from the shaft while installing the pipe umbrella.

The supply hole was located in the center of drift 2. From there it was easy to supply shotcrete via hoses to drifts 1 and 3. After completion of these two drifts, the shotcrete supply line was re-routed through drift 1 for the excavation of drift 2.

This set-up mitigated the dust problem due to the fact that the main source of dust, the shotcrete plant, was positioned on the surface. Access to the shotcrete plant from the tunnel was possible via a stair tower installed inside the shaft. However, because the shaft was located within railroad property individuals were only granted access after having passed an Amtrak certified safety training course.

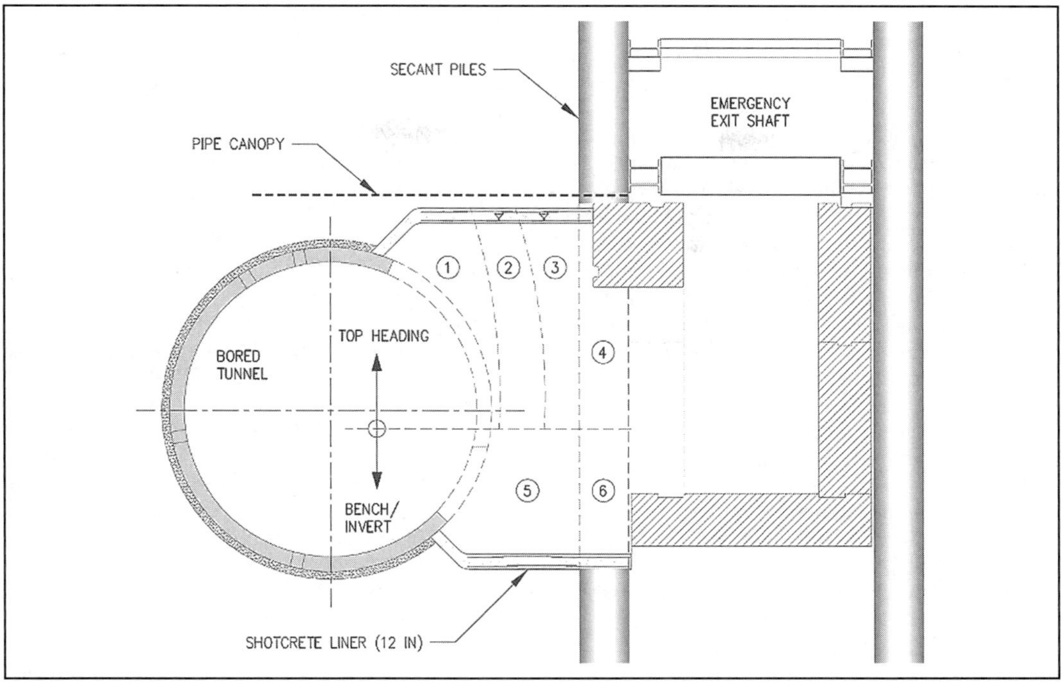


**Figure 10. Excavation sequence cross passages [2]**

**Figure 11. Shotcrete plant at ground surface**

The oven dry shotcrete material was delivered to the job-site in supersacks where they were emptied into a vertical 5-cy silo, using a telehandler. During shotcreting, the dry-mix was transferred from the silo via a pre-dampener to the shotcrete pot from where it was conveyed pneumatically through the short piping system (around 150 ft long) down the shaft and to the heading. Additional water and liquid alkali-free accelerator were supplied through separate lines and added directly at the nozzle. The combination of using dry material and short transportation length further reduced the risk of line plugs.

In order to keep the dust levels below the allowable limits, a bulkhead with two box fans was installed at the end of the TBM tunnel for ventilation. In addition, a secondary ventilation system was provided with a rigid duct reaching into the cross passage all the way to the face to exhaust dust generated from shotcreting and excavation through a wet scrubber back into the TBM tunnel.

### Shotcrete Trials

Shotcrete trials were conducted a couple of weeks prior to the start of cross passage excavation. The contract called for overhead and vertical sprayed test panels. Spraying of the panels had to be performed using materials and equipment that was intended to be used later during execution of the work. Spraying of the test panels was also part of the program for certifying the nozzlemen in accordance with ACI guidelines, which was a requirement in the contract.

During construction, the average shotcrete compressive strength obtained from three cores taken from a panel sprayed in the heading for every 8-hour shift had to be, at a minimum:

- 1,400 psi at 24 hours,
- 3,500 psi at 7 days (single core not less than 3,200 psi)
- 5,000 psi at 28 days (single core not less than 4,670 psi)

This is equal to the J2 line according to the Austrian Sprayed Shotcrete Guideline [1].

In addition to the cores, readings using the Meyco Penetration Needle were taken from the test panel for monitoring the early age strength development of the shotcrete. These readings were then compared with the baseline data obtained during the trials, and used as an early warning indicator.

The results obtained in the field during production testing confirmed that the material selected was capable of fulfilling the design requirements.

Exposed ground had to be sealed immediately with an unreinforced, minimum 2 inch flashcoat of shotcrete. The design also included special requirements for face support to cover situations where excavation had to be interrupted for a prolonged period of time. For stoppages of more than 24 hours but less than 3 days, the face had to be supported with a 5-inch layer of shotcrete. For a stoppage of 3 or more days, 10 inches of shotcrete had to be applied.

Extra care had to be taken when excavating around the perimeter of the opening in the segmental liner of the TBM tunnel, to prevent damages to the surfaces of the edges created by the saw-cut. Keeping these surfaces intact and smooth was imperative, as they were used to accommodate the termination detail for the waterproofing sheet membrane, prior to the placing of the final cast-in-place liner. In addition, a hydrophilic gasket, along with a re-injectable grout hose were attached to this surface around the perimeter of the opening as a precaution, in case the waterproofing termination leaked.

## WATERPROOFING

The contract specifications called for a post-applied composite waterproofing membrane, consisting of a polyethylene film (HDPE) and a self-adhesive rubberized asphalt, that bonds with the cast-in-place concrete once it has fully set. Given the lack of experience available with such membranes in SEM-mined tunnels with shotcrete support, the JV considered proposing a non-adhering PVC membrane instead. However, since the shaft had already been lined with the product specified in the contract, a complex mechanical connection at the interface between the two types of materials, i.e., PVC and HDPE, would have been required if the cross passage would have been lined with PVC. Because of warranty issues and the fact that such interfaces are prone to leaking, the JV decided to proceed with the specified material. This decision was further supported by the fact that the manufacturer of the adhesive HDPE membrane had just introduced a thicker and more ductile product specifically developed for shotcrete lined underground structures. The waterproofing was installed by a specialized subcontractor whose workers had undergone a comprehensive certification program for waterproofing installation by the manufacturer. In order to smooth out the rough surface of the shotcrete liner, and to prevent localized hydrostatic pressure build-up, a layer of

**Figure 12. Drift 1—Wall rebar and waterproofing**

**Figure 13. Drift 3—Form for crown pour of CIP liner**

geotextile was placed between the shotcrete and the membrane. The individual membrane sheets were joined together using a special tape developed by the membrane manufacturer. At the interface with the precast concrete segmental liner of the TBM tunnel, the membrane was terminated with a gasketed and galvanized steel strip, bolted to the concrete of the segments at 6 in. centers. To prevent water pressure build-up behind the membrane prior to the pouring of the cast-in-place (CIP) liner, sumps with sacrificial electric submersible pumps were installed in the invert of each cross passage. The pumps were turned off and the sumps grouted tight after the CIP concrete liner had reached its design strength.

## CAST-IN-PLACE LINING

The cast-in-place (CIP) liner for each drift was installed in three lifts, i.e., invert, walls, and crown. Due to the complicated geometry of the structure, the form consisted of a combination of a modular formwork system, pre-fabbed steel panels, and custom fabricated panels made from plywood and lumber. The concrete was supplied to the jobsite ready-mixed and pumped from the surface down the shaft into the tunnel via a 5-inch slickline. The crown pours proved to be the most challenging because the concrete had to be displaced into a fully closed form. Therefore, compaction of the concrete had to be accomplished using pneumatically driven form-mounted vibrators. To avoid over-pressuring or under-filling of the form, bleeder pipes were installed in the crown through the soffit of the form. These pipes were fitted with a ball-valve at the bottom end, which was immediately closed when concrete started to appear during the pour. Upon closure of the last bleeder pipe, the Shifter usually turned on the vibrators for a last time and ordered the pump operator to pump two more strokes before calling the pour complete.

Approximately 4 hours later, the bleeder pipes were pulled and the remaining holes were used for contact grouting later on. To ensure that the concrete would flow into every corner of the form, an 8-inch slump, pea-gravel mix was used. The rebar was supplied to the job already cut to the required length and bent, as needed, on the surface, by members of Local 46 prior to being sent underground on flatcars. The rebar cages were stick-built inside the cross passages by the Sandhogs. After the CIP liner had attained its design strength, all remaining temporary pump sumps inside the cross passages and shaft were abandoned and grouted tight before performing the contact grouting.

Contact grouting was accomplished using a neat cement grout that was batched on the surface using a compact grout plant and pumped through a hose, down the shaft into the cross passage. The end of the hose was equipped with a pressure gauge and a mechanical rubber packer that fit inside the hole left in the crown by the bleeder pipes. Grouting pressures were typically in the range of 30 to 40 psi. A few minor leaks in the CIP liner at the interface with the TBM tunnel and the shaft had to be repaired later with two-component polyurethane grout.

## SCHEDULE

Construction of the 3 cross passages, including the cast-in-place liner, took approximately 30 weeks. As expected, there was a learning curve involved, which resulted in significantly shorter construction periods for cross passage drifts 3 and 2, although the technical difficulties during construction of the center drift, such as demolition of the shotcrete side walls and connecting of the shotcrete shell to neighboring shotcrete shells, were harder to overcome. A very positive impact on the schedule resulted from the change of excavation equipment. Excavation and

**Figure 14. Finished cross passages—View from the TBM tunnel**

installation of the temporary support took 7 weeks for drift 1, 3 weeks for drift 3, and 2.5 weeks for drift 2. The durations for waterproofing and CIP liner installation were 10 weeks for drift 1, 7 weeks for drift 3, and 6 weeks for drift 2. These durations are based on two 8-hour shifts per day, except for the first 8 weeks, where work was performed in three 8-hour shifts. Saturday work was occasionally performed during installation of the CIP liner for the purpose of concrete pouring.

## SUMMARY

Applying the 3-cell approach in conjunction with the principles of the Sequential Excavation Method (SEM) for the construction of the cross passage at the Yard Lead Emergency Exit Structure proved to be the right choice. Besides keeping ground deformations below allowable tolerance, and maintaining stable face conditions at all times, thereby safeguarding the railroad above, the JV was able to benefit from a significantly reduced construction period.

Changing of the excavation equipment from a traditional mini-excavator with hoe ram attachment to a remote controlled and electrically powered demolition breaker after completion of the first drift resulted in improved production rates in the jet-grout and a safer working environment for the operator.

## REFERENCES


[1] Österreichischer Betonverein: Sprayed Shotcrete Guideline (current edition).

[2] Robinson, B. and Wehrli, J.M. 2013, East Side Access—Queens Bored Tunnels Case Study, 2013 Rapid Excavation and Tunneling Conference Proceedings. Littleton, CO: SME. pp. 1,014–1,041.

[3] Max John, Factors of shotcrete lining design: Tunnels & Tunnelling International, October, 2003.

[4] East Side Access Project, Geotechnical Interpretative Report, Contract CQ031, Queens Bored Tunnels and Structures, Rev B, March 2009.

# Rock Mass Characterization and Initial Support Performance of the Caldecott Tunnel Fourth Bore, San Francisco Bay Area, California

**Sean Harvey, Bill Zietlow, David Kwietnewski, and Alan L. Howard**
Brierley Associates

**ABSTRACT:** The Caldecott Tunnel Fourth Bore was mined using Sequential Excavation Method (SEM) tunneling. Rock mass characterization was based on observations made during excavation and initial support performance was determined through daily surveying of monitoring points. Rock mass conditions observed during excavation were often better than anticipated by the GBR and the design, and often led to classifications at least one Ground Class better than anticipated in the design distribution. This paper provides a summary of the rock mass characterization as encountered during excavation as well as comparison of the performance of the rock mass and initial support systems relative to that anticipated by the design.

## INTRODUCTION

The recently completed Caldecott Tunnel Fourth Bore located on Highway 24 in the San Francisco Bay Area of California is an approximately 990 meter long horseshoe-shaped tunnel that extends through the Oakland-Berkeley Hills and connects Oakland with the Walnut Creek area. It consists of a two-lane tunnel that lies north of the existing three bores that were built between 1934 and 1937 (Bore Nos. 1 and 2), and between 1960 and 1964 (Bore No. 3). The Fourth Bore was excavated with dimensions of approximately 16 meters wide feet wide by 12 meters tall. In addition to the main tunnel, seven cross passages which connect to the existing Bore 3 were excavated in a horseshoe shape with approximate dimensions of 4 meters wide by 4 meters tall.

The Fourth Bore project was designed based on the principles of the New Austrian Tunneling Method (NATM) also known as the Sequential Excavation Method of tunneling (SEM). The SEM tunnel design is a modified version of European NATM to account for U.S. conditions, and requirements including the "limited experience with NATM construction, the prevailing contractual environment, and preferences for contractual simplicity" (Thapa et al., 2009). The resulting design was "prescriptive" and detailed the initial support to include shotcrete, rock dowels, spiles, lattice girders, face dowels, and pipe canopies. The initial support was divided into seven Support Categories, which were defined by each item of support including advance lengths, shotcrete thickness, and rock dowel layout. Final support consisted of reinforced cast-in-place concrete.

The seven Support Categories (labeled I through IV with subcategories "A" and "B") were "...designed to address seven anticipated ground behaviors" which were labeled "Ground Classes" in the project documents. Accordingly, the Ground Classes and Support Categories were expected to have a one-to-one correlation. For example, if a rock mass was identified as Ground Class 1a, the corresponding initial support would be as prescribed by Support Category IA.

The seven Ground Classes were defined based on 18 rock mass types (RMT's) identified during the design phase. Each RMT was intended to have distinct geological and / or geotechnical characteristics. Overall RMT rock mass descriptions used Geological Strength Index (GSI) system terminology, with the design descriptions for each RMT intended to correspond to a rock mass volume that controlled rock mass behavior at tunnel-scale. Rock mass conditions for each RMT and GSI description were further described using United States Bureau of Reclamation (USBR) terminology. Special RMTs were developed for the faults, dikes and sills that occur along the alignment. The GBR generally described the rock mass on the project as ranging from "blocky" to "disintegrated." The anticipated behavior of each RMT was then used to define the Ground Classes based on the behaviors defined in Table 1. Table 2 shows the anticipated primary behavior defined for each Ground Class.

A certain distribution of Ground Classes and corresponding required Support Categories were included in the contract for bidding purposes only. The contract included unit prices per unit length of tunnel for excavation and support of each installed Support Category, with additional units and pricing for various additional support measures.

Based on observed geologic and rock mass conditions, and measurements from geotechnical

**Table 1. Ground behaviors (from Thapa et al., 2013*)**

| Behavior | Description of Failure Modes and Manifestations in an Unsupported Tunnel |
|---|---|
| Block Failure | Discontinuity-controlled, gravity-induced failure of rock blocks that manifests as falling and sliding of blocks. |
| Raveling | The progressive, discontinuity-controlled failure of small rock blocks within the general rock mass at or near the excavation surface. Manifested as successive fallout of small rock blocks and can ultimately result in significant overbreak. |
| Shallow Shear Failure | Results from overstressing of the ground within 0.25D to 0.5D of the tunnel perimeter (D = tunnel diameter) and may be enhanced by the potential for discontinuity and gravity-controlled failure modes. Manifested by moderate inward movement of the tunnel perimeter, including invert heave, and possibly by movement of rock into the tunnel opening along discontinuities. |
| Deep Shear Failure | Results from overstressing of the ground beyond 0.25D to 0.5D from the tunnel perimeter. Deep-seated shear failure manifests as large radial convergence of the tunnel perimeter, including invert heave. |
| Slaking / Softening | The deterioration and breakdown of intact rock upon exposure by excavation and manifests as slabbing of material from the crown and sidewalls. The severity of this behavior is assessed on the basis of slake durability tests performed according to ASTM Test Method D4644. Softening, which is dependent on wetting and exposure by excavation, is the reduction of intact rock strength at the invert or elsewhere and manifests as the development of a muddy or unstable invert or sloughing along segments of the tunnel perimeter elsewhere. |
| Swelling | Occurs because of absorption of water by clay minerals in rock upon excavation-induced unloading. Swelling manifests as movement of the ground into the tunnel opening or additional tunnel support loading. |
| Crown instability due to low cover | Excessive crown geological overbreak and chimney-type failure will occur because of lack of confinement under low-cover reaches at portals. It manifests as block fallout and raveling above the crown. |

* As modified from Austrian Society for Geomechanics, 2004.

instrumentation, Support Category recommendations were advanced at daily SEM meetings that involved the Contractor, Owner, Designer, and Construction Manager. Agreement between the parties generally resulted; however, occasional disagreements about Support Category could not be resolved, in which case the Owner directed the installed initial support.

Because of generally better-than-anticipated ground conditions and behavior, and through the collaborative efforts between the Contractor and Owner during construction, the as-built ground classification and support recommendations resulted in a significantly different distribution of Ground Classes (and Support Categories), with virtual elimination of one of the four main classes (Harvey, et al., 2012).

During bidding and initial submittal preparation, the Contractor Team developed an alternative approach to the design that captured the major design considerations for the initial support as well as simplified the design into three support categories with modifications. However, the Contractor was directed to implement the prescriptive design as bid because of risk concerns and contractual requirements.

The following discussion will review the Contractor Team's estimation of the expected behavior and compare the baseline Ground Classes with the as-built Ground Classes. A brief discussion of the Contractor Team's expected initial support performance compared with the as-built initial lining performance will be presented.

**Table 2. Ground classes and behavior**

| Ground Class | Primary Ground Behavior |
|---|---|
| 1a | Block failure |
| 1b | Block failure |
| 2a | Shallow shear failure |
| 2b | Shallow shear failure |
| 3a | Deep shear failure |
| 3b | Deep shear failure |
| 4 | Crown instability |

## PRE-CONSTRUCTION GROUND BEHAVIOR ESTIMATE

During the bid preparation phase the Contractor's Team reviewed and performed statistical analyses of the field investigation data and laboratory test data provided in the Geotechnical Data Report (GDR) and studied the ground classifications discussed in the Geotechnical Baseline Report (GBR). This initial investigation led the Contractor's Team to begin thinking that the ground could potentially behave better than described in the GBR. In order to test this theory, a finite element model (FEM) was developed using the Ground Class properties presented in the GDR and GBR.

Because of the complicated nature of differentiating between 18 different RMT's which included much overlap in the geologic and geotechnical parameters, the Ground Classes were simplified into two alternate Ground Classes which were labeled TS-1 and TS-2.

Alternate ground class TS-1 was developed using statistical analyses of the RMT's included in baseline Ground Classes 1a, 1b, and 2a. Alternate ground class TS-2 was developed using statistical analyses of the RMT's included in baseline Ground Classes 2b, 3a, and 3b.

For simplicity, the ground for the FEM was simulated numerically with a Mohr-Coulomb constitutive model. Mohr-Coulomb strength parameters include cohesion (c), internal angle of friction of the rock mass (φ), modulus of elastic deformation of the rock mass ($E_{rm}$), and Poisson's ratio (υ). Poisson's ratio was based on values given in the GBR. The calculation used to develop the values for c, φ, and $E_{rm}$ was introduced by Hoek and Brown and determines Mohr-Coulomb strength parameters from the Hoek-Brown failure criterion via a series of UCS test values, simulating field scale tests, and a statistical curve fitting process (Hoek et al., 1998, Hoek et al., 2002, and Hoek, 2007). The values for c, φ, and $E_{rm}$ were calculated using data presented in the GBR and GDR.

These FEM numerical analyses were performed with Plaxis 2D, with a two-dimensional, plane strain configuration. Plane strain analyses are used to model problems with consistent geometry and material properties along one axis. The two-dimensional analysis was used for simplicity relative to three-dimensional modeling and because it is conservative in relation to stresses in the lining. Three-dimensional modeling would allow stresses in the rock mass to be redistributed ahead of and around the excavation. Because this isn't possible in a two-dimensional analysis, more stress is carried by the lining.

The staging of the FEM calculation steps also had a conservative effect. At each stage of the numerical tunnel excavation, the lining segment was installed prior to the excavation. This "wishing" the lining in place, while obviously not physically possible, conservatively results in attracting more stress to the lining by ignoring real world displacements that occur before the lining is installed and reducing the arching effect that would redistribute stresses in the rock mass around the tunnel opening.

The results of these analyses are shown in the moment-thrust diagrams below, Figures 1 and 2. Figure 1 shows the moment-thrust diagram for the TS-1 ground class. Figure 2 shows the moment-thrust diagram for the TS-2 ground class. Both Figures 1 and 2 show the allowable thrust-moment interaction in the shotcrete for each calculation phase and each lining segment (due to symmetry in the lining, all moments are shown as positive), including plastic deformation of the shotcrete. The minimum compressive and tensile strength required for all of the moment-thrust combination for 8 in. of fiber reinforced shotcrete are 600 psi and 96 psi, respectively. This is a factor of safety of 2.5 relative to the early strength shotcrete and a factor of safety of approximately 8 relative to the 28-day minimum specified strength.

Additionally, to check the ground properties against the ground behavior assumptions described in the GBR, the tunnel was modeled as a bald excavation. If the excavation was performed in alternate ground class TS-1 without installing the lining, a thin layer of relatively widely spaced, sporadically distributed plastic points develops in the rock mass around the bench excavation. This layer reached a maximum thickness of approximately 0.5 meters. The lack of a thick, dense, uniform zone of plastic points in the rock mass confirmed that the primary ground behavior was "block failure," consistent with behavior defined for Ground Classes 1a and 1b.

This same procedure was used to check the ground behavior assumptions for alternate ground class TS-2. The simulated unsupported excavation resulted in a uniform layer of plastic points developed around the bench excavation which extended beyond the excavation perimeter approximately 2.5 meters. This size zone of plastic points indicates that the primary ground behavior is consistent with "shallow shear failure" as defined in the project documents.

The FEM analyses did not show ground behavior that would have been consistent with "deep shear failure" as defined in the project documents, which further led the Contractor Team to believe that the ground would behave better than predicted in the GBR.

Tables 3 and 4 present the FEM rock mass input data for the alternate ground classes TS-1 and TS-2, respectively.

## AS-BUILT GROUND CLASSES

Geologic mapping was required by the project specifications for the purpose of selecting and installing the prescribed initial support in a systematic manner. Both the Contractor and the Owner provided teams of experienced geologists to perform tunnel mapping. The Contractor maps for each round comprised daily submittals that formed the basis of Support Category selection.

Field mapping of each excavated round allowed the geologists the opportunity to observe ground

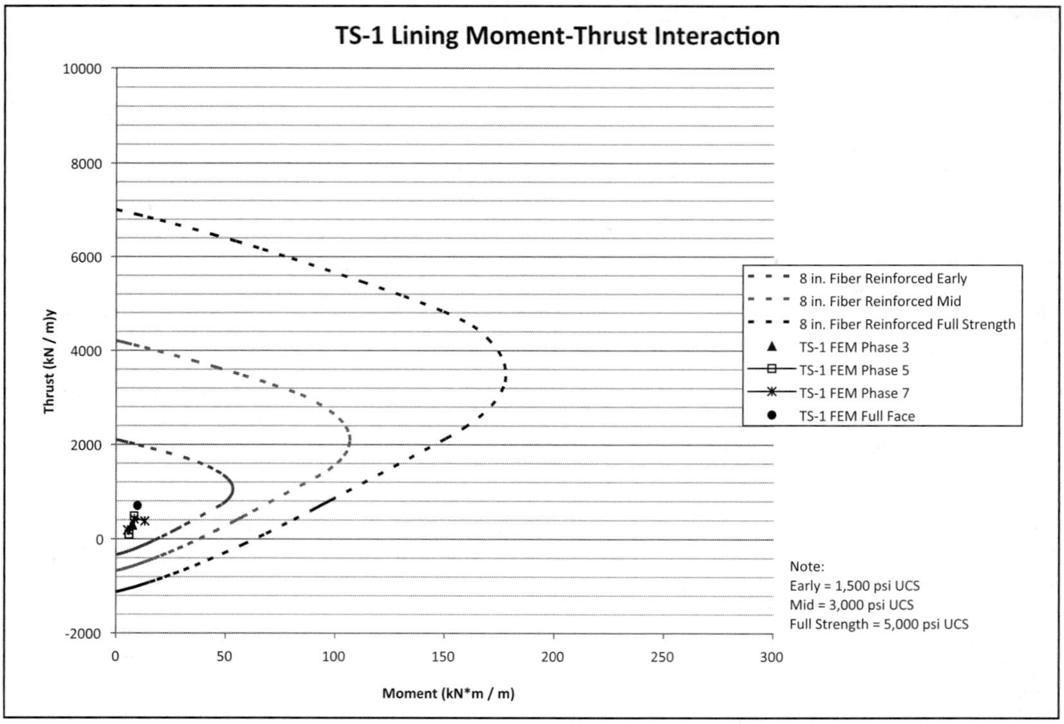


**Figure 1. Moment-thrust for alternative ground class TS- 1**

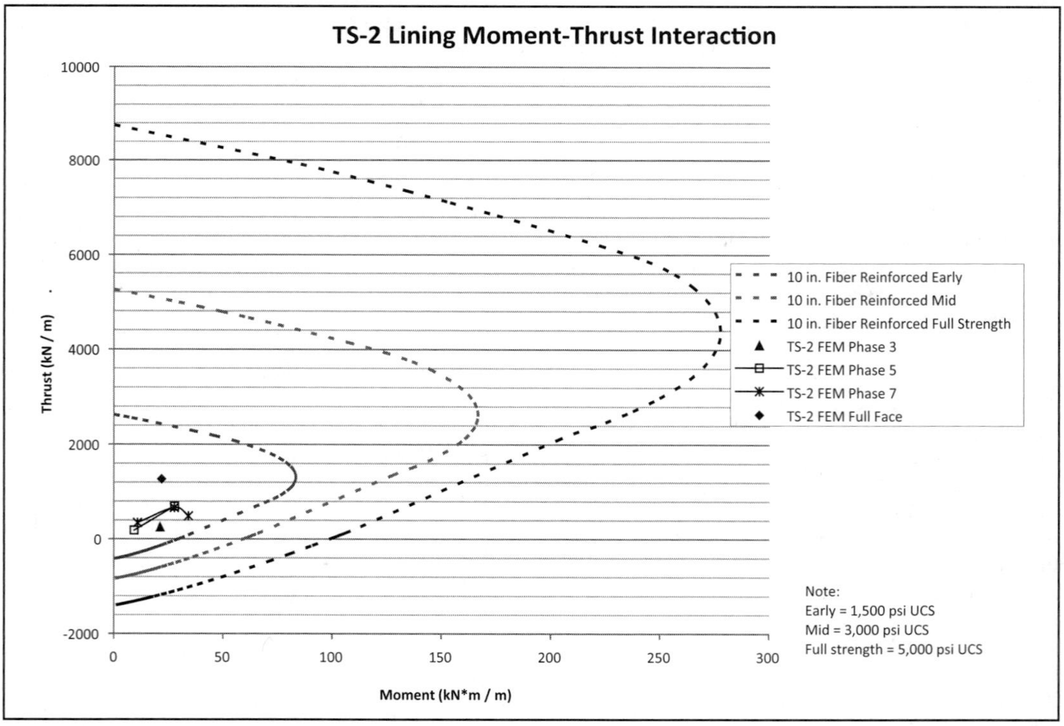


**Figure 2. Moment-thrust for alternative ground class TS-2**

**Table 3. Original TS-1 FEM parameters**

| Data | Value |
|---|---|
| UCS | 6.6 MPa |
| $m_i$ | 7.5 |
| GSI | 53 |
| Unit wt. | 22.5 kN/m$^3$ |
| **Resulting FEM Input** | |
| C | 0.283 MPa |
| Φ | 30° |
| $E_{rm}$ | 3,054 MPa |

**Table 4. Original TS-2 FEM parameters**

| Data | Value |
|---|---|
| UCS | 5.9 MPa |
| $m_i$ | 11 |
| GSI | 36 |
| Unit wt. | 21.7 kN/m$^3$ |
| **Resulting FEM Input** | |
| C | 0.193 MPa |
| Φ | 29° |
| $E_{rm}$ | 1,082 MPa |

**Table 5. As-built ground class distribution**

<table>
<tr><th>Ground Class</th><th>Primary Ground Behavior</th><th colspan="2">Bid Tunnel Length (%)</th><th>Mapped Tunnel Length (%)</th></tr>
<tr><td>1a</td><td>Block failure</td><td>9</td><td rowspan="2">27</td><td rowspan="2">36</td></tr>
<tr><td>1b</td><td>Block failure</td><td>18</td></tr>
<tr><td>2a</td><td>Shallow shear</td><td colspan="2">4</td><td>36</td></tr>
<tr><td>2b</td><td>Shallow shear</td><td colspan="2">38</td><td>23</td></tr>
<tr><td>3a</td><td>Deep shear</td><td colspan="2">22</td><td>0</td></tr>
<tr><td>3b</td><td>Deep shear</td><td colspan="2">4</td><td>0</td></tr>
<tr><td>4</td><td>Crown instability</td><td colspan="2">5</td><td>5</td></tr>
<tr><td>Total</td><td></td><td colspan="2">100</td><td>100</td></tr>
</table>

behavior during and immediately after excavation. The data collected for each round of excavation included detailed rock mass and discontinuity data. Additional mapping information included groundwater and rock mass behavior observations. Primary behaviors defined in the GBR included block failure, shallow shear, deep shear, and crown instability (corresponding respectively to major Ground Classes 1, 2, 3, and 4). Secondary behaviors included raveling, slaking, swelling and softening. Following the data collection and interpretation, the geologist finalized each map by recommending a prevailing Ground Class, which in turn determined the recommended Support Category.

Rock mass conditions observed during excavation were often better than anticipated by the GBR and the design, and often led to classifications at least one Ground Class better than anticipated in the design distribution. This, in turn, led to installation of more Support Categories for "better" ground than the basis of bidding. For example, where the design anticipated Ground Class 2b, actual conditions were often classified in the field as Ground Class 2a, or even 1a or 1b. This example proved to be particularly important because Support Category IIA had no spiles, except as additional measures and Support Category IIB had a full canopy of spiles for standard pre-support (Harvey, et al., 2012).

Additionally, although some limited Support Category III was installed at the direction of the Owner, observations made by the Contractor's geologists of the rock mass conditions and behavior during excavation never led to a classification of Ground Class 3. In total quantities, the entire Support Category III was essentially unused and replaced by Support Category IIA. Examples of the better-than-predicted rock mass conditions and behavior occurred all along the tunnel in every geologic formation encountered on the project, except for the First Shale. The First Shale was poor quality rock with minimal standup time requiring Support Category IV, as designed.

Table 5 summarizes the Ground Classes for the project and compares the anticipated conditions to the actual conditions encountered during construction as mapped by the Contractor's geologists (Harvey, et al., 2012).

The initial support consisting of various round lengths, shotcrete thicknesses, lattice girders, and rock dowels generally performed very well. Initial support performance was measured by systematic installation of radial optical survey convergence. The contract also included multiple point borehole extensometers (MPBX) at each portal to measure performance of the pillar between Bore No. 3 and Bore No. 4. Systematic MPBX installations to attempt to measure shallow and deep shear were ultimately abandoned and interpretation of the survey data was the main means to evaluate initial support performance.

Performance of the initial lining was measured using optical survey monitoring arrays. Monitoring arrays were required to be installed and initial readings taken within six hours of completion of excavation and initial support of the first advance length behind the face. Measured deformations occurring during top heading excavation typically ranged from about 10 mm to 30 mm, not including deformations that naturally occur during excavation and support prior to it being feasible to install the monitoring arrays.

Instances where measured deformation exceeded warning levels, which ranged from 20 mm to 50 mm, were extremely rare and higher alarm levels were never reached. This confirms that the initial support installed performed as intended for the ground conditions encountered.

## AS-BUILT BACK ANALYSES

During construction, there was much debate between the Contractor and the Owner about the observed behavior of the rock mass and measured deformation of the initial lining and whether shallow shear failure or deep shear failure was occurring in the rock mass. Shallow shear failure and deep shear failure behaviors were defined in the GBR and are presented above in Table I. However, MPBX's that were included in the contract were not long enough to adequately allow for measurement of the deepest extents of the deep shear zone. The Contractor proposed to install MPBX's and shotcrete strain gauges on a number of occasions where there was disagreement with the Owner about the observed behavior of the rock mass. These proposals were turned down by the Owner each time.

For our own research, Brierley has performed FEM back-analyses using the as-built data for two different locations which were mapped as Ground Class 2a and 2b by the Contractor. These locations correspond to approximately 575 linear meters ("tunnel meter 575") and approximately 777 linear meters ("tunnel meter 777") west of the east portal.

The baseline Ground Class for tunnel meter 575 was 55% Ground Class 3a/3b, 45% Ground Class 2b indicating the predominant primary behavior was expected to be deep shear failure and, to a lesser extent, shallow shear failure. The Contractor's geologists mapped this location as Ground Class 2b and the resulting initial support scheme installed was Support Category IIB. The maximum radial deformation measured during the top heading excavation was approximately 12 millimeters as shown on Figure 3. Additional deformation of approximately 8 to 10 millimeters occurred as the bench was excavated in this location.

The baseline Ground Class for tunnel meter 777 was Ground Class 3a indicating an expected primary behavior of deep shear failure. The Contractor's geologists mapped this location as Ground Class 2b. Because of raveling behavior in this location, a hybrid support scheme was installed based on certain Support Category IIB and certain Support Category IIIA initial support elements which included 305 mm of shotcrete. As shown on Figure 4, little to no deformation was measured.

It can be seen that deformation did not exceed the warning level, confirming that the initial support installed performed as intended for the ground encountered, which ranged from one to two Ground Classes better than anticipated by the baseline.

The FEM back-analyses were performed using the as-built data and baseline data for tunnel meter 575 and tunnel meter 777. The as-built GSI data for tunnel meter 571.6 to tunnel meter 578.6 ranged from 25 to 40. A GSI value of 32 was used in the FEM back analyses. The as-built GSI data for tunnel meter 773.2 to tunnel meter 779.1 ranged from 15 to 40. A GSI value of 29 was used in the FEM back analyses. Baseline values for rock mass type Tc-2 were used for tunnel meter 575 and baseline values for rock mass type Tcp were used for tunnel meter 777. Additional FEM input values are presented in Tables 6 and 7.

There was also some discussion during construction that the use by the contractor of higher strength than specified shotcrete allowed the use of lower than designed initial Support Categories. To investigate this possibility, we have performed the FEM back-analyses with both the specified and as-built shotcrete strengths. The as-built shotcrete strength data is shown in Table 8. The plastic zones resulting from the FEM back-analyses are shown in Figures 5 and 6 for tunnel meter 575 and tunnel meter 777. These figures show the plastic zones that develop when using the design strength shotcrete in the FEM. Adjusting the shotcrete strength using the as-built shotcrete data did not result in a significantly different plastic zone in either model.

The resulting plastic zone for tunnel meter 575, which occurs in the lower corner of the tunnel perimeter, extends approximately 4 meters into the rock mass. As previously shown, the measured radial convergence for tunnel meter 575 did not approach the warning level and is considered minimal. This is consistent with the definition provided for shallow shear failure (refer to Table 1).

The resulting plastic zones for tunnel meter 777 occur in the crown and in the lower corner of the tunnel perimeter. The plastic zone in the crown extends approximately 4 meters into the rock mass and the plastic zone in the lower corner extends approximately 8 to 10 meters into the rock mass. While 10 meters extends just beyond the definition for shallow shear of 0.25D to 0.5D, the measured convergence at tunnel meter 777 was negligible. Therefore, it can be concluded that the observed behavior is consistent with the definition provided for shallow shear failure.

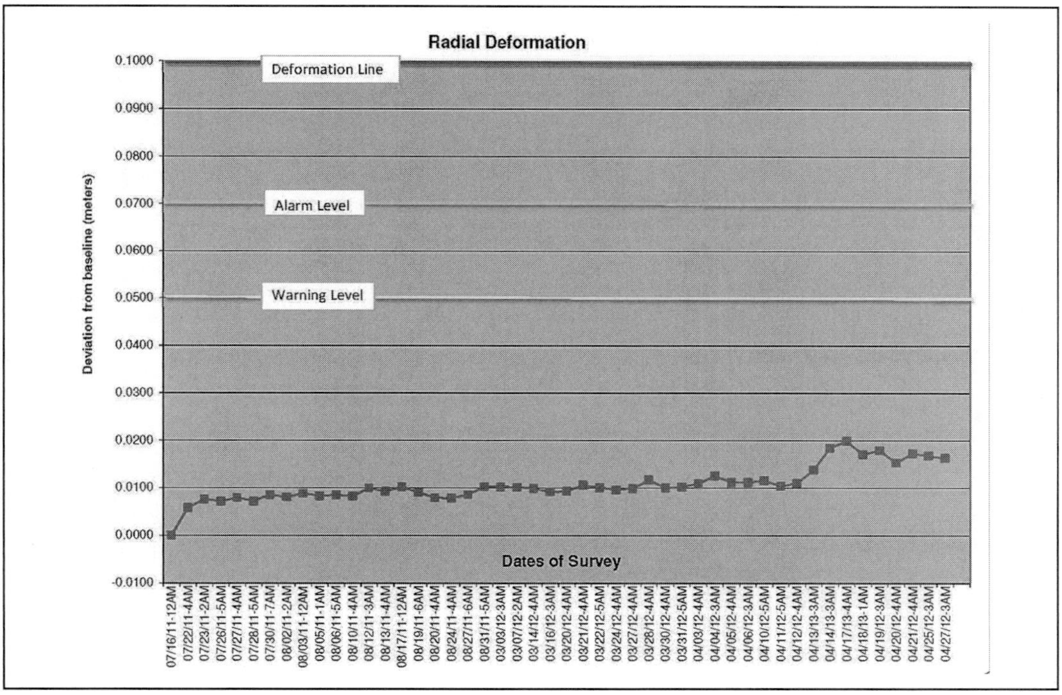


**Figure 3. Radial deformation, tunnel meter 575**

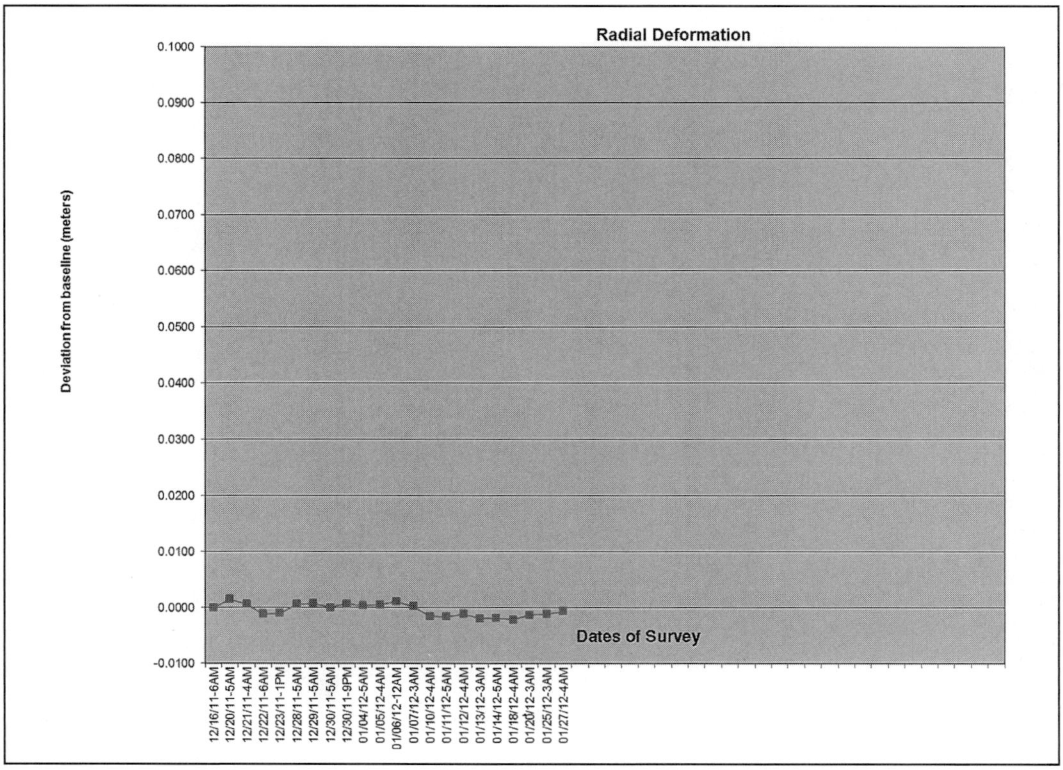


**Figure 4. Radial deformation, tunnel meter 777**

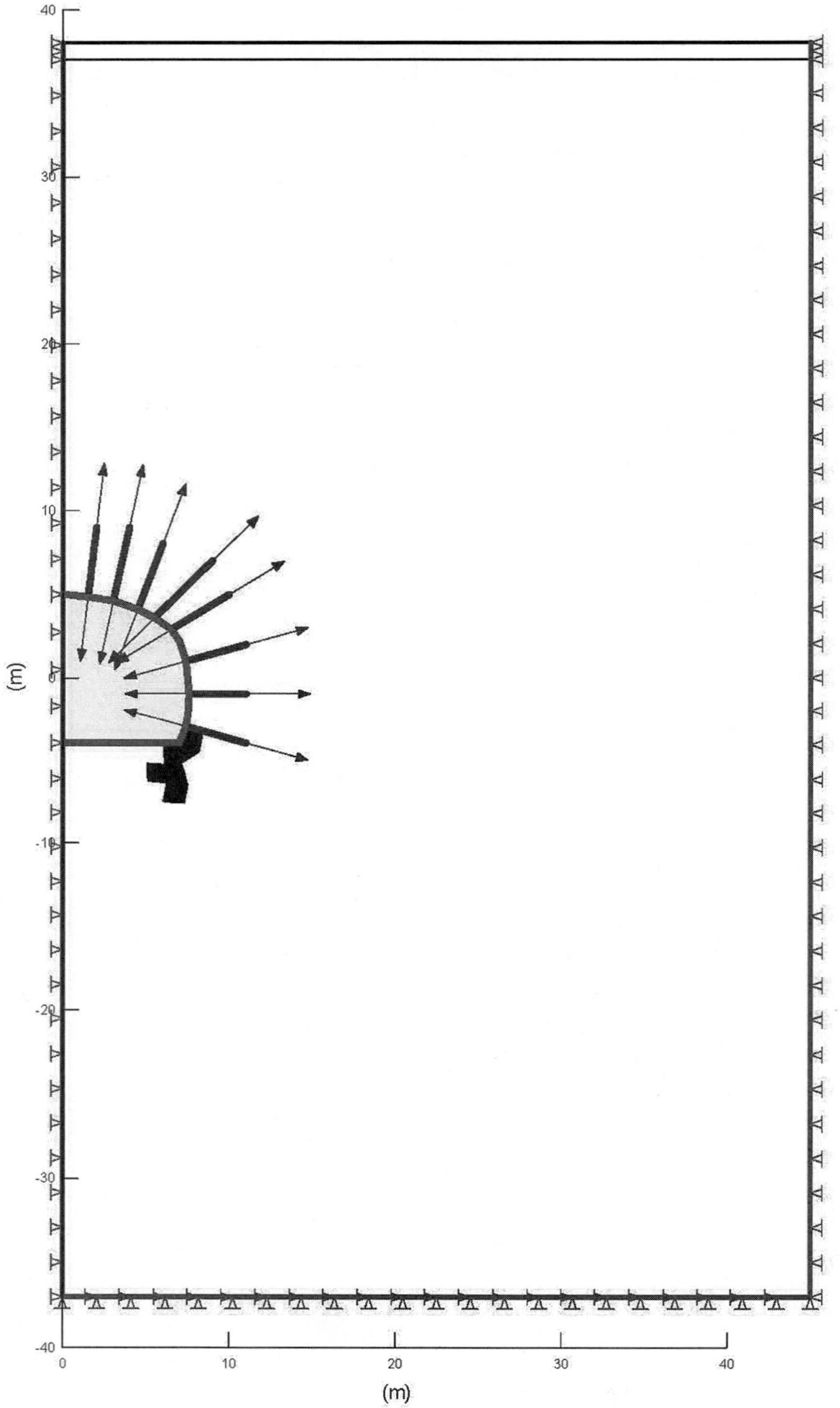


**Figure 5. Tunnel meter 575 FEM model showing plastic zone**

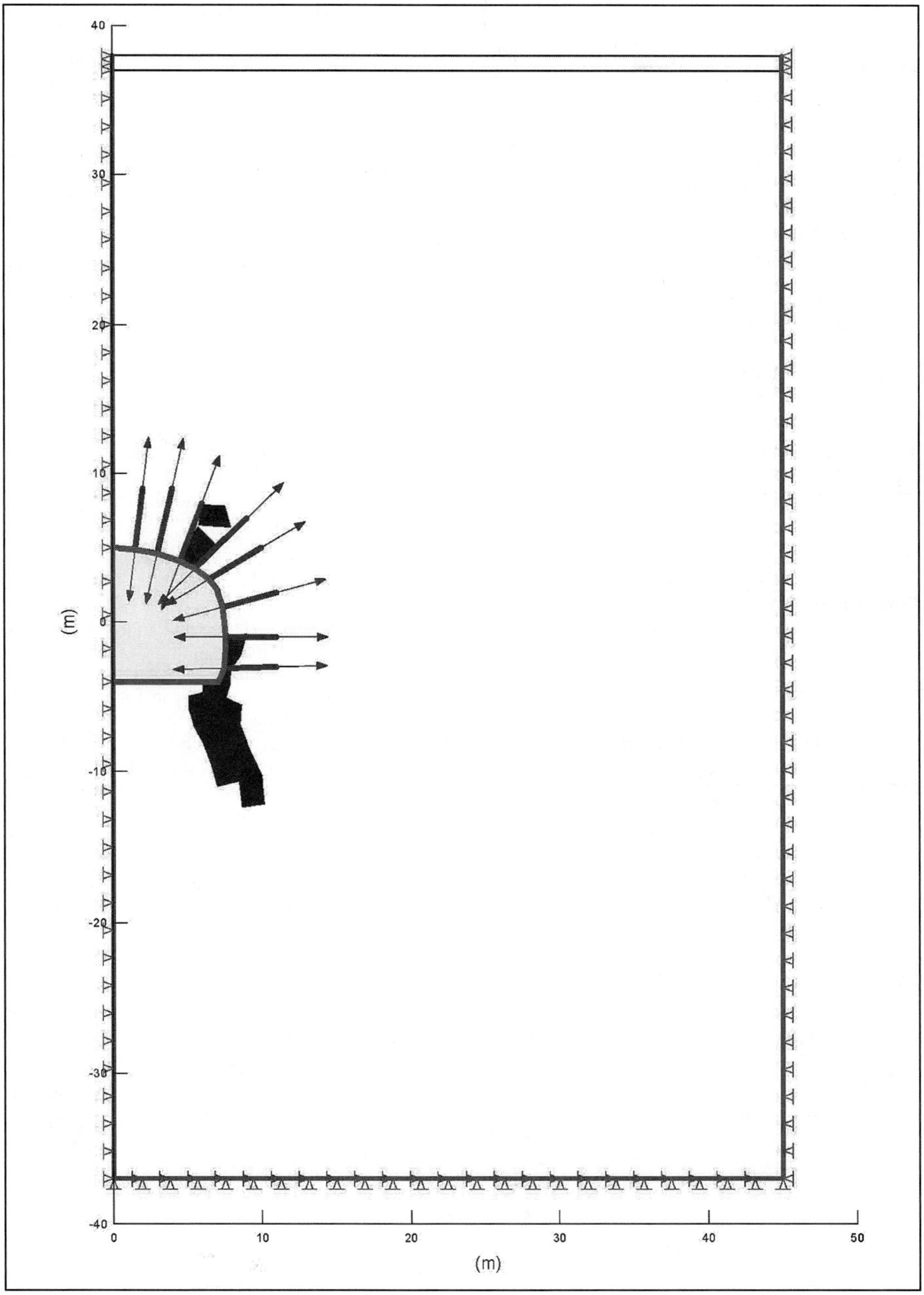


**Figure 6. Tunnel meter 777 FEM model showing plastic zones**

## CONCLUSIONS

Four conclusions can be drawn from the FEM back analyses:

1. The Contractor Team's initial estimate of ground class and ground behavior is validated in the as-built Ground Class distribution as well as the virtual elimination of Support Category III;
2. Ground behavior at locations that were baselined to behave as deep shear failure instead appears to be consistent with shallow shear failure, as observed during construction and verified through FEM back-analyses;
3. Using FEM input data that most closely resembles the rock mass (versus applying safety factors to the rock mass data) leads to good correlation between anticipated ground behavior and actual ground behavior; and
4. The difference between specified and as-built shotcrete strength has very little impact on ground behavior.

**Table 6. Tunnel meter 575 FEM parameters**

| Data | Value |
|---|---|
| UCS | 21.7 MPa |
| $m_i$ | 8 |
| GSI | 32 |
| Unit Wt. | 24.3 kN/m$^3$ |
| **Resulting FEM Input** | |
| C | 0.648 MPa |
| Φ | 27° |
| $E_{rm}$ | 1,655 MPa |

**Table 7. Tunnel meter 777 FEM parameters**

| Data | Value |
|---|---|
| UCS | 17.2 MPa |
| $m_i$ | 7.5 |
| GSI | 29 |
| Unit Wt. | 24.3 kN/m$^3$ |
| **Resulting FEM Input** | |
| C | 0.407 MPa |
| Φ | 27° |
| $E_{rm}$ | 1,241 MPa |

In general, based on mapped ground conditions, installed initial support, measured convergence, and back analyses of these variables, the Contractor Team's pre-bid estimation of ground class distribution and ground behavior appears to have been accurate. The original design appears to have been conservative which resulted in a significantly different distribution of Ground Classes than bid due to more favorable geologic conditions and ground behavior encountered during construction. The installed support varied significantly even resulting in the virtual elimination of one major Support Category and significant quantity overruns for others. Tunnel construction was completed with a different distribution of design Support Categories plus a few Owner-directed, new design configurations.

## REFERENCES


Harvey, S.; Howard, A.; Baldwin, J.; Monahan, W.; "Contractor Implementation of SEM Tunneling and Comparison Between Design and Observed Ground Conditions, Caldecott Tunnel Fourth Bore, San Francisco Bay Area, California"; Proceedings, North American Tunneling (NAT) Conference; June 2012.

Thapa, B.; Marcher, T.; McRae, M.; Skovajsova, Z.; and Momenzadeh, M.; "NATM Strategies in the U.S.—Lessons Learned from the Initial Support Design for the Caldecott 4th Bore"; Proceedings, Rapid Excavation and Tunneling Conference (RETC); June 2009.

Thapa, B; Nitschke, A.; Ramirez, I.; McRae, M.; Gall, V.; "Lessons Learned From NATM Design and Construction of the Caldecott Fourth Bore"; Proceedings, Rapid Excavation and Tunneling Conference (RETC); June 2013.


**Table 8. As-built shotcrete test data**

| Date | Station (tunnel meter) | 10 min Needle Avg. (MPa) Required 0.275 | 60 min Needle Avg. (MPa) Required 0.485 | Compression 1 day Avg. (MPa) Required 9.7 | Compression 7 day Avg. (MPa) Required 22.1 | Compression 28 day Avg. (MPa) Required 28.0 |
|---|---|---|---|---|---|---|
| 7/14/11 | 571 | 0.40 | 0.83 | 29.1 | 31.4 | 46.1 |
| 7/19/11 | 578 | 0.38 | 0.79 | 16.7 | 47.1 | 64.4 |
| 10/31/11 | 761 | 0.44 | 0.98 | 34 | 37.6 | 51.5 |

# McCook MTS Bifurcation—A Paradigm of Craftsmanship

**Bryan Hansen**
National Welding Corporation

**Clay Haynes, Faruk Oksuz, and Miguel Sanchez**
Black & Veatch

**Dave Schiemann**
USACE Chicago District

**ABSTRACT:** The Metropolitan Water Reclamation District of Greater Chicago's (MWRDGC) Tunnel and Reservoir Plan (TARP) and the McCook Reservoir will further reduce flood damages and combined sewer overflows (CSOs) for the city of Chicago and Cook County, Illinois. The McCook Reservoir will receive approximately 37.8 billion liters (10 billion gallons) of water via the Mainstream Tunnel System, which connects the TARP Mainstream Tunnel to the McCook Reservoir and the Distribution Tunnel System, which connects the TARP Des Plaines Tunnel to the McCook Reservoir. The U. S. Army Corps of Engineers (USACE) tasked Black & Veatch to design the Main Tunnel System. This paper describes the Main Tunnel System project components; how this final piece fits into the Chicago's TARP program; and provides an update on the construction progress to date with a specific emphasis on the bifurcation liner installation.

## PROJECT DETAILS

The Main Tunnel System design includes a 27.5m (90 ft) diameter and 92m (300 ft) deep Main Gate and tunnel Construction Access Shaft and associated wet-well shaft arrangements to house six high-head and large 4.4m by 9m (14.5 ft by 29.5 ft) wheel gates; a 10m (33 ft) diameter and 490m (1,600 ft) long Main Tunnel in rock including a tunnel bifurcation (with steel and concrete lining) at the Main Gate/Access Shaft; a live tap connection to the existing Mainstream Tunnel; and energy dissipation and portal structures.

The construction of the Main Tunnel System has been divided into the Main Gate/Access Shaft contract and the Main Tunnel System contract. Construction of the Main Gate/Access Shaft was completed in August 2011 and the Main Tunnel System construction is currently ongoing. Construction progress updates and discussion of key construction issues to date are presented in this paper. Kiewit Infrastructure Co. was awarded the construction of the CUP McCook Main Tunnel System with a Notice to Proceed on January 1, 2012.

## HISTORY

The MWRDGC has been dealing with CSOs and flooding problems since the late 1960s and formally adopted the Tunnel and Reservoir Plan (TARP) in 1972. Phase I of TARP, which included construction of 175 km (109 miles) of deep storage and conveyance tunnels with diameters up to 10m (33 ft), was completed in 2006. Phase I resulted in substantial improvements in surface water quality enhancing the Chicago riverfront. Additional improvements are expected as Phase II comes on-line, including three large reservoir systems, as shown in Figure 1.

Phase II includes a series of storage reservoirs to increase flood storage capacity and further reduce CSO discharges, with additional storage capacity projected to come on-line over the next several years.

## MCCOOK RESERVOIR OVERVIEW

Authorized in the Water Resources Development Act of 1999, the McCook Reservoir Project is a key component of Chicago's ongoing TARP Project. The McCook Reservoir will provide approximately 37.8 billion liters (10 billion gallons) of additional CSO and flood water storage for TARP. The reservoir will store excess CSO and floodwater from TARP's Mainstream and Des Plaines deep tunnel systems during periods of wet-weather peak flows. This stored volume will be pumped to the MWRDGC's Stickney Wastewater Treatment Plant (WWTP) for treatment prior to discharge to Des Plaines River.

The McCook Reservoir is currently under construction and being excavated in dolomite limestone. An aerial image of the reservoir is shown in Figure 2.

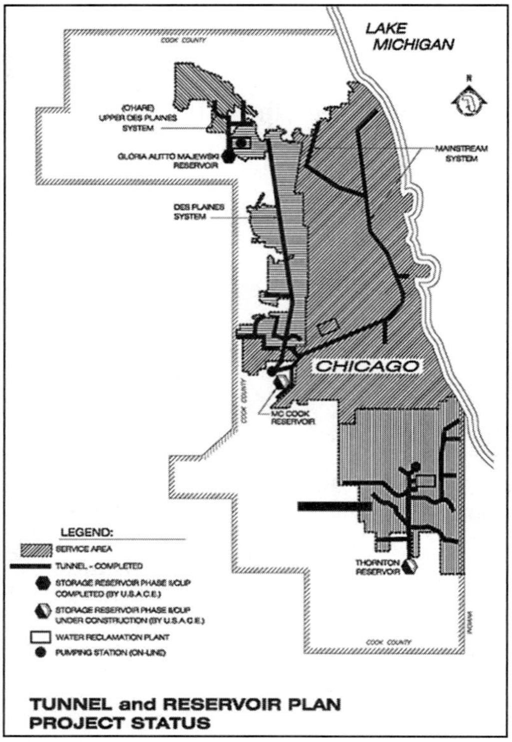


**Figure 1. A schematic layout of Chicago's TARP (Source: MWRDGC)**

The reservoir walls are nearly vertical (Figure 3) and excavated to depths up to 107m (350 ft) below grade. The McCook Reservoir development as part of the Phase II of TARP includes tunnels for TARP connections servicing the Des Plaines and Mainstream systems. The subject of this paper is the McCook Main Tunnel System (MTS) connecting the TARP Mainstream tunnel to the McCook Reservoir.

## OVERVIEW OF THE MTS PROJECT

The MTS tunnel is the primary inlet for CSOs and floodwater from the TARP tunnels into the McCook Reservoir. The tunnel will be concrete lined for long-term stability and to minimize infiltration and exfiltration. The MTS extends from the existing Mainstream Tunnel through the gate shaft and connects to the McCook Reservoir. The MTS is approximately 490m (1,600 ft) long with a finished inside diameter of 10m (33 ft). The MTS will be bifurcated through the gate shaft. Kiewit is using drill-and-blast construction methods on this relatively short tunnel. A rock plug will be left in place in the MTS until the installation of gates and related structures have been completed and the reservoir is ready to receive flows. The MTS is characterized by the following components:

- Main Tunnel—an approximately 490m (1,600 ft) long, 10m (33 ft) inside diameter

**Figure 2. An aerial view of McCook Reservoir and quarry**

(ID) drill-blast tunnel connecting the existing Mainstream Tunnel and the McCook Reservoir, bifurcated for approximately 88m (290 ft) through the Main Gate/Access Shaft.

- Main Gate/Access Shaft—a 27m (88 ft) ID, 90m (295 ft) below-grade circular shaft located near the Main Tunnel midpoint to house the gate system. The outer shell of the shaft has been designed and constructed under a separate contract by the USACE. This shaft will be used for construction of the MTS and eventually will contain the high-head wheel gates for controlling flow between the TARP Mainstream Tunnel and McCook Reservoir.
- Construction Shaft (contractor option)—an optional, 7.6-m (25 ft) ID, approximately 87m (285 ft) below-grade shaft to be located at approximately 91m (300 ft) downstream of the Mainstream Tunnel Connection. Kiewit elected to build the construction shaft. The construction shaft is not a requirement for operation of the system.

**Figure 3. McCook Reservoir and limestone quarry**

- Gates—installation of six rectangular, wheel gates and the associated gate controls. Each bifurcation of the Main Tunnel contains one main gate and two guard gates–one upstream and one downstream of the main gate. The gates and associated gate hydraulic cylinders and controls were manufactured under a separate contract and will be provided to Kiewit as government furnished items. The gates were designed by Black & Veatch and were fabricated by Oregon Iron Works.
- Main Tunnel/Mainstream Tunnel Connection—the live connection from the MTS to the existing Mainstream Tunnel. The connection geometry was analyzed and evaluated to minimize potential turbulence and cavitation using computational fluid dynamics (CFD).
- MTS/McCook Reservoir Connection—the Main Tunnel portal connection to the McCook Reservoir, including the construction of an Energy Dissipation Structure. The portal will be excavated in rock, with long-term support provided by rock bolts and shotcrete. Hydraulic structures have been designed to minimize erosion during reservoir filling and emptying.
- Control Building—a surface facility to house gate operating controls, hydraulic power units and provide limited storage (Figure 4).

The MTS design, construction, operation, and commissioning will be coordinated with the overall McCook Reservoir water control plan, as well as the reservoir excavation, high wall stabilization, groundwater protection system construction, and Distribution Tunnel connections.

The MTS includes a live connection to the Mainstream Tunnel. Operating Mainstream Tunnel

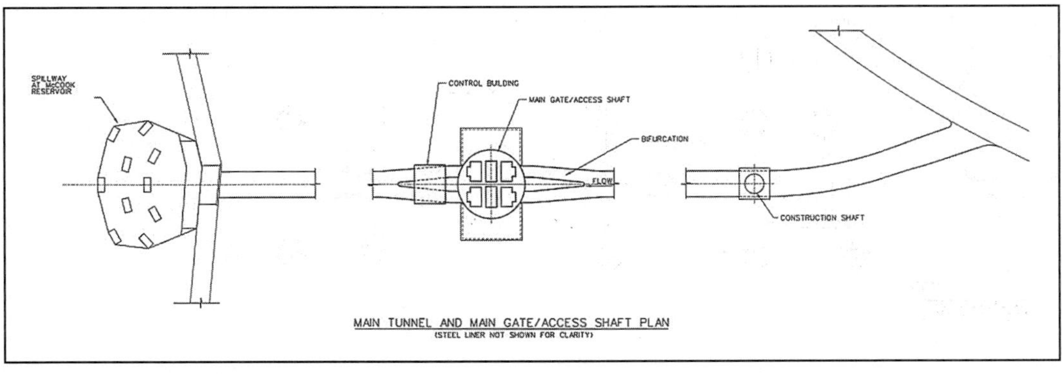


**Figure 4. Plan view of project components**

Figure 5. Main gate shaft during excavation

Figure 6. Main gate shaft at final depth

disruptions will be minimized as part of the live connection, construction planning and all other MTS facilities must be completed and ready to receive CSO water before the connection can be completed. This connection will be one of the more challenging aspects of the construction project.

### Future Components of McCook Reservoir

Reservoir excavations, distribution tunnel connections, and the final reservoir preparation work is now under design. Rock excavated from the reservoir is hauled to a nearby quarry and the market demand for aggregate influences the rate of reservoir excavations. Final reservoir preparation will include removal of rock plugs between the distribution tunnel and McCook Reservoir, installation of inlet/outlet works, completion of ongoing grout curtain installation around the reservoir perimeter, and reservoir slope stabilization.

## MAIN GATE SHAFT

The excavation and initial support of the Main Gate Shaft was designed by USACE and constructed by McHugh Construction with Notice to Proceed in October 2009 and construction completion in August 2011. Figure 5 and Figure 6 demonstrate various stages of construction of the shaft.

The design of the gates and operating cylinders was performed by Black & Veatch under a separate contract and the component fabrication is complete. The gate shaft will have a 27m (88 ft) finished diameter and a 0.9m (3 ft) thick concrete liner to a depth of approximately 73m (240 ft).

The shaft will house the gates and all the components necessary to operate the gates. At the base of the shaft, the Main Tunnel will be split into two sections (bifurcation) so that the flow can be regulated into one or both of the bifurcations. The flow in each bifurcation will be regulated by a set of three gates, one main gate and two guard gates (total of six gates). Provisions have been made to provide man-basket access to the Main Tunnel in order to perform general maintenance on the gates and gate slots.

### Gate Design

The design includes six gates (i.e., two main gates and four guard gates). The main gates were designed to resist flow in both directions, whereas the guard gates were designed to resist flow in only one direction. Each gate is operated by a hydraulic cylinder and operates in a guide slot by means of wheels. The main gates are 5.49m (18 ft) wide and 9.52m (31.23 ft) in height. The guard gates are 4.98m (16.33 ft) wide and 9.21m (30.23 ft) in height. Each main gate weighs approximately 104 mt (230 kips). Each gate was designed to resist a static load of 79m (260 ft) of hydraulic head. Figure 7 shows one example of the gate analysis deformation results using ANSYS. Figure 8 shows the gates during fabrication.

## BIFURCATION DESIGN CONSIDERATIONS

A bifurcation was included in the design for redundancy and to reduce the required size of the closure gates (Figure 9). By having two conduits, the gate actuators and appurtenances could be maintained and repaired in a shut position while the other conduit's gates could be operated in accordance with the operational needs of the MTS. However, scheduled maintenance and repair would be planned in the dry season because the full hydraulic capacity of the MTS is not realized with one side of the bifurcation out of service. In summary, both sides of the bifurcation are needed to meet the operational needs of the MTS.

The bifurcation was steel lined for two primary reasons. The first reason was that the steel gates could not make a seal against reinforced concrete. Secondly, excessive velocity gradients develop in the

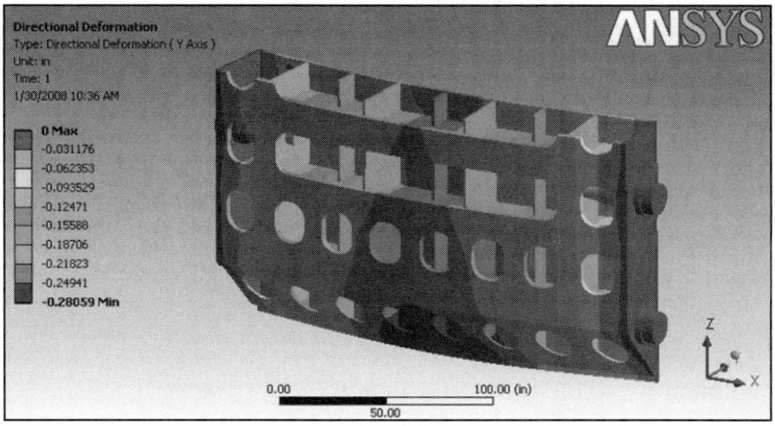


**Figure 7. Gate ANSYS Model (created from 3-D CAD file)**

**Figure 8. Gate fabrication in progress**

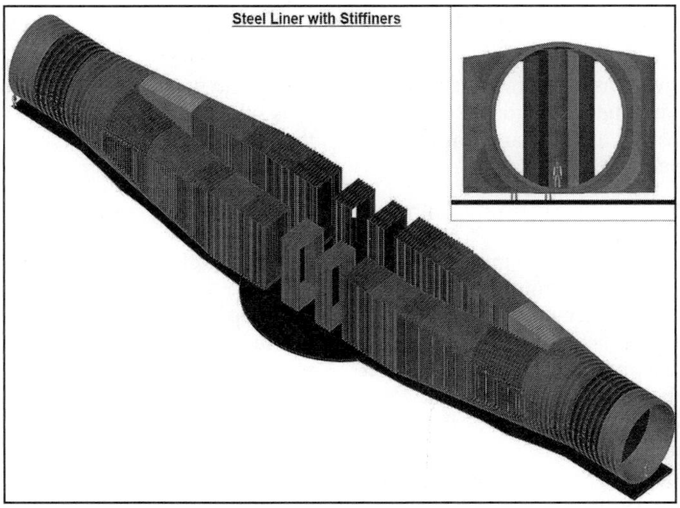


**Figure 9. Rendering of the bifurcation**

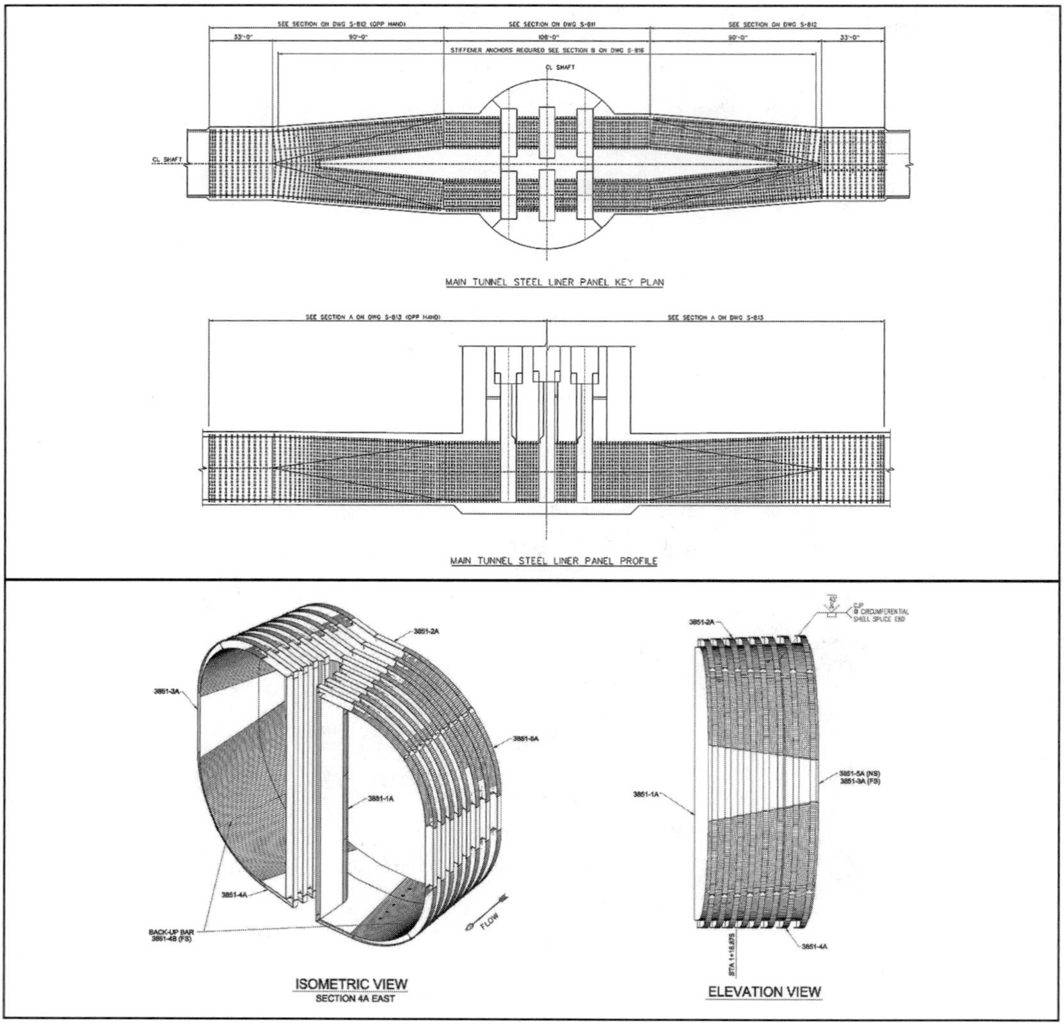


**Figure 10. Tunnel liner plan, section and isometric view**

bifurcation as the flow transitions from a single circular section to two rectangular sections. Reinforced concrete cannot withstand the wear and tear of these high velocity gradients.

## BIFURCATION DETAILS

Approximately 108m (354 ft) of the tunnel includes a steel liner which bifurcates from a 4m × 9.8m (19 ft × 32 ft) rectangular cross section to a 10m (33 ft) diameter round section. The bifurcation legs were designed to accommodate six high-head wheel gates located at the bottom of the shaft. These gates serve as flow control and are an integral feature of the McCook Reservoir system. The body of the liner utilizes "T" Steel rings wrapped around the liner shell on (0.4m–0.9m (16 in.–36 in.) centers which allows the use of a relatively thin liner shell thickness of 19mm (0.75") considering the diameter of these components. This system was designed by Black and Veatch (Figure 10).

## THE CONSTRUCTION TEAM

National Welding Corporation was responsible to assemble, fit and weld the tunnel liner as a subcontractor to Kiewit Infrastructure. Kiewit Infrastructure was the General Contractor of the overall project and self-performed most of the other key project elements including tunnel excavation, tunnel concrete lining and oversight of all other activities. Selway Corporation prepared the shop drawings, performed all the shop fabrication, and shipping to the project site.

## PREASSEMBLY PLANNING

As these large sections could not be fabricated in shippable sized rings, the components to the 10m (33 ft) diameter sections and bifurcations were required to be fabricated in transportable sized pieces. An initial task for the construction team was to maximize the piece sizes which would minimize the onsite assembly requirements. The team decided the 10m (33 ft) round sections would be most manageable in quarter round sections which would be feasible to ship but still require oversized permits (Figure 11).

Prior to shipping, the manufacturer was required to preassemble all sections into complete rings, then connect the girth joints. This was performed as a precaution to minimize any onsite rework (Figure 12). The construction team elected to use bolted connections for the preassembly which could be reused on site to expedite the initial subassembly. These quarter sections were transported from Montana to Chicago, Illinois using 48 specially permitted loads.

The quarter sections were received onsite by Kiewit Construction and delivered to National Welding Corporation for onsite subassembly, fitting, and welding. Subassembly proved to be challenging with large pieces, inclement weather, and the changing geometry of the liner sections. We concluded each ring assembly would be built on the surface in the same manner as if manufacturing a tank to minimize the difficult handling of such heavy pieces and awkward shapes in a tunnel or in a horizontal orientation. The quarter round sections were initially erected on a concrete pad which was within the crane radius range (Figure 13).

The rigging and section support was all custom designed for each pair of tunnel liner sections as the liner geometry was constantly changing. Wind conditions and weather effected surface subassembly but the bolted connections used for the subassembly

Figure 11. Quarter sections of 10m (33 ft) diameter liner

Figure 12. Preassembly of sections

**Figure 13. Assembly of sections**

**Figure 14. Rigging and connections**

**Figure 15. Bracing subassembly and shelter**

minimized this effect as there was no need for preheating welds and the connection time was expedited (Figure 14).

Large cross-bracing was installed during the initial subassembly to maintain liner section shape. The assembled liner sections were then transported by rollers into a temperature-controlled temporary fabrication facility for fitting and welding of the longitudinal joints and other attachments. The temporary facility provided the environmental controls needed for fitting and welding longitudinal seams and attachments during the cold Chicago winter. The liner shell was manufactured with ASTM A537 Class 2 which is a high strength material mandating controlled preheat and interpass temperatures which required the shelter's controlled environment to assure meeting those needs (Figure 15).

Once in the shelter, Fitting of the liner seams was accomplished using tank tools and hydraulic rams to bring the sections into alignment and conformance with the project specifications (Figure 16). The root openings of the axial seams were carefully controlled to avoid irregularities in the liner diameter and assure meeting the Ultrasonic Testing (UT) requirements. The roundness of the as-fit round section was measured and documented prior to any permanent welding as the 13mm (½ in. on radius and 0.8mm (1/32 in.) deviation at the joints (Figure 16).

The weldment location was preheated as dictated by the submittals and FCAW (Flux Cored Arc Welding) Welding Procedure Specification (WPS) (Figure 17) utilizing Induction Heating methods. The FCAW process allowed maximizing the welding production while maintaining a predictable high quality of the finished work product. The production welding of the axial seams was then started utilizing automatic and semi-automatic welding methods (Figure 17).

**Figure 16. Shelter, hydraulic fit-up, and documentation**

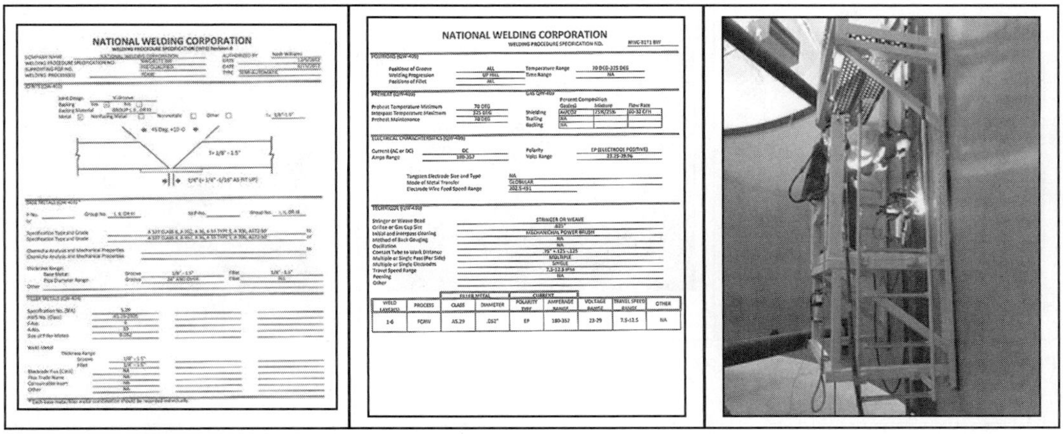


**Figure 17. WPS, longitudinal seam weld in process**

**Figure 18. "J" anchor welding and anchor layout**

All welded liner seams required complete joint penetration (CJP) butt welds. Weld integrity and quality was verified by using phased array Ultrasonic Testing (UT) of 100% of longitudinal welds and 50% of circumferential girth seams. Fillet welds were used for some fixtures and temporary attachments and received Magnetic Particle (MT) inspection.

After completion of the subassembly in the shelter, many of the sections utilized "J" shaped anchors on the outside of the piece to secure the finished liner to the concrete backfill after installation in the tunnel (Figure 18). Over 16,000 of these anchors were field installed on this project (Figure 18). The large anchor count was primarily due to the sections sizes which became too large for shipping with the anchors attached at by the fabricator. The anchor weldment was a full perimeter fillet weld.

The most challenging pieces to assemble were the approach to the bifurcations and the actual bifurcation sections. These sections were very complicated shapes to accommodate the changing geometry of each ring assembly (Figure 19).

After completion of the subassembly the pieces were then transported out of the shelter and staged for installation. Sections were stockpiled at the tunnel location where the liner section was prepared and

**Figure 19. Bifurcation and bull nose assembly**

**Figure 20. Bifurcation and bull nose assembly**

**Figure 21. Liner section rigging**

rigged up for a 90 degree rotation to the horizontal orientation. This maneuver was particularly difficult as the weight of some sub assembled pieces was in excess of 164 mt (180 tons) (Figure 20). Support legs were then attached to support the new orientation and the piece was readied for the tunnel installation.

The liner sections were then rigged for picking and lowering down the shaft (Figure 21). The tunnel crew along with the surface crew spotted the clearances as the pieces were lowered.

Sections were lowered approximately 91m (300 ft) down the shaft to the tunnel floor where a roller mechanism was utilized to transport the liner section into the tunnel (Figure 22). The liner section elevation, clocking, and orientation were carefully verified prior to securing to assure the next adjoining section would fit correctly. This operation was meticulously performed due to the consequence of error.

The liner was placed into the final tunnel location and support legs were secured to the tunnel floor (Figure 22). The subsequent liner sections could then be installed and this sequence continued. Scaffolding had to be custom designed and was installed at this point to facilitate accessing the joints, performing fit-up, and circumferential seam welding. As the liner grew the piece shapes were constantly changing

**Figure 22. Lowering liner sections**

**Figure 23. Support legs and various cross sections**

**Figure 24. Circumferential seam welding and technical welding team**

which made it difficult to utilize a repeatable method for rigging, fitting, scaffolding, and general handling issues (Figure 23).

As with the surface subassembly, after final placement of each liner section, the seams were adjusted to meet the Project Specifications and the location and tolerances were verified and documented. The weldment location was preheated as dictated by the submittals using induction heating methods and the FCAW production welding procedures were utilized. Production welding of the circumferential seams was then started utilizing automatic welding methods by a technical welding team to assure a high quality result (Figure 24).

## CONCLUSION

This tunnel liner was exceptionally difficult to construct due to the constantly changing geometry and unusually large size of the pieces. The key elements to the success of this project were the team approach and careful planning and development prior to and during all construction activities. The owner and managing engineer of the project are of equal importance as team members as their cooperation in timely resolution of design or construction issues was invaluable. The project was fortunate to have multiple companies involved that understood and were attentive to the needs for successful project. Many activities required unique and creative solutions for difficult conditions and time constraints. Project components and methods used can be applied to future projects. The importance of a partnering/team approach with competent members was crucial to the success of the project. These challenges reinforce and justify the benefit of negotiated contracts between parties based on the following priorities: (a) construction capability, (b) experience, (c) key personnel, d) references, (e) financial strength, (f) scheduling, and (g) cost. The proactive cooperation and communications between all parties involved, including the USACE and MWRDGC, the general contractor, Kiewit Infrastructure, the specialty contractor, National Welding Corp, and the engineer, Black and Veatch, contributed significantly to the resolution of challenging problems and have made the project a success.

# NATM Excavation and Support Design and Construction of the Caldecott Fourth Bore

**Axel G. Nitschke and Vojtech Gall**
Gall Zeidler Consultants

**Mike McRae**
Jacobs Associates

**Ivan Ramirez**
Contra Costa Transportation Authority

**ABSTRACT:** The two-lane, 1,036 m long (3,399 ft) Caldecott Fourth Bore opened to traffic in November 2013 and along with the original three Caldecott bores provides a key transportation link between Alameda and Contra Costa counties. Seven cross passages connect the new Fourth Bore to the existing Third Bore. This paper describes the design of the initial ground support and final lining for the tunnel. The paper also documents the construction of the tunnel, including the daily support selection process, organization of the construction support team, examples of predicted versus observed ground behaviors, contractual considerations regarding support selection criteria, and the management of community impacts.

## PROJECT DESCRIPTION

The original Caldecott Tunnel consist of three bores along State Route 24 (SR 24) through the Berkeley Hills in Oakland, California. The California Department of Transportation (Caltrans) and the Contra Costa Transportation Authority (CCTA) proposed construction of a Fourth Bore to provide two additional traffic lanes to address congestion on SR 24 near the original three Caldecott Tunnels. The horseshoe-shaped Fourth Bore is 1,036 m long (3,399 ft), 15.2 m wide (50 ft), and 9.7 m high (32 ft). The project included short sections of cut-and-cover tunnel at each portal, seven cross passages between the Fourth Bore and the original Third Bore, and a new Operations and Maintenance Control (OMC) building. The Fourth Bore provides two 3.6 m (11.8 ft) traffic lanes and two shoulder areas that are 3 m (9.8 ft) and 0.6 m (2 ft) wide, respectively (Figure 1). The tunnel includes a jet fan ventilation system, a wet standpipe fire protection system, and various operation and control systems, including closed circuit television (CCTV), heat and pollutant sensors, and traffic monitoring.

### Ground Conditions

The geology along the alignment is characterized by Middle to Late Miocene-age marine and nonmarine sedimentary rocks, which strike northwest with high dip angles and are locally overturned. The western end of the alignment traverses marine shale and sandstone of the Sobrante Formation, which includes the First Shale, Portal Sandstone, and Shaley Sandstone members. The middle section of the alignment traverses chert, shale, and sandstone of the Claremont Formation, which consists of the Preliminary Chert, Second Sandstone, and Claremont Chert and Shale members (Page 1950). The eastern end of the alignment traverses nonmarine claystone, siltstone, sandstone, and conglomerate of the Orinda Formation. The limits of the major formations along the tunnel are shown in Figure 2. The Fourth Bore alignment encountered four major inactive faults, which occur at the contacts between the geologic units. These faults strike northwest, perpendicular to the tunnel alignment. In addition to the major faults, many other zones of weak ground were encountered, such as smaller faults, shears, and crushed zones. The active Hayward fault, located 1.4 km (0.9 mi) west of the project area, is the closest regional active fault. Additional details on the geologic conditions along the tunnel alignment are presented in Thapa et al. (2008a, 2009).

## DESIGN

The following is a brief summary of the design of the Caldecott Fourth Bore.

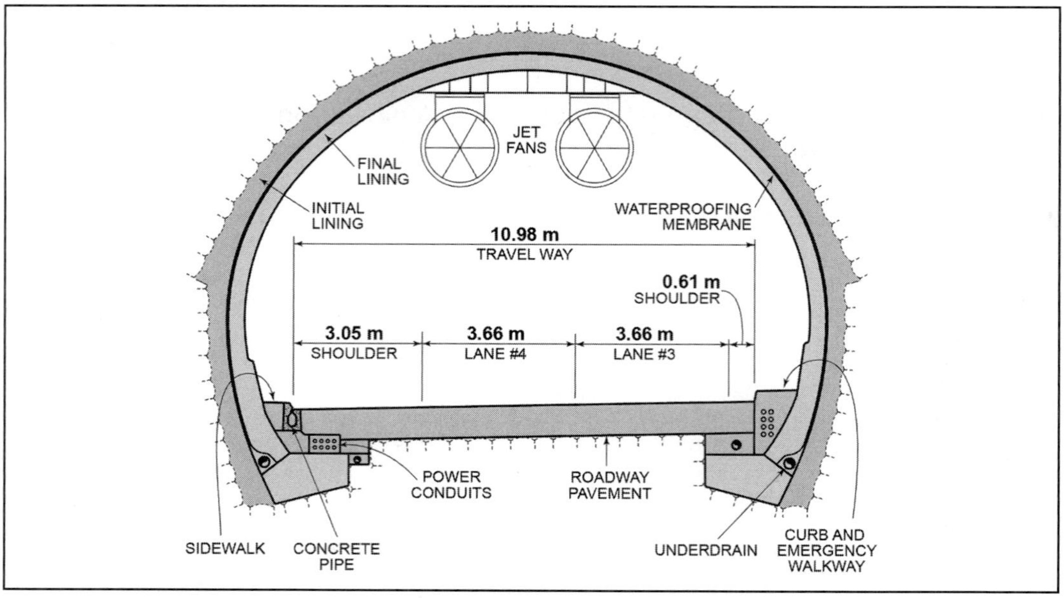


**Figure 1. Typical cross section of the Caldecott Fourth Bore Tunnel**

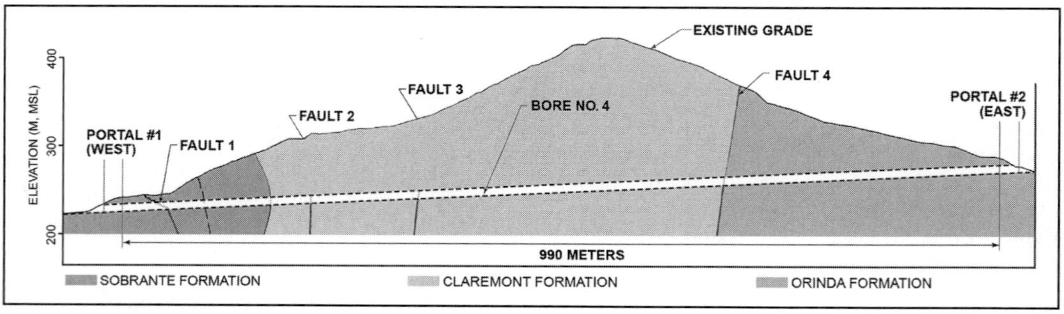


**Figure 2. Simplified geological longitudinal profile of the Caldecott Fourth Bore**

## Design of Initial Ground Support

The excavation and support design followed the principles of the New Austrian Tunnel Method (NATM), also commonly referred to as Sequential Excavation Method (SEM). Design drawings for the project included detailed requirements for the excavation sequence and initial ground support systems for the anticipated range of ground conditions including restrictions on advance length for each stage of excavation and the arrangement, dimensions, and capacity requirements for the support elements. The design included four major initial support systems, referred to as support categories (SC), SC I through SC IV, and three subvariations of the support categories, labeled with A and B. Table 1 summarizes the components by support categories, and Figure 3 presents a typical design drawing showing arrangement and installation requirements for the support elements for one of the major support categories. The design also included a toolbox of 20 additional support measures, consisting of shotcrete used as face sealing, initial lining, or temporary inverts, different types of rock dowels and spiles, lattice girders, face dowels, as well as probe and drain holes. The toolbox measures were used to augment the standard support category, if required by the encountered ground conditions.

## Design of Final Lining

The final lining for the Caldecott Fourth Bore consists of cast-in-place reinforced concrete placed against a PVCsheet waterproofing geomembrane backed by a drainage geotextile. The waterproofing geomembrane extends only over the arch and sidewalls of the tunnel and drains into a drainage system located at invert level. The lining is 381 mm (15 in.)

**Table 1. Summary of systematic support measures per support categories**

<table>
<tr><th>SC</th><th>Max. Advance Length</th><th>Systematic Presupport</th><th>Face Support (FRS=fiber reinf. shotcrete)</th><th>Min. Shotcrete Thickness</th><th>Aver. Radial Dowel Spacing</th><th>Temporary Shotcrete Invert Arch</th></tr>
<tr><td>IA</td><td rowspan="2">1.8 m<br>~5′–11″</td><td rowspan="2">None</td><td>Face dowels, sealing FRS, as required</td><td rowspan="2">~20 cm (8″)</td><td rowspan="2">1.8 m<br>~5′–11″</td><td rowspan="2">None</td></tr>
<tr><td>IB</td><td>Systematic face dowels, sealing FRS</td></tr>
<tr><td>IIA</td><td rowspan="2">1.4 m<br>~4′–7″</td><td>None</td><td rowspan="2">Face dowels, sealing FRS OR sloping core, sealing FRS</td><td rowspan="2">~25 cm (10″)</td><td rowspan="2">1.5 m<br>~4′–11″</td><td rowspan="2">None</td></tr>
<tr><td>IIB</td><td>Spiles</td></tr>
<tr><td>IIIA</td><td rowspan="2">1.0 m<br>~3′–3″</td><td rowspan="2">Spiles</td><td rowspan="2">Sloping core, sealing FRS</td><td rowspan="2">~30 cm (12″)</td><td rowspan="2">1.2 m<br>~3′–11″</td><td>None</td></tr>
<tr><td>IIIB</td><td>Top heading and bench</td></tr>
<tr><td>IV</td><td>1.0 m<br>~3′–3″</td><td>Pipe arch canopy</td><td>Sloping core, sealing FRS</td><td>~30 cm (12″)</td><td>None</td><td>Top heading and bench</td></tr>
</table>

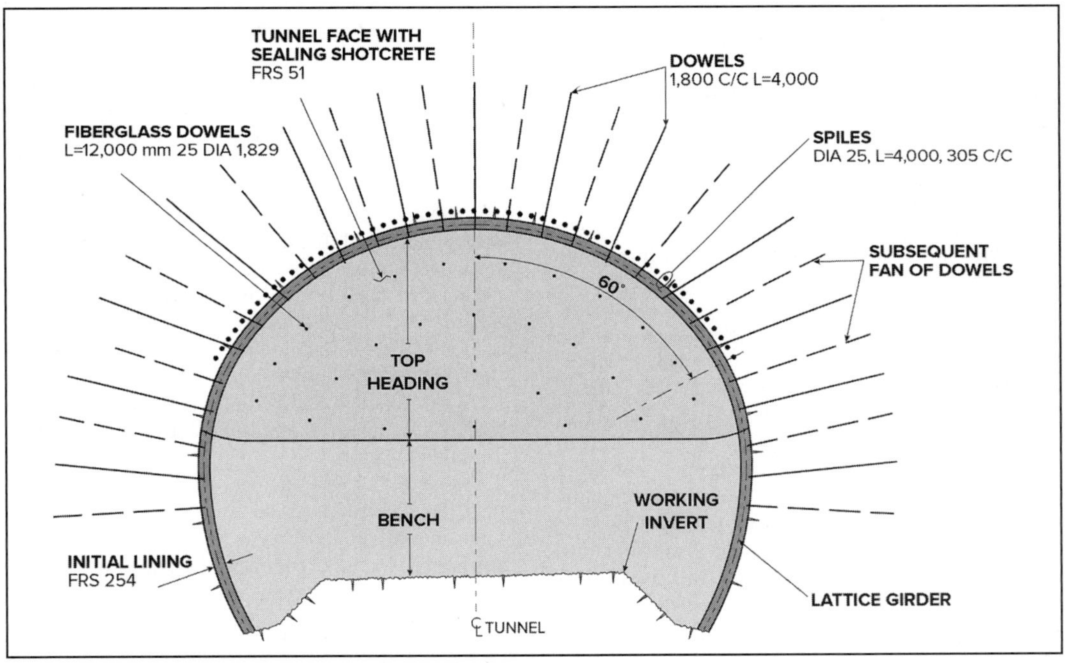


**Figure 3. Example of support category requirements, typical excavation cross section for SC IIB**

thick and includes two layers of reinforcing steel. The inner layer consists of 19 mm (No. 6) bars at 200 mm (7.9 in.) in the hoop direction and 13 mm (No. 4) bars at 200 mm (7.9 in.) in the longitudinal direction; the outer layer of steel consists of 16 mm (No. 5) bars at 400 mm (15.7 in.) in the hoop direction and 13 mm (No. 4) bars at 400 mm in the longitudinal direction. The design strength of the concrete is 35 MPa (5,000 psi).

The initial shotcrete lining and final concrete lining were designed as a combined support system, under the assumption that a portion of the ground load, initially carried by the initial support system, will be transferred to the final lining because of deterioration of the rock dowels and shotcrete comprising the initial support. Analyses indicate that approximately 50% of the load ground loads carried by the initial support system will be transferred to the final lining (Thapa et al. 2008b). However, in the design the final lining was conservatively assumed to carry two-thirds of the ground loads originally carried by the initial support system.

Extensive seismic analyses were performed to evaluate performance of the tunnel lining and ensure that the tunnel structure would meet Caltrans serviceability criteria for lifeline routes (Thapa et al. 2008a).

These analyses indicated that a single layer of reinforcing steel on the inside face of the lining would satisfy the seismic performance criteria. However, Caltrans decided to include a second layer of reinforcing steel to improve ductility during a seismic event (Thapa et al. 2013).

### Quantitative Risk Analysis

A quantitative risk analysis was performed at several points during the design phase to evaluate and manage the risks to both project cost and schedule. Results of the risk studies were used to proactively plan risk mitigations and to help differentiate among design alternatives, and ultimately to establish the project budget and completion schedule. A series of facilitated workshops were attended by Caltrans design, construction, and legal personnel and members of the consultant design team, as well as external experts, focused first on identifying and quantifying all the project risks, and then on developing strategies to mitigate the most important risks.

The risks were grouped into five major categories: environmental, design, right-of-way, bid, and construction. Once the risks had been identified and quantitatively assessed, a custom probabilistic risk-based integrated cost and schedule model was built and used to evaluate the possible range in project cost and schedule, explicitly considering both: (1) the possible variation in the baseline cost and schedule; and (2) the various risks. One of the major risks identified for the project was a significant delay (and thereby cost) associated with either a challenge to the environmental documents or a legal action by one of the community groups opposed to the project. Based on accepted precedent for large infrastructure projects, the 80th percentile of mitigated cost and schedule was used for planning purposes. Identifying, quantifying, and then mitigating the risks allowed Caltrans and CCTA to establish and ultimately meet a realistic and stable budget and schedule for the project.

## CONSTRUCTION

### Construction Milestones

The contract was advertised to bidders by the Caltrans in May 2009 and was awarded to the low bidder, Tutor Saliba Corporation (TSC), on November 20, 2009. Tunnel construction was preceded by portal excavation and support, which began concurrently on the east and west sides of the alignment in May 2010. The Contractor elected to drive the top heading from both ends of the alignment concurrently to expedite the schedule. Approximately 80% of the top heading (800 m [2,625 ft]) was excavated from the East Portal, and the remaining 200 m (656 ft) were excavated from the West Portal. Break-in occurred in August 2010 at the East Portal by TSC, and in March 2011 at the West Portal by subcontractor FoxFire Constructors. Breakthrough of the top heading occurred at the end of November 2011 from the East Portal heading after tunneling from the west side was completed to the breakthrough location roughly two weeks earlier.

Benching followed completion of the full top heading. TSC's bench excavation sequence consisted of a center cut excavation followed by excavation of remaining side berms and installation of the tunnel sidewall support. TSC performed the center cut bench excavation working eastward from the breakthrough point for the majority of this reach. Foxfire excavated the full face of the bench from the West Portal towards the breakthrough point. Invert excavation and support followed benching, where required. Bench and invert excavation were completed in September 2012.

Final lining construction used a 15 m long (49 ft) form that was advanced uphill from the west to the east from April to October 2012. Typically, it took 8 to 10 hours to move, set, and place the concrete and another 8 to 10 hours for the concrete to set sufficiently to allow form removal, resulting in 4 to 5 form advances per week over a 6-day workweek.

Several life-safety systems were installed in the Fourth Bore and Third Bore, including linear heat detectors, smoke detectors systems, gas detection systems, message signs, fire suppression systems, jet fans. All of the systems are monitored and can be controlled at the control center located in the Operations and Maintenance Center (OMC), which also controls the systems for the three existing bores and it is planned to monitor all Bay Area tunnels from this location.

The tunnel was opened to traffic on November 16, 2013 (Figure 4).

### Support Selection Process and Organization of the Construction Team

The Contract Documents described the design basis for the support categories and the criteria for selecting the appropriate support category based on the ground conditions and ground behaviors observed in the tunnel. Each support category was developed to support a defined ground condition that, along with the in situ conditions, resulted in certain ground behaviors. Defined ground behaviors included block failure, raveling, shallow shear failure, deep shear failure, slaking and softening, swelling, and crown instabilities due to low cover. (Thapa et al. 2008a,b, 2013). In addition, the design documents included the applicable toolbox support measures required for different observed or measured behaviors of the tunnel excavation.

The construction management team consisted of Caltrans personnel, augmented by Parsons

**Figure 4. Caldecott Fourth Bore Ribbon Cutting Ceremony on November 15, 2013**

Brinckerhoff and Gall Zeidler Consultants. Jacobs Associates as the designer provided the Design Representative during tunnel excavation and installation of the final lining. Gall Zeidler Consultants provided the NATM Engineer, who was in charge of the NATM tunnel–related technical construction management and led the team of engineers, geologists, and inspectors on-site.

Daily meetings were held during the mining phase between the Contractor's and Engineer's tunnel experts to select the appropriate standard support category and any required toolbox support measures based on observed ground conditions, observed support performance, and measured lining deformation.

Encountered ground conditions and behaviors were mapped by both the Contractor's and Engineer's geologists on a daily basis during all phases of excavation of the tunnel for each face. Probe holes were instrumented using an automatic data logger that recorded feed pressure, torque, and advance rate, and this information was interpreted to predict the ground conditions ahead of the tunnel face. Convergence monitoring was carried out across the tunnel arch and bench walls at instrumentation stations spaced approximately 15 m (49 ft) apart that were typically monitored within 100 m (328 ft) of the tunnel heading. In one area, long-term convergences in the top heading footing area were observed additional rock dowels were installed and successfully controlled the ongoing convergence. However, in general the monitored movements stayed well below the warning levels defined by the design.

All the information described above was reviewed at daily meetings between the Contractor's and the Engineer's tunneling experts and provided the basis for a joint ground classification and support selection for each tunnel advance. The selected standard support categories and associated excavation were paid for on a per meter basis, whereas the toolbox support measures were paid for on a unit price basis.

## CONTRACT

### Predicted Versus Observed Ground Behaviors and Support Requirements

The encountered ground conditions along the alignment were generally consistent with the design prognosis, with the exception of two tunnel reaches that total 87 m (285 ft), or 9% of the alignment. These two reaches of differing site conditions occurred within the Second Sandstone between Tunnelmeter (TM) 241 and 322 (79 m [259 ft]) and within the Claremont Chert and Shale between TM 386 and 394 (8 m [26 ft]). In the Second Sandstone, the rock structure between TM 241 and 322 was blocky to massive, in contrast to the predicted blocky structure, and the intact rock strength was approximately 25% higher on average than indicated from strength tests performed during the design stage. The sandstone dikes in the Claremont Chert and Shale encountered in the tunnel between TM 386 and 394 exhibited a blocky to massive structure, in contrast to the predicted very blocky rock structure in the best rock mass in this formation. The Contractor and the Engineer negotiated a modified compensation for this differing site condition based on the item price for the original line item and documented effort.

### Quantity Deviations

The major deviation from the design prognosis is the lesser quantity of SC III that was actually installed. While ground conditions anticipated to require SC III based on GSI, UCS data, and ground cover were encountered, SC II could be used in these reaches. This was because the strength of the fiber reinforced shotcrete as installed was higher than specified in the design. The higher than specified shotcrete strength

Table 2. Predicted versus installed support for Support Class II, including subtypes IIA and IIB

| Support Category | Predicted Quantity | Installed Quantity |
|---|---|---|
| II | 412 m | 568 m |
| IIA | 35 m | 380 m |
| IIB | 377 m | 188 m |

allowed for support selection of a thinner shotcrete lining, while still maintaining the required lining performance (Thapa et al. 2013). The predicted total quantity of SC III was 257 m (843 ft), compared to the installed quantity of 60 m (196 ft).

Another significant deviation from the design prognosis was the extent and payment for spiling in SC IIA and SC IIB. Spiling was an additional support measure in SC IIA, whereas SC IIB included systematic spiling (54 spiles total) over the entire arch (Table 1). The design intent was that SC IIB would be utilized where spiling was necessary around the majority of the arch and that SC IIA would be utilized where spiling was required over a limited portion of the arch. The Contractor's interpretation of the contract was to apply the pay item for additional spiles applicable to SC IIA unless the full number of 54 spiles, as prescribed for SC IIB, was required. Negotiations between the Contractor and Engineer established a payment mechanism that compensated the Contractor for SC IIB when more than 37 spiles were required at a particular location and compensated the Contractor for SC IIA plus the unit price for the number of spiles when less than 37 spiles were required. This deviation from the design intent resulted in differences between the predicted support and as-installed support (Table 2).

### Contractual Considerations and Unit Price

A key advantage of NATM is the flexibility of the method to adapt to the observed ground conditions with suitable systematic and additional support measures. Support selection decisions were made by the on-site team in the daily meetings as described above. Based on this joint review and considering potential operational constraints, the group would decide on the support class and support measures required for the day's advances.

The flexibility of the NATM tunneling method with the commensurate frequent variation in excavation sequence and support requirements can result in contractual challenges related to fair risk sharing between the Contractor and the Owner and equitable compensation mechanisms. One approach to addressing the issue would be to break all different elements of the excavation and support classes of the design into numerous separate unit price line items, for example, shotcrete, lattice girders, spiles, and rock bolts. Typically, only the line item for excavation is tied to a specific support category. The support measures, on the other hand, become independent from the support classes. ITA's Working Group 19 addresses such an approach and provides design and contractual guidelines for the execution of NATM, referred to as Conventional Tunneling by the ITA (ITA 2009 and 2013). A detailed discussion of the different support measures is also provided in the Federal Highway Administration's Design Manual (FHWA 2009).

The application of NATM in Austria and Germany typically allows selection of excavation sequence and initial support elements in combinations appropriate to variations in encountered ground behaviors, so as to achieve the most efficient tunnel support system possible. This approach often results in a highly variable excavation and support process that requires different pay items for each support element such that they can be combined as needed. Mostly, unexpected ground conditions do not become the basis for a differing site condition claim if the designed systematic and additional support measures are applied, even if they are modified from the standard support classes. However, using this approach can result in significant variations between predicted and actual quantities, and this quantity variation has to be appropriately addressed in the contract.

The Caldecott Tunnel used a detailed and prescriptive design of the excavation and support sequence, which was developed to minimize the number of support categories and pay items with the goal of simplifying the construction operations and avoiding an overly complex and cumbersome contractual payment process. Standard support categories were measured and paid on a per meter basis, with the pay item covering all associated excavation and support requirements. This approach was judged to be more conducive to promoting competitive and responsive bids. The payment approach for each support category was successful (except in the case of the spiling that is part of SC II, described above). Based on the divergence of the Contractor's interpretation from the design intent, and the variability in the number of spiles required per advance, it may have been more advantageous to remove a prescriptive design for the spiling from systematic support measures and pay for the spiles as additional support (including time-dependent costs such as impacts on advance rates).

## THIRD-PARTY INVOLVEMENT

### California Division of Occupational Safety and Health Administration

After evaluating the information gathered during the ground investigation phase, the California Division of Occupational Safety and Health Administration (Cal/OSHA) classified the Caldecott Tunnel as "Gassy with Special Conditions." This classification imposed stringent requirements on the Contractor with regard to equipment, operation, and health and safety precautions. During mining activities, the Contractor was required to measure and record gas readings every hour at the face of the main tunnel and the cross passages. The records were available at the site for review by Cal/OSHA engineers during their bimonthly visit. Additionally, the excavation equipment had to be fitted with a measuring device continuously measuring for traces of gas.

At the conclusion of the top heading excavation, and before the bench and invert excavation was completed, Caltrans requested that Cal/OSHA relax the classification to "Potentially Gassy with Special Conditions." After reviewing all the records from the top heading excavation and finding that there were no significant traces of gas, Cal/OSHA reclassified the tunnel to "Potentially Gassy with Special Conditions." This allowed the use of more standard equipment, without the stringent requirements imposed by a classification as gassy, allowing for expedited excavation.

### Emergency Response Plan

A significant construction risk often overlooked is the integration of the electrical and mechanical systems. For road tunnels, NFPA Code 502 (NFPA 2011) requires preparation of an Emergency Response Plan (ERP). For the Caldecott Fourth Bore, the ERP was developed under the supervision of the office of the State Fire Marshall, the California Highway Patrol, and the Oakland and Orinda-Moraga Fire Departments. In compliance with the NFPA code, seven cross-passages were constructed between Bore 4 and the existing Bore 3.

Prior to the tunnel being opened to traffic, the emergency scenarios in the ERP were tested with the use of the tunnel safety-life systems. Experience gained at the Caldecott Fourth Bore project and other road tunnels recently completed in California reveals that integration of the system is complicated, difficult, and time consuming. Hardware and software issues occurred at any time during the integration process. Even though a particular system had passed during the individual testing, it could develop issues when integrated with other systems. In the Fourth Bore, there was the additional complication of integrating existing systems of the Third Bore with the newer systems in the Fourth Bore. Thus, sufficient schedule time was necessary for systems integration and testing, as well as continuous coordination with vendors, subcontractors, integrators, emergency responders and the operators.

### Community Outreach

An extensive community outreach was initiated prior to construction and continued throughout to keep key stakeholders, taxpayers, and the motoring public well informed. A comprehensive strategic communications plan served as a blueprint for project-specific messaging and community outreach protocols, and helped to standardize communications among partner agencies, including Contra Costa Transportation Authority (CCTA), Caltrans, Metropolitan Transportation Commission (MTC), and Alameda County Transportation Commission (Alameda CTC). Prior to start of construction, CCTA and project partners launched a project website, www.caldecott-tunnel.org, which provided ongoing information. In addition, a full-time Public Information Officer provided regular updates to the many stakeholders and the public.

### Blasting and Community Impact

The need for blasting and its potential impact on the nearby residents and structures were assessed in detail during the design phase. These assessments indicated that all encountered rock types along the tunnel alignment could be excavated by a large roadheader. However, in order to provide additional flexibility to the contractor in the event of encountering stronger, more massive rock than anticipated, blasting was permitted. Given the concerns of the public living in close proximity to the project, Caltrans required close controls on all blasting operations per the project specifications, including requirements for a 24-hour notice prior to blasting, prohibiting blasting during evening hours, submittal of detailed blasting plans, monitoring of ground vibrations and air overpressures, and strict limits on peak particle velocity and air overpressures at specified locations. As anticipated during the design phase, blasting was not required because of the utilization of a very powerful roadheader.

### Noise Reduction

Managing noise impacts during construction on nearby residents was also a key consideration. The project plans included a detailed design for a sound wall adjacent to the West Portal to shield a large residential development from construction noise. In addition, the Contract Documents required the Contractor to prepare a detailed sound control plan. Monitoring was performed prior to construction to measure ambient noise levels. The ambient levels were recorded by placing recording devices in close proximity to the construction area at locations designated by the specifications. During the construction period, the Contractor was required to continuously monitor and record ambient noise levels and compare them to the ambient baseline levels. In addition, the Contractor was required to install four monitoring and recording devices near the construction areas, with locations approved by the Engineer, and monitor for noise levels exceeding 86 dBA. If an event occurred that exceeded the noise levels, a notification was immediately sent to the Engineer, and the Contractor was required to determine the cause of the elevated sound level within 20 minutes of the occurrence. If the noise exceedance was caused by the Contractor's activities, the Contractor was required to suspend operations and take measures to mitigate the sound.

## CONCLUSION

The design and construction of the Caldecott Fourth Bore was based on the principles of the New Austrian Tunneling Method (NATM). The support systems as designed and implemented during construction were successful in supporting the tunnel opening, controlling ground behaviors, and limiting tunnel convergence to below the predicted thresholds. The NATM approach provided the required flexibility to adapt the support for the wide-span tunnel to the encountered weak and variable ground conditions. The experience with construction of the Fourth Bore indicates that the simplified contract structure minimizes the potential for misinterpretation of the contract as related to a multitude of support variations. With growing experience with NATM execution, it will be possible to develop designs with more flexibility that will require more refined contractual payment structures.

The successful completion of the Caldecott Fourth Bore on schedule and under budget demonstrates that large NATM tunneling is a cost-effective approach to tunnel construction.

## ACKNOWLEDGMENTS

The authors acknowledge the major contribution of Dr. Bhaskar Thapa to the successful completion of the Caldecott Fourth Bore. He worked on the project through all phases, from the initial investigation to the completion of the final lining. Dr. Thapa passed away unexpectedly in June 2013, just four months prior to the opening of the Fourth Bore. The successful completion of this landmark project serves as a tribute to his hard work, dedication, and passion for tunnel design and construction. The high-quality analyses, reports, and construction documents that Dr. Thapa prepared will serve as models for future generations of tunnel engineers.

Dr. Bill Roberds with Golder Associates led the risk workshops and performed the risk analyses.

## REFERENCES

National Fire Protection Association (NFPA). 2011. NFPA 502—Standard for Road Tunnels, Bridges, and Other Limited Access Highways. 2011 Edition.

Page, B.M. 1950. Geology of the Broadway Tunnel, Berkeley Hills, California. *Economic Geology,* 45(2).

Thapa, B.B., McRae, M.T., and Van Greunen, J. 2007. Preliminary design of the Caldecott Fourth Bore. In *2007 RETC Proceedings. Toronto, Ontario, Canada, June 10–13.* Littleton, CO: SME.

Thapa, B.B., Van Greunen, J., Sun, Y., McRae, M.T., and Law, H. 2008a. Design analyses for a large-span tunnel in weak rock subject to strong seismic shaking. In *2008 NAT Proceedings. San Francisco, CA, June 8–11.* Littleton, CO: SME.

Thapa, B.B., Marcher, T., McRae, M.T., Sander, H.J. 2008b. Design of NATM initial support on the Caldecott Fourth Bore. In *Proceedings of the American Rock Mechanics Symposium. San Francisco, California.*

Thapa, B.B., Marcher, T., McRae, M.T., John, M., Skovajsova, Z., and Momenzadeh, M. 2009. NATM strategies in the U.S.—Lessons learned from the initial support design for the Caldecott 4th Bore. In *2009 RETC Proceedings, Toronto, Ontario, Canada, June 10–13.* Littleton, CO: SME.

Thapa, B.B., Nitschke, A., Ramirez, I., McRae, M.T., and Gall, V., 2013. Lessons learned from NATM design and construction of the Caldecott Fourth Bore. In *2013 RETC Proceedings,. Washington, D.C., July 24–26.* Littleton, CO: SME.

The International Tunnelling and Underground Space Association (ITA). April 2009. ITA Working Group No. 19—Conventional Tunnelling: General Report on Conventional Tunnelling.

The International Tunnelling and Underground Space Association (ITA). May 2013. ITA Working Group No. 19—Conventional Tunnelling: Guidelines on Contractual Aspects of Conventional Tunnelling.

U.S. Department of Transportation Federal Highway Administration (FHWA). March 2009. Technical Manual for Design and Construction of Road Tunnels—Civil Elements. NHI Course 130090. Publication No. FHWA-NHI-09-010.

# Digging Deep for Cost Savings: Mill Creek Regional Effluent Tunnel

**Brian Glynn, Brian Gettinger, and Cary Hirner**
Black & Veatch

**Dick Alexander**
Johnson County Wastewater

**Lizan Gilbert**
SJ Louis Construction of Texas

**ABSTRACT:** The Mill Creek Regional Effluent Tunnel parallels the Kansas River for nearly 3,050 m (10,000 ft) through gassy shale to provide 680 ML/D (180 MGD) capacity for the Mill Creek Regional Waste Water Treatment Plant Outfall. Proximity to the river required water tight support of excavation through over 15 m (50 ft) of alluvial soils, and bedrock excavation methods were limited at the retrieval shaft due to its close proximity to the railroad. This paper will focus on design decisions and lessons learned throughout construction including dealing with gassy conditions and groundwater, shaft and tunnel excavation, and coordination with the busiest railroad tracks in the region.

## INTRODUCTION

Johnson County Wastewater's (JCW) Mill Creek Regional Wastewater Treatment Plant (WWTP) is located in the City of Shawnee, Kansas, just south of the Kansas River and west of Interstate I-435. The WWTP serves portions of the Mill Creek, Tooley Creek and Cedar Creek watersheds and discharges treated effluent via a series of diffusers located in the Kansas River, just downstream of an existing raw water intake.

The Mill Creek Regional WWTP was built in the 1990s and had an original capacity of 34 ML/D (9 MGD) and was later upgraded in 2006 to a capacity of 72 ML/D (19 MGD). The WWTP was also designed to treat higher, short-term flows caused by wet weather events through a series of aerated lagoons. The original capacity of the effluent pumping station, which pumps the treated effluent from the WWTP to a series of diffusers in the Kansas River, was 212 ML/D (56 MGD). At the time of the WWTP upgrade in 2006, the pump station peak flows were modeled to be 303 ML/D (80 MGD) for the upgrade project and 469 ML/D (124 MGD) for the future; however peak flows of either size had not been encountered at the time of the upgrade project at the WWTP and the improvements to the effluent pump station were deferred in order to reduce costs.

Soon after the WWTP upgrade project was completed in 2006, the WWTP began to receive larger wet weather events than previously encountered, which nearly exceeded the discharge capacity of the effluent pumping station. A pre-design study for the effluent discharge capacity improvements was then conducted by Black & Veatch, which included determining wastewater flow projections through the year 2034. The study recommended that the effluent discharge capacity of the WWTP needed to be increased in order to meet development occurring in the watersheds that were served by the WWTP and that the WWTP should be capable of handling wet weather events of up to 680 ML/D (180 MGD) in the future.

Fourteen different alternatives were evaluated, which comprised of different combinations of the following: a parallel force main, a pump station upgrade, a shallow tunnel, an open channel overflow to nearby tributary stream, and a deep gravity flow tunnel. A triple bottom line approach was used to evaluate the alternatives. Key factors used in the evaluation included capital cost, operating cost and complexity, 50 year net present value, as well as community and environmental impacts. A deep gravity tunnel flowing as an inverted siphon was selected as the preferred alternative as it eliminated a pump station and it was the simplest to operate with no additional operating expenses despite it not having the lowest initial capital cost.

## DESIGN PHASE

The first priority of the design phase of the project was the design of a temporary discharge line from the WWTP to Mill Creek, which would be utilized if the plant encountered any wet weather flows

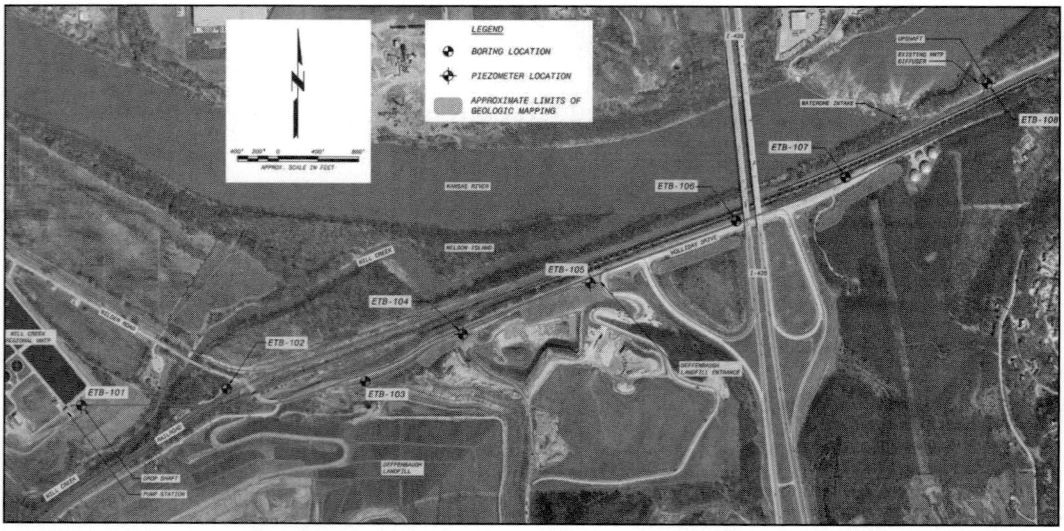


**Figure 1. Tunnel alignment (plan view)**

that exceeded the original discharge capacity of the effluent pump station of 212 ML/D (56 MGD). The temporary discharge line would help protect the WWTP and prevent manholes, located upstream of the WWTP, from overflowing, until the permanent discharge capacity improvements could be made to the WWTP.

Once the design of the temporary discharge was completed and construction was ongoing, the project team began work on the design of the deep tunnel and associated surface work. Black & Veatch had previously completed the design and construction services for a nearby tunnel crossing the Kansas River, for the installation of a 1.5 m (60 in) diameter water main in the Tacket Shale Formation. The Tacket Shale proved to be a favorable tunneling medium with minimal groundwater inflow even when situated directly below the Kansas River. Based on this experience, the Mill Creek Regional Effluent Tunnel was also planned to be excavated within the Tacket Shale.

## Alignment

The tunnel alignment as shown in Figure 1 follows Holiday Drive from the WWTP to the existing diffuser in a northeasterly direction and is nearly 3,000 m (10,000 ft) in length. Following beneath an existing right of way for Holiday Drive eliminated acquiring additional easements for most of the alignment. The tunnel is bordered on the north by the Kansas River and Burlington Northern Santa Fe (BNSF) railroad tracks and on the south by Deffenbaugh Industries Landfill, which is one of the largest Subtitle D landfills in the region. The alignment crosses beneath the BNSF railroad tracks twice as well as beneath Interstate I-435. The proximity of the landfill caused concerns for encountering contaminated groundwater and methane gas in the tunnel excavation. Naturally occurring methane gas in the Tacket Shale Formation was also a concern.

## Geotechnical Investigation

Twelve borings were drilled along the alignment to identify soil and rock properties and to confirm the vertical alignment of the tunnel as shown in Figure 1. Ten borings were drilled vertically and two were drilled at an inclined angle of 20 degrees from vertical to identify geologic features and discontinuities. Four piezometers were installed in borings to collect water samples and monitor the groundwater level prior to and during construction. Soil and rock sampling was performed in the borings drilled at each shaft site and rock samples were collected and geologically logged from borings along the alignment.

Geotechnical lab testing for rock included: Moduli in Uniaxial Compression, Unconfined Compressive Strength (UCS), Brazilian Tensile Strength (BTS), Slake Durability, Cerchar Abrasivity and Dry Unit Weight. Geotechnical lab testing for soil included: Grain Size Analysis, Moisture Content, UCS and Atterberg Limits. Sampling was also performed of the groundwater, which was tested for the presence of dissolved methane.

## Geologic Setting

Significant thickness of overburden was encountered at both shaft sites. The overburden at the down shaft was comprised of mostly stiff, high plasticity clay to a depth of nearly 15 m (50 ft) underlain by a 1.5 m (5 ft) thick layer of well graded gravel with

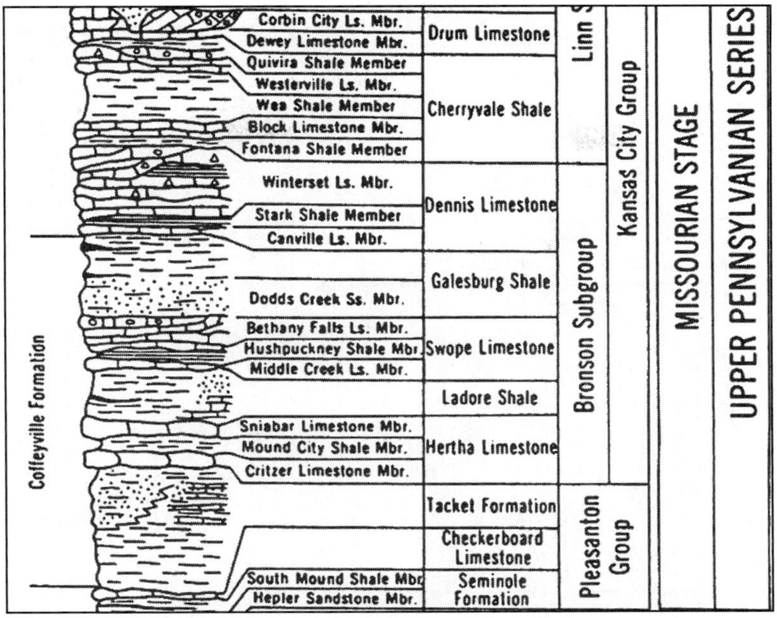


*Source: Kansas Geological Survey (www.kgs.ku.edu)*

**Figure 2. Local stratigraphic column**

cobbles. Overburden at the upshaft site consisted of 4 m (13 ft) of silt underlain by 8.5 m (28 ft) of alluvial sand and 0.9 m (3 ft) of highly weathered shale. Water encountered during drilling and in monitoring wells installed in borings confirmed that groundwater in the overburden at both the down shaft and upshaft is hydraulically connected to the nearby Mill Creek and Kansas River. Water inflow at the down shaft through the well graded gravel was estimated to be 760 L/min (200 gpm) and inflow at the upshaft through the alluvial sand and silt was estimated to be 15,140 L/min (4,000 gpm).

Bedrock encountered in the shaft and tunnel excavations is a part of the Kansas City and Pleasanton Group of the Upper Pennsylvanian Series as shown in Figure 2. Formations encountered at the down shaft and upshaft begin with the Winterset Limestone through the Tacket Shale, as shown in the tunnel profile in Figure 3. The tunnel alignment is entirely in the Tacket Shale formation, while the shafts pass through interbedded shale and limestone formations.

### Shaft Sites and Surface Work

The down shaft site is located on the property of the WWTP near the existing plant lagoons. The down shaft was designed to be 6 m (20 ft) in finished diameter and 41 m (135 ft) deep in order to reach the tunnel invert. To convey flow from the lagoon to the down shaft, three morning glory weirs and a 2.4 m (96 in) discharge pipe will be installed in the lagoon as shown in Figure 4. The 2.4 m (96 in) discharge pipe will connect to a new junction box which will collect flow from the WWTP's existing mechanical treatment process. All plant effluent is then conveyed through a flow metering and shaft inlet structure prior to entering the down shaft.

The upshaft site is located along a bank of the Kansas River, just downstream of a raw water intake. The property is owned by the BNSF Railroad, but JCW has an existing utility easement for the existing diffuser and junction box already located at the upshaft site. The new tunnel will flow by gravity as an inverted siphon up the 3 m (10 ft) diameter, 35 m (115 ft) deep upshaft. After exiting the upshaft, flow will be conveyed through a new junction box into an existing 1.4 m (54 in) diameter pipe to the existing diffusers located in the Kansas River as shown in Figure 5. In addition to being constructed on BNSF property, the upshaft is located within 45 m (150 ft) of four active railroad tracks, which carry some of the highest volume of rail traffic in the Kansas City area.

### Baseline Conditions and Anticipated Rock Behavior

At the conclusion of the geotechnical investigation, a Geotechnical Data Report (GDR) and a Geotechnical Baseline Report (GBR) were published to summarize the anticipated geotechnical conditions that could be expected. The GBR and GDR were included in the

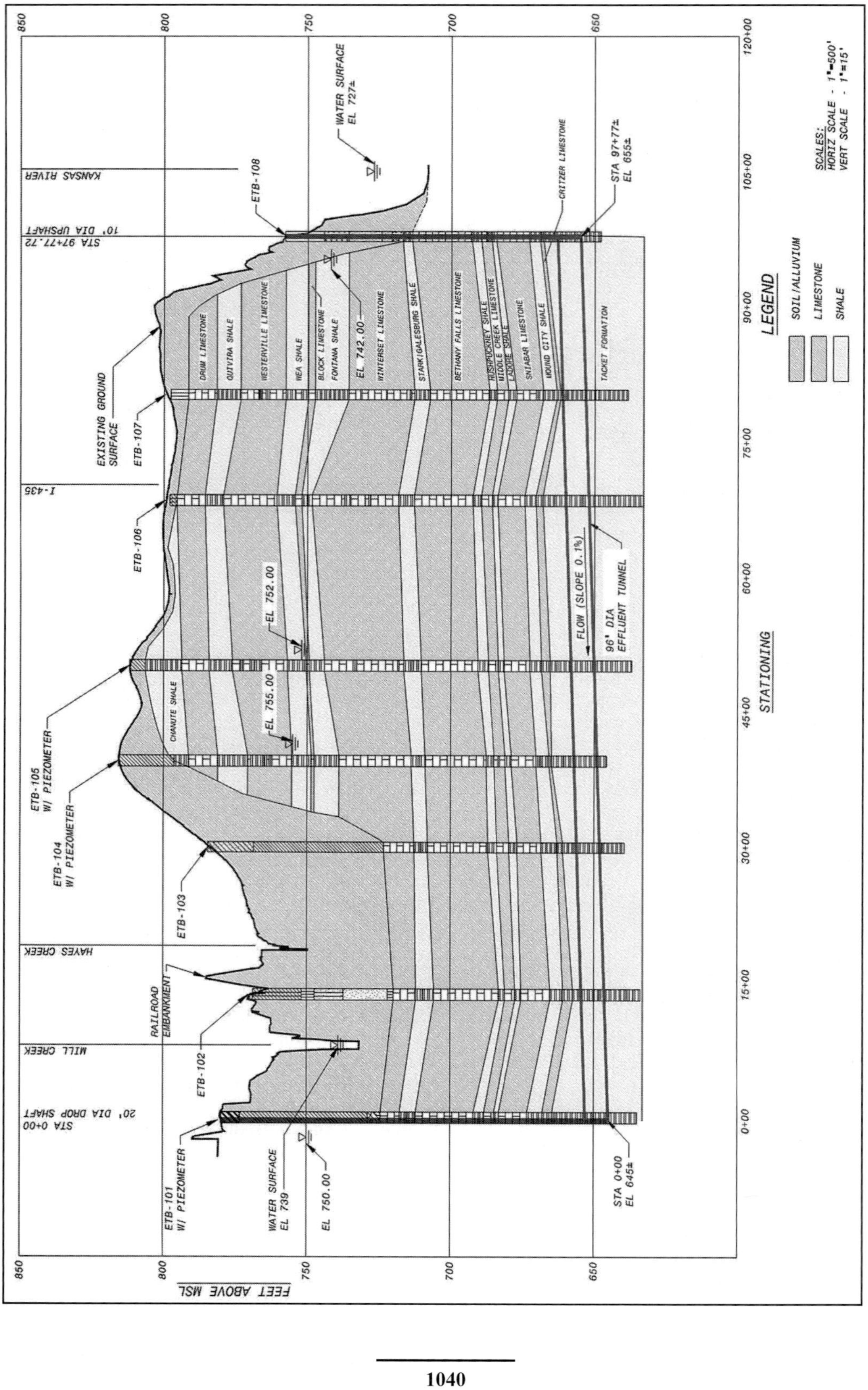


Figure 3. Tunnel geologic profile

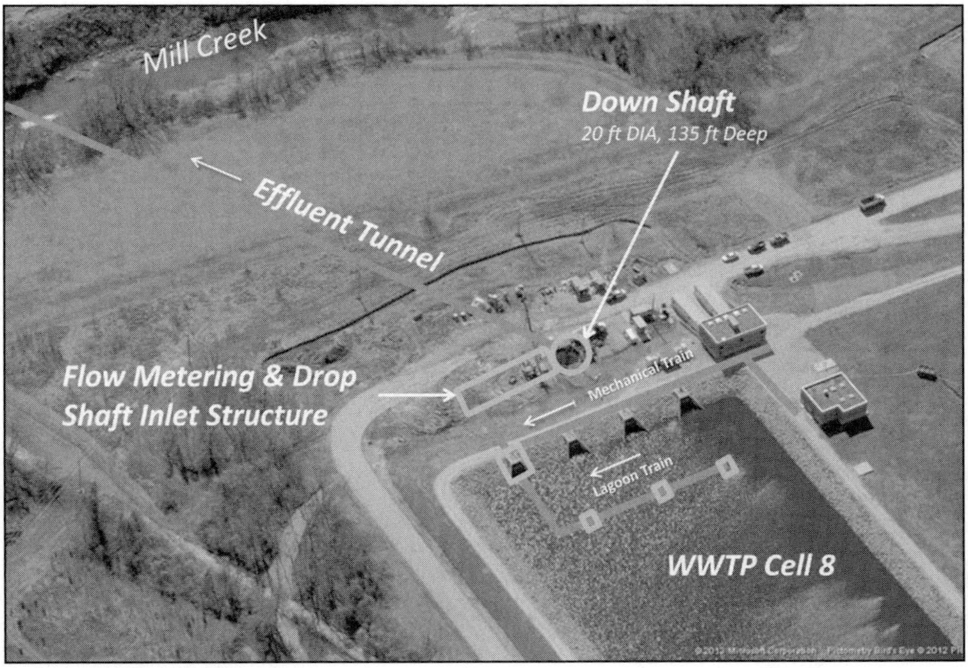


**Figure 4. Down shaft site plan**

**Figure 5. Upshaft site plan**

**Table 1. Baseline properties of limestone and shale units**

| Location | Rock Type | Baseline Criteria | Splitting Tensile Strength, MPa (psi) | Unconfined Compression Strength, MPa (psi) | Cerchar Abrasivity | Slake Durability, % |
|---|---|---|---|---|---|---|
| Shafts | Limestone Units | Average | 5.9 (850) | 124.1 (18,000) | No Baseline | No Baseline |
| | | Maximum | 12.8 (1,860)* | 213.7 (31,000) | No Baseline | No Baseline |
| Tunnel | Shale Units | Average | 1.5 (210) | 34.5 (5,000) | 0.8 | 70 |
| | | Maximum | 2.0 (290) | 68.9 (10,000) | 1.5 | No Baseline |
| | | Minimum | No Baseline | No Baseline | No Baseline | 50 |

* Based on a correlated value of 0.06 times the maximum unconfined compressive strength.

Contract Documents to assist contractors in pricing their bid. The baseline properties for the limestone and shale to be encountered in shaft excavations and the shale to be encountered in tunnel excavation are shown in Table 1.

In addition to presenting quantitative baselines of geotechnical properties, the anticipated ground behavior was also discussed, key points are summarized below:

**Shafts**

- Alluvial sand and gravel will flow
- Clay and silt will ravel
- Saturated silt and sandy clay will flow
- Detached blocks of shale will slake and could fall
- Detached blocks of limestone could fall

**Tunnel**

- Shale in crown will ravel and slab due to gravity and in-situ horizontal stress
- Failures of detached blocks in tail and starter tunnel
- Tunnel invert deterioration due to presence of water and softness of shale

### Dictated Construction Methods

In order to mitigate risk associated with construction methods, preferred methods were specified in the Contract Documents for use on both the shaft and tunnel excavations. Since the upshaft is located on BNSF property and it is in close proximity to active railroad tracks, no blasting was allowed at the upshaft site in order to comply with a BNSF permit. At both shaft sites, water tight initial support keyed into bedrock was required through the overburden prior to excavating soil. Sheet piling was not allowed since it could not be socketed into competent limestone.

The tunnel was required to be mined with an intrinsically safe tunnel boring machine (TBM) with a slotted partial shield. Intrinsic safety is required due to the potential for methane gas to be encountered within the shale. Dissolved methane was detected in groundwater samples taken during the geotechnical investigation. A slotted partial shield was required on the TBM to provide a safe working area for installation of initial rock support in order to mitigate spalling and slabbing of the shale in the crown of the excavated tunnel.

### Minimum Required Initial Support for Excavation in Rock

In addition to specifying construction methods, minimum initial support requirements were included in the Contract Documents. Minimum initial support requirements were based on previous experience in the tunneling medium as well as field and lab testing. These requirements were included to assist the contractors in pricing their bid, however, additional support required to maintain a safe excavation remained the responsibility of the Contractor.

For shaft excavations, the initial support is dependent on the rock formation exposed. In exposed limestone formations, 1.5 m (5 ft) long No. 8 rock dowels, evenly spaced around the shaft perimeter, were specified every 1.5 m (5 ft) vertically with each row offset as shown in Figure 6. In exposed shale formations, rock dowels were also required and were supplemented by W4×W4 welded wire fabric (WWF) with a 10 cm (4 in) openings and a minimum of 5 cm (2 in) of shotcrete was to be applied over the WWF. The rock dowels were installed to prevent block failure and the WWF and shotcrete were installed to prevent spalling and weathering of the shale in the shafts.

Minimum initial support for the tunnel excavation was specified as a set of two 1.5 m (5 ft) long, No. 8 rock dowels installed vertically every 1.5 m (5 ft) along the tunnel length and W4×W4 WWF with 5 cm (2 in) openings in the top 60 degrees of the circular tunnel excavation as shown in Figure 7. WWF and rock dowels were installed to prevent spalling and slabbing of the shale in the crown of the tunnel.

## CONSTRUCTION PHASE

The project was bid on May 17, 2011 and six contractors submitted bids for the project. The bids ranged from approximately $32 million to $40 million. The engineer's estimate was $37 million. The project was

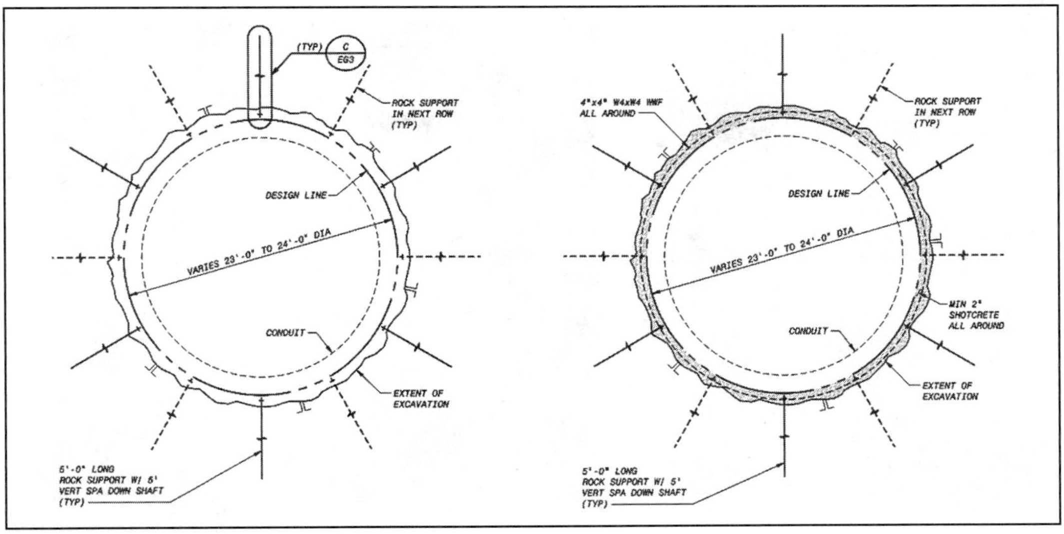


**Figure 6. Down shaft initial support minimum requirements**

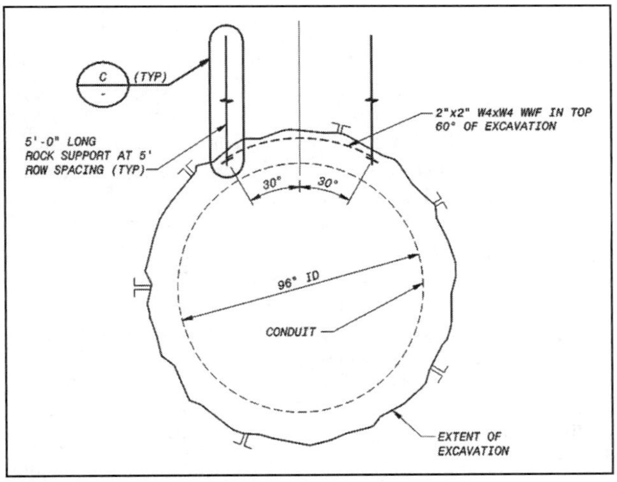


**Figure 7. Tunnel initial support minimum requirements**

awarded to the low bidder, S.J. Louis Construction of Texas, LTD. (Contractor). The notice to proceed was issued for September 19, 2011.

### Shaft Excavation

The project began with the construction of a circular secant pile wall at the down shaft site and later at the upshaft site to allow for the excavation of the shafts through the overburden. The overburden was excavated for the secant piles using the segmented casing method down to bedrock. A rock socket was drilled into the top of the limestone bedrock layer. Primary piles were installed first, followed by the secondary piles. Each secant pile was 90 cm (35 in) in diameter. The inside diameter of the circular sheet pile wall was 8 m (26 ft) at the down shaft site and 4.5 m (15 ft) at the upshaft site. The secant piles were approximately 18 m (60 ft) deep at the down shaft site and 15 m (50 ft) deep at the upshaft site.

Once the circular secant pile walls were installed and cured, the excavation of the overburden inside the circular secant pile walls was performed using a mini track excavator, down to bedrock. A crane and muck box was used to remove the excavated materials. The depth of the overburden was approximately 17 m (56 ft) at the down shaft site and 12.5 m (41 ft) at the upshaft site. After the overburden was excavated, rock excavation began inside the shafts. At the down shaft site, rock excavation was performed mainly by drilling and blasting. At the upshaft site, blasting was not permitted, due to restrictions from BNSF, therefore a mini track excavator with a breaker attachment and jack hammers were used for rock excavation. The depth of rock excavation was approximately 24 m (79 ft) at the down shaft site and 19 m (62 ft) at the upshaft site. The excavation of the overburden at each shaft site only took a few weeks. The rock excavation took approximately 15 weeks at the down shaft site and approximately 48 weeks at the upshaft site. The vertical rate of rock excavation was approximately 1.6 m (5.25 ft) per week at the down shaft site and approximately 0.4 m (1.30 ft) per week at the upshaft site. The rock consisted of layers of limestone and shale as mentioned previously. The minimum initial

**Figure 8. Secant pile installation and completed circular secant pile wall at down shaft**

support was installed as required in the shafts as the excavation progressed. Additional rock dowels were installed in some areas of the shafts, where additional support was deemed necessary by the Contractor for safety reasons (Figure 8).

### Starter and Tail Tunnel Excavation

After the down shaft was completely excavated, the excavation of the starter and tail tunnels began. The starter and tail tunnels were excavated through the use of a roadheader machine as well as by drilling and blasting. The starter and tail tunnels were excavated over a period of approximately 15 weeks. The roadheader allowed for precise rock excavation of the starter and tail tunnels; however the Contractor experienced mechanical problems with their roadheader machine and reverted to drilling and blasting at times. Drilling and blasting within the shale created significant over break, which resulted in a larger starter tunnel than desired in some areas. When complete, the starter tunnel was approximately 70 m (225 ft) in length, 4.5 m (15 ft) in width and 4.5 m (15 ft) in height and allowed for the assembly and launch of the TBM underground. The tail tunnel was approximately 23 m (75 ft) in length, 3 m (10 ft) in width and 3 m (10 ft) in height and allowed for the efficient removal of muck cars from the down shaft for disposal of the tunnel muck on the surface.

### Tunnel Boring Machine

The TBM used on this project was a remanufactured machine provided by the Robbins Company of Solon, Ohio. The machine was a double shield type of TBM. The diameter of the cutterhead was 3.15 m (124 in). The cutterhead contained 24, 38 cm (15 in) diameter disc cutters. The cutterhead was driven by four, 186 kW (250 hp) water cooled VFD electric motors, which could produce up to 1,077 kN-m (794,100 lb-ft) of torque. The maximum stroke of the TBM was 1.27 m (50 in). The maximum recommended operating thrust was 2,670 kN (600,000 lbf). The TBM could produce a maximum of 8,943 kN (2,010,619 lb) of thrust via the eight main thrust cylinders. The TBM required a 4,160 volt electrical service run to the site. The overall weight of all of the components of the TBM was approximately 235 metric tons (260 tons).

The TBM arrived on site in late September 2012, after a significant delay in the refurbishing process by the TBM manufacturer. The TBM took approximately seven weeks to assemble, including a few weeks of troubleshooting of the electrical and hydraulic connections. Portions of the TBM were pre-assembled on the surface as shown in Figure 9 that could be lowered into the shaft, but the majority of the assembly took place underground.

Once assembled, the TBM began the tunnel excavation phase of the project. As the TBM advanced, the crown of the tunnel was supported with the minimum required initial support. Additional rock dowels, mine straps and welded wire fabric were installed in the crown of the tunnel in some areas where additional support was deemed necessary by the Contractor for safety reasons. As anticipated during design, overbreak of up to approximately two to three feet in the crown was experienced in areas where the quality of rock was lower.

**Figure 9. TBM above ground during pre-assembly at down shaft site**

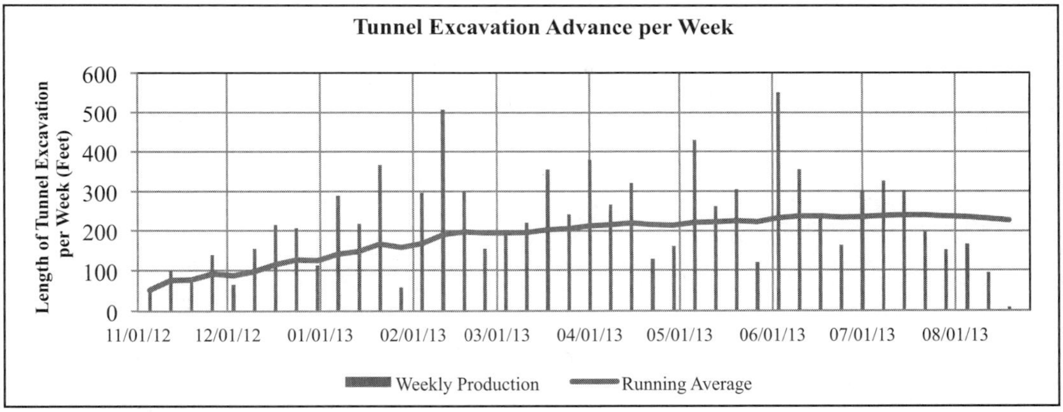


**Figure 10. Tunnel excavation advance by week**

### Tunnel Excavation Advance Rates

Tunnel excavation began on November 7, 2012 and ended August 19, 2013, a period of approximately 40 weeks. The overall average excavation rate was approximately 10 m (33 ft) per day or 72.5 m (238 ft) per week. The highest advance of the TBM in a single day was nearly 51 m (168 ft), which occurred on February 13, 2013 and the highest advance of the TBM in a single week was nearly 168 m (550 ft), which occurred in the first week of June 2013. Operating issues which impacted the advance rate included problems with the guidance system on board the TBM, hydraulic hose ruptures on the TBM, buildup of muck inside the cutterhead of the TBM and higher than expected wear of disc cutters on the cutterhead (Figure 10).

## LESSONS LEARNED

The soft invert of the excavated tunnel created problems for the Contractor, specifically problems with the tracks for man and material transport, which was installed in the invert of the tunnel. The soft invert allowed the tracks to shift and/or sink as the locomotives, muck boxes and utility cars would pass over them, which over time would cause derailments. The Contract Documents did not require the Contractor to install any temporary lining over the invert of the tunnel during the excavation phase or prior to the final lining phase of this project. The Contract Documents only required the Contractor to not allow water to pond in the invert of the tunnel.

The Contractor attempted to place ballast, concrete, steel ribs and larger railroad ties in order to support the rail, in areas of the tunnel where the invert was especially problematic. These efforts temporarily improved conditions, but delayed tunnel excavation operations, as excavated materials from the TBM could not be removed from the tunnel while materials were brought in to improve the invert. Over time the ballast placed in portions of the invert allowed water to pond in some sections of the tunnel. The materials placed in the invert of the tunnel also had to be removed before the final lining of the tunnel could be installed.

In future tunnels, especially in this medium, possible additions to the Contract Documents would include, requiring the Contractor to line the invert with pre-cast concrete panels, cast-in-place concrete or steel ribs and lagging. These methods would improve working conditions underground, but would also add a somewhat significant cost to the project.

## FUTURE DISCUSSION

At the time this paper was written, in November 2013, the project was about a year away from completion and the final lining of the tunnel had yet to begin. A future paper will discuss the tunnel and shaft lining processes.

# TRACK 4: CASE STUDIES

## Session 4: Rock Tunnels, Caverns, and Shafts

Mike Stokes, Chair

# Successful Fast-Track Construction Sequencing of a Complex Station and Tunnels Project in a Highly Developed Urban Environment

**Amitabha Mukherjee and Thomas Peyton**
Parsons Brinckerhoff

**Anthony Del Vescovo**
Schiavone Construction Company

**Anil Parikh**
NYC-MTACC

**ABSTRACT:** New York City's extensive transit system had not been expanded for several decades until construction of the Second Avenue Subway program was re-started in 2006. Many of the construction contracts were performed at an ambitious pace to meet this projected completion date. Phase 1 of the overall program includes three new stations and connecting tunnels over a distance of almost two miles, with restrictive surface work sites and significant operational constraints in a densely developed portion of New York City's Upper East Side.

This paper will describe the highly coordinated methods to plan, sequence, schedule and implement the complex 72nd Street Station and Tunnels Project on a fast-tracked 37-month schedule. Construction work included extensive rock excavation (blasting) for several large caverns, tunnels and shafts as well as final concrete lining before a sequenced turn-over to follow-on finishes and system-wide contractors. Close coordination and effective communications between the Owner, Contractor, Construction Manager and the Design Engineer during the construction phase contributed to meeting schedule and quality goals that initially seemed extremely ambitious and possibly out of reach. Challenges and solutions will be presented with examples that confirm that the concentrated collaborative approach between the parties was essential for the critical success on this very complex and risky project located in a highly developed urban environment in New York City's Upper East Side.

## SUMMARY AND INTRODUCTION

The 72nd Street Station and Tunnels Project in New York City was a difficult undertaking with restrictive site conditions and an aggressive construction schedule. Nonetheless, it was completed successfully, on time and provided MTA with an excellent opportunity to meet its interim goals within the Second Avenue Phase 1 subway construction program schedule (Figure 1).

This project was one of the key core contracts in Phase 1 of the Second Avenue Subway construction program and as such, received considerable attention from all parties including the MTACC, Designer Aecom-Arup, the Consultant Construction Manager, PB and the civil works contractor, SSK Constructors JV formed by Schiavone Construction, JF Shea Construction and Kiewit Infrastructure Co.

This paper will describe the key operations and conditions that contributed to the high quality, on-time completion of this fast-tracked subway project. It is not too common to find projects of this nature and complexity performed under rigorous site and environmental conditions completed per plan and within a tightly controlled schedule. This paper will summarize the important aspects of cooperation, collaboration, planning and expeditious problem resolution as well as other unique features of the work including innovative means and methods within the design and the construction operations. In that the work was performed within the densely developed Upper East Side of New York City, the project stands as an accomplishment to be recognized by Owners, planners and builders of similar work.

The civil works portion of the station and tunnels contract was built on a reliable and well

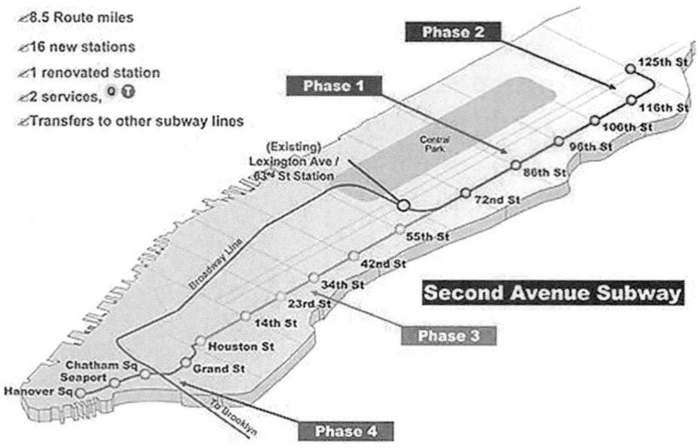


Phases of Construction
Courtesy of MTA Capital Construction Company

**Figure 1.**

thought-out detailed design that it itself, recognized the difficult site conditions in the performance of the work. The MTA's designers were frequently present at the site and had participation in all phases of construction progress. Their day-to-day involvement made a significant positive difference in the project outcome.

Overall, the MTA's expectations for construction performance were reasonable and attainable but not without exceptional collaborative efforts by the parties, not only for the fundamental construction aspects, but to address the concurrent side issues and concerns largely from local neighborhood residents who both wanted the subway built but resisted its construction and challenges to the quality of their lives throughout the multi-year construction program. Initially, the work seemed daunting and the construction schedule out-of-reach but as an aggressive excavation and final concrete lining program was developed, interim and final milestones were deemed achievable. Figure 2 illustrates the general scope and complexity of the project.

## CIVIL CONSTRUCTION SCOPE AND SCHEDULE

The scope of the civil work contract for the 72nd Street Station and Tunnels project was substantial and included rock excavation and final concrete lining of all station and cross-over caverns as well as the 2,000 LF south tunnels and caverns. Additional work included the excavation and shotcrete lining of two Ancillary areas and three Entrances as well as connecting adits and inclines. Under a prior contract, two TBM-bored tunnels were excavated but left unlined in the 72nd Street project area. Figures 3 and 4 illustrate the primary portions of the scope in the civil work contract. There were multiple concurrent work areas requiring careful planning and execution to meet schedule goals shown in Figures 5, 6 and 7. Completion of the civil works would precede the follow-on station finishing and system-wide contracts that were so critical to MTA's Second Avenue Subway program schedule.

Overall, the civil work contract would require the excavation of over 175,000 BCY of rock (granite, gneiss and Manhattan schist) by blasting methods and placing in excess of 50,000 CY of reinforced concrete final lining. It was an extensive construction program performed on a three-shift-per-day basis for the entire construction program of 37 months, plus three months approved Extensions-of-Time for directed Extra Work. Except for interruptions due to severe weather including hurricanes Irene and Sandy, the work was performed continuously following Notice-of-Award (NOA) and Notice-to-Proceed (NTP) on 01 Oct 10. Milestone 1 for all work located north of Grid Line 17 (the northerly 250 LF of the station cavern and the North Cross-Over cavern) was to be completed by the end of Month 31 (30 Apr 13).

## JOB SITE CONDITIONS AND WORK RESTRICTIONS

Owing to the location of the work, in the Upper East Side, there were inherent site conditions that were adverse to heavy construction activities. Additionally, local residents had come to enjoy their peace and quality of life without the clamor and nuisances from construction equipment, many construction workers

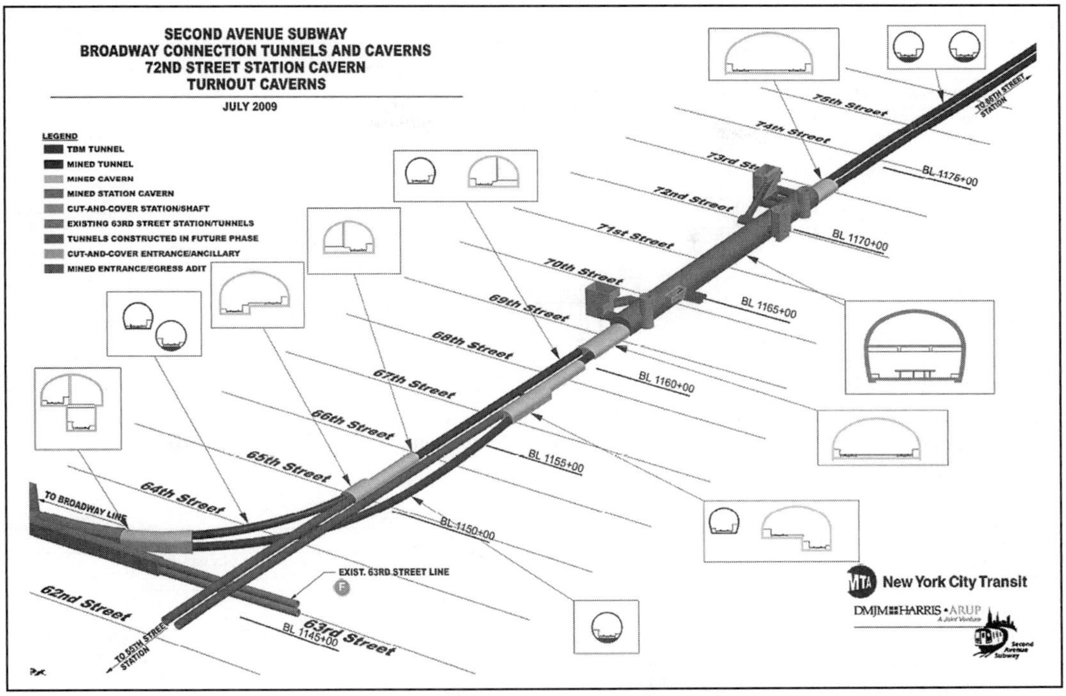


**Figure 2. Overall site plan for the 72nd Street Station and Tunnels Project for the MTA. The project extended from 73rd Street in the north to the tie-in point at the existing 63rd Street Station for a distance of about 3,000 LF.**

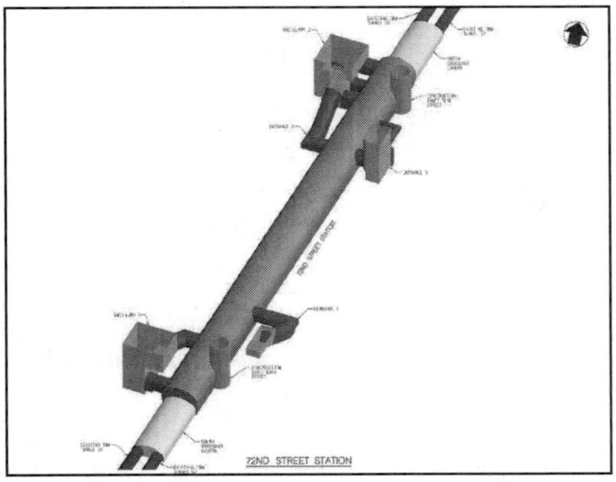


**Figure 3. General arrangement of the station cavern, cross-overs, ancillaries and entrances**

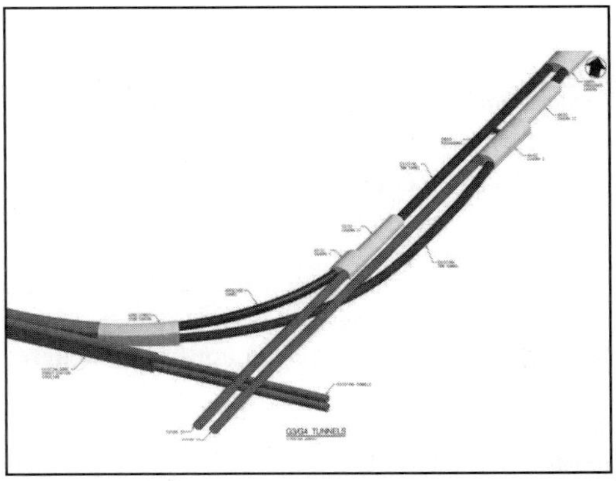


**Figure 4. General arrangement of the south tunnels and caverns connecting to the existing 63rd St. Station**

as well as changes to traffic patterns and business disruptions. Apart from high rise construction, there was little street-level construction activity common to the Second Avenue corridor including nearby side streets until October 2010 and the start of the 72nd Street Project that would continue for more than three years for the civil work plus an additional three years for station finishing and system-wide facilities. The early start to the two Construction Shafts under a separate contract will be described below.

In general terms, the neighborhood favored having a new subway station at 72nd Street and extending south to 69th Street, but was annoyed with lengthy construction related activities, traffic issues,

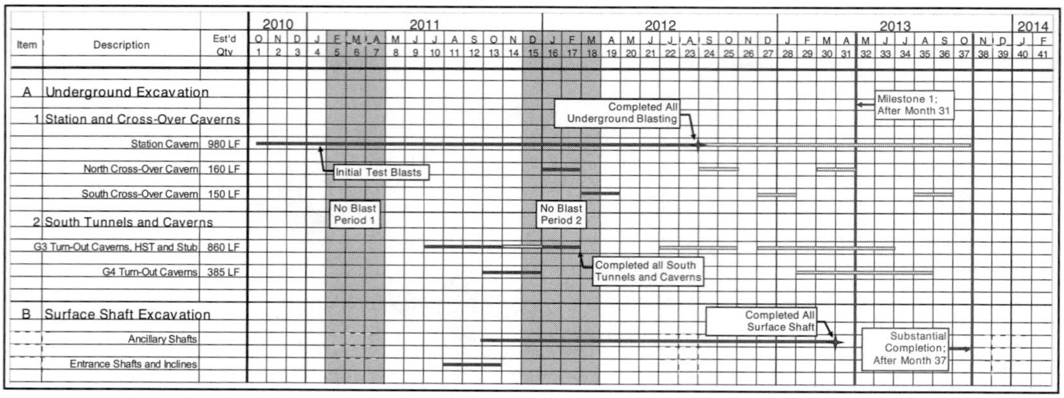


**Figure 5. Overall excavation and final concrete lining schedule for the project—including all caverns, tunnels, adits, surface shafts and entrances. Additional construction activities are detailed in the following figures.**

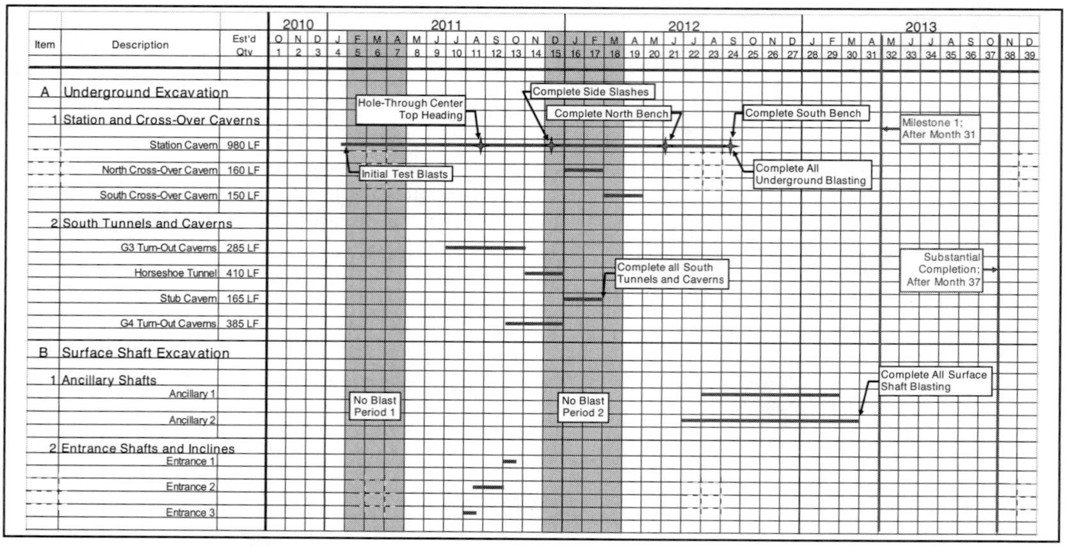


**Figure 6. Overall excavation schedule—including all caverns, tunnels, adits, surface shafts and entrances. The work required multiple concurrent heading operations while constantly mucking to two Construction Shafts.**

noise, dust, odors, blast vibrations and unfamiliar personnel in the streets at all hours that the work was underway. Nonetheless, and after the initial year of blasting the station cavern and connecting tunnels, the neighborhood acquiesced to the ongoing work—largely focused at the two Construction Shafts—located at 72nd and 69th Streets on Second Avenue. This acceptance was due in part from an effective community a relations and outreach program by the MTA that included monthly meetings, frequent underground public tours, a store-front Community Center and a monthly newsletter, among other services and information systems. Additionally, the Contractor maintained a tidy, well organized secure work site.

## Site Conditions

The site conditions were typical for a densely developed urban setting but were further complicated by the heavy use of Second Avenue as one of the primary vehicle access corridors to mid-town Manhattan as well as points further south. The character of the site included the following, for example:

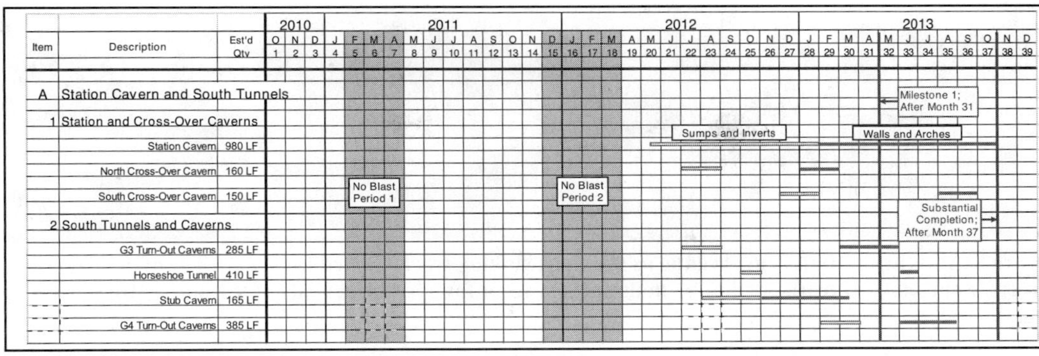


**Figure 7. Overall final concrete lining schedule—including all caverns, tunnels, adits, surface shafts and entrances. The work required many concurrent operations while constantly supplying concrete from drop shafts.**

- **Road Way and Sidewalk Conditions**
  - Narrow, one-way side streets with no truck access or parking available
  - Prescribed truck routes to approach and depart from the site
  - Enormous numbers of pedestrians present on the streets throughout the day
- **Construction Operating Conditions**
  - Restrictive loading and unloading zones on Second Avenue
  - No storage areas available at the site
  - Extremely limited areas for the use of heavy equipment; e.g., crawler cranes
- **Surface and Underground Conditions**
  - Roadway paving conditions that were far from "smooth" until an extensive repaving program was initiated by the MTA in 2011 (good investment)
  - Antiquated underground utility systems on Second Avenue
  - Limited quantities of temporary electrical power available
  - Difficult temporary sewer and water connections for construction operations

### Local Interest Groups and Agencies

- Neighborhood groups; both in favor of and opposed to MTA's extensive multi-year Second Avenue Subway construction plan
- Stakeholder interest groups having special needs and requirements
- Numerous city and state agencies having involvement in the project

### Work Restrictions

Work restrictions were onerous and confining but still, they were in keeping with the needs of the local residents and neighborhood character. Several of the more important work restrictions included the following:

- **Surface Work Operations**
  - 7:00 AM start for all surface work (e.g., equipment units), Monday to Friday
  - 10:00 PM end to all surface operations, Monday to Friday
  - Saturday surface work hours restricted from 10:00 AM to 7:00 PM
- **Blasting Restrictions**
  - Various vibration limits (PPV) prescribed for different classes of buildings
  - Blast related dust, smoke and odor concerns
  - Prescribed muck truck haul routes
  - Ventilation fan noise limits
- **Final Lining Operations**
  - Concrete age restrictions before pumping and placing
  - Ready-mix truck queuing restrictions on Second Avenue
- **Construction Schedule and Key Dates**
  - **Milestone 1**—after 31 months; completion of all work north of Grid Line 17
  - Access to the existing Bellmouth area at 63rd Street—after 31 months
  - **Substantial Completion**—after 37 months; completion of all work

## SELECTION OF THE RIGHT APPROACH FOR JOB CONSTRUCTION

While the site conditions and work restrictions were formidable and the design had considerable embedded construction practicalities, choosing the right construction approach was essential for the project to be successful and competed on time. The selection of resources, construction facilities, work sequence, equipment, means and methods fell to the Contractor

**Figure 8. Aerial view along 2nd Avenue as the shaft sinking work started in late 2010 at 69th and 72nd Streets. Four lanes of traffic had to be maintained except for specific periods of the day—Monday to Friday.**

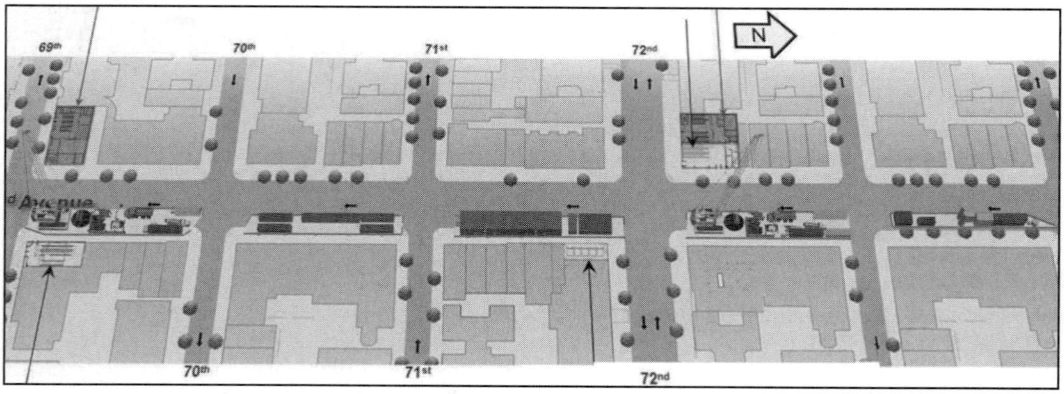


**Figure 9. Aerial view along 2nd Avenue as the shaft sinking work got underway in late 2010 at 69th and 72nd Streets. Large cranes were used to service the shafts with marginal productivity due to the physical confinements of the site.**

to select and implement in a manner that would meet safety, quality and schedule goals under the contract. This was done well and with a highly productive and innovative muck handling facility that was later adapted to handle large quantities of concrete forms and reinforcing steel for the final lining.

The following describes many of the key facilities and resources brought to the project by the MTA, its Designer, Construction Manager and Contractor that contributed to the critical success of the 72nd Street Project and the ongoing Second Avenue Subway construction program.

### Construction Shafts and Utility Relocations

There were potentially several means to access the underground areas and construction operations. Only the two Construction Shafts located on Second Avenue at 69th and 72nd streets, however, provided the best overall solution for access to the work from street level directly to station invert and were used throughout the excavation and final lining phases. The MTA's original design for the station incorporated two temporary Construction Shafts since it was apparent that alternate means of access were less appealing, meanwhile reliable access to the underground work areas was critically important to meet schedule goals.

The initial portion (35' deep) of each Construction Shaft was excavated under a prior contract and lined to full depth at 30' diameter. Please refer to Figure 8. While they provided a good starting point for the follow-on station and tunnels contract, they also established construction work sites on Second Avenue and relocated several underground utility systems well in advance of the primary station excavation phase. The advance utility relocations also provided a substantial benefit to the job and avoided predictable delays to the crucial early phase work.

The Construction Shafts were located approximately 60 to 80 feet from the station end walls and were, therefore, ideally suited for excavation and final lining operations. They were ultimately excavated in multiple stages to full depth to the station invert level as station cavern excavation advanced. As such, they provided full-time access to the caverns and tunnels. Please refer to Figures 9 and 10 that illustrate the work site phased sequences.

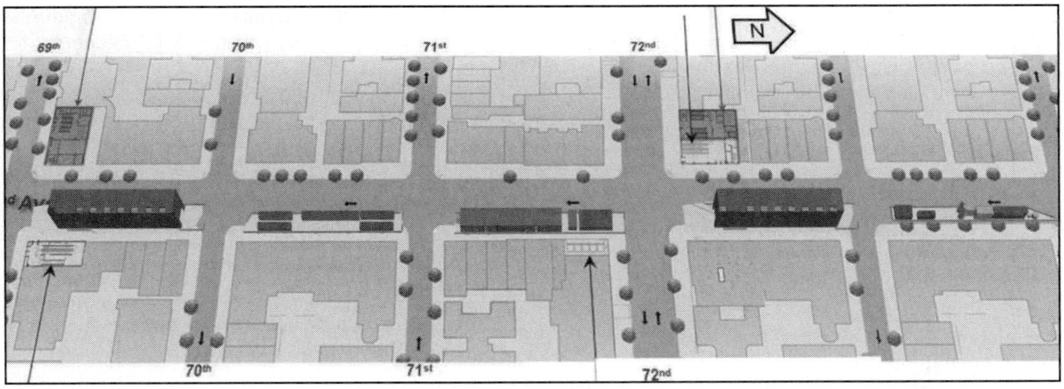


**Figure 10. Aerial view along 2nd Avenue after Muck Houses had been installed in mid-2011 at 69th and 72nd Streets. The integrated mucking systems were critical to the schedule and, therefore, overall success of the project.**

### Muck Houses at Construction Shafts

At the location of the two Construction Shafts, temporary structural steel-framed enclosures were built primarily to support overhead gantry cranes, but also to store large daily quantities of blasted materials from the underlying caverns and tunnels. This innovative approach was loosely envisioned in the Environmental Impact Statement (EIS) for the project but was much further refined and optimized by the Contractor to handle his operations in a manner that minimized construction nuisance issues while efficiently loading-out large quantities of muck on a daily basis. These Muck Houses became the operations centers for all underground work.

The Muck House concept was initially resisted by the MTA and local resident groups until it was realized that alternate means for crucial continuous muck handling were far more egregious and would require considerably more Critical Path time in the construction schedule. The Muck House design was refined, successfully built and provided numerous benefits to the project and the local community throughout a period of use of approximately two years at each shaft while still allowing natural light and air to circulate to the neighboring buildings. Construction noise, smoke and odors were minimized and the all-weather enclosures proved invaluable for continuous construction operations through both the excavation and final lining phases. The enclosures were tidy, well-constructed and maintained and blended well into the local area while providing lighting for sidewalks and streets. Please see Figures 11 and 12 that illustrate the construction features and finishes. Exterior features were closely coordinated with local aspects.

The Muck Houses were built to provide efficient flow of construction traffic in and out. Additionally,

**Figure 11. Construction of one of two Muck Houses erected at the site—for efficient materials handling**

**Figure 12. Operational Muck House at 72nd Street site. Ventilation and electrical systems are also enclosed.**

they housed the following essential systems and equipment from the elements that would otherwise have been located around the site:

- Muck storage bins (12 × 25 CY each)
- Underground ventilation systems and controls
- Construction shaft hoisting equipment
- Power distribution and communications equipment
- Concrete and shotcrete pumps (at 72nd Street only)
- Centers for material handling and management
- Gathering point for emergency drills and safety meetings

### TBM Bored Tunnels—Previous Construction Contract

Two TBM-bored tunnels underlying the station and cross-over caverns were built beforehand under a separate contract and were integrated into the design of the station and south tunnels. These tunnels originate at the 96th Street Station (Launch Box) and run through the 86th Street and 72nd Street stations terminating at the existing 63rd Street Station Bellmouth cavern. They became the south running tunnels on the 72nd Street project with portions enlarged to become turn-out caverns and the stub cavern adjacent to the 63rd Street Station Bellmouth. One tunnel was extended 400 LF (Horseshoe Tunnel) and connected to the 63rd Street Stub Cavern.

While there were several friendly debates over the merits and delays from the construction of the TBM bored tunnels, they nevertheless provided the following significant benefits to the project as well as to the overall Second Avenue Subway construction program and should, therefore, be considered on other similar applications.

- Running tunnels completed early and in advance of the follow-on cavern excavation
- Reduced quantity of tunnel and cavern rock excavation to be removed through the 69th and 72nd Street Construction Shafts
- Provided advance geotechnical assessment of the ground and water conditions for cavern and tunnel excavation—throughout the 3,000 LF length of the project
- Provided potential muck haulage passageways between the Construction Shafts
- Initial unobstructed connection from the Launch Box at 96th St to 63rd St Bellmouth

Notwithstanding the benefits achieved for the overall Second Avenue Subway construction program as a result of the pre-excavation of the TBM-bored tunnels, there were also potential issues to address and resolve in advance and to avoid potential conflict from the shared use of space for the 72nd Street project (on its own) and particularly in light of the fast-track construction schedule prescribed for the civil work contract. Figures 13 and 14 show the confluence of the initial TBM bored tunnels with the follow-on tunnels and cavern excavations.

### Multiple Drop Holes for Concrete and Gravel Supply

Although not a new concept for underground construction, the use of well-placed drop holes for concrete and gravel supply proved to be extremely beneficial to maintaining progress in the final lining operations, including the preceding drainage system installation in the caverns. The drop holes were drilled 60' to 80' deep on Second Avenue at 66th and

**Figure 13. Final stage in the station cavern—for the removal of the rock surrounding the TBM tunnel**

**Figure 14. Completion of the G3 and G4 tunnels at the Stub Cavern—with a final separation of only 6 feet**

68th Streets and fitted with slicklines, compressed air, water lines as well as power and communications systems to link the underground operations to large concrete pumps and the expanded work sites above. This arrangement immediately "decongested" the Construction Shaft areas and provided critically important alternate points of delivery using ready-mix trucks for all final lining concrete. Gravel for drainage layers was dropped through the 68th Street pipe to an underground storage bin.

A third concrete pumping station was established at 73rd Street for the final lining in the station, North Cross-Over and nearby adits. At this location, additional slicklines were installed in the 72nd Street Construction Shaft and used until well after the shaft was backfilled to grade. The majority of all station area concrete was pumped through the 72nd Street Construction Shaft. Please see Figures 15 and 16 that show two of the essential concrete slickline installations.

Use of the drop holes and more particularly, the use of the *expanded work site areas* along Second Avenue were extremely beneficial to the job and especially to public safety in the congested Construction Shaft areas. Consider the following for example, as direct outcomes;

- Traffic flows greatly improved along Second Avenue; to and from side streets also
- Improved pedestrian safety by minimizing truck maneuvering at intersections
- Fewer concurrent operations located at the two Construction Shafts
- Flexibility in material supply—including gravel for the drainage layers
- Separate concrete pumping sites and operations allowed for less congestion and more focus on safety and quality
- Expanded and separate work sites (pumping stations) were well laid-out, maintained and easily accessible by heavy trucks.

## Prescribed Truck Routes for Muck Haulage Off-Site

The City of New York established specific truck haul routes for muck leaving the site. This was important since all prescribed routes efficiently led to bridges and tunnels and onto various dump and fill sites outside of Manhattan in New Jersey, Queens and Brooklyn. Use of these designated routes resulted in the least impact to traffic and general public safety, while at the same time, minimizing haul times that restricted daily production to between two to four loads per truck. Over 15,000 loads of muck were needed to handle the excavated quantities that at peak rates resulted in over 100 loads per work day

**Figure 15. Concrete slicklines in G3 running tunnel connected to the 66th Street pumping station**

**Figure 16. Concrete slicklines in the 72nd Street shaft connected to the 73rd Street pumping station**

from the two Construction Shafts/Muck House operations combined. The majority of the muck was carried over the George Washington Bridge to New Jersey.

Additional trucking was required for more than 50,000 CY for final lining concrete. These deliveries also followed pre-determined inbound and outbound truck haul routes that utilized Second Avenue as the primary north-south corridor. All concrete was sourced from Queens or Brooklyn and, therefore, had to be transported through tunnels and bridges to Manhattan.

### Local Truck Queuing Areas

In consideration of the scope of the work and particularly the delivery and removal of construction materials, it was easy to appreciate the dilemma of a small linear construction site located in a densely

Table 1. Summary of the overall truck delivered quantities. While these quantities were approximate, they illustrate that approximately 22,000 loads were transported to and from the site during the construction period of 39 months.

| Item | Description | Overall Data | | Average | Peak Rates | | |
|---|---|---|---|---|---|---|---|
| | | Qty | Time | /Month | /Week | /Day | /Hour |
| 1 | Muck from excavation | 15,000 lds | 20 mon | 750 | 550 | 120 | 8 |
| 2 | Concrete for final lining | 6,000 lds | 16 mon | 375 | 150 | 80 | 12 |
| 3 | Rebar for final lining | 300 lds | 16 mon | 20 | 8 | 2 | |
| 4 | Construction materials | 350 lds | 16 mon | 20 | 12 | 4 | |
| 5 | Construction debris | 300 lds | 16 mon | 20 | 10 | 2 | |

Figure 17. Construction site on 2nd Avenue showing the field office (blue) and Muck House (white)

Figure 18. Construction site on 2nd Avenue showing features of field office (blue) and Muck House (white)

developed urban setting. Hence, truck delivery and material removal strategies and routes had to be carefully considered in a manner to avoid impacts to the day-to-day construction operations. These strategies included the following, for example:

- Limited working hours for surface operations—Monday to Friday and Saturdays.
- NYC DOT load dimension restrictions and specified Manhattan access points.
- Seasonal embargoes on heavy and oversize vehicle traffic.
- Total quantities delivered and removed and rates per hour.

In very general terms, the truck delivery and removal quantities included the following as listed in Table 1.

At peak production during the rock excavation phase, over 100 muck haul trucks per day (15 hours/day) were at the site. In the same period, additional trucking for construction materials and concrete were needed. In order to accommodate the flow and availability of trucks, special provisions were made to temporarily use travel and parking lanes on Second Avenue for as many as three full city blocks (200' per block). In this manner and with truck coordinators present, 15 to 20 trucks could be queued and rapidly dispatched to the Construction Shafts or any of the three concrete pumping stations. This was an invaluable arrangement needed for the reliable and timely supply of deliveries to the site. It was apparent that in order to meet the construction schedule, the quantity and flow of delivery trucks to the site had to be well managed with use of nearby queuing areas, without impacting local residents or businesses.

### Tidy Industrial Park Concept for All Temporary On-Site Surface Facilities

While most construction sites are set-up around the direct construction needs and supplied with readily available equipment, temporary buildings and facilities, much of the 72nd Street project area was planned in advance, laid-out and deliberately mobilized with "appearance" in mind. To do so included choice of temporary facility locations, dimensions, colors and surface finishes in a manner best suited for the local environment and generally pleasing but still fully functional for continuous construction needs. Figures 17 and 18 illustrate the size and conformity.

The block-to-block layout of dozens of facilities was orchestrated to match existing utility systems, Construction Shafts, truck routes, fencing as well as vehicle and pedestrian flows. Careful attention was paid to fencing and gate locations to minimize the neighborhood impacts. Ultimately, fences were covered with decorative panels and announcement boards promoting local businesses and MTA's "Shop Second Avenue" campaign.

While the overall appearance of the linear site both from street level and above was a dense functional construction operation, the tidy, color-coordinated and enclosed appearance was orderly, secure and in many respects, nested well into the local neighborhood—albeit as a tidy "industrial park" on a temporary basis.

### Re-Sequencing of Work in the South Tunnels and Caverns

The original concept and work sequence of the job was integrated into the initial approved CPM schedule. After agreements were in place to continue uninterrupted blasting through two separate "No Blasting" periods, a thorough review of the work sequence in the south tunnels and caverns was initiated. This resulted in a proposed overhaul to the work sequence that simplified and streamlined the excavation and final lining operations throughout the 2,000 LF of south tunnels and caverns. It resulted in multiple concurrent operations in a retreating direction from the 63rd Street Stub Cavern northward to the station and the 69th Street Construction Shaft. This shaft area was the final exit point from the job and was key to the re-sequencing plan for the south tunnels and caverns. The original work flow was not so sequential.

The MTA reviewed and quickly endorsed the re-sequencing plan since it made full use of the "No Blasting" periods and eliminated possible conflicts at the 63rd Street Station contract interface area. There was also a potential for earlier turn-over of the south tunnels and caverns to the follow-on System-Wide contract, well before the Substantial Completion milestone.

One issue that had to be addressed with the re-sequencing of the work involved a significant overhaul of the CPM schedule, not only for the south tunnels and caverns but also to untie integrated and interlocked station area work activities. The re-sequenced work provided more flexibility to perform tasks as two independent and potentially competing operations, whereas, formerly one area was largely conditional and dependent on the other. Schedule risks declined.

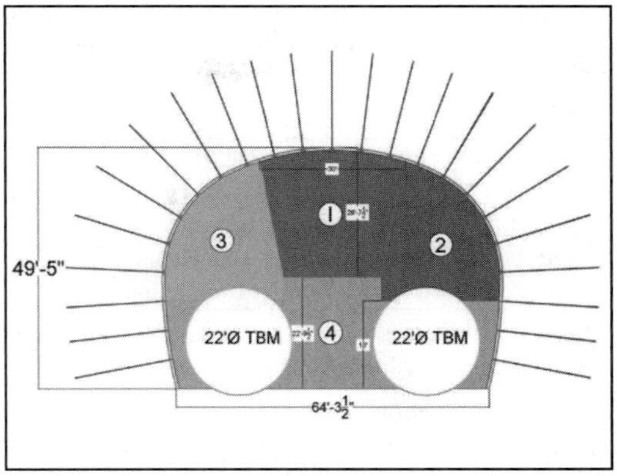


**Figure 19. Section through the station cavern showing the Initial Support layout for various excavation phases**

### Universal Initial Ground Support Design

Initial ground support in the caverns, tunnels and adits was prescribed in the Contract Documents and included fully resin-grouted rock bolts and dowels of different lengths—depending on cavern or tunnel location. All rock bolts and dowels were specified with the same diameter rods, steel grade and resin encapsulation. The concept worked well with the bolt lengths as the only variable for all prescribed locations. Spacing patterns were also prescribed for each tunnel, adit or cavern location. Figure 19 illustrates the *Initial Support* in the station.

Rock dowels were similarly detailed for the station side walls. Whereas rock bolts were installed with a prescribed pre-tension (load), rock dowels were not tensioned. The prescribed *Initial Support* design was beneficial to the extent that it could be procured well in advance, materials stockpiled and was readily available.

## COMPREHENSIVE PLANNING FOR THE WORK

Detailed planning for the work was deliberately incorporated into the final design and illustrated on the Contract Drawings. Construction planning for implementation of the work was performed by the Contractor and recorded on technical submittals. Additional submittals that addressed detailed planning and procurements were provided by subcontractors as the work progressed.

The detailed planning process was well embraced by the Contractor, the Construction Manager, the MTA and its Designer. This concept and work process was evident in all major and

**Table 2. Summary contract durations of the primary work in the stations and tunnels for Phase 1 of the Second Avenue Subway Program schedule***

| | | Contract Time Allocations | | | |
|---|---|---|---|---|---|
| **Item** | **Contract Location** | **Civil** | **Finish** | **Systems** | **Totals** |
| 1 | 63rd Street Station (Rebuild) | 46 | 0 | 6* | **52** |
| 2 | 72nd Street Station and Tunnels | 37+2.5 | 31* | 6* | **74.5** |
| 3 | 86th Street Station and Tunnels | 37 | 36* | 6* | **79** |
| 4 | 96th Street Station and Tunnels | 38 | 30* | 6* | **74** |
| 5 | **Second Avenue Target completion date** | | | | **Dec 2016** |

* Many durations overlapped with one another but none could be delayed.

modest construction operations and was frequently the focus of many meetings and correspondence between the parties. Clearly the commitment to through, innovative and comprehensive construction operations planning was one of many activities that contributed to the successful completion of the work. But good planning had to be communicated and fully implemented in the field to be successful. Between all parties, this commitment and follow-though was demonstrated through highly productive meetings and frequent (daily) discussions.

## Construction Planning in the Detailed Design Phase

The design phase for the project integrated the program schedule that addressed and balanced the construction needs and work sequence and durations of the initial four concurrent work sites. Table 2 summaries the overall planned durations of the contracts on Second Avenue.

A significant amount of construction planning was integrated in advance into the initial design of the work for the 72nd Street project. This was clearly evident starting with the pre-excavation of the underlying TBM tunnels. The Contract Documents also included detailed phasing diagrams for all excavation and ground support for all tunnels and caverns. The final lining operations were defined by construction joint locations and highly repeatable forming and poring sequence for all inverts, walls and arches in all areas. Additionally, drainage and waterproofing designs were common for all areas. Later, some highly beneficial Value Engineering concepts further simplified the work and helped improve the construction schedule in specific areas.

## Construction Planning for the Excavation Phase

The initial stage of the work involved a massive multi-phase rock excavation that at times included as many as seven separate but concurrent operations—in the first 12 months (approximately) of operation. These included the following, for example:

- **Underground Rock Excavation**
  - Station cavern top heading and bench—north end
  - Station cavern top heading and bench—south end
  - Stub cavern top heading and G3 cavern enlargement
- **Ancillary and Entrance Shafts**
  - ANC-1 shaft excavation
  - ANC-2 shaft excavation
  - ENT-3 shaft excavation

Some of the above-listed work areas involved multiple drills and concurrent working faces—all sequenced with blasting, mucking and installation of *Initial Support* materials. In all work areas, careful planning was needed and included attention to the following activities, for example:

- Detailed excavation work sequence and blasting plans
- Work schedules and safety systems planning
- Rock reinforcement sequencing for shotcrete, rock bolts and dowels
- Quality Control plans for rock reinforcement
- Materials handling logistics for muck removal

In order for all concurrent rock excavation operations to progress well and without delays and interruptions, many other supporting activities had to be carefully addressed, including:

- Dust control and particulate measurements
- Blast vibration control and measurements
- Traffic flow adjustments and detailed safety provisions for the works and the public

Overall, underground blasting finished as scheduled but due to the sequence of the final stages, (in the station cavern area), a successful early start to the final lining operations was achieved concurrently in two distinct areas. This early start was critical to the

**Table 3. Summary of the principle work elements comprising the final tunnel liner in the tunnels, caverns and adits throughout the project. All elements had to be carefully sequenced to avoid conflicts and work interruptions.**

| | | Final Concrete Liner Locations | | | | |
|---|---|---|---|---|---|---|
| | | Caverns | | | Tunnels | |
| Item | Work Description | Invert | Walls | Arch | Invert | Arch |
| 1 | Drainage system materials | ● | | | ● | |
| 2 | Waterproofing membrane | ● | ● | ● | ● | ● |
| 3 | Reinforcing steel and embedded metals | ● | ● | ● | fibers | fibers |
| 4 | Formwork and shoring systems | ● | ● | ● | ● | ● |
| 5 | Custom support systems | | | ● | | ● |
| 7 | Concrete placing | ● | ● | ● | ● | ● |
| 8 | Architectural finishing in designated areas | | | ● | | |

success of the final lining operations that assumed prominence on the Critical Path schedule until the end of the project.

### Construction Planning for the Final Concrete Lining

Initial planning for the final lining started in earnest at about Month 4 (January 2011). Initial pours were made in the 63rd Street Stub Cavern in Month 18 (March 2012) and the station cavern in Month 21 (June 2012). Detailed planning work for the final lining was intensive and involved dozens of professional staff from the MTA, the Designer, Construction Manager, Contractor as well as specialty suppliers and subcontractors. Table 3 lists the work scopes.

Careful and comprehensive planning for the final liner was essential for successful completion of the work within the remaining time on the schedule after the completion of rock excavation. The work would require as many as seven concurrent operational work areas, each actively placing elements of the final liner that included the following.

In order to accomplish the work with high quality results, an extensive arrangement of personnel, submittals, schedules, materials and equipment procurement was integrated into an intensive planning program. Planning for the work, took over 2.5 years while concurrently, portions of the lining was constructed. Much of the final liner was constructed on the Critical Path and therefore, controlled the completion date of the project. Detailed and thorough planning was, therefore, critical to meet interim Milestone and Substantial Completion dates.

The strategy for installation of the final concrete liner included five major arch forming systems, two curved wall forming systems, one tunnel invert and a separate arch forming system, in addition to numerous flat panel braced forms. Concrete was placed from drop holes connected to large concrete pumps on the surface. Additionally, site logistics were extensively modified from muck handling to materials and concrete handling at multiple locations.

## OWNER, DESIGNER, AND CONSTRUCTION MANAGER FIELD OFFICE

The MTA fully staffed their field office with experienced construction personnel in addition to the Construction Management staff from the start of the work to the end of the project. It also added specific Designer personnel for extensive periods to augment their on-site staff to efficiently deal with construction and occasional design-related issues. This was a good solution overall since it responded well to the nature and rapid pace of the schedule with prompt responses to inquires, RFI's, submittals and Technical Meeting follow-up discussions.

Attendance at frequent on-site meetings was particularly beneficial in the first 18 months of the project. The entire Construction Management team stayed fully informed and engaged in the day-to-day progress of the works and was always available for information and discussions. Very effective lines of communication were established early that endured throughout the job for the benefit of not only the parties but also the project, and the Second Avenue Subway program as a whole. It cannot be understated, the great importance to the success of the project of proper and sufficient staffing in the MTA's field office—with experienced Construction Manager and Designer professionals.

## MANAGEMENT OF ISSUES AND CHANGES TO THE SCOPE OF WORK

On fast-track work, issues that arose needed to be handled and finalized very promptly. In the event of lingering delays to issue resolutions, the progress of the work would suffer. This problem is also evident for directed changes to the work. The project staff must in all cases, maintain a clear vision of the work while being equipped to handle potential issues, delays, work interruptions and periodic changes

**Table 4. Summary of meeting types and frequencies as well as the usual attendees. All site meetings were held at MTA's field office, the most convenient for all attendees and located very proximate to the site at 66th Street.**

| | | Meeting Frequency | | | | Meeting Attendees | | | | |
|---|---|---|---|---|---|---|---|---|---|---|
| **Item** | **Description of Meetings** | **Wky** | **BWy** | **Mon** | **A/N** | **MTA** | **DHA** | **CCM** | **SSK** | **FTA** |
| 1 | Progress Meetings | ❍ | ● | ❍ | | ● | ● | ● | ● | |
| 2 | Technical Meetings | | ● | | | ● | ● | ❍ | ● | |
| 3 | "Seniors" Meetings | | | ● | | ● | ● | ● | ● | |
| 4 | Quality Control Meetings | | ● | ❍ | | ● | ● | ● | ● | ● |
| 5 | Special Safety Meetings | | ❍ | | ● | ● | | ● | ● | |
| 6 | Special Schedule Meetings | | | | ● | ● | ● | ● | ● | |
| 7 | Change Order Discussions | | | | ● | ● | | ● | ● | |

in scope. The MTA, the Designer, Construction Manager and the Contractor fully recognized that problems would result from tardy resolution and had in place, sufficient experienced personnel to handle these events. Additionally, directed changes had to be priced and fully agreed before the changed work could proceed. This process kept contract administration staffs working quickly to avoid delays while settling scope, quantity and pricing matters. They were all mindful of the project schedule and the discreet work sequences that may be impacted from modest changes in scope or delays that may occur.

### Prompt Resolution of Design and Construction Issues

The fast-track construction schedule inherently meant that technical and administrative issues and procedures would be equally promptly handled to avoid impacts to the work. In the case of technical and design issues that periodically arose—both from the Owner and Contractor's perspective, most were handled promptly and in a collaborative manner. To do so, involved frequent and productive discussions, meetings and exchange of comprehensive information. Few issues required lengthy resolution times and if any were identified as such, additional early attention was provided. The means for communications included the following:

- Requests for Information (RFI) submitted electronically and responded in the same mode. Some RFI's involved meeting discussions and informal resolutions
- Submittal of shop and working drawings as well as supporting technical information through an electronic portal established for this purpose

Additional communications included biweekly Technical Meetings and joint site inspections as well as special meetings with peripheral groups. Summaries, schedules and logs of discussions and resolutions to technical issues were also maintained to record progress and closure.

### Periodic Meeting Arrangements for Design and Construction Issues

Time for review and resolution of issues was always limited, hence the establishment of a multi-level meeting structure and frequency to identify, discuss and resolve issues, as well as to review progress of the work, planning, schedules and other matters. These meetings were generally held as listed in Table 4. Preparation, discussions and follow-up were essential.

It was very evident at each meeting and in all levels of the on and off-site organizations, the pace of the work was fast and essential to meet the key schedule goals. Virtually everyone had a keen sense of time and accomplishment related to progress of the work, hence the strong universal sense of urgency, issue resolution as well as periodic reviews, alternate work sequences and remedies, all focused to achieve schedule advantages.

## MANAGING THE CONSTRUCTION SCHEDULE

The Second Avenue Program schedule was extremely important for the MTA to maintain and to closely monitor since the four individual station schedules were nested within this program schedule. Considerable visibility, discussion and analysis of the 72nd Street schedule including progress of the work and planning for future operations, were frequent topics of discussion at formal and informal meetings. While the extensive CPM schedule for the 72nd Street project idetntified large and small work activities, the more defined 6-Week Look-Ahead schedules were more frequently used for effective day-to-day and week-to-week operations planning.

### Schedule Reviews and Updates

There were broad and diverse audiences for various scheduling work products. Each was carefully reviewed on a weekly, biweekly or monthly basis with periodic in-depth projections made to gage progress, delays and conflicts in specific work areas. Managing the schedule process and updates fell to several individuals on the MTA and Contractor staffs. See Table 5.

Overall, the CPM schedule, 6-Week Look-Ahead and other special schedules were very closely monitored and utilized in planning the work. Please see Figure 20. Careful preparation was needed for many of the short-term schedules shared between the MTA and the Contractor since many outside factors often had an influence on the progress and outcome. It was these plans and time-sequenced schedule work products that when fully used, made a very positive impact on the job. They communicated well among the parties, avoided mis-understandings while identifying critical "hot spots," Hold Points, and "hard targets" for many competing, conflicting and concurrent operations, often using shared resources in several areas of the site.

### "No Blast" Periods on the Critical Path for Station Cavern Excavation

The contract included two lengthy "No Blast" periods in the first 18 months of the construction schedule. These periods were intended to accommodate the safe passage of the TBM below the station cavern excavation (only) and required separate, three and four

**Table 5. Summary of the scheduling work products and the primary persons involved in this important process. Meetings were frequently held on-site to address and evaluate progress updates, issues and potential delays.**

| Item | Schedule Work Product | On-Site Staff Personnel | |
|---|---|---|---|
| | | MTA | SSK |
| 1 | **CPM Schedule Management** | | |
| | Periodic updates | Schedule Manager | Schedule Engineer |
| | Logic changes and Extra Work changes | | |
| 2 | **6-Week Look Ahead Schedules** | | |
| | Weekly updates and distribution | Resident Engineer | Operations Manager |
| | Planning, submittals and new work | | |
| 3 | **Special Work Operation Schedules** | | |
| | New work plans | Resident Engineer and | Operation Manager and |
| | Special operations, integration and tasks | Construction Manager | Engineering Manager |
| 4 | **Time Impact Analyses** | | |
| | Directed Extra Work scopes | Resident Engineer and | Project Manager and |
| | Unanticipated delays & work interruptions | Schedule Manager | Schedule Engineer |

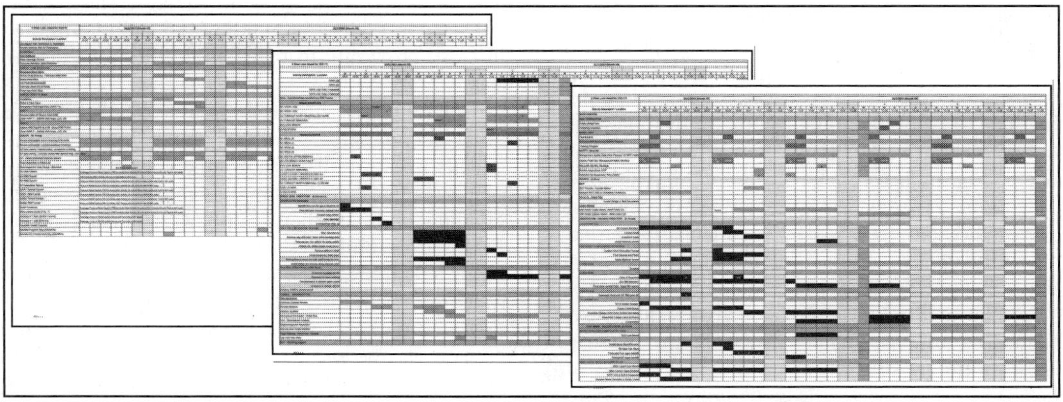

**Figure 20. Typical 6 Week Look-Ahead construction schedule during the final lining operations. Considerable detail and linkage between multiple work areas, material supply tasks and forward planning activities were typical.**

**Table 6. Summary of key dates and data related to the excavation and final concrete lining operations in the station cavern, south tunnels and caverns—within the initial and adjusted contract schedule times**

| | | Key Date, Month, and Data | | |
|---|---|---|---|---|
| **Item** | **Description** | **Date** | **Month** | **Data** |
| **A** | **Key Contractual Dates** | | | |
| 1 | Notice-of-Award and Notice-to-Proceed | 01 Oct 10 | 0 | Start of 37 month schedule |
| 2 | **Milestone 1:** | | | |
| | Initial date per contract | 30 Apr 13 | 31 | After 31 months |
| | Adjusted date with Extensions-of-Time | 15 Jul 13 | 33.5 | with 2.5 months EOT |
| 3 | **Substantial Completion:** | | | |
| | Initial date per contract | 30 Oct13 | 37 | After 37 months |
| | Adjusted date with Extensions-of-Time | 15 Jan 14 | 39.5 | with 2.5 months EOT |
| **B** | **Key Operations Dates** | | | |
| 1 | **Rock Excavation** | | | |
| | First test blast at 69th Street Shaft | 21 Jan 11 | 4 | Small round in construction shaft |
| | Final underground blast—at 69th area | 08 Sep 12 | 24 | Final bench/invert blast |
| 2 | **Final Lining Operations** | | | |
| | Initial final lining (mud mat & sand walls) | 15 Mar 12 | 18 | 63rd Street Stub Cavern |
| | Initial final lining (reinforced concrete) | 19 Jun 12 | 33 | 63rd Street Stub Cavern |
| | Initial station cavern pour (sump invert) | 26 Jul 12 | 34 | Slab at 105' below street |
| | Final station cavern pour (arch No.33) | 03 Dec 13 | 39 | 30' long at south end wall |

month periods of no blasting in the station cavern. The "No Blasting" periods did not apply to the south tunnels and caverns, leaving these areas free to continue unhindered excavation (or final lining) operations—to the extent that these were underway and accessible at the time of the "No Blasting" periods.

While the "No Blasting" periods were determined to be at specific times in the construction schedule, there were still large variables with respect to the actual impact periods and what other productive work could be performed during these periods. On closer examination, the "No Blasting" periods were intended to facilitate "blast vibration free" periods during the TBM passage through the station cavern area; i.e., below the ongoing top heading excavation. As such, the "No Blast" periods were deemed to be extremely disruptive to the work and would delay the station Critical Path schedule during the critically important station excavation phase but still had to be accommodated in the schedule of operations. Table 6 lists key dates.

A solution was achieved between the adjoining contractors and endorsed by the FDNY (governing authority in New York City for blasting matters) that allowed for continuous blasting through both "No Blasting" periods provided that special work sequences were used and that two emergency escape shafts were constructed to connect the station top heading excavation to the underlying TBM bored tunnel (east side only). The result was continuous TBM mining concurrently with continuous drill and blast excavation in the station top headings—including the side slashes. Overall, this arrangement greatly benefited the two projects in the Second Avenue Subway construction program without delaying the Critical Path of the 72nd Street project. In the absence of the joint accommodation for concurrent, operations, there would have been a *seven month delay* to the station cavern excavation and subsequent final lining operations. Please see Figures 5, 6 and 7 that highlight the "No Blast" periods.

## Blast Vibration Controls and Their Influence on the Construction Schedule

Several specific blast vibration limits were set in the Contract Documents to minimize public disturbances and potential property damage. These limits are listed in Table 7 for building and facilities located in close proximity to the site and the Construction Shafts.

The blast vibration limits (Peak Particle Velocity) measured in inches per second, were frequently a point of discussion largely due to the means of measurement since they were intended to measure ground vibrations (energy) only without the dynamic effects from buildings or utility systems. The PPV limits had a direct link to the blasting quantities per round and, therefore, *set the pace of the rock excavation work*—especially in the early stages when the excavation was located closest to the surface. Many factors also influenced the reliable measurement of blast vibrations. Recently, the FDNY has recommended the use of a strain gauge measurement system to replace error-prone seismographs in order to properly measure PPVs. This system would improve the quality of measurements and provide a more reliable indicator of the ground transmitted blast energy imparted to a structure. The excavation schedule would also benefit considerably, with longer blast round lengths *within the PPV limits*.

## CONSTRUCTION SUPPORT PROVISIONS AND SYSTEMS

Although not directly involved with day-to-day construction operations, MTA engaged numerous on and off-site personnel and outside organizations for support of all Second Avenue Subway construction activities. The support systems and organizations were *absolutely critical* to maintaining public awareness and support for a complex project located within a densely developed neighborhood, where the "NIMBY" concept had to be frequently addressed in a manner that was patiently overcome. Please refer to Table 8 for a summary. Ultimately, confidence in the MTA and the Second Avenue Subway were successfully achieved.

The MTA engaged several specialized and experienced personnel from its staff, outside consultants and the Construction Management staff for the following tasks, services, and functions, all focused on construction support systems and structured organizational approach. Please see Table 8 that summarizes the programs and service established.

Several other groups had significant involvement in the project from time-to-time that contributed

**Table 7. Summary of the prescribed (and adjusted) ground-born allowable blast vibrations; velocity and frequency data, as measured at the location of various buildings and utility systems along the tunnel and cavern route**

| | | Blast Vibration Data | | | | | Affected Work Areas | | | | | |
|---|---|---|---|---|---|---|---|---|---|---|---|---|
| | | Original Values | | | Revised Values | | Caverns | | Tunnels and Adits | | | |
| **Item** | **Description** | **Velocity** | **Freq.** | **Distance** | **Velocity** | **Freq.** | **Station** | **T/Os** | **Stub** | **G3** | **G4** | **Adits** |
| 1 | Normal Buildings | 1.92 ips | >40 hz | | 1.92 ips | >40 hz | ● | ● | ● | ● | ● | ● |
| 2 | Fragile Buildings | 0.50 ips | >40 hz | None | 1.20 ips | >40 hz | ● | ● | ● | ● | ● | ● |
| 3 | Sensitive Buildings | 0.50 ips | >40 hz | None | 1.20 ips | >40 hz | ● | | | | | ● |
| 4 | Historic Buildings | 0.50 ips | >40 hz | None | 1.20 ips | >40 hz | ● | ● | ● | ● | ● | ● |
| 5 | Landmark Buildings | 0.50 ips | >40 hz | | 0.50 ips | >40 hz | | ● | | ● | | |
| 6 | Underground Utility Systems | 0.50 ips | >40 hz | | 0.50 ips | >40 hz | ● | | ● | | | ● |

**Table 8. Summary of MTA established Community Relations and Outreach Programs for the 72nd Street Station and Tunnels Project. Ultimately, a storefront Community Relations Center was established at the 86th Street Station.**

| | | Prime | Responsible Party | | | |
|---|---|---|---|---|---|---|
| **Item** | **Description** | **Location** | **MTA** | **Consul** | **CCM** | **Contr** |
| 1 | Community Relations Program | Site + Web | ● | ❍ | ❍ | |
| 2 | Community Outreach Program | Site + Web | ● | ❍ | ❍ | |
| 3 | Media relations organization | HQ + Web | ● | ❍ | | |
| 4 | Public tours—underground areas | Site | ● | ❍ | ● | ● |

**Table 9. Summary listing of the principle New York City agencies having influence and significant jurisdiction on portions of the work performed at the project. Many agency representatives were very supportive of the project.**

| | Description | | Specific Operations | |
|---|---|---|---|---|
| **Item** | **Abb.** | **Department** | **Responsibility** | **Work Phase** |
| 1 | FDNY | New York City—Fire Department | Blasting controls | All rock excavation |
| 2 | DOB | New York City—Department of Buildings | Building vibrations | All rock excavation |
| 3 | DOB | New York City—Cranes and Derricks | Crane inspections | All rock excavation |
| 4 | DOT | New York City—Dept of Transportation | Oversize loads | All rock excavation |

**Table 10. Summary of issues and activities that needed special attention over the course of the work. These items were generally initiated by neighborhood concerns and were handled professionally by the MTA and others as listed.**

| Item | Description | Construction Operation | Specialists Acquired From | | | | |
|---|---|---|---|---|---|---|---|
| | | | MTA | PB | DHA | CCM | SSK |
| **A** | **Excavation Operations** | | | | | | |
| 1 | Odor, fumes & smoke from blasting | Excavation | ○ | ● | | ● | |
| 2 | Vibrations from blasting | Excavation | ○ | ● | ○ | ● | ● |
| 3 | Fugitive dust from blasting | Excavation | ○ | ● | | ● | ● |
| 4 | Truck traffic flows through the site | Excavation | ○ | | | ● | |
| 5 | Truck queuing on Second Avenue | Excavation | ○ | ● | | ● | |
| 6 | Additional site work areas | Excavation | ○ | | | ● | |
| **B** | **Final Lining Operations** | | | | | | |
| 1 | Truck traffic flows through the site | Final Lining | ○ | | | ● | |
| 2 | Truck queuing on 2nd Avenue | Final Lining | ○ | | | ● | |
| 3 | Additional site work areas | Final Lining | ○ | ● | | ● | |

to its visibility and momentum for on-time completion. Please see Table 9.

There were numerous occasions when very specialized services were needed to aid in the construction process and for which the MTA was best suited to address—for information management and contractual responsibility. Please see Table 10.

## CONCLUSIONS AND RECOMMENDATIONS

The 72nd Station and Tunnels Project was a critical success to MTA, the Designer, Construction Manager and Contractor for several reasons. It demonstrated that an aggressive fast-tracked construction schedule could be achieved amongst difficult site and working conditions. But in order for this to occur, many factors had to be considered and deliberately built into the detailed design, construction planning and execution—all managed around an attainable construction schedule while attending to multiple concurrent operations and Critical Paths. On many occasions, many construction operations competed or conflicted with one another. Numerous compromises, risk analyses, and comparative studies were made. Ultimately, three separate but concurrent excavation operations were underway, followed by nominally seven separate concurrent final lining operations. Absent the multiple concurrent, well sequenced, equipped and staffed construction operations, the project would have not met key schedule goals.

Successful multiple concurrent operations were the result of the factors, conditions and activities, listed in Table 11—all dedicated to achieving timely results.

There were several outstanding aspects of the work that contributed to fulfillment of the fast-track schedule. While many seem ordinary and fundamental to the work, they were nonetheless provided and/or maintained in manner that greatly benefited the work while keeping pace with the progress and all challenges encountered.

- Open and transparent communications at all levels throughout the construction
- Frequent face-to-face meetings that were focused and productive
- Clear mutual understanding of project goals, obstacles and schedules
- Trust between the parties with a strong sense of urgency and mutual confidence
- Problem identification and resolution amongst the parties
- Deliberately shared resources and benefits between the parties

## ACKNOWLEDGMENTS

The authors would like to acknowledge the considerable contributions to this paper (and corresponding oral presentation) from the MTACC (Program Manager), NYCTA (Project Owner), Aecom-Arup (Project Designers), Parsons Brinckerhoff (Consultant Construction Manager) and SSK (Contractor). In particular, very helpful information was provided and included in this paper from Bill Goodrich (MTACC), Jaroslav Prokop (PB) as well as Jaidev Sankar and Verya Nasri from Aecom-Arup. Brian Fulcher, SSK's Underground Operations Manager responsible for all excavation and final lining operations, initially developed the concept for this paper provided timely review and comments throughout its development.

**Table 11. Summary of activities and services brought to the project for the benefit of the parties. These items can be traced to the critical success of the fast-tracked schedule and multiple concurrent work operations and schedules.**

| | | Credit Going to the Parties | | | | |
|---|---|---|---|---|---|---|
| **Item** | **Description of Activity or Service** | **MTA** | **DHA** | **CM** | **SSK** | **Subs** |
| **A** | **Initial Design and Construction Approach** | | | | | |
| 1 | Final design—solid and comprehensive | ● | ● | ○ | ○ | ○ |
| 2 | Access to the work—multiple points | ○ | | ○ | ● | ● |
| 3 | Expanded work areas to suit compressed schedules | ○ | | ○ | ● | ● |
| 4 | Construction approach with concurrent operations | ○ | | ○ | ● | ○ |
| **B** | **Personnel and Experience** | | | | | |
| 1 | Experienced field staff | ● | ○ | ● | ● | ● |
| 2 | Field coordination | | | ● | ● | ○ |
| 3 | Supporting systems and technical expertise | ○ | ● | ○ | ● | |
| **C** | **Planning and Scheduling Commitment** | | | | | |
| 1 | Detailed planning and technical submittals | ○ | ○ | ○ | ● | ○ |
| 2 | Schedules and work sequenced plans | ○ | | ○ | ● | |
| 3 | Alternate work schemes and sequences | ○ | | ○ | ● | |
| **D** | **Resources Applied to the Work** | | | | | |
| 1 | Plant and equipment quantities and capacities | ○ | | ○ | ● | |
| 2 | Labor relations—problem-less and productive | ● | | ● | ● | ○ |
| 3 | Community Relations and Outreach programs | ● | | ○ | ○ | |

## REFERENCES AND ADDITIONAL INFORMATION

There have been many recent papers and publications written on the design and construction of the 72nd Street Station and Tunnels project as part of MTA's Second Avenue Subway. While many predate the events of the site work described in this paper, they should be considered as useful preconstruction references.


Desai, D., Nail, M., Rossler, K., Stone, C., New York Subway Caverns and Cross-Overs—a Tale of Trials and Tribulations, 2005, *Proceedings of the Rapid Tunnels and Excavation Conference, 2005*, Seattle, Washington. Littleton: Society of Mining Engineers.

Desai, D., Lagger, H., Stone, C., New York Subway Stations and Cross-Over Caverns—Update on Initial Support Design, 2007, *Proceedings of the Rapid Tunnels and Excavation Conference, 2007*, Toronto, Canada, Littleton: Society of Mining Engineers.

Fulcher, B., Menge, S., Grillo, J. 2013; Construction of a Large Span Station Cavern, Running Tunnels, Cross-Over and Turn-Out Caverns, Shafts and Entrances, 2013; *Proceedings of the Rapid Tunnels and Excavation Conference, 2013*, Washington, DC. Littleton: Society of Mining Engineers.

Nasri, V., Hoffman, S., Parikh, A., 2013; Construction of 86th Street Station—Large and Shallow Rock Cavern in New York; *Proceedings of the World Tunnel Congress 2013*, Geneva, Switzerland, International Tunnelling Association.

Nasri, V., Fulcher, B., Redmond, R., 2008; Design of the Second Avenue Subway in New York, 2008; *Proceedings of the World Tunnel Congress 2008*, Agra, India, International Tunnelling Association.

Nasri, V., 2008; Design of Large and Shallow Cavern of the New York Second Avenue Subway, 2008; *Proceedings of the North American Tunneling Conference, 2008*, San Francisco, California. Littleton: Society of Mining Engineers.

Nasri, V., Bergeson, W., 2010; Continuum and Discontinuum Modelling for Design of Shallow Caverns in Jointed Manhattan Schist New York, 2010; *Proceedings of the World Tunnel Congress, 2010*, Vancouver, Canada, International Tunnelling Association.

Nasri, V., Fulcher, B., Redmond R., and Parikh, A. 2011, Geotechnical Investigation for the Second Avenue Subway in New York. *Proceedings of Planning and Development of Underground Space*, 2011. Hong Kong, China.

Nasri, V., Fulcher, B., Redmond, R., and Parikh, A. 2012, Design and Construction of 72nd Street Large and Shallow Rock Cavern Station in New York City. *Proceedings of the North American Tunneling Conference, 2012*, Indianapolis, Indiana. Littleton: Society of Mining Engineers.

Nasri, V., Fulcher, B., and Redmond, R. 2012, Design and Construction of 72nd Street Station Rock Cavern in New York. *Proceedings of the World Tunnel Congress, 2012*, Bangkok, Thailand, International Tunneling Association.

Parikh, A., Fosbrook, G., Phillips, D., 2005; Second Avenue Subway—Tunnelling Beneath Manhattan, 2005; *Proceedings of the Rapid Excavation and Tunnelling Conference, 2005,* Seattle, Washington. Littleton: Society of Mining Engineers.

Sankar, J., Mukherjee, A., Delle, A., 2013; Designer / Construction Manager / Contractor Perspective of Design and Construction of the New 72nd Street Station of the Second Avenue Subway; *Proceedings of the Rapid Tunnels and Excavation Conference, 2013*, Washington, D.C. Littleton: Society of Mining Engineers.

Snee, C.P., Ponti, M.A., Shah, A.N., 2004; Investigation of Complex Geologic Conditions for the Second Avenue Subway Tunnel Alignment in New York City, *Proceedings of the North American Tunnelling Conference, 2004,* Atlanta, Georgia. Littleton: Society of Mining Engineers.

Snee, C.P., 2008; Engineering Geology and Cavern Design for New York City, *Proceedings of the North American Tunnelling Conference, 2008,* San Francisco, California. Littleton: Society of Mining Engineers.

## SELECTED ADDITIONAL READINGS

Garavito-Bruhn, E., Napoli, A., Towell, N., 2012; Second Avenue Subway Project, New York: Refurbishment of an Existing Underground Station, *Proceedings of the Structures Congress 2012.* American Society of Civil Engineers.

Parikh, A., Phillips, D., Sykes, M., Second Avenue Subway Project—History and Construction Challenges, *Proceedings of the Geotechnical Engineering for Transportation Projects*, American. Society of Civil Engineers.

# Manhattan Underground—Rapid SEM Mining by Drill and Blast of the Second Avenue Subway's 86th Street Station Cavern

**John Kiernan and Michael Stokes**
Skanska USA Civil Northeast

**Julie Freitas**
Parsons Brinckerhoff

**ABSTRACT:** As part of the Second Avenue Subway Project, rapid sequential excavation methods (SEM) by drill and blast were used to mine the station cavern, two open cut vertical shafts, and ancillary structures for the 86th Street Station Cavern Mining and Heavy Civil/Structural Contract now under construction. The top heading, intermediate and bottom benches and sumps were blasted under low rock cover in the densely urbanized, Manhattan Upper East Side neighborhood. This paper describes the means and methods employed to coordinate expedited drill, excavation and rock support production with minimal impact to the community and surrounding buildings. Overcoming challenges such as observing stringent vibration criteria, coordinating with the public, working with utilities in low headroom, operating within very tight surface staging, and mitigating environmental impacts, proved critical. Regular coordination among representatives of the contractor, designer, construction manager and owner demonstrated to be essential for safely managing adverse ground conditions, maintaining the project schedule, and mitigating difficult ground.

## INTRODUCTION

### Phase 1 Overview

New York City Transit (NYCT), for the first time in over sixty years, is expanding their subway system with the four-phase Second Avenue Subway (SAS) Project. The first phase of the project, includes new tunnels from 105th Street to 63rd Street, with new stations at 96th, 86th, and 72nd Streets, and new entrances to the existing Lexington Avenue/63rd Street Station at 63rd Street and Third Avenue (MTA). The 86th Street Station Cavern Mining and Heavy Civil / Structural Contract now under construction includes the removal of approximately 197,357,097 $m^3$ (160,000 BCY) of rock, and the placement of the permanent Station concrete lining (Figure 1).

The Metropolitan Transportation Authority Capital Construction (MTACC) is the client, and the design engineer is the joint venture AECOM/Arup. The consultant construction manager is Parsons Brinckerhoff (PB/CCM), and the contractor is Skanska/Traylor joint venture (STJV). The estimated daily ridership for Phase 1 is expected to be 213,000, with a target completion date of December 2016, at a cost of $4.45 billion (MTA 2013).

During Phase 1, there are four concurrent station cavern construction contracts in progress at 96th, 86th, 72nd, and 63rd Streets. The first Phase 1 contract consisted of two TBM (tunnel boring machine) mined 6.7 m (22 ft)-diameter tunnels under Second Avenue running parallel from roughly 92nd to 65th Streets. The total mined length was 3900 m (12,800 ft); the S1 (West) tunnel was 2,375 m (7,800 ft), and the S2 (East) tunnel was 1,524 m (5,000 ft) in length. TBM mining completed on September 22, 2011, and the 86th Street Station Cavern contract notice to proceed (NTP) was issued August 4, 2011.

### Excavation Limits

Just below Second Avenue, the 86th Street Station Cavern spans from 87th to 83rd Streets. The project contains two ancillary and a public cavern, several adits and cross passages, three open cuts, two shafts and two inclines in the heart of Manhattan's Upper East Side neighborhood (Figure 2). The cavern measures approximately 286 m (938 ft) long, 21 m (69 ft) wide and 14.6 m (49 ft), and 18.5 m (60 ft) high in the Public and Ancillary Caverns measures, respectively. At an average invert depth of 30.5 m (100 ft) below ground surface, the rock cover above the cavern varied from 7.6 to 19.8 m (25 to 65 ft). At 83rd and 87th Streets, there are construction shafts, and two corresponding Ancillary structures. The shaft at the north end of the cavern measures roughly 12 m (40 ft) by 9 m (30 ft) and the open cut at the south end of the cavern measures 61 m (200 ft) by

12 m (40 ft). Two entrances and an elevator shaft will be constructed; Entrance 1 will be incorporated into the basement and first floor of an existing high rise residential building and Entrance 2 will be adjacent to one of the largest luxury residential buildings in Manhattan.

## HITTING THE GROUND RUNNING

### Construction Power and Utilities Logistics

The New York Metro area, especially Manhattan is often referred to the 'city that never sleeps,' and being such, requires a substantial utility network for support.

#### *Protection of Utilities at the North End (87th Street)*

In the early planning stages of the contract, the contractor needed to decide on how to obtain 480 V power for all mining and surface operations. The typical method used in Manhattan is to obtain feeders from Con-Edison (Con Ed) to a service box and then disperse power throughout the project. This method can be timely and could have caused delay to the construction of surface support facilities needed to begin the project. Skanska, joint ventured with Shea and Schiavone (S3) on the previous TBM mining contract, and was still utilizing the TBM substation to feed temporary power into the 96th Street Launch Box* and the running tunnels. The contractor and S3 set up a memorandum of understanding (MOU) to turn over the substation and use this for temporary construction power on the 86th Street Station project.

After the TBM operations were completed, the sub-station was re-purposed to power the works at the 86th Street Station Cavern, allowing the construction to commence in a timely manner. This was completed with the same cable utilized for the TBM to three transformer stations which converted the 13.2 kV to 480V AC power. Electric power was necessary to run two Atlas Copco E2C Drill Jumbos, two Meyco Potenza Shotcrete Robots, two Meyco Suprema Shotcrete Pumps, two overhead 39,916 kg (44 ton) gantry cranes, two muck house dump cranes, two Ingersol Rand 25 $m^3/m$ (900 CFM) compressors, and all other temporary power for the project.

The sub-station was designed to be completely self-contained and modular to allow for accelerated installation and removal. A team of engineers worked with the utility to allow for this original design. Special heavy haul trailers transported the sub-station from Ohio and was set with a heavy lift crane. As this is a fully certified Con Ed accepted sub-station, the unit can be re-utilized for future projects instead of sold for parts or salvage as done on previous projects. This was key to shortening the duration of time needed to kick off the 86th Street Station Cavern construction.

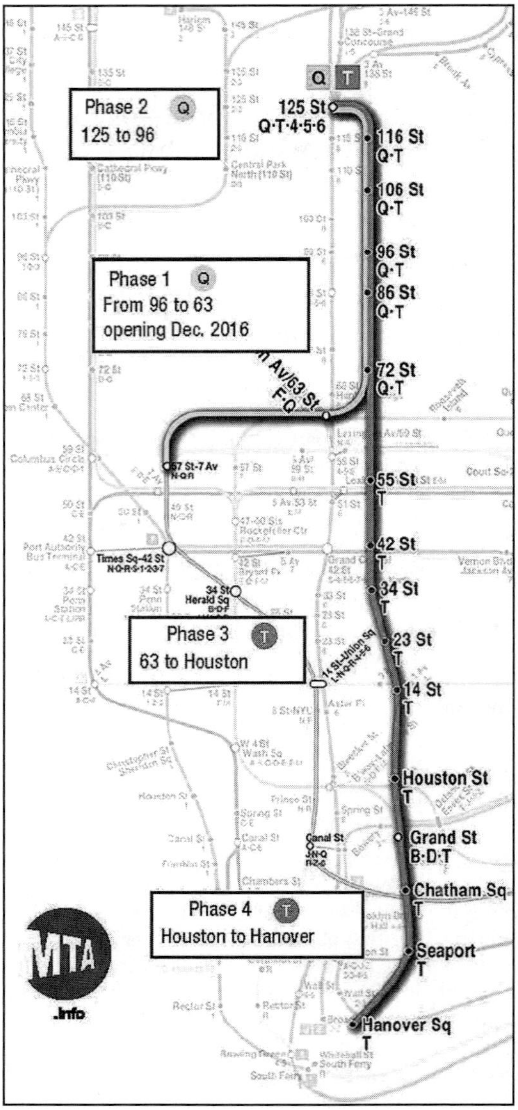


Source: Metropolitan Transportation Authority (MTA).

**Figure 1. Project phases for the Second Avenue Subway**

In order to convert the 13.2 kV power from the substation at 96th Street to 480 V AC power at 86th Street, a 13.2 kV cable was stretched through the 96th Launch Box and hung under the temporary deck installed by a previous contractor. The cable was then secured along the crown of the previously

*The TBM was assembled and launched southward from an open cut excavation on Second Avenue located between 92nd and 93rd Streets.

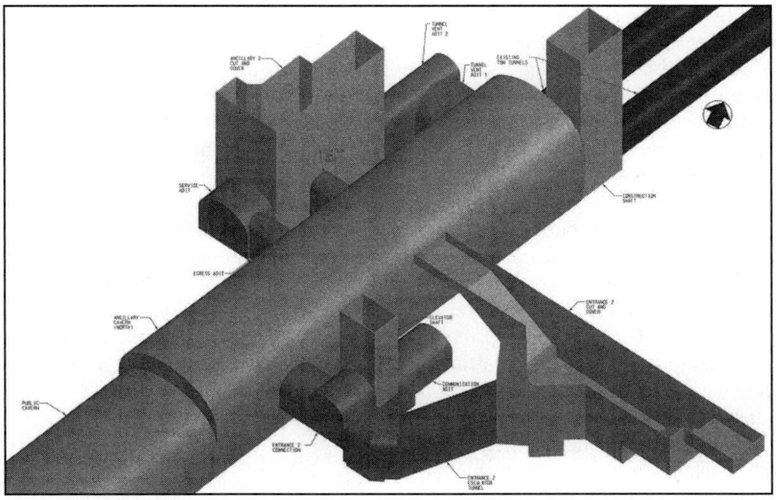


Source: Adapted from DMJM Harris/Arup (JV) 2011.

**Figure 2. Isometric view of station (North end)**

bored west TBM tunnel and terminated 4.6 m (15 ft) north of the 86th Street Station cavern north endwall. From the middle lane of Second Avenue, a 25.4 cm (10in)-diameter hole was drilled from street level and penetrated the west TBM tunnel. From the hole, the cable was fed 30.5 m (100 ft) vertically and brought to switchgear which transformed it into 480 V AC power for the North Shaft located at 87th Street.

In order to begin top down mining activities, temporary utilities were needed at five locations. Power, water, and compressed air were needed for subsurface mining equipment and drill and blast operations, the powering of the gantry and muck system, shotcrete, and temporary site power. In order to feed power to all the locations from 88th to 82nd Streets, a temporary utility trench was excavated along Second Avenue to house all construction utilities. The trench was approximately 0.9 m (3 ft) wide by 0.8 m (2.5 ft) deep and 457 m (1,500 lf) in total length. Due to the dense utilities present in Manhattan, special care was taken to plan and excavate the trench without damaging or disrupting services to the buildings in the neighborhood. Thankfully, the contractor was fortunate enough to have a clear path along the east side of Second Avenue within the designated work zones to bury the utilities. Also, work was able to be sequenced without closing any of the cross streets and without incident.

### *Protection of Utilities at the South End (83rd Street)*

At the south end of the cavern (near 83rd Street) in the middle of the open cut, was a tremendous amount of utilities. Water and gas mains, fiber optics, electrical duct banks were all located in the excavation footprint and had a major influence on how blasting would commence. Utility protection needed to be designed and installed to shield these facilities from any potential fly rock.

Numerous gas, electrical power, water, fiber optics and drainage lines were in the way, or extremely close to the south end open cut. These utilities were overhead and their location proved to be a great challenge. These utilities which included two gas lines, electrical duct banks, and two water mains were hung from the deck beams and were not permitted to be relocated. The protection plan formulated by the contractor included securing the utilities in place and monitoring their operational status during the excavation phase.

Since the rock to deck beam clearance was only 4.3 m (14 ft) not only would these utilities need to be protected, but also a spotter would constantly have to be aware of swinging booms from excavators and drills, and the matting of shots would have to be exact. A 0.9 m (36 in) line located directly over the excavation limits was evaluated by the team as being particularly vulnerable to penetration by potential fly rock during a blast if shots were not matted correctly. For this reason, this gas line was completely encased in a protection shield formed out of multiple layers of 2.5 cm (1 in) plywood with heavy bracing every 30.4 m (12 in) inside the encasement.

## GEOLOGIC SETTING AND GROUND CONDITIONS

### Project Geology

New York City is situated at the extreme southern terminus of the Manhattan Prong, part of the New

England Upland Physiographic Province. The geologic history of the area spans from the Precambrian to the Holocene, during which it experienced several major mountain building events, and Pleistocene glaciation. Within New York City, the Precambrian to Ordovician crystalline rocks are separated by the regional NE-SW trending Cameron's thrust fault. The Manhattan and Hartland Formations, lie respectively to the west and east of the fault, and exhibit structural elements such as faults, shears, and joint systems formed during past thrust movement. The regional tectonic history has left a complex stress imprint on the New York City bedrock; principal horizontal stress is compressional in the northeast-southwest direction (Figure 3).

The Manhattan Schist, member of the Manhattan Formation, underlies the majority of the project site, and has an anisotropic metamorphic fabric. Typically crystalline, Cambro-Ordovician schist predominantly quartz and mica in composition, is typically a garnetiferous biotite and muscovite mica schist, often with a gneissic fabric (DMJM Harris-Arup JV 2011). Also encountered were localized, but irregular and unpredictable pegmatite bodies, and thin amphibolite lenses typically conformable to foliation. The entire excavation was mined through the Manhattan schist.

Foliation and its corresponding joint set typically dips 5° to 30° to the southeast, and are the dominant and most structurally controlling factors within the rock mass. Foliation generally strikes perpendicular to the tunnel alignment. The alignment, which trends N29°E, is commonly referred to as 'Manhattan north.' Cross foliation joint sets typically have moderate and high angle dips to the southeast and northwest. During TBM mining, the machine was driven down-dip. However, during cavern mining, headings were drilled and blasted both down, against and perpendicular to foliation dip.

### Rock Mass Properties

Typical to this part of Manhattan, the rock mass was highly competent, fresh to slightly weathered, and strong to very strong. Some base lined (2011) rock mass properties include unit weight of 2485 to 3,124 kg/m$^3$ (155 to 195 pcf), uniaxial compressive strength of 15.8 to 99.9 MPa (2.3 to 14.5 Ksi), tensile strength (Brazilian method) of 6.2 to 24.8 MPa (0.9 to 3.6 Ksi), Poisson's Ratio of 0.10 to 0.40, and

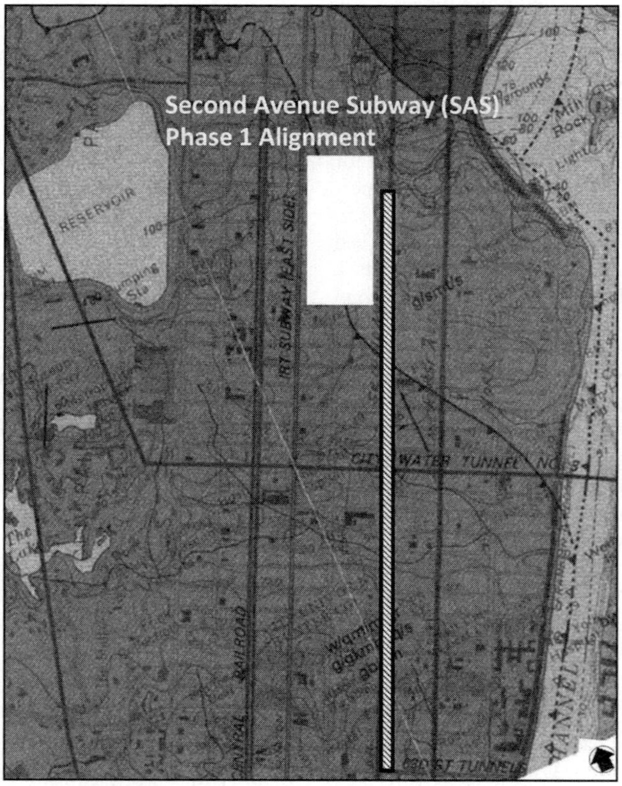


Source: Adapted from Baskerville1994.

**Figure 3. Manhattan engineering bedrock, and engineering geologic map**

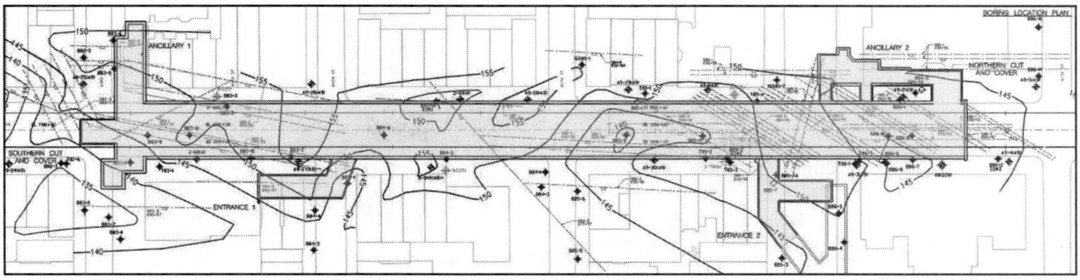

Source: Adapted from DMJM Harris/Arup (JV) 2011.

**Figure 4. Plan of interpreted geologic structures and bedrock topography**

Cerchar abrasivity index of 2.4 to 4.9. In order to identify ground or water infiltration conditions, probe holes were drilled in the crown of the center pilot for at least 9 m (30 ft) ahead of the face for the entire length of the cavern excavation. The probe holes, along with the rock mass itself was relatively dry, with very few water inflows, ranging from dry to very low flows (<3,785 $cm^3/m$). Although the excavation was situated below the organic groundwater levels, inflows were particularly higher in areas of pegmatite-schist contacts. The overall low flows may in part be due to the extensive deep building foundations, existing underground infrastructure, and few recharge areas in this part of Manhattan.

The Geotechnical Baseline Report (GBR) (2011) indicated that up to forty-six geologic structures* ranging between 1.2 to 3.6 m (4 to 12 ft) thick. Interpreted from acoustic televiewer (ATV) and boring logs, and a cross hole tomography survey, these structures were projected to encounter the cavern alignment (Figure 4). To record these and other discontinuities, the CCM performed geologic mapping for the faces and crown of the entire top heading, and walls of open cut excavations.

## UP FOR THE CHALLENGE

Prior to starting any excavation began, it was necessary to take into account the need of both the client and residents of this close knit, highly connected community. Since operations would run 24 hours a day, five days a week, it was critical that the 86th Street excavation not impact pedestrians, traffic, or the overall harmony of the neighborhood. In order to keep traffic flowing, the excavation had to be broken into phases, each one coordinated with a specific Maintenance and Protection of Traffic Plan (MPT).

*Geologic structures are defined as rocks that have been affected by faults, shears and alteration.

### Protecting the Public

As in any construction project, the need to protect the public is always the highest of priorities. In most cases, pedestrians can be rerouted away from the work zone or sometimes an area of work can be completely closed down to pedestrians. However, due to the high population density and high volume traffic in this part of Manhattan, higher level of protection measures were implemented. In order to safeguard and reroute pedestrians, the following preventative actions were taken:

- New walkways were painted
- Construction notifications were posted throughout the neighborhood
- Sound blankets were hung to reduce the impact of construction noise
- Dampening blankets were hung to reduce the air overpressure generated by a blast event
- Water mister were employed to control dust from the excavation

### Job-Specific Gantry and Muck System

Due to the constraints of working in New York City, and the lack of lay down area for equipment and trucks, the contractor was well aware that mucking would be a tremendous challenge. In order to "hog out" the blasted rock and to maintain the schedule, a new way of mucking would have to be considered and come to fruition. The size of the area around the north shaft and south open cut prevented the use of conventional muck conveyors, or the general "big crane, big box, big dump" system. In order to overcome this challenge, the project team along with MCT Murer Inc. designed a gantry and an elevated carousel with 11.5 $m^3$ (15 cy.) muck storage containers for mucking operations. This system had the speed to keep up with blasting, an elevated carousel that allowed trucks to pull in underneath from Second Avenue, and was quiet which was welcomed by the neighborhood residents. This mechanized

Source: MTACC 2013.

**Figure 5. 86th Street station mucking system**

mucking and shaft hoisting system was placed at the north shaft (at 87th Street) and at the south open cut (at 83rd Street) and served as the primary muck locations. Each muck system was less than one block long (approximately 61 m) and only 1½ traffic lanes wide (Figure 5).

### Sequential Excavation Method (SEM)

The cavern, adits, and open cuts were mined using sequential excavation method (SEM) by drill and blast. Main elements of SEM include a defined round length, support measures (especially shotcrete), multiple drifts/headings with support installed every round, pre or localized support, and instrumentation. The main benefit is that SEM allows for field adjustment, and typically benefits from a unit-price contract form, which the project had in place.

#### *Excavation by Drill and Blast*

For the excavation of the cavern, the northern end of the alignment was the first to commence. Mining from the south open cut started shortly after the main north heading was developed. The excavation started with the sinking of a 9.1 m (30 ft) by 12.1 m (40 ft) shaft roughly 30.5 m (100 ft) deep. For the mining operations, the contractor chose to use the Atlas Copco D3 with a 7.3 m (24 ft) mast and the ECM 350 air track. At the start of the excavation, the top of rock was only 7.3 m (24 ft) deep from the street surface, so the shaft was drilled and shot in halves. These first half shots were only 1.8 m (6 ft) deep and had a powder factor of 0.9 m (3 lbs) per delay. However, the contractor had a distinct advantage. Prior to any excavation, they decided to drill a 35.5 cm (14 in) relief hole to facilitate the blasting and to help minimize vibration and fly rock. Once into sequence with the shaft operations, the work crews began to open up the production. After two 1.8 m (6 ft) lifts, one half at a time, the contractor opened up to full width shaft shots at 3.6 m (12 ft) rounds dramatically increasing production. In the shaft sequence, the following ten steps were followed:

1. Drill round
2. Pull equipment (drill rigs) out of shaft
3. Load explosives and wire
4. Cover shaft with steel covers and rubber blast mats
5. Shoot
6. Excavate blasted rock
7. Install rock bolts
8. Shotcrete
9. Complete excavation
10. Blow bottom

This basic procedure became the rhythm of the crew. It was not long until a second and third shift was added to the sequence to enhance the 24-hour blast cycle.

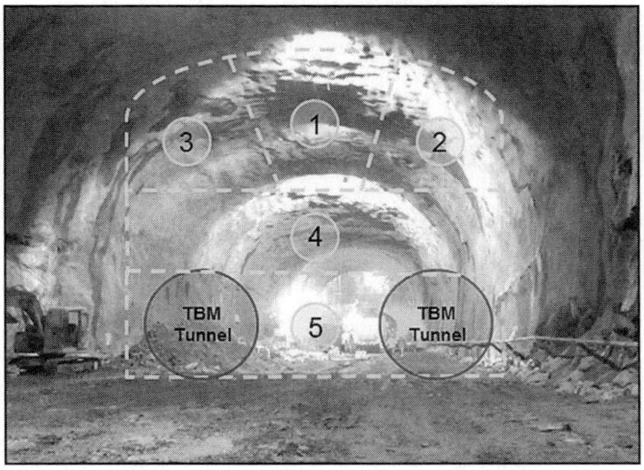


Source: MTACC 2013.

**Figure 6. Cavern excavation sequence**

### *Blasting Parameters*

The contractor established shaft excavation schematics which included 0.5 m (1.5 ft) spacing on the perimeter holes, 0.8 m (2.5 ft) spacing on the productions holes, with the burn at 0.2 to 0.3 m (6 in to 1 ft) spacing on the diamond and box. Each blast hole was 4.8 cm (1⅞ in) and prior to loading was blown clean, and stemmed with pea stone or clay. The main type of explosives used on the 86th Street Station Cavern was Emmulex 927, weighing 400 g (0.88/lbs) per stick, and Red-d-lite trim powder, weighing 272 g (0.60/lbs) per stick. The primers were dual delays made by Austin Powder and were 200/5,000 ms. The actual shot delays used were 9 ms, 17 ms, 25 ms and 42 ms quick relays also made by Austin Powder, and supplied by Explo Inc. These explosives are the only explosives allowed in Manhattan by the Fire Department of New York (FDNY). They also control all blasting within the five boroughs of New York City.

After sinking the construction shafts at 87th and 83rd Streets, the top heading was drilled and blasted. The top heading was divided into a center pilot, and east and west slashes (Nos. 1–3), followed by an intermediate (No. 4) and bottom bench (No. 5) (Figure 6).

Two main top headings were mined from the north and south, trending towards one another. The intermediate bench was stoped into the previously mined tunnel below. The project has to adhere to strict blasting criteria in the top heading such as a maximum 3.6 m (12 ft) round and a 0.9 m (36 in) stagger between faces. Each heading blast consisted of an array of approximately 150 holes and approximately 980 kg (2,160 lbs) of total powder. While setting up the top heading the blasters we able to wire in what was called a "triple." A triple would be three faces wired and shot together, shot under the allowable vibration, and shot under the 5 seconds allowed in the total delay time, minimizing cut-offs.

### *Blasting Protocols and Project Schedule*

Along with demanding site conditions, the project faced a challenging project schedule. Rock excavation predominantly by drill and blast, and some mechanical methods, was estimated at 197,357,097 $m^3$ (160,000 BCY) (240,000 BCY swelled). Since blasting operations were allowed only between the hours of 08:00–20:00, Monday through Friday the contractor incorporated a tight blasting schedule, sometimes up to three individual blasts a day to maintain the most efficient blasting cycle. Blasting started at the beginning of April 2012 and was completed in November 2013, with the main cavern taking just over a year to mine. Some impacts to the schedule not included in the baseline consisted of noise and work restrictions during *Rosh Hashanah* and Yom Kippur holidays, untimely Local Law 11* work at an adjacent high rise residential building, and damage and delays incurred by Hurricane Sandy. The project team coordinated with New York City's largest public agencies, such as the MTACC, New York City Department of Transportation (NYCDOT), New York City Department of Environmental Protection (NYCDEP), and FDNY to ensure safe and successful operations.

The contractor would submit a blast plan for each individual blast which was reviewed and approved by the engineer prior to blasting. When,

*Local Law 11/98 is a law enacted by the City of New York, which requires landlords to hire an architect or engineer to perform an inspection of the exterior walls of buildings greater than six-stories, every five years.

on occasion, vibration readings were exceeded, the responsible parties were notified (by email) with real-time instrumentation alerts from the Argus monitoring system. The engineer, CCM, and the FDNY shared the contractor's technical interpretation of any exceedences, and what measures would be undertaken to ensure that vibration threshold values would not be exceeded in the future. Limiting response values of 0.013 m/s (0.5 in/sec) peak particle velocity (PPV) were established for historical, sensitive, and fragile buildings, some of which are over 100 years old, and 0.049 m/s (1.92 in/sec) PPV for all other buildings.

Though there are limits to every operation, the blasting operations were carried out effectively and efficiently by coordinating with local agencies, stakeholders and the CCM. A comprehensive plan to notify personnel and position all personnel prior to the blast was established by the contractor and CCM and the contractor was able to consistently alert the public so no one was alarmed when warning horns, or rumbling at the street was heard.

An additional, indispensable element to maintaining a strong partnership was communication. Weekly coordination meetings during the blasting operation provided a forum to discuss and settle issues brought to the table and these meetings went a long way in creating a sense of partnership between the different parties involved including other contractors on adjacent contracts of the Second Avenue Subway project.

## Rock Reinforcement

### *Open Cut*

Several different initial rock reinforcement methods were considered by the contractor and its design consultants for the open cut. The guiding parameters for the design required a support method that would utilize the positive in situ rock properties while ensuring a system that would coincide with expected field production and the planned blasting scheme. The engineer ultimately selected a traditional rock stability analysis design method that produced a support scheme that could stabilize theoretical "worst case" rock wedges in the side walls of the excavation.

A combination of #10 Dywidag and Swellex PM24 rock bolts in conjunction with a 7.6 cm (3 in) thick layer of fiber reinforced shotcrete were selected as the excavation support system, fulfilling the numerous design prerequisites. The bolt pattern in the walls was based on a staggered 1.8 m (6 ft) horizontal by 1.8 m (6 ft) vertical spacing with 3.0 m (10 ft) long pre-tensioned Dywidag bolts as the levels of support. The initial ground support in the portal faces of the cavern and the C.I.R. room was designed exclusively with a lighter Swellex pattern due to the previously noted reduced surcharge in these areas and the straightforward means by which these could be removed in subsequent construction phases to complete the new excavations. Based on field logistical and engineering support reasons it was determined that the excavation would progress in sync with the rock reinforcement installation in six lifts of approximately 1.8 to 2.4 m (6 to 8 ft) each.

### *Cavern*

Initial rock reinforcement in the public and ancillary caverns, and adits was established by the designer at predetermined station ranges. The two initial rock reinforcement types for the cavern consist of pattern bolts and dowels consisting of #10 Dywidag, Grade 75, 6.1 m (20 ft) in length:

- Type I
  - Arch: 1.8 by 1.8 m (6 by 6 ft) pattern rock bolts, pre-stressed 133,446 N (30 (kips)
  - Wall: 1.8 by 3.6 m (6 by 12 ft) pattern rock dowels, un-tensioned
- Type II
  - Arch: 1.5 by 1.5 m (5 by 5 ft) pattern rock bolts, pre-stressed 133,446 N (30 (kips)
  - Wall: 1.5 by 3.0 m (5 by 10 ft) pattern rock dowels, un-tensioned

A 13 cm (5 in) layer of fiber reinforces shotcrete was applied to the rock surface, followed by a 5 cm (2 in) layer of smoothing shotcrete (Figure 7).

## Equipment

A wide variety of equipment was used to excavate the cavern, which was mined entirely within rock. For drilling, the contractor used several Atlas Copco D3's, and 2 E2C 2 boomer drills with a spare on hand. At peak production, the cavern was running four CAT 321's, two CAT 980s, a CAT 960, and a CAT 963. For the shotcrete operations in the cavern the contractor used two Meyco Potenza and one Meyco Aruga, along with three Suprema pumps topside. Also crawling around the cavern at peak production were four (60 ft) track man lifts (Figure 8).

### *Hard Rock Drilling Using Tunnel Manager*

The contractor utilized two Atlas Copco Boomer E2C jumbo rig for drilling production holes, rock reinforcement, and probe holes. The two boom jumbo is fully automated, and the booms can be programmed to a hole location by using the Rig Control System (RCS 5) supported by Tunnel Manager software (Atlas Copco). Due to New York City labor agreements, the contractor opted not to use the fully automated system but still used the Tunnel Manager

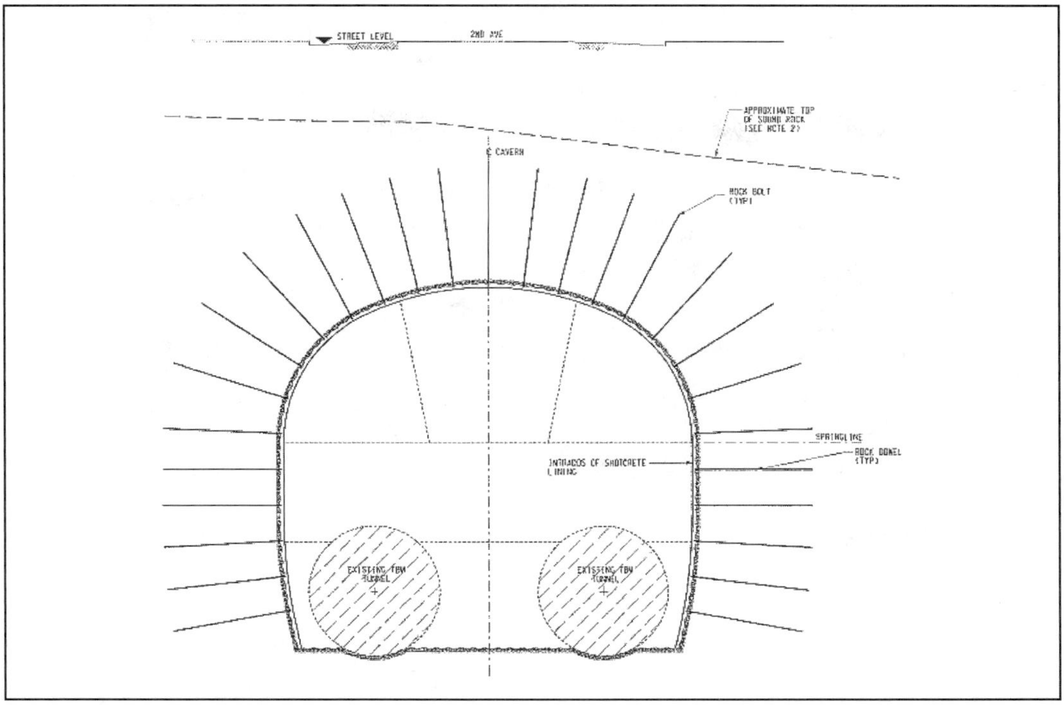


Source: DMJM Harris/Arup (JV) 2011.

**Figure 7. Initial ground support**

Source: MTACC 2013.

**Figure 8. Cavern equipment**

Source: MTACC 2013.

**Figure 9. Excavation and geometry—CIR adit and TBM tunnels at South Open Cut**

software. This software was used to record drill data, and to drill hole locations according to the blast plan. However, establishing the blast plans required a series of detailed steps. The geometry of the shots were drafted first and consisted of a top heading with a center pilot and two side slashes. Once the geometries were drawn in Tunnel Manager, the rounds could be developed. Each hole was designated with an individual number and hole type (i.e., contour, burn, lifter, and easer). In the program, each hole is given an angle of lookout and a starting coordinate and ending coordinate to set the hole depth. The angle of lookout is designated by a tail array attached to the hole. After the pattern has been drawn, it is loaded onto a USB memory stick and uploaded into the computer in the cab of the drill rig. After the pattern is uploaded, it is displayed on the cab's monitor, and the driller uses a joystick control to line up the hole and matches the tail of the curser with the tail of the hole to gain the correct drilling angle.

Unlike traditional blasting, this method required minimal layout and survey, and helped reduce the overall blast cycle time. To start drilling the production array, just one (starter) hole was needed to be laid out by survey on the face, and the drill was lined up using the true alignment of the cavern. This helped save time by not having to paint out every face with a hole pattern. Another major advantage of the software was being able to control over break throughout the cavern. By setting a 3% to 5% hole lookout, the over break in most locations was limited to less than 0.46 m (18 in). This helped to keep concrete material cost down by not having to pour additional concrete during the subsequent concrete lining stage (Figure 7).

### Trades

Integrating the work between the cavern and street levels also required coordination between the different trades, as they all play an important part to the project success. When it comes to the work force, New York City is one-of-a kind because numerous union trades take part in the project. Below is a list of the trades that played a major role in the project.

- Local 147 Sandhogs
- Local 29 Drill runners
- Local 15 Operating engineers
- Local 14 operating engineers
- Local 731 laborers
- Local 3 Electricians
- Local 40 Iron workers
- Local 456 Timbermen
- Local 157 Carpenters
- Local 46 Lathers
- Local 282 Teamsters
- Local 1453 Dock builders

The project was fortunate to have twelve different trades working together safely!

### Cavern Geometry

With so many underground excavations and tunnels already mined and constructed in Manhattan, you may wonder, what could pose a risk? Similar to other underground excavations, the 86th Street Station Cavern design also consisted of complicated geometry, with numerous adits, open cuts, and shafts intersecting a large horse shoe-shaped cavern, tapered in at the toe. This was constructed under relatively low rock cover, and close to the street surface, where the maximum depth from street level to the cavern crown is approximately 16.7 m (55 ft) (Figure 9). The geometry of the designed structure, in combination with in situ joint orientation, and occasional geological over break during blasting created some localized areas of instability, notably:

1. Vertical, relatively thin pillar between the cavern and an ancillary structure
2. Numerous adit and shaft intersections at staggered elevations in an open cut, adjacent to the existing mined tunnels
3. Vertical walls of an open cut structure with a few re-entrant corners

In addition, a significant number of concrete building foundations for are located within 0.6 m (2 ft) of the excavation limits. The close proximity of these major structural elements to the vertical face of the cut created challenges for the support of excavation design.

### Environmental Measures

The contract specification for dust monitoring on site required that particulate matter would be measured using monitoring stations. These stations utilize dust track monitoring devices with the ability to measure particle sizes less than 10 μm (0.0004 in) in size (PM-10). The stations would be located in the upwind and downwind perimeters of the work zone. The contract limited particulate levels to remain below 100 mcg/m$^3$. If this limit was exceeded, the dust track monitors would send out an alert and dust suppression measures would have to be implemented.

The idea was to engineer a system that would utilize two fans at each shaft location to the cavern excavation below. One fan would supply fresh air to the workers at the heading during all construction activities, including drilling, shotcrete, loading, bolting, and mucking operations, the second fan was dedicated to exhausting the cavern after a controlled blast to capture the air and channel it through the scrubber system prior to exiting into the atmosphere. Two specific and differently engineered scrubber systems were evaluated for the project.

The system designed and engineered by Shauenberg, was determined to be the preferred method with physical data proving it would satisfy the requirements for our application. This system is most commonly associated with TBM mining and treats the air at the heading so that it passes through and meets all criteria for breathable air. The air monitoring data showed that the Shauenberg units were able to provide a particulate concentration of 1 mcg/m$^3$, while the contract specification required readings to remain below 100 mcg/m$^3$.

The exhausted air is sent into a venturi which is outfitted with nozzles around the perimeter that spray a fine mist of water onto the air and dust mixture. The air is mixed with the fine mist from the nozzles that saturates the dust in the air, and then is transmitted to a secondary treatment. The secondary treatment consists of a filter bank consisting of PVC spin filters which mechanically separate the saturated dust particles created in the initial treatment. The spin filters consist of a plastic tube with a stationary spinner mounted at the intake end. As air carrying the saturated dust particles is forced through the inlet of the filter tube, it engages the stationary spinner, which creates a centrifugal force sufficient to send the dust particles to the periphery of the tube, thereby separating them from the air.

The scrubber systems were installed at the north and south shafts in the summer of 2012. After a blast, when the scrubber system is in operation, visible dust or smoke cannot be detected in the area immediately surrounding the access shafts, and cannot be seen in the exhaust stream.

## CLOSING REMARKS

### Alleviating Congestion and Demand

As New York City, as well as many other major US cities, is faced with not only updating and/or maintaining aging infrastructure, but also with preparing the City for sustainable transportation in the future. Though the future can be uncertain, ensuring the public can get to where they need to go–safely, efficiently and cost effectively, is a fundamental transportation aspect that does not change with time. Continually, we are challenged at all levels, from regional planning and urban development down to the geologic conditions encountered during mining.

The Second Avenue Subway Project is an essential step for New York City to alleviate congestion on the Lexington Avenue Line, meet systemwide demand, and ensure sustainable growth for the New York Metro region. According to the MTA (2012), though New York City's population grew by one million in the previous three decades, the system did not expand a single new mile of highway or track. Furthermore, PlaNYC2030 (2011) states that the City's population is still growing, and that by 2030 is projected to increase to more than nine million. The resilience of the subway system is tested daily, especially in the Upper East Side where the 4, 5, 6 subway lines carry 1.8 million passengers every weekday–more than the weekday rail ridership of Boston, Chicago and San Francisco Muni combined (MTA 2013). Furthermore, MTACC (2013) estimates that Phase 1 will decrease crowding on the Lexington Avenue Line by as much as 13% or 23,500 fewer riders on an average weekday and travel time will be reduced by 10 minutes or more for many riders. Construction of the 86th Street Station

Cavern will serve as a fundamental transportation node essential for sustainable urban growth.

## Mitigating Risks While Maintaining Design

Though SEM mining by drill and blast was an overall success, difficult geological conditions within the excavation were often encountered. In addition, overhead utilities not only dictated how contractor could implement the contractor's means and methods but also were a constant reminder of the potential dangers during the excavation. However, as Snee (2008) describes, though the construction process just imparts a new stress-strain regime on the rock mass, there still requires a bridge between geology, analysis and design that is established early in the project life, maintained, and continually strengthened. This philosophy was successfully implemented as representatives from the designer, contractor, and CCM were readily available onsite, and timely responses were received for any geotechnical or blasting related concerns.

### *Geology and Geometry*

Though regional geologic trends, such as foliation striking on line with the Appalachian metamorphic trend, and geologic structures oriented comparably to Cameron's thrust fault, were generally anticipated, in situ conditions varied and required continual assessment. In the center pilot, and western and eastern slashes of the main cavern, a ubiquitous sub-vertical, southeast dipping geologic structure was mapped; typically 0.6 to 0.9 m (2 to 3 ft) wide, moderately to severely weathered, had a combination of closely to widely spaced joints, breccia, mylonite, gouge, clay, stilbite, epidote, calcite, limonite, iron oxide staining, and slickensides. Within each heading, this dominant structure was intermittently mapped between 47 to 85 m (155 to 280 ft) in total length. This structure strikes typically N29°E, parallel to the cavern alignment, or "Manhattan north."

Since the structure's orientation was parallel to the direction of mining and due to its pervasive nature, the presence of this feature proved challenging. Within the east heading, it was often present behind the sidewall, but not visible, and in the center and west headings it was often visible and created difficult ground. For similar reasons, this structure also threatened the stability of the pillars, and open cut walls discussed in the Geology and Cavern Geometry Section (Nos. 1–3). To mitigate these concerns, additional ground support measures were implemented where needed. Approximately 1,161 m (3,810 ft) of additional ground support steel was installed and over 76 $m^3$ (100 $yd^3$) of shotcrete was applied and paid for under contract bid items for additional "as directed" work.

### *The Keys to Success*

Over 440 blasts were successfully executed in open cuts, cavern, shafts and adits in this Upper East Side neighborhood during normal business hours, and at one location in close proximity to an Olympic-sized swimming pool! Public outreach, or "community engineering" (G. Almeraris, personal communication) as it was coined, was a major component in the success of the project at the street level. Underground, in addition to having a coordinated project team effort, having blaster(s)-in-charge with extensive experience, knowledge of the local geology, and continuity between mining contracts also proved essential in executing safe and efficient blasting. For instance, local knowledge played a key role in maintaining a smooth cavern profile even when headings trended along, against and perpendicular to foliation dip, conditions that typically result in a saw-tooth profile. Also, implementing the Tunnel Manager software played a major part in saving blast cycle time. Once the cavern was developed and in full production mining, advance rates approximated 12 m (40 ft) per week in each direction. Not losing sight of the overall goal, the team safely and efficiently addressed geologic conditions that contributed to instability.

A simple, but fundamental, lessons learned was the important role communication plays during mining and construction; daily coordination between the designer, contractor and CCM proved essential. Though the contract included unit priced items for additional ground support measures, field issues were addressed immediately, and for everyone, safety was always paramount. In closing, the success of SEM by drill and blast on this project demonstrated it was an appropriate approach for cavern excavation in a densely urbanized area at shallow depths.

## REFERENCES


City of New York. 2011. *PlaNYC2030*. New York, NY: City of New York, Mayor Michael R. Bloomberg.

DMJM Harris-Arup JV. 2011. Geotechnical Baseline Report. New York, NY. Unpublished work.

DMJM Harris-Arup JV. 2011. Geotechnical Interpretive Report. New York, NY. Unpublished work.

Second Avenue Subway. 2013. Building for the future. http://web.mta.info/capital/sas_alt.html. Accessed January 2013.

Snee, Chris. 2008. Engineering Geology in Manhattan–Building a Bridge Between Geology and Design. Presented at the "Manhattan on the Rocks" A Specialty Seminar by ASCE Metropolitan Section Geotechnical Group, New York, NY, May 8.

# Blue Lake Hydroelectric Expansion Project—Tunnel Excavations

**Kellan Osborne**
J.S. Redpath Corporation

**ABSTRACT:** Conventional drill and blast methods were used to successfully excavate three new tunnels and two new shafts, employing typical rubber tired equipment and a mechanized raise climber, at the Blue Lake Expansion Project.

The underground excavations were one phase of the ongoing $88 million Blue Lake Expansion Project owned by the City and Borough of Sitka, Alaska, a primarily fishing and tourism based community of approximately 10,000 residents located in southeast Alaska. Working in a remote island community accessible only by plane or boat presented many challenges.

Underground excavations consisted of a 42.67 m (140 ft) long exploratory tunnel located 60.96 m (200 ft) from the base of the existing Blue Lake Dam, accessible only by crane; a 143.26 m (470 ft) long access adit developed to facilitate excavation of a 99.67 m (327 ft) surge shaft driven from the bottom up, using a Mechanized Raise Climber (MRC) breaking through in a remote surface location accessible by helicopter only; a 256.03 m (840 ft) long intake tunnel which portaled in 9.14 m (30 ft) above Blue Lake, Sitka's drinking water supply; and a 33.22 m (109 ft) gate shaft driven from the new intake tunnel, excavated with an MRC.

This paper will provide a description of the logistical and engineering challenges involved with performing underground excavations in this isolated and demanding environment. Work areas accessible by crane and helicopter only, working directly above Sitka's drinking water supply, and excavating to within 9.14 m (30 ft) of the Blue Lake Dam and existing penstock tunnel are a few of the many factors which challenged this critical project.

## INTRODUCTION

To keep up with the demand for clean, renewable, eco-friendly energy, the City and Borough of Sitka, Alaska (CBS) is undertaking an expansion of the existing Blue Lake Hydroelectric facility. Sitka is a remote southeast Alaskan town with no roads connecting it to other communities. The community is 100% self-sufficient in water and energy supply. The Blue Lake Expansion Project (BLX) was undertaken to prevent future energy shortages, decrease dependency on oil produced heating, and diminish power supplemented by diesel generators.

BLX consists of raising the height of the existing Blue Lake Dam by 25.3 m (83 ft), excavating new waterways, modifying penstocks, and constructing a new powerhouse with three new 5.3MW (nominal capacity) generating units. Blue Lake Tunnelers (BLT) scope of work consisted of excavating three new tunnels and two new shafts using conventional drill and blast methods and Mechanized Raise Climber (MRC).

In the fall of 2012, the CBS contracted Barnard Construction Company, Inc. (BCC) as prime contractor to execute the project. BCC awarded underground excavation work to BLT, a joint venture partnership between underground mine contractor J.S. Redpath Corporation (Redpath) and heavy civil contractor Frontier-Kemper Constructors, Inc. Redpath provided expertise in underground mining while Frontier-Kemper Constructors offered experience in heavy civil construction project management.

BLX tunnels will be completed in two phases. Phase One, completed in 2013, consisted of excavating an adit tunnel, surge raise and breakthrough location, drainage tunnel, intake tunnel and gate shaft. Connections to existing workings, left as rock plugs, will be removed during a critical 2014 generation outage, completing Phase Two.

## LOGISTICS

Sitka, Alaska is located in a chain of islands in Southeast Alaska. Unlike many cities, Sitka is located on a small island accessible only by boat or air. Normal freight deliveries by truck or rail are not options. Products that cannot be flown must be shipped via barge from elsewhere in Alaska or from Seattle, WA. Barges arrive once per week from Seattle in the winter months and twice weekly during summer months. An additional two week lead time was required on any deliveries from outside Alaska. As underground excavation supplies and consumables were not readily available locally, thorough pre-project planning was critical to maintain the project schedule. Delays associated with material

**Figure 1. Yarder unloading**

and equipment procurement were among the greatest project risks.

Equipment and materials received by barge or air were commonly delivered to site in trucks or on lowboy trailers. However, power lines between the barge terminal and site restricted delivery of BLT's Mini-Madill log yarder. A yarder was set up and used like an aerial tram to transport supplies, small equipment, steel structures and concrete to the remote surge shaft breakthrough location. The yarder required delivery on a small barge equipped with an unloading ramp. A neighboring fish processing facility had an area suitable to drive the 35 ton machine onto land (Figure 1).

## ENVIRONMENTAL CONSIDERATIONS

The proximity of BLT's work areas to Blue Lake, Sawmill Creek, and Alaskan Forest Service land required special attention to hydrocarbon runoff and reclamation of impacted land. The intake tunnel portal, located 9.14 m (30 ft) above Blue Lake, had zero tolerance for discharge water seepage into the lake. Discharge water was pumped to a location with no probability of contacting the lake. To eliminate spills in Sawmill Creek, a spawning ground for salmon, a Baker settling tank and geotextile filter were installed near the drainage tunnel portal to clean sediment from drill cuttings. A sump was established at the adit tunnel portal to eliminate sediment in discharge water.

The remote location of work areas combined with an abundant population of brown bears was also a concern for project employees. Sawmill Creek was hiked daily by drainage tunnel crews and being a spawning stream for salmon, the possibility of human and bear interaction was a reality. Daily waste removal from work areas and precautionary bear spray were the only feasible solutions. No interaction with bears occurred throughout the project by BLT crews.

Access to the drainage tunnel and dam staging area (intake tunnel and gate shaft) was by Blue Lake Road only. During winters with heavy snowfall two chutes commonly produce avalanches that would block the road. Signage and visual checks before crossing avalanche zones mitigated this risk. One avalanche occurred covering a 15.24 m (50 ft) stretch of road with 3.05 m (10 ft) of hard packed snow. Fortunately, no personnel or equipment were present during the slide.

The breakthrough location for the surge shaft was remotely located and monitored by the Fish and Game Department. Upon completion of surface excavation and steel erection, the breakthrough excavation area was vegetated with transplanted trees and native grasses. Water runoff was monitored to ensure drain patterns followed natural courses.

## DRAINAGE TUNNEL

To measure competency of the rock supporting the dam's left abutment and to measure water being produced by the abutment, a 3.05 m × 3.96 m (10 ft × 13 ft), 42.67 m (140 ft) long inclined tunnel was excavated. At completion, the drainage tunnel allowed core hole samples to be drilled from under the abutment and seepage from the lake to be quantified by a weir installed at the tunnel portal.

The location chosen for the drainage tunnel portal presented many challenges. Portal location was 60.96 m (200 ft) from the Blue Lake Dam, approximately 7.62 m (25 ft) from a 18.29 m (60 ft) deep plunge pool and directly below a 60.96 m (200 ft) cliff which was plagued with falling debris due to constant rain, freezing, and thawing (Figure 2). With no access roads available to the portal and construction of a road unfeasible due to steep canyon walls

**Figure 2. Drainage tunnel portal location**

and Sawmill Creek, equipment and supplies had to be lowered by crane. A Leibherr LF 1600-2, with a capacity of 660 tons and boom length of 124.97 m (410 ft), was positioned on the right abutment providing equipment and personnel transportation to the portal. Equipment required for excavation included a Tamrock Quasar single boom drill jumbo, 9,070 kg (20,000 lb), R-1300 Caterpillar LHD, 20,865 kg (46,000 lb), and small forklift for loading and hanging utilities. Meticulous attention given to the gear and rigging ensured appropriately rated slings and straps, balanced loads, and successful lifts.

Initial planning called for workers to transport in and out of the canyon via a crane man basket. Due to crane availability, crews hiked to and from the portal on a mountain trail ½ mile below the dam. At a safe conservative pace, travel time was 15 minutes. This allowed greater effective working hours while still having the crane available as secondary egress in emergency situations.

Running power, air, and water from the dam staging area to the portal was not an option. As an alternative, 12.19 m (40 ft) shipping flats were assembled prior to portal availability equipped with the necessary utilities and supplies to complete the work. One 2.44 m × 12.19 m (8 ft × 40 ft) flat was equipped with a 375cfm Ingersoll Rand diesel air compressor and shipping container; one 2.44 m × 12.19 m (8 ft × 40 ft) flat was equipped with a 275kW Atlas Copco Generator and electrical panel; and one 2.44 m × 12.19 m (8 ft × 40 ft) flat equipped with a warming hut, outhouse, and ground support materials.

To eliminate contaminated drill water from entering Sawmill Creek a settling tank was established outside the portal to collect construction water. Geotextile settling bags attached to tank overflow valves ensured water produced during excavation was filtered before being released into Sawmill Creek.

Before tunneling could begin, precautionary blasting and scaling of the highwall was performed. Highwall scaling would be mandatory following each blast until portal steel sets and canopy were installed to provide worker and equipment safety.

Rock conditions for the drainage tunnel were highly competent. Preliminary reports called for class I ground support, spot bolting as required, for a majority of the tunnel, which was encountered as predicted. A comprehensive work place setup, timely crane scheduling for equipment and supply transport as well as excellent ground conditions allowed drainage tunnel development to finish ahead of schedule. Working dayshift only, advance averaged 3.05 meters per day (10 feet per day) following portal socket establishment and steel set installation. At the ending tunnel station of 0+42.67 m (1+40 ft), Blue Lake Dam was only 9.14 m (30 ft) away.

## INTAKE TUNNEL

Excavating a new intake tunnel and gate shaft was required as the existing intake gate was deteriorating. The intake tunnel was designed as a 3.05 m × 3.66 m × 256.03 m (10 ft × 12 ft × 840 ft) long, 17% decline and would have a 33.22 m (109 ft) gate shaft located 16.46 m (54 ft) into the tunnel. The chosen location for the intake tunnel portal was on a cliff bordering Blue Lake which required a surface contractor to bench down the highwall and establish a portal access road. Portal access was delayed 65 days making on-schedule tunnel completion a necessity as rising lake levels would ultimately exceed the tunnel invert. Had lake levels exceeded the portal invert elevation prior to tunnel and shaft completion, all work would be delayed until water levels receded again in

**Figure 3. Drilling in intake tunnel**

2014, potentially delaying substantial completion of the project.

The limited area available around the portal socket posed a challenge for establishing utilities and mucking. Water was pumped directly from Blue Lake with a submersible pump. Air and power supply were established at the top of the highwall some 24.38 m (80 ft) above the portal face. A power cable and 2 inch air lines were run down the highwall and tied off intermittently to eye-pins for support. Tunnel discharge water was pumped up and over the highwall to an approved drainage location. Tunnel discharge water was closely monitored as Blue Lake is the source of the community of Sitka's drinking water.

Tunnel excavation averaged 10.06 meters per day (33 feet per day) with minimal ground support required. Preliminary ground support studies called for two fault zones which would require steel sets and shotcrete. These zones were never encountered. Mud seams were commonly intercepted producing up to 15gpm of water inflow but requiring minimal additional ground support. Approximately 152.4 m (500 ft) into tunnel excavation, accumulated water from the multiple mud seams caused several hours of delay per round to pump the face dry prior to drilling. This time was used to advance utilities and maintain surface muck piles. Had tunnel length been substantially greater a small sump station would have been required to prevent tunnel water from accumulating at the face (Figure 3).

The existing intake tunnel remained "live" during all excavations to continue supplying power and water to the city. To ensure that tunnel alignment and surveys were accurate BLT chose to drill a probe hole into the live tunnel. This would increase probability of a successful rock plug removal during the 2014 shutdown. Surveys showed the distance between existing and new intake tunnels to be 6.1 m (20 ft). A one meter (3.28 ft) long, two inch diameter hole was drilled in the rock plug and reamed to a 3.5 inch diameter. A mechanical packer was installed with an inner diameter of two inches allowing the jumbo, equipped with extension steel, to freely drill through it and record the actual distance between tunnels. The existing intake tunnel was successfully "tapped" and valve closed shutting off the inflow of water. Following the probe hole during tunnel plug excavation will minimize survey risk and ensure an accurate connection between tunnels during the critical 2014 generation outage.

## ADIT TUNNEL

To mitigate the risk of a surge or water hammer forming upstream of the powerhouse and damaging turbines, a surge shaft was required. This shaft allows a relief route away from the powerhouse in the event of a surge. The designed location for the surge chamber was 143.26 m (470 ft) into a mountain. One option involved sinking a shaft, top down, but accessibility made this impractical. A 3.66 m × 3.96 m × 143.26 m (12 ft × 13 ft × 470 ft) long tunnel would be excavated to access the bottom of the surge shaft allowing for shaft excavation from the bottom up using an MRC.

Adit tunnel portal access posed minimal challenge in accessibility. A location 60.96 m (200 ft) from shop facilities and accessible by truck was a welcome relief in relation to drainage, intake, and breakthrough locations. Conventional drill and blast methods in extremely competent rock yielded higher than expected advance rates. Drilling 3.66 m (12 ft) rounds with a Tamrock 205 two boom jumbo and mucking with a Caterpillar R-1300 LHD, tunnel advance averaged 9.14 meters per day (30 feet per day), many days advancing 12.19 m (40 ft) plus. Initial ground control plans called for type II ground

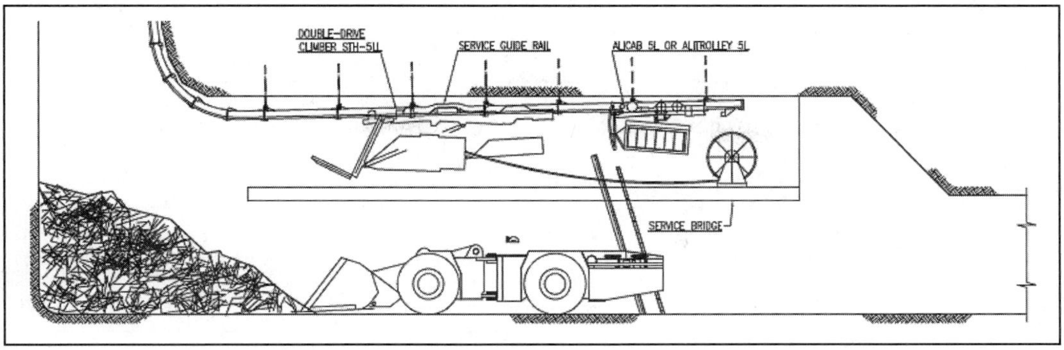


**Figure 4. Elevated nest arrangement**

throughout the tunnel, however, field inspection by Jacobs Associates determined spot bolting as necessary was sufficient for approximately 106.68 m (350 ft) of the tunnel.

Approaching the last 24.38 m (80 ft) of tunnel excavation, the cross-section of the drift was increased to 4.88 m × 6.1 m (16 ft × 20 ft) to accommodate an elevated MRC nest (Figure 4). Since the shaft was located at the end of the tunnel, an elevated nest setup was required for mucking.

Connecting the adit tunnel and surge shaft to existing tunnel workings is a 1.83 m (six ft) diameter shaft 9.14 m (30 ft) in length to be excavated during the 2014 generation outage. To confirm the location of the existing workings and establish a pilot hole to follow for the 2014 season, a two inch vertical probe hole was drilled through a mechanical packer, similar to the intake tunnel tap. To minimize work required in 2014, a 1.83 m (six ft) diameter, 1.52 m (five ft) deep pre-sink was excavated. Tapping into the live tunnel was initially a concern due to proximity to the existing powerhouse and the possibility of breakthrough debris reaching active turbines. Analysis by McMillen concluded that breakthrough debris posed no threat to powerhouse activities. The existing tunnel workings were successfully probed into confirming adit tunnel location accuracy and establishing a pilot hole to follow in 2014.

## SURGE SHAFT BREAKTHROUGH PREPARATION

To optimize surge shaft collar ground control and for safety concerns, surface preparation was required before the shaft reached surface. The location for the breakthrough was inaccessible via road and constructing a road was not feasible. Mud, dense forest and inclines as great as 45% eliminated realistic road construction options. Helicopters were required to fly initial equipment and supplies. To safely land helicopter loads, crews established a landing zone by felling trees and prepping the area with handheld tools and equipment.

**Figure 5. Helicopter lift**

The initial helicopter flight utilized an A-star 220 B-Series which lifted one Ingersoll Rand LM-100 surface drill, one Hitachi ZX-35 excavator, two air compressors and miscellaneous supplies (Figure 5). With a lifting capacity of only 998 kg (2200 lb), equipment would need to be disassembled prior to flight. At the breakthrough site, equipment would be reassembled. Another flight, using a larger Bell 214 helicopter with 2495 kg (5500 lb) lifting capacity followed. A John Deere 120 excavator was

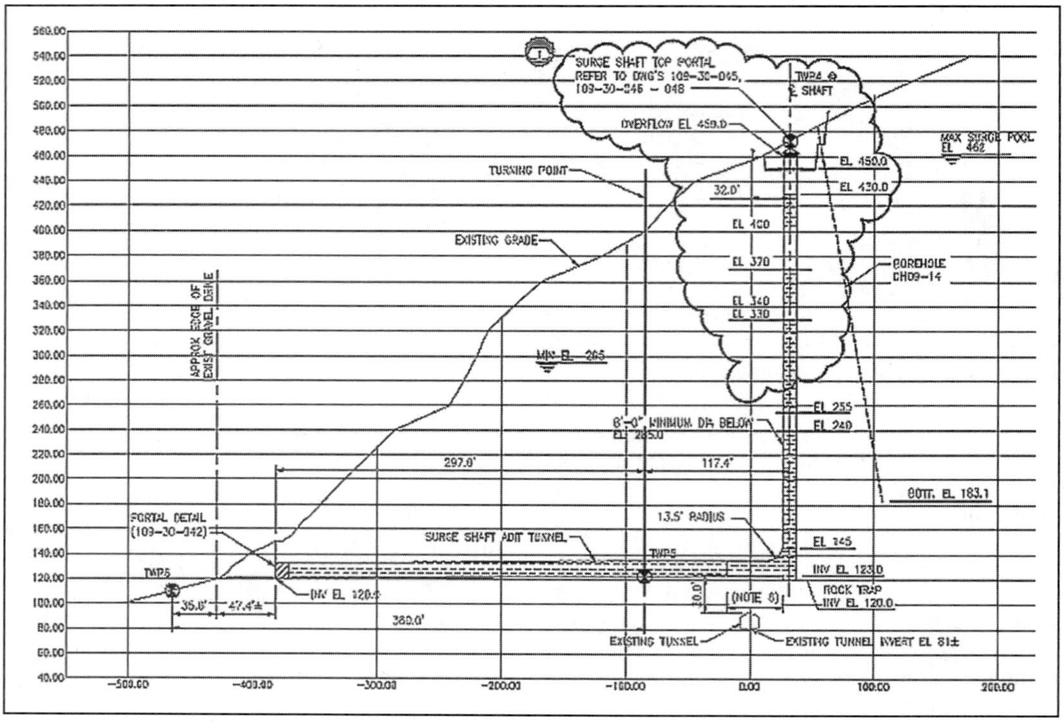


**Figure 6. Adit tunnel and surge shaft profile**

flown in, along with supplies, and reassembled at site.

Surface drilling and blasting were performed to establish a basin for the shaft to break into (Figure 6). In the event a significant surge takes place, the basin will regulate the amount of water released, eliminating the risk of significant runoff. A total of 1980 cubic meters (2590 cubic yards) of rock, in situ, had to be removed to establish the breakthrough basin. Typical benches were 3.05 m (10 ft) in depth, drilled with the LM-100, and muck removed by excavators. 3.05 m (10 ft) before reaching elevation 450 (feet above sea level) the surge basin bottom elevation, excavation was halted until the raise breakthrough. A pilot hole was surveyed and drilled 9.14 m (30 ft) in depth. This allowed MRC crews to accurately measure the distance between the shaft working face and surface, and aided final raise alignment.

Prior to breaking to surface, line holes were drilled from surface around the raise perimeter to eliminate collar over break. Approximately 80 holes were drilled allowing the final round relief points to break to. This method proved effective and necessary to accurately install a bearing plate for the steel can structure.

## SURGE SHAFT

Elevating an MRC nest allows for muck removal to take place under the nest and eliminates additional tunnel development to accommodate a nest. LHDs have adequate room to access blasted raise muck without restricting nest access. An elevated nest eliminates workers' exposure to unsupported ground.

Initial shaft rounds were drilled with a two boom jumbo allowing adequate room to establish the curve rail. Subsequent rounds are drilled with handheld stoper drills. The MRC runs on two air powered motors which are supplied to the ascending or descending machine via 2 inch bull hose on a retractable reel. Typical rounds consisted of 80, 1⅜ inch holes loaded with 1¼ inch packaged emulsion explosive. Lead wire was run down the rail and to the adit portal to safely initiate each blast. MRC rail has air and water lines built in. Air and water headers are turned on prior to the blast to expedite blast fume ventilation. Rail is advanced as necessary, generally one 2m rail per round. Ground conditions in the surge raise were deemed extremely competent. No ground support besides scaling was required in the lower 91.44 m (300 ft) of shaft. BLT chose to install 1.83 m (six ft) galvanized split sets on 2.44 m

**Figure 7. Completed surge shaft breakthrough steel structure**

(eight ft) centers throughout the shaft as a safety precaution. Typical advance throughout the raise was 2.44 m (8 ft) per shift exceeding estimates. In total, 99.66 m (327 ft) of raise was excavated at a diameter of 3.35 m (11 ft).

## SURGE SHAFT BREAKTHROUGH STEEL STRUCTURE

The surface breakthrough presented exposure of Sitka's drinking water to outside contaminants. To minimize water contamination risk, a 3.35 m (11 ft) diameter steel can was installed in the collar and grouted in. Covering the can was an engineered canopy restricting animal or foreign object entrance. Corrugated steel "window" flaps allowed water discharge in the event of a surge while prohibiting unwanted entrance (Figure 7).

The steel can sections were fabricated using ¼ inch steel coated with Lifelast Durashield 310 to prevent weathering and deterioration. The canopy was galvanized for this same purpose. Transporting the 907 kg (2000 lb) cans 91.44 vertical m (300 ft) up a mountain required the yarder and fabricated skids to prevent damage. After arriving at the yarder landing zone, the JD-120 excavator would maneuver cans into the surge basin and align in the shaft collar.

A final heavy lift would be performed by a Columbia Vertol helicopter demobilizing all equipment and materials. With a lifting capacity of 4,535 kg (10,000 lb), only the JD-120 excavator would need to be disassembled prior to flight.

## CONCLUSION

Successful completion of the underground excavation in this remote, challenging region required precision, continuous project planning and scheduling. Work areas accessible by crane or helicopter exclusively provided constant logistical operational challenges. The ability of BLT crews to adapt to new challenging work areas regularly and safely greatly increased project success. With diligent planning, competent crews and innovative site establishments a quality product was produced.

# Applying High-Strength Shotcrete to 86th Street Underground Cavern While Maintaining the Highest Quality Control Standards

**Ivan Djordjevic and Robert Begonja**
Skanska USA Civil Northeast

**Hitesh Shah**
Parsons Brinckerhoff

**ABSTRACT:** As part of the ongoing 2nd Avenue Subway program, which will relieve major congestion for subway riders on Manhattan's east side, MTA Capital Construction (MTA CC) contracted Skanska/Traylor Joint Venture (STJV) to excavate and concrete line the station cavern for the 86th St Station. Excavation was done using the drill and blast method. The approximately 1,000 ft long by 60 ft high and 60 ft wide hard rock cavern was excavated and then reinforced using rock bolts and fiber reinforced shotcrete.

This paper will describe the means and methods used to safely and productively apply quality reinforced shotcrete to the cavern walls and arch with minimal impact to the community and other excavation operations. The obstacles that were overcome include continuous public observation, very tight surface staging, controlling environmental impacts (noise, dust, air quality) and safety challenges when working in one of the most densely populated and influential areas of Manhattan, the Upper East Side.

## QUALITY CONTROL (PRE-PRODUCTION AND PRODUCTION TESTING)

### Mix Development

The shotcrete mix design used for the 86th street project was previously used and developed on the Skanska/Shea/Schiavone contract of the No 7 line extension. This mix contained cementitious material, sand, stone, water, water reducer, and set retarder. Also, at the spray nozzle accelerator was added to speed up the setting process.

### Pre-Construction Testing

The shotcrete program was set up prior to production to check for capability of equipment, workmanship and quality of material under field conditions. Due to the lack of in-situ area due to the fact that the excavation had not yet begun, shotcrete was applied to the test panels. These panels were prefabricated out of ply wood and measured 3 feet by 3 feet by 6" (Figure 1). Shotcrete was tested for compressive and flexural strength.

### Compressive Strength

As per specifications the compressive strength of Shotcrete must reach following values

1. At 10 hrs: 1500 psi minimum.
2. At 24 hrs: 2700 psi minimum
3. At 28 days: 6000 psi minimum

A series of 5 overhead and 5 vertical panels were sprayed (see Figure 2). Three of each were tested after 10 hrs, 24 hrs and 28 days. One of each of these panels was used for compressive strength testing after 3, 7 and 14 days respectively to give an early indication of the development of the final strength and two panels. An additional panel in each position was sprayed to give the possibility to replace any specimen that was damaged during coring or handling. Average 28 day Compressive Strength was 7,682 psi.

### Flexural Strength

As per specifications the Steel Fiber Reinforced Shotcrete (SFRS) must meet following strength and toughness requirements:

1. Average first crack flexural strength at seven days greater than 700 psi
2. Residual strength factor R20,50 at seven days of 40.

To determine the optimal mix for Steel Fiber Reinforced Shotcrete the same base mix design was used with addition of steel fibers. For testing purposes we used steel fiber by two different companies. We sprayed 3 panels for each supplier with varying amounts of fiber. These panels were then cut in 4" wide strips and tested for flexural strength as per ASTM C1609. Concurrently another 9, 4" × 4" × 14", beams (see Figure 2) were poured inside molds for use in the same test. After the testing was completed it was decided that 53 lb/cy will give us the desired results.

### Nozzleman Evaluation

Pre-production testing was used to evaluate the proposed nozzleman for proper shotcrete application technique on vertical and overhead surfaces. Shotcrete panels were checked for honeycombing and segregation and cores were checked for leasing and voids (see Figure 3).

### Production Testing

During production, shotcrete was checked at point of mixing in the concrete plant, delivery tickets were checked at the job site, shotcrete panels were periodically sprayed and cored for compressive strength and applied shotcrete was checked for soundness.

**Figure 1. Test panel**

At the plant an independent testing laboratory, employed by Construction Managers office, checked all the trucks prior to them being released.

At the job site shotcrete delivery tickets were checked and shotcrete was checked for lumps or bleeding. To test shotcrete for compressive strength a test panel was sprayed for every 25 CY applied. After it was demonstrated that the quality of shotcrete supplied and applied met and exceeded contract requirements this frequency was changed to one panel per 100 CY sprayed and no less than one per week. Test panels were left in situ for at least 24 hrs after which they were taken to the lab for coring and subsequent testing of specimens.

Shotcrete applied was visually inspected for any visible imperfections and checked with a hammer for soundness to confirm good bonding between shotcrete and rock surface.

## QUALITY ASSURANCE

### Qualifications of Applicator

The minimum experience for a nozzlemen shall be at least three years experience applying Plain and Fiber Reinforced Shotcrete. Each nozzleman had to demonstrate the acceptable proficiency with uniform application of shotcrete on vertical and overhead test panels and confirm the laboratory testing of cores taken from the test panels.

### Shotcrete Foreman

Shotcrete was applied under the immediate supervision of a foreman with at least five years of shotcrete operations experience. At least one year shall include operations experience involving the application of shotcrete on to rock.

**Figure 2. Test panels before and after shotcrete application**

**Figure 3.**

### Confirmation to Regulatory Requirements

The operations conducted were complied to Code of Federal Regulations (CFR)—29 CFR 1910 Occupational Safety and Health Standards (OSHA), safety requirements for working platforms or lifting equipment and personal protective equipment.

### Mix Design and Testing Prior to Production

Mix design was submitted and field trials were held prior to the actual application of shotcrete. The shotcrete mix design called for 6000 psi. The accelerator admixture to develop a quick set mix that was incorporated was: Initial setting time of shotcrete was 3 minutes maximum from time of placement. Final setting was approximately at 12 minutes. Accelerator percentage was adjusted to account for changes in rock condition, height, ambient temperature and material slump. The test panels for the compressive strength was of 3 ft × 3 ft & 6" high square wood panel . Shotcrete beams were also made to test the flexural strength. The panel for the flexural test was 45" × 15" to make 9 samples of 14" × 4." The flexural test was conducted as per ASTM C1609. For both compression test and flexural test samples were made with both types of shotcrete—with and without fibers.

### Testing

- Standard Concrete Core Testing: Panels will made for every 100 cubic yards. Cylinders of size 3" diameter and 6" long were cored from the test panels. Core samples were tested at 28 days for compressive strength test for 6000 psi.
- Independent test agency Tectonic Engineering was utilized for sampling and testing of shotcrete. Personnel conducting the field and laboratory test were ACI certified.
- The test results for the shotcrete core samples were tabulated in a log form and submitted to the client on a monthly basis.

## FIELD APPLICATIONS

### Equipment

Atlas Copco MEYCO products were used for shotcrete applications at the 86th Street Cavern project. This fleet of products included three MEYCO Suprema shotcrete pumps, two MEYCO Potenza Spraying units, and one MEYCO Oruga spraying unit.

MEYCO Suprema pumps were located at both the North shotcrete zone and South shotcrete zone. These pumps were used for all conveying of both shotcrete material and accelerator from the surface to the application site. The third pump was used as a spare and as a mobile pump for other onsite shotcrete needs not associated with the main cavern excavation.

MEYCO Potenza spraying units were located in the North cavern and South cavern. Each unit was responsible for fulfilling the shotcrete needs at its site. Each of these units averaged the application of 50–60 CY of shotcrete per shift. The Potenza allowed the project team to successfully apply overhead shotcrete at a height of more than 35 feet while allowing the nozzleman and crew to be a safe distance behind and unexposed to freshly applied overhead shotcrete. (see Figure 4).

MEYCO Oruga spraying unit was a smaller track unit which was used during earlier stages of shotcrete application. Both in the North shaft and South shaft the Oruga was used since it had a much smaller footprint and application height was only in the 15–18 foot range. The Oruga also came in handy for smaller adits and drifts which were limited in size and made bringing the large Potenza unit difficult. (see Figure 5).

### Field Constraints

Several constraints had to be overcome in order to successfully create and operate the shotcrete program for the 86th Street Cavern Project.

#### *Off-site Batching*

Since an onsite batching plant was not feasible, and off-site supplier was used and ready-mix trucks would travel to the pumping stations on the project.

**Figure 4.**

**Figure 5.**

The ready mix trucks had to travel 11 miles from the batching plant over two bridges through Manhattan traffic. This created an environment where you could not consistently rely on travel time of the ready mix trucks. This uncertainty posed issues when trying to productively pump 50–60 CY per shift. To help combat this, retarder was used in the ready mix truck to slow down set time of the shotcrete material. Retarder levels were constantly being adjusted to help combat traffic, weather conditions, or other field delays. Traditionally the material was able to maintain a workability level of 4–5 hours from batching.

Once at site ready mix trucks had fiber added to them when reinforced shotcrete was being used for the day. Fiber was delivered onsite in truckloads and manually added to the trucks before discharge at pump. Fiber dosage was strictly monitored to ensure adequate quantity of fiber and proper mixing was achieved. It was also vital to maintain proper fiber storage onsite. Protecting fiber from weather and preventing rust was critical to maintain high quality results.

### *Long Distance Conveying by Means of Pump*

Due to the underground aspect of the project all shotcrete material had to be conveyed from the surface to application site using long distance pumping. Traditional discharge from ready mix trucks into spraying units at application site was not feasible. Due to the linear layout of the project and the fact that access shafts were on the northern and southern most ends of the project, two shotcrete work zones were required to be outfitted. Each zone would be required to independently field the shotcrete needs of the SEM teams at each end.

The main work zones posed their own constraints due to the congested urban environment of the project. Neither the North Shaft nor the South shaft had adequate work space to allow pump setup. Therefore adjacent ancillary work zones were used to setup the shotcrete pump, bulk accelerator, fiber storage, and allow for ample space for ready mix trucks to discharge shotcrete material. Again due to the congested nature of the project site setting several issues had to be overcome in order to setup the shotcrete work zones.

Steel conveying pipe (slickline) used to pump the shotcrete material had to cross underneath busy city streets. This required street closures in order to safely dig a trench for pipe installation while avoiding a web of existing utilities. This slick line was then encased in concrete and the street reopened. Since the slick line was considered inaccessible the project team decided to install spare pipe runs in the event a plug in shotcrete material developed and was unable to be removed. Therefore the pipe trench consisted of two 2.5" slickline, two 4" slickline, and two ¾" CPVC line used for accelerator.

Accelerator used to activate the shotcrete at the nozzle was also conveyed from the surface. Using the MEYCO Suprema pump accurate dosing was achieved based on material flow rate. This dosing was done topside at the pump. CPVC lines were used to carry the accelerator material from the surface down the shaft and to the application site. During the winter months several precautionary measures were put in place to prevent these lines and the accelerator material from freezing. All CPVC piping at the surface, in the trench, and down the shaft was heat treated to limit the temperature exposure to the material. In addition after every application, CPVC lines were flushed out with water, blown out with compressed air, and then anti-freeze was blown through the lines to prevent any moisture from freezing and possibly creating a blockage for the next application.

The aboveground pump site conveyed the shotcrete material and accelerator down to application site a distance of approximately 800–1,000 feet. This posed great risks which had to be mitigated as best as possible. Due to that distance, both conveying pump on MEYCO Suprema and the accelerator pump were pushed to their limits. This further stressed a need for consistent shotcrete material. Shotcrete material needed to have a high slump in order to limit the pressure on the pump, plasticizer was added onsite in order to keep material consistent and workable. Material flowed through the 4" slick line and into hoses on spraying unit. Clean, well maintained slick line and hoses was constantly being stressed to the work crews, again in order to mitigate the risk of blockage.

### *Safe Field Application*

Both the fleet of equipment and the above ground shotcrete setup culminated underground with the proper and safe application of shotcrete. Shotcrete crews were well trained and well disciplined in their craft. The shotcrete foreman ran two crews, each crew consisted of nozzle man and two miners. Each of these crews were responsible for all shotcrete field activities on their site. In the eight hour shift, each crew was able to setup their equipment and work area for application, spray 50–60 CY and cleanup and demobilize application site.

Application site setup required driving in the MEYCO Potenza spraying unit to the application site. Shotcrete material conveying pipe (slickline) had to be run to unit along with CPVC accelerator line, water line, and compressed air line. Once set up was complete the exposed rock face had to be well washed with water and compressed air, Potenza nozzle was used for this as it allowed nozzle man to be clear from exposed rock face. This was critical to maintain proper adhesion to rock surface and minimize any fallout from the overhead application. Exclusion zones were then put in place where shotcrete application was to take place, barrels, cones, and caution tape were used to cordon off the areas. Only shotcrete crews who had intimate knowledge of the days progress were allowed in these areas. The project team did not want to risk other workers walking under freshly applied overhead shotcrete. While rapid setup times were achieved some fallout did occur and personnel safety could not be risked.

During shotcrete operations dust is always an ongoing issue. In order to combat this dust for the safety of craft and prevent dust exposure from leaving the cavern several mitigation items were used by the project team. Ventilation fans were used on each end of the cavern for intake of fresh air for the men. Exhaust fans were also equipped to maintain a steady air exchange flow. However, due to the crowded urban environment in which the project took place and the high power community presence the project team did not wish to exhaust tunnel air onto the busy Manhattan sidewalks. In order to prevent this, a scrubber system was installed. This scrubber system took all exhaust air and filtered it through an internal filter, this filtered air then entered a chamber of spraying misters which knocked down any remaining particles. These fine misters allowed the air to exit the scrubber system clean and hazard free.

Within the cavern smaller ventilation fans were also in place to help alleviate dust occurrence during shotcrete operations. These small rotational fans also had a mister ring installed in front of them which would knock down dust particles from shotcrete in the immediate area. The project team was able to maintain a very clean dust free environment during excavation operations. However, to further ensure the safety of the craft and management all personnel working in or around shotcrete areas were required to wear half face respirators equipped with filter cartridges. All these precautionary measure allowed for safe shotcrete application within the 86th street cavern.

## CONCLUSION

Completing the 86th street cavern required a high pace, highly focused approach to all excavation operations. The densely populated urban environment of Manhattan's Upper East Side allowed little room for error. The crowded staging areas and high powered community in the surrounding apartment buildings created a jobsite environment unlike any other. The excavation operations required precise blasting, high production mucking, and quality shotcrete application. The shotcrete program was able to be maintained with strict adherence to quality control, quality assurance, and was completed in a safe productive manner. In total the project team was able to apply approximately 12,000 CY of fiber reinforced shotcrete and 3,000 CY of unreinforced shotcrete in a short eight month period. This was productively completed without any environmental or safety incidents and left a happy community with little disturbance to surrounding areas.

# Successfully Constructing Deep Shafts Through High-Yield Glacial Outwash Aquifer Overburden, Shale, and Limestone Rock for the Deep Rock Tunnel Connector Project

**Alston M. Noronha**
Black & Veatch

**John Morgan**
Citizens Energy Group

**Stuart Lipofsky**
Shea-Kiewit Joint Venture

**ABSTRACT:** Construction of the Deep Rock Tunnel Connector (DRTC) combined sewer overflow (CSO) tunnel more than 60 m (200 ft) below ground in Indianapolis used several shaft construction techniques. Slurry wall construction for two 15 m (50 ft) diameter shafts through approximately 30 m (100 ft) of high yielding aquifer was successful. Conventional Drill & Blast excavation of shale and limestone completed these shafts. Drilled steel casing through overburden and raise-bored rock excavation was employed for nine utility, drop, and ventilation shafts, 2.3–3.6 m (7.5–12 ft) in diameter. This paper describes the shaft construction operations including challenges and solutions thereto.

## PROJECT OVERVIEW

The city of Indianapolis, like a lot of developing cities across America, has older sewer infrastructure in the central part of the city, which is unable to keep up with the growing population and its needs. Both sanitary and storm water are carried to two main treatment plants through the same sewer infrastructure. During precipitation events, these pipes get overwhelmed and CSO water discharges into the river. In an attempt to clean up the city's water ways, a Long-Term Control Plan (LTCP) has been put in place which includes construction of approximately 40 km (25 mi) of 5.5 m (18 ft) diameter tunnel in rock, between 70 and 80 m (230 and 260 ft) below ground.

The DRTC is the southernmost part of this LTCP. Its main components are:

- Two large shafts varying in diameter from 13.4 m (44 ft) at the surface to 10.7 m (35 ft) in rock
- 12 km (7.5 mi) of 5.5 m (18 ft) finished diameter tunnel (6.1 m (20 ft) excavated diameter)
- 975 m (3200 ft) of adits
- Three 2.4 m (8 ft) diameter utility shafts
- Three drop shafts, 1.4 to 2.1 m (4.5 to 7 ft) diameter
- Three vent shafts, 1.1 to 1.8 m (3.5 to 6 ft) diameter
- Three diversion structures.

Along with providing storage capacity for the CSO water, the DRTC will convey flows from all tunnel branches upstream of it, to the Southport water treatment facility, as seen in Figure 1.

The local geology consists of overburden between 24 and 34 m (80 and 110 ft) thick, which overlies the bedrock. Construction challenges through the overburden include high groundwater table and boulders.

## GEOLOGY

### Regional Overburden Soils

The project site is located in the City of Indianapolis which entirely encompasses Marion County, Indiana. The overburden soils in Marion County, Indiana are unconsolidated glacial deposits averaging 30 m (100 ft) in thickness. The overburden along the DRTC alignment consists primarily of sand and gravel with a few discontinuous till layers. The presence of sub-rounded cobbles and boulders was noted when drilling the DRTC borings. Their depths and locations were generally random in nature, although they were more frequently encountered immediately

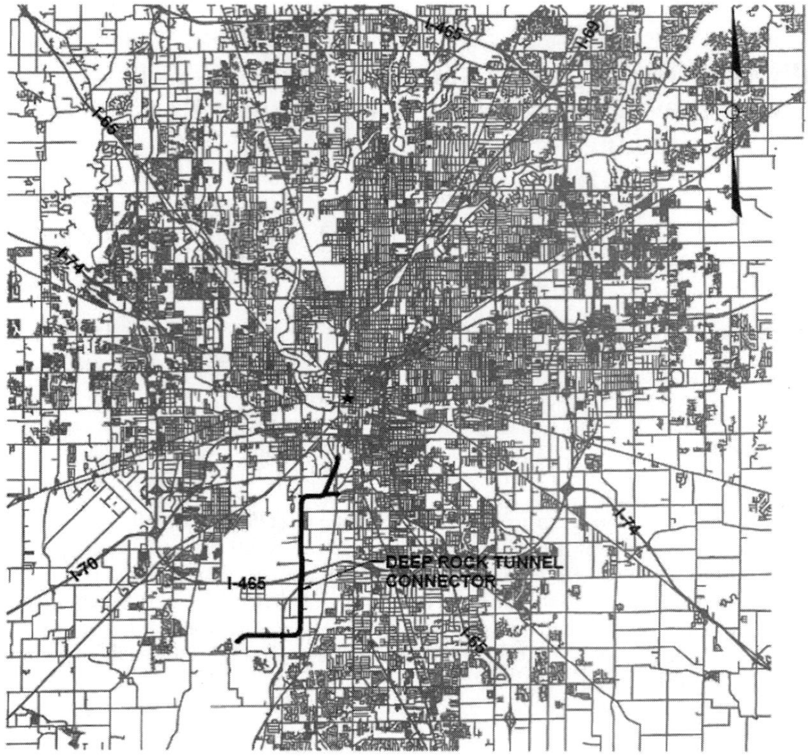


**Figure 1. DRTC project plan and location**

above the top of bedrock. They tend to be made up of hard rock such as granite and gneiss from the Canadian Shield, transported by glacial action. Boulders ranging in sizes from about 0.3 to 1.5 m (1 to 5 ft) in diameter can be seen around a local quarry.

### Regional Bedrock

Devonian and Silurian carbonate rock underlie most of Indianapolis and the White River valleys where the DRTC is located. The bedrock formations encountered in boreholes, in descending order are as below.

- New Albany shale of Devonian-Mississippian geologic age;
- North Vernon Formation limestone of Devonian Muscatatuck Group;
- Vernon Fork and Geneva Members of the Jeffersonville Formation, also of Muscatatuck Group;
- Mississinewa or Liston Creek Member of the Silurian Wabash Formation.

The shallowest bedrock formation along the DRTC alignment is predominantly New Albany shale, encountered between elevations El. 550 and 625 feet mean sea level (msl) (NAVD 88). It is a black to brownish green shale, slightly weathered to fresh, and soft. This formation is very thin towards the northern portions of the tunnel alignment.

The North Vernon Formation consists of moderately hard, brown to gray limestone, with scattered fossiliferous zones. It is approximately 9 to 15 m (30 to 50 ft) thick in the project area, with the shallowest contact at the alignment's north end and getting progressively deeper toward the alignment's southern end.

The Vernon Fork Member of the Jeffersonville Formation consists of moderately hard to hard, gray to white limestone with clear calcite grains, stylolites, and scattered pyrite. The lower part of this unit becomes dolomitic. It has been observed to be approximately 12 to 24 m (40 to 80 ft) thick in the project area.

The Geneva Member of the Jeffersonville Formation consists of tan to brown dolomite which is granular, moderately hard to hard, with scattered fossil casts. It is approximately 12 to 18 m (40 to 60 ft) thick in the project area. The Wabash Formation, where encountered in the boreholes, consists of greenish-gray dolomite which is moderately hard to hard, and encountered at depths between 63 and 90 m (205 and 295 ft) below ground. No DRTC borings have fully penetrated the Wabash formation.

### Local Hydrogeology

Groundwater in the DRTC alignment vicinity predominantly flows through an outwash aquifer system primarily consisting of coarse sand and gravel with scattered cobbles, which is generally unconfined within the project area. The unconsolidated soils, including the outwash aquifer along the DRTC alignment range between about 15 and 40 m (50 and 130 ft) thick. Regional piezometric maps indicate shallow groundwater is between about El. 650 and 675 feet msl along the DRTC alignment (IDNR, 2002).

The most productive aquifers in the region are typically associated with sand and gravel outwash deposits located in the White River Valley and flood plains. Production wells completed within this aquifer reportedly yield over 2,000 gallons per minute (gpm) (Shrewsberry, 2007). The bedrock aquifer generally provides less groundwater than the outwash. IDNR *(2*002) reports most wells drilled into the upper 100 feet of bedrock produce less than 5 gpm. Productive bedrock aquifer zones are reportedly in areas where the New Albany shale is absent (Black and Veatch, 2007).

## LAUNCH/RETRIEVAL SHAFTS

The two largest shafts on the project are located at each end of the tunnel. The launch shaft is located at the downstream end, close to the Southport treatment plant. It serves at the main access shaft to the tunnel and all tunnel boring machine (TBM) mining related activities go through the launch shaft. The retrieval shaft is located at the upstream end of the tunnel, north of the intersection of West Street and White River Parkway East Drive. As the name suggests, it will be used to retrieve the TBM and other equipment and materials once construction of the tunnel is complete. This shaft will also serve as the launch shaft for the White River Tunnel which will connect to it from the North.

The launch and retrieval shafts were constructed using similar techniques. Given the large size of the shaft, the high groundwater table (few feet below ground), and the high hydraulic conductivity of the overburden soil, dewatering was not an option and a water tight shaft construction method had to be employed. A ring of reinforced concrete walls was used through the overburden and a few feet into top of bedrock. This was further supported by a reinforced concrete ring at the soil-rock interface. Below the reinforcing ring, rock was excavated using drill and shoot technique down to tunnel invert. The excavated rock portion was supported by pattern rock bolting and shotcrete. This was later concrete lined.

### Slurry Wall Construction Through Overburden

The general contractor, Shea-Kiewit Joint Venture (S-K JV), utilized the expertise of Bencor to construct the slurry walls for both shafts.

Bencor laid out the slurry wall pattern on site, which consisted of ten interlocking panels, five primary panels and five closing panels as seen in Figure 2. The sequence involved constructing all primary panels first followed by the closing panels.

The slurry wall layout was excavated down to a few feet below the surface and precut Styrofoam blocks were placed, forming the outline for the slurry wall. Concrete was then placed around the Styrofoam blocks, creating a slurry guidewall system to help keep the hydromill vertical during the start of each panel excavation. Bencor utilized a Bauer BC 40 hydromill (Figure 3) to excavate the overburden and a few feet into top of rock. The hydromill was 0.9 m (3 ft) wide, 3.1 m (10.3 ft) long, and weighed around 40 tons. Each panel was filled with bentonite slurry during and after excavation to stabilize the opening. For this purpose a bentonite batching plant and three slurry ponds were setup on site. As hydromill excavation progressed, cuttings were pumped out with the bentonite slurry and transported to a desanding unit which separated the cuttings out and the slurry was pumped back to the ponds where it was reused. Several slurry tests, including viscosity, density, pH and filtration were run a few times each day to make sure the slurry was within specifications and working as desired.

The primary panels were excavated in three bites, one on each side and the third bite in the center. The hydromill was then run along the bottom of the panel to de-sand the excavation, until the sand content was less than 5%. Following this, Bencor would check the internal profile of the excavated panel using a Koden device, which would print out the excavation extents of each panel as the Koden sensor was lowered and raised in the excavation.

The panels had a reinforced steel cage as part of the design which was tied together on site as two halves, each around 15.2 m (50 ft) long. The rebar cages had concrete block spacers on the inside and outside faces of the shaft, but on the two sides adjacent to other panels, U-shaped PVC pipes were used. These spacers help keep the steel reinforcing in the center of the excavation. Additionally, 10 cm (4 in) diameter schedule 80 vertical PVC bedrock grout

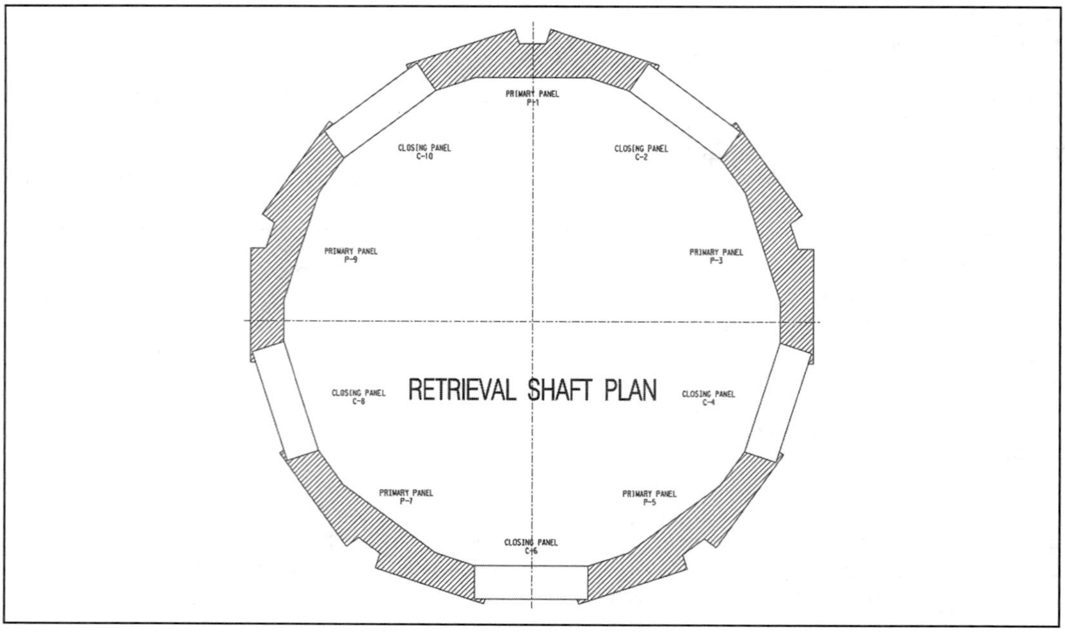


**Figure 2. Slurry wall plan**

tubes were tied into the steel cage, 4 and 2 per primary and closing panel respectively, (total 30) to act as conduits for pressure grouting at a later stage.

Excavation of the closing panels was done in one bite. This involved excavating a little into the concrete from primary panels on either side. Hence the U-shaped PVC pipe spacers were used on the sides so that the hydromill could easily go through the PVC rather than steel or some harder material.

The 34.5 MPa (5,000 psi) concrete for the primary panels was placed using three equally spaced tremie pipes simultaneously, to avoid differential pressure in the excavation, which might move the rebar cage. The closing panels used one tremie pipe. The primary panels were excavated opposite each other to give the recently concreted panel time to gain strength and minimize differential pressures.

Construction of the launch and retrieval shaft slurry walls was similar with a few differences as noted here. The retrieval shaft slurry walls were around 4.6 m (15 ft) shorter than those of the launch shaft. The retrieval shaft had high voltage overhead power lines running close to the shaft along the north-west section. Hence excavation in this area took place with the crane and hydromill working from the inside of the shaft footprint rather than the outside, like at the launch shaft site. Slurry wall construction for the launch shaft took 5 weeks and began in May 2012. For the retrieval shaft it took 7 weeks and started in July 2012.

Once slurry wall construction was completed, Layne Christensen was brought in to pressure grout below the slurry wall through the PVC grout tubes placed in the rebar cages of the slurry wall. Most of the grout holes did not take any grout, with a few of them taking very little. In this case, the post grouting exercise turned out to be more of a precautionary step.

## Obstacles

The geotechnical investigation showed that there were boulders present in the overburden. To deal with potential boulders, Bencor had a mechanical clamshell and a 7,710 kg (17,000 lb) chisel on site. Fortunately these were never used at either of the shaft sites. Most of the boulders encountered were either broken up by the hydromill or got wedged between the two cutting wheels of the hydromill, at which point, the hydromill was retrieved from the excavation, its cutting wheels reversed to drop out the boulder, and excavation was resumed.

Over excavation near the panel top was another obstacle that Bencor encountered initially, but overcame quickly by limiting the depth of excavation with a track hoe, prior to hydromill excavation.

Majority of the retrieval shaft overburden excavation challenges came as a result of work that took place nearly 100 years ago in this area. At that point in time the White River flow path was directly through this shaft location. However, due to a flood in downtown Indianapolis in 1913 the river was realigned to help prevent future floods.

On July 3, 2012 human bones were found during retrieval shaft slurry wall excavation of the initial

**Figure 3. Hydromill**

panel. The Police Department and Crime Lab representatives were on location. Within three hours of arrival of public safety personnel all operations were permitted to resume. The bones were clearly old and were identified as such, thankfully, creating minimal disruption to the activities on the project site.

The biggest obstruction during slurry wall construction was encountered at the retrieval shaft, around 4.6 m (15 ft) below ground, while excavating panel P5. Not knowing what the obstruction was, and dealing with sand and gravel type soil, a trench box had to be used to get down to the obstruction which turned out to be a tree trunk approximately 46 cm (18 in) in diameter and 6.1 m (20 ft) long. This caused a two week delay in construction of slurry walls at the retrieval shaft. Given the history at this project location, overall it was found to be a successful operation with minimal downtime.

### Shaft Excavation

The launch and retrieval shafts used a track-hoe lowered into the shaft to excavate the overburden. Excavated spoil was removed from the shaft in muck buckets using a crane. A sump with a pump was setup in the shaft during excavation to remove water, mostly trapped within the slurry wall itself with some water coming from below the slurry wall. Once excavated to top of rock, a reinforced concrete ring 0.8 m (2.5 ft) thick by 1.2 m (4 ft) deep was constructed at the soil-rock interface, towards the bottom of the slurry wall, to provide structural support.

In rock, drill and shoot excavation technique was used. A 30.5 m (12 in) diameter burn hole was drilled at the shaft center, down to tunnel invert, prior to shaft excavation. Each shot averaged 3.7 m (12 ft) in depth. Shot rock was mucked out of the shaft using an excavator and muck buckets. The shaft rock face was supported by a 1.8 × 1.8 m (6 × 6 ft) rock bolting pattern, 2.4 m (8 ft) deep, followed by shotcrete. This was later concrete lined up to the reinforced concrete ring at the soil-rock interface.

Depths of the launch and retrieval shafts were approximately 76 m (250 ft) and 65.5 m (215 ft) respectively. Rock blasting for the launch shaft began in July 2012. Overall, the contractor made two shots per week during the launch shaft construction which took approximately 7 weeks.

Work at the retrieval shaft began in June 2013. Here work has been performed by a crew that works around the overall project site as needed. The retrieval shaft has not yet been a critical path item and as such helped to utilize extra manpower as they become available. TBM hole-through is expected to take place April 2014 barring any problems with the tunnel drive.

## UTILITY, DROP, AND VENT SHAFTS

Aside from the larger launch and retrieval shafts, the other nine shafts on the DRTC project will be excavated using the below procedure. The project has three utility shafts to aid in air release from the tunnel as it fills up with water and to provide access for future inspection and maintenance of the tunnel. Three sites have been designed to incorporate diversion structures which divert CSO flows to drops shafts. Hence there are three drops shafts on the DRTC project which will convey near surface flows to the tunnel. Each of the drop shafts have a de-aeration chamber at the tunnel elevation which force entrapped air out of the water and to the surface through a vent shaft at its downstream end. See Table 1 for designed shaft diameters.

### Steel Casing Through Overburden

S-K JV used the services of Case Foundation to sink steel casings through the overburden for all nine shafts (utility, drop, and vent shafts). Prior to Case Foundation starting their work at any of the shafts, a 31 cm (12¼ in) diameter pilot hole, at the shaft center was drilled, from the surface down to the tunnel elevation, and filled with pea gravel. Excavation of these nine shafts followed a similar construction sequence as explained below.

Case Foundation first spun in a temporary casing at the surface, which ranged in diameter from 3.2 to

**Table 1. Shaft design and excavation diameters**

| Location | Shaft Design Diameter m (ft) | Overburden Permanent Casing Diameter m (ft) | Overburden Depth m (ft) | Raise Bore Reamer Diameter Through Rock m (ft) | Raise Bore Depth m (ft) |
|---|---|---|---|---|---|
| Utility Shaft No. 1 | 2.4 (8.0) | 3.7 (12.0) | 35.1 (115) | 3.0 (10.0) | 27.4 (90) |
| Utility Shaft No. 2 | 2.4 (8.0) | 3.7 (12.0) | 35.1 (115) | 3.0 (10.0) | 28.7 (94) |
| Utility Shaft No. 3 | 2.4 (8.0) | 3.7 (12.0) | 18.3 (60) | 3.0 (10.0) | 46.3 (152) |
| CSO 008 Drop Shaft | 2.1 (7.0) | 3.7 (12.0) | 30.5 (100) | 2.7 (9.0) | 32.6 (107) |
| CSO 008 Vent Shaft | 1.8 (6.0) | 3.4 (11.0) | 30.5 (100) | 2.7 (9.0) | 32.6 (107) |
| CSO 117 Drop Shaft | 1.4 (4.5) | 2.9 (9.5) | 30.2 (99) | 2.1 (7.0) | 29.6 (97) |
| CSO 117 Vent Shaft | 1.1 (3.5) | 2.1 (7.0) | 30.2 (99) | 1.8 (6.0) | 29.6 (97) |
| CSO 118 Drop Shaft | 1.7 (5.5) | 3.2 (10.5) | 24.1 (79) | 2.4 (8.0) | 36.9 (121) |
| CSO 118 Vent Shaft | 1.2 (4.0) | 2.3 (7.5) | 24.1 (79) | 1.8 (6.0) | 36.9 (121) |

4.6 m (10.5 to 15 ft) and in depth from 4.9 to 6.7 m (16 to 22 ft). After excavating material from inside the temporary casing, it was filled with bentonite slurry, and this level was maintained through the entire excavation. Excavation then commenced through the overburden using smaller, 2.1 m (7 ft) diameter drill tools, all the way to top of rock. The drilling tools used for excavation included open augers, hollow stem augers, and drilling buckets. Once the central portion of the shaft was excavated to top of rock, the excavation was reamed out using progressively larger drilling tools to open up the excavation.

Case Foundation then used a desanding unit to removed sand and suspended materials from the bentonite slurry, until it was less than 5%. The next phase was to place the permanent casing (see Table 1 for sizes) and spin it a couple of feet into top of rock.

If the permanent casing was less than 90 feet long, it was placed as one piece (see Figure 4) while casings longer than 90 feet were placed as two separate halves, welded together on site. A crane lowered the permanent can approximately in the center of the excavated hole, followed by a float can placed inside the permanent casing. The float can was essentially a steel can open at the bottom and closed at the top, with a few valves on the top to pump in and release air. This setup helped twist the can into top of rock at the surveyed in location.

A grout fill was tremie placed in the annular space between the permanent casing and excavation. Once completed to the surface, the temporary oversized surface casing was removed and the opening covered up and secured. On average Case Foundation took one week per shaft.

## Obstacles

While most of the shaft excavations and casing placements went on smoothly, there were a few sites with issues associated during the excavation process.

One of the biggest challenges was at Utility Shaft #2 where the contractor encountered what seemed to be a zone of boulders just above bedrock. After trying to get through them with soil excavating tools proved to be inefficient, Case Foundation brought in their set of rock drilling tools which included rock augers, scrapers, and smaller core buckets to tackle this obstruction. It took them nearly twice as long to excavate this shaft compared to the others. But the rock drilling tools were successful in breaking up the boulders and getting them out of the excavation.

## Raise Bore Excavation of Rock

Prior to starting raise bore operations; any groundwater inflow into the bottom of the shaft was controlled by pre-excavation grouting through four ports located in the permanent casing, six feet from the bottom of the casing. DMC Mining Services were to perform raise bore excavation of the shafts. To date, only one raise bore (Utility Shaft #2) has been completed.

Raise bore operations were slightly different depending on the location of the shaft. The three utility shafts are located above the tunnel while the three drop and vent shafts are off the tunnel alignment. For the utility shafts that were located in the corner of the crown, the tunnel was squared off using drill and shoot from tunnel crown to springline.

The surface setup included, raise bore beams aligned and pinned to the concrete pad spanning the caisson. The Subterranean 009 raise drill used two jacks to pull the raise bore head while excavating the rock. Drill rods complete with 31 cm (12¼ in) diameter × 1.5 m (5 ft) raise bore stabilizers and a pilot tricone roller bit was added to the raise drill and lowered into the pilot hole. Once down to tunnel elevation, a hydraulic Makeup/Breakout tool was used to break the bottom rod/stabilizer joint, which was removed along with the pilot bit.

The reamer was transported through the tunnel using a railroad flat car and positioned directly below the pilot hole. The stem of the reamer was

**Figure 4. Case foundation placing permanent casing into excavation at Utility Shaft #3**

**Figure 5. Raise bore reamer, as seen in the tunnel**

aligned with the bottom of the stabilizer (Figure 5) using voice communication (mine phones from tunnel to top of launch shaft, and cell phones from there to raise bore operator). The Makeup/Breakout tool was used to torque the joint to the proper value. See Table 1 for reamer diameters and depths of rock excavation for the various shafts.

To start reaming, the reamer is collared into the rock face until all of the cutters are in contact with the rock. After collaring is complete, rpm and thrust values are increased to appropriate values. For shafts not above the tunnel, the material is loaded into muck cars positioned in the TBM tunnel. The muck cars are then transported to the shaft and hoisted to the surface and dumped. The cuttings from the utility shafts (located directly above the tunnel) land on a muck chute which was installed after the raise bore reamer was pulled to the rock face. The muck chute diverts the cuttings to the TBM tunnel belt.

Once the raise bore reamer holed through, at the soil rock interface, it was brought up through the permanent overburden casing and retrieved from the surface. Following which the excavation was immediately covered and secured.

## CONCLUSIONS

A detailed geotechnical investigation is one of the first steps in identifying potential challenges and anomalies that may be encountered during construction of a tunnel project. This coupled with knowledge of similar projects in the area help to set baselines so that bidding parties are aware of what they might encounter and be prepared for during construction.

Due to high groundwater, high permeability soils, and boulders in the overburden, several shaft construction methods like shaft sinking, and use of sheet piles were not allowed on this project. Employing specialized contractors to perform various parts of the project was key to keep the project construction on schedule.

Maintaining good communication, via daily discussions and weekly progress meetings, with the owner, and a good relationship between the Contractor, Construction Inspection Team, and the Owner made tackling issues that much easier.

## REFERENCES


AECOM and Christopher B Burke Engineering, October 2011. Prepared for Citizens Energy Group. *Conformed Plans for Project No. CS-38-010C Deep Rock Tunnel Connector and Project No. LD-38-003 Southport AWT Levee Improvements and Extension.*

AECOM, April 2011. Prepared for Indianapolis Department of Public Works (DPW) / Citizens Energy Group. *Deep Rock Tunnel Connector Geotechnical Baseline Report.*

Black & Veatch, 2008. Prepared for DPW and USACE. *Fall Creek/White River Tunnel Groundwater Management Plan—Final Report.*

Indiana Department of Natureal Resources (IDNR), 2002. *Groundwater Resources in the White River Basin.* (http://www.in.gov/dnr/water/files/WFWR_web1-26.pdf).

Shrewsberry and Associates, 2007. Prepared for IDEM. *Conceptual Plan—Tunnel Spoils Disposal: Former Tibbs-Banta Disposal Site—Belmont/Southport Interplant Connection*, Marion County, Indianapolis, Indiana.

# Widening of the Twin Tunnels

**Raymond Henn, David Kwietnewski, and Gina Ross**
Brierley Associates

**Todd Kilduff**
T.M. Kilduff Consulting, Inc.

**ABSTRACT:** The Twin Tunnels, operated by the Colorado State Department of Transportation (CDOT), are located on interstate I-70 near Idaho Springs, Colorado. The existing tunnels, both two lanes, were constructed from 1960 through 1961. Since their original construction, traffic volume along the I-70 corridor has steadily increased. In 2012 CDOT awarded a CM/GC contract to widen the eastbound tunnel from two to three lanes. The widening project included detouring eastbound traffic onto the nearby county road, high scaling and rock support at the portals, demolition of the existing tunnel liner, drill and blast excavation, placement of new concrete tunnel lining, contact grouting, and construction of portal structures.

## INTRODUCTION

The Twin Tunnels are located along interstate I-70 approximately one mile east of the city of Idaho Springs, Colorado. I-70 is the primary east-west highway through Colorado. The existing Twin Tunnels consists of two tunnels, one westbound and one eastbound, each with two travel lanes. During the winter months I-70 sees very heavy traffic volume during ski season as well as in the summer months during tourist season. The tunnels experience both auto and commercial traffic traveling between Denver and Vail, Colorado.

This paper talks about the construction activities involved in widening the eastbound tunnel from two travel lanes to three. The portal developments started in late March 2013 and the enlarged tunnel was opened to traffic in December 2013. The project was awarded as a CM/GC contract with the Kramer/Obayashi JV as contractors. Atkins and Parsons Brinkerhoff were the tunnel designers. Brierley Associates performed constructability reviews, cost estimating and scheduling, and provided CM services during construction. Plans are underway to enlarge the westbound tunnel during 2014 construction season.

## HISTORY

The existing Twin Tunnels were constructed between 1960 and 1961 as part of a widening of US 6 & 40 through Clear Creek Canyon. The original US 6 & 40 paralleled I-70 on the south side of Clear Creek, through the Twin Tunnels area. Portions of it still exist today as County Road 314. They were excavated using the drill and blast method and lined with reinforced concrete. The finished cross section of each tunnel was approximately 32ft wide by 23ft high. The tunnels have performed well over the past 50 plus years. Figure 1 shows the existing west portals of the eastbound tunnel as it appeared just prior to the start of the 2013 widening project.

## DETOUR

The existing eastbound tunnel of the Twin Tunnels had to be completely closed to traffic during the entire widening construction. The I-70 eastbound tunnel traffic was detoured to the south and onto County Road 314 (Figure 2). This gave the contractor several hundred feet of yard/laydown area to the east and west of the existing eastbound tunnel portals. The detour was constructed under a separate contract during 2012/2013. The work included rock excavation, rock reinforcement, construction of retaining walls, strengthening of an existing bridge over Clear Creek and repaving of approximately two miles of the existing County Road 314.

## PORTAL DEVELOPMENT

Once the detour was in-place portal development started on both the east and west portal at the same time. The portal work as well as tunnel rock excavation and initial support, and placement of reinforcing steel and final tunnel liner concrete placement were done concurrently from each portal utilizing two separate but equal size tunnel crews and equipment spreads (Figure 3).

**Figure 1. West side portals before the start of the widening with the eastbound tunnel portal on the right**

**Figure 2. I-70 Detour to the south**

## EXISTING TUNNEL LINER REMOVAL

The existing tunnel liner consisted of 18 inches of concrete with No. 5 and No. 6 reinforcing steel. Also within the concrete liner were steel tunnel sets, W 10 × 39, on 4 foot centers which were installed during tunnel excavation as initial support.

Liner removal, which went on concurrently with portal development and tunnel excavation, started with cutting the liner longitudinally and radially using a 40 inch diameter diamond impregnated steel rotating saw blade (Figure 4). After saw cutting the liner concrete was in approximately 6ft by 6ft size pieces which remained in place even after saw cutting.

The concrete was removed utilizing a Liebheir R932 Litronic Tunneling Excavator equipped with a hydraulic impact hammer on a 45 degree swivel attachment (excavator arm). The swivel attachment provided greater range of motion for the excavator arm in the tight quarters of the existing tunnel.

Part way through tunnel excavation, the contractor elected to remove the south wall and arch of the old tunnel liner with the excavation blasts. Radial cuts were no longer made and a new longitudinal saw cut at the bottom of the south wall for the

**Figure 3. Portal development on east side**

**Figure 4. Diamond saw cutting of existing liner**

full tunnel length facilitated a clean cut during each blast. The north pillar wall of the old tunnel was left in place from the invert up to spring line for the full length of the tunnel.

## TUNNEL EXCAVATION AND INITIAL SUPPORT

The contractor project schedule called for three 8-hour shifts per day, 6-days a week for the tunnel work. Shortly after the start of tunnel excavation a structural steel framed canopy was erected at each portal (Figure 5).

The canopy served three purposes. First, it provided rock fall protection throughout the course of the construction. Timber crane mats and concrete barriers were set on top of the canopy structure for this purpose. Second, it supported the steel blasting mats, hung from the frame. The blasting mats were supported from the frame by rollers attached to the flanges of the trolley beams that were installed with the canopy girders. This allowed the mats to be moved across the face, like a curtain, for each blast. They were moved (rolled) to the sides during the drilling and mucking operations. Third, after the

**Figure 5. Structural steel framed canopy with blast mats attached**

completion of the tunnel excavation the same frame was used to lift the final tunnel lining rebar cage off the rebar gantry and onto the final tunnel liner formwork. This is discussed in more detail in the section on Final Lining.

### Geology

The bedrock at the Twin Tunnels is metamorphic gneiss, biotite gneiss, and hornblende gneiss. Locally, variations in the orientation in the rock structure are attributed to the numerous folds and minor faults. Igneous intrusions of pink granite and pegmatite occur at various locations. A zone containing fault gouge, soft seams, platy crushed rock and some veins of pyrite was encountered in the first 100 to 150 feet of the tunnels from the West portal including a zone of weaker mineralized rock is present in the area.

Discontinuities (joints and fractures) observed in the tunnel show a moderate to high spacing frequency depending on location within the tunnel. The quality of the rock improves from west to east. The area around the west portal is lower quality than the rock comprising the east portal. A description of the ground classes, as designed is given in Table 1.

The following is the initial tunnel support required by ground class.

- TTP–West
  - 8″ Shotcrete at portal brow extending 5′ above brow
  - 2 Rows spiling, #11 grade 60 groutable hollow bar; 20 LF embedment
  - First layer FRS shotcrete 4″ thick
  - W 12×65 Steel sets spaced on 4′ centers
  - Second layer FRS shotcrete 6″ thick

- TTP–East
  - 8″ Shotcrete at portal brow extending 5′ above brow
  - 1 Row spiling, #11 grade 60 groutable hollow bar; 20 LF embedment
  - First layer FRS shotcrete 4″ thick
  - W 12×65 Steel sets spaced on 4′ centers
  - Pillar bolts (outside of turn-under point), #9 Grade 75 threaded bar; resin grouted, 16 LF embedment; tensioned

- TT1 (per row, longitudinal spacing, 5′ centers)
  - 15 EA, Arch dowels, #9 Grade 75 threadbar; 16 LF embedment; 5′ transverse spacing
  - 2 EA, Sidewall dowels, #9 Grade 75 threadbar; 12 LF embedment; 5′ spacing
  - 1 EA Pillar dowels, #9 Grade 75 threaded bar; resin grouted, 16 LF embedment, tensioned
  - WWF: 6″×6″, W4.0×W4.0
  - Headings I, II: 15′ excavation allow 75′ lag between headings

- TT2 (per row, longitudinal spacing, 5′ centers)
  - 15 EA, Arch dowels, #9 Grade 75 threadbar; 16 LF embedment; 4′ transverse spacing

**Table 1. Ground Classes**

| Ground Class | Typical Rock Characteristics | Ground Behavior |
|---|---|---|
| TTP–West | Typically slightly weathered moderately to extremely fractured gneiss. Rock cover is thinner to the south. | Raveling |
| TTP–East | Typically dark gray-white to dark gray-black moderately fractured gneiss with a close joint spacing. Joint surface is typically planar, smooth. | Stucturally controlled block instability |
| TT1 | Typically fresh intact dark gray gneiss with a joint /fracture spacing in excess of 6 feet. Joints/fractures are very widely spaced, with very widely spaced clusters of very closely to closely to controlled spaced fractures. Slickensided fractures are rare. Mineralization along joints/fractures is rare and few in-filled joints/fractures are observed. Joint surfaces range from planar, rough and irregular, undulating, smooth; undulating, rough or irregular. | Stucturally controlled block instability |
| TT2 | Typically slightly weathered gray gneiss and light gray to white pegmatite with a joint spacing from 2 to 6 feet. Shear/fault planes, joint/fracture weathering, and alteration products are typical for this rock class. Open, in-filled and slickensided fractures are present. Observed shear/fault planes slow raveling may contain disintegrated rock between rock surfaces, with a thickness of alteration products generally less than 6 inches. The rock mass contains distinct sub-domains of lower quality rock characterized by clusters of very closely to closely spaced fractures and persistent in-filled fractures. | Slow raveling |
| TT2S Mixed Face TT2/TT3 | TT3 ground on the northern side and TT2 on the southern side. | |
| TT3 | Typically moderately weathered light gray pegmatite and dark gray gneiss with a joint /fracture spacing of less than 2 feet, or multiple and random joint/fracture sets with smooth or slickensided fast surfaces, irrespective of joint/fracture spacing, or multiple zones of brecciated and heavily fractured rock with clay or disintegrated between rock surfaces, or one or more shear/fault planes with a filling thickness greater than 6 inches. | Fast raveling |

- 2 EA, Sidewall dowels, #9 Grade 75 threadbar; 12 LF embedment; 4′ Spacing
- 1 EA Pillar dowels, #9 Grade 75 threaded bar; resin grouted, 16 LF embedment, tensioned
- First layer FRS shotcrete 3″ thick
- Second layer FRS shotcrete 3″ thick
- Headings I, II: 10′ excavation allow 40′ lag between headings

- TT2s (per row, longitudinal spacing, 4′ centers)
  - 19 EA, Arch dowels, #9 Grade 75 threadbar; 16 LF embedment; 4′ transverse spacing
  - 2 EA, Sidewall dowels, #9 Grade 75 threadbar; 12 LF embedment; 4′ spacing
  - 1 EA Pillar dowels, #9 Grade 75 threaded bar; resin grouted, 16 LF embedment, tensioned
  - First layer FRS shotcrete 3″ thick
  - MC12×50 Channel through Heading I
  - 23 EA Spiling bars, #9 Grade 60 bar; 14 LF embedment; 2′ transverse spacing; only in Heading I, 8 LF longitudinal spacing
  - Second layer FRS shotcrete 3″ thick
  - Heading I,II: 5′ excavations allow 25′ lag between headings

- TT3 (per row, longitudinal spacing, 4′ centers)
  - 19 EA, Arch dowels, #9 Grade 75 threadbar;16 LF embedment; 4′ transverse spacing
  - 3 EA, Sidewall dowels, #9 Grade 75 threadbar; 16 LF embedment; 4′ spacing
  - 1 EA Pillar bolts, #11 Grade 150 threaded bar; epoxy coated, 25 LF embedment, tensioned; Package 1B
  - First layer FRS shotcrete 4″ thick
  - MC12×50 Channel through Headings I, II
  - 38 EA Spiling bars, #9 Grade 60 bar; 14 LF embedment; 2′ transverse spacing; Headings I, II, 8 LF longitudinal spacing
  - Second layer FRS shotcrete 6″ thick
  - Heading I, II (Heading III not used): 5′ excavation, 35′ lag between three headings
- Equipment
  - Bolting: Fletcher, J-251-LS, Single Boom Jumbo; Atlas Copco E2C Boomer Drill

- Shotcrete: Reed B20 Shotcrete Pump; Shotcrete Technologies Shot-Tech 32.3 Robotic arm

***Blast Hole Drilling Equipment***

- Atlas Copco E2C Boomer Drills

***Blast Hole Lengths***

- Depths
  - TTP: 5 LF rounds
  - TT1: 13.5 LF rounds
  - TT2: 10 LF rounds
  - TT2S: 5 LF round in Heading I, 10 LF round in Heading II
  - TT3: 5 LF rounds
  - Invert: 6′ deep for footings, 5′ deep for main invert

- Number of holes in each round (1⅞″ diameter holes unless otherwise noted)
  - TTP (Full Face Heading): 100 Production Holes, 47 Perimeter Holes
  - TT1 (Heading I, Heading II): Heading I—20 Production, 18 Perimeter; Heading II—70 Production, 26 Perimeter
  - TT2: Same as TT1
  - TT2S: Mechanically excavate Heading I; Heading II—70 Production, 26 Perimeter
  - TT3: Contractor never excavated as TT3, used TT2S blast plans
  - Invert: 3 Footing Holes spaced 4′ apart, 5 Production Holes spaced 4′ apart holes on 5′ centers through tunnel east to west (2½″ dia. holes)

***Type of Explosives***

- Dynosplit: Trim Holes, + ½ stick Dyno-AP Booster at bottom of loading column
- Dyno-AP: Production Holes
- Nonel Lead Line: Ignition cord, tied into first hole, and ignited outside tunnel
- EZTL, 25 ms, 42 ms surface delays, at heading face
- EZDET 700/25, Blasting caps, down hole

***Mucking Equipment***

- 2 EA Volvo L250 Loaders
- 1 EA Cat D9R Dozer

### Shotcrete and Drain Board

Per the design, Geocomposite Drain Board was installed throughout the tunnel. The Geocomposite drain board material used was J-Drain 200, a dimpled impermeable polymeric sheet with a layer of non- woven filter fabric to retain smaller materials so that they may not pass into the drainage core. There were two types of drain board coverage. Continuous drain board coverage was used inside both portals of the tunnel, with the addition of 130 LF of continuous drainage added during construction when heavy rain fall occurred and water was very apparent through a section of tunnel near the West Portal. The second type of drain board coverage was the use of 3' wide drain board strips that covered the perimeter of the tunnel profile. The strips were spaced every 20 LF through the remainder of the tunnel. All drain board was fastened to the walls using Hilti Soft Material Attachment fasteners which were pinned through the board and into the smoothing shotcrete wall. The drain board material was tied into the installed formation drain located within the footings of the tunnel and covered with porous concrete material.

To install the Geocomposite drain board, the design asked for a smoothness criteria of shotcrete so that the drain board could be fastened tight to the walls and to prevent anything from protruding through the drain board material. Each Drain board strip throughout the tunnel required a smoothed shotcrete surface to be placed to cover rock, dowels, and WWF. Where continuous drain board was to be installed smoothing shotcrete was also required to cover installed dowels, channels, and steel sets.

Supplemental smoothing shotcrete was initially intended to be used if additional drain board was to be added to the project. During construction there were areas where overbreak was encountered due to blasting and ground that was less stable. After installation of MC channels there were large voids behind the channels that needed to be filled prior to final liner concrete installation, thus the Supplemental smoothing shotcrete was also used in these areas.

Smoothing shotcrete was placed by hand out of a man lift instead of utilizing the shotcrete robot.

### Average Cycle Time

Tunnel driven from the East Portal (driving west) proceeded much faster than the West Portal (driving east) because of the more competent rock resulting in longer rounds and less time installing initial support. The East Portal crew drove 460 feet of tunnel in 102 calendar days resulting in 4.5 feet per calendar day. Average cycle time was a little less than 2 days per round. The best cycle being just under 24 hours. The West Portal crew drove 175 feet of tunnel in 98 calendar days resulting in 1.8 feet per calendar day. Average cycle time was about 3.5 days per round. The best cycle being 36 hours.

## FINAL TUNNEL LINING

The final tunnel liner was cast-in-place double mat reinforced concrete. Two sets of rebar gantries and two sets of concrete formwork were utilized. Each form was 40 ft long. The tunnel walls and arch were 18 inches and 24 inches thick, as-designed depending on location along the tunnel. For the 18 inch profiles the reinforcing steel in the arch consisted of #5 longitudinal bars and #9 radial bars, and the wall consisted of #5 longitudinal bars and #7 radial bars placed in an inner and outer mat. For the 24 inch profiles the longitudinal bars were changed to #6, but the radial bars remained the same as the 18 inch mats. The inner layer of reinforcing was epoxy coated while the outer layer was black bar. In the 18 inch mats longitudinal bars were spaced on 12-inch centers and the radial bars were on 9-inch centers, and in the 24 inch mats the radial bars were spaced on 6-inch centers and the longitudinal bars remained at 12-inch centers. The required 28 day concrete compressive strength was 4,500 psi. The finished tunnel cross section is 53 ft wide and 40 ft high.

The complete rebar cage for each pour was constructed outside the tunnel portals on the rebar gantry. Rail had been laid on a mud mat through the tunnel and extended outside the concrete portals to allow for a rebar gantry and a tunnel form to be assembled outside the portals and then moved throughout the tunnel. To ready a form for a pour the rebar gantry, with the completed rebar cage was moved under the structural steel frame canopy at the portal where the rebar cage was lifted off the gantry utilizing chain falls, the gantry was moved out from under the frame. The concrete form was then moved under the frame and the rebar lowered onto the form. The form was then pulled on the rail into place for the next pour (Figure 6).

The first 40 ft long concrete pour was made near the center of the tunnel with each subsequent pour moving eastward and westward from the first pour. The concrete form had rows of four inspection/placement doors and four guillotine valves. Each 40 ft long form had 24 doors and 20 guillotine valves. The forms were also fitted with external form vibrators. Additionally the forms were fabricated with holes near the crown to allow for future contact grouting. The forms were pre-plumbed (piped) with concrete delivery piping to allow the concrete to be placed starting at the lowest level of placement doors and moving upward utilizing the rows of placement doors and guillotine valves. Figure 7 shows concrete form in tunnel. Two concrete boom pump trucks were used for placement. The boom pump truck discharge lines were connected directly to the pre-installed piping system on the formwork.

## CONTACT GROUTING

After form stripping, each pour of the cast-in-place concrete tunnel liner could be contact grouted. Since the minimum required strength to allow form stripping was 1,000 psi, usually achieved in 24 hours, contact grouting could not start any sooner than this. In reality because of the logistics of moving and resetting up the form coupled with leaving a safe and workable distance between the concreting operations and the contact grouting operations, contact grouting

**Figure 6. Rebar cage waiting to be lowered onto form**

**Figure 7. Concrete form in tunnel**

**Figure 8. Tunnel concrete liner with grout packers installed**

normally occurred a minimum of a week after a pour was made. Contact grout holes were located at the approximate 11 o'clock, noon, and 1 o'clock positions near the crown of the tunnel. There were four contact grout holes at each clock position spaced on 5 ft centers along the tunnel. There were a total of 12 contact grout holes per 40 ft pour.

The grout mix was Portland cement and water mixed at a 0.45 water cement ratio. The grout was batched at an off-site batch plant and delivered to the tunnel in 4½ CY loads in a concrete ready mix truck. A Haney grout plant was used to agitate and pump the grout. The grout was discharged from the ready mix truck directly into the plant's agitator tank, then transferred to the plant's mixer tank and into the piston pump.

Before the start of contact grouting each of the 12 contact grout holes had a mechanical packer installed in it (Figure 8). The grout was pumped through each packer, one at time until refusal was reached. Refusal was defined as one gallon or less of grout for one minute at full injection pressure (25 psi) measured at the packer.

**Figure 9. Final paving and tunnel entrance**

## PORTAL STRUCTURES

The portal structures were cast-in-place reinforced concrete. They extended from the last final tunnel lining pour to some distance outside the tunnel. The east portal structure extended 40 feet from the last tunnel pour and the west portal structure extend 135 feet from the last tunnel pour. The tunnel form was used as the inside formwork while the outside formwork was "stick built." The electrical, lighting, drainage, and mechanical work started within the tunnel prior to the last several final tunnel lining pours.

## CONCLUSION

At the start of construction in March 2013 the end of Colorado's winter weather held back the project pace. Likewise the start of winter weather in November slowed the project's completion a little. The project was completed on December 12, 2013 and was opened to three lanes of traffic for the beginning of the ski season. Figure 9 shows the completed tunnel.

## ACKNOWLEDGMENTS

The authors would like to thank the Colorado Department of Transportation (CDOT) and especially, Mike Keleman CDOT's Tunnel Project Engineer on the Twin Tunnels Project for providing several of the pictures used in the paper.

# TRACK 4: CASE STUDIES

## Session 5: Mechanical Excavation

Brian Zalenko, Chair

# A History of Tunneling in Los Angeles

**Jason M. Buenker and Robert A. Robinson**
Shannon & Wilson, Inc.

**ABSTRACT:** Tunneling is booming in Los Angeles. The surge in recent tunneling projects, coupled with the arrival of the North American Tunneling Conference in downtown Los Angeles presents an ideal opportunity to look back on the local history of tunneling. Our timeline begins in the early 1900s with the hand mined roadway tunnels on 2nd and 3rd Streets and extends forward to the earth-pressure balance TBM tunnels constructed by LA Metro in the last decade. We reexamine local tunneling obstacles (settlement, gassy ground) and preview the next great tunneling projects to come.

## INTRODUCTION

As a city built into the foothills of the Santa Monica Mountains, Los Angeles has long depended on tunnels as a solution to transportation and infrastructure problems. However, rocky local tunneling experiences, subsurface obstacles, and the multi-lane highway alternative caused Angelinos to periodically sour on tunneling as a low impact solution. Following a surge of transit, water, and wastewater tunnels from 1900 to 1940, tunneling in Los Angeles nearly ceased for the next 30 years. Local municipalities returned to tunneling in the 1970s to expand public transportation networks and utility systems; however, gassy ground and excessive settlements nearly derailed the effort. Finally, with the advent of pressurized-face tunneling and gas-resistant liners, tunneling in Los Angeles appears here to stay, with ever large tunnels on the horizon.

The following history of tunneling in Los Angeles is far from comprehensive, but instead represents a re-opening of the rich tunneling experiences of Los Angeles and the often costly lessons learned from previous tunneling ventures. This paper naturally bends toward projects that received the greatest volume of print or available references (typically transit projects). The authors hope this document functions chiefly as a starting point for the student of tunneling history in Los Angeles and a background for future tunneling projects.

## SUBSURFACE FRAMEWORK FOR TUNNELING

Geology and other subsurface issues played an outsized role in design and construction of tunnels in the Los Angeles Basin. Difficult and variable geologic conditions such as running sand, sticky clay, abrasive granular soils, liquefiable soils, mixed face conditions, and multiple perched groundwater tables are tunneling issues found in many geologic settings. Challenges specific to Los Angeles include gassy ground, tar laden deposits, highly corrosive soils, and numerous active fault zones.

Soils and rock units recognized in the Los Angeles Basin include:

- Young Alluvium—Holocene-age surficial sediments covering most of the basin as loose to medium dense granular, and soft to stiff fine grained soils deposited by streams, rivers, and lakes.
- Dune Sand—Clean to silty sand and gravelly sand with some clay layers. Recent dune sand is loose to medium dense, and older dune sand is dense to very dense and may be lightly cemented.
- Old Alluvium—Interbedded late-Pleistocene stiff, fine grained and dense, coarse grained alluvium dissected by stream channel and floodplain deposits. Includes the Lakewood Formation. Deposit may contain cobbles and boulders, specifically near the Los Angeles River, and interbeds of marine clay and estuary soils near the coast.
- San Pedro Formation—Fine grained sand, silty sand, and sandy silt with interbeds of dense, medium to coarse grained sand and stiff to hard silt and clay layers. Occasional cobbles and gravelly sand layers.
- Fernando Formation—Massive, weak siltstone and claystone, weathered in the upper 3 meters to stiff to hard clay, with thin sandstone interbeds, cobbles, and boulder sized concretions that are moderately to strongly cemented with calcium carbonate.
- Puente Formation—Marine sedimentary rock with interbedded weak sandstone, shale, siltstone, and conglomerate, which have

generally been a favorable tunneling medium on several local projects.

- Topanga Formation—Interbedded weak sandstone, shale, siltstone, and conglomerate, which have generally been a favorable tunneling medium.
- Crystalline basement rocks—Granitic, basaltic and metamorphic massive to highly fractured and faulted bedrock, generally found in the cores of the mountain ranges around the Los Angeles Basin.

Groundwater levels have dropped significantly over the last 100 years since historic high groundwater levels. Groundwater depths measured since the 1930s and 1940s have dropped by up to 6 meters or more. Due to interlayering of fine grained soils and coarse granular soils, numerous perched and confined groundwater tables occur locally.

Manmade subsurface hazards including tieback anchors, mapped or unmapped oil well casings, unmapped water and sewer tunnels, and abandoned foundation elements also represent obstacles to new tunneling. Substantial effort continues to be expended during tunnel design to qualify risk associated with these obstacles.

## HISTORICAL SUMMARY

Over 80 tunnels have been constructed in the Los Angeles region in the past 140 years with a combined length of over 220 kilometers. Table 1 lists selected tunnels and includes brief design and construction details. Figure 1 maps the approximate locations of selected local tunnels. As summarized in the table, tunneling in Los Angeles generally occurred in three phases. Phase 1 encompasses the beginnings of transit and utility infrastructure using hand mining or drill and blast methods. Phase 2 follows after a 30 year period of dormancy with the arrival of digger shields and early rock tunnel boring machines (TBM). Phase 3 builds on the successes and failures of Phase 2, as digger shields yield to pressure-faced TMB tunneling. The table also includes a Phase 4 of proposed tunneling projects.

Note that the table has several relevant omissions, including most tunnels associated with the Los Angeles Aqueduct, the Colorado River Aqueduct, and the California Aqueduct. Railroad tunnels excavated north of the Newhall Pass or west of the San Fernando Valley are also omitted. Although these tunnels had great impacts on the Los Angeles area, they fall outside the scope of tunnels having direct roots in the Los Angeles Basin and are well covered in other publications.

### Phase 1: Hand Mined Tunneling—1875 to 1967

Los Angeles tunnel construction began in the late 19th century as the Southern Pacific Railroad traversed the San Gabriel Mountains via the San Fernando Tunnel. Transit tunneling followed in downtown Los Angeles in the early 1900s to relieve the now familiar problem of traffic congestion. Trolley and highway tunnels breached hilly topography in the early urban core at Broadway, Third, Hill, and Second Streets. The Broadway Tunnel, constructed in 1901, had the greatest width, at 12.1m, of any tunnel in the U.S. These developments cleared the way for growth at the city's business and manufacturing core within reasonable commute time from residential areas. Mass-transit tunnels followed shortly thereafter. In the early 20th century, the Los Angeles Railway provided urban transportation and the Pacific Electric Railway connected to outlying communities with the Hollywood Subway Tunnel, the first subway tunnel constructed in Los Angeles (Photo 1). With the rise of the automobile came additional tunneling through the Elysian Park Hills to connect the San Fernando Valley with downtown Los Angeles in the 1930s.

The growth of Southern California prompted further construction of major utility tunnels. The City of Los Angeles built 142 tunnels totaling 69 kilometers for the Los Angeles Aqueduct from 1908 to 1913. The Metropolitan Water District of Southern California (MWD) followed with construction of 29 tunnels totaling 148 kilometers for the Colorado River Aqueduct from 1932 to 1938 and another 26 tunneling kilometers of system expansion in the San Gabriel Valley during that period. Los Angeles also expanded the sanitary system during this era with the North Outfall Sewer and White Point Outfall Sewer Tunnels.

These early tunnels were mined by hand using stacked drift construction, with picks and shovels, later augmented with pneumatic spaders, steam shovels, and front-end loaders without the benefit of a protective shield. Muck was removed with wheelbarrows or rail-mounted cars pushed by men, pulled by mules, or hauled by electric or diesel locomotives. Tunnel support started with timber and masonry, transitioning into steel ribs, timber lagging, and steel liner plates; final liners consisted of cast-in-place concrete or masonry (Photo 2). Most of these Phase 1 tunnels are still in use today.

Los Angeles nearly ceased tunneling in the 1940s and 1950s, possibly due to resources and labor diverted to World War II efforts. Transit tunneling also likely decreased due to the rise of the automobile and utility tunneling likely decreased because of lessened utility needs due to increased capacity provided by new infrastructure. Tunneling resumed with a major MWD system expansion in the late 1960s

**Table 1. Selected Los Angeles area tunnels**

| Tunnel (use) | Date | Width (m) | Length (m) | Excavation Method | Support Method | Ground Conditions | Salient Features |
|---|---|---|---|---|---|---|---|
| **Phase 1: Hand Mined Tunneling—1875 to 1967** | | | | | | | |
| San Fernando (railroad) | 1876 | 4.8 | 2,124 | Drill & blast | Timber | Sandstone | Several cave-ins and deaths. |
| Santa Susana (railroad) | 1904 | | 2,247 | | Timber, CIP | Sandstone,siltstone | |
| Chatsworth (railroad) | | | 303 | | Timber, shotcrete, rock bolts | Sandstone | Rehabbed in 1980s |
| No. 28 (railroad) | | | 183 | | Timber, CIP, steel ribs, shotcrete, rock bolts | | |
| Broadway (road) | 1901 | 12.1 | 232 | Stacked drift | Timber, CIP | Fernando | Demolished 1950s. |
| 3rd Street (road) | 1901 | 9.7 | 378 | Stacked drift | Steel, timber, brick | Fernando | 13 workers buried. 6 deaths. |
| Elizabeth (water) | 1911 | 3 | 8,192 | Drill & blast | Timber, CIP | Sandstone. | 200 m/mo. rock world record |
| Hill Street (trolley & road) | 1909 | ~8 | 300 | Stacked drift | Timber, CIP | Fernando | Demolished 1955 |
| Newhall Highway (road) | 1910 | 5.3 | 133 | Drill & blast | Steel, CIP, masonry | Shale, sandstone | Open cut 1939 |
| Broadway (rebuild) | 1916 | 12.1 | 232 | Steam shovel | Timber, brick, CIP | Fernando | Demolished 1949 |
| 2nd Street (road) | 1921 | ~12 | 458 | Stacked drift | Timber, CIP | Fernando | |
| North Outfall Sewer (sewer) | 1924 | 4.8 | 1,500 | Hand mining | Timber, pipe | Alluvium | 3 tunnel segments |
| Hollywood Subway | 1925 | 8.5 | 1,319 | Blasting, steam shovels | Timber, CIP | Fernando Formation | 1st LA subway—Pacific Electric Railroad. Aka Belmont Tunnel |
| Elysian Park No. 3 (road) | 1931 | 14.1 | 141 | Stacked drift, rail mounted mucking | Timber, CIP | Sandstone | 3 cave-ins totalling 800 $m^3$ |
| Elysian Park No. 1 (road) | | 14.1 | 123 | | | | |
| Sepulveda Canyon (road) | 1930 | 12.7 | 200 | Stacked drift, Steam shovels | Timber, CIP | Slate | |
| Pasadena (water) | 1937 | 3 | 3,701 | Spaders, augers | Steel ribs & spiles | Alluvium | 1st use of Conway mucker in LA |
| Monrovia No. 1 (water) | 1938 | 3 | 2,399 | Drill & blast, stacked drift | Timber ribs, light steel supports. CIP | Gneiss and granite <10 gpm water flow | Crosses Sierra Madre Fault |
| Monrovia No. 2 (water) | | | 287 | | | | |
| Monrovia No. 3 (water) | | | 10,330 | | | | |
| Pasadena Ext. (water) | 1936 | 3 | 1,709 | Spaders, auger drills, blasting | Steel spiling & ribs, pipe | Alluvium | |
| White Point Outfall (sewer) | 1936 | 3 | 10,000 | Drill & blast | Timber, pipe (?) | Shale, mudstone, sand. | Replaced Harbor City outfall |
| Elysian Park No. 4 (road) | 1936 | 14.1 | 230 | Stacked drift | Timber, CIP | Sandstone | On Arroyo Seco Parkway |
| San Rafael No. 1 (water) | 1938 | 3 | 1,234 | Drill & blast Spaders | Timber & steel ribs. Welded steel pipe backed with concrete and lined with gunite | Gneiss | |
| San Rafael No. 2 (water) | | 3 | 1,728 | | | Gneiss | |
| Monrovia No. 4 (water) | | 3 | 2,480 | | | Gneiss | |
| Sierra Madre (water) | 1936 | 3 | 2043 | Spaders | Steel ribs & spiling, CIP | Alluvium | |
| Hollywood Reservoir (water) | 1941 | 2 | 1,128 | Drill & blast | Steel ribs, grouted pipe | Topanga Fm, basalt | 600 gpm inflow along rock contact |
| Cajalco (water) | 1954 | 3 | 2,299 | | Concrete | | |

(table continues)

**Table 1. Selected Los Angeles area tunnels (continued)**

| Tunnel (use) | Date | Width (m) | Length (m) | Excavation Method | Support Method | Ground Conditions | Salient Features |
|---|---|---|---|---|---|---|---|
| San Juan (water) | 1956 | 3 | 2,072 | Drill & blast. | Timber sets, CIP | Rock | |
| La Cienega–San Fernando (sewer) | 1956 | 3 | 12,621 | Drill & blast | Steel ribs, liner plate, CIP | Topanga Fm | <800 gpm inflow |
| Glendora (water) | 1970 | 4.7 | 9,906 | Drill & blast. | Steel ribs. CIP. | | Plus 503 m long adit |
| Balboa Outlet (water) | 1970 | 3.8 | 1,146 | | Concrete pipe | | |
| Castic No. 1 (water) | | | 1,689 | | | | |
| Castic No. 2 (water) | 1971 | 6.2 | 3,953 | | | | |
| Saugus (water) | | | 1,913 | | | | |
| Placerita (water) | | | 366 | | | | |
| **Phase 2: Shield Tunneling—1967 to 2000** | | | | | | | |
| Newhall (water) | 1971 | 6.2 | 5,576 | Digger shield | Steel ribs. CIP. | Sandstone | 1st digger shield in LA area |
| Balboa Inlet (water) | | 4.3 | 1,524 | | Steel ribs, CIP | | |
| Sepulveda (water) | 1971 | 2.4 | 2,233 | Rock TBM | Steel ribs, CIP | | Large settlement |
| Sylmar (water) | 1971, 1974 | 6.7 | 8,854 | Digger shield | Timber, CIP | Saugus & Sunshine Ranch Fm | Explosion June 24, 1971 killed 16. Led to modern gas test regs. |
| Tonner No. 1 (water) | 1976 | | 1,402 | Rock TBM | Liner plates, CIP. Grouted concrete pipe | | |
| Tonner No. 2 (water) | | | 5,610 | | | | |
| Sacatella Flood Control | 1977 | 5.5 | 967 | Digger shield | Concrete segments | Puente Fm | Gassy but high ventilation rates, oil |
| Red Line (A171–A130) | | | | | | | |
| A171 (7th/ Flower to MacArthur Park) | 1991 | Twin 5.8 | 1,529 | Digger shield | Precast segments, CIP | Alluvium, Fernando Fm | Ground settlement of 6 to 25 mm |
| A146 (5th/Hill to 7th/Flower) | | | 655 | | Ribs and lagging, CIP | Alluvium, dry | Chemical & compaction grouting |
| A141 (Union Station to Pershing Square) | 1991 | Twin 5.8 | 1,582 | Digger shield | Steel ribs, CIP | Alluvium, Fernando Fm. | Chemical & compaction grouting |
| A130 (Yard leads) | | | 223 | | Steel ribs, CIP | Alluvium | Tunnel fire July 13, 1990 collapsed portion not chemically grouted |
| North Outfall Replacement Sewer | 1992 | 2.8 to 3.8 | 12,805 | Digger shields & partial EPBM | Single gasket precast segments | Dune Sand, Young & Old Alluvium, San Pedro Fm | Inglewood Fault. 1st partial EPBM in LA. 46 cavities & 102 loose zones. |

(table continues)

**Table 1. Selected Los Angeles area tunnels (continued)**

| Tunnel (use) | Date | Width (m) | Length (m) | Excavation Method | Support Method | Ground Conditions | Salient Features |
|---|---|---|---|---|---|---|---|
| Red Line (B201–C331) | | | | | | | |
| B201 (MacArthur Park to Wilshire/Vermont) | 1995 | Twin 5.9 | 1,383 | Digger shield | Precast concrete segments, CIP | Puente Fm | Settlement of 10 to 40mm |
| B221 (Wilshire/Vermont to Wilshire/Western) | 1995 | Twin 5.9 | 1,273 | Digger shield | Precast concrete segments, CIP | Puente Fm | Settlement < 25 mm |
| B251 (Wilshire/Vermont to Hollywood/Highland) | 1995 | Twin 5.9 | 9,514 | Digger shield | Precast concrete segments, CIP | Young & Old Alluvium over Puente Fm, | 3 areas out of alignment and remined. Hollywood Blvd sinkhole |
| C311 (Hollywood/High-land to Universal City) | 1996 | | 4,018 | Rock TBMs (2) | Chemical grout. CIP | Granite, Chico, Simmi & Topanga | <180m water, Hollywood & Benedict Canyon Faults |
| C331 (Universal City to North Hollywood) | 1997 | | 3,206 | Digger shield | Precast segments, PVC membrane, CIP | < 10m water, alluvium, Topanga Formation | Alluvium with boulders over soft bedrock. >75 mm.settlement. |
| **Phase 3: Pressure Face TBM—2000 to 2013** | | | | | | | |
| Hollywood Reservoir Bypass & Utility (water) | 2000 | 2.9<br>1.8 | 1829<br>366 | EPBM | Grouting, precast segments, steel pipe | Topanga Fm, basalt | Tunneling method eliminated probe drilling and face grouting |
| Riverside Badlands (water) | 2002 | 3.7 | 12,878 | Shielded TBM | Grouting ahead, precast segments, steel pipe | Alluvium, sedimentary & metamorphic bedrock | Multiple faults, depth of over 200 m and groundwater over 100 m |
| North Outfall Sewer–East Central Interceptor Sewer (ECIS) | 2004 | 4.1 | 20,000 | 4 EPBMs | Gasketed precast segment liner. PVC-lined carrier pipe. | Alluvium, San Pedro & Lakewood Fm. Dry. Inglewood Fault | Excessive settlement mitigated with compaction and 2.9M gallons chemical grout. |
| Northeast Interceptor Sewer (NEIS) | 2004 | 3.3 | 8532 | 3 rock TBMs, 2 EPBMs | | Alluvium, Puente Fm. ~ 10 m groundwater | Permeation and chemical grouting in sands |
| Gold Line Extension (MGLEE) | 2007 | Twin 5.9 | 2,127 | EPBM (2) | Double-gasketed, precast segments. | Alluvium. 0 to 12m groundwater. | 1st specified closed-face TBM by Metro. Boyle Heights Oil Field. |
| Arrowhead (water) | 2008 | 3.7 | 15,400 | Rock TBM (2) | Grouting, precast segments, steel pipe | Multiple faults, gniss, marble & granite | 600 m groundwater, forest fire (2003) and flashflood (2004) |

(table continues)

**Table 1. Selected Los Angeles area tunnels (continued)**

| Tunnel (use) | Date | Width (m) | Length (m) | Excavation Method | Support Method | Ground Conditions | Salient Features |
|---|---|---|---|---|---|---|---|
| **Phase 4: Future Tunneling in Los Angeles Region** | | | | | | | |
| Silver Lake | 2014 | 3 | 1,400 | Rock TBM | Grouted steel pipe | Fernando Fm | |
| Clearwater (sewer) | ca. 2016 | 4.9 | 10,109 | Closed-face TBM | Gasketed, precast segments | Shale, mudstone, and marine sand | |
| Crenshaw–LAX Transit Corridor | ca. 2019 | | 2,740 | Closed-face TBM | | Alluvium, Inglewood Fault. <7 m water | 1st design build tunnel contract by LA Metro. Potero Oil Field |
| Regional Connector | ca. 2020 | | 1,450 | EPBM & 108m SEM | | Alluvium, Fernando Fm. | Thru LA Financial District |
| Westside Subway Extension | | | | | | | |
| Seg. 1—Western to La Cienega | ca. 2023 | Twin 5.9 | 6,280 | | Double-gasketed, Precast segments | Alluvium, San Pedro & Fernando Fm. 0 to 80 ft water | Tar laden. Gassy. Abandoned oil wells |
| Seg. 2—La Cienega to Century City | ca. 2026 | | 3,700 | Closed-face TBM | | Alluvium, San Pedro Fm. 0 to 50 ft water | Tar laden. Gassy. Abandoned oil wells. Inglewood Fault. |
| Seg. 3—Century City to Westwood | ca. 2035 | | 4,350 | | | Alluvium. 0 to 80 ft groundwater | Tar laden. Gassy. Abandoned oil wells. Santa Monica Fault |
| SR-710 | ca. 2030 | 11 to 16 | 8,000 to 18,000 | TBD | TBD | Alluvium, Fernando, Puente, Topanga Fm | Ryamond, Alhambra, Eagle Rock, San Rafael Faults |
| Sepulveda Pass | ca. 2023 | ~20 | 16,000 | Closed-face soil and rock TBM | TBD | TBD—Soil and rock | Mixed soil & rock. Private-public partnership. $10 billion estimate |

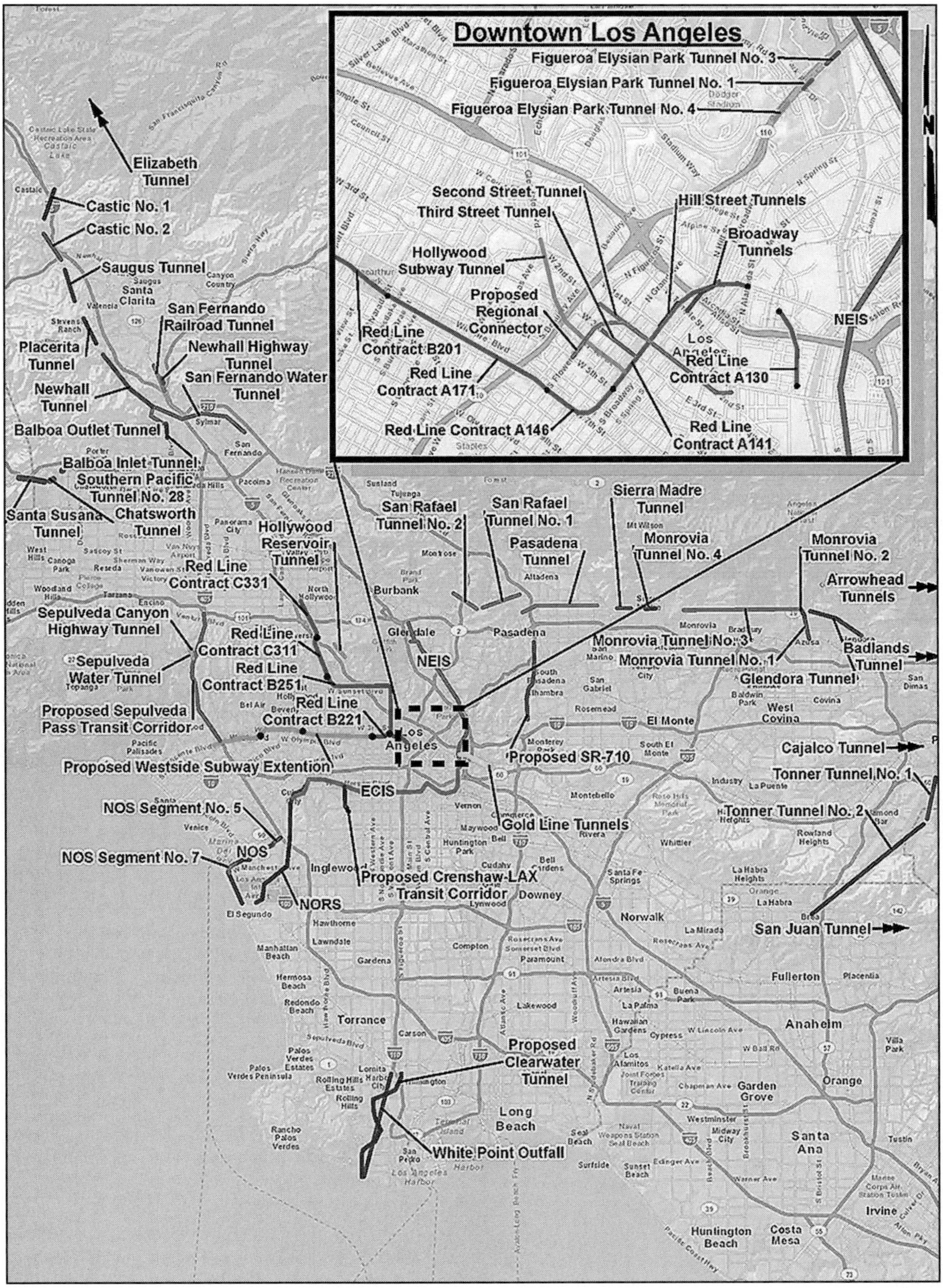


Figure 1. Selected Los Angeles tunnel projects

connected with Los Angeles Aqueduct feeders in the San Gabriel Mountains.

Despite relatively little tunnel construction near the end of Phase 1, political and public developments during this era set the table for rapid tunneling development in the late 1970s and 1980s. The California Legislature established the Los Angeles Metropolitan Transit Authority in 1951, which became the Southern California Rapid Transit District in 1964 and later the Los Angeles County Metropolitan Transportation Authority (Metro) in 1994. None of those agencies generated much interest in renewing public transportation until Los Angeles voters approved a 240 kilometer subway system in November 1980 (Murthy and Monsees 1989).

### Phase 2: Shield TBM Tunneling—1967 to 2000

The hiatus in Los Angeles tunneling ended in the mid-1970s as MWD constructed a major system expansion through the Newhall Pass. The MWD tunnels made the first known use of digger shields and TBMs in the Los Angeles area. This tunneling effort also encountered the first documented gas obstacles of any tunnel in Los Angeles. These gas issues culminated in the deadly June 24, 1971 explosion in the unfinished San Fernando Water Tunnel. The explosion led to greater subsurface regulations concerning gassy ground, which continues to impact local and national tunneling to this day.

Metro officials began planning a new 29.8 kilometer starter Red Line subway in the early 1980s. Planning centered on construction in tar laden, gassy ground conditions through the old Salt Lake Oil Fields and the search for tunnel liners capable of excluding gas from diffusing into the new subway tunnels (Proctor and Monsees 1985). Construction centered on the use of digger shields, similar to that shown on Photo 3, augmented with compressed air and extensive grouting in an attempt to exclude groundwater and gas from the advancing tunnels.

The effort hit a roadblock on March 24, 1985, when the buildup of methane gas in a West Los Angeles department store basement caused a powerful explosion that injured 23 people and damaged the building interior. A City of Los Angeles task force and other researchers connected the methane explosion to the nearby Salt Lake Oil Field and buildup of pressurized subsurface methane gas (City of Los Angeles 1985; Hamilton and Meehan 1992). Congressional legislation responded by prohibiting use of federal subway funding for tunneling through "potential risk zones" for methane gas. In response, Metro established alignment alternatives that rerouted the Red Line to the San Fernando Valley via Vermont Avenue, thus bypassing the gas issues in the Fairfax district (Murthy and Monsees 1989).

**Photo 1. Hollywood Subway Tunnel (Courtesy of University of Southern California, on behalf of the USC Libraries Special Collections)**

**Photo 2. Stacked drift construction for the Second Street Tunnel, ca. 1921 (Courtesy of University of California, Los Angeles, on behalf of the UCLA Libraries Special Collections)**

Tunneling began on the Red Line at Minimum Operating Segment (MOS)-1 in November 1987 using Mitsubishi or Robbins digger shields that typically included hinged breasting doors with hydraulic jacks, a backhoe-type excavator, and overcutters at the shield periphery (Escandon et al. 1989). The digger shields performed well in the stable Fernando Formation, as surface settlements ranged from 1 to 1.5 millimeters (Robison et al. 1989); however, digger shield excavation in alluvial soils prompted compaction and chemical grouting programs from the ground surface (Gularte et al. 1991, Mahar 1994) and through the tunnel heading (Robison and Wardwell,

Photo 3. Digger shield with hinged hydraulic breasting doors (Courtesy of The Robbins Co.)

Photo 4. A130 tunnel fire and collapse segment (Courtesy of F. Gularte)

Photo 5. Hollywood sinkhole (Courtesy of R. Sage)

1991) to control ground loss. Chemical grouting proved highly effective, as a fire on July 13, 1990 in the unfinished A130 tunnel consumed nearly all temporary wood lagging in the tunnel. Despite the sudden lack of ground support, chemically grouted ground under the freeway continued to stand with only local spalling. In contrast, ungrouted ground beneath a vacant lot adjacent to the highway suffered complete collapse and required remedial excavation and repair (Photo 4, Gularte et al. 1991). In general, although the Red Line MOS-1 tunnels experienced peripheral construction issues (misaligned segmental liner, inadequate liner thicknesses, change order litigation), tunneling excavation was largely successful (Mahar 1994).

Tunneling turned to MOS-2 in 1993, with contract B251 pushing the Red Line northward into Hollywood via Vermont Avenue and Hollywood Boulevard. Problems and delays plagued tunneling from the outset (Gordon et al. 1995). Groundwater inflows, flowing granular soils, and liner compressive failure eventually led to up to 4.3 centimeters of subsidence and stoppage of work in August 1994 (Roth and Stirbys 2006). Tunneling resumed in January 1995, but on June 22, 1995 an 80-foot section of Hollywood Boulevard collapsed during remining of an improperly aligned tunnel segment (Photo 5, Gordon and Kennedy 1995). Political ill-will and cleanup costs mounted in the following months leading to contractor dismissal and lawsuits (Wallis 1995). The segment was later finished by other contractors and opened in 1999 amid the ongoing settlement of several claims (Roth and Stirbys 2006).

Red Line MOS-3 began as MOS-2 construction drew to a close. The C311 contract alignment passed under the Santa Monica Mountains through hard rock, with mixed face conditions under the Hollywood Freeway that prompted a dewatering and chemical grouting program (Taylor et al. 1997). The program limited settlement to about 1.6 centimeters (Kramer and Albino 1997) and tunneling progressed through the rock segment using extensive face grouting to control high groundwater inflow and protect natural springs (Kramer et al. 1998). In contrast to the success of C311, the C331 contract from Universal City to North Hollywood experienced issues from the outset. Clean sand and gravels raveled and flowed into the digger shield due to insuffienct face support, causing settlements to propagate upward to Lankershim Boulevard. Contractor-owner disputes followed, leading C311 to follow a similar path as B251 (Roth and Stirbys 2006).

At this juncture, Metro commissioned a study regarding the ongoing problems experienced during Red Line tunnel construction. The study concluded that the geologic and subsurface conditions in Los Angeles are compatible (if not favorable) to tunnel construction and that Red Line construction was generally successful. Technical problems were isolated to liner thickness issues during MOS-1, subsidence along Hollywood Boulevard during B251 tunneling, the Hollywood Sinkhole created while realigning liner segments of B251, and the settlements under Lankershim Boulevard on C331 (Eisenstein et al.

**Photo 6. Earth pressure balance TBM (Courtesy of J. Critchfield)**

1995). The study argued that many of these issues could be solved by specifying laser guided alignment systems and positive face control tunneling with earth pressure balance TBMs (EPBM).

The Red Line projects were not alone in their struggles with digger shields that led to settlement and sinkhole issues. Digger shields used on the North Outfall Replacement Sewer (NORS) created ground losses by failing to maintain forward pressure that led to a 4.8 meter deep sinkhole near Taxiway 49 at LAX and other sinkholes on airport property. In response, over 900 borings (augmented with ground penetrating radar and cone penetration tests) were drilled at LAX in two phases to find 46 cavities and 102 loosened ground zones (Gordon and Sherry 1993). However, despite the settlement issues, the NORS project offered a preview into the next phase of local tunneling through the first use of an EPBM in Los Angeles.

NORS used a Lovat EPBM capable of operating in full EPBM mode or in partial-EPB mode with "pressure relieving gates" to regulate pressure. The EPBM used full-circle steel ribs and timber lagging or prefabricated steel liner plates for initial support and thrust reaction. However, the liner plates often buckled, requiring stiffening with T-ring stiffeners to accommodate the EPBM thrusts (Gordon and Sherry 1993). The EPBM pressure relieving gates were unable to control ground loss when the machine encountered cobbles and boulders. The ground loss resulted in settlement and sink holes (Roth and Kamine 1997).

### Phase 3: Pressure Face TBM Tunneling—2000 to 2013

The introduction of pressure face tunneling to Los Angeles marked a major turning point in the fortunes of local tunneling efforts. Full EPBM tunneling (Photo 6) on the East Central Interceptor Sewer (ECIS) and Northeast Sewer (NEIS) in dry to wet granular alluvium, combined with gasketed, bolted, precast concrete segment single-pass liners (Photo 7) resulted in minimal settlement over the full alignment (Crow and Holzhouser 2003, Keller and Crow 2004, Varley et al. 2004). An advanced grout injection system compatible with the new liner further stabilized and sealed the tunnels (Zernich et al. 2005). This type of integrated excavation and support system is capable of excavating water-laden flowing silt and sand, and clay, with groundwater heads over 30 meters, and high methane and hydrogen sulfide concentrations. The ECIS and NEIS projects effectively demonstrated that EPBMs could tunnel competently where digger shields could not.

Further advances included the addition of rock disc cutters on EPBMs to grind up boulders and excavate through bedrock, such as the sandstone and basalt at the Hollywood Reservoir Bypass Tunnel (Colzani et al. 2001). Improvements in grouting capabilities and the understanding of groundwater impacts helped control groundwater inflows at the Arrowhead Tunnels (Fulcher et al. 2007, Shamma et al. 2003). Changes in contracting formats also allowed for greater sharing of risk between owners and contractors (escrowed bid documents, labor and energy escalation clauses, design summary and geotechnical baseline reports, and disputes review boards) that contributed to fairer and more equitable tunnel contracting. Specifically, Metro's Gold Line Extension (MGLEE) was successfully completed in 2007 with less than 1.1 centimeters of surface settlement versus an allowable 5 centimeters. MGLEE was successful in part due to contract requirements for continuous EPBM operation in closed face mode, with a screw auger and minimum face pressure of 0.6 bar above ambient groundwater pressure (Choueiry et al. 2007, Robinson and Bragard 2007).

## PHASE 4: FUTURE TUNNELING IN LOS ANGELES

Several major tunneling projects are in various phases of design or construction contract negotiation that will again expand the limits of local tunneling experience. Metro will use design-build contracting

**Photo 7: Initial gasketed bolted precast segmental liner and final carrier pipe liner. East Central Interceptor portion of North Outfall Sewer. (Courtesy of J. Critchfield)**

methods to construct the Crenshaw/LAX Transit Corridor, Regional Connector, and Westside Subway Extension projects. Major highway tunnels are also in the early conceptual planning stages including the Sepulveda Pass Tunnel beneath I-405, and the SR-710 tunnels between Pasadena and Los Angeles. Table 1 indicates that over 46 kilometers of new tunnels could be constructed in the next 20 years.

## CONCLUSIONS

Tunneling in the Los Angeles region has evolved significantly over the last 140 years enabling the construction of more than 80 tunnels, totaling over 220 kilometers of construction, in ever more difficult ground conditions. In the next 20 years, an additional 46 kilometers of tunnels are already in various stages of planning, design, and initial construction.

The current phase of tunneling typically involves full EPBM tunneling with a gasketed precast concrete segmental lining, which has proven to be very effective in:

- Variable, granular and cohesive soil conditions, with multiple perched groundwater and cobbles and scattered boulders
- Mixed-face conditions with soils overlying relatively soft siltstone and sandstone bedrock
- Groundwater pressures of 0 to over 200 ft of head
- High hydrogen sulfide and methane levels
- Abrasive granular soils
- Sticky, clogging cohesive soils
- Fractured bedrock with high water pressures

Other tunneling methodologies including large openings constructed by the sequential excavation method (SEM) and large diameter closed-face TBMs over 12 meters in diameter have yet to be tried in the LA soils and bedrock, but have potential applications on future tunneling projects such as the Westside Subway Extension, SR 710 highway, and through Sepulveda Pass on I-405.

## ACKNOWLEDGMENTS

The authors thank Amanda Elioff for her insightful review of this paper, the Metropolitan Water District's archive staff for access to their extensive compilation of old tunnel records, and especially Judy Bloch for her assistance in assembling these references.

## REFERENCES

Choueiry, E., Elioff, A., Richards, J., and Robinson, B., 2007, Planning and Construction of the Metro Gold Line Eastside Extension Tunnels, Los Angeles, CA, *Proceedings of the Rapid Excavation and Tunneling Conference,* pp. 808–829.

City of Los Angeles, 1985, Task Force Report on the March 24, 1985 Methane Gas Explosion and Fire in Fairfax Area, *Metropolitan Transit Authority Library,* Los Angeles, California, June 10, 1985. CF85-0563-53.

Colzani, G, Strid, J., Cole, S., and Olsen, D., 2001, Tunneling at Hollywood Reservoir, *Proceedings of the Rapid Excavation and Tunneling Conference*, pp. 421–431.

Crow, M. and Holzhauser, J., 2003 Performance of Four EPB-TBMs Above and Below the Groundwater Table on the ECIS Project, Los Angeles, CA. *Proceedings of the Rapid Excavation and Tunneling Conference*, pp. 905–926.

Eisenstein, D., Martin, G., and Parker, H., 1995, Los Angeles Metro Rail Project: Report on Tunneling Feasibility and Performance, Los Angeles County Metropolitan Transportation Authority, Los Angeles, CA, 1995.

Escandon, R.F., Lindsay, R.C., Navid, S.S., and Piepenburg, M.A., 1989, Tunneling Performance and Geotechnical Aspects of Los Angeles Metro Rail Contract A-171, *Proceedings of the Rapid Excavation and Tunneling Conference*, pp. 641–660.

Fulcher, B., Bednarski, J., Bell, M., Tzobery, S., and Burger, W., 2007, Piercing the Mountain and Overcoming Difficult Ground and Water Conditions with Two Hybrid Hard Rock TBMs, *Proceedings of the Rapid Excavation and Tunneling Conference,* pp. 229–254.

Gordon, L., and Kennedy, M. (1995, June 23). Street Vanishes in Subway Sinkhole: MTA: Burst water main weakens earth above tunnel, causing half a block of Hollywood Boulevard to collapse. *Los Angeles Times*. Retrieved from www.articles.latimes.com.

Gordan, R. and Sherry, G., 1993. NORS Tunnel Project Case History, *Proceedings of the Rapid Excavation and Tunneling Conference,* pp. 223–239.

Gordon, R.B., Poulton, D.K., and Burg, S.M., 1995, Vermont/Hollywood Tunnel Project, Los Angeles Metro Redline Contract B251, Development of the Main Work Site, *Proceedings of the Rapid Excavation and Tunneling Conference,* pp. 111–126.

Gularte, F.B., Taylor, G.E., Monsees, J.E., Whyte, J.P., 1991, Tunneling Performance of Chemically Grouted Alluvium and Fill, Los Angeles Metro Rail, Contract A-130, *Proceedings of the Rapid Excavation and Tunneling Conference,* pp. 291–307.

Hamilton, D.H. and Meehan, R.L., 1992, Cause of the 1985 Ross Store Explosion and Other Gas Ventings, Fairfax District, Los Angeles, *Engineering Geology Practice in Southern California, Association of Engineering Geologists, Special Publication No. 4*, Belmont, California, pp. 145–158.

Keller, E. and Crow, M., 2004, Tunneling through an operational oilfield and active faults on the ECIS Project, Los Angeles, CA, USA, *Proceedings of the 2004 North American Tunneling Conference*, pp. 441–448.

Kramer, G. and Albino, J., 1997, Los Angeles Metro—TBM Starter Tunnels—Mixed Face Conditions, *Proceedings of Construction Congress V: Managing Engineering Construction in Expanding Local Markets*, Sponsored by the American Society of Civil Engineers, Minneapolis, Minnesota, pp. 241–250.

Kramer, G., Roach, M., Townsend, J., and Warren, S., 1998, Grouting of TBM Rock Tunnels for the Los Angeles Subway, *Grouts and Grouting: A Potpourri of Projects*, *Geotechnical Special Publication No. 80*, American Society of Civil Engineers, Reston, VA, pp. 100–110.

Mahar, J., 1994, Field Observation and Evaluation of Final Lining Conditions in L.A. Metro Red Line Segment 1 Tunnels, *Report for Tunnel Review Panel appointed by Los Angeles County Metropolitan Transportation Authority*, Los Angeles County Metropolitan Transportation Authority, Los Angeles, CA.

Murthy, K.N. and Monsees, J.E., 1989, Los Angeles Subway—A Triumph Despite Tribulations, *Proceedings of the Rapid Excavation and Tunneling Conference,* pp. 585–603.

Proctor, R.J. and Monsees, J.E., 1985, Los Angeles Metro Rail Project, Design Issues Related to Gassy Ground, *Proceedings of the Rapid Excavation and Tunneling Conference*, pp. 488–505.

Robinson, B. and Bragard, C., 2007, Los Angeles Metro Goldline Eastside Extension—Tunnel Construction Case History, *Proceedings of the 2007 Rapid Excavation and Tunneling Conference*, pp. 472–494.

Robison, M.J., Lyman, T.J., and Ruemmele, W.A., 1989, Ground Response to Tunneling, Los Angeles Metro Rail Contract A171, *Proceedings of the Rapid Excavation and Tunneling Conference,* pp. 641–660.

Robison, M.J. and Wardwell, S.R., 1991, Chemical Grouting to Control Ground Losses and Settlements on Los Angeles Metro Rail Contract A146, *Proceedings of the Rapid Excavation and Tunneling Conference,* pp. 179–195.

Roth, W.H. and Haber Kamine, M., 1997, Two Construction Claims on the NORS Project, *Proceedings of the 23*rd *World Tunnel Congress,* Vienna, Austria, pp. 743–748.

Roth, W. and Stirbys, A., 2006, Claims and Forensic Engineering in Tunneling, Forensic Engineering, pp. 591–604.

Shamma, J., Tempelis, D., Duke, S., Fordham, E., and Freeman, T., 2003, Arrowhead Tunnels: Assessing Groundwater Control Measures in a Fractured Hard Rock Medium, *Proceedings of the Rapid Excavation and Tunneling Conference,* pp. 296–305.

Strosnider, J.A., 1987, Los Angeles Metro Rail Project, *Proceedings of the Rapid Excavation and Tunneling Conference,* pp. 1093–1111.

Taylor, G.E., Cavey, J.K., and Roach, M.F., 1997, Horizontal Chemical Grouting Beneath the Hollywood Freeway, Los Angeles Metrorail, C-311, *Proceedings of the Rapid Excavation and Tunneling Conference,* 1997.

Varley, Z., Patel, R., and McDonald, J., 2004. *City of Los Angeles Northeast Interceptor Sewer Tunnel,* Proceedings of the North American Tunneling Conference, pp. 461–468.

Wallis, S. (June 1995). What is Going on in Los Angeles? *Tunnel Talk, Direct by Design.* Retrieved from www.tunneltalk.com on December 12, 2013.

Zernich, B., Robinson, B., and Krulc, M, 2005, Northeast Interceptor Sewer Case History, *Proceedings of the 2005 Rapid Tunneling and Excavation Conference,* pp. 357–392.

# The Learning Curve—North American Microtunnel Industry Leaps Forward with Recent Curved Drives

**Rory P.A. Ball and Marc Gelinas**
Hatch Mott MacDonald

**ABSTRACT:** The tunneling industry in North America recently completed both horizontal and vertical curved microtunnel drives in a variety of soil conditions. These successful installations have provided designers with flexibility along alignments while saving projects both time and money. North American contractors, engineers, and suppliers are learning what is required for the design of successful curved microtunnel drives. This paper offers direct experience from four curved microtunnel alignments including lessons learned from the design and construction management of an S-curve tunnel and a simultaneous vertical and horizontal curve tunnel. Important considerations for alignments, machine and pipe design are discussed for curved microtunneling.

## INTRODUCTION

This paper focuses on curved microtunnels advanced using conventional jacking frames. Some of the design and construction aspects of these tunnels that are discussed are also applicable to curved microtunnels advanced using pipe thrusters. This type of tunneling includes Herrenknecht's Direct Pipe technology and MTS's System2 technology. These technologies require additional considerations beyond standard microtunneling projects. However, because a microtunnel boring machine (MTBM) is used for all these technologies and they have similar benefits, project references are included for all these technologies.

There has been confusion in the industry regarding the proper terminology for tunneling alignments that incorporate curves. In the recent past, some authors have used the word "compound" to refer to alignments with vertical and horizontal curves. However, this word becomes confusing for those who are familiar with highway terminology. This paper will instead refer to a "spatial" curve. A spatial curve is an alignment incorporating a vertical and horizontal curve. Some alignments, such as the Keswick Project discussed later in this paper, have navigated the vertical and horizontal curves simultaneously. These drives will be characterized as spatial curves as well, but with an extra note about the curves being done simultaneously.

## NORTH AMERICAN MICROTUNNELING INDUSTRY

The history of microtunneling in North America has traditionally been comprised of straight alignments beginning when the first microtunnel launched in 1984 under the I-95 in Miami, Florida. Since then, hundreds of microtunnels have been completed in most types of soil and rock scattered across the USA, Canada, and Mexico. This paper focuses on the USA and Canada. This seasoned industry is always striving to improve and involves experienced stakeholders in several fields such as planning, design, construction, construction management, supply/manufacturing, and surveying. With decades of implementing projects constructing straight drives, the experienced contractors can frequently deliver a project introducing innovative solutions thereby managing the project risk for a competitive price.

Until recently, owners have normally been reluctant to design curved drives or accept value engineering proposals from contractors who offer curved alignments. In some cases, the owners were not willing to consent to the higher risks associated with curved alignments, even if the net savings of time and costs were a benefit to the project. Some contractors state the extra risks would be covered solely on themselves, but it is rare that a major tunneling stoppage does not somehow affect the owner's domain. In the case histories mentioned in this paper, the majority of owners who have allowed curved alignments have experienced the benefits in electing to incorporate them.

## BENEFITS OF INCORPORATING CURVES

In planning and design, the option to incorporate simple curves or spatial curves should, at a minimum, be placed under consideration. By taking the time to review the option of integrating curves into the project's alignment, the net benefit in terms of cost and time can be weighed against added risks. After reviewing curved microtunnel projects from around the world, the authors created the following list of potential benefits:

- Reduce the number of shafts
- Reduce the need to move the jacking frame and ancillary equipment
  - Some shafts, necessary for manholes, can be "push-thru" shafts
- Reduce shaft depths
  - Savings can be significant in locations where hydrostatic pressure is reduced
- Minimize environmental impacts (by eliminating shafts)
  - Community
  - Wildlife
- Reduce project schedule
  - Especially if Environmental Impact Reports (EIRs) are eliminated for excluded shafts
- Flexibility along alignment / ROW
  - "Thread the needle" through right-of-way alignments
  - Eliminate the need for larger laydown space at excluded shafts
- Avoid obstacles and sensitive zones
- Hydraulic optimization
  - Removing unnecessary manholes/transitions

Unique characteristics of each project will determine which potential benefits are relevant for consideration in the design. Some owners place more weight on certain aspects (environmental, community impact, etc.) even if the net savings might be minimal compared to the contract price. The designer must use their past project experience, supplemented by their knowledge of the worldwide industry, to help the owner to weight the potential benefits against any risks added to their job.

## WORLDWIDE PROJECTS

The worldwide MTBM industry has numerous projects to showcase what types of curves are possible. The first curved microtunneling project was completed in 1982 in Japan (Camp 2001). Since then, curved microtunnel alignments have routinely been completed in multiple continents with varying degrees of difficulty. North America is only now starting to adopt techniques used on worldwide projects for over *three* decades.

The following list includes some notable curved MTBM drives around the world (some aspects are in bold where significant or milestone historic achievements have been attained):

- (**1982**, soil) Edo River Crossing; Chiba, Japan
  - 847-m (2,779-ft) long; 1800-mm (5.9-ft) OD
  - 300-m (984-ft) horizontal ("H") radius
- (1987, soil) Tokyo Electric Power Transmission Main, Japan
  - 202.5-m (664-ft) long; 3200-mm (12.1-ft) OD
  - **Two 15-m (50-ft) radii H curves; two 30-m (98-ft) radii vertical ("V") curves; 29% maximum grade**
- (1997, soil) Nokendai Railway Station; Kanagawa, Japan
  - 524-m (1,719-ft) long; 1,200-mm (3.9-ft) OD
  - **Five sequential curves** of the following radii: 240-m (787-ft) H, 300-m (984-ft) H, 500-m (1,640-ft) V, 50-m (164-ft) H, 100-m (328-ft) H
- (1998, **rock**) Horden Outfall; Horden, England
  - 553-m (1,814-ft) long; 2,420-mm (7.9-ft) OD
  - **Two vertical curves** w/ grade up to 14%
- (2004, **rock**) Korean Electric Power Conduit; Seoul, Korea
  - 800-m (2,625-ft) long; 3060-mm (10.0-ft) OD
  - **Five s-curves** with H radii from 200-m (656-ft) to 250-m (820-ft)
- (2007, soil, pipe thruster) Rhine River Crossing; Worms, Germany
  - 464-m (1,522-ft) long; 1,326-mm (4.4-ft) OD
  - 1450-m (4,757-ft) V radius
  - **First use of Herrenknecht Direct Pipe technology**
- (2007) Zug/CH; Sewer Zugerbergstrasse
  - 270-m (886-ft) long;
  - 110-m (361-ft) H radius
  - **First use of Jackcontrol joint system**
- (2013, soil, pipe thruster) MosStrojTransGas Pipeline; Khimki, Russia
  - 357-m (1,171-ft) long; 1,220-mm (4.0-ft) OD
  - 217.4-m (713.3-ft) V radius
  - **First use of MTS System2 pipe thruster technology**

The above list highlights projects where contractors have successfully negotiated installations with challenging attributes that provided net benefits to Owners. So how do North American projects stack up against worldwide curved MTBM drives?

## NORTH AMERICAN PROJECTS

At the time of publication, seven planned curved microtunnels have been completed in the USA and

Canada. A matrix of these seven projects is included in Table 1 along with two additional non-microtunneling projects; see notes below the table for further details.

Within the industry's history, there are numerous microtunnel drives with some degree of curvature due to normal steering, unintended steering, unintended deflection (due to obstructions or geologic stratigraphy), or unforeseen guidance issues. However, only the planned microtunnel drives are discussed herein.

A majority of curved projects listed in Table 1 were bid as straight drive microtunnels, and later became curved drives through value engineering proposals from the contractors. Only the Beachwalk Forcemain was designed as a curved alignment and sent out to bid by the owner. To the best of the authors' knowledge, North American highlights of curved projects are listed below with additional details found in Table 1 (Rush 2013):

- (1990) Hochtief Inc. completed a 1,041-m (3,415-ft) pipe jacking drive in Sayreville, NJ for the Middlesex County Outfall Project. Although the 4.2-m (13.7-f)t diameter hydroshield TBM was pipe jacked, guided, and provided continuous face-support, the operator's station was located in the shield instead of at the surface and therefore does not meet the ASCE definition of a microtunnel. The alignment included planned spatial (vertical and horizontal) curves and utilized the largest precast concrete pipe in the world at the time. It was also the first use of a large diameter hydroshield in the USA.
- (2010) Northeast Remsco completed the first planned curved drive in the United States as part of a value engineering proposal– a 183-m (600-ft) drive in Hartford for the Metropolitan District Commission's Clean Water Project. (Palmer et al. 2010).
- (2010) REM Directional Inc. installed the first use of Herrenknecht's Direct Pipe system in the USA—a 215-m (705-ft) drive in Arcadia, Florida. Direct Pipe is a hybrid technology combining the benefits of microtunneling and horizontal directional drilling (HDD). The spatial (vertical and horizontal) curve drive was completed in only three days of drilling.
- (2012) Frank Coluccio Construction Co. completed a double curve as part of the Beachwalk Forcemain Project in Honolulu. The S-curve drive spanned 378 m (1,241 ft) and was the first microtunnel project bid as a curve in the United States.
- (2013) James W. Fowler Co. completed the longest S-curve drive as part of the Santa Ana River Interceptor (SARI) Relocation Project in California. The curve drive spanned 478 m (1,567 ft) and was the first use of the Jackcontrol joint system in North America. A second single-curve drive was also performed.
- (2013) Ward and Burke completed four microtunnel drives as part of the Keswick WPCP Effluent Outfall Expansion Project. One drive marked the first curved microtunnel drive in Canada (March), another drive marked the first spatial (simultaneous vertical and horizontal) microtunnel curve in North America (May), and the last drive marked the first underwater MTBM extraction in Canada (June; a straight drive).
- (2013) Ward and Burke completed the longest curved microtunnel in North America for the Elgin Mills Watermain Project. The 739-m (2,425-ft) drive included three horizontal curves with the following radii: 450-m (1,476-ft), 800-m (2,624-ft), and 8,000-m (26,247-ft).

Each of these projects provided benefits to Owners and Contractors to varying degrees. The elimination of shafts on a couple of the projects saved time and money by eliminating lengthy environmental impact studies. Some projects found benefits from avoiding construction of shafts in high risk (and costly) areas. Others used curves to stay within tight right-of-way alignments.

## DESIGN CONSIDERATIONS

Extensive planning is necessary for every microtunnel. As tighter curves and more complicated geometry are incorporated, the designer's level of effort escalates as does the risk to the project. The design team's evaluation of the cost for risk mitigation measures is critical to help the owner weigh the net benefit from integrating curves. After all, why go through the headache if no appreciable cost or time savings come with it? This section presents several design considerations regarding curved microtunnels.

### Ground Conditions

Knowing the characteristics of the ground is critical to evaluating the interaction with almost every component of the microtunneling process. Specifically for curved drives, the designer must identify geologic interfaces between layers of significantly different strengths. Operators will struggle to steer a machine when skimming a bedrock surface or trying to transition from a soft to hard layer. Conditions like this may lead to unfavorable joint rotations.

Depending on the accuracy and grade of the system, ground conditions might also dictate whether a two pass or one pass tunnel is necessary. Nearly flat gravity installations in ground containing cobbles and/or boulders may lead to reverse grades. In this case, a two-pass tunnel allows the contractor to smooth out the grade (to an extent) when installing the carrier pipe. The casing must be designed to be pipe jacked along the curved alignment. Also, the carrier pipe must be designed to be transported and connected on the curve as well.

### Geometry

Whatever the intention may be for incorporating curves, designers must be careful to not develop an alignment with incompatible or competing geometry. For example, if the designer limits the tunnel's overcut size to reduce the chance of surface settlement, the ability of the MTBM to steer into a curve is diminished. A smaller initial overcut will also reduce the allowable length of the tunnel in abrasive ground if the MTBM head cannot be accessed along the way to refurbish the gauge tooling. On the other hand, too large of an overcut can waste grout and lube, cause settlement, and risk flotation of the pipe string. The designer must consider whether the MTBM's face will need to be accessed during the tunnel drive. Compressed air interventions are one method to do this, but must be planned out before the MTBM is ordered for the job. This way of accessing the head also causes a prolonged stoppage of the pipe string and creates safety hazards for the miners. Backloading tooling can be changed from within the cutterhead, but the contractor will not be able to add any hardfacing material around the circumference of the cutterhead. Therefore, another option is to incorporate a "push-thru" intermediate shaft where the MTBM can be temporarily refurbished under free-air conditions. This method is also preferred by crew if the machine's inside diameter is tight. An example of this method is included later in the paper.

Undersized interior diameters of casing pipe constrain a person's movement in the tunnel and prevent in-tunnel operations from being performed efficiently (or at all). Tight curves may require frequent control surveys to confirm MTBM position and update guidance systems. The designer must consider the safety of personnel who will need to enter the tunnel while the pipe string is stationary; some of the subsequent questions should be asked during design. Will surveyors be able to enter and exit the tunnel safely amongst all the tunneling system's equipment at the maximum length of the tunnel? Can survey equipment be used effectively and efficiently to not hold up operations? Are miners able to reach the MTBM head to do maintenance if necessary? Questions like these may lead the designer to consider larger MTBM sizes or reduce the tunnel's length. With the interior dimensions in mind, the layout of the curves on the alignment is concurrently considered.

Although starting a tunnel immediately on curve is not a show stopper, it does have drawbacks. Starting a tunnel on a straight section allows the crew to get a feel for the ground conditions, the machine's controls, the steering interaction with the ground, and to deal with slight frame or shaft exit seal misalignments. Depending on the location along the alignment, each section of pipe may need its travel path considered as well as its final orientation (Camp 2001).

If the curve(s) can be placed further from the jacking pit, fewer pipes then travel through the curve and may allow more force to be used from the main jacks. The exact allowable geometry of the curves may be dictated by what the pipe is capable of handling.

### Pipe and Joints

To reduce risk of damage from eccentric loading, pipe must be specifically designed for pipe jacking with high loading capabilities. For example, Reinforced Concrete Pipe (RCP) is commonly Class V (ASTM C76). Once the options for pipe is made, the pipe designer can consider the geometry of the tunnel, the anticipated ground conditions, the anticipated MTBM equipment and the plan of operation (namely, their working hours) to estimate thrust forces. Curves in the alignment lead to additional eccentric loading between pipe segments which will decrease the allowable jacking load.

The other consequence of curved alignments is the joints opening during driving which may lead to fluid infiltration (lubrication, grout, or water) if they are not properly designed. The expected joint opening widths must be evaluated to determine what type of joint design, packer and gasket material are necessary. Maximum joint opening widths are dependent on the tightest radius combined with normal steering tolerances through that zone. The pipe manufacturer should certify the compatibility of any joint packer used with their pipe. For example, joint packers consisting of full-circumference hydraulic hoses filled with fluid may need special block-outs in the pipe joints.

The pipe designer may be tempted to require shorter pipe segment lengths to minimize joint rotation, but caution is recommended. If the pipe segment length is shorter than the outside diameter, unfavorable pitch and yaw may be experienced along the tunnel. Shorter pipe segments also lead to higher costs from the additional pipe changes, more expensive pipe, and more joint material. But can shorter pipes lead to savings from smaller shafts? Not if the limiting length comes from the MTBM. Therefore,

**Table 1. Curved drives in soil completed with MTBMs in the USA and Canada**

| Year | Project | Location | Contractor | Curve Geometry | Machine / Guidance | Pipe | Ref |
|---|---|---|---|---|---|---|---|
| 1990 | Middlesex County Outfall Project* | Sayreville, NJ | Hochtief Inc. | 1,041 m (3,415 ft) long; unknown radii for spatial curves (both H&V) | Hochtief 4.16-m (13.7-ft) OD mixshield / ZED Guidance | 3.4-m (11-ft) ID RCP | 3,4 |
| 2010 | Homestead Avenue Interceptor Extension | Hartford, CT | Northeast Remsco Construction | 183 m (600 ft) long; 414-m (1,359-ft) radius | HK AVND-1800AB / VMT SLS-LT | 1.8-m (5.9-ft) ID Vianini RCP | 5,6 |
| 2010 | Florida Gas Phase VIII Expansion† | Arcadia, FL | REM Directional Inc. | 215 m (705 ft) long;<br>Spatial curves—radii of 914-m (3,000-ft) V and 1,829-m (6,000-ft) H | HK AVN-600 plus pipe-thruster; 914-mm (36-in.) OD / VMT | 762-mm (30-in.) steel | 7,8,9 |
| 2012 | Beachwalk Forcemain | Honolulu, HI | Coluccio Construction Co. | 378 m (1,241 ft) long;<br>S-curve 2×274-m radius | Rasa DH-1800 / Tokyo Keiki SV Corporation gyrocompass and hydro-level | 2,159-mm (85-in.) OD, 1,829-mm (72-in.) ID polymer concrete by US Composite Pipe South | 10 |
| 2013 | SARI Relocation | Yorba Linda, CA | JW Fowler | Drive #1:<br>478 m (1,567 ft) long S-curve radii of 4,838-m (15,873-ft) and 4,730-m (15,518-ft)<br>Drive #2:<br>190 m (622 ft) long, 1,676-m (5,500-ft) radius | HK AVND-2000AB / VMT SLS-LT | 2.58-m (8.46-ft) OD Ameron RCP w/ Jackcontrol p-system | 11,12 |
| 2013 | Keswick WPCP Outfall | Keswick, Ontario | Ward and Burke | Drive #1:<br>210 m (689 ft) long spatial curve—simultaneous radii of 875-m (2,870-ft) H and 6,600-m (21,653-ft) V<br>Drive #2:<br>340 m (1,115 ft) long, 6,500-m (21,325-ft) radius | HK AVND- / VMT SLS-LT | 1,219-mm (48-in.) OD Munro RCP | 13 |
| 2013 | Elgin Mills Road Watermain | Richmond Hill, Ontario | Ward and Burke | Triple horizontal curves—739 m (2,425 ft) long; radii—450-m (1,476-ft), 800-m (2,624-ft), and 8,000-m (26,247-ft) | HK AVND- / VMT SLS-LT | 1,219-mm (48-in.) OD Munro RCP | 14 |

* The Middlesex County Outfall Project does not meet the ASCE definition of a microtunnel because the machine was not operated remotely. However, this fact is trivial compared to the complexities successfully negotiated by the project team to complete a multiple curve, guided, pipe-jacked alignment with continuous face support from a hydroshield TBM.

† The Florida Gas Phase VIII Expansion Project was drilled using Herrenknecht's Direct Pipe thruster and not a traditional microtunnel setup.

the key is to keep the pipes as long as feasible, and protect them from damage or fluid infiltration.

### MTBM System

The allowable minimum curve radius may come from the MTBM as opposed to the jacking pipe. The MTBM must be designed to excavate the tightest planned curve of the project. As previously mentioned, the overcut of the cutterhead and the expected gauge wear are extremely important for a successful drive. On very tight curves, some MTBMs might need to be outfitted with more than one articulation joint, an adjustable bend angle on the cutterhead, overexcavation devices or copy cutters. The articulation joints must be self-cleaning with flushing ports to prevent the buildup of soil that may prevent steering. Additionally, over-skinning of an MTBM head must be kept to a minimum on curved drives. The amount of flexing an MTBM can undergo is greatly reduced as the depth of skinning increases. The designer must bear in mind every component of the MTBM system which may limit the radius of the curves.

Intermediate Jacking Stations (IJS) must also be designed to pass through the curves, depending on the position of each of them. These segments of the tunnel are rigid and must be closed up at the end of excavation once the hydraulic jacks are removed. IJS are considered backup thrust devices to be employed only when the friction along the pipe string builds (possibly from sitting still too long). Once they are used to get the pipe string moving, friction typically decreases enough to rely solely on the main jacks. This is why IJS must be installed before thrust forces reach too close to the allowable force on the pipe segments. Some contracts require an IJS to be installed at regular intervals and when thrust forces reach 70% of the maximum allowable thrust force. On some alignments, contractors may place IJS to wind up in final positions that rest before a curve in the alignment. By doing so, the main jacks can be used to close up the IJS without pushing the normal pipe segments further through a curve.

Lubrication of the entire pipe string is critical to reduce friction and to maintain the overcut to reduce settlement. The ingredients in the lubrication must be specifically designed for the soil and groundwater environment. Injection of the lubrication must take place up and down the entire alignment at regular intervals to keep friction low and reduce the ground loss in the overcut. Automatic lubrication devices are preferred as way to make this process more thorough and efficient. The machine's operator can create software programs to place lubrication at variable intervals and locations along the pipe string. This method is also safer as it reduces the number of personnel needing to enter the tunnel.

### Guidance and Surveying

An experienced specialist in underground surveying is critical to the accuracy of a microtunnel. Small errors originating from conveying control points from the surface into the shaft can magnify up the tunnel. Temperature fluctuations lead to inversion layers within the tunnel and shaft, causing further errors. Only a closed loop survey within the tunnel allows the survey team to identify high and low points, and must be done with traditional surveying equipment (not GPS based). To protect against unwanted disturbance, guidance systems and survey equipment must be protected within the jobsite and tunnel at all times. Inexperienced personnel cause control surveys to create significant, unanticipated delays in the production cycle. These delays can lead to cost and time increases not only from the time the survey takes, but from the increase in jacking forces when the need for using IJS is triggered. These examples are meant to stress the importance of requiring rigorous survey company and personnel requirements.

Quality in tunneling accuracy and efficiency will improve if the surveyor is intimately familiar with the guidance system. Alternatively, the guidance equipment manufacturer's representative can be on-site to take the surveyor's coordinates to update the system. It is also critical for the surveyor to understand the need to use geodetic surveys for the guidance system as opposed to planar surveys (Camp 2013). Geodetic surveys take into account the curvature of the Earth and provide a more accurate survey. But once the tunnel is in place, how can the owner locate it from the surface?

Since the pipeline can no longer be located on the surface by the tangent between manholes, the owner may want to consider installing monuments located on the surface above the tunnel alignment. One option includes electronic monuments buried longitudinally a couple meters below the surface to help locate the tunnel for future connections.

## CONSTRUCTION CONSIDERATIONS

Before construction can begin, the prime contractor assembles all shaft and tunneling information into (typically) a comprehensive submittal for review by the engineer. This stage is critical to ensure coordination is occurring between the various subcontractors and suppliers involved with the whole tunneling process. Contingency plans to deal with construction issues should be well planned and prewritten before construction.

One of the most common construction issues on MTBM drives is high thrust force. Pipes being jacked on curves will have a lower allowable thrust capacity, and therefore a contingency plan to protect the pipes must include several components. Assuming IJS are installed at regular intervals and at prescribed levels of jacking capacity (say 70% of maximum allowable), some contractors elect (or are required) to have an additional IJS on hand if it is needed. As previously mentioned, the IJS are for backup jacking capacity and are only used until thrust forces are reduced enough to fall back to solely on the main jacks. If the IJS are unable to lower the overall jacking capacity on their own, the jacking operation may need to continue 24/7 to keep the pipe string moving. Longer hours of production usually lead to a drop in overall jacking force. Additionally, the lubrication mixture may need to be adjusted with different ingredients to better suit the ground conditions.

Joint rotation should be monitored at the zones of curvature along the tunnels to check for over-rotation. If a problematic zone is found, every subsequent pipe that travels through the zone will experience nearly the same rotation (if they are the same length). To prevent damage to pipes, the contractor may elect to use an IJS behind this zone to push the pipe with a lower force than would be needed to push the entire MTBM and pipe string. Although rarely done, a few past projects have remediated excessive joint rotations by realigning the pipe string during jacking. In this scenario, the contractor would use a port (existing or newly drilled) along a pipe segment to pump a fluid or grout to try and heave the pipe in a desired direction. This example is a time-consuming (emergency) process that should be prevented ahead of time using good construction techniques.

Long distance drives with multiple IJS may require the contractor to install ventilation, slurry or lubrication booster pumps within the pipe string. Electrical transformers may also be necessary. As line losses increase, the spoil laden slurry will lose velocity and particles may fall out of suspension. Additionally, abrasion of the pumps may further reduce the efficiency of the system and velocity of the slurry. Contractors and designers must keep these considerations in mind both from the standpoint of preventing work stoppages and preventing face stability issues. If pumps do not supply or flush slurry correctly, or pressures become difficult to maintain, overexcavation at the face may occur in certain MTBM systems. This is true for even straight drives, but is especially true for drives with vertical curves.

Elevation differences along the tunnel alignment can cause problems within the slurry system if booster pumps are not carefully considered. If the tunnel is being driven uphill compared to the shaft depth, the contractor must be mindful of the slurry system when pipe changes occur. Slurry will tend to run out of the system into the shaft when this occurs and air will enter the slurry pipes. Valves may need to be placed on the slurry lines within every pipe segment. If leaks occur within the tunnel, such as at the IJS location, where will the slurry travel to? If the alignment is convex, such as driving a tunnel underneath a river, fluids will tend to run to the face of the tunnel while driving downwards and pool in the middle of the tunnel afterwards. Drain lines may need to be installed to remediate this condition.

Ground conditions must be evaluated throughout the tunneling operation. Negotiating the curves was preplanned based on the anticipated ground conditions before excavation begins. If unexpected soft or hard ground is encountered at points of curvature, the operator may struggle to maintain the planned radius. Hopefully, the MTBM was outfitted ahead of time with some of the components mentioned in the design considerations. Copy cutters can begin excavating the interior of the curve several meters before the curve to help the MTBM steer into the curve. Past worldwide projects have installed jet grout columns in very soft soils to support the MTBM turning through curves.

## SPECIFIC AUTHOR EXPERIENCE

### Santa Ana River Interceptor (SARI) Microtunnels

Out of five microtunnels included in two SARI contracts, two of the drives were curved on the "Mainline" contract. The Mainline Contractor, WA Rasic, and their microtunneling subcontractor, Fowler, proposed a value engineering change in the project alignment to add three curves, thus eliminating a tunnel shaft and converting another to a push-through (intermediate) shaft. This revised alignment included one S-curve tunnel (Drive #4) and one single-curve tunnel (Drive #5), both completed in mid-year 2013. The changes yielded a project cost savings over $1M that was to be split between the Owner and the Contractor. Details of the curved drive are the following:

1. Drive #4—Mainline Coal Canyon 2-Curve Drive
   a. 477.6 m (1,567 ft) long; 2,578-mm (101.5-in.) OD; 2,134-mm (84-in.) ID; RCP casing
   b. S-Curved; 4,838-m (15,873-ft) radius curve first, then a 4,730-m (15,518-ft) radius curve
   c. Launch shaft: soldier piles and lagging
   d. Intermediate shaft (cutterhead inspection): lattice girder reinforced shotcrete
   e. Receiving shaft (same receiving shaft will be used in other Coal Canyon Drive): lattice girder reinforced shotcrete
2. Drive #5—Mainline Coal Canyon 1-Curve Drive

**Figure 1. SARI "push-thru" intermediate shaft for MTBM inspection**

a. 190 m (622 ft) long; 2,578-mm (101.5-in.) OD; 2,134-mm (84-in.) ID; RCP casing
b. Curved; 1,676-m (5,500-ft) radius
c. Launch shaft: soldier piles and lagging
d. Receiving shaft (same receiving shaft will be used in other Coal Canyon Drive): lattice girder reinforced shotcrete

The alluvial ground along the SARI alignment was tested during design for abrasivity. Based on Miller testing, mineral types, and grain sizes, the geotechnical baseline report indicated a very high potential for abrasion of the MTBM system. Fowler was proposing to combine two of the straight drives into one longer, double curved drive. One of the added risks for a longer drive is the increased potential of having the various components on the tunnel face wear out and either slow or stop production. If this was to occur, the machine would need to be retrofitted, done via a costly and time consuming compressed air intervention entry to the tunnel face. Therefore, Fowler proposed to use a "push-thru" intermediate shaft where the cutterhead face could be accessed. The circular shaft was sunk before pipe jacking started, and included a low-strength slurry block poured in the MTBM's path along the tunnel. As shown in Figure 1, the slurry included a block-out portion that allowed Fowler to inspect the cutterhead in quadrants. This allowed for an efficient way to change tooling and weld addition hard-facing material as needed. After the MTBM was retrofitted, the shaft was partially backfilled above the tunnel crown with fluid. This fluid allowed the slurry and lubrication fluid pressures to be maintained within the annulus during the remainder of the drive.

To reduce eccentric loading on the Class V RCP, Fowler utilized the Jackcontrol joint system. This system, debuted in North America on the SARI project, effectively distributes stresses through neoprene hoses installed in the joints that are filled with hydraulic fluid. The system also provides the operator with feedback concerning joint pressures and joint rotations. To successfully provide control survey and navigate the machine through the curves, Fowler retained the full-time services of VMT who utilized their SLS-Microtunnel LT system.

### Keswick Outfall Microtunnels

The initial design of the Keswick Outfall microtunnels provided for a 762-mm (30-in.) diameter outfall pipeline, mirroring the design of the existing effluent outfall. In order to limit drive lengths and thereby maintain jacking forces at acceptable levels, the initial design called for six microtunnel drives and six tunnel shafts. The microtunnel drives between all of the shafts were designed to be straight.

Following award, the microtunnel subcontractor, Ward and Burke, proposed to eliminate two of the six shafts from the design and reduce the number of microtunnel drives from six to four. However, two of the four resulting drives would need to be curved. A critical condition of this change was that the diameter of the outfall had to be upsized from 762-mm (30-in.) to 1219-mm (48-in.). Two key reasons why the increase in diameter was necessary were:

1. The increased performance characteristics of the larger diameter MTBM which made it more appropriate for long-distance drives, and
2. Increased ease of installation and removal of intermediate jacking stations.

Other factors which played into this decision included the availability of microtunnel jacking pipe, and the fact that the larger 1219-mm (48-in.) MTBM was more neutrally buoyant than the smaller 762-mm (30-in.) MTBM. This last consideration was considered advantageous as alignment soils consisted predominantly of very soft (SPT "N" = 0), silts and clays and maintaining MTBM stability was of prime concern.

For the first curved drive, the required curve would be a plan curve only with a radius of 6500 m (21,325 ft). This curve was necessitated by the need to keep the plan alignment of the microtunnel from crossing directly below power poles along the alignment. Because of the large radius, this curved drive was considered to be of low risk. In addition, the use of a curved drive, combined with the increased drive length capability of the larger MTBM allowed a mid-drive shaft to be eliminated. This shaft was to be located in close proximity to existing private residences, and was considered to pose a significant risk for third-party damage claims. Therefore, the nominal risk taken on by adding the plan curve was considered to be more than offset by the elimination of the mid-drive shaft.

For the second curved drive, the required curve would be a spatial curve with simultaneous plan and profile curves with radii of 875 m (2,870 ft) horizontal and 6,600 m (21,653 ft) vertical respectively. The plan curve was necessitated by the need to avoid existing steel sheet pile shoring while maintaining the outfall within the public right of way. The profile curve was required to avoid an existing utility, to keep the upstream reception shaft at its design depth (i.e., to avoid deepening the shaft) and maintain the hydraulics within the manhole planned at that location. The combination of factors driving the plan and profile curves gave rise to the need for the final design of a simultaneous spatial curve. As was the case for the first curved drive, the use of a curved drive and the larger MTBM allowed a mid-drive shaft to be eliminated. This shaft was to be located immediately adjacent to a major intersection, and would have required temporary lane restrictions in order to provide truck access for spoils off-haul. In this case, because of the need for the curve to be a simultaneous spatial curve, the risk in adding the curve was considered to be only partially offset by elimination of the mid-drive shaft. However, the Region agreed to share this risk with the contractor for two main reasons:

1. The Contractor would have the opportunity to demonstrate their ability to jack a curved drive on the first, plan-curve only drive, before attempting the spatial curve drive, and
2. The contract carried provisional allowances for emergency rescue shafts, ground modification and utility support should issues be encountered during the drive.

The design and contractor teams, along with York Region, worked together to evaluate the technical risks and benefits of these changes. Key factors that were taken into account when evaluating the changes included the experience of the microtunnel subcontractor, Ward and Burke. Ward and Burke had completed curved drives in the past on European installations, and owned the specialized survey equipment necessary for long distance curved drives (VMT system). In addition, Ward and Burke's proposed MTBM operators and survey specialist all had curved drive experience.

## CONCLUSIONS

The North American MTBM industry now has several contractors, owners, designers and construction managers who have gone through the learning curve of curved alignments. Design and construction lessons learned must continue to be shared by the various project stakeholders within the industry for us to continue building the list of successful projects with curves. As the complexity of North American projects increase, experience from the worldwide MTBM industry will help guide what it takes to complete these projects. The industry will begin to see longer drives, sharper radii, curves in rock, and alignments with more consecutive or spatial curves. This can be accomplished with solid risk management strategies, experienced industry expertise, providing greater benefits and resulting in more successful projects for the owners.

Owners must build a team of experts who can thoroughly evaluate curved drive risks, and find ways to implement mitigations to minimize them. Designers must create robust contract documents with detailed experience requirements for the contractor, operators, machinery, pipe supplier, and surveyors. Contractors must take advantage of equipment that will help their operation be safe and efficient, along with personnel experienced in navigating and driving on curves. Construction management professionals experienced with microtunneling and curved microtunneling must be full-time in their quality assurance role with detailed construction records. Collectively the various stakeholders all having an important and critical project role will help to deliver a successful project.

## REFERENCES

[1] Camp, C. (2001)—Microtunneling through Designed Curves. Society for Mining, Metallurgy & Exploration (SME), North American Tunneling (NAT), USA.

[2] Rush, J., and Ball, R. (2013)—Advances in Microtunneling. Trenchless Technology, Benjamin Media, USA.

[3] Babendererde, S. (1991)—Long Distance Pipe Jacking in Sayreville, New Jersey. SME, Rapid Excavation &Tunneling Conference (RETC), USA.
[4] Essex, Randy (2013)—Personal Interview. Hatch Mott MacDonald.
[5] Palmer, R., Martin, L., and Seilert, A. (2010)—The First Planned Microtunnel in the United States. NASTT, No-Dig Show, USA.
[6] Seilert, A. (2010)—First Planned, Curved Microtunnel in the United States. SME, NAT, USA.
[7] Pfeff, D. and Lang, G. (2010)—Direct Pipe as Alternative to HDD in Permeable Soil Conditions. North American Society for Trenchless Technology (NASTT), No-Dig Show, USA.
[8] Direct Pipe Method Proves Out in First Use in U.S. Pipeline & Gas Journal, May 2011, Vol. 238 No. 5.
[9] Premiere Performance for Direct Pipe in USA. TunnelTalk, November 2010.
[10] Gibbs, S. (2013)—No Longer a Pipe Dream. Trenchless Technology, January.
[11] Ball, R. et al. (2013)—Microtunneling through Abrasive Soil with Cobbles and Boulders: Construction of the SARI Relocation. NASTT, No-Dig Show, USA.
[12] Young, D. et al. (2013)—Construction of the Santa Ana River Interceptor (SARI) Relocation. SME, NAT, USA.
[13] Gelinas, M. el al. (2014)—Microtunnelling Overcomes Design and Construction Challenges. NASTT, No-Dig Show, USA.
[14] Knoll, M. (2013)—Personal Interview. VMT-US.
[15] Camp, C. (2013)—Surveying and Guidance for Microtunnel Construction. NASTT, No-Dig Show, USA.

# Pushing the Limit—Shallow Cover Slurry TBM Mining Between Active Commuter Rail Tracks

**Eric Prantil and Frank Perrone**
Hatch Mott MacDonald

**David Smith**
Parsons Brinckerhoff

**Jean-Marc Wehrli**
Traylor Bros., Inc.

**ABSTRACT:** Slurry TBM evolution has allowed for increased control of face pressure and ground stability which correlates to increased success for shallow cover mining. This was a necessary component of the B/C Tunnel Extension performed in Queens, NY for the East Side Access Project. Through a coordinated effort, the designer, contractor, and construction management teams pushed a 6.86m (22'-6") diameter Herrenknecht slurry TBM beyond contract expectations, culminating in less than ½ shield diameter of cover. This paper explores the operation strategy with details regarding the surface and underground coordination, instrumentation plan, and program wide cost and schedule savings associated with its success.

## BACKGROUND

The East Side Access project being constructed by the Metropolitan Transportation Authority Capital Construction (MTACC) in New York City will, when completed provide a new direct connection to for Long Island Railroad from Queens to a new terminal located beneath the existing historical Grand Central Terminal on the east side of Manhattan, thereby helping relieve overcrowding at New York Penn Station and connecting subways. To achieve this, extensive underground work is required in the boroughs of Manhattan and Queens including Contract CQ031: Queens Bored Tunnels and Structures. This contract included the construction of four separate running tunnels, A, Yard Lead (YL), B/C and D; totaling approx. 10,500 ft using 2 Herrenknecht Slurry TBM's together with pre cast segmental concrete linings, a first in New York. These tunnels were constructed through varying ground conditions which included everything from a full face of rock, glacial till and clays, and mixed face conditions and beneath the busiest railroad interlocking in the US handling over 800 passenger trains a day in and out of Penn Station.

Following the minimal impact to the overlying railroad operations resulting from the initial A and Yard Lead TBM drives, a feasibility study to investigate the possibility of extending either of the two remaining tunnel drives was performed. The Construction Management (CM) team was responsible for coordinating this effort with the Designer, the General Engineering Consultants (GEC), and the Contractor, Granite Northeast—Traylor Brothers, Inc.—Frontier Kemper (GTF).

It was recognized early on in the study that geographical location of the proposed extensions would be a key driver in determining the feasibility of the extensions. Figure 1 shows the original planned termini and the surrounding Long Island Railroad (LIRR) and Amtrak railroad infrastructure. The transition of the future ESA revenue tracks from the existing LIRR mainline tracks on the surface into the bored tunnels requires the construction of open cut and cut and cover approach structures. These structures require extensive piling work for the support of excavation, ground excavation, and cast in place concreting operations. Performing such work adjacent to railroad tracks requires significant coordination with the railroads to secure any necessary track outages, service revisions and access and protection staff to provide a safe work zone. The replacement of surface works with bored tunnel that utilized the existing slurry TBM and treatment plant offered the program cost and schedule benefits and significant reduction in future schedule disruption to the railroads.

As shown in Figure 1, the B/C approach structure was to be constructed between the LIRR Westward Freight Track to the North, the LIRR

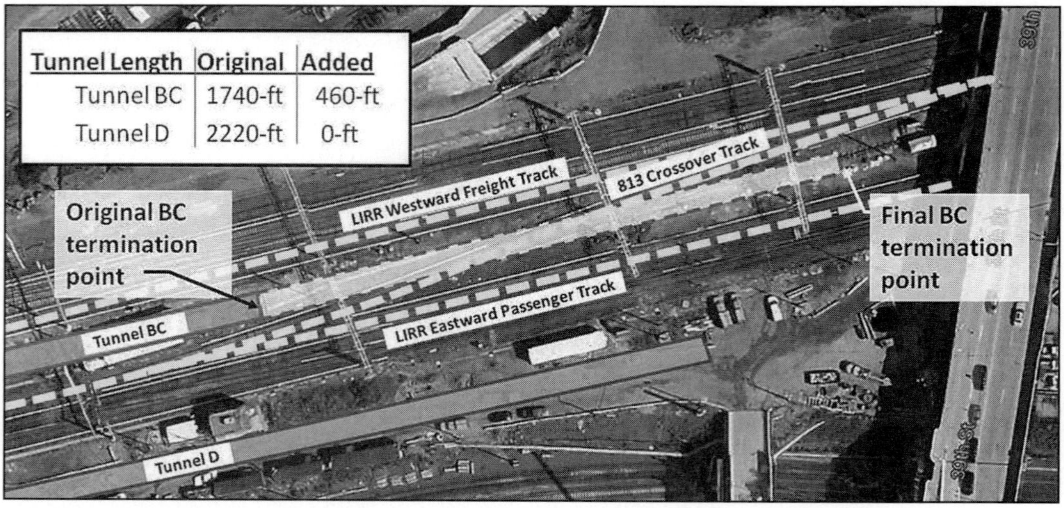


**Figure 1. B/C and D Base Contract tunnel lengths and in place railroad infrastructure**

Eastward Passenger Line to the South, and through the LIRR 813 Crossover Switch all of which would require to be taken out of service for considerable durations during open cut/cut and cover construction. Replacement of these structures by bored tunnel would therefore show schedule benefits, reduce future railroad resource requirements and reduce risk to the program In addition the duration of the installation of a new switch above the B/C approach to replace the existing 813 switch and associated track outage and service diversions would be reduced by some 8 months. Although new risks and challenges exist when tunneling with shallow cover beneath these rail lines, it was determined that on balance the proposed TBM extension offered sufficient benefits to out weigh these risks and detailed studies, engineering and preparation then started involving the Construction Manager, GEC, Contractor and Railroad teams.

## PARTIES INVOLVED—MAIN CONCERNS AND RESPONSIBILITIES

### Construction Manager

During the planning and execution phases of the B/C Extension work it was critical to identify all major stakeholders that could become affected by this operation in order to have the full support of all parties. This included approaching both LIRR and Amtrak to develop a dialogue on the intent of the extension work, while protecting railroad assets and operation. This coordination was the responsibility of the Construction Management team. Additional coordination was required with adjacent ESA Contracts including ESA Contract CH053, which would be working concurrently within the Harold Site with the B/C Extension operation.

The Geotechnical evaluation of the proposed extension was undertaken by the CM team Lead Geotechnical Engineer in conjunction with the GEC, and was based on existing geotechnical borings in the area. This analysis was used by the GEC to determine the loading of the finished tunnel structure, and by GTF to plan the operation of the slurry TBM. This information was also used to develop an instrumentation and monitoring plan which was to be implemented along the alignment extension.

Given the risk of the potential impacts to the surrounding railroad infrastructure during the mining operation, comprehensive risk informed contingency action plans were developed with input from the Contractor, GEC and Railroads.

### GEC

The B/C Extension plan was constrained by several design factors. The primary factor was to increase the length of the bored tunnel while maintaining the proper grade for the future train operations. This was coupled with the acceptable limits of the construction of the approach structure. Additional design consideration was given to the structural integrity of the precast concrete tunnel lining with regards to the decreasing ground cover and proximity to the active rail loading, especially beneath the critical LIRR 813 Switch. All redesign was still subject to AREMA design loading and for these rail lines, where Cooper E80 loading plus impact governed.

### Contractor

The focus of the planning by the Contractor was to ensure that the 22'-6" diameter Herrenknecht Slurry TBM would function as planned with cover of less than half a diameter and above the water table. Specifically, the extent of the functionality of the slurry circuit feed and return lines, face pressure, bubble pressure and tail shield grouting pressure all had to be determined. Equally important as the underground work, the contractor was also responsible for establishing a program of surface contingency measures in the event of slurry loss to the surface or excessive ground settlement or heave. These plans were executed both prior to and during the tunneling effort.

### Railroad

Concurrence from both Amtrak and LIRR was necessary to move forward with the B/C Extension. The railroads would gain several operational benefits in agreeing to perform the extension. These benefits included: avoiding a cut and cover operation in the middle of Harold Interlocking, preserving critical crossover infrastructure longer duration than originally planned, and reducing the amount of railroad support required to execute the cut and cover work. From a contractual standpoint, this work would reduce the construction duration of the CH058 contract, responsible for the construction of the cut and cover structure.

The critical factor for the railroads was the status of the 813 Switch and LIRR Freight Track. The railroads weighed the options regarding a short term versus long term outage and developed a systematic approach that would allow the work to be undertaken without overly impacting the daily train operations. The resulting coordination identified single points of failure which were a major concern for the railroads. These areas were reinforced with additional operational resources and the staging of stand by equipment in the field in order to mitigate any potential issues that could arise.

## PRE-EXTENSION EFFORT

Planning and coordination efforts began in September 2011 when, the first two tunnels were still being mined. The A TBM, which would relaunch as the D TBM was approximately 40% complete with the 1,919 foot A drive, and the Yard Lead Drive, with a total length of 4,320 feet, was approximately 22% complete. Upon breakthrough, the Yard Lead TBM would be relaunched to mine the B/C Tunnel. This therefore gave approximately six to seven months for the planning effort to be performed prior to the launch of the B/C Tunnel.

The GEC was tasked to determine the feasibility of the B/C Extension with regard to the segmental lining capability and the operational perspective. Initially the existing tunnel vertical alignment was simply projected forward from the designed end point STA.BC 1199+86, to determine how far the TBM drive could potentially extend. This resulted in a revised end point that remained within the influence zone of the 813 switch. To extract maximum benefit and remove the TBM stopping point from beneath the 813 switch the alignment was dropped by roughly 4'-0". Attention had to be paid also to the future surface tie in and signal locations. See Figure 2.

As can be seen the future B/C Approach Structure also ties in structurally with the piers of the 39th Street Bridge to the east. In order to smoothly transition from the bored tunnel to the cut and cover approach structure, the limits of mining were set at STA. BC 1204+25. Therefore, the revised design allowed for a potential extension of 439 feet. Taking into account the railroad infrastructure in the area, Figure 1 shows the physical relocation of the 813 Crossover track during construction of the B/C approach could be completely eliminated by mining underneath it. Additional cost and schedule delays in future ESA contracts could be avoided, provided that the CQ031 mining effort was successful.

With the revised design and tunneling limits in play, input from GTF was necessary in order to fully understand the capabilities of the slurry TBM for shallow cover mining (Figure 3).

Based on the developed geotechnical profile, the proposed 439 feet extension would have ground cover ranging from 17'-0" at the start station to approximately 6'-6" at the revised end station, measured from the ground surface to the TBM crown. Working from the ground surface down, following a layer of track ballast, the composition of the overburden consisted of a top layer of uncontrolled man made Fill ($\gamma = 125$ pcf) anticipated to be 6-ft to 10-ft thick, which transitioned to Glacial Till ($\gamma = 135$ pcf) extending well below the tunnel. The variable and unknown depth of the fill created one of the main concerns of the extension work. The slurry TBMs had performed well under similar conditions on the job where shallow mining was performed beneath overburden composed of 100% Glacial Till, but its performance in Fill was unknown. The Fill was a mix of non-compacted, loose material which was placed using overburden excavated for the construction of adjacent passenger rail lines and known from previous experience to contain railroad debris such as pieces of rail, ties, spikes, and splice plates.

As the crown of the TBM encountered increasing amounts of Fill, the probability of achieving the full target extension decreased. The TBM manufacturer,

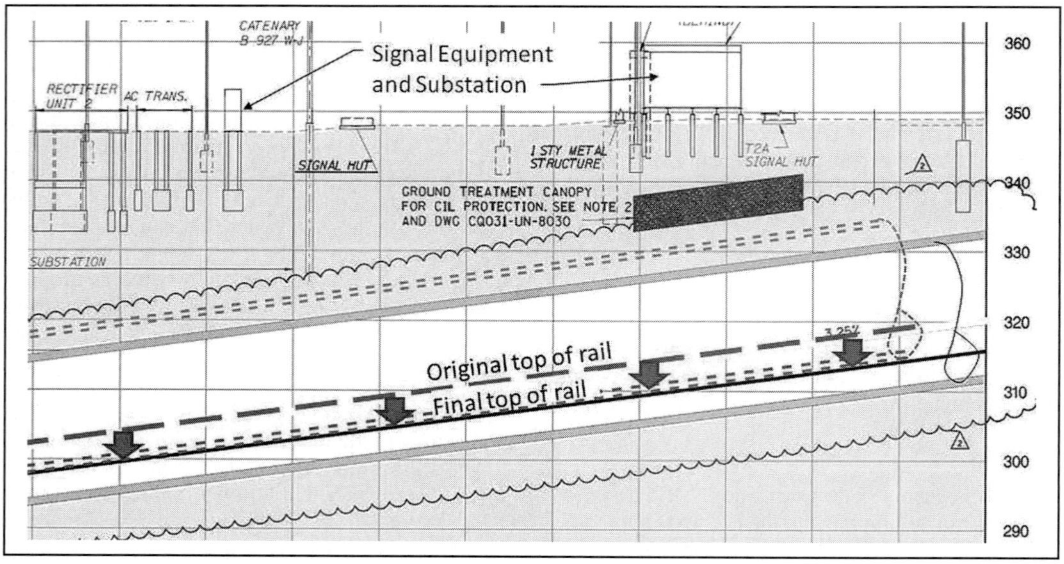


**Figure 2. Comparison of B/C tunnel contract profile with revised extension profile**

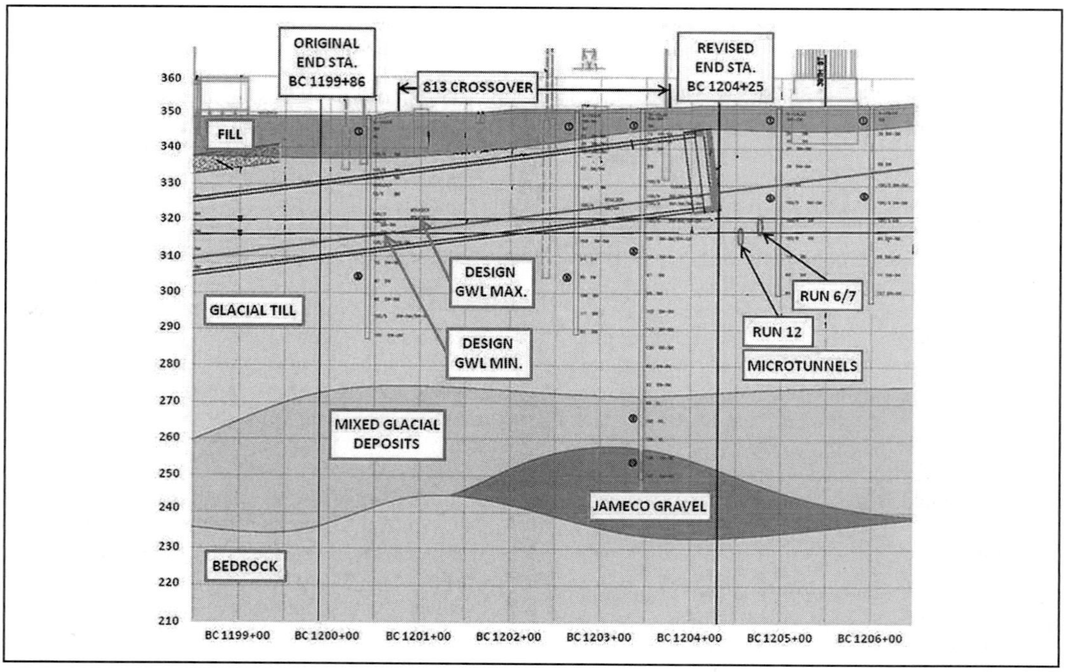


**Figure 3. Detailed geotechnical profile of the B/C extension**

Herrenknecht (HK), advised the Contractor of the minimum operating pressure needed to effectively run the slurry circuit was 0.6 Bar (8.70 psi) This pressure was measured at the springline of the shield and resulted in approximately 8.45 feet of required cover above the TBM crown (19.7 feet from springline). The pressure directly influenced the main slurry return pump (P2.1 Pump) located underneath the operators cab within the TBM shield. If the necessary back pressure on the pump could not be developed, it would be rendered inoperable halting forward progress. Based on this analysis, 8.45 feet of cover equated roughly to STA. BC 1202+65, resulting in an extension length of 279 feet, 64% of what was governed by the design. By maintaining the required pressure past this point the risk of a slurry frack to the surface increased markedly. Soil friction was considered to play a major role in providing a necessary safeguard against this occurrence. Overall, the performance of the slurry return pump was the major concern regarding the mechanical ability of the TBM to perform this work.

The information gathered from the technical meetings with the GEC and the Contractor enabled the CM team to lay the groundwork for the required discussions with the railroads. At this point in the planning process the railroad management team was engaged and informed of the scope and context of the extension work that was to be undertaken. The primary objective was to establish realistic expectations for the work and identify the potential effects this operation could have on the railroad infrastructure. This subsequently led to the development of jointly prepared contingency plans to safeguard personnel and railroad infrastructure.

The CM team also worked with LIRR and Amtrak to develop a track outage window to decrease the risk to railroad operations. Due to the shallow nature of the tunnel drive the area of influence was projected to include the LIRR Westward Freight Track as well as the LIRR 813 Switch, thus requiring a window where both tracks would be out of service The outlying tracks which paralleled the alignment, including the LIRR Eastward Passenger and Westward Passenger Lines would remain in service throughout the operation. Based on concurrent railroad construction operations, an outage window was secured from Friday, July 6th 2012 to Friday August 17th 2012. This outage took into consideration the expected mining production as well as several key tasks which the railroad required. At this stage in the planning, it was established that the TBM would mine and erect lining until either one of two events occurred; the TBM completed the proposed 439 feet of extension or slurry broke through to the surface. The risk of slurry loss to the surface was anticipated, planned for and accepted by all parties involved.

It was agreed that if slurry were to communicate to the surface and foul adjacent tracks, these lines would have to be refurbished. This would include the removal of the existing ballast, installation of new ballast, tamping of the material, and running of test trains over these tracks. The timeframe for track restoration work was built into the railroad outage period. With the concerns of the railroads addressed, and the collective team all working towards the same goal, all the items required for the construction of the necessary surface works were in place and the preparatory work could begin. This work started approximately at the beginning of June 2012.

The surface preparation plan implemented by GTF focused on three main areas: slurry containment measures in case of frack-out, installation of the monitoring instruments, and protection of an existing signal trough located immediately above the tunnel alignment. The scope of the instrumentation included track, surface, and structural monitoring points (Figure 4).

The track monitoring program included 32 rail prisms (SP Points) installed across the LIRR Westward Freight Track, Eastward Passenger Track, and the 813 Switch. Each was used to determine any settlement or heave of the rail. These prisms were read by a pair of Automated Total Stations (AMTS Units) mounted on catenary poles outside the zone of influence to provide readings in real time. The prisms were complemented by the use of nine (9) rail monitoring points which were manual points designated to measure the cross track tilt, or the differential between the top of rail height of the north and south rails.

The ground and structure monitoring programs were just as extensive. A deep monitoring borehole extensometer (BX Unit) was installed at the designated stoppage location of STA. BC 1200+00 where the TBM was instructed to hold if it arrived prior to the designated outage date of July 6th 2012. The purpose of this instrument was to monitor the ground immediately above the cutterhead, in the event that the restart of mining from the holding position caused any surface settlement. A full program of 39 manually read surface settlement points were installed along the proposed 439 feet of the extension. Nine (9) tiltmeters were also installed on four (4) multi-span catenary structures which had foundations that could potentially be influenced by the TBM mining.

Review and Alert limits for track work and catenary poles, consistent with Federal Railroad Authority (FRA) standards, were established per the already in place railroad infrastructure monitoring plan. Installation of all monitoring devices was undertaken by GTF's subconsultant Wang Technologies. The monitoring and data analysis was performed by Geocomp Corporation, which was overseen by the Construction Management Team's lead geotechnical engineer.

The slurry containment and mitigation measures that were undertaken by GTF had the primary goal of preventing slurry from migrating onto adjacent tracks and having the ability remove it from the designated area as quickly and efficiently as possible if leakage did occur (Figure 5).

The limits of this plan encompassed the same area as the instrumentation plan. The slurry containment zone was established between the LIRR

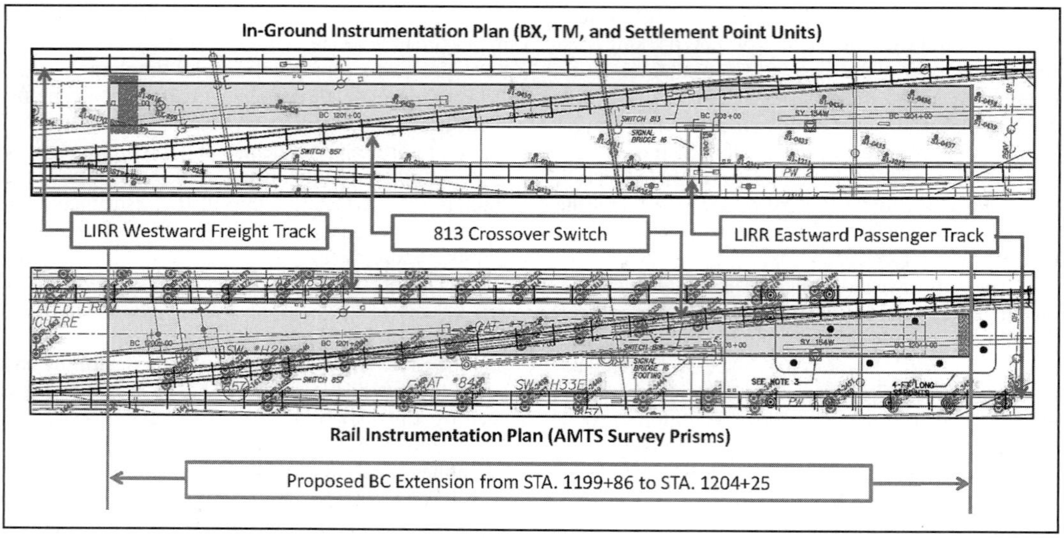


**Figure 4. Plan view of B/C extension instrumentation**

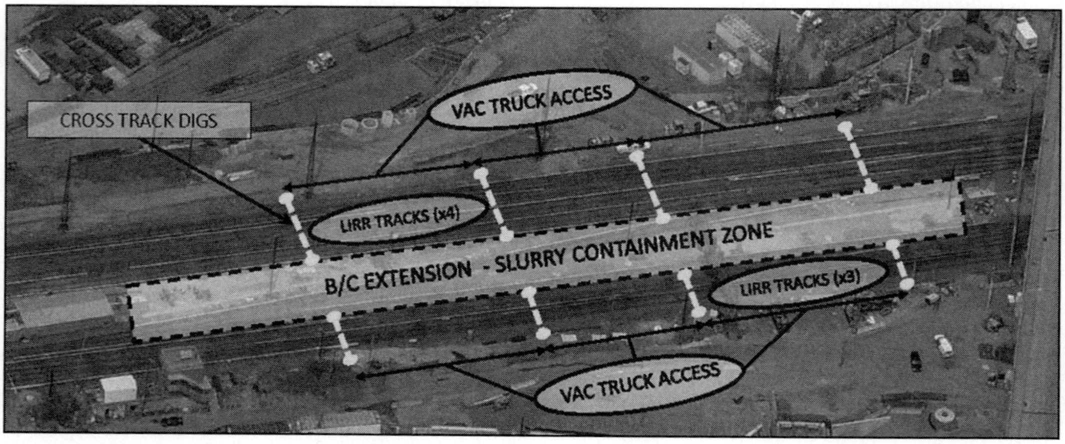


**Figure 5. Plan view of B/C extension slurry contingency plan**

Westward Freight Track to the North and the LIRR Eastward Passenger Track to the South. A perimeter was established using a silt fence of 30" tall nylon filter fabric held in place by wooden stakes driven into the ground. This was considered adequate to contain the slurry to the interior of the tracks and delineate a safe zone for access above the tunnel during mining. In plan, the area of the B/C Extension is flanked by four rail tracks to the North and three to the south. This arrangement does not allow easy access for equipment. So to facilitate the potential need for removal of the spilled slurry, 10 inch diameter PVC piping was installed underneath the active rail tracks which extended from the B/C Extension area to adjacent railroad property which could be easily accessed. A total of eight "cross track digs" as they came to be known were installed. These would allow Screwsucker Pumps and heavy duty Vac-Trucks to be placed and manned during the mining effort, to allow for immediate response if slurry were to escape to the surface. Hoses were installed through the cross track digs to reach the mining area. As the TBM would mine forward, the surface equipment would be moved forward, essentially keeping up with the TBM cutterhead.

One of the more critical pieces of railroad infrastructure above the tunnel alignment was a concrete LIRR Signal Cable trough containing 96 signal power and communication cables that controlled the entirety of Harold Interlocking. This 16" deep trough ran directly above the centerline of the tunnel for the length of the extension. The risk was that

**Figure 6. Signal trough support system + slurry breakthrough and containment**

if slurry was lost to the surface, TBM face support would be lost and a sinkhole could form, potentially causing the cables to sag and malfunction. At the recommendation of the GEC, a trough support system was installed by hand (Figure 6). This included a longitudinal support beam that would span any potential sinkhole, and would support and safeguard the cables.

With the completion of all necessary preparation work, a final coordination and readiness meeting was held on Thursday July 5th 2012 between the Construction Management Team, GTF, GEC, LIRR, Amtrak, and ESA Program Management. All parties who were involved over the course of planning and execution attended. The final plan was laid out and any remaining questions regarding scope and execution were addressed. With the outage scheduled for Friday July 6th, 2012, there was nothing to impede the start of the work.

## MINING EFFORT

The plan established between the Construction Management Team, GTF, and GEC, and agreed upon by all parties involved several key factors which were in place prior to the start of mining. Based on the cover to the TBM crown and overburden composition, the target bubble pressure was set at a maximum of 0.9 Bar (13.05 psi) and the tailshield grouting pressure set at a maximum of 2.9 Bar (43.22 psi). This 2.0 Bar differential between the bubble pressure and grouting pressure was the target set on all previous TBM drives up to this point on the CQ031 Contract and was already proven effective.

A secondary grouting program was also in place at key grouting ports on the segmental lining (Figure 7). This was recommended by the GEC in order to densify the ground at the springline and prevent potential ring squat due to the shallow cover as well as to handle future effects of the railroad loading from tracks adjacent and above the bored tunnel. There were two types of segmental lining rings used on the CQ031 contract. Both were identical with the exception that the "specialty" rings had additional secondary grouting ports. These were used in specialized mining conditions on the contract where additional ground support from inside the tunnel was required. The "normal" rings had six (6) total grout ports or one per segment. While the "specialty" rings had a total of 16 ports, three per segment except at the key which retained one. These "specialty" rings were to be installed when the TBM mined beneath the 813 Switch, from STA. BC 1201+90 to 1203+20.

The secondary grouting operation consisted of the installation of non return valves in the segmental lining grout ports near the springline of the built ring prior to the ring installation. Once the ring was installed, and shoved out of the TBM tail shield, secondary grouting commenced approximately two to three rings behind the tailshield. Grout hoses were hooked up to the TBM onboard grouting system, which controlled the pressure and volume criteria. The secondary grouting pressure criteria were set at 2.9 Bar (43.22 psi) and a volume criteria of 400 liters was established. This data was carefully monitored and reviewed for possible adjustments on a daily basis.

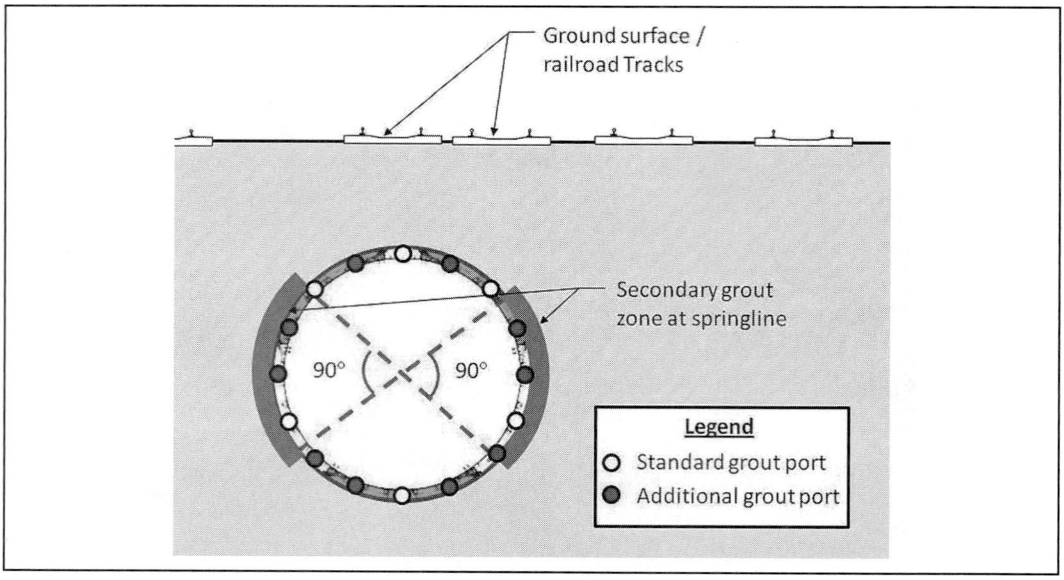


**Figure 7. Secondary grouting through the segmental lining**

The TBM reached the designated holding location of STA. BC 1200+00 on Friday June 29th, 2012. This allowed for a week of maintenance and preparation of the interior of the TBM components prior to beginning the extension effort. This included work to the slurry inlet pipes, segment erector, and grease pump system. All equipment was to be in the best condition possible, since it was agreed that once mining began; operations would be performed 24 hours / 7 days a week. The continuous mining requirement reduced the risk of potential ground movements during a TBM restart.

On Friday July 6th 2012, the TBM resumed mining and commenced the BC extension at 10:00 AM. The first day of mining the extension was productive with the TBM meeting its targets over three shifts, however production began to slow shortly thereafter. Mining rates reached approximately 6 mm/min to 10 mm/min due to high boulder and clay content which was encountered in the ground. Daily meetings between the CM, GTF, and GEC drew conclusions that the crusher chamber and grizzly bars were beginning to clog due to the insufficient back pressure in the bubble chamber to assist the movement of larger particles through the primary slurry return pump. After five (5) days of averaging approximately 15 feet per day, the team agreed to challenge conventional practice and increase the bubble pressure to 1.1 Bar (15.95 psi). This decision had immediate positive effects, with the TBM achieving 25 feet per day due to the increase.

One week into the B/C Extension, the TBM had mined 140.5 feet past the STA. 1200+00 holding point and a total of 154.5 feet from the original contract end station of STA. BC 1199+86. At this point in the operation 35% of the intended goal was completed and there were no surface, rail, or infrastructure movement or slurry leakage.

The second week of mining continued at an average of 27 feet per day without any issues arising on the surface. Both primary and secondary grouting takes were as expected, and ring installation without issue. The TBM slurry circuit however was plagued with multiple problems during this time period. These included the breakage of outgoing slurry pipes due to the high boulder content and low back pressure and leaks developing in the return lines on the TBM itself. Impacts to the slurry circuit led to hours of downtime to remove replace or repair components. By the 14th day into the extension, the TBM had mined 322 feet beyond the original contractual end station attaining 73% of the target extension.

On Thursday July 19th, 2012, the surface instrumentation began showing heave in response to the TBM tailshield grouting pressure. Hairline cracks developed at the surface in the soil. At this stage, the TBM was located at STA. BC1203+17 with approximately seven (7) to eight (8) feet of cover at the crown. In response to this, the bubble pressure was lowered to 0.9 Bar (13.05 psi) and the grouting pressure lowered to 2.5 Bar (36.25 psi).

Upon entering the third weekend of the extension, mining continued, but issues ensued due to worn slurry return pipes, specifically on the telescoping lines located at the rear of the TBM. However the biggest impact came during Swing Shift on Sunday

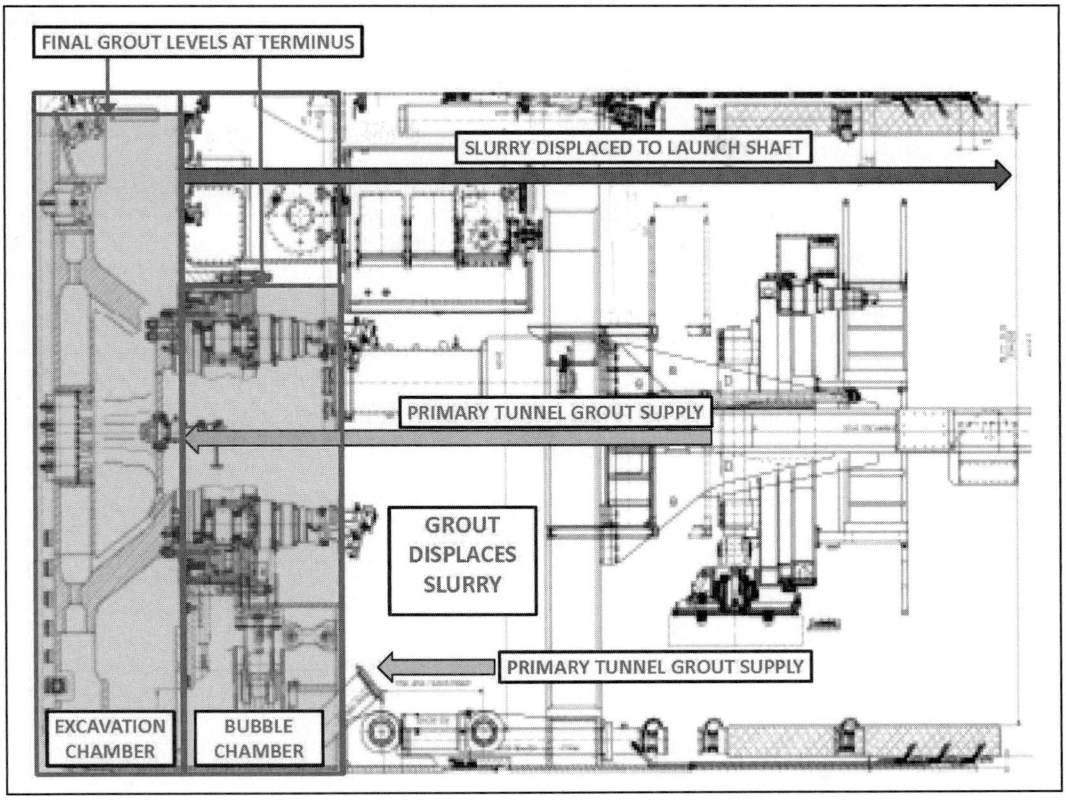


**Figure 8. Terminating the TBM cutterhead**

July 22nd, 2012, when the Contractor lost the functionality of the TBM crusher. It did however stall in a position which did not block the grizzly screen bars, and mining continued into the following day.

At approximately 7:45 AM on Monday, July 23rd 2012, a slurry leak propagated to the surface just outside the containment barrier adjacent to the out of service LIRR Westward Freight Line. The Team responded in accordance with the established plan. The slurry was diverted utilizing pre-staged sand bags into the central area of the extension and extracted using the cross track digs, Vac Trucks and Screwsucker pumps (Figure 6). Even though this event occurred at the peak of the morning rush hour for trains heading into New York Penn Station, no delays were experienced. The response plan proved successful and no railroad infrastructure was damaged throughout the process.

Overall the B/C TBM cutterhead reached an end station of STA. BC1204+15 and successfully mined an additional 429.4 feet of bored tunnel, achieving 98% of the intended project goal. It is estimated that the amount of ground cover at the terminus of the cutterhead was approximately 6.5 feet which equates to roughly 28% of shield diameter.

## POST-MINING EFFORT

Upon completion of the mining, there were several tasks which needed to be completed prior to returning the 813 switch back into service. This included the remainder of the surface cleanup, reballasting the tracks, as well as the stabilization of the ground in front of the cutterhead.

Due to the challenges described herein the BC TBM could not be recovered and was planned to be abandoned in place for future incorporation into the future structure. In order to ensure that he ground ahead of the TBM would not experience any settlement once the slurry pressure was eliminated and the TBM abandoned, GTF developed a scheme of pumping the TBM cutterhead full of tailshield grout to stabilize the face and fill the voids within the cutterhead and forward bulkhead. This method had already been implemented on the terminus of the D Tunnel, and had been executed without issue (Figure 8).

Through the rerouting of grout lines and utilizing access ports between the bubble chamber and the free-air side of the TBM bulkhead wall, tailshield grout was able to be pumped into the bubble chamber and eventually into the excavation chamber as well. This displaced slurry through access ports near the crown at the free air side of the TBM. The bubble and excavation chamber were filled with approximately 50,000 liters of grout in a process which took approximately one shift on Tuesday, July 24th 2012.

With the face stabilized and surface work concluding, the LIRR was able to run a Test Train on the 813 Switch on the night of Wednesday, July 25th 2012. As the Test Train was being run, convergence monitoring was performed inside the tunnel using prisms installed on the segmental lining throughout the length of the B/C Extension. Negligible movement was recorded. The GEC was satisfied with these results and deemed the rings stable and structurally sound for normal rail operations.

The LIRR 813 Switch and LIRR Westward Freight Lines went back into service on Friday, July 27th. This was a full three weeks earlier than anticipated.

## CONCLUSION

The successful execution of the BC extension shows that shallow cover slurry TBM mining is possible given the proper planning and ground conditions. Through the efforts and coordination of the Construction Management Team, Contractor, and GEC, with almost nine months of planning and 18 days of execution, several hundred feet of cut and cover structure was eliminated from future East Side Access Program saving time and money and allowing the railroads to maintain critical infrastructure during the process.

# Slurry on the Rocks: Trials and Tribulations of ESA Frozen STBM Safe Havens

**J. Ike Isaacson and Frank Perrone**
Hatch Mott MacDonald

**Brett Robinson**
Traylor Bros., Inc.

**Greg Ziegler**
Moretrench American Corporation

**David Smith**
Parsons Brinckerhoff

**ABSTRACT:** TBM safe havens were incorporated as part of the segmentally lined Slurry Tunnel Boring Machine (STBM) tunnels beneath Sunnyside Yard on the East Side Access Project in Queens, New York, and included one frozen safe haven. Considerations were given to providing a good seal behind the TBM, keeping it from freezing into the block, and the slurry procedures for mining into and out of the block. However, the frozen ground did not provide the safe haven that was hoped. This paper will discuss the experience with the safe haven and the lessons learned by multiple parties for application to future projects.

## BACKGROUND

A Joint Venture of Granite, Traylor Brothers, and Frontier Kemper (GTF) proposed on and was awarded the CQ031 contract to construct 4 new segmentally lined rail tunnels beneath the Sunnyside Yard train storage facility in Queens, New York (Figure 1). During the bidding phase, GTF expressed concerns about whether the ground conditions were suitable for compressed air cutterhead interventions, and so presented a plan to install safe zones along the alignment to mitigate the risk. The successful proposal and project scope were modified to incorporate the provision of "safe havens" within which free air cutterhead interventions could be utilized to change worn cutting tools and conduct repairs. In studying the alignment, the most logical zones to construct the safe havens were located at the emergency exit shaft structure areas. However, shortly after Contract award, a change to the scope occurred, eliminating two shafts; the Three Tunnel Emergency Exit, and the Tunnel D Emergency Exit. The Three Tunnel Emergency Exit structure had originally included a perimeter secant piled wall that fully encapsulated three tunnels; Yard Lead (YL), A, and B/C Tunnels. This perimeter would have provided a major component of the initial safe havens design for the first driven tunnels. In order to provide the intended safe havens, alternatives were examined at the Three Tunnel Emergency Exit site that would still encapsulate two tunnels, YL and A, into a single safe haven. Proximity to a major sewer and revenue passenger mainline tracks, site access, schedule, slurry chemistry complications, and the challenges of creating angled jet grout columns led to the consideration of ground freezing alternatives. Freeze pipe drilling was considered a cleaner operation, more favorable to daily railroad operations, and the plant had a smaller, more confined footprint compared to grouting. Thus the concept of parking a STBM in frozen ground to provide for free air interventions was born. Figure 2 illustrates the STBM locations with respect to the LIRR mainline tracks and Amtrak's maintenance of way tracks.

## GROUND, STBM, AND LINER PREPARATION

### Ground Freeze at the Three Tunnel Location

The configuration of the freeze zone consisted of a battered and twisted rectangular box that would encapsulate a portion of both the YL and A Tunnel alignments to form the safe haven. Figure 3 illustrates the twisted geometry of the freeze zone.

**Figure 1. Freeze plant and connection piping at the Three Tunnel Safe Haven site, Sunnyside Yard, East Side Access, NY**

This required the drilling of 90 steel freeze pipes, 8 Temperature Pipes and 11 Heat Pipes between 75 and 110 feet deep averaging a 35-degree angle inclination from vertical.

Moretrench (MTAC), the ground freezing specialty subcontractor, performed the thermal design of the frozen block (Safe Haven). MTAC utilized Finite Element Modeling with TEMP/W software to confirm the design and installation of the mass freeze. The theoretical freeze pipe layout and spacing was modeled utilizing a brine temperature of –30 degrees Celsius with in the 3.5" O.D. steel freeze pipes. The goal of the modeling was to accommodate free-air access to the face of the cutterhead via a 3ft × 11ft × 11ft hand excavated work zone and to ensure that groundwater within the limits of the safe haven was completely frozen. The modeling confirmed that an average temperature of -10 Celsius would be achieved for the frozen block (a requirement for structural analysis). In addition to modeling the frozen block Safe Haven (approximately 10,000 cubic yards of frozen soil), thermal modeling was conducted to establish a baseline for freeze formation and determine how STBM heat loading would impact the sustainability of the freeze zone following the removal of selected freeze pipes. The frozen block was modeled with the TBM parked in the block for repair. This analysis incorporated a TBM internal temperature of 70 degrees Fahrenheit.

The freeze pipe installation process went quickly and smoothly. Moretrench monitored the temperatures of the frozen block through eight temperature monitors that were strategically located. These monitors were 3.5 inch O.D. steel pipes with thermocouple wires installed to take temperature readings to depths below the invert of the deeper YL tunnel. Several temperature monitors were located on the perimeter of the freeze to confirm growth external to the last row of freeze pipes. Other temperature monitors were installed internal to the frozen block between the freeze pipes to confirm that all soils inside the planned frozen block were frozen. Moretrench ran a refined thermal finite element analysis (model) of the frozen block based on as-built location of freeze pipes, actual brine temperatures and actual field temperatures from the temperature monitors.

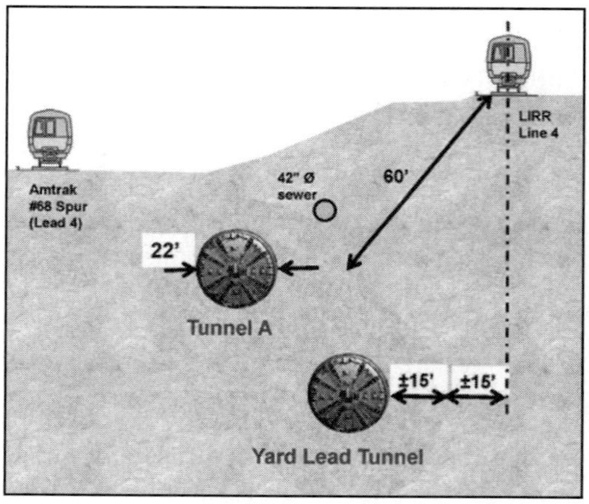


**Figure 2. Cross section at safe zone**

Based on these parameters, the finite element analysis modeling showed a completely frozen block after freezing day 63 through the circulation of chilled brine supplied by two refrigeration plants during the active freezing phase. Once this confirmation of the frozen block was made the commencement of pipe retractions could begin. The initial intent was to reduce from two refrigeration plants to

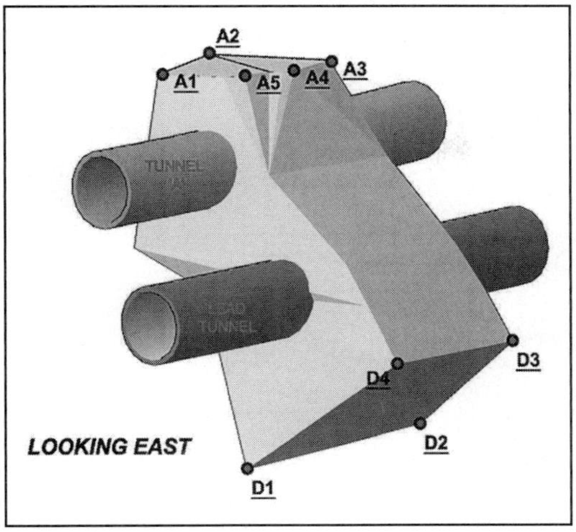


Figure 3. Rendering of freeze zone

Figure 4. Stainless-steel ring beam

one, going into a freeze maintenance phase once the target temperature had been achieved.

In order to accommodate access to the face of the cutterhead via a 3 ft × 11ft × 11 ft hand excavated work zone and ensure groundwater within the limits of the safe haven was completely frozen, a target average temperature of –10 degrees Celsius was established, based on structural modeling of the freeze to accomplish safe free-air cutterhead.

### TBM Modifications

Using frozen ground as a safe haven raised concerns that if the STBM were parked long enough, it could become stuck if the gap between the ground and the shield iced up. In order to address that concern, a low viscosity, oxidation inhibiting lubricant designed for extreme low temperatures was identified, which would remain fluid in the frozen ground. The lubricant, herein referred to as arctic grease could be injected into the annulus around the shield as it mined in and through the ice, replacing the slurry that would be susceptible to freezing. Herrenknecht designed and GTF modified both STBMs with eight rows of grease ports drilled through the shield to inject the grease. The forward most rows contained 24 ports, whereas the remaining rows contained roughly 16 ports with the final configuration varying by location.

### Special Precast Concrete Segments

As the TBM mines into the frozen block, the cutterhead creates an excavation with a diameter greater than the precast tunnel liner. Typically the annulus is filled with grout injected from the TBM. However, to protect the two component tailshield injection system from freezing, and to allow greater ability to perform additional grouting in the frozen zone if required, it was decided to use special precast segments with additional grout ports, and to grout the annulus between the segments and frozen ground through the segments. To achieve this 72 precast concrete rings were designed and manufactured with 16 grout ports compared with the standard 6 per ring.

### Sealing the Annulus

After the TBM mined into the frozen ground, in order to obtain a positive seal between the segmental liner and the frozen soil outside the lining, GTF designed a ring beam seal. The seal was installed between two precast concrete rings, and was composed of a stainless steel ring beam (to overcome long-term durability concerns) that matched the interior and exterior diameter of the concrete segments, with the web oriented longitudinally. Exterior to the web, a Bullflex bag (similar to a fire hose), was installed, which was protected by a sheet metal plate. The ring beam was assembled inside the TBM in a similar manner to precast concrete segmental rings, and when pushed out of the STBM shield, the Bullflex bag would be inflated with grout to create a positive seal in the exterior annular space. Two ring beams were fabricated prior to the Yard Lead (YL) tunnel machine and A tunnel TBM arriving at the safe haven. Figure 4 illustrates the final configuration of the stainless steel ring beam prior to installation in the YL tunnel.

## STBM APPROACH AND FREEZE ENTRY

### Annular Grease Purge of Slurry

To ensure that the slurry within the annular gap around the shield of the TBM was not trapped behind the installed segmental liner by the grease, a sequence was developed to fill the annular space surrounding the shield, thereby forcing the migration of slurry toward the cutterhead chamber. The STBM

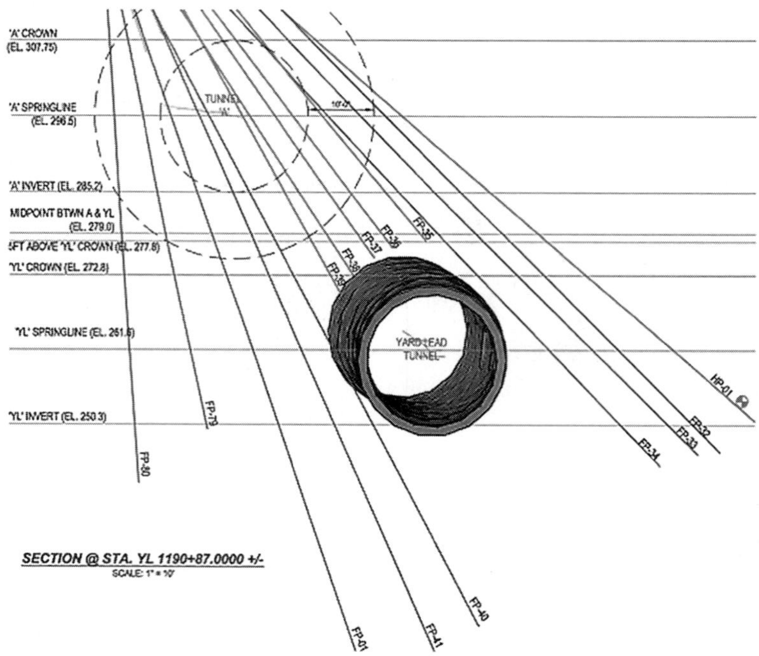


**Figure 5. Cross section showing freeze pipe removal for YL tunnel**

was driven to a face station that placed a given portion of frozen ground behind the cutterhead, at which time the grease purge was initiated starting from the grease ports closest to the tail seals. Once the estimated purge grease volume had been injected, injections were then made at the forward shield ports as the STBM was advanced into its temporary parking position. Affirmation that the annular space had been purged came in the form of grease fragments in the return slurry at the surface plant. In addition, arctic grease was also used to replace the Condat grease in the tail brushes, as it was felt the Condat would not provide an adequate seal at freezing temperatures.

### Retraction of Freeze Pipes Within Excavation Perimeter

The freeze pipes extended through the excavation line of both YL and A tunnels in the safe haven. Steel pipes were used for the freeze as these were deemed strong enough to retract out of the way of the TBM path. Figure 5 illustrates the freeze pipes that were retracted within the YL tunnel horizon. Heated brine was circulated in the pipes just until the exterior bond holding them in place was broken, whereupon the pipes were pulled just above the mining envelope. Chilled brine was again circulated and the pipes froze back into the ground. The retraction process for the YL tunnel, which included 27 pipes, took approximately 1 week to accomplish.

### Ring Beam Seal Installation and Grouting

At a point where the STBM position indicated that the rear of the tailshield was at the estimated boundary of the safe haven ice, one of the Bullflex ring beams was installed. Once the ring beam was pushed back beyond the shield, the Bullflex bag was inflated with Portland grout. After it was allowed to set, additional grouting into the annular space aft of the Bullflex seal was implemented through the special segments to further tighten the annular space.

### Tail-Shield Grout System Purge and Switchover

Due to concerns that had arisen during the testing of the two component grout setup in freezing conditions, a switchover was made to Portland cement grout once the annulus reached the estimated edge of the safe haven ice. With this switchover of grout type, grout injection also shifted from the tailshield ports to grouting through the segments. To protect the two component tailshield injection system in the tailshield, which would be subjected to freezing temperatures, the STBM tailshield system was purged of grout components and replaced with arctic grease so that it could be reinstated during freeze exit with minimal difficulty.

Once the STBM had reached the predetermined parking station within the freeze, the slurry within the cutterhead was drained to a level just above the invert to prevent it from becoming a block of ice that would prevent cutterhead rotation. With the removal of the slurry, pressurized air now filled the bulk of the cutterhead chamber. The cutterhead was rotated at pre-determined intervals to also keep the cutterhead free before the maintenance work in the cutterhead began. As a safety measure, anti-freeze was injected into the excavation chamber to ensure any remnants of slurry did not freeze, whilst the machine was parked.

## LOSS OF SEAL AND RESULTING ISSUES

### Depressurization and Inflow of Groundwater

The seal formed between the extrados of the segmental lining and the excavated soil by the grout-inflated Bullflex Bag and stainless steel ring beam was allowed to set over a 3 day weekend to allow time for cure. After this, the first attempt to go to free air for an intervention was made. Depressurization was undertaken at a slow, controlled rate, and entry was then made to observe the cutterhead and face conditions. During the initial entry, while the TBM engineers where assessing cutterhead wear and setting up to begin the planned cutter tool change-out, groundwater inflow was first noted, which rapidly increased in a short period of time from above the right shoulder of the STBM shield coming from the rear of the TBM. The cutterhead was evacuated, airlock sealed, and face pressure re-established with compressed air, but the damage had been done, and sufficient soil had migrated to the bottom of the excavation chamber blocking the slurry circuit and connection to the bubble chamber.

### Soil Migration and Fouling of Excavation Chamber

A subsequent free air entry into the cutterhead was made. At this time, engineers attempted to understand the source of the leak and estimated the water inflow to be approximately 175 gpm. At this time it was not clear whether seal behind the machine had failed, or whether the freeze was compromised. A pumping system was established, and water level in the excavation chamber was lowered almost to a level where the nature and extent of the blockage could be visibly understood. Unfortunately, a mistake was made whilst managing the pumping system, and the level with the excavation chamber began to rise, causing concerns to all parties that perhaps the inflow was increasing. Exit and re-pressurization was made, and all parties met to begin formulating strategies to clear the apparent blockage. During this time, the loss of compressed air through the ground to the surface from the cutterhead chamber was monitored and it was found to be increasing with time at an alarming rate, indicating a second potential flaw in the frozen safe zone and creating a extremely time sensitive issue. If the compressor plant could not keep up with the air flow loss, then compressed air interventions would not be possible due to the known defect in the frozen safe zone.

Upon notification of the issue, Moretrench conducted a full depth temperature profiling of all freeze pipes. This process incorporates a systematic shutdown of each pipe for a specific time period and retrieval of temperature data at very close spatial intervals. This process took approximately one week and failed to shed new light on the suspected defect as the results confirmed the intended frozen ground temperatures were being maintained.

## COMPLETION OF INTERVENTION AND MINE-OUT

At this stage, further free-air interventions were discounted, and the focus was to re-establish mining capability. Consideration was given to abandoning this location to undertake cutter change-out and drive the STBM out of the frozen safe haven into unmodified ground. A three pronged strategy was established to accomplish this new goal.

### Mother Mud Re-cakes

First, the increasing amounts of compressed air being lost to the ground had to be addressed as well as establishing a reserve compressed air capacity to potentially perform compressed air interventions. There was no such safety margin left with compressed air system, so the team agreed to inject thick, freshly batched bentonite "mother mud" slurry around the shield in the frozen safe haven, knowing all too well that it was not understood how long the team had until this slurry would freeze. After several days of allowing the thick slurry to penetrate under pressure to the possible sources of the leak, essentially "caking" the ground the cutterhead was drained and air losses were checked. Fortunately, it was determined that air losses had been reduced to levels which could accommodate manned compressed air entry into the cutterhead.

### Clearing Excavation Chamber and Return Pump Inlet

Once manned compressed air entries began, the initial focus was to remove the soil material blocking the inlet of the slurry return pump. This meant excavating by hand with shovels, filling bags of material which were removed from the cutterhead through the tool lock. At the same time, TBM engineers assessed

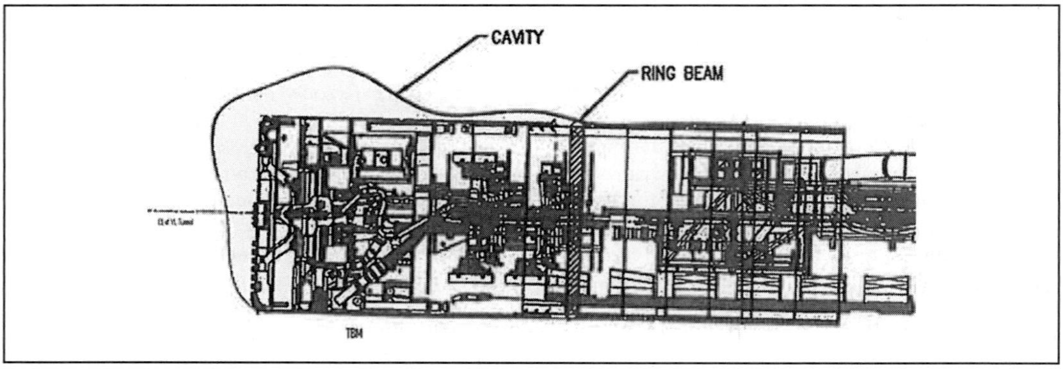


**Figure 6. Cross section of TBM at over break location**

cutterhead tools and replaced any that were necessary to allow the machine to mine out of the safe haven.

With the passage of time during this work, the mother mud dried and increasing air loss thru the ground was monitored by engineers. After some study of the compressed air system, rules were established to evacuate crews when air loss approached an unsafe level. At the established cutoff point the cutterhead would be evacuated, and another cycle of mother mud was injected, to re-cake the ground. During this work, observations were made to try to determine the area of the excavation chamber where air loss was occurring. These observations revealed that material filling the cutterhead chamber had been eroded from above the STBM shield, where an 8'×6'×15' area of overbreak had been noted. Figure 6 shows a graphic representation of the eroded cavity identified above the YL STBM. This area would have to be filled-in after STBM exit in order to prevent propagation of the void to the surface after the ground thawed.

### Planned Cutterhead Maintenance and Exit Strategy

Once the slurry return pump inlet had been cleared and the ability of the machine to resume mining re-established, the situation was again re-assessed. Continual mother-mud re-cakes were undertaken, and additional grout injections were made at the ring beam in an attempt to improve the seal. Attempts to augment the slurry with sawdust to provide a better seal resulted in a slurry polymer reaction between the bentonite slurry and the sawdust, which produced a high viscosity that nearly clogged the slurry circulation circuit. Thus augmentation attempts using sawdust was aborted. The Contractor (GTF) at this point felt the overall situation was stable enough to attempt cutter tool replacement as originally planned, but under compressed air instead of free air conditions. Tool change-out commenced while the team formulated a plan to drive the STBM out of the frozen safe haven and grout the crown void that had been noted above the machine.

Mapping of the void above the shield was conducted and two of the freeze pipes were selected which appeared to coincide with the position of the highest point of the void. These pipes would be extracted to serve as vents to allow air to escape the crown void while it was filled with grout.

Observations were showing that with each successive re-cake attempt using the mother mud, the intervals between successive re-cakes was reducing, meaning the re-cakes were experiencing diminishing returns. It was believed that the warm compressed air itself, in flowing through the leak, was further eroding and enlarging the defect(s) in the frozen block.

### Resumption of Mining with Minimal Annular Grouting

In order to exit the safe haven, the remaining freeze pipes left in front of the STBM within the excavation line had to first be retracted. The remaining twelve pipes were retracted above the excavation line, and mining was resumed with the YL STBM, 20 days after the original free air attempt was made.

During mine-out, a minimal amount of Portland grouting was conducted to establish a cradle of annular grout beneath approximately 30% of the ring circumference. Mine-out continued until the tailshield was clear of the freeze and void area, at which time grouting was switched back to the two component grout, which was injected through the segments to form a second lift which cradled approximately 60% of the rings.

At that time, two of the freeze pipes were totally extracted from above the void to serve as vents. The slurry within the cutterhead was maintained and provided face pressure via the column of grout within the open vent holes to the surface. Two-component grout was injected at the machine to fill the remaining annular space outside the segmental liner behind the machine back to the ring beam seal, as well as to fill the void above the crown created by the inflow. Upon exiting the freeze zone there the two component grout communicated briefly to the surface through one of the freeze pipe holes prior to either gelling or freezing in the hole.

At this point, the use of modified rings was discontinued, the tailshield grouting system was reactivated, and normal mining operations resumed. A compressed air intervention was made in unmodified ground to check the cutterhead condition, which was found to be satisfactory.

Meanwhile, the A STBM had by this time caught up with the YL machine and had temporarily halted just outside the safe haven to prevent the two machines from being in the compromised safe haven at the same time. It was decided that the A STBM would not be allowed into the safe haven block until the void had been fully grouted. Due to the success of the compressed air intervention on the YL machine under less than favorable conditions the decision was made to attempt the A STBM cutter tool change-out under compressed air prior to entering the safe haven. A compressed air intervention was conducted to assess the condition of cutter tools and air loss to the ground. The frozen ground of the safe haven would be considered a 'last resort', if air loss was too great. The initial intervention was not successful. The face began to collapse, and air loss was very high. The A tunnel machine mined a little further along, where interventions were tried again, and this time the ground was suitable, and work began on re-tooling the cutterhead. Mother mud re-cakes were required to control air losses. All freeze pipes in the Tunnel A alignment were heated and retracted to allow the TBM to mine through the frozen block. Upon completion of cutterhead refurbishment, the A STBM recommenced mining and mined directly through the frozen safe haven without stopping.

## LIKELY FAILURE MODES AND CONTRIBUTING FACTORS

As with all things, hindsight offers the chance to piece information and events together to allow the analysis of what happened and what the possible causes were. Some of these were:

- Heat generated by the TBM, grout, and grease used may have been higher than anticipated.
- The freeze block may not have been sufficiently long to ensure the ring beam and grout bag were far enough into the fully frozen ground.
- Prolonged and multiple decompressions facilitated enough water flow to exacerbate the loss of seal. The use of brine during the thawing and retraction of freeze pipes above TBM may have introduced leakage pathways to the cutterhead chamber along the annulus of retracted freeze pipes.
- Groundwater chemistry may have been a contributing factor

## LESSONS LEARNED

Teamwork and close coordination were vital during both the planning and recovery phases of this project. Key lessons learned from this experience are:

- Close coordination between all parties was crucial to the successful recovery of the STBM.
- Need multiple methods to verify that the ground is truly frozen
- If groundwater flow is noted during free-air intervention, immediately seal and go to compressed air to minimize further seal loss
- A lower heat of hydration grout mix should be tested.
- Allow more time for retracted freeze pipes to refreeze and seal with ground before attempting mine-in.

## ACKNOWLEDGMENTS

The authors would like to express their gratitude to the extraordinary effort of members of the GTF, the General Engineering Consultant, and MTACC team to accomplish this work and for the opportunity to discuss it with the tunneling community.

## REFERENCES

Robinson, Brett and Jean-Marc Wehrli (2013). East Side Access—Queens Bored Tunnels Engineering Challenges. In M.A. DiPonio and C. Dixon (Ed.) *Rapid Excavation and Tunneling Conference—2013 Proceedings* (pp. 1,086–1,118), Englewood, Colorado, United States: Society for Mining, Metallurgy, and Exploration.

Robinson, Brett and Jean-Marc Wehrli (2013). East Side Access—Queens Bored Tunnels Case Study. In M.A. DiPonio and C. Dixon (Ed.) *Rapid Excavation and Tunneling Conference—2013 Proceedings* (pp. 1,014–1,041), Englewood, Colorado, United States: Society for Mining, Metallurgy, and Exploration.

# A Change in Ground Conditions Necessitates Fast-Track Design Changes

**Stephen O'Connell and Kyle White**
Black & Veatch

**Steven Kirk**
City of Charleston

**Paul Kassouf**
Triad Engineering

**Seth Pollak**
Arup

**ABSTRACT:** Construction had started on a 4,000 feet long, 12 feet diameter storm water tunnel below Charleston, SC when a geotechnical investigation revealed an abandoned sewer tunnel was not in the condition or the location originally believed. Concurrent with sinking of the access shaft caisson, the project team worked to engineer a solution to mitigate the risks of construction near the abandoned tunnel. The resulting design changes increased the depth of the shaft and flattened out the grade of the tunnel maintaining the full length of the storm water tunnel safely below the existing tunnel's profile.

## INTRODUCTION

The Market Street Drainage Improvements Project, part of the City of Charleston's overall Master Plan to relieve flooding on the Charleston peninsula, is a three phase initiative to remediate frequent flooding within the main tourist district in the downtown Charleston, SC peninsula. Construction of the first phase of the project was completed in 2007 and included enhancements to the currently operating Concord Street Pump Station to allow for future acceptance of flows from the Market Street Tunnel. Phase two construction, which includes the tunnels and shafts, began in September 2012 and is scheduled for completion in August 2014, at which time phase three of the project, which includes the near surface network of pipelines and vortex structures can begin.

## BACKGROUND

Given Charleston peninsula's minimal topographic relief and proximity to the Atlantic Ocean, adequate stormwater drainage has been an ongoing challenge for the city since its inception. While the most severe flooding occurs during rain events of moderate to severe intensity within two hours of high tide, it is not uncommon for several areas of the Charleston peninsula to flood during high tidal events without any rain. This can result in hours of standing water on the streets and in the neighborhoods (Figure 1).

As a result of frequent flooding, in 1837 the Mayor of Charleston offered a $100 gold coin to anyone who could engineer a feasible solution to the City's stormwater problems. Numerous ideas were submitted in pursuit of this gold coin; however, no design stood above the rest and ultimately the Mayor developed a solution by combining several of the best ideas.

The early solution was to construct a network of interconnected brick arches that discharged stormwater by gravity to either the Cooper River or Ashley River, which flank the east and west sides of the peninsula (Figure 2). Gates were installed on the outfalls to control the tidal waters. In addition the system was slightly undersized to help facilitate scouring velocities to reduce sedimentation during flood events. Unfortunately, the gravity system was never very efficient in conveying stormwater flows to the rivers, particularly during high tide events. Although the system provided some minor flood relief, the frequent flooding could not be overcome. Faulty gate valves and years of siltation have further clogged the system of brick arches. Today, many

Figure 1. Flooding along Market Street

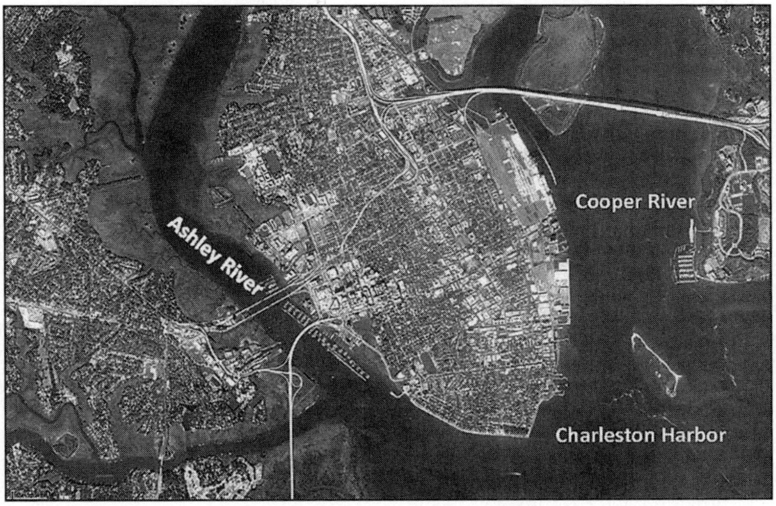


Figure 2. Charleston Peninsula

of the gates have since been removed as they were causing restrictions in an already undersized system.

In an effort to properly address the growing city's concern over stormwater drainage a Master Plan was developed in 1984 by Davis & Floyd, Inc., a regional engineering firm with an office in Charleston. This Master Plan divided the Charleston peninsula into discrete drainage basins from which the City of Charleston engineers began to plan improvements projects based on order of community need and the severity of the flooding.

Due to the urban and historic nature of the Charleston peninsula it was logical to seek stormwater relief via a deep tunnel system which inherently limits the amount of surface disruption during construction and can generally increase conveyance capacity. In addition to the now well-known advantages of tunneling in urban settings, City engineers and their consultants planned to introduce pumping stations into the system in an effort to overcome the Charleston area's lack of topographic relief and move the maximum amount of water possible in the shortest amount of time. These pump stations would work in conjunction with the deep conveyance tunnels and discharge to either the Cooper or Ashley River.

The City of Charleston first implemented tunnel techniques on their Meeting Street/Calhoun Street tunnel. The project was actually designed as a major open-cut endeavor to fix the frequent flooding issues. However, after input from various contractors, the project was converted into a deep underground conveyance tunnel in an effort to minimize impacts to existing utilities and avoid public disruption. The tunnel project incorporated a new pump station to discharge the tunnel flows out into the Cooper River.

This concept has proven highly successful for the City on past projects and is currently being implemented on the Spring/Fishburne US 17 Drainage Improvements Project, which is currently being constructed.

### Basic Geology

The geology in Charleston can be categorized into two basic classifications for purposes of this paper: Surficial soils (shallow sedimentary deposits) and Cooper Group, known locally as the Cooper Marl. The Charleston peninsula is part of an estuary, and the surficial and shallow geology is influenced by a combination of marine and continental geomorphological processes. The Surficial soils were deposited in a wide range of sedimentary facies including fluvial, overbank, tidal marsh, tidal channel, tidal flat, lagoon, beach, barrier island, and shallow marine deposits and consist primarily of highly plastic organic silt and clay with interbedded sand lenses. The typical engineering characteristics of the material are very high moisture content, very low shear strength generally with a soft to very-soft consistency.

Lying below the surficial soils is the Cooper Marl, a thick sequence of marine sediments which is generally characterized as a relatively massive, homogenous, olive green, highly calcareous, phosphatic, fossiliferous, clayey sand and silt. On the Charleston peninsula, the Cooper Marl generally lies 30–75 feet below the ground surface. An irregular erosional contact surface often separates the Cooper Marl from the surficial soils and the farther inland one travels from the Atlantic Ocean the closer the Cooper Marl is to the surface. The Cooper Marl is a remarkably homogenous formation and exhibits consistent engineering properties with very little variation with depth or along profile making it an excellent tunneling medium exhibiting sufficient standup time for erection of initial support yet soft enough to excavate by shovel and air spade if desired.

The Cooper Marl's strength and standup time can be primarily attributed to the calcareous bonds which give the soil formation "rock-like" properties in its natural state. Cooper Marl, while composed of clayey sand and silt, cannot easily be defined by the Tunnelman's Ground Classification system (Heuer, 1974). However, once the calcareous bonds are broken they do not re-mold and the material completely loses its strength. For this reason, the more the Cooper Marl is handled and broken down, the more difficult it becomes to work with turning into a sticky, sloppy mess which makes it difficult to use as any type of structural fill.

### Topography

The coastal environment has helped Charleston prosper over the years as it is ideal for commerce and travel. However, its location has also played a part in some of its most difficult engineering challenges. The coastal region of South Carolina is commonly referred to as the "Lowcountry," and with good reason as the average elevation of the City is only a few feet above sea level with little to no topographic relief. Engineers have struggled with these elevation constraints since the City began installing pipelines for movement of water services. Gravity systems have been particularly difficult to construct as even moderate slopes will push the infrastructure at or below the tidal zones and render the system ineffective. This has facilitated the need for Charleston to divide the peninsula into a series of smaller basins that help limit the runs of gravity lines and incorporate pump stations to increase velocities for purposes of scouring and discharge.

## A HISTORY OF TUNNELING IN CHARLESTON

Tunneling within the Cooper Marl is nothing new for Charleston where the City of Charleston's stormwater services division and Charleston Water System's wastewater and water supply divisions have been tunneling for decades to provide critical infrastructure services to its residents. Tunnels for "water" conveyance have roots in Charleston dating back to 1928 when a system of water supply tunnels were constructed to bring water from the Edisto River and Foster Creek to the Hanahan Water Treatment Plant to supplement groundwater sources (Figure 3). This trend of constructing tunnels for "water" conveyance has continued throughout the years until present day where tunnels are still being designed and constructed. To date nearly 50 miles of tunnels have been constructed or designed in the Charleston region by these two entities (Swartz, et al., 2012).

## MARKET STREET DRAINAGE IMPROVEMENTS PROJECT DETAILS

Phase two of the Market Street Drainage Improvements Project consists of two tunnels excavated and supported from a single 20 foot ID access shaft and lined 140 feet below ground surface: the 2,725 linear feet Concord Street tunnel and the 1,235 linear feet Market Street tunnel (Figure 4). The tunnels were designed to be concrete lined with a finished internal diameter of 10 feet. Three 54-inch steel drop shafts, located between the historic Market Street sheds (Figure 5), will eventually drop stormwater from a near surface network of pipelines to the deep tunnel system which will convey the stormwater to its termination point at the Concord

**Figure 3. Tunnel constructed in Charleston circa 1920**

Street Pump Station where the stormwater will be discharged into the Cooper River via the previously constructed river outfall.

Davis & Floyd, Inc. began the original design work for the Market Street Drainage Improvements Project in 1998 and subcontracted design of the tunnels and shafts to URS, Corp. Due to numerous issues including funding and obtaining permits from the South Carolina Ports Authority for the proposed access shaft site, bidding of the project did not occur until March 2012 with NTP given to the joint venture of Triad-Midwest Mole in July of that year. Triad-Midwest Mole is utilizing Arup for design services during construction.

In addition to iterative changes to the design including turnover of the responsible engineering staff the approximately 14 year delay from design to construction presented the owner, engineer, construction manager, and contractor with several issues which arose during construction including.

- The presence of new high rise condominiums in close proximity to the tunnel alignment founded on piles.
- Changes in land ownership including the need for acquiring an easement for locating the projects only access shaft within State Port's Authority property.
- Further deterioration to the adjacent CWS wastewater tunnel which led to sizable voids in the ground surrounding the tunnel.

During the bidding phase for the project, several contractor-driven alternate design proposals within the specific guidelines established by the design team were allowed. The alternates were submitted with the bid by prequalified contractors with subconsultants engaged. Among the alternates, a reduction in tunnel diameter from 10 feet to 9 feet was accepted. The reduction was allowed in an effort to accommodate a range of proposed concrete formwork and TBM sizes. Following award, a reduction in finished access shaft diameter from 25 feet to 20 feet was accepted. The Market Street Drainage system, like the majority of the City's stormwater system, is designed for conveyance only and not for storage and as a result the reduction in shaft and tunnel diameters did not affect the hydraulics or required level of service.

Construction engineering and inspection services for the project are being provided by Black & Veatch, Corp. who is working as a subconsultant to Davis & Floyd, Inc. with URS, Corp., the original designer of record for the tunnels and shafts.

## WASTEWATER TUNNELS IN THE VICINITY OF THE PROPOSED MARKET STREET TUNNEL

The original unlined deep tunnel wastewater system on the Charleston peninsula was constructed in the late 1960s to early 1970s. Tunnel branches run down both the east and west sides of the Charleston peninsula and eventually terminate at the Plum Island Wastewater Treatment plant across the Charleston Harbor. This system served the City well until the 1990s when CWS began discovering extensive corrosion in the tunnels after commercial divers were hired to inspect the tunnel system. What they found was unsettling: collapsed ribs, which were used for primary support of the tunnel excavation, gaping holes in the carrier pipe inside the excavated tunnel, large sections of sloughing Cooper Marl, and accumulation of sludge (Figure 6). Engineers feared the severity of deterioration could cause a blockage in the tunnel, which would result in sewer overflows in downtown Charleston. The divers made temporary repairs, but their findings made it clear that Charleston needed a new sewer tunnel system. CWS immediately began an aggressive replacement program to build a new tunnel system. The fast-tracked project was divided into several phases with the first phase completed in 2001 and the final replacement phase currently under construction. The sewer tunnel

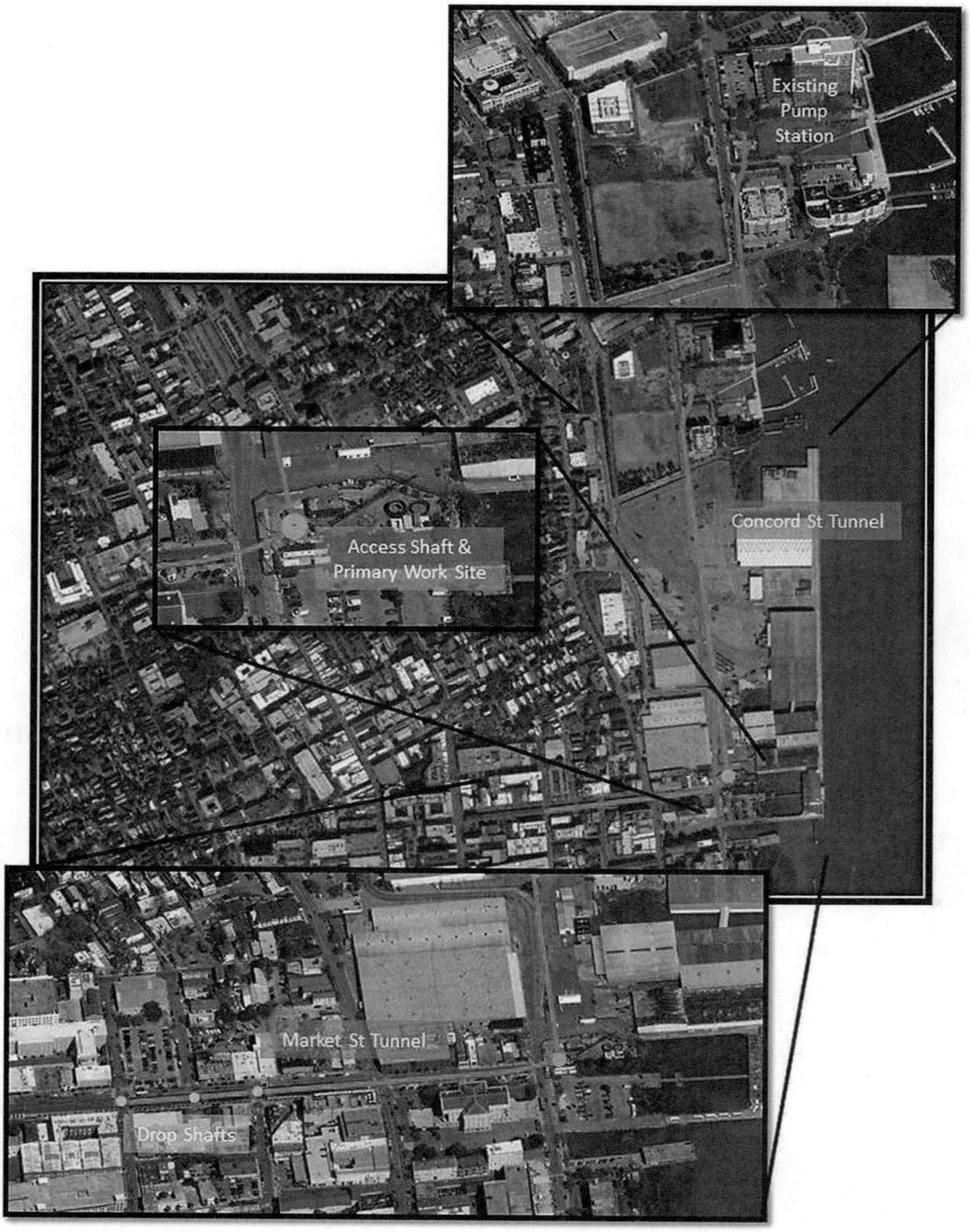


**Figure 4. Market Street project alignment**

replacement program is the single largest infrastructure program in the utility's history.

The relatively small size of the Charleston peninsula (approximately 5 square miles) in conjunction with the location of the stormwater drainage basins created a conflict in tunnel alignments to the point that the Market Street tunnels would need to cross beneath and parallel to the abandoned CWS wastewater tunnel and cross beneath the new CWS wastewater tunnel, which was designed and constructed after the original design of the Market Street Project.

**Figure 5. Drop shaft installation along Market Street**

**Figure 6. Deteriorating condition of the existing wastewater tunnels**

## CHANGE ORDER

The original depth of the Market Street Tunnel was driven by an effort to keep construction costs down by reducing the depth of the access shaft and drop shafts. When the project was first being designed in the 1990s, the original CWS sewer tunnel was still functioning and the extent of the tunnel's deterioration along with the resultant voids in the surrounding ground were not yet known. In the fifteen years that passed between design of the project and commencement of construction, many factors had changed that would ultimately have an effect on the tunnel's design: a building with a pile foundation was installed along the alignment, a new wastewater tunnel was installed at the same general depth in the area, and additional time elapsed on existing historical structures. The realization of these unknowns led to the need for additional geotechnical investigation that was to be performed by the Contractor.

Leading up to the Contractor's mobilization, Cone Penetration Tests (CPT) were performed by the Contractor's geotechnical subconsultant in an effort to better determine the location of an existing, potentially unlined sewer tunnel that was abandoned by CWS in 2004 as previously noted. Existing documentation and an early investigation by the Owner, which was included in the Contract, provided an approximate location for the existing tunnel with a minimum design separation of approximately 10 feet. The results of the additional CPT testing led by the Contractor concluded the following:

- The existing sewer tunnel may be closer to the proposed tunnel alignment than originally anticipated
- The tunnel may not have been constructed in its exact design location (no as-builts were available for the tunnel)

- The primary tunnel excavation support, carrier pipe, and Cooper Marl failures along the tunnel could be more extensive than originally thought, and
- Subsurface failures may be progressively expanding outward.

Discussion among the owner, design team, and contractor centered on safety concerns during the Construction phase and long term structural and stability concerns of the proposed tunnel. Concurrent with Change Order discussions, the access shaft caisson was being sunk and a decision was quickly required so as to not impact or hold up progress.

## DESIGN CHANGES

As part of the Change Order package, the Contractor utilized their design subconsultant, Arup, to provide a design for the changed components. In general, the Change Order resulted in deepening of the access shaft and three drop shafts by 60 feet and reducing the tunnel grade.

While safety was the driving factor of the Change Order, the Project also benefitted in other aspects including a full diameter connection to an existing pump station (as opposed to a reduced-diameter connection to an existing tunnel) and a flatter grade that proved beneficial to both construction and maintenance.

The drop in alignment meant that a potential re-design in the 20 feet ID caisson shaft would be necessary, possibly requiring a thicker lining to deal with increased hoop forces. In order to avoid this, it was decided that the caisson would terminate approximately 20 feet above the proposed shaft invert and that traditional mining and temporary support utilizing steel ribs and liner plate lagging would be utilized below the caisson to invert level. Design checks were made to ensure that the caisson would "hang up" once undermined. These checks considered the side friction and bentonite slurry placed around the caisson annulus to assist in sinking. Using force equilibrium analyses, it was determined that the caisson needed to be supported by a bearing pad, excavated as a trench under the cutting shoe. This 3 feet wide × 1 foot 7 inch thick pad was designed to support the excess dead weight of the caisson not accounted for through side friction. The pad was reinforced with 12 lb/yd$^3$ of macrosynthetic fiber in order to provide adequate shear strength. Additional steel hooks were emplaced to provide reinforcement continuity between the pad and future cast-in-place final lining below the pad. Lastly, as this joint represented a "weak" point in the system, and possible source of water inflow/outflow, a PVC injection hose system was installed along the underside of the cutting shoe. Upon full cure of the bearing pad, the hose system was fully grouted with microfine cement to fill any voids and provide a strong bond between the pad and underside of the caisson.

The W14×48 steel rib and liner plate support installed below the bearing pad was complicated by the fact that three simultaneously open tunnels at 12 feet 6 inches. diameter each were required in order to meet the project schedule. While one tunnel was being lined by concrete following excavation, the TBM could be re-launched in the adjacent tunnel, which required a separate tail tunnel behind. Traditionally, large vertical beams are installed around either side of the portal, providing full load transfer around the opening through bending. These beams intrude on the final profile, and must be either removed or accommodated by a larger excavated diameter. In the case of the Market Street Shaft, diagonal bracing was designed around each opening to transfer the cut steel rib hoop thrust in compression to the adjacent ring beams—similar to a truss. Within the rib supported ground, an active condition was assumed to mobilize for the Cooper Marl based on the calculated amount of movement during each 4 ft. advance prior to rib installation. This was confirmed via convergence-confinement and longitudinal displacement profile calculations, similarly carried out for tunnels. A value of Ka=0.4 was ultimately adopted based on Peck, 1969.

By utilizing the bracing in compression only, the beam size was vastly reduced compared to the traditional vertical member scheme. The diagonal beams were sized to fit within the horizontal flanges of the ring beams and transferred their horizontal load component through welds. The vertical component was accommodated by a vertical member which either carried the load up to the underside of the bearing pad, or down to the invert mud slab. The scheme meant that no enlargement in shaft diameter was necessary, nor was removal of the temporary bracing prior to casting of the final lining.

The tunnel temporary support design was analyzed using a closed form convergence-confinement approach and validated using two dimensional finite difference code software. A relaxation factor of 55% was adopted based on the closed form solution and applied in the calculations. The stiffness of Cooper Marl was derived from SPT-N blow counts and dilatometer testing and assumed to be 350 ksf to a depth of 120 feet increasing to 475 ksf below this point. Upon completion of relaxation, a structural liner representing W4×13 steel ribs at 5 feet. spacing was installed and the model solved to equilibrium. The extracted thrust and moment forces were checked in an elastic interaction diagram which assumes the maximum supporting pressure of the circular rib (at zero blocking angle) is:

$$Ps, max = (A_s \times \sigma_y) / S \times r$$

Where

$A_s$ = area of steel
$\sigma_y$ = yield stress of steel
S = set spacing
r = radius of rib

The maximum moment at zero thrust is determined by the simple expression $S_y \times \sigma_y$, where $S_y$ is the section modulus of the rib. A second stage was added to the model in which the final 10 inch thick concrete lining was added and hydrostatic pressure applied to the gap in between the temporary and final lining. The extracted forces were then factored by 1.4 and verified in an interaction diagram. Further checks on the final lining design under various load combinations were made using a simple beam-spring model, in which the spring represents the passive reaction of the lining against the ground. The 10 in. unreinforced lining with f'c=4500psi satisfied all criteria, including a seismic check Hashash, 2002 which considered an Operating Basis Earthquake (OBE) and Maximum Design Earthquake (MDE) of 5.3 and 7.3, respectively. The induced strains from each case were significantly smaller than the limiting level of 0.003 established for design purposes. It is noted that 1.5 lb/yd$^3$ of microsynthetic fibers were included in the final lining concrete in order to reduce plastic shrinkage cracks during the first few hours of curing. Further curing measures were specified during the first 7 days after stripping, utilizing either wet or membrane curing compound methods. Stripping strength was set at a value of 600 psi based on a simple model loaded under dead weight only, which was typically achieved after 12 hours.

The final design hurdle was to assess the impact that the Concord Street Tunnel penetration on the existing Calhoun Street Shaft. The shaft was sunk by open end caisson in 1995 to a depth of 145 feet. The revised Concord Street Tunnel alignment was advantageous in that it breaks into the shaft a few feet above invert level. To assess the structural impact and potential need for retrofit, a 3D structural beam-spring model using computer software was developed. The applied loading on the existing shaft was assumed from the original design drawings, which gave an average uniform lateral earth pressure (as an equivalent fluid density). Two openings were modeled in the 30 feet. ID shaft, a circular one for the new Concord Street Tunnel, as well as a square one for the existing 8 feet diameter circular tunnel, which was constructed within a 12 feet. square "break-out" opening eye in the caisson wall. Included in the structural model was an internal baffle structure composed of 8in. thick walls, fortuitously positioned in between the two tunnels. The result of the modeling showed that the internal restraint of the baffle walls provided significant rigidity and support to the stressed wall in between the two openings, limiting the bending stresses which would have been excessive had the wall not been where it was. The cut in the shaft wall was made 8in. larger than the tunnel radius so that a reinforced circular collar could be constructed around the opening using #5 bars at 6in. spacing. Continuity between the new tunnel lining and existing shaft wall was made by drilling and grouting in #5 standard hooks and extending these into the tunnel lining some distance away from the tie-in. This "closure pour" between the completed tunnel unreinforced lining and the shaft wall was set at a distance of 8 feet. 6 inches based on the structural 3D model. As in the bearing pad connection joint, a PVC injection hose system was installed in the center of the tie-in connection joint to ensure good bond and water tightness at the junction.

## LESSONS LEARNED

The project was initially designed to be excavated and supported by sequential excavation methods (SEM) as a result of the relatively short tunnel lengths and absence of a retrieval shaft. However, prior to bidding an alternative option to utilize a TBM with steel ribs and timber lagging as temporary support was added to the design. If chosen the TBM would have to be dismantled and pulled back to the access shaft within the tunnel for retrieval purposes. This bid option that allowed the contractor to provide an alternative plan supported and designed by the contractor's consultant proved beneficial to the owner in this case. Of the eight bids submitted on the project four opted for the TBM option and four for the SEM option, however, all four TBM bids came in below the SEM bids. In addition the lowest bid submitted was significantly below the second lowest bid and below the engineer's estimate. The contractor's methods and execution utilizing the TBM option with steel rib and timber lagging allowed the project to finish a few months early excepting the delays caused by the change order negotiation.

In addition, the elimination of the originally planned 2.4% slope on the Concord tunnel provided a safer, more efficient working environment. The change of the outlet connection from the Calhoun Tunnel to the pump station wetwell is a more efficient hydraulic design and provides easier maintenance access for the owner.

As the Change Order was discussed and executed, it became apparent that the lengthy delay between initial design and bidding resulted in some inherent problems for the project. For similar scenarios on future projects, the authors would suggest an additional period of time be provided to the design and construction management team prior to the bidding period to provide adequate time for a more thorough review of the Project's design and its relation to changes in industry standards and locality conditions. During the years elapsed from design completion to the start of construction numerous advancements in the tunneling industry including TBMs becoming more cost effective occurred. In addition the project site settings, including the extreme deterioration of CWS's wastewater tunnel, which was in close proximity to the proposed Market Street Tunnel, changes in land ownership, and new a high rise condominium being built along the alignment, may have dictated some upfront changes to the contract.

In hind-sight leaving the task of pinpointing the exact location of the abandoned CWS tunnel to the Contractor after contract award may have been rethought. In general it has become standard practice of care in the underground industry to place the risk on the party best able to handle it, which in the case of completing additional geotechnical investigations to locate an existing tunnel along the tunnel alignment would likely fall to the Owner and their Engineer. While the ultimate solution resulting in the change order solved the problem, the Client may have been able to save money in the long run if they had engaged the designer to conduct additional studies prior to finalizing the contract documents so that the project would have been bid based on the eventual tunnel depth. However, each project and owner has unique drivers and constraints to negotiate during a projects life-cycle with regards to budgetary issues, political considerations, and social impacts. In most cases the owners understand the benefits of proceeding with design and construction in a linear fashion without major delays and with properly investigating and mitigating potential risk to the project as early as possible, however, their unique situations often dictate a course of action which is in conflict with industry standards. It is the duty of the engineer to properly inform and educate owners with regards to project risk and to suggest proper mitigation measures over the life of a project.

## ACKNOWLEDGMENTS

The authors wish to thank the City of Charleston for their permission to publish this paper and in whose work provided the basis for this paper.

## REFERENCES

Cabiness, L.S., Kirk, S.A., O'Connell, S.A., and Swartz, J.T.: *Sustainable Underground Solutions for an Above Ground Problem*, North America Tunneling Conference, June 2010, Seattle, Washington.

Hashash, Y.M.A and Park D., "*Viscous Damping Formulation and High-Frequency Components in Deep Deposits,*" Soil Dynamics and Earthquake Engineering, V.22 (7), pp. 611–624.

Heuer, R., *"Important Ground Parameters in Soft Ground Tunnels,*" Proceedings of Engineering Foundation Conference on Subsurface Exploration for Underground Excavation and Heavy Construction, August 1974, Henniker, NH, ASCE.

Peck, R.B., *"Deep Excavations and Tunnels in Soft Ground,"* Proceedings of the 7th International Conference on Soil Mechanics and Foundation Engineering, Mexico City, State of the Art Volume, pp. 225–290.

Swartz, J., Cabiness, L., O'Connell, S., and Kirk, S., *Below these Crowded Streets: Charleston's Network of Underground Infrastructure*, APWA Reporter, June 2010.

Swartz, J., Drolet, L., Cabiness, L., O'Connell, S., *Boring Old Charleston—Tunneling and Trenchless Technology Key to Charleston's Infrastructure*, North American Tunneling Conference, June 2012, Indianapolis, Indiana.

# Challenges of Fully Assembled TBM Back-up Through Completed Tunnel Bore

**E.C. Wang**
Jacobs Engineering Group, Inc.

**J. Velez**
Judlau/OHL

**E. Jordan**
Hatch Mott MacDonald

**ABSTRACT:** To expedite the excavation schedule of the Metropolitan Transportation Authority Capital Construction (MTACC)'s East Side Access (ESA) Project (Contracts CM009 and CM019), a plan was developed to prepare a newly-mined tunnel for retrieval of a fully-assembled Tunnel Boring Machine (TBM) thereby avoiding dis-assembly/re-assembly delays. Engineering analyses and constructability evaluations were performed to assess and address challenges of pulling back the double-shield TBM for subsequent re-launch from the newly-excavated wye cavern. Following a comprehensive geotechnical evaluation and risk assessment, measures including replacement of the existing initial support with equivalent minimal profile support systems and enlargement/pre-support of the TBM bore within adverse ground conditions were implemented. This paper presents lessons learned.

## BACKGROUND

The East Side Access (ESA) project provides a critical rail link for Long Island and eastern Queens commuters traveling to Manhattan's east side with a connection into a new LIRR terminal beneath Grand Central Terminal (GCT). See Figure 1.

Two contracts facilitated eight tunnel drives, six wye caverns, two main station caverns, several cross-overs and cross-passages, as well as nine shafts, both vertical and inclined. The planning of an efficient excavation sequence that included advancing multiple headings simultaneously by tunnel boring machines (TBMs), roadheaders, and conventional drill and blast methods, was critical to maintaining construction schedule.

## UNIQUE CHALLENGES

Two contracts, CM-009 and CM-019, involved mining a series of upper and lower level eastbound TBM tunnels and caverns under Midtown Manhattan. The excavation sequence created challenges to the construction schedule, one of which included optimizing TBM back-up operations through a completed tunnel bore (Figure 2). Specifically, the initial Eastbound-4 (EB4) bore that was comprised of five tunnel zones with existing steel rib supported sections (Figure 3), would have required lengthy delays associated with the disassembly of a double-shield TBM to accommodate the "backing up" for a subsequent re-launch. This daunting task, completed between late 2009 and early 2010, presented several challenges in terms of excavation sequence, TBM limitations, overlying critical infrastructure, as well as difficult ground conditions. In addition, several value engineering initiatives were incorporated. Extensive coordination and planning, including workshop sessions were initiated with the Contractor, the construction manager (PM-CM), and the chief tunnel designer (GEC) staff and included several site visits to evaluate the feasibility of time-saving alternatives.

### Excavation Sequence

The key steps for EB-4 construction of the combined CM-009 and CM-019 contracts are summarized below:

- Commencing from an existing concrete bulkhead under 63rd Street and Second Ave, a TBM launch chamber was created by controlled drill-&-blast excavation and pre-cast gripper walls within the south terminus of the existing two-track lower level 63rd Street tunnel box.

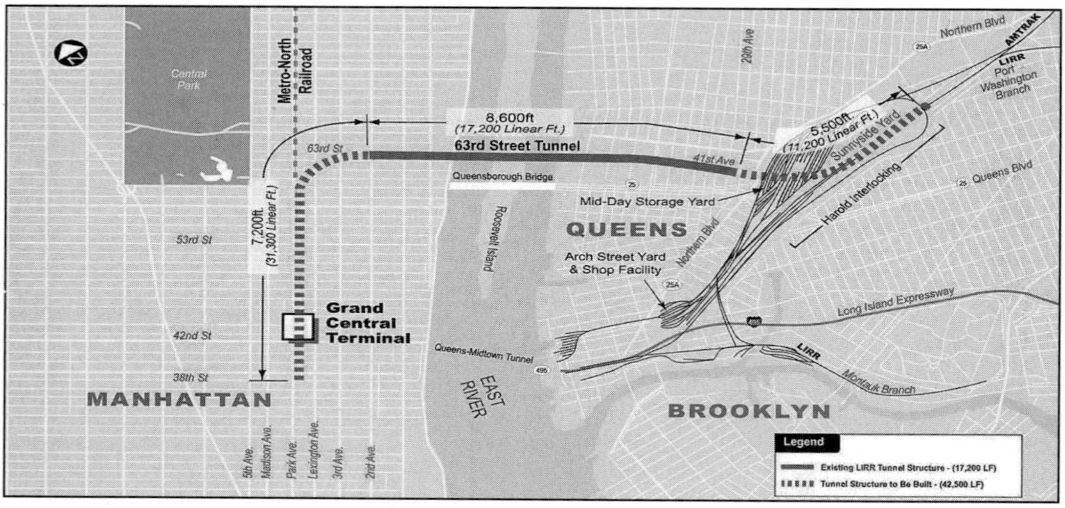


**Figure 1. East Side Access project site plan**

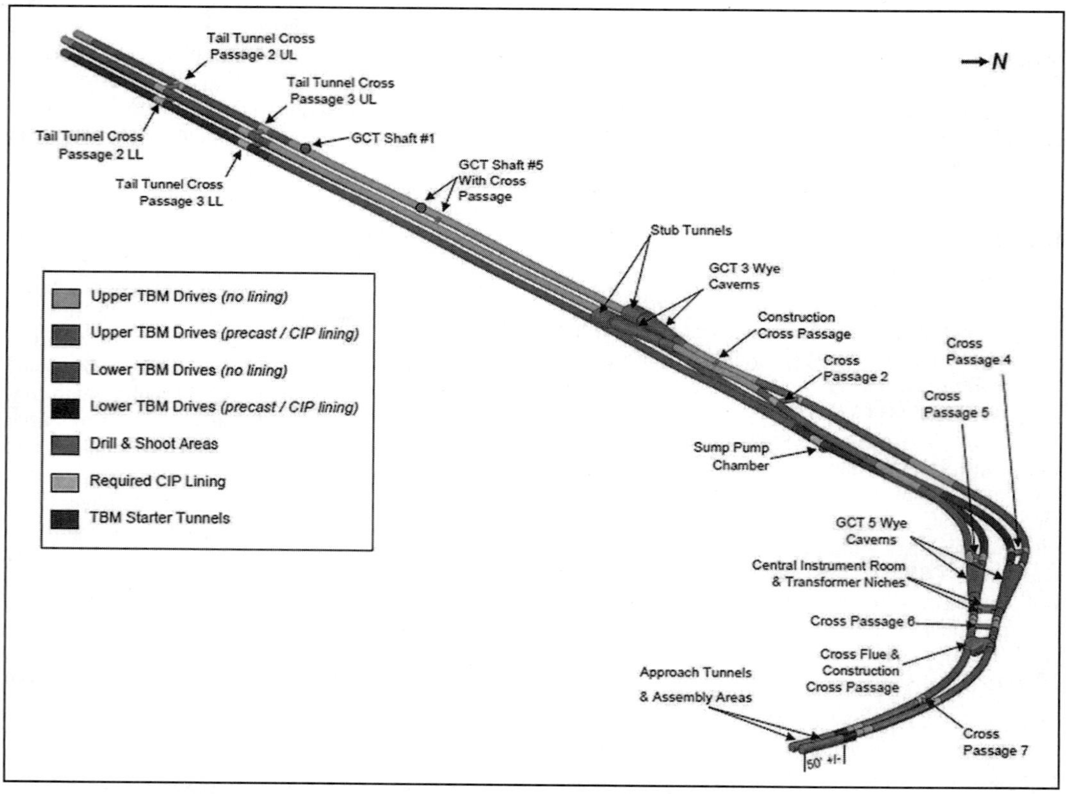


**Figure 2. General configuration of combined Contracts CM-009 and CM-019**

- TBM Tunnel drive 7,400 linear ft. (2,256 m) (initial—upper EB tunnel from 63rd Street to 38th Street)
- Drill-&-Blast Construction Cross-over (upper cavern at GCT 3 from 50th to 52nd Streets/Park Avenue)
- Pull back SELI TBM 5,200 ft (1,585 m) length through initial completed EB bore.
- Partially backfilled the initial bore within the GCT 5 EB wye cavern limits with a cellular concrete plug to expedite TBM re-launch.
- Re-launch TBM from GCT 5 EB Wye.
- TBM Tunnel drive 5,200 linear ft. (1,585 m) (second—lower EB tunnel from 59th Street & Lexington Avenue to 38th Street)

Following the TBM re-launch, the initial bore was enlarged to a 465-foot (142 m) long Wye Cavern from Station EB4-1069+50 to EB4-1064+85, by excavating and supporting the ground to final GCT5 EB Wye Cavern limits.

**Figure 3. Full circumference ring sets to be removed for TBM back-up**

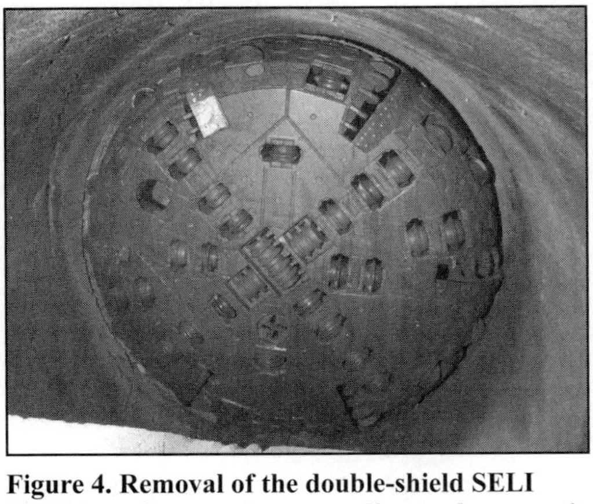

**Figure 4. Removal of the double-shield SELI TBM (no tolerance for protruding rock support)**

The SELI-TBM was to be pulled back through several areas with restrictive profile rock support, including mesh, mine straps, steel channels and full circumference ring sets. The steel ring set sections (5) and surrounding ground, presented the greatest challenge for retrieving the double-shield TBM. The existing steel rib supported sections, included a total of 349 linear feet (106-m) along the EB4 alignment, were located at the following five areas:

- Set #1—"Cathedral Void" set. 19 full-circumference ring sets on 5-foot (1.5-m) centers for 75 feet (23 m), from Station EB4-1062+13 to EB4-1061+38
- Set #2—'Intermediate shear zone" set. 15 ring sets on 5-foot (1.5-m) centers for 70 feet (21 m), from Station EB4-1064+93 to EB4-1064+23
- Set #3—"Christmas Shear Zone" set. 13 ring sets on four to five-foot (1.2 to 1.5-m) centers for 42 feet (13 m), from Station EB4-1067+09 to EB4-1066+49.
- Set #4—Curve set. A total of 15 ring sets for 65 feet (20 m), from Station EB4-1057+71 to EB4-1057+6 and 1058+57 to 1058+07 (15 ea.) steel rib sets.
- Set #5—53rd Street Subway Crossing set. 16 ring sets on four to five-foot (1.2 to 1.5-m) centers for 77 feet (23.5-m), from Station EB4-1050+89 to EB4-1050+12.

### TBM Limitations

The SELI double-shield TBM provided flexibility for mining through various ground conditions ranging from hard rock (gripper mode) to localized poor rock such as shear zones (shield). However, as shown in Figure 4, the cutter head was not designed for partial disassembly to facilitate retrieval through the rib-supported tunnel bore.

### Urban Environment and Overlying Infrastructure

Mining beneath Midtown Manhattan is a challenging endeavor as this environment features skyscrapers, subway lines, and subsurface utilities. As such, addressing ground control risk is critically important.

For example, the Rib Set #5 area extending from Station EB4-1050+89 to EB4-1050+12 was located below the existing New York City Transit IND 53rd Street subway tunnel crossing. Similar to the other four rib set areas, a site reconnaissance was conducted to re-evaluate the actual ground conditions.

### Adverse Ground Conditions

The SELI-TBM initially encountered a variety of ground conditions and rock types. The most adverse ground conditions were encountered at locations of major fault shear zones, some of which consisted of intersecting fault zones.

The initial EB tunnel was mined in the Manhattan Formation, a metamorphic rock formation of Lower Cambrian to Middle Ordovician age. The Formation was subjected to several tectonic events that resulted in folded and faulted ground.

The tunnel drive presented a variety of geotechnical challenges ranging from excellent quality rock to very poor quality with very short stand-up times. Most of the rock formation consisted of Mica Schistose Gneiss. Rock strength ranged from weak (UCS, 4,000 psi [30 MPa] to very strong (UCS 36,000 psi [250 MPa]).

The formation along the tunnel drive included minor and major folds and faults, some of which had measurable thicknesses up to 100 feet (30.5 m) of tunnel length.

While the majority of the tunnel drive was mined in what was considered very good quality rock, there were several localized areas of fair quality rock and a few lengthy zones with poor quality rock. These poor quality rock zones presented the greatest challenges in terms of tunnel arch support condition, rock quality and corresponding alternate rock support design. Moreover, the greater challenge, arguably, was implementing the alternative design and installing the alternate support in a timely manner to accommodate the assembled TBM pull-back, as well as enlarging the tunnel to a large wye cavern in poor quality rock.

## ENGINEERING EVALUATION

The systematic approach to assessing the feasibility of replacing existing steel rib support along initial EB4 drive included: (1) initial site visit by Contractor, GEC and PM/CM staff to field verify actual ground conditions and evaluate support requirements; (2) workshop attended by Contractor, GEC, and PM-CM staff to evaluate site findings and identify hazards/ risks, as well as develop engineered methodology for alternative rock support; and (3) develop risk register.

### Feasibility Study: Schedule Impact Evaluation

The schedule evaluation focused on the two critical path activities: TBM dis-assembly and re-assembly, and cavern development for TBM re-launch. A minimum delay of eight months was estimated for TBM dis-assembly and re-assembly considering the required enlargement of the GCT 3EB and GCT5 EB wye caverns to accommodate dis-assembly and re-assembly, respectively. Furthermore, considering that this option precluded use of more cost-effective TBM mining for approximately 60% of the overall wye cavern rock excavation than conventional drill-and-blast, a better solution was virtually mandated.

### Geotechnical Assessment (Competent Ground to Fault Shear Zone)

The geotechnical characteristics of the SELI-mined tunnel presented a number of concerns in terms of rock quality, existing rock support, and trends of fault zones. The contract for the SELI-TBM mined tunnel did not classify the rock per se, rather it identified or segregated the tunnel drive by a range of values for certain geotechnical parameters and corresponding rock support classes (SCI, SCII, or SCIII).

A contract Geotechnical Baseline Report (GBR) described locations with poor quality rock in terms of the Rock Quality designation (RQD), Q System, and Rock Mass Rating (RMR) System and baselined tunnel lengths with a range of anticipated geotechnical parameter values and where certain rock support schemes or classes could be expected, which was based on findings from a typical pre-construction exploration phase of the ESA project. In order to supplement and verify anticipated conditions in terms of rock quality described in the GBR, a full-time team of geotechnical engineers and geologists prepared geological and geotechnical maps with a 'map as you mine' approach as shown in Figure 5.

The collected mapping data was logged, tabulated and interpreted, in terms of RQD and the Q-system. RMR values were calculated using the following equation (Barton, 1995) that was provided in the contract geotechnical document.

$$RMR = 15 \mathrm{Log}\, Q + 50 \qquad (1)$$

In addition to geotechnical mapping, traditional geological structure mapping was performed based on trends of dominant features, like fault zones. Projection mapping assisted with developing an understanding of the nature and trend of fault zones (i.e., geologic structures) to better determine the effectiveness of alternative rock support designs.

Both the GBR and subsequent mapping indicated rock quality along the tunnel drive included a full range of conditions in terms of rock quality, RQD and Q values. Although most of the tunnel drive was in very good quality rock, certain lengthy locations, included major fault shear zones that exhibited very short stand-up times, including arch failure and deformation when the tunnel was initially mined. The range of project geotechnical analytical results is provided in Table 1.

### Risk Assessment (Schedule & Shallow Cover Considerations)

In preparation for the upcoming SELI TBM back-up, a site reconnaissance was conducted by DJ/JV in

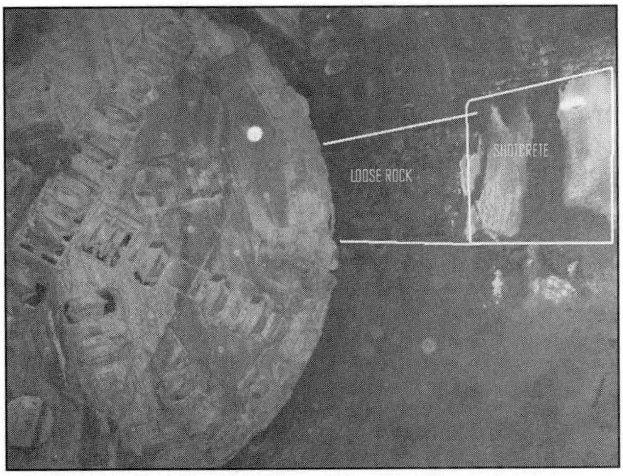


**Figure 5. SELI TBM with location of poor quality rock at springline identified**

**Table 1. Summary of rock properties, ESA, GBR, March 2006**

| | Most Probable Range of Values | | Minimum and Maximum Test Values | |
|---|---|---|---|---|
| **Property** | **Failure Type** | **Most Probable Range** | **Minimum Test Value** | **Maximum Test Value** |
| | | **Density and Strength Properties** | | |
| Density (air-dried) | | 170–1,180 pcf | 158 pct | 184 pcf |
| Uniaxial Compressive | Structural failure | 4,000–16,000 psi | 2751 psi | 19,686 psi |
| Strength (UCS) | Non-structural failure | 7,000–22,000 psi | 6,540 psi | 28,177 psi |
| Brazilian Tensile | Structural failure | 600–1,700 psi | 490 psi | 1,764 psi |
| Strength (BTS) | Non-structural failure | 800–2,300 psi | 357 psi | 2,550 psi |
| Point Load Strength | Structural failure | 100–750 psi | 71 psi | 1,242 psi |
| Index Failure (PLSI) | Non-structural | 150–1,280 psi | 64 psi | 1,281 psi |

conjunction with the CM and GEC (Figure 6). The purpose of this site visit was to inspect and evaluate the ground conditions behind the rib-supported areas (Set #1 through #5) located between GCT 3 and GCT 5. The field observation data collected was subsequently discussed during a September 2, 2009 workshop attended by Contractor, GEC, and PM/CM staff. Site findings and risks associated with the Contractor's proposed rib-replacement support systems were identified during this brainstorming session with involved parties. This transparent and collaborative approach was beneficial in both identifying potential hazards as well as vetting proposed solutions. The resulting Risk Register which was developed listed risks/hazards associated with each of the Contractor's proposed replacement support schemes at each of the five existing rib-supported areas (Sets #1 through #5). The Risk Register also provided a framework for evaluating the Contractor's Excavation and Support submittal. The following general notes included in the SELI TBM back-up risk assessment demonstrate the level of coordination in addressing ground control concerns:

- Continuity is needed from shift to shift.
- Survey—TBM clearance must be ensured, install convergence monitoring points, as directed by Construction Manager.
- Contingency plan is needed for re-installing sets, in case one is needed as supplemental support.
- Post back-up contingency is needed for monitoring and supplemental support installation.
- Install rock surface protection as required or as directed by Construction Manager, once the TBM passes.
- Maintain supplies of material and equipment at the immediate work site.
- Designate full-time experienced Project Superintendents and qualified Contractor's Geotechnical Engineers to be in direct control of the work, in accordance with section 02407.
- Submit a Safe Work Plan for these specific activities, in accordance with section 01540-3.04A.
- Provide organization chart showing shift assignments for the Contractor's personnel in direct control of the work.

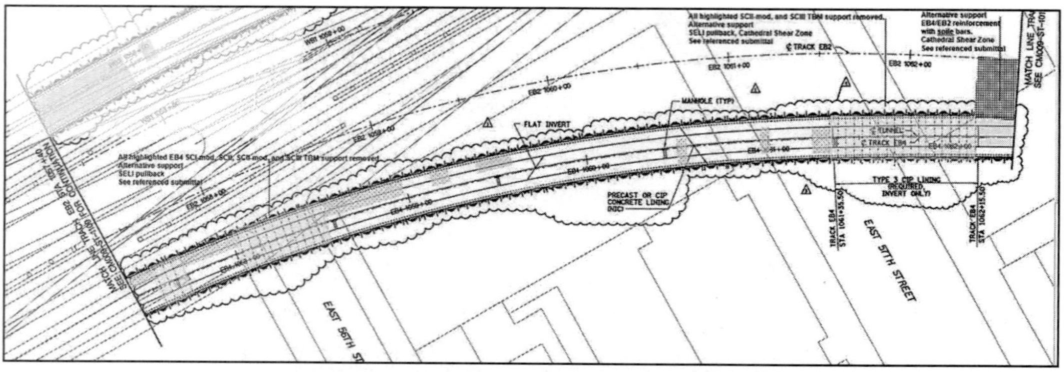


**Figure 6. Eastbound SELI-mined, tunnel drive with highlighted areas of concern. Yellow-hatched locations depict zones with full circumference rings (removed).**

The comprehensive Risk Register provided detailed identification of risks and mitigation measures associated with each area including the following headings under Specific Notes:

- Set/Description,
- Rock Support,
- Proposed Support,
- Proposed Work Steps (as described in DJ/JV's submittal),
- Risks/ Hazards,
- Mitigation, and finally
- DJ/JV's Comments

The risk matrix jointly developed by the Contractor, GEC, and PM/ CM staff helped to clarify the critical issues to be addressed. The risk matrix development process also demonstrated a sound understanding of the Contractor's proposed means and methods for replacing existing steel ribs with an alternative ground support system which would allow the fully-assembled TBM to back-up.

### Constructability

Once the associated risks and hazards were identified and mitigation strategies were developed, detailed Construction Work Plans (CWP) for each major operation were prepared by the Contractor. Specifically, the CWP for Rib Removal for SELI TBM Backup identified the five areas where existing steel rib removal was required including the observed field conditions and replacement support and testing requirements. Figure 7 shows a critical location within Rib Set #1 "Cathedral Fallout Zone" where a Self-Drilling Anchor (MAI bolt) system was selected. A decision-making flowchart (Figure 8) was developed to accommodate critical bolting and rib-removal activities associated with this area. As shown in Figure 7, an initial larger diameter countersunk hole was cored through the existing lagging/mesh/shotcrete. The bolt orientation was selected based on geologic mapping data prepared by CM's Senior Project Geologist. During drilling of the SDA MAI bolt, a DJ/JV geologic engineer performed detailed logging of the varying penetration rates between grout and host rock to ensure a minimum 10-ft embedment into sound rock was achieved per design.

From the counter-sunk opening, drilling of the 1.25-inch (32 mm) diameter MAI bolt was extended through the cellular grout and into a minimum of 10-ft (3-m) of sound rock with continuous grouting to obtain a fully-encapsulated friction bolt.

Figure 8 outlines the key steps involved in comprehensive contact grouting procedure (Figure 9) which was prepared to address proof drilling concerns to verify location of and depth to competent rock for pressure grouting in order to fully backfill any existing voids.

## PLANNING AND IMPLEMENTATION OF PREPARATORY MEASURES

As previously mentioned, a series of integrated Construction Work Plans were developed by DJ/JV which considered the overall operation objectives. Specifically, these plans included individual work components such as the sequence of preparatory steps including relocation of existing conveyor and ventilation systems to permit efficient excavation and support of existing rib supported areas.

For example at the GCT5EB Wye, preparatory measures included installation of support for the future wye cavern enlargement prior to pouring the concrete plug due to drilling access constraints. Similarly, removal of existing ribs located within the future narrow pillar area was necessary to allow TBM back-up through the initial bore, Subsequent re-installation of steel ribs to be combined with concrete backfill plug required placement of a PVC bond breaker membrane covering prior to commencing the

**Figure 7. MAI bolts installed through cellular grout into minimum 10-ft (3-m) sound bedrock**

concrete plug pour to facilitate post-pour stripping without damaging the ribs. Furthermore, the associated relocation of ventilation and conveyor components as well as construction of temporary bulkheads to direct the flow of fresh air to and block construction debris exhaust air from active work zones was mandatory.

## Alternate Ground Support

Alternative rock support was designed for a total of 59 ring sets and approximately 360 feet along the tunnel drive.

The existing tunnel conditions were highly variable due to adverse geology. Procedures for replacement, alternative support that were adopted, included the following.

For Set #1 area where existing overbreak in crown and arch sustained during TBM mining had been initially stabilized with cellular grout and timber lagging as shown in Figure 7, a ring replacement bolting scheme was developed. Specifically, the rock support system was divided into three distinct pre-support zones as follows: (a) Ribs #1 through #3 which had been determined to be a "Transition zone" corresponding to Support Class II (SCII) requiring (7 ea.) 12-ft (3.65-m) Swellex PM24 extending over the upper 120° of the arch; (b) "Main Fall-out zone" extending from Ribs #4 to #12 per Dr. Sauer Corporation's approved design consisted of Swellex PM24 combined with a Self-Drilling Anchor system, MAI bolts drilled through the existing cellular grout backfill and embedded a minimum of 10-ft (3-m) into sound competent rock above pre-existing backfilled voids and/or weak rock; and (c) A SCII "Transition zone" supported by (7 ea.) 12-ft (3.65-m) long Swellex PM24 bolts. To monitor potential ground movement within this area, instrumentation including convergence points and monitoring targets were installed between the ribs at the crown and springline locations prior to commencing any drilling activities.

Upon completion of bolt installation, all bolts were pull-tested to 110% of design load corresponding to 30 kips (132 kN). Replacement bolts were installed for any failed bolt test locations.

After the TBM was backed up through the areas where rings were removed, the Contractor and CM field verified actual ground condition, evaluated support requirements, and instrumentation data. Preparatory activities included scaling to remove any loose rock present until sound competent rock encountered. Any voids detected which needed to be backfilled prior to TBM backup, were completely backfilled with shotcrete to accommodate TBM retrieval. For remaining void-free areas, any additional surface protection (flashcrete, mine straps, and mesh) per agreement between Contractor and CM and ground support was installed.

For Set #2 located at a future narrow pillar of GCT 5 EB bifurcation, the existing 15 ea. steel ribs were within the proposed concrete plug limits and would need to be removed to facilitate TBM relaunch through. Based on mapping and site visits to verify actual conditions, the ground conditions were reclassified as SCI- SCII. The proposed replacement support system consisted of six (6 ea.) PM24, 10-foot long Swellex bolts within the upper 120° of tunnel arch installed at 6-ft (1.8-m) longitudinal spacing. To address stability concerns of the narrow pillar DJ/JV proposed reduced gripper pressures combined with the concrete plug placement. Contingency measures such as pillar through bolts (i.e., tie-rods) were also provided in the updated GCT 5 EB wye cavern design drawings prepared by the GEC.

Set #3 known as "Christmas Shear Zone" included 13 existing ribs and the initial support was re-classified based on mapping and site visits

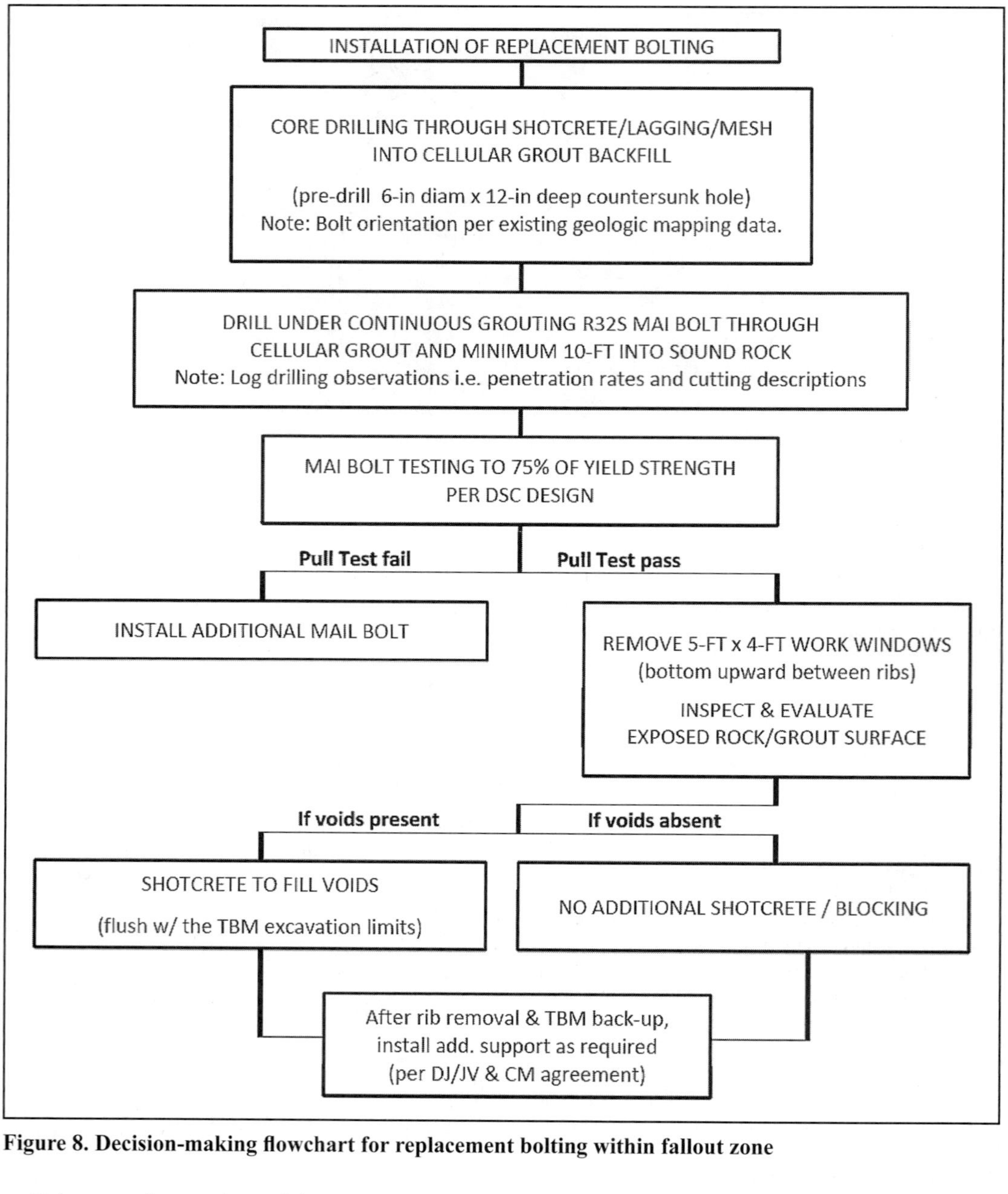


**Figure 8. Decision-making flowchart for replacement bolting within fallout zone**

verifying actual ground conditions. Revised support classes were as follows: (a) Ribs #1 to #3 re-evaluated as "transition zone" corresponding to SCII. Similar to Set #1 transition zone, rib replacement support included (7 ea.) 12-ft (3.65-m) Swellex PM24 extending over the upper 120° of the arch; (b) Ribs #4 to #9 (Shear Zone). Per the approved design prepared by the contractor's designer (ILF), where excavation to final wye cavern limits was necessary, the design called for spiling, installed at 1-ft (0.3-m) center-to-center spacing over the steel rings. The enlargement cycle and sequence of replacement support included removal of one ring at a time, apply 2-inch (50-mm) flashcrete plus install (7 ea.) 12-ft (3.65-m) long Dywidag dowels extending over upper 120° of tunnel arch covered with flashcrete plus 6-inch (150-mm) shotcrete layer; and (c) Ribs #10 to #13 were re-evaluated as a "transition zone" corresponding to SCII support, with (7 ea.) 12-ft

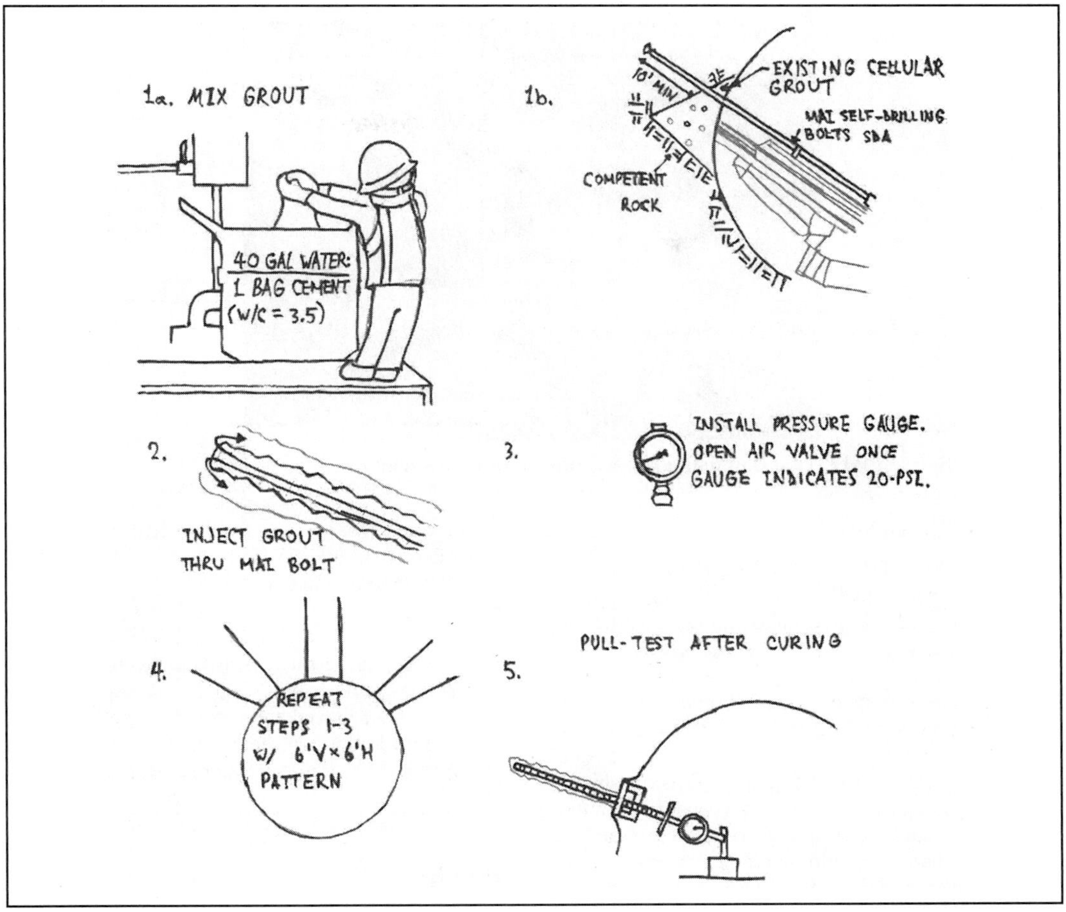


**Figure 9. Contact grouting steps**

(3.65-m) Swellex PM24 extending over the upper 120° of the arch.

The initial support (12-ft (3.65-m) Dywidag on 6-ft (1.8-m) radial by 5-ft (1.5-m) longitudinal spacing on a staggered pattern combined with shotcrete) within the shear zone is shown in Figure 10.

For Set #4A (Sta. EB4-1058+57 to EB4-1058+07) the existing 11 ea. steel ribs were replaced by a pre-support bolting system based on mapping and field inspection that was performed by DJ/JV and CM geotechnical staff which led to re-classification of actual ground conditions from SCIII to Support Class I/II. The corresponding pre-support system included installing (4 ea.) 10-ft (3-m) Swellex PM24 spaced 6-ft (1.8-m) longitudinally adjacent to existing rib immediately after last. Contingency measures consisted of installing grouted self-drilling anchors MAI bolts as required by actual ground conditions encountered during bolting and rib removal. A total of 3% of installed bolts were subjected to pull-testing to 80% of yield strength.

For Set #4B (Sta. EB4-1057+71 to EB4-1057+56) consisting of 4 ea. existing ribs, based on re-evaluation of ground conditions based on mapping and observation during site visit to inspect actual ground conditions, re-classification of support class to SCI/II was determined. A similar pre-support bolting system, testing and contingency approach was used for Set #4A.

The fifth set of rib supported area was located below the 53rd Street crossing and extended from Station EB4-1050+89 to EB4-1050+12. This is an area where massive and competent bedrock is present, but cover is shallow. The pre-support bolting system consisting of (7 ea.) = 10-ft (3-m) Swellex within the upper 120° of the arch. After the TBM back-up cleared this area, re-installation of ribs was required to account for risks and hazards associated with shallow cover conditions and the overlying subway crossing.

**Figure 10. Sequential excavation using pre-support spiling in shear zone**

## LESSONS LEARNED

Fully-assembled TBM pullbacks are possible even when poor rock quality requiring steel supports under limited access and challenging construction conditions exist.

- Rock quality concerns can be overcome through sound technical oversight, geological mapping, well conceived alternate rock support designs, and quality craftsmanship.
- Logistical issues such as preparatory measures and construction sequencing can be resolved by careful planning and comprehensive preparation of construction work plans.
- Use of a risk register with input and active participation from all parties involved (Contractor, PM/CM and GEC) is a beneficial tool for identifying potential hazards/risks and developing effective mitigation strategies.

## ACKNOWLEDGMENTS

The authors gratefully acknowledge the Metropolitan Transportation Authority Capital Construction (MTACC) for permission to present this paper. Additionally, the authors wish to acknowledge the Dragados-Judlau joint venture (Contractor), URS (PMC), Hatch Mott MacDonald (Tunnel Construction Manager), Jacobs-LiRo (Consultant Construction Manager) and Parsons Brinckerhoff (General Engineering Consultant) as well as the contractor's design consultants Dr. Sauer Corporation and ILF Consultants for their support and contributions during this critical phase of the project. The authors wish to thank Michael Goldemberg and Kevin Tse for editorial and graphic support.

The information presented in this paper is based on observations, opinion, interpretation and knowledge of the authors. It may not be representative their companies, representatives, or subsidiaries.

## REFERENCES

Barton, N. (1995), The Influence of joint properties in modeling jointed rock masses. Keynote Lecture, Eighth International Society for Rock Mechanics Congress of ISRM, Tokyo, Japan, Vol. 3:1,023–1,032. Rotterdam, Netherlands: Balkema.

Dr. Sauer Corporation, CM009 Manhattan Tunnels—East Side Access Project, New York, Support Class III Replacement for TBM Pull Back—Calculation No. 405DC020A-00 dated November 21, 2009.

ILF Consultants, Inc., East Side Access Project—GCT 5 Wye Cavern Finite Element Analysis Section III dated October 08, 2008.

# Design for the Ballard Siphon Replacement Project

**Russell Berends, Gregg Sherry, and Mohamed Gamal**
Brierley Associates

**John Fowler**
James W. Fowler Company

**ABSTRACT:** The purpose of the Ballard Siphon Replacement Project is to upgrade the aged wooden sewer lines under the Salmon Bay in Seattle, WA. A new 2.85-m-inner-diameter (9.4 ft), 604-m-long (1,980 ft) tunnel was constructed approximately 18 m (60 ft) underneath the bay bottom using EPB TBM methods. The 9.0-m-inner-diameter (29.5 ft), 45.4-m (149 ft) excavated depth launching shaft was excavated in the wet using a vertical shaft machine, which lowers the shaft liner from the top while excavating using roadheader methods and removing the cuttings in a slurry, the first time this technology has been used in North America. This paper discusses the design and construction of the temporary support for the launching shaft and the tunnel.

## INTRODUCTION AND BACKGROUND

In 2011, the James W. Fowler Company began work on the Ballard Siphon Replacement Project in Seattle, Washington. Part of the system that transports wastewater and overflow from North Seattle beneath the Salmon Bay to the West Point Treatment Plant at Discovery Park consists of two 0.91-m-diameter (36 in) wood stave pipes that lay approximately 6 m (20 ft) below the bottom of the bay and were constructed circa 1935. The Ballard Siphon Replacement Project was initiated to increase the capacity of the CSO (combined sewer overflow) system by sliplining the existing, aged siphon and constructing a new siphon consisting of two vertical access shafts and a 2.85-m-inner-diameter (9.4 ft) tunnel. Brierley Associates designed the temporary support for the launching shaft and tunnel for the general contractor, James W. Fowler Company. King County is the owner of the new wastewater siphon structures, and the project was designed by Tetra Tech, Landau Associates and Staheli Trenchless Consultants. This paper discusses the design and construction of the tunnel and the south shaft of the new siphon.

The tunnel for the new siphon runs approximately 18 m (60 ft) below the bottom of the Salmon Bay, roughly 12 m (40 ft) below the elevation of the existing wood stave pipe siphon, for a length 604 m (1,980 ft). A site plan view is shown in Figure 1. The tunnel was constructed using an EPB (earth pressure balance) TBM (tunnel boring machine) to resist approximately 35 m (116 ft) of water pressure near the shaft and 29 m (95 ft) of water pressure underneath the Salmon Bay. The TBM was launched from the shaft on the south side of the bay.

The specifications of the launching shaft allowed for three possible methods for its construction: (1) slurry diaphragm walls, (2) caisson methods, or (3) ground freezing with a structural liner. Due to delays in the early phases of construction and scheduling issues, the contractor ruled out the use of slurry diaphragm walls. Additionally, the particular groundwater and geologic conditions at the project site were not favorable for ground freezing, eliminating that option. The third and final option, traditional caissons, typically would not be acceptable for such a deep shaft because of the increased risk of the caisson lodging in the excavation from the lateral soil pressure on the caisson and the issue of maintaining proper caisson alignment. To solve these problems, the contractor chose to use a vertical shaft machine (VSM)—the first time this method has been used in North America. The VSM excavation method is similar to traditional caisson methods in that a concrete lining is constructed from the surface and sunk into the excavation; however, the VSM method, developed by Herrenknecht, is different in that the precast concrete liner is hung from cables and lowered in a controlled manner, alleviating the alignment issue, and a roadheader that can excavate ground below the concrete liner is setup at the bottom of the shaft liner, solving the lodging issue.

## SUBSURFACE CONDITIONS

Much of Seattle's topography and geology has been shaped by glacial action, and the majority of the material encountered while excavating the siphon

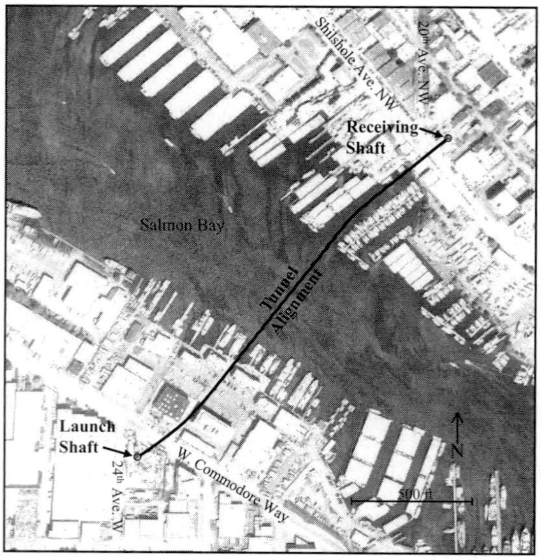


**Figure 1. Project site plan**

shafts and tunnel consisted of glacial deposits. Fill material comprises the surficial 2.4 m (8 ft) at the launch shaft. The fill is underlain by approximately 3 m (10 ft) of Vashon advance outwash–alluvial deposits from glacial streams consisting of sand with silt, gravel, and cobbles. The Vashon advance outwash is underlain by approximately 21 m (70 ft) of Pre-Fraser interglacial deposits consisting of very stiff to hard silt and clay interbedded with medium dense to very dense sand. The material encountered in the bottom 19 m (61 ft) of the launch shaft excavation is Pre-Fraser slickensided deposits consisting of very stiff to hard, low to highly plastic silt and clay. The horizontal tunnel bore passes through Pre-Fraser slickensided deposits for roughly half its length, followed by Pre-Fraser interglacial deposits for the second half. A geologic profile is shown in Figure 2.

There is potential to encounter cobbles and boulders throughout the Vashon advance outwash, Pre-Fraser interglacial deposits and the Pre-Fraser slickensided deposits, causing potential delays and problems for both the TBM and VSM. Additionally, the Pre-Fraser interglacial deposits and the Pre-Fraser slickensided deposits have been subjected to high overburden loads from overlying glaciers and can potentially have high stresses locked in the medium—stresses which result in loads that the shaft and tunnel liners must support.

## TUNNEL DESIGN AND CONSTRUCTION

The tunnel originates at the south launching shaft and continues at an upward grade ranging from 0.1% to 3% for a length of 604 m (1,980 ft), terminating at a receiving shaft on the north side of the bay. Due to the high water pressure resulting from working 29 m (95 ft) below the surface of the Salmon Bay and the job specifications, an EPB TBM was required for the bore. The EPB TBM counteracts the high water pressure by pressurizing soil conditioning foam in front of the cutter head as it bores through the soil. The Herrenknecht EPB2850AH TBM was selected because of its record of success in dealing with high water pressure and because of the efficiencies of using the same brand as the VSM that was being utilized for the launch shaft construction. The cutter head overcut the liner segments by 36 mm (1.4 in) and used both cutter disks and drag teeth to excavate the stiff to hard clay, coarse-grained soils and possible boulders of the glacial deposits. The cuttings were removed from the face of the TBM through a screw conveyor, transported to the launching shaft in muck carts and lifted to an onsite slurry separation plant.

As the TBM advanced, the precast concrete segments were assembled within the TBM. Segment rings have an internal diameter of 2.85 m (9.4 ft), a wall thickness of 200 mm (7.9 in), a length of 1.0 m (39.4 in) and consist of six segments. Longitudinal joints use bolts to connect the segments, and circumferential joints are kept in place with Buclock brand dowels. Concrete with a minimum 28-day compressive strength of 41 MPa (6000 psi) and 39 kg/m$^3$ (65 lb/yd$^3$) of steel fiber reinforcement was selected for the precast segments. Packing with a thickness of 3.2 mm (one-eighth in) was placed between the segments to limit concentrated concrete bearing stresses while allowing for adequate compression of the gaskets to resist the water pressure of the bay above.

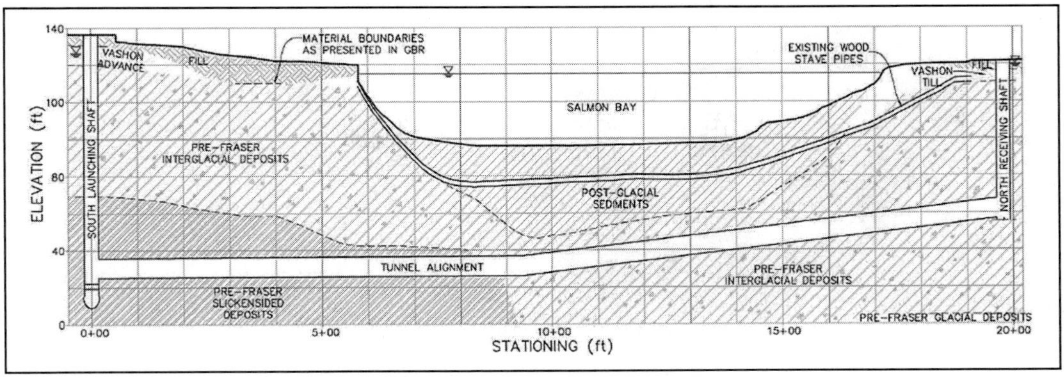


**Figure 2. Geologic profile**

To understand the segmented concrete liner behavior and to estimate stresses in the segments, the liner was analyzed as a 3-D model in the finite element program SAP2000, as shown in Figure 3. The liner was designed to resist earth and water loads of 1125 kPa (23.5 ksf) at the springline, 718 kPa (15.0 ksf) at the crown and 780 kPa (16.3 ksf) at the invert, per the project specifications. The soil surrounding the tunnel was modeled as a series of radial springs. The lateral loads are large compared to the vertical loads because of the locked-in stresses in the overconsolidated glacial material. Shear, moment and axial stresses from the numerical model were checked against the capacity of the fiber reinforced concrete to ensure that the segments could handle the loads. The stress-strain flexural behavior of the steel fiber reinforced concrete was estimated from flexural testing using methods by Barros (2005). Thrust-moment failure envelopes were determined by assuming many different linear strains across the segmental liner to determine the outer boundary of the envelope, and the combined axial and flexural loads were plotted with the envelope to determine the adequacy of the fiber reinforced concrete, as shown in Figure 4.

Segment ring thrust and moment loads were also estimated using Szechy's methods for tunnels (1966) and Roark's solid mechanics tables (Young and Budynas, 2002), and moments were reduced to account for the presence of the joints using methods by Muir Wood (1975). The thrusts and reduced moments of Roark's solid mechanics tables compared favorably to the numerical analysis. A second loading scenario was analyzed using the finite element model where 70% of the earth loads are applied to the liner as well as 340 kpa (50 psi) to a 60° arc of the liner to simulate grouting outside of the concrete liner. In addition to checking the soil and grout loading onto the liner, the capacity of the segmented liner was checked against the form-stripping, handling, storage and TBM jacking loads.

## LAUNCHING SHAFT DESIGN AND CONSTRUCTION

The shaft at the south end of the tunnel was used as the launching shaft which allowed tunneling to proceed at an upward grade, easing muck removal and tunnel dewatering efforts. VSM methods were chosen to excavate the 9.0-m-inner-diameter (29.5 ft), 45.4-m (149 ft) excavated depth TBM launching shaft. VSM excavation methods include lowering a concrete shaft liner composed of precast segments into the shaft in a controlled manner with cable hoists at the surface, as shown in Figure 5. A telescoping roadheader breaks up material below the bottom of the shaft liner. The use of the VSM requires excavating the shaft in the wet and then placing bentonite slurry in the annulus of the concrete liner to help keep the hole open and to act as a lubricant. Cuttings are pumped out of the shaft off of the cutterhead in slurry form. For this project, the cuttings were removed from the slurry in an onsite separation plant. In addition to the ability of the VSM to excavate the shaft on the limited timeline as discussed above, the VSM also works well in a confined construction space, is well suited to excavation with a high water table, and runs quietly off of city power, thereby limiting construction noise. The Herrenknecht VSM 9000 model was used for this project because it was compatible with the shaft dimensions and ground conditions, and it was available for the required time window.

To begin excavation using VSM methods, a reinforced concrete ring beam was constructed at the surface around the location that the concrete segmented liner was to be sunk. The purpose of the ring beam is to aid in guiding the concrete liner as it is lowered and to act as a foundation for the four winches that lower the concrete liner. Next, five liner rings, including the bottom cutting ring, were assembled and set within the ring beam. Four additional liner rings were assembled on top of the existing segmented liner assembly, and the VSM roadheader was

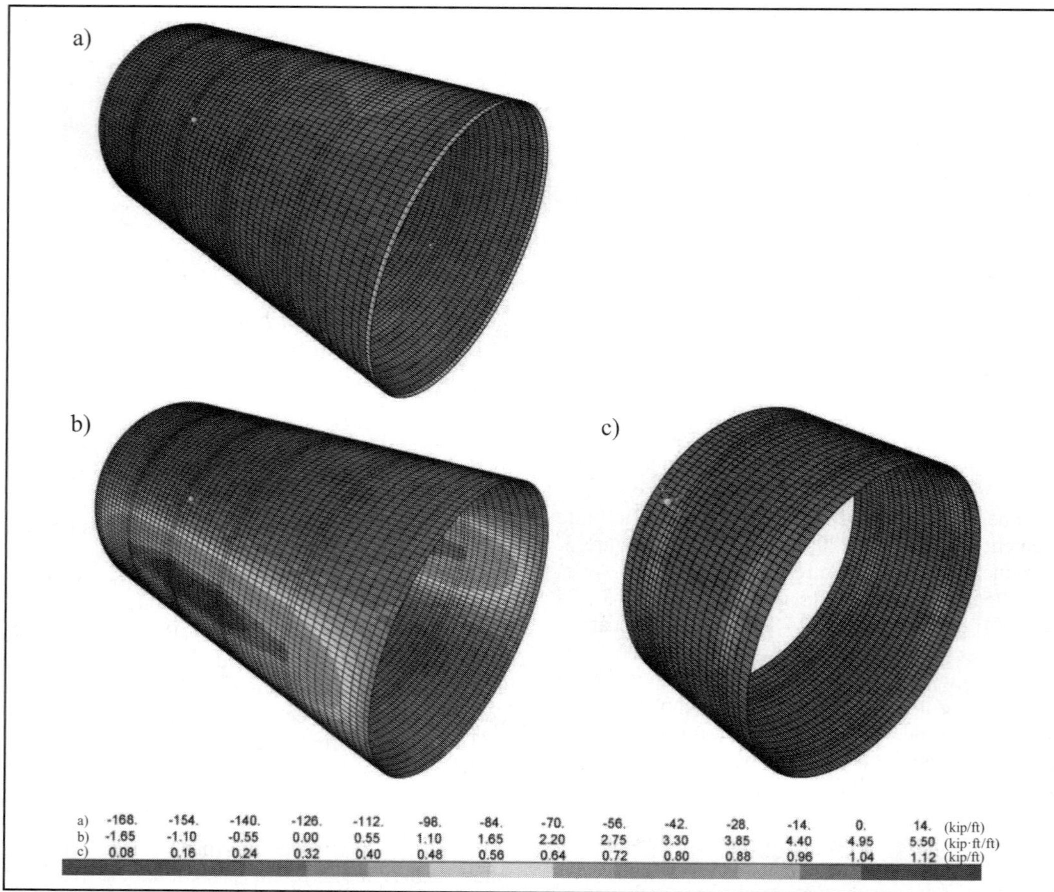


**Figure 3. Finite element modeling results for tunnel lining including: (a) axial hoop forces; (b) circumferential moment; and (c) out-of-plane shear**

set inside the concrete liner, as shown in Figure 6. Finally, the winch lines were connected to the concrete liner, the liner was flooded and excavation using the roadheader began. The shaft was excavated to the design liner depth from April 12, 2012 to May 16, 2012, and the production rate of the VSM averaged approximately 1.8 m (6 ft) per day. After the shaft lining was advanced to the design elevation, the material was excavated to a depth about 2.1 m (7 ft) below the tip of the liner and 0.6 m (1 ft) beyond the outer radius of the liner for placement of the tremie slab concrete. The VSM was then removed from the shaft. The 0.6-m (1-ft) overcut of the tremie slab allowed the tremie slab to engage the weight of the overburden ground, thus eliminating the need for dowels or an excessively thick tremie slab to resist shaft uplift. The 344 $m^3$ (450 $yd^3$) tremie slab was completed in one pour on June 1, 2012. A 0.9-m (3-ft) reinforced concrete slab was installed at the bottom of the shaft up to a finished shaft depth of 41.5 m (136 ft), the liner annulus was grouted, and then finally the shaft was dewatered.

As the VSM was advanced, segmented concrete liner rings were assembled on top of the sinking caisson. The liner rings have an internal diameter of 9.0 m (29.5 ft), a wall thickness of 400 mm (15.75 in), a length of 1.0 m (39.4 in), and consist of four segments. Longitudinal joints are connected with two bolts and have guiding rods to help align the segments. The circumferential joint of each liner ring has twelve dowels and is connected to the ring below with bolts that are placed all the way through the segments and thread into the segment ring below. Hardboard packing with a thickness of 2.0 mm (0.08 in) was placed in the circumferential joints. The shaft segments used the same steel fiber reinforced concrete mix that was used on the tunnel segments. The segments are also reinforced with rebar at select locations around the lifting anchors and on the inner side of the segments so that the segments could be lifted out of the segment forms. Additional rebar reinforcement was used

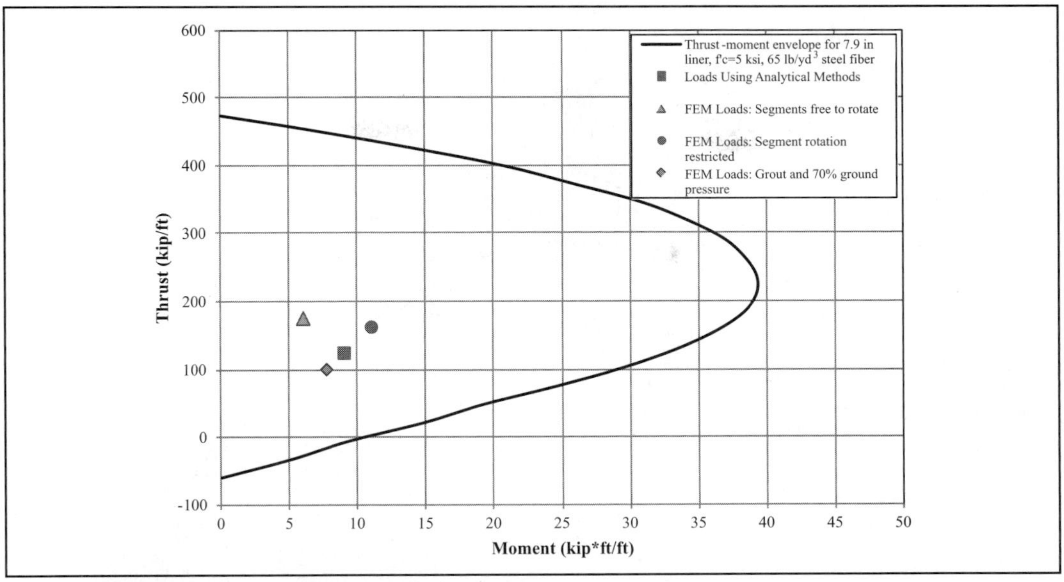


**Figure 4. Estimated loads on tunnel lining plotted on thrust-moment diagram**

**Figure 5. Vertical shaft machine system viewed from the surface including: (a) a shaft liner hoist; (b) a vertical shaft machine hoist; and (c) the utility line guidance tower**

in the third and fifth rings from the bottom of the shaft liner, where the VSM roadheader was mounted. The cutting ring also has additional rebar installed, especially around the mounting plates where shaft liner winch cables attach.

The shaft segmented concrete liner was designed to resist hydrostatic loads and the soil loads that were prescribed in the project specifications. The prescribed earth loads ranged from 10 kPa (0.2 ksf) at the surface to over 177 kPa (3.7 ksf) at the shaft floor elevation. The shaft liner was modeled using finite element methods similar to those used for the tunnel liner. This model used no tangential springs, because the annulus was filled with slurry for a long period of time, unlike the tunnel lining. This numerical model assumed that the shaft was fully excavated with the floor slab installed in order to determine the ring compression loads in the shaft after it was dewatered. Buckling of the shaft was also checked with Euler buckling equations.

Finite element methods were also used to model the TBM break-out hole in the side of the shaft lining. The TBM break-out hole was modeled with no tangential springs, and the horizontal movement of the base of the liner was restrained due to the presence of the tremie slab. A second model of the TBM break-out hole allowed horizontal deflection of the bottom of the liner, but tangential springs were used to simulate the annulus grout. High concentrated stresses around the hole required that a portal frame consisting

Figure 6. Lowering the vertical shaft machine into the shaft liner

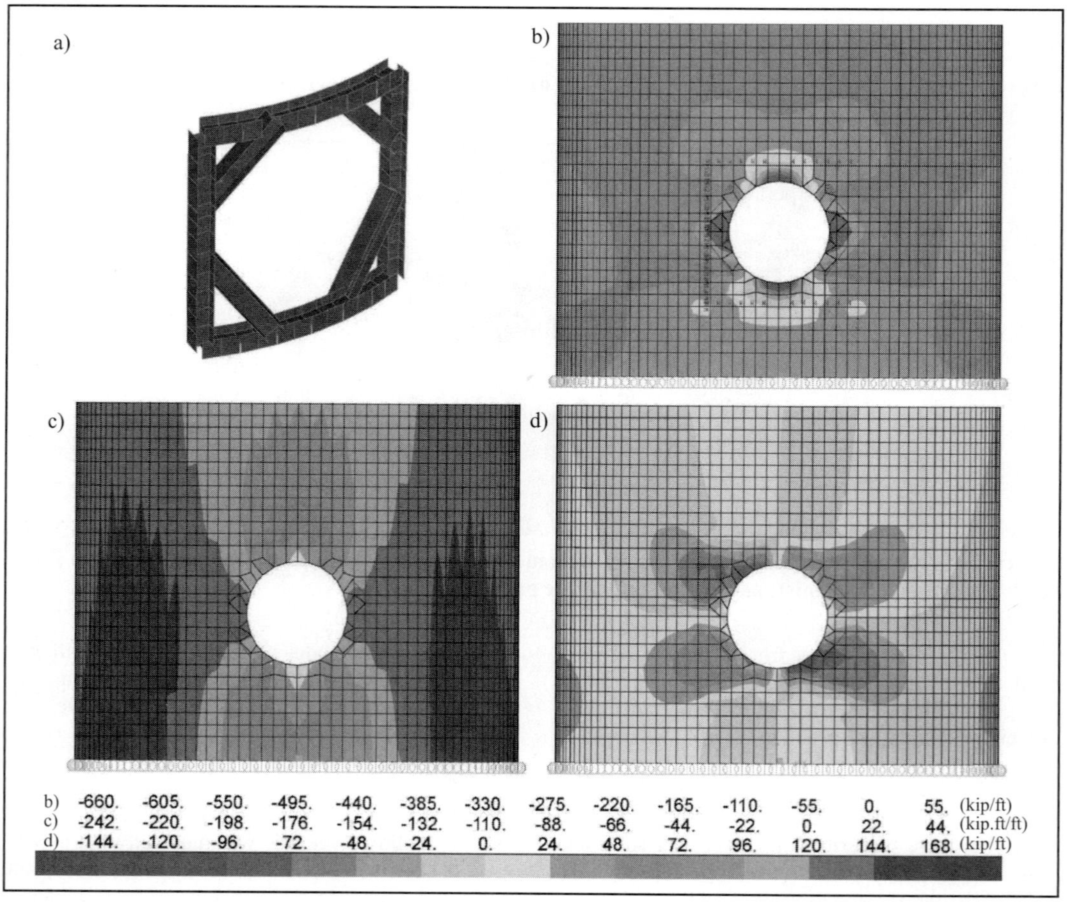


Figure 7. Finite element modeling results of the framed tunnel break-out hole in the shaft liner assuming no tangential springs and a horizontally restrained liner base: (a) the modeled portal frame; (b) circumferential thrust; (c) circumferential moment; and (d) in-plane shear

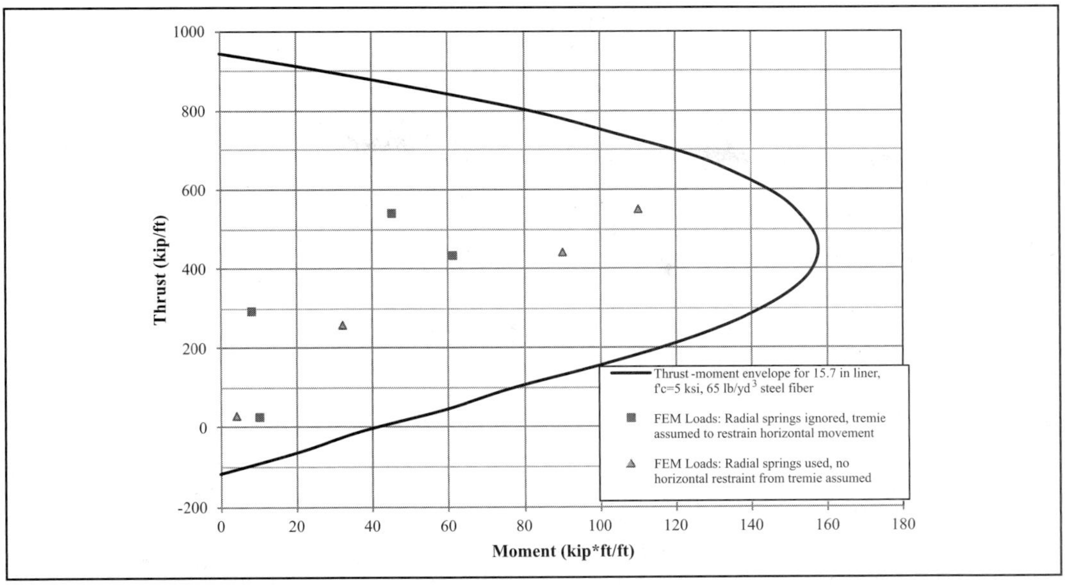


**Figure 8. Estimated loads on shaft liner around tunnel break-in plotted on thrust-moment**

of W250×80 (W10×54 imperial) members encased in reinforced concrete be installed around the opening. This portal frame was incorporated into the finite element models; the results of the model that assumed no tangential springs and a restrained liner bottom are shown in Figure 7. Thrust-moment failure envelopes were developed in the same manner as for the tunnel segments to make sure the shaft segments could resist the combined axial-flexural loading around the tunnel break-out, as shown in Figure 8.

After the shaft was excavated and dewatered, the TBM was lowered into the shaft to begin tunnel construction. The tunneling work was performed from the week of October 29, 2012 to the week of August 19, 2013.

## CONCLUSION

The Ballard Siphon Replacement Project marked the first use of a Vertical Shaft Machine in North America. The project increased the capacity of the CSO system in Seattle by sliplining the existing wood stave pipe siphon and constructing a new siphon composed of two vertical shafts and a tunnel under the Salmon Bay. High water pressure and variable ground conditions required the use of an Earth Pressure Balance TBM. Potentially high lateral loads resulting from the overconsolidated glacial material required that special attention be given to loading of the tunnel segments. The timeline and depth of the launch shaft for the EPB TBM made the VSM the right choice for the job. This project demonstrated that the VSM is capable of handling varying conditions, such as those found in glacial material, and is well suited to urban construction sites where space is at a premium.

## REFERENCES


Barros, J.A.O., Cunha, V.M.C.F., Ribeiro, A.F., & Antunes, J.A.B. (2005). Post-cracking behaviour of steel fibre reinforced concrete. *Materials and Structures, 38*(275), 47–56.

Fowler, J. (printed 2012). First Use of Vertical Shaft Machine Technology in North America. *NASTT Pacific Northwest Trenchless Review, 2013 ed.* 11–14.

King County. (2009). Contract documents for Ballard Siphon Replacement Project, Contract number C00507C10. Seattle, WA.

Muir Wood, A.M. (1975). The circular tunnel in elastic ground. *Géotechnique, 25*(1), 115–127.

Szechy, K. (1966). *The art of tunneling.* Budapest: Hungary: Akadémiai Kiadó.

Young, W.C., & Budynas, R.G. (2002). *Roark's formulas for stress and strain* (7th ed.). New York, NY: McGraw-Hill.

# Index

**D**

**E**